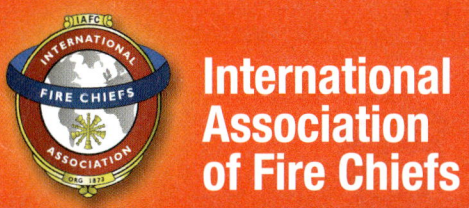

International
Association
of Fire Chiefs

Fundamentals of
Fire Fighter Skills
and Hazardous
Materials Response

CANADIAN FOURTH EDITION

JONES & BARTLETT
LEARNING

Jones & Bartlett Learning
World Headquarters
5 Wall Street
Burlington, MA 01803
978-443-5000
info@jblearning.com
www.psglearning.com

National Fire Protection Association
1 Batterymarch Park
Quincy, MA 02169-7471
www.NFPA.org

International Association of Fire Chiefs
4025 Fair Ridge Drive
Fairfax, VA 22033
www.IAFC.org

Jones & Bartlett Learning books and products are available through most bookstores and online booksellers. To contact the Jones & Bartlett Learning Public Safety Group directly, call 800-832-0034, fax 978-443-8000, or visit our website, www.psglearning.com.

17230-0

Production Credits

General Manager and Executive Publisher: Kimberly Brophy
VP, Product Development: Christine Emerton
Senior Managing Editor: Donna Gridley
Executive Editor: Bill Larkin
Senior Editor: Amanda J. Mitchell
Senior Editor: Barbara Scotese
VP, Sales, Public Safety Group: Phil Charland
Business Development Manager, Fire: Janet E. Maker
Director of Canadian Sales: Rob Rosenitsch
Project Manager: Kristen Rogers
Project Specialist: John Fuller
Digital Project Specialist: Angela Dooley

Director of Marketing Operations: Brian Rooney
VP, Manufacturing and Inventory Control: Therese Connell
Composition: S4Carlisle Publishing Services
Cover Design: Kristin E. Parker
Text Design: Scott Moden
Director, Content Services and Licensing: Joanna Gallant
Rights & Media Manager: Shannon Sheehan
Rights & Media Specialist: John Rusk
Media Development Editor: Troy Liston
Cover Image: Courtesy of Brent Thomas
Printing and Binding: LSC Communications
Cover Printing: LSC Communications

Library of Congress Cataloging-in-Publication Data unavailable at time of printing

6048

Printed in the United States of America
23 22 21 20 19   10 9 8 7 6 5 4 3 2 1

Brief Contents

Contents

CHAPTER **3**

Personal Protective Equipment 65

CHAPTER **4**

Fire Service Communications 131

Skill Drills

Acknowledgments

Jones & Bartlett Learning, the National Fire Protection Association, and the International Association of Fire Chiefs would like to thank the current editors, contributors, and reviewers of *Fundamentals of Fire Fighter Skills, Canadian Fourth Edition.*

Editorial Board for the US Edition

Dave Schottke
Fairfax Volunteer Fire Department
Fairfax, Virginia

Rob Schnepp
Division Chief of Special Operations (retired)
Alameda County Fire Department
Alameda, California

Contributors and Reviewers

Kent Abbott
Toromont Power Systems
Newfoundland and Labrador

Brian Byrnell
Lakeland College
Vermilion, Alberta

Steve Harnum
Conception Bay South
Newfoundland and Labrador

Alan Jones
Lakeland College
Vermilion, Alberta

Jeff Kirk
Orillia Fire Department
Orillia, Ontario

Steve Knight
Lambton College Sarnia
Lambton, Ontario

Kevin C. Lambert
Ottawa Fire Training Centre and Ontario Fire College
Ottawa, Ontario

Roy Langmead
St. John's International Airport
Newfoundland and Labrador

Jeremy Parkin
Rama First Nation
Rama, Ontario

Kenny Roche
St. John's Regional Fire Department
Newfoundland and Labrador

Trevor Shea
Mississauga Fire and Emergency Services
Burlington, Ontario

Brent Thomas
Orillia Fire Department
Orillia, Ontario

Dave Wallack
Justice Institute of British Columbia
New Westminster, British Columbia

Richard While
Justice Institute of British Columbia
New Westminster, British Columbia

Photographic Contributions

We would like to extend a huge "thank you" to Glen E. Ellman, the photographer for this project. Glen is a commercial photographer and fire fighter based in Fort Worth, Texas. His expertise and professionalism are unmatched!

Thank you to the following organizations that opened up their facilities for these photo shoots:

Tucson Fire Department, Tucson, Arizona
Jim Critchley, Fire Chief
Mike Garcia, Assistant Chief
Mike Fischback, Deputy Chief, Training
Brian Stevens, Battalion Chief
Casey Justen, Battalion Chief
Vinny McGurk, Senior Housing Tech

Stephen Scionti, Captain
Andy Yeoh, Captain
Patrick Bunker, Captain
Nate Weber, Captain
Mike Wintrode, Engineer
Jason West, Engineer
Brian Barrett, Engineer
Thomas Campuzano, Engineer
Matthew Walter, Fire Fighter
Jason Greenawalt, Fire Fighter
Mario Carrasco, Fire Fighter
Ryan Vaughan, Fire Fighter
Cole Mayfield, Fire Fighter

CAL FIRE Academy, Ione, California

Brent Stangeland, Department Training Chief
Jeremy Lawson, Deputy Chief Training Operations
Ann Johnston Rosales, Division Chief
Mike Olivarria, Division Chief
Ty LaBelle, Battalion Chief
Robert Wheatly, Battalion Chief
Trisha Reis, Facility Coordinator
Anthony Wuester, Captain
Darrin Lehman, Captain
Keeton Brookins, Fire Apparatus Engineer
Kevin Seymour, Fire Apparatus Engineer
Luke Manton, Fire Apparatus Engineer
Jason Hawley, Fire Apparatus Engineer
Bronson Wilcox, Fire Apparatus Engineer

Atlantic County Fire Academy, Egg Harbor Township, New Jersey

Mike Corbo, Director
Marylynn Graham, Instructor
Eric Grace, Instructor
Casey Hunt, Instructor

The Mississippi State Fire Academy, Jackson, Mississippi

Reggie Bell, Executive Director
Michael Word, Instructor Advanced
Shannon Sandridge, Curriculum/Public Relations
Luke Smith, Staff Instructor
Donny Collins, Instructor
Kris Koziol, Senior Instructor
Robert Wells

Fort Worth Fire Department, Fort Worth, Texas

Paul Macken, Fire Fighter
Alex Cacho, Engineer
Chad Self, Fire Fighter
Frank Becerra, Fire Fighter
Jacky Haggard, Fire Fighter
Scott Smith, Fire Fighter
Kurt Howard, Fire Fighter
Jimmy Carter, Lieutenant
Brian Oates, Fire Fighter
Joe Salazar, Fire Fighter

Boca Raton Fire Rescue Services, Boca Raton, Florida

Burleson Fire Department, Burleson, Texas

Chicago Fire Department, Chicago, Illinois

Des Plains Fire Department, Des Plains, Illinois

Fairfax City Fire Department, Fairfax, Virginia

San Antonio College, Regional Fire Academy, San Antonio, Texas

New for the *Canadian Fourth Edition*

The Canadian fourth edition of *Fundamentals of Fire Fighter Skills and Hazardous Materials Response* was significantly updated to better serve Fire Fighter I and II instructors and students. Not only were the underpinnings of the textbook and curriculum shifted from the NFPA 472 standard to the professional qualifications standard of NFPA 1072, but the overall organization and design of the material were also updated accordingly. The program now has five distinct sections: Fire Fighter I, Fire Fighter II, Hazardous Materials Awareness, Hazardous Materials Operations, and Hazardous Materials Operations: Mission Specific. This new organization will allow you the flexibility to teach your Fire Fighter I and II course(s) exactly the way you wish. The first 21 chapters set the foundation for Fire Fighter I knowledge and understanding. The next 7 chapters encompass the Fire Fighter II section. With this core Fire Fighter I and II knowledge covered, the final 9 chapters address awareness level, operations level, and operations mission-specific requirements.

Updated Design and Online Learning

The new interior design was developed to embrace online learning. Throughout the textbook, additional online learning opportunities can be accessed by redeeming the corresponding Navigate 2 access code in the front of the printed textbook. These activities are earmarked with the Navigate 2 logo and include:

Fire Alarm—Reinforce comprehension of course materials and hone critical thinking skills.

Knowledge Check—Access the interactive eBook to evaluate understanding of course materials with knowledge check questions and an end-of-chapter quiz. To make the most of the eBook, create personal and collaborative notes, highlight key concepts, and bookmark selected content for easy reference, even when you're offline!

On Scene—Review the On Scene case studies, and check out other valuable resources, such as flashcards, audiobook, progress and performance analytics, and more.

Beyond enhancing the student learning experience, Navigate offers a full suite of instructor tools that will help ensure a successful program. In addition to comprehensive lecture materials, instructors can assign and customize homework and assessments, access a robust gradebook, and tailor course plans using the real-time, actionable data found in the analytics dashboard.

Improve learning outcomes by taking full advantage of these engaging, accessible, and effective solutions and tools!

Fire Fighter I

Fire Fighter I

The Fire Service

KNOWLEDGE OBJECTIVES

After studying this chapter, you will be able to:

- List five guidelines for successful fire fighter training. (p. 5)
- Describe the mission of the fire service. (**NFPA 1001: 4.1.1**, p. 5)
- Describe the culture of the fire service. (pp. 5–6)
- Describe the general qualifications for becoming a fire fighter. (pp. 6–7)
- Outline the roles and responsibilities of a Fire Fighter I. (**NFPA 1001: 4.1.1**, pp. 8–9)
- Describe the common roles of fire fighters within the fire department. (pp. 9–10)
- Describe the specialized response roles within the fire department. (p. 10)
- List the Five E's of community risk reduction. (p. 11)
- Describe the characteristics of a Community Risk Reduction (CRR) program. (pp. 10–11)
- Identify common safety hazards in the home. (pp. 12–13)
- Describe the basic types of residential smoke alarms. (pp. 13–15)
- Describe a situation in which you will interact with other organizations within your community. (pp. 15–16)
- Explain the concept of governance, and describe how regulations, standards, policies, and standard operating procedures affect it. (**NFPA 1001: 4.1.1**, pp. 16–17)
- Locate information in departmental documents and standard operating procedures. (**NFPA 1001: 4.1.2**, pp. 16–17)
- Describe the organization of the fire service. (**NFPA 1001: 4.1.1**, pp. 17–21)
  - List the different types of fire department companies, and describe their functions. (pp. 17–18)
  - Describe how to organize a fire department in terms of staffing, function, and geography. (p. 19)
  - Explain the basic structure of the chain of command within the fire department. (pp. 19–20)
  - Define the four basic management principles used to maintain organization within the fire department. (pp. 20–21)
- Explain the evolution of the methods and tools of firefighting from colonial days to the present. (pp. 21–23)
- Explain how building codes prevent the loss of life and property. (p. 23)
- Describe the evolution of training and education for fire department services. (pp. 23–24)
- Describe the evolution of fire equipment for fire department services. (pp. 24–25)
- Describe the evolution of communications for fire department services. (pp. 25–26)
- Describe the evolution of funding for fire department services. (p. 26)

SKILLS OBJECTIVES

After studying this chapter, you will be able to perform the following skill:

- Locate information in departmental documents and standard operating procedures. (NFPA 1001: 4.1.2, pp. 16–17)

ADDITIONAL STANDARDS

- NFPA 72, *National Fire Alarm and Signaling Code*
- NFPA 1500, *Standard on Fire Department Occupational Safety, Health, and Wellness Program*
- NFPA 1582, *Standard on Comprehensive Occupational Medical Program for Fire Departments*

Fire Alarm

It is your first day in the station after having completed 4 months of recruit academy. Your new crew welcomes you. As you are being shown around the station, the doorbell rings—and you are told your first official duty is to answer it. At the door, you find a young woman and her young son. She sheepishly says that her son has been begging her to meet the fire fighters because he wants to be one when he grows up.

As you begin the tour, you show the pair the living quarters and then move to the apparatus floor. Like most youngsters, the visiting boy is full of questions and asks, "How old do you have to be to be a fire fighter? Do you drive the ladder truck? Why does it have a big ladder on top?" The mother tells him to give you a chance to explain before asking any more questions. You smile as you ponder your answers.

1. What is the age requirement to become a fire fighter?

2. How do the duties of the captain, the fire apparatus driver/operator, and the fire fighters differ?

3. What are the various types of apparatus used in the fire service, and how do they function?

 *Access Navigate for more practice activities.*

Introduction

Training to become a fire fighter is challenging. The art and science of extinguishing fires is much more complex than most people imagine. You will be challenged—both physically and mentally—during this course. You must keep your body in excellent condition so you can complete your assignments; you must also remain mentally alert to cope with the various conditions you will encounter. Fire fighter training will expand your understanding of fire suppression—that is, the various activities involved in controlling and extinguishing fires.

When you join the fire service, you join a profession with a long and noble history of protecting and serving the community. By the time you complete this course, you will be well equipped to continue a centuries-old tradition of preserving lives and property threatened by fires (**FIGURE 1-1**). You will have an understanding of the many duties and responsibilities of the fire service, and you should have a feeling of personal satisfaction about your accomplishment.

This chapter offers five guidelines for you to remember as you progress through this course and take your place as a fire fighter in your community. Next, it presents the responsibilities of a Fire Fighter I and a Fire Fighter II. The chapter then explains the organization of the fire service and describes the chain of command that connects all members of a department. Finally, it reviews the history and development of the modern-day fire service.

FIGURE 1-1 When you join the fire service, you join a profession with a long and noble history of protecting and serving the community.
© Vince Talotta/Toronto Star/Getty.

Fire Fighter Guidelines

During this course of study, you will need to practise and work hard. Do your best. The five guidelines listed here will keep you on target to become a proud and accomplished fire fighter.

1. *Be safe.* Safety should always be uppermost in your mind. Keep yourself safe. Keep your teammates safe. Keep the public you serve safe.
2. *Follow orders.* Your supervisors have more training and experience than you do. If you can be counted on to follow orders, you will become a dependable member of the team.
3. *Work as a team.* Fighting fires requires the coordinated efforts of all department members. Teamwork is essential to success.
4. *Think!* Lives will depend on the choices you make. Put your brain in gear. Think about what you are studying.
5. *Follow the Golden Rule.* Treat each team member, victim, or citizen as an important person or as you would treat a member of your own family. Everyone is an important person or family member to someone, and everyone deserves your best efforts.

The Mission of the Fire Service

The primary mission of the fire department is to save lives and protect property and the environment through prevention, education, suppression, and rescue activities. Some of the actual activities may vary from one fire department to another depending on the location of the department and what other community organizations are present, but the primary mission of the fire department remains the same. The mission statement of your department should be short enough that you can easily remember it. One example of a fire department mission statement is as follows, *"The Phoenix Fire Department is committed to providing the highest level of public safety services to our community. We protect lives and property through fire suppression, emergency medical and transportation services, disaster services, fire prevention, and public education. Our members will Prevent Harm, Survive, and Be Nice!"*

Many fire departments post the mission statement of their fire department in a visible place where department members will see it regularly.

The Culture of the Fire Service

The culture of an organization can be defined as "the set of shared attitudes, values, goals and practices that characterize that institution or organization" (merriam-webster.com, 2017). This definition identifies some of the characteristics upon which the fire service culture is based. Earlier, we talked about the goal/mission of the fire service—to save lives and protect property and the environment. By looking at the culture of the fire service we can better understand the values that are held by the fire service. There are many ways to list these values. The following list represents some of these cultural characteristics:

- **Courage** can be described as the physical, moral, and mental strength of a person. While it takes tremendous courage to run into a burning building, it takes a different type of courage to treat all people with respect and to stand up for the rights of people who do not have the ability to stand up for their rights. It takes many types of courage to be a competent and caring fire fighter. Fire fighters are respected for these many types of courage.
- **Honour** guides a person to demonstrate ethical and moral behaviour. We honour those who have lost their lives in the middle of a fire fight. We also need to honour those who maintain their competency in order to be ready to answer a call to help any citizen at any time. Fire fighters are highly respected. Work hard to earn that respect.
- **Duty** may be described as an act or course of action that is required of a fire fighter. It is a fire fighter's duty to adhere to standard operating procedures, show technical proficiency, follow established best practices,

assess situations, weigh possible options, make sound decisions, and initiate actions that are required to create a positive outcome. Become a lifelong learner.

- **Respect** can be described as a deep admiration for something or someone elicited by his or her abilities, achievements, or qualities. To gain respect from people, we must first respect them. We can do this by getting to know the people we interact with and looking out for the well-being of everyone.
- **Integrity** may be described as a measure of where a person stands in times of challenge and controversy. We are accountable to all members of the public and to our department. As fire fighters, we should learn to take advantage of our strengths and improve on our weaknesses. As fire fighters, we should seek responsibility as well as accept responsibility for our actions.
- **Character** is the combination of emotional, intellectual, and moral qualities that constitute a person. Character is based on values, and it is a combination of our actions and words. We are ultimately responsible for the decisions we make and for the actions we take. We can set the example by showing optimism and encouraging others.
- **Moral courage** is required to take action when you have doubts or fears about the consequences. Moral courage requires deliberation and careful thought. Moral courage may also require physical courage when the consequences are bodily peril or punishment. Moral courage is required when a tough decision requires you to think and then take an unpopular position.

FIRE FIGHTER TIP

Parts of the organizational structure of our organization are built from military traditions.

Fire Fighter Qualifications

Not everyone can become a fire fighter. Those who do succeed in achieving this status understand the vital mission of the fire department. A fire fighter must be healthy and in good physical condition and assertive enough to enter a dangerous situation yet mature enough to work as a member of a team (**FIGURE 1-2**). The job requires a person who has the desire to learn, the will to practise, and the ability to apply the skills of the trade. A fire fighter is constantly learning as the body

FIGURE 1-2 Those who succeed in achieving fire fighter status understand the vital mission of the fire department: to save lives and protect property and the environment through prevention, education, and suppression.

© Jones & Bartlett Learning. Photographed by Glen E. Ellman.

of knowledge about fires expands and the technology used in fighting fires evolves.

The training and performance qualifications for fire fighters are specified in National Fire Protection Association (NFPA) 1001, *Standard for Fire Fighter Professional Qualifications*. Age, education requirements, medical requirements, and other criteria are established locally.

Age Requirements

Most career fire departments require that candidates be at least 18 years of age, although some have specified a minimum age of 21. Candidates should possess a valid driver's license, have a clean driving record, have no criminal record, and be drug free.

Volunteer fire departments usually have different age requirements, which often depend on insurance considerations. Some volunteer fire departments allow "junior" members to join but restrict their activities until they reach age 18.

Education Requirements

Most career fire departments require that applicants have at least a high school diploma or equivalent. Some require candidates to have completed college-level courses in a field related to fire or emergency medical services (EMS). All departments require fire fighters who wish to be considered for promotion or extra responsibility to take additional courses.

Medical Requirements

Because firefighting is both stressful and physically demanding, fire fighters must undergo a medical evaluation before their training begins. This medical evaluation identifies most medical conditions or physical limitations that could increase the risk of injury or illness either to the candidate or to other fire fighters. Part of the medical requirement is to successfully pass a drug screening test. Medical requirements for fire fighters are specified in NFPA 1582, *Standard on Comprehensive Occupational Medical Program for Fire Departments*.

Physical Fitness Requirements

Physical fitness requirements are established to ensure that fire fighters have the strength and stamina needed to perform the tasks associated with firefighting and emergency operations. Several performance-testing scenarios have been validated by qualified organizations. NFPA 1001 allows individual fire departments to choose the fitness-testing method that will be used for fire fighter candidates. It is important that fire fighters maintain a high level of fitness throughout their careers in order to meet the physical rigours required of fire fighters and to help prevent injuries and illness.

Many departments require applicants for firefighting to complete a standardized physical ability test. These tests are designed to measure the strength and endurance required by active fire fighters. One of the most widely used tests is the Candidate Physical Ability Test (CPAT). This testing process was developed by the International Association of Fire Fighters (IAFF) and the International Association of Fire Chiefs (IAFC) (IAFC, 2017). The test consists of the following eight scenarios, which are completed while wearing a 50-pound vest to simulate the weight of equipment worn during a fire.

- Stair climb: Climbing stairs while carrying an additional 11-kg (25-lb) simulated hose pack
- Ladder raise and extension: Placing a ground ladder at the fire scene and extending the ladder to the roof or a window
- Hose drag: Stretching uncharged hose lines and advancing hose lines
- Equipment carry: Removing and carrying equipment from fire apparatus to fire ground
- Forcible entry: Penetrating a locked door and breaching a wall

- Search: Crawling through dark, unpredictable areas to search for victims
- Rescue: Removing a victim or partner from a fire scene
- Ceiling breach and pull: Locating fire and checking for fire extension

Firefighting candidates have 10 minutes and 20 seconds to complete all eight events. Many fire departments require fire fighters to complete this series of events every year to ensure that they maintain the level of physical fitness required for active firefighting.

Emergency Medical Care Requirements

Delivering emergency medical care is an important function of most fire departments. NFPA 1001 allows individual fire departments to specify the level of emergency medical care training required for entry-level personnel. At a minimum, you will need to understand infection control procedures and be able to perform cardiopulmonary resuscitation (CPR), control bleeding, and manage shock. Some departments require fire fighters to become certified as an Emergency Medical Responder (EMR), Emergency Medical Technician (EMT), Advanced Emergency Medical Technician (AEMT), or a Paramedic.

Testing and Interview Requirements

The steps in the application process to become a fire fighter are decided by each fire department and approved by the municipality. There are many variations in this process. Some departments have a much simpler process than others. Smaller municipal departments and volunteer departments often have a less formal screening process than large metropolitan departments. Most candidate screening processes consist of a written exam and an oral interview.

Written exams can be either a general knowledge test that measures oral and written communications skills, basic mathematics, and mechanical aptitude or a test of fire and EMS knowledge. Your department usually will provide you with an instruction sheet to help prepare you for the written test and the oral interview.

Some fire departments conduct oral interviews to measure the candidate's ability to organize thoughts, communicate clearly, and follow instructions under the stress of an interview. There is a wide variation to the manner in which each department conducts interviews. If you will be interviewed, the department usually will give you some instructions to help you prepare for the interview.

Roles and Responsibilities of Fire Fighter I and Fire Fighter II

The first step in understanding the organization of the fire service is to learn your roles and responsibilities as a Fire Fighter I or Fire Fighter II. As you progress through this text, you will learn what to do and how to do it so that you can take your place confidently among the ranks. As discussed previously, the training and performance qualifications for fire fighters are outlined in NFPA 1001. The following sections describe the roles and responsibilities of a Fire Fighter I and a Fire Fighter II. In some cases, the same or similar roles and responsibilities may appear under Fire Fighter I and Fire Fighter II. This enables fire fighters to learn the fundamentals of those concepts as a Fire Fighter I and increase their knowledge of the same concepts as a Fire Fighter II.

Training programs vary. You may take one course that covers both Fire Fighter I and Fire Fighter II, or you may take two separate courses.

Roles and Responsibilities for Fire Fighter I

- Don (put on) and doff (take off) personal protective equipment (PPE) properly.
- Hoist hand tools using appropriate ropes and knots.
- Understand and correctly apply appropriate communication protocols.
- Use self-contained breathing apparatus (SCBA).
- Respond on apparatus to an emergency scene.
- Establish and operate safely in emergency work areas.
- Force entry into a structure.
- Exit a hazardous area safely as a team.
- Set up and use ground ladders safely and correctly.
- Attack a passenger vehicle fire, an exterior Class A fire, and an interior structure fire.
- Conduct search and rescue in a structure.
- Perform ventilation of an involved structure.
- Overhaul a fire scene.
- Conserve property with salvage tools and equipment.
- Connect a fire department pumper to a water supply.
- Extinguish incipient Class A, Class B, and Class C fires.
- Illuminate an emergency scene.
- Turn off utilities.
- Combat ground cover fires.
- Perform fire safety surveys.
- Clean and maintain equipment.
- Locate information in departmental documents and standard operating procedures.
- Operate as part of a team.

Roles and Responsibilities for Fire Fighter II

- Perform scene size-up.
- Determine the need for the incident command system (ICS).
- Arrange and coordinate ICS until command is transferred.
- Prepare reports.
- Communicate the need for assistance.
- Coordinate an interior attack line team.
- Extinguish an ignitable liquid fire.
- Control a flammable gas cylinder fire.
- Protect evidence of fire cause and origin.
- Assess and disentangle victims from motor vehicle accidents.
- Assist special rescue team operations.
- Perform a fire safety survey.
- Present fire safety information.
- Maintain fire equipment.
- Perform annual service tests on fire hose.

NFPA 1001 also requires the Fire Fighter I to meet the requirements in Chapter 4, Awareness; Chapter 5, Operations; and the mission-specific responsibilities in Section 6.2, Personal Protective Equipment and Section 6.6, Product Control of NFPA 1072, *Standard for Hazardous Materials/Weapons of Mass Destruction Emergency Response Personnel Professional Qualifications*.

Every member of the fire service will interact with the public. In addition to interacting with citizens at the scene of incidents, fire fighters will encounter people who visit the fire station requesting a tour or asking questions about specific fire safety issues (**FIGURE 1-3**). Fire fighters should be prepared to assist these visitors and use this opportunity to provide them with additional fire safety information. Keep in mind that interaction with the public occurs both on and off duty. Use every instance to deliver positive public relations and send educational messages.

FIGURE 1-3 Any contact with the public should be used as an educational opportunity. Whenever the public may come in contact with a fire fighter's PPE during demonstrations, the PPE should be clean and free of contaminants.
Courtesy of Marsha Giesler.

Roles Within the Fire Department

General Roles

A fire fighter may be assigned any task from placing hose lines to extinguishing fires. Generally, the fire fighter is not responsible for any command functions and does not supervise other personnel, except on a temporary basis when promoted to an acting officer. Through additional training and promotion, a fire fighter may fulfill many roles during a career, particularly in smaller departments. Some of the more common positions include the following:

- **Fire apparatus driver/operator**: Often called an engineer or technician, the driver is responsible for getting the fire apparatus to the scene safely, as well as setting up and running the pump or operating the aerial device once it arrives on the scene. This function is a full-time role in some departments; in other departments, it is rotated among the fire fighters.
- **Company officer**: The company officer is usually a lieutenant or captain who is in charge of an apparatus. This person leads the company both on scene and at the station. The company officer is responsible for initial firefighting strategy, personnel safety, and the overall activities of the fire fighters on their apparatus. Once command is established, the company officer focuses on tactics.
- **Incident safety officer**: The incident safety officer watches the overall operation for unsafe practices. He or she has the authority to halt any firefighting activity, allowing the activity to resume only when it can be done safely and correctly. The senior

ranking officer may act as an incident safety officer until the appointed incident safety officer arrives or until another officer is delegated those duties.
- **Training officer**: The training officer is responsible for updating the training of current fire fighters and for training new fire fighters. He or she must be aware of the most current techniques of firefighting, EMS, and other specialized services provided by the fire department. The training officer must coordinate various aspects of the fire department's training program and maintain documentation of all training activities.
- **Incident commander (IC)**: The incident commander is responsible for the management of all incident operations. This position focuses on the overall strategy of the incident and is often assumed by the district/battalion chief.
- **Fire marshal**: The fire marshal delivers, manages, and/or administers fire protection and life-safety–related codes and standards, investigations, education, and/or prevention services.
- **Fire inspector**: The fire inspector inspects businesses and enforces public safety laws and fire codes. In some departments, this may also be referred to as a fire prevention officer.
- **Fire investigator**: The fire investigator responds to fire scenes to help investigate the cause of a fire. The investigator may have full police powers to investigate and arrest suspected arsonists and people causing false alarms.
- **Fire and life-safety educator (FLSE)/fire prevention officer (FPO)**: The FLSE/FPO educates the public about fire safety and injury prevention and presents juvenile fire safety programs. The FLSE/FPO and the public information officer may be part of a larger community risk reduction program within the department.
- **911 dispatcher/telecommunicator**: From the communications centre, the dispatcher takes calls from the public, sends appropriate units to the scene, assists callers with emergency medical information (treatment that can be performed until the medical unit arrives), and assists the IC with obtaining needed resources.
- **Emergency vehicle technician (EVT)**: Emergency vehicle technicians repair and service fire and EMS vehicles, keeping them ready to respond to emergencies. These individuals are usually trained by equipment manufacturers to repair vehicle engines, lights, and all parts of the fire pump and aerial ladders.

- **Fire police officer**: Fire police officers are usually fire fighters who control traffic and secure the scene from public access. Many fire police officers are also sworn peace officers.
- **Information management**: "Info techs" are fire fighters or civilians who take care of a department's computer and networking systems. Maintaining computerized fire reports, e-mail, and a functional computer network requires the services of properly trained computer personnel.
- **Public information officer (PIO)**: The PIO serves as a liaison between the IC and the news media.
- **Fire protection engineer**: The fire protection engineer reviews plans and works with building owners to ensure that their fire suppression and detection systems meet the applicable codes and function as needed. Some fire protection engineers design these systems, and most hold a degree in fire engineering.

Specialized Response Roles

Many assignments require specialized training. Indeed, most large departments have teams of fire fighters who respond to specific types of calls. Members of these teams are usually required to be fire fighters before they begin additional training. Specialist positions include the following:

- **Aircraft/crash rescue fire fighter (ARFF)**: ARFFs are based at military and civilian airports and receive specialized training in aircraft fires, extrication of victims on aircraft, and extinguishing agents. They wear special PPE and respond in specialized fire apparatus that protects them from high-temperature fires caused by substances such as jet fuel.
- **Hazardous materials technician**: Hazardous materials ("Hazmat") technicians have training and certification in chemical identification, leak control, decontamination, and clean-up procedures.
- **Technical rescuer**: "Tech rescuers" are trained in special rescue techniques for incidents involving structural collapse, trench rescue, swiftwater rescue, confined-space rescue, high-angle rescue, and other unusual situations. The units they work in are sometimes called urban search and rescue teams.
- **SCUBA dive rescue technician**: Many fire departments, especially those located near waterways, lakes, or an ocean, use SCUBA technicians who are trained in rescue, recovery, and search procedures in both water and under-ice situations. SCUBA stands for self-contained underwater breathing apparatus.

- **Emergency medical services (EMS) personnel**: EMS personnel administer prehospital care to people who are sick or injured. Prehospital calls account for the majority of responses in many departments, so fire fighters are often cross-trained with EMS personnel. EMS training levels are normally divided into four categories: Emergency Medical Responder, Emergency Medical Technician, Advanced Emergency Medical Technician, and Paramedic.
  1. **Emergency Medical Responder (EMR)**: The first trained professional, such as a police officer, fire fighter, lifeguard, or other rescuer, to arrive at the scene of an emergency to provide initial medical assistance. EMRs have basic training and often perform in an assistant role within the ambulance.
  2. **Emergency Medical Technician (EMT)**: Most EMS providers are EMTs. They have training in basic emergency care skills, including oxygen therapy, bleeding control, CPR, automated external defibrillation, basic airway devices, and assisting patients with certain medications.
  3. **Advanced Emergency Medical Technician (AEMT)**: AEMTs can perform more procedures than EMTs, but they are not yet Paramedics. They have training in specific aspects of advanced life support, such as intravenous (IV) therapy, interpretation of cardiac rhythms, defibrillation, and airway intubation.
  4. **Paramedic**: Paramedics have completed the highest level of training in EMS. They have extensive training in advanced life support, including IV therapy, administering drugs, cardiac monitoring, inserting advanced airways (endotracheal tubes), manual defibrillation, and other advanced assessment and treatment skills.

LISTEN UP!

From 2014 to 2016, the NFPA reported 44 percent of fire departments provided no emergency medical services, 51 percent provided basic life support (BLS), and 5 percent provided advanced life support (ALS) (NFPA, Canadian Fire Department Profile, 2018).

Working Within the Community

Community Risk Reduction (CRR) is a comprehensive, all-hazard, unifying approach that includes programs, actions, and services used by a community, which prevent or mitigate the loss of life, property, and resources associated with life safety, fire, and other disasters within a community (NFPA, 1035, 2015). CRR is different from the traditional ideas of fire prevention because it is concerned with a comprehensive approach to reducing the overall incidence and impact of emergencies within the community.

The concept of CRR began as a pilot program in Merseyside, England. The quantifiable success of these efforts to significantly reduce fire fatalities defined the program as an effective model for providing risk reduction in communities. CRR and similar programs are gaining momentum in Canada and the United States. Because a single "canned" prevention program will not effectively mitigate the risks for all communities, CRR assists communities by identifying and developing a combination of unique strategic interventions to address individual problems with the most efficient use of resources.

As fire prevention is the responsibility of each province or territory, it has been difficult to nationalize fire prevention initiatives. However, many fire services are adopting approaches similar in nature to the CRR.

A basic overview of CRR is provided here for insight into the methods of risk reduction. CRR is addressed through six basic steps:

1. Identify risks.
2. Prioritize risks.
3. Develop strategies and tactics to mitigate risks.
4. Prepare the CRR plan.
5. Implement the CRR plan.
6. Monitor, evaluate, and modify the plan.

The first step in developing a CRR program is to identify risks. A risk can be defined as any factor, human or otherwise, that could lead to an emergency that would negatively affect the people, places, or resources of that community. A logical place to begin identifying community risks is to look at what the department experiences in terms of the types of incidents, call volumes, locations, and causes. These data are available in many departments through the records management system (RMS). Another good source for data concerning risks in the community is likely available in a geographic information system (GIS). A GIS understands location and is particularly well suited to pinpoint risks. It can analyze preparedness capabilities and build layers of data, which can help members of the department visualize the relationships between variables impacting risk and preparedness. Your GIS might be located in the fire department or in another government agency.

The second step in developing a CRR program is to prioritize risks. When building a model for risk reduction, evaluate each of the identified risks based on factors such as the severity, frequency, and duration of an event. It is also important to consider the likelihood of occurrence and the department's capacity to respond. One particularly important risk factor to consider is people. Often, identifying the relationship between the most vulnerable populations and high call volume areas will help you to identify risk reduction priorities.

The third step in developing a CRR program is to develop strategies and tactics to mitigate risks or reduce the occurrence of preventable incidents. To do this effectively, one must utilize all of the available resources within the community. The Five E's of fire prevention—education, engineering, enforcement, economic incentives, and emergency response—provide the basic strategies and tactics to address this third step. All risks should be examined through the lens of each of the following strategies:

- **Education:** Changing behaviour by teaching people about fire and emergency prevention and response. This is the traditional approach to fire safety.
- **Engineering:** Using technology to make buildings and products safer. Smoke alarms, child-resistant medication caps, and car seats are all examples of using technology to improve safety.
- **Enforcement:** Fire and building codes are used to create and maintain safety in the built environment. Enforcing these codes reduces the risk in the community in both the short and long term.
- **Economic incentives:** Providing financial motivation can encourage beneficial behaviours and choices. Financial motivation can be in the form of reward or punishment. For example, providing tax credits for sprinkler systems or fines for the absence or removal of smoke alarms.
- **Emergency response:** A well-trained and prepared emergency response force is essential. Deployment should be responsive to the unique aspects of the communities served.

These first three steps of developing a CRR program comprise a risk assessment. A risk assessment is the identification and prioritization of potential and likely risks within a particular community.

Steps four through six of developing a CRR program are related to the development of a CRR plan. The fourth step is to prepare the CRR plan. This plan should be developed with input from personnel within the fire department, as well as from community groups and people outside the fire department. Once finished, the plan should be communicated broadly to build understanding and buy-in. The fifth step is the implementation of the plan. The sixth step is monitoring, evaluating, and modifying the plan as needed.

CRR is an exciting and important trend in the fire service. Our goal should be to prevent an incident from occurring rather than trying to mitigate an incident that has occurred.

Educating the Public about the Risks

A fire fighter must be able to identify hazards and understand how to rectify those hazards. Many fire departments offer to conduct fire and life-safety surveys in private dwellings. These surveys identify fire and life-safety hazards and provide the occupants with recommendations on how to make their home safer.

During a fire and life-safety survey, be on the lookout for potential fire hazards and non-fire safety hazards. Poisoning, slip and fall hazards, drowning, burns, and toppling furniture are all common causes of injuries in the home. Simple precautions can prevent some of these emergencies:

- Pools must be properly enclosed.
- Furniture, televisions, and appliances must be properly secured to prevent tip-overs.
- Medicines and chemicals must be stored where children cannot access them.
- Stairs and balconies should have handrails and guards.
- Domestic hot water temperature should not exceed 49°C (120°F).

During the survey, point out hazards, explain the reasons for making recommendations, and answer any questions the occupants may have. Always look for fire protection equipment, such as smoke alarms, carbon monoxide alarms, and fire extinguishers. Make sure there are working smoke alarms on each floor, in the corridor outside each sleeping area, and in each bedroom. Make sure that any fire alarm or suppression equipment in the home is properly installed, maintained, and operational. Take the time to test all smoke alarms and recommend adding additional alarms if some areas are not protected. Some fire departments install one or more smoke alarms at no cost in any dwelling that lacks this important protection. Your goal should be to educate the occupant about common fire hazards.

Stress the importance of keeping ignition sources away from combustible materials and keeping bedroom doors closed to limit fire growth and spread. It is especially important to discuss the risk of children playing with lighters and matches, leaving food cooking on the stove unattended, and storing combustibles such as towels, potholders, and cooking oils too close to the stove. Many fires have resulted from someone inadvertently turning on the wrong burner and igniting something that was left on the stovetop. Every kitchen should be equipped with an approved ABC-rated fire extinguisher.

If the room contains a fireplace, a wood stove, or a portable heater, ensure that no combustible materials are stored nearby. A fireplace should have a screen to keep sparks and hot embers from escaping. If solid fuels are used, the chimney or flue pipe should be professionally inspected at least once each year.

Conduct the survey in a systematic fashion, checking both the inside and the outside of the home. Some fire fighters begin by inspecting the outside of the house, whereas others prefer to start with the interior. Outside the house, make sure that the house number, or address, is clearly visible from the street, exits are not obstructed, and recommend that fireplaces and chimneys be inspected annually. Inside the house, systematically inspect each room for fire hazards, and explain why different situations are considered potential fire hazards. A person who understands why an overloaded electrical circuit is a fire hazard will be more likely to avoid overloading circuits in the future. Stress the importance of using power strips that contain circuit breakers to prevent overloading of extension cords. Help the occupant identify escape routes.

Explain the importance of good housekeeping and the need to clear junk out of garages, basements, and storage areas. Furnaces and water heaters that operate with an open flame are often located in a basement or garage; they can ignite flammable materials that are stored too close to them.

Storage of gasoline and other flammable substances is a major concern because an open flame or pilot light can easily ignite flammable vapours. Gasoline and other flammable liquids should be stored only in approved containers and in outside storage areas or outbuildings. Propane tanks, such as those used in gas grills, should also be stored outside or in outbuildings. Small quantities of flammable and combustible liquids (such as paint, thinners, varnishes, and cleaning fluids) should be stored in closed metal containers away from heat sources. Oily or greasy rags should also be stored in closed metal containers. It is recommended that homeowners maintain fully charged fire extinguishers in basements and garages.

Listen carefully to the occupants, because their questions will enable you to address any special concerns or correct inaccurate information. Talk to the occupants, and take the time to answer their questions fully. Continue to emphasize the importance of smoke alarms, home exit plans, and fire drills. Smoke alarms and carbon monoxide detectors are mandatory in most jurisdictions. Many fire departments provide educational materials that address safety issues, such as the installation and maintenance of smoke alarms, the role of residential fire sprinklers in preventing fire deaths, and the selection and use of portable fire extinguishers.

After you have completed a home fire safety survey or any other type of inspection, you must file your report according to your department's policies. If you identified hazards that require further action or follow-up, discuss them with your company officer or a designated representative from the code enforcement office. For more information on conducting a safety survey, see Chapter 27, *Fire and Life-Safety Initiatives.*

Residential Smoke Alarms

Fire alarm and detection systems range from simple, single-station smoke alarms for private homes to complex fire detection and control systems for high-rise buildings. This section discusses residential smoke alarms.

The most common type of residential fire alarm is the single-station smoke alarm (**FIGURE 1-4**). This life-safety device includes a smoke detection sensor, an automatic control unit, and an audible alarm within a single device. It alerts occupants when a fire occurs. Millions of single-station smoke alarms have been installed in private dwellings and apartments.

Smoke alarms can be battery-powered, hard-wired to a 120-volt electrical system, or both. Current building codes require smoke alarms in all newly constructed dwellings to be hard-wired and powered by the building's electrical system. These smoke alarms also must be interconnected, so that all of the alarms will sound if one alarm detects smoke. Hard-wired smoke alarms also must contain a backup battery in case electrical service is disrupted. Many older residences have single-station battery-operated smoke alarms that are not hard-wired and not connected to other smoke alarms. It is imperative that smoke alarm batteries be replaced on a regular basis. Some communities recommend replacing smoke alarm batteries when Daylight Saving Time begins and ends. Some smoke alarms come with a long-life lithium battery that may last up to 10 years. NFPA 72, *National Fire Alarm and Signaling Code,* recommends that all smoke alarms be replaced after 10 years from the date of manufacture.

NFPA 72 requires new residences to have at least one smoke alarm on each level of the house, one in the corridor or hallway outside each sleeping area, and one in each bedroom. Older residences built before these requirements were established may not have enough smoke alarms to provide early detection and notification. Fire fighters should recommend that the occupants install additional smoke alarms to improve their protection. Reference your department's policies before making recommendations on fire protection equipment.

Ionization and Photoelectric Smoke Alarms.

Two types of fire detection technology may be used in a smoke alarm to detect combustion (**TABLE 1-1**):

- Ionization alarms are activated by the smaller, invisible products of combustion.
- Photoelectric alarms are activated by the larger, visible products of combustion.

Ionization smoke detection works on the principle that burning materials release many different products of combustion, including electrically charged microscopic particles. An ionization alarm senses the presence of these invisible charged particles (ions) (**FIGURE 1-5**).

An ionization smoke alarm contains a very small amount of radioactive material inside its inner chamber. This radioactive material releases charged particles into the chamber, and a small electric current flows between two plates. When smoke particles enter the chamber, they neutralize the charged particles and interrupt the current flow. The alarm senses this interruption and activates.

Photoelectric smoke alarms use a light beam and a photocell to detect larger, visible particles of smoke (**FIGURE 1-6**). When visible particles of smoke enter the inner chamber, they reflect some of the light onto the photocell, thereby activating the alarm.

Ionization smoke alarms react more quickly than photoelectric smoke alarms to fast-burning fires, such

FIGURE 1-4 A single-station smoke alarm.
Courtesy of Kidde Residential and Commercial Division.

TABLE 1-1 Comparison of Ionization and Photoelectric Smoke Alarms*

Ionization Smoke Alarms	Photoelectric Smoke Alarms
Use a small amount of radioactive material to ionize air in a sensing chamber, allowing current to flow between two charged plates. Smoke entering the chamber reduces the current flow and causes the detector to go into alarm.	Detect smoke based on how it affects a projected light beam. The light-scattering type is more common. Smoke entering the chamber scatters the light to the sensor. When sufficient light is reflected into the sensor, it goes into alarm.
Somewhat better response to small particles from flaming fires than photoelectric alarms. Can detect invisible smoke in the form of fire gases.	Considerably faster response to smouldering fires than ionization alarms.
May give a false alarm if improperly placed too close to the kitchen stove or hot, steamy bathrooms.	Less likely to produce nuisance alarms during cooking and in areas subject to changes in humidity and atmospheric pressure.
More common than photoelectric alarms.	Less common than ionization alarms.
Inexpensive.	Somewhat more expensive then ionization alarms.

*When properly installed and maintained, both types of smoke alarms are effective for detecting fires and alerting occupants early enough for them to escape safely. It is recommended that home owners have both types of smoke alarms.

FIGURE 1-5 An ionization smoke alarm.
© Brendan Byrne/age fotostock.

FIGURE 1-6 A photoelectric smoke alarm.
© Serov Aleksei/Shutterstock.

as a fire in a trash can, which may, at first, produce little visible smoke. Ionization alarms are more susceptible to nuisance alarms from common activities, such as light smoke from cooking and steam from a shower.

By comparison, photoelectric smoke detectors are more responsive to slow-burning or smouldering fires, such as a fire caused by a cigarette caught in a couch, which usually produces a large quantity of visible smoke. They are less prone to false alarms from steam than are ionization smoke detectors. Recent studies indicate that both ionization and photoelectric smoke detectors are acceptable life-safety devices. These studies determined that an adequate number of properly spaced detectors of either type detected smoke within acceptable time limits. Combination ionization/photoelectric smoke alarms are also available. These alarms quickly react to both fast-burning and smouldering fires. They are not suitable for use near kitchens or bathrooms because they are prone to the same nuisance alarms as regular ionization smoke detectors.

Combination ionization and photoelectric alarms are available (**FIGURE 1-7**). The NFPA Educational Messages Advisory Committee currently recommends using both ionization and photoelectric alarms and placing them in the most appropriate setting or using combination alarms (photoelectric and ionization) for residential settings.

Be sure to explain the importance of properly installing and maintaining smoke alarms as recommended by NFPA 72, *National Fire Alarm and Signaling Code*. Remind homeowners of the following:

- Test smoke alarms once a month using the test button.
- Change alkaline batteries in smoke alarms every 6 months.
- Replace all smoke alarms every 10 years, or if they fail the monthly test.
- Clean smoke alarms regularly to prevent false alarms.

FIGURE 1-7 Combination ionization and photoelectric smoke alarm.

Courtesy of Kidde Fire Safety.

Working with Other Organizations

Fire departments are a vital part of their communities. To fulfill its mission, each fire department must interact with other organizations in the community. Consider the response to a motor vehicle crash (**FIGURE 1-8**). In an area with a centralized 911 call centre, the fire department, EMS, law enforcement officials, and tow-truck operators are all notified of this event. If the crash affects utility services or highway structures, the utility company or highway department is also notified. If EMS is not part of the fire department, a separate EMS provider is contacted. In addition, fire fighters frequently interact with hospital personnel who assume care for ill or injured victims who are transported to emergency departments.

When multiple agencies work together at an incident, a unified command must be established as part of the **incident command system (ICS)**. A unified command system eliminates multiple command posts, establishes a single set of incident goals and objectives, and ensures mutual communication and cooperation. Although there can be only one IC at each scene, each agency must have input in handling an emergency. The ICS is discussed more thoroughly in Chapter 22, *Establishing and Transferring Command*.

Other organizations may be involved in different types of incidents. For example, utility companies, the Salvation Army, and the Red Cross may be part of the team responding to structure fires. Wildland fires may involve representatives from various government agencies and jurisdictions. Large-scale incidents may call upon several agencies, including but not limited to:

- Public works
- School administrators

FIGURE 1-8 Many agencies work together at the scene of a motor vehicle crash.

© Patrick Kane/AP Photos.

- Medical examiners or coroners
- Funeral directors
- Government officials (mayor, city manager)
- Federal and provincial/territorial investigators
- Public Safety Canada
- Royal Canadian Mounted Police (RCMP)
- Military
- Local, provincial/territorial, and federal emergency management agencies
- Highway departments
- Search and rescue teams
- Fire investigators
- Various provincial/territorial agencies

Communications centres keep a list of contact names and phone numbers so these parties may be notified without delay when their presence at an incident is required.

Fire Department Governance

Governance is the process by which an organization exercises authority and performs the functions assigned to it. The governance of a fire department depends on regulations, standards, policies, and standard operating procedures (SOPs). Each of these concepts is discussed in this section.

Regulations are developed by various government or government-authorized organizations to implement a law that has been passed by a government body. For example, federal occupational health and safety laws are adopted as regulations by some provinces or territories. These regulations may apply to activities within the fire department.

Standards are issued by nongovernmental entities and are generally consensus based. A standard may be voluntary, meaning that the standards can be adopted by an **authority having jurisdiction (AHJ)** as a requirement for that area. The AHJ is the governing body that sets operational policy and procedures for the jurisdiction in which you operate. For example, organizations such as the **National Fire Protection Association (NFPA)** and the International Code Council (ICC) issue voluntary consensus-based codes and standards that set the expected construction, performance, and operation of many aspects of fire service operations. NFPA 1001, *Standard for Fire Fighter Professional Qualifications*, sets the parameters of the Fire Fighter I and Fire Fighter II courses. Other NFPA standards set requirements for the construction and performance of fire apparatus, PPE, and SCBA.

It is important to understand that these standards are designed to help you become a more efficient, competent, and safe fire fighter. You will see many NFPA standards referenced throughout this course. Most fire departments have access to these standards.

Policies are developed to provide definitive guidelines for present and future actions. Fire department policies, for example, outline what is expected in stated conditions. These policies often require personnel to make judgments and to determine the best course of action within the stated policy. Policies governing parts of a fire department's operations may be developed by other government agencies, such as personnel policies that cover all employees of a city or county.

Standard operating procedures (SOPs) provide specific information on the actions that should be taken to accomplish a certain task (**FIGURE 1-9**). SOPs are developed within the fire department, are approved by the chief of the department, and ensure that all members of the department perform a given task in the same manner. They provide a uniform way of dealing with emergency situations, enabling fire fighters from different stations or companies to work together smoothly, even if they have never encountered one another before. These procedures are vital because they enable everyone in the department to function properly and know what is expected for each task. Fire fighters must learn and frequently review departmental SOPs. An example would be your policy and procedure manuals.

In some fire departments, standard operating guidelines (SOGs) are utilized. SOGs are similar to SOPs; however, SOGs may vary due to circumstances surrounding a particular incident. SOGs are not as strict as SOPs, because conditions may dictate that the fire fighter or officer uses his or her personal judgment in completing the procedure. This flexibility allows the responder to deviate from a set procedure yet still be held accountable for that action.

A practical way to organize a department's SOP manual is with removable pages collected in a three-ring binder, which ensures that updates can be made easily. Each department member should have an SOP manual and must update it as needed. This manual should be organized in sections, such as administration, safety, scene operations, apparatus and equipment, station duties, uniforms, and miscellaneous, and use a simple numbering system based on section and policy numbers. Many fire departments maintain their SOPs on a computer network, which simplifies the process of providing all employees with up-to-date SOPs.

Anytown Fire Department

Standard Operating Procedure
Date: 1/1/18
Section: Administration SOP 01-01   Page 1 of 1
Maintaining Station Logbooks

Purpose
This guideline is provided to ensure that information documented in station logbooks is maintained in a consistent manner, station to station, shift to shift, throughout the department. The station commander shall have discretion in formatting the information.

Scope
This guideline shall be followed by all authorized department personnel entering information into the station logbooks. Any deviations from this guideline will be the responsibility of the individual making the deviation.

Policy
The station log is an important component of the fire department's record-keeping system. All entries must be legible, written in black ink, and the writer identified by name and computer identification number. All members should treat the logbook as a legal document and record all of the station's activities as well as other information deemed important by the station commander.

Procedure
The following guidelines should be used when making logbook notations:
1. The day, date, and shift noted on the top line of the page.
2. A list of each person assigned to that shift to include their computer identification numbers, and the apparatus to which they are assigned for the shift. Personnel on leave should be identified with the type of leave (ie, annual, sick, leave without pay, funeral, military, training) noted. A notation should be made by the name of any member temporarily assigned to another station during that shift.
3. Daily safety talks should be recorded prior to the section dedicated to emergency responses. For more information on safety talks, refer to SOP 31.3
4. The section of the log dedicated to recording chronological activity should have two (2) columns on the side of the book: the far left should note time and the second column the incident number. A column to the far right should be used to note the identification number of the person making the entry.
5. Incident numbers for medic unit responses should be noted in blue ink.
6. Incident numbers for fire suppression responses should be noted in red ink.
7. Units responding, the address to which the response is being made, and the type of situation found should be noted behind the message number (ie, E-5 responded to 304 Albemarle Drive/Gas Leak).
8. The term fill-in should be used in the logbook to denote one station standing by another station to cover a company that has responded to an incident.
9. Entries for medic units should include the hospital to which the patient was transported.
10. Station maintenance/repairs and apparatus maintenance should be documented in the section of the logbook dedicated to that purpose. This documentation may also be maintained in a separate logbook dedicated to those types of entries.

   Although the information requirements provided here must be maintained in the station's logbook, the format used for ensuring that documentation is left to the station commander's discretion.

Approved:_____   Date: _____
                                  Fire Chief

FIGURE 1-9 A sample standard operating procedure.
© Jones & Bartlett Learning.

The Organization of the Fire Service

Company Types and Apparatus

A fire department includes many different types of companies, each of which has its own job at the scene of an emergency. Companies may be composed of various combinations of people and equipment; in smaller departments, a company may fill many roles. The most common types of companies and apparatus are described here:

- **Engine or pump company**: An engine or pump company is responsible for securing a water source, deploying handlines, conducting search and rescue operations, and putting water on the fire (**FIGURE 1-10**). Each pumper, also referred to as a fire engine, has a pump, carries hose, and maintains a booster tank of water. Pumpers also carry a limited quantity of ladders and hand tools.

- **Truck or ladder company**: A truck or ladder company specializes in forcible entry, ventilation, roof operations, search and rescue, and deployment

FIGURE 1-10 A pump company secures a water source and extinguishes the fire.
© Jones & Bartlett Learning. Photographed by Glen E. Ellman.

of ground ladders (**FIGURE 1-11**). Each ladder company, also referred to as fire truck, tower ladder company, ladder truck, or aerial ladder, carries several ground ladders, ranging from 2.4 to 15 metres (8 to 50 feet) in length, as well as an extensive quantity of tools. Ladder trucks are also equipped with an aerial device, such as an aerial ladder, tower ladder, or ladder/platform. These aerial devices can be raised and positioned above a roof to provide fire fighters with a stable, safe work zone.

- **Rescue company**: A rescue company usually is responsible for rescuing victims from fires, confined spaces, trenches, and high-angle situations. Rescue companies carry an extensive array of regular and specialized tools.
- **Wildland/brush company**: A wildland/brush company is dispatched to vegetation fires, and they are specifically designed to work in this type of environment. Because they often work in rough terrain, wildland/brush companies

use four-wheel drive vehicles. They carry a tank of water and a pump that enables them to pump water while the ladder is moving. These companies also carry special firefighting equipment such as portable pumps, McLeod rakes, shovels, and chainsaws.

- **Hazardous materials company**: A hazardous materials company responds to and controls scenes involving spilled or leaking hazardous chemicals. These companies have special equipment, PPE, and training to handle most emergencies involving chemicals.
- **Emergency medical services company**: An EMS company may include medical units such as ambulances or first-response vehicles. These companies respond to and assist in transporting medical and trauma victims to medical facilities for further treatment. They often carry medications, defibrillators, and other equipment that can stabilize a critical patient during transport. Pump or ladder companies may be staffed with EMS providers who act as first responders until a transport ambulance arrives (**FIGURE 1-12**).

Additional apparatus may include the following:

- **Quint apparatus**: "Quint" is short for "quintuple," meaning five. The quint apparatus has five functions associated with it: pump, water tank, fire hose storage, aerial, and ground ladders.
- **Initial attack apparatus**: The primary purpose of the initial attack apparatus is to initiate a fire suppression attack on structural, vehicular, or vegetation fires. Also referred to as the quick attack apparatus.
- **Mobile water supply apparatus**: The mobile water supply apparatus is designed to transport water to emergency scenes. It may or may not have a pump. This apparatus is also referred to as a tanker or water tender.

FIGURE 1-11 A ladder company provides aerial support at the fire.
© Jones & Bartlett Learning. Photographed by Glen E. Ellman.

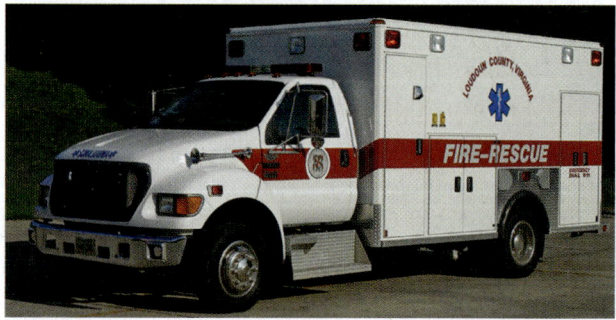

FIGURE 1-12 An EMS company delivers medical services.
© Jones & Bartlett Learning. Photographed by Glen E. Ellman.

Other Views of Fire Service Organization

There are several other ways to look at the organization of a fire department, including in terms of staffing, function, and geography.

Staffing

A fire department must have sufficient trained personnel available to respond to a fire at any hour of the day, every day of the year. Staffing issues affect all fire departments—career departments, composite departments, and volunteer departments. In volunteer departments, it is especially important to ensure that enough responders are available at all times, particularly during the day. In the past, when many people worked in or near the communities where they lived, volunteer response time was not an issue. Today, however, many people have longer commutes and work longer hours, so the number of people available to respond during the day may be limited. Some volunteer departments have been forced to hire full-time fire fighters during daytime hours to ensure sufficient personnel are available to respond to an incident. These organizations are referred to as composite departments.

Function

Fire departments can be organized along functional lines. For example, the training division is responsible for leading and coordinating department-wide training activities; ladder companies have certain functional responsibilities at a fire; and hazardous materials squads have different functional responsibilities that support the overall mission of the fire department.

Geography

Each fire agency is responsible for a specific geographic area, and each station in the department is assigned the primary responsibility for a geographic area within the community. Fire stations are distributed throughout a community in a manner intended to ensure a rapid response time to every location in that community. This design enables the fire department to distribute and use specialized equipment efficiently throughout the community.

Chain of Command

The organizational structure of a fire department consists of a chain of command. Although the precise ranks may vary in different departments, the basic concept remains the same across the fire service. The chain of command creates a structure for managing the department and the fire-ground operations.

Fire fighters usually report to a lieutenant, who is responsible for a single fire company (usually a ladder or rescue company) on a single shift. Lieutenants can provide a number of practical skills and tips to new recruits.

The next level in the chain of command is the captain. Captains are responsible not only for managing a fire company on their shift but also for coordinating the company's activities with other shifts. A captain is generally in charge of a station when his/her shift is on duty.

Captains report directly to chiefs. Several levels of chiefs are designated. District/battalion chiefs are responsible for coordinating the activities of several fire companies in a defined geographic area or district. A district/battalion chief usually assumes command at all large incidents within his or her district.

Above district/battalion chiefs are platoon chiefs. Platoon chiefs are in charge of a single platoon or shift and are responsible for the daily activity of all members on duty.

Assistant or division chiefs are usually in charge of a functional area, such as training, within the department. These officers report directly to the deputy or assistant deputy chief of the department.

The chief of the department has overall responsibility for the administration and operations of the fire department. The chief can delegate responsibilities to other members of the department but is still responsible for ensuring that these activities are carried out properly.

The fire service's chain of command is used to implement department policies. This organizational structure enables a fire department to determine the most efficient and effective way to fulfill its mission and to communicate this information to all members of the department (**FIGURE 1-13**). Adhering to the chain of command ensures that a given task is carried out in a uniform manner. A variety of documents, including the previously discussed regulations, policies, procedures, and SOPs, are used to achieve this goal.

Source of Authority

Governments—whether municipal, provincial/territorial, or federal—are charged with protecting the welfare of the public against common threats. Fire is one such peril, because an uncontrolled fire threatens everyone in the community. Citizens accept certain restrictions on their behaviour and pay taxes to protect themselves and the common good. As a consequence,

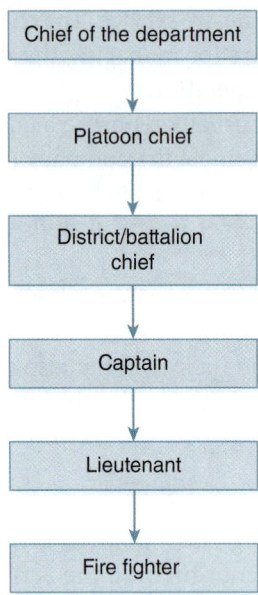

FIGURE 1-13 The chain of command ensures that the department's mission is carried out effectively and efficiently.
© Jones & Bartlett Learning.

people charged with protecting the public may be given certain privileges to enable them to perform effectively. For example, fire departments can legally enter a locked home without permission to extinguish a fire and protect the public.

The fire service draws its authority from the governing entity responsible for protecting the public from fire—whether it is a town, a city, a county, a municipality, or a special fire district. Federal and provincial/territorial governments also grant authority to fire departments. In some provinces/territories, private corporations have contracts to provide fire protection to municipalities or government agencies. The head of the fire department (the fire chief) is accountable to the leaders of the governing body, such as the city council, the municipality, the mayor, or the city manager. Because of the relationship between a fire department and local government, fire fighters should consider themselves to be civil servants, working for the tax-paying citizens who fund the fire department. The ultimate customers for a fire department are the citizens of a community.

Basic Principles of Organization

The fire department uses a paramilitary style of leadership. In other words, fire fighters operate under a rank system, which establishes a chain of command (discussed earlier in this chapter) in which the fire chief serves as head of the department. Most fire departments use four basic management principles:

- Discipline
- Division of labour
- Unity of command
- Span of control

Discipline

Discipline is guiding and directing fire fighters to do what their fire department expects of them. Positive discipline consists of providing guidelines for the right way of doing things. Examples of positive discipline are policies, SOPs, training, and education. Corrective discipline consists of actions taken to discourage inappropriate behaviour or poor performance. Examples of corrective discipline are counselling sessions, formal reprimands, or suspension from duty.

Division of Labour

Division of labour is a way of organizing an incident by breaking down the overall strategy into a series of smaller tasks. Some fire departments are divided into units based on function. For example, pump companies establish water supplies and pump water; ladder companies perform forcible entry, rescue, and ventilation functions. Each of these functions can be divided into multiple assignments, which can then be assigned to individual fire fighters. With division of labour, the specific assignment of a task to an individual makes that person responsible for completing the task and prevents duplication of job assignments.

Unity of Command

Unity of command is the concept that each fire fighter answers to only one supervisor, each supervisor answers to only one boss, and so on (**FIGURE 1-14**). In this way, the chain of command ensures that everyone is answerable to the fire chief and establishes a direct route of responsibility from fire chief to fire fighter.

At a fire ground, all functions are assigned according to incident priorities. A fire fighter with more than one supervising officer during an emergency may be overwhelmed with various assignments, and the incident priorities may not be accomplished in a timely and efficient manner. The concept of unity of command is designed to avoid such conflicts and lead to more effective firefighting.

Span of Control

Span of control is the number of people that one person can supervise effectively. According to the Federal Emergency Management Agency (FEMA), the span of control of personnel is between three and seven (FEMA, 2013). In complex or rapidly changing environments this number may be lower to maintain tighter accountability of personnel.

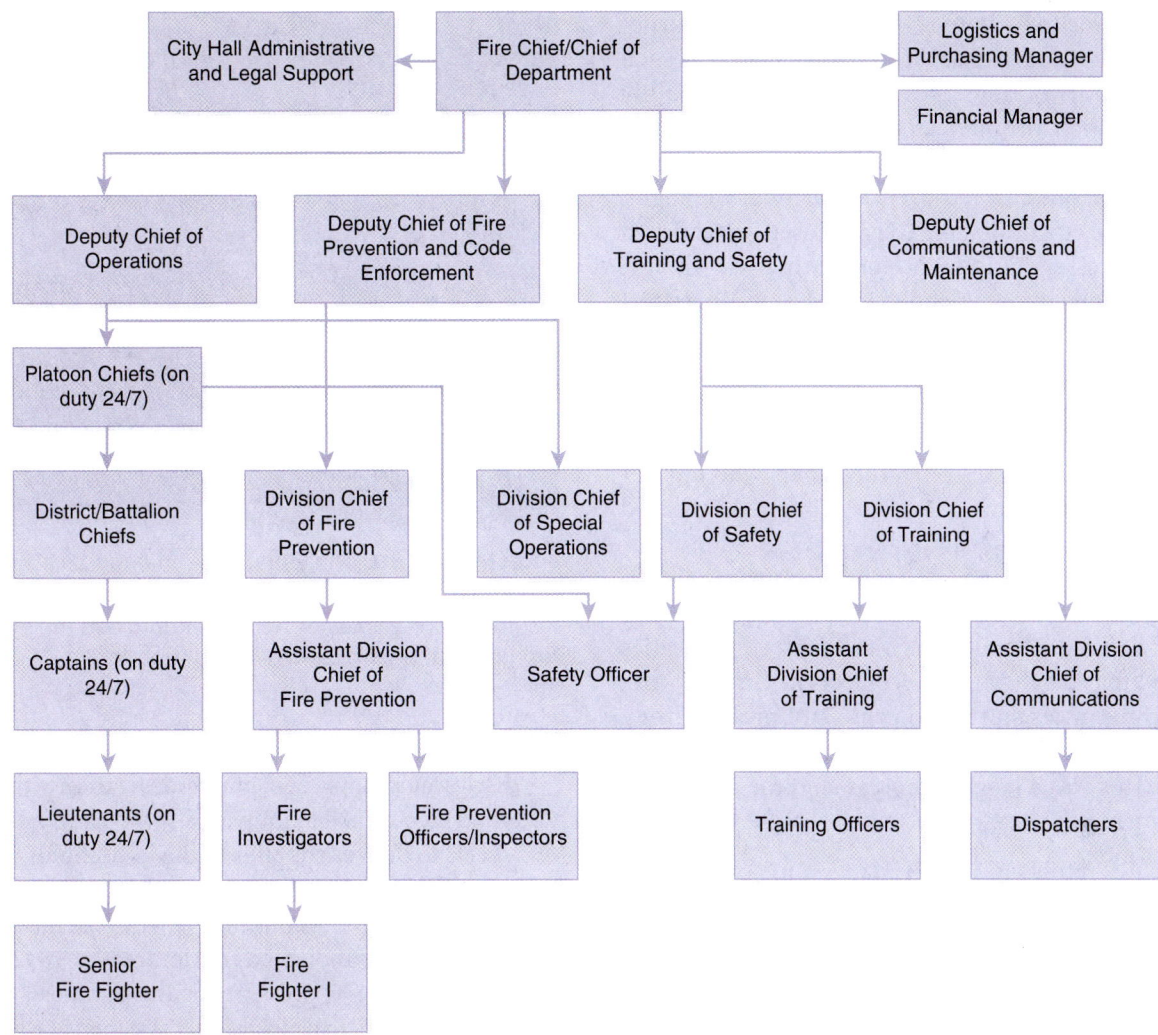

FIGURE 1-14 The organization of a typical fire department.
© Jones & Bartlett Learning.

The History of the Fire Service

Since prehistoric times, controlled fire has been a source of comfort and warmth, but uncontrolled fire has brought death and destruction. Historical accounts from the ancient Roman Empire describe community efforts to suppress uncontrolled fire. In 24 B.C., the Roman emperor Augustus Caesar created what was probably the first fire department. Called the Familia Publica, it was composed of approximately 600 slaves who were stationed around the city and charged with watching for and fighting fires. Of course, because the Familia Publica consisted of slaves, these conscripts had little interest in preserving the homes of their masters and little desire to take risks, so fires continued to be a problem.

In about 60 A.D., under the emperor Nero, the Corps of Vigiles was established as the Roman Empire's fire protectors. This group of 7000 free men was responsible for firefighting, fire prevention, and building inspections. The Corps of Vigiles adopted the formal rank structure of the Roman military, which continues to be used by today's fire departments.

The Canadian Fire Service

When early French and British settlers came to Canada, they established communities in forested areas that were vulnerable to fire and did not consider fire

---

FIRE FIGHTER TIP

The Maltese cross is an international symbol for the fire service. You will see it on most fire apparatus, on fire fighter uniforms, and on fire fighter badges. The Maltese cross is a symbol of honour and protection. There are many versions of the stories of the brave Maltese knights at different periods of time. It is important to remember that this symbol has been passed down through the ages and continues to embody lives of service, bravery, and sacrifice. When you wear this cross on your uniform or on your badge, wear it proudly, and strive to uphold the long-standing and brave traditions of the fire service.

protection for their homes. At that time, most structures were built entirely of combustible materials such as straw and wood, as these materials were readily available. Local ordinances soon required the use of less flammable building materials and mandated that fires be banked (covered over) throughout the night. In 1630, the city of Boston, Massachusetts established the first fire regulations in North America when it banned wood chimneys and thatched roofs. In a short time, these regulations were being applied in Canada as well. There were many fires caused by build-up in chimneys, and because of this, the new fire regulations also required that chimneys be swept out regularly. Fire wardens imposed fines on homeowners who did not obey these regulations; the money collected was used to pay for firefighting equipment.

It was not until the late 1700s that organized firefighting methods began to develop. After major fire losses in Montreal and in the Maritimes, communities decided that they wanted to control the threat of fire. As communities started to grow and prosper, they would organize local councils to address the fire safety issues. This was a lengthy process, and the merchants that suffered the losses determined that they could do a better job.

The earliest firefighting measures required households to have several buckets filled with water close to their front door, and homeowners would respond to fires with these buckets when a fire was identified in the community. They also used axes, ladders, ropes, and hook devices to help with the prevention of fire spread.

FIRE FIGHTER TIP

Early insurance companies marked the homes of their policy holders with a plaque or fire mark that showed the name or logo of the insurance company. The insurance company paid fire companies to respond to those buildings that displayed their fire mark; sadly, others were left to burn.

The Great Fire of London in 1666 prompted the organization and development of fire insurance companies, which would provide their own fire protection with on-call firefighting teams (**FIGURE 1-15**). In the late 1700s, insurance policies became available in Canada, and at this time, the British built fire pumpers to replace some of the primitive tools and equipment from earlier periods. The Royal Navy and the British Army showed the cities in Canada how to use pumping engines, which were already in use on warships. With this technology available, the older communities started to purchase their own fire pumpers. Some of the pumpers were donated by British fire insurance companies to help with the fire prevention effort.

FIGURE 1-15 The Great Fire of London in 1666 prompted the organization and development of fire insurance companies, which would provide their own fire protection with on-call firefighting teams.
© North Wind Picture Archives/Alamy Stock Photo.

Early water supply in communities started with basic wells and then turned to individuals who had containers on wheels to deliver supplies to the households. Soon, the municipal officials began paying these individuals to show up to fires in the community. The first municipal piped-water system opened in Montreal in 1832. This system provided low pressure for the pumpers. Other cities opened municipal piped-water systems shortly after.

In many countries, the military or police were the individuals who participated in early firefighting efforts. In Canada, the local citizens that had the most property to lose would band together and protect each other's property. They organized into what would ultimately be known as fire companies, and they would each have their own hand pumper and a hose reel or a ladder on a cart, which carried salvage materials. This organized group became well known in the various colonies of New England and was looked upon as a positive promotion of volunteer fire companies. Many people who fled the American Revolution came to Canada to settle in its established communities, and they brought this enthusiasm for volunteer fire companies with them. When the gold rush arrived on the west coast, it brought people to Victoria, British Columbia, including many ex-volunteer fire fighters from the United States, who helped to create very strong fire companies.

During the 1800s, the volunteer fire company model would be developed in all Canadian cities and towns. These volunteer fire fighters took their jobs very seriously and were often involved in disputes with municipal officials. These individuals would respond to calls quickly and would take extreme pride in the work they were providing. During this turbulent political time in Canada, these volunteer fire companies were involved in the riots over the Rebellion Losses Bill in Montreal—the then capital city. These fire companies participated in the riots and would ultimately be held responsible for burning the Parliament buildings in Montreal.

> ### FIRE FIGHTER TIP
>
> Alfred Perry was the captain of the volunteer corps of fire fighters and was active in the promotion of fire protection in the 18th century for areas such as Montreal. Perry could be considered the Canadian version of Benjamin Franklin, who was the first leader of the volunteer fire company in the United States. Perry might also be considered the father of the Canadian fire service.

Building Codes

Throughout history, the threat of fire has served as an impetus for communities to establish building codes. Although the first building codes—which were developed in ancient Egypt—focused on preventing building collapse, building codes were quickly recognized as an effective means of preventing, limiting, and containing fires.

Colonial communities had few building codes. The first settlers had a difficult time erecting even primitive shelters, which often were constructed of wood with thatched straw roofs. The fireplaces used for cooking and heating may have had chimneys constructed of smaller logs. The all-wood construction and use of open fires meant that fires were a constant threat to early settlers. As communities developed, they enacted codes restricting the hours during which open fires were permitted and the materials that could be used for roofs and chimneys. After British forces burned Washington in 1814 (during the War of 1812), codes prohibited the building of wooden houses in the area. Some building codes required the construction of a fire-resistive wall, or fire wall, of brick or mortar between two buildings.

Today's building codes not only govern construction materials but also frequently require built-in fire prevention and safety measures. Required fire detection equipment notifies both building occupants and the fire department. Built-in fire suppression and sprinkler systems help contain fires to a small area and prevent small fires from growing into major fires. Fire escapes, stairways, and doors that unlock when the alarms sound and that open outward enable occupants to escape a burning building safely. Without modern building code requirements, high-rise buildings and large shopping centres could not be built safely.

> ### FIRE FIGHTER TIP
>
> Modern building codes help to ensure that fire departments receive prompt notification of fires and limit or extinguish fires that might otherwise overwhelm local firefighting resources.

At first, each community established its own building codes. As larger government jurisdictions were established, however, a more uniform code was adopted across Canada, reflecting a minimum standard. Local communities were permitted to make this code stricter, as needed.

Today, Canadian codes and standards are written by national organizations such as the National Research Council (NRC), Canadian Standards Association (CSA), the International Codes Council (ICC), and the National Fire Protection Association (NFPA). Volunteer committees of citizens and representatives of businesses, insurance companies, and government agencies research and develop proposals, which are in turn debated and reviewed by various groups. The final document, known as a **consensus document**, is then presented to the public. Many Canadian jurisdictions adopt selected NFPA codes and standards as law.

Training and Education

Fire fighter training and education have also come a long way over the years. The first fire fighters simply needed sufficient muscular strength and endurance to pass buckets or operate a hand pumper. As equipment became more complex, however, the importance of formalized training and good judgment increased.

Today's fire fighters operate high-tech, costly equipment, including million-dollar apparatus, radios, thermal imaging devices (**FIGURE 1-16**), and **self-contained breathing apparatus (SCBA)**. These tools, as well as better fire detection devices, have greatly increased the safety and effectiveness of modern-day fire fighters. The most important "machines" on the fire scene, however, remain the intelligent, knowledgeable, well-trained, physically capable fire fighters, who have the ability and determination to attack the fire. A thermal imaging device may be able to find someone trapped in a burning building, but it takes a smart, able-bodied fire fighter to remove the victim safely.

The increasing complexity of both the world and the science of firefighting requires that fire fighters continually sharpen their skills and increase their knowledge of potential hazards. That need for ongoing education is why training courses such as this

FIGURE 1-16 A thermal imaging device is one of many tools available to modern-day fire fighters.
© Jones & Bartlett Learning. Courtesy of MIEMSS.

one are just as important as good physical fitness to today's fire fighters.

Fire Equipment

Today's equipment and apparatus evolved over a number of years, as new inventions were adapted to the needs of the fire service. Colonial-era fire fighters, for example, had only buckets, ladders, and **fire hooks** (tools used to pull down elements in burning structures) at their disposal. Homeowners were required to keep buckets filled with sand or water and to bring them to the scene of the fire. Some towns also required that ladders be available so that fire fighters could access the roof to extinguish small fires. If all else failed, the fire hook was used to pull down burning elements inside buildings and prevent the fire from spreading to nearby structures. The "hook-and-ladder truck" evolved from this early equipment. The town of York (present day Toronto) created a bucket "regulation" in the early 19th century requiring homes to have at least two 8-L (2-gal) buckets and two ladders.

Buckets gave way to hand-powered pumpers in 1720, when Richard Newsham developed the first such pumper

in London, England. As many as 16 strong men were used to power the pump, making it possible to propel a steady stream of water from a safe distance. In 1829, more powerful steam-powered pumpers replaced the hand-powered pumpers. Unfortunately, many volunteer fire fighters felt threatened by these steam pumpers and fought against their use. Steam pumpers were heavy machines that were pulled to the fire by a trained team of horses. They required constant attention, which limited their use to larger cities that could bear the costs of maintaining the horses and the steamers.

The advent of the internal combustion engine in the early 1900s greatly changed the fire service and enabled even small towns to have machine-powered pumpers. Today, both staffed and unstaffed firehouses keep pump companies ready to respond at any hour of the day or night. Although they require regular maintenance, current equipment does not require the constant attention that horses or steam pumpers did. Modern fire apparatus carry water, a pumping mechanism, hose, equipment, and personnel. In this sense, a single apparatus has replaced several single-function vehicles from the past.

The progress in fire protection equipment extends beyond trucks. Without an adequate water supply, modern-day apparatus would be helpless. The advent of municipal water systems provided large quantities of water to extinguish major fires.

Romans developed the first municipal water systems, just as they had developed the first fire companies. It was not until the 1800s, however, that water distribution systems were applied to fire suppression efforts. Frederick Graff Sr., a fire fighter in New York City, developed the first fire hydrants in 1817. He realized that using a valve to control access to the water in the pipes would enable fire fighters to tap into the system whenever a fire occurred. These valves, or **fireplugs** (hydrants), were used with both above-ground and below-ground piping systems. In Canada, the first municipal piped-water system was opened in Montreal in 1832. In 1861, Toronto and Montreal installed more powerful water main systems that were equipped with hydrants.

Because small fires are more easily controlled, the sooner a fire department is notified, the more likely it will be able to extinguish the fire and minimize losses. The introduction of public call boxes in Washington, D.C., in 1860 represented a major advance in this direction. The call boxes, which were placed around the city, enabled citizens to send a coded telegraph signal to the fire department, which received the message as a series of bells. The fire department could determine the location of the fire alarm box being used by the number of bells in the signal. When fire fighters arrived at the alarm box, the caller could then direct them to the exact location of the fire. Similar units are still used in many areas but are being replaced by more immediate and effective communications systems.

Communications

Good communication is vital for effective firefighting. When a fire is discovered, fire fighters must be summoned and citizens alerted to the danger. During the firefight, officers must be able to communicate with fire fighters or summon additional resources. Not surprisingly, improvements in communication systems are tied to improvements in the fire service.

During the colonial period, fire wardens or night watchmen patrolled neighbourhoods and sounded the alarm if a fire was discovered. Some towns, including Charleston, South Carolina, built a series of fire towers from which wardens could watch for fires. In many towns, ringing the community fire bell or church bells alerted citizens to a fire.

In the late 1800s, telegraph fire alarm systems were installed in large cities. These systems enabled more rapid reporting of fires and made it possible for officers to request additional resources or to let dispatchers know that the fire was extinguished. In small towns, telegraph fire alarm systems were gradually replaced by community sirens mounted on poles or tall rooftops to signal a fire. Today, most fire departments rely on pagers or two-way radios to summon part-time or volunteer fire fighters to emergencies.

Many of the early communications systems have been replaced by hard-wired and cellular telephones that enable citizens to report an emergency from almost anywhere. Use of telephones has greatly reduced the time and difficulty of reporting a fire. The introduction of computer-aided dispatch facilities has likewise improved response times, because the closest available fire units can quickly be sent to the site of the emergency.

Communications during the firefight also have improved over the years, from simply shouting loud enough to be heard over the chaos to using two-way radios. Before electronic amplification became available, the chief officer shouted commands through his trumpet. The **chief's trumpet**, or bugle, eventually became a symbol of authority (**FIGURE 1-17**). Although chief officers no longer use trumpets for communicating, the use of multiple trumpets to symbolize the rank of chief signifies this person's need to communicate as well as to lead (**FIGURE 1-18**). Today's two-way radios enable fire units and individual fire fighters to remain in contact with one another at all times. In addition, voice amplification systems, SCBA alarms, and personal alert safety system (PASS) devices communicate potential safety issues to the fire fighter and to the crew.

FIGURE 1-17 The chief's trumpet was once used to amplify the commander's voice. Today it serves as a symbol of authority in the fire service.

© Jones & Bartlett Learning.

FIGURE 1-18 The historic symbol of the chief officer's trumpet is still used on a chief's badge. This series of crossed trumpets is one of the cherished traditions of the fire service.
© Jones & Bartlett Learning. Photographed by Glen E. Ellman.

Paying for Fire Service

As previously noted, volunteer fire departments were common in colonial times and are still used today in many areas. The first fire wardens were employed by communities and paid from community funds. The introduction of hand-powered pumpers hastened the development of a permanent fire service. The question of who would pay for the equipment and the fire fighters, however, was not settled for many years.

Fire insurance companies were established in England soon after the Great Fire of London in 1666 to help victims cope with the financial loss from fires. Such companies collected fees (premiums) from homeowners and businesses, in return pledging to repay the owner for any losses resulting from fire. Fire insurance policies first became available in Canada in the late 1700s. "An ounce of prevention is worth a pound of cure"—an apt motto for fire safety.

Because the insurance companies could save money if a fire was put out before much damage was done, they agreed to pay fire companies for trying to extinguish fires. Early insurance companies marked the homes of their policy holders with a plaque or **fire mark** displaying the name or logo of the insurance company (**FIGURE 1-19**). Most fire companies were loosely governed and organized, and more than one company might show up to fight a fire. If two fire companies arrived at a fire, however, a dispute might arise over which company would collect the money.

FIGURE 1-19 Fire marks were originally symbols affixed to the front of a building designating the insurance company responsible for covering that fire.
© Jones & Bartlett Learning. Photographed by Glen E. Ellman.

Consequently, municipalities began assuming the role of providing fire protection.

Today almost all the fire protection in Canada is funded directly or indirectly through tax dollars. Some jurisdictions fund the fire department as a public service; others contract with an independent fire department. Volunteer fire departments frequently conduct fund-raising activities. Ultimately, however, all funds for operating fire departments come from the citizens of the community.

Fire Service in Canada Today

Today, the fire service in North America is the product of evolution occurring over the past 400 years. As a novice fire fighter, it is helpful for you to learn from the past as well as to study the modern-day Canadian fire service.

According to the NFPA, there were 152,650 local fire fighters in Canada from 2014 to 2016. Of this number, approximately 17 percent are career fire fighters, and 83 percent are volunteers. Approximately 3672 fire departments protected Canada from 2014 to 2016 (NFPA, Canadian Fire Department Profile, 2014-2016). These departments may be composed of all career fire fighters, all volunteer fire fighters, or a mix of career and volunteer fire fighters, depending on the community's needs and available resources (**FIGURE 1-20**). According to the Canadian Fire Department Profile, most career fire fighters worked in communities that protected 50,000 or more people and most volunteer fire fighters worked in departments that protected fewer than 50,000 people.

- All career—All of the members of the department are paid, full-time fire fighters.
- Composite departments:
  - Composite—These departments include both paid, full-time fire fighters and either on-call fire fighters or volunteers.
  - Mostly volunteer—More than half of the members in these departments are volunteers, although they do include paid, full-time fire fighters as well.
- All volunteer—All of the members of the fire department are volunteer fire fighters.

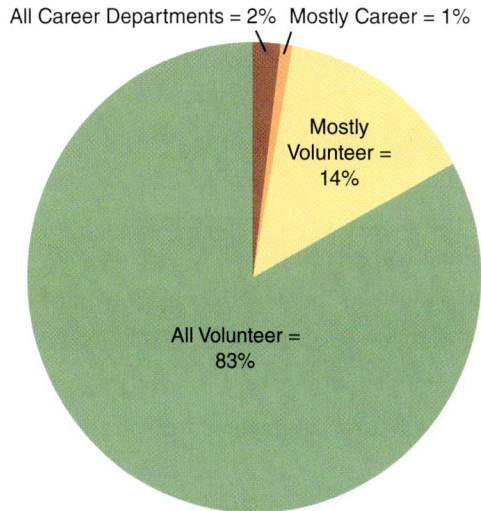

Total Fire Departments = 3,672

All Career Departments = 2%   Mostly Career = 1%

Mostly Volunteer = 14%

All Volunteer = 83%

FIGURE 1-20 The majority of fire departments consist of fire fighters who are volunteers.
© Jones & Bartlett Learning.

After-Action REVIEW

IN SUMMARY

- Throughout your training and career, keep in mind the five fire fighter guidelines:
  - *Be safe.*
  - *Follow orders.*
  - *Work as a team.*
  - *Think!*
  - *Follow the Golden Rule.*
- It is imperative that you understand the importance of the mission statement of your fire department.
- The fire service culture is based on many characteristics, including courage, honour, duty, respect, integrity, character, and moral courage.
- The training and performance qualifications for fire fighters are specified in NFPA 1001, *Standard for Fire Fighter Professional Qualifications*. Age, education, physical fitness, medical, and interviewing requirements are established locally.
- Many departments require firefighting applicants to complete an annual physical ability test. The Candidate Physical Ability Test (CPAT) is the most widely used. During this test, candidates have to complete eight scenarios within a set time limit.
- A Fire Fighter I works in a team under direct supervision to suppress fires.
- A Fire Fighter II works in a team under general supervision. A Fire Fighter II may assume command, transfer command, and coordinate command within the incident command system.
- Throughout your career, you may assume several roles in the fire department. Each role requires additional training:
  - The fire apparatus driver/operator is responsible for getting the fire apparatus to the scene safely, as well as setting up and running the pump or operating the aerial device.
  - The company officer leads the company both on the scene and at the station.
  - The safety officer watches the overall operation for unsafe practices.

- The training officer is responsible for updating the training of current and new fire fighters.
- The incident commander (IC) is responsible for the management of all incident operations.
- The fire marshal delivers, manages, and/or administers fire protection and life-safety–related codes and standards, investigations, education, and/or prevention services.
- The fire inspector inspects businesses and enforces public safety laws and fire codes.
- The fire investigator responds to fire scenes to help investigate the cause of a fire.
- The fire and life-safety educator (FLSE)/fire prevention officer educates the public about fire safety and injury prevention.
- The 911 dispatcher/telecommunicator takes calls from the public and dispatches appropriate units.
- Emergency vehicle technicians (EVTs) repair and service fire and EMS vehicles.
- Fire police officers control traffic and secure the scene from the public.
- Information management professionals are fire fighters or civilians who take care of a fire department's computer network system.
- The public information officer (PIO) serves as a liaison between the incident commander and the news media.
- The fire protection engineer reviews plans and works with building owners to ensure that their fire suppression and detection systems will meet the applicable codes and function as needed.

- Many emergencies require specialized skills. Most large fire departments have teams of specialized fire fighters who can respond to specific emergencies.

- Community Risk Reduction (CRR) is a comprehensive unifying approach to prevent or mitigate the loss of life, property, and resources associated with life safety, fire, and other disasters within a community. The six steps of CRR are:
  1. Identify risks.
  2. Prioritize risks.
  3. Develop strategies and tactics to mitigate risks.
  4. Prepare the CRR plan.
  5. Implement the CRR plan.
  6. Monitor, evaluate, and modify the plan.

- The Five E's of fire prevention—Education, Engineering, Enforcement, Economic incentives, and Emergency response—provide the basic strategies and tactics to mitigate risks.

- Basic types of residential smoke alarms include ionization smoke alarms, photoelectric smoke alarms, and a combination of ionization and photoelectric smoke alarms.

- Common nonfire safety hazards found in the home include poisoning, slip and fall hazards, drowning, burns, and toppling furniture.

- When multiple agencies, such as police, fire, and EMS, work together at an incident, a unified command must be established as part of the incident command system. A unified command establishes a single set of incident goals under a single leader and ensures mutual communication and cooperation.

- Governance is the process by which an organization exercises authority and performs the functions assigned to it. The governance of a fire department depends on regulations, standards, policies, and standard operating procedures (SOPs).

- A fire department includes many different types of companies and equipment to perform specific tasks at the scene of an emergency:
  - Engine or pump company
  - Truck or ladder company
  - Rescue company

- Wildland/brush company
- Hazardous materials company
- Emergency medical services (EMS) company

- The chain of command may differ from fire department to fire department, but the basic concept remains the same across the fire service. The chain of command, from lowest rank to highest, is:
  - Fire fighter
  - Lieutenant
  - Captain
  - District/battalion chief
  - Platoon chief
  - Chief of the department

- Four basic management principles apply to the fire service:
  - Discipline comprises the set of guidelines that a fire department establishes for fire fighters. Regulations, policies, and procedures are all forms of discipline.
  - Division of labour is a way of organizing an incident by breaking down an overall strategy into a series of smaller tasks.
  - Unity of command is the concept that each fire fighter answers to only one supervisor.
  - Span of control is the number of people that one person can supervise effectively.

- Highly destructive fires spurred communities to enact strict building and fire codes in an effort to prevent large loss of life and property. Today's building codes not only govern construction materials but also frequently require built-in fire prevention and safety measures such as sprinkler systems.

- The fire service in North America is the product of an evolution over the past 400 years. As a beginning fire fighter, it is helpful for you to learn from the past and to study the fire service today.

KEY TERMS

Access Navigate for flashcards to test your key term knowledge.

911 dispatcher/telecommunicator From the communications centre, the dispatcher takes the calls from the public, sends appropriate units to the scene, assists callers with treatment instructions until the EMS unit arrives, and assists the incident commander with needed resources.

Advanced Emergency Medical Technician (AEMT) A member of EMS who can perform limited procedures that usually fall between those provided by an EMT and those provided by a Paramedic, including IV therapy, interpretation of cardiac rhythms, defibrillation, and airway intubation.

Aircraft/crash rescue fire fighter (ARFF) An individual who takes firefighting actions to prevent, control, or extinguish fire involved or adjacent to an aircraft for

the purpose of maintaining maximum escape routes for occupants using normal and emergency routes for egress. (NFPA 414)

Assistant or division chief A midlevel chief who often has a functional area of responsibility, such as training, and who answers directly to the fire chief.

Authority having jurisdiction (AHJ) An organization, office, or individual responsible for enforcing the requirements of a code or standard or for approving equipment, materials, an installation, or a procedure. (NFPA 1)

Banked Covering a fire to ensure low burning.

Captain The second rank of promotion in the fire service, between the lieutenant and the district/battalion chief. Captains are responsible for managing a fire company and for coordinating the activities of that company among the other shifts.

Chain of command A rank structure, spanning the fire fighter through the fire chief, for managing a fire department and fire-ground operations.

Chief of the department The top position in the fire department. The fire chief has ultimate responsibility for the fire department and usually answers directly to the mayor or other designated public official.

Chief's trumpet An obsolete amplification device that enabled a chief officer to give orders to fire fighters during an emergency. Also called a bugle, it was a precursor to a bullhorn and portable radios.

Community Risk Reduction (CRR) Programs, actions, and services used by a community, which prevent or mitigate the loss of life, property, and resources associated with life safety, fire, and other disasters within a community. (NFPA 1035)

Company officer The individual responsible for command of a company, a designation not specific to any particular fire department rank (can be a fire fighter, lieutenant, captain, or chief officer, if responsible for command of a single company). (NFPA 1026)

Consensus document A code document jointly developed by people representing various organizations and interests. NFPA codes and standards are consensus documents.

Discipline The guidelines that a department sets for fire fighters to work within.

District/battalion chief Usually the first level of fire chief. These chiefs are often in charge of running calls and supervising multiple stations or districts within a city. A district/battalion chief is usually the officer in charge of a single-alarm working fire.

Division of labour Breaking down an incident or task into a series of smaller, more manageable tasks and assigning personnel to complete those tasks.

Doff To take off an item of clothing or equipment.

Don To put on an item of clothing or equipment.

Emergency Medical Responder (EMR) The first trained professional, such as a police officer, fire fighter, lifeguard, or other rescuer, to arrive at the scene of an emergency to provide initial medical assistance. EMRs have basic training and often perform in an assistant role within the ambulance.

Emergency medical services (EMS) company A company that may be made up of medical units and first-response vehicles. Members of this company respond to and assist in the transport of medical and trauma victims to medical facilities. They often have medications, defibrillators, and Paramedics who can stabilize a critical patient.

Emergency medical services (EMS) personnel Personnel who are responsible for administering prehospital care to people who are sick and injured. Prehospital calls make up the majority of responses in most fire departments, and in some organizations, EMS personnel are crossed-trained as fire fighters.

Emergency Medical Technician (EMT) EMS personnel who account for most of the EMS providers in the United States. An EMT has training in basic emergency care skills, including oxygen therapy, bleeding control, CPR, automated external defibrillation, use of basic airway devices, and assisting patients with certain medications.

Emergency vehicle technician (EVT) The individual who performs maintenance, diagnosis, and repair on emergency vehicles.

Engine or pump company A group of fire fighters who work as a unit and are equipped with one or more pumping engines that have rated capacities of 2839 L/min (750 gpm) or more. (NFPA 1410).

Fire and life-safety educator (FLSE/fire prevention officer (FPO)) The individual who has demonstrated the ability to coordinate, create, administer, prepare, deliver, and evaluate educational programs and information.

Fire apparatus driver/operator A fire department member who is authorized by the authority having jurisdiction to drive, operate, or both drive and operate fire department vehicles. (NFPA 1451)

Fire Fighter I A person, at the first level of progression as defined in Chapter 4 of NFPA 1001, who has demonstrated the knowledge and skills to function as an integral member of a firefighting team under direct supervision in hazardous conditions. (NFPA 1001)

Fire Fighter II A person, at the second level of progression as defined in Chapter 5 of NFPA 1001, who has demonstrated the skills and depth of knowledge to function under general supervision. (NFPA 1001)

Fire hooks Tools used to pull down burning elements in structures; also called pike poles.

Fire inspector An individual who conducts fire code inspections and applies codes and standards. (NFPA 1037)

Fire investigator An individual who has demonstrated the skills and knowledge necessary to conduct, coordinate, and complete an investigation. (NFPA 1037)

Fire mark Historically, an identifying symbol on a building informing fire fighters that the building was insured by a company that would pay them for extinguishing the fire.

Fire marshal A person designated to provide delivery, management, and/or administration of fire protection and life-safety–related codes and standards, investigations, education, and/or prevention services for local, county/provincial, federal, tribal, or private sector jurisdictions as adopted or determined by that entity. (NFPA 1037)

Fireplug Historically speaking, a plug installed to control water accessed from wooden pipes. Today, this is a slang term used to describe a fire hydrant.

Fire police officer An individual officially deployed who provides scene security, directs traffic, and conducts other duties as determined by the authority having jurisdiction. (NFPA 1091)

Fire protection engineer A member of the fire department who is responsible for reviewing plans and working with building owners to ensure that the design of and systems for fire detection and suppression will meet applicable codes and function as needed. They also may be employed by an architectural firm to assure that buildings are constructed in a fire-safe manner.

Fire wardens Individuals who were charged with enforcing fire regulations in the colonial period.

Governance The process by which an organization exercises authority and performs the functions assigned to it.

Hazardous materials company A fire company that responds to and controls scenes where hazardous materials have spilled or leaked. Responders wear special suits and are trained to deal with most chemicals.

Hazardous materials technician A person who responds to hazardous materials/weapons of mass destruction incidents using a risk-based response process by which he or she analyzes the problem at hand, selects applicable decontamination procedures, and controls a release while using specialized protective clothing and control equipment. (NFPA 472)

Incident commander (IC) The individual responsible for all incident activities, including the development of strategies and tactics and the ordering and release of resources. (NFPA 1500)

Incident command system (ICS) The combination of facilities, equipment, personnel, procedures, and communications operating within a common organizational structure that has responsibility for the management of assigned resources to effectively accomplish stated objectives pertaining to an incident or training exercise. (NFPA 1670)

Incident safety officer A member of the command staff responsible for monitoring and assessing safety hazards and unsafe situations and for developing measures for ensuring personnel safety. (NFPA 1500)

Information management Fire fighters or civilians who take care of the computer and networking systems that a fire department needs to operate.

Initial attack apparatus Fire apparatus with a fire pump of at least 946 L/min (250 gpm) capacity, water tank, and hose body, whose primary purpose is to initiate a fire suppression attack on structural, vehicular, or vegetation fires and to support associated fire department operations. May also be referred to as quick attack apparatus.

Ionization smoke detection The principle of using a small amount of radioactive material to ionize the air between two differentially charged electrodes to sense the presence of smoke particles. Smoke particles entering the ionization volume decrease the conductance of the air by reducing ion mobility. The reduced conductance signal is processed and used to convey an alarm condition when it meets preset criteria. (NFPA 72)

Lieutenant A company officer who is usually responsible for a single fire company on a single shift; the first in line among company officers.

Mobile water supply apparatus A vehicle designed primarily for transporting (pickup, transporting, and delivering) water to fire emergency scenes to be applied by pumping equipment or by other vehicles.

National Fire Protection Association (NFPA) The association that develops and maintains nationally recognized minimum consensus standards on many areas of fire safety and specific standards on hazardous materials.

Paramedic EMS personnel with the highest level of training in EMS, including cardiac monitoring, administering drugs, inserting advanced airways, manual defibrillation, and other advanced assessment and treatment skills.

Photoelectric smoke alarm A detector that uses a light beam and a photocell to detect larger visible particles of smoke. When visible particles of smoke enter the inner chamber they reflect some of the light onto the photocell, thereby activating the alarm.

Platoon chief Manages the on duty shift. In most career departments there are four platoons working a 24-hour shift. The platoon chief answers to either the assistant deputy chief or the deputy chief of operations.

Policies Formal statements that provide guidelines for present and future actions. Policies often require personnel to make judgments.

Public information officer (PIO) An individual who has demonstrated the ability to conduct media interviews and prepare news releases and media advisories. (NFPA 1035)

Quint apparatus Fire apparatus with a permanently mounted fire pump, a water tank, a hose storage area, an aerial ladder or elevating platform with a permanently mounted waterway, and a complement of ground ladders.

Regulations Mandates issued and enforced by governmental bodies such as the Canadian Standards Association.

Rescue company A group of fire fighters who work as a unit and are equipped with one or more rescue vehicles. (NFPA 1410)

SCUBA dive rescue technician A responder who is trained to handle water rescues and emergencies, including recovery and search procedures, in both water and under-ice situations. (SCUBA stands for self-contained underwater breathing apparatus.)

Self-contained breathing apparatus (SCBA) An atmosphere-supplying respirator that supplies a respirable air atmosphere to the user from a breathing air source that is independent of the ambient environment and designed to be carried by the user. (NFPA 1981)

Single-station smoke alarm A detector comprising an assembly that incorporates a sensor, control components, and an alarm notification appliance in one unit operated from a power source either located in the unit or obtained at the point of installation. (NFPA 72)

Span of control The maximum number of personnel or activities that can be effectively controlled by one individual (usually three to seven). (NFPA 1006)

Standard operating procedure (SOP) A written organizational directive that establishes or prescribes specific operational or administrative methods to be followed routinely for the performance of designated operations or actions. (NFPA 1521)

Standards Documents, the main text of which contain only requirements and which are in a form generally suitable for mandatory reference by another standard or code or for adoption into law. Nonmandatory provisions shall be located in an appendix or annex, footnote, or fine-print note and are not to be considered a part of the requirements of a standard. (NFPA 1)

Technical rescuer A person who is trained to perform or direct a technical rescue. (NFPA 1006)

Training officer The person designated by the fire chief with authority for overall management and control of the organization's training program. (NFPA 1401)

Truck or ladder company A group of fire fighters who work as a unit and are equipped with one or more pieces of aerial fire apparatus. (NFPA 1410)

Unity of command The concept by which each person within an organization reports to one, and only one, designated person. (NFPA 1026)

Wildland/brush company A fire company that is dispatched to vegetation fires where larger pumpers cannot gain access. Wildland/brush companies have four-wheel drive vehicles and special firefighting equipment.

REFERENCES

Federal Emergency Management Agency. ICS Management: Span of Control. https://emilms.fema.gov/IS200b/ICS0102370.htm. Accessed October 1, 2018.

Giesler, M. 2016. *Fire and Life Safety Educator: Principles and Practice*, 2nd Edition. Jones & Bartlett Learning.

International Association of Fire Fighters. "Candidate Physical Ability Test Program Summary." http://www.iaff.org/hs/CPAT/cpat_index.html. Accessed October 1, 2018.

Merriam-Webster.com. 2017. "Culture." https://www.merriam-webster.com/dictionary/culture?utm_campaign=sd&utm_medium=serp&utm_source=jsonld. Accessed October 1, 2018.

National Fire Protection Association. 2018. "Canadian Fire Department Profile 2014-2016." https://www.nfpa.org/News-and-Research/Fire-statistics-and-reports/Fire-statistics/The-fire-service/Administration/Canada-Fire-Department-Profile. Accessed September 19, 2018.

National Fire Protection Association. NFPA 1035, *Standard on Fire and Life Safety Educator, Public Information Officer, Youth Firesetter Intervention Specialist and Youth Firesetter Program Manager Professional Qualifications*, 2015. www.nfpa.org. Accessed October 1, 2018.

National Fire Protection Association. NFPA 1500, *Standard on Fire Department Occupational Safety, Health, and Wellness Program*, 2018. www.nfpa.org. Accessed October 1, 2018.

National Fire Protection Association. NFPA 1582, *Standard on Comprehensive Occupational Medical Program for Fire Departments*, 2018. www.nfpa.org. Accessed October 1, 2018.

On Scene

You have just completed your first week of your Fire Fighter I and II course. Your class consists of fire fighters from different sizes of fire departments. Your instructor has announced that there will be a quiz covering the first chapter of your textbook on Friday. You and several of your friends have decided to join a small informal study group to review the material before the quiz. Your discussion generates a number of questions.

1. Which of the following is *not* typically a type of incident to which fire departments respond?

A. Motor vehicles collisions

B. Hazardous materials spills

C. Vicious dogs on the loose

D. Emergency medical requests

2. Which of the following is *not* a common staffing method of fire departments?

A. Safety officer

B. Volunteer

C. Career

D. Composite

3. Which role would most likely fall under the fire prevention division?

A. Fire police officer

B. Fire marshal

C. Incident commander

D. Company officer

4. If there are an equal number of fire fighters assigned to each station at your department, what is generally considered to be the number of people one person can directly supervise effectively while maintaining normal span of control?

A. 1

B. 2–3

C. 3–7

D. 5–15

(continued)

On Scene Continued

5. If a fire fighter is responsible for reporting to both a captain and a lieutenant, which organizational principle does this violate?

A. Span of control

B. Unity of command

C. Division of labour

D. Chain of command

6. Which position is between the rank of fire fighter and district/battalion chief?

A. Chief of the department

B. Lieutenant

C. Fire fighter

D. Platoon chief

7. The incident commander is responsible for which of the following?

A. Leading the company both on the scene and at the station

B. Watching the overall operation for unsafe practices

C. Management at an emergency incident

D. Updating the training of current fire fighters and new fire fighters

8. Which of the following is the highest rank in the fire department chain of command?

A. Captain

B. District/battalion chief

C. Lieutenant

D. Chief

Access Navigate to find answers to this On Scene, along with other resources such as an audiobook.

Fire Fighter Health and Safety

KNOWLEDGE OBJECTIVES

After studying this chapter, you will be able to:

- List the major causes of death and injury in fire fighters. (pp. 37–38)
- Describe the 16 fire fighter life safety initiatives. (**NFPA 1001: 4.1.1**, p. 39)
- Describe some of the organizations that set the regulations, standards, and procedures intended to ensure a safe working environment for the fire service. (**NFPA 1001: 4.1.1**, pp. 39–40)
- Describe the connection between physical fitness and fire fighter safety. (**NFPA 1001: 4.1.1**, p. 41)
- Describe the components of a well-rounded physical fitness program. (**NFPA 1001: 4.1.1**, pp. 41–43)
- Explain the practices fire fighters should take to promote optimal physical and mental health and well-being. (**NFPA 1001: 4.1.1**, pp. 41–46)
- Explain the role of a critical incident stress debriefing in preserving the mental well-being of fire fighters. (**NFPA 1001: 4.1.1**, pp. 44–45)
- List signs and symptoms of behavioural and emotional distress. (**NFPA 1001: 4.1.1**, p. 44)
- Describe the purpose of an employee assistance program. (**NFPA 1001: 4.1.1**, p. 46)

- Explain how fire fighter candidates, instructors, and veteran fire fighters work together to ensure safety during training. (p. 46)
- Describe how to safely mount an apparatus. (**NFPA 1001: 4.3.2**, pp. 47–48)
- Describe how to safely ride a fire apparatus to an emergency scene. (**NFPA 1001: 4.3.2**, pp. 47–49)
- Describe how to safely dismount an apparatus. (**NFPA 1001: 4.3.2**, pp. 48–49)
- Describe hazards and safety measures associated with riding apparatus. (**NFPA 1001: 4.3.2**, pp. 47–49)
- List the NFPA standards that require fire fighters to wear safety belts while riding in a fire apparatus (**NFPA 1001: 4.3.2**, p. 48).
- List the prohibited practices when riding in a fire apparatus to an emergency scene. (**NFPA 1001: 4.3.2**, pp. 48–49)
- Describe how to manage traffic safely at an emergency scene. (**NFPA 1001: 4.3.3**, pp. 49–50)
- List the four general principles that govern emergency vehicle operation. (p. 51)
- Explain how the teamwork concept is applied during every stage of an emergency incident to ensure the safety of all fire fighters. (p. 55)
- Describe how the personnel accountability system is implemented during an emergency incident. (pp. 55–56)

- Explain considerations for hazard and scene control. (**NFPA 1001: 4.3.3**, p. 56)
- List the common hazards at an emergency incident. (**NFPA 1001: 4.3.3**, pp. 57–59)
- Explain how to shut off a structure's electrical service. (**NFPA 1001: 4.3.3, 4.3.18**, pp. 57–58)
- Describe the measures fire fighters follow to ensure electrical safety at an emergency incident. (**NFPA 1001: 4.3.3**, pp. 57–58)
- Explain how to shut off a structure's gas service. (**NFPA 1001: 4.3.3, 4.3.18**, pp. 58–59)
- Explain how to shut off a structure's water service. (**NFPA 1001: 4.3.3, 4.3.18**, p. 59)
- Describe how to lift and move objects safely. (p. 59)
- Explain how rehabilitation is used to protect the safety of fire fighters during an emergency incident. (p. 59)
- Describe how to ensure safety at the fire station. (p. 60)
- Describe how to ensure safety outside of the workplace. (p. 60)

SKILLS OBJECTIVES

After studying this chapter, you will be able to perform the following skills:

- Mount an apparatus safely. (**NFPA 1001: 4.3.2**, p. 48)
- Dismount from an apparatus safely. (**NFPA 1001: 4.3.2**, p. 49)

ADDITIONAL STANDARDS

- **NFPA 1250**, *Recommended Practice in Fire and Emergency Services Organization Risk Management*
- **NFPA 1451**, *Standard for a Fire and Emergency Service Vehicle Operations Training Program*
- **NFPA 1500**, *Standard on Fire Department Occupational Safety, Health, and Wellness Program*
- **NFPA 1582**, *Standard on Comprehensive Occupational Medical Program for Fire Departments*

Fire Alarm

You and your crew are at a large fire in an apartment building that started on the third floor, extended into the attic, and is now venting through the roof. Your captain gave the incident commander (IC) the accountability tags, and you were assigned to Division D to confine the fire on the third floor. You quickly ran through an air cylinder as you pulled ceilings after having stretched the 65-mm (2½-in.) hose to the third floor. Your crew quickly changes cylinders and gets ready to continue the arduous task at hand. Then you get the word from the IC: "Pumper 213, report to rehab."

1. What is the leading cause of fire fighter injury and death?

2. What are the safety measures taken during this incident?

3. What are the potential hazards during this incident?

 *Access Navigate for more practice activities.*

Introduction

This chapter covers the topics of injury and illness prevention, means of reducing fire fighter deaths, and safety and health measures needed during all activities performed by fire fighters, including training, response, fire-ground operations, and fire station duties. Firefighting, by its very nature, is dangerous. However, many fire fighters are injured off of the fire ground, as well.

Every fire fighter must be aware of the risks inherent in all job responsibilities and activities and must learn safe methods of confronting all of the risks.

Every fire department must do everything it can to reduce the hazards and dangers of the job and help prevent fire fighter injuries, illness, and deaths. Each fire department must have a strong commitment to fire fighter health and safety, with fire fighters taking the lead for their own health and safety. When fire fighters understand the various risks associated with their job responsibilities and activities, and the actual threat that they present, safety measures can become routine, consistent, and fully integrated into every activity, procedure, and job description.

Advances in standards, technology, and equipment require fire departments to review and revise their health and safety policies and procedures regularly. Safety officers are responsible for evaluating the hazards of various situations and recommending appropriate safety measures to the incident commander (IC). Each accident, injury, or near miss must be thoroughly investigated to learn why it happened and how it can be avoided in the future. After-incident reviews and research by designated health and safety officers can identify new hazards as well as appropriate safety measures. In addition, reports of accidents, fatalities, and near misses from other fire departments can help identify common problems and lead to the development of effective preventive actions.

Because fire fighters must be ready to react immediately to an alarm, preparations for response begin long before the alarm is sounded. These preparations include physical and mental readiness, checking personal equipment, ensuring that the fire apparatus is ready, and making sure that all equipment carried on the apparatus is ready for use. Fire fighters also should be familiar with their response district, know the buildings under their protection, and understand their department's standard operating procedures (SOPs).

Response actions for the apparatus driver also include considering road and traffic conditions, determining the best route to the incident, identifying nearby hydrant locations or water sources, and selecting the best position for the apparatus at the incident scene.

Causes of Fire Fighter Deaths and Injuries

While Canada does not currently maintain a national database of statistics on fire fighter injuries and line-of-duty deaths, individual provincial/territorial workers' compensation statistics are maintained. Thankfully, the statistical trend shows that the number of annual fire fighter deaths in Canada is quite low. Comparatively, the National Fire Protection Association (NFPA) reports that in the United States, 69 fire fighters were killed in the line of duty in 2016 (NFPA, Fire Fighter Fatalities in the United States, 2016). These deaths occurred during fire-ground operations, at nonemergency incident scenes, in nonemergency situations, such as in fire stations, during training, and while responding to or returning from emergency situations (**FIGURE 2-1**).

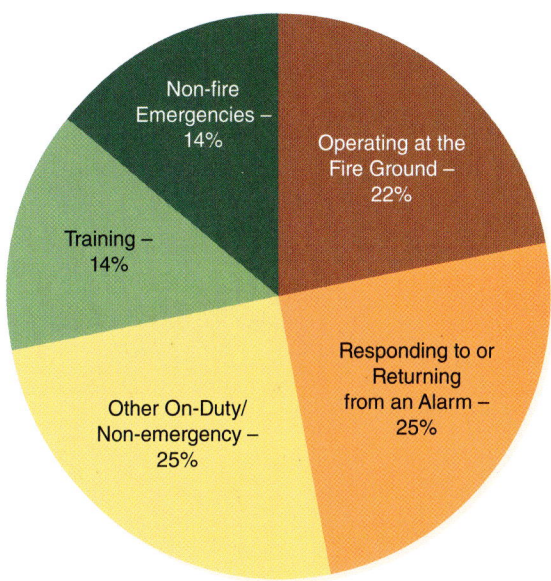

FIGURE 2-1 Fire fighter deaths in the United States by type of duty. Sixty-nine fire fighters died in the line of duty in 2016.
Source: National Fire Protection Association, Fire Fighter Fatalities in the United States, 2016.

The largest share of deaths occurred from stress, overexertion, and medical issues (**TABLE 2-1**). Cardiac events accounted for 38 percent of these deaths. The second leading cause of death was vehicle crashes. In 2016, 19 fire fighters died in vehicle accidents, a majority of whom died responding to or returning from incidents.

The NFPA estimates that in the United States, 62,085 fire fighters were injured in the line of duty in

TABLE 2-1 Fire Fighter Deaths by Cause of Injury	
Overexertion, stress, medical	42%
Vehicle accidents	25%
Falls	10%
Struck by objects	6%
Other	6%
Fatal assault	4%
Structural collapse	4%
Struck by vehicles	3%

Source: National Fire Protection Association, Fire Fighter Fatalities in the United States, 2016.

2016, an 8.8 percent decrease from the previous year. **FIGURE 2-2** shows the breakdown of injuries by type of duty.

Fewer than half of the injuries occurred on the fire ground. The leading cause of fire-ground injuries was overexertion or strain (**TABLE 2-2**).

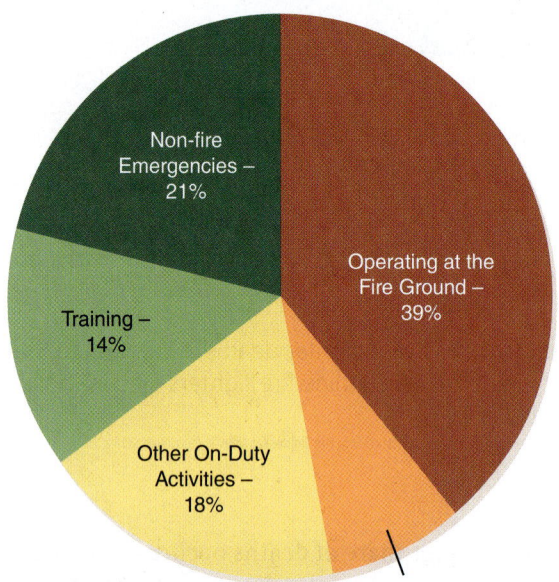

FIGURE 2-2 Fire fighter injuries in the United States by type of duty. In 2016, 62,085 fire fighters were injured in the line of duty.
Source: National Fire Protection Association, Fire Fighter Injuries in the United States, 2016.

TABLE 2-2 Fire-Ground Injuries by Cause	
Type of Injury	**Percentage of Total Injuries**
Overexertion/strain	27.1%
Fall, jump, slip	21%
Other	16.4%
Exposure to fire products	13.6%
Contact with an object	9.7%
Struck by an object	5.9%
Exposure to chemicals or radiation	3.7%
Extreme weather	3.1%

Source: National Fire Protection Association, Fire Fighter Injuries in the United States, 2016.

Reducing Fire Fighter Deaths and Injuries

Reducing fire fighter injuries and deaths requires every fire fighter, every fire department, and the entire fire community to work together. In 2003, the Canadian Fallen Firefighters Foundation (CFFF) was incorporated to construct a national memorial to Canada's fallen fire fighters, conduct an annual national memorial service, and support the families of the fallen.

Most fire fighter injuries and deaths are the result of preventable situations. Recognizing this fact, organizations such as the International Association of Fire Chiefs (IAFC) and the National Fallen Firefighters Foundation have developed programs with the goal of reducing line-of-duty deaths. For example, developed in 2005, the Near-Miss Reporting System provides a method for reporting situations that could have resulted in injuries or deaths. This system, which is accessible over the Internet, provides a means for all fire fighters to learn from situations that occur both rarely and frequently.

In an effort to do more to prevent line-of-duty deaths and injuries, the first National Firefighter Life Safety Summit convened in 2004, uniting fire service leaders and organizations across the United States. The result was the creation of the Everyone Goes Home program and a set of key initiatives, the 16 Firefighter Life Safety Initiatives. The goal of the Everyone Goes Home program is to raise awareness of life safety issues, improve safety practices, and allow everyone to return home at the end of his or her shift. The 16 Firefighter Life Safety Initiatives describe steps that need to be taken to change the current culture of the fire service to help make it a safe work environment. They are listed in **TABLE 2-3**.

Reducing fire fighter injuries and deaths requires the dedicated efforts of every fire fighter, of every fire department, and of the entire fire community working together. It also requires a safety program that integrates important components such as regulations, standards, procedures, personnel, training, and equipment. These components are discussed next.

SAFETY TIP

Injury and illness prevention is a responsibility shared by each member of the firefighting team. Fire fighters must always consider three groups when ensuring safety at the scene:

- Their personal safety
- The safety of other team members
- The safety of everyone present at an emergency scene

TABLE 2-3 16 Firefighter Life Safety Initiatives

1. **Cultural change:** Define and advocate the need for a cultural change within the fire service relating to safety and incorporating leadership, management, supervision, accountability, and personal responsibility.
2. **Accountability:** Enhance the personal and organizational accountability for health and safety throughout the fire service.
3. **Risk management:** Focus greater attention on the integration of risk management with incident management at all levels, including strategic, tactical, and planning responsibilities.
4. **Empowerment:** All fire fighters must be empowered to stop unsafe practices.
5. **Training and certification:** Develop and implement national standards for training, qualifications, and certification (including regular recertification) that are equally applicable to all fire fighters based on the duties they are expected to perform.
6. **Medical and physical fitness:** Develop and implement national medical and physical fitness standards that are equally applicable to all fire fighters based on the duties they are expected to perform.
7. **Research agenda:** Create a national research agenda and a data collection system that relates to the 16 Firefighter Life Safety Initiatives.
8. **Technology:** Utilize available technology whenever it can produce higher levels of health and safety.
9. **Fatality, near-miss investigation:** Thoroughly investigate all fire fighter fatalities, injuries, and near misses.
10. **Grant support:** Grant programs should support the implementation of safe practices and procedures and/or mandate safe practices as an eligibility requirement.
11. **Response policies:** National standards for emergency response policies and procedures should be developed and championed.
12. **Violent incident response:** National protocols for response to violent incidents should be developed and championed.
13. **Psychological support:** Fire fighters and their families must have access to counseling and psychological support.
14. **Public education:** Public education must receive more resources and be championed as a critical fire and life safety program.
15. **Code enforcement and sprinklers:** Advocacy must be strengthened for the enforcement of codes and the installation of home fire sprinklers.
16. **Apparatus design and safety:** Safety must be a primary consideration in the design of apparatus and equipment.

Source: National Fallen Firefighters Association. Everyone Goes Home. 16 Fire Fighter Life Safety Initiatives.

Regulations, Standards, and Procedures

Safety is the highest priority. Ensuring a safe working environment for the members of the fire service is undertaken by several professional organizations and results in various regulations, standards, and procedures.

As discussed in Chapter 1, *The Fire Service*, standards are issued by nongovernmental entities and are generally consensus based. The NFPA develops standards that help to standardize training courses, apparatus, equipment, and operations. Their mission is focused on safety: to save lives and reduce loss with information, knowledge, and passion. It is important to understand that these standards are designed to help you become a safe, efficient, and competent fire fighter. You will see many NFPA standards referenced throughout this course. Most fire departments have access to these standards.

To reduce the risks of accidents, injuries, occupational illnesses, and fatalities, a successful health and safety program that complies with NFPA 1500, *Standard on Fire Department Occupational Safety, Health, and Wellness Program*, must be established.

NFPA 1500 includes guidance on several key aspects of health and safety including policies, training and education, apparatus operation, personal protective equipment (PPE), emergency operations, station safety, medical and physical requirements, and health and safety programs. This chapter addresses each of these areas briefly.

NFPA 1500 provides a template for implementing a comprehensive health and safety program. Additional NFPA standards focus on specific subjects directly related to health and safety—for example, NFPA 1582, *Standard on Comprehensive Occupational Medical Program for Fire Departments*.

In the United States, the federal **Occupational Safety and Health Administration (OSHA)**, along with a variety of health and safety agencies, develops and enforces government regulations on workplace safety. In Canada, health and safety agencies within each province/territory develop and enforce workplace safety regulations. It is critical that fire departments ensure that they are cognizant of and compliant with the specific regulations regarding safety in each respective province/territory in Canada. NFPA standards are often incorporated by reference in government regulations.

Every fire department should have a set of SOPs, or standard operating guidelines (SOGs), that provide specific information on the actions that should be taken to accomplish a certain task. These procedures are vital because they enable everyone in the department to function properly and know what is expected for each task. Each fire fighter is responsible for understanding and following these procedures. This enables fire fighters from different stations or companies to work together safely and smoothly.

The fire department chain of command also enforces safety goals and procedures. In particular, the command structure keeps everyone working toward common goals in a safe manner. An incident command system (ICS) is a nationally recognized plan to establish command and control of emergency incidents. The ICS is flexible enough to meet the needs of any emergency situation, so it should be implemented at every emergency scene—from a routine auto accident to a major disaster involving responders from numerous agencies. More information on the ICS is presented in Chapter 22, *Establishing and Transferring Command*.

Most provincial and territorial governments have mandated that employers establish joint occupational health and safety committees. These committees are generally responsible for guiding the creation of health and safety policies and the monitoring of fire fighter health and wellness. Members of this committee should include representatives from every area, component, and level within the department, from fire fighters to chief officers. The health and safety officer and the fire department physician also should be members of the committee.

Personnel

A health and safety program is only as effective as the individuals who implement it. As discussed in Chapter 1, *The Fire Service*, safety officers are members of the fire department whose primary responsibility is safety. At the emergency scene, the designated safety officer reports directly to the IC and has the authority to correct or stop any action that is judged to be unsafe. Safety officers observe operations and conditions, evaluate

risks, and work with the IC to identify hazards and ensure the safety of all personnel. They also determine when fire fighters can work without self-contained breathing apparatus (SCBA) after a fire is extinguished. Safety officers can enhance safety in the workplace, at emergency incidents, and at training exercises. Even so, it is important to remember that each and every member of the fire department share the responsibility for promoting safety, both as an individual and as a member of the team.

Teamwork is an essential element of safe emergency operations. On the fire ground and during any hazardous activity, fire fighters must work together to get the job done. Freelancing has no place on the fire ground; it poses a danger to the fire fighter who acts independently and every other fire fighter on the emergency scene. Freelancing is acting independently of a superior's orders or the fire department's SOPs. Freelancing is discussed in more detail later in the chapter.

Training

Adequate training is essential for fire fighter safety. The initial fire fighter training covers the potential hazards of each skill and outlines the steps necessary to avoid injury. Fire fighters must avoid sloppy practices or shortcuts that might potentially contribute to injuries. They also must learn how to identify hazards and unsafe conditions.

The knowledge and skills developed during training classes are essential to maintain safety at actual emergency scenes. The initial training course is just the beginning—fire fighters must continually seek out additional courses to keep their skills current.

Equipment

A fire fighter's equipment ranges from power and hand tools to PPE and electronic instruments. Fire fighters must know how to use equipment in the correct manner and then operate it safely at all times. Equipment also must be properly maintained. Poorly maintained equipment can create additional hazards to the user or fail to operate when needed.

Manufacturers usually supply operating instructions and safety procedures for their equipment. These instructions cover proper use of the equipment, its limitations, and warnings about potential hazards. Fire fighters must read and heed these warnings and instructions. In addition, new equipment must meet applicable standards to ensure that it can perform under the difficult and dangerous conditions often encountered on the fire ground.

Personal Health and Well-Being

Safety and well-being are directly related to personal health and physical fitness. Although fire departments regularly monitor and evaluate the health of fire fighters, each department member is responsible for his or her own personal health, conditioning, and nutrition. To be an effective fire fighter, you must exercise regularly, eat a healthy diet, get an adequate amount of sleep, and take preventive measures to avoid illnesses such as heart disease and cancer.

Physical Fitness

All fire fighters—whether career or volunteer—should spend at least an hour each day in physical fitness training (**FIGURE 2-3**). Fire fighters should be examined by either a personal or departmental physician before beginning any new workout routine. An exercise routine that includes weight training, cardiovascular workouts, and stretching with a concentration on job-related exercises is ideal. For example, many fire fighters use a stair-climbing machine to focus on the muscle groups used for climbing. This type of exercise builds cardiovascular endurance for the fire ground; however, other muscle groups should not be neglected. Physical fitness must be a career-long activity; it is not something to be left at the fire academy when you graduate. Firefighting is a stressful activity that demands that you maintain a good fitness level throughout your career.

Nutrition

Diet is another important aspect of personal health. A healthy menu includes fruits, vegetables, healthy fats, whole grains, and lean protein. Pay attention to portion sizes. Unfortunately, most people eat larger portions than their bodies need. Substitute high-calorie desserts with healthy choices (such as fruit).

LISTEN UP!

Maintain your physical fitness, because your life and the lives of your crew depend on it!

Hydration

Hydration is an important part of staying healthy (**FIGURE 2-4**). A good guideline is to consume 237 to 296 mL (8 to 10 oz) of water for every 5 to 10 minutes of physical exertion. Do not wait until you feel thirsty to start hydrating. In fact, fire fighters should drink up to 4L (1 gal) of water each day to keep properly hydrated. Being adequately hydrated before an emergency occurs will enable you to better maintain adequate hydration and continue as an effective and healthy fire fighter. The

FIGURE 2-3 Regular exercise will help you to stay healthy and perform your job effectively.
© Jones & Bartlett Learning. Photographed by Glen E. Ellman.

FIGURE 2-4 Consume 237 to 296 mL (8 to 10 oz) of water for every 5 to 10 minutes of physical exertion.
Courtesy of Adam Ferrari.

amount of water needed to maintain adequate hydration will depend on the type of work you are doing and the ambient temperature. Remember, any time you are working in full PPE, your internal environment will rapidly become hot and you cannot dissipate body heat to the outside environment. Proper hydration enables muscles to work longer and reduces the risk of illness and injuries at the emergency scene. Some recent studies have indicated that maintaining a good level of hydration while engaged in firefighting activities also may reduce your chance of having a heart attack. More information on hydration is presented in Chapter 20, *Fire Fighter Rehabilitation*.

SAFETY TIP

Maintaining proper hydration is essential to performing at your peak physical level. Proper hydration is ongoing.

Sleep

Good health requires that you get an adequate amount of uninterrupted sleep to maintain alertness, prevent stress, and avoid illnesses and injuries. Because it is not always possible to get adequate sleep during long duty shifts, it is important that you get adequate amounts of uninterrupted sleep during your off-duty time. Establish a consistent sleep schedule and sleep routine, such as turning off all electronic devices one-half hour before bedtime to allow your mind to wind down and prepare for sleep.

Heart Disease

Heart disease is a leading cause of death in Canada as a whole and a leading cause of death among fire fighters in particular. A healthy lifestyle that includes a balanced diet, weight training, and cardiovascular exercise can help reduce many risk factors for heart disease and enable fire fighters to meet the physical demands of the job. It is also important that you have regular physical examinations to identify heart disease at an early stage.

Research projects are under way that utilize advanced technology to improve the health and safety in the fire service. For example, the SMARTER project (Science, Medicine, Research, Technology for Emergency Responders) uses specially designed, wearable monitoring equipment to measure the physiological stress of fire fighters while they perform on-the-job duties. Participants' heart rates, core body temperatures, respiratory rates, and other inputs are recorded, allowing for early detection of abnormalities and real-time monitoring of atmospheric conditions (Smith, 2017).

Cancer

Today's modern furnishings are composed of petroleum-based synthetic materials that result in the creation of polycyclic aromatic hydrocarbons (PAH) under fire conditions. Cancer represents more than 86 percent of all fatality claims in Canada, with an annual rate of 50 fatalities per 100,000 firefighters (CTV Vancouver). Cancer, now considered to be the leading cause of death among fire fighters, can be caused by a wide variety of cancer-causing substances (carcinogens) entering the body (International Association of Fire Fighters, 2017). These include exhaust from diesel engines, poisonous gases in smoke, and a wide variety of chemical particles. The dirt and soot that attach themselves to a fire fighter's protective gear and uniform contain large quantities of substances known to cause cancer. Fire fighters' hoods and gloves are thought to contain especially high concentrations of carcinogens. Carcinogens can be ingested through the mouth, injected into the body, absorbed through the respiratory system, or absorbed through the skin. Fire fighters are most likely to absorb carcinogens through their skin and through their respiratory systems.

The Firefighter Cancer Support Network estimates that fire fighters have a 9 percent higher risk of being diagnosed with cancer than the general U.S. population (Firefighter Cancer Support Network, 2017). Along the same lines, a study conducted by the National Institute for Occupational Safety and Health (NIOSH) concluded that the nearly 30,000 participants had a greater number of cancer diagnoses and cancer-related deaths than the general U.S. population. Most were digestive, oral, respiratory, and urinary cancers. When comparing fire fighters in this study to each other, they found that the chance of lung cancer increased with the amount of time spent at fires, and the chance of death from leukemia increased with the number of fire incidents (Centers for Disease Control and Prevention, 2016). Individual provinces and territories in Canada may have this information available as well.

Remember, contaminated objects that are placed in the cab of a pumper or in the trunk of a fire fighter's personal vehicle continue to release cancer-causing substances into the area around them. The longer contaminated objects are present, the more these objects continue to release toxic substances. Remove bunker gear and all other contaminated clothing as soon as possible. Bunker gear should be transported away from the riding compartment in a fire apparatus and be thoroughly washed immediately according to the manufacturer's instructions (**FIGURE 2-5**).

FIGURE 2-5 Special washing machines are available to launder personal protective clothing.
© Jones & Bartlett Learning. Photographed by Glen E. Ellman.

The Firefighter Cancer Support Network (FCSN)[1] suggests actions you can take to protect yourself. We will list some of them here.

1. Use SCBA from initial attack to finish of overhaul. (Not wearing SCBA in both active and post-fire environments is the most dangerous voluntary activity in the fire service today.)
2. Do gross field decon of PPE to remove as much soot and particulates as possible.
3. Use cleansing wipes to remove as much soot as possible from head, neck, jaw, throat, underarms, and hands immediately and while still on the scene.
4. Change your clothes, and wash them immediately after a fire.
5. Shower thoroughly after a fire.
6. Clean your PPE, gloves, hood, and helmet immediately after a fire.
7. Do not take contaminated clothes or PPE home or store them in your vehicle.
8. Decon fire apparatus interior after fires.
9. Keep bunker gear out of living and sleeping quarters.
10. Stop using tobacco products.
11. Use sunscreen or sun block.

The importance of annual medical examinations cannot be overstated—early detection and early treatment are essential to increasing survival. These 11 lifesaving actions are excerpted from FCSN's "Taking Action Against Cancer" white paper; it is available free of charge at firefightercancersupport.org.

Some cancers do not present for 20 years or more after exposure to a carcinogen. Therefore, it is important to reduce your exposure to cancer-causing substances starting on your first day of service as a fire fighter.

More information on performing field reduction of contaminants can be found in Chapter 3, *Personal Protective Equipment*.

Tobacco, Alcohol, and Illicit Drugs

Many fire departments have adopted policies that prohibit the use of tobacco products by fire fighters, both on duty and off duty. Smoking is a major risk factor for cardiovascular disease, it reduces the efficiency of the body's respiratory system, and it increases the risk of lung and other types of cancer. Fire fighters should avoid tobacco products entirely for both health and insurance reasons.

Alcohol is another substance that fire fighters should avoid. Alcohol is a mood-altering substance that can be abused. Excessive alcohol use can damage the body and affect performance. Fire fighters who have consumed alcohol within the previous 8 hours must not be permitted to engage in training or emergency operations. In addition, alcohol use increases the risk of mouth, throat, larynx, esophagus, liver, colon, and breast cancers (American Cancer Society, 2017).

Illicit drug use has absolutely no place in the fire service. Many fire departments have drug-testing programs to ensure that fire fighters do not use or abuse drugs. The illegal use of drugs endangers your life, the lives of your team members, and the public you serve.

SAFETY TIP

Everyone is subject to an occasional illness or injury. You should not try to work when ill or injured. Operating safely as a member of a team requires both fitness and concentration. Do not compromise the safety of the team or your personal health by trying to work while you are ill or injured.

Counselling and Critical Incident Stress Management

Fire fighters are often exposed to stressful situations and work in an environment that is subject to unscheduled

1. Material in the Cancer section on this page is courtesy of Firefighter Cancer Support Network.

and unexpected emergency events. Many fire fighters see more traumatic situations in a short time than most other citizens see in their entire lifetime. Firefighting involves not only the stresses directly connected to fighting the fire and to rendering emergency medical care but also the added burdens of disrupted sleep patterns, rotating work schedules, unscheduled overtime, and interrupted meal schedules. High levels of stress can produce a variety of symptoms. Some people are not able to sleep well; others tend to gain or lose weight. Many people become irritable when stressed. Overeating, increased consumption of alcoholic beverages, and the use of non-prescribed drugs may be the result of stress. Stress may produce depression or suicidal thoughts in some people.

LISTEN UP!

Find time for yourself and your family as a mental buffer from the stressors of the job.

To diminish the effect of these stressors, it is important to get adequate sleep, maintain a healthy, balanced diet, and get adequate exercise. It is also important for you to balance your work schedule with other activities and monitor your behavioural health. Finally, it is essential to identify and utilize the resources that are available to assist in maintaining behavioural health. Some of these resources are provided by employers, such as the employee assistance programs that will be discussed later. Other resources and assistance are provided by professional organizations. For example, the International Association of Fire Fighters (IAFF) offers information, available through their website, as well as the in-patient IAFF Center of Excellence for Behavioral Health Treatment and Recovery.

Critical Incident Stress Management

Critical incidents challenge the capacity of most individuals to deal with stress. It is important to understand what the stressors in this job are and to learn how to work to diminish their effect. Examples of critical incidents include the following:

- Line-of-duty deaths (police, fire/rescue, emergency medical services)
- Suicide of a colleague
- Serious injury to a colleague
- Situations that involve a high level of personal risk to fire fighters
- Events in which the victim is known to the fire fighters
- Multiple-casualty/disaster/terrorism incidents

- Events involving death or life-threatening injury or illness to a victim, especially a child
- Events that are prolonged or end with a negative or unexpected outcome

This list is not complete, nor is it necessarily a fact that any of these situations will seriously trouble every individual. Normal coping mechanisms help many fire fighters to handle many situations. Some individuals have a high capacity to deal with stressful situations through exercise, talking to friends and family, or turning to their religious beliefs. These are healthy, nondestructive ways to deal with the pressures of being exposed to a critical incident.

Sometimes individuals react to critical incidents in ways that are not healthy, such as alcohol or drug abuse, depression, an inability to function normally, or a negative attitude toward life and work. These symptoms can occur in anyone, even individuals who normally have healthy coping skills. Reactions vary from one individual to the next, both in type and severity. Many times fire fighters do not realize they are affected in a deeply negative manner. A somewhat routine incident, however, may trigger negative reactions from a critical incident that occurred in the past. Critical incident stress also can be cumulative, building up over time. This condition, which is called burnout, cannot be traced to any one incident.

The recognized stages of emotional reaction experienced by fire fighters and other rescue personnel after a stressful incident can include the following:

- Anxiety
- Denial/disbelief
- Frustration/anger
- Inability to function logically
- Remorse
- Grief
- Reconciliation/acceptance

These stages can occur within minutes or hours, or they can take several days or months to unfold. Not all of the steps will occur for every event, and they do not necessarily occur in the order given here.

Fire fighters should understand the resources available to them and know how they can access them. Maintaining good emotional health is a simple but important part of fire fighter survival. The aim of counselling, peer support teams, employee assistance programs, and **critical incident stress management (CISM)** programs is to prevent these emotional reactions from having a negative impact on the fire fighter's work and life, over both the short term and the long term (**FIGURE 2-6**). The major difference between CISM and peer support teams is that CISM teams usually respond immediately after a crisis incident,

FIGURE 2-6 Group stress debriefings are sometimes used to alleviate stress reactions generated by high-stress emergency situations.
© Tom Carter/Getty Images.

and peer support teams provide continuous, ongoing support. For example, a **critical incident stress debriefing (CISD)** is held as soon as possible after a traumatic call. It provides a forum for firefighting and emergency medical services (EMS) personnel to discuss the anxieties, stress, and emotions triggered by a difficult call. Follow-up sessions can be arranged for individuals who continue to experience stressful or emotional responses after a challenging incident. With peer support teams, specially trained department members help those struggling with day-to-day issues by communicating with them regularly and recommending resources to assist them.

Everyone handles stress differently, and new fire fighters need time to develop the personal resources necessary to deal with difficult situations. If any incident proves mentally or emotionally disturbing to you, and it is beyond your ability to deal with it alone or with those closest to you, ask your supervisor for assistance. Your supervisor should be familiar with the resources available to you and may refer you to a qualified professional. Signs that you may need assistance include having trouble sleeping or having difficulty dealing with your thoughts or feelings. Some departments are proactive and include training in things such as stress first aid.

Firefighting is a job that requires close teamwork. It is the responsibility of all fire fighters to monitor themselves and other members of their team for signs of mental stress. We are all interdependent on the members of our team in order to accomplish our goal. The team needs to function well, not only at emergency scenes but also when performing routine tasks or relaxing between runs. It is important that you respect each member of your team. Bullying or discrimination should never be tolerated. It is everyone's job to be on the lookout for signs of unhealthy behaviour or stress. If you think a fellow fire fighter is exhibiting signs of stress or suffering from depression, there are several ways in which you may be able to help. Although the ways in which this outcome is accomplished vary, the purpose remains the same.

Sometimes it may be helpful to sit down with a co-worker and ask him or her how he or she is doing or to let him or her know that you are concerned about him or her. You may be able to encourage him or her to seek help. If you see signs of stress, talk with your officer or someone who can help that person receive assistance.

Suicide Awareness and Prevention

In recent years there have been more fire fighter suicides than line-of-duty deaths. In 2016, there were 99 reported fire fighter deaths as a result of suicide in the United States (Firefighter Behavioral Health Alliance, 2017). Recent estimates reveal that a fire department is three times more likely to experience a suicide in any given year than a line-of-duty death (The National Fallen Firefighters Association, 2017). This fact reminds us that stressful occupations may experience a higher rate of suicides than the general population. Many people are concerned about the incidence of suicide in fire fighters and other emergency care providers.

While fire fighters are trained to be the best they can be, many are not prepared for the ill effects or aftermath of stress or a traumatic situation. It is important to understand that by recognizing signs of stress or depression in a team member you may be able to help him or her get assistance for his or her problem before it becomes worse.

> **FIRE FIGHTER TIP**
>
> According to the Firefighter Behavioral Health Alliance (FBHA), the top five warning signs of job-related stress include:
>
> - Recklessness/impulsiveness
> - Anger
> - Isolation
> - Loss of confidence in abilities or skills
> - Sleep deprivation
>
> If you believe you are suffering from these issues please seek help from your employee assistance program, chaplain, peer support team, or a qualified counsellor in your community.

Employee Assistance Programs

There are many formal programs to support fire fighter behavioural health. These programs are maintained by fire departments, unions, local governments, and even charitable organizations. Support is provided through special events, projects, peer support, chaplain programs, and education. A traditional and common form of support for fire fighters is the employee assistance program. **Employee assistance programs (EAPs)** provide confidential help with a wide range of problems that might affect performance. Many fire departments have established EAPs so that fire fighters can get counselling, support, or other assistance in dealing with physical, financial, emotional, or substance abuse problems. EAPs include a variety of helpful resources, including access to qualified counsellors and chaplains who have a working knowledge of the fire service. Some fire departments have qualified counsellors available 24 hours a day. The initial counselling may consist of a group session for all fire fighters and rescuers; alternatively, it can also be done on a one-on-one basis or in smaller groups. A fire officer may refer a fire fighter to an EAP if a problem starts to affect the individual's job performance. Fire fighters who take advantage of an EAP can do so with complete confidentiality and without fear of retribution.

According to the FBHA, many do not seek help, mainly due to confidentiality or job promotion concerns or simply because counsellors may not know the culture of the fire service. The FBHA recommends inviting counsellors to the station, including them on ride time, in training, and during meals. Another suggestion is to have the EAP counsellors create a video biography so department members and their families can see who they are and get to know them.

Safety During Training

During training, fire fighters learn and practise the skills that they will use later under emergency conditions. Typically, the patterns that develop during training continue during actual emergency incidents. Thus, developing the proper working habits during training courses helps ensure safety later. Use of proper protective gear and good teamwork are as important during training as they are on the fire ground.

Instructors and veteran fire fighters are more than willing to share their experiences and advice. They can explain and demonstrate every skill and point out the safety hazards involved because they have performed these skills hundreds of times and know what to do. But here, too, safety is a shared responsibility. Do not attempt anything you feel is beyond your ability or knowledge. If you see something that you believe is

an unsafe practice, bring it to the attention of your instructors or a designated safety officer.

Avoid freelancing. Wait for specific instructions or orders before beginning any task. Do not assume that something is safe and act independently. Follow instructions and learn to work according to the proper procedures.

Teamwork is also important during training exercises. Assignments are given to firefighting teams during most live fire exercises. Teams must stay together. If any member of the team becomes fatigued, is in pain or discomfort, or needs to leave the training area for any reason, notify the instructor or safety officer. EMS personnel should be available to perform an examination and to transport ill or injured personnel to further treatment if necessary. A fire fighter who is injured during training should not return until medically cleared for duty.

Safety During the Emergency Response

Safety during an emergency response must begin before there is an actual response. Safety begins with fire fighters being prepared for emergencies. Fire fighters must be ready to respond to an emergency at any time during their tour of duty. This process begins by ensuring that your PPE is complete, ready for use, and in good condition. At the beginning of each tour of duty, place your PPE in its designated location, which depends on your assigned riding position on the apparatus (**FIGURE 2-7**).

You should also conduct a daily inspection of the SCBA for your riding position at the beginning of each tour of duty (**FIGURE 2-8**). The air cylinder should be full, the face piece clean, and the personal alert safety system (PASS) operable. Also check the availability and operation of your hand light and any hand tools you might require, based on your assigned position on the apparatus (**FIGURE 2-9**). Recheck your PPE and tools thoroughly when you return from each emergency response, and clean them whenever they are used. Personal protective gear should be properly positioned so that you can don it quickly. This routine will help ensure that your gear and equipment are fully functional when the next alarm comes.

Every response to an incident has a high potential for accidents, injuries, and death. **Response** actions include receiving the alarm, donning protective clothing and equipment, mounting and dismounting the apparatus, and transporting equipment and personnel to and from the emergency incident quickly and safely.

FIGURE 2-7 Protective clothing should be properly positioned so that you can quickly don it.
© Jones & Bartlett Learning.

FIGURE 2-8 Conduct the daily check of your SCBA at the beginning of each tour of duty.
© Jones & Bartlett Learning. Photographed by Glen E. Ellman.

Alarm Receipt

When an alarm is received, the response should be prompt and efficient. Responding fire fighters should walk briskly to the apparatus. There is no need to run; the objective is to respond quickly, without injuring anyone

FIGURE 2-9 Check the tools assigned to your riding position.
© Jones & Bartlett Learning. Photographed by Glen E. Ellman.

or causing any damage. Follow established procedures to ensure that stoves, faucets, and other appliances at the station are shut off. Wait until the apparatus bay doors are fully open before leaving the station.

SAFETY TIP

Volunteer fire fighters who are not assigned to specific tours of duty or riding positions should check their PPE, SCBA, and associated tools and equipment on a regular basis to ensure that these items are ready for use whenever an emergency response becomes necessary. After each use, all PPE should be carefully cleaned and checked before it is put away.

Riding the Apparatus

Expectations for donning PPE should be referenced in your department SOPs. A common practice is for fire fighters to don PPE prior to mounting the apparatus. While a fire apparatus is in motion, all crew members should be wearing seat belts properly. Do not attempt to don PPE while the apparatus is on the road. Wait until you dismount at the incident scene to don any protective clothing that was not donned prior to mounting the apparatus. Don an SCBA only after the apparatus stops at the scene, unless the SCBA is seat mounted.

All equipment should be properly mounted, stowed, or secured on the fire apparatus. Unsecured equipment in the crew compartment can prove dangerous if the apparatus must stop or turn quickly, because a flying tool, map book, or PPE can seriously injure a fire fighter.

Be careful when mounting apparatus, because the steps on fire apparatus are often high and can be slippery. Use handrails when mounting or dismounting the apparatus. Follow the steps in **SKILL DRILL 2-1** to mount an apparatus properly.

SKILL DRILL 2-1
Mounting Apparatus Fire Fighter I, NFPA 1001: 4.3.2

1 When mounting (climbing aboard) fire apparatus, always have at least one hand firmly grasping a handhold and at least one foot firmly placed on a foot surface. Maintain the one hand and one foot placement until you are seated.

2 Fasten your seat belt, and leave it fastened until the apparatus is stopped at its destination. Don any other required safety equipment for the response, such as hearing protection and intercom systems.

© Jones & Bartlett Learning. Photographed by Glen E. Ellman.

All fire fighters must be seated in their assigned riding positions with seat belts and/or harnesses fastened before the apparatus begins to move. NFPA 1500, *Standard on Fire Department Occupational Safety, Health, and Wellness Programs,* and NFPA 1001, *Standard for Fire Fighter Professional Qualifications,* require all fire fighters to be in their seats, with seat belts secured, whenever the vehicle is in motion. Do not unbuckle your seat belt to don any clothing or equipment while the apparatus is en route to an incident. Vehicle occupants who are not wearing seat belts are much more likely to suffer serious injuries or death than are occupants who have their seat belts properly fastened. Air bags are most effective when seat belts are properly fastened; they are not effective without properly applied seat belts.

Seat belts also greatly reduce the possibility that vehicle occupants will be ejected from the vehicle.

For safety's sake, SOPs often prohibit specific actions during response. As noted previously, fire fighters must remain seated with their seat belts securely fastened while the emergency vehicle is in motion. Never unfasten your seat belt to retrieve or don equipment. Do not dismount the apparatus until the vehicle comes to a complete stop. In addition, you should never stand up while riding on apparatus. Do not hold on to the side of a moving vehicle or stand on the rear step. When a vehicle is in motion, everyone aboard must be seated and belted in an approved riding position.

The noise produced by sirens and air horns can have long-term, damaging effects on a fire fighter's hearing.

SKILL DRILL 2-2
Dismounting a Stopped Apparatus Fire Fighter I, NFPA 1001: 4.3.2

1 Become familiar with your riding position and the safest way to dismount.

2 Maintain the one hand and one foot placement when leaving the apparatus, especially on wet or potentially icy roadway surfaces.

© Jones & Bartlett Learning. Photographed by Glen E. Ellman.

For this reason, your department should provide hearing protection for personnel riding on fire apparatus. Some of these devices include radio and intercom capabilities so that fire fighters can talk to one another and hear information from the dispatcher or IC.

During response, limit conversation to the exchange of pertinent information. Listen for instructions from the IC, for instructions from your company officer, and for additional information about the incident over the radio. The vehicle operator's attention should be focused on driving the apparatus safely to the scene of the incident.

The ride to the incident is a good time to consider any relevant factors that could affect the situation. These factors could include the time of day or night, the temperature, the presence of precipitation or wind, the type of occupancy, the type of construction, and the location and type of incident. Using this time to think ahead will help you mentally prepare for the various possibilities that you might encounter at the scene.

When the apparatus arrives at the incident scene, the driver/operator will park it in a location that is both safe and functional. Wait until the vehicle comes to a complete stop before dismounting. Always check for traffic before opening the doors or stepping out of the apparatus. During the dismount, watch for other hazards—for example, ice and snow, downed power lines, uneven terrain, or hazardous materials—that could be present.

Be careful when dismounting apparatus. The increased weight of PPE and adverse conditions can contribute to slips, strains, and sprains. Follow the steps in **SKILL DRILL 2-2** to dismount an apparatus safely.

Traffic Safety on the Scene

An emergency incident scene presents several risks to fire fighters in addition to the hazards of fighting fires and performing other duties. One of these dangers is traffic, particularly when the incident scene is on a street or highway. Traffic safety should be a major concern for the first-arriving units because approaching drivers might not see emergency workers or realize how much room fire fighters need to work safely.

The first unit or units to arrive at the incident scene have a dual responsibility. Not only must the fire fighters focus on the emergency situation facing them, but they also must consider approaching traffic, including other emergency vehicles, and other, less obvious hazards. Always check for traffic before opening doors and dismounting the apparatus, and watch out for traffic when working in the street.

One of the most dangerous work areas for fire fighters is on a roadway, where traffic can be approaching at high speeds. Follow departmental SOPs to close streets quickly and to block access to areas where operations are being conducted. Place traffic cones, flares, emergency scene signage, and other warning devices far enough away from the incident to slow approaching traffic and direct it away from the work area (**FIGURE 2-10**). Police officers should assist by diverting traffic at a safe distance outside the hazard area. It is not uncommon to set large control zones at the onset of an incident, only to discover that the zones may have been established too liberally. At the same time, control zones should not be defined too narrowly. As the IC gets more information about the incident, the control zones may be expanded or reduced. Wind shifts are a common reason why control zones are modified during the incident. If there is a prevailing wind pattern in your area, then factor that information into your decision-making process when it comes to control zones.

Placement of emergency vehicles on the scene is also critical. With proper placement, such vehicles can act as a barrier between oncoming traffic and the scene. All fire fighters working at highway incidents should wear high-visibility safety vests in addition to their normal PPE. These vests should meet the ANSI 207 standard for public safety vests. Many fire departments have specific SOPs covering required safety procedures for these incidents.

FIGURE 2-10 The scene of a crash should be marked properly, and traffic should be diverted so that responders have enough room to work.
© Mike Legeros. Used with permission.

Safe Driving Practices

In the United States in 2016, the NFPA estimated there were 15,425 collisions involving fire department emergency vehicles responding to or returning from incidents (NFPA, U.S. Firefighter Injuries, 2016). As discussed earlier, in 2016, 19 fire fighters died in vehicle-related incidents. Eleven of the 19 died responding to or returning from an incident (NFPA, Firefighter Fatalities in the United States, 2016).

Drivers of fire apparatus have a great responsibility to get the fire apparatus and the crew members to the emergency scene without having or causing a traffic accident en route. Drivers must always exercise caution when driving to an incident. Fire apparatus can be large, heavy, and difficult to maneuver, and they must know the streets in their first-due area and any target hazards. They must be able to operate the vehicle skillfully and keep it under control at all times. In addition, they must anticipate all responses from other drivers who might not see or hear an approaching emergency vehicle or know what to do when confronted by one. Operating an emergency vehicle without the proper regard for safety can endanger the lives of both the fire fighters on the vehicle and any civilian drivers and pedestrians encountered along the way.

Prompt response is a goal, but safe response is a much higher priority. It often has been stated that the biggest cause of vehicle crashes is "a loose nut behind the wheel." In other words, a major factor in crashes is the attitude and ability of the vehicle operator. A competent emergency vehicle operator needs to have a confident, but not cocky, attitude. Aggressive driving has no place in the fire service. The emergency vehicle operator should have good judgment and reactions, mental fitness, maturity, physical fitness, alertness, and good driving habits. It is important to maintain a clean driving record, because a person who has been cited for multiple moving violations in his or her personal vehicle will usually be an increased risk as the driver of an emergency vehicle. To be a good driver, it is important to know the provincial/territorial and local laws relating to motor vehicle operations. In addition, an emergency vehicle operator needs to understand the reaction time, braking distance, and stopping distance of the vehicle.

Driving while impaired is simply asking for trouble. Emergency driving, even when properly performed, increases risks. Impaired driving dramatically increases those risks. Impairment can result from many different sources—for example, using some prescribed medicines, using some over-the-counter medicines, and being overly fatigued. Anyone who has been drinking alcoholic beverages should not drive. Eating, texting, and talking on a communications device while driving are all forms of distracted driving. Even cell phones with navigation technology should not be used in such a way that the driver is distracted. Distracted driving during routine driving or when responding to an emergency is to be avoided—it is the cause of many collisions and deaths. In some cases, this has led to criminal prosecution and conviction.

Fire fighters who drive emergency vehicles must have special driver training and know the laws and regulations that apply to emergency responses. Many jurisdictions require a special driver's license to operate fire apparatus. The rules that apply to emergency vehicles are very specific, and the driver/operator is legally responsible for the safe operation of the vehicle and the safety of the vehicle occupants at all times. These driving skills are not required for the Fire Fighter I and Fire Fighter II courses and are beyond the scope of this textbook.

Although most provinces/territories permit drivers of emergency vehicles to take exception to specific traffic regulations when responding to emergency incidents, driver/operators must always consider the potential actions of other drivers before making such a decision. For example, traffic laws require other drivers to yield the right of way to an emergency vehicle. There is no assurance, however, that other drivers will do so when an emergency vehicle approaches. The driver/operator also must anticipate which routes other units responding to the same incident will take. All passengers and drivers of emergency vehicles should wear seat belts on routine and emergency responses.

Many collisions occur when the motor vehicle operator loses control of the vehicle. These crashes may be caused by driving too fast for the prevailing conditions, braking inappropriately, changing directions too abruptly, or tracking around a curve too fast. In addition, many collisions involving emergency vehicles occur on open roads, where excessive speed is often cited as a primary cause. Intersections are common sites of collisions involving emergency vehicles; most of these collisions occur when the emergency vehicle

operator fails to stop at an intersection to ensure that other traffic has stopped before proceeding through the intersection.

A motor vehicle collision actually consists of a series of separate collision events. The first collision occurs when the vehicle collides with a second vehicle or with a stationary object; the second collision occurs when the occupants of the vehicle collide with the interior of the vehicle.

SAFETY TIP

Always fasten your seat belt each and every time you get into a motor vehicle. It is vital that you follow this rule each and every time.

Laws and Regulations Governing Emergency Vehicle Operation

Four general principles govern emergency vehicle operation:

- Emergency vehicle operators are subject to all traffic regulations unless a specific exemption is made. A specific exemption is a statement that appears in a statute, such as "The driver of an authorized vehicle may exceed the maximum speed limits so long as he or she does not endanger life or property."
- Exemptions are legal only when the vehicle is operating in emergency mode.
- Even with an exemption, the emergency vehicle operator can be found criminally or civilly liable if involved in a crash.
- An exemption does not relieve the operator of an authorized emergency vehicle from the duty to drive with reasonable care for all persons using the highway.

Laws governing emergency vehicle operation may vary from one province/territory to another. You must follow the laws and regulations of your province/territory regarding emergency vehicle operations. Some provinces/territories permit private vehicles to operate as emergency vehicles when responding to a fire station or to the scene of an emergency; other provinces/territories outline a limited use of emergency equipment and dictate certain restrictions. You must understand and follow the specific laws and regulations of your province/territory as well as your department's SOPs.

In many cases, one of the best predictors of future performance is past behaviour. Your driving record is an important consideration in your career in a fire department. A person who has past moving violations for speeding, reckless operation, aggressive driving, chargeable crashes, or driving under the influence of alcohol or drugs may be excluded from driving any emergency vehicles. Your driving record, both on duty and off duty, is important to your career as a fire fighter.

SAFETY TIP

Developing situational awareness on the emergency scene is essential to your safety and survival. This concept can be summed up by saying that you must maintain a "big picture" view of the emergency scene and avoid "tunnel vision." Situational awareness requires that fire fighters observe, not just see. Observation is deliberate, and seeing is casual. This ability develops through experience and with well-developed competency in basic fire fighter skills. The less you have to concentrate on the components of your task, the more you are able to observe your surroundings. This quality emerges through hours and hours of practice and skill repetition. Do not stop practicing until your basic skills become procedural memories. If you can do your job with your eyes closed, you will be able to observe much more when they are open!

Standard Operating Procedures for Personal Vehicles

In some departments where fire fighters are not on duty at the fire station, it may be necessary for members to respond to the emergency scene in their private vehicles. For you to use your personal vehicle in a situation requiring an emergency response, this use must be permitted by provincial or territorial laws and regulations, and this operation must also be permitted by your local fire department. If your local fire department does not permit the use of personal vehicles for emergency responses, then you are not permitted to engage in this behaviour regardless of what is permitted by your laws.

Fire fighters who respond to emergency incidents in their personal vehicles must follow the specific laws and regulations of their province/territory and follow departmental SOPs related to this issue. Some jurisdictions require fire fighters to equip their privately owned vehicles with warning devices and to operate them as emergency vehicles; others grant no special status to privately owned vehicles driven during an emergency response. In some areas, volunteer fire fighters responding to an emergency incident use colored lights to request the right of way from other drivers.

Your fire department should have SOPs explaining the requirements for personal vehicle use for emergency responses. These procedures will likely address which kind of training you must complete before you are permitted to use your personal vehicle for emergency response. This training usually includes a course in defensive driving and functioning as an emergency vehicle operator.

Vehicle Collision Prevention

Engaging in safe driving practices will likely prevent vehicle collisions. If you are using your personal vehicle to respond to an emergency, it is important to take the characteristics of your vehicle into account while driving. For example, four-wheel-drive pickup trucks and large sport utility vehicles have a higher center of gravity than sleek sports cars, and sudden changes in direction are more likely to result in roll-over crashes when a vehicle has a high center of gravity. For this reason, it is critical to learn the characteristics and limitations of your vehicle.

Anticipate the road and road conditions. If you regularly travel the same roads to report to your fire station, learn the characteristics of that strip of roadway, and be able to anticipate when you need to slow down or stop. Always drive at a safe speed for that road: Travelling on a limited-access highway is much different from travelling through a school zone, for example. Observe the traffic conditions, and slow down when traffic congestion is present. Expect that motorists around you may do anything at any time. Upon the approach of an emergency vehicle, for example, some motorists will slow down, some will stop, some will speed up, some will pull to the right, and others will turn left in front of you. Expect the unexpected.

Make allowances for weather conditions. On a clear day, it might seem that you can see forever, but, really, you cannot. In conditions of rain, snow, fog, dust, or darkness, the distance over which you can see is greatly reduced; reduce your speed to compensate for the limited visibility. There also may be vehicle collisions, downed trees or power lines, or debris in the road. At night, your vision is limited by the distance your headlights reach, so reduce your speed accordingly. Recognize that you cannot see objects outside the projection of the headlight beams. Understand that the goal of an emergency response is not to drive as fast as you can but rather to arrive on the scene as quickly as you can while maintaining safety.

Adjust your speed of response to accommodate any storm conditions. Rainstorms reduce visibility and produce slippery road surfaces. When water collects on a roadway, your vehicle can hydroplane on the thin film of water that separates your tires and the road surface. You have no control over your vehicle when hydroplaning occurs. Be aware of high winds from tornadoes, hurricanes, or other windstorms that can topple trees onto the roadway. During heavy rainstorms, watch for water flowing over the road. You cannot see how deep the water is, and you cannot tell whether the floodwater has washed out part of the roadway. Slow down when driving following an earthquake or tornado because you will not know where the roadway, bridges, and overpasses have been blocked, damaged, or destroyed.

When operating an emergency vehicle, you are not exempt from the laws of physics. The laws of physics tell us that when the speed of a vehicle doubles, the force exerted by that vehicle increases by a factor of four. Thus, when the speed of a vehicle increases from 32 to 64 kilometres per hour (20 to 40 miles per hour), the force exerted by that vehicle is increased by a factor of

four. Higher speeds require more braking power and a longer distance to bring the vehicle to a stop.

To drive safely, you need to understand where the greatest risks are located. As mentioned earlier, studies have shown that many crashes involving emergency vehicles occur at intersections (**FIGURE 2-11**). Most fire departments require all emergency vehicles to come to a full stop when the emergency vehicle operator encounters a stop sign or a red traffic light. If you enter an intersection without stopping, you may not be able to see approaching traffic, and many motorists do not drive defensively. They assume that if there is no stop sign

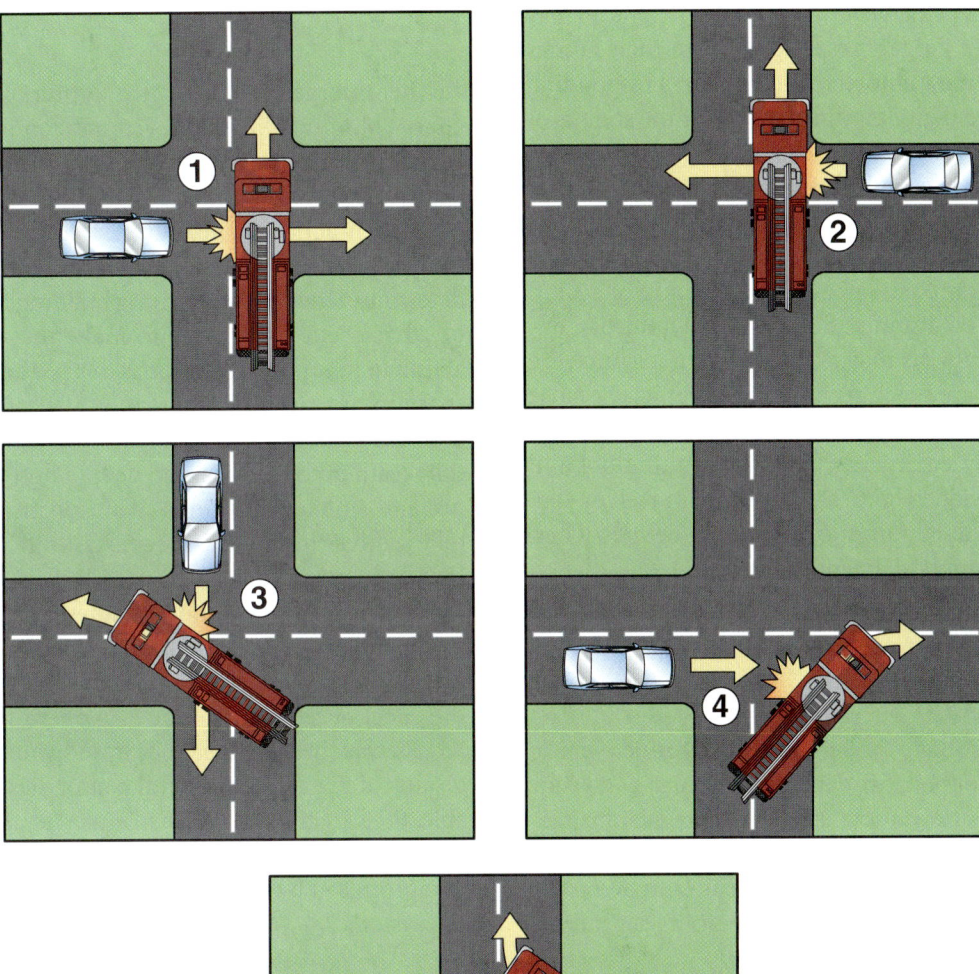

FIGURE 2-11 Every two-lane intersection has five potential crash points, making intersections very dangerous.

© Jones & Bartlett Learning.

or red light, they can proceed through the intersection without looking for oncoming vehicles.

In addition, many emergency vehicle crashes occur on open stretches of straight roads during daylight hours. The primary cause of these crashes is excessive speed. Most of us have limited experience with high-speed driving, so it is important to keep within the limits of your driving skills and within the limits of the vehicle you are operating. Many departments limit the speed during emergency responses to 8 kilometres per hour (5 miles per hour) over the posted speed limit. Multiple studies have shown that little time is saved by driving at high rates of speed.

During an emergency response, be alert for the presence of other emergency vehicles. These vehicles may carry other fire fighters responding from their homes or fire, EMS, or law enforcement personnel responding to the same incident. It is difficult to hear the sounds of other apparatus when you are on an emergency response.

Drive with a cushion of safety around you—that is, in such a manner that even if a motorist does the unexpected, you still have enough time and distance to avoid a collision. Remember that the sound of a siren carries a limited distance. Sometimes in an urban environment motorists may not be able to determine the location of an approaching emergency vehicle because of surrounding buildings, which distort the source of the siren. Similarly, emergency lights have limitations and cannot be depended on to clear a path of travel for you.

TABLE 2-4 summarizes the guidelines when driving to respond to an emergency call.

The Importance of Vehicle Maintenance

Never underestimate the importance of proper vehicle maintenance for both fire department vehicles and your private vehicle. It is important to perform regular maintenance of the engine, transmission, and other equipment to ensure dependable starting and running. Keep the brakes and tires in proper condition to ensure dependable

stopping. In addition, note that tire tread depth for emergency vehicles should exceed the minimum required for nonemergency vehicles. The suspension system and steering system need to be properly maintained to ensure safe handling of the vehicle. Likewise, the windshield wipers and windshield washers must be kept in good condition to ensure clear vision. Regularly check all headlights, taillights, and turn signals for proper operation. Private vehicles that are used for emergency response should be maintained at a higher level than vehicles that are not subject to the stresses of emergency operation.

Safety at the Incident

At the emergency scene, fire fighters should never charge blindly into action. From the moment that fire fighters arrive at an emergency incident scene, their department's SOPs and the structured ICS must guide all of their actions. The officer in command will "size up" the situation, carefully evaluating the conditions to determine whether the burning building is safe to enter or what needs to be done to make the scene safe for action to take place. Your job is to pay attention to your surroundings and to report things that do not match your expectations. For example, observe the time of day, the weather conditions, the people on the scene, and fire and smoke conditions. Wait for instructions and follow directions for the specific tasks to be performed. Fire fighters should always work in assigned teams (companies or crews) and be guided by a strategic plan for the incident. Teamwork and disciplined action are essential to provide for the safety of all fire fighters and the effective, efficient conduct of operations.

Freelancing, whether done by individual fire fighters, groups of fire fighters, or full companies, is unacceptable. Freelancing cannot be tolerated at any emergency incident. The safety of every person on the scene can be compromised by fire fighters who do not work within the system.

TABLE 2-4 Guidelines for Safe Emergency Vehicle Response
1. Drive defensively.
2. Follow agency policies in regard to posted speed limits.
3. Always maintain a safe distance. Use the "4-second rule." Stay at least 4 seconds behind another vehicle in the same lane.
4. Maintain an open space or cushion in the lane next to you as an escape route in case the vehicle in front of you stops suddenly.
5. Always assume that other drivers will not hear your siren or see your emergency lights.
6. Select the shortest and least congested route to the scene at the time of dispatch.
7. Visually clear all directions of an intersection before proceeding.
8. Go with the flow of traffic.
9. Watch carefully for bystanders and pedestrians. They may not move out of your way or could move the wrong way.

For example, a fire fighter who enters a burning structure without informing a superior may be trapped by rapidly changing conditions. By the time the fire fighter is missed, it may be too late to perform a rescue. Searching for a missing fire fighter exposes others to unnecessary risk.

Personnel must not respond to an emergency incident scene unless they have been dispatched or have an assigned duty to respond. Unassigned units and individual personnel arriving on the scene can overload the IC's ability to manage the incident effectively. Individuals who simply show up and find something to do are likely to compromise their own safety and create more problems for the command staff. All personnel must operate within the established system, reporting to a designated supervisor under the direction of the IC.

Teamwork

On the fire ground or emergency scene, a firefighting team should always consist of at least two fire fighters who work together and are in constant communication with each other. Some departments call this scheme the buddy system (**FIGURE 2-12**). In some cases, fire fighters work directly with the company officer, and all crew members function as a team. In other situations, two individual fire fighters may form a team that is assigned to carry out a specific task. In either case, the company officer must always know where teams are and what they are doing.

Partners or assigned team members should enter together, work together, and leave together. If one member of a team must leave the fire building for any reason, the entire team must leave together, regardless of whether it is a two-person team or an entire crew working as a team. Once one SCBA alarm activates, the entire crew should exit the structure together. The officer should be the last to leave to ensure everyone has evacuated the scene. The officer reports the incident through the accountability system.

Before entering a burning building to perform interior search and rescue or fire suppression operations, fire fighters must be properly equipped with approved and appropriate PPE. Partners should check each other's PPE to ensure it has been donned and is working correctly before they enter a hazardous area. Team members should maintain visual, vocal, or physical contact with one another at all times. At least one member of each team should have a portable radio to maintain contact with the IC or a designated individual in the chain of command who remains outside the hazardous area. Two fire fighters must remain outside the hazardous area, properly equipped to respond immediately if the entry team needs to be rescued. This rapid intervention crew/company (also known as a rapid intervention team) must be able to communicate with the entry team, either by sight or by radio, and should be ready to provide assistance. For more information about the importance of teamwork and a rapid intervention crew/company, see Chapter 18, *Fire Fighter Survival.*

Personnel Accountability

Every fire department must have a personnel accountability system to track personnel and assignments on the emergency scene. This system should record the individuals assigned to each company, crew, or entry team; the assignments for each team; and the team's current activities. If any fire fighters are reported missing, lost, or injured, the accountability system can identify who is missing and what his or her last assignment was. Several kinds of accountability systems are acceptable, ranging from paper assignments or display boards to laptop computers and electronic tracking devices.

Some departments use a "passport" system. In this scheme, each company officer carries a small magnetic board called a passport. Each crew member on duty with that company has a magnetic name tag on the board (**FIGURE 2-13**). At the incident scene, the company's passport is given to a designated

FIGURE 2-12 A firefighting team should consist of at least two members who work together.
© Jones & Bartlett Learning.

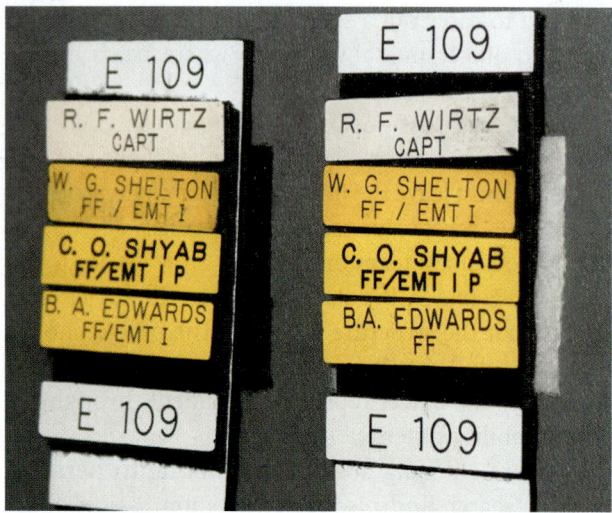

FIGURE 2-13 A passport lists each fire fighter assigned to a crew.

individual, who uses it to track the assignment and location of every company at the incident.

Other fire departments employ an accountability-tag system. With this approach, each fire fighter carries a name tag and turns it in or places it in a designated location on the apparatus when he or she is on the scene. The tags are then collected and used to track crews who are working together as a unit. This system works well when crews are organized based on the available personnel.

Accountability systems provide an up-to-date accounting of everyone who is working at the incident and how he or she is organized. At set intervals, an accountability check, or personnel accountability report (PAR), is completed to account for everyone. A personnel accountability report is also completed when the operational strategy changes, a significant fire event or collapse occurs, or when a situation occurs that could endanger fire fighters. Usually, a company officer reports on the status of each crew. The company officer should always know exactly where each crew is and what it is doing. If a crew splits into two or more teams, the company officer should be in contact with at least one member of each team. A list of personnel and assignments should be readily available at the command post.

Fire fighters must learn which kind of accountability system their department uses, how to work within it, and how this system works within the ICS. They are responsible for complying with this system and staying in contact with a company officer or assigned supervisor at all times. Most importantly, fire fighters must always remember that teams must stay together.

Chapter 18, *Fire Fighter Survival*, covers many of these accountability systems, self-survival skills, and the steps for calling for help ("mayday").

Scene Hazards

Fire fighters must be aware of their surroundings when performing their assigned tasks at an emergency scene. At an incident, make a safe exit from the fire apparatus, and look at the building or situation for safety hazards such as traffic, downed utility wires, and adverse environmental conditions. If arriving at a fire call, dismounting firefighters should look up at the roof line to determine the roof structure and begin evaluating the structure type and occupancy. The introduction of new technologies requires fire fighters to be familiar with their run areas and observant of new hazards. This includes the increasingly common photovoltaic power systems (solar power systems) and battery energy storage systems within electrical power grids. An incident on a street or highway must first be secured with proper traffic- and scene-control devices. A variety of traffic-incident management techniques may be appropriate. Flares, traffic cones, and barrier tape are all measures that can help keep the public at a safe distance from the scene and to divert traffic around the area. Emergency vehicles can be placed to block traffic and protect the incident scene. Always operate within established boundaries and protected work areas.

Changing fire conditions also will affect safety. Through situational awareness, fire fighters monitor the changing conditions and maintain safety. This should continue even during the overhaul phase and while picking up equipment—watch out for falling debris, smouldering areas of fire, and sharp objects. If a safety officer is not on the scene, another qualified person should be assigned to monitor the atmosphere to ensure it is safe to remove SCBA or enter the area without SCBA. Because the chance for injury increases when you are tired, do not let down your guard even though the main part of the firefighting operation is over. For more information about hazard recognition, see Chapter 18, *Fire Fighter Survival*.

Utilities

Controlling utilities is one of the first tasks that must be accomplished at many working structure fires. Once utilities have been controlled, this information should be communicated to the IC. If additional utility hazards are found while conducting operations, the IC should be notified immediately. Everyone working on the fire ground should know the status of the utilities. If control of the utilities cannot be completed, the IC should determine the safest course of action. Most departments have written SOPs that define when the utilities are to be shut off. Although this responsibility is often assigned to a particular company or crew, all fire fighters should know how to shut off a building's electrical, gas, and water service.

Controlling utilities is particularly important if fire fighters need to open walls or ceilings to look for fire in the void spaces, to cut ventilation holes in roofs, or to penetrate through floors. Because these spaces often contain both electrical lines and gas pipes, the danger of electrocution or explosion exists unless these utilities are disconnected.

Electrical, gas, and water utilities at a burning building may be disconnected for any of several reasons. For example, if faulty electrical equipment or a gas appliance caused the fire, shutting off the supply will help alleviate the problem. It also will reduce the risk of injury to fire fighters.

Electrical Service

Electricity is an invisible emergency scene hazard that must always be respected. Electricity is present in most structure fires, many outdoor emergencies, and in many traffic collisions. Disconnecting electrical service reduces the risk of electrical injury to fire fighters. For example, a fire fighter using an axe to open a void space could be electrocuted if the tool contacts energized electrical wires. Disconnecting electrical service can prevent problems such as short circuits and electrical arcing that could result from fire or water damage. In addition, disconnecting electrical service eliminates potential ignition sources that might cause an explosion if leaking gas has accumulated inside the building. The electrical supply must be disconnected at a location outside the area where gas might be present. Interrupting the power from a remote location alleviates the risk of an electrical arc that could cause a gas explosion.

The type of electrical service delivery depends on the utility company providing that service and the age of the system. The most common installation in older areas is a service drop from above-ground utility wires to the electric meter, which is typically mounted on the outside of the building (**FIGURE 2-14**). Some newer developments include underground distribution systems that include underground cables and transformers. Electrical meters may be located inside or outside of a building. There may be separate meters for each tenant or one meter for multiple occupancies.

Electrical wires may be energized yet nearly invisible when lying on the ground or when dangling from a pole or a building. This is especially true in conditions of limited visibility such as darkness, fog, and smoke. Always check for overhead power lines before raising ladders. During any major fire, the electrical power supply to the building should be turned off as part of the fire-ground task called "controlling the utilities."

FIGURE 2-14 The electrical service often has an exterior meter connected to above-ground utility wires.
© Jones & Bartlett Learning. Photographed by Glen E. Ellman.

In addition to structure fires, fire departments are often called to electrical emergencies, such as those involving downed power lines, fires, arcing wires, and transformer fires. Whenever there is a possibility that there are live power lines, have the power disconnected or turn off the power if you are trained and permitted by department policy.

Fire apparatus should be parked outside the area and away from power lines when responding to a call for an electrical emergency. A downed power line should be considered energized until the power company confirms that it is dead. Secure the area around the power line, and keep the public, and other first responders, at a safe distance. Never drive fire apparatus over a downed line or attempt to move it using tools. If a sparking power line causes a brush fire, and it is safe to do so, attempt to contain the fire from a safe distance, but do not use water near the power line.

Fire departments must work with their local electrical utility companies to identify the different types of electrical services and arrangements used in the area and to determine the proper procedures for dealing with each type. Often, a main disconnect switch can be operated to interrupt the power. In general, to turn off an electrical service to a structure, you must first identify the circuit breaker box that supplies the building or part of the building in which you need to disconnect the power. Then you can turn off the main circuit breaker that supplies the whole breaker box. It is important to leave a fire fighter at the circuit breaker box, or to use a lockout system that tags the box, to ensure that the power does not get turned on accidentally. Most utility companies will train fire fighters to identify and operate shut-off devices on typical installations. Other utility companies request that the fire department call a company representative to shut off the power. Shutting off large systems that involve high-voltage equipment should be done by a company technician or a trained individual from the premises.

Sometimes a utility company representative needs to be called to interrupt power from a remote location, such as a utility pole. This step may be necessary if the outside wires have been damaged by fire, if fire fighters are working with ladders or aerial apparatus, or if an explosion is possible. In urban areas, the electrical utility company can usually dispatch a qualified technician or crew to respond quickly to a fire or emergency incident. In many cases, qualified technicians are dispatched automatically to all working fires. Some utility companies can interrupt the service to an area from a remote location in response to a fire department request.

Gas Service

Shutting down gas service eliminates the potential for explosion due to damaged, leaking, or ruptured gas piping. For example, a power saw slicing through a gas pipe can cause a rapid release of gas and create the potential for an explosion. A structural collapse can rupture gas lines or create leaks in the stressed piping.

Both natural gas and liquefied petroleum gas (LPG) are used for heating and cooking. Generally, natural gas is delivered through a network of underground pipes. By contrast, LPG is usually delivered by a tank truck and stored in a container on the premises. In some areas, LPG is distributed from one large storage tank through a local network of underground pipes to several customers.

A single valve usually controls the natural gas supply to a building. This valve is generally located outside the building at the entry point of the gas piping; natural gas service has a distinctive piping arrangement. In older buildings, the shut-off valve for a natural gas system may be located in the basement. The shut-off valve for a natural gas system is usually a quarter-turn valve with a locking device so it can be secured in the off position (**FIGURE 2-15**). When the handle is in line with the pipe, the valve is open; when the handle is at a right angle to the pipe, the valve is closed. A special key, an adjustable wrench, or a spanner wrench can be used to turn the handle.

The shut-off valve for an LPG (propane) system is usually located at the storage tank (**FIGURE 2-16**).

FIGURE 2-15 The natural gas shut-off valve is generally located outside the building at the entry point of the gas piping. In older buildings, the shut-off valve for a natural gas system may be located in the basement.
© lucag_g/Shutterstock.

FIGURE 2-16 The shut-off valve for an LPG system is usually located at the storage tank.
Courtesy of Amanda Mitchell.

The more common type of LPG valve, however, has a distinctive handle that indicates the proper rotation direction to open or close the valve. To shut off the flow of gas, rotate the handle to the fully closed position.

A gas valve that has been shut off must not be reopened until the system piping has been inspected by a qualified person. Air must be purged from the system, and any pilot lights must be reignited to prevent gas leaks.

Water Service

If a serious water leak has occurred inside the building, shutting off the water supply may help to minimize additional water damage to the structure and contents.

Water service to a building can usually be shut off by closing one valve at the entry point. Many communities permit the water service to be turned off at the connection between the utility pipes and the building's system. This underground valve, which is often accessed through a curb box, is located outside the building and can be operated with a special wrench or key. In most cases, another valve is found inside the building, usually in the basement (if there is one), where the water line enters. In warmer climates, water supply valves are sometimes located above ground.

Lifting and Moving

Lifting and moving objects are part of a fire fighter's daily duties. Do not try to move something that is too heavy alone—ask for help. Never bend at the waist to lift an object; instead, bend at the knees, and use your legs to lift the weight. If you must move objects over a long distance, use equipment such as handcarts, hand trucks, and wheelbarrows.

In addition to moving inanimate objects, fire fighters must often move sick or injured victims. Do not move an injured person, unless his or her life is in danger, until the appropriate medical personnel are on scene to stabilize the victim. Discuss and evaluate the options before moving a victim, and then proceed very carefully. If necessary, request help. Never be afraid to call for additional resources, such as another pump or ladder company, to assist in lifting and moving a heavy victim.

Adverse Weather Conditions

In adverse weather conditions, fire fighters must dress appropriately. A structural firefighting protective coat, pants, boots, and helmet can keep you warm and dry in rain, snow, or ice. Firefighting gloves and knit caps also help retain body heat and keep you warm. If conditions are icy, make smaller movements, watch your step, and keep your balance. During the hot summer months, it is difficult to stay cool in PPE. Many departments have SOPs covering PPE use in summer months. Check your department's SOPs for guidance on acceptable PPE use in adverse weather.

Rehabilitation

Rehabilitation is a systematic process that provides longer periods of rest and recovery for emergency workers during an incident. It is usually conducted in a designated area away from the hazards of the emergency scene. The rehabilitation area, or "rehab," is usually staffed by fire and EMS personnel.

Never be afraid or embarrassed to admit you need a break when on the emergency scene. Typically, breaks are in the form of recycling and rehabilitation. Recycling is brief and consists of taking a few minutes to rest, hydrate, and change the SCBA bottle before getting back to work.

While in rehabilitation, fire fighters should take advantage of the opportunity to rest, rehydrate, have their vital signs checked, and receive treatment for minor injuries (**FIGURE 2-17**). Rehabilitation gives fire fighters the chance to cool off in hot weather and to warm up in cold weather. Rehabilitation time can also be used to replace SCBA cylinders, obtain new batteries for portable radios, and make repairs or adjustments to tools or equipment. Firefighting teams can discuss recently completed assignments and plan their next work cycle. When a crew is released from rehabilitation, its members should be rested, refreshed, and ready for another work cycle. If members of the crew are too exhausted or unable to return to work or have not been medically cleared to return to work, they should be replaced and released from the incident. More information on fire fighter rehabilitation is presented in Chapter 20, *Fire Fighter Rehabilitation*.

FIGURE 2-17 In the rehabilitation area, fire fighters can rest and rehydrate.
© Jones & Bartlett Learning. Photographed by Glen E. Ellman.

Violence

Fire fighter safety is jeopardized by violence. Fire fighters are sometimes dispatched to calls involving domestic disputes, active shooters, injuries from an assault, or other violent scenes. In such incidents, the staging area and the fire apparatus should be located at some distance from the scene until the police arrive, investigate, and declare the scene safe. Only then should fire fighters proceed to enter the emergency scene. Recently, there have been violent incidents where fire fighters have been targeted, and in some cases, they have been mistaken for intruders. This increases the need for situational awareness at all times. A fire fighter's personal safety should always be paramount. If there is any threat to your personal safety, retreat as quickly as possible from the emergency scene to a safe distance, and request the police to secure the scene. Do not become a victim.

If you are confronted with a potentially violent situation, do not respond violently. Remain calm, speak quietly, and attempt to gain the person's trust and call for police assistance. You might consider taking additional classes to increase your understanding and develop appropriate skills for these situations.

Safety at the Fire Station

The fire station is just as much a workplace as the fire ground. Indeed, fire fighters spend much of their time during a shift at their fire station. The length of each shift varies, from 12 to 24, to 48, to 72 hours and, in some cases, to a 96-hour shift. Some may be longer. The need for safety around the fire station is as important as it is on the fire ground or at other emergency scenes. While in the fire station, be careful when working with power tools and equipment, ladders, electrical appliances, pressurized cylinders, and hot surfaces. Practise using tools and equipment properly and safely before attempting to use them at an actual emergency incident. Follow the proper procedures and safety precautions both in training and at an incident scene.

Equipment should always be in excellent condition and ready for use. Proper maintenance of tools includes sharpening, lubricating, and cleaning each tool. All fire fighters should be able to do basic repairs such as changing a saw blade or a hand-light battery. Practise using and maintaining tools and equipment at the fire station until you can perform these tasks quickly and safely. Remember, injuries that occur at the firehouse can be just as devastating as those that occur at an emergency incident scene.

Safety Outside Your Workplace

Continue to follow safe practices when you are off duty. An accident or injury, regardless of where it happens, can end your career as a fire fighter. For example, if you are using a ladder while off duty, follow the same safety practices that you would use while on duty. Keep your seat belt fastened in your personal vehicle, just as you are required to do when you are on duty (**FIGURE 2-18**).

FIGURE 2-18 Always use your seat belt.
© Jones & Bartlett Learning. Photographed by Glen E. Ellman.

After-Action REVIEW

IN SUMMARY

- Every fire fighter and fire department must have a strong commitment to safety and health. Safety must be fully integrated into every activity, procedure, and job description.

- In the United States, the majority (42 percent) of all fire fighter deaths are caused by stress, overexertion, and medical issues. Cardiac events account for 38 percent of these deaths.

- Motor vehicle accidents are the second leading cause of fire fighter fatalities in the United States. Always wear your seat belt each and every time you are in a motor vehicle.

- The majority of all fire-ground injuries are caused by overexertion or strain.

- Fire fighters must always consider three groups when ensuring safety at the scene:
  - Their personal safety
  - The safety of other team members
  - The safety of everyone at the emergency scene

- A successful safety program must have four major components:
  - Standards and procedures—Regulations, ranging from NFPA standards, to OSHA regulations, to local laws, govern the fire department's SOPs and SOGs on ensuring a safe work environment.
  - Personnel—From teams working together to safety officers, personnel ensure that a fire department's safety program is implemented correctly.
  - Training—Training does not end with this course. To ensure that you attain and maintain a high level of skills and knowledge, you must seek out additional courses and training opportunities throughout your career.
  - Equipment—To ensure safety, equipment must be properly maintained and fire fighters must be trained how to use it correctly.

- The National Fallen Firefighters Foundation's "16 Firefighter Life Safety Initiatives" describes the steps that need to be taken to change the current culture of the fire service to help make it a safer place for all fire fighters.

- Safety and well-being are directly related to personal health and fitness. You should eat a healthy diet, maintain a healthy weight, exercise regularly, stay hydrated, and sleep for 7 to 8 hours whenever possible.

- A well-rounded fitness routine includes weight training, cardiovascular workouts, and stretching. A well-rounded diet includes appropriate portions of fruits, vegetables, healthy fats, whole grains, and lean protein.

- Employee assistance programs are available to provide fire fighters with confidential counselling, support, or assistance in dealing with a physical, financial, emotional, or substance abuse problem.

- The aim of counselling, peer support teams, employee assistance programs, and critical incident stress management programs is to prevent emotional reactions from having a negative impact on the fire fighter's work and life, over both the short term and the long term.

- Many fire fighter fatalities occur during training exercises. Do not attempt anything you feel is beyond your ability or knowledge. If you see something you feel is unsafe, say something.

- Response actions include receiving the alarm, donning protective clothing and equipment, mounting the apparatus, and transporting equipment and personnel to the emergency incident quickly and safely.

- The response process begins when the alarm is received at the fire station. Dispatch messages will include the location of the incident, the type of emergency, and the units that are due to respond.

- Don your PPE before mounting the apparatus or after arriving at the emergency scene. While riding in the apparatus, your seat belt should be secure, and you should be concentrating on mentally preparing for the emergency. Consider any relevant factors that could affect the emergency, such as the weather and the time of day.

- Upon arrival at the scene, traffic safety should be a major concern. Always check for traffic before exiting the apparatus. Follow departmental SOPs to close streets quickly and block access for civilian vehicles to the incident.
- Four general principles govern emergency vehicle operation:
  - Emergency vehicle operators are subject to all traffic regulations unless a specific exemption is made.
  - Exceptions are legal only when operating in emergency mode.
  - Even with an exemption, the emergency vehicle operator can be found criminally or civilly liable if involved in a crash.
  - An exemption does not relieve the operator of an authorized emergency vehicle from the duty to drive with reasonable care for all persons using the highway.
- Emergency driving requires good reactions and alertness. Driving while impaired or distracted is unacceptable. Your driving record, both on duty and off duty, will impact your career as a fire fighter.
- Safe driving practices will prevent most vehicle collisions. Anticipate the road and road conditions. Make allowances for weather conditions. Learn where the greatest risks are located on your routes. Be alert for the presence of other emergency vehicles.
- Fire fighters should always work in assigned teams and be guided by a strategic plan for the incident. Teamwork and disciplined action are essential to provide for the safety of all fire fighters and the effective, efficient conduct of operations.
- Upon arriving at the scene, check in to the personnel accountability system. This system tracks personnel and assignments at the emergency scene. Several types of systems are acceptable, ranging from paper assignments, to display boards, to laptop computers and electronic tracking devices.
- To protect the safety of fire fighters, controlling utilities is one of the first tasks to be accomplished. The electrical service and gas supply should be shut off and locked.
- Do not try to move a heavy object alone—ask for help.
- The fire station is just as much a workplace as the fire ground. Be careful when working with power tools, ladders, electrical appliances, pressurized cylinders, wet floors, and hot surfaces.
- An accident or injury, regardless of when or where it happens, can end your firefighting career. Continue to follow safe practices when you are off duty.

KEY TERMS

Access Navigate for flashcards to test your key term knowledge.

Critical incident stress debriefing (CISD) A post-incident meeting designed to assist rescue personnel in dealing with psychological trauma as the result of an emergency. (NFPA 1006)

Critical incident stress management (CISM) A program designed to reduce acute and chronic effects of stress related to job functions. (NFPA 450)

Employee assistance programs (EAPs) An employee-sponsored service designed for personal or family problems, including mental health, substance abuse, various addictions, marital problems, parenting problems, emotional problems, or financial or legal concerns. (NFPA 450)

Freelancing The dangerous practice of acting independently of command instructions.

Occupational Safety and Health Administration (OSHA) The U.S. federal agency that regulates worker safety and, in some cases, responder safety. OSHA is part of the U.S. Department of Labor.

Response Immediate and ongoing activities, tasks, programs, and systems to manage the effects of an incident that threatens life, property, operations, or the environment. (NFPA 1600)

REFERENCES

American Cancer Society. 2017. "Alcohol Use and Cancer." https://www.cancer.org/cancer/cancer-causes/diet-physical-activity/alcohol-use-and-cancer.html. Accessed October 11, 2018.

Centers for Disease Control and Prevention. 2016. "Findings from a Study of Cancer Among U.S. Fire Fighters." https://www.cdc.gov/niosh/firefighters/health.html. Accessed October 11, 2018.

CTV Vancouver. "Cancer the Leading Cause of Death for Canadian Fire Fighters: Study, 2018". https://bc.ctvnews.ca/cancer-the-leading-cause-of-death-for-canadian-firefighters-study-1.3862986. Accessed February 11, 2019.

Firefighter Behavioral Health Alliance. www.ffbha.org. Accessed October 11, 2018.

Firefighter Cancer Support Network. "Who We Are." https://firefightercancersupport.org/Who.We.Are/. Accessed October 11, 2018.

International Association of Fire Fighters. IAFF Cancer Summit, 2018. www.iaff.org. Accessed October 11, 2018.

National Fire Protection Association. "Firefighter Fatalities in the United States - 2016." Figure 2. http://www.nfpa.org/news-and-research/fire-statistics-and-reports/fire-statistics/the-fire-service/fatalities-and-injuries/firefighter-fatalities-in-the-united-states. Accessed October 11, 2018.

National Fire Protection Association. "U.S. Firefighter Injuries - 2016." http://www.nfpa.org/News-and-Research/Fire-statistics-and-reports/Fire-statistics/The-fire-service/Fatalities-and-injuries/Firefighter-injuries-in-the-United-States. Accessed October 11, 2018.

National Fire Protection Association. NFPA 1250, *Recommended Practice in Fire and Emergency Services Organization Risk Management*, 2015. www.nfpa.org. Accessed October 11, 2018.

National Fire Protection Association. NFPA 1451, *Standard for a Fire and Emergency Service Vehicle Operations Training Program*, 2018. www.nfpa.org. Accessed October 11, 2018.

National Fire Protection Association. NFPA 1500, *Standard on Fire Department Occupational Safety, Health, and Wellness Program*, 2018. www.nfpa.org. Accessed October 11, 2018.

National Fire Protection Association. NFPA 1582, *Standard on Comprehensive Occupational Medical Program for Fire Departments*, 2018. www.nfpa.org. Accessed October 11, 2018.

The National Fallen Firefighters Association. "Everyone Goes Home. Confronting Suicide in the Fire Service." https://www.everyonegoeshome.com/2014/12/02/suicides-preventable-reaching-vulnerable/. Accessed October 11, 2018.

Smith, Denise. 2017. "Can SMARTER Technology Reduce Firefighter Injuries and Fatalities?" http://horizons.globeturnoutgear.com/can-smarter-technology-reduce-firefighter-injuries-and-fatalities/. Accessed October 11, 2018.

On Scene

You live in a small town, so it is common for the fire fighters to personally know their elected officials. One day, you are eating at the only diner in town when the mayor comes over and asks if you mind if he sits down with you. You readily say yes to the opportunity to eat with the mayor.

The conversation quickly turns to the fire department. From his line of questioning, you quickly recognize that the mayor does not understand the risks of firefighting and why safety has become so important. He says he is curious why the fire fighters need new safety gear and equipment. You explain that he should contact your chief to discuss the matter further. As the conversation shifts gears, you think about how your chief will answer this question.

1. How many fire fighters were killed in the United States in the line of duty in 2016?

A. 50

B. 69

C. 130

D. 98

2. What is the cause of most fire fighter deaths in the United States?

A. Burns

B. Building collapse

C. Stress, overexertion, and medical issues

D. Vehicle accidents

(continued)

On Scene Continued

3. Out of all fire fighter injuries in the line of duty in the United States, what percentage occurred on the fire ground?

   A. 75 percent

   B. 39 percent

   C. 65 percent

   D. More than 50 percent

4. Which program was created at the National Firefighter Life Safety Summit in 2004 to help describe the steps that need to be taken to change the current culture of the fire service?

   A. U.S. Fire Administration's Firefighter Health, Wellness and Fitness program

   B. 16 Firefighter Life Safety Initiatives

   C. Firefighter Close Calls Resources

   D. SAFENET

5. According to the Firefighter Cancer Support Network, what is the first step to helping to protect yourself from cancer?

   A. Change your clothes, and wash them immediately after a fire.

   B. Decontaminate the interior of the fire apparatus after a fire.

   C. Use self-contained breathing apparatus from initial attack to finish of overhaul.

   D. Keep structural firefighting protective gear out of living quarters.

6. What is the most important action you can take to ensure your safety while en route to an emergency incident?

   A. Don personal protective clothing.

   B. Fasten your seat belt.

   C. Fasten your self-contained breathing apparatus straps.

   D. Wear your helmet.

7. Another reason for needing new safety PPE would be that the leading cause of firefighter injury in the United States in 2016 was which of the following?

   A. Training

   B. Responding to or returning from an alarm

   C. Non-fire emergencies

   D. Operating at the fire ground

 *Access Navigate to find answers to this On Scene, along with other resources such as an audiobook.*

Personal Protective Equipment

KNOWLEDGE OBJECTIVES

After studying this chapter, you will be able to:

- List the components of personal protective equipment (PPE). (pp. 67–73)
- Explain the role of the fire fighter's work clothing as part of the PPE ensemble. (p. 67)
- Describe the type of protection provided by structural firefighting PPE. (pp. 68–74)
- Explain how each design element of a fire helmet works to protect the head, face, and eyes. (pp. 69–70)
- Explain why protective hoods are a part of structural firefighting PPE. (p. 70)
- Explain how each design element of a structural firefighting protective coat works to protect the upper body. (pp. 70–71)
- Explain how each design element of structural firefighting protective pants works to protect the lower body. (p. 71)
- Describe how each design element of boots works to protect the feet. (pp. 71–72)
- Describe how each design element of gloves works to protect the hands and wrists. (p. 72)
- Explain how a personal alert safety system (PASS) helps to ensure fire fighter safety. (**NFPA 1001: 4.3.1**, p. 72)

- List the limitations of PPE. (p. 74)
- Describe the procedure for donning personal protective clothing. (**NFPA 1001: 4.1.2**, pp. 74–76)
- Describe the procedure for doffing personal protective clothing. (**NFPA 1001: 4.1.2**, pp. 77–79)
- Describe how to inspect the condition of PPE. (**NFPA 1001: 4.1.2**, pp. 79–80)
- Describe how to properly maintain PPE. (**NFPA 1001: 4.1.2, 4.5.1**, pp. 79–80)
- Describe why thoroughly cleaning PPE immediately after it has been exposed to smoke or fire conditions is an important step in reducing your chance of developing cancer. (**NFPA 1001: 4.1.2, 4.5.1**, pp. 77–79)
- Describe the specialized protective equipment required for vehicle extrication and wildland fires. (p. 80)
- List the respiratory hazards posed by smoke and fire. (**NFPA 1001: 4.3.1**, pp. 81–82)
- List the conditions that require respiratory protection or self-contained breathing apparatus (SCBA). (**NFPA 1001: 4.3.1**, p. 82)
- Describe the types of breathing apparatus. (pp. 82–83)
- Describe the differences between open-circuit self-contained breathing apparatus and closed-circuit self-contained breathing apparatus. (p. 84)
- Describe the limitations of SCBA. (**NFPA 1001: 4.3.1**, pp. 85–86)

- Describe the physical and psychological limitations of an SCBA user. (**NFPA 1001: 4.3.1**, pp. 86–87)
- List and describe the major components of SCBA. (**NFPA 1001: 4.3.1**, pp. 87–89)
- Describe the devices on an SCBA that can assist the user in air management. (**NFPA 1001: 4.3.1**, pp. 90–91)
- Describe the pathway that air travels through an SCBA. (p. 91)
- Explain the breathing techniques used to conserve air supply. (**NFPA 1001: 4.3.1**, p. 91)
- List the complete sequence of donning PPE. (**NFPA 1001: 4.1.2, 4.3.1**, p. 104)
- Describe the importance of SCBA inspections and SCBA operational testing. (**NFPA 1001: 4.5.1**, pp. 107–110)
- Explain how to inspect an SCBA to ensure that it is operation ready. (**NFPA 1001: 4.5.1**, pp. 107–110)
- Explain the procedures for refilling SCBA air cylinders. (pp. 116, 120)

SKILLS OBJECTIVES

After studying this chapter, you will be able to perform the following skills:

- Don approved personal protective clothing. (**NFPA 1001: 4.1.2**, pp. 74–76)
- Doff approved personal protective clothing. (**NFPA 1001: 4.1.2**, pp. 77–79)
- Don an SCBA from an apparatus seat mount. (**NFPA 1001: 4.3.1**, pp. 92–95)
- Don an SCBA from an apparatus compartment mount. (**NFPA 1001: 4.3.1**, p. 96)
- Don an SCBA from a storage case using the over-the-head method. (**NFPA 1001: 4.3.1**, pp. 97–98)
- Don an SCBA from a storage case using the coat method. (**NFPA 1001: 4.3.1**, pp. 99–101)
- Don a face piece. (**NFPA 1001: 4.3.1**, pp. 101, 102–103)
- Doff an SCBA. (**NFPA 1001: 4.3.1**, pp. 104, 105–107)
- Perform a visible inspection of an SCBA. (**NFPA 1001: 4.5.1**, pp. 107–110)
- Perform an operational inspection of an SCBA. (**NFPA 1001: 4.5.1**, pp. 108–110, 111–113)
- Replace an SCBA air cylinder. (**NFPA 1001: 4.3.1**, pp. 110, 114–116)
- Replace an SCBA air cylinder on another fire fighter. (**NFPA 1001: 4.3.1**, pp. 114, 116, 117–119)
- Refill an SCBA air cylinder from a compressor or a cascade system. (**NFPA 1001: 4.5.1**, pp. 116, 120, 121)
- Clean an SCBA. (**NFPA 1001: 4.5.1**, pp. 120, 122–123)

ADDITIONAL STANDARDS

- Canadian Standards Association (CSA), Z94.4-18, *Selection, Care, and Use of Respirators*
- Canadian Standards Association (CSA), Z180.1-13, *Compressed Breathing Air and Systems*
- Canadian Standards Association (CSA), CAN/CSA Z96-15, *High-Visibility Safety Apparel*.
- **NFPA 1404**, *Standard for Fire Service Respiratory Protection Training*
- **NFPA 1500**, *Standard on Fire Department Occupational Safety, Health, and Wellness Program*
- **NFPA 1582**, *Standard on Comprehensive Occupational Medical Program for Fire Departments*
- **NFPA 1851**, *Standard on Selection, Care, and Maintenance of Protective Ensembles for Structural Fire Fighting and Proximity Fire Fighting*
- **NFPA 1852**, *Standard on Selection, Care, and Maintenance of Open-Circuit Self-Contained Breathing Apparatus (SCBA)*
- **NFPA 1971**, *Standard on Protective Ensembles for Structural Fire Fighting and Proximity Fire Fighting*
- **NFPA 1975**, *Standard on Emergency Services Work Clothing Elements*
- **NFPA 1977**, *Standard on Protective Clothing and Equipment for Wildland Fire Fighting*
- **NFPA 1981**, *Standard on Open-Circuit Self-Contained Breathing Apparatus (SCBA) for Emergency Services*
- **NFPA 1982**, *Standard on Personal Alert Safety Systems (PASS)*

Fire Alarm

One cool fall evening, you are dispatched for a report of a smoke odour in the area. It seems like a fairly routine call. The residents are beginning to use their fireplaces for the first time of the year, and the weather conditions tend to cause the smoke to hang low. As the pumper drives slowly through the neighbourhood, you notice the distinctive smell of a working house fire rather than that of firewood.

As you turn the corner, you see smoke drifting from the basement window of a house. Your captain completes a quick size-up and requests a full assignment from dispatch. She then tells you and another fire fighter to pull a cross-lay and attack the fire.

1. Which personal protective equipment would you need for this situation?

2. What is the most efficient way to don a self-contained breathing apparatus?

3. Once the fire is over, what is the proper way to care for your protective equipment?

 *Access Navigate for more practice activities.*

Introduction

Two safety components used by fire fighters require special consideration: personal protective equipment (PPE) and self-contained breathing apparatus (SCBA). Fire fighter safety and individual job performance depend upon a complete understanding of all safety equipment, including its limitations. PPE is designed to protect fire fighters from heat and physical injury. SCBA provides respiratory protection for an established period of time depending on the type of air supply being used. The use of PPE, in combination with SCBA, provides fire fighters with protection to enter smoke-filled and toxic environments. Improper use of either PPE or SCBA can result in serious injury or death.

Personal Protective Equipment

Personal protective equipment (PPE) is an essential component of a fire fighter's safety system. It enables a person to survive under conditions that might otherwise result in death or serious injury. Different PPE are designed for specific hazardous conditions, such as structural firefighting, wildland firefighting, airport rescue and firefighting, hazardous materials operations, and emergency medical operations.

PPE provides protection based on specifications that clearly define the design, application, and limitations of the gear. It is the responsibility of every fire fighter to be familiar with these specifications to ensure that the equipment is used within the manufacturer's

recommended guidelines. The following information focuses on standard fire service PPE, work clothing, and SCBA use.

Work Clothing

At the lower end of the PPE spectrum is normal street clothing or work uniforms, which offer the least amount of protection. Fire fighters (who are not responding to a call), police officers, and emergency medical services (EMS) providers typically wear this level of "protection."

Keep in mind that certain synthetic fabrics can melt at relatively low temperatures and cause severe burns, even when worn under PPE. Clothing containing nylon or polyester, even if these materials are blended with natural fibres, should not be worn in a firefighting environment. Clothing made of natural fibres, such as cotton or wool, is generally safer to wear. Nylon and polyester will melt, but cotton and wool may char or even burn without melting. Special synthetic fibres such as Nomex® and PBI®, which are used in structural firefighting protective clothing, have excellent resistance to high temperatures. NFPA 1975, *Standard on Emergency Services Work Clothing Elements*, defines criteria for selecting appropriate fabrics for work uniforms.

SAFETY TIP

Do not mix and match different brands and styles of personal protective clothing. Some styles are not compatible with others and may leave gaps that will expose the wearer.

Structural Firefighting PPE

Structural firefighting PPE provides the highest level of protection for personnel engaged in structural firefighting (**FIGURE 3-1**). A fire fighter's PPE must provide full-body coverage and protection from a variety of hazards. To be effective, the entire ensemble must be worn whenever potential exposure to those hazards exists. Structural firefighting PPE enables fire fighters to enter burning buildings and work in areas with elevated temperatures and concentrations of toxic gases. Without PPE, fire fighters would be unable to conduct search and rescue operations or perform fire suppression activities.

Structural firefighting PPE consists of a protective coat, pants, coveralls, a helmet, a protective hood, boots, and gloves. The clothing can be worn with a **self-contained breathing apparatus (SCBA)** and a **personal alert safety system (PASS)** device. In order to achieve the maximum level of protection required for structural firefighting activities, all of these components must be worn together (**FIGURE 3-2**). Structural firefighting PPE is designed to cover every inch of the body. It provides protection from heat and fire, protects against water saturation, and helps reduce trauma from cuts or falls. There are a number of different manufacturers and styles of structural firefighting PPE.

PPE must be cleaned, maintained, and inspected regularly to ensure that it will continue to provide the intended degree of protection when it is needed.

FIGURE 3-2 Structural firefighting PPE consists of a helmet, coat, pants, protective hood, gloves, boots, an SCBA, and a PASS device.
© Jones & Bartlett Learning. Photographed by Glen E. Ellman.

SAFETY TIP

Structural firefighting protective ensemble is the term used to identify standard response PPE designed to protect the wearer from the wide range of interior and exterior fires to which a fire department responds. The ensemble is not designed for other functions, such as hazardous materials, water rescue, high-angle rescue, or wildland firefighting. Instead, each of these specialized functions requires a specific set of PPE. If your department performs specialized operations, you need the appropriate PPE for that activity.

Thorough cleaning of PPE immediately after it has been exposed to a fire situation is important to remove the cancer-causing chemicals that are part of smoke. PPE should also be cleaned after it has been exposed to biological contaminants, petroleum products, or chemicals. It is equally important to remove all of your clothing and wipe down your skin at the emergency scene. All gear should be bagged for cleaning immediately upon removal at the emergency scene. Shower as soon as possible to reduce your exposure to these carcinogens. Remember to repair or replace worn or damaged articles.

Protection Provided

Structural firefighting PPE is designed to provide full-body coverage and offers several different types of protection:

- Provides thermal protection
- Provides protection from water
- Provides impact protection
- Protects against cuts and abrasions
- Provides padding against injury
- Provides respiratory protection

The coat and pants have outer shells that can withstand elevated temperatures for a limited time, repel water, and provide protection from abrasions and sharp objects. The knees may be reinforced with pads for

Heads-up air pressure display

Integrated PASS device

Remote pressure gauge

Face piece

Air saver/donning switch

Purge valve

Regulator

FIGURE 3-1 Structural firefighting PPE provides protection from multiple hazards.
© Jones & Bartlett Learning. Photographed by Glen E. Ellman.

greater protection when crawling. Reflective trim adds visibility in dark or smoky environments. Insulating layers of fire-resistant materials protect the skin from high temperatures. A moisture barrier between different layers of PPE keeps liquids and vapours, such as hot water or steam, from reaching the skin. Helmets help provide head protection from falling debris, along with the face shield providing eye protection. A fire-retardant protective hood covers any exposed skin between the coat collar and the helmet. Gloves protect the hands from heat, cuts, and abrasions. Boots protect the feet and ankles from the fire, keep them dry, prevent puncture injuries, and protect the toes from crushing injuries.

Each item of PPE clothing is designed to overlap from one item to the next. For example, boots overlap the pants, coats overlap the pants, and coats overlap the gloves. This ensures that the body is protected at all times.

An SCBA is a **respirator** that provides an independent, limited air supply. In this way, the fire fighter's respiratory system remains safe from toxic products and hot gases present in the atmosphere.

PPE for fire fighters is manufactured according to exacting standards. According to NFPA 1971, *Standard on Protective Ensembles for Structural Fire Fighting and Proximity Fire Fighting*, each item must have a permanent label verifying that the particular item meets the requirements of a certification organization (**FIGURE 3-3**). Labels also include the appropriate term for the piece of PPE and manufacturer information, including date of manufacture, model name or number, size or size range, and principal material of construction. Usage limitations as well as cleaning and maintenance instructions should also be provided.

The requirements for firefighting PPE are outlined in two standards:

- NFPA 1971, *Standard on Protective Ensembles for Structural Fire Fighting and Proximity Fire Fighting*
- NFPA 1977, *Standard on Protective Clothing and Equipment for Wildland Fire Fighting*

The requirements for SCBA and PASS devices are outlined in these standards:

- Canadian Standards Association (CSA), Z94.4-18, *Selection, Care, and Use of Respirators*
- Canadian Standards Association (CSA), Z180.1-13, *Compressed Breathing Air and Systems*
- NFPA 1981, *Standard on Open-Circuit Self-Contained Breathing Apparatus (SCBA) for Emergency Services*
- NFPA 1982, *Standard on Personal Alert Safety Systems (PASS)*
- NFPA 1852, *Standard on Selection, Care, and Maintenance of Open-Circuit Self-Contained Breathing Apparatus (SCBA)*
- NFPA 1984, *Standard on Respirators for Wildland Fire Fighting Operations*

Helmet

Fire helmets are manufactured in several designs and shapes using a variety of materials. Nevertheless, each design must meet the requirements specified in NFPA 1971. The hard outer shell of the helmet is lined with energy-absorbing material and has a suspension system to provide impact protection against falling objects (**FIGURE 3-4**). The helmet shell also repels water, protects against steam, and creates a thermal barrier against heat and cold. Its shape helps to deflect water away from the head and neck.

Face and eye protection can be provided by a **face shield**, goggles, or both. These components must be

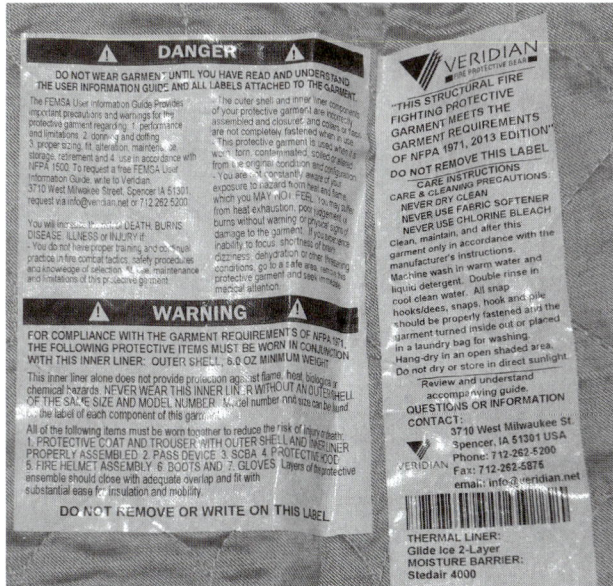

FIGURE 3-3 PPE labels provide important information about each item of PPE.
© Jones & Bartlett Learning. Photographed by Glen E. Ellman.

FIGURE 3-4 A helmet is constructed with multiple layers and components.
© Jones & Bartlett Learning. Photographed by Glen E. Ellman.

attached to the helmet and are used when an SCBA is not needed or when the SCBA face piece is not in place. A chin strap is required and must be worn to maintain the helmet in the proper position. This strap helps to keep the helmet on the fire fighter's head during an impact.

Fire helmets have an inner liner for added thermal protection. This liner also provides protection for the wearer's ears and neck. When entering a burning building, the fire fighter should pull down the ear tabs to ensure maximum protection.

Helmets are manufactured with adjustable inner suspension systems that hold the head away from the shell and cushion it against impacts. This suspension system must be adjusted to fit the individual, who should try on the helmet with the SCBA face piece and protective hood in place.

In many departments, helmet shells are colour coded according to the fire fighter's rank and function. In addition, they often identify the function of the company and carry company numbers, rank insignia, or other markings. Some fire departments use a shield mounted on the front of the helmet to identify the fire fighter's rank and company. Bright, fluorescent reflective materials or decals are applied to make the fire fighter more visible in all types of lighting conditions. Some decals are luminescent (glow in the dark).

Protective Hood

Although the helmet's ear tabs cover the ears and neck, this area is still at risk for burns when the head is turned or the neck is flexed. Protective hoods provide additional thermal protection for these areas. Such a hood, which is constructed of a flame-resistant material such as Nomex, PBI, or carbon fibre, covers the whole head and neck, except for that part of the face protected by the SCBA face piece (FIGURE 3-5). The lower part of the protective hood, which is called the bib, drapes down inside the protective coat. Because of the danger

from the contaminants in smoke itself, some protective hoods provide particulate protection to reduce the amount of contaminants that reach the skin.

Protective hoods are designed to be worn over the edges of the SCBA face piece but under the fire helmet. After securing the face piece straps, the fire fighter should carefully fit the protective hood around the face piece so that no areas of bare skin are left exposed. The protective hood must fit snugly around the clear area of the face piece so that vision is not compromised and hot gases cannot leak between the face piece and the protective hood. Protective hoods are available in various styles, lengths, materials, and colours.

Structural Firefighting Protective Coat

The structural firefighting protective coats meet NFPA 1971 requirements. These coats may be referred to as bunker coats or turnout coats (FIGURE 3-6).

Structural firefighting protective coats consist of three layers. The outer layer or shell is constructed of a sturdy, flame-resistant material such as Nomex, Kevlar®, or PBI. Fluorescent reflective material applied to the outer shell makes the fire fighter more visible in smoky conditions and at night. When light-coloured fabrics are used for protective coats, it makes it easier to identify contaminants such as hydrocarbons, blood, and body fluids on the coat.

The second layer of the protective coat is the moisture barrier, which usually consists of a flexible membrane attached to a thermal barrier material (the third layer). The moisture barrier helps prevent the transfer of water, steam, and other fluids to the skin. It is critical because water applied to a fire generates large amounts of super-heated steam, which can engulf fire fighters and burn unprotected skin.

The third layer of the protective coat comprises a thermal barrier, which is made of a multilayered or quilted material that insulates the body from external

FIGURE 3-5 A protective hood.
© Jones & Bartlett Learning. Photographed by Glen E. Ellman.

FIGURE 3-6 A structural firefighting protective coat.
© Jones & Bartlett Learning. Photographed by Glen E. Ellman.

temperatures. It enables fire fighters to operate in elevated temperatures generated by a fire and keeps the body warm during cold weather.

Structural firefighting PPE will protect you under limited conditions. The longer you remain in an environment with elevated temperatures, the more heat energy the clothing absorbs. After a certain period of time in a high-temperature environment, your gear will start to transfer more heat through the coat, and you will start to feel this stored heat energy transmitted to your body. This can happen very quickly in a fire situation.

The front of the protective coat has an overlapping flap to provide a secure seal. The inner closure is secured first, and then the outer flap is secured, creating a double seal. Several different combinations of zippers, Velcro, snaps, and D-rings can be used to secure the inner and outer closures.

The collar of the protective coat works with the protective hood to protect the neck. The collar has snaps or a Velcro closure system in front to keep it in a raised position. The coat's sleeves feature thumb wristlets (straps or reinforced openings). Wristlets prevent the sleeves from riding up on the wrists so that no skin is exposed between the gloves and the sleeves, which could result in wrist burns. It is important that the wearer use the thumb wristlets.

An integrated safety component is the drag rescue device (DRD). The DRD is a fabric handle integrated within the protective coat that is designed to aid in the rescue of a fire fighter. A rescuer can grab onto the drag rescue device and drag the incapacitated person to safety. Most manufacturers measure fire fighters for the appropriate-sized coats and pants. If equipment is being transferred to another fire fighter it is imperative to ensure that the protective coat cover the person's body without any area becoming exposed with movement.

Pockets in the coat can be used for carrying small tools or extra gloves. Additional pockets or loops can be installed to hold radios, microphones, flashlights, or other accessories.

Structural Firefighting Protective Pants

The **structural firefighting protective pants** or trousers included as part of structural firefighting PPE are constructed in a waist-length design or a bib-overall configuration (**FIGURE 3-7**). Like protective coats, protective pants must satisfy the conditions specified in NFPA 1971. These pants may be referred to as bunker pants or turnout pants.

Also like protective coats, protective pants are constructed with three layers. The outer shell resists abrasions and repels water. The second layer is a moisture barrier that protects the skin from liquids and steam burns. The third, inner layer is a quilted, thermal barrier that

FIGURE 3-7 Structural firefighting protective pants.
© Jones & Bartlett Learning. Photographed by Glen E. Ellman.

protects the body from elevated temperatures. Protective pants are reinforced around the ankles and knees with leather or extra padding.

Protective pants are manufactured with a double-fastener system at the waist, similar to the front flap of the protective coat. Fluorescent or reflective stripes around the ankles provide added visibility. Suspenders hold the pants up. Pants should be large enough that you can don them quickly. They should be big enough to allow you to crawl and bend your knees easily, but they should not be bigger than necessary.

Boots

Structural firefighting boots can be constructed of rubber or leather; they are available in a variety of styles. Rubber firefighting boots come in a step-in style without laces (**FIGURE 3-8**). Leather firefighting boots are available in pull-on style or in a shorter version with laces (**FIGURE 3-9**). Taller boots provide shin protection. Many fire fighters install a zipper on the laced boots to aid in quick donning and doffing of this component of PPE.

FIGURE 3-8 Rubber firefighting boots.
© Jones & Bartlett Learning. Photographed by Glen E. Ellman.

FIGURE 3-9 Leather firefighting boots.
© Jones & Bartlett Learning. Photographed by Glen E. Ellman.

Both boot styles must meet the requirements specified in NFPA 1971. The outer layer of these boots repels water and must be flame and cut resistant. Boots also must have a heavy sole with a slip-resistant design, a puncture-resistant sole, and a reinforced toe to prevent injury from falling objects. An inner liner constructed of a material such as Nomex or Kevlar adds thermal protection.

Boots must be the correct size for the fire fighter's feet. The foot should be secure within the boot to prevent ankle injuries and enable secure footing on ladders or uneven surfaces. Improperly sized boots will cause blisters and other problems.

Gloves

Gloves are an important part of firefighting PPE, because most fire suppression tasks require the use of the hands (**FIGURE 3-10**). Gloves must provide adequate protection yet permit the manual dexterity needed to accomplish tasks. NFPA 1971 specifies that gloves must be resistant to heat, liquid absorption, vapours, cuts, and penetration. A liner adds thermal protection and serves as a moisture barrier.

Many fire fighters carry a second set of gloves in their gear or on the fire apparatus so they can change

FIGURE 3-10 Firefighting gloves.
© Jones & Bartlett Learning. Photographed by Glen E. Ellman.

gloves when one pair becomes dirty, wet, or damaged. Do not wring or twist wet gloves, as this motion can tear or damage the inner liners.

Although gloves furnish needed protection, they inevitably reduce manual dexterity. For this reason, fire fighters need to practise manual skills while wearing gloves to become accustomed to them and to learn to adjust their movements accordingly.

SAFETY TIP

Plain leather work gloves or plastic-coated gloves should never be used for structural firefighting. Gloves used for firefighting must be labelled as such and meet the specifications given by NFPA 1971, *Standard on Protective Ensembles for Structural Fire Fighting and Proximity Fire Fighting*. Latex or nitrile gloves are required for rendering emergency medical care.

Respiratory Protection

The SCBA is an essential component of the PPE used for structural firefighting. Without adequate respiratory protection, fire fighters would be unable to mount an attack. The design, parts, operation, use, and maintenance of this complex piece of equipment are covered later in this chapter.

Personal Alert Safety System

A personal alert safety system (PASS) is an electronic device that emits a loud audible signal when a fire fighter becomes trapped or injured. This device sounds automatically if a fire fighter is motionless for a set time period. It also can be manually activated to notify other fire fighters that the user needs assistance.

Integrated PASS devices turn on automatically when the SCBA is activated (**FIGURE 3-11**). Non-integrated PASS devices must be turned on manually and will be found with older SCBA. Some PASS devices may be used by fire fighters not wearing SCBA. All fire fighters must confirm that their PASS devices are on and working properly before they enter a burning building or hazardous area. PASS devices are discussed in more detail later in this chapter.

Additional PPE

Some eye protection is provided by the face shield mounted on the fire helmet. When additional eye protection is needed, such as when using power saws or hydraulic rescue tools, fire fighters can use approved goggles or safety glasses. Goggles can be carried easily in the pocket of the protective coat.

Fire fighters are often exposed to loud noises such as sirens and engines. Because hearing loss is cumulative,

FIGURE 3-11 An integrated PASS device can save a fire fighter's life.
© Jones & Bartlett Learning. Photographed by Glen E. Ellman.

FIRE FIGHTER TIP

Fire fighters have personal preferences about the tools and equipment they carry in their pockets. Observe which items seasoned members of your department carry in their pockets. Their choices will help you decide whether you need to carry similar equipment. Some common items:

- Spare flashlight
- Multi-tool
- Lineman's cutters
- Wedges
- Webbing
- Screwdriver with an interchangeable head
- Shove knife
- Folding knife

FIGURE 3-12 Some fire apparatus are equipped with a headset and intercom systems. These systems can provide hearing protection, permit communications between crew members, and enable everyone to hear radio communications.
© Jones & Bartlett Learning. Photographed by Glen E. Ellman.

it is important to limit exposure to loud sounds. A headset and intercom system on the apparatus provides hearing protection, permits communications between crew members, and enables everyone to hear radio communications (**FIGURE 3-12**). A small speaker is incorporated into large earmuffs located at each riding and operating position. Fire fighters don the earmuffs, which reduce engine and siren noise. At the same time, the speaker enables crew members to listen, and a boom microphone allows them to talk to each other in a normal tone of voice and to hear the apparatus radio. Flexible ear plugs are useful in other situations involving loud sounds. Fire fighters should use the hearing protection supplied by their departments to prevent hearing loss when using power equipment or other equipment that produces noise.

A fire fighter should always carry a **hand light**, given that most interior firefighting takes place in near-dark, zero-visibility conditions. A good working hand light can illuminate your surroundings, mark your location, and help you find your way under difficult conditions. In addition to carrying a hand light, hands-free options such as helmet-mounted lights or other lights that clip to PPE or SCBA may be useful.

Two-way radios link the members of a firefighting team. At least one member of each team working inside a burning building or in any hazardous area should always have a radio. Some fire departments provide a radio for every on-duty fire fighter. Follow your department's standard operating procedure (SOP) on radio use. A radio should be considered part of PPE and carried with you whenever it is appropriate.

Any time you are operating in a roadway or are close to traffic, you are required to wear a bright-coloured reflective safety vest that meets the latest iteration of the CSA Z96-15, *High-Visibility Safety Apparel*.

Limitations of Structural Firefighting PPE

Structural firefighting PPE protects a fire fighter from the hostile environment of a fire. To provide complete protection, each component must be properly donned and worn continuously. Unfortunately, even today's advanced PPE has drawbacks and limitations. Understanding those limitations will help you avoid situations that could result in serious injury or death.

PPE tends to be challenging to don as it includes several individual components, all of which must be put on in the proper order and correctly secured. You must be able to don your equipment quickly and correctly, either at the fire station before you respond to an emergency or after you arrive at the scene. Practise donning your protective clothing until you can do so quickly and smoothly (**FIGURE 3-13**).

PPE is heavy, accounting for nearly 23 kg (50 lb) of extra weight. This increased weight means that everything you do—even walking—requires more energy and strength. Another drawback of PPE is that it limits mobility. Full PPE not only limits the range of motion but also makes movements awkward and difficult. Tasks such as advancing an attack line up a stairway or using an axe to ventilate a roof can be difficult, even for a fire fighter in excellent physical condition.

Because PPE retains body heat and perspiration, wearing this equipment makes it difficult for the body to cool itself. Perspiration is retained inside the protective clothing rather than evaporating to cool you. As a consequence, fire fighters in full protective gear can rapidly develop elevated body temperatures, even when the

ambient temperature is cool. The problem of overheating is more acute when surrounding temperatures are high, which is one reason why fire fighters must undergo regular rehabilitation and adequate fluid replacement. Remove as much of your PPE as possible when you arrive at rehabilitation, and place the PPE in an area away from direct contact with yourself and others. Removing your PPE will help you to cool down quickly. More important, PPE will continue to off-gas toxins from the hazardous conditions from which they just came.

Wearing PPE also decreases normal sensory abilities. Wearing heavy gloves, for example, reduces the sense of touch. Structural firefighting protective coats and pants protect skin but reduce its ability to determine the temperature of hot air. Sight is restricted when you wear an SCBA. The plastic face piece reduces peripheral vision, and the helmet, protective hood, and coat make turning the head difficult. Both the earflaps and the protective hood over the ears limit hearing. Speaking becomes muffled and distorted by the SCBA face piece, even if it is equipped with a special voice amplification system.

For these reasons, fire fighters must become accustomed to wearing and using PPE. Practicing skills while wearing PPE will help you become comfortable with its operation and limitations.

Donning Personal Protective Clothing

Don (i.e., put on) protective clothing in a specific order to obtain maximum protection. Donning must also be done quickly; however, your goal should be to don PPE properly and utilize all fasteners.

> ### FIRE FIGHTER TIP
>
> Sometimes referred to as a proximity suit, high temperature–protective equipment allows a properly trained fire fighter to work in extreme fire conditions, such as those posed by aircraft fires. This type of PPE shields the wearer during short-term exposures to high temperatures.

Following a set pattern of donning PPE helps reduce the time it takes to dress. This exercise does not include donning an SCBA, which will be discussed later. First become proficient in donning personal protective clothing, and then add the SCBA. Either don protective clothing in the firehouse before mounting the apparatus, or put it on at the scene after you arrive. *Do not* don protective clothing inside the apparatus while en route to an emergency incident. Instead, stay in your assigned seat, properly secured by a seat belt or safety harness, while the vehicle is in motion. Follow the steps in **SKILL DRILL 3-1** to don personal protective clothing.

FIGURE 3-13 It takes practice to don the full set of PPE.

© Jones & Bartlett Learning. Photographed by Glen E. Ellman.

SKILL DRILL 3-1
Donning Personal Protective Clothing Fire Fighter I, NFPA 1001: 4.1.2

1 Place your equipment in a logical order for donning.

2 Place your protective hood over your head and down around your neck.

3 Put on your boots, and pull up your protective pants. Place the suspenders over your shoulders, and secure the front of the pants.

4 Put on your protective coat, and close the front of the coat.

(continued)

SKILL DRILL 3-1 Continued
Donning Personal Protective Clothing Fire Fighter I, NFPA 1001: 4.1.2

5 Place your helmet on your head, and adjust the chin strap securely. Turn up your coat collar, and secure it in front.

6 Put on your gloves.

7 Have your partner check your clothing.

Doffing Personal Protective Clothing

To **doff** (i.e., remove) your personal protective clothing, reverse the procedure used in getting dressed. Follow the steps in **SKILL DRILL 3-2**.

PPE should be cleaned any time it becomes dirty—for example, by being exposed to smoke, blood or body fluids, or chemicals—to ensure your safety and health.

Smoke contains **carcinogens**, substances capable of causing cancer. Body fluids can transmit diseases, and chemicals can cause burns.

Doff the PPE at the scene and then use a brush or water to remove as much dirt and debris as possible. Do not ride back to the station wearing or sitting next to contaminated PPE. Thoroughly clean the PPE as soon as you return to the station. Once PPE has been properly cleaned, it should be placed in a convenient

---

SKILL DRILL 3-2
Doffing Personal Protective Clothing Fire Fighter I, NFPA 1001: 4.1.2

1 Remove your gloves.

2 Open the collar of your protective coat.

3 Release the helmet chin strap, and remove your helmet.

(continued)

SKILL DRILL 3-2 Continued
Doffing Personal Protective Clothing Fire Fighter I, NFPA 1001: 4.1.2

4 Remove your protective coat.

5 Remove your protective pants and boots.

6 Remove your protective hood.

© Jones & Bartlett Learning. Photographed by Glen E. Ellman.

location until the next response. Current NFPA standards call for decontaminated gear to be stored in a separate room of the fire station that has good ventilation and limited UV lighting. The separation and ventilation aid in keeping cancer-causing agents away from fire fighters, and reduced UV sunlight helps improve the life of PPE. PPE should not be kept in living areas, sleeping areas, or vehicles because even when cleaned it may still contain some carcinogens. Wherever it is stored, personal protective clothing must be properly maintained, organized, and ready for the next emergency response.

Inspection and Maintenance of PPE

Approved personal protective clothing is built to exacting standards. It requires proper care if it is to continue to afford maximum protection. Avoid unnecessary wear on clothing. A complete set of approved clothing (excluding SCBA) costs more than $3000. Keep this expensive equipment in good shape for its intended use—structural firefighting.

Inspect the condition of your personal protective clothing on a regular basis. Look for excessive wear, contamination, dirt, and damage. Repair worn or damaged clothing at once. If the fabric is damaged, it must be properly repaired to retain its protective qualities. Clothing that is worn or damaged beyond repair must be replaced immediately because it will not be able to protect you.

Avoid unnecessary cuts or abrasions on the outer material. This material already meets NFPA standards—do not look for new opportunities to test its effectiveness. Follow the manufacturer's instructions for repairing or replacing this PPE.

In addition, you must keep personal protective clothing clean to reduce your exposure to carcinogens and to maintain its protective properties. Dirt will build up in the clothing fibres from routine use and exposure to fire environments. Smoke particles also will become embedded in the outer shell and will continue to off-gas if not properly cleaned. The interior layers will frequently be soaked with perspiration. Allowing fire by-products to stay embedded on PPE will break down the protective components of the gear, causing the potential for the material to fail.

Other contaminants are formed from the by-products of burnt plastics and synthetic products. These residues, which are flammable, can become trapped between the fibres or build up on the outside of PPE, damaging the materials and reducing their protective qualities. A fire fighter who is wearing contaminated PPE is, in essence, bringing additional fuel into the fire on the clothing.

PPE that has been badly soiled by exposure to smoke, other products of combustion, melted tar, petroleum products, or other contaminants needs to be cleaned as soon as possible. Regular and routine cleaning should remove most of these contaminants. Items that have been exposed to chemicals or hazardous materials may have to be cleaned by a professional cleaning company or be impounded for decontamination or disposal.

Cleaning instructions are listed on the label or tag attached to the clothing. Always follow the manufacturer's cleaning instructions. Failure to do so may reduce the effectiveness of the garment and create an unsafe situation for the wearer.

Fire departments should have special washing machines that are approved for cleaning protective clothing. If they do not have approved cleaning devices, they can contract with an outside firm to clean and repair protective clothing. In either case, the manufacturer's instructions for cleaning and maintaining the garment must be followed. If you wash your protective clothing in house, follow the manufacturer's instructions for drying these garments properly. Make sure your PPE is dry before using it on the fire ground. If wet protective clothing is exposed to the high temperatures of a structural fire, the water trapped in the liner materials will turn into steam and be trapped inside the moisture barrier. The result may be painful steam burns. Many fire departments issue two sets of gear to each fire fighter so that each always has a set of clean, dry gear available while one set is being cleaned.

Other PPE also may require regular cleaning and maintenance. For example, the outer shell of your helmet should be cleaned with a mild soap as recommended by the manufacturer. The inner parts of helmets should be removed and cleaned according to the manufacturer's instructions. The chin strap and suspension system must be properly adjusted and all parts of the helmet kept in good repair.

Protective hoods and gloves get dirty quickly and should be cleaned according to the manufacturer's instructions. Most protective hoods can be washed with appropriate soaps or detergents. Dirty protective hoods may contain large quantities of carcinogens following a fire. Repair or discard gloves or protective hoods that have holes in them; do not use them. Even a small cut or opening in these items can result in a burn injury.

Boots should be maintained according to the manufacturer's instructions. Rubber boots need to be kept in a place that does not result in damage to the boot. Leather boots must be properly maintained to keep them supple and in good repair. Any type of boot needs to be repaired or replaced if the outer shell is damaged. Do not store these items or any other PPE in direct sunlight.

All PPE that has been exposed to hazardous conditions should be cleaned and removed as soon as possible following the fire (**FIGURE 3-14**). Perform a field reduction of contaminants to remove as much dirt and debris as possible. Remove all surface contaminants. Doff the PPE at the scene. Thoroughly clean the PPE as soon as you return to the station. As discussed, correctly storing your PPE is another important step in the proper care and maintenance of your PPE. PPE should be stored only in the proper PPE storage areas.

Specialized Protective Equipment

Vehicle Extrication

Firefighters are often required to respond to motor vehicle collisions and should wear PPE appropriate for the tasks they will be performing. The officer in charge also may designate one or more members to don SCBA and stand by with a charged hose line. PPE can protect against sharp metal edges, cuts, and abrasions.

Some protective clothing, such as special gloves and coveralls or jumpsuits, has been specifically designed for vehicle extrication. These items are generally lighter in weight and more flexible than structural firefighting PPE, although they may be constructed of the same basic materials.

Fire fighters performing vehicle extrication must always be aware of the possibility of contact with blood or other body fluids. Latex or nitrile gloves should be worn when providing patient treatment. Eye protection should be worn, given the possibilities of breaking glass, contact with body fluids, metal debris, and accidents with tools. Proper cleaning of extrication gear is especially important because many of the fluids encountered during vehicle extrication are oil based—for example, gasoline, diesel fuel, transmission fluid, and brake fluid. Because of the flammability of these fluids, they pose a hazard for fire fighters who enter a burning building while wearing oil-contaminated gear. Always wear a high-visibility safety vest if you will be operating on a roadway.

Wildland Fires

Gear that is designed specifically for fighting wildland or brush fires must meet the requirements outlined in NFPA 1977, *Standard on Protective Clothing and Equipment for Wildland Fire Fighting*. In this type of firefighting ensemble, the jacket and pants are made of fire-resistant materials, such as Nomex or specially treated cotton, which are designed for comfort and maneuverability while working in the wilderness. Wildland fire fighters wear a helmet of a thermoresistant plastic, eye protection, and pigskin or leather gloves. The boots are designed to provide comfort and sure footing while hiking to a remote fire scene and to provide protection against crush and puncture injuries. Structural firefighting gear is not designed for extended wildland firefighting. It may produce high body temperatures resulting in heat exhaustion or heat stroke.

SAFETY TIP

Some recent studies have shown that protective hoods can transfer large amounts of carcinogens to your skin. When doffing PPE, *do not* leave a contaminated protective hood or SCBA mask around your neck and below your nose and mouth. You will absorb or inhale additional toxins. To remove as much dirt from your face and neck as possible, use wet wipes as soon as possible in the field. Follow this by taking a shower as soon as possible. It is also important to thoroughly clean your protective hood after each use. Follow your fire department's policies in regard to cleaning contaminated PPE. Your fire department may require that the PPE be bagged in a biohazard bag and professionally cleaned. Never clean your PPE or work uniform in home washers to prevent cross-contamination with your home laundry.

FIGURE 3-14 Perform a field reduction of contaminants to remove dirt and debris before returning to the station.
© Jones & Bartlett Learning. Photographed by Glen E. Ellman.

Respiratory Protection

Respiratory protection equipment is an essential component of the structural firefighting protective ensemble. Respiratory protection is both expensive and complicated; using one confidently requires practice. You must be proficient in using it before you engage in fire suppression activities.

The atmosphere of a burning building is considered to be **IDLH (immediately dangerous to life and health)**. Attempting to work in such an atmosphere without proper respiratory protection can cause serious injury or death. Never enter or operate in a hazardous atmosphere without first ensuring that you have appropriate respiratory protection.

Respiratory Hazards of Fires

Fire fighters need respiratory protection because a fire involves a complex series of chemical reactions that can rapidly affect the atmosphere in unpredictable ways. The most readily evident by-product of a fire is smoke. Some of the harmful substances found in smoke include soot (carcinogen), carbon monoxide, cyanide compounds, formaldehyde, and the oxides of nitrogen. These substances are dangerous if inhaled, and repetitive exposures to these materials may have negative health effects.

Products of Combustion

Collectively, the airborne products of combustion are called smoke. Smoke contains three highly toxic products of combustion or physical states of matter: particles (solids), vapours (finely suspended liquids), and gases.

Smoke particles consist of unburnt, partially burnt, and completely burnt substances. Many smoke particles are so small that they can pass through the natural protective mechanisms of the respiratory system and enter the lungs or pass through the skin, move into the bloodstream, and be transported throughout the body. Some are toxic to the body and can result in severe injuries or death if they are inhaled. These particles also can prove extremely irritating to the eyes and digestive system.

Smoke vapours, which are finely suspended liquids—that is, aerosols—are similar to fog, which consists of small water droplets suspended in the air. When oil-based compounds burn, they produce small hydrocarbon droplets that become part of the smoke. If inhaled or ingested, these compounds can affect the respiratory and circulatory systems. Some of the toxic droplets in smoke can cause poisoning if they are absorbed through the skin.

A fire also produces several types of gases. The amount of oxygen available to the fire and the type of fuel being burned determine precisely which gases are produced. Many of the gases produced by residential or commercial fires are highly toxic. Carbon monoxide (CO), hydrogen cyanide, and phosgene are three of many gases often present in smoke. These gases are discussed in more detail in Chapter 5, *Fire Behaviour*.

The harmful products produced by smoke require fire fighters to use respiratory protection in all fire environments, regardless of whether the environment is known to be contaminated, suspected of being contaminated, or could possibly become contaminated without warning. The use of respiratory protection allows fire fighters to enter and work in a hazardous atmosphere with a safe, independent air supply.

LISTEN UP!

Research conducted by Underwriters Laboratories demonstrated that residual smoke on gear retains many heavy metals, including arsenic, cobalt, chromium, lead, mercury, phosphorus, and phthalates. Exposure to these chemicals has been linked to increased risk of cancer. Dirty PPE is a danger to your health!

Oxygen Deficiency

Oxygen is required to sustain human life. Normal outside or room air contains approximately 21 percent oxygen. A decrease in the amount of oxygen in the air, however, may drastically affect an individual's ability to function (**TABLE 3-1**). An atmosphere with an

TABLE 3-1 Physiological Effects of Reduced Oxygen Concentration

Oxygen Concentration	Effect
21%	Normal breathing air
17%	Judgment and coordination impaired; lack of muscle control
12%	Headache, dizziness, nausea, fatigue
9%	Unconsciousness
6%	Respiratory arrest, cardiac arrest, death

oxygen concentration of 19.5 percent or less is considered oxygen deficient (OSHA, *Confined or Enclosed Spaces or Other Dangerous Atmospheres,* 2017). If the oxygen level drops below 17 percent, people can experience disorientation, an inability to control their muscles, and irrational thinking, which can make escaping a fire much more difficult.

LISTEN UP!

Smoky environments and oxygen-deficient atmospheres are deadly! It is essential to use an SCBA at all times when operating in a smoky environment. Also, keep in mind that smoke is not just a respiratory danger; toxins can enter your body through the skin.

During compartment fires (i.e., fires burning within enclosed areas), oxygen depletion occurs in two ways. First, the fire consumes large quantities of the available oxygen, thereby decreasing the concentration of oxygen in the atmosphere. Second, the fire produces large quantities of other gases, which decrease the oxygen concentration by displacing the oxygen that would otherwise be present inside the compartment.

Increased Temperature

Heat is also a respiratory hazard. The temperature of the smoke (particles, vapours, or gases) varies, depending on the fire conditions and the distance travelled. Inhaling superheated smoke can cause severe burns of the respiratory tract immediately. If the smoke is hot enough, a single inhalation can cause fatal respiratory burns. More information about fire behaviour and products of combustion appears in Chapter 5, *Fire Behaviour*.

Other Toxic Environments

Not all hazardous atmospheric conditions are caused by fires. Indeed, fire fighters may encounter oxygen-deficient atmospheres in numerous types of emergency situations. Respiratory protection is just as important in these situations as it is in a fire suppression operation. For example, toxic gases may be released at hazardous materials incidents from leaking storage containers or industrial equipment, from chemical reactions, or from the normal decay of organic materials. Internal combustion engines or improperly operating heating appliances can produce carbon monoxide.

Always approach the scene from a safe location and direction. Take time to assess the scene and interpret clues. Once you have determined the dangers that may be present, you can formulate a plan for addressing the situation.

Conditions That Require Respiratory Protection

More fire deaths are caused by smoke inhalation than burns. Clearly, adequate respiratory protection is essential to your safety as a fire fighter. In fact, the products of combustion from house fires and commercial fires may be so toxic that just a few breaths can result in death. This is why you always must take respiratory protection precautions before entering the fire. When you first arrive at the scene of a fire, you do not have any way to measure the immediate danger to your life and your health posed by that fire. You must use full PPE and approved breathing apparatus if you enter and operate within this atmosphere.

This includes response to outside fires, such as vehicle and dumpster fires. Continue to wear full PPE and approved breathing apparatus during overhaul until the air has been tested and proven to be safe by the safety officer. Do not remove your respiratory protection just because a fire has been knocked down.

Respiratory protection and PPE also must be used in any situation where there is a possibility of toxic gases or oxygen deficiency, such as in a confined space. Always assume that the atmosphere is hazardous until it has been tested and proven to be safe.

Types of Breathing Apparatus

There are two major types of breathing apparatus, atmosphere-supplying respirators and air-purifying respirators.

Atmosphere-supplying respirators (ASRs) use air that is supplied from a source independent of the ambient (room) air. This may be air carried in a compressed gas cylinder such as the fire service self-contained breathing apparatus (SCBA), or it may be a supplied-air respirator that receives air through a hose. The construction and operation of SCBA will be more fully described in the following sections.

A **supplied-air respirator (SAR)** uses an external source for the breathing air (**FIGURE 3-15**). In this type of device, a hose line is connected to a breathing-air compressor or to compressed air cylinders located outside the hazardous area. The user breathes air through the line and exhales through a one-way valve, just as with an open-circuit SCBA. Supplied-air respirators are more frequently used in industrial settings and in some hazardous materials situations. Their use will be described in more detail as part of your hazardous materials training.

The second major type of breathing apparatus is an air-purifying respirator. An **air-purifying respirator (APR)** is supplied with an air-purifying filter, cartridge, or a canister that removes specific (known) air contaminants

FIGURE 3-15 A supplied-air respirator (SAR) may be needed for special rescue operations. A SAR may also be referred to as a *positive-pressure air-line respirator* (with escape unit).
© Jones & Bartlett Learning. Photographed by Glen E. Ellman.

by passing ambient (room) air through the air-purifying element (**FIGURE 3-16**). APRs are designed to remove particulate matter or specified gases or vapours. They are used in some industrial settings where the air is monitored carefully to determine the precise quantity of contaminants that are present. APRs are not suitable for firefighting operations because it is not possible to know the type and concentration of contaminants present in any given situation. Furthermore, they do not supply oxygen in an oxygen-deficient atmosphere.

A **powered air-purifying respirator (PAPR)** is similar in function to the standard APR described above but includes a small fan to help circulate air into the mask (**FIGURE 3-17**). The fan draws outside air through the filters and into the mask via a low-pressure hose. PAPRs are not considered to be true positive-pressure units like an SCBA because it is possible for the wearer to "outbreathe" the flow of supplied air, thereby creating a negative-pressure situation inside the mask, possibly

FIGURE 3-17 Powered air-purifying respirators (PAPRs) have a small fan to help circulate air into the mask, diminishing the work of breathing of the wearer, helping to reduce fogging in the mask, and providing a constant flow of cool air across the face.
Courtesy of Chris Hawley.

allowing contaminants to enter the face mask due to a poor seal with the face.

APRs and PAPRs are typically worn to respond to a hazardous materials situation where the type and quantity of contaminants are known. These respirators should not be used in structural firefighting situations.

Standards and Regulations

In Canada, the Canadian Standards Association (CSA) sets the requirements for the selection, use, and care of respirators in the workplace (CAN/CSA Z94.4, 2016). CSA is a federal agency that researches, develops, and implements occupational safety and health programs. Canadian fire services also refer to the **National Institute for Occupational Safety and Health (NIOSH)** in the United States. NIOSH is a federal agency that researches, develops, and implements occupational health and safety programs that are relevant to the North American fire service. It also investigates fire fighter fatalities and serious injuries and makes recommendations on how to prevent accidents from recurring. In addition, NFPA 1981, *Standard on Open-Circuit Self-Contained Breathing Apparatus (SCBA) for Emergency Services*, requires that all SCBA equipment used in the emergency service

FIGURE 3-16 Air-purifying respirators (APRs) offer specific degrees of protection if the hazard present is known and the appropriate filter canister is used.
Courtesy of Sperian Respiratory Protection.

meet the design, testing, and certification requirements established by NIOSH.

Although the CSA standard Z94.4 must be followed in Canadian departments, the NFPA has developed three standards directly related to SCBA in particular:

- NFPA 1500, *Standard on Fire Department Occupational Safety, Health, and Wellness Program*, includes the basic requirements for SCBA use and program management.
- NFPA 1404, *Standard for Fire Service Respiratory Protection Training*, sets the requirements for an SCBA training program within a fire department.
- NFPA 1981, *Standard on Open-Circuit Self-Contained Breathing Apparatus (SCBA) for Emergency Services*, includes requirements for the design, performance, testing, and certification of open-circuit SCBA for the fire service.

These standards specify approved SCBA equipment, initial training for each fire fighter, and annual retraining.

> **FIRE FIGHTER TIP**
>
> Do not confuse SCBA and SCUBA. An SCBA is used by fire fighters during fire suppression activities. A **self-contained underwater breathing apparatus (SCUBA)** is used by divers while swimming underwater.

The CSA and provincial and territorial agencies are also responsible for the administration of an effective respiratory protection program in the workplace (CSA Z94.4, 2016). Fire departments in all provinces and territories must follow the CSA Z94.4 standard, which covers the fit testing, medical screening, and training of all fire fighters in the use of SCBA. In some Canadian provinces and territories, individual occupational safety and health agencies establish and enforce these regulations. In the United States the **Occupational Safety and Health Administration (OSHA)** is responsible for establishing and enforcing regulations for respiratory protection programs (OSHA, *Personal Protective Equipment*, 2017). Each fire department must follow applicable standards and regulations to ensure safe working conditions for all personnel.

Self-Contained Breathing Apparatus

The basic respiratory protection hardware used by the fire service is the self-contained breathing apparatus, or SCBA. The term "self-contained" in the acronym for this equipment refers to the requirement that the apparatus be the sole source of the fire fighter's air supply. In other

words, an SCBA is an independent air supply that will last for a predictable duration.

The two main types of SCBA are open-circuit and closed-circuit devices. An **open-circuit self-contained breathing apparatus** is typically used for structural firefighting. A cylinder of compressed air provides the breathing air supply for the user, and exhaled air is released into the atmosphere through a one-way valve (**FIGURE 3-18**). Approved open-circuit SCBA come in several different models, designs, and options.

A **closed-circuit self-contained breathing apparatus** recycles the user's exhaled air. The air passes through a mechanism that removes carbon dioxide; then it is supplemented with oxygen and "rebreathed" by the wearer. No exhaled air is released to the outside environment, making this type of unit a closed-circuit system (**FIGURE 3-19**).

Many closed-circuit SCBA units include a small oxygen cylinder as well as chemically generated oxygen. A closed-circuit SCBA is often used for extended operations, such as mine rescue work, and operations in long tunnels where breathing apparatus must be worn for a long time.

An SCBA is designed to provide clean breathing air (the same as environmental air) to fire fighters who are working in the hostile environment of a fire. It must meet rigid manufacturing specifications so that it can function in the increased temperatures and smoke-filled environments that fire fighters encounter. When properly maintained, this equipment will provide sufficient quantities of air to enable fire fighters to perform rigorous tasks.

FIGURE 3-18 Open-circuit SCBA.
© Jones & Bartlett Learning. Photographed by Glen E. Ellman.

FIGURE 3-19 A closed-circuit SCBA is commonly referred to as a "rebreather."
© Jones & Bartlett Learning. Photographed by Glen E. Ellman.

Using an SCBA requires that fire fighters develop unique skills, including different breathing techniques. You should always remember that this equipment limits normal sensory awareness—the senses of smell, hearing, and sight are all affected by the apparatus. Proficiency in the use of SCBA and other PPE requires ongoing training and practice.

Limitations of SCBA Equipment

Like any type of equipment, an SCBA also has its limitations. Some of these limitations apply to the equipment; others apply to the user's physical and psychological abilities. Because an SCBA carries its own air supply in a pressurized air cylinder, its use is limited by the amount of air in the cylinder. SCBA air cylinders for structural firefighting must carry enough air for a minimum of 30 minutes; air cylinders rated for 45 minutes, 60 minutes, and 75 minutes are also available. These duration ratings, however, are based on ideal laboratory conditions. The realistic useful life of an SCBA air cylinder for firefighting operations is usually much less than the rated duration, and actual use time will depend on the size of the user, his or her physical fitness and conditioning, the amount of physical exertion, and the user's degree of calm. An SCBA air cylinder generally has a realistic useful life of no more than 50 percent of the rated time. For example, an SCBA air cylinder rated for 30 minutes can be expected to last for a maximum of 15 minutes during strenuous firefighting.

Fire fighters must manage their working time while using SCBA so that they have enough time to exit from the hazardous area before exhausting the air supply. To properly manage the air supply, a fire fighter must consider the following factors:

- The time and the effort it will take to reach the task destination. Climbing stairs will take more energy and air than walking across a flat floor, for example.
- The amount of air that will be available upon reaching the task destination.
- The amount of time necessary to complete the task and the air that will be used during that period. Some tasks will take more energy and air.
- The amount of time it will take to reach a safe area. At the end of this time, the fire fighter must have a reserve of air for unexpected emergencies.

An SCBA provides a very limited window of time for firefighting and a safe exit from the hazardous conditions of the fire. It is essential that you have a margin of safety built into your air supply for the unexpected. In some cases, you may need to begin exiting from the fire scene before half of your air supply is exhausted.

The weight of an SCBA varies, based on the manufacturer and the type and size of the air cylinder. Generally, an SCBA weighs at least 11 kg (25 lb). The size and shape of an SCBA may make it difficult for you to fit through tight openings when wearing this equipment (**FIGURE 3-20**). Several techniques may help you navigate these spaces:

- Change your body position: Rotate your body by 45 degrees and try again.
- Loosen one shoulder strap and change the location of the SCBA on your back.
- If you have no other choice, you may have to remove your SCBA. In this case, do not let go of the harness for any reason. Keep the unit in front of you as you navigate through the tight space. Reattach the SCBA harness as soon as you are through the restricted space. *Note*: This is a last resort procedure!

The added weight and bulk of the SCBA decrease the user's flexibility and mobility and shift the user's center of gravity.

As previously noted, the design of the SCBA face piece limits the fire fighter's vision—particularly his or her peripheral vision. The face piece lens may fog up under some conditions, further limiting visibility. Face pieces also can fail if the temperature and radiant heat are too high. NIST studies show that the face piece lens can soften at approximately 150°C (300°F), with the lens

FIGURE 3-20 An SCBA expands a user's profile, making it more difficult to pass through tight spaces.
© Jones & Bartlett Learning. Photographed by Glen E. Ellman.

FIGURE 3-21 This SCBA face piece shows damage from heat.
© Jones & Bartlett Learning. Photographed by Dave Casey.

starting to "craze" or bubble at higher temperatures and then actually fail at temperatures above 215°C (419°F) (NIST, 2011) (**FIGURE 3-21**). SCBA may affect the user's ability to communicate, depending on the type of face piece and any additional hardware provided, such as voice amplification and radio microphones. The equipment can be noisy during inhalation and exhalation, which may limit the user's hearing as well, especially when coupled with a protective hood and helmet.

Structural fire fighting in extremely cold environments is both hard on fire personnel and the equipment they depend on to remain safe. It is not uncommon for SCBA to freeze under extremely cold conditions, causing an immediate loss of air supply. There are two primary causes for this type of failure. The first is caused by not opening the primary air supply valve at the base of the air cylinder fully. If the valve is only partially opened, ice can form on the needle component of the valve, occluding the opening and preventing a free flow of air. The second cause is allowing moisture to enter the regulator. This causes the valve to freeze (primarily on regulators that attach to the face piece itself). Firefighters must be mindful to both fully open the air cylinder shut-off valve and prevent water from accumulating in an exposed face piece. Also, when exiting the warm environment of a fire to change air cylinders or go to

rehabilitation, be aware of the impact the drastic change in temperature has on your face piece. Avoid breathing on ambient air when you exit the structure under cold weather conditions, as the temperature differential can cause the regulator valve to freeze. Always ensure that your SCBA is functioning properly prior to reentry into an IDLH environment.

Physical Limitations of the SCBA User

Fire fighters are required to maintain good physical conditioning. A person with ideal body weight and in good physical condition will be able to perform more work per cylinder of air than a person who is overweight or out of shape. An out-of-shape or obese fire fighter will consume the air supply from this equipment more quickly and will have to exit the fire building long before a well-conditioned fire fighter does so. Overweight or poorly conditioned fire fighters also are at greater risk for heart attack due to physical stress.

Altogether, the protective clothing and SCBA that must be worn when fighting fires can weigh more than 23 kg (50 lb). Moving with this extra weight requires additional energy, which, in turn, increases air consumption and body temperature. Taken collectively, this activity places additional stress on a fire fighter's body.

The weight and bulk of the complete PPE limit a fire fighter's ability to walk, climb ladders, lift objects, and crawl through restricted spaces. Fire fighters must become accustomed to these limitations and learn to alter their movements accordingly. Practice and conditioning are key to becoming proficient in wearing and using PPE while fighting fires.

Psychological Limitations of the SCBA User

In addition to the physical limitations, the user must make mental adjustments when wearing an SCBA. Breathing through an SCBA is different from normal breathing, and it can be stressful. Covering your face with a face piece, hearing the air rushing in, hearing valves open and close, and exhaling against positive pressure are foreign sensations. The surrounding environment, which is often dark and filled with smoke, is foreign as well.

Fire fighters must adjust so that they can operate effectively under these stressful conditions. Practice in donning PPE, breathing through SCBA, and performing firefighting tasks in darkness help to build confidence, not only in the equipment but also in the fire fighter's personal skills.

Training generally introduces one skill at a time. Practise each skill as it is introduced, and try to become proficient in that skill. As your skills improve, you will be able to tackle tasks characterized by increasing levels of difficulty.

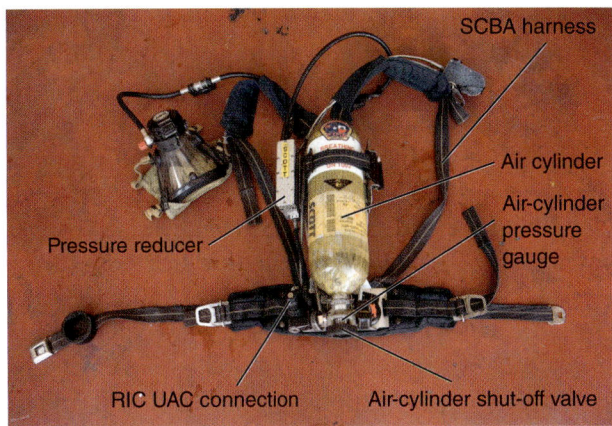

FIGURE 3-22 The components of self-contained breathing apparatus.
© Jones & Bartlett Learning. Photographed by Glen E. Ellman.

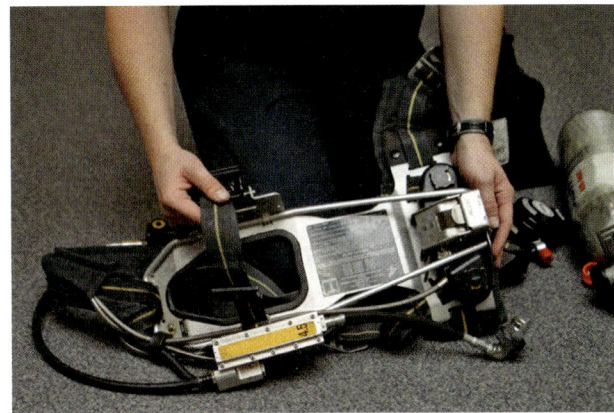

FIGURE 3-23 SCBA harnesses come in a variety of models.
© Jones & Bartlett Learning. Photographed by Glen E. Ellman.

FIRE FIGHTER TIP

The goal of training is to bring you to a level of comfort and proficiency in using your equipment. Start with a friendly environment until you are used to your equipment. Then add stressors such as darkness, smoke, and heat, one at a time.

Components of SCBA

An SCBA consists of four main parts: the harness, the air cylinder assembly, the regulator assembly, and the face piece assembly. Although the basic features and operations of all models are similar, you need to become familiar with the specific model of SCBA used by your department. The following sections will discuss the components of an SCBA (**FIGURE 3-22**).

SCBA Harness

The **SCBA harness** consists of the backpack or frame for mounting the working parts of the SCBA and the straps and fasteners used to attach the SCBA to the fire fighter (**FIGURE 3-23**). It is usually constructed of a lightweight metal or composite material.

Most SCBA harnesses have two adjustable shoulder straps and a waist belt. These must be constructed of a material such as Kevlar that is strong and able to withstand elevated temperatures. Depending on the specific model of SCBA, the waist belt and shoulder straps carry different proportions of the pack's weight. The procedures for tightening and adjusting the straps also vary based on the model. The SCBA harness must be secure enough to keep the SCBA firmly fastened to the user but not so tight that it interferes with breathing or movements. The waist belt must be tight enough to keep the SCBA from moving from side to side or getting caught on obstructions. Some SCBA models are equipped with a reinforced harness that can be used to help drag a fallen fire fighter out of danger.

Air Cylinder Assembly

A compressed **air cylinder** holds the breathing air for an SCBA. This removable air cylinder is attached to the SCBA harness and can be changed quickly in the field.

An experienced fire fighter should be able to remove and replace the air cylinder in complete darkness.

Fire fighters should be familiar with the type of air cylinders used in their departments. Air cylinders are marked with the materials used in their construction, the working pressure, and the rated duration.

The air pressure in filled SCBA air cylinders ranges from 2200 to 5500 **pounds per square inch (psi)** (15,168 to 37,921 kilopascals [kPa]). The greater the air pressure, the more air that can be stored in the cylinder.

Low-pressure air cylinders, which are pressurized at 15,168 kPa (2200 psi), can be constructed of steel or aluminum and are usually rated for 30 minutes of use. Composite air cylinders are generally constructed of an aluminum shell wrapped with carbon, Kevlar, or glass fibres. They are significantly lighter in weight, can be pressurized up to 37,921 kPa (5500 psi), and are rated for 30, 45, 60, or 75 minutes of use.

As previously noted, the rated duration times are established under laboratory conditions. A working fire fighter can quickly use up the air because of exertion, so the ratings should be viewed with caution. Generally, the working time available for a particular air cylinder is half the rated duration.

The neck of an air cylinder is equipped with a hand-operated valve. The air cylinder valve controls the flow of air leaving the air cylinder. A **pressure gauge** is located near the air cylinder valve; it shows the amount of air currently in the cylinder. This is referred to as the air-cylinder pressure gauge. A second pressure gauge is located on the shoulder strap or in another location where it can be seen while the SCBA is being used. This is referred to as the remote pressure gauge. Be careful not to damage the threads or let any dirt get into the outlet of the air cylinder.

FIGURE 3-24 SCBA regulator.
© Jones & Bartlett Learning. Photographed by Glen E. Ellman.

the air supply. Exhaling opens a second valve (the exhalation valve), thereby expelling the exhaled air into the atmosphere. SCBA regulators are capable of delivering large volumes of air to support the strenuous activities required in firefighting. Most SCBA units are equipped with a **dual-path pressure reducer**, a feature that automatically provides a backup method for air to be supplied to the regulator if the primary passage malfunctions.

SCBA regulators maintain a slight positive air pressure to the face piece in relation to the ambient air pressure outside the face piece. This feature helps to prevent the hazardous atmosphere outside the face piece from leaking into the face piece during inhalation. If any leakage occurs in the area where the face piece and the face make the seal, the positive-pressure breathing air inside the face piece will keep the hazardous atmosphere from entering the device. Regardless of the positive pressure, a proper face-piece-to-face seal must always be maintained. Breathing with this slight positive pressure may require some practice. New fire fighters often report that it takes more energy to breathe when first using positive-pressure SCBA, but this sensation gradually decreases.

To activate the SCBA, it is necessary to open the air cylinder valve, don the SCBA harness, don the face piece, and depending on the type of regulator you are using, attach the regulator to the face piece or connect the low-pressure air supply hose to the regulator, and breathe. If the air supply is partially or completely cut off during use, you can fully open the **regulator purge/bypass valve** to secure an emergency supply of air. This will create a constant flow of air into the face piece; however, this action will rapidly deplete the remaining air supply in the air cylinder. If it is necessary to open the regulator purge/bypass valve, you should immediately exit from the IDLH area. During normal use, the regulator purge/bypass valve can be momentarily opened to remove condensation or residual air from the respirator after the cylinder valve is turned off.

FIRE FIGHTER TIP

The air-cylinder pressure gauge is important for checking the amount of air in the cylinder. This gauge cannot be viewed by the user while wearing the SCBA. A second pressure gauge can be seen by the user. This remote pressure gauge is mounted directly on the regulator, or it may be located on a separate hose so it can be attached to a shoulder strap for easier viewing.

Regulator Assembly

SCBA regulators may be mounted on the waist belt or shoulder strap of the SCBA harness or attached directly to the face piece (**FIGURE 3-24**). The regulator controls the flow of air to the user. Inhaling decreases the air pressure in the face piece. This change in pressure opens the regulator, which in turn releases air from the cylinder. When inhalation stops, the regulator shuts off

As discussed, a visible remote pressure gauge enables the user to monitor the amount of air remaining in the cylinder. The gauge can be mounted directly on the regulator, or it may be located on a separate hose so it can be attached to a shoulder strap for easier viewing. If the pressure gauges (air-cylinder pressure gauge and remote pressure gauge) are working correctly, the readings should be within 10 percent of each other. A pressure gauge must be visible to the user during SCBA use.

The regulator may have an air saver/donning switch that prevents the rapid loss of the air supply if the cylinder valve is open and the face piece is removed from the face or the regulator is removed from the face piece. In other words, if you open the air cylinder valve, air will not flow unless you depress the air saver/donning switch.

Although many models of SCBA regulators exist, fire fighters must learn how to operate only the particular model that is used in their department. They should be able to operate the regulator in the dark and while wearing firefighting gloves.

Face Piece Assembly

The **face piece** delivers breathing air to the fire fighter and protects the face from high temperatures and smoke (**FIGURE 3-25**). It consists of a face mask with a clear lens, an exhalation valve, and—on models with a harness-mounted regulator—a flexible low-pressure supply hose. In some models, the regulator is attached directly to the face piece.

Full face pieces cover the nose, mouth, and eyes. Half face pieces cover the nose and mouth. The part that comes in contact with the skin is made of special rubber or silicone, because these materials provide for a tight seal. The clear lens allows for better vision. Exhaled air is expelled from the face piece through the one-way exhalation valve, which has a spring mechanism to maintain positive pressure inside the face

FIGURE 3-25 SCBA face pieces come in several sizes.
© Jones & Bartlett Learning. Photographed by Glen E. Ellman.

piece. Because it is difficult to communicate through a face piece, a voice amplification device or mechanical diaphragm is used to facilitate communication. A mechanical diaphragm uses a vibrating airtight membrane to transmit the fire fighter's voice without the use of electricity.

Face pieces are equipped with **nose cups** to help prevent fogging of the clear lens. In addition, nose cups prevent the build-up of carbon dioxide (CO_2) by directing exhaled air toward the exhalation valve. Fogging is a greater problem in colder climates. It occurs because the compressed air you breathe is dry, but the air you exhale is moist. The regulator purge/bypass valve can be opened slightly for a second or two to clear condensation from the eyepieces. The flow of dry air from the regulator helps to prevent fogging of the lens.

The face piece is held in place with a weblike series of straps or a net and straps. Face pieces should be stored with the straps in the longest position to make them easier to don. Pull the end of the straps toward the back of the head (not out to the sides) to tighten them and to ensure a snug fit.

A leak in the face piece seal may result from an improperly sized face piece, an improper donning procedure, or facial hair around the edge of the face piece. In particular, the following factors can affect the seal on an SCBA face piece:

- Facial hair, sideburns, or beard
- A low hairline that interferes with the sealing surface
- Ponytails or buns that interfere with the smooth and close fit on the head harness
- A skull cap that projects under the face piece or temple pieces
- The absence of teeth
- Improper size face piece

An improperly fitted face piece may lead to exposure to a hazardous environment. Such leaks are dangerous for two reasons. First, a large leak could overcome the positive pressure in the face piece and allow contaminated air to enter the face piece. Second, a leak of any size will deplete the breathing air and reduce the amount of time available for firefighting.

Face pieces are manufactured in several sizes. CSA Z94.4 (2016) and NFPA 1500 requires that all fire fighters have their face pieces fit-tested annually to ensure that they are wearing the proper size. Some departments issue individual face pieces to each fire fighter; others provide a selection of sizes on each apparatus. CSA Z94.4 (2016) and NFPA 1500 also require that the sealing surface of the face piece be in direct contact with the user's skin; that is, hair or a beard cannot be in the seal area.

Additional Features

An SCBA is also required to have a **heads-up display (HUD)**, which must be visible to the user while wearing the face piece, enabling the user to constantly monitor the amount of air in the air cylinder. Most SCBA face pieces contain **light-emitting diodes (LEDs)** that indicate the amount of air remaining in the cylinder. These indicate whether the air cylinder is full, three-fourths full, one-half full, or one-fourth full. Other LEDs on the display may provide additional information—for example, low batteries or other problems (**FIGURE 3-26**).

NFPA standards require that SCBA include two **end-of-service-time indicators (EOSTIs)** or low-air alarms, that operate independently of each other and activate different senses. For example, one alarm might ring a bell, whereas the second alarm might vibrate, whistle, or flash an LED.

This warning device tells the user that the end of the breathing air supply is approaching. Currently, the EOSTI or low-air warning alarm is constructed to sound when the pressure in the air cylinder is down to 35 percent of its capacity. At this pressure level you must begin to exit from the IDLH environment (NFPA 1981, 2019). Most fire departments require fire fighters to exit from the IDLH area before the EOSTI alarm sounds because the low-air alarm does not sound until two-thirds of the air supply has been exhausted. This reserve of one-third of the air supply is not always adequate for a safe escape (NFPA 1982, 2018).

As discussed earlier, many SCBA models are manufactured with an integrated PASS device. A PASS device is designed to help colleagues locate a downed fire fighter by sending out a loud audible signal. The PASS device combines an electronic motion sensor with an alarm system. If the user remains motionless for 30 seconds, it will produce a low warning tone before sounding a full alarm. The user can reset the device by moving during this warning period. A fire fighter in distress also can manually activate this device. Turning on the air supply in an SCBA with an integrated PASS device automatically activates the PASS device; this ensures that a fire fighter does not forget to turn the PASS device on when entering a hazardous area. Newer PASS devices also may include a radio transmitter that sends a signal to the command post when the alarm sounds. It is important to learn how to activate and deactivate the PASS device on your SCBA.

Communications among fire fighters wearing SCBA are difficult. To facilitate communication, an SCBA is required to be equipped with a voice communication system. This functionality may be as simple as a mechanical voice diaphragm, or it may be as sophisticated as an electronic system. Some SCBA are also equipped with voice communications systems that utilize wireless communication interoperability to interface with mobile radios. Recent changes have been made in the NFPA 1981 and 1982 standards to improve the face piece voice communication intelligibility and volume.

In 2013 the NFPA allowed for open-circuit SCBA to have an **emergency breathing safety system (EBSS)** integrated. The EBSS allows for SCBA users to "buddy breathe" or share their available air supply when one is low or out of air (NFPA 1981, 2019). An SCBA is required to be equipped with a **rapid intervention crew/company universal air connection (RIC UAC)** (**FIGURE 3-27**).

FIGURE 3-26 A heads-up display enables the user to constantly monitor the amount of air in the air cylinder.
© Jones & Bartlett Learning. Photographed by Glen E. Ellman.

RIC UAC connection

FIGURE 3-27 The rapid intervention crew/company (RIC) uses the RIC UAC to refill the SCBA cylinder of a trapped fire fighter who is running out of air. This connection is used only for emergency refilling of an air cylinder.
© Jones & Bartlett Learning. Photographed by Glen E. Ellman.

The rapid intervention crew/company (RIC) uses this connection to refill the SCBA cylinder of a trapped fire fighter who is running out of air. A universal air connection (UAC) is attached to a hose from a full air cylinder and brought to the downed fire fighter by an RIC, who refills the downed fire fighter's cylinder. This connection should not be used for the routine refilling of a cylinder. The RIC UAC is designed only to be used for emergency refilling of an air cylinder. It is important that the dust cap is in place over the UAC to prevent damage. This universal connection can be used between SCBA packs of any manufacturer.

Fire department SCBA are required to be certified to provide protection against certain chemical, biological, radiological, and nuclear agents—that is, agents that could be released as a result of a terrorist attack. An SCBA provides protection against these agents by preventing chemical fumes, disease-causing biological organisms, and radioactive particles from being inhaled and entering the fire fighter's respiratory system. However, an SCBA does not protect from contamination by other means of transmission.

Many accessories are available for SCBA, including data logging, unit IDs, tracking devices, corrective eye lenses for face pieces, thermal imaging capabilities, and electronic communications devices (NFPA 1981, 2019). Data logging and unit IDs aid in keeping track of the use of each SCBA and checking the proper functioning of each unit. Some of these tracking devices work by sending out sounds of varying pitch. Corrective lenses for face pieces enable SCBA users to use their face pieces as corrective eyewear. Electronic communications systems integrate radio communications between SCBA users. Some SCBA offer a thermal imaging device that can be added into the unit to provide each fire fighter with the ability to detect heat sources. If your SCBA is equipped with any of these accessories, you must become competent in the operation of these extra devices. To ensure proper functioning of your SCBA, do not add any devices or accessories that are not approved by the manufacturer of the equipment.

Pathway of Air Through an SCBA

In an SCBA, the breathing air is stored under pressure in the air cylinder. This air passes through the cylinder valve into the high-pressure **air line** (hose), which then takes it to the regulator. The regulator reduces the high pressure to low pressure.

The regulator opens when the user inhales, reducing the pressure on the downstream side. In an SCBA unit with a face piece–mounted regulator, the air goes directly into the face piece. In units with a harness-mounted regulator, the air travels from the regulator through a low-pressure hose into the face piece. From the face piece, the air is inhaled through the user's air passages and into the lungs.

When the user exhales, used air is returned to the face piece. The exhaled air is exhausted from the face piece through the exhalation valve. This cycle repeats with every inhalation. As the pressure in the face piece drops, the exhalation valve closes, and the regulator opens.

Breathing Techniques

Many breathing techniques can be used to conserve air. The skip-breathing technique helps conserve air while using an SCBA in a firefighting situation. In this technique, the fire fighter takes a short breath, holds it, takes a second short breath (without exhaling in between breaths), and then relaxes with a long exhale. Each breath should take 5 seconds.

A simple drill can demonstrate the benefits of skip breathing. In this exercise, one fire fighter dons PPE and an SCBA with a full air cylinder and walks in a circle around a set of traffic cones, the track at the local school, or, if safety permits, the parking lot at the fire station. A second fire fighter times how long it takes for the fire fighter to completely deplete the air in the SCBA. After the first fire fighter is completely rested, the air cylinder is replaced, and the same drill is repeated using the skip-breathing technique. A comparison of the times after completion of both cycles should confirm that skip breathing conserves the air in the cylinder.

Another method that can be used is the Reilly (humming) technique. Humming allows for a longer exhalation, and this slows the breathing rate. This technique is accomplished by slowly inhaling and then exhaling while making a humming sound.

Keep in mind that when you are able to decrease the rate and depth of your breathing, you increase the amount of time you can breathe from a single cylinder of air. Keep calm, perform your assignments efficiently, and do not start breathing from your air supply until you need to.

The controlled breathing technique also helps extend the SCBA air supply. It consists of a conscious effort to inhale naturally through the nose and to force exhalation from the mouth. Practicing controlled breathing during training will help you to maximize the efficient use of air. It is important to note that fire fighters should always follow SOPs regarding the use of various breathing techniques.

Donning an SCBA

Donning an SCBA is an important skill. Fire fighters should be able to don and activate an SCBA in 1 minute; both the fire fighter's personal safety and the effectiveness of the firefighting operation depend on mastery of this skill. Fire fighters must already be wearing personal protective clothing before they don an SCBA (NFPA 1001, 2019). Before beginning the actual donning process, fire fighters must carefully check the SCBA to ensure it is ready for operation:

- Check whether the air-cylinder pressure gauge is at least 90 percent of its rated pressure.
- If the SCBA has an air saving/donning switch, confirm that it is activated to prevent air from flowing.
- Open the air cylinder valve two or three turns, listen for the low-air alarm to sound, and then open the valve fully.
- Test the PASS device.
- Check the remote pressure gauge and the air-cylinder pressure gauge. The reading for these gauges should be within 10 percent of each other.
- Check all harness and face piece straps to ensure that they are fully extended.
- Check all valves to ensure that they are in the correct position. (An open regulator purge/bypass valve will waste air.)

Donning an SCBA from an Apparatus Seat Mount

An SCBA should be located so that fire fighters can don it quickly when they arrive at the scene of a fire. Seat-mounted brackets enable fire fighters to don SCBA en route to an emergency scene, without unfastening

their seat belts or otherwise endangering themselves (**FIGURE 3-28**). This enables fire fighters to begin work as soon as they arrive on scene.

Several types of apparatus seat-mounting brackets are available. Some hold the SCBA with the friction of a clip. Others are equipped with a mechanical hold-down device that must be released to remove the SCBA. Regardless of which mounting system is employed, it must hold the SCBA securely in the bracket; a collision or sudden stop should not dislodge the SCBA from the brackets, because a loose SCBA can be a dangerous projectile. The fire fighter who dons an SCBA from a seat-mounted bracket should not tighten the shoulder straps while seated, so as not to dislodge the SCBA in a sudden-stop situation. Fire fighters must remain securely restrained by a seat belt or combination seat belt/shoulder harness any time the apparatus is moving.

Donning an SCBA while en route to an emergency can save valuable time. This maneuver requires that

FIGURE 3-28 Seat-mounted brackets hold the SCBA securely in place and enable fire fighters to don an SCBA en route to an emergency scene.

© Jones & Bartlett Learning. Photographed by Glen E. Ellman.

you don your protective pants, boots, and coat before mounting the apparatus. Place your protective hood around your neck. Keep your helmet and gloves close by (**FIGURE 3-29**). Confirm that the SCBA has been inspected and is ready for service.

When you exit the apparatus, be sure to take a face piece with you. Face pieces should be kept either in a storage bag close to each seat-mounted SCBA or attached to the SCBA harness.

Follow the steps in **SKILL DRILL 3-3** to don an SCBA from a seat-mounted bracket. Before beginning this skill drill, inspect the SCBA to ensure it is ready for service.

FIGURE 3-29 Do not don your helmet until you arrive at the scene.
© Jones & Bartlett Learning. Photographed by Glen E. Ellman.

SKILL DRILL 3-3
Donning an SCBA from an Apparatus Seat Mount Fire Fighter I, NFPA 1001: 4.3.1

1 Don your protective hood, pants, boots, and coat. Safely mount the apparatus, and sit in the seat. Place your arms through the SCBA shoulder straps. Partially tighten the shoulder straps; do not fully tighten them.

2 Fasten your SCBA waist belt.

(continued)

SKILL DRILL 3-3 Continued

Donning an SCBA from an Apparatus Seat Mount Fire Fighter I, NFPA 1001: 4.3.1

3 Fasten your seat belt. When the apparatus stops at the emergency scene, release the seat belt, and release the SCBA from its brackets. If the apparatus has an SCBA locking device, detach the SCBA from the locking device.

4 Carefully exit the apparatus. Maintain three points of contact with the vehicle while exiting.

5 Cinch down the SCBA waist belt.

6 Adjust shoulder straps until they are snug.

SKILL DRILL 3-3 Continued
Donning an SCBA from an Apparatus Seat Mount Fire Fighter I, NFPA 1001: 4.3.1

7 Open the main cylinder valve. Activate the air saver/donning switch to prevent the flow of air, if needed.

8 Don the face piece, and check for leaks. Pull the protective hood up over your head, put the helmet on, and secure the chin strap.

9 If necessary, connect the regulator to the face piece or attach the low-pressure air supply hose to the regulator. Activate the air flow and ensure that the PASS device alarm is operating.

Donning an SCBA from an Apparatus Compartment Mount

Compartment-mounted SCBA units may be mounted on fire apparatus and can be donned quickly. These units are used by fire fighters who arrive in vehicles without seat-mounted SCBA or whose seats were not equipped with them. Driver/operators and fire fighters who arrive in their private vehicles often use these types of SCBA. The mounting brackets should be positioned high enough to allow for easy donning of the equipment. Some mounting brackets allow the fire fighter to lower the SCBA without removing it from the mounting bracket. Older apparatus may have the brackets mounted on the exterior of the vehicle. Any exterior-mounted SCBA should be protected from weather and dirt by a secure cover.

The SCBA should be stored on the apparatus in ready-for-use condition, with the air cylinder valve closed.

Follow the steps in **SKILL DRILL 3-4** to don an SCBA from a side-mounted compartment or bracket.

Before starting, don your protective pants, boots, and coat. Place your protective hood around your neck. Place your helmet and gloves close by. Confirm that the SCBA has been inspected and is ready for service. Examine the equipment to see how it is mounted in the compartment. Check the release mechanism, and determine how it operates. If the SCBA is mounted on an exterior bracket, remove the protective cover before beginning the donning sequence.

Donning an SCBA from the Ground, the Floor, or a Storage Case

Fire fighters sometimes must don an SCBA that is on the ground, or the SCBA may be kept in a storage case. Storage cases are usually used for transporting extra SCBA units. It should not be used to transport SCBA that will be used during the initial phase of operations at a fire scene as it can add considerable time to remove the unit from the case and prepare it for use.

SKILL DRILL 3-4
Donning an SCBA from an Apparatus Compartment Mount Fire Fighter I, NFPA 1001: 4.3.1

1. Stand in front of the SCBA bracket, and fully open the air cylinder valve. Ensure that the regulator is in the off position.

2. Turn your back toward the SCBA, slide your arms through the shoulder straps, and partially tighten the straps.

3. Release the SCBA from the bracket, and step away from the apparatus.

4. Attach the waist belt, and tighten it.

5. Adjust the shoulder straps. Fully open the main air cylinder valve. Activate the air saver/donning switch to prevent the flow of air, if needed.

6. Don the face piece, and check for an adequate seal.

7. Pull the protective hood into position, don your helmet, and secure the chin strap. Never hook the chin strap onto the helmet chin strap or onto the face piece or regulator itself.

8. If necessary, connect the regulator to the face piece or attach the low-pressure air supply hose to the regulator. Activate the air flow, and ensure that the PASS device alarm is operating.

Two methods of donning can be used in these situations: the over-the-head method and the coat method.

Over-the-Head Method. Follow the steps in **SKILL DRILL 3-5** to don an SCBA using the over-the-head method. Before starting, confirm that the SCBA has been inspected and is ready for service. Don your protective pants, boots, and coat. Place your protective hood around your neck. Place your helmet and gloves close by.

SKILL DRILL 3-5
Donning an SCBA Using the Over-the-Head Method Fire Fighter I, NFPA 1001: 4.3.1

1 Lay out the SCBA so that the cylinder is resting on the floor or ground, the backplate is facing up, and the air cylinder valve is facing away from you. Move the shoulder straps to the sides.

2 Fully open the main air cylinder valve. Activate the air saver/donning switch to prevent the flow of air, if needed.

3 Bend down and grasp the SCBA backplate with both hands. Using your knees to support and lift the extra weight, lift the SCBA up and over your head. Once the SCBA clears your head, rotate it 180 degrees so that the waist belt straps are pointed toward the ground.

(continued)

SKILL DRILL 3-5 Continued
Donning an SCBA Using the Over-the-Head Method Fire Fighter I, NFPA 1001: 4.3.1

4 Slowly slide the pack down your back. Make sure that your arms slide into the shoulder straps. Once the SCBA is in place, tighten the shoulder straps, and secure the waist belt.

5 Don the face piece, and check for an adequate seal. Pull your protective hood into position on your head, don your helmet, and secure the chin strap.

6 If necessary, connect the regulator to the face piece, or attach the low-pressure air supply hose to the regulator. Activate the air flow, and ensure that the PASS device alarm is operating.

Coat Method. Follow the steps in **SKILL DRILL 3-6** to don an SCBA using the coat method. Before starting, confirm that the SCBA has been inspected and is ready for service. Don your protective pants, boots, and coat.

Place your protective hood around your neck. Place your helmet and gloves close by.

These instructions will have to be modified for different SCBA models. In particular, the sequence

SKILL DRILL 3-6
Donning an SCBA Using the Coat Method Fire Fighter I, NFPA 1001: 4.3.1

1 Lay out the SCBA so that the cylinder is resting on the floor or ground, the backplate is facing up, and the air cylinder valve is facing toward you. Move the shoulder straps to the sides. Fully open the air cylinder valve. Activate the air saver/donning switch to prevent the flow of air, if needed. Place your dominant hand on the opposite shoulder strap. For safety reasons, be sure to grasp the strap as close to the backplate as possible.

2 Lift the SCBA, and swing it over your dominant shoulder, being mindful of people or objects around you.

3 Slide your other hand between the SCBA cylinder and the corresponding shoulder strap.

(continued)

SKILL DRILL 3-6 Continued
Donning an SCBA Using the Coat Method Fire Fighter I, NFPA 1001: 4.3.1

4 Tighten the shoulder straps.

5 Attach the waist belt, and adjust its tightness.

6 Don the face piece, and check for an adequate seal.

7 Pull the protective hood into position on your head, don your helmet, and secure the chin strap. If necessary, connect the regulator to the face piece, or attach the low-pressure air supply hose to the regulator. Activate the air flow, and ensure that the PASS device alarm is operating.

for adjusting the shoulder straps and the waist belt varies with different models. Modifications also must be made for SCBA models with waist-mounted regulators. Refer to the specific manufacturer's instructions supplied with each unit. Follow the SOPs for your department.

Donning the Face Piece

Your face piece keeps contaminated air outside and pure breathable air inside. To perform properly, it must be the correct size and must be adjusted to fit your face. Make sure you have been tested to determine the proper size for you. The requirements for face piece fit testing are described in CSA Z94.4 (2016) and in NFPA 1500. Fit testing needs to be performed before fire fighters are permitted to use SCBA. To ensure a proper fit, testing should be repeated every year.

No facial hair can be present in the seal area. Eyeglasses that pass through the seal area cannot be worn with a face piece, because they can cause leakage between the face piece and your skin. Your face piece must match your SCBA—you cannot interchange a face piece from a different SCBA model.

Face pieces for various brands and models of SCBA may differ slightly. Some have the regulator mounted on the face piece; others have it mounted on the harness straps. Fire fighters must learn about the specific face pieces used by their departments.

Follow the steps in **SKILL DRILL 3-7** to don a face piece. Before beginning this procedure, make sure you have donned your PPE and the protective hood is around your neck. Remove your helmet.

SAFETY TIP

The following method is one way to check the seal of a face mask:

1. Don the SCBA, and begin breathing cylinder air.
2. Completely close the air cylinder valve.
3. Breathe through the face piece until all air stops flowing from the breathing regulator.
4. Inhale slowly, and hold your breath. Your face piece should be slightly drawn to your face.
5. Listen and feel for air leakage around the face piece seal. Check that the negative pressure in the face piece does not change.
6. If no change occurs, no leaks are present.

Safety Precautions for SCBA

As you practise using your SCBA, remember that this equipment is your protection against serious injury or death in hazardous conditions. Practise safe procedures from the beginning.

Before you enter a hazardous environment, make sure that you have an adequate supply of air in your air cylinder and that your PASS device is activated. Be sure that you are properly entered into your personnel accountability system. Always work in teams of two in hostile environments. In addition, always have at least two fire fighters outside at the ready whenever two fire fighters are working in a hostile environment.

FIRE FIGHTER TIP

Many fire departments need to take precautions to ensure proper operation of SCBA in cold temperatures. SCBA should be stored in a location that is kept at a temperature greater than 0°C (32°F). After use, the equipment should be cleaned and dried thoroughly to avoid retained moisture that might freeze moving parts. When SCBA units are doffed in below-freezing temperatures, the face piece and the regulator should be placed inside the protective coat to keep it warm and prevent any moisture from freezing.

SCBA Use During Emergency Operations

As a fire fighter, your job is to practise using SCBA until you are confident that you can carry out a variety of tasks in hazardous conditions while depending on your equipment to supply you with safe air to breathe. Because hostile environments are often unpredictable, fire fighters must be prepared to react if an emergency situation occurs while they are using an SCBA. In emergencies, follow simple guidelines. Most important, keep calm, stop, and think. Panic increases air consumption. Try to control your breathing by maintaining a steady rate of respirations. A calm person has a greater chance of surviving an emergency.

If you experience a problem with your SCBA, try to exit the IDLH area to a safe environment. If your cylinder contains air but no air comes out of your regulator, many SCBA have a regulator purge/bypass valve that can be opened slightly to release a constant supply of air. This measure will rapidly empty your cylinder, however. In this circumstance, you must immediately exit from the hazardous environment. If you are in danger, follow the steps for self-survival and calling a "mayday." This topic is discussed in Chapter 18, *Fire Fighter Survival*.

You will be using your SCBA in a variety of conditions that most people would consider to be emergencies. Because you are a fire fighter, these activities are predictable for you. Therefore, you need to master a wide variety of firefighting skills while wearing SCBA.

SKILL DRILL 3-7

Donning a Face Piece Fire Fighter I, NFPA 1001: 4.3.1

1 Fully extend the straps on the face piece.

2 Rest your chin in the chin pocket at the bottom of the face mask. Fit the face piece to your face, bringing the straps or webbing over your head.

3 Tighten the lowest two straps. To tighten them, pull the straps straight back, rather than out and away from your head. Check the head harness net to make sure it is lying flat against the back of your head. Tighten the pair of straps at your temple, if these straps are present. If your model has additional straps, tighten the top strap(s) last.

4 Check for a proper seal. This process depends on the model and type of face piece you use. Confirm that your nose fits in the nose cup.

SKILL DRILL 3-7 Continued
Donning a Face Piece Fire Fighter I, NFPA 1001: 4.3.1

5 Pull the protective hood into position on your head. Make sure it does not get under your face piece or obscure your vision.

6 Don your helmet, and secure the chin strap.

7 If needed, attach the regulator to your face piece, or attach the low-pressure air supply hose to the regulator.

Practise these tasks first in conditions of good visibility, and then progress to doing the same tasks in conditions of limited visibility. You need to practise most of the skills you will perform as a fire fighter—for example, advancing hose lines, climbing ladders, crawling through windows, performing rescues, providing medical assistance, and crawling through confined spaces—while wearing SCBA.

The following are three exercises that may be practised in order to assist in the mastery of using SCBA properly.

The Conducting a Primary Search Using the Standard Search Method skill drill in Chapter 12, *Search and Rescue*, describes how to perform search and rescue while wearing SCBA. Another skill you need to practise is passing through a restricted space while wearing an SCBA. This activity is illustrated in the Opening a Wall to Escape skill drill in Chapter 18, *Fire Fighter Survival*. A third activity to familiarize you with your SCBA simulates the activities you would perform when ventilating a roof. This activity is illustrated in Chapter 13, *Ventilation*. Once you have mastered donning and doffing your PPE ensemble, you may perform this activity to help you master the use of your SCBA. In doing so, you will work with at least one other team member.

LISTEN UP!

Refilling a cylinder using the rapid intervention crew/company universal air connection (RIC UAC) in below-freezing conditions presents two concerns. The first concern relates to the protective cap on the RIC UAC coupling. Water on the cover of the coupling may freeze, preventing a connection between the filling hose and the universal air connection. The second concern is that if the universal air connection is used to fill an SCBA cylinder in below-freezing temperatures, and the SCBA is then taken into a warmer temperature, the difference in temperature could cause the pressure in the cylinder to expand to unsafe levels. If there is excess pressure, open the air cylinder valve and the regulator purge/bypass valve to release the excess pressure.

Doffing an SCBA

The procedure for doffing an SCBA depends on which model you use and whether it has a face piece–mounted regulator or a harness-mounted regulator. Follow the procedures recommended by the manufacturer and your department's SOPs. Your department may teach some variation of these steps. Follow the procedure taught by your department.

Follow the steps in **SKILL DRILL 3-8** to doff your SCBA. Do not leave a contaminated hood or face piece

around your neck, below your nose and mouth, after you have taken your PPE off. You will inhale the toxins.

Putting It All Together: Donning the Entire PPE Ensemble

The complete PPE ensemble consists of both personal protective clothing and respiratory protection. Although donning personal protective clothing and respiratory protection can be learned and practised separately, you must be able to integrate these skills to have a complete PPE ensemble. Each component of the ensemble must be in the proper place to provide whole-body protection.

The steps for donning complete PPE (personal protective clothing and SCBA) are summarized here:

1. Place the protective hood over your head, and bring it down around your neck.
2. Put on your protective pants and boots. Adjust the suspenders, and secure the front flap of the pants.
3. Put on your protective coat, and secure the front.
4. Open the air cylinder valve on your SCBA, and check the air pressure. Press the air saver/donning switch to prevent air flow, if needed.
5. Put on your SCBA harness.
6. Tighten both shoulder straps of the SCBA harness.
7. Attach the waist belt of the SCBA harness, and tighten it. Tighten the chest straps, if present.
8. Fit the face piece to your face.
9. Tighten the face piece straps, beginning with the lowest straps.
10. Check the face piece for a proper seal. (Follow the manufacturer's instructions.)
11. Pull the protective hood into position over the face piece straps so that it covers all bare skin but does not obscure your vision.
12. Place your helmet on your head with the ear tabs extended, and secure the chin strap.
13. Turn up your coat collar, and secure it in front.
14. Put on your gloves.
15. Check your clothing to be sure it is properly secured.
16. Make sure your PASS device is turned on (if it is not integrated).
17. Attach your regulator, or open the air cylinder valve to start the flow of breathing air.
18. Work safely!

SKILL DRILL 3-8
Doffing an SCBA Fire Fighter I, NFPA 1001: 4.3.1

1 Remove the regulator from your face piece, or disconnect the low-pressure air supply hose from the regulator.

2 Close the air cylinder valve, or fully depress the air saver/donning switch to stop the flow of air.

3 Release your waist belt.

(continued)

SKILL DRILL 3-8 Continued
Doffing an SCBA Fire Fighter I, NFPA 1001: 4.3.1

4 Loosen the shoulder straps, and remove the SCBA harness. If you have not already done so, close the air cylinder valve.

5 Bleed the air pressure from the regulator by opening the regulator purge/bypass valve.

6 Ensure that the PASS device is turned off. Place the SCBA in a safe location. Clean your SCBA as soon as possible, following the manufacturer's instructions.

SKILL DRILL 3-8 Continued
Doffing an SCBA Fire Fighter I, NFPA 1001: 4.3.1

7 Remove your gloves. Remove your helmet, and pull your protective hood down around your neck. Loosen the straps on your face piece.

8 Remove your face piece.

© Jones & Bartlett Learning. Photographed by Glen E. Ellman.

SCBA Inspection and Maintenance

An SCBA must be properly cleaned in accordance with the manufacturer's maintenance requirements. As a general rule, SCBA should be cleaned after every use and inspected on a daily basis to ensure that it is ready for use at all times. The air cylinder must be changed or refilled, the face piece and regulator must be sanitized according to the manufacturer's instructions, and the unit must be cleaned, inspected, and checked for proper operation. After operating in a fire environment, your SCBA will be coated with dirt and soot, which contains many dangerous and carcinogenic substances. It is very important to thoroughly clean all parts of your SCBA as soon as possible. Follow the instructions of the SCBA manufacturer and of your department. It is the user's responsibility to ensure that the SCBA is in good working order and in ready-to-use condition before it is returned to the fire apparatus.

Inspection and operational testing should be conducted both after a unit has been used and on a regular schedule. In career departments, inspection and testing are done at the beginning of each shift. In volunteer departments, this step is commonly performed on a weekly schedule. A complete annual inspection and maintenance procedure must be performed on each SCBA unit. The annual inspection must be performed by a certified manufacturer's representative or a person who has been trained and certified to perform this work.

If an SCBA inspection reveals any problems that cannot be remedied by routine maintenance, the SCBA must be removed from service for repair. Only properly trained and certified personnel are authorized to repair SCBA.

FIRE FIGHTER TIP

Follow the manufacturer's recommendations, and avoid the use of aerosol cleaners or any alcohol-containing cleaner, as these materials degrade the rubber material in the face piece. Face pieces should be air dried or wiped with a soft, nonabrasive cloth to avoid scratching the lens.

The purpose of an SCBA inspection is to identify any parts of the SCBA that are visibly damaged and need to be repaired or replaced to ensure continued safe operation. This visual inspection can be done in conjunction with the operational testing sequence discussed next. Follow the steps in **SKILL DRILL 3-9** for the SCBA inspection. A more detailed inspection is required if a cylinder has been exposed to excessive heat, has come into contact with flame, has been exposed to chemicals, or has been dropped.

Operational Testing

A pressurized SCBA cylinder contains a tremendous amount of potential energy. Not only does the air within the cylinder exert considerable pressure on its walls, but the cylinder itself is used under extreme conditions on the fire ground. If the cylinder ruptures and suddenly releases this energy, it can cause serious injury or death. For this reason, cylinders must be regularly inspected

SKILL DRILL 3-9
Visible SCBA Inspection Fire Fighter I, NFPA 1001: 4.5.1

1 Visually inspect the air cylinder and valve assembly for dents and gouges. Look for black or discoloured areas that indicate exposure to flame.

2 Check the cylinder for the current hydrostatic test date and date of manufacture. Check the air-cylinder pressure gauge to be sure it is full.

SKILL DRILL 3-9 Continued
Visible SCBA Inspection Fire Fighter I, NFPA 1001: 4.5.1

3 Inspect hose and rubber parts for damage or deterioration.

4 Inspect the SCBA harness, webbing, buckles, fasteners, and cylinder retention system for damage.

5 Verify that the SCBA has been cleaned according to the manufacturer's and department's recommendations. Inspect the regulator for intact gaskets and visible damage.

6 Inspect the face piece for damage and worn components. Look for damage to the lenses, and check for the presence of a nose cup.

(continued)

SKILL DRILL 3-9 Continued
Visible SCBA Inspection Fire Fighter I, NFPA 1001: 4.5.1

7 Inspect the head harness to confirm that all parts are present and working properly.

8 Check the quick disconnects and the RIC UAC to make sure they are not damaged, they are operating properly, and the dust cap is in place.

© Jones & Bartlett Learning. Photographed by Glen E. Ellman.

and tested to ensure they are safe. The operational testing sequence is designed to check the function of the many parts of the SCBA to ensure safe use of the device. It concentrates on the working parts of the SCBA. Follow the steps in **SKILL DRILL 3-10** for operational testing of an SCBA.

The standard on *Compressed Breathing Air and Systems* (CAN/CSA-Z180) requires **hydrostatic testing** for SCBA cylinders on a periodic basis and limits the number of years that a cylinder can be used. Hydrostatic testing seeks to identify any defects or damage that might render the cylinder unsafe. Any cylinder that fails a hydrostatic test should be immediately taken out of service and cannot be used.

Cylinders constructed of different materials have different testing requirements. Aluminum, steel, and carbon-fibre cylinders must be hydrostatically tested every 5 years (CSA Z180, *Compressed Breathing Air*

and Systems). Cylinders constructed of composite materials such as Kevlar or fibreglass fibres must be tested every 3 years (CSA Z180, *Compressed Breathing Air and Systems*). Fire fighters must know which types of cylinders are used by their departments and must check each cylinder for a current hydrostatic test date before filling it.

Replacing SCBA Cylinders

A used air cylinder can be quickly replaced with a full cylinder in the field to enable you to continue firefighting activities. A fire fighter who is working alone must doff his or her SCBA harness to replace the air cylinder; two fire fighters who are working together can change each other's cylinders without removing their SCBA harness. The steps listed in this section outline how a single person makes a cylinder change. This procedure may

SKILL DRILL 3-10
SCBA Operational Inspection Fire Fighter I, NFPA 1001: 4.5.1

1 Check the regulator purge/bypass valve to be sure it is closed.

2 Depress the air saver/donning switch, if present, to start the flow of air.

3 Slowly open the air cylinder valve. Check for proper operation of the heads-up display and of the low-battery indicator. Confirm that the low-air alarm and PASS devices are working.

4 Check the remote pressure gauge for proper operation.

(continued)

SKILL DRILL 3-10 Continued

SCBA Operational Inspection Fire Fighter I, NFPA 1001: 4.5.1

5 Don the face piece. Adjust it to obtain a good seal. Inhale sharply to start the flow of air. Breathe normally to check for proper operation.

6 Remove the regulator or face piece; air should flow freely.

7 Depress the air saver/donning switch to stop the flow of air.

8 Open the regulator purge/bypass valve to check for air flow.

SKILL DRILL 3-10 Continued
SCBA Operational Inspection Fire Fighter I, NFPA 1001: 4.5.1

9 Close the regulator purge/bypass valve to stop the flow of air.

10 Rotate the air cylinder valve to close it.

11 Open the regulator purge/bypass valve slightly to vent residual air pressure from the system. Watch the heads-up display to verify its proper operation as the air pressure is exhausted.

12 Once the air flow stops, close the regulator purge/bypass valve. Complete any reporting that is required.

© Jones & Bartlett Learning. Photographed by Glen E. Ellman.

vary slightly depending on the model of SCBA being used. Follow the procedure recommended by the SCBA manufacturer and by your department's SOPs.

Practise changing air cylinders until you become proficient at this task. A fire fighter should be able to change an air cylinder in the dark and while wearing gloves if necessary. Follow the steps in **SKILL DRILL 3-11** to replace an SCBA air cylinder.

Replacing an SCBA Cylinder on Another Fire Fighter

If you need to quickly reenter the fire scene, a second person can replace your air cylinder while you continue to wear your SCBA harness. Be sure that you are physically able to fight a second round with the fire. It is better to allow yourself a few minutes in rehab than to get back

SKILL DRILL 3-11
Replacing an SCBA Cylinder Fire Fighter I, NFPA 1001: 4.3.1

1 Place the SCBA on the floor or a bench.

2 Close the air cylinder valve.

3 Open the regulator purge/bypass valve to bleed off the pressure.

SKILL DRILL 3-11 Continued
Replacing an SCBA Cylinder Fire Fighter I, NFPA 1001: 4.3.1

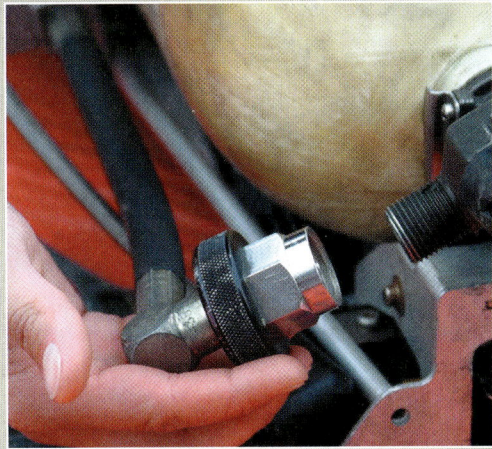

4 Disconnect the high-pressure supply hose. Keep the ends clean.

5 Release the air cylinder from the SCBA harness, and remove the depleted air cylinder.

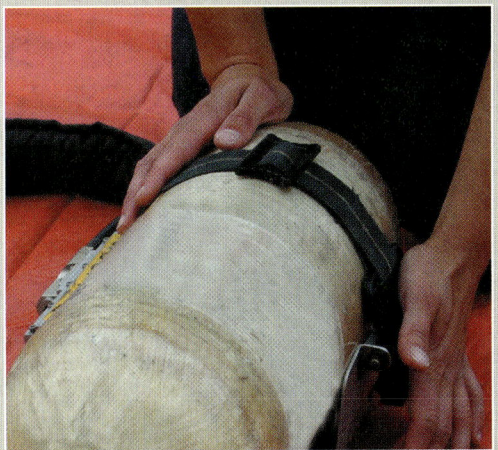

6 Slide a full air cylinder into the SCBA harness. Align the outlet to the supply hose. Lock the air cylinder in place.

(continued)

SKILL DRILL 3-11 Continued
Replacing an SCBA Cylinder Fire Fighter I, NFPA 1001: 4.3.1

7 Check that the "O" ring is present and in good shape.

8 Connect the high-pressure hose to the air cylinder. Hand-tighten only.

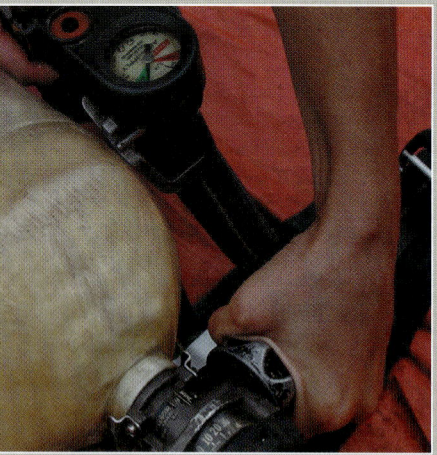

9 Open the air cylinder valve. Check the air-cylinder pressure gauge and the remote pressure gauge.

© Jones & Bartlett Learning. Photographed by Glen E. Ellman.

into the fire immediately and require rescue from other crew members. Do not overtax yourself by replacing the air cylinder and going back to work without adequate rest when you need it.

Follow the steps in **SKILL DRILL 3-12** to replace an SCBA air cylinder on another fire fighter.

Refilling SCBA Cylinders

Compressors and **cascade systems** are used to refill SCBA air cylinders. Compressors and cascade systems can be permanently located at a maintenance facility or at a firehouse, or they can be mounted on a truck or

SKILL DRILL 3-12
Replacing an SCBA Cylinder on Another Fire Fighter Fire Fighter I, NFPA 1001: 4.3.1

1 Remove the regulator from the face piece, or remove the face piece so that you can breathe ambient air.

2 Close the air cylinder valve on the used air cylinder.

3 Open the regulator purge/bypass valve to bleed off pressure from the high-pressure supply line.

4 Disconnect the high-pressure supply line.

(continued)

SKILL DRILL 3-12 Continued
Replacing an SCBA Cylinder on Another Fire Fighter Fire Fighter I, NFPA 1001: 4.3.1

5 Release the SCBA air cylinder from the SCBA harness, and set it aside.

6 Slide the full SCBA air cylinder into the SCBA harness.

7 Check for the presence and satisfactory condition of the "O" ring.

SKILL DRILL 3-12 Continued
Replacing an SCBA Cylinder on Another Fire Fighter Fire Fighter I, NFPA 1001: 4.3.1

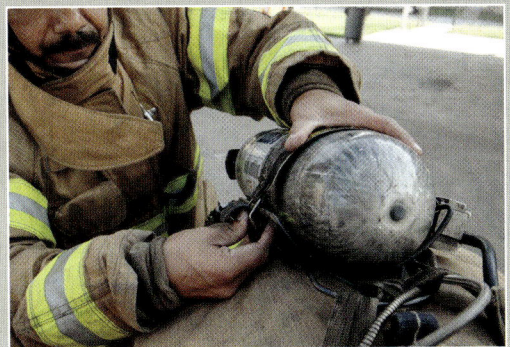

8 Lock the SCBA air cylinder into place.

9 Connect the high-pressure line to the SCBA cylinder.

10 Open the air cylinder valve, and notify the fire fighter that the SCBA cylinder change is complete. State the pressure reading of the SCBA air cylinder to the fire fighter.

a trailer for mobile use. Mobile filling units often are brought to the scene of a large fire.

Compressors filter atmospheric air, compress it to a high pressure, and transfer it to the SCBA air cylinders (**FIGURE 3-30**). Cascade systems have several large storage cylinders of compressed breathing air connected by a high-pressure manifold system (**FIGURE 3-31**). The empty SCBA air cylinder is connected to the cascade system, and compressed air is transferred from the storage cylinders to the SCBA air cylinder. The storage cylinder valves must be opened and closed, one at a time, to fill the SCBA air cylinder to the recommended pressure. This is done by opening the lowest pressure SCBA air cylinder first and then filling it from cylinders containing more pressure. The specific steps for filling SCBA air cylinders vary with different systems.

Proper training is required to fill SCBA cylinders. Refilling SCBA cylinders requires special precautions because of the high pressures involved in this procedure. The SCBA cylinder must be placed in a shielded container while it is being refilled. Such a container is designed to prevent injury if the cylinder ruptures. In addition, the hydrostatic test date must be checked before the cylinder is refilled to ensure that its certification has not expired. Special procedures must be followed to ensure that the air used to fill the SCBA cylinder is not contaminated.

Whether your department has an air compressor or a cascade system, only those fire fighters who have been trained on the safe use of this equipment should refill air cylinders. Follow the manufacturer's recommendations. Follow the steps in **SKILL DRILL 3-13** to safely fill SCBA air cylinders.

Fire fighter breathing air replenishment systems (FBARS) are required in some high-rise buildings based on the local jurisdiction's code. FBARS offer the option to refill SCBA from inside the building on any floor. A fire department air connection is installed outside the building, and plumbing within the structure carries the air throughout the building. Fill stations are located in designated areas throughout the building.

FIGURE 3-30 Compressors filter and compress atmospheric air before transferring it to an SCBA air cylinder.

© Jones & Bartlett Learning. Photographed by Glen E. Ellman.

FIGURE 3-31 Cascade systems store filtered and compressed breathing air in storage cylinders and transfer it to SCBA air cylinders.

© Jones & Bartlett Learning. Photographed by Glen E. Ellman.

Cleaning and Sanitizing SCBA

Most SCBA manufacturers provide specific instructions for the care and cleaning of their models. The first step in cleaning the SCBA is to rinse the entire unit with clean water using a hose. It is important to clean the SCBA *before* taking it apart. This prevents foreign substances getting into SCBA connectors and internal parts. The SCBA harness assembly and air cylinder can be cleaned with a mild detergent/soap-and-water solution. If additional cleaning is needed, the unit can be scrubbed with a stiff brush. After scrubbing, the SCBA harness and air cylinder should be rinsed with clean water. Follow the manufacturer's cleaning instructions and the protocols of your department.

SKILL DRILL 3-13
Refilling SCBA Cylinders Fire Fighter I, NFPA 1001: 4.5.1

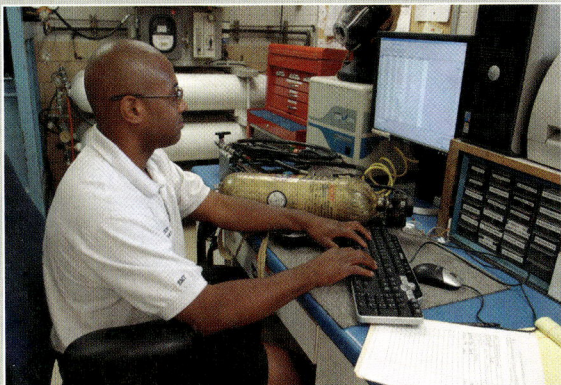

1 Complete the fill record form, including the date, hydrostatic test date, and air cylinder serial number.

2 Ensure the air cylinder is safe to fill by checking its date of manufacture and the hydrostatic test date.

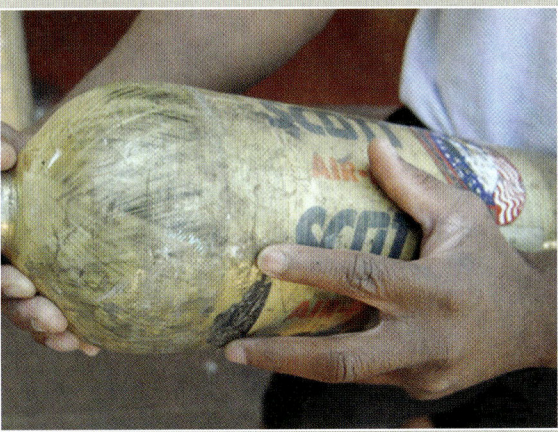

3 Check the air cylinder for visible damage. Follow the compressor or cascade system filling procedures. Secure the system after use according to system procedures.

After a fire, face pieces and regulators can be cleaned with a mild detergent/soap-and-water solution or with a disinfectant cleaning solution. The face piece should be fully submerged in the cleaning solution. If additional cleaning is needed, a soft brush can be used to scrub the face piece. During the cleaning process, avoid scratching the lens or damaging the exhalation valve. The regulator can be cleaned with the same solution, but it should not be submerged. The face piece and regulator should then be rinsed with clean water.

Allow the SCBA time to dry completely before returning it to service. Also check for any damage before returning the equipment to service. Follow the steps in **SKILL DRILL 3-14** to clean and sanitize an SCBA.

SKILL DRILL 3-14
Cleaning an SCBA Fire Fighter I, NFPA 1001: 4.5.1

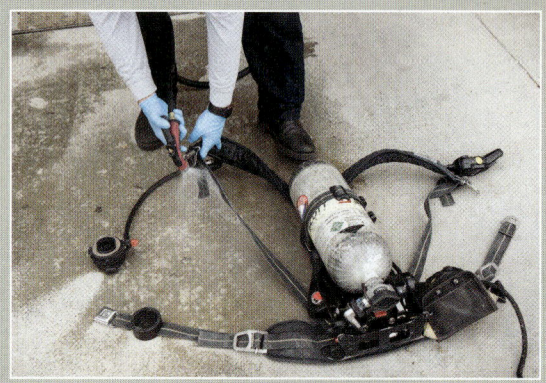

1 Rinse the entire unit using a hose with clean water. Inspect the SCBA for any damage that might have occurred before cleaning.

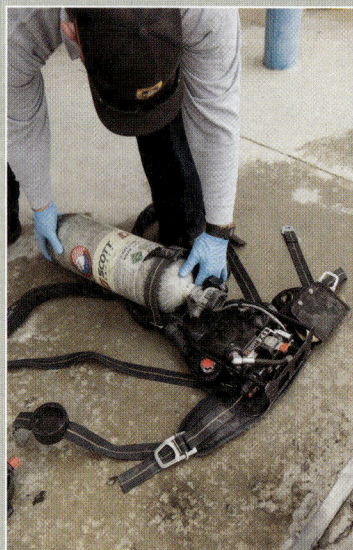

2 On some models, the regulator also can be removed from the SCBA harness. Detach the SCBA air cylinder from the harness.

3 Using a stiff brush, along with a mild detergent/soap-and-water solution, scrub the SCBA air cylinder and harness. Rinse and set these pieces aside to dry.

SKILL DRILL 3-14 Continued
Cleaning an SCBA Fire Fighter I, NFPA 1001: 4.5.1

4 In a 19-L (5-gal) bucket, make a mixture of mild detergent/soap-and-water solution; alternatively, use the manufacturer's recommended cleaning and disinfecting solution and water. Submerge the SCBA face piece in the soapy water or cleaning solution. For heavier cleaning, allow the face piece to soak.

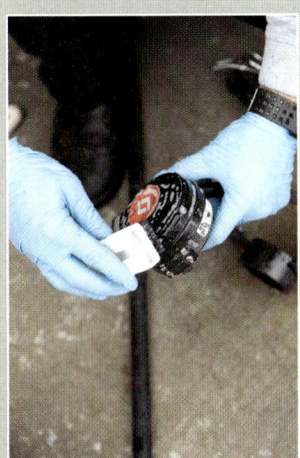

5 Clean the regulator with the soapy water or cleaning solution, following the manufacturer's instructions. Use a soft brush, if necessary, to scrub contaminants from the face piece and regulator.

6 Completely rinse the face piece and the regulator with clean water. Do not submerge the regulator. Set them aside and allow them to dry. Reassemble and inspect the entire SCBA before placing it back in service.

After-Action REVIEW

IN SUMMARY

- Personal protective equipment is an essential component of a fire fighter's safety system. It enables a person to survive under conditions that might otherwise result in death or serious injury.
- Structural firefighting PPE enables fire fighters to enter burning buildings and work in areas with elevated temperatures and concentrations of toxic gases. This type of PPE is designed to be worn with an SCBA.
- The components of structural PPE include the following items:
  - Structural firefighting protective coat and pants—Have tough outer shells to withstand temperature extremes and repel water and insulating layers to protect the skin.
  - Helmet—Protects the entire head and face from trauma; includes ear coverings and a face shield.
  - Protective hood—Constructed of a flame-resistant material to provide additional protection to the ears and neck.
  - Boots—May be constructed of rubber or leather; designed to protect the feet from heat and trauma.
  - Gloves—Constructed of heat-resistant leather to protect the hands from heat, water, vapours, cuts, and puncture wounds.
  - Work uniform—Must be made of natural fibres or special synthetic fibres to prevent burns.
  - SCBA—Provides essential respiratory protection that enables fire fighters to enter the fire atmosphere with a safe, independent air supply.
  - PASS device—An alarm that will automatically or manually sound if a fire fighter is in need of help.
  - Additional equipment—Goggles, earmuffs, ear plugs, hand light, headset and intercom system, and two-way radios.
- Structural PPE has several limitations:
  - Weight—PPE adds 23 kg (50 lb) of extra weight to the fire fighter.
  - Overheating—PPE retains body heat and perspiration, which raises the risk of overheating.
  - Mobility—Full structural firefighting gear limits the range of motion and makes movements awkward and difficult.
  - Senses—PPE limits touch, sight, hearing, and speech.
- Fire fighters should be able to quickly don PPE. Following a set pattern of donning PPE helps reduce the time it takes to dress.
- Check the condition of PPE on a regular basis. Clean it when necessary, and get it repaired at once if it is worn or damaged. Damaged PPE will not protect you.
- Keep PPE clean to maintain its protective properties and to reduce your chance of developing cancer. The flammable by-products of burnt plastics and synthetic products can become trapped between the fibres, damaging the PPE and increasing your chance of developing cancer.
- Cleaning instructions are listed on the tag attached to the clothing. Always follow the manufacturer's cleaning instructions.
- The specialized protective equipment used during vehicle extrication includes special gloves and coveralls or jumpsuits. Latex or nitrile gloves may be used if the fire fighter provides emergency medical care.

- The PPE for wildland fires is designed for comfort and maneuverability. It includes a jacket and pants made of fire-resistant materials, a helmet, eye protection, and pigskin or leather gloves. Wildland boots are designed to provide comfort and sure footing in uneven terrain.

- Respiratory hazards associated with fires include smoke, smoke particles, smoke vapours, toxic gases, oxygen deficiency, and increased temperatures.

- The two main types of SCBA are open-circuit and closed-circuit devices. An open-circuit self-contained breathing apparatus is typically used for structural firefighting. A tank of compressed air provides the breathing air supply for the user, and exhaled air is released into the atmosphere through a one-way valve. A closed-circuit self-contained breathing apparatus recycles the user's exhaled air and is often used for extended operations, such as mine rescue work.

- One of the biggest limitations of an SCBA is the limited amount of air in the cylinder. Fire fighters must manage their working time while using an SCBA so that they have enough time to exit from the hazardous area before exhausting the air supply.

- Physical conditioning is important for all SCBA users. A fire fighter with an ideal body weight and in good physical condition will be able to perform more work per cylinder of air than a person who is overweight or out of shape.

- Breathing through an SCBA is different from normal breathing and can be stressful. Covering your face with a face piece, hearing the air rushing in, hearing valves open and close, and exhaling against a positive pressure are foreign sensations. Practicing breathing through an SCBA in darkness helps to build confidence in the equipment and your skills.

- An SCBA consists of four main parts: the harness, the air cylinder assembly, the regulator assembly, and the face piece assembly. A compressed air cylinder is removable from the SCBA harness and can be changed quickly in the field. SCBA regulators may be mounted on the waist belt or shoulder strap of the harness or attached directly to the face piece; they control the flow of air to the user. The face piece delivers breathing air to the fire fighter and protects the face from high temperatures and smoke.

- The pathway of air through an SCBA begins in the air cylinder. The air passes through the air cylinder valve into the high-pressure air line that takes it to the regulator. The regulator opens when the user inhales, reducing the pressure on the downstream side. In an SCBA unit with a face piece–mounted regulator, the air goes directly into the face piece. In units with a harness-mounted regulator, the air travels from the regulator through a low-pressure hose into the face piece. From the face piece, the air is inhaled through the user's air passages and into the lungs. When the user exhales, used air is returned to the face piece. The exhaled air is exhausted from the face piece through the exhalation valve.

- Several breathing techniques can be used to conserve air:
  - The skip-breathing technique helps conserve air while using an SCBA in a firefighting situation. In this technique, the fire fighter takes a short breath, holds it, takes a second short breath without exhaling in between breaths, and then relaxes with a long exhale. Each breath should take 5 seconds.
  - The Reilly (humming) technique allows for a longer exhalation, and this slows the breathing rate. In this technique, the fire fighter slowly inhales and then exhales while making a humming sound.
  - The controlled breathing technique helps extend SCBA air supply. In this technique, the fire fighter makes a conscious effort to inhale naturally through the nose and then force exhalation from the mouth.

- Each SCBA must be checked on a regular basis to ensure that it is ready for use. Inspection and operational testing should be conducted both after a unit has been used and on a regular schedule.

- Compressors and cascade systems are used to refill SCBA air cylinders. A compressor or a cascade system can be permanently located at a maintenance facility or at a firehouse, or it can be mounted on a truck or a trailer for mobile use. Mobile filling units are often brought to the scene of a large fire.

- The steps for donning complete PPE are as follows:

  1. Place the protective hood over your head, and bring it down around your neck.
  2. Put on your protective pants and boots. Adjust the suspenders, and secure the front flap of the pants.
  3. Put on your protective coat, and secure the front.
  4. Open the air cylinder valve on your SCBA, and check the air pressure. Press the air saver/donning switch to prevent air flow, if needed.
  5. Put on your SCBA harness.
  6. Tighten both shoulder straps of the SCBA harness.
  7. Attach the waist belt of the SCBA harness and tighten it. Tighten the chest straps, if present.
  8. Fit the face piece to your face.
  9. Tighten the face piece straps, beginning with the lowest straps.
  10. Check the face piece for a proper seal. (Follow the manufacturer's instructions.)
  11. Pull the protective hood into position over the face piece straps so that it covers all bare skin but does not obscure your vision.
  12. Place your helmet on your head with the ear tabs extended, and secure the chin strap.
  13. Turn up your coat collar, and secure it in front.
  14. Put on your gloves.
  15. Check your clothing to be sure it is properly secured.
  16. Make sure your PASS device is activated.
  17. Attach your regulator, and open the air cylinder valve to start the flow of breathing air.
  18. Work safely!

KEY TERMS

Access Navigate **for flashcards to test your key term knowledge.**

Air cylinder The pressure vessel or vessels that are an integral part of the SCBA and that contain the breathing gas supply; can be configured as a single cylinder or other pressure vessel or as multiple cylinders or pressure vessels. (NFPA 1981)

Air line The hose through which air flows, either within an SCBA or from an outside source to a supplied air respirator.

Air-purifying respirator (APR) A respirator that removes specific air contaminants by passing ambient air through one or more air purification components. (NFPA 1984)

Atmosphere-supplying respirator (ASR) A respirator that supplies the respirator user with breathing air from a source independent of the ambient atmosphere and includes self-contained breathing apparatus (SCBA) and supplied air respirators (SAR). (NFPA 1981)

Canadian Standards Association (CSA) The federal agency responsible for the regulation of respirator fit testing, training, and breathing air systems.

Carcinogen A cancer-causing substance that is identified in one of several published lists, including, but not limited to, *NIOSH Pocket Guide to Chemical Hazards,*

Hazardous Chemicals Desk Reference, and the *ACGIH 2007 TLVs* and *BEIs*. (NFPA 1851)

Cascade system A method of piping air tanks together to allow air to be supplied to the SCBA fill station using a progressive selection of tanks, each with a higher pressure level. (NFPA 1901)

Closed-circuit self-contained breathing apparatus Self-contained breathing apparatus designed to recycle the user's exhaled air. This system removes carbon dioxide and generates fresh oxygen.

Compressor A device used for increasing the pressure and density of a gas. (NFPA 853)

Doff To take off an item of clothing or equipment.

Don To put on an item of clothing or equipment.

Dual-path pressure reducer A feature that automatically provides a backup method for air to be supplied to the regulator of an SCBA if the primary passage malfunctions.

Emergency breathing safety systems (EBSS) A device on an SCBA that allows users to share their available air supply in an emergency situation. (NFPA 1981)

End-of-service-time indicator (EOSTI) A warning device on an SCBA that alerts the user that the reserved air supply is being utilized. (NFPA 1981)

Face piece Describes both full face pieces that cover the nose, mouth, and eyes and half face pieces that cover the nose and mouth. (NFPA 1404)

Face shield A protective device commonly intended to shield the wearer's face, or portions thereof, in addition to the eyes from certain hazards, depending on face shield type. (NFPA 1500)

Fire helmet Protective head covering worn by fire fighters to protect the head from falling objects, blunt trauma, and heat.

Hand light A small, portable light carried by fire fighters to improve visibility at emergency scenes; it is often powered by rechargeable batteries.

Heads-up display (HUD) Visual display of information and system conditions status that is visible to the wearer. (NFPA 1981)

Hydrostatic testing A test performed by filling pressure-containing components completely with water or other incompressible fluid while expelling all contained air, closing or capping all open ports of the pressure-containing components, and then raising and maintaining the contained pressure to pressurize the pressure-containing components to a prescribed value through an externally supplied pressure-generating device. (NFPA 1901)

Immediately dangerous to life and health (IDLH) Any condition that would pose an immediate or delayed threat to life, cause irreversible adverse health effects, or interfere with an individual's ability to escape unaided from a hazardous environment. (NFPA 1670)

Light-emitting diodes (LEDs) Electronic semiconductors that emit a single-colour light when activated. LEDs are used for operational displays in SCBA.

National Institute for Occupational Safety and Health (NIOSH) The U.S. federal agency responsible for research and development on occupational safety and health issues.

Nose cups An insert inside the face piece of an SCBA that fits over the user's mouth and nose.

Occupational Safety and Health Administration (OSHA) The U.S. federal agency that regulates worker safety and, in some cases, responder safety. It is part of the U.S. Department of Labor.

Open-circuit self-contained breathing apparatus An SCBA in which the exhaled air is released into the atmosphere and is not reused.

Personal alert safety system (PASS) A device that continually monitors for lack of movement of the wearer and automatically activates an alarm signal, indicating the wearer is in need of assistance; can also be manually activated to trigger the alarm signal. (NFPA 1982)

Personal protective equipment (PPE) Consists of full personal protective clothing, plus a self-contained breathing apparatus (SCBA) and a personal alert safety system (PASS) device. (NFPA 1001)

Pounds per square inch (psi) The standard unit for measuring pressure.

Powered air-purifying respirator (PAPR) An air-purifying respirator that uses a powered blower to force the ambient air through one or more air-purifying components to the respiratory inlet covering. (NFPA 1984)

Pressure gauge A device that measures and displays pressure readings. In an SCBA, the pressure gauges indicate the quantity of breathing air that is available at any time.

Protective hood A part of a fire fighter's personal protective equipment that is designed to be worn over the head and under the helmet; it provides thermal protection for the neck and ears.

Rapid intervention crew/company universal air connection (RIC UAC) A system that allows emergency replenishment of breathing air to the SCBA of disabled or entrapped fire or emergency services personnel. (NFPA 1407)

Regulator purge/bypass valve A device or devices designed to bypass a regulator.

Respirator The complete assembly, including the respiratory inlet covering air purification components, electronics, batteries, harness, cables, and hose where applicable; designed to protect the wearer from inhalation of atmospheres containing harmful gases, vapors, or particulate matter. (NFPA 1994)

SCBA harness The backpack or frame for mounting the working parts of the SCBA and the straps and fasteners used to attach the SCBA to the fire fighter.

SCBA regulator The part of the SCBA that reduces the high pressure in the cylinder to a usable lower pressure and controls the flow of air to the user.

Self-contained breathing apparatus (SCBA) A respirator worn by the user that supplies a respirable atmosphere, that is either carried in or generated by the apparatus and that is independent of the ambient environment. (NFPA 350)

Self-contained underwater breathing apparatus (SCUBA) A respirator with an independent air supply that is used by underwater divers.

Structural firefighting protective coat The protective coat worn by a fire fighter for interior structural firefighting.

Structural firefighting protective pants The protective trousers worn by a fire fighter for interior structural firefighting.

Supplied-air respirator (SAR) An atmosphere-supplying respirator for which the source of breathing air is not designed to be carried by the user. Also known as an air line respirator. (NFPA 1989)

Two-way radios Portable communication devices used by fire fighters. Every firefighting team should carry at least one radio to communicate distress, progress, changes in fire conditions, and other pertinent information.

REFERENCES

Canadian Standards Association (CSA). CAN/CSA Z94.4, 2016. https://www.csagroup.org/codes-standards/. Accessed September 26, 2018.

Canadian Standards Association (CSA). CAN/CSA Z180, Compressed Breathing Air and Systems, 2018. https://www.csagroup.org/codes-standards/. Accessed September 26, 2018.

Canadian Standards Association (CSA). CAN/CSA Z96-15, 2015. High-Visibility Safety Apparel. https://www.csagroup.org/codes-standards/. Accessed September 26, 2018.

National Fire Protection Association. NFPA 1001, *Standard on Fire Fighter Professional Qualifications*, 2019. www.nfpa.org. Accessed September 26, 2018.

National Fire Protection Association. NFPA 1404, *Standard for Fire Service Respiratory Protection Training*, 2018. www.nfpa.org. Accessed September 26, 2018.

National Fire Protection Association. NFPA 1500, *Standard on Fire Department Occupational Safety, Health, and Wellness Program*, 2018. www.nfpa.org. Accessed September 26, 2018.

National Fire Protection Association. NFPA 1582, *Standard on Comprehensive Occupational Medical Program for Fire Departments*, 2018. www.nfpa.org. Accessed September 26, 2018.

National Fire Protection Association. NFPA 1851, *Standard on Selection, Care, and Maintenance of Protective Ensembles for Structural Fire Fighting and Proximity Fire Fighting*, 2014. www.nfpa.org. Accessed September 26, 2018.

National Fire Protection Association. NFPA 1852, *Standard on Selection, Care, and Maintenance of Open-Circuit Self-Contained Breathing Apparatus (SCBA)*, 2019. www.nfpa.org. Accessed September 26, 2018.

National Fire Protection Association. NFPA 1901, *Standard for Automotive Fire Apparatus,* 2016. http://www.nfpa.org/. Accessed September 26, 2018.

National Fire Protection Association. NFPA 1971, *Standard on Protective Ensembles for Structural Fire Fighting and Proximity Fire Fighting,* 2018. www.nfpa.org. Accessed September 26, 2018.

National Fire Protection Association. NFPA 1975, *Standard on Emergency Services Work Clothing Elements,* 2014. www.nfpa.org. Accessed September 26, 2018.

National Fire Protection Association. NFPA 1977, *Standard on Protective Clothing and Equipment for Wildland Fire Fighting,* 2016. www.nfpa.org. Accessed September 26, 2018.

National Fire Protection Association. NFPA 1981, *Standard on Open-Circuit Self-Contained Breathing Apparatus for Emergency Services,* 2019. www.nfpa.org. Accessed September 26, 2018.

National Fire Protection Association. NFPA 1982, *Standard on Personal Alert Safety Systems,* 2018. www.nfpa.org. Accessed September 26, 2018.

National Institute of Standards and Technology (NIST). 2011. "Fire Exposures of Fire Fighter Self-Contained Breathing Apparatus Facepiece Lenses." NIOST Technical Note 1724. http://ws680.nist.gov/publication/get_pdf.cfm?pub_id=909917. Accessed September 26, 2018.

Occupational Safety and Health Administration (OSHA). "Confined or Enclosed Spaces or Other Dangerous Atmospheres." https://www.osha.gov/SLTC/etools/shipyard/shiprepair/confinedspace/oxygendeficient.html. Accessed September 27, 2018.

Occupational Safety and Health Administration (OSHA). "Occupational Safety and Health Standards: Personal Protective Equipment." https://www.osha.gov/pls/oshaweb/owadisp.show_document?p_table=standards&p_id=12716. Accessed September 27, 2018.

U.S. Department of Transportation. 2017. 49 *Code of Federal Regulations* 180.205 https://www.ecfr.gov/cgi-bin/text-idx?SID=475de645c98062d4043077d9794a583f&mc=true &node=se49.3.180_1205&rgn=div8. Accessed September 27, 2018.

On Scene

Your crew was first in to the fully involved house, and you were the lead on the nozzle as you worked to put down the flames. After completing rehab, you were tasked with overhaul, but you do not mind; you are getting to fight fire. You think to yourself, "This is the greatest job in the world."

You are still pumped from the excitement as you pull back into the station. As you step off the pumper, you notice the soot covering your gear and your SCBA. You think to yourself that it will be a while before you get to bed. Your captain, who is also acting as your mentor, asks you the following questions.

1. Which of the following is *not* a component of the structural firefighting PPE?

A. Protective hood

B. Self-contained breathing apparatus (SCBA)

C. Mobile radio

D. Structural firefighting protective coat

2. Which method should routinely be used to clean the SCBA?

A. Mild acid solvent

B. Clean water only

C. Mild detergent/soap-and-water solution

D. Solvent-based degreaser

(continued)

On Scene Continued

3. Pressure readings on the air cylinder pressure gauge and the remote pressure gauge should be _____ each other.

   A. the same as

   B. within 20 percent of

   C. within 50 percent of

   D. within 10 percent of

4. Full PPE weighs nearly:

   A. 11 kg (25 lb)

   B. 16 kg (35 lb)

   C. 23 kg (50 lb)

   D. 32 kg (70 lb)

5. A(n) _____ is an SCBA designed to recycle the user's exhaled air.

   A. supplied-air respirator

   B. open-circuit self-contained breathing apparatus

   C. closed-circuit self-contained breathing apparatus

   D. cascade system

Access Navigate to find answers to this On Scene, along with other resources such as an audiobook.

Fire Service Communications

KNOWLEDGE OBJECTIVES

After studying this chapter, you will be able to:

- Describe the role of the communications centre in the fire service. (pp. 133–134)
- Describe the role and responsibilities of a telecommunicator. (p. 134)
- List the requirements of a communications centre. (pp. 134–135)
- Describe the equipment used in a communications centre. (pp. 135–136)
- Describe how computer-aided dispatch assists in dispatching the correct resources to an emergency incident. (pp. 135–136)
- Describe the steps in processing an emergency incident. (pp. 137–142)
- Explain methods of receiving emergency and nonemergency fire department communications. (**NFPA 1001: 4.2.1**, pp. 143–144)
- Describe how telecommunicators conduct a telephone interrogation. (p. 138)
- Describe how location validation systems operate. (pp. 140–141)
- Determine if a communication is emergent or nonemergent. (**NFPA 1001: 4.2.1**, pp. 141, 143)

- Explain procedures for transmitting the emergency information to a dispatch centre. (**NFPA 1001: 4.2.1**, p. 142)
- Describe procedures for handling nonemergency calls. (**NFPA 1001: 4.2.2**, pp. 143–144)
- Describe procedures for handling emergency calls. (**NFPA 1001: 4.2.1**, pp. 143–144)
- Explain the importance of following department SOPs for receiving and processing communications. (**NFPA 1001: 4.2.1**, pp. 138, 141, 143)
- Identify information to be obtained when taking a report of an emergency to enable necessary assistance to be dispatched. (**NFPA 1001: 4.2.1**, p. 138)
- Describe the three types of fire service radios. (pp. 145–146)
- Identify other modes of fire service communication. (**NFPA 1001: 4.2.1**, p. 135)
- Describe how two-way radio systems operate. (p. 145)
- Explain how a repeater system works to enhance fire service communications. (pp. 147–148)
- Explain how digital radios and trunked systems work to enhance fire service communications. (p. 148)
- Describe the basic principles of effective radio communication. (**NFPA 1001: 4.2.3**, pp. 148–149)

- Identify radio departmental procedures and codes for using fire department radios. (**NFPA 1001: 4.2.1; 4.2.3**, pp. 148–151)
- Describe when to use plain language and how ten-codes are implemented in fire service communications. (**NFPA 1001: 4.2.3**, p. 148)
- Outline the information provided in size-up and progress reports. (pp. 149–150)
- Recognize routine traffic, emergency traffic, and emergency evacuation signals. (**NFPA 1001: 4.2.3**, pp. 150–151)

- Send and receive messages over the fire department radio. (**NFPA 1001: 4.2.2, 4.2.3**, pp. 147–148)
- Determine if a radio communication is routine or emergency traffic. (**NFPA 1001: 4.2.3**, pp. 150–151)

ADDITIONAL STANDARDS

- **NFPA 901**, *Standard Classifications for Incident Reporting and Fire Protection Data*
- **NFPA 902**, *Fire Reporting Field Incident Guide*
- **NFPA 1061**, *Standard for Public Safety Telecommunications Personnel Professional Qualifications*
- **NFPA 1221**, *Standard for the Installation, Maintenance, and Use of Emergency Services Communications Systems*

SKILLS OBJECTIVES

After studying this chapter, you will be able to perform the following skills:

- Receive a phone call, and obtain, route, and document information according to department procedures. (**NFPA 1001: 4.2.2**, pp. 143–144)
- Observe the operation of a communications centre. (pp. 136–137)

Fire Alarm

At 3:04 AM, you are dispatched to a report of a fire at 3256 Queen Street West. Your captain asks if any additional details are available, and the dispatcher says that the caller said there was smoke coming from the eaves of the building and then hung up. You are getting excited about the prospect of a working fire. Then the dispatcher comes back and says that this may be a false report, because the caller used a cell phone from the east side of town. Your aggravation rises at the prospect of being woken up for another false call—at least until the dispatcher comes back with a frantic sound in her voice and says that a home security company just reported a fire alarm at 3256 Queen Street *East*. Thoughts plow through your mind as you wonder what is going on.

1. What is the process for receiving 911 calls in an emergency communications centre?

2. How would the dispatcher know where the cell phone call originated?

3. How are alarm systems monitored and then reported to the communications centre?

 *Access Navigate for more practice activities.*

Introduction

Technology and advanced communications systems are having a tremendous effect on fire department communications. As a fire fighter, you must be familiar with the communications systems, equipment, and procedures used in your department. This chapter provides a basic guide to help you understand how fire department communications systems work and how commonly used systems are configured.

Every fire department depends on a functional communications centre. When a citizen requests assistance or an alarm sounds, the communications centre dispatches the appropriate units to the incident, and it continues to maintain communication with those units throughout the duration of the incident. Communications personnel (telecommunicators) play an integral role in ensuring constant and reliable communication for on scene personnel, thereby improving firefighter safety. The communications centre should always know which units can be dispatched to an incident and be able to dispatch units accordingly. Another important responsibility of the telecommunicator is to ensure adequate coverage for the response area. This may include shuffling apparatus from one district to another or calling for mutual aid. The actions taken by the communications centre will be mandated by the size of the coverage area and established guidelines.

At the scene, fire fighters need to communicate with one another so that the incident commander (IC) can manage the operation efficiently based on progress reports or requests for assistance from fire fighters. The incident command system (ICS) depends on the presence of a functional onsite communications system. During incidents, fire fighters must be able to communicate not only with one another but also with other emergency response agencies.

In addition to these special communications requirements, a fire department must have a communications infrastructure that allows it to function as an effective organization. Basic administration and day-to-day management require an efficient communications network that includes, but is not limited to, telephone and data links with every fire station and work site.

The Communications Centre

Most requests for fire department response are made by either landline telephone or cellular phone. Emergency and nonemergency calls are then directed to a **public safety communications centre** for that community or jurisdiction. The communications centre may be a designated **public safety answering point (PSAP),** which serves and dispatches multiple agencies (fire, EMS, and law enforcement), or a stand-alone communications centre, which serves and dispatches a single agency (e.g., fire department). Some communications centres work with several fire departments, and some work with all of the fire departments in a county or region. In Canada, the 911 operations centres are located in the municipal police dispatch centre. 911 dispatchers will answer calls, ascertain the nature of the emergency, and send the call to the appropriate agency. In remote regions within large geographic areas that have small populations, the Royal Canadian Mounted Police and/or the provincial policing agency, as in Ontario and Quebec, are typically responsible for dispatching emergency vehicles, including fire services.

A joint facility may house separate personnel and independent systems for each service delivery agency; alternatively, the entire operation may be integrated, with all employees cross-trained to receive calls and dispatch responders to any type of emergency incident. When fire, EMS, and law enforcement communications are located in the same facility, the call can be answered, processed, and dispatched immediately. If these agencies operate in separate facilities, those calls must be transferred to the appropriate telecommunicator. Some agencies also have a mobile communications centre that allows dispatchers to be onsite and run communications for a larger incident. The call-taking process is described in more detail later in this chapter.

Regardless of how or where the call is transferred, the public safety communications centre is the hub of the fire department emergency response system. It serves as the central processing point for all information relating to an emergency incident and all of the information relating to the location, status, and activities of fire department units. It connects and controls all of the department's communications systems or serves as the PSAP. The communications centre functions much like the human brain—that is, information comes in via the nerves, is processed, and then is sent back out to be acted upon by the various parts of the body.

The communications centre is a physical location, whose size and complexity vary depending on the needs of the department. The communications centre for one department may be a small room in a fire station, whereas another department may have a specially designed, highly sophisticated facility with advanced technological equipment. Fire departments in smaller communities may only require a simple system whereas public safety agencies in a larger metropolitan area may require a larger facility. There tends to be distinct differences between rural and urban dispatch centres—specifically,

FIGURE 4-1 All communications centres perform the same basic functions.
© Jones & Bartlett Learning. Photographed by Glen E. Ellman.

the number of dispatchers needed at any given time, the type of training that is required, and, to some extent, the type of equipment used. Regardless of their size, location, and configuration, communications centres perform the same basic functions (**FIGURE 4-1**).

Telecommunicators

The employees who staff a communications centre are known as call takers, dispatchers, or **telecommunicators**. The job of a telecommunicator can be complicated, demanding, and extremely stressful. Even in the face of these challenges, the successful telecommunicator must be able to receive, process, or disseminate information, understand and follow complicated procedures, perform multiple tasks effectively, memorize information, and make decisions quickly.

Telecommunicators who have been professionally trained to work in a public safety communications environment and who have completed advanced training and professional certification programs are critical to the overall effectiveness of the communications system.

One of the telecommunicator's most important skills is the ability to communicate effectively with citizens to obtain critical information, even when the caller is highly stressed or in extreme personal danger. Telecommunicators must always maintain a calm, professional demeanor even if a caller is overly emotional or insulting. Voice control and the ability to maintain composure under pressure are important qualities; the telecommunicator must always be in control.

A telecommunicator must be skilled in operating all of the systems and equipment in the communications centre. He or she must understand and follow the fire department's operational procedures, particularly those relating to dispatch policies and protocols, radio communications, and incident management. The telecommunicator must keep track of the status and location of each unit at all times and must monitor the overall deployment and availability of resources throughout the system. NFPA 1061, *Standard for Public Safety Telecommunications Personnel Professional Qualifications*, contains a complete list of qualifications for telecommunicator candidates in the United States. In Canada, there are no established nationwide requirements for the certification of telecommunicators. Most municipalities train their personnel in house to perform the duties required for their specific fire service. In municipalities east of of the Ottawa River, telecommunicators must be bilingual in French and English. In most areas within Ontario and the western provinces and territories, bilingualism is not mandatory but is considered an asset.

Communications Facility Centre Requirements

The fire service communications centre must be designed and operated to ensure that its critical mission can be performed with a high degree of reliability. The performance requirements in NFPA 1221, *Standard for the Installation, Maintenance, and Use of Emergency Services Communications Systems*, govern the design and construction of a fire department or public safety communications centre. These requirements apply whether the communications centre serves a small community with only one or two fire stations or a metropolitan area with dozens of stations.

The communications centre should be well protected against natural threats, and it should be able to withstand predictable damaging forces such as floods, earthquakes, snowstorms, tornadoes, hurricanes, and other severe storms. It should be located so that it can continue to function in times of civil unrest. In addition, this centre should be able to operate at maximum capacity, without interruption, even when other community services are severely affected. The building should be equipped with emergency generators and other systems so that it can continue to operate for several days in even the most challenging conditions.

NFPA 1221 also requires backup systems for all of the critical equipment in a communications centre so that the failure of a single component or system will not disable the entire operation. For example, there must be more than one way of transmitting a dispatch message from the communications centre to each fire station. A backup radio transmitter should be available, and the telephone system must be able to receive calls even if part of the system is damaged. The design of the facility should minimize its vulnerability to a fire originating inside the building as well as to nearby fires. The centre also must be secured to prevent unauthorized entry.

A backup communications centre at a different location should be established. If some unanticipated situation makes it impossible to operate from the primary location, this backup location can be activated to ensure ongoing operation of the communications function.

Plans need to be in place for the reporting of emergencies in the event that the emergency communications centre fails. Some communities have plans in place to locate fire department and law enforcement vehicles at strategic locations so citizens can report emergencies. Fire station personnel may plan to staff a watch desk to report emergencies from citizens who walk into the fire station.

Communications Centre Equipment

A communications centre is usually equipped with several types of communications systems and equipment. Although the specific requirements depend on the size of the operation and the configuration of the local communications systems, most centres will have the following equipment:

- Dedicated 911 telephones
- Public telephones
- Direct-line phones to other agencies
- Equipment to receive alarms from public or private fire alarm systems
- Computers utilizing **geographic information systems (GIS)** and/or hard-copy files and maps to locate addresses and select units to dispatch; the GIS stores and integrates data gathered from GPS
- Equipment for alerting and dispatching units to emergency calls
- Two-way radio system(s)
- Recording devices to record phone calls and radio traffic
- Backup electrical generators
- Records and record management system
- TTY/TDD capabilities

Computer-Aided Dispatch

Computers are used in almost all communications centres. Most large communications centres use sophisticated **computer-aided dispatch (CAD)** systems that perform many functions; many smaller centres have smaller-scale, less sophisticated versions of CAD systems.

A CAD system helps meet the most important objective in processing an emergency call: sending the appropriate units to the correct location as quickly as possible. As the name suggests, a CAD system is a combination of hardware and software designed to assist a telecommunicator by performing specific functions more quickly and efficiently than they can be done manually.

FIGURE 4-2 A CAD system enables a telecommunicator to work more quickly and efficiently.
© Jones & Bartlett Learning. Photographed by Glen E. Ellman.

Once the address is determined and the incident description is in the CAD system, the system quickly makes a recommendation on the appropriate units to dispatch. Although all of the functions performed by a CAD system can be performed manually by trained and experienced telecommunicators, the system can make a recommendation on the appropriate units to dispatch in less than a second (**FIGURE 4-2**). Data links that provide the location of the caller and automatically enter the address can also save time. If duplicate addresses exist or if the address entered is not a valid location, the CAD system will prompt the telecommunicator to ask the caller for more information.

The most advanced type of CAD system utilizes **global positioning system (GPS)** data devices to confirm the location of the caller. The CAD system determines the location of each unit, which units are available to respond to a call, which units are assigned to incidents, which units are temporarily assigned to cover different areas, and the closest fire stations in order of response for any location. The CAD then indicates the units that can respond quickly to an alarm, even if some of the units that would normally respond to the call are currently unavailable. A major advantage of a CAD system is that it automatically captures and stores every event as it occurs. Before such systems were developed, someone had to write down and time-stamp every transaction, including every radio transmission, and all of these hard-copy logs had to be retained for future reference.

Some CAD systems transmit dispatch information directly to **mobile data terminals (MDTs)** on the apparatus or to computers that are located in the fire station (**FIGURE 4-3**). In addition, CAD systems can provide immediate access to information such as preincident plans, hazardous materials lists, lockbox locations, and information such as whether persons with limited mobility reside at the address of the emergency call.

FIGURE 4-3 Some CAD systems transmit dispatch information directly to terminals in fire stations and mobile data terminals in the apparatus.
© Jones & Bartlett Learning. Photographed by Glen E. Ellman.

FIGURE 4-4 Telecommunicators must obtain information and relay it accurately to the appropriate responders.
© Jones & Bartlett Learning. Photographed by Glen E. Ellman.

By linking the CAD system to other data files, fire fighters gain access to even more useful information such as travel route instructions, maps, and reference materials.

Voice Recorders and Activity Logs

Almost everything that happens in a communications centre is recorded, either by a **voice recording system** or by an **activity logging system**. Most communications centres can automatically record everything that is said over the telephone or radio, 24 hours a day. In addition, most have an instant playback unit that allows the telecommunicator to replay conversations for the previous 10–15 minutes at the touch of a button. This feature is particularly valuable if the caller talks quickly, has an unusual accent, hangs up, or is disconnected. The telecommunicator can replay the message several times, if necessary, to determine exactly what was said.

The logging system keeps a detailed record of every incident and activity that occurs. These records include every call that is entered, every unit that is dispatched, and every significant event that occurs in relation to an emergency incident. Times are recorded when a call is received, when the units are dispatched, when they report that they are en route, when they arrive at the scene, when the incident is under control, and when the last unit leaves the scene.

Voice recorders and activity logs are maintained for several reasons. First, they serve as legal records of the official delivery of a government service by a public agency (i.e., the fire department). These records may be required for legal proceedings, sometimes years after the incident occurred. They may be needed to defend the fire department's actions when questions are raised about an unfortunate outcome. The records provided by voice recorders and activity logs accurately document the events and often can demonstrate that the organization and its employees performed ethically, responsibly, and professionally. They also make it difficult to hide an error, if a mistake was made.

Second, records are valuable for reviewing and analyzing information about department operations. Good record keeping enables a fire department to examine what happened on a particular call as well as to measure workloads, system performance, activity trends, and other factors as part of its planning and budget preparation. For example, analysts and planners can use data from CAD systems to study deployment strategies and to make the most efficient use of fire department resources.

The telecommunicator's first responsibility is to obtain information that is required to dispatch the appropriate units to the correct location (**FIGURE 4-4**). At that point, the incident must be processed according to standard protocols. Using the CAD system information, the telecommunicator must decide which agencies or units should respond and transmit the necessary information to them.

Communications Centre Operations

Several different functions are performed in a fire department or public safety communications centre. All of these activities must be performed accurately and efficiently, even in the most challenging circumstances. For example, the activity level in a communications centre can increase rapidly in a short period of time, but the chaos outside must not affect the operations inside. If the communications centre fails to perform its mission, the fire department will not be able to deliver much-needed emergency services.

A communications centre performs the following basic functions:

- Receiving calls for emergency incidents and dispatching fire department units
- Supporting the operations of fire department units delivering emergency services
- Coordinating fire department operations with other agencies
- Tracking the location of all operational vehicles and special operational assets
- Monitoring the level of coverage and managing the deployment of available units
- Notifying designated individuals and agencies of particular events and situations
- Maintaining records of all emergency-related activities
- Maintaining information required for dispatch purposes
- Assisting with on scene accountability and supporting incident command

Most fire department responses begin when a call comes in to the communications centre, which then dispatches one or more units. If a citizen reports an emergency directly to a fire station or a crew discovers a situation, it is imperative that the information and request for help be immediately relayed to the communications centre. The communications centre can then initiate the emergency response process. Most emergency response services in Canada generate an incident number for every call for service; this usually includes emergency and nonemergency incident response. The assignment of an incident call number enables effective referencing from the records management system. Responding to a reported emergency without using the proper dispatch procedures can result in confusion and delay in the necessary resources reaching the emergency.

The communications centre should have both a primary method and a backup method of transmitting alarms to stations. Although radio, telephone, and public address systems often are used to transmit information to fire stations, the use of fixed or mobile computer terminals and printers to transmit dispatch messages is increasing. Some fire departments still use a system of bells to transmit alarms. Although volunteer or rural departments may use outdoor sirens or horns to summon fire fighters to an emergency, most volunteer fire fighters receive dispatch messages over pagers or cell phones.

Processing an emergency incident includes the following steps:

1. Call receipt
2. Location validation
3. Classification and prioritization
4. Unit selection
5. Dispatch

Call Receipt

Call receipt refers to the process of receiving a call for service and obtaining the necessary information to initiate a response. It includes telephone calls from the general public as well as other notification methods, such as automatic fire alarm systems, municipal fire alarm boxes, requests from other agencies, calls reported directly to fire stations, and calls initiated via radio from fire department units or other public safety agencies.

LISTEN UP!

Dispatchers must be careful about which information is given to the public. Information such as availability of individual fire departments, apparatus placement, and available equipment should be monitored. A standard operating procedure should clarify which information is allowed to be broadcast and which information is not.

Telephones

The public generally uses landline telephones or cellular phones to report emergency incidents. Most emergencies that occur outside a building are reported using a cellular phone. Some communities estimate that about 70 percent of their emergency calls are reported with cellular phones (Federal Communications Commission [FCC], 2016). In most communities, calls to 911 connect the caller with a PSAP. The PSAP can take the information immediately or transfer the call to the appropriate agency, based on the nature of the emergency. In some systems, calling 911 connects the caller directly with a telecommunicator, who obtains the required information.

LISTEN UP!

Some communities have implemented a 311 system to handle nonemergency calls. This system often is used to link calls with health and human services or other community resources that can assist with a problem when it might not necessarily be life threatening in nature. The underlying premise of the 311 system is to reduce the number of nonemergency calls to 911.

The communications centre has a seven-digit nonemergency telephone number. Some people may use that number to report an emergency because they are not sure that their particular situation is serious enough to be considered a "true emergency."

Any number that is published as a fire department telephone number should be answered at all times, including nights and weekends. The 911 emergency number should always be pronounced as "nine-one-one"; it should not be pronounced as "nine-eleven," because there is no "11" on a telephone.

The telecommunicator who takes the call must conduct a **telephone interrogation**, asking the caller questions to obtain the required information. Initially, the telecommunicator needs to know the location of the emergency and the nature of the situation. As discussed, many 911 systems can automatically provide the location of the telephone where a call originates. However, this information may be unavailable or inaccurate if the call was made from a cellular phone or from a location other than the site of the emergency incident. The exact location must be obtained so that units can respond directly to the incident scene.

The telecommunicator also must interpret the nature of the problem from the caller's description. The caller may be distressed, excited, and unable to organize his or her thoughts. There may be language barriers; the caller might speak a different language or might not know the right words to explain the situation. Telecommunicators must follow standard operating procedures (SOPs) and use active listening to interpret the information. Many communications centres use a structured set of questions to obtain and classify information on the nature of the situation. The telecommunicator must remember that the caller thinks the situation is an emergency, and he or she must treat every call as such until it is determined that no emergency exists.

Telecommunicators cannot allow gaps of silence to occur while questioning the caller. If the caller suddenly becomes silent, something may have happened to him or her; the caller might be in personal danger or extremely upset. Conversely, if the telecommunicator is silent, the caller might think that the telecommunicator is no longer on the line or no longer listening.

Disconnects are another problem, because callers to 911 might hang up accidentally or be disconnected after initially reaching the communications centre. If the caller's location is known, a telecommunicator who is unable to return the call and reach the original caller usually dispatches a police officer to the location to determine if more help is needed.

With just two critical pieces of information—the location and nature of the problem—the telecommunicator can initiate a response. Local SOPs, however, may require the telecommunicator to obtain additional information. Getting the caller's name and contact phone number is useful in case it is necessary to call back for additional information. If the caller is in danger or distress, the telecommunicator will try to keep the line open and remain in contact with the caller until help arrives. In many communities, telecommunicators are trained to provide self-help instructions for callers, advising them on what to do until the fire department arrives. For example, if a building fire is being reported, the telecommunicator will advise the occupants to evacuate and wait outside; if the caller is reporting a medical incident, the telecommunicator may provide first-aid instructions, based on the patient's symptoms.

TTY/TDD/Text. Speech- and/or hearing-impaired citizens can communicate by telephone using special relay devices that display text rather than transmitting audio (**FIGURE 4-5**). These include the TTY (teletype), the TDD (telecommunications device for the deaf), and text devices. **TTY/TDD systems** and text devices have a screen and a keyboard for exchanging words. The **Canadian Radio-television and Telecommunications Commission (CRTC)** is a regulatory agency with a mandate as an overseer for broadcasting and telecommunications. TTYs/TDDs and computer modems are all standard equipment in Canada's numerous communications centres. Telecommunicators must know how to use this equipment to communicate with people who are speech and hearing impaired. Many communications centres have a special telephone line with a seven-digit number set aside for such calls.

While not all communications centres accept text-to-911 messages at this time, text service providers continue to develop capabilities. The CRTC is promoting an agenda that includes the next generation of 911 communications, referred to as NG911.

> **LISTEN UP!**
>
> Next-generation dispatch systems permit dispatch centres to receive data, texts, video clips, and other communications from the general public (Dolcourt, 2014).

Direct-Line Telephones. Even though cellular devices often are used for most functions, it is important to have a second means to communicate if the cellular system or radio

FIGURE 4-5 TTY/TDD/text devices enable speech-and/or hearing-impaired persons to communicate over telephone lines or wireless networks using text communication.
© Jones & Bartlett Learning.

system becomes inoperable. A **direct-line** (or ring-down) telephone connects two predetermined points. Picking up the phone at one end causes an immediate ring at the other end. Direct-line connections often link police and fire communications centres or two fire communications centres that serve adjacent areas. Direct lines also may connect hospitals, private alarm companies, utility companies, airports, and similar facilities with the communications centre. These lines often are used in both directions so that the communications centre can both receive calls for assistance and send notifications and requests for response.

Direct lines may connect the communications centre to each fire station in its jurisdiction. A direct line also can be linked to the station's public address speakers to announce dispatch messages.

Municipal Fire Alarm Systems

Some communities have fire alarm boxes or emergency telephones on street corners or in public places. A fire alarm box transmits a coded signal to the communications centre and, in some cases, directly to individual fire departments (**FIGURE 4-6**). By manually pulling a lever, a signal transmits an alarm and identifies the location of the box; however, it does not indicate the type of emergency that is occurring. Municipal fire alarm boxes also may be connected to automatic fire alarm systems in private commercial, industrial, and residential buildings. When the building's fire alarm system is activated, the fire alarm box transmits an alarm. Municipal fire alarm boxes are usually connected by networks of dedicated cables.

In recent years, most communities have eliminated their municipal fire alarm systems due to the inability to determine the nature of the call, an increasing number of false alarms, high maintenance costs, and development of alternative notification systems, including private and public hard-wired and cellular telephones. In many cases, radio-operated fire alarm boxes have replaced these hard-wired boxes. A radio-operated fire alarm box uses radio waves to transmit information (**FIGURE 4-7**). Other communities have converted their municipal fire alarm systems into call box systems.

The use of municipal fire alarm systems has decreased since the advent of cellular communication devices. Private alarm companies, described in the next section, also receive a large number of fire alarms.

A **call box** connects a person directly to a telecommunicator. The caller can request a full range of emergency assistance—from police, fire, or emergency medical assistance to a tow truck. Call boxes can be directly connected to telephone networks or can operate on a radio system. Emergency call boxes often are located along major highways and near bridges, tunnels, subways, large complexes such as universities, and in other locations where nearby

FIGURE 4-6 Fire alarm boxes are still in use in some parts of the country.
© Beschi Mauro/Shutterstock.

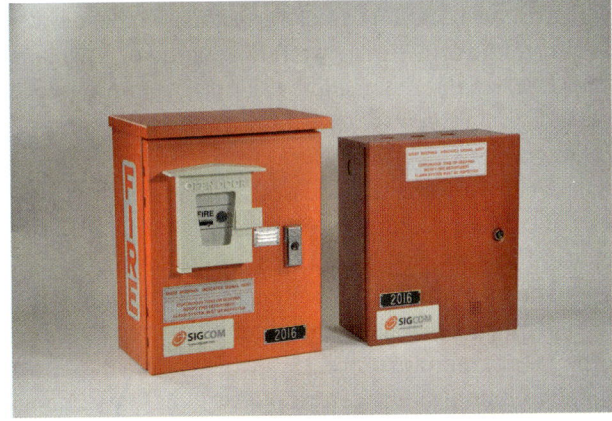

FIGURE 4-7 A radio-operated fire alarm box.
Courtesy of Signal Communications Corporation (Sigcom).

phones are not available and where cellular phones do not work. In locations without electric or telephone service, call boxes may be solar powered (**FIGURE 4-8**).

Private Automatic Fire Alarm Systems

As discussed, many private commercial, industrial, and residential buildings have automatic fire alarm systems,

FIGURE 4-8 Wireless call boxes are sometimes found in places without other kinds of telephone service.
© Jones & Bartlett Learning. Photographed by Glen E. Ellman.

which use heat detectors, smoke detectors, or other devices to initiate an alarm, such as water flow alarms or automatic sprinkler systems. Automatic fire alarm systems use several types of arrangements to transmit alarms to the communications centre. Chapter 26, *Fire Detection, Suppression, and Smoke Control Systems*, discusses these systems in detail.

The connection used to transmit an alarm from a private fire alarm system to a communications centre depends on many factors. These systems may be monitored by a privately operated central station alarm service or other alarm monitoring service, which relays the alarm to the communications centre. A direct line may connect a central alarm service and the communications centre. Some communications centres provide monitoring services. No matter how the alarm reaches the communications centre, the result is the creation of an incident report and the dispatch of fire department apparatus.

Walk-ins

Although most emergencies are reported by telephone, sometimes people actually come to a fire station seeking assistance. When a walk-in occurs, the station should contact and advise the communications centre of the

situation immediately. The communications centre creates an incident report and dispatches any needed additional assistance. Even if the units at the station can handle the situation, they should notify the communications centre that they are occupied with an incident.

Citizens should be able to come to a fire station and report an emergency at all times, even when the station is unoccupied. Many departments install a direct-line telephone to the communications centre just outside each fire station. These phones should be marked with a simple sign stating "If the station is vacant, pick up the telephone in the red box to report an emergency." This enables citizens to contact the dispatch centre quickly when the station is not occupied or when all units are out on emergency calls.

Location Validation

In order to process the request for service and before dispatching units, the telecommunicator must validate that the location information received is adequate. For example, potential duplicate addresses—such as two streets with the same or similar names—must be eliminated from consideration. For example, it is important to differentiate 123 East Main Street from 123 West Main Street. This process of elimination is done through the GIS system in CAD. The information must point to a valid location on a map or in a street index system, and that location must be within the geographic jurisdiction of the potential dispatch units. Without a valid address, a communications centre will be unable to send units to the proper location.

Enhanced 911 Systems

Most 911 systems currently in operation are called enhanced 911 systems. Enhanced 911 systems have features that can help the telecommunicator obtain identifying information. For example, **automatic number identification (ANI)** shows the telephone number where the call originated. **Automatic location identification (ALI)** queries a database to show the location of the telephone, the subscriber's name, and other details (**FIGURE 4-9**). Enhanced 911 is designed to improve dispatching time and accuracy. When combined with an accurate database, enhanced 911 generates a computer display listing all of the necessary information when the call is received from a landline telephone. Another feature in an enhanced 911 system ensures that each call is directed to the appropriate PSAP for that location.

The telecommunicator should always confirm that the information that is generated is correct and refers to the actual location of the emergency. If the caller has recently moved, the database may contain the old address. Also, sometimes a billing address for the telephone service is provided rather than the physical location of the telephone.

Although cellular phones have certainly enhanced public access to emergency services (a majority of the

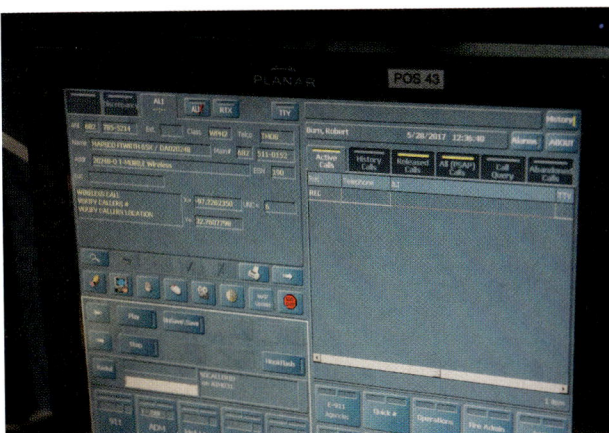

FIGURE 4-9 An automatic location identification (ALI) system provides information about the origin of a 911 call.
© Jones & Bartlett Learning. Photographed by Glen E. Ellman.

calls are coming from cellular devices), they have also created challenges. Efforts to provide location validation for calls received from cellular devices are advancing; however, in some areas the dispatcher cannot determine the location of many 911 calls. This includes a large number of calls made from cellular devices or those that use the Internet for phone transmission using **Voice over Internet Protocol (VoIP)**. VoIP technology converts a person's voice into a digital signal that can be sent via the Internet to another computer, a VoIP phone, or a traditional phone with a specialized adapter.

Keep in mind that many people calling 911 may not know their exact location. In addition, the use of cellular phones often results in numerous calls for the same incident, making it hard to determine the exact location of the emergency and overloading the communications centre.

To address these challenges, next-generation 911 systems (NG911) allow digital information, including voice, photos, text messages, and video, to flow from any communications device through the network to a PSAP. It relies on high quality mapping and the use of GIS to determine the location from which a call is being made. Although the technology to transition to NG911 is available, it is still evolving. PSAP technology is also advancing; PSAPs will be able to receive data and alerts

from safety devices, medical devices, and sensors, and they will be able to issue emergency alerts to wireless devices and highway alert systems.

Call Classification and Prioritization

Classification and prioritization is the process of assigning a response category, based on the nature of the reported problem. Most fire departments respond to many kinds of situations, ranging from outside ground cover fires and fires in high-rise buildings to heart attacks and multiple-casualty incidents. The nature of the call dictates the urgency of the call and its priority. Although most fire department calls are dispatched immediately, it is sometimes necessary to delay dispatch to a lower-priority call if a more urgent situation arises or if several calls come in at the same time. Some calls qualify for emergency response (red lights and siren); others are considered nonemergency.

The call classification and prioritization is an important part of processing a request for service because it will help you select the appropriate units to respond to the incident. The standard response to different incidents is established by the fire department's SOPs and dispatch protocols and can range from a single unit to an initial assignment of 10 or more units and dozens of fire fighters.

Unit Selection

Unit selection is the process of determining exactly which unit or units to dispatch, based on the location and classification of the incident. The usual policy is to dispatch the closest available unit that can provide the necessary assistance. When all units are available and in their fire stations, this determination can usually be made quickly based on preprogrammed information. Generally, the standard assignment to each type of incident in each geographic area is stored on a system of **run cards**. They list units in the proper order of response, based on response distance or estimated response time, and often specify the units that would be dispatched through several levels of multiple alarms. Run cards are prepared in advance and are usually entered into the CAD system. In the past, departments stored these in hard-copy form.

Unit selection becomes more complicated when some units are out of position or unavailable. This step often requires quick decision-making skills, even when the CAD system is programmed to follow set policies for various situations. Some communications centres are equipped with automatic vehicle locator systems that track apparatus using GPS devices. These systems enable telecommunicators to dispatch the closest units that are in service, thereby increasing the efficiency of the system. Many departments have entered into automatic mutual aid agreements that specify that the

FIRE FIGHTER TIP

As cellular and radio technology advances, FirstNet was created by the U.S. Department of Commerce to build a nationwide broadband network for communication among first responders and 911 centres. The goal of FirstNet is to have a designated wireless network for emergency responses that will allow a national coordinated response to emergencies (United States Department of Transportation, 2017).

closest available units will be sent even if that requires dispatching units from several different jurisdictions.

Most CAD systems are programmed to select the units for an incident automatically, based on the location, call classification, and actual status of all units. Such a system will recommend a dispatch assignment, which the telecommunicator can then accept or adjust, based on circumstances or special information. The same process is used to dispatch any additional units to an incident, whether it is a multiple alarm or a request for particular units or capabilities.

Many large departments are adding the ability for the CAD system to determine the closest available apparatus for response through the use of an Automatic Vehicle Response Recommendation. Through GPS, the system identifies the closest vehicle to a call and automatically dispatches the vehicle through the vehicle's MDT.

Dispatch

Dispatch is the important step of actually communicating the request for service by quickly and accurately alerting the selected units to respond and transmitting the information to them. Fire departments use a variety of dispatch systems, ranging from telephone lines to radio systems. The communications centre must have at least two separate methods of sending a dispatch message from the communications centre to each fire station.

The primary connection can be a hard-wired circuit, a telephone line, a computer-based data link, a microwave transmission system, or a radio system. Most fire departments dispatch verbal messages to the appropriate fire stations. The dispatch message is broadcast over speakers in the fire station so that everyone immediately knows the location and nature of the incident and the units that should respond. Radio transmissions are used to contact a unit that is out of the station. The fire station or each vehicle must confirm that the message was received and the units are responding. If the communications centre does not receive the confirmation within a set time period, it must dispatch substitute units.

A CAD system can be programmed to alert the appropriate fire stations automatically. Such a system can send the dispatch information to computer terminals or printers, sound distinctive tones, turn on lights and public address speakers, turn off the stove, and perform additional functions. The dispatch message also can go directly to each individual vehicle that has a computer terminal as well as to the station. The CAD notification often is accompanied by a verbal announcement over the radio and the fire station speakers.

Fire departments with volunteer responders must be able to reach them with a dispatch message. Some departments issue pagers or use cellular phones or use online or mobile apps to reach volunteers. Some CAD systems permit text messages to be sent to cellular phones with incident information. Volunteer fire departments in some communities also may rely on outdoor sirens, horns, or whistles to notify their members of an emergency. These audible devices usually can be activated by remote control from the communications centre. Volunteers then call the communications centre by radio or telephone to receive specific instructions.

Operational Support and Coordination

After the communications centre dispatches the units, it begins to provide incident support and coordination. Someone in the communications centre must remain in contact with the responding units throughout the entirety of the incident. The telecommunicator must confirm that the dispatched units actually received the alarm, record their en route times, provide any additional or updated information, and record their on-scene arrival times.

Operational support and coordination encompass all communications between the units and the communications centre during an entire incident. Two-way radios, laptop computers, mobile communications devices, telephones, and mobile computer terminals all may be used to exchange information.

Generally, the IC will communicate with a telecommunicator operating a radio in the communications centre. In larger departments with large communications centres, a dedicated telecommunicator is usually assigned to deal with a large incident or fire. This ensures operational continuity and allows the telecommunicator to focus on one incident and not be distracted by other incidents. Progress and incident status reports, requests for additional units or release of extra units, notifications, and requests for information or outside resources are examples of incident communications. The communications centre closely monitors the radio and provides any needed support for the incident.

Each part of the public safety network—fire, EMS, and police—must be aware of what other agencies are doing at the incident. The communications centre serves as the hub of the network that supports units operating at emergency incidents, and it coordinates the fire department's activities and requirements with other agencies and resources. For example, the telecommunicator might need to notify nonemergency resources such as gas, electric, or telephone companies of the incident and request that they take certain actions. The communications centre should have accurate, current telephone numbers and contact information for every relevant agency.

Status Tracking and Deployment Management

The communications centre must know the location and status of every fire department unit at all times. Units should never get "lost" in the system, whether they are available for dispatch, assigned to an incident, or in the repair shop. As previously stated, the communications centre must always know which units are available and which units are not available for dispatch. The changing conditions at an incident may require frequent reassignment of units, including ambulances. In addition, units may be needed from outside the normal response area or from other districts.

Tracking the status of the various units is difficult if the telecommunicator must rely on radio reports and coloured magnets or tags on a map or status board. CAD systems using GPS data make this job much easier, because status changes can be entered through digital status units or computer terminals.

Communications centres must continually monitor the availability of units in each geographic area and redeploy units when an area has insufficient coverage. Many fire departments list both unit relocations and multiple response units on the run cards. This information is only valid, however, if major incidents occur one at a time and when all units are available. If a department has a large volume of routine incidents, it may need to redeploy units to balance coverage, even when no major incidents are in progress.

Usually, a supervisor in the communications centre is responsible for determining when and where to redeploy units as well as for requesting coverage from surrounding jurisdictions. For example, units from many different jurisdictions may be redeployed to respond to large-scale incidents under regional or province-/territory-wide plans. These plans must include a system for tracking every unit and a designated communications centre for maintaining contact with all units.

Taking Calls: Emergency and Nonemergency Calls

One of the first things you should learn when you are assigned to a fire station is how to use the telephone and intercom/radio systems. You must be able to use them to answer a call or to initiate a response. Keep nonemergency calls to a minimum, so incoming phone lines are open to receive emergency calls.

A fire fighter who answers the telephone in a fire station, fire department facility, or communications centre is a representative of the fire department. Use your department's standard greeting when you answer the phone: "Pleasant Town Fire. Captain Smith speaking. How may I help you?" Be prompt, polite, professional, and concise.

Detailed SOPs for obtaining information and processing calls should be provided to any fire fighter assigned to answer incoming emergency telephone lines. An emergency call can come in on any fire department telephone line. As mentioned earlier, someone may call your fire station directly instead of dialing 911 or another published emergency number. If this happens, you are responsible for ensuring that the caller receives the appropriate emergency assistance. Your department's SOPs should outline exactly which steps you should take in this situation, such as whether you should take the information yourself or connect the caller directly to the communications centre, if possible.

Fire fighters should understand the steps followed by the personnel who staff the communications centre as they receive calls, solicit needed information from a caller, and initiate a response to an emergency. Follow the steps in **SKILL DRILL 4-1** to receive a call, obtain the essential information from a caller, and initiate a response to an emergency.

Touring the Communications Centre

It is helpful for new fire fighters to tour the emergency communications centre. Observing the actual operation of this centre provides a much better understanding of the role of the telecommunicator. **SKILL DRILL 4-2** lists the steps for touring your local communications centre.

SAFETY TIP

Vehicle locator systems and mobile tracking devices are valuable tools to accurately and effectively keep command staff and telecommunicators up to date on the status of all units. These systems are especially useful in large departments and at large emergency incidents.

SKILL DRILL 4-1
Receiving a Call and Initiating a Response to an Emergency
Fire Fighter I, NFPA 1001: 4.2.1 and 4.2.2

1 Answer promptly and professionally. Identify yourself, your agency, and your location. Determine immediately whether there is an emergency. If the call involves an emergency, follow your department SOPs. Organize your questions to get the following information:

- Incident location (including cross streets and identifying landmarks)
- Type of incident/situation
- Scene safety information
- When the incident occurred
- Caller's name
- Location of the caller, if different from the incident location
- Caller's callback number

Always terminate the call in a courteous manner, and let the caller hang up first.

2 Record the information needed, including the date and time of the call. Initiate a response following the protocols of your communications centre. The protocols in your department may vary from the steps listed here. Follow the protocols of the agency having jurisdiction for your department's communications.

© Jones & Bartlett Learning. Photographed by Glen E. Ellman.

SKILL DRILL 4-2
Touring the Communications Centre

1 Arrange for a tour. Conduct yourself in a professional manner. Observe the use of equipment.

2 Observe the receipt of a reported emergency. Differentiate the needs of fire, police, and emergency medical services (EMS) personnel. Understand the telecommunicator's job.

© Jones & Bartlett Learning. Photographed by Glen E. Ellman.

Radio Systems

Fire department communications systems depend on two-way radio systems. Radios link the communications centre and individual units; they also link units at an incident scene. A radio system is an integral component of the ICS because it links all of the units on an incident—both up and down the chain of command and across the organization chart. A radio may be the fire fighter's only link to the incident organization and the only means to call for help in a dangerous situation. Radios also are used to transmit dispatch information to fire stations, to page fire fighters, and to link mobile computer terminals.

Fire departments use many different types of radios and radio systems. Technological advances are rapidly adding new features and system configurations, making it impossible to describe all of the possible features, principles of operation, and systems here. Instead, this section describes common systems and operating features. As a fire fighter, you must know how to operate your assigned radio and learn your own department's radio procedures.

Radio Equipment

There are three main types of fire service radios: the portable radio, the mobile radio, and the base station.

A **portable radio** is a hand-held two-way radio that is small enough for a fire fighter to carry at all times (**FIGURE 4-10**). The radio body contains an integrated speaker and microphone, an on/off switch or knob, a volume control, channel select switch, and a "push-to-talk" (PTT) button.

A portable radio must have an antenna to receive and transmit signals. A popular optional attachment is an extension microphone/speaker unit that can be clipped to a collar or shoulder strap, while the radio remains in a pocket or pouch.

A portable radio is usually powered by a rechargeable battery, which should be checked at the beginning of each shift or prior to each use. Because the battery has a limited capacity, it must be recharged or replaced after extended operations. Battery-operated portable radios also have limited transmitting power. The signal can be heard only within a certain range and is easily blocked or overpowered by a stronger signal.

Mobile radios are more powerful two-way radios that are permanently mounted in vehicles and powered by the vehicle's electrical system (**FIGURE 4-11**). Both mobile and portable radios share similar features, but mobile radios usually have a fixed speaker and an attached, hand-held microphone on a coiled cord. The PTT button is on the microphone, and the antenna is usually mounted on the exterior of the vehicle. Fire apparatus

FIGURE 4-10 A portable radio should be carried by each individual fire fighter.
© Jones & Bartlett Learning. Photographed by Glen E. Ellman.

often include headsets with a combined intercom/radio system that enables crew members to talk to each other and hear the radio.

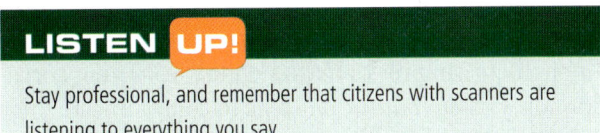

LISTEN UP!

Stay professional, and remember that citizens with scanners are listening to everything you say.

Base station radios are permanently mounted in a building, such as a fire station, communications centre, or remote transmitter site. These kinds of radios are more powerful than either portable or mobile radios. The antenna of the base station is often mounted on a radio tower so the transmissions have a wide coverage area (**FIGURE 4-12**). Public safety radio systems often use multiple base stations at different locations to cover large geographic areas. The communications centre operates these stations by remote control.

As discussed, MDTs are computers that are located in the fire station or on the apparatus that transmit data by radio or cellular signal. An MDT allows for greatly expanded communication capabilities. Instead of having

FIGURE 4-11 A mobile radio is permanently mounted in a vehicle.
© Jones & Bartlett Learning. Photographed by Glen E. Ellman.

FIGURE 4-12 A base station radio antenna is often mounted on a radio tower to provide maximum coverage.
© Jones & Bartlett Learning. Photographed by Glen E. Ellman.

to listen to the telecommunicator and determine whether he or she said 11345 Main Street or 11354 Main Street, you can look at the terminal where the address is displayed and obtain directions. An MDT also may allow for communication without the use of the radio. For example, fire fighters can press a button on the MDT that corresponds with *en route* or *on scene* to communicate their acknowledgement to dispatch. Satellite communications can track progress to the scene via GPS mapping and can provide important scene information. Some departments use MDTs to track unit status, transmit dispatch messages, and exchange different types of information.

FIRE FIGHTER TIP

To enable fire fighters to communicate more effectively while wearing an SCBA mask, several manufacturers have developed electronic communication systems that interface a portable radio with SCBA face pieces. These devices enable fire fighters to hear messages more clearly while wearing SCBA and to transmit information over their portable radios more easily, clearly, and simply. If your department uses this type of device, learn how to use it effectively.

Radio Operation

Every radio system in Canada must be licensed and operated within the guidelines as established by the CRTC.

Radios work by broadcasting electronic signals on certain frequencies, which are designated in units of megahertz (MHz). A **frequency** is the number of cycles (oscillations) per second of a radio signal; only those radios tuned to that specific frequency can hear the message. The CRTC licenses an agency to operate on one or more specific frequencies. These frequencies often are programmed into the radio and can be adjusted only by a qualified technician.

A radio **channel** is an assigned frequency or frequencies used to carry voice and/or data communications. Assigned radio frequencies may be used in a variety of systems. With a **simplex channel** (push to talk, release to listen), each radio transmits signals and receives signals on the same frequency, so a message goes directly from one radio to every other radio that is set to receive that frequency. When one party transmits, the other can only receive. With a **duplex channel**, two different frequencies are used at the same time to permit simultaneous transmission (talk) and reception (listen), like a telephone. Duplex channels are used with repeater systems, which are described later in this chapter. **Multiplex channels** combine both analog and digital signals and can simultaneously transmit two or more different types of information, such as voice and telemetry, in either or both directions over the same frequency.

Frequency bands are portions of the radio frequency (RF) spectrum assigned for specific uses. The most commonly used bands for public safety communications are the **very high-frequency (VHF) band** and the **ultrahigh-frequency (UHF) band**. The VHF band extends from roughly 30 to 300 MHz. VHF has been arbitrarily divided into a low band (30–50 MHz) and a high band (132–174 MHz).

Canadian fire service frequencies are located in several different ranges, including 33–46 MHz (VHF low band); 150–174 MHz (VHF high band); 450–460 MHz (UHF band); 700 MHz; and 800–900 MHz. Additional groups of frequencies are allocated in specific geographic areas with a high demand for public safety agency radio channels. Each band has certain advantages and disadvantages relating to geographic coverage, topography (hills and valleys), and penetration into structures. Some bands have a large number of different users, which can cause interference problems, particularly in densely populated metropolitan areas.

Generally, one radio can be programmed to operate on several frequencies in a particular band. This limitation may become a problem if neighbouring fire departments or police, fire, and EMS systems within the same jurisdiction operate on different bands. When different agencies working at the same incident communicate via different bands, they must make complicated arrangements with cross-band repeaters and multiple radios in command vehicles so that they can communicate with one another. Currently, public safety frequencies are being reallocated and changed. This will result in many departments operating on new frequencies and is designed to improve the operation of public safety radios and data transmission.

Repeater Systems

Radio messages can be broadcast over a limited distance for two reasons: (1) The signal weakens as it travels farther from the source; and (2) buildings, tunnels, and topography create interference. These problems are most significant for hand-held portable radios, which have limited transmitting power. Mobile radios that operate in systems covering large geographic areas face similar problems. To compensate for these shortcomings, fire departments may use radio repeater systems (**FIGURE 4-13**).

A **repeater** is a special base station that uses two separate frequencies—one to receive messages and one to transmit messages. A repeater has a large antenna that is able to receive lower-power signals, such as those from a portable radio, from a long distance away. The signal is then rebroadcast to all of the radios set on the designated channel with all the power of the base station. Communications systems that use repeaters usually have outstanding systemwide communications and are able to get the best signal from portable radios. This approach enables the transmission to reach a wider coverage area.

Many public safety radio systems have multiple receivers (voting receivers), which are geographically distributed over the service area to capture weak signals. The individual receivers forward their signals to a device that selects the strongest signal and rebroadcasts it over the system's base station radio(s) and transmission tower(s). With this configuration, messages that originate from a mobile or hand-held portable radio have as strong a signal as a base station radio at the communications centre. As long as the original signal is strong enough to reach one of the voter receivers, the system is effective. If the signal does not reach a repeater, however, no one will hear it.

Some fire departments switch from a duplex channel to a simplex channel for on scene communications. This configuration is sometimes called a **talk-around channel**, because it bypasses the repeater system. In this case, the radios transmit and receive on the same frequency, and the signal is not repeated. A talk-around channel often works well for short-distance communications, such as from the command post to crews inside a house fire or from one unit to another inside a building. Conversations on a talk-around channel cannot be monitored by the communications centre; instead, the IC must use a more powerful radio on a repeater channel to maintain contact with the communications centre.

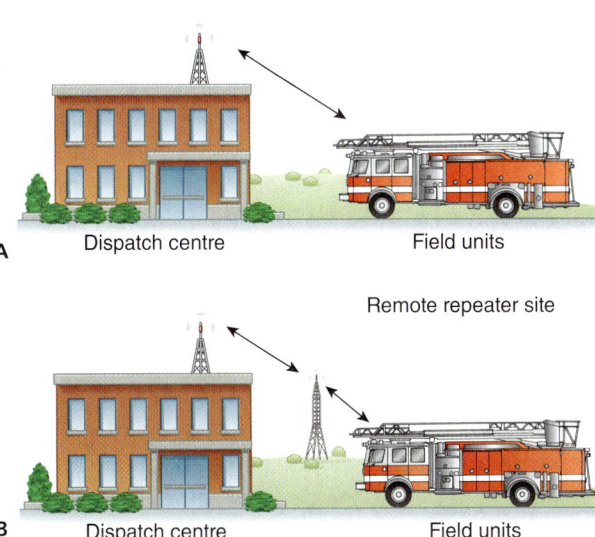

FIGURE 4-13 Direct and repeater channels send and receive transmissions in different ways. **A.** Direct channel (simplex). **B.** Repeater channel (duplex).
© Jones & Bartlett Learning.

Mobile repeater systems also can boost signals at the incident scene by creating a localized, onsite repeater system. The mobile repeater can be permanently mounted in a vehicle or set up at the command post. It captures the weak signal from a portable radio for rebroadcast. Some fire departments use cross-band repeaters, which boost the power of a weak signal and transmit it over a different radio band. In this way, a fire fighter using a UHF portable radio inside a building can communicate with the IC who is outside on a VHF radio. Similar systems may be used in large buildings or underground structures so crews working inside can communicate with units on the outside or with the communications centre.

Digital Radio and Trunked Systems

Radio systems based on newer technology and digital communications are being introduced to take advantage of the additional frequencies being allocated by the CRTC for public safety use. Unfortunately, these new systems are generally incompatible with older radios and require expensive infrastructure changes.

Digital radio systems allow the transmission of digital (computer) signals or analog (voice) signals that have been digitized and compressed by a computer. With digital trunked radios, either 800- or 900-MHz systems, instead of being assigned to one or two frequencies, many frequencies are assigned to a group. These groups can be thought of as virtual channels that appear and disappear as conversations occur. As a radio conversation begins, a computer selects or scans for the next open frequency and directs all of the radios in the talk group to receive the message on that frequency. As the conversation continues, you will likely be speaking on a different frequency because the computer is constantly monitoring for frequency load and reassigning transmissions to unused frequencies. A trunking system makes more efficient use of available frequencies and can provide coverage over extensive networks. Therefore, you do not need to worry about being able to transmit or receive. In a trunking system, the computer switches you to another channel without your knowledge, and you operate the radio as you normally do. Trunking systems also allow different radios to be tied together for a given incident. With a trunking system, it is possible to link different agencies on the same system. For instance, public works, the highway department, fire department, and the police department could all be linked for a specific incident if they needed to share radio communications.

Using digital technology also is more efficient than using analog technology, and it enables more users to communicate at the same time in a limited portion of the radio spectrum.

Using a Radio

As a fire fighter, you must learn how to operate any radio assigned to you and how to work with the particular radio systems used by your fire department. You must know when to use "Channel 3" or "Tac 5" or "Charlie 2," which buttons to press, and which knobs to turn. Study the training materials and SOPs provided by your department to learn this information.

When a radio is assigned to you, make sure you know how to operate and maintain it. Check the battery to be sure it is fully charged. Learn the protocol for changing and recharging portable radio batteries. On the fire ground, your radio is your link to the outside world. It must work properly—because your life depends on it.

If you hold a portable radio perpendicular to the ground with the antenna pointing toward the sky, you will get better transmission and reception. Range and transmission quality also improve if you remove the radio from the radio pocket or belt clip before you use it.

You should be familiar with all departmental SOPs governing the use of radios. All radio transmissions should be pertinent to the situation, with unnecessary radio traffic being avoided. Remember that radio communications are automatically recorded. Anyone with a radio scanner also may hear them, including the news media and the general public. Given this fact, you should avoid saying anything of a sensitive nature or anything you might later regret. Radio tapes provide a complete record of everything that is said and can be admissible as evidence in a legal case.

Most fire departments use plain English for radio communications, but some use codes for standard messages. A few departments have developed intricate systems of radio codes to fit any situation. Ten-codes, a system of coded messages that begin with the number 10, were once widely used. They can be problematic, however, when units from different jurisdictions have to communicate with one another. To be effective, codes must be understood by all parties and mean the same thing to all users. The National Incident Management System (NIMS) and NFPA standards recommend using plain English rather than codes. If your department uses radio codes, you must know all of the codes and understand how they are used. Even plain English may include some generally accepted terminology within a system, such as "code blue" for a cardiac arrest. The objective is to be clearly and easily understood, without having to explain every message in detail. If your department has a few codes or special words for special situations, they should be specified in the SOPs.

To use a radio, follow the steps in **SKILL DRILL 4-3**.

LISTEN UP!

Ten-codes are not approved by the NIMS. Using plain English is the preferred method of communication.

SKILL DRILL 4-3
Using a Radio Fire Fighter I, NFPA 1001: 4.2.3

1 Listen to determine that the channel is clear of any other traffic. Depress the "push-to-talk" (PTT) button, and wait at least 2 seconds before speaking. This delay enables the system to capture the channel without cutting off the first part of the message. Some systems sound a distinctive tone when the channel is ready.

2 Speak across the microphone at a 45-degree angle, and hold the microphone 2.5 to 5 centimetres (1 to 2 inches) from your mouth. Never speak on the radio if you have something in your mouth.

3 Know what you are going to say before you start talking. Speak clearly, and keep the message brief and to the point.

4 Release the PTT button only after you have finished speaking.

© Jones & Bartlett Learning. Photographed by Glen E. Ellman.

Size-up and Progress Reports

Fire fighters and telecommunicators must know how to transmit clear, accurate size-up and progress reports. The first-arriving unit at an incident—whether a fire fighter, a company officer, or a chief—should always give

LISTEN UP!

Some departments have adopted a regional numbering system for apparatus to avoid confusion at major incidents. This numbering often involves assigning a two- or three-digit number to each piece of apparatus. Most municipalities have their own numbering system.

a brief initial size-up report and establish command as described in Chapter 22, *Establishing and Transferring Command*. This initial report should convey a preliminary assessment of the situation and give other units a sense of what is happening so that they can anticipate what their assignments may be when they arrive at the scene. The communications centre should repeat the initial size-up information.

For example, on arrival at the scene, a pump company might make the following report:

Dispatch from Pump 35. We are on location at 123 Main Street. This is a two-storey detached type V residential structure. Nothing showing from side A. Pump 35 will be 123 Main Street command.

The communications centre would respond by announcing:

Pump 35 is on location, nothing showing from side A. Pump 35 will be Main Street command.

This initial report should be followed by a more detailed report after the IC has had time to gather more information and make initial assignments. An example of such a report follows:

IC: Communications, this is High Street command.
Communications Centre: Go ahead, High Street Command, this is High Street command.
IC: At 176 High Street, we are using all companies on the first alarm for a fire on the second floor, side A, of a two-storey, wood-frame dwelling, approximately 9 m by 12 m (30 ft by 40 ft). Exposure Alpha is the street, Exposure Baker is a similar dwelling, Exposure Charlie is an alley, and Exposure Delta is a vacant lot. Companies are engaged in offensive operations, and the fire appears to be confined. We have an all clear at this time. Platoon Chief 406 is in Command.
Communications Centre: Copied High Street Command. You have an all-clear for a fire in a dwelling at 176 High Street. The fire appears to be confined to the second floor. Platoon Chief 406 is IC.

For the duration of an incident, the IC should provide regular updates to the communications centre. Information about certain events—such as command transfer, "all clear," and "fire under control"—should be provided when they occur. Reports also should be made when the situation changes significantly and at least every 10–20 minutes during a working incident.

In some fire departments, the communications centre prompts the IC to report at set intervals, known as **time marks**. Time marking allows the IC to assess the progress of the incident and decide whether the strategy or tactics should be changed. All progress reports are noted in the activity log to document the incident.

If additional resources, such as apparatus, equipment, or personnel, are needed, the IC transmits these requests to the communications centre. The communications centre will advise the IC of additional resources that are dispatched.

Emergency Messages

Sometimes the telecommunicator must interrupt normal radio transmissions for **emergency traffic**. Emergency traffic is an urgent message that takes priority over all other communications. When a unit needs to transmit emergency traffic, the telecommunicator generates a distinctive alert tone to notify everyone on the frequency to stand by so that the channel is available for the emergency communication. Once the emergency message is complete, the telecommunicator notifies all units to resume normal radio traffic.

Some radio systems also have emergency buttons located on portable and mobile units that allow fire fighters in trouble to push the button to transmit an emergency signal to the telecommunicator. This button alerts the communications centre that the ID assigned to that radio has activated an emergency need and, in some cases, identifies the location of the unit in trouble. If your radios are equipped with this feature, learn the procedure for using it.

The most important emergency traffic is a fire fighter's call for help. Most departments use **mayday** to indicate that a fire fighter is lost, is missing, or requires immediate assistance. If a mayday call is heard on the radio, all other radio traffic should stop immediately. The fire fighter making the mayday call should describe the situation, location, and help needed. Fire fighters should study and practise the procedure for initiating and responding to a mayday call.

Some agencies utilize the acronym LUNAR to report a mayday. LUNAR stands for: **Location**, your location in the building/incident; **Unit**, the unit you are assigned to; **Name**, who you are; **Air**, the amount of air you have in your cylinder or **Assignment**, where you were last assigned; **Resources**, what you need to get you out of the mayday situation.

An example of a mayday call follows:

Fire fighter: MAYDAY . . . MAYDAY . . . MAYDAY. [All radio traffic stops.]

IC: Unit calling MAYDAY, go ahead.

Fire fighter: This is Pump 403. We are on the second floor and running out of air. Fire has cut off our escape route. We request a ladder to the window on the Charlie side of the building so we can evacuate.

IC: Command copied. Pump 403, your escape route cut off by fire. I am sending the rapid intervention crew to Charlie side with a ladder.

The procedure for responding to a mayday call—which is used when fire fighters are in imminent danger—should be studied and practised frequently. Ensure that you are familiar with your department's SOPs regarding mayday calls. The steps for initiating a mayday are explained more fully in Chapter 18, *Fire Fighter Survival*.

Another emergency traffic message is "evacuate the building," which warns all units inside a structure to abandon the building immediately. Most fire departments have specified a standard **evacuation signal** to warn all personnel to pull back to a safe location. The evacuation signal that is commonly used is a sequence of three blasts on an apparatus air horn, repeated several times, or sirens sounded on "high-low" for 15 seconds. An evacuation warning should be announced at least three times to ensure that everyone hears it; the IC also should announce the warning on the radio. Because no universal evacuation signal has been established, fire fighters must learn their department's SOP for emergency evacuation.

After the evacuation, the radio airwaves should remain clear so the IC can request a Personnel Accountability Report (PAR). A PAR is usually requested following an evacuation by the IC. This step ensures that all personnel have safely exited the building. After the PAR, the IC will allow the resumption of normal radio traffic.

After-Action REVIEW

IN SUMMARY

- Every fire department depends on a functional communications system.
- The communications centre is the central processing point for all information relating to an emergency incident and all information relating to the location, status, and activities of fire department units. Depending on the size of the fire department, the communications centre may be a designated public safety answering point, which serves and dispatches multiple agencies, or a stand-alone communications centre, which serves only one agency.
- Telecommunicators receive information from citizens and process that information to correctly dispatch resources to an emergency incident.
- The following equipment is found in communications centres:
  - Dedicated 911 telephones
  - Public telephones
  - Direct-line phones to other agencies
  - Equipment to receive alarms from public or private fire alarm systems
  - Computers utilizing geographic information systems (GIS) and/or hard-copy files and maps to locate addresses and select units to dispatch
  - Equipment for alerting and dispatching units to emergency calls
  - Two-way radio system(s)
  - Recording devices to record phone calls and radio traffic
  - Backup electrical generators
  - Teletype/telecommunication (TTY/TDD) capabilities
- Computer-aided dispatch (CAD) enables telecommunicators to work more effectively. It tracks the status of units and assists the telecommunicator in quickly dispatching units to an emergency incident. Some CAD systems transmit dispatch information directly to mobile data terminals (MDTs). The most advanced type of CAD system utilizes global positioning system (GPS) data devices to confirm the location of the caller. A CAD system automatically captures and stores every event as it occurs.
- Most communications centres automatically record everything that is said over the telephone or radio. This feature can prove valuable if the caller talks very quickly, has a strong accent, hangs up, or is disconnected. The recordings also serve as legal records for the fire department. They can assist in reviewing and analyzing information about department operations.
- A communications centre performs the following basic functions:
  - Receiving calls for emergency incidents and dispatching fire department units
  - Supporting the operations of fire department units delivering emergency services
  - Coordinating fire department operations with other agencies
  - Keeping track of the status of each fire department unit at all times
  - Monitoring the level of coverage and managing the deployment of available units
  - Notifying designated individuals and agencies of particular events and situations

- Maintaining records of all emergency-related activities
- Maintaining information required for dispatch purposes

- Communications centres should have both primary and backup methods of transmitting alarms to stations. Despite the method being used, proper dispatch procedures must always be followed to avoid confusion and delay of resources. There are five major steps in processing an emergency incident:
  - Call receipt—the process of receiving an initial call and gathering information
  - Location validation—ensuring that the address is valid
  - Classification and prioritization—assigning a response category based on the nature of the problem
  - Unit selection—the process of determining exactly which unit or units to dispatch based on the location and classification of the incident
  - Dispatch—the alerting of the selected units to respond and transmit information to them

- Calls may be received via telephone, municipal fire alarm systems, private and automatic fire alarm systems, and walk-ins.

- It is important for the telecommunicator to validate that the location information received is correct.

- Enhanced 911 systems are able to display information about where a landline call originated and even the name and address of the caller.

- The next generation of 911 systems is being designed to enable telecommunicators to pinpoint the location of an emergency call that is made with a cellular device.

- Fire department communications systems depend on two-way radio systems. A radio system is an integral component of the ICS because it links all of the units on an incident—both up and down the chain of command and across the organization chart.

- Three types of fire service radios may be used:
  - Portable radio—a hand-held two-way radio that the fire fighter carries at all times. The battery of such a radio should be checked at the beginning of each shift.
  - Mobile radio—two-way radios permanently mounted in vehicles and powered by the vehicle's electrical system.
  - Base station—radios permanently mounted in a building, such as a fire station, communications centre, or remote transmitter site. Public safety radio systems often use multiple base stations at different locations to cover large areas.

- Radios work by broadcasting electronic signals on certain frequencies. These frequencies often are programmed into the radio and can be adjusted only by a qualified technician.

- A radio channel uses one simple frequency (simplex channel), two frequencies (duplex channel), or one complex frequency (multiplex channel). With a simplex channel, each radio transmits and receives signals on the same frequency, so a message goes directly from one radio to every other radio set to that frequency (push to talk, release to listen). With a duplex channel, each radio transmits signals on one frequency and receives messages on another frequency. This allows for simultaneous transmission and reception, like a telephone. Multiplex channels simultaneously transmit two or more different types of information in either or both directions over the same frequency.

- In a repeater system, each radio channel uses two separate frequencies—one to transmit and the other to receive. When a low-power radio transmits over the first frequency, the signal is received by a repeater unit that automatically rebroadcasts it on the second frequency over a more powerful radio. All radios set on the designated channel receive the boosted signal on the second frequency. This approach enables the transmission to reach a wider coverage area.

- Unlike digital radios, trunked radios allow many frequencies to be assigned to a group. When a user presses the transmit button, a computer assigns a frequency for that message and directs all of the radios in the talk group to receive the message on that frequency. A trunking system makes more efficient use of available frequencies and can provide coverage over extensive networks.

- The first-arriving unit at an incident should always give a brief initial radio report. This report should convey a preliminary assessment of the situation and give other units a sense of what is happening so that they can anticipate what their assignments may be when they arrive at the scene.

- Emergency traffic is an urgent message that takes priority over all other communications.

- When a unit needs to transmit emergency traffic, the telecommunicator generates a distinctive alert tone to notify everyone on the frequency to stand by so that the channel is available for the emergency communication. Once the emergency message is complete, the telecommunicator notifies all units to resume normal radio traffic.

- The most important emergency traffic is a fire fighter's call for help. Most departments use mayday to indicate that a fire fighter is lost, is missing, or requires immediate assistance. If a mayday call is heard on the radio, all other radio traffic should stop immediately. The fire fighter making the mayday call should describe the situation, location, and help needed.

- From a legal standpoint, records and reports are vital parts of the emergency. Information must be complete, clear, and concise because these records can become admissible evidence in a court case.

- A fire fighter who answers the telephone in a fire station, fire department facility, or communications centre is a representative of the fire department. Use your department's standard greeting when you answer the phone. Be prompt, polite, professional, and concise.

KEY TERMS

Access Navigate for flashcards to test your key term knowledge.

Activity logging system A device that keeps a detailed record of every incident and activity that occurs.

Automatic location identification (ALI) A series of data elements that informs the recipient of the location of the alarm. (NFPA 1221)

Automatic number identification (ANI) A series of alphanumeric characters that informs the recipient of the source of the alarm. (NFPA 1221)

Base station A stationary radio transceiver with an integral AC power supply. (NFPA 1221)

Call box A system of telephones connected by phone lines, radio equipment, or cellular technology to a communications centre or fire department.

Canadian Radio-television and Telecommunications Commission (CRTC) The authority that regulates and supervises broadcasting and telecommunications in Canada.

Channel An assigned frequency or frequencies used to carry voice and/or data communications.

Computer-aided dispatch (CAD) A combination of hardware and software that provides data entry, makes resource recommendations, and notifies and tracks those resources before, during, and after fire service alarms,

preserving records of those alarms and status changes for later analysis. (NFPA 1221)

Digital radio The transmission of information via radio waves using native digital (computer) data or analog (voice) signals that have been converted to a digital signal and compressed.

Direct line A telephone that connects two predetermined points.

Dispatch To send out emergency response resources promptly to an address or incident location for a specific purpose. (NFPA 450)

Duplex channel A radio system that is able to simultaneously use two frequencies per channel; one frequency transmits and the other receives messages. Such a system uses a repeater site to transmit messages over a greater distance than is possible with a simplex system.

Emergency traffic An urgent message, such as a call for help or evacuation, transmitted over a radio that takes precedence over all normal radio traffic.

Evacuation signal A distinctive signal intended to be recognized by the occupants as requiring evacuation of the building. (NFPA 72)

Frequency The number of cycles (oscillations) per second of a radio signal.

Geographic information systems (GIS) A system of computer software, hardware, data, and personnel to describe information tied to a spatial location. (NFPA 450)

Global positioning systems (GPS) A satellite-based radio navigation system comprised of three segments: space, control, and user. (NFPA 414)

Mayday A verbal declaration indicating that a fire fighter is lost, missing, or trapped and requires immediate assistance.

Mobile data terminals (MDTs) Technology that allows fire fighters to receive data while in the fire apparatus or at the station.

Mobile radio A two-way radio that is permanently mounted in a fire apparatus.

Multiplex channel Simultaneous transmission of multiple data streams, most often voice signals, in either or both directions over the same frequency.

Portable radio A battery-operated, hand-held transceiver. (NFPA 1221)

Public safety answering point (PSAP) A facility equipped and staffed to receive emergency and nonemergency calls requesting public safety services via telephone and other communication devices. (NFPA 1061)

Public safety communications centre A building or portion of a building that is specifically configured for the primary purpose of providing emergency communications services or public safety answering point (PSAP) services to one or more public safety agencies under the authority or authorities having jurisdiction. (NFPA 1061)

Repeater A special base station radio that receives messages and signals on one frequency and then automatically retransmits them on a second frequency.

Run cards Cards used to determine a predetermined response to an emergency.

Simplex channel A radio system that uses one frequency to transmit and receive all messages; transmissions can occur in either direction but not simultaneously in both; when one party transmits, the other can only receive, and the party that is transmitting is unable to receive.

Talk-around channel A simplex channel used for onsite communications.

Telecommunicator An individual whose primary responsibility is to receive, process, or disseminate information of a public safety nature via telecommunication devices. (NFPA 1061)

Telephone interrogation The phase in a 911 call during which the telecommunicator asks questions to obtain vital information such as the location of the emergency.

Ten-codes A system of predetermined coded messages, such as "What is your 10–20?" used by responders over the radio.

Time marks Status updates provided to the communications centre every 10–20 minutes. Such an update should include the type of operation, the progress of the incident, the anticipated actions, and the need for additional resources.

Trunked radios A radio system that uses a computerized shared bank of frequencies to make the most efficient use of radio resources.

TTY/TDD systems User devices that allow speech- and/or hearing-impaired citizens to communicate over a telephone system. TTY stands for teletype, and TDD stands for telecommunications device for the deaf; the displayed text is the equivalent of a verbal conversation between two hearing persons.

Ultrahigh-frequency (UHF) band Radio frequencies between 300 and 3000 MHz.

Very high-frequency (VHF) band Radio frequencies between 30 and 300 MHz; the VHF spectrum is further divided into high and low bands.

Voice over Internet Protocol (VoIP) Technology that converts a person's voice into a digital signal that can be sent via the Internet to another device.

Voice recording system Recording devices or computer equipment connected to telephone lines and radio equipment in a communications centre to record telephone calls and radio traffic.

REFERENCES

Americans with Disabilities Act. Title II Regulations (35.162). https://www.ada.gov/regs2010/titleII_2010/titleII_2010_regulations.htm#a35162. Accessed October 24, 2018.

Dolcourt, Jessica. Text to 9-1-1: What you need to know (FAQ). 2014. https://www.cnet.com/news/text-to-911-what-you-need-to-know-faq/. Accessed October 24, 2018.

Federal Communications Commission. 911 Wireless Services. 2016. https://www.fcc.gov/consumers/guides/911-wireless-services. Accessed October 24, 2018.

National Fire Protection Association. NFPA 901, *Standard Classifications for Incident Reporting and Fire Protection Data*, 2016. www.nfpa.org. Accessed October 24, 2018.

National Fire Protection Association. NFPA 902, *Fire Reporting Field Incident Guide*, 1997. www.nfpa.org. Accessed October 24, 2018.

National Fire Protection Association. NFPA 1061, *Standard for Public Safety Telecommunications Personnel Professional Qualifications*, 2018. www.nfpa.org. Accessed October 24, 2018.

National Fire Protection Association. NFPA 1221: *Standard for the Installation, Maintenance, and Use of Emergency Services Communications Systems*. 2016. www.nfpa.org. Accessed October 24, 2018.

United States Department of Transportation. NG911 & FirstNet. https://www.911.gov/911connects/issue-4/New-NG911-and-FirstNet-Guide-for-State-and-Local-Authorities.html. Accessed October 24, 2018.

On Scene

Today is the day you study dispatching in your recruit class. You see that 4 hours is allocated for this topic. You wonder how it can take 4 hours to show how a dispatcher answers a telephone, asks some questions, and then pushes a button to dispatch the fire units.

The director of the emergency communications centre arrives and tells the class that you will be going over to the communications centre. You are asked to sit next to a man with a headset on and five computer screens in front of him. He hands you a headset so you can listen in. Within minutes, you are amazed at how hard this job really is.

1. _____ is an enhanced 911 service feature that displays where the call originated or where the phone service is billed.

   A. Automatic location identification (ALI)

   B. Automatic number identification (ANI)

   C. Call box

   D. TTY/TDD/text phone system

2. A _____ is a radio system that uses a shared bank of frequencies to make the most efficient use of radio resources.

   A. trunked system

   B. simplex system

   C. radio repeater system

   D. mobile data system

3. A _____ is a radio system that uses one frequency to transmit and receive all messages.

   A. trunked system

   B. simplex system

   C. radio repeater system

   D. mobile data system

4. A(n) _____ is a verbal declaration indicating that a fire fighter is lost, missing, or trapped and requires immediate assistance.

   A. emergency evacuation signal

   B. ten-code

   C. box call

   D. mayday

(continued)

On Scene Continued

5. Two-way radios that are permanently mounted in vehicles and powered by the vehicle's electrical system are called:

A. portable radios.

B. mobile radios.

C. base radios.

D. simplex radios.

6. Which step comes first in the processing of an emergency incident in a communications centre?

A. Classification and prioritization

B. Unit selection

C. Location validation

D. Dispatch

Access Navigate to find answers to this On Scene, along with other resources such as an audiobook.

Fire Behaviour

KNOWLEDGE OBJECTIVES

After studying this chapter, you will be able to:

- Describe the chemistry of fire. (pp. 159–167)
- List the three states of matter. (**NFPA 1001: 4.3.10**, pp. 159–160)
- List the five forms of energy. (pp. 160–161)
- Explain the concept of the fire triangle. (**NFPA 1001: 4.3.11**, p. 161)
- Explain the concept of the fire tetrahedron. (**NFPA 1001: 4.3.11**, pp. 161–162)
- Describe the chemistry of combustion. (**NFPA 1001: 4.3.11**, p. 162)
- Describe the by-products of combustion. (**NFPA 1001: 4.3.11**, pp. 162–163)
- Explain how fires are spread by conduction, convection, and radiation. (**NFPA 1001: 4.3.12**, pp. 164–166)
- Define flow path, and describe how it influences the growth of a building fire. (**NFPA 1001: 4.3.11**, p. 165)
- Describe the four methods of extinguishing fires. (pp. 166–167)
- Define Class A, B, C, D, and K fires. (pp. 167–169)
- Describe the importance of the following characteristics in solid-fuel fires: composition of fuel, amount of fuel, and configuration of fuel. (**NFPA 1001: 4.3.11**, pp. 169–170)
- Describe the four stages of fire development: incipient stage, growth stage, fully developed stage, and decay stage. (**NFPA 1001: 4.3.11**, pp. 170–176)
- Define the following terms: thermal layering, neutral plane, roll-over, flashover, backdraft, fuel-limited fires, ventilation-limited fires, and smoke explosion. (pp. 171–175, 176)
- Describe the conditions that cause thermal layering. (**NFPA 1001: 4.3.12**, pp. 171–172)
- Describe the conditions that lead to roll-over. (**NFPA 1001: 4.3.12**, p. 172)
- Describe the conditions that lead to flashover. (**NFPA 1001: 4.3.12**, pp. 172–173)
- Describe the conditions that lead to a backdraft. (**NFPA 1001: 4.3.11**, pp. 173–174)
- Describe the conditions that lead to rapid fire growth. (p. 174)
- Describe the conditions that lead to a fuel-limited fire and a ventilation-limited fire. (pp. 174–175)
- Describe the conditions that lead to a smoke explosion. (p. 176)
- Describe how fire behaves in modern structures. (pp. 176–177)
- Describe how the wind effect impacts fire behaviour. (p. 177)
- Describe the characteristics of liquid-fuel fires. (pp. 177–179)
- Define the following terms: boiling point, flash point, and fire point. (pp. 178–179)
- Define the characteristics of gas-fuel fires. (pp. 179–180)

- Explain the concept of vapour density. (p. 179)
- Explain the concept of flammable range. (p. 179)
- Define the following terms: lower explosive limit (LEL) and upper explosive limit (UEL). (p. 179)
- Describe the causes and effects of a boiling liquid/expanding vapour explosion (BLEVE). (pp. 179–180)
- Describe the process of reading smoke. (pp. 180–183)
- Describe the four key attributes of smoke. (pp. 180–182)

SKILLS OBJECTIVES

There are no skills objectives for Fire Fighter I candidates. NFPA 1001 contains no Fire Fighter I Job Performance Requirements for this chapter.

Fire Alarm

It is 3:46 AM and you arrive at a house for the report of a fire called in by a neighbour. You pull a preconnected 45-mm (1¾-in.) hose line and advance it to the front door. Even though there is little visible smoke, you smell the familiar scent of fire. You notice the windows are stained black with soot. As you prepare to force entry, you notice light brown smoke pushing through between the threshold and the door. There is a crack across the living room window. The hair on the back of your neck rises as you get an unsettled feeling.

1. Why would this situation make you uneasy?

2. What type of conditions may be present in this structure?

3. What type of event may be likely to occur?

 *Access Navigate for more practice activities.*

Introduction

This chapter describes the behaviour of fire.

Since prehistoric times, when humans discovered how to use fire for cooking food and keeping warm, fire has fuelled our lives. Although we have progressed from using open fires for cooking and for warmth, we continue to depend on fires to provide the energy for modern society. Burning fossil fuels creates most of our electric power. Our gasoline- and diesel-powered vehicles depend on small explosions (internal combustion) to propel them. Open flames in well-designed furnaces heat most of our homes.

Unfortunately, destruction of lives and property by uncontrolled fires also has been occurring since ancient times. Despite our advanced technology for detecting and combating fires, we continue to wage an ongoing battle with fire. From 2012-2016, the number of total calls reported in the Province of Ontario (Canada's largest province comprising 38.7% of the overall population), including fire calls and non-fire calls, increased from 462,542 in 2012 to 494,811 in 2016. While the number of total calls increased, loss fires, or fires with injury, fatality, or dollar loss, declined from 11,294 in 2012 to 10,844 in 2016 (Ontario Ministry of Community Safety and Correctional Services).

This chapter explores the relationship between various types of fuels, oxygen, heat, and the combustion process. It describes the four stages of fire growth. It emphasizes research studies that have demonstrated how fire growth is dramatically influenced by changes in ventilation and how it can be controlled by limiting ventilation or the quick application of water. It describes how the composition of the fuel, the amount of fuel, and the configuration of the fuel influence the combustion process with solid fuels. In addition, the following terms are defined: *thermal layering, roll-over, neutral plane, backdraft, flashover, ventilation-limited fires, fuel-limited fires*, and *smoke explosions*. This chapter describes how to assess the volume, velocity, density, and colour of smoke. By relating the key attributes of smoke to the size of the building, weather conditions, and the rate of change of the smoke, it may assist you in determining the location of the fire and its stage of development. This chapter also describes the characteristics of liquid fuel fires.

This text book uses the metric system and provides imperial unit (British) conversion factors or equivalent values as necessary. In the metric system, distance is measured in metres, centimetres, and millimetres, liquid volume is measured in litres, and millilitres, temperature is measured in degrees Celsius, and pressure is measured in pascals or kilopascals. In the imperial system, distance is measured in feet and inches, liquid volume is measured in gallons, temperature is measured in degrees Fahrenheit, and pressure is measured in pounds per square inch.

The Chemistry of Fire

To operate in a more safe and effective manner on the fire ground, you need to understand the conditions needed for a fire to ignite and grow. A well-trained and experienced fire fighter can put out more fire with less water because of his or her understanding of the chemistry of fire.

What Is Fire?

Fire is a rapid chemical process that produces heat and usually light. Everyone can identify fire when they see it, but few can explain the process that is taking place to produce it. Fire is characterized by the production of a flame, which can be seen in many different colours, depending on what is burning and how much heat is being produced. Fire requires fuel that is in the form of combustible gas or vapour in order for the chemical process of combustion to occur. Materials that remain in a solid or liquid state will not burn while in that state. As these materials are heated they produce a gas or vapour that is ultimately ignited. Solid fuels such as wood, liquid fuels such as gasoline, and gaseous fuels such as propane will all burn after they have changed into a vapour.

States of Matter

Matter is made up of atoms and molecules. Matter exists in three states: solid, liquid, and gas (**FIGURE 5-1**). In Chapter 3, *Personal Protective Equipment*, we used smoke as the example of matter. As a by-product of combustion, smoke is known to contain high levels of toxicity including, but not limit to, polycyclic aromatic hydrocarbons (PAHs) and known carcinogens. These substances adhere to personal protective equipment (PPE) in such a manner as to require fire fighters to undergo decontamination and hygiene procedures before leaving the scene of any fire where they have been exposed to smoke or other gases. Fire fighters are exposed to these products in three **states of matter**: particles (solids), vapours (finely suspended liquids or aerosols), and gases. An understanding of these states of matter is helpful in understanding fire behaviour.

We all know what makes an object solid: It has a definite size and shape. Most uncontrolled fires are fed by solid fuels. In structure fires, the building and most of the contents exist as **solids**. Some solids, such as wax or plastic, may change to a liquid or a gaseous state when heated. Cold makes most solids more brittle, whereas heat makes them more flexible. Because a solid is rigid, only a limited number of molecules are present on its surface; the majority of molecules are cushioned or insulated by the outer surface of the solid.

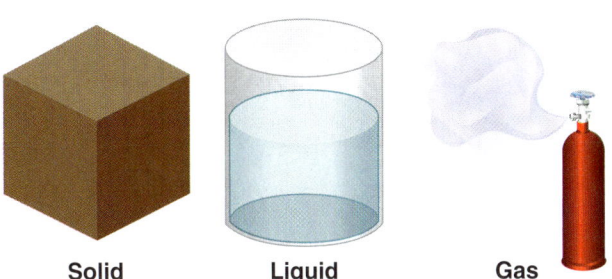

Solid   Liquid   Gas

FIGURE 5-1 States of matter, or the physical state of a substance, can be classified as solids, liquids, or gases.
© Jones & Bartlett Learning.

A **liquid** will assume the shape of the container in which it is placed. Most liquids expand when heated and will turn into gases when sufficiently heated. Liquids, for all practical purposes, cannot be compressed. This characteristic allows fire fighters to pump water for long distances through pipelines or hose. A liquid has no independent shape, but it does have a definite volume. The liquid with which fire fighters are most concerned is water.

A **gas** is a type of fluid that has neither independent shape nor independent volume but rather tends to expand indefinitely. The gas we most commonly encounter is air, the mixture of invisible, odourless, tasteless gases that surrounds the earth. The mixture of gases in air maintains a constant composition—21 percent oxygen, 78 percent nitrogen, and 1 percent argon and trace amounts of other gases such as carbon dioxide, neon, methane, helium, krypton, hydrogen, and xenon. Not only is oxygen required for us to live, but it also is required for fires to burn. We will explore the reaction of fuels with oxygen as we look at the chemistry of burning.

Fuels

Fuels are materials that store energy. Think of the vast amount of heat that is released during a large fire. The energy released in the form of heat and light has been stored in the fuel before it is burned. The release of the energy in a gallon of gasoline, for example, can move a car many miles down the road (**FIGURE 5-2**). Many common household materials, such as computers, TVs, furniture, and carpets, consist of fuels that will burn under the right conditions.

Types of Energy

Energy exists in the following forms: chemical, mechanical, electrical, light, and nuclear. Regardless of the form

in which the energy is stored, it can be changed from one form to another. For example, electrical energy can be converted to heat or to light. Similarly, mechanical energy can be converted to electrical energy through a generator.

Chemical Energy

Chemical energy is the energy created by a chemical reaction. Some chemical reactions produce, or give off, heat (**exothermic**); others absorb heat (**endothermic**). The combustion process (fire) is an example of an exothermic reaction, because it releases heat energy. Ice cubes melting (absorbing heat) is an example of an endothermic reaction. Most chemical reactions occur because bonds are established between two substances or bonds are broken as two substances are chemically separated. Heat is produced whenever oxygen combines with a combustible material. If the reaction occurs slowly in a well-ventilated area, the heat is released harmlessly into the air. If the reaction occurs very rapidly or within an enclosed space, the mixture can be heated to its ignition temperature and can begin to burn. **Ignition temperature** is the minimum temperature at which a fuel, when heated, will ignite in the presence of air and continue to burn. Fire is an example of energy being released as a result of a chemical reaction. An example of this occurs when a bundle of rags soaked with linseed oil releases enough heat through oxidization, causing the rags to ignite spontaneously.

Mechanical Energy

Mechanical energy is converted to heat when two materials rub against each other and create friction. For example, a fan belt rubbing against a seized pulley or vehicle tires spinning on pavement produce heat. Heat also is produced when mechanical energy is used to compress air in a compressor. Water falling over a dam is another example of mechanical energy.

Electrical Energy

Electrical energy is converted to heat energy in several different ways. For example, electricity produces heat when it flows through a wire or any other conductive material. The greater the flow of electricity and the greater the resistance of the material, the greater the amount of heat produced. Examples of electrical energy that can produce enough heat to start a fire include electric heating elements, overloaded wires, electrical arcs, batteries and energy storage systems, and lightning. Electrical energy is carried through the electrical wires inside homes and can be stored in batteries that convert chemical energy to electrical energy.

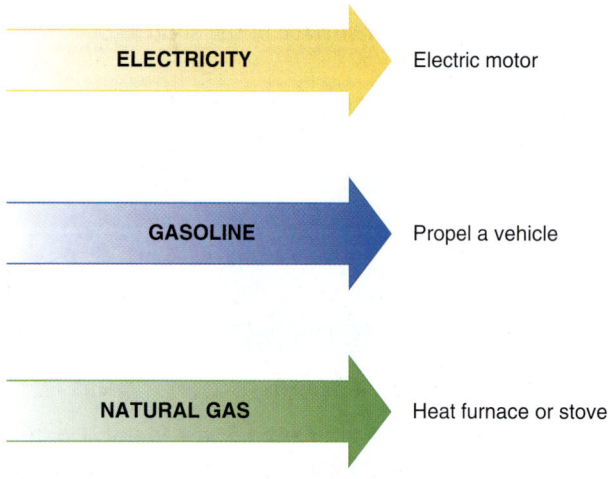

ELECTRICITY → Electric motor

GASOLINE → Propel a vehicle

NATURAL GAS → Heat furnace or stove

FIGURE 5-2 Energy being converted to work.
© Jones & Bartlett Learning.

Light Energy

Light energy is produced by electromagnetic waves packaged in discrete bundles called photons. This energy travels as thermal radiation, a form of heat. When light energy is hot enough, it can sometimes be seen in the form of visible light. One example of light energy is the radiant energy we receive from the sun. We think of candles, fires, light bulbs, and lasers as forms of light energy. We should recognize that, while these do produce light energy, they also produce heat. They emit both heat and light. They transfer most of their heat via convection and radiation. If they are touching something, they also transfer heat using conduction. If light energy is of a frequency that we cannot see, the energy may be felt as heat but not seen as visible light.

Nuclear Energy

Nuclear energy is created by splitting the nucleus of an atom into two smaller nuclei (nuclear fission) or by combining two small nuclei into one large nucleus (fusion). Nuclear reactions release large amounts of energy in the form of heat. These reactions can be controlled, as in a nuclear power plant, or uncontrolled, as in an atomic bomb explosion. In a nuclear power plant, the nuclear reaction releases carefully controlled amounts of heat, which then heat water to produce steam, which powers a steam turbine generator. Both uncontrolled explosions and controlled reactions release radioactive material, which can cause injury or death. Nuclear energy is stored in radioactive materials and converted to electricity by nuclear power generating stations.

Conservation of Energy

The law of conservation of energy states that energy cannot be created or destroyed by ordinary means. Energy can, however, be converted from one form to another. Think of an automobile. Chemical energy in the gasoline is converted to mechanical energy when the car moves down the road. When you apply the brakes to stop the car, the mechanical energy is converted to heat energy by the friction between the wheel rotators and the brake pads. Similarly, in a house fire, the stored chemical energy in the wood structure and plastic-based contents of a house are converted into heat and light energy during the fire.

Conditions Needed for Fire

To understand the behaviour of fire, you need to consider the three basic elements needed for combustion to occur: fuel, oxygen, and heat. First, a combustible fuel must be present. Second, oxygen must be available in sufficient quantities. Third, a source of ignition (heat) must be present. If we graphically place these three components together, the result is the fire triangle (**FIGURE 5-3**).

A fourth factor, an uninhibited chemical chain reaction, must result from the first three elements to produce and maintain a self-sustaining fire. That is, the fuel, oxygen, and heat must interact in such a way as to create a self-sustaining chemical chain reaction to keep the combustion process going.

One way of visualizing this process is to show the chemical chain reaction joining the elements of the fuel, oxygen, and heat. This depiction reflects the central role that the chemical reaction plays in maintaining the process of flaming combustion. This relationship is sometimes characterized as the fire tetrahedron (**FIGURE 5-4**).

FIGURE 5-3 The fire triangle consists of fuel, oxygen, and heat.
© Jones & Bartlett Learning.

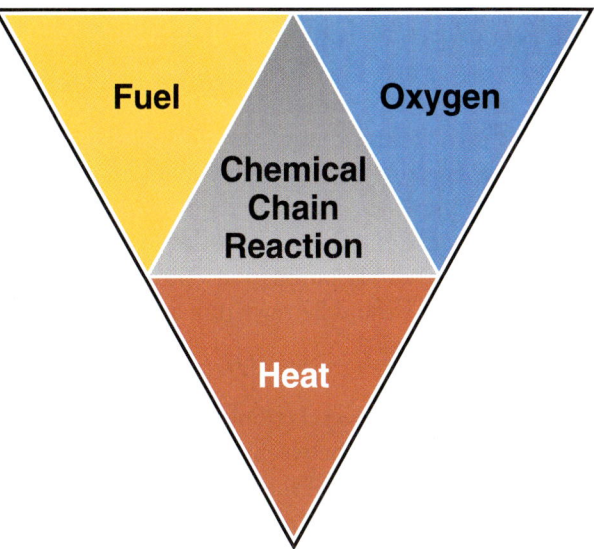

FIGURE 5-4 A chemical chain reaction unites fuel, oxygen, and heat in the fire tetrahedron.
© Jones & Bartlett Learning.

A tetrahedron is a four-sided, three-dimensional figure. Each side of the fire tetrahedron represents one of the four elements needed for a fire to occur.

The key point to remember is that fuel, oxygen, and heat must be present for a fire to start and to continue burning. If you remove any of these elements, the fire will go out.

Chemistry of Combustion

The smallest unit of matter is an **atom**. Some atoms exist in nature as pairs. For example, two atoms of oxygen are paired up in air. The formula for the oxygen pair is written O_2.

When atoms of one element combine chemically with atoms of another element, they produce a compound that is made up of molecules. For example, when an atom of oxygen (O) combines chemically with two atoms of hydrogen (H_2), the resulting compound is water (H_2O).

Almost all fuels consist of hydrogen (H) and carbon (C) atoms; hence, they are called hydrocarbons. Fuels may contain a wide variety of other elements, but for our purpose we can describe the self-sustaining chemical reaction of combustion simply by considering the fuel to contain carbon and hydrogen. When this fuel combines with oxygen, it produces two by-products, water and carbon dioxide (CO_2):

Hydrogen–Carbon Fuel + Oxygen = Water and Carbon Dioxide

Graphically, this reaction is depicted as follows:

$$\text{H–C (hydrocarbon fuels)} + O_2 = H_2O + CO_2$$

Of course, the reactions that occur in most actual fires are not as simple as this basic reaction suggests. First, hydrocarbon fuels are very complex molecules that contain complex chains of atoms in addition to hydrogen and carbon. As a consequence, the combustion process produces numerous toxic by-products. Second, most structure fires encountered by fire fighters burn in the presence of a limited amount of oxygen. The resulting **incomplete combustion** produces significant quantities of deadly gases and compounds and a variety of by-products, which are released into the atmosphere. Also, the production of large quantities of heat and light is an integral part of combustion. Thus, the reaction for combustion should be rewritten as follows:

$$\text{H–C (hydrocarbon fuels)} + O_2 = H_2O + CO_2 + \text{Light} + \text{Heat}$$

To understand the reactions involved in fire, it is helpful to differentiate between oxidation, combustion, and pyrolysis. **Oxidation** is the process in which oxygen combines chemically with another substance to create a new compound. For example, steel that is exposed to oxygen results in rust. The process of oxidation can be very slow; indeed, it can take years for oxidation

to become evident. Slow oxidation does not produce easily measurable heat. **Combustion**, by contrast, is a rapid chemical process in which the combination of a substance with oxygen produces heat and light. For fire fighters' purposes, the terms "combustion" and "fire" can be used interchangeably. **Pyrolysis** is the process that liberates gaseous fuel vapours due to the heating of a solid fuel. Large quantities of flammable vapours greatly increase the size of a fire. Pyrolysis is evident when wood, heated sufficiently, breaks down into vapours and char (**FIGURE 5-5**).

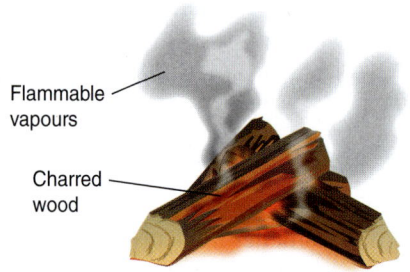

Flammable vapours

Charred wood

FIGURE 5-5 Pyrolysis is evident when wood, heated sufficiently, breaks down into vapours and char.
© Jones & Bartlett Learning.

SAFETY TIP

Smoke consists mainly of unburned forms of hydrocarbon fuels. As a consequence, when smoke is combined with adequate oxygen and heat, it will burn. Smoke contains large amounts of potential energy that can be released violently under the right conditions. Regardless of whether or not smoke is heated or cooled, all smoke should be considered dangerous.

Products of Combustion

The classic chemical reaction for burning that results in only water and carbon dioxide does not occur in most fire situations that fire fighters encounter. As discussed, most structure fires burn with a limited supply of oxygen, result in incomplete combustion, and produce a variety of toxic by-products. These by-products are collectively called **smoke**. Smoke includes three major components: particles (solids), vapours (finely suspended liquids—that is, aerosols), and gases.

Smoke particles include unburned, partially burned, and completely burned substances. The unburned particles are lifted in the **thermal column** produced by the fire and are usually readily visible. Some partially burned particles become part of the smoke because inadequate oxygen is available to allow for their complete combustion. Completely burned particles are primarily ash; the unburned and partially burned smoke particles can include a variety of substances. Some particles in smoke are small enough that they can get past the protective mechanisms of the respiratory system and enter the lungs, and most of them are toxic to the body. The concentration of unburned or partially burned particles depends on the amount of oxygen that was available to fuel the fire.

Smoke vapours, which are finely suspended liquids (aerosols), are similar to fog, which consists of small water droplets suspended in the air. When water is applied to a fire, additional small water droplets also may be suspended in the smoke or haze that forms. Similarly, when oil-based compounds burn, they produce small oil-based droplets that become part of the smoke. Oil-based or lipid compounds can cause great harm when they are inhaled. In addition, some toxic droplets cause poisoning if absorbed through the skin.

Smoke contains a wide variety of gases. The composition of gases in smoke varies greatly, depending on the amount of oxygen available to the fire at that instant. The composition of the substance being burned also influences the composition of the smoke. In other words, a fire fuelled by wood will produce a different composition of gases than a fire fuelled by petroleum-based fuels, such as plastics.

Because smoke is the product of incomplete combustion and it contains unburned hydrocarbons, we need to remember that it is a form of fuel. A column of hot black smoke coming in contact with an adequate supply of oxygen and an ignition source can ignite suddenly and violently.

Almost all of the gases produced by a fire are toxic to the body, including carbon monoxide, hydrogen cyanide, and phosgene. **Carbon monoxide (CO)** is deadly in small quantities and has been used to kill people in gas chambers. When inhaled, this gas quickly replaces the oxygen in the bloodstream because it binds with hemoglobin molecules (which normally carry oxygen to the body's cells) in the blood 200 times more readily than oxygen does. Even a small concentration of CO can quickly disable and kill a fire fighter.

Hydrogen cyanide is formed when plastic and foam products—such as furniture and the polyvinyl chloride pipe used in residential construction—burn. This poisonous gas is quickly absorbed by the blood and interferes with cellular respiration. Just a small amount of hydrogen cyanide can easily render a person unconscious. Recent studies have indicated that cyanide poisoning in fire fighters may be more common than previously recognized. Hydrogen cyanide also was used to kill convicted criminals in gas chambers.

Phosgene gas is formed from incomplete combustion of many common household products, including vinyl materials. This gas can affect the body in several ways. At low levels, it causes itchy eyes, a sore throat, and a burning cough. At higher levels, phosgene gas can cause pulmonary edema (fluid retention in the lungs) and death. Phosgene gas was used as a weapon in World War I to disable and kill soldiers.

Fires produce other gases that may affect the human body in harmful ways. Among these toxic gases are hydrogen chloride and several compounds containing different combinations of nitrogen and oxygen. Many of the compounds cause cancer, years after exposure. Higher rates of exposure to smoke increase the possibility of fire fighters contracting cancer.

Carbon dioxide, which is produced when sufficient oxygen is available to allow for complete combustion, is not toxic, but it can displace oxygen from the atmosphere and cause hypoxia (inadequate oxygenation of the blood) and asphyxiation. Carbon dioxide (CO_2) is a gas that is commonly used in restaurants to make carbonated beverages. Some restaurants have large CO_2 storage cylinders located in their basements. If a CO_2 tank is leaking in a basement, it will displace the normal room air, because the CO_2 is heavier than the air. This creates an atmosphere that may appear entirely normal but is immediately dangerous to life and health (IDLH) because of the lack of sufficient oxygen. You must use self-contained breathing apparatus (SCBA) in such a setting. Inhalation of and contact with these substances should be avoided.

A discussion of the by-products of combustion would be incomplete without considering heat. Because smoke is the result of fire, it is hot. The temperature of smoke varies depending on the conditions of the fire and the distance the smoke travels from the fire. Injuries from smoke may occur because of the inhalation of the particles, droplets, and gases that make up smoke. The inhalation of hot gases in smoke also may cause injuries in the form of severe burns of the skin and the respiratory tract.

SAFETY TIP

Toxic gases can quickly fill confined spaces or below-grade structures. Any confined space or below-grade area must be treated as a hazardous atmosphere until it has been tested to ensure that the concentration of oxygen is adequate and that no hazardous or dangerous gases are present (Underwriters Laboratories, Inc., 2010).

Fire Spread and Heat Transfer

Fire is a process of combustion that follows the laws of physics. In order to predict how and where the fire may spread and the impacts of our chosen tactics, we need to understand the basics of fire behaviour and fire spread and the influences of fire spread.

The primary principle needed to understand fire spread is the principle of heat transfer. By studying heat transfer, fire fighters can understand how a fire grows, how to predict the size and growth rate of the fire, and determine the most effective means by which to attack the fire. The exchange of thermal energy between materials is known as **heat transfer** and is measured as energy flow per unit of time. When there is a difference in temperature between two objects, heat will transfer from a hotter object to a cooler object until they reach equal temperatures. When two objects have the same temperature heat transfer does not occur.

SAFETY TIP

Smoky environments are deadly! It is essential to use an SCBA whenever you are operating in a smoky environment. This rule applies whether you are dealing with a structure fire, a dumpster fire, a car fire, or any other type of fire that generates smoke. Toxic gases are still present during the overhaul stage of a fire, even when there is little smoke visible. An Underwriters Laboratories (UL) smoke study determined that 97 percent of a fire's particulates are too small to be visible to the human eye.

The rate of heat transfer is dependent on two factors: the difference in temperature between the two substances and the ability of the materials to conduct heat. The rate of transfer increases when the differences between the objects are greater. **Heat flux** is the measure of the rate of heat transfer to or from one surface to another, which is commonly expressed as kilowatts per square metre (kW/m^2) or British thermal units per square foot (Btu/ft^2). For example, if a sofa is actively burning and heat is moving to the ceiling, the heat flux would indicate how much heat was being transferred to the ceiling.

Most people use the terms *heat* and *temperature* interchangeably; however, they have different meanings. **Temperature** is a measurement of the amount of molecular activity when compared with a reference or standard, whereas *heat* refers to a form of energy. An increase or decrease in the amount of energy transferred to an object can affect the molecular activity within an object and result in a corresponding change in an object's temperature. Higher temperatures reflect greater energy of a material. Temperature is measured on the Celsius scale or the Fahrenheit scale.

There are three primary mechanisms by which heat is transferred: conduction, convection, and radiation.

Conduction

Conduction is the process of transferring heat to and through one solid to another. This transfer occurs when two solids are in contact with each other and one has a higher temperature than the other. This heat transfer occurs because of the kinetic energy of the molecules within the solid materials (**FIGURE 5-6**). Conduction transfers energy directly from one molecule to another, much the same way as a billiard ball transfers energy from one billiard ball to the next. Objects vary in their ability to conduct energy. Metals generally have a greater ability to conduct heat than wood does. Objects that have more tightly packed molecules are more efficient in conducting heat than objects that are less densely constructed. For example, heat applied to a copper pipe will be readily conducted along the pipe. **TABLE 5-1** shows a comparison of the thermal conductivity of some common materials.

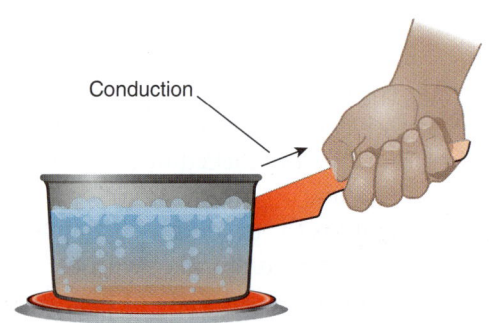

Conduction

FIGURE 5-6 Conduction through a solid object.
© Jones & Bartlett Learning.

TABLE 5-1 Thermal Conductivity of Common Building Materials and Air	
Material	**Thermal Conductivity (W/m × K)**
Copper	385
Carbon steel	36
Concrete	0.8
Brick	0.5
Air	0.024
Mineral insulation	0.04
Wood, oven dried	0.04

Measured at 20°C; W/m×K: Watts per metre Kelvin.

Heat transfer by conduction is dependent on three separate factors. The first factor is the thermal conductivity of the material being heated—better conductors transfer more heat. The second factor is the size of the area being heated. If all other factors remain the same, heating a small area results in less heat transfer than heating a larger area. The third factor is the difference in temperature between the heated object and the object that is being heated. As discussed previously, the greater the temperature difference or energy difference between the two materials, the greater the rate of heat transfer.

Insulating materials, on the other hand, are designed to limit the transfer of heat. Insulating materials have little ability to conduct heat due to the air trapped within the insulating material. The air or other gas trapped in small pockets prevents the degree of heat transfer that occurs in more densely constructed materials, and gases are not efficient conductors of heat from one solid to another because the gas molecules are farther apart than they are in denser materials. As a result, most insulating materials limit the rate of heat transfer by conduction. It is important to note, however, that although some materials are poor conductors, such as polyurethane foam, they can be highly combustible. They pyrolyze when heat is applied to them. Because they are poor conductors, the surface absorbing the heat will pyrolyze faster than if the material conducted more heat away from the surface being heated. As more combustible material is pyrolyzed, more fuel is available to the fire. The result is that some materials that are poor heat conductors may ignite and spread fire rapidly as heat is applied to them.

Convection

Convection is the circulatory movement that occurs in a gas or fluid such as air or water, with areas of differing temperatures being present in the same medium (**FIGURE 5-7**).

A simple definition of convection is the transfer of heat by the flow of gases or fluids from hotter areas to cooler areas. Convection in a fire primarily involves smoke and hot gases generated by the fire. This transfer occurs from a hotter gas to a cooler surface. The heat of the fire warms the gases and particles in the smoke. The hotter and less dense column of gases rises and displaces cooler, denser gases downward. A large fire burning in the open can generate a **plume**, an elongated and moving column of heated gases and smoke that rises high in the air. This convection stream can carry smoke and large bands of burning fuel for several blocks before the gases cool and fall back to the earth. If winds are present during a large fire, they will influence the direction in which the convection currents travel.

When a fire occurs in a building, the convection currents generated by the fire rise in the room and travel along the ceiling. This is referred to as the ceiling jet. These currents carry hot gases, which may ultimately heat combustible room surfaces and contents enough to ignite them. Fire fighters may take advantage of convection currents by using the cooler, lower layer to advance a hose line or conduct a search for occupants.

If the fire continues to grow, increasing the volume of hot gases and smoke, and if the pressure in the fire **compartment** is sufficient, the hot gases will push laterally outside the room where the fire started. Because the pressure in the lower parts of the fire compartment is lower, cooler air will be drawn into the lower levels of the fire compartment. The volume of space in which the cooler air is being drawn into the fire and the volume of space where the hotter gases and smoke are escaping from the fire are called the **flow path**. A flow path is the area(s) within a structure where heat, smoke, and air flow from areas of higher pressure to lower pressure. It comprises at least one intake vent, one exhaust vent, and the connecting volume between the vents. In other words, the flow path transfers hot gases and smoke from the higher pressure within the fire area toward the lower pressure areas accessible through doorways, window openings, and roof openings and may be influenced by external factors such as wind direction and speed. The fire growth will progress along the exhaust portion of the flow path. In addition, convection currents are influenced by the layout of the building. For example, air will flow quickly through a structure with an open floor plan versus a structure made up of small, closed rooms.

As the hot gases exit the fire compartment at the upper levels and the cooler air enters the fire compartment at a lower level, there is a level in the compartment opening where the pressure exerted by the lighter hot gases flowing out and the inward pressure of the cooler air entering the compartment will be equal. This is called the neutral plane and is discussed in more detail later.

Chapter 13, *Ventilation*, contains more information on flow paths.

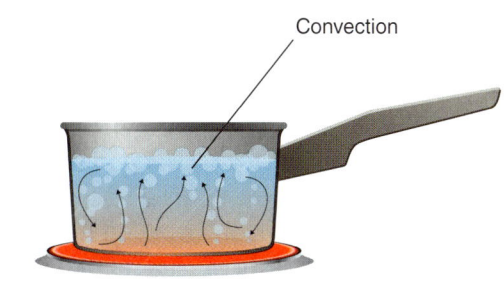

Convection

FIGURE 5-7 Convection.
© Jones & Bartlett Learning.

Radiation

Radiation is the transfer of heat through the emission of energy in the form of invisible electromagnetic waves (**FIGURE 5-8**). The sun radiates energy to the earth over the vast miles of outer space; this electromagnetic radiation readily travels through the vacuum of space. When this energy is absorbed, it is converted to heat—that is, the heat you feel as the sun's energy touches your body on a warm day. The direction in which the radiation travels can be changed or redirected, as when a mirror reflects the sun's rays and bounces the energy in another direction. The electromagnetic waves increase the temperature of any substance or object that is capable of absorbing the radiation. For example, heat from a fire is transferred in a direct line away from the source object and absorbed by cooler objects, including liquids and gases. Most exposure fires, that is, fires that start at a location that is remote from the area being protected and grow to expose that which is being protected, are the result of thermal radiation. Radiation also occurs in a fire compartment when fire fighters encounter heat from the fire and heat that is radiated back from the surfaces of the fire compartment.

Thermal radiation from a fire travels in all directions. The effect of thermal radiation, however, is not seen or felt until the radiation strikes an object and heats the surface of the object. Thermal radiation is a significant factor in the growth of a campfire from a small flicker of flame to a fire hot enough to ignite large logs. The growth of a small fire in a wastebasket to a full-blown room-and-contents fire is due in part to the effect of thermal radiation. A building that is fully involved in fire radiates a tremendous amount of energy in all directions. Indeed, the radiant heat from a large building fire can travel several hundred feet to ignite an unattached building.

Thermal radiation has an important influence on the performance of the fire fighter's PPE. The typical modern fuel load contains more petroleum-based materials, and fire fighters are encountering situations where the burning material is producing a large amount of energy. These increased **heat release rates (HRR)**, the rate at which heat energy is generated (measured in joules per second or watts), create conditions where fire fighters are exposed to higher levels of energy in a shorter amount of time—potentially saturating their PPE. This energy, in the form of heat, is then transferred to the operating fire fighter potentially leading to dangerous conditions.

> **FIRE FIGHTER TIP**
>
> Radiant heat can travel through the air gaps in a water spray.

Methods of Extinguishment

There are many variations on the methods used to extinguish fires; however, they boil down to four main methods: cooling the burning material, excluding oxygen from the fire, removing fuel from the fire, and interrupting the chemical reaction with a flame inhibitor (**FIGURE 5-9**). There are many variations on the way that these methods can be implemented, and sometimes a combination of these methods is used to achieve suppression of fires.

The method most commonly used to extinguish fires is to cool the burning material. The actions fire fighters take to set up a water supply, lay hose lines to the fire, and apply water to the fire are all steps in implementing this method of fire extinguishment.

A second method of extinguishing a fire is to exclude oxygen from the fire. One simple way to do this is to place the lid on an unvented charcoal grill or to close the door to a fire room. Reducing the amount of air reaching a fire will retard growth or extinguish the fire. Applying foam to a petroleum fire covers flammable vapours.

Removing fuel from a fire is the third method to extinguish the fire. For example, shutting off the supply of natural gas to a fire being fuelled by this gas will extinguish the natural gas fire (but it will not necessarily extinguish any resulting fires caused by the gas). Reducing or cooling hot gases and smoke also diminishes the supply of fuel available to a fire. In wildland fires, a firebreak cut around a fire puts further fuel out of reach of the fire.

The fourth method of extinguishing a fire is to interrupt the chemical reaction with a flame inhibitor. Halon- or Halotron-type fire extinguishers work in this way. These clean, fire-extinguishing agents can be applied with portable extinguishers or through a fixed system designed to flood an enclosed space. They leave no residue and are often used to protect electronics and computer systems. In Canada, Halon systems are not permitted for use. Their use is contrary to components of the Canadian Environmental Protection Act, 1999. A list of Halon alternatives for use in fire suppression can be obtained through your provincial/territorial authority or through Underwriters Laboratories of Canada (ULC).

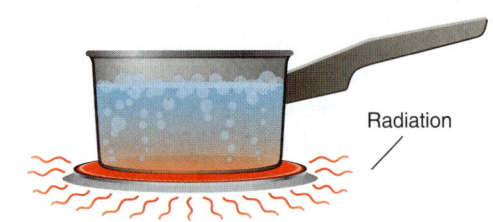

FIGURE 5-8 Radiation.
© Jones & Bartlett Learning.

A

B

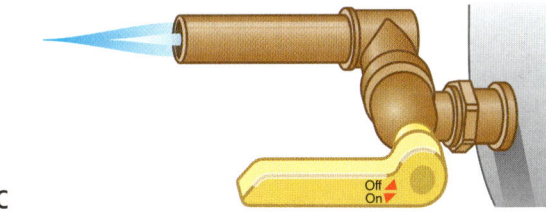

C

D

FIGURE 5-9 The four basic methods of fire extinguishment. **A.** Cool the burning material. **B.** Exclude oxygen from the fire. **C.** Remove fuel from the fire. **D.** Interrupt the chemical reaction with a flame inhibitor.
© Jones & Bartlett Learning.

Classes of Fire

Fires are generally categorized into one of five classes: Class A, Class B, Class C, Class D, and Class K.

Class A Fires

Class A fires involve ordinary solid combustible materials such as wood, paper, plastics, and cloth (**FIGURE 5-10**). Natural vegetation such as the grass that burns in ground cover fires also is considered to be part of this group of materials. The method most commonly used to extinguish Class A fires is to cool the fuel with water to a temperature that is below the ignition temperature or a combination of limiting ventilation and applying water.

Class B Fires

Class B fires involve flammable or combustible liquids such as gasoline, kerosene, diesel fuel, grease, tar, lacquer, oil-based paints, and motor oil (**FIGURE 5-11**). Fires

FIGURE 5-10 A Class A fire involves wood, paper, or other ordinary combustibles.
© schankz/Shutterstock.

FIGURE 5-11 A Class B fire involves flammable liquids such as gasoline.
© thaloengsak/iStock/Getty Images.

involving gases such as propane or natural gas are also classified as Class B fires. These fires can be extinguished by shutting off the supply of fuel or by using foam to exclude oxygen from the fuel.

Class C Fires

Class C fires involve energized electrical equipment (**FIGURE 5-12**). A Class C fire could involve building wiring and outlets, fuse boxes, circuit breakers, transformers, generators, or electric motors. Power tools, lighting fixtures, household appliances, and electronic devices such as televisions, radios, and computers could be involved in Class C fires as well. They are listed in a separate class because of the electrical hazard they present. Incorrectly attacking a Class C fire with an extinguishing agent that conducts electricity can result in injury or death to the fire fighter. Once the power is cut to a Class C fire, the fire is treated as a Class A or Class B fire depending on the type of material that is burning.

Class D Fires

Class D fires involve combustible metals such as sodium, magnesium, zirconium, lithium, potassium, and titanium (**FIGURE 5-13**). These fires are assigned to a special class because the application of water to fires involving these metals will result in violent explosions. Instead, these fires must be attacked with special agents to prevent explosions and to smother the fire and remove the supply of oxygen. Many automobile manufacturers use magnesium components to reduce the weight and increase the strength of vehicles.

Class K Fires

Class K fires involve combustible cooking oils and fats in kitchens (**FIGURE 5-14**). Heating vegetable or animal fats or oils in appliances such as deep-fat fryers

FIGURE 5-13 A Class D fire involves metals such as magnesium, sodium, or titanium. Many automobile manufacturers use magnesium components to reduce the weight and increase the strength of vehicles.
© Andrew Lambert Photography/Science Source.

FIGURE 5-14 A Class K fire involves combustible cooking oils and fats.
© Kathie Nichols/Shutterstock.

can result in serious fires that are difficult to fight with ordinary fire extinguishers. Special Class K extinguishers are available to handle this type of fire. These extinguishers contain a wet agent that combines with the cooking oils to produce a soap-like substance. These agents also absorb heat from the fire.

Some fires may fit into more than one class. For example, a fire involving a wood building also could involve petroleum-based contents. Likewise, a fire involving energized electrical circuits—a Class C fire—also might

FIGURE 5-12 A Class C fire involves energized electrical equipment.
© PhotoStock-Israel/Alamy Stock Photo.

involve Class A or Class B materials. If a fire involves live electrical sources, it should be treated as a Class C fire until the source of electricity has been disconnected.

Characteristics of Solid-Fuel Fires

Most building fires that fire fighters encounter involve solid fuels. Solid fuels have a definite size and shape. A variety of solid fuels are found in most buildings. These include wood and wood-based products, fabrics, paper, carpeting, and many petroleum-based fuels, such as plastics and petroleum-based foams. Wood is a commonly encountered building material. Older furniture was commonly constructed of solid-wood frames, and natural materials such as cotton were used for padding and upholstery. Wall coverings in older houses were often paper. Modern furniture is more frequently constructed from petroleum-based foams and fabrics and uses press board and plastic frames. Today, wall coverings are more likely to be latex-based paint or other petroleum-based materials. Solid fuels have certain characteristics that influence how a fire progresses. The structural building materials and the building contents will influence how a fire burns.

Solid fuels burn when they are heated sufficiently to change them into flammable vapours. When fuels are heated, they begin to pyrolyze, thereby releasing flammable vapours. Some fuels change from a solid form directly into a vapour. Other solid fuels first change into a liquid before becoming a vapour. Plastics and petroleum-based materials pyrolyze faster than wood-based products. This means that it requires less heating to decompose and ignite many of these fuels than it does to ignite wood.

Heating wood products causes them to pyrolyze. This decomposition results in the release of flammable vapours. Wood contains varying amounts of moisture. When wood is heated, the first change that occurs is that the water is vapourized and escapes. This causes the wood to begin to char. This charring is a natural mechanism of the wood in an attempt to protect itself. As the wood chars, it begins to lose mass, which can compromise its load-carrying capabilities. When the wood is heated to about 218°C (425°F), pyrolysis begins.

As the wood is heated, the energy required to vapourize the water delays the beginning of pyrolysis.

The characteristics of solid fuels that influence their combustion can be broken into three major categories: the composition of the fuel, the amount of the fuel, and the configuration of the fuel.

Composition of Fuel

The chemical composition of the fuel has a significant impact on how the fuel burns. Wood is composed primarily of a natural fibre called cellulose. Cellulose is combustible. Most of the modern room contents, including synthetic foams, upholstery coverings, various types of wall coverings, and plastic furniture, are manufactured from petroleum products. Petroleum-based products generally contain more potential heat energy than products made from natural products. This means that fires fuelled by modern petroleum-based products have a much higher HRR than fires fuelled by natural products such as wood and cotton.

Another factor that influences how the fire burns and the amount of heat that is released is the amount of moisture contained in the fuel. The water contained in a fuel serves as a cooling agent. It requires heat to vapourize the water from the fuel before the fuel can be pyrolyzed to create a flammable vapour. For example, green firewood from a recently cut tree contains a lot of moisture and is difficult to burn in a fireplace, whereas firewood that has been allowed to dry will ignite more easily and burn faster with a higher HRR.

Amount of Fuel

The second characteristic of fuel that determines how it burns is the amount of fuel available to the fire. In a **fuel-limited fire**, the fire has sufficient oxygen for fire growth but has a limited amount of fuel available for burning. If all other factors are the same, when more fuel is available, there is a higher HRR than when less fuel is available. For example, under identical conditions, a fire burning with four wooden pallets will have a much lower HRR than a fire fuelled with 40 wood pallets.

Configuration of Fuel

The third characteristic of fuel that determines how it burns is the configuration of the fuel. There are three factors that contribute to configuration. These are the surface-to-mass ratio, the orientation of the fuel, and the continuity of the fuel.

Surface-to-Mass Ratio

As discussed, solid fuels have a definite size and shape. The size and shape of the fuel greatly impact the ability of the fuel to ignite, the time it takes the fuel to be consumed, and the HRR of the burning fuel. Using wood as an example, consider a large log that is 3 m (10 ft) long. This log has a large mass or weight; however, the surface area of this wood is relatively small compared to its mass. If we cut this log into 102-mm by 102-mm (4-in. by 4-in.) beams, the total mass remains the same as that of the log, but the surface area is much greater than it was when in the form of a single log. If the original log were cut into thin shingles, it would maintain the same original total mass as the log but would have a far greater total surface area than the original log or the 102-mm by 102-mm (4-in. by 4-in.) beams. The energy required for ignition of the low surface-to-mass ratio log is much greater than the energy required to ignite the higher surface-to-mass ratio shingles. The higher the surface-to-mass ratio, the less energy that is required to ignite the fuel. This is because less energy is required to heat the material to a temperature sufficient to burn. For example, holding a match to a large log usually will not result in ignition of the log, but holding a match to a thin piece of shingle from the same source usually will result in ignition of the shingle.

Orientation of the Fuel

The second factor that affects the configuration of the fuel is the orientation of the fuel. For example, a board that is placed in a horizontal position will burn more slowly than an equal-sized board placed in a vertical position. A board in a horizontal position that is ignited on one corner will produce hot fire gases that rise away from the surface of the board. As convection currents carry heat upward, heat from the fire is carried away from the surface of the board. Convection will contribute little to the spread of the fire to other parts of the board. Because the heat from the convection currents rises, it has little effect on the spread of the fire. However, if a similar board is placed in a vertical position and the board is ignited on one of the lower corners, the convection created by the hot fire gases will transfer some of the heat along the vertical surface of the board. The heat absorbed by convective heat transfer will be much greater than the heat that is absorbed by the board in a horizontal position.

Continuity of the Fuel

The third factor that impacts the configuration of the fuel is the continuity of the fuel. Continuity refers to the closeness of one piece of fuel to another. The closer the fuel is, the easier and more quickly it will ignite. Continuity can occur in a horizontal direction along floors, ceilings, or horizontal surfaces of building contents. It also can occur in a vertical direction along walls or the vertical surfaces of building contents. Fuel

that is in contact with other fuel will ignite faster and reach a peak HRR more quickly. This rate of growth is partly due to spread by radiation, convection, and, in some cases, close proximity conduction.

Solid-Fuel Fire Development

A ventilated fire progresses through four classic stages of growth: the incipient stage, the growth stage, the fully developed stage, and the decay stage (**FIGURE 5-15**).

1. Incipient Stage

The first stage of fire development is called the **incipient stage**. *Incipient* means beginning to come into being. The incipient stage occurs when there is an adequate supply of fuel, oxygen, and heat or ignition (**FIGURE 5-16**). **Ignition** is defined as the action of setting something on fire. The ignition of a fire can occur in many different ways. One simple way to ignite a fire is by lighting some paper with a match. At the incipient stage a fire is small and confined to the initial fuel that was ignited. Because the fire is small, the initial growth is largely dependent on the type of fuel and how much fuel can be pyrolyzed into a vapour. Radiant heat from the fire begins to pyrolyze the fuel, increasing the amount of flammable gases. A fire at

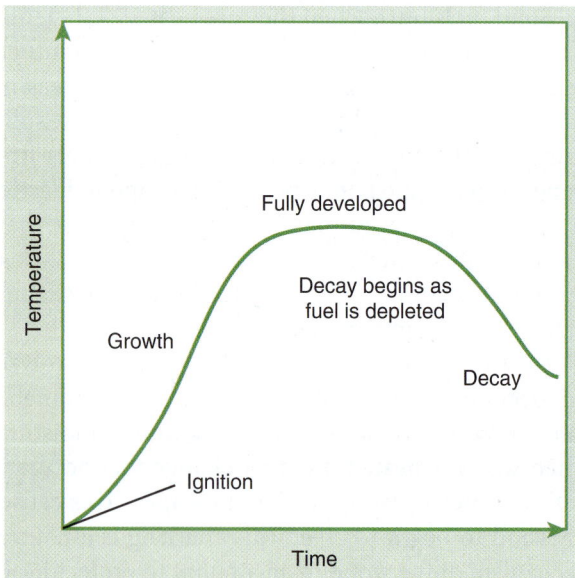

FIGURE 5-15 Illustration of a traditional fire growth curve with a fuel-limited fire.
Courtesy of NIST.

FIGURE 5-16 The incipient stage of a fire.
Courtesy of NIST.

FIGURE 5-17 The growth stage of a fire.
Courtesy of NIST.

this stage consumes relatively small amounts of oxygen, so at this stage the fire is fuel dependent.

As the fire begins to produce more heat, a plume of hot gases and smoke begins to rise from the fire. As this plume rises, it is cooled by the surrounding cooler air. As increasing amounts of heat are generated, the plume carries more heat and hot gases upward. If it reaches the ceiling, it will spread out in a circle that increases in size. These hot gases begin to heat the surfaces with which they make contact. In the incipient stage, because the amount of heat is limited, there is little increase in the temperature of the room. The amount of smoke and combustion gases in the compartment usually is not high enough to pose a serious threat to occupants if they exit quickly. Many fires in the incipient stage can be extinguished safely with a portable fire extinguisher.

It is important to understand, however, that the stages of fire growth do not have a clear dividing line between them. A fire can progress rapidly from the incipient stage to a growth stage in a very short period of time. Because of the rapid fire growth, building occupants and fire fighters who are not equipped with a complete ensemble of PPE, including SCBA, should exit from even a small fire.

2. Growth Stage

The second stage of fire development is the **growth stage**. A fire in the growth stage produces more interaction and is more dependent on the environment in the compartment around it than a fire in the incipient stage (**FIGURE 5-17**). Especially important are the composition of the compartment surfaces and contents and the placement and configuration of the compartment contents. The amount of ventilation is also an important factor in continued growth or the fire becoming ventilation-limited, which we will discuss shortly.

As a fire begins to grow it produces larger amounts of heat, producing a higher HRR. The hot gases and smoke start to rise into a plume because they become lighter as they are heated. The location of the fuel in a compartment influences the fire development. If the fuel is in the center of the compartment, it can draw air from four sides. The hot gases and smoke in the plume mix with cooler air. This cooling reduces the rate of growth, the length of the flame, and the length of the plume, thereby reducing vertical extension of the fire. These factors help to slow the rate of fire growth. If the burning fuel is located close to a corner, the plume can entrain (draw along and transport) air from only two sides. This has the effect of drawing in less cool air at the base of the fire. As a result, the combustion zone increases in height. In this situation, the growth of the fire will be faster than if the burning fuel were in the middle of the room. If the fuel is located close to a wall, it can entrain air on three sides. This situation will result in a growth rate slower than a fire located in a corner but faster than a fire with fuel in the middle of the room. A higher fire plume will increase the hot gases at the ceiling level of the compartment. More hot gases at the ceiling will radiate more heat back to the room surfaces, the room contents, and to the burning fuel, thereby increasing the rate of fire growth.

Specific types of fire conditions are described next. These include thermal layering, roll-over, flashover, backdraft, rapid fire growth, and the behaviour of ventilation-limited fires. These conditions are not limited to the growth stage of fire development but are introduced here because many of them are likely to occur in the growth stage.

Thermal Layering

As a fire continues to grow, more hot gases are generated. Hot gases are lighter and buoyant and therefore tend to seek a higher level in the compartment. Because cooler gases are heavier, they tend to occur at lower levels in the compartment. The phenomenon of gases forming into layers according to temperature is called **thermal layering**. This is also referred to as *heat stratification*. Any time the ventilation of the fire is changed, it changes the thermal layering.

As the hot fire gases rise, they increase the rate of heat transfer back to the contents and walls of the fire compartment. The layer of hot gases in the upper level of the compartment creates an increase in pressure. This layer of hot gases tends to push out of the upper level through any available openings such as doors or windows. Because the cooler gases at the bottom of the room exert less pressure than the hotter gases, cooler air is drawn into the fire compartment, usually at lower levels.

The common boundary or interface between the hot gases and the cooler gases is called the **neutral plane**. The neutral plane is the volume of space in a compartment opening at which the pressure of the hot gases and smoke leaving the compartment and the pressure of the cooler air entering the compartment are equal. In order for a neutral plane to exist, there must be a flow of cooler air entering the compartment and a flow of hot gases exiting the compartment. Fire fighters should avoid working in a layer of hot gases. Stay low whenever possible to remain under the layer of hot gases in the upper levels of the fire compartment. You will learn more about steps that can be taken to ventilate and control the fire to reduce the amount of hot gases and to increase the level of the neutral plane.

As the fire continues to grow and the amount of heat generated increases, you may notice small flames "dancing" in the hot gas layer. These isolated flames are an indication that the gases in the hot layer are within their flammable range. This indicates that this layer is hot enough to cause ignition. Flammable range is discussed in more detail later. Often, these flames will be around the edges of the hot plume because this area has sufficient oxygen for combustion. These isolated flames may indicate that it is a **ventilation-limited fire**. A ventilation-limited fire is restricted because there is not enough oxygen available for the fire to burn as rapidly as it would with an unlimited supply of oxygen. This also may indicate that conditions are approaching a point where a fire may soon flash over, or ignite all exposed surfaces.

Roll-Over

Roll-over (also known as flameover) is the spontaneous ignition of hot gases in the upper levels of a room or compartment (**FIGURE 5-18**). During the growth stage of the fire, the hottest gases rise to the top of the room. If some of these gases reach ignition temperature,

FIGURE 5-18 Roll-over is a sign that flashover is imminent unless actions are taken to change the fire conditions.
© Jones & Bartlett Learning. Photographed by Glen E. Ellman.

they will ignite from the radiation of heat coming from the flaming fire or from direct contact with open flames. In other words, the upper layer of flammable vapour catches fire. These flames can flicker across the ceiling and then go out. Alternatively, when they get a few degrees hotter, they can extend throughout the room at ceiling level. Roll-over is a sign that the temperature is rising—and if it continues to rise, the temperature throughout the room will soon reach the point where the room and contents spontaneously and rapidly ignite. Roll-over is a sign that flashover is imminent unless actions are taken to change the fire conditions.

FIRE FIGHTER TIP

Flashover can be prevented by cooling the fuel, removing the fuel, or reducing the amount of oxygen present by controlling ventilation.

Flashover

Flashover is not a specific moment but rather the transition from a fire that is growing by igniting one type of fuel to another, to a fire where all of the exposed surfaces have ignited. It is a rapid change or transition from the growth stage to the fully developed stage. In a fire, it may be hard to determine which stage a fire is in. Fires do not always follow the four classic stages of development. Fire fighters should be aware that a flashover can occur any time temperatures and sufficient oxygen are available to support combustion (**FIGURE 5-19**). Flashovers are less likely to occur if the fire is in a large room or compartment that will supply more air during fire development. Larger areas entrain more air in the fire plume, which tends to cool

FIGURE 5-19 Flashover is the rapid change or transition from the growth stage to the fully developed stage.
Courtesy of Dave Casey.

the plume and reduce the radiation and convection of heat to the room surface and contents.

The critical temperature for a flashover to occur is approximately 538°C (1000°F). Once this temperature is reached, all fuels in the room are involved in the fire, including the floor coverings. This means that the temperature at the floor level may be as high as the temperature of the ceiling just before flashover. Fire fighters, even with full PPE, cannot survive for more than a few seconds in a flashover.

Flashover may not occur if the fire remains ventilation-limited. A ventilation-limited fire will produce a limited amount of heat energy. If the supply of oxygen is not increased, the fire may enter a decay stage. During this decay stage, heat will continue to pyrolyze fuels in the compartment. This pyrolysis produces additional vapourized fuel in the form of smoke. A fire that is in the decay stage will flash over quickly, however, if temperatures are maintained and sufficient oxygen is added. Keeping doors and windows to the fire compartment closed will help to reduce the amount of oxygen available to the fire and may help prevent a flashover. It is important to note that even with increased ventilation,

FIRE FIGHTER TIP

It is important to be able to recognize when there are signs that may indicate a flashover. These conditions include off-gassing or smoke from furniture and carpeting, a sudden increase in heat, or zero visibility. Even when you recognize these conditions are present, you often will not have time to escape from danger. Work to avoid placing yourself in these conditions by either cooling the environment as you go or avoiding locations where flashover indicators are present.

many compartment fires will remain ventilation-limited rather than advancing to a free-burning state because there is insufficient oxygen to produce a free-burning state. The ventilation-limited stage of a fire is illustrated in (**FIGURE 5-20**).

Backdraft

A **backdraft** is caused by the introduction of oxygen, a change of the ventilation profile, into an enclosure where the superheated gases and contents are already hot enough for ignition but do not have sufficient oxygen to combust. The development of a backdraft requires a unique set of conditions. When a fire generates quantities of combustible gases, these gases can become heated above their ignition temperatures. Backdrafts require a "closed box"—that is, a room, compartment, or building that is ventilation-limited. When the fire compartment has a limited amount of oxygen, combustion is reduced, yet superheated fuel in the form of smoke still fills the compartment (**FIGURE 5-21**). If a supply of oxygen is introduced into the room by ventilation, such as opening the front door or an accidental event such as a window breaking because of excess heat, sudden and explosive combustion can occur owing to the presence of the superheated flammable gases. This kind of explosive combustion may exert enough force to cause severe injury or death to fire fighters. Signs and symptoms of an impending backdraft include the following:

- Any confined fire with a large heat build-up
- Little or no visible flame from the exterior of the building
- A "living fire," where the building appears to be breathing due to smoke puffing out of and then being sucked back into the building
- Smoke that seems to be pressurized

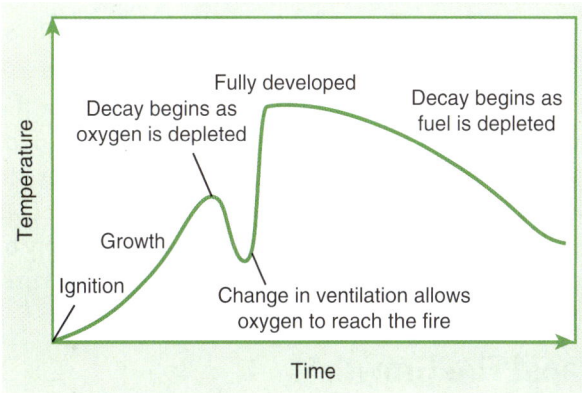

FIGURE 5-20 Illustration of a ventilation-limited fire, which often occurs in houses with lightweight construction and petroleum-based contents.
Courtesy of NIST.

A

B

FIGURE 5-21 These photos show the rapid transition to a backdraft. The photos were taken only seconds after the compartment door was opened.
© Jones & Bartlett Learning. Photographed by Glen E. Ellman.

- Smoke-stained windows (an indication of a significant fire)
- No smoke showing
- Turbulent smoke
- Thick yellowish/dark brown and black smoke exiting eves or around doors and windows (containing sulfur compounds)
- A rapid and consistent inflow of air as fire fighters undertake their initial entry

It is important for fire fighters to stay alert for the conditions that signal a possible backdraft and to reduce the possibility of a backdraft developing. Chapter 13, *Ventilation*, and Chapter 16, *Supply Line and Attack Line Evolutions*, explain specific ways to limit the supply of oxygen to the fire and to cool a ventilation-limited fire.

Rapid Fire Growth

Introducing air into a ventilation-limited fire can result in explosive rapid fire growth. Research has demonstrated that fire fighters making entry through the front door can introduce enough air into the fire area to produce rapid

fire growth and flashover. This change can occur so fast that it is not possible to escape this deadly environment. Fire fighters need to consider carefully what constitutes ventilation. Opening any door, window, skylight, or roof introduces oxygen into a burning building. Repeated experiments have been conducted that produced rapid fire growth after the front door was opened. Rapid fire growth can be prevented by cooling the fuel, removing the fuel, or controlling the amount of oxygen present. During rapid growth, a fire can progress to its maximum HRR unless actions are taken to limit the supply of oxygen or apply water to the fire.

Behaviour of Ventilation-Limited Fires

Experiments conducted by UL, the National Institute of Standards and Technology (NIST), and the New York City Fire Department (FDNY) have continuously demonstrated that many building fires become ventilation-limited because of a limited supply of oxygen. Newer houses are tightly sealed, with added insulation and caulking, double-paned windows, and storm windows. A fire in a building during the growth stage can consume enough oxygen to reduce the concentration of oxygen below the level needed to maintain a growing fire. When this occurs, the fire is said to have become ventilation-limited. Assessing a ventilation-limited fire presents a challenge to fire fighters. A fire in this condition initially may appear to be a small fire in the incipient stage, but it contains a large amount of energy in the form of hot gases and smoke, which are vapourized fuel. Often flames are not visible. If the supply of oxygen is increased, a ventilation-limited fire can progress suddenly to a flashover in less than 2 minutes. Research indicates that fires in modern residential occupancies are likely to enter a ventilation-limited decay stage prior to the arrival of the first due company (Underwriters Laboratories, Inc., 2010).

From the outside, a ventilation-limited fire may be deceiving—it may look like a small fire. All that is needed is more oxygen to produce explosive fire growth. Be aware that a fire that appears to be "small and relatively harmless" may in fact be a ventilation-limited fire containing enormous amounts of energy that just needs a healthy dose of oxygen to grow rapidly.

3. Fully Developed Stage

The third stage of fire development is the **fully developed stage**. During this stage, the fire is consuming the maximum amount of fuel possible, and it is achieving the maximum HRR possible for the fuel supply and oxygen present (**FIGURE 5-22**). During the fully developed stage, the fire may be ventilation-limited or fuel-limited. When a fire has an unlimited supply

FIGURE 5-22 The fully developed stage of a fire.
Courtesy of NIST.

of oxygen, the fire is fuel-limited. Consider a building under construction that has been framed with wood and combustible sheathing but lacks windows and doors. If this building catches fire, the fully developed fire may be fuel-limited because it has an abundant supply of oxygen. In contrast, most structure fires that fire fighters encounter have limited openings from the fire compartment to the outside. These fires are limited by the amount of oxygen available, because there are limited openings available to admit oxygen and to exhaust the hot fire gases and smoke.

During the fully developed stage, as the fire releases large quantities of heat, the energy pyrolyzes large amounts of fuel, generating large quantities of smoke and fire gases. The smoke and fire gases generate increased pressure in the fire compartment and force hot fire gases from the fire compartment. The HRR will be dependent on the size of the compartment openings, which provide a supply of oxygen and openings for hot fire gases.

When the hot fire gases are exhausted from the fire building, if they are above the ignition temperature of the gases, they may ignite upon mixing with a fresh supply of oxygen. This is what produces flames from compartment openings during this active burning state. If these hot fire gases travel to another room or compartment, they can transfer enough heat to spread the fire to rooms or compartments adjacent to the fire compartment. The supply of oxygen may be greater in adjacent areas. As hot vapourized fuel travels to adjacent areas, it can ignite when it reaches a new supply of oxygen. The large quantities of heat transferred to adjacent areas will pyrolyze fuel in these areas, increasing the supply of vapourized fuel available to the fire. This provides an additional source of fuel for the fire.

Not all fires reach the fully developed stage. If the fire compartment has limited openings that result in the fire being ventilation-limited, the fire probably will not

reach the conditions needed to achieve the maximum rate of heat production of a fully developed fire that has an unlimited supply of oxygen.

4. Decay Stage

The fourth and final stage in fire development is the **decay stage**. This condition can occur because of a decreasing fuel supply or because of a limited oxygen supply (**FIGURE 5-23**).

One example of a fuel-limited fire occurs in cases where an outdoor fire consumes the fuel supply. It also can occur in a room or compartment fire when the fuel supplied by the compartment contents is being consumed and the fire has not extended beyond the compartment because of fire-resistant construction. During the decay stage, active flaming combustion decreases or stops. This change is fuel-limited and not ventilation-limited. The heat will continue to pyrolyze the fuel and create flammable gases and vapours. These flammable fuel products will continue to burn, but the rate of combustion will continue to decrease and eventually stop when the fuel supply is exhausted.

A second type of situation occurs when a fire with an available fuel supply becomes oxygen-limited. This type of fire also can enter a decay stage but for a different reason. In oxygen-limited decay, the rate of combustion slows, and visible flames decrease or disappear. Because there is still significant heat in and around the fire and

FIGURE 5-23 The decay stage of a fire.
Courtesy of NIST.

the fire compartment, fuels will continue to pyrolyze and create additional flammable vapours and gases. The rate of pyrolysis will slow as the rate of combustion slows, but large quantities of flammable fuel may still be present. This decay is caused by a ventilation-limited fire. If additional oxygen is introduced into the fire compartment, rapid or violent fire growth can develop quickly.

Smoke Explosion

A smoke explosion occurs when a mixture of flammable gases and oxygen is present, usually in a void or other area separate from the fire compartment. Smoke may travel some distance from the fire. When it comes in contact with a source of ignition, the flammable mixture ignites—often in a violent manner. The conditions needed to produce a smoke explosion often include the presence of void spaces, combustible building materials, and a ventilation-limited fire that produces unburned fuel. Smoke explosions can occur during the decay stage of a fire, but they are not limited to that stage. They sometimes occur during final extinguishment or overhaul. In a smoke explosion, there is no change to the ventilation profile such as an open door or window; rather, it occurs from the travel of smoke within the structure to an ignition source.

Fire Behaviour in Modern Structures

Modern construction techniques and synthetic contents have altered fire behaviour in today's fires. Houses built in the last 20 or 30 years are constructed to be energy and cost efficient. They contain insulation to prevent heat loss in the winter and loss of cool air in the summer. In addition, they are sealed to prevent air exchange through the walls and around the doors and windows. As a consequence, when a fire occurs in a modern house, only a limited quantity of air will enter the structure when the doors and windows are closed. In many parts of the country, doors and windows are kept closed most of the year because buildings are being heated or cooled by heating and cooling systems.

Many newer houses are constructed with lightweight manufactured building components. These manufactured building components contain many highly combustible products, such as glues and coatings, most of which are made from petroleum-based products. Modern homes constructed with lightweight components contain less material (less mass) than houses built with dimensional lumber. Petroleum-based products contain more potential heat energy. This means that fires in newer houses have a higher HRR than fires in older houses constructed primarily of wood. These building products are not

only combustible, but they also burn hotter and faster than a house that is constructed primarily of wood. The lightweight and new-growth frames of newer houses fail also much faster than the heavier wood framing of older homes. The lightweight construction often provides greater surface area of structural members and the potential for greater void spaces between floors.

The second major change in houses relates to the widespread use of plastics and petroleum-based products for furniture and accessories. These products catch fire easily, reach high temperatures quickly, and have a high HRR. When burning, these products release huge quantities of thick, dark smoke. This rapid combustion requires huge amounts of oxygen to maintain the growth stage (i.e., to remain burning) (**FIGURE 5-24**).

These changes in building construction materials and techniques and changes in the building contents have, in turn, changed the dynamics of fire behaviour inside a structure. Tests by UL have consistently documented a pattern of fire behaviour that had been previously observed by veteran fire fighters.

Fires in modern structures progress to the fully developed stage quickly. In most cases, these fires enter a ventilation-limited stage, because there is a limited supply of oxygen available. During this ventilation-limited stage, the heat from the fire continues to pyrolyze available fuels, thereby increasing the supply of vapourized fuel. All that is needed is an increased supply of oxygen to change the fire into a rapid growth stage, often in the form of a flashover.

FIRE FIGHTER TIP

The intensity and speed with which a fire burns depend on the concentration of oxygen available to the fire. During the ignition stage, there is usually sufficient oxygen available for the fire to grow. In the growth, fully developed, and decay stages, a fire in an enclosed compartment may be regulated by the amount of oxygen available to it. If the atmosphere contains 16 to 21 percent oxygen, the fire typically will burn with an open flame. If the concentration of oxygen drops to less than 16 percent, open-flame burning will decrease. If the concentration of oxygen drops to less than 10 percent of the atmosphere, burning of most substances will stop (Gann and Friedman, 2015). If oxygen is introduced into a dying fire, however, the conditions can change quickly and result in a dangerous situation for fire fighters. Properly coordinated ventilation and fire attack procedures are designed to prevent these conditions from occurring.

Often, the fire department arrives on the scene while the fire is in a ventilation-limited decay stage. One of the first actions taken by some fire departments is to open the front door to gain access to the building for an interior attack. Unfortunately, opening the door has

Today's Fire Environment

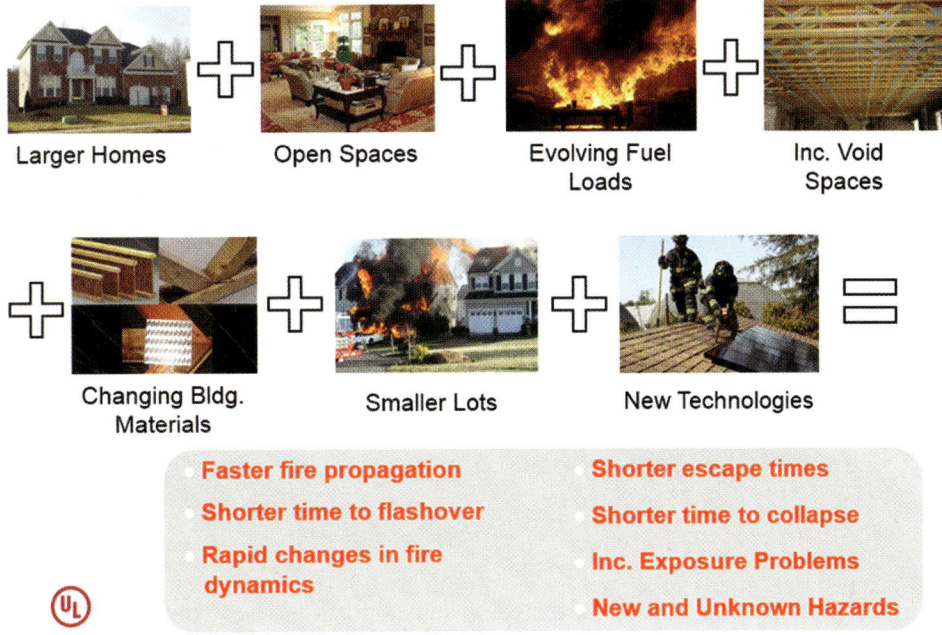

FIGURE 5-24 Modern fire environment.
Courtesy of UL.

the unintentional effect of introducing a fresh supply of oxygen to the fire. As discussed previously, when a fresh supply of oxygen comes in contact with superheated flammable fuel, the result is rapid growth of the fire, often in the form of a flashover.

To prevent this scenario from occurring, fire fighters must limit the amount of airflow to the fire. This requires coordinated timing in implementing building entry, ventilation, and application of water to the fire. More information on these topics is presented in Chapter 10, *Forcible Entry*; Chapter 13, *Ventilation*; Chapter 16, *Supply Line and Attack Line Evolutions*; and Chapter 17, *Fire Suppression*.

Wind Effect

One factor that greatly influences the behaviour of a fire is the wind. When wind is present, the behaviour of the fire changes drastically. For example, when the air is relatively still, the best approach to a fire may be through the "A" side of a building. Conversely, if the wind is blowing at 32 to 40 km/h (20 to 25 mi/h) from the "C" side of the same building, entering on the "A" side may be a deadly miscalculation because the wind is pushing the fire to the "A" side of the building (**FIGURE 5-25**). The impact of the wind is similar to placing several large ventilation fans on the "C" side of the building and pushing huge quantities of air and fire toward the "A" side of the building.

The potential effect of the wind may not be evident during the initial size-up when the structure is tightly closed. However, because the wind can have a huge effect on fire behaviour, the initial size-up of a structure fire must always include an evaluation of the impact of the wind. This evaluation will sometimes change the plan for attacking the fire. For example, if the seat of the fire is located on the "C" side of the building and a strong wind is blowing from that direction, the incident commander might choose not to begin ventilation operations on the "C" side windows if the rescue crew needs to enter through the "A" side of the building. Opening up a window on the "C" side could cause the fire to spread rapidly toward the "A" side. NIST research has shown that wind speeds as low as 16 km/h (10 m/h) can create wind-driven fire conditions inside the structure (NIST, 2008).

Characteristics of Liquid-Fuel Fires

Fires involving liquid or gaseous fuels have some different characteristics from fires that involve solid fuels. To understand the behaviour of these fires, you need to understand the characteristics of liquid and gaseous fuels.

Recall that solid fuels do not burn in the solid state but instead must be converted to a vapour before they will burn. Liquids share the same characteristic: They must be converted to a vapour and be mixed with

Wind travels from the rear of the house

FIGURE 5-25 An interior structure fire can be affected by an exterior wind. Note the lack of visible smoke exiting the window on the windward side. This indicates a high pressure flow path of fresh air.
© Jones & Bartlett Learning.

oxygen in the proper concentration before they will burn. Three conditions must be present for a vapour and air mixture to ignite:

- The fuel and air must be present at a concentration within a flammable range.
- There must be an ignition source with enough energy to start ignition.
- The ignition source and the fuel mixture must make contact for long enough to transfer the energy to the air–fuel mixture.

As liquids are heated, the molecules in the material become more active, and the speed of vapourization of the molecules increases. Most liquids eventually reach their boiling point during a fire. **Boiling point** is the temperature at which a liquid will continually give off vapours in sustained amounts and, if held at that temperature long enough, will turn completely into a gas. The boiling point of water, for example, is 100°C (212°F). As the boiling point is reached, the amount of flammable vapour generated increases significantly (**FIGURE 5-26**). Because most liquid fuels are a mixture of compounds (e.g., gasoline contains approximately 100 different compounds), however, the fuel does not have a single boiling point. Instead, the compound with the lowest ignition temperature determines flammability of the mixture—the temperature at which that fuel will spontaneously ignite.

The amount of liquid that vapourizes is related to the **volatility** of the liquid. Liquids that have a lower

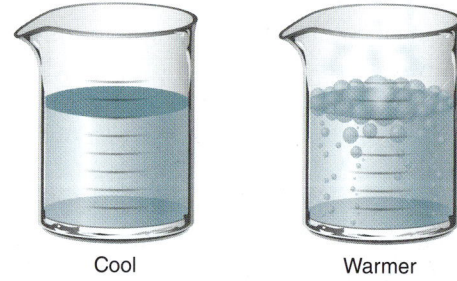

Cool   Warmer

FIGURE 5-26 As liquids are heated, the molecules become more active, and the speed of vapourization increases.
© Jones & Bartlett Learning.

molecular weight tend to vapourize more readily than liquids with a higher molecular weight. In addition, the higher the temperature, the more the liquid will evaporate. As more of the liquid vapourizes, it may reach a point where enough vapour is present in the air to create a flammable vapour–air mixture.

Two additional terms are used to describe the flammability of liquids: flash point and fire point. The **flash point** is the lowest temperature at which a liquid or solid produces a flammable vapour. It is measured by determining the lowest temperature at which a liquid will produce enough vapour to support a small flame or flash fire for a short period of time until the fuel is consumed (the flame may go out quickly) (**TABLE 5-2**). The **fire point** (also known as the flame point) is the lowest temperature at which a liquid produces enough vapour to sustain a continuous fire. For most materials, the fire point is only slightly higher than the flash point.

Characteristics of Gas-Fuel Fires

By learning about the characteristics of flammable gas fuels, you can help to prevent injuries or deaths in emergency situations and work to mitigate the conditions causing the problem. Two terms are used to describe the characteristics of flammable vapours: vapour density and flammable range.

Vapour Density

Vapour density refers to the weight of a gas fuel and measures the weight of the gas compared to air (**TABLE 5-3**). The weight of air is assigned the value of 1. A gas with a vapour density less than 1 will rise to the top of a confined space or rise in the atmosphere. For example, hydrogen gas, which has a vapour density of 0.07, is a very light gas. Conversely, a gas with a vapour density greater than 1 is heavier than air and will settle close to the ground. For example, propane gas has a vapour density of 1.55 and will settle to the ground. By comparison, carbon monoxide has a vapour density of

TABLE 5-2 Flash Point	
Liquid	**Flash Point (°C/°F)**
Water	N/A
Gasoline	−43°C/−45°F
Acetone	−20°C/−4°F
#2 grade diesel	52°C/125°F

From *Principles of Fire Protection Chemistry and Physics.* Used by permission.

TABLE 5-3 Vapour Density of Common Gases	
Gaseous Substance	**Vapour Density**
Gasoline	>3.0
Ethanol	1.6
Methane	0.55
Propane	1.55

0.97—almost the same as that of air—so it mixes readily with all layers of the air. In situations where a flammable gas is present, fire fighters need to recognize the vapour density of the escaping fuel so that they can take actions to prevent the ignition of the fuel and allow the gaseous fuel to safely escape into the atmosphere.

Flammable Range

Mixtures of flammable gases and air will burn only when they are mixed in certain proportions. If too much fuel is present in the mixture, there will not be enough oxygen to support the combustion process; if too little fuel is present in the mixture, there will not be enough fuel to support the combustion process. The range of gas–air mixtures that will burn varies from one fuel to another. Carbon monoxide will burn when mixed with air in concentrations between 12.5 percent and 74 percent. By contrast, natural gas will burn only when it is mixed with air in concentrations between 4.5 percent and 15 percent.

The **flammable range (explosive limit)** is the range in concentration between the lower and upper flammable limits. The terms *flammable range* and *explosive limit* are used interchangeably because under most conditions, if the flammable gas–air mixture will not explode, it will not ignite. The **lower explosive limit (LEL)**, also referred to as the lower flammable limit (LFL), refers to the minimum amount of gaseous fuel that must be present in a gas–air mixture for the mixture to be flammable. In the case of carbon monoxide, the LEL is 12.5 percent. The **upper explosive limit (UEL)**, also referred to as the upper flammable limit (UFL), of carbon monoxide is 74 percent. Test instruments are available to measure the percentage of fuels in gas–air mixtures and to determine when an emergency scene is safe.

Boiling Liquid/Expanding Vapour Explosions

One potentially deadly set of circumstances involving liquid and gaseous fuels is a **boiling liquid/expanding vapour explosion (BLEVE)**. A BLEVE occurs when a liquid fuel is stored in a vessel under pressure. If the vessel is filled with propane, for example, the bottom part of the vessel would contain liquid propane, and the upper part of the vessel would contain vapourized propane (**FIGURE 5-27**).

If this sealed container is subjected to heat from a fire, the pressure that builds up from the expansion of the liquid will prevent the liquid from evaporating. Normally, evaporation cools the liquid and allows it to maintain its temperature. If heating continues, the temperature inside the vessel reaches a level that exceeds the boiling point of the liquid. At some point, the vessel will fail,

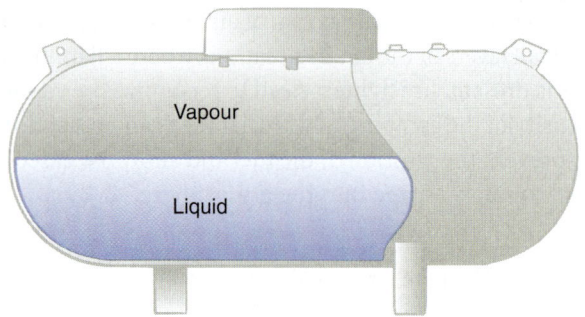

FIGURE 5-27 A propane tank contains both liquid and vapour.
© Jones & Bartlett Learning.

FIGURE 5-28 This photo shows a fireball formed from an ignited BLEVE.
© Ivan Cholakov/Shutterstock.

releasing all of the heated fuel in a massive explosion. The released fuel instantly becomes vapourized and ignites as a huge fireball (**FIGURE 5-28**).

The key to preventing a BLEVE is to cool the top of the tank, which contains the vapour. This action will prevent the fuel from building up enough pressure to cause a catastrophic rupture of the container. Prevention of BLEVEs is covered in more detail in Chapter 23, *Advanced Fire Suppression*.

> **FIRE FIGHTER TIP**
>
> Flames indicate what is happening now, whereas smoke gives a more complete picture of the characteristics of the fire and where it is going.

Smoke Reading

Learning the principles underlying fire behaviour helps fire fighters to understand the rules that govern the way fires burn. One practical application of these principles

is learning how to "read" the smoke at a fire. Being able to read smoke may enable you to learn where the fire is, how big it is, and where it is moving. At a fire, most untrained people look at flames. Flames indicate where the fire is now, but they do not tell you how big the fire is and where it is moving. Fires are dynamic events—what you see this minute probably will change quickly. Before developing a plan of attack, fire fighters want to anticipate where the fire will be in a few minutes. At a structure fire, often there are no visible flames, because the fire is occurring inside the building. The ability to read smoke gives fire fighters information that they need to mount a more effective attack on a fire, and this may help to save either their lives or the lives of the building's occupants.

It is helpful to think about smoke as being a fuel. Fuels behave in predictable ways, when they are combined with the right mixture of oxygen and heat. For example, flashovers occur when fuels in a room or confined space are heated until they ignite. Most of the fuel in a flashover is in the form of smoke. Recall that a backdraft is a sudden, explosive ignition of fire gases that occurs when oxygen is introduced into a superheated space with limited oxygen. As in a flashover, these fire gases are really smoke. Both flashovers and backdrafts are fed by superheated fuel, in the form of smoke. Thus, as you study the art of smoke reading, think of this exercise as studying the fuel that is all around you at a fire.

As mentioned earlier, smoke is composed of three major components: particles (solids), vapours (finely suspended liquids or aerosols), and gases. The solids consist of carbon, soot, dust, and fibres; the vapours contain suspended hydrocarbons and water; and the gases consist of carbon monoxide and a wide variety of other gases. Taken together, these three components of smoke represent fuels that will burn when heated sufficiently in the presence of oxygen. Hot smoke is extremely flammable and will ultimately dictate the fire's behaviour. In addition, smoke today is not the smoke of yesterday. Many of today's building materials and building contents are plastics, which are made from petroleum-based products. As a consequence, these materials give off more toxic gases and burn at higher temperatures than the materials used for construction and contents in earlier times.

The best place to observe patterns of smoke is from the outside of the fire building. Compare the smoke coming out of openings in different parts of the building. It will tell you more than looking at the flames.

Step 1: Determine the Key Attributes of Smoke

The first step in reading smoke is to consider four key attributes of the smoke: **smoke volume, smoke velocity** or speed, **smoke density**, and **smoke colour**. You also need to factor in the size of the building, the layout of the building, the wind, and any other characteristics of the building that might change the appearance of these key attributes.

The volume of smoke coming from the fire provides some idea of how much fuel is being heated to the point that it gives off gases (off-gassing). Think of these materials as preheating. As you assess the smoke volume, also consider the size of the burning building (the box). It takes relatively little smoke to fill up a small building, but a lot of smoke is required to fill up a large building. Thus, you must consider the size of the box when considering the amount of smoke coming from it. Assessing the volume of smoke alone will not provide a complete picture of the fire, but it sets the stage for better understanding the fire.

The velocity (speed) at which smoke is leaving the building suggests how much pressure is accumulating inside the building. Smoke is pushed both by heat and by volume. When smoke is pushed by heat, it will rise and then slow down gradually. When smoke is pushed by volume, it will slow down immediately. Assess the smoke velocity to determine whether it is being pushed by heat or by volume.

Consider whether the smoke has a laminar flow or a turbulent flow. **Laminar smoke flow** is a smooth or streamlined flow. It indicates that the box and its contents are absorbing heat and the pressure in the box is not too high (**FIGURE 5-29**). By contrast, **turbulent smoke flow** is agitated, boiling, or angry. This type of flow is caused by rapid molecular expansion of the gases within the smoke and the restrictions created by the building (the box) (**FIGURE 5-30**). This expansion occurs when the box cannot absorb any more heat. In such circumstances, the radiant heat that the box has absorbed is radiated back into the smoke. Turbulent smoke contains an immense amount of energy. When

FIGURE 5-30 Turbulent smoke flow.
Courtesy of Keith Muratori.

this energy reaches a point where the smoke is heated to its ignition temperature in the presence of sufficient oxygen, flashover will occur. Whenever you see turbulent smoke, be aware that flashover is likely to occur very soon.

By comparing the speed and type of flow from similar-sized openings, fire fighters can get a good idea of where the fire is located. A similar-sized opening with a higher velocity of smoke is closer to the location of the fire. Look for the fastest smoke coming from the most restrictive opening. Studying the velocity of smoke provides valuable information as fire fighters prepare to attack the fire.

Evaluate the thickness of the smoke. Smoke density suggests how much fuel is contained in the smoke. The denser the smoke is, the more fuel it contains. Dense smoke can produce a powerful flashover when conditions are right. In fact, thick smoke can flash over even without turbulent flow because the supply of fuel reaches from the heart of the fire to the location where the smoke is issuing from the building.

When you think of smoke as fuel, you quickly realize that fire fighters who are surrounded by dense smoke are working in a pool of flammable fuel. While no one would send a fire fighter into a pool of petroleum, we are doing just that when we send a fire fighter into dense smoke.

Moreover, dense smoke contains many poisonous and cancer-causing substances. When fire fighters are in an environment of dense smoke that has banked down to the floor, they are in an environment that will not support life for a building occupant. Just a few breaths of this toxic brew will result in death. Also, if a flashover occurs, this environment will immediately change to a deadly environment for fire fighters—even those in full PPE with SCBA. There is no sense in a fire fighter risking his or her life to try to save someone who is already dead.

The colour of smoke may give you some indication of which stage the fire is in and which substances

FIGURE 5-29 Laminar smoke flow.
Courtesy of Dave Dodson.

are burning. When a single substance is burning, the colour of the smoke may provide valuable clues about what that substance is. Of course, most fires involve multiple substances burning, so using smoke colour alone to determine what is burning may not always yield a clear answer.

When first-arriving fire fighters evaluate smoke colour, they can determine the stage of heating and obtain information about the location of the fire in the building. Some solid materials such as wood will emit a white-coloured smoke when they are first heated. This white colour is primarily a result of moisture being released from the material. The same effect is apparent when fire fighters apply water to a fire: The smoke goes from black to white. As material dries out, however, the colour of the smoke emitted changes. Smoke from burning wood, for example, changes to tan or brown. Plastics and painted or stained surfaces emit a gray smoke, which is a combination of black from the hydrocarbons and white from the escaping moisture. When synthetic materials are heated, the white vapour created is uncombusted fuel.

The colour of smoke also can help fire fighters determine the location of the fire. As smoke travels, its heat evaporates moisture from some of the materials it passes through. This added moisture tends to change black smoke to lighter smoke the farther it travels from the fire. In addition, the carbon suspended in the black smoke settles out and is filtered by the materials it passes through. Therefore, carbon-rich black smoke often becomes lighter in colour as it travels farther from the initial fire site, because of the loss of carbon from the smoke.

How do you differentiate between white smoke caused by early heating and white smoke from a hot fire that has travelled a distance? White smoke that is lazy or slow indicates early heating. White smoke that has its own pressure indicates smoke from a hot fire that has travelled for some distance.

Black Fire

Black fire is a high-volume, high-velocity, turbulent, ultra-dense, black smoke. You can think of it as a form of fire, because in many ways it is. Black fire is so hot (temperatures up to 538°C [1000°F]) that it creates a deadly environment. Black fire can produce charring as well as heat damage to steel and concrete. Its presence indicates impending autoignition and flashover. Thus, black fire is often just seconds away from becoming a deadly environment even for fire fighters wearing full PPE. Because black fire is not tenable for fire fighters in full protective equipment, it certainly is not tenable for unprotected occupants of a building. When black fire occurs, there are usually no lives to be saved in that location.

Step 2: Determine What Is Influencing the Key Attributes

Consider the size of the structure that is on fire. How big is the box in which the fire is contained? A small amount of smoke rapidly fills up a small building, so it does not take a big fire to fill a small building with smoke. Conversely, in a large building (such as a "big box" store or a warehouse), it takes a large fire to produce enough smoke to fill the whole building and pressurize the smoke coming out of the building. A fire fighter who pulls up to a large building that has signs of smoke from the outside should realize that only a significant fire would produce enough smoke to vent to the outside. Always consider the size of the box when examining smoke for clues about the fire.

Also consider how wind, thermal balance, fire streams, ventilation openings, and an operating fire sprinkler system might affect the normal characteristics of smoke. Wind can change the direction in which the smoke travels. The normal thermal balance in a room or building can be affected by the heating and air-conditioning systems, especially in a high-rise building. Ventilation openings are made with the intention of changing the normal flow of smoke, so the characteristics of the smoke would naturally be expected to change as the building is ventilated. An operating sprinkler system causes smoke to be cooled; this cool smoke will hang close to the bottom of the room, making it difficult to see.

Step 3: Determine the Rate of Change

To complete the smoke reading, determine the rate of change for the event. Remember that flames indicate what is happening now, whereas smoke gives a more complete picture of the characteristics of the fire and where it is going. Are the volume of smoke, the velocity of smoke, the density of smoke, and the colour of smoke changing? How are they changing? How rapidly are these changes occurring? What do these changes suggest about the progression of the fire?

Step 4: Predict the Event

To assess the size and location of a fire, it is necessary to first consider the four key attributes of smoke—volume, velocity, density, and colour. Then, fire fighters need to consider what might be influencing these key attributes. Next, fire fighters should try to determine the rate of change: Is it getting better, or is it getting worse? Finally, it is time to put these pieces of the puzzle together. The previously mentioned information should help fire fighters to better determine the location of the fire, the size of the fire, and the potential for a hostile fire event such as a backdraft or a flashover. Communicate the key parts of these observations to the company officer. With practice, this assessment process will help you systematically evaluate the information the smoke signals are giving you about the fire.

One way to become more proficient in smoke reading is to review videos of fires. Assess the smoke at the beginning of the tape, and try to identify the location of the fire and the stage of burning. You will be surprised at how much information you can obtain from a short video clip. This information on smoke reading might not have been taught to fire fighters who completed their initial training in the past, but today's smoke is deadly. If you want to stay alive, you need to understand it.

Smoke Reading Through a Door

When you see indications of a hot fire, such as darkened windows, but little visible smoke is coming from around closed doors and from around windows, you may be dealing with a fire that is in a decay stage because it does not have sufficient oxygen (ventilation-limited). This is a sign of great danger: As soon as this fire receives a fresh supply of oxygen, it will likely produce a rapid fire growth event and possibly backdraft. Fires can be dangerous even when little smoke is showing.

When you open a door, watch what the smoke does, and identify the neutral plane. If smoke exits through the top half of the door and clean air enters through the bottom half of the opening, then the fire is probably on the same level. This phenomenon is sometimes called "smoke that has found balance."

When you open a door, if smoke rises and the opening clears out, fresh air is being pulled into the building. This indicates the fire is probably above the level of the opening.

If smoke thins when the door is opened, but smoke still fills the door, the fire may be below the level of the opening.

After-Action REVIEW

IN SUMMARY

- Fire is a rapid chemical process that produces heat and light.
- Matter is made up of atoms and molecules.
- Matter exists in three states: solid, liquid, and gas.
  - A solid has definite capacity for resisting forces and, under ordinary conditions, retains a definite size and shape.
  - A liquid assumes the shape of the container in which it is placed.
  - A gas is a type of liquid that has neither independent shape nor independent volume but rather tends to expand indefinitely.
- Fuels are materials that store energy. Energy exists in many forms, including chemical, mechanical, electrical, light, or nuclear.
  - Chemical energy is the energy created by a chemical reaction.
  - Mechanical energy is converted to heat when two materials rub against each other and create friction.
  - Electrical energy is converted to heat energy when it flows through a conductive material.
  - Light energy is produced by electromagnetic waves packaged in photons and travels as thermal radiation.
  - Nuclear energy is stored in radioactive materials and is converted to electricity by nuclear power generating stations.

- The three basic conditions needed for a fire to occur are fuel, oxygen, and heat. A fourth factor, a chemical chain reaction, is required to maintain a self-sustaining fire.

- A by-product of fire is smoke. Smoke includes three major components: small solids (particles), vapours (aerosols), and gases. Smoke consists mainly of unburned forms of hydrocarbon fuels.

- Fire may be spread by conduction, convection, and radiation.

- Direct contact is a flame touching a fuel.

- Conduction is the transfer of heat through matter, like heat travelling up a metal spoon.

- Convection is the circulatory movement that occurs in a gas or fluid. Convection currents in a fire involve hot gases generated by the fire that rise because they are lighter, creating a higher pressure. Cooler gases are denser and move to the lower areas of the compartment. Convection pushes hot gases from the fire compartment to other areas of the building and is instrumental in spreading the fire beyond the room of origin.

- Radiation is the transfer of heat through the emission of energy in the form of electromagnetic waves. Thermal radiation has an important influence on the performance of the fire fighter's PPE.

- The four principal methods of fire extinguishment are cooling the fuel, excluding oxygen, removing the fuel, and interrupting the chemical reaction.

- Fires are categorized as Class A, Class B, Class C, Class D, and Class K. These classes reflect the type of fuel that is burning and the type of hazard that the fire represents.

  - Class A fires involve ordinary solid combustible materials such as wood, paper, and cloth.
  - Class B fires involve flammable or combustible liquids such as gasoline, kerosene, diesel fuel, and motor oil.
  - Class C fires involve energized electrical equipment.
  - Class D fires involve combustible metals such as sodium, magnesium, and titanium.
  - Class K fires involve combustible cooking oils and fats in kitchens.

- Most fires encountered by fire fighters involve solid fuels. Solid fuels do not actually burn in a solid state. Instead, they must be heated or pyrolyzed to decompose into a vapour before they will burn.

- Three primary factors that influence the combustion of solid fuel fires are the composition of the fuel, the amount of fuel, and the configuration of the fuel. The factors that influence the fuel configuration are the surface-to-mass ratio, the orientation of the fuel, and the continuity of the fuel.

- There are four stages of fire development: the incipient stage, the growth stage, the fully developed stage, and the decay stage.

  - The incipient stage occurs as the fuel starts to burn; the fire is confined to the area of origin and does not significantly affect the atmosphere in the fire compartment.
  - The growth stage occurs as the fire begins to involve fuels beyond the ignition point. During this stage, the fire begins to create a plume of hot gases that begins to act on the fire compartment. The temperature within the fire compartment begins to rise and to pyrolyze fuels close to the fire. When pressure from the burning gases becomes higher, the fire may begin to spread beyond the room of origin.
  - The fully developed stage occurs when all available fuel has ignited and heat is being produced at the maximum rate. Some fires will not reach a fully developed stage because they become ventilation-limited.
  - During the decay stage, the rate of burning slows down because less fuel is available or the oxygen supply is limited.

- A ventilation-limited fire is in a state of decay because there is a limited supply of oxygen available to the fire.

- The growth of room-and-contents fires depends on the characteristics of the room and the contents of the room. Synthetic products are widely used in today's homes. The by-products of heated plastics are not only flammable but also toxic.

- Special conditions within the fire compartment include thermal layering of gases, roll-over, backdraft, flashover, rapid fire growth, and ventilation-limited fires.

- Thermal layering is the property of gases in an enclosed space in which they form layers according to their temperature. The hottest gases travel by convection currents to the top level of the room.

- Roll-over is the ignition of the hot, unburned gases that have accumulated at the top of the fire compartment; this can be a precursor to a flashover.

- Flashover is the near-simultaneous ignition of most of the exposed combustible materials in an enclosed area.

- Backdraft is caused by a change of the ventilation profile, permitting the introduction of oxygen into an enclosure where superheated gases and contents are hot enough for ignition, but the fire does not have sufficient oxygen to cause their combustion.

- Modern structures tend to be more tightly sealed, be constructed of lighter-weight materials, and contain more plastics. These characteristics can lead to a greater risk of backdrafts when a fire occurs in such a structure.

- Liquid-fuel fires require the proper mixture of fuel and air, an ignition source, and contact between the fuel mixture and the ignition.

- The characteristics of flammable vapours can be described in terms of vapour density and flammability limits. Vapour density reflects the weight of a gas compared to that of air. Flammability limits vary widely for different fuels.

- A boiling liquid/expanding vapour explosion (BLEVE) is a catastrophic explosion in a vessel containing both a boiling liquid and a vapour.

- Assessment of smoke volume, velocity, density, and colour assists fire fighters to potentially predict the location of a fire and its stage of development.

- Smoke reading requires fire fighters to evaluate the effect of the building, the weather, and ventilation on the smoke.

KEY TERMS

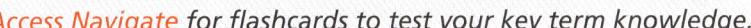

Access Navigate for flashcards to test your key term knowledge.

Atom The smallest particle of an element, which can exist alone or in combination.

Backdraft A deflagration (explosion) resulting from the sudden introduction of air into a confined space containing oxygen-deficient products of incomplete combustion. (NFPA 1403)

Black fire A hot, high-volume, high-velocity, turbulent, ultra-dense black smoke that indicates an impending flashover or autoignition.

Boiling liquid/expanding vapour explosion (BLEVE) An explosion that occurs when pressurized liquefied materials (e.g., propane or butane) inside a closed vessel are exposed to a source of high heat.

Boiling point The temperature at which the vapour pressure of a liquid equals the surrounding atmospheric pressure. (NFPA 1)

Carbon dioxide A colourless, odourless, electrically nonconductive inert gas that is a suitable medium for extinguishing Class B and Class C fires. (NFPA 10)

Carbon monoxide (CO) A toxic gas produced through incomplete combustion.

Chemical energy Energy that is created or released by the combination or decomposition of chemical compounds.

Class A fire A fire in ordinary combustible materials, such as wood, cloth, paper, rubber, and many plastics. (NFPA 1)

Class B fire A fire in flammable liquids, combustible liquids, petroleum greases, tars, oils, oil-based paints, solvents, lacquers, alcohols, and flammable gases. (NFPA 1)

Class C fire A fire that involves energized electrical equipment. (NFPA 1)

Class D fire A fire in combustible metals, such as magnesium, titanium, zirconium, sodium, lithium, and potassium. (NFPA 1)

Class K fire A fire in a cooking appliance that involves combustible cooking media (vegetable or animal oils and fats). (NFPA 1)

Combustion A chemical process of oxidation that occurs at a rate fast enough to produce heat and usually light in the form of either a glow or a flame. (NFPA 1)

Compartment A space completely enclosed by walls and a ceiling. Each wall in the compartment is permitted to have openings to an adjoining space if the openings have a minimum lintel depth of 200 mm (8 in.) from the ceiling and the total width of the openings in each wall does not exceed 2.4 m (8 ft). A single opening of 900 mm (36 in.) or less in width without a lintel is permitted when there are no other openings to adjoining spaces. (NFPA 13)

Conduction Heat transfer to another body or within a body by direct contact. (NFPA 921)

Convection Heat transfer by circulation within a medium such as a gas or a liquid. (NFPA 921)

Decay stage The stage of fire development within a structure characterized by either a decrease in the fuel load or available oxygen to support combustion, resulting in lower temperatures and lower pressure in the fire area. (NFPA 1410)

Electrical energy Heat that is produced by electricity.

Endothermic Reactions that absorb heat or require heat to be added.

Exothermic Reactions that result in the release of energy in the form of heat.

Fire A rapid, persistent chemical reaction that releases both heat and light.

Fire point The lowest temperature at which a liquid will ignite and achieve sustained burning when exposed to a test flame in accordance with ASTM 92, Standard Test Method for Flash and Fire Points by Cleveland Open Cup Tester. (NFPA 1)

Fire tetrahedron A geometric shape used to depict the four components required for a fire to occur: fuel, oxygen, heat, and chemical chain reactions.

Fire triangle A geometric shape used to depict the three components of which a fire is composed: fuel, oxygen, and heat.

Flammable range (explosive limits) The range in concentration between the lower and upper flammable limits. (NFPA 67)

Flashover A transition phase in the development of a compartment fire in which surfaces exposed to thermal radiation reach ignition temperature more or less simultaneously, and fire spreads rapidly throughout the space, resulting in full room involvement or total involvement of the compartment or enclosed space. (NFPA 921)

Flash point The minimum temperature at which a liquid or a solid emits vapour sufficient to form an ignitable mixture with air near the surface of the liquid or the solid. (NFPA 115)

Flow path The movement of heat and smoke from the higher pressure within the fire area toward the lower pressure areas accessible via doors, window openings, and roof structures. (NFPA 1410)

Fuel A material that will maintain combustion under specified environmental conditions. (NFPA 53)

Fuel-limited fire A fire in which the heat release rate and fire growth are controlled by the characteristics of the fuel because there is adequate oxygen available for combustion. (NFPA 1410)

Fully developed stage The stage of fire development where heat release rate has reached its peak within a compartment. (NFPA 1410)

Gas The physical state of a substance that has no shape or volume of its own and will expand to take the shape and volume of the container or enclosure it occupies. (NFPA 921)

Growth stage The stage of fire development where the heat release rate from an incipient fire has increased to the point where heat transferred from the fire and the combustion products are pyrolyzing adjacent fuel sources and the fire begins to spread across the ceiling of the fire compartment (roll-over). (NFPA 1410)

Heat flux The measure of the rate of heat transfer to a surface, typically expressed in kilowatts per metre squared (kW/m^2) or Btu/ft^2. (NFPA 268)

Heat release rates (HRR) The rates at which heat energy is generated by burning. (NFPA 921)

Heat transfer The movement of heat energy from a hotter medium to a cooler medium by conduction, convection, or radiation.

Hydrogen cyanide An extremely toxic gas produced by the combustion of many common plastic-based materials. Low-level exposure can cause cyanosis, headache, dizziness, unsteady gait, and nausea.

Ignition The action of setting something on fire.

Ignition temperature Minimum temperature a substance should attain in order to ignite under specific test conditions. (NFPA 402)

Incipient stage The early stage of fire development where the fire's progression is limited to a fuel source and the thermal hazard is localized to the area of the burning material. (NFPA 1410)

Incomplete combustion A burning process in which the fuel is not completely consumed, usually due to a limited supply of oxygen.

Laminar smoke flow Smooth or streamlined movement of smoke, which indicates that the pressure in the building is not excessively high.

Liquid A fluid (such as water) that has no independent shape but has a definite volume and does not expand indefinitely and that is only slightly compressible.

Lower explosive limit (LEL) The minimum concentration of a combustible vapour or combustible gas in a mixture of the vapour or gas and gaseous oxidant, above which propagation of flame will occur on contact with an ignition source. (NFPA 115)

Mechanical energy A form of potential energy that can generate heat through friction.

Neutral plane The interface at a vent, such as a doorway or a window opening, between the hot gas flowing out of a fire compartment and the cool air flowing into the compartment where the pressure difference between the interior and exterior is equal.

Oxidation Reaction with oxygen either in the form of the element or in the form of one of its compounds. (NFPA 53)

Phosgene A chemical agent that causes severe pulmonary damage; it is a by-product of incomplete combustion.

Plume The column of hot gases, flames, and smoke rising above a fire; also called convection column, thermal updraft, or thermal column. (NFPA 921)

Pyrolysis A process in which material is decomposed, or broken down, into simpler molecular compounds by the effects of heat alone; pyrolysis often precedes combustion. (NFPA 921)

Radiation The combined process of emission, transmission, and absorption of energy travelling by electromagnetic wave propagation (e.g., infrared radiation) between a region of higher temperature and a region of lower temperature. (NFPA 550)

Roll-over The condition in which unburned fuel (pyrolysate) from the originating fire has accumulated in the ceiling layer to a sufficient concentration (i.e., at or above the lower flammable limit) so that it ignites and burns. This can occur without ignition of, or prior to the ignition of, other fuels separate from the origin; also known as flameover. (NFPA 921)

Smoke The airborne solid and liquid particulates and gases evolved when a material undergoes pyrolysis or combustion, together with the quantity of air that is entrained or otherwise mixed into the mass. (NFPA 1404)

Smoke colour The attribute of smoke that reflects the stage of burning of a fire and the material that is burning in the fire.

Smoke density The thickness of smoke. Because it has a high mass per unit volume, smoke is difficult to see through.

Smoke explosion A violent release of confined energy that occurs when a mixture of flammable gases and oxygen is present, usually in a void or other area separate from the fire compartment, and comes in contact with a source of ignition. In this situation, there is no change to the ventilation profile, such as an open door or window; rather, it occurs from the travel of smoke within the structure to an ignition source.

Smoke particles The unburned, partially burned, and completely burned substances found in smoke.

Smoke velocity The speed of smoke leaving a burning building.

Smoke volume The quantity of smoke, which indicates how much fuel is being heated.

Solid One of the three stages of matter; a material that has three dimensions and is firm in substance.

States of matter The physical state of a material—solid, liquid, or gas.

Temperature The degree of sensible heat as measured by a thermometer or similar instrument.

Thermal column A cylindrical area above a fire in which heated air and gases rise and travel upward.

Thermal layering The stratification (heat layers) that occurs in a room as a result of a fire.

Thermal radiation The means by which heat is transferred to other objects.

Turbulent smoke flow Agitated, boiling, angry-movement smoke, which indicates great heat in the burning building. It is a precursor to flashover.

Upper explosive limit (UEL) The maximum amount of gaseous fuel that can be present in the air if the air/fuel mixture is to be flammable or explosive.

Vapour density The weight of an airborne concentration (vapour or gas) compared to an equal volume of dry air.

Ventilation-limited fire A fire in which the heat release rate and fire growth are regulated by the available oxygen within the space. (NFPA 1410)

Volatility The ability of a substance to produce combustible vapours.

REFERENCES

Gann, Richard, & Raymond Friedman. 2015. *Principles of Fire Behaviour and Combustion*, 4th edition. Jones & Bartlett Learning.

National Institute of Standards and Technology (NIST). Videos for Wind-Driven Fires: Governors Island and Laboratory Experiments. 2008. https://www.nist.gov/el/fire-research-division-73300/firegov-fire-service/videos-wind-driven-fires-governors-island. Accessed September 25, 2018.

Ontario Ministry of Community Safety and Correctional Services. Firc Loss in Ontario, 2012-2016. https://www.mcscs.jus.gov.on.ca/english/FireMarshal/MediaRelationsand Resources/FireStatistics/OntarioFires/FireLossesCauses TrendsIssues/stats_causes.html. Accessed October 25, 2018.

Underwriters Laboratories (UL). Firefighter Exposure to Smoke Particulates, 2010. https://ulfirefightersafety.org/research-projects/firefighter-exposure-to-smoke-particulates.html. Accessed September 25, 2018.

Underwriters Laboratories (UL). Impact of Ventilation on Fire Behaviour in Legacy and Contemporary Residential Construction. 2010. https://ulfirefightersafety.org/research-projects/impact-of-ventilation-on-fire-behaviour-in-legacy-and-contemporary-residential-construction.html. Accessed September 25, 2018.

On Scene

You are spending your evening searching the Internet for firefighting videos. Hours pass by as you watch video clip after video clip. Then you run across a video clip that shows two fire fighters entering a two-storey house with fire coming out of a side window. Smoke is pouring out of the upper half of the front door, and you can see that the upper floor is full of smoke.

As the clip continues, the smoke on the upper floor gets darker and darker and begins boiling out of the window. To your surprise, a fire fighter emerges in the turbulent smoke of the window. He is clearly in trouble.

1. What stage is this fire in?

A. Incipient stage

B. Growth stage

C. Fully developed stage

D. Decay stage

2. How is the heat from this fire travelling from one room to another on the second floor?

A. Convection

B. Direct contact

C. Conduction

D. Radiation

3. Which of the following is *not* a form in which energy exists?

A. Chemical

B. Mechanical

C. Digital

D. Electrical

4. A fire that involves energized electrical equipment is considered which of the following?

A. Class K fire

B. Class B fire

C. Class C fire

D. Class A fire

5. The introduction of additional oxygen into or a change in the ventilation profile of an enclosure where the superheated gases and contents are already hot enough for ignition can cause which of the following?

A. Roll-over

B. Thermal layering

C. Backdraft

D. BLEVE

6. Which of the following is the speed of the smoke leaving a burning building?

A. Smoke velocity

B. Smoke density

C. Smoke volume

D. Smoke colour

7. Which of the following is a hot, high-volume, high-velocity, turbulent, ultra-dense black smoke that indicates an impending flashover or autoignition?

A. BLEVE

B. Laminar flow

C. Black fire

D. Pyrolysis

Access Navigate to find answers to this On Scene, along with other resources such as an audiobook.

CHAPTER 6

Fire Fighter I

Building Construction

KNOWLEDGE OBJECTIVES

After studying this chapter, you will be able to:

- Explain how occupancy classifications affect fire suppression operations. (p. 191)
- Explain how the contents of a structure affect fire suppression operations. (p. 191)
- Describe the characteristics of masonry building materials. (pp. 192–193)
- Describe the characteristics of concrete building materials. (p. 193)
- Describe the characteristics of steel building materials. (pp. 193–194)
- Describe the characteristics of glass building materials. (pp. 194–195)
- Describe the characteristics of gypsum building materials. (pp. 195–196)
- Describe the characteristics of wood building materials. (pp. 196–197)
- Describe the characteristics of engineered wood building materials. (p. 197)
- Describe the characteristics of plastic building materials. (pp. 197–198)
- List the five types of building construction. (pp. 198–204)
- Describe the characteristics of Type I construction. (pp. 199–200)
- Describe the effects of fire on Type I construction. (pp. 199–200)

- Describe the characteristics of Type II construction. (pp. 200–201)
- Describe the effects of fire on Type II construction. (pp. 200–201)
- Describe the characteristics of Type III construction. (p. 201)
- Describe the effects of fire on Type III construction. (p. 201)
- Describe the characteristics of Type IV construction. (p. 202)
- Describe the effects of fire on Type IV construction. (p. 202)
- Describe the characteristics of Type V construction. (pp. 202–203)
- Describe the effects of fire on Type V construction. (pp. 202–205)
- Describe the characteristics of balloon-frame construction. (pp. 203–204)
- Describe the effects of fire on balloon-frame construction. (pp. 203–204)
- Describe the characteristics of platform-frame construction. (pp. 204–205)
- Describe the effects of fire on platform-frame construction. (pp. 204–205)
- Describe the challenges associated with fighting a fire in a hybrid building. (pp. 204–205)
- Describe the purpose of a foundation in a structure. (p. 205)

- List the warning signs of foundation collapse. (p. 205)
- Explain how floor construction affects fire suppression operations. (pp. 205–206)
- Describe the characteristics of fire-resistive floors. (p. 206)
- Describe the characteristics of wood-supported floors. (pp. 206–207)
- Describe the characteristics of ceiling assemblies. (p. 207)
- List the three primary components of roof assemblies. (p. 207)
- List the three primary types of roofs. (pp. 207–210)
- Describe the characteristics of trusses. (p. 210)
- List the types of trusses. (pp. 210–211)
- Describe the effects of fires on trusses. (pp. 211–212)
- Describe the characteristics of walls. (p. 212)
- List the common types of walls in structures. (pp. 212–213)
- Describe the characteristics of door assemblies. (p. 214)
- Describe the characteristics of window assemblies. (p. 214)
- Describe the characteristics of fire doors. (pp. 214–215)
- Describe the characteristics of fire windows. (pp. 214–215)
- Explain the effect that interior finishes have on fire suppression operations. (p. 215)

- Explain the effect that exterior finishes and siding have on fire suppression operations. (p. 216)
- Describe the hazards that buildings under construction or demolition pose to fire fighters. (pp. 216–217)
- Describe the factors that increase the chance of building collapse. (pp. 217–219)
- Describe how building construction factors into preincident planning and incident size-up. (pp. 219–220)

SKILLS OBJECTIVES

There are no skills objectives for Fire Fighter I candidates. NFPA 1001 contains no Fire Fighter I Job Performance Requirements for this chapter.

ADDITIONAL STANDARDS

- **NFPA 80**, *Standard for Fire Doors and Other Opening Protectives*
- **NFPA 220**, *Standard on Types of Building Construction*

Fire Alarm

Your crew is driving around your district looking at various buildings. The Driver/Operator drives through a historic residential district and points to a two-storey house with a wraparound porch, tall narrow windows, and a stone foundation. He then cuts over to the old main street and points to a commercial building. It is three storeys tall and made of brick. Every seventh layer of brick is laid crosswise. It has a flat roof and is about 9 m (30 ft) wide and 27 m (90 ft) deep. It has large plate glass windows on the first floor, and there are normal size windows on the second and third floors.

Next, he drives to a residential part of town built in the 1940s. Each house is one storey, is about 12 m (40 ft) square, and has a roof with asphalt shingles. Each home has wooden clapboard siding and a small porch on the front.

Finally, he drives to a newer commercial area where there is a stand-alone convenience store. It was built in 1998 and has a flat roof and metal siding. He turns to you and says, "We just looked at four different buildings. Tell me about each one."

1. Why do fire fighters need to understand building construction?

2. What effect does building construction have on fires?

3. Which building would most likely collapse first if exposed to fire?

 *Access Navigate for more practice activities.*

Introduction

Knowing the basic types of building construction is vital for fire fighters because building construction affects how fires grow and spread. Fire fighters must be able to recognize and understand different types of building construction. They must understand how each type of building construction reacts when exposed to the effects of heat and fire. This knowledge helps them anticipate the fire's behaviour and allows them to respond accordingly. An understanding of building construction also helps fire fighters determine when it is safe to enter a burning building and when it is necessary to evacuate a building. Your safety, the safety of your team members, and the safety of the building's occupants all depend on your knowledge of building construction.

Construction, however, is merely one component of a complex relationship. Fire fighters also must consider the occupancy of the building and the building contents. Fire risk factors vary depending on how a building is used. To establish a course of action at a fire, it is necessary to consider the interactions between three factors: the building construction, the occupancy of the building, and the contents of the building. The following sections relate to occupancy and contents, and the rest of this chapter focuses on those aspects of building construction that are important to a fire fighter.

Occupancy

The term **occupancy** refers to how a building is used. Based on a building's occupancy classification, a fire fighter can predict who is likely to be inside the building, how many people are likely to be there, and what they are likely to be doing. Occupancy also suggests the types of hazards and situations that might be encountered in the building.

Occupancy classifications are used in conjunction with building and safety codes to establish regulatory requirements, including the types of construction that can be used to construct buildings of a particular size, use, or location. Regulations classify building occupancies into major categories—such as residential, health care, business, and industrial—based on common characteristics.

Occupancy classifications can be used to predict the number of occupants who are likely to be at risk in a fire. For example, hospitals and nursing homes are occupied 24 hours a day by persons who will probably need assistance to evacuate. An office building probably has a large population during the day, but most of the workers should be able to evacuate the site without assistance. A daycare centre might be filled with young children during the day but is likely to be empty at night. A nightclub is typically empty during the day and crowded at night. Fire fighters must consider these factors when responding to a particular building.

Building codes require that certain types of building construction be used for specific types of occupancies. In most communities, however, buildings that were originally built for one reason often have been repurposed to another use. Buildings constructed before modern building codes were adopted may present significant hazards for fire fighters. In addition, a building may be used for a variety of occupancies. For example, a single commercial building may contain many different types of businesses with vastly different contents and hazards. A mixed-use building may include several levels of underground parking, multiple commercial occupancies on the first floor, and residential apartment units on the upper floors. This type of building presents a wide variety of challenges for responding fire fighters.

Contents

Contents also must be considered when responding to a building. Building contents vary widely but are usually closely related to the occupancy of a building. A computer repair shop has different characteristics from an auto transmission shop, for example. Some buildings have noncombustible contents that would not feed a fire, whereas others could be so dangerous that fire fighters could not safely attack any fire at the site. For example, a factory that manufactures cast-iron pipe fittings or airplanes may contain many toxic and flammable substances. Whenever possible, fire fighters should prepare a prefire or preincident plan that accounts for these possibilities when buildings present special hazards.

Even similar occupancies can pose different levels of risk, however. For example, three warehouses, although identical from the exterior, could have different contents that would either increase or reduce the risks to fire fighters. One might contain ceramic floor tiles, the second might be filled with wooden furniture, and the third might serve as a storehouse for swimming pool chemicals. The tiles will not burn, the furniture will burn easily, and the chemicals will create toxic products of combustion and contaminated runoff.

When evaluating the risks at any response, consider the occupancy, the contents, and the building construction. Together, they represent an interwoven package of risks to fire fighters.

Types of Construction Materials

An understanding of building construction begins with the materials used. Building components are usually made of different materials. The properties of these materials and the details of their construction determine the basic fire characteristics of the building.

Function, appearance, price, and compliance with building and fire codes are all considerations when selecting building materials and determining construction methods. Architects often place a priority on functionality and aesthetics when selecting materials, whereas builders are often more concerned about price and ease of construction. For their part, building owners are usually interested in durability and maintenance expenses, as well as the materials' initial cost and aesthetics.

Fire fighters, however, use a completely different set of factors to evaluate construction methods and materials. Their chief concern is the behaviour of the building under fire conditions. In particular, a building material or construction method that is attractive to architects, builders, and building owners might have the potential to create serious problems or deadly hazards for fire fighters.

The most commonly used building materials are masonry, concrete, steel, aluminum, glass, gypsum board, wood, engineered wood products, and plastics. Within these basic categories, hundreds of variations are possible. The key factors that affect the behaviour of each of these materials under fire conditions are outlined here:

- **Combustibility**: Whether or not a material will burn determines its combustibility. Materials such as wood will burn when they are ignited, releasing heat, light, and smoke, until they are completely consumed by the fire. Concrete, brick, and steel are noncombustible materials that cannot be ignited and are not consumed by a fire.
- **Thermal conductivity**: This characteristic describes how readily a material will conduct heat. Heat flows very readily through metals such as steel and aluminum. By contrast, brick, concrete, and gypsum board are poor conductors of heat.
- Decrease in strength at elevated temperatures: Many materials lose strength at elevated temperatures. For example, steel loses strength and will bend or buckle when exposed to the high temperatures associated with fires, and aluminum melts in a fire. Some engineered building products are constructed of wood products held together by glue or plastics; the glue in these products begins to fail long before they are hot enough to burn. By contrast, bricks and concrete can generally withstand high temperatures for extended periods of time.
- Thermal expansion when heated: Some materials—steel, in particular—expand significantly when they are heated. A steel beam exposed to a fire will stretch (elongate); if it is restrained so that it cannot elongate, it will sag, warp, or twist. As a general rule, a steel beam will elongate at a rate of 25.4 mm (1 in.) per 3 m (10 ft) of length at a temperature

of 538°C (1000°F). This expansion can push out the walls of buildings and cause collapse. When water or cooling is applied to a heated steel beam, the beam contracts and can cause collapse. It is important to realize that from the moment of its construction, a building is under constant stress from gravity, wind, and natural elements.

Masonry

Masonry includes stone, concrete blocks, and brick. The individual components are usually bonded together into a solid mass with mortar, which is produced by mixing sand, lime, water, and Portland cement. Brick or concrete blocks embedded with steel reinforcing rods (known as concrete–masonry units [CMU]) may be preassembled into a wall section. A complete concrete masonry wall can be delivered to a construction site, ready to be erected without further preparation.

Masonry materials are inherently fire resistive (**FIGURE 6-1**); that is, they do not burn or deteriorate at the temperatures normally encountered in building fires. Masonry is also a poor conductor of heat, so it will limit the passage of heat from one side through to the other side. For these reasons, masonry is often used to construct fire walls to protect vulnerable materials.

A

B

FIGURE 6-1 Masonry materials are inherently fire resistive. **A.** Solid masonry wall. **B.** Brick veneer wall.
A: © AbleStock; B: © Jones & Bartlett Learning. Photographed by Glen E. Ellman.

A **fire wall** helps prevent the spread of a fire from one side of the wall to the other (**FIGURE 6-2**). Although masonry materials do not conduct heat well, they can act as a heat sink or reservoir, causing excessive heat to be held in the building after the fire is extinguished.

Not all masonry walls are necessarily fire walls, however. A single layer of masonry may be placed over a wood-framed building to create a robust weather-resistant aesthetic. Unprotected openings in a masonry wall can provide a conduit for fire spread. If the mortar breaks down or the wall has been exposed to fire for a prolonged time, a masonry wall can collapse during a fire.

Masonry structures can collapse under fire conditions if the roof or floor assembly fails. In such a case, the masonry fails because of the mechanical action of the collapse. Likewise, aged, weakened mortar can contribute to a collapse. Regardless of the cause, a collapsing masonry wall can be a deadly hazard to fire fighters.

Concrete

Like masonry, concrete is a naturally fire-resistive material. Because it does not burn or conduct heat well, it is often used to insulate other building materials from fire. Concrete does not have a high degree of thermal expansion—that is, it does not expand greatly when exposed to heat and fire—nor will it lose strength when exposed to high temperatures.

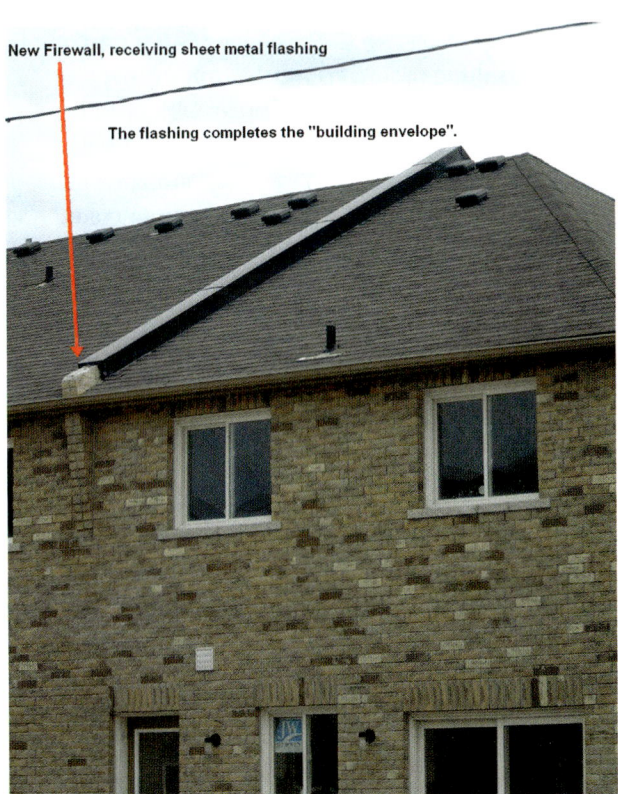

FIGURE 6-2 Fire walls.
Courtesy of Achim Hering.

Concrete is made from a mixture of Portland cement and aggregates such as sand and gravel. Different formulations of concrete can be produced for specific building purposes. Concrete is inexpensive and easy to shape and form. It is often used for foundations, columns, floors, walls, roofs, and exterior pavement.

Under compression, concrete is strong and can support a great deal of weight. Under tension, however, it is weak. When this material is used in building construction, steel reinforcing rods are often embedded into the concrete to strengthen it under tension. In turn, the concrete insulates the steel reinforcing rods from heat.

Although it is inherently fire resistive, concrete can be damaged by exposure to a fire. For example, a fire may convert trapped moisture in the concrete to steam. As the steam expands, it creates internal pressure that can cause sections of the concrete surface to break off in a process called **spalling**. Severe spalling in reinforced concrete can expose the steel reinforcing rods to the heat of the fire. If the fire is hot enough, the steel might weaken, resulting in a structural collapse, although this is a rare event. In many Canadian cities, salt is used on roads as a de-icing agent in winter months. Vehicles carry salt into and onto concrete constructed parking structures resulting in premature degradation of the reinforced steel in the concrete. Fire fighters should be extra vigilant of collapse when fighting fires in older parking structures.

SAFETY TIP

Heat transfer can occur through exposed steel reinforcing rods. This mode of heat transfer is known as conduction.

Steel

Steel is the strongest building material in widespread use, in terms of both tension and compression. It is an alloy of iron and carbon. Other metals may be added to the mix to produce steel with special properties, such as stainless steel or galvanized steel. Steel can be produced in a wide variety of shapes and sizes, ranging from heavy beams and columns to thin sheets. This material is often used in the structural framework of a building to support floor and roof assemblies. Additionally, many residential occupancies use steel studs instead of wood studs. Steel is resistant to aging and does not rot, although most types of steel will rust unless they are protected from exposure to air and moisture.

When considered by itself, steel is not fire resistive. It will melt at extremely high temperatures, although these extraordinary temperatures are rarely encountered at structure fires (**FIGURE 6-3**). Steel conducts heat well, so it tends to expand and lose strength as the temperature

FIGURE 6-3 Steel will lose strength at temperatures of more than 538°C (1000°F).

© Jones & Bartlett Learning. Photographed by Glen E. Ellman.

increases. For this reason, other materials—such as masonry, concrete, or layers of gypsum board—are often used to protect steel from the heat of a fire. Sprayed-on coatings of mineral- or cement-like materials are also used to insulate steel members. The amount of heat absorbed by steel depends on the mass of the object and the amount of protection surrounding it. Smaller, lighter pieces of steel absorb heat more easily than larger and heavier pieces.

Failure of a steel structure depends on three factors: the mass of the steel components, the loads placed upon them, and the methods used to connect the steel pieces. Heated steel beams often sag and twist, whereas steel columns tend to buckle as they lose strength. The uneven heating that occurs in actual fire situations causes the bending and distortion. Any sign of bending, sagging, or stretching of steel structural members should be considered a warning of an immediate risk of failure. If this sign is noticed, all fire fighters should immediately clear the area because of the potential for imminent collapse.

Other Metals

A variety of other metals, including copper, zinc, iron, and aluminum, is also used in building construction. Copper is used primarily for electrical wiring and piping, and it is sometimes used for decorative roofs, gutters, and downspouts. Zinc is used primarily as a coating to protect metal parts from rust and corrosion. Two types of iron are used in building construction: wrought iron and cast iron. Wrought iron is used for water pipes, rivets, and other decorative work, and cast iron is used primarily for columns and support beams.

Aluminum is occasionally used as a structural material in building construction. It is more expensive and not as strong as steel, so its use is generally limited to light-duty applications such as awnings, siding, window frames, door frames, roof panels, and sunshades. Aluminum expands more extensively than steel when

heated and loses strength quickly when exposed to a fire. This material has a lower melting point than steel, so it often melts and drips in a fire.

Glass

Almost all buildings contain glass—in windows, doors, skylights, and sometimes walls. Ordinary glass breaks easily, but glass can be manufactured to resist breakage and to withstand impacts or high temperatures.

Although ordinary glass usually breaks when exposed to fire, specially formulated fire-rated glass can be used as a fire barrier in certain situations. The thermal conductivity of glass is rarely a significant factor in the spread of fire.

Many different types of glass are available for use in building construction:

- Ordinary window glass usually breaks with a loud pop when heat exposure to one side causes it to expand and creates internal stresses that fracture the glass. The broken glass forms large shards, which usually have sharp edges.
- **Tempered glass** is much stronger than ordinary glass and more difficult to break (**FIGURE 6-4**). Some tempered glass can be broken with a spring-loaded centre punch; it will shatter into small pieces that do not have the sharp edges of ordinary glass. Tempered or laminated glass usually is required in installations such as storm doors to prevent injuries from shards of broken glass.
- **Laminated glass** is made by placing a thin sheet of plastic between two sheets of glass (**FIGURE 6-5**). The resulting product is much stronger than ordinary glass, is difficult to break with ordinary hand tools, and will usually deform instead of breaking. When exposed to a fire, laminated glass windows are likely to crack but remain in place. Laminated glass is sometimes used in buildings to help soundproof areas.

FIGURE 6-4 Tempered glass will shatter into small pieces that do not have the sharp edges of ordinary glass.

© Sylvie Bouchard/Shutterstock.

FIGURE 6-5 Laminated glass is much stronger than ordinary glass and will usually deform instead of breaking.
Courtesy of David Sweet.

FIGURE 6-6 Glass blocks are thick pieces of glass that are designed to be built into a wall. They are similar to bricks or tiles.
© Jones & Bartlett Learning. Photographed by Glen E. Ellman.

- **Glass blocks** are thick pieces of glass similar to bricks or tiles (**FIGURE 6-6**). They are designed to be built into a wall with mortar so that light can be transmitted through the wall. Glass blocks have limited strength and are not intended to be used as part of a load-bearing wall, but they can usually withstand a fire. Some glass blocks are approved for use with fire-rated masonry walls.
- **Wired glass** is made by molding tempered glass with a reinforcing wire mesh (**FIGURE 6-7**). When wired glass is subjected to heat, the wire holds the glass together and prevents it from breaking. This material often is used in fire doors and in windows designed to prevent fire spread.

FIGURE 6-7 Wired glass.
© Jones & Bartlett Learning.

Gypsum Board

Gypsum is a naturally occurring mineral composed of calcium sulfate and water molecules and used to make plaster of Paris. Gypsum is a good insulator and noncombustible; it will not burn even in atmospheres of pure oxygen.

Gypsum board (also called drywall, sheetrock, or plasterboard) is commonly used to cover the interior walls and ceilings of residential living areas and commercial spaces (**FIGURE 6-8**). It is manufactured in

FIGURE 6-8 Gypsum board consists of a layer of gypsum between two layers of specially produced paper.
© Jones & Bartlett Learning. Photographed by Glen E. Ellman.

large sheets consisting of a layer of compacted gypsum sandwiched between two layers of specially produced paper. Although gypsum board is a good finishing material, it is not a strong structural material by itself. In a building, gypsum board is nailed or screwed in place on a framework of wood or metal studs, and the edges are secured with a special tape. The nail or screw heads and tape are then covered with a thin layer of plaster.

Gypsum board has limited combustibility, because the paper covering will burn slowly when exposed to a fire. It does not conduct or release heat to an extent that would contribute to fire spread, so this material is often used to create a firestop or to protect building components from fire. Notably, gypsum blocks are sometimes used as protective insulation around steel members or to create fire-resistive enclosures or interior fire walls. Gypsum board is required in some applications, such as residential garages, to create a rated fire-resistant barrier. When mounted on wood framing, gypsum board will protect the wood from fire for a limited time.

When gypsum board is heated, some of the water present in the calcium sulfate will evaporate, causing the board to deteriorate. If it is exposed to a fire for a long time, this material will fail. Even if the gypsum board retains its integrity after a fire, the sections that were directly exposed to the fire will probably be unable to serve as effective fire barriers in the future and should be replaced after the incident. Water also weakens and permanently damages gypsum board.

Gypsum board can be reinforced with other materials when additional strength is needed. The resulting product is commonly referred to as impact-resistant drywall. Impact-resistant drywall can be difficult to breach in a fire situation. Methods commonly used for reinforcement include heavy paper backing, fibre reinforcement, and fibreglass matting. Gypsum may be mixed with Portland cement to produce a cement board that can be used for exterior sheathing or as an underlay for ceramic floor or wall tile systems. Another type of impact-resistant wallboard uses a thin sheet of Lexan backing laminated to a paper-faced gypsum board. When exposed to fire, this plastic will smoulder and melt and may burn as a combustible liquid.

Wood

Wood is one of the most commonly used building materials in today's environment. It is inexpensive to produce, is easy to use, and can be shaped into many different forms, ranging from heavy structural supports to thin strips of exterior siding. Both soft woods (such as pine) and hard woods (such as oak) are used for building construction.

FIRE FIGHTER TIP

Older buildings were built with lumber that was true to its stated dimensions. A two-by-four stud actually measured 51 mm by 102 mm (2 in. by 4 in.). In buildings built today, a two-by-four stud measures 38 mm by 86 mm ($1\frac{1}{2}$ in. by $3\frac{3}{8}$ in.) or less. In a fire, the older, true two-by-four stud resists fire longer than the newer stud. The type of lumber used in the past to construct buildings is often referred to as full dimensional lumber.

Solid lumber is squared and cut into uniform lengths. Examples of solid lumber range from the heavy timbers used in mills and barns, to smaller two-by-four studs, to wood siding, interior wood trim, wood floors, and wood shingles on roofs. Solid wood can be used in a wide variety of applications in a building. This type of lumber may be found in multiple sizes in either older, or **legacy construction**, or newer, or **contemporary construction**.

For fire fighters, the most important characteristic of wood is its high combustibility. Wood acts as fuel in a structural fire and can provide a path for the fire to spread. It ignites at fairly low temperatures and gradually becomes consumed by the fire. A burning wood structure becomes weaker every minute and eventually collapses, making it imperative that fire fighters understand building construction and maintain situational awareness when combating fires. The wood burning process creates great quantities of heat and hot gases, until eventually all that remains after the fuel is consumed is a small quantity of residual ash.

The rate at which wood ignites, burns, and decomposes depends on several factors:

- Ignition: A small ignition source contains less energy, and it takes a longer time to ignite a fire. Using an accelerant greatly speeds this process.
- Moisture: Damp or moist wood takes longer to ignite and burn. New lumber usually contains more moisture than lumber that has been in a building for many years. A higher relative humidity in the atmosphere also makes wood more difficult to ignite.
- Density: Heavy, dense wood is harder to ignite than lighter or less dense wood.
- Preheating: The more the wood is preheated, the faster it ignites.
- Size and form: The rate of combustion is directly related to the surface area of the wood. During the combustion process, high temperatures release flammable gases from the surface of the wood. A large solid beam is difficult to ignite and burns relatively slowly; by contrast, a lightweight truss ignites more easily and is consumed rapidly.

Exposure to the high temperatures generated by a fire can decrease the strength of wood through the process of pyrolysis. This chemical change occurs when wood or other materials are heated to a temperature high enough to release some of the volatile compounds in the wood, albeit without igniting these gases. The wood then begins to decompose without combustion.

Engineered Wood Products

Engineered wood products include plywood, fibreboard, **oriented strand board (OSB)**, and particle board—materials that are used in roofs, walls, floors, countertops, doors, I-joists, trusses, and cabinets. These products, which are also referred to as manufactured board, manmade wood, or composite wood, are manufactured from small pieces of wood that are held together with glue or adhesive. The adhesives include urea-formaldehyde resins, phenol-formaldehyde resins, melamine-formaldehyde resins, and polyurethane resins.

Engineered wood products are preferred over natural wood for a variety of reasons. In particular, they are usually less expensive than large, solid planks of natural wood. Such products can be fabricated in longer lengths and for specific applications that are not possible with natural wood. They can maximize the natural characteristics of wood, thereby producing greater strength and stiffness. These products are easy to work with using ordinary tools, and they can be readily shaped into curved surfaces. With all of these advantages, it is not surprising that the use of engineered wood products has greatly increased in recent years and is expected to increase another 19–25 percent by 2019 (APA, 2015).

Unfortunately, some drawbacks arise with engineered wood products. One disadvantage is that some of these products are prone to warping in conditions of high humidity. A second disadvantage is that these products may contain toxic products, such as formaldehyde, which are introduced during manufacturing. The biggest disadvantage—at least from fire fighters' perspective—is that these products contribute to faster fire spread, shorter time to flashover and collapse, and rapid changes in fire behaviour.

SAFETY TIP

The glue that is used to create **laminated wood** is flammable.

Fire-Retardant–Treated Wood

Wood cannot be treated to make it completely noncombustible. Nevertheless, impregnating wood with mineral salts makes it more difficult to ignite and slows the rate of burning. Fire-retardant treatment can significantly reduce the fire hazards of wood construction. On the downside, the treatment process can reduce the strength of the wood.

In some cases, fire-retardant–treated wood can pose a danger for fire fighters. Some fire-retardant chemicals used to treat plywood roofing panels, for example, cause the wood to rapidly deteriorate and weaken. The plywood may fail, necessitating the replacement of these panels. Fire fighters should know whether fire-retardant–treated plywood is used in the community. If it is, they should avoid standing on roofs made with these panels during a fire, because the extra weight could cause the plywood to collapse. Fire fghters should be engaged in systemic and consistent prefire planning. Prefire planning provides fire fighters with vital information concerning building construction and collapse potential.

SAFETY TIP

Avoid standing on roofs that have been constructed with fire-retardant–treated wood or with trusses. Both types of roofs can collapse in a short period of time.

Plastics

Plastics are synthetic materials that are used in a wide variety of products (**FIGURE 6-9**). These materials may be transparent or opaque, stiff or flexible, and tough or brittle. Plastics are rarely used for structural support; however, they may be found throughout a building. When used to produce building products, they are often combined with other materials.

Building exteriors may include vinyl siding, plastic window frames, plastic panel skylights, and clear plastic panes. Two plastics that are commonly used in place

FIGURE 6-9 Plastic building materials can be used for siding, window frames, and interior finishes.
© AbleStock.

of glass are polycarbonates and acrylics. These plastics may be used for added strength and resistance against breakage. Foam plastic materials may be used as exterior or interior insulation. Plastic pipe and fittings, plastic tub and shower enclosures, and plastic lighting fixtures are commonly used. Even carpeting and floor coverings often contain plastics.

The combustibility of plastics varies greatly. Some plastics ignite easily and burn quickly, whereas others burn only when an external source of heat is present. Some plastics can withstand high temperatures and fire exposure without igniting.

Many plastics produce quantities of heavy, dense, dark smoke and release high concentrations of toxic gases as they burn. This type of smoke resembles smoke from a petroleum fire—which should not be surprising given that most plastics are made from petroleum products.

Thermoplastic materials melt and drip when exposed to high temperatures, even those as low as 260°C (500°F). Although heat can be used to shape these materials into a variety of desired forms, the dripping, burning plastic can rapidly spread a fire. By contrast, **thermoset materials** are fused by heat and will not melt as the plastic burns, although their strength will decrease dramatically.

Types of Construction

Canadian building construction is regulated by the federal government in all federally regulated industries, provinces/territories, and local municipal governments. The National Building Code of Canada (NBC) and National Fire Code of Canada (NFC) are developed by the National Research Council Canada (NRC) with input from provincial and municipal building officials. The NBC and NFC are model codes that, everywhere except Prince Edward Island, are either adopted by provincial/territorial governments or used as the basis for provincial/territorial codes and standards. Provinces/territories that adopt the national codes have regulatory mechanisms to vary the requirements of the code to meet the needs of their jurisdictions. National and provincial/territorial codes are minimum construction standards. Variations on the code by municipal governments generally only increase the minimum code requirements; they are not allowed to reduce requirements.

Since 2005, the Canadian code system has been an objective-based system, which allows for multiple building solutions to deliver on its four objectives. The objectives under which the codes have been developed are safety, health, accessibility for persons with disabilities,

and fire and structural protection of buildings. Prior to that time, the codes were prescriptive in nature. Building codes specify the type of construction to be used based on the height, area, occupancy classification, and location of the building. Other factors, such as the presence of automatic sprinklers, will also affect the required construction classification. National Fire Protection Association (NFPA) 220, *Standard on Types of Building Construction*, provides detailed requirements for each type of building construction. Building construction is dictated by the occupancy classification of the building. The NBC has nine sections: Two sections dictate the fire safety requirements of buildings. Buildings that are required to be designed by professional architects and engineers, such as assembly occupancies and large buildings, generally fall under sections 1–8. Section 9 of the code relates specifically to housing, small buildings of three storeys or less that do not exceed 600 m², and major occupancies (Groups C, D, E, and F [Divisions 2 and 3]). The two main sections of the code that deal with fire-safe construction are sections 3 and 9. Individual provinces and territories may have their own building code and that should be your reference document.

Fire fighters need to understand and recognize five types of building construction—designated as Types I through V. These classifications are directly related to fire protection and fire behaviour in the building.

Buildings are classified based on the combustibility of the structure and the fire resistance of its components. Buildings using Type I or Type II construction are assembled with primarily noncombustible materials and limited amounts of wood and other materials that will burn. This does not mean, however, that Type I and Type II buildings cannot be damaged by a fire. In fact, if the contents of the building burn, the fire can seriously damage or destroy the structure.

In buildings using Type III, Type IV, or Type V construction, both the structural components and the building contents will burn. Wood and wood products are used to varying degrees in these buildings. If the wood ignites, the fire will weaken and consume both the structure and its contents. Structural elements of these buildings can be damaged or destroyed in a very short time.

As mentioned earlier, a building is classified based on the fire resistance of its components. **Fire resistance** refers to the length of time that a building or building component can withstand a fire before igniting. Fire-resistance ratings are stated in hours, based on the time that a test assembly withstands or contains a standard test fire. For example, walls are rated based on whether they stop the progress of the test fire for periods ranging from 20 minutes to 4 hours. A floor assembly is rated based on whether it supports a load for 1 hour or

2 hours. Because ratings are based on a standard test fire, an actual fire could be more or less severe than the test fire. Fire ratings are considered merely guidelines; that is, an assembly with a 2-hour rating will not necessarily withstand an actual fire for 2 hours.

Building codes specify the type of construction to be used, based on the height, area, occupancy classification, and location of the building. Other factors, such as the presence of automatic sprinklers, also affect the required construction classification. NFPA 220, *Standard on Types of Building Construction*, provides detailed requirements for each type of building construction.

Type I Construction: Fire Resistive

Type I construction (fire-resistive) is the most fire-resistive category of building construction. It is used for buildings designed for large numbers of people, buildings with a high life-safety hazard, tall buildings, large-area buildings, and buildings containing special hazards. Type I construction is commonly found in schools, hospitals, and high-rise buildings (**FIGURE 6-10**).

Buildings with Type I construction can withstand and contain a fire for a specified period of time. The fire resistance and combustibility of all construction materials are carefully evaluated, and each building component must be engineered to contribute to fire resistance of the entire building to warrant this classification.

All of the structural members and components used in Type I construction must be made of noncombustible materials, such as steel, concrete, or gypsum board. In addition, the structure must be constructed or protected so that it has at least 2 hours of fire resistance. Building codes specify the fire-resistance requirements for different components. For example, columns and load-bearing walls in multistorey buildings could have

a fire-resistance requirement of 3 or 4 hours, whereas a floor might be required to have a fire-resistance rating of only 2 hours. Some codes allow Type I buildings to use a limited amount of combustible materials as part of the interior finish.

If a Type I building exceeds specific height and area limitations, codes generally require the use of fire-resistive walls and/or floors to subdivide it into compartments. A compartment might consist of a single floor in a high-rise building or a part of a floor in a large-area building. In any event, a fire in one compartment should not spread to any other parts of the building. To ensure that fire is contained, stairways, elevator shafts, and utility shafts should be enclosed in construction that prevents fire from spreading from floor to floor or from compartment to compartment.

Type I buildings typically use reinforced concrete and protected steel-frame construction. As mentioned earlier, concrete is noncombustible and provides thermal protection around the steel reinforcing rods. Reinforced concrete can fail, however, if it is subjected to a fire for a long period of time or if the building contents create an extreme fire load. Given its tendency to lose strength in the face of high temperatures, structural steel framing must be protected from the heat of a fire. In Type I construction, the structural steel members are generally encased in concrete, shielded by a fire-resistive ceiling, covered with multiple layers of gypsum board, or protected by a sprayed-on insulating material (**FIGURE 6-11**). An unprotected steel beam exposed to a fire can fail and cause the entire building to collapse.

By themselves, Type I building materials should not provide enough fuel to create a serious fire. Thus it is the contents of the building that determine the severity of a Type I building fire. In theory, a fire could consume all of the combustible contents of a Type I building yet leave the structure basically intact.

FIGURE 6-10 Type I construction is commonly found in schools, hospitals, and high-rise buildings.
© John Foxx/ Stockbyte/Getty Images.

FIGURE 6-11 Sprayed-on fireproofing materials often are used to protect structural steel.
© Michael Doolittle/Alamy Stock Photo.

Although Type I construction is designed to give fire fighters time to conduct an interior attack, it can be very difficult to extinguish a fire in a Type I building. A fire that is fully contained within fire-resistive construction can be very hot and difficult to ventilate because the burning contents may produce copious quantities of heat and smoke. For this reason, fire-resistive construction in modern high-rise buildings is typically combined with an automatic fire detection system and automatic sprinklers.

Sometimes, the burning contents of a building provide sufficient fuel and generate enough heat to overwhelm the fire-resistant properties of the construction. In these cases, the fire can escape from its compartment and damage or destroy the structure. Similarly, inadequate or poorly maintained construction can undermine the fire-resistive properties of the building materials. Under extreme fire conditions, a Type I building can collapse. Even so, these buildings have a much lower potential for structural failure than do other buildings.

Type II Construction: Noncombustible

Type II construction (noncombustible) is also referred to as noncombustible construction (**FIGURE 6-12**). All of the structural components in a Type II building must be made of noncombustible materials. The fire-resistive requirements, however, are less stringent for Type II construction than for Type I construction. In some cases, no fire-resistance requirements apply to the building. In other cases, known as protected noncombustible construction, a fire-resistance rating of 1 or 2 hours may be required for certain elements.

Noncombustible construction is most common in single-storey warehouse or factory buildings, where vertical fire spread is not an issue. Unprotected construction of this type is generally limited to a maximum of two to four storeys in non-residential construction. Some multistorey buildings are constructed using protected noncombustible construction. In Canada, a good rule of thumb for identifying Type II structures is visible exposed steel structural elements, such as roof trusses and columns. Type II buildings include, but are not limited to, big box stores, stand-alone fast food restaurants, grocery stores, warehouses, industrial buildings and retail outlets.

Steel is the most common structural material in Type II buildings. Insulating materials can be applied to the steel when fire resistance is required. A typical example of Type II construction is a large-area, single-storey building with a steel frame, metal or concrete block walls, and a metal roof deck. Fire walls are sometimes used to subdivide these large-area buildings and prevent catastrophic losses. Undivided floor areas must be protected with automatic sprinklers to limit the fire risk.

Fire severity in a Type II building is determined by the contents of the building, because the structural components contribute little or no fuel, and interior finish materials are limited. If the building contents provide a high fuel load, a fire can collapse and destroy the structure. In this kind of construction, automatic sprinklers should be used to protect combustible and

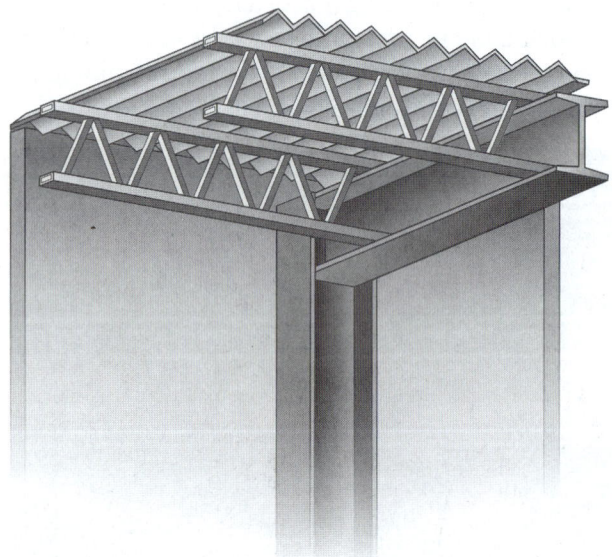

FIGURE 6-12 Type II construction is also referred to as noncombustible construction.
© Jones & Bartlett Learning.

LISTEN UP!

It is increasingly common to find commercial and residential structures, including single-family residences, constructed with lightweight steel (Havel, 2008). Lightweight steel may be present in the structural components (e.g., studs, trusses, joists) of Type I, Type II, and even some Type III buildings. In these buildings, the steel components are fastened together with screws and then covered with sheathing material. Interior and exterior walls can then be finished in any manner, such as with siding or masonry finishes. Once the finishes are in place, there is no easy way to tell that a building is constructed of lightweight steel. Fire fighters should be aware of buildings constructed with lightweight steel in their response area because this type of construction presents additional dangers that affect firefighting operations. For example, lightweight steel components are noncombustible and fire resistant when properly sheathed or covered, but, when the sheathing is breached, exposure to heat can cause rapid failure. Additionally, steel framing members may have sharp edges and projecting screws and fasteners that can damage firefighting gear and equipment and injure fire fighters. Pay attention to buildings under construction in your response area so that you are prepared to respond appropriately.

valuable contents. It is sound practice for pump operators to supply the sprinkler system as soon as possible after establishment of a water supply. Fires in buildings equipped with monitored fire detection systems are more likely to be reported earlier than fires in buildings without monitored fire detection systems.

Type III Construction: Ordinary

Type III construction (ordinary) is also referred to as ordinary construction because it is used in a wide variety of buildings, ranging from commercial strip malls to small apartment buildings (**FIGURE 6-13**). Ordinary construction is usually limited to buildings of no more than four storeys, but it can sometimes be found in buildings as tall as six or seven storeys.

Type III buildings have masonry load-bearing exterior walls, which support the floors and the roof structure. This type of construction was far more prevalent in most Canadian towns and cities during the 19th century and the first half of the 20th century. The interior structural and nonstructural members—including the walls, floor, and roof—are primarily constructed of wood. In most ordinary construction, gypsum board or plaster is used as an interior finish material, covering the wood framework and providing minimal fire protection.

Fire-resistance requirements for interior construction of Type III buildings are limited. The gypsum or plaster coverings over the interior wood components provide some fire resistance. Key interior structural components may be required to have fire-resistance ratings of 1 or 2 hours, or there may be no requirements at all. Some Type III buildings use interior masonry load-bearing walls to meet the requirements for fire-resistant structural support.

A building of Type III construction has two separate fire loads: the contents and the combustible building materials used. Even a vacant building contains a sufficient quantity of wood and other combustible components to produce a large fire. Given the combustibility of these materials, a fire involving both the contents and the structural components can quickly destroy the building.

Fortunately, most fires originate with the building contents and are extinguished before the flames ignite the Type III structure. Once a fire extends into the structure and begins to consume the fuel within the walls and above the ceilings, it becomes much more difficult to control and prone to roof or floor collapse.

Type III construction presents several problems for fire fighters. For example, an electrical fire can begin inside the void spaces within the walls, floors, and roof assemblies and extend to the contents. The void spaces allow a fire to extend vertically and horizontally, spreading from room to room and from floor to floor. In such a case, fire fighters have to open the void spaces to fight the fire. An uncontrolled fire within the void spaces is likely to destroy the building. Fire fighters need to be especially aware that fires in void spaces under floors or under the roof can quickly destroy the strength of these components. Do not be surprised by these "hidden" fires; instead, anticipate them.

The fire resistance of interior structural components often depends on the age of the building and on local building codes. Older Type III buildings were built from solid dimensional lumber, which can contain or withstand a fire for a limited time. Newer or renovated buildings might contain lightweight assemblies that can be damaged much more quickly and are prone to early failure. Floors and roofs in these buildings collapse suddenly and without warning. Because fire fighters cannot always determine the type of structure during a fire, you should assume that lightweight components may be present in any Type III building.

Exterior walls can collapse if a fire causes significant damage to the interior structure. Because the exterior walls, the floors, and the roof are connected in a stable building, the collapse of the interior structure could make the free-standing masonry walls unstable and prone to collapse.

FIGURE 6-13 Type III construction is referred to as ordinary construction because it is used in a wide variety of buildings.
© Ken Hammon/USDA.

Type IV Construction: Heavy Timber

Type IV construction (heavy timber) is also known as heavy timber construction (**FIGURE 6-14**). A heavy timber building has exterior load-bearing walls that consist of masonry construction and interior walls, columns, beams, floor assemblies, and roof structure that are made of wood. The exterior walls are usually brick and are extra thick to support the weight of the building and its contents.

The wood used in Type IV construction is much heavier than that used in Type III construction. It is more difficult to ignite and will withstand a fire for a much longer time before the building collapses. A typical heavy timber building might have 203-mm-square wood posts (8-in.) supporting 356-mm-deep wood beams (14-in.) and 203-mm floor joists (8-in.). The floors could be constructed of solid wood planks, 51 or 76 mm (2 or 3 in.) thick, with a top layer of wood serving as the finished floor surface.

Heavy timber construction, if it meets building codes, has no concealed spaces or voids. This structure helps reduce the horizontal and vertical fire spread that often occurs in ordinary construction buildings. Unfortunately, many heavy timber buildings do contain vertical openings for elevators, stairs, or machinery, which can provide a path for a fire to travel from one floor to another.

In the past, heavy timber construction was used to construct buildings as tall as six to eight storeys with open spaces suitable for manufacturing and storage occupancies. Today, modern buildings of this style usually are built using either fire-resistive (Type I) or noncombustible (Type II) construction. New buildings of Type IV construction are rare, except for special structures that feature the construction components as architectural elements.

The heavy, solid wood columns, support beams, floor assemblies, and roof assemblies used in heavy timber construction will withstand a fire much longer than the smaller wood members used in ordinary and lightweight combustible construction. Nevertheless, once involved in a fire, the structure of a heavy timber building can burn for many hours. A fire that ignites the combustible portions of heavy timber construction is likely to burn until it runs out of fuel and the building is reduced to a pile of rubble. As the fire consumes the heavy timber support members, the masonry walls will become unstable and collapse.

Mill construction was common during the 1800s in the United States and Canada, especially in the U.S. northeast and central Canada (Ontario and Quebec). In particular, large mill buildings were used as factories and warehouses. In many communities, dozens of these large buildings were clustered together in industrial areas, creating the potential for huge fires involving multiple multistorey buildings. The radiant heat from a major fire in one of these buildings could spread the fire to nearby buildings, resulting in the loss of several surrounding structures.

Mill construction was state of the art for its time. Automatic sprinkler systems were developed to protect buildings of mill construction; as long as the sprinklers are properly maintained, these buildings have a good safety record. Without sprinklers, however, a heavy timber mill building becomes a major fire hazard.

Only a few of these original mill buildings are still used for manufacturing. Many more have been converted into small shops, galleries, office buildings, and residential occupancies. These conversions tend to divide the open spaces into smaller compartments and create void spaces within the structure. The revamped buildings may contain either noncombustible contents or highly flammable contents, depending on their use. Occupancies may range from unoccupied storage facilities to occupied loft apartments and office space. Appropriate fire protection and life-safety features, such as modern sprinkler systems, must be built into the conversions to ensure the safety of occupants.

FIGURE 6-14 Type IV construction was once used for buildings suitable for manufacturing and storage occupancies. Today, new buildings of Type IV construction are rare.
© Helen Filatova/Shutterstock.

Type V Construction: Wood Frame

In **Type V construction (wood frame)**, all of the major components are constructed of wood or other combustible materials (**FIGURE 6-15**). Type V construction is often called wood-frame construction. Wood-frame construction is used not only in one- and two-family dwellings and small commercial buildings but also in larger structures such as apartment and condominium complexes and office buildings up to four storeys in height.

Many wood-frame buildings do not have any fire-resistive components. Some codes require a 1-hour fire-resistance rating for limited parts of wood-frame buildings, particularly those of more than two storeys. Plaster or gypsum board barriers often are used to achieve

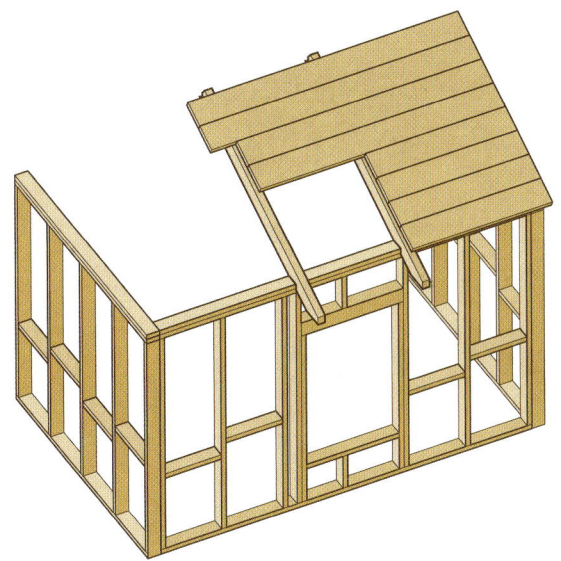

FIGURE 6-15 Type V (wood-frame) construction is the most common type of construction used today.
© Jones & Bartlett Learning.

this rating, but fire fighters should not assume that these barriers are present or effective in all Type V buildings.

Because all of the structural components are combustible, wood-frame buildings can rapidly become totally involved in a fire. In addition, Type V construction usually creates voids and channels that allow a fire within the structure to spread quickly. A fire that originates in a Type V building can easily extend to other nearby buildings.

Wood-frame buildings often collapse and suffer major destruction from fires. Smoke detectors are essential to warn building residents early if a fire occurs. Although compartmentalization can help limit the spread of a fire, fire detection devices and automatic sprinklers are the most effective way of protecting lives and property in Type V buildings.

Wood-frame buildings are constructed in a variety of ways. Older wood-frame construction was assembled from solid lumber, which relied on its mass to provide strength. To reduce costs and create the largest building with the least amount of material, lightweight construction makes extensive use of engineered wood products such as wooden I-beams and wooden trusses. With these construction techniques, structural assemblies are engineered to be just strong enough to carry the required load. As a result, there is little built-in safety margin, and these buildings can collapse early, suddenly, and completely during a fire.

The presence of flammable building contents may greatly increase the severity of a fire and accelerate the speed with which the fire destroys a wood-frame building. Indeed, the structural elements of such a building can be damaged or destroyed in a very short time. Whenever a fire has been burning under a floor or roof, you should assume that the structure is unsafe and may collapse. Fires that start to consume structural parts of a Type V

building present a high risk of building collapse. From a fire fighter's point of view, lightweight wood-frame construction is a campfire waiting to happen.

> **SAFETY TIP**
>
> Lightweight wood-frame construction is a campfire waiting to happen. A fire in this type of construction can spread rapidly throughout the building and to adjacent exposures.

Wood-frame buildings might be covered with wood siding, vinyl siding, aluminum siding, brick veneer, or stucco. However, the covering on a building does not reflect the type of construction of that building. Just because you see a brick covering does not mean that the building is constructed of solid brick walls. During a structural fire, the brick veneer is likely to collapse or peel away as the wall behind it burns or collapses. A single thickness of bricks is rarely strong enough to stand independently. Fire fighters should be aware of buildings constructed in this manner and maintain a safe distance from potentially falling bricks. For this reason, it is imperative that a collapse zone be set up around a burning building.

Two systems are used to assemble wood-frame buildings: balloon-frame construction and platform-frame construction. Because these systems developed in different eras, fire fighters often can anticipate the type of construction based on the age of a building.

> **SAFETY TIP**
>
> Many buildings have a brick veneer on the exterior walls of wood-frame construction. A thin layer of brick or stone might be applied to enhance the appearance of the building or to reduce the risk of fire spread. Unfortunately, this practice can give the impression that a building is Type III (ordinary) construction when it is really Type V (wood-frame) construction.

Balloon-Frame Construction

Balloon-frame construction was popular between the late 1800s and the mid-1900s. In a balloon-frame building, the exterior walls are assembled with wood studs that run continuously from the basement to the roof (**FIGURE 6-16**). In a two-storey building, the floor joists that support the first and second floors are nailed to these continuous studs. As a result, an open channel between each pair of studs extends from the foundation to the attic. Each of these channels provides a path that enables a fire to spread from the basement to the attic without being visible on the first- or second-floor levels. Fire fighters must anticipate that a fire originating on a lower level will quickly extend through these voids, and

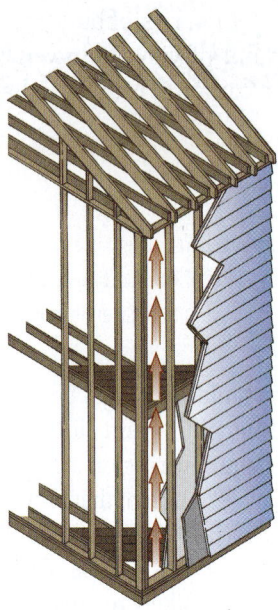

FIGURE 6-16 In a balloon-frame building, the exterior walls create channels that enable a fire to spread from the basement to the attic.
© Jones & Bartlett Learning.

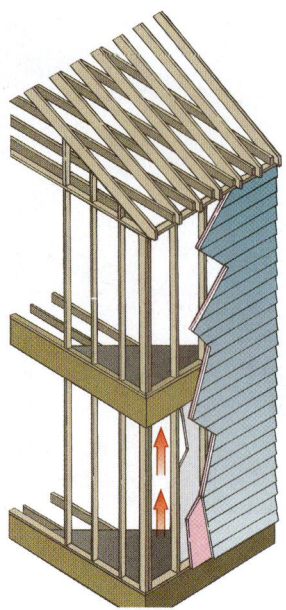

FIGURE 6-17 In a platform-frame building, the floor platform blocks the path of fire, preventing it from spreading from one floor to another.
© Jones & Bartlett Learning.

they should open the void spaces to check for hidden fires and to prevent rapid vertical extension.

Platform-Frame Construction

Platform-frame construction is used for almost all modern wood-frame construction. In a building with platform construction, the exterior wall studs are not continuous. Instead, the first floor is constructed as a platform, and the studs for the exterior walls are erected on top of it. The first set of studs extends only to the underside of the second-floor platform, which blocks any vertical void spaces. The studs for the second-floor exterior walls are erected on top of the second-floor platform. At each level, the floor platform blocks the path of any fire rising within the void spaces (**FIGURE 6-17**). Platform framing prevents fire from spreading from one floor to another through continuous stud spaces. Although a fire can eventually burn through the wood platform, the platform will slow the fire spread.

> ### LISTEN UP!
>
> Buildings may be designed specifically for the local climate and uses of a community. Additionally, building materials may vary from one region to another depending on cost and what is readily available. Learning about the unique types of buildings that exist in a community will help you predict the fire behaviour of those buildings.

Hybrid Building Construction

Learning about the different types of construction is important in understanding the stability of each type of construction and how each type of building will react under fire conditions. It is equally important to understand that a single building may contain more than one type of construction. **Hybrid buildings** are those that do not fit entirely into any of the five construction types because they incorporate building materials of more than one type, such as wooden beams and steel columns.

A hybrid building may result from initial construction or from renovations made to the structure. Some buildings are initially constructed with one part of the structure adhering to one type of construction and other parts of the structure adhering to a different type of construction. For example, a building may be constructed with a parking garage on its first level and residential apartments or condo units on its upper levels. In this case, local building codes may require that the first level be constructed with fire-resistive construction (Type I) but permit the upper levels to be constructed with contemporary lightweight wood framing (Type V), provided that the building is protected with a fire sprinkler system. Until the fire detection and sprinkler systems are installed and fully operational, this type of building is at high risk of being completely destroyed by a fire in a short period of time. Because there is such a huge fire load with no fire stops, a fire can rapidly engulf and destroy the building in spite of aggressive firefighting efforts.

Extensive renovations to a structure also may result in a hybrid building. For example, a structure initially constructed of legacy construction (Type III) may be renovated to include an addition constructed using contemporary lightweight construction (Type V). Renovated buildings can be dangerous because they can give

FIGURE 6-18 This figure shows a residence that was initially constructed with Type III ordinary legacy construction in the 1920s but was renovated with contemporary Type V lightweight construction during 2015-2016. The current house (as shown) consists of Type III ordinary construction on the left half of the house and Type V lightweight wood-frame construction on the right half of the house. Fire spread and collapse will occur much faster in the right half (Type V) than in the left half (Type III).
Courtesy of Jennifer Schottke.

the impression that they are one type of construction when they are really a mixture of construction types. The original part of a renovated building may consist of a stronger or more fire-resistive construction, while the newer part may consist of lightweight construction that will quickly fail under fire conditions (**FIGURE 6-18**). The result is a building in which fire spread and collapse will occur much faster in one section.

Building Components

The construction classification system gives fire fighters a general idea of the materials used in building and an indication of how the materials will react to fire. In addition, each building has several different components that should be recognized. A fire fighter who understands how these components function will have a better understanding of the risks involved with a burning building. Some building construction features are safer for fire fighters, as this section explains.

The following major components of a building are discussed in this section:

- Foundations
- Floors and ceilings
- Roofs
- Trusses
- Walls
- Doors and windows
- Interior finishes and floor coverings
- Exterior finishes and siding

Foundations

The primary purpose of a building foundation is to transfer the weight of the building and its contents to the footings and then to the ground (**FIGURE 6-19**). The weight of the building itself is called the **dead load**; the weight of the building's contents is called the **live load**. The foundation ensures that the base of the building is planted firmly in a fixed location, which helps keep all other components connected.

Modern foundations are usually constructed of concrete or masonry, although wood pilings may be used in some areas. Foundations can be either shallow or deep, depending on the type of building and the soil composition. Some buildings are built on concrete footings or piers; others are supported using steel piles or wooden posts driven into the ground.

As long as the foundation remains stable and intact, it usually is not a major concern for fire fighters. Most foundation problems are caused by circumstances other than a fire, such as improper construction, shifting soil conditions, or earthquakes. Insulated concrete forms are becoming more prevalent in Canada as a means to reduce heat loss and reduce the incidents of cracked walls. Although fires can damage timber post and other wood foundations, most foundations remain intact even after a severe fire has damaged the rest of the building. If the foundation shifts or is in poor repair, however, it can become a critical concern. A burning building with a weak foundation or inadequate lateral bracing could be in imminent danger of collapse.

When examining a building, take a close look at the foundation. Look for cracks that indicate movement of the foundation. If the building has been modified or remodelled, look for areas where the support could be compromised.

Floors and Ceilings

Floor construction is important to fire fighters for three major reasons. First, fire fighters who are working inside a building must rely on the integrity of the floor to support their weight. A floor that fails could drop

FIGURE 6-19 The foundation supports the entire weight of the building.
© Dorn1530/Shutterstock.

the fire fighters into a fire on a lower level. Second, in a multistorey structure, fire fighters may be working below a floor (or roof) that would fall on them if it collapsed. Third, the floor system influences whether a fire spreads vertically from floor to floor within a building or whether it remains contained on a single level. Some floor systems are designed to resist fires, whereas others have no fire-resistant capabilities.

Fire-Resistive Floors

In multistorey buildings, floors and ceilings are generally considered as a combined structural system—that is, the structure that supports a floor also supports the ceiling of the storey below. In a fire-resistive building, this system is designed to prevent a fire from spreading vertically and to prevent a collapse when a fire occurs in the space below the floor–ceiling assembly. Fire-resistive floor–ceiling systems are rated in terms of hours of fire resistance based on a standard test fire.

Concrete floors are common in fire-resistive construction. The concrete can be cast in place or assembled from panels or beams of precast concrete, which are made at a factory and then transported to the construction site for placement. The thickness of the concrete floor depends on the load that the floor needs to support.

Concrete floors can be either self-supporting or supported by a system of steel beams or trusses. In the latter case, the steel can be protected by sprayed-on insulating materials or covered with concrete or gypsum board. If the ceiling is part of the fire-resistive rating, it provides a thermal barrier to protect the steel members from a fire in the area below the ceiling.

The ceiling below a fire-resistant floor can be constructed from plaster or gypsum board, or it can be a system of tiles suspended from the floor structure. In many cases, a void space between the ceiling and the floor above it contains building systems and equipment such as electrical or telephone wiring, heating and air-conditioning ducts, and plumbing and fire sprinkler system pipes (**FIGURE 6-20**). If the space above the ceiling is not subdivided by fire-resistant partitions or protected by automatic sprinklers, a fire can quickly extend horizontally across a large area.

Wood-Supported Floors

Wood floor structures are common in non–fire-resistive construction. Wooden floor systems range from heavy timber construction, which is often found in old mill buildings, to modern, lightweight truss or engineered wooden I-beam construction.

Although heavy timber construction provides a huge fuel load for a fire, it also can contain and withstand a fire for a considerable length of time without

FIGURE 6-20 The space between the ceiling and the underside of the floor above it often holds electrical and communications wiring and other building systems.
© Lourens Smak/Alamy Stock Photo.

collapsing. Heavy timber construction uses posts and beams that are at least 203 mm (8 in.) on the smallest side and often as large as 356 mm (14 in.) in depth. The floor decking is often assembled from solid wood boards, 51 to 76 mm (2 or 3 in.) thick, which are covered by an additional 25 mm (1 in.) of finished wood flooring. The depth of the wood in this system often will contain a fire for an hour or more before the floor fails or burns through.

Conventional wood flooring, which was widely used for many decades, is much lighter than heavy timber but uses solid lumber as beams, floor joists, decking, and finished flooring. It burns readily when exposed to a fire but generally takes approximately 20 minutes to burn through or reach structural failure. This time estimate is only a general, unscientific guideline; the actual burning rate depends on many other factors.

Modern lightweight construction uses structural elements that are much less substantial than full dimensional lumber. For example, lightweight wooden trusses or engineered wooden I-beams are often used as supporting structures. Thin sheets of plywood or OSB boards are used as decking, and the top layer often consists of a thin layer of concrete or wood covered by carpet. This floor construction provides little resistance to fire. The lightweight structural elements can fail, or the fire can burn through the decking quickly. One study demonstrated that while it took 19 minutes for a traditional unprotected residential floor assembly to fail, an unprotected lightweight engineered I-joist failed in only 6 minutes (NIOSH, 2009).

It is impossible to tell how a floor is constructed by looking at it from above. The important information about a floor can be observed only from below, if it is visible at all. The building's age and local construction methods can provide significant clues to a floor's stability, but you should be aware that many older buildings have been renovated using modern lightweight systems and materials. Fire fighters should use preincident surveys and other planning activities to gather essential structural information about buildings. In some reconstructions of office or computer rooms, the floor has been replaced with removable floor panels to hide the different types of cables under the floor.

Ceiling Assemblies

From a structural standpoint, the ceiling is considered to be part of the floor assembly. Ceilings are mainly included in buildings for appearance reasons. Their primary function is to hold light fixtures and to diffuse light. Ceilings also conceal heating, ventilation, and air-conditioning (HVAC) distribution systems. In addition, they conceal wiring and fire sprinkler systems.

Ceilings can be part of the fire-resistive package. For example, they can be covered with plaster or gypsum board, or they can consist of dropped ceilings covered with mineral tiles. Some ceilings are fire rated as part of the floor assembly.

Fire fighters should be aware that hollow, void spaces between the floor and the ceiling can contribute to the horizontal spread of fire. Penetrations in the floor can contribute to the vertical spread of fire.

Roofs

The primary purpose of a roof is to protect the inside of a building from the weather. In some cases, the roof is also vital to the stability of the building. Roof assemblies consist of three major components: the supporting structure, the roof decking, and the roof covering (**FIGURE 6-21**).

The most common types of roof supports are solid beams of wood, steel, or concrete or a system of trusses. Solid-beam construction uses solid components, such as girders, beams, and **rafters** (solid wood joist mounted in an inclined position), to support the roof. A **truss** is a collection of smaller, lightweight structural components joined to form a triangular configuration or a system of triangles. Trusses used to support roofs may be made of wood, steel, or combinations of wood and steel.

The roof decking is the portion of roof between the roof supports and the roof covering. It is composed of a rigid layer of wooden boards, plywood sheets, or metal panels that is anchored to the supporting structure and then treated or covered to prevent water penetration. The roof covering is applied on top of the deck; it is the part of the roof exposed to the weather.

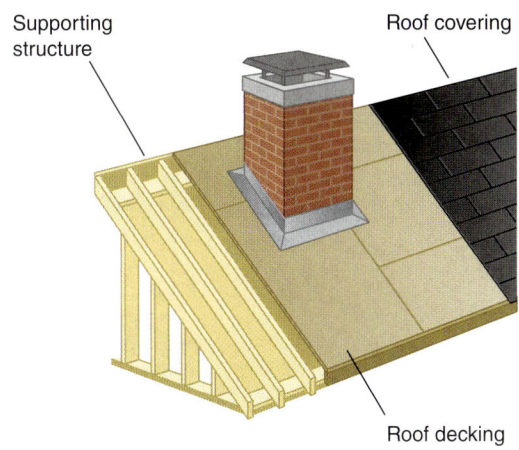

FIGURE 6-21 Roof assemblies consist of a supporting structure, decking, and covering.
© Jones & Bartlett Learning.

Several materials and methods are used for roof construction. Roof materials may be combustible, noncombustible, or both. Roofs are constructed in three primary designs: pitched, curved, and flat. A building may include a combination of these designs. Each type of roof is different and presents different challenges.

Pitched Roofs

A **pitched roof** has sloping or inclined surfaces. Pitched roofs are used on many houses and some commercial buildings. The pitch or angle of the roof can vary depending on local architectural styles and climate. Variations of pitched roofs include gable, hip, mansard, gambrel, and lean-to roofs (**FIGURE 6-22**).

The slope of a pitched roof can present either a minor inconvenience or a complete lack of secure footing, depending on its angle. Fire fighters working on a pitched roof are always in danger of falling, but the risk is particularly high when the roof is wet or icy, when the roof covering is not secure, or when smoke or darkness obscures vision. Roof ladders and aerial apparatus are used to provide a secure working platform for fire fighters working on a pitched roof.

Pitched roofs are usually supported by either rafters or trusses. Pitched roofs supported by rafters usually have solid wood boards as the roof decking.

Modern lightweight construction uses manufactured wood trusses to construct most pitched roofs. Many lightweight trusses are manufactured using metal **gusset plates** that penetrate the wood no more than 10 mm (⅜ in.) and tie the chords and members of the truss together. The decking usually consists of thin plywood or a sheeting material such as wood particleboard. When trusses and decking are exposed to heat and fire, they often fail after a short period of time with no prior warning. Fire fighters cannot work safely on this type of roof when the supporting structures become involved in a fire.

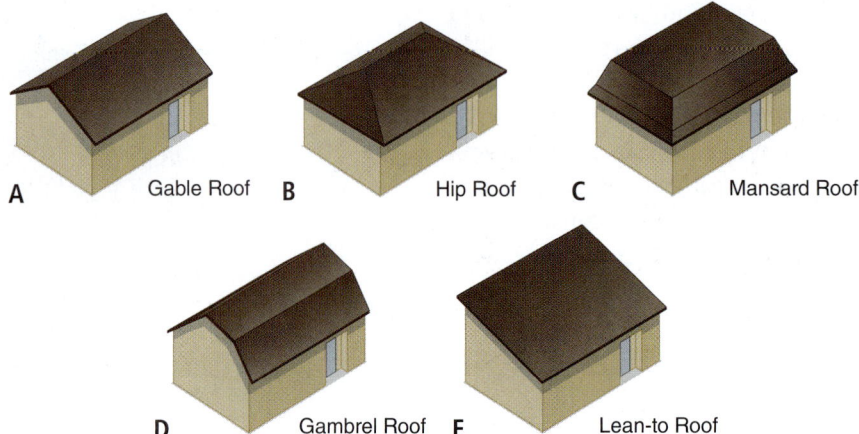

FIGURE 6-22 Examples of pitched roofs: **A.** A gable roof. **B.** A hip roof. **C.** A mansard roof. **D.** A gambrel roof. **E.** A lean-to roof.
© Jones & Bartlett Learning.

Steel trusses also are used to support pitched roofs. The fire resistance of steel trusses is directly related to how well the steel is protected from the heat of the fire.

Several roof-covering materials are used on pitched roofs, usually in the form of shingles or tiles (**FIGURE 6-23**). Shingles usually are made from felt or mineral fibres impregnated with asphalt, although metal and fibreglass shingles also are used. Wood shingles, often made from cedar, are popular in some areas. In older construction, wood shingles often were mounted on individual wood slats instead of continuous decking; this configuration of shingles burns rapidly in dry weather conditions.

Shingles are generally durable, economical, and easily repaired. A shingle roof that has aged and deteriorated should be removed completely and replaced. Some older buildings may have newer layers of shingles on top of older layers, which can make it difficult to cut an opening for ventilation.

Roofing tiles are usually made from clay or concrete products. Clay tiles, which have been used for roofing since ancient times, can be either flat or rounded. Rounded clay tiles are sometimes called mission tiles. Clay tiles are both durable and fire resistant.

Slate tiles are produced from thin sheets of rock. Slate is an expensive, long-lasting roofing material, but it is brittle and becomes slippery when wet. Slate tiles also pose a threat to fire fighters below when a pitched roof is being ventilated, because falling tiles can become airborne missiles.

Metal panels of galvanized steel, copper, and aluminum also may be used on pitched roofs. Expensive metals such as copper often are used for their decorative appearance, whereas lower-cost galvanized steel is used on barns and industrial buildings. Corrugated, galvanized metal panels are strong enough to be mounted on a roof without roof decking. Metal roof coverings will not burn; however, they can conduct heat to the roof decking.

Curved Roofs

Curved roofs are often used for supermarkets, warehouses, industrial buildings, arenas, auditoriums, bowling alleys, churches, airplane hangars, and other buildings that require large, open interiors. Steel or wood bowstring trusses or arches usually are used to support these kinds of roofs.

The decking on curved roof buildings can range from solid wooden plywood and OSB boards to corrugated steel sheets. The type of material in use must be identified before ventilation openings can be made. Often, the roof covering consists of layers that include felt, mineral fibres, and asphalt, although some curved roofs are covered with foam plastics or plastic panels.

Flat Roofs

Flat roofs are found on houses, apartment buildings, shopping centres, warehouses, factories, schools, and hospitals (**FIGURE 6-24**). Most flat roofs have a slight slope so that water can drain off the structure. If the roof does not have the proper slope or if the drains are not maintained, water may pool on the roof, overloading the structure and causing a collapse.

The support systems for most flat roofs are constructed of either wood or steel. A wood support structure uses solid **wooden beams** and joists; laminated wood beams may be included to provide extra strength or to span long distances. Lightweight construction uses wood trusses or wooden I-beams as the supporting members. Such lightweight assemblies are much less fire resistant and collapse more quickly than do solid-wood systems.

Steel also may be used to support flat roofs. Open-web steel trusses (sometimes called bar joists) can span long

A: Paul Springett/Up the Resolution/Alamy Images; B: © Thinkstock/Stockbyte/Getty Images; C: © bildagentur-online.com/th-foto/Alamy Stock Photo.

FIGURE 6-23 Several roof-covering materials are used on pitched roofs. **A.** A tile roof. **B.** A wood shingle roof. **C.** A slate roof.

FIGURE 6-24 Most flat roofs have a slight slope so that water can drain off.
Courtesy of Captain David Jackson, Saginaw Township Fire Department.

SAFETY TIP

It is inadvisable to cut ventilation holes in lightweight roof structures. Studies show that this action unnecessarily places fire fighters at risk, as lightweight components will burn more rapidly and degrade the roof upon which fire fighters are standing. Cutting a vent hole above the fire creates a flow path that entrains more, not less, oxygen into an attic space, promoting rapid fire growth and exponential fire spread, in most circumstances. Determine the type of roof construction before beginning roof ventilation operations. If the roof is constructed with lightweight trusses, ventilate the roof from an aerial device.

distances and remain stable during a fire if they are not subjected to excessive heat.

Flat roof decks can be constructed of wood planking, plywood, corrugated steel, gypsum, or concrete. The material used depends on the building's age and size, the climate, the cost of materials, and the type of roof covering.

Most flat roofs are covered with multiple-layer, built-up roofing systems that help prevent leakage. A typical built-up roof covering contains five layers: a vapour barrier, thermal insulation, a waterproof membrane, a drainage layer, and a wear course. The outside of the flat roof reveals only the top layer, which is often gravel; it serves as the wear course and protects the underlying layers from wear and tear. Cutting through all of the layers to open a ventilation hole can be quite a challenge.

Most flat roof coverings contain highly combustible materials, including asphalt, roofing felt, tarpaper, rubber or plastic membranes, and plastic insulation layers. These materials can be difficult to ignite. Once lit, however, they burn readily and release great quantities of heat and black smoke.

Flat roofs can present unique problems during vertical ventilation operations. A roof with a history of leaks could have several patches where additional layers of roofing material make for an extra-thick covering. Sometimes a whole new roof is constructed on top of the old one. After opening a hole in the new roof, fire fighters might discover the old roof and have to cut into that as well. Fire fighters also might discover fire burning in the void space between the two roofs.

A similar problem may be encountered in remodelled buildings or buildings with additions. An old flat roof may be found under a new, pitched section, resulting in two separate void spaces. The old roof may provide a large supply of fuel for the fire. Such a situation is difficult to predict unless the incident commander has the building plan in advance.

SAFETY TIP

Flat roofs with parallel chord trusses and corrugated metal decks have been known to roll down into ventilation holes during roof operations. Some fire departments no longer perform vertical ventilation in buildings with this type of roof.

Trusses

As previously discussed, a truss is a structural component that is composed of smaller pieces joined to form a triangular configuration or a system of triangles. Trusses are used extensively in support systems for both floors and roofs. They are common in modern construction for several reasons. The triangular geometry creates a strong, rigid structure that can support a load much greater than its own mass. For example, both a solid beam and a simple truss with the same overall dimensions can support the same load. The truss has several advantages over the solid beam, however. It requires much less material than the beam, is much lighter, and can span a long distance without supports. Trusses often are prefabricated and transported to the building location for installation.

Trusses are widely used in new construction and often replace heavier solid beams and joists in renovated or modified older buildings. They are frequently used in residential construction, apartment buildings, small office buildings, commercial buildings, warehouses, fast-food restaurants, airplane hangars, churches, and even firehouses. They may be clearly visible, or they may be concealed within the construction.

The strength of a truss depends on both its members and the connections between them. A properly assembled and installed truss is strong. If any member or connector begins to fail, however, the strength of the entire truss will be compromised.

Trusses can be used for many different purposes. In building construction, those made of wood, steel, or combinations of wood and steel are used primarily to support roofs and floors. Trusses are everywhere, so it is important that you learn how to recognize them. In some buildings, the trusses are exposed and readily visible from the inside of the building. In other buildings, the floor or roof trusses are not visible because they are enclosed in attic or floor spaces (**FIGURE 6-25**).

FIGURE 6-25 This building may have hidden trusses.
© imac/Alamy Stock Photo.

Three of the most commonly used types of trusses are the parallel chord truss, the pitched chord truss, and the bowstring truss.

A **parallel chord truss** has two parallel horizontal members connected by a system of diagonal and sometimes vertical members (**FIGURE 6-26**). The top and bottom members are called the chords, and the connecting pieces are called the web members. Parallel chord trusses are typically used to support flat roofs or floors. In lightweight construction, an engineered **wood truss** is often assembled with wood chords and either wood or light steel web members (**FIGURE 6-27**). A steel bar joist is another example of a parallel chord truss.

A **pitched chord truss** is typically used to support a sloping roof (**FIGURE 6-28**). Most modern residential

FIGURE 6-26 A parallel chord truss.
© Jones & Bartlett Learning.

FIGURE 6-27 A parallel chord truss assembled with wood chords and wood web members.
© Jones & Bartlett Learning. Photographed by Glen E. Ellman.

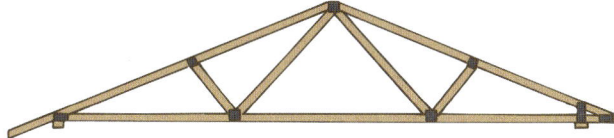

FIGURE 6-28 A pitched chord truss.

© Jones & Bartlett Learning.

construction uses a series of prefabricated wood pitched chord trusses to support the roof. The roof deck is supported by the top chords, and the ceilings of the occupied rooms are attached to the bottom chords. In this way, the trusses define the shape of the attic.

A **bowstring truss** has the same shape as an archery bow, where the top chord represents the curved bow and the bottom chord represents the straight bowstring (**FIGURE 6-29**). The top chord resists compressive forces, and the bottom chord resists tensile forces. Bowstring trusses are usually quite large and widely spaced in the structure. They were once popular in warehouses, supermarkets, and similar buildings with large, open floor areas but have since faded from popularity (Jakubowski, 2013). The roof of a building with bowstring trusses has a distinctive curved shape (**FIGURE 6-30**).

Under normal circumstances, a properly engineered and installed truss will do a satisfactory job of supporting the roof or floor assembly for which it was designed. However, like many other building components, trusses display different characteristics when exposed to heat and fire. Because these components are engineered to work with the least amount of wood or steel needed to support a given weight, they provide little margin of safety for fire fighters operating during a fire.

FIGURE 6-29 A bowstring truss.

© Jones & Bartlett Learning.

FIGURE 6-30 A building with a bowstring truss.

Courtesy of Captain David Jackson, Saginaw Township Fire Department.

SAFETY TIP

Trusses can fail without warning early in a fire. Fire fighters should stay off of roofs any time fire is known or suspected to be in the truss loft.

Trusses consist of a small cross-section of wood with a high surface-to-mass ratio. As a consequence, a fire burning for a very short period of time can destroy enough wood to weaken a truss. All it takes is a failure at one point in a truss to produce a failure in the entire truss. Wooden trusses constructed with metal gusset plates are especially prone to failure. Because the gusset plates are embedded in the wood at a depth of only 10 mm (⅜ in.), heating the metal gusset plate will decompose the wood fibres, quickly leading to failure of the truss. Likewise, wooden trusses held together with steel pins can fail in a fire. When heated, the steel pins conduct heat into the wood, which in turn causes decomposition of the wood. When the wood weakens, failure initially occurs at that point, followed by failure of the entire truss. Fire also can weaken the glue in glued trusses, causing the truss to weaken.

Steel trusses are prone to failure during fires. Automatic sprinklers can protect the steel from the heat; without such protection, however, the steel over a fire will eventually elongate and twist. A 31-m-long (100-ft-long) beam or truss can elongate by as much as 229 mm (9 in.) when heated to 538°C (1000°F) (National Fire Protection Association, 2015). Elongated steel beams can rotate and drop wooden trusses—with deadly consequences. Elongation also can push down the exterior walls of the building. Cracks in the outside walls at the roof level are a warning indicator of this condition.

Truss spaces can contribute to the spread of fires. Spaces under floors occupied by trusses can serve as conduits for the spread of fires. These fires often are difficult to detect and may spread rapidly, much as fires spread quickly through any open spaces. Because truss-supported floors can be covered with many types of material (including cement), it may be difficult to determine that a fire burning under the floor has weakened the floor. Be alert for the possibility of a fire burning under you. Fires that occur in the space above a truss-supported ceiling are challenging to detect. Whenever a truss-supported ceiling has fire above it, be prepared for the possibility of a rapid collapse of the trusses supporting the roof.

Because trusses exist in many types of buildings and give fire fighters few signs of impending failure, some jurisdictions have passed laws requiring that buildings constructed with trusses be marked with a specified sign. Identifying the presence of trusses enables you to better predict the behaviour of a particular building under fire conditions and to evaluate the risk and benefit of different approaches to fire suppression in that building.

Walls

Walls are the most visible parts of a building because they shape the exterior and define the interior. Walls may be constructed of masonry, wood, steel, aluminum, glass, or other materials.

Walls are either load bearing or nonbearing. **Load-bearing walls** provide structural support and can be either exterior or interior walls (**FIGURE 6-31**). A load-bearing wall supports a portion of both the building's weight (dead load) and its contents (live load), transmitting that load down to the building's foundation.

Damaging or removing a load-bearing wall can result in a partial or total collapse of the building.

Nonbearing walls, by contrast, support only their own weight (**FIGURE 6-32**). For this reason, most nonbearing walls can be breached or removed without compromising the structural integrity of the building. Many nonbearing walls are interior partitions that divide the building into rooms and spaces. The exterior walls of a building can be nonbearing when a system of columns supports the building.

In addition to load-bearing and nonbearing walls, several specialized walls may be used in construction:

- **Party walls** are constructed on the line between two properties and are shared by a building on each side of the line. They are almost always load-bearing walls. A party wall is often—but not always—constructed as a fire wall between the two properties.
- Fire walls are designed to limit the spread of fire from one side of the wall to the other side. A fire wall might divide a large building into sections or separate two attached buildings. Fire walls usually extend from the foundation up to and through the roof of a building. They are constructed of fire-resistant materials and may be fire rated.
- **Fire barrier walls** are interior walls that extend from a floor to the underside of the floor above. They often enclose fire-rated interior corridors or divide a floor area into separate fire compartments.
- **Fire separations** are fire-rated assemblies that enclose interior vertical openings, such as stairwells, elevator shafts, and chases for building

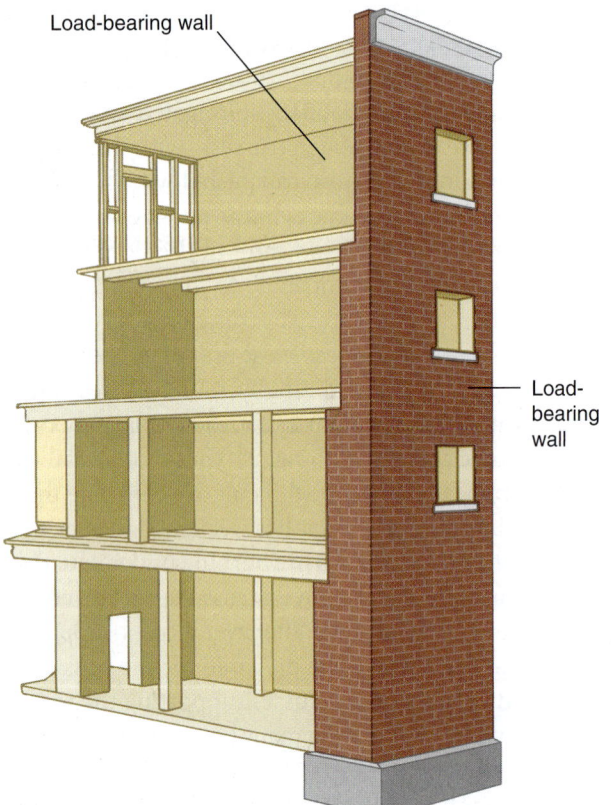

FIGURE 6-31 A load-bearing wall provides structural support.
© Jones & Bartlett Learning.

FIGURE 6-32 A nonbearing wall supports only its own weight.
© Jones & Bartlett Learning.

utilities (**FIGURE 6-33**). A fire separation prevents fire and smoke from spreading from floor to floor via the vertical opening. In multistorey buildings, fire separations also protect the occupants when they are using the exit stairways.

- **Curtain walls** are nonbearing exterior walls attached to the outside of the building. These walls often serve as the exterior skin on a steel-frame high-rise building (**FIGURE 6-34**).

Wood framing is used to construct the walls in most houses and many small commercial buildings. The wood framing used for exterior walls can be covered with a variety of materials, including wooden siding, vinyl siding, aluminum siding, stucco, and masonry veneer. Moisture barriers, wind barriers, and other types of insulation are usually applied to the outside of the vertical studs before the outer covering is applied.

Vertical wooden studs support the walls and partitions inside the building. If fire resistance is critical, steel studs are used to frame walls. The wood framing usually is covered with gypsum board and any of a variety of interior finish materials. The space between the two wall surface coverings may be empty, it may contain thermal and sound insulating materials, or it may contain electrical wiring, telephone wires, and plumbing. These spaces often provide pathways through which fire can spread.

FIGURE 6-33 Fire separation.
© Jones & Bartlett Learning. Photographed by Glen E. Ellman.

LISTEN UP!

The United States Fire Administration (USFA) has partnered with the American Forest and Paper Association to develop educational materials to enhance fire fighter awareness of the fire performance of different types of lightweight construction components. The fire fighter educational material developed under this partnership is available, free of charge from the USFA website.

Walls may be constructed of masonry. Solid, load-bearing masonry walls, at least 152–203 mm (6–8 in.) thick, can be used for buildings up to six storeys high. In contrast, nonbearing masonry walls can be almost any height. Older buildings often have masonry load-bearing walls that are several feet thick at the bottom and then decrease in thickness as the building's height increases. Modern masonry walls are typically reinforced with steel rods or concrete to provide a more efficient structural system.

Masonry walls provide a durable, fire-resistant outer covering for a building, so they are often used as fire walls. A well-designed masonry fire wall is often completely independent of the structures on either side. Even if the building on one side burns completely and collapses, the fire wall should prevent the fire from spreading to the building on the other side. When properly constructed and maintained, masonry walls are strong and can withstand a vigorous assault by fire. If the interior structure begins to collapse and exerts unanticipated forces on the exterior walls, however, solid masonry walls may fail during a fire.

Doors and Windows

Doors and windows are important components of any building. Although they generally have different functions—doors provide entry and exit, whereas windows provide light and ventilation—in an emergency, doors

FIGURE 6-34 Curtain walls.
Courtesy of Glenn Corbett.

and windows are almost interchangeable. A window can serve as an entry or an exit, for example, whereas a door can provide light and ventilation.

Hundreds of door and window designs exist, with many different applications. Of particular concern to fire fighters are fire doors and fire windows. Even wood doors provide some barrier to fire spread. Closing a door during a fire may give you a few additional minutes to search a room.

Door Assemblies

Most doors are constructed of either wood or metal. Hollow-core wooden doors are often used inside buildings. A typical hollow-core door has an internal framework whose outer surfaces are covered by thin sheets of wood. Although a fire will burn through a hollow-core door, these types of door can act as a fire barrier for a short period of time. Additionally, these doors can be easily opened with simple hand tools; they should not be used where security is a concern.

Solid-core wooden doors are used where a more substantial door is required. These types of doors are manufactured from solid panels or blocks of wood and are more difficult to force open. A solid-core door provides some fire resistance and often can keep a fire contained within a room for 20 minutes or more, giving fire fighters a chance to arrive on the scene and mount an effective attack.

Metal doors are more durable and fire resistant than wooden doors. Some metal doors have a solid wood core that is covered on both sides by a thin sheet of metal; others are constructed entirely of metal and reinforced for added security. The interior of a metal door can be either hollow or filled with wood, sound-deadening material, or thermal insulation. Although most approved fire doors are metal, you should not assume that all metal doors are fire doors.

FIRE FIGHTER TIP

Manufactured (mobile) homes use lightweight building components throughout the structure to reduce weight. As a result, most parts of these structures are combustible. Such homes typically have few doors and small windows, making entry for fire suppression or rescue difficult. Once a fire starts, especially in an older manufactured (mobile) home, it can destroy the entire structure within a few minutes.

Window Assemblies

Fire fighters frequently use windows as entry points to attack a fire and as emergency exits. Opening a window can greatly change the path of a fire and can provide ventilation, allowing smoke and heat to escape the building and cooler, fresh air to enter it.

Windows come in many shapes, sizes, and designs for different buildings and occupancies (**FIGURE 6-35**). Fire fighters must become familiar with the specific types

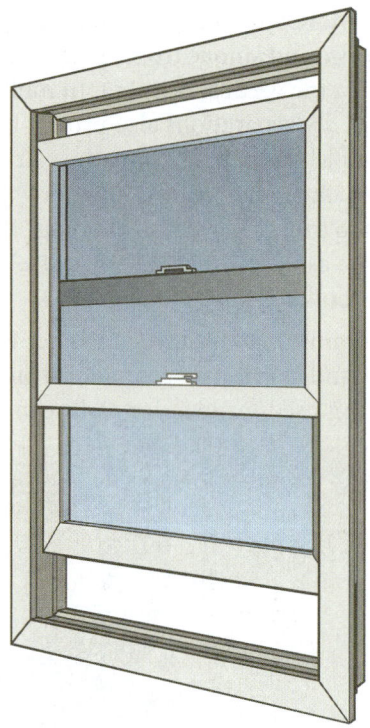

FIGURE 6-35 Windows can open in many different ways. The double-hung window shown here is just one example.
© Jones & Bartlett Learning.

of windows found in local occupancies, learn to recognize them, and understand how they operate. It is especially important to know if a particular type of window is difficult to open or cannot be opened. Today's windows are being built with more energy-conserving features (U.S. Department of Energy, 2017). As part of this trend, double- and triple-pane windows are common in some parts of the country. Window coatings and more durable types of glass may make forcible entry more difficult during firefighting operations. Chapter 10, *Forcible Entry*, contains more information on window construction.

Fire Doors and Fire Windows

Fire door assemblies and **fire windows** are constructed to prevent the passage of flames and heat through an opening during a fire. They must be tested and meet the standards specified in NFPA 80, *Standard for Fire Doors and Other Opening Protectives*. Fire doors and fire windows come in many different shapes and sizes, and they provide different levels of fire resistance. For example, fire doors can swing on hinges, slide down or across an opening, or roll down to cover an opening.

The fire rating on a door or window covers the actual door or window, the frame, the hinges or closing mechanism, the latching hardware, and any other equipment that is required to operate the door or window. All of these items must be tested and approved as a combined system. All fire doors must have a mechanism that keeps the door closed or automatically closes the door when a fire occurs. Doors that are normally open can be closed by the release of a fusible link or electromagnet. The closing of these doors is triggered by the activation of a smoke detector, heat detector, or fire alarm system.

Fire doors and fire windows are rated for a particular duration of fire resistance to a standard test fire. This is similar to the fire-resistance rating system used for building construction assemblies. A 1-hour rating, however, does not guarantee that the door will resist any fire for 60 minutes; rather, it simply establishes that the door will resist the standard test fire for 60 minutes. In any given fire situation, a door rated at 1 hour will probably last twice as long as a door rated 30 minutes.

Fire windows are used when a window is needed in a required fire-resistant wall. These windows often are made of wired glass, which is designed to withstand exposures to high temperatures without breaking. Fire-resistant glass without wires is available for some applications; in such cases, special steel window frames are required to keep the glass firmly in place.

Wired glass is used to provide vision panels in or next to fire doors (**FIGURE 6-36**). Vision panels allow a person to view conditions on the opposite side of a door without opening the door or to make sure no one is standing in front of the door before opening

it. In these configurations, the components making up the entire assembly—including the door, the window, and the frame—must be tested and approved together.

When a window is required only for light passage, glass blocks can sometimes be used instead of wired glass. Glass blocks resist high temperatures and remain in place during a fire (assuming they are properly installed). The size of the opening is limited and depends on the fire-resistance rating of the wall.

Interior Finishes and Floor Coverings

The term **interior finish** is typically used to refer to the exposed interior surfaces of a building. These materials affect how a particular building or occupancy reacts when a fire occurs. Considerations related to interior finishes include whether a material will ignite easily or resist ignition, how quickly a flame will spread across the material surface, how much energy the material will release when it burns, how much smoke it will produce, and which components the smoke will contain (e.g., toxic particles or gases).

A room with a bare concrete floor, concrete block walls, and a concrete ceiling has no interior finishes that will increase the fire load. But when the same room has an acrylic carpet with rubber padding on the floor, wooden baseboards, varnished wood panelling on the walls, and foam plastic acoustic insulation panels on the ceiling, the situation is vastly different: These interior finishes ignite quickly, spread flames quickly across other surfaces, and release significant quantities of heat, flames, and toxic smoke.

Different interior finish materials contribute in various ways to a building fire. Each individual material has certain characteristics, and fire fighters must evaluate the particular combination of materials in a room or space. Typical wall coverings include painted plaster or gypsum board, wallpaper or vinyl wall coverings, wood panelling, and many other surface finishes. Floor coverings might include different types of carpet, vinyl floor tiles, or finished wood flooring. All of these products will burn to some extent, and each one involves a different set of fire risk factors. Fire fighters must understand the hazards posed by different interior finish materials if they are to operate safely at the scene of a fire. Keep in mind that most of the furnishings and interior finishes in a house consist of some type of plastic that is derived from petroleum products.

FIGURE 6-36 Wired glass is used to provide vision panels in or next to fire doors.
Courtesy of Securalldoors.com.

LISTEN UP!

Remember, after occupants have been removed from the building, no building is worth a fire fighter's life.

Exterior Finishes and Siding

The outside of buildings may be covered with a wide variety of materials, such as stucco, brick, stone, metal, wood, and plastics. Depending on the material used, siding may or may not ignite or burn under ordinary circumstances. Some siding materials actually help to prevent the spread of fire.

Stucco is a thin concrete surface that can be used over any material, such as brick, tile, or wood. Compared to wood siding, stucco offers better protection against fire exposure; however, it may be more difficult to make openings in walls covered with this type of siding.

Brick veneer siding is popular for wood-frame residences, garden apartments, and smaller commercial buildings in areas where brick is economical. This siding consists of a single layer of brick or stone applied to the exterior walls of the building to give the appearance of a durable and architecturally desirable outer covering. This brick is not structural, and it is held up by the house. Natural or artificial stone can be used in the same manner (**FIGURE 6-37**). Even though a building has an outer layer of brick or stone, fire fighters should never assume that the exterior walls are masonry. Many buildings that look like masonry are actually constructed using wood-frame techniques and materials. If the wood structure is damaged during a fire, this veneer is likely to collapse.

Metal siding, when used on residences, is usually made up to resemble some other material. For example, aluminum siding is made to look like clapboards, and embossed sheet metal may be made to look like stone. Metal siding can present severe electrical hazards, both from stray electrical currents and from lightning. Although bonding and grounding of the siding may prevent such hazards, this is rarely done, and fire fighters need to be aware of the possibility that the siding is energized.

Wood siding may consist of thick or thin boards or shake shingles. Some forms of wood siding are more flammable than others. For example, thick boards on a

FIGURE 6-37 A stone veneer wall under construction.
Courtesy of Glenn Corbett.

barn require more heat to ignite; however, they contain large amounts of energy that will result in increased fire growth and sustainability. Thin boards, on the other hand, require less heat to ignite and burn up quickly. Wooden shake shingles also ignite easily and spread flames quickly.

Plastic or vinyl siding is made to look like wood siding and may be used to cover up old wood siding. This type of siding melts at relatively low temperatures and will deform, burn, and drip when on fire.

Some types of siding are bonded with insulation. These insulation materials may be flammable. The use of flammable siding and insulation contributed to the rapid fire spread and many fire deaths in the Grenfell Tower Fire in London in 2017 (BBC, 2017). If siding catches fire, the fire can rapidly spread across the outside of the building, through soffit vents or open doors or windows, and into the interior of the building.

Buildings Under Construction or Demolition

Buildings that are under construction or renovation or in the process of demolition present a variety of problems and extra hazards for fire fighters. In many cases, the fire protection features found in finished buildings are missing from these structures. For example, automatic sprinklers might not have been installed yet or might have been disconnected. Smoke detectors might not have been installed or might be inoperable. There may be no coverings on the walls, leaving the entire wood framework fully exposed. Missing doors and windows can provide an unlimited supply of fresh air to feed the fire and allow the flames to spread rapidly. Fire-resistive enclosures may be missing, leaving critical structural components unprotected. Sprinkler and standpipe systems may be inoperative. In addition, construction sites and demolition sites often have large quantities of building materials stored close to buildings. These materials can add a huge additional fire load and contribute to the rapid spread of fire, often from one building to another. All of these factors can enhance fire spread, leading to quick structural failure.

Many fires at construction and demolition sites are inadvertently caused by workers using torches to weld, cut, or take apart pieces of the structure. Tanks of flammable gases and piles of highly combustible construction materials might be left in locations where they could add even more fuel to a fire. Buildings under construction or demolition often are left unoccupied for many hours, resulting in delayed discovery and reporting of fires. In some cases, it might prove difficult for fire apparatus to approach the structure or for fire fighters to access working hydrants. All of these problems must be anticipated when considering the fire risks associated with a construction or demolition site.

FIGURE 6-38 A building collapse can kill or injure building occupants, fire fighters, and rescuers.
Courtesy of Glenn Corbett.

Building Collapse

Partial or complete building collapse has the potential to kill or injure building occupants, fire fighters, and rescuers (**FIGURE 6-38**). Each department should have information and knowledge about the local types of building construction and hazards. Often, this information is obtained through preincident planning, and it should be available to fire fighters who are dispatched to an emergency scene. It should include characteristics of the building that might lead to building collapse. Preincident planning is discussed later in this chapter and in Chapter 27, *Fire Prevention and Community Risk Reduction.*

To fully understand the potential for building collapse, you must first understand the forces that act on every building. Gravity exerts downward force on a building, which, in turn, exerts stress on the sides of a building. Buildings are constructed to resist the force of gravity. Structural elements and the connections that tie the building together help to transfer the load and contents of the building to another part of the structure. For example, the roof is supported by the walls of a building. When fire or lateral forces, such as wind or an explosion, destroy or collapse the walls, a heavier load is placed on the roof, and a collapse is more likely to occur. The failure of a building component, such as a floor collapsing, can exert enough lateral force to bring down a brick or masonry wall.

Once a collapse begins, the results are unpredictable. It is important that fire fighters do everything possible to anticipate collapse. Although we have discussed some of the indicators of collapse throughout this chapter, a review of these signs is important given their critical nature. Signs of collapse may include the following:

- Cracks in walls, especially cracks that develop or grow during a fire (**FIGURE 6-39**)
- Leaning walls
- Pitched or sagging doors
- Doors stuck in shifted frames
- Moaning/groaning or cracking sounds
- Any type of movement or vibrations
- Movement or shifting of water on the floor
- Smoke pushing through cracks in the wall
- Lack of water runoff from firefighting operations

FIGURE 6-39 A crack indicating serious building instability.
© monbibi/Shutterstock.

In addition to recognizing the signs, it is important to be able to identify the factors that increase the likelihood of building collapse so you can take the steps needed to reduce the risk of injury or death. These factors include the environment, building occupancy, existing structural instability, fire and explosion damage, and lightweight construction.

Environmental Factors

Building codes require structures be constructed to withstand heavy snow, rain, winds, and earthquakes in a distributed system (e.g., heavy snow load spread over a roof); however, under fire conditions, these environmental factors can quickly lead to building collapse. As fire begins to alter the structural stability of the building, the additional stress from the snow, rain, or wind can cause the already weakened structural members to collapse prematurely.

Building Occupancy

The occupant of the building may contribute to building collapse by overloading the floors and/or roofs. Often, roofs are not designed to be as strong as floors, especially in warmer climates that do not experience heavy snowfall. If the space under the roof is used for storage, or if extra HVAC equipment is mounted on the roof, the load can exceed the designer's expectations. Adding the weight of several fire fighters to an overloaded roof can be disastrous. Also, this space may contain different types of insulation to help protect the structure from climate changes. In older homes, for example, this insulation often contains ground-up paper that can lead to smouldering and fire spread.

Water-saturated building contents can contribute to building collapse. The contents of a building may become soaked during firefighting operations and retain the water that was used to extinguish the fire. Water weighs 1 kg/L (8.34 lb/gal), so a nozzle that flows 379 L/min (100 gal/min) adds about 376 kg (830 lb) of weight to the building each minute. It is important that fire fighters calculate the total water load on the building. Waterlogged conditions may cause a floor to collapse, which, in turn, may cause a wall to collapse.

Existing Structural Instability

The stability of a building is dependent on all structural components being attached to each other. Structural components may be weakened or broken due to a number of causes, including previous fire damage, rotting, or corrosion from water leaks.

During construction of a building, structural components may be unstable. For example, buildings constructed with poured concrete floors are initially supported with shoring (**FIGURE 6-40**). If this shoring is removed before the concrete is sufficiently cured, a collapse of the top floor can occur, resulting in the progressive collapse of each floor below. Precast concrete panels are constructed in one location and then hoisted into place onsite. They are unstable until all connections have been completed. Supporting members, such as joists, beams, or trusses, may fail during

FIGURE 6-40 These shores serve as a temporary structure that supports the load of the concrete during the course of construction.
© photoslb/Shutterstock.

construction if too many holes are cut into them. A failure of one member can result in a building collapse.

Buildings that are being renovated present similar concerns as those under construction. Load-bearing members may be structurally compromised during renovation. Additionally, if a building is renovated from one occupancy type to another, the new contents of the building can exceed the load that the building was initially designed to support. Damage to any structural component can cause a building to collapse.

Fire and Explosion Damage

Sustained moderate to heavy fire conditions can result in building collapse. Interior fires generate hot gases that accumulate under the roof. Fire fighters who are performing ventilation activities on the roof to release the heated gases should be supported on an aerial apparatus. This is still risky because opening a ceiling, floor, or attic void space could produce a backdraft, flashover, or explosion, resulting in a sudden roof collapse. Additionally, the collapse itself can result in a sudden blast of fire. Thus, after switching to a defensive firefighting mode, fire fighters should stay away from doors and windows in the event that a collapse occurs.

Building collapse can occur during overhaul activities. Use caution when switching from active firefighting to overhaul activities. This is not a time when speed is paramount. A building may fail because of the added weight from the water used during firefighting activities. A fire also may decrease the mass of wooden structural components. Even post-fire cooling, the contraction of structural components and building contents can cause movements that lead to collapse. If there is any doubt about the structural integrity of the building, it may be necessary to overhaul the building from outside and have it inspected by an engineer.

Lightweight Construction

Modern lightweight construction poses perhaps the greatest collapse danger to fire fighters today. This type of construction collapses more frequently and approximately three times faster than other types of construction (Underwriters Laboratories [UL], 2008). The trusses used in these buildings are constructed with a small margin of safety; failure of one wooden truss usually results in overload and subsequent failure of adjoining trusses. For this reason, failure of a single wooden truss usually leads to catastrophic failure of the floor or roof it supports.

Steel trusses also are prone to failure during fires. When thin tubular steel is heated enough to cause a tube in the truss to weaken, the entire truss can fail. If steel trusses are protected from heat with thermal insulation, they may withstand the heat from a fire for a longer period of time. Many steel trusses do not have any protection from heat, however, so they often fail during a major fire. Steel has a tendency to sag before it fails completely, which may give a slight warning of its impending failure.

In some cases, trusses may fail because of overload. As buildings are modified, or when the use of buildings changes, more weight may be placed on floors or roofs than they were intended to carry. This extra weight might exceed the weight for which the trusses were designed. In the event of a fire, the added weight can contribute to the rapid failure of a truss-supported floor or roof.

In the past, fire fighters were taught that sounding the floor or checking the roof to see if it was "spongy" would help determine if that part of the structure was safe. This technique has proved unreliable when dealing with structural members built from trusses. Fire fighters have reported many cases where roofs and floors have collapsed with no warning signs. For this reason, department regulations often require fire fighters to evacuate the building any time lightweight wood trusses or wooden I-beams are subjected to fire.

SAFETY TIP

Checking the floor or roof to see if it is spongy is not a valid test to determine the safety of a floor or roof built using modern lightweight construction. In older Type III buildings, be mindful of water pooling in rooms and hallways, and note changes as work progresses. Pooling water may be an indicator that the structural integrity of the building has been compromised. When in doubt, exit immediately.

Preincident Planning and Incident Size-Up

When a building is on fire, how do you determine which type of construction was used to build the building? Even experienced fire fighters need to assess a building carefully to make this determination.

The best way to gather this information is to conduct preincident planning. Preincident planning allows fire department personnel to identify the type of building construction as well as to survey the specific characteristics of that building. This type of investigation helps you prepare to quickly, safely, and effectively attack a fire in that building. Preincident planning allows you the luxury of being at the building before a fire breaks out.

During preincident planning, you can document the characteristics of a building before a fire starts there (**FIGURE 6-41**). In your examination of the building, consider the following questions: Which type of building construction is present? Which type of occupancy is this building? What type of contents does it contain? Which

FIGURE 6-41 Preincident inspections help you determine the type of building construction before a fire occurs.
© Jones & Bartlett Learning.

FIGURE 6-42 A thorough incident size-up can help identify hazards at a fire.
© Steven Townsend/Code 3 Images.

type of support do the floors have? Is it safe to climb onto the roof? Does the roof include photo-voltaic solar panels or grow areas? Does the building contain roof trusses, floor trusses, or manufactured wood I-joists that might collapse suddenly? Which types of fire detection and fire suppression systems are installed in the building? The importance of preplanning cannot be overemphasized. An excellent way to identify building construction is to inspect the building multiple times while it is being built. Your preplanning will be greatly advanced by this exercise. More detailed information about preplanning is presented in Chapter 27, *Fire Prevention and Community Risk Reduction.*

Of course, it is not possible to do preincident planning for every property in your department. In cases where you have not conducted a preincident inspection, it will be necessary to rely on an incident size-up to determine the type of construction present in the fire building (**FIGURE 6-42**). By learning the general characteristics of the building types in your first-due area, you will get some idea of the type of construction

in a given neighbourhood. In a neighbourhood where the houses were built in the 1920s, for example, you are likely to find balloon construction. In a neighbourhood populated by recently completed houses, you are likely to encounter lightweight construction with wooden I-beams and wooden trusses.

Use the information you have learned about building construction to make firefighting safer both for you and for building residents. Your knowledge of building construction, preincident inspections, and incident size-up will help you gather the information you need to remain a safe and effective fire fighter.

Over the course of your career as a fire fighter, it is important to continue to keep up with the changes in building construction. One of the most important effective safety tools a fire fighter can utilize is his or her own mind. Fire fighters must be open minded to new ideas and be students of their job. It is commonly known that medical professionals such as doctors and lawyers are practitioners of their profession. This is because they continually practise, do not remain intellectually stagnant, and they are continually learning and applying their knowledge to their work. Fire fighters should consider themselves practitioners of their trade. This attitude will be their greatest safety tool. Spend some time at construction sites. Review the latest fire studies from places like the UL Firefighter Safety Research Institute and the National Institute of Standards and Technology (NIST), which are available online. The information you gain may be valuable at your next fire scene.

FIRE FIGHTER TIP

It is important to know which types of construction are found in your response area. When a building is under construction, contact the builder or site foreperson and arrange for a tour. Ask questions about the structure and its construction. Inquire about any special features and its fire protection system.

After-Action REVIEW

IN SUMMARY

- It is vital for fire fighters to understand the basic types of building construction and to recognize how building construction affects fire growth and spread.

- Based on a structure's occupancy classification, a fire fighter can predict who is likely to be inside the building. For example, a bank is unlikely to have occupants at 3 AM.

- Building contents usually are related to the occupancy of the structure. For example, a warehouse is likely to store goods or chemicals, whereas an apartment building is likely to be filled with human occupants and furniture.

- The most commonly used building materials are wood, engineered wood products, masonry, concrete, steel, aluminum, glass, gypsum board, and plastics. The key factors that affect the behaviour of each of these materials under fire conditions are combustibility, thermal conductivity, decrease in strength at elevated temperatures, and thermal expansion when heated.

- Masonry includes stone, concrete blocks, and brick. It is fire resistive and a poor conductor of heat. Even so, a masonry structure can cause excessive heat to be held in a building after a fire is extinguished or can collapse under fire conditions if the roof or floor assembly collapses.

- Concrete is a fire-resistive material that does not conduct heat well. It often is used to insulate other building materials from fire.

- Steel is used in the structural framework of buildings. It is strong and resistant to aging, but it is not fire resistive and conducts heat. The risk of failure of a steel structure depends on the mass of the steel components, the loads placed on them, and the methods used to connect the components.

- Glass is noncombustible but not fire resistive. Ordinary glass will usually break when exposed to fire. Tempered glass is stronger and more difficult to break. Laminated glass is likely to crack and remain in place when exposed to fire. Glass blocks have limited strength and are not intended to be used as part of a load-bearing wall. Wired glass incorporates wires that hold the glass together and prevent it from breaking when exposed to heat.

- Gypsum board is commonly used to cover the interior walls and ceilings of residences. It has limited combustibility and does not conduct or release heat to contribute to fire spread. It may fail if exposed to fire for extended periods.

- The most important characteristic of wood is its high combustibility. The rate at which wood ignites, burns, and decomposes depends on several factors: ignition, moisture, density, preheating, and the size and form.

- Engineered wood products are also called manufactured board, manmade wood, or composite wood. These products are manufactured from small pieces of wood that are held together with glue or adhesives. Engineered wood products may warp under high humidity and release toxic fumes during a fire.

- Wood cannot be treated to make it completely noncombustible, but it can be made to be more difficult to ignite and burn through application of a fire-retardant treatment.

- Plastics are rarely used for structural support but may be found throughout a building. The combustibility of plastics varies greatly. Many plastics produce quantities of heavy, dense, dark smoke and release high concentrations of toxic gases.

- There are five types of building construction:
  - Type I construction: fire resistive—commonly used in schools, hospitals, and high-rise buildings.
  - Type II construction: noncombustible—commonly used in single-storey warehouses or factory buildings where fire spread is not an issue.
  - Type III construction: ordinary—used in a wide variety of buildings, ranging from strip malls to small apartment buildings.
  - Type IV construction: heavy timber—used only for special structures that feature construction components such as architectural elements.

- Type V construction: wood frame—used in one- and two-family houses, commercial and office buildings, and apartment and condominium complexes. Type V building constructions are assembled using either balloon-frame construction or platform-frame construction.
- Hybrid buildings consist of more than one type of construction. This may occur during initial construction or because of later renovations.
- Every building contains the following major components:
  - Foundation
  - Floors and ceilings
  - Roof
  - Trusses
  - Walls
  - Doors
  - Windows
  - Interior finishes and floor coverings
  - Exterior finishes and siding
- Buildings under construction or demolition often lack fire protection features and present additional hazards to fire fighters.
- Factors related to building collapse include the environment, building occupancy, existing structural instability, fire and explosion damage, and lightweight construction. Once a collapse begins, the results are unpredictable.
- The preincident plan should contain information on the structure's type of building construction.

KEY TERMS

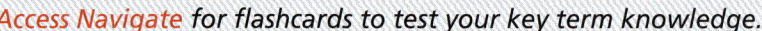

Access Navigate for flashcards to test your key term knowledge.

Balloon-frame construction An older type of wood frame construction in which the wall studs extend vertically from the basement of a structure to the roof without any fire stops.

Bowstring truss A truss that is curved on the top and straight on the bottom.

Combustibility The property describing whether a material will burn and how quickly it will burn.

Contemporary construction Buildings constructed since about 1970 that incorporate lightweight construction techniques and engineered wood components. These buildings exhibit less resistance to fire than older buildings.

Curtain wall Nonbearing walls that separate the inside and outside of the building but are not part of the support structure for the building.

Curved roof A roof with a curved shape.

Dead load Dead loads consist of the weight of all materials of construction incorporated into the building including but not limited to walls, floors, roofs, ceilings, stairways, built-in partitions, finishes, cladding and other similarly incorporated architectural and structural

items, and fixed service equipment including the weight of cranes. (NFPA 5000)

Fire barrier wall A wall, other than a fire wall, having a fire-resistance rating. (NFPA 5000)

Fire door assembly Any combination of a fire door, a frame, hardware, and other accessories that together provide a specific degree of fire protection to the opening. (NFPA 80)

Fire resistance The measure of the ability of a material, product, or assembly to withstand fire or give protection from it. (NFPA 251)

Fire separation A horizontal or vertical fire resistance–rated assembly of materials that have protected openings and are designed to restrict the spread of fire. (NFPA 45)

Fire wall A wall separating buildings or subdividing a building to prevent the spread of fire and having a fire-resistance rating and structural stability. (NFPA 5000)

Fire window A window assembly rated in accordance with NFPA 257 and installed in accordance with NFPA 80. (NFPA 5000)

Flat roof A horizontal roof; often found on commercial or industrial occupancies.

Glass blocks Thick pieces of glass that are similar to bricks or tiles.

Gusset plate Connecting plate made of a thin sheet of steel used to connect the components of the truss.

Gypsum A naturally occurring material consisting of calcium sulfate and water molecules.

Gypsum board The generic name for a family of sheet products consisting of a noncombustible core primarily of gypsum with paper surfacing. (NFPA 5000)

Hybrid building A building that does not fit entirely into any of the five construction types because it incorporates building materials of more than one type.

Interior finish The exposed surfaces of walls, ceilings, and floors within buildings. (NFPA 5000)

Laminated glass Safety glass; the lamination process places a thin layer of plastic between two layers of glass so that the glass does not shatter and fall apart when broken.

Laminated wood Pieces of wood that are glued together.

Legacy construction An older type of construction that used sawn lumber and was built before about 1970.

Live load The load produced by the use and occupancy of the building or other structure, which does not include construction or environmental loads such as wind load, snow load, rain load, earthquake load, flood load, or dead load. Live loads on a roof are those produced (1) during maintenance by workers, equipment, and materials and (2) during the life of the structure by movable objects such as planters and by people. (NFPA 5000)

Load-bearing wall A wall that supports any vertical load in addition to its own weight or any lateral load.

Manufactured (mobile) home A structure, transportable in one or more sections, which, in the travelling mode, is 2.4 m (8 body-ft) or more in width or 12 m (40 body-ft) or more in length or, when erected on site, is 29.7 m² (320 ft²) or more and which is built on a permanent chassis and designed to be used as a dwelling, with or without a permanent foundation, when connected to the required utilities, and includes plumbing, heating, air-conditioning, and electrical systems contained therein. (NFPA 5000)

Masonry Built-up unit of construction or combination of materials such as clay, shale, concrete, glass, gypsum, tile, or stone set in mortar. (NFPA 5000)

Nonbearing wall Any wall that is not a bearing wall. (NFPA 5000)

Occupancy The purpose for which a building or other structure, or part thereof, is used or intended to be used. (NFPA 5000)

Oriented strand board (OSB) An engineered wood product manufactured from small pieces of wood that are held together with glue or adhesives. The adhesives include urea-formaldehyde resins, phenol-formaldehyde resins, melamine-formaldehyde resins, and polyurethane resins.

Parallel chord truss A truss in which the top and bottom chords are parallel.

Party wall A wall constructed on the line between two properties.

Pitched chord truss A type of truss typically used to support a sloping roof.

Pitched roof A roof with sloping or inclined surfaces.

Platform-frame construction Construction technique for building the frame of the structure one floor at a time. Each floor has a top and bottom plate that acts as a firestop.

Rafters Joists that are mounted in an inclined position to support a roof.

Spalling Chipping or pitting of concrete or masonry surfaces. (NFPA 921)

Tempered glass A type of safety glass that is heat treated so that, under stress or fire, it will break into small pieces that are not as dangerous.

Thermal conductivity A property that describes how quickly a material will conduct heat.

Thermoplastic material Plastic material capable of being repeatedly softened by heating and hardened by cooling and, that in the softened state, can be repeatedly shaped by molding or forming. (NFPA 5000)

Thermoset material Plastic material that, after having been cured by heat or other means, is substantially infusible and cannot be softened and formed. (NFPA 5000)

Truss A collection of lightweight structural components joined in a triangular configuration that can be used to support either floors or roofs.

Type I construction (fire resistive) The type of construction in which the fire walls, structural elements, walls, arches, floors, and roofs are of approved noncombustible or limited-combustible materials that have a specified fire resistance.

Type II construction (noncombustible) The type of construction in which the fire walls, structural elements, walls, arches, floors, and roofs are of approved noncombustible or limited-combustible materials without fire resistance.

Type III construction (ordinary) The type of construction in which exterior walls and structural elements that are portions of exterior walls are of approved noncombustible or limited-combustible materials and in which fire walls, interior structural elements, walls, arches, floors, and roofs are entirely or partially of wood of smaller dimensions than required for Type IV construction or are of approved noncombustible, limited-combustible, or other approved combustible materials. (NFPA 14)

Type IV construction (heavy timber) The type of construction in which fire walls, exterior walls, and interior bearing walls and structural elements that are portions of such walls are of approved noncombustible or limited-combustible materials. Other interior structural elements, arches, floors, and roofs are constructed of solid or laminated wood or cross-laminated timber without concealed spaces within allowable dimensions of the building code. (NFPA 14)

Type V construction (wood frame) The type of construction in which structural elements, walls, arches, floors, and roofs are entirely or partially of wood or other approved material. (NFPA 14)

Wired glass A glazing material with embedded wire mesh.

Wooden beam Load-bearing member assembled from individual wood components.

Wood truss An assembly of small pieces of wood or wood and metal.

REFERENCES

APA. Market outlook: industry expects increasing demand for engineered wood products. July 2, 2015. https://www.apawood.org/market-outlook-2015. Accessed May 10, 2018.

BBC. London fire: What happened at Grenfell Tower? July 19, 2017. http://www.bbc.com/news/uk-england-london-40272168. Accessed September 27, 2018.

Havel, Gregory. Building Construction: Lightweight Steel Framing. January 1, 2008. http://www.fireengineering.com/articles/print/volume-161/issue-1/features/building-construction-lightweight-steel-framing.html. Accessed September 27, 2018.

Jakubowski, Greg. Bowstring Truss Roof Construction Hazards. January 10, 2013. http://www.firerescuemagazine.com/articles/print/volume-8/issue-3/firefighter-safety/bowstring-truss-roof-construction-hazards.html. Accessed September 27, 2018.

National Fire Protection Association (NFPA). 2015. *Brannigan's Building Construction for the Fire Service*, 5th edition. Burlington, MA: Jones & Bartlett Learning.

National Fire Protection Association. NFPA 80, *Standard for Fire Doors and Other Opening Protectives*. www.nfpa.org. Accessed September 27, 2018.

National Fire Protection Association. NFPA 220, *Standard on Types of Building Construction*. www.nfpa.org. Accessed September 27, 2018.

National Institute for Occupational Safety and Health (NIOSH). Preventing Deaths and Injuries of Fire Fighters Working Above Fire-Damaged Floors. February 2009. DHHS (NIOSH) Publication No. 2009–114. https://www.cdc.gov/niosh/docs/wp-solutions/2009-114/default.html. Accessed September 27, 2018.

National Institute for Occupational Safety and Health (NIOSH). Two Career Fire Fighters Die Following a Seven-Alarm Fire in a High-Rise Building Undergoing Simultaneous Deconstruction and Asbestos Abatement—New York. August 5, 2010. https://www.cdc.gov/niosh/fire/reports/face200737.html. Accessed September 27, 2018.

Underwriters Laboratories (UL). Report on Structural Stability of Engineered Lumber in Fire Conditions. September 30, 2008. https://www.ul.com/global/documents/offerings/industries/buildingmaterials/fireservice/NC9140-20090512-Report-Independent.pdf. Accessed September 27, 2018.

U.S. Department of Energy. Energy-Efficient Windows. https://energy.gov/energysaver/energy-efficient-windows. Accessed September 27, 2018.

On Scene

A tornado ripped through your community in the early hours of the morning, damaging numerous buildings. Your crew has been tasked with conducting a damage assessment of a two-block area and reporting your findings to the incident commander.

The first house is slightly damaged with no risk of collapse. The next house has significant damage. It is a wood-frame house with several collapsed load-bearing walls. It is unsafe to enter without shoring. The third wood-frame home is completely destroyed. The fourth building is a noncombustible commercial structure with block walls and steel trusses. It is partially collapsed. As you continue your search, you realize the magnitude of this event.

1. What type of construction is the commercial structure?

 A. Type II

 B. Type III

 C. Type IV

 D. Type V

2. What type of construction were the houses?

 A. Type II

 B. Type III

 C. Type IV

 D. Type V

3. A _____ wall is a wall that is designed to provide structural support for a building.

 A. nonbearing

 B. load-bearing

 C. curtain

 D. fire

4. While evaluating the structures, you recall that the buildings' _____, which are designed to transfer the weight of the building and its contents to the ground, may have been affected by the storm.

 A. foundations

 B. roofs

 C. beams

 D. columns

5. The damage assessment includes evaluating the structural stability of the buildings. In order to complete this evaluation, you must determine the loads that may have been in place for the building. Which type of load can be difficult to evaluate because the items that the occupants place within the buildings affect it?

 A. Dead load

 B. Occupancy load

 C. Live load

 D. Fire load

6. Which of the following are mounted in an inclined position to support the roof of the houses and have most likely been affected by the tornadoes?

 A. Trusses

 B. Rafters

 C. Joists

 D. Beams

(continued)

On Scene Continued

7. While evaluating the damaged houses that were constructed with trusses, you recall that if the structure utilized _____, only a thin sheet of metal embedded in the wood at a depth of 10 mm (⅜ in.) was holding the trusses together before the strong winds from the tornadoes affected their capability.

A. steel pins

B. lag bolts

C. gusset plates

D. tongue and groove

8. Which of the following types of construction usually would be most resistive to the effects of a tornado?

A. Fire-resistive construction

B. Noncombustible construction

C. Ordinary construction

D. Wood-frame construction

Access Navigate to find answers to this On Scene, along with other resources such as an audiobook.

Portable Fire Extinguishers

KNOWLEDGE OBJECTIVES

After studying this chapter, you will be able to:

- State the primary purposes of fire extinguishers. (**NFPA 1001: 4.3.16**, pp. 229–230)
- Define Class A fires. (**NFPA 1001: 4.3.16**, p. 231)
- Define Class B fires. (**NFPA 1001: 4.3.16**, p. 231)
- Define Class C fires. (**NFPA 1001: 4.3.16**, pp. 231–232)
- Define Class D fires. (**NFPA 1001: 4.3.16**, p. 232)
- Define Class K fires. (**NFPA 1001: 4.3.16**, p. 232)
- Explain the classification and rating system for fire extinguishers. (**NFPA 1001: 4.3.16**, pp. 232–233)
- Explain the labelling system for fire extinguishers. (**NFPA 1001: 4.3.16**, pp. 233–234)
- Describe the three risk classifications for area hazards. (**NFPA 1001: 4.3.16**, pp. 235–236)
- Describe the types of agents and operating systems used in fire extinguishers. (**NFPA 1001: 4.3.16**, pp. 238–246)
- Select the proper class of fire extinguisher. (**NFPA 1001: 4.3.16**, p. 247)
- Describe the basic steps of fire extinguisher operation. (**NFPA 1001: 4.3.16**, p. 246)
- Explain the basic steps of inspecting, maintaining, recharging, and hydrostatic testing of fire extinguishers. (**NFPA 1001: 4.3.16**, pp. 256–257)

SKILLS OBJECTIVES

After studying this chapter, you will be able to perform the following skills:

- Transport the fire extinguisher to the location of the fire. (**NFPA 1001: 4.3.16**, pp. 247–248)
- Extinguish a Class A fire with a stored-pressure water-type fire extinguisher. (**NFPA 1001: 4.3.16**, pp. 249–250)
- Extinguish a Class A fire with a multipurpose dry-chemical fire extinguisher. (**NFPA 1001: 4.3.16**, p. 250)
- Extinguish a Class B flammable liquid fire with a dry-chemical fire extinguisher. (**NFPA 1001: 4.3.16**, pp. 251–252)
- Extinguish a Class B flammable liquid fire with a stored-pressure foam fire extinguisher. (**NFPA 1001: 4.3.16**, pp. 252–253)
- Operate a carbon dioxide fire extinguisher. (**NFPA 1001: 4.3.16**, p. 254)
- Operate a halogenated agent-type fire extinguisher. (**NFPA 1001: 4.3.16**, p. 255)
- Operate a dry-powder fire extinguisher. (**NFPA 1001: 4.3.16**, p. 255)
- Operate a wet-chemical fire extinguisher. (**NFPA 1001: 4.3.16**, p. 256)

ADDITIONAL STANDARDS

- **NFPA 10**, *Standard for Portable Fire Extinguishers*
- **NFPA 11**, *Standard for Low-, Medium-, and High-Expansion Foam*

Fire Alarm

You are the designated on-duty cook for the day, and the kitchen is bare. You go into the grocery store while the rest of your crew waits in the apparatus. As you make your way through the produce section, an employee runs up to you and says that there is a fire in the back room. You quickly radio your officer to inform her of the situation and then immediately proceed to the back room. You locate a small fire in a trash can that is just starting to spread. You quickly scan the room looking for a fire extinguisher. Not seeing one, you ask the employee where the closest extinguisher is located.

1. What type of fire extinguisher would be most effective for this fire? Why?

2. As the fire grows, at what point will using a fire extinguisher become ineffective?

3. How do fire inspectors ensure the fire extinguisher will work when needed?

 JONES & BARTLETT LEARNING **NAVIGATE 2** *Access Navigate for more practice activities.*

Introduction

Portable fire extinguishers are required in many types of occupancies as well as in commercial vehicles, boats, aircraft, and various other locations. Fire prevention efforts encourage citizens to keep fire extinguishers in their homes, particularly in their kitchens. Fire extinguishers are used successfully to put out hundreds of small fires every day, preventing millions of dollars in property damage as well as saving lives. Most fire extinguishers are easy to operate and can be used effectively by an individual with only basic training.

Fire extinguishers range in size from models that can be operated with one hand to large, wheeled models that contain several hundred pounds of **extinguishing agent** (material used to stop the combustion process) (**FIGURE 7-1**). Extinguishing agents include water, water with different additives, dry chemicals, wet chemicals, dry powders, and gaseous agents. Each agent is suitable for specific types of fires.

Fire extinguishers are designed for different purposes and involve different operational methods. This chapter discusses the most appropriate kind of extinguisher to use for different types of fires and which kinds must not be used for certain fires. The selection of the proper fire extinguisher builds on the information presented in Chapter 5, *Fire Behaviour*.

This chapter also covers the operation and simple inspection and maintenance of the most common types of portable fire extinguishers. These principles

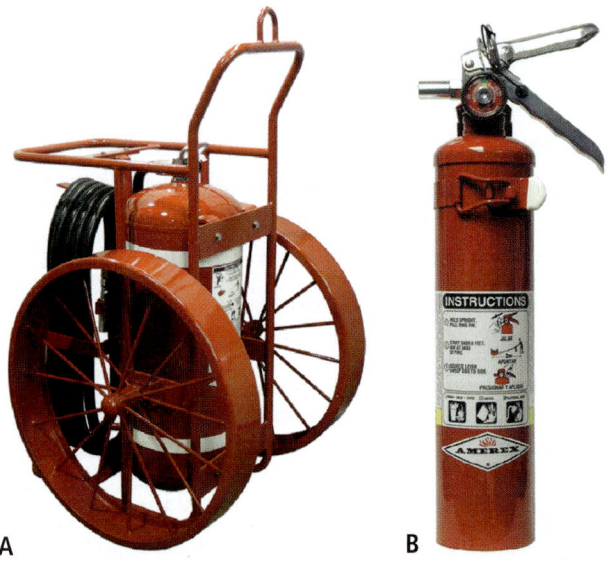

A   **B**

FIGURE 7-1 Portable fire extinguishers can be large or small. **A.** A wheeled fire extinguisher. **B.** A hand-held fire extinguisher.
Courtesy of Amerex Corporation.

FIRE FIGHTER TIP

Fire fighters are often called out to provide fire extinguisher training for the public. It is essential that you understand the characteristics and operations of each type of fire extinguisher.

will enable you to use fire extinguishers correctly and effectively, thereby reducing the risk of personal injury and property damage.

Purposes of Fire Extinguishers

Portable fire extinguishers have two primary uses: to extinguish incipient stage fires (those that have not spread beyond the area of origin) and to control fires where traditional methods of fire suppression are not recommended.

Extinguishing Incipient Stage Fires

Fire extinguishers are placed in many locations so that they will be available for immediate use on small, incipient stage fires, such as a fire in a wastebasket. A trained individual with a suitable fire extinguisher can usually control this type of fire (**FIGURE 7-2**). As flames spread beyond the wastebasket to other contents of the room, however, such a fire becomes increasingly difficult and dangerous to control.

One advantage of fire extinguishers is their portability. It may take less time to control a fire with a portable fire extinguisher than it does to advance and charge a hose line. Most fire department vehicles carry at least one fire extinguisher, and many vehicles carry two or more extinguishers of different types. Fire department vehicles that are not equipped with water or fire hose usually carry at least one multipurpose fire extinguisher. Fire fighters often use these portable fire extinguishers to control incipient stage fires quickly. At times, a fire fighter may even use a fire extinguisher from the fire-site premises to control an incipient stage fire.

The primary disadvantage of fire extinguishers is that they are "one-shot" devices. In other words, once the contents of a fire extinguisher have been discharged, the device is no longer effective in fighting fires until it is recharged or replaced. If the fire extinguisher does not control the fire, some other device or method must be employed. This is a serious limitation when compared to a fire hose with a continuous water supply. When using a portable fire extinguisher to control an incipient stage fire, make sure you are not placing yourself in a dangerous situation.

SAFETY TIP

Do not place yourself in a dangerous situation by trying to fight a large fire with a small fire extinguisher. You cannot fight a fire or protect yourself with an empty extinguisher, an undersized extinguisher, or the inappropriate type of extinguisher. Never let the fire come between you and a means of escape. This is a general rule of firefighting.

Extinguishing Fires with the Appropriate Extinguishing Agent

Fire extinguishers are also used to control fires in situations where water is not recommended. For example, using water on fires that involve energized electrical equipment increases the risk of electrocution to fire fighters and can cause extensive damage to the electrical equipment. In these cases, it would be better to use a fire extinguisher containing an appropriate extinguishing agent. The appropriate extinguishing agent also must be used for fires that involve flammable liquids, cooking oils, and combustible metals.

As a fire fighter, you must know which fires require certain extinguishing agents, which type of fire extinguisher should be used, and how to operate the different types of fire extinguishers. Using an improper type of fire extinguisher or extinguishing agent can spread burning material, cause unnecessary damage, and may pose a danger to the fire extinguisher operator.

Portable fire extinguishers are sometimes used in combination with other extinguishing techniques. For example, with a pressurized-gas fire, water may be used to cool hot surfaces and prevent reignition, while a dry-chemical agent is used to extinguish the flames (**FIGURE 7-3**). Certain types of portable fire extinguishers can be helpful in overhauling a fire. The extinguishing agents contained in these devices break

FIGURE 7-2 A trained individual with a suitable fire extinguisher can usually control an incipient stage fire.
© Jones & Bartlett Learning. Photographed by Glen E. Ellman.

FIGURE 7-3 Water can cool hot surfaces.
© Jones & Bartlett Learning. Photographed by Glen E. Ellman.

down the surface tension of the water, allowing water to penetrate the materials and reach deep-seated fires.

Methods of Fire Extinguishment

Understanding the nature of fire is key to understanding how fire extinguishers work and how they differ from one another. As discussed in Chapter 5, *Fire Behaviour*, all fires require three basic ingredients: fuel, heat, and oxygen (fire triangle). In scientific terminology, burning is called *oxidation*. This chemical process occurs when another substance such as a fuel is combined with oxygen, resulting in the formation of ash or other waste products and the release of energy as heat and light.

The combustion process begins when the fuel is heated to its ignition temperature—the temperature at which it begins to burn. The energy that initiates this process can come from many different sources, including a spark or flame, friction, electrical energy, or a chemical reaction. Once a substance begins to burn, it will generally continue burning as long as adequate supplies of oxygen and fuel to sustain the chemical reaction are present, unless something interrupts the process. This relationship is sometimes characterized as the fire tetrahedron.

Most fire extinguishers stop the burning by cooling the fuel below its ignition temperature, by cutting off the supply of oxygen, or by performing both actions. Some extinguishing agents interrupt the complex chemical reactions that occur between the heated fuel and the oxygen. Modern portable fire extinguishers contain agents that use one or more of these methods.

Cooling the Fuel

If the temperature of the fuel falls below its ignition temperature, the combustion process will stop. Water extinguishes a fire using this method.

Cutting Off the Supply of Oxygen

Creating a barrier that interrupts the flow of oxygen to the flames will extinguish a fire. Putting a lid on a pan of burning food is an example of this technique (**FIGURE 7-4**); applying a blanket of foam to the surface of a burning liquid is another example. Similarly, surrounding the fuel with a layer of carbon dioxide (CO_2) can cut off the supply of oxygen necessary to sustain the burning process.

Interrupting the Chain of Reactions

Some extinguishing agents work by interrupting the molecular chain reactions required to sustain combustion.

FIGURE 7-4 Covering a pan of burning food with a lid will extinguish a fire by cutting off the supply of oxygen.
© Jones & Bartlett Learning.

Halon-type extinguishing agents and some dry-chemical agents work in this way. In some cases, just a very small quantity of the agent can accomplish this objective.

Classes of Fires

It is essential to match the appropriate type of fire extinguisher to the type of fire. Fires and fire extinguishers are grouped into classes according to their characteristics. Some extinguishing agents work more efficiently than others on certain types of fires. In some cases, selecting the proper extinguishing agent will mean the difference between extinguishing a fire and being unable to control it.

Before selecting a fire extinguisher, ask yourself, "Which class of fire am I fighting?" There are five classes of fires, each of which affects the choice of extinguishing equipment.

Class A Fires

Class A fires involve ordinary solid combustible materials such as wood, paper, cloth, rubber, household rubbish, and some plastics (**FIGURE 7-5**). Natural vegetation, such as grass and trees, is also Class A material. Water is the most commonly used extinguishing agent for Class A fires, although several other agents can be used effectively.

LISTEN UP!

Most residential fires involve Class A materials. Even though the source of the ignition could be Class B or Class C in nature, the resulting fire frequently involves ordinary combustibles (NFPA, *Home Structure Fires*, 2017). A fire that begins with an unattended pot of grease on the stove, for example, might ignite the wooden cabinets and other kitchen contents. A short circuit in an electrical outlet might ignite combustible material in the immediate vicinity. In both of these cases, the fire would involve primarily Class A materials.

FIGURE 7-5 Most ordinary combustible materials are included in the definition of Class A fires.
© Jones & Bartlett Learning.

Class B Fires

Class B fires involve flammable or combustible liquids, such as gasoline, oil, grease, tar, lacquer, oil-based paints, and some plastics (**FIGURE 7-6**). Fires involving flammable gases, such as propane or natural gas, are also categorized as Class B fires. Examples of Class B fires include a fire in a pot of molten roofing tar, a fire involving splashed fuel on a hot lawnmower engine, and burning natural gas that is escaping from a gas meter struck by a vehicle. Several different types of extinguishing agents are approved for use in Class B fires.

Class C Fires

Class C fires involve energized electrical equipment, which includes any device that uses, produces, or delivers electrical energy (**FIGURE 7-7**). A Class C fire could involve building wiring and outlets, fuse boxes, circuit breakers, transformers, generators, or electric motors. Power tools, lighting fixtures, household appliances, and electronic devices such as televisions, radios, and computers could be involved in Class C fires as well. The equipment must be plugged in or connected to an

FIGURE 7-6 Class B fires involve flammable liquids and gases.
© thaloengsak/iStock/Getty Images.

FIGURE 7-7 Class C fires involve energized electrical equipment or appliances.
© PhotoStock-Israel/Alamy Stock Photo.

electrical source, but not necessarily operating, for the fire to be classified as Class C. Once the power is cut to a Class C fire, the fire is treated as a Class A or Class B fire depending on the type of material that is burning.

Electricity does not burn, but electrical energy can generate tremendous heat that could ignite nearby Class A or B materials. As long as the equipment is energized, the incident must be treated as a Class C fire. Agents that will not conduct electricity, such as dry chemicals or CO_2, must be used on Class C fires. Incorrectly attacking a Class C fire with an extinguishing agent that conducts electricity, such as water, can result in injury or death to the fire fighter. Once the power is cut to a Class C fire, the fire is treated as a Class A or Class B fire depending on the type of material that is burning.

Class D Fires

Class D fires involve combustible metals such as magnesium, titanium, zirconium, sodium, lithium, and potassium. Special techniques and extinguishing agents are required to fight combustible metal fires (**FIGURE 7-8**). Normally used extinguishing agents can react violently—even explosively—if they come in contact with burning metals. Violent reactions also can occur when water strikes burning combustible metals.

Class D fires are most often encountered in industrial occupancies, such as machine shops, repair shops, and metal recycling plants, as well as in fires involving aircraft and certain models of automobiles. Magnesium and titanium—both combustible metals—are used to produce automotive and aircraft parts because they combine high strength with light weight. Sparks from cutting, welding, or grinding operations could ignite a Class D fire, or the metal items could become involved in a fire that originated elsewhere.

FIGURE 7-8 Combustible metals in Class D fires require special extinguishing agents.
© Andrew Lambert Photography/Science Source.

FIGURE 7-9 Class K fires involve combustible cooking oils and fats.
© Kathie Nichols/Shutterstock.

Because of the chemical reactions that could occur during a Class D fire, it is important to select the proper extinguishing agent and application technique for this kind of incident. Choosing the correct fire extinguisher for a Class D fire requires expert knowledge and experience.

Class K Fires

Class K fires involve combustible cooking oils and fats (**FIGURE 7-9**). Cooking oil fires were once classified as Class B combustible liquid fires; however, the introduction of modern, high-efficiency deep-fat fryers and the trend toward using vegetable oils instead of animal fats to fry foods have resulted in higher cooking temperatures, which required the development of a new class of wet-chemical extinguishing agents. Some restaurants continue to use extinguishing agents that were approved for Class B fires, however. Class B fire extinguishers are not as effective for cooking oil fires as are Class K fire extinguishers.

Classification and Rating of Fire Extinguishers

Portable fire extinguishers are classified and rated based on their extinguishing properties and capabilities. This information is important for selecting the proper fire extinguisher to fight a particular fire (**TABLE 7-1**). It is also used to determine which types of fire extinguishers should be placed in a given location so that incipient stage fires can be controlled quickly.

In Canada, the **Underwriters Laboratory of Canada (ULC)** is the organization that developed the standards, classification, and rating system for portable fire extinguishers. Each fire extinguisher has a specific rating that identifies the classes of fires for which it is both safe and effective.

The classification system for fire extinguishers uses both letters and numbers. The letters indicate the

TABLE 7-1 Types of Fires

Class A	Ordinary combustibles
Class B	Flammable or combustible liquids
Class C	Energized electrical equipment
Class D	Combustible metals
Class K	Kitchen fires involving oils and fats

classes of fire for which the fire extinguisher can be used, and the numbers indicate its relative effectiveness. Fire extinguishers that are safe and effective for more than one class will be rated with multiple letters. For example, a fire extinguisher that is safe and effective for Class A fires will be rated with an "A"; one that is safe and effective for Class B fires will be rated with a "B"; and one that is safe and effective for both Class A and Class B fires will be rated with both an "A" and a "B."

Class A and Class B fire extinguishers also include a number, indicating the relative effectiveness of the fire extinguisher in the hands of a nonexpert user. On Class A fire extinguishers, this number reflects the amount of water it contains. A fire extinguisher that is rated 1-A contains the equivalent of 5 L (1.25 gal) of water. A typical Class A fire extinguisher contains 2.5 gal (9.5 L) of water and has a 2-A rating. The higher the number, the greater the extinguishing capability of the fire extinguisher.

The effectiveness of Class B fire extinguishers is based on the approximate area (measured in square meters [m^2]) of burning fuel that these devices are capable of extinguishing. Certification testing of these fire extinguishers is performed by trained experts who are able to control a larger fire than a nonexpert user; therefore, the numerical rating is about 40 percent of the area of burning fuel that an expert can consistently extinguish (Underwriters Laboratories, 2017). For example, a 10-B rating indicates that a nonexpert user should be able to extinguish a fire in a pan of flammable liquid that is 0.9 m^2 (10 ft^2) in surface area, while an expert user should be able to extinguish a fire that is 2.3 m^2 (25 ft^2) in surface area. A nonexpert user should be able to use a fire extinguisher rated 40-B to control a flammable liquid pan fire with a surface area of 3.7 m^2 (40 ft^2), while an expert user should be able to extinguish a fire with a surface area of 9 m^2 (97 ft^2).

Numbers are used to rate a fire extinguisher's effectiveness only for Class A and Class B fires. If the fire extinguisher can also be used for Class C fires, it contains an agent proven to be nonconductive to electricity and safe for use on energized electrical equipment. For instance, a fire extinguisher that carries a 2-A:10-B:C rating can be used on Class A, Class B, and Class C fires. It has the extinguishing capabilities of a 2-A fire extinguisher

when applied to Class A fires, has the capabilities of a 10-B fire extinguisher when applied to Class B fires, and can be used safely on energized electrical equipment.

Standard test fires are used to rate the effectiveness of fire extinguishers. Such testing may involve different agents, amounts, application rates, and application methods. Fire extinguishers are rated for their ability to control a specific type of fire as well as for the extinguishing agent's ability to prevent rekindling. Some agents can successfully suppress a fire but are unable to prevent the material from reigniting. A rating is given only if the fire extinguisher completely extinguishes the standard test fire and prevents rekindling.

Labelling of Fire Extinguishers

Fire extinguishers that have been tested and approved by an independent laboratory are labelled to clearly designate the classes of fire the unit is capable of extinguishing safely. This traditional lettering system has been used for many years and is still found on many fire extinguishers. More recently, a universal pictograph system has been developed. This system does not require the user to be familiar with the alphabetic codes for the different classes of fires.

Traditional Lettering System

The traditional lettering system uses the following labels (**FIGURE 7-10**):

- Fire extinguishers suitable for use on Class A fires are identified by the letter "A" on a solid green triangle. The triangle has a graphic relationship to the letter "A."

ORDINARY

COMBUSTIBLES

FLAMMABLE

LIQUIDS

ELECTRICAL

EQUIPMENT

COMBUSTIBLE

METALS

FIGURE 7-10 Traditional letter labels on fire extinguishers often incorporate a shape as well as a letter. The Class K icon is a black hexagon with a white "K."

- Fire extinguishers suitable for use on Class B fires are identified by the letter "B" on a solid red square. Again, the shape of the letter mirrors the graphic shape of the box.
- Fire extinguishers suitable for use on Class C fires are identified by the letter "C" on a solid blue circle, which also incorporates a graphic relationship between the letter "C" and the circle.
- Fire extinguishers suitable for use on Class D fires are identified by the letter "D" on a solid yellow five-pointed star.
- Fire extinguishers suitable for use on Class K fires are identified by a pictograph showing a fire in a frying pan. Because the Class K designation is new, there is no traditional-system alphabet graphic for it.

Pictograph Labelling System

The pictograph system, such as described for Class K fire extinguishers, uses symbols rather than letters on the labels. This system clearly indicates whether a fire extinguisher is inappropriate for use on a particular class of fire. The pictographs are all square icons, each of which is designed to represent a certain class of fire (**FIGURE 7-11**). The icon for Class A fires is a burning trash can beside a wood fire. The Class B fire extinguisher icon is a flame and a gasoline can; the Class C icon is a flame and an electrical plug and socket; and the Class D icon is a flame and a metal gear. As noted previously, fire extinguishers rated for fighting

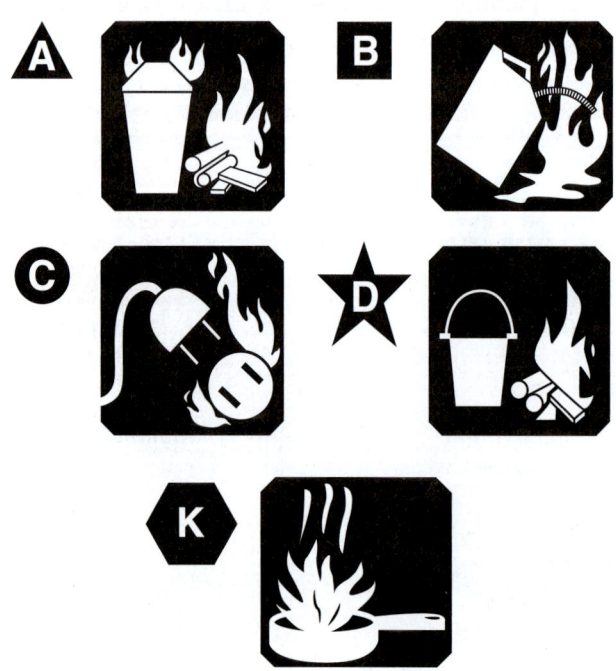

FIGURE 7-11 The icons for Class A, B, C, D, and K fires.
© Jones & Bartlett Learning.

Class K fires are labelled with an icon showing a fire in a frying pan.

Under this pictograph labelling system, the presence of an icon indicates that the fire extinguisher has been rated for that class of fire. A missing icon indicates that the fire extinguisher has not been rated for that class of fire. A red slash across an icon indicates that the fire extinguisher must not be used on that type of fire, because doing so would create additional risk. For example, a fire extinguisher rated for Class A fires only would show all three icons, but the icons for Class B and Class C would have red diagonal lines through them. This three-icon array signifies that the fire extinguisher contains a water-based extinguishing agent, making it unsafe to use on flammable liquid or electrical fires.

Certain fire extinguishers labelled as appropriate for Class B and Class C fires do not include the Class A icon but may be used to put out small Class A fires. The fact that they have not been rated for Class A fires indicates that they are less effective in extinguishing a common combustible fire than a comparable Class A fire extinguisher would be.

SAFETY TIP

The safest and surest way to extinguish a Class C fire is to turn off the power and treat it like a Class A or B fire. If you are unable to turn off the power, you should be prepared for reignition, because the electricity could reignite the fire after it has been extinguished.

Fire Extinguisher Placement

Fire and building codes and regulations require the installation of fire extinguishers in many areas so that they will be available to fight incipient stage fires. NFPA 10, *Standard for Portable Fire Extinguishers*, lists the requirements for placing and mounting portable fire extinguishers as well as the appropriate mounting heights.

The regulations for each type of occupancy specify the maximum floor area that can be protected by each fire extinguisher, the maximum travel distance from the closest fire extinguisher to a potential fire, and the types of fire extinguishers that should be provided. Two key factors must be considered when determining which type of fire extinguisher should be placed in each area: the class of fire that is likely to occur and the potential magnitude of an incipient stage fire.

Fire extinguishers should be mounted so they are readily visible and easily accessible (**FIGURE 7-12**). Heavy fire extinguishers should not be mounted high on a wall. If the fire extinguisher is mounted too high,

FIGURE 7-12 Fire extinguishers should be mounted in locations with unobstructed access and visibility.
© Jones & Bartlett Learning.

a smaller person might be unable to lift it off its hook or could be injured in the attempt.

The recommended mounting heights for the placement of fire extinguishers are as follows (NFPA, *NFPA 10, Standard for Portable Fire Extinguishers*, 2017):

- Fire extinguishers weighing up to 18 kg (40 lb) should be mounted so that the top of the extinguisher is not more than 1.5 m (5 ft) above the floor.
- Fire extinguishers weighing more than 18 kg (40 lb) should be mounted so that the top of the extinguisher is not more than 1 m (3.5 ft) above the floor.
- The bottom of an extinguisher should be at least 102 mm (4 in.) above the floor.

Classifying Area Hazards

Areas are divided into three risk classifications—light, ordinary, and extra hazard—based on the amount and types of combustibles that are present, including building materials, contents, decorations, and furniture. The quantity of combustible materials present is sometimes called a building's **fire load** and is measured as the average weight of combustible materials per square foot (or per square metre) of floor area. The larger the fire load, the larger the potential fire.

The occupancy use category does not necessarily determine the building's hazard classification. The recommended hazard classifications for different types of occupancies are simply guidelines based on typical situations. The hazard classification for each area should be based on the actual amount and types of combustibles that are present. It is important to note that each provincial and territorial jurisdiction will have established regulations pertaining to the requirement for fire extinguishers.

Light (Low) Hazard

Light (low) hazard locations are areas where the quantity, combustibility, and heat release of the materials are low, and the majority of materials are arranged so that a fire is not likely to spread. Light hazard environments usually contain limited amounts of Class A combustibles, such as wood, paper products, cloth, and similar materials. A light hazard environment might also contain some Class B combustibles (flammable liquids and gases), such as copy machine chemicals or modest quantities of paints and solvents, but all Class B materials must be kept in closed containers and stored safely. Examples of common light hazard environments include most offices, classrooms, churches, assembly halls, and hotel guest rooms (**FIGURE 7-13**).

Ordinary (Moderate) Hazard

Ordinary (moderate) hazard locations contain more Class A and Class B materials than do light hazard locations. The combustibility and heat release rate of the materials are moderate. Typical examples of ordinary hazard locations include retail stores with on-site storage areas, light manufacturing facilities, auto showrooms, parking garages, research facilities, and workshops or service areas that support light hazard locations, such as hotel laundry rooms or restaurant kitchens (**FIGURE 7-14**).

FIGURE 7-13 Light hazard areas include offices, churches, and classrooms.
Courtesy of Bill Larkin.

FIGURE 7-14 Auto showrooms, hotel laundry rooms, and parking garages are classified as ordinary hazard areas.
© Jones & Bartlett Learning.

Ordinary hazard areas also include warehouses that contain Class I and Class II commodities. Class I commodities include noncombustible products stored on wooden pallets or in corrugated cartons that are shrink-wrapped or wrapped in paper. Class II commodities include noncombustible products stored in wooden crates or multilayered corrugated cartons.

Extra (High) Hazard

Extra (high) hazard locations contain more Class A combustibles and/or Class B flammables than do ordinary hazard locations. The combustibility and heat release rate of the materials are high. Typical examples of extra hazard areas include woodworking shops; service and repair facilities for cars, aircraft, or boats; and many kitchens and other cooking areas that have deep fryers, flammable liquids, or gases under pressure (**FIGURE 7-15**). In addition, areas used for manufacturing processes such as painting, dipping, or coating, as well as facilities used for storing or handling flammable liquids, are classified as extra hazard environments. Warehouses containing products that do not meet the definitions of Class I and Class II commodities also are considered extra hazard locations.

Determining the Most Appropriate Placement of Fire Extinguishers

Several factors must be considered when determining the numbers and types of fire extinguishers that should be placed in each area of an occupancy. Among these factors are the types of fuels found in the area and the quantities of those materials.

FIGURE 7-15 Kitchens, woodworking shops, and auto repair shops are considered possible extra hazard locations.
© Jones & Bartlett Learning.

Some areas may need more than one type of fire extinguisher or fire extinguishers with more than one rating. Environments that include Class A combustibles require a fire extinguisher rated for Class A fires; those with Class B combustibles require a fire extinguisher rated for Class B fires. Areas that contain both Class A and Class B combustibles, however, require either a fire extinguisher that is rated for both types of fires or a separate fire extinguisher for each class of fire.

Most buildings require fire extinguishers that are suitable for fighting Class A fires because ordinary combustible materials—such as furniture, partitions, interior finish materials, paper, and packaging products—are so common. Even where other classes of products are used or stored, there is still a need to defend the facility from a fire involving common combustibles.

A single multipurpose fire extinguisher is generally less expensive than two individual fire extinguishers and eliminates the problem of selecting the proper fire extinguisher for a particular fire. However, it is sometimes more appropriate to install Class A fire extinguishers in general-use areas and to place fire extinguishers that are especially effective in fighting Class B or Class C fires near those specific hazards.

In some facilities, a variety of conditions are present. In these occupancies, each area must be individually evaluated and the extinguisher installation tailored to its particular circumstances. A restaurant is a good example of this situation. The dining areas contain common combustibles, such as furniture, tablecloths, and paper products, that would require a fire extinguisher rated for Class A fires. In the restaurant's kitchen, where the risk of fire involves cooking oils, a Class K fire extinguisher would provide the best defense.

Similarly, within a hospital, fire extinguishers for Class A fires would be appropriate in hallways, offices, lobbies, and patient rooms. Class B fire extinguishers should be mounted in laboratories and areas where flammable anesthetics are stored or handled. Electrical rooms should have fire extinguishers that are approved for use on Class C fires, whereas hospital kitchens would need Class K fire extinguishers.

Fire Extinguisher Design

Portable fire extinguishers contain a variety of extinguishing agents—the substances that put out a fire (**TABLE 7-2**). These extinguishing agents are expelled from the fire extinguisher through the use of pressure, and many fire extinguishers rely on pressurized gas to eject the extinguishing agent. This gas can be stored either with the extinguishing agent in the body of the fire extinguisher or externally in a separate cartridge or cylinder. With external storage, the extinguishing agent is put under pressure only as it is used.

Some extinguishing agents, such as CO_2, are called **self-expelling agents**. Most self-expelling agents are normally gases but are stored as liquids under pressure in the fire extinguisher. When the confining pressure is released, the agent rapidly expands, causing it to self-discharge. Hand-operated pumps are used to expel the agent when water or water with additives is the extinguishing agent.

TABLE 7-2 Basic Types of Extinguishing Agents
- Water
- Dry chemicals
- CO_2
- Foam
- Halogenated agents
- Dry powder
- Wet chemicals

Portable Fire Extinguisher Components

Most hand-held portable fire extinguishers have six basic parts (**FIGURE 7-16**):

- A cylinder that holds the extinguishing agent
- A carrying handle
- A nozzle or horn
- A trigger and discharge valve assembly
- A locking mechanism to prevent accidental discharge
- A pressure indicator

Cylinder

The body of the fire extinguisher, known as the **cylinder** or container, holds the extinguishing agent. Nitrogen, compressed air, or CO_2 can be used to pressurize the cylinder to expel the agent. **Stored-pressure fire extinguishers** hold both the extinguishing agent, in wet or dry form, and the expeller gas under pressure in the cylinder. **Cartridge/cylinder-operated fire extinguishers** rely on an external cartridge of pressurized gas, which is released only when the fire extinguisher is to be used.

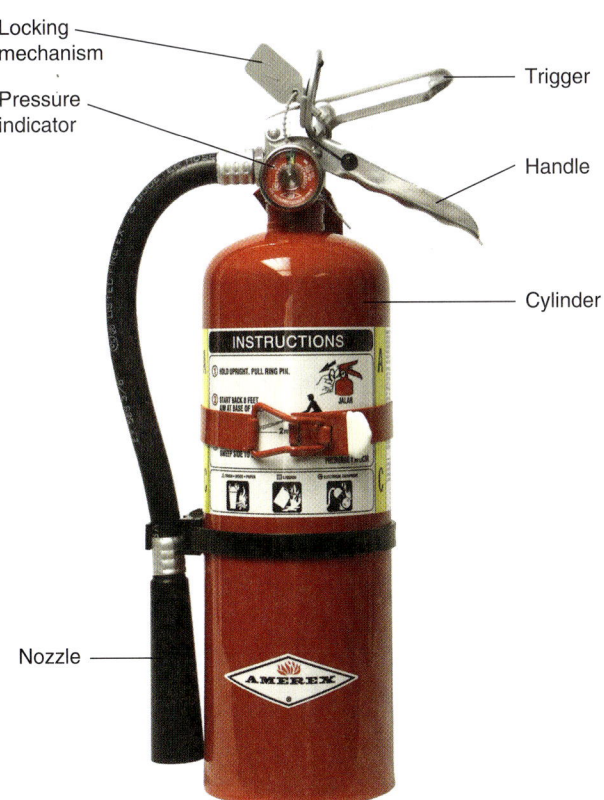

Locking mechanism
Pressure indicator
Trigger
Handle
Cylinder
Nozzle

FIGURE 7-16 Portable fire extinguishers have six basic parts.
Courtesy of Amerex Corporation.

Handle

The handle is used to carry the portable fire extinguisher and, in many cases, to hold it during use. The actual design of the handle varies from model to model, but all fire extinguishers that weigh more than 1.4 kg (3 lb) have handles. In many cases, the handle is located just below the trigger mechanism.

Nozzle or Horn

The extinguishing agent is expelled through a nozzle or horn. In some fire extinguishers, the nozzle is attached directly to the valve assembly at the top of the extinguisher. In other models, the nozzle is found at the end of a short hose.

Trigger

The trigger is the mechanism that is squeezed or depressed to discharge the extinguishing agent. In some models, the trigger is a button positioned just above the handle. On most portable fire extinguishers, however, the trigger is a lever located above the handle. To release the extinguishing agent, the operator lifts the fire extinguisher by the handle and simultaneously squeezes down on the discharge lever.

Cartridge/cylinder-operated fire extinguishers usually have a two-step operating sequence. First, a handle or lever is pushed to pressurize the stored agent; then, a trigger-type mechanism incorporated in the nozzle assembly is used to control the discharge.

Locking Mechanism

The locking mechanism is a simple quick-release device that prevents accidental discharge of the extinguishing agent. The simplest form of locking mechanism is a stiff pin, which is inserted through a hole in the trigger to prevent it from being depressed. The pin usually has a ring at the end so that it can be removed quickly.

A special plastic tie, called a tamper seal, is used to secure the pin. This seal is designed to break easily when the pin is pulled. Removing the pin and tamper seal is best accomplished with a twisting motion. The tamper seal makes it easy to see whether the fire extinguisher has been used and not recharged. It also discourages people from playing or tinkering with the fire extinguisher.

Pressure Indicator

The pressure indicator or gauge shows whether a fire extinguisher has sufficient pressure to operate properly. Over time, the pressure in a fire extinguisher may dissipate. Checking the gauge first will tell you whether the fire extinguisher is ready for use. Not all fire extinguishers

have pressure indicators; therefore measures such as weighing may be required to determine if the extinguisher is ready to operate.

Pressure indicators vary in terms of both design and sophistication. Most fire extinguishers use a needle gauge. Pressure may be shown in pounds per square inch (psi) or on a three-step scale (too low, proper range, too high). Pressure gauges are usually colour-coded, with a green area indicating the proper pressure zone.

Some disposable fire extinguishers intended for home use have an even simpler pressure indicator—that is, a plastic pin built into the cylinder. Pressing on the pin tests the pressure within the fire extinguisher. If the pin pops back up, the fire extinguisher has enough pressure to operate; if it remains depressed, the pressure has dropped below an acceptable level.

Wheeled Fire Extinguishers

Wheeled fire extinguishers are large units that are mounted on wheeled carriages and typically contain between 68 and 159 kg (150 and 350 lb) of extinguishing agent. The wheeled design lets one person transport the fire extinguisher to the fire. If a wheeled fire extinguisher is intended for indoor use, doorways and aisles must be wide enough to allow for its passage to every area where it could be needed.

Wheeled fire extinguishers usually have a long delivery hose, so the unit can stay in one spot as the operator moves around to attack the fire from more than one side. Usually, a separate cylinder containing nitrogen or some other compressed gas provides the pressure necessary to operate the fire extinguisher.

Wheeled fire extinguishers are most often installed in special hazard areas, such as at military bases, airports, and industrial settings. Models with rubber tires or wide-rimmed wheels are available for outdoor installations.

Types of Fire Extinguishers

Portable fire extinguishers vary according to their extinguishing agent, capacity, effective range, and the time it takes to completely discharge the extinguishing agent. They also have different mechanical designs. This section describes the basic types of extinguishing agents used in portable fire extinguishers, as well as the basic characteristics of the different types of extinguishers. The seven types of fire extinguishers discussed in this section are organized by type of extinguishing agent:

- Water-type fire extinguishers
- Dry-chemical fire extinguishers

- CO₂ fire extinguishers
- Class B foam fire extinguishers
- Halogenated-agent fire extinguishers
- Dry-powder fire extinguishers
- Wet-chemical fire extinguishers

Water-Type Fire Extinguishers

Water is an efficient, plentiful, and inexpensive extinguishing agent. When it is applied to a fire, it is quickly converted from a liquid into steam, absorbing great quantities of heat in the process. As the heat is removed from the combustion process, the fuel cools below its ignition temperature, and the fire stops burning.

Water-type fire extinguishers are intended for use primarily on Class A fires. Many Class A fuels will absorb liquid water, which further lowers the temperature of the fuel. This also prevents rekindling. Water is a much less effective extinguishing agent for other classes of fires. Using a water-type fire extinguisher on hot cooking oil, for example, can cause explosive splattering, which can spread the fire and endanger the operator of the fire extinguisher. Many burning flammable liquids will simply float on top of water. Because streams of water conduct electricity, it is dangerous to apply a stream of water to any fire that involves energized electrical equipment. If a water-type fire extinguisher is used on a burning combustible metal, a violent reaction can occur. Because of these limitations, plain water is used only in Class A fire extinguishers. Class B foam fire extinguishers, which are a specific type of water extinguisher, are intended for fires involving flammable liquids.

Stored-Pressure Water-Type Fire Extinguishers

The most popular kind of **stored-pressure water-type fire extinguisher** is the 9.5-L (2.5-gal) model with a 2-A rating (**FIGURE 7-17**). Many fire department vehicles carry this type of fire extinguisher for use on incipient stage Class A fires. Such a fire extinguisher expels water in a solid stream with a range of 9 to 12 m (30 to 40 ft) through a nozzle at the end of a short hose. The discharge time is approximately 55 seconds if the fire extinguisher is used continuously. A full fire extinguisher weighs about 14 kg (30 lb) (NFPA, *Standard for Portable Fire Extinguishers*, 2017).

Because water can freeze, stored-pressure water-type fire extinguishers should not be installed in areas where the temperature is expected to drop below 0°C (32°F). Antifreeze models of stored-pressure water-type fire extinguishers, called loaded-stream fire extinguishers, are available. The loaded-stream agent will not freeze at temperatures as low as −40°C (−40°F) (NFPA, *Standard for Portable Fire Extinguishers*, 2017).

FIGURE 7-17 Most fire departments use stored-pressure water-type fire extinguishers.
© Jones & Bartlett Learning.

The recommended procedure for operating a stored-pressure water-type fire extinguisher is to set it on the ground, grasp the handle with one hand, and pull out the ring pin or release the locking latch with the other hand. At this point, the fire extinguisher can be lifted and used to douse the fire. Use one hand to aim the stream at the fire, and squeeze the trigger with the other hand. The stream of water can be turned into a spray by putting a thumb at the end of the nozzle; this technique is often used after the flames have been extinguished as part of the effort to thoroughly soak the fuel.

Stored-pressure water-type fire extinguishers can be recharged at any location that provides water and a source of compressed air. It is recommended that recharging be performed by a trained individual or a company that specializes in fire extinguishers. If that is not possible, however, be sure to follow the manufacturer's instructions to ensure proper and safe recharging.

Water Mist Fire Extinguishers

Another method of applying water is through the use of water mist. **Water mist fire extinguishers** are constructed in a manner similar to stored-pressure water-type extinguishers; however, they are usually more easily identifiable because they are typically white in colour (**FIGURE 7-18**). Instead of the discharge

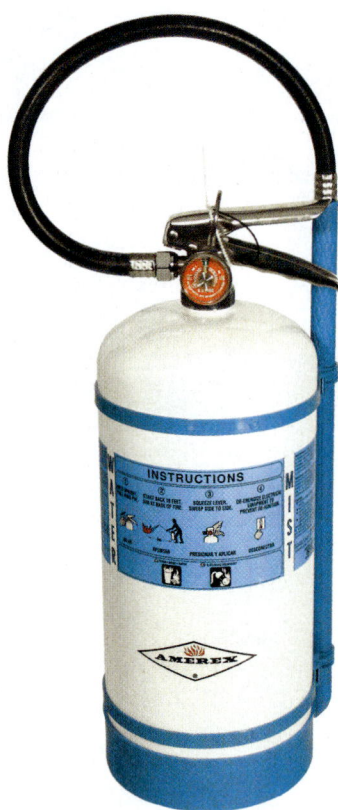

FIGURE 7-18 Water mist fire extinguishers are easily identifiable because they are typically white in colour.
Courtesy of Amerex.

hose and nozzle assembly of a stored-pressure water-type extinguisher, they have a discharge hose that is connected to an applicator wand and a misting nozzle. They are commonly available in 6.6-L (1.75-gal) and 9.5-L (2.5-gal) sizes.

Water mist fire extinguishers contain distilled or de-ionized water, which conducts less electricity than tap water. The water is discharged from the misting nozzle as a fine spray that provides safety from electrical shock. For this reason, water mist fire extinguishers are safe to use on Class A and Class C fires. They are rated at 2-A:C. The operator must be within 1.5 to 3.7 m (5 to 12 ft) of the fire in order for this type of fire extinguisher to be effective (NFPA, *Standard for Portable Fire Extinguishers*, 2018).

These fire extinguishers are used where regular fire extinguishers might cause excessive damage. Typical uses include museums, rare book collections, hospital environments, telecommunication facilities, and "clean room" manufacturing facilities.

Loaded-Stream Fire Extinguishers

One notable disadvantage of water as an extinguishing agent is that it freezes at 0°C (32°F). In areas that are subject to below-freezing temperatures, **loaded-stream fire extinguishers** can be used to counteract this limitation.

Loaded-stream fire extinguishers discharge a solution of water containing an alkali metal salt that prevents freezing at temperatures as low as −40°C (−40°F). Pressure for these fire extinguishers is supplied by a separate cylinder of CO_2.

The most common model is the 9.5-L (2.5-gal) unit, which is identical to a typical stored-pressure water-type extinguisher. Hand-held models are available with capacities of 4 to 9.5 L (1 to 2.5 gal) of water and are rated from 1-A to 3-A. Larger units, including a 64-L (17-gal) unit rated 10-A and a 125-L (33-gal) unit rated 20-A, are also available.

Wetting-Agent and Class A Foam Fire Extinguishers

Wetting-agent fire extinguishers expel water that contains a solution intended to reduce surface tension (the physical property that causes water to bead or form a puddle on a flat surface). Reducing the surface tension allows water to spread over the fire and penetrate more efficiently into Class A fuels.

Class A foam fire extinguishers contain a solution of water and Class A foam concentrate. This extinguishing agent has foaming properties as well as the ability to reduce surface tension.

Both wetting-agent and Class A foam fire extinguishers are available in the same configurations as water extinguishers, including hand-held stored-pressure models and wheeled units. These fire extinguishers should not be exposed to temperatures below 4°C (40°F).

Pump Tank Fire Extinguishers

Pump tank fire extinguishers come in sizes ranging from 1-A-rated, 6-L (1.5-gal) units to 4-A-rated, 19-L (5-gal) units. The water in these devices is not stored under pressure. Instead, the pressure needed to expel the water is provided by a hand-operated, double-acting, vertical piston pump, which moves water out through a short hose on both the up and the down strokes. The manually operated pump may be mounted directly on the cylinder of the fire extinguisher, or it may be part of the nozzle assembly. This type of fire extinguisher sits upright on the ground during use. A small bracket at the bottom allows the operator to steady the fire extinguisher with one foot while pumping.

Pump tank fire extinguishers can be used with antifreeze agents. The manufacturer should be consulted for details, because some antifreeze agents (such as common salt) can corrode the fire extinguisher or damage the pump. Fire extinguishers with steel shells corrode more easily than those with copper or nonmetallic shells.

Backpack Fire Extinguishers

Backpack fire extinguishers are used primarily outdoors for fighting brush and grass fires (**FIGURE 7-19**).

FIGURE 7-19 A backpack fire extinguisher can be refilled easily from a natural water source such as a lake or river.

© Jones & Bartlett Learning.

Most of these units have a tank capacity of 19 L (5 gal) and weigh approximately 23 kg (50 lb) when full. Backpack fire extinguishers are listed by UL but do not carry numeric ratings.

The water tank can be made of fibreglass, stainless steel, galvanized steel, nylon, canvas, or brass. Backpack fire extinguishers are designed to be refilled easily in the field, such as from a lake or a stream, through a wide-mouth opening at the top. A filter keeps dirt, stones, and other contaminants from entering the tank. Antifreeze agents, wetting agents, or other special water-based extinguishing agents can be used with backpack fire extinguishers.

Most backpack fire extinguishers are operated via hand pumps. The most common design has a trombone-type, double-acting piston pump located at the nozzle, which is attached to the tank by a short rubber hose. To discharge the fire extinguisher, the operator holds the pump in both hands and moves the piston back and forth.

Some models have a compression pump built into the side of the tank. On these devices, it takes about 10 strokes of the pump handle to build up the initial pressure, which is maintained through continuous slow strokes. The operator uses the other hand to control the discharge. A lever-operated shut-off nozzle is provided at the end of a short hose.

Dry-Chemical Fire Extinguishers

Dry-chemical fire extinguishers deliver a stream of finely ground particles onto a fire. Different chemical compounds are used to produce **dry chemicals** of varying capabilities and characteristics. These extinguishing agents work in several ways. First, the finer particles of the chemical vaporize when they reach the high temperature of the flame and release a vapour that interrupts the chemistry of the flame. Second, the particles of the dry chemical shield the fuel surface from the flame radiation, thereby reducing the rate at which the burning fuel is being vaporized. And third, when extensive dry chemical is applied and reaches the surface of the fuel, it can smother the fire by forming an insulating blanket.

Five compounds are used as the primary dry-chemical extinguishing agents:

- Sodium bicarbonate (rated for Class B and C fires only)
- Potassium bicarbonate (rated for Class B and C fires only)
- Urea-based potassium bicarbonate (rated for Class B and C fires only)
- Potassium chloride (rated for Class B and C fires only)
- Ammonium phosphate (rated for Class A, B, and C fires)

Sodium bicarbonate is often used in small household fire extinguishers. Potassium bicarbonate, potassium chloride, and urea-based potassium bicarbonate all have greater fire-extinguishing capabilities (per unit volume) for Class B fires than does sodium bicarbonate. Potassium chloride is more corrosive than the other dry-chemical extinguishing agents.

Ammonium phosphate is the only dry-chemical extinguishing agent that is rated as suitable for use on Class A fires. Although ordinary dry-chemical fire extinguishers can be used against Class A fires, a water dousing is also needed to extinguish any smouldering embers and prevent rekindling.

Dry-chemical fire extinguishers offer several advantages over water-type fire extinguishers:

- They are effective on Class B (flammable liquid and gas) fires.
- They can be used on Class C (energized electrical equipment) fires, because the chemicals are nonconductive.
- They can be stored and used in areas where temperatures fall below the freezing point.

Although dry-chemical fire extinguishers can be used on Class C fires that involve energized electrical equipment, the chemicals can be very damaging to computers, electronic devices, and electrical equipment. The fine particles are carried in air and settle like a fine dust inside the equipment. Over a period of months, this residue can corrode metal parts, causing considerable damage. Another disadvantage of dry-chemical fire extinguishers is that the chemicals may make breathing more difficult when discharged in an enclosed environment.

SAFETY TIP

Discharging a dry-chemical fire extinguisher in a confined space can create a cloud of fine dust that can impair vision and cause difficulty breathing. Self-contained breathing apparatus (SCBA) should be used to protect fire fighters from both toxic gases from the fire and the dry-chemical dust discharged from the fire extinguisher.

Stored-pressure units expel the dry-chemical agent in the same manner as stored-pressure water-type fire extinguishers. The dry chemical in a cartridge/cylinder-operated fire extinguisher is not stored under pressure. Instead, these fire extinguishers have a sealed, pressurized cartridge connected to the storage cylinder. They are activated by pushing down on a lever that punctures the cartridge and pressurizes the cylinder.

The trigger on this type of fire extinguisher allows it to be discharged intermittently, starting and stopping the agent flow. Releasing the trigger stops the flow of agent; however, this does not mean that the fire extinguisher can be put aside and used again later. These fire extinguishers do not retain their internal pressure for extended periods, because the granular dry chemical causes the valve to leak.

Depending on the fire extinguisher's size, the horizontal range of the discharge stream can be from 1.5 to 14 m (5 to 45 ft; NFPA, *Standard for Portable Fire Extinguishers*, 2018). Some models have special nozzles that allow for a longer range. The long-range nozzles are useful when the fire involves burning gas or a flammable liquid under pressure or when the operator is working in a strong wind.

Most small, hand-held dry-chemical fire extinguishers are available with capacities ranging from 0.5 to 14 kg (1 to 30 lb) of agent and are designed to discharge their contents completely in as little as 8 to 20 seconds. Larger units may discharge for as long as 25 seconds (NFPA, *Standard for Portable Fire Extinguishers*, 2018). Wheeled fire extinguishers are available with capacities up to 159 kg (350 lb) of agent. Large dry-chemical fire extinguishers also may be mounted on fire apparatus to deal with special risks.

The selection of which dry-chemical fire extinguisher to use depends on the compatibility of different agents with one another and with any products that they might contact. Some dry-chemical extinguishing agents cannot be used in combination with particular types of foam.

Ordinary Dry-Chemical Fire Extinguishers

The first dry-chemical fire extinguishers were introduced during the 1950s and were rated only for Class B and C fires. The industry term for these B:C-rated units is "ordinary dry-chemical" extinguishers. Ordinary dry-chemical fire extinguishers are available in hand-held models with ratings up to 160-B:C. Larger wheeled units carry ratings up to 640-B:C.

LISTEN UP!

A dry-chemical fire extinguisher usually will continue to lose pressure after a partial discharge. Pressure loss can occur even when only a small amount of agent has been discharged and the pressure gauge still indicates that the fire extinguisher is properly charged. This loss occurs because the agent leaves residue in the valve assembly that allows the stored pressure to leak out slowly.

Multipurpose Dry-Chemical Fire Extinguishers

During the 1960s, **multipurpose dry-chemical fire extinguishers** were introduced. These fire extinguishers contain an ammonium phosphate–based extinguishing agent and are rated for Class A, B, and C fires. The chemicals in these fire extinguishers take the form of fine particles that are treated with other chemicals to prevent the particles from absorbing moisture, which could cause packing or caking and interfere with the extinguisher's discharge and flow. When discharged, the chemicals in these fire extinguishers form a crust over Class A combustible fuels, thereby preventing rekindling (**FIGURE 7-20**).

Multipurpose dry-chemical fire extinguishers should never be used on cooking oil (Class K) fires, such as those involving deep-fat fryers located in commercial kitchens. The ammonium phosphate–based extinguishing agent is acidic and will not react with cooking oils to produce the smothering foam needed to extinguish this type of fire. Even worse, the acid will counteract the foam-forming properties of any alkaline extinguishing agent that is applied to the same fire.

Multipurpose dry-chemical fire extinguishers are available as stored-pressure, cartridge, or nitrogen cylinder–type hand-held models with ratings ranging from 1-A to 20-A and from 10-B:C to 120-B:C. Larger, wheeled models have ratings ranging from 20-A to 40-A and from 60-B:C to 320-B:C.

FIGURE 7-20 Multipurpose dry-chemical fire extinguishers can be used for Class A, B, and C fires.
© Jones & Bartlett Learning.

Carbon Dioxide Fire Extinguishers

Carbon dioxide is a gas that is 1.5 times heavier than air, colourless, odourless, and nontoxic. When CO_2 is discharged on a fire, it forms a dense cloud that displaces the air surrounding the fuel. This effect interrupts the combustion process by reducing the amount of oxygen that can reach the fuel. The placement of a blanket of CO_2 over the surface of a liquid fuel also can disrupt the fuel's ability to vaporize.

CO_2 is both an expelling agent and an extinguishing agent. In portable **carbon dioxide (CO_2) fire extinguishers**, CO_2 is stored under a pressure of 5674 kPa (823 psi), which keeps the CO_2 in liquid form at room temperature. When the pressure is released, the liquid CO_2 rapidly converts to a gas, and the expansion of the gas forces the agent out of the container.

The CO_2 is discharged through a siphon tube that reaches to the bottom part of the storage cylinder, forced through a hose, and expelled through a horn or cone-shaped applicator that directs the flow of the agent on the fire. When discharged, the agent is very cold and contains a mixture of CO_2 gas and solid CO_2, which is quickly converted to a gas. The gas forms a visible cloud of "dry ice" that helps to cool the burning materials and

surrounding areas when moisture in the air freezes as it comes into contact with the CO_2.

CO_2 fire extinguishers are rated for Class B and C fires only. This extinguishing agent does not conduct electricity and has two significant advantages over dry-chemical agents: It is not corrosive, and it does not leave any residue because it dissipates into the air. These factors are important in areas where costly electronic components or computer equipment must be protected. CO_2 fire extinguishers are also used around food preparation areas and in laboratories.

CO_2 fire extinguishers have several limitations and disadvantages:

- Weight: CO_2 fire extinguishers are heavier than similarly rated extinguishers that use other extinguishing agents (**FIGURE 7-21**).
- Range: CO_2 fire extinguishers have a short discharge range (1 to 2.4 m [3 to 8 ft]), which requires the operator to be close to the fire, increasing the risk of personal injury (NFPA, *Standard for Portable Fire Extinguishers*, 2018).

FIGURE 7-21 Carbon dioxide fire extinguishers are heavy due to the weight of the container and the large quantity of agent needed to extinguish a fire. They also have a large discharge nozzle, making them easily identifiable.
© Jones & Bartlett Learning. Photographed by Glen E. Ellman.

- Weather: CO_2 fire extinguishers do not perform well at temperatures below –18°C (0°F) or in windy or drafty conditions, because the extinguishing agent dissipates before it reaches the fire.
- Confined spaces: When CO_2 fire extinguishers are used in confined areas, the extinguishing agent dilutes the oxygen in the air. If the air is diluted enough, people in the space can begin to suffocate.
- Suitability: CO_2 fire extinguishers are not suitable for use on fires involving pressurized fuel or on cooking grease fires.

Depending on their size, CO_2 fire extinguishers can discharge completely in 8 to 30 seconds (NFPA, *Standard for Portable Fire Extinguishers*, 2018). The trigger mechanism can be operated intermittently to preserve any remaining agent. The pressurized CO_2 will remain in the fire extinguisher, but the extinguisher must be recharged after use. The fire extinguisher can be weighed to determine how much agent is left in the storage cylinder.

The smaller CO_2 fire extinguishers contain from 1 to 2 kg (2.5 to 5 lb) of agent. These units are designed to be operated with one hand. The horn is attached directly to the discharge valve on the top of the fire extinguisher by a hinged metal tube. In larger models, the horn is attached at the end of a short hose. These models require two-handed operation.

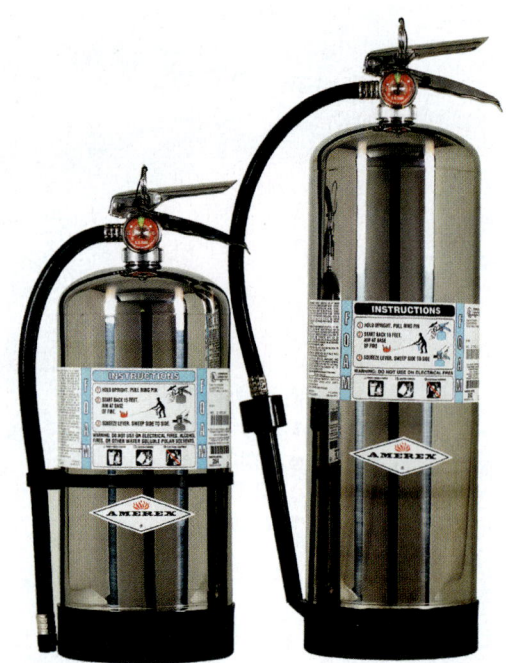

FIGURE 7-22 An AFFF fire extinguisher produces an effective foam for use on Class B fires.
Courtesy of Amerex.

////////////////////////////////////

SAFETY TIP

Do not aim the CO_2 fire extinguisher discharge at anyone or allow it to come in contact with exposed skin, because frostbite could result. CO_2 discharged into a confined space will reduce the oxygen level in that space. For this reason, anyone entering the confined area must wear SCBA.

Class B Foam Fire Extinguishers

Class B foam fire extinguishers are very similar in appearance and operation to water-type fire extinguishers. Instead of plain water, however, they discharge a solution of water and either aqueous film-forming foam (AFFF) or film-forming fluoroprotein (FFFP) foam (**FIGURE 7-22**). The agent is discharged through an aspirating nozzle, which mixes air into the stream. The result is a foam solution that floats over the surface of a burning liquid, creating a blanket that separates the fuel from oxygen. This blanket prevents the fuel from vaporizing. It forms a barrier between the fuel and the oxygen, extinguishing the flames and preventing reignition.

Class B foam fire extinguishers are very effective in fighting Class A fires but are not suitable for Class C fires or for fires involving flammable liquids or gases under pressure. They are not intended for use on fires that involve pressurized fuels or cooking oils, and only specifically labelled foam fire extinguishers can be used on fires involving polar solvents—that is, water-soluble flammable liquids such as alcohols, acetone, esters, and ketones. These fire extinguishers also are not effective at freezing temperatures. Consult the fire extinguisher manufacturer for information on using foam agents effectively at low temperatures. It is important to note that many brands of "fluoroprotein" are harmful to the environment. Runoff should be controlled when using these types of foam products.

AFFF and FFFP hand-held stored-pressure fire extinguishers are available in two sizes: 6 L (1.6 gal), rated 2-A:10-B, and 9.5 L (2.5 gal), rated 3-A:20-B. A wheeled model with a 125-L (33-gal) capacity and a rating of 20-A:160-B is also available.

Both AFFF and FFFP concentrates produce highly effective foams. Which one should be used depends on the product's compatibility with a particular flammable liquid and other extinguishing agents that could be used on the same fire. Detailed information on the use of AFFF and FFFP is available in NFPA 11, *Standard for Low-, Medium-, and High-Expansion Foam*.

Wet-Chemical Fire Extinguishers

Wet-chemical fire extinguishers are the only type of extinguisher to protect Class K (kitchen) installations, which are equipped with deep-fat fryers, cooking oils, and grills. They use wet-chemical extinguishing agents, which include aqueous solutions of potassium acetate, potassium carbonate, and potassium citrate, either singly or in various combinations. These wet agents convert the fatty acids in cooking oils or fats to a soap or foam, a process known as saponification.

Before Class K extinguishing agents were developed, most fire-extinguishing systems for kitchens used dry chemicals. The minimum requirement for a commercial kitchen was a 40-B–rated sodium bicarbonate or potassium bicarbonate fire extinguisher. These systems required extensive clean-up after their use, which often resulted in serious business interruptions. All fixed extinguishing systems installed in restaurants and commercial kitchens after 1994 use wet-chemical extinguishing agents (Griffin, 2014). These extinguishing agents are discharged as a fine spray, which reduces the risk of splattering. Clean-up afterward is much easier, allowing a business to reopen sooner.

Wet-chemical extinguishing agents are specifically formulated for use in commercial kitchens and food-product manufacturing facilities, especially where food is cooked in a deep fryer. The fixed, automatic fire-extinguishing systems discharge the agent directly over the cooking surfaces. Portable Class K wet-chemical fire extinguishers are available in three sizes: 3 L (0.8 gal), 6 L (1.5 gal), and 9.5 L (2.5 gal). There are no numerical ratings for these fire extinguishers.

SAFETY TIP

Before deciding to use a fire extinguisher, size up the fire to ensure that the extinguisher has an adequate capacity and holds the proper extinguishing agent.

Halogenated-Agent Fire Extinguishers

Halogenated-agent fire extinguishers use halogenated agents, which are produced from a family of liquefied gases, known as halogens, that includes fluorine, bromine, iodine, and chlorine. Hundreds of different formulations can be produced from these elements; these myriad versions have many different properties and potential uses. Although several of these formulations are effective for extinguishing fires, only a few of them are commonly used as extinguishing agents.

Halogenated agents leave no residue and are ideally suited for areas that contain computers or sensitive electronic equipment. Per pound, they are approximately twice as effective at extinguishing fires as is CO_2 (Conroy, 2003).

Two categories of halogenated agents are distinguished: Halons and halocarbons. A 1987 international agreement, known as the Montreal Protocol, limited Halon production and importation because these agents damage the earth's ozone layer. Since then, Halons have been replaced by halocarbons (clean agents), which are not subject to the same environmental restrictions. Both types of agents are available in hand-held fire extinguishers rated for Class B and C fires. Larger capacity models are also rated for use on Class A fires.

In this type of fire extinguisher, the agent is stored as a liquid and discharged under relatively high pressure. The fire extinguisher releases a mist of vapour and liquid droplets that disrupts the molecular chain reactions within the combustion process, thereby extinguishing the fire. Halogenated-agent fire extinguishers have a horizontal discharge stream range of 2 to 11 m (6 to 35 ft) (NFPA, *Standard for Portable Fire Extinguishers*, 2018). These agents dissipate rapidly in windy conditions, as does CO_2, so their effectiveness is limited in outdoor locations. Because halogenated agents also displace oxygen, they should be used with care in confined areas.

Currently, four types of halocarbon agents are used in portable fire extinguishers: hydrochlorofluorocarbon, hydrofluorocarbon, perfluorocarbon, and fluoroiodocarbon. Halon 1211 (bromochlorodifluoromethane) is available in hand-held stored-pressure fire extinguishers with capacities that range from 1 kg (2 lb), rated 2-B:C, to 10 kg (22 lb), rated 4-A:80-B:C. These fire extinguishers are unsuited for use on fires involving pressurized fuels or cooking grease. Wheeled Halon 1211 models are available with capacities up to 68 kg (150 lb) with a rating of 30-A:160-B:C. The wheeled fire extinguishers use a nitrogen booster charge from an auxiliary cylinder to expel the agent.

Dry-Powder Fire Extinguishers and Extinguishing Agents

Dry powder is a chemical compound used to extinguish fires involving combustible metals (Class D fires). These agents are stored in fine granular or powdered form and are applied to smother the fire. They form a solid crust over the burning metal, which both blocks out oxygen (the fuel for the fire) and absorbs heat.

The extinguishing agents and techniques required to put out Class D fires vary greatly, depending on the specific fuel, the quantity involved, and the physical form of the fuel (e.g., grindings, shavings, or solid objects). Each dry-powder extinguishing agent is listed for use on specific combustible-metal fires. The agent and the application method must be suited to the particular situation.

The most commonly used dry-powder extinguishing agent is formulated from finely ground sodium chloride (table salt) plus additives to help it flow freely over a fire. A thermoplastic material mixed with the agent binds the sodium chloride particles into a solid mass when they come into contact with a burning metal. **Dry-powder fire extinguishers** using sodium chloride–based agents are available with 14-kg (30-lb) capacity in either stored-pressure or cylinder/cartridge models. Wheeled models are available with 68- to 159-kg (150- and 350-lb) capacities.

Dry-powder fire extinguishers usually are carried on specialty apparatus such as hazardous materials units; they are typically carried on frontline apparatus only if there is a known hazard within the primary response area. They also are found in businesses or manufacturing plants where specific hazards are present. The fire extinguishers are easily identified by their yellow cylinder (**FIGURE 7-23**). They have adjustable nozzles that allow the operator to vary the flow of the extinguishing agent. When the nozzle is fully opened, the hand-held models have a range of 2 to 2.4 m (6 to 8 ft; NFPA, *Standard for Portable Fire Extinguishers*, 2018). Extension wand applicators are available to direct the discharge from a more distant position. Class D agents must be applied carefully so that the molten metal does not splatter. No water should come in contact with the burning metal, because even a trace quantity of moisture can cause a violent reaction.

The same sodium chloride–based agent that is used in portable fire extinguishers can be stored in bulk form and applied by hand. Another dry-powder extinguishing agent for Class D fires, graded granular graphite mixed with phosphorus-containing compounds, cannot be expelled from a portable fire extinguisher. Instead, this agent is produced in bulk and must be applied manually from a pail or other container using a shovel or scoop. When applied to a metal fire, the phosphorus compounds release gases that blanket the fire and cut off its supply of oxygen; the graphite absorbs heat from the fire, allowing the metal to cool below its ignition temperature. Bulk dry-powder agents are available in 18- and 23-kg (40- and 50-lb) pails and 159-kg (350-lb) drums.

Other specialized dry-powder extinguishing agents are available for fighting specific types of metal fires. For information about Class D agents and fire extinguishers, consult the manufacturer's recommendations and NFPA's *Fire Protection Handbook*.

LISTEN UP!

Dry-chemical agents and dry-powder agents have very different meanings in relation to fire extinguishers. Dry-chemical fire extinguishers are rated for Class B and C fires or for Class A, B, and C fires. Dry-powder fire extinguishers are designed for use in suppressing Class D fires. Do not confuse these two terms.

Use of Fire Extinguishers

Fire extinguishers should be simple to operate. Indeed, an individual with only basic training should be able to use most fire extinguishers safely and effectively. Every portable fire extinguisher should be labelled with printed operating instructions.

There are six basic steps in extinguishing a fire with a portable fire extinguisher:

1. Locate the fire extinguisher.
2. Select the proper classification of fire extinguisher.
3. Ensure your personal safety by having an exit route.
4. Transport the fire extinguisher to the location of the fire.
5. Activate the fire extinguisher to release the extinguishing agent.
6. Apply the extinguishing agent to the fire for maximum effect.

Although these steps are not complicated, practice and training are essential for effective fire suppression with a fire extinguisher. In particular, tests have shown that the effective use of Class B portable fire extinguishers depends heavily on user training and expertise. A trained expert can extinguish a larger fire than a nonexpert can, using the same extinguisher.

As a fire fighter, you should be able to operate any fire extinguisher that you might be required to use, whether it is carried on your fire apparatus, hanging on the wall of your firehouse, or placed in some other location in your community.

FIGURE 7-23 Dry-powder fire extinguishers are easily identified by their yellow cylinder.
© Jones & Bartlett Learning. Photographed by Glen E. Ellman.

Locating a Fire Extinguisher

As a fire fighter, you should know which types of fire extinguishers are carried on your department's apparatus and where each type of extinguisher is located. You also should know where fire extinguishers are located in and around the fire station and other workplaces. You should have at least one fire extinguisher in your home and another in your personal vehicle, and you should know exactly where they are located. Knowing the exact locations of fire extinguishers can save valuable time in an emergency.

Selecting the Proper Fire Extinguisher

Selecting the proper fire extinguisher requires an understanding of the classification and rating system for fire extinguishers. Knowing which types of agents are available, how they work, which ratings the fire extinguishers carried on your fire apparatus have, and which extinguisher is appropriate for a particular fire situation is also important.

Fire fighters should be able to assess a fire quickly, determine whether the fire can be controlled by a fire extinguisher, and identify the appropriate extinguisher. Using a fire extinguisher with an insufficient rating may not completely extinguish the fire, which can place the operator in danger of being burned or otherwise injured. If the fire is too large for the fire extinguisher, consider other options, such as obtaining additional extinguishers or making sure that a charged hose line is ready to provide backup.

Understanding the fire extinguisher rating system and the different types of agents will enable a fire fighter who must use an unfamiliar extinguisher to determine whether it is suitable for a particular fire situation. A quick look at the label should be all that is needed in such a case.

In most municipalities in Canada, fire prevention officers or fire marshals enforce the fire code within their jurisdiction and offer advice on fire extinguisher types and placement. In volunteer departments, it is usually the fire chief who enforces the fire code. Fire fighters should be able to determine the most appropriate type of fire extinguisher to place in a given area, based on the types of fires that could occur and the hazards that are present. In some cases, one type of fire extinguisher might be preferred over another. An

extinguishing agent such as CO_2 or a halogenated agent is better for a fire involving electronic equipment, for example, because it leaves no residue. A dry-chemical fire extinguisher is generally more appropriate than a CO_2 fire extinguisher to fight an outdoor fire, because wind will quickly dissipate CO_2. A dry-chemical fire extinguisher also would be the best choice for a fire involving a flammable liquid leaking under pressure from a pipe, whereas foam would be a better choice for a fire involving a liquid spill on the ground.

Ensuring Your Personal Safety

Before using a fire extinguisher, be sure to approach the fire with an exit behind you. If the fire suddenly expands or the fire extinguisher fails to control it, you must have a planned escape route. Never let the fire get between you and a safe exit. After suppressing a fire, do not turn your back on it. Always watch it, and be prepared for it to rekindle until the fire has been fully overhauled.

When ordinary civilians use fire extinguishers on incipient stage fires, they are probably wearing their normal clothing. As a fire fighter, however, you should wear your personal protective clothing and use appropriate personal protective equipment (PPE). Take advantage of the protection they provide.

If you must enter an enclosed area where a fire extinguisher has been discharged, wear full PPE, and use SCBA. The atmosphere within the enclosed area will probably contain a mixture of combustion products and extinguishing agents, and the oxygen content within the space may be dangerously low.

Transporting a Fire Extinguisher

The best method of transporting a hand-held portable fire extinguisher depends on the extinguisher's size, weight, and design. Hand-held portable models can weigh as little as 0.45 kg (1 lb) or as much as 23 kg (50 lb). The ability to handle the heavier fire extinguishers depends on an individual operator's personal strength.

Fire extinguishers with a fixed nozzle should be carried in the favoured or stronger hand. This approach enables the operator to depress the trigger and direct the discharge easily. Fire extinguishers that have a hose between the trigger and the nozzle should be carried in the weaker or less favoured hand so that the favoured hand can grip and aim the nozzle.

Heavier fire extinguishers may have to be carried as close as possible to the fire and placed upright on the ground. The operator can then depress the trigger with one hand, while holding the nozzle and directing the stream with the other hand.

To transport a fire extinguisher, follow the steps in **SKILL DRILL 7-1.**

SKILL DRILL 7-1
Transporting a Fire Extinguisher Fire Fighter I, NFPA 1001: 4.3.16

1 Locate the closest fire extinguisher.

2 Assess that the fire extinguisher is safe and effective for the type of fire being attacked. Release the mounting bracket straps.

3 Lift the fire extinguisher using good body mechanics. Lift small fire extinguishers with one hand and large extinguishers with two hands.

4 Walk briskly—do not run—toward the fire. If the fire extinguisher has a hose and nozzle, carry the extinguisher with one hand, and grasp the nozzle with the other hand.

Operating a Fire Extinguisher

Activating a fire extinguisher to apply the extinguishing agent is a single operation that involves four steps. The **PASS** acronym is a helpful way to remember these steps:

1. **Pull** the safety pin.
2. **Aim** the nozzle at the base of the flames.
3. **Squeeze** the trigger to discharge the agent.
4. **Sweep** the nozzle across the base of the flames.

Most fire extinguishers have very simple operation systems. Practise discharging different types of fire extinguishers in training situations to build confidence in your ability to use them properly and effectively.

To extinguish a Class A fire with a stored-pressure water-type fire extinguisher, follow the steps in **SKILL DRILL 7-2**.

Note: Stored-pressure water-type fire extinguishers are safe and effective to use on Class A fires. They are not effective on Class B fires, and they are not safe to use on Class C fires.

SKILL DRILL 7-2
Extinguishing a Class A Fire with a Stored-Pressure Water-Type Fire Extinguisher Fire Fighter I, NFPA 1001: 4.3.16

1 Size up the fire to determine whether a stored-pressure water-type fire extinguisher is safe and effective for the fire. Ensure the fire extinguisher is large enough to be safe and effective. Ensure your safety. Make sure you have an exit route from the fire. Do not turn your back on a fire.

2 Remove the hose and nozzle. Quickly check the pressure gauge to verify that the fire extinguisher is adequately charged.

3 Pull the pin to release the fire extinguisher control valve. You must be within 9 to 12 m (30 to 40 ft) of the fire to be effective.

4 Aim the nozzle, and sweep the water stream at the base of the flames.

(continued)

SKILL DRILL 7-2 Continued
Extinguishing a Class A Fire with a Stored-Pressure Water-Type Fire Extinguisher Fire Fighter I, NFPA 1001: 4.3.16

© Jones & Bartlett Learning. Photographed by Glen E. Ellman.

5 Overhaul the fire; take steps to prevent rekindling, break apart tightly packed fuel, and summon additional help if needed.

SKILL DRILL 7-3
Extinguishing a Class A Fire with a Multipurpose Dry-Chemical Fire Extinguisher Fire Fighter I, NFPA 1001: 4.3.16

1 Size up the fire to determine whether a multipurpose dry-chemical fire extinguisher is safe and effective for this fire. Ensure the fire extinguisher is large enough to be safe and effective. Ensure your safety. Make sure you have an exit route from the fire. Do not turn your back on a fire.

2 Remove the hose and nozzle. Quickly check the pressure gauge to verify that the fire extinguisher is adequately charged.

3 Pull the pin to release the fire extinguisher control valve. Depending on the size of the fire and fire extinguisher, you must be within 1.5 to 14 m (5 to 45 ft) of the fire to be effective.

4 Aim the nozzle, and sweep the dry-chemical discharge at the base of the flames. Coat the burning fuel with dry chemical.

5 Overhaul the fire; take steps to prevent rekindling, break apart tightly packed fuel, and summon additional help if needed.

To extinguish a Class A fire with a multipurpose dry-chemical fire extinguisher, follow the steps in **SKILL DRILL 7-3**. If outdoors, approach the fire from the upwind side to improve visibility and ensure that the extinguishing agent does not blow back in your face. Full PPE should be worn with your SCBA activated and face piece on.

Note: Multipurpose dry-chemical fire extinguishers are safe and effective on Class A, B, and C fires.

To extinguish a Class B flammable liquid spill or pool fire with a dry-chemical extinguisher, follow the steps in **SKILL DRILL 7-4**.

Note: Dry-chemical fire extinguishers are safe to use on Class B and C fires. They are effective on Class B fires, but they are not rated for Class A fires.

To extinguish a Class B flammable liquid spill or pool fire with a stored-pressure fire extinguisher containing an AFFF or FFFP extinguishing agent, follow the steps in **SKILL DRILL 7-5**.

Note: Stored-pressure foam fire extinguishers are safe to use on Class A, B, and C fires. They are effective on Class B fires by forming a foam blanket. They are effective on Class A fires by cooling and penetrating the fuel.

To operate a CO_2 fire extinguisher, follow the steps in **SKILL DRILL 7-6**.

Note: CO_2 fire extinguishers are most effective when used in areas without wind. They are designed for Class B and C fires only.

To extinguish a fire in an electrical equipment room with a halogenated stored-pressure fire extinguisher, follow the steps in **SKILL DRILL 7-7**.

Note: Halogenated stored-pressure fire extinguishers are safe to use on Class A, B, and C fires. They are effective on Class B fires but require large quantities to be effective on Class A fires.

To extinguish an incipient stage fire involving combustible metal filings or shavings (Class D fire) with a dry-powder fire extinguisher, follow the steps in **SKILL DRILL 7-8**.

Note: Dry-powder fire extinguishers are safe to use only on Class D fires. They are effective only on the metals for which they are intended.

To extinguish a fire in a deep-fat fryer with a wet-chemical fire extinguisher, follow the steps in **SKILL DRILL 7-9**.

Note: Wet-chemical fire extinguishers are designed to be used on Class K fires—that is, fires in deep-fat fryers and on grills. They are not rated for other classes of fire.

SKILL DRILL 7-4
Extinguishing a Class B Flammable Liquid Fire with a Dry-Chemical Fire Extinguisher Fire Fighter I, NFPA 1001: 4.3.16

1 Size up the fire, and ensure your safety. Check the pressure gauge.

2 Pull the pin to release the fire extinguisher valve. Depending on the size of the fire and fire extinguisher, you must be within 1.5 to 14 m (5 to 45 ft) of the fire to be effective.

(continued)

SKILL DRILL 7-4 Continued

Extinguishing a Class B Flammable Liquid Fire with a Dry-Chemical Fire Extinguisher Fire Fighter I, NFPA 1001: 4.3.16

3   Aim the nozzle, and sweep the dry-chemical discharge across the surface of the burning liquid. Start at the near edge of the fire, and work toward the back.

© Jones & Bartlett Learning. Photographed by Glen E. Ellman.

4   Overhaul the fire; take steps to prevent rekindling, keep a blanket of dry chemical over the fuel, and summon additional help if needed.

SKILL DRILL 7-5

Extinguishing a Class B Flammable Liquid Fire with a Stored-Pressure Foam Fire Extinguisher (AFFF or FFFP) Fire Fighter I, NFPA 1001: 4.3.16

1   Size up the fire, and ensure your safety. Remove the hose and nozzle. Quickly check the pressure gauge to verify that the fire extinguisher is adequately charged. Pull the pin to release the fire extinguisher control valve. You must be within 6 to 8 m (20 to 25 ft) of the fire to be effective.

SKILL DRILL 7-5 Continued
Extinguishing a Class B Flammable Liquid Fire with a Stored-Pressure Foam Fire Extinguisher (AFFF or FFFP) Fire Fighter I, NFPA 1001: 4.3.16

2 Aim the nozzle, and discharge the stream of foam so the foam drops gently onto the surface of the burning liquid at the front or the back of the container. Let the foam blanket flow across the surface of the burning liquid. Avoid splashing foam on the burning liquid, because it can cause the burning fuel to splatter.

3 Overhaul the fire; keep a thick blanket of foam intact over the hot liquid, reapply foam over any hot spots, and summon additional help if needed.

SKILL DRILL 7-6
Operating a Carbon Dioxide Fire Extinguisher Fire Fighter I, NFPA 1001: 4.3.16

1 Size up the fire to determine whether CO_2 is a safe and effective agent for this fire. Ensure the fire extinguisher is large enough to be safe and effective. Ensure your safety. Make sure you have an exit route from the fire. Do not turn your back on a fire. Remove the horn or nozzle. Pull the pin to release the fire extinguisher control valve.

2 Quickly squeeze to verify that the fire extinguisher is charged; CO_2 fire extinguishers do not have pressure gauges. You must be within 1 to 2.4 m (3 to 8 ft) of the fire to be effective.

3 Aim the horn or nozzle, and sweep at the base of the flames.

4 Overhaul the fire; take steps to prevent rekindling, and summon additional help if needed.

SKILL DRILL 7-7
Operating a Halogenated Stored-Pressure Fire Extinguisher
Fire Fighter I, NFPA 1001: 4.3.16

1. Size up the fire to determine whether a halogenated stored-pressure fire extinguisher is safe and effective for this fire. Ensure the fire extinguisher is large enough to be safe and effective.

2. Ensure your safety. Turn off electricity if possible. Make sure you have an exit route from the fire. Do not turn your back on a fire.

3. Remove the hose and nozzle. Quickly check the pressure gauge to verify that the fire extinguisher is adequately charged.

4. Pull the pin to release the fire extinguisher control valve. Depending on the size of the fire and fire extinguisher, you must be within 1 to 11 m (3 to 35 ft) of the fire to be effective.

5. Aim the nozzle at the base of the flames to sweep the flames off the surface starting at the near edge of the flames.

6. Overhaul the fire; take steps to prevent rekindling, continue to apply the extinguishing agent to cool the fuel, and summon additional help if needed.

SKILL DRILL 7-8
Operating a Dry-Powder Fire Extinguisher Fire Fighter I, NFPA 1001: 4.3.16

1. Size up the fire to determine whether a dry-powder fire extinguisher is safe and effective for this fire. Ensure the fire extinguisher is large enough to be safe and effective. Avoid water or other extinguishing agents that might react with the combustible metals.

2. Ensure your safety. Make sure you have an exit route from the fire. Do not turn your back on a fire.

3. Remove the hose and nozzle. Quickly check the pressure gauge to ensure the fire extinguisher is charged.

4. Pull the pin to release the control valve. You must be within 2 to 2.4 m (6 to 8 ft) of the fire to be effective.

5. Aim the nozzle, and fully open the valve to provide a maximum range, and then reduce the valve to produce a soft, heavy flow down to completely cover the burning metal. The method of application may vary depending on the type of metal burning and the extinguishing agent.

6. Overhaul the fire; take steps to prevent rekindling, continue to place a thick layer of the extinguishing agent over the hot metal to form an airtight blanket, allow the hot metal to cool, and summon additional help if needed.

SKILL DRILL 7-9
Operating a Wet-Chemical Fire Extinguisher Fire Fighter I, NFPA 1001: 4.3.16

1 Size up the fire, and ensure your safety.

2 Remove the hose and nozzle. Quickly check the pressure gauge to verify that the fire extinguisher is adequately charged.

3 Pull the pin to release the fire extinguisher control valve. You must be within 2.4 to 3.7 m (8 to 12 ft) of the fire to be effective.

4 Aim the nozzle, and discharge so the stream of wet chemical drops gently into the surface of the burning liquid at the front or the back of the deep-fat fryer.

5 Let the deep foam blanket flow across the surface of the burning liquid. Avoid splashing foam on the burning liquid.

6 Continue to apply the agent until the foam blanket has extinguished all of the flames.

7 Do not disturb the foam blanket even after all of the flames have been suppressed. If reignition occurs, repeat these steps.

The Care of Fire Extinguishers

Fire extinguishers must be regularly inspected and properly maintained to ensure that they will be available for use in an emergency. Records must be kept to confirm that the required inspections and maintenance have been performed on schedule. The individuals assigned to perform these functions must be properly trained and must always follow the manufacturer's recommendations for inspecting, maintaining, recharging, and testing the equipment. Persons performing maintenance or recharging of portable fire extinguishers should be qualified according to NFPA 10, *Standard for Portable Fire Extinguishers*.

Inspection

According to NFPA 10, an inspection is a "quick check" to verify that a fire extinguisher is available and ready for immediate use. Fire extinguishers on fire apparatus should be inspected as part of the regular equipment checks mandated by your department. Inspection should take place each month. The fire fighter charged with inspecting the fire extinguishers should perform the following tasks:

- Ensure that all tamper seals are intact.
- Determine fullness by weighing or "hefting" the fire extinguisher.
- Examine all parts for signs of physical damage, corrosion, or leakage.
- Check the pressure gauge to confirm that it is in the operable range.
- Ensure that the fire extinguisher is properly identified by type and rating.
- Check the hose and nozzle for damage or obstruction by foreign objects.
- Check the hydrostatic test date of the fire extinguisher.

If an inspection reveals any problems, the fire extinguisher should be removed from service until the required maintenance procedures are performed. Spare fire extinguishers should be used until the problem is corrected.

The pressure gauge on a stored-pressure fire extinguisher indicates whether the pressure is sufficient to expel the entire agent. Some stored-pressure fire extinguishers use compressed air, whereas others use compressed nitrogen. The weight of the fire extinguisher and the presence of an intact tamper seal should indicate that the unit is full of extinguishing agent.

Cartridge-type fire extinguishers contain a predetermined quantity of a pressurizing gas that expels the agent. This gas is released only when the cartridge is punctured. A properly charged fire extinguisher will be full of extinguishing agent and will have a cartridge that has not been punctured.

Self-expelling agents, such as CO_2 and some halogenated agents, do not require a separate gas cartridge. The only way to determine whether a CO_2 fire extinguisher is properly charged is to weigh it. The proper weight should be indicated on the outside of the fire extinguisher.

Maintenance

The maintenance requirements and intervals for various types of fire extinguishers are outlined in NFPA 10. Maintenance includes an internal inspection as well as any repairs that may be required. These procedures must be performed periodically, depending on the type of fire extinguisher. An inspection also may reveal the need to perform maintenance procedures.

Maintenance procedures must always be performed by a qualified person. Some procedures can be performed only at a properly licensed facility. The specific qualifications and training requirements are determined by the manufacturer and the jurisdictional authority. Untrained personnel should never be allowed to perform fire extinguisher maintenance.

Common indications that a fire extinguisher needs maintenance include the following findings:

- The pressure gauge reading is outside the normal range.
- The inspection tag is out-of-date.
- The tamper seal is broken, especially in fire extinguishers with no pressure gauge.
- The fire extinguisher does not appear to be full of extinguishing agent.
- The hose or nozzle assembly is obstructed.
- There are signs of physical damage, corrosion, or rust.
- There are signs of leakage around the discharge valve or nozzle assembly.

Recharging

All performance standards assume that a fire extinguisher will be fully charged when it has to be used. A rechargeable fire extinguisher must be taken out of service immediately and recharged after each and every use, even if it was not completely discharged. This guideline also applies to any fire extinguisher that leaks or has a pressure gauge reading above or below the proper operating range. Most rechargeable fire extinguishers must be recharged by qualified personnel. Nonrechargeable fire extinguishers should be replaced after any use.

When a fire extinguisher is recharged, the extinguishing agent is refilled, and the system that expels the agent is properly pressurized. Both the quantity of the agent and the pressurization must be verified. After recharging is complete, a tamper seal is installed; this seal provides assurance that the fire extinguisher has not been fully or even partially discharged since the last time it was recharged.

It is recommended that fire extinguishers be recharged by a trained individual with the proper equipment or a fire extinguisher company that maintains fire extinguishers. If that is not possible, however, typical 9.5-L (2.5-gal) stored-pressure water-type fire extinguishers can be recharged by fire fighters using water and a source of compressed air. Before the fire extinguisher is refilled, all remaining stored pressure must be discharged so that the valve assembly can be safely removed. Water is then added up to the water-level indicator, and the valve assembly is replaced. Finally, compressed air is introduced to raise the pressure to the level indicated on the gauge.

Hydrostatic Testing

Most fire extinguishers are pressurized vessels, meaning that they are designed to hold a steady internal pressure. The ability of a fire extinguisher to withstand this internal pressure is measured by periodic **hydrostatic testing**. In most provinces and territories, the hydrostatic testing requirements for fire extinguishers are established by the Standards Council of Canada (CAN/ULC-S508) and can be found in NFPA 10. Such tests are conducted in a special test facility and involve filling the fire extinguisher with water and applying above-normal pressure.

Each fire extinguisher has an assigned maximum interval between hydrostatic tests, usually 5 or 12 years, depending on the construction material and vessel type. The date of the most recent hydrostatic test must be indicated on the outside of the fire extinguisher. A fire extinguisher may not be refilled if the date of the most recent hydrostatic test is not within the prescribed limit. Any fire extinguisher that is out-of-date should be removed from service and sent to the appropriate maintenance facility for hydrostatic testing.

After-Action REVIEW

IN SUMMARY

- Fire extinguishers are used successfully to put out hundreds of fires every day, preventing millions of dollars in property damage and saving uncounted lives.

- Portable fire extinguishers have two primary uses: to extinguish incipient stage fires and to control fires where traditional methods of fire suppression, such as use of a hose line, are not recommended.

- The primary disadvantage of fire extinguishers is that they are "one-shot" devices.

- The five classes of fires affect the choice of extinguishing equipment:
  - Class A fires involve ordinary combustibles such as wood, paper, cloth, rubber, household rubbish, and some plastics.
  - Class B fires involve flammable or combustible liquids, such as gasoline, oil, grease, tar, lacquer, oil-based paints, and some plastics.
  - Class C fires involve energized electrical equipment, which includes any device that uses, produces, or delivers electrical energy.
  - Class D fires involve combustible metals such as magnesium, titanium, zirconium, sodium, lithium, and potassium. Special techniques and extinguishing agents are required to fight combustible-metal fires.
  - Class K fires involve combustible cooking oils and fats.

- Portable fire extinguishers are classified and rated based on their extinguishing properties and capabilities.

- The classification system for fire extinguishers uses both letters and numbers. The letters indicate the classes of fire for which the fire extinguisher can be used, and the numbers indicate its relative effectiveness. Fire extinguishers that are safe and effective for more than one class are rated with multiple letters. Numbers are used to rate a fire extinguisher's effectiveness only for Class A and Class B fires.

- There are two fire extinguisher labelling systems: the traditional lettering system and the pictograph system.
  - Fire extinguishers suitable for use on Class A fires are identified by the letter "A" on a solid green triangle or by an icon showing a burning trash can beside a wood fire.
  - Fire extinguishers suitable for use on Class B fires are identified by the letter "B" on a solid red square or by an icon showing a flame and gasoline can.
  - Fire extinguishers suitable for use on Class C fires are identified by the letter "C" on a solid blue circle or by an icon showing a flame, electrical plug, and socket.
  - Fire extinguishers suitable for use on Class D fires are identified by the letter "D" on a solid yellow, five-pointed star or by an icon showing a flame and a metal gear.
  - Fire extinguishers suitable for use on Class K (combustible cooking oil) fires are identified by an icon showing a fire in a frying pan. Because the Class K designation is new, there is no traditional-system alphabet graphic for it.

- NFPA 10 lists the requirements for placing and mounting portable fire extinguishers of different weights. Following these requirements ensures safe fire extinguisher storage and accessibility.

- Areas are divided into three risk classifications based on the fire risks associated with the materials in those areas:
  - Light (low) hazard locations
  - Ordinary (moderate) hazard locations
  - Extra (high) hazard locations

- Most fire extinguishers stop fires by cooling the fuel below its ignition temperature, by cutting off the supply of oxygen, or by combining the two techniques.

- Portable fire extinguishers use seven basic types of extinguishing agents:
  - Water—Used to extinguish Class A fires.
  - Dry chemicals—Used mainly to extinguish Class B and C fires. This type of fire extinguisher can be used to suppress Class A fires, but water is also needed to fully extinguish any embers.
  - Carbon dioxide—Used to extinguish Class B and C fires only.
  - Foam—Used to extinguish Class A or B fires. Be sure to double-check the type of foam in the fire extinguisher. Most Class B foams can be used on Class A fires, but Class A foams are not effective on Class B fires.
  - Halogenated agents—Include Halons and halocarbons. Halons are limited-use chemicals due to their propensity to damage the environment. These agents are used to extinguish Class B and C fires.
  - Dry powders—Used to extinguish Class D fires.
  - Wet chemicals—Used to extinguish Class K fires.
- Most hand-held portable fire extinguishers have six basic parts:
  - A cylinder or container that holds the extinguishing agent
  - A carrying handle
  - A nozzle or horn
  - A trigger and discharge valve assembly
  - A locking mechanism to prevent accidental discharge
  - A pressure indicator or gauge that shows whether a stored-pressure fire extinguisher has sufficient pressure to operate properly
- Portable fire extinguishers vary according to their mechanical design, extinguishing agent, capacity, effective range, and the time it takes to completely discharge the extinguishing agent.
  - Stored-pressure water-type fire extinguishers—Expel water in a solid stream with a range of 11 to 12 m (35 to 40 ft) through a nozzle at the end of a short hose.
  - Water mist fire extinguishers—Discharge a fine spray of water mist that cools and soaks the fuel and provides safety from electrical hazard.
  - Loaded-stream fire extinguishers—Discharge a solution of water containing a chemical to prevent freezing.
  - Wetting-agent fire extinguishers—Expel water that contains a solution to reduce its surface tension, which enables water to penetrate more efficiently into Class A fuels.
  - Pump tank fire extinguishers—Do not store water under pressure, but instead expel it through use of a hand-operated, double-acting, vertical piston pump.
  - Backpack fire extinguishers—Five-gallon water tanks used primarily outdoors for fighting brush and grass fires.
  - Ordinary dry-chemical fire extinguishers—Available in hand-held models and larger, wheeled units to extinguish Class B and C fires.
  - Multipurpose dry-chemical fire extinguishers—Available in hand-held models and larger, wheeled units to extinguish Class A, B, and C fires.
  - Carbon dioxide fire extinguishers—Include a siphon tube and horn or cone-shaped applicator that is used to direct the flow of the agent. These fire extinguishers have a relatively short discharge range of 1 to 2.4 m (3 to 8 ft).
  - Class B foam fire extinguishers—Very similar in appearance and operation to water-type fire extinguishers, except these extinguishers have an aspirating nozzle.
  - Wet-chemical fire extinguishers (Class K)—Available in three sizes, 3 L (0.8 gal), 6 L (1.5 gal), and 9.5 L (2.5 gal).

- Halogenated-agent fire extinguishers—Available in hand-held fire extinguishers for Class B and C fires and larger capacity models for use on Class A fires.

- Dry-powder fire extinguishers—Rely on sodium chloride–based agents. Models with 14-kg (30-lb) capacity are available in either stored-pressure or cylinder/cartridge versions. Wheeled models are available with 68- and 159-kg (150- and 350-lb) capacities.

- There are six basic steps in extinguishing a fire with a portable fire extinguisher:

  - Locate the fire extinguisher.

  - Select the proper classification of fire extinguisher.

  - Ensure your personal safety by having an exit route.

  - Transport the fire extinguisher to the location of the fire.

  - Activate the fire extinguisher to release the extinguishing agent.

  - Apply the extinguishing agent to the fire for maximum effect.

- Activating a fire extinguisher to apply the extinguishing agent is a single operation that involves four steps. The PASS acronym is a helpful way to remember these steps:

  1. **P**ull the safety pin.

  2. **A**im the nozzle at the base of the flames.

  3. **S**queeze the trigger to discharge the agent.

  4. **S**weep the nozzle across the base of the flames.

- When using a fire extinguisher, always approach the fire with an exit behind you.

- Fire extinguishers must be regularly inspected and properly maintained to ensure that they will be available for use in an emergency. The fire fighter charged with inspecting the fire extinguishers should perform the following tasks:

  - Ensure that all tamper seals are intact.

  - Determine fullness by weighing or "hefting" the fire extinguisher.

  - Examine all parts for signs of physical damage, corrosion, or leakage.

  - Check the pressure gauge to confirm that it is in the operable range.

  - Ensure that the fire extinguisher is properly identified by type and rating.

  - Check the hose and nozzle for damage or obstruction by foreign objects.

  - Check the hydrostatic test date of the fire extinguisher.

- Common indications that a fire extinguisher needs maintenance include the following findings:

  - The pressure gauge reading is outside the normal range.

  - The inspection tag is out-of-date.

  - The tamper seal is broken, especially in fire extinguishers with no pressure gauge.

  - The fire extinguisher does not appear to be full of extinguishing agent.

  - The hose or nozzle assembly is obstructed.

  - There are signs of physical damage, corrosion, or rust.

  - There are signs of leakage around the discharge valve or nozzle assembly.

- A fire extinguisher must be recharged after each and every use, even if it was not completely discharged. The only exceptions are nonrechargeable extinguishers, which should be replaced after any use.

- Most fire extinguishers are pressurized vessels, meaning that they are designed to hold a steady internal pressure. The ability of a fire extinguisher to withstand this internal pressure is measured by periodic hydrostatic testing. These tests, which are conducted in a special test facility, involve filling the fire extinguisher with water and applying above-normal pressure.

KEY TERMS

Access Navigate for flashcards to test your key term knowledge.

Ammonium phosphate An extinguishing agent used in dry-chemical fire extinguishers that can be used on Class A, B, and C fires.

Aqueous film-forming foam (AFFF) A solution based on fluorinated surfactants plus foam stabilizers to produce a fluid aqueous film for suppressing liquid fuel vapours. (NFPA 10)

Carbon dioxide A colourless, odourless, electrically nonconductive inert gas that is a suitable medium for extinguishing Class B and Class C fires. (NFPA 10)

Carbon dioxide (CO_2) fire extinguisher A fire extinguisher that uses carbon dioxide gas as the extinguishing agent. It is rated for use on Class B and C fires.

Cartridge/cylinder-operated fire extinguisher A fire extinguisher in which the expellant gas is in a separate container from the agent storage container. (NFPA 10)

Class A fire A fire in ordinary combustible materials, such as wood, cloth, paper, rubber, and many plastics. (NFPA 10)

Class B fire A fire in flammable liquids, combustible liquids, petroleum greases, tars, oils, oil-based paints, solvents, lacquers, alcohols, and flammable gases. (NFPA 10)

Class C fire A fire that involves energized electrical equipment. (NFPA 10)

Class D fire A fire in combustible metals, such as magnesium, titanium, zirconium, sodium, lithium, and potassium. (NFPA 10)

Class K fire A fire in a cooking appliance that involves combustible cooking media (vegetable or animal oils and fats). (NFPA 10)

Clean agent Electrically nonconducting, volatile, or gaseous fire extinguishant that does not leave a residue upon evaporation. (NFPA 10)

Cylinder The body of the fire extinguisher where the extinguishing agent is stored.

Dry chemical A powder composed of very small particles, usually sodium bicarbonate, potassium bicarbonate, or ammonium phosphate based with added particulate material supplemented by special treatment to provide resistance to packing, resistance to moisture absorption (caking), and the proper flow capabilities. (NFPA 10)

Dry-chemical fire extinguisher A fire extinguisher that uses a powder composed of very small particles, usually sodium bicarbonate, potassium bicarbonate, or ammonium phosphate, based with added particulate

material supplemented by special treatment to provide resistance to packing, resistance to moisture absorption (caking), and the proper flow capabilities. These fire extinguishers are rated for use on Class B and C fires, although some are also rated for Class A fires.

Dry powder Solid materials in powder or granular form designed to extinguish Class D combustible metal fires by crusting, smothering, or heat-transferring means. (NFPA 10)

Dry-powder fire extinguisher A fire extinguisher that uses solid materials in powder or granular form to extinguish Class D combustible metal fires by crusting, smothering, or heat-transferring means.

Extinguishing agent A material used to stop the combustion process. Extinguishing agents may include liquids, gases, dry-chemical compounds, and dry-powder compounds.

Extra (high) hazard locations Occupancies where the total amounts of Class A combustibles and Class B flammables are greater than expected in occupancies classed as ordinary (moderate) hazards. The combustibility and heat release rate of the materials are high.

Film-forming fluoroprotein (FFFP) foam A protein-foam solution that uses fluorinated surfactants to produce a fluid aqueous film for suppressing liquid fuel vapours. (NFPA 10)

Fire load The total energy content of combustible materials in a building, space, or area including furnishing and contents and combustible building elements expressed in MJ. (NFPA 557)

Halocarbon Halocarbon agents include hydrochlorofluorocarbon (HCFC), hydrofluorocarbon (HFC), perfluorocarbon (PFC), fluoroiodocarbon (FIC) types of agents, and other halocarbons that are found acceptable under the Environmental Protection Agency Significant New Alternatives Policy program. (NFPA 10)

Halogenated agent A liquefied gas extinguishing agent that extinguishes fire by chemically interrupting the combustion reaction between fuel and oxygen. Halogenated agents leave no residue. (NFPA 402)

Halogenated-agent fire extinguisher A fire extinguisher that uses a halogenated extinguishing agent; also called a clean agent fire extinguisher.

Halons Halons include bromochlorodifluoromethane (Halon 1211), bromotrifluoromethane (Halon 1301), and mixtures of Halon 1211 and Halon 1301 (Halon 1211/1301). (NFPA 10)

Halon 1211 A halogenated agent whose chemical name is bromochlorodifluoromethane ($CBrClF_2$) and that is a multipurpose, Class ABC–rated agent effective against flammable liquid fires. (NFPA 408)

Handle The grip used for holding and carrying a portable fire extinguisher.

Horn The tapered discharge nozzle of a carbon dioxide fire extinguisher.

Hydrostatic testing Pressure testing of a fire extinguisher to verify its strength against unwanted rupture. (NFPA 10)

Light (low) hazard locations Occupancies where the quantity, combustibility, and heat release of the materials is low, and the majority of materials are arranged so that a fire is not likely to spread.

Loaded-stream fire extinguisher A water-based fire extinguisher that uses an alkali metal salt as a freezing-point depressant.

Locking mechanism A device that locks a fire extinguisher's trigger to prevent its accidental discharge.

Multipurpose dry-chemical fire extinguisher A fire extinguisher that uses an ammonium phosphate–based extinguishing agent that is effective on fires involving ordinary combustibles, such as wood or paper, and fires involving flammable liquids. It is rated to fight Class A, B, and C fires.

Nozzle A device for use in applications requiring special water discharge patterns, directional spray, or other unusual discharge characteristics. (NFPA 13)

Ordinary (moderate) hazard locations Occupancies that contain more Class A and Class B materials than are found in light hazard locations. The combustibility and heat release rate of the materials is moderate.

PASS Acronym for the steps involved in operating a portable fire extinguisher: Pull pin, Aim nozzle, Squeeze trigger, Sweep across burning fuel.

Polar solvent A water-soluble flammable liquid such as alcohol, acetone, ester, and ketone.

Pressure indicator A gauge on a pressurized portable fire extinguisher that indicates the internal pressure of the expellant.

Pump tank fire extinguisher A nonpressurized, manually operated water-type fire extinguisher that is rated for use on Class A fires. Discharge pressure is provided by a hand-operated, double-acting piston pump.

Saponification The process of converting the fatty acids in cooking oils or fats to soap or foam; the action caused by a Class K fire extinguisher.

Self-expelling agent An agent that has sufficient vapour pressure at normal operating temperatures to expel itself from a fire extinguisher.

Stored-pressure fire extinguisher A fire extinguisher in which both the extinguishing agent and expellant gas are kept in a single container and that includes a pressure indicator or gauge. (NFPA 10)

Stored-pressure water-type fire extinguisher A fire extinguisher in which water or a water-based extinguishing agent is stored under pressure.

Tamper seal A retaining device that breaks when the locking mechanism is released.

Trigger The button or lever used to discharge the agent from a portable fire extinguisher.

Underwriters Laboratory of Canada (ULC) The organization that tests and certifies that fire extinguishers (among many other products) meet established standards.

Water mist fire extinguisher A fire extinguisher containing distilled or de-ionized water and employing a nozzle that discharges the agent in a fine spray. (NFPA 10)

Wet-chemical extinguishing agent Normally an aqueous solution of organic or inorganic salts or a combination thereof that forms an extinguishing agent. (NFPA 10)

Wet-chemical fire extinguisher A fire extinguisher containing a wet-chemical extinguishing agent for use on Class K fires.

Wetting-agent fire extinguisher A fire extinguisher that expels water combined with a concentrate to reduce the surface tension and increase its ability to penetrate and spread.

Wheeled fire extinguisher A portable fire extinguisher equipped with a carriage and wheels intended to be transported to the fire by one person. (NFPA 10)

REFERENCES

Conroy, Mark. *NFPA Guide to Portable Fire Extinguishers*. Burlington, MA: Jones & Bartlett Learning, 2003.

Griffin, Bill. "60 Years of Commercial Kitchen Fire Suppression." *ASHRAE Journal*. June 2014.

National Fire Protection Association. *Home Structure Fires*. Quincy, MA: 2017.

National Fire Protection Association. NFPA 10, *Standard for Portable Fire Extinguishers*, 2018. www.nfpa.org. Accessed November 5, 2018.

National Fire Protection Association. NFPA 11, *Standard for Low-, Medium-, and High-Expansion Foam*, 2016. www.nfpa.org. Accessed November 5, 2018.

Underwriters Laboratories. "Extinguishers and Extinguishing System Units, FWFZ." 2017. http://productspec.ul.com /document.php?id=FWFZ.GuideInfo. Accessed November 5, 2018.

On Scene

You and your crew are at a local high school giving a tour of the pumper. You get to the compartment that has three fire extinguishers in it. You explain that the first is a wetting-agent fire extinguisher, the second is a dry-chemical fire extinguisher, and the third is a CO_2 fire extinguisher. One of the students asks if there are other types of fire extinguishers, and you respond that there are. The teacher then tells you that they have been studying the chemistry of fire over the past 2 weeks and asks you to explain each type of fire extinguisher and how each works to extinguish a fire. The other crew members chuckle to themselves as they watch you get tested over fire extinguishers.

1. Which class of fire involves combustible cooking media such as vegetable oils, animal oils, and fats?

   A. Class A

   B. Class B

   C. Class J

   D. Class K

2. Which fire extinguisher would you use on a Class C fire?

   A. Dry powder

   B. Film-forming fluoroprotein (FFFP) foam

   C. Halon 1211

   D. Wet chemical

3. Which fire extinguisher can typically be used to extinguish fires in ordinary combustibles, combustible liquids, and charged electrical equipment?

   A. Wet-chemical fire extinguisher

   B. Multipurpose dry-chemical fire extinguisher

   C. Wetting-agent fire extinguisher

   D. CO_2 fire extinguisher

4. What does the acronym "PASS" stand for?

   A. Press, Aim, Squeeze, Sweep

   B. Pull, Aim, Squeeze, Sweep

   C. Peek, Aim, Squeeze, Sweep

   D. Poke, Aim, Squeeze, Sweep

(continued)

On Scene Continued

5. Hydrostatic testing confirms that the fire extinguisher:

A. has sufficient strength to withstand the internal pressures.

B. will properly expel the propellant when activated.

C. can be used on electrical fires.

D. contains a water-based extinguishing agent.

6. Which of the following is a polar solvent?

A. Alcohol

B. Gasoline

C. Diesel fuel

D. Water

7. What is the term for the retaining device that breaks when the locking mechanism is released?

A. Handle

B. Trigger

C. Tamper seal

D. Horn

8. How do you determine if a carbon dioxide fire extinguisher is completely filled?

A. Examine the fire extinguisher.

B. Check the pressure gauge to assure adequate pressure.

C. Weigh the fire extinguisher.

D. Shake the fire extinguisher.

Access Navigate to find answers to this On Scene, along with other resources such as an audiobook.

CHAPTER **8**

Fire Fighter I

Fire Fighter Tools and Equipment

KNOWLEDGE OBJECTIVES

After studying this chapter, you will be able to:

- Describe the general purposes of tools and equipment. (pp. 266–267)
- Describe the safety considerations for the use of tools and equipment. (p. 266)
- Describe why it is important to use tools and equipment effectively. (p. 267)
- Describe why it is important for you to know where tools are stored. (**NFPA 1001: 4.5.1**, p. 267)
- List and describe tools and equipment that are used for rotating. (pp. 267, 269–270)
- List and describe tools and equipment that are used for pushing or pulling. (pp. 270–271)
- List and describe tools and equipment that are used for prying or spreading. (pp. 271–273)
- List and describe tools and equipment that are used for striking. (pp. 273–275)
- List and describe tools and equipment that are used for cutting. (pp. 275–277, 278)
- Describe the tools used in response and size-up activities. (pp. 278–279)
- Describe the tools used in a forcible entry. (pp. 279–280)

- Describe the tools used during an interior attack. (pp. 280–281)
- Describe the tools used during search and rescue operations. (p. 281)
- Describe the tools used during ventilation operations. (pp. 281–282)
- Describe the tools used during salvage and overhaul operations. (p. 282)
- Describe the importance of properly maintaining tools and equipment. (**NFPA 1001: 4.5.1**, p. 283)
- Describe the supplies needed to clean and inspect hand tools. (**NFPA 1001: 4.5.1**, p. 283)
- Explain the importance of replacing tools in their assigned locations. (**NFPA 1001: 4.5.1**, p. 283)
- Identify procedures, including reporting requirements, for removing a damaged tool from service. (**NFPA 1001: 4.5.1**, p. 283)

SKILLS OBJECTIVES

After studying this chapter, you will be able to perform the following skill:

- Clean and inspect hand tools. (**NFPA 1001: 4.5.1**, p. 283)

Fire Alarm

You are excited about getting to travel to another station to work on a pump company. Ever since you were in recruit class, pump work has appealed to you. You have seen the crew dismount the pumper with their tools in hand, heading to the structure to force open a door. You are excited knowing that today that fire fighter could be you.

It isn't long before the alarm sounds, "Person trapped, motor vehicle collision, Fourth and Elm," directs the dispatcher. You slip into your gear, take your seat, and put on your seat belt. As the pump company pulls out from the station, you mull over how this changes your plan for what you thought you would be doing today.

1. How will your role be different for this particular call?

2. What kinds of tools will you need for this job?

3. What other kind of knowledge will you need?

 *Access Navigate for more practice activities.*

Introduction

Fire fighters use tools and equipment to perform a wide range of activities. A fire fighter must know how to use tools effectively, efficiently, and safely, even when it is dark or visibility is limited. This chapter provides an overview of the general functions of the most commonly used tools and equipment and discusses how they are used during fire suppression and rescue operations. As you will see, the same tools may be used in different ways during each phase of fire suppression or rescue operations. This chapter also explains how to maintain tools and equipment so that they will always be ready for emergency use.

General Considerations

Tools and equipment are used in almost all fire suppression and rescue operations. As you progress through your training, you will learn how to operate the different types of tools and equipment used by your department.

Hand tools are used to extend or multiply the actions of your body and to increase your effectiveness in performing specific functions. Most of these tools operate using simple mechanical principles: A pike pole extends your reach, allows you to penetrate through a ceiling, and enables you to apply force to pull down ceiling material; an axe multiplies the cutting force you can exert on a given area. By contrast, power tools and equipment use an external source of power, such as an electric motor or an internal combustion engine. In certain cases, they are faster and more efficient than hand tools.

Safety

Safety is a prime consideration when using any kind of tool or equipment. You need to operate the equipment so that you, your fellow fire fighters, victims, and bystanders are not accidentally injured. Personal safety also requires the use of the proper personal protective equipment (PPE):

- Approved helmet
- Protective hood
- Eye protection
- Face shield
- Firefighting gloves
- Structural firefighting protective coat
- Structural firefighting protective pants
- Boots
- Self-contained breathing apparatus (SCBA) with personal alert safety system (PASS)

Conditions of Use/Operating Conditions

The best way to learn how to use tools and equipment properly is under optimal conditions of visibility and safety. In the beginning, you must be able to see what you are doing and practise without endangering yourself and others. As you become more proficient, you should practise using tools and equipment under more difficult working conditions. Eventually, you should be able to use tools and equipment safely and effectively even when darkness or smoke decreases visibility. You must be able to work safely in hazardous areas, from a ladder or pitched roof, where you are surrounded by noise and other activities, while wearing all of your protective clothing, and while using your SCBA. For this reason, some departments require fire fighters to practise certain skills and evolutions in total darkness or with their face masks covered to simulate the darkness of actual fires.

Effective Use

The first rule of effective tool use is to properly understand the problem and select the right tool for the job. The second rule is to fully understand the capabilities and limitations of each tool so that you can maximize its effectiveness. Time and energy can be wasted if you begin utilizing the wrong tool for the work required. A perfect example is when a fire fighter uses an axe to force a steel door when a Halligan tool or sledge hammer would be more effective. When you are assigned a task on the fire ground, your objective is to complete that task safely and quickly. If you waste energy by working inefficiently, you will not be able to perform additional tasks. If you know which tools and equipment are needed for each phase of firefighting, you will be able to achieve the desired objective quickly and have the energy needed to complete the remaining tasks. That being said, it is also important to pace yourself. Exceeding your personal limits will prevent you from completing the required tasks.

New fire fighters are often surprised by the strength and energy required to perform many tasks. The best tool on the fire ground is a firefighter who is fit, thinks his or her way through problems, and understands how to use the tools required to perform his or her duty. An aggressive, continuous program of physical fitness will enable you to maintain your body in the optimal state of readiness. It does little good to practise using a tool for hours if you are not in good physical condition.

As your training continues, you will learn which tools and equipment are used during different phases of fire-ground operations. For example, the tools needed for forcible entry differ from the tools needed for overhaul. Knowing which tools are needed for the work that must be done will help you prepare for the different tasks that unfold on the scene of a fire. The specific steps for properly using tools and equipment are presented in this chapter and throughout this text; your fire department instructors will demonstrate them as well.

Most fire departments have standard operating procedures (SOPs) or standard operating guidelines (SOGs) that specify which tools and equipment are needed in various situations. As a fire fighter, you must know where every tool and piece of equipment are carried on your apparatus. Knowing how to use a piece of equipment does you no good if you cannot find it quickly. Your company officer and your department's SOPs will assist you in knowing which tools to bring to different situations.

Some fire fighters carry a selection of small tools and equipment in the pockets of their coats or pants. Check whether your department requires you to carry certain tools and equipment at all times, and ask senior fire fighters for recommendations about which tools and equipment to carry.

Functions

Certain firefighting tools are carried by pump companies, ladder companies, and rescue squads. Other tools are carried only by specific types of companies. In the beginning, it may seem like the number of different types of tools is overwhelming. Often, the easiest way to learn and remember these tools is to group them by the function that each performs (**TABLE 8-1**). Most of the tools carried by fire departments fit into the following functional categories:

- Rotating (assembly or disassembly) (**FIGURE 8-1**)
- Prying or spreading (**FIGURE 8-2**)
- Pushing or pulling (**FIGURE 8-3**)
- Striking (**FIGURE 8-4**)
- Cutting (**FIGURE 8-5**)
- Multiple use (**FIGURE 8-6**)

Rotating Tools

Rotating tools apply a rotational force to make something turn. The most commonly used rotating tools are screwdrivers, wrenches, and pliers, which are used to assemble (fit together) or disassemble (take apart) parts that are connected with threaded fasteners (**TABLE 8-2**).

Examples of rotating tools include the following items (**FIGURE 8-7**):

- **Box-end wrench**: A hand tool with a closed end that is used to tighten or loosen bolts. Some styles of box-end wrenches have ratchets (a mechanism that will rotate only one way) for easier use.
- **Gripping pliers**: A hand tool with a pincer-like working end that can also be used to bend wire or hold smaller objects.
- **Hydrant wrench**: A hand tool used to open or close a hydrant by rotating the valve stem; also

TABLE 8-1	Tool Functions
Function	**Tool**
Rotate	Box-end wrench Gripping pliers Hydrant wrench Open-end wrench Pipe wrench Screwdriver Socket wrench Spanner wrench
Push or pull	Ceiling hook Clemens hook Drywall hook K tool Multipurpose hook Pike pole Plaster hook Roofman's hook Shove knife San Francisco hook
Pry or spread	Claw bar Crowbar Flat bar Halligan tool Hux bar Hydraulic spreader Kelly tool Pry bar Rabbit tool
Strike	Battering ram Chisel Flat-head axe Hammer Mallet Maul Pick-head axe Sledgehammer Spring-loaded center punch
Cut	Axe Bolt cutter Chainsaw Circular saw Cutting torch Hacksaw Handsaw Hydraulic shears Reciprocating saw Rotary saw Seat belt cutter

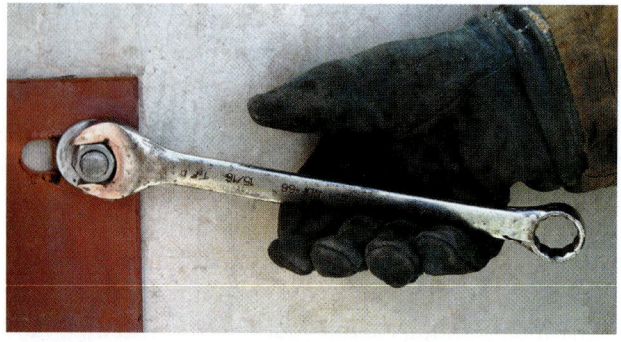

FIGURE 8-1 Rotate, assemble, or disassemble.
© Jones & Bartlett Learning.

FIGURE 8-2 Pry or spread.
© Jones & Bartlett Learning.

FIGURE 8-3 Push or pull.
© Jones & Bartlett Learning. Photographed by Glen E. Ellman.

FIGURE 8-4 Strike.
© Jones & Bartlett Learning.

FIGURE 8-5 Cut.
© Jones & Bartlett Learning. Photographed by Glen E. Ellman.

used to remove the caps from the hydrant outlets. Some versions are plain wrenches, whereas others have a ratchet feature (**FIGURE 8-8**) (also referred to as a hydrant key).

- **Open-end wrench**: A hand tool with an open end that is used to tighten or loosen bolts.
- **Combination wrench**: A hand tool with an open-end wrench on one end and a box-end wrench on the other.

FIGURE 8-6 A combination of functions.
© Jones & Bartlett Learning. Photographed by Glen E. Ellman.

TABLE 8-2 Tools for Assembly and Disassembly

- Screwdrivers
- Box-end wrenches
- Pipe wrenches
- Hydrant wrenches
- Socket wrenches
- Open-end wrenches
- Spanner wrenches
- Gripping pliers

FIGURE 8-7 Hydrant wrench (hydrant key), two spanner wrenches, pipe wrench, open-end/box-end wrench, gripping pliers, screwdriver, and socket wrench.
© Jones & Bartlett Learning.

FIGURE 8-8 A hydrant wrench in action.
© Jones & Bartlett Learning. Photographed by Glen E. Ellman.

- **Pipe wrench**: A wrench with one fixed grip and one movable grip that can be adjusted to fit securely around pipes and other tubular objects.
- **Screwdriver**: A tool that is used to turn screws.
- **Socket wrench**: A wrench that fits over a nut or bolt and uses the action of an attached handle to tighten or loosen the nut or bolt.
- **Spanner wrench**: A special wrench used to tighten or loosen fire hose **couplings** (a set or pair of connection devices that allow a fire hose to be interconnected with additional lengths of hose). Spanner wrenches come in several sizes.

Assembling and disassembling are basic mechanical skills that are routinely employed by fire fighters to solve problems. Most fire apparatus carries a tool kit with a selection of screwdrivers with different heads, open-end wrenches, box wrenches, socket wrenches, adjustable wrenches, pipe wrenches, and pliers.

Various sizes and types of screw heads are available, including slotted head, Phillips head, Roberts head, and others. A screwdriver with interchangeable heads is sometimes more useful than a selection of different screwdrivers. Pliers and wrenches come in a variety of shapes and sizes for different applications.

Pushing/Pulling Tools

A second group of tools are those used for pushing or pulling. These tools can extend the reach of the fire fighter as well as increase the power the fire fighter can exert upon an object (**TABLE 8-3**). Pushing and pulling tools have many different uses in fire department operations.

An example of a tool that extends the fire fighter's reach in order to poke through and hook material is a **pike pole** (**FIGURE 8-9**). Pike poles have a wide variety of uses. Pike poles can be used for overhauling ceilings and walls and pulling apart garbage and stuffing from furniture or mattresses, and they can also be used in the construction of water chutes. This wood or fibreglass pole has a metal head attached to one end. It is

TABLE 8-3 Tools for Pushing and Pulling
- Ceiling hook
- Multipurpose hook
- Roofman's hook
- Clemens hook
- Pike pole
- San Francisco hook
- Drywall hook
- Plaster hook
- Shove knife

FIGURE 8-9 A pike pole in action.
© Jones & Bartlett Learning.

used primarily to pull down a ceiling in an effort to get to the seat of a fire burning above. The metal head has a sharpened point that can be punched through the ceiling and a hook that can grab and pull the ceiling down. A common mistake made by new fire fighters when they first learn to use a pike pole is to punch through and pull without turning the tool 90 degrees to ensure the hook properly grabs the material. A fire fighter who is proficient in using a pike pole can be extremely effective on the fire ground. The proper use of a pike pole is discussed in depth in Chapter 19, *Salvage and Overhaul*.

Pike poles come in several different sizes and with a variety of heads (**FIGURE 8-10**). The most common length of 1 to 2 m (4 to 6 ft) enables a fire fighter to stand on a floor and pull down a 3-m-high (10-ft) ceiling. **Closet hooks**, which are intended for use in tight spaces, are commonly 0.6 to 1 m (2 to 4 ft) long. Some pike poles are equipped with handles as long as 3.7 to 4 m (12 or 14 ft) for use in rooms with very high ceilings; others may have a "D"-type handle for better pulling power. It is important to bring the right-sized pike pole to a fire. If the pike pole is too short, you will not be able to reach the ceiling. If it is too long, you will not be able to use it in a room with a low ceiling.

The different head designs of pike poles are intended for different types of ceilings and come in a variety of configurations. Many fire departments use one type of pike pole for plaster ceilings and another type for drywall ceilings.

Commonly encountered types of pulling poles and hooks include the following:

- **Ceiling hook**: A tool consisting of a long wood or fibreglass pole and a metal point with a spur at right angles that can be used to probe ceilings and pull down plaster lath material.
- **Clemens hook**: A multipurpose tool that can be used for forcible entry and ventilation applications because of its unique head design.
- **Drywall hook**: A specialized version of a pike pole designed to remove drywall.
- **Multipurpose hook**: A long pole with a wooden or fibreglass handle and a metal hook.
- **Plaster hook**: A long pole with a pointed head and two retractable cutting blades on the side.

- **Roofman's hook**: A long pole with a solid metal hook.
- **San Francisco hook**: A multipurpose tool used for forcible entry and ventilation applications. It includes a built-in gas shut-off and directional slot.

A **K tool** is another type of pushing or pulling tool (**FIGURE 8-11**). It is used to pull the lock cylinder out of a door, exposing the locking mechanism so it can be unlocked easily. A **shove knife** is also used to release the latch on a door, specifically outward-swinging doors. Pulling the tool down and outward releases the locking mechanism of the door (**FIGURE 8-12**). You should not use this technique to try to gain entry through inward-swinging doors. It is not effective on doors with metal frames and only works on a limited number of doors with wooden frames.

Prying/Spreading Tools

A third group of tools are those used for prying or spreading. These tools may be as simple as a pry bar or as mechanically

FIGURE 8-11 Components of a K tool.
Courtesy of Fire Hooks Unlimited.

FIGURE 8-12 A shove knife.
© Jones & Bartlett Learning. Photographed by Glen E. Ellman.

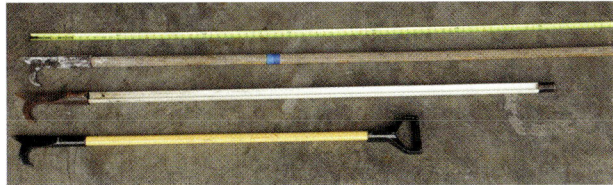

FIGURE 8-10 Pike poles come in several different sizes.
© Jones & Bartlett Learning. Photographed by Glen E. Ellman.

complex as a hydraulic spreader (**TABLE 8-4**). They come in several sizes and with different features that are designed for different applications.

One of the most popular prying tools is a **Halligan tool**, which was designed in 1948 by a New York City fire fighter (Firefighter Nation, 2007). Many variations of this tool exist, and many names are used to describe it and its parts. The tool incorporates a sharp tapered pick, a blade (either a wedge or an adze), and a fork or claw (**FIGURE 8-13**). It can be used for forcible entry applications, as can several other prying tools.

TABLE 8-4 Tools for Prying and Spreading
■ Claw bar   ■ Halligan tool   ■ Kelly tool   ■ Crowbar   ■ Hux bar   ■ Pry bar   ■ Flat bar   ■ Hydraulic spreader   ■ Rabbit tool

The specific uses and applications of these tools are covered in the chapters on the various phases of fire suppression.

In addition to the Halligan tool, the following hand tools are used for prying (**FIGURE 8-14**):

- **Claw bar**: A tool with a pointed claw-hook on one end and a forked- or flat-chisel pry on the other end that can be used for forcible entry.
- **Crowbar**: A straight bar made of steel or iron with a forked chisel on the working end.
- **Flat bar**: A specialized prying tool made of flat steel with prying ends suitable for performing forced entry.
- **Hux bar**: A multipurpose tool that can be used for forcible entry and ventilation applications because of its unique design. A Hux bar also can be used as a hydrant wrench.
- **Kelly tool**: A steel bar with two main features—a large pick and a large chisel or fork.
- **Pry bar**: A specialized prying tool made of a hardened steel rod with a tapered end that can be inserted into a small area. The bar acts as a lever to multiply the force that a person can exert to bend or pry objects apart. A properly positioned pry bar can apply an enormous amount of force.

Hydraulic spreaders are an example of machine-powered rescue tools that can be used for prying and spreading (**FIGURE 8-15**). The use of hydraulic power enables you to apply several tons of force on a very small area. You must have special training to operate these machines safely, however. Fire and rescue departments most commonly use this equipment for extrication of victims from motor vehicles and machinery (**FIGURE 8-16**).

Hand-powered hydraulic tools also are used for prying and spreading. One of these, the **rabbit tool**, is designed to quickly open doors (**FIGURE 8-17**).

FIGURE 8-13 A Halligan tool in action.
© Jones & Bartlett Learning. Photographed by Glen E. Ellman.

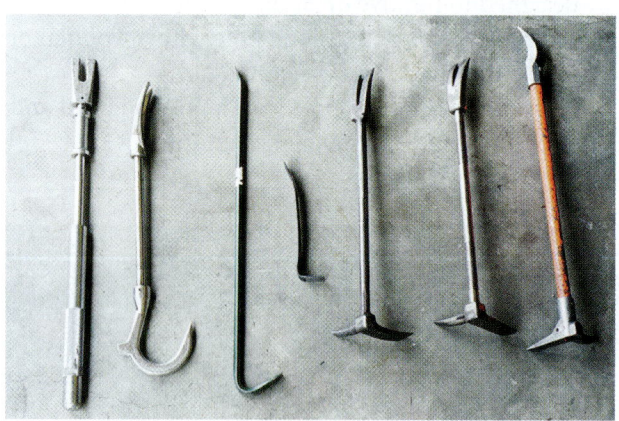

FIGURE 8-14 Tools used for prying and spreading.
© 2003, Berta A. Daniels.

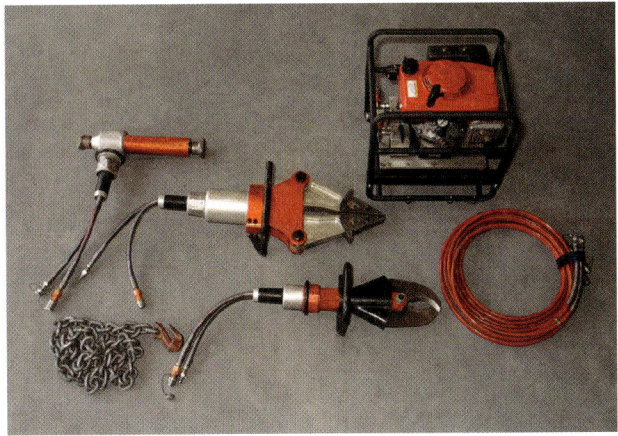

FIGURE 8-15 The components of a hydraulic rescue tool.
© Jones & Bartlett Learning.

FIGURE 8-16 A hydraulic spreader often is used for extrication of victims from vehicles.
© Jones & Bartlett Learning. Photographed by Glen E. Ellman.

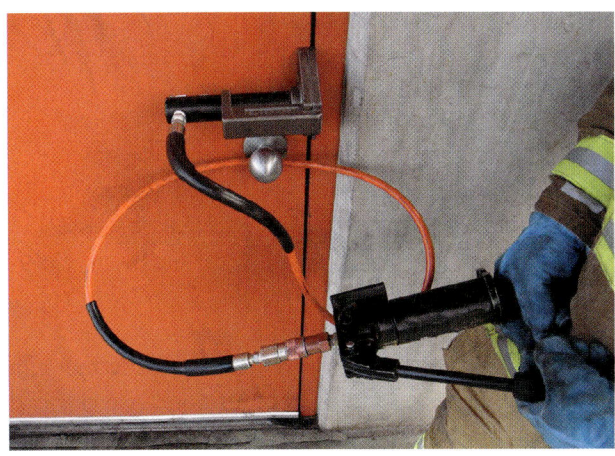

FIGURE 8-17 A rabbit tool can open doors quickly.
© Jones & Bartlett Learning.

TABLE 8-5 Striking Tools
■ Battering ram
■ Hammer
■ Pick-head axe
■ Chisel
■ Mallet
■ Sledgehammer
■ Flat-head axe
■ Maul
■ Spring-loaded center punch

A

B

FIGURE 8-18 Two types of axes: **A.** Flat-head axe. **B.** Pick-head axe.
© Jones & Bartlett Learning.

Striking Tools

Striking tools are used to apply an impact force to an object (**TABLE 8-5**). They often are employed to gain entrance to a building or a vehicle or to make an opening in a wall or roof. This equipment also can be used to force the end of a prying tool into a small opening. Specific use of these tools is covered in Chapter 10, *Forcible Entry*.

Among the most frequently used tools in the fire service are axes, including both **pick-head axes** and **flat-head axes** (Fire Rescue, 2014) (**FIGURE 8-18**).

Both types of axes have a wide cutting blade that can be used to chop into a wall, roof, or door. A pick-head axe has a point or pick that can be used for puncturing, pulling, and prying (**FIGURE 8-19**). A flat-head axe has a flat head that can be used for striking objects. It can be used as a striking tool for forcible entry, usually in combination with a prying tool, such as a Halligan tool. Together, the flat-head axe and the Halligan tool are sometimes referred to as "the **irons**"; this combination is highly effective in most forcible-entry situations (**FIGURE 8-20**).

FIGURE 8-19 A pick-head axe can be used to pry up boards.
© Jones & Bartlett Learning. Photographed by Glen E. Ellman.

FIGURE 8-20 A flat-head axe can be used with a Halligan tool to force open a door.
© Jones & Bartlett Learning. Photographed by Glen E. Ellman.

FIRE FIGHTER TIP

Many fire departments carry a striking tool and a prying bar strapped together, a combination sometimes referred to as "the irons." A flat-head axe and a Halligan tool are carried together on many vehicles and are taken into a fire building by one of the crew members.

In addition to axes, striking tools include the following items (**FIGURE 8-21**):

- **Hammer**: A hand tool constructed of solid material with a long handle and a head affixed to the top of the handle, with one side of the head used for striking and the other side used for prying.
- **Mallet**: A short-handled hammer with a round head.
- **Maul**: A specialized striking tool (weighing 3 kg [6 lb] or more) with an axe on one side of the head and a sledgehammer on the other side.
- **Sledgehammer**: A hammer that can be one of a variety of weights and sizes. The head of the hammer can weigh from 1 to 9 kg (2 to 20 lb), and the handle may be short like a carpenter's hammer or long like an axe handle.
- **Battering ram**: A heavy metal bar used to break down doors and breach walls (**FIGURE 8-22**).
- **Chisel**: A metal tool with one sharpened end that can be used to break apart material when used in conjunction with a hammer, mallet, or sledgehammer.
- **Spring-loaded center punch**: A spring-loaded punch that is used to break tempered automobile glass.

The spring-loaded center punch is used primarily on cars (**FIGURE 8-23**). It can exert a large amount of force on a pinpoint-size portion of tempered automobile glass. This action disrupts the integrity of tempered glass and causes the window to shatter into small, uniform-sized pieces. A spring-loaded center

FIGURE 8-21 Striking tools (from top): hammer, maul, mallet, sledgehammer.
© Jones & Bartlett Learning. Photographed by Glen E. Ellman.

FIGURE 8-22 A battering ram is used to break down doors and breach walls.
© Jones & Bartlett Learning. Photographed by Glen E. Ellman.

Spring-loaded punch

FIGURE 8-23 A spring-loaded center punch can be used to break a car window safely.
© Jones & Bartlett Learning. Photographed by Glen E. Ellman.

SAFETY TIP

Some tools require a considerable amount of movement and room to operate. Always confirm that no one is in danger of being injured before you use these tools. Look around and make sure you can operate an axe or sledgehammer safely and effectively.

punch is used in vehicular crashes to gain access to a victim who needs care. Vehicle extrications are covered in depth in Chapter 24, *Vehicle Rescue and Extrication*.

Cutting Tools

Cutting tools have sharp edges that sever objects. They come in several forms and are used to cut a wide variety of substances (**FIGURE 8-24**). Cutting tools used by fire fighters range from knives and wire cutters that can be carried in the pockets of protective coats to seat belt cutters, bolt cutters (a scissors-like tool used to cut through items such as chains or padlocks), saws, hydraulic shears, and cutting torches (a torch that produces a high-temperature flame capable of melting metal) (**TABLE 8-6**). Each of these tools is designed to work on certain types of materials. Fire fighters can be injured and cutting tools can be ruined if the tools are used incorrectly.

Bolt cutters are most often used to cut through chains or padlocks to open doors or gates. By concentrating the cutting force on a small area, it is possible to break through many chains in just a few seconds.

Fire departments often carry several different types of saws. This equipment can be classified into two main categories, based on the power source. Handsaws are manually powered, whereas mechanical saws usually are powered by electric motors or gasoline-powered engines. Handsaws include hacksaws, carpenter's handsaws, keyhole saws, and coping saws.

Hacksaws are designed to cut metal (**FIGURE 8-25**). Different blades can be used, depending on the type of metal being cut. Hacksaws are useful when metal needs to be cut under closely controlled conditions.

Carpenter's handsaws are designed for cutting wood. Saws with large teeth are effective in cutting large timber or tree branches; they are often useful at motor vehicle crashes where tree limbs may hamper the rescue effort. Saws with finer teeth are designed for cutting finished lumber. A coping saw is used to cut curves in wood; it consists of a handsaw with a narrow blade set between the ends of a U-shaped frame. A keyhole saw, a specialty saw, is narrow and slender and can be used to cut keyholes in wood and drywall.

Although handsaws have a valuable role in fire-fighting operations, power saws have the advantage of accomplishing more work in a shorter period of time. They also enable fire fighters to conserve energy, resulting in less fatigue. Because mechanical saws are powerful, only trained operators should use them. Nevertheless, they offer some distinct disadvantages relative to handsaws. Specifically, power saws are heavy to carry and sometimes can be difficult to start. They may require an electrical connection, although battery-powered cordless models are available.

A

B

C

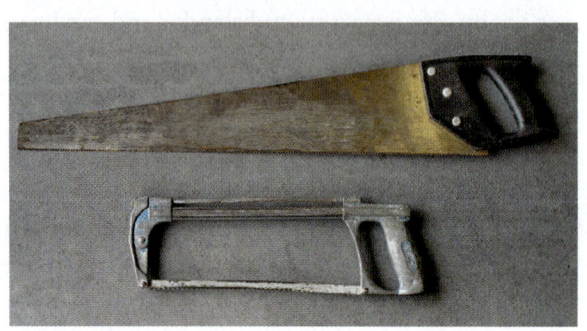

D

FIGURE 8-24 Cutting tools. **A.** Combination tool. **B.** Seat belt cutter. **C.** Bolt cutters. **D.** Handsaws.

© Jones & Bartlett Learning.

TABLE 8-6 Cutting Tools
▪ Axes
▪ Bolt cutters
▪ Chainsaws
▪ Circular saws
▪ Cutting torches
▪ Hacksaws
▪ Handsaws
▪ Hydraulic shears
▪ Reciprocating saws
▪ Rotary saws
▪ Seat belt cutters

FIGURE 8-25 A fire fighter using a hacksaw to cut through metal.

© Jones & Bartlett Learning. Photographed by Glen E. Ellman.

Three primary types of mechanical saws are chainsaws, rotary saws, and reciprocating saws. Most people are familiar with the gasoline-powered or electric **chainsaws** commonly used to cut wood, particularly trees. Fire fighters often use saws with special cutting chains to cut ventilation openings in roofs constructed of wood, metal, tar, gravel, or insulating materials (**FIGURE 8-26**).

Rotary saws may be powered by either electric motors or gasoline engines (**FIGURE 8-27**). In some rotary saws, the cutting part of the saw is a round metal blade with teeth. Different blades are used depending on the type of material being cut. Other rotary saws use a flat, abrasive disk for cutting. The disks are made of composite materials and are designed to wear down as they are used. It is important to match the appropriate saw blade or saw disk to the material being cut.

LISTEN UP!

Before forcing any door or window, check to see if it is unlocked. Always try before you pry!

FIGURE 8-26 A fire fighter can use a chainsaw to cut through a roof.
© Jones & Bartlett Learning. Photographed by Glen E. Ellman.

FIGURE 8-27 A fire fighter using a rotary saw.
© Jones & Bartlett Learning. Photographed by Glen E. Ellman.

Reciprocating saws are powered by either an electric motor or a battery motor that rapidly pulls a saw blade back and forth (**FIGURE 8-28**). As with rotary saws, different blades are used to cut different materials. Reciprocating saws are most commonly used to cut metal during extrication of a victim from a motor vehicle.

FIRE FIGHTER TIP

Ventilation saws may have a depth gauge on the bar and are specifically designed for roof ventilation.

The cutting tools that require the most training are hydraulic shears and cutting torches. **Hydraulic shears** often are used along with hydraulic spreaders and rams

A

B

FIGURE 8-28 **A.** The components of a reciprocating saw. **B.** A reciprocating saw being used during the extrication of a victim from a vehicle.
© Jones & Bartlett Learning.

in extrication of victims from motor vehicles; the same hydraulic power source can be used with all three types of tools. Hydraulic shears can cut quickly through metal posts and bars.

Cutting torches produce an extremely high-temperature flame and are capable of heating steel until it melts, thereby cutting through an object (**FIGURE 8-29**). Because these torches produce such high temperatures (3149°C [5700°F]), operators must be specially trained before using them. Cutting torches are sometimes used for rescue situations and for cutting through heavy steel objects. One drawback of this equipment is that it cannot be used in situations where flammable fuels are present.

Multiple-Function Tools

Certain tools are designed to perform multiple functions, thereby reducing the number of tools needed to achieve a goal. For example, a flat-head axe can be used as either a cutting tool or a striking tool. Some combination tools can be used to cut, to pry, to strike, and to turn off utilities.

FIGURE 8-29 A cutting torch can be used to cut through a metal door.
© Jones & Bartlett Learning.

FIGURE 8-30 Heavy-duty air bags can be used to lift vehicles in rescue situations.
© Jones & Bartlett Learning. Photographed by Glen E. Ellman.

Special-Use Tools

Some fire situations require special-use tools that perform other functions. For example, fire departments located in areas where brush and ground fires occur frequently may need to carry fire rakes, firefighting brooms, shovels, and combination tools that can be used for raking, chopping, cutting, and leaf blowing. These tools are described in Chapter 21, *Wildland and Ground Fires*.

Rescue squads, such as trench/collapse rescue, heavy vehicle rescue, elevator rescue, and confined space rescue units, also may use specialized equipment such as jacks and air bags for lifting heavy objects, **come alongs** (small, hand-operated winches) for lifting or moving heavy objects or bending objects, and tripods (**FIGURE 8-30**). You can learn more about the proper use of this special equipment by taking special rescue courses or during in-service training.

Phases of Use

The process of extinguishing a fire usually involves a sequence of steps or stages. Each phase of a fire-ground operation may require the use of certain types of tools and equipment.

The basic steps of fire suppression are summarized here:

- **Response** and **size-up**: This phase begins when the emergency call is received and continues as the units travel to the incident scene. The last part of this phase involves the initial observation and evaluation of factors used to determine which strategy and tactics will be employed.
- **Forcible entry**: This phase begins when entry to buildings, vehicles, aircraft, or other confined areas is locked or blocked, requiring fire fighters to use special techniques to gain access.
- **Interior attack**: During this phase, a team of fire fighters is assigned to enter the fire structure, locate the fire, and extinguish the fire.
- **Search and rescue**: This phase involves searching for any victims trapped by the fire and extricating them from the building.
- Rapid intervention: During this phase, a **rapid intervention crew/company (RIC)** provides immediate assistance to injured or trapped fire fighters.
- **Ventilation**: This step involves changing air within a compartment by natural or powered means.
- **Salvage**: This phase includes bagging and removing items such as pictures, computers, and documents and protecting property from further loss. The efficient removal of water can also be considered part of salvage operations.
- **Overhaul**: The final phase is to ensure that all hidden fires are extinguished after the main fire has been suppressed.

You may need to use different tools in each of these phases, or you may need to use the same tools to complete different tasks.

Response and Size-Up Tools and Equipment

The response and size-up phase enables you to anticipate emergency situations. At this time, you should consider the information from the dispatcher along with preincident plan information about the location. This information can provide you with an idea of the nature and possible gravity of the situation, as well as the types of problems that might arise during firefighting operations. For example, an automobile fire on the highway presents different problems and requires different tools than a call for smoke coming from a single-family house. A different thinking process occurs when you are dispatched at midnight to a house fire that may have trapped a family inside than when you respond to a report of a kitchen stove fire at suppertime. Even though information about the incident is limited at this point, the response and size-up phase is the time to start thinking about the types of tools and equipment that you might need.

Most fire departments have SOPs or SOGs that specify the tools and equipment required for different types of fires. Each crew member is expected to bring specific tools and equipment from the apparatus. These requirements take into account the roles that different units have within the fire department. A member from a pump company, for example, will not bring the same tools that a member from a ladder company will bring. Likewise, the tools carried for an interior fire attack differ from those carried for an outside or defensive attack.

Upon arrival at the scene, the company officer will perform a size-up and develop the action plan for each company. As part of individual self awareness, every fire fighter should engage in a size-up. You should identify the building construction, occupancy, height, and exact nature of the problem immediately before taking any action. This will help you to make important decisions regarding hazards, collapse potential, and means of egress. This action plan often will determine the tools that will be needed for each assigned task.

Forcible Entry Tools and Equipment

Gaining entrance to a locked building or structure can present a challenge to even the most seasoned fire fighter. Most buildings are equipped with security devices designed to keep unwanted people out, but these same devices can make it difficult for fire fighters to gain access to the building. Forcible entry is the process of entering a building by overcoming these barriers. The skills involved in this phase are discussed in depth in Chapter 10, *Forcible Entry*.

Several types of tools can be used for forcible entry, including an axe, a prying tool, a K tool, or a shove knife. A flat-head axe and a Halligan tool, collectively called the irons, are used in combination to pry open a door, although they may permanently damage both the door and the frame. Prying tools used for gaining access include pry bars, crowbars, Halligan tools, Hux bars, and the hydraulic-powered rabbit tool.

A K tool can be used to pull out a cylinder lock mounted in a wood or heavy metal door so that the lock can be released (**FIGURE 8-31**). This comparatively nondestructive process leaves the door and most of the locking mechanism undamaged. The building owner can have the lock cylinder replaced at a relatively low cost. A shove knife also leaves the door and locking mechanism undamaged. This tool is used to open outward-swinging doors. The knife is inserted between the door and door frame and is then pulled down and outward.

A variety of striking tools can be used for forcible entry when brute force is needed to break into a building. These items include flat-head axes, hammers, sledgehammers, and battering rams.

Sometimes the easiest—or only—way to gain access is to use cutting tools. An axe can be used to cut out a door panel. A power saw can be used to cut through a wood wall. Bolt cutters can be used to remove a padlock. Cutting torches or rotary power saws can be used to cut through metal security bars, hinges, or padlocks on metal security doors. Although cutting is a destructive process, it is justified when required to save lives or property.

Many techniques and types of tools can be used to gain entry into secured structures (**TABLE 8-7**). The exact tool needed will depend on the method of entry and the type of obstacle. Because having greater experience usually suggests the best way to gain entry in each situation, rely on the orders and advice of your officer and fellow fire fighters. Additional techniques used for forcible entry are described in Chapter 10, *Forcible Entry*.

FIGURE 8-31 Use a K tool to pull out and release a cylinder lock.
© Jones & Bartlett Learning.

TABLE 8-7 Forcible Entry Tools

Type of Tool	Use in Forcible Entry
Prying tools	Prying tools can include a Halligan tool, a flat bar, and a crowbar, among other options. They can be used to break windows or force open doors.
Axe, flat-head axe, pick-head axe	An axe can be used to cut through a door or break a window. A flat-head axe can be used in conjunction with a Halligan tool to force open a door.
Sledgehammer	A sledgehammer can be used to breach walls or break a window. It also can be used in conjunction with other tools, such as a Halligan tool, to force open a door or break off a padlock.
Hammer, mallet	These tools may be used in conjunction with other tools such as chisels or punches to force entry through windows or doors.
Chisel, punch	These tools can be used to make small openings through doors or windows.
K tool	The K tool provides a "through the lock" method of opening a door, thereby minimizing damage to the door.
Shove knife	A shove knife is used to open outward-swinging doors.
Rotary saw, chainsaw, reciprocating saw, circular saw	Power saws can cut openings through obstacles including doors, walls, fences, gates, security bars, and other barriers.
Bolt cutter	A bolt cutter can be used to cut off padlocks or cut through obstacles such as fences.
Battering ram	A battering ram can be used to breach walls.
Hydraulic door opener	A hydraulic door opener can be used to force open a door.
Hydraulic rescue tool	A hydraulic rescue tool can be used in a variety of situations to break, breach, or force openings in doors, windows, walls, fences, or gates.

Interior Firefighting Tools and Equipment

The process of fighting a fire inside a building involves several tasks that are usually performed simultaneously or in rapid succession by teams of fire fighters. While one crew is advancing a hose line to attack the fire, another crew may be searching for occupants, and a third crew may be performing ventilation tasks. This is referred to as combined operations (combined ops). An additional company may be standing by as an RIC, in case a fire fighter needs to be rescued.

Every crew working inside a burning building should carry some basic tools and equipment. These basic tools enable them to solve problems they may encounter while performing interior operations. For example, crew members may encounter obstacles such as locked doors, or they may need to open an emergency escape route. They may need to establish horizontal ventilation by forcing, opening, or breaking a window. They may have to gain access to the space above the ceiling by using a pike pole or making a hole in a wall or floor with an axe. A powerful light is also important, because smoke can quickly reduce interior visibility to just a few inches.

The basic set of tools for interior firefighting includes the following items:

- A prying tool, such as a Halligan tool
- A striking tool, such as a flat-head axe or a sledgehammer
- A cutting tool, such as an axe
- A pushing/pulling tool, such as a pike pole
- A **hand light** or portable light

Crews also may carry specialized tools and equipment needed for their particular assignment.

The specific tools that must be carried by each crew usually are defined in the fire department's training manuals and SOPs. These requirements are based on local conditions and preferences and may differ depending on the type of company and the assignment. See Chapter 17, *Fire Suppression*, for more detailed information on interior attack operations.

Search and Rescue Tools and Equipment

Search and rescue needs to be carried out quickly, shortly after arrival on the fire ground. See Chapter 12, *Search and Rescue*, for more detailed information on search and rescue operations. A search team should carry the same basic hand tools as the interior attack team including at least one forcible entry tool, such as an axe, sledgehammer, or Halligan tool. These tools can be used both to open an area for a search and, if necessary, to open an emergency exit path.

In addition to being equipped for forcible entry and emergency exit, a search and rescue team may use hand tools to probe under beds for unconscious victims. A short pike pole is relatively light and may reduce the time needed to search an area by extending the fire fighter's reach.

The following tools are also used during search and rescue operations:

- Hand light or flashlight
- Portable radio
- **Thermal imaging devices**
- Hose lines
- Ladders
- Search rope(s)
- Tubular webbing or short rope
- Chalk, crayons, felt tip markers, spray paint, or masking tape (for marking the doors of rooms that have been searched)

Rapid Intervention Tools and Equipment

The RIC is designated to stand by to provide immediate assistance to any fire fighters who become lost, trapped, or injured during an incident or training exercise. Members of this team should carry the standard set of tools for interior firefighting as well as extra tools and equipment that are particularly important for search and rescue tasks. The extra tools and equipment should help the RIC find and gain access to a fire fighter who is in trouble, extricate a fire fighter who is trapped under debris, provide breathing air for a fire fighter who has experienced an SCBA failure or run out of air, and remove an injured or unconscious fire fighter from the building. This equipment should be gathered and staged with the RIC, so that it will be immediately available if needed.

An RIC should carry the following special equipment:

- Thermal imaging devices
- Additional portable lighting
- Search ropes
- Prying tools
- Striking tools
- Cutting tools, including a power saw
- SCBA and spare air cylinders for use with **rapid intervention crew/company universal air connection (RIC UAC)**
- Litter or patient packaging device

Ventilation Tools and Equipment

The objective of ventilation is to provide openings so that fresh air can enter a burning structure and hot gases and products of combustion can escape from the building. See Chapter 13, *Ventilation*, for more detailed information on ventilation operations. Many of the same tools used for forcible entry also are applied to provide ventilation. For example, power saws and hand tools are commonly used to create vent openings. Axes, Halligan tools, pry bars, tin cutters, pike poles, and other hooks can all be used to remove coverings from existing openings, cut through roof decking, remove sections of the roof, and punch holes in the interior ceiling.

To provide adequate ventilation, unlocked or easily released windows and doors should be opened normally. Locked or jammed windows and doors may have to be broken or forced open using a hand tool (Halligan tool, axe, or pike pole) or ladder. It also may be necessary to create interior openings within the building so that contaminated air can reach the exterior openings.

In addition to making or controlling openings to influence ventilation, fire fighters can use mechanical ventilation to direct the flow of combustion gases. Large high-powered fans are used to remove smoke from a building or to introduce fresh air into a structure. Negative-pressure fans (known as smoke ejectors) draw contaminated air *out of* a building. Positive-pressure fans blow clean air *into* a building to force contaminated air out.

Ventilation fans can be powered either by electric or gasoline motors or by water pressure (**FIGURE 8-32**). A series of electrically powered fans may be placed throughout a large structure to help move smoke in the desired direction. A gasoline-powered fan may not be suitable in some situations, because it can introduce carbon monoxide into a structure if an exhaust hose is not available. If the building atmosphere contains potentially dangerous levels of flammable gases, a water-powered fan may be the best choice.

FIGURE 8-32 Different types of fans used in ventilation.
© Jones & Bartlett Learning. Photographed by Glen E. Ellman.

Salvage and Overhaul Tools and Equipment

The purpose of salvage and overhaul is to protect property and belongings from damage and ensure that all hidden fires are extinguished. During this phase, burned debris must be removed, and potential hot spots in enclosed spaces behind walls, above ceilings, and under floors are exposed. Both tasks can be accomplished using simple hand tools.

The following tools are used during salvage and overhaul operations:

- Salvage covers and floor runners
- Sprinkler shut-off tools
- Water-removal equipment (wet vacuums, submersible pumps and hose)
- Ventilation equipment (fans, power blowers)
- Pushing tools (pike poles and ceiling hooks of varying lengths)
- Prying tools (crowbars and Halligan tools)
- Cutting tools (axes, power saws)
- Debris-removal tools (pitchforks, shovels, brooms, rakes, buckets, wheelbarrows, carryalls)
- Thermal imaging devices

Salvage efforts at residential fires often focus on protecting personal property. Salvage covers are used to shield and cover building contents, while floor runners can be used to cover a section of carpet or hardwood flooring. In buildings equipped with sprinkler systems, fire fighters should shut off any sprinkler heads that have been activated by the fire to reduce excess water damage and to restore the fire protection system in the building. A single sprinkler head can be shut off with a wooden sprinkler wedge—a technique that leaves the

sprinkler system with an inoperable sprinkler head when it is turned on. A sprinkler head also can be shut off with a sprinkler stop, a special tool that contains a fusible link and allows an activated sprinkler head to be placed back in service until a replacement head can be installed.

After sprinkler heads have been shut off, water-removal equipment may be needed to help remove water that has accumulated within a building. Special vacuum cleaners that suck up water are available. Drainage pumps may also be used to remove water that has accumulated in basements or below ground level.

During overhaul, pike poles and ceiling hooks are commonly used for pulling down ceilings and opening holes in walls. Prying tools are used to open closed spaces and remove baseboards and window or door casings. Axes, and sometimes power saws, are used to open walls, floors, and ceilings. Shovels, pitchforks, brooms, rakes, buckets, wheelbarrows, and carryalls are used to clear away debris.

Thermal imaging devices are used to "see" hot spots behind walls without physically cutting into them (**FIGURE 8-33**). This technology has reduced the risk of missing dangerous hot spots as well as curtailed the amount of time and effort it takes to overhaul a fire scene. The use of thermal imaging devices is also discussed in Chapter 12, *Search and Rescue*. The process of overhauling a fire scene and the tools used to accomplish this task are covered more fully in Chapter 19, *Salvage and Overhaul*.

SAFETY TIP

There is a valuable saying in the fire service: "Hand tools—never leave your apparatus without them."

FIGURE 8-33 Thermal imaging devices can see hot spots.
© Jones & Bartlett Learning.

Maintenance

All tools and equipment must be properly maintained so that they will be ready for use when they are needed. This means you must keep equipment clean and free from rust, keep cutting blades sharpened, and keep fuel tanks filled. Every tool and piece of equipment must be ready for use before you respond to an emergency incident. Follow the manufacturer's instructions for cleaning and maintaining each piece of equipment. In addition, use equipment only for its intended purposes. For example, a pike pole is made for pushing and pulling; it is not a lever and will break if used inappropriately. Use the right tool for the job.

SAFETY TIP

Keep the manufacturers' manuals for all of the department's tools and equipment in a safe and easily accessible location. Follow the manufacturers' advice on the use and maintenance of tools and equipment, and refer to the manuals whenever you have questions about a device.

Cleaning and Inspecting Hand Tools

All hand tools should be cleaned completely and inspected and their conditions documented after each use. Remove all dirt and debris from the tools. If appropriate, use water streams to remove the debris and mild soap to clean the equipment thoroughly. Learn how to safely use the cleaning solutions that your department and the manufacturer specify for cleaning tools and equipment.

Before any tool is placed back into service, it should be inspected for damage. Avoid painting tools, because paint may hide defects or visible damage. Keep the number of markings on a tool to a minimum.

To clean and inspect hand tools, follow the steps in **SKILL DRILL 8-1**.

Immediately after cleaning and inspecting tools, place them back in their proper location, ready for use. If tools are damaged and need repair or replacement, they must be removed from service. Follow department procedures for documenting the damage and removing the tool from service.

SKILL DRILL 8-1
Cleaning and Inspecting Hand Tools Fire Fighter I, NFPA 1001: 4.5.1

- Clean and dry all metal parts. Metal tools must be dried completely, either by hand or by air, before being returned to the apparatus. Remove rust with steel wool. Coat unpainted metal surfaces with a light film of lubricant to help prevent rusting. Do not oil the striking surface of metal tools, as this treatment may cause them to slip.

- Inspect wood handles for damage such as cracks and splinters. Repair or replace any damaged handles. Sand the handle if necessary. Do not paint or varnish a wood handle; instead, apply a coat of boiled linseed oil. Check that the tool head is tightly fixed to the handle.

- Clean fibreglass handles with soap and water. Inspect for damage. Repair or replace any damaged handles. Check that the tool head is tightly fixed to the handle.

- Inspect cutting edges for nicks or other damage. Cutting tools should be sharpened after each use. File and sharpen as needed. Power grinding may weaken some tools, so hand sharpening may be required.

© Jones & Bartlett Learning. Photographed by Glen E. Ellman.

After-Action REVIEW

IN SUMMARY

- Tools and equipment are used in almost all fire suppression and rescue operations.
- Hand tools are used to extend or multiply the actions of your body and to increase your effectiveness in performing specific functions.
- Power tools and equipment use an external source of power and are faster and more efficient than hand tools.
- Always wear appropriate PPE when using tools or equipment.
- Most tools fit into the following functional categories:
  - Rotating tools—Apply a rotational force to make something turn and are used to assemble and disassemble items. Rotating tools include screwdrivers, wrenches, and pliers.
  - Prying or spreading tools—Used to pry and spread. May be as simple as a pry bar or as mechanically complex as a hydraulic spreader. Prying and spreading tools include the claw bar, crowbar, and Halligan tool.
  - Pushing or pulling tools—Extend your reach and increase the power you can exert upon an object. Pushing and pulling tools include pike poles, closet hooks, K tools, shove knives, and ceiling hooks.
  - Striking tools—Used to apply an impact force to an object. Striking tools include hammers, mallets, axes, and spring-loaded center punches.
  - Cutting tools—Tools with sharp edges that sever objects. Cutting tools include axes, bolt cutters, chainsaws, cutting torches, handsaws, and power saws.
  - Multiple-use tools—Designed to perform multiple functions, thereby reducing the total number of tools needed to achieve a goal. Some combination tools can be used to cut, to pry, to strike, and to turn off utilities.
- The following forcible entry tools are used to gain entrance to a locked building:
  - Prying tools—Used to break windows or force open doors.
  - Axe—Used to cut through a door or break a window.
  - Sledgehammer—Used to breach walls or break a window.
  - Hammer or mallet—May be used with other tools such as chisels or punches to force entry through windows or doors.
  - Chisel or punch—Used to make small openings through doors or windows.
  - K tool—Provides a "through the lock" method of opening a door, minimizing damage to the door.
  - Shove knife—Used to open outward-swinging doors.
  - Saw—Can cut openings through obstacles, including doors, walls, fences, gates, security bars, and other barriers.
  - Bolt cutter—Used to cut off padlocks or cut through obstacles such as fences.
  - Battering ram—Used to breach walls.
  - Hydraulic door opener—Used to force open a door.
  - Hydraulic rescue tool—Used in a variety of situations to break, breach, or force openings in doors, windows, walls, fences, or gates.
- The interior attack team is responsible for advancing a hose line, finding the fire, and applying water to extinguish the flames. The members of this team need the following basic tools that will allow them to reach the seat of the fire:
  - Prying tool
  - Striking tool
  - Cutting tool
  - Pushing/pulling tool
  - Strong hand light or portable light
  - Thermal imaging device

- A search and rescue team should carry the same basic hand tools as the interior attack team, as well as the following:
  - Short pike pole
  - Hand light or flashlight
  - Portable radio
  - Thermal imaging devices
  - Hose lines
  - Ladders
  - Search rope(s)
  - Tubular webbing or short rope
  - Chalk, crayons, felt tip markers, spray paint, or masking tape
- An RIC should carry the following special equipment:
  - Thermal imaging devices
  - Portable lighting
  - Search ropes
  - Prying tools
  - Striking tools
  - Cutting tools, including a power saw
  - SCBA or spare air cylinders for use with rapid intervention crew/company universal air connection (RIC UAC)
- The objective of ventilation is to provide openings so that fresh air can enter a burning structure and hot gases and products of combustion can escape from the building. The following special equipment is needed for ventilation:
  - Pulling and pushing tools
  - Cutting tools
  - Negative-pressure fans (smoke ejectors)
  - Positive-pressure fans
- The purpose of salvage and overhaul is to protect property and belongings from damage and ensure that all hidden fires are extinguished. The following tools are used during salvage and overhaul operations:
  - Salvage covers and floor runners
  - Sprinkler shut-off tools
  - Water-removal equipment
  - Ventilation equipment
  - Pushing tools
  - Prying tools
  - Cutting tools
  - Debris-removal tools
  - Thermal imaging devices
- All tools and equipment must be properly maintained so that they will be ready for use when they are needed. This means you must keep equipment clean and free from rust, keep cutting blades sharpened, and keep fuel tanks filled.

KEY TERMS

Access Navigate for flashcards to test your key term knowledge.

Battering ram A tool made of hardened steel with handles on the sides used to force doors and to breach walls. Larger versions may be used by as many as four people; smaller versions are made for one or two people.

Bolt cutter A cutting tool used to cut through thick metal objects such as bolts, locks, and wire fences.

Box-end wrench A hand tool used to tighten or loosen bolts. The end is enclosed, as opposed to an open-end wrench. Each wrench is a specific size, and most have ratchets for easier use.

Carpenter's handsaw A saw designed for cutting wood.

Ceiling hook A tool with a long wooden or fibreglass pole that has a metal point with a spur at right angles at one end. It can be used to probe ceilings and pull down plaster lath material.

Chainsaw A power saw that uses the rotating movement of a chain equipped with sharpened cutting edges. It is typically used to cut through wood.

Chisel A metal tool with one sharpened end that is used to break apart material in conjunction with a hammer, mallet, or sledgehammer.

Claw bar A tool with a pointed claw-hook on one end and a forked- or flat-chisel pry on the other end. It is often used for forcible entry.

Clemens hook A multipurpose tool that can be used for several forcible entry and ventilation applications because of its unique head design.

Closet hook A type of pike pole intended for use in tight spaces, commonly 0.6 to 1 m (2 to 4 ft) in length.

Combination wrench A hand tool with an open-end wrench on one end and a box-end wrench on the other.

Come along A hand-operated tool used for dragging or lifting heavy objects that uses pulleys and cables or chains to multiply a pulling or lifting force.

Coping saw A saw designed to cut curves in wood.

Coupling One set or pair of connection devices attached to a fire hose that allow the hose to be interconnected to additional lengths of hose or adapters and other firefighting appliances. (NFPA 1963)

Crowbar A straight bar made of steel or iron with a forked chisel on the working end that is suitable for performing forcible entry.

Cutting torch A torch that produces a high-temperature flame capable of heating metal to its melting point, thereby cutting through an object. Because of the high temperatures (3149°C [5700°F]) that these torches produce, the operator must be specially trained before using this tool.

Drywall hook A specialized version of a pike pole that can remove drywall more effectively because of its hook design.

Flat bar A specialized type of prying tool made of flat steel with prying ends suitable for performing forcible entry.

Flat-head axe A tool that has a head with an axe on one side and a flat head on the opposite side.

Forcible entry Techniques used by fire personnel to gain entry into buildings, vehicles, aircraft, or other areas of confinement when normal means of entry are locked or blocked. (NFPA 402)

Gripping pliers A hand tool with a pincer-like working end that can be used to bend wire or hold smaller objects.

Hacksaw A cutting tool designed for use on metal. Different blades can be used for cutting different types of metals.

Halligan tool A prying tool that incorporates a sharp tapered pick, a blade (either an adze or wedge), and a fork; it is specifically designed for use in the fire service.

Hammer A striking tool.

Hand light A small, portable light carried by fire fighters to improve visibility at emergency scenes. It is often powered by rechargeable batteries.

Handsaw A manually powered saw designed to cut different types of materials. Examples include hacksaws, carpenter's handsaws, keyhole saws, and coping saws.

Hux bar A multipurpose tool that can be used for several forcible entry and ventilation applications because of its unique design. It also may be used as a hydrant wrench.

Hydrant wrench A hand tool that is used to operate the valves on a hydrant; it also may be used as a spanner wrench. Some models are plain wrenches, whereas others have a ratchet feature.

Hydraulic shears A lightweight, hand-operated tool that can produce up to 4536 kg (10,000 lb) of cutting force.

Hydraulic spreader A lightweight, hand-operated tool that can produce up to 4536 kg (10,000 lb) of prying and spreading force.

Interior attack The assignment of a team of fire fighters to enter a structure and attempt fire suppression.

Irons A combination of tools, usually consisting of a Halligan tool and a flat-head axe, that are commonly used for forcible entry.

Kelly tool A steel bar with two main features: a large pick and a large chisel or fork.

Keyhole saw A saw designed to cut keyhole circles in wood and drywall.

K tool A tool that is used to remove lock cylinders from structural doors so the locking mechanism can be unlocked.

Mallet A short-handled hammer.

Maul A specialized striking tool, weighing 3 kg (6 lb) or more, with an axe on one side of the head and a sledgehammer on the other side.

Mechanical saw A saw that usually is powered by an electric motor or a gasoline engine. The three primary types of mechanical saws are chainsaws, rotary saws, and reciprocating saws.

Multipurpose hook A long pole with a wooden or fibreglass handle and a metal hook on one end used for pulling.

Open-end wrench A hand tool that is used to tighten or loosen bolts. The end is open, as opposed to a box-end wrench. Each wrench is a specific size.

Overhaul The process of final extinguishment after the main body of a fire has been knocked down. All traces of fire must be extinguished at this time. (NFPA 402)

Pick-head axe A tool that has a head with an axe on one side and a pointed end ("pick") on the opposite side.

Pike pole A pole with a sharp point ("pike") on one end coupled with a hook. It is used to make openings in ceilings and walls. Pike poles are manufactured in different lengths for use in rooms of different heights.

Pipe wrench A wrench having one fixed grip and one movable grip that can be adjusted to fit securely around pipes and other tubular objects.

Plaster hook A long pole with a pointed head and two retractable cutting blades on the side.

Pry bar A specialized prying tool made of a hardened steel rod with a tapered end that can be inserted into a small area.

Rabbit tool A hydraulic spreading tool designed to pry open doors that swing inward.

Rapid intervention crew/company (RIC) A minimum of two fully equipped personnel on site, in a ready state, for immediate rescue of disoriented, injured, lost, or trapped rescue personnel. (NFPA 1500)

Rapid intervention crew/company universal air connection (RIC UAC) A system that allows emergency replenishment of breathing air to the SCBA of disabled or entrapped fire or emergency services personnel. (NFPA 1407)

Reciprocating saw A saw that is powered by an electric motor or a battery motor and whose blade moves back and forth.

Response Immediate and ongoing activities, tasks, programs, and systems to manage the effects of an incident that threatens life, property, operations, or the environment. (NFPA 1600)

Roofman's hook A long pole with a solid metal hook used for pulling.

Rotary saw A saw that is powered by an electric motor or a gasoline engine and that uses a large rotating blade to cut through material. The blades can be changed depending on the material being cut.

Salvage A fire-fighting procedure for protecting property from further loss following an aircraft accident or fire. (NFPA 402)

San Francisco hook A multipurpose tool that can be used for several forcible entry and ventilation applications because of its unique design, which includes a built-in gas shut-off and directional slot.

Screwdriver A tool used for turning screws.

Search and rescue The process of searching a building for a victim and extricating the victim from the building.

Seat belt cutter A specialized cutting device that cuts through seat belts.

Shove knife A forcible entry tool used to trip the latch on outward-swinging doors.

Size-up The process of gathering and analyzing information to help fire officers make decisions regarding the deployment of resources and the implementation of tactics. (NFPA 1410)

Sledgehammer A hammer that can be one of a variety of weights and sizes.

Socket wrench A wrench that fits over a nut or bolt and uses the ratchet action of an attached handle to tighten or loosen the nut or bolt.

Spanner wrench A type of tool used to couple or uncouple hose by turning the rocker lugs or pin lugs on the connections.

Spring-loaded center punch A spring-loaded punch used to break automobile glass.

Thermal imaging device An electronic device that detects differences in temperature based on infrared energy and then generates images based on those data. It is commonly used in smoke-filled environments to locate victims as well as to search for hidden fire during size-up and overhaul.

Ventilation The controlled and coordinated removal of heat and smoke from a structure, replacing the escaping gases with fresh air. (NFPA 1410)

REFERENCES

Firefighter Nation. "History of the Halligan Tool". December 24, 2007. http://my.firefighternation.com/group/firefighting historymyths/forum/topics/889755:Topic:233830. Accessed September 25, 2018.

Fire Rescue. "Effective and Safe Use of Axes on the Fireground". May 1, 2014. http://www.firerescuemagazine.com/articles /print/volume-9/issue-5/training-0/effective-and-safe-use-of -axes-on-the-fireground.html. Accessed September 25, 2018.

On Scene

You have really settled into your role as a volunteer fire fighter. You know you are helping your community, and you feel satisfaction every time you make a good stop on a fire. It doesn't take long for you to get noticed by others in the department. In fact, at a recent training course, the training officer approached you on break and asked you to teach a class on basic tools and equipment. You feel great about his confidence in your skills, but you wonder whether you know enough about the tools to teach the subject to others.

1. A _____ is a tool that is used to remove lock cylinders from structural doors so the locking mechanism can be unlocked.

   A. K tool
   B. Kelly tool
   C. rabbit tool
   D. keyhole saw

2. A _____ is a prying tool that incorporates a pick, a blade, and a fork, designed for multiple uses in the fire service.

   A. ceiling hook
   B. pry bar
   C. Halligan tool
   D. crowbar

3. A _____ is a tool made of hardened steel with handles on the sides used to force doors and to breach walls.

   A. battering ram
   B. Halligan tool
   C. maul
   D. mallet

4. The "irons" usually consists of what two tools?

   A. Pick-head axe and Halligan tool
   B. Flat-head axe and Halligan tool
   C. Flat-head axe and Clemens hook
   D. Pick-head axe and Clemens hook

5. A hacksaw is included in which functional category?

   A. Cutting
   B. Pushing or pulling
   C. Prying or spreading
   D. Rotating

6. A _____ is a short hooked pole designed for use in tight places.

   A. pike pole
   B. claw bar
   C. crowbar
   D. closet hook

7. A hydrant wrench may be used for which of the following functions?

   A. To reduce water pressure
   B. To shut off electric power
   C. As an alternative to a spanner wrench
   D. To perform forcible entry

8. A pike pole is included in which functional category?

   A. Cutting
   B. Pushing or pulling
   C. Prying or spreading
   D. Rotating

Access Navigate to find answers to this On Scene, along with other resources such as an audiobook.

Ropes and Knots

KNOWLEDGE OBJECTIVES

After studying this chapter, you will be able to:

- Describe the four primary types of fire service rope. (**NFPA 1001: 4.3.20**, pp. 291–293)
- List the two types of life safety rope and their minimum breaking strength. (**NFPA 1001: 4.3.20**, p. 291)
- Describe the characteristics of escape rope and fire escape rope. (**NFPA 1001: 4.3.20**, pp. 291–292)
- Describe the characteristics of water rescue throwlines. (**NFPA 1001: 4.3.20**, p. 292)
- Describe the characteristics of utility ropes. (**NFPA 1001: 4.3.20**, pp. 292–293)
- Describe the characteristics of webbing. (**NFPA 1001: 4.3.20**, p. 293)
- List the disadvantages of natural fibre ropes. (**NFPA 1001: 4.3.20**, pp. 293–294)
- List the advantages of synthetic fibre ropes. (**NFPA 1001: 4.3.20**, pp. 293–294)
- List the disadvantages of synthetic fibre ropes. (**NFPA 1001: 4.3.20**, p. 294)
- List the types of synthetic fibres that are used in fire service rope. (**NFPA 1001: 4.3.20**, pp. 293–294)
- Describe how twisted ropes are constructed. (**NFPA 1001: 4.3.20**, pp. 294–295)
- Describe how braided ropes are constructed. (**NFPA 1001: 4.3.20**, p. 295)
- Describe how kernmantle ropes are constructed. (**NFPA 1001: 4.3.20**, pp. 295–296)
- Explain the differences between dynamic kernmantle rope and static kernmantle rope. (**NFPA 1001: 4.3.20**, pp. 295–296)
- List the four components of the rope maintenance formula. (**NFPA 1001: 4.3.20**, p. 296)
- Describe how to preserve rope strength and integrity. (**NFPA 1001: 4.3.20**, pp. 296–297)
- Describe how to clean rope. (**NFPA 1001: 4.5.1**, pp. 297–298)
- Describe how to inspect rope. (**NFPA 1001: 4.5.1**, pp. 297–299)
- Describe how to keep an accurate rope record. (**NFPA 1001: 4.5.1**, pp. 299–300)
- Describe how to store rope properly. (**NFPA 1001: 4.5.1**, pp. 300–301)
- List the terminology used to describe the parts of a rope when tying knots. (**NFPA 1001: 4.3.20**, pp. 301–302)
- List the terminology used to describe the bends in rope that are formed when a knot is tied. (**NFPA 1001: 4.3.20**, p. 302)
- List the common types of knots that are used in the fire service. (**NFPA 1001: 4.3.20**, p. 302)
- Describe the characteristics of a safety knot. (**NFPA 1001: 4.3.20**, pp. 303–304)

- Describe the characteristics of a half hitch. (**NFPA 1001: 4.3.20**, pp. 303, 305)
- Describe the characteristics of a clove hitch. (**NFPA 1001: 4.3.20**, pp. 303, 306–307)
- Describe the characteristics of a figure eight knot. (**NFPA 1001: 4.3.20**, p. 308)
- Describe the characteristics of a bowline knot. (**NFPA 1001: 4.3.20**, pp. 311–312)
- Describe the characteristics of a sheet bend. (**NFPA 1001: 4.3.20**, pp. 311, 313)
- Describe the characteristics of a water knot. (**NFPA 1001: 4.3.20**, pp. 311, 314)
- Describe the methods used to hoist a tool. (**NFPA 1001: 4.3.20**, pp. 314–320)

SKILLS OBJECTIVES

After studying this chapter, you will be able to perform the following skills:

- Care for life safety ropes. (**NFPA 1001: 4.5.1**, pp. 296–297)
- Clean fire department ropes. (**NFPA 1001: 4.5.1**, pp. 297–298)
- Inspect fire department ropes. (**NFPA 1001: 4.5.1**, pp. 297–299)
- Place a life safety rope in a rope bag. (**NFPA 1001: 4.5.1**, pp. 300–301)
- Tie a safety knot. (**NFPA 1001: 4.1.2, 4.3.20**, pp. 303–304)
- Tie a half hitch. (**NFPA 1001: 4.1.2, 4.3.20**, pp. 303, 305)
- Tie a clove hitch in the open. (**NFPA 1001: 4.1.2, 4.3.20**, pp. 303, 306)

- Tie a clove hitch around an object. (**NFPA 1001: 4.1.2, 4.3.20**, pp. 303, 307)
- Tie a figure eight knot. (**NFPA 1001: 4.1.2, 4.3.20**, p. 308)
- Tie a figure eight on a bight. (**NFPA 1001: 4.1.2, 4.3.20**, p. 309)
- Tie a figure eight follow-through. (**NFPA 1001: 4.1.2, 4.3.20**, p. 310)
- Tie a figure eight bend. (**NFPA 1001: 4.1.2, 4.3.20**, p. 311)
- Tie a bowline. (**NFPA 1001: 4.1.2, 4.3.20**, pp. 311–312)
- Tie a sheet or Becket bend. (**NFPA 1001: 4.1.2, 4.3.20**, pp. 311, 313)
- Tie a water knot. (**NFPA 1001: 4.1.2, 4.3.20**, pp. 311, 314)
- Hoist an axe. (**NFPA 1001: 4.1.2, 4.3.20**, pp. 314–316)
- Hoist a pike pole. (**NFPA 1001: 4.1.2, 4.3.20**, p. 316)
- Hoist a ladder. (**NFPA 1001: 4.1.2, 4.3.20**, pp. 316–317)
- Hoist a charged hose line. (**NFPA 1001: 4.1.2, 4.3.20**, p. 318)
- Hoist an uncharged hose line. (**NFPA 1001: 4.1.2, 4.3.20**, pp. 318–319)
- Hoist an exhaust fan or power tool. (**NFPA 1001: 4.1.2, 4.3.20**, pp. 318, 320)

ADDITIONAL STANDARDS

- **NFPA 1858**, *Standard of Selection, Care, and Maintenance of Life Safety Rope and Equipment for Emergency Services*
- **NFPA 1983**, *Standard on Life Safety Rope and Equipment for Emergency Services*

Fire Alarm

You and your partner are on the fourth floor of a multistorey apartment building and find yourself trapped by a rapidly progressing fire. You get to a bedroom at the front of the building and know your only escape route is out the window. You don't have time to wait for the aerial ladder to get to you, so you quickly consider your training for self-rescue. You pull out your rope and begin the most important task you have ever done in your life.

1. What type of rope would you want to use for this type of activity?

2. What rope construction method would be used for this rope?

3. What type of knot would you use?

 *Access Navigate for more practice activities.*

Introduction

In the fire service, ropes are widely used to hoist or lower tools, appliances, or people; to perform rescues; or to serve as a life line in an emergency. A rope might be your only means of accessing a trapped person or your only way of escaping from a fire.

Learning about ropes and knots is an important part of your training as a fire fighter. This chapter gives you a basic understanding of the importance of ropes and knots in the fire service. You can build on this foundation as you develop skills in handling ropes and tying knots. You must be able to tie simple knots accurately without hesitation or delay.

This chapter discusses different types of rope construction and the materials used in making ropes. It covers the care, cleaning, inspection, and storage of ropes. It also shows how to tie 10 essential knots and explains how to secure tools and equipment so that they can be raised or lowered using ropes.

Types of Rope

Four primary types of rope are used in the fire service, each of which is dedicated to a distinct function. **Life safety rope** is used solely for supporting people. It must be used whenever a rope is needed to support a person, whether during training or during firefighting, rescue, or other emergency operations. **Escape rope** and **fire escape rope** are single-purpose, emergency self-escape, self-rescue ropes. Water rescue **throwline** is a floating rope that is used during water rescue. **Utility rope** is used in most other cases, when it is not necessary to support the weight of a person, such as when hoisting or lowering tools or equipment.

Life Safety Rope

Life safety rope is a critical tool used *only* for life-saving purposes; it must never be used for other purposes (**FIGURE 9-1**). Life safety rope must be used in every situation where the rope must support the weight of one or more persons. In these situations, rope failure could result in serious injury or death.

Because life safety rope must be extremely reliable, the criteria for design, construction, and performance of this type of rope and its related equipment are specified in NFPA 1983, *Standard on Life Safety Rope and Equipment for Emergency Services.* This standard also requires rope manufacturers to provide detailed instructions for the proper use, maintenance, and inspection of the life safety rope, including the conditions for removing the rope from service. In addition, the manufacturer must supply a list of criteria that must be reviewed before a life safety rope that has been used in the field can be

FIGURE 9-1 A life safety rope.
© Jones & Bartlett Learning. Photographed by Glen E. Ellman.

used again. If the rope does not meet all of the criteria, it must be retired from service.

Types of Life Safety Ropes

NFPA 1983 defines the performance requirements for two types of life safety rope: **technical use life safety rope** and **general use life safety rope**. Technical use life safety rope, by definition, has a diameter that is 9.5 mm (⅜ in.) or greater but is less than 13 mm (½ in.; NFPA, 2016). It has a minimum breaking strength of lbf [20 kN] (4496 pounds force). Highly trained rescue teams that deploy to technical environments, such as mountainous and/or wilderness terrain, use technical use life safety ropes (**TABLE 9-1**).

General use life safety rope is the most common life safety rope carried by the fire service in Canada. A general use life safety rope can be no larger than 16 mm (⅝ in.) and no smaller than 11 mm (⁷⁄₁₆ in.; NFPA, 2016). In addition to its larger diameter, general use life safety rope is also stronger than technical use life safety rope. With a minimum breaking strength of 40 kN (8992 lbf), the general use life safety rope allows the technical rescuer a much greater margin of safety.

Escape Rope

When you are responding to an emergency, you should always have a safe way to get out of a situation and reach a safe location. You might be able to go back through the door that you entered, or you might have another exit route, such as through a different door, through a window, or down a ladder. If conditions suddenly change for the worse, having an escape route can save your life.

Sometimes, however, you can find yourself in a situation where conditions deteriorate so quickly that

TABLE 9-1 Life Safety Rope Strength	
Rope Classification	**Life Safety Rope Strength**
Technical use life safety rope	20 kN (4496 lbf)
General use life safety rope	40 kN (8992 lbf)

Source: National Fire Protection Association. NFPA 1983, *Standard on Life Safety Rope and Equipment for Emergency Services.* Quincy, MA: 2016.

you cannot use your planned exit route. For example, the stairway you used might collapse behind you, or the room you are in might suddenly flash over (a phase in the development of a contained fire in which exposed surfaces reach ignition temperature more or less simultaneously and fire spreads rapidly throughout the space), blocking your planned route out. In such a situation, you might need to take extreme measures to get out of the building. Two types of ropes were developed for this type of situation. The escape rope is intended for emergency self-rescue situations, while the fire escape rope is intended specifically for emergency self-rescue from an immediately hazardous environment in which fire or fire products are involved.

Both types of escape rope are designed to carry the weight of only one person and to be used only one time (**FIGURE 9-2**). They can fit easily into a small packet or pouch and are easy to carry. Many of these ropes are integrated within a harness system, similar to an SCBA pack. Their purpose is to provide the fire fighter with a method of escaping from a life-threatening situation. The escape rope or fire escape rope should be replaced with a new rope if it is exposed to an immediate danger to life and health environment.

> **FIRE FIGHTER TIP**
>
> Escape ropes are not classified as life safety ropes.

Water Rescue Throwline

Water rescue throwlines are used in water rescue operations. They range from 15 to 31 m (50 to 100 ft) in length and are designed to be kept in a special throw bag that can be thrown to a person in the water (**FIGURE 9-3**).

FIGURE 9-2 An escape rope is designed to be used by only one person.
© Jones & Bartlett Learning. Photographed by Glen E. Ellman.

A

B

FIGURE 9-3 A. Water rescue throwlines are kept in special throw bags. **B.** To use, the rescuer holds onto one end of the rope and throws the bag to the victim in the water.
© fotomy/Alamy Stock Photo.

The rescuer holds on to one end of the rope and throws the bag—which is attached to the other end of the rope—to the victim to catch. The rescuer then uses the rope to pull the victim to shore. Throwlines also can be used as a tether for rescuers entering the water.

This type of rope is made of material that floats, such as polypropylene, so the rope does not snag on underwater hazards or get entangled in motorboat propellers. It does not have the strength or abrasion resistance required to be used as a life safety rope.

Utility Rope

Utility rope is used when it is not necessary to support the weight of a person. Fire department utility rope is used for hoisting or lowering tools or equipment, for making ladder **halyards** (rope used on extension ladders to raise a fly section), for marking off areas, and for stabilizing objects (**FIGURE 9-4**). This type of rope requires regular inspection.

Utility ropes must not be used in situations where life safety rope is required. Conversely, life safety rope must not be used for utility applications. A fire fighter

FIGURE 9-4 Utility ropes are used for hoisting and lowering tools.
© Jones & Bartlett Learning. Photographed by Glen E. Ellman.

SAFETY TIP

Many fire departments use colour coding or other visible markings to identify different types of rope. This allows a fire fighter to determine quickly if a rope is a life safety rope or a utility rope. The length of each rope should be clearly marked by a tag or a label on the rope bag.

must be able to recognize the type of rope instantly from its appearance and markings.

Webbing

Webbing is a woven material of flat or tubular weave in the form of a long strip (**FIGURE 9-5**). Although it is not actually considered a type of rope, webbing is often used in conjunction with rope. Because of its special characteristics, webbing may even be preferable to rope in certain situations. For example, webbing commonly is used for anchoring because it is less expensive than rope.

There are hundreds of types and sizes of webbing. Most webbing is made of nylon or polyester and ranges in width from 13 mm (½ in.) up to 152 mm (6 in.).

FIGURE 9-5 Webbing is available in flat or tubular construction.
Courtesy of Skedco, Inc.

Flat webbing is constructed of a single layer of material; tubular webbing consists of a flattened tube of material. Although flat webbing is less expensive, tubular webbing is more supple and easier to work with.

Rope Materials

Ropes can be made from many different types of materials. The earliest ropes were made from naturally occurring vines or fibres that were woven together. Today, ropes are made of synthetic materials such as nylon or polypropylene. Because ropes have many different uses, certain materials can work better than others in particular situations.

Natural Fibres

In the past, fire departments used ropes made from natural fibres, such as manila or cotton, because there were no alternatives. In these ropes, the natural fibres are twisted together to form strands. A strand can contain hundreds of individual fibres of different lengths. Today, ropes made from natural fibres rarely are used as utility ropes and are no longer acceptable as life safety ropes (**TABLE 9-2**). Natural fibre ropes can be easily weakened by mildew, and they deteriorate with age, even when properly stored. A manila rope absorbs water when wet, making it very susceptible to deterioration. Once wet, a natural fibre rope is difficult to dry.

Synthetic Fibres

Since the introduction of nylon in 1938, synthetic fibres have been used to make ropes. In addition to nylon, several newer synthetic materials—such as polyester,

TABLE 9-2 Drawbacks to Using Natural Fibre Ropes

- Lose their load-carrying ability over time
- Subject to mildew
- Absorb water when wet
- Degrade quickly

polypropylene, and polyethylene—have been used in rope construction (**FIGURE 9-6**). Synthetic fibres have several advantages over natural fibres (**TABLE 9-3**). For example, synthetic fibres are generally stronger than natural fibres, so it is possible to use a smaller diameter rope without sacrificing strength. In addition, synthetic materials can produce very long fibres that run the full length of a rope to provide greater strength and added safety.

Synthetic ropes are more resistant to rotting and mildew than natural fibre ropes and do not degrade as rapidly. Depending on the material, they might also provide more resistance to melting and burning than natural fibre ropes. In addition, synthetic ropes absorb much less water when wet, and they can be washed and dried. Some types of synthetic rope can float on water, which is a major advantage in water rescue situations.

However, ropes made from synthetic fibres do have some drawbacks. Prolonged exposure to ultraviolet light as well as exposure to strong acids or alkalis can damage some synthetic ropes and decrease their life expectancy. Additionally, synthetic ropes need to be protected from abrasion and sharp objects that can damage or cut the rope fibres.

FIGURE 9-6 Synthetic fibres are generally stronger than natural fibres.
© Jones & Bartlett Learning. Photographed by Glen E. Ellman.

TABLE 9-3 Advantages to Using Synthetic Fibre Ropes

- Strength-to-diameter ratio
- Difficult to meet block creel construction requirement with natural fibres
- Longevity over natural fibres

FIGURE 9-7 Polypropylene rope is often used in water rescues.
© Jones & Bartlett Learning. Photographed by Glen E. Ellman.

Life safety ropes are always made from synthetic fibres. A variety of synthetic fibres are used to manufacture this type of rope; a common material is nylon. It has a high melting temperature compared to other synthetic materials, good abrasion resistance, and is strong and lightweight. Nylon ropes are also resistant to most acids and alkalis. Polyester is another common synthetic fibre used for life safety ropes. Some life safety ropes are made of a combination of nylon and polyester or other synthetic fibres.

Polypropylene is the lightest of the synthetic fibres. Because it does not absorb water and floats, polypropylene rope is often used for water rescue situations (**FIGURE 9-7**). Polypropylene, however, is not as suitable as nylon for fire department life safety uses because it is not as strong, it is hard to knot, and it has a low melting point compared to other synthetic materials.

Rope Construction

There are several different types of rope construction. The best choice of rope construction depends on the specific application.

Twisted Rope

Twisted ropes, which are also called laid ropes, are made of individual fibres twisted into strands. The strands are then twisted together to make the rope (**FIGURE 9-8**). This method of rope construction has been used for hundreds of years. Both natural and synthetic fibres can be used to make twisted rope.

This method of construction does have a disadvantage in that it exposes all of the fibres to the outside of the rope, where they are subject to abrasion. Abrasion can damage the rope fibres and reduce rope strength. Twisted ropes tend to stretch and are prone to unraveling when a load is applied.

FIGURE 9-8 Twisted rope.
© Jones & Bartlett Learning. Photographed by Glen E. Ellman.

Braided Rope

Braided ropes are constructed by weaving, or intertwining strands—typically synthetic fibres—together in the same way that hair is braided (**FIGURE 9-9**). This method of construction also exposes all of the strands to the outside of the rope where they are subject to abrasion. Braided rope stretches under a load, but it is not prone to twisting. A double-braided rope has an inner braided core covered by a protective braided sleeve so that only the fibres in the outer sleeve are exposed; the inner core remains protected from abrasion in this type of construction (**FIGURE 9-10**).

Kernmantle Rope

Kernmantle rope consists of two distinct parts: the kern and the mantle. The kern is the centre or core of the rope. The mantle, or sheath, is a braided covering that protects the core from dirt and abrasion. Although both parts of a kernmantle rope are made with synthetic fibres, different types of synthetic fibres can be used for the kern and the mantle.

Each fibre in the kern extends for the entire length of the rope without knots or splices. This block creel construction is required under NFPA 1983 for all life safety ropes. The continuous filaments produce a core

FIGURE 9-9 Braided rope.
© Jones & Bartlett Learning. Photographed by Glen E. Ellman.

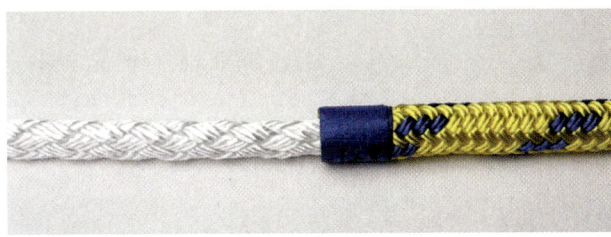

FIGURE 9-10 Double-braided rope.
Courtesy of Steve Hudson.

that is stronger than one constructed of shorter fibres that are twisted or braided together.

Kernmantle construction produces a strong and flexible rope that is relatively thin and lightweight (**FIGURE 9-11**). This construction is well suited for rescue work and is very popular for life safety rope.

Kernmantle ropes can be either dynamic or static. A dynamic rope is designed to be elastic and stretch when it is loaded. A dynamic kernmantle rope is constructed with overlapping or woven fibres in the core (**FIGURE 9-12**). When the rope is loaded, the core fibres are pulled tighter, which gives the rope its elasticity. NFPA 1983 specifies the limits of stretch or elongation of life safety ropes and the testing requirements for life safety ropes with moderate elongation.

A static rope has a limited range of elasticity. In the core of a static kernmantle rope, all of the fibres are laid parallel to each other (**FIGURE 9-13**). Such a rope has little elasticity and limited elongation under an applied load. Most fire department life safety ropes use static kernmantle construction. This type of rope is well suited for lowering a person and can be used with a pulley system for lifting individuals. It can also be used to create a bridge between two structures. Static kernmantle rope includes an embedded trailer from the

FIGURE 9-11 Kernmantle construction.
Courtesy of Steve Hudson.

FIGURE 9-12 Typical dynamic rope core.
Courtesy of Pigeon Mountain Industries.

FIGURE 9-13 Typical static rope core.
Courtesy of Pigeon Mountain Industries.

manufacturer. This trailer may include the name of the manufacturer, model number, make, serial number, and date of manufacture.

Rope Maintenance

All ropes—but especially life safety ropes—need proper care to perform in an optimal manner. Maintenance is necessary for all kinds of equipment and all types of rope, and it is absolutely essential for life safety ropes. Your life and the lives of others depend on the proper maintenance of your life safety ropes. For more information on the selection, care, and maintenance of life safety rope, see NFPA 1858, *Standard of Selection, Care, and Maintenance of Life Safety Rope and Equipment for Emergency Services*.

There are four parts to the maintenance formula:

- Care
- Clean
- Inspect
- Store

Care for the Rope

You must follow certain principles to preserve the strength and integrity of rope:

- Protect the rope from sharp and abrasive surfaces. Use edge protectors when the rope must pass over a sharp or unpadded surface.
- Protect the rope from rubbing against another rope or webbing. Friction generates heat, which can damage or destroy the rope.
- Protect the rope from heat, chemicals, and flames.
- Protect the rope from prolonged exposure to sunlight. Ultraviolet radiation can damage rope.
- Do not step on the rope! Your footstep could force shards of glass, splinters, or abrasive

---

FIRE FIGHTER TIP

A **shock load** can occur when a rope is suddenly placed under unusual tension—for example, when someone attached to a life safety rope falls until the length of the rope or another rescuer stops the drop. A utility rope can be shock loaded in a similar manner if a piece of equipment that is being raised or lowered drops suddenly.

Any rope that has been shock loaded should be inspected and might have to be removed from service. Although there might not be any visible damage, shock loading can cause damage that is not immediately apparent. Repeated shock loads can severely weaken a rope so that it can no longer be used safely. Keeping accurate rope records helps identify potentially damaged rope.

particles into the core of the rope, damaging the rope fibres.

- Follow the manufacturer's recommendations for rope care.

To care for life safety ropes properly, follow the steps in **SKILL DRILL 9-1**.

Clean the Rope

Many ropes made from synthetic fibres can be washed with a mild soap and water. In addition, a special rope washer can be attached to a garden hose (**FIGURE 9-14**). Some manufacturers recommend placing the rope in a mesh bag and washing it in a front-loading washing machine. Top-loading washing machines should not be used to wash ropes, because they can bind or wind the rope and produce unpredictable or undetectable stresses on the rope.

When washing a rope, use a mild detergent. Do not use bleach because it can damage rope fibres. Follow the manufacturer's recommendations for specific care of your rope. Do not pack or store wet or damp rope. Air drying is usually recommended. The drying rope should be suspended and should not lie on the floor. The use of mechanical drying devices is usually not recommended. Rope should never be dried or stored in direct sunlight.

Follow the steps in **SKILL DRILL 9-2** to clean fire department ropes.

Inspect the Rope

Life safety ropes must be inspected after each use, whether the rope was used for an emergency incident or in a training exercise (**TABLE 9-4**). Unused rope also should be inspected on a regular schedule. Some departments inspect all rope, including life safety and utility ropes, every 3 months. Obtain the recommended inspection criteria from the rope manufacturer.

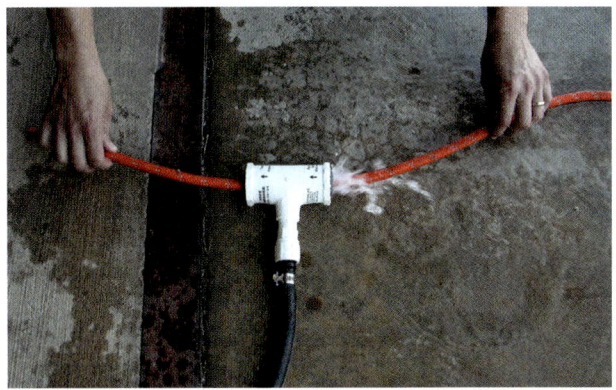

FIGURE 9-14 Some fire departments use a rope washer to clean their ropes.
Courtesy of Captain David Jackson, Saginaw Township Fire Department.

SKILL DRILL 9-1
Caring for Life Safety Ropes Fire Fighter I, NFPA 1001: 4.5.1

Courtesy of Pigeon Mountain Industries.

1. Protect the rope from sharp and abrasive edges; use edge protectors.

2. Protect the rope from rubbing against other ropes.

3. Protect the rope from heat, chemicals, flames, and sunlight.

4. Avoid stepping on the rope.

SKILL DRILL 9-2
Cleaning Fire Department Ropes Fire Fighter I, NFPA 1001: 4.5.1

1 Wash the rope with mild soap and water.

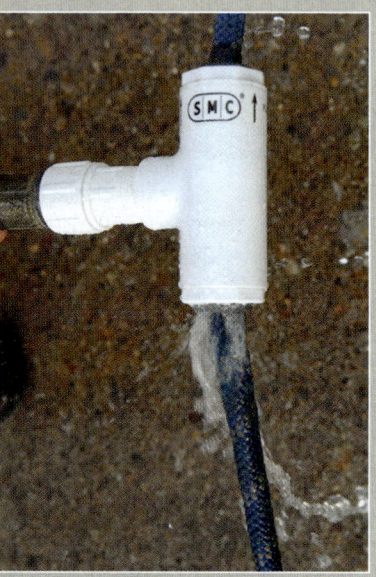

2 Use a rope washer or machine if recommended by the rope's manufacturer.

3 Air dry the rope out of direct sunlight. Inspect the rope, and replace it in the rope bag so that it is ready for use.

© Jones & Bartlett Learning. Photographed by Glen E. Ellman.

TABLE 9-4 Questions to Consider When Inspecting Life Safety Ropes

- Has the rope been exposed to heat or flame?
- Has the rope been exposed to abrasion?
- Has the rope been exposed to chemicals?
- Has the rope been exposed to shock loads?
- Are there any depressions, discolourations, or lumps in the rope?

Inspect the rope visually, looking for cuts, frays, or other damage as you run it through your fingers. Make sure that your grasping hand is gloved to protect yourself from sharp objects embedded in the rope. Because you cannot see the inner core of a kernmantle rope, feel for any depressions (flat spots or lumps on the inside). Examine the sheath for any discolourations, abrasions, or flat spots (**FIGURE 9-15**). If you have any doubt about whether the rope has been damaged, consult your company officer (**TABLE 9-5**).

FIGURE 9-15 Rope inspection is a critical step.
© Jones & Bartlett Learning. Photographed by Glen E. Ellman.

TABLE 9-5 Signs of Possible Rope Deterioration
• Discolouration
• Shiny markings from heat or friction
• Damaged sheath
• Core fibres poking through the sheath
• Inconsistencies in the thickness of the "feel" of the rope

A life safety rope that is no longer usable must be pulled from service and either destroyed or marked as a utility rope. A downgraded rope must be clearly marked so that it cannot be confused with a life safety rope.

Follow the steps in **SKILL DRILL 9-3** to inspect fire department ropes.

Rope Record (Log)

Each inspection of a utility rope or life safety rope should be recorded in a **rope record**. An individual record,

SKILL DRILL 9-3
Inspecting Fire Department Ropes Fire Fighter I, NFPA 1001: 4.5.1

1 Inspect the rope after each use and at regular intervals.

■ Examine the core by looking for depressions, flat spots, or lumps.

■ Examine the sheath by looking for discolourations, abrasions, flat spots, and embedded objects.

2 Remove the rope from service if it is damaged.

Courtesy of Steve Hudson.

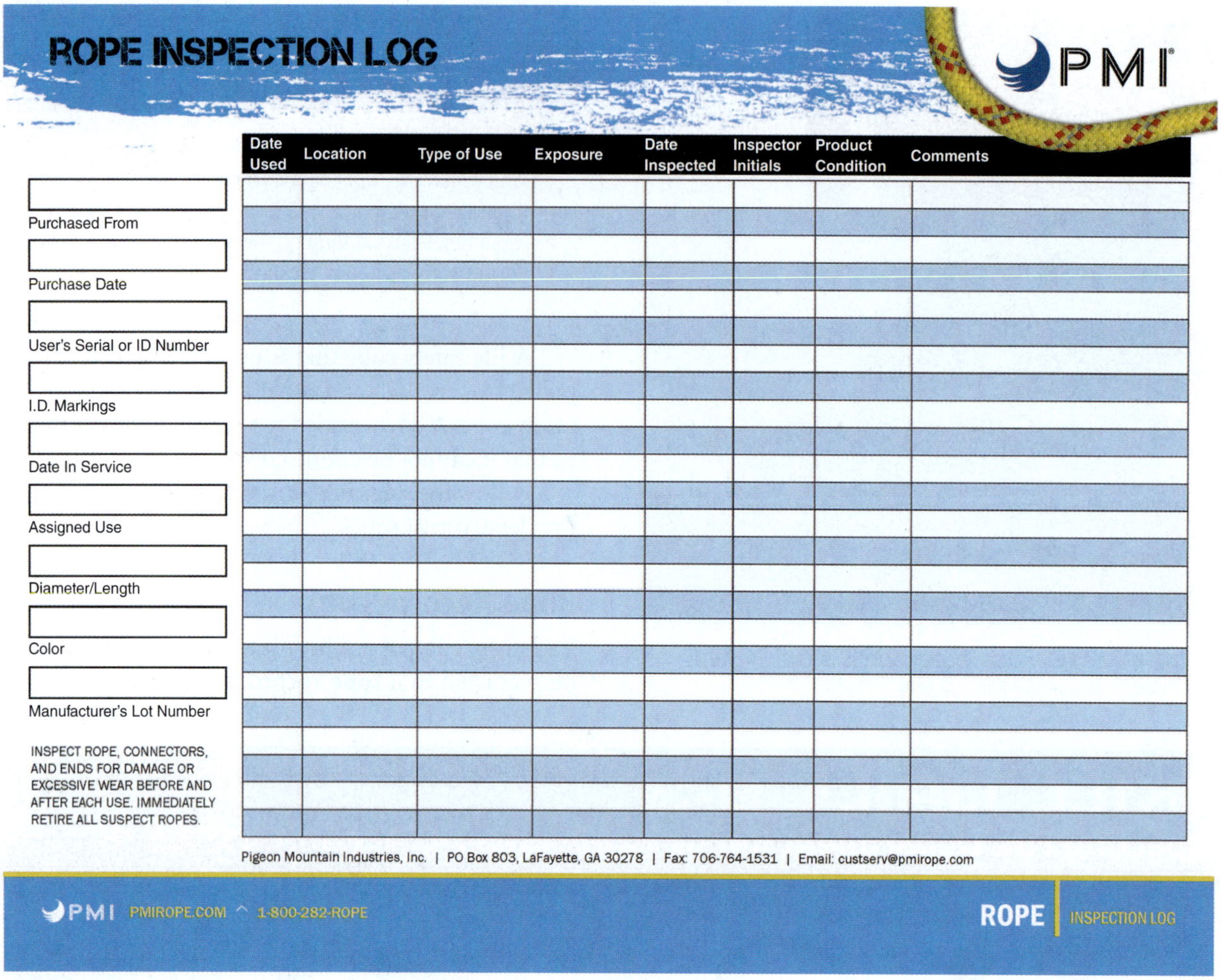

ROPE INSPECTION LOG

PMI

Purchased From

Purchase Date

User's Serial or ID Number

I.D. Markings

Date In Service

Assigned Use

Diameter/Length

Color

Manufacturer's Lot Number

INSPECT ROPE, CONNECTORS, AND ENDS FOR DAMAGE OR EXCESSIVE WEAR BEFORE AND AFTER EACH USE. IMMEDIATELY RETIRE ALL SUSPECT ROPES.

Date Used	Location	Type of Use	Exposure	Date Inspected	Inspector Initials	Product Condition	Comments

Pigeon Mountain Industries, Inc. | PO Box 803, LaFayette, GA 30278 | Fax: 706-764-1531 | Email: custserv@pmirope.com

PMI   PMIROPE.COM   1-800-282-ROPE   **ROPE** | INSPECTION LOG

FIGURE 9-16 Rope inspection should be recorded in a rope record.
Courtesy of Pigeon Mountain Industries.

which must be kept for each piece of life safety rope, should include the purchase date, dates of use, and the types of loads imposed on the rope. The record could also include the expected retirement date of the rope (**FIGURE 9-16**).

Store the Rope

Proper care ensures a long life for your rope and reduces the chance of equipment failure and accident. Store ropes in areas where there is some air circulation, away from temperature extremes, and out of sunlight. Avoid placing ropes where fumes from gasoline, oils, hydraulic fluids, or other petroleum products might damage them. Apparatus compartments used to store ropes should be separated from compartments used to store any oil-based products or machinery powered by gasoline or diesel fuel. Do not place any heavy objects on top of the rope (**FIGURE 9-17**).

Rope bags are used to protect and store ropes. Each bag should hold only one rope. Rope may also

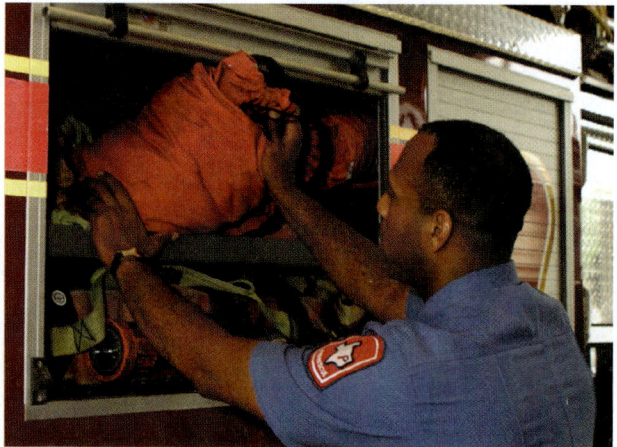

FIGURE 9-17 Ensure that ropes are stored safely.
© Jones & Bartlett Learning. Photographed by Glen E. Ellman.

be coiled for storage (**FIGURE 9-18**). Very long pieces of rope are sometimes stored on spools.

Follow the steps in **SKILL DRILL 9-4** to place a life safety rope into a rope bag.

FIGURE 9-18 Ropes may be coiled for storage.
© AbleStock.

Knots

Knots are prescribed ways of fastening lengths of rope or webbing to objects or to each other. As a fire fighter, you must know how to tie certain knots and when to use each type of knot. Knots can be used for one or more particular purposes. Safety knots, such as the overhand knot, are used to secure the ends of ropes to prevent them from coming untied. Hitches, such as the clove hitch, are used to attach a rope around an object. Loop knots, such as the figure eight and the bowline, are used to form loops. Bends, such as the sheet bend or Becket bend, are used to join two ropes together.

Any knot reduces the load-carrying capacity of the rope in which it is placed by a certain percentage (**TABLE 9-6**). You can avoid an unnecessary reduction in rope strength if you know which type of knot to use and how to tie the knot correctly.

Terminology

Specific terminology is used to refer to the parts of a rope in describing how to tie knots (**FIGURE 9-19**):

- The **working end** is the part of the rope used for forming the knot.
- The **running end** is the part of the rope used for lifting or hoisting.
- The **standing part** is the part of the rope between the working end and the running end.

SKILL DRILL 9-4
Placing a Life Safety Rope in a Rope Bag Fire Fighter I, NFPA 1001: 4.5.1

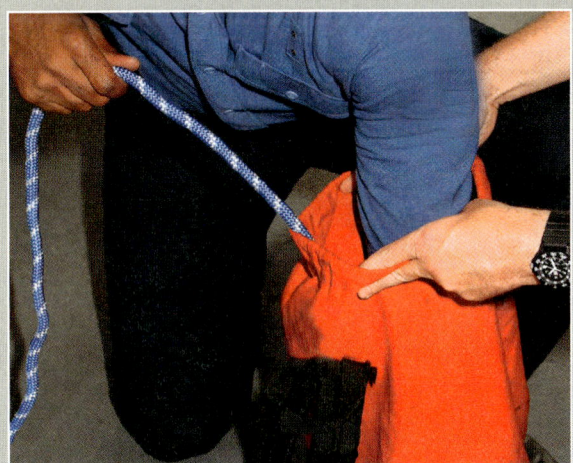

1 Inspect the rope to be certain it is fit for service. Carefully load the life safety rope into the rope bag.

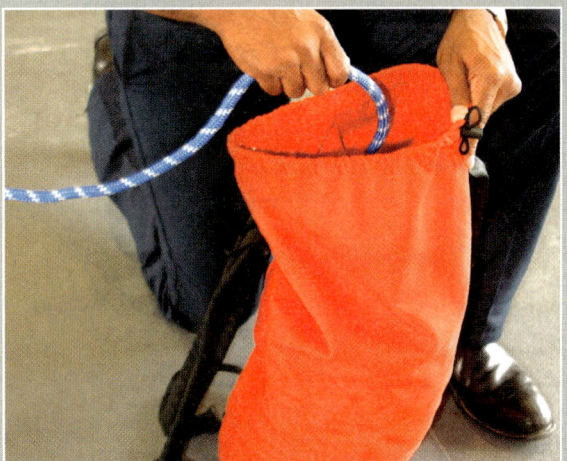

2 Do not try to coil the rope in the bag because this action will cause the rope to kink and become tangled when it is pulled out. Return the rope and the rope bag to its proper location for storage or service.

© Jones & Bartlett Learning. Photographed by Glen E. Ellman.

TABLE 9-6 Effect of Loop Knots on Rope Strength	
Knot	**Reduction in Strength**
Figure eight	25 percent
Figure eight on a bight	22 percent
Bowline	37 percent

Source: Richards, Dave. 2005. "Knot Break Strength vs Rope Break Strength." http://caves.org/section/vertical/nh/50/knotrope-hold.html. Accessed June 18, 2018.

FIGURE 9-20 A bight.
© Jones & Bartlett Learning. Photographed by Glen E. Ellman.

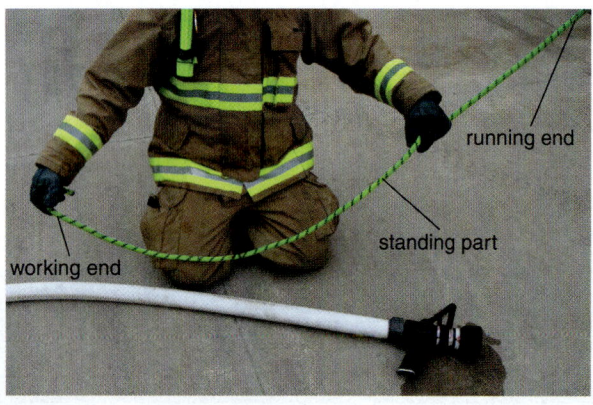

FIGURE 9-19 The sections of a rope used in tying knots.
© Jones & Bartlett Learning. Photographed by Glen E. Ellman.

FIGURE 9-21 A loop.
© Jones & Bartlett Learning. Photographed by Glen E. Ellman.

Specific terminology is also used when describing the bends in rope that are formed when a knot is tied:

- A **bight** is formed by reversing the direction of the rope to form a U bend with two parallel ends (**FIGURE 9-20**).
- A **loop** is formed by making a circle in the rope (**FIGURE 9-21**).
- A **round turn** is formed by making a loop and then bringing the two ends of the rope parallel to each other (**FIGURE 9-22**).

A fire fighter should know how to tie and properly use the following simple knots:

- Safety knot
- Half hitch
- Clove hitch
- Figure eight
- Figure eight on a bight
- Figure eight follow-through

FIGURE 9-22 A round turn.
© Jones & Bartlett Learning. Photographed by Glen E. Ellman.

- Figure eight bend
- Bowline
- Sheet bend (Becket bend)
- Water knot (ring bend)

There are many ways to correctly tie each of these knots. Find one method that works for you, and use it all the time. Your department might require that you learn how to tie knots in addition to those listed here.

A knot should be properly "dressed" by tightening and removing twists, kinks, and slack from the rope. The finished knot is then firmly fixed in position. The configuration of a properly dressed knot should be evident so that it can be easily inspected.

It is important to become proficient in tying knots. Knot-tying skills can be quickly lost without adequate practice. Practise tying knots while you are on the telephone or watching television (**FIGURE 9-23**). You never know when you will need to use these skills in an emergency situation. For added practice, try tying these knots with your gloves on or in darkness. These are conditions you will encounter often as a fire fighter.

Safety Knot

A safety knot is used to secure the leftover working end of the rope. It should always be used to finish other basic knots, as it provides a degree of safety to ensure that the primary knot will not become undone. When used to finish another knot, a safety knot must be properly dressed (the strands aligned and uncrossed), compacted (all ends pulled down so the knot is compact), and tied close to the primary knot.

A safety knot is a knot in the loose end of the rope that is made around the standing part of the rope. It secures the loose end and prevents it from slipping back through the primary knot.

Follow the steps in **SKILL DRILL 9-5** to tie a safety knot.

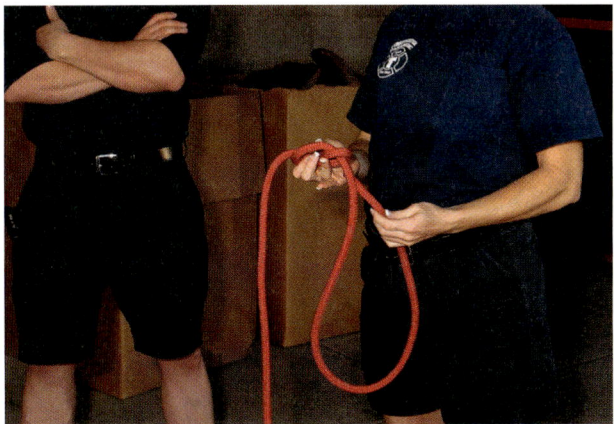

FIGURE 9-23 To maintain your knot-tying skills, practise tying different knots frequently.
© Jones & Bartlett Learning. Photographed by Glen E. Ellman.

Hitches

Hitches are knots that wrap around an object, such as a pike pole or a fencepost. They are used to secure the working end of a rope to a solid object or to tie a rope to an object before hoisting it.

Half Hitch

The half hitch is not a secure knot by itself, which is why it is used only in conjunction with other knots. For example, when hoisting an axe or pike pole, you use the half hitch to keep the hoisting rope aligned with the handle. On long objects, you might need to use several half hitches.

Follow the steps in **SKILL DRILL 9-6** to tie a half hitch.

Clove Hitch

A clove hitch is used to attach a rope firmly to a round object, such as a tree or a fencepost. It can also be used to tie a hoisting rope around an axe or pike pole. A clove hitch can be tied anywhere in a rope and will hold equally well if tension is applied to either end of the rope or to both ends simultaneously.

There are two different methods of tying this knot. A clove hitch tied in the open is used when the knot can be formed and then slipped over the end of an object, such as an axe or pike pole. It is tied by making two consecutive loops in the rope. Follow the steps in **SKILL DRILL 9-7** to tie a clove hitch in the open.

If the object is too large or too long to slip the clove hitch over one end, the same knot can be tied around the object. Follow the steps in **SKILL DRILL 9-8** to tie a clove hitch around an object.

Loop Knots

Loop knots are used to form a loop in the end of a rope. Loops can be used for hoisting tools, for securing a person during a rescue, for securing a rope to a fixed object, or for identifying the end of a rope stored in a rope bag. When tied properly, these knots will not slip yet are easy to untie.

> **FIRE FIGHTER TIP**
>
> When tying a safety knot in a clove hitch, the end of the rope should pass around the object the clove hitch is being tied to. The safety knot is then tied onto the rope.

SKILL DRILL 9-5
Tying a Safety Knot Fire Fighter I, NFPA 1001: 4.1.2, 4.3.20

1 Take the loose end of the rope, beyond the knot, and form a loop around the standing part of the rope.

2 Pass the loose end of the rope through the loop.

3 Tighten the safety knot by pulling on both ends at the same time. Test whether you have tied a safety knot correctly by sliding it on the standing part of the rope. A knot that is tied correctly will slide. Note: When complete, the safety knot should sit directly next to any accompanying knots.

SKILL DRILL 9-6
Tying a Half Hitch Fire Fighter I, NFPA 1001: 4.1.2, 4.3.20

1 Grab the rope with your palm facing away from you.

2 Rotate your hand so that your palm is facing you. This will make a loop in the rope.

3 Pass the loop over the end of the object.

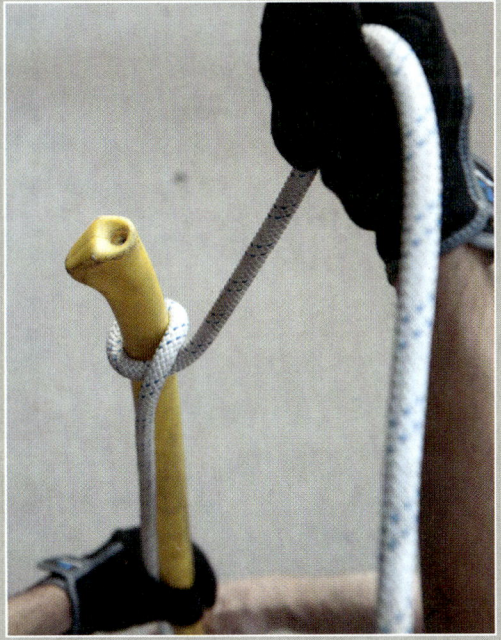

4 Finish the half-hitch knot by positioning it and pulling tight.

SKILL DRILL 9-7
Tying a Clove Hitch in the Open Fire Fighter I, NFPA 1001: 4.1.2, 4.3.20

1 Starting from left to right on the rope, grab the rope with crossed hands with the left positioned higher than the right.

2 Holding on to the rope, uncross your hands. This will create a loop in each hand.

3 Slide the right-hand loop behind the left-hand loop.

4 Slide both loops over the object.

5 Pull in opposite directions to tighten the clove hitch. Tie a safety knot in the working end of the rope.

SKILL DRILL 9-8
Tying a Clove Hitch Around an Object Fire Fighter I, NFPA 1001: 4.1.2, 4.3.20

1 Place the working end of the rope over the object.

2 Make a complete loop around the object, working end down.

3 Make a second loop around the object a short distance above the first loop. Pass the working end of the rope under the second loop, above the point where the second loop crosses over the first loop.

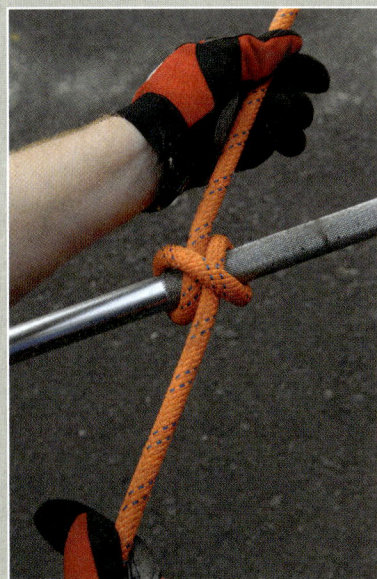

4 Tighten the knot, and secure it by pulling on both ends.

5 Tie a safety knot in the working end of the rope.

Figure Eight Knot

A figure eight is a basic knot used to produce a family of other knots, including the figure eight on a bight, the figure eight follow-through, and the figure eight bend. Follow the steps in **SKILL DRILL 9-9** to tie a figure eight knot.

1 Form a bight in the rope.

2 Loop the working end of the rope completely around the standing part of the rope.

3 Thread the working end back through the opening of the bight.

4 Tighten the knot by pulling on both ends simultaneously. When you pull the knot tight, it will have the shape of a figure eight.

Figure Eight on a Bight

The figure eight on a bight knot creates a secure loop at the working end of a rope. This loop can be used to attach the end of the rope to a fixed object or a piece of equipment or to tie a life safety rope around a person. Follow the steps in **SKILL DRILL 9-10** to tie a figure eight on a bight.

SKILL DRILL 9-10
Tying a Figure Eight on a Bight Fire Fighter I, NFPA 1001: 4.1.2, 4.3.20

1 Double over a section of the rope to form a bight. The closed end of the bight becomes the working end of the rope.

2 Hold the two sides of the bight together as if they were one rope. Form a loop in the doubled section of the rope.

3 Feed the working end of the bight back through the loop.

4 Pull the knot tight to lock the neck of the bight and form a secure loop.

5 Use a safety knot to secure the loose end of the rope to the standing part. Follow your department SOPs (SOGs) regarding the regular use of safety knots.

Figure Eight Follow-Through

A figure eight follow-through knot creates a secure loop at the end of the rope when the working end must be wrapped around an object or passed through an opening before the loop can be formed. It is very useful for attaching a rope to a fixed ring or a solid object with an "eye." Follow the steps in **SKILL DRILL 9-11** to tie a figure eight follow-through.

SKILL DRILL 9-11
Tying a Figure Eight Follow-Through Fire Fighter I, NFPA 1001: 4.1.2, 4.3.20

1 Tie a simple figure eight in the standing part of the rope, far enough back to make a loop. Leave this knot loose.

2 Thread the working end through the opening or around the object, and bring it back through the original figure eight knot in the opposite direction.

3 Once the working end has been threaded through the knot, pull the knot tight.

4 Secure the loose end with a safety knot. Follow your department SOPs (SOGs) regarding the regular use of safety knots.

SKILL DRILL 9-12
Tying a Figure Eight Bend Fire Fighter 1, NFPA 1001: 4.1.2, 4.3.20

1 Tie a figure eight near the end of one rope.

2 Thread the end of the second rope completely through the knot from the opposite end. Pull the knot tight.

3 Tie a safety knot on the loose end of each rope to the standing part of the other.

© Jones & Bartlett Learning. Photographed by Glen E. Ellman.

Figure Eight Bend

The figure eight bend or tracer 8 is used to join two ropes together. Follow the steps in **SKILL DRILL 9-12** to tie a figure eight bend.

Bowline

A bowline knot also can be used to form a loop. This type of knot is frequently used to secure the end of a rope to an object or anchor point. It also can be used to hoist equipment.

When a bowline knot in a loaded rope is pulled over an edge, the knot can deform. In addition, a bowline knot may become loose when it is not under load. For these reasons, bowline knots always need to be secured with a safety knot. Follow the steps in **SKILL DRILL 9-13** to tie a bowline.

Bends

Bends are used to join two ropes together. These knots also can be used to join rope to a chain or cable.

Sheet Bend

The sheet bend or Becket bend can be used to join two ropes of unequal size. A sheet bend knot also can be used to join rope to a chain or to pull up a rope of a different diameter. It should not be relied upon to support a human load. Follow the steps in **SKILL DRILL 9-14** to tie a sheet bend.

Water Knot

The water knot or ring bend is used to join webbing of the same or different sizes together. When a single

SKILL DRILL 9-13
Tying a Bowline Fire Fighter I, NFPA 1001: 4.1.2, 4.3.20

1 Make the desired sized loop, and bring the working end back to the standing part.

2 Form another small loop in the standing part of the rope with the section close to the working end on top. Thread the working end up through this loop from the bottom.

3 Pass the working end over the loop and around and under the standing part.

4 Pass the working end back down through the same opening.

5 Tighten the knot by holding the working end and pulling the standing part of the rope backward.

6 Tie a safety knot in the working end of the rope.

SKILL DRILL 9-14
Tying a Sheet or Becket Bend Fire Fighter I, NFPA 1001: 4.1.2, 4.3.20

1 Using your left hand, form a bight at the working end of one rope. If the ropes are of unequal size, the bight should be made in the larger rope.

2 Thread the working end of the second rope up through the opening of the bight, between two parallel sections of the first rope.

3 Loop the second rope completely around both sides of the bight. Pass the working end of the second rope between the original bight and under the second rope.

4 Tighten the knot by holding the first rope firmly while pulling back on the second rope.

5 Tie a safety knot in the working end of each rope.

© Jones & Bartlett Learning. Photographed by Glen E. Ellman.

SKILL DRILL 9-15
Tying a Water Knot Fire Fighter I, NFPA 1001: 4.1.2, 4.3.20

1 In one end of the webbing, approximately 152 mm (6 in.) from the end, tie an overhand knot.

2 With the other end of the webbing, start retracing from the working end through the knot until approximately 152 mm (6 in.) is left on the other end.

3 Tie an overhand knot on each tail as a safety.

© Jones & Bartlett Learning. Photographed by Glen E. Ellman.

piece of webbing is used and the opposite ends are tied to each other, a loop or sling is created. Follow the steps in **SKILL DRILL 9-15** to tie a water knot.

Hoisting

Tying knots is not an idle exercise but rather a practical skill that you will use frequently on the job. In emergency situations, you might have to raise or lower a tool to other fire fighters. It is essential to ensure that the rope is tied securely to the object being hoisted so that the tool does not fall. Additionally, your co-workers must be able to remove the rope and place the tool into service quickly.

It is important for you to learn how to raise and lower an axe, a pike pole, a ladder, a charged hose line, an uncharged hose line, and a fan. When you are hoisting or lowering a tool, make sure no one is standing under

the object. Keep the scene clear of people to avoid any chance of an accident. Attach a **tag line** (a separate rope that ground personnel can use to guide an object that is being hoisted or lowered) on the equipment being raised or lowered to help control its path and prevent the tool or equipment from getting caught on an obstruction, such as an electrical line, utility pole, or structural feature. The tag line should be attached in a way that allows it to be removed easily and dropped back down to ground personnel.

Hoisting an Axe

An axe should be hoisted in a vertical position with the head of the axe down. After receiving the end of the utility rope from the fire fighter above, follow the steps in **SKILL DRILL 9-16** to hoist an axe.

SKILL DRILL 9-16
Hoisting an Axe Fire Fighter I, NFPA 1001: 4.1.2, 4.3.20

1 Tie the end of the hoisting rope around the handle of the axe near the head using either a figure eight on a bight or a clove hitch. Slip the knot down the handle from the end to the head.

2 Loop the standing part of the rope under the head.

3 Place the standing part of the rope parallel to the axe handle.

4 Use one or two half hitches along the axe handle to keep the handle parallel to the rope.

5 Communicate with the fire fighter above that the axe is ready to raise.

To release the axe, hold the middle of the handle, release the half hitches, and slip the knot up and off.

Hoisting a Pike Pole

A pike pole should be hoisted in a vertical position with the head at the top. After receiving the end of the utility rope from the fire fighter above, follow the steps in **SKILL DRILL 9-17** to hoist a pike pole.

Hoisting a Ladder

A ladder should be hoisted in a vertical position. A tag line should be attached to the bottom to keep the ladder under control as it is hoisted. If it is a roof ladder, the hooks should be in the retracted position. After receiving the end of the utility rope from the fire fighter above, follow the steps in **SKILL DRILL 9-18** to hoist a ladder.

SKILL DRILL 9-17
Hoisting a Pike Pole Fire Fighter I, NFPA 1001: 4.1.2, 4.3.20

1 Place a clove hitch over the bottom of the handle, and secure it close to the bottom of the handle. Leave enough length of rope below the clove hitch for a tag line while raising the pike pole.

2 Place a half hitch around the handle above the clove hitch to keep the rope parallel to the handle.

3 Slip a second half hitch over the handle, and secure it near the head of the pike pole.

4 Communicate with the fire fighter above that the pike pole is ready to raise.

SKILL DRILL 9-18
Hoisting a Ladder Fire Fighter I, NFPA 1001: 4.1.2, 4.3.20

1 Tie a figure eight on a bight to create a loop.

2 The loop should be approximately 0.9–1.2 m (3–4 ft) in diameter and large enough to fit around both ladder beams.

3 Pass the rope between two beams of the ladder, three or four rungs from the top. Pull the end of the loop under the rungs and toward the tip at the top of the ladder.

4 Place the loop around the top tip of the ladder.

5 Pull on the running end of the rope to remove the slack from the rope. Attach a tag line to the bottom rung of the ladder to stabilize it as it is being hoisted. Communicate with the fire fighter above that the ladder is ready to be raised. Hold on to the tag line to stabilize the bottom of the ladder as it is being hoisted.

SKILL DRILL 9-19
Hoisting a Charged Hose Line Fire Fighter I, NFPA 1001: 4.1.2, 4.3.20

1 Make sure that the nozzle is completely closed and secure. An unsecured handle (known as the bail) could get caught on something as the hose is being hoisted, and the nozzle could open suddenly. Use a clove hitch, 0.3–0.6 m (1–2 ft) behind the nozzle, to tie the end of the hoisting rope around a charged hose line. Use a safety knot to secure the loose end of the rope below the clove hitch.

2 Make a bight in the rope even with the nozzle shut-off handle, which must be in the forward (completely off) position. Insert the bight through the handle opening, and slip it over the end of the nozzle. When the bight is pulled tight, it will create a half hitch and secure the handle in the off position while the charged hose line is hoisted. Communicate with the fire fighter above that the hose line is ready to hoist.

© Jones & Bartlett Learning. Photographed by Glen E. Ellman.

Hoisting a Charged Hose Line

It is almost always preferable to hoist a dry hose line because water adds considerable weight to a hose line. Water weighs 3.78 kg (8.33 lb) per gallon, which can make hoisting a charged line much more difficult. However, there can be occasions when it is necessary to hoist a charged hose line. After receiving the end of the utility rope from the fire fighter above, follow the steps in **SKILL DRILL 9-19** to hoist a charged hose line.

The knot can be released after the line is hoisted by removing the tension from the rope and slipping the bight back over the end of the nozzle.

Hoisting an Uncharged Hose Line

Before hoisting a dry hose line, you should fold the hose back on itself and place the nozzle on top of the hose. This action ensures that water will not reach the nozzle if the hose is accidentally charged while being hoisted. It also eliminates an unnecessary stress on the couplings by ensuring that the rope pulls on the hose and not directly on the nozzle. After receiving the end of the utility rope from the fire fighter above, follow the steps in **SKILL DRILL 9-20** to hoist an uncharged hose line.

Hoisting a Fan or Power Tool

Several types of tools and equipment—including a fan, a chainsaw, a circular saw, or any other object that has a strong closed handle—can be hoisted using the same technique. In this situation, the hoisting rope is secured to the object by passing the rope through the opening in the handle. A figure eight follow-through knot is then used to close the loop.

Some types of equipment require that you use additional half hitches to balance the object in a particular

SKILL DRILL 9-20
Hoisting an Uncharged Hose Line Fire Fighter I, NFPA 1001: 4.1.2, 4.3.20

1 Fold about 1 m (3 ft) of hose back on itself, and place the nozzle on top of the hose.

2 Tie a clove hitch around both the nozzle and the hose, and tighten it securely. The clove hitch should hold both the nozzle and the hose.

3 Tie a half hitch around the hose about 31 cm (1 ft) from the nozzle. Tie a second half hitch around the hose about 31 cm (1 ft) above the first half hitch.

4 Tie a safety knot around the nozzle and the hose below the clove hitch.

5 Communicate with the fire fighter above that the hose line is ready to hoist. Hoist the hose with the fold at the top and the nozzle pointing down. Before releasing the rope, the fire fighters at the top must pull up enough hose so that the weight of the hanging hose does not drag down the hose.

position while it is being hoisted. For example, power saws are hoisted in a level position to prevent the fuel from leaking out.

You should practise hoisting the actual tools and equipment used in your department. You should be able to perform this task automatically and in adverse conditions.

Remember that you always use utility rope for hoisting tools. You do not want to get oil or grease on designated life safety ropes. If a life safety rope becomes oily or greasy, it should be taken out of service and destroyed so that it will not be used again mistakenly as a life safety rope. The damaged rope can be cut into short lengths and used for utility rope.

After receiving the end of the utility rope from the fire fighter above, follow the steps in **SKILL DRILL 9-21** to hoist an exhaust fan.

SKILL DRILL 9-21
Hoisting an Exhaust Fan or Power Tool Fire Fighter I, NFPA 1001: 4.1.2, 4.3.20

1 Tie a figure eight knot in the rope about 3 ft (1 m) from the working end of the rope.

2 Loop the working end of the rope around the fan handle and back to the figure eight knot.

3 Secure the rope by tying a figure eight follow-through by threading the working end back through the first figure eight in the opposite direction.

4 Attach a tag line to the fan for better control. Communicate with the fire fighter above that the exhaust fan is ready to hoist.

© Jones & Bartlett Learning. Photographed by Glen E. Ellman.

After-Action REVIEW

IN SUMMARY

- The four primary types of rope used in the fire service are:
  - *Life safety rope.* Used to support people during a rescue. This type of rope is used only for life-saving purposes; it is rated as either technical or general use life safety rope.
  - *Escape rope and fire escape rope.* Used as a single-purpose emergency self-escape rope.
  - *Throwline.* A floating rope used during water rescue operations.
  - *Utility rope.* Used to perform all other tasks, such as hoisting equipment.
- Ropes can be made of natural or synthetic fibres. Natural fibres can be weakened by abrasion, mildew, and age, so they cannot be used as life safety ropes. Synthetic fibres are generally stronger than natural fibres and are used in life safety ropes.
- Synthetic fibres can be damaged by ultraviolet light and abrasion.
- Nylon, polyester, and polypropylene are common synthetic fibres used in life safety ropes. Polypropylene does not absorb water, so it is often used for water rescues.
- Common types of rope construction include the following:
  - Twisted ropes
  - Braided ropes
  - Kernmantle ropes
- All ropes need proper care to perform in an optimal manner. The rope maintenance formula consists of four parts:
  - Care
  - Clean
  - Inspect
  - Store
- The principles of caring for a rope include protecting the rope from:
  - Sharp and abrasive surfaces
  - Rubbing against another rope or webbing
  - Heat, chemicals, and flames
  - Prolonged exposure to sunlight
- When inspecting life safety rope, consider these questions:
  - Has the rope been exposed to heat or flame?
  - Has the rope been exposed to abrasion?
  - Has the rope been exposed to chemicals?
  - Has the rope been exposed to shock loads?
  - Are there any depressions, discolourations, or lumps in the rope?
- A rope record for a life safety rope includes a history of when the rope was purchased, when it was used, how it was used, each inspection, and what kinds of loads were applied to it.
- Ropes should be stored away from temperature extremes, out of sunlight, and in areas with good air circulation. Rope bags can be used to protect and store ropes.
- Knots can be used for one or more particular purposes.
  - Safety knots, such as the overhand knot, are used to secure the ends of ropes to prevent them from coming untied.
  - Hitches, such as the clove hitch, are used to attach a rope around an object.

- Loop knots, such as the figure eight and the bowline, are used to form loops.
- Bends, such as the sheet bend or Becket bend, are used to join two ropes together.
- Specific terminology is used to refer to the parts of a rope when describing how to tie knots:
  - The working end is the part of the rope used for forming the knot.
  - The running end is the part of the rope used for lifting or hoisting.
  - The standing part is the rope between the working end and the running end.
- Specific terminology is used when describing the bends in rope that are formed when a knot is tied:
  - A bight is formed by reversing the direction of the rope to form a U bend with two parallel ends.
  - A loop is formed by making a circle in the rope.
  - A round turn is formed by making a loop and then bringing the two ends of the rope parallel to each other.
- The knots that a fire fighter should know how to tie are as follows:
  - Safety knot
  - Half hitch
  - Clove hitch
  - Figure eight
  - Figure eight on a bight
  - Figure eight follow-through
  - Figure eight bend
  - Bowline
  - Sheet bend
  - Water knot
- In emergency situations, a fire fighter might need to hoist a tool to other fire fighters. Fire fighters should practise hoisting the tools and equipment used in their departments.

KEY TERMS

Access Navigate for flashcards to test your key term knowledge.

Bend A knot that joins two ropes or webbing pieces together. (NFPA 1670)

Bight The open loop in a rope or piece of webbing formed when it is doubled back on itself. (NFPA 1006)

Block creel construction Rope constructed without knots or splices in the yarns, ply yarns, strands or braids, or rope. (NFPA 1983)

Braided rope Rope constructed by intertwining strands in the same way that hair is braided.

Dynamic rope A rope generally made from synthetic materials that is designed to be elastic and stretch when loaded. Mountain climbers often use dynamic rope.

Escape rope A single-purpose, emergency self-escape (self-rescue) rope; not classified as a life safety rope. (NFPA 1983)

Fire escape rope An emergency self-rescue rope used to escape an immediately hazardous environment involving fire or fire products; not classified as a life safety rope. (NFPA 1983)

General use life safety rope A life safety rope that is no larger than 16 mm (⅝ in.) and no smaller than 11 mm (⁷⁄₁₆ in.), with a minimum breaking strength of 40 kN (8992 lbf).

Halyard Rope used on extension ladders for the purpose of raising a fly section(s). (NFPA 1931)

Hitch A knot that attaches to or wraps around an object so that when the object is removed, the knot will fall apart. (NFPA 1670)

Kernmantle rope Rope made of two parts—the kern (interior component) and the mantle (the outside sheath).

Knot A fastening made by tying rope or webbing in a prescribed way. (NFPA 1670)

Life safety rope Rope dedicated solely for the purpose of supporting people during rescue, firefighting, other emergency operations, or during training evolutions. (NFPA 1983)

Loop A piece of rope formed into a circle.

Rope bag A bag used to protect and store rope so that the rope can be easily and rapidly deployed without kinking.

Rope record A record for each piece of rope that includes a history of when the rope was placed in service, when it was inspected, when and how it was used, and which types of loads were placed on it.

Round turn A piece of rope looped to form a complete circle with the two ends parallel.

Running end The part of a rope used for lifting or hoisting.

Safety knot A knot used to secure the leftover working end of the rope.

Shock load An instantaneous load that places a rope under extreme tension, such as when a falling load is suddenly stopped as the rope becomes taut.

Standing part The part of a rope between the working end and the running end.

Static rope A rope generally made out of synthetic material that stretches very little under load.

Tag line A separate rope that ground personnel can use to guide an object that is being hoisted or lowered.

Technical use life safety rope A life safety rope with a diameter that is 9.5 mm (⅜ in.) or greater but is less than 13 mm (½ in.), with a minimum breaking strength of 20 kN (4496 lbf). Used by highly trained rescue teams that deploy to technical environments such as mountainous and/or wilderness terrain.

Throwline A floating rope that is intended to be thrown to a person during water rescues or as a tether for rescuers entering the water. (NFPA 1983)

Twisted rope Rope constructed of fibres twisted into strands, which are then twisted together.

Utility rope Rope used for securing objects, for hoisting equipment, or for securing a scene to prevent bystanders from being injured. Utility rope must never be used in life safety operations.

Water knot A knot used to join the ends of webbing together.

Webbing Woven material of flat or tubular weave in the form of a long strip. (NFPA 1983)

Working end The part of the rope used for forming a knot.

REFERENCES

National Fire Protection Association. NFPA 1858, *Standard of Selection, Care, and Maintenance of Life Safety Rope and Equipment for Emergency Services*, 2018. www.nfpa.org. Accessed November 8, 2018.

National Fire Protection Association. NFPA 1983, *Standard on Life Safety Rope and Equipment for Emergency Services*, 2017. www.nfpa.org. Accessed November 8, 2018.

On Scene

You are normally assigned to a ladder company, but today you are assigned to the rescue company at another station. The captain is scheduled to teach the Introduction to Technical Rescue course to the fire fighters from several departments, so you quickly familiarize yourself with the location of the equipment, climb in the ladder, and head to the training centre. It immediately becomes apparent that one of the students lacks basic rope skills that the remainder of the class already possesses. The lead instructor asks you to review the basics with the fire fighter while the rest get everything laid out.

1. What type of rope is designed to be used for securing objects, hoisting equipment, and blocking off scenes?

  A. Technical use life safety rope

  B. General use life safety rope

  C. Utility rope

  D. Escape rope

2. What is the part of a rope used for lifting or hoisting?

  A. Standing end

  B. Running end

  C. Bight

  D. Round turn

(continued)

On Scene Continued

3. The knot used to secure the leftover working end of the rope is called a _____ knot.

A. safety

B. figure eight

C. bowline

D. half hitch

4. Which of the following is a rope that is generally made out of synthetic material and stretches very little under load?

A. Dynamic rope

B. Twisted rope

C. Braided rope

D. Static rope

5. Once rope has gotten wet, what should be done with it?

A. Lay it in the sunlight so that it does not get moldy.

B. Air dry it out of direct sunlight.

C. Place it in a clothes dryer on high heat.

D. The rope can no longer be used.

6. Which knot is used to create a secure loop at the end of the rope when the working end must be wrapped around an object or passed through an opening before the loop can be formed?

A. Figure eight follow-through

B. Figure eight on a bight

C. Half hitch

D. Becket bend

7. Which of the following is a piece of rope looped to form a complete circle with the two ends parallel?

A. Bight

B. Hitch

C. Round turn

D. Loop

8. Which of the following types of rope should be able to float?

A. Technical life safety rope

B. Utility rope

C. Throwline

D. Kernmantle rope

Access Navigate to find answers to this On Scene, along with other resources such as an audiobook.

Forcible Entry

KNOWLEDGE OBJECTIVES

After studying this chapter, you will be able to:

- Describe the situations and circumstances that require forcible entry into a structure. (**NFPA 1001: 4.3.4**, pp. 327–328)
- List the general safety rules to follow when utilizing forcible entry tools. (**NFPA 1001: 4.3.4**, pp. 328–329)
- List the general carrying tips when utilizing forcible entry tools. (**NFPA 1001: 4.3.4**, p. 329)
- List the general maintenance tips when utilizing forcible entry tools. (p. 329)
- List the types of tools used in forcible entry. (**NFPA 1001: 4.3.4**, pp. 329–330)
- List the rotating tools used in forcible entry. (**NFPA 1001: 4.3.4**, p. 330)
- Describe the tasks that rotating tools are used for in forcible entry. (**NFPA 1001: 4.3.4**, p. 330)
- List the striking tools used in forcible entry. (**NFPA 1001: 4.3.4**, p. 330)
- Describe the tasks that striking tools are used for in forcible entry. (**NFPA 1001: 4.3.4**, p. 330)
- List the prying and spreading tools used in forcible entry. (**NFPA 1001: 4.3.4**, pp. 330–332)
- Describe the tasks that prying and spreading tools are used for in forcible entry. (**NFPA 1001: 4.3.4**, pp. 330–332)
- List the cutting tools used in forcible entry. (**NFPA 1001: 4.3.4**, pp. 332–333)

- Describe the tasks that cutting tools are used for in forcible entry. (**NFPA 1001: 4.3.4**, pp. 332–333)
- List the pushing and pulling tools used in forcible entry. (**NFPA 1001: 4.3.4**, pp. 333–334)
- Describe the tasks that pushing and pulling tools are used for in forcible entry. (**NFPA 1001: 4.3.4**, pp. 333–334)
- List the special-use and lock tools used in forcible entry. (**NFPA 1001: 4.3.4**, pp. 334–336)
- Describe the tasks that special-use and lock tools are used for in forcible entry. (**NFPA 1001: 4.3.4**, pp. 334–336)
- Describe the basic components of a door. (**NFPA 1001: 4.3.4**, p. 336)
- Explain the differences between a solid-core and a hollow-core door. (**NFPA 1001: 4.3.4**, p. 337)
- Describe the basic classifications of doors by opening type. (**NFPA 1001: 4.3.4**, pp. 337–345)
- Explain how the door classification affects forcible entry operations. (**NFPA 1001: 4.3.4**, pp. 337–345)
- Describe the basic configurations of window construction. (**NFPA 1001: 4.3.4**, pp. 343, 345)
- Describe the common styles of window frames. (**NFPA 1001: 4.3.4**, pp. 346–352)
- Explain how the style of window frame affects forcible entry operations. (**NFPA 1001: 4.3.4**, pp. 346–352)
- Describe the major components of a door lock. (**NFPA 1001: 4.3.4**, pp. 352–353)
- Describe the major components of a padlock. (**NFPA 1001: 4.3.4**, p. 353)

- Describe the four major types of locks. (**NFPA 1001: 4.3.4**, pp. 353–355)
- Explain how the type of lock affects forcible entry operations. (**NFPA 1001: 4.3.4**, pp. 353–356)
- Describe the tools used to force entry through locks. (**NFPA 1001: 4.3.4**, pp. 356, 357, 358)
- Describe how to force entry through doors with drop bars. (**NFPA 1001: 4.3.4**, p. 357)
- Describe how to force entry through security gates and windows. (**NFPA 1001: 4.3.4**, pp. 357, 358–359)
- Explain the differences between load-bearing and nonbearing walls. (**NFPA 1001: 4.3.4**, p. 359)
- Describe the materials used in exterior and interior walls. (**NFPA 1001: 4.3.4**, pp. 359–361, 362)
- Describe the materials used in floors. (pp. 361, 363–364)
- List the basic steps and considerations in forcible entry operations. (pp. 364–365)

SKILLS OBJECTIVES

After studying this chapter, you will be able to perform the following skills:

- Force entry into an inward-opening door. (**NFPA 1001: 4.3.4**, pp. 338, 339–340)

- Force entry into an outward-opening door. (**NFPA 1001: 4.3.4**, pp. 338, 340–341)
- Open an overhead garage door using the triangle method. (**NFPA 1001: 4.3.4**, p. 344)
- Open an overhead garage door using the hinge method. (**NFPA 1001: 4.3.4**, p. 345)
- Force entry through a wooden double-hung window. (**NFPA 1001: 4.3.4**, pp. 347–348)
- Force entry through a casement window. (**NFPA 1001: 4.3.4**, pp. 350–351)
- Force entry through a projected window. (**NFPA 1001: 4.3.4**, pp. 351–352)
- Force entry using a K tool. (**NFPA 1001: 4.3.4**, pp. 356–357)
- Force entry using an A tool. (**NFPA 1001: 4.3.4**, pp. 356, 358)
- Force entry by unscrewing a lock. (**NFPA 1001: 4.3.4**, pp. 356, 358)
- Breach a wall frame. (**NFPA 1001: 4.3.4**, p. 360)
- Breach a masonry wall. (**NFPA 1001: 4.3.4**, p. 361)
- Breach a metal wall. (**NFPA 1001: 4.3.4**, p. 362)
- Breach a floor. (**NFPA 1001: 4.3.4**, pp. 363–364)

Fire Alarm

There is a working fire on the second floor of a local hotel. Your crew is assigned to search for occupants who may still be in the building. The hotel is a three-storey, Type III ordinary construction building with a hallway down the center. It was built in 1908. As you dismount the fire apparatus, your captain tells you to grab the tools for forcible entry.

1. What kind of access issues would you expect for this type of structure?

2. What kind of tools would you want to take with you?

3. How would the forcible entry challenges for this structure differ from other types of structures?

 *Access Navigate for more practice activities.*

Introduction

One of a fire fighter's most dynamic and challenging tasks, forcible entry, is defined as gaining access to a structure when the normal means of entry are locked, secured, obstructed, blocked, or unable to be used for some other reason. This chapter examines forcible entry into structures.

Forcible entry requires strength, knowledge, proper techniques, and skill. Because it often causes damage to the property, fire fighters must consider the results of using different forcible entry methods and select the one appropriate for the situation. If rapid entry is needed to save a life or prevent a more serious loss of property, it is appropriate to use maximum force. When the situation is less urgent, fire fighters can take more time and use less force.

Another factor to consider when making forcible entry is the need to secure the premises after operations are completed. Fire fighters must never leave the premises in a condition that would allow unauthorized entry.

A fire fighter who is skilled in forcible entry should be able to get the job done quickly, with as little damage as possible. As part of this task, the fire fighter must consider the type of building construction, possible entry points, the types of securing devices that are present, and the best tools and techniques for the specific situation. Selecting the right tool and using the proper technique save valuable time—and could save lives.

Doors, windows, locks, and security devices can be combined in countless variations to prevent easy entry. Given this diversity, fire fighters must be familiar with new technology, including new styles of windows, doors, locks, and security devices that are common in the local response area. In addition, they must understand how these devices operate. Learn to recognize different types of doors, windows, and locks. The best time to examine these components is during inspection and preincident planning visits. Touring buildings when they are under construction or being renovated is another excellent way to learn about building construction and to examine different devices. Talking with construction workers and locksmiths can provide valuable information about how to best approach a particular lock, window, or door.

This chapter reviews door, window, and lock construction. In addition, it discusses the tools and techniques used for forcible entry through doors, windows, walls, and floors. The skill drills cover the most common forcible entry techniques.

Forcible Entry Situations

Forcible entry is usually required at emergency incidents where time is a critical factor. A rapid forcible entry, using the right tools and methods, might result in a successful rescue or allow a pump company to make an interior attack and control a fire before it extends farther into the structure or to adjoining structures. Forcible entry also may be required in cases of medical emergencies when a person needs immediate medical care and is unresponsive or unable to unlock the door. Fire fighters must study and practise forcible entry techniques so that their proficiency increases with every experience.

Company officers usually select both the point of entry and the method to be used. They ensure that the efforts of different companies are properly coordinated for safe, effective operations. For example, forcible entry actions must be coordinated with hose teams because the entry must be made before pump companies can get inside the structure to attack a fire. Opening a door before the hose lines are in place and ready to advance allows fresh air to enter the structure, possibly resulting in rapid fire spread or a backdraft. Making an opening at the wrong location can undermine a well-planned attack. Any introduction of fresh air can create a flow path that can rapidly accelerate fire growth. Prior to making entry into any fire-involved structure, personnel must determine the impact this will have on fire growth. In a ventilation-limited fire, forcing entry of a door on the windward side of a structure may cause backdraft or another significant fire event. Consider forcible entry to be part of ventilation, and make sure it is coordinated with the rest of the fire attack. Remember that the flow path transfers hot gases and smoke from the higher pressure within the fire area toward the lower pressure areas accessible through doorways, window openings, and roof openings and may be influenced by external factors such as wind direction and speed.

Securing entry into a building is a vital part of the firefighting process. In most cases you must gain entry in order to perform rescue and to extinguish the fire. Work to achieve entry as simply as possible. Before beginning forcible entry, remember this simple rule: "Try before you pry." In other words, always check doors or windows to confirm that forcible entry is truly needed. Be absolutely sure that you are at the address of the emergency incident. There is nothing harder to explain than why you broke down the door of a house located three doors down from the structure that is on fire. Check for the presence of a key. An unlocked door requires no force; a window that can be opened does not need to be broken. Maintaining the structural integrity of doors and windows is an important component with regards to tactical fire suppression. Verifying the status of a closed door or window only takes a few seconds and can prevent unnecessary damage. It is equally important to look for alternative entry points. Do not spend time working on a locked door when, for example, a nearby window provides easy access to the same room.

If forcible entry is required, you need to consider several factors. First, evaluate the seriousness of the situation. If there are indications of an active fire, the report of people trapped, or a medical emergency, then a rapid entry may be required. On the other hand, if there are no signs of fire or other emergency, the speed with which you make your entry may be less important. Consider the nature of the call. If possible, consider using a method of entry that results in the least amount of damage to the door and building. It is easier to replace a door lock than to replace an expensive door and door frame. Next evaluate the type of occupancy and construction. Determine the type of door or window. Is it wood, metal, or glass? Does the door open outward or inward? What type of hinges does it have? Determine what type of lock is present. After considering these factors you will be able to determine which type of forcible entry is best suited for that situation.

Unusually difficult forcible entry situations may require the expertise of specialty companies such as technical rescue teams. These companies have experience working with specialized tools and equipment to gain access to almost any area.

FIRE FIGHTER TIP

Many communities have installed lockbox systems. With these systems, a master key carried by fire department members opens a small lockbox at the incident structure. This lockbox contains the keys needed to open the structure's door. If your community has a lockbox system, always look for a lockbox before beginning any type of forcible entry. In some communities, lockboxes are mounted next to the fire department connection (FDC). The fastest way to gain access to a building is with a key.

Forcible Entry Tools

Choosing the right forcible entry tool can be a very important decision. Fire departments use many different types of forcible entry tools, ranging from basic cutting, prying, and striking tools to sophisticated mechanical, electrical, and hydraulic equipment. Fire fighters must be familiar with the following information:

- Which tools the department uses
- What the uses and limitations of each tool are
- How to select the proper tool for the job
- How to safely operate each tool
- How to carry the tools safely
- How to inspect and maintain each tool, thereby ensuring its readiness for service and safe operation

General Tool Safety

Any tool that is used incorrectly or is maintained improperly can be dangerous to both the operator and other persons in the vicinity. Anyone who uses a tool should understand the proper operating procedures before beginning the task. Always follow the manufacturer's recommendations for operation and maintenance.

General safety tips for using tools include the following points:

1. Always wear the appropriate personal protective equipment (PPE) (**FIGURE 10-1**). When conducting forcible entry during fire suppression operations, this includes a full set of structural firefighting PPE.
   - Approved helmet
   - Protective hood
   - Eye protection
   - Face shield
   - Firefighting gloves
   - Protective coat

FIGURE 10-1 When conducting forcible entry during fire suppression operations, always wear a full set of structural firefighting PPE.

© Jones & Bartlett Learning. Photographed by Glen E. Ellman.

- Protective pants
- Boots
- Self-contained breathing apparatus (SCBA) with personal alert safety system (PASS)

2. Learn to recognize the materials used in building and lock construction, and become familiar with the appropriate tools and techniques used for each kind of material. For example, many locks are made of **case-hardened steel**, which combines carbon and nitrogen to produce a hard outer coating that cannot be penetrated by most ordinary cutting tools. Using the wrong tool to cut case-hardened steel could break the tool and potentially injure the user and other fire fighters.

3. Keep all tools clean, properly serviced according to the manufacturer's guidelines, and ready for use. Immediately report any tool that is damaged or broken, and take it out of service.

4. Do not leave tools lying on the ground or floor; instead, return them to a tool staging area or the apparatus. Tools that are left lying around are tripping hazards that may increase the number of injuries on the fire ground.

General Carrying Tips

Tools can cause injuries even when they are not in use. Carrying a tool improperly, for example, can result in muscle strains, abrasions, or lacerations. The following general carrying tips apply to all tools:

- Do not try to carry a tool or piece of equipment that is too heavy or designed to be used by more than one person. Instead, request assistance from another fire fighter to carry this type of tool.
- Always use your legs—not your back—when lifting heavy tools.
- Keep all sharp edges and points away from your body at all times. Cover or shield them with a gloved hand to protect those around you.
- Carry long tools with the head of the tool down toward the ground. Be aware of overhead obstructions and wires, especially when using pike poles and ladders.

General Maintenance Tips

All tools should be kept in a "ready state"—that is, the tool should be in proper working order, in its proper storage place, and ready for immediate use. Forcible entry tools are generally stored in a designated compartment on the apparatus. Hand tools should be clean, and cutting blades should be sharp. Power tools should be completely fuelled and treated with a fuel-stabilizing

product, if necessary, to ensure easy starting. Remember that metal cutting blades can be damaged by gasoline or other petroleum vapours. Every crew member should be able to locate the right tool immediately and be confident that it is ready for use (**FIGURE 10-2**).

All tools require regular maintenance and cleaning to ensure that they will be ready for use in an emergency. Thorough, conscientious daily or weekly checks of these items should be performed, particularly with infrequently used tools. Always follow the manufacturer's instructions and guidelines, and store maintenance manuals in an easily accessible location. Keep proper records to track maintenance, repairs, and any warranty work that is performed on tools.

> ### FIRE FIGHTER TIP
>
> The forcible entry tools covered in this chapter are not the only tools available. Different fire departments use different tools, and regional preferences for a particular piece of equipment may influence the items that are made available. You must become familiar with all of the tools used in your department.

Types of Forcible Entry Tools

Several types of tools can be used for forcible entry, including an axe, a prying tool, a K tool, or a shove knife. Forcible entry tools include hand tools and power tools. As discussed in Chapter 8, *Fire Fighter Tools and Equipment*, tools can be classified into the following functional categories:

- Rotating tools
- Striking tools
- Prying/spreading tools
- Cutting tools

FIGURE 10-2 All tools should be kept in a neat and ready state.
© Jones & Bartlett Learning.

- Pushing/pulling tools
- Multiple-function tools
- Special-use tools/lock tools

Rotating Tools

Rotating tools apply a rotational force to make something turn. The most commonly used rotating hand tools are screwdrivers, wrenches, and pliers, which are used to assemble (fit together) or disassemble (take apart) parts that are connected with threaded fasteners. Power tools used for forcible entry include rotating power saws that can create openings through obstacles including doors, walls, fences, gates, security bars, and other barriers.

Assembling and disassembling are basic mechanical skills that are routinely employed by fire fighters to solve problems. Most fire apparatus carry a tool kit with a selection of screwdrivers with different heads, open-end wrenches, box wrenches, socket wrenches, adjustable wrenches, pipe wrenches, and pliers.

Various sizes and types of screw heads are available, including slotted head, Phillips head, Robertson head, and others. A screwdriver with interchangeable heads is sometimes more useful than a selection of different screwdrivers. Pliers and wrenches also come in a variety of shapes and sizes for different applications. A description of the most common rotating tools is presented in Chapter 8, *Fire Fighter Tools and Equipment*.

Striking Tools

Striking tools generate an impact force directly on an object or another tool. These items include flat-head axes, hammers, sledgehammers, and battering rams.

They are often employed to gain entrance to a building or a vehicle or to make an opening in a wall or roof. This equipment can also be used to force the end of a prying tool into a small opening. Striking tools are generally hand tools that are powered by human energy. The head of a striking tool is usually made of hardened steel.

SAFETY TIP

Before swinging a striking tool, make sure there is a clear area of at least the length of the tool handle around you. Station another fire fighter outside the swing area as a safety person to keep others from entering the zone. Make sure that you have checked for potential signs of backdraft before using any striking tool.

Flat-Head Axe. The flat-head axe was one of the first tools developed by humans and is still widely used for a variety of purposes. One side of the axe head is a cutting blade; the other side is a flat striking surface. Fire fighters often use the flat side to strike a Halligan tool and drive a wedge into an opening. Most fire apparatus carry both flat-head axes and pick-head axes.

Battering Ram. The battering ram is a heavy metal bar used to force doors and breach walls. Originally, it was a large log used to smash through enemy fortifications. Today's battering rams are usually made of hardened steel and have handles; two to four people are needed to use a battering ram. Currently, battering rams are more commonly used by law enforcement agencies than by fire departments.

Sledgehammer. Sledgehammers are hammers that come in a variety of weights and sizes. The head of the hammer can weigh from 1 to 9 kg (2 to 20 lb). The handle may be short like a carpenter's hammer or long like an axe handle. A sledgehammer can be used by itself to break down a door or in conjunction with other tools such as the Halligan tool.

Prying/Spreading Tools

Prying and spreading tools are often used by fire fighters to force entry into buildings. These tools may be simple or mechanically complex. Prying tools used for gaining access include pry bars, crowbars, Halligan tools, Hux bars, and the hydraulic-powered rabbit tool. They come in several sizes and with different features that are designed for different applications. This section describes commonly used prying and spreading tools used in forcible entry.

Halligan Tool. The Halligan tool is widely used by the fire service (**FIGURE 10-3**). The tool incorporates three different tools: a sharp tapered pick, a blade (either a wedge or an adze), and a fork or claw. The adze or blade end is used to pry open doors and windows; the pick end is used to make holes, drive through locks and shackles, or break glass; and the fork or claw end is primarily used to gain a mechanical advantage when prying and forcing doors, windows, and boards.

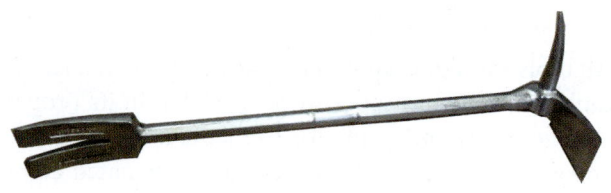

FIGURE 10-3 A Halligan tool has multiple purposes.
© Jones & Bartlett Learning. Photographed by Glen E. Ellman.

Pairing a Halligan tool with a flat-head axe creates a tool set often referred to collectively as the **irons**. These two tools are commonly used to perform forcible entry. A fire fighter can use a flat-head axe (or a sledgehammer) to strike the Halligan tool into place, thereby creating a **purchase point** (a small opening that allows better tool access in forcible entry) between a door or window and its frame. There are many variations of this tool available throughout the country, and different terms are used to describe the various parts of the tool.

Pry Bar/Hux Bar/Crowbar. Pry bars, Hux bars, and crowbars are made from hardened steel cast into a variety of shapes and sizes. These tools are most commonly used to force doors and windows, to remove nails, or to separate building materials. The different shapes allow fire fighters to exert different amounts of leverage in diverse situations.

Pry Axe. The **pry axe** is a multipurpose tool that can be used both to cut and to force open doors and windows (**FIGURE 10-4**). It includes an adze, a pick, and a fork or claw. The tool consists of two parts: the body, which has the adze and pick and looks similar to a miniature pick axe, and a handle with a fork or claw at the end, which slides into the body. The handle of the pry axe can be extended to provide extra leverage when prying, or it may be removed and inserted into the head of the adze to provide rotational leverage. Extreme caution should be used when handling this tool. Over time, the mechanism that locks the handle into position may become worn, allowing the handle to slip out and potentially injure fire fighters.

Hydraulic Tools. Hydraulic-powered tools such as spreaders, cutters, and rams often are used for forcible entry. These tools require hydraulic pressure, which can be provided by a high-pressure, motor-operated pump or a hand pump.

There are two types of hydraulic tools that are designed to force open doors that open inward. The rabbit tool is a three-piece tool that comprises a hand-held hydraulic pump connected to a hydraulic jaw or spreader by a high-pressure hydraulic hose (**FIGURE 10-5**). The pumping action (manual or powered) forces hydraulic fluid through the hose and causes the spreader to expand and force the door.

The second tool, the Hydra-Ram, is a one-piece unit that contains a manual pump and jaw in a single unit (**FIGURE 10-6**). These tools work best when the door has a metal frame. They also can be used to force sliding elevator doors.

FIGURE 10-5 The rabbit tool is a hydraulic tool comprising a hand-held hydraulic pump, a hydraulic jaw or spreader, and, in some cases, a high-pressure hydraulic hose. The pumping action (manual or powered) forces hydraulic fluid through the hose and causes the spreader to expand and force a door.
© Jones & Bartlett Learning.

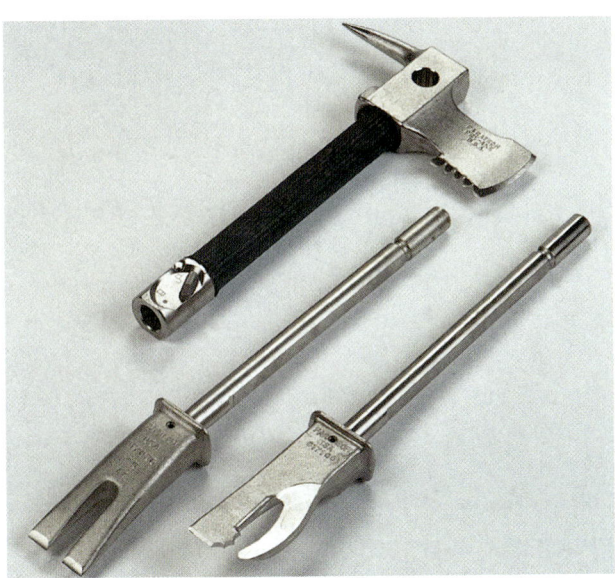
FIGURE 10-4 The pry axe is a multipurpose tool that can be used to cut and pry.
Courtesy of Paratech Inc.

FIGURE 10-6 The Hydra-Ram is a one-piece hydraulic tool that contains both the pump and jaw or spreader in a single unit.
Courtesy of Fire Hooks Unlimited.

To force an inward-opening door using a hydraulic tool, open a gap between the door and the door frame using a prying tool such as a Halligan tool. Insert the closed hydraulic jaws of the spreader in the door frame about halfway between the lock and the doorknob. Force the door by pumping the hydraulic pump to open the jaws. If this does not work the first time, reposition the jaws to a place closer to the door lock.

Larger hydraulic cutters and spreaders are typically used in vehicle extrication but can also be used in some forcible entry situations (**FIGURE 10-7**). Hydraulic rams can be used to apply a powerful force in one direction, and they come in a variety of lengths and sizes. The operation of hydraulic tools used in vehicle extrication is covered in Chapter 24, *Vehicle Rescue and Extrication*.

Cutting Tools

Cutting tools are primarily used for cutting doors, roofs, walls, and floors. Although cutting is a destructive process, it is justified when required to save lives or property. Although they do not work as quickly as power tools, hand-operated cutting tools are proven and reliable in many situations. For example, an axe can be used to cut out a door panel, and bolt cutters can be used to remove a padlock. Because they do not require a power source, these items often can be deployed more quickly than power tools.

Several power cutting tools are available for different applications. For example, cutting torches or rotary power saws can be used to cut through metal security bars, hinges, or padlocks on metal security doors. A power saw can be used to cut through a wood wall. These tools can be powered by batteries, electricity, gasoline, or hydraulics, depending on the amount of power required.

Each type of tool and power source has advantages and disadvantages. For example, battery-powered tools are portable and can be placed in operation quickly but have limited power and operating times.

Axe. Many different types of axes are available, including flat-head, pick-head, pry, and multipurpose axes. The cutting edge of an axe can be used to break into plaster and wood walls, roofs, and doors (**FIGURE 10-8**).

Although a flat-head axe was described earlier in this chapter as a striking tool, it is also classified as a cutting tool. The pick-head axe is similar to a flat-head axe but has a pick instead of a striking surface opposite the blade. This pick can have many uses as a striking tool, including making an entry point or a small hole if needed.

Specialty axes such as the pry axe and the multipurpose axe also can be used for purposes other than cutting. The multipurpose axe, for example, can be used for cutting, striking, or prying. It includes a pick, a nail puller, a hydrant wrench, and a gas main shutoff wrench.

FIGURE 10-8 The cutting edge of an axe can be used to break into plaster and wood walls, roofs, and doors.
© Jones & Bartlett Learning.

FIGURE 10-7 A hydraulic spreader.
© Jones & Bartlett Learning.

Bolt Cutters. Bolt cutters are used to cut metal components such as bolts, padlocks, chains, and chain-link fences. These tools are available in several different sizes, based on the blade opening and the handle length. The longer the handle, the greater the cutting force that can be applied. Note that bolt cutters may not be able to cut into some heavy-duty padlocks that are constructed from case-hardened steel.

Rotary Saws. Rotary saws may be powered by either electric motors or gasoline engines. In some rotary saws, the cutting part of the saw is a round metal blade with teeth, with different blades being used depending on the type of material being cut. Other rotary saws use a flat, abrasive disk for cutting. The disks are made of composite materials and are designed to wear down as they are used. It is important to match the appropriate saw blade or saw disk to the material being cut.

Reciprocating Saws. Reciprocating saws are powered by either an electric motor or a battery motor that rapidly pulls a saw blade back and forth. As with rotary saws, different blades are used to cut different materials. Reciprocating saws are most commonly used to cut metal during extrication of a victim from a motor vehicle.

Circular Saw. Most fire departments use gasoline-powered circular saws for making forcible entry and for cutting ventilation holes. These tools are light, powerful, and easy to use, with blades that can be changed quickly. The different types of blades enable the saw to cut many materials.

Fire departments generally carry several different types of circular saw blades so that the proper blade can be used in each situation.

- Carbide-tipped blades are specially designed to cut through hard surfaces or wood. They stay sharp for long periods, so more cuts can be made before the blade needs to be changed.
- Metal-cutting blades consist of a composite material made with aluminum oxide. They are used to cut metal doors, locks, or gates.
- Masonry-cutting blades are abrasive and made of a composite material that includes silicon carbide or steel. They can cut concrete, masonry, and similar materials. Because masonry-cutting blades resemble metal-cutting blades, the operator must check the label on the blade before using it. Blades with missing labels should be discarded. Do not store masonry-cutting blades near gasoline or petroleum products because petroleum vapours

SAFETY TIP

Almost all cutting tools have built-in safety features. Always take advantage of the safety features on every tool. To avoid injury, never remove safety guards from tools—not even for "just one cut."

will cause the composite materials in the blade to decompose. In such a case, when the blade is later used, it could disintegrate.

Pushing/Pulling Tools

Pushing and pulling tools can extend the reach of the fire fighter as well as increase the power the fire fighter can exert upon an object. Pushing and pulling tools have many different uses in forcible entry. For example, the K tool and the shove knife are both pulling tools that are designed to release the locking mechanisms on doors.

K Tool, A Tool, and J Tool. There are a few tools that are used to force the locking mechanisms on doors (FIGURE 10-9). The K tool is designed to remove a lock cylinder. To remove the lock cylinder, the cutting edge of the K tool is placed over the cylinder lock, a pry bar is inserted into the slot on the face of the K tool, and a flat-head axe is used to strike the pry bar with light blows, until the wedging blades or notches take a bite into the cylinder body of the lock, pulling the cylinder from the door. Once the lock cylinder is removed, another tool can then be used to open the locking mechanism. If the lock has a protective ring, the K tool might not be able to cut through it. This comparatively nondestructive process leaves the door and most of the locking mechanism undamaged.

The A tool is similar to the K tool except that the pry bar is built into the cutting part of the tool. This tool gets its name from its A-shaped cutting edges. To use an A tool, put the cutting head over the lock cylinder, and use a striking tool to force it down into the cylinder until the lock cylinder can be forced out of the door. Once the lock cylinder is removed, you then use another tool to open the locking mechanism.

A J tool will fit between double doors that have push bars or panic bars. Slide the J tool between the doors, and pull to engage the panic bars.

Shove Knife. A shove knife is an old tool yet is frequently used by modern-day fire fighters. A shove knife is a pulling tool that is used to trip the latch on outward-swinging doors (FIGURE 10-10). The knife is inserted between the door and the door jamb

FIGURE 10-9 There are a few tools that are used to force the locking mechanisms on doors. **A.** K tool. **B.** A tool. **C.** J tool.
© Jones & Bartlett Learning. Photographed by Glen E. Ellman.

FIGURE 10-10 A shove knife.
© Jones & Bartlett Learning. Photographed by Glen E. Ellman.

and is then pulled down and outward to release the locking mechanism of the door. You should not use this technique to try to gain entry through inward-swinging doors because it is not effective on doors with metal frames and only works on a limited number of doors with wood frames. Some newer doors have latches that do not respond to this tool.

Multiple-Function Tools

Certain tools are designed to perform multiple functions, thereby reducing the total number of tools needed to achieve a goal. For example, a flat-head axe can be used as either a cutting tool or a striking tool. Some combination tools, such as the Halligan tool, can be used to pry and to strike.

In addition, some tools are used in combination with other tools. For example, a flat-head axe and a Halligan tool, collectively called the irons, are used in combination to pry open a door, although they may permanently damage both the door and the frame.

Special-Use/Lock Tools

Special-use tools used in forcible entry include lock tools that are used to disassemble locking mechanisms such as padlocks. These devices cause minimal damage to the door and the door frame. If they are used properly, the door and frame should be undamaged, although the lock generally must be replaced. An experienced user usually can gain entry in less than a minute with these tools.

Duck-Billed Lock Breakers.

The duck-billed lock breaker has a large metal wedge attached to a handle (**FIGURE 10-11**). The narrow end of the wedge is driven through the center of the shackle (the U-shaped part of the padlock) using a striking tool. The increasing size of the wedge forces the shackles apart until they break.

Locking Pliers and Chain.

The locking pliers and chain is a pair of pliers with a chain attached to the handle. This tool is used to clamp a padlock securely in place so that the shackles can be cut safely with a circular saw or cutting torch (**FIGURE 10-12**). To use this tool, one fire fighter clamps the pliers to the lock body while a second fire fighter maintains a steady tension on the chain as the lock is being cut. Securing the padlock is necessary because it is too dangerous to hold the lock with a gloved hand or to cut into it while it is loose.

Bam-Bam Tool.

The bam-bam tool is a sliding hammer that contains a case-hardened screw, which is inserted, secured, and driven into the keyway of a lock. Once the screw is set, the sliding hammer pulls the tumblers out of the lock so that the trip-lever mechanism inside the lock can be opened manually.

Many techniques and types of tools can be used to gain entry into secured structures (**TABLE 10-1**). The exact tool needed depends on the method of entry and the type of obstacle. Rely on the orders and advice of your officer and fellow fire fighters. The techniques used for forcible entry are described next.

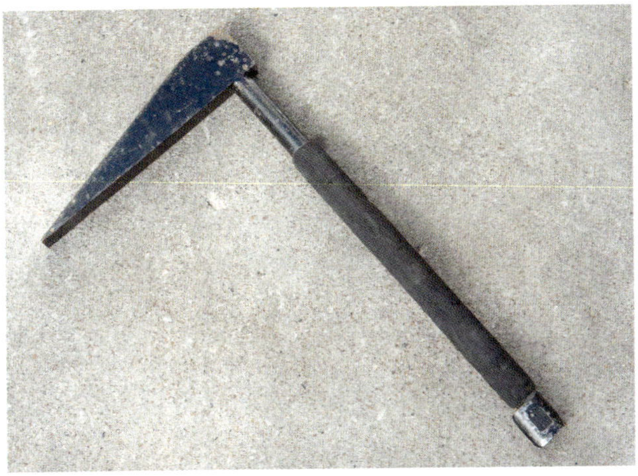

FIGURE 10-11 Duck-billed lock breakers force padlock shackles apart until they break.
© Jones & Bartlett Learning. Photographed by Glen E. Ellman.

FIGURE 10-12 A locking pliers and chain holds a padlock securely in place so shackles can be cut safely. The pliers clamp on to the lock body, and the chain is used to create tension while the shackles are cut.
© Jones & Bartlett Learning. Photographed by Glen E. Ellman.

TABLE 10-1 Forcible Entry Tools	
Type of Tool	**Use in Forcible Entry**
Prying tools	Prying tools can include a Halligan tool, a flat bar, and a crowbar, among other options. They can be used to break windows or force open doors.
Axe, flat-head axe, pick-head axe	An axe can be used to cut through a door or break a window. A flat-head axe can be used in conjunction with a Halligan tool to force open a door.
Sledgehammer	A sledgehammer can be used to breach walls or break a window. It also can be used in conjunction with other tools, such as a Halligan tool, to force open a door or break off a padlock.
Hammer, mallet	These tools may be used in conjunction with other tools such as chisels or punches to force entry through windows or doors.

(continued)

TABLE 10-1 Forcible Entry Tools *(Continued)*	
Type of Tool	**Use in Forcible Entry**
Chisel, punch	These tools can be used to make small openings through doors or windows.
K tool	The K tool provides a "through the lock" method of opening a door, thereby minimizing damage to the door.
Shove knife	A shove knife is used to open outward-swinging doors.
Rotary saw, chainsaw, reciprocating saw	Power saws can cut openings through obstacles including doors, walls, fences, gates, security bars, and other barriers.
Bolt cutter	A bolt cutter can be used to cut off padlocks or cut through obstacles such as fences.
Battering ram	A battering ram can be used to breach walls.
Hydraulic door opener	A hydraulic door opener can be used to force open a door.
Hydraulic rescue tool	A hydraulic rescue tool can be used in a variety of situations to break, breach, or force openings in doors, windows, walls, fences, or gates.

Doors

In most cases, the best point to attempt forcible entry to a structure is the door or a window. Doors and windows are constructed as entry points, so they are made generally of weaker materials than are walls or roofs. An understanding of the basic construction of doors will help you select the proper tool and increase your likelihood of successfully gaining entry. Windows will be discussed in the next section.

Basic Door Construction

Doors can be categorized both by their construction material and by the way they open. Both interior and exterior doors have the same basic components (**FIGURE 10-13**):

- **Door** (the entryway itself)
- **Door jamb** (the upright or vertical parts of a door frame onto which a door is secured)
- **Hardware** (the handles, hinges, and other components)
- **Locking mechanism**

Construction Materials

Doors are generally constructed of wood, metal, or glass. The design of the door and the construction material used determine the difficulty in forcing entry through this point.

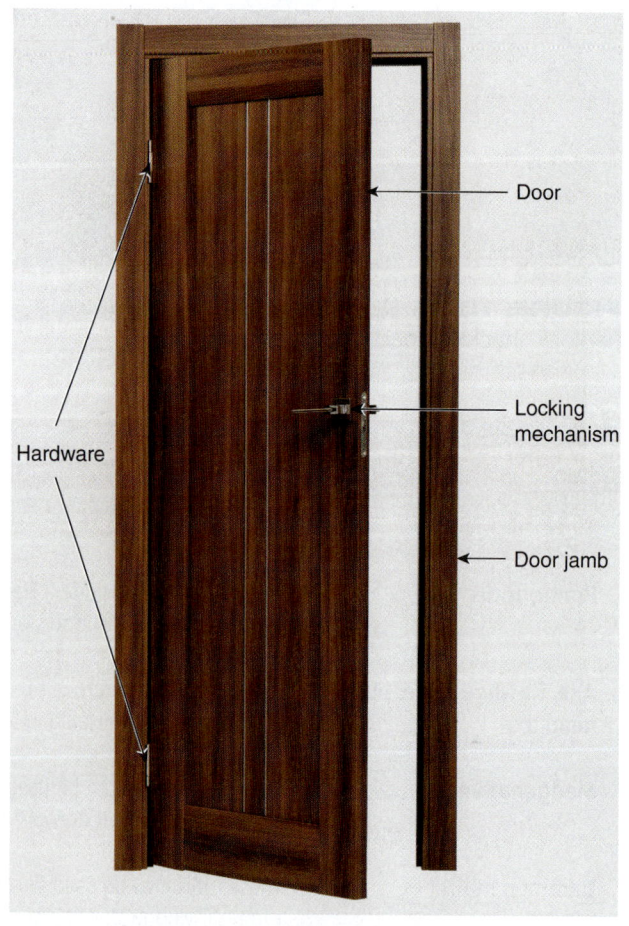

FIGURE 10-13 The parts of a door.
© Ivan Nakonechnyy/Shutterstock.

Wood

Wood doors are commonly used in residences and may be found in some commercial buildings. Three types of wood swinging doors are distinguished: slab, ledge, and panel. Slab doors come in solid-core or hollow-core designs and are attached to wood-frame construction with normal hardware (**FIGURE 10-14**).

Solid-core doors are constructed of solid wood core blocks covered by a face panel. Some have a fire rating. Such doors are typically used for entrance doors. Solid-core doors are heavy and may be difficult to force, but their construction enables them to contain fire better than hollow-core doors do.

Hollow-core doors have a lightweight, honeycomb interior, which is covered by a face panel. They are often used as interior doors, such as for bedrooms. Hollow-core doors are easy to force. These doors will serve as a fire barrier for a short period of time, but they burn through more quickly than a solid-core door.

Ledge doors are simply wood doors with horizontal bracing. These doors, which are often constructed of tongue-and-groove boards, may be found on warehouses, sheds, and barns.

Panel doors are solid wood doors that are made from solid planks to form a rigid frame with solid wood panels set into the frame. Panel doors are used as both exterior and interior doors and may be made from a variety of types of wood. These doors resist fire longer than hollow-core slab doors do and are typically easier to breach than solid-core slab doors if entry is attempted at the panels.

Metal

Many Canadian jurisdictions require closures (doors) in Type I residential and commercial high-rise buildings to be metal (steel) doors equipped with self-closing and self-locking mechanisms. Like wood doors, they may have either a hollow-core or a solid-core construction. Hollow-core metal doors have a metal framework interior so they are as lightweight as possible. By contrast, solid-core metal doors have a foam or wood interior that is intended to reduce the door's weight without affecting its strength. Metal doors may be set in either a wood frame or a metal frame. If fire rated, metal (steel) doors need to be set in a steel frame in order to maintain the required fire rating. Residential metal doors may appear to be panel doors and are often used as entry doors. Metal doors are more durable and fire resistant than wooden doors.

Glass

Glass doors generally have a steel frame with tempered glass; alternatively, they may be simply tempered glass and not require a frame but have metal supports to attach hardware. Glass doors are easy to force but can be dangerous owing to the large number of small broken pieces that are produced when glass is broken.

LISTEN UP!

Opening a door or window can allow air to enter the structure, potentially resulting in rapid fire spread, ventilation-induced flashover, or a backdraft. Forcing doors and windows should be considered ventilation. Be mindful of the flow path and listen to instructions when performing forcible entry. Everything needs to be coordinated!

Types of Doors

Doors are also classified based on how they open. The five most common ways that doors open are inward, outward, sliding, revolving, and overhead. Overhead, sliding, and revolving doors are the most readily identified. Inward-opening and outward-opening doors can be differentiated based on whether the hinges are visible (**FIGURE 10-15**). If you can see the hardware, the door will swing toward you (outward). If the hinges are not visible, the door will swing away from you (inward).

Door frames may be constructed of either wood or metal. Wood-framed doors come in two styles: stopped and rabbet. Stopped door frames have a piece of wood attached to the frame that stops the door from swinging

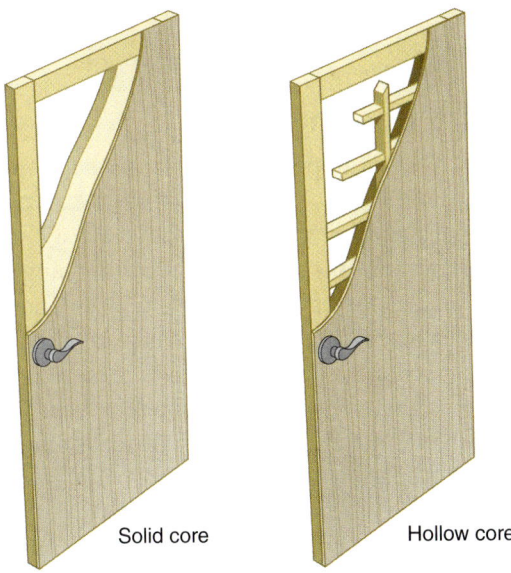

Solid core   Hollow core

FIGURE 10-14 A slab door may have either solid-core or hollow-core construction.

© Jones & Bartlett Learning.

FIGURE 10-15 Inward-opening and outward-opening doors can be differentiated based on whether the hinges are visible. **A.** An outward-opening door will have hinges showing. **B.** A door with the hinges not showing will be inward opening.

A: © Jones & Bartlett Learning; B: © Jones & Bartlett Learning. Photographed by Glen E. Ellman.

past the latch. Rabbeted door frames are constructed with the stop cut built into the frame so it cannot be removed.

Metal-framed doors are more difficult to force than wood-framed doors. Metal frames have little flexibility, so when a metal frame is used with a metal door, forcing entry can be difficult. Metal frames look like rabbeted door frames.

Inward-Opening Doors

Design. Inward-opening doors of wood, steel, or glass are found in most structures. They have an exterior frame with a stop or rabbet that keeps the door from opening past the latch. Types of locking mechanisms range from standard doorknob locks, to deadbolt locks, to sliding latches. From the inside, an inward-opening door opens with a pull.

Forcing Entry. A simple solution to gaining entry through inward-opening doors may be to break a small window on the door or adjacent to it, reach inside, and operate the locking mechanism. Remember—"try before you pry."

If no window is available and the door is locked, stronger measures are required. To force entry into an inward-opening door, follow the steps in **SKILL DRILL 10-1**.

Outward-Opening Doors

Design. Outward-opening doors are used in commercial occupancies and for most exits (**FIGURE 10-16**). They are designed so that people can leave a building quickly during an emergency. Outward-opening doors may be constructed of wood, metal, or glass. They usually have exposed hinges, which may present an entry opportunity. More frequently, however, these hinges will be sealed so that the pins cannot be removed. Several types of locks, including handle-style locks and deadbolts, may be used with these doors.

Forcing Entry. Before forcing entry to an outward-opening door, check the hinges to see if they can be disassembled or the pins removed. If that would take too long or cannot be done, place the adze end of a prying tool such as the Halligan into the space between the

SKILL DRILL 10-1
Forcing Entry into an Inward-Opening Door Fire Fighter I, NFPA 1001: 4.3.4

1 Look for any safety hazards as you evaluate the door. Inspect the door for the location and number of locks and their mechanisms.

2 Insert the adze end of a prying tool (such as a Halligan tool) into the space between the door and the door jamb about 15 cm (6 in.) above or below the door lock. Push up or down on the tool to rotate the adze and to create a gap (or crease) in the door. To capture your progress, insert a wedge or tool before removing the adze end of the tool from the door.

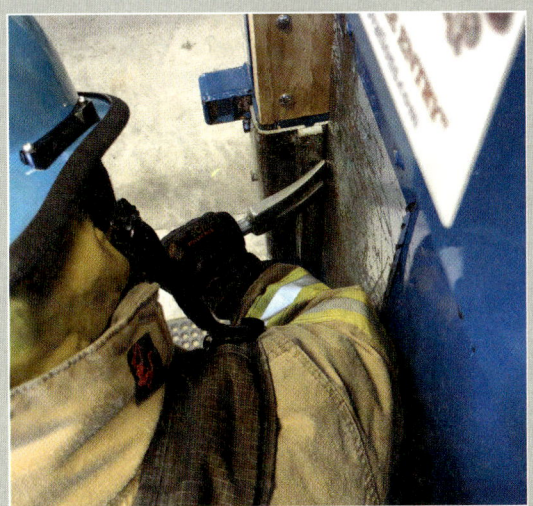

3 Insert the fork end of the tool into the gap between the door and the door frame.

4 Instruct a second fire fighter to drive the fork in forcefully with a flat-head axe. Keep constant pressure on the tool and move the tool away from the door as it is being driven in. The tool is set when the arch of the fork is even with the inside edge of the door/doorstop and is perpendicular to the door.

(continued)

SKILL DRILL 10-1 Continued
Forcing Entry into an Inward-Opening Door Fire Fighter I, NFPA 1001: 4.3.4

5 When the Halligan tool is set, push the tool sharply inward toward the center of the door to create maximum force.

Courtesy of Jessica Holmes.

6 Maintain control of the door when it opens by hooking the door with the adze end of the Halligan or by attaching a strap to the knob. Limit ventilation of the fire.

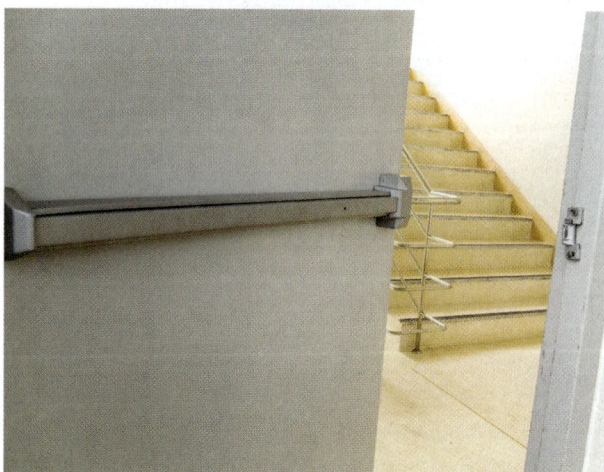

FIGURE 10-16 The outward-opening door swings open with a push.

© thexfilephoto/iStock/Getty Images.

door and the door jamb near the locking mechanism. Use a striking tool to drive the adze farther into the door jamb until it has a good bite on the door. Then, leverage the tool to force the door outward away from the jamb.

FIRE FIGHTER TIP

Self-closing doors may work via a means as simple as a mechanism above the door or as complex as a magnet or electrically charged release that functions by closing the doors when the fire alarm sounds. Either way, when these doors are forced, the mechanism that closes the doors should be disabled, along with using some type of door chock or wedge to keep the doors from closing behind you once entry is made.

One method to force entry into an outward-opening door is shown in **SKILL DRILL 10-2**.

SKILL DRILL 10-2
Forcing Entry into an Outward-Opening Door Fire Fighter I, NFPA 1001: 4.3.4

1 Size up the door, looking for any safety hazards. Determine the number and location of locks and the locking mechanism. Place the adze end of the Halligan tool in the space between the door and the door jamb. Gap the door by rocking the tool up and down to spread the door from the door jamb.

2 Set the tool, and pry the door out by pulling on the Halligan tool so the adze can be driven in past the door jamb. Be careful not to bury the tool in the door jamb. Strike the Halligan tool to drive the adze end of the tool past the door jamb.

3 Force the door by pulling the Halligan tool away from the door. Control the door to limit ventilation of the fire and to prevent it from closing behind you.

© Jones & Bartlett Learning. Photographed by Glen E. Ellman.

Sliding Doors

Design. Most sliding doors are constructed of tempered glass in a wooden or metal frame. Such doors are commonly found in residences and hotel rooms that open onto balconies or patios (**FIGURE 10-17**). Sliding doors generally have two sections and a double track; one side is fixed in place, while the other side slides. A weak latch on the frame of the door secures the movable side. Many people prop a wood or metal rod in the track to provide additional security. If this has been done, look for another means of entry if possible to avoid destroying the door and leaving the premise open.

FIGURE 10-17 A reinforced sliding glass door may be difficult to open without breaking the glass.
© Jones & Bartlett Learning. Photographed by Glen E. Ellman.

FIGURE 10-18 Building codes may require outward-opening doors next to a revolving door.
Courtesy of Amanda Mitchell.

Forcing Entry. Forcing open sliding glass doors may be either very easy or very difficult. If the doors are not reinforced with a rod in the track, they should be easy to force. The locking mechanisms are not strong, and any type of prying tool can be used. Place the tool into the space between the door and the door jamb near the locking mechanism, and force the door away from the mechanism. If a rod in the track prevents the door from moving and there is no other way to enter, the only choice is to break the glass to gain forcible entry. This may create a hazard if fire fighters must drag hose lines or victims through the sharp glass debris. Clean all broken glass from the opening to avoid such problems.

Revolving Doors

Design. Revolving doors are most commonly found in large commercial or residential buildings. These are commonly found in large cities (**FIGURE 10-18**). They are usually made of four glass panels with metal frames. The panels are designed to collapse outward with a certain amount of pressure to allow for rapid escape during an emergency. Revolving doors are generally secured by a standard cylinder lock or slide latch lock.

Because of Life Safety Code®, fire, and building code requirements, standard outward-opening doors

are often found adjacent to the revolving doors. It may be easier to attempt to force entry through these doors instead of going through the revolving doors. Repairing any damage to the outward-opening doors will be less expensive for the building owner, too.

Forcing Entry. Forcible entry through revolving doors should be avoided whenever possible. Even if the doors can be forced open, the opening created will not be large enough to allow many people to exit quickly and easily. Forcible entry through a revolving door can be achieved by attacking the locking mechanism directly or by breaking the glass.

Overhead Doors

Design. Overhead doors come in many different designs, ranging from standard residential garage doors to high-security commercial roll-up doors (**FIGURE 10-19**). Most residential overhead garage doors have three or four panels, which may or may not include windows. Some residential overhead garage doors come in a single section and tilt rather than roll up. These doors, which may be made of wood or metal, usually have a hollow core that is filled with insulation or foam. By contrast, commercial security overhead doors are made of metal panels or hardened steel rods. They may use solid-core or hollow-core construction, depending on the amount of security needed. Overhead doors can be secured with cylinder-style locks, padlocks, or automatic garage door openers.

A

B

FIGURE 10-19 Overhead doors come in many different designs, ranging from standard residential garage doors to high-security commercial doors. **A.** Commercial overhead doors are made of metal panels or steel rods. **B.** Residential overhead garage doors usually have a hollow core that is filled with insulation or foam.

A: © Dave White/iStockphoto.com; B: © Photodisc/Creatas.

Forcing Entry. Before forcing entry through an overhead door, perform a careful size-up of the door. Most residential garage doors are not very sturdy; breaking a window or panel and manually operating the door lock or pulling the emergency release on the automatic opener from the inside may be all that is needed. If the fire is located behind the overhead door, however, it might have weakened the door springs, making it impossible to raise the door. In such a case, either the door must be cut or another means of entry found.

Remember to secure any raised door with a pike pole or other support to ensure that it does not close on either fire fighters or their attack lines. Put the prop under the door near the track. If the door has an emergency release cord, it might be able to serve as a second safety measure. Other options include bending the track with the fork of a Halligan tool or securing the door with a set of vice grips.

There are many ways to cut an overhead door to gain access, and there are many theories to justify each of these cuts. For the sake of simplicity, we will demonstrate how to make a cut using the triangle method and the hinge cut. The triangle method requires only two cuts to open the door, and it can be used on doors constructed of a variety of materials. Some other cuts, such as the box cut, require three separate cuts. Others, such as the hinge cut, cannot be used with doors constructed of certain types of materials. You should learn how to perform the cut recommended by your department.

To open an overhead garage door or a roll-up security door using the triangle method, follow the steps in **SKILL DRILL 10-3**.

The hinge cut is another method of cutting an overhead garage door. This method has an advantage in that you can make a large opening in the door, as well as open and close the cut section of the door as needed to control the air exchanged through the opening. The disadvantages to this technique are that it requires more time and more cuts to produce the opening. This technique is also called the inverted L cut and the West Coast cut.

To open an overhead garage door or a roll-up security door using the hinge method, follow the steps in **SKILL DRILL 10-4**.

SAFETY TIP

Remember to control doors to limit ventilation of the fire and to prevent them from closing behind you.

Windows

Windows provide both air flow and light to the inside of buildings; they also can provide emergency entry or exit. Windows often are easier to force than doors. Understanding how to force entry into a window requires an understanding of both window-frame construction and glass construction. Window frames are made of the same materials used in doors—wood, metal, vinyl, or combinations of these materials—and will often match the door construction of a building.

SKILL DRILL 10-3
Opening an Overhead Garage Door Using the Triangle Method
Fire Fighter I, NFPA 1001: 4.3.4

1 Before cutting, check for any safety hazards during the size-up of the garage door. Select the appropriate tool to make the cut. (The best choice is a power saw with a metal-cutting blade.) Wearing full protective gear and eye protection, start the saw, and ensure it is in proper working order.

2 Be aware of the environment behind the door. Check the outside of the door for heat. Try to lift the door to assure that it is not unlocked. If necessary, cut a small inspection hole, large enough to insert a hose nozzle.

3 Starting at a center high point in the door, make a diagonal cut to the right, down to the bottom of the door.

4 Next to the same starting point, make a second diagonal cut to the left, down to the bottom of the door. Fold the door down to form a large triangle.

5 If necessary, pad or protect the cut edges of the triangle and the bottom panel to prevent injuries as fire fighters enter or leave the premises. If you have time, you can remove the cut portion of the door and the bottom rail by raising the bottom rail with a prying tool and then cutting the bottom rail in two places.

SKILL DRILL 10-4
Opening an Overhead Garage Door Using the Hinge Method
Fire Fighter I, NFPA 1001: 4.3.4

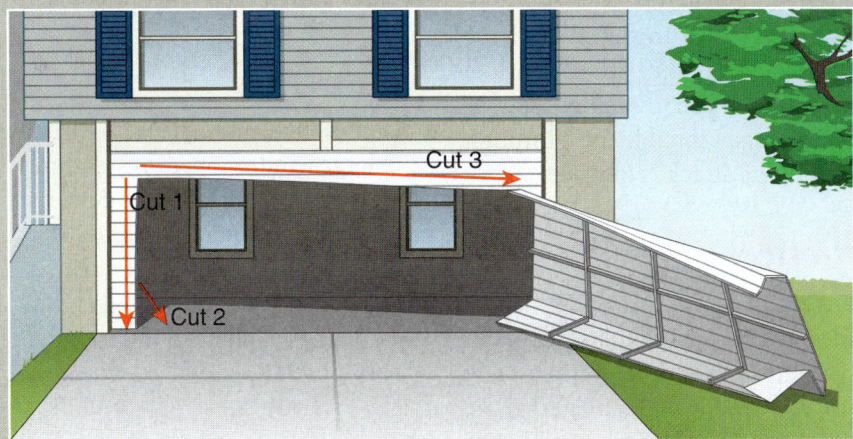

1 Before cutting, check for any safety hazards during the size-up of the garage door. Select the appropriate tool to make the cut. (The best choice is a power saw with a metal-cutting blade.) Wearing full protective gear and eye protection, start the saw, and ensure it is in proper working order.

2 Be aware of the environment behind the door. Check the outside of the door for heat. Try to lift the door to assure that it is not unlocked. If necessary, cut a small inspection hole, large enough to insert a hose nozzle.

3 Make a vertical cut down the left side of the door starting as high as you can comfortably reach.

4 Make a diagonal cut about 45 cm (18 in.) from the bottom of the door starting at the first cut. Make a cut to the bottom of the door. This will make a small triangle. Kick or push in the triangle formed by these two cuts.

5 Insert the saw blade into the opening made by the triangle cuts at the bottom of the door. Cut through the angle iron, or L-shaped steel, at the bottom of the door.

6 Resting the saw on your shoulder, make a horizontal cut the width of the door, starting at the top of the first cut. Continue cutting to the right side of the door. Do not extend the cut all the way to the vertical edge as this will make the door panels unstable and cause the door to collapse on that side. The uncut side of the door will act as the hinge.

7 Open the door flap outward. This will produce a large opening that can be opened and closed to control the flow of air to the fire. Secure the door in an open position anytime fire fighters are inside the building.

© Jones & Bartlett Learning.

This section reviews window construction, glass construction, and forcible entry techniques. The goal of forcible entry through windows is to gain entry with minimal or no damage to the window. As with doors, you should try to open the window normally before using any force.

Safety

Always take proper protective measures, including wearing full PPE, when forcing entry through windows.

Ensure that the area around the window, both inside and outside the building and below the window, is clear of other personnel.

Although breaking the glass is the easiest way to force entry, it is also dangerous. Broken pieces of wood, shattered glass, and sharp metal can cause serious injury. Always stand to the windward side, with your hands higher than the breaking point, when breaking windows. This positioning ensures that the broken glass will fall away from your hands and body. Placing the tip of the tool in the corner of the window will give you more control in breaking the window. After the window is broken, clear glass from the entire frame so that no glass shards will stick out and cause injury to fire fighters who enter or exit through the window. Glass may be the least expensive part of the window, but replacement costs are increasing as energy-efficient windows are being used more widely.

Remember that forcing entry through windows during a fire situation will change the path of fire flow and cause a ventilation-limited fire to rapidly grow or change direction. Do not attempt forcible entry through a window unless a proper fire attack is in place.

Glass Construction

The glazing is the transparent part of the window and is most commonly made of glass. Window glass comes in several configurations: regular glass, double/triple-pane glass, plate glass (for large windows), laminated glass, and tempered glass. Plastics such as Plexiglas and Lexan may also be used in windows. The window may contain one or more panes of glass. Insulated glass, for example, usually consists of two or more pieces of glass in the window and will be discussed in more detail later.

FIRE FIGHTER TIP

Lexan is used as a security measure because it cannot be broken with conventional tools and methods. It looks like glass, but if you take an axe to it, the axe will bounce back. Some carbide saw blades will cut Lexan. Another method is to freeze the Lexan with a CO_2 fire extinguisher and then hit it with the pick of an axe. In some cases, it may be easier to cut the window frame in order to gain entry.

Regular or Annealed Glass

Single-pane regular or annealed glass often is used in construction because it is relatively inexpensive; larger pieces are called plate glass. This type of glass is easily broken with a pike pole. When broken, plate glass creates long, sharp pieces called shards, which can penetrate helmets, boots, and other protective gear, causing severe lacerations and other injuries.

Double/Triple-Pane Glass (Insulated Windows)

Double/triple-pane glass is used in many homes because it improves home insulation by using two panes of glass with an air pocket between them. Some double/triple-pane windows may include an inert gas such as argon between the panes for additional insulation value. These windows

are sealed units, which makes them more expensive to replace. However, replacing the glass alone is less expensive than replacing the entire window assembly. Forcing entry through insulated windows is basically the same as forcing entry through single-pane windows, except that the two panes may need to be broken separately. These kinds of windows produce dangerous glass shards.

Plate Glass

Commercial plate glass is a stronger, thicker glass used in large window openings. Although it is being replaced by tempered glass in modern construction for safety reasons, commercial plate glass can still be found in older large buildings, storefronts, and residential sliding doors. It can be broken easily with a sharp object such as a Halligan tool or a pike pole. When broken, commercial plate glass windows produce dangerous glass shards.

Laminated Glass

Laminated glass, also known as safety glass, is used to prevent windows from shattering and causing injury. Laminated glass is molded with a sheet of plastic between two sheets of glass. This type of glass is most commonly used in vehicle windshields, but it may also be found in other applications such as doors or building windows.

Tempered Glass

Tempered glass is specially heat treated, making it four times stronger than regular glass. This type of glass is commonly found in side and rear windows in vehicles, in commercial doors, in newer sliding glass doors, and in other locations where a person might accidentally walk into the glass and break it. Tempered glass breaks into small pellets without sharp edges to help prevent injury during accidents. The best way to break tempered glass is by using a sharp pointed object in the corner of the frame. During vehicle extrication, a center punch is often used to break a vehicle window.

Wired glass is tempered glass that has been reinforced with wire. This kind of glass may be clear or frosted, and it is often used in fire-rated doors that require a window or sight line from one side of the door to the other. Wired glass is difficult to break and force.

Frame Designs

Window frames come in many styles. This chapter covers the most common ones, but fire fighters must be familiar with the types in their response area and the forced entry techniques for each type.

Double-Hung Windows

Design. Double-hung windows contain two movable panels, or sashes, usually made of wood or vinyl, that slide up and down (**FIGURE 10-20**). These kinds of windows are frequently found in residences and have wood, plastic, or metal tracks. Newer double-hung window sashes may be removed or swung in for cleaning. They may have either one locking mechanism that is found in the center of the window or two locks on each side of the lower sash that prevent the sashes from moving up or down.

Forcing Entry. Forcing entry through double-hung windows involves opening or breaking the locking mechanism. Place a prying tool between the windowsill and the lower sash, and force it up to break the lock or remove it from the track. This technique must be done carefully; otherwise, it may cause the glass to shatter. Because forcing the locks causes extensive damage to the window, breaking the glass and

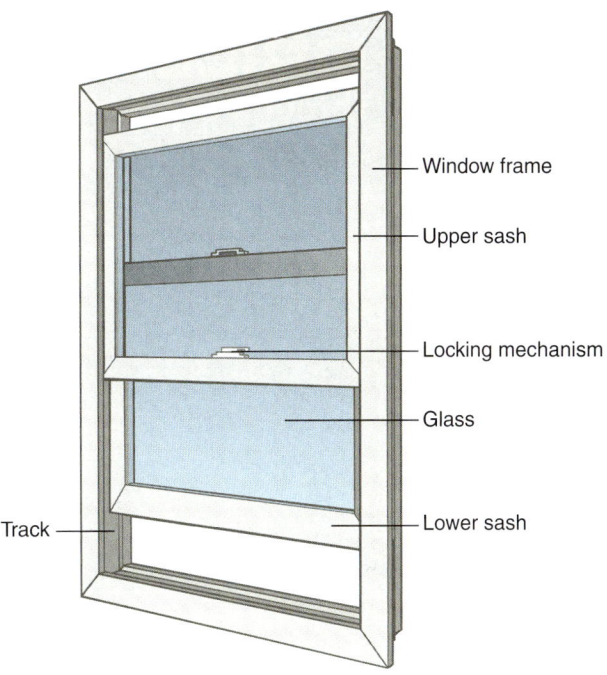

FIGURE 10-20 Double-hung windows allow the inner and outer sashes to move freely up and down.
© Jones & Bartlett Learning.

opening the locks may be a less expensive method of gaining entry.

To force entry through a wooden double-hung window, follow the steps in **SKILL DRILL 10-5**.

SKILL DRILL 10-5
Forcing Entry Through a Wooden Double-Hung Window
Fire Fighter I, NFPA 1001: 4.3.4

1 Size up the window for any safety hazards, and locate the locking mechanism.

2 Place a prying tool such as the Halligan tool between the windowsill and the bottom sash in line with the locking mechanism.

(continued)

SKILL DRILL 10-5 Continued
Forcing Entry Through a Wooden Double-Hung Window
Fire Fighter I, NFPA 1001: 4.3.4

© Jones & Bartlett Learning. Photographed by Glen E. Ellman.

3 Pry the bottom sash upward to displace the locking mechanism. Secure the window in the open position so that it cannot close.

FIRE FIGHTER TIP

When forcing entry through older double-hung windows, try sliding a hacksaw blade or similar blade up between the sashes. The teeth on the blade will grip the locking mechanism and can often move it into the unlocked position. This technique will not work on newer double-hung windows because of their newer latch design.

Single-Hung Windows

Design. Single-hung windows are similar to double-hung windows, except that the upper sash is fixed and only the lower sash moves (**FIGURE 10-21**). The locking mechanism is the same as on double-hung windows. It may be difficult to distinguish between single-hung and double-hung windows from the exterior of a building.

Forcing Entry. Forcing entry through a single-hung window involves the same technique as forcing entry through a double-hung window. Place a prying tool between the windowsill and the lower sash, and force it up. This will usually break the lock or remove the sash from the track. Be careful because the glass may shatter. Breaking the glass and opening the window are generally easier with a single-hung window.

Jalousie Windows

Design. A jalousie window is made of adjustable sections of tempered glass encased in a metal frame that overlap each other when closed (**FIGURE 10-22**).

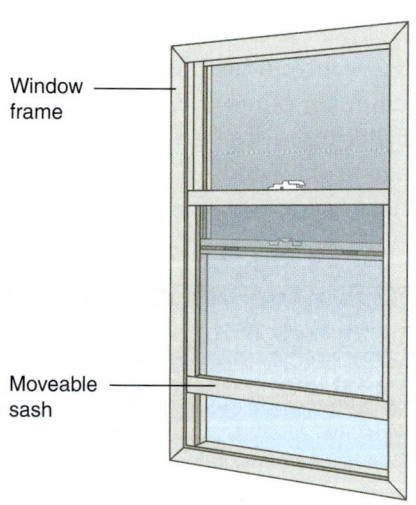

Window frame

Moveable sash

FIGURE 10-21 A single-hung window has only one movable sash.
© Jones & Bartlett Learning.

This type of window is often found in mobile homes and is operated by a small hand-wheel or crank located in the corner of the window.

Forcing Entry. Forcing entry through jalousie windows can be difficult because they are made of tempered glass and have several panels. Forcing open the panels or breaking a lower one to operate the crank is possible but does not leave a large-enough opening for a person to enter the building. Removing or breaking the panels one at a time is time consuming and does not always leave a clean entry point, with the framing in place. The best strategy is to avoid these windows if possible.

FIGURE 10-22 Jalousie windows are opened and closed with a small hand crank.
© Jones & Bartlett Learning. Photographed by Bill Larkin.

Awning Windows

Design. Awning windows are similar in operation to jalousie windows, except that they usually have one large or multiple medium-sized glass panels that do not overlap when they are closed (**FIGURE 10-23**). The hinge is on the top, and they open outward. Awning windows are operated by a hand crank located in the corner or in the center of the window (**FIGURE 10-24**). These kinds of windows can be found in residential, commercial, and industrial structures. Residential awning windows may be framed in wood, vinyl, or metal, whereas commercial and industrial windows usually have metal frames.

Commercial and industrial awning windows often use a lock and a notched bar to hold the window open, rather than a crank.

Forcing Entry. Forcing entry through an awning window requires the same technique that is used for jalousie windows—that is, break or force the lower panel, and operate the crank, or break out all the panels. Depending on the size of the panels and the window frame, it may be easier to access an awning window than a jalousie window.

Horizontal-Sliding Windows

Design. Horizontal-sliding windows are similar to sliding doors (**FIGURE 10-25**). In older windows, the locking mechanisms attach to the window frame. Newer sliding windows have latches between the windows, similar to those found on double-hung windows.

FIGURE 10-23 An awning window has larger panels than a jalousie window.
© Jones & Bartlett Learning. Photographed by Bill Larkin.

FIGURE 10-24 An awning window is operated by a hand crank located in the corner or in the center of the window.
© Jones & Bartlett Learning. Photographed by Bill Larkin.

People often place a rod or pole in the track to prevent break-ins.

Forcing Entry. Forcing entry through sliding windows is just like forcing entry through sliding doors: Place a pry bar near the latch to break the latch or the plate. If a rod or pole has been inserted into the track, look for another entry point, or break the glass, although the latter step should be the last resort.

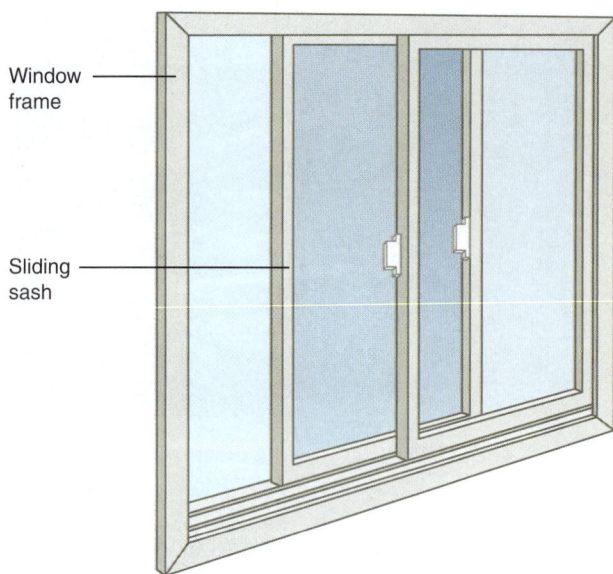

FIGURE 10-25 Horizontal-sliding windows work like sliding doors.
© Jones & Bartlett Learning.

Casement Windows

Design. Casement windows have a steel or wood frame and open away from the building with a crank mechanism (**FIGURE 10-26**). Although they are similar to awning windows that have large glass panels, casement windows have a side hinge, rather than a top hinge. Several types of locking mechanisms can be used with these windows. Like jalousie windows, these windows should be avoided when forcible entry is necessary because they are difficult to force open.

Forcing Entry. The best way to force entry through a casement window is to break out the glass from one or

FIGURE 10-26 Casement windows open to the side with a crank mechanism.
© Jones & Bartlett Learning. Photographed by Bill Larkin.

more of the panes, locate the locking mechanisms, and crank them manually. To force entry through a casement window, follow the steps in **SKILL DRILL 10-6**.

SKILL DRILL 10-6
Forcing Entry Through a Casement Window Fire Fighter I, NFPA 1001: 4.3.4

1 Size up the window to check for any safety hazards, and locate the locking mechanism. Select an appropriate tool to break out a windowpane. Stand to the windward side of the window, and break out the pane closest to the locking mechanism.

(continued)

SKILL DRILL 10-6 Continued
Forcing Entry Through a Casement Window Fire Fighter I, NFPA 1001: 4.3.4

2 Remove all of the broken glass in the pane to prevent injuries.

3 Reach in and unlock the window, then manually operate the window crank to open the window.

© Jones & Bartlett Learning. Photographed by Glen E. Ellman.

Projected Windows

Design. **Projected windows** (also called factory windows) are usually found in older warehouse or commercial buildings (**FIGURE 10-27**). They are awning-type windows that can project inward or outward on a top or bottom hinge. Screens are rarely used with these windows, but forcing entry may not be easy, depending on the integrity of the frame, the type of locking mechanism used, and the window's distance off the ground. These windows may have fixed metal-framed wire glass panes above them.

Forcing Entry. Avoid forcing entry through a projected window when possible. These windows are often difficult to force open, and they may be difficult for a

FIGURE 10-27 Projected windows may open inward or outward on a top or bottom hinge.
© Jones & Bartlett Learning. Photographed by Bill Larkin.

person to enter. To force entry, break out the glass from one pane, unlock the mechanism, and open the window by hand. If the opening created is not large enough, break out the entire window assembly.

FIRE FIGHTER TIP

In areas that are prone to hurricanes, special hardened glass may be installed in windows. This product, which is not easy to recognize in a fire situation, consists of a shatter-proof membrane between two layers of tempered or specially treated glass. These windows are harder to remove; however, they can be shattered with the sharp end of an axe or a Halligan tool.

To force entry through a projected or factory window, follow the steps in **SKILL DRILL 10-7**.

Locks

Locks have been used for hundreds of years. They range from the very simple push-button locks found in most homes to the complex computer-operated locks found in banks and other high-security areas.

Parts of a Door Lock

Most door locks have the same basic parts. Gaining entry is easier when you have an understanding of what these parts are and how they operate (**FIGURE 10-28**). Door locks have three major components:

- **Latching device**—The part of the lock that "catches" and holds the door frame

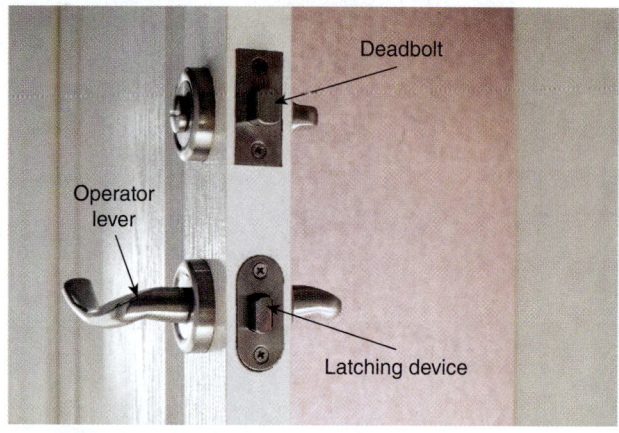

FIGURE 10-28 Most door locks have the same basic components.
© Maxal Tamor/Shutterstock.

SKILL DRILL 10-7
Forcing Entry Through a Projected Window Fire Fighter I, NFPA 1001: 4.3.4

1. Size up the window to check for any safety hazards, and locate the locking mechanism. Select an appropriate tool, such as a pike pole, to break a pane of glass. Stand to the windward side of the window, and break the pane closest to the locking mechanism.

2. Remove all of the broken glass in the window frame to prevent injuries.

3. Reach in and manually open the locking mechanism and the window. If necessary, use a cutting tool, such as a torch, to remove the window frame and enlarge the hole.

© Jones & Bartlett Learning. Photographed by Glen E. Ellman.

- **Operator lever**—The handle, doorknob, or keyway that turns the latch to lock it or unlock it
- **Deadbolt**—A second, separate latch that locks and reinforces the regular latch

Parts of a Padlock

Padlocks are portable or detachable locks. Most padlocks, like most door locks, have similar components (**FIGURE 10-29**):

- **Shackle**—The U-shaped top of the padlock that slides through a hasp and locks in the padlock itself
- **Unlocking mechanism**—The keyway, combination wheel, or combination dial used to open the padlock
- **Lock body**—The main part of the padlock that houses the locking mechanisms and the retention part of the lock

Padlocks are discussed in more detail later.

Safety

Tool safety is an important consideration when executing forcible entry through a door lock or a padlock. The tools used for removing the cylinders in a door lock must be kept sharp, and fire fighters must be careful to avoid being cut when using these tools. Care also must be taken when using striking tools when forcing a lock to avoid pinching or crushing fingers or hands. Fire fighters must wear proper eye protection and gloves and use tools carefully to minimize the risk of injury.

Types of Locks

The four major lock categories are cylindrical locks, padlocks, mortise locks, and rim locks.

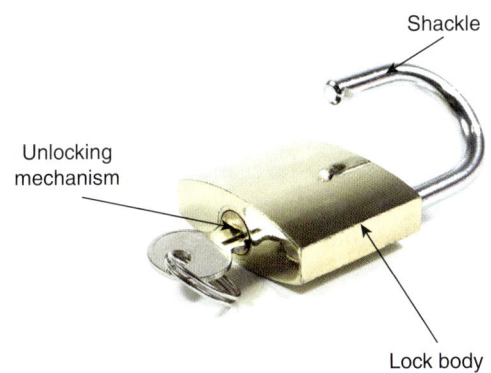

FIGURE 10-29 Padlocks have the same basic parts.
© Lik Studio/Shutterstock.

Cylindrical Locks

Design. **Cylindrical locks** are the most common fixed lock in use today (**FIGURE 10-30**). They are relatively inexpensive and are standard in most doors. The locks and handles are set into a predrilled hole that goes straight through the width of the door. The outside of the doorknob usually has a key-in-the-knob lock; the inside of the doorknob will have either a keyway, a button, or another type of locking/unlocking mechanism.

Forcing Entry. There are several ways to force entry when a cylindrical lock is present. One method is to use a type of lock puller such as a K tool. If this first method does not work, an alternative method is to place a pry

A

B

FIGURE 10-30 A cylindrical lock. **A.** The outside of the doorknob usually has a key-in-the-knob lock. **B.** The inside of the doorknob will have either a keyway, a button, or another type of locking mechanism.
Courtesy of Amanda Mitchell.

bar near the locking mechanism and lever it to force the lock.

Padlocks

Design. Padlocks are one of the most common locks fire fighters may encounter, and they come in many designs and strengths. Both regular-duty and heavy-duty padlocks are available. Padlocks come with a variety of unlocking mechanisms, including keyways, combination wheels, or combination dials. Operating the unlocking mechanism opens one side of the lock to release the shackle and allow entry. Shackles for regular padlocks generally have a diameter of 6 mm (¼ in.) or less and are not made of case-hardened metal. Shackles for heavy-duty padlocks are 6 mm (¼ in.) or larger in diameter and made of case-hardened steel or some other case-hardened metal. The design of some padlocks hides the shackle, making forced entry difficult.

One of these locks is referred to as the hockey puck lock (**FIGURE 10-31**). These types of locks have hidden shackles and cannot be forced through conventional methods. A large pipe wrench with a cheater bar can be used to twist the locks from the mounting tabs. However, if the lock has a mounting bracket, this technique will be ineffective. In such a case, you will need a rotary saw or a cutting torch to force entry. When using this method, you should focus your effort approximately two-thirds of the way up from the keyway in order to cut the pin in half.

FIGURE 10-31 The hockey puck lock has hidden shackles.
Courtesy of Master Lock Company LLC.

Forcing Entry. Several techniques can be used to force entry through padlocks without causing extensive property damage. Before breaking the padlock, consider cutting the hardware (hasp and shackle that the padlock is attached to) first. This approach saves the padlock and makes securing the building easier for the building owner. Otherwise, breaking the shackle on the padlock may be the best method for forcing entry through padlocked doors. If the padlock is made of case-hardened steel, however, many conventional methods of breaking the lock will prove ineffective.

The tools most commonly used to force entry through a padlock are bolt cutters, duck-billed lock breakers, a bam-bam tool, locking pliers and chain, and a rotary saw or a torch:

- Bolt cutters can quickly and easily break regular-duty padlocks but cannot be used on heavy-duty, case-hardened steel padlocks. To use bolt cutters on a padlock, open the jaws as wide as possible. Close the jaws around one side of the lock shackle to cut through the shackle. Once one side of the shackle of a regular-duty lock is cut, the other side will spin freely and allow access.

- The duck-billed lock breaker has a large metal wedge attached to a handle. To open a padlock using a duck-billed lock breaker, place the narrow end of the wedge into the center of the shackle, and force it through with another striking tool. The wedge will spread the shackle until it breaks, freeing the padlock and allowing access to the building. While this causes no damage to the lock body, the lock will be inoperable after use of the duck-billed lock breaker.

- The bam-bam tool is a sliding hammer that can pull the lock cylinder out of a regular-duty padlock. It will not work on higher-end padlocks that include security rings to hold the cylinder in place. This tool contains a case-hardened screw that is placed in the keyway. To open a padlock using the bam-bam tool, insert and secure the screw into the keyway. Once the screw is set, use the sliding hammer to pull the tumblers out of the padlock so that the trip-lever mechanism inside the lock can be opened manually.

- The locking pliers and chain is used to clamp a padlock securely in place so that the shackles can be cut safely. To use a locking pliers and chain to secure a padlock, clamp the pliers to the lock body. One fire fighter maintains tension on the chain while another fire fighter uses a rotary saw with a metal-cutting blade or a torch to cut the padlock.

Mortise Locks

Design. Mortise locks are designed to fit in predrilled openings on the side of a door; they are commonly found in hotel rooms (**FIGURE 10-32**). Most mortise locks have both a locking or nonlocking latch and a deadbolt built into the same mechanism, each of which operates independently of the other. While the latch may lock the door, the deadbolt can be deployed for added security. Mortise locks are known for their strength.

Forcing Entry. The design and construction of mortise locks make them difficult to force. A door with a mortise lock may be forced with conventional means such as those described but will probably require the use of a through-the-lock technique.

Rim Locks and Deadbolts

Design. Rim locks and deadbolts are locks that can be surface mounted on the interior of the door frame (**FIGURE 10-33**). They are commonly found in residences as secondary locks to support the through-the-handle locks. Such locks may be identified from the outside by the keyway that has been bored into

FIGURE 10-33 Rim locks and deadbolts are mounted on the inside of a residence door.
© seele/iStock/Getty Images.

the door. These locks have a bolt that extends at least 25 mm (1 in.) into the door frame, making the door more difficult to force.

Forcing Entry. Rim locks and deadbolts are the most challenging locks to break. They can be very difficult to force with conventional methods, such that through-the-lock methods may be the only option. An additional option is to use a powered circular saw with a metal cutting blade to cut through the portion of the lock between the door and the door frame. This is sometimes called the throw of the lock.

Electromagnetic Locks

Design. Electromagnetic locks use powerful electromagnets and an armature plate that are energized by electricity (**FIGURE 10-34**). These locks may be operated with keyless lock systems, discussed next. The electromagnets exert as much as 544 kg (1200 lb) of force to hold the door shut. These devices can be either "fail safe" or "fail secure." Fail-safe locking devices are unlocked when deenergized. A fail-secure locking device remains locked when power is lost.

Forcing Entry. Doors with electromagnetic locks may be forced open by inserting prying tools between the two sides of the electromagnetic plate. Some locks may require hydraulic tools. Once these locks are open, tape a small object, such as a nail, on the electromagnetic plate to prevent the door from locking again. Follow the SOPs of your fire department.

Keyless Door Locks

Design. Also called electronic locking systems, keyless door locks operate without a traditional metal key.

FIGURE 10-32 Mortise locks usually have a latch and a deadbolt, which operate independently of each other.
Courtesy of Amanda Mitchell.

FIGURE 10-34 Many electromagnet locks release with the loss of electric power.
Courtesy of Amanda Mitchell.

FIGURE 10-35 Keyless locking systems are widely used in hotels, offices, public buildings, secured installations, gated communities, and parking garages.
Courtesy of Amanda Mitchell.

They are widely used in hotels, offices, public buildings, secured installations, gated communities, and parking garages. They can be operated by entering a numerical code, inserting or swiping a key card, or by using a fingerprint or eye scan (**FIGURE 10-35**). They can be accessed directly or from a remote location using a computer or a smartphone. Some electronic door lock systems are powered by batteries installed in the locking system. Others are powered by the electrical system in the building. When the lock receives the correct electronic input, such as a code or a key card, it activates a small motor or actuator that releases the locking mechanism. Keyless locking systems can be coupled with a wide variety of locking mechanisms.

Forcing Entry. A wide variety of techniques may be needed to access keyless locking systems. Whenever possible, obtain the master key card from the facilities manager, security personnel, or alarm company personnel. This will enable you to quickly enter any room you need to access to mitigate the emergency situation in a setting such as a hotel. Alternatively, the facilities manager, security personnel, or alarm company personnel may be able to release the locks remotely. Some fire departments carry access key cards in the apparatus for facilities to which they regularly respond. If these actions are not successful, it may be necessary to force each individual door lock. The method you use to gain entry will depend on the type of lock that is coupled with the keyless locking system and will use the techniques covered throughout this chapter.

Through-the-Lock Techniques

The following Skill Drills present various through-the-lock techniques.

The K tool is designed to remove a lock cylinder. To force entry using a K tool, follow the steps in **SKILL DRILL 10-8**.

FIRE FIGHTER TIP

If the K tool does not fit over a residential lock, try an A tool.

As discussed, the A tool is similar to the K tool except that the pry bar is built into the cutting part of the A tool. To force entry using an A tool, follow the steps in **SKILL DRILL 10-9**.

To force entry by unscrewing the lock, follow the steps in **SKILL DRILL 10-10**.

SKILL DRILL 10-8
Forcing Entry Using a K Tool Fire Fighter I, NFPA 1001: 4.3.4

1 Size up the door and lock area, checking for any safety hazards. Determine which type of lock is used, whether it is regular or heavy-duty construction, and whether the lock cylinder has a case-hardened collar, which may hamper proper cutting. Place the K tool over the face of the lock cylinder, noting the location of the keyway.

2 Place the adze end of the Halligan tool or similar prying tool into the slot on the face of the K tool. Using a flat-head axe or similar striking tool, strike the end of the prying tool to drive the K tool farther onto the lock cylinder.

3 Pry up on the tool to remove the lock cylinder and expose the locking mechanism. Using the small tools that come with the K tool, turn the locking mechanism to open the lock. The lock should release, allowing you to open the door.

© Jones & Bartlett Learning. Photographed by Glen E. Ellman.

Forcing Entry Through Doors with Drop Bars

Many commercial buildings have rear single or double doors that are used for receiving deliveries and removing refuse. These doors are usually constructed of metal and may have no windows or a very small window. To prevent unauthorized entry into the building, occupants often equip the doors with removable dropdown bars that fit into holders or brackets mounted on the inside of the door. The dropdown bar is placed in the brackets and extends beyond the sides of the door frame. You may recognize the presence of these bars from the outside by looking for the heads of carriage bolts that are holding the brackets on the inside of the door. Gaining entry through these doors requires you to disable these devices. There are many techniques that can be used. One of the simplest is to cut the heads off the carriage bolts and drive these bolts back through the door.

Forcing Entry Through Security Gates and Windows

Many homes are equipped with metal security gates and security bars over the windows, which are intended to provide protection from break-ins for the building

SKILL DRILL 10-9
Forcing Entry Using an A Tool Fire Fighter I, NFPA 1001: 4.3.4

1 Size up the door and lock area, checking for any safety hazards. Determine which type of lock is used, whether it is regular or heavy-duty construction, and whether the lock cylinder has a case-hardened collar, which may hamper proper cutting. Place the cutting edges of the A tool over the lock cylinder, between the lock cylinder and the door frame.

2 Using a flat-head axe or similar striking tool, drive the A tool onto the lock cylinder. Pry up on the A tool to remove the lock cylinder from the door. Insert a key tool into the hole to manipulate the locking mechanism and open the door.

© Jones & Bartlett Learning. Photographed by Glen E. Ellman.

SKILL DRILL 10-10
Forcing Entry by Unscrewing the Lock Fire Fighter I, NFPA 1001: 4.3.4

1 Size up the door and lock area, checking for any safety hazards. Using a set of vise grips, lock the pliers onto the outer housing of the cylinder lock with a good grip.

2 Unscrew the housing until it can be fully removed. Manipulate the locking mechanism with lock tools to disengage the arm, and then open the door.

© Jones & Bartlett Learning. Photographed by Glen E. Ellman.

residents. Security gates may be equipped with locks on the outside and the inside of the gate. As a consequence, if a key is not available to the building occupants, they can become trapped in their own house. Many security bars on windows are equipped with a locking mechanism that enables the residents to quickly open the security bars from the inside and use the window for an exit in case of a fire. Unfortunately, some security bars on windows are permanently mounted into the window frame and cannot be removed quickly.

Thus these security devices present a dilemma: The same devices that were designed to keep criminals out can, in the event of a fire, make it difficult for the building occupants to escape from the fire and difficult for fire fighters to access the fire. Such devices can complicate fire fighters' efforts to gain access to the building to perform search and rescue operations and fire suppression activities. Given these facts, it is important for fire fighters to check for the presence of a lockbox and know how to rapidly force entry when these security devices are in place. Some gates that secure driveways may be siren activated. In such a case, no forcible entry technique other than sounding the siren on the fire apparatus is required to open the gates. Other gates may be opened using a key card that is kept in the fire apparatus.

One method of breaching a security gate is to remove the lock cylinder using a K tool, an A tool, or a bam-bam tool. If the lock cylinder can be accessed and removed, it is a simple matter to turn the locking mechanism and open the gate. If the lock cylinder has been protected to minimize the chance of break-in, however, it may be necessary to cut through part of the gate to open it. This can be done by using a circular saw that is equipped with an appropriate blade or, in some cases, by using a hydraulic cutter. An alternative is to use a hydraulic spreader to force the anchor from the masonry or wood to which it is attached. Be sure to wear PPE, including adequate eye protection, when operating these tools. Plan your cuts to minimize the amount of metal that needs to be cut to open the gate.

Breaching security bars on a window can sometimes be accomplished by removing the lock cylinder, if one is present on the outside of the security bars. If no lock cylinder is present, it will be necessary to cut through the bars using either a circular saw or hydraulic cutters. If the bars are not well anchored, you might be able to pry them from the window. Sometimes a hydraulic spreader can be used to force the anchor from the masonry or wood to which it is attached. This method will not always work, however, when the bars are securely anchored in masonry or concrete.

Look for the types of security gates and bars that are present in your community. Talking with the installers of these products will give you more knowledge about how to gain emergency access or egress.

Breaching Walls and Floors

On occasion, forced entry through a window, lock, or door might not be possible or might take too long. In these cases, consider the option of breaching a wall or floor. Breaching a wall can be an option for removal of injured people or for emergency escape. An understanding of basic construction concepts is required to execute this technique safely and quickly.

Load-Bearing/Nonbearing Walls

Walls may be constructed of masonry, wood, steel, aluminum, glass, and other materials. Before breaching a wall, first consider whether the interior or exterior wall is load bearing. A load-bearing wall supports a portion of the building's weight and its contents, so removing or damaging such a wall could result in partial or total collapse. By contrast, a nonbearing wall supports only its own weight and can be removed safely and without danger. Many nonbearing walls are interior walls or **partitions**. Refer to Chapter 6, *Building Construction*, for a review of wall construction.

Exterior Walls

Exterior walls form the perimeter of the building and can be constructed of one or more materials. Exterior walls are often load bearing. Many residences have both wood and brick, vinyl or aluminum siding, or masonry block construction (**FIGURE 10-36**). Commercial buildings usually have concrete, masonry, or metal exterior walls. Commercial buildings also may have steel I-beam construction; older commercial structures may have heavy timber construction. Because there are exceptions to every rule, it is important to know the construction methods and materials used for buildings in your specific response area.

Deciding whether to breach an exterior wall is a difficult choice. Masonry, metal, and brick are formidable materials, and breaking through them can be very

FIGURE 10-36 Residences may have exterior walls of multiple materials, such as wood and masonry block.
Courtesy of DOE/NREL, Credit-Ed Hancock.

difficult. The best tools to use in breaching a concrete or masonry exterior wall are a battering ram, a sledgehammer, and a rotary saw with a concrete blade.

Interior Walls

Interior walls in residences are usually constructed of wood or metal studs covered by plaster, gypsum, or sheetrock. Some newer residential construction contains a laminate sheetrock, which is extremely difficult to penetrate. Commercial buildings may have concrete block interior walls. Breaching an interior wall can be dangerous to fire fighters for several reasons. For example, many interior walls contain electrical wiring, plumbing, cable wires, and phone wires—all of which present hazards to fire fighters. Interior load-bearing walls can be breached without undue hazard as long as studs are not removed. Extreme care should be taken, however, if any studs are removed.

After determining whether the wall is load bearing, sound it to locate a stud away from any electrical outlets or switches. Tap on the wall; the area between studs will make a hollow sound compared to the solid sound directly over the stud.

After locating an appropriate site, make a small hole (which can also serve as an inspection hole to check for fire) to check for any obstructions. If the area is clear, expand the opening to reveal the studs. Walls should be breached as close as possible to the studs because this choice makes a large opening, and cutting is easier. If possible, enlarge the opening by removing at least one stud to enable quick escapes.

To breach a wall frame, follow the steps in **SKILL DRILL 10-11**.

SKILL DRILL 10-11
Breaching a Wall Frame Fire Fighter I, NFPA 1001: 4.3.4

1 Size up the wall, checking for safety hazards such as electrical outlets, wall switches, or any signs of plumbing. Inspect the overall scene to ensure that the wall is not load bearing. Using a tool, sound the wall to locate any studs. Make a hole between the studs.

2 Cut the sheetrock as close to the studs as possible.

3 Enlarge the hole by extending it from stud to stud and as high as necessary.

To breach a masonry wall, create an upside-down V in the wall. Follow the steps in **SKILL DRILL 10-12**.

To breach a metal wall, follow the steps in **SKILL DRILL 10-13**.

Floors

The two most popular floor materials found in residences and commercial buildings are wood and poured

---

SKILL DRILL 10-12
Breaching a Masonry Wall Fire Fighter I, NFPA 1001: 4.3.4

1 Size up the wall, checking for any safety hazards such as electrical outlets, wall switches, and plumbing. Inspect the overall area to ensure that the wall is not load bearing. Select a row of five masonry blocks at 61–91 cm (2–3 ft) above the floor. Using a sledgehammer, knock two holes in each masonry block. Each hole should pierce into the hollow core of the masonry blocks.

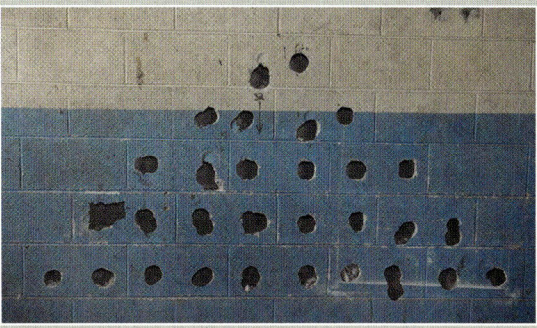

2 Repeat the process on four masonry blocks above the first row. Repeat the process on three masonry blocks above the second row, on two masonry blocks above the third row, and on one masonry block above the fourth row. An inverted V-cut has now been created.

3 Knock out the remaining portion of the masonry blocks by hitting the masonry blocks parallel to the wall rather than perpendicular to it.

4 Clear any reinforcing wire using bolt cutters, and enlarge the hole as needed.

© Jones & Bartlett Learning. Photographed by Glen E. Ellman.

SKILL DRILL 10-13
Breaching a Metal Wall Fire Fighter I, NFPA 1001: 4.3.4

1 Size up the wall, checking for any safety hazards such as electrical outlets, wall switches, or plumbing. Inspect the overall area to ensure that the wall is not load bearing. Select an appropriate tool, such as a rotary saw with a metal-cutting blade or a torch. Open the wall using a diamond-shaped or inverted V-cut.

2 Make the cut so the cut piece of the wall falls into the building and no sharp bottom edges are exposed.

concrete. Both are very resilient and may be difficult to breach. Unfortunately, it is impossible to tell how a floor is constructed by looking at it from above. The important information about a floor can be observed only from below, if it is visible at all. The building's age and local construction methods can provide significant clues to a floor's stability. Conduct a thorough size-up of the site before considering forcible entry through this route; breaching a floor should be a last-resort measure. A rotary saw with an appropriate blade is the best tool to breach a floor; a chainsaw may be a better choice when used on a wood floor.

To breach a floor, follow the steps in **SKILL DRILL 10-14**.

FIRE FIGHTER TIP

Fire departments are regularly called because people are locked out of their cars. Many fire departments do not respond to these calls unless a child or a pet has been locked in a vehicle. This type of situation represents an emergency situation that requires a fire department response.

Some fire departments receive training from a locksmith in the use of slim jims and other devices to unlock vehicles. If you are not trained and equipped to gain entry into cars, in an emergency you can quickly gain entry by breaking a side window as far away from the trapped child or pet as possible. This is easily done using a spring-loaded center punch or other sharp object as described in Chapter 24, *Vehicle Rescue and Extrication*.

SKILL DRILL 10-14
Breaching a Floor Fire Fighter I, NFPA 1001: 4.3.4

1 Size up the floor area, checking for hazards in the area to be cut. Sound the floor with an axe or similar tool to locate the floor joists.

2 Use an appropriate cutting tool to cut one side of the hole. Cut the opposite side. Make two additional cuts at right angles to the first cuts. This will form a rectangle. Avoid cutting into the floor joists.

(continued)

SKILL DRILL 10-14 Continued
Breaching a Floor Fire Fighter I, NFPA 1001: 4.3.4

3 Remove any flooring, such as carpet, tile, or floorboards that have been loosened. Cut a similar size opening into the subfloor, until the proper size hole is achieved.

4 Secure the area around the hole to prevent others working in the area from falling through the hole.

© Jones & Bartlett Learning. Photographed by Glen E. Ellman.

Systematic Forcible Entry

The first step in forcible entry is to think. Several issues need to be evaluated before deciding on a course of action. Double-check the address for the incident—it is embarrassing to try to explain to homeowners why you have destroyed their back door when the burning food on the stove was reported three houses down the street. Finding the right address can be a challenge, especially if rear entrances off an alley are not clearly marked.

Second, look for the presence of a lockbox. Lockboxes are attached to buildings, usually close to a main entrance (**FIGURE 10-37**). These boxes can be unlocked with

FIGURE 10-37 A lockbox contains the keys needed to open the structure's door.
Courtesy of Wayland, Massachusetts.

a master key that is carried by the fire department. The master key is coded to open all lockboxes in that jurisdiction. Once entry is made into the lockbox, the fire department officer can remove keys that will open the specific building. With a lockbox system, rapid entry can be obtained by the fire department into any building equipped with a lockbox, and the building can be secured after the incident by the fire department. To maintain security over the lockbox master keys, some systems are equipped with a device that keeps the master key locked in the fire apparatus unless released by the emergency dispatcher. This approach creates a record of each time the master key is used, allows the dispatcher to record the time and place the key was used, and identifies the officer requesting the use of the master key. You need to determine if your department has a lockbox program and, if it does, how it works.

Third, evaluate the threat level of the incident. When flames and smoke are visible, the need for forcible entry is much higher than when a fire alarm activation occurs in the middle of a summer thunderstorm with no sign of fire. Life safety is a major consideration. A fire alarm activation in a closed business in the middle of the night is much less potentially life threatening than a call where people are reportedly in distress or trapped.

Fourth, consider how to make entry with the least amount of damage. This objective must be balanced with the severity of the emergency. Generally speaking, you want to make entry with the least damage in the shortest amount of time, but this assessment must be balanced with the type of emergency present.

Remember that forcible entry must be coordinated with other fireground activities. Significant fire events can occur if forcible entry is conducted without considering the consequences of actions. Try not to damage doors or windows to the point that they are no longer usable. Functional doors and windows can prove helpful in controlling flow paths and air intake, resulting in increased tactical efficiency and improved safety.

After-Action REVIEW

IN SUMMARY

- Forcible entry is required at emergency incidents where time is a critical factor. Company officers usually select both the point of entry and the method to be used.

- Before beginning forcible entry operations, remember to "try before you pry."

- It is important to wear the appropriate PPE when conducting forcible entry.

- Tools should be kept clean, properly serviced according to the manufacturer's guidelines, and ready for use at all times.

- Most tools fit into the following functional categories:

  - Rotating tools—Apply a rotational force to make something turn and use to assemble and disassemble items. Rotating tools used in forcible entry include screwdrivers, wrenches, and pliers.

  - Striking tools—Generate an impact force directly on an object or another tool. Striking tools used in forcible entry include the flat-head axe, battering ram, and sledgehammer.

  - Prying/spreading tools—Designed for prying and spreading. Prying/spreading tools used in forcible entry include the Halligan tool, pry bar/hux bar/crowbar, pry axe, spreaders, cutters, and rams.

  - Cutting tools—Primarily used for cutting doors, roofs, walls, and floors. Cutting tools used in forcible entry include the axe, bolt cutters, reciprocal saw, and circular saw.

  - Pushing/pulling tools—Extend your reach and increase the power you can exert upon an object. Pushing and pulling tools used in forcible entry include the K tool, A tool, J tool, and shove knife.

  - Lock tools and specialty tools—Used to disassemble or release the locking mechanism on a door. They include the duck-billed lock breaker, locking pliers and chain, and bam-bam tool.

- Every door contains four major components:
  - Door—The entryway itself
  - Door jamb—The upright or vertical sides of a window or door frame onto which a door is secured
  - Hardware—The handles, hinges, and other components
  - Locking mechanism
- Doors are generally constructed of wood, metal, or glass.
- Doors are classified by how they open:
  - Inward—Can be made of wood, steel, or glass and are found in most structures. The locking mechanisms range from standard doorknob locks to deadbolt locks or sliding latches.
  - Outward—Used in commercial occupancies and for most exits. The hinges are often exposed. Several types of locks, including handle-style locks and deadbolts, may be used with these doors.
  - Sliding—Constructed of tempered glass in a wooden or metal frame. A weak latch on the frame of the door secures the movable side.
  - Revolving—Made of four glass panels with metal frames. Generally secured by a standard cylinder lock or slide latch lock.
  - Overhead—Range from standard residential garage doors to high-security commercial roll-up doors. May be secured with cylinder-style locks, padlocks, or automatic garage door openers.
- Window construction includes glazed glass, regular or annealed glass, double/triple-pane glass, plate glass, laminated glass, and tempered glass.
- Window frame designs include the following types:
  - Double-hung windows—Made of wood or vinyl and two movable sashes.
  - Single-hung windows—Similar to double-hung windows except that the upper sash is fixed and only the lower sash moves.
  - Jalousie windows—Made of adjustable sections of tempered glass encased in a metal frame that overlap each other when closed.
  - Awning windows—Operate like jalousie windows except that awning windows are one large or multiple medium-sized glass panels. These windows open outward.
  - Horizontal-sliding windows—Similar to sliding doors.
  - Casement windows—Have a steel or wood frame and open away from the building with a crank mechanism.
  - Projected windows—Usually found in older warehouses and project either outward or inward on a top or bottom hinge.
- Locks range in sophistication from simple push-button locks to complex computer-operated locks.
- There are three major parts of a door lock:
  - Latching device—The part of the lock that "catches" and holds the door frame
  - Operator lever—The handle, doorknob, or keyway that turns the latch to lock it or unlock it
  - Deadbolt—A second, separate latch that locks and reinforces the regular latch
- Most padlocks, like most door locks, have similar parts:
  - Shackle—The U-shaped top of the lock that slides through a hasp and locks in the padlock itself
  - Unlocking mechanism—The keyway, combination wheel, or combination dial used to open the padlock
  - Lock body—The main part of the padlock that houses the locking mechanisms and the retention part of the lock

- Locks can be classified into four major categories:
  - Cylindrical locks—Most common fixed lock in use. One side of the door usually has a key-in-the-knob lock; the other side will have a keyway, a button, or another type of locking/unlocking mechanism.
  - Padlocks—Most common locks on the market. Available with a variety of unlocking mechanisms, including keyways, combination wheels, or combination dials.
  - Mortise locks—Designed to fit in predrilled openings and have both a latch and a deadbolt built into the same mechanism, each of which operates independently of the other.
  - Rim locks—Include deadbolts that can be surface mounted on the interior of the door frame. These locks have a bolt that extends at least 25 mm (1 in.) into the door frame.
- Electromagnetic and keyless door locks operate without a traditional metal key.
- On occasion, breaching a wall or floor may be necessary. Before breaching a wall, first consider whether the wall is load bearing. Removing or damaging this type of wall could cause the building or wall to collapse.
- The steps in systematic forcible entry are as follows:
  - Think.
  - Look for a lockbox.
  - Evaluate the situation.
  - Make entry with the least amount of damage possible.

KEY TERMS

Access Navigate for flashcards to test your key term knowledge.

Adze The blade or wedge part of a tool such as the Halligan tool.

Annealed The process of forming standard glass.

A tool A cutting tool with a pry bar built into the cutting part of the tool.

Awning windows Windows that have one large or multiple medium-size panels that do not overlap when they are closed. The window is operated by a hand crank from the corner of the window. The hinge is on the top.

Bam-bam tool A sliding hammer with a case-hardened screw, which is inserted, secured, and driven into the keyway of a lock to remove the keyway from the lock.

Battering ram A tool made of hardened steel with handles on the sides used to force doors and to breach walls. Larger versions may be used by as many as four people; smaller versions are made for one or two people.

Bolt cutter A cutting tool used to cut through thick metal objects, such as bolts, locks, and wire fences.

Case-hardened steel Steel created in a process that uses carbon and nitrogen to harden the outer core of a steel component, while the inner core remains soft. Case-hardened steel can be cut only with specialized tools.

Casement windows Windows in a steel or wood frame that open away from the building via a crank mechanism. These windows have a side hinge.

Cylindrical locks The most common fixed locks in use today. The locks and handles are placed into a predrilled hole in the door. The outside of the doorknob will usually have a key-in-the-knob lock; the inside will usually have a keyway, a button, or another type of locking/unlocking mechanism.

Deadbolt Surface- or interior-mounted lock on or in a door with a bolt that provides additional security.

Door An entryway; the primary choice for forcing entry into a vehicle or structure.

Door jamb The upright or vertical parts of a door frame onto which a door is secured.

Double-hung windows Windows that have two movable panels or sashes that can move up and down.

Double/triple-pane glass A window design that traps air or inert gas between two pieces of glass to help insulate a house.

Duck-billed lock breaker A tool with a point that can be inserted into the shackles of a padlock. As the point is driven farther into the lock, it gets larger and forces the shackles apart until they break.

Exterior wall A wall—often made of wood, brick, metal, or masonry—that makes up the outer perimeter of a building. Exterior walls are often load bearing.

Forcible entry Techniques used by fire personnel to gain entry into buildings, vehicles, aircraft, or other areas of confinement when normal means of entry are locked or blocked. (NFPA 402)

Fork The fork or claw end of a tool.

Glazing Glass or transparent or translucent plastic sheet used in windows, doors, skylights, or curtain walls. [ASCE/SEI 7:6.2] (NFPA 5000)

Halligan tool A prying tool that incorporates a sharp tapered pick, a blade (either an adze or wedge), and a fork or claw; it is specifically designed for use in the fire service.

Hardware The parts of a door or window that enable it to be locked or opened.

Hockey puck lock A type of padlock with hidden shackles that cannot be forced open through conventional methods.

Hollow-core door A door made of panels that are honeycombed inside, creating an inexpensive and lightweight design.

Horizontal-sliding windows Windows that slide open horizontally.

Interior wall A wall inside a building that divides a large space into smaller areas.

Irons A combination of tools, usually consisting of a Halligan tool and a flat-head axe, that is commonly used for forcible entry.

Jalousie windows Windows made of small slats of tempered glass, which overlap each other when the window is closed. Often found in trailers and mobile homes, jalousie windows are held together by a metal frame and operated by a small hand wheel or crank found in the corner of the window.

J tool A tool that is designed to fit between double doors equipped with push bars or panic bars.

K tool A tool that is used to remove lock cylinders from structural doors so the locking mechanism can be unlocked.

Laminated glass Safety glass. The lamination process places a thin layer of plastic between two layers of glass so that the glass does not shatter and fall apart when broken.

Latching device A spring-loaded latch bolt or a gravity-operated steel bar that, after release by physical action, returns to its operating position and automatically engages the strike plate when it is returned to the closed position. (NFPA 80)

Lock body The part of a padlock that holds the main locking mechanisms and secures the shackles.

Locking mechanism A standard doorknob lock, deadbolt lock, or sliding latch.

Mortise locks Door locks with both a latch and a bolt built into the same mechanism; the two locking mechanisms operate independently of each other. Mortise locks often are found in hotel rooms.

Operator lever The handle, doorknob, or keyway of a door that turns the latch to open it.

Padlocks One of the most common types of locks on the market today, portable locks built to provide regular-duty or heavy-duty service. Several types of locking mechanisms are available, including keyways, combination wheels, and combination dials.

Partition A nonstructural interior wall that spans horizontally or vertically from support to support. The supports may be the basic building frame, subsidiary structural members, or other portions of the partition system. [ASCE/SEI 7:11.2] (NFPA 5000)

Pick The pointed end of a tool, which can be used to make a hole or purchase point in a door, floor, or wall.

Plate glass A type of glass that has additional strength so it can be formed in larger sheets but will still shatter upon impact.

Projected windows Windows that project inward or outward on a top or bottom hinge; also called factory windows. They are usually found in older warehouses or commercial buildings.

Pry axe A specially designed hand axe that serves multiple purposes. Similar to a Halligan tool, it can be used to pry, cut, and force doors, windows, and many other types of objects. Also called a multipurpose axe.

Purchase point A small opening made to enable better tool access in forcible entry.

Reciprocating saw A saw that is powered by an electric motor or a battery motor and whose blade moves back and forth.

Rim locks Surface-mounted, interior locks located on or in a door with a bolt that provide additional security.

Rotary saw A saw that is powered by an electric motor or a gasoline engine and that uses a large rotating blade to cut through material. The blades can be changed depending on the material being cut.

Shackle The U-shaped part of a padlock that runs through a hasp and then is secured back into the lock body.

Shove knife A forcible entry tool used to trip the latch of outward swinging doors.

Solid-core door A door design that consists of wood filler pieces inside the door. This construction creates a stronger door that may be fire rated.

Tempered glass A type of safety glass that is heat treated so that it will break into small pieces that are not as dangerous.

Unlocking mechanism A keyway, combination wheel, or combination dial used to open a padlock.

On Scene

You are on the first-in pump for an alarm in a large industrial building. The company manufactures rubber belts and hose. It is early Sunday morning, so you are pretty sure this is going to be an accidental alarm trip. Dispatch advises that the key holder is en route with a 25-minute estimated time of arrival. There is no evidence of fire showing from the outside of the structure. Your captain has you set a ground ladder to a high first-floor window to take a look across the factory floor. From the ladder, you can see a thick haze inside.

1. The window you peer through is called a factory window or a _____ window.

  A. casement

  B. projected

  C. double-hung

  D. single-hung

2. The front door to the office area is a glass door. What type of glass would most likely be used in this door?

  A. Tempered

  B. Plate

  C. Double-pane

  D. Laminate

3. Which type of door is most commonly used as a front exit?

  A. Inward swinging

  B. Outward swinging

  C. Sliding

  D. Revolving

4. On a(n) _____ door, you can check the hinges to see if they can be disassembled or the pins removed before more destructive methods are used.

  A. inward-swinging

  B. outward-swinging

  C. sliding

  D. revolving

(continued)

On Scene Continued

5. Which tool is designed to fit between double doors equipped with panic bars so it can be forced open?

A. Halligan tool

B. Bam-bam tool

C. K tool

D. J tool

6. A _____ is a second, separate latch that locks and reinforces the regular latch.

A. deadbolt

B. jamb

C. latch

D. rabbet

7. Bolt cutters can be used to cut through all of the following *except*:

A. Padlocks

B. Bolts

C. Wire fences

D. Mortise locks

8. Which of the following is a combination of forcible entry tools called the irons?

A. Hydraulic pump and cutter used to cut security bars

B. J tool and K tool

C. Halligan tool and flat-head axe

D. Pick and pry axe

Access Navigate to find answers to this On Scene, along with other resources such as an audiobook.

Ladders

KNOWLEDGE OBJECTIVES

After studying this chapter, you will be able to:

- List and describe the parts of a ladder. (**NFPA 1001: 4.3.6**, pp. 373–375)
- Categorize the different types of ladders. (**NFPA 1001: 4.3.6**, pp. 376–380)
- Inspect ladders. (**NFPA 1001: 4.5.1**, pp. 380–381, 382–383)
- Maintain ladders. (**NFPA 1001: 4.5.1**, pp. 381, 382–383)
- Clean ladders. (**NFPA 1001: 4.5.1**, pp. 381–383)
- Describe when, where, and who performs service testing on ladders. (**NFPA 1001: 4.5.1**, pp. 383, 384)
- Specify the hazards associated with ladders. (**NFPA 1001: 4.3.6**, p. 384)
- Itemize the measures fire fighters should take to ensure safety when working with and on ladders. (**NFPA 1001: 4.3.6**, pp. 385–386)
- Cite the factors and guidelines used to select the appropriate ladder from the fire apparatus. (**NFPA 1001: 4.3.6**, pp. 387–388)
- Determine appropriate ladder placement for common fire-ground tasks. (**NFPA 1001: 4.3.6**, pp. 398, 399–400)
- Describe how to remove a ladder from the apparatus. (pp. 388–389)
- Describe how to lift ladders. (pp. 389–390)

SKILLS OBJECTIVES

After studying this chapter, you will be able to perform the following skills:

- Inspect, clean, and maintain a ladder. (**NFPA 1001: 4.5.1**, pp. 380–383)
- Carry a ladder using the one-fire fighter shoulder carry. (**NFPA 1001: 4.3.6**, pp. 390–391)
- Carry a ladder using the two-fire fighter shoulder carry. (**NFPA 1001: 4.3.6**, p. 391)
- Carry a ladder using the three-fire fighter shoulder carry. (**NFPA 1001: 4.3.6**, pp. 391, 392)
- Carry a ladder using the two-fire fighter suitcase carry. (**NFPA 1001: 4.3.6**, pp. 392, 393)
- Carry a ladder using the three-fire fighter suitcase carry. (**NFPA 1001: 4.3.6**, pp. 392, 394)
- Carry a ladder using the three-fire fighter flat carry. (**NFPA 1001: 4.3.6**, pp. 393, 395)
- Carry a ladder using the four-fire fighter flat carry. (**NFPA 1001: 4.3.6**, pp. 393, 395–396)
- Carry a ladder using the three-fire fighter flat-shoulder carry. (**NFPA 1001: 4.3.6**, pp. 396–397)
- Carry a ladder using the four-fire fighter flat-shoulder carry. (**NFPA 1001: 4.3.6**, pp. 398, 399)
- Raise a ladder using the one-fire fighter flat raise for ladders less than 14 feet. (**NFPA 1001: 4.3.6**, p. 401)

- Raise a ladder using the one-fire fighter flat raise for ladders more than 14 feet. (**NFPA 1001: 4.3.6**, pp. 402–403)
- Tie the halyard. (**NFPA 1001: 4.3.6**, pp. 402, 403, 404)
- Raise a ladder using the two-fire fighter beam raise. (**NFPA 1001: 4.3.6**, pp. 404–406)
- Raise a ladder using the two-fire fighter flat raise. (**NFPA 1001: 4.3.6**, pp. 407–408)
- Raise a ladder using the three-fire fighter flat raise. (**NFPA 1001: 4.3.6**, pp. 407, 409–410)
- Raise a ladder using the four-fire fighter flat raise. (**NFPA 1001: 4.3.6**, pp. 407, 411–412)
- Climb a ladder. (**NFPA 1001: 4.3.6**, pp. 413–414)
- Use a leg lock to work from a ladder. (**NFPA 1001: 4.3.6**, p. 416)

- Deploy a roof ladder. (**NFPA 1001: 4.3.6, 4.3.12**, pp. 417–418)
- Inspect a chimney. (pp. 418–419)

ADDITIONAL STANDARDS

- **NFPA 1901**, *Standard for Automotive Fire Apparatus*
- **NFPA 1931**, *Standard for Manufacturer's Design of Fire Department Ground Ladders*
- **NFPA 1932**, *Standard on Use, Maintenance, and Service Testing of In-Service Fire Department Ground Ladders*
- **NFPA 1983**, *Standard on Life Safety Rope and Equipment for Emergency Services*

Fire Alarm

Along with your crew, you find yourself at the scene of a fast-moving fire in an apartment building. When your crew reports to the command post, the incident commander (IC) tells your captain that a ladder company was sent inside to conduct a primary search so he needs your crew to set ground ladders to the building. The IC tells you that he wants you and the other fire fighters to ladder the first, second, and third floors of the B and D sides of the building while he does a 360-degree walk-around. You head to the ladder truck, quickly making estimates of the ladders you will need.

1. Which types of ladders will work best for this situation?

2. How would you estimate the ladder lengths you will need?

3. What concerns do you have when you are laddering the building?

 **Access Navigate for more practice activities.**

Introduction

Fire department ground ladders are among the most useful, versatile, durable, easy-to-use, and rapidly deployable tools used by fire fighters. Portable ladders are carried on nearly every piece of structural firefighting apparatus and serve a wide variety of functions for firefighting and for other rescue situations (**FIGURE 11-1**). Every fire fighter must be proficient in the basic skills of working with ladders and physically capable of performing a variety of ladder evolutions.

Primary Uses of Ground Ladders

Ladders can be used to gain access to a window or to the roof for purposes of search and rescue, ventilation, fire suppression, and overhaul. They also can be raised to a window or to the roof as an emergency exit for fire fighters or trapped victims. They can provide access to and egress from a cockloft or an attic. Ladders also can be used to provide a safe pathway between floors, enabling fire fighters to avoid a damaged or unsafe stairway. Roof ladders can be used to provide stable

FIGURE 11-1 Ladders can provide access to or exit from a structure.
© Jones & Bartlett Learning.

footing and distribute the weight of fire fighters during roof operations. Ladders can be used to support a hose line over an opening. They also can be deployed across a small opening to create a bridge. They can be placed to enable a fire fighter to climb over a fence or obstruction. In certain cases the best escape route from a fire may be to travel horizontally across a ladder used as a bridge to a nearby building. During salvage operations, a ladder covered with a waterproof tarp can be used as a chute to funnel water out of a building.

Other Uses of Ground Ladders

Ladders have a wide variety of uses in non-firefighting situations. A ladder may be used to reach an injured person in a ravine or down a steep highway embankment. In trench rescue situations, fire fighters can use ladders to access a victim or to aid in bringing him or her to the surface. They can also be used to provide access or escape from other below-grade locations such as a manhole. A ladder A-frame (an A-shaped structure formed with two ladder sections) can also be used to lower rescuers into a trench or manhole or to raise or lower victims safely. A specialized rescue team can use a ladder to provide access to and exit from buses, railcars, or airplanes. Ladders can be used for a variety of functions during a hazardous materials incident. In addition, ladders can be used as ramps to assist in moving equipment during emergencies.

Ladder Construction and Components

Ladders must meet the performance requirements of NFPA 1931, *Standard for Manufacturer's Design of Fire Department Ground Ladders*. The standard requires manufacturers to adhere to basic requirements such

as specifications for the diameter and spacing of rungs; the inclusion of specific labels and markings, including ladder positioning stickers, heat sensor labels, and electrocution hazard warning labels; and length and width requirements. Additional specific requirements are outlined for the various types of ladders.

Most commonly used ladders are constructed of one of three materials: metal, wood, or fibreglass. Each material has advantages and disadvantages. The vast majority of fire service ladders are constructed of metal, primarily aluminum. Metal ladders are relatively light in weight, inexpensive, and easy to repair. Disadvantages include electrical conductivity and failure when exposed to high heat. A few departments use wooden ladders. Wood ladders are less likely to conduct electricity, retain more strength when exposed to heat, and are more flexible than metal ladders. Disadvantages may include a higher cost, more maintenance, and heavier weight. Fibreglass is less frequently used in the fire service. Similar to wood ladders, fibreglass ladders are less likely to conduct electricity and have increased strength. However, these ladders have an increased chance of failure if overloaded and may be flammable if exposed to flames. It is important to follow the specific manufacturer's instructions and cautions for the use of your ladders.

In its most basic design, a ladder consists of two beams connected by a series of parallel rungs. Fire service ladders, however, are specialized tools with several different parts. To use and maintain ladders properly, fire fighters must be familiar with the various types of ladders, as well as with the different parts and terms used to describe them.

Basic Ladder Components

The basic components of a ladder are discussed next (**FIGURE 11-2**).

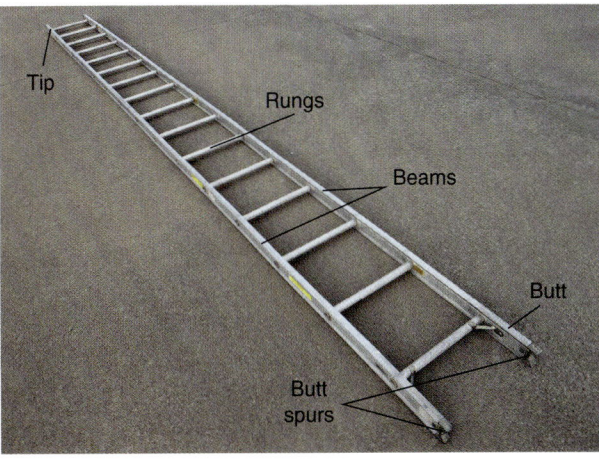

FIGURE 11-2 The basic components of a ladder.
Courtesy of David Schottke.

Beams

A **beam** is one of the two main structural components that runs the entire length of most ladders or ladder sections. The beams support the rungs and carry the load of a person.

Three basic types of ladder beam construction are used (**FIGURE 11-3**):

- **Trussed beam**: A trussed beam ladder has a top rail and a bottom **rail**, which are joined by a series of smaller pieces called **truss blocks**. The rungs are attached to the truss blocks. Trussed beams are usually constructed of aluminum or wood. This construction is often used for longer extension ladders. The term "rail" can also be used to refer to the top and bottom surfaces of an I-beam.
- **I-beam**: An I-beam ladder has thick sections at the top and the bottom of the beams, which are connected by a thinner section. The rungs are attached to the thinner section of the beam. This type of beam is usually made from fibreglass.
- **Solid beam**: A solid beam ladder has a simple rectangular cross-section. Many wooden ladders have solid beams. Rectangular aluminum beams, which are usually hollow or C shaped, are also classified as solid beams.

Rung

A **rung** is a crosspiece that spans the two beams of a ladder. The rungs serve as steps and transfer the weight of the user to the beams. Most ground ladders used by fire departments have aluminum rungs, but wooden ladders are still constructed with wood rungs. Per NFPA 1931, the surface area of rungs is required to be skid resistant (NFPA 1931, 2015).

Tip

The **tip** is the very top of the ladder.

Butt

The **butt** is the end of the ladder that is placed against the ground when the ladder is raised. It is sometimes called the heel or base.

Butt Spurs

Butt spurs are metal spikes that are attached to the butt of a ladder. They prevent the butt from slipping out of position.

Butt Plate

A **butt plate** is an alternative to a simple butt spur. This swiveling plate is attached to the butt of the ladder and incorporates both a spur and a cleat or pad.

Roof Hooks

Roof hooks are spring-loaded, retractable, curved metal pieces

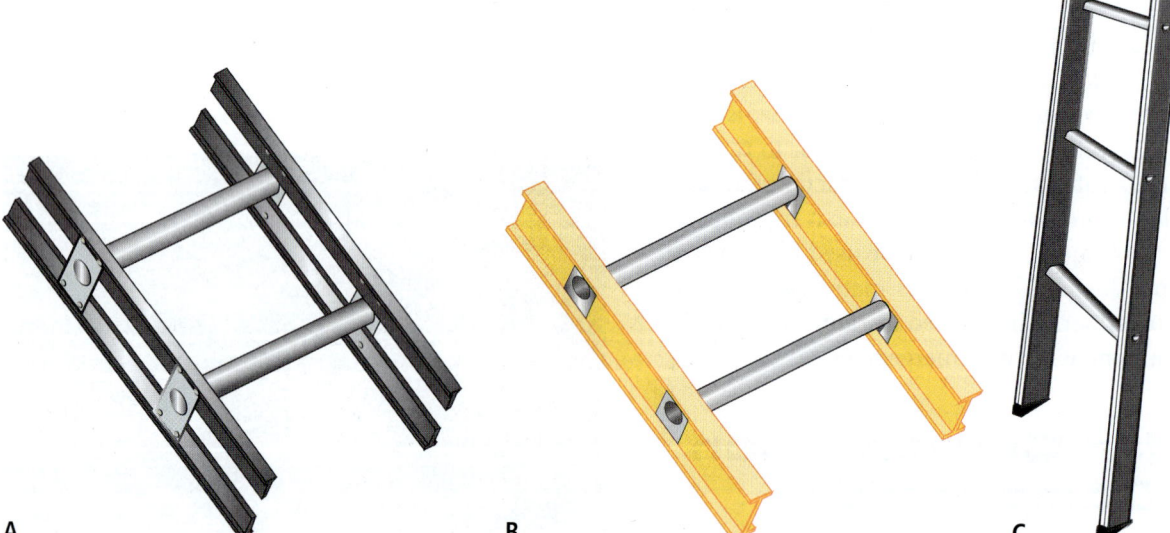

A   **B**   **C**

FIGURE 11-3 Three types of beam construction are found in ladders: **A.** Trussed beam. **B.** I-beam. **C.** Solid beam.

© Jones & Bartlett Learning.

that are attached to the tip of a roof ladder. These hooks are used to secure the tip of the ladder to the peak of a pitched roof.

Heat Sensor Label

A **heat sensor label** identifies when the ladder has been exposed to a specific amount of heat conditions that could damage its structural integrity; such a label changes colours when it is exposed to a particular temperature (**FIGURE 11-4**).

Protection Plates

Protection plates are reinforcing pieces that are placed on a ladder at chafing and contact points to prevent damage from friction or contact with other surfaces.

Tie Rod

A **tie rod** is a metal bar under each rung that runs from one beam of the ladder to the other. The tie rods help to keep the beams from separating from the rungs. Tie rods are typically found in wooden ladders.

Extension Ladder Components

An extension ladder is an assembly of two or more ladder sections that fit together and can be extended or retracted to adjust the ladder's length. Extension ladders have additional parts (**FIGURE 11-5**) that are not found on straight ladders.

FIGURE 11-4 The heat sensor label indicates if the ladder has been exposed to excessive heat. All metal and fibreglass ground ladders include heat sensors that are preset for 149°C (300°F) ± 5 percent and include an expiration date.
© Jones & Bartlett Learning. Photographed by Glen E. Ellman.

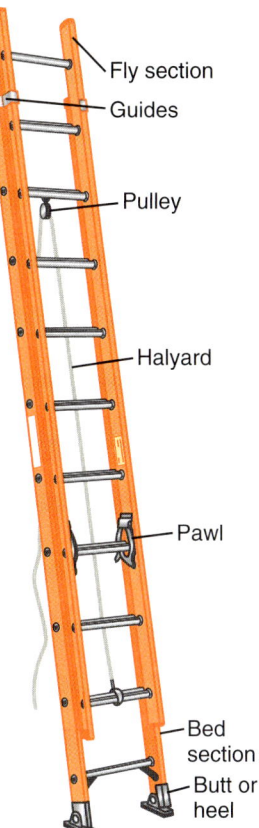

FIGURE 11-5 Extension ladders have additional parts that are not found in straight ladders.
© Jones & Bartlett Learning.

Bed (Base) Section

The **bed (base) section** is the widest section of an extension ladder. It serves as the base; all other sections are raised from the bed section. The bottom of the bed section rests on the supporting surface.

Fly Section

A **fly section** is the part of an extension ladder that is raised or extended from the bed section. Extension ladders often have more than one fly section, each of which extends from the previous section.

Guides

Guides are strips of metal or wood that guide a fly section of an extension ladder as it is being extended. Channels or slots in the bed or fly section may also serve as guides.

Halyard

The **halyard** is the rope or cable used to extend or hoist the fly sections of an extension ladder. The halyard runs through the pulley.

Pawls

Pawls are the mechanical locking devices that are used to secure and hold the extended fly sections of an extension ladder in place. They are sometimes called ladder locks, dogs, or rung locks.

Pulley

The **pulley** is a small grooved wheel that is used to change the direction of the halyard pull. A downward pull on the halyard creates an upward force on the fly sections, extending the ladder.

Stops

Stops are pieces of wood or metal that prevent the fly sections of a ladder from overextending and collapsing the ladder. They are also referred to as stop blocks.

Staypole

Staypoles (also called tormentor poles) are long metal poles that are attached to the top of the bed section and that help stabilize an extension ladder as it is being raised and lowered. Staypoles are required for ladders that are 12 m (40 ft) or longer; however, they can be found on some 11-m (35-ft) ladders. One pole is attached to each beam of an extension ladder with a swivel joint. Each pole has a spur on the other end. Ladders with staypoles are typically referred to as Bangor ladders.

Types of Ladders

Ladders used in the fire service can be classified into two broad categories: aerial and ground. **Aerial ladders** are permanently mounted and operated from a piece of motorized aerial fire apparatus. Ground ladders are carried on fire apparatus but are designed to be removed and used in other locations.

Aerial Apparatus

Aerial ladders vary greatly in terms of their design and function. This discussion is limited to a brief overview of the basic styles of aerial fire apparatus as it relates to ladders. Detailed requirements for aerial fire apparatus are documented in NFPA 1901, *Standard for Automotive Fire Apparatus*. This standard sets minimum performance requirements for fire apparatus. Aerial fire apparatus may be classified as aerial ladders, elevated platforms, and water towers (**FIGURE 11-6**).

A

B

FIGURE 11-6 Aerial apparatus. **A.** Aerial ladder. **B.** Elevated platform.
© Jones & Bartlett Learning. Photographed by Glen E. Ellman.

Aerial ladders are permanently mounted, power-operated ladders with a working length of between 15 m (50 ft) and 42 m (137 ft). The apparatus that aerial ladders and elevated platforms are mounted to may be referred to as ladder companies, tower ladder companies, ladder trucks, trucks, or aerial ladders. Aluminum or steel aerial ladders contain two or more sections that telescope into one another. The most common construction of an aerial ladder is truss construction (**FIGURE 11-7**). Most aerial ladders have a permanently mounted waterway and monitor nozzle.

Those built on a straight **chassis** are often referred to as straight-stick aerials; those built on a tractor-trailer chassis are called tillered aerials or tiller trucks. A tillered ladder truck requires a second operator at the back of the trailer to steer the back of the trailer. Tillered ladder

CHAPTER 11 Ladders   **377**

FIGURE 11-7 A steel aerial ladder.
© Jones & Bartlett Learning. Photographed by Glen E. Ellman.

trucks can be maneuvered into tight places that straight chassis trucks cannot reach.

The **elevated platform** apparatus includes a passenger-carrying platform (bucket) attached to the tip of a boom or ladder. The boom may or may not serve as a ladder, although the ladder-type boom is most common. The ladder-type boom allows for continuous access to the platform. An elevated platform that is 33.5 m (110 ft) or less in length must have a prepiped waterway and a permanently mounted monitor nozzle. Some aerial ladder apparatus are equipped with a fire pump and hose; other aerial apparatus are constructed without a fire pump.

Water towers have two or more booms that may telescope or articulate. They are used to deliver elevated streams of water and contain high capacity prepiped waterways. Water towers do not have the capacity to carry fire fighters or rescue victims.

Ground Ladders

Most fire apparatus carry ground ladders. Ground ladders are designed to be removed from the apparatus and used in different locations. They may take the form of either general-purpose ladders, such as straight or extension ladders, or specialized ladders such as folding ladders (attic ladders) or roof ladders.

Fire department pumpers or engines are required to carry at least one straight ladder with roof hooks, one extension ladder, and one folding ladder. While minimum length specifications are not included in NFPA 1901, most fire department pumpers carry 7- or 8-m (24- or 28-ft) ladders, which can reach the roof of a typical two-storey building.

Ladder companies are required to carry a total of 35 m (115 ft) of ground ladders. They must carry at least two straight ladders with roof hooks, two extension ladders, and one folding ladder. These ladder trucks usually carry 11- or 12-m (35- or 40-ft) ladders, which can reach the roofs of most three-storey buildings.

Initial attack fire apparatus, also referred to as quick attack apparatus, is required to have one combination or extension-type ladder that is 3.7 m (12 ft) or longer.

Quint fire apparatus has a permanently mounted fire pump, a water tank, a hose storage area, an aerial ladder or elevated platform with a permanently mounted waterway, and a complement of ground ladders. The quint carries a minimum total complement of 26 m (85 ft) of ground ladders including one extension ladder, one straight ladder equipped with roof hooks, and one folding ladder.

Generally, fire service ground ladders are limited to a maximum length of 15 m (50 ft). If a building's height exceeds the capabilities of these ladders, aerial apparatus must be used.

> **FIRE FIGHTER TIP**
>
> Fire service ground ladders should be constructed and certified as compliant with the most recent edition of NFPA 1931, *Standard for Manufacturer's Design of Fire Department Ground Ladders*.

Straight Ladder

A straight ladder is a single-section, fixed-length ground ladder. Straight ladders may also be called wall ladders or single ladders. These lightweight ladders can be raised quickly, and they can reach the windows and roofs of one- and two-storey structures. Straight ladders are commonly 3.7 to 4 m (12 to 14 ft) long but can be as much as 6 m (20 ft) long.

Roof Ladder

A **roof ladder** (sometimes called a hook ladder) is a straight ground ladder that is equipped with retractable roof hooks at one end. The roof hooks secure the tip of the ladder to the peak of a pitched roof, when the ladder lies flat on the roof. A roof ladder provides stable footing and distributes the weight of fire fighters and their equipment, thereby helping to reduce the risk of structural failure in the roof assembly. Roof ladders also can be used on flat roofs as a work platform to distribute weight. The roof ladders used in the fire service are usually 4 to 5.5 m (12 to 18 ft) long (**FIGURE 11-8**). In addition to being used on

FIGURE 11-8 The roof ladders used in the fire service are commonly 12 to 18 ft (3.7 to 5 m) long.
© Jones & Bartlett Learning. Photographed by Glen E. Ellman.

roofs, roof ladders can be used in situations where a straight ladder is needed.

> **LISTEN UP!**
>
> Roof ladders are not free-hanging ladders. The roof hooks will not support the full weight of the ladder and anyone on it when the ladder is in a vertical position.

Extension Ladder

An **extension ladder** is an adjustable-length ground ladder with two or more sections that fit together. The bed (base) section supports one or more fly sections. Pulling on the halyard extends the fly sections along a system of guides, or channels or slots, in the bed or fly section (**FIGURE 11-9**). The fly sections, in turn, lock in place at the pawls so that the rungs are aligned to facilitate climbing.

An extension ladder is usually heavier than a straight ladder of the same length and normally requires more than one person to set it up. Because the length is adjustable, an extension ladder can replace several straight ladders, and it can be stored in places where a longer straight ladder would not fit.

Bangor Ladder

Bangor ladders are extension ladders with staypoles that offer added stability during raising, lowering, and climbing operations. As discussed, staypoles are required on ladders of 12 m (40 ft) or greater length but can be found on some 11-m (35-ft) ladders. The poles help keep these heavy ladders under control while fire fighters are maneuvering them into place

FIGURE 11-9 An extension ladder can be used at any length, from fully retracted to fully extended.
© Jones & Bartlett Learning. Photographed by Glen E. Ellman.

(**FIGURE 11-10**). When the ladder is positioned correctly, the staypoles are planted in the ground on either side for additional stability.

Combination Ladder

A **combination ladder** can be used as a stepladder (A-frame) or an extension ladder. Such ladders are convenient for indoor use and for maneuvering in tight spaces. They are generally 2 to 3 m (6 to 10 ft) in length in the A-frame configuration and 3 to 5 m (10 to 15 ft) in length in the extension configuration. A multipurpose ladder can be used as a step ladder or a straight ladder (**FIGURE 11-11**).

Folding Ladder

A **folding ladder** (also called an attic ladder) is a narrow, collapsible ladder that is designed to allow access to attic scuttle holes and confined areas (**FIGURE 11-12**). The two beams fold in to enhance the ladder's portability. Folding ladders are commonly available in 2.4- to 4-m (8- to 14-ft) lengths.

FIGURE 11-10 Staypoles are used to stabilize a Bangor ladder while it is being raised or lowered.
Courtesy of Duo-Safety Ladder Corporation.

FIGURE 11-11 Multipurpose ladders can be used in several different configurations.
© Jones & Bartlett Learning.

FIGURE 11-12 Folding ladders are designed to be used in narrow spaces or restrictive passages. This photo shows a folding ladder in the portable position.
© Jones & Bartlett Learning. Photographed by Glen E. Ellman.

Fresno Ladder

A **Fresno ladder** is a narrow, two-section extension ladder that is designed to provide attic access. The Fresno ladder is generally short—just 3 to 4 m (10 to 14 ft)—so it has no halyard; it is extended manually (**FIGURE 11-13**). A Fresno ladder can be used in tight space applications, such as bridging over a damaged section of an interior stairway.

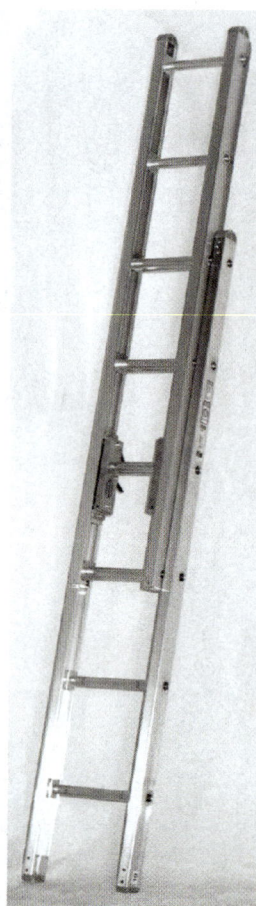

FIGURE 11-13 A Fresno ladder is extended manually.
Courtesy of Duo-Safety Ladder Corporation.

Inspection, Maintenance, and Service Testing of Ground Ladders

Ground ladders used by the fire service must be able to withstand extreme conditions. They can be dropped, overloaded, or exposed to temperature extremes during use. As discussed previously, NFPA 1931, *Standard for Manufacturer's Design of Fire Department Ground Ladders*, establishes requirements for the construction of new ground ladders. These requirements are based on probable conditions during emergency operations.

NFPA 1932, *Standard on Use, Maintenance, and Service Testing of In-Service Fire Department Ground Ladders*, provides general guidance for the use of ground ladders. Ladders must be regularly inspected, maintained, and service tested, following the NFPA standard. In addition, ladders should always be inspected and maintained in accordance with the manufacturer's recommendations.

Inspection

Ladders must be visually inspected at least monthly. A ladder should also be inspected after each use. A general visual inspection should, at a minimum, address the following:

- Beams: Check the beams for cracks, splintering, breaks, gouges, checks, wavy conditions, or deformations.
- Rungs: Check all rungs for snugness, tightness, punctures, wavy conditions, worn serrations, splintering, breaks, gouges, checks, or deformations.
- Halyard: Check the halyard for fraying or kinking; ensure that it moves smoothly through the pulleys.
- Wire-rope halyard extensions: Check the wire-rope halyard extensions on three- and four-section ladders for snugness. This check should be performed when the fly section is fully bedded to ensure that the upper sections will align properly during operation.
- Ladder guides: Check the guides for chafing. Also check for adequate wax, if the manufacturer requires wax.
- Pawls: Check the pawls for proper operation.
- Butt spurs: Check the butt spurs for excessive wear or other defects.
- Heat sensors: Check the heat sensor labels to see whether the sensors indicate that the ladder has been exposed to excessive heat.
- Bolts and rivets: Check all bolts and rivets for tightness; bolts on wood ladders should be snug and tight without crushing the wood.
- Welds: Check all welds on metal ladders for cracks or apparent defects.
- Roof hooks: Check the roof hooks for sharpness and proper operation.
- Metal surfaces: Check metal surfaces for signs of surface corrosion.
- Fibreglass and wood surfaces: Check fibreglass ladders for loss of gloss on the beams. Check for damage to the varnish finish on wooden ground ladders. Check for signs of wood rot—that is, dark, soft spots.

If the inspection reveals any deficiencies, the ladder must be removed from service until repairs are made. Properly trained fire fighters can often perform minor repairs that require simple maintenance at the fire station. By contrast, only qualified personnel at a properly equipped repair facility should perform repairs involving the structural or mechanical components of a ladder.

Ladders that have been exposed to excessive heat must be removed from service and pass a service test before being used again. A ladder that fails the structural stability test should be permanently removed from service.

Maintenance

All fire fighters should be able to perform routine ladder maintenance. Maintenance is simply the regular process of keeping the ladder in proper operating condition. Fundamental maintenance tasks include the following:

- Clean and lubricate the pawls, following the manufacturer's instructions (**FIGURE 11-14**).
- Clean and lubricate the guides on extension ladders in accordance with the manufacturer's recommendations.
- Replace worn halyards and wire rope halyard extensions on extension ladders when they become frayed or kinked (**FIGURE 11-15**).
- Clean and lubricate roof hooks. Remove rust and other contaminants, and lubricate the folding roof hook assemblies on roof ladders to keep them operational (**FIGURE 11-16**).
- Check the heat sensor labels. Replace the heat sensor labels when they reach their expiration date. When a ladder has been exposed to high temperatures, remove it from service until it has passed a service test.

FIGURE 11-14 Extension ladder pawls must operate smoothly.
© Jones & Bartlett Learning. Photographed by Glen E. Ellman.

FIGURE 11-15 Replace the halyard if it is worn or damaged.
© Jones & Bartlett Learning. Photographed by Glen E. Ellman.

FIGURE 11-16 Roof hooks must operate smoothly.
© Jones & Bartlett Learning.

- Maintain the finish on fibreglass and wooden ladders in accordance with the manufacturer's recommendations.
- Ensure that ladders are *not* painted except for the top and bottom 0.5 m (18 in.) of each section, because paint can hide structural defects in the ladder. The tip and butt are painted for purposes of identification and visibility.
- Keep records of all maintenance.

Ladders that are in storage should be placed on racks or in brackets and protected from the weather. Fibreglass ladders can be damaged by prolonged exposure to direct sunlight (**FIGURE 11-17**).

Cleaning

Ladders must be cleaned regularly to remove any road grime and dirt that have built up on the apparatus during storage. Ladders should be cleaned before each inspection to ensure that any hidden faults can be observed. They should also be cleaned after each use to remove dirt and debris.

FIGURE 11-17 Ladders should be stored on racks or in brackets, out of the weather or direct sunlight.
© Jones & Bartlett Learning.

Use a soft-bristle brush and water to clean ladders. A mild, diluted detergent may be used, if allowed by the manufacturer's recommendations. Remove any tar, oil, or grease deposits with a safety solvent as recommended by the manufacturer.

Rinse and dry the cleaned ladder before placing it back on the apparatus.

To inspect, clean, maintain, and store ladders, follow the steps in **SKILL DRILL 11-1**.

SAFETY TIP

Do not get any solvent on the halyard of an extension ladder. Contact with solvents can damage halyard ropes.

SKILL DRILL 11-1
Inspect, Clean, and Maintain a Ladder Fire Fighter I, NFPA 1001: 4.5.1

1 Clean all components following manufacturer and national standards. Visually inspect the ladder for wear and damage.

2 Lubricate the ladder pawls, guides, and pulleys using the recommended material. Ensure you do not get lubricant on the halyard or rungs of the ladder.

SKILL DRILL 11-1 Continued
Inspect, Clean, and Maintain a Ladder Fire Fighter I, NFPA 1001: 4.5.1

3   Perform a functional check of all components.

4   Complete the maintenance record for the ladder. Tag and remove the ladder from service if deficiencies are found. Return the ladder to the apparatus or storage area.

© Jones & Bartlett Learning. Photographed by Glen E. Ellman.

Service Testing

Service testing is performed periodically to evaluate the continued usefulness of a ladder during its life. Service tests should be performed on both new ladders and ladders that have been in use for some time. Such tests measure the structural integrity of a ladder, ensuring that ladders are safe to use for their intended purposes. Service testing of ground ladders must follow NFPA 1932.

Service tests should be conducted before a new ladder is used and annually while the ladder remains in service. A ladder that has been exposed to extreme heat, has been overloaded, has been impact or shock loaded, is visibly damaged, or is suspected of being unsafe for any other reason must be removed from service until it has passed a service test. A repaired ladder must also undergo a service test before it can be returned to service, unless a halyard replacement was the only repair.

One part of the ladder service test is the horizontal-bending test, which evaluates the structural strength of a ladder. To perform this test, the ladder is placed in a horizontal position across a set of supports. A weight is then put on the ladder, and the amount of deflection or bending caused by the weight applied is measured to evaluate the ladder's strength (**FIGURE 11-18**). Additional tests are performed on the extension hardware of extension ladders. The hooks on roof ladders are tested as well.

Service testing of ladders requires special training and equipment and must be conducted only by qualified personnel. Many fire departments use outside contractors to perform these tests. The results of all service tests must be recorded and kept for future reference.

FIGURE 11-18 A horizontal-bending test evaluates the structural strength of a ladder.
Courtesy of UL.

Ladder Safety

Several potential hazards are associated with ladder use, but unfortunately these risks are easily overlooked during emergency operations. As a consequence, many fire fighters have been seriously injured or killed in ladder accidents, both on the fire ground and during training sessions.

Ladders must be used with caution; follow standard operating procedures and regularly reinforce your skills through training.

In any firefighting operation, personal protective equipment (PPE) is essential for working with ladders. Protective clothing provides protection from mechanical injuries as well as from fire and heat injuries. Helmets, coats, pants, gloves, and protective footwear provide protection from falling debris, impact injuries, and pinch injuries. Fire fighters must be able to work with and on ladders while wearing self-contained breathing apparatus (SCBA). You should practise working with ladders while wearing all of your PPE, including helmets, gloves, and eye protection.

Many safety precautions should be followed from the time a ladder is removed from the apparatus until it is returned to the apparatus. Basic safety issues include the following concerns:

- Lifting and moving ladders
- Placement of ground ladders
- Working on a ladder
- Rescue operations
- Ladder damage

These issues are discussed here and will be revisited in the discussions of using specific types of ladders.

Lifting and Moving Ladders

Teamwork is essential when working with ladders. Some ladders are heavy and can be very awkward to maneuver, particularly when they are extended. Crew members must use proper lifting techniques and coordinate all movements.

Never attempt to lift or move a ladder that weighs more than you are capable of lifting safely. Because of their shape, ladders may be awkward to carry, especially over uneven terrain, through gates, or over snow and ice. It is better to ask for help in moving a ladder than to risk injury or delay the placement of the ladder.

SAFETY TIP

When raising or lowering ground ladders, personnel must ensure they look up at the tip of the ladder regularly. This practice will minimize the chance of ladder contact with overhead obstructions.

Placement of Ground Ladders

Fire fighters working on the fire ground should survey the area where a ladder will be used before placing the ladder or beginning to raise it. If possible, inspect the area before retrieving the ladder from the apparatus. If you note any hazards, consider changing the position of the ladder. Work in the safest available location.

Sometimes a ladder must be placed in a potentially hazardous location. If you are aware of the hazard, you can take corrective actions and implement special precautions to avoid accidents.

The most important check assesses the location of overhead utility lines. If a ladder comes in contact with or close to an electrical wire, the fire fighters who are handling it can be electrocuted. If the line falls, it can electrocute other fire fighters as well. Avoid placing a ladder against a surface that has been energized by a damaged or fallen power line.

Do not assume that it is safe to use wood or fibreglass ladders around power lines. Metal ladders will certainly conduct electricity, but a wet or dirty wood or fibreglass ladder can also conduct electricity.

If possible, do not raise ladders anywhere near overhead wires. At a minimum, keep ladders at least 3 m (10 ft) away from any power lines while using or raising them. If a power line is nearby, make sure that enough fire fighters are present to keep the ladder under control as it is raised. One fire fighter should watch to make sure that the ladder does not come too close to the line. If the ladder falls into a power line, the results can be deadly.

Other types of overhead obstructions also may be hazards during a ladder raise. If the ladder hits something as it is being raised, the weight and momentum of the ladder will shift, which can cause fire fighters to lose control and drop the ladder.

Ladders must be placed on stable and relatively level surfaces (**FIGURE 11-19**). The shifting weight of a

FIGURE 11-19 Ladders must be placed on stable and relatively level surfaces.
© Jones & Bartlett Learning. Photographed by Glen E. Ellman.

climber can easily tip a ladder placed on a slope or unstable surface. Always evaluate the stability of the surface before placing the ladder, and check it again during the course of the firefighting operations. A ladder placed on snow or mud, for example, may become unstable as the snow melts or as the base of the ladder sinks into the mud. In addition, water from the firefighting operations can make soft ground even more unstable.

It is also important to ensure that the tip of the ladder is placed against a stable surface. Placing the tip of a ladder against an unstable surface can result in the ladder falling forward into a fire or being pushed backward. The failure of an unstable wall, for example, can occur during firefighting operations or during overhaul. Avoid injury or death—do not place the tip of a ladder against any unstable surface.

Do not place the ladder where it will be exposed to direct flames or extreme temperatures, because excessive heat can cause permanent damage to or catastrophic failure of such equipment. Check the heat sensor labels immediately after a ladder has been exposed to high temperatures.

SAFETY TIP

Always maintain at least 3 m (10 ft) of clearance between a ladder and utility lines to prevent electrocution.

Working on a Ladder

Before climbing a ladder, make sure it is set at the proper climbing angle (approximately 75 degrees) for

maximum load capacity and safety. To accomplish this, position the base of the ladder out from the building at a distance equal to one-fourth of the ladder's height (**FIGURE 11-20**). For example, if a ladder is being raised to a window 6 m (20 ft) above the ground, the base of the ladder should be placed 1.5 m (5 ft) from the building. Achieving a proper angle is important to your safety and the safety of people being rescued.

When using an extension ladder, ensure that the pawls are locked and the halyard tied before anyone climbs the ladder.

To prevent slipping, another fire fighter must secure the base of the ladder by heeling (footing/butting) it (**FIGURE 11-21**). This technique uses the fire fighter's weight to keep the base of the ladder from slipping. The base of the ladder can also be mechanically secured to a solid object. An alternative to

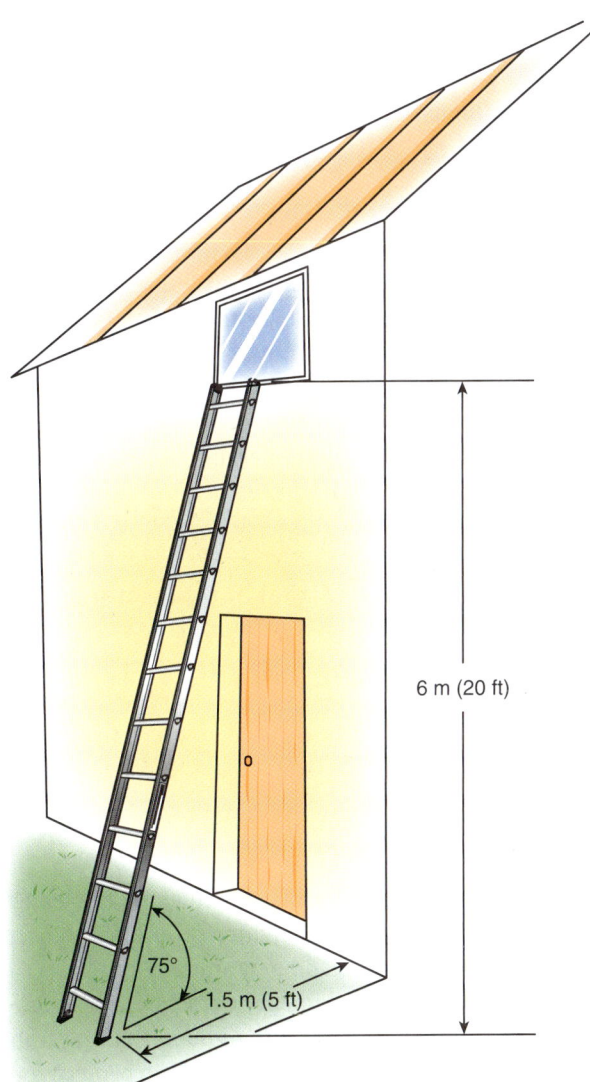

FIGURE 11-20 Place the base of a ladder out from the building at a distance equal to one-fourth of the ladder's height to achieve a climbing angle of 75 degrees.
© Jones & Bartlett Learning.

FIGURE 11-21 To prevent a ladder from slipping, fire fighters should secure the base by heeling (butting) the ladder.

© Jones & Bartlett Learning. Photographed by Glen E. Ellman.

securing the base is to use a rope or strap to secure the tip of the ladder to the building.

All ladders have weight limits. Most fire department ground ladders are designed to support a weight of 340 kg (750 lb) (NFPA 1932, 2015). In a rescue situation, this weight limit is equivalent to one victim and two fire fighters wearing full protective clothing and equipment. It would also accommodate two fire fighters climbing or working on the ladder along with their equipment. In contrast, ground ladders designed for construction and home use may be rated for as little as 102 kg (225 lb). Given the potential difference in ratings, it is important to avoid using unapproved ladders for fire department operations.

> ### LISTEN UP!
>
> Many ladders used for construction work and home repair are rated for a working load of only 102 to 136 kg (225 to 300 lb)—significantly less than the 340-kg (750-lb) rating of approved fire department ladders. Remember the difference between these two types of ladders. If you encounter a situation where a construction ladder is already present, you still need to set up and use a fire department–approved ladder to avoid the possibility of a catastrophic ladder failure.

The weight should be distributed along the length of the ladder. Only one fire fighter should be on each section of an extension ladder at any time.

While climbing the ladder, be prepared for falling debris, misguided hose streams, or people falling from the building. Your bunker gear will help to protect you.

If fire conditions change rapidly while you are working on a ladder, you must be prepared to climb down quickly. For example, if the fire suddenly flashes over or if flames break through a window located near

the ladder, you must be prepared to move quickly out of danger. Your gear will not protect you from direct exposure to flames for more than a few seconds.

A fire fighter who is working from a ladder is in a less stable position than one who is working on the ground. While on a ladder, you must constantly adjust your balance, especially when swinging a hand tool or reaching for a trapped occupant. There is always the danger of falling as well as a risk to people below if something falls or drops. To minimize your risk of falling, secure yourself to the ladder. Methods for securing yourself to a ladder are discussed in detail later.

Rescue

Ladders can be an essential tool for rescue. For example, ground ladders are often used to reach and remove trapped occupants from the upper storeys of a building. When positioning a ladder for rescue, place the tip of the ladder at or slightly below the windowsill. This position provides a direct pathway for the fire fighter entering the window and a direct route for exiting the structure. The ladder should not occlude any part of the window opening. Fire fighters need to address several important safety concerns before going up a ladder to rescue someone.

A person who is in extreme danger may not wait to be rescued. Jumpers not only risk their own lives but may also endanger the fire fighters trying to rescue them. Fire fighters have been seriously injured by persons who jumped before a rescue could be completed.

A trapped person might try to jump onto the tip of an approaching ladder or to reach out for anything or anyone nearby. As a fire fighter, you might be pulled or pushed off the ladder by the person you are trying to rescue. If several people are trapped, they might all try to climb down the ladder at the same time.

It is important to make verbal contact with any person you are trying to rescue. You must remain in charge of the situation and not allow the individual to panic. Maintain eye contact and tell the person to remain calm and wait to be rescued. If enough fire fighters are present, one could maintain contact with the person while others raise the ladder.

After you reach the person, you still have to get him or her down to the ground safely. Speak to the person in a calm, steady voice and ask him or her not to make any erratic movements. It is often helpful to have one fire fighter assist the person, while a second fire fighter acts as a guide and backup to the rescuer. Although modern fire ladders are designed to support this load, a ladder that is overloaded during a rescue operation must be removed from service for inspection and service testing. See Chapter 12, *Search and Rescue*, for specific ladder rescue techniques.

Ladder Damage

Ladders are easily damaged while in use. For example, a ladder might be overloaded or used at a low angle during a rescue. An unexpected shift in fire conditions could bring the ladder in direct contact with flames. Whenever a ladder is used outside of its recommended limits, it should be taken out of service for inspection and testing, even if there is no visible damage.

Using Ground Ladders

Ground ladders are often urgently needed during emergency incidents. Because an accident or error in handling or using a ladder can result in death or serious injury, all fire fighters must know how to work with ladders.

Using a ladder requires that fire fighters complete a series of consecutive tasks. The first step is to select the best ladder for the job from those available. Fire fighters must then remove the ladder from the apparatus and carry it to the location where it will be used. The next step is to raise and secure the ladder. At the end of the operation, the ladder must be lowered and returned to the apparatus. Each of these tasks is important to the safe and successful completion of the overall objective.

Selecting the Ladder

The first step in using a ladder is to select the appropriate ladder for the required task. What will the ladder be used for? Selecting the right ladder is dependent upon its purpose. Fire fighters must be familiar with all of the ladders carried on their apparatus. Pump and ladder company apparatus usually carry several ground ladders of various lengths. Many other types of apparatus, such as tankers (water tenders) and rescue trucks, carry additional ladders. NFPA 1901, *Standard for Automotive Fire Apparatus*, identifies minimum requirements for ground ladders for each type of apparatus (**TABLE 11-1**).

Selecting an appropriate ladder requires that fire fighters estimate the heights of windows and rooflines. General guidelines have been developed for estimating the heights for residential and commercial buildings (**TABLE 11-2**).

Ladder placement will affect the size and length of the ladder needed. When the ladder is used to access a roof, for example, the tip of the ladder should extend several feet above the roofline to provide a convenient handhold and secure footing for fire fighters as they mount and dismount. This extra length also makes the tip of the ladder visible to fire fighters who are working on the roof (**FIGURE 11-22**). A general

TABLE 11-1 Minimum Ladder Complement for Apparatus

Type of Apparatus	Minimum Ladder Complement
Pumper/Engine	1 folding ladder 1 straight ladder equipped with roof hooks 1 extension ladder
Quick attack/initial attack	1 combination or extension-type ladder 3.7 m (12 ft) or longer
Aerial/ladder	Ground ladders that have a total length of 35 m (115 ft) or more and contain a minimum of: ■ 1 folding ladder ■ 2 straight ladders equipped with roof hooks ■ 2 extension ladders
Quint	Ground ladders that have a total length of 26 m (85 ft) or more and contain a minimum of: ■ 1 folding ladder ■ 1 straight ladder equipped with roof hooks ■ 1 extension ladder

Source: NFPA 1901, *Standard for Automotive Fire Apparatus*, 2016.

TABLE 11-2 Approximate Distances for Residential and Commercial Construction

Type of Construction	Approximate Height
Residential floor-to-floor height	2.4–3 m (8–10 ft)
Residential floor-to-windowsill height	1 m (3 ft)
Commercial floor-to-floor height	3–3.7 m (10–12 ft)
Commercial floor-to-windowsill height	1 m (4 ft)

FIGURE 11-22 The tip of the ladder should extend above the roofline during roof operations.
© Jones & Bartlett Learning. Photographed by Glen E. Ellman.

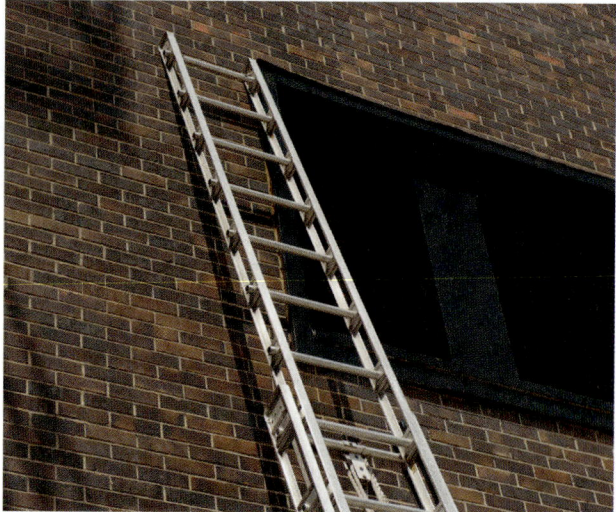

FIGURE 11-23 To provide access through a window, the tip of the ladder should be on the upwind side of the window and even with the top of the window opening.
© Jones & Bartlett Learning. Photographed by Glen E. Ellman.

FIGURE 11-24 For rescue operations, the tip of the ladder should be just below the windowsill.
© Jones & Bartlett Learning. Photographed by Glen E. Ellman.

rule is to ensure that at least five ladder rungs show above the roofline.

Ladders used to provide access to windows must be longer than those used in rescue operations. During access operations, the ladder is placed next to the window, with the ladder tip even with the top of the window opening (**FIGURE 11-23**). During rescue operations, however, the tip of the ladder should be immediately below the windowsill, which prevents the ladder from obstructing the window opening while a trapped occupant is removed (**FIGURE 11-24**).

The final factor in determining the correct length of ladder is the angle formed by the ladder and the placement surface (ground). A ground ladder should be placed at an angle of approximately 75 degrees for maximum strength and stability. To satisfy this criterion, the ladder must be slightly longer than the vertical distance between the ground and the target point. Generally, a ladder requires an additional 1 ft (0.5 m) in length for every 5 m (15 ft) of vertical height.

Thus, to reach a window that is 9 m (30 ft) above grade level, placed at an angle of approximately 75 degrees, a ladder would have to be at least 10 m (32 ft) long. Because ladders used in roof operations need to extend above the roofline, accessing a roof that is 9 m (30 ft) above grade level requires a ladder at least 11 m (35 ft) long.

Removing the Ladder from the Apparatus

Ladders are mounted on apparatus in various ways. Such equipment should be stored in locations where it will not be exposed to excessive heat, engine exhaust, or mechanical damage. Fire fighters must be familiar with how ladders are mounted on their apparatus and practise removing them safely and quickly.

Ladders are often nested one inside another and mounted on brackets on the side of the pump/engine. Fire fighters should note the nesting order and location of the ladders relative to the brackets when removing or replacing them. Ladders that are not needed or put aside should not be placed on the ground in front of an exhaust pipe where hot exhaust could damage them.

Ladders also can be stored on overhead hydraulic lifts (**FIGURE 11-25**). The hydraulic mechanism keeps them out of the way until they are needed, when they can be lowered to a convenient height. It is important to review the operation of the hydraulic lift used on your fire department's apparatus so that you can operate it efficiently and safely during an emergency situation.

On some vehicles, ground ladders are stored in compartments under the hose bed or aerial device. The ladders may lie flat or vertically on one beam (**FIGURE 11-26**). To remove or replace them, fire

FIGURE 11-25 This hydraulic mechanism lowers the ladders to a convenient height when they are needed.
Courtesy of Dennis Wetherhold, Jr.

FIGURE 11-26 Ladders in rear compartments can be stored either flat or vertically. This configuration requires adequate space behind the truck to remove ladders at an emergency scene.
© Jones & Bartlett Learning.

fighters slide the ladders out of the rear of the vehicle. For this reason it is important to leave adequate room behind the apparatus so that ladders can be removed at a fire scene.

Lifting Ladders

Many ladders are heavy and awkward to handle. To prevent lifting injuries while handling ladders, fire fighters must work together to lift and carry long or heavy ladders.

When fire fighters are working as a team to lift or carry a ladder, one fire fighter must act as the leader, providing direction and coordinating the actions of all team members. The lead fire fighter needs to call out the intended movements using two-step commands. For example, when lowering a ladder, the leader should say, "Prepare to lower," followed by the command "Lower." Team members must communicate in a clear and concise fashion, using the same terminology for commands; there should be no confusion about the specific meaning of each command. Because terminology may differ among jurisdictions, fire fighters who are newly or temporarily assigned to an apparatus should verify which commands are used and what they mean.

A prearranged method should be in place for determining the lead fire fighter for ladder lifts. In some fire departments, the fire fighter on the right side at the butt end of the ladder is always the leader. In other departments, the fire fighter at the tip of the ladder on the right side may be the leader. The departmental policy should be consistent for all ladder operations.

Additionally, you must use good lifting techniques when handling ladders. When bending to pick up a ladder, bend at the knees and keep the back straight (**FIGURE 11-27**). Lift and lower the load with the legs, rather than the back. Take care to avoid twisting motions during lifting and lowering, as these motions often lead to back strains.

FIGURE 11-27 Bend your knees and keep your back straight when lifting or lowering a ladder.
© Jones & Bartlett Learning. Photographed by Glen E. Ellman.

Carrying Ladders

Once a ladder has been removed from the apparatus or lifted from the ground, it must be carried to the placement site. Ladders can be carried at shoulder height or at arm's length. They can be carried either on edge or flat. The most common carries are explained in the following sections.

One-Fire Fighter Shoulder Carry

Most straight ladders and roof ladders less than 5.5 m (18 ft) long can be safely carried by a single fire fighter. To perform the steps involved in the one-fire fighter shoulder carry, follow the steps outlined in **SKILL DRILL 11-2**.

SKILL DRILL 11-2
One-Fire Fighter Shoulder Carry Fire Fighter I, NFPA 1001: 4.3.6

1 Start with the ladder mounted in a bracket or standing on one beam. Locate the center of the ladder. Grasp the two rungs on either side of the middle rung.

2 Lift the ladder, and rest it on your shoulder.

3 Walk carefully with the butt end first and pointed slightly downward.

© Jones & Bartlett Learning. Photographed by Glen E. Ellman.

Most ladders are carried with the butt end forward, because the placement of the butt determines the positioning of the ladder. For example, a straight ladder is carried with the butt end first and pointed slightly downward. Roof ladders are carried tip first with the hooks toward the front. A roof ladder usually must be carried up another ladder to reach the roof. The roof ladder is then usually pushed up the pitched roof slope from the butt end until the hooks engage the peak.

Two-Fire Fighter Shoulder Carry

The two-fire fighter shoulder carry is generally used with extension ladders up to 11 m (35 ft) long. It also can be used to carry straight ladders or roof ladders that are too long for one person to handle. To perform the two-fire fighter shoulder carry, follow the steps in **SKILL DRILL 11-3**.

Three-Fire Fighter Shoulder Carry

Three fire fighters may be needed to carry a heavy ladder. This carry is similar to the two-fire fighter shoulder carry, with the additional fire fighter positioned at the middle of the ladder. All three fire fighters stand on the same side of the ladder. To perform the three-fire fighter shoulder carry, follow the steps in **SKILL DRILL 11-4**.

SKILL DRILL 11-3
Two-Fire Fighter Shoulder Carry Fire Fighter I, NFPA 1001: 4.3.6

1 Start with the ladder mounted in a bracket or standing on one beam. Both fire fighters are positioned on the same side of the ladder. Facing the butt end of the ladder, one fire fighter is positioned near the butt end of the ladder, and a second fire fighter is positioned near the tip of the ladder.

2 Both fire fighters place one arm between two rungs and, on the leader's command, lift the ladder onto their shoulders. The ladder is carried butt end first.

3 The fire fighter closest to the butt end covers the sharp butt spur with a gloved hand to prevent injury to other fire fighters.

© Jones & Bartlett Learning. Photographed by Glen E. Ellman.

SKILL DRILL 11-4
Three-Fire Fighter Shoulder Carry Fire Fighter I, NFPA 1001: 4.3.6

1 All three fire fighters are positioned on the same side of the ladder. Facing the butt end of the ladder, one fire fighter is positioned at the butt end of the ladder, one fire fighter is at the middle of the ladder, and one fire fighter is at the tip of the ladder.

2 Each fire fighter places an arm between two rungs, and, on the leader's command, all fire fighters hoist the ladder onto their shoulders.

3 The ladder is carried butt end first. The fire fighter closest to the butt end covers the sharp butt spur with a gloved hand to prevent injury to other fire fighters.

© Jones & Bartlett Learning. Photographed by Glen E. Ellman.

Two-Fire Fighter Suitcase Carry

The two-fire fighter suitcase carry is commonly used with straight and extension ladders. With this technique, the ladder is carried on its edge by a beam at arm's length. To perform the two-fire fighter suitcase carry, follow the steps in **SKILL DRILL 11-5**.

Three-Fire Fighter Suitcase Carry

A three-fire fighter suitcase carry can be used for heavier ladders. This technique is similar to the two-fire fighter suitcase carry, with the addition of a third fire fighter at the center of the ladder. All three fire fighters remain on the same side of the ladder. To perform

SKILL DRILL 11-5
Two-Fire Fighter Suitcase Carry Fire Fighter I, NFPA 1001: 4.3.6

1 Start with the ladder resting on the ground on one beam. Both fire fighters are positioned on the same side of the ladder. Facing the butt end of the ladder, one fire fighter is positioned at the butt end of the ladder, and the other fire fighter is positioned at the tip of the ladder.

2 Each fire fighter reaches down and grasps the upper beam of the ladder.

3 On the leader's command, both fire fighters pick the ladder up from the ground and carry it with the butt end forward. The fire fighter closest to the butt end covers the sharp butt spur with a gloved hand to prevent injury to other fire fighters.

© Jones & Bartlett Learning. Photographed by Glen E. Ellman.

the three-fire fighter suitcase carry, follow the steps in **SKILL DRILL 11-6**.

Three-Fire Fighter Flat Carry

The three-fire fighter flat carry is typically used with extension ladders up to 11 m (35 ft) long. To perform the three-fire fighter flat carry, follow the steps in **SKILL DRILL 11-7**.

Four-Fire Fighter Flat Carry

The four-fire fighter flat carry is similar to the three-fire fighter flat carry described in the previous section, except that two fire fighters are positioned on each side of the ladder. On each side, one fire fighter is at the tip and the other is at the butt end of the ladder. To perform the four-fire fighter flat carry, follow the steps in **SKILL DRILL 11-8**.

SKILL DRILL 11-6

Three-Fire Fighter Suitcase Carry Fire Fighter I, NFPA 1001: 4.3.6

1 Start with the ladder resting on the ground on one beam. All three fire fighters are positioned on the same side of the ladder. Facing the butt end of the ladder, one fire fighter is positioned at the butt end of the ladder, one fire fighter is at the middle of the ladder, and one fire fighter is at the tip of the ladder.

2 All three fire fighters reach down and grasp the upper beam of the ladder.

3 On the leader's command, the fire fighters pick the ladder up from the ground and carry it at arm's length.

4 The fire fighter closest to the butt end covers the sharp butt spur with a gloved hand to prevent injury to other fire fighters.

SKILL DRILL 11-7
Three-Fire Fighter Flat Carry **Fire Fighter I, NFPA 1001: 4.3.6**

1 The carry begins with the bed section of the ladder flat on the ground. Two fire fighters stand on one side of the ladder, and the third fire fighter stands on the opposite side of the ladder. Facing the butt end of the ladder, one fire fighter is positioned at the butt end of the ladder with the second fire fighter at the tip end of the ladder. The third fire fighter is positioned at the center of the ladder on the opposite side.

2 All three fire fighters kneel down and grasp the closer rung at arm's length.

3 On the leader's command, all three fire fighters pick up the ladder and carry it butt end forward. The fire fighter closest to the butt end covers the sharp butt spur with a gloved hand to prevent injury to other fire fighters.

© Jones & Bartlett Learning. Photographed by Glen E. Ellman.

SKILL DRILL 11-8
Four-Fire Fighter Flat Carry **Fire Fighter I, NFPA 1001: 4.3.6**

1 The carry begins with the bed section of the ladder flat on the ground. Two fire fighters stand on each side of the ladder. All four fire fighters face the butt end of the ladder. One fire fighter is positioned at each corner of the ladder, with two fire fighters at the butt end of the ladder and two fire fighters at the tip end of the ladder.

(continued)

SKILL DRILL 11-8 Continued
Four-Fire Fighter Flat Carry Fire Fighter I, NFPA 1001: 4.3.6

2 On the leader's command, all four fire fighters kneel down and grasp the closer rung at arm's length.

3 On the leader's command, all four fire fighters pick up the ladder and carry it butt end forward.

4 The fire fighters closest to the butt end cover the sharp butt spurs with gloved hands to prevent injury to other fire fighters.

© Jones & Bartlett Learning. Photographed by Glen E. Ellman.

Three-Fire Fighter Flat Shoulder Carry

The three-fire fighter flat shoulder carry is similar to the three-fire fighter flat carry, except that the ladder is carried on the shoulders instead of at arm's length. A minimum of three fire fighters should perform the flat shoulder carry. The fire fighters face the tip of the ladder as they lift it but then pivot into the ladder as they raise it to shoulder height.

This carry is useful when fire fighters must carry the ladder over short obstacles. Raising the ladder to shoulder height increases the potential for back strain, however, so fire fighters must follow proper lifting techniques.

To perform the three-fire fighter flat shoulder carry, follow the steps in **SKILL DRILL 11-9**.

SKILL DRILL 11-9
Three-Fire Fighter Flat Shoulder Carry Fire Fighter I, NFPA 1001: 4.3.6

1 The carry begins with the bed section of the ladder flat on the ground. Two fire fighters are positioned on one side of the ladder, and the third fire fighter is on the opposite side of the ladder. Facing the tip end of the ladder, one fire fighter is positioned at the butt end of the ladder, and the second fire fighter is at the tip end of the ladder. The third fire fighter is at the center of the ladder on the opposite side.

2 All three fire fighters kneel and grasp the closer beam. On the leader's command, the fire fighters stand, raising the ladder.

3 As the ladder approaches chest height, the leader instructs the fire fighters to pivot into the ladder.

4 The fire fighters place the beam of the ladder on their shoulders. The ladder is carried in this position with the butt moving forward. The fire fighter closest to the butt end covers the sharp butt spur with a gloved hand to prevent injury to other fire fighters.

Four-Fire Fighter Flat Shoulder Carry

The four-fire fighter flat shoulder carry is similar to the three-fire fighter flat shoulder carry described in the previous section, except that two fire fighters are positioned on each side of the ladder. On each side, one fire fighter is at the tip and the other at the butt end of the ladder. To perform the four-fire fighter flat shoulder carry, follow the steps in **SKILL DRILL 11-10**.

Placing a Ladder

The first step in placing a ladder is selecting the proper location for the ladder. Generally, an officer or senior fire fighter will select the general area for ladder placement, and the fire fighter at the butt of the ladder will determine the exact site at which to position it. Both the officer ordering the ladder and the fire fighter

///////////////////////////////

SAFETY TIP

When fire fighters of different heights are working together and carrying ladders, you can make these operations safer and smoother by considering the placement of fire fighters. When performing three-person carries, place shorter fire fighters at the butt end of the ladder and taller fire fighters at the tip end of the ladder. This strategy prevents a shorter fire fighter from being placed between two taller fire fighters and having to carry the ladder at an awkward and unsafe height. When performing four-person carries, placing shorter fire fighters at the butt end of the ladder prevents the ladder from being tipped to one side while it is being carried.

placing the ladder need to consider several factors in their decisions.

SKILL DRILL 11-10
Four-Fire Fighter Flat Shoulder Carry Fire Fighter I, NFPA 1001: 4.3.6

1 The carry begins with the bed section of the ladder flat on the ground. Two fire fighters are positioned on each side of the ladder. All four fire fighters face the tip end of the ladder. One fire fighter is positioned at each corner of the ladder, with two fire fighters at the butt end and two fire fighters at the tip end of the ladder.

2 On the leader's command, all four fire fighters kneel and grasp the closer beam.

SKILL DRILL 11-10 Continued
Four-Fire Fighter Flat Shoulder Carry Fire Fighter I, NFPA 1001: 4.3.6

3 On the leader's command, the fire fighters stand, raising the ladder.

4 As the ladder approaches chest height, the fire fighters all pivot toward the ladder.

5 The fire fighters place the beam of the ladder on their shoulders. All four fire fighters face the butt of the ladder. The ladder is carried in this position, with the butt moving forward. The fire fighters closest to the butt end cover the sharp butt spurs with gloved hands to prevent injury to other fire fighters.

© Jones & Bartlett Learning. Photographed by Glen E. Ellman.

As discussed, a raised ladder should be positioned at an angle of approximately 75 degrees to provide the best combination of strength, stability, and vertical reach. This angle also creates a comfortable climbing angle. When a fire fighter is standing on a rung, the rung at shoulder height will be approximately one arm's length away. The ratio of the ladder height (vertical reach) to the distance from the structure should be 4:1. For example, if the vertical reach of a ladder is 6 m (20 ft), the butt of the ladder should be placed 1.5 m (5 ft) out from the wall. Most new ladders have an inclination or level guide sticker on their beams such that the bottom line of the indicator will be parallel to the ground when the ladder is positioned properly. Standing at the base of the ladder and extending the arms straight out is another way to check the climbing angle (**FIGURE 11-28**). The hands should comfortably reach the beams or rungs if the angle is appropriate. Finally, the angle can be measured by ensuring that the butt of the ladder is one-fourth of the working height out from the base of the structure.

When calculating vertical reach, a ladder generally requires an additional 31 cm (1 ft) in length for every 5 m (15 ft) of vertical height. Also, remember to add extra footage for rooftop operations. Ladders used in roof operations need to extend above the roofline. If the butt of the ladder is placed too close to the building, the climbing angle will be too steep. In such a case, the tip of the ladder could pull away from the building as the fire fighter climbs, making the ladder unstable. Conversely, if the butt is too far from the structure, the climbing angle will be too shallow, reducing the load capacity of the ladder and increasing the risk that the butt could slip out from under the ladder.

FIGURE 11-28 The climbing angle can be checked by standing on the bottom rung and holding the arms straight out.

© Jones & Bartlett Learning. Photographed by Glen E. Ellman.

FIGURE 11-29 A ladder placed on an uneven surface is unsafe because it can easily tip over.
© Jones & Bartlett Learning.

The ladder should be placed on a stable and relatively level surface, as ladders placed on uneven surfaces are prone to tipping (**FIGURE 11-29**). As a fire fighter climbs the ladder, the moving weight will cause the load to shift back and forth between the two beams. Unless both butt ends are firmly placed on solid ground, this uneven load can start a rocking motion that will tip the ladder over.

Ladders should not be placed on top of unstable structures such as manholes or trap doors. The combined weight of the ladder, fire fighters, and equipment could cause the cover or door to fail, injuring the fire fighters.

As mentioned earlier, ladders should be placed in areas with no overhead obstructions. If a ladder comes into contact with overhead utility lines—and particularly electric power lines—both the fire fighters working on the ladder and those stabilizing it could be killed or injured. This is true for all ladders, regardless of their composition (fibreglass, wood, or metal).

Be aware that a ladder can become energized even if it does not actually touch an electric line. For example, a ladder that enters the electromagnetic field surrounding a power line can become energized. For this reason, ladders should be placed at least 3 m (10 ft) away from energized power lines. Always maintain eye contact with the tip of the ladder when raising and or lowering it.

Finally, ground ladders should not be placed in high-traffic areas unless no other alternative is available. For example, because the main entrance to a structure is heavily used, a ground ladder should not be placed where it would obstruct the door.

SAFETY TIP

Ladders can become energized without making contact with overhead electric lines. An electric force field surrounds electric lines, and the size of this field increases as the voltage is increased. Conductivity of the air is also increased as the relative humidity increases. Always keep ladders at least 3 m (10 ft) away from electric lines.

Raising a Ladder

Once the position has been selected, the ladder must be raised for climbing. Two commonly used techniques for raising ground ladders are the beam raise and the flat raise, also known as the rung raise. A beam raise is typically used when the ladder must be raised parallel to the target surface. A flat raise (rung raise) is often used when the ladder can be raised from a position perpendicular to the target surface. There are many different names used to describe these ladder raises. This text will use the terms *beam raise* and *flat raise*.

The number of fire fighters required to raise a ladder depends on the length and weight of the ladder, as well as on the available clearance from obstructions. A single fire fighter can safely raise many straight ladders and lighter extension ladders. Two or more fire fighters are required to raise longer and heavier ladders.

One-Fire Fighter Flat Raises

There are two variations of the one-fire fighter flat raise. One is used by a single fire fighter to raise a small straight ladder, typically 4 m (14 ft) or less in length. Always check for overhead hazards before raising a ladder. To perform the one-fire fighter flat raise for ladders less than 4 m (14 ft), follow the steps in SKILL DRILL 11-11.

The other variation of the one-fire fighter flat raise is generally used with straight ladders longer than

SKILL DRILL 11-11
One-Fire Fighter Flat Raise for Ladders Less Than 4 Metres (14 Feet) Long
Fire Fighter I, NFPA 1001: 4.3.6

1 Place the butt of the ladder on the ground directly against the structure so that both butt spurs contact the ground and the structure. Lay the ladder on the ground. If the ladder is an extension ladder, place the bed (base) section on the ground. Stand at the tip of the ladder, and check for overhead hazards. Take hold of a rung near the tip, bring that end of the ladder to chest height, and then step beneath the ladder. Raise the ladder using a hand-over-hand motion as you walk toward the structure until the ladder is vertical and against the structure. If an extension ladder is being used, hold the ladder vertical against the structure, and extend the fly section by pulling the halyard smoothly, with a hand-over-hand motion, until the desired height is reached and the pawls are locked.

2 Pull the butt of the ladder out from the structure to create a 75-degree climbing angle. To move the butt away from the structure, grip a lower rung, and lift slightly while pulling outward. At the same time, apply pressure to an upper rung to keep the tip of the ladder against the structure. If the ladder is an extension ladder, it will be necessary to rotate the ladder so the fly section is out. The halyard should be tied as described in the Tying the Halyard skill drill. Check the tip and the butt of the ladder to ensure safety before climbing.

4 m (14 ft) and with extension ladders that can be safely handled by one fire fighter. Each fire fighter will have a different safety limit, depending on his or her strength and the weight of the ladder. To perform the one-fire fighter flat raise for ladders more than 4 m (14 ft), follow the steps in **SKILL DRILL 11-12**.

With practice, it is possible to move the ladder from the carrying position to placing the butt of the ladder directly into the position where it should be when the ladder is raised. However, in the beginning it is generally easier and more efficient to lay the ladder down flat with the butt against the building before moving the ladder into the vertical position.

Tying the Halyard

The halyard of an extension ladder should always be tied after the ladder has been extended and lowered into place. A tied halyard stays out of the way and provides a safety backup to the pawls for securing the fly section. One method of tying a halyard is shown in **SKILL DRILL 11-13**.

SKILL DRILL 11-12
One-Fire Fighter Flat Raise for Ladders More Than 4 Metres (14 Feet) Long
Fire Fighter I, NFPA 1001: 4.3.6

1 Place the butt of the ladder on the ground directly against the structure so that both spurs contact the ground and the structure. Lay the ladder on the ground. If the ladder is an extension ladder, place the base section on the ground. Stand at the tip of the ladder, and check for overhead hazards.

2 Take hold of a rung near the tip, bring that end of the ladder to chest height, and then step beneath the ladder.

3 Raise the ladder using a hand-over-hand motion as you walk toward the structure until the ladder is vertical and against the structure.

4 If an extension ladder is being used, hold the ladder vertical against the structure, and extend the fly section by pulling the halyard smoothly, with a hand-over-hand motion, until the desired height is reached and the pawls are locked.

SKILL DRILL 11-12 Continued
One-Fire Fighter Flat Raise for Ladders More Than 4 Metres (14 Feet) Long
Fire Fighter I, NFPA 1001: 4.3.6

5 Pull the butt of the ladder out from the structure to create a 75-degree climbing angle. To move the butt away from the structure, grip a lower rung, and lift slightly while pulling outward. At the same time, apply pressure to an upper rung to keep the tip of the ladder against the structure.

© Jones & Bartlett Learning. Photographed by Glen E. Ellman.

6 If the ladder is an extension ladder, it will be necessary to rotate the ladder so the fly section is out. The halyard should be tied as described in the Tying the Halyard skill drill. Check the tip and the butt of the ladder to ensure safety before climbing. Always secure the base by heeling the ladder.

SKILL DRILL 11-13
Tying the Halyard Fire Fighter I, NFPA 1001: 4.3.6

1 Wrap the excess halyard rope around two rungs of the ladder, and pull the rope tight across the upper of the two rungs.

(continued)

SKILL DRILL 11-13 Continued
Tying the Halyard Fire Fighter I, NFPA 1001: 4.3.6

2   Tie a clove hitch around the upper rung and the vertical section of the halyard. Refer to Chapter 9, *Ropes and Knots*, to review how to tie a clove hitch.

3   Pull the clove hitch tight.

4   Place an overhand safety knot as close to the clove hitch as possible to prevent slipping.

© Jones & Bartlett Learning. Photographed by Glen E. Ellman.

Two-Fire Fighter Beam Raise

If the ladder is small, this raise can be performed by one fire fighter. In many cases the beam raise should be performed with two or more fire fighters. The two-fire fighter beam raise is used with midsized extension ladders up to 11 m (35 ft) long. To perform the two-fire fighter beam raise, follow the steps in **SKILL DRILL 11-14**.

SKILL DRILL 11-14
Two-Fire Fighter Beam Raise Fire Fighter I, NFPA 1001: 4.3.6

1 The two-fire fighter beam raise begins with a shoulder or suitcase carry. One fire fighter stands near the butt end of the ladder, and the other fire fighter stands near the tip.

2 The fire fighter at the butt of the ladder places that end of the ladder on the ground, while the fire fighter at the tip of the ladder holds that end.

3 The fire fighter at the butt of the ladder places a foot on the butt of the beam that is in contact with the ground and grasps the upper beam.

4 The fire fighter at the tip of the ladder checks for overhead hazards and then begins to walk toward the butt, while raising the lower beam in a hand-over-hand fashion until the ladder is vertical.

(continued)

SKILL DRILL 11-14 Continued
Two-Fire Fighter Beam Raise Fire Fighter I, NFPA 1001: 4.3.6

5 The two fire fighters pivot the ladder into position as necessary.

6 The fire fighters face each other, one on each side of the ladder, and heel the ladder by each placing the toe or instep of one boot against the opposing beams of the ladder.

7 One fire fighter extends the fly section by pulling the halyard smoothly with a hand-over-hand motion until the fly section is at the height desired and the pawls are locked. The other fire fighter stabilizes the ladder by holding the outside of the base section beams so that if the fly comes down suddenly it will not strike the fire fighter's hands.

8 The fire fighter facing the structure places one foot against one beam of the ladder, and then both fire fighters lean the ladder into place. The halyard is tied. The fire fighters check the ladder for a 75-degree climbing angle and check for stability at the tip and at the butt end of the ladder.

Two-Fire Fighter Flat Raise

The two-fire fighter flat raise is also commonly used with midsized extension ladders up to 11 m (35 ft) long. To perform the two-fire fighter flat raise, follow the steps in **SKILL DRILL 11-15**.

Three- and Four-Fire Fighter Flat Raises

The three- and four-fire fighter flat raises are used for very heavy ladders. The basic steps in these raises are similar to the two-fire fighter flat raise. In a three-fire

---

SKILL DRILL 11-15
Two-Fire Fighter Flat Raise Fire Fighter I, NFPA 1001: 4.3.6

1 The two-fire fighter flat raise begins from a shoulder carry or suitcase carry, with one fire fighter near the butt of the ladder and the other near the tip. The fire fighter at the butt of the ladder places the butt of the lower beam on the ground, while the fire fighter at the tip holds the other end. The fire fighter at the tip rotates the ladder so that both butts are in contact with the ground.

2 The fire fighter at the butt of the ladder stands on the bottom rung, grasps a higher rung with both hands, crouches down, and leans backward to heel the ladder. The fire fighter at the tip of the ladder checks for overhead hazards, then swings under the ladder and walks toward the butt, advancing down the ladder and lifting the rungs in a hand-over-hand fashion until the ladder is vertical.

3 The two fire fighters stand on opposite sides of the ladder and pivot it into position as necessary.

(continued)

SKILL DRILL 11-15 Continued
Two-Fire Fighter Flat Raise Fire Fighter I, NFPA 1001: 4.3.6

4 The fire fighters face each other, one on each side of the ladder, and heel the ladder by each placing the toe or instep of one boot against the opposing beams of the ladder.

5 If using an extension ladder, one fire fighter extends the fly section by pulling the halyard smoothly with a hand-over-hand motion until the tip is at the desired height and the pawls are locked. The other fire fighter stabilizes the ladder by holding the outside of the base section beams so that if the fly comes down suddenly it will not strike the fire fighter's hands.

6 The fire fighter facing the structure places one foot against one beam of the ladder, and then both fire fighters lean the ladder into place. The halyard is tied. The fire fighters check the ladder for a 75-degree climbing angle and for stability at the tip and at the butt end of the ladder.

fighter flat raise, the third fire fighter assists with the hand-over-hand raising of the ladder. When four fire fighters are raising the ladder, two anchor the butt of the ladder while the other two raise the ladder. Each of the two fire fighters at the butt places his or her inside foot on the bottom rung of the ladder and bends over, grasping a rung in front of him or her.

To perform the three-fire fighter flat raise, follow the steps in **SKILL DRILL 11-16**.

SKILL DRILL 11-16
Three-Fire Fighter Flat Raise Fire Fighter I, NFPA 1001: 4.3.6

1 The raise begins from a shoulder carry or suitcase carry, with one fire fighter near the butt of the ladder, one fire fighter in the middle, and one fire fighter near the tip. The fire fighter at the butt of the ladder places the butt of the lower beam on the ground, while the fire fighter at the tip holds that end. The fire fighter in the middle moves to the tip. The fire fighters at the tip rotate the ladder so that both butts are in contact with the ground. The fire fighter at the butt of the ladder stands on the bottom rung, grasps a higher rung with both hands, crouches down, and leans backward to heel the ladder.

2 The fire fighters at the tip of the ladder check for overhead hazards, then begin to walk toward the butt, advancing down the ladder and lifting the rungs in a hand-over-hand fashion until the ladder is vertical.

3 All three fire fighters pivot the ladder into position as necessary.

(continued)

SKILL DRILL 11-16 Continued
Three-Fire Fighter Flat Raise Fire Fighter I, NFPA 1001: 4.3.6

4 Two fire fighters face each other, one on each side of the ladder, grasp the outsides of the beams, and heel the ladder by each placing the toe or instep of one boot against the opposing beams of the ladder.

5 The third fire fighter extends the fly section by pulling the halyard smoothly with a hand-over-hand motion until the tip is at the desired height and the pawls are locked.

6 One fire fighter heels the ladder while the other two lean the ladder into place. The halyard is tied. The fire fighters check the ladder for a 75-degree climbing angle and for security at the tip and at the butt end of the ladder.

To perform the four-fire fighter flat raise, follow the steps in **SKILL DRILL 11-17**.

Fly Section Orientation

When raising extension ladders, fire fighters must know whether the fly section should be facing toward the building (fly in) or away from the building (fly out). The ladder manufacturer and department standard

SKILL DRILL 11-17
Four-Fire Fighter Flat Raise Fire Fighter I, NFPA 1001: 4.3.6

1 The raise begins with a four-fire fighter flat carry. Two fire fighters are at the butt of the ladder, and two fire fighters are at the tip.

2 The two fire fighters at the butt of the ladder place the butt end of the ladder on the ground, while the two fire fighters at the tip hold that end.

3 The two fire fighters at the butt of the ladder stand side by side, facing the ladder. Each fire fighter places the inside foot on the bottom rung and the other foot on the ground outside the beam. Both crouch down, grab a rung (preferrably the same rung), and lean backward.

4 The two fire fighters at the tip of the ladder check for overhead hazards and then begin to walk toward the butt of the ladder, advancing down the rungs in a hand-over-hand fashion until the ladder is vertical.

(continued)

SKILL DRILL 11-17 Continued
Four-Fire Fighter Flat Raise Fire Fighter I, NFPA 1001: 4.3.6

5 The fire fighters pivot the ladder into position, as necessary.

6 Two fire fighters heel the ladder by placing a boot against each beam. Each fire fighter places the toe or instep of one boot against one of the beams. The third fire fighter stabilizes the ladder by holding it on the outside of the rails.

7 The fourth fire fighter extends the fly section by pulling the halyard smoothly with a hand-over-hand motion until the tip reaches the desired height and the pawls are locked.

8 The two fire fighters facing the structure each place one foot against one beam of the ladder while the other two fire fighters lower the ladder into place. The halyard is tied. The fire fighters check the ladder for a 75-degree climbing angle and for security at the tip and at the butt end of the ladder.

operating procedures (SOPs) will specify whether the fly should be facing in or out. In general, manufacturers of fibreglass and metal ladders recommend that the fly sections are facing away (fly out) from the structure. In contrast, wooden fire-service ladders often are designed to be used with the fly facing in. With some ladders, it is not possible to raise the fly section of the ladder from the fly side of the ladder.

Securing the Ladder

Several approaches may be used to prevent a ladder from moving once it is in place. There are a few methods used to heel or foot a ladder. One option is to have a fire fighter stand on the outside of the ladder, facing the structure, and heel the ladder by placing a boot against the butt end of the ladder (**FIGURE 11-30**).

FIGURE 11-30 A fire fighter can stand on the outside of the ladder, facing the structure, and heel the ladder by placing a boot against the butt end of the ladder.

© Jones & Bartlett Learning. Photographed by Glen E. Ellman.

This method has advantages and disadvantages. This technique allows the heeling fire fighter to survey the building and warn the climber of possible hazards, but this approach can cause injury to the heeling fire fighter if there is falling debris.

A rope, a rope–hose tool, or webbing also can be used to secure a ladder in place. The lower part of the ladder can be tied to any solid object to keep the base from kicking out; in fact, the base should always be secured if the ladder is used at a low angle. The tip of the ladder can be tied to a secure object near the top to keep it from pulling away from the building. The best method is to secure both the tip and the base.

Climbing the Ladder

Climbing a ladder on the fire or emergency scene should be done in a deliberate and controlled manner. Always make sure the ladder is secure (tied or heeled). Before climbing an extension ladder, verify that the pawls are locked and the halyard is secured.

The proper climbing angle should be checked as well. See the section Placing a Ladder for more information.

Keep any bouncing and shifting to a minimum while climbing the ladder. Your eyes should be focused forward, with only occasional glances upward. This approach will prevent debris such as falling glass from injuring your face or eyes. Lower the protective face shield for additional protection.

Use a hand-over-hand motion on the rungs of the ladder, or slide both hands along the underside of the beams while climbing. Sliding the hands along the beams is more secure because it maintains three contact points with the ladder at all times (both hands and one foot).

If tools must be moved up or down, it is better to hoist them by ropes than to carry them up a ladder. Carrying tools on a ladder reduces the fire fighter's grip and increases the potential for injury if a tool slips or falls. Sometimes you will need to carry a tool up a ladder, however, so you should practise doing it safely. To climb a ladder while holding a tool, hold the tool against one beam with one hand, and maintain contact with the opposite beam with the other hand. Tools such as pike poles can be hooked onto the ladder and moved up every few rungs. Be sure not to overload the ladder. No more than two fire fighters should be on a ladder at the same time. As mentioned earlier, a properly placed ladder should be able to support two rescuers (with their protective clothing and equipment) and one victim.

One method of climbing a ladder while carrying a tool is described in **SKILL DRILL 11-18**.

SKILL DRILL 11-18
Climbing a Ladder While Carrying a Tool Fire Fighter I, NFPA 1001: 4.3.12

1 Check the ladder to ensure that it is at a safe angle to climb and is secure. Grasp the tool comfortably and securely in one hand, and hold that hand and the tool against the beam of the ladder.

2 Wrap the other hand around the underside of the beam, and begin climbing.

3 Climb smoothly and safely by maintaining contact between your free hand and the beam and by sliding the tool along the opposite beam. Climb with confidence, with your back as straight as possible and your arms extended. Refrain from "hugging" the ladder as you ascend or descend.

© Jones & Bartlett Learning. Photographed by Glen E. Ellman.

Dismounting the Ladder

When a ladder is used to reach a roof or a window entry, the fire fighter will have to dismount the ladder. Fire fighters can minimize their risk of slipping and falling by making sure that the surface is stable—that is, testing its stability with a tool—before dismounting. For example, fire fighters who dismount from a ladder onto a roof typically sound the roof with an axe before stepping onto the surface (**FIGURE 11-31**).

Try to maintain contact with the ladder at three points when dismounting. For example, a fire fighter who steps onto a roof should keep two hands and one foot on the ladder while checking the footing. This consideration is particularly important if the surface slopes or is covered with rain, snow, or ice. Do not shift your weight onto the roof until you have tested the footing.

It is also important to check the footing when entering a window from a ladder positioned to the side of the window. Before stepping into the window, confirm that the interior surface is structurally sound and offers secure footing. The three points of contact rule also applies in this situation.

Under heavy fire and smoke conditions, fire fighters sometimes dismount from a ladder by climbing over the tip and sliding over the windowsill into the building. Sound the floor inside the window before entering to confirm that it is solid and stable. To remount a ladder under these conditions, back out the window feet first, and rest your abdomen on the sill until you can feel the ladder under your feet. Under better conditions, sit on the windowsill with legs out and roll onto the ladder.

Working from a Ladder

Fire fighters must often work from a ladder. To avoid falling while doing so, the fire fighter must be secured to

FIGURE 11-31 Check the stability of the roof before dismounting from the ladder.
© Jones & Bartlett Learning. Photographed by Glen E. Ellman.

the ladder. Fire fighters use different methods to secure themselves to a ladder.

The first method involves the use of a **ladder belt**—a piece of equipment specifically designed to secure a fire fighter to a ladder or elevated surface (**FIGURE 11-32**). Fire fighters should use only ladder belts designed and certified under the requirements specified by NFPA 1983, *Standard on Life Safety Rope and Equipment for Emergency Services*. Utility belts designed to carry tools should never be used as ladder belts.

There are a few methods that do not require special equipment that can be used to secure a fire fighter to a ladder. One method is to secure yourself by spreading your legs outward, securing yourself against the beams. Another method is the leg lock.

To apply a leg lock, follow the steps in **SKILL DRILL 11-19**.

Placing a Roof Ladder

Several methods can be used to properly position a roof ladder on a sloping roof. One commonly used method is described in **SKILL DRILL 11-20**.

Descending a Ladder

As you prepare to descend a ladder, apply the same safety practices as when climbing a ladder. Even though you are descending, you are still in an elevated position and at risk of slipping and falling. Before starting to

FIGURE 11-32 A ladder belt has a large hook that is designed to secure a fire fighter to a ladder.
© Jones & Bartlett Learning. Photographed by Glen E. Ellman.

SKILL DRILL 11-19
Using a Leg Lock to Work from a Ladder Fire Fighter I, NFPA 1001: 4.3.6

1 Climb to the desired work height, and step up to one more rung.

2 Note the side of the ladder where the work will be performed. Extend your leg between the rungs on the side opposite the side you will be working.

3 Once your leg is between the rungs, bend your knee, and bring your foot back under the rung and through to the climbing side of the ladder.

4 Secure your foot against the next lower rung or the beam of the ladder. Use your thigh for support, and step down one rung with the opposite foot.

5 The use of a leg lock enables you to have two hands free for a variety of tasks, including placing a roof ladder or controlling a hose stream.

descend, take time to ensure that the ladder is at an appropriate angle for climbing and descending. Be sure it is on stable footing and that someone is heeling (or footing) the ladder. Communicate to the person heeling the ladder that you are about to descend. Remember that you cannot see the ladder rungs during descent, so first check that they are not slippery from water or, in the case of cold weather, from ice. If you are holding a

SKILL DRILL 11-20
Deploying a Roof Ladder Fire Fighter I, NFPA 1001: 4.3.12

1 Carry the roof ladder to the base of the climbing ladder that is already in place to provide access to the roofline.

2 Place the butt end of the roof ladder on the ground, and rotate the hooks of the roof ladder to the open position.

3 Raise and lean the roof ladder against one beam of the other ladder with the hooks oriented outward, away from you. Climb the lower climbing ladder until you reach the midpoint of the roof ladder that is positioned next to you, and then slip one shoulder between two rungs of the roof ladder, and shoulder the roof ladder.

4 Climb to the roofline of the structure, carrying the roof ladder on one shoulder. Secure yourself to the ladder.

(continued)

SKILL DRILL 11-20 Continued
Deploying a Roof Ladder Fire Fighter I, NFPA 1001: 4.3.12

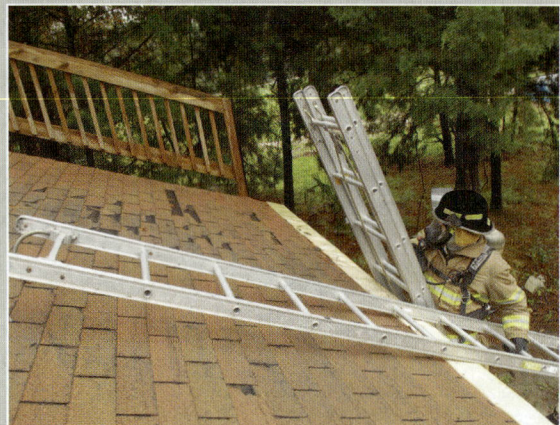

5 Place the roof ladder on the roof surface with hooks down. Push the ladder up toward the peak of the roof with a hand-over-hand motion.

6 Once the hooks have passed the peak, pull back on the roof ladder to set the hooks, and check that they are secure. To remove a roof ladder from the roof, reverse the process just described. After releasing the hooks from the peak, it may be necessary to turn the ladder on one of its beams or turn it so the hooks are pointing up to slide the ladder down the roof without catching the hooks on the roofing material. Carrying a roof ladder in this manner requires strength and practice.

© Jones & Bartlett Learning. Photographed by Glen E. Ellman.

tool, establish a secure grip on it that will allow you to maintain good contact with the ladder as you descend. Consider lowering tools using a utility rope.

As you begin to descend, face the ladder, keeping your back perpendicular with the ground and your arms almost fully extended. Maintain three-point contact as you descend, keeping two hands and one foot or two feet and one hand on the ladder. Have the arches of your feet on the rungs. Descend slowly, avoid sudden movements, and stay in the middle of the ladder to increase stability. Minimize bouncing and swaying back and forth. Do not overload the ladder. Unless it is an emergency, one person descends at a time.

Inspect a Chimney

Each year, fire departments receive many calls for chimney fires and other chimney-related problems.

Many times it is necessary to inspect the chimney to determine whether a problem is truly present. To access the chimney, it is necessary to safely ladder the building and proceed to the roof. The steps for inspecting a chimney are described in SKILL DRILL 11-21.

As you work on learning the skills needed to perform vital ladder operations, do not lose sight of the dangers inherent in working with ladders. As you practise the skills described in this chapter, follow these safety rules:

- Wear appropriate PPE. This equipment should include SCBA when you are operating on the fire ground.
- Choose the proper ladder for the job.
- Choose the proper number of people for the job.
- Move smoothly, safely, and efficiently, keeping the ladder under your control at all times.
- Lift with your legs, keeping your back straight.

SKILL DRILL 11-21
Inspecting a Chimney

1 Make sure you are equipped with a complete set of PPE, including SCBA, and are operating under the direction of your officer.

2 Remove the proper-size extension ladder and the roof ladder from the apparatus.

3 Carry the ladders to the building, and raise the extension ladder so that five rungs project above the roofline.

4 Confirm that the roof is safe to climb on. If it is not, use an aerial ladder.

5 When possible, place the roof ladder on the upwind side of the chimney.

6 Working with a partner, climb to the chimney while carrying the appropriate equipment.

7 Inspect the chimney, looking for holes, cracks, and structural instability. Report your findings to the incident commander.

8 Remove the tools and roof ladder when the task is complete.

© Jones & Bartlett Learning.

- Use proper hand and foot placement during carries, raises, climbing, and lowering of ladders.
- Look overhead for wires and obstructions before raising or lowering a ladder. Communicate the results of this check by stating "Clear overhead" to your partner.
- Allow the heel person to take charge of the processes of raising and lowering the ladder.
- After extending an extension ladder, check the pawls to confirm that they are set, and make sure the halyard is tied off.

- Check the ladder for the proper 75-degree angle.
- Ensure that the wall or roof will support the ladder before climbing it.
- Remain aware of hazards to yourself and to others.
- Maintain hand contact with the ladder. Watch your hands when climbing up, and watch your legs when climbing down.
- Use an appropriate method to secure yourself to the ladder.

After-Action REVIEW

IN SUMMARY

- The primary function of ladders is to provide safe access to and egress from otherwise inaccessible areas.
- Ladders have a wide variety of uses in non-firefighting situations, including ravines or steep highway embankments, trench rescue situations, below-grade locations such as manholes, vehicle extrication situations, or hazardous materials incidents.
- Ladders are made of metal, wood, and fibreglass.

- The two main structural components of the ladder are the beam and the rung. The beams support the rungs and carry the load of the fire fighter from the rungs down to the ground.

- There are three basic types of ladder beam construction: trussed beam, I-beam, and solid beam. Trussed beams are usually constructed of metal or wood. I-beam ladders are usually made from fibreglass.

- Many wooden ladders have solid beams. Rectangular aluminum beams, which are usually hollow or C shaped, are also classified as solid beam ladders.

- The rung is a crosspiece that spans the two beams of the ladder. Most ground ladders in fire departments have aluminum rungs, but wooden ladders are still constructed with wood rungs.

- Additional parts of a typical ground ladder include the following components:
  - Rail
  - Truss block
  - Tie rod
  - Tip
  - Butt
  - Butt spurs
  - Butt plate
  - Roof hooks
  - Heat sensor label
  - Protection plates

- An extension ladder is an assembly of two or more ladder sections that fit together and can be extended or retracted to adjust the length of the ladder. Additional parts of an extension ladder include the following components:
  - Bed (base) section
  - Fly section
  - Guides
  - Halyard
  - Pawls
  - Pulley
  - Stops
  - Staypoles

- There are two general categories of ladders: aerial and ground. Aerial ladders are permanently mounted and operated from a fire apparatus. Ground ladders are carried on fire apparatus and removed from the apparatus to be used in other locations.

- Ground ladders include general-purpose ladders such as straight or extension ladders and specialized ladders such as folding (attic) or roof ladders.

- Most ladder companies are required to carry at least two straight ladders with roof hooks, two extension ladders, and one folding ladder.

- A straight ladder is a single-section, fixed-length ground ladder.

- A roof ladder is a straight ground ladder with retractable hooks at one end.

- An extension ladder is an adjustable-length ground ladder with multiple sections.

- Bangor ladders are extension ladders with staypoles for added stability.

- A combination ladder can be converted from a straight ladder to a stepladder configuration or from an extension ladder to a stepladder configuration.

- A folding (attic) ladder is a narrow, collapsible ladder designed to allow access to attic scuttle holes and confined areas.

- A Fresno ladder is a narrow, two-section extension ladder designed to provide attic access.

- A ground ladder should be inspected visually after every use. If the inspection reveals any deficiencies, the ladder must be removed from service until repairs are made.

- All fire fighters should be able to perform routine ladder maintenance, including cleaning the ladders to remove any road grime and dirt. Ladders should be cleaned after every use.

- Service testing of ground ladders must follow NFPA 1932.

- Safety precautions should be followed from the time a ladder is removed from the apparatus until it is returned to the apparatus. Full turnout gear and PPE are essential for working with ground and aerial ladders. Additional general safety rules include the following:
  - Lift with your legs, keeping your back straight.
  - Look overhead for wires and obstructions before raising or lowering a ladder.
  - After extending an extension ladder, check the pawls (ladder locks) to confirm that they are set, and make sure the halyard is tied off.
  - Check the ladder placement for the proper 75-degree angle. Ensure that the wall or roof will support the ladder before climbing it.

- When positioning a ladder for rescue, place the tip of the ladder at or slightly below the windowsill. This provides a direct pathway to enter and exit the window.

- Communication is key to coordinating efforts when working with ladders. One leader should be designated, and his or her commands should be followed during carries and raises.

KEY TERMS

Access Navigate for flashcards to test your key term knowledge.

Aerial ladder A self-supporting, turntable-mounted, power-operated ladder of two or more sections permanently attached to a self-propelled automotive fire apparatus and designed to provide a continuous egress route from an elevated position to the ground. (NFPA 1901)

Bangor ladder A ladder equipped with tormentor poles, or staypoles, that stabilize the ladder during raising and lowering operations.

Base (bed) section The lowest or widest section of an extension ladder. (NFPA 1931)

Beam The main structural side of a ground ladder. (NFPA 1931)

Beam raise A ladder raise used to raise a ladder perpendicular to a building.

Butt The end of the beam that is placed on the ground, or other lower support surface, when ground ladders are in the raised position. (NFPA 1931)

Butt plate An alternative to a simple butt spur; a swiveling plate with both a spur and a cleat or pad that is attached to the butt of the ladder.

Butt spurs That component of ground ladder support that remains in contact with the lower support surface to reduce slippage. (NFPA 1931)

Chassis The basic operating motor vehicle, including the engine, frame, and other essential structural and mechanical parts, but exclusive of the body and all appurtenances for the accommodation of driver, property, passengers, appliances, or equipment related to other than control. Common usage might, but need not, include a cab (or cowl).

Combination ladder A ground ladder that is capable of being used both as a stepladder and as a single or extension ladder. (NFPA 1931)

Elevated platform An apparatus that includes a passenger-carrying platform (bucket) attached to the tip of a boom or ladder.

Extension ladder A non–self-supporting ground ladder that consists of two or more sections travelling in guides, brackets, or the equivalent arranged so as to allow length adjustment. (NFPA 1931)

Fire department ground ladder Any portable ladder specifically designed for fire department use in rescue, firefighting operations, or training. (NFPA 1931)

Flat raise (rung raise) A ladder raise used to position a ladder parallel to a building. Also called a rung raise.

Fly section Any section of an aerial telescoping device beyond the base section. (NFPA 1901) This definition applies to aerial ladder devices and ground ladders.

Folding ladder A single-section ladder with rungs that can be folded or moved to allow the beams to be brought into a position touching or nearly touching each other. (NFPA 1931)

Fresno ladder A narrow, two-section extension ladder that has no halyard. Because of its limited length, it can be extended manually.

Guides Strips of metal or wood that serve to guide a fly section during extension. Channels or slots in the bed or fly section may also serve as guides.

Halyard Rope used on extension ladders for the purpose of raising a fly section(s). (NFPA 1931)

Heat sensor label A label that changes colour at a preset temperature to indicate a specific heat exposure. (NFPA 1931)

I-beam A ladder beam constructed of one continuous piece of I-shaped metal or fibreglass to which the rungs are attached.

Ladder A-frame An A-shaped structure formed with two ladder sections. It can be used as a makeshift lift when raising a trapped person. Sometimes referred to as an A-frame hoist.

Ladder belt A compliant equipment item that is intended for use as a positioning device for a person on a ladder. (NFPA 1983)

Pawls Devices attached to a fly section(s) to engage ladder rungs near the beams of the section below for the purpose of anchoring the fly section(s). Also called locks or dogs. (NFPA 1931)

Protection plates Reinforcing material placed on a ladder at chafing and contact points to prevent damage from friction and contact with other surfaces.

Pulley A device with a free-turning, grooved metal wheel (sheave) used to reduce rope friction. Side plates are available for a carabiner to be attached. (NFPA 1670)

Rail The top or bottom piece of a trussed beam assembly used in the construction of a trussed ladder. Also, the top and bottom surfaces of an I-beam ladder. Each beam has two rails.

Roof hooks The spring-loaded, retractable, curved metal pieces that allow the tip of a roof ladder to be secured to the peak of a pitched roof. The hooks fold outward from each beam at the top of a roof ladder.

Roof ladder A single ladder equipped with hooks at the top end of the ladder. (NFPA 1931)

Rung The ladder crosspieces, on which a person steps while ascending or descending. (NFPA 1931)

Solid beam A ladder beam constructed of a solid rectangular piece of material (typically wood), to which the ladder rungs are attached.

Staypoles Poles attached to each beam of the base section of extension ladders, which assist in raising the ladder and help provide stability of the raised ladder. (NFPA 1931)

Stop A piece of material that prevents the fly sections of a ladder from becoming overextended, leading to collapse of the ladder.

Tie rod A metal rod that runs from one beam of the ladder to the other to keep the beams from separating. Tie rods are typically found in wood ladders.

Tip The very top of the ladder.

Truss block A piece of wood or metal that ties the two rails of a trussed beam ladder together and serves as the attachment point for the rungs.

Trussed beam A ladder beam constructed of top and bottom rails joined by truss blocks that tie the rails together and support the rungs.

REFERENCES

National Fire Protection Association. NFPA 1901, *Standard for Automotive Fire Apparatus*, 2016. www.nfpa.org. Accessed September 28, 2018.

National Fire Protection Association. NFPA 1931, *Standard for Manufacturer's Design of Fire Department Ground Ladders*, 2015. www.nfpa.org. Accessed September 28, 2018.

National Fire Protection Association. NFPA 1932, *Standard on Use, Maintenance, and Service Testing of In-Service Fire Department Ground Ladders*, 2015. www.nfpa.org. Accessed September 28, 2018.

National Fire Protection Association. NFPA 1983, *Standard on Life Safety Rope and Equipment for Emergency Services*, 2017. www.nfpa.org. Accessed September 28, 2018.

On Scene

Your pump company is the first to arrive at a large, old, two-and-a-half-storey wood frame house that has been converted into apartments. You find people trapped on the second floor with heavy smoke conditions throughout. One victim will need to be rescued down the 11-m (35-ft) extension ladder. With the help of your crew member, you start by pulling the extension ladders off of the apparatus.

1. When possible, which is the preferred method of cleaning the ladder?

   A. Stiff-bristle brush, mild detergent, and water

   B. Soft-bristle brush and water

   C. Stiff-bristle brush and water

   D. Soft-bristle brush, mild solvent, and water

2. How should wire-rope halyard extensions be checked?

   A. With the fly section fully extended

   B. With the fly section partially extended

   C. With the fly section fully bedded

   D. The position of the fly section does not matter

3. In addition to after each use, when should a ground ladder normally be visually inspected?

   A. Daily

   B. Weekly

   C. Monthly

   D. Yearly

4. What are the spring-loaded, retractable, curved metal pieces that allow the tip of a roof ladder to be secured to the peak of a pitched roof?

   A. Halyards

   B. Tie rods

   C. Pawls

   D. Roof hooks

5. What are the two main structural components that run the entire length of most ladders or ladder sections?

   A. Rails

   B. Truss blocks

   C. Beams

   D. Guides

(continued)

On Scene Continued

6. If a ladder is raised 6 m (20 ft) to gain access to a roof, how far away from the building should you position the butt of the ladder from the building?

A. Approximately 1 m (4 ft)

B. Approximately 5 m (5 ft)

C. Approximately 2.4 m (8 ft)

D. Approximately 3 m (10 ft)

7. What are the devices attached to the fly section of a ladder that engage ladder rungs to anchor the fly sections of the ladder called?

A. Tie rods

B. Truss blocks

C. Butt spurs

D. Pawls

Access Navigate to find answers to this On Scene, along with other resources such as an audiobook.

CHAPTER 12

Fire Fighter I

Search and Rescue

KNOWLEDGE OBJECTIVES

After studying this chapter, you will be able to:

- Describe the mission of search operations. (**NFPA 1001: 4.3.9**, pp. 428–432)
- Describe the mission of rescue operations. (**NFPA 1001: 4.3.9**, pp. 428–432)
- Explain how fire fighters maintain safety through risk management. (**NFPA 1001: 4.3.9**, pp. 427–428)
- Explain how search and rescue operations are coordinated with other fire suppression operations. (**NFPA 1001: 4.3.9**, pp. 428–429)
- Identify the factors to evaluate during a search and rescue size-up. (**NFPA 1001: 4.3.9**, pp. 429–431)
- Explain how search operations are coordinated. (**NFPA 1001: 4.3.9**, pp. 431–432)
- Describe the role of a fire officer during search operations. (**NFPA 1001: 4.3.9**, pp. 433, 438)
- List the priorities of search operations. (**NFPA 1001: 4.3.9**, p. 432)
- List the tools and equipment used in search and rescue operations. (**NFPA 1001: 4.3.9**, pp. 432–433)
- Describe the methods fire fighters use to determine whether an area is tenable. (**NFPA 1001: 4.3.9**, pp. 433–434)
- Describe the objectives of a primary search. (**NFPA 1001: 4.3.9**, pp. 434–435)

- Describe the search patterns commonly used in primary search operations. (**NFPA 1001: 4.3.9**, p. 434)
- Explain how thermal imaging devices are used during search operations. (**NFPA 1001: 4.3.9**, p. 436)
- Describe a primary search using the standard search method. (**NFPA 1001: 4.3.9**, pp. 437, 438)
- Describe a primary search using the oriented search method. (**NFPA 1001: 4.3.9**, pp. 437, 438–441)
- Describe a primary search using the oriented-vent-enter-isolate-search (O-VEIS) method. (**NFPA 1001: 4.3.9**, pp. 441–442)
- Describe a primary search using the team search method. (**NFPA 1001: 4.3.9**, pp. 442–443)
- Describe the objectives of a secondary search. (**NFPA 1001: 4.3.9**, pp. 443–444)
- List the major types of rescue methods. (**NFPA 1001: 4.3.9**, p. 445)
- Describe the concept of sheltering-in-place. (**NFPA 1001: 4.3.9**, pp. 445–446)
- Describe how to assist a victim to an exit. (**NFPA 1001: 4.3.9**, pp. 446–447)
- List the common types of simple victim carries performed during rescue operations. (**NFPA 1001: 4.3.9**, pp. 448–451)
- List the six emergency drags performed during rescue operations. (**NFPA 1001: 4.3.9**, pp. 452–458)
- Describe the conditions that may require a ground ladder rescue. (**NFPA 1001: 4.3.9**, pp. 458–462)
- Describe the advantages of using aerial ladders and platforms during rescue operations. (**NFPA 1001: 4.3.9**, pp. 462, 465)

SKILLS OBJECTIVES

After studying this chapter, you will be able to perform the following skills:

- Conduct a primary search using the standard search method. (**NFPA 1001: 4.3.9**, pp. 437, 438)
- Conduct a primary search using the oriented search method. (**NFPA 1001: 4.3.9**, pp. 437, 438–441)
- Conduct a primary search using the oriented-vent-enter-isolate-search (O-VEIS) method. (**NFPA 1001: 4.3.9**, pp. 441–442)
- Conduct a primary search using the team search method. (**NFPA 4.3.9**, pp. 442–443)
- Conduct a secondary search. (**NFPA 4.3.9**, pp. 443–444)
- Perform a one-person walking assist. (**NFPA 4.3.9**, p. 446)
- Perform a two-person walking assist. (**NFPA 4.3.9**, p. 447)
- Perform a two-person extremity carry. (**NFPA 4.3.9**, p. 448)
- Perform a two-person seat carry. (**NFPA 4.3.9**, p. 449)
- Perform a two-person chair carry. (**NFPA 4.3.9**, pp. 449, 450)
- Perform a cradle-in-arms carry. (**NFPA 4.3.9**, pp. 450, 451)
- Perform a clothes drag. (**NFPA 4.3.9**, p. 452)
- Perform a blanket drag. (**NFPA 4.3.9**, pp. 452, 453, 454)
- Perform a standing drag. (**NFPA 4.3.9**, pp. 453, 454, 455)
- Perform a webbing sling drag. (**NFPA 4.3.9**, pp. 455, 456)
- Perform a fire fighter drag. (**NFPA 4.3.9**, pp. 455, 457)
- Perform a one-person emergency drag from a vehicle. (**NFPA 4.3.9**, pp. 455, 458)
- Rescue a conscious victim from a window. (**NFPA 4.3.9**, pp. 459–461)
- Rescue an unconscious victim from a window. (**NFPA 4.3.9**, pp. 461, 462)
- Rescue an unconscious child or small adult from a window. (**NFPA 4.3.9**, pp. 461, 462, 463)
- Rescue a large adult from a window. (**NFPA 4.3.9**, pp. 462, 464)

ADDITIONAL STANDARDS

- **NFPA 1500**, *Standard on Fire Department Occupational Safety, Health, and Wellness Program*

Fire Alarm

You arrive on the scene of a fast-moving apartment fire. The structure is a three-storey, center-hallway, Type III ordinary-constructed apartment building with fire on the second floor. The second- and third-floor hallways are smoke charged. Dispatch is reporting occupants trapped. You are assigned to conduct a primary search on the third floor while another crew begins the fire attack. This is a call that will tax the resources of your department.

1. How will you go about making the search?

2. What will you do if you find a victim in the first unit you search?

3. Will you take a hose line with you on the search?

 *Access Navigate for more practice activities.*

Introduction

The primary mission of the fire department is to save lives and protect property. More recently, many fire departments include "protection of the environment" as a part of their overall duty. Saving lives is the highest strategic priority at a fire scene. Search and rescue operations are complementary tasks that are usually performed immediately upon arrival. The first fire fighters to arrive must verify that there are no people trapped within the involved structure.

The safety of fire fighters and members of the public must remain the highest strategic priority throughout the incident. All search-related activities and tactical objectives must be carried out in the safest possible manner. Unfortunately, there may be circumstances that make it impossible to perform a rescue due to the degree of structural fire involvement.

Search and rescue are almost always performed in tandem, yet they are actually separate actions. The purpose of search operations is to locate living victims. The purpose of rescue operations is to physically remove an occupant or victim from a dangerous environment. Rescue occurs when a fire fighter leads an occupant to an exit, carries an unconscious victim out of a burning building, or rescues the victim using a ladder. All three are examples of rescues because both the occupant and the victim were physically removed from imminent danger through the actions of a fire fighter.

Any fire department company or unit can be assigned to search and rescue operations. All fire fighters must be trained and prepared to perform search and rescue functions.

Search and rescue operations must be conducted quickly and efficiently. A systematic approach will ensure that every occupant who can possibly be saved is successfully located and removed from danger. Fire fighters should practise the specific search and rescue procedures used by their departments.

Often, the first-arriving fire fighters will not know whether any occupants are inside a burning building or how many occupants might be inside. Fire fighters should never assume that a building is unoccupied. The only way to confirm that every occupant has safely evacuated a building is to conduct a thorough search. The decision to enter a burning building to search for victims is one that may be made by the incident commander (IC), the officer in charge, or a fire fighter and is based on a risk/benefit analysis. However, fire fighters need to recognize that a room that has flashed over is not survivable even for a fire fighter in full personal protective equipment (PPE) for more than a few seconds. A room that contains a ventilation-limited fire may, at first glance, appear to be a small incipient fire, but in reality it may be a larger fire in a ventilation-limited state.

Risk Management

Search and rescue situations require a very special type of risk management. Although every emergency operation involves a degree of unavoidable inherent risk, fire fighters will probably encounter situations during search and rescue operations that involve a significantly higher degree of personal risk. The risk involved in conducting search and rescue operations must always be weighed against the probability of finding someone who is savable—that is, who is still alive to be rescued.

The IC is responsible for managing the level of risk during emergency operations. He or she must perform a risk/benefit analysis to determine which actions will be taken in each situation. Actions that present a high level of risk to the safety of fire fighters are justified only if victims are known or believed to be in immediate danger and there is a reasonable probability that lives can be saved. In contrast, only a limited risk level is acceptable to save property. The risk/benefit analysis should consider the stage of the fire, the condition of the building, and the presence of any other hazards.

When there is no possibility of saving either lives or property, no risk is acceptable, and search and rescue operations cannot be performed at all. For example, if a building is fully involved in flames, a room has flashed over, or a building demonstrates potential backdraft conditions, the risk to fire fighters might be too high and the possibility of saving any occupants inside the structure might be too low to justify sending a search and rescue team into the building (**FIGURE 12-1**). A similar decision might be made if the fire occurs in

FIGURE 12-1 The IC must weigh the risk to the fire fighters against the possibility of saving any occupant inside before authorizing an interior search.

Courtesy of District Chief Chris E. Mickal/New Orleans Fire Department, Photo Unit.

an abandoned building or a lightweight construction building in danger of structural collapse. The IC may be able to identify these conditions from the exterior, or he or she may learn of them from a team assigned to conduct a search. A search team that encounters conditions that make entry impossible should report their findings back to the IC. In these situations, it might not be possible to conduct an interior search until the fire has been extinguished.

A risk/benefit analysis of the scene is often guided by established policies. For example, many fire departments have policies that prohibit fire fighters from entering abandoned structures known to be in poor structural condition. This kind of policy reflects the fact that such a building has a high risk of collapse and the low probability that anyone would be inside.

FIRE FIGHTER TIP

The IC must always balance the risks involved in an emergency operation with the potential benefits:

- Actions that present a high level of risk to the safety of fire fighters are justified only if there is a potential to save lives.
- Only a limited level of risk is acceptable to save valuable property.
- It is not acceptable to risk the safety of fire fighters when there is no chance of saving lives or property.

In order to assess the degree of risk and to minimize the chance of injuries, it is important to do the following:

- Conduct a thorough 360-degree size-up.
- Use a thermal imaging device.
- Evaluate the possibility of a ventilation-limited fire.
- Practise good door control at the point of entry into the building to limit the supply of oxygen to the fire.
- Consider the use of a transitional fire attack to cool the fire before starting the search process.

Search and Rescue Operations

Coordination with Fire Suppression Operations

Although search and rescue is a top priority, it is rarely the only action taken by first-arriving fire fighters. Instead, the IC and fire fighters must plan and coordinate all fire suppression operations—from ventilation to fire attack—to support the search and rescue priority.

All rescue efforts must take into account the flow path of the fire. Any operation conducted in the flow path increases the risks posed to the fire fighter. Getting caught in a flow path that is carrying hot gases away from the fire can result in a deadly situation in a matter of seconds. Any change in any opening in the fire building, whether it is the intentional opening of a door, window, or roof or an unintentional change produced by glass failure or a change in the wind direction, can be deadly to both rescuers and persons trapped in the fire.

There may be instances where the most effective tactical approach to performing rescue is an aggressive and sustained fire attack. This tactic will be dependent on staffing levels and should ensure protection of means of egress and entry. Sometimes, the most effective way to perform search and rescue is to minimize or eliminate the life safety threat. Because smoke is fuel, the initial cooling of hot gases, particles, and aerosols in the room of origin can reduce temperatures and limit fire spread. This technique—which you will learn how to conduct in Chapter 17, *Fire Suppression*—is called a **transitional attack**. It is an offensive fire attack initiated by a quick, indirect, exterior attack into the fire compartment to initiate cooling and darken the fire, allowing fire fighters to quickly transition to an interior attack for final suppression. The use of a transitional attack can be an important part of an effective search and rescue operation. It also makes conditions safer for fire fighters conducting interior operations.

Other fire scene activities also must be coordinated with search and rescue. For example, forcible entry is sometimes needed to provide entrances and exits for search and rescue teams. Sometimes, well-placed ventilation may be indicated. At other times, it may be best to limit ventilation and protect trapped occupants by closing doors or by not opening windows. Ladders may be raised to second-storey windows to provide an escape route if needed. Portable lighting can also provide valuable assistance to interior search crews.

Often, the search for potential victims provides valuable information about the location and extent of the fire within a building. In essence, the searchers act as a reconnaissance team to determine which areas are involved and where a fire might spread. They report this information to the IC, who develops the overall fire suppression plan.

The proper identification of the building construction type should always be the first step in the size-up process. Firefighters should never enter into a structure without establishing the most probable building construction type. Identifying the construction type informs the following with regard to safely executing search and rescue activities:

- Means of entry and egress
- Room location and configuration
- Vent, enter, and search potential

- Floor stability
- Collapse potential
- Flow path locations
- Likely avenues of fire spread and propagation—i.e., Type V structures support fire growth and spread and become part of the overall fire load. Type I structures do not contribute to the fire load. Awareness of this fact will guide the amount of time and resources required for search and rescue.
- Forcible entry tools—i.e., steel doors in steel frames are harder to defeat than wood doors in wood frames.

Search and Rescue Size-Up

The size-up process at every fire should include a specific evaluation of the critical factors for search and rescue—namely, building construction, occupant information, such as the number of occupants in the building, their location, the degree of risk to life safety, and their ability to evacuate by themselves. Usually, this information is not immediately available, so fire fighters' actions must be based on a combination of observations and expectations. The type of building construction; the size, occupancy, and arrangement of the building; the visible smoke and fire conditions; weather conditions; and the time of day and the day of the week are all important observations that are required to guide decision making.

A search and rescue plan can then be developed based on this information. Such a plan identifies the areas to be searched, the priorities for searching different areas, the number of search teams required, and any additional actions needed to support the search and rescue activities. One search team might be sufficient for a small building, whereas multiple teams might be needed to search a large building.

Building Size, Construction, and Arrangement

The size, construction, and arrangement of the building are important factors to consider in planning and conducting a search. A small, one-storey house is simpler to search than a larger building with multiple units or a large commercial building. A large building with many rooms must be searched in a systematic fashion. Even though the complexity of the search increases as a building gets larger, searchers need to be aware of the fire flow path in the building and avoid actions that could place them in a dangerous position.

The risk posed to victims and fire fighters may also increase depending on the building construction: Occupants of an unprotected, wood-frame building are in greater danger than those in a building with fire-resistive construction and automatic sprinklers.

Access to an interior layout or floor plan is often helpful when planning and assigning teams to search a building. To ensure that each area is searched promptly and that no areas are omitted, specific areas must be assigned to different search teams in an organized manner. Assignments are often based on the stairway locations, corridor arrangements, and the apartment or room numbering system. Teams must practise good door control at the point of entry into the building and report when they have completed searching each area.

Because it is difficult to determine a building's layout or floor plan in the midst of an emergency, fire fighters often conduct preincident surveys of buildings such as apartment complexes and offices. The information assembled during the preincident survey is used to prepare preincident plans, so fire fighters will be prepared when an emergency occurs. In some departments this information is available to responding companies electronically or through paper forms (**FIGURE 12-2**).

Preincident plans can include a variety of valuable information about a building:

- Corridor layouts
- Exit locations
- Stairway locations
- Apartment layouts
- Number of bedrooms in apartments
- Locations of handicapped residents' apartments
- Special function rooms or areas

Fire fighters should note how the floors of a building are numbered. Buildings constructed on a slope may appear to have a different number of levels when viewed from the "A" or front side than when viewed from the "C" or back side. For example, a fire may appear to be on the third floor to a fire fighter standing in front of the building, while a fire fighter at the rear of the building might report that it is on the fifth floor. Without knowing

FIGURE 12-2 Preincident plans may be available electronically or through paper forms.
© Jones & Bartlett Learning.

how floors are numbered, fire fighters can be sent to work above a fire instead of below it. This situation resulted in multiple fire fighter deaths in Pittsburgh, Pennsylvania, in 1995 (Routley, 1995).

Occupant Information

The initial size-up of an incident can provide valuable information on the number of occupants in the building, their most likely location, the degree of risk to their life safety, and the probability of locating them. Fire fighters should never automatically assume that a building is unoccupied.

An observant fire fighter notices clues that indicate whether a building is occupied and how many occupants are likely to be present. For example, the first-responding unit at a residential fire might see two cars in the driveway and numerous toys on the front lawn. These signs indicate that the house is probably occupied by a family with children. In such a situation, the IC would likely emphasize the importance of quickly searching every room where victims might be located. Conversely, an absence of cars, locked doors, and an

overflowing mailbox might suggest that the house is vacant or unoccupied. If there is a chance that savable victims might be present, the IC will always order a search in case someone is inside (**FIGURE 12-3**).

Similar observations can be made for other structures. Although an empty parking lot does not guarantee that a building is empty, the presence of a car probably indicates that at least one maintenance worker or security guard is inside—but it might also mean that a car is simply parked in the lot. Boarded-up windows and doors on a building surrounded by a chain-link fence are indications that the building may be unoccupied.

When a building is known or believed to be occupied, fire fighters should first rescue the occupants who are in the most immediate danger, followed by those who are in less danger. Search teams should be assigned on the basis of these priorities. Several factors determine the level of risk faced by the occupants of a burning building—namely, the location of the fire within the building, the direction of the fire spread, the speed and direction of the wind, the volume and intensity of the fire, and smoke conditions in different areas of the structure.

A

B

FIGURE 12-3 Exterior observations may provide a good indication of whether a building is occupied. **A.** Parked cars indicate the residence may be occupied. **B.** Boarded-up windows indicate the residence may be unoccupied.
A: © Jones & Bartlett Learning. Photographed by Glen E. Ellman; B: Courtesy of Captain David Jackson, Saginaw Township Fire Department.

Its is essential to establish a search priority based on where occupants are most likely to be located. Occupants who are close to the fire, above the fire, or in the fire flow path are usually at greater risk than occupants who are located farther away from the fire. An occupant in the apartment where the fire started is probably in immediate danger, whereas the occupants of a lower or upper floor may be relatively safe. This decision will be somewhat dependent on the type of construction used in the fire building.

Building occupants who are at the windows or on balconies calling for help obviously realize that they are in danger and want to be rescued. There may be other occupants in the building who cannot be seen from the exterior, however—perhaps because they are asleep, unconscious, incapacitated, or trapped. Fire fighters must make sure that everyone is accounted for. There may be instances whereby rescue can only be determined through a primary and secondary search.

It is important to verify information and consider the source of the information. People may identify the term "baby" in reference to a pet rather than a child. It is important to confirm and verify all information.

Occupancy Type

Young children and elderly people, who are more likely to need assistance to evacuate, are at greater risk than adolescents and adults. Individuals who are confined to a bed or wheelchair are at greater risk than occupants who can move freely. Consequently, more fire fighters and resources will be needed for search and rescue operations in certain occupancies.

A fire in a single-family dwelling generally places the members of one family at risk, whereas a fire in a large apartment building may endanger multiple families. A warehouse fire probably presents a direct risk to few occupants.

Visible Smoke and Fire Conditions

Visible smoke and fire conditions can provide clues to the location and intensity of the fire (**FIGURE 12-4**). Observe the volume and density of the smoke and the velocity of the smoke movement. More volume, darker colour, and greater speed usually indicate more severe fire conditions.

Assess the direction of the wind, and determine where it may push the fire. Look for the amount of visible flames. Determine the number of openings where flames are visible. All of these factors will help you to determine the location and size of the fire, the potential for growth, and the direction of travel. Careful assessment of these factors will help you to develop a rescue plan.

FIGURE 12-4 The visible smoke and fire conditions at the scene can help fire fighters determine the location and intensity of the fire.
© Irene Teesalu/Shutterstock, Inc.

Time of Day and Day of Week

Generally, the life risk in residential occupancies is higher at night and on weekends, when more people are at home. Risk is greatest late at night, when the occupants are probably asleep. A fire that occurs in a residential occupancy on a weekday afternoon would present a significantly different rescue problem from a fire in the same location at 2:00 AM.

An office building that is fully occupied on a weekday afternoon is typically unoccupied or has a low occupancy at midnight. Conversely, a nightclub fire on a Friday or Saturday night could endanger hundreds of lives.

SAFETY TIP

Search and rescue size-up considerations include the following factors:

- Building construction type, size, and orientation
- Occupancy information (i.e., number of occupants, degree of risk to occupants, and ability of occupants to exit on their own)
- Occupancy type
- Visible smoke and fire conditions
- Time of day and day of week

Search Coordination

Life safety must be at the forefront of every decision made throughout the incident. The safety of all personnel and members of the public is paramount. Generally speaking, the IC will assign a search and rescue sector/division within the command system and assign an officer to assume the radio designation. As its primary tactical objective, the goal of the search and rescue sector/division is to ensure that all occupants have been removed from danger. As the searchers complete a search of each area, the search officer notifies the IC of the status and results of the search effort. An "All clear" report indicates that an area has been searched, and all victims have been removed.

Another critical aspect of search coordination is keeping track of everyone who was rescued or who escaped without assistance. This information should be tracked at the incident command post so that reports of missing occupants can be matched to reports of rescued victims. Fire fighters should also conduct an exterior search for any missing occupants. An occupant who escaped from the fire may be lying unconscious on the ground nearby or in the care of neighbours, for example. An occupant who jumped from a window could be injured and unable to move.

Search Priorities

A search begins in the areas where victims are at the greatest risk. Search teams must work together closely and coordinate their searches to ensure that all areas are covered. One or two search teams can usually go through all of the rooms in a single-family dwelling in 15 minutes or less (Coleman, 2011). Multiple search teams and a systematic division of the building are needed to cover larger structures such as apartment buildings and high-rise buildings. Area search assignments should be based on a system of priorities:

- The first priority is to search those areas where live victims may be located immediately around the fire and then the rest of the fire floor.
- The second priority is to search the area directly above the fire and the rest of that floor.
- The third priority is higher-level floors, typically working from the top floor down, because smoke and heat are likely to accumulate in these areas.
- Generally, areas below the fire floor are a lower priority unless there is danger of collapse.

These priorities are guidelines and need to be modified based on occupancy, construction type, and the circumstances at that incident.

Search and Rescue Safety

Search and rescue operations present a high risk to fire fighters. During searches, fire fighters are exposed to the same hazards that may endanger the lives of potential victims. Even though fire fighters have the advantages of protective clothing, protective equipment, training, teamwork, and SOPs, they can still be seriously or fatally injured during these operations. Safety is an essential consideration in all search and rescue operations.

Search and Rescue Tools and Equipment

To perform search and rescue properly, fire fighters must have the appropriate tools and equipment (**TABLE 12-1**). Full PPE, which is always required for structural firefighting, is essential. The proper attire includes a helmet, protective hood, protective coat, protective pants, boots, and gloves. Each fire fighter must use self-contained breathing apparatus (SCBA) and carry a flashlight or hand light. Before entering the building, fire fighters must verify their personal alert safety system (PASS) devices are activated.

At least one member of each search team should be equipped with a portable radio. Ideally, each individual should have a radio. If a fire fighter gets into trouble, the radio is the best means to obtain assistance.

Thermal imaging devices can also be very helpful during search and rescue operations. This tool, which is similar to a video camera, can be used in smoke-filled buildings to locate victims and search for hidden fires

---

LISTEN UP!

When addressing the third priority of area search, fire fighters typically work from the top floor down. High-rise buildings are an exception to this rule, however, due to the impact of the stack effect. The stack effect—which is discussed more thoroughly in Chapter 13, *Ventilation*—is the vertical air flow within a building caused by the differences in temperature inside and outside the building. This phenomenon may result in large quantities of products of combustion being present in the upper levels of the building. In the case of high-rise buildings, officers must take the stack effect into account and initiate search operations from the highest floor impacted by the by-products of fire. Unaffected floors can be searched after these floors have been swept.

TABLE 12-1 Search and Rescue Tools and Equipment

- PPE
- Hand light or flashlight
- Portable radio
- Thermal imaging devices
- Forcible entry (exit) tools
- Hose lines
- Ladders
- Search rope(s)
- A piece of tubular webbing or short rope (5–7 m; 16–24 ft)
- Chalk, crayons, felt tip markers, or masking tape

during size-up. The use of thermal imaging devices is discussed later in this chapter.

Each fire fighter assigned to search and rescue should carry the same basic hand tools as the interior attack team, including at least one forcible entry tool, such as an axe, sledgehammer, Halligan tool, or short pike pole. These tools can be used both to open an area for a search and, if necessary, to open an emergency exit path. A hand tool can also be used to extend the fire fighter's reach during a sweep for unconscious victims.

A search team that is working close to the fire should carry a hose line or be accompanied by another team with a hose line. A hose line can protect the fire fighters and enable them to search a structure more efficiently. This measure is essential when a search team is working close to or directly above the fire. The hose line can be used to knock down the fire, to protect a means of egress (stairway or corridor), and to protect the victims as they are escaping.

A search team may need to use ground ladders to gain access to a search site that is located on a second or higher floor. Ladders may also be used to rescue victims from dangerous areas if regular exits cannot be used. Using a ground ladder to rescue a trapped occupant is a stressful and demanding task that carries a considerable risk of injury to both fire fighters and victims. Proper ladder rescue techniques are discussed later in this chapter.

Search ropes may be used during search operations to provide searchers with an anchor point that will help them maintain their orientation while searching for victims. Search ropes are especially useful when search is being conducted in large, open areas. Webbing can be used when it is necessary to drag or carry a victim from a hazardous environment.

Once a room or apartment is searched, personnel should use chalk, felt tip markers, or tape to mark the door. This marking lets other personnel know that the room has already been searched.

Methods to Determine Whether an Area Is Tenable

Fire fighters must make rapid, accurate, and ongoing assessments about the safety of the building while working at a structure fire. When they arrive, fire fighters must quickly determine the type of structure involved, the stability of the structure and the possibility of collapse, and the life-safety risk involved. Try to determine how long the structure has been on fire. Are there reports of fire showing before your arrival? Consider the possibility that if no fire is showing, this could indicate a ventilation-limited fire condition. Dark black and turbulent smoke, blackened windows, the appearance of a "breathing" building, soot around windows or doors indicating smoke travel, puffing or little to no smoke showing, and intense heat are signs of possible flashover or backdraft. Cracks in walls, leaning walls, pitched or sagging doors, and any partial collapse are significant indicators of an impending collapse. Additional signs of collapse are discussed in Chapter 6, *Building Construction*.

Even after the decision has been made to enter a burning structure, fire fighters must continually re-evaluate the safety of the operation. There should be a continual assessment of the building and fire conditions, and this information should be regularly communicated to personnel on an ongoing basis. Fire officers must be alert for changing conditions inside the building and for any information about changing conditions that is from other parts of the fire scene. The IC may know or see things that fire fighters working inside the building might not know, for example, and may call for an evacuation as the fire situation changes. For their part, fire fighters in the interior of the building should check and recheck the surfaces they are working on. Floors that have burned through, that are sagging, or that are warm; furniture or carpeting off-gassing; or flame "rollover" should be avoided. A rapid rise in the amount of heat or flame "rollover" may indicate the potential for flashover.

Evaluate the floor before entering the building. A size-up is essential to determine where the fire is located. Sounding the floor is not a valid measure of stability and is a dangerous practice to be discouraged. Floors that are supported by lightweight wooden trusses and other manufactured components fail quickly when exposed to fire. Entering a burning building when there is a fire burning beneath you is an act of great risk. All residential floor systems are at risk of sudden catastrophic collapse without warning (UL FSRI, 2006). Extreme caution should be used when operating on any floor above a fire.

SAFETY TIP

During search and rescue operations, fire fighters should follow these guidelines:

- Work from a single plan.
- Maintain radio contact with the IC, both through the chain of command and via portable radios.
- Monitor fire conditions and building stability during the search.
- Coordinate ventilation operations with search and rescue activities.
- Adhere to the personnel accountability system.
- Stay with a partner.
- Maintain constant situational awareness, listen to radio reports from other sectors, and close doors while searching rooms, to reduce the possibility of being caught in a flow path.

Only in imminent life-threatening situations where immediate action can prevent the loss of life or serious injury should fire fighters take actions at a higher level of risk. The initial IC must evaluate the situation and determine whether this level of risk is justified. The IC also must be prepared to explain his or her decision.

Primary Search

Two types of searches are performed in buildings: primary and secondary. A **primary search** is an immediate and quick attempt to locate any potential victims who are in danger. This search should be as thorough as time permits and should cover any places where victims are likely to be found. Fire conditions might make it impossible to conduct a primary search in some areas, or they may limit the time available for an exhaustive search. A primary search should be completed in 15 minutes or less (Coleman, 2011).

A **secondary search** is conducted after the fire has been suppressed. At this point, remaining occupants may have been exposed to conditions that cannot support life, such as flashover. The main purpose of the secondary search is to find any occupants or victims who were not found in the primary search. If possible, it should be conducted by a different search team so that each area of the building is examined with a fresh set of eyes. The secondary search is covered in more detail later in this chapter.

During the primary search, fire fighters rapidly search the accessible areas of a burning structure where conditions may still permit human life to exist. The objective of this search is to find any potential victims as quickly as possible and to remove them from danger. When fire fighters complete the primary search, they have gone as far as they can and have removed anyone who was rescued. The phrase "Primary search is all clear" is used to report that the primary search has been completed.

By necessity, the primary search is conducted quickly and gives priority to those areas where victims are most likely to be located. Time is a critical concern, because fire fighters must reach potential victims before they are burned, overcome by smoke and toxic gases, or trapped by a structural collapse. Search teams may have only a few minutes to conduct a primary search. In that limited amount of time, fire fighters must try to find anyone who could be in danger and remove those individuals to a safe area. Active fire conditions may limit the areas that can be searched quickly as well as the time that can be spent in each area.

Fire fighters should try to check all of the areas where victims might be, such as beds, cribs, and sofas. Adults who tried to escape on their own are often found near doors or windows. Some people—particularly children—may try to hide in a closet, in a bathtub or shower, or behind or under a piece of furniture. Be aware that young children may be frightened by the appearance of a fire fighter wearing full PPE.

The primary search is frequently conducted in conditions that expose both fire fighters and victims to the risks presented by heavy smoke, heat, structural collapse, and entrapment by the fire. Search teams must often work in conditions of zero visibility and may have to crawl along the floor to stay below layers of hot gases. Because the beam from a powerful hand light might be visible only a few inches inside a smoke-filled building, search teams must practise searching for victims in total darkness. They must know how to keep track of their location and how to get back to their entry point. Practising these skills in a controlled environment will enable fire fighters to perform confidently under similar conditions at a real emergency incident.

Fire fighters must rely on their senses when they search a building:

- Sight: Can you see anything?
- Sound: Can you hear someone calling for help, moaning, or groaning?
- Touch: Do you feel a victim's body?

If smoke and fire restrict visibility, fire fighters must use sound and touch to find victims. Every few seconds a member of the search team should yell out "Is anyone in here? Can anyone hear me?" Then listen. Hold your breath to quiet your breathing regulator, and stop moving. People who have suffered burns or are semiconscious may not be able to speak intelligibly, so you will need to listen for guttural sounds, cries, faint voices, groans, and moans. Focus on the direction of any sounds you hear.

As you search, feel around for the hands, legs, arms, and torsos of potential victims. Use side crawls to extend your reach sideways; sweep ahead and to the sides as you crawl. Practise until you can tell the difference between a piece of furniture and a human body. The type of furniture you encounter may also provide clues for finding victims. Spindle-type legs in a bedroom may indicate a crib that needs to be searched. An unusually low bed may be the bottom berth of a children's bunk bed, indicating that both the bottom bed and top bed need to be searched.

Zero-visibility conditions can be disorienting. In such circumstances, fire fighters must follow walls and note turns and doorways to avoid becoming lost. The use of a thermal imaging device can be beneficial in improving the speed and efficiency of a search. After locating a victim or completing a search of an area, the search team must be able to retrace its path and return to its entry point.

Search teams must also identify a secondary escape route in case fire conditions change and block their planned exit route. Thus, during the search, fire fighters should note the locations of stairways, doors, and windows. Always be aware of the nearest exit and an alternate exit. It is also important to know the location of interior doors in the case that fire fighters need to isolate themselves from rapid fire growth. If the situation deteriorates rapidly, a window can serve as an emergency exit from a room. Fire fighters should keep track of the exterior walls and window locations, even reaching up to feel for the windows if conditions force them to crawl.

The search team must remain in voice contact with one another and in radio contact with the IC or someone outside the building. The search team members must be able to give their location, including the building section and floor, to the IC if they need assistance. If the search team needs a ladder for the rescue, the IC will need to know where the ladder should be placed (which side of the building) and how long a ladder will be needed (which floor). Fire fighters should use standard incident command system terminology to describe their location over the radio.

FIGURE 12-5 Search teams always include at least two members.
© Jones & Bartlett Learning. Photographed by Glen E. Ellman.

SAFETY TIP

While searching for victims, you also need to focus on your safety. Continuously monitor your air supply and the conditions of the fire and the structure. You should also be aware of the status of other crews, such as the fire attack crew and the ventilation crew.

General Search Techniques

Fire fighters should employ standard techniques to search assigned areas quickly, efficiently, and safely. They should operate in teams of two or more and should always stay close together (**FIGURE 12-5**). Partners must remain in direct visual, voice, or physical contact with each other.

As discussed previously, at least one member of each search team should also have a radio to maintain contact with the command post or someone outside the building. The search team can use this radio to call for help if they become disoriented, trapped by fire, or need assistance. If the search team finds a victim, they can notify the IC so that help will be available to remove the victim from the building and begin medical treatment. The search team must notify the IC when the search of each area has been completed so the IC can make informed decisions about which steps to take next.

Remember, the primary search is a quick attempt to locate any potential victims who are in danger. The primary search should be efficient and as thorough as possible. It should cover any places that can be safely searched where a victim might be located. Fire conditions may make it impossible or unsafe to search in some parts of a building. Changing fire conditions will limit the time available for a primary search. Team members must keep themselves safe. They need to monitor changes in the fire conditions to make sure their position is not in danger of an immediate flashover and to monitor changes in the smoke level and characteristics, the heat level, and the location of the flames. Possible indications of structural changes, such as unusual sounds or sagging, must be noted and reported to the IC. They also need to carefully monitor their air supplies to ensure they know when to exit in order to be out of the IDLH zone before their low-air alarm sounds.

Changes in the fire suppression activities are very important, too. The search team needs to know if the fire attack team is making progress or if they are having difficulty in establishing a water supply or advancing the attack line. The search team members must have a plan for conducting a search that will maximize their efforts. Perhaps most importantly, they need to keep track of the progress they are making in completing the search.

Using Thermal Imaging Devices

A **thermal imaging device** is a valuable tool for conducting a search in a smoke-filled building. This piece of equipment is similar to a video camera, except that it displays images of heat emitted from surfaces instead of visible light images. The images appear on a display screen and show the relative amounts of heat being radiated by different objects. It is important that you learn how to properly set any controls on your thermal imaging device and practise operating it in order to be proficient in its use during an emergency situation.

A major benefit of using a thermal imaging device is that it can "see" the heat signature of a person in conditions of total darkness or through smoke that totally obscures normal vision (**FIGURE 12-6**). Fire fighters will be able to identify the shape of a human body with a thermal image scan because the body will be either warmer or cooler than its surroundings.

Temperature differences also mean that the thermal imaging device can show furniture, walls, doorways, and windows. This information may enable a fire fighter to navigate through the interior of a smoke-filled building. The thermal imaging device can sometimes be used to locate a fire in a smoke-filled building or behind doors, walls, or ceilings.

One of the limitations of thermal imaging devices is that the user of the device is the only person who can see the images. The person operating the thermal imaging device must communicate the information seen on the camera screen to the search team members. Additionally, fire fighters can find themselves in a difficult situation if the thermal imaging device fails and leaves the team members without this special set of eyes. It is important to use good search methods and to encourage all team members to remain aware of their surroundings despite the use of a thermal imaging device.

Remember that the thermal imaging device does not "see" fire. It measures the heat flux—that is, the heat that is given off by an object—and shows differences in the amount of heat given off by different objects or from different parts of the room. By itself, it does not show you the integrity of a structure. Be alert for areas that show small heat signatures (areas of higher temperature), which may indicate that a floor is not safe to walk on or an area of roof that is unsafe to be under. If the thermal imaging device gives any indication of an unsafe condition, make sure that a thorough exterior size-up has been conducted.

Several types of thermal imaging devices are available. Some are hand-held devices, similar to a hand-held computer with a built-in monitor. Others are helmet-mounted devices that produce an image in front of the face piece of the user's SCBA. Each type of thermal imaging device has different controls and functions, and the images may be displayed in a variety of ways. It is important to carefully read and understand the instructions for the thermal imaging devices used by your department. Fire fighters need training and practice to become proficient in using these devices, interpreting the images, and maneuvering confidently through a building while using a thermal imaging device. Regular equipment checks are required to ensure adequate battery life and proper operation.

FIRE FIGHTER TIP

Not all rooms have just four walls; some may have five or more. To determine how many walls a room has, check for walls on each side of the door as you enter the room. If there is a wall parallel to the door, on the right of the door, and to the left of the door, then you will need to search five wall surfaces. If there is a parallel wall only to the right or to the left of the door, then the room has four walls. Noting the number of walls in a room about to be searched is a good way to ensure that the entire room has been searched and to maintain orientation.

Four Methods of Conducting a Primary Search

Four methods of conducting a primary search are described here. These are the standard search method, the oriented search method, the oriented-vent-enter-isolate-search (O-VEIS) method, and the team search method. It is important to understand how each is conducted and where each is most appropriately used. All four methods require a thorough training program that includes sufficient practice in order for you to perform efficient and effective searches. And you need regular practise in simulated fire conditions to maintain proficiency.

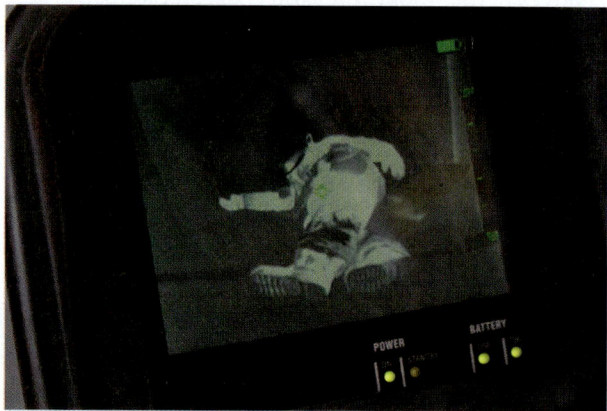

FIGURE 12-6 A thermal imaging device can capture an image of a person through thick smoke.
© Jones & Bartlett Learning. Photographed by Glen E. Ellman.

The Standard Search Method

A primary search using the standard search method is conducted with a search team of two or more members. All members of the search team cover the same area at the same time.

The search pattern will be dictated by the size of the room, scene conditions, potential number of victims, and number of available team members. Searchers conduct a quick and systematic search by staying on an outside wall of the room. They may choose to use a left-handed or a right-handed search pattern. The fire fighter closest to the wall keeps his or her left or right shoulder close to the wall at all times (**FIGURE 12-7**). Fire fighters must follow walls and note turns and doorways to avoid becoming lost. Searchers should use the most efficient movement based on the hazard encountered and should maintain team integrity during the search by using visual, voice, or direct contact.

Some departments advocate that each searcher carry a tool to extend his or her reach. Others believe that the tool reduces the searcher's ability to feel the texture and composition of an object. The advantages

A

B

FIGURE 12-7 The standard search method. **A.** A left-handed search pattern. **B.** A right-handed search pattern.
© Jones & Bartlett Learning. Photographed by Glen E. Ellman.

and disadvantages of using a tool when searching are discussed later in this chapter. If the room is large, it may be necessary for searchers to perform side crawls to check the center parts of the room. To do this, one searcher stays close to the wall at all times, and a partner may move closer to the center of the room with two or three side crawls.

Depending on the department's procedures, searchers can mark the door as searched or verbally report their progress to the command post after completing the search.

FIRE FIGHTER TIP

When search and rescue efforts are coordinated with fire suppression and ventilation operations, the operational plan may include closing the door to the room as it is being searched. When properly coordinated, this action can reduce smoke and heat from entering the room and allows the fire fighters to open a window to clear smoke from the room if needed. After completing the search, the searchers should close any open windows before opening the door of the room, reentering the hallway, closing the door behind them, and proceeding to the next room. The use of this technique will depend on department procedures and the overarching operational plan.

The standard search method is the most commonly taught method of primary search used by fire fighters. The main use of the standard search method is in residential fires, where two fire fighters should be able to search an average-sized house in 15 minutes or less (Coleman, 2011).

A primary search using the standard search method keeps searchers together and in voice contact of each other. A search officer or team leader may not be available to help monitor the searchers' air supply and safety, the stability of the structure, and other fire-ground operations. The searchers must be careful to pay attention to their surroundings and watch out for deteriorating conditions. If a victim is located, the searchers must exit with that victim; for this reason, the continuity of the search is more likely to be lost than if using the oriented search method.

Follow the steps in **SKILL DRILL 12-1** to perform a primary search of a residential structure and rescue victims using the standard search method.

The Oriented Search Method

Depending on the size of the room, scene conditions, potential number of victims, and number of available team members, a primary search using the oriented search method is conducted using a search team consisting of an officer or team leader and one to three searchers. This

SKILL DRILL 12-1
Conducting a Primary Search Using the Standard Search Method
Fire Fighter I, NFPA 1001: 4.3.9

1 Don your PPE, including SCBA, and enter the personnel accountability system. Bring hand tools, a hand light, a radio, a thermal imaging device, and search ropes if indicated. Notify command that the search is starting, and indicate the area to be searched and the direction of the search. Use hand tools or ground ladders if needed to gain access to the site. Conduct a quick and systematic search by staying on an outside wall and searching from room to room. Maintain contact with an outside wall.

2 Maintain team integrity using visual, voice, or direct contact. Use the most efficient movement based on the hazard encountered: duck walk, crawl, stand only when you can see your feet and it is not hot. Use tools to extend your reach if recommended by your department. Clear each room visually or by touch, and then close the door. Search the area, including stairs up to the landing on the next floor.

3 Periodically listen for victims and sounds of fire. Communicate the locations of doors, windows, and inside corners to other team members. Observe fire, smoke, and heat conditions; update command on this information. Locate and remove victims; notify the IC. When the search is complete, conduct a personnel accountability report. Report the results of the search to your officer.

© Jones & Bartlett Learning. Photographed by Glen E. Ellman.

method of search should only be conducted if approved by your department.

With the oriented search method, the officer remains just outside or in the entranceway to the individual rooms that are being searched, and a single searcher or multiple searchers systematically search one room at a time using the right-handed or left-handed search method. At first glance this seems like it would decrease the efficiency of the search; however, in this scenario the safety of the crew is maintained by the officer.

The search officer is responsible for maintaining the searchers' safety, monitoring their air supplies, monitoring the progression of the fire and the fire suppression efforts, developing a systematic search plan, assessing the progress of the search effort, monitoring the activities on the rest of the fire scene, and coordinating activities with the IC. This provides a greater degree of safety for the search team and allows the searchers to focus on the search itself. The officer also keeps track of which rooms have been searched and which are yet to be searched.

By staying just outside or in the entranceway of the rooms being searched, the officer is in a better position to note these changes than the search team who is searching inside individual rooms. The officer must maintain voice contact with each member of the search team. This method enables the fire fighters who are searching to concentrate all their attention on the task of searching their assigned area. The officer controls the door and monitors the means of exit for the search team. If possible, there should be a primary exit and a second one that can be used if the first one becomes untenable.

Follow the steps in **SKILL DRILL 12-2** to conduct a primary search using the oriented search method. The oriented search method takes less time than using the standard search method (Coleman, 2011).

Depending on department procedures, the searched room or apartment can be marked so other personnel will know that it has been searched. Chalk, crayons, felt tip markers, or masking tape can be used to mark the door for this purpose. Some fire departments use a two-part marking system to indicate when a search is in progress and when it has been completed: In this system, a slash ("/") indicates that a search is in progress, and an "X" indicates that the search has been completed (**FIGURE 12-8**). Other fire departments may place an object in the doorway or attach a tag or latch strap to the doorknob to indicate that the room has been searched.

Continuity of the search can be easily lost when a victim is found and removed from the hazardous environment by a single search team. When this happens, it is hard to resume the search from the same location. By the time a second team has been assembled and arrives at the location where the victim was found, it is hard to pick up the search without duplication of efforts and wasting valuable time. In switching from one team to another, it is easy to miss areas that need to be searched. The most efficient way to maintain the continuity of the search is for the search officer to notify command that a victim has been found, move toward the exit with the victim, and be met by a team that can remove the victim the rest of the way out of the building and begin emergency medical care. This approach

SKILL DRILL 12-2
Conducting a Primary Search Using the Oriented Search Method
Fire Fighter I, NFPA 1001: 4.3.9

1 A search team consisting of one officer or team leader and one to three searchers is assembled.

2 Searchers don their PPE, including SCBA and hand tools, and enter the personnel accountability system.

(continued)

SKILL DRILL 12-2 Continued
Conducting a Primary Search Using the Oriented Search Method
Fire Fighter I, NFPA 1001: 4.3.9

3 The officer notifies the IC that search is starting and directs the searchers to the area to be searched.

4 The officer remains outside the rooms to be searched to monitor safety conditions, air supplies, and the status of the fire. The officer maintains a systematic search pattern and coordinates activities with the IC.

5 Searchers use a left-handed or a right-handed search pattern and perform two to three side crawls as necessary to extend the search toward the center of the room. Upon completion of the search of each room, the officer directs searchers to the next rooms to be searched and closes the door. If a victim is found, the officer notifies the IC and requests a second team to help remove the victim. The search team moves the victim toward the exit and turns over the care and removal of the victim to the second team.

6 The search team then returns to the last location searched and continues the systematic search of the building. When the search is complete, the officer conducts a personnel accountability report. The officer reports the results of the search and the personnel accountability report to the IC.

FIGURE 12-8 In large buildings, doors should be marked after the rooms are searched.
© Jones & Bartlett Learning. Photographed by Glen E. Ellman.

enables the search team to return to the location they were searching and continue the search for additional victims without losing much time and without losing the continuity of the search. This operation requires planning, a lot of practise, and a well-coordinated fire suppression operation.

> ### FIRE FIGHTER TIP
>
> Specific safety requirements for search and rescue operations are defined in NFPA 1500, *Standard on Fire Department Occupational Safety, Health, and Wellness Program,* and in regulations enforced by the Occupational Safety and Health Administration. The NFPA requirements state that a team of at least two fire fighters must enter together, and at least two other fire fighters must remain outside the danger area, ready to rescue the fire fighters who are inside the building (NFPA, 2018). This policy is sometimes called the **two-in/two-out rule**.

The Oriented-Vent-Enter-Isolate-Search (O-VEIS) Method

The original vent-enter-search (VES) method of conducting a primary search was developed for situations in which there is a porch located in front of a bedroom window and a person in the bedroom needs to be rescued. This method of primary search is meant to be used in extreme situations in which there is a shortage of personnel. One fire fighter can quickly place a ladder to the porch roof, quickly open or break the bedroom window, enter the bedroom, perform a quick search of the room, and exit back onto the porch roof. If it is possible to access a second bedroom from the same porch, this method allows for search to be conducted in two bedrooms in a short period of time. This evolution is very dangerous

for several reasons, however. Opening a bedroom window is a type of ventilation, which adds oxygen to the fire, contributing to a flashover or directing the hot fire gases toward the open window. Additionally, the VES method violates the two-in/two-out rule, and because it is often done before any hose line is in place, the fire may expand rapidly. VES is a very dangerous method and should not be used. A safer modification of VES is the oriented-vent-enter-isolate-search (O-VEIS) method. This method stresses the need to quickly close the door between the room to be searched and the rest of the house.

> ### SAFETY TIP
>
> There are two additions to the classic vent-enter-search method that can make this procedure safer:
>
> 1. Isolate the room to be searched. Be sure to close the door between the room and the hallway as soon as you enter the room. According to studies conducted by the National Institute of Standards and Technology, this step will help to keep the room more tenable for a few minutes while you perform a quick search of the room (UL Firefighter Safety Research Institute, 2013).
> 2. Use a team of two fire fighters. One fire fighter enters the room to isolate it and perform the search. The second fire fighter remains outside the window through which entry was made; this fire fighter monitors the fire conditions and can warn the inside fire fighter of changing fire conditions. The second fire fighter can also assist in rescuing any victims who are found.

A primary search using the O-VEIS method is conducted using a search team consisting of an officer or team leader and one searcher. The team places a ladder in front of the window leading to the room or to the porch in front of the room to be searched. They then climb to the window or porch roof and assess the situation. After determining the room appears to be tenable for a victim and for the fire fighter, they open or remove the window. The officer remains outside—on the porch roof or, if no porch is present, on the ladder—to continue to assess the situation in the room, maintain radio contact, and monitor activities in other parts of the fire ground. The officer should have a hand light to assist in orienting the searcher back to the point of entry (**FIGURE 12-9**).

The searcher quickly enters the room. As soon as possible, this fire fighter locates the door to the hallway and immediately closes it. This action isolates the room to be searched from the fire flow path (UL Firefighter Safety Research Institute, 2013). Even a thin hollow core door will provide some protection from the flow path for a short period of time. This isolation reduces the heat and smoke in that room for a few minutes and

FIGURE 12-9 In the oriented-vent-enter-isolate-search (O-VEIS) method of primary search, the search officer remains outside to monitor the fire and safety conditions and to assist the inside searcher.
© Jones & Bartlett Learning. Photographed by Glen E. Ellman.

decreases the chance that the room will quickly become untenable for the rescuer and any victims present. It prevents unplanned ventilation, which may rapidly accelerate the fire growth; it also improves visibility in the bedroom and reduces the chance for flashover.

If a victim is found, the officer can assist in removing the victim to the porch roof or ladder. If a victim is found and there is another bedroom to be searched from the same porch, it may be more efficient for the officer to request a second rescue crew to remove the victim from the porch roof so the original search team can begin to search the second bedroom.

A primary search using the O-VEIS method is somewhat safer and more efficient than the original VES method. This is not to say that it is a safe procedure. It should be considered only in dire emergencies in which a sizeable risk has a large potential benefit. It should not be attempted if the room is fully charged with smoke and in danger of flashing over when additional oxygen is introduced.

The Team Search Method

The team search (also known as the rope search) method of conducting a primary search uses a **search rope** or hose line to provide searchers with an anchor point to keep them in contact with each other and with the egress (exit) route from a burning building. The search team includes an anchor person who ties off the search rope to a stationary object that is located outside the building, about 3 m (10 ft) outside the exit door. As the searchers progress through the building, the anchor person feeds the line out while keeping slight tension on it at a comfortable height above the floor (about 0.5 m). This method gives the searchers an identifiable route out of the building if the direction of travel inside the building needs to be changed. Additionally, the anchor person

can let the searchers know how far they have progressed into the building.

This method of conducting a primary search is more complicated than the three methods previously discussed. It is time consuming and requires a lot of training and coordination, but it is well suited for locating and removing lost occupants or injured fire fighters. The team search method is an especially valuable tool for a rapid intervention crew/company working to locate a missed, downed, or trapped fire fighter in a large space. Its use also enables other fire fighters to locate the search team quickly. The team search method is not often used in a residential fire unless the house is very large or has large open spaces; however, it can be a valuable method to locate an occupant in a large warehouse or other large open structure when visibility is severely limited.

FIRE FIGHTER TIP

The team search method of conducting a primary search helps fire fighters keep in contact with the exit in a building that contains cold smoke. Cold smoke is generated when a fire sprinkler system is activated during a fire. The sprinkler cools the smoke and causes it to bank down to the floor. Cool smoke and condensed water droplets severely limit visibility.

With the team search method, the officer of the search team is responsible for developing a search plan, keeping the team safe, and communicating with the IC. The officer stays in contact with the search rope or hose line and maintains voice contact with the searchers. This helps to keep the searchers oriented. The officer also ensures that each member is monitoring his or her air supply. This officer is usually responsible for operating the thermal imaging device and directing the team based on the images seen on the screen. The officer must also assess any changing conditions around the search team and communicate these changes to the IC.

The members of the search team are responsible for following the search rope or hose line into the building and conducting the primary search off the rope or hose line. This is done by placing one searcher on each side of the search rope. When searching large areas, each of the searchers performs two or three side crawls at a right angle to the rope. The searcher on the left side of the rope side crawls to the left of the rope, and the searcher on the right side of the rope side crawls to the right of the rope. Some departments use a short lateral rope that clips to the main rope. This enables them to move a limited distance from the main search rope.

When the search team is ready to exit, the officer should assume the anchor position on the search line and maintain tension on the line as the team works its way along the line to the building exit. The officer assumes this position in order to remain accountable for the other members of the team.

The team search method keeps searchers together, preventing individuals from getting lost or separated. It also keeps searchers oriented to the exit and provides a safe, quick, and direct route out of the building. This method is useful when searching large areas, especially where there are limited landmarks and limited visibility to provide directional guidance. It also allows a rescue team to find the searchers if they need to enter to remove a victim who has been located by the search team.

However, because the team search method relies solely on the search rope, the search team spends more time holding on to the rope than engaging in active searching. It is also hard to maintain the continuity of the search if a victim is found and removed. If the search team locates a victim and leaves to remove the victim, it will be hard for a new crew to return to the same place to continue a systematic search.

FIGURE 12-10 A secondary search is conducted after the fire is under control.
© Jones & Bartlett Learning. Courtesy of MIEMSS.

FIRE FIGHTER TIP

When a rope is used as a search tool, knots are sometimes placed every 20 ft (6 m) to enable searchers to know how far they have progressed into the building (Coleman, 2011). The number of knots indicates the number of feet from the end of the rope. For example, four knots would equal 24 m (80 ft) of travel into the building. Some fire departments prefer to use a fire hose in place of a search rope because it is larger, always available, and easier to locate when visibility is limited. In cases of emergency, a fire fighter can determine which direction leads to the exit by identifying the male and female couplings on the hose. Additionally, some departments drop flashing hand lights at intervals leading toward the exit.

Secondary Search

The purpose of the primary search is to locate victims and remove them from the hazardous environment. The purpose of the secondary search is to locate any occupants or victims who are unaccounted for and were not found and removed during the primary search. The secondary search is conducted as soon as the fire is under control or fully extinguished and there are sufficient resources available (**FIGURE 12-10**). The secondary search is conducted slowly and methodically to ensure that no areas are overlooked. Areas that could not be covered during the primary search are covered during the secondary search. The time needed to complete a secondary search will be dependent on the building size, the type of occupancy, the type of building construction, the fire severity, the amount of contents or clutter in the building, and the probability of collapse.

The secondary search should be conducted by a different team than the one that conducted the primary search. Bringing in a fresh crew helps to lessen the possibility that a victim will be missed a second time.

There are fewer risks to fire fighters during the secondary search than during the primary search. Once the fire is under control, conditions during the secondary search should be much different, ventilation should have cleared out most of the smoke, and visibility should be greatly improved. Lights can be brought and set up to make it easier to see the area that is being searched. It is important to remain alert for conditions indicating an unstable structure that may result in a collapse. Continue to watch for signs of a fire **rekindle**.

The secondary search team needs to wear proper PPE, including SCBA, helmets and gloves, and structural firefighting protective gear. They should continue to use SCBA until the air quality has been thoroughly tested. Air that looks clear can contain high concentrations of poisonous gases and small suspended particles of toxic by-products of the fire. Approximately 97% of these particulates are invisible to the human eye. In addition, the by-products of combustion (soot, smoke, and ash) contain cancer-causing substances that can be absorbed through the skin (Underwriters Laboratories, 2010). Air monitoring is discussed in Chapter 19, *Salvage and Overhaul*.

The secondary search should be conducted in an orderly, systematic manner to ensure that all areas have been searched. In some instances it may be most efficient to start the secondary search in the areas less affected by the fire. This is especially true when overhaul is still occurring in the burned parts of the building. The search team can then systematically work their way toward the area affected by the fire. In other instances it is important to begin the secondary search in the area affected by the fire—for example, when the fire is small or when you have a strong reason to believe a victim is located close to the fire. Pay special attention to any witnesses, especially if they have knowledge of the building occupants. Be methodical and coordinated. Move furniture away from walls, and carefully look under beds, furniture, boxes, and clothes that might hide a child or adult. Be especially thorough in checking all places where a child could hide, including closets, bathtubs, shower enclosures, behind doors, under windows, and inside toy chests. If debris is present from the fire, remove it to thoroughly search all areas of the room.

When searching a large commercial or residential building, the secondary search should include the floor above the fire, the floor below the fire, stairwells, elevator shafts, the roof, the outside perimeter of the building, under trees and shrubs around the building, and even the roof tops of exposed buildings. As you are searching, be alert for fire rekindle and any signs of the origin of the fire that need to be reported to the IC. A systematic search will be helpful in ensuring that all areas are searched.

When the secondary search has been completed, the search officer or team leader should notify the IC by reporting, "Secondary search is clear."

Follow the steps in **SKILL DRILL 12-3** to perform a secondary search.

SAFETY TIP

Even though the fire is under control, safety remains a top consideration during a secondary search. SCBA should be used until the air is tested and pronounced safe to breathe because levels of carbon monoxide and other poisonous gases often remain high for a long time after a fire is extinguished.

SKILL DRILL 12-3
Performing a Secondary Search Fire Fighter I, NFPA 1001: 4.3.9

1 ▪ Don PPE, including SCBA, and enter the personnel accountability system.

▪ Bring hand tools, a hand light, a radio, a thermal imaging device, and search ropes if indicated.

▪ Notify command that the search is starting, and indicate the area to be searched and the direction of the search.

2 ▪ Conduct a systematic, thorough search for any possible victims.

▪ In addition to checking for victims, be alert for hidden fire, salvage needs, and indicators of the fire's cause, its origin, or arson.

▪ Protect any evidence.

▪ Maintain team integrity using visual, voice, or direct contact.

▪ Notify command of the results of the search and personnel accountability report.

Rescue Techniques

Rescue is the removal of a located person who is unable to escape from a dangerous situation. Fire fighters rescue people not only from fires but also from a wide variety of accidents and mishaps. Although this section refers primarily to rescuing occupants from burning buildings, many of the techniques discussed here can be used for other types of situations. As a fire fighter, you must learn and practise the various types of assists, carries, drags, and other techniques used to rescue people from fires. After mastering these techniques, you will be able to rescue victims from life-threatening situations.

Rescue is the second component of search and rescue. When you locate a victim during a search, you must direct, assist, or carry that person to a safe area. Rescue can be as simple as verbally directing an occupant toward an exit, or it can be as demanding as extricating a trapped, unconscious victim and physically carrying that person out of the building. The term "rescue" is generally applied to situations where the rescuer physically assists or removes the victim from the dangerous area.

Most people who realize that they are in a dangerous situation will attempt to escape on their own. Children and elderly persons, physically or developmentally handicapped persons, impaired people, and ill or injured persons, however, may be unable to escape and will need to be rescued. People who are sleeping, are under the influence of alcohol or drugs, or have diminished mental capacity may not become aware of the danger in time to escape from the structure on their own. Toxic gases may incapacitate even healthy individuals before they can reach an exit. Victims also may become trapped when a rapidly spreading fire, an explosion, or a structural collapse cuts off potential escape routes.

Because fires threaten the lives of both victims and rescuers, the first priority is to remove the victim from the fire building or dangerous area as quickly as possible. It is sometimes better to move the victim to a safe area first and then provide any necessary medical treatment. The assists, lifts, and carries described in this chapter should not be used if you suspect that the victim has a spinal injury, however, unless there is no other way to remove the person from the life-threatening situation.

Always use the safest and most practical means of egress when removing a victim from a dangerous area. A building's normal exit system, such as interior corridors and stairways, should be used if it is open and safe. If the regular exits cannot be used, an outside fire escape, a ladder, or some other method of egress must be found. Ladder rescues, which are covered in this chapter, can be both difficult and dangerous, whether the victim

is conscious and physically fit or is unconscious and injured (**FIGURE 12-11**).

Shelter-in-Place

In some situations, the best option is to shelter the occupants in place instead of trying to remove them from the building. This option should be considered when the occupants are conscious and are found in a part of the building that is adequately protected from the fire by fire-resistive construction or fire suppression systems. If smoke and fire conditions block the exits, victims might be safer staying in the sheltered location than attempting to evacuate through a hazardous environment.

Such a situation occurs in a high-rise apartment building when a fire is confined to one apartment or one floor. In this scenario, the stairways and corridors may be filled with smoke, but the occupants who are remote from the fire would be very safe on their balconies or in their apartments with the windows open to provide fresh air. They would be exposed to greater levels of risk if they attempted to exit than if they remained in their apartments until the fire is extinguished. When occupants remain in place, it is important for them to isolate the room where they are from the fire. This can be done by practicing good door control or by

FIGURE 12-11 Bringing an unconscious victim or a person who needs assistance down a ladder can be difficult and dangerous.

determining whether they are safer with the windows open or with the windows closed. Open windows can serve as an entry point for smoke from the outside or actually draw the fire to the occupant. The decision to follow a shelter-in-place strategy in this case must be made by the rescue officer or IC. Potential locations to shelter-in-place should be identified on the preincident plan.

Exit Assist

The simplest rescue is the exit assist, in which the victim is responsive and able to walk without assistance or with very little assistance. The fire fighter may simply need to guide the person to safety or provide a minimal level of physical support.

Two types of assists can be used to help responsive victims exit a fire situation:

- One-person walking assist
- Two-person walking assist. Even if the victim can walk without assistance, the fire fighter should take the person's arm or use the one-person walking assist to make sure that the victim does not fall or become separated from the rescuer.

One-Person Walking Assist

The one-person walking assist can be used if the person is capable of walking. To perform a one-person walking assist, follow the steps in **SKILL DRILL 12-4**.

SKILL DRILL 12-4
Performing a One-Person Walking Assist Fire Fighter I, NFPA 1001: 4.3.9

1 Help the victim stand next to you, facing the same direction.

2 Have the victim place his or her arm around your neck, and hold on to the victim's wrist, which should be draped over your shoulder. Put your free arm around the victim's waist, and help the victim to walk.

Two-Person Walking Assist

The two-person walking assist is useful if the victim cannot stand and bear weight without assistance. With this assist technique, the two rescuers completely support the victim's weight. It may be difficult to walk through doorways or narrow passages using this type of assist. To perform a two-person walking assist, follow the steps in **SKILL DRILL 12-5**.

SKILL DRILL 12-5
Performing a Two-Person Walking Assist Fire Fighter I, NFPA 1001: 4.3.9

1 Two fire fighters stand facing the victim, one on each side of the victim. Both fire fighters assist the victim to a standing position.

2 Once the victim is fully upright, place the victim's right arm around the neck of the fire fighter on the right side. Place the victim's left arm around the neck of the fire fighter on the left side. The victim's arms should drape over the fire fighters' shoulders. The fire fighters hold the victim's wrist in one hand.

3 Both fire fighters put their free arms around the victim's waist for added support. Both fire fighters slowly assist the victim to walk. Fire fighters must coordinate their movements and move slowly.

Simple Victim Carries

Four simple carry techniques can be used to move a victim who is conscious and responsive but incapable of standing or walking:

- Two-person extremity carry
- Two-person seat carry
- Two-person chair carry
- Cradle-in-arms carry

Two-Person Extremity Carry

The two-person extremity carry requires no equipment and can be performed in tight or narrow spaces, such as the corridors of mobile homes, small hallways, and narrow spaces between buildings. The focus of this carry is on the victim's extremities. To perform a two-person extremity carry, follow the steps in **SKILL DRILL 12-6**.

SKILL DRILL 12-6
Performing a Two-Person Extremity Carry **Fire Fighter I, NFPA 1001: 4.3.9**

1 Two fire fighters help the victim to sit up.

2 The first fire fighter kneels behind the victim, reaches under the victim's arms, and grasps the victim's wrists.

3 The second fire fighter backs in between the victim's legs, reaches around, and grasps the victim behind the knees.

4 The first fire fighter gives the command to stand and carry the victim away, walking straight ahead. Both fire fighters must coordinate their movements.

Two-Person Seat Carry

The two-person seat carry is used with victims who are disabled or paralyzed. This type of carry requires the assistance of two fire fighters, and moving through doors and down stairs may be difficult. To perform a two-person seat carry, follow the steps in **SKILL DRILL 12-7**.

Two-Person Chair Carry

The two-person chair carry is particularly suitable when a victim must be carried through doorways, along narrow corridors, or up or down stairs. In this technique, two rescuers use a chair to transport the victim. A folding chair should not be used for this purpose, because folding chairs can collapse when used in this

SKILL DRILL 12-7
Performing a Two-Person Seat Carry **Fire Fighter I, NFPA 1001: 4.3.9**

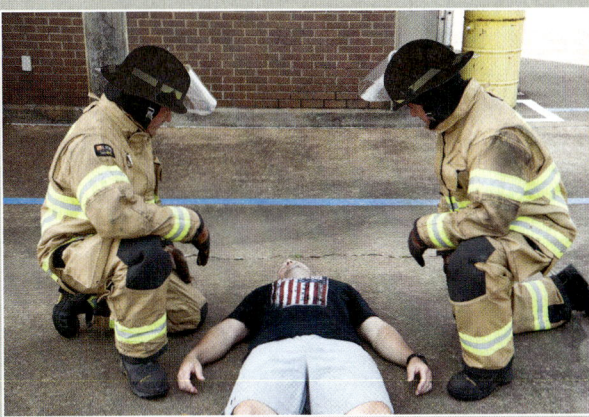

1 Two fire fighters kneel near the victim's hips, one on each side of the victim.

2 Both fire fighters raise the victim to a sitting position and link arms behind the victim's back.

3 The fire fighters place their free arms under the victim's knees. If possible, the victim puts his or her arms around the fire fighters' necks for additional support. The rescuers stand and move toward safety.

© Jones & Bartlett Learning. Photographed by Glen E. Ellman.

way to carry a victim. The chair must be strong enough to support the weight of the victim while he or she is being carried. The victim should feel much more secure with this carry than with the two-person seat carry, and he or she should be encouraged to hold on to the chair. To perform a two-person chair carry, follow the steps in **SKILL DRILL 12-8**.

Cradle-in-Arms Carry

The cradle-in-arms carry can be used by one fire fighter to carry a child or a small adult. With this technique, the fire fighter should be careful of the victim's head when moving through doorways or down stairs. To perform the cradle-in-arms carry, follow the steps in **SKILL DRILL 12-9**.

SKILL DRILL 12-8
Performing a Two-Person Chair Carry Fire Fighter I, NFPA 1001: 4.3.9

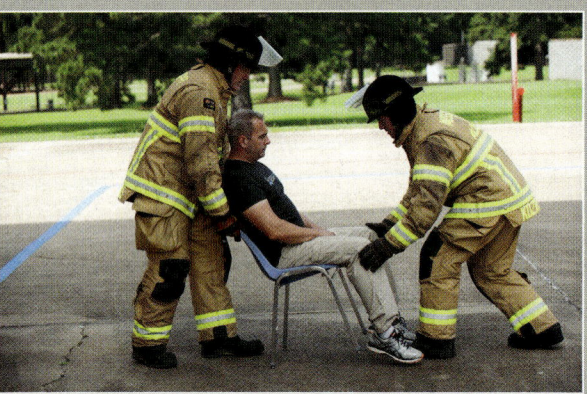

1 Tie the victim's hands together, or have the victim grasp his or her hands together. This prevents the victim from reaching for a stationary object while you are moving him or her. One fire fighter stands behind the seated victim, reaches down, and grasps the back of the chair.

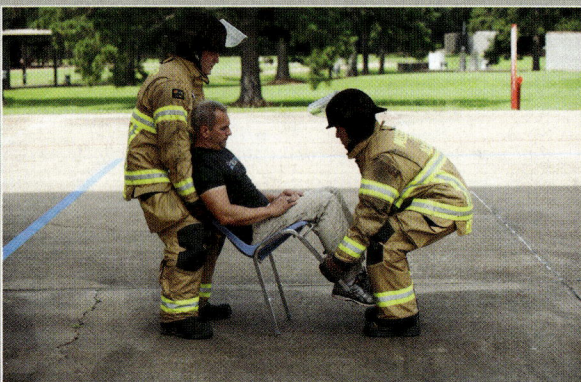

2 The fire fighter tilts the chair slightly backward on its rear legs so that the second fire fighter can step back between the legs of the chair and grasp the tips of the chair's front legs. The victim's legs should be between the legs of the chair.

3 When both fire fighters are correctly positioned, the fire fighter behind the chair gives the command to lift and walk away. Because the chair carry may force the victim's head forward, watch the victim for airway problems.

© Jones & Bartlett Learning. Photographed by Glen E. Ellman.

SKILL DRILL 12-9
Performing a Cradle-in-Arms Carry Fire Fighter I, NFPA 1001: 4.3.9

1 Kneel beside the child, and place one arm around the child's back and the other arm under the thighs.

2 Lift slightly and roll the child into the hollow formed by your arms and chest.

3 Be sure to use your leg muscles to stand.

Emergency Drags

The most efficient method to remove an unconscious or unresponsive victim from a dangerous location is a drag. Six types of emergency drags can be used to remove unresponsive victims from a fire situation:

- Clothes drag
- Blanket drag
- Standing drag
- Webbing sling drag
- Fire fighter drag
- Emergency drag from a vehicle

When using an emergency drag, the rescuer should make every effort to pull the victim in line with the long axis of the body to provide as much spinal protection as possible. The victim should be moved head first to protect the head.

Clothes Drag

The clothes drag is used to move a victim who is on the floor or the ground and is too heavy for one rescuer to lift and carry alone. In this technique, the rescuer drags the person by pulling on the clothing in the neck and shoulder area. The rescuer should grasp the clothes just behind the collar, use the arms to support the victim's head, and drag the victim away from danger. To perform the clothes drag, follow the steps in **SKILL DRILL 12-10**.

Blanket Drag

The blanket drag can be used to move a victim who is not dressed or who is dressed in clothing that is too flimsy for the clothes drag (e.g., a nightgown). This procedure requires the use of a large sheet, blanket, curtain, or rug.

SKILL DRILL 12-10
Performing a Clothes Drag   Fire Fighter I, NFPA 1001: 4.3.9

1 Crouch behind the victim's head, grab the shirt or jacket around the collar and shoulder area, and support the head with your arms.

2 Lift with your legs until you are fully upright. Walk backward, dragging the victim to safety.

Place the item on the floor, and roll the victim onto it, and then pull the victim to safety by dragging the sheet or blanket. To perform a blanket drag, follow the steps in **SKILL DRILL 12-11**.

Standing Drag

The standing drag can be used to move a victim who is unconscious or disoriented. To perform the standing drag, follow the steps in **SKILL DRILL 12-12**.

SKILL DRILL 12-11
Performing a Blanket Drag Fire Fighter I, NFPA 1001: 4.3.9

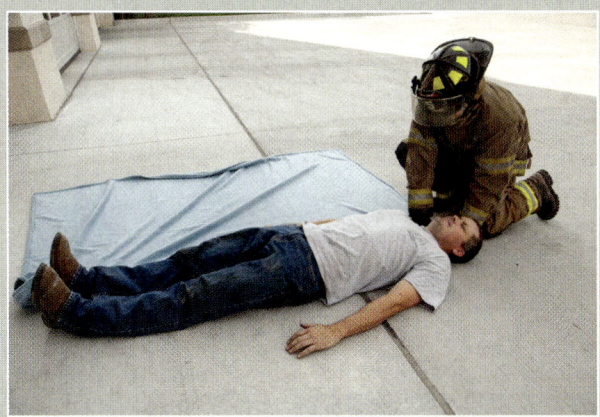

1 Lay the victim supine (face up) on the ground. Stretch out the material you are using next to the victim.

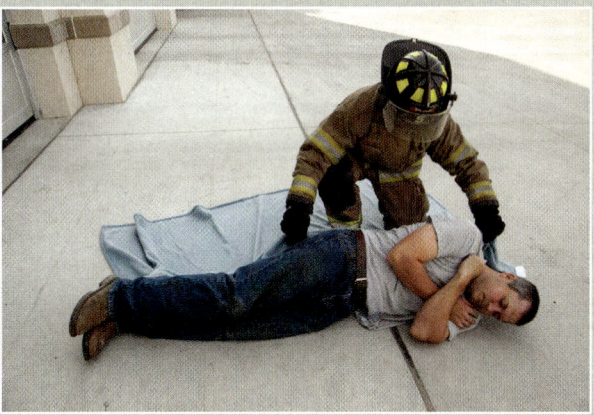

2 Roll the victim onto one side. Neatly bunch one-third of the material against the victim's body so the victim will lie approximately in the middle of the material.

3 Lay the victim back down (supine). Pull the bunched material out from underneath the victim, and wrap it around the victim.

(continued)

SKILL DRILL 12-11 Continued
Performing a Blanket Drag Fire Fighter I, NFPA 1001: 4.3.9

© Jones & Bartlett Learning. Photographed by Glen E. Ellman.

4 Grab the material at the head, and drag the victim backward to safety.

SKILL DRILL 12-12
Performing a Standing Drag Fire Fighter I, NFPA 1001: 4.3.9

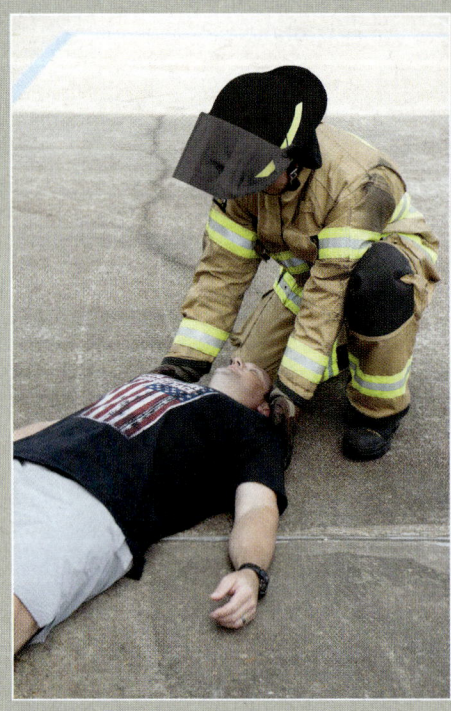 

1 Kneel at the head of the supine victim.

2 Raise the victim's head and torso by 90 degrees so that the victim is leaning against you.

SKILL DRILL 12-12 Continued
Performing a Standing Drag Fire Fighter I, NFPA 1001: 4.3.9

3 Reach under the victim's arms, wrap your arms around the victim's chest, and lock your arms.

4 Stand straight up using your legs. Drag the victim out.

© Jones & Bartlett Learning. Photographed by Glen E. Ellman.

Webbing Sling Drag

The webbing sling drag provides a secure grip around the upper part of a victim's body, allowing for a faster removal from the dangerous area. In this drag, a sling is placed around the victim's chest and under the armpits and then used to drag the victim. The webbing sling helps support the victim's head and neck. A webbing sling can be rolled and kept in a bunker coat pocket. A carabiner can be attached to such a sling to secure the straps under the victim's arms and provide additional protection for the victim's head and neck.

To perform the webbing sling drag, follow the steps in **SKILL DRILL 12-13**.

Fire Fighter Drag

The fire fighter drag can be used if the victim is heavier than the rescuer because it does not require lifting or carrying the victim. To perform the fire fighter drag, follow the steps in **SKILL DRILL 12-14**.

One-Rescuer Emergency Drag from a Vehicle

An emergency drag from a vehicle is performed when the victim must be quickly removed from a vehicle to save his or her life. The drag described in this section might be used, for example, if the vehicle is on fire or if the victim requires cardiopulmonary resuscitation.

There is no effective way for one person to remove a victim from a vehicle without some movement of the neck and spine. Preventing excess movement of the victim's neck, however, is important. To perform an emergency drag from a vehicle with only one rescuer, follow the steps in **SKILL DRILL 12-15**.

SKILL DRILL 12-13
Performing a Webbing Sling Drag Fire Fighter I, NFPA 1001: 4.3.9

1 Using a prepared webbing sling, place the victim in the center of the loop so the webbing is behind the victim's back in the area just below the armpits.

2 Take the large loop over the victim, and place it above the victim's head. Reach through, grab the webbing behind the victim's back, and pull through all the excess webbing. This creates a loop at the top of the victim's head and two loops around the victim's arms.

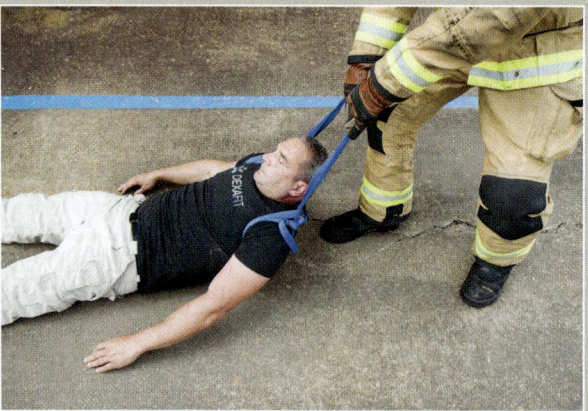

3 If necessary, adjust your hand placement to protect the victim's head while dragging.

SKILL DRILL 12-14
Performing a Fire Fighter Drag Fire Fighter I, NFPA 1001: 4.3.9

1 Tie the victim's wrists together with anything that is handy—a cravat (a folded triangular bandage), gauze, a belt, or a necktie.

2 Get down on your hands and knees, and straddle the victim.

3 Pass the victim's tied hands around your neck, straighten your arms, and drag the victim across the floor by crawling on your hands and knees.

© Jones & Bartlett Learning. Photographed by Glen E. Ellman.

SKILL DRILL 12-15
Performing a One-Person Emergency Drag from a Vehicle Fire Fighter I, NFPA 1001: 4.3.9

1 Grasp the victim under the arms, and cradle his or her head between your arms.

2 Gently pull the victim out of the vehicle.

3 Lower the victim down into a horizontal position in a safe place.

© Jones & Bartlett Learning. Photographed by Glen E. Ellman.

Assisting a Person Down a Ground Ladder

Using a ground ladder to rescue a trapped occupant is one of the most critical, stressful, and demanding tasks performed by fire fighters. Assisting someone down a ladder carries a considerable risk of injury to both fire fighters and victims. Whenever it is possible to use a stairway, fire escape, or aerial tower for rescue, these options should be considered before using ground ladders. Fire fighters must use proper techniques to safely accomplish a ladder rescue. In addition, they must have the physical strength and stamina needed to accomplish the rescue without causing injury to anyone involved. Although circumstances could require an individual fire fighter to work alone, at least two fire fighters should work as a team to rescue a victim whenever possible.

Time is a critical factor in many rescue situations—someone waiting to be rescued is often in immediate danger and may be preparing to jump. As a consequence, fire fighters may have only a limited time to work. They must quickly and efficiently raise a ladder and assist the victim to safety.

Ladder rescue begins with proper placement and securement of the ladder. As described in Chapter 11, *Ladders*, a ladder used to rescue a person from a window should have its tip placed just below the windowsill. This positioning makes it easier for the victim to mount the ladder. If possible, one or more fire fighters in the interior of the building should help the victim onto the ladder, and one fire fighter should stay on the ladder to assist the individual down.

Any ladder used for rescue should be heeled or tied in. The weight of an occupant and one or two fire fighters, all moving on the ladder at the same time, can easily destabilize a ladder that is not adequately secured. Additional personnel will be needed to secure the ladder and to assist in bringing the victim down the ladder.

It is not often that fire fighters need to perform a rescue using a ground ladder. This type of rescue is physically taxing and requires carefully executed skills. It is important to practise the skills to be prepared for this event, as those skills may be needed at any time.

Rescuing a Conscious Person from a Window

A ladder rescue is often frightening to conscious victims. When a rescue involves a conscious person, fire fighters should establish verbal contact as quickly as possible to reassure the victim that help is on the way. Maintain eye contact, and communicate in a calm, measured tone. People have jumped to their deaths just seconds before a ladder could be raised to a window.

All ladder rescues should be performed with two fire fighters whenever possible. **SKILL DRILL 12-16** presents a technique that could, if necessary, be

SKILL DRILL 12-16
Rescuing a Conscious Victim from a Window Fire Fighter I, NFPA 1001: 4.3.9

1 The rescue team places the ladder into the rescue position, with the tip of the ladder just below the windowsill, and secures the ladder in place.

2 The first fire fighter climbs the ladder, makes contact with the victim, and climbs inside the window to assist the victim. The fire fighter should make contact as soon as possible to calm the victim and encourage the victim to stay at the window until the rescue can be performed.

(continued)

SKILL DRILL 12-16 Continued
Rescuing a Conscious Victim from a Window Fire Fighter I, NFPA 1001: 4.3.9

3 The second fire fighter climbs up to the window, leaving at least one rung available for the victim. When ready, the fire fighter advises the victim to slowly come out onto the ladder, feet first and facing the ladder.

4 The second fire fighter forms a semi-circle around the victim, with both hands on the beams of the ladder.

5 The second fire fighter and victim proceed slowly down the ladder, one rung at a time, with the fire fighter always staying one rung below the victim. If the victim slips or loses his or her footing, the fire fighter's legs should keep the victim from falling. The fire fighter can take control of the victim at any time by leaning in toward the ladder and squeezing the victim against the ladder. The fire fighter should verbalize each step and talk to the person being rescued to help reassure and calm him or her and encourage the person to keep his or her gaze forward.

performed by a single fire fighter. Using a live person to practise this skill can be dangerous; for this reason, many departments use a manikin when performing this drill.

Rescuing an Unconscious Person from a Window

Rescuing an unconscious victim by ladder is dangerous and difficult, but it may be the best way to save a life. If the trapped person is unconscious, one or more fire fighters will have to climb inside the building and pass the person out of the window to a fire fighter on a

ladder. Caution should be used when lowering an unconscious victim down a ladder, because it is very easy for the victim's arms or legs to get caught in the ladder. To rescue an unconscious person via a ladder, follow the steps in **SKILL DRILL 12-17**.

Rescuing an Unconscious Child or a Small Adult from a Window

Small adults and children can be cradled across a fire fighter's arms during a rescue. To use this rescue technique, the child must be light enough that the fire fighter can descend safely, using only arm strength to

SKILL DRILL 12-17
Rescuing an Unconscious Victim from a Window Fire Fighter I, NFPA 1001: 4.3.9

1 The rescue team sets up and secures the ladder in rescue position with the tip of the ladder just below the windowsill.

2 One fire fighter climbs up the ladder and enters the window to rescue the victim. The second fire fighter climbs up to the window opening and waits for the victim.

(continued)

SKILL DRILL 12-17 Continued
Rescuing an Unconscious Victim from a Window Fire Fighter I, NFPA 1001: 4.3.9

3 The second fire fighter places both hands on the rungs of the ladder, with one leg straight and the other horizontal to the ground with the knee at an angle of 90 degrees. The foot of the straight leg should be one rung below the foot of the bent leg. When both fire fighters are ready, the first fire fighter passes the victim out through the window and onto the ladder, keeping the victim's back toward the ladder.

4 The victim is lowered so that he or she straddles the second fire fighter's leg. The fire fighter's arms should be positioned under the victim's arms, holding on to the rungs. The fire fighter keeps the balls of both feet on the rungs of the ladder to make it easier to move his or her feet. The fire fighter climbs down the ladder slowly, one rung at a time, transferring the victim's weight from one leg to the other. The victim's arms can also be secured around the fire fighter's neck.

© Jones & Bartlett Learning. Photographed by Glen E. Ellman.

support the victim. To carry an unconscious child down a ladder, follow the steps in SKILL DRILL 12-18.

Rescuing a Large Adult from a Window

Three or four fire fighters using two ladders may be needed to rescue very tall or heavy adults. To rescue a large adult using a ladder, follow the steps in SKILL DRILL 12-19.

Removing a Victim by Aerial Ladder or Platform

An aerial ladder or platform can be used for rescue operations. The same basic rescue techniques are used

SKILL DRILL 12-18
Rescuing an Unconscious Child or a Small Adult from a Window
Fire Fighter I, NFPA 1001: 4.3.9

1 The rescue team sets up and secures the ladder in rescue position, with the tip below the windowsill.

2 The first fire fighter climbs the ladder and enters the window to assist the victim. The second fire fighter climbs the ladder to the window opening and waits to receive the victim. Both of the second fire fighter's arms should be level with his or her hands on the beams.

3 When ready, the first fire fighter passes the victim to the second fire fighter so the victim is cradled across the second fire fighter's arms.

4 The second fire fighter climbs down the ladder slowly, with the victim being held in his or her arms. The fire fighter's arms should stay level, and his or her hands should slide down the beams.

SKILL DRILL 12-19
Rescuing a Large Adult from a Window Fire Fighter I, NFPA 1001: 4.3.9

1 The rescue team places and secures two ladders, side by side, in the rescue position. The tips of the two ladders should be just below the windowsill.

2 Multiple fire fighters may be required to enter the window to assist from the inside.

3 Two fire fighters, one on each ladder, climb up to the window opening and wait to receive the victim.

4 When ready, the victim is lowered down across the arms of the fire fighters, with one fire fighter supporting the victim's legs and the other supporting the victim's arms. Once in place, the fire fighters can slowly descend the ladder, using both hands to hold on to the ladder rungs.

© Jones & Bartlett Learning. Photographed by Glen E. Ellman.

with both aerial and ground ladders, but aerial ladders offer several advantages over ground ladders. Aerial ladders are much stronger and have a longer reach. In addition, they are wider and more stable than ground ladders, with side rails that provide greater security for both rescuers and victims (**FIGURE 12-12**).

Aerial platforms are even more suitable than aerial ladders for rescue operations. These devices reduce the risk of slipping and falling because the victim is lowered to the ground mechanically. An aerial platform is usually preferred for rescue work if one is available.

FIGURE 12-12 An aerial ladder is stronger and more stable than a ground ladder.
© Jones & Bartlett Learning. Photographed by Glen E. Ellman.

After-Action REVIEW

IN SUMMARY

- Search and rescue are almost always performed in tandem, yet they are separate operations. The purpose of search operations is to locate living victims. The purpose of rescue operations is to physically remove an occupant or victim from a dangerous environment.

- The IC and fire fighters must plan and coordinate all fire suppression operations to support the search and rescue priority. Sometimes the most effective initial step is to slow down the growth of the fire or to extinguish it before search and rescue operations can begin.

- The size-up process at every fire should include a specific evaluation of the following critical factors:
  - Building construction type and size
  - Occupancy information
  - Occupancy type
  - Visible smoke and fire conditions
  - Time of day and day of week
  - Weather conditions

- Plans for search and rescue can be developed based on the information gathered during the size-up. The plans must consider the risks and benefits of the operations. In some cases, such as when conditions are ripe for a backdraft, search and rescue operations may not be performed.

- Never assume that a building is occupied or unoccupied. Look for clues, such as cars parked in the driveway or toys on the front lawn, that indicate the presence of an occupant or occupants. However, the decision to perform search and rescue operations should not be made solely on these observations.

- A search begins in the areas where victims are at the greatest risk. Area search assignments should be based on a system of priorities:
  - First priority—Search the areas where live victims may be located immediately around the fire and then the rest of the fire floor.
  - Second priority—Search the area directly above the fire and the rest of that floor.
  - Third priority—Search the higher-level floors, typically working from the top floor down, because smoke and heat are likely to accumulate in these areas.
  - Lowest priority—Search areas below the fire floor unless there is a danger of collapse.

- Although every emergency operation involves a degree of unavoidable, inherent risk to fire fighters, during search and rescue operations fire fighters will probably encounter situations that involve a significantly higher degree of personal risk. Assuming this level of risk is acceptable only if there is a reasonable probability of saving a life. In some cases, the IC may decide that the risk to fire fighters is too great, and a primary search may not be conducted.

- Dark black turbulent smoke, blackened windows, the appearance of a "breathing" building, and intense heat are signs of possible flashover or backdraft. Cracks in walls, leaning walls, pitched or sagging doors, and any partial collapse are significant indicators of an impending collapse.

- A primary search is a quick attempt to locate victims who are in danger. Fire fighters must rely on their senses such as touch, hearing, and sight as they search for victims in smoky conditions.

- When searching a room, fire fighters should use the walls to orient themselves and as guides around the room. The room should be searched in a systematic pattern.

- Thermal imaging devices may be used to search for victims in darkness or smoke-obscured conditions. A thermal imaging device detects the temperature differential of the victim's body from the room.

- Four methods of conducting a primary search are the standard search method, the oriented search method, the oriented-vent-enter-isolate-search (O-VEIS) method, and the team search method.

- A secondary search is conducted after the fire is under control or fully suppressed. It is a slower search that is designed to locate any occupants or victims who were not found during the primary search.

- Rescue is the removal of a located person who is unable to escape from a dangerous situation; it is the second component of search and rescue. When you locate a victim during a search, you must direct, assist, or carry that person to a safe area. Rescue can be as simple as verbally directing an occupant toward an exit, or it can be as demanding as extricating a trapped, unconscious victim and physically carrying that person out of the building.

- The following may be unable to escape from a structure and therefore may need to be rescued:
  - Children and elderly persons
  - Physically or developmentally handicapped persons
  - Impaired people
  - Ill or injured persons
  - People who are sleeping
  - People under the influence of alcohol or drugs
  - People who become trapped due to blocked escape routes

- Because fires threaten the lives of both victims and rescuers, the first priority is to remove the victim from the fire building or dangerous area as quickly as possible.

- In some situations, the best option is to shelter the occupants in place instead of trying to remove them from a fire building. This option should be considered when the occupants are conscious and are found in a part of the building that is adequately protected from the fire by fire-resistive construction and/or fire suppression systems. Such a situation occurs in a high-rise apartment building when a fire is confined to one apartment or one floor.

- The simplest rescue is the exit assist, in which the victim is responsive and able to walk without assistance or with very little assistance. Two types of assists can be used to help responsive victims exit a fire situation:
  - One-person walking assist
  - Two-person walking assist

- Four simple carry techniques can be used to move a victim who is conscious and responsive but incapable of standing or walking:
  - Two-person extremity carry
  - Two-person seat carry
  - Two-person chair carry
  - Cradle-in-arms carry

- The most efficient method to remove an unconscious or unresponsive victim from a dangerous location is a drag. Six emergency drags can be used to remove unresponsive victims from a fire situation:
  - Clothes drag
  - Blanket drag

- Standing drag
- Webbing sling drag
- Fire fighter drag
- Emergency drag from a vehicle

- Assisting someone down a ground ladder carries a considerable risk of injury to both fire fighters and victims. Whenever it is possible to use a protected stairway, fire escape, or aerial tower for rescue, these options should be considered before using ground ladders.

KEY TERMS

Access Navigate for flashcards to test your key term knowledge.

Primary search An immediate and quick search of the structures likely to contain survivors. (NFPA 1670)

Rekindle A return to flaming combustion after apparent but incomplete extinguishment. (NFPA 921)

Rescue Those activities directed at locating endangered persons at an emergency incident, removing those persons from danger, treating the injured, and providing for transport to an appropriate healthcare facility. (NFPA 1500)

Search Land-based efforts to find victims or recover bodies. (NFPA 1951)

Search rope A guide rope used by fire fighters that allows them to maintain contact with a fixed point.

Secondary search A detailed, systematic search of an area that is conducted after the fire has been suppressed. (NFPA 1670)

Thermal imaging device An electronic device that detects differences in temperature based on infrared energy and then generates images based on those data. It is commonly used in smoke-filled environments to locate victims as well as to search for hidden fire during size-up and overhaul.

Transitional attack An offensive fire attack initiated by an exterior, indirect handline operation into the fire compartment to initiate cooling while transitioning into interior direct fire attack in coordination with ventilation operations.

Two-in/two-out rule A safety procedure that requires a minimum of two personnel to enter a hazardous area and a minimum of two backup personnel to remain outside the hazardous area during the initial stages of an incident.

REFERENCES

Coleman, John "Skip." *Searching Smarter*. Tulsa, Oklahoma: 2011.

National Fire Protection Association. NFPA 1500, *Standard on Fire Department Occupational Safety, Health, and Wellness Program*. 2018. www.nfpa.org. Accessed October 3, 2018.

Routley, J. Gordon. "Three Firefighters Die in Pittsburgh House Fire." US Fire Administration: 1995. https://www.usfa.fema.gov/downloads/pdf/publications/tr-078.pdf. Accessed October 3, 2018.

UL Firefighter Safety Research Institute. "Governors Island Experiments." May 21, 2013. https://ulfirefightersafety.org/research-projects/governors-island-experiments.html. Accessed October 3, 2018.

UL Firefighter Safety Research Institute. "Improving Fire Safety by Understanding the Fire Performance of Engineered Floor Systems." August 1, 2006. https://ulfirefightersafety.org/research-projects/engineered-floor-systems.html. Accessed October 3, 2018.

Underwriters Laboratories. "Firefighter Exposure to Smoke Particulates." April 1, 2010. https://ulfirefightersafety.org/research-projects/firefighter-exposure-to-smoke-particulates.html. Accessed October 3, 2018.

On Scene

Shortly after 1:15 AM, you are called to a fire in a house located just down the street from the station. You don your PPE and quickly mount the apparatus when the dispatcher indicates that neighbours are reporting they hear screaming coming from the house. When you arrive, you find a two-storey, wood-frame house with heavy smoke showing. The first arriving company begins pulling handlines. Two members of your crew are assigned ventilation duties, and you and the other fire fighter are assigned primary search responsibility. You begin to mentally review some of the information you were recently taught about search and rescue.

1. Where should your search begin?

A. Areas where live victims may be located immediately around the fire

B. Area directly above the fire and the rest of that floor

C. Higher-level floors, working from the top floor down

D. Fire floor, working toward your way toward the fire

2. During the search, you locate a victim who needs to be rescued from a window. Where should you place the tip of the ladder to rescue this victim?

A. Just below the windowsill

B. Just above the windowsill

C. Beside the window

D. Above the window header

3. The minimum number of rescuers required to rescue an unconscious person from a window is _____.

A. 2

B. 3

C. 4

D. 5

4. Your crew member locates an unconscious victim. The most efficient method to remove this victim from a dangerous location is a _____.

A. two-person walking assist

B. clothes drag

C. two-person extremity carry

D. two-person chair carry

5. You need to search for a victim in a second-storey bedroom and are unable to gain access to the bedroom from the hallway. What type of primary search method should you consider?

A. Standard search method

B. Oriented search method

C. Oriented-vent-enter-isolate-search (O-VEIS) method

D. Team search method

6. If this house was a very large building with a lot of open spaces and poor visibility, which of the following primary search methods might be most suitable?

A. Standard search method

B. Oriented search method

C. O-VEIS method

D. Team search method

7. You locate a heavy victim in the second-storey bedroom who needs to be removed through the bedroom window. What is the minimum number of fire fighters needed to remove this victim?

A. 2

B. 3

C. 6

D. 4

8. Following your search, another team of fire fighters will conduct a(n) _____ to find victims who were not found during the primary search.

A. oriented search

B. rescue

C. team search

D. secondary search

Access Navigate to find answers to this On Scene, along with other resources such as an audiobook.

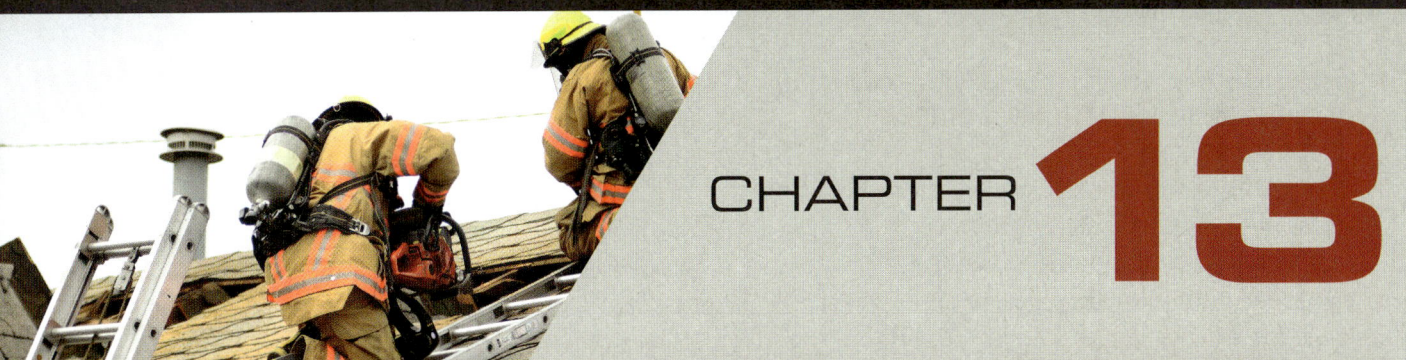

CHAPTER 13

Fire Fighter I

Ventilation

KNOWLEDGE OBJECTIVES

After studying this chapter, you will be able to:

- Describe the characteristics of a ventilation-limited fire. (p. 472)
- Describe the impact of door control on ventilation. (pp. 472–473)
- Describe the impact of ventilation location. (p. 473)
- Describe the impact of ventilation hole size. (p. 474)
- Describe the impact of wind on fire behaviour. (pp. 474–475)
- Describe the impact of exterior suppression on fire behaviour. (p. 475)
- Describe the importance of including ventilation considerations in a size-up. (pp. 475–479)
- Describe how the location, size, and stage of fire affect ventilation operations. (**NFPA 1001: 4.3.11**, pp. 476–477)
- Describe how the characteristics of different construction types affect ventilation operations. (**NFPA 1001: 4.3.11**, pp. 477–479)
- Describe the importance of the timing and coordination of ventilation and suppression. (**NFPA 1001: 4.3.11**, p. 480)
- Describe steps that can be taken to minimize backdrafts and flashovers. (**NFPA 1001: 4.3.11**, pp. 480–481)

- List the two basic types of ventilation. (**NFPA 1001: 4.3.11**, p. 481)
- Explain how horizontal ventilation removes contaminated atmosphere from a structure. (**NFPA 1001: 4.3.11**, pp. 482–491)
- List the two methods of horizontal ventilation. (**NFPA 1001: 4.3.11**, p. 482)
- Explain how natural ventilation removes contaminated atmosphere from a structure. (**NFPA 1001: 4.3.11**, pp. 483–486)
- Describe the techniques used to provide natural ventilation to a structure. (**NFPA 1001: 4.3.11**, pp. 483–486)
- Explain how mechanical ventilation removes contaminated atmosphere from a structure. (**NFPA 1001: 4.3.11**, pp. 486–491)
- Describe the techniques used to provide mechanical ventilation to a structure. (**NFPA 1001: 4.3.11**, pp. 486–491)
- Describe how negative-pressure ventilation removes contaminated atmosphere from a structure. (**NFPA 1001: 4.3.11**, pp. 486–488)
- Describe the techniques used to provide negative-pressure ventilation to a structure. (**NFPA 1001: 4.3.11**, pp. 486–488)
- Describe how positive-pressure ventilation removes contaminated atmosphere from a structure. (**NFPA 1001: 4.3.11**, pp. 488–490)

- Describe the techniques used to provide positive-pressure ventilation to a structure. (**NFPA 1001: 4.3.11**, pp. 488–489)

- Describe how hydraulic ventilation removes contaminated atmosphere from a structure. (**NFPA 1001: 4.3.11**, pp. 489, 491)

- Describe the techniques used to provide hydraulic ventilation to a structure. (**NFPA 1001: 4.3.11**, pp. 489, 491)

- Describe how vertical ventilation removes contaminated atmosphere from a structure. (**NFPA 1001: 4.3.12**, pp. 491–493)

- Describe how to ensure fire fighter safety during vertical ventilation operations. (**NFPA 1001: 4.3.12**, pp. 492–493)

- Identify the warning signs of roof collapse. (**NFPA 1001: 4.3.12**, pp. 493–494)

- Describe the components and characteristics of roof assemblies. (**NFPA 1001: 4.3.12**, pp. 493–497)

- List the differences in solid-beam construction and lightweight construction in roofs. (**NFPA 1001: 4.3.12**, pp. 494–495)

- Explain how roof construction affects fire resistance. (**NFPA 1001: 4.3.12**, pp. 495–496)

- List the basic types of roof design. (**NFPA 1001: 4.3.12**, pp. 496–497)

- Describe the characteristics of flat roofs. (**NFPA 1001: 4.3.12**, p. 496)

- Describe the characteristics of pitched roofs. (**NFPA 1001: 4.3.12**, pp. 496–497)

- Describe the characteristics of curved roofs. (**NFPA 1001: 4.3.12**, p. 497)

- Describe the techniques of vertical ventilation. (**NFPA 1001: 4.3.12**, pp. 497–507)

- List the tools utilized in vertical ventilation. (**NFPA 1001: 4.3.12**, pp. 498–501)

- List the types of roof cuts utilized in vertical ventilation operations. (**NFPA 1001: 4.3.12**, pp. 501–509)

- Describe the characteristics of a rectangular or square cut. (**NFPA 1001: 4.3.12**, pp. 501–502)

- Describe the characteristics of a seven, nine, eight (7, 9, 8) rectangular cut. (**NFPA 1001: 4.3.12**, pp. 502–504)

- Describe the characteristics of a louver cut. (**NFPA 1001: 4.3.12**, pp. 503, 505)

- Describe the characteristics of a triangular cut. (**NFPA 1001: 4.3.12**, pp. 505, 506)

- Describe the characteristics of a peak cut. (**NFPA 1001: 4.3.12**, pp. 506–508)

- Describe the characteristics of a trench cut. (**NFPA 1001: 4.3.12**, pp. 507–509)

- Describe the special considerations in ventilating basements. (**NFPA 1001: 4.3.12**, pp. 509–510)

- Describe the special considerations in ventilating concrete roofs. (**NFPA 1001: 4.3.12**, p. 510)

- Describe the special considerations in ventilating metal roofs. (**NFPA 1001: 4.3.12**, pp. 510–511)

- Describe the special considerations in ventilating high-rise buildings. (**NFPA 1001: 4.3.12**, pp. 511–512)

- Describe the special considerations in ventilating windowless buildings. (**NFPA 1001: 4.3.12**, p. 512)

- Describe the special considerations in ventilating large buildings. (**NFPA 1001: 4.3.12**, p. 513)

- Explain how to ensure that ventilation equipment is in a state of readiness. (pp. 513–515)

SKILLS OBJECTIVES

After studying this chapter, you will be able to perform the following skills:

- Break glass with a hand tool. (**NFPA 1001: 4.3.11**, p. 484)

- Break a window with a ladder. (**NFPA 1001: 4.3.11**, p. 485)

- Deliver negative-pressure ventilation. (**NFPA 1001: 4.3.11**, pp. 486–487)

- Deliver positive-pressure ventilation. (**NFPA 1001: 4.3.11**, pp. 488–490)

- Perform hydraulic ventilation. (**NFPA 1001: 4.3.11**, pp. 489, 491)

- Operate a power saw. (**NFPA 1001: 4.3.12**, pp. 499–501)

- Make a rectangular cut to deliver vertical ventilation. (**NFPA 1001: 4.3.12**, p. 502)

- Make a seven, nine, eight (7, 9, 8) rectangular cut to deliver vertical ventilation. (**NFPA 1001: 4.3.12**, pp. 502–504)

- Make a louver cut to deliver vertical ventilation. (**NFPA 1001: 4.3.12**, p. 505)

- Make a triangular cut to deliver vertical ventilation. (**NFPA 1001: 4.3.12**, p. 506)

- Make a peak cut to deliver vertical ventilation. (**NFPA 1001: 4.3.12**, pp. 507–508)

- Make a trench cut to deliver vertical ventilation. (**NFPA 1001: 4.3.12**, p. 509)

- Perform a readiness check on a power saw. (**NFPA 1001: 4.5.1**, pp. 513–514)

- Maintain a power saw. (**NFPA 1001: 4.5.1**, p. 515)

Fire Alarm

It is a routine day at the grocery store where you work full time. You are busy stocking shelves when your pager goes off, alerting you to a structure fire only a few blocks away. On your arrival, you find a one-storey, wood-frame house with dirty brown smoke pushing out through the eaves, siding, and other cracks. The windows are stained black. The neighbours say they do not think anyone is home because the occupants work during the day.

1. What is the best way to ventilate this structure?

2. What concerns do you have about ventilating this structure?

3. When would you want to ventilate this structure?

Introduction

Ventilation is the controlled and coordinated removal of heat and smoke from a structure. Effective ventilation not only removes heat and smoke, it also replaces the escaping gases with cooler, cleaner, and oxygen-rich air. This firefighting operation must be planned and systematic. Understanding how and when to perform ventilation is vital to conducting safe and effective fire suppression. When properly performed and coordinated, ventilation can help to remove hot gases from the fire compartment, make it easier to locate the seat of the fire, improve visibility, and contribute to faster and safer knockdown, more effective fire suppression, and improved efficiency of searching. Poorly performed and poorly timed ventilation can quickly add oxygen to a fire and, if not coordinated with fire attack and other fire-ground operations, can result in increased hazards and an actively growing fire. Recently, tactical considerations for ventilation have changed because of the findings from multiple live fire experiments conducted by Underwriters Laboratories (UL) and the National Institute of Standards and Technology (NIST). The information and techniques presented in this chapter incorporate the results of those experiments.

Fire service ventilation can be looked at as a two-step process. The first step, which can be done during size-up, is a determination of the need for ventilation, an assessment of the location and amount of ventilation needed, and the coordination of ventilation operations with other fire-suppression operations. Planning and assessment of ventilation needs are discussed in the first sections of this chapter.

The second step of ventilating a fire is the mechanical operations of actually opening or closing doors and windows, opening skylights, and cutting openings in the roof. The information on conducting ventilation operations is discussed in the latter sections of this chapter.

Before fire fighters open a door, take out a window, or cut a hole in a roof, they should ask themselves the following questions:

- Why am I ventilating? Is it to provide a tenable atmosphere for a potential occupant? Am I creating a more tenable atmosphere for possible occupants and fire fighters?
- Where do I want to accomplish the ventilation? Do I want to perform horizontal ventilation or vertical ventilation? Where is the fire in relation to my anticipated vent?
- When do I want to perform the ventilation? Have I coordinated my efforts with the actions of the suppression team and the search and rescue teams?

Ventilation and Fire Behaviour

Before ventilating, it is important to understand how a given action will affect the behaviour of the fire. First, fire fighters must determine if they are dealing with a ventilation-limited fire or a fuel-limited fire. Chapter 5, *Fire Behaviour*, contains more information on these fire conditions and the different stages of fire growth.

SAFETY TIP

Effective ventilation removes smoke, gases, and heat, but it also brings in cooler, cleaner, and oxygen-rich air. If ventilation is not coordinated with fire attack operations, the hazards from the fire can increase.

A **ventilation-limited fire** contains large quantities of thermal energy and hot, unburned fire gases in an oxygen-deficient environment. Once sufficient oxygen is introduced into the fire compartment, flashover will occur, and the fire will transition to the fully developed stage. It is at the fully developed stage that the fire releases its maximum energy. Well-trained fire fighters manage ventilation-limited fires by controlling flow paths and using specific cooling techniques. Many building fires today involve modern synthetic contents, which cause fires to grow much faster than fires involving older (natural-based) contents; therefore, many of the building fires that fire fighters encounter will be in a ventilation-limited decay stage upon arrival (Kerber, 2012). Any action that results in the introduction of oxygen to the fire is likely to change the fire from a ventilation-limited decay stage to a **fuel-limited fire**—that is, the fire will have a limited amount of fuel available for burning, but it will have sufficient oxygen for fire growth. The action can be as simple as opening a door of the building. Any opening establishes a potential **flow path** within the structure.

Chapter 5, *Fire Behaviour*, describes the flow path as the area(s) within a structure where heat, smoke, and air flow from areas of higher pressure to areas of lower pressure. Hot gases and smoke are transferred from the higher pressure within the fire area toward the lower pressure areas accessible through doorways, window openings, and roof openings. The flow path can function as a single unidirectional flow between the fire and an opening to the outside (exhaust). It also may act as a bidirectional flow, acting both as an inlet for fresh air and an outlet, or exhaust. In a bidirectional flow, a fire fighter may identify the neutral plane that defines the inlet and outlet. The **neutral plane** is the interface at a ventilation opening, such as a doorway or a window, between the hot gas flowing out of a fire compartment and the cool air flowing into the compartment where the pressure difference between the interior and exterior is equal.

The flow path will be determined by the building design and which doors and windows are open to the outside. Every new ventilation opening may provide a new flow path to the fire and vice versa. Any operations that are conducted in the flow path place fire fighters at significant risk for injury or death because of the significant amount of heat and smoke that travel along the flow path. Fire fighters should avoid placing

---

SAFETY TIP

Fire fighters operating in the flow path are at a significantly increased risk for injury or death due to the flow of hot fire gases and smoke moving over them.

---

themselves in the flow path. However, when it becomes unavoidable, proper water application and ventilation techniques can be used to manage the flow path and minimize risk.

Several factors influence the effectiveness of ventilation and the speed with which the ventilation changes the flow path. These include the impact of door control, the impact of the ventilation location, the impact of the ventilation hole size, the impact of wind, and the impact of exterior suppression on fire behaviour.

SAFETY TIP

Interior fire attack is still the most common and important fire attack method, but emphasis needs to be placed on controlling ventilation and cooling fire gases. Apply water to the smoke to cool it as you move through it.

The Impact of Door Control

Following a thorough size-up, one of the first actions taken by most fire departments at a structure fire is to gain access to the building by opening a door. Although forcing entry may be necessary to fight the fire, this action changes the ventilation profile of the fire by supplying additional oxygen to the fire. As previously discussed, a ventilation-limited fire can rapidly change into a fuel-limited fire. Once a door is opened, the clock is ticking before either the fire gets extinguished or it grows until an untenable condition exists, potentially jeopardizing the safety of everyone in the structure and immediate fire ground.

Limiting the air inlet will limit the fire's ability to grow. The simple act of closing the door after forcing entry will limit the air supply to the fire and slow the fire growth until a crew is ready to make access to the building and perform a coordinated fire attack. Sometimes the most effective act in ventilation is to limit ventilation until a coordinated fire attack can be made. Leaving the door closed as long as possible aids significantly in controlling a fire by limiting the amount of oxygen supplied to it. When making an interior fire attack or rescue effort, try to maintain control of the door, keeping it closed as much as possible and opening it just enough to allow the hose line to be advanced to the fire. It may be necessary to station a fire fighter at the door to maintain this partial door closure (**FIGURE 13-1**). This tactic is known as "manning the door." This fire fighter can also assist in advancing the hose line as he or she controls the door.

Fire growth also may be limited by closing interior doors. Closing the door after searching a room potentially limits the supply of air available to the fire. It also helps to reduce the risk of the fire transitioning

FIGURE 13-1 When making an interior fire attack or rescue effort, try to maintain control of the exterior door, keeping it closed as much as possible to limit the flow of air.
© Jones & Bartlett Learning. Photographed by Glen E. Ellman.

into that room or other areas connected to that room. Closing the door to the room of origin or the involved compartment aids in reducing the oxygen flow, which in turn reduces room temperature, until a suppression agent can be applied.

LISTEN UP!

Forcing the door must be considered an act of ventilation. While ventilation is a critical part of fire-suppression activities, it should always be conducted in coordination with effective and sustained suppression efforts. Once a door is opened, attention must be given to the flow through the door. A rapid rush of air or a tunneling effect could indicate a ventilation-limited fire. Effective forcible entry techniques that reduce damage to the door will ultimately make the door more effective for flow path management.

SAFETY TIP

If you limit air inlets to the fire, you can limit the ability of the fire to grow. Sometimes the best ventilation is less ventilation.

The Impact of the Ventilation Location

Ventilation openings have a significant impact on fire growth. In many instances, fires are vented by the time firefighters arrive on scene. However, in those instances where the fire has not vented, fire fighters must be mindful of how their ventilation tactics can impact fire growth.

Ventilation should occur as close to the fire as possible. The most desirable options are to provide a ventilation opening directly over the seat of the fire or to open a door or window that will release heat and smoke from the fire directly to the exterior. Ventilating directly over the fire produces the fastest impact on the behaviour of the fire and will exhaust the greatest amount of combustion products. However, if this ventilation is done without coordinated water application, the result will most likely be added fire growth because of the added supply of oxygen to the fire. In other words, ventilating over the fire is helpful only if ventilation is coordinated with fire attack. Proper water application reduces the overall temperature within a fire compartment, reducing the energy created by the main body of the fire. Once the fire growth has been slowed, a ventilation opening can be more effective in removing hot gases and smoke. This will improve the conditions in the fire compartment, reducing the air temperature and improving visibility.

If it is not possible to create a ventilation opening in the immediate area of the fire, fire fighters must be able to predict how ventilation from another location will affect the fire. In simplistic terms, fire compartments should be viewed as high-pressure environments. Even if the opening is in a remote location from the fire, a low-pressure opening is created. The energy generated by the fire will move from a high-pressure environment to a low-pressure environment. It is important to note that in winter months, heavier cold air will impact ventilation differently than in summer months. Fire travels toward a ventilation opening, unless influenced by outside forces such as strong winds along the flow path. Additionally, if the ventilation opening is farther from the fire, it will take longer for any changes in ventilation to impact the fire. These distant openings are less effective in removing smoke and hot gases from the fire.

Other factors that affect how fast the fire responds to oxygen include the stage the fire is in, whether the fire is ventilation limited or fuel limited, the number and size of the ventilation inlets and outlets, the shape of the ventilation openings, the temperature of the fire room, the configuration of the walls, and the amount and type of contents in the flow path.

The Impact of the Ventilation Hole Size

Fire studies conducted by UL demonstrate that larger sized vertical ventilation openings do not localize the growth of a fire. In addition, vertical ventilation alone does not reduce the temperatures in a fire building. The studies compared the impact of cutting a 1.2-m by 1.2-m (4-ft by 4-ft) roof ventilation hole with the effect of cutting a 1.2-m by 2.4-m (4-ft by 8-ft) roof ventilation hole. Neither the small hole nor the larger hole had the effect of lowering the temperature in the fire area when only ventilation was performed (Kerber, 2013).

LISTEN UP!

The information in the following list is taken from the "Fire Service Summary Report: Study of the Effectiveness of Fire Service Vertical Ventilation and Suppression Tactics in Single Family Homes," published by the UL Firefighter Safety Research Institute:

- The fire does not react to additional oxygen instantaneously. A ventilation action may appear to be positive at first, as the air is entrained into the ventilation-limited fire; 2 minutes later, conditions could become deadly without water application.
- The higher the interior temperatures, the faster the fire reacts. If the fire is showing on arrival, the interior temperatures are higher than if the house is closed. This means that additional ventilation openings are going to create more burning in a shorter period of time.
- The closer the air is to the fire, the faster the fire reacts. Venting the fire room will cause burning to increase quickly; it will also let the hot gases out faster after water is applied.
- The higher the ventilation, the faster the fire reacts. Faster and more efficient ventilation means faster air entrainment, which means more burning and higher temperatures. It also means better ventilation after water is applied.
- The more air, the faster the fire reacts. Also, the more exhaust, the more air that can be entrained into the fire. A bigger ventilation hole in the roof means that more air will be entrained into the fire. If the fire is fuel limited, this is good, but if the fire is ventilation limited, this could be bad.

SAFETY TIP

Limiting the flow path by limiting inlet and exhaust openings until water is ready to be applied limits the supply of oxygen to the fire and, in turn, reduces the amount of heat release and temperature rise within the structure.

Although vertical ventilation by itself does not appear to have a positive effect on a ventilation-limited fire, vertical ventilation in coordination with an exterior or interior application of water as close to the fire as possible has the effect of improving visibility, reducing the temperature in the fire compartment, and temporarily limiting the growth of the fire. When a coordinated attack is conducted in conjunction with water application, a larger ventilation hole is more effective in lowering the temperature of the fire compartment and other parts of the building (Kerber, 2013). However, it is important to understand that with today's fires, the fire service simply cannot vent enough to vent the enormous amounts of smoke and particulates.

The Impact of Wind

Wind is a powerful force that can change the direction and speed of a fire and its flow path rapidly. Any time a window or door is opened on the side of a building that faces the wind, known as the upwind side, an unlimited amount of oxygen under high pressure is introduced into the fire. This action may be unintentional, such as a glass window or patio slider door breaking because of heat from the fire, or it can be an intentional act of ventilation. A sudden increase in oxygen, combined with hot flammable smoke (fuel) from the fire, will result in rapid fire growth and can produce a sudden change in the direction of the flow path due to the pressure imposed by the wind. Think of the wind as a giant positive-pressure fan forcing huge quantities of oxygen into a ventilation-limited fire (**FIGURE 13-2**). Fire fighters should remember to keep the wind at their backs during a fire attack and avoid ventilating on the upwind or downwind side of a fire unless it is part of a well-organized suppression effort. NIST has conducted experiments on wind-driven fires. More information

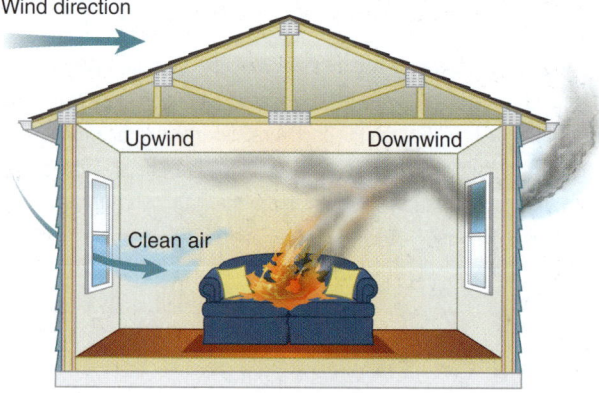

FIGURE 13-2 On a windy day, proper ventilation helps to remove smoke and heat from a structure.

© Jones & Bartlett Learning.

about these studies is available at the NIST website. These tests were conducted using high-rise structures, but wind is also a very real threat in one- and two-family structure fires.

The Impact of Exterior Suppression

UL and NIST fire research studies have demonstrated that our traditional assumptions about the effects of ventilation need to be modified. Ventilation has often been defined as the systematic removal of heat, smoke, and fire gases from a building and replacement of those elements with cool, fresh air. Although this definition is and always has been technically true, it can lead to a misapplication of the principle. What research has repeatedly demonstrated is that when dealing with ventilation-limited fires, any cooling effect produced by ventilation alone is minimal and short lived (UL Firefighter Safety Research Institute [FSRI], 2013). Because of the extremely fuel-rich environment found on today's fire ground, ventilation that is not preceded by and concurrent with or immediately followed by effective suppression will introduce enough oxygen to rapidly bring the fire area to flashover. **TABLE 13-1** shows the conclusions for residential fire behaviour that were drawn from the fire experiments conducted by NIST, UL, and the New York City Fire Department (FDNY).

For the greatest reduction in production of hot gases and smoke, use a coordinated fire attack by performing ventilation and applying water to the fire from a safe location as close to the fire as possible. Applying water to the fire as quickly as possible from either the exterior or interior can help to improve conditions in the entire fire structure. Even a small amount of water improves the conditions within the structure. Well-planned ventilation, when coordinated with the proper application of water, has several benefits. It removes large amounts of highly flammable fuels in the form of superheated gases and smoke and cools the fire building, which

greatly reduces the chance for flashovers or backdrafts. Ventilation also helps to improve the visibility within the fire building, thereby helping fire fighters maintain their orientation within the building and making it easier to search for fire victims. Reduced temperatures improve fire fighter safety and increase the potential for the survival of building occupants.

An offensive fire attack initiated by a quick, indirect, exterior attack into the fire compartment is called a transitional attack. This technique—which you will learn how to conduct in Chapter 17, *Fire Suppression*—initiates cooling and darkens the fire, allowing fire fighters to quickly transition into an interior attack for final suppression. When used, this technique should be performed prior to entry, search, and suppression. Surprisingly, the UL/NIST research demonstrated the application of water from the outside, when properly applied, did *not* push the fire from one place to another (Kerber, 2013). The water introduced into the fire from the outside produced cooling, not just in the fire compartment but also in other areas of the fire building. Significant cooling was achieved even when the water was applied to a body of fire in the interior of the structure at a distance from the seat of the fire. While this application of water did not totally extinguish the fire, it cooled the fire gases throughout most parts of the fire building, making it safer for fire fighters to enter and complete the suppression of the fire. This greatly increased the chance of survival for building occupants.

It is important to stress that the transitional fire attack does not reduce the importance of proper ventilation and that ventilation remains a crucial part of fire suppression. Close coordination between the fire attack team and the ventilation crew is required at a fire. Providing ventilation without this coordination can contribute to the growth of the fire and may lead to very dangerous conditions (**FIGURE 13-3**). More information about mounting an effective and safe transitional fire attack is presented in Chapter 17, *Fire Suppression*.

Size-Up and Ventilation

A thorough size-up of each fire is important to develop a plan not only for rescue and suppression but also for the tactics needed to ventilate the fire structure. Ventilation is not a one-size-fits-all type of operation, where fire fighters perform exactly the same actions at every fire. The size-up will help to determine which actions need to be taken and in which order these actions need to be initiated. It should include consideration of the following factors: the location, size, and stage of the fire; the fire department arrival time; the size of

TABLE 13-1 Residential Fire Behaviour

- Increasing the air flow to a ventilation-limited structure fire by opening doors, windows, or roof openings will increase the fire hazard from the fire for both building occupants and fire fighters.
- Increasing the air flow to a ventilation-limited structure fire may lead to a rapid transition to flashover.
- There is a need for systematic fire attack to coordinate the activities of size-up, entry, ventilation, search and rescue, and application of water.

Courtesy of NIST.

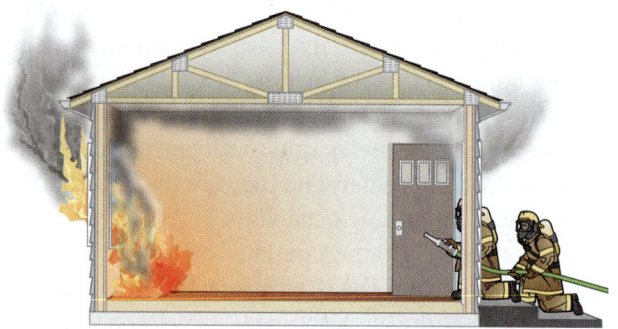

A

B

FIGURE 13-3 Proper ventilation enables fire fighters to control a fire rapidly. **A.** Vented structure. **B.** Unvented structure.
© Jones & Bartlett Learning.

FIGURE 13-4 The colour, location, and amount of smoke can provide valuable clues for fire fighters.
Courtesy of District Chief Chris E. Mickal/New Orleans Fire Department, Photo Unit.

the building; the type of building construction; the geometry or shape of the building; the potential for rescue; the potential for building collapse; the amount of fuel in the building contents; the impact of modern (synthetic) contents versus traditional (natural-based) contents; and the type of fire attack that can be used. Fire fighters also need to assess possible places to ventilate and evaluate how these ventilation options will impact the flow path and the spread of hot fire gases and smoke.

Location, Size, and Stage of Fire

Fire fighters must be able to recognize when ventilation is needed and where it should be provided, based on the circumstances of each fire situation. One of the most important aspects of size-up is the determination of where the fire is located, how big the fire is, and what stage the fire is in. An experienced fire fighter learns to recognize these significant factors immediately and to interpret each situation quickly.

The colour, location, movement, and amount of smoke can provide valuable clues about the fire's size, intensity, and fuel (**FIGURE 13-4**). A very hot fire produces smoke that moves quickly, rolling and forcing its way out through an opening. The hotter the fire, the

more the gases expand, the greater the pressure in the structure, the faster the movement of the smoke. Cooler smoke moves more slowly and gently. On a cool, damp day with very little wind, this type of smoke might hang low to the ground (a phenomenon known as **smoke inversion**).

FIRE FIGHTER TIP

The seemingly benign appearance of light and light-colored smoke should not fool fire fighters: Such smoke can be just as dangerous as thick black smoke. Always remember that these unburned products of combustion produce toxic gases and vapours, no matter what color the smoke is.

A different phenomenon occurs in buildings with automatic sprinkler systems. The water discharged from the sprinklers cools the smoke and produces a cold smoke that may hardly move within the building. This type of smoke can fill a large warehouse space from floor to ceiling with an opaque mixture that behaves much like fog on a damp day. Because this cold smoke has a lower temperature than the surrounding atmosphere, it sits at the lower levels of the room, making it hard to see and hard to locate the seat of a fire. Mechanical ventilation is often needed to clear this type of smoke from the building. A similar situation occurs when smoke becomes trapped within a building long enough for it to cool to the ambient temperature.

A fire where there is little or no smoke showing may indicate a small fire in the incipient stage (a fire that can easily be extinguished with a small amount of water and little fire damage) or a fire that has exhausted most of its fuel supply. However, a fire that initially has no visible flames and very little or no smoke also may be a ventilation-limited fire.

Type of Building Construction

The way a building is constructed affects ventilation operations. The different types of construction are discussed in more detail in Chapter 6, *Building Construction*. Each construction type presents its own set of problems in regard to ventilation operations.

Home construction in Canada has changed significantly over the past several years. New home construction is primarily comprised of engineered wood products. These products are often left unprotected in basements and in areas where the building code does not require them to be covered over with fire-resistant materials. When exposed to flames or extreme heat, these elements support fire growth, becoming part of the overall fire load. They are also known to fail quickly, resulting in localized or catastrophic collapse. The heat release rate of these materials is challenging the ability of fire fighters to engage in safe ventilation practices. Affordability, rapid urbanization, and the need to create more energy-efficient homes has created challenging circumstances for fire fighters. Additionally, most recently built residences are sealed from the outside with tight windows, vapour barriers, and insulation. The wide use of heating and air conditioning systems means that in many parts of the country houses are tightly closed for most of the year. The result is that in a fire event today, many buildings develop higher interior temperatures much more quickly. If the doors and windows remain closed, the fire will reach a point where it has such a limited supply of oxygen that it will not sustain open combustion. When this happens, the fire, which is still very hot and has an abundance of fuel, becomes a ventilation-limited fire.

Type I Construction (Fire-Resistive)

Type I construction (fire-resistive) refers to a building in which all of the structural components are made of noncombustible materials, such as steel or concrete. This type of construction generally has spaces divided into compartments, which limit potential fire spread. A fire is most likely to involve the combustible contents of an interior space within such a building.

Although Type I construction is intended to confine a fire, it does contain potential avenues for fire spread. Openings for mechanical systems such as heating and cooling ducts, plumbing and electrical **chases**, elevator shafts, and stairwells all provide paths for fire spread. In all types of construction, fire can also spread from one floor to another through exterior windows, a phenomenon called *vertical fire extension* (or auto-exposure) (**FIGURE 13-5**).

Smoke can travel through a Type I building using the same routes. Stairways; elevator shafts; and heating, ventilation, and air-conditioning (HVAC) systems are

FIGURE 13-5 Vertical fire extension can occur when the fire jumps from floor to floor through exterior windows. This figure shows fire spread in a Type V (wood frame) structure.
© Jones & Bartlett Learning.

common pathways. Because the windows in these buildings are often sealed and are not broken easily, there can be limited opportunities for creating ventilation openings. In some cases, one stairway can be used as an exhaust shaft, while other stairways are cleared for rescue and access. Ventilation fans are often required to direct the flow of smoke. Some windowless buildings contain a number of glass panels with a symbol (sometimes a Maltese cross) in a lower corner, indicating that the panel is made of tempered safety glass and can be broken easily with a sharp, pointed object.

In many of these buildings, engineered designs are installed to assist fire fighters with ventilation or controlling the movement of hot gases and smoke throughout the building. Smoke and heat vents are installed in the ventilation systems; stairwells and elevator shafts are designed with positive pressure to limit smoke migration; and HVAC systems are designed to be used by arriving fire fighters to pressurize floors or areas to limit fire spread. In some cases, the most valuable person on scene may be the building engineer, who can assist in the operations of many of these systems from the fire command centre. It is never advisable for the incident commander to turn off air handling systems in a Type I structure, unless the system is contributing to fire or smoke spread. In Canada, Type I structures have been engineered to resist the spread of fire and smoke through a combination of compartmentalization and pressurization.

The roof on a Type I building usually is supported by concrete or steel roof decking. In these types of roofs, which are discussed later in this chapter, it can be difficult or impossible to make vertical ventilation holes. These roofs might, however, have openings for skylights or for HVAC ducts.

Type II Construction (Noncombustible)

Type II construction (noncombustible) refers to a building in which all of the structural components are made of noncombustible materials. A typical example of Type II construction is a large-area, single-storey building with a steel frame, metal or concrete block walls, and a metal roof deck.

Fire walls are sometimes used to subdivide these large-area buildings and limit the potential fire spread. Additionally, Type II construction is most common in single-storey warehouse or factory buildings, where vertical fire spread is not an issue. Similar to Type I construction, a fire in this type of building is most likely to involve the combustible contents of an interior space within the building, as the structural components contribute little or no fuel, and interior finish materials are limited.

Many Type II buildings have few window openings; therefore, horizontal ventilation is often limited to existing doors. Vertical ventilation of these buildings should be attempted only with aerial devices. If vertical ventilation is attempted, remember that the building contents may provide a high fuel load, and a fire in this type of building can result in failure of the building walls and collapse of the roof at any time. Additionally, the roof on a Type II building usually is supported by metal roof decking. It can be difficult or impossible to make vertical ventilation holes in these types of roofs.

Type III Construction (Ordinary)

Buildings constructed of Type III construction (ordinary) have exterior walls made of noncombustible or limited-combustible materials that support the roof and floor assemblies. The interior walls and floors are usually wood construction; the roof usually has wood decking and a wood structural support system. In this type of construction, the roof can be cut with power saws or axes to open vertical ventilation holes. Type III construction buildings usually have windows and doors that can be used for horizontal ventilation.

In this type of construction, the walls and floors might include numerous openings for plumbing and electrical chases. If a fire breaks through the interior finish materials (plaster or drywall), it can spread undetected within these void spaces to other portions of the structure. Interior stairwells generally allow a fire to extend vertically within the building. Often, Type III construction buildings have been remodelled and have multiple roofs, such as when a newer roof is constructed on top of an older roof, making ventilation more difficult. Although ventilation can be performed either horizontally or vertically in such a case, it can be hard to ventilate in the area close to where the fire is burning.

The vertical openings in Type III construction frequently provide a path for fire to extend into an attic or cockloft (open space between the ceiling of the top floor and the underside of the roof). Heat and smoke can accumulate within these spaces and then spread laterally between sections of a large building or into attached buildings. Coordinated suppression and vertical ventilation are essential for extinguishing attic fires (**FIGURE 13-6**).

FIRE FIGHTER TIP

Ventilation should not take place until the suppression team is in place and is in a position to apply water or foam. A well-coordinated, disciplined fire attack requires constant communication between the fire attack team and the ventilation team.

FIGURE 13-6 Coordinated suppression and vertical ventilation are essential for extinguishing attic fires. Incident commanders should exercise extreme caution when considering roof ventilation in Type V lightweight construction.

Courtesy of District Chief Chris E. Mickal/New Orleans Fire Department, Photo Unit.

Type IV Construction (Heavy Timber)

Type IV construction (heavy timber) has exterior walls that consist of masonry construction and interior walls, columns, beams, floor assemblies, and roof structure that are made of wood. The exterior walls are usually brick and are extra thick to support the weight of the building and its contents.

The heavy, solid wood used in Type IV construction is difficult to ignite, but once involved in a fire, the structure of these types of buildings can burn for many hours. As the fire consumes the heavy timber support members, the masonry walls will become unstable and collapse.

Heavy timber construction should have no concealed spaces or voids, reducing the risk of horizontal and vertical fire spread that often occurs in ordinary construction buildings. However, many Type IV buildings have been converted into small shops, galleries, office buildings, and residential occupancies. These conversions tend to divide the open spaces into smaller compartments and create void spaces within the structure. Additionally, many heavy timber buildings contain vertical openings for elevators, stairs, or machinery, which can provide a path for a fire to travel from one floor to another.

Type IV buildings usually contain a large number of windows. A large fire in this type of construction will often self-ventilate as the windows break from the heat of the fire. Vertical ventilation, however, may be difficult because of the thick layers of wood sheathing and the buildup of multiple layers of roofing materials.

Type V Construction (Wood Frame)

Type V construction (wood frame) has many of the same features as Type III construction. The primary difference is that the exterior walls in a Type V building are not required to be constructed of masonry or noncombustible materials.

Type V buildings often contain many void spaces where fire can spread, including attics or cocklofts. Modern, fast-growth lumber; lightweight wood-truss roofs; and manufactured I-beam floors, which can fail quickly under fire conditions, are common in newer buildings.

Older buildings of Type V construction were often assembled with balloon-frame construction. This type of construction includes direct vertical channels within the exterior walls, so a fire can spread very quickly from a lower level to the attic or cockloft in such a building. Heated smoke and gases can accumulate under the roof of these buildings, requiring rapid vertical roof ventilation.

Modern Type V construction typically uses platform-frame construction. In these buildings, the structural frame is built one floor at a time. Between each floor a plate at the floor and the ceiling acts as a fire stop. Its presence limits the fire from spreading upward and helps contain and limit the fire on a single floor.

Modern Type V construction typically uses lightweight components to save money and provide longer and more open areas, which presents problems when a fire occurs. Engineered structural components used in residential construction, such as trusses and wooden I-beams, contain much less wood than the solid beams used in older construction. These products burn more rapidly and might fail quickly. The fire service cannot become complacent with modern dimensional lumber as new lumber is engineered to "grow quickly." In some cases, a tree may be only 18 years of age when harvested. UL studies have shown that modern dimensional lumber (e.g., 0.6 m by 2.4 m [2 ft by 8 ft], 0.6 m by 3 m [2 ft by 10 ft], 0.6 m by 3.7 m [2 ft by 12 ft]) may fail as quickly as engineered products (Kerber, 2012).

LISTEN UP!

As building construction materials and techniques have changed and the fuel load of building contents is increased, fire behaviour in residential fires has changed. This requires modification of some of the ventilation tactics that have been used in the past. These updates are based on the observations of fire fighters at today's fires and the controlled tests conducted by NIST and UL. These studies are available at www.nist.gov/fire or at www.ulfirefightersafety.org.

Timing and Coordination of Ventilation and Suppression

Limiting the amount of air entering the fire compartment limits the fire's ability to grow. This can be achieved by adopting a disciplined approach to ventilation practices. Fire fighters must resist the urge to break windows and force open doors. A methodical approach includes keeping the doors and windows closed until the fire suppression team is ready to apply water into the fire compartment. Once the water has begun to cool the fire, more energy is being absorbed by the water than is being generated by the fire. At this point temperatures will likely be reduced to enable traditional ventilation methods to eliminate the residual heat and smoke. When thinking about ventilation, think about the Three Ws—when, where, and why—which includes timing. Effective ventilation requires an incident command function to coordinate timing and a fire attack.

SAFETY TIP

Many structure fires today are ventilation-limited and are discovered during the ventilation-limited stage.

Minimizing Backdrafts and Flashovers

Ventilation is a major factor in two significant fire-ground phenomena: backdraft and flashover. Chapter 5, *Fire Behaviour*, contains more information on these conditions. Both can be deadly situations, and fire fighters should exercise great caution when conditions indicate that either is possible. A transitional fire attack reduces the chance of a backdraft or flashover. Apply water to the fire from a safe location as close to the fire as possible. This will reduce heat and fuel in the form of hot gases and smoke from the fire compartment. It will also reduce the temperature of the space, ideally converting the ventilation-limited fire to a fuel-limited fire. Then, ventilation will be more effective in removing hot gases and smoke from the fire space, and this will allow entry into the interior space for final fire suppression (**FIGURE 13-7**).

Backdrafts and Ventilation

A backdraft can occur when a building is charged with superheated gases and most of the available oxygen has

A

B

Air

FIGURE 13-7 A. Building structure and fire conditions should be evaluated for signs of backdraft before any type of ventilation opening is made. **B.** Improperly venting a fire in an oxygen-deficient atmosphere can cause a backdraft.
© Jones & Bartlett Learning.

been consumed. Few flames might be apparent, but the hot gases can contain large amounts of unburned or partially burned fuel. If oxygen is introduced to the mixture, the fuel can ignite and explode. In some cases, a backdraft can occur in as little as 10 seconds after ventilation (Kerber, 2012). To reduce the chance of a backdraft, fire fighters must release as much heat and unburned products of combustion as possible.

Placing a ventilation opening as high as possible within the building or area can help to eliminate potential backdraft conditions. A roof opening can draw the hot gases up and relieve the interior pressure. As the mixture rises into the open atmosphere and is exposed to oxygen, it might ignite. The ventilation crew should not ventilate until the attack crew is ready to advance the hose line.

Before entering the building, the attack crews should charge their hose lines and check for proper hose function by flowing water and checking the nozzle outside the building. It is important to remember that opening the front door of a structure to begin a fire attack is an act of ventilation. The door should remain closed until the attack team is ready to advance the hose line into the building.

Once the attack team sees flames or encounters high temperatures inside the structure, they should open their hose streams to extinguish the fire and cool the interior atmosphere as quickly as possible. In cases where flames are beginning to move across the ceiling, a narrow-patterned fog stream of water directed at the ceiling helps to cool the upper area of the room without upsetting the thermal balance. "Cool as you go" is a good phrase to help remember this concept.

SAFETY TIP

The flow of hot gases in a flow path has been measured at speeds as high as 24 kph (15 mph) with no wind present (Norwood and Ricci, 2014). This is faster than a fire fighter can run to try to escape the path of the flashover.

Flashovers and Ventilation

Flashover is the transition from a fire that is growing by igniting one type of fuel to another to a fire where all of the exposed surfaces have ignited (**TABLE 13-2**). The critical temperature for a flashover to occur is approximately 538°C (1000°F). In a typical room-and-contents fire, the ceiling temperature can reach 1204°C (2200°F; Kerber, 2012). Because the potential harm from a flashover is so great, fire fighters need to be able to recognize any conditions that might indicate the possibility of a flashover and take steps to prevent it whenever possible. These conditions include off-gassing or smoke from furniture and carpeting, a sudden increase in heat, or zero visibility. Do not enter an environment if signs of an impending flashover are present. Use a transitional attack to cool the fire compartment, vent, and then enter.

TABLE 13-2 Chemical Properties of Fire Gases		
Chemical Name	Ignition Temperature	Flammable Range
Propane	480°C/896°F	2.5 to 9.0 percent
Natural Gas	628°C/1163°F	5.0 to 15 percent
Hydrogen	538°C/1000°F	4.0 to 75 percent

SAFETY TIP

There are no reliable signs or symptoms to warn you of an impending flashover in a ventilation-limited fire. If you recognize the potential signs of pending flashover, either exit from the nearest egress or apply water to the upper areas of the compartment immediately and monitor conditions closely.

Types of Ventilation

To remove the products of combustion and other airborne contaminants from a structure, fire fighters use two basic types of ventilation: horizontal and vertical. The type of ventilation used depends on several factors, including the ease and effectiveness of executing (performing) that type of ventilation.

Horizontal ventilation takes advantage of the doors, windows, and other openings at the same level as the fire. Doors are often the easiest of these to use for ventilation. They often need to be opened anyway to provide access to or egress from the building. It is possible to open some doors without tools. Additionally, the door may have been left open by the building residents. If the door is locked, it is usually not too difficult to perform forcible entry with simple tools. Windows are the next easiest to open. Remember, *try before you pry*—that is, try manually opening a window before removing the glass and frame. Second-floor windows are usually readily accessible using ground ladders. In some cases, fire fighters might make additional openings in a wall to provide horizontal ventilation.

Vertical ventilation involves making openings in roofs or floors so that heat, smoke, and toxic gases can escape from the structure in a vertical direction. Pathways for vertical ventilation can include ceilings, stairwells, exhaust vents, and roof openings such as skylights, scuttles (small openings or hatches with moveable lids), or monitors. Ventilating skylights and roofs usually requires gaining access using ground ladders or aerial devices. The safety of the fire fighter should always be the first consideration when making these tactical decisions. Generally, it is inadvisable to place fire fighters on a lightweight constructed roof. Additional openings can be created by cutting holes in the roof or the floor and making sure that the opening extends through every layer of the roof or floor. Vertical ventilation typically incorporates horizontal openings, such as doors and windows, to provide an intake for the clean air to replace the smoke that is being exhausted through the vertical vent.

Ventilation can be either natural or mechanical. Natural ventilation depends on convection currents and other natural forces, such as the wind, to move heat and smoke out of a building and to allow clean air to enter. Mechanical ventilation uses fans or other powered equipment to introduce clean air or to exhaust heat and smoke.

Ventilation can also be classified as intentional or unintentional. Intentional ventilation occurs when the ventilation is planned and done on purpose. Unintentional ventilation occurs when a window or door fails because of the fire or a window or door is left open by mistake. Any ventilation may cause a rapid change in the direction or intensity of a fire, and unexpected ventilation may jeopardize fire fighters operating inside a structure due to sudden changes in the fire's behaviour.

The effectiveness of each type of ventilation depends on the size of the opening and its location. Opening a door usually creates a larger ventilation opening than opening a window. Some windows are opened easily while others are constructed with reinforced materials that are difficult to remove or break. Opening a skylight is often easier and faster than cutting a hole in the roof. Roofing materials vary greatly as noted in Chapter 6, *Building Construction*. Even after a roof has been cut open, there may be additional layers of ceiling materials that need to be removed to create an exhaust opening.

FIGURE 13-8 Windows are frequently used in horizontal ventilation.
Courtesy of District Chief Chris E. Mickal/New Orleans Fire Department, Photo Unit.

LISTEN UP!

When referring to ventilation techniques, fire fighters use the term *contaminated atmosphere* to describe the products of combustion that must be removed from a building and the term *clean air* to refer to the outside air that replaces them. The contaminated atmosphere can include any combination of heat, smoke, and gases produced by combustion, including toxic gases. A contaminated atmosphere also can be defined as a dangerous or undesirable atmosphere that was not caused by a fire. For example, natural gas, carbon monoxide, ammonia, and many other products can contaminate the atmosphere as a result of a leak, spill, or similar event. The same basic techniques of ventilation can be applied to many of these situations. Contaminated atmospheres must be considered environments that are immediately dangerous to life and health (IDLH).

Horizontal Ventilation

Horizontal ventilation uses horizontal openings in a structure, such as windows and doors, and can be employed in many situations, particularly in small fires (**FIGURE 13-8**). Horizontal ventilation is commonly used in residential fires, room-and-contents fires, and fires that can be controlled quickly by the attack team.

Horizontal ventilation can be a rapid, generally easy way to clear a contaminated atmosphere. Often,

outlets for horizontal ventilation can be made simply by opening a door or window. In other cases, fire fighters might need to break a window or to use forcible entry techniques. Search and attack teams operating inside a structure, as well as fire fighters operating outside the building, may use horizontal ventilation to reduce heat and smoke conditions. As previously mentioned, this action must be conducted in coordination with fire attack and other fire-ground operations.

Horizontal ventilation is most effective when the opening goes directly into the room or space where the fire is located. Opening an exterior door or window, for example, allows heat and smoke to flow directly outside and eliminates the problems that can occur when these products travel through the interior spaces of a building.

Horizontal ventilation is more difficult if there are no direct openings to the outside or if the openings are inaccessible. In these situations, it might be necessary to direct the air flow through other interior spaces or to use vertical ventilation.

Horizontal ventilation can be used in less urgent situations if the building can be ventilated without additional structural damage. Low-urgency situations might include residual smoke caused by a stovetop fire or a small natural gas leak.

Horizontal ventilation tactics include both natural and mechanical methods.

Natural Ventilation

Natural ventilation depends on convection currents, wind, and other natural air movements to allow a contaminated atmosphere to flow out of a structure. Convection currents, which are created by the fire, move smoke and gases up toward the roof or ceiling and out and away from the fire source. Opening or breaking a window or door allows these products of combustion to escape through natural ventilation.

Natural ventilation can be used only when the natural convection air currents or wind is adequate to move the contaminated atmosphere out of the building and replace it with clean air. In winter months, the positive effects of natural ventilation may be more discernible due to the presence of cold, heavy air. Mechanical devices (discussed later in this chapter) can be used if natural forces do not provide adequate ventilation.

Natural ventilation is often used when quick ventilation is needed, such as when attacking a room-and-contents fire or a first-floor residential fire with people trapped on the second floor. For search and attack teams to enter these buildings and act quickly, smoke and heat must be ventilated immediately.

Wind speed and direction play important roles in natural ventilation. If possible, windows on the downwind side of a building should be opened first so that the contaminated atmosphere flows out. Openings on the upwind side can then be used to bring in clean air. Conversely, opening a window on the upwind side first could push the fire into uninvolved areas of the structure, particularly on a windy day.

Breaking Glass

Natural ventilation uses existing or created openings in a building. The methods used to create horizontal openings employ a variety of tools and techniques. For example, some windows can simply be opened by hand. If the window cannot be opened and the need for ventilation is urgent, however, fire fighters should not hesitate to radio their officer and ask permission to break the glass. Although breaking a window is a fast and simple way to create a ventilation opening, it is important to communicate with the officer before employing this technique because the fire fighter may not be able to see how the building is reacting to the fire or where other crews are working.

When breaking glass, fire fighter should always use a hand tool (Halligan tool, axe, or pike pole) and keep their hands above or to the side of the falling glass. This tactic prevents pieces of glass from sliding down the tool and potentially causing injury. The tool should then be used to clear the entire opening of all remaining pieces of glass, thereby creating the largest opening possible and providing a way for fire fighters to enter or exit through the window in the event of an emergency.

Clearing the broken glass also reduces the risk of injury to anyone who might be in the area and eliminates the danger of sharp pieces falling out later. Before breaking glass—particularly if the window is above the ground floor—fire fighters must look out to ensure that no one will be struck by the falling glass.

To break glass on the ground floors with a hand tool, follow the steps in **SKILL DRILL 13-1**.

Breaking a Window from a Ladder. A similar technique can be used to break a window on an upper floor, with the fire fighter working from a ladder. The ladder should be positioned to the upwind side of the window so that smoke is carried away from the fire fighter. The tip of the ladder should be even with the top of the window. The fire fighter should climb to a position level with the window and lock into the ladder for safety. Using a hand tool, the fire fighter should strike and break the window and then clear the opening completely. The fire fighter should not be positioned below the window, where glass could slide down the handle of the tool (**FIGURE 13-9**).

Breaking a Window with a Ladder. Fire fighters on the ground can use the tip of a ladder to break and clear a window when immediate ventilation is needed on an upper floor. This technique requires proper ladder selection. Usually, second-floor windows can be reached by 5- (16-ft) or 6-m (20-ft) roof ladders, as well as by bedded 7- (24-ft) or 8.5-m (28-ft) extension ladders. Extension ladders are needed to break windows on the third or higher floors.

To perform this maneuver, the ladder can be raised directly into the top half of the window, or it can be raised next to the window to determine the proper height for the tip. The ladder is then rolled into the window, drawn back at the tip, and forcibly dropped into the top third of the window. The objective is to push the broken glass into the window opening, although there is always

SKILL DRILL 13-1
Breaking Glass with a Hand Tool Fire Fighter I, NFPA 1001: 4.3.11

1 Wear full personal protective equipment (PPE), including eye protection and self-contained breathing apparatus (SCBA). Select a hand tool, and position yourself to the side of the window.

2 With your back facing the wall, swing backward forcefully with the tip of the tool striking the top one-third of the glass.

3 Clear the remaining glass from the opening with the hand tool.

© Jones & Bartlett Learning. Photographed by Glen E. Ellman.

FIGURE 13-9 When breaking an upper-storey window, the fire fighter should place the ladder to the side of the window.
© Jones & Bartlett Learning.

a risk that some of the glass will fall outward. For this reason, fire fighters working below the window must wear full PPE to shield themselves from falling glass. A word of caution: The higher the window, the greater the danger of falling glass. It is also important to remember that this form of ventilation must still be performed in coordination with suppression operations and the incident's tactical objectives.

When a ladder is used to break a window, shards of glass might be left hanging from the edges of the opening. If the opening will be used for access, it needs to be cleared with a tool before anyone enters through it.

FIRE FIGHTER TIP

When breaking a window from above, watch for glass sliding down the beams of the ladder. Wear full protective clothing and SCBA (on air) and gloves when performing this type of ventilation because this procedure can cause injury to an unwary fire fighter. Coordinate operations with the incident's tactical objectives.

To break a window with a ladder, follow the steps in **SKILL DRILL 13-2**.

SKILL DRILL 13-2
Breaking Windows with a Ladder Fire Fighter I, NFPA 1001: 4.3.11

1 Wear full PPE, including eye protection. Select the proper size ladder for the job. Check for overhead lines. Use standard procedures for performing a ladder raise.

2 Raise the ladder next to the window. Extend the tip so that it is even with the top third of the window. If a roof ladder is used, extend the hooks toward the window.

3 Coordinate operations with the incident's tactical objectives. Position the ladder in front of the window.

4 Forcibly drop the ladder into the window. Exercise caution—falling glass can cause serious injury.

5 Raise the ladder from the window, and move it to the next window to be ventilated. Either carry the ladder vertically or pivot the ladder on its feet.

Opening Doors

Like windows, door openings can be used for natural ventilation operations. Doorways have certain advantages when used for ventilation. Specifically, they are large openings—often twice the size of an average window. In fact, some doorways can be large enough to drive a truck through. A locked door can be opened using forcible entry techniques.

The problem with using doorways as ventilation openings is that their use for this purpose compromises entry and exit for interior fire attack teams as well as search teams. If heat and smoke are pouring through a doorway, fire fighters cannot enter or exit through that portal. Because protecting means of egress is a priority, doorways are more suitable as entries for clean air. The contaminated atmosphere should, therefore, be discharged through a different opening.

Fire fighters often approach a fire through the interior of a building to attack from the unburned side. Opening an exterior door that leads directly to the fire can create an opening for oxygen-rich air to enter the building or a vent that pushes the heat and smoke out of the building. Treat the opening of a door as a major act of ventilation. It should be opened only when the hose line is charged and the attack team is ready to advance.

Before opening an exterior door, examine the area around the door. A small amount of smoke can be the sign of a small fire or a sign of a ventilation-limited fire. Proceed with extreme caution. Once the exterior door is opened, note how the air moves through the exterior door. A rapid movement of air into the structure can indicate a ventilation-limited fire. If the fire is actively burning, clean air may be seen entering through the bottom half of the exterior door and smoke escaping through the top half (bidirectional flow).

Doorways are good places to put mechanical ventilation devices when natural ventilation needs to be augmented. Fans can be used to push clean air in (positive-pressure ventilation) or to pull contaminated air out (negative-pressure ventilation).

SAFETY TIP

Opening an exterior door is an act of ventilation. In a ventilation-limited fire, this act can quickly produce a backdraft.

Mechanical Ventilation

In addition to making or controlling openings to influence ventilation, fire fighters can use mechanical ventilation to direct the flow of combustion gases. **Mechanical ventilation** uses large high-powered fans or other powered equipment to augment natural ventilation and direct the flow of combustion gases. There are three different methods of mechanical ventilation. **Negative-pressure ventilation** uses fans called **smoke ejectors** to pull smoke and heat through openings of a structure. **Positive-pressure ventilation** uses fans to introduce clean air into a structure and push the contaminated atmosphere out. **Hydraulic ventilation** moves hot gases and smoke from a room by using fog or broken-pattern fire streams to create a pressure differential behind and in front of the nozzle.

In some buildings, HVAC systems can be used to provide mechanical ventilation. To perform this function, such a system needs to be configured so that the zones can be controlled correctly and effectively. The building engineer or maintenance staff might have the specific knowledge required to assist in this effort. Some HVAC systems can be used to exhaust smoke or supply clean air if a fire occurs, whereas others should be shut down immediately. Information about HVAC systems should be obtained during preincident planning surveys.

Negative-Pressure Ventilation

The basic principles of air flow are used in negative-pressure ventilation. Fire fighters locate the source of the fire and, using a smoke ejector, exhaust the products of combustion out through a window or door. The smoke ejector creates a negative pressure that draws the heat, smoke, and fire gases out of the building. In turn, fresh, clean air must enter the structure to replace the exhausted contaminated air.

Smoke ejectors are usually 40 to 60 cm (16 to 24 in.) in diameter, although larger models are available. They can be powered by electricity, gasoline, or water pressure (**FIGURE 13-10**).

FIGURE 13-10 A smoke ejector pulls products of combustion out of a structure. Gasoline-powered smoke ejectors might have problems running because of reduced oxygen or carbon particles in the smoke.

Courtesy of Super Vacuum Mfg. Co., Inc.

Negative-pressure ventilation can be used to clear smoke out of a structure after a fire, particularly if natural ventilation would be too slow or if no natural cross-ventilation exists. In large buildings, negative-pressure ventilation can be used to pull heat and smoke out while fire fighters are working, although it is not generally used before or during a fire attack. This type of ventilation is usually associated with horizontal ventilation; however, a smoke ejector can also be used in a roof-vent operation to pull heated gases out of an attic.

Negative-pressure ventilation has several limitations related to its positioning, power source, maintenance, and air flow control. Often, fire fighters must enter the heated and smoke-filled environment to set up the smoke ejector, and they might need to use braces and hangers to position it properly. Because most smoke ejectors run on electricity, a power source and a cord are required for them to function. Smoke ejectors also get very dirty while they are running, so they must be taken apart and cleaned after each use.

Air flow control with a negative-pressure system is difficult. The smoke ejector must be completely sealed so that the exhausted air is not immediately drawn back into the building. This phenomenon, which is called *recirculation*, reduces the effectiveness of the smoke ejector. Recirculation can be eliminated by completely blocking the opening around the ejector.

Smoke ejectors usually have explosion-proof motors, which make them excellent choices for venting flammable or combustible gases and other hazardous environments.

To perform negative-pressure ventilation, follow the steps in **SKILL DRILL 13-3**.

SKILL DRILL 13-3
Delivering Negative-Pressure Ventilation Fire Fighter I, NFPA 1001: 4.3.11

1 Determine the area to be ventilated and the direction of the outside wind. Place the smoke ejector so that it is not on the upwind side of the building.

2 Hang the smoke ejector in the upper part of the ventilation opening.

3 Use a salvage cover to prevent recirculation if the structure's windows are not intact. Provide an opening on the upwind side of the structure to provide cross-ventilation.

© Jones & Bartlett Learning. Photographed by Glen E. Ellman.

Positive-Pressure Ventilation

Large, powerful fans known as positive-pressure fans can be used to force fresh air into a structure. The fans create a positive pressure inside the structure, which displaces the contaminated atmosphere and pushes heat and other products of combustion out. Positive-pressure fans can be used to reduce interior temperatures and smoke conditions in coordination with an initial fire attack or used to clear a contaminated atmosphere after a fire has been extinguished. When positive-pressure fans are used to control the flow of products of combustion prior to fire control, the technique is known as positive-pressure attack. A **positive-pressure attack** can provide increased visibility and tenability for fire fighters and potential occupants while fire suppression efforts are under way. Positive-pressure fans can also be used to remove products of combustion after a structure fire has been controlled. This technique is known as positive-pressure ventilation.

Positive-pressure fans are usually set up at exterior doorways, often at the same opening used by the attack team (**FIGURE 13-11**). This creates a cone of fresh air that is pushed in the same direction as the advancing hose line so that fire fighters are enveloped in a stream of clean, cool air while they move toward the fire.

When using positive-pressure ventilation, it is necessary to provide an outlet or exhaust opening—such as an open window or door—to release the positive pressure created by the fan. This opening must be located in the fire room to effectively allow the heat and smoke to escape (**FIGURE 13-12**). The size of the exhaust opening should be a 2:1 ratio (or, at minimum, a 1:1 ratio) to the entry opening to create the desired positive pressure within the room. It is also important that the interior door of the fire room—not the exterior door of the structure—dictate the size of the exhaust opening. A higher pressure must be maintained at this opening in order to retain an efficient flow. If there is no exhaust opening or if the opening is too small or located in other areas of the structure, the heat and smoke can flow into the other areas of the building or back toward the attack team. In addition, this kind of system will not work effectively if the building is not intact.

To increase the efficiency of positive-pressure attack or ventilation and to prevent smoke and fire from

FIGURE 13-11 Positive-pressure fans are usually set up at exterior doorways, often at the same opening used by the attack team.
© Jones & Bartlett Learning.

FIGURE 13-12 When using positive-pressure ventilation, there must be an opening in the fire room to allow heat and products of combustion to escape.
© Jones & Bartlett Learning.

reaching other parts of the building, fire fighters should close doors to unaffected areas of the structure. This is an effective action to maintain a tenable environment for people trapped in other parts of the building.

When planning positive-pressure attack or ventilation, the force and direction of the wind need to be taken into consideration. Positive-pressure fans should never be placed in a position where they are blowing against the wind. Check the direction and speed of the wind before placing fans, and monitor the wind for changes in direction or intensity.

Recent studies by UL and NIST have demonstrated that opening all windows immediately after fire suppression operations is most effective in ventilating a single-family or two-family residence. Note, this recommendation only applies to positive-pressure ventilation—that is, after the fire has been extinguished. During positive-pressure attack, on the other hand, only openings in the fire room have been proven to be effective. Openings other than those in the fire room permit the fire to propagate to others areas of the structure. In a high-rise building, positive-pressure fans can blow fresh air up through a stair shaft. By opening the doors, fire fighters can clear one floor at a time. Very large structures can be ventilated by placing multiple fans side by side.

FIRE FIGHTER TIP

In high-rise fires, a large number of people need to be safely evacuated, often using a single stairway. The attack stairway and evacuation stairways must be established and announced early in the event. The evacuation stairway must be free of smoke and toxic gases. Positive-pressure fans can be used to keep smoke and products of combustion from entering the stairwell. In emergencies, they can also be used to introduce clean, oxygen-rich air into the potentially toxic atmosphere of the stairwell. When using positive-pressure fans for this purpose, keep in mind that the engines can increase carbon monoxide levels, so ensure that the fans are properly ventilated.

Positive-pressure ventilation has several advantages over negative-pressure ventilation. For instance, a single fire fighter can set up a fan very quickly. Because the fan is positioned outside the structure, the fire fighter does not have to enter a hazardous environment, and the

fan does not interfere with interior operations. Positive pressure is both quick and efficient because the entire space within the structure is under pressure.

When used with a well-placed exhaust opening, positive-pressure ventilation can help keep a fire confined to a smaller area and increase safety for interior operating crews, as well as any building occupants. In addition, positive-pressure fans do not require as much cleaning and maintenance as negative-pressure fans because the products of combustion never move through the fan.

Positive-pressure ventilation does have some disadvantages, however. The most important concern is the potential for improper use of positive pressure to spread a fire. If the fire is burning in a structural space, such as in a pipe chase or above a ceiling, positive pressure can push it into unaffected areas of the structure. If the fire is located in structural void spaces, positive-pressure ventilation should not be used until access to these spaces is available and attack crews are in place. Ineffective positive-pressure ventilation occurs if positive pressure is used and there are no openings for exhaust gases to release the positive pressure from the building.

Positive-pressure fans operate at high velocity and can be very noisy. Most are powered by internal combustion engines, so they can increase carbon monoxide levels if they are run for significant periods of time after the fire is extinguished. Natural ventilation should be used after the structure is cleared to prevent carbon monoxide build-up.

SKILL DRILL 13-4 outlines the steps to performing positive-pressure ventilation.

FIRE FIGHTER TIP

Doors used for ventilation must be kept open to ensure that ventilation is sustained. Depending on the time available and the type of door and lock, fire fighters might remove the door, disable the closure device, or place a wedge in the hinge side or under the door. Overhead doors must be locked in the open position, thereby ensuring that they cannot accidentally roll down and close. Check the manufacturer's recommendations.

On some fans, if the fan is tipped down while removing it from the apparatus, the oil sensor will prevent the fan from being started. Fire fighters need to know the procedure to use to get the fan to start. If the ventilation causes conditions to worsen, the fan should be turned off.

Hydraulic Ventilation

Hydraulic ventilation uses the water stream from the hose line to exhaust smoke and heated gases from a structure. To use this ventilation technique, the fire

SKILL DRILL 13-4
Delivering Positive-Pressure Ventilation Fire Fighter I, NFPA 1001: 4.3.12

1 Place the fan approximately 1 m (3 ft) in front of the opening to be used for the fire attack. The exact position depends on the size of the opening, the size of the fan, and the direction of the wind.

2 Provide an exhaust opening in the fire room. This opening can be made either before the fan is started or when the fan is started.

3 Start the fan.

4 Check the cone of air produced; it should completely cover the opening. This can be checked by running a hand around the door frame to feel the direction of air currents. Monitor the exhaust opening to ensure there is a unidirectional flow. Allow the smoke to clear.

fighter working the hose line directs a narrow fog stream or a broken-pattern stream from a fog-stream nozzle out of the building through an opening, such as a window or doorway. The contaminated atmosphere is drawn into a low-pressure area behind the nozzle. An induced draft created by the high-pressure stream of water then pulls the smoke and gas out through the opening (**FIGURE 13-13**). A well-placed fog stream or broken-pattern stream can move a tremendous volume of air through an opening.

Hydraulic ventilation can move several thousand cubic feet of air per minute, and it is effective in clearing both heat and smoke from a fire room. It is most

FIGURE 13-13 Hydraulic ventilation can be used to draw smoke out of a building after the fire is controlled.
© Jones & Bartlett Learning.

useful after the fire is under control. Because hydraulic ventilation does not require any specialized equipment, it can be performed by the attack team with the same hose line that was used to control the fire.

Hydraulic ventilation has some disadvantages. Fire fighters must enter the heated, toxic environment and remain in the path of the products of combustion as they are being exhausted. If the water supply is limited, using water for ventilation must be balanced with using water for fire attack. This technique can cause excessive water damage if it is used improperly or for long periods of time. Also, in cold climates, ice build-up can occur on the ground, creating a safety hazard.

Ideally, the ventilation opening should be created before the hose line advances into the fire area. The hose stream itself can be used to break a window before it is directed on the fire. The steam created when the water hits the fire will push the heat and smoke out through the vent. Hydraulic ventilation can then be used to clear the remaining heat and smoke from the room.

To perform hydraulic ventilation, follow the steps in **SKILL DRILL 13-5**.

Vertical Ventilation

Vertical ventilation refers to the release of smoke, heat, and other products of combustion into the atmosphere in a vertical direction. Vertical ventilation occurs naturally, owing to the fire's flow path (convection), if an opening

SKILL DRILL 13-5
Delivering Hydraulic Ventilation Fire Fighter I, NFPA 1001: 4.3.12

1 Enter the room, and remain close to the ventilation opening. Place the nozzle through the opening, and open the nozzle to a narrow fog or broken-pattern spray.

2 Keep directing the stream outside and back into the room until the stream almost fills the opening. The nozzle should be 0.6 to 1.2 m (2 to 4 ft) inside the opening.

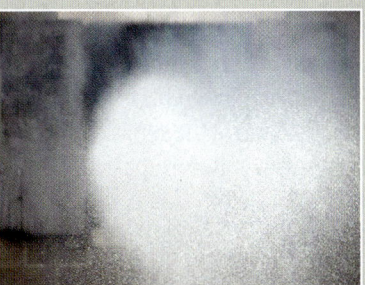

3 Stay low, out of the heat and smoke, or to one side to keep from partially obstructing the opening.

© Jones & Bartlett Learning. Photographed by Glen E. Ellman.

is available above a fire. Convection causes the products of combustion to rise and flow through the opening. In some situations, vertical ventilation can be assisted by mechanical means such as fans or hose streams.

Although vertical ventilation may involve any opening that allows the products of combustion to travel up and out, this term is most often applied to describe operations on the roof of a structure. The ventilation openings can be existing features, such as skylights or bulkheads, or holes created by fire fighters who cut through the roof covering. The choice of ventilation openings depends primarily on the building's roof construction.

A vertical ventilation opening should be made as close as possible to the seat of the fire. Smoke issuing from the roof area, melted asphalt shingles, or steam coming from the roof surface are all signs that fire fighters can use to identify the hottest point. A thermal imaging device also may be helpful in locating the part of the roof closest to the fire. In order for a vertical exhaust vent to be effective, there must be a horizontal intake vent to admit air. In addition, if horizontal vents are made on a floor above the fire, they can produce the effect of vertically venting the fire.

> ### FIRE FIGHTER TIP
>
> "Taking the lid off" does not guarantee positive results. Vertical ventilation is the most efficient type of natural ventilation. It allows for most hot gases to exit the structure; however, it also allows the most air to be entrained into the structure. Vertical ventilation must be coordinated with fire attack—just like with horizontal ventilation.

Safety Considerations in Vertical Ventilation

Vertical ventilation should be performed only when it is necessary and can be done safely. Rooftop operations involve significant inherent risks. If horizontal ventilation can sufficiently control the situation, vertical ventilation might be unnecessary. Before performing vertical ventilation, fire fighters must evaluate all pertinent safety issues and avoid unnecessary risks. In particular, they should assess the roof for skylights, roof scuttles, heat vents, plumbing vents, louver ventilation, translucent roof panels, cell phone towers, photovoltaic panels (solar panels), and fan shafts.

The most obvious risk in a rooftop operation is the possibility that the roof will collapse if the fire compromises the structural integrity of the roof support system. If the structural system supporting the roof fails, both the roof and the fire fighters standing on it will fall into the fire. Fire fighters also can fall through an area of the roof decking that has been weakened by the fire.

Determine the type of roof construction before beginning roof ventilation operations. If the roof is constructed with lightweight trusses, ventilate the roof from an aerial device (**FIGURE 13-14**). Roofs built using lightweight manufactured trusses can fail quickly and without warning (UL FSRI, 2006).

> ### SAFETY TIP
>
> Ventilate roofs constructed with lightweight construction from an aerial device. Do not attempt to stand on these roofs if there are signs of heat or fire underneath.

Falling from a roof—either off of the building or through a ventilation opening, open shaft, or skylight into the building—is a risk fire fighters encounter when operating on roofs. Smoke reduces visibility and can cause disorientation, increasing the risk of falling off the edge of a roof. Factors such as darkness, snow, ice, or rain can create hazardous situations. Not surprisingly, a pitched roof presents even greater risks than a flat roof. Safe operations on pitched roofs require fire fighters to operate from an aerial device or to use a roof ladder.

Effective vertical ventilation may assist with prompt fire control and reduces the risks to fire fighters operating inside the building. In some situations, prompt, proper, and coordinated vertical ventilation can save the lives of building occupants by relieving interior conditions, opening exit paths, and allowing fire fighters to enter and perform search and rescue operations.

Vertical ventilation should always be performed as quickly, efficiently, and safely as possible. If they provide an adequate opening, skylights, ventilators, bulkheads,

FIGURE 13-14 Fire fighters operating from the safety of an aerial device.

Courtesy of Captain David Jackson, Saginaw Township Fire Department.

and other existing openings should be opened first. If fire fighters must cut through the roof covering to create a hole, the location should be identified and the operation should be performed promptly and efficiently. It must be done in coordination with the other fire fighters operating on the fire scene.

Fire fighters who are working on the roof should always have two safe exit routes. A second ground ladder or aerial device should be positioned to provide a quick alternative exit, in case conditions on the roof deteriorate rapidly. The two exit routes should be separate from each other, preferably in opposite directions from the operation site. The team working on the roof should always know where these exit routes are located, and they should plan ventilation operations to work toward these locations.

The ventilation opening should never be located between the fire fighters and their exit routes. A ventilation opening releases smoke and hot gases from the fire. If this exhaust flow path of hot gases and smoke separates the fire fighters from their exit from the roof, they will be in serious danger (**FIGURE 13-15**).

SAFETY TIP

Fire fighters operating on roofs must be alert for the presence of skylights, translucent panels, and photovoltaic panels (solar panels). Fire fighters have been injured or killed due to stepping on and/or falling through these types of panels. These panels are not required to support the weight of a fire fighter or to be marked so fire fighters can identify them during visual limited operations. Additionally, photovoltaic panels are energized and can result in electrocution. Preincident planning and careful operations are required.

Once a ventilation opening has been made, fire fighters should withdraw to a safe location. There is no good reason to stand around the vent hole and look at it; this behaviour exposes the fire fighters to an unnecessary risk.

Fire fighters should operate on a roof only if absolutely necessary; this technique should not be standard practice. Roof operations should be conducted from a roof ladder or an aerial device. Roofs—pitched or flat—can fail suddenly and without warning due to either fire damage or modern lightweight construction. Modern lightweight construction frequently uses thin plywood for roof decking, with a variety of waterproof and insulating coverings applied to the plywood. A relatively minor fire under this type of roof can cause the plywood to delaminate, with no visible indication of damage or weakness from above. The weight of a fire fighter on a damaged area can be enough to open a hole and cause the fire fighter to fall through the roof.

With pitched roofs, vertical ventilation operations should be conducted from aerial devices or at least from a well-secured roof ladder. Any rooftop operation should be conducted only after a complete size-up and assessment have been conducted and only when it is specifically needed to achieve a positive outcome.

A fire fighter's path to a proposed vent-hole site should follow the areas of greatest support and strength. These locations include the roof edges, which are supported by bearing walls, and the hips and valleys of the roof, where structural materials are doubled for strength. These areas should be stronger than interior sections of the roof.

Fire fighters should always be aware of their surroundings when working on a roof. Knowing where the roof's edge is located, for example, can help avoid accidentally walking or falling off the roof. Maintaining this knowledge is especially important—although difficult—in darkness, adverse weather, or heavy smoke conditions.

The order of the cuts in making a ventilation opening should be planned carefully. Fire fighters should be upwind, have clear exit routes, and be standing on a firm section of the roof or using a roof ladder. If fire fighters are upwind from the open hole, the wind will push the heat and smoke away from them; thus, the escaping heat and smoke should not block their exit route. To prevent an accidental fall into the building, fire fighters should stand on a portion of the roof that is firmly supported or work from a roof ladder, particularly when making the final cuts.

Basic Indicators of Roof Collapse

Roof collapse poses the greatest risk to fire fighters performing vertical ventilation. Some roofs—particularly truss roofs—give little or no warning that they are about to collapse.

Fire fighters who are assigned to vertical ventilation tasks should always be aware of the condition and construction of the roof. Avoid vertical ventilation

FIGURE 13-15 Be prepared for smoke and flames that are released when the roof is opened.
© craig robinson/iStock/Getty Images Plus/Getty Images.

operations that require fire fighters to be placed on a roof suspected to be comprised of lightweight materials. They should immediately retreat from the roof if they notice any of the following signs:

- Visible indication of sagging roof supports
- Any indication that the roof assembly is separating from the walls, such as the appearance of fire or smoke near the roof edges
- Structural failure of any portion of the building, even if it is some distance from the ventilation operation
- A sudden increase in the intensity of the fire from the roof opening (e.g., fire showing around roof vents, melting snow, and evaporating or steaming water)
- High heat indicators on a thermal imaging device

Roof Construction

Roofs can be constructed of many different types of materials in several configurations. Nevertheless, all roofs have three major components: the supporting structure, the roof decking, and the roof covering (**FIGURE 13-16**). Roof construction and design are discussed in Chapter 6, *Building Construction*.

The roof support system provides the structural strength to hold the roof in place. It must also be able to bear the weight of any rain or snow accumulation, any rooftop machinery or equipment, and any other loads placed on the roof, such as fire fighters or other people walking on the roof. The support system can be constructed of solid beams of wood, steel, or concrete or a system of lightweight construction.

The roof decking is the portion of roof between the roof supports and the roof covering. It is composed of

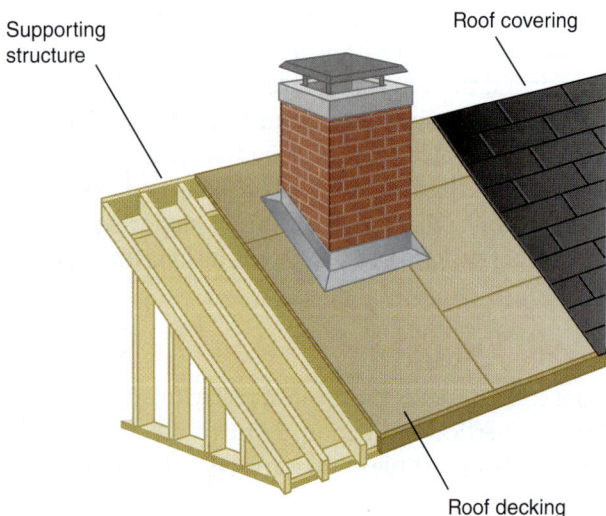

FIGURE 13-16 Roof assemblies consist of a supporting structure, decking, and covering.
© Jones & Bartlett Learning.

a rigid layer made of wooden boards, plywood sheets, or metal panels. The roof covering is applied on top of the decking and can have several layers. It constitutes the weather-resistant surface of the roof. It typically includes a waterproof membrane and insulation to retain heat in winter and limit solar heating in the summer. Roof covering materials include shingles and composite materials, tar and gravel, rubber, foam plastics, and metal panels.

Vertical ventilation operations often involve cutting a hole through the roof covering. The selection of tools, the technique used, the time required to make an opening, and the personnel requirements all depend on the roofing materials and the layering configurations.

Solid-Beam vs. Lightweight Construction

The two major structural support systems for roofs use either solid-beam or lightweight construction. It can be impossible to determine which system is used simply by looking at the roof from the exterior of the building, particularly when the building is on fire. Given this fact, information on the roof support system should be obtained during preincident planning surveys. Some provinces and territories require building owners to place a sign near the building entrance to indicate that the roof is constructed using lightweight construction.

The basic difference between solid-beam and lightweight construction is in the way individual load-bearing components are made. Solid-beam construction uses solid components, such as girders, beams, and rafters, to support the roof. Lightweight construction is assembled from smaller, individual components and includes trusses or engineered systems such as I-joists. In most cases, it makes no difference whether solid-beam or lightweight construction is used. For fire fighters, however, there is an important difference: Lightweight construction can collapse quickly and without warning when it is exposed to a fire even for a short period of time.

Trusses are constructed by assembling relatively small and lightweight components of wood, steel, or a combination of wood and steel in a series of triangles (**FIGURE 13-17**). Engineered systems are constructed from composite materials that are usually glued together (**FIGURE 13-18**). The resulting systems can efficiently span long distances while supporting a load. In most cases, a solid beam that spans an equal distance and carries an equivalent load will be much larger and heavier than a truss or engineered system.

The disadvantage of lightweight construction is that some systems fail completely when only one of the smaller components is weakened or when only one of the connections between components fails. If only one component fails, other components in the system might be able to absorb the additional load, but the

FIGURE 13-17 Truss construction uses smaller, individual components to support the roof.
© Jones & Bartlett Learning. Photographed by Glen E. Ellman.

FIGURE 13-18 Engineered systems, such as these I-joists, are constructed from composite materials that are usually glued together.
© Jones & Bartlett Learning. Photographed by Glen E. Ellman.

overall strength of the system is still compromised. A fire that causes one component to fail probably weakens additional components in the same area. These components can also fail, sometimes as soon as the additional load is transferred. In most cases, a series of trusses or engineered components fail in rapid succession, resulting in a total collapse of the roof. Sometimes just the weight of a fire fighter walking on the roof is enough to trigger a collapse.

Lightweight construction is not necessarily bad or inherently weak construction. Indeed, some lightweight construction is very strong and is assembled with sufficient protection from fire-retardant materials, so that it can resist fire just as well as solid beams. Many of these systems, however, are made of lightweight materials and are designed to carry as much load as possible with as little mass as possible. The individual components of lightweight construction used for roof support are often 51-mm by 102-mm (2-in. × 4-in.) wood sections or a combination of wood and lightweight steel bars or

tubes. Because these components are small, they can be weakened quickly if not protected from a fire.

Even more critical, however, are the points of the lightweight construction where the individual components join together. The components of a truss constructed with 51-mm × 102 mm (2-in. × 4-in.) wood pieces can be connected with heavy-duty staples or with **gusset plates** (connecting plates made of a thin sheet of steel) (**FIGURE 13-19**). The components of an engineered system must be properly installed and are usually glued together. These connection points are often the weakest portion of the lightweight construction and the most common point of failure in a fire. When one connection fails, the system loses its ability to support a load.

Trusses also may be made of individual steel bars or angle sections that are welded together. Lightweight steel trusses, known as bar joists, often support flat roofs on commercial or industrial buildings. When exposed to the heat of a fire, these metal trusses expand and lose strength. A roof supported by bar joists will probably sag before it collapses, a warning that fire fighters should heed. The expansion can elongate the trusses, which may cause the supporting walls to collapse suddenly. Horizontal cracks in the upper part of a wall may indicate that the steel roof supports are pushing outward.

Lightweight construction can be found in almost any type of roof or floor, including flat roofs, pitched roofs, or curved roofs. Fire fighters should assume that any modern construction uses lightweight construction for the roof support system until proven otherwise.

Effects of Roof Construction on Fire Resistance

Fire can quickly burn through some roofs, whereas others retain their integrity for long periods. The difference arises because each type of material used in roof construction is affected differently by fire. These

FIGURE 13-19 Gusset plates, sometimes referred to as gang nails, will eventually fail in extreme heat, causing the entire roof truss to fail.
Courtesy of Captain David Jackson, Saginaw Township Fire Department.

reactions either increase or decrease the time available to perform roof operations. For example, solid beams are generally more fire resistant than truss systems because solid beams are larger and heavier (more mass). Wooden beams are affected by fire more quickly than concrete beams. Steel, unlike wood, does not burn but will elongate and lose strength when heated.

Most roofs eventually fail as a result of fire exposure—some quickly and others more gradually. In some situations, the fire weakens or burns through the supporting structure, which collapses even though the roof covering remains intact. However, failure of the supporting structure usually results in sudden and total collapse of the roof. A structural failure involving other building components could also cause the roof to collapse.

In other cases, the fire burns through the roof covering, even though the supporting structure is still sound. This type of roof failure usually begins with a "burn through" close to the seat of the fire or at a high point above the fire. As combustible products in the roof covering become involved in the fire, the opening spreads, eventually causing the roof to collapse.

The inherent strength and fire resistance of a roof are determined by local climate conditions and building code requirements. In areas that receive winter snow loads, roof supporting structures can usually support a substantial weight and might include multiple layers of insulating material. A serious fire could burn under this type of roof with very little visible evidence from above. Heavily constructed roofs, supported by solid structural elements, can burn for hours without collapsing.

In warmer climates, however, roof construction is often very light. Roofs in these geographic areas might simply function as umbrellas. The supporting structure can be just strong enough to support a thin roof decking with a waterproof, weather-resistant membrane. This type of roof can burn through quickly, and the supporting structure can collapse after a short fire exposure.

Roof Design

Fire fighters must be able to identify the various types of roof designs and the materials used in their construction. Creating ventilation openings requires the use of many different tools and techniques, depending on the type of design and roof decking.

Flat Roofs. Flat roofs can be constructed with many different support systems, roof decking systems, and materials. Although they are classified as flat, most roofs have a slight slope so that water can drain off the structure.

Flat roof construction is generally very similar to floor construction. The roof structure can be supported by solid components, such as wooden beams and rafters, or by trusses. The horizontal beams or trusses often run from one exterior load-bearing wall (a wall that supports the weight of a floor or roof) to another load-bearing wall. In some buildings, the roof is supported by a system of vertical columns and/or interior load-bearing walls.

The roof decking is usually constructed of multiple layers, beginning with wood planking, plywood, or metal panels. Then come one or more layers of roofing paper, insulation, tar and gravel, rubber, gypsum, lightweight concrete, or foam plastic. Some roof decking has only a single layer, consisting of metal panels or precast concrete sections.

Flat roofs often have vents, skylights, scuttles, or other features that penetrate the roof decking. Removing the covers from these openings provides vertical ventilation without the need to make cuts through the roof decking.

Flat roofs also can have parapet walls, freestanding walls that extend above the normal roofline (**FIGURE 13-20**). A **parapet** can be an extension of a firewall, a division wall between two buildings, or a decorative addition to the exterior wall.

Pitched Roofs. Pitched roofs have a visible slope that provides for rain, ice, and snow runoff. The pitch or angle of the roof can vary depending on local climate and architectural styles. A pitched roof is usually supported by either trusses or rafters. The rafters usually run from one load-bearing wall up to a center ridge pole and back to another load-bearing wall.

Most pitched roofs have a layer of thin plywood or a sheeting material such as wood particleboard (oriented strand board) as decking. This layer is often covered by a weather-resistant membrane and an outer covering such as shingles or tiles. Some pitched roofs have a system of **laths** (thin, parallel strips of wood) instead of solid sheeting to support the outer covering.

FIGURE 13-20 Flat roofs can have parapet walls that extend above the normal roofline.

© Jones & Bartlett Learning. Photographed by Glen E. Ellman.

As with a flat roof, the type of roof construction material will dictate how to ventilate a pitched roof. For example, a tile pitched roof can be opened by breaking the tiles and pushing them through the supporting laths. A tin roof can be cut and "peeled" back like the lid on a can. Opening a wooden roof usually requires cutting, chopping, or sawing.

Roof ladders should be used to provide a stable support while working on a pitched roof. A ground ladder or aerial device can be used to access the lower part of the roof; in such a case, the roof ladder is placed on the sloping surface and hooked over the center ridge of the roof.

Curved Roofs. Curved roofs are generally found in commercial structures because they create large open spans without requiring the use of columns (**FIGURE 13-21**). This type of roof is common in supermarkets, warehouses, industrial buildings, arenas, auditoriums, bowling alleys, churches, airplane hangars, and other similar buildings.

Steel or wood bowstring trusses or arches usually are used to support curved roofs. These trusses are usually spaced 1.8 to 6 m (6 to 20 ft) apart and are what give the roof its distinctive curved shape. The decking on curved roof buildings can range from solid wooden boards or plywood to corrugated steel sheets. Often, the roof covering consists of layers that include felt, mineral fibres, and asphalt, although some curved roofs are covered with foam plastics or plastic panels.

Although these roofs are quite distinctive when seen from above or outside, their structure might not be evident from inside the building because a flat ceiling is attached to the bottom chords of the trusses. This creates a huge attic space that is often used for storage. A hidden fire within this space can severely and quickly weaken the bowstring trusses. Fire fighters should be wary of any fire involving the truss space in these buildings.

FIGURE 13-21 A curved roof is commonly used for warehouses, supermarkets, and other similar buildings.
© NicVW/Alamy Stock Photo.

The collapse of a bowstring truss roof is usually very sudden. Fire fighters have been killed when a bowstring truss roof collapsed (Jokubowski, 2013).

Buildings with this type of roof construction should be identified and documented during preincident planning surveys. Parapet walls, smoke, and other obstructions make it difficult to recognize the distinctive curved shape of a bowstring truss roof during an incident.

Vertical Ventilation Techniques

The objective of any vertical ventilation operation is simple: to provide the largest opening in the appropriate location, using the least amount of time and the safest technique. Among the different roof openings that can provide vertical ventilation are built-in roof openings, examination openings to locate the optimal place to vent, primary expandable openings located directly over the fire, and defensive secondary openings intended to prevent fire spread.

Before starting any vertical ventilation operation, fire fighters must make an initial assessment. They should note construction features and indications of possible fire damage, establish safety zones and exit paths, and identify built-in roof openings that can be used immediately.

Ventilation operations must not be conducted in unsafe locations. A less efficient opening in a safe location is better than an optimal opening in a location that jeopardizes the lives of the ventilation team. In some cases, it might not be safe to conduct any rooftop operations, unless they can be performed from the safety of an aerial device. If vertical ventilation cannot be performed, fire fighters have to rely on horizontal ventilation techniques.

Information on the fire's location as well as visible clues from the roof can be used to pinpoint the best location to vent. The incident commander or attack crews might be able to provide some direction. Examination openings can be used to determine how large an area is involved, whether a fire is spreading, and in which direction it is moving (**FIGURE 13-22**). These openings allow the ventilation team to evaluate conditions under the roof and to verify the proper location for a ventilation opening. Sometimes a single slit cut (a **kerf cut**) into the roof with a power saw gives fire fighters some indication of the location of the fire. If a single cut is not effective, a triangular examination opening can be created very quickly by making three small cuts in the roof.

Once the optimal place to vent is located, the ventilation team should determine the most appropriate type of opening to make. Built-in rooftop openings provide readily available ventilation openings

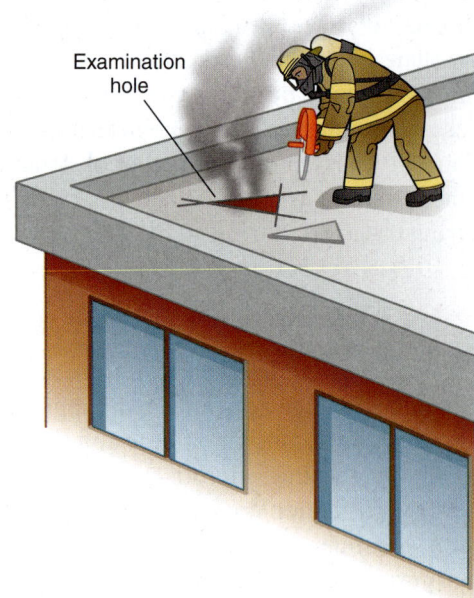

FIGURE 13-22 A triangular cut can be used as an examination opening.
© Jones & Bartlett Learning.

FIGURE 13-23 Built-in rooftop openings often can provide vertical ventilation quickly.
© Jones & Bartlett Learning. Photographed by Glen E. Ellman.

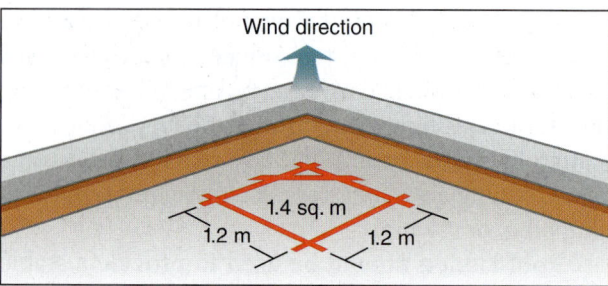

Total 1.4 sq. m

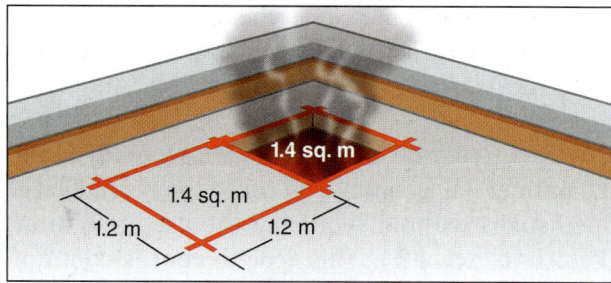

Total 2.8 sq. m

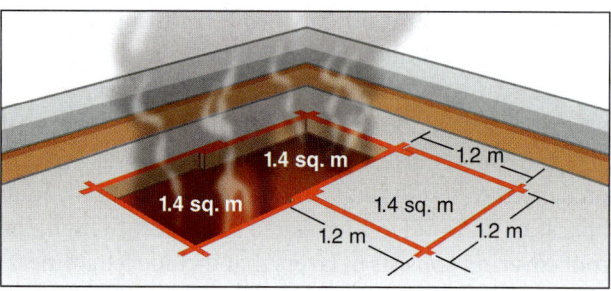

Total 4.2 sq. m

FIGURE 13-24 A ventilation hole can be quickly expanded by extending the cuts.
© Jones & Bartlett Learning.

(**FIGURE 13-23**). Skylights, rooftop stairway exit doors, louvers, and ventilators can quickly be transformed into ventilation openings simply by removing a cover or an obstruction. This approach also results in less property damage to the building. Many residential roofs have a vent installed along the ridge line of the roof to keep the attic space ventilated. Often, these can be removed quickly because they are attached with small nails.

If the roof must be cut to provide a ventilation opening, cutting one large hole is better than making several small ones. Some departments recommend starting with a square opening about 1.2 m long by 1.2 m wide (4 ft long by 4 ft wide). If necessary, this hole can be expanded by continuing the cuts to make a larger opening (**FIGURE 13-24**). If the crews inside the building report that smoke conditions are lifting and temperature levels are dropping, the vent is effective and probably large enough. If interior crews do not see a difference, the opening might be obstructed or need to be expanded.

Once the roof opening is made, a hole of the same size should be made in the ceiling material below to allow heat and smoke to escape from the interior of the building. If the ceiling is not opened, only the heat and smoke from the space between the ceiling and the roof will escape through the roof opening. The blunt end of a pike pole or hook can be used to push down as much of the ceiling material as possible. A roof opening is much less effective if there is no ceiling hole or if the ceiling hole is too small.

Tools Used in Vertical Ventilation

Several tools can be used in roof ventilation. Fire fighters commonly use power saws to cut vent openings, but they also use many hand tools. For example, axes, Halligan tools, pry bars, tin cutters, pike poles, and other hooks can all be used to remove coverings from existing openings, cut through the roof decking, remove sections of the roof, and punch holes in the interior ceiling. Usually the specific tools needed can be determined by looking at the building and the roof construction features.

Ventilation team members should always carry a standard set of tools as well as a utility rope for hoisting additional equipment if needed. All personnel involved in roof operations should use SCBA and wear full PPE. Ground ladders or aerial devices can provide access to the roof.

Power saws, such as a rotary saw with a wood- or metal-cutting blade or a chain saw, effectively cut through most roof coverings. Special carbide-tipped saw blades are especially useful when cutting through typical roof construction materials. To operate a power saw, follow the steps in **SKILL DRILL 13-6**.

SKILL DRILL 13-6
Operating a Power Saw Fire Fighter I, NFPA 1001: 4.3.12

1 Confirm that the cutting device or blade is appropriate for the material anticipated. Briefly inspect the blade or chain for obvious damage. Ensure that proper protective gear is in place.

2 Before going to the roof, start the saw to ensure that it runs properly. Set the choke to the halfway position or as recommended by the manufacturer.

(continued)

SKILL DRILL 13-6 Continued
Operating a Power Saw Fire Fighter I, NFPA 1001: 4.3.12

3 Stay clear of any moving parts of the saw, use a foot or knee to anchor the saw to the ground, and pull the starter cord as recommended to start the saw.

4 Run the saw briefly at full throttle to verify its proper operation.

5 Shut down the saw, wait for the blade or chain to stop completely, and then carry the saw to the roof.

6 Whenever possible, work off of a roof ladder or aerial device for added safety.

SKILL DRILL 13-6 Continued
Operating a Power Saw Fire Fighter I, NFPA 1001: 4.3.12

7 Start the saw in an area slightly away from where you intend to cut.

8 Always run the saw at maximum throttle when cutting. The saw should be running at full speed before the blade or chain touches the roof decking. Keep the throttle fully open while cutting and removing the blade or chain from the cut to reduce the tendency for the blade or chain to bind.

© Jones & Bartlett Learning. Photographed by Glen E. Ellman.

Types of Roof Cuts

Roof construction is a major consideration in determining which type of cut to use. Some roofs are thin and easy to cut with an axe or a power saw; others have multiple layers that are difficult to cut. Being familiar with the types of roof construction in the local area and the best methods for opening them increases the efficiency and effectiveness of ventilation operations.

Rectangular or Square Cut. A 1.2-m by 1.2-m (4-ft by 4-ft) square cut or a rectangular cut is the most common vertical ventilation opening. It requires the fire fighter to make four cuts completely through the roof decking, using an axe or power saw. When using a power saw, the fire fighter must carefully avoid cutting through the structural supports. To do this, the fire fighter should sound the roof to locate the structural supports before starting to cut. **Sounding** is the process of striking a roof with a tool to locate support members. Once the structural supports are located, the fire fighter should cut beside the supports rather than through them. After the four cuts are completed, a section of the roof decking can be removed.

The fire fighter should stand upwind of the opening, with two unobstructed exit routes. The first and last

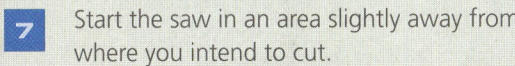

SAFETY TIP

Always carry and handle tools and equipment in a safe manner. When carrying tools up a ladder, hold the beam of the ladder instead of the rungs, and run the hand tool up the beam. This technique works for most hand tools, including axes, pike poles and hooks, and tin roof cutters. Hoist tools using a rope with a tag line (a separate rope that ground personnel can use to guide the tool that is being hoisted) to keep them away from the building. Carry power saws and equipment in a sling across your back. Always turn off power equipment before you carry it. For more information on hoisting tools and equipment, see Chapter 11, *Ladders*.

cuts should be made parallel to and just inside the roof supports. The fire fighter making the cuts must always stand on a solid portion of the roof or on a roof ladder. A triangular cut in one corner of the planned opening can be used as a starting point for prying the decking up with a hand tool.

Depending on the roof construction, it might be possible to lift out the entire section at once. If several layers of roofing material are present, fire fighters might have to peel them off in layers. The decking itself could consist of plywood sheets or individual boards that have to be removed one at a time.

To perform the rectangular or square cut, follow the steps in **SKILL DRILL 13-7**.

Seven, Nine, Eight (7, 9, 8) Rectangular Cut.
Large commercial buildings require larger ventilation openings to exhaust the heat and smoke from a fire. When these buildings have flat roofs, the seven, nine, eight (7, 9, 8) rectangular cut is an effective ventilation technique. The steps to create this cut are similar to the steps for creating a square or rectangular cut. Seven cuts are produced, resulting in a 1.2-m by 2.4-m (4-ft by 8-ft) ventilation hole.

SKILL DRILL 13-7
Making a Rectangular or Square Cut Fire Fighter I, NFPA 1001: 4.3.12

1 Locate the roof supports by sounding. Make the first cut parallel to the roof support. The support should be outside the area that will be opened.

2 Make a triangular cut at the first corner of the opening. Then make two cuts perpendicular to the roof supports.

3 Make the final cut parallel to and slightly inside another roof support. Be sure to stand on the solid portion of the roof when making this cut.

4 Use a hand tool to pull out or push in the triangle cut to create a small starter hole. If possible, pull the entire roof section free, and flip it over onto the solid roof. It might be necessary to pull the decking out one board at a time.

5 Use a pike pole to punch out the ceiling below. The hole should be the same size as the opening in the roof decking. Be wary of a sudden updraft of hot gases or flames.

To perform the seven, nine, eight (7, 9, 8) cut, follow the steps in **SKILL DRILL 13-8**.

Louver Cut. Another common cut is the louver cut, which is particularly suitable for flat or sloping roofs with plywood decking. To make a louver cut, fire fighters use a power saw or axe to make two parallel cuts, approximately 1.2 m (4 ft) apart, perpendicular to the roof supports. The fire fighters then make cuts parallel to the roof supports, approximately halfway between

1 Locate the roof supports by sounding. Make the first cut parallel to the roof supports. The cut should be 1 to 1.2 m (3.5 to 4 ft) long.

2 Make the second cut at a 45-degree angle to the first cut. The length of this knock-out cut should be 0.3 to 0.6 m (1 to 2 ft) in length.

3 Make the third cut perpendicular to the first cut from the corner where the knock-out cut was made. This cut should be about 2.4 m (8 ft) long. The three cuts should produce the shape of the number "7."

(continued)

SKILL DRILL 13-8 Continued
Making a Seven, Nine, Eight (7, 9, 8) Rectangular Cut Fire Fighter I, NFPA 1001: 4.3.12

4 Make the fourth cut perpendicular to the first cut on the opposite side of the knock-out cut (the second cut). The cut should be about 1.2 m (4 ft) long.

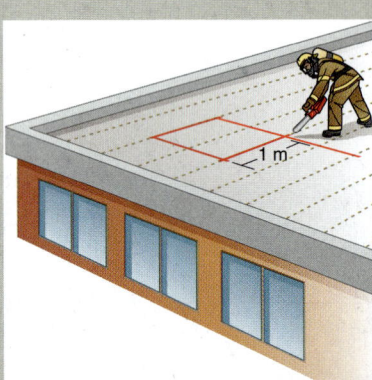

5 Make the fifth cut parallel to the first cut. Starting on the third-cut side of the hole, make a cut that connects with the fourth-cut side of the hole. These five cuts should produce the shape of the number "9."

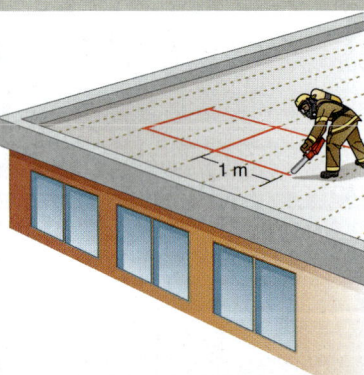

6 Make the sixth cut perpendicular to the first cut. Extend the fourth cut approximately 1.2 m (4 ft) until it is even with the third cut.

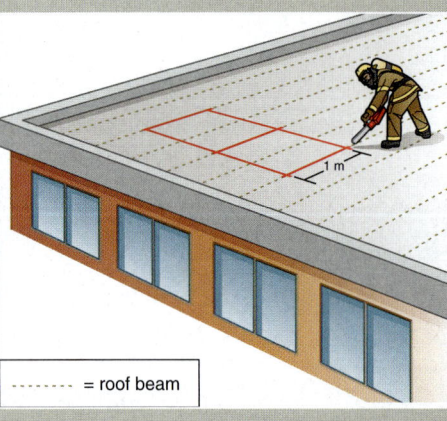

7 Make the seventh cut parallel to the first cut and the fifth cut. Start on the side of the sixth cut, and cut toward the side of the third cut, until the cuts are connected. These seven cuts should produce the shape of the number "8."

-------- = roof beam

each pair of supports. Using each roof support as a fulcrum, the cut sections are tilted to create a series of louvered openings.

Louver cuts can quickly create a large opening. Continuing the same cutting pattern in any direction creates additional louver sections. To make a louver cut, follow the steps in **SKILL DRILL 13-9**.

Triangular Cut. The triangular cut works well on metal roof decking because it prevents the decking from rolling away as it is cut. Using saws or axes, the fire

fighter removes a triangle-shaped section of decking. Smaller triangular cuts can be made between supports (so that the decking falls into the opening) or over

---

FIRE FIGHTER TIP

Ventilating a steel-decked roof can be labour-intensive work. Removing thick insulation covered in tar and gravel often requires a significant amount of staffing. Incident commanders must be certain that they have the resources to undertake and support this type of operation.

SKILL DRILL 13-9
Making a Louver Cut Fire Fighter I, NFPA 1001: 4.3.12

1 Locate the roof supports by sounding.

2 Make two parallel cuts perpendicular to the roof supports.

3 Cut parallel to the supports and between pairs of supports in a rectangular pattern.

4 Tilt the panel to a vertical position. Open the interior ceiling area below the opening by using the butt end of a pike pole. This hole should be the same size as the opening made in the roof decking.

supports (to create a louver effect). Because triangular cuts are generally smaller than other types of roof ventilation openings, several might be needed to create an adequately sized vent.

To make a triangular cut, follow the steps in **SKILL DRILL 13-10**.

Peak Cut. **Peak cuts** are limited to pitched roofs sheeted with plywood. In these structures, the 1.2-m by 2.4-m (4-ft by 8-ft) sheets of plywood are applied starting at the bottom edge of the roof and working toward the top. A partial sheet is usually placed at the top, along the roof peak. A small gap is left between the two sides of the roof to allow the plywood sheeting to expand and contract with temperature changes.

To make a peak cut, the fire fighter first uses a hand tool to clear the outer covering along the peak (roof cap), which reveals the roof supports through the gap in the sheeting. Using a power saw or axe, the fire fighter makes a series of vertical cuts between the supports from the

SKILL DRILL 13-10
Making a Triangular Cut Fire Fighter I, NFPA 1001: 4.3.12

1 Locate the roof supports.

2 Make the first cut from just inside a support member in a diagonal direction toward the next support member.

3 Begin the second cut at the same location as the first, and make it in the opposite diagonal direction, forming a V shape.

4 Make the final cut along the support member to connect the first two cuts. Cutting from this location allows fire fighters the full support of the member directly below them while performing ventilation.

top to the bottom of the plywood sheet. The individual panels are then struck with an axe and louvered or pried up with a hook.

Repeating this technique on both sides of the roof essentially removes the entire peak. Unless multiple layers of roof decking are present, there is no need to make horizontal cuts. The opening can be expanded either vertically or horizontally with a minimum of cutting.

To make a peak cut, follow the steps in **SKILL DRILL 13-11**.

Trench Cut. A trench cut (or strip cut) is used to stop fire spread in long narrow buildings, such as strip

---

SKILL DRILL 13-11
Making a Peak Cut Fire Fighter I, NFPA 1001: 4.3.12

1 Locate the roof supports.

2 Clear the roofing materials away from the roof peak.

3 Make the first cut vertically, at the farthest point away. Start at the roof peak in the area between the support members, and cut down to the bottom of the first plywood panel.

4 Make parallel downward cuts between supports, moving horizontally along the roofline to make additional ventilation openings.

(continued)

SKILL DRILL 13-11 Continued
Making a Peak Cut Fire Fighter I, NFPA 1001: 4.3.12

5 Strike the nearest side of the roofing material with an axe or maul, pushing it in, using the support located at the center as a fulcrum. This causes one end of the roofing material to go downward into the opening and the other to rise up. If necessary, repeat this process on both sides of the peak, horizontally across the peak, or vertically toward the roof edge.

© Jones & Bartlett Learning. Photographed by Glen E. Ellman.

6 Open the interior ceiling area below the opening by using the butt end of a pike pole. This hole should be the same size as the vent opening made in the roof decking.

malls and small storage complexes. A trench cut creates a large opening ahead of the fire, removing a section of fuel and letting heated smoke and gases flow out of the building. Essentially, it is a firebreak in the roof. Water is sometimes applied through the trench cut.

Trench cuts are a defensive ventilation tactic intended to stop the progress of a large fire, particularly one that is advancing through an attic or cockloft. The incident commander who chooses this tactic is "writing off" part of the building and identifying a point where crews will be able to stop the fire.

A trench cut is made from one exterior wall across to the other. It begins with two parallel cuts, spaced 0.6 m to 1.2 m (2 ft to 4 ft) apart. Approximately every 1.2 m (4 ft), fire fighters make short perpendicular cuts between the two parallel cuts. They can then lift the roof covering out in sections, completely opening a section of the roof. On a pitched roof, the trench should run from the peak down.

A trench cut is a secondary cut, used to limit the fire spread, rather than a primary cut, located over the seat of the fire. A primary cut should still be made before crews start working on the trench cut.

A trench cut must be made far in advance of the fire. Incident commanders must anticipate adequate staffing is available to perform the task. The ventilation crew must be able to complete the cut, and the interior crew must be able to get into position before the fire passes the trench. Examination openings should be made to ensure that the fire has not already passed the chosen site before the trench cut is completed. The crew members' lives will be in danger, and the tactic is useless if the fire advances beyond the trench before it is opened.

Although trench cuts are effective, they require both time and personnel to complete. As with all types of vertical ventilation, careful coordination between the ventilation crew and the interior attack crew is essential. While the trench cut is being made, attack teams should be deployed inside the building to defend the area in front of the cut.

To make a trench cut, follow the steps in **SKILL DRILL 13-12.**

SKILL DRILL 13-12
Making a Trench Cut Fire Fighter I, NFPA 1001: 4.3.12

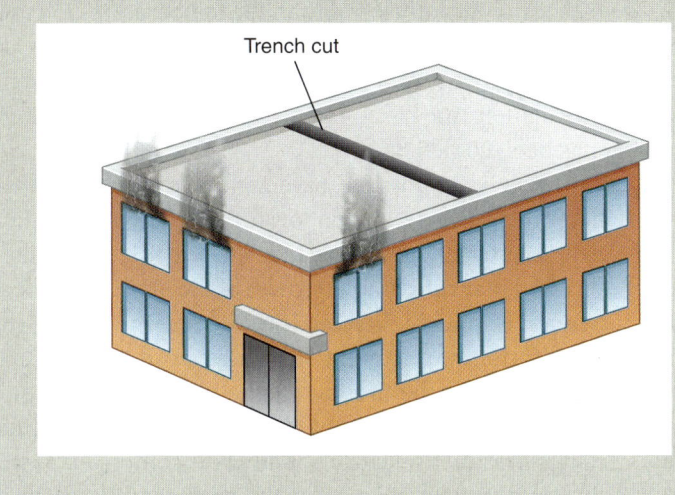

Trench cut

© Jones & Bartlett Learning.

1 After a primary cut has been made over the seat of the fire, cut a number of small inspection holes to identify a point sufficiently far ahead of the fire travel.

2 Make two parallel cuts, 0.6 to 1.2 m (2 to 4 ft) apart, across the entire roof, starting at the ridge pole (for pitched roofs) or a load-bearing wall (for flat roofs).

3 Cut between the two long cuts to make a row of rectangular sections.

4 Remove the rectangular panels to open the trench.

Special Considerations

Many obstacles can be encountered during ventilation operations. Fire fighters must be creative and remember the basic objectives of ventilation—to create high openings as rapidly as possible so that hose lines can be advanced into the building.

Poor access or obstructions such as trees, fences, electric lines, or tight exposures can prevent fire fighters from getting close enough to place ladders. Steel bars, shutters, and other security features can also hinder ventilation efforts. Window openings in abandoned buildings might be boarded or sealed. A building that requires forcible entry is also likely to present ventilation challenges. To enhance their security, some commercial buildings have roofs made of steel plating. In these buildings, fire fighters should not attempt vertical ventilation through the roof but rather should use alternative ventilation techniques.

Many residential and commercial roofs have multiple layers; some even have a new roof built on top of an older one. Buildings that have two roofs present a challenge for fire fighters. A well-executed roof-ventilation evolution may not result in effective ventilation for these buildings. In this case, it is necessary to determine the reason the ventilation is ineffective. Additional holes may need to be cut in the second roof, or another type of ventilation may need to be performed. Whenever possible, buildings with two roofs should be identified during preincident surveys.

Ventilating Basements

Research involving basement fires has provided valuable information about safe and effective ways to ventilate and extinguish basement fires. Many vertical voids within a building originate or terminate in the basement, providing ample opportunities for fire gases to spread throughout the building. In many parts of the country, especially in newer subdivisions with lightweight construction, basement ceilings are left unfinished when the home is sold. Under these conditions, fires originating in the basement will quickly involve the floor system, resulting in a failure of the floor over the fire and early

SAFETY TIP

Remember that basement fires often do not look like basement fires. Because of the many voids between the basement and other parts of the building, a fire that started in the basement may at first appear to be a kitchen fire or a fire on the first or second floor of the building. As a safety measure, operate as though the floor, your working platform, has been weakened by a fire underneath until you can prove that there is no fire beneath you.

collapse, with the potential for causing injury or death to fire fighters entering the building with conditions of limited visibility.

The traditional tactic of applying water down the interior stairway while advancing to the seat of the fire should no longer be considered a safe operation. One of the findings of recent research on basement fires was that water applied via the interior stairs had a limited effect on cooling the basement or extinguishing the fire. However, water applied through an external window or external door quickly darkened the fire and reduced temperatures throughout the building (**FIGURE 13-25**). This technique did not "push" fire or hot gases up the interior stairs. The cooling effect lasted for several minutes before the fire grew back to the size it was before the application of the water (UL FSRI, 2013) (**TABLE 13-3**).

When attacking basement fires, the use of interior stairs for ventilation should not be considered a safe option because of the potential for failure of the floor system

FIGURE 13-25 Applying water to a basement fire through a basement window can quickly darken the fire and reduce temperatures throughout the building.
Courtesy of NIST.

TABLE 13-3 Basement Fire Findings
■ Basement fires spread to the floors above from routes other than the stairs, as the floor assembly often failed close to the location where the fire started. ■ Flowing water at the top of the interior stairs had limited impact on basement fires. ■ Offensive exterior attack using a straight or solid stream of water through a basement window was effective in cooling the fire compartment. ■ Offensive attack using a straight or solid stream of water through an exterior door was effective in cooling the fire compartment.

Courtesy of NIST.

and the danger of fire fighters being in the exhaust flow path of hot gases from the fire. In some basement fires, the temperature reaches untenable levels even for fire fighters in full PPE. Recent UL tests have shown that in some basement fires, the temperature at the floor level is not significantly lower than the temperatures in other parts of the basement (UL FSRI, 2013). This seems to indicate that it may not always be a safe idea to send fire fighters into basement fires before the compartment has been cooled.

Whenever possible, basement fires should be ventilated through exterior windows or doors. This ventilation will result in oxygen being added to a ventilation-limited fire and may cause rapid growth of the fire unless a hose stream is quickly directed into the volume of fire. The application of a straight stream as close to the fire as possible until the fire is darkened is effective in cooling the fire compartment and the rest of the building. Once this offensive exterior attack is completed, it will be safer and easier to enter the basement to complete extinguishment of the remaining fire. More information about extinguishing basement fires is presented in Chapter 17, *Fire Suppression*.

Ventilating Concrete Roofs

Some commercial or industrial structures have concrete roofs. Concrete roofs can be constructed with "poured-in-place" concrete, with precast concrete sections of roof decking placed on a steel or concrete supporting structure, or with T-beams (a load-bearing structural beam with a T-shaped cross section). Such roofs are generally flat and difficult to breach. The roof decking is usually very stable, but fire conditions underneath could weaken the supporting structural components or load-bearing walls, leading to failure and collapse.

There are few options for ventilating concrete roofs. Even special concrete-cutting saws are generally ineffective against these structures. In such cases, fire fighters should use alternative ventilation openings such as vents, skylights, and other roof penetrations or horizontal ventilation.

Ventilating Metal Roofs

Metal roofs and metal roof decking present many challenges for the ventilation crew. Ventilation on these types of roofs should only be deployed when necessary and when staffing and time permits. Because metal conducts heat more quickly than other roofing materials, discoloration and warping of this material can indicate the seat of the fire. Tin-cutter hand tools can be used to slice through thin metal coverings, whereas special saw blades might be needed to cut through metal roof decking. In many cases, the metal is on the bottom and supports a built-up or composite roof covering.

Metal roof decking is often supported by lightweight steel bar joists, which can sag or collapse when exposed to a fire. Because the metal decking is lightweight, the supporting structure can be relatively weak, with widely spaced bar joists. The resulting assembly can fail quickly with only limited fire exposure.

As the fire heats the metal decking, the tar roof covering can melt and leak through the joints into the building, where it releases flammable vapours. When this sequence of events occurs, it can quickly spread the fire over a wide area under the roof decking. Fire fighters should look for indications of dripping or melting tar and begin rapid ventilation to dissipate the flammable vapours before they can ignite. Hose streams should be used to cool the roof decking from below to stop the tar from melting and producing vapours.

When metal roof decking is cut, the metal can roll down and create a dangerous slide directly into the opening. The triangular cut prevents the decking from rolling away as easily, so it is the preferred option, even though several cuts can be needed to create an adequately sized vent.

Ventilating High-Rise Buildings

Ventilating a high-rise building can be challenging. A high-rise building resembles a stack of individual floor compartments connected by stairways, elevator shafts, and other vertical passages. Most high-rise buildings have sealed windows that are difficult to break. In addition, high-rise buildings have unique patterns of smoke movement, such that smoke might be trapped on individual floors or it may move up or down within the vertical shafts.

Many newer high-rise buildings have incorporated smoke management capabilities into their HVAC systems. Use of this type of system enables different areas to be pressurized with fresh air and contaminated air to be exhausted directly to the outside. If the HVAC system does not have this capacity, it can complicate problems by circulating smoke to different areas of the building.

A phenomenon called the **stack effect** can occur in high-rise structures. The stack effect is a response to the differences in temperature inside and outside a building. A cold outer atmosphere and a heated interior cause smoke to rise quickly through stairways, elevator shafts, and other vertical openings, filling the upper levels of the building (**FIGURE 13-26**).

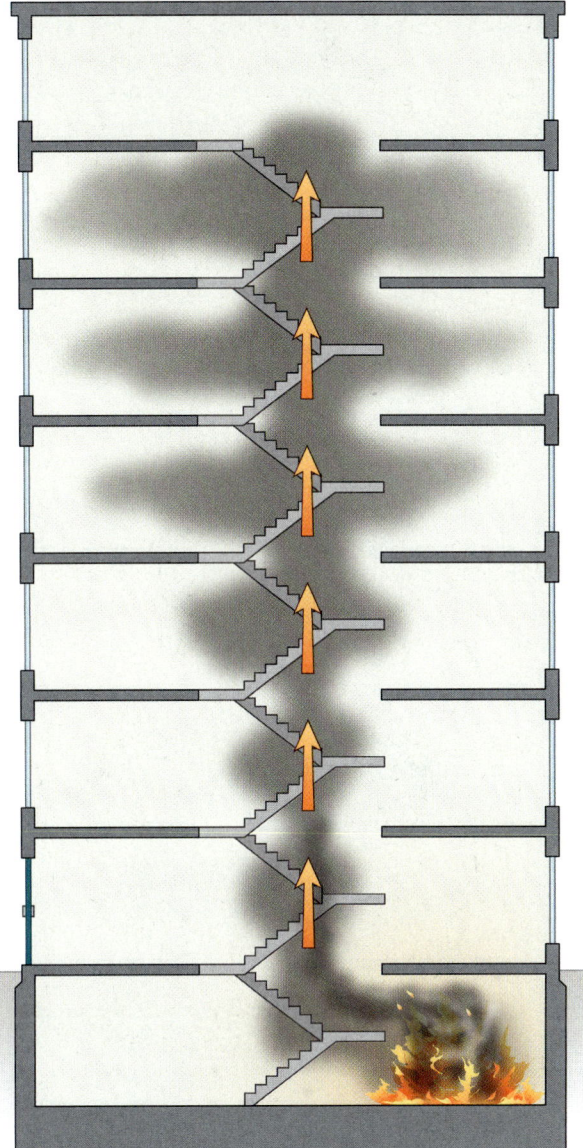

Winter stack

FIGURE 13-26 A winter stack effect occurs when the outside air is much cooler than the interior temperature.
© Jones & Bartlett Learning.

The opposite situation can occur on a hot summer day, when the interior temperature is much cooler than the outside atmosphere. In this scenario, the heavy cooler air pushes the smoke down the vertical openings, toward a lower level exit (**FIGURE 13-27**).

The situation can change if the fire produces enough heat to alter the temperature profile within the building. The air currents within a tall building might, for example, be stronger than the convection currents that normally govern the flow of smoke and other products of combustion. A strong wind through an open or broken window can change the direction of

Summer stack

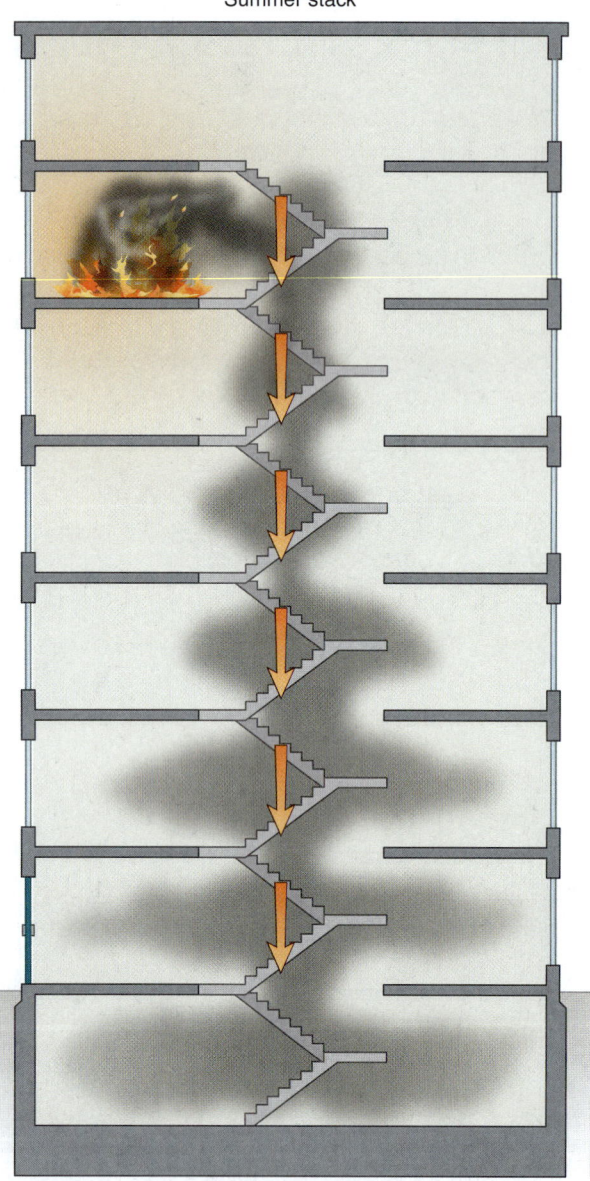

FIGURE 13-27 A summer stack effect occurs when the interior temperature is much cooler than the outside air.
© Jones & Bartlett Learning.

smoke. Contaminated air might suddenly move toward the opening or be pushed back as a strong draft of fresh air enters the building.

After smoke mixes with fresh air or is hit by water from sprinklers or hose streams, it cools and can "sit" in one location. This cold smoke can fill several floors and usually needs to be cleared by mechanical means.

A key objective in ventilating a high-rise building is to manage the air movement in stairways and elevator shafts. Positive-pressure ventilation can help keep stairways clear of smoke and ventilate individual floors in high-rise structures. Incident commanders should

assign a ventilation officer to coordinate all ventilation-related activities.

At least one stairwell should be designated as the evacuation stairwell and announced to the operating companies. This stairwell should be kept well ventilated and designated as an occupant and rescue route. These stairs must be used by rescuers and escaping victims only. Some newer buildings have smoke-proof stair towers or pressurized stair shafts that are designed to keep smoke out of the stairway. Otherwise, positive-pressure fans can be used in a stairway to keep smoke out. Place the positive-pressure fans at a ground-floor doorway, positioning them so that fresh air blows into the stairway. Make certain that the engines of the positive-pressure fans are well ventilated because they can increase carbon monoxide levels.

A pressurized stairway also can be used to clear smoke from a floor. Opening a door from the stairway allows fresh air to enter the floor. The contaminated atmosphere can then be vented out through a window or another stairway.

Ventilating Windowless Buildings

Many structures do not have windows (**FIGURE 13-28**). Some buildings are designed without windows; others have bricked-up or covered windows. These buildings pose two significant risks to fire fighters: Heat and products of combustion are trapped, and fire fighters have no secondary exit route.

Windowless buildings are similar to basements in terms of the ventilation approach to be used. Any ventilation needs to be as high as possible and probably requires mechanical assistance. Using existing rooftop openings, cutting openings in the roof, reopening boarded-up windows or doors, and making new openings in exterior walls are all possible ways to ventilate windowless structures.

FIGURE 13-28 Structures without windows pose significant problems in ventilation.
© olaf schlueter/Shutterstock.

Ventilating Large Buildings

Providing adequate ventilation is more difficult in large buildings than in smaller ones. In a large building, a ventilation hole placed in the wrong location can draw the fire toward the opening, spreading the fire to an area that had not previously been involved. This underscores the importance of coordinating ventilation operations with the overall fire attack strategy.

Smoke cools as it travels into unaffected portions of a large building. A fire sprinkler system also cools the smoke, causing it to stratify. As the cold smoke fills the area, it becomes more difficult to clear.

If possible, fire fighters should use interior walls and doors to create several smaller compartments in a large building, thereby limiting the spread of heat and smoke. The smaller areas can be cleared one at a time with positive-pressure fans. Several fans can be used in a series or in parallel lines to clear smoke from a large area.

Equipment Maintenance

It is important that all equipment used for ventilation is kept in good repair and ready to operate at peak efficiency. It is important to read and follow the manufacturer's instructions for maintaining power equipment at regular and properly documented intervals. Fuel must be rotated regularly if not used. Fuel tanks need to be filled to the recommended levels. All fire fighters who use ventilation tools need to practise using them regularly to ensure safe and effective operation when needed.

To perform a readiness check on a power saw, follow the steps in **SKILL DRILL 13-13**.

To maintain a power saw, follow the steps in **SKILL DRILL 13-14**.

Perform periodic checks of the power saw, and document the maintenance on the departmental log.

SKILL DRILL 13-13
Performing a Readiness Check on a Power Saw Fire Fighter I, NFPA 1001: 4.5.1

1 Make certain that the fuel tank is full. Make certain that the bar and chain oil reservoirs are full.

2 Check the throttle trigger for smooth operation.

(continued)

SKILL DRILL 13-13 Continued
Performing a Readiness Check on a Power Saw Fire Fighter I, NFPA 1001: 4.5.1

3 Ensure that the saw, blade, air filter, and chain brake are clean and working. Inspect the blade for even wear, and lubricate the sprocket tip if needed.

4 Check the chain for wear, missing teeth, or other damage. Check the chain end for proper tension. Check the chain catcher.

5 Check for loose nuts and screws, and tighten them if needed. Check the starter and starter cord for wear.

6 Inspect the spark plugs.

7 Start the saw. Make certain that both the bar and the chain are being lubricated while the saw is running. Check the chain brake.

8 Make certain that the stop switch functions. Record the results of the inspection.

SKILL DRILL 13-14
Maintaining a Power Saw Fire Fighter I, NFPA 1001: 4.5.1

1 Remove, clean, and inspect the clutch cover, bar, and chain for damage and wear. Replace if necessary.

2 Inspect the air filter and clean/replace as needed.

3 Lubricate components as recommended by the manufacturer. Reinstall the bar and chain, flipping the bar over each time to help wear the bar evenly. Replace the clutch cover.

4 Adjust the chain tension (make sure the bar and chain cool before adjusting).

5 Fill the power saw with fuel. Fill the bar and chain oil reservoirs.

© Jones & Bartlett Learning. Photographed by Glen E. Ellman.

After-Action REVIEW

IN SUMMARY

- Ventilation is the controlled and coordinated removal of heat and smoke from a structure. Effective ventilation not only removes heat and smoke, it also replaces the escaping gases with cooler, cleaner, and oxygen-rich air.

- Fire size-up needs to include the type of ventilation that is appropriate for the fire conditions and a plan to implement ventilation operations in coordination with other fire-suppression activities.

- Before ventilating, fire fighters must first determine if they are dealing with a ventilation-limited fire or a fuel-limited fire. If the fire is ventilation-limited, supply oxygen. If the fire is fuel-limited, establish a flow path. Uncoordinated ventilation can cause rapid fire growth.

- Factors that influence the effectiveness of ventilation operations include door control, the ventilation location relative to the fire, the ventilation hole size, the impact of wind, and the impact of exterior suppression on fire behaviour.

- Modern construction practices contribute to ventilation-limited fires and rapid heat build-up.

- Backdrafts and flashovers can be minimized through the use of a transitional fire attack that is coordinated with appropriate ventilation.

- Horizontal ventilation takes advantage of the doors, windows, and other openings at the same level as the fire. In some cases, fire fighters make additional openings in a wall to provide horizontal ventilation. Horizontal ventilation is commonly used in residential fires, room-and-contents fires, and fires that can be controlled quickly by the attack team.

- Vertical ventilation refers to any opening that allows the products of combustion to travel up and out. It involves making openings in roofs or floors so that heat, smoke, and toxic gases can escape from the structure in a vertical direction. Pathways for vertical ventilation can include ceilings, stairwells, exhaust vents, and roof openings such as skylights, scuttles, or monitors. Additional openings can be created by cutting holes in the roof or the floor and making sure that the opening extends through every layer of the roof or floor. The choice of roof openings depends primarily on the building's roof construction.

- Mechanical ventilation includes negative-pressure ventilation, positive-pressure ventilation, and hydraulic ventilation.

  - Negative-pressure ventilation uses smoke ejectors to exhaust smoke and heat from a structure. It can be used to move smoke out of a structure after a fire.

  - Positive-pressure ventilation uses fans to introduce clean air into a structure and push the contaminated atmosphere out. It can be used to reduce interior temperatures and smoke conditions in coordination with a fire attack or clear a contaminated atmosphere after a fire has been extinguished.

  - Hydraulic ventilation moves air using fog or broken-pattern fire streams to create a pressure differential behind and in front of the nozzle. It is most useful in clearing smoke and heat out of a room after the fire is under control.

- Before performing vertical ventilation, fire fighters must evaluate all pertinent safety issues and avoid unnecessary risks. The biggest risk is roof collapse. Assess the roof for roof scuttles, heat vents, plumbing vents, louver ventilation, solar panels, and fan shafts to prevent tripping or falling from the roof.

- When working on a roof, have two safe exit routes. A second ground ladder or aerial device should be positioned to provide a quick escape route. The ventilation opening should never be located between the exit route and the ventilation crew.

- Fire fighters who are assigned to vertical ventilation tasks should always be aware of the condition of the roof. They should immediately retreat from the roof if they notice any of the following signs:

  - Visible indication of sagging roof supports

  - Any indication that the roof assembly is separating from the walls, such as the appearance of fire or smoke near the roof edges

- Structural failure of any portion of the building, even if it is some distance from the ventilation operation
- A sudden increase in the intensity of the fire from the roof opening
- High heat indicators on a thermal imaging device

■ The three major components of roof construction are roof support structures, roof decking, and roof coverings. The roof support system provides the structural strength to hold the roof in place. The structural system is either solid-beam or lightweight construction. The roof covering is the weather-resistant surface and may consist of many layers. The roof decking is a protective layer between the support structures and the coverings.

■ Roof designs include flat roofs, pitched roofs, and curved roofs.

- Flat roof construction is similar to floor construction. It can be supported by solid components or by trusses. Flat roofs often have vents, skylights, scuttles, or other features that penetrate the roof deck. Removing the covers from these openings provides vertical ventilation without the need to make cuts through the roof deck.

- Pitched roofs have a visible slope. They can be supported by trusses or a system of rafters and beams. Many of these roofs have a layer of solid sheeting covered by a weather-resistant membrane and outer covering. The roof construction material dictates how to ventilate the roof.

- Curved roofs create large open spans without the use of columns. They are often supported with bowstring trusses or arches. The collapse of a bowstring truss is usually very sudden.

■ The types of vertical ventilation openings include the following:

- Built-in roof openings
- Inspection openings
- Primary (expandable) openings
- Secondary (defensive) openings

■ Some commercial or industrial structures have concrete roofs. There are few options for ventilating these structures. Use alternative ventilation openings such as vents or skylights.

■ Metal roofs conduct heat and are often supported by lightweight steel metal joists.

■ Both horizontal and vertical ventilation can be required to vent a basement.

■ HVAC systems may be used to ventilate high-rise buildings.

■ To ensure successful ventilation operations, all equipment and tools must be in a ready state and properly maintained.

KEY TERMS

Access Navigate for flashcards to test your key term knowledge.

Chase Open space within walls for wires and pipes.

Cockloft The concealed space between the top-floor ceiling and the roof of a building.

Flow path The movement of heat and smoke from the higher pressure within the fire area toward the lower pressure areas accessible via doors, window openings, and roof structures. (NFPA 1410)

Fuel-limited fire A fire in which the heat release rate and fire growth are controlled by the characteristics of the fuel because there is adequate oxygen available for combustion. (NFPA 1410)

Gusset plate Connecting plate made of a thin sheet of steel used to connect the components of a truss. May also be referred to as a gang nail.

Horizontal ventilation The opening or removal of windows or doors on any floor of a fire building to create flow paths for fire conditions. (NFPA 1410)

Hydraulic ventilation Ventilation that relies on the movement of air caused by a fog stream that is placed 0.6 m to 1.2 m (2 ft to 4 ft) in front of an open window.

Kerf cut A cut that is the width and depth of the saw blade. It is used to inspect cockloft spaces from the roof.

Lath Thin strips of wood used to make the supporting structure for roof tiles.

Louver cut A cut that is made using power saws and axes to cut along and between roof supports so that the sections created can be tilted into the opening.

Mechanical ventilation A process of removing heat, smoke, and gases from a fire area by using exhaust fans, blowers, air-conditioning systems, or smoke ejectors. (NFPA 402)

Natural ventilation The flow of air or gases created by the difference in the pressures or gas densities between the outside and inside of a vent, room, or space. (NFPA 853)

Negative-pressure ventilation Ventilation that relies on electric fans to pull or draw the air from a structure or area.

Neutral plane The interface at a vent, such as a doorway or a window opening, between the hot gas flowing out of a fire compartment and the cool air flowing into the compartment where the pressure difference between the interior and exterior is equal.

Parapet The part of a wall entirely above the roofline. (NFPA 5000)

Peak cut A ventilation opening that runs along the top of a pitched roof.

Positive-pressure attack The use of positive-pressure fans to control the flow of products of combustion while fire suppression efforts are under way.

Positive-pressure ventilation Ventilation that relies on fans to push or force clean air into a structure after a structure fire has been controlled.

Primary cut The main ventilation opening made in a roof to allow smoke, heat, and gases to escape.

Roof covering The membrane, which may also be the roof assembly, that resists fire and provides weather protection to the building against water infiltration, wind, and impact. (NFPA 5000)

Roof decking The rigid portion of roof between the roof supports and the roof covering.

Secondary cut An additional ventilation opening made for the purpose of creating a larger opening or limiting fire spread.

Seven, nine, eight (7, 9, 8) rectangular cut A ventilation opening that is usually about 1.2 m by 2.4 m (4 ft by 8 ft) in size; it is primarily used for large commercial buildings with flat roofs.

Smoke ejectors A mechanical device, similar to a large fan, that can be used to force heat, smoke, and gases from a post-fire environment and draw in fresh air. (NFPA 402)

Smoke inversion The condition in which smoke hangs low to the ground because of the presence of cold air.

Sounding The process of striking a roof with a tool to determine where the roof supports are located.

Stack effect The vertical air flow within buildings caused by the temperature-created density differences between the building interior and exterior or between two interior spaces. (NFPA 92)

Strip cut Another term for a trench cut.

Transitional attack An offensive fire attack initiated by an exterior, indirect handline operation into the fire compartment to initiate cooling while transitioning into interior direct fire attack in coordination with ventilation operations.

Trench cut A roof cut that is made from one load-bearing wall to another load-bearing wall and that is intended to prevent horizontal fire spread in a building.

Triangular cut A triangle-shaped ventilation cut in the roof decking that is made using a saw or an axe.

Type I construction (fire resistive) The type of construction in which the fire walls, structural elements, walls, arches, floors, and roofs are of approved noncombustible or limited-combustible materials that have a specified fire resistance.

Type II construction (noncombustible) The type of construction in which the fire walls, structural elements, walls, arches, floors, and roofs are of approved noncombustible or limited-combustible materials without fire resistance.

Type III construction (ordinary) The type of construction in which exterior walls and structural elements that are portions of exterior walls are of approved noncombustible or limited-combustible materials and in which fire walls, interior structural elements, walls, arches, floors, and roofs are entirely or partially of wood of smaller dimensions than required for Type IV construction or are of approved noncombustible, limited-combustible, or other approved combustible materials. (NFPA 14)

Type IV construction (heavy timber) The type of construction in which fire walls, exterior walls, and interior bearing walls and structural elements that are portions of such walls are of approved noncombustible or limited-combustible materials. Other interior structural elements, arches, floors, and roofs are constructed of solid or laminated wood or cross-laminated timber without concealed spaces within allowable dimensions of the building code. (NFPA 14)

Type V construction (wood frame) The type of construction in which structural elements, walls, arches, floors, and roofs are entirely or partially of wood or other approved material. (NFPA 14)

Ventilation The controlled and coordinated removal of heat and smoke from a structure, replacing the escaping gases with fresh air. (NFPA 1410)

Ventilation-limited fire A fire in which the heat release rate and fire growth are regulated by the available oxygen within the space. (NFPA 1410)

Vertical ventilation The vertical venting of structures involving the opening of bulkhead doors, skylights, scuttles, and roof cutting operations to release smoke and heat from inside the fire building. (NFPA 1410)

REFERENCES

Jakubowski, Greg. 2013. "Bowstring Truss Roof Construction Hazards." *Fire Rescue*, January 10, 2013. http://www .firerescuemagazine.com/articles/print/volume-8/issue-3 /firefighter-safety/bowstring-truss-roof-construction -hazards.html. Accessed November 12, 2018.

Kerber, Stephen. 2012. "Analysis of Changing Residential Fire Dynamics and Its Implications on Firefighter Operational Timeframes." Underwriters Laboratories. https://newscience.ul.com/wp-content/uploads/2014/04 /Analysis_of_Changing_Residential_Fire_Dynamics _and_Its_Implications_on_Firefighter_Operational _Timeframes.pdf. Accessed November 12, 2018.

Kerber, Stephen. 2013. "Study of the Effectiveness of Fire Service Vertical Ventilation and Suppression Tactics." UL Firefighter Safety Research Institute. https://ulfirefightersafety.org /docs/2010-DHS-FD-Summary.pdf. Accessed November 12, 2018.

Norwood, P. J., and Frank Ricci. 2014. "Ventilation Limited Fire: Keeping It Rich and Other Tactics Based Off Science." *Fire*

Engineering, January 24, 2014. http://www.fireengineering .com/articles/2014/01/ventilation-limited-fire-keeping-it -rich-and-other-tactics-based-off-science.html. Accessed November 12, 2018.

UL Firefighter Research and Safety Institute. 2006. "Structural Stability of Engineered Lumber in Fire Conditions." https://ulfirefightersafety.org/research-projects/structural -stability-of-engineered-lumber-in-fire-conditions.html. Accessed November 12, 2018.

UL Firefighter Research and Safety Institute. 2011. "Effectiveness of Fire Service Vertical Ventilation and Suppression Tactics." https://ulfirefightersafety.org/research-projects/vertical -ventilation-and-suppression-tactics.html. Accessed November 12, 2018.

UL Firefighter Research and Safety Institute. 2013. "Governors Island Experiments." https://ulfirefightersafety.org /research-projects/governors-island-experiments.html. Accessed November 12, 2018.

On Scene

You arrive at work and your captain barely has time to say good morning before you are dispatched to a structure fire. The structure is a newer three-storey commercial Type II structure with metal cladding. The caller indicated that a construction crew was digging a trench when it hit a gas main. The gas ignited, impinged on the building, and caused the fire to spread up the wall, igniting combustible roofing material above one of the units. The incident commander indicated that this would be an offensive attack with an objective to limit fire spread from the current suite that is involved. Your crew is assigned to ventilate the structure to support fire control.

1. You are assigned to vertically ventilate the structure. What should be your greatest concern?

A. Falling from the roof

B. The roof collapsing

C. Cutting the hole in the wrong location

D. Crews putting a hose line through the ventilation hole

2. When cutting a vertical ventilation hole using a power saw, when should you initially start the power saw?

A. Just before you start up the ladder

B. At the edge of the roof before you get to the location of the vent hole

C. Just before you start cutting on the roof

D. Before going to the roof

(continued)

On Scene Continued

3. You need to make a cut using a power saw and axe along and between roof supports so that the sections created can be tilted into the opening. What is this cut called?

   A. Louver cut

   B. Kerf cut

   C. Triangle cut

   D. Secondary cut

4. As the fire begins to spread from one end of the building to the other, the incident commander orders your company to make a roof cut from one load-bearing wall to another load-bearing wall to try to prevent the horizontal spread of the fire. What is this cut called?

   A. Louver cut

   B. Kerf cut

   C. Seven, nine, eight cut

   D. Trench cut

5. Which of the following tools will most likely be used to push down the ceiling after you cut a vertical ventilation hole?

   A. Axe

   B. Pike pole

   C. Power saw

   D. Halligan tool

6. When setting up positive-pressure ventilation, where should the fan be placed?

   A. Inside the building, 0.9 to 1.2 m (3 to 4 ft) from the doorway

   B. In the doorway

   C. 0.9 to 1.2 m (3 to 4 ft) in front of the doorway

   D. 2.4 to 3 m (8 to 10 ft) in front of the doorway

7. You are assigned to perform hydraulic ventilation. What is this?

   A. The process of making openings on the same level as the fire so that smoke, heat, and gases can escape horizontally from a building through openings such as doors and windows

   B. The process of removing heat, smoke, and gases from a fire area by using exhaust fans, blowers, air-conditioning systems, or smoke ejectors

   C. Ventilation that relies on fans to push or force clean air into a structure

   D. Ventilation that relies on the movement of air caused by a fog stream that is placed 0.6 to 1.2 m (2 to 4 ft) in front of the open window

8. Your captain says you are experiencing smoke inversion. What is this?

   A. The condition in which smoke hangs low to the ground due to the presence of cold air

   B. Vertical air flow within buildings caused by temperature-created density differences between the building interior and exterior or between two interior spaces

   C. The process in which rising smoke, heat, and gases encounter a horizontal barrier such as a ceiling and begin to move out and back down

   D. Recirculation of exhausted air that is drawn back into a negative-pressure fan in a circular motion

Access Navigate to find answers to this On Scene, along with other resources such as an audiobook.

Water Supply Systems

KNOWLEDGE OBJECTIVES

After studying this chapter, you will be able to:

- Describe how municipal water systems supply water to communities. (pp. 523–526)
- Describe the common guidelines that govern the location of fire hydrants. (p. 526)
- List the types of fire hydrants. (pp. 526–527)
- Describe the characteristics of dry-barrel hydrants. (**NFPA 1001: 4.3.15**, pp. 526–527)
- Describe the characteristics of wet-barrel hydrants. (**NFPA 1001: 4.3.15**, p. 527)
- Explain the principles of fire hydraulics. (pp. 528–534)
- Describe how water flow is measured. (pp. 528–534)
- Describe how water pressure is measured. (pp. 528–534)
- Compare potential and kinetic energy. (p. 532)
- Describe the similarities between static pressure and normal operating pressure of a system. (p. 533)
- Explain how friction loss affects water pressure. (p. 534)
- Explain how elevation pressure affects water pressure. (p. 534)
- Describe how to prevent water hammer. (p. 535)
- Describe how to inspect a fire hydrant. (p. 535)
- Describe how to test a fire hydrant. (pp. 536–537)
- Describe the equipment and procedures that are used to access static sources of water. (**NFPA 1001: 4.3.15**, pp. 538–541)
- Describe the characteristics of a mobile water supply apparatus. (**NFPA 1001: 4.3.15**, pp. 541–542)
- Describe the advantages of a portable tank system. (**NFPA 1001: 4.3.15**, pp. 542–543)

SKILLS OBJECTIVES

After studying this chapter, you will be able to perform the following skills:

- Operate a dry-barrel fire hydrant. (**NFPA 1001: 4.3.15**, pp. 528–530)
- Shut down a dry-barrel fire hydrant. (**NFPA 1001: 4.3.15**, p. 531)
- Operate a wet-barrel fire hydrant. (**NFPA 1001: 4.3.15**, pp. 528, 532)
- Shut down a wet-barrel fire hydrant. (**NFPA 1001: 4.3.15**, p. 533)
- Conduct a hydrant flow test. (pp. 537–538)
- Assist the pump driver/operator with drafting. (**NFPA 1001: 4.3.15**, pp. 539–540)
- Set up a portable tank. (**NFPA 1001: 4.3.15**, pp. 542–544)

ADDITIONAL STANDARDS

- **NFPA 24**, *Standard for the Installation of Private Fire Service Mains and Their Appurtenances*
- **NFPA 1142**, *Standard on Water Supplies for Suburban and Rural Fire Fighting*

Fire Alarm

You arrive at a fire in a barn that is fully involved. The attack is clearly going to be a defensive operation, but with a house and other buildings nearby, aggressive actions will be required to prevent their loss. The house and other buildings are located one-quarter of a mile away from the barn, down a narrow, tree-lined lane. While the other crews are setting up exposure protection, you are assigned to be water supply officer. The incident commander says he needs a 1893 litres per minute [L/min] (500 gallons per minute [gpm]) flow to be established.

1. What water sources might be available?

2. Which type of water supply would be most dependable?

3. Which type of water source would require the most resources to implement?

 *Access Navigate for more practice activities.*

Introduction

This chapter covers several topics related to **water supply**. Generally speaking, water supply refers to how water is supplied to the fire scene. In municipalities, water is usually supplied through fire hydrants. In rural areas, water sources may include static sources of water such as ponds or rivers or stored sources of water such as storage tanks or cisterns. This chapter covers the water flow during the first half of its journey—from the water source to the fire pumper or engine.

Water supply for the purposes of firefighting activities must be adequate, reliable, and continuous. Fire fighters must have confidence in their water supply in order to effectively undertake their duties.

If the water supply is interrupted while crews are working inside a building, fire fighters can be injured or killed. Fire fighters entering a burning building need to be confident that their water supply is both reliable and adequate to operate hose lines—for their protection and to extinguish the fire. As will be elaborated on in Chapter 15, *Fire Hose, Appliances, and Nozzles*, hose lines are the primary means by which fire fighters apply water to control, confine, and ultimately extinguish fires.

Fire fighters can obtain water from one of two sources. **Municipal-type water systems** furnish water under pressure through fire hydrants (**FIGURE 14-1**). Fire hydrants make the municipal water supply available to the fire department. In addition, most automatic sprinkler systems and many standpipe systems are connected directly to a municipal water source. In contrast, rural areas may depend on **static sources of water** such

FIGURE 14-1 The water that comes from a hydrant is provided by a municipal or private water system.
© Jones & Bartlett Learning. Photographed by Glen E. Ellman.

as lakes, ponds, or municipal storage tanks designed for fire service use. These sources serve as drafting sites for fire department apparatus to obtain and deliver water to the fire scene.

FIGURE 14-2 Mobile water supply apparatus can deliver limited quantities of water to the scene of a fire.
© Jones & Bartlett Learning. Photographed by Glen E. Ellman.

Often, the water that is carried in a tank on one of the first-arriving vehicles is used for the initial attack (**FIGURE 14-2**). Although many fires are successfully controlled using tank water, this does not ensure the adequacy and reliability of the water supply. The establishment of an adequate, reliable, and continuous water supply is an essential tactical objective, which needs to be accomplished soon after the first fire fighters arrive on scene. The first arriving pump driver/operator must take action to back up the water supply or begin laying supply lines for use by subsequent arriving apparatus. Without a continuous water supply, any fire attack on a well-involved structure fire will be limited. However, in the hands of a well-trained and knowledgeable attack crew, the water on board a pump can go a long way toward controlling a compartment fire in a limited space.

Municipal Water Systems

Municipal water systems in Canada are regulated by provincial and territorial governments. These systems provide clean water for drinking, cooking, cleaning, and fire protection. As the name suggests, most municipal water systems are owned and operated by a local government agency, such as a city, county, or special water district. Although some municipal water systems are privately owned, the basic design and operation of both private and municipal systems are very similar. Municipal water is supplied to homes, commercial establishments, and industries.

A municipal water system has three major components: the water source, the treatment facility, and the distribution system.

Water Sources

Municipal water systems can draw water from wells, rivers, streams, lakes, or human-made water storage facilities called reservoirs. The source depends on the geographic and hydrologic features of the area. Many municipal water systems draw water from several sources to ensure a sufficient supply. Underground pipelines or open canals supply some cities with water from sources that are many miles away.

The water source for a municipal water system needs to be large enough to meet the total demands of the service area. Most of these systems include large storage facilities, thereby ensuring they will be able to meet the community's water supply demands if the primary water source becomes unavailable for some reason. The backup supply for some systems can provide water for several months or years. In other systems, however, the supply may last only a few days.

Water Treatment Facilities

Municipal water systems usually include a water treatment facility, where impurities are removed from the water (**FIGURE 14-3**). The nature of the treatment system depends on the quality of the untreated source water. Water that is clean and clear from the source requires little treatment. Other systems must use extensive filtration to remove impurities and foreign substances. Some treatment facilities use chemicals to remove impurities and improve the water's taste. All of the water in the system must be suitable for drinking.

Chemicals and ultraviolet (UV) radiation are used to kill bacteria and harmful organisms and to keep the water pure as it moves through the distribution system to individual homes or businesses.

FIGURE 14-3 Impurities are removed at the water treatment facility.
© Hundley Photography/Shutterstock.

Water Distribution System

The distribution system delivers water from the treatment facility to the end users and fire hydrants through a complex network of underground pipes, known as **water mains**. In most cases, the distribution system also includes pumps, storage tanks, control valves, reservoirs, and other necessary components to ensure that the required volume of water can be delivered, where and when it is needed, at the required pressure.

The pressure required in any given system will depend on various factors including, but not limited to, population size, elevation, industrial demand, and overall usage. Generally, water pressure ranges from 138 to 552 kPa (20 pounds per square inch [psi] to 80 psi) at the delivery point. The recommended minimum pressure for water coming from a fire hydrant is 138 kPa (20 psi), but it is possible to operate with lower hydrant pressures under some circumstances. In some locations with high fire risk, higher pressure systems may be present.

Most water distribution systems rely on an arrangement of pumps to provide the required water pressure, either directly or indirectly. In systems that use pumps to supply direct pressure, if the pumps stop operating, the pressure will be lost, and the system will be unable to deliver adequate water to the end users or to hydrants. Most municipal systems maintain multiple pumps and backup power supplies to reduce the risk of a service interruption due to a pump failure. The extra pumps can be used to supplement the water pressure when the pressures in the system drop below a predetermined setting, such as for a major fire or during other high-demand periods.

In a pure **gravity-feed system**, the water source, treatment facility, and storage facilities are located on high ground while the end users live in lower lying areas, such as a community in a valley (**FIGURE 14-4**). This type of system may not require any pumps, because gravity, through the elevation differentials, provides the necessary pressure to deliver the water. In some systems,

FIGURE 14-5 Water that is stored in an elevated tank can be delivered to the end users under pressure.
© Jones & Bartlett Learning. Photographed by Glen E. Ellman.

the elevation pressure is so high that pressure-control/reduction devices are needed to keep parts of the system from being subjected to excessive pressures.

Most municipal water supply systems use a combination of pumps and gravity to deliver water. Pumps can be used to deliver water from the treatment facility to **elevated water storage towers** or to reservoirs located on hills or high ground. These elevated water storage facilities maintain the desired water pressure in the distribution system, ensuring that water can be delivered under pressure even if the pumps are not operating (**FIGURE 14-5**). When the elevated water storage facilities need refilling, large supply pumps are used. Additional pumps may be installed to increase the pressure in particular areas, such as a booster pump that provides extra pressure for a hilltop neighbourhood.

A combination pump and gravity-feed system must maintain enough water in the elevated storage tanks and reservoirs to meet anticipated demands. If more water is being used than the pumps can supply, or if the pumps are out of service, some systems will be able to operate for several days by relying solely on their elevated storage reserves. Other systems may be able to function for only a few hours in such scenarios.

FIGURE 14-4 A gravity-feed system can deliver water to a low-lying community without the need for pumps.
© Jones & Bartlett Learning.

The underground distribution system that delivers water to the end users is made up of three different sizes of water mains (**FIGURE 14-6**). Large mains, known as **primary feeders** or trunk lines, carry large quantities of water to a section of a town or city. Connected to the primary feeders are smaller mains, called **secondary feeders** or branch lines, that distribute water to a smaller area. The smallest mains, called **distributors**, carry water to the users and to fire hydrants along individual streets.

The size of the water mains required depends on the amount of water needed for normal consumption and for fire protection in each location. Most jurisdictions specify the minimum-size main that can be installed in a new municipal water system to ensure an adequate flow. Some municipal water systems, however, may have undersized water mains in older areas of the community. In addition, the volume of water delivered through a water main may decrease if the pipe becomes corroded or partly filled with sediment. Fire fighters must know the arrangement and capacity of the water systems in their response areas.

Water mains in a well-designed distribution system will follow a grid pattern. A grid arrangement provides water flow to a fire hydrant from two or more directions and establishes multiple paths from the source to each area. This kind of organization helps to ensure an adequate flow of water for firefighting. The grid design also helps to minimize downtime for the other portions of the system if a water main breaks or needs maintenance work. With a grid pattern, the water flow can be diverted around the affected section.

Older water distribution systems may have dead-end water mains, which supply water from only one direction. Such dead-end water mains may still be found in the outer reaches of a municipal system. Fire hydrants on this type of main will have a limited water supply. If two or more hydrants on the same dead-end main are used to fight a fire, the upstream hydrant will have more water and water pressure than the downstream hydrants.

Control valves installed at intervals throughout a water distribution system allow different sections of the water supply to be turned off or isolated. These valves are used when a water main breaks or when work must be performed on a section of the system.

Shut-off valves are located at the connection points where the underground mains meet the distributor mains. These valves control the flow of water to individual customers or to individual fire hydrants (**FIGURE 14-7**). If the water system in a building or

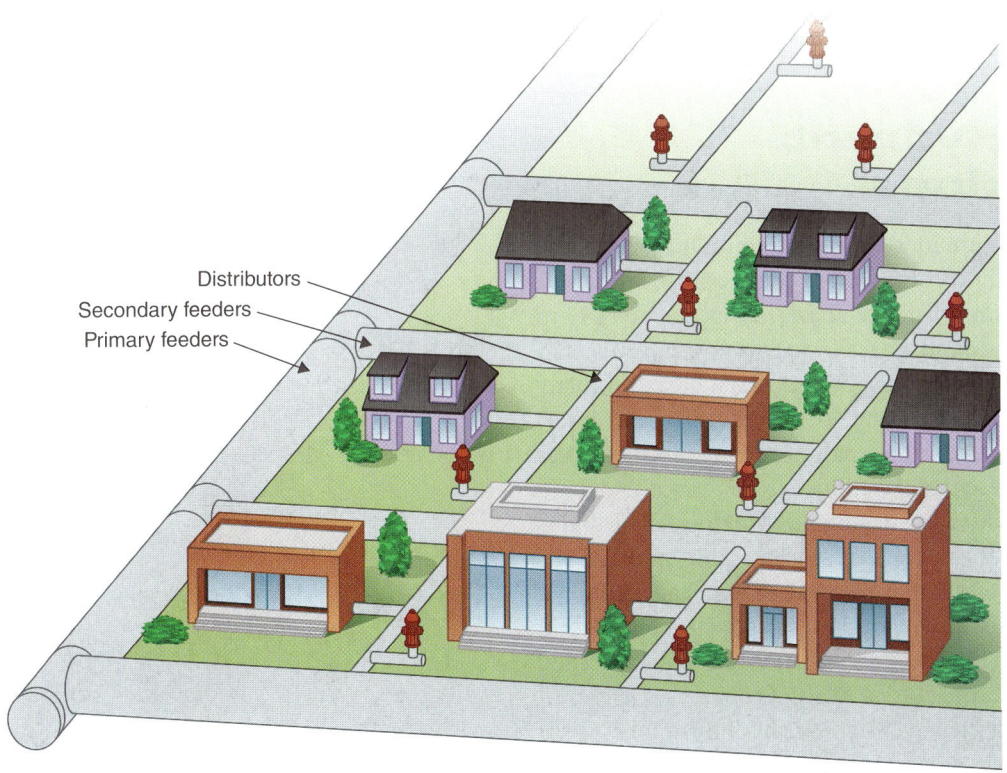

FIGURE 14-6 The underground distribution system includes three different sizes of water mains: primary feeders, secondary feeders, and distributors.

FIGURE 14-7 A shut-off valve controls the water supply to an individual user or fire hydrant. The shut-off valve is located under the cover.

© Jones & Bartlett Learning. Photographed by Glen E. Ellman.

to a fire hydrant becomes damaged, the shut-off valve can be closed to prevent further water flow.

The fire department should notify the water department when fire operations will require prolonged use of large quantities of water. The water department may, in turn, be able to increase the normal volume and/or pressure by starting additional pumps. In some systems, the water department can open valves to increase the flow to a certain area in response to fire department operations at major fires.

Fire Hydrant Locations

Fire hydrants are located according to local standards and nationally recommended practices. They may be placed a certain distance apart, perhaps every 152 m (500 ft) in residential areas and every 91 m (300 ft) in high-value commercial and industrial areas. In many communities, hydrants are located at every street intersection, with mid-block hydrants being installed if the distance between intersections exceeds a specified limit.

In some cases, the requirements for locating hydrants are based on the occupancy, construction, and size of a building. A builder may be required to install additional hydrants when a new building is constructed so that no part of the building will be more than a specified distance from the closest hydrant.

Knowing the location of fire hydrants makes them easier to find in emergency situations. Fire department preincident plans should identify the locations and flow rates of nearby fire hydrants for each building or group of buildings.

Types of Fire Hydrants

Fire hydrants provide water for firefighting purposes. Public fire hydrants are part of the municipal water distribution system and draw water directly from the public water mains. Fire hydrants are also installed on **private water systems** and may be supplied by the municipal water system or from a separate source. The water source, as well as the adequacy and reliability of the supply to private hydrants, must be noted and identified to ensure that sufficient water supplies will be available when fighting fires.

Most fire hydrants consist of an upright steel casing (barrel) attached to the underground water distribution system. The two major types of municipal fire hydrants are the dry-barrel hydrant and the wet-barrel hydrant. Fire hydrants are equipped with one or more valves to control the flow of water through the hydrant. In addition, one or more outlets are provided to connect fire department hose to the hydrant. The outlets may be of various sizes depending on the local jurisdiction, although 65-mm (2½-in.) connections and one larger connection (114, 127, or 152 mm [4½, 5, or 6 in.]) are common. Threads utilized on fire hydrants are usually of the national standard type, although some jurisdictions have their own thread type.

Dry-Barrel Hydrants

Dry-barrel hydrants (frostproof hydrants) are used in climates where temperatures fall below freezing (**FIGURE 14-8**). These fire hydrants are connected

FIGURE 14-8 A dry-barrel hydrant.

Courtesy of American AVK Company.

to a pressurized municipal water system. The valve that controls the flow of water into the barrel of the hydrant is located at the base, several feet below ground (below the frost line), to keep the hydrant from freezing (**FIGURE 14-9**). The length of the barrel depends on the climate and the depth of the valve. Water enters the barrel of the hydrant only when it is opened. Turning the stem nut on the top of the hydrant rotates the operating stem; this opens the valve so that water flows up into the barrel of the hydrant.

Whenever this type of hydrant is not in use, the barrel must remain dry. If the barrel contains standing water, it will freeze in cold weather and render the hydrant inoperable. After each use, the water drains out through an opening at the bottom of the barrel. This drain is fully opened when the hydrant valve is fully closed. When the hydrant valve is opened, the drain closes, thereby preventing water from being forced out of the drain when the hydrant is under pressure. If the drain becomes clogged, however, the hydrant may not drain. In this case, it may be necessary to pump water out of the hydrant before it freezes.

A partially opened hydrant valve means that the drain is also partially open, and pressurized water can flow out. This leakage can erode (undermine) the soil around the base of the hydrant and may damage the hydrant. For this reason, a hydrant valve should be always fully opened or fully closed. A fully opened hydrant valve also makes the maximum flow available to fight a fire.

Most dry-barrel hydrants contain only one large valve that controls the flow of water to all outlets. Each outlet must be connected to a hose or an outlet valve, or it must have a hydrant cap firmly in place before the valve is turned on. Many fire departments use special hydrant valves (gate valves) that allow additional connections to be made after the hydrant is charged and water is flowing through the hose attached to the first outlet. If a hydrant valve is not used when initially connecting to a hydrant, the hydrant would have to be shut off before additional lines could be added. If additional outlets may be needed later, separate hydrant valves should be connected before the hydrant is opened.

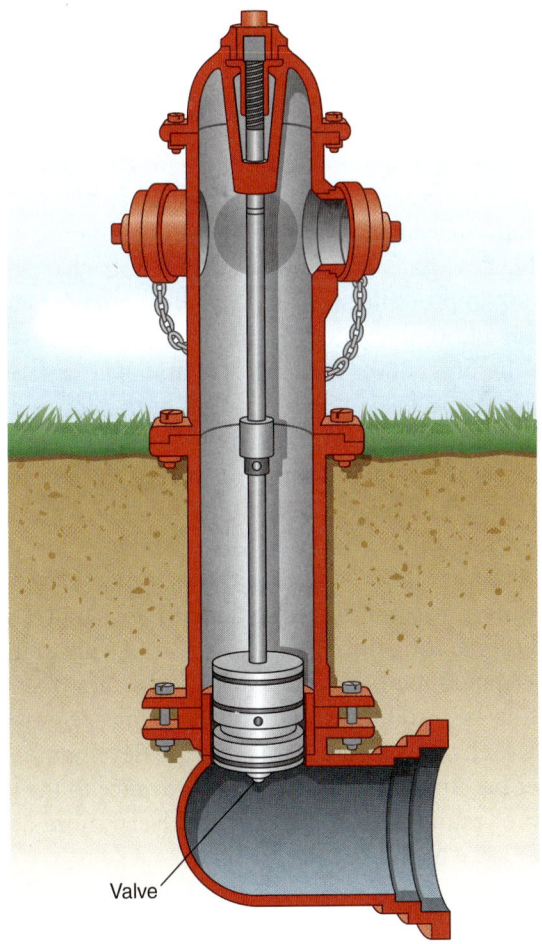

FIGURE 14-9 A dry-barrel hydrant is controlled by an underground valve.

© Jones & Bartlett Learning.

Valve

FIRE FIGHTER TIP

The terms *dry-barrel hydrant* and *dry hydrant* are easy to confuse, but they are very different. Dry-barrel hydrants are connected to a pressurized municipal water system in areas where the temperature drops below freezing. Dry hydrants (also known as drafting hydrants) are a permanent piping system that is connected to a static water source such as a stream, pond, or lake. To get water from a dry or drafting hydrant, you must draft (pump) the water using a pumper/engine or a portable pump.

Wet-Barrel Hydrants

Wet-barrel hydrants are used in locations where temperatures do not drop below freezing. These fire hydrants always have water in the barrel and do not have to be drained after each use.

Wet-barrel hydrants usually have separate valves that control the flow of water to each individual outlet (**FIGURE 14-10**). A fire fighter can hook up one hose line and begin flowing water and later attach a second hose line and open the valve for that outlet, without shutting down the fire hydrant.

Fire Hydrant Operation

Fire fighters must be proficient in operating fire hydrants. First, always ensure that the hydrant is operational before use. Debris in the water distribution system may be forced into the barrel of a hydrant, a leaking ground

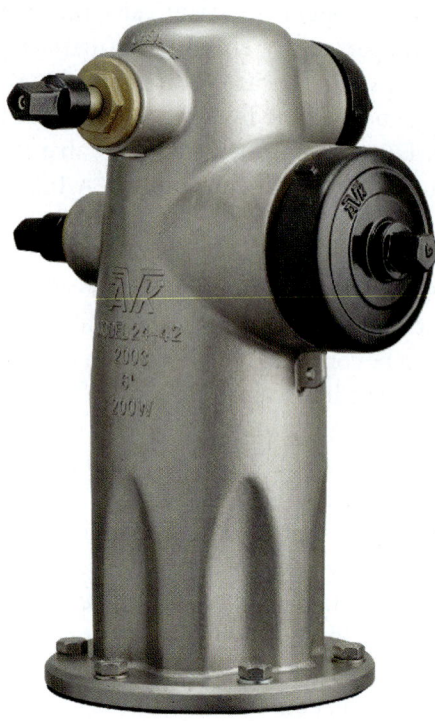

FIGURE 14-10 A wet-barrel hydrant has separate valves that control the flow of water to each outlet.
Courtesy of American AVK Company.

valve may allow water to enter the hydrant barrel and freeze during inclement weather, or vandals may remove caps and place trash or foreign objects into the hydrant. These materials can obstruct the water flow or damage a fire department pumper if they are drawn into the pump.

Regardless of the hydrant type, it is a good practise to check the operation of the hydrant and flush out debris before connecting a hose to a hydrant. To do so, the fire fighter making the connection opens the large outlet cap and then opens the hydrant valve just enough to ensure that water flows into the hydrant and flushes out any foreign matter. This step takes a few seconds. The fire fighter then closes the hydrant valve, connects the hose, and reopens the valve all the way. Water distribution and supply are the responsibilities of each municipality. Therefore, the responsibility for the regular maintenance of the water supply system is likely to be that of a water or municipal works department.

Operating a Dry-Barrel Fire Hydrant

Dry-barrel fire hydrants are located in areas where the temperatures dip below freezing. **SKILL DRILL 14-1** outlines the steps for getting water from a dry-barrel fire hydrant efficiently and safely.

Individual fire departments may have their own variations on this procedure. For example, some departments specify that the wrench be left on the hydrant. Other departments require that the wrench be removed and returned to the fire apparatus so that an unauthorized person cannot interfere with the operation. Always follow the standard operating procedures (SOPs) for your department.

Shutting a hydrant down properly is just as important as opening a hydrant properly. If the hydrant is damaged during shutdown, it cannot be used until it has been repaired. To shut down a dry-barrel hydrant efficiently and safely, follow the steps in **SKILL DRILL 14-2**.

Operating a Wet-Barrel Fire Hydrant

Wet-barrel fire hydrants are located in areas where the temperatures do not normally dip below freezing. These hydrants maintain water in their barrels and do not require draining. The steps for operating them are different from the steps required to operate a dry-barrel hydrant. **SKILL DRILL 14-3** lists the steps for getting water from a wet-barrel fire hydrant efficiently and safely.

Follow the steps in **SKILL DRILL 14-4** to shut down a wet-barrel fire hydrant.

Fire Hydraulics

The amount of water available to fight a fire at a given location is a crucial factor in planning an attack. Will the hydrants deliver enough water at the needed pressure to enable fire fighters to control a fire? If not, what can be done to improve the water supply? How can fire fighters obtain additional water if a fire does occur?

The procedures for testing hydrants are relatively simple, but a basic understanding of the concepts of hydraulics and careful attention to detail are required. This section explains some of the basic theory and terminology of hydraulics and describes how the tests are conducted and the results are recorded.

Fire hydraulics deals with the properties of energy, pressure, and water flow as related to fire suppression. When operating hose lines at a fire, it is important to understand some basic principles of hydraulics—namely, friction loss in different sizes of hose lines, elevation-related changes in pressure, and the development of water hammers. Fire fighters who advance to the position of pump driver/operator will learn more about fire service hydraulics.

Water Flow and Pressure

The **water flow**, or quantity of water moving through a pipe, hose, or nozzle, is measured in terms of its **volume**,

SKILL DRILL 14-1
Operating a Dry-Barrel Fire Hydrant Fire Fighter I, NFPA 1001: 4.3.15

1 Remove the cap from the outlet you will be using.

2 Look inside the hydrant opening for debris.

3 Check that the remaining caps are snugly attached.

4 Attach the hydrant wrench to the stem nut located on top of the hydrant. Check the top of the hydrant for an arrow indicating the direction to turn to open.

(continued)

SKILL DRILL 14-1 Continued
Operating a Dry-Barrel Fire Hydrant Fire Fighter I, NFPA 1001: 4.3.15

5 Open the hydrant valve enough to verify flow of water and to flush out any debris that may be in the hydrant.

6 Close the hydrant valve to stop the flow of water. Attach the hose or valve to the hydrant outlet.

7 When instructed to do so by your officer or the pump driver/operator, start the flow of water by turning the hydrant wrench to fully open the valve. This may take 12 or more turns depending on the type of hydrant.

8 Open the hydrant slowly to avoid a pressure surge. Once the flow of water has begun, you can open the hydrant valve more quickly. Make sure that you open the hydrant valve completely. If the valve is not fully opened, the drain hole will remain open.

SKILL DRILL 14-2
Shutting Down a Dry-Barrel Fire Hydrant Fire Fighter I,
NFPA 1001: 4.3.15

1 Turn the hydrant wrench until the stem valve is closed. Always stand behind the hydrant when opening or closing the valve.

2 Allow the hose to drain by opening a drain valve or disconnecting a hose connection downstream. Slowly disconnect the hose from the hydrant outlet, allowing any remaining pressure to escape.

3 Leave one hydrant outlet open until the hydrant is fully drained.

4 Replace the hydrant cap. Do not leave or replace the caps on a dry-barrel hydrant until you are sure that the water has completely drained from the barrel. If you feel suction on your hand when you place it over the opening, the hydrant is still draining. In very cold weather, you may have to use a hydrant pump to remove all of the water and prevent freezing.

© Jones & Bartlett Learning. Photographed by Glen E. Ellman.

usually specified in units of gallons per minute (gpm). Canadian fire departments use the metric system and measure flow in litres per minute (L/min). The conversion factor is 3.785 L = 1 gal.

Water pressure refers to the amount of energy or force per unit area of water and is measured in kilopascals (kPa) or in units of pounds per square inch (psi). The conversion factor is 6.894 kPa = 1 psi. Water

SKILL DRILL 14-3
Operating a Wet-Barrel Fire Hydrant Fire Fighter I, NFPA 1001: 4.3.15

1. Remove the cap from the outlet you will be using.

2. Look inside the hydrant opening for debris.

3. Check that the remaining caps are snugly attached.

4. Attach the hydrant wrench to the stem nut located behind the outlet you will be using. Check the hydrant for an arrow indicating the direction to turn to open.

5. Open the hydrant valve enough to verify flow of water and to flush out any debris in the hydrant.

6. Close the hydrant valve to stop the flow of water.

7. Attach the hose or hydrant valve to the hydrant outlet.

8. When instructed to do so by your officer or the pump driver/operator, start the flow of water by turning the hydrant wrench to fully open the valve. This may take 12 or more turns, depending on the type of hydrant.

9. Open the hydrant slowly to avoid a pressure surge. Once the flow of water has begun, you can open the hydrant valve more quickly. Make sure that you open the hydrant valve completely.

Courtesy of Bill Larkin.

pressure is required to push water through a hose, to expel water through a nozzle, or to lift water up to a higher level. A pump adds energy to a water stream, causing an increase in pressure. Water flow and water pressure are two different, but mathematically related, measurements.

Water that is not moving has **potential energy**, or energy that it has stored up as a result of its position or condition. When a stream of water flows out through an opening (known as an orifice), the pressure is converted to kinetic energy. **Kinetic energy** is the energy possessed by an object as a result of its motion. When the water is moving, it has a combination of potential (stored) energy and kinetic (in motion) energy. Both the water flow and the water pressure under a specific set of conditions must be measured as part of testing any water system, including fire hydrants. To calculate the volume of water flowing, fire fighters measure the

SKILL DRILL 14-4
Shutting Down a Wet-Barrel Fire Hydrant Fire Fighter I,
NFPA 1001: 4.3.15

1 Turn the hydrant wrench until the valve opposite the outlet you are using is closed.

2 Allow the hose to drain by opening a drain valve or disconnecting a hose connection downstream. Slowly disconnect the hose from the hydrant outlet, allowing any remaining pressure to escape.

3 Replace the hydrant cap.

Courtesy of Bill Larkin.

water pressure at the center of the water stream as it passes through the opening and then factor in the size and flow characteristics of the orifice. A **Pitot gauge** is used to measure water pressure in kPa (or psi) and to calculate the volume of water in litres (or gallons) per minute.

Static pressure is the amount of pressure in a system when the water is not moving. Static pressure is potential energy, because it would cause the water to move if there were some place the water could go. This kind of pressure causes the water to flow out of an opened fire hydrant. If no static pressure is present, the fire hydrant is inoperable.

To measure static pressure in a water distribution system, place a pressure gauge on a hydrant outlet, and open the hydrant valve. No water can be flowing out of the hydrant when a static pressure measurement is being taken.

When measured in this way, the static pressure reading assumes that there is no flow in the system. Of course, because municipal water systems deliver water to

hundreds or thousands of users, there is almost always some water flowing within the system. Thus, in most cases, a static pressure reading actually measures the normal operating pressure of the system.

Normal operating pressure refers to the amount of pressure in a water distribution system during a period of normal consumption. In a residential neighbourhood, for example, people are constantly using water to care for lawns, wash clothes, bathe, and perform other normal household activities. In an industrial or commercial area, normal consumption occurs during a normal business day as water is used for various purposes. The water distribution system uses some of the static pressure to deliver this water to residents and businesses. A pressure gauge connected to a hydrant during a period of normal consumption will indicate the normal operating pressure of this system.

Fire fighters need to know how much pressure will be in the system when a fire occurs. Because the regular users of the system will be drawing off a normal amount of water even during firefighting operations, the normal operating pressure is sufficient for measuring available water. (*Note*: The normal operating pressure may change according to the time of day in some areas owing to high demand during certain hours.)

Residual pressure is the amount of pressure that remains in the system when water is flowing. For example, when fire fighters open a hydrant and start to draw large quantities of water out of the system, some of the potential energy of still water has converted into the kinetic energy of moving water. However, not all of the potential energy turns into kinetic energy—some of it is used to overcome friction in the pipes. The pressure remaining after this loss is the residual pressure.

Residual pressure is important because it provides the best indication of how much more water is available in the system. As more water flows, there is less residual pressure in the system. In theory, when the maximum amount of water is flowing, the residual pressure is zero, and there is no more potential energy to push more water through the system. In reality, 138 kPa (20 psi) is considered the minimum usable residual pressure necessary to reduce the risk of damage to underground water mains or pumps.

At the scene of a fire, the pump driver/operator uses the difference between static pressure and residual pressure to determine how many more attack lines or appliances can be operated from the available water supply. The pump driver/operator can refer to a set of tables to calculate the maximum amount of available water; these tables are based on the static and residual pressure readings taken during hydrant testing. Knowing the static pressure, the water flow in litres (gallons) per minute, and the residual pressure enables fire fighters to calculate the amount of water that can be obtained from a hydrant or a group of hydrants on the same water main. The procedure for conducting a fire hydrant flow test is described later.

Friction Loss

Friction loss is the decrease in pressure that occurs as water moves through a pipe or hose due to the resistance it meets as it travels. This loss of pressure represents the energy required to push the water through the hose. Friction loss is influenced by the diameter of the hose, valves or hose appliances applied to the hose, the volume of water travelling through the hose, and the distance the water travels. In a given size hose, a higher flow rate produces more friction loss. In a 65-mm (2½-in.) hose, for example, a 1136-L/min (300-gpm) flow causes much more friction loss than a 757-L/min (200-gpm) flow. At a given flow rate, the smaller the diameter of the hose, the greater the friction loss. At a flow of 1893 L/min (500 gpm), the friction loss in a 65-mm (2½-in.)

hose is much greater than the friction loss in a 102-mm (4-in.) hose. With any combination of flow and diameter, the friction loss is directly proportional to the distance. At a flow of 946 L/min (250 gpm) in a 65-mm hose (2½-in.), for example, the friction loss in 61 m (200 ft) of hose is double the friction loss in 31 m (100 ft) of hose.

Elevation Pressure

Static pressure is generally created by **elevation pressure**, pump pressure, or both. An elevated water tank, for example, creates elevation pressure in the water mains due to the difference in height between the water in the water tank and the underground delivery pipes. Similarly, if a fire hose is laid down a hill, the water at the bottom will have additional pressure because of the change in elevation. Conversely, if a fire hose is advanced upstairs to the third floor of a building, it will lose water pressure as a result of the energy that is required to lift the water. The pump driver/operator must take elevation changes into account when setting the discharge pressure on the pumper/engine.

Gravity creates elevation pressure (sometimes referred to as head pressure) in a water system as the water flows from a hilltop reservoir to the water mains in the valley below. Pumps create pressure by applying energy from an external source into the system.

Water Hammer

Water hammer is a surge in pressure caused by suddenly stopping the flow of a stream of water. A fast-moving stream of water has a large amount of kinetic energy. If the water suddenly stops moving when a valve is closed, all of the kinetic energy is converted to an instantaneous increase in pressure. Because water cannot be compressed, the additional pressure is transmitted along the hose or pipe as a shock wave. Water hammer can rupture a hose, cause a coupling to separate, or damage the plumbing on a piece of fire apparatus. Severe water hammer can damage an underground piping system. Fire fighters have been injured by the effects of water hammer.

A similar situation can occur if a valve is opened too quickly and a surge of pressurized water suddenly fills a hose. The surge in pressure can damage the hose or cause the fire fighter at the nozzle to lose control of the stream.

To prevent water hammer, always open and close fire hydrant valves slowly. Similarly, pump driver/operators need to open and close the valves on the pumper slowly. When operating the nozzle on an attack line, the fire fighter must open the nozzle slowly. Just as important, when closing the shut-off valve on an attack line, the fire fighter must do it slowly.

Maintaining Fire Hydrants

Inspecting Fire Hydrants

Because fire hydrants are essential to fire-suppression efforts, fire fighters must understand how to inspect and maintain them. Depending on the jurisdiction, inspections of fire hydrants may fall to the fire department that utilizes the system or the water department that maintains it. Fire hydrants should be checked on a regular schedule—no less than once a year—to ensure that they are in proper operating condition. If required to inspect fire hydrants, fire fighters should receive proper training in identifying issues. It is unlikely that large municipalities will require fire fighters to regularly inspect or test fire hydrants or water systems.

The first factors to check when inspecting fire hydrants are visibility and accessibility. Fire hydrants should always be visible from every direction, so they can be easily spotted. A fire hydrant should not be hidden by tall grass, brush, fences, debris, dumpsters, or any other obstructions (**FIGURE 14-11**). In winter, fire hydrants must be clear of snow. It is a universal norm that it is illegal to park in front of a fire hydrant.

In many communities, fire hydrants are painted in bright reflective colours for increased visibility. The bonnet (the top of the fire hydrant) may be colour coded to indicate the available flow rate of the fire hydrant (**TABLE 14-1**). While some jurisdictions use their own colour-coding system, it is recommended that NFPA 291, *Recommended Practice for Fire Flow Testing and Marking of Hydrants*, be followed. Coloured reflectors are sometimes mounted next to fire hydrants or placed in the pavement in front of them to make them more visible at night. In municipalities with

Class	Flow Available at 20 psi (138 kPa)	Colour
Class C	Less than 1893 L/min (500 gpm)	Red
Class B	1893–3782 L/min (500–999 gpm)	Orange
Class A	3785–5674 L/min (1000–1499 gpm)	Green
Class AA	5678 L/min and higher (1500 gpm)	Light blue

TABLE 14-1 Fire Hydrant Colours

NFPA 291 recommends that fire hydrants be colour coded to indicate the water flow available from each hydrant at 138 kPa (20 psi). It is recommended that the top bonnet and the fire hydrant caps be painted according to the above system. The colours give you an idea of how much water can be obtained from a hydrant during a fire.

high accumulations of snow, fire hydrants will likely be equipped with a winter fire hydrant marker, making them visible in winter conditions.

Fire hydrants should be installed at an appropriate height above the ground. Their outlets should not be so high or so low that fire fighters have difficulty connecting hose lines to them. NFPA 24, *Standard for the Installation of Private Fire Service Mains and Their Appurtenances*, requires a minimum of 450 mm (18 in.) from the center of a hose outlet to the finished grade. Fire hydrants that experience snow accumulations may require placement of hydrants farther above the finish grade. Fire hydrants should be positioned so that the connections—especially the large-diameter port or steamer port on the hydrant—face the street.

During a fire hydrant inspection, check the exterior of the hydrant for signs of damage. Open the steamer port of dry-barrel hydrants to ensure that the barrel is dry and free of debris. Make sure that all caps are present and that the outlet hose threads are in good working order (**FIGURE 14-12**).

FIGURE 14-11 Fire hydrants should not be hidden or obstructed.
Courtesy of Captain David Jackson, Saginaw Township Fire Department.

LISTEN UP!

Before you leave the fire scene, make sure that dry-barrel hydrants are completely drained, even if the weather is warm. During winter, any water left in the hydrant can freeze. Fire fighters may lose valuable time connecting a hose to a frozen hydrant, only to discover that it will not operate. If this happens, fire fighters will be without water until they can locate a working hydrant and can reposition and reconnect the hose lines.

FIGURE 14-12 All hydrants should be checked at least annually.
© Jones & Bartlett Learning. Photographed by Glen E. Ellman.

The second part of the inspection ensures that the hydrant works properly. Open the hydrant valve just enough to confirm that water flows out and flushes any debris out of the barrel. After flushing, shut down the hydrant. A properly draining hydrant will create suction against a hand placed over the outlet opening. When the hydrant is fully drained, replace the cap.

If the threads on the discharge ports need cleaning, use a steel brush and a small triangular file to remove any burrs in the threads. Also check the gaskets in the caps to make sure they are not cracked, broken, or missing. Replace worn gaskets with new ones, which should be carried on each apparatus. Follow the manufacturer's recommendations for maintaining any parts that require lubrication.

Testing Fire Hydrants

Generally speaking, fire fighters are not responsible for maintaining, testing, or repairing fire hydrants in Canada. The only exception to this might be under exceptional circumstances such as a massive snowstorm, when fire personnel may be utilized to clear snow from around fire hydrants. The procedure for testing hydrant flows requires two adjacent hydrants, a Pitot gauge, and an outlet cap with a pressure gauge. As part of testing, fire fighters measure static pressure and residual pressure at one hydrant and then open the other hydrant to let water flow out. The two hydrants should be connected to the same water main and preferably at approximately the same elevation (**FIGURE 14-13**).

The cap gauge is placed on one of the outlets of the first hydrant. The hydrant valve is then opened to allow water to fill the hydrant barrel. The initial pressure reading on this gauge is recorded as the static pressure (the pressure in a system when the water is not moving).

At the second hydrant, fire fighters remove one of the discharge outlet caps and open the fire hydrant. They put the Pitot gauge into the middle of the stream and take a reading, which is recorded as the Pitot pressure. At the same time, fire fighters at the first fire hydrant record the residual pressure reading (the amount of pressure that remains in the system when water is flowing).

Using the size of the discharge opening (usually 65 mm [2½ in.]) and the Pitot pressure, fire fighters calculate the flow in L/min (or gpm) or look it up in a table. Such a table usually incorporates factors to adjust for the shape of the discharge opening. Fire fighters can use special graph paper or computer software to plot the static pressure and the residual pressure at the test flow rate. The line defined by these two points shows the number of litres (gallons) per minute that is available at 138 kPa (20 psi) residual pressure.

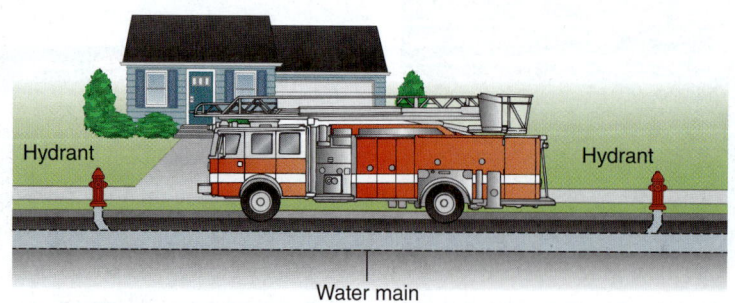

FIGURE 14-13 Testing hydrant flow requires two hydrants on the same water main.
© Jones & Bartlett Learning.

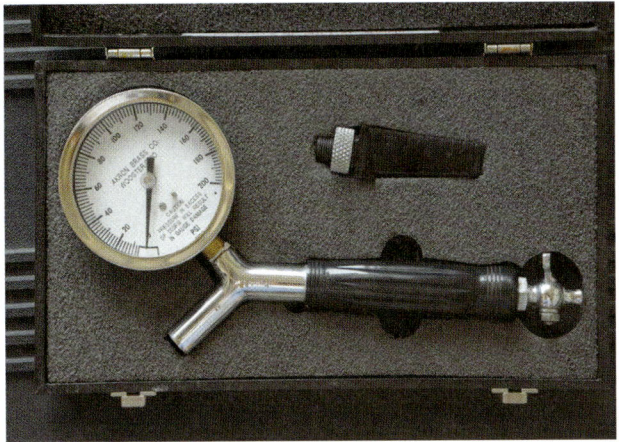

FIGURE 14-14 The Pitot gauge.
© Jones & Bartlett Learning. Photographed by Glen E. Ellman.

Several devices are available to simplify the process of taking accurate Pitot readings. Outlet attachments have smooth tips and brackets that hold the Pitot gauge in the exact required position (**FIGURE 14-14**). Or,

the flow can be measured with an electronic flow meter instead of a Pitot gauge.

To test the operability and flow of a fire hydrant, follow the steps in **SKILL DRILL 14-5**.

Rural Water Supplies

Many fire departments protect areas that are not serviced by municipal water systems or private water systems.

SKILL DRILL 14-5
Conducting a Fire Hydrant Flow Test Fire Fighter I

1 Remove the cap from the hydrant port, open the hydrant, and allow water to flow until it runs clear. Close the hydrant valve. Place a cap gauge on one of the outlets of the first hydrant.

2 Open the hydrant valve to fill the hydrant barrel. No water should be flowing. Record the initial pressure reading on the gauge. This is the static pressure.

(continued)

SKILL DRILL 14-5 Continued
Conducting a Fire Hydrant Flow Test Fire Fighter I

3 Move to the second hydrant, remove one of the discharge caps, and open the second hydrant.

4 Place the Pitot gauge one-half the diameter of the orifice away from the opening, and record this pressure as the Pitot pressure. At the same time, fire fighters at the first hydrant should record a second pressure reading. This is the residual pressure. Use the recorded pressure readings to calculate or look up the flow rates at 138 kPa (20 psi) residual pressure. Document your findings.

© Jones & Bartlett Learning. Photographed by Glen E. Ellman.

In these areas, residents usually depend on individual wells or cisterns to supply water for their domestic uses. Because there are no fire hydrants in these areas, fire fighters must depend on water from other (static) sources for firefighting activities. Fire fighters in rural areas must know how to get water from the sources that are available.

In Canada, many municipalities attempt to maintain sufficient water volume to meet the Fire Underwriters Survey (FUS) requirement for Accredited Superior Tanker Shuttle Service, a recognized equivalency to hydrant protection. A fire service's ability to provide a continuous and reliable water supply is directly related to the insurance ratings in the municipality: The better the water supply, the better the insurance rates, and the safer the community.

Static Sources of Water

Several potential static water sources can be used for fighting fires in rural areas. Both natural and human-made bodies of water such as rivers, streams, lakes, ponds, oceans, canals, reservoirs, swimming pools, and cisterns can be used to supply water for fire suppression (**FIGURE 14-15**). Some areas have many different static sources, whereas others have few or none at all.

Water from a static source can be used to fight a fire directly, if it is close enough to the fire scene. Otherwise, it must be transported to the fire using long hose lines, engine relays, or mobile water supply apparatus.

To be useful, static water sources must be accessible to a pumper or portable pump. If a road or hard surface

FIGURE 14-15 Any accessible body of water can be used as a static source.

© Jones & Bartlett Learning. Photographed by Glen E. Ellman.

is located within 6 m (20 ft) of the static water source, a pumper can drive close enough to draft water directly from the water source into the pump through a hard suction hose. Some fire departments construct special access points so pumpers can approach the water source. Rural fire departments should identify static water sources in the area and practise establishing drafting operations at all of these locations. Once the drafting location is identified, fire fighters should inspect the swivel gaskets. Next, the appropriate length of hose is selected and connected, and a strainer is placed on the end of the hose. The strainer prevents large debris such as trash, rocks, weeds, small twigs, and animals/fish from entering the pump. When the strainer is in place, drafting can begin.

SKILL DRILL 14-6 shows the steps to take to assist the pump driver/operator with assembling the equipment needed to draft water from a static water supply.

SKILL DRILL 14-6
Assisting the Pump Driver/Operator with Drafting Fire Fighter I, NFPA 1001: 4.3.15

1 After the pump driver/operator has positioned the pumper at the draft site, inspect the swivel gaskets on the female coupling for damage or debris.

2 Connect each section of suction hose together, and connect the strainer to the end of the hose that will be placed in the water.

(continued)

SKILL DRILL 14-6 Continued
Assisting the Pump Driver/Operator with Drafting Fire Fighter I, NFPA 1001: 4.3.15

3 Connect the other end of the suction hose to the fire pump.

4 Advance the suction hose assembly into position with the strainer in the water.

5 Ensure that the strainer assembly has at least 61 cm (24 in.) of water in all directions around the strainer.

Dry hydrants provide an alternative means of accessing static water sources in areas that are inaccessible to fire apparatus. Dry hydrants, also called drafting hydrants, provide quick and reliable access to static water sources. A dry hydrant is a fixed piping system with a strainer on one end and a connection for a hard suction hose on the other end (**FIGURE 14-16**). The strainer end should be located below the water's surface and away from any silt or potential obstructions. The other end of the pipe should be accessible to fire apparatus, with the connection at a convenient height for hook-up to a pumper or a portable pump (**FIGURE 14-17**). When a hard suction hose is connected to a drafting hydrant, the pumper can draft water from the static source.

Dry hydrants are often installed in lakes and rivers and close to clusters of buildings where there is a recognized need for fire protection. They may also be installed in farm cisterns or connected to swimming pools on private property to make water available for the local fire department. In some areas, dry or drafting hydrants are used to enable fire fighters to reach water under the frozen surface of a lake or river. NFPA 1142, *Standard on Water Supplies for Suburban and Rural Fire Fighting*, has more information about dry hydrants.

The portable pump can assist fire fighters in accessing water in areas that are inaccessible to fire apparatus (**FIGURE 14-18**). The portable pump can be hand-carried or transported using an off-road vehicle to the water source. Portable pumps can deliver as much as 1893 L/min (500 gpm).

Water Shuttle Operations

Mobile water supply apparatus can be used to deliver water to fight fires. When a large volume of water is needed for an extended period at a fire in an area without an adequate municipal or private water supply, mobile water supply apparatus can be used to deliver water from a fill site to the fire scene. These trucks are commonly referred to as tankers or water tenders. In the western part of the United States where aerial water tankers are used commonly in fighting wildland fires, the term *water tender* is used to describe truck-mounted mobile water supply apparatus. In parts of the country where aerial apparatus is not used commonly, the term *tanker* is used to describe truck-mounted mobile water supply apparatus. This text will use the term *tanker* to mean mobile water supply apparatus.

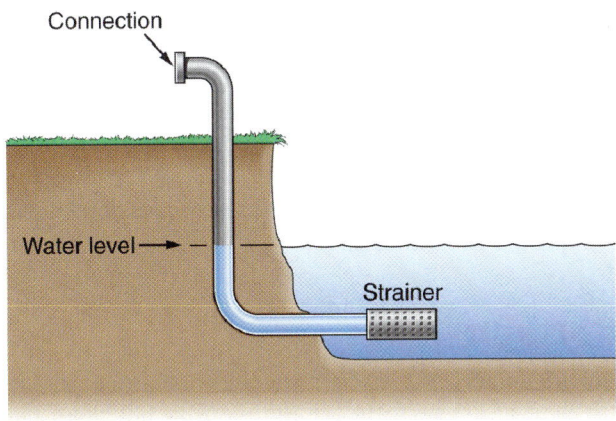

FIGURE 14-16 A dry hydrant or drafting hydrant can be placed at an accessible location near a static water source.
© Jones & Bartlett Learning.

FIGURE 14-17 The height of the dry hydrant or drafting hydrant connection is convenient for hook-up to a pumper or a portable pump.
Courtesy of Ryan Van Buskirk.

FIGURE 14-18 A portable pump can be used if the water source is inaccessible to a fire department pumper.
Courtesy of Castle Rock Fire, Ketchum, ID.

These trucks are designed to conduct **water shuttle operations**. That is, they transport large volumes of water from a water source to a fire scene using mobile water supply apparatus.

Fire department pumpers usually carry at least 1893 L (500 gal) of water in the booster tank, whereas fire department tankers generally carry 3785 to 13,249 L of water (1000 to 3500 gal) (**FIGURE 14-19**). Some tankers can transport as much as 18,927 L (5000 gal) of water.

The number of tankers needed will depend on the distance between the fill site and the fire scene, the time it takes the tanker to dump and refill, and the flow rate required at the fire scene. In rural areas, several tankers may be dispatched for a structural fire.

All of the components of a water shuttle operation must be set up so that water moves efficiently from the fill site to the fire scene. If a mobile water supply apparatus is the only source of water for fighting a structural fire, the attack must be carefully planned. Enough water must be available on the scene to supply the required hose lines. Some fire departments begin the attack using water from the booster tank of the first-arriving

unit. Tankers then pump additional water directly into the attack pumper to keep it full. At the fill site, the tankers must be refilled without delay. The routes in both directions must be planned so that tankers make efficient round trips, without having to back up or make U-turns. At the fire scene, the tankers should be able to drive up, dump their water into the portable tanks, and immediately return to the fill site. The portable tanks, in turn, must be large enough to receive the full load of each tanker as it arrives. An effective water shuttle operation can deliver several hundred litres of water per minute without interruption. If your department uses mobile water supply apparatus, you need to learn the specific system used by your area.

Departments must practise water shuttle operations to ensure proper coordination and effective water delivery. An effective operation requires preplanning and excellent leadership. There are many components to the process, and they must be practised in order to be effective. If the water supply is exhausted before the fire is extinguished, the attack team will be in serious danger, and the building will probably be lost. Conversely, if the use of water is too conservative, fire-suppression efforts will probably be unsuccessful.

Portable Tanks

Portable tanks carried on fire apparatus can be quickly set up at a fire scene. These tanks typically hold between 2271 and 18,927 L (600 and 5000 gal) of water and should be placed so that they can be accessed from multiple directions. With this supply method, one pumper drafts water from the portable tank, using a hard suction hose just as it would if the water came from any other static source (**FIGURE 14-20**). A tanker is used to fill the portable tank. The pump

FIGURE 14-19 A fire department tanker or water tender.
© IAN MARLOW/Alamy Stock Photo.

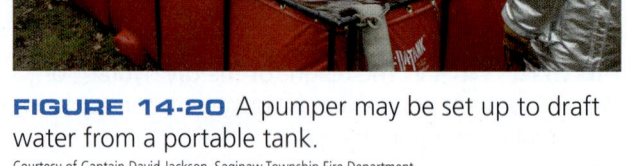

FIGURE 14-20 A pumper may be set up to draft water from a portable tank.
Courtesy of Captain David Jackson, Saginaw Township Fire Department.

driver/operator primes the pump and begins drafting water out of the portable tank, while the tanker leaves to get another load of water.

Speed is a primary advantage when using a portable tank system because tankers do not have to hook up to the pumper to transfer the water into the portable tank. Instead, a dump valve enables the tankers to offload as much as 11,356 L (3000 gal) of water in 1 minute into a portable tank (**FIGURE 14-21**). Some apparatus may use a pump system to offload the water even faster. The faster the tanker can offload its water, the more quickly it can return to the fill site for another load.

Another advantage of the portable tank system is its ability to expand rapidly. Additional portable tanks can be set up and linked together to increase water storage capacity, additional pumpers can be used to draft water from the portable tanks, and additional tankers can be used to deliver more water at a faster rate. A series of tankers may be used as shuttles, dumping water either simultaneously or in sequence. When using multiple portable tanks, a jet siphon assists in keeping the water level at the maximum capacity in the tank closest to the pumper.

To set up a portable tank, follow the steps in **SKILL DRILL 14-7**.

FIGURE 14-21 A dump valve allows a tanker to discharge water into the portable tank quickly.
© Jones & Bartlett Learning. Photographed by Glen E. Ellman.

SAFETY TIP

Never get between two apparatus or tankers!

SKILL DRILL 14-7
Setting Up a Portable Tank Fire Fighter I, NFPA 1001: 4.3.15

1 Two fire fighters lift the portable tank off the apparatus. This tank may be mounted on a side rack or on a hydraulic rack that lowers it to the ground. Place the portable tank on as level ground as possible beside the pumper. The pump driver/operator will indicate the best location.

(continued)

SKILL DRILL 14-7 Continued
Setting Up a Portable Tank Fire Fighter I, NFPA 1001: 4.3.15

2 Expand the tank (metal-frame type), or lay it flat (self-expanding type).

3 One fire fighter helps the pump driver/operator place the strainer on the end of the suction hose, put the suction hose into the tank, and connect it to the pumper.

4 The second fire fighter helps the tanker driver discharge water into the portable tank. If the tank is self-expanding, the fire fighters may need to hold the collar until the water level is high enough for the tank to support itself.

After-Action REVIEW

IN SUMMARY

- The primary objective at a fire is to get water on the fire—the action that cools the fire and extinguishes it.

- Municipal water systems draw water from wells, rivers, streams, lakes, or reservoirs; they carry the water via pipelines or canals to a water treatment facility and then to the water distribution system, a complex network of underground pipes.

- Municipal water systems make clean water available to people in populated areas and provide water for fire protection. Fire hydrants make this water supply available to the fire department.

- Most municipal water supply systems use both pumps and gravity to deliver water.

- Underground water mains come in several sizes. Large mains, or primary feeders, carry large quantities of water to a section of a town or city. Smaller mains, called secondary feeders or branch lines, distribute water to a smaller area. The smallest mains, or distributors, carry water to the end users and to fire hydrants along individual streets.

- Shut-off valves are located at the connection points where the underground mains meet the distributor mains. These valves can be used to prevent water flow if the water system in the building or the fire hydrant is damaged.

- Fire hydrants are located according to local standards and nationally recommended practices. In many communities, fire hydrants are located at every street intersection.

- The two types of fire hydrants are dry-barrel hydrants and wet-barrel hydrants.

- Dry-barrel hydrants are used in areas where temperatures drop below freezing. When this type of hydrant is not in use, the barrel must be dry.

- Wet-barrel hydrants are used in areas where temperatures do not drop below freezing. These hydrants do not have to be drained after each use.

- Fire fighters must be proficient in operating fire hydrants, including the tasks of turning on the hydrant, shutting off the hydrant, and inspecting the hydrant.

- To understand fire hydrant testing procedures, fire fighters must understand some basic concepts of fire hydraulics, which deal with the properties of energy, pressure, and water flow as related to fire suppression.

- The water flow, or the quantity of water moving through a pipe, hose, or nozzle, is described in terms of its volume; it is usually specified in units of litres (or gallons) per minute.

- Water pressure refers to the amount of energy or force per unit area of the water, and it is measured in units of kilopascals (kPa) or in pounds per square inch (psi). Water flow and water pressure are two different, but mathematically related, measurements.

- Water that is not moving has potential (stored) energy. When the water is moving, it has a combination of potential energy and kinetic (in motion) energy. Both the water flow and the water pressure under a specific set of conditions must be measured when testing any water system, including fire hydrants.

- A Pitot gauge is used to measure water pressure in kPa (or psi) and to calculate the volume of water in litres per minute (gallons per minute).

- Static pressure is the pressure in a system when the water is not moving.

- Because municipal water systems deliver water to many people, there is almost always some water flowing within the system. In most cases, a static pressure reading actually measures the normal operating pressure of the system. Normal operating pressure refers to the amount of pressure in a water distribution system during a period of normal consumption.

- Residual pressure is the amount of pressure that remains in the system when water is flowing.

- Knowing the static pressure, the water flow in litres (gallons) per minute, and the residual pressure enables fire fighters to calculate the amount of water that can be obtained from a hydrant or a group of hydrants on the same water main.

- Static pressure is generally created by elevation pressure, pump pressure, or both.

- Fire hydrants should be checked on a regular schedule of no less than once per year to ensure that they are in proper operating condition.

- As part of fire hydrant testing, fire fighters measure static pressure and residual pressure at one hydrant and then open another hydrant to let water flow out.

- In rural areas, fire departments often depend on water from static sources to maintain their water supply.

- Water may be used from the static source directly or transported via a mobile water supply apparatus (tanker or tender). These trucks are designed to carry large volumes of water, ranging from 1893 L (500 gal) to 18,927 L (5000 gal).

- Dry or drafting hydrants and portable pumps provide alternative means of accessing static water sources in areas that are inaccessible to fire apparatus.

- Portable tanks are carried on fire apparatus and can hold between 2271 L (600 gal) and 18,927 L (5000 gal) of water. These tanks can be set up quickly and linked together to increase the water storage capacity.

KEY TERMS

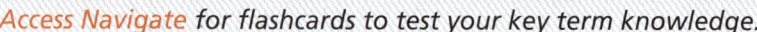

Access Navigate for flashcards to test your key term knowledge.

Distributors The smallest diameter underground water main pipes in a water distribution system that deliver water to local users within a neighbourhood.

Dry-barrel hydrant (frostproof hydrant) A type of hydrant used in areas subject to freezing weather. The valve that allows water to flow into the hydrant is located underground below the frost line, and the barrel of the hydrant is normally dry.

Dry hydrant An arrangement of pipe permanently connected to a water source other than a piped, pressurized water supply system that provides a ready means of water supply for firefighting purposes and that utilizes the drafting (suction) capability of a fire department pump. (NFPA 1142)

Dump valve A large opening from the water tank of a mobile water supply apparatus for unloading purposes. (NFPA 1901)

Elevated water storage tower An above-ground water storage tank that is designed to maintain pressure on a water distribution system.

Elevation pressure The amount of pressure created by gravity. Also known as head pressure.

Fire hydraulics The physical science of how water flows through a pipe or hose.

Friction loss The reduction in pressure resulting from the water being in contact with the side of the hose. This contact requires force to overcome the drag that the wall of the hose creates.

Gravity-feed system A water distribution system that depends on gravity to provide the required pressure. The system storage is usually located at a higher elevation than the end users of the water.

Kinetic energy The energy possessed by an object as a result of its motion.

Mobile water supply apparatus A vehicle designed primarily for transporting (pickup, transporting, and delivering) water to fire emergency scenes to be applied by other vehicles or pumping equipment. (NFPA 1901)

Municipal-type water system A system having water pipes servicing fire hydrants and designed to furnish, over and above domestic consumption, a minimum of 950 L/min (250 gpm) at 138 kPa (20 psi) residual pressure for a 2-hour duration. (NFPA 1141)

Normal operating pressure The observed static pressure in a water distribution system during a period of normal demand.

Pitot gauge A type of gauge that is used to measure the velocity pressure of water that is being discharged from an opening. It is used to determine the flow of water from a hydrant or nozzle.

Portable tanks Folding or collapsible tanks that are used at the fire scene to hold water for drafting.

Potential energy The energy that an object has stored up as a result of its position or condition. A raised weight and a coiled spring have potential energy.

Primary feeders The largest diameter water main pipes in a water distribution system that carry the greatest amounts of water.

Private water system A privately owned water system that operates separately from the municipal water system.

Reservoir A water storage facility.

Residual pressure The pressure that exists in the distribution system, measured at the residual hydrant at the time the flow readings are taken at the flow hydrants. (NFPA 24)

Secondary feeders Smaller diameter water main pipes in the water distribution system that connect the primary feeders to the distributors.

Shut-off valve Any valve that can be used to shut down water flow to a water user or system.

Static pressure The pressure that exists at a given point under normal distribution system conditions measured at the residual hydrant with no hydrants flowing. (NFPA 24)

static sources of water A water source such as a pond, river, stream, or other body of water that is not under pressure.

Steamer port The large-diameter port on a fire hydrant.

Volume The quantity of water flowing; usually measured in litres (gallons) per minute.

Water flow The amount of water flowing through pipes, hose, and fittings, usually expressed in litres (gallons) per minute (L/min or gpm).

Water hammer The surge of pressure that occurs when a high-velocity flow of water is abruptly shut off. The pressure exerted by the flowing water against the closed system can be seven or more times that of the static pressure. (NFPA 1962)

Water main A generic term for any underground water pipe.

Water pressure The application of force by one object against another. When water is forced through the distribution system, it creates water pressure.

Water shuttle operations A method of transporting water from a source to a fire scene using a number of mobile water supply apparatus.

Water supply A source of water for firefighting activities. (NFPA 1144)

Wet-barrel hydrant A hydrant used in areas that are not susceptible to freezing. The barrel of the hydrant is normally filled with water.

REFERENCES

National Fire Protection Association. NFPA 24, *Standard for the Installation of Private Fire Service Mains and Their Appurtenances*, 2019. www.nfpa.org. Accessed November 13, 2018.

National Fire Protection Association. NFPA 1142, *Standard on Water Supplies for Suburban and Rural Fire Fighting*, 2017. www.nfpa.org. Accessed November 13, 2018.

On Scene

You are part of a study group at your class at the fire academy. In preparation for your test on water supply next week, you have prepared the following questions for discussion with your study group.

1. Which of the following is *not* a major component of a municipal water system?

A. Water source

B. Water treatment system

C. Waste disposal system

D. Water distribution system

2. Which of the following are the largest diameter pipes in a water distribution system that carry the greatest amount of water?

A. Distributors

B. Primary feeders

C. Secondary feeders

D. Large-diameter hose

3. Which type of fire hydrant is connected to a static water source?

A. Dry hydrant

B. Pressurized hydrant

C. Dry-barrel hydrant

D. Wet-barrel hydrant

4. When shutting down a dry-barrel hydrant, why is it important to not replace the caps until you are sure that the water has completely drained from the barrel?

A. To make it easier to access the hydrant if necessary

B. To handle any water leaking from the hydrant

C. To prevent water hammer

D. To allow the hydrant to drain through the bottom drain

(continued)

On Scene Continued

5. To follow NFPA recommendations, what colour should you paint a fire hydrant that flows 4732 L/min (1250 gpm)?

A. Orange

B. Light blue

C. Green

D. Red

6. _____ is the pressure in a system when there is no water moving.

A. Static pressure

B. Flow pressure

C. Residual pressure

D. Elevation pressure

7. A portable pump is used to secure water from a(n):

A. municipal water system.

B. static water source.

C. elevated water system.

D. gravity-feed water system.

8. Mobile water supply apparatus is used when:

A. round trips to a water source are short.

B. a hydrant does not work.

C. large quantities of water are needed and hydrants are not available.

D. fire fighters want to avoid laying out hose lines.

Access Navigate to find answers to this On Scene, along with other resources such as an audiobook.

CHAPTER **15**

Fire Fighter I

Fire Hose, Appliances, and Nozzles

KNOWLEDGE OBJECTIVES

After studying this chapter, you will be able to:

- List the two types of fire hose. (**NFPA 1001: 4.3.15**, p. 551)
- Describe the various sizes of fire hose and how they are used. (**NFPA 4.3.15**, pp. 551–552)
- Describe the characteristics of attack hose. (**NFPA 1001: 4.3.10**, pp. 551–552)
- Explain how fire hose is constructed. (**NFPA 1001: 4.3.15**, pp. 552–553)
- Describe the characteristics of single-jacket hose. (**NFPA 1001: 4.3.15**, p. 552)
- Describe the characteristics of multiple-jacket hose. (**NFPA 1001: 4.3.15**, p. 552)
- Describe the characteristics of rubber-covered hose. (**NFPA 1001: 4.3.15**, p. 552)
- Describe the characteristics of couplings. (**NFPA 1001: 4.3.10**, pp. 553–556)
- List the common types of couplings. (**NFPA 1001: 4.3.10**, pp. 553–556)
- Describe supply hose. (**NFPA 1001: 4.3.15**, pp. 556, 561)
- Describe the two types of suction hose. (**NFPA 1001: 4.3.15**, p. 562)
- List the common types of hose damage. (**NFPA 1001: 4.5.2**, pp. 563–564)

- Describe how to clean and maintain hose. (**NFPA 1001: 4.5.2**, pp. 564–566)
- Describe the importance of a hose inspection. (**NFPA 1001: 4.5.2**, p. 566)
- List the common types of hose rolls used to organize supply hose. (**NFPA 1001: 4.5.2**, pp. 567–570)
- List the common hose appliances used in conjunction with fire hose. (**NFPA 1001: 4.3.15**, pp. 570–571, 574–578)
- Describe the characteristics of wyes. (**NFPA 1001: 4.3.10**, pp. 572, 574)
- Describe the characteristics of water thieves. (**NFPA 1001: 4.3.15**, p. 574)
- Describe the characteristics of Siamese connections. (**NFPA 1001: 4.3.15**, pp. 574–575)
- Describe the characteristics of adaptors and reducers. (**NFPA 1001: 4.3.15**, p. 575)
- Describe the characteristics of hose jackets. (**NFPA 1001: 4.3.10**, pp. 575–576)
- Describe the characteristics of hose rollers. (**NFPA 1001: 4.3.10**, p. 576)
- Describe the characteristics of hose bridges. (**NFPA 1001: 4.3.15**, p. 576)
- Describe the characteristics of hose clamps. (**NFPA 1001: 4.3.10**, p. 576)
- Describe the types of valves used to control water in pipes or hose lines. (**NFPA 1001: 4.3.15**, pp. 576–578)

- Describe the different types of master stream appliances. (pp. 578–579)
- Discuss the differences between smooth-bore nozzles and fog-stream nozzles. (**NFPA 1001: 4.3.10**, pp. 579–583)

SKILLS OBJECTIVES

After studying this chapter, you will be able to perform the following skills:

- Replace the swivel gasket on a fire hose. (**NFPA 1001: 4.5.2**, p. 555)
- Perform the one-fire fighter foot-tilt method of coupling a fire hose. (**NFPA 1001: 4.3.10**, p. 557)
- Perform the two-fire fighter method of coupling a fire hose. (**NFPA 1001: 4.3.10**, p. 558)
- Perform the one-fire fighter knee-press method of uncoupling a fire hose. (**NFPA 1001: 4.3.10**, p. 559)
- Perform the two-fire fighter stiff-arm method of uncoupling a fire hose. (**NFPA 1001: 4.3.10**, p. 560)
- Uncouple a hose with a spanner wrench (hose key). (**NFPA 1001: 4.3.10**, pp. 560–561)
- Clean and maintain hose. (**NFPA 1001: 4.5.2**, pp. 564–565)
- Mark a defective hose. (**NFPA 1001: 4.5.2**, p. 567)
- Perform a straight hose roll. (**NFPA 1001: 4.5.2**, p. 568)
- Perform a single-doughnut hose roll. (**NFPA 1001: 4.5.2**, pp. 569–570)
- Perform a twin-doughnut hose roll. (**NFPA 1001: 4.5.2**, p. 571)
- Perform a self-locking twin-doughnut hose roll. (**NFPA 1001: 4.5.2**, pp. 572–573)
- Use a hose jacket. (**NFPA 1001: 4.3.10**, p. 576)
- Open and close nozzles slowly to prevent water hammer. (**NFPA 1001: 4.3.10**, p. 577)
- Operate a smooth-bore nozzle. (**NFPA 1001: 4.3.10**, pp. 580–581)
- Operate a fog-stream nozzle. (**NFPA 1001: 4.3.10**, pp. 582–583)

ADDITIONAL STANDARDS

- **NFPA 24**, *Standard for the Installation of Private Fire Service Mains and Their Appurtenances*
- **NFPA 1142**, *Standard on Water Supplies for Suburban and Rural Fire Fighting*
- **NFPA 1961**, *Standard on Fire Hose*
- **NFPA 1962**, *Standard for the Care, Use, Inspection, Service Testing, and Replacement of Fire Hose, Couplings, Nozzles, and Fire Hose Appliances*
- **NFPA 1963**, *Standard for Fire Hose Connections*
- **NFPA 1964**, *Standard for Spray Nozzles*
- **NFPA 1965**, *Standard for Fire Hose Appliances*

Fire Alarm

Shortly after beginning fire training you are assigned to ride along with a pump company for a shift. After a rather uneventful day, your company is dispatched for a working fire at a warehouse. When you arrive there are flames coming from the front of the building. You hear the incident commander's instruction to lay a 102-mm (4-in.) supply line and to prepare for an attack with your deck gun.

1. Why is it important to understand the function and proper uses of each type of hose carried on your apparatus?

2. What are the advantages of using a master stream appliance on this fire?

Access Navigate for more practice activities.

Introduction

One of the primary functions of firefighting is to quickly and effectively apply sufficient quantities of water for fire control and extinguishment. In order to achieve this objective it is necessary to understand the tools that enable you to complete this task. This chapter discusses fire hose, appliances, and nozzles. It explains the functions, sizes, and construction of fire hose and hose couplings. It covers the maintenance and inspection of hose and describes some simple hose rolls. Hose appliances are used in conjunction with fire hose to expand the versatility of hose. This chapter describes the function and use of many hose appliances, including wyes, water thieves, Siamese connections, adaptors, reducers, hose jackets, hose rollers, hose bridges, hose clamps, and some valves. This chapter also explains the difference between smooth-bore and fog-stream nozzles. Special purpose nozzles, such as cellar nozzles and water curtain nozzles, are also described.

Fire Hose

Fire hose are used for two main purposes: as supply hose and as attack hose. **Supply hose** (or supply lines) are used to deliver water to an attack pump, aerial device, or tanker from a pressurized source such as a fire hydrant or from a water supply pump that may be operating from a fire hydrant or from a static (unpressurized) water source. Supply hose are designed to carry larger volumes of water at lower pressures than attack hose.

    Attack hose (or attack lines) are used to discharge water from an attack pump onto the fire. Most attack hose carry water directly from the attack pump to a nozzle that is used to direct the water onto the fire. In some cases, an attack hose is attached to a master stream appliance, which can be a truck-mounted deck gun, an aerial device, or a portable master stream appliance.

    Supply hose and attack hose are discussed in more detail later. NFPA 1961, *Standard on Fire Hose*, defines the specifications, inspection, and testing for the design and construction of new fire hose.

Sizes of Hose

Fire hose range in size from 19 mm to 152 mm (¾ in. to 6 in.) in diameter (**FIGURE 15-1**). The nominal hose size refers to the inside diameter of the hose when it is filled with water.

38-mm (1½-Inch) and 45-mm (1¾-Inch) Attack Hose

Most fire departments use either 38-mm (1½-in.) or 45-mm (1¾-in.) hose as the primary attack hose for

FIGURE 15-1 Fire hose comes in a wide range of sizes for different uses and situations.
© Jones & Bartlett Learning. Photographed by Glen E. Ellman.

most fires. Both sizes use the same 38-mm (1½-in.) couplings. This hose is the primary hose used in training due to its size and maneuverability. Hose of this size can usually be operated by one fire fighter, although a second fire fighter usually assists with line advancement. Initial attack lines are usually preconnected for immediate deployment and range in length from 46 to 107 m (150 to 350 ft).

    The primary difference between 38-mm (1½-in.) and 45-mm (1¾-in.) hose is the amount of water that can flow though the hose. Depending on the pressure in the hose and the type of nozzle used, a 38-mm (1½-in.) hose can generally flow between 189 and 473 L/min (50 and 125 gpm). An equivalent 45-mm (1¾-in.) hose can flow between 454 and 681 L/min (120 and 180 gpm). A 45-mm (1¾-in.) hose can deliver much more water and is only slightly heavier and more difficult to advance than a 38-mm (1½-in.) hose.

Booster Hose. A **booster hose (booster line)** is 25-mm (1-in.) diameter hose usually carried on a hose reel that holds 45 m or 61 m (150 ft or 200 ft) of hard rubber hose. Booster hose contains a reinforced material that gives it a rigid shape. This rigidity allows the hose to flow water without pulling all of the hose off the reel. Booster hose is light in weight and can be advanced quickly by one person.

    The disadvantage of booster hose is its limited flow. The normal flow from a 25-mm (1-in.) booster hose is in the range of 151 to 189 L/min (40 to 50 gpm). As a consequence, the use of booster hose is typically limited to small outdoor fires and trash dumpster fires. This type of hose should not be used for structural or vehicle firefighting, unless you are specifically trained to use it for these circumstances.

Forestry Fire Hose. A lightweight, collapsible, 19-mm (¾-in.), 25-mm (1-in.), or 38-mm (1½-in.) diameter hose, known as **forestry fire hose**, is often used to fight wildland and ground cover fires. Large volumes of water are not usually required to control wildland fires. The single jacket construction of forestry fire hose makes it light to carry and easy to maneuver in rough terrain comprised of brush, trees, rocks, and slopes. The hose usually comes in 15 m (50 ft) rolls that are connected together to stretch long distances. This type of hose can also be found in the hose cabinets of Type I buildings.

65-mm (2½-Inch) to 76-mm (3-Inch) Hose

The 65-mm (2½-in.) hose can be used as either attack hose or supply hose, but it is most often used as attack hose. A 65-mm (2½-in.) hose is used as an attack hose for fires that are too large to be controlled by a 38-mm (1½-in.) or 45-mm (1¾-in.) hose. A 65-mm (2½-in.) hose is generally considered to deliver a flow of approximately 946 L/min (250 gpm). It takes at least two fire fighters to safely control a 65-mm (2½-in.) hose due to the weight of the hose, the water, and the nozzle reaction force. A 15-m (50-ft) length of dry 65-mm (2½-in.) hose weighs about 14 kg (30 lb). A charged 65-mm (2 ½-in.), 31-m (100-ft) hose contains approximately 45 kg (100 lb) of water. A 65-mm (2½-in.) hose is most often used in large structural fire situations with large volumes of fire. These lines can be used in both defensive and offensive fire control operations.

While not as prevalent as in the past, the 76-mm (3-in.) hose can be used to deliver water to portable monitors, deck guns, and aerial master stream appliances. These hose sizes usually come in 15-m (50-ft) lengths.

> ### FIRE FIGHTER TIP
>
> Never use LDH to back up either a sprinkler or standpipe system by attaching the LDH hose directly to a Fire Department Connection (FDC). The FDC is not designed to accommodate a 65-mm (2½-in.) hose connection. The weight of the hose, combined with the pressure, will likely result in damaging the inlet connection or breaking it off completely.

Large-Diameter Hose

Large-diameter hose (LDH) has a diameter of 88 mm (3½ in.) or more. Standard LDH sizes include 102 mm (4 in.) and 127 mm (5 in.). LDH is widely used throughout various Canadian fire services. While most LDH is constructed as supply hose, some fire departments use

special LDH that can withstand higher pressures. The largest LDH size is 152 mm (6 in.) in diameter. Standard lengths of either 15 m (50 ft) or 31 m (100 ft) are available for LDH. These commonly produce a water flow between 1325 L/min (350 gpm) and 5678 L/min (1500 gpm). As with the 76-mm (3-in.) hose described earlier, LDH is also used to deliver water to master stream appliances.

> ### FIRE FIGHTER TIP
>
> Small-diameter hose (SDH) are used as attack lines; large-diameter hose (LDH) are almost always used as supply lines. Medium-diameter hose (MDH) can be used as either attack lines or supply lines.

Hose Construction

Most fire hose is constructed with an inner waterproof liner surrounded by either one or two outer layers or reinforcements. Regardless of the number of layers, the outer layer is commonly referred to as the jacket. The jacket is a woven mesh made from high-strength synthetic fibres (such as nylon) that are resistant to high temperatures, mildew, and many chemicals. These fibres can also withstand some mechanical abrasion.

A **single-jacket** hose is constructed with one layer of woven fibre. A **multiple-jacket** hose is constructed with two or more layers of woven fibres. In a multiple-jacket hose with two layers (**double-jacket hose**), the outer layer is bonded to the inner woven layer during a heating process. Without the outer layer, the rubber inner layer would simply expand with the introduction of pressurized water, much like a balloon expands when filled with air.

Some double-jacket fire hose are constructed with a durable rubber-like compound on the outer layer or jacket. This material is bonded to a single layer of strong woven fibres. This type of construction is called **rubber-covered hose** (**FIGURE 15-2**). In a rubber-covered hose, the inner layer and outer jacket are usually bonded together, and the woven fibres are contained within these layers.

Both types of hose are designed to be stored flat and to fold easily when there is no water inside the hose. This allows a much greater length of hose to be stored in the hose compartments on a fire apparatus.

The **hose liner** is the inner part or layer of the hose (**FIGURE 15-3**). This liner prevents water from leaking out of the hose and provides a smooth inside surface for water to move against. Without this smooth surface, excessive friction would arise between the moving water and the inside of the hose, reducing the amount of pressure that could reach the nozzle. The hose liner

FIGURE 15-2 Rubber-covered hose.
© Jones & Bartlett Learning. Photographed by Glen E. Ellman.

FIGURE 15-3 The hose liner inside a fire hose can be made from synthetic rubber or a variety of other membrane materials.
© Jones & Bartlett Learning.

is usually made of a synthetic rubber compound or a thin flexible membrane material that can be flexed and folded without developing leaks.

Hose Couplings

Hose couplings are used to connect, or couple, individual lengths of fire hose together. They are also used to connect a hose to a fire hydrant, to an inlet and outlet connection on a pumper, or to a variety of nozzles, fittings, and appliances. A coupling is permanently attached to each end of a section of fire hose. NFPA 1963, *Standard for Fire Hose Connections*, defines the performance requirements for hose couplings and adaptors, as well as the specifications for the connections of the couplings and adaptors.

SAFETY TIP

Gloves should always be worn when handling hose. Metal shavings, glass shards, or other sharp objects may potentially become embedded in the fibres of the hose. These objects could easily sever a muscle or tendon in a bare hand—and end a career.

The two most common types of supply hose and attack hose couplings are threaded hose couplings and Storz-type (nonthreaded) hose couplings.

Threaded Hose Couplings

Threaded hose couplings are generally used on hose up to 76 mm (3 in.) in diameter. A length of fire hose has a male hose coupling with threads on the outside on one end of the hose and a female hose coupling with threads on the inside on the other end of the hose (**FIGURE 15-4**). The female hose coupling has a swivel, enabling a female end to be secured around a male end without twisting the hose. Because a length of fire hose has a male and a female coupling, the use of double male and double female adaptors is necessary in some situations.

Male hose couplings are manufactured in a single piece, whereas female hose couplings are manufactured in two pieces. Occasionally, when fire fighters need to attach a smaller hose to a larger hose or when they need to connect two hose that possess different threads, a reducer or adaptor can be added to both the male and female hose couplings. This arrangement is called a five-piece coupling set. Reducers and adaptors are discussed later. Most fire departments use standardized hose threads, which allows fire hose from different departments to be connected together. The use of standardized hose threads is not universal among all jurisdictions, however, so most fire departments carry special adaptors on their apparatus for use during mutual aid firefighting operations. For example, the provinces of Ontario and Quebec adhere to different thread standards for fire hydrants.

When connecting fire hose with threaded hose couplings, the threads must be properly aligned so that the male and female hose couplings will engage fully with minimal resistance. Threaded hose couplings provide a secure connection between two sections of hose when properly coupled. These couplings are not likely to become disconnected during proper use. The swivel

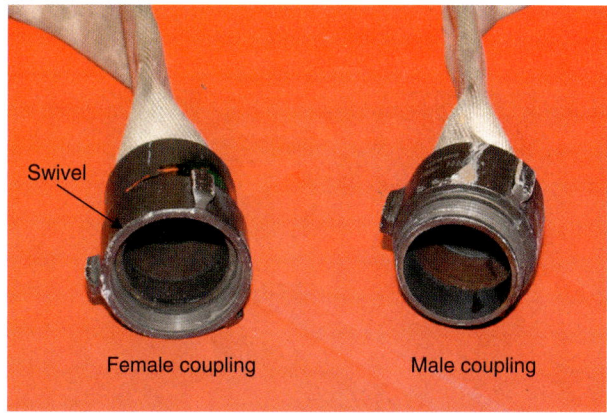

Swivel

Female coupling   Male coupling

FIGURE 15-4 A set of threaded hose couplings includes one male coupling and one female coupling. The female coupling has threads on the inside of the coupling and a swivel. The male coupling has exposed threads on the outside of the coupling.
© Jones & Bartlett Learning. Photographed by Glen E. Ellman.

on the female hose coupling should be turned until the connection is snug, but only hand tight, so that the hose couplings can be easily disconnected.

A disadvantage of threaded hose couplings is they are prone to cross threading, which can result in leakage and possible separation. If any leakage occurs after the hose is filled with water, further tightening may be needed. You can use a **spanner wrench (hose key)**, a rotating tool used to couple (connect) and uncouple (disconnect) hose couplings, to gently tighten the couplings until the leakage is stopped (**FIGURE 15-5**). In particular, you may need to use a spanner wrench (hose key) to uncouple the hose after it has been pressurized with water. Normally, two spanner wrenches (hose keys) are used together to rotate the two hose couplings in opposing directions to couple or uncouple the hose couplings.

Threaded hose couplings are constructed with lugs (extensions or indentations) that provide leverage to aid in the coupling and uncoupling of hose couplings. Three types of lugs are used: pin lugs, rocker lugs, and recessed lugs (**FIGURE 15-6**). Pin lugs look like small cylinders that extend outward from the hose coupling. They are rarely found on fire hose today because the pins tend to snag as the hose is being pulled over rough surfaces. Instead, pin lugs have largely been replaced by rocker lugs. **Rocker lugs** (or rocker pins) have rectangular-shaped extensions. The edges of rocker lugs are beveled to prevent them from catching on objects as the hose is dragged across a surface. Recessed lugs have circular indentations and require a specially designed spanner wrench (hose key) to engage them. These lugs are usually found on the couplings for 19-mm (¾-in.) or 25-mm (1-in.) booster hose.

A **Higbee indicator** is a notch or cut on the outside of one of the lugs that indicates the position of the first thread on a coupling. Aligning the Higbee indicators

FIGURE 15-6 Lugs are extensions or indentations that provide leverage to aid in the connection and disconnection of hose couplings.
© Jones & Bartlett Learning. Photographed by Glen E. Ellman.

on two couplings helps determine exactly where the first thread is on a pair of couplings and will help you to couple hose more quickly. When the indicators on the male and female couplings are aligned, the two couplings should connect quickly and easily (**FIGURE 15-7**).

An important part of a threaded hose coupling is the rubber swivel gasket. The swivel gasket is an O-shaped piece of rubber that sits inside the swivel section of the female hose coupling. When a male hose coupling is tightened against it, the swivel gasket forms a seal that stops water from leaking. If the swivel gasket is damaged or missing, the hose coupling will leak. Swivel gaskets can deteriorate with time. Also, using a wrench to tighten couplings on an empty hose or to overtighten couplings

FIGURE 15-5 Spanner wrenches (hose keys) are used to couple or uncouple hose couplings.
© Jones & Bartlett Learning. Photographed by Glen E. Ellman.

FIGURE 15-7 Higbee indicators are notches or cuts that show the position where the threads on a pair of couplings properly align with each other. The notches or cuts are especially helpful when the threads are not visible, such as at night.
© Jones & Bartlett Learning. Photographed by Glen E. Ellman.

on a filled hose can damage the swivel gaskets and cause them to leak. The swivel gaskets must be changed periodically as part of hose maintenance.

Although a leaking hose coupling is not a critical problem during most firefighting operations, it can result in unnecessary water damage. During cold weather, a leaking hose coupling can cause ice to form and create a significant safety hazard. The best way to prevent leaks is to make sure the swivel gaskets are in good condition and to replace any swivel gaskets that are missing or damaged. To replace the swivel gasket in a female hose coupling, follow the steps in **SKILL DRILL 15-1**.

SKILL DRILL 15-1
Replacing the Swivel Gasket Fire Fighter I, NFPA 1001: 4.5.2

1 Fold the new swivel gasket in half by bringing the thumb and the forefinger together to create two loops.

2 Place either of the two loops inside the hose coupling, and position it against the gasket seat.

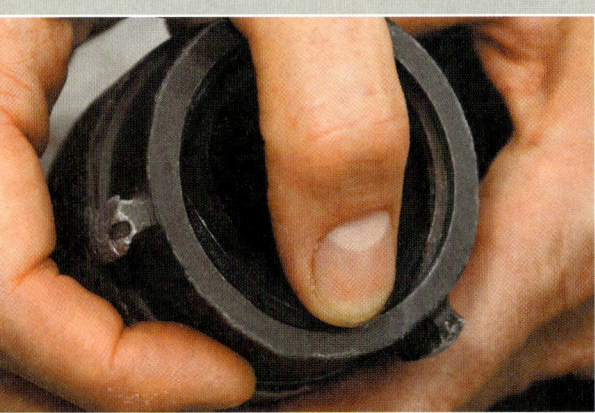

3 Using the thumb, push the remaining unseated portions of the swivel gasket into the hose coupling until the entire swivel gasket is properly positioned against the gasket seat inside the coupling.

Storz-Type (Nonthreaded) Hose Couplings

Storz-type (nonthreaded) hose couplings are designed so that the couplings on both ends of a length of hose are the same. In other words, there is no male or female end to the hose. When this system is used, each coupling can be attached to any other coupling of the same diameter (**FIGURE 15-8**). Storz-type hose couplings are made for all hose sizes; however, in North America they are most often used on LDH.

Since there are no threads to make a connection, Storz-type hose couplings are connected by mating the two couplings face-to-face and then turning them clockwise into a locking position. To disconnect a set of Storz-type hose couplings, the two parts are rotated counterclockwise until they release. They may require spanner wrenches (hose keys) for complete coupling and uncoupling. Some fire hydrant manufacturers place Storz-type hose connections on fire hydrants—a practice that allows fire departments to make hydrant connections rapidly.

If fire hydrants are not equipped with Storz-type hose couplings, connection to a fire hydrant will require the attachment of an adaptor in order to connect the Storz-type hose coupling on the hose to the fire hydrant. These couplings are more prone to accidental disconnect if they are not completely coupled. Some Storz-type hose couplings have a small pin or lever that must be released before uncoupling the hose—a design that prevents accidental uncoupling of the hose.

Adaptors are used to connect Storz-type couplings to threaded couplings or to connect couplings of different sizes together. Many fire departments use LDH with Storz-type hose couplings as a supply line between a fire hydrant and a pumper.

Storz-type hose couplings can also be found on fire department connections (FDCs). FDCs are fire hose connections that fire departments use to pump water into a standpipe or sprinkler system.

Coupling and Uncoupling Hose

Several techniques may be used for connecting and disconnecting hose couplings. Depending on the circumstances, one technique may be more effective than another. A fire fighter should learn how to perform each skill. To perform the one-fire fighter foot-tilt method of coupling fire hose, follow the steps in **SKILL DRILL 15-2**.

To perform the two-fire fighter method for coupling a fire hose, follow the steps in **SKILL DRILL 15-3**.

Charged hose lines should never be disconnected while the hose is under pressure. Forcing the separation of couplings under pressure can result in serious injury and cause a hose to flail about in an uncontrolled manner. Prior to decoupling a hose line, verify that the hose you wish to disconnect has been relieved of pressure. To verify that the pressure has been released, gently crack open the bale on the nozzle. You should see and feel a reduction in both water flow and back pressure. The hose should collapse and become limp if there is no pressure in the line. Ideally, the hose should be disconnected from the pump prior to breaking the line. This will prevent unexpected accidental water flow and line pressurization. If the coupling resists an attempt to uncouple it, check to make sure the pressure is relieved before using spanner wrenches (hose keys) to loosen the coupling.

To perform the one-fire fighter knee-press method of uncoupling a fire hose, follow the steps in **SKILL DRILL 15-4**.

To perform the two-fire fighter stiff-arm method of uncoupling a hose, follow the steps in **SKILL DRILL 15-5**.

To uncouple a hose with spanner wrenches (hose keys), follow the steps in **SKILL DRILL 15-6**.

SAFETY TIP

Never attempt to uncouple a charged hose line.

Supply Hose

As previously discussed, supply hose is used to deliver water to an attack pump from a pressurized source, such as a fire hydrant or from a water supply pump that may be operating from a hydrant or from a static

FIGURE 15-8 Storz-type hose couplings are designed so that the couplings on both ends of a length of hose are the same.

© Jones & Bartlett Learning. Photographed by Glen E. Ellman.

SKILL DRILL 15-2
Performing the One-Fire Fighter Foot-Tilt Method of Coupling a Fire Hose
Fire Fighter I, NFPA 1001: 4.3.10

1 Place one foot on the hose behind the male coupling. Push down with your foot to tilt the male coupling upward.

2 Place one hand behind the female coupling, and grasp the hose.

3 Place the other hand on the swivel of the female coupling. Bring the two couplings together, and align the Higbee indicators. Turn the female coupling counterclockwise until it clicks, which indicates that the threads are aligned. Rotate the swivel in a clockwise direction to connect the hose.

SKILL DRILL 15-3
Performing the Two-Fire Fighter Method of Coupling a Fire Hose
Fire Fighter I, NFPA 1001: 4.3.10

1 Pick up the male coupling. Grasp it directly behind the coupling, and hold it tightly against the body.

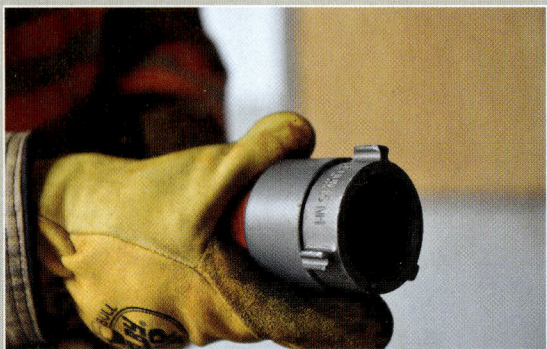

2 The second fire fighter holds the female coupling firmly with both hands.

3 The second fire fighter brings the female coupling to the male coupling and aligns the female coupling with the male coupling, using the Higbee indicators for easy alignment.

4 The second fire fighter turns the female coupling counterclockwise until it clicks, which indicates that the threads are aligned, and then turns the female coupling swivel clockwise to couple the hose.

© Jones & Bartlett Learning. Photographed by Glen E. Ellman.

SKILL DRILL 15-4
Performing the One-Fire Fighter Knee-Press Method of Uncoupling a Fire Hose Fire Fighter I, NFPA 1001: 4.3.10

1 Pick up the female coupling.

2 Turn the connection upright, resting the male coupling on a firm surface.

3 Place a knee on the female coupling, and press down on it with your body weight (which compresses the swivel gasket). Turn the female coupling swivel counterclockwise, and loosen the coupling.

SKILL DRILL 15-5
Performing the Two-Fire Fighter Stiff-Arm Method of Uncoupling a Fire Hose Fire Fighter I, NFPA 1001: 4.3.10

© Jones & Bartlett Learning. Photographed by Glen E. Ellman.

1 Two fire fighters face each other and firmly grasp their respective coupling.

2 With elbows locked straight, they push toward each other, compressing the swivel gasket.

3 While pushing toward each other, the fire fighters turn the coupling counterclockwise, loosening the coupling.

SKILL DRILL 15-6
Uncoupling a Hose with Spanner Wrenches (Hose Keys) Fire Fighter I, NFPA 1001: 4.3.10

1 With the connection on the ground, straddle the connection above the female coupling.

2 Place one spanner wrench (hose key) on the swivel of the female coupling, with the handle of the spanner wrench (hose key) to the left.

SKILL DRILL 15-6 Continued
Uncoupling a Hose with Spanner Wrenches (Hose Keys) Fire Fighter I, NFPA 1001: 4.3.10

3 Place the second spanner wrench (hose key) on the male coupling, with the handle of the spanner wrench (hose key) to the right.

4 Push both spanner wrench (hose key) handles down toward the ground, loosening the connection.

© Jones & Bartlett Learning. Photographed by Glen E. Ellman.

water source. Supply hose are designed to carry larger volumes of water at lower pressures than attack hose. They are 65 mm (2½ in.), 76 mm (3 in.), 102 mm (4 in.), 127 mm (5 in.), and 152 mm (6 in.) in diameter. The choice of diameter is based on the preferences and operating requirements of each fire department. It also depends on the amount of water needed to supply the attack pump, the distance from the source to the attack pump, and the pressure that is available at the source.

Fire department pumpers are normally loaded with at least one bed of hose that can be laid out as a supply hose. When threaded couplings are used, this hose can be laid out from the fire hydrant or water supply pump to the fire (known as a forward hose lay) or from the fire to the hydrant or water supply pump (known as a reverse hose lay). Sometimes pumpers are loaded with two beds of hose so they can easily drop a supply hose in either direction. If Storz-type couplings are used or the necessary adaptors are provided, hose from the same bed can be laid in either direction.

A 65-mm (2½-in.) hose may be used as either a supply hose or an attack hose. Such hose has a limited flow capacity as supply hose, but it can be effective at low to moderate flow rates and over short distances. Sometimes two parallel lines of 65-mm (2½-in.) hose are used to provide a more effective water supply.

LDH is much more efficient than 65-mm (2½-in.) hose for moving larger volumes of water over longer distances. Given this fact, many fire departments use 102-mm (4-in.) or 127-mm (5-in.) hose as their standard supply hose. A single 127-mm (5-in.) supply hose can deliver flows exceeding 5678 L/min (1500 gpm) under some conditions. LDH is heavy and difficult to move after it has been charged with water, however. This hose comes in 15-m (50-ft) and 31-m (100-ft) lengths. The NFPA recommends that a fire pumper carry at least 244 m (800 ft) of 65-mm (2½-in.) or larger supply hose (NFPA 1901, 2016). A typical pumper may carry anywhere from 244 m (800 ft) to 381 m (1250 ft) of supply hose.

Supply hose must be tested annually at a pressure of at least 1379 kilopascals [kPa] (200 psi) or at a pressure not to exceed the service test pressure marked on the hose.

Suction Hose

Suction hose is used to supply water to the suction (or intake) side of the fire pump. There are two different types of suction hose, soft sleeve hose and hard suction hose. Soft sleeve hose is used to transport water from a pressurized source, such as a fire hydrant, to the suction side of the fire pump. Hard suction hose is designed to draft water from a static water source, such as a pond, river, or portable tank, to the suction side of a fire pump. Each type of hose is more fully described next.

A soft sleeve hose (also known historically as a soft suction hose) is a short section of large-diameter supply hose that is used to provide water from the large steamer outlet (the large-diameter port) on a fire hydrant or other pressurized water source to the suction side of the fire pump (**FIGURE 15-9**). The soft sleeve hose is used to allow as much water as possible to flow from the water source to the suction side of the fire pump through a single line. Soft sleeve hose may be as small as 65 mm (2½ in.) or as large as 152 mm (6 in.) in diameter. A soft sleeve hose may have a Storz or similar connection on both ends, or it may have threaded female connections on each end, with one end matching the local fire hydrant threads and the other end matching the threads on the suction side of the fire pump. If it has two Storz connections, an adaptor is required to connect the large diameter inlet to the pumper and the fire hydrant. The couplings have large handles to allow for quick tightening by hand. The hose is usually between 3 m (10 ft) and 7.6 m (25 ft) in length.

A hard suction hose is a short section of rigid hose that is primarily used to draft water from a static source such as a river, lake, or portable drafting basin to the suction side of the fire pump on a fire department pumper or into a portable pump (**FIGURE 15-10**). It can also be used to carry water from a fire hydrant to the pumper.

Hard suction hose normally comes in 3-m (10-ft) or 6-m (20-ft) sections. The diameter is based on the

FIGURE 15-10 Hard suction hose.
© Jones & Bartlett Learning. Photographed by Glen E. Ellman.

capacity of the pump but can be as large as 152 mm (6 in.). Hard suction hose can be made from either rubber or plastic; however, the plastic versions are much lighter and more flexible.

Long handles are provided on the female couplings of hard suction hose to assist in tightening the hose. To draft water, it is essential to have an airtight connection at each coupling. Sometimes it may be necessary to tap the handles on the female couplings with a rubber mallet to tighten the hose or to disconnect it. Tapping these handles with anything metal, however, could cause damage to the handles or the coupling. To create an airtight seal between couplings, it is important to have a flexible gasket to facilitate the seal. These gaskets need to be inspected and replaced regularly. Pump operators should have spare gaskets on hand for this purpose.

Attack Hose

Attack hose is designed to be used for fire suppression and control. This hose carries the water from the attack pump to the fire or to a FDC or from a standpipe system to the fire. The common diameters for attack hose are 38-mm (1½-in.) or 45-mm (1¾-in.) lines, 25-mm (1-in.) booster lines, and 25-mm (1-in.) or 38-mm (1½-in.) forestry fire hose (**FIGURE 15-11**). Some fire departments also use 65-mm (2½-in.) attack hose. Each section of attack hose is usually 15 m (50 ft) long.

Attack hose must withstand high pressure and is designed to be used during fire suppression where it can be subjected to high temperatures, sharp surfaces, abrasion, and other potentially damaging conditions. For this reason, attack hose must be tough yet flexible and light in weight. Attack hose must be tested annually at a pressure of at least 2068 kPa (300 psi) and is intended to be used at pressures up to 1896 kPa (275 psi).

Attack hose can be either multiple-jacket or rubber-covered construction.

FIGURE 15-9 Soft sleeve hose.
© Jones & Bartlett Learning. Photographed by Glen E. Ellman.

FIGURE 15-11 Attack hose are used during fire suppression.
© Jones & Bartlett Learning. Photographed by Glen E. Ellman.

Hose Care, Maintenance, and Inspection

Fire hose should be regularly inspected and tested following the procedures in NFPA 1962, *Standard for the Care, Use, Inspection, Service Testing, and Replacement of Fire Hose, Couplings, Nozzles, and Fire Hose Appliances.* Hose that are not properly maintained can deteriorate over time and eventually burst. In addition, the swivel gaskets in female couplings need to be checked regularly and replaced when they are worn or damaged.

Causes and Prevention of Hose Damage

Fire hose is a life line for fire fighters. Every time fire fighters respond to a fire, they rely on fire hose to deliver the water needed to attack the fire and protect themselves from it. Fire hose is a highly engineered product designed to perform well under adverse conditions. Fire fighters must be careful to prevent damage to the hose that could result in premature or unexpected failure. The factors that most commonly damage fire hose include mechanical causes, heat, cold, UV radiation, chemicals, and mildew.

Mechanical Damage

Mechanical damage can occur from many sources. Hose that is dragged over rough objects or along a roadway can be damaged by abrasion, for example. Broken glass

and sharp objects can cut through the hose. Laying hose lines through broken windows and glass can cause a puncture in the line. Ensure that all the glass is cleared from the window opening. Particles of grit caught in the fibres can damage the jacket or puncture holes in the liner. Reloading dirty hose can cause damage to the fibres in the jacket of the hose.

Fire hose is likely to be damaged if it is run over by a vehicle. For this reason, hose ramps or bridges should be used if traffic must drive over a hose that is in the roadway. Hose couplings can be damaged by dropping them on the ground. In particular, the exposed threads on male couplings are easily damaged if they are dropped. Avoid dragging hose couplings, as this practice can cause damage to the threads and to the swivels.

Heat and Cold

Hose can be damaged by heat and cold as well as by prolonged exposure to sunlight. Heat is an obvious concern when fighting a fire. A hose that is directly exposed to a fire can burn through and burst quickly. Burning embers and hot coals can also damage the hose, causing small leaks or weakening the hose so that it is likely to burst under pressure. Always visually inspect any hose that has come in direct contact with a fire.

Avoid storing a hose in places where it will come in contact with hot surfaces, such as a heating unit or the exhaust pipe on a vehicle. If the apparatus is parked outside, use a hose cover to protect the hose from sunlight. It is important to keep the hose from becoming brittle.

In cold weather, freezing is a threat to hose. Freezing can rupture the hose liner and break fibres in the jacket. When fire fighters are working in below-freezing temperatures, water should be kept flowing through the hose to prevent freezing. If a line must be shut down temporarily, the nozzle should be left partly open to keep the water moving, and the stream should be directed to a location where it will not cause additional water damage. When a line is no longer needed, the hose should be drained and rolled before it freezes. In extremely cold weather, it may be difficult to roll hose. In these instances, hose can be piled. However, firefighters must be mindful not to bend the hose too much, as this may cause the frozen fibres within the jacket to wear or break.

Hose that is frozen or encased in ice often can be thawed out with a steam generator. Another option is to carefully chop the hose out using an axe, being careful not to cut the hose itself. The hose can be transported back to the fire station to thaw. Do not attempt to bend a section of frozen hose. In situations where the hose is frozen solid, it may be necessary to transport the hose back to the fire station on a flatbed truck.

Chemicals

Many chemicals can damage fire hose. Such chemicals may be encountered at incidents in facilities where chemicals are manufactured, stored, or used as well as in locations where their presence is not anticipated. Most vehicles contain a wide variety of chemicals that can damage fire hose, including battery acid, gasoline, diesel fuel, antifreeze, motor oil, and transmission fluid. Hose may come in contact with these chemicals at vehicle fires or at the scene of a collision where chemicals are spilled on the roadway. In particular, supply hose often come in contact with residues from these chemicals when lines are laid in the roadway. Fire fighters should remove chemicals from the hose as soon as possible and wash the hose with an approved detergent, thoroughly rinse it, and let it dry completely.

Mildew

Mildew is a type of fungus that can grow on fabrics and materials in warm, moist conditions. A fire hose that has been packed away while it is still wet and dirty is a natural breeding ground for mildew. This fungus feeds on nutrients found in many natural fibres, which can cause the fibres to rot and deteriorate. In the days when cotton fibres were used in hose jackets, mildew was a major problem. Hose had to be washed and completely dried after every use before it could be placed back on the apparatus.

Modern fire hose is made from synthetic fibres that are resistant to mildew, and most types can be repacked without drying. Nevertheless, mildew may still grow on exposed fibres if they are soiled with contaminants that will provide mildew with the necessary nutrients. The fibres in rubber-covered hose are protected from mildew.

SAFETY TIP

Whenever a fire hose has suffered possible damage, it should be thoroughly inspected and tested according to NFPA 1962, *Standard for the Care, Use, Inspection, Service Testing, and Replacement of Fire Hose, Couplings, Nozzles, and Fire Hose Appliances*, before it is returned to service.

Cleaning and Maintaining Hose

It is important to properly clean and maintain fire hose. Because hose can be made from different materials, the exact steps needed to clean and maintain it will vary. In general, try to prevent hose from coming in contact with petroleum and abrasive substances whenever possible. When hose becomes dirty it is important to clean it as soon as possible. Follow the manufacturer's recommendations. For a mild cleaning, cool water and a soft brush may be adequate. For dirty hose, it may be necessary to use a mild detergent, especially if the hose has come in contact with petroleum-based products.

To clean hose that is dirty or contaminated, follow the steps in **SKILL DRILL 15-7**.

SKILL DRILL 15-7
Cleaning and Maintaining Hose Fire Fighter I, NFPA 1001: 4.5.2

1 Lay the hose out flat. Rinse the hose with water.

SKILL DRILL 15-7 Continued
Cleaning and Maintaining Hose Fire Fighter I, NFPA 1001: 4.5.2

2 Gently scrub the hose with mild detergent, paying attention to soiled areas.

3 Turn over the hose, and repeat steps 1 and 2. Give a final rinse to the hose with water. Make sure the hose is dry before storing it.

© Jones & Bartlett Learning. Photographed by Glen E. Ellman.

FIRE FIGHTER TIP

Modern fire hose is made from synthetic fibres that are resistant to mildew, and most types can be repacked without drying. Always follow manufacturer recommendations for washing and drying fire hose.

Some fire departments use specially manufactured equipment for washing and drying fire hose. The equipment ranges from simple to complex (**FIGURE 15-12**).

More complex hose washers consist of a large cabinet-style mechanical hose washer. These may contain an automatic feed that moves hose through a power washing cycle and then squeegees some of the water off the hose before it leaves the washing machine. This type of machine enables one person to wash a large quantity of hose in a fairly short period of time. **FIGURE 15-13** shows one type of mechanical hose washer.

Though some fire hose can be replaced on fire apparatus while it is wet, it is preferable to dry the hose prior to reloading it onto an apparatus. Fire hose can be dried in several different ways. Some fire stations have angled racks with slats on which wet hose can be placed. The angled construction allows water to drain from the inside of the hose, and the slats allow the outside of the hose to dry. Other fire stations are

FIGURE 15-12 A simple hose washer.
© Jones & Bartlett Learning. Photographed by Glen E. Ellman.

FIGURE 15-13 A mechanical hose washer.
Courtesy of Circul-Air Corp.

FIGURE 15-14 A hose dryer.
Courtesy of Circul-Air Corp.

FIRE FIGHTER TIP

To prevent UV radiation damage, do not dry hose in sunlight.

built with hose towers. To dry hose in a hose tower, one end of the clean and wet hose is hoisted to the top of the tower. This allows water to drain from the inside of the hose and allows the outside of the hose to dry. A third method of drying hose is the use of a heated hose drying cabinet. These cabinets are built to hold wet hose that has been loosely folded or coiled. Once the wet hose is placed into the cabinet, heated air is circulated through the cabinet. Heated hose dryers may dry hose faster than other methods in some climates. They can be used in any season. **FIGURE 15-14** shows one type of heated hose dryer. Some heated hose dryers can also be used for drying personal protective equipment.

If your department has hose washing and drying equipment, you need to follow the manufacturer's instructions to be sure the hose is being properly cleaned. Some hose washing equipment will not completely clean the hose couplings. Hose couplings may require special attention to be sure they are clean and will operate easily and effectively under emergency conditions.

Hose Inspections

Visual hose inspections should be performed at least quarterly. A visual inspection should also be performed after each use, either while the hose is being cleaned and dried or when it is reloaded onto the apparatus. If any defects are found, that length of hose should be immediately removed from service and tagged with a description of the problem. Hose that has not been used in 30 days should be unpacked, inspected, cleaned, and reloaded. The appropriate notifications must be made to have the hose repaired.

To clearly mark a defective hose, follow the steps in **SKILL DRILL 15-8**.

Hose Testing and Records

Each length of hose should be tested at least annually, according to the procedures listed in NFPA 1962, *Standard for the Care, Use, Inspection, Service Testing, and Replacement of Fire Hose, Couplings, Nozzles, and Fire Hose Appliances*. The procedures are complicated and require special equipment. This equipment must be operated according to the manufacturer's instructions.

A hose record is a written history of each individual length of fire hose. Each length of hose should be identified with a unique number stenciled or painted on

SKILL DRILL 15-8
Marking a Defective Hose Fire Fighter I, NFPA 1001: 4.5.2

1 Inspect the hose for defects.

2 Upon finding a defect, mark the area on the hose, and remove the hose from service. Tag the hose as defective, provide a description of the defect, and notify your superiors.

© Jones & Bartlett Learning. Photographed by Glen E. Ellman.

it. A hose record will contain information such as hose size, type, manufacturer, date of manufacture, date of purchase, and testing dates. Hose testing and recording hose records are Fire Fighter II tasks. To perform an annual service test on a fire hose, record the results of the service test, and maintain records, see Chapter 23, *Advanced Fire Suppression*.

Hose Rolls

An efficient way to transport a single section of fire hose is in the form of a roll. Rolled hose is both compact and easy to manage. A fire hose can be rolled many different ways, depending on how it will be used. Follow the standard operating procedures (SOPs) of your department when rolling hose.

Straight or Storage Hose Roll

The straight or storage hose roll is a simple and frequently used hose roll. It is used for general handling and transportation of hose as well as for rack storage of hose (**FIGURE 15-15**). With this arrangement, the male coupling is at the center of the roll, and the female coupling is on the outside of the roll.

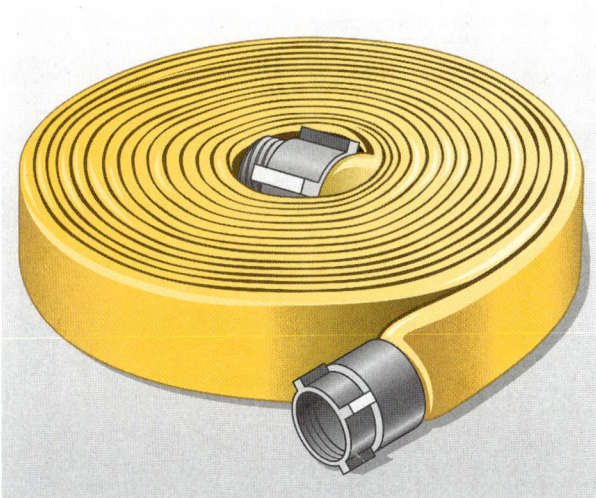

FIGURE 15-15 A straight or storage hose roll is used for transporting and storing hose. The exposed threads of the male coupling stay protected at the center of the roll.
© Jones & Bartlett Learning.

To perform a straight hose roll, follow the steps in **SKILL DRILL 15-9**.

SKILL DRILL 15-9
Performing a Straight or Storage Hose Roll Fire Fighter I,
NFPA 1001: 4.5.2

1 Lay the hose flat and in a straight line.

2 Fold the male coupling over on top of the hose.

3 Roll the hose to the female coupling.

4 Set the hose roll on its side, and tap any protruding hose flat with a foot. With this arrangement, the male coupling is at the center of the roll, and the female coupling is on the outside of the roll.

Single-Doughnut Hose Roll

The single-doughnut roll is used when the hose will be put into use directly from the rolled state. With this arrangement, and for easy access, both couplings are on the outside of the roll (**FIGURE 15-16**). The hose can be connected and extended by one fire fighter. The hose unrolls as it is extended.

To perform a single-doughnut roll, follow the steps in **SKILL DRILL 15-10**.

Twin-Doughnut Hose Roll

The twin-doughnut roll is used primarily to make a small compact roll that can be carried easily (**FIGURE 15-17**).

To perform a twin-doughnut hose roll, follow the steps in **SKILL DRILL 15-11**.

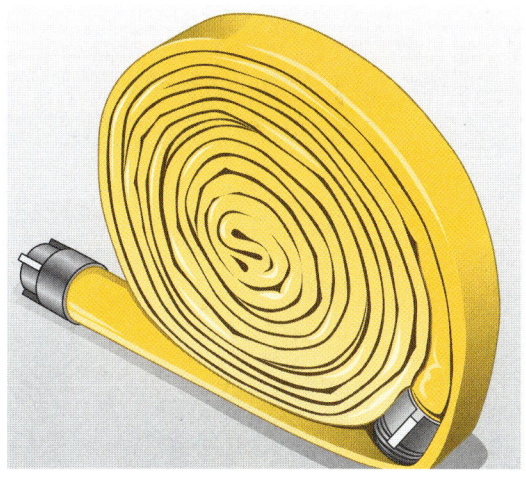

FIGURE 15-16 A doughnut hose roll is used when the hose will be put into use directly from the rolled state. Both couplings are easily accessible.
© Jones & Bartlett Learning.

SKILL DRILL 15-10
Performing a Single-Doughnut Hose Roll Fire Fighter I, NFPA 1001: 4.5.2

1 Lay the hose flat and in a straight line.

2 Locate the midpoint of the hose.

(continued)

SKILL DRILL 15-10 Continued
Performing a Single-Doughnut Hose Roll Fire Fighter I, NFPA 1001: 4.5.2

3 From the midpoint, move 1.5 m (5 ft) toward the male coupling end. From this point, start rolling the hose toward the female coupling.

4 At the end of the roll, wrap the excess hose of the female end over the male coupling to protect the threads on the male coupling. With this arrangement, and for easy access, both couplings are on the outside of the roll.

© Jones & Bartlett Learning. Photographed by Glen E. Ellman.

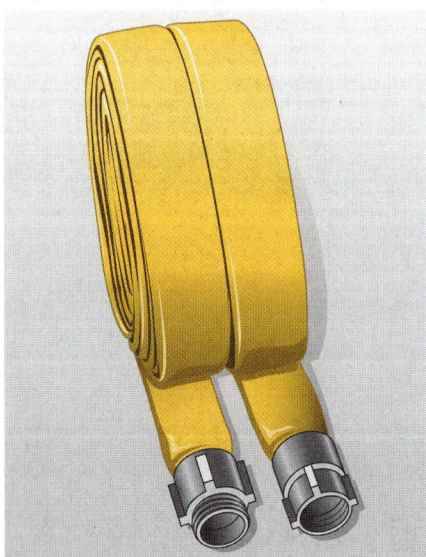

FIGURE 15-17 A twin-doughnut hose roll is used to make a small compact roll that can be carried easily.
© Jones & Bartlett Learning.

Self-Locking Twin-Doughnut Hose Roll

The self-locking twin-doughnut hose roll is similar to the twin-doughnut roll, except that it forms its own carry loop (**FIGURE 15-18**).

To perform a self-locking twin-doughnut roll, follow the steps in **SKILL DRILL 15-12**.

Hose Appliances

A hose appliance is any device used in conjunction with a fire hose for the purpose of delivering water, including wyes, water thieves, Siamese connections, adaptors, reducers, hose jackets, hose rollers, hose bridges, hose clamps, and some valves. It is important to learn how to use the hose appliances and tools provided by your fire department. In particular, you should understand

SKILL DRILL 15-11
Performing a Twin-Doughnut Hose Roll Fire Fighter I, NFPA 1001: 4.5.2

1 Lay the hose flat and in a straight line.

2 Bring the male coupling alongside the female coupling.

3 Fold the far end over, and roll both sections of hose toward the couplings, creating a double roll.

4 The roll can be carried by hand, by a rope, or by a hose strap.

© Jones & Bartlett Learning. Photographed by Glen E. Ellman.

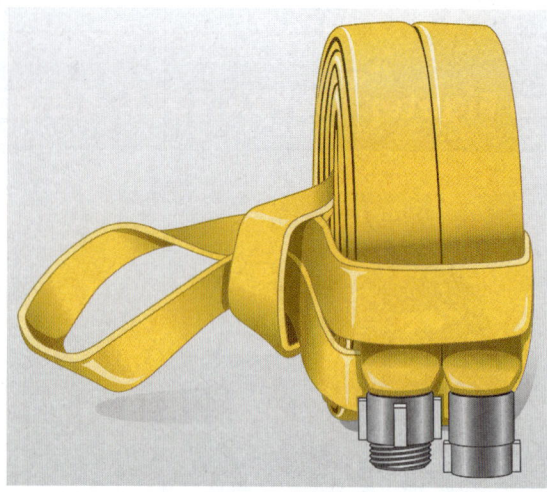

FIGURE 15-18 A self-locking twin-doughnut hose roll forms its own carry loop.
© Jones & Bartlett Learning.

the purpose of each device and be able to use each appliance correctly. Some hose appliances are used primarily with supply hose, whereas others are most often used with attack hose; many appliances have applications with both supply hose and attack hose. Both types are discussed in this section.

Wyes

A **wye** is a device that splits one hose into two or more separate lines. The word *wye* refers to a Y-shaped part or object. When threaded couplings are used, a wye has one female connection and two or more male connections. The wye that is most commonly used in the fire service splits one 65-mm (2½-in.) hose into two 38-mm (1½-in.) hose lines. This appliance is used primarily on attack hose.

SKILL DRILL 15-12
Performing a Self-Locking Twin-Doughnut Hose Roll Fire Fighter I, NFPA 1001: 4.5.2

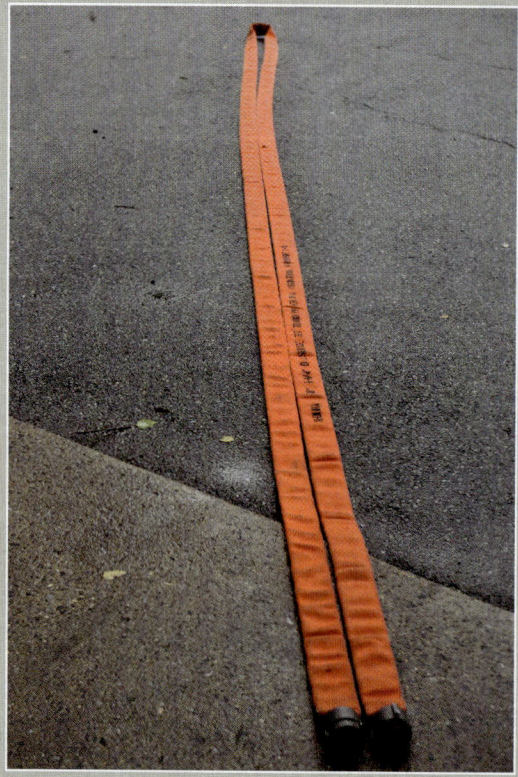

1 Lay the hose flat, and bring the couplings alongside each other.

2 Move one side of the hose over the other, creating a loop. This creates the carrying shoulder loop.

SKILL DRILL 15-12 Continued
Performing a Self-Locking Twin-Doughnut Hose Roll Fire Fighter I, NFPA 1001: 4.5.2

3 Bring the loop back toward the couplings to the point where the hose crosses.

4 From the point where the hose crosses, begin to roll the hose toward the couplings.

5 Position the loops so that one is larger than the other. Pass the larger loop over the couplings and through the smaller loop, which secures the rolls together and forms the shoulder loop.

6 The finished result is the self-locking twin-doughnut roll.

A **gated wye** is equipped with two quarter-turn ball valves so that the flow of water to each of the split lines or outlets can be controlled independently (**FIGURE 15-19**). A gated wye enables fire fighters to initially attach and operate one hose and then add a second hose later if necessary. The use of a gated wye avoids the need to shut down the hose supplying the wye to attach the second hose. Some gated wyes are used on LDH. These wyes have an inlet for LDH (102 mm [4 in.] or 127 mm [5 in.]) and several 65-mm (2½-in.) outlets.

Water Thief

A **water thief** is similar to a gated wye but includes a 65-mm (2½-in.) inlet, a 65-mm (2½-in.) outlet, and two 38-mm (1½-in.) outlets (**FIGURE 15-20**). It is used to supply many hose from a single supply line. A water thief is used to supply attack hose. Larger water thiefs are available and designed for use with LDH. The water that comes from a single 65-mm (2½-in.) inlet can be directed to two 38-mm (1½-in.) outlets or one 65-mm (2½-in.) outlet. Under most conditions, it is not possible to supply all three outlets at the same time because the capacity of the supply hose is limited.

A water thief can be placed near the entrance to a building to provide the water for an interior attack line or lines. One or two 38-mm (1½-in.) attack hose can be used in such a case. If necessary, they can then be shut down and a 65-mm (2½-in.) hose substituted for them. When using any appliance, you should follow your departments SOPs/SOGs and ensure you are well versed in their application.

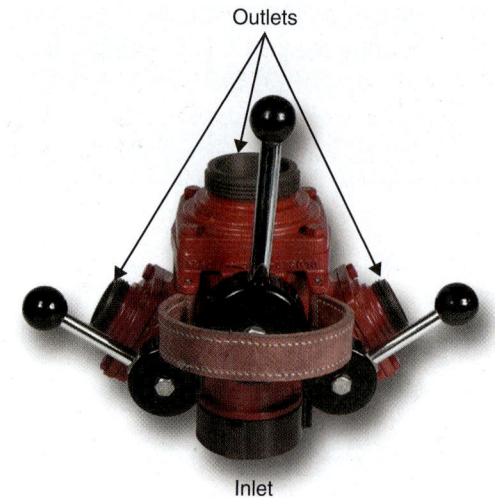

FIGURE 15-20 A water thief is similar to a gated wye, but it includes a 65-mm (2½-in.) inlet, a 65-mm (2½-in.) outlet, and two 38-mm (1½-in.) outlets.
Courtesy of Akron Brass Company.

Siamese Connection

A **Siamese connection** is a hose appliance that combines two or more hose lines into one. The most commonly used type of Siamese connection combines two 65-mm (2½-in.) hose lines into a single 65-mm (2½-in.) line (**FIGURE 15-21**). This scheme increases the flow of water on the outlet side of the Siamese connection. A Siamese connection that is used with threaded couplings has two female inlets and one male outlet. A Siamese connection may be equipped with quarter-turn ball valves or clapper valve mechanisms. Clapper mechanisms are discussed in more detail later. These connections can be used with supply lines and attack lines.

A Siamese connection can be attached to a pumper inlet to allow water supply from two different supply

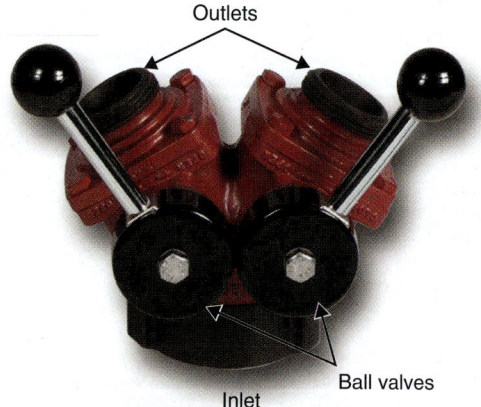

FIGURE 15-19 A wye splits one hose stream into two or more hose streams. On the gated wye, two quarter-turn ball valves allow the flow of water to each outlet to be controlled independently.
Courtesy of Akron Brass Company.

FIGURE 15-21 A typical Siamese connection combines two or more hose lines into one (two female inlets and a single male outlet).
Courtesy of Akron Brass Company.

lines. These kinds of connections are also used to supply master stream appliances and ladder pipes. Siamese connections are commonly installed on the FDCs that are used to supply water to standpipe and sprinkler systems in buildings. Fire fighters should review their department's policies on the correct procedures for supplying fire department Siamese connections.

Adaptors

Adaptors are devices that allow fire hose couplings to be safely interconnected with couplings of different sizes, threads, or mating surfacing or that allow fire hose couplings to be safely connected to other appliances. Dissimilar threads could be encountered when different fire departments are working together or in industrial settings where the hose threads of the building's equipment do not match the threads of the municipal fire department. Some private fire hydrants may have different threads from the municipal system and require an adaptor. Adaptors are also used to connect threaded couplings to Storz-type (unthreaded) couplings. They are useful for both supply hose and attack hose.

Adaptors can also be used when it is necessary to connect two female couplings or two male couplings. A **double-female adaptor** is used to join two male hose couplings. A **double-male adaptor** is used to join two female hose couplings. The use of double-male and double-female adaptors is not as prevalent today as in the years prior to the adoption of LDH supply lines. In the past, these adapters played a vital role in establishing reverse hose lays for water supply (**FIGURE 15-22**).

Reducers

A **reducer** is a type of adaptor used to attach a smaller-diameter hose to a larger-diameter hose

(**FIGURE 15-23**). Usually the larger end has a female connection and the smaller end has a male connection. One type of reducer is used to attach a 65-mm (2½-in.) hose to a 38-mm (1½-in.) hose. Many 65-mm (2½-in.) nozzles are constructed with a built-in reducer so that a 38-mm (1½-in.) hose can be attached for overhaul. Reducers are also used to attach a 65-mm (2½-in.) supply hose to a larger suction inlet on a fire pumper.

Hose Jacket

A **hose jacket** is a device that is placed over a leaking section of hose to stop a leak (**FIGURE 15-24**). The best way to handle a leak in a section of hose is to replace the defective section of hose. However, a hose jacket can provide a temporary fix until the section of hose can be replaced. This device should be used only in cases where it is not possible to quickly replace the leaking section of hose. It can be used for both supply hose and attack hose.

FIGURE 15-23 A reducer is used to connect a smaller-diameter hose to a larger-diameter hose.
© 2003, Berta A. Daniels.

Double-female adaptor   Double-male adaptor

FIGURE 15-22 Double-female and double-male adaptors are used to join two couplings of the same configuration.
© Jones & Bartlett Learning. Photographed by Glen E. Ellman.

FIGURE 15-24 A hose jacket is used to repair a leaking hose.
© 2003, Berta A. Daniels.

SKILL DRILL 15-13
Using a Hose Jacket Fire Fighter I, NFPA 1001: 4.3.10

1 Open the hose jacket, and place the damaged coupling in one end.

2 Place the second coupling in the other end of the jacket.

3 Close the hose jacket, ensuring that the latch is secure. Slowly bring the hose line up to pressure, allowing the gaskets to seal around the hose ends.

The hose jacket consists of a split metal cylinder that fits tightly over the outside of a hose line. This cylinder is hinged on one side to allow it to be placed over the leak. A fastener is then used to clamp the cylinder tightly around the hose. Gaskets on each end of the hose jacket prevent water from leaking out the ends of the hose jacket.

To use a hose jacket, follow the steps in **SKILL DRILL 15-13**.

Hose Roller

A **hose roller** is a device used to prevent chafing or kinking over a sharp edge as hose is being hoisted over the edge of a roof or over a windowsill (**FIGURE 15-25**). A hose roller is sometimes called a hose hoist because it has rollers to assist in raising (i.e., hoisting) a hose over the edge of the building. Hose rollers are also used to protect ropes when hoisting an object over the edge of a building and during rope rescue operations. They are typically used with attack hose.

FIGURE 15-25 A hose roller is used to protect a hose when it is hoisted over a sharp edge of a roof or a windowsill.
© 2003, Berta A. Daniels.

Hose Bridge

A **hose bridge** protects a hose when it is necessary to drive a vehicle over a hose line. With rubber hose bridges, the bridge is placed under the hose, and the hose is placed into a trough in the center of the hose bridge. With metal hose bridges, the bridge is placed over the hose, and the hose is placed in an opening in the center of the hose bridge. Some hose bridges are made to accommodate 65-mm (2½-in.) hose. Other bridges can accommodate up to 127-mm (5-in.) diameter hose. Whenever possible, it is best to avoid driving over hose even with hose bridges because this practice may damage hose.

Hose Clamp

A **hose clamp** is used to temporarily stop the flow of water in a hose line. These devices are often applied to supply hose, allowing a hydrant to be opened before the line is attached to the suction side of the attack pump. Placement of a hose clamp on the supply line allows the fire fighter who is assigned to the hydrant to charge the line before it has been attached to the pump, saving time and improving efficiency. As soon as the hose is connected, the clamp is released. A hose clamp can also be used to stop the flow in a line if a hose ruptures or if an attack hose needs to be connected to a different appliance (**FIGURE 15-26**). Hose clamps use a screw mechanism, a long-handled lever, or hydraulic power to clamp off the flow of water. Fire fighters need to learn how to operate the type of hose clamp used by their department.

Valves

Valves are used to control the flow of water in a pipe or hose. Several different types of valves are used on fire hydrants, fire apparatus, standpipe and sprinkler systems, and attack hose. The important thing to remember

FIGURE 15-26 A hose clamp is used to temporarily interrupt the flow of water in a hose.
© Jones & Bartlett Learning. Photographed by Glen E. Ellman.

when opening and closing any valve or nozzle is to do it s-l-o-w-l-y to prevent water hammer.

Commonly encountered valves include the following types:

- **Ball valves**: Ball valves are used on nozzles, gated wyes, and pump discharge gates. They consist of a ball with a hole in the middle. When the hole is lined up with the inlet and the outlet, water flows through it. As the ball is rotated, the flow of water is gradually reduced until it is shut off completely (**FIGURE 15-27A**).
- **Clapper mechanisms**: Clapper mechanisms or valves prevent water from flowing in a backwards direction. They close automatically if water flows against them. Some Siamese valves are equipped with clapper mechanisms. The most common use of a clapper mechanism is in FDCs for standpipe systems and automatic sprinkler systems. They allow the fire fighter to connect a hose to one side of the FDC and flow water while the other side of the FDC is uncapped. These are discussed in more detail in Chapter 26, *Fire Detection, Suppression, and Smoke Control Systems*.
- **Gate valves**: Gate valves are found on fire hydrants and on sprinkler systems. Rotating a spindle causes a gate to move slowly across the opening. The spindle is rotated by turning it with a wrench or a wheel-type handle (**FIGURE 15-27B**).
- **Butterfly valves**: Butterfly valves are often found on the large pump intake connections where a hard suction hose or soft sleeve hose is connected. They are opened or closed by rotating a handle one-quarter turn (**FIGURE 15-27C**).

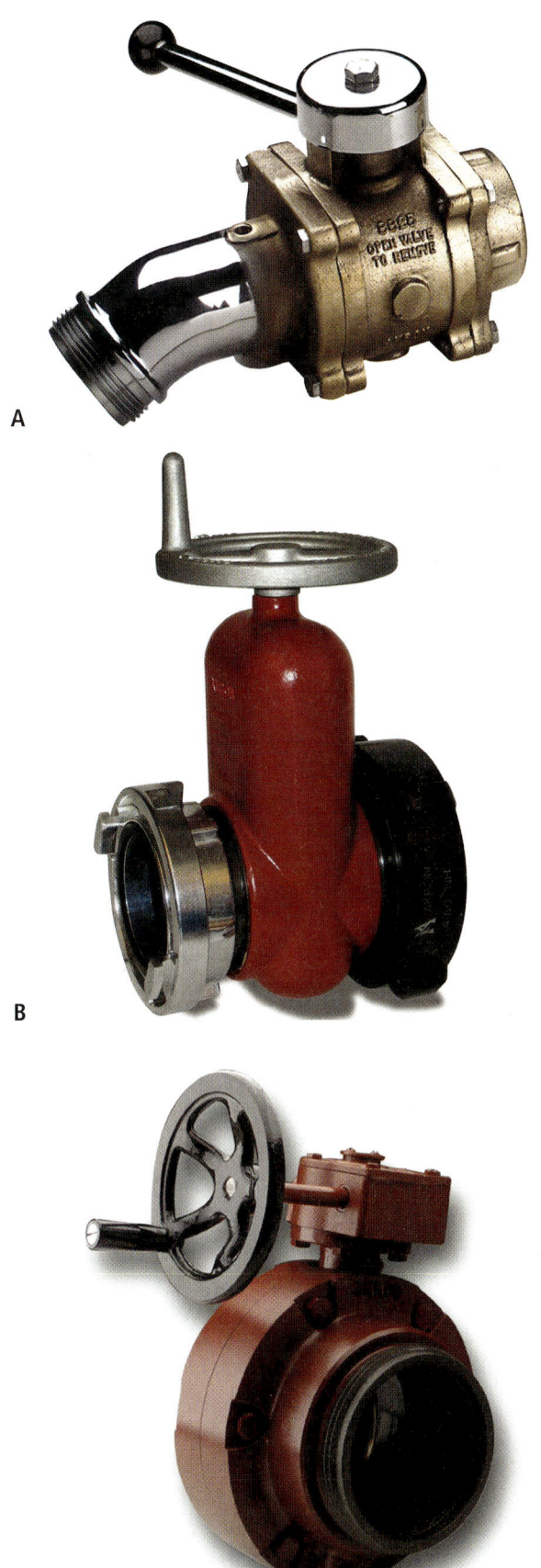

A

B

C

FIGURE 15-27 Valves are used to control the flow of water in a pipe or a hose. **A.** Ball valve. **B.** Gate valve. **C.** Butterfly valve.
Courtesy of Akron Brass Company.

- **Four-way hydrant valves:** This appliance, which is attached to a fire hydrant, enables water to flow directly from the fire hydrant to the attack pump close to the fire. By placing a second fire pumper at the fire hydrant, it is possible to boost the pressure in the supply hose line by changing the position of the four-way hydrant valve to enable the second-arriving fire pumper to pump from the hydrant into the supply line. It can boost pressure in a supply hose without interrupting the flow of water. The use of a four-way hydrant valve is covered in detail in Chapter 16, *Supply Line and Attack Line Evolutions.*
- **Remote-controlled hydrant valves:** Remote-controlled hydrant valves are attached to a fire hydrant. They allow the operator to turn the hydrant on without flowing water into the hose line. These valves are operated by the pump operator using a radio control. They free up the personnel assigned to hydrant duty for other tasks and give the pump driver/operator control over the hydrant from a distance.

Master Stream Appliances

Master stream appliances or devices are used to produce high volume water streams for large fires. Most master stream appliances discharge between 1325 L/min (350 gpm) and 5678 L/min (1500 gpm), though much larger capacities are available. Master stream appliances can discharge a stream of water farther than a handline. A master stream appliance can be either manually or remotely operated from a fixed position. There are three main types of master stream appliances.

A **deck gun** is permanently mounted on and operated from a vehicle, and it is equipped with a piping system that delivers water to the gun (**FIGURE 15-28**). These devices are also called turret pipes or wagon pipes. Sometimes a hose must be connected to the deck gun in order to place it in operation.

A **portable monitor** is a master stream appliance that can be carried on apparatus and removed from the apparatus when needed (**FIGURE 15-29**). It is placed on the ground, and hose lines are connected to the portable monitor to supply the water. Most of these devices come equipped with either one, two, or three inlets. Portable monitors are built with 65-mm (2½-in.) inlets or with an LDH inlet. Smaller portable monitors are sometimes set up attached to preconnected hose. This allows the monitor to be quickly placed in service by one fire fighter.

Elevated master stream appliances are mounted on aerial apparatus such as aerial ladders, tower ladders,

FIGURE 15-28 A deck gun is permanently mounted on a vehicle and equipped with a piping system that delivers water to the device.
© Jones & Bartlett Learning. Photographed by Glen E. Ellman.

FIGURE 15-29 A portable monitor is placed on the ground and supplied with water from one or more hose lines. Note the manner in which the hose lines enter the appliance from the front, forming an anchor for each supply line.
© Jones & Bartlett Learning. Photographed by Glen E. Ellman.

elevated platforms, or special hydraulically operated master stream booms (**FIGURE 15-30**). A **ladder pipe** is a removable elevated master stream device that is mounted close to the tip of an aerial ladder or tower ladder and supplied by a hose. Aerial apparatus are usually equipped with permanently mounted waterways that supply the master stream appliances that are mounted close to the top of the aerial device.

Additional information on the uses and operation of these devices is presented in Chapter 17, *Fire Suppression.*

FIGURE 15-30 Elevated master stream devices can be mounted on aerial apparatus.
© SteveStone/iStockphoto/Getty Images.

Nozzles

Nozzles are attached to the discharge end of attack hose to give shape and direction to fire streams (streams of water or extinguishing agents such as foam). Without a nozzle, the water discharged from the end of a hose would extend only a short distance. Nozzles are used on all sizes of hose and on master stream appliances.

Nozzles can be classified into three groups based on the size of the fire stream:

- **Low-volume nozzles** flow 151 L/min (40 gpm) or less. In structural firefighting, low-volume nozzles are primarily used for booster hose; their use is limited to small outdoor fires. In wildland firefighting, some nozzles use as little as 57 L/min (15 gpm).
- **Handline nozzles** are used on hose ranging from 38 mm (1½ in.) to 65 mm (2½ in.) in diameter. Handline nozzle streams usually flow between 189 L/min and 1325 L/min (50 and 350 gpm).
- **Master stream nozzles** are used on deck guns, portable monitors, and ladder pipes that flow more than 1325 L/min (350 gpm).

Low-volume and handline nozzles incorporate a nozzle shut-off valve that is used to control the flow of water. The shut-off valve for a master stream appliance is usually separate from the nozzle itself.

Nozzle Shut-Off

The nozzle shut-off valve or bale enables the fire fighter at the nozzle to control (start or stop) the flow of water. The handle that controls this valve is called a bale. Some nozzles incorporate a rotary control valve that is operated by rotating the nozzle in one direction to open (turn on) and in the opposite direction to close (shut off) the flow of water.

Some nozzles are made so that the tip of the nozzle can be separated from the shut-off valve. Such a break-away fire nozzle allows fire fighters to shut off the flow, unscrew the nozzle tip, and then add more lengths of hose to extend the hose without shutting off the valve at the pumper.

Types of Nozzles

Nozzles can also be classified by type. All nozzles direct the water stream into a certain shape or pattern. Some nozzles incorporate a mechanism that can automatically adjust the flow based on the water's volume and pressure.

Two different types of nozzles are manufactured for the fire service: smooth-bore nozzles and fog-stream nozzles (combination nozzles). Smooth-bore nozzles produce a solid stream, or a solid column of water, whereas fog-stream nozzles can be adjusted to produce a straight stream or to separate the water into droplets to produce a variety of fog streams. The size of the water droplets and the discharge pattern can be varied by adjusting the nozzle setting. Nozzles must have an adequate volume of water and an adequate pressure to produce a good fire stream. The volume and pressure requirements vary according to the type and size of nozzle.

Smooth-Bore Nozzles

The simplest smooth-bore nozzle consists of a shut-off valve and a smooth-bore tip that gradually decreases the diameter of the solid stream to a size smaller than the hose diameter (**FIGURE 15-31**). Smooth-bore nozzles are manufactured to fit both handlines and master stream appliances. Those that are used

FIGURE 15-31 A smooth-bore nozzle.
© Jones & Bartlett Learning. Photographed by Glen E. Ellman.

for master stream appliances often consist of a set of stacked tips, where each successive tip in the stack has a smaller-diameter opening. Tips can be quickly added or removed to provide the desired stream size. This approach allows different sizes of streams to be produced under different conditions.

Using a smooth-bore nozzle offers several advantages. For example, a smooth-bore nozzle has a longer reach than a fog-stream nozzle operating at a straight stream setting. Smooth-bore nozzles also operate at lower pressures than fog-stream nozzles do. Most smooth-bore nozzles are designed to operate at 345 kPa (50 psi), whereas fog-stream nozzles generally require pressures of 517 to 690 kPa (75 to 100 psi). Lower nozzle pressure makes it easier for a fire fighter to handle the nozzle.

A straight or solid stream extinguishes a fire with less air movement and less disturbance of the thermal layering than does a fog stream, which in turn makes the heat conditions less intense for fire fighters during an interior attack. It is also easier for the operator to see the pathway of a solid stream than a fog stream.

There are some disadvantages associated with smooth-bore nozzles. These nozzles are not effective for hydraulic ventilation. (Hydraulic ventilation is discussed in Chapter 13, *Ventilation*.) A fire fighter cannot change the setting of a smooth-bore nozzle to produce a fog stream; in contrast, a fog-stream nozzle can be set to produce a straight stream. In addition, smooth bore nozzles are not as effective when used for compartment-based firefighting as they generate large water droplets that tend not to be as effective at heat absorption as those produced by a fog pattern.

To operate a smooth-bore nozzle, follow the steps in **SKILL DRILL 15-14**.

SKILL DRILL 15-14
Operating a Smooth-Bore Nozzle Fire Fighter I, NFPA 1001: 4.3.10

1 Select the desired tip size, and attach it to the nozzle shut-off valve. Attain a stable stance (if standing).

2 Slowly open the valve, allowing water to flow.

SKILL DRILL 15-14 Continued
Operating a Smooth-Bore Nozzle Fire Fighter I, NFPA 1001: 4.3.10

3 Open the valve completely to achieve maximum effectiveness.

4 Direct the stream to the desired location.

© Jones & Bartlett Learning. Photographed by Glen E. Ellman.

Fog-Stream Nozzles

Fog-stream nozzles are capable of producing either a straight stream or various fog patterns (**FIGURE 15-32**). These are sometimes called combination nozzles, spray nozzles, or adjustable fog-stream nozzles. The advantage of creating these droplets of water is that they absorb heat much more quickly and efficiently than does a solid column of water. A fog-stream nozzle can be used to reduce the room temperature to prevent a flashover. The smaller droplets absorb more heat per gallon than water from a straight stream. Discharging 4 litres of water in 3 cubic metres (1 gallon of water in 100 cubic feet) of involved interior space can extinguish a fire in 30 seconds. Fog-stream nozzles can be adjusted to produce

FIGURE 15-32 A fog-stream nozzle.
© Jones & Bartlett Learning. Photographed by Glen E. Ellman.

a variety of stream patterns, ranging from a straight stream to a narrow fog cone of less than 45 degrees to a wide-angle fog pattern that is close to 90 degrees. The straight streams produced by fog-stream nozzles have openings in the center. Therefore, a fog-stream nozzle cannot produce a solid stream. The straight stream from a fog-stream nozzle breaks up faster and does not have the reach of a solid stream. A straight stream from a fog-stream nozzle is also affected more dramatically by wind than a solid stream is.

The use of fog-stream nozzles offers several advantages. First, these nozzles can be adjusted to produce a variety of stream patterns by rotating the tip of the nozzle. In addition, fog streams are effective at absorbing heat and can be used to create a water curtain to protect fire fighters from extreme heat.

Fog-stream nozzles move large volumes of air along with the water. This can be an advantage or a disadvantage, depending on the situation. A fog stream can be used to exhaust smoke and gases through hydraulic ventilation. Unfortunately, this type of air movement can result in sudden heat inversion in a room, which then pushes hot steam and gases down onto the fire fighters. Research has shown that if used incorrectly, some applications of a fog pattern can push the fire into unaffected areas of a building by the addition of air flow. Fire fighters must be mindful of the effect the hose stream has on the fire environment and make adjustments accordingly.

To produce an effective stream using a fog-stream nozzle, nozzles must be operated at the pressure recommended by the manufacturer. For many years, the standard operating pressure for fog-stream nozzles was 690 kPa (100 psi). Some manufacturers have produced low-pressure nozzles that are designed to operate at 345 or 517 kPa (50 or 75 psi). The advantage of low-pressure fog-stream nozzles is that they produce less reaction force, which makes them easier to control and advance. Lower nozzle pressure also decreases the risk that the nozzle will get out of control. Some manufacturers adjust the bale reaction according to the required specifications of individual departments so that firefighters can produce a finer mist for maximum control.

To operate a fog-stream nozzle, follow the steps in **SKILL DRILL 15-15**.

Three types of fog-stream nozzles are available. The difference among the types is related to the water delivery capability:

- A **fixed-gallonage fog nozzle** delivers a preset flow at the rated discharge pressure. The nozzle could be designed to flow 114, 227, or 379 L/min (30, 60, or 100 gpm).
- An **adjustable-gallonage fog nozzle** allows the operator to select a desired flow from several settings by rotating a selector bezel to adjust the size of the opening. For example, a nozzle

SKILL DRILL 15-15
Operating a Fog-Stream Nozzle Fire Fighter I, NFPA 1001: 4.3.10

1 Select the desired nozzle. Attain a stable stance (if standing).

2 Slowly open the valve, and allow water to flow.

SKILL DRILL 15-15 Continued
Operating a Fog-Stream Nozzle Fire Fighter I, NFPA 1001: 4.3.10

3 Open the valve completely.

4 Select the desired water pattern by rotating the bezel of the nozzle. Apply water where needed.

© Jones & Bartlett Learning. Photographed by Glen E. Ellman.

could have the options of flowing 227, 360, or 473 L/min (60, 95, or 125 gpm). Once the setting is chosen, the nozzle delivers the rated flow for as long as the rated pressure is provided at the nozzle.

- An automatic-adjusting fog nozzle can deliver a wide range of flows. As the pressure at the nozzle increases or decreases, an internal spring-loaded piston moves in or out to adjust the size of the opening. The amount of water flowing through the nozzle is adjusted to maintain the rated pressure and produce a good stream. A typical automatic nozzle could have an operating range of 341 to 852 L/min (90 to 225 gpm) while maintaining 690 kPa (100 psi) discharge pressure.

For more information about smooth-bore nozzles and fog-stream nozzles, see Chapter 17, *Fire Suppression*.

Other Types of Nozzles

Several other types of nozzles are used for special purposes. If your fire department has other types of specialty nozzles, you need to become proficient in their use and operation. Piercing nozzles, for example, are used to make a hole in automobile sheet metal, aircraft, or building walls, roofs, or floors to extinguish fires behind these surfaces (**FIGURE 15-33**).

FIGURE 15-33 A piercing nozzle can pierce holes in certain materials such as sheet metal.
© Goodman Photography/Shutterstock.

Cellar nozzles and **Bresnan distributor nozzles** are used to fight fires in cellars or basements and other inaccessible places such as attics and cocklofts (**FIGURE 15-34**). These nozzles discharge water in a wide circular pattern as the nozzle is lowered vertically through a hole into the cellar. They work like a large sprinkler head.

Water curtain nozzles are used to deliver a flat screen of water that then forms a protective sheet (curtain) of water on the surface of an exposed building (**FIGURE 15-35**). The water curtains must be directed onto the exposed building because radiant heat can pass through the air gaps in the water curtain.

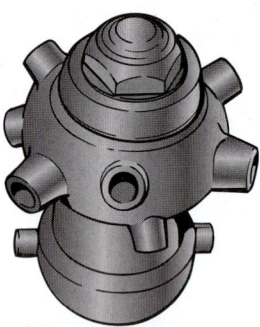

FIGURE 15-34 A Bresnan distributor nozzle is used to fight fires in inaccessible places such as cellars or basements and attics.
Courtesy of Akron Brass Company.

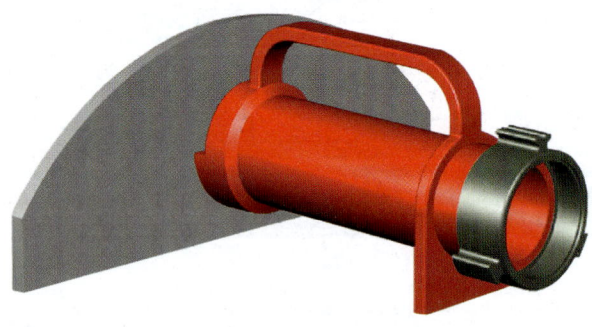

FIGURE 15-35 A water curtain nozzle delivers a flat screen of water that then forms a protective sheet (curtain) of water.
Courtesy of POK of North America, Inc.

Nozzle Maintenance and Inspection

Nozzles should be inspected on a regular basis, along with all of the equipment on every fire department vehicle. In particular, nozzles should be checked after each use before they are placed back on the apparatus. They should be kept clean and clear of debris. Debris inside the nozzle will affect the performance of the nozzle, possibly reducing flow. Dirt and grit can also interfere with the valve operation and prevent the nozzle from opening and closing fully. A light grease on the surface of the ball valve keeps it operating smoothly.

On fog-stream nozzles, inspect the fingers and teeth on the face of the nozzle (**FIGURE 15-36**). Make sure all fingers are present and that the moveable part of the nozzle spins freely. Any missing fingers or failure of the ring to spin will drastically affect the fog pattern. Any problems noted should be referred to a competent technician for repair.

FIGURE 15-36 On fog-stream nozzles, inspect the fingers and teeth on the face of the nozzle. Make sure all fingers are present and that the moveable part of the nozzle spins freely.
© Jones & Bartlett Learning. Photographed by Glen E. Ellman.

After-Action REVIEW

IN SUMMARY

- Fire hose range in size from 19 mm to 152 mm (¾ in. to 6 in.) in diameter.

  - Forestry fire hose is 19 mm (¾ in.), 25 mm (1 in.), or 38 mm (1½ in.) in diameter.

  - Booster hose is 25 mm (1 in.) in diameter. Although it is classified as attack hose, it should not be used for structural or vehicle firefighting. The normal flow from a 25-mm (1-in.) booster hose is in the range of 151 to 189 L/min (40 to 50 gpm).

  - Most fire departments use either 38-mm (1½-in.) or 45-mm (1¾-in.) diameter hose as the primary attack hose for most fires.

  - A 65-mm (2½-in.) hose can be used as either attack hose or supply hose, but it is most often used as attack hose for large fires. It generally delivers a flow of approximately 946 L/min (250 gpm).

  - A 76-mm (3-in.) hose is often used as supply hose. Most large-diameter hose (89 mm [3½ in.] or more) is constructed as supply hose. The largest large-diameter hose size is 152 mm (6 in.) in diameter.

- Fire hose is constructed with an inner waterproof liner surrounded by either one or two outer layers.

- Hose couplings are used to connect or couple individual lengths of fire hose together. They are also used to connect a hose line to a hydrant, to a suction or discharge valve on a pumper, or to a variety of nozzles, fittings, and appliances. A coupling is permanently attached to each end of a section of fire hose.

- The two types of couplings are threaded hose couplings and Storz-type hose couplings.

  - A set of threaded hose couplings includes a male hose coupling with threads on the outside on one end of the hose and a female hose coupling with threads on the inside on the other end of the hose. Threaded hose couplings are used on most hose up to 76 mm (3 in.) in diameter. Threaded hose couplings provide a secure connection between two sections of hose when properly coupled. A disadvantage is that they are prone to cross threading.

  - Storz-type couplings are designed so that the couplings on both ends of a length of hose are the same. When this system is used, each coupling can be attached to any other coupling of the same diameter. They are connected by mating the two couplings face-to-face and then turning them clockwise into a locking position. One disadvantage is that these couplings are more prone to accidental disconnect if they are not completely coupled.

- Several techniques are used for connecting and disconnecting hose couplings. Depending on the circumstances, one technique may be more effective than another.

- Fire hose are used as supply hose and as attack hose.

- Supply hose are used to deliver water from a static water source or from a fire hydrant to an attack pumper. They are designed to carry larger volumes of water at lower pressures than attack hose. Supply hose can either be soft sleeve or hard suction.

  - Soft sleeve hose is a short section of large-diameter supply hose that is used to connect a fire department pumper directly to the large steamer outlet on a hydrant. A soft sleeve hose may have a Storz or similar connections on both ends, or it may have threaded female connections on each end, with one end matching the local fire hydrant threads and the other end matching the threads on the suction side of the fire pump. If it has two Storz connections, an adaptor is required to connect the large diameter inlet to the pumper and the fire hydrant.

  - A hard suction hose is a special type of supply hose that is used to draft water from a static source such as a river, lake, or portable drafting basin to the suction side of the fire pump on a fire department pumper or into a portable pump. It can also be used to carry water from a fire hydrant to the pumper. It is designed to remain rigid and will not collapse when a vacuum is created in the hose to draft the water into the pump.

- Attack hose is designed to be used for fire suppression. This hose carries the water from the attack pumper to the fire or to an FDC or from a standpipe system to the fire. Attack hose usually operate at higher pressures than supply lines do.

- Fire hose should be inspected often and tested annually following the procedures in NFPA 1962, *Standard for the Care, Use, Inspection, Service Testing, and Replacement of Fire Hose, Couplings, Nozzles, and Fire Hose Appliances.*

- The most common causes of hose damage are mechanical damage, heat and cold damage, UV radiation, chemical damage, and mildew damage.

- Follow the manufacturer's recommendations for cleaning and maintaining hose. For a mild cleaning, cool water and a soft brush may be adequate. For dirty hose, it may be necessary to use a mild detergent.

- Visual hose inspections should be performed at least quarterly. A visual inspection of hose should be performed after each use. If any defects are found, the length of hose should be marked, and the hose should be removed from service.

- Rolled hose is compact and easy to manage and transport. A fire hose can be rolled different ways, depending on how it will be used. Follow the SOPs of your department when rolling hose.

- Hose appliances are used in conjunction with a fire hose for the purpose of delivering water. Appliances include wyes, water thieves, Siamese connections, adaptors, reducers, hose jackets, hose rollers, hose bridges, hose clamps, and some valves.

- Master stream appliances or devices are used to produce high volume water streams for large fires. Most master stream appliances discharge between 1325 L/min (350 gpm) and 5678 L/min (1500 gpm). These include deck guns, portable monitors, and elevated master stream appliances.

- Nozzles produce streams of water or extinguishing agents such as foam, giving them shape and direction.

- Nozzles can be classified based on the size of the fire stream:
  - Low-volume nozzles flow 151 L/min (40 gpm) or less. In structural firefighting, they are primarily used for booster hose; their use is limited to small outside fires.
  - Handline nozzles are used on hose lines ranging from 38 mm (1½ in.) to 65 mm (2½ in.) in diameter. Handline streams usually flow between 189 and 1325 L/min (50 and 350 gpm).
  - Master stream nozzles are used on deck guns, portable monitors, and ladder pipes that flow more than 1325 L/min (350 gpm).

- Nozzle shut-off valves enable the fire fighter at the nozzle to start or stop the flow of water.

- Nozzles can be classified by type. Two types, the smooth-bore nozzle and the fog-stream nozzle, are manufactured for the fire service.
  - Smooth-bore nozzles produce a solid stream, or a solid column of water. Smooth-bore nozzles are manufactured to fit both handlines and master stream appliances.
  - Fog-stream nozzles can be adjusted to produce a straight stream or to separate the water into droplets to produce a variety of fog streams.

- Several other types of nozzles are used for special purposes.
  - Piercing nozzles are used to make a hole in sheet metal or building walls to extinguish fires.
  - Cellar nozzles and Bresnan distributor nozzles are used to fight fires in cellars or basements and other inaccessible places such as attics and cocklofts.
  - Water curtain nozzles are used to deliver a flat screen of water that forms a protective sheet of water on the surface of an exposed building.

- Nozzles should be inspected on a regular basis, along with all of the equipment on every fire department apparatus.

KEY TERMS

Access Navigate **for flashcards to test your key term knowledge.**

Adaptor A device that allows fire hose couplings to be safely interconnected with couplings of different sizes, threads, or mating surfacing or that allows fire hose couplings to be safely connected to other appliances.

Adjustable-gallonage fog nozzle A nozzle that allows the operator to select a desired flow from several settings.

Attack hose Hose designed to be used by trained fire fighters and fire brigade members to combat fires beyond the incipient stage. (NFPA 1961)

Automatic-adjusting fog nozzle A nozzle that can deliver a wide range of water stream flows. As the pressure at the nozzle increases or decreases, an internal spring-loaded piston moves in or out to adjust the size of the opening.

Ball valves Valves used on nozzles, gated wyes, and pumper discharge gates. They consist of a ball with a hole in the middle of the ball.

Booster hose (booster line) A noncollapsible hose used under positive pressure having an elastomeric or thermoplastic tube, a braided or spiraled reinforcement, and an outer protective cover. (NFPA 1962)

Breakaway type nozzle A nozzle with a tip that can be separated from the shut-off valve.

Bresnan distributor nozzle A nozzle that can be placed in confined spaces such as cellars or basements. The nozzle spins, spreading water over a large area.

Butterfly valves Valves that are found on the large pump intake connections where the suction hose connects to the suction side of the fire pump.

Cellar nozzle A nozzle used to fight fires in cellars or basements and other inaccessible places. The device works by spreading water in a wide pattern as the nozzle is lowered through a hole into the cellar.

Clapper mechanism A mechanical device installed within a piping system that allows water to flow in only one direction.

Deck gun A device that is permanently mounted on and operated from a vehicle and equipped with a piping system that delivers water to the gun.

Double-female adaptor A hose adaptor that is used to join two male hose couplings.

Double-jacket hose A hose constructed with two layers of woven fibres.

Double-male adaptor A hose adaptor that is used to join two female hose couplings.

Elevated master stream appliance A nozzle mounted on the end of an aerial device that is capable of delivering large amounts of water into a fire or exposed building from an elevated position.

Fire streams Streams of water or extinguishing agents.

Fixed-gallonage fog nozzle A nozzle that delivers a set number of gallons per minute (litres per minute) as per the nozzle's design, no matter what pressure is applied to the nozzle.

Fog-stream nozzle A nozzle that is placed at the end of a fire hose and can be adjusted to produce a straight stream or to separate the water into droplets to produce a variety of fog streams. May also be referred to as a combination nozzle, spray nozzle, or adjustable fog-stream nozzle.

Forestry fire hose A hose designed to meet specialized requirements for fighting wildland fires. (NFPA 1961)

Gate valves Valves found on hydrants and sprinkler systems. Rotating a spindle causes a gate to move slowly across the opening.

Gated wye A valved device that splits a single hose into two separate hose, allowing each hose to be turned on and off independently.

Handline nozzle A nozzle with a rated discharge of less than 350 gpm (1325 L/min).

Hard suction hose A short section of supply hose that is used to draft water from a static source such as a river, lake, or portable drafting basin to the suction side of the fire pump on a fire department pumper or into a portable pump.

Higbee indicators Indicators on the male and female threaded couplings that indicate where the threads start. These indicators should be aligned before fire fighters start to thread the couplings together.

Hose appliance A piece of hardware (excluding nozzles) generally intended for connection to fire hose to control or convey water. (NFPA 1962)

Hose bridge A device that protects a hose when it is necessary for a vehicle to drive over a hose.

Hose clamp A device used to compress a fire hose to stop water flow.

Hose jacket A device used to stop a leak in a fire hose or to join hose that have damaged couplings.

Hose liner The inside portion of a hose that is in contact with the flowing water; also called the hose inner jacket.

Hose roller A device that is placed on the edge of a roof and is used to protect hose as it is hoisted up and over the roof edge.

Ladder pipe A monitor that attaches to the rungs of a vehicle-mounted aerial ladder. (NFPA 1965)

Large-diameter hose (LDH) A hose 89 mm (3.5 in.) or larger that is designed to move large volumes of water to supply master stream appliances, portable hydrants, manifolds, standpipe and sprinkler systems, and fire department pumpers from hydrants and in relay. (NFPA 1410)

Low-volume nozzle A nozzle that flows 151 L/min (40 gpm) or less.

Master stream appliance Devices used to produce high volume water streams for large fires. Most master stream appliances discharge between 1325 L/min (350 gpm) and 5678 L/min (1500 gpm), though much larger capacities are available. These devices include deck guns and portable ground monitors.

Master stream nozzle A nozzle with a rated discharge of 1325 L/min (350 gpm) or greater. (NFPA 1964)

Mildew A fungus that can grow on hose if the hose is stored wet. Mildew can damage the jacket of a hose.

Multiple-jacket A construction consisting of a combination of two separately woven reinforcements (double jacket) or two or more reinforcements interwoven. (NFPA 1962)

Nozzle A device for use in applications requiring special water discharge patterns, directional spray, or other unusual discharge characteristics. (NFPA 13)

Nozzle shut-off valve A device that enables the fire fighter at the nozzle to start or stop the flow of water.

Piercing nozzle A nozzle that can be driven through sheet metal or other material to deliver a water stream to that area.

Portable monitor A monitor that can be lifted from a vehicle-mounted bracket and moved to an operating position on the ground by not more than two people. (NFPA 1965)

Reducer A fitting used to connect a small hose line or pipe to a larger hose line or pipe. (NFPA 1142)

Remote-controlled hydrant valve A valve that is attached to a fire hydrant to allow the operator to turn the hydrant on without flowing water into the hose line.

Rocker lugs Fittings on threaded couplings that aid in coupling the hose. Also referred to as rocker pins.

Rubber-covered hose Hose whose outside covering is made of rubber, which is said to be more resistant to damage. Also referred to as a rubber-jacket hose.

Siamese connection A hose appliance that allows two hose to be connected together and flow into a single hose.

Single-jacket A construction consisting of one woven jacket. (NFPA 1962)

Smooth-bore nozzle A nozzle that produces a solid stream, or a solid column of water.

Smooth-bore tip A nozzle device that is a smooth tube, which is used to deliver a solid column of water. Different-sized tips can be attached to a single nozzle or appliance.

Soft sleeve hose A short section of large-diameter supply hose that is used to provide water from the large steamer outlet (the large-diameter port) on a fire hydrant or other pressurized water source to the suction side of the fire pump.

Solid stream A solid column of water.

Spanner wrench (hose key) A type of tool used to couple or uncouple hose by turning the rocker lugs or pin lugs on the connections.

Storz-type (nonthreaded) hose coupling A hose coupling that has the property of being both the male and the female coupling. It is connected by engaging the lugs and turning the coupling a one-third turn.

Straight stream A stream made by using an adjustable nozzle to provide a straight stream of water.

Suction hose A short section of supply hose that is used to supply water to the suction side of the fire pump.

Supply hose Hose designed for the purpose of moving water between a pressurized water source and a pump that is supplying attack lines. (NFPA 1961)

Threaded hose coupling A type of coupling that requires a male fitting and a female fitting to be screwed together.

Water curtain nozzle A nozzle used to deliver a flat screen of water that forms a protective sheet of water to protect exposures from fire.

Water thief A device that has a 65-mm (2½-in.) inlet and a 65-mm (2½-in.) outlet in addition to two 38-mm (1½-in.) outlets. It is used to supply many hose from one source.

Wye A device used to split a single hose into two or more separate lines.

REFERENCES

National Fire Protection Association. NFPA 24, *Standard for the Installation of Private Fire Service Mains and Their Appurtenances*, 2019. www.nfpa.org. Accessed October 4, 2018.

National Fire Protection Association. NFPA 1142, *Standard on Water Supplies for Suburban and Rural Fire Fighting*, 2017. www.nfpa.org. Accessed October 4, 2018.

National Fire Protection Association. NFPA 1901, *Standard for Automotive Fire Apparatus*, 2016. www.nfpa.org. Accessed October 4, 2018.

National Fire Protection Association. NFPA 1961, *Standard on Fire Hose*, 2013. www.nfpa.org. Accessed October 4, 2018.

National Fire Protection Association. NFPA 1962, *Standard for the Care, Use, Inspection, Service Testing, and Replacement of Fire Hose, Couplings, Nozzles, and Fire Hose Appliances*, 2018. www.nfpa.org. Accessed October 4, 2018.

National Fire Protection Association. NFPA 1963, *Standard for Fire Hose Connections*, 2014. www.nfpa.org. Accessed October 4, 2018.

National Fire Protection Association. NFPA 1964, *Standard for Spray Nozzles*, 2018. www.nfpa.org. Accessed October 4, 2018.

National Fire Protection Association. NFPA 1965, *Standard for Fire Hose Appliances*, 2014. www.nfpa.org. Accessed October 4, 2018.

On Scene

You arrive as part of a third alarm response to a fire at a large industrial complex. A huge column of black smoke is billowing into the sky—it is visible for miles. Your company is directed to a fire hydrant two blocks away to set up master streams to cool the fire. The engineer weaves the apparatus through the hose that is laid throughout the road. Once you reach your final destination, you step out and break out the large-diameter hose before setting up a master stream appliance.

1. Large-diameter hose is _____ or larger and is designed to move large volumes of water to supply master stream appliances, portable hydrants, manifolds, standpipe and sprinkler systems, and fire department pumpers from hydrants and in relay.

A. 65 mm (2½ in.)

B. 152 mm (6 in.)

C. 89 mm (3½ in.)

D. 102 mm (4 in.)

2. Which of the following is *not* an advantage of Storz-type hose couplings?

A. They fit all fire hydrants.

B. They have no male or female end to the hose.

C. They cannot be cross threaded.

D. They can be quickly connected.

3. Which of the following sizes of hose is *not* used as a supply hose?

A. 76 mm (3 in.)

B. 65 mm (2½ in.)

C. 45 mm (1¾ in.)

D. 102 mm (4 in.)

4. A reducer is a type of adaptor that is used to:

A. connect two hose with different threads.

B. change a supply hose to an attack hose.

C. attach a smaller-diameter hose to a larger-diameter hose.

D. control the flow of water.

5. A _____ is a device that has a 65-mm (2½-in.) inlet and a 65-mm (2½-in.) outlet in addition to two 38-mm (1½-in.) outlets. It is used to supply many hose from one source.

A. water thief

B. gated wye

C. four-way hydrant valve

D. dump valve

6. A _____ is used to join two female hose couplings.

A. hose jacket

B. double-male adaptor

C. double-female adaptor

D. Siamese connection

(continued)

On Scene Continued

7. A hose roller is used:

   A. for rolling large-diameter hose.

   B. for rolling suction hose.

   C. to prevent chafing or kinking of hose at an edge of a roof or windowsill.

   D. for storing booster hose.

8. Which of the following is *not* a type of nozzle?

   A. Attic nozzle

   B. Water curtain nozzle

   C. Cellar nozzle

   D. Fog-stream nozzle

Access Navigate to find answers to this On Scene, along with other resources such as an audiobook.

CHAPTER **16**

Fire Fighter I

Supply Line and Attack Line Evolutions

KNOWLEDGE OBJECTIVES

After studying this chapter, you will be able to:

- Describe the procedures used to connect supply lines to a fire hydrant. (**NFPA 1001: 4.3.15**, p. 593)
- Describe the common types of supply line evolutions. (**NFPA 1001: 4.3.15**, pp. 593–601)
- Describe the common techniques used to load supply hose. (**NFPA 1001: 4.5.2**, pp. 599, 601–607)
- Describe the common technique used to attach hose to a fire hydrant. (**NFPA 1001: 4.3.15**, p. 607)
- Describe the common techniques used to carry and advance supply hose. (**NFPA 1001: 4.3.15**, pp. 607, 610–613)
- Describe the two types of standpipe systems. (p. 614)
- Describe the general procedures that are followed during attack line evolutions. (**NFPA 1001: 4.3.10**, p. 615)
- Describe the types of loads used to organize attack hose. (**NFPA 4.5.2**, pp. 615–623)
- Describe the procedures to follow when advancing attack hose. (**NFPA 1001: 4.3.10**, pp. 623, 625–631)
- Describe how to extend an attack line. (**NFPA 1001: 4.3.10**, pp. 631–632)
- Describe how to connect and advance an attack line from a standpipe outlet. (**NFPA 1001: 4.3.10**, pp. 632–634)
- Describe how to replace a damaged section of attack hose. (**NFPA 1001: 4.3.10**, pp. 634–635)
- Describe why hose should be unloaded and reloaded on a regular basis. (p. 637)
- Describe how to unload a fire hose. (p. 637)

SKILLS OBJECTIVES

After studying this chapter, you will be able to perform the following skills:

- Perform a forward hose lay. (**NFPA 1001: 4.3.15**, pp. 595–596)
- Attach a fire hose to a four-way hydrant valve. (**NFPA 1001: 4.3.15**, pp. 597–598)
- Perform a reverse hose lay. (**NFPA 1001: 4.3.15**, p. 600)
- Perform a split hose lay. (**NFPA 1001: 4.3.15**, p. 601)
- Perform a flat hose load. (**NFPA 1001: 4.5.2**, pp. 602–603)
- Perform a horseshoe hose load. (**NFPA 1001: 4.5.2**, p. 604)
- Perform an accordion hose load. (**NFPA 1001: 4.5.2**, pp. 605–606)
- Attach a soft sleeve hose to a fire hydrant. (**NFPA 1001: 4.3.15**, pp. 608–609)
- Perform a working hose drag. (**NFPA 1001: 4.3.10**, p. 610)

- Perform a shoulder carry. (**NFPA 1001: 4.3.10**, p. 611)
- Advance an accordion load. (**NFPA 1001: 4.3.10**, pp. 612–613)
- Connect a hose line to supply a fire department connection. (**NFPA 1001: 4.3.15**, p. 614)
- Perform a minuteman hose load. (**NFPA 1001: 4.5.2**, pp. 616–617)
- Advance a minuteman hose load. (**NFPA 1001: 4.3.10**, p. 618)
- Perform a preconnected flat hose load. (**NFPA 1001: 4.5.2**, pp. 619–620)
- Advance a preconnected flat hose load. (**NFPA 1001: 4.3.10**, pp. 620–621)
- Perform a triple-layer hose load. (**NFPA 1001: 4.5.2**, p. 622)
- Advance a triple-layer hose load. (**NFPA 1001: 4.3.10**, p. 623)
- Unload and advance wyed lines. (**NFPA 1001: 4.3.10**, p. 624)
- Advance an attack line from an attack pump to the door. (**NFPA 1001: 4.3.10**, pp. 625–626)
- Advance an attack line from the door to the fire. (**NFPA 1001: 4.3.10**, pp. 626–627)

- Advance an attack line up a stairway. (**NFPA 1001: 4.3.10**, pp. 627–628)
- Advance an attack line down a stairway. (**NFPA 1001: 4.3.10**, pp. 628–629)
- Advance an uncharged attack line up a ladder. (**NFPA 1001: 4.3.10**, p. 630)
- Operate an attack line from a ladder. (**NFPA 1001: 4.3.10**, pp. 631–632)
- Connect and advance an attack line from a standpipe outlet. (**NFPA 1001: 4.3.10**, pp. 633–634)
- Replace a damaged section of hose. (**NFPA 1001: 4.3.10**, p. 635)
- Drain a fire hose. (**NFPA 1001: 4.5.2**, pp. 635–636)

ADDITIONAL STANDARDS

- **NFPA 1142**, *Standard on Water Supplies for Suburban and Rural Fire Fighting*
- **NFPA 1961**, *Standard on Fire Hose*
- **NFPA 1962**, *Standard for the Care, Use, Inspection, Service Testing, and Replacement of Fire Hose, Couplings, Nozzles, and Fire Hose Appliances*
- **NFPA 1963**, *Standard for Fire Hose Connections*

Fire Alarm

It is midmorning when your pump company is dispatched to a strip mall for the report of a working fire in a fast-food restaurant. You are dispatched as the first-due pumper. Your company is assigned to advance the first attack line into the fire. As you are responding, your dispatcher informs you that a closer pump company will be first due, and your company is now assigned to establish a supply line for the first-due company.

1. Why is it important to understand the difference between supply line evolutions and attack line evolutions?

2. What is the primary purpose of supply line evolutions?

3. What is the primary purpose of establishing an attack line?

 *Access Navigate for more practice activities.*

Introduction

Fire hose evolutions are standard methods of working with fire hose to accomplish different objectives in a variety of situations. Most fire departments set up their equipment and conduct regular training so that fire fighters will be prepared to perform a set of standard evolutions. Evolutions involve specific actions that are assigned to specific members of a crew, depending on their riding positions on the apparatus. Every fire fighter should know how to perform all of the standard evolutions quickly and proficiently. When an officer calls for a particular evolution to be performed, each crew member should know exactly what to do.

As discussed in Chapter 14, *Water Supply Systems*, ensuring a dependable water supply is a critical fire-ground operation that must be accomplished as soon as possible. A water supply should be established at the same time as other initial fire-ground operations are conducted. At many fire scenes, size-up, forcible entry, raising ladders, search and rescue, ventilation, and establishing a water supply occur concurrently.

As discussed in Chapter 15, *Fire Hose, Appliances, and Nozzles*, fire hose is used for two main purposes: as supply hose and as attack hose. These also may be referred to as supply lines and attack lines. A hose line (one or more lengths of hose) is a single component in the overall water delivery system; a supply line or an attack line is the name of the system with all of the included appliances. Supply hose (or supply line) is used to either deliver water from a pressurized source, such as a fire hydrant, to an attack pump, or to deliver water from another water supply pump that is being used to provide a water supply for the attack pump. Most attack lines carry water directly from the attack pump to a nozzle that is used to direct the water onto the fire. For this reason, hose evolutions can be divided into supply line evolutions and attack line evolutions. This chapter discusses both.

SAFETY TIP

When performing a forward hose lay, the fire fighter who is connecting the supply hose to the fire hydrant must not stand between the hose and the fire hydrant. When the apparatus starts to move off, the hose could become tangled and suddenly be pulled taut. Anyone standing between the hose and the fire hydrant could be seriously injured.

Supply Line Evolutions

Supply line operations involve laying hose lines and making connections between a water supply source, such as a fire hydrant, and an attack pump.

Laying Supply Hose

The objective of laying supply hose is to deliver water from a water source, such as a fire hydrant, to an attack pump. This can be done using a forward hose lay, a reverse hose lay, or a split hose lay (**FIGURE 16-1**). In most cases, this operation involves using the forward hose lay, or laying a continuous hose line out of the bed of the fire apparatus as the vehicle drives forward. Each fire department will determine its own preferred methods and procedures for supply line operations.

FIRE FIGHTER TIP

When connecting supply hose to a fire hydrant, wait until you get the signal from the pump driver/operator before you charge the hose line. If the fire hydrant is opened prematurely, the hose bed could become charged with water or a loose hose line could discharge water at the fire scene. Either situation will disrupt the water supply operation and could cause serious injuries. Make sure that you know your department's signal to charge a hose line, and do not become so excited or rushed that you make a mistake.

Forward Hose Lay

The **forward hose lay** is most often used by the first-arriving pump company at the scene of a fire. Using this method, also referred to as the straight hose lay, the hose is laid out from the water supply source, such as a fire hydrant, and unfolds as the apparatus approaches the fire. This allows the pump company to establish a water supply without assistance from another company and places the attack pump close to the fire, allowing access to additional hose, tools, and equipment that are carried on the apparatus. To perform the forward hose lay, the apparatus stops near the fire hydrant, and a fire fighter steps off of the apparatus and secures the hose around the fire hydrant. Once the hose is secured, the apparatus proceeds to the fire. As the apparatus moves forward, a length of supply hose unfolds from the apparatus and onto the ground.

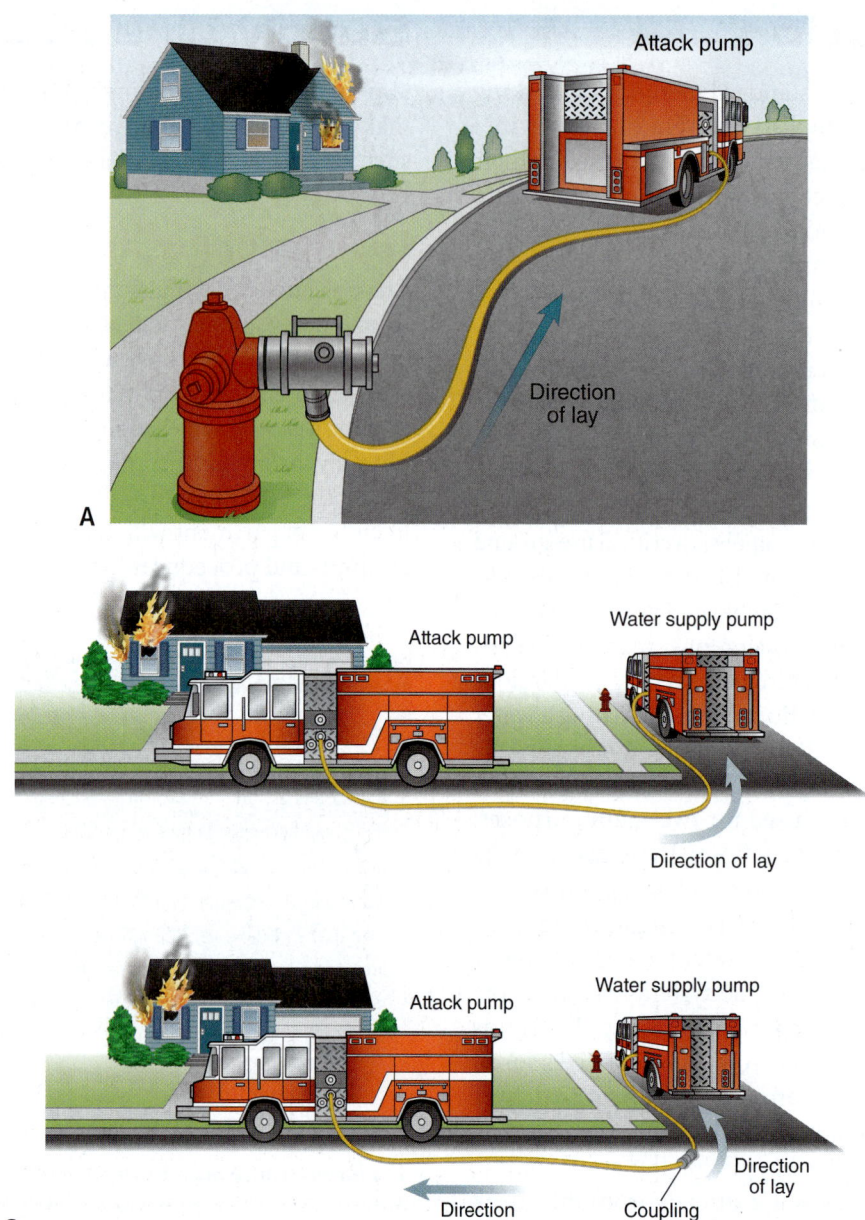

FIGURE 16-1 Laying supply hose can be done using a forward hose lay, a reverse hose lay, or a split hose lay. **A.** A forward hose lay is made from the water source, such as a fire hydrant, to the attack pump or fire. **B.** A reverse hose lay is made from the attack pump or fire to a water source such as a fire hydrant. **C.** Two pumpers perform a split hose lay. The attack pump performs a forward hose lay from the corner of the street to the fire. The water supply pump performs a reverse hose lay from the hose at the corner of the street to a water source such as a fire hydrant.

© Jones & Bartlett Learning.

A forward hose lay can be performed using 65-mm (2½-in.), 76-mm (3-in.), or 89-mm (3½-in.) hose or larger. The larger the diameter of the hose, the more water can be delivered through a single supply hose. When 65-mm (2½-in.) or 76-mm (3-in.) hose is used and the beds are arranged to lay dual hose lines, a company can lay two parallel lines from the fire hydrant to the fire.

If the fire hydrant is located close to the fire, it may supply a sufficient quantity of water from the fire hydrant alone. For example, a 127-mm (5-in.) hose can supply 2650 L/min (700 gpm) over a distance of 152 m (500 ft), losing only about 138 kPa (20 lb) of pressure due to friction loss.

To perform a forward hose lay, follow the steps in **SKILL DRILL 16-1**.

SKILL DRILL 16-1
Performing a Forward Hose Lay Fire Fighter I, NFPA 1001: 4.3.15

1 The pump driver/operator stops the fire apparatus 3 m (10 ft) from the fire hydrant.

2 Step off of the apparatus carrying the hydrant wrench and all necessary tools. Grasp enough hose to reach to and loop around the fire hydrant. Loop the end of the hose around the fire hydrant, or secure the hose as specified in the local standard operating procedure (SOP). Do not stand between the hose and the fire hydrant.

3 Signal the pump driver/operator to proceed to the fire once the hose is secured.

4 Once the apparatus has moved off and a length of supply hose has been removed from the apparatus and is lying on the ground, remove the appropriate-size fire hydrant cap from the outlet nearest to the fire. Follow the local SOP for checking the operating condition of the fire hydrant. Always flush the hydrant before attaching a hose. This practice will ensure that the water supply is free of any debris that can inhibit the smooth flow of water.

(continued)

SKILL DRILL 16-1 Continued
Performing a Forward Hose Lay Fire Fighter I, NFPA 1001: 4.3.15

5 Attach the supply hose to the outlet on the fire hydrant. An adaptor may be needed if a large-diameter hose with Storz-type couplings is used.

6 Attach the hydrant wrench (hydrant key) to the stem nut on the fire hydrant. Check the top of the hydrant for an arrow indicating the direction to turn to open. The pump driver/operator uncouples the hose and attaches the end of the supply hose to the suction side of the pump on the attack pump or clamps the hose closed to the fire pump, depending on the local SOP.

7 When the pump driver/operator signals to charge the hose by prearranged hand signal, radio, or air horn, open the hydrant valve slowly and completely.

8 Follow the hose back to the pump, and remove any kinks from the supply hose.

© Jones & Bartlett Learning. Photographed by Glen E. Ellman.

Four-Way Hydrant Valve

In cases where long supply lines are needed or when the flow pressure from the hydrant is not adequate, it is often necessary to place a water supply pump at the fire hydrant to provide additional pressure in the supply line. Some departments use a **four-way hydrant valve** to achieve this (**FIGURE 16-2**).

When the four-way hydrant valve is placed on the fire hydrant, the water flows initially from the fire hydrant through the valve to the supply hose, which delivers the

FIGURE 16-2 A four-way hydrant valve.
© Jones & Bartlett Learning. Photographed by Glen E. Ellman.

water to the attack pump. The water supply pump can hook up to the four-way hydrant valve and redirect the flow of water by changing the position of the four-way hydrant valve. The water then flows from the fire hydrant to the water supply pump. The water supply pump increases the pressure by using a pump and discharges the water into the supply hose, boosting the flow of water to the attack pump. A four-way hydrant valve ensures that the supply hose can be charged with water immediately using the pressure from the hydrant yet still allows for a second pump, the water supply pump, to connect to the hose line later to provide pump pressure.

This operation can be accomplished without uncoupling any lines or interrupting the flow of water.

To use a four-way hydrant valve, follow the steps in **SKILL DRILL 16-2**.

SKILL DRILL 16-2
Attaching a Fire Hose to a Four-Way Hydrant Valve Fire Fighter I, NFPA 1001: 4.3.15

1 Stop the attack pump 3 m (10 ft) past the fire hydrant to be used. Grasp the four-way hydrant valve, the attached hose, and enough hose to reach to and loop around the fire hydrant. Carry the four-way hydrant valve from the apparatus, along with the hydrant wrench and any other needed tools. Loop the end of the hose around the fire hydrant, or secure the hose with a rope as specified in the local SOP. Do not stand between the fire hydrant and hose. Signal the attack pump driver/operator to proceed to the fire.

2 Once enough hose has been removed from the apparatus and is lying on the ground, remove the steamer port (large-diameter port) from the fire hydrant. Follow the local SOP for checking the operating condition of the fire hydrant. Attach the four-way hydrant valve to the fire hydrant outlet (an adaptor may be needed). Attach the hydrant wrench to the fire hydrant. The attack pump driver/operator uncouples the hose and attaches the end of the supply line to the suction side of the pump on the attack pump. The attack pump driver/operator signals by prearranged hand signal, radio, or air horn to charge the supply line. Open the hydrant valve slowly and completely.

(continued)

SKILL DRILL 16-2 Continued
Attaching a Fire Hose to a Four-Way Hydrant Valve Fire Fighter I, NFPA 1001: 4.3.15

3 Initially, the attack pump is supplied with water from the fire hydrant. When the water supply pump arrives at the fire scene, the water supply pump driver/operator stops at the fire hydrant that has the four-way valve.

4 The water supply pump driver/operator attaches a hose from the four-way hydrant valve outlet to the suction side of the pump on the water supply pump.

5 The water supply pump driver/operator attaches a second hose to the inlet side of the four-way hydrant valve and connects the other end to the discharge side of the pump on the water supply pump.

6 Change the position of the four-way hydrant valve to direct the flow of water from the fire hydrant through the water supply pump and into the supply line.

© Jones & Bartlett Learning. Photographed by Glen E. Ellman.

Reverse Hose Lay

The reverse hose lay is the opposite of the forward hose lay. In the reverse hose lay, the hose is laid out from the fire to the water source, such as a fire hydrant, in the direction opposite to the flow of the water. It may be a standard tactic in areas where sufficient fire hydrants are available and additional companies that can assist in establishing a water supply will arrive quickly. In this scenario, one of the additional companies will be assigned to lay a supply line from the attack pump to a fire hydrant. In some cases, a reverse lay is used to establish a supply line.

FIRE FIGHTER TIP

When laying out supply hose with threaded couplings, you may find that the wrong end of the hose is on top of the hose bed. Double-male adaptors and double-female adaptors will allow you to join two couplings of the same type or configuration. A set of adaptors (one double-male and one double-female) should be easily accessible for these situations. Some departments place a set of adaptors on the end of the supply hose for this purpose.

To perform the reverse hose lay, the water supply pump stops close to the attack pump on arrival, and supply hose is pulled from the bed of the water supply pump and connected to the suction side of the pump on the attack pump. The water supply pump then drives to the fire hydrant (or alternative water source), connects the supply hose to the water source, and pumps water back to the attack pump. Usually the water supply pump parks in such a way that hose can easily be pulled from the water supply pump to the suction side of the pump on the attack pump.

To perform a reverse hose lay, follow the steps in **SKILL DRILL 16-3**.

Split Hose Lay

A split hose lay is performed by two pump companies in situations where hose must be laid in two different directions to establish a water supply. This evolution could be used when the attack pump must approach a fire either along a dead-end street with no hydrant or down a long driveway. To perform a split hose lay, the attack pump drops the end of its supply hose at the corner of the street and performs a forward lay toward the fire. The water supply pump stops at the same intersection, pulls off enough hose to connect to the end of the supply line that is already there, and then performs a reverse hose lay to the hydrant or static water source. With the two lines connected together, the water supply pump can pump water to the attack pump.

A split hose lay often requires coordination by two-way radio, because the attack pump must advise the water supply pump of the plan and indicate where the end of the supply line is being dropped. In many cases, the attack pump is out of sight when the water supply pump arrives at the split point.

A split hose lay does not necessarily require split hose beds (hose beds that are divided into two or more sections). It can be performed with or without split hose beds if the necessary adaptors are used.

SAFETY TIP

When loading hose on an apparatus, always use caution in climbing up and down on the apparatus. If you are loading hose at a fire scene, watch out for wet, slippery surfaces, ice, or other hazards. Also, wet hose can be heavy, and you may need to reach, stretch, or lift the hose to get it into the hose bed. Use caution!

To perform a split hose lay, follow the steps in **SKILL DRILL 16-4**.

Loading Supply Hose

This section describes commonly utilized procedures for loading supply hose into the hose beds on a fire apparatus. Hose can be loaded in several different ways, depending on how it will be laid out at the fire scene. The hose must be easily removable from the hose bed, without kinks or twists, and without the possibility of becoming caught or tangled. The ideal hose load would be easy to load, avoid wear and tear on the hose, have few sharp bends, and allow the hose to lay out of the hose bed smoothly and easily.

You must learn the specific hose loads used by your fire department. When loading hose, always remember that the time and attention that go into loading the hose properly will prove valuable when it becomes necessary to use the hose at a fire. You should wear appropriate personal protective equipment (PPE) when loading hose—at a minimum, your helmet, gloves, and boots. Remember that many modern hose beds are a metre (several feet) above the ground and that a fall from a hose bed could result in significant injuries. When referring to the hose bed, the end closest to the cab is called the front of the hose bed. The end closest to the tailboard is called the rear of the hose bed.

SKILL DRILL 16-3
Performing a Reverse Hose Lay Fire Fighter I, NFPA 1001: 4.3.15

1 Pull sufficient hose to reach from the water supply pump to the suction side of the fire pump on the attack pump. Anchor the hose to a stationary object if possible. Do not stand between the hose and the stationary object. The water supply pump driver/operator drives away, laying out hose from the attack pump to the fire hydrant or static water source.

2 The attack pump driver/operator connects the supply hose to the suction side of the pump on the attack pump.

3 The water supply pump driver/operator uncouples the supply hose from the hose remaining in the hose bed and attaches the supply hose to the discharge side of the pump on the water supply pump.

4 The water supply pump driver/operator connects the water supply pump to the fire hydrant or water source. Upon the signal from the attack pump, the water supply pump driver/operator charges the supply line.

SKILL DRILL 16-4
Performing a Split Hose Lay Fire Fighter I, NFPA 1001: 4.3.15

1 The attack pump driver/operator stops at the intersection or driveway entrance.

2 A fire fighter from the attack pump removes the end of the supply hose from the hose bed and anchors it.

3 The attack pump driver/operator lays out a hose line while proceeding slowly toward the fire.

4 The attack pump fire fighter either proceeds by foot to the fire or waits at the intersection for the water supply pump, according to local SOP.

5 When the water supply pump arrives at the intersection, it stops and connects the end of its supply hose to the hose end laid in the street by the attack pump. If threaded couplings are used, a double-male adaptor may be required.

6 A fire fighter anchors the end of the supply line from the water supply pump company and lays out a hose line to the fire hydrant or static water source.

7 The attack pump driver/operator can start pumping from the booster tank on the pump and then switch over to the supply hose when the water supply pump is ready to pump from the fire hydrant or static water source.

8 The water supply pump driver/operator positions the apparatus at the water supply according to the local SOP.

9 The water supply pump driver/operator pulls off hose from the hose bed until the next coupling. The hose is uncoupled at this connection and is connected to the discharge side of the pump on the water supply pump or the four-way hydrant valve according to the local SOP.

10 Upon the signal from the attack pump, the water supply pump driver/operator charges the supply line.

© Jones & Bartlett Learning.

Three basic hose loads are commonly used for loading supply hose: the flat hose load, the horseshoe hose load, and the accordion hose load (**FIGURE 16-3**). Any one of these methods can be used to load hose for either a forward hose lay or a reverse hose lay.

Flat Hose Load

The flat hose load is the easiest loading technique to implement and can be used for any size of attack hose or supply hose, including large-diameter hose (LDH). Because the hose is placed flat in the hose bed, it should lay out flat without twists or kinks. With this type of hose load, wear and tear on the edges of the hose from the movement and vibration of the vehicle during travel are minimized. The flat hose load can be used with a single hose bed or a split hose bed. Many variations of the flat hose load exist, so follow your department's SOPs.

To perform a flat hose load, follow the steps in **SKILL DRILL 16-5**.

Horseshoe Hose Load

The horseshoe hose load is accomplished by placing the hose on its edge and positioning it around the perimeter of the hose bed in a U-shape. At the completion of the first U-shape, the hose is folded inward to form another U-shape in the opposite direction. This continues until a complete layer is filled; then another layer is started above the first. When the hose load is completed, the hose in each layer is in the shape of a horseshoe. A major advantage of the horseshoe hose load is that it contains fewer sharp bends than the other hose loads.

A horseshoe hose load cannot be used for LDH because the hose tends to fall over when it stands on edge. This kind of load leads to more wear on the hose

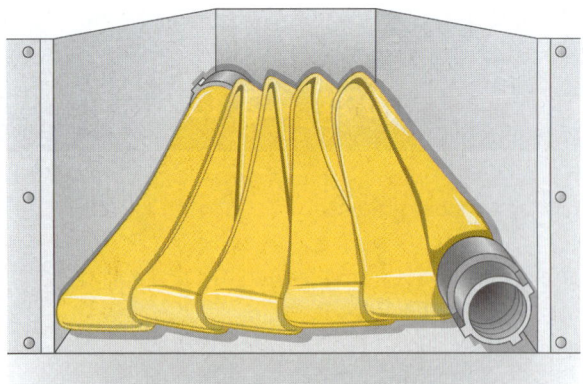

A

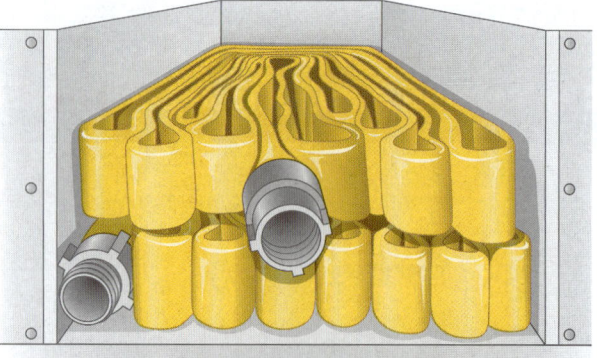

B

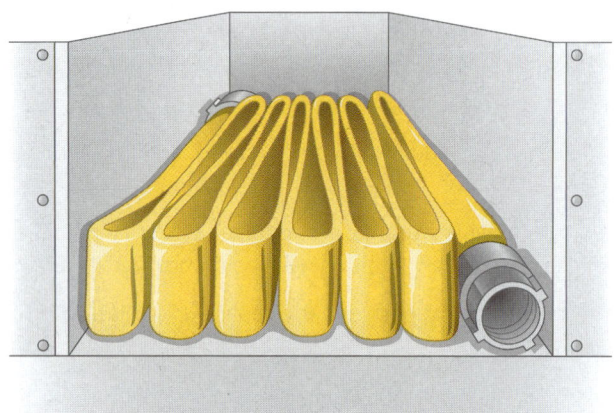

C

FIGURE 16-3 Three basic hose loads are used to load supply hose onto the apparatus. **A.** The flat hose load. **B.** The horseshoe hose load. **C.** The accordion hose load.

© Jones & Bartlett Learning.

SKILL DRILL 16-5
Performing a Flat Hose Load Fire Fighter I, NFPA 1001: 4.5.2

1 If you are loading supply hose with threaded couplings, determine whether the hose will be used for a forward hose lay or a reverse hose lay. To set up the hose for a forward hose lay, place the male hose coupling in the hose bed first. To set up the hose for a reverse hose lay, place the female hose coupling in the hose bed first. Start the hose load with the coupling at the front end of the hose bed.

SKILL DRILL 16-5 Continued
Performing a Flat Hose Load Fire Fighter I, NFPA 1001: 4.5.2

2 Fold the hose back on itself at the rear of the hose bed.

3 Run the hose back to the front end of the hose bed on top of the previous length of hose. Fold the hose back on itself so that the top of the hose is on the previous length.

4 While laying the hose back to the front of the hose bed, angle the hose to the side of the previous fold.

5 Continue to lay the hose in neat folds until the whole hose bed is covered with a layer of hose. To make this hose load neat, make every other layer of hose slightly shorter, or alternate the folds. This keeps the ends from getting too high at the folds. Continue to load the layers of hose until the required amount of hose is loaded.

because the weight of the hose is supported by the edges. Also, when laying out a horseshoe hose load, the hose tends to lay out in a wave-like manner from one side of the street to the other.

Many variations of the horseshoe hose load exist, so follow your department's SOPs. To perform a horseshoe hose load, follow the steps in **SKILL DRILL 16-6**.

SKILL DRILL 16-6
Performing a Horseshoe Hose Load Fire Fighter I, NFPA 1001: 4.5.2

1 If you are loading supply hose with threaded couplings, determine whether the hose will be used for a forward hose lay or a reverse hose lay. For a forward hose lay, start with the male coupling in the rear of the hose bed. For a reverse hose lay, start with the female coupling in the rear of the hose bed.

2 Lay the first length of hose on its edge against the right or left wall of the hose bed.

3 At the front of the hose bed, lay the hose across the width of the bed, and continue down the opposite side toward the rear of the hose bed.

4 When the hose reaches the rear of the hose bed, fold the hose back on itself, and continue laying it back toward the front of the hose bed. Keep the hose tight to the previous row of hose around the hose bed until it is back to the rear on the starting side. Fold the hose back on itself again, and continue packing the hose tight to the previous row.

5 Continue to pack the hose on the first layer. Each fold of hose will decrease the amount of space available inside the horseshoe. Once the center of the horseshoe is filled in, begin a second layer by bringing the hose from the rear of the hose bed and laying it around the perimeter of the hose bed. Complete additional layers using the same pattern as used for the first layer. Finish the hose load with any appliances used by your department.

Accordion Hose Load

The accordion hose load is performed with the hose placed on its edge. In this loading technique, the hose is laid side to side in the hose bed. This load is easy to implement and makes the hose easy to carry. Fire fighters can carry multiple folds per person from the hose bed. One layer is loaded from left to right, and then the next layer is loaded above it from left to right.

The accordion hose load also has some disadvantages. Because the hose is stacked on its side, the hose experiences more wear than with a flat load. The accordion hose load is not recommended for LDH because this kind of hose tends to collapse when placed on its side.

Many variations of the accordion hose load exist, so follow your department's SOPs. To perform the accordion hose load, follow the steps in **SKILL DRILL 16-7**.

SKILL DRILL 16-7
Performing an Accordion Hose Load Fire Fighter I, NFPA 1001: 4.5.2

1 If you are loading hose with threaded couplings, determine whether the hose will be used for a forward hose lay or a reverse hose lay. For a forward hose lay, place the male hose coupling in the rear of the hose bed first. For a reverse hose lay, place the female hose coupling in the rear of the hose bed first. Lay the first length of hose on its edge against the right or left wall of the hose bed.

2 Double the hose back on itself at the front of the hose bed. Leave the female end extended and accessible so that two hose lines can be cross-connected later, if needed, in order to create a single hose line from both sides of a split hose bed.

(continued)

SKILL DRILL 16-7 Continued
Performing an Accordion Hose Load Fire Fighter I, NFPA 1001: 4.5.2

3 Lay the hose next to the first length, and bring it to the rear of the hose bed. Fold the hose at the rear of the hose bed so that the bend is even to the edge of the hose bed. Continue to lay folds of hose across the hose bed. Alternate the length of the hose folds at each end to allow more room for the folded ends.

4 When the bottom layer is completed, angle the hose upward to begin the second tier. Continue the second layer by repeating the steps used to complete the first layer.

© Jones & Bartlett Learning. Photographed by Glen E. Ellman.

Split Hose Beds

A **split hose bed** is a hose bed that is divided into two or more sections. This division is made for several purposes:

- One compartment in a split hose bed can be loaded for a forward hose lay (with the female coupling hanging below the hose bed), and the other side can be loaded for a reverse hose lay (with the male coupling hanging below the hose bed). This allows a line to be laid in either direction without adaptors.
- Two parallel hose lines can be laid at the same time (called "laying dual hose lines"). Dual hose lines are beneficial if the situation requires more water than one hose line can supply.
- The split hose beds can be used to store hose of different size. For example, one side of the hose bed could be loaded with 65-mm (2½-in.) hose that can be used as supply hose or as attack

hose. The other side of the hose bed could be loaded with 127-mm (5-in.) hose for use as a supply hose. This setup enables the use of the most appropriate-size hose for a given situation.

- All of the hose from both sides of the hose bed can be laid out as a single hose line. This is done by coupling the end of the hose in one bed to the beginning of the hose in the other bed.

A combination hose load is a hose load used when one long hose line is needed. When a split hose bed is loaded for a combination load, the end of the last length of a hose in one bed is coupled to the beginning of the first length of hose in the opposite bed. All of the hose plays out of one hose bed first, and then the hose continues to play out from the second hose bed. To lay dual hose lines, the connection between the two hose beds is uncoupled, and the hose can play out of both beds simultaneously. When the two sides of a split bed are loaded with the hose in opposite directions, either a double-female or double-male adaptor is used to make the connection between the two hose beds.

Connecting a Pump to a Water Supply

When a pump is located at a fire hydrant, a supply hose must be used to deliver the water from the fire hydrant to the pump. Supply hose is used to deliver as much water as possible over a short distance. In most cases, a soft sleeve hose (also known historically as a soft suction hose) is used to transport water from a pressurized source, such as a fire hydrant, to the suction side of the fire pump. Although it is uncommon, the connection can also be made with a hard suction hose or with a short length of large-diameter (soft suction) supply hose.

Attaching a Soft Sleeve Hose to a Fire Hydrant

Large-diameter soft sleeve hose is normally used to connect an attack pump directly to a fire hydrant. To attach a soft sleeve hose to a fire hydrant, follow the steps in **SKILL DRILL 16-8**.

Supply Hose Carries and Advances

Several different techniques are used to carry and advance supply hose. The best technique for a particular situation will depend on the size of the hose, the distance over which it must be moved, and the number of fire fighters available to perform the task. The same techniques can be used for supply lines and attack lines.

Whenever possible, a hose line should be laid out and positioned as close as possible to the location where it will be operated before it is charged with water. A charged hose line is much heavier and more difficult to maneuver than a dry hose line. A suitable amount of extra hose should be available to allow for maneuvering after the hose line is charged.

Working Hose Drag

The working hose drag technique is used to deploy hose from a hose bed and advance the hose line over a relatively

---

FIRE FIGHTER TIP

The following tips will help you to do a better job when loading hose:

- Drain all of the water out of the hose before loading it.
- Rolling the hose first will result in a flatter hose load, because there will be no air in the hose.
- Do not load hose too tightly. Leave enough room so that you can slide a hand between the folds of hose. If hose is loaded too tightly, it may not lay out properly.
- Load hose so that couplings do not have to turn around as the hose is pulled out of the hose bed. Make a short fold in the hose close to the coupling to keep the hose properly oriented. This short fold is called a **Dutchman**. The Dutchman fold allows the hose to deploy or unload smoothly; it prevents the coupling from turning and becoming stuck as the hose is laid out (**FIGURE 16-4**).
- Couple sections of hose with the flat sides oriented in the same direction.
- Check swivel gaskets before coupling hose.
- Tighten couplings so that they are hand tight only. With a good gasket, the hose should not leak.

FIGURE 16-4 A Dutchman fold prevents the coupling from turning and becoming stuck as the hose is laid out.
Courtesy of Scott Dornan, ConocoPhillips Alaska.

SKILL DRILL 16-8
Attaching a Soft Sleeve Hose to a Fire Hydrant Fire Fighter I, NFPA 1001: 4.3.15

1 The pump driver/operator positions the apparatus so that the suction side of the pump on the attack pump is the correct distance from the fire hydrant. Remove the hose from the hose bed along with any needed adaptors and the hydrant wrench.

2 Attach the soft sleeve hose to the suction side of the pump on the attack pump if it is not already attached. In some departments, this end of the hose is preconnected. It may be necessary to use an adaptor.

3 Unroll the hose.

4 Remove the large fire hydrant cap. Check the fire hydrant for proper operation.

SKILL DRILL 16-8 Continued
Attaching a Soft Sleeve Hose to a Fire Hydrant Fire Fighter I, NFPA 1001: 4.3.15

5 Attach the soft sleeve hose to the fire hydrant.

6 Ensure that there are no kinks or sharp bends in the hose that might restrict the flow of water.

7 Open the fire hydrant valve slowly when indicated by the driver/operator. Check all connections for leaks. Tighten the couplings if necessary.

8 Where required, place chafing blocks under the hose where it contacts the ground to prevent mechanical abrasion.

© Jones & Bartlett Learning. Photographed by Glen E. Ellman.

short distance to the desired location. Depending on the size and length of the hose, several fire fighters may be required to perform this task.

To perform a working hose drag, follow the steps in **SKILL DRILL 16-9**.

Shoulder Carry

The shoulder carry is used to transport full lengths of hose over a longer distance than it is practical to drag the hose. It is also useful when a hose line must be advanced around obstructions. For example, this technique can be used to stretch an attack line from the front of a building, around to the rear entrance, and up to the second floor. The shoulder carry can also be employed to stretch a supply line to an attack pump in a location where the hose cannot be laid out by another pump company.

This technique requires practice and good teamwork to be successful. By working together to complete tasks efficiently, fire fighters can achieve their goal of extinguishing the fire in the shortest period of time. To perform a shoulder carry, follow the steps in **SKILL DRILL 16-10**.

To advance an accordion load using a shoulder carry, follow the steps in **SKILL DRILL 16-11**.

SKILL DRILL 16-9
Performing a Working Hose Drag Fire Fighter I, NFPA 1001: 4.3.10

1 Place the end of the hose over one shoulder. Hold on to the coupling with the alternate hand.

2 Walk in the direction you want to advance the hose.

3 As the next hose coupling is ready to come off the hose bed, a second fire fighter grasps the coupling and places the hose over the shoulder.

4 Continue this process until enough hose has been pulled out of the hose bed.

© Jones & Bartlett Learning. Photographed by Glen E. Ellman.

SKILL DRILL 16-10
Performing a Shoulder Carry Fire Fighter I, NFPA 1001: 4.3.10

1 Stand at the tailboard of the apparatus. Grasp the end of the hose, and place it over your shoulder so that the coupling is at chest height. Have a second fire fighter place additional hose on your shoulder so that the ends of the folds reach about knee level. Continue to place folds on your shoulder, but only as much as you can safely carry.

2 Hold the hose to prevent it from falling off your shoulder. Continue to hold the hose, and move forward about 4.5 m (15 ft).

3 A second fire fighter then stands at the tailboard to receive a load of hose and then moves forward. When enough fire fighters have received hose loads, the hose can be uncoupled from the hose bed. The pump driver/operator connects the coupling to the pump.

4 All of the fire fighters start walking toward the fire. The fire fighter closest to the apparatus starts offloading hose from his or her shoulder. Once the fire fighter closest to the apparatus has laid out all his or her hose, the next fire fighter in line starts laying out hose from his or her shoulder. Each fire fighter lays out his or her supply of hose until the entire length is laid out.

© Jones & Bartlett Learning. Photographed by Glen E. Ellman.

SKILL DRILL 16-11
Advancing an Accordion Load Fire Fighter I, NFPA 1001: 4.3.10

1 Find the end of the accordion load, whether it is a nozzle or a coupling. Using two hands, grasp the end of the load and the number of folds it will take to make an adequate shoulder load.

2 Pull the accordion load about one-third of the way off the apparatus.

3 Twist the folds so that they become flat, with the end of the accordion load (nozzle or coupling) on the bottom of the now flat shoulder load.

4 Transfer the hose to your shoulder while turning so that you face in the direction you will walk.

SKILL DRILL 16-11 Continued
Advancing an Accordion Load Fire Fighter I, NFPA 1001: 4.3.10

5 Place the shoulder load over your shoulder, and grasp it tightly with both hands.

6 Walk away from the apparatus, pulling the shoulder load out of the hose bed. If additional fire fighters are needed, they may follow steps 1–4 to assist in removing the required amount of hose.

© Jones & Bartlett Learning. Photographed by Glen E. Ellman.

Connecting Supply Hose Lines to Standpipe and Sprinkler Systems

Another water supply evolution is furnishing water to standpipe and sprinkler systems. Fire department connections (FDCs) on buildings are provided so that the fire department can pump water into standpipe and sprinkler systems. This setup is considered to be a supply line because it supplies water to the standpipe and sprinkler systems. What is different is that this supply line is connected to the discharge side of the pump on the attack pump.

The function of the hose line in this case is to provide either a primary or secondary water supply for the sprinkler or standpipe system. The same basic techniques are used to connect the hose lines to either type of system.

Standpipe systems are used to provide a water supply for attack lines that will be operated inside the building. Outlets are provided inside the building where

fire fighters can connect attack lines. The fire fighters inside the building must then depend on fire fighters outside the building to supply the water to the FDC (**FIGURE 16-5**).

FIGURE 16-5 A standpipe connection.
© Jones & Bartlett Learning. Photographed by Glen E. Ellman.

Two types of standpipe systems exist:

- A dry standpipe system depends on the fire department to provide all of the water.
- A wet standpipe system has a built-in water supply, but the FDC is provided to deliver a higher flow or to boost the pressure.

The pressure requirements for standpipe systems depend on the height where the water will be used inside the building.

The FDC for a sprinkler system is also used to supplement the normal water supply. The required pressures and flows for different types of sprinkler systems can vary significantly. As a guideline, sprinkler systems should be fed at 1034 kPa (150 psi) unless more specific information is available.

To connect a hose line to supply an FDC, follow the steps in **SKILL DRILL 16-12**.

SKILL DRILL 16-12
Connecting a Hose Line to Supply a Fire Department Connection
Fire Fighter I, NFPA 1001: 4.3.15

1 Locate the fire department connection (FDC) to the standpipe or sprinkler system. Extend a hose line from the discharge side of the pump on the apparatus to the FDC using the size of hose required by the fire department's SOPs. Some fire departments use a single hose line, whereas others call for two or more lines to be connected.

2 Remove the caps on the standpipe inlet. Some caps are threaded into the connections and must be unscrewed. Other caps are designed to break away when struck with a tool such as a hydrant wrench or spanner wrench (hose key).

3 Visually inspect the interior of the connection on the FDC to ensure that it does not contain any debris that might obstruct the water flow. Never stick your hand or fingers inside the connections; fire fighters have been injured from sharp debris left inside these connections. Attach the hose line to the FDC. Notify the pump driver/operator when the connection has been completed.

Attack Line Evolutions

Attack line operations deliver water from an attack pump to a handline (a hose and nozzle that can easily be held and maneuvered by hand), which discharges the water onto the fire. Attack line evolutions are standard methods of working with attack lines to accomplish different objectives in a variety of situations.

Attack lines are usually stretched from an attack pump to the fire. The attack pump is usually positioned close to the fire, and attack lines are stretched manually by fire fighters. In some situations, an attack pump will drop an attack line at the fire and drive from the fire to a fire hydrant or water source. This procedure is similar to the reverse hose lay evolution for supply lines, except that the hose will be used as an attack line.

Most pumpers are equipped with preconnected attack lines, which provide a predetermined length of attack hose that is already equipped with a nozzle and connected to a pump discharge outlet. The outlet connection may be in the hose bed or in another location, depending on the individual configuration of the pumper. An additional supply of attack hose that is not preconnected is usually carried in another hose bed or compartment. To create an attack line with this hose, the desired length of hose is removed from the bed, disconnected from the next coupling in the hose bed, and attached to the pump discharge outlet. This hose can also be used to extend the length of a preconnected attack line or to attach to a wye or a water thief.

Attack hose is loaded in such a manner that it can be quickly and easily deployed. There are many ways to load attack lines into a hose bed, and this section presents only a few of the most common hose loads. Your department might use a variation of one of these techniques. It is important for you to master the hose loads used by your department.

Loading Preconnected Attack Lines

Preconnected hose lines are intended for immediate use as attack lines. A preconnected hose line has a predetermined length of hose with the nozzle already attached, and it is connected to the pump discharge outlet on the attack pump. Because every fire situation is different, most departments load attack lines of different lengths. A pumper could be equipped with a 46-m (150-ft) preconnected line and a 76-m (250-ft)

preconnected line, for example. Fire fighters should pull an attack line that is long enough to reach the fire but not so long that an excess of hose might slow down the operation and become tangled. The most commonly used attack lines are 45-mm (1¾-in.) hose. Many pumps are also equipped with preconnected 65-mm (2½-in.) hose lines to enable them to make a quick attack on larger fires.

Preconnected hose lines can be placed in several different locations on the apparatus. For example, a section of a divided hose bed at the rear of the apparatus can be loaded with a preconnected attack line. Transverse hose beds are installed above the pump and loaded so that the hose can be pulled off from either side of the apparatus. Preconnected lines can be loaded into special trays that are mounted on the side of fire apparatus. Many pumps also include a special compartment in the front bumper that can store a short preconnected hose line. This hose line is often used for vehicle fires and dumpster fires, where the apparatus can drive up close to the incident and a longer hose line is not needed. This hose line is sometimes referred to as the "trash line."

Booster hose is another type of preconnected attack line. Booster hose reels holding 25-mm (1-in.) hose can be mounted in a variety of locations on fire apparatus. Booster hose should not be used for structural or vehicle firefighting.

Attack lines should be loaded in the hose bed in a manner that ensures they can be quickly stretched from the attack pump to the fire. It should be possible for one or two fire fighters to remove the hose quickly from the hose bed and advance the hose to the fire. Whichever hose load technique is used, the hose should not become tangled as it is being removed from the hose bed and advanced. Laying out the hose should not require multiple trips between the pumper and the fire attack point. It should be easy to lay the hose around obstacles and corners. If additional fire fighters are available, they can remain behind to help move hose line around any obstacles. It should also be possible to repack the hose quickly and with minimal personnel. There is no perfect hose load that works well for every situation. The three most common hose loads for preconnected attack lines are the minuteman hose load, the flat hose load, and the triple-layer hose load. Variations in circumstances can make one type of hose load preferable for your community.

Minuteman Hose Load

The minuteman hose load allows a single fire fighter to flake out, or drop, one fold or loop of hose at a time

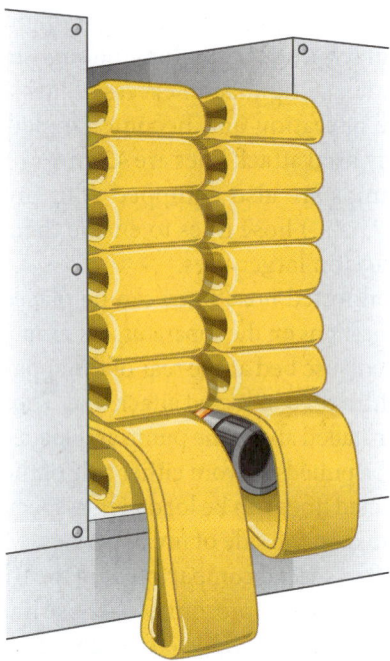

FIGURE 16-6 The minuteman hose load allows a single fire fighter to flake out, or drop, one fold or loop of hose at a time from the shoulder, as he or she advances toward the fire.
© Jones & Bartlett Learning.

from the shoulder, as he or she advances toward the fire (**FIGURE 16-6**). This load avoids having to maneuver the hose lines around obstacles and helps to prevent sharp kinks. It maintains sufficient hose for entry into the building at the entrance to the building. The minuteman hose load controls the amount of hose the nozzle person can deploy, reducing the chances of a fire fighter taking too much hose. This load requires some knowledge to load but is easier to load than the triple-layer hose load.

To perform a minuteman hose load, follow the steps in **SKILL DRILL 16-13**.

To advance a minuteman hose load, follow the steps in **SKILL DRILL 16-14**.

Preconnected Flat Hose Load

The **preconnected flat hose load** is loaded in a similar manner to the flat hose load used for supply lines, except the female end of the hose line is preconnected to a pump discharge outlet, and a nozzle is attached to the male end of the hose line (**FIGURE 16-7**). The flat hose load is easy to load and easy to deploy. It can be deployed on the shoulder

SKILL DRILL 16-13
Performing a Minuteman Hose Load Fire Fighter I, NFPA 1001: 4.5.2

1 Connect the female end of the first length of hose to the preconnected discharge outlet. Flat load the hose even to the edges of the hose bed. At the second fold, make a loop or ear. This loop will create a handle to be used when advancing the hose load.

2 Flat load the rest of the 31 m (100 ft) of hose. Place the male hose coupling to the side of the hose bed.

SKILL DRILL 16-13 Continued
Performing a Minuteman Hose Load Fire Fighter I, NFPA 1001: 4.5.2

3 Couple the remaining two hose sections, and attach the nozzle.

4 Place the nozzle in the hose bed facing outward. Feed one layer of hose under the nozzle. Wrap the first layer of hose around the tip of the nozzle.

5 Continue flat loading the remaining two sections of hose.

6 Connect the female coupling from the last section of hose to the male coupling from the first section of hose.

SKILL DRILL 16-14
Advancing a Minuteman Hose Load Fire Fighter I, NFPA 1001: 4.3.10

1 Grasp the nozzle and the folds on top of it. Pull this top part of the load approximately one-third out of the hose bed.

2 Turn away from the hose bed, and place this top part of the hose load on your shoulder. Walk away from the apparatus until the rest of the top section of hose drops from the hose bed.

3 Turn back around toward the hose bed, and grasp the loop or ear from the first section of hose.

4 Walk away from the apparatus until all hose is clear from the hose bed. Continue walking away, allowing the rest of the hose to deploy from the top of the load on your shoulder.

FIGURE 16-7 The preconnected flat hose load is loaded in a similar manner to the flat hose load used for supply lines, except the female end of the hose line is preconnected to a pump discharge outlet, and a nozzle is attached to the male end of the hose line.
© Jones & Bartlett Learning.

or by grasping the loop of the hose line with your arm. It can be used in jurisdictions that have a variety of types of structures, including single family and multifamily buildings. It is also suited to taller buildings with stairs.

To perform a preconnected flat hose load, follow the steps in **SKILL DRILL 16-15**.

To advance a preconnected flat hose load, follow the steps in **SKILL DRILL 16-16**.

Triple-Layer Hose Load

The triple-layer hose load is suited for departments that generally respond to fires in one- or two-storey single family dwellings (**FIGURE 16-8**). It is a hose loading method in which the hose is folded back onto itself to reduce the overall length to one-third before loading the hose in the bed. This method is used by pump companies with minimal personnel. One fire fighter can clear the hose bed and deploy this type of hose load. It can be deployed and charged before entering the building. The triple-layer hose load is more difficult to load. It requires practice and space in which to lay out the hose beside the apparatus.

SKILL DRILL 16-15
Performing a Preconnected Flat Hose Load Fire Fighter I,
NFPA 1001: 4.5.2

1 Attach the female end of the hose to the preconnect discharge outlet.

2 Begin flat loading the hose in the hose bed.

(continued)

SKILL DRILL 16-15 Continued
Performing a Preconnected Flat Hose Load Fire Fighter I,
NFPA 1001: 4.5.2

3 Load the first layer of hose, and make the first fold even with the edge of the hose bed. Then, make a loop, or ear, on the second or higher fold. This loop will be used as a pulling handle.

4 Flat load the remainder of the hose in the hose bed. Attach the nozzle to the male coupling at the end of the hose.

© Jones & Bartlett Learning. Photographed by Glen E. Ellman.

SKILL DRILL 16-16
Advancing a Preconnected Flat Hose Load Fire Fighter I,
NFPA 1001: 4.3.10

1 Grasp the top section of hose, and place it over your shoulder.

2 Turn away from the apparatus until the top load drops onto your shoulder.

SKILL DRILL 16-16 Continued
Advancing a Preconnected Flat Hose Load Fire Fighter I,
NFPA 1001: 4.3.10

3 Turn back around toward the hose bed, and grasp the lower loop, or ear. Pull the remainder of the hose from the hose bed.

4 Turn away from the apparatus, and continue walking, letting the hose lay out off of your shoulder.

© Jones & Bartlett Learning. Photographed by Glen E. Ellman.

FIGURE 16-8 The triple-layer hose load method is a hose loading method in which the hose is folded back onto itself to reduce the overall length to one-third before loading the hose in the bed.

© Jones & Bartlett Learning.

To perform a triple-layer hose load, follow the steps in **SKILL DRILL 16-17**.

To advance a triple-layer hose load, follow the steps in **SKILL DRILL 16-18**.

Wyed Lines

To reach a fire that is some distance from the pump, it can be necessary first to advance a larger-diameter supply line, such as a 65-mm (2½-in.) hose line, and then split it into two 45-mm (1¾-in.) attack lines. This is accomplished by attaching a gated wye or a water thief to the end of the 65-mm (2½-in.) line and then attaching the two attack lines to the gated outlets.

FIRE FIGHTER TIP

To prepare to advance attack hose, form an upright loop about 1.2 m (4 ft) in diameter outside the door to the fire building. If more hose is needed, one person can roll this loop toward the nozzle. This will yield several additional feet of hose at the nozzle.

SKILL DRILL 16-17
Performing a Triple-Layer Hose Load Fire Fighter I, NFPA 1001: 4.5.2

1 Attach the female end of the hose to the preconnect discharge outlet.

2 Connect the sections of hose together.

3 Extend the hose directly from the hose bed. Pick up the hose two-thirds of the distance from the preconnect discharge outlet to the hose nozzle.

4 Carry the hose back to the apparatus, forming a three-layer loop.

5 Pick up the entire length of folded hose. (This will take at least two fire fighters.)

6 Lay the triple-folded hose in the hose bed in an S-shape with the nozzle on top.

SKILL DRILL 16-18
Advancing a Triple-Layer Hose Load Fire Fighter I, NFPA 1001: 4.3.10

1 Grasp the nozzle and the top fold.

2 Turn away from the hose bed, and place the hose on the shoulder. Walk away from the apparatus until the entire load is out of the bed.

3 When the load is out of the bed, drop the fold.

4 Extend the nozzle the remaining distance.

© Jones & Bartlett Learning. Photographed by Glen E. Ellman.

One method of unloading and advancing wyed lines is shown in **SKILL DRILL 16-19**.

Advancing Attack Lines

The purpose of this section is to help fire fighters build upon the hose handling skills they have already been taught and to integrate these new skills with those they already know. Remember that the overall goal is to get the attack line from the apparatus to the fire as quickly, efficiently, and safely as possible. This section consists of two main parts: The first part focuses on techniques and tips for getting the attack line from the attack pump to the building entry point. The second part of this section covers techniques for advancing the attack line from the entry point of the building to the fire.

Attack lines are used for three different types of hose evolutions: offensive, defensive, and transitional operations. When fire fighters advance hose lines (one

SKILL DRILL 16-19
Unloading and Advancing Wyed Lines Fire Fighter I, NFPA 1001: 4.3.10

1 Grasp the wye that is attached to the end of a 65-mm (2½-in.) attack line, and pull it from the bed.

2 Advance the 65-mm (2½-in.) attack line toward the fire.

3 Attach the female end of a 45-mm (1¾-in.) attack line to one outlet on the gated wye.

4 Attach the female end of a second 45-mm (1¾-in.) attack line to the second outlet of the gated wye. The individual 45-mm (1¾-in.) attack lines can now be extended to the desired positions.

or more lengths of hose) toward a building to attack a fire, the operation is termed *offensive*. By contrast, a *defensive* operation is conducted from the exterior, or while moving away from the building or fire, by directing water streams toward the fire from a safe distance. A third type of fire attack is a combined, offensive, exterior and interior operation, sometimes referred to as a *transitional attack*, a blitz attack, softening the target, or resetting the fire. As a new fire fighter, you are not responsible for determining the type of fire attack that will be used. More information about these evolutions is presented in Chapter 17, *Fire Suppression*.

Advancing Attack Lines from the Attack Pump to the Door

There are many techniques used to advance attack lines from the attack pump to the entry point of the fire building. Regardless of the technique used, the objective is to lay out the hose in a single pass using limited personnel so that the hose line can be charged quickly in a manner that permits the attack crew to advance the hose line from the entry point to the seat of the fire. The technique used will depend on the distance from the apparatus to the door, the number of obstacles between the apparatus and the entry point, the size of the fire building, and the distance from the entry point to the seat of the fire. All of these factors will impact the type of hose load fire fighters will use

and the evolution they will use to advance the hose to the entry point. The evolutions presented earlier in this chapter provide a preliminary framework of skills. When the attack line has been laid out to the entry point, the hose should be flaked out in a serpentine pattern with lengths of hose running parallel to the front of the fire building so that it can be easily advanced into the building and it will not become tangled when it is charged (**FIGURE 16-9**). It should also be set back from the doorway so that it does not obstruct the entry and exit path.

Make sure that you flake out the hose before it is charged with water. Once the hose line is charged, the hose becomes much more difficult to maneuver and advance. It is also important to verify that the hose length will reach the location where the hose line is needed inside the building.

While fire fighters are flaking out the hose and preparing to enter the building, the company officer will be completing the size-up. Other members of the team will be carrying out other tasks to support the operation. For example, they might be forcing entry, getting into position for ventilation, and establishing a water supply. All of these tasks must be performed in a coordinated manner to maintain a safe environment and to extinguish the fire efficiently.

Once the hose is flaked out, signal the driver/operator to charge the hose line. Open the nozzle slowly to bleed out any trapped air and to make sure the hose is operating properly. If you are using an adjustable nozzle,

FIGURE 16-9 Before advancing an attack line into a building, the hose should be flaked out in a serpentine pattern outside the building entrance. This ensures that the hose will not become tangled once it is charged.

make sure the nozzle is set to deliver the appropriate stream. Once this is done, slowly close the nozzle.

Quickly recheck all parts of your PPE. Make sure your coat is fastened and your collar is turned up and fastened in front. Check your gloves, and make sure you have a hand light. Check your partner's equipment, and have your partner check your equipment. If you have time, catch your breath while you are breathing ambient air. In cold weather conditions, refrain from breathing on ambient air as frost can build up on the inside of your SCBA face piece. Be ready to start breathing air from your SCBA and to advance the charged hose line as soon as your officer directs you to do so.

When you are given the command to advance the hose, maintain situational awareness. Your safety and that of your fellow fire fighters are paramount. Make sure the other members of the nozzle team are ready. Do not stand in front of the door as it is opened: You do not know what might happen.

Advancing Attack Lines from the Door to the Fire

As you move inside the building, stay low to avoid the greatest amount of heat and smoke. If you cannot see because of the dense smoke, use your hands to feel the pathway in front of you so that you do not fall into a hole or other opening. If visibility is low, stop, take a breath, and listen for the sounds created by the fire. Look for the glow of fire, and check for the sensation of heat coming through your face piece. Communicate with the other members of the nozzle team as you advance.

Advancing a hose line from the door of the building to the fire requires a team effort. A hose team may consist of as few as two members. Whenever possible, placing more people on this team will result in a more efficient and smoother operation. Ideally, a hose line crew consists of at least three members on the hose line near the nozzle and a fourth member at the door. Charged hose lines are not easy to advance through a house or other building. As resistance is encountered in advancing the hose line, the fire fighter at the nozzle can help to pull more hose, while the fire fighter at the door is responsible for feeding more hose into the building and controlling the door, thereby controlling the air flow into the building. This door control will limit the ability of the fire to grow rapidly as fire fighters get into position to apply water.

FIRE FIGHTER TIP

Hose advancement is seldom successful when you rely on brute strength alone.

As you advance the hose line, you need to have enough hose to enable you to move forward. Successful hose advancement is based on anticipation of the need and teamwork. Move the hose using the large muscles of your legs rather than tiring the smaller, weaker muscles of your arms. Use your arms for small movements.

The goal should be to have surplus hose to reach the target in one pass, rather than having to go back to get more hose. If everyone on the hose line initially moves an appropriate amount of hose, you should have enough to reach the target with a little surplus. The objective of moving hose is to move the amount that is needed, not to move as much hose as possible. Too much hose placed close to the nozzle can cause as much of a problem as having too little hose at the nozzle. The amount of hose that each member of the team takes will be dependent on several factors. These factors include the number of people on the team, the length of the hose, and the distance between friction points.

Friction points are those places along the hose line where the hose is changing directions, turning a corner, going through a doorway, or going up or down stairs. Friction points are those places where you can predict that the hose will become caught and will slow or stop the advance of the hose line. All members of the hose line team should work to avoid problems at friction points and to provide the best angle from which to apply water to the fire. When approaching inward-opening doors, approach from the hinge side to allow the nozzle person to direct the stream into the overhead without obstruction. When approaching outward-opening doors, the unhinged side provides clear access to the room and allows the nozzle team to make a bight (a U-bend with two parallel ends) outside the door and advance the hose line on the unhinged side of the door to avoid the friction point on the hinged side of the door.

When you encounter a corner, pull the hose to the outside of the corner to create surplus hose and to avoid contact with the inside corner friction point. This surplus hose allows the nozzle person enough surplus hose that he or she can choose the best angle of attack for the stream. In cases where there may be roll-over, it is possible to direct a stream of water down the hallway in the overhead. This angle gives the nozzle person access to that area while providing him or her with cover. Create surplus hose by pulling extra hose in front of a friction point. This hose can be pulled into a bight or a loop. The bight or loop can be flat on the ground, or it can be placed against the wall. When moving hose forward between two friction points, it is usually best to work in the middle between the two friction points. When you are located at a corner, position yourself to the outside of the hose line. The nozzle person must communicate continuously with other team members so they will know how much more hose will be needed to reach the fire.

It is a good idea to have someone tending each corner and to be ready to move in an organized fashion. Each member on the hose line takes control over the span of hose between him or her and the next member. The length of the hose span is based on the distance between the friction points. Members must space out on the hose line. Try to maximize the amount of hose carried by each person. This will help to maintain a fast, efficient stretch. The use of bights or loops will help to keep surplus hose available and ready for any needed advancement. Helpful tips for successful hose movement include the following:

- Communicate hose needs, distance, and direction.
- Move hose line with your legs; feed with your arms from the corners.
- Create S bights or loops for surplus spans.

- Position team members in corners.
- Overshoot corners and load off hose in the direction of travel.

Advancing an Attack Line up a Stairway

When advancing a hose line up stairs, arrange to have an adequate amount of extra hose close to the bottom of the stairs. Utilizing a shoulder carry with the minuteman hose load allows for easy advancement. Make sure all members of the team are ready to move on command. It is difficult to move a charged hose line up a set of stairs while flowing water through the nozzle; shutting down the hose line while you are moving up the stairs will often allow you to get to the top of the stairs more quickly and safely. Follow the directions of your officer.

To advance an uncharged attack line up a stairway, follow the steps in **SKILL DRILL 16-20**.

SKILL DRILL 16-20
Advancing an Attack Line up a Stairway Fire Fighter I, NFPA 1001: 4.3.10

1 Use a shoulder carry to advance up the stairs.

2 When ascending the stairway, lay the hose against the outside of the stairs to avoid sharp bends and kinks and to reduce tripping hazards.

(continued)

SKILL DRILL 16-20 Continued
Advancing an Attack Line up a Stairway Fire Fighter I, NFPA 1001: 4.3.10

© Jones & Bartlett Learning. Photographed by Glen E. Ellman.

3 Arrange excess hose so that it is available to fire fighters entering the fire floor.

Advancing an Attack Line down a Stairway

Advancing a charged hose line down a stairway is difficult and dangerous. The smoke and flames from the fire tend to travel up the stairway in a residential structure. This means fire fighters will be operating in the exhaust portion of the flow path. Due to the risks of operating in a potential flow path, fire officers should consider cooling the fire environment from the safest location possible, the exterior. Applying a straight or solid stream into an exterior window prior to entry will cool the fire environment, limit fire damage to the floor system, and reduce the thermal assault to the operating members. All of these actions improve the operating environment for the fire company. If the application of an exterior attack line is not possible and an interior attack must be made, get down the stairway, and position yourself below the heat and smoke as quickly as you can, making sure the hose does not get stuck or kink on a friction point, as this can restrict the flow of water and potentially compromise safety. Keep as low as possible to avoid the worst of the heat and smoke. When you advance a hose line down a stairway, you have a major advantage on your side: Gravity is working with you to bring the hose line down the stairs. You

should never advance toward a fire unless your hose line is charged and ready to flow water. Wearing PPE and SCBA changes your center of gravity. If you try to crawl down a stairway headfirst, you are likely to find yourself tumbling head over heels. Instead, move down the stairway feet first, using your feet to feel for the next step. Move carefully but as quickly as possible to get below the heat and smoke.

To advance a charged attack line down a stairway, follow the steps in **SKILL DRILL 16-21**.

Advancing an Attack Line up a Ladder

If a hose line has to be advanced up a ladder, it should be done before the hose line is charged. The hose line should be placed across your chest, with the nozzle draped over your shoulder. With the hose arranged in this way, the hose line will not push you away from the ladder if it is mistakenly charged while you are climbing.

Additional fire fighters should pick up the hose about every 7.6 m (25 ft) and help to advance it up the ladder. Additional hose should be fed up the ladder until sufficient hose is inside the building to reach the fire.

To advance an attack line up a ladder, follow the steps in **SKILL DRILL 16-22**.

SKILL DRILL 16-21
Advancing an Attack Line down a Stairway Fire Fighter I,
NFPA 1001: 4.3.10

1 Advance forward with the charged hose line.

2 Descend stairs feet first. If there is smoke, position yourself underneath it.

3 Position fire fighters at areas where hose lines could snag.

© Jones & Bartlett Learning. Photographed by Glen E. Ellman.

SKILL DRILL 16-22

Advancing an Uncharged Attack Line up a Ladder Fire Fighter I, NFPA 1001: 4.3.10

1 If a hose line needs to be advanced up a ladder, it should be advanced before it is charged. Advance the hose line to the ladder. Pick up the nozzle; place the hose across the chest, with the nozzle draped over the shoulder. Climb up the ladder with the uncharged hose line.

2 Once the first fire fighter reaches the first fly section of the ladder, a second fire fighter shoulders the hose to assist advancing the hose line up the ladder. To avoid overloading of the ladder, enforce a limit of one fire fighter per fly section. The nozzle is placed over the top rung of the ladder and advanced into the fire area.

3 Additional hose can be fed up the ladder until sufficient hose is in position. The hose can be secured to the ladder with a hose strap to support its weight and keep it from becoming dislodged.

Operating an Attack Line from a Ladder

A hose stream can be operated from a ladder and directed into a building through a window or other opening. To operate an attack line from a ladder, follow the steps in **SKILL DRILL 16-23**.

Extending an Attack Line

When choosing a preconnected hose line or assembling an attack line, it is better to have too much hose than not enough. With a hose that is longer than necessary, you can flake out the excess hose. With a hose that is too short,

---

SKILL DRILL 16-23
Operating an Attack Line from a Ladder Fire Fighter I, NFPA 1001: 4.3.10

1 Climb the ladder with a hose line to the height at which the hose line will be operated. Apply a leg lock, or use a ladder belt.

2 Secure the hose to the ladder with a hose strap or rope hose tool, rope, or piece of webbing.

(continued)

SKILL DRILL 16-23 Continued
Operating an Attack Line from a Ladder Fire Fighter I, NFPA 1001: 4.3.10

3 Carefully operate the hose stream from the ladder. Be careful when opening and closing nozzles and redirecting the stream because of the nozzle backpressure. This force could destabilize the ladder.

© Jones & Bartlett Learning. Photographed by Glen E. Ellman.

you cannot advance it to the seat of the fire without shutting down the hose line and taking the time to extend it.

In some circumstances, it can be necessary to extend a hose line by adding additional lengths of hose—for example, if the fire is farther from the apparatus than initially estimated. More hose line could also be needed to reach the burning area or to complete extinguishment.

There are two basic ways to extend a hose line. First, fire fighters can disconnect the hose from the discharge outlet on the attack pump and add the extra hose at that location. This method requires advancing the full length of the attack line to take advantage of the extra hose, which could take time and considerable effort.

Alternatively, fire fighters can add the extra hose to the discharge end of the attack line. This addition can be achieved if the nozzle is a **breakaway type nozzle** that can be separated from the shut-off valve. With this type of nozzle, the valve is closed to stop the flow of water, the nozzle tip is removed, and the extra hose is attached to the shut-off valve. The nozzle tip can then be installed on the male end of the added hose. This evolution, which should occur in a protected area, allows fire fighters to lengthen the attack hose without shutting off the valve at the pump on the apparatus. A standpipe kit can be used to supply the added section of hose. Standpipe

kits include the appropriate hose and nozzle, a spanner wrench (hose key), and any required adaptors.

> **FIRE FIGHTER TIP**
>
> Upon arrival at a high-rise fire, note the location of the standpipe stations (hose cabinets) on an uninvolved floor so you can orient yourself to its location. These cabinets will be aligned with one another on each floor.

Advancing an Attack Line from a Standpipe Outlet

Standpipe outlets inside a building are provided for fire fighters to connect attack hose lines. Their availability eliminates the need to advance hose lines all the way from the attack pump on the street outside to an upper floor or to a fire deep inside a large building. In tall buildings, it would be impossible to advance hose lines up the stairways in a reasonable time and to supply sufficient pressure to fight a fire on an upper floor. Instead, fire fighters carry hose inside the building and connect to the standpipe outlet.

In most Canadian jurisdictions, building codes require standpipe connections for attack lines to be located in hose stations (hose cabinets) in the hallways on each floor. If smoke conditions on the fire floor allow, it is preferable to hook up on the same floor as the fire to prevent smoke contamination of the stairwell(s). Keeping the stairwell doors closed also reduces the potential for a flow path to be created in the hallway, if the fire compartment is vented on the windward side of the structure.

It is safe practice to initially access the fire floor from a stairwell and not the elevator. The officer should carefully open the door leading onto the fire floor to assess the conditions. If the fire floor is filled with smoke, it may be preferable to hook onto the hose station on the floor below the fire.

When hooking up on the floor below the fire, it is essential to ensure you have enough hose to make it to the involved unit. Many fire services carry an additional 15-m (50-ft) length of 65-mm (2½-in.) hose for this purpose. Most hose pack configurations include at least 31 m (100 ft) of 45 mm (1¾ in.) lightweight attack line. A fire fighter will attach the 65-mm (2½-in.) hose to the closest hose station on the floor below the fire and flake the hose up the stairwell to where it is attached to the attack line. If attaching a 38-mm (1½-in.) or 45-mm (1¾-in.) line to the 38-mm (1½-in.) connection, be sure to reach into the male connection and remove any pressure-reducing device by gently pulling down. The pressure-reducing device is usually a cylindrical disk with a hole in the middle designed to reduce water flow in the hose line. This will enable sufficient water flow for the hose line.

The second firefighter will flake the hose line up past the stairwell landing to the next landing and back down ensuring that the hose line is not fouled or kinked. Once the officer and lead fire fighter are in a position to advance onto the fire floor, they will request that the hose line be charged. It is sound practice to flush all standpipe connections before charging any hose line. When the hose line is charged and advanced into the fire floor, gravity will help to move the hose line forward. This technique is much easier than trying to pull the charged hose line up the stairs.

It is imperative that the officer designates the stairwell as the *fire attack stairwell* over the air. This is to ensure that the attack stairwell is not used for evacuation, as it may become contaminated with heat and smoke. The officer should be equipped with a thermal imaging device, if possible, and a hand light. Once the hose line has been charged, the water supply fire fighter should assist in advancing the hose line so that the officer is free to direct operations and assess conditions.

To connect and advance an attack line from a standpipe outlet, follow the steps in **SKILL DRILL 16-24**.

LISTEN UP!

Each length of hose should be tested at least annually, according to the procedures listed in NFPA 1962, *Standard for the Care, Use, Inspection, Service Testing, and Replacement of Fire Hose, Couplings, Nozzles, and Fire Hose Appliances.*

SKILL DRILL 16-24
Connecting and Advancing an Attack Line from a Standpipe Outlet
Fire Fighter I, NFPA 1001: 4.3.10

1 Carry a standpipe hose bundle to the standpipe outlet that is one floor below the fire.

2 Remove the cap from the standpipe outlet. Open the standpipe valve to flush the standpipe. Attach the proper adaptor or an appliance such as a gated wye to the standpipe outlet.

(continued)

SKILL DRILL 16-24 Continued
Connecting and Advancing an Attack Line from a Standpipe Outlet
Fire Fighter I, NFPA 1001: 4.3.10

3 Flake the hose up the stairs to the floor above the fire floor or along a hallway outside the fire compartment. It is better to have too much hose than not enough hose.

4 Extend the hose back down to the fire floor, and prepare for the fire attack.

© Jones & Bartlett Learning. Photographed by Glen E. Ellman.

Replacing a Defective Section of Hose

With proper maintenance and testing, the risk of fire hose failure should be low, but even so, it is always a possibility. Fire fighters need to know what to do if a section of supply or attack hose bursts or develops a major leak while it is being used. Every fire fighter should know how to quickly replace a length of defective hose and restore the flow.

A burst hose line should be shut down as soon as possible. The fire fighters operating the hose line should be removed from the area if hazardous. If the hose line cannot be shut down at the fire pump, at the fire hydrant, or at a control valve, a hose clamp can be used to temporarily stop the flow of water in a hose line. Place the hose clamp on an undamaged section of hose upstream from the damaged section. After the water flow has been shut off, quickly remove the damaged section of hose, mark it as defective, and

replace it with two sections of hose. Using two sections of hose will ensure that the replacement hose is long enough to replace the damaged section.

To replace a damaged section of hose, follow the steps in **SKILL DRILL 16-25**.

Many fire department SOPs require two hose lines to be connected to an FDC. If one line breaks or if the

SAFETY TIP

Some departments use LDH with Storz-type couplings to connect to FDCs. Hose that is rated for use as attack hose should be used to connect to a standpipe system.

Large-diameter supply hose is rated for a safe working pressure of 1276 kPa (185 psi). Standpipe and sprinkler systems generally require at least 1034 kPa (150 psi) to be provided at the FDC, but a water hammer can cause a spike of much higher pressure. The excess pressure could burst the hose and place fire fighters who are working inside the building in danger.

SKILL DRILL 16-25
Replacing a Damaged Section of Hose Fire Fighter I, NFPA 1001: 4.3.10

1 Shut down or clamp off the damaged hose line.

2 Remove the damaged section of hose.

3 Replace the damaged section of hose with two new sections to ensure that the length of the hose will be adequate.

4 Restore the water flow.

© Jones & Bartlett Learning.

flow is interrupted, water will still be delivered to the system through the other line.

Draining and Picking Up Hose

To put a hose back into service, it must be drained of water. To do so, lay the hose straight on a flat surface, and then lift one end of the hose to shoulder level. Gravity will allow the water to flow to the lower portion of the hose and eventually out of the hose. As you proceed down the length of hose, fold the hose back and forth over your shoulder. When you reach the end of the section, you will have the whole section of hose on your shoulder.

To drain a hose, follow the steps in **SKILL DRILL 16-26**.

SKILL DRILL 16-26
Draining a Hose Fire Fighter I, NFPA 1001: 4.5.2

1 Lay the section of hose straight on a flat surface.

(continued)

SKILL DRILL 16-26 Continued
Draining a Hose Fire Fighter I, NFPA 1001: 4.5.2

2 Starting at one end of the hose, lift the hose to shoulder level.

3 Move down the length of hose, laying it on the ground or folding it back and forth over the shoulder.

4 Continue down the length until the entire hose is on the shoulder.

Unloading Hose

There are times other than a fire when fire fighters will need to unload the hose from an apparatus. For example, hose should be unloaded and reloaded on a regular basis to place the bends in different portions of the hose, because leaving bends in the same locations for long periods of time is likely to cause weakened areas. In addition, hose might have to be unloaded to change out apparatus. Sometimes all of the equipment from one vehicle must be transferred to another vehicle. It could also be necessary to offload the hose for annual testing to be conducted.

The following procedure should be used to unload a hose from a hose bed:

1. Use a large area such as a parking lot for this procedure. Be sure it is clean.
2. Disconnect any gate valves or nozzles from the hose before you begin.
3. Grasp the end of the hose, and pull it off the apparatus in a straight line.
4. When a coupling comes off the apparatus, uncouple the hose, and pull off the next section of hose.
5. When all of the hose has been removed from the hose bed, use a broom to brush off any dirt or debris on both sides of the hose, or jacket of the hose.
6. Sweep out any debris or dirt from the hose bed.
7. Store hose rolls off the floor on a rack in a cool dry area.

After-Action REVIEW

IN SUMMARY

- The objective of laying a supply line is to deliver water from a water source, such as a fire hydrant or a static water source, to an attack pump. In most cases, this operation involves laying a continuous hose line out of the bed of the fire apparatus as the vehicle drives forward. It can be done using either a forward hose lay, a reverse hose lay, or a split hose lay.
  - A forward hose lay starts at the water source, such as a fire hydrant, and proceeds toward the fire; the hose is laid in the same direction as the water flows—from the fire hydrant to the fire.
  - A reverse hose lay involves laying the hose from the fire to the water source, such as a fire hydrant; the hose is laid in the opposite direction to the water flow.
  - A split hose lay is performed by two pump companies in situations where hose must be laid in two different directions to establish a water supply. The attack pump drops the end of its supply hose at the corner of the street and performs a forward hose lay toward the fire. The water supply pump stops at the same intersection, pulls off enough hose to connect to the end of the supply line that is already there, and then performs a reverse hose lay to the fire hydrant or static water source. With the two lines connected together, the water supply pump can pump water to the attack pump.
- Hose can be loaded into the hose bed in several different ways, depending on how it will be laid out at the fire scene. The hose must be easily removable from the hose bed, without kinks or twists, and without the possibility of becoming caught or tangled. Three basic hose loads are commonly used for loading supply hose: the flat hose load, the horseshoe hose load, and the accordion hose load.
  - The flat hose load: The hose is placed flat in the hose bed.
  - The horseshoe hose load: The hose is placed on its edge and positioned around the perimeter of the hose bed in a U-shape.
  - The accordion hose load: The hose is placed on its edge and positioned side to side in the hose bed.
- In most cases, a soft sleeve hose is used to transport water from a pressurized source, such as a fire hydrant, to the suction side of the fire pump.
- Several different techniques are used to carry and advance supply hose. The best technique for a particular situation will depend on the size of the hose, the distance over which it must be moved, and the number of fire fighters available to perform the task. The same techniques can be used for supply lines and attack lines.
- The working hose drag technique is used to deploy hose from a hose bed and advance the hose line over a relatively short distance to the desired location.

- The shoulder carry is used to transport full lengths of hose over a longer distance than it is practical to drag the hose.
- Attack line operations deliver water from an attack pump to a handline, which discharges water onto the fire. The attack pump is usually positioned close to the fire, and attack lines are stretched manually by fire fighters.
- Attack lines should be loaded so that they can be quickly and easily deployed. The three most common hose loads used for preconnected attack lines are the minuteman hose load, the flat hose load, and the triple-layer hose load.
  - The minuteman hose load allows a single fire fighter to flake out, or drop, one fold or loop of hose at a time from the shoulder, as he or she advances toward the fire.
  - The flat hose load is loaded in a similar manner to the flat hose load used for supply lines, except the female end of the hose line is preconnected to a pump discharge outlet, and a nozzle is attached to the male end of the hose line.
  - The triple-layer hose load is a hose loading method in which the hose is folded back onto itself to reduce the overall length to one-third before loading the hose into the bed. This method is used by pump companies with minimal personnel.
- To reach a fire that is some distance from the pump, it can be necessary first to advance a larger-diameter supply line, such as a 65-mm (2½-in.) hose line, and then split it into two 45-mm (1¾-in.) attack lines. This is accomplished by attaching a gated wye or a water thief to the end of the 65-mm (2½-in.) hose line and then attaching the two attack lines to the gated outlets.
- To attack an interior fire, an attack line is usually advanced in two stages: The first stage involves laying out the hose to the building entrance. The second stage is to advance the hose line into the building to the location where it will be operated.
- The objective of moving hose is to move the amount that is needed, not to move as much hose as possible.
- Every fire fighter should know how to quickly replace a length of defective hose and restore the flow.
- Every fire fighter should know how to put a hose back into service.

KEY TERMS

Access Navigate for flashcards to test your key term knowledge.

Accordion hose load A method of loading hose on a vehicle that results in a hose appearance that resembles accordion sections. This is achieved by standing the hose on its edge and laying it side to side in the hose bed.

Attack line operations The delivery of water from an attack pump to a handline, which discharges the water onto the fire.

Breakaway type nozzle A nozzle with a tip that can be separated from the shut-off valve.

Combination hose load A hose loading method used when one long hose line is needed; the end of the last length of a hose in one bed of a split hose bed is coupled to the beginning of the first length of hose in the opposite bed.

Dutchman A short fold placed in a hose when loading the hose into a hose bed; the fold keeps the hose properly oriented and prevents the coupling from turning in the hose bed.

Flat hose load A hose loading method in which the hose is laid flat and stacked on top of the previous section.

Forward hose lay A method of laying a supply line where the hose line starts at the water source and ends at the attack pump. This is also referred to as the straight hose lay.

Four-way hydrant valve A specialized type of valve that can be placed on a hydrant and that allows another pump to increase the supply pressure without interrupting flow.

Friction points Those places where a hose line encounters resistance, such as corners, doorways, and stairs.

Handline A hose and nozzle that can be held and directed by hand. (NFPA 11)

Horseshoe hose load A hose loading method in which the hose is laid on its edge around the perimeter of the hose bed so that it resembles a horseshoe.

Minuteman hose load A hose loading method that allows a single fire fighter to flake out the necessary amount of hose from the shoulder while advancing toward the fire;

this load avoids having to maneuver the hose lines around obstacles and helps to prevent sharp kinks.

Preconnected flat hose load A hose loading method similar to the flat hose load used for supply lines, except the female end of the hose line is preconnected to a pump discharge outlet, and a nozzle is attached to the male end of the hose line.

Reverse hose lay A method of laying a supply line where the supply line starts at the attack pump and ends at the water source.

Split hose bed A hose bed arranged to enable the pump to lay out either a single supply line or two supply lines simultaneously.

Split hose lay A scenario in which the attack pump forward lays a supply line from an intersection to the fire, and the water supply pump reverse lays a supply line from the hose left by the attack pump to the water source.

Supply line operations The delivery of water from a water supply source, such as a fire hydrant or a static water source, to an attack pump.

Triple-layer hose load A hose loading method in which the hose is folded back onto itself to reduce the overall length to one-third before loading the hose into the hose bed. This load method reduces deployment distances.

REFERENCES

National Fire Protection Association. NFPA 1142, *Standard on Water Supplies for Suburban and Rural Fire Fighting*, 2017. www.nfpa.org. Accessed November 16, 2018.

National Fire Protection Association. NFPA 1961, *Standard on Fire Hose*, 2013. www.nfpa.org. Accessed November 16, 2018.

National Fire Protection Association. NFPA 1962, *Standard for the Care, Use, Inspection, Service Testing, and Replacement of Fire Hose, Couplings, Nozzles, and Fire Hose Appliances*, 2018. www.nfpa.org. Accessed November 16, 2018.

National Fire Protection Association. NFPA 1963, *Standard for Fire Hose Connections*, 2014. www.nfpa.org. Accessed November 16, 2018.

On Scene

You are dispatched on the second alarm for a working fire at a residential two-storey duplex. You are assigned to establish an initial supply line for Pump 403 and then to report to the captain to assist them with the attack on the fire.

1. If you are asked to lay the supply line from Pump 403 to the hydrant, what type of hose lay would you use?

  A. Forward hose lay

  B. Reverse hose lay

  C. Triple hose lay

  D. Split hose lay

2. If your department uses 102-mm (4-in.) supply hose, what type of hose load would you use?

  A. Split hose load

  B. Accordion hose load

  C. Flat hose load

  D. Horseshoe hose load

3. When connecting your soft sleeve hose to the fire hydrant, which of the following is *not* a recommended step?

  A. Checking the other hydrant caps to be sure they are attached snugly

  B. Opening the hydrant quickly to get water to the fire as soon as possible

  C. Assuring that the pump is located within reach of the soft sleeve hose

  D. Checking the hydrant for proper operation before attaching the soft sleeve hose to the hydrant

(continued)

On Scene Continued

4. Which size attack hose is most commonly used when fighting a fire in a single family or duplex structure?

   A. 102-mm (4-in.) hose

   B. 65-mm (2½-in.) hose

   C. 45-mm (1¾-in.) hose

   D. 76-mm (3-in.) hose

5. Which of the following produces fewer sharp bends than other hose loads?

   A. Horseshoe hose load

   B. Minuteman hose load

   C. Triple-layer hose load

   D. Flat hose load

6. The amount of hose that each member of the team takes will be dependent on several factors. These include the number of people on the team, the length of the hose, and the:

   A. distance between friction points.

   B. amount of brute strength used.

   C. position of team members.

   D. diameter of the hose.

Access Navigate to find answers to this On Scene, along with other resources such as an audiobook.

CHAPTER 17

Fire Fighter I

Fire Suppression

KNOWLEDGE OBJECTIVES

After studying this chapter, you will be able to:

- Describe the objectives of a defensive operation. (**NFPA 1001: 4.3.10**, p. 644)
- Describe the operations performed during a defensive operation. (**NFPA 4.3.10**, p. 644)
- Describe the objectives of an offensive operation. (**NFPA 1001: 4.3.10**, p. 644)
- Describe the operations performed during an offensive operation. (**NFPA 1001: 4.3.10**, p. 644)
- Describe the objectives of a transitional attack. (**NFPA 1001: 4.3.10**, pp. 644–649)
- Describe the operations performed during a transitional attack. (**NFPA 1001: 4.3.10**, pp. 644–649)
- Describe the characteristics of a solid stream. (**NFPA 1001: 4.3.8**, p. 650)
- Describe the characteristics of a straight stream. (**NFPA 1001: 4.3.8**, p. 650)
- Describe the characteristics of a fog stream. (**NFPA 1001: 4.3.8**, pp. 650–651)
- Describe the objectives of a direct attack. (**NFPA 1001: 4.3.10**, pp. 651, 652–653)
- Describe the objectives of an indirect attack. (**NFPA 1001: 4.3.10**, pp. 651, 654, 655)
- Describe the objectives of a combination attack. (**NFPA 1001: 4.3.10**, pp. 654, 656–657)

- Describe the techniques used to operate large handlines. (**NFPA 1001: 4.3.8**, pp. 656, 658–661)
- Describe the characteristics of a master stream appliance. (**NFPA 1001: 4.3.8**, p. 661)
- Describe the characteristics of a deck gun. (**NFPA 1001: 4.3.8**, p. 661)
- Describe the characteristics of a portable monitor. (**NFPA 1001: 4.3.8**, p. 662)
- Describe the characteristics of elevated master stream appliances. (**NFPA 1001: 4.3.8**, pp. 662–663)
- Describe the characteristics of concealed-space fires. (**NFPA 1001: 4.3.10**, p. 663)
- Describe the tactics used to suppress concealed-space fires. (**NFPA 1001: 4.3.10**, p. 663)
- Describe the characteristics of basement fires. (**NFPA 1001: 4.3.10**, pp. 663–665)
- Describe the tactics used to suppress basement fires. (**NFPA 1001: 4.3.10**, pp. 663–665)
- Describe the characteristics of fires above ground level. (**NFPA 1001: 4.3.10**, p. 665)
- Describe the tactics used to suppress fires above ground level. (**NFPA 1001: 4.3.10**, p. 665)
- Describe the characteristics of attic fires. (**NFPA 1001: 4.3.10**, pp. 665–667)
- Describe the tactics used to suppress attic fires. (**NFPA 1001: 4.3.10**, pp. 665–667)
- Describe the characteristics of fires in large buildings. (**NFPA 1001: 4.3.10**, p. 667)

- Describe the tactics used to suppress fires in large buildings. (NFPA 1001: 4.3.10, p. 667)
- Describe the characteristics of fires in buildings under construction, renovation, or demolition. (NFPA 1001: 4.3.10, p. 667)
- Describe the tactics used to suppress fires in buildings under construction, renovation, or demolition. (NFPA 1001: 4.3.10, p. 667)
- Describe the characteristics of fires in lumberyards. (NFPA 1001: 4.3.8, pp. 667–668)
- Describe the tactics used to suppress fires in lumberyards. (NFPA 1001: 4.3.8, pp. 667–668)
- Describe the characteristics of fires in stacked or piled materials. (NFPA 1001: 4.3.8, p. 668)
- Describe the tactics used to suppress fires in stacked or piled materials. (NFPA 1001: 4.3.8, p. 668)
- Describe the characteristics of fires in trash containers. (NFPA 1001: 4.3.8, pp. 668, 669–670)
- Describe the tactics used to suppress fires in trash containers. (NFPA 1001: 4.3.8, pp. 668, 669–670)
- Describe the characteristics of fires in confined spaces. (NFPA 1001: 4.3.8, pp. 668, 670)
- Describe the tactics used to protect exposures. (NFPA 1001: 4.3.8, pp. 670–671)
- Describe the characteristics of fires on buildings with solar photovoltaic systems. (pp. 671–672)
- Describe the tactics used to suppress fires on buildings with solar photovoltaic systems. (pp. 671–672)
- Describe the characteristics of chimney fires. (NFPA 1001: 4.3.10, pp. 672–673)
- Describe the tactics used to suppress fires in chimneys. (NFPA 1001: 4.3.10, pp. 672–673)
- Describe the characteristics of vehicle fires. (NPFA 1001: 4.3.7, p. 673)
- Describe the types of motor vehicles. (NFPA 4.3.7, p. 673)
- Describe the different types of alternative fuels that power motor vehicles. (NFPA 4.3.7, pp. 673–677)
- Describe the tactics used to suppress vehicle fires. (NPFA 1001: 4.3.7, pp. 678, 679–680)
- Describe the tactics used to suppress fires in the passenger area of a vehicle. (NPFA 1001: 4.3.7, p. 678)
- Describe the tactics used to suppress fires in the engine compartment of a vehicle. (NPFA 1001: 4.3.7, pp. 678, 680–681)
- Describe the tactics used to suppress fires in the trunk of a vehicle. (NPFA 1001: 4.3.7, p. 681)
- Describe how to overhaul a vehicle fire. (NPFA 1001: 4.3.7, p. 681)

- Describe when gas service should be shut off. (NFPA 1001: 4.3.18, p. 682)
- Describe when the electrical service should be shut off. (NFPA 1001: 4.3.18, pp. 682–684)
- Describe the hazards posed by electrical fires. (NFPA 1001: 4.3.18, pp. 683–684)
- Describe the tactics used to suppress an electrical fire. (pp. 683–684)
- Describe when water service should be shut off. (NFPA 1001: 4.3.18, p. 684)

SKILLS OBJECTIVES

After studying this chapter, you will be able to perform the following skills:

- Perform a transitional attack. (NFPA 1001: 4.3.10, pp. 644–649)
- Perform a direct attack. (NFPA 1001: 4.3.10, pp. 651, 652–653)
- Perform an indirect attack. (NFPA 1001: 4.3.10, pp. 651, 654, 655)
- Perform a combination attack. (NFPA 1001: 4.3.10, pp. 654, 656–657)
- Perform the one-fire fighter method for operating a large handline. (NFPA 1001: 4.3.8, pp. 658, 659)
- Perform the two-fire fighter method for operating a large handline. (NFPA 4.3.8, pp. 658, 660)
- Operate a deck gun. (NFPA 1001: 4.3.8, p. 661)
- Deploy and operate a portable monitor. (NFPA 1001: 4.3.8, p. 662)
- Locate and suppress concealed-space fires. (NFPA 1001: 4.3.10, p. 663)
- Extinguish an outside trash fire or other outside Class A fire. (NFPA 1001: 4.3.8, pp. 668, 669–670)
- Extinguish a vehicle fire. (NPFA 1001: 4.3.7, pp. 678–681)
- Shut off gas utilities. (NFPA 1001: 4.3.18, p. 682)
- Shut off electric utilities. (NFPA 1001: 4.3.18, pp. 682–684)

ADDITIONAL STANDARDS

- NFPA 1006, *Standard for Technical Rescue Personnel Professional Qualifications*
- NFPA 1403, *Standard on Live Fire Training Evolutions*

Fire Alarm

This is it—your first night at the fire station. You lie in your bed wide awake, just hoping for a call. Just past midnight, the alarm sounds for a fire in a house. You slide down the pole and quickly slip on your gear before climbing into the cab. Your heart is racing as the other crew members mount the apparatus like it is just another day at the office. As you get close to the scene, you can see the orange glow: The incident is a working fire.

1. What is the difference between an offensive operation and a defensive operation?

2. When would you use an indirect attack rather than a direct attack?

3. Which type of nozzle would you use for an indirect attack?

 *Access Navigate for more practice activities.*

Introduction

The term "fire suppression" refers to the strategies, tactics, and tasks involved in extinguishing fires. There are several methods used by fire fighters to extinguish fires. All of these methods involve removal of one of the three basic elements needed for combustion to occur—fuel, oxygen, or heat. Together, these three elements form what is known as the fire triangle. A fourth factor, a chemical chain reaction, must result from the first three elements to produce and maintain a self-sustaining fire. This is known as the fire tetrahedron. If you remove any of these four elements, the fire will be extinguished.

This chapter presents the basic methods that fire fighters most frequently use to extinguish fires. Common fire terminology is also presented. The apparatus and equipment employed by most fire departments are designed to apply large volumes of water to a fire in an attempt to cool the fuel below its ignition temperature. Although fire departments typically deploy a variety of extinguishing agents for different situations, water is used more often than any other agent. Fire experiments conducted by National Institute of Standards and Technology (NIST) and Underwriters Laboratories (UL) have increased our understanding of fire behaviour. They have also demonstrated safer and more efficient ways in which to attack certain fires. The results of these studies are incorporated throughout this chapter and the rest of this text. Many of these changes are included in the sections pertaining to transitional fire attack, attic fires, and basement fires. See Chapter 23, *Advanced Fire Suppression*, for advanced fire suppression techniques.

Defensive, Offensive, and Transitional Attack Operations

One of the most important aspects of fire suppression is a methodical evaluation of the incident. In the vernacular of the fire service, this is called "size-up." In order to make sound decisions, personnel require information. A proper size-up involves, but is not limited to, identifying the nature of the emergency, the life-safety risk, and the need for additional resources. An initial size-up is often conducted by the first-arriving company officer, who serves as the incident commander (IC) until a higher ranking officer arrives on scene to assume command, usually a District Chief. The initial IC will identify the building construction type, the occupancy, the location of the fire, the need for rescue, and the the overall scope of the situation. In most instances a proper size-up will include a complete 360-degree reconnaissance of the involved structure to determine rescue requirements, identify exposures, collapse potential, and other safety concerns. Generally the decisions made in the first initial 5–10 minutes of an incident will determine the outcome of the incident. The initial size-up is often conducted by the first-arriving company officer, who serves as the IC until a higher-ranking officer arrives on the scene and assumes command. The IC will use the information derived from the initial size-up to formulate an initial incident action plan (IAP). In most instances the initial IAP will be a mental plan that establishes an initial strategy and first tactical objectives.

The first two types of fire suppression operations are offensive and defensive. Generally most structure fires are suppressed by fire fighters using an offensive strategy that is based on an aggressive interior fire attack. By contrast, a defensive operation is conducted from the exterior of the building or fire compartment, or while moving away from the building or fire, by directing water streams toward the fire from a safe distance. A third type of fire attack is a brief exterior, indirect attack into the fire compartment to initiate cooling, followed by an interior attack to suppress the fire (in coordination with ventilation operations). This is sometimes referred to as a transitional attack, a blitz attack, softening the target, or resetting the fire. All three of these methods will be discussed in this section.

As a new fire fighter, you are not responsible for determining the type of fire attack that will be used. Nevertheless, you should understand the various factors that go into making these decisions and understand why one type would be favoured over another. The IC determines the overall strategy and assigns tactical objectives based on the situation and available resources. The IC continually re-evaluates the situation as the incident progresses, reassigning tactical objectives and personnel until a "loss stopped" is declared.

If the IC decides to switch from one strategy to another at any point during an operation, this change must be clearly communicated and understood by all fire fighters. A change from offensive to defensive operations could be warranted if an interior attack is unsuccessful or if the risk factors are determined to be too great to justify having fire fighters work inside the building. Sometimes the strategy switches from defensive to offensive after an exterior attack has reduced the volume of fire inside a building to the point at which fire fighters can enter and complete extinguishment. Specific strategies and tactics are usually driven by consideration of fire fighter safety and victim survivability.

LISTEN UP!

As a Fire Fighter I, you are not responsible for determining the type of fire attack that will be used. NFPA 1001 states that the Fire Fighter I works under direct supervision in hazardous conditions.

Defensive Operations

A defensive operation or a defensive attack is chosen for those instances when interior operations are not feasible, which may include a fully involved structure or a ventilation-controlled fire where occupant survivability is highly unlikely. The primary objective of a defensive operation is to prevent the fire from spreading. Defensive operations are typically conducted from the exterior. Defensive operations often use high-volume handlines or master streams from a safe position to control the fire. Water is directed into the building through doorways, windows, and openings in the roof or onto exposures to keep the fire from spreading. This is known as an indirect attack. Exposure protection is a high priority during a defensive operation. Sometimes a defensive operation can be made inside a large structure if the fire is contained to one area. Fire streams can be directed toward the involved area from a safe location to keep the fire from spreading to other parts of the building.

Offensive Operations

An offensive operation or an offensive attack typically exposes fire fighters to heat and smoke inside the building. Interior operations are considered immediately dangerous to life or health (IDLH) environments and as such fire fighters will be at greater risk of injury. The objective in an offensive operation is for fire fighters to get close enough to the fire to perform a direct attack or to apply extinguishing agents directly onto the fire. This allows the extinguishing agent to overpower and extinguish the fire. An offensive operation is used in situations where the fire is not too large or too dangerous to be extinguished by applying water using handlines. When an offensive attack is successful, the fire can be controlled with the least amount of property damage. An offensive operation requires well-planned coordination among crews performing different tasks, such as ventilating, operating hose lines, and conducting aggressive search and rescue.

SAFETY TIP

It is imperative to extinguish all fire as you proceed to the seat of the fire. Failure to do so may allow the fire to grow behind you, entrapping you by cutting off your escape route.

Transitional Attack

The primary means of suppressing fires is to apply water directly on the involved area or areas on the structure. Traditionally, it was thought that attacking the fire from the outside may push the fire through other parts of the building. This has led to the assumption that offensive operations must be made exclusively from the inside and from the unburned side of the structure, effectively "pushing the fire" away from the uninvolved portion of the building and any occupants who might be behind the advancing hose team. This technique requires fire fighters to enter a burning building, often with low visibility and little idea of where the fire is located. NIST and UL research demonstrates that these assumptions are not always accurate.

In 2013, UL released its *Innovating Fire Attack Tactics* report, which outlined the results of a series of tests conducted with NIST and the New York City Fire Department (FDNY) (UL, 2013). During the structure fire experiments, researchers found that applying a limited amount of water through an open window or door into the fire compartment before entering the structure may dramatically improve interior conditions for both occupants and fire fighters (**FIGURE 17-1**).

As NIST and UL measured the movement of the fire, they determined that a properly applied straight or solid stream does not "push" fire. However, creating a higher pressure with improper stream applications may cause a high-pressure point at the exhaust, causing an internal flow path. They determined that the movement of the fire is dependent on the flow path—that is, the

movement of heat and smoke in a structure from the higher pressure within the fire area toward the lower-pressure areas accessible via doorways, halls, stairs, and window openings. After verifying these results with multiple experiments, they realized that if water was introduced into the fire from outside, it produced cooling not just in the fire compartment but also in other areas of the fire building (**FIGURE 17-2**). Temporary cooling was achieved even when the water entered the structure at a distance from the seat of the fire. In one experiment, a straight stream of water flowing 681 L/min (180 gpm) was applied for 28 seconds and flowed 318 L (84 gal) of water into the house at some distance from the fire. The water reduced the temperature in the front of the living room from 649°C (1200°F) to 149°C (300°F). (*Note:* These numbers apply only to the Modern Fire Behavior experiments conducted at Governor's Island by FDNY, NIST, and UL.)

These experiments showed that an intentional, brief offensive exterior attack—introducing water from the outside to a body or volume of fire—reduced temperatures in other parts of the house at some distance from the fire, but it did not completely extinguish the fire. It slowed the growth of the fire by cooling huge quantities of very hot gaseous fuel and solid fuel below its ignition temperature. The effect of this offensive exterior attack is significant but short-lived. These gains are only temporary unless additional water can be applied to the seat of the fire.

To prevent the fire from regaining its former state, to complete extinguishment, and to search for and remove any occupants, it is necessary to quickly enter the building and to complete extinguishment of the remaining fire immediately following the initial knockdown from the exterior.

When the goal is to get inside quickly, and although there is no consensus on a term for this type of operation, the tactics remain the same. A **transitional attack** is an offensive operation initiated by a brief exterior, indirect attack into the fire compartment to initiate cooling and to stop the progress of the fire. This exterior attack quickly transitions to an interior attack to suppress the fire, in coordination with ventilation operations. The transitional attack "softens the target" (resets the fire) or cools the fire gases, reduces the risk of flashover, improves visibility, and allows fire fighters to enter the fire compartment quickly. This method can often effectively place water on the seat of the fire sooner, thus making conditions tenable for any trapped occupants (**FIGURE 17-3**). With the fire knocked down, more aggressive ventilation can follow without

FIGURE 17-1 Prior to making entry, the nozzle fire fighter directs a straight stream through the top of the door or window and deflects the stream off of the ceiling. Evaluate fire conditions before transitioning to an interior fire attack.

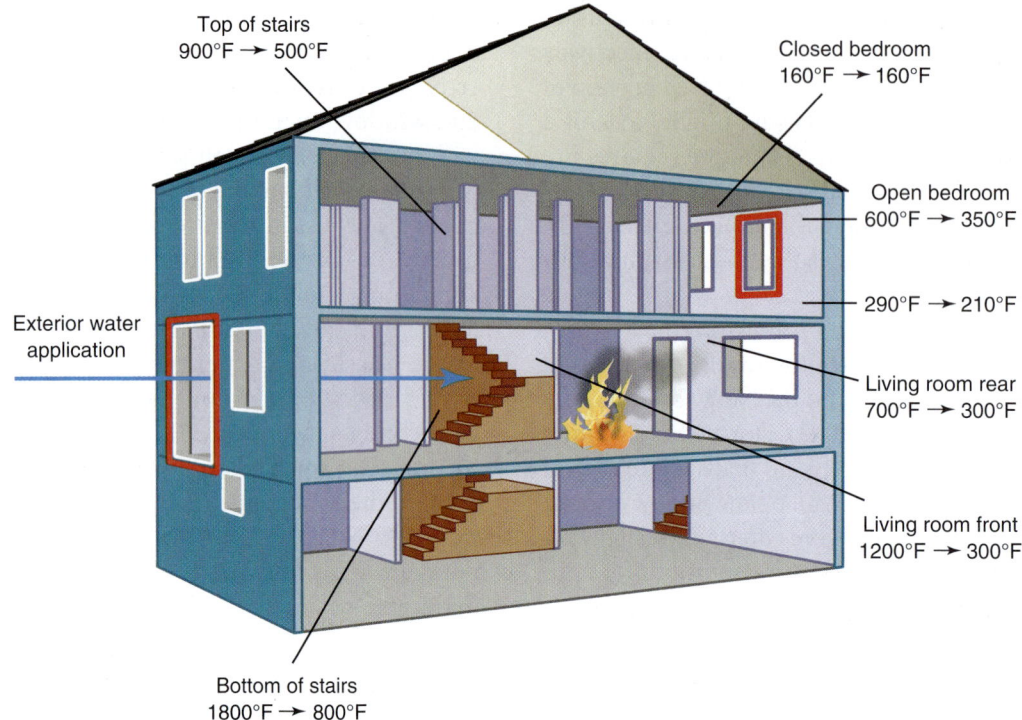

Top of stairs
900°F → 500°F

Closed bedroom
160°F → 160°F

Open bedroom
600°F → 350°F

290°F → 210°F

Exterior water
application

Living room rear
700°F → 300°F

Living room front
1200°F → 300°F

Bottom of stairs
1800°F → 800°F

FIGURE 17-2 An offensive exterior attack some distance from the fire reduces temperatures throughout the house and partially suppresses the fire. The temperatures listed are from before and after the exterior application of water.
Courtesy of NIST.

the risk of inducing flashover. As visibility continues to improve, fire fighters can move through the building more effectively, allowing them to conduct rescue operations, final suppression, and overhaul more safely and efficiently.

FIGURE 17-3 Fire fighters can use a transitional attack to reduce the burning and quickly cool the interior before entering a burning building.
Courtesy of UL.

SAFETY TIP

The acronym SLICE-RS has been developed by fire service instructors to help first-arriving company officers remember the steps to be taken for initial pump company operations upon arrival at a structure fire scene. The steps are as follows:

Sequential Actions

S Size-up

L Locate

I Identify and control the fire flow path

C Cool the space from the safest location

E Extinguish the fire

Actions of Opportunity

R Rescue

S Salvage

When using any acronym, keep in mind the objectives of your actions. It may be necessary to modify your actions to fit a specific situation.

To avoid unnecessary water damage, only enough water to control the fire should be used. Although it is not always possible to use a transitional attack, fire fighters should consider using it, especially if the fire has already self-vented through a door or window.

An initial exterior attack can be started using a preconnected hose line or a master stream appliance. Because this attack is conducted from the outside, it does not place fire fighters in a dangerous position in the fire compartment. Therefore, it can be started using water from the tank on the apparatus even before a permanent water supply has been established.

The results of the experiments conducted by NIST and UL require us to reconsider our long-standing approaches to ventilation and the application of water into the fire building. **TABLE 17-1** lists the conclusions regarding exterior fire attacks that were drawn from the fire experiments conducted by NIST, UL, and FDNY.

To perform a transitional attack, follow the steps in **SKILL DRILL 17-1**.

FIRE FIGHTER TIP

Deflecting a straight- or solid-stream nozzle off a ceiling to create water droplets does not work as sufficiently on drop-down ceilings or if the ceiling is comprised of ceiling panels. In these circumstances use of a narrow fog pattern would likely yield better results.

TABLE 17-1 Exterior Fire Attack

- A properly applied offensive exterior fire attack through a window or door, even when it is the only exterior vent, will not push fire.
- Water application is most effective if a straight stream is aimed through the smoke into the ceiling of the fire compartment. This technique allows heated gases to continue to vent from the fire compartment while cooling the hot fuel inside. Fog patterns should not be used in this application. The fog pattern entrains large volumes of air and pushes air into the building. A fog stream can also block a ventilation opening, effectively changing the fire flow path.
- Applying a straight or solid hose stream through a window or door into a room involved in a fire resulted in improved conditions throughout the structure.
- Even in cases where the front and rear doors were open and windows had been vented, application of water through one of the vents improved conditions throughout the structure.
- Applying water directly into the compartment as soon as possible resulted in the most effective means of suppressing the fire.
- Transitional attack is an offensive exterior fire attack that occurs just prior to entry, search, and tactical ventilation. This technique is also known as a blitz attack or softening the target.
- The transitional attack should begin from the outside, but it is necessary to finish it from the inside.
- Coordinate the fire attack with ventilation—do not ventilate before an attack stream is ready.

Courtesy of NIST.

SKILL DRILL 17-1
Performing the Transitional Attack Fire Fighter I, NFPA 1001: 4.3.10

1 Don full personal protective equipment (PPE) and self-contained breathing apparatus (SCBA). Select the proper handline to be used to attack the fire based on the fire's size, location, and type. Advance the hose line from the apparatus to the entry point of the structure. Flake out excess hose in front of the entry point.

(continued)

SKILL DRILL 17-1 Continued
Performing the Transitional Attack Fire Fighter I, NFPA 1001: 4.3.10

2 Don the face piece, and activate the SCBA and personal alert safety system (PASS) device prior to entering the building.

3 Notify the pump driver/operator that you are ready for water. Open the nozzle to purge air from the system, and make sure water is flowing. If using an adjustable nozzle, ensure that it is set to the proper nozzle pattern for entry. Shut down the nozzle until you are in a position to apply water.

4 If fire has vented from a door or window, apply a straight stream through the top of the opening from a safe, exterior location so it deflects off the ceiling. Evaluate the effectiveness of the hose stream before transitioning to an interior fire attack.

SKILL DRILL 17-1 Continued
Performing the Transitional Attack Fire Fighter I, NFPA 1001: 4.3.10

5 Check the entry door for heat before making entry.

6 Control the flow path as crews advance toward the fire.

7 If it is safe to do so, advance into the fire compartment and apply water to the base of the fire. If it is not safe to advance into the fire compartment, apply water from a safe location such as a hallway, adjoining room, or doorway, until the room begins to darken.

8 Shut the nozzle off, and reassess the fire conditions. Confirm that ventilation has been completed.

9 Locate and extinguish hot spots until the fire is completely extinguished.

Operating Handlines

Some of the most basic skills that must be mastered by every fire fighter involve the use of handlines to apply water onto a fire. Put simply, a fire fighter must be able to advance and operate a handline effectively to extinguish a fire. The proper operation of a handline is also essential to protect yourself, your crew, and any trapped victims from the fire. Fire attack operations are often conducted under extremely stressful conditions, including high-heat conditions, limited or zero visibility, and unfamiliar surroundings. Care should be taken not to have opposing handlines, such that two crews are working "against" each other, due to potential injury from the water streams and debris.

Fire fighters must learn how to operate both large and small handlines, as well as master stream appliances. Small handlines can be as large as 51 mm (2 in.) in diameter. The most frequently used handlines for interior fire attack are 38- or 45-mm (1½-in. or 1¾-in.) handlines.

Larger handlines may be 65 mm (2½ in.) in diameter or more. Because water can flow through these hose lines at a rate of more than 946 L/min (250 gpm), larger handlines are heavier and less maneuverable than smaller handlines. Master stream appliances are used when large quantities of water are needed to control a large fire. Such a stream can deliver water at a rate of at least 1325 L/min (350 gpm), and some master stream appliances can flow more than 7570 L/min (2000 gpm). The most commonly used master stream appliances deliver flows between 1325 and 5678 L/min (350 and 1500 gpm). Master stream appliances are operated from a fixed position—either on the ground (portable master stream device), on top of a piece of fire apparatus (deck gun), or on an aerial device (ladder or elevated platform). They are typically used for defensive operations, although master streams can be used to "blitz" a fire before beginning an interior offensive attack. This fire suppression method knocks the main body of fire down with a heavy stream; crews can then stretch handlines into the site and extinguish the remaining fire.

Fire Streams

Different types of fire streams are produced by using different types of nozzles. As described in Chapter 15, *Fire Hose, Appliances, and Nozzles*, the nozzle defines the shape and direction of the water or extinguishing agent that is discharged onto the fire. Fire streams are also affected by pump pressure, friction loss (length, diameter, and type and condition of hose), elevation (A 30.5-cm [1-ft] rise in the elevation of a hose line

equates to 3 kPa [.434 psi] pressure loss due to the weight of water), gravity, wind, and distance from the nozzle to the fire. A fire stream can be produced with a smooth-bore nozzle or a combination nozzle. Fire department policies and standard operating procedures (SOPs) usually dictate the types of nozzles that are used with different types of hose lines. The nozzle operator must know which type of nozzle should be used in the specific situation at hand.

A combination or smooth-bore nozzle produces a **solid stream** (**FIGURE 17-4**). A solid stream of water has a greater reach and penetrating power than a straight stream of water, because it is discharged as a continuous column of water.

A fog-stream nozzle adjusts to produce either a straight stream or multiple fog-stream patterns (**FIGURE 17-5**).

A **straight stream** has a greater reach than a fog stream, given similar water flow rates and nozzle pressures. A straight stream also keeps the water concentrated in a small area, so it can penetrate through a hot atmosphere to reach and cool the burning surfaces. To produce a straight stream, the fire fighter sets the adjustable fog-stream nozzle to the narrowest pattern it can discharge. This type of stream is made up of a highly concentrated pattern of droplets that are all discharged in the same direction.

Solid stream

FIGURE 17-4 A solid stream is produced by a smooth-bore nozzle.
© Jones & Bartlett Learning.

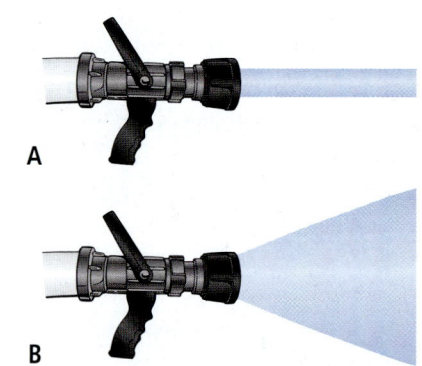

A

B

FIGURE 17-5 Fog-stream nozzles produce straight streams or multiple fog streams. **A.** Straight stream. **B.** Fog stream.
© Jones & Bartlett Learning.

A fog stream divides water into droplets, which have a very large surface area and can absorb heat efficiently. When heat levels in a building need to be lowered quickly, a combination of ventilation and a fog stream may be the fire suppression method of choice. A fog stream can also be used to protect fire fighters from the radiant heat of a large fire. Most adjustable nozzles can be adjusted from a straight stream, to a narrow fog pattern, to a wide fog pattern, depending on the reach that is required and how the stream will be used. When a fog-stream nozzle is used, the nozzle operator must know how to set the discharge pattern to produce different kinds of streams.

One consideration when selecting and operating nozzles is the amount of air that is moved along with the water. A fog stream naturally moves a large quantity of air along with the mass of water droplets. This air flows into the fire area along with the water. When this air movement is combined with steam production as the water droplets encounter a heated atmosphere, the thermal balance is likely to be disrupted quickly. In such a case, the hot fire gases and steam may be displaced back toward the nozzle operator. Straight and solid streams move little air in comparison with a fog stream, so fewer concerns with displacement and disruption of the thermal balance arise when these types of streams are used for fire suppression. A very narrow fog stream applied in short pulses into the upper smoke layer can aid in cooling the atmosphere without disrupting the thermal balance. These techniques are very specific, and fire fighters should be properly trained in both fire dynamics and compartment-based firefighting tactics before using them in a real fire situation.

When applied correctly, the air movement created by a fog stream can be used for ventilation. Discharging

FIRE FIGHTER TIP

Interior fire attack can be conducted on many different scales. In many cases, such an attack is geared toward a fire that is burning in only one room; this kind of fire may be controlled quickly by one small handline. Larger fires require more water, which could be provided by two or more small handlines working together or by one or more larger handlines. Fires that involve multiple rooms, large spaces, or concealed spaces are more complicated and require more extensive coordination; nevertheless, the basic techniques for attacking these fires are similar to the techniques used when extinguishing smaller fires. A trained Fire Fighter II should be able to understand and coordinate an interior fire attack. See Chapter 23, *Advanced Fire Suppression*, for more information about interior fire attack.

a fog stream out through a window or doorway, for example, will draw smoke and heat out in the same manner as an exhaust fan. This operation must be performed carefully to prevent accidentally drawing hidden fire toward the nozzle operator. The use of a water stream to provide ventilation is called hydraulic ventilation and is discussed further in Chapter 13, *Ventilation*.

Stream Placement

Direct attack and indirect attack are two different methods of applying water onto a fire. A combination attack is performed in two stages, beginning with an indirect attack and continuing with a direct attack. Small handlines are typically used for these types of attacks.

Direct Attack

An effective means of fire suppression in some situations is a **direct attack**. This kind of attack delivers water directly onto the base of the fire. The water cools the burning surface until it is below its ignition temperature. Water should be applied until flame is no longer visible. To perform a direct attack, follow the steps in **SKILL DRILL 17-2**.

Indirect Attack

The indirect application of water can be used in situations where the interior temperature is increasing and it appears that the room or space is ready to flash over. This application can be accomplished from a safer location, such as a doorway or hallway, to reduce the thermal exposure to operating fire fighters. With this fire suppression method, the fire fighter aims water at the ceiling to cool the superheated gases in the upper levels of the room or space. This action can prevent or delay flashover long enough for fire fighters to apply water directly to the seat of the fire (perform a direct attack) or to make a safe exit. Follow your department's SOPs regarding the application of water.

The objective of an **indirect attack** is to apply water to the hot gases by sweeping the ceiling with water to cool the gases and droplets. A straight stream directed toward the ceiling has been shown to be effective in cooling the fire gases in a room. This action may also hit and cool some of the solid fuel and burning surfaces. This cooling only lasts for a short period of time unless the stream is sustained. The intention of this indirect attack is to absorb the heat radiating from the walls and ceiling and to cool the hot gases in the fire compartment. As soon as the water has reduced the temperature, the fire stream can transition into a direct attack. Cooling the walls, ceiling, and hot gases

SKILL DRILL 17-2
Performing a Direct Attack Fire Fighter I, NFPA 1001: 4.3.10

1 Don full PPE and SCBA. Select the proper handline to be used to attack the fire based on the fire's size, location, and type. Advance the hose line from the apparatus to the entry point of the structure. Flake out excess hose in front of the entry point.

2 Don the face piece, and activate the SCBA and PASS device prior to entering the building.

3 Notify the pump driver/operator that you are ready for water.

4 Open the nozzle to purge air from the system, and make sure water is flowing. If using an adjustable nozzle, ensure that it is set to the proper nozzle pattern for entry. Shut down the nozzle until you are in a position to apply water.

SKILL DRILL 17-2 Continued
Performing a Direct Attack Fire Fighter I, NFPA 1001: 4.3.10

5 Check the entry door for heat before making entry.

6 Control the flow path as crews advance toward the fire.

7 If it is safe to do so, advance into the fire compartment, and apply water to the base of the fire.

8 Shut the nozzle off, and reassess the fire conditions. Confirm that ventilation has been completed. Locate and extinguish hot spots until the fire is completely extinguished.

© Jones & Bartlett Learning. Photographed by Glen E. Ellman.

as you progress toward a fire is also an effective way to reduce the possibility of a flashover.

Today's indirect attack is different from the indirect method that Lloyd Laymen advocated starting in the 1950s. In Layman's indirect attack, the idea was to keep the fire compartment closed, introduce a low flow of water fog (95–114 L/min [25–30 gpm]) into a window, and fill the compartment with steam. Today's indirect attack uses a straight stream or a solid stream (or a narrow fog stream applied in bursts), flowing between 454 and 681 L/min (120 and 180 gpm), to cool everything quickly, and does not generate significant amounts of steam.

To perform an indirect attack, follow the steps in **SKILL DRILL 17-3**.

Combination Attack

A **combination attack** employs both indirect attack and direct attack methods in a sequential manner. This strategy should be used when a room's interior has been heated to the point that it is nearing a flashover condition. Fire fighters should first use an indirect attack method from the safest location, such as a hallway, adjoining room, or doorway, to cool the fire gases down to safer temperatures and prevent flashover from occurring. This operation is followed with a direct attack on the main body of fire. In a combination attack, the fire fighter who is operating the nozzle should be given plenty of space to maneuver.

FIRE FIGHTER TIP

Check for high heat behind a door by feeling with the back of your gloved hand or aiming a straight stream at the top of the door. If the water converts to steam, the door is extremely hot (**FIGURE 17-6**).

FIGURE 17-6 Steam from a quick burst of water from the nozzle onto the entry door can indicate high heat.
© Jones & Bartlett Learning.

SKILL DRILL 17-3
Performing an Indirect Attack Fire Fighter I, NFPA 1001: 4.3.10

1 Don full PPE and SCBA. Select the proper handline to be used to attack the fire based on the fire's size, location, and type. Advance the hose line from the apparatus to the entry point of the structure. Flake out excess hose in front of the entry point.

2 Don the face piece, and activate the SCBA and PASS device prior to entering the building.

SKILL DRILL 17-3 Continued
Performing an Indirect Attack Fire Fighter I, NFPA 1001: 4.3.10

3 Notify the pump driver/operator that you are ready for water.

4 Open the nozzle to purge air from the system, and make sure water is flowing. If using an adjustable nozzle, ensure that it is set to the proper nozzle pattern for entry. Shut down the nozzle until you are in a position to apply water.

5 Check the entry door for heat before making entry.

6 Control the flow path as crews advance toward the fire.

7 From the safest location, such as a hallway, adjoining room, or doorway, apply water to the superheated gases at the ceiling, and move the stream back and forth. Flow water until the room begins to darken.

8 Shut the nozzle off, and reassess the fire conditions. Confirm that ventilation has been completed. Locate and extinguish hot spots until the fire is completely extinguished.

To perform a combination attack, follow the steps in **SKILL DRILL 17-4**.

Operating Large Handlines

Larger handlines may be 65 mm (2½ in.) in diameter or more and can be used either for offensive or defensive operations. In an offensive attack situation, a 65-mm (2½-in.) attack line can be advanced into a building to apply a heavy stream of water onto a large volume of fire.

The same direct and indirect attack techniques that were described for small handlines can also be used with large handlines. A 65-mm (2½-in.) handline can overwhelm a substantial interior fire if it can be discharged directly into the involved area. The extra reach of the stream can also prove valuable when making an interior attack in a large building.

It is difficult for fire fighters to advance and maneuver a large handline inside a building, particularly in tight

SKILL DRILL 17-4
Performing a Combination Attack Fire Fighter I, NFPA 1001: 4.3.10

1 Don full PPE and SCBA. Select the proper handline to be used to attack the fire based on the fire's size, location, and type. Advance the hose line from the apparatus to the entry point of the structure. Flake out excess hose in front of the entry point.

2 Don the face piece, and activate the SCBA and PASS device prior to entering the building.

3 Notify the pump driver/operator that you are ready for water.

4 Open the nozzle to purge air from the system, and make sure water is flowing. If using an adjustable nozzle, ensure that it is set to the proper nozzle pattern for entry. Shut down the nozzle until you are in a position to apply water.

SKILL DRILL 17-4 Continued
Performing a Combination Attack Fire Fighter I, NFPA 1001: 4.3.10

5 Check the entry door for heat before making entry.

6 Control the flow path as crews advance toward the fire.

7 From the safest location, such as a hallway, adjoining room, or doorway, apply water to the superheated gases at the ceiling, and move the stream back and forth. Flow water until the room begins to darken.

8 If it is safe to do so, advance into the fire compartment, and apply water to the base of the fire.

9 Shut the nozzle off, and reassess the fire conditions. Confirm that ventilation has been completed. Locate and extinguish hot spots until the fire is completely extinguished.

quarters or around corners. Although one fire fighter can control a large handline if it is firmly anchored, at least two team members are needed to advance and maneuver a 65-mm (2½-in.) handline inside a building. These fire fighters must contend with the nozzle reaction force as well as the combined weight of the hose and the water. In situations where the hose line must be advanced over a considerable distance into a building, additional fire fighters will be required to move the line. The extra effort required to deploy such a line, however, is balanced by the powerful fire suppression capabilities of a large handline.

SAFETY TIP

Indicators of possible structural collapse include:

- Cracks in walls, especially cracks that develop or grow during a fire
- Leaning walls
- Pitched or sagging doors, roofs, or floors
- Doors stuck in shifted frames
- Any type of movement or vibrations
- Smoke emitting from cracks in the wall
- Moaning, groaning, creaking, or cracking sounds
- Lack of runoff from firefighting operations

In addition to recognizing the signs, it is important to be able to identify the factors that increase the likelihood of structural collapse so you can take the steps needed to reduce the risk of injury or death. These factors include the environment, building occupancy, existing structural instability, fire and explosion damage, and lightweight construction.

Large handlines are often used in defensive situations to direct a heavy stream of water onto a fire from an exterior position. In these cases, the nozzle is usually positioned so that it can be operated from a single location by one or two fire fighters. The stream can be used to attack a large exterior fire or to protect exposures. It can also be directed into a building through a doorway or window opening to knock down a large volume of

LISTEN UP!

Although a fire fighter's primary concern is saving lives and property, you will inevitably make observations and gather information as you perform your duties. Note the time of day, the weather conditions, and any route obstructions. As the incident progresses, make a mental note of the building, fire and smoke conditions, bystanders, and any other information you observe. The information could help the fire investigator determine the origin and cause of the fire.

fire inside. If the exterior attack is successful in reducing the volume of fire, the IC might decide to switch to an interior attack to complete extinguishment of the fire.

One-Fire Fighter Method for Operating a Large Handline

One fire fighter can operate a large handline if it is firmly anchored. This task can be accomplished by utilizing a webbing strap or by forming a large loop of hose about 0.6 m (2 ft) behind the nozzle and sitting down on it. When the loop is placed over the top of the nozzle, the weight of the hose stabilizes the nozzle and reduces the nozzle reaction. To add more stability, lash the hose loop to the hose behind the nozzle where they cross. This technique reduces the energy needed to control the hose line if it is necessary to maintain the water stream for a long period of time. Although this method does not allow the hose to be moved while water is flowing, it is a good choice for protecting exposures when fire fighters are operating in a defensive situation.

To perform the one-fire fighter method for operating a large handline, follow the steps in **SKILL DRILL 17-5**.

Two-Fire Fighter Method for Operating a Large Handline

When two fire fighters are available to operate a large handline, one should act as the nozzle operator, while the other serves as a backup. The nozzle operator grasps the nozzle with one hand and holds the hose behind the nozzle with the other hand. The hose should be cradled across the fire fighter's hip for added stability. The backup fire fighter should be positioned approximately 0.9 m (3 ft) behind the nozzle operator, on the same side of the hose as the nozzle operator. The backup fire fighter grasps the hose with both hands and holds the hose against a leg or hip. The backup fire fighter can also use a hose strap to maintain a better hand grip on a large handline. When the handline is operated from a fixed position, the second fire fighter can kneel on the hose with one knee to stabilize it against the ground.

To perform the two-fire fighter method for operating a large handline, follow the steps in **SKILL DRILL 17-6**.

If it is necessary to advance a flowing 65-mm (2½-in.) handline over a short distance and only two fire fighters are available, be aware of the large reaction force exerted by the flowing water. It is much easier to shut down the nozzle momentarily and move it to the new position than to relocate a flowing line. If the hose line must be moved while water is flowing, both fire fighters must brace the hose against their bodies to

SKILL DRILL 17-5
Performing the One-Fire Fighter Method for Operating a Large Handline
Fire Fighter I, NFPA 1001: 4.3.8

1 Select the correct size of fire hose for the task to be performed. While wearing full PPE and SCBA, advance the hose into the position from which you plan to attack the fire. Signal that you are ready for water. Open the nozzle to allow air to escape and to ensure that water is flowing, and then close the nozzle.

2 Make a loop with the hose. Ensure that the nozzle is under the hose line that is coming from the fire apparatus. Using rope or a strap, secure the hose sections together where they cross, or use your body weight to kneel or sit on the hose line at the point where the hose crosses itself.

3 Allow enough hose to extend past the section where the hose line crosses itself for maneuverability.

4 Open the nozzle, and direct water onto the designated area.

SKILL DRILL 17-6
Performing the Two-Fire Fighter Method for Operating a Large Handline
Fire Fighter I, NFPA 1001: 4.3.8

1 Don all PPE and SCBA. Select the correct hand-line for the task at hand. Stretch the hose line from the fire apparatus into position.

2 Signal that you are ready for water, and open the nozzle a small amount to allow air to escape and to ensure water is flowing. Advance the hose line as needed.

3 Before attacking the fire, the fire fighter on the nozzle should cradle the hose on his or her hip while grasping the nozzle with one hand and supporting the hose with the other hand. The second fire fighter should stay 0.9 m (3 ft) behind him or her and grasp the hose, securing it with two hands.

4 Open the nozzle in a controlled fashion, and direct water onto the fire or designated exposure.

keep it under control. Three fire fighters can stabilize and advance a large handline more comfortably and safely than can two fire fighters.

Operating Master Stream Appliances

Master stream appliances are used to produce high-volume water streams for large fires. Several types of master stream appliances exist, including portable monitors, deck guns, ladder pipes, and other elevated stream devices. Most master streams discharge between 1325 and 5678 L/min (350 and 1500 gpm), although much larger capacities are available for special applications. In addition, the stream that is discharged from a master stream appliance has a greater range than the stream from a handline, so it can be effective from a greater distance.

A master stream appliance can be either manually operated or directed by remote control. Many of these devices can be set up and then left to operate unattended. This capability may prove extremely valuable in a high-risk situation, because it eliminates the need to leave a fire fighter in an unsafe location or a hazardous environment to operate the device.

Master streams are used mainly during defensive operations. They should never be directed into a building while fire fighters are operating inside the structure, because these streams can push heat, smoke, or fire onto the fire fighters. The force and impact of the stream can also dislodge loose materials or cause a structural collapse.

SAFETY TIP

A nozzle flowing 3785 L/min (1000 gpm) is adding 3.81 tonnes (4.2 tons, 8400 lb, or 3810 kg) of weight to the building every minute. Be aware of the increasing risk of structural collapse in such a scenario.

Deck Guns

A deck gun is permanently mounted on a vehicle and equipped with a piping system that delivers water to the device. These devices are sometimes called turret pipes or wagon pipes. If the vehicle is equipped with a pump, the pump driver/operator can usually open a valve to start the flow of water. Sometimes, however, a hose must be connected to a special inlet to deliver water to the deck gun. If your fire apparatus is equipped with a deck gun, you need to learn your role when placing it in operation. To operate a deck gun, follow the steps in **SKILL DRILL 17-7**.

SKILL DRILL 17-7
Operating a Deck Gun Fire Fighter I, NFPA 1001: 4.3.8

1 Make sure that all firefighting personnel are out of the structure. Place the deck gun in the correct position. Aim the deck gun at the fire or at the target exposure. Signal the pump driver/operator that you are ready for water.

2 Once water is flowing, adjust the angle, aim, or water flow as necessary.

© Jones & Bartlett Learning. Photographed by Glen E. Ellman.

Portable Monitors

A portable monitor is a master stream appliance that can be positioned wherever a master stream is needed. It is placed on the ground, and hose lines are then connected to the portable monitor to supply the water. Most of these devices come equipped with either one, two, or three inlets. Portable monitors can be built with 65-mm (2½-in.) inlets or with a large-diameter hose inlet. Smaller portable monitors are sometimes set up attached to a preconnected hose. This arrangement enables one fire fighter to step away from the pumper with the portable monitor and attached hose, walk to an assigned area, and quickly place the portable monitor in operation. Some master stream appliances can be both used as deck guns or taken off the fire apparatus and used as portable monitors.

To deploy a portable monitor, remove it from the apparatus, and carry it to the location where it will be used. Advance an adequate number of hose lines from the pumper to the monitor. The number of hose lines needed depends on the volume of water to be delivered and the size of the hose lines. Form a large loop in the end of each hose line in front of the monitor, and then attach the male coupling to the inlets of the monitor. The loops serve to counteract the force created by the flow of the water through the nozzle.

To set up and operate a portable monitor, follow the steps in **SKILL DRILL 17-8**.

If the portable monitor is not adequately secured, the nozzle reaction force can cause it to move from the position where it was originally placed. A moving portable monitor poses a danger to anyone in its path. Many of these master stream appliances are equipped with a strap or chain that may be secured to a fixed object to prevent the monitor from moving. Pointed feet on the base also help to keep a portable monitor from moving. If the stream is operated at a low angle, the reaction force will tend to make the monitor unstable. For this reason, a safety lock is usually provided to prevent the monitor from being lowered beyond a safe angle of 35 degrees. When setting up any portable monitor, always follow the manufacturer's instructions and your department's SOPs to ensure its safe and effective operation.

Elevated Master Stream Appliances

Elevated master stream appliances are mounted on aerial ladders, aerial platforms, or special hydraulically operated booms. A ladder pipe is an elevated master

SKILL DRILL 17-8
Operating a Portable Monitor Fire Fighter I, NFPA 1001: 4.3.8

1. Remove the portable monitor from the fire apparatus, and move it into the desired position.

2. Attach the necessary hose lines to the monitor as per your department's SOPs or the manufacturer's instructions.

3. Loop the hose lines in front of the monitor to counteract the force created by water flowing out of the nozzle.

4. Signal the pump driver/operator that you are ready for water.

5. Aim the water stream at the fire or onto the designated exposure, and adjust the stream as necessary.

© Jones & Bartlett Learning. Photographed by Glen E. Ellman.

stream appliance that is mounted at the tip of an aerial ladder or tower ladder. On many older aerial ladders, the ladder pipe is attached to the top of the ladder only when it is actually needed, and a hose is run up the ladder to deliver water to the device. Secure the hose to the ladder with ropes or straps before flowing water from the ladder pipe. Most newer aerial ladders and tower ladders are equipped with a fixed piping system to deliver water to a permanently mounted master stream appliance at the top. This arrangement saves valuable setup time at a fire scene. If your apparatus is equipped with a ladder pipe, you need to learn how to assist in its setup.

Specific Fire-Ground Operations

Concealed-Space Fires

Fires in ordinary and wood-frame construction can burn in combustible void spaces behind walls and under subfloors and ceilings. To prevent the fire from spreading, these fires must be found and suppressed. To locate and suppress fires behind walls and under subfloors, follow the steps in **SKILL DRILL 17-9**.

Basement Fires

Fires in basements or below grade level present several different challenges. First and foremost they may be

---

SAFETY TIP

Many fires today are likely to be ventilation-limited when you arrive. Remember that even the act of opening the front door can introduce additional oxygen to the fire, contributing to a violent flashover in a very short period of time!

SKILL DRILL 17-9
Locating and Suppressing Concealed-Space Fires Fire Fighter I, NFPA 1001: 4.3.8

1 Locate the area of the building where a hidden fire is believed to exist. Look for signs of fire such as smoke coming from cracks or openings in walls, charred areas with no outward evidence of fire, and peeling or bubbled paint or wallpaper. Listen for cracks and pops or hissing steam. Use a thermal imaging device to look for areas of heat that may indicate a hidden fire. Use the back of your hand to feel for heat coming from a wall or floor.

2 If a hidden fire is suspected, use a tool such as an axe or Halligan tool to remove the building material over the area. If fire is found, expose the area as much as possible without causing unnecessary damage, and extinguish the fire using conventional firefighting methods.

© Jones & Bartlett Learning. Photographed by Glen E. Ellman.

hard to recognize as basement fires. Basement fires may spread to areas above. This means that when a fire is discovered by a neighbour or when fire fighters arrive on the scene, the fire may have spread from the basement to the first floor or above. Fire fighters may assume that a basement fire is a fire in one of the main rooms of the house, since that is where the fire is most visible.

Many vertical voids within a building originate or terminate in the basement, providing ample opportunities for fire gases to spread throughout the building. In many parts of the country, especially in subdivisions with lightweight construction built prior to the adoption of the 2012 International Residential Code, basement ceilings are most likely left unfinished when the house is constructed. Under these circumstances, fires originating in the basement will quickly involve the floor and support system, resulting in a failure of the floor over the fire and early collapse. These floor failures have the potential for causing injury or death to fire fighters entering the building with conditions of limited visibility. If fire fighters do not identify the basement fire prior to entering the building above the basement fire, they are at an increased risk for falling through the damaged floor and ending up in the burning basement below. Also remember that basement fires can spread to upper floors in houses with balloon-frame construction. In a balloon-frame building, the exterior walls are assembled with wood studs that run continuously from the basement to the roof. As a result, an open channel between each pair of studs extends from the foundation to the attic. Each channel provides a path that enables a fire to spread from the basement to the attic without being visible on the first- or second-floor levels.

Basements are difficult and dangerous spaces to access because they have limited routes for entering and exiting. Many older buildings have only one interior stairwell to the basement, so they may be difficult to ventilate because of the limited openings to the outside. Basements are often used for storage, so fire fighters may find narrow, disorganized, or cluttered spaces. Many basements contain a high volume of flammable materials, which results in a high fuel load and the production of large quantities of heat. All of these risks highlight the importance of completing a 360-degree size-up prior to entering a structure below grade.

One of the findings of research on basement fires is that water applied via the interior stairs has a limited effect on cooling the basement or extinguishing the fire. Researchers also discovered that the temperatures at the bottom of the basement stairs are not sufficiently cool to enable fire fighters to operate safely in that environment. Therefore the traditional tactic of advancing down the interior stairs to attack the seat of a fire may no longer be considered a safe and effective option for suppressing basement fires. If the only point of entry into a basement is an interior stairway, fire fighters must look for alternative openings to apply an exterior stream to the basement, reducing temperatures and the fire's impact on the floor system (UL FSRI, 2016).

Research investigating the characteristics of basement fires has provided us with valuable information about how we can use a safer and more effective approach to attack basement fires. If the basement has exterior windows or an external door, these should be considered as points for ventilation and suppression. Once an external window or basement door is opened, it impacts the flow path of the fire. The growth of these fires can be slowed and the basement temperatures can be reduced through the use of a transitional attack. A straight stream of water applied through an external window or door quickly darkens down a fire and greatly reduces temperatures throughout the building. Furthermore, fire experiments have shown that this transitional attack, when applied properly, does not push fire or hot gases up the interior stairs, nor does it create dangerous steam conditions. The darkening of the fire lasts for several minutes before the fire may experience significant regrowth. If the basement does not have external windows or doors, the fire can be attacked by cutting a small opening above the fire and inserting a fog-stream nozzle, cellar nozzle, or a Bresnan distributor nozzle to temporarily cool the basement compartment.

To conduct a safe and effective attack on a basement fire, identify any external openings from the basement to the outside and advance a charged hose line to those openings. At the same time, make sure a crew is

---

FIRE FIGHTER TIP

Some points to remember regarding basement fires are as follows:

- Thermal imaging devices may help indicate whether there is a basement fire but cannot be used to assess structural integrity or fire severity from above.
- Attacking a basement fire from a stairway places fire fighters in a high-risk location due to being in the path of hot gases flowing up the stairs and working over the fire on a flooring system that has the potential to collapse due to fire exposure.
- Coordinating ventilation is extremely important. Ventilating the basement may create a flow path up the stairs and out through the front door of the structure, almost doubling the speed of the hot gases and increasing the temperature of the gases to levels that could cause injury or death to a fully protected fire fighter.

prepared to ventilate through one or more openings. As soon as the ventilation crew is ready to ventilate, apply a straight stream through the opening. Following this, fire fighters can advance into the basement to complete the knockdown, ventilation, final extinguishment, and overhaul of the fire. **TABLE 17-2** shows the conclusions drawn from the fire experiments conducted by NIST, UL, and the FDNY regarding basement fires.

Fires Above Ground Level

Advancing charged handlines up stairs and along narrow hallways requires much more physical effort than advancing a charged handline on a level surface. It is important to protect stairways and other vertical openings between floors when fighting a fire in a multilevel structure. Specifically, handlines must be placed to keep the fire from extending vertically and to ensure that exit paths remain available.

Interior fire crews must always look for a secondary exit path in case their entry route becomes blocked by the fire or by a structural collapse. This secondary exit could be a second interior stairway, an outside fire escape, a ground ladder placed to a window, or an aerial device.

In high-rise buildings, the standpipe system is typically used to supply water for hose lines. Fire fighters must practise connecting hose lines to standpipe outlets and extending lines from stairways into remote floor areas. Additional hose lines, tools, air cylinders, and emergency medical services (EMS) equipment should be staged one or two floors below the fire.

SAFETY TIP

Be alert to the risk of structural instability and collapse. Remember that modern construction does not always provide clear signs of weakness before collapse. Studies have shown fire fighters cannot rely on sounding the floor for accurate stability indicators.

TABLE 17-2 Basement Fires

- Fire flows from basement fires developed in locations other than the stairs, as the floor assembly often failed close to the location where the fire started.
- Flowing water at the top of the interior stairs had limited impact on basement fires.
- Offensive exterior attack through a basement window was effective in cooling the fire compartment.
- Offensive fire attack through an exterior door was effective in cooling the fire compartment.

Courtesy of NIST.

Attic Fires

To understand how to extinguish attic fires, it is helpful to know where these fires start, how they receive a supply of oxygen, and how the fire flow path contributes to the spread of these fires. Fires that involve attics can start in several different locations. Some start in the attic area. An example of this is a fire that starts from an electrical short circuit in the attic. A second location where attic fires can start is in a location apart from the attic. An example of this is an older house built using balloon-frame construction. In this case, a fire that starts in the basement can quickly spread from the basement to the attic through the openings between the studs, which extend from the basement to the attic. A third location where attic fires can start is outside. These fires often start as a ground cover fire or a grilling adventure gone awry. They then impinge on the vinyl or other combustible siding and spread up the side of the house. When the fire impinges on the vinyl siding, it will cause the siding to melt and fall away from the exterior wall and, in many cases, ignite the combustible foam insulation. These insulation fires can transition into the attic space through the exposed soffit very rapidly.

When a fire enters the attic space, it reaches a place where there are large quantities of wood and possibly other combustible materials, open air spaces, and limited or absent fire stops. Convection currents within the attic and radiated heat contribute to the growth of the fire. The attic spaces in most houses are designed to have a limited yet constant exchange of air to prevent excess moisture from building up in the attic and to remove excess heat during hot weather. In modern construction, fresh air enters the attic space through vents mounted in the soffit. These soffit vents usually run the length of the house on both sides and supply cooler fresh air to the attic under normal conditions. These same soffit vents serve as a limited source of oxygen during a fire that involves the attic space. In other words, soffit vents serve as a limited inlet during an attic fire, although a new trend in the building industry is to create temperature-controlled attic spaces with the use of spray polyurethane foam to seal the attic and soffit.

During normal conditions, convection currents exhaust warmer air from the attic through vents, which are usually mounted along the peak or ridge of the roof. In newer houses, this vent runs in a continuous pathway along the peak of the roof. When there is a fire in the attic, this ridge vent serves as the exhaust for the fire gases. Most modern construction uses plastic soffit vents and plastic ridge vents. Older houses may be constructed with vents at each end of the house instead of using ridge vents. In older houses constructed using balloon-frame construction, part of the flow path inlet will be through the open spaces running from the basement to the attic, which are located between the wall studs. It is important to understand how flow paths operate in attic fires. It is also important to understand that there are many different types of building construction and vents (**FIGURE 17-7**). Each type will affect how fires grow and spread.

Attic fires contain large quantities of fuel; however, because of the limited air supply available through the soffit vents and stud spaces, they usually become ventilation-limited fires. Further limiting the exhaust flow path is the fact that the ridge vent is often made of plastic. Heat from the fire will often melt the ridge vent and seal this opening, thereby limiting the exhaust of hot fire fuel gases from the fire. This means that ventilation without coordinated suppression will usually result in rapid fire growth and spread. Treat attic fires as ventilation-limited fires, and use a transitional fire attack, if possible. Apply water through the vent openings to cool the compartment in coordination with vertical ventilation.

To initiate a transitional attack on an attic fire, it is necessary to apply a stream of water into the attic space to remove fuel gases and to cool the compartment. Remember, if air can flow into the attic, water can be applied through the same openings. Because the air is entering the fire compartment through the soffit vents, it is logical that water applied through these openings should be effective. Live fire experiments have indicated

that this is true. When performing a transitional attack on an attic fire, quickly remove the soffit vents from the underside of the soffit. In most modern-constructed houses, these are made of plastic and, in the case of a single-storey house, can be easily removed with a pike pole. Once the vents are removed, use a straight-stream nozzle to spray a stream of water into the soffit-vent opening, directing water all the way along one side of the house.

This will wet the underside of the sheathing on that side, deflect off the ridge line, and run down the other side of the interior roof sheathing. This application of water serves several purposes. It removes large quantities of the fire fuel gases, cools the fire compartment, and wets the fuel in the attic. All of these actions serve to reduce the growth of the fire and partially extinguish it. Following this initial exterior attack, it will be necessary to ventilate the attic space and transition into an interior attack for final suppression. In the case of a two-storey house, the initial attack may be accomplished by using aerial devices to apply a stream of water through the soffit vent openings.

Older houses are sometimes built with vent openings at each gable end of the building. Application of water through these openings on the end may be an effective way to perform a transitional fire attack.

Remember that these fire flow paths change if there is wind blowing to enhance or retard the air flow. Understanding flow paths is an important step in understanding how to extinguish attic fires.

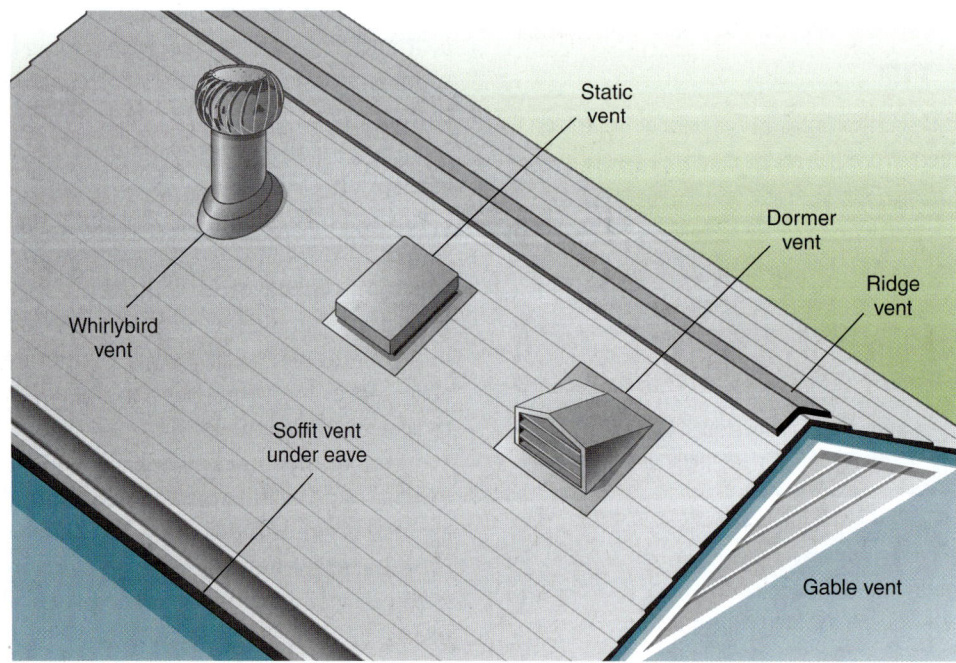

Static vent

Dormer vent

Ridge vent

Whirlybird vent

Soffit vent under eave

Gable vent

FIGURE 17-7 Roof vents will affect how fires grow and spread.
© Jones & Bartlett Learning.

Fires in Large Buildings

Large buildings have many different uses and layouts. Big-box stores contain one very large, open space surrounded by smaller rooms and storage areas. These stores are usually one storey tall. Other large buildings are constructed for multiple offices and may be many storeys tall. The fire protection for most large buildings is dependent on fire-resistive construction, fire detection, and automatic sprinkler systems. In large buildings equipped with automatic sprinkler systems, it is important to augment the water supply by using a fire apparatus to pump into the fire department connection to increase the water volume and pressure in the sprinkler system.

The fire load in large buildings varies greatly depending on the building contents. Large stores and home improvement centres contain an extensive amount of flammable contents ranging from lumber, to flammable fuels, to large quantities of home furnishings and clothing. Recognize that the fire load may be high in some of these buildings. Consider the need for long hose lays if there is not an adequate number of standpipe connections. Remember that big buildings may produce big fires and require the use of large-diameter hose lines with high-volume water supplies.

Many large buildings have floor plans that can cause fire fighters to become lost or disoriented while working inside, particularly in low-visibility or zero-visibility conditions. In such a setting, search rope may be necessary to provide searchers with an anchor point to keep them in contact with each other and with the egress (exit) route from a burning building. A well-organized

preincident plan of the structure can be essential when fighting this type of fire. Knowing the occupancy and the other hazards beforehand will help in determining the best strategy and tactics.

Fires in Buildings During Construction, Renovation, or Demolition

Buildings that are under construction, renovation, or demolition are all at increased risk for destruction by fire. These buildings often have large quantities of combustible materials exposed, while lacking the fire-resistant features of a finished building. If the building lacks intact windows and doors, an almost unlimited supply of oxygen is available to fuel a fire. Fire detection, fire alarm, and automatic fire suppression systems often are not installed or are inoperable. In addition, construction workers using torches and other flame-producing devices pose a notable fire risk. Moreover, these buildings are often unoccupied and can be easy targets for arsonists.

Under these conditions, a fire in a large building can be difficult to extinguish. If no life-safety hazards are involved, fire fighters may need to use a defensive operation for this type of fire. In such a case, no fire fighters should enter the building, and a collapse zone should be established. A defensive exterior operation should be conducted using master streams, aerial streams, and large handlines to protect exposures. Inspections of construction and demolition sites can give fire fighters indications if construction is following code.

Fires in Lumberyards

Lumberyard fires are often prime candidates for a defensive firefighting operation. A typical lumberyard contains large quantities of highly combustible materials that are stored in the open or in sheds where plenty of air is available to support combustion. Given this rich fuel supply, a lumberyard fire will usually produce tremendous quantities of radiant heat and release burning embers that will cause the fire to spread quickly from stack to stack. Lumberyards may be sited in locations where there is an inadequate water supply. Such fires may quickly extend to nearby buildings and other structures. Large buildings at lumberyards are often constructed using trusses and other lightweight building techniques, so you must be aware of the potential for rapid structural collapse at these incidents.

Protecting exposures is often the primary objective at lumberyard fires, when there is little life-safety risk. Exposure protection should be dealt with early in the operation by placing large handlines and master stream appliances where they can be most effective. A collapse zone must be established around any stacks of burning material and buildings to keep fire fighters out of dangerous positions.

Fires in Stacked or Piled Materials

Fires occurring in stacked or piled materials, such as plastics, tires, and packaging materials, can present a variety of hazards. The greatest danger to fire fighters is the possibility that a stack of heavy material, such as rolled paper or baled rags, will collapse without warning. Such an event might occur, for example, if a fire has damaged the stacked materials or if water has soaked into them. Absorbed water can increase the weight of many materials and weaken cardboard and paper products. In fact, the water discharged by automatic sprinklers can be sufficient to make some stacked materials unstable. A tall stack of material that falls on top of a team of fire fighters can cause injury or death.

Given these risks, fires in stacked materials should be approached cautiously. All fire fighters must remain outside potential collapse zones. Mechanical equipment should be used to move material that has been partially burned or water soaked.

Conventional methods of fire attack can often be used to gain control of the fire; however, water must penetrate into the stacked material to fully extinguish the residual combustion. Class A foam and wetting agents can be applied to extinguish smouldering fires in tightly packed combustible materials. See Chapter 23, *Advanced Fire Suppression*, for more information about firefighting foam application. Overhaul will require fire fighters to separate the materials to expose any remaining deep-seated fire. This operation can be a labour-intensive process unless mechanical equipment can be used to separate the damaged material.

Trash Container and Rubbish Fires

Trash container (dumpster) fires usually occur outside of a structure and appear to present fewer challenges than fires inside buildings. Even so, fire fighters must be vigilant in wearing full PPE and using SCBA when fighting trash container fires or trash pile fires, because there is no way of knowing what might be included in a collection of trash. Some trash containers may contain hazardous materials or materials that are highly flammable or explosive. Some residential dumpsters are now constructed from polyethylene, which can burn very rapidly, and the fire service needs to be aware of their presence.

If the fire is deep seated, fire fighters will have to overhaul the trash to make sure that the fire is completely extinguished. Manual overhaul involves pulling the contents of a trash container apart with pike poles and other hand tools so that water can reach the burning material. This process can be labour intensive and involves considerable risk to fire fighters. The fire fighters are exposed to any contaminants in the container as well as to the risks of injury from burns, smoke, or other causes. Considering the low value of the contents of a trash container, it is difficult to justify any risk to fire fighters' safety.

Class A foam is useful for extinguishing many trash container fires, because it allows water to soak into the materials and, therefore, can eliminate the need for manual overhaul. Some fire departments use the deck gun on the top of a pumper to extinguish large trash container fires and then complete extinguishment by filling the container with water. This is done by pointing the deck gun at the dumpster and slowly opening the discharge gate to lob water into the container.

Trash containers are often placed behind large buildings and businesses. If the container is close to the structure, be sure to check for fire extension. Also look around the container for the presence of telephone, cable, and power lines that might have been damaged by the fire.

To extinguish an outside trash fire or other outside class A fire, follow the steps in **SKILL DRILL 17-10**.

Confined Spaces

Fires and other types of emergencies can occur in confined spaces. Fires in underground vaults and utility rooms (such as transformer vaults) are too dangerous to enter. In such cases, fire fighters should summon the utility company and keep the area around manhole covers and other openings

SKILL DRILL 17-10
Extinguishing an Outside Class A Fire Fire Fighter I, NFPA 1001: 4.3.8

1 Don full PPE, including SCBA; enter the personnel accountability system; and work as a team. Perform size-up, and give an arrival report. Call for additional resources if needed. Ensure that apparatus is positioned uphill and upwind of the fire and that it protects the scene from traffic.

2 Deploy an appropriate attack line (at least 38 mm [1½ in.] in diameter).

3 Open the nozzle to purge air from the system, and make sure water is flowing. If using an adjustable nozzle, ensure that it is set to the proper nozzle pattern. Shut down the nozzle until you are in a position to apply water.

4 Direct the crew to attack the fire in a safe manner—uphill and upwind from the fire.

(continued)

SKILL DRILL 17-10 Continued
Extinguishing an Outside Class A Fire Fire Fighter I, NFPA 1001: 4.3.8

5 Break up compact materials with hand tools or hose streams.

6 Overhaul the fire, and notify command when the fire is under control. Identify obvious signs of the origin and cause of the fire. Preserve any evidence of arson. Return the equipment and crew to service.

© Jones & Bartlett Learning. Photographed by Glen E. Ellman.

clear while awaiting the arrival of utility company personnel. To manage emergencies in these areas the CSA (Canadian Standards Association) Z1006 standard requires that you follow all of the OHS (Occupational Health and Safety) regulations and you understand the standard (CCOHS, 2018). Never attempt to enter a confined space for fire suppression or rescue operations unless you have been trained as a technical rescuer. Refer to your department's SOPs for confined space emergencies.

LISTEN UP!

Fire fighters must identify obvious signs of the origin and cause of the fire, and they should attempt to preserve evidence related to the cause of the fire, particularly when working near the suspected area of fire origin and when the fire cause is not obvious. In general, disturb as little as possible in or near the area of origin until a fire investigator determines the cause of the fire and gives permission to clear the area.

Protecting Exposures

Protecting exposures refers to actions that are taken to prevent the spread of a fire to areas that are not already burning. Exposure protection is a consideration at every fire; it becomes even more important with a large fire. If the fire is relatively small and contained within a limited area, the best way to protect exposures is usually to extinguish the fire; when the fire is extinguished, the exposure problem ceases to exist. Modern fires produce heat flux levels that may affect exposures earlier in the event. Heat flux is the rate of heat transfer to or from one surface to another. In cases where the fire is too large to be controlled, exposure protection becomes a priority. In some cases, the best outcome that fire fighters can hope to obtain is to stop the fire from spreading.

The IC must consider the size of the fire and the risk to exposures in relation to how much firefighting capability is available and how quickly those resources can be assembled. In some cases, the IC will direct the first-arriving companies to protect exposures while a second group of companies prepares to attack the fire. At other times, the IC must identify a point where the progress of the fire can be stopped and direct all fire-fighting efforts toward that objective.

Protecting exposures involves different tactics from offensive fire attacks. At a large free-burning fire, the first priority is to protect exposed buildings and property from the three primary mechanisms by which heat is transferred: conduction, convection, and radiation (**FIGURE 17-8**). The best option in these circumstances is usually to direct the first hose streams at the exposures rather than at the fire itself. Wetting the exposures will keep the fuel from reaching its ignition temperature. Master stream appliances are excellent tools for protecting exposures. They also ensure that large volumes of water can be directed onto the exposures from a safe distance without putting fire fighters in the path of excess heat or in danger of structural collapse.

Fog streams can sometimes be used to absorb some of the heat coming from the main body of fire.

Solar Photovoltaic Systems

Solar photovoltaic (PV) systems generate electrical power from sunlight by converting particles of sunlight (photons) from light energy into electrical energy. These systems consist of individual solar cells that are mounted together to form thin **solar panels**. Multiple solar panels are usually mounted close together on the roof of a building. Each solar panel is interconnected by electrical cables. The electrical current generated by solar panels is direct current (DC). The electrical current is transmitted through a conduit to an inverter that changes the DC into alternating current (AC). The energy created by the solar panels can be used in that building, sold back to the power company, or stored in a battery storage system on the premises. A battery storage system is a system that stores energy through the use of a battery technology so that the energy can be used at a later time. These systems are also called battery energy storage systems. A PV system generates electrical current any time light strikes the solar panels. Usually, there is no single switch that will disable the entire PV system. Electrical shut-offs may not be adequately marked, so fire fighters

FIGURE 17-8 Protecting an exposure from radiant heat. Take care that the water applied to an exposure does not damage the structure (flood the basement or damage siding).
© Jones & Bartlett Learning.

should assume that all components of these systems are energized.

Because PV systems generate electrical energy, it is important to learn how to operate safely around them. Almost all parts of the PV system pose potential electrocution hazards. Fire fighters must be aware of the electrical shock hazard posed by rooftop solar panels and by battery storage systems. The electrical shock hazard due to the application of water is dependent on the voltage of the system, water conductivity, distance, and spray pattern. Tests conducted by UL have demonstrated that operating a hose stream at a distance of at least 6.1 m (20 ft) from the solar panels using a fog pattern of at least 10 degrees is safe (UL, 2011). Because salt water conducts electrical current more readily than fresh water, it should not be used around PV systems. Fire fighters should be aware that electrical enclosures are not resistant to water penetration from the force of a fire hose stream. Fire fighter's gloves and boots offer limited protection against electrical shock, and the amount of protection from electric shock is reduced if the gloves or boots are wet.

It is hard to completely mitigate the electrocution hazard from PV systems because most PV systems cannot be turned off with a single disconnect switch; therefore, fire fighters need to exercise care during all interior and exterior operations. Be alert for metal roofs where solar panels are present. Metal roofs can conduct electrical hazards to other parts of the building. Exercise care during ventilation and overhaul operations. Fire damage or damage from tools may sever part of a conductor in the system, which may pose a shock hazard or a fire hazard. Fire fighters must treat all electrical panels as charged electrical equipment, even at night.

Other hazards related to PV systems include fall and collapse hazards associated with solar roof panels. During fire conditions it is important to stay away from the roof line underneath solar roof panels. Roof panels may become dislodged from the roof and slide off, posing a dangerous fall hazard. Because solar panels are heavy, there is an increased chance for roof failure during a fire. By understanding the hazards posed by PV systems and following your department's SOPs, you should be able to operate safely and effectively around solar PV systems.

Chimney Fires

Fireplaces and wood-burning stoves are designed for open flames inside buildings, so they are often the ignition source and area of origin for structure fires. Chimney fires can present several different firefighting challenges because they are usually built into the structure, and they pass from floor to floor, through the attic, and above the roof line. Newer chimneys are usually made from double-wall or triple-wall metal, and there is sufficient clearance between the hot metal and the combustible structural members of the building. Older chimneys were made from brick and mortar or clay tiles, which often deteriorate over time. The mortar in old chimneys may crumble, leaving cracks and openings between bricks where flames and embers can escape into void spaces in the wall or attic. Creosote is a by-product of burning wood, and it coats the inside of the chimney and builds up over time. Creosote is highly combustible, and many chimney fires are caused by the lack of maintenance and cleaning. Chimney fires involving creosote are often extremely hot and can cause flames to shoot out of the top of the chimney.

When responding to a reported chimney fire, look for smoke or flames coming from the top of the chimney. If fire fighters see smoke coming from the eaves or attic vents, they should suspect that the fire has extended into void spaces or into the attic itself. Avoid using water to extinguish the fire unless it is necessary because the water may crack or damage the hot flue. Dry-chemical extinguishers are commonly used to extinguish chimney fires. Remember that fire moves up from the source, so it is important to extinguish the fire in the firebox, as well as in the chimney. While wearing full PPE and SCBA, discharge the extinguisher into the firebox and up into the chimney. If the fire is not severe, use salvage tarps to cover the floor and contents near the fireplace to reduce damage from the extinguisher discharge. Sometimes, it is possible to discharge the extinguisher from the top of the chimney down into the firebox, but the convective movement of the heat may prevent the agent from reaching the firebox. It may be necessary to attack the fire from both the top and the bottom. Some fire departments carry small plastic kitchen bags filled with dry-chemical powder. As these are dropped from the top of the chimney, they melt and release the contents of the bag. When operating on a roof, be sure to use safe roof operation techniques, including using a roof ladder. Because chimney fires usually occur in cold weather, watch for snow and ice on the roof.

Use a thermal imaging device to check for hot spots behind walls and in the attic. Visually check the attic for extension, and overhaul any areas where fire has breached the chimney. It may be necessary to remove drywall or siding the entire length of the chimney if smoke is still being produced after the fire is extinguished in the firebox and chimney. Remove all debris, and ensure that all burning embers are extinguished. It is important to thoroughly overhaul chimney fires. Many chimney fires have rekindled with disastrous results several hours after the fire department has left the scene.

Vehicle Fires

Vehicle fires are one of the most common types of fires handled by fire departments. In 2016, the NFPA reported that approximately 173,000 highway vehicle fires occurred in the United States (NFPA, Fire Loss in the United States, 2016). These fires may result from a variety of causes. For example, discarded smoking materials can cause fires in upholstery. Electrical short circuits may cause fires in many different parts of a vehicle. Friction caused by dragging brakes or defective wheel bearings may cause fires. Motor vehicle accidents (MVAs) may lead to ruptured fuel lines, resulting in fires.

Modern vehicles contain a variety of gas-filled, pressurized cylinders and containers that may explode when exposed to fire. Hydraulic pistons are used to support hatchbacks, trunks, tailgates, and automobile hoods. Energy-absorbing bumper systems contain hydraulic pistons. In addition, many modern vehicles use a MacPherson strut suspension system to absorb road shocks. When these gas-filled components are quickly heated to high temperatures in a fire, they can release pressure explosively, sending metal parts hurtling away from the vehicle. In addition, the supplemental restraint systems (SRS) found in most vehicles include air bags and air curtains containing chemicals that can ignite explosively during a fire.

Modern automobiles are constructed from hundreds of pounds of plastics, which produce large quantities of heat and toxic smoke when they burn. They also contain a variety of petroleum products, including gasoline or diesel fuel, motor oil, brake fluid, and automatic transmission fluid. These products ignite easily, burn with high intensity, and produce large quantities of toxic gases. In addition to common passenger vehicles, fires occur in transport trucks, passenger buses, and recreational vehicles. These fires present additional hazards and operational challenges. For this reason, it is important to always wear full PPE, including SCBA, when fighting a vehicle fire.

It is important to understand the hazards involved in fighting vehicle fires and to mount a safe attack on the fire. Only when a viable victim is trapped in a burning vehicle does the scenario become a life-or-death situation. If a victim is visible in the vehicle, immediate rescue is the priority if it is safe to approach the vehicle. All fire fighters must wear full PPE and SCBA during this operation. One or more fire fighters should attempt to rapidly remove the victim from the vehicle while another fire fighter provides protection with a fog stream.

If the driver is outside of the burning vehicle, ask about any specific hazards that may be present in the vehicle, such as portable propane cylinders, propane torches, medical oxygen equipment, cans of spray paint, and other hazardous materials. If no driver or occupant is present, do not assume that the vehicle is safe; always be cautious as you approach a vehicle fire. Perform a risk/benefit analysis—that is, look at the big picture and weigh the options. Do not risk injury when fighting a vehicle fire.

> ### FIRE FIGHTER TIP
>
> To reduce confusion and minimize the potential for mistakes at these scenes, it is important to use standardized terminology when referring to specific parts of vehicles. For example, in the United States and Canada, the driver's seat is on the left side of the vehicle. The right side of a vehicle is where the passenger's seat is located. Also, in general, vehicles consist of three main components or compartments: the engine compartment, the passenger compartment, and the trunk (cargo area).

Types of Vehicles

Most of the vehicles on the road today are **conventional vehicles** that use internal combustion engines for power. Internal combustion engines burn gasoline or diesel fuel to produce power. Fuel creates a hazard if it leaks after an MVA. Other hazards associated with conventional vehicles include short circuits and battery acid leaks, as well as a wide variety of combustibles in the passenger compartment or trunk. The 12-volt electrical systems in these vehicles pose a minimal threat to rescuers.

Alternative-fuel vehicles use anything other than a petroleum-based motor fuel (gasoline or diesel fuel) to propel a motorized vehicle. These vehicles may be powered by alternative fuels such as propane, natural gas, methanol, or hydrogen. Some vehicles are powered by electricity (battery electric vehicles) or a combination of electricity and fuel (hybrid electric vehicles). Following an MVA or fire, these vehicles present hazards that are not encountered in incidents involving conventional vehicles. It is important for rescuers to recognize the hazards these vehicles pose both to rescuers and to victims and to be familiar with the additional steps needed to mitigate these hazards.

Alternative-Fuel Vehicles

Alternative-fuel vehicles may be powered by electricity, electricity and liquid fuel (hybrid), blended liquid fuels, compressed gases, and fuel cells.

Alternative-fuel vehicles are usually identified by markings on the vehicle (**FIGURE 17-9**). In addition to the hazards from conventional vehicles, vapours from battery electric vehicle or hybrid electric vehicle batteries can be carried in smoke or steam. Never approach these vehicles without proper PPE and SCBA.

Battery Electric and Hybrid Electric Vehicles

Battery electric vehicles are propelled solely by an electric motor that is powered by batteries. Usually these batteries must be recharged from an external source. Hazards posed by these vehicles include the potential for electrical shock or arcing due to their powerful batteries and high-voltage electrical systems and vehicle movement. Additional hazards include off-gassing and delayed-reaction electrical fire from damaged batteries. Be aware that these vehicles use a high-voltage system in addition to the 12-volt system found in conventional vehicles.

Hybrid electric vehicles use both a battery-powered electric motor and a liquid-fuelled engine (**FIGURE 17-10**). The batteries power the electric motor that drives the wheels, much like a train locomotive. The gasoline-powered engine is used for auxiliary power and to recharge the batteries. Hybrid electric vehicles contain all of the same hazards associated with conventional gasoline-powered vehicles, plus additional hazards linked to the electrical system.

This chapter will refer to battery electric vehicles and hybrid electric vehicles as electric drive vehicles. As discussed, these vehicles are powered by electricity or a combination of electricity and fuel. The batteries that power these vehicles have a higher voltage than traditional automotive batteries, and it may take several minutes for a high-voltage system to de-energize after the

FIGURE 17-10 A hybrid sticker indicates that the vehicle is powered by both a liquid fuel engine and an electric motor, and it contains a high-voltage battery or batteries.
© Jones & Bartlett Learning.

main battery is turned off. Although each type of electric drive vehicle has its own unique features, one feature is common throughout—the need for properly trained responders to disable the vehicle's electrical system to prevent accidental starting, electrical shock, or further fire or explosion. Most of these vehicles are designed such that the high-voltage system is isolated once the 12-volt electrical system is disconnected. The batteries in electric drive vehicles may not be located in the engine compartment but in other areas, such as the trunk or under the seats or floorboards (**FIGURE 17-11**). Furthermore, there may be more than one battery present.

In addition to the battery pack and as part of the electrical system, brightly coloured (typically orange) high-voltage cables are routed from the high-voltage battery pack to the inverter and electric motor, which are located in the engine compartment (Vangelder, 2018). These cables are usually located beneath the passenger compartment toward the center of the vehicle, so they

FIGURE 17-9 Alternative-fuel vehicles are usually identified by markings on the vehicle.

FIGURE 17-11 A battery pack installed in a Toyota Prius.

do not interfere with most extrication procedures. However, in many vehicles, high voltage is used to operate a variety of accessories, such as power steering and air conditioning. Be aware that these high-voltage cables can be located nearly anywhere in the vehicle (**FIGURE 17-12**) (Vangelder, 2018). Because these cables carry high voltage, they pose a risk to rescuers. Look for placards and high-voltage labels that identify a potential shock hazard. Avoid contact, and remain vigilant and aware of the dangers.

There are some considerations to keep in mind when dealing with fires involving electric drive vehicles:

- When responding to fires involving these types of vehicles, you need to don your PPE, including SCBA.
- Apply chock blocks to the wheels to prevent the vehicle from moving forward or in reverse.
- Determine whether the vehicle is a hybrid electric vehicle or a battery electric vehicle. If possible, turn the power off by activating the vehicle shut-off switch, removing the key from its switch, or removing the proximity key from the vehicle. Some vehicles have a disconnect switch or fuses that can be removed to disconnect the motors from the power source.

- Avoid direct contact with the batteries. Watch for increased voltage in the battery compartments. High-voltage DC cables connect the batteries to the electric motors that power the wheels. Avoid cutting these cables.

Electric drive vehicles have automatic shut-offs that are designed to disable the high-voltage electrical system in the event of an MVA. It is important to establish a water supply, because fires in these vehicles require more water to extinguish than fires in conventional vehicles. According to the current manufacturers' recommendations, water remains the extinguishing agent of choice in case of a fire involving an electric drive vehicle. Controlled fires in hybrid electric vehicles and battery electric vehicles have not shown any electrical hazards to fire fighters while they are applying water to fires in these vehicles (NFPA, 2015).

It will take more water and a longer period of time to extinguish these fires. Apply water even after the flames are no longer visible; this is necessary to continue to cool the batteries. Batteries can reheat and reignite for a long period of time after the flames are extinguished. Listen for popping sounds, which may indicate that the battery is heating up. In cases where you can see the battery, you can use a thermal imaging device to see if the battery is cooling or heating up. After a fire, electric vehicles and hybrids should be stored at least 15 m (50 ft) from any exposures because of the possibility of a late rekindle.

The components of these vehicles continue to change, and the design and location of each component vary from one manufacturer to another. It is important to keep up with changes in vehicle architecture, as they affect the response to incidents involving the vehicles. Most manufacturers offer a wealth of information that can be viewed or downloaded from the manufacturer's website. It is a good idea for fire fighters to receive additional training on alternative-fuel fires. For more information on electric drive vehicles, including how to disable the electrical system, see Chapter 24, *Vehicle Rescue and Extrication*.

FIGURE 17-12 High-voltage cable.
© Jones & Bartlett Learning.

Blended Liquid Fuel–Powered Vehicles

Two types of alcohol—methanol and ethanol—are commonly mixed with gasoline to produce a blended fuel (**FIGURE 17-13**). Most gasoline-powered vehicles can run on a blend consisting of gasoline and up to 10 percent ethanol, which is available at some regular services station across Canada (Natural Resources Canada, 2016). Diesel fuel also may be blended with other combustibles to produce a blended diesel fuel.

Biofuels are derived from agriculturally produced products such as grains, grasses, or recycled animal oils. They are processed to produce a type of alcohol that is added to gasoline to produce a blended fuel.

Vehicles operating on these blended fuels are usually identified by special designations. They are singled out for attention primarily because of their environmental benefits.

MVAs involving vehicles that rely on blended fuels are handled in the same manner as MVAs involving vehicles powered by conventional fuels.

Compressed Gas–Powered Vehicles

Three forms of compressed gas are used to power vehicles. Compressed natural gas (CNG) is commonly stored in steel, aluminum, or carbon fibre–wound tanks, under pressures ranging from 20,684 to 24,821 kilopascal [kPa] (3000 to 3600 psi). Many cities power some of their municipal bus fleets with CNG. In particular, buses, delivery vans, and fleet sedans may be powered by CNG (**FIGURE 17-14**). Automobiles powered by CNG contain storage cylinders that are very similar to SCBA cylinders. These cylinders are usually located in the trunk of smaller vehicles or on the roof of larger vehicles and contain CNG at high pressures (**FIGURE 17-15**). These vehicles can be identified by the "pregnant shape" of the rooftop. When a fire occurs, they must be cooled and protected just like

FIGURE 17-13 A flex fuel identification badge.
© David R. Frazier Photolibrary, Inc./Alamy Images.

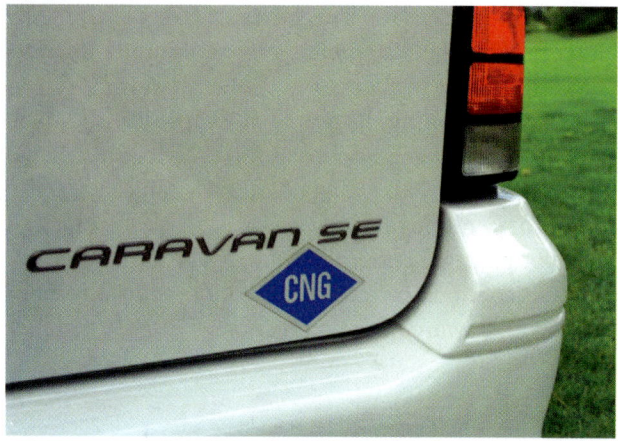

FIGURE 17-14 The CNG sticker on this van indicates the that it is powered by compressed natural gas.
Courtesy of DOE/NREL, Credit - Warren Gretz.

FIGURE 17-15 These CNG tanks are mounted to the roof of a bus.
© Jones & Bartlett Learning.

any gas cylinder. CNG is a nontoxic, lighter-than-air gas that will rise and dissipate if it is released into the atmosphere. Any sign of fire in a vehicle powered by CNG should be treated with extreme caution.

Liquefied natural gas (LNG) is stored at very low temperatures. Storage tanks for LNG are double walled and must be insulated to keep the liquefied gas cold. The use of LNG is not as common as the use of CNG.

Liquefied petroleum gas (LPG), also called propane, is the third type of compressed gas used to power vehicles. LPG is popular for light-duty vehicles, medium-duty trucks, and shuttle vans used in high-mileage fleet applications, for example, taxis and delivery vehicles (**FIGURE 17-16**). Propane is stored in a liquid state at a pressure of 345 kPa (50 psi).

Look for signage or placards on the vehicle identifying it as being powered by a compressed gas. If a compressed gas is present, check for the presence of

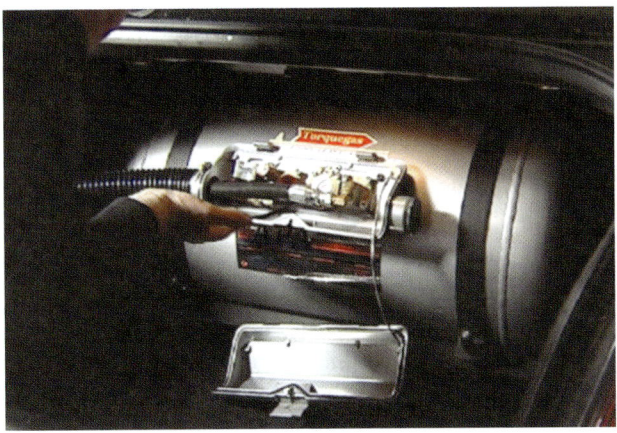

FIGURE 17-16 The LPG tanks used in automobiles such as taxis are usually found in the trunk of the vehicle.
© Jones & Bartlett Learning.

leaking cylinders or fuel lines. Damage to the gas cylinders or fuel lines could result in the escape of flammable natural gas. In addition, a fire in a gas-powered vehicle poses the threat of a boiling liquid/expanding vapour explosion (BLEVE). In the event of a fire, keep the gas cylinders cool to avoid a BLEVE. If you suspect that a leak is present, keep all sources of ignition away from the vehicle and take measures to stop the leak if possible. Fully involved fires in vehicles powered by compressed gas should be fought with a handline from a protected location or with an unmanned master stream to prevent injuries from exploding gas cylinders. If LNG is leaking, avoid spraying a stream of water on the escaping liquid, as this action heats the very cold liquid; it may then convert to a gaseous state with explosive force. It is acceptable to spray a fog stream on an escaping cloud of vapour, however. Ensure that a fire fighter in full protective equipment is standing by with a charged hose line that is at least 38 mm (1½ in.) in diameter. Refer to Chapter 23, *Advanced Fire Suppression*, for more information about compressed gas cylinder fires.

Fuel Cell–Powered Vehicles

Fuel cells generate electricity through a chemical reaction between hydrogen gas and oxygen gas to produce water and in the process produce electricity and heat. The purpose of fuel cells is not to produce water, however, but rather to produce electricity that can be used to drive an electric motor. Fuel cell–powered vehicles are essentially electric vehicles with a different device for generating the electricity. In a vehicle powered by a fuel cell, the electric motor is powered by electricity generated by the fuel cell. The major components needed for a fuel cell–powered vehicle are a hydrogen gas storage tank, a high-output battery, a group of fuel cells, and an electric motor. The use of fuel cells is being advocated because they produce less pollution than internal combustion engines.

The first production line vehicle with a dedicated hydrogen fuel cell platform was made available for public sale in 2008 (**FIGURE 17-17**) (Vangelder, 2018).

Although only a few fuel cell–powered vehicles are being produced today, new technology is continually being developed, so you should expect to see more of them on the road in the future.

Responding to the Scene

Many hazards are associated with vehicle fires—traffic hazards, electrical hazards, fuel, pressurized cylinders, fire, and toxic smoke. Be sure to wear full PPE to guard against these hazards.

Because vehicle fires usually occur on streets and highways, one of the biggest hazards faced by fire fighters is the danger posed by traffic. Drivers are easily distracted by the sight of a burning vehicle, which may lead to subsequent MVAs. To counteract this risk, you should use your apparatus to block traffic. Do not hesitate to work with law enforcement to shut down or divert traffic flow if necessary to ensure safety for fire fighters. Fire apparatus operators often place their vehicles 30 m (100 ft) behind the burning vehicle to stop traffic and to position the apparatus a safe distance from the burning vehicle. Any time Canadian workers are operating in a roadway, they are required to wear high-visibility safety clothing that meets the CSA (Canadian Standards Association) Standard Z96-15 High-Visibility Safety Apparel (CCOHS, 2018). This is not a requirement, however, if the fire fighter is wearing SCBA or conducting firefighting operations. Follow the steps outlined by your department for guarding against traffic hazards.

FIGURE 17-17 The Honda Clarity fuel cell.
© VDWI Automotive/Alamy Stock Photo.

Attacking Vehicle Fires

Conduct a quick size-up, and look for any people that may be in the vehicle. Take immediate action to protect and remove them if it is safe to do so. As you prepare to approach a vehicle fire, make sure that you have created a safe perimeter around the vehicle, and remove all bystanders from the area. Use a hose line at least 38 mm (1½ in.) in diameter; such a hose will provide sufficient cooling power to suppress the fire and provide protection from a sudden flare-up. Charge the hose line from a safe location, and slowly open the nozzle to bleed the air from the hose. Set the adjustable fog-stream nozzle to initially deliver a straight stream from as far away as possible. Approach the vehicle from an uphill and upwind position, moving in at a 45-degree angle (**FIGURE 17-18**). Some vehicles have compressed gas pistons in the front and rear bumpers. Approaching the vehicle from a 45-degree angle will help protect you if the fire causes these pistons to explode.

Open the nozzle, and sweep the ground and the bottom part of the vehicle using a horizontal motion. Extinguish all visible fire while advancing toward the vehicle. Sweeping along the undercarriage helps to cool the bumper pistons, shock absorbers, and hydraulic struts; cool the tires before they explode; and cool the fuel tank before it fails. By applying a sufficient quantity of water to the lower part of the vehicle, you reduce the chance of an explosive event. Observe the area under the car during the approach for any sign of leaking flammable liquids. Adjust the nozzle to a narrow fog pattern as you get closer to the vehicle. Continue to widen the fog pattern as you approach the vehicle.

Once this sweep of the exterior is complete, fire fighters can begin to attack any fire in the passenger area, the engine compartment, and the cargo compartment. Some departments may use compressed air foam or Class B foam hose lines to attack vehicle fires or if a large quantity of fuel is burning.

To extinguish a vehicle fire, follow the steps in **SKILL DRILL 17-11**.

Fire in the Passenger Area

Fires in the passenger area of a vehicle are more visible and accessible than fires in the engine compartment or the cargo area. Usually it is logical to extinguish the fire in the passenger compartment before moving on to extinguish any fire in the engine and cargo compartments. Often the windows are broken or a door is ajar, presenting an opening through which fire fighters can direct a stream of water. If the doors will not open, stand upwind from the window, and use a striking tool to break out one or more windows. Be cautious, because a backdraft may occur.

As you get closer to the vehicle, change the nozzle to produce a wider pattern that will cool a wider area and give you some protection from the heat of the fire. Pay special attention to cooling areas such as the steering column and the dashboard on the passenger side. Cool areas that contain side-curtain air bags; cooling will greatly reduce the chance for accidental deployment of the SRS.

Fire in the Engine Compartment

The engine compartment of a vehicle is filled with a variety of devices that use petroleum products to power or lubricate them. It also contains components made from plastics and rubber. As a consequence, these devices produce a large amount of smoke when they burn. Other hazards present in the engine compartment include suspension struts and hydraulic lift cylinders for the hood. Vehicle batteries contain sulfuric acid, which can cause serious burns. Some vehicles have magnesium parts within the engine compartment; when magnesium is present, you must use a Class D extinguishing agent.

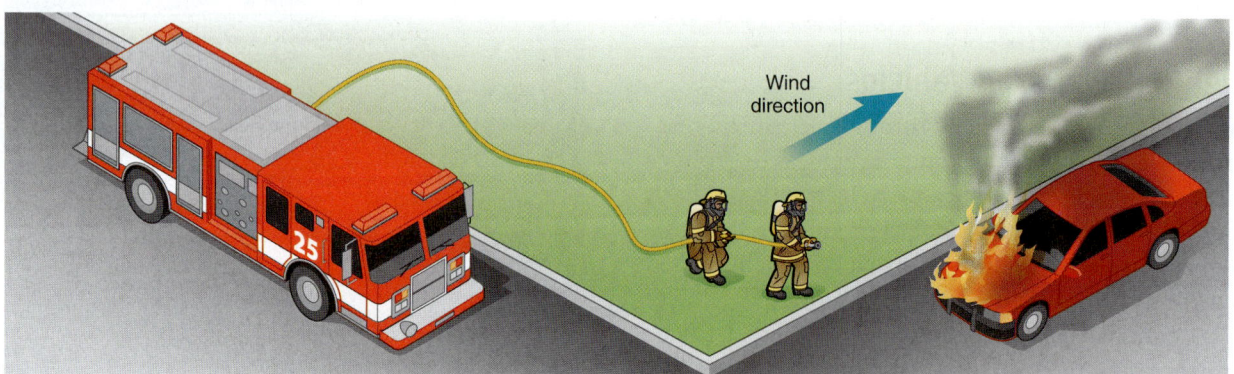

FIGURE 17-18 When possible, approach vehicle fires in the hood or engine area from a 45-degree angle and from the uphill and upwind side.

SKILL DRILL 17-11
Extinguishing a Vehicle Fire Fire Fighter I, NFPA 1001: 4.3.7

1 Don full PPE, including SCBA; enter the personnel accountability system; and work as a team. Perform size-up, and give an arrival report. Call for additional resources if needed. Ensure that apparatus is positioned uphill and upwind of the fire and that it protects the scene from traffic.

2 Deploy an appropriate attack line (at least 38 mm [1½ in.] in diameter).

3 Open the nozzle to purge air from the system, and make sure water is flowing. If using an adjustable nozzle, ensure that it is set to the proper nozzle pattern. Shut down the nozzle until you are in a position to apply water.

4 Direct the crew to attack the fire in a safe manner. Attack from uphill and upwind of the fire and at a 45-degree angle, and extinguish any fire under the vehicle.

(continued)

SKILL DRILL 17-11 Continued
Extinguishing a Vehicle Fire Fire Fighter I, NFPA 1001: 4.3.7

5 Carefully approach the vehicle, and completely suppress the fire.

6 Overhaul all areas of the vehicle, including the passenger compartment, engine compartment, and cargo area (trunk).

7 Notify command when the fire is under control. Identify obvious signs of the origin and cause of the fire. Preserve any evidence of arson. Return the equipment and crew to service.

© Jones & Bartlett Learning. Photographed by Glen E. Ellman.

A challenging part of extinguishing a fire in the engine compartment is gaining access to the fire. An initial attack can be made through the wheel well. The plastic liner between the engine compartment and the wheel well is often consumed during such a fire, so it may be possible to spray water into the engine compartment through this opening. An alternative initial approach is to spray water through the grille after thoroughly cooling the area around the bumper. When using this tactic, avoid standing directly in front of the bumper until this area is thoroughly cooled. Although neither of these two methods is effective at totally extinguishing the fire, they will help to diminish the volume of fire while fire fighters are gaining access to the engine compartment.

The fastest way to gain access to the engine compartment is to pull the hood-release lever inside the passenger compartment to open the hood. Unfortunately, this method rarely works during a vehicle fire, so other methods usually need to be tried. An alternative method is to insert a pry bar along the side of the hood between the edge of the hood and the fender. Pry the side of the hood away from the fender to produce an opening big enough to apply a stream of water into the engine compartment.

A second method of gaining access to the engine compartment is to manually open the hood of the vehicle. In most fires, the cable that normally opens the hood is damaged and will not release from the inside of the vehicle. To open the hood of a burning vehicle, break out the plastic grille, and find the hood-release cable. With a gloved hand, pull on the cable to release the hood, or use the forked end of a Halligan tool to twist the cable until the hood releases. Fire fighters can also use pliers to grasp and pull the cable.

Once the hood is raised, fire fighters should have good access to any remaining fire and hot spots. Use plenty of water or foam to cool this area. Once the vehicle's hood has been opened and any fire in the engine compartment extinguished, disconnect the power to the vehicle by cutting the battery cables. If the battery is not under the hood, check the wheel wells or trunk. First, remove a section of the negative cable at least 15 cm (6 in.) long by making two cuts with wire cutters. Then, remove a similar section of the positive cable. This excision will disrupt the flow of power to the vehicle and ensure that nothing is powered accidentally.

Fire in the Cargo Area (Trunk)

A fire in the trunk of a vehicle may present unknown hazards because it is not possible to know what the trunk contains. In addition, fires in this area are challenging to access. Initial access can sometimes be made by knocking out the tail light assembly on one side, which enables fire fighters to direct a hose stream into the trunk to cool down and partially extinguish fire in this area. A fire in the trunk area of an automobile can also be accessed by first using the pick of a Halligan tool to force the lock into the trunk and then using a screwdriver or key tool (K-tool lock puller) to turn the lock cylinder in a clockwise direction. A charged hose line must be ready when the trunk lid is raised.

Fires in the rear of light trucks and vans must always be approached cautiously. In addition, vehicles using alternative fuels (discussed earlier in this chapter) may contain compressed natural gas or propane cylinders

in the trunk. Vans are often used by couriers and could contain medical waste, laboratory specimens, and radioactive material.

Overhauling Vehicle Fires

Overhaul of vehicle fires is just as important as overhaul of structure fires. As soon as it is safe to approach the vehicle, chock the wheels to prevent the vehicle from moving. If a vehicle fire erupts quickly, the driver may not have time to set the brake and parking gear. Also, fire can damage cables and wires that control the operation of parking and braking mechanisms. After all visible fire has been knocked down, allow a few minutes for the steam and smoke to dissipate before starting overhaul. This delay will allow visibility to improve so that overhaul can be completed safely.

During overhaul of interior fires, remember that air bags can deploy without warning in a burning automobile. Never place any part of your body in the path of a front or side air bag.

As you overhaul the vehicle, be systematic and thorough. Do not miss areas that may contain lingering sparks. Direct the hose stream under the dashboard, and soak and remove smouldering upholstery. Apply water over and under all parts of the engine compartment. Confirm that no fluids are leaking from the vehicle. If contents in the car might potentially be salvaged, treat them with respect. Continue to use your SCBA as long as smoke or fumes are present.

SAFETY TIP

Some vehicle components (engines or body) may be constructed from magnesium or other flammable or explosive metals that can react violently when water is applied during fire suppression. If possible, a Class D extinguishing agent should be used in these cases instead of water. All areas of the vehicle must be checked for fire extension and for potential victims. There have been numerous cases of vehicle fires that were used to cover a crime.

Shutting Off Building Utilities

Because most structures use electricity or gaseous fuel to operate lighting, heating, cooling, cooking, or manufacturing equipment, it is important to shut off these utilities as early as possible during fire suppression activities. The IC should identify which utilities are present during size-up and request that the utility companies respond to disconnect service to the building. If it is safe to do so, fire fighters can often shut off these services before the utility company arrives.

Shutting Off Gas Service

Many structures use either natural gas or propane gas for heating or cooking. In addition, these two energy sources have many industrial applications. If a gas line inside a structure becomes compromised during a fire, the escaping gas can add fuel to the fire. The means by which the gas is supplied to the structure must be located to stop the flow.

Most residential gas supplies are delivered through a gas meter connected to an underground utility network or from a storage tank located outside the building. If the gas is supplied by an underground distribution system, the flow can be stopped by closing a quarter-turn valve on the gas meter. If the gas is supplied from an outside LPG storage cylinder, closing the cylinder valve will stop the flow.

After the gas service has been shut off, use a lockout or shut-off tag to ensure that it is not turned back on. Only a professional can reestablish the flow of gas to a structure.

To shut off gas utilities, follow the steps in **SKILL DRILL 17-12**.

Gas Fire Suppression

Fires involving gaseous fuels such as natural gas or propane can be very dangerous. Gaseous fuel that escapes from the system is called fugitive gas. It can travel between floors, inside walls or other void spaces, and even underground. If the correct mixture of fuel and air comes into contact with an ignition source, an explosion can occur. Because the fuel comes from the piping system under pressure, these fires can ignite Class A combustibles and cause those fires to burn more rapidly. If the gas has not ignited, immediately shut off the gas valve (see Skill Drill 17-12), and evacuate the area. If the gas is already burning, shut off the meter or main valve before fully extinguishing the fire. Once the fuel source is shut off, completely extinguish and overhaul any remaining Class A fire. Refer to Chapter 23, *Advanced Fire Suppression*, for information on compressed gas cylinder fires.

Shutting Off Electrical Service

The greatest danger with most fires involving electrical equipment is electrocution. For this reason, only Class C extinguishing agents should be used when energized equipment is involved in a fire. All electrical equipment should be considered as potentially energized until the power company or a qualified electrical professional confirms that the power is off. Once the electrical service has been disconnected, most fires in electrical equipment can be controlled using the same tactics and procedures that are used for a Class A fire.

When a fire occurs in a building, the electrical service should be turned off quickly to reduce the risks of injury or death to fire fighters, even if there is no involvement of electrical equipment in the fire. If possible, this shut-off should take place at the main circuit breaker box, with a lockout tag being used to prevent someone from accidentally turning the electrical current back on. Some buildings have an exterior main disconnect

SKILL DRILL 17-12
Shutting Off Gas Utilities Fire Fighter I, NFPA 1001: 4.3.18

1. Don PPE, including SCBA; enter the personnel accountability system.

2. Acknowledge the assignment.

3. Locate the exterior gas shut-off valve.

4. Close the gas shut-off valve.

5. Attach a shut-off or lockout tag and lock if required.

6. Notify command that the gas is shut off.

© Jones & Bartlett Learning. Photographed by Glen E. Ellman.

switch that shuts off electrical power to the building. These can be identified during preincident planning. Fire fighters should never attempt to remove an electric meter to disconnect power.

If it is not possible to turn the electricity off at the breaker box or the main disconnect switch, fire fighters must notify the electrical utility to send a representative to the fire scene to disconnect the service from a location outside the building. In many cases, the local utility company is automatically notified of any working fire.

To shut off the electrical utilities, follow the steps in **SKILL DRILL 17-13**.

Electrical Fire Suppression

Fire suppression methods for fires involving electrical equipment vary according to the type of equipment and the power supply. In many cases, the best approach is to wait until the power is disconnected and then use the appropriate extinguishing agents to control the fire. If the power cannot be disconnected or the situation requires immediate action, only Class C extinguishing agents—such as Halon agents, CO_2 or dry chemicals, or water mist—should be used.

When delicate electronic equipment is involved in a fire, Halons, CO_2, or water mist should be used to limit the damage as much as possible. These agents cause less damage to computers and sensitive equipment than do water or dry-chemical agents.

When power distribution lines or transformers are involved in a fire, special care must be taken to ensure the safety of both emergency personnel and the public. No attempt should be made to attack these fires until the power has been disconnected. In some cases, fire fighters may need to protect exposures or extinguish a fire that has spread to other combustible materials, if this can be accomplished without coming in contact with the electrically energized equipment. If a hose stream comes in contact with the energized equipment, the current can flow back through the water to the nozzle and electrocute fire fighters who are in contact with the hose line.

Many electrical transformers contain a cooling liquid that includes polychlorinated biphenyls (PCBs), a cancer-causing material. Do not apply water to a burning transformer. Water can cause the transformer's cooling liquid to spill or splash, contaminating both fire fighters and the environment. If the transformer is located on a pole, it should be allowed to burn until electrical utility professionals arrive and disconnect the power. Dry-chemical extinguishers can then be used to control the fire. Fires in ground-mounted transformers can also be extinguished with dry-chemical agents after the power has been disconnected. Fire fighters should stay out of the smoke and away from any liquids that are discharged from a transformer, and they must wear full PPE and SCBA to attack the fire.

Some very large transformers contain large quantities of cooling oil and require foam for fire extinguishment.

SKILL DRILL 17-13
Shutting Off Electrical Utilities Fire Fighter I, NFPA 1001: 4.3.18

1. Don PPE, including SCBA; enter the personnel accountability system.

2. Acknowledge the assignment.

3. Locate the appropriate disconnect switch or main breaker.

4. Shut off the power.

5. Attach a lockout tag and lock, or post a fire fighter at the shut-off until the utility company permanently disconnects electrical power.

6. Notify command that the electricity is shut off.

© Jones & Bartlett Learning. Photographed by Glen E. Ellman.

This foam can be applied only after the power has been disconnected. Until the power is off, fire fighters can still protect exposures while taking care to avoid contamination.

Underground power lines and transformers are often located in underground vaults. Explosive gases can build up within these vaults. If a spark ignites these gases, the resulting explosion can lift a manhole cover from a vault and hurl it for a considerable distance. Products of combustion can also leak into buildings through the underground conduits. Fire fighters should never enter an underground electrical vault while the equipment is energized. Even after the power has been disconnected, these vaults should be considered to be confined spaces containing potentially toxic gases, explosive atmospheres, or oxygen-deficient atmospheres. Special precautions are required to enter this type of confined space.

Large commercial and residential structures often have high-voltage electrical service connections and interior rooms containing transformers and distribution equipment. These areas should be clearly marked with electrical hazard signs, and fire fighters should not enter them unless there is a rescue to be made. Until the power has been disconnected, fire suppression efforts should be limited to protecting exposures. Fire fighters must wear full PPE and SCBA owing to the inhalation hazards presented by the burning of the plastics and cooling liquids that are often used with equipment of this size.

Shutting Off Water Service

If a serious water leak has occurred inside the building, shutting off the water supply may help to minimize additional water damage to the structure and contents. Water service to a building can usually be shut off by closing one valve at the entry point. Many communities permit the water service to be turned off at the connection between the utility pipes and the building's system. This underground valve, which is often accessed through a curb box, is located outside the building and can be operated with a special wrench or key. In most cases, another valve is found inside the building, usually in the basement (if there is one), where the water line enters. In warmer climates, water supply valves are sometimes located above ground.

After-Action REVIEW

IN SUMMARY

- Fire suppression refers to all of the tactics and tasks performed on the fire scene to extinguish a fire.
- All fire suppression operations are either offensive or defensive:
  - Offensive—Tasks performed while moving toward the fire or inside the fire structure. Used when the fire is not too large or dangerous to be extinguished using interior handlines.
  - Defensive—Tasks typically performed outside the fire structure. Used when the fire is too large to be controlled by an offensive attack or when the level of risk is too high.
- A combination attack involves both offensive and defensive suppression operations.
- The IC constantly evaluates conditions to determine the type of attack that should be used. An offensive attack may be switched to a defensive attack if necessary and vice versa.
- A fire fighter must be able to advance and operate a handline effectively to extinguish a fire.
- Care should be taken not to have opposing handlines. Opposing handlines may harm fire fighters with water streams or debris.
- The most frequently used sizes for interior fire attack are 38- or 45-mm (1½- or 1¾-in.) handlines.
- Different types of fire streams are produced by different types of nozzles:
  - A fog stream divides water into droplets, which can absorb heat efficiently. Fog streams can be used to protect fire fighters from the intense heat of the fire and to assist in ventilation.
  - A straight stream can hit a fire from a longer distance away; it also keeps the water concentrated in a small area.
  - A solid stream has an even greater reach and penetrating power than a straight stream and is charged as a continuous stream of water.

- The most effective means of fire suppression in most situations is a direct attack. This kind of attack delivers water directly onto the base of the fire. The water cools the fuel until it is below its ignition temperature.

- The objective of an indirect attack is to quickly absorb as much heat as possible from the fire atmosphere. Such an approach is particularly effective at preventing flashover from occurring. This method of fire suppression should be used when a fire has produced a layer of hot gases at the ceiling level.

- A combination attack employs both indirect attack and direct attack methods in a sequential manner. This strategy should be used when a room's interior has been heated to the point that it is nearing a flashover condition. Fire fighters first use an indirect attack method to cool the fire gases, then make a direct attack on the main body of fire.

- Large handlines can be used either for offensive or for defensive operations.

- Master stream appliances are used to produce high-volume water streams for large fires. Several types of master stream appliances exist, including portable monitors, deck guns, ladder pipes, and other elevated stream devices. Most master streams discharge between 1325 and 5678 L/min (350 and 1500 gpm) of water.

- Fires in ordinary and wood-frame construction can burn in combustible void spaces behind walls and under subfloors and ceilings.

- Fires in basements or below grade level may cause the floor on the ground level to collapse.

- During an interior fire attack, protect stairways and other vertical openings between floors when fighting a fire in a multilevel structure. Handlines must be placed to keep the fire from extending vertically and to ensure that exit paths remain available to fire fighters and victims.

- Many large buildings have floor plans that can cause fire fighters to become lost or disoriented while working inside, particularly in low-visibility or zero-visibility conditions. Search rope may be necessary in these settings.

- Buildings that are under construction, renovation, or demolition are all at increased risk for destruction by fire. These buildings often have large quantities of combustible materials exposed, while lacking the fire-resistant features of a finished building.

- A lumberyard fire will usually produce tremendous quantities of radiant heat and release burning embers that can cause the fire to spread quickly from stack to stack.

- Fires in stacked materials should be approached cautiously. All fire fighters must remain outside potential collapse zones. Mechanical equipment should be used to move material that has been partially burned or water soaked.

- Fires in trash containers may be fought with foam or a deck gun.

- Fires in confined spaces such as utility rooms are too dangerous to enter. Fire fighters should summon the utility company and keep the area around manhole covers and other openings clear while awaiting arrival of utility personnel.

- Electrical shut-offs may not be adequately marked for a solar PV system, so fire fighters should assume that all components of these systems are energized. Gloves and boots offer limited protection against electrical shock, especially when wet. Fire fighters should operate a hose stream at a distance of at least 6.1 m (20 ft) from the solar panels.

- Use dry-chemical extinguishers to extinguish chimney fires when possible, not water. Be sure to extinguish the fire in the firebox as well as in the chimney.

- Vehicle fires are one of the most common types of fires handled by fire departments. These fires may result from a variety of causes, including discarded smoking materials, electrical short circuits, friction caused by dragging brakes or defective wheel bearings, and ruptured fuel lines due to MVAs.

- Many hazards related to vehicle fires are possible—traffic hazards, fuel, pressurized cylinders and containers that can explode, fire, and toxic smoke.

- Overhaul of vehicle fires is just as important as overhaul of structure fires. As soon as it is safe to approach the vehicle, chock the wheels to prevent the vehicle from moving. After all visible fire has been knocked down, allow a few minutes for the steam and smoke to dissipate before starting overhaul. As you overhaul the vehicle, be systematic and thorough.

- Always be alert for signs that a burning vehicle could be powered by an alternative fuel, such as compressed natural gas (CNG), liquefied natural gas (LNG), or liquefied petroleum gas (LPG). Fully involved fires in vehicles powered by either type of fuel should be fought with an unmanned master stream to prevent injuries from exploding gas cylinders.

- A fire in a gas-powered vehicle poses the threat of a boiling liquid/expanding vapour explosion (BLEVE). In the event of a fire, keep the gas cylinders cool to avoid a BLEVE.

- The greatest danger with most fires involving electrical equipment is electrocution. Only Class C extinguishing agents should be used when energized equipment is involved in a fire. All electrical equipment should be considered as potentially energized until the power company or a qualified electrical professional confirms that the power is off.

- Once the electrical service has been disconnected, most fires in electrical equipment can be controlled using the same tactics and procedures as are used for Class A fires.

KEY TERMS

Access Navigate for flashcards to test your key term knowledge.

Alternative-fuel vehicles A vehicle that uses anything other than a petroleum-based motor fuel (gasoline or diesel fuel) to propel a motorized vehicle.

Battery electric vehicles Vehicles that are powered by electricity.

Combination attack A type of attack employing both direct attack and indirect attack methods.

Conventional vehicles Vehicles that use internal combustion engines for power.

Defensive operation Actions that are intended to control a fire by limiting its spread to a defined area, avoiding the commitment of personnel and equipment to dangerous areas. (NFPA 1500)

Direct attack Firefighting operations involving the application of extinguishing agents directly onto the burning fuel. (NFPA 1145)

Fuel cells Cells that generate electricity through a chemical reaction between hydrogen gas and oxygen gas to produce water and in the process produce electricity to propel a vehicle.

Hybrid electric vehicles Vehicles that use both a battery-powered electric motor and a liquid-fuelled engine to propel a vehicle.

Indirect attack Firefighting operations involving the application of extinguishing agents to reduce the build-up of heat released from a fire without applying the agent directly onto the burning fuel. (NFPA 1145)

Ladder pipe A monitor that attaches to the rungs of a vehicle-mounted aerial ladder. (NFPA 1964)

Offensive operation Actions generally performed in the interior of involved structures that involve a direct attack on a fire to directly control and extinguish the fire. (NFPA 1500)

Soffit The material covering the gap between the edge of the roof and the exterior wall of the house. The soffits in most modern houses contain a system of vents to maintain air flow to the attic area. Some attics may have sealed soffits to create a conditioned air space for energy conservation.

Solar panels A general term for thermal collectors or photovoltaic (PV) modules. (NFPA 780)

Solar photovoltaic (PV) systems A power system designed to convert solar energy into electrical energy; may also be referred to as a PV system or a solar power system.

Solid stream A solid column of water.

Straight stream A stream made by using an adjustable nozzle to provide a straight stream of water.

Transitional attack An offensive fire attack initiated by an exterior, indirect handline operation into the fire compartment to initiate cooling while transitioning into interior direct fire attack in coordination with ventilation operations.

REFERENCES

CCOHS (Canadian Centre for Occupational Health and Safety). Confined spaces. https://www.ccohs.ca/topics/hazards/workplace/confinedspaces/. Accessed November 21, 2018.

CCOHS (Canadian Centre for Occupational Health and Safety). High-visibility safety apparel. https://www.ccohs.ca/oshanswers/prevention/ppe/high_visibility.html. Accessed November 21, 2018.

National Fire Protection Association, 2015. "Electric Vehicle Emergency Field Guide, Classroom Edition." http://www.ncdoi.com/OSFM/RPD/PT/Documents/Coursework/EV_SafetyTraining/EV%20EFG%20Classroom%20Edition.pdf. Accessed June 14, 2018.

National Fire Protection Association. 2016. "Fire Loss in the United States." https://www.nfpa.org/News-and-Research/Fire-statistics-and-reports/Fire-statistics/Fires-in-the-US/Overall-fire-problem/Fire-loss-in-the-United-States. Accessed June 14, 2018.

National Fire Protection Association. NFPA 1006, *Standard for Technical Rescue Personnel Professional Qualifications*, 2017. www.nfpa.org. Accessed November 19, 2018.

National Fire Protection Association. NFPA 1403, *Standard on Live Fire Training Evolutions*, 2018. www.nfpa.org. Accessed November 19, 2018.

Natural Resources Canada. What is ethanol? https://www.nrcan.gc.ca/energy/alternative-fuels/fuel-facts/ethanol/3493#a1. Accessed November 21, 2018.

UL. 2011. "Firefighter Safety and Photovoltaic Installations Research Report." https://ulfirefightersafety.org/docs/PV-FF_SafetyFinalReport.pdf. Accessed June 14, 2018.

UL. 2013. "Innovating Fire Attack Tactics." https://newscience.ul.com/wp-content/themes/newscience/library/documents/fire-safety/NS_FS_Article_Fire_Attack_Tactics.pdf. Accessed June 14, 2018.

UL Fire Safety Research Institute. 2014. "Study of Residential Attic Fire Mitigation Tactics and Exterior Fire Spread Hazards on Fire Fighter Safety." https://ulfirefightersafety.org/docs/Attic-Final-Report-Online.pdf. Accessed June 14, 2018.

UL Fire Safety Research Institute. 2016. "Understanding and Fighting Basement Fires." https://ulfirefightersafety.org/research-projects/understanding-and-fighting-basement-fires.html. Accessed June 14, 2018.

On Scene

You are just returning to the station from a medical emergency when your pumper is dispatched to the report of a car on fire in a garage. The battalion chief arrives first and establishes command. The IC gives a size-up indicating that the structure is a single-storey, wood-frame house with dark, turbulent smoke coming from the garage area. He then assigns your crew to fire attack.

1. Which of the following is used to cool the ceiling temperatures and darken the fire if conditions are unsafe for fire fighters to enter the fire compartment?

A. Interior attack

B. Indirect attack

C. Direct attack

D. Combination attack

2. Which of the following is *not* a master stream appliance?

A. Deck gun

B. Portable ground monitor

C. Elevated stream

D. Solid stream

(continued)

On Scene Continued

3. A _____ is made by using a smooth-bore nozzle to produce a penetrating stream of water.

A. solid stream

B. fog stream

C. straight stream

D. master stream

4. A _____ is an offensive operation initiated by a brief exterior indirect attack into the fire compartment to initiate cooling and to stop the progress of the fire.

A. master stream

B. transitional attack

C. direct attack

D. defensive attack

5. What should be done once the gas has been shut off to a building?

A. Attach a shut-off or lockout tag.

B. Report to staging.

C. Report to rehabilitation.

D. Notify the gas company.

6. Large handlines will flow at least _____ L/min (gpm).

A. 568 (150)

B. 757 (200)

C. 946 (250)

D. 1136 (300)

7. Which challenge do fire fighters frequently face when extinguishing a vehicle engine compartment fire?

A. Confining the fire

B. Vehicle movement

C. Gaining access

D. Locating the seat of the fire

 *Access Navigate to find answers to this On Scene, along with other resources such as an audiobook.*

CHAPTER 18

Fire Fighter I

Fire Fighter Survival

KNOWLEDGE OBJECTIVES

After studying this chapter, you will be able to:

- Describe how to apply a risk/benefit analysis to an emergency incident. (pp. 690–691)
- List the common hazard indicators that should alert fire fighters to a potentially life-threatening situation. (**NFPA 1001: 4.3.5**, pp. 691–692)
- List the 11 Rules of Engagement for Fire Fighter Survival. (pp. 692–693)
- Explain how to maintain team integrity during emergency operations. (**NFPA 1001: 4.3.5**, pp. 693–694)
- Define personnel accountability system. (**NFPA 1001: 4.3.5**, p. 694)
- Describe the types of personnel accountability systems and how they function. (**NFPA 1001: 4.3.5**, pp. 694–695)
- Explain how a personnel accountability report is taken. (**NFPA 1001: 4.3.5**, p. 695)
- Describe how to initiate emergency communications procedures. (**NFPA 1001: 4.3.5**, pp. 695–696)
- Describe the information that should be included in a mayday call for emergency assistance. (**NFPA 1001: 4.2.4**, pp. 696–698)
- Define rapid intervention crews/companies. (p. 698)
- Describe the methods used for maintaining orientation. (p. 699)

- Describe common self-rescue techniques. (**NFPA 1001: 4.3.5**, pp. 699–708)
- Describe how to find a safe location while awaiting rescue. (**NFPA 1001: 4.3.5**, pp. 706, 708)
- Describe air management procedures. (**NFPA 1001: 4.3.1, 4.3.5**, pp. 708–709)
- Describe common techniques for rescuing a downed fire fighter. (**NFPA 1001: 4.3.9**, pp. 709–714)
- Describe how a rapid intervention pack can provide an emergency air supply to a downed or trapped fire fighter. (**NFPA 1001: 4.3.9**, pp. 712, 714–718)
- Explain the importance of the rehabilitation process. (p. 718)

SKILLS OBJECTIVES

After studying this chapter, you will be able to perform the following skills:

- Initiate a mayday call for emergency assistance. (**NFPA 1001: 4.2.4**, pp. 696–698)
- Perform a self-rescue using a hose line. (**NFPA 1001: 4.3.5**, pp. 700, 701)
- Locate a door or window for an emergency exit. (pp. 700, 701–702)
- Use the backhanded swim technique to escape through a wall. (**NFPA 1001: 4.3.1, 4.3.9**, pp. 703–704)
- Use the forward swim technique to escape through a wall. (**NFPA 1001: 4.3.1, 4.3.9**, pp. 703, 705–706)

- Escape from an entanglement. (pp. 707–708)
- Rescue a downed fire fighter using the fire fighter's SCBA straps. (**NFPA 1001: 4.3.9**, pp. 710–711)
- Rescue a downed fire fighter using a drag rescue device. (**NFPA 1001: 4.3.9**, pp. 711–712)
- Rescue a downed fire fighter as a two-person team. (**NFPA 1001: 4.3.9**, pp. 713–714)
- Supply air to a downed fire fighter using the low-pressure hose from a rapid intervention pack. (**NFPA 1001: 4.3.9**, pp. 715–716)

- Supply air to a downed fire fighter using the high-pressure hose from a rapid intervention pack. (**NFPA 1001: 4.3.9**, pp. 717–718)

ADDITIONAL NFPA STANDARDS

- **NFPA 1407**, *Standard for Training Fire Service Rapid Intervention Crews*
- **NFPA 1500**, *Standard on Fire Department Occupational Safety, Health, and Wellness Program*

Fire Alarm

You are dispatched as the third-due pump company to a large three-storey, Type V, single-family dwelling. All of the houses in the part of the district to which you have been dispatched were built over the last 10 years using light-weight construction techniques and materials. As you respond, the dispatcher states that he has received multiple calls reporting smoke showing from the outside of the residence. Shortly after, the first-due unit arrives on the scene and reports a working fire.

1. Why is a personnel accountability system so important for fire fighter survival when working at a fire such as this one?

2. How does the type of fire attack influence fire fighter survival techniques?

3. Why should rehabilitation be considered a valuable part of ensuring fire fighter survival at a working fire?

 *Access Navigate for more practice activities.*

Introduction

Fire fighter survival is the most important consideration in any successful fire department operation. Fire fighters often perform their duties in environments that are inherently dangerous. The most desirable outcome, for any incident, is for the situation to be resolved with no injuries sustained. This can be achieved by working and training in a manner that recognizes risks and hazards and consistently applies safe operating policies and procedures. "Everybody goes home" should be the goal of every member of the team. This chapter introduces the actions, attitudes, and systems that are important in achieving that goal.

Fire fighters often encounter situations where their survival depends on making the right decisions and taking the appropriate actions. Many of the factors that cause fire fighter deaths and injuries appear repeatedly in postincident studies as risks that are often present at fires and other emergency incidents. Fire fighters must learn to recognize dangerous situations and to take actions to prevent further dangerous situations from occurring.

Risk/Benefit Analysis

A **risk/benefit analysis** approach to emergency operations can limit the risk of fire fighter deaths and injuries. This kind of analysis weighs the positive results that can be achieved against the probability and severity of potential negative consequences. A standard approach to risk/benefit analysis should be incorporated into the standard operating procedures (SOPs) for all fire departments. NFPA 1500, *Standard on Fire Department Occupational Safety, Health, and Wellness Program*, is an excellent reference guide to this approach as it applies to fire department operations.

In the fire service, risk/benefit analysis should be practised and implemented by everyone both at the station and on the fire ground. At incidents the incident commander (IC) is responsible for the analysis of the overall scene and assigning tactical objectives to fire companies within risk management guidelines. Officers and fire fighters are responsible for executing tactical assignments. Subsequently, personnel on the scene of an emergency must maintain situational awareness and be mindful of subtle and or rapidly changing events that may compromise their safety.

The risk/benefit method attempts to anticipate potential risks that fire fighters may face when entering a burning structure and weighs them against the results that might be achieved by undertaking such a mission. If a building is known to be unoccupied and has no value, there is no justification for risking fire fighters' lives to save it. Similarly, if a fire has already reached the stage where no occupants could survive and there is little or no chance that property can be saved, there is no justification for risking the lives of fire fighters in an interior attack (**FIGURE 18-1**).

Firefighting activities are inherently dangerous. However, many instances require fire fighters to enter a structure fire to perform rescue and effect extinguishment. These circumstances are evaluated by the IC as part of a risk management strategy. Most fire services in Canada have a risk management policy to guide decision making. However, the dynamic and fast-moving nature of the fire ground often results in interior attacks that are marginal. Officers must have the confidence, training, and experiential knowledge to be able to terminate an attack if they feel the risks outweigh any potential benefit. All standard approaches to operational safety must be followed without compromise. No property is ever worth the life of a fire fighter.

FIGURE 18-1 If a fire has already reached the stage where no occupants could survive and no property of value can be saved, there is no justification for risking the lives of fire fighters in an interior attack.
© Kent Weakley/Shutterstock.

Actions that present a high level of risk to the safety of fire fighters are justified only if victims are known or believed to be in immediate danger and there is a reasonable probability that lives can be saved. The determination that a risk is acceptable in a particular situation does not justify taking unsafe actions, however; it merely justifies taking actions that involve a higher level of risk.

The IC must continually assess the risks and benefits before committing crews to the interior of a burning structure. If the potential benefits are marginal and the risks are too high, fire fighters should not be engaged to conduct an interior attack. Company officers and safety officers should engage in risk/benefit analysis on an ongoing basis. If they perceive that the risk/benefit balance has shifted, it is their responsibility to report their observations and recommendations to the IC. In addition, each individual fire fighter involved in the operation should conduct a risk/benefit analysis from his or her own perspective. A report from an observant fire fighter to his or her company officer can be crucial to the safety of an operation. The IC must reevaluate the risk/benefit balance whenever conditions change and decide whether the strategy must be changed. There should be no hesitation to terminate an interior attack and move to an exterior defensive operation at any time if the risks associated with an offensive operation change so that they outweigh the benefits of this strategy.

Hazard Indicators

As you progress in your career as a fire fighter, it will become evident that fire suppression and other types of emergency operations involve many different types of hazards. Fire fighters must be capable of working safely in an environment that includes a wide range of inherent hazards. While the danger of firefighting should never be taken for granted or perceived as routine, a fire fighter must learn to routinely follow safe operating procedures. You should recognize the various types of hazardous conditions and react to them appropriately.

An example of a commonly encountered hazard is the presence of smoke inside a structure. Fires produce smoke, and breathing smoke is recognized as dangerous. Fire fighters entering any immediately dangerous to life or health (IDLH) environment must ensure that they wear personal protective equipment (PPE) appropriate for the hazard. At the same time, it should be recognized that black fire (high-volume, high-velocity, turbulent, ultra-dense, black smoke) indicates that the atmosphere may soon prove deadly even for fire fighters who are in complete PPE including self-contained breathing apparatus (SCBA). (Chapter 5, *Fire Behaviour*, has more information on smoke reading.)

Hazardous conditions may or may not be evident by simple observation. For example, smoke and flames are often visible, whereas hazards related to the construction details of a building might not be so easy to detect. You should be aware of the indicators of less obvious hazards. Identifying known or recognized hazards with less obvious hazards is part of a continual size-up process that informs decision making and critical thinking. Hazard recognition becomes easier through study and experience:

- Building construction: A complete understanding of building construction and how fires are likely to propagate is an essential component in anticipating fire behaviour in different types of buildings and recognizing the potential for structural collapse. Is the building designed to withstand a fire for a period of time, such as in Type I structures, or is it likely to weaken quickly? Has the building been modified from its original use? Does the structure incorporate void spaces that would allow a fire to spread quickly, as is the case in modern Type V construction? Does the building incorporate truss roofs or floors or lightweight construction that could fail in a short period of time? (Chapter 6, *Building Construction*, has more information on hazard indicators related to building construction.)
- Weather conditions: Inclement weather can turn a routine incident into a hazardous one. Rain, snow, and freezing temperatures all challenge normal firefighting operations. These conditions usually result in less fire fighter mobility, slip and fall hazards, reduced visibility, and equipment failure, such as frozen hydrants, hose lines, and SCBA. Raising a ladder close to power lines during windy weather involves an increased level of risk. Fires that are driven by strong winds affect fire growth and intensity. Changing wind patterns can place fire fighters at more risk. It is always preferable to conduct fire operations on the windward side of the building, reducing the potential of fire fighters conducting operations in a flow path.
- Occupancy: The occupancy type is another important factor in determining risk, tactical objectives, and overall strategy. For example, Type I buildings may have residential or commercial occupancies. Exterior signage can be one of the first indicators of occupancy. Furniture, paint, and hardware occupancies can represent an increased risk for fire fighters due to a high fire load, chemicals, and shelved storage. High-risk occupancies such as assembly occupancies, senior residences, and group homes present their own unique challenges. These types of occupancies will likely require a need for additional human and physical resources (**FIGURE 18-2**).

FIGURE 18-2 The NFPA 704 diamond indicates that hazardous materials are present.
© Jones & Bartlett Learning.

These are just a few examples of observable factors that might indicate a hazard. At each and every incident, fire fighters should look for what could be critical indications of a hazard.

Standard Operating Procedures

SOPs define the manner in which a fire department conducts operations at an emergency incident. The adoption of SOPs is designed to improve fire ground efficiency, fire ground discipline, command and control activities, and safety.

SOPs must be learned and practised before they can be implemented. Training is the only way for individuals to become proficient at performing any set of skills. To be effective, the fire fighter must apply the same skills routinely and consistently at every training session and at every emergency incident. Under stressful conditions, most people will automatically default to actions they are most familiar. Ideally, when faced with a stressful circumstance or dangerous situation, well-trained fire fighters will react in a manner consistent with their training.

Rules of Engagement for Fire Fighter Survival

To help define safe practices for structural firefighting, the International Association of Fire Chiefs (IAFC) and National Volunteer Fire Council (NVFC) developed

Rules of Engagement for Firefighter Survival. These 11 rules are summarized here:

1. **Size up your tactical area of operation.** Fire fighters and company officers need to pause for a moment and look over their area of responsibility. During this size-up, you need to evaluate your individual risk exposure and determine a safe approach to complete your assigned objectives.

2. **Determine the occupant survival profile.** Fire fighters and company officers need to consider the fire conditions to determine if an occupant could be alive in the conditions that present themselves. They must maintain an ongoing assessment of these conditions.

3. ***Do not* risk your life for lives or property that cannot be saved.** Do not engage in high-risk search and rescue and firefighting activities when the fire conditions prevent occupant survival or if the property destruction is inevitable.

4. **Extend *limited* risk to protect *savable* property.** Limit your risk to a reasonable, cautious, and conservative level when trying to save a building.

5. **Extend *vigilant* and *measured* risk to protect and rescue *savable* lives.** Engage in search and rescue and firefighting operations in a calculated, controlled, and safe manner, while remaining alert to changing conditions during high-risk search and rescue operations, only in conditions where lives can be saved.

6. **Go in together, stay together, and come out together.** Ensure that you and your fellow fire fighters always enter burning buildings as a team of two or more and that no fire fighter is allowed to be alone at any time while entering, operating in, or exiting a hazardous building.

7. **Maintain continuous awareness of your air supply, situation, location, and fire conditions.** Maintain constant situational awareness of your SCBA air supply, your location in the building, and all that is happening in your area of operation and elsewhere on the fire ground that may affect your risk and safety. Monitor your air supply, and ensure that you are outside the hazard zone before your low-air alarm activates.

8. **Constantly monitor fire-ground communications for critical radio reports.** Maintain constant awareness of all fire-ground radio communications on your assigned channels for progress reports, critical messages, and other information that may affect your risk and safety.

9. **Report unsafe practices or conditions that can harm you. Stop, evaluate, and decide.** Report unsafe practices and exposure to unsafe conditions to your supervisor. There should be no penalty for this reporting.

10. **Abandon your position and retreat before deteriorating conditions can harm you.** Be aware of unsafe fire conditions, and exit immediately to a safe area when you are exposed to deteriorating conditions, unacceptable risk, or a life-threatening condition.

11. **Declare a mayday as soon as you *think* you are in danger.** Declare a mayday as soon as you are faced with a life-threatening situation; declare a mayday as soon as you think you are in trouble. Failure to declare a mayday as soon as possible puts your life and the lives of fellow fire fighters in even more danger. The process of initiating a mayday is discussed later in this chapter.

Team Integrity

Teamwork is an essential requirement in firefighting (**FIGURE 18-3**). A team is defined as two or more individuals attempting to achieve a common goal. Given the intense physical labour necessary to pull hose or raise ladders, team efforts are imperative in firefighting. The most important reason for maintaining team integrity in an emergency operation is for safety: Team members keep track of one another and provide immediate assistance to other team members when needed. The standard operational team in firefighting operations is a company, usually consisting of three to five fire fighters working under the direct supervision of a company officer. Career fire fighters may arrive as a company on the same apparatus, whereas on-call or volunteer fire fighters should assemble in companies upon arrival.

FIGURE 18-3 A team attempts to achieve a common goal.
© Jones & Bartlett Learning.

Individual companies are assigned as a group to perform tasks at the scene of an emergency incident. Chapter 1, *The Fire Service*, discusses the various types of companies. The IC keeps track of the companies involved in the response and communicates with company officers. Every company officer must have a portable radio to maintain contact with the IC. The company officer is responsible for knowing the location of every company member; he or she should know exactly what each individual fire fighter is doing at all times.

Team integrity means that a company arrives at a fire together, works together, and leaves together. The members of a company should always be oriented to one another's location, activities, and condition. In a hazardous atmosphere, such as inside a structure, a fire fighter should always be able to contact his or her company members by voice, sight, or touch. When air cylinders need to be replaced, all of the company members should leave the building together, change cylinders together, and return to work together. If any members require rehabilitation, the full company should go to rehabilitation together and return to action together.

In some cases, a company will be divided into teams of two to perform specific tasks. For example, a ladder company might be split into a roof team and an interior search team. The initial attack on a room-and-contents fire in a single-family dwelling might be made by two fire fighters, while two other fire fighters remain outside, following the two-in/two-out rule. For safety and survival reasons, fire fighters should always use a buddy system, working in teams of at least two. If one team member should become incapacitated, his or her partner can provide help immediately and call for assistance. The two team members must remain in direct contact with each other. At least one member of every team must have a portable radio to maintain contact with the IC; in many fire departments, every member carries a portable radio.

Personnel Accountability System

Every fire department must have a **personnel accountability system**. This system provides a systematic way to keep track of every company, crew, or team of fire fighters operating at the scene of an incident, from the time they arrive until the time they are released (**FIGURE 18-4**). In addition, it must be able to identify each individual fire fighter who is assigned to a particular company, crew, or team; the location of each team; the assignments for each team; and the team's current activities. The system assumes that all of the company members will be together—that is, working as a unit under the supervision of the company officer—unless the IC has been advised of a change. At any point in the operation, the IC should

FIGURE 18-4 Personnel accountability system.
Courtesy of Mike Legeros.

be able to identify the location and function of each company and, within seconds, determine which individual fire fighters are assigned to that company. If any fire fighters are reported missing, lost, or injured, the accountability system can identify who is missing and what his or her last assignment was.

> **LISTEN UP!**
>
> It is important that volunteer fire departments have a method to keep track of all members responding to an incident and to track which team each member is working with. While career departments have established shifts and riding positions, some volunteer departments will not know who is able and available to respond to the emergency until they arrive to the emergency scene. This makes an accountability system harder to implement but also especially needed.

The personnel accountability systems used by fire departments take many different forms. Fire fighters must learn which kind of accountability system their department uses, how to work within it, and how this system works within the incident command system (ICS). Whatever type of system is used, the only way for it to be truly effective is if every fire fighter takes responsibility for its use at all incidents. This rule includes even the smallest of incidents. Some fire departments start with a written roster of the team members assigned to each company; others use a computer database tied to a daily staffing program. Lists, roll calls, or electronic tracking devices should be carried on each piece of apparatus and should be used to account for all personnel at the beginning of every tour of duty or when changes are made to assignments throughout the tour.

While some departments may use electronic tracking devices for their personnel accountability systems, others may use simpler systems consisting of Velcro or magnetic nametags to account for each company member assigned to an apparatus. With this system, the tags are affixed to a special display board carried in the cab. This board, which is sometimes called a passport, must be kept up-to-date throughout the tour of duty and must denote any change in the assigned team. In the case of a volunteer company, the tags can be affixed to the board at the time of an alarm.

At the scene of an incident, these boards are physically left with a designated person at the command post or at the entry point to a hazardous area. Depositing the board and the tags indicates that the company and those individual fire fighters have entered the hazardous area. Later, when the company leaves the hazardous area or is released from the scene, the company officer must retrieve the board—an action that indicates the entire company has safely left the hazardous area.

As part of the personnel accountability system, a **personnel accountability report (PAR)** may be conducted. A PAR is an accountability check or roll call taken by each supervisor at an emergency incident. Many departments conduct PARs at regular set intervals. The IC should also request a PAR at tactical benchmarks, such as when the operational strategy changes or when a situation occurs that could endanger fire fighters. When the IC requests a PAR, each company officer physically verifies that all assigned members are present and confirms this information to the IC.

The company officer should always know exactly where each crew is and what it is doing. He or she must be in verbal, visual, or physical contact with all crew members to verify their status. If a crew splits into two or more teams, the company officer should be in contact with at least one member of each team. Once the company officer verifies the status of the crew, he or she then communicates with the IC by radio to report a PAR. Whenever a fire fighter cannot be accounted for, he or she is considered missing until proven otherwise. A report of a missing fire fighter always becomes the highest priority at the incident scene.

A PAR should always be performed if unusual or unplanned events occur at an incident. For example, if there is a report of an explosion, a structural collapse, or a fire fighter missing or in need of assistance, a PAR should immediately take place so that command can determine if any personnel are missing, identify any missing individuals, and establish their last known locations and assignments. These locations would then serve as the starting points for any search and rescue crews.

Emergency Communications Procedures

Communication is a critical part of any emergency operation, but it is especially important in relation to fire fighter safety and survival (**FIGURE 18-5**).

FIGURE 18-5 Communication is a critical part of any emergency operation.
© Mark C. Ide.

Breakdowns in the communication process may contribute to fire fighter injuries and deaths. Conversely, good communications can save the lives of fire fighters in a dangerous situation.

Chapter 4, *Fire Service Communications*, discusses the proper radio communications procedures. The following is a model of good radio communication:

"Central IC to Ladder 203."

"This is Ladder 203. Go ahead, IC."

"Ladder 203 ventilate the roof."

"Ladder 203 copies. Beginning roof ventilation."

Standard communications such as this one ensure that the message is clearly stated and then repeated back as confirmation that it has been received and understood. Most fire services have established their own procedures for radio communication. It is important for personnel to know the procedures and strictly adhere to them.

In circumstances that involve an imminent safety hazard or a missing or injured fire fighter, it is vital that all communication is clear and concise. Safety is the primary reason for standard radio communication protocol. However, in order to achieve efficient, effective fire-ground operations, a standardized approach to radio communication is essential. Many fire services have SOPs that contain specific standard emergency communication phrases. Likewise, a separate set of standard phrases should be used. These phrases should be known and practised by everyone in the department. The same procedures and terminology should be used by all of the fire departments that can be expected to work together at emergency incidents. In many areas, these procedures are coordinated regionally. The following represent some, but not all, of the standard messages fire fighters can expect to hear.

Emergency traffic is an urgent message that is transmitted to communicate that all messages to follow are prioritized, with the exception of a mayday call. It is used by many departments to indicate an imminent fire-ground hazard, such as a potential explosion or structural collapse. This kind of message may also be used to order fire fighters to immediately withdraw from interior offensive attack positions and switch to a defensive strategy.

The word *mayday* is used to indicate that a fire fighter is in trouble and requires immediate assistance. It could be used if a fire fighter is lost, is trapped, has an SCBA failure, or is running out of air. A fire fighter can transmit a mayday message to request assistance or to report that a fire fighter is missing or in trouble. Any use of the term *mayday* should bring an immediate response and take precedence over all other radio communications.

In many fire departments, the communications centre sounds a special tone over the radio to alert all members that an emergency situation is in progress. It then repeats the information to be certain that everyone at the incident scene heard the message correctly. Some departments sound a horn in such circumstances. Many departments have adopted a three-tone radio report combined with three air horn activations on the closest apparatus to the scene to indicate an "emergency bail" or immediate withdrawal from a structure.

Similar procedures are followed for emergency traffic situations. All imminent hazards and emergency instructions should be announced in a manner that captures the attention of everyone at the incident scene and ensures that the message is clearly understood.

Initiating a Mayday

A mayday message, as described previously, is used to indicate that a fire fighter is in trouble and needs immediate assistance. Analysis of fire fighter fatalities and serious injuries has shown that fire fighters often wait too long before calling for assistance. Instead of initiating a mayday call when they first get into trouble, fire fighters often wait until the situation is absolutely critical before requesting help. This delay could reflect a fear of embarrassment if the situation turns out to be less severe than anticipated, combined with a hope that the fire fighter will be able to resolve the problem without assistance.

A failure for the IC to act promptly on hearing a reported *mayday* may result in serious injury or death. Systems have been designed to do everything possible to protect fire fighters and to rescue fire fighters from dangerous situations—but those systems are effective only when fire fighters act appropriately when they find themselves in trouble. Do not hesitate to call for help when you think you need it.

To initiate a mayday call, follow your fire department's SOPs. In most cases, this will involve the fire fighter in trouble transmitting the message "Mayday, mayday, mayday" over the radio to initiate the process. The IC will interrupt any other communications and direct the person reporting the mayday to proceed. That person will then give a LUNAR report. LUNAR stands for: **Location**, the fire fighter's location in the building/incident; **Unit**, the unit the fire fighter is assigned to; **Name**, the name of the fire fighter; **Air**, the amount of air the fire fighter has in his or her cylinder, or **Assignment**, where the fire fighter was last assigned; **Resources**, what the fire fighter needs to get out of the mayday situation. Other departments may report using "who," "what," and "where" instead of providing a LUNAR report.

The IC will repeat this information back and then initiate procedures to rescue the fire fighter. The next actions could include committing the rapid intervention crew/company (RIC) to the rescue effort, calling for additional resources, and redirecting other teams to support a search and rescue operation. An example of a mayday call follows:

Fire fighter: MAYDAY . . . MAYDAY . . . MAYDAY.

[All radio traffic stops.]

IC: Unit calling MAYDAY, go ahead.

Fire fighter: This is Pump 403. We are on the second floor and running out of air. Fire has cut off our escape route. We request a ladder to the window on the Charlie side of the building so we can evacuate.

IC: Command copied. Pump 403, your escape route cut off by fire. I am sending the rapid intervention crew to Charlie side with a ladder.

In addition to the mayday call, the fire fighter in trouble should manually activate his or her personal alert safety system (PASS) device so that other fire fighters can hear the signal and come to the fire fighter's location. The fire fighter should also activate the emergency alert button on his or her portable radio, if it has this feature. Both of these alert signals will make it difficult for the downed fire fighter to be heard over the radio; therefore, it is important to deactivate the PASS device and emergency alert button before talking into the radio. Reactivate these alerts once you have transmitted your message. All of these actions will increase the probability that the RIC or other fire fighters will be able to locate the fire fighter in trouble. It is also important that the fire fighter continue to work to free himself or herself from the dangerous situation.

To initiate a mayday call, follow the steps in **SKILL DRILL 18-1.**

SKILL DRILL 18-1
Initiating a Mayday Call for Emergency Assistance Fire Fighter I, NFPA 1001: 4.2.4

1 Use your radio to call "MAYDAY, MAYDAY, MAYDAY." Give a LUNAR report (your location, unit number, name, air or assignment, and resources needed) or report who, what, where.

2 Activate your PASS device. Attempt self-rescue. If you are able to move, identify a safe location where you can await rescue.

3 If unable to self-rescue, lie on your side in a fetal position with your PASS device pointing out so that it can be heard.

(continued)

SKILL DRILL 18-1 Continued
Initiating a Mayday Call for Emergency Assistance Fire Fighter I, NFPA 1001: 4.2.4

4 Point your flashlight toward the ceiling. Slow your breathing as much as possible to conserve your air supply.

© Jones & Bartlett Learning. Photographed by Glen E. Ellman.

Rapid Intervention Crews/Companies

Rapid intervention crews/companies (RICs) are established for the sole purpose of rescuing fire fighters who are operating at emergency incidents. In some fire departments, the term *rapid intervention team* or *fire fighter assistance and safety team* is used to describe this assignment. An RIC is a company or crew that is assigned to stand by at the incident scene, fully dressed in PPE and SCBA and equipped for action, and that is ready to deploy immediately when assigned to do so by the IC.

An RIC is an extension of the two-in/two-out rule. During the early stages of an interior fire attack, a minimum of two fire fighters are required to establish an entry team, and a minimum of two additional fire fighters are required to remain outside the hazardous area. The two fire fighters who remain outside the hot zone can perform other functions, but they must have donned full PPE and SCBA so that they can immediately enter and assist the entry team if the interior fire fighters need to be rescued. The two fire fighters who remain outside the hazardous area fulfill the first stage of an RIC procedure. The second stage involves a dedicated RIC, specifically assigned to stand by for fire fighter rescue assignments.

NFPA 1407, *Standard for Training Fire Service Rapid Intervention Crews*, includes training procedures for rapid intervention operations, and NFPA 1500, *Standard on Fire Department Occupational Safety, Health, and Wellness Program*, includes a complete description of RIC procedures and provides examples of their use at typical scenes. A number of options are available for establishing and organizing RICs, including dispatching one or more additional companies to incidents with this specific assignment. Some small fire departments rely on an automatic response from an adjoining community to provide a dedicated RIC at working incidents.

SOPs should dictate when an RIC is required, how it is assigned, and where it should be positioned at an incident scene. An RIC should be in place at any incident where fire fighters are operating in conditions that are IDLH, which includes any structure fire where fire fighters are operating inside a building and using SCBA. Many fire departments specify a special set of rescue equipment, including extra SCBA air cylinders, that is brought to the standby location. All RIC members must wear full PPE and SCBA, and they must be ready for action at an instant.

The IC should quickly deploy the RIC to any situation where a fire fighter needs immediate assistance. Situations that call for the use of the RIC would include a lost or missing fire fighter, an injured fire fighter who has to be removed from a hazardous location, and a fire fighter trapped by a structural collapse.

Fire Fighter Survival Procedures

Some of the most important procedures you need to learn are directly related to your personal safety and to the safety of the fire fighters who will be working with you. These procedures are designed to keep you from entering into dangerous situations and to guide you in situations where you could be in immediate danger. Your personal safety will depend on learning, practising, and consistently following these procedures. However, never lose sight of the fact that preventing life-threatening situations is preferable to utilizing survival procedures.

Maintaining Orientation

As you practise working with SCBA, you will discover that it is easy to become disoriented in a dark, smoke-filled building. Sometimes the visibility inside a burning building can be measured in inches. In these conditions, it is extremely important to stay oriented. If you become lost, you could run out of air before you find your way out. The atmosphere outside your face piece could be fatal in one or two breaths. Even if you call for assistance, rescuers might not be able to find you. For these reasons, safety and survival inside a fire building are directly related to remaining oriented within the building.

Several methods can help fire fighters stay oriented inside a smoke-filled building. Before entering a building, look at it from the outside to get an idea of the size, shape, arrangement, and number of storeys. After entering the structure, paint a picture in your mind of your surroundings. Given that most of your senses will be disrupted by dense smoke, touch may become your only means of moving through a structure. By feeling out your surroundings, you will be able to tell which type of room you have entered. Feel for changes in floor coverings, and always look for alternative means of egress.

Preplanning helps with orientation. By reviewing the preincident plan before entering the structure you will have a general idea of the structure's floor plan. See Chapter 27, *Fire and Life Safety Initiatives,* for more information.

One of the most basic methods to remain oriented is to always stay in contact with a hose line—for example, by always keeping a hand or foot on the hose line. To find your way out, feel for the couplings. Place one hand on the male coupling and the other hand on the female coupling. To exit the building, travel in the direction of the hand holding the male coupling.

SAFETY TIP

To exit a building by following a hose line, you should travel in the direction of the male hose coupling. To remember which coupling to follow, some fire fighters use the phrase, "Smooth, bump, bump—back to the pump."

Team integrity is an important factor in maintaining orientation. Everyone works as a team to stay oriented.

When the team members cannot see one another, they must stay in direct physical contact or within the limits of verbal contact. Standard procedures are employed to ensure that fire fighters work efficiently as a team, such as having one member remain at the doorway of the room while the other members perform a systematic search inside. The searchers then work their way around the room in one direction until they get back to the fire fighter at the door.

A **guideline** can be used for orientation when inside a structure. A guideline is a rope that is attached to an object on the exterior or a known fixed location; it is stretched out as a team enters the structure. The fire fighters can follow the guideline back out, or another team can follow the guideline in to find them. Some fire departments use a series of knots to mark distances along the rope to indicate how far it is stretched. Be aware that using the guideline technique requires intense practice.

Training in low visibility will assist in building confidence and competence. This type of training can be conducted inside any building, while using an SCBA face piece with the lens covered. More realistic conditions can be simulated by using smoke machines inside training facilities. Fire fighters should practise navigating around furniture, following charged hose lines, climbing and descending stairs, and fitting through narrow spaces. Distracting noises, such as operating fans, can add a realistic feel. Team integrity must always be maintained in these practice sessions.

Self-Rescue

The most important step in ensuring your safety in a dangerous circumstance is to remain calm. Take a second to collect your thoughts and focus on the issue. Panicking will likely result in exacerbating the situation and is likely to deplete your air supply more quickly. If you become separated from your team, or if you become lost, disoriented, or trapped inside a structure, you can use several techniques for **self-rescue**. The first step is always to call for assistance by issuing a mayday. Do not wait until it is too late for anyone to find you before making it known that you are in trouble. As previously discussed, time is critical when your safety is at risk. You need to initiate the process as soon as you *think* you are in trouble not when you are absolutely *sure* you are in grave danger.

If you are simply separated from your team, the best option may be to take a normal route of egress to the exterior. Follow a hose line back to an open doorway, descend a ladder, or locate a window or door—all of these options can provide a direct exit and require no special tools or techniques. You must immediately communicate with your company officer or the IC to inform him or her that you have exited safely.

To rescue yourself by locating and following a hose line to an exit, follow the steps in **SKILL DRILL 18-2**.

To locate a door or window that can be used for an exit, follow the steps in **SKILL DRILL 18-3**. In addition to these actions, fire fighters can use many more complicated techniques to escape from dangerous predicaments, such as breaching a wall. These methods must be learned and practised regularly so that you will be ready to perform them without hesitation if you are ever in a situation that requires them.

To escape through a wall that is constructed with wood or metal studs, you need to expose the studs and then reduce the size of your profile to get your body and your SCBA through the narrow opening. In addition to reducing your profile, sometimes you can get through a narrow opening by loosening your waist strap and one strap of your SCBA and then repositioning your SCBA as you twist through the limited opening. If these methods fail, you may have to remove your SCBA backpack while keeping your face piece on. You will then have to push your SCBA in front of you. Hold on to your backpack carefully to prevent it from becoming separated from your face piece. This technique should not be used unless more simple techniques will not work. The technique that

SKILL DRILL 18-2
Performing Self-Rescue Using a Hose Line Fire Fighter I, NFPA 1001: 4.3.5

1 Initiate a mayday. Stay calm, and control your breathing.

2 Systematically search the room to locate a hose line. Follow the hose line to a hose coupling. Identify the male and female ends of the coupling. Move from the female coupling to the male coupling.

SKILL DRILL 18-2 Continued
Performing Self-Rescue Using a Hose Line Fire Fighter I, NFPA 1001: 4.3.5

3 Follow the hose out. Exit the hazard area. Notify command of your location.

© Jones & Bartlett Learning. Photographed by Glen E. Ellman.

SKILL DRILL 18-3
Locating a Door or Window for Emergency Exit Fire Fighter I, NFPA 1001: 4.3.5

1 Initiate a mayday. Stay calm, and control your breathing.

(continued)

SKILL DRILL 18-3 Continued
Locating a Door or Window for Emergency Exit Fire Fighter I, NFPA 1001: 4.3.5

2   Systematically locate a wall.

3   Use a sweeping motion on the wall to locate an alternative exit. Identify the opening as a window, interior door, or external door. Beware of closets, bathrooms, and other openings without egress.

4   If the first opening identified is not adequate for an exit, continue to search. Maintain your orientation, and stay low. Exit the room safely if possible. If unable to exit, assume the downed fire fighter position by lying on your stomach or side on the floor in a safe location, or find refuge. Keep command informed of your situation.

works for you will be dependent on your size. A small person may be able to squeeze through a limited-size opening without removing SCBA; a larger person trying to escape through the same small opening may have to remove the SCBA backpack to fit through it.

One method of opening a wall and using a reduced profile to escape through the opening is shown in **SKILL DRILL 18-4**.

Another method of escaping through a wall opening is the forward swim technique. To use the forward swim technique to reduce your SCBA profile and escape through a wall opening, follow the steps shown in **SKILL DRILL 18-5**. The success of this technique relies on several factors, including the size of the fire fighter. As

previously mentioned, a small fire fighter may be able to squeeze through a limited-size opening without removing SCBA; a larger fire fighter trying to escape through the same small opening may have to remove the SCBA backpack to fit through it. Also, remember that the distance between the studs will vary depending on the date of construction, the type of construction materials used, and the size of the building.

Disentanglement is an important skill that fire fighters need to learn and practise. When crawling in a smoke-filled building, a fire fighter can easily become entangled in fallen wires, cables, and debris. Above most dropped ceilings is a huge and varied mass of electrical wires and conduit, computer and TV cables, plumbing and sewage

SKILL DRILL 18-4
Using the Backhanded Swim Technique to Escape Through a Wall
Fire Fighter I, NFPA 1001: 4.3.5

1 Identify deteriorating conditions that require exiting through a wall. Initiate a mayday. Use a hand tool or your feet to open a hole in the wall between two studs. If using a hand tool, drive the tool completely through the wall to check for obstacles on the other side.

2 Enlarge the hole. Enter the hole head first to check the floor and fire conditions on the other side of the wall.

(continued)

SKILL DRILL 18-4 Continued
Using the Backhanded Swim Technique to Escape Through a Wall
Fire Fighter I, NFPA 1001: 4.3.5

3 Loosen the SCBA waist strap, and remove one shoulder strap. Sling the SCBA to one side to reduce your profile.

4 Escape through the opening in the wall. Adjust the SCBA straps to their normal position.

5 Assist others through the opening in the wall. Report your status to command.

© Jones & Bartlett Learning. Photographed by Glen E. Ellman.

SKILL DRILL 18-5
Using the Forward Swim Technique to Escape Through a Wall
Fire Fighter I, NFPA 1001: 4.3.5

1 Identify deteriorating conditions that require exiting through a wall. Initiate a mayday. Use a hand tool or your feet to open a hole in the wall between two studs. If using a hand tool, drive the tool completely through the wall to check for obstacles on the other side.

2 Enlarge the hole. Check the floor and fire conditions on the other side of the wall. Loosen the SCBA shoulder straps if necessary, but do not remove them.

3 Lie prone, with your stomach flat on the ground and your head pointed toward the wall opening. Stretch your arms toward your head and attempt to touch your ears with your upper arms to reduce your profile.

4 Extend your arms, and head into the wall opening first.

(continued)

SKILL DRILL 18-5 Continued
Using the Forward Swim Technique to Escape Through a Wall
Fire Fighter I, NFPA 1001: 4.3.5

5 Rotate your shoulders, SCBA, and waist as necessary to escape through the opening. Do not lift yourself up on your elbows, as this will raise your profile. Assist others through the opening in the wall. Report your status to command.

© Jones & Bartlett Learning. Photographed by Glen E. Ellman.

pipes, telephone cables, and heating and air-conditioning tubing and ducts. Because many dropped ceilings are held up only with wires, it does not take much fire or water damage to bring this huge tangle of utilities crashing down. If these wires fall on your back while you are wearing an SCBA, you could be easily trapped.

Many fire fighters carry small tools in the pockets of their PPE that can be used to cut through wires or small cables. This procedure can be very difficult if poor-visibility conditions prevent the fire fighter from seeing and identifying the entangling material.

To escape from an entanglement, follow the steps in **SKILL DRILL 18-6**.

Some additional self-rescue methods involve using tools and equipment in ways in which they were not designed. These so-called last-resort methods should be taught only by experienced and qualified instructors and must be practised with strict safety measures in place. Some of these methods are controversial. Some fire departments consider them unacceptable owing to the high risk of injury when they are used or even practised; other fire departments provide training in them because these measures could save the life of a fire fighter in an extreme situation. This policy decision must be made by each fire department.

Safe Locations

A **safe location** is a temporary location that provides refuge for fire fighters who are awaiting rescue or finding a method of self-rescue from an extremely hazardous situation. The term *safe* in this case is relative: The location may not be particularly safe, but it is less dangerous than the alternatives.

Safe locations are important when situations become critical and few alternatives are available. When a critical situation occurs, fire fighters should know where to look for a safe location and recognize one when it presents itself.

If fire fighters become trapped inside a burning building, a room with a door and a window could be a safe location. Ensure that the door to the room of refuge is closed. This will prevent the creation of a flow path when you break the window to escape. Ideally, the room of refuge should be parallel with the hallway and not at the end of the hallway. Also, the area of refuge should not be over the main body of a fire, particularly in Type V construction. In this case, the safe location ideally provides time for a rescue team to reach the fire fighters, a ladder to be raised to the window, or another group of fire fighters to control the fire.

A roof or floor collapse often leaves a void adjacent to an exterior wall. If trapped fire fighters can reach this void, it could be the best place to go. Breaching the wall might be the only way out of the void or the only way for a rescue team to reach the trapped personnel.

Maintaining team integrity is important when escaping to a safe location. In some cases, fire fighters have survived in safe locations long enough to work together to create an opening. The ability to use a radio

SKILL DRILL 18-6
Escaping from an Entanglement Fire Fighter I, NFPA 1001: 4.3.5

1 Initiate a mayday. Stay calm, and control your breathing.

2 Change your position—back up and turn on your side to try to free yourself.

3 Use the swimmer stroke to try to free yourself.

4 Loosen the SCBA straps, remove one arm, and slide the air pack to the front of your body to try to free the SCBA.

(continued)

SKILL DRILL 18-6 Continued
Escaping from an Entanglement Fire Fighter I, NFPA 1001: 4.3.5

5 Cut the wires or cables causing the entanglement. Be aware of any possible electrocution risk. If you are unable to disentangle yourself, notify command of your situation. If you are able to exit, notify command that you are out of danger.

© Jones & Bartlett Learning. Photographed by Glen E. Ellman.

to call for help and to describe the location where fire fighters are trapped can be critical.

Like all of the emergency procedures described in this chapter, these activities require good instruction and practice. Always follow your department's operating guidelines when it comes to self-rescue techniques.

Air Management

Air management is important to all fire fighters when using SCBA. Air equals time. How long a fire fighter can survive in a dangerous atmosphere depends on how much air is in the SCBA and how quickly the fire fighter uses it. Fire fighters have to manage their air supply to manage their time in a hazardous atmosphere.

The time in the hazardous atmosphere must take into account the time it takes to enter and get to the location where the fire fighter plans to operate and the time it takes to exit safely. The work time is limited by the amount of air in the SCBA cylinder, after allowing for entry and exit time. If it takes 6 minutes to reach a safe atmosphere and the fire fighter has only 3 minutes of air left, the results could be disastrous.

Recall that the time rating for an SCBA is based on a standard rate of consumption for a typical adult under low-exertion conditions in a test laboratory. Fire fighters typically consume air much more rapidly than occurs in the standard test. In fact, fire fighters often consume a 45-minute-rated air supply in just 20 to 25 minutes.

The actual rate of air consumption varies significantly among individual fire fighters and depends on the activities that they are performing. Chopping holes with an axe or pulling ceilings with a pike pole, for example, will cause a fire fighter to consume air more rapidly than will conducting a survey with a thermal imaging device. Even given the same workload, no two individuals will use exactly the same amount of air from their SCBA cylinders. Differences in body size, fitness levels, and even heredity factors cause air consumption rates to be faster or slower than the average.

Air management must be a team effort as well as an individual responsibility. Fire fighters will always be working as part of a team when using SCBA, and team members inevitably use air at different rates. The member who uses the air supply most rapidly determines the working time for the team.

A good way to determine your personal air usage rate is to participate in an SCBA consumption exercise. This drill puts fire fighters through a series of exercises and obstacles to learn how quickly each fire fighter uses air when performing different types of activities. Repeated practice can help you learn to use air more efficiently by building your confidence and learning to control your breathing rate. Keeping physically fit results in a more efficient use of oxygen and allows you to work for longer periods with a limited air supply. Chapter 3, *Personal Protective Equipment*, discusses breathing techniques that can be used to conserve air.

Practice sessions should take place in a nonhazardous atmosphere. A typical skill-building exercise might include the following components:

- While wearing full PPE and breathing air from SCBA, perform typical activities such as carrying hose packs up stairs, hitting an object with an axe or sledgehammer, or crawling along a hallway.

- Record the following information: how much air is in the cylinder at the start, how long you work, whether the physical labour is light or heavy, when the low-air alarm sounds, and when the cylinder is completely empty.

This type of exercise will allow you to determine your personal rate of consumption and work on improving your times. With practice, you will be able to predict how long your air supply should last under a given set of conditions.

Knowledge of your team members' physical conditions and their workloads can help you look out for their safety as well. Because many tasks are divided up in any given operation, one member may be working to his or her maximum level while another team member is acting only in a supportive role. This difference in exertion levels could cause the first team member to use up his or her air supply much faster without realizing it.

When using SCBA at an emergency incident, be aware of the limitations of the device. Do not enter a hazardous area unless your air cylinder is full. Keep track of your air supply by using the heads-up display (if equipped) to determine how much air you have left. Always follow the rule of air management: Know how much air you have in your SCBA before you go in and manage that amount so that you leave the hazardous environment *before* your SCBA low-air alarm activates. Do not wait until the low-pressure alarm sounds to start thinking about leaving the hazardous area; you should be out of the hazardous area before this alarm sounds.

SAFETY TIP

Follow the rule of air management: Know how much air you have in your SCBA before you go in, and manage that amount so that you leave the hazardous environment *before* your SCBA low-air alarm activates.

Even when fire fighters employ the best possible air management practices, emergency situations can occur. An SCBA can malfunction, or fire fighters may be trapped inside a building by a structural collapse

---

SAFETY TIP

Before entering a hazard zone:
- Ensure that all crew members have portable radios and appropriate hand tools.
- Keep crews together.
- Work under the incident command system.
- Follow guidelines for managing your air supply.

---

or an unexpected situation. In these circumstances, remaining calm and knowing how to use all of the emergency features of the SCBA can be critically important. Remaining calm, controlling your breathing rate, and taking shallow breaths can slow air consumption. All of these techniques require practice to build confidence.

Rescuing a Downed Fire Fighter

One of the most critical and demanding situations in the fire service arises when fire fighters have to rescue another fire fighter who is in trouble. Air management has critical implications in a fire fighter rescue situation. Specifically, air management must be considered for both the rescuers and the fire fighter who is in trouble.

When you reach a downed fire fighter, the first step is to assess the fire fighter's condition. Is the fire fighter conscious and breathing? Does he or she have a pulse? Is the fire fighter trapped or injured? Make a rapid assessment, and notify the IC of your situation and location. Have the RIC deployed to your location, and quickly determine the additional resources needed.

The most critical decision at this point is whether the fire fighter can be moved out of the hazardous area quickly and easily or whether considerable time and effort will be required to bring the fire fighter to safety. If it will take more than a minute or two to remove the fire fighter from the building, air supply will be an important consideration. You need to provide enough air to keep the fire fighter breathing for the duration of the rescue operation.

Fire fighters must know and understand all aspects of their SCBA in this type of situation. If the downed fire fighter has a working SCBA, determine how much air remains in the cylinder. A fire fighter who is breathing and has an adequate air supply is not in immediate life-threatening danger. If the SCBA contains very little air or no air, however, air management is a critical priority. In this case, rescuers need to move the fire fighter out of the hazardous area immediately or provide an additional air supply. A spare SCBA air cylinder or rapid intervention pack (discussed later in this chapter) should be brought to the location and connected to the downed fire fighter's SCBA through the rapid intervention crew/company universal air connection (RIC UAC). This procedure provides for rapid emergency refilling of breathing air to an SCBA for the downed or trapped fire fighter. If your agency does not use RIC UAC, follow local policies and guidelines.

If the fire fighter's SCBA is not working, check whether a simple fix can make it operational. Did a hose or regulator come off? Was the cylinder valve accidentally closed? If regulator failure is a possibility, open the bypass valve.

When it is necessary to quickly remove a downed fire fighter, you can use the fire fighter's SCBA straps to create a rescue harness. To rescue a downed fire fighter by making a rescue harness from the fire fighter's SCBA straps, follow the steps in **SKILL DRILL 18-7**.

You can also use a downed fire fighter's **drag rescue device** to remove him or her from the hazardous area.

The drag rescue device is a strap- or harness-type component that is attached to the fire fighter's structural firefighting protective coat. To rescue a downed fire fighter who is wearing PPE and SCBA from the immediate hazard using a drag rescue device, follow the steps in **SKILL DRILL 18-8**.

As a member of a two-person team, you can also remove a downed fire fighter with PPE and SCBA from a hazardous area by having one fire fighter pull on the

SKILL DRILL 18-7
Rescuing a Downed Fire Fighter Using SCBA Straps as a Rescue Harness
Fire Fighter I, NFPA 1001: 4.3.5

1 Locate the downed fire fighter. Activate the mayday procedure, if that step has not already been taken. Quickly assess the condition of the downed fire fighter and the situation. Shut off the PASS device as needed to aid in communication.

2 Position yourself at the legs of the downed fire fighter. Lift one leg of the downed fire fighter up onto your shoulder.

3 Locate and loosen the waist straps and shoulder straps (if needed) of the downed fire fighter's SCBA harness.

SKILL DRILL 18-7 Continued
Rescuing a Downed Fire Fighter Using SCBA Straps as a Rescue Harness
Fire Fighter I, NFPA 1001: 4.3.5

4 Unbuckle the downed fire fighter's waist strap, and buckle the waist strap under the lifted leg. Tighten the straps. Remove the downed fire fighter from the hazard area to a safe area.

© Jones & Bartlett Learning. Photographed by Glen E. Ellman.

SKILL DRILL 18-8
Rescuing a Downed Fire Fighter Using a Drag Rescue Device
Fire Fighter I, NFPA 1001: 4.3.5

1 Locate the downed fire fighter. Activate the mayday procedure, if that step has not already been taken. Shut off the PASS device to aid in communication. Assess the situation and the condition of the downed fire fighter.

2 Use the RIC UAC to fill the downed fire fighter's air supply cylinder, if needed.

(continued)

3 Access the fire fighter's drag rescue device.

4 Remove the downed fire fighter from the hazard area to a safe area.

© Jones & Bartlett Learning. Photographed by Glen E. Ellman.

downed fire fighter's SCBA straps and the second fire fighter support the legs of the downed fire fighter. To do so, follow the steps in **SKILL DRILL 18-9**.

Rapid Intervention Pack

A **rapid intervention pack** (also referred to as an RIT pack) is a portable air supply intended for use by an RIC (**FIGURE 18-6**). It provides an emergency source of breathing air for a single fire fighter who has run out of air or whose air supply is insufficient to safely exit from an IDLH atmosphere. There are many situations that may call for the use of a rapid intervention pack. These include a downed fire fighter who cannot be quickly moved because he or she is injured or trapped and a fire fighter who has suffered some type of SCBA failure. The steps for using a rapid intervention pack are described

in the following sections. Follow the manufacturer's recommendations for use.

The rapid intervention pack contains the same basic components as an SCBA. See Chapter 3, *Personal Protective Equipment*, for a more detailed discussion of these components.

Like SCBA, an air cylinder holds the breathing air for a rapid intervention pack. Different packs may contain cylinders of different sizes and pressures; the cylinder must be compatible with the downed fire

SKILL DRILL 18-9
Rescuing a Downed Fire Fighter as a Two-Person Team Fire Fighter I, NFPA 1001: 4.3.9

1 Locate the downed fire fighter. Activate the mayday procedure, if that step has not already been taken. Shut off the PASS device to aid in communication. Assess the situation and the condition of the downed fire fighter. Use the RIC UAC to fill the downed fire fighter's air supply cylinder, if needed.

2 The second fire fighter converts the downed fire fighter's SCBA harness into a rescue harness.

3 The first fire fighter grabs the shoulder straps or uses the webbing to create a handle to pull the downed fire fighter.

4 The second fire fighter stays at the downed fire fighter's legs and supports the legs of the downed fire fighter.

(continued)

SKILL DRILL 18-9 Continued
Rescuing a Downed Fire Fighter as a Two-Person Team Fire Fighter I, NFPA 1001: 4.3.9

5 Remove the downed fire fighter from the hazard area to a safe area.

© Jones & Bartlett Learning. Photographed by Glen E. Ellman.

FIGURE 18-6 A rapid intervention pack is a portable air supply system used by rapid intervention crews/companies.
© Jones & Bartlett Learning. Photographed by Glen E. Ellman.

fighter's SCBA equipment. The neck of the air cylinder is equipped with a valve that controls the flow of air leaving the air cylinder. A pressure gauge is located near the air cylinder valve; a set of LEDs identical to those located on SCBAs shows the amount of air currently in the cylinder. Two steady green LEDs indicate that the air cylinder is full, one steady green LED indicates that the air cylinder is ¾ full, one yellow blinking LED indicates that the cylinder is ½ full, and a rapidly flashing red LED indicates that the cylinder is ¼ full. The pack also contains a regulator assembly to control the flow of air to the user and a low-pressure and a high-pressure

hose through which the emergency supply of air passes through from the air cylinder.

An emergency face piece is also included in the rapid intervention pack. This face piece is different from that used on SCBA—it is not equipped with a nose piece assembly and may not have voice transmitting capability. It is designed to be used only as an emergency face piece. Most rapid intervention packs are supplied with a heavy duty container to store all the components needed. Some packs have other compartments to carry additional supplies needed by the RIC. The key to successful use of RIC UAC is continuous training.

Use of the Low-Pressure Hose

The low-pressure hose on the rapid intervention pack can be used to supply air to a trapped or downed fire fighter in three different ways. The first is by connecting it to the downed or trapped fire fighter's SCBA regulator. This method should only be used if the fire fighter's SCBA face piece and regulator are in place and functioning properly. The second way the low-pressure hose can be used is by connecting it to the rapid intervention pack regulator and substituting the rapid intervention pack regulator for the fire fighter's SCBA regulator. This method is used if the fire fighter's SCBA regulator is missing, displaced, or malfunctioning. The third way the low-pressure hose can be used is by connecting it to the regulator and emergency face piece found in the rapid intervention pack and replacing the downed fire fighter's SCBA regulator and face piece. This method is used if any part of the fire fighter's face piece or regulator is damaged or displaced. This method requires time and additional movement of the fire fighter. You will need to remove the downed fire fighter's helmet, hood, face piece, and regulator, leaving the fire fighter without air for a period of time. For these reasons, this method should only be used if the other two methods will be ineffective.

When it is necessary to supply air to a fire fighter using the low-pressure hose from the rapid intervention pack, follow the steps in **SKILL DRILL 18-10**.

Use of the High-Pressure Hose

The high-pressure hose on the rapid intervention pack is used when a downed or trapped fire fighter has a limited supply of air that will not last long enough for the fire fighter to be removed from the hazardous

SKILL DRILL 18-10
Supplying Air to a Downed Fire Fighter Using the Low-Pressure Hose from a Rapid Intervention Pack Fire Fighter I, NFPA 1001: 4.3.9

1 Inspect the rapid intervention pack for proper operation, and make sure that the air cylinder is full. Turn on the air cylinder valve.

2 Check the downed fire fighter's SCBA to determine which parts of the fire fighter's SCBA are not operating properly.

3 If the problem is that there is no air being supplied to the fire fighter's low-pressure SCBA hose, and the rest of the SCBA is operational, detach the low-pressure hose from the fire fighter's SCBA regulator, and attach the low-pressure hose from the rapid intervention pack to the SCBA regulator.

(continued)

SKILL DRILL 18-10 Continued

Supplying Air to a Downed Fire Fighter Using the Low-Pressure Hose from a Rapid Intervention Pack Fire Fighter I, NFPA 1001: 4.3.9

4 If the problem is that the fire fighter's SCBA regulator is not operating properly, attach the low-pressure hose of the rapid intervention pack to the pack's regulator. Detach the fire fighter's SCBA regulator, and attach the rapid intervention pack regulator and low-pressure hose.

5 If the problem is that the fire fighter's face piece has been displaced, is missing, or is damaged, remove the fire fighter's helmet, hood, and face piece. Place the emergency face piece from the rapid intervention pack on the fire fighter, and attach the rapid intervention pack regulator and low-pressure hose to the emergency face piece.

6 Continue to monitor the air pressure and operation of the downed fire fighter's SCBA and the pressure in the rapid intervention pack air cylinder.

atmosphere. It should be used only when the fire fighter has an operational SCBA with a properly functioning and attached regulator and face piece. It should not be used if the fire fighter's face piece has been detached or if the fire fighter's SCBA is not functioning properly.

When the high-pressure hose is connected to the fire fighter's SCBA through an RIC UAC, air is transferred from the higher pressure in the rapid intervention pack air cylinder to the lower pressure in the fire fighter's air cylinder. Once the pressure in the two cylinders is equalized, the high-pressure hose should be disconnected

from the fire fighter's RIC UAC. If necessary, a second transfill can be done; however, there will be significantly less air pressure available the second time than there was initially. During this operation it is important for one person to monitor the downed fire fighter's SCBA for proper operation and remaining air.

Follow the steps in **SKILL DRILL 18-11** to fill a downed fire fighter's SCBA air cylinder using a high-pressure hose from a rapid intervention pack. A fire fighter should at no time remove his or her own SCBA mask to share air with a downed fire fighter.

SKILL DRILL 18-11
Supplying Air to a Downed Fire Fighter Using the High-Pressure Hose from a Rapid Intervention Pack Fire Fighter I, NFPA 1001: 4.3.9

1 Check the rapid intervention pack to assure that the air cylinder is full and the high-pressure hose is in proper condition. Turn on the air cylinder valve.

2 Check the downed fire fighter's SCBA to be sure it is operating properly, that the face piece and regulator are attached, and that the air supply is low.

3 Remove the high-pressure hose from the rapid intervention pack, and inspect the fire fighter's RIC UAC to ensure it is clean and undamaged.

(continued)

SKILL DRILL 18-11 Continued
Supplying Air to a Downed Fire Fighter Using the High-Pressure Hose from a Rapid Intervention Pack Fire Fighter I, NFPA 1001: 4.3.9

4 Attach the rapid intervention pack high-pressure hose to the RIC UAC of the fire fighter's SCBA, and leave it in place until the pressure is equalized between the two air cylinders.

5 Disconnect the high-pressure hose from the RIC UAC, and monitor the air pressure and operation of the downed fire fighter's SCBA.

© Jones & Bartlett Learning. Photographed by Glen E. Ellman.

Rehabilitation

Chapter 20, *Fire Fighter Rehabilitation*, covers fire fighter rehabilitation in detail. The purpose of rehabilitation is to reduce the effects of fatigue during an emergency operation (**FIGURE 18-7**). Firefighting involves demanding physical labour. When combined with extremes of weather and the mental stresses associated with emergency incidents, it can even challenge the strength and endurance of fire fighters who are physically fit and well prepared. Rehabilitation helps fire fighters retain the ability to perform at the current incident and restores their capacity to work at later incidents.

At the simplest of incidents, a rehabilitation area can be set up on the tailboard of an apparatus with a water cooler. At larger incidents, a complete rehabilitation operation should be established, including personnel to monitor the vital signs of fire fighters and provide first aid. In whatever size or form it takes, rehabilitation is integral to fire fighter safety and survival. The personnel accountability system

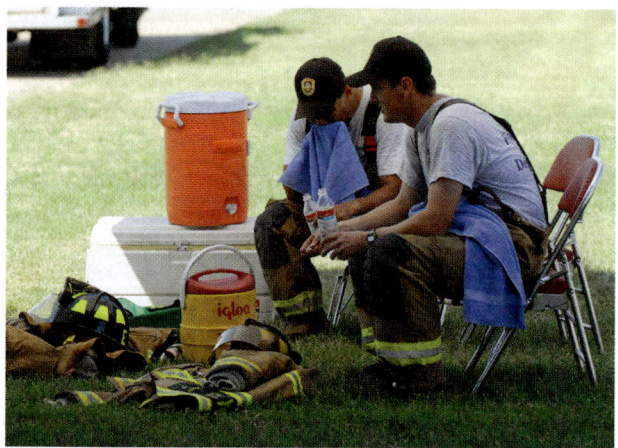

FIGURE 18-7 Rehabilitation reduces the effects of fatigue produced by an operation.
© Jones & Bartlett Learning. Photographed by Glen E. Ellman.

must continue to track fire fighters who are assigned to report to rehabilitation and note when they are released from rehabilitation for another assignment.

After-Action REVIEW

IN SUMMARY

- Risk/benefit analysis weighs the positive results that can be achieved against the probability and severity of potential negative consequences. It should be practised and implemented by everyone on the fire ground. While the IC is responsible for the analysis of the overall scene, company officers and fire fighters need to process the risks and benefits of the strategies implemented by command.

- No property is ever worth the life of a fire fighter.

- Smoke and flames are obvious hazards of fire. You should also be familiar with the less obvious hazard indicators, including inclement weather conditions and the likelihood that certain occupancies might hold hazardous materials.

- Safe operating procedures define the manner in which a fire department conducts operations at an emergency incident.

- The 11 Rules of Engagement identified by the IAFC and NVFC are as follows:
    - Size up your tactical area of operation.
    - Determine the occupant survival profile.
    - Do not risk your life for lives or property that cannot be saved.
    - Extend limited risk to protect savable property.
    - Extend vigilant and measured risk to protect and rescue savable lives.
    - Go in together, stay together, and come out together.
    - Maintain continuous awareness of your air supply, situation, location, and fire conditions.
    - Constantly monitor fire-ground communications for critical radio reports.
    - Report unsafe practices or conditions that can harm you. Stop, evaluate, and decide.
    - Abandon your position and retreat before deteriorating conditions can harm you.
    - Declare a mayday as soon as you think you are in danger.

- Team integrity means that a company arrives at a fire together, works together, and leaves together.

- Personnel accountability systems help to ensure team integrity by tracking the location and function of every company, crew, or team of fire fighters operating at the scene of an incident.

- A personnel accountability report is a roll call taken by each supervisor at an emergency incident. These head-counts are made at regular set intervals throughout the incident.

- Communication is important in relation to fire fighter safety and survival. There are standard methods to transmit emergency information, standard information items to convey, and standard responses to emergency messages.

- Emergency traffic is used to indicate an imminent fire-ground hazard. This kind of message would also be used to order fire fighters to immediately withdraw from interior offensive attack positions and switch to a defensive strategy. An emergency traffic message takes precedence over all other radio communications except a mayday call.

- Mayday is used if a fire fighter is lost, is trapped, has an SCBA failure, or is running out of air. A fire fighter can transmit a mayday message to request assistance, or another fire fighter can transmit a mayday message to report that a fire fighter is missing or in trouble.

- A mayday call takes precedence over all other radio communications.

- The sole purpose of RICs is to rescue fire fighters operating at emergency incidents. An RIC is an extension of the two-in/two-out rule. This team should be in place at any incident where fire fighters are operating in conditions that are immediately dangerous to life and health and should be deployed to any situation where a fire fighter needs immediate assistance.

- One of the most basic methods to remain oriented in a dark, smoke-filled environment is to always stay in contact with a hose line or guideline.

- The first step in attempting a self-rescue is to initiate a mayday.

- A safe location is a temporary location that provides refuge while awaiting rescue or finding a method of self-rescue. In this case, "safe" means less hazardous.

- Always follow the rule of air management: Know how much air you have left in your SCBA before you enter a building, and manage that amount so that you leave the hazardous environment before your low-air alarm sounds.

- SCBA straps or a drag rescue harness may be used to help drag a downed fire fighter to a safe location.

- A rapid intervention pack can be used to provide an emergency source of breathing air for a fire fighter who has run out of air or whose air supply is insufficient to safely exit from an IDLH atmosphere.

- Rehabilitation reduces the effects of fatigue during an emergency operation and helps fire fighters retain the ability to perform at the incident.

KEY TERMS

Access Navigate for flashcards to test your key term knowledge.

Air management The use of a limited air supply in such a way as to ensure that it will last long enough to enter a hazardous area, accomplish needed tasks, and return safely.

Drag rescue device (DRD) A component integrated within the protective coat element to aid in the rescue of an incapacitated fire fighter. (NFPA 1851)

Guideline A rope used for orientation when fire fighters are inside a structure where there is low or no visibility. The line is attached to a fixed object outside the hazardous area.

Personnel accountability report (PAR) Periodic reports verifying the status of responders assigned to an incident or planned event. (NFPA 1026)

Personnel accountability system A system that readily identifies both the location and function of all members operating at an incident scene. (NFPA 1500)

Rapid intervention crew/company A minimum of two fully equipped personnel on site, in a ready state, for immediate rescue of disoriented, injured, lost, or trapped rescue personnel. (NFPA 1500)

Rapid intervention pack A portable air supply that provides an emergency source of breathing air for a single fire fighter who has run out of air or whose air supply is insufficient to safely exit from an IDLH atmosphere.

Risk/benefit analysis An assessment of the risk to rescuers versus the benefits that can be derived from their intended actions. (NFPA 1006)

Safe location A location remote or separated from the effects of a fire so that such effects no longer pose a threat. (NFPA 101)

Self-rescue Escaping or exiting a hazardous area under one's own power. (NFPA 1006)

REFERENCES

International Association of Fire Chiefs (IAFC). *Rules of Engagement for Firefighter and Incident Commander Survival.* http://iafcsafety.org/rules-of-engagement-training-poster-released/. Accessed November 8, 2018.

National Fire Protection Association. NFPA 1407, *Standard for Training Fire Service Rapid Intervention Crews*, 2015. www.nfpa.org. Accessed November 8, 2018.

National Fire Protection Association. NFPA 1500, *Standard on Fire Department Occupational Safety, Health, and Wellness Program*, 2018. www.nfpa.org. Accessed November 8, 2018.

Occupational Safety and Health Administration (OSHA). *Respiratory Protection, 29 CFR 1910.134.* https://www.osha.gov/pls/oshaweb/owadisp.show_document?p_table=standards&p_id=12716. Accessed November 8, 2018.

On Scene

It is 2:00 AM and you are on the scene of a fire in an apartment complex. The building is a five-storey, ordinary construction, with a center hallway design. The fire is on the second floor, and there are heavy smoke conditions on the second, third, fourth, and fifth floors. The IC assigns your captain to the search group; you and your crew are assigned to the primary search of the third floor. The following questions go through your head as you review the possible situations ahead.

1. Who is responsible for conducting a risk/benefit analysis during this fire?

   A. Incident commander

   B. Company officer

   C. Fire fighter

   D. The first arriving company

2. Which of the following is *not* a hazard indicator?

   A. Time of day

   B. Type of occupancy

   C. Weather conditions

   D. Building construction

3. Which of the following is *not* part of the 11 Rules of Engagement?

   A. Constantly monitor fire-ground communications for critical radio reports.

   B. Report unsafe practices or conditions that can harm you.

   C. Do not abandon your position and retreat before deteriorating conditions can harm you unless directed to do so by the IC.

   D. Declare a mayday as soon as you *think* you are in danger.

4. If you become lost during your search, what should your first action be?

   A. Continue searching until you find something that is familiar.

   B. Switch the hand that is in contact with the wall, and try to find your way out.

   C. Transmit a mayday communication over the radio.

   D. Transmit an "emergency traffic" communication over the radio.

5. The time rating for an SCBA is based on a standard rate of consumption for a typical:

   A. fire fighter under high exertion conditions in a test laboratory.

   B. adult under high exertion conditions in a test laboratory.

   C. adult under low exertion conditions in a test laboratory.

   D. fire fighter under low exertion conditions in a test laboratory.

6. When is a personnel accountability report not necessary?

   A. When changing from an offensive to a defensive strategy

   B. When an emergency evacuation has been declared

   C. Every 10 minutes

   D. Upon transfer of command

(continued)

On Scene Continued

7. Which of the following rapid intervention pack evolutions is accomplished using the high-pressure hose?

A. Replacing the regulator and air supply on a downed fire fighter's SCBA

B. Filling a downed fire fighter's SCBA air cylinder to increase the pressure in his or her air cylinder

C. Replacing the hose that supplies the SCBA regulator on the downed fire fighter's SCBA

D. Replacing the face piece and air supply on a downed fire fighter's SCBA

8. What is the first step you should take if you are trapped in an entanglement during an emergency situation?

A. Change your position to try to free yourself.

B. Use the swimmer stroke to try to free yourself.

C. Cut the wires or cables causing the entanglement.

D. Initiate a mayday.

Access Navigate to find answers to this On Scene, along with other resources such as an audiobook.

CHAPTER **19**

Fire Fighter I

Salvage and Overhaul

KNOWLEDGE OBJECTIVES

After studying this chapter, you will be able to:

- Describe the types of lights used to illuminate exterior and interior scenes. (**NFPA 1001: 4.3.17**, pp. 725–726)
- Describe the equipment used to illuminate an emergency scene. (**NFPA 1001: 4.3.17**, pp. 726–727)
- Describe the safety precautions to take when working with lighting equipment. (**NFPA 1001: 4.3.17**, p. 727)
- Describe how to operate lighting equipment to light exterior and interior scenes. (**NFPA 1001: 4.3.17**, pp. 725–728)
- Explain the purpose of salvage operations. (**NFPA 1001: 4.3.14**, pp. 728–729)
- List the tasks involved in a salvage operation. (pp. 728–729)
- Describe how forcible entry operations affect salvage operations. (**NFPA 1001: 4.3.14**, p. 729)
- Describe the safety precautions that need to be considered when performing salvage. (p. 729)
- List the tools used to perform salvage operations. (pp. 729–730)
- Describe the salvage techniques commonly used to prevent water damage. (pp. 730–738)
- Describe the general procedures for preventing excess water damage from fire sprinklers. (**NFPA 1001: 4.3.14**, pp. 730–734)
- List the equipment used to shut down fire sprinklers. (**NFPA 1001: 4.3.14**, pp. 730–734)

- Describe the identifying characteristics of a main control valve of a fire sprinkler system. (**NFPA 1001: 4.3.14**, pp. 733–734)
- Describe the general procedures and equipment used to remove excess water from a structure. (**NFPA 1001: 4.3.14**, pp. 734–738)
- Describe the general procedures and equipment used to limit smoke and heat damage. (pp. 738–745)
- Describe how to maintain salvage covers. (**NFPA 1001: 4.5.1**, pp. 745, 746)
- Explain when fire investigators should become involved in salvage operations. (**NFPA 1001: 4.3.14**, p. 746, 750)
- Describe the purpose of overhaul operations. (**NFPA 1001: 4.3.13, 4.3.14**, p. 746)
- List the concerns that must be addressed to ensure the health of fire fighters who are performing overhaul. (**NFPA 1001: 4.3.13**, pp. 746–747)
- Describe the common methods of air monitoring at the fire scene. (**NFPA 1001: 4.3.21**, pp. 747–749)
- List the concerns that must be addressed to ensure the safety of fire fighters who are performing overhaul. (**NFPA 1001: 4.3.13**, pp. 749–750)
- List the indicators of possible structural collapse. (**NFPA 1001: 4.3.13**, p. 729)
- Explain how to preserve structural integrity during overhaul. (**NFPA 1001: 4.3.13**, p. 750)

- Describe how to preserve evidence during overhaul operations. (**NFPA 1001: 4.3.13**, p. 750)
- Explain how fire fighters determine overhaul locations. (**NFPA 1001: 4.3.13**, pp. 750–753)
- List the tools that are used for overhaul operations. (**NFPA 1001: 4.3.13**, p. 753)
- Describe the general techniques used in overhaul operations. (**NFPA 1001: 4.3.13, 4.3.14**, pp. 753–755)

- Construct a water chute. (**NFPA 1001: 4.3.14**, pp. 735–736)
- Construct a water catch-all. (**NFPA 1001: 4.3.14**, pp. 736–737)
- Fold a salvage cover for one-fire fighter deployment. (**NFPA 1001: 4.3.14**, pp. 739–740)
- Fold a salvage cover for two-fire fighter deployment. (**NFPA 1001: 4.3.14**, p. 741)
- Fold and roll a salvage cover. (**NFPA 1001: 4.3.14**, p. 742)
- Perform a one-fire fighter salvage cover roll. (**NFPA 1001: 4.3.14**, p. 743)
- Perform a salvage cover shoulder toss. (**NFPA 1001: 4.3.14**, p. 744)
- Perform a salvage cover balloon toss. (**NFPA 1001: 4.3.14**, p. 745)
- Use a multi-gas air monitoring device. (**NFPA 1001: 4.3.21**, pp. 747–748)
- Open a ceiling to check for fire using a pike pole. (**NFPA 1001: 4.3.13**, p. 754)
- Open an interior wall to check for fire. (**NFPA 1001: 4.3.13**, p. 755)

SKILLS OBJECTIVES

After studying this chapter, you will be able to perform the following skills:

- Illuminate an emergency scene. (**NFPA 1001: 4.3.17**, p. 728)
- Use a sprinkler wedge to shut down a sprinkler head. (**NFPA 1001: 4.3.14**, p. 731)
- Use a sprinkler stop to shut down a sprinkler head. (**NFPA 1001: 4.3.14**, p. 732)
- Close and reopen a main OS&Y valve. (**NFPA 1001: 4.3.14**, p. 733)
- Close and open a main post indicator valve. (**NFPA 1001: 4.3.14**, p. 734)

Fire Alarm

After a bitterly cold and snowy winter with several sub-zero days with strong winds, the sun is finally shining. Today the temperature is climbing to near record highs. The sun and unexpected warmth have brightened the collective mood of the fire station when your company is dispatched on a public service call. As you are responding, your dispatcher gives you additional information. This is a call for a residence that has a water leak. You are not surprised—when the temperature rises, a building's water starts to flow through pipes that may have been ruptured by the extremely cold weather.

As you arrive on the scene, you note a large three-storey residence. You also note that there is water flowing from the bottom of the front door.

1. How would you go about protecting the contents of this house?

2. How are the skills needed on this call related to the salvage and overhaul skills presented in this chapter?

3. What equipment will you need to prevent further damage to this house and its contents?

 *Access Navigate for more practice activities.*

Introduction

The basic goals of firefighting are first to save lives, second to control the fire, and third to protect property. Salvage and overhaul operations help meet the third goal by limiting and reducing property losses resulting from a fire. Salvage efforts protect property and belongings from damage, particularly from the effects of smoke and water—although some salvage techniques are useful for types of calls other than fire calls, such as a call involving a burst pipe. Overhaul ensures that a fire is completely extinguished by finding, exposing, and suppressing any smouldering or hidden pockets of fire in an area that has been burned.

Salvage and overhaul are usually conducted in close coordination with each other and have a lower priority than search and rescue operations or fire suppression. If a department has enough personnel on scene to address more than one priority, salvage can be performed concurrently with fire suppression. Depending on the building construction type, such as Type V balloon frame construction, fire suppression activities will include opening void spaces for hidden fires to prevent the fire from enveloping the structure from the inside. In this situation overhauling is considered synonymous with suppression activities.

Fire fighters performing salvage and overhaul must be able to see where they are going, what they are doing, and be aware of potential hazards. The smoke might have cleared, but electrical power might still be unavailable. Portable lights should be set up to fully illuminate areas during salvage and overhaul.

Throughout the salvage and overhaul process, fire fighters must attempt to preserve evidence related to the cause of the fire, particularly when working near the suspected area of fire origin and when the fire cause is not obvious. In general, disturb as little as possible in or near the area of origin until a fire investigator determines the cause of the fire and gives permission to clear the area. Knowing what caused a fire can help prevent future fires and enable victims to recoup losses under their insurance coverage. This subject is covered in greater depth in Chapter 28, *Fire Origin and Cause*.

Lighting

Lighting is an important concern at many fires and other emergency incidents. Because many emergency incidents occur at night, lighting is required to illuminate both the exterior scene and the interior of the building, as it is likely the electrical supply has been disconnected. Quite often flood lights powered by a generator or generators will need to be deployed to allow for personnel to complete various tasks safely.

Fire departments may use different types of lighting equipment depending on the particular situation. Spotlights, for example, project a narrow, concentrated beam of light. Floodlights project a more diffused light over a wide area. Lights can be portable or permanently mounted on fire apparatus.

In most departments it is the responsibility of a ladder company to supply lighting; however, any company can be assigned this responsibility. Lighting equipment can be mounted on any apparatus, and some fire departments equip special vehicles with high-powered lights to illuminate incident scenes.

Lighting Methods

Effective lighting improves the efficiency and safety of all crews at an emergency incident, but it requires practice to set up quickly and efficiently. Departmental standard operating procedures (SOPs) or standard operating guidelines (SOGs), or the incident commander (IC), will dictate the responsibility for setting up lighting.

Exterior Lighting

Exterior lighting should be provided at all incident scenes whenever visibility is reduced or personnel are working in darkness. This has been explained earlier. Scene lighting also makes fire fighters more visible to drivers who are approaching the scene or maneuvering emergency vehicles. Powerful exterior lights can often be seen through the windows and doors of a dark, smoke-filled building and can guide disoriented fire fighters or victims to safety.

SAFETY TIP

Personnel closest to the scene of a structure fire should keep the emergency lights on as they often flash off the interior walls of the structure for better orientation to side A and help determine if the room they are in has a window.

Apparatus operators should turn on their apparatus-mounted floodlights and position them for maximum effectiveness so that as much of the area as possible is illuminated. The entire area around a fire-struck building should be illuminated. If some areas cannot be covered with apparatus-mounted lights, use portable lights.

Interior Lighting

Portable lights can provide interior lighting at a fire scene. During interior operations, quickly extend a power cord, and set up a portable light at the entry point to the structure. Fire fighters looking for the building exit can then use this light at the door as a beacon. Extend additional lights into the building to illuminate interior areas as needed and as time permits.

Interior operations often progress through the fire suppression and search and rescue phases quickly; sometimes these steps are completed before effective interior lighting can be established. If interior operations continue for a lengthy period, portable lighting should be used to provide as much light as possible in the areas where fire fighters are working.

When operations reach the salvage and overhaul phase, time should be taken to provide adequate interior lighting in all areas so that crews can work safely. Proper lighting enables fire fighters to see what they are doing and to observe any dangerous conditions that need to be addressed. Although exterior lighting is generally needed only at night, interior lighting may be needed even when it is a bright day outside.

Lighting Equipment

Portable lights can be taken into buildings to illuminate the interior. They can also be set up outside the structure to illuminate the fire or emergency incident scene (**FIGURE 19-1**). Portable lights usually range from 300 watts to 1500 watts and can use several types of bulbs, including quartz bulbs, halogen bulbs, and light-emitting diodes (LEDs).

The electricity for portable lights is supplied by a generator, an inverter, or a building's electrical system.

Chapter 25, *Assisting Special Rescue Teams*, provides further detail about these sources of electricity. Portable light fixtures are connected to generators and inverters with electrical cords. These electrical cords should be stored neatly coiled or on permanently mounted reels attached to the fire apparatus. The cord is then pulled from the reel to the place where the power is needed. Portable reels can be taken from the apparatus into the fire scene.

Junction boxes are used as mobile power outlets. They are placed in convenient locations so that cords for individual lights and electrical equipment can be attached to them. Junction boxes used by fire departments are protected by waterproof covers and are often equipped with small lights so that they can be easily located in the dark.

The connectors and plugs used for fire department lighting have special connectors that attach with a slight clockwise twist. This type of plug is considered *intrinsically safe*, reducing the possibility of a spark being created when the the cord is disconnected from a power source. This is an important safety measure so that electrical equipment can be used in areas where the by-products of combustion may be present. This setup keeps the power cords from becoming unplugged during fire department operations.

Most modern fire apparatus come equipped with spotlights or LED lighting that enables the operator to illuminate an emergency scene. In many cases these lights run off an onboard generator; however, due to the low voltage required by LED lighting, some may run off the engine of the vehicle itself (**FIGURE 19-2**). Some

FIGURE 19-1 Portable emergency lights come in various sizes and levels of brightness.
© StockPhotosArt/Shutterstock, Inc.

FIGURE 19-2 A side apparatus-mounted light.
Courtesy of Bill Larkin.

vehicle-mounted lights can be manually raised to illuminate a larger area. Mechanically operated light towers, which can be raised and rotated by remote control, can create near-daylight conditions at an incident scene (**FIGURE 19-3**).

Battery-powered lights generally are used by individual fire fighters to find their way in dark areas or to illuminate their immediate work area. These kinds of lights are most often used during the first few critical minutes of an incident because they are lightweight, are easily transported, do not require power cords, and can be used immediately (**FIGURE 19-4**). The lights are powered by either disposable or rechargeable batteries; thus, they have a limited operating time before the batteries need to be recharged or replaced.

In order to be effective in smoke conditions, fire fighters need to use powerful hand lights. Large hand lights that project a powerful beam of light are preferred for search and rescue activities and interior fire suppression operations when fire fighters must penetrate smoke-filled areas quickly. These lights can be equipped with a shoulder strap for easy transport. Every crew member entering a fire building should be equipped with a high-powered hand light.

A personal flashlight is another type of battery-operated light used by fire fighters. Always carry a flashlight as part of your personal protective equipment (PPE); it could be a lifesaver if your primary light source fails. Flashlights are not as powerful as the larger hand lights, however, and they will not operate for as long on one set of batteries. A fire fighter's flashlight should be rugged and project a strong light beam.

Safety Principles and Practices

The lighting and power equipment used at a fire scene generally operates on 110-volt AC (alternating current), which is the same as standard household current. Some systems require higher voltage. All electrical cords, junction boxes, lights, and power tools must be maintained properly and handled carefully to avoid electrical shocks. In addition, all electrical equipment must be properly grounded. Electrical cords must be well insulated, without cuts or defects, and properly sized to handle the required amperage.

Generators should have ground-fault interrupters (GFI) to prevent a fire fighter from receiving a potentially fatal electric shock. A GFI senses when there is a problem with an electrical ground and interrupts the current in such a case, shutting down both the power source and the equipment it is feeding.

Some portable generators are equipped with a grounding rod that must be inserted into the ground. Always use a grounding device if one is provided. Avoid areas of standing or flowing water when placing power cords and junction boxes at a fire scene, and place electrical equipment on higher ground whenever possible.

Follow the steps in **SKILL DRILL 19-1** to illuminate an emergency scene.

FIGURE 19-3 An apparatus-mounted light tower can provide near-daylight conditions at an incident scene.
© Glen E. Ellman.

FIGURE 19-4 Hand lights manufactured for use by fire fighters have light outputs ranging from 3500 to 75,000 candlepower.
© 2003, Berta A. Daniels.

SAFETY TIP

A good rule to remember: Light early, light often, and light safely.

SKILL DRILL 19-1
Illuminating an Emergency Scene Fire Fighter I, NFPA 1001: 4.3.17

1 Wear PPE. Depending on scene conditions, you may or may not wear self-contained breathing apparatus (SCBA).

2 Inspect all equipment while setting it up.

3 Start the portable generator, engage the inverter, or check that there is electrical power in the building.

4 Connect cords, plug adaptors, GFIs, and lighting equipment.

5 Ensure proper grounding and GFI use.

6 Ensure that the scene is adequately and safely illuminated.

© Jones & Bartlett Learning. Photographed by Glen E. Ellman.

Salvage Overview

Salvage operations are conducted to protect property from fire, heat, water, and smoke damage. Many fire departments are placing a higher priority on salvage operations. As most modern homes are constructed of lightweight construction materials, the time fire fighters have to properly salvage belongings is greatly reduced. Often, the damage caused by smoke and water can be more extensive and costly to repair or replace than the property that is burned. Salvage operations are also conducted in non-fire situations that endanger property, such as floods or water leaking from burst pipes.

At the scene of a fire or other type of emergency, the victims' property is entrusted to the care of fire fighters, who are expected to protect that property from avoidable loss or damage to the best of their ability. Salvage operations are usually aimed at preventing or limiting secondary losses that result from smoke and water damage, fire suppression efforts, and other causes. These efforts must be performed promptly to preserve the building and protect the contents from avoidable damage. Salvage operations begin by focusing on bagging and tagging personnel objects and removing them to a place of safety. Fire fighters will also use a tarp over furniture and ensure that much of the water has been removed. Depending on circumstances, fire fighters may cover windows with plastic or place traps over holes in the roof to prevent further damage.

Salvage efforts at residential fires focus on protecting personal property. These items can have personal implications for the victims' lives. Some of the most valuable items personnel can protect and remove are pictures, passports, wallets, purses, medication, jewelry, antiques, and computer terminals. While conducting fire operations, fire fighters should be trained to bring an item out of the structure every time they leave the hot zone to retrieve a tool or replenish their air supply. In this manner salvage can be undertaken as a matter of course, maximizing time and efficiency.

LISTEN UP!

One way many fire departments help homeowners after a fire is by having a list of board-up and fire restoration companies available that they can give to the homeowner. Board-up companies are prepared to respond quickly and secure a property to protect it from rain, snow, or damage from vandals. Fire restoration companies are skilled in removing smoke and water from a building and its contents before more damage occurs. Because many building owners are not aware of these companies, recommendations are greatly appreciated by the owner or occupant of a damaged building at this highly stressful time.

At commercial or industrial fires, the salvage priorities are different from those at residential fires. Preincident planning should identify critical items, enabling fire fighters to efficiently focus their salvage efforts. For example, a law firm might identify its files as the most valuable property, whereas a library might place more importance on books and computer files than on furniture. In a factory, machinery may be a higher priority than the finished products stored on premises. A preincident plan should note whether the contents of a warehouse consist of valuable electronic components or industrial steel pipe. After the operation, the building must be secured to prevent break-ins or theft. Once fire operations are completed it becomes the responsibility of the owner to ensure the building is secure. If the fire is suspicious the police service with jurisdiction is responsible for ensuring security and the continuity of evidence. Do everything possible to reduce the amount of damage caused by both the fire and your own actions. Whether boarding a window or providing contact information for nongovernmental aid agencies, people are appreciative of any effort fire fighters make during and after an incident. Fire fighters should never take the appreciation the public has for their profession for granted. Engaging in systemic salvage operations is an important part of the service you provide and should be performed with as much enthusiasm as you would extinguishing the fire. Chapter 10, *Forcible Entry*, discusses forcible entry techniques in more detail.

Safety Considerations During Salvage Operations

Safety is a primary concern during salvage operations. The PPE required will depend on the site of the salvage operations, the stage of the fire, and any hazards present. Salvage operations often begin while the fire is still burning and continue for several hours after it has been extinguished. During firefighting operations and salvage operations, fire fighters should wear a full set of protective clothing and equipment, including SCBA.

If salvage is conducted in a different part of the building than where the fire occurred, several floors below the fire, or in a neighbouring building, fire fighters may have other protective clothing requirements. Hazards in these areas might be different from those in the fire area. Fire fighters conducting salvage operations in remote areas can wear partial PPE (boots, turnout pants, helmet, and gloves), but only with the approval of the safety officer.

Structural collapse is always a possibility during salvage operations because of fire damage to the building's structural components. For example, lightweight trusses can collapse quickly when damaged by a fire. Heavy objects, such as air-conditioning units and appliances, can crash through roofs or ceilings weakened by fire damage.

Even the water used to douse the fire can contribute to structural collapse. Recall that each gallon of water weighs 4 kg (8.34 lb). Extinguishing a major fire can, therefore, load the structure with tons of extra water weight. If the structure is already weakened by the fire, the extra water weight can cause a catastrophic structural collapse, with the potential to injure or kill fire fighters. For example, a ladder pipe discharging 3028 L/min (800 gpm) of water into a building is adding 3.2 tons [2.9 tonnes] (6400 lb) of weight to the building every minute.

Sometimes this danger may be masked by absorbent materials such as bolts of cloth or bales of rags stored in the building. Uneven absorption of water also presents a danger. For example, stacked rolls of newsprint in a printing factory can collapse if the bottom layer becomes waterlogged and, therefore, unable to support the stack.

Water creates potential hazards for fire fighters conducting salvage operations on floors below the fire. Water used to extinguish a fire inevitably leaks down through the building and can cause major damage. If it becomes trapped in the void space above a ceiling, it may cause sections of the ceiling to collapse. Fire fighters should watch for warning signs such as water leaking through ceiling joints and sagging areas of ceilings. Water that accumulates in a basement can hide a stairway and present additional hazards if utility connections are also located in the basement.

Damage to the building's utility systems presents significant risks to fire fighters who are conducting salvage operations. Gas and electrical services should always be shut off to eliminate the potential hazards of electrocution or explosion stemming from gas leaks.

Salvage Tools

There are a variety of salvage tools used in salvage operations (**TABLE 19-1**). Salvage covers (large sheets of heavy canvas or plastic material), floor runners, and other protective items shield and cover building contents from water runoff, falling debris, soot, and particulate matter in smoke residue. Other salvage tools remove smoke, water, and other products that can damage property (**FIGURE 19-5**).

TABLE 19-1 Salvage Tools

- Salvage covers or rolls of polyethylene film
- Box cutter (for cutting plastic)
- Floor runners
- Wet/dry vacuums
- Squeegees
- Drainage pumps and hose
- Sprinkler shut-off kit
- Ventilation fans, power blowers
- Small tool kit
- Pike poles (to construct water chutes to catch dripping water)

FIGURE 19-5 Salvage tools.
© Jones & Bartlett Learning. Photographed by Glen E. Ellman.

Using Salvage Techniques to Prevent Water Damage

The best way to prevent water damage at a fire scene is to limit the amount of water used to fight the fire. In other words, use only enough water to cool surfaces and knock down the fire quickly; do not use more water than is needed. Do not continue to douse a fire that is already out, because this action merely creates unnecessary water damage. Turn off hose nozzles when they are not in use. If a nozzle must be left partly open to prevent freezing during cold weather, direct the flow out through a window. Take the time to tighten any leaking hose couplings after the fire has been knocked down, because this measure prevents additional water damage during overhaul. Look for leaking water pipes in the building; if you find any, shut off the building's water supply.

Deactivating Sprinklers

Buildings equipped with sprinkler systems require special steps to limit water damage. Sprinklers can control a fire efficiently, but they will keep flowing until the activated sprinklers are shut off or the entire system is shut down. This ongoing flow can cause significant water damage. Responding fire companies should know where and how to shut down any automatic sprinkler system, if needed. If the system is accidentally activated, it should be shut down as soon as possible to avoid excessive water damage.

In an actual fire, the order to shut off the sprinkler system should come only from the IC. Generally, the sprinkler system should not be shut down until the fire is completely extinguished. If sprinklers are shut down prematurely, the fire may rekindle, causing major damage. Between the time when sprinklers are shut down and when overhaul is completed, a fire fighter with a portable radio should stand by and be ready to reactivate the sprinklers if necessary.

Sprinkler systems are usually designed so that only those sprinkler heads directly over or close to the fire will be activated. Indeed, most fires can be contained or fully extinguished by the release of water from only one or two sprinkler heads (**FIGURE 19-6**). Although movies may show all sprinkler heads activating at once, this does not occur in real life except in specially designed deluge sprinkler systems. More information on sprinkler systems is presented in Chapter 26, *Fire Detection, Suppression, and Smoke Control Systems*.

If only one or two sprinklers have been activated, inserting a sprinkler wedge or a sprinkler stop can quickly stop the flow. This action keeps the rest of the system operational during overhaul.

A **sprinkler wedge** is a small triangular piece of wood. Inserting two wedges on opposite sides, between the orifice and the deflector of the sprinkler, and pushing them together plugs the opening effectively, stopping the flow from standard upright and pendant sprinkler heads.

A **sprinkler stop** is a more sophisticated mechanical device with a rubber stopper that can be inserted into a sprinkler head (**FIGURE 19-7**). Several types of sprinkler stops are available, including some that work only with specific sprinkler heads. Some types of sprinkler stops such as a Shutgun™ are manufactured with a fusible link so the sprinkler head remains functional until it can be permanently replaced (**FIGURE 19-8**).

To stop the flow of water from a sprinkler using a pair of sprinkler wedges, follow the steps in **SKILL DRILL 19-2**. To stop the flow of water from a sprinkler using a sprinkler stop, follow the steps in **SKILL DRILL 19-3**.

FIGURE 19-6 Most fires are controlled by the activation of only one or two sprinkler heads.
Courtesy of Tyco Fire and Building Products.

FIGURE 19-7 A sprinkler stop can be used to stop the flow of water from a sprinkler head that has been activated.
© Jones & Bartlett Learning.

FIGURE 19-8 This sprinkler stop shuts off the sprinkler and provides a temporary fusible link to keep the sprinkler head in service.
Courtesy of Scott Dornan, ConocoPhillips Alaska.

SKILL DRILL 19-2
Using Sprinkler Wedges Fire Fighter I, NFPA 1001: 4.3.14

1 Hold one wedge in each hand.

2 Insert the two wedges, one from each side, between the discharge orifice and the sprinkler deflector.

3 Bump the wedges securely into place to stop the water flow.

© Jones & Bartlett Learning.

SKILL DRILL 19-3
Using a Sprinkler Stop Fire Fighter I, NFPA 1001: 4.3.14

1 Have a sprinkler stop in hand.

2 Place the flat-coated part of the sprinkler stop over the sprinkler orifice and between the frame of the sprinkler.

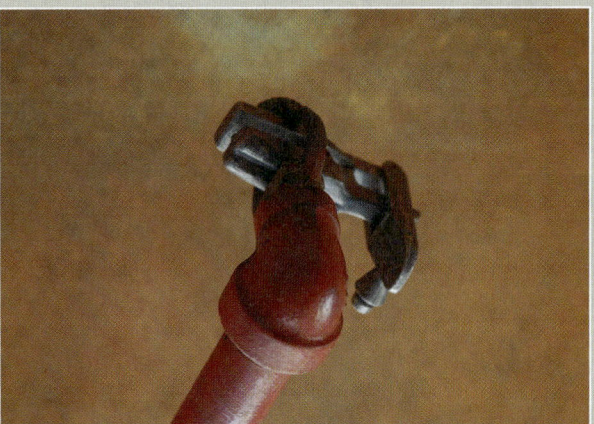

3 Push the lever to expand the sprinkler stop until it snaps into position.

Not all sprinklers can be shut off with a sprinkler wedge or a sprinkler stop. Recessed sprinklers, which are often installed in buildings with finished ceilings, are usually difficult to shut off. If the individual heads cannot be shut off or if several sprinklers have been activated, you can stop the flow by closing the main sprinkler control valve. After the valve is closed, the water trapped in the piping system will drain out through the activated sprinklers for several minutes. If a drain valve is found near the control valve, open it to drain the system quickly. This step directs the water flow to a location where it will not cause additional damage.

The main control valve for a sprinkler system is usually an **outside screw and yoke (OS&Y) valve** or a **post indicator valve (PIV)**. An OS&Y valve is typically found in a mechanical room in the basement or on the ground-floor level of a building. A PIV is located outside the building or on an exterior wall. Some sprinkler systems also have zone valves that control the flow of water to sprinklers in different areas of the building. High-rise buildings, for example, usually have a zone valve for each floor. Closing a zone valve stops the flow to sprinklers in the zone; in the rest of the building, however, the sprinklers remain operational. The locations of the main control valves and zone valves should be identified during preincident planning visits to a building.

Sprinkler control valves should always be locked in the open position, usually with a chain and padlock, and should be equipped with a tamper alarm that is tied into the fire alarm system. Fire fighters who are sent to shut off a sprinkler control valve should take a pair of bolt cutters to remove the lock if the key is not available.

To close and reopen an OS&Y valve, follow the steps in **SKILL DRILL 19-4**.

To close and reopen a PIV, follow the steps in **SKILL DRILL 19-5**.

SKILL DRILL 19-4
Closing and Reopening a Main OS&Y Valve Fire Fighter I, NFPA 1001: 4.3.14

1 Locate the OS&Y valve as indicated on the preincident plan. Identify the valve that controls sprinklers in the fire area. If the valve is locked in the open position with a chain and padlock and the key is readily available, unlock and remove the chain. If no key is available, cut the lock or the chain with a pair of bolt cutters. Cut a link close to the padlock so that the chain can be reused.

2 Turn the valve handle clockwise to close the valve. Keep turning until resistance is strong and little of the valve stem is visible.

3 To reopen the OS&Y valve, turn the handle counterclockwise until resistance is strong and the valve stem is visible again. Lock the valve in the open position.

© Jones & Bartlett Learning. Photographed by Glen E. Ellman.

SKILL DRILL 19-5
Closing and Reopening a Main Post Indicator Valve Fire Fighter I, NFPA 1001: 4.3.14

1 Locate the PIV as indicated on the preincident plan. Unlock the padlock with a key, or cut the lock with a pair of bolt cutters.

2 Remove the handle from its storage position on the PIV, and place it on top of the valve, similar to the use of a hydrant wrench. Turn the valve stem in the direction indicated on top of the valve to close the valve. Keep turning until resistance is strong and the visual indicator changes from "Open" to "Shut."

3 To reopen the PIV, turn the valve stem in the opposite direction until resistance is strong and the indicator changes back to "Open." Lock the valve in the open position.

© Jones & Bartlett Learning. Photographed by Glen E. Ellman.

Before a sprinkler system can be restored to normal operation, the sprinkler heads that have been activated must be replaced. Every sprinkler system should have spare heads stored somewhere, usually near the main control valves. An activated sprinkler head must be replaced with another head of the same design, size, and temperature rating. It is not generally accepted practice for fire service personnel to replace sprinkler heads for liability reasons. The building owner is ultimately responsible for ensuring the building meets the requirements of the fire code in any jurisdiction. Many fire services require a form to be filled out and signed by the owner or manager acknowledging the need to service the alarm or standpipe system immediately to ensure fire code compliance.

The main sprinkler control valve or the appropriate zone valve must be closed, and the system must be drained before a sprinkler head can be changed. Special wrenches must be used when replacing sprinklers to prevent damage to the operating mechanism.

After the activated heads are replaced, the sprinkler system can be restored to service. Restoring a sprinkler system to service takes special training and should be performed only by qualified individuals.

SAFETY TIP

Do not shut down the water to a sprinkler system until ordered to do so by the IC. The fire must be completely out, and hose lines must be available should the fire reignite. A fire fighter should be stationed at the main sprinkler control valve with a portable radio, ready to reopen the valve if necessary.

Removing Water

Water that accumulates within a building or drips down from higher levels should be channeled to a drain or to the outside of the building to prevent or limit water damage to the structure. A salvage pump may be needed to help remove the water in some cases.

Some buildings have floor drains that funnel water into a below-ground sewer system. These floor drains should be kept free from debris so that the water can drain freely. Fire fighters can use squeegees to direct the water into the drain.

In buildings or houses without floor drains, it may be possible to shut off the water supply valve to a floor-mounted toilet, remove the nuts that hold the toilet bowl to the floor, and remove the toilet from the floor flange. This maneuver creates a large drain capable of handling large quantities of water, as long as someone keeps the opening from becoming clogged with debris.

Water on a floor at ground level can often be channeled to flow outside of the structure through a doorway or other opening. Water on a floor above ground level can sometimes be drained to the outside by making an opening at floor level in an exterior wall of the building.

Water chutes or water catch-alls are often used to collect water leaking down from firefighting operations on higher floor levels and to help protect property on calls involving burst pipes or leaking roofs. A water chute catches dripping water and directs it toward a drain or to the outside through a window or doorway. A water catch-all is a temporary pond that holds dripping water in one location. The accumulated water must then be drained to the outside of the building. Water chutes and catch-alls can be constructed quickly using salvage covers or a roll of polyethylene film.

A water chute can be constructed with a single salvage cover or with a cover and two pike poles for support. To construct a water chute using a salvage cover, follow the steps in **SKILL DRILL 19-6**.

To construct a water catch-all, follow the steps in **SKILL DRILL 19-7**.

SKILL DRILL 19-6
Constructing a Water Chute Fire Fighter I, NFPA 1001: 4.3.14

1 Fully open a large salvage cover flat on the ground.

2 If using pike poles, lay one pole on one edge of the cover, and roll the cover around the handle. Roll the cover tightly toward the middle.

3 Repeat the actions in Step 2 on the opposite edge of the cover, rolling the opposite edge tightly toward the middle until the two rolls are 31–91 cm (1–3 ft) apart.

(continued)

SKILL DRILL 19-6 Continued
Constructing a Water Chute Fire Fighter I, NFPA 1001: 4.3.14

4 Turn the cover upside down. Position the chute so that it collects dripping water and channels it toward a drain or outside opening. Place the chute on the floor, with one end propped up by a chair or other object.

5 Use a stepladder or other tall object to support chutes constructed with pike poles.

© Jones & Bartlett Learning. Photographed by Glen E. Ellman.

SKILL DRILL 19-7
Constructing a Water Catch-All Fire Fighter I, NFPA 1001: 4.3.14

1 Open a large salvage cover on the ground, and roll each edge of the cover toward the opposite side.

SKILL DRILL 19-7 Continued
Constructing a Water Catch-All Fire Fighter I, NFPA 1001: 4.3.14

2 Fold each corner over at a 90-degree angle, starting each fold approximately 91 cm (3 ft) in from the edge.

3 Roll the remaining two edges inward approximately 61 cm (2 ft).

4 Lift the rolled edge over the corner flaps, and tuck it in under the flaps to lock the corners in place.

Wet Vacuum

Special vacuum cleaners that suck up water can be used during salvage operations. Two types of **wet vacuums** (also known as water vacuums) are available: a small-capacity backpack type and a larger wheeled unit. The backpack vacuum cannot be used by someone who is wearing SCBA. Wet/dry shop vacuums may be used as a low-cost alternative to wet vacuums.

Drainage Pumps

Drainage pumps remove water that has accumulated in basements or below ground level. Portable electric submersible pumps can be lowered directly into the water to pump it out of a building. Gasoline-powered portable pumps must be placed outside a building because they exhaust poisonous carbon monoxide gas. These pumps use a hard-suction hose, which is lowered into the basement through a window and drafts water out of the building.

SAFETY TIP

You cannot operate a gasoline-powered pump safely inside a structure because the operation of the pump generates deadly carbon monoxide gas.

Using Salvage Techniques to Limit Smoke and Heat Damage

One way to reduce property loss is to keep heat and smoke out of areas that are not involved in the fire. A closed door can effectively keep smoke and heat out of a room or area. Given this fact, fire fighters doing search and rescue operations should remember to close doors after searching a room.

Properly timed and effective ventilation practices using exhaust fans and natural ventilation can be used to limit smoke and heat damage. Rapid ventilation will often reduce smoke damage in an area already filled with smoke. The visible components of smoke are mostly soot particles and other products of combustion, including corrosive chemicals that will settle on any horizontal surface when the smoke cools. Blowing the smoke out of the building is a better option than allowing these contaminants to settle on the contents. All members of the fire suppression team—not just the

salvage crews—should recognize opportunities to prevent smoke and heat damage. Ventilation is discussed in detail in Chapter 13, *Ventilation*.

Salvage Covers

The most common method of protecting building contents is to cover them with salvage covers. As previously mentioned, salvage covers are large square or rectangular sheets of heavy canvas or plastic material that are used to protect furniture and other items from water runoff, falling debris, soot, and particulate matter in smoke residue.

Salvage crews usually begin their work on the floor immediately below the fire, with the goal being to prevent water damage to furniture and other contents on lower floors. If the fire is in an attic, fire fighters may have enough time to spread covers over the furniture in the rooms directly below the fire before pulling the ceilings to attack the flames.

The most efficient way to protect a room's contents is to move all the furniture to the center of the room, away from the walls, where water could damage the backs of the furniture. This approach reduces the total area that must be covered, enabling one or two fire fighters to cover the pile quickly and move on to the next room (**FIGURE 19-9**).

Remove any pictures from the walls and place them with the furniture. Put smaller pictures and valuable objects in drawers or wherever they will be protected from breakage. If you have enough time, roll up any rugs, and place them on the pile.

Some departments use rolls of construction-grade polyethylene film for protecting a structures' contents instead of salvage covers. This material comes in rolls as

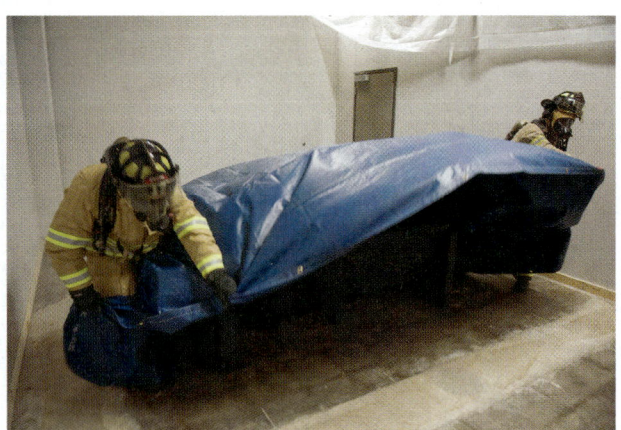

FIGURE 19-9 Move furniture to the center of the room so that a single salvage cover can protect all of the contents.

© Jones & Bartlett Learning. Photographed by Glen E. Ellman.

long as 37 m (120 ft), so it can be unrolled over the room's contents and cut to the correct length with a box cutter. Polyethylene film is particularly useful for covering long surfaces, such as retail display counters. This material is disposable and can be left behind after the fire; in contrast, traditional salvage covers must be picked up, washed, dried, and properly folded for storage in fire apparatus compartments after each use. Occasionally, salvage covers may be left behind as protection for the building contents, but they should be picked up when they are no longer needed.

Special folding and rolling techniques are used to store salvage covers so that they can be deployed quickly by one or two fire fighters. Be certain to follow your fire department's SOPs/SOGs when folding or rolling a salvage cover.

To fold a salvage cover to prepare it for one fire fighter to deploy, follow the steps in **SKILL DRILL 19-8**. To

SKILL DRILL 19-8
Performing a Salvage Cover Fold for One-Fire Fighter Deployment
Fire Fighter I, NFPA 1001: 4.3.14

1 Spread the salvage cover flat on the ground with a partner facing you.

2 On the right side of the salvage cover, make a fold at the quarter point of the cover. Next, make a second fold that ends in the middle of the cover.

3 On the left side of the salvage cover, make a fold at the quarter point of the cover. Next, make a second fold that ends in the middle of the cover.

(continued)

SKILL DRILL 19-8 Continued
Performing a Salvage Cover Fold for One-Fire Fighter Deployment
Fire Fighter I, NFPA 1001: 4.3.14

4 Fold the two halves together, and flatten the salvage cover to remove any trapped air.

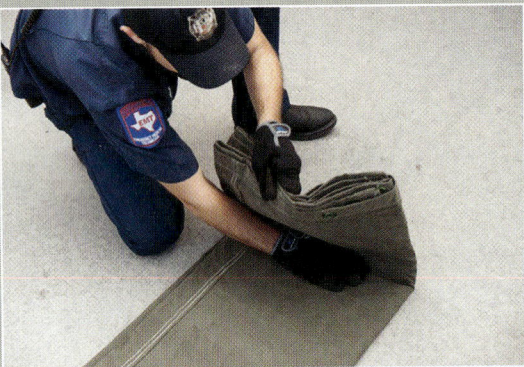

5 Make 31-cm (1-ft) folds from each end of the cover until you reach the center of the salvage cover.

6 Fold the two halves together.

© Jones & Bartlett Learning. Photographed by Glen E. Ellman.

fold a salvage cover to prepare it for two fire fighters to deploy, follow the steps in **SKILL DRILL 19-9**.

To fold and roll a salvage cover, follow the steps in **SKILL DRILL 19-10**.

Although it often takes two people to fold and roll a salvage cover for storage, this technique enables one person to unroll the cover easily and quickly. To perform the one-person salvage cover roll, follow the steps in **SKILL DRILL 19-11**.

A single fire fighter can also use a shoulder toss to spread a salvage cover. To perform a shoulder toss, follow the steps in **SKILL DRILL 19-12**.

SKILL DRILL 19-9
Performing a Salvage Cover Fold for Two-Fire Fighter Deployment
Fire Fighter I, NFPA 1001: 4.3.14

1 Spread the salvage cover flat on the ground with a partner facing you. Together, fold the cover in half.

2 Together, grasp the unfolded edge, and fold the cover in half again. Flatten the salvage cover to remove any trapped air.

3 Move to the newly created narrow ends of the salvage cover, and fold the salvage cover in half lengthwise.

4 Fold the salvage cover in half lengthwise again. Make certain that the open end is on top.

5 Fold the cover in half a third time.

SKILL DRILL 19-10
Folding and Rolling a Salvage Cover Fire Fighter I, NFPA 1001: 4.3.14

1 Spread the salvage cover flat on the ground with a partner facing you.

2 Together, fold the outside edge in to the middle of the cover, creating a fold at the quarter point.

3 Fold the outside fold in to the middle of the cover, creating a second fold.

4 Repeat Step 2 from the opposite side of the cover.

5 Repeat Step 3 from the opposite side of the cover so the folded edges meet at the middle of the cover, with the folds touching but not overlapping.

6 Tightly roll up the folded salvage cover from the end.

SKILL DRILL 19-11
Performing a One-Person Salvage Cover Roll Fire Fighter I, NFPA 1001: 4.3.14

1 Stand in front of the end of the object that you are going to cover.

2 Start to unroll the cover over one end of the object.

3 Continue unrolling the cover until you reach the top of the object. Allow the remainder of the cover to unroll and settle at the end of the object.

4 Spread the cover, unfolding each side outward over the object to the first fold.

5 Unfold the second fold on each side, and drape the cover completely over the object.

6 Tuck in all loose edges of the cover around the object.

© Jones & Bartlett Learning. Photographed by Glen E. Ellman.

SKILL DRILL 19-12
Performing a One-Fire Fighter Salvage Cover Shoulder Toss
Fire Fighter I, NFPA 1001: 4.3.14

1 Place the folded salvage cover over one arm.

2 Toss the cover over the salvaged object with a straight-arm movement.

3 Unfold the cover until it completely drapes the object.

Two fire fighters can use a balloon toss to cover a pile of building contents quickly. To perform a balloon toss, follow the steps shown in **SKILL DRILL 19-13**.

Salvage Cover Maintenance

Salvage covers must be adequately maintained to preserve their shelf life. Salvage cover maintenance depends on the type of cover used. A canvas cover can usually be cleaned with a scrub brush and clean water. If the cover becomes particularly dirty, however, cleaning with a mild detergent may be necessary. Covers should be adequately rinsed when a detergent is used.

Canvas covers must be properly dried before being returned to service. Effectively drying a canvas cover will reduce mildewing. Vinyl type covers are easily

SKILL DRILL 19-13
Performing a Two-Fire Fighter Salvage Cover Balloon Toss
Fire Fighter I, NFPA 1001: 4.3.14

1 Place the salvage cover on the ground beside the object.

2 Unfold the cover so that it runs along the entire base of the object. Each fire fighter grabs one edge of the cover and brings it up to waist height.

3 Together, lift the cover quickly so that it fills with air like a balloon.

4 Move quickly to the other side of the item, using the air to help support the cover, and spread the entire cover over the object.

maintained by rinsing and do not mildew as easy as canvas covers.

Once dried, salvage covers should be inspected for tears and holes. Any damage found can be patched by using duct tape or a sewn-on patch.

Floor Runners

As previously mentioned, floor runners protect carpets or hardwood floors from water, debris, fire fighters' boots, and firefighting equipment. Fire fighters entering an area for salvage operations should unroll this protective material ahead of themselves and stay on the floor runner while working in the area.

Other Salvage Operations

The best way to protect the contents of a building may be to move them to a safe location. The IC will make this determination. Any items removed from the building should be placed in a dry, secure area—preferably a single location. In some cases, salvaged items can be moved to a suitable location within the same building. If items are moved outside, they should be protected from further damage caused by firefighting operations or the weather. Valuable items should be placed in the care of a law enforcement officer if the property owner is not present.

Sometimes the building contents can provide clues about the cause or spread of the fire. In these situations, fire investigators should be consulted, and they should supervise the removal process.

Salvage operations sometimes extend outside the building to include valuable property such as vehicles or machinery. Use a salvage cover to protect outside property or move vehicles if this action can be done without compromising the fire suppression efforts.

Overhaul Overview

Overhaul is the process of searching for and extinguishing any pockets of fire that remain after a fire has been brought under control. A new fire remains a possibility even if 99 percent of a fire is out and just 1 percent is smouldering. Indeed, a single pocket of embers can rekindle after fire fighters leave the scene and cause even more damage and destruction than the original fire. A fire cannot be considered fully extinguished until the overhaul process is complete.

The process of overhaul begins after the fire is brought under control. Overhaul can be a time-consuming, physically demanding process. The greatest challenge during this phase of fire operations is to identify and open any void spaces in a building where the fire might be burning undetected. If the fire extends into any void spaces, fire fighters must open the walls and ceilings to expose the burned area. Any materials that are still burning must be soaked with water or physically removed from the building. This process must continue until all of the burned material is located and unburned areas are exposed. Overhaul is also required for nonstructure fires such as those occurring in automobiles, bulk piles (tires or woodchips), vegetation, and even garbage.

Health Considerations During Overhaul

During overhaul you need to be aware of the threats to your health. Overhaul, even after ventilation, is conducted in an environment that has unknown amounts of smoke, soot, and other partly burned substances. Smoke and soot are particles from burned materials. The atmosphere during overhaul contains poisonous gases such as carbon monoxide and hydrogen cyanide in unknown quantities. In cases of an enclosed compartment, the atmosphere may contain elevated levels of carbon dioxide or decreased levels of oxygen. In addition to poisonous gases, smoke and soot contain toxic compounds such as ammonia, hydrogen chloride, sulfur dioxide, hydrogen sulfide, hydrogen cyanide, the oxides of nitrogen, formaldehyde, acrolein, and polycyclic aromatic hydrocarbons. These compounds present two different types of health hazards to fire fighters. Some of these gases, such as carbon monoxide and hydrogen cyanide, can result in immediate and deadly illness. Others are carcinogens, which present long-term health hazards by increasing your chance of contracting a variety of cancers later in life.

During overhaul, poisonous substances can enter the body in three different ways: absorption through the skin, inhalation, and accidental ingestion. To prevent the poisonous particles in smoke and soot from being absorbed through your skin, wear full PPE, including SCBA, boots, a structural firefighting protective coat and pants, a protective hood, a helmet, and gloves. In addition to protecting your face from absorption, SCBA prevents poisonous gases and particles from being inhaled into the lungs. Carefully remove and clean contaminated PPE as soon as possible. Be sure to thoroughly shower to remove any toxic substances from your hair and skin, and clean your hands and body before eating to prevent inadvertent ingestion of toxic products. All of these steps will go a long way in preventing immediate and long-term health problems.

An additional health consideration is adequate rehabilitation. Working for extended periods in full PPE creates heat stress and dehydration; overhaul crews should work for short periods and take frequent breaks. Fresh crews or crews that have been properly rehabilitated should replace fatigued crews.

Air Monitoring

Because there is such a wide variety of toxic substances present at the fire scene, there is no single best practice when it comes to detection and monitoring in the fire environment, specifically during overhaul. Chapter 35, *Hazardous Materials Responder Health and Safety*, discusses the common methods of post-fire detection and monitoring and the use of certain detection devices at the fire scene. Although most fire gases and particulates can be detected and measured by one type of technology or another, there is no single device that can detect all of them.

The most common method of gas detection involves the use of single-gas detection devices; however, detection of a single gas does not determine whether the entire building is safe or unsafe to operate in without SCBA. For example, testing the air for carbon monoxide levels does nothing to measure the amount of complex carcinogens present. Additional methods of detection and monitoring include using multi-gas meters configured with a variety of sensors (e.g., oxygen, carbon monoxide, hydrogen sulfide, flammable gas) and using more than one technology to look for multiple gases. To use a multi-gas meter to provide atmospheric monitoring, follow the steps in **SKILL DRILL 19-14**. The authority having jurisdiction will determine the exact monitoring

SKILL DRILL 19-14
Using a Multi-gas Meter Fire Fighter I, NFPA 1001: 4.3.21

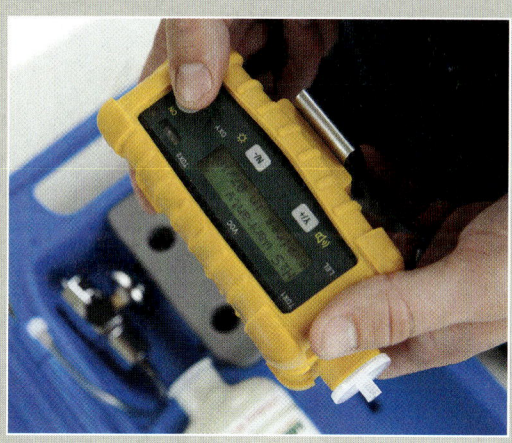

1 Turn the unit on and let it warm up (usually 5 minutes is sufficient) in a well-ventilated area, away from the fire scene and any vehicles. Ensure the battery has sufficient life for the operational period. Identify the installed sensors (e.g., oxygen, flammability, hydrogen sulfide, carbon monoxide), and verify that the sensors are not expired. Review alarm limits and the audio and visual alarm notifications associated with those limits. Review and understand the types of gases and vapours that could harm or destroy the sensors. Use other methods to check for those substances (e.g., pH paper) to ensure they are not present in the atmosphere to be sampled. Care must be taken to avoid pulling liquids into the device—it is designed to sample air, not liquid!

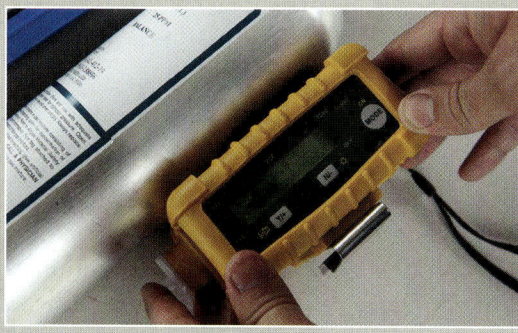

2 Perform a test on the pump by occluding the inlet and ensuring the appropriate alarm sounds.

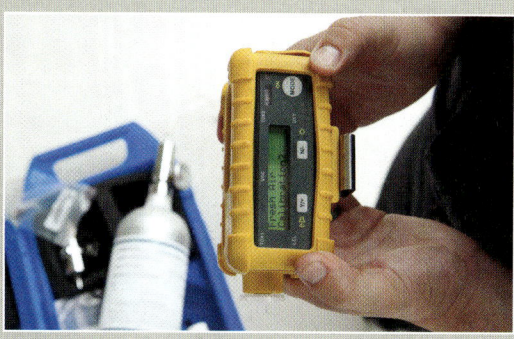

3 Perform a fresh air calibration, and "zero" the unit.

(continued)

SKILL DRILL 19-14 Continued
Using a Multi-gas Meter Fire Fighter I, NFPA 1001: 4.3.21

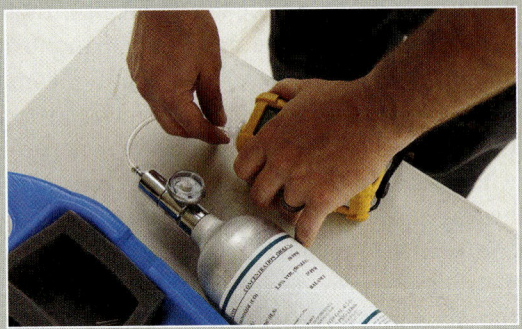

4 Ensure the meter is operating correctly by exposing the unit to a substance or substances that the unit should detect and react to accordingly. In essence, you are making sure the unit will "see" what it is supposed to see before it is called upon to see it in a real situation.

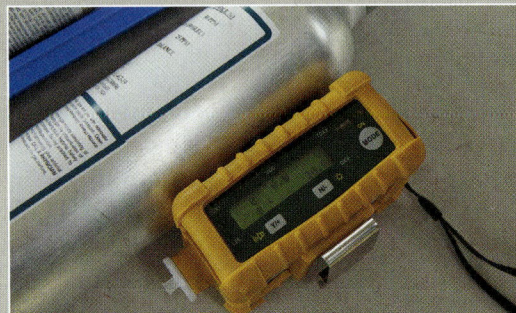

5 Allow the device to reset, or return to fresh air calibration state, then review the alarm levels and resetting procedures for addressing sensors that become saturated, or exposed to too much gas or vapour.

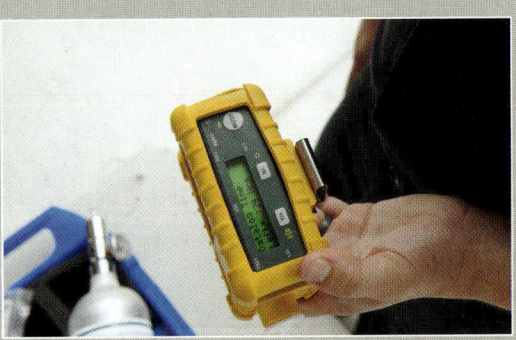

6 Review other device functions such as screen illumination, data logging (if available), and low-battery alarm.

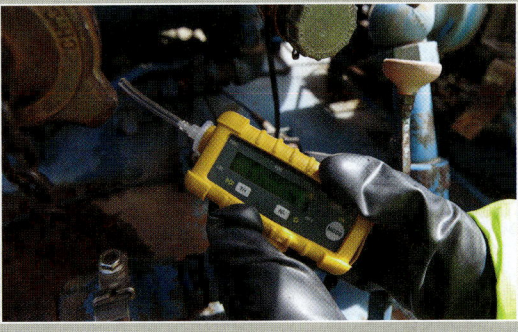

7 Review decontamination procedures. Carry out monitoring and detection per the device manufacturer's instructions. If the high- or low-level alarm activates, follow the standard operating procedures established by the authority having jurisdiction.

© Jones & Bartlett Learning. Photographed by Glen E. Ellman.

device to be used for the particular situation, the alarm levels and configuration of that device, and the standard operating procedures to follow if the device's high- or low-level alarm activates.

Electrochemical sensors are one of the most common technologies used in post-fire detection and monitoring devices. These sensors are filled with a chemical reagent—a substance that is designed to react when it encounters another specific substance. Typical electrochemical sensors used in post-fire detection and monitoring test for oxygen, hydrogen cyanide, carbon monoxide, hydrogen sulfide, ammonia, and chlorine. When the reagent reacts with the target gas, a meter reading results (**FIGURE 19-10**). If the amount of gas present is above a preset level, the low-level or high-level alarm will sound. For example, a detection device with an electrochemical sensor used to detect carbon monoxide may have a low-level alarm

FIGURE 19-10 Electrochemical sensors ready to be installed and used.
Courtesy of Rob Schnepp.

FIGURE 19-11 A photoionization detector (PID) in service during a session of fire investigation research.
Courtesy of Rob Schnepp.

setting of 35 ppm. A low-level alarm sounding during post-fire detection would indicate that the level of carbon monoxide in the area of detection has reached 35 ppm, and personnel should take protective action, such as donning an SCBA mask or performing additional ventilation. Electrochemical sensors are commonly found in single-gas devices.

Another type of sensor is a photoionization detector (PID). A PID is available as a stand-alone unit, or it may be incorporated into a multi-gas meter (**FIGURE 19-11**). In either case, the PID uses ultraviolet light to ionize the gases that move through the sensor. When the gases are ionized, sensors are able to determine the ionization of the molecules. The PID detects organic materials (materials that contain carbon as part of their molecular structure), such as benzene, acetone, toluene, ethanol, butane, and many others. It will not detect the presence of carbon monoxide, hydrogen cyanide, or compounds such as natural gas. Additionally, the PID does not identify which toxic gas is in the air—it only alerts the user that a toxic gas is present.

Another option for the detection of toxic gases is colorimetric tubes, which can be used to detect the presence and/or levels of specific chemical vapours and/or to confirm the readings of the electrochemical sensors or other technologies. The glass tubes are filled with reagents that react to a unique substance at a particular concentration when a certain amount of sample air

moves through the tube. The tubes are set up to detect single substances and/or chemical families (groups).

Regardless of the technology used, air monitoring does not address airborne contamination in total. It is up to the user to understand the alarm levels of each type of device used by his or her department and to know the established emergency actions to take if the low- or high-level alarm on the device sounds. Ultimately, the safest way for a fire fighter to reduce the possibility of inhalation exposures is to keep his or her SCBA on!

Safety Considerations During Overhaul

Many injuries can occur during overhaul. As mentioned earlier, overhaul is strenuous work conducted in an area already damaged by fire. Fire fighters who were involved in the fire suppression efforts may be physically fatigued and overlook hazards. As previously mentioned, adequate breaks for rehabilitation need to be provided during the overhaul process. Overhaul should occur at a steady, not rushed, pace. It should not be as hurried as the pace of initial firefighting.

Several additional hazards may be present in the overhaul area. Notably, the structural safety of the building is often compromised. Catastrophic building collapses have occurred during overhaul. Heavy objects could lead to roof or ceiling collapse, debris could litter the area, and there could be holes in the floor. Visibility is often limited, so fire fighters may have to depend on portable lighting. The presence of wet or icy surfaces makes falls more likely. In addition, during overhaul operations, dangerous equipment—including axes, pike poles, and power tools—is used in close quarters.

Fire fighters must be aware of these hazards and proceed with caution. When necessary, take extra time to evaluate the hazards and determine the safest way to proceed. There is no need to rush during overhaul, and there is no excuse for risking injury during this phase of operations.

A charged hose line must always be ready for use during overhaul operations because flare-ups may occur. For example, a smouldering fire in a void space could reignite suddenly when fire fighters open the space and allow fresh air to enter; such a fire could quickly become intense and dangerous. In addition, fine-grained combustible materials such as sawdust can smoulder for a long time and then ignite explosively when they are disturbed and oxygenated.

A safety officer should always be present during overhaul operations to note any hazards and ensure that operations are conducted safely. Company officers should supervise operations, look for hazards, and make sure that all crew members work carefully. The work should proceed at a steady pace, with an appropriate number of crew members: Too many people working in a small area creates chaos.

Always evaluate the structural condition of a building before beginning overhaul. Chapter 6, *Building Construction*, discusses building collapse. Signs of collapse may include the following:

- Cracked or leaning walls
- Pitched or sagging doors or doors stuck in shifted frames
- Moaning/groaning or cracking sounds
- Any type of movement or vibrations
- Movement or shifting of water on the floor
- Smoke pushing through cracks in the wall
- Lack of water runoff from firefighting operations

During the overhaul phase, do not compromise the structural integrity of the building. When opening walls and ceilings, remove only the outer coverings, and leave the structural members in place. If more invasive overhaul is required, be careful not to compromise the structure's load-bearing members. Avoid cutting lightweight wood trusses, load-bearing wall studs, and floor or ceiling joists, especially when using power tools.

If fire damages the structural integrity of a building, the IC may call for a "hydraulic overhaul" rather than a standard overhaul. In this situation, larger-diameter hose streams are used to completely extinguish a fire from the exterior. This strategy is appropriate if the site poses excessive risks to fire fighters and damage to the property is so extensive that it has no salvage value. Heavy mechanical equipment may be used to demolish unsafe buildings and expose any remaining hot spots. A condemned or abandoned building that is going to be torn down is not a place to risk the life or health of a fire fighter.

If a complete overhaul cannot be conducted, the IC can establish a fire watch. The fire watch team remains at the fire scene with a charged hose line and watches for signs of rekindling. This team can request additional help if the fire reignites.

Preserving Evidence During Overhaul

Overhaul crews must work with fire investigators to ensure that important evidence that could indicate the cause or **area of origin** of a fire is not lost or destroyed as a result of their efforts to extinguish all remnants of the fire. Ideally, a fire investigator should examine the area before overhaul operations begin, identifying evidence and photographing the scene before it is disturbed.

If a fire investigator is not immediately available, the overhaul crews should make careful observations and report to the investigator later. When performing overhaul in or near the suspected area of origin, note fire patterns and smoke residue on the walls or ceilings that could indicate the exact site of origin or the path of fire travel. Often the point of origin is found directly below the most damaged area on the ceiling, where the heat of the fire was most intensely concentrated. Fire patterns and damage often spread outward from the room or area where the damage is most severe. A piercing nozzle can be used to extinguish hidden fire in these void areas; its use limits the damage in and around the suspected area of origin. When moving appliances or other electrical items, note whether they were plugged in or turned on. Always look for evidence that the fire investigator can use to determine the cause of the fire.

If you discover anything suspicious—particularly indications of arson—you should leave it in place, make sure that no one interferes with the item or the surrounding area, and notify a fire officer or fire investigator immediately. Ensure that the fire will not rekindle, but do not do anything to unnecessarily compromise the scene once the fire is under control. Never discard debris until the fire investigator gives his or her approval to do so. Chapter 28, *Fire Origin and Cause*, provides additional information on preservation of evidence in case of fire.

Where to Overhaul

Determining when, where, and how much property needs to be overhauled requires good judgment. Overhaul must be thorough and extensive enough to ensure that the fire is completely out. At the same time, fire fighters should try not to destroy more property than is necessary.

Generally, it is better to make sure that the fire is definitely out than it is to be too careful about damaging property. If the fire rekindles, more property damage will occur, and the fire department could be held responsible for the additional losses.

The area that must be overhauled depends on the building's construction, its contents, and the size of the fire. All areas directly involved in the fire must be overhauled. If the fire was confined to a single room, all of the furniture in that room must be checked for smouldering fire. The overhaul process must ensure that the fire did not extend into any void spaces in the walls, above the ceiling, or into the floor. If signs indicate that the fire spread into the structure itself or into the void spaces, all suspect areas must be opened to expose any hidden fire.

If the fire involved more than one room, overhaul must include all possible paths of fire extension. The paths available for a fire to spread within a building are directly related to the type of construction. As a fire fighter, you must learn to anticipate where and how a fire is likely to spread in different types of buildings to find any hidden pockets of fire.

Fires in Type I *fire-resistive* construction can be considered contents-only fires due to the type and nature of the materials used in their construction. Generally, Type I buildings in Canada are constructed using reinforced load-bearing concrete and protected steel elements. In these buildings fire can spread through pipe and electrical shafts, electrical conduit, or finishings such as interior combustible walls and floor coverings. It is important for fire fighters to always check for fire extension on the floor above the fire unit.

Wood-frame and ordinary construction buildings may contain several areas where a hidden fire could be burning. In particular, these structures often have void spaces under the wall-covering materials. When a serious fire strikes a building of ordinary or wood-frame construction, fire fighters may need to open every wall, ceiling, and void space to check for fire extension. Neighbouring buildings may need to be overhauled if the fire could have spread into them.

In balloon-frame construction, a fire can extend directly from the basement to the attic, without obvious signs of fire on any other floor (**FIGURE 19-12**). For this reason, these buildings require a thorough floor-by-floor overhaul, as the main body of the fire is still being attacked. Any delay in opening up the walls nearest and above the seat of the fire can result in a total loss fire. Ideally, personnel should be directed to investigate fire spread in the roof area if it is suspected that the fire has found its way into any wall. As these buildings were primarily built in the late 19th and early 20th century, the cock-loft areas tend to be less tall than in Type III ordinary constructed buildings. If fire has extended into the roof area, smoke will likely exit through small grill-like vents just under the roof

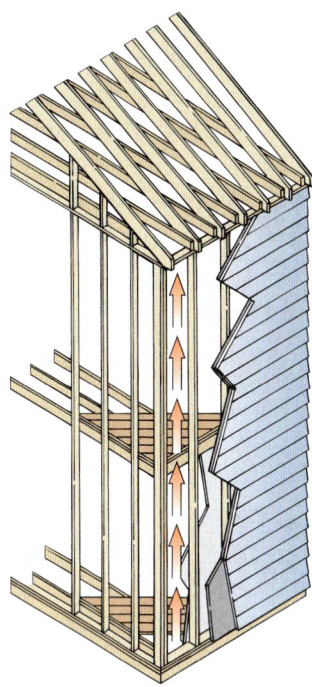

FIGURE 19-12 A fire in a balloon-frame construction building can spread directly from the basement to the attic.
© Jones & Bartlett Learning.

eaves. Preincident planning will aid in identifying these buildings, which are most likely located in the oldest areas of large Canadian cities, particularly in Eastern Canada. The problems could be compounded if the attic insulation consists of blown-in cellulose materials—these materials can smoulder for a long time. If cellulose fibre insulation has been exposed to fire, it can present a challenge for fire fighters. If you are in doubt about whether or not the insulation is safe, there is no doubt it all should be removed to the outside and washed down. Be careful not to use fans as this can cause embers to travel into inaccessible areas. A fire watch should be posted for several hours post-fire to ensure that a rekindle does not occur. Additional information on how building construction may affect fire spread is found in Chapter 6, *Building Construction*.

FIRE FIGHTER TIP

One method used to help you determine where you need to overhaul is to think of a fire compartment as having seven "sides." These sides are the four walls of the compartment, the ceiling, the floor, and any void spaces. Thoroughly check each of these areas to be certain that the fire is fully extinguished.

If a building has been extensively remodelled, overhaul presents special challenges. Fire can hide in the space between a dropped ceiling and the original ceiling, or it may extend into a different section of the building through doors and windows covered by new construction. Some buildings have two roofs—one original and one added later—with a void space between them. A fire in this void space presents especially difficult overhaul problems.

The cause of the fire can indicate the extent of necessary overhaul. A kitchen stove fire may extend into the kitchen exhaust duct and could ignite combustible materials in the immediate vicinity. Follow the path of the duct, and open the areas around it to locate any residual fire. A lightning strike releases enormous energy through the wiring and piping systems, which can start multiple fires in different parts of the building. Overhaul after this type of incident must be extensive.

Using Your Senses

Efficient and effective overhaul requires the use of all of your senses. Look, listen, and feel to detect signs of potential burning.

Look for these signs:

- Smoke seeping from cracks or from around doors and windows
- Fresh or new smoke
- Red, glowing embers in dark areas
- Burned areas
- Discoloured material
- Peeling paint or cracked plaster
- "Hot spots" shown on a thermal imaging device

Listen for these sounds:

- Crackling sounds that indicate active fire
- Hissing sounds that indicate water has touched hot objects

Feel for this condition:

- Heat, shown on a thermal imaging device or felt on the back of your hand (only if it is safe to remove your glove)

Experience is a valuable trait during overhaul situations. By using his or her senses in a systematic manner, an experienced fire fighter can often determine which parts of the structure need to be opened and which areas were untouched by the fire.

Thermal Imaging

The thermal imaging device is a valuable high-tech tool that is used during overhaul. The same type of thermal imaging device used for search and rescue (as discussed in Chapter 12, *Search and Rescue*) can be used to locate hidden hot spots or residual pockets of fire. It can quickly differentiate between unaffected areas and areas that need to be opened. Using a thermal imaging device can decrease the amount of time needed to overhaul a fire scene and reduce the amount of physical damage to the building.

The thermal imaging device distinguishes between objects or areas with different temperatures and displays hot areas and cold areas as different colours on the video screen. Because it is sensitive, this device can "see" a hot spot even if the heat source is located behind a wall. It can show the pattern of fire within a wall, even if there is only a few degrees' difference in the wall's surface temperature.

Interpreting the readings from a thermal imaging device requires practice and training (**FIGURE 19-13**). This device displays relative temperature differences, so an object will appear to be warmer or cooler than its surroundings. If the room has been superheated by fire, all of the contents—including the walls, floor, and ceiling—will appear hot. In these circumstances, an extra-hot spot behind a wall could be difficult to distinguish. In this situation, it might be better to look at the wall from the side that was not exposed to the fire to identify hot spots.

SAFETY TIP

Before overhaul operations begin, the safety officer should confirm that electric power and gas service in the overhaul area are shut off. Be careful: Even though the gas supply may be turned off, gas under pressure might still be present in the lines and will need to be bled off.

Do not rely solely on the information from a thermal imaging device to identify areas of persistent fire, because anything that acts as an insulator can hide hot spots. Hot spots behind heavy padding and carpeting or in insulated walls, for example, might not show up as hot spots on the thermal imaging device screen.

The fire fighter must consider the thermal imaging device as just another tool to complement the rest of the

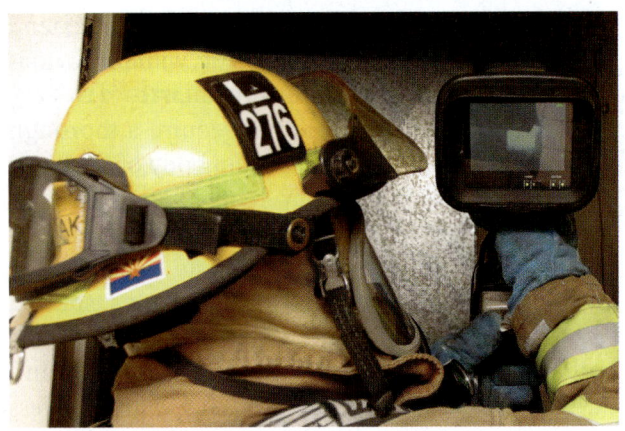

FIGURE 19-13 Thermal imaging devices can help locate hot spots.

© Jones & Bartlett Learning.

overhaul operation; it cannot completely rule out the possibility of concealed fire. The only way to ensure that no hidden fire exists is to open a ceiling or wall and perform a direct inspection.

Overhaul Techniques

The objective of overhaul is to find and extinguish any fires that could still be burning after a fire is brought under control. Any fire uncovered during overhaul must be thoroughly extinguished. Overhaul operations should continue until the IC is satisfied that no smouldering fires are left.

During overhaul operations, a charged hose line should be available to douse any sudden flare-ups or exposed pockets of fire. If necessary, use a direct stream from the hose line but avoid unnecessary water damage. Extinguish small pockets of fire or smouldering materials with the least possible amount of water by using a short burst from the hose line or simply drizzling water from the nozzle directly onto the fire.

Extinguish smouldering objects that can be safely picked up by dropping them into a bathtub or bucket filled with water. Remove materials prone to smouldering, such as mattresses and cushioned furniture, from the building and thoroughly soak them outside. Roll mattresses, and secure them with rope or webbing to move them out of the building and decrease the possibility of rekindling while being moved out of the building.

Place debris that is moved outside far enough away from the building to prevent any additional damage if it reignites. Do not allow debris to block entrances or exits. In some cases, a window opening can be enlarged so that debris can be removed more easily. Heavy machinery, such as a front-end loader, may be used to remove large quantities of debris from commercial buildings.

Overhaul Tools

Tools used during overhaul are designed for cutting, prying, and pulling, thereby ensuring that fire fighters can access spaces that might contain hidden fires. Many of the tools used for overhaul are also used for ventilation and forcible entry. The following tools are frequently required in overhaul operations:

- Pike poles and ceiling hooks: for pulling ceilings and removing gypsum wallboard
- Crowbars and Halligan-type tools: for removing baseboards and window or door casings
- Axes: for chopping through wood, such as floor boards and roofing materials
- Power tools such as battery-powered saws: for opening up walls and ceilings
- Pitchforks and shovels: for removing debris

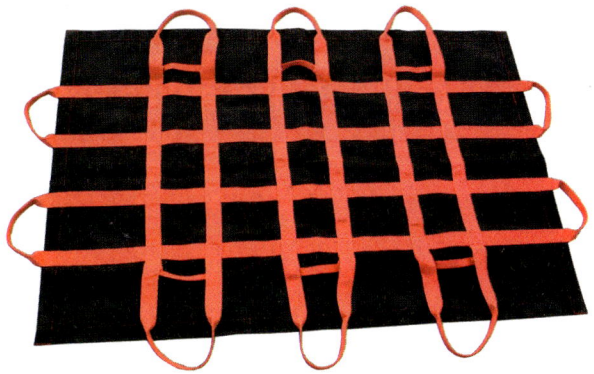

FIGURE 19-14 A carryall is used to remove debris during overhaul.
Courtesy of Cascade Fire Equipment Company.

- Rubbish hooks and rakes: for pulling things apart
- Thermal imaging devices: for identifying hot spots

Because overhaul situations usually do not require high pressures or large volumes of water, a 38- or 45-mm (1½- or 1¾-in.) hose line is usually sufficient to extinguish hot spots. Follow your department's SOPs when choosing a hose line during overhaul.

Buckets, tubs, wheelbarrows, and carryalls are used to remove debris from a building. A **carryall** is a 1.8-m-square (6-ft-square) piece of heavy canvas material, with rope handles in each corner, used to carry rubbish (**FIGURE 19-14**).

> **FIRE FIGHTER TIP**
>
> Overhaul is physically demanding work. Pace yourself, and use proper technique to increase your efficiency. Ask experienced members of your team for tips on improving your technique.

Opening Walls and Ceilings

Pike poles are used to open ceilings and walls to expose hidden fire. To pull down a ceiling with a pike pole, follow the steps in **SKILL DRILL 19-15**.

Use the pike pole to break through and pull down large sections of ceilings made with gypsum board. Pulling down laths and breaking through plaster ceilings require more force. Power saws may be required to cut through ceilings made with plywood or solid boards.

Pike poles, axes, power saws, and handsaws can be used to open a hole in a wall. Use the same technique to open a wall with a pike pole as you do to open a ceiling. When using an axe, make vertical cuts with the blade of the axe, and then pull the wallboard away from the studs by hand or with the pick end of the axe. A power saw also can be used to make vertical cuts. Pull the wall section away with another tool or by hand.

To open an interior wall with a pick-head axe, follow the steps in **SKILL DRILL 19-16**.

SKILL DRILL 19-15
Pulling a Ceiling Using a Pike Pole Fire Fighter I, NFPA 1001: 4.3.13

1 Select the appropriate length of pike pole based on the height of the ceiling. Determine which area of the ceiling will be opened. Typically, the most heavily damaged areas are opened first, followed by the surrounding areas. Position yourself to begin work with your back toward a door so the debris you pull down will not block your access to the exit.

2 Using a strong, upward-thrusting motion, penetrate the ceiling with the tip of the pike pole. Face the hook side of the tip away from you. Once the pike has penetrated the material, turn the pike pole 180 degrees; this allows for the hook, on the pike, to grab more material. This technique is particularly useful when pulling lath and plaster ceilings, as the hook will grab the slats and allow you to make a larger hole with each pull.

3 Pull down and away from your body, so the ceiling material falls away from you.

4 Continue pulling down sections of the ceiling until the desired area is opened. Pull down any insulation, such as rolled fibreglass, found in the ceiling.

SKILL DRILL 19-16
Opening an Interior Wall Fire Fighter I, NFPA 1001: 4.3.13

1 Determine which area of the wall will be opened. Open those areas most heavily damaged by the fire first, followed by the surrounding areas.

2 Use the axe blade to begin cutting near the top of the wall. Cut downward between wall studs. Be alert for electrical switches or receptacles, as they indicate the presence of electrical wires behind the wall.

3 Make two vertical cuts, using the pick end of the axe to pull the wall material away from the studs and open the wall. Work from top to bottom. Remove items such as baseboards or window and door trim with a Halligan tool or axe.

4 Continue opening additional sections of the wall until the desired area is open. Pull out any insulation, such as fibreglass, found behind the wall.

After-Action REVIEW

IN SUMMARY

- Salvage and overhaul limit and reduce property losses from a fire. Salvage efforts protect property and belongings from damage, particularly from the effects of smoke and water. Overhaul ensures that a fire is completely extinguished by finding, exposing, and suppressing any smouldering or hidden pockets of fire in an area that has been burned.

- Because many emergency incidents occur at night, lighting is required to illuminate the scene and enable safe, efficient operations. Both interior and exterior lighting are essential for fire fighters to conduct search and rescue, ventilation, fire suppression, salvage, and overhaul operations.

- All electrical equipment must be properly grounded. Electrical cords must be well insulated, without cuts or defects, and properly sized to handle the required amperage.

- Lights can be portable or mounted on apparatus. Large hand lights that project a powerful beam of light are preferred for search and rescue activities and interior fire suppression operations when fire fighters must penetrate smoke-filled areas quickly.

- Salvage efforts usually are aimed at preventing or limiting secondary losses. Salvage operations include ejecting smoke, removing heat, controlling water runoff, removing water from the building, securing a building after a fire, covering broken windows and doors, and providing temporary patches for ventilation openings in the roof to protect the structure and contents.

- Safety is a primary concern during salvage operations. Structural collapse is always a possibility because of fire damage to the building's structural components. In addition, the water used to douse the fire can add extra weight to the damaged structural components.

- The best way to prevent water damage at a fire scene is to limit the amount of water used to fight the fire. One way to do so is to shut down the sprinkler system. Sprinklers should not be shut down until ordered to do so by the IC.

- Sprinklers can be shut down by using sprinkler wedges or sprinkler stops or by shutting off the main control valve.

- Water may be removed from a structure with a salvage pump, water chutes, water catch-alls, wet vacuums, or drainage pumps.

- Salvage techniques to limit smoke and heat damage include closed doors, ventilation, salvage covers, and floor runners.

- The best way to protect an object from water and smoke damage is to remove it from the structure and place it in a safe location.

- Overhaul is the process of searching for and extinguishing any pockets of fire that remain after a fire has been brought under control. A single pocket of embers can rekindle after fire fighters leave the scene and cause even more damage and destruction than the original fire.

- Threats to health are an important consideration during overhaul:
  - Use full PPE, including SCBA, to prevent contamination from carcinogen agents and poisonous gases that are present during overhaul.
  - Use air monitoring technology to avoid toxic substances present at the fire scene.

- Many injuries can occur during overhaul. During this time, fire fighters may be physically fatigued and require rehabilitation. In addition, the structure may be compromised and visibility limited. Be aware of these hazards, and proceed with caution.

- During all phases of fire-ground operations, be alert for unusual signs or activities that might help the fire investigator during the fire investigation.

- Be sure to coordinate your activities with those of the fire investigator.

- A charged hose line must always be ready for use during overhaul operations to suppress flare-ups and explosions.

- Always evaluate the structural condition of a building before beginning overhaul. Look for the following indicators of possible structural collapse:
  - Cracked or leaning walls
  - Pitched or sagging doors or doors stuck in shifted frames
  - Moaning/groaning or cracking sounds
  - Any type of movement or vibrations
  - Movement or shifting of water on the floor
  - Smoke pushing through cracks in the wall
  - Lack of water runoff from firefighting operations
- During overhaul operations, do not compromise the structural integrity of the building.
- The area that must be overhauled depends on the building's construction, its contents, and the size of the fire. All areas directly involved in the fire must be overhauled.
- Use your senses of sight, hearing, and touch to determine where overhaul is needed.
- A thermal imaging device is helpful in locating hot spots during overhaul.

KEY TERMS

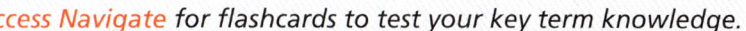

Access Navigate for flashcards to test your key term knowledge.

Area of origin A structure, part of a structure, or general geographic location within a fire scene in which the "point of origin" of a fire or explosion is reasonably believed to be located. (NFPA 901)

Carryall A piece of heavy canvas with handles, which can be used to tote debris, ash, embers, and burning materials out of a structure.

Floodlight A light that can illuminate a broad area.

Floor runner A piece of canvas or plastic material available in various lengths that is used to protect flooring from dropped debris and dirt from shoes and boots.

Junction box A device that attaches to an electrical cord to provide additional outlets.

Outside screw and yoke (OS&Y) valve A sprinkler control valve with a valve stem that moves in and out as the valve is opened or closed.

Overhaul The process of final extinguishment after the main body of a fire has been knocked down. All traces of fire must be extinguished at this time. (NFPA 402)

Post indicator valve (PIV) A sprinkler control valve with an indicator that reads either "open" or "shut" depending on its position.

Salvage A firefighting procedure for protecting property from further loss following an aircraft accident or fire. (NFPA 402)

Salvage cover A large square or rectangular sheet made of heavy canvas or plastic material that is spread over furniture and other items to protect them from water runoff and falling debris.

Secondary loss Property damage that occurs due to smoke, water, or other measures taken to extinguish the fire.

Spotlight A light designed to project a narrow, concentrated beam of light.

Sprinkler stop A mechanical device inserted between the deflector and the orifice of a sprinkler head to stop the flow of water.

Sprinkler wedge A piece of wedge-shaped wood placed between the deflector and the orifice of a sprinkler head to stop the flow of water.

Water catch-all A salvage cover that has been folded to form a container to hold water until it can be removed.

Water chute A salvage cover that has been folded to direct water flow out of a building or away from sensitive items or areas.

Wet vacuum A device similar to a wet/dry shop vacuum cleaner that can pick up liquids. It is used to remove water from buildings.

On Scene

Dusk is rapidly turning to evening. You have responded to a fire in a three-storey, wood-frame apartment building. The fire appears to have begun on the third-floor balcony and spread into the attic through the soffit. The fire destroyed four units on one end of the building, and about two-thirds of the roof is burned away.

The fire is now extinguished, and there is a tremendous amount of water on the upper floors due to the use of master stream appliances. You have been assigned to perform salvage operations. The IC has determined that it is safe to re-enter the building in order to knock down any remaining hot spots and to protect the personal property of the tenants.

1. A _____ is a piece of canvas or plastic material available in various lengths that is used to protect flooring from dropped debris and/or dirt from shoes and boots.

   A. floor runner

   B. water chute

   C. water catch-all

   D. carryall

2. A _____ is a salvage cover that has been folded to direct water flow out of a building or away from sensitive items or areas.

   A. floor runner

   B. water chute

   C. water catch-all

   D. carryall

3. What is your role related to fire investigation during salvage and overhaul?

   A. Contact the fire investigator for your jurisdiction.

   B. Attempt to preserve evidence related to the cause of a fire.

   C. Delay overhaul until the fire investigator arrives on the scene.

   D. Work to identify the point of origin of the fire.

4. Which is the first choice for stopping water from flowing through a sprinkler system that has activated?

   A. Close the OS&Y valve.

   B. Close the PIV.

   C. Use a sprinkler wedge to deactivate the sprinklers.

   D. Close the sprinkler system control valve.

5. Which of the following is a second method of turning a sprinkler system off if you cannot shut off an individual sprinkler head?

   A. Shut off the tamper switch.

   B. Close the OS&Y valve.

   C. Close the hydrant valve.

   D. Close the municipal water system valve.

6. Which of the following should you perform during overhaul operations?

   A. Clear off all floors so that when the fire investigator arrives, he or she can check for fire patterns.

   B. Open all interior walls in the structure.

   C. Remove all furniture from the room to prevent rekindling.

   D. Preserve evidence, and note fire patterns near the suspected areas of origin.

7. When is it safe to remove SCBA during overhaul operations?

   A. When the smoke has been cleared from the fire compartment

   B. After the air has been tested with a multi-gas meter

   C. It is not safe to remove SCBA while overhauling a fire scene

   D. When carbon monoxide levels are safe

8. Which of the following tools is *not* commonly used during overhaul operations?

   A. Axes

   B. Battery-powered saws

   C. Vise grips

   D. Pike poles

Access Navigate to find answers to this On Scene, along with other resources such as an audiobook.

CHAPTER 20

Fire Fighter I

Fire Fighter Rehabilitation

KNOWLEDGE OBJECTIVES

After studying this chapter, you will be able to:

- Define rehabilitation. (p. 760)
- Describe the factors and causes that require rehabilitation for fire fighters. (pp. 761–762)
- Explain how heat stress and personal protective equipment tax the fire fighter's body. (pp. 762–763)
- Describe the hazards of dehydration, and explain how dehydration can be prevented. (pp. 762–763)
- List the signs of dehydration. (pp. 763–764)
- Describe the types of extended fire incidents during which fire fighters need rehabilitation. (pp. 765–766)
- Describe the other types of incidents during which fire fighters need rehabilitation. (pp. 766–767)
- Explain why the body needs rehabilitation during severe weather conditions. (pp. 767–768)
- List the steps in rehabilitation. (p. 768)
- Describe the types of fluids that are ideal for fire fighters to drink during rehabilitation. (p. 770)
- Describe the types of food that are ideal for fire fighters to eat during rehabilitation. (pp. 771–772)
- Explain what the individual fire fighter's personal responsibilities are in rehabilitation. (p. 773)

SKILLS OBJECTIVES

There are no skills objectives for Fire Fighter I candidates. NFPA 1001 contains no Fire Fighter I Job Performance Requirements for this chapter.

ADDITIONAL STANDARDS

- **NFPA 1500**, *Standard on Fire Department Occupational Safety, Health, and Wellness Program*
- **NFPA 1561**, *Standard on Emergency Services Incident Management System and Command Safety*
- **NFPA 1584**, *Standard on the Rehabilitation Process for Members During Emergency Operations and Training Exercises*

Fire Alarm

It is shaping up to be a hot day, so you decide to get your workout in early. You have just completed 5 km (3 mi) on the treadmill when you are dispatched to a house fire in an old section of town that dates back to the late 1890s. Upon arrival, your crew is assigned to fire attack and are told that the fire is in the basement. Your crew has trouble locating the seat of the fire in the dirty brown smoke. When you finally reach it, you realize it is a large fire that is spreading quickly to the upper floors. Just as you are able to get a knockdown of the fire in the basement, you notice you are getting low on air. You notify command of your status, and you are asked to get your cylinder changed out quickly and go to the first floor to begin opening up walls.

You change out your cylinder and head back in. You begin pulling walls and notice that you are breathing much harder than usual. You begin getting mild cramps in your arms as you tire from swinging the axe. You strive to keep up with your partner. You are relieved when you notice you are getting low on air, knowing you will be sent to rehabilitation.

1. What is the purpose of rehabilitation?

2. Should an incident commander consider sending personnel to rehabilitation sooner than after two cylinders of air have been consumed?

3. Which actions could you have taken to reduce the heat stress on your body?

JONES & BARTLETT LEARNING
NAVIGATE 2   *Access Navigate for more practice activities.*

Introduction

A common saying within the fire service is "Take care of yourself first, take care of the rest of your team second, and take care of the people involved in the incident third." At first glance, this statement might seem self-centered. After all, fire fighters have a deep commitment to help others. If you were to put your personal comfort first, you probably would not want to get up in the middle of a cold winter night to answer a call. But this statement actually has a deeper meaning: You must take care of yourself physically and mentally so you can continue to help others. You must place a high priority on the health and well-being of your fellow fire fighters for the good of the fire department as a whole. Only when you and your teammates are physically fit and mentally alert will you be able to perform the tasks necessary to save lives and protect property.

To **rehabilitate** means to restore someone or something to a condition of health or to a state of useful and constructive activity. Fighting fires is an occupation that requires excellent physical conditioning to combat the rigours of heat, cold, smoke, flames, physical exertion, and emotional stress that are routinely encountered by team members. Even a seasoned, well-conditioned fire fighter can quickly become fatigued when battling a tough fire (**FIGURE 20-1**). New fire fighters are frequently amazed at the amount of energy expended

FIGURE 20-1 Firefighting often involves extreme physical exertion and results in rapid fatigue.
© Jones & Bartlett Learning.

during fire suppression activities. This exertion takes its toll on your body, resulting in heat build-up, dehydration, hunger, and fatigue. Rehabilitation provides an opportunity for fire fighters to rehydrate and rest so that they can be reassigned to operational duties as quickly and as safely as possible. Rehabilitation is a critical factor in maintaining your health and well-being. It is so critical, in fact, that it is the focus of several NFPA standards, including NFPA 1500, *Standard* on *Fire Department Occupational Safety, Health, and Wellness Program*; NFPA 1561, *Standard on Emergency Services Incident Management System and Command Safety*; and NFPA 1584, *Standard on the Rehabilitation Process for Members During Emergency Operations and Training Exercises*.

Emergency incident rehabilitation is part of the overall emergency effort. This aspect of the emergency response ensures that fire fighters who are exhausted, dehydrated, hungry, ill, injured, or emotionally upset can take a break for rest, fluids, food, medical monitoring, and treatment of illnesses or injuries (**FIGURE 20-2**). Without the opportunity to rest and recover, fire fighters may develop physical symptoms, such as fatigue, headaches, lightheadedness, gastrointestinal problems, or collapse. Rehabilitation gives them the opportunity to evaluate and address these issues.

SAFETY TIP

A tired or dehydrated fire fighter is more likely to be injured and runs the risk of collapsing. Rehabilitation is essential to correct imbalances in the body that, if left untreated, could endanger you, your crew, and others at the scene.

Rehabilitation enables fire fighters to continue to perform more safely and effectively at an emergency scene. A tired or dehydrated fire fighter is not able to accomplish as much work and is more likely to become ill or injured. The effort that is required to rescue an ill or injured fire fighter takes time and resources away from fire suppression activities. For all these reasons, rehabilitation is essential to fight fires safely and effectively.

FIGURE 20-2 Rehabilitation provides fire fighters with an opportunity to take a break for rest, cooling or rewarming, rehydration, calorie replacement, medical monitoring, and ensuring accountability.
© Jones & Bartlett Learning. Courtesy of MIEMSS.

FIRE FIGHTER TIP

The amount of rest needed to recover from physical exertion is directly related to the intensity of the work performed. Fire fighters who have expended a tremendous amount of energy will require a longer recovery period than those who have performed moderate work.

The Need for Rehabilitation

Many conditions come together during a firefighting operation to produce a stressful environment. Consider the stresses involved in a typical middle-of-the-night call for a working fire. The loud, jarring sound of the alarm jolts your sleeping body awake. Without hesitation, you must immediately get up, get dressed, and put on your personal protective equipment (PPE). You do not have time to eat or get something to drink. You may have to drive an emergency vehicle, haul hose, position a ladder, and climb to the roof to cut a ventilation hole. All of these tasks require a significant amount of energy and concentration, and you must be able to move into action quickly with no time to warm up your muscles as athletes do before an event.

You are required to respond under adverse conditions and circumstances, day or night. As emergency calls can never be anticipated, on-duty fire fighters are under constant stress. Radio messages reporting people trapped or injured often result in increased awareness, anxiety, and fear as personnel respond to a dangerous situation. Feelings of anxiety and fear can adversely affect a fire fighter's performance. It is incumbent on all personnel to recognize their triggers for fear and anxiety and address them accordingly. Fire fighters must ensure that they manage their stress properly. Rehabilitation, at the scene of a fire, provides an opportunity for fire fighters to alleviate both physical and emotional stress.

Studies have shown that proper rehabilitation is one way to prevent fire fighters from collapsing or suffering injuries during fire suppression activities and emergency operations. Taking short breaks, replacing fluids, ingesting healthy food, and cooling or rewarming are

FIRE FIGHTER TIP

The SAID principle—Specific Adaptation to Imposed Demands—identifies a need for training that mimics the type of work to be performed. In other words, the best way to prepare for physical work is to perform activities that match the type, intensity, and duration of the work. The optimal training program should include both high- and low-intensity cardiovascular training and muscular work that involves strength, endurance, and power.

all measures that reduce the risks of injury and illness (**FIGURE 20-3**). Rehabilitation also helps to improve the quality of decision making, because people who are tired tend to make poor decisions.

Personal Protective Equipment

The construction and nature of PPE add to metabolic heat build-up and exacerbate the threat of heat stress and dehydration (**FIGURE 20-4**). PPE can weigh 18 kg (40 lb) or more, and the extra weight increases the amount of energy needed to simply move around. PPE creates a protective envelope around a fire fighter that protects his or her body from the smoke, flames, heat, and steam of a fire. At the same time, however, it traps almost all body heat inside the protective envelope.

When a fire fighter is wearing PPE, perspiration soaks the inner clothing. Normally, evaporating perspiration helps cool your body so that it does not overheat, but with PPE, the same vapour barrier that keeps hot liquids and steam from getting to the body also results in limited evaporation. Evaporative cooling—such as might take place from bare skin—does not occur. Because evaporative cooling cannot occur, a fire fighter's core temperature quickly rises to dangerous levels.

Regardless of the temperature and relative humidity in the ambient air, the temperature and relative humidity inside your PPE will quickly rise to high levels. These high temperatures and high relative humidity produce heat stress. Heat stress reduces your ability to do work, and excessive amounts can lead to collapse if not treated. Because your body continues to perspire in an attempt to cool your body, you can quickly become dehydrated in addition to being overheated under conditions of heat stress.

FIGURE 20-3 Taking short breaks to rehabilitate lowers your risk of injury and illness.
© Jones & Bartlett Learning.

FIGURE 20-4 The PPE that is worn to protect a fire fighter contributes to heat stress.
© Jones & Bartlett Learning.

SAFETY TIP

The heat stress index combines temperature and relative humidity to measure the degree of heat stress. **TABLE 20-1** shows the heat stress index. In Canada, the humidex rating measures the combined effects of warm temperatures and humidity. Environment Canada uses humidex ratings to inform the general public when conditions of heat and humidity are possibly uncomfortable.

Dehydration

Dehydration is a state in which the body's fluid losses are greater than fluid intake. If left untreated, this imbalance can lead to electrolyte imbalance, shock, and death. Fighting fires is a very strenuous activity, and the large amounts of muscular energy required during firefighting activities produce a significant amount of heat. During this exertion, the body loses a substantial amount of water through perspiration. In fact, fire fighters

TABLE 20-1 Heat Stress Index

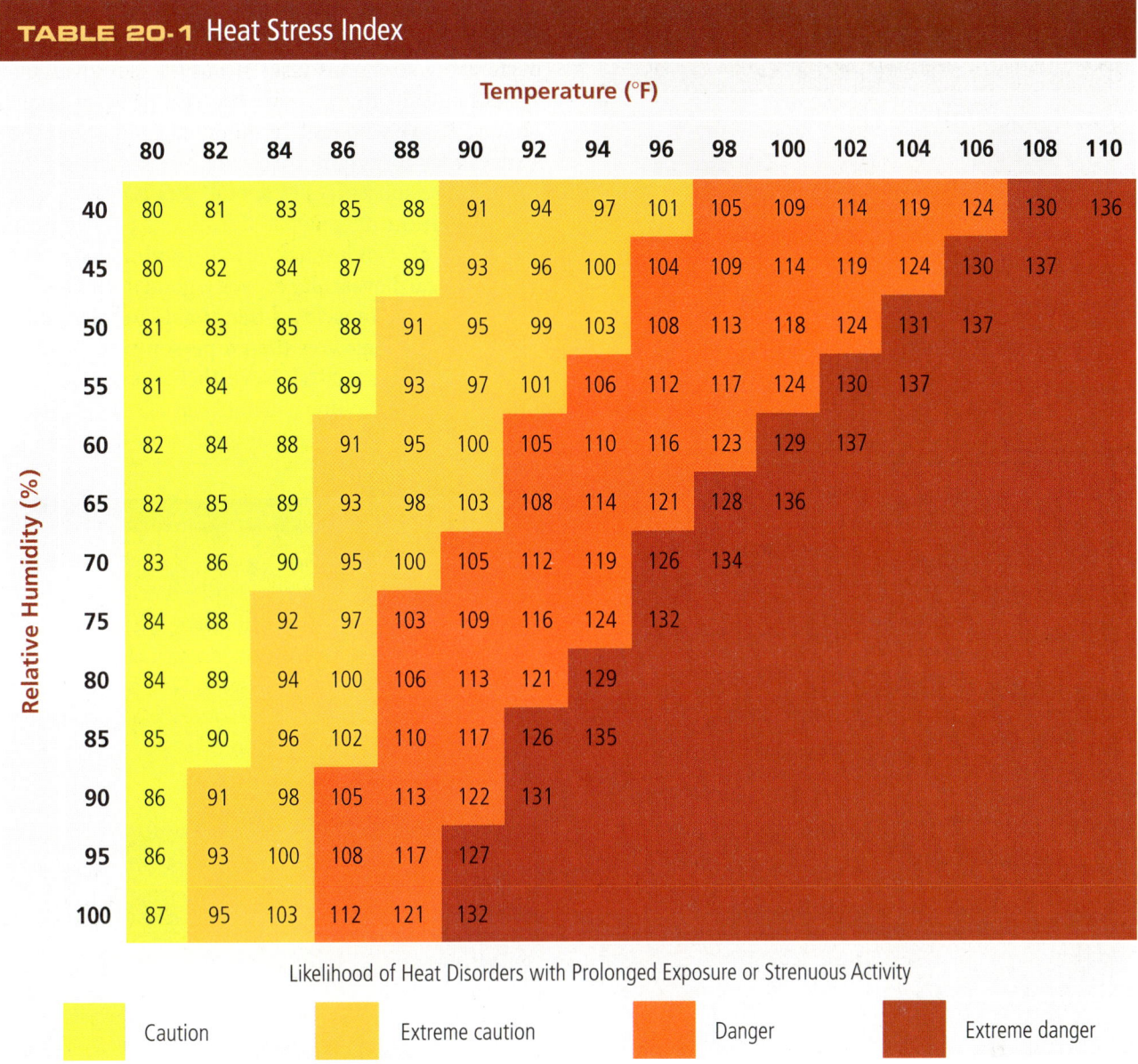

Temperature (°F)

Relative Humidity (%)	80	82	84	86	88	90	92	94	96	98	100	102	104	106	108	110
40	80	81	83	85	88	91	94	97	101	105	109	114	119	124	130	136
45	80	82	84	87	89	93	96	100	104	109	114	119	124	130	137	
50	81	83	85	88	91	95	99	103	108	113	118	124	131	137		
55	81	84	86	89	93	97	101	106	112	117	124	130	137			
60	82	84	88	91	95	100	105	110	116	123	129	137				
65	82	85	89	93	98	103	108	114	121	128	136					
70	83	86	90	95	100	105	112	119	126	134						
75	84	88	92	97	103	109	116	124	132							
80	84	89	94	100	106	113	121	129								
85	85	90	96	102	110	117	126	135								
90	86	91	98	105	113	122	131									
95	86	93	100	108	117	127										
100	87	95	103	112	121	132										

Likelihood of Heat Disorders with Prolonged Exposure or Strenuous Activity

☐ Caution ☐ Extreme caution ☐ Danger ☐ Extreme danger

National Weather Service (NWS), http://www.nws.noaa.gov/om/heat/heat_index.shtml.

in action can lose as much as 1 L (1 qt) of fluid in less than 1 hour (U.S. Fire Administration [USFA], 2008).

Dehydration reduces strength, endurance, and mental judgment, as evidenced by a variety of signs and symptoms (**TABLE 20-2**). Extreme dehydration can lead to confusion and total collapse. It is important to prevent dehydration or correct it as quickly as possible. Drink at least a litre (quart) of water per hour during periods of heavy physical exertion. It is simple to plan for and accomplish this fluid intake when you have a scheduled physical conditioning session or in-service drill. Fires are unscheduled events, however, so you must try to keep your body pre-hydrated when you are on duty. In addition, replenishing fluids during rehabilitation is essential to correct any fluid imbalance in your body. It is better to prevent dehydration by drinking adequate amounts of water before an incident than it is to try to correct severe dehydration during or after an incident.

Energy Consumption

Food provides the fuel your muscles need to work properly. In times of strenuous activity, the body burns carbohydrates and fats for energy. These energy sources need to be replenished. Without a sufficient supply of the right types of food for energy, your body cannot continue to perform at peak levels for extended periods. The number of calories you need depends on the duration of the activity, the time since your last meal, and your general physical condition. The type of calories that you consume is also important—it is important to refuel the body with a balanced diet consisting of nutritious food.

TABLE 20-2 Signs and Symptoms of Dehydration

Percentage of Body Weight (Lost)	Signs and Symptoms
1%	Increased thirst
2%	Loss of appetite Dry skin Dark urine Fatigue Dry mouth
3%	Increased heart rate
4–5%	Decreased work capacity by up to 30%
5%	Increased respiration Nausea Decreased sweating Decreased urine output Increased body temperature Markedly increased fatigue Muscle cramps Headache
10%	Muscle spasms Markedly elevated pulse rate Vomiting Dim vision Confusion Altered mental status

Source: IAFC. *A Guide for Best Practices: An Introduction to NFPA 1584 (2008 Standards)*.

SAFETY TIP

You should always sanitize your hands before you eat.

Tolerance for Stress

Every individual has a different tolerance level for the stresses encountered when fighting fires. For example, younger people tend to have greater endurance and can tolerate higher levels of stress. Likewise, a person who is well rested and in good condition will have a greater level of endurance than someone who is tired and in poor condition. In addition, carrying extra weight and performing strenuous tasks strain the cardiovascular system and greatly increase the risk of heart attack. Finally, people who take certain medications and substances may lose more fluids than those who are not taking these substances (**TABLE 20-3**).

Conditioning plays a significant role in a fire fighter's level of endurance. A well-conditioned person with good cardiovascular capacity, good flexibility, and well-developed muscles is better able to tolerate the stresses of fighting fires than a person who is out of shape (**FIGURE 20-5**). Nevertheless, even the most impressive conditioning will not keep a fire fighter from becoming exhausted under physically stressful situations.

FIRE FIGHTER TIP

Your department should engage in a variety of activities to help develop optimal stair-climbing endurance. Whether at a health club or fire station, fire fighters may participate in step aerobics classes or use stair-climbing devices (with or without gear) to develop their leg muscles and lung capacity. When exercise equipment is not available, fire fighters can use step boxes or stairwells to enhance their climbing capacity.

TABLE 20-3 Common Heat Sensitizing Medications

Drug	Brand Name	Use
Diphenhydramine	Benadryl Genahist	Allergies
Pseudoephedrine	Sudafed Dimetapp	Congestion
Timolol	Blocadren	Heart disease, hypertension
Selegiline	Eldepryl	Parkinson's disease
Amitriptyline Doxepin	Elavil Sinequan	Depression
Chlorothiazide Hydrochlorothiazide	Diuril HydroDIURIL	Hypertension, heart failure
Prochlorperazine Atropine	Compazine Donnatal	Nausea

FIGURE 20-5 A well-conditioned fire fighter will have a greater tolerance for the stresses encountered when fighting fires.
© Jones & Bartlett Learning.

When Is Rehabilitation Needed?

It is important to know when rehabilitation is needed during an emergency incident. The following guidelines are based on self-contained breathing apparatus (SCBA) use. A fire or emergency crew should perform self-rehabilitation, resting with fluid replacement, for a minimum of 10 minutes following the depletion of one 30-minute SCBA cylinder or 20 minutes of intense work without SCBA. A fire or emergency crew should rest with fluid replacement for a minimum of 20 minutes following the depletion of two 30-minute SCBA cylinders, the depletion of one 45- or 60-minute SCBA cylinder, whenever a full encapsulating suit is worn, or following 40 minutes of intense work without SCBA (USFA, 2008).

Follow the rehabilitation guidelines established by your department. After a period of rehabilitation has been completed, company officers or other crew members should assure that members are fit to return to duty. During larger incidents it may be necessary to designate a person to serve as a time recorder to assure that all crew members are rotating out to rehabilitation as often as necessary.

The concept of rehabilitation needs to be addressed at all incidents, but it will not be necessary to implement all components of a rehabilitation centre for every incident. For example, fire fighters who put out a small fire in a single room might require only water for rehydration, whereas those who are involved in extinguishing a major wildland fire may need a full-fledged rehabilitation station.

Extended Fire Incidents
Structure Fires

Large emergency incidents require full-scale rehabilitation efforts. Major structure fires that involve extended time on the scene can be hard on crews (**FIGURE 20-6**). Personnel working on the interior of a structure fire will become dehydrated and fatigued quickly because of the intense heat and stressful conditions brought on by working in full PPE. They will require rehabilitation in order to continue working at a healthy and safe level. Most major fire services require that a rehabilitation sector be established at the outset of a working fire. The responsibility for ensuring that a rehabilitation sector is established is that of the IC. Generally speaking, the IC will evaluate the need for a rehabilitation sector and assign personnel and resources accordingly. Rotating

FIGURE 20-6 Major fires often require a strong effort that lasts for an extended period of time. Rehabilitation and crew rotation are important measures that limit the risk of exhaustion and injuries to fire fighters.
© Jerry S. Mendoza/AP Images.

crews off the fire ground and bringing in fresh crews promote health and safety and help get the job done in an efficient manner.

High-Rise Fires

High-rise fires place increased stress on fire fighters at the same time that they drain crew members' energy resources. It takes considerable energy to walk up many flights of stairs and then work in PPE and SCBA. Hose packs, hand tools, and extra SCBA cylinders must be carried up staircases that may be crowded with people exiting the building. Just getting everyone and everything up to the location of the fire may require a marathon effort, even before the attack on the fire begins. Many large departments have established standard operating procedures (SOPs) for high rise fires. While it is generally accepted practice for elevators to return to a predesignated floor in the event of a fire alarm, fire personnel can override this feature in an effort to capture elevators for use in ferrying personnel and equipment safely. This reduces the fatigue associated with lugging equipment up several floors, increasing operational fire attack efficiency.

During a high-rise fire, the incident commander (IC) may assign three companies to do the work that is normally done by one company. This strategy enables the companies to rotate between attacking the fire, replenishing their air supplies, and meeting their needs for rest, fluids, and nutrition. At high-rise fires, the rehabilitation centre is often located two or three floors below the level of the fire to reduce the time and energy expended walking up and down the stairs.

FIRE FIGHTER TIP

During active operations, fire fighters may be reluctant to admit that they need a rest. Asking an obviously exhausted fire fighter if he or she needs to go to rehabilitation almost invariably brings the response, "No—I'm okay." All company members need to watch out for one another, and company officers must monitor their crews for indications of fatigue. It is better to go to rehabilitation a few minutes early than to wait too long and risk the consequences of injury or exhaustion. The IC should always plan ahead so that a fresh or rested crew is ready to rotate with a crew that needs rehabilitation.

Wildland Fires

Emergency incident rehabilitation was first used on a widespread basis during firefighting operations carried out in wildland areas such as grasslands and forests. Given the size and intensity of major wildland fires, crews need to work in shifts so that their bodies can recover from the heat, smoke, and stresses of the fire. Minimal

rehabilitation efforts may be all that are needed for a small ground fire that is easily extinguished. By contrast, a large forest fire may require hundreds of fire fighters and take several weeks to control and extinguish; such a fire will require a significant and ongoing rehabilitation effort for all crews involved (**FIGURE 20-7**).

Proper nutrition and rehydration are essential for keeping the crews healthy and ready to resume duty. Some studies have shown that wildland fire fighters may require as much as 21 to 25 J (5000 to 6000 calories) per day (Sharkey, Ruby, and Cox, 2002).

Training Exercises

Training exercises, including live-burn exercises in training towers or actual structures, simulate many of the conditions of actual firefighting. Fire fighters may be encapsulated in full PPE, including SCBA. The exercises may be conducted on days when the ambient temperature is hot or cold and may last for several hours or an entire day.

Adequate pre-hydration and rehabilitation should be incorporated in the planning for these activities (**FIGURE 20-8**). Training exercises conducted over a full day should be set up to allow the participants to go to rehabilitation between strenuous activities. Rehabilitation is no less important during training activities than it is during emergency incidents.

Other Types of Incidents Requiring Rehabilitation

Other types of incidents may require extensive rehabilitation efforts to maintain the health and safety of fire fighters. For example, hazardous materials incidents that require responders to wear a **fully encapsulated suit** (a protective suit that fully covers the responder,

FIGURE 20-7 Wildland fires often involve hundreds of fire fighters who work for days or weeks at the site.
Courtesy of Michael Rieger/FEMA News Photo.

FIGURE 20-8 The same type of rehabilitation procedures should be implemented for training activities as are used for actual emergency incidents.
© Jones & Bartlett Learning.

FIGURE 20-9 Responders wearing fully encapsulated suits must be carefully monitored for symptoms of heat stress.
© Jones & Bartlett Learning. Photographed by Glen E. Ellman.

including the breathing apparatus) are especially stressful (**FIGURE 20-9**). These kinds of incidents can expose responders to strenuous conditions for extended periods of time. In addition, they may require that the staging area be located at significant distance from the hazardous materials, requiring responders to walk for long distances while wearing the heavy PPE. For these

reasons, hazardous materials incidents require adequate rehabilitation for responders.

On other occasions, the fire department may be involved in long-duration search and rescue activities or other incidents that require the presence of public safety agencies for extended periods of time. These situations can be both mentally and physically taxing. The establishment of a rehabilitation centre where fire fighters can recuperate during these stressful incidents is essential.

The need for rehabilitation is not limited to emergency situations. Athletic events and stand-by assignments may require rehabilitation. Whenever fire fighters are required to be ready for action for an extended period of time, some provision should be made for providing replenishing fluids, nourishing foods, and warming or cooling for crew members.

Severe Weather Conditions

Weather conditions are an important consideration with regards to determining both the need for rehabilitation and the nature of rehabilitation.

The rehabilitation sector should be located in an area that it out of the public view and free of vehicle exhaust. The space must be large enough to accommodate more than one crew at a time, have adequate drinking water and food supply, be located in a shaded or air-conditioned area, and have medical personnel available for those who require attention.

Emergencies that occur when the temperature is very hot increase the need to rotate crews and allow extensive rehabilitation. Crews working on the interior fire attack may not notice much difference, because their environment during the attack is hot regardless of the external temperature. Hot weather, however, will definitely affect crews working on the outside. These fire fighters tend to become dehydrated and fatigued much faster than they would if the ambient temperature was in a more comfortable range.

Another factor that must be considered is the humidity of the air, because humidity plays an important role in evaporative cooling. High humidity reduces evaporative cooling, making it more difficult for the body to regulate its internal temperature. If the body

SAFETY TIP

PPE is designed to protect fire fighters from hazards encountered in emergency operations. Fire fighters who are overheated should remove their PPE as soon as possible to permit evaporative cooling. PPE must be removed only in a safe environment, outside of the hazard area.

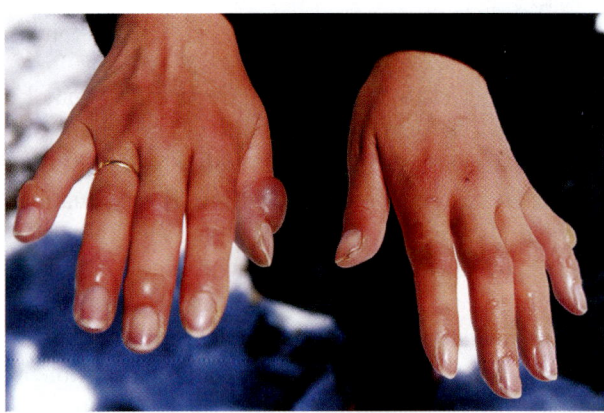

FIGURE 20-10 Prolonged exposures to freezing weather can result in severe frostbite.
Courtesy of Neil Malcom Winkelmann.

is unable to disperse heat because it is covered by PPE, serious short- and long-term medical situations, such as **heat rash**, **heat cramps**, **heat exhaustion**, and **heat stroke** could result.

Cold weather also increases the need for rehabilitation and crew rotation. **Hypothermia**, a condition in which the internal body temperature falls below 35°C (95°F), can lead to loss of coordination, muscle stiffness, coma, and death. **Frostbite** (damage to tissues resulting from prolonged exposure to cold) can also occur (**FIGURE 20-10**). Even in cold weather, the weight of PPE and the physical exertion of fighting a fire cause the body to sweat inside the protective clothing. The combination of damp clothing and cold temperatures can quickly lead to hypothermia.

FIRE FIGHTER TIP

At most fire department incidents, only one rehabilitation centre will be established. In contrast, at large-scale incidents, it may be necessary to set up multiple rehabilitation centres. In some communities, rehabilitation is initiated entirely by fire department personnel. In other communities, establishing the rehabilitation facility may be a combined effort between the Salvation Army, the Canadian Red Cross, the fire department auxiliary unit, or a separate emergency medical services (EMS) agency.

How Does Rehabilitation Work?

One way to understand how an emergency incident rehabilitation centre works is to look at the functions it is designed to perform. A widely used model of rehabilitation comprises seven parts (Bledsoe, 2009):

- Relief from climatic conditions
- Rest and recovery
- Active or passive cooling or warming

- Rehydration and calorie replacement
- Medical monitoring
- Member accountability
- Release and reassignment

Although these seven activities are not necessarily completed in the order listed, it is important that each step be addressed. In addition to these steps, a fire fighter entering a rehabilitation centre must perform a field reduction of contaminants.

Field Reduction of Contaminants

Before entering a rehabilitation centre, it is important to minimize exposure to toxins through off-gassing of PPE and skin contamination. The most effective way to perform this is by performing a field reduction of contaminants. Wash down your PPE (weather conditions permitting), and remove it. Wash or wipe down your face, neck, hands, and arms, or wash using mild soap and water. This process will vary depending on the fire fighter's proximity to the fire scene, the temperature of the rehabilitation centre, and the amount of time needed at the rehabilitation centre. PPE should be stored only in the proper PPE storage area to avoid continued exposure to possible carcinogens. Follow your department's protocols.

Relief from Climatic Conditions

Few days provide the perfect climatic conditions for rehabilitation. When the temperature is between approximately 20°C and 22°C (68°F and 72°F), the humidity is low, and plenty of shade is available, indoor rehabilitation may not be needed. Conversely, on days when the temperature is warmer or colder than 20°C to 22°C (68°F to 72°F), or when there is high humidity or wind, rehabilitation needs to take place in a climate-controlled location. In that case, a rehabilitation centre may be set up in a nearby building such as a church, school, business, or shopping centre, or rehabilitation services may be provided in a transit bus that is brought to the emergency scene.

When entering the rehabilitation centre, it is important to take off your PPE to help your body warm up or cool down as needed (**FIGURE 20-11**). If your inner clothing is wet, it may be necessary to change clothing. Many fire fighters carry an extra pair of socks and a second pair of gloves to handle such emergencies. Use a towel to dry off, if needed.

Rest and Recovery

Rest should begin as soon as the fire fighter arrives for rehabilitation. The rehabilitation centre should be located away from the central activity of the emergency, so fire fighters can disengage from all other stressful activities

FIGURE 20-11 In the rehabilitation centre, fire fighters should be able to remove personal protective clothing and get some rest before returning to action.
© Tom Carter/PhotoEdit, Inc.

and remove personal protective clothing. Rest continues as fire fighters go through each aspect of revitalization. Rehabilitation centres are equipped with chairs or cots so fire fighters can sit or lie down and relax in a comfortable position.

Passive or Active Cooling or Warming

The third step of rehabilitation is stabilizing the body's internal temperature. Interior firefighting requires wearing full PPE, including SCBA. As previously discussed, this encapsulating gear and the rigourous exercise required during firefighting generate huge amounts of body heat that cannot be dissipated by normal means. Even without the use of SCBA, performing strenuous external firefighting activities, especially during times of hot temperatures and high relative humidity, generates huge amounts of body heat. At other times, fire fighters may suffer from cold temperatures. Mounting a defensive attack on a large outdoor lumberyard in –15°C (5°F) temperatures, for example, can lead to hypothermia. Rehabilitation centres must be ready to provide means to both warm cold fire fighters and cool overheated fire fighters.

Passive and Active Cooling

For the overheated fire fighter, PPE should be removed as soon as possible to allow the body to cool. Two types of cooling are possible: passive and active.

Passive cooling is performed by removing PPE and other clothing and moving to a cooler environment. These measures allow the body to cool through effective sweating. Passive cooling works best when both the ambient temperature and the relative humidity are low, which is why stabilizing a fire fighter's body temperature through passive cooling becomes more complicated during periods of hot weather or high humidity. During these periods, rehabilitation centres need to be climate controlled so that fire fighters can achieve a normal body temperature before resuming active duty. Some fire departments have air-conditioned vehicles available for rehabilitation. Others use buses or establish rehabilitation centres in an already-cooled area such as a shopping centre.

If passive cooling is not adequate to produce a reduction in body temperature, active cooling may be required. Active cooling is performed by placing something cool on the fire fighter. This can be done by engaging a misting fan, by placing cool, wet towels around the head and neck, or by dipping both forearms in cool water. Active cooling can produce a more rapid reduction in the body temperature.

Passive and Active Warming

Fighting fires in cold temperatures can cause problems as well. As previously mentioned, fire fighters responding to incidents during cold weather are subject to both hypothermia and frostbite (**FIGURE 20-12**). In these

FIGURE 20-12 Winter weather presents a different set of problems for fire fighters than summer weather. Rehabilitation procedures must be conducted in a warm space, and fire fighters should be given hot liquids and food as well as dry clothing.
© Jack Dagley Photography/Shutterstock.

cases, the rehabilitation centre needs to be heated so that fire fighters can warm up before returning to the cold environment. Passive warming occurs when the fire fighter gets out of the cold and the body restores itself to an equilibrium temperature.

Active heating may be needed when a fire fighter's body is unable to restore itself to a normal temperature balance. Active warming is achieved by medical professionals through the use of heat packs, warming blankets, or warmed IV fluids. Fire fighters who are wet or severely chilled should be wrapped in warming blankets and moved into a well-heated area before they remove their PPE. As soon as possible, all wet clothing should be removed and replaced with warm, dry clothing.

Rehydration and Calorie Replacement

Rehydration

When the body becomes overheated, it sweats so that evaporative cooling can reduce body temperature. Perspiration is composed of water and other dissolved substances such as salt. During fire suppression activities, a fire fighter can perspire enough to lose 2 L (2 qt) of water in the time it takes to go through two cylinders of air (Bledsoe, 2009). Because 1 L if water weighs 1 kg (1 pt weighs 1 lb), losing 2 L (2 qt) of water is approximately equivalent to a 2 percent loss in body weight for a 91-kg (200-pound) fire fighter. The loss of that much body fluid can result in impaired body temperature regulation. Further dehydration can result in reduced muscular endurance, reduced strength, and heat cramps. In addition, severe cases of dehydration can contribute to heat stroke and death. Because a fire fighter can lose such a large quantity of water during fire suppression activities, it is important to replace fluids early—before severe dehydration takes place (**FIGURE 20-13**).

When the body perspires, it loses **electrolytes** (certain salts and other chemicals that are dissolved in body fluids and cells) as well as water. If a fire fighter loses large amounts of water through perspiration, his or her electrolytes must be replenished in addition to the water. Sports-type drinks are often used in rehabilitation for this purpose; they contain water, balanced electrolytes, and some sugars.

Rehydrating the body is not as simple as drinking lots of fluids quickly. Drinks such as colas, coffee, and tea should be avoided because they contain caffeine; caffeine acts as a diuretic that causes the body to excrete more water. Sugar-rich carbonated beverages are not tolerated or absorbed as well as straight water or sports drinks because too much sugar is difficult to digest and

FIGURE 20-13 The rehabilitation centre should have plenty of fluids available to rehydrate fire fighters. Plain water or diluted sports drinks are preferred.
© Jones & Bartlett Learning.

causes swings in the body's energy level. In addition, drinks that are too cold or too hot may be difficult to consume and may prevent the fire fighter from ingesting enough liquids.

An additional concern is the rate at which fluids are absorbed from the stomach. Drinking too much too quickly can cause bloating, a condition in which air fills the stomach. It causes a feeling of fullness and can lead to discomfort, nausea, and even vomiting.

Studies have demonstrated that the stomach can absorb approximately 1 L (1 qt) of fluid per hour, but, as previously noted, the body can also lose up to 1 L (1 qt) of fluid per hour during intense activity. Thus the body may lose fluids as rapidly as they can be replaced. Once the initial 1 L (1 qt) of water is lost, the body will require 1 to 2 hours to recover.

One key to remaining hydrated while fighting fires is to make sure you are hydrated before you reach the fire ground. To do so, try to drink enough fluids to keep your body properly hydrated whenever you are on duty, particularly during hot weather. This habit will decrease your risk of dehydration and the need to play catch-up with fluids when an emergency incident occurs. Adequate hydration also ensures peak physical and mental performance. More information on hydration is presented in Chapter 2, *Fire Fighter Health and Safety*.

Calorie Replacement

A fire fighter performs intense physical work and burns many calories during active firefighting, and it is necessary to replace those calories during rehabilitation. In much the same way that an engine runs on diesel fuel, the human body runs on **glucose**. Glucose (also known as blood sugar) is carried throughout the body by the bloodstream and is needed to burn fat efficiently and release energy.

For the body to work properly, its glucose levels need to be in balance. If blood sugar levels drop too low, the body becomes weak and shaky. If blood sugar levels are too high, the body becomes sluggish. Blood sugar levels can be balanced by eating a proper diet of carbohydrates, proteins, and fats.

Carbohydrates—a major source of fuel for the body—can be found in grains, vegetables, and fruits. The body converts carbohydrates into glucose, so "carbs" are an excellent energy source. A common dietary myth is that carbohydrates are fattening and should be avoided. In fact, carbohydrates are needed for a balanced diet. Carbohydrates have the same number of calories per gram as proteins and fewer calories than fat. Additionally, carbohydrates are the only fuel that the body can readily use during high-intensity physical activities such as fighting fires.

Proteins perform many vital functions within the body. Most protein comes from meats and dairy products, but smaller amounts are found in grains, nuts, legumes, and vegetables. The body uses proteins (amino acids) to grow and repair tissues; proteins are used as a primary fuel source only in extreme conditions, such as during starvation. Like other nutrients, excess proteins are converted and stored in the body as fat.

Fats are also essential for life. These nutrients are used as a source of energy, for insulating and protecting organs, and for breaking down certain vitamins. Some fats, such as those found in fish and nuts, are healthier and more beneficial to the body than others, such as the fats found in margarine. However, excess fat consumption—particularly of saturated fats (which come mostly from animal products)—is linked to high cholesterol, high blood pressure, and cardiovascular disease.

Both candy and soft drinks contain sugar (unless they are specifically labelled as "sugar-free"), so the body can quickly absorb and convert these foods to fuel. However, simple sugars also stimulate the production of insulin, which reduces blood glucose levels. As a consequence, eating a lot of sugar can actually result in lower energy levels.

For fire fighters to sustain peak performance levels, it is necessary for them to "refuel" during rehabilitation. During short-duration incidents, low-sugar, high-protein sports bars can be used to keep the glucose balance steady. During extended-duration incidents, however, fire fighters should eat a more complete meal (**FIGURE 20-14**). The proper balance of carbohydrates, proteins, and fats will maintain energy levels throughout the emergency. To ensure peak performance, the meal should include complex carbohydrates such as whole-grain breads, whole-grain pasta, rice, and vegetables. It is also better to eat a series of smaller meals over time, rather than one or two large meals, because larger meals can increase glucose levels and slow down the body.

Healthy, balanced eating is the recommended lifestyle for all fire fighters. Proper nutrition reduces stress, improves health, and provides more energy. It is just as

FIGURE 20-14 Plans for rehabilitation at extended-duration incidents should include complete, balanced meals.
© Photodisc/Creatas.

important to keep blood sugar levels balanced throughout the day as it is to remain hydrated. If you follow this regimen, you will be ready to react to an emergency at any time. In Canada, the federal government has created the Canada Food Guide. It is suggested that you review the Canada Food Guide in order to understand what is recommended for a balanced diet. The Canada Food Guide is regularly reviewed and amended to ensure that it remains current.

> **FIRE FIGHTER TIP**
>
> During rehabilitation, it is important to eat foods that contain carbohydrates. Foods such as energy bars or meal replacement bars are readily available and can be stored for extended periods of time, as opposed to perishable foods such as fruit. Fruit, however, provides carbohydrates and contains a lot of water, which helps maintain a good degree of hydration.

Medical Monitoring

Medical monitoring during rehabilitation is important to identify potentially dangerous health problems. Checking vital signs is one way to identify potential problems. Many rehabilitation protocols specify the following types of medical monitoring:

- Checking the fire fighter's heart rate, respiratory rate, blood pressure, and temperature
- Determining how much oxygen is in the fire fighter's bloodstream by taking a pulse oximetry reading
- Performing a carbon monoxide assessment

If the fire fighter's vital signs or test results are abnormal upon entering the rehabilitation centre, the fire fighter should be reexamined before being permitted to return to firefighting.

In addition, signs of illness or injury should be assessed in the rehabilitation centre before the fire fighter resumes active duty. As a fire fighter, you should be alert for any of the following signs and symptoms:

- Chest pain
- Dizziness
- Shortness of breath
- Weakness
- Headache
- Cramps
- Mental status changes
- Behavioural changes
- Changes in speech
- Changes in gait
- Abnormal body temperature

Report these signs to medical personnel at the rehabilitation centre if you notice any of them in yourself or in one of your crew members.

> **FIRE FIGHTER TIP**
>
> All fire fighters entering rehabilitation should be quickly checked by medical personnel before returning to active firefighting.

Transportation to a Hospital

Another function of the rehabilitation centre is to arrange for transportation of ill or injured fire fighters. An ambulance should be available to ensure that ill or injured fire fighters receive prompt transport.

> **FIRE FIGHTER TIP**
>
> Many fire departments have found that assigning a fire or EMS officer to be in charge of the rehabilitation centre ensures that fire fighters stay in rehabilitation as long as needed and that fire fighters who need to be transported to a hospital for further evaluation are transported rapidly.

Member Accountability

Crew integrity should be maintained during rehabilitation. Crew members should be released to rehabilitation together and should be released from rehabilitation together. This practice makes accountability on the fire ground much easier for the IC to control. It also ensures that all members of the crew report to rehabilitation to receive the rest, cooling or warming, hydration, food, and medical monitoring that they require. In the event that one member of a crew is not able to return to the active fire scene, the IC must determine how to handle the situation.

Release and Reassignment

Once they are rested, rehydrated, refuelled, and rechecked to make certain that they are fit for duty, fire fighters can be released from rehabilitation and reassigned to active duty. Release and reassignment are the

> **FIRE FIGHTER TIP**
>
> Depending on their levels of exertion, the type of work that they have performed, and how they feel, fire fighters entering rehabilitation should be quickly checked by medical personnel before returning to active firefighting. It is important to note that Canadian law prohibits forced medical procedures except under exceptional circumstances, such as diminished mental ability to understand the consequences and specific circumstances involving minors. Therefore, one cannot be forced to submit to medical evaluation in a rehabilitation unit.

last step in the rehabilitation process. Reassignments may be to the same job performed before the trip to rehabilitation or to a different task, depending on the decision of the IC.

Personal Responsibility in Rehabilitation

The goal of every firefighting team is to save lives first and property second. To achieve this goal, you need to take care of yourself first, take care of the rest of your crew second, and then take care of the people involved in the incident. Another way of looking at this issue is to remember that safety begins with you. The other members of your crew depend on you to bear your share of the load. To meet their expectations, you need to maintain your body in peak condition.

Part of your responsibility is to know your own limits. No one else can know what you ate, whether you are lightheaded or dehydrated, whether you are feeling ill, or whether you need a breather. You are the only person who knows these things. Therefore, you may be the only person who knows when you need to request rehabilitation. It may be difficult to say, "I need a break," while the rest of the crew is still hard at work. Even so, it is better to take a break when you need it than to push yourself too far and have to be rescued by other members of your crew.

Regular rehabilitation enables fire fighters to accomplish more work during a major incident, and it decreases the risk of stress-related injuries, illness, and death. Remember—safety begins and ends with you.

SAFETY TIP

Stress management skills are crucial for high-quality performance, regardless of the type of stressor involved and the role that the person plays in emergency assistance. A fire fighter needs to implement basic stress management skills to deliver high-quality performance. Exercise, proper nutrition, and emotional support are all helpful in managing stress.

FIRE FIGHTER TIP

You must take care of yourself first if you are to be able to help others.

After-Action REVIEW

IN SUMMARY

- Rehabilitation is an important part of firefighting. It is designed to ensure that emergency personnel can rest, cool off or warm up, receive fluids and nourishment, and be evaluated for potential medical problems.

- Conditions such as stress, physical exertion, and weather extremes take a toll on the body.

- Rehabilitation is essential to correct physical imbalances, such as overheating and dehydration, that could endanger members of the firefighting crew.

- Due to their protective qualities and weight, PPE and SCBA add heat stress on the body.

- Dehydration is a state in which the body's fluid losses are greater than fluid intake. If left untreated, it can lead to electrolyte imbalance, shock, and even death. To avoid dehydration, frequently drink small amounts of nonsugary and noncaffeinated beverages throughout the shift.

- To work at peak capacity, the body needs to be fuelled with the right amount of nutrient-rich foods. The proper balance of carbohydrates, proteins, and fats will maintain energy levels throughout an emergency.

- Rehabilitation provides fire fighters with periods of rest and time to recover from the physical fatigue and mental stresses of fighting fires. Taking short breaks, replacing fluids, ingesting healthy food, and cooling or rewarming are all measures that can reduce the risk of injury and illness. They also improve mental focus.

- Rehabilitation needs to be addressed at all incidents, but establishing a rehabilitation centre may not be necessary for every incident. Types of incidents that require a dedicated rehabilitation centre include structure fires, high-rise fires, and wildland fires. Rehabilitation is also needed for hazardous materials incidents, search and rescue operations, and training exercises.

- Emergencies that occur during severe weather conditions require a rehabilitation centre for cooling or rewarming.

- Rehabilitation includes seven components:
  - Relief from climatic conditions
  - Rest and recovery
  - Active or passive cooling or warming
  - Rehydration and calorie replacement
  - Medical monitoring
  - Member accountability
  - Release and reassignment
- Before entering a rehabilitation centre, you must minimize exposure to carcinogens and other contaminants by performing a field reduction of contaminants.
- Report the following signs to medical personnel at the rehabilitation centre if you notice any of them in yourself or in one of your crew members:
  - Chest pain
  - Dizziness
  - Shortness of breath
  - Weakness
  - Headache
  - Cramps
  - Mental status changes
  - Behavioural changes
  - Changes in speech
  - Changes in gait
  - Abnormal body temperature
- Your responsibilities in rehabilitation are to know your limits, to listen to your body, and to use the rehabilitation facilities when needed.

KEY TERMS

Access Navigate for flashcards to test your key term knowledge.

Dehydration A state in which the body's fluid losses are greater than fluid intake. If left untreated, dehydration may lead to shock and even death.

Electrolytes Certain salts and other chemicals that are dissolved in body fluids and cells. Proper levels of electrolytes need to be maintained for good health and strength.

Emergency incident rehabilitation A function on the emergency scene that cares for the well-being of the fire fighters. It includes relief from climatic conditions, rest, cooling or warming, rehydration, calorie replacement, medical monitoring, member accountability, and release.

Frostbite A localized condition that occurs when the layers of the skin and deeper tissue freeze. (NFPA 704)

Fully encapsulated suit A protective suit that completely covers the fire fighter, including the breathing apparatus, and does not let any vapour or fluids enter the suit. It is commonly used in hazardous materials emergencies.

Glucose The source of energy for the body. One of the basic sugars, it is the body's primary fuel, along with oxygen.

Heat cramps Painful muscle spasms that occur suddenly during or after physical exertion; usually involve muscles in the leg or abdomen.

Heat exhaustion A mild form of shock that occurs when the circulatory system begins to fail because of the body's inadequate effort to give off excessive heat.

Heat rash Itchy rash on skin that is wet from sweating; seen after prolonged sweating.

Heat stroke A severe, sometimes fatal condition resulting from the failure of the body's temperature-regulating capacity. Reduction or cessation of sweating is an early symptom; body temperature of 40.5°C (105°F) or higher, rapid pulse, hot and dry skin, headache, confusion, unconsciousness, and convulsions may occur as well.

Hypothermia A condition in which the internal body temperature falls below 35°C (95°F), usually a result of prolonged exposure to cold or freezing temperatures.

Rehabilitate To restore someone or something to a condition of health or to a state of useful and constructive activity.

REFERENCES

Bledsoe, Bryan E. *Rehabilitation and Medical Monitoring: A Guide for Best Practices, an Introduction to NFPA 1584, 2008 Standards.* Midlothian, TX: Cielo Azul Publications, 2009.

National Fire Protection Association. NFPA 1500, *Standard on Fire Department Occupational Safety, Health, and Wellness Program,* 2018. www.nfpa.org. Accessed November 13, 2018.

National Fire Protection Association. NFPA 1561, *Standard on Emergency Services Incident Management System and Command Safety,* 2014. www.nfpa.org. Accessed November 13, 2018.

National Fire Protection Association. NFPA 1584, *Standard on the Rehabilitation Process for Members During Emergency Operations and Training Exercises,* 2015. www.nfpa.org. Accessed November 13, 2018.

Sharkey, Brian, Brent Ruby, and Carla Cox. "Feeding the Wildland Firefighter." 2002. https://www.fs.fed.us/t-d/pubs/htmlpubs/htm02512323/. Accessed November 13, 2018.

U.S. Fire Administration (USFA). "Emergency Incident Rehabilitation." 2008. https://www.usfa.fema.gov/downloads/pdf/publications/fa_314.pdf. Accessed November 13, 2018.

On Scene

You arrive at your station for your tour of duty but are told by your captain that you need to report to Station 8 and will be assigned to the rehabilitation unit for the shift. As you drive to Station 8, you dread the thought of being assigned to rehab. Then you wonder, "Do if I remember the steps necessary to set up rehab?" For the remainder of your drive, you have a nagging feeling that you are not prepared for this assignment. When you finally arrive, you pull the captain aside and express your concerns. He begins to review the basics of your rehabilitation plan by asking you the following questions.

1. Which of the following describes a state in which the body's fluid losses are greater than fluid intake?

   A. Homeostasis

   B. Dehydration

   C. Hypothermia

   D. Hyperthermia

2. Which of the following fluids are recommended for use during rehabilitation?

   A. Water, coffee, and sports drinks

   B. Carbonated cola drinks and water

   C. Sports drinks and water

   D. Coffee and sports drinks

3. Where should the rehabilitation centre be located?

   A. Near the SCBA bottle changing station

   B. Away from the central activity of the emergency

   C. Close to the ancillary activities of the emergency

   D. Close to the central activity of the emergency

4. Which of the following are recommended parts of the protocol for medical monitoring during rehabilitation?

   A. Vital signs, EKG, and pulse oximetry

   B. Glucometry, carbon monoxide assessment, and vital signs

   C. Pulse oximetry, EKG, and carbon monoxide assessment

   D. Carbon monoxide assessment, pulse oximetry, and vital signs

(continued)

On Scene Continued

5. Which of the following is *not* usually one of the seven parts of rehabilitation?

  A. Active or passive heating or cooling

  B. Member accountability

  C. Rest and recovery

  D. PPE replacement

6. Removing PPE and other clothing and moving to a cooler environment are considered:

  A. active cooling.

  B. passive cooling.

  C. active warming.

  D. passive warming.

7. How long should a fire or emergency crew perform self-rehab and rest with fluid replacement following the depletion of one 30-minute SCBA cylinder or 20 minutes of intense work without SCBA?

  A. A maximum of 10 minutes

  B. A minimum of 10 minutes

  C. A minimum of 30 minutes

  D. The same amount of time as spent in the fire

8. What is the last step in the rehabilitation process?

  A. Rest and recovery

  B. Active cooling or warming

  C. Release and reassignment

  D. Rehydration

Access Navigate to find answers to this On Scene, along with other resources such as an audiobook.

CHAPTER 21

Fire Fighter I

Wildland and Ground Cover Fires

KNOWLEDGE OBJECTIVES

After studying this chapter, you will be able to:

- Define the terms *wildland fires* and *ground cover fires*. (NFPA 1001: 4.3.19, p. 778)
- Explain how the three elements of the wildland fire triangle affect each side of the fire triangle. (pp. 779–781)
- Describe subsurface fuels, surface fuels, and aerial fuels. (NFPA 1001: 4.3.19, p. 779)
- Explain the relationship between a fuel's properties and the speed at which the fuel ignites and the ensuing fire spreads. (NFPA 1001: 4.3.19, pp. 779–780)
- Describe how weather conditions and topography influence the growth of wildland fires. (NFPA 1001: 4.3.19, pp. 781–782)
- Label the parts of a wildland fire. (NFPA 1001: 4.3.19, p. 782)
- Describe the methods and tools used to cool a fuel with water. (NFPA 1001: 4.3.19, pp. 783–784)
- Describe the methods and tools used to remove a fuel from wildland fires. (NFPA 1001: 4.3.19, pp. 784–785)
- Describe the methods and tools used to smother wildland fires. (NFPA 1001: 4.3.19, p. 785)
- Itemize the characteristics of the fire apparatus used to suppress wildland fires. (NFPA 1001: 4.3.19, pp. 785–787)
- Describe how a direct attack is mounted on wildland fires. (NFPA 1001: 4.3.19, pp. 788–789)
- Describe how an indirect attack is mounted on wildland fires. (NFPA 1001: 4.3.19, p. 789)
- Describe how a parallel attack is mounted on wildland fires. (NFPA 1001: 4.3.19, p. 789)
- Describe how the ten standard firefighting orders can be used to prevent future fire fighter tragedies. (p. 790)
- Describe how the eighteen watch out situations can be used to determine if an assignment is safe. (p. 790)
- Describe the parts of the LCES mnemonic. (pp. 790–791)
- Describe the hazards associated with wildland and ground cover firefighting. (NFPA 1001: 4.3.19, pp. 791–792)
- Describe the personal protective clothing and equipment needed for wildland firefighting. (NFPA 1001: 4.3.19, pp. 792–796)
- Explain the problems created by the wildland/urban interface. (pp. 796–797)

SKILLS OBJECTIVES

After studying this chapter, you will be able to perform the following skills:

- Suppress a ground cover fire. (NFPA 1001: 4.3.19, p. 786)
- Deploy a fire shelter. (NFPA 1001: 4.3.19, pp. 793–795)

ADDITIONAL STANDARDS

- NFPA 1051, *Standard for Wildland Firefighting Personnel Professional Qualifications*
- NFPA 1901, *Standard for Automotive Fire Apparatus*
- NFPA 1906, *Standard for Wildland Fire Apparatus*
- NFPA 1977, *Standard on Protective Clothing and Equipment for Wildland Fire Fighting*

Fire Alarm

You have been watching the national news with awe as you see homes with only their foundations remaining as another major wildland fire sweeps through the formerly pristine forest. Lately it seems that these types of fires have become a regular nightly news item. You think about your own community and whether it is possible that you would ever have to face a wildland fire because you are part of a large municipal fire department that is not surrounded by forests.

1. How is wildland firefighting similar to structural firefighting?

2. Are urban/suburban fire fighters likely to face wildland fires? Why or why not?

3. How are wildland vehicles and equipment similar to the equipment used in structural firefighting?

 *Access Navigate for more practice activities.*

Introduction

Wildland is land in an uncultivated natural state that is covered by timber, woodland, brush, or grass. Most fire departments are responsible for at least some areas that can be classified as wildland. For example, most large cities have parklands that are covered with natural vegetation. In some cases, fire has occurred in these large urban forested areas. **Wildland fires** are defined by the National Fire Protection Association (NFPA) as unplanned fires burning in vegetative fuel (grass, leaves, crop fields, and trees). Some fire departments respond more often to wildland fires than to structural fires.

Wildland fires can consume grasslands, brush, and trees of all sizes. These fires burn vegetation on a variety of terrain types and threaten not only vegetative fuels but also buildings and other improvements that have been made on the land. The term **wildland/urban interface** is used to describe the area where wildland meets with human-made structures. The wildland/urban interface and its associated challenges will be discussed later in the chapter.

The incidence of wildland fires varies from season to season and from one year to the next. Wildland fires occur naturally and are not always caused by human activity. Fire has played an important ecological role in natural evolution of forests and certain tree species. In Canada, forest management is primarily a provincial and or territorial responsibility. Forests under federal ownership account for a small portion of Canada's forests (Natural Resources Canada, 2017). The threat and intensity of wildland fires are closely tied to the moisture content and the combustibility of the vegetative fuels, as well as to weather conditions and topography. Although many calls for wildland fires involve small incidents, some calls will escalate into major incidents.

Wildland fires are sometimes referred to by different terminology, including brush fires, forest fires, grass fires, ground cover fires, ground fires, natural cover fires, and wildfires. While these names provide descriptions of specific types of wildland fires, all are properly classified as wildland fires.

Ground cover fires are a type of wildland fire that burns loose debris on the surface of the ground. This debris includes vegetation such as grass, as well as dead leaves, needles, and branches that have fallen from shrubs and trees. In some fire departments, ground cover fires are the most common type of wildland fire encountered.

Most fire fighters will be called on to extinguish smaller wildland and ground cover fires at some point in their careers. The information in this chapter is intended to teach you how to extinguish wildland and ground cover fires safely. It will not train you to be a wildland fire fighter, however. Fire fighters whose primary responsibility is suppressing wildland fires must complete separate and comprehensive wildland firefighting training.

Large wildland fires are handled by agencies that specialize in this type of firefighting. Each province/territory has an agency designated to coordinate wildland firefighting. Federal government agencies such as the U.S. Forest Service, Bureau of Land Management, National Wildfire Coordinating Group, Bureau of Indian Affairs, National Park Service, and U.S. Fish and Wildlife Service are also responsible for coordinating firefighting activities at large incidents and incidents that occur on federal lands. Due to the size and nature of wildland fires, a great deal of organization is required. Initiation of the incident command system (ICS) is crucial immediately upon the arrival of the first fire personnel. See Chapter 22, *Establishing and Transferring Command*, for more information on working within the ICS during a wildland fire.

Wildland and Ground Cover Fires and the Fire Triangle

To safely and effectively extinguish wildland and ground cover fires, fire fighters need to understand the various factors that cause fire ignition and affect the growth and spread of wildland fires. The fire triangle provides a good model for explaining the behaviour of wildland and ground cover fires (**FIGURE 21-1**).

The fire triangle, as discussed in Chapter 5, *Fire Behaviour*, consists of three elements: fuel, oxygen, and heat. Wildland and ground cover fires require the same three elements to start and propagate as any other type of fire. However, unlike structure fires, wildland fires can burn for days, weeks, and months, depending on the fuel source, wind, topography, and weather. Uncontrolled wildland fires can spread rapidly, burning kilometres of forest and ground cover in a very short period of time.

Fuel

The first side of the fire triangle is fuel. The primary fuel for wildland and ground cover fires is the vegetation (grasslands, brush, and trees) in the area. The amount of fuel in an area may range from sparse grass to heavy underbrush and large trees. Some fuels ignite readily and burn rapidly when dry, whereas others are more difficult to ignite and burn more slowly.

Vegetative fuels can be classified according to their location. Subsurface fuels are those fuels located under the ground. Roots, moss, duff, and decomposed stumps are examples of subsurface fuels. Fires involving subsurface fuels are difficult to locate and challenging to extinguish.

Surface fuels, which are sometimes called ground fuels, are located close to the surface of the ground. They include grass, fallen leaves, twigs, needles, small trees, and slash (the pieces of logs, branches, bark, stumps, and other vegetative debris left over from logging and land-clearing operations). Brush less than 2 m (6 ft) above the ground is also classified as a surface fuel. Surface fuels are involved in ground cover fires.

Aerial fuels (also called canopy fuels) are located more than 2 m (6 ft) above the ground. They usually consist of trees, including tree limbs, leaves and needles on limbs, and moss attached to the tree limbs.

Fuel Characteristics

The characteristics of wildland fuels may determine how quickly the fuel ignites, how rapidly it burns, and how readily it spreads to other areas. These properties include the fuel's size and shape, its compactness, its continuity, its volume, its moisture level, and its orientation.

The size and shape of a fuel influence how it burns (**TABLE 21-1**). For example, fine fuels have a large surface area relative to their volume, which causes them to ignite easily, burn quickly, and produce a lot of heat. Fine fuels include dried vegetation such as twigs, leaves, needles, grass, moss, and light brush. Ground duff—the partly decomposed organic material on a forest floor—is another type of fine fuel. Such fuels are usually the main type of fuel present in ground cover fires. They facilitate the ignition of heavier fuels and cause fire to burn and spread with great intensity. In addition, they burn rapidly and unpredictably. As a consequence of these characteristics, fires spread more quickly in fine fuels than in heavy timber and brush.

Heavy fuels are of a larger diameter than fine fuels. They include slash, large brush, heavy timber, stumps, branches, and dead timber on the ground. Although fires involving heavy fuels do not spread as rapidly as fires involving fine fuels do, heavy fuels can burn with a high intensity.

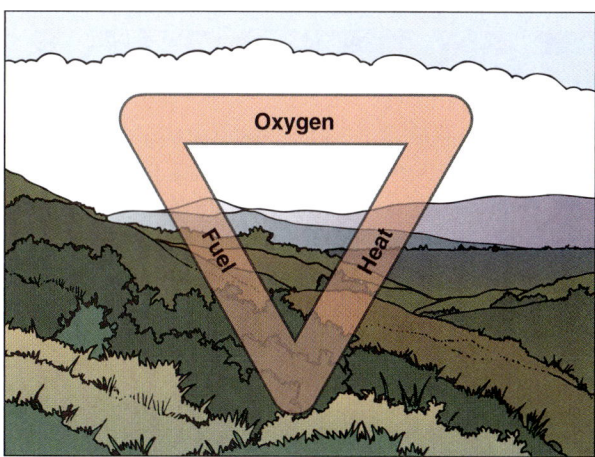

FIGURE 21-1 The wildland fire triangle.
© Jones & Bartlett Learning.

SAFETY TIP

Heavy fuels such as trees and stumps require close attention. Spraying water on the outside of these materials will not extinguish the fire, because heavy fuels have a tendency to continue to burn on the inside. Heavy fuels need to be cut out to confirm that fire fighters have extinguished all of the fire and to prevent rekindling.

TABLE 21-1 Fuel Classification

	Examples	Fuel Size	Fuel Moisture Content	Fuel Burn Time
Fine fuels	Grass, leaves, tree moss, and loose surface litter	6 mm (¼ in.) in diameter or less	Absorb or lose moisture very rapidly; can change quickly from no flammability to high flammability	1 hour
Light fuels	Small twigs and stems, both living and dead	6–25 mm (¼–1 in.) in diameter	Dry out very quickly and can become highly flammable	10 hours
Medium fuels	Sticks, branch wood, and medium duff	25–76 mm (1–3 in.) in diameter	Gain and lose moisture relatively slowly and cumulatively; add to wildfire intensity	100 hours
Heavy fuels	Large brush, heavy timber, stumps, branches, dead timber on the ground, and slash	76 mm (3 in.) or more in diameter	Change slowly and cumulatively over long periods of time	1000 hours

© Jones & Bartlett Learning.

Fuel compactness influences the rate at which a wildland fire burns. When fuels are tightly compressed, adequate oxygen and heat cannot get to large surfaces of the fuel in a short period of time. Therefore, compact fuels often burn more slowly and over a longer period of time. For example, a compact fuel such as a large tree trunk burns more slowly than a less compact fuel such as dry grass.

Fuel continuity refers to wildland fuels that have uninterrupted connections. For example, intertwined tree branches provide for fuel continuity. Fuels with continuity touch one another or are located in such close proximity that fire can spread from one source to the next rapidly. Fire spread may occur laterally, or vertically, such as between **ladder fuels** that carry fire from the surface of the ground into the tops of shrubs and trees. In situations where there is no fuel continuity, a fire may not extend past the initial area of origin.

Fuel volume refers to the quantity of fuel available in a specific area. The amount of fuel in a given area influences the growth and intensity of the fire.

Fuel moisture refers to the amount of moisture contained in a fuel. This characteristic influences the ignition temperature, speed of ignition, the rate of spread, and the intensity of the fire. Fuels with high moisture content will not ignite and burn as readily as fuels with low moisture content. Some types of vegetation naturally contain more moisture than others. Fuel moisture content is affected by changes in weather and season effects such as rainfall and relative humidity. One of the driest points of the year is often after the spring runoff,

when trees and foliage are still dry and new growth is just beginning. The forest floor may still be tinder dry if spring rains have not occurred. In a year with decreased rainfall, less moisture will be present in vegetative fuels compared with the same vegetation during a rainy year. The capacity of fuels to vary in fuel moisture content depends on the type and condition of the fuels.

Fuel orientation refers to the position of the fuel relative to the ground. Fuel that is parallel to the ground will have different burning characteristics than fuel that is in a more upright or vertical position. Because the heat from burning fuel rises, fuel that is oriented in a vertical position will catch fire and burn more readily than fuel that is in a different configuration.

SAFETY TIP

All fires involving fine or heavy fuels should be treated with the same caution and safety. Many fire fighters have died because they underestimated fire behaviour. Treat all wildland and ground cover fires with respect.

Oxygen

The second side of the fire triangle is oxygen. Oxygen is needed to initiate and support the process of combustion. In a structure fire, a limited supply of oxygen results in a slow-burning fire. If additional oxygen becomes available to a fire, it will begin to burn more rapidly.

In wildland and ground cover fires, which are usually free burning, oxygen is rarely an important variable in the ignition or spread of the fire. Unlike in structure fires, oxygen is available in unlimited quantities in these open-air fires. Nevertheless, air movement around wildland and ground cover fires will influence the speed with which the fire moves. Wind blowing on a wildland and ground cover fire brings more oxygen to the fire, accelerates the process of combustion, and influences the direction in which the fire travels.

Heat

The third side of the fire triangle is heat. Sufficient heat must be applied to fuel in the presence of adequate oxygen to produce a fire. Wildland and ground cover fires may be ignited by natural causes, accidental causes, and intentional causes.

Lightning is the source of almost all naturally caused wildland and ground cover fires. Some storms produce dry lightning, in which lightning strikes occur without rain. A single lightning storm can produce many lightning strikes and start many fires.

A variety of accidental causes may result in wildland and ground cover fires. These causes may include discarded smoking materials, improperly extinguished campfires, and downed electrical wires.

Finally, some wildland and ground cover fires are the result of human activity. These activities may result in unintentional or intentional fires. Carelessness or intentional fires can result in criminal charges such as arson. Arson and the motives behind this crime are discussed in more detail in Chapter 28, *Fire Origin and Cause.*

Other Factors That Affect Wildland Fires

Once a wildland and ground cover fire has started, the growth of the fire is influenced by weather factors and the topography of the land involved in the fire. Weather conditions and topography have a greater effect on wildland and ground cover fires than they do on structure fires.

Weather

Weather conditions have a major impact on the course of a wildland fire. The two most critical aspects of weather that influence a wildland fire are moisture and wind.

Moisture

Moisture can be present either in the form of relative humidity or precipitation. **Relative humidity** is the ratio of the amount of water vapour present in the air

compared to the maximum amount of water vapour that the air can hold at a given temperature. Warm air has a higher capacity for moisture content than cool air does.

Relative humidity is expressed as a percentage. If the air contains the maximum amount of water vapour possible, the relative humidity is said to be 100 percent. On a rainy day, the relative humidity is usually 100 percent. In dry climates, the relative humidity may be 20 percent or less, which means the water vapour contained in the air is only 20 percent of the maximum amount that the air is capable of holding.

Relative humidity is a major factor in the behaviour of wildland and ground cover fires. When relative humidity is low, fire behaviour increases because vegetative fuels dry out, making them more susceptible to ignition. Conversely, when relative humidity is high, fire behaviour decreases because the moisture from the air is absorbed by the vegetative fuels, making them less susceptible to ignition. Changes in relative humidity affect fine fuels more dramatically than they affect heavy fuels. If the relative humidity drops and stays at 10 percent, for example, eventually all fuels will equalize at that level—but fine fuels will reach that level much more quickly than heavy fuels.

Relative humidity typically varies with the time of day. As the temperature increases throughout the day, the relative humidity generally drops. Conversely, as the temperature decreases during the evening and nighttime hours, the relative humidity rises. A rise in the relative humidity can help to slow the spread of a wildland fire and aid the work of fire crews who are trying to control the fire.

Average relative humidity varies from region to region and season to season. These changes contribute to variations in the moisture content of vegetative fuels—a factor that partly explains why many regions of Canada demonstrate seasonal patterns of wildland and ground cover fires. The seasons with the most wildland and ground cover fires are often the seasons with the lowest relative humidity.

The second type of moisture that influences the behaviour of wildland and ground cover fires is precipitation. Rainfall is absorbed by plants, increasing their moisture content and renders the plants less susceptible to combustion. In seasons with adequate precipitation, the incidence of wildland and ground cover fires is typically much lower than in years with below-average precipitation.

Wind

On May 1, 2016, a fire that started approximately 15 km (9 miles) from Fort McMurray, Alberta, grew rapidly, aided by strong winds and dry conditions. The fire's intensity and sheer size earned it the nickname

"The Beast." The fire created a micro environment that sucked in air for hundreds of metres adding to its overall wind-driven effect. The fire jumped rivers and highways and rampaged through small communities. The fire destroyed tens of thousands of hectares of forest and over 3000 homes and structures. It took the combined efforts of thousands of fire fighters, over 80 helicopters, several water bomber aircraft, and a little over a year to completely extinguish. Wind was a definitive contributing factor to both the size and intensity of this fire (Alberta, Canada, Government, 2017).

Topography

The movement of wildland and ground cover fires also reflects the topography of the land. **Topography**, which refers to the changes of elevation in the land as well as the positions of natural and human-made features, has a major effect on the behaviour of wildland and ground cover fires. Heat from these fires rises just like the heat in a structure fire. When a wildland or ground cover fire burns on any sort of slope or on flat land that is influenced by wind, much of the heat from the fire will carry into the air, thereby preheating the fuels above the main body of fire. For example, when a fire is burning on a hillside, convective and radiant heat from the fire will preheat the fuels on the slope above, causing them to dry out and be more flammable. Consequently, fire will spread more rapidly moving up a slope than on flat land.

Other topographical features may affect fire growth and spread. Rivers, lakes, and highways can prevent barriers to fire spread.

Extinguishing Wildland Fires

To safely attack and effectively extinguish wildland and ground cover fires, fire fighters must consider how the factors of fuels, weather, and topography come together

to influence the behaviour of the fire. Firefighters need to know the terms that are commonly used to describe different parts of the fire so that they can both understand directions from command and report information back to fire officers.

Anatomy of a Wildland Fire

The location where a wildland or ground cover fire begins is called the **area of origin**. The **head of the fire** is the leading edge of the fire as it moves forward consuming new fuel (**FIGURE 21-2**). As the fire grows, the area closest to the area of origin is referred to as the **heel of the fire** or the **rear of the fire**. The **flank of the fire** is the edge between the head and heel of the fire that runs parallel to the direction of the fire spread.

As the fire grows, a change in weather, topography, or fuel may cause it to move in such a way that it projects out into a long, narrow extension called a **finger**. A finger can produce a secondary direction of travel for the fire. The unburned area between a finger and the main body of the fire is called a **pocket**. A pocket is a dangerous place for fire fighters because this area of unburned fuel is surrounded on three sides by fire. An area of land that is left untouched by the fire, but is surrounded by burned land, is called an **island**.

A **spot fire** is a new fire that starts outside the perimeter of the main fire. Spot fires usually begin when flaming vegetation, in the form of an ember or spark, is picked up by the convection currents generated by the fire and dropped some distance away from the original fire.

Green and **black** are terms often used by wildland fire fighters. "Green" describes areas containing unburned fuels; these areas are more likely to become involved in the fire than are black areas. "Black" describes areas

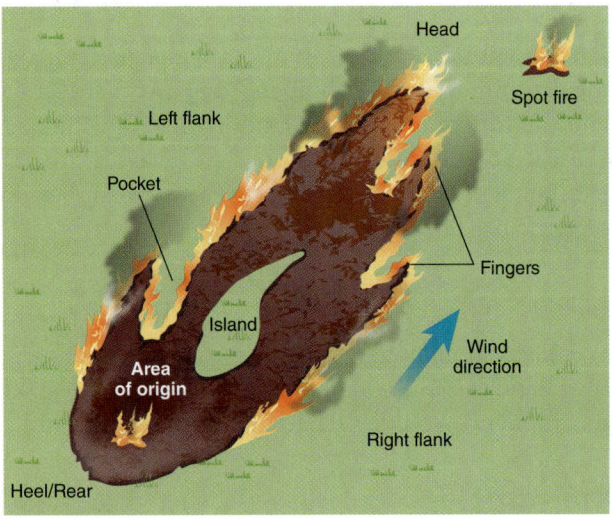

FIGURE 21-2 The anatomy of a wildland fire.
© Jones & Bartlett Learning.

that have already been burned. Because black areas contain few unburned fuels, they are often relatively safe for fire fighters.

Methods of Extinguishment

The methods used to extinguish wildland and ground cover fires focus on disrupting the conditions represented by the fire triangle. Wildland and ground cover fires can be controlled and extinguished by cooling the fuel, by removing fuel from the fire, or by smothering the fire. Methods used to extinguish fires are further discussed in Chapter 5, *Fire Behaviour*.

Cooling the Fuel

Water is used to cool the fuels that drive wildland and ground cover fires. Although the principles involved in cooling wildland and ground cover fires are the same as those used for structural fires, the methods used to apply the water may be different.

For small fires with a light fuel load, backpack fire extinguishers may be effective (**FIGURE 21-3**). These devices, which are discussed in further detail in

Chapter 7, *Portable Fire Extinguishers*, can be transported to the fire and quickly put into operation before the fire has an opportunity to grow. Most portable backpack fire extinguishers hold 19 L (5 gal) of water and are designed to be refilled easily in the field, such as from a lake or stream. Fire fighters expel the water by using a hand pump.

Water from booster tanks carried on structural fire apparatus or on special wildland fire trucks is another means of cooling a wildland fire's fuel. (The types of fire apparatus used for wildland fires are discussed later in this chapter.) Unfortunately, it is not always possible to get trucks into a position where they can attack a wildland fire. Structural fire apparatus is not designed to be taken off the road, and wildland fire apparatus can reach only limited parts of wildland. Forestry fire hose is sometimes extended for hundreds of feet. This type of hose is typically single-jacket, lightweight, collapsible, and smaller in diameter than the hose required for structural firefighting. The water supply may be limited to the amount of water carried by the apparatus, unless static water sources can be tapped by fire pumpers or portable pumps. Backpack fire extinguishers and forestry fire hose are used to extinguish most ground cover fires.

A third means of applying water to a wildland fire is from aircraft. Fixed-wing aircraft can take on a load of water from a lake or river and deliver it to fire fighters at the scene. Fire retardant can also be delivered by fixed-wing aircraft (**FIGURE 21-4**). Canada forestry fire personnel will commonly use the Canadair CL-415 "Water Bomber." The specialized aircraft, made in Canada, is designed to scoop water from a lake, large river, reservoir, or other large body of water and dump it onto wildland fires. The water bomber design is extremely efficient and takes only seconds to fill. Rotary-wing aircraft (helicopters) can either

FIGURE 21-3 Backpack fire extinguishers may be effective for small fires with a light fuel load.
© Jones & Bartlett Learning.

FIGURE 21-4 Aircraft can be used for delivering water or fire retardant to wildland fires.
© STRINGER/AFP/Getty Images.

carry water in an internal tank and drop it directly on the fire, or they can use a long cable and bucket system to transport the water to the fire fighters working at the scene (**FIGURE 21-5**). Helicopters can also carry fire retardant or foam extinguishing agents. The use of aircraft can be an effective means to control wildland fires, but it requires close communication and coordination between ground and air resources.

A

B

FIGURE 21-5 Rotary-wing aircraft can be used to apply water to a wildland fire by either **A.** carrying the water in an internal tank and dropping it directly on the fire or **B.** by using a long cable and bucket system to transport the water to the scene.
Courtesy of Jeff Pricher.

Removing the Fuel

The second method of controlling and extinguishing wildland and ground cover fires is to remove the fuel from the fire. This can be accomplished using hand tools or mechanized equipment.

In terms of hand tools, a fire broom or steel fire rake is used to sweep away fine fuels such as light grass or weeds. Other special-use tools that are designed for cutting, digging, and scraping can be used to create a **fire control line** (a constructed or natural barrier used to control the fire) in light brush (**FIGURE 21-6**). Examples of tools used for this purpose are the **McLeod** (a combination hoe and rake tool) and the **Reinhart** (a scraping tool that looks like an oversized garden hoe). When working in heavier brush, a **Hazel hoe** (also known as an adze hoe) can be used to grub out brush to create a fire control line. The **Pulaski axe** combines an adze and an axe for brush removal. A **council rake** is a long-handled rake constructed with hardened, triangular-shaped steel teeth that is used for digging, for rolling burning logs, for cutting grass and small brush, and for raking a fire control line down to soil with no subsurface fuel (mineral soil).

In some situations, saws are used to remove heavy brush and trees from a fire's path. The types of saws employed for this purpose range from handsaws to gasoline-powered chainsaws. Other powered equipment used to control wildland and ground cover fires includes tractors, plows, and bulldozers.

Setting a fire is another technique that can be used to remove the fuel from a wildland fire. A **backfire** is a fire that is intentionally set by fire fighters along the inner edge of a fire control line in an attempt to burn an area of vegetation, thereby creating a buffer area in which fuel has already been burned. This forces the fire

FIGURE 21-6 Special-use hand tools can be used to create a fire control line in light brush.
Courtesy of Jeff Pricher.

to cease or change direction. Backfires must be set at the right time and at the proper place to work correctly. The technique is potentially dangerous and should be conducted only by trained personnel when deemed necessary by the incident commander (IC) and when a coordinated plan has been established.

Firing out is another method of setting a fire to remove fuel (**FIGURE 21-7**). Firing out is used to extend a flank or other part of the fire to a preestablished fire control line or to clean up or strengthen a fire control line. This technique is typically a smaller operation and involves burning less fuel than a backfire. To accomplish this technique, the fire fighter uses a drip torch, fusee, or flare gun. As with setting a backfire, firing out should be conducted only by trained personnel when deemed necessary by the crew supervisor or IC.

Once a fire has been set on the ground, it cannot be undone. When improperly set, this technique may worsen a wildland fire, create a new fire, or even cause fire fighter injuries or fatalities.

Smothering the Fire (Removing Oxygen)

The third method that can be used to control and extinguish wildland fires is to remove oxygen from the fire. The smothering technique is most commonly used when fire fighters are overhauling the last remnants of a wildland and ground cover fire. In these circumstances, water and mineral soil (dirt with no organic material) that is dug up at the scene are often thrown on smouldering vegetation until it is cool to the touch to prevent flare-ups. Smothering is more useful during the less active phases of a fire.

Compressed air foam systems (CAFS) combine foam concentrate, water, and compressed air to produce a foam that will stick to vegetation and structures that

FIGURE 21-7 Firing out is a method of setting a fire to remove fuel from a wildland fire.
Courtesy of Jeff Pricher.

are in the path of a wildland and ground cover fire. The foam absorbs the fire's heat and breaks down the surface tension of brush and wood fibres, thereby penetrating deeper into the woody material, cooling the fuel, and preventing oxygen from combining with the fuel. CAFS can be delivered from a backpack fire extinguisher or from an appliance that is mounted on an apparatus. Such systems help to extinguish fires with less water because the compressed air foam achieves better penetration of fuels and clings to the fuel, thereby increasing the effectiveness of the water used. The use of CAFS reduces the incidence of fires rekindling.

In some locations, gel is used in place of foam to pretreat fuels in the path of a wildland or ground cover fire. Like foam, the gel sticks to vegetation and structures, absorbs the fire's heat, and prevents the fuel from burning. Unlike foam, gel cannot be pumped through a fire hose, so it must be introduced at the nozzle of a hose or backpack fire extinguisher. Depending on the type of gel, an application will remain effective for several hours.

Suppressing a Ground Cover Fire

To suppress a ground cover fire, follow the steps in **SKILL DRILL 21-1**.

Fire Apparatus Used for Wildland Fires

Specialized wildland fire and structural fire apparatus are commonly used for wildland fire suppression. Structural fire apparatus are often used because most small wildland fires are fought by departments whose primary mission is structural firefighting. For small wildland fires, structural fire pumpers carry enough water to mount an attack, and booster lines provide enough water for attacking. For larger fires, preconnected 45-mm (1¾-in.) structural hose lines can be used.

Using structural fire apparatus for wildland fires has several drawbacks. In particular, most structural fire apparatus are not designed to pump water at the same time that the truck is moving forward. The lack of a pump-and-roll capacity means that the apparatus must stop and be shifted into pump gear before a stream of water can be pumped. It also makes the apparatus less efficient when a continuous stream of water is needed while the truck moves along the flank of a fire. In addition, most structural fire apparatus is designed for on-the-road use. As a consequence, trucks that lack all-wheel drive and underside protection may become stuck or damaged when used in off-road situations. Structural fire apparatus that

SKILL DRILL 21-1
Suppressing a Ground Cover Fire Fire Fighter I, NFPA 1001: 4.3.19

1 Don appropriate personal protective equipment (PPE). Identify safety and exposure risks. Protect exposures if necessary.

2 Construct a fire control line by removing fuel with hand tools.

3 As an alternative to Step 2, extinguish the fire with a backpack pump extinguisher or a handline.

4 Overhaul the area completely to ensure complete extinguishment of the ground cover fire.

© Jones & Bartlett Learning. Photographed by Glen E. Ellman.

meet the requirements of NFPA 1901, *Standard for Automotive Fire Apparatus*, are not primarily designed for fighting wildland fires.

By contrast, specialized wildland fire apparatus are designed for fighting wildland fires. Because the terrain and conditions vary from one geographic location to another, many variations in wildland fire apparatus exist. The design of any such apparatus needs to be suited to the local wildland conditions. Wildland fire apparatus range from small pick-up trucks or Jeep-type vehicles to large trucks. Most of this apparatus is designed with all-wheel drive, underside protection, and high road clearance, which enables the vehicles to travel off-road. Wildland fire apparatus normally pump between 38 L/min and 2082 L/min (10 gpm and 550 gpm). They are equipped with a water tank that holds from 189 to 2839 L (50 to 750 gal) of water, and they are able to operate a water pump while moving. This pump-and-roll capability is achieved by powering the pump with a separate engine or by powering the pump from a **power take-off shaft**. Wildland fire apparatus should meet the requirements of NFPA 1906, *Standard for Wildland Fire Apparatus*.

Small wildland fire apparatus typically carry 757 to 1136 L (200 to 300 gal) of water. Most such small apparatus have a pump that is powered by a small gasoline or diesel engine, an arrangement that gives the vehicle a pump-and-roll capability. These smaller brush trucks can respond quickly. Often they can contain a fire before it grows into a larger blaze (**FIGURE 21-8**).

Larger wildland fire apparatus are often built on a midsized or large truck chassis (**FIGURE 21-9**). These trucks carry more water than small apparatus, which enables them to operate for a longer period of time without being refilled. Many of these vehicles have crew cabs that enable an entire crew of fire fighters to travel to the fire scene.

A third type of wildland fire apparatus is a tanker or water tender (**FIGURE 21-10**). These trucks are designed to transport water to the scene of a wildland fire. They are usually equipped with a pump that enables them to fill from a drafting site and to offload water when needed. Many tenders are equipped with a dump valve for rapidly unloading water into a portable tank. Some tenders are suited for off-road use, whereas others must remain on a road.

Wildland fire apparatus also include tractors, plows, and bulldozers (**FIGURE 21-11**). These vehicles are used to create fire control lines.

As mentioned earlier, fixed-wing and rotary-wing aircraft are valuable resources for dropping water or fire retardant where the terrain makes it dangerous or impossible to fight a fire with ground-based fire fighters. Aircraft can be used to attack the head of a fire, thereby slowing the growth and spread of the fire.

FIGURE 21-8 A smaller brush truck can respond quickly to wildland fires.
Courtesy of Jeff Pricher.

FIGURE 21-10 A tanker or water tender is used to transport water to wildland fires.
© Steven Townsend/Code 3 Images.

FIGURE 21-9 Larger wildland fire apparatus carry more water than small apparatus.
© Michael Routh/Alamy Stock Photo.

FIGURE 21-11 A wildland bulldozer is used to create fire control lines.
Courtesy of Jeff Pricher.

Types of Attacks

Wildland and ground cover fires can advance quickly and can change directions quickly. Given their unpredictability, they present a very hazardous environment for fire fighters. For these reasons, it is important for the IC to match the type of firefighting attack to the conditions present at the fire. The general types of attacks used to contain and extinguish wildland fires are direct attack, indirect attack, and parallel attack. Depending on the situation, you may need to use a combination of these methods when fighting the fire.

Before determining which method of attack to use, the IC must assess and evaluate the priorities for preserving lives and property. The top priority, of course, is to ensure the safety of both fire fighters and citizens. All fire fighters should make sure that they know the escape route and escape plan before attacking the fire. In the wildland/urban interface environment, having good situational awareness is especially crucial. Preestablished plans will ensure that everyone knows when to use escape routes and safety zones. The second priority is to minimize the property losses from the fire as much as possible. Due to the unpredictable nature of wildland fires, the IC must not hesitate to call in additional resources early in the life of any wildland fire incident.

Fire attacks should always start from an **anchor point**—that is, an area close to the place where the fire started and from which the vegetation has already been burned. The anchor point provides an area of safety for fire fighters. Ideally, it should be located close to a highway or road that can be used as an escape route. This strategy helps prevent a situation in which the fire might change direction, circle back on fire fighters, and trap them in a point of danger.

Direct Attack

A **direct attack** on a wildland and ground cover fire is mounted by containing and extinguishing the fire at its burning edge. A direct attack can be made with a fire crew and hand tools, a structural pump company, a wildland pump company, a bulldozer, or an aircraft. Fire fighters might smother the fire with dirt, use hose lines to apply water or compressed air foam to cool the fire, or remove the fuel. A direct attack is typically used on a small wildland fire before it has grown to a large size.

Any direct attack needs to be coordinated through the ICS. Before a direct attack is started, the IC will assess the resources available and determine the best way to attack the fire.

A direct attack is dangerous because fire fighters must work in smoke and heat close to the fire. However, this type of attack has the advantage of accomplishing quick containment of a fire. By extinguishing the fire while it is small, fire fighters reduce the risk posed by the fire. Most wildland and ground cover fires are kept small because the initial crews used an aggressive direct attack on the fire.

Two major types of direct attacks are used on wildland and ground cover fires: the anchor, flank, and pinch attack and the flanking attack.

Anchor, Flank, and Pinch Attack. The **anchor, flank, and pinch attack** (also known as the pincer attack) is a method of direct attack that requires two or more teams of fire fighters. This is the safest and most reliable method of attack. The key to this method is the establishment of two solid anchor points. One team establishes an anchor point on the left side of the fire near the point of origin and mounts a direct attack along the left flank of the fire, working toward the head of the fire. A second team of fire fighters establishes an anchor point on the right side of the fire near the point of origin and mounts a direct attack along the right flank of the fire, also working toward the head of the fire. As the teams advance, the fire gets "pinched" between them, which reduces its growth. By starting at a safe anchor point, the risk to fire fighters is minimized. A successful attack using this method requires the availability of sufficient personnel to be able to mount a two-pronged attack (**FIGURE 21-12**).

Flanking Attack. The **flanking attack** is a method of direct attack that requires only one team of fire fighters.

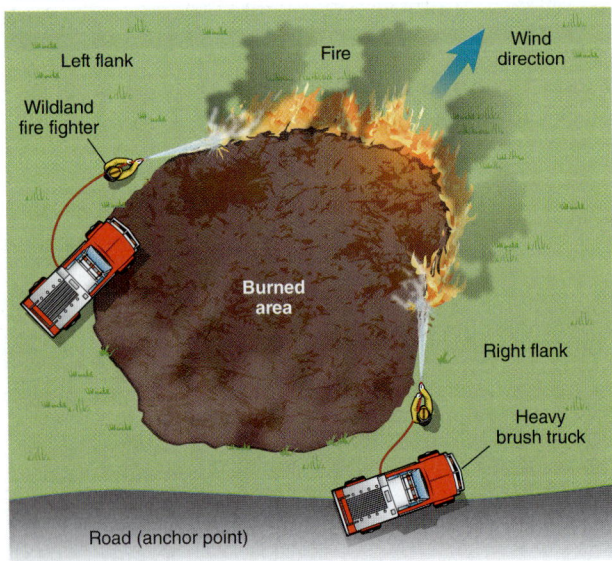

FIGURE 21-12 The anchor, flank, and pinch attack involves engagement on both flanks of a wildland fire.

© Jones & Bartlett Learning.

This team starts at an anchor point and attacks along one flank of the fire. Such an attack is used when sufficient resources are not available to mount two simultaneous attacks or when the risks posed by the fire are high on one flank of the fire and lower on the other flank of the fire. The decision regarding which flank of the fire to attack is based on the determination of which side of the fire poses the greatest risk (**FIGURE 21-13**).

Indirect Attack

An **indirect attack** is most often used for large wildland and ground cover fires that are too dangerous to approach through a direct attack because there is not enough equipment or personnel available. It is also appropriate when the topography or weather conditions pose a high risk to fire fighter safety.

An indirect attack is mounted by building a fire control line (fire break) along a predetermined route, based on natural fuel breaks or favourable breaks in the topography, and then burning out the intervening fuel (**FIGURE 21-14**). This strategy is similar to using a defensive attack on a structure fire. The fire control line may be constructed fairly close to the fire, or it may be constructed several miles away. An indirect attack requires only one team of fire fighters and can be mounted using either hand tools or mechanized equipment.

Parallel Attack

As with the indirect attack, a **parallel attack** is used when the fire edge is too hot to approach through a direct attack. To maintain fire fighter safety, a fire

FIGURE 21-14 An indirect attack is made along natural fuel breaks, at favourable breaks in the topography, or at a considerable distance from the fire.
© Jones & Bartlett Learning.

control line is constructed parallel to the fire edge, with a distance of about 1.5 to 15 m (5 to 50 ft) between the control line and the fire edge—far enough from the fire that workers and equipment can continue to work effectively if the intensity of the fire grows (**FIGURE 21-15**). In some cases, the fire control line can be shortened by being constructed across the fire's unburned fingers. The unburned fuel usually burns out as the fire moves alongside the control line, but it can also burn out unassisted with the main fire as long as the weather and/or terrain does not present a threat to the fire control line.

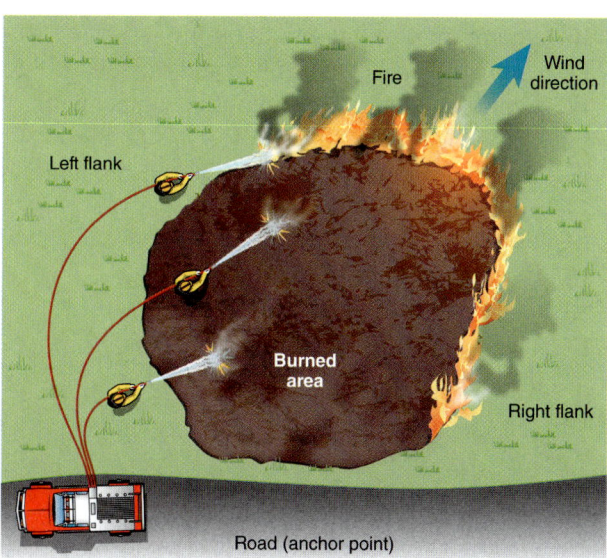

FIGURE 21-13 A flanking attack is made on the left flank or right flank of a wildland fire.
© Jones & Bartlett Learning.

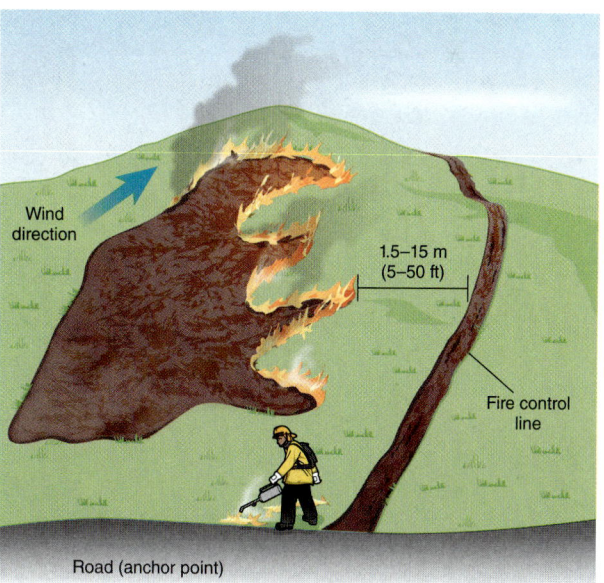

FIGURE 21-15 A parallel attack is made parallel to the fire's edge.
© Jones & Bartlett Learning.

Safety in Wildland Firefighting

Fighting wildland and ground cover fires is hazardous duty. Responses for wildland and ground cover fires involve all of the hazards encountered by personnel responding to structural fires, in addition to hazards unique to the wildland environment.

Ten Standard Firefighting Orders

The **ten standard firefighting orders** were developed by the U.S. Department of Agriculture Forest Service Chief after a task force investigated 16 tragic wildland fires that resulted in multiple fire fighter fatalities. The ten orders, which are meant to prevent further fire fighter tragedies from occurring during these types of fires, are broken into three categories as follows:

Fire behaviour
1. Keep informed on fire weather conditions and forecasts (**FIGURE 21-16**).
2. Know what your fire is doing at all times.
3. Base all actions on the current and expected behaviour of the fire.

Fireline safety
4. Identify escape routes and safety zones and make them known.

5. Post lookouts when there is possible danger.
6. Be alert. Keep calm. Think clearly. Act decisively.

Organizational control
7. Maintain prompt communications with your forces, your supervisor, and adjoining forces.
8. Give clear instructions, and ensure they are understood.
9. Maintain control of your forces at all times.

If 1 to 9 are considered, then . . .
10. Fight fire aggressively, having provided for safety first.

The Eighteen Watch Out Situations

The **eighteen watch out situations** are an expansion of the ten standard firefighting orders. The eighteen watch out situations can be used as a checklist to determine if an assignment is safe. If five or more of the situations are encountered, the fire fighter should reconsider engaging in the assignment.

The eighteen watch out situations are as follows:

1. Fire not scouted and sized up
2. In country not seen in daylight
3. Safety zones and escape routes not identified
4. Unfamiliar with weather and local factors influencing fire behaviour
5. Uninformed on strategy, tactics, and hazards
6. Instructions and assignments not clear
7. No communication link with crew members or supervisor
8. Constructing line without safe anchor point
9. Building fireline downhill with fire below
10. Attempting frontal assault on fire
11. Unburned fuel between you and the fire
12. Cannot see main fire and not in contact with anyone who can
13. On a hillside where rolling material can ignite fuel below
14. Weather becoming hotter and drier
15. Wind increases or changes direction
16. Getting frequent spot fires across line
17. Terrain and fuels make escape to safety zones difficult
18. Taking a nap near the fireline

FIGURE 21-16 When responding to a wildland fire, it is important to keep informed on fire weather conditions and forecasts.
Courtesy of Jeff Pricher.

Lookouts, Communications, Escape Routes, and Safety Zones (LCES)

Before attacking a wildland fire, fire fighters should ensure that lookouts, communications, escape routes, and safety zones (**LCES**) are in place. The LCES mnemonic is derived from the core safety principle outlined in the ten standard firefighting orders and the eighteen watch out situations previously discussed.

Ensuring the components of LCES are in place before attacking a wildland fire reduces the risk associated with fighting these types of fires:

- *Lookouts* need to be stationed in a location from which they can see the fire fighters working at the scene (**FIGURE 21-17**). If the lookout recognizes that the fire fighters are becoming endangered by the fire or the hazards of the environment, he or she notifies the fire fighters so that they can reposition themselves to a safer area. Lookouts must be knowledgeable about the wildland environment and must be able to anticipate changes in the behaviour of the fire and the weather.
- *Communications* on the wildland fire scene must be clear, concise, and prompt (**FIGURE 21-18**). All personnel should be on the same radio frequencies. Effective communications ensure that everyone is aware of approaching hazards.
- *Escape routes* provide fire fighters with a path from a dangerous location to an area away from the fire. The effectiveness of an escape route is dependent on the behaviour of the fire. Fire fighters should make sure that more than one escape route is available to them at all times.
- *Safety zones* are preestablished areas of refuge to which fire fighters can retreat where they will not need to use a fire shelter. The size of the zone will be based on the behaviour of the fire and the fire fighter's location relative to the fire. A basic guideline is to choose an area that provides a

FIGURE 21-18 Communications on the wildland fire scene must be clear, concise, and prompt to ensure that everyone is aware of approaching hazards.
Courtesy of Jeff Pricher.

separation of four times the flame height. This guideline is based on a three-person pump company crew on flat ground with no wind. The size of the safety zone would need to be adjusted to take into account wind, location on a slope, and the number of resources.

LACES is an adaptation of the LCES mnemonic that is used in various parts of the country. The "A" component may have three different meanings depending on the area in which it is being used: Anchor points, Attitude, or Awareness.

Remember to plan for the components of LCES/LACES before firefighting in the wildland environment. Everyone must be informed of the escape routes and safe zones, and everyone should adhere to the operational plans.

Hazards of Wildland Firefighting

As previously mentioned, responses for wildland and ground cover fires involve all of the hazards encountered by personnel responding to structural fires. In addition, they may involve some unique hazards associated with driving fire apparatus on poorly maintained roads or trails and driving apparatus off-road. Driving on unimproved roads and steep terrain greatly increases the chance of fire apparatus rollovers. Given these dangers, drivers must thoroughly understand the operating characteristics of their fire apparatus and operate the apparatus within the safe limits specific to its design. All fire fighters must keep their seat belts fastened whenever the apparatus is moving.

When working in rough terrain, wildland and ground cover fire fighters are at increased risk for falls. Rough ground often contains holes that are difficult to see in

FIGURE 21-17 Lookouts need to be stationed in a location from which they can see the fire fighters working at the scene.
Courtesy of Jeff Pricher.

smoky conditions. Steep terrain also increases the likelihood of falls. It is important to prevent injuries caused by the sharp tools involved in wildland and ground cover firefighting. Carry these tools with the sharp edge facing the ground. If a tool has sharp edges on both ends, be sure to use a tool guard. Tool awareness is the key to injury prevention.

Other hazards of fighting wildland and ground cover fires include burns, smoke inhalation, dehydration, and heat stroke. Because the PPE worn by wildland and ground cover fire fighters provides less protection than the PPE worn by structural fire fighters, fire fighters must keep far enough from the heat of the fire to prevent burns. Also, because wildland and ground cover firefighting is done without self-contained breathing apparatus (SCBA), fire fighters must avoid inhaling poisonous gases and suspended smoke particles.

When engaged in wildland and ground cover firefighting, be alert for the hazards posed by falling trees. During a fire, the lower parts of trees may burn away and weaken the support for the rest of the tree. Another area of concern is the tops of trees. In some instances, fire may burn inside a tree, from the base to the top, causing the top to break off. As a result, trees of all sizes can fall with little warning. One of the most serious and often overlooked hazards is the use of chainsaws on trees. After vehicle accidents and medical-related occurrences, chainsaw use and hazard trees are the next leading cause of wildland fire fighter fatalities and injuries (Wildland Fire Lessons Learned Center, 2017). In the wildland/urban interface environment, it is important to look up in wooded areas to identify fire-weakened trees, loose branches (sometimes referred to as widow-makers), and your proximity to them before using a chainsaw. Only specially trained personnel should operate chainsaws. If you are not qualified for chainsaw use, flag or mark the tree and area so that others will not work around or underneath them.

Also be alert for the presence of electrical hazards. Electrical transmission lines and other electrical wires may be present in the location of a wildland fire. Wires that drop on vegetation may ignite a wildland and ground cover fire and pose an electrical hazard to fire fighters. Many of these safety hazards can be difficult to see at night and in smoky conditions. Electricity from a downed electrical wire, however, may spark a ground cover fire. Be alert for electrical hazards even in the middle of an area covered with vegetation with no buildings around.

Additionally, be alert for conductivity caused by smoke and heat. Heat and smoke from a fire burning under a high-voltage transmission line can sometimes ionize above the fire, creating a conductor to the ground or across other conductors in the transmission lines. If possible, do not construct escape routes or anchor points underneath high-voltage transmission or power lines.

Because many wildland fires occur during hot weather and because combating them involves strenuous physical activity, all fire fighters must be alert to the possibility of dehydration and heat stroke. Make sure you are drinking enough water, and take breaks when needed to give your body time to recover.

LISTEN UP!

Serious and often overlooked hazards are chainsaws and hazard trees. It is important to look up in wooded areas to identify fire-weakened trees, loose branches, and your proximity to them.

SAFETY TIP

An ash-covered pit filled with hot embers can be left behind when stumps and large logs burn out. The ash can look like stable ground and ensnare unsuspecting fire fighters. As you navigate through a stump-laden area, watch your step!

Personal Protective Clothing and Equipment

It is important for wildland fire fighters to wear appropriate PPE (**FIGURE 21-19**). Specifically, they should be equipped with a one-piece jumpsuit or a coat, shirt, and trousers that meet the requirements of NFPA 1977, *Standard on Protective Clothing and Equipment for Wildland Fire Fighting*. These garments should be constructed of a fire-resistant material, such as Nomex. Such construction ensures that the clothing will stop burning as soon as the heat and flame are removed, which reduces the chance of fire fighters being burned. Wildland fire fighters should also wear an approved helmet with a protective shroud, eye protection, gloves, and protective footwear. All of this equipment should meet the provisions of NFPA 1977.

Respiratory protection for wildland and ground cover firefighting is usually limited to a filter mask. These masks are designed to filter out small particles of smoke; however, they do not protect against inhalation of heavy smoke or poisonous gases. Check the manufacturer's instructions to determine the limits of each filter mask.

FIGURE 21-19 PPE for a wildland fire fighter.
Courtesy of Jeff Pricher.

Such masks should not be relied upon beyond the limits for which they are approved.

The PPE that is provided for wildland and ground cover firefighting varies from one location to another, depending on the types of wildland fires encountered and on the climate in a given location. Follow the local fire department's standards regarding PPE. The use of structural PPE for wildland fires can quickly result in rapid body overheating and dehydration. Given this risk, fighting wildland fires in structural PPE for long periods of time is dangerous and not recommended.

SAFETY TIP

Avoid the use of structural PPE for wildland firefighting. Wearing structural PPE for long periods of time will result in the retention of body heat and will greatly increase the chance of dehydration, heat exhaustion, and heat stroke.

Fire Shelters

One of the most important pieces of PPE for wildland and ground cover fire fighters is the fire shelter. This life-saving piece of equipment should be issued

to all wildland and ground cover fire fighters. Fire shelters are made of a thin, reflective-foil layer that is attached to a layer of fibreglass (**FIGURE 21-20**). They are designed to reflect approximately 95 percent of a fire's radiant heat for a short period of time (NWCG, 2007). The reflective property allows a rapidly moving fire to pass over a fire fighter who has deployed a fire shelter.

Fire shelters are carefully folded and carried in a protective pouch on the fire fighter's belt. When fire fighters are in danger of being overrun by rapidly moving fire, they should try to get to a safe location. Only when they cannot escape should they use their fire shelters. To use a fire shelter, the fire fighter finds the largest available clearing, scrapes away flammable litter, opens the shelter, lies face down on the shelter, and covers himself or herself with the shelter. When properly used by well-trained personnel, this equipment can save lives. As with all equipment, however, it is important to receive proper training to use a fire shelter safely. To use a fire shelter, follow the steps in **SKILL DRILL 21-2**. Note that the fire shelter pictured here is a practice fire shelter, as indicated by its orange case. Fire shelters with orange cases are intended for training purposes only. Fire shelters intended for real fire purposes come in a blue case.

SAFETY TIP

Following training with a fire shelter, take the time to properly repack the practice fire shelter so that it is ready to deploy realistically during the next training session. Follow the repacking instructions provided by the manufacturer.

FIGURE 21-20 A wildland fire shelter.
Courtesy of Anchor Industries, Inc.

SKILL DRILL 21-2
Using a Fire Shelter Fire Fighter I, NFPA 1001: 4.3.19

1 Pick the largest available clearing. Wear gloves and a hard hat, and cover your face and neck if possible. Scrape away flammable litter if you have time.

2 Pull the red ring to tear off the plastic bag covering the shelter.

3 Grasp the handle labelled "Left Hand" in your left hand and the handle labelled "Right Hand" in your right hand.

4 Shake the shelter until it is unfolded. If it is windy, lie on the ground to unfold the shelter.

SKILL DRILL 21-2 Continued
Using a Fire Shelter Fire Fighter I, NFPA 1001: 4.3.19

5 Slip your feet through the hold-down straps.

6 Slip your arms through the hold-down straps.

7 Lie down in the shelter with your feet toward the oncoming fire. Push the sides out for more protection, and keep your mouth near the ground.

The Wildland/Urban Interface

As previously mentioned, the wildland/urban interface is the area where undeveloped land with vegetative fuels meets with human-made structures. Another similar term is the wildland/urban intermix. This term describes an area where vegetative fuels intermingle with human-made structures, with no clear boundary between them.

The mixing of wildlands and developed areas may occur for a number of reasons. For example, people may want a house that is surrounded by the beauty of nature—that is, they may want a house that has modern conveniences but is set in the middle of the woods. Additionally, overcrowding and the high expense of living in a big city have caused some people to find homes in remote areas that are located next to forests and uninhabited areas. The mixing of wildlands and developed areas in this way has created massive problems in many parts of Canada. Wildland fires regularly ignite buildings and become structure fires. Structure fires, conversely, regularly ignite vegetative fuels and become wildland fires. In many cases, it is not possible to clearly separate one type of fire from the other. This phenomenon explains why most fire departments are involved in fighting some types of wildland fires.

The wildland/urban interface presents a major challenge for fire fighters. Because many people live in the wildland/urban interface, fires in these areas present a significant life-safety hazard. Also, because so many structures are built in these areas, there is a huge potential for property loss whenever a fire occurs. Finally, many parts of the wildland/urban interface do not have adequate municipal water systems.

Much of the effort geared toward reducing the loss from wildland fires is best directed at prevention. Many communities work to involve homeowners by educating them about the risks their homes pose and the steps that they can take to reduce these risks. One important measure that has been implemented in many jurisdictions is to outlaw the use of flammable cedar shake roofs in houses in the wildland/urban interface. Another important step is to encourage or require homeowners to maintain a defensible space around their structures. A defensible space is an area around a structure where combustible vegetation that can spread fire has been cleared, reduced, or replaced. This includes ensuring gutters are cleaned, moving woodpiles are away from structures, and cleaning up all leaf litter and fire-resistive shingles and siding. These acts reduce the fire load. A defensible space provides a barrier between a structure and an advancing fire. The best offense against the huge problem of fire in the wildland/urban interface is often vigourous prevention.

Strategic Considerations for the Wildland/Urban Interface

The wildland/urban interface emphasizes the importance of having structural fire fighters learn the basic principles of fighting wildland fires. While the material presented in this chapter will help fire fighters understand the basic principles of fighting wildland fires, it is not intended to provide all of the skills and knowledge needed by wildland fire fighters. These types of fires require special training on what to look for during a size-up and which tactical and strategic considerations to take into account. If your department is regularly involved in fighting wildland fires, you should complete a separate course on wildland firefighting based on NFPA 1051, *Standard for Wildland Firefighting Personnel Professional Qualifications*.

Firefighting strategies for the wildland/urban interface are not based on containing and extinguishing the fire; instead, they are based on protecting structures in the path of the fire. As you approach a structure, evaluate its location and surrounding area. All structures in the path of the wildland fire should be assigned a triage category, similar to the way medical triage is conducted in mass casualty situations.

The triage process should determine if the structure is not threatened, is threatened and needs protection, or is threatened and cannot be protected. When making this determination, consider how much fuel is located close to the structure, whether the structure's roofing material and siding are flammable, and the intensity of the approaching fire. Additionally, consider how many resources are available and the amount of time necessary for them to arrive at the scene. Keep your own safety in mind. Do you have a safe evacuation route? Is the fire behaviour too extreme, or are there other hazards, such as liquid propane gas (LPG) tanks, that threaten your safety? If you do not have access to enough water, it may be necessary to move out. Once you have evaluated these factors, you can determine the level of threat to the structure.

As you approach a threatened structure, back in your apparatus to permit a speedy exit if needed. Leave the apparatus running and the lights on. Move any debris or vegetation away from the structure. Lay hose lines to protect your crew and the structure, and keep your water tank filled. Remember that water systems may fail because of a loss of power. Use fire hydrants to fill your tank, but do not hook up to a hydrant for an ongoing water supply; instead, seek static sources of water such as lakes and swimming pools. Do not waste water wetting down surfaces. Use water to control the fire or to reduce radiant heat that may ignite the structure. Because you may need to move on short notice, keep your evolutions mobile and your hose lines as short as possible. Do not use booster lines for these types of evolutions; if you

need to leave the scene quickly, it will take too long to roll up the booster line.

Carefully monitor the movement of the fire. If possible, control the fire at the edge of the yard to prevent serious damage to the threatened structure. In cases in which this is not possible, you may be able to knock down or redirect the fire that is moving toward the structure. If the fire is too intense and/or moving too quickly, you may not be able to protect the structure and will need to evacuate the area and move to a safe location.

Remain mobile and be ready for a quick exit. Above all, keep safe and follow your department's SOPs.

LISTEN UP!

Many areas that are prone to wildland fires have limited numbers of roads leading into and out of them. Be aware of the routes by which you can escape a rapidly moving fire or a fire that changes direction quickly. Do not let yourself get caught in a wildland fire with no escape route.

After-Action REVIEW

IN SUMMARY

- Wildland is land in an uncultivated natural state that is covered by timber, woodland, brush, or grass.
- Wildland fires are unplanned fires burning in vegetative fuel that sometimes includes structures.
- Ground cover fires are a type of wildland fire that burns loose debris on the surface of the ground. This debris includes vegetation such as grass, dead leaves, needles, and branches that have fallen from shrubs and trees.
- Large wildland fires are handled by specialized agencies, such as Parks Canada.
- The wildland fire triangle consists of fuel, oxygen, and heat. These three elements affect the intensity of the fire.
- The primary fuel for wildland fires is vegetation, such as grasslands, brush, and trees. Vegetative fuels can be classified according to their location and include subsurface fuels, surface fuels, and aerial fuels.
- The size, shape, compactness, continuity, volume, moisture level, and orientation of a fuel influence how it burns. Fine fuels burn more quickly than heavy fuels and require less heat to reach their ignition temperature.
- In wildland fires, air movement influences the speed and direction in which the fire moves. Wind blowing on a wildland fire brings more oxygen to the fire and speeds the process of combustion.
- Wildland and ground cover fires may be ignited by natural causes, accidental causes, and intentional causes.
- Lightning is the source of almost all naturally caused wildland fires. Discarded smoking materials and campfires are common accidental causes.
- The two most critical weather conditions that influence a wildland fire are moisture and wind. When the relative humidity is low, fire behaviour increases. When relative humidity is high, fire behaviour decreases. Wind can push the fire into additional fuels.
- Topography can influence the spread of the fire and the speed of the fire. A wildland fire will spread more quickly over dry, flat land than over a river valley.
- The location where the wildland fire begins is called the area of origin.

- The most rapidly moving area of a wildland fire is called the head of the fire.

- As a wildland fire gets bigger, the area close to the area of origin is called the heel or rear of the fire.

- The edge between the head and heel of the wildland fire is called the flank of the fire.

- A long, narrow extension of fire that projects from the head of the fire is called a finger.

- A pocket is an unburned area between a finger and the head of the wildland fire.

- An area of land that is left untouched by the wildland fire, but is surrounded by burned land, is called an island.

- A new fire that starts outside the perimeter of the main wildland fire is called a spot fire.

- Green areas contain unburned fuels, whereas black areas have already been burned. Green areas are more likely to become involved in a wildland fire.

- Wildland fires can be controlled and extinguished by cooling the fuel, by removing fuel from the fire, or by smothering the fire.

- Backpack fire extinguishers may be effective against small wildland fires with light fuel loads. Water may also be applied by hose lines on special wildland fire trucks or by aircraft.

- Hand tools, such as rakes, and power tools, such as chainsaws, may be used to remove fuel from the fire. Setting a fire is another technique that can be used to remove the fuel from a wildland fire.

- Compressed air foam systems or gel may be used to smother wildland fires.

- Both structural fire apparatus and specialized wildland fire apparatus are used for wildland fire suppression:
  - Structural fire apparatus are commonly used because most small wildland fires are fought by departments whose primary mission is structural firefighting. Structural fire apparatus are not designed to pump water at the same time the truck is moving forward, which makes the apparatus less efficient in larger wildland fire suppression.
  - Specialized wildland fire apparatus vary in design and size based on local wildland conditions. Specialized wildland fire apparatus have a pump-and-roll capability.

- Other types of wildland fire apparatus include tankers, tractors, plows, bulldozers, and aircraft.

- All fire fighters should ensure that they know the escape route and escape plan before attacking a wildland fire.

- The three general types of attacks used to contain and extinguish wildland fires are direct attack, indirect attack, and parallel attack.
  - A direct attack is mounted by containing and extinguishing the fire at its burning edge.
  - An indirect attack is mounted by building a fire control line along natural fuel breaks, at favourable breaks in the topography, or at a considerable distance from the fire and then burning out the intervening fuel.
  - A parallel attack is used when the fire edge is too hot to approach through a direct attack. During a parallel attack, a fire control line is constructed parallel to the fire edge.

- The ten standard firefighting orders are meant to prevent fire fighter tragedies from occurring during wildland fires. They are broken into three categories: fire behaviour, fireline safety, and organizational control.

- The eighteen watch out situations can be used as a checklist to determine if an assignment is safe. If five or more of the situations are encountered, the fire fighter should reconsider engaging in the assignment.

- Ensuring the components of LCES/LACES are in place before attacking a wildland fire reduces the risk associated with wildland firefighting.

- The hazards of wildland fires include fire apparatus rollovers, falls, burns, smoke inhalation, dehydration, and heat stroke.

- The fire shelter is one of the most important pieces of PPE for wildland fire fighters. Fire shelters are designed to reflect 95 percent of a fire's radiant heat for a short period of time.

- The wildland/urban interface is the mixing of wildland and developed areas. Because many people live in the wildland/urban interface, fires in this zone present a significant risk to both lives and structures.

KEY TERMS

Access Navigate for flashcards to test your key term knowledge.

Aerial fuels Fuels located more than 2 m (6 ft) off the ground, usually part of or attached to trees.

Anchor, flank, and pinch attack A direct method of suppressing a wildland or ground cover fire that involves two teams of fire fighters establishing anchor points on each side of the fire and working toward the head of the fire until the fire gets "pinched" between them; also known as the pincer attack.

Anchor point A strategic and safe point from which to start constructing a fire control line. An anchor point is used to reduce the chance of fire fighters being flanked by fire.

Area of origin A structure, part of a structure, or general geographic location within a fire scene, in which the "point of origin" of a fire or explosion is reasonably believed to be located. (NFPA 921)

Backfire A fire set along the inner edge of a fire control line to consume the fuel in the path of a wildland fire or change the direction of force of the fire's convection column. (NFPA 901)

Backpack fire extinguisher A portable fire extinguisher usually consisting of a 19-L (5-gal) water tank that is worn on the user's back and features a hand-powered piston pump for discharging the water.

Black An area that has already been burned.

Compressed air foam system (CAFS) A foam system that combines air under pressure with foam solution to create foam. (NFPA 1901)

Council rake A long-handled rake constructed with hardened triangular-shaped steel teeth that is used for raking a fire control line down to soil with no subsurface fuel, for digging, for rolling burning logs, and for cutting grass and small brush.

Defensible space An area, as defined by the authority having jurisdiction [typically a width of 9 m (30 ft) or more], between an improved property and a potential wildland fire where combustible materials and vegetation have been removed or modified to reduce the potential for fire on improved property spreading to wildland fuels or to provide a safe working area for fire fighters protecting life and improved property from wildland fire. (NFPA 1051)

Direct attack A method of wildland fire attack in which fire fighters focus on containing and extinguishing the fire at its burning edge.

Eighteen watch out situations A list of situations published by the National Wildfire Coordinating Group (NWCG) and used to assess whether or not a wildland firefighting assignment is safe to conduct.

Fine fuels Fuels that ignite and burn easily, such as dried twigs, leaves, needles, grass, moss, and light brush.

Finger A narrow point of fire whose extension is created by a shift in wind or a change in topography.

Fire control line Comprehensive term for all constructed or natural barriers and treated fire edges used to control a fire. (NFPA 901)

Fire shelter An item of protective equipment configured as an aluminized tent utilized for protection, by means of reflecting radiant heat, in a fire entrapment situation. (NFPA 1500)

Firing out A wildland firefighting technique that involves setting a fire along the inner edge of a fire control line to consume the fuel between a fire control line and the fire's edge.

Flanking attack A direct method of suppressing a wildland or ground cover fire that involves placing a suppression crew on one flank of a fire.

Flank of the fire The edge between the head and heel of the fire that runs parallel to the direction of the fire spread.

Fuel compactness The extent to which fuels are tightly packed together.

Fuel continuity The relative closeness of wildland fuels, which affects a fire's ability to spread from one area of fuel to another.

Fuel moisture The amount of moisture present in a fuel, which affects how readily the fuel will ignite and burn.

Fuel orientation The position of a fuel relative to the ground.

Fuel volume The amount of fuel present in a given area.

Green An area of unburned fuels.

Ground cover fire A fire that burns loose debris on the surface of the ground.

Ground duff Partly decomposed organic material on a forest floor; a type of light fuel.

Hazel hoe A hand tool used to grub out heavy brush to create a fire control line; also known as an adze hoe.

Head of the fire The main or running edge of a fire; the part of the fire that spreads with the greatest speed.

Heavy fuels Fuels of a large diameter, such as large brush, heavy timber, snags, stumps, branches, and dead timber on the ground. These fuels ignite and are consumed more slowly than light fuels.

Heel of the fire The side opposite the head of the fire, which is often close to the area of origin.

Incident Response Pocket Guide (IRPG) A job and training reference for personnel operating at a wildland fire. It may also be used for all-hazard incident response.

Indirect attack A method of wildland fire attack in which the control line is located along natural fuel breaks, at favourable breaks in the topography, or at considerable distance from the fire, and the intervening fuel is burned out.

Island An unburned area surrounded by fire.

Ladder fuels Fuels that provide vertical continuity between the ground and the tops of trees or shrubs, thereby allowing fire to move with relative ease.

LCES A mnemonic that stands for Lookouts, Communications, Escape routes, and Safety zones. Fire fighters should ensure that the components of LCES are in place before attacking a wildland fire to reduce the risk associated with fighting these types of fires.

McLeod A hand tool used for constructing fire control lines and overhauling wildland fires. One side of the head consists of a five-toothed to seven-toothed fire rake; the other side is a hoe.

Parallel attack A method of attack in which the control line is located parallel to the fire edge, at a distance of about 1.5 to 15 m (5 to 50 ft) from the fire. The intervening fuel usually burns out as the fire control line moves alongside the fire but can also burn out with the main fire.

Pocket A deep indentation of unburned fuel along the fire's perimeter, often found between a finger and the head of the fire.

Power take-off shaft A supplemental mechanism that enables a pump apparatus to operate a pump while the apparatus is still moving.

Pulaski axe A hand tool that combines an adze and an axe for brush removal.

Rear of the fire The side opposite the head of the fire. Also called the heel of the fire.

Reinhart A hand tool used for constructing fire control lines and overhauling wildland fires. The tool is similar to an oversized garden hoe.

Relative humidity The ratio between the amount of water vapour in the gas at the time of measurement and the amount of water vapour that could be in the gas when condensation begins, at a given temperature. (NFPA 79)

Slash Debris resulting from natural events such as wind, fire, snow, or ice breakage; or from human activities such as building or road construction, logging, pruning, thinning, or brush cutting. (NFPA 1144)

Spot fire A new fire that starts outside areas of the main fire, usually caused by flying embers and sparks.

Subsurface fuels Partially decomposed matter that lies beneath the ground, such as roots, moss, duff, and decomposed stumps.

Surface fuels Fuels that are close to the surface of the ground, such as grass, leaves, twigs, needles, small trees, logging slash, and low brush. Also called ground fuels.

Ten standard firefighting orders A set of systematically organized rules developed by the U.S. Department of Agriculture Forest Service task force to reduce danger to firefighting personnel.

Topography The land surface configuration. (NFPA 1051)

Wildland Land in an uncultivated, more or less natural state and covered by timber, woodland, brush, and/or grass. (NFPA 901)

Wildland fire An unplanned fire burning in vegetative fuels. (NFPA 1051)

Wildland/urban interface The line, area, or zone where structures and other human development meet or intermingle with undeveloped wildland or vegetative fuels. (NFPA 5000)

Wildland/urban intermix An area where improved property and wildland fuels meet with no clearly defined boundary. (NFPA 5000)

REFERENCES

Alberta, Canada, Government. "Home Again: Recovery After the Wood Buffalo Wildfire." https://web.archive.org/web/20170131021144/https://www.alberta.ca/documents/Wildfire-Home-Again-Report.pdf#. Accessed November 30, 2018.

National Fire Protection Association. NFPA 1051, *Standard for Wildland Firefighting Personnel Professional Qualifications,* 2016. www.nfpa.org. Accessed November 21, 2018.

National Fire Protection Association. NFPA 1901, *Standard for Automotive Fire Apparatus,* 2016. www.nfpa.org. Accessed November 21, 2018.

National Fire Protection Association. NFPA 1906, *Standard for Wildland Fire Apparatus,* 2016. www.nfpa.org. Accessed November 21, 2018.

National Fire Protection Association. NFPA 1977, *Standard on Protective Clothing and Equipment for Wildland Fire Fighting,* 2016. www.nfpa.org. Accessed November 21, 2018.

National Wildfire Coordinating Group. "S-130, Firefighter Training," 2007. https://training.nwcg.gov/classes/s130/508%20Files/071231_s130_m3_508.pdf. Accessed November 21, 2018.

Natural Resources Canada. "Canada's Forest Laws." https://www.nrcan.gc.ca/forests/canada/laws/17497. Accessed November 30, 2018.

Wildland Fire Lessons Learned Center. "2017 Incident Review Summary," 2017. https://www.wildfirelessons.net/HigherLogic/System/DownloadDocumentFile.ashx?DocumentFileKey=0a9d9baf-5eaf-dae1-0914-dcbe02f2da09&forceDialog=0. Accessed November 21, 2018.

On Scene

The alarm squawks out, "Grass Fire. Two miles east of Highway 125." You jump to your feet and scramble to the pump. Your captain directs you to the brush unit.

Highway 125 is a hilly road that winds through the countryside. The road is lined with pastures that are interspersed with groves of trees. You notice a large plume of billowing white smoke in the distance. This is your first real wildland fire, and your mind begins to race. To calm yourself, you quickly cycle through what you learned in your basic training.

1. A grass fire is characterized by what category of fuel?

A. Fine

B. Light

C. Medium

D. Coarse and heavily compacted

2. Which of the following would be considered surface fuels?

A. Roots

B. Decomposed stumps

C. Brush

D. Large trees

3. Which of the following is one leg of the fire triangle?

A. Weather

B. Fuel

C. Humidity

D. Slash

4. In the context of a wildland fire, which of the following best describes a "pocket"?

A. A safe place for fire fighters

B. A dangerous place for fire fighters

C. The area at the heel of the fire

D. The area at the head of the fire

(continued)

On Scene Continued

5. A McLeod tool is a:

   A. tool that combines an axe with an adze.

   B. tool that combines a hoe and an adze.

   C. tool that combines a hoe and a rake.

   D. long-handled rake with hardened, triangular-shaped steel teeth.

6. A backfire is a(n):

   A. pincer attack.

   B. flanking attack.

   C. direct attack.

   D. indirect attack.

7. The basic guideline for a safety zone should include a minimum separation area of at least ____ times the height of the fire.

   A. two

   B. three

   C. four

   D. five

8. The ten standard firefighting orders are broken into which categories?

   A. Safety, fire behaviour, and LCES

   B. Fire behaviour, fireline safety, and organizational control

   C. Organizational control, LCES, and fire behaviour

   D. LACES, fire behaviour, and ICS

Access Navigate to find answers to this On Scene, along with other resources such as an audiobook.

SECTION 2

Fire Fighter II

Establishing and Transferring Command

KNOWLEDGE OBJECTIVES

After studying this chapter, you will be able to:

- Outline the roles and responsibilities of a Fire Fighter II. (**NFPA 1001: 5.1.1**, pp. 807–808)
- Explain the importance of communicating crew (or team) progress on an assigned task to the crew leader. (**NFPA 1001: 5.2.2**, p. 808)
- Describe the National Incident Management System (NIMS). (p. 809)
- Explain the organization of the incident command system. (**NFPA 1001: 5.1.1**, pp. 809–814)
- Describe how to function within an assigned role in the incident command system. (**NFPA 1001: 5.1.2**, pp. 815–819)
- Describe the characteristics of the incident command system. (**NFPA 1001: 5.1.1**, pp. 819–823)
- Describe the process of performing an initial size-up. (**NFPA 1001: 5.1.2**, pp. 823–827)
- List the two basic categories of information used in the size-up process. (**NFPA 1001: 5.1.2**, pp. 824–826)
- Explain how the size-up process determines the resources required at the emergency incident. (**NFPA 1001: 5.1.2**, pp. 826–827)
- Explain how the size-up process can be used to determine whether additional resources are needed. (**NFPA 1001: 5.2.2**, pp. 826–827)
- Explain the need for requesting additional resources to complete a task. (**NFPA 1001: 5.2.2**, pp. 826–827)
- Organize and coordinate an incident command system. (**NFPA 1001: 5.1.2**, pp. 827–831)
- Establish command of an incident command system until command of the incident is transferred. (**NFPA 1001: 5.1.1**, pp. 828–829)
- Transfer command of a scene within an incident command system. (**NFPA 1001: 5.1.1**, pp. 829–831)
- List the three incident priorities from which an incident action plan is based. (**NFPA 1001: 5.1.2**, p. 831)
- Describe the acronym RECEO-VS and how it provides a general guideline for incident commanders to systematically address the incident priorities. (pp. 832–835)
- Describe the acronym SLICE-RS and how it provides initial pump company operations a short list of objectives prior to the arrival of additional resources. (p. 835)
- Explain the importance of an incident report. (**NFPA 1001: 5.2.1**, pp. 835–836)
- Describe how to collect the necessary information for a thorough incident report. (**NFPA 1001: 5.2.1**, p. 836)
- Describe the resources that list the codes utilized in incident reports. (**NFPA 1001: 5.2.1**, p. 836)
- Explain the consequences of an incomplete or inaccurate incident report. (**NFPA 1001: 5.2.1**, p. 836)
- Describe the goals of crew resource management in the fire service. (p. 836)

SKILLS OBJECTIVES

After studying this chapter, you will be able to perform the following skills:

- Operate within the incident command system. (**NFPA 1001: 5.1.2**, p. 828)
- Establish or assume command of an incident. (**NFPA 1001: 5.1.1**, p. 829)
- Transfer command of an incident to another fire fighter. (**NFPA 1001: 5.1.1**, p. 831)
- Complete an incident report. (**NFPA 1001: 5.2.1**, pp. 835–836)

Note: To be NIMS compliant, in this chapter we will refer to the incident management system as the incident command system.

ADDITIONAL STANDARDS

- **NFPA 1026**, *Standard for Incident Management Personnel Professional Qualifications*
- **NFPA 1500**, *Standard on Fire Department Occupational Safety, Health, and Wellness Program*
- **NFPA 1521**, *Standard for Fire Department Safety Officer Professional Qualifications*
- **NFPA 1561**, *Standard on Emergency Services Incident Management System and Command Safety*

Fire Alarm

Your pump company has been dispatched to the smell of smoke in a rural area near the edge of your jurisdiction. As you get closer to the location, you see smoke drifting across the road. When you turn the corner, you see flames in the second-storey window of an old farm house. While you advance an attack line from the hose bed, your company officer does a 360° walk around the building. He gives a size-up over the radio, requests additional resources (including a tanker from the neighbouring department), and establishes command. The driver/operator charges the attack line, and you hear a neighbour ask your officer if all of the occupants are outside.

1. What are the three incident priorities you must address at all emergency scenes?

2. What is the incident command system, and how will the command structure expand as additional resources arrive on the scene?

3. What are the primary benefits of operating under the incident command system during mutual aid incidents?

 *Access Navigate for more practice activities.*

Introduction

Most fire department operations involve the management of resources. Routine station duties, fire prevention activities, training, and emergency incident responses all require the safe, effective, and efficient use of fire department personnel, apparatus, and equipment. Most fire departments operate within a paramilitary-style command structure with various ranks of authority and responsibility. Ultimately, the fire chief is answerable for both the operations and the services provided by the fire department. However, many operational responsibilities are delegated to lower-ranking officers within the organizational structure. Routine tasks and duties are performed by the lower ranks; decision making and strategic planning are carried out by officers of higher ranks. For example, as discussed in Chapter 1, *The Fire Service*, personnel with a **Fire Fighter I** designation perform basic fire fighter duties, such as the proper use of personal protective equipment (PPE), tools, and hose, and performing forcible entry, ventilation, search and rescue operations, and basic firefighting methods, under direct supervision. Personnel with a **Fire Fighter II** designation perform more advanced skills such as size-up and establishing and transferring command to a more experienced fire fighter, under general supervision. The assumption of command by an individual fire fighter is a rare occurrence in practice. It would most likely occur in small volunteer departments where an individual fire fighter would arrive at an incident first. The systems employed by career municipal departments ensure that an officer is the first to assume command at most incidents.

This **chain of command** results in the upward and downward flow of information and assignments needed to effectively manage both routine operations and complex incidents. Fire services have evolved their daily operational activities in a more structured manner incorporating the **incident command system (ICS)** into every call. The first arriving member is to assume command and maintain command until it is transferred or no longer required. An ICS is an incident management tool that identifies a single **incident commander (IC)** who is responsible for all aspects of the operation. All personnel, apparatus, and resources work under the authority and direction of the IC.

All fire department emergency operations and training exercises should be conducted within the framework of an ICS. Using an ICS ensures that operations are coordinated and conducted safely and effectively. This type of system provides a standard approach, structure, terminology, and operational procedure to organize and manage any operation—from a training session to an emergency scene. The same principles apply whether the situation involves a single pump company or hundreds of emergency responders from dozens of different agencies. These principles also can be applied to many nonemergency events, such as large-scale public events. Planning, delegation, supervision, and communication are key components of an effective ICS.

Roles and Responsibilities of the Fire Fighter II

As discussed in Chapter 1, *The Fire Service*, the first step in understanding the organization of the fire service is to learn your roles and responsibilities. Although this chapter is specific to the Fire Fighter II, the following list provides a review of the roles and responsibilities of the Fire Fighter I and the Fire Fighter II. Remember, in some cases, the same or similar roles and responsibilities may appear under Fire Fighter I and Fire Fighter II. This enables fire fighters to learn the fundamentals of those concepts as a Fire Fighter I and increase their

knowledge of the same concepts as a Fire Fighter II. In addition, training programs vary. You may take one course that covers both Fire Fighter I and Fire Fighter II, or you may take two separate courses.

As discussed earlier, personnel with a Fire Fighter I designation work under *direct* supervision. The roles and responsibilities required for personnel with a Firefighter I designation include the following:

- Don and doff PPE properly.
- Hoist hand tools using appropriate ropes and knots.
- Understand and correctly apply appropriate communication protocols.
- Use self-contained breathing apparatus (SCBA).
- Respond on apparatus to an emergency scene.
- Establish and operate safely in emergency work areas.
- Force entry into a structure.
- Exit a hazardous area safely as a team.
- Set up and use ground ladders safely and correctly.
- Attack a passenger vehicle fire, an exterior Class A fire, and an interior structure fire.
- Conduct search and rescue in a structure.
- Perform ventilation of an involved structure.
- Overhaul a fire scene.
- Conserve property with salvage tools and equipment.
- Connect a fire department pump to a water supply.
- Extinguish incipient Class A, Class B, and Class C fires.
- Illuminate an emergency scene.
- Turn off utilities.
- Combat ground cover fires.
- Perform fire safety inspections.
- Clean and maintain equipment.
- Locate information in departmental documents and SOPs.
- Operate as part of a team.

As a Fire Fighter II, you will receive a higher level of training that will allow you to more fully assist in mitigating emergency situations. For example, the Fire Fighter II coordinates and directs other fire fighters and assumes a greater level of responsibility for incident management. The Fire Fighter II works under *general* supervision to do the following:

- Perform scene size-up.
- Determine the need for ICS.
- Arrange and coordinate ICS until command is transferred.
- Prepare reports.
- Communicate the need for assistance.
- Coordinate an interior attack line team.
- Extinguish an ignitable liquid fire.

- Control a flammable gas cylinder fire.
- Protect evidence of fire cause and origin.
- Assess and disentangle victims from motor vehicle collisions.
- Assist special rescue team operations.
- Perform a fire safety survey.
- Present fire safety information.
- Maintain fire equipment.
- Perform annual service tests on fire hose.

Many of these duties involve interacting with the public, working with special rescue teams, and performing basic scene assessment and decision-making functions in the initial stages of an emergency incident.

Communications

The ICS establishes a clear administrative structure built from the top down. The system, when used properly, provides for clear and concise communication and fire-ground discipline by breaking the overall incident down into more manageable components. Each team, division, sector branch, and staff member, including command, is assigned a specific radio designation. These designations may be transferred from one company to the next, but they survive for the duration of the incident or until command eliminates them, as goals and strategic objectives are achieved. The IC can provide initial and updated situational information, make assignments, and alert personnel of safety concerns down the chain of command. Personnel performing at the task level can notify the IC of changing fire conditions, provide task progress reports, or request additional resources through that same chain of command.

For example, fire fighters may be assigned to perform search and rescue operations on the third floor of an apartment building. The IC gives the assignment to the company officer or to the Fire Fighter II leading a crew, either face-to-face or over the radio. The crew leader repeats the assignment information to confirm that he or she has received and understands the assignment. It is important to keep the IC informed about crew location, safety conditions, and if fire or building conditions change while working on an assignment. The IC may ask for periodic personnel accountability report (PAR) updates, especially if you are working in a hazardous environment.

Most ICS follow a very similar template; however, many fire services adopt their own system of management. As a new fire fighter you must ensure that you are well versed in the requirements of your specific department's SOPs and safety requirements. When you are assigned to a specific team or task force, your officer will communicate progress toward completion of your assigned task or assignment. Depending on the scope of the incident, your officer may report directly to command

or to a branch officer. Once the assignment has been completed or if you are unable to complete the assignment, notify the IC and wait for further instructions. The IC may send additional resources, reassign your crew, or assign your crew to rehabilitation or staging.

History of the Incident Command System

Prior to the 1970s, each individual fire department had its own method for commanding and managing incidents. Often, the organizational structure established to direct operations at an incident scene depended on the style of the chief on duty. Not surprisingly, this individualized approach did not work well when **units** from different districts or mutual aid companies responded to a major incident. In structural firefighting, the basic units are companies. What was adequate for routine incidents became ineffective and confusing with large-scale incidents, rapidly changing situations, and units that did not normally work together.

Such a fragmented approach to managing emergency incidents is no longer considered acceptable. Over the past 30 years, formal incident management systems have been developed and refined. Today's ICS structures comprise an organized system of roles, responsibilities, and SOPs that are widely used to manage and direct emergency operations. As a result, the same basic approaches, organizational structures, and terminology are used by thousands of fire departments and emergency response agencies across Canada.

The move to develop a standard system began in the United States in the early 1970s after several large-scale wildland fires in southern California proved disastrous for both the fire service and residents of the region. A number of fire-related agencies at the local, state, and federal levels decided that better organization was necessary to effectively combat these costly fires. Collectively, these agencies established an organization known as **FIRESCOPE** (**FI**re **RES**ources of **C**alifornia **O**rganized for **P**otential **E**mergencies) to develop solutions for a variety of problems associated with large, complex emergency incidents during wildland fires. These problems related to the following issues:

- Command and control procedures
- Resource management
- Terminology
- Communications

FIRESCOPE developed the first standard ICS. Originally, ICS was intended only for large, multijurisdictional or multiagency wildland incidents involving more than 25 resources or operating units. It proved so successful, however, that it was applied to structural firefighting and eventually became an accepted system for managing all emergency incidents in the United States (**FIGURE 22-1**). It was not long after many U.S. fire departments adopted the ICS in the early 1980s that Canadian fire services also adopted the system. In Canada, variations of the original command system have been adopted. The ICS has become an essential component of modern fire services emergency management.

At the same time, the **fire-ground command (FGC)** system was developed and adopted by many fire departments. The basic concepts of the FGC system were similar to those that formed the foundation of the ICS, with some differences in terminology and organizational structure. The FGC system was initially designed for day-to-day fire department incidents involving fewer than 25 fire suppression companies, but it could be expanded to meet the needs of much larger incidents.

During the 1980s, the ICS developed by FIRESCOPE was adopted by all American federal and most state wildland firefighting agencies. The National Fire Academy (NFA) also used the FIRESCOPE ICS as the model fire service ICS for all of its courses. Additional federal agencies, including the Federal Emergency Management Agency (FEMA) and the Federal Bureau of Investigation, adopted the same model for use during major disasters or terrorist events. All federal agencies in the United States could now learn and use the same basic system.

Several federal regulations and consensus standards—including NFPA 1500, *Standard on Fire Department Occupational Safety, Health, and Wellness Program*—were adopted during the 1980s. These standards mandated the use of an ICS at emergency incidents. NFPA 1561, *Standard on a Fire Department Incident Management System*, which was issued in 1990, identified the key components of an effective system and described the importance of using such a system at all emergency incidents. Fire departments could use either the FIRESCOPE ICS or the FGC system to meet the requirements of NFPA 1561.

In the years that followed, users of different systems from across the United States formed the National Fire Service Incident Management System Consortium to develop "model procedure guides" for implementing effective incident management systems at various types of incidents. The resulting system, which blended the best aspects of both FIRESCOPE'S ICS and the FGC system, is now known formally as ICS—the incident command system. An ICS can be used at any type or size of emergency incident and by any type or size of department or agency. Reflecting this change, NFPA 1561 is now called *Standard on Emergency Services Incident Management System and Command Safety*. NFPA 1026, *Standard for Incident Management Personnel Professional Qualifications*, was developed to define the various job performance requirements for each of the positions classified within the National Incident Management System (NIMS) model. The NIMS model is discussed next.

Incident Command and the NIMS Model

In 2004, the U.S. Department of Homeland Security (DHS) developed a standard approach to incident management that can be used by many different agencies under a wide range of emergency and nonemergency situations. The **National Incident Management System (NIMS)** provides a consistent, nationwide framework for incident management, enabling federal, state, and local governments, private sector and nongovernmental organizations, and all other organizations who assume a role in emergency management to work together effectively and efficiently.

Like the ICS, the NIMS model can be used regardless of the cause, size, complexity, or type of incident, including acts of terrorism and natural disasters. Along with the basic concepts of flexibility and standardization, the NIMS principles are now taught in every incident management course. You can learn more about NIMS on FEMA's website.

The ICS Organization

The ICS is an important component of the NIMS. At the Fire Fighter II level, you will frequently work within the framework of the NIMS ICS model, so it is imperative

FIGURE 22-1 ICS was first developed to coordinate efforts during large-scale wildland fires.
© Kathryn Capaldo/Alamy Stock Photo.

that you understand the basic terminology, structure, and roles commonly used on emergency scenes.

The ICS structure identifies a full range of duties, responsibilities, and functions that are performed at emergency incidents; it also defines the relationships among those components. One of the key principles of the ICS is its adaptability. While some components are used at almost every incident, others are not. The strength of the ICS is its ability to expand or contract according to the needs of the incident. The five major components of an ICS organization are command, operations, planning, logistics, and finance/administration. These five components are discussed in this chapter.

The requirements for logistics, finance administration, and planning are usually reserved for extraordinary emergency incidents, such as an earthquake or other significant event, that involve a multiagency response. Command is established by the first-arriving **company officer** who performs an initial size-up, identifies safety hazards, releases or requests additional resources, and begins formulating an **incident action plan (IAP)**. The initial IAP is a mental exercise that establishes an adopted strategy based on risk management principles, develops the incident action plan, requests resources, and oversees all operational aspects of the incident. As the incident becomes more complex, however, the IC formally delegates other duties, such as operations, planning, logistics, and finance/administration, to other qualified individuals, under the principle of span of control.

An ICS organization chart may be quite simple or very complex. Each block on an ICS organization chart refers to a specific function or job description. Positions are staffed as they are needed. The only position that must be filled at every incident is command. Command decides which additional components are needed for the specific situation and activates those positions by assigning someone to perform those tasks. To help clarify roles within the ICS organization, standard position titles are used for personnel within the organization, most of which will be discussed in this chapter. Each level of the organization has a different designator for the individual in charge. The position title typically includes the functional or geographic area of responsibility, followed by a specific designator (**TABLE 22-1**).

All fire fighters must understand the overall structure of ICS as well as the basic roles and responsibilities of each position within the ICS organization. As an emergency develops, a fire fighter could start in logistics, move to operations, and eventually assume a command staff position. Knowing how ICS works enables fire fighters to see how different roles and responsibilities work together and relate to one another. This allows them to focus on their specific roles without being overwhelmed by the entire incident.

TABLE 22-1 Levels of an ICS Organization		
ICS Level	**ICS Function/ Location**	**Position Designator**
Command	Command and control	Incident commander
Command staff	Safety, liaison, information	Officer
General staff	Operations, planning, logistics, finance/ administration	Section chief
Branch	Varies (e.g., EMS)	Director
Division/group	Varies (e.g., Division A)	Supervisor
Unit/crew/strike	Varies (e.g., Rehab)	Leader
Team/task force		Company officer

Command

On an ICS organization chart, the first component is **command** (**FIGURE 22-2**). As has been mentioned, command resides in a single individual. Command cannot be transferred to anyone who has not arrived on the scene of the incident. The IC has the authority to undertake any action deemed necessary to save lives and stabilize the incident. Command is established when the first unit arrives on the scene and is maintained until the last unit leaves the scene. Each of the other positions may be filled at the IC's discretion as the incident becomes more complex.

In the ICS structure, command (either single or unified) is ultimately responsible for managing an incident and has the authority to direct all activities at the incident scene. Initially, command is directly responsible for the following tasks:

- Establishing the incident action plan, including strategies and tactics to mitigate the incident
- Establishing command and expanding the ICS organization as needed
- Managing resources
- Coordinating resource activities
- Providing for scene safety
- Releasing information about the incident
- Coordinating with outside agencies

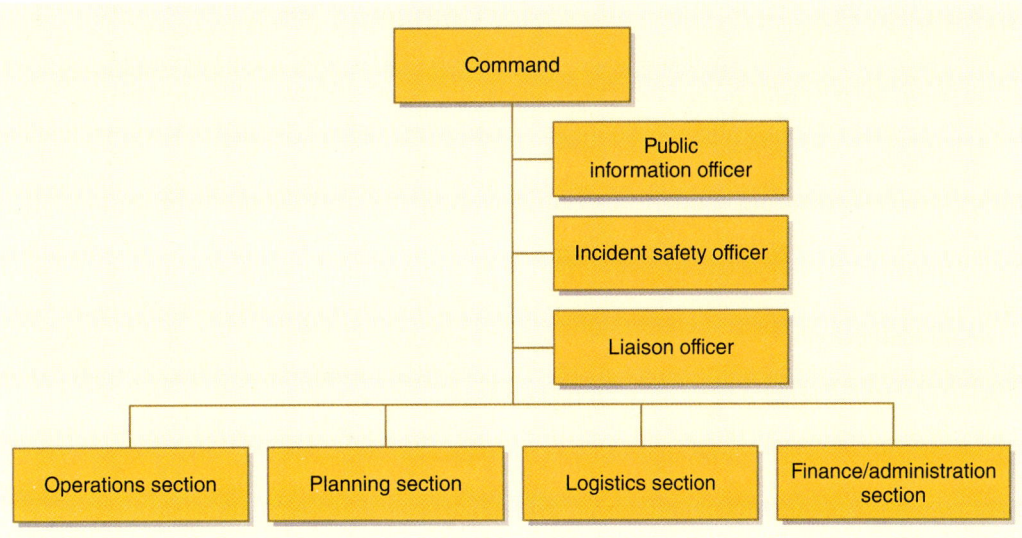

FIGURE 22-2 The ICS organization chart.
© Jones & Bartlett Learning.

As the incident becomes more complex, the IC may delegate some of these activities to other ICS positions, but he or she is still responsible for ensuring that all necessary activities are completed.

Single Command

Single command is the most traditional perception of the command function and is the genesis of the term *incident commander*. When an incident occurs within a single jurisdiction, and when there is no jurisdictional or functional agency overlap, a single IC should be identified and designated as having overall incident management responsibility by the appropriate jurisdictional authority. This does not mean that other agencies do not respond or do not have a role in supporting the management of the incident; it simply means that they report to and must work under the direction and authority of the IC.

Single command is also appropriate in the later stages of an incident that was initially managed by a unified command, or a shared command authority. Over time, as many incidents become stabilized, the strategic objectives become increasingly focused on a single jurisdiction or discipline. For example, a mass-casualty incident resulting from an explosion will likely use a unified command structure (such as shared command between fire, law enforcement, and EMS). Once all victims have been treated and transported from the scene, command will shift to a law enforcement officer while the crime scene is processed. In this situation, it is appropriate to transition from a unified command structure to a single command structure under the direction of a single IC.

It is also acceptable, if all agencies and jurisdictions agree, to designate a single IC in multiagency and multijurisdictional incidents. In this situation, however, command personnel should be carefully chosen. The IC is responsible for developing the strategic incident objectives on which the incident action plan (IAP) will be based. IAPs are oral or written plans containing general objectives reflecting the overall strategy for managing an incident. IAPs are discussed in more detail in this chapter. The IC is responsible for developing the IAP and ordering and releasing incident resources.

Unified Command

When multiple agencies with overlapping jurisdictions or legal responsibilities respond to the same incident, a **unified command** provides several advantages. Under a unified command structure, representatives from each agency cooperate to share command authority. They work together and are directly involved in the decision-making process. Operating under a unified command helps to ensure cooperation, avoid confusion, and guarantee agreement on strategies and tactics needed to mitigate the emergency. Information about overlapping jurisdictions and unified command is discussed in greater detail later. As the discipline of emergency management has evolved, major Canadian cities have established Emergency Management Departments, whose sole responsibility is to identify risk and ensure cooperation among various agencies and stakeholders. The principal officers within a unified command structure will either be located in a mobile command post or at an emergency command centre.

Incident Command Post

The incident command post (ICP) is the headquarters location for the incident. Command functions are centered in the ICP, so command and all direct support staff should always be located at this site. The location of the ICP should be clearly marked and announced to all personnel as soon as it is established. This is especially critical for incidents involving large structures or large geographic areas.

The ICP should be located in a protected location near the incident scene; however, it should not be in the immediate vicinity of emergency operations. Often, the ICP for a major incident is located in a special vehicle or building (**FIGURE 22-3**). This choice of location enables the command staff to function without needless distractions or interruptions. Possible locations for the ICP can be determined during preincident planning of large structures. For incidents involving large geographic areas, the ICP may be remote from some parts of the operational area.

FIGURE 22-3 A unified command involves many agencies directly in the decision-making process for a large incident.
© Jones & Bartlett Learning.

Command Staff

Individuals who are assigned to the command staff perform functions that report directly to command, and they cannot be delegated to work in other major sections of the organization (**FIGURE 22-4**). The incident safety officer, liaison officer, and public information officer are always part of the command staff. In addition, aides, assistants, and advisors may be assigned to work directly for members of the command staff. An aide is a fire fighter (sometimes an officer) who serves as a direct assistant to a member of the command or general staff.

Incident Safety Officer

The incident safety officer is responsible for ensuring that safety issues are managed effectively at the incident scene. He or she serves as the eyes and ears of command, identifying and evaluating hazardous conditions, preventing and correcting unsafe practices, and ensuring that safety procedures are followed. For departments without a full-time safety officer, the incident safety officer (ISO) should be appointed early during an incident, especially if personnel are exposed to hazardous conditions or high-risk operations. As the incident becomes more complex and the number of resources present at the scene increases, additional qualified personnel can be assigned as assistant incident safety officers.

Although the incident safety officer is an advisor to command, he or she has the authority to stop or suspend operations when unsafe situations occur. This authority is clearly stated in national standards, including NFPA 1500, *Standard on Fire Department Occupational Safety, Health, and Wellness Program*; NFPA 1521, *Standard for Fire Department Safety Officer Professional Qualifications*; NFPA 1561, *Standard on Emergency Services Incident Management System and Command Safety*; and NFPA 1026, *Standard for Incident Management Personnel Professional Qualifications*. Several municipalities and fire departments require the assignment of an ISO at all complex incidents.

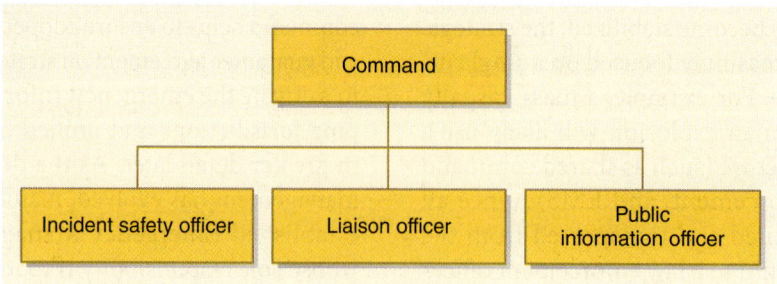

FIGURE 22-4 The command staff report directly to command.
© Jones & Bartlett Learning.

At a fire scene, the incident safety officer should be an individual who is knowledgeable in fire behaviour, building construction and collapse potential, firefighting strategy and tactics, hazardous materials, rescue practices, and departmental safety rules and regulations. He or she should also have considerable experience in incident response and specialized training in occupational safety and health. Many larger fire departments have full-time incident safety officers who perform administrative functions relating to health and safety when they are not responding to emergency incidents. Fire fighters interested in serving in this capacity should pursue training and certification under NFPA 1521, *Standard for Fire Department Safety Officer Professional Qualifications*.

Liaison Officer

The liaison officer is command's point of contact for representatives from assisting and cooperating agencies. An assisting agency or organization is one that assists in operations and provides personnel or other tactical resources to the agency with jurisdictional authority over the incident. Fire departments, law enforcement agencies, EMS providers, and heavy equipment contractors are examples of assisting organizations. Cooperating agencies such as the American Red Cross and the Salvation Army provide supplies and support but do not participate in the actual operations component of incident management. The liaison officer is responsible for coordinating with the representatives from those agencies. During an active incident, command may not have time to meet directly with everyone who comes to the ICP. Under these circumstances, the liaison officer functions as the representative of command, obtaining and providing information or directing people to the proper location or authority. The liaison area should be adjacent to, but not inside, the ICP.

Public Information Officer

The public information officer (PIO) is responsible for gathering incident information and releasing it to the news media, other agencies, and through social media (**FIGURE 22-5**). At a major incident, the public will want to know what has happened and what is being done to mitigate the incident. Because command must make managing the incident the top priority, the PIO serves as the contact person for media requests, allowing command to concentrate on the incident. A media headquarters should be established near, but not in, the ICP. The information presented to the media by the PIO must be approved by the IC before it is released.

FIGURE 22-5 The public information officer (PIO) is responsible for gathering and releasing incident information to the media and other appropriate agencies.
Courtesy of Captain David Jackson, Saginaw Township Fire Department.

General Staff Functions

Command has the overall responsibility for the entire incident command organization, although some elements of command's responsibilities can be handled by the command staff. When the incident is too large or too complex for just one person to manage effectively, command may appoint someone to oversee parts of the operation. Everything that occurs at an emergency incident can be divided among the major functional components—referred to as "sections"—within the ICS structure (**FIGURE 22-6**):

- Operations
- Planning
- Logistics
- Finance/administration

Resources operating within these four sections are led by section chiefs who form the ICS general staff. Command decides which, if any, of these four positions need to be activated, when to activate them, and who should be placed in each position. As discussed previously, only the IC position must be established early in the incident. The other command staff or general staff positions may be filled when the IC determines the need for additional personnel. As each section chief is appointed, additional personnel and resources may then be assigned to work under his or her direction.

The four section chiefs on the general staff, when assigned, may perform their duties from the main ICP, although this structure is not required. At a large incident, the four functional organizations may operate from different locations, but they will always be in direct contact with command.

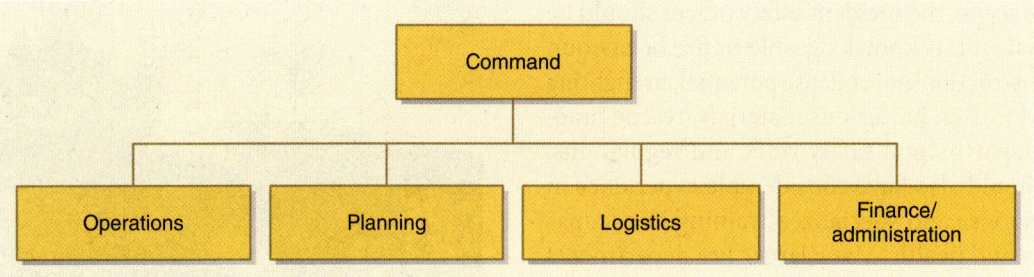

FIGURE 22-6 Major functional components, or sections, of ICS.
© Jones & Bartlett Learning.

Operations

The **operations section** is responsible for taking direct action to control the incident. Personnel functioning within the operations section perform tactical operations such as fighting the fire, rescuing any trapped individuals, treating any injured victims, and doing whatever else is necessary to alleviate the emergency situation.

For most structure fires, command directly supervises the functions of the operations section. At complex incidents, a separate **operations section chief** assumes this responsibility so that command can focus on the overall strategy while the operations section chief focuses on the tactics that are required to get the job done.

Operations are conducted in accordance with an IAP that outlines what the strategic objectives are and how emergency operations will be conducted. At most incidents, the IAP is relatively simple and can be expressed in a few words or phrases. For a large-scale incident, however, it can be a lengthy document that is regularly updated and used for daily briefings of the command staff.

Planning

The **planning section** is responsible for the collection, evaluation, dissemination, and use of information relevant to the incident. The planning section works with preincident plans, building construction drawings, maps, aerial photographs, diagrams, reference materials, and status boards. It is also responsible for developing, disseminating, and updating the IAP. The planning section develops what needs to be done and by whom, and it identifies which resources are needed.

Command activates the planning section when information needs to be obtained, managed, and analyzed. The **planning section chief** reports directly to command. Individuals assigned to planning examine the current situation, review available information, predict the probable course of events, and prepare recommendations for strategies and tactics. The planning section also keeps track of resources at large-scale incidents and provides command with regular situation and resource status reports.

Logistics

The **logistics section** is responsible for providing supplies, services, facilities, and materials during the incident. The **logistics section chief** reports directly to command and serves as the supply officer for the incident. Among the responsibilities of this section are keeping apparatus fuelled; providing food, refreshments, and rehabilitation facilities for fire fighters; obtaining the foam concentrate needed to fight a large flammable liquid fire; and arranging for specialized equipment to remove a large pile of debris.

In many fire departments, these logistical functions are routinely performed by permanent support services personnel. These groups work in the background to ensure that the members of the operations section have the resources they need to get the job done. Resource-intensive or long-duration situations may require the formal assignment of a logistics section chief, however, because service and support requirements may be so complex or so extensive that they need their own management component.

Finance/Administration

The **finance/administration section** is the fourth major ICS component managed directly by command. This section is responsible for the accounting and financial aspects of an incident, as well as any legal issues that may arise in its aftermath. This function is not staffed at most incidents because cost and accounting issues are typically addressed after the incident. Nevertheless, a **finance/administration section chief** may be assigned at large-scale and long-term incidents that require immediate fiscal management, particularly when outside resources must be procured quickly. A finance/administration section may also be established during a natural disaster or during a hazardous materials incident where reimbursement may come from the shipper, carrier, chemical manufacturer, or insurance company.

Standard ICS Concepts and Terminology

To ensure consistency at emergency incidents, fire departments develop SOPs and then train and practise using ICS. This approach increases safety and efficiency. Emergency scenes tend to be chaotic, so organizing operations at an incident often poses a serious challenge, particularly if the agencies involved use different terms to describe certain concepts and resources. As mentioned earlier, one of the strengths of ICS is its use of standard terminology. The system applies specific terms to various parts of an incident organization. As a fire fighter you must learn the basic concepts and terminology of the ICS.

Some fire departments may use slightly different terminology. For this reason, fire fighters must ensure that they know and understand their ICS departmental SOPs. This section defines important ICS terms and examines their use in organizing and managing an incident. For more information about ICS terminology, see NFPA 1026, *Standard for Incident Management Personnel Professional Qualifications*.

Single Resources and Crews

A single resource is an individual, a vehicle and its assigned personnel, or a crew or team of individuals with an identified supervisor that can be used on an incident or planned event (**FIGURE 22-7**). For example, a pump and its crew would be a single resource; a ladder company would be a second single resource. A company officer is the individual in charge of a company. A company operates as a work unit, with all crew members working under the supervision of the company officer. Companies are assigned to perform tasks such as search and rescue, attacking the fire with a hose line, forcible entry, and ventilation.

A crew is a team of two or more fire fighters who are working without apparatus. For example, members of a pump company or ladder company who are assigned to operate inside a building would be considered a crew. Additional personnel at the scene of an incident who are assembled to perform a specific task may also be called a crew. A crew must have an assigned leader or company officer.

A small-scale incident can often be handled successfully by one single resource. The organizational structure implemented at an emergency incident starts small, with the arrival of the first unit or units. When a limited number of resources are involved, command can often manage the organization personally or with the assistance of an experienced aide. Subsequently, a typical command structure for a one-alarm structure fire is usually of command and two or three reporting resources (**FIGURE 22-8**). A larger incident requires a more complex command structure to ensure that no details are overlooked and that personnel safety is not compromised.

Groups and Divisions

The most frequently used ICS components in structural firefighting, outside of command and reporting resources, are divisions and groups (**FIGURE 22-9**). These components place several single resources under one supervisor. Command can also assign individuals to special jobs, such as incident safety officer and liaison officer, to establish a more effective organization for the incident.

The primary reason for establishing groups and divisions is to maintain an effective span of control.

- A group usually refers to companies and/or crews working on the same task or function, although not necessarily in the same location.

FIGURE 22-7 A single resource is an individual, a vehicle and its assigned personnel, or a crew or team of individuals with an identified supervisor who is assigned to perform specific tasks at an incident.
© Jones & Bartlett Learning. Courtesy of MIEMSS.

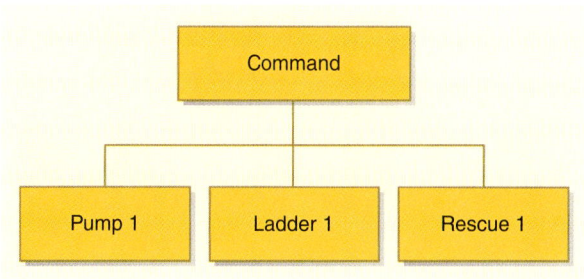

FIGURE 22-8 At a small-scale incident, the typical ICS command structure may consist of command and reporting resources.
© Jones & Bartlett Learning.

- A **division** (or sector) usually refers to companies and/or crews working in the same geographic area.

In the early stages of an incident, individual companies are often assigned to work in different areas or perform different tasks. As the incident grows and more resources arrive on scene, command can establish groups and divisions.

These organizational units are particularly useful when several resources are working near one another in the warm and hot zones. One officer can oversee the activities of more than one company in a specific geographic area or according to a specific function.

Designating divisions ensures that all activities on the scene of an emergency are coordinated and integrated. An IC can maximize and optimize efficiency by specifying areas of responsibility through division assignments. International coordination will occur between divisional supervisors without the need to engage command, which will enable the IC to focus on the macro needs of the incident (**FIGURE 22-10**).

An alternative way of organizing resources is by function (rather than by location). A group includes those resources assigned to a specific function, such as ventilation, search and rescue, or water supply. Groups are responsible for performing an assignment, wherever it may be required, and often work across division lines. For example, the officer assigned to supervise several crews performing ventilation operations would use the radio designation "Ventilation Group."

Division and group supervisors have the same rank within ICS. They are usually chief officers, but company officers also may be assigned to these positions. Divisions do not report to groups, and groups do not report to divisions. Instead, the officers are required to work together—that is, to coordinate their actions and activities with one another. For example, a **group supervisor** (officer) must coordinate with a **division supervisor** (officer) when the group enters the division's geographic area, particularly if the group's assignment will affect the division's personnel, operations, or safety. The division supervisor, in turn, must be aware of everything that is happening within that area. Effective communication between divisions and groups is critical during emergency operations.

Branches

A **branch** is a higher level of combined resources than divisions and groups. At a major incident, several different activities may occur in separate geographic areas or involve distinct functions. The span of control might still be a problem, even after the establishment of divisions and groups. For example, a structural collapse during a major fire may result in several trapped victims and many victims who need medical treatment and transportation. In such a situation, the operations section chief would have multiple responsibilities that could exceed his or her span of control. Activating a fire suppression branch, a rescue branch, and an EMS branch would address this problem (**FIGURE 22-11**). One officer would be responsible for each branch and report to the operations

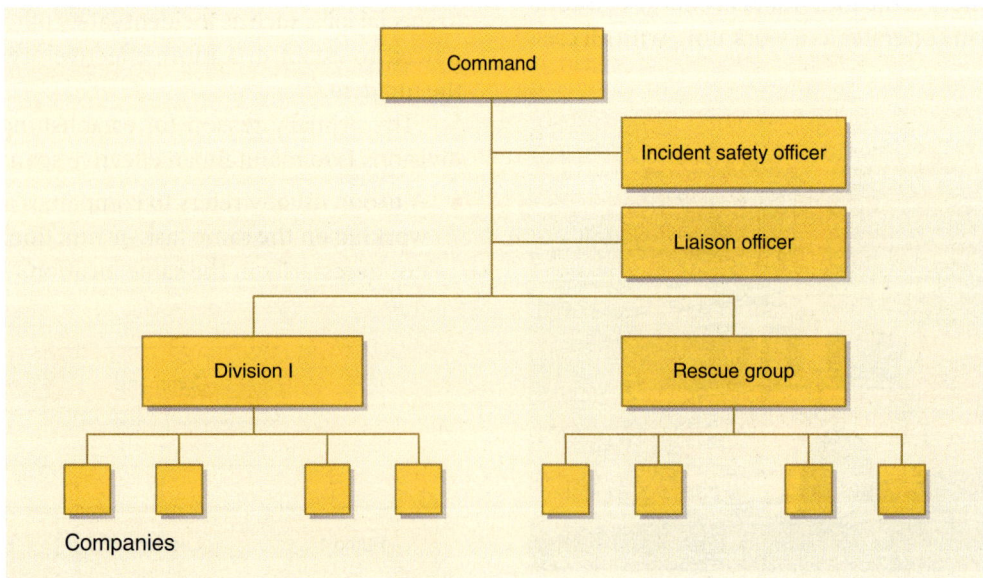

FIGURE 22-9 Divisions and groups are organized to manage the span of control and to supervise and coordinate units working together.

© Jones & Bartlett Learning.

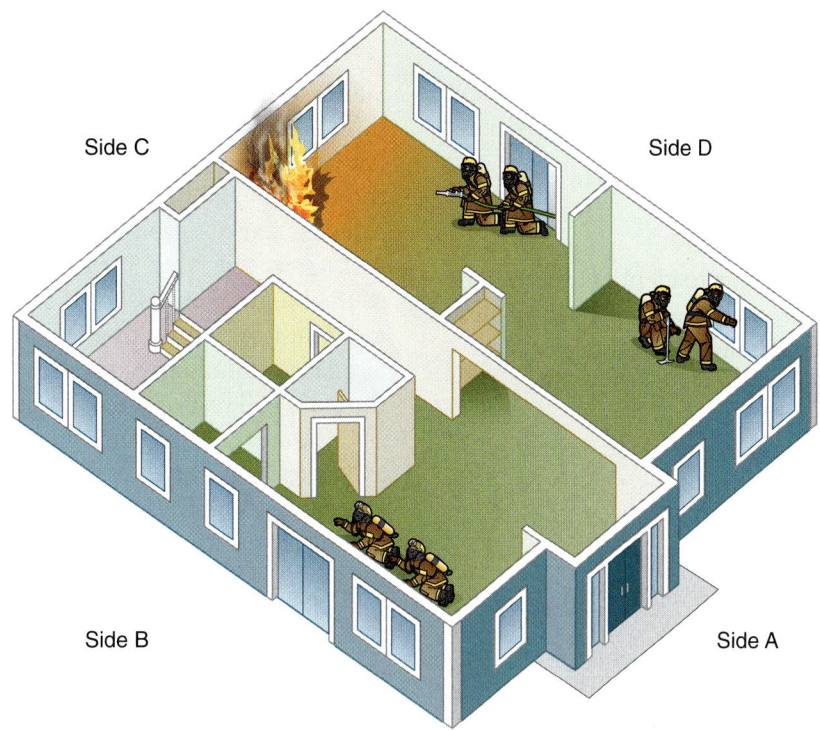

Side C

Side D

Side B

Side A

FIGURE 22-10 A division (or sector) refers to companies or crews working in the same geographic area.
© Jones & Bartlett Learning.

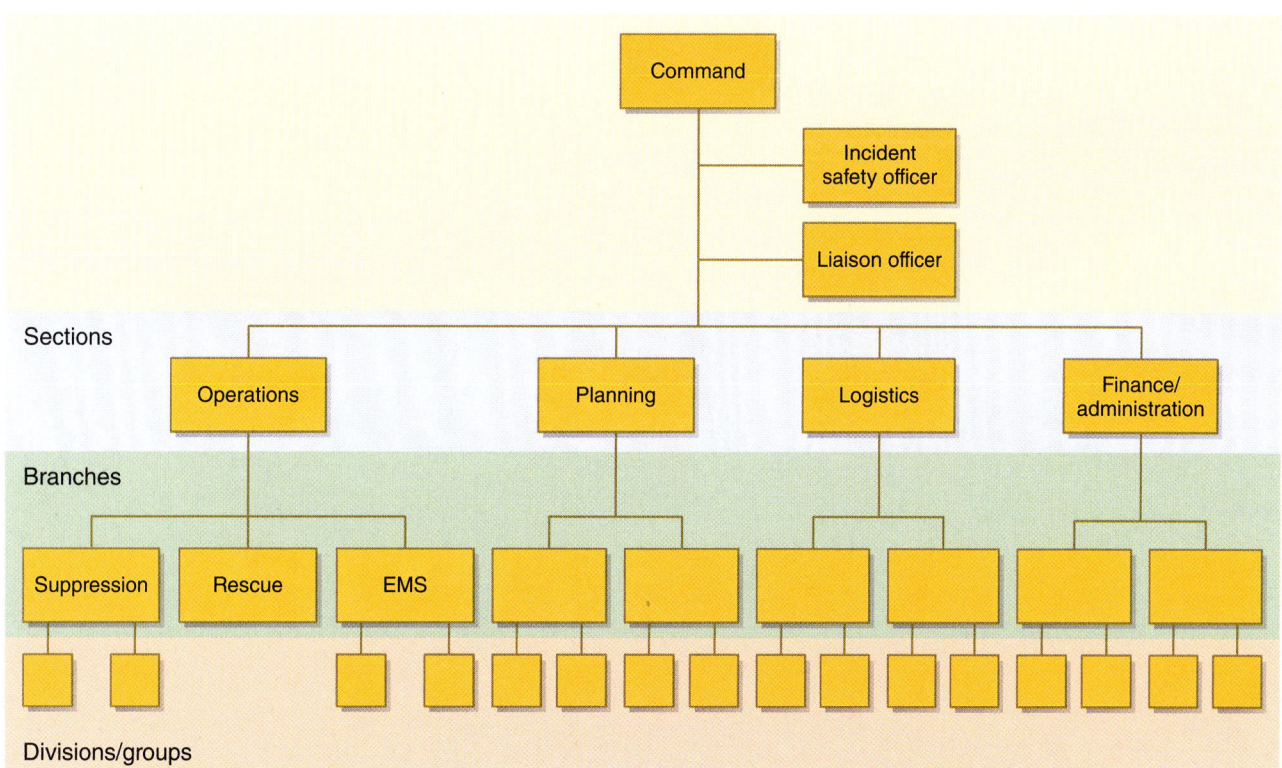

FIGURE 22-11 Creating branches within the operations section is one way to manage the span of control during a large incident.
© Jones & Bartlett Learning.

section chief. Within each branch, several divisions or groups would report to the branch director.

In these situations, a higher-level supervisor (a **branch director**) is in charge of a number of divisions or groups.

Location Designators

The ICS uses a standard system to identify the different parts of a building or a fire scene. Every fire fighter must be familiar with this terminology.

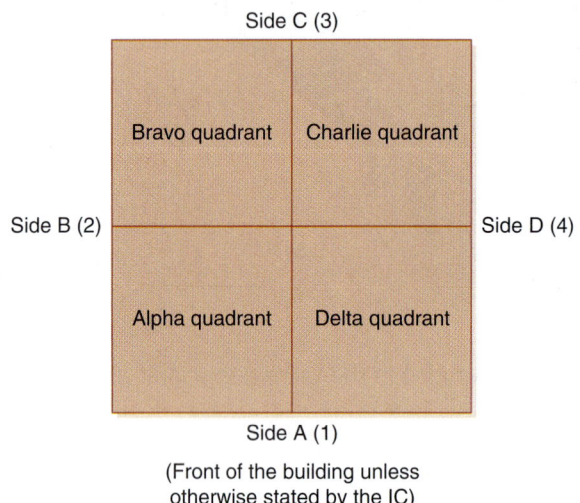

FIGURE 22-12 Location designators in an ICS.
Courtesy of Kevin Lambert.

The exterior sides of a building are generally known as Sides A, B, C, and D. The front of the building is Side A, with Sides B, C, and D following in a clockwise direction around the building. However, many Canadian fire services use a numerical system in which the front of the building is designated Side 1, with sides 2, 3, and 4 designated in a clockwise direction (**FIGURE 22-12**). The companies working in front of the building are assigned as "Division A," and the radio designation for their supervisor is "Division A." Similar terminology is used for companies working at the sides and rear of the building.

The areas adjacent to a burning building are called exposures. Exposures take the same letter as the adjacent side of the building. A fire fighter facing Side A can see the adjacent building on the left (Exposure B) and the building on the right (Exposure D). If the burning building is part of a row of buildings, the buildings to the left are called Exposures B, B1, B2, and so on. The buildings to the right are Exposures D, D1, D2, and so on (**FIGURE 22-13**).

Within a building, divisions are commonly assigned with the number of the floor on which they are working. For example, fire fighters working on the fifth floor would be in "Division 5," and the radio designation for the officer assigned to that area would be "Division 5." Crews doing different tasks on the fifth floor would all be part of this division.

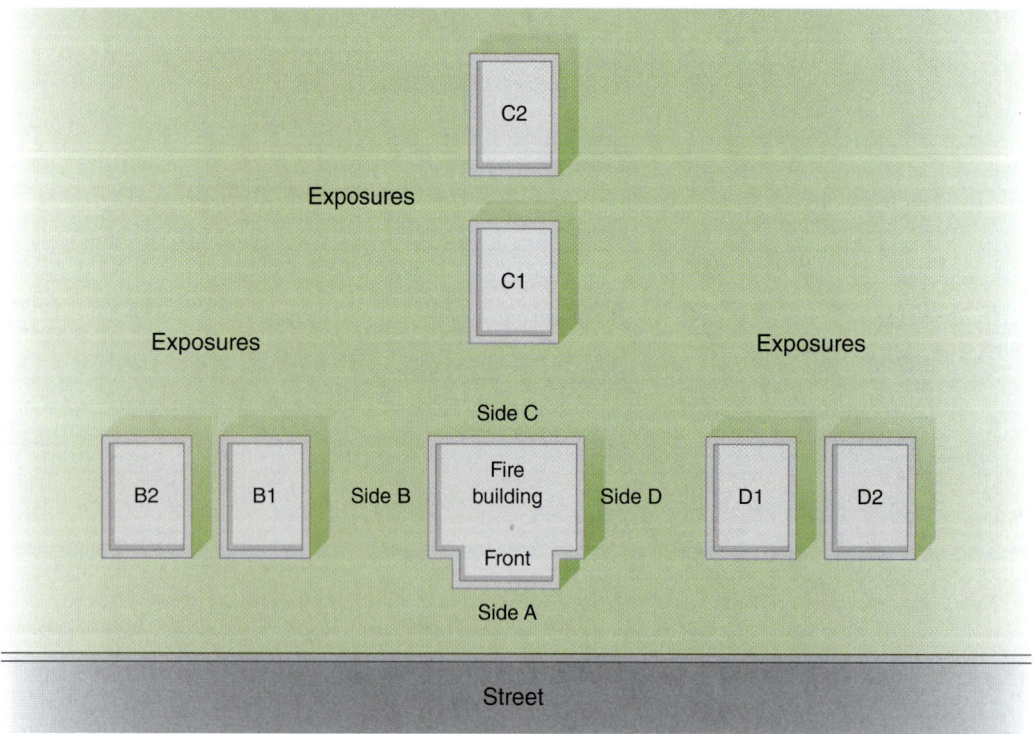

FIGURE 22-13 Location designators in ICS.
© Jones & Bartlett Learning.

Task Forces and Strike Teams

Task forces and strike teams are groups of single resources that have been assigned to work together for a specific purpose or for a certain period of time. Combining resources reduces the span of control by placing several units under a single supervisor.

A **task force** includes two or more single resources, such as different types of units assembled to accomplish a specific task. For example, a task force may be composed of two pumps and one ladder company, two pumps and two brush units, or one rescue company and four ambulances. A task force uses a common communications system and operates under the supervision of a task force **leader**. All communications for the separate units in the task force are directed to the task force leader, who then communicates up the chain of command within the ICS.

Task forces may be assembled for a specific incident need, but they are often part of a fire department's standard dispatch philosophy. For example, a fire department in a large, urban city might dispatch two pump companies, two ladder companies, a rescue company, and a chief officer to all reported structure fires in a high-rise district. Some departments create task forces consisting of one pump and one brush unit for responses during wildland fire season. The brush unit responds with the pump company wherever it goes.

A **strike team** consists of two or more single resources of the same type with an assigned leader. A strike team could be five pumps (pump strike team), five ladders (ladder strike team), or five ambulances (EMS strike team). A strike team operates under the supervision of a strike team leader and uses a common communications system.

Strike teams are commonly used to combat wildland fires—incidents to which dozens or hundreds of companies may respond. During wildland fire season, many departments establish strike teams consisting of two or more pump companies that will be dispatched and work together on major wildland fires (**FIGURE 22-14**). The assigned companies rendezvous at a designated location and then respond to the scene together. Each pump company has a company officer and fire fighters, but only one of the company officers is designated as the strike team leader for each strike team. All communications for the strike team are directed to the strike team leader.

EMS strike teams, which comprise two or more ambulances and a supervisor, are often organized to respond to multiple-casualty incidents or disasters. For example, rather than requesting 15 ambulances and establishing an organizational structure to supervise

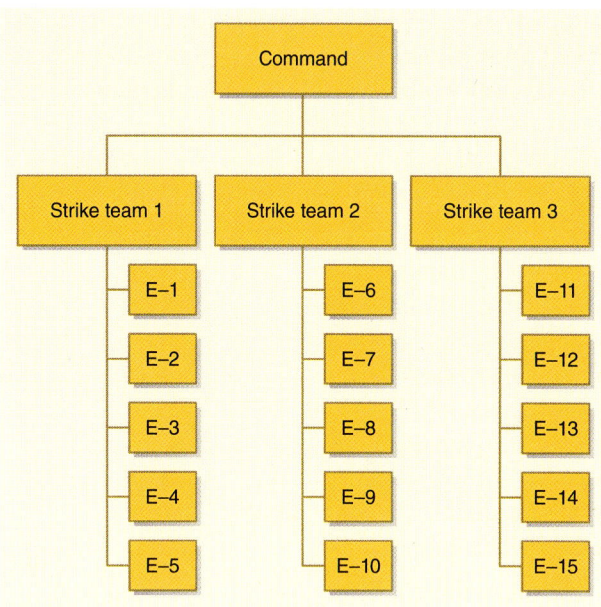

FIGURE 22-14 The organization of a strike team.
© Jones & Bartlett Learning.

15 single resources, command can request three EMS strike teams and can coordinate with three strike team leaders.

FIRE FIGHTER TIP

Strategy is the overall aim or goal used to achieve a desired end result. In the fire service, strategies are usually defined as either offensive or defensive. A tactical objective would be a step or action used to achieve the overall goal or aim of the operation. Tactics are short term and are applied procedures used to complete a tactical objective. Tasks are the actual actions taken to accomplish a tactical objective. For example, aggressively attacking an interior fire is an offensive strategy. The tactical objective may be to prevent the fire's spread while the manner in which the suppressing agent is applied is a tactic, such as a coordinated attack. Forcing a door and stretching hose lines are all tasks associated with the tactical objective. *Tasks* are the specific assignments that will get the job done: "Ladder 1 will go to the roof to make the trench cut; Pumps 3 and 5 will take the hose lines to the third floor; and Ladder 4 will pull ceilings to provide access into the cockloft" are all tasks.

Characteristics of the Incident Command System

Effective management of incidents requires an organizational structure to provide both a hierarchy of authority and responsibility and formal channels for communications. Under the ICS, the specific responsibilities and authority of everyone in the organization are clearly stated, and all relationships are well defined. Important characteristics of an ICS include the following:

- Recognized jurisdictional authority and responsibility
- Applicable to all risk and hazard situations
- Applicable to both day-to-day operations and major incidents
- Unity of command
- Span of control
- Modular organization
- Common terminology
- Integrated communications
- Incident action plans
- Designated incident facilities
- Resource management

Each of these characteristics is discussed next.

Jurisdictional Authority

An IC must be identified at every emergency incident, even if a single pump company or EMS crew responds to and mitigates the emergency. In an effective ICS, everyone working at the incident knows who is in charge of the incident and is responsible for all of the operations required to mitigate the emergency situation. Usually, the role of IC is filled by the ranking fire department, law enforcement, or EMS officer on the scene, depending on the type of emergency. Local protocols determine which agency has jurisdictional authority for various types of incidents. For example, the ranking law enforcement officer may serve as the IC during a localized civil disturbance, while a fire department company officer may fill that role during a routine structural fire incident.

As discussed earlier, the command function is structured in one of two ways: single or unified. When an incident occurs within a single jurisdiction, and when there is no jurisdictional or functional agency overlap, a single IC should be identified and designated as having overall incident management responsibility by the appropriate jurisdictional authority. Identifying the appropriate IC becomes more complicated, however, when several jurisdictions are involved or when multiple agencies within a single jurisdiction have authority for various aspects of the incident. For example, suppose a military aircraft crashes in a national park and ignites a wildland fire that spreads across the park's boundaries into an adjoining province or territory. Such a situation involves both military and civilian agencies as well as multiple levels of government. Chaos would ensue if each affected jurisdiction claimed to be in charge of the incident and attempted to issue orders to the other jurisdictions.

When responsibilities of responders overlap, a unified command may be used instead of a single IC.

The unified command brings representatives of different agencies together to work on one plan and ensures that all actions are fully coordinated. The function of command, whether conducted by an individual IC or through a team of individuals functioning as the unified command, is a management and leadership position. This role is responsible for setting strategic objectives, maintaining a comprehensive understanding of the impact of an incident, and identifying the strategies required to manage that impact effectively.

The introduction of NIMS sparked tremendous discussion and debate related to the broader emphasis given to the concept of unified command. Unified command is a critical evolution of the ICS system that recognizes an important reality of incident management: A multiagency or multijurisdictional response to an incident is a routine occurrence. Unified command provides a framework that allows agencies with different legal, geographic, and functional responsibilities to coordinate, plan, and interact effectively. Unified command, through a consensus-based approach, addresses the challenges that were encountered nationwide when multiple organizations responding to the same incident established separate, but concurrent, incident management structures. This approach, which is still routine in many jurisdictions, leads to inefficiency and duplication of effort, resulting in ineffective incident management, on scene conflict, and substantial safety issues. Unified command is crafted specially to address these issues.

The concept of unified command represents a clear departure from the traditional view of a single IC. Unfortunately, it is frequently misunderstood and difficult for some organizations to implement. To be effective in implementing unified command, agencies and individuals need to worry less about the structural change and more about the changes required in organizational culture and interagency relationships. As incidents become increasingly complex, it is unlikely that any one individual has the knowledge and expertise required to effectively develop, implement, and evaluate strategic objectives for a major incident.

The lesson of unified command is to be concerned less with who is in charge and more with what is required to safely and effectively manage the incident. For unified command to be effective, the management of the incident depends on a collaborative process to establish incident objectives and designate priorities that accommodate those objectives. This approach must yield a single IAP that clearly defines the various agency and jurisdictional responsibilities for incident management. Unified command also yields a single integrated set of incident objectives. Moreover, the use of unified command reinforces those other aspects of ICS that make it effective for managing incidents. For example, because command is provided using the unified command model, the incident is managed using a single organizational structure, thereby limiting duplication and freeing more resources for overall incident management duties. In addition, unified command ensures that the incident will be managed from a single ICP. Finally, unified command allows for a single planning section, which helps prevent multiple, redundant requests for similar items and the inefficient use of limited supplies and resources. If a unified command structure is to be effectively implemented and used, all responding agencies must understand and support the purpose of functioning under this type of ICS.

FIRE FIGHTER TIP

Incoming units and personnel need to be patient and give the IC enough time to perform a proper and thorough size-up. This step is essential in creating an effective IAP. If all personnel are not patient and do not wait for the IAP, then freelancing could occur. **Freelancing** is acting independently of a superior's orders or the fire department's SOPs. Freelancing can disrupt the operations, lead to confusion, and endanger the lives of both fire fighters and civilians.

All-Risk, All-Hazard System

ICS has evolved into an all-risk, all-hazard system that can be applied to manage resources at fires, floods, tornadoes, plane crashes, earthquakes, hazardous materials incidents, mass-casualty incidents, or any other type of emergency situation (**FIGURE 22-15**). This kind of system has also been used to manage many nonemergency events, such as large-scale public events, that have similar requirements for command, control, and communication. The flexibility of the ICS enables the management structure to expand as needed, using whatever components are necessary. The operations of multiple agencies and organizations can be integrated smoothly in the management of the incident.

Everyday Applicability

An ICS can and should be used for everyday operations, as well as for major incidents. Command should be established at every incident, whether it is a trash bin fire, an automobile collision, or a building fire (**FIGURE 22-16**). Regular use of the system ensures familiarity with standard procedures and terminology. It also increases the users' confidence in the system. Frequent use of ICS for routine situations makes it easier to apply to larger incidents.

Unity of Command

Unity of command is a management concept in which each person reports to only one direct supervisor. All orders and assignments come directly from that supervisor, and all reports are made to the same supervisor. This approach eliminates the confusion that can result when a person receives orders from multiple supervisors. Unity of command reduces delays in solving problems as well as the potential for life and property losses. By ensuring that each person has only one supervisor, unity of command can increase overall accountability, prevent freelancing, improve the flow of communication, both up and down the chain of command, assist with the coordination of operational issues, and enhance the safety of the entire situation.

FIGURE 22-15 ICS can be used to manage different types of emergency incidents involving several agencies.
© Mark C. Ide.

FIGURE 22-16 Command should be established at every incident.
© Rick McClure/AP Images.

Span of Control

Span of control refers to the number of subordinates who report to one supervisor at any level within the organization. Span of control relates to all levels of ICS—from the strategic level, to the operational/tactical level, to the task level (individual companies or crews).

In most situations, one person can effectively supervise only three to seven people or resources (FEMA, 2013). Because of the dynamic nature and rapid pace of emergency incidents, an individual who has command or supervisory responsibilities in an ICS normally should not directly supervise more than five people. The actual span of control should depend on the complexity of the incident and the nature of the work being performed. For example, at a complex incident involving hazardous materials, the span of control might be only three; during less intense operations, the span of control could be as high as seven.

Modular Organization

As discussed, the ICS structure and roles are predefined and ready to be staffed and made operational as needed. The ICS has often been characterized as an organizational toolbox, where only the tools needed for the specific incident are used. In an ICS, these tools consist of position titles, job descriptions, and an organizational structure that defines the relationships between positions. Some positions and functions are used frequently, whereas others are needed only for complex or unusual situations. Any position can be activated simply by assigning someone to it.

For example, a small structure fire can usually be managed by the IC, in the form of a person who directly supervises four or five company officers. Each company officer supervises his or her respective company fire fighters. At a larger fire, when more companies respond, the ICS structure would expand to include functional groups or geographic divisions of up to five companies under the direction of a supervisor. At the same time, company officers would be assigned to perform specific ICS functions, such as safety and planning. At even more complex incidents, the additional levels and positions within the ICS structure would be filled in the same manner.

An ICS is not necessarily a rank-oriented system. The best qualified person should be assigned at the appropriate level for each situation, even if that means a lower-ranking individual is temporarily assigned to a higher-level position in the command structure. For complex incidents, however, fire fighters should have experience and special training if they function in ICS positions, such as command, or as section chiefs in the operations, planning, logistics, and finance/administration sections.

Common Terminology

The ICS promotes the use of common terminology both within an organization and among all agencies involved in emergency incidents. Common terminology means that each word has a single definition, and no two words used in managing an emergency incident have the same definition. Everyone uses the same terms to communicate the same thoughts, so everyone understands what is meant. Each job comes with one set of responsibilities, and everyone knows who is responsible for each duty.

Common terminology is particularly important for radio communications. In most Canadian jurisdictions the term "tanker" refers to a vehicle that transports a large volume of water to a fire scene. In some jurisdictions, the term "pump tanker" is used in reference to a tanker with a pump capable of supplying multiple hose lines. The term "water bomber" refers to a plane designed to scoop water from a static source and drop it on a wildland or brush fire. It is important for fire fighters to understand the vernacular of both the fire service and for those with whom they have mutual aid agreements.

> **FIRE FIGHTER TIP**
>
> FEMA's *Resource Typing Library Tool* lists the proper terminology for the common resources used at large-scale emergencies. This tool is available through the FEMA website.

Integrated Communications

Under the NIMS model, all ICS participants must be able to communicate within the command structure. Using an integrated communications system ensures that everyone operating at an emergency can communicate with both supervisors and subordinates. The ICS must support communication up and down the chain of command at every level. A message must be able to move efficiently through the system from command down to the lowest level and from the lowest level up to command (**FIGURE 22-17**). Within a fire department, the primary means of communicating at the incident scene is by radio.

Incident Action Plans

An ICS ensures that everyone involved in the incident is following one overall plan. Different components of the organization may perform different functions, but all of their efforts contribute to meeting the same goals and objectives. At smaller incidents, the IC develops an IAP and verbally communicates the incident priorities,

FIGURE 22-17 Integrated communications are essential for successful operations during an emergency.
© Jones & Bartlett Learning.

LISTEN UP!

Emergency traffic is an urgent message that takes priority over all other communications. The most important emergency traffic is a fire fighter's call for help. Most departments use *mayday* to indicate that a fire fighter is lost, is missing, or requires immediate assistance. If a mayday call is heard on the radio, all other radio traffic should stop immediately. Some agencies utilize the acronym LUNAR to report a mayday. LUNAR stands for: **Location**, your location in the building/incident; **Unit**, the unit to which you are assigned; **Name**, who you are; **Air**, the amount of air you have in your cylinder or **Assignment**, where you were last assigned; **Resources**, what you need to get you out of the mayday situation. The procedure for responding to a mayday call should be studied and practised frequently. The steps for calling a mayday are explained more fully in Chapter 18, *Fire Fighter Survival*.

objectives, strategies, and tactics to all of the operating units. For larger, complex incidents using either a single command or a unified command, a formal, written plan is used to record important incident information. Key representatives from all participating agencies meet regularly to develop and update this plan. In both large and small incidents, all personnel involved in the incident understand what their specific roles are and how they fit into the overall plan.

Designated Incident Facilities

The ICS designates the locations where specific functions are to be performed. For example, command will always be based at the ICP. The staging area, incident base, and helispot are all designated areas where particular functions take place. The facilities required for each incident are established according to the specific IAP or a predefined ICS plan.

Resource Management

The ICS provides a standard method of **resource management** for assigning and keeping track of the resources involved in the incident. Pump companies, ladder companies, and other units are dispatched to the incident and assigned specific functions. The ICS structure is designed to keep track of the various company assignments through an appropriate chain of command and span of control.

At large-scale incidents, units are often dispatched to a **staging area**, rather than going directly to the incident location. The staging area should be close to the incident scene. A number of units can be held in reserve at this location, ready to be assigned if needed.

Personnel are the most vital resource the fire service has. Under the ICS, the IC uses a personnel accountability system to track each member at the scene. Chapter 18, *Fire Fighter Survival*, discusses the various personnel accountability systems in detail.

Size-Up

A formal part of establishing the ICS is **size-up**, the process of initially evaluating an emergency situation to determine which actions need to be taken and which actions can be undertaken safely. It is always the first step in making plans to bring the situation under control. The initial size-up of an incident is conducted when the first unit arrives on the scene and identifies the appropriate actions for that unit or units. Size-up is also used to determine whether additional resources are needed. Size-up is one of the most important steps that needs to be taken at an emergency scene. It is often the key factor in determining whether an emergency operation is a success or a failure.

The initial size-up is often conducted by the first-arriving company officer, who serves as the IC until a higher-ranking officer arrives on the scene and assumes command. The IC uses the size-up information to develop an initial IAP and to set the stage for the actions that follow.

As more complete and detailed information becomes available, the IC will evaluate and update both the initial size-up and the initial IAP. At a major incident, the size-up process might continue through several stages. The ongoing size-up must consider the effectiveness of the initial plan, the impact that fire fighters are having on the problem, and any changing circumstances at the incident.

Although officers usually perform the size-up, all fire fighters must understand how to gather and process information, how to formulate an operational plan, and how to use this information to change plans during the operation. If no officer is present on the first-arriving unit, a fire fighter is responsible for establishing command and conducting the preliminary size-up until an officer arrives. Individual fire fighters are often asked to obtain information and to report their observations to command for ongoing size-up. They should routinely make observations during incidents to maintain their personal awareness of the situation and to develop their personal competence.

Managing Information

The size-up process requires a systematic approach to managing information. Emergency incidents are often complicated, chaotic, and rapidly changing. The IC must look at a complicated situation, identify the key factors that apply to the specific incident, and develop an action plan based on known facts, observations, realistic expectations, and certain assumptions. As the operation unfolds, the IC must continually evaluate, revise, update, supplement, and reprocess the size-up information to ensure that the plan is still valid or to identify when the plan needs to be changed.

Size-up relies on two basic categories of information: facts and probabilities. Facts include information that we know to be true, either through observation or knowledge. In the context of sizing up a fire or emergency situation, the first arriving personnel will analyze the situation in an effort to obtain information. In the case of a fire, the facts can be obtained through visual observation such as flames showing from a second-floor window. Facts can also be based on knowledge, such as properly identifying a structure as Type V lightweight construction. Facts can also be determined through interaction with occupants or victims. Sound fire-ground decision making is predicated on factual information. The Oxford English Dictionary defines probability as "*The quality or state of being probable; the extent to which something is likely to happen or be the case.*" Due to the nature of fire, commanders must make decisions based on probability. Quite often, from the moment the IC arrives on scene, he or she will be constantly reacting to rapidly unfolding events. For fire fighters and officers, a properly performed size-up by thinking in terms of anticipating events is a superior way to manage the fire ground. Information filtered through a balance of probabilities allows decision makers to allocate and place resources before they are needed, thereby assigning tactical objectives intended to mitigate further damage. It is often the case that the initial arriving IC will be faced with the need to make several decisions within the first few minutes of arrival. Some of these decisions will be made through experiential knowledge based on facts. Other decisions, such as establishing hose lines to protect exposures, will be

based on the anticipatory probability that an event may occur. As the incident evolves, the IAP evolves as well. When time and circumstance permit, the IC establishes a permanent command post and begins taking notes and formalizing plans in writing. Many Canadian fire services have developed tactical worksheets that provide guidance for this process. Experienced and knowledgeable fire-ground commanders will make decisions based on their analysis of the situation. Effectively sizing up a situation involves skills that can only be obtained through constant practice, training, education, and experience.

FIRE FIGHTER TIP

Personnel should view firefighting as a profession that is continually changing. No one is an expert fire fighter because our understanding of fire dynamics, building construction, and tactical approaches to fire fighting are constantly changing. Like many professions, firefighting requires continual practice and sincere dedication to learning.

Facts

Facts are obtained from various sources. For example, the communications centre will provide some facts about the incident during dispatch. A **preincident plan** will contain many facts about the structure. Maps, manuals, and other references could provide additional information. A preincident plan generally provides details about a building's construction, layout, contents, special hazards, and fire protection systems. Chapter 27, *Fire and Life Safety Initiatives*, contains specific information about the contents and development of such plans. As you review the information, consider how preincident plans impact the size-up process.

Mobile computer data terminals, laptop computers, and tablets have the ability to call up previously entered facts at a touch of a few buttons. Individuals with specific training, such as a building engineer or a utility representative, can add even more specific information. In addition, a seasoned officer will have built up a knowledge bank of valuable information based on experience, training, and direct observations.

The initial dispatch information will contain facts such as the location and nature of the situation. This information could be very general ("a reported building fire in the vicinity of Central Avenue and Main Street") or very specific ("a smoke alarm activation in room 3102 of the First National Bank Building at 173 South Main Street").

The time of day, temperature, and weather conditions are other factors that can easily be determined and incorporated into the initial size-up. Based on these basic facts, an officer might have certain expectations about the incident.

For example, whether a building is likely to be occupied or unoccupied, whether the occupants are likely to be awake or sleeping, and whether traffic will delay the arrival of additional units may all be inferred from the time of day. Weather conditions such as snow and ice might delay the arrival of fire apparatus, create operational problems with equipment, and require additional fire fighters to perform basic functions such as stretching hose lines and raising ladders. Strong winds can cause rapid extension or spread of a fire to exposed buildings or push a fire in unexpected ways. High heat and humidity will affect fire fighters' performance and may cause heat injuries.

A well-trained, experienced fire officer will have a basic knowledge of the community and the department's available resources. The officer may remember the types of occupancies, construction characteristics of typical buildings, and specific information about particular buildings from previous incidents or preincident planning visits.

Basic facts about a building's size, layout, construction, and occupancy can often be observed upon arrival, if they are not known in advance. In particular, the officer must consider the size, height, and construction of the building during the size-up. The approach taken for developing an IAP for a two-storey Type V detached residential dwelling is quite different from one developed for a Type I fire-resistive residential high-rise.

A plan for rescuing occupants and attacking a fire must also incorporate information about the building layout, such as the number, locations, and construction of stairwells. The plan for a building with an open stairwell that connects several floors might identify keeping the fire out of the stairwell as a priority. This focus enables fire fighters to use the stairwell for rescue and prevents the fire from spreading to other floors. Ladder placement, use of aerial or ground ladders, and emergency exit routes all depend on the building layout.

Any special factors that will assist or hinder operations must also be identified during size-up. For example, bars on the windows will limit access and complicate the rescue of trapped victims. Fire detection and suppression systems play a vital part in the initial size-up. A fire in a building that contains a properly designed and installed fire detection system with external notification will be reported sooner to the fire department than an unoccupied building with no alarm system. Likewise, an operational sprinkler system may confine or extinguish the fire. The presence of a standpipe system makes getting a hose line in operation on an upper floor of a burning building much quicker and easier.

The occupancy of a building also represents critical information. An apartment building, a restaurant, an automotive repair garage, an office building, a warehouse filled with concrete blocks, and a chemical manufacturing plant all present fire fighters with different sets of problems that must be addressed.

The IC should initially perform a size-up by walking or driving completely around all four sides of the building, a process often referred to as a "walk-around" or "doing a 360." By completing a loop around a building, the IC can observe fire and smoke conditions on all sides of the building and determine if immediate rescue is required. The IC can look for signs of occupancy, locations of windows and doors, building construction, fire suppression devices, exposures, and locations of utility entrances such as electrical and gas. It will not always be possible for the IC to complete this. In those cases, the companies assigned to the sides and back can give a limited size-up report to the IC, who will usually be located on the front (Side A or Side 1) of the building.

Properly identifying the size and scope of the problem is a crucial component of the size-up. Determining the fire's exact location can be challenging when there are large volumes of smoke and heat but little fire showing from the exterior. The size, type, and location of attack lines are wholly dependent on the situation and the need for immediate rescue.

Direct visual observations will give the best information about the size and location of a fire, particularly when combined with information about the building. Although visible flames can indicate where the fire is located and how intensely it is burning, it may not be an accurate portrayal of the overall situation. Flames issuing from only one window suggest that the fire is confined to that room, but it could also be spreading through void spaces to other parts of the building.

An experienced fire fighter will observe where smoke is visible, how much is apparent, what colour it is, and how it moves. Smoke reading, as described in Chapter 5, *Fire Behaviour*, is a vital part of scene size-up because this process helps the IC predict where the fire might go.

Fire fighters quickly learn that no two fires are the same. Fire conditions within a given structure may be so intense that visibility is negligible. Deprived of sight, fire fighters must feel their way to the main body of the fire. If smoke limits visibility, a crackling sound or the sensation of heat coming from one direction may indicate the location of the fire. Blistering paint and smoke seeping through cracks can lead fire fighters to a hidden fire burning within a wall. Thermal imaging devices are effective devices for locating fires because they can sometimes pinpoint the location of a fire that otherwise would be hard to find. In this way, thermal imaging devices can help fire fighters determine the location of a fire quickly and without putting fire fighters at much risk.

The IC needs to gather as much factual information as possible about a fire. Because the IC often is located at a fixed ICP outside the building, company officers will be requested to report their observations from different locations. A company that is operating a hose line inside the building can report on interior conditions, whereas the ventilation crew on the roof will have a very different perspective. The IC may request an officer or a fire fighter to gather specific information that will help him or her develop or modify the IAP.

Regular progress reports from companies working in different areas will provide updated information about the situation. Progress reports enable an IC to judge whether an operational plan is effective or needs to be changed.

Probabilities

Probabilities refer to events and outcomes that can be predicted or anticipated, based on facts, observations, common sense, and previous experiences. Fire fighters frequently use probabilities to anticipate or predict what is likely to happen in various situations. The IAP is also based on probabilities, predicting where the fire is likely to spread and anticipating potential problems.

An IC must be able to quickly identify the probabilities that apply to a given situation. For example, a fire in an apartment building in the middle of the night will probably involve occupants who need to be rescued. In such circumstances, the IC would assign additional crews to search for potential victims, even if no factual information indicates that any occupants need to be rescued. Similarly, a fire burning on the top floor of a structure in a row of attached buildings has a high probability of spreading to adjoining buildings through the cockloft. In this case, the IC's plan would include opening the roof above the fire and sending crews into the exposed occupancies to check for fire extension in the cockloft.

The concepts of convection, conduction, and radiation enable an IC to predict how a fire will extend in a particular situation. By observing a particular combination of smoke and fire conditions in a particular type of building, the IC can identify a range of possibilities for fire extension within the structure or to other exposed buildings. Using these probabilities, the IC can predict what is likely to happen and develop a plan to control

the situation effectively. Effective size-up requires a good knowledge of fire behaviour to evaluate the probabilities for the spread of a fire.

The IC must also evaluate the potential for collapse of a burning structure. The building's construction, the location and intensity of the fire, and the length of time the structure has been burning are all factors that must be considered in making this determination. Houses built in the last 30 years using lightweight construction techniques, for example, tend to collapse in a much shorter period of time than houses built in earlier eras. When the possibility of collapse exists, risk management dictates that the IC not permit fire fighters to enter the building or immediately order all fire fighters to exit the building.

Resources

Resources include all of the means that are available to fight a fire or conduct emergency operations at any other type of emergency incident. Resource requirements depend on the size and type of incident. Resource availability depends on the capacity of a fire department to deliver fire fighters, fire apparatus, equipment, water, and other items that can be used at the scene of an incident.

A fire department's basic resources consist of its personnel and apparatus. Firefighting resources are usually defined as the number of pump companies, ladder companies, special units, and command officers required to control a particular fire. An IC should be able to request the required number of companies and know that each unit will arrive with the appropriate equipment and the necessary fire fighters to perform a standard set of functions at the emergency scene. The IC should know how many and which types of companies are available to respond, how they are staffed and equipped, and how long they should take to arrive. This information might have to be updated at the incident, particularly in areas served by volunteer fire fighters, because the number of fire fighters available to respond may vary at different times.

Water supply is another critical resource. It is rarely a problem if the fire occurs in a district that has fire hydrants and a strong, reliable municipal water system with adequate-sized water mains. In an area without fire hydrants, however, water supply may be severely limited. Even when a static water supply is available, it takes time to establish a water supply from such a source. If the water supply is limited to tankers or tenders, the amount of water that can be delivered is limited. In this situation, the IC would need to call for an additional supply of tankers.

Sometimes, resources include more than fire fighters, fire apparatus, and a water source. A fire in a flammable liquids storage facility will require large quantities of foam and the equipment to apply it effectively. A hazardous materials incident might require special monitoring equipment, chemical protective clothing, and bulk supplies to neutralize or absorb a spilled product. A building collapse might require heavy equipment to move debris and a structural engineer to determine where fire fighters can work safely. The fire department must have these supplies available or be able to obtain them quickly when they are needed. Resources for a large-scale incident must include food and fluids for the rehydration of fire fighters, fuel for the apparatus, and other supplies. The Red Cross, Salvation Army, and law enforcement agencies often provide support resources.

Sizing-up enables the IC to determine which resources will be needed to control the specific situation and to ensure their availability at an appropriate time. An IAP to control an incident can be effective only if the necessary resources can be assembled quickly. If a delay occurs, the IC must anticipate how much the fire will grow and where it will spread. If the desired resources cannot be obtained, the IC must develop a realistic plan using the available resources to gain eventual control of the situation.

Ideally, a fire department will be able to dispatch enough fire fighters and apparatus to control any situation within its jurisdiction. In reality, however, many fire departments do not have an adequate number of apparatus or personnel on duty to meet the emergency response needs of their communities. Most departments have established mutual aid agreements to assist surrounding jurisdictions if a situation requires more resources than the individual community can provide. In some areas, hundreds of fire fighters and apparatus can be assembled to respond to a large-scale incident. In more remote areas, the resources available to fight a fire can be very limited.

If resources are insufficient or delayed, a fire can become too large to be controlled by available resources. For example, if only one or two apparatus and a few fire fighters respond to a structure fire, the IC will have to determine whether they will offensively attack the fire, defensively flow available water on the fire, or protect the exposures while the original building burns. The IC must always determine which actions can be taken safely with the resources that are currently available.

Resources, in turn, must be organized to support efficient emergency operations. Fire fighters must be properly equipped and organized in companies. Equipment and procedures must be standardized. The ICS must be used to manage all of the resources that could be used at a large-scale incident, and the communications system must enable the IC to coordinate operations effectively.

Implementing the Incident Command System

To an outsider, the ICS might appear to be a large, complicated organizational model that involves a complex set of SOPs. By contrast, to the individual fire fighter working within this system, the ICS is simple and uncomplicated.

This section discusses how every emergency incident is conducted. Certain components of the ICS are used on every incident and at every training exercise. Fire fighters generally have specific responsibilities and procedures to follow in most situations.

Three basic components always apply:

- Command is established at every incident and is maintained from the time that the first unit arrives until the time that the last unit leaves. The identity of command may change, but there is always one individual (or group of individuals from different agencies or jurisdictions) who is in charge of the incident and responsible for everything that happens. SOPs may dictate who will be command at any time.
- Each fire fighter always reports to one supervisor. A fire fighter's supervisor will usually be a company officer. The company officer directly supervises a small crew of fire fighters, such as a pump company or ladder company, who work together. At an incident scene, the company officer provides instructions and must always know where each fire fighter is and what he or she is doing. If the company officer assigns two fire fighters to work together away from the rest of the company, both fire fighters remain under the supervision of the company officer. The company officer could be an acting officer (a fire fighter temporarily designated as a "fill-in" officer), or a fire fighter could be assigned to work temporarily under the supervision of a different officer.
- The company officer reports to command. If only one company is present at the scene, the company officer is command, at least until someone else arrives and assumes that role. At a small incident, the company officer may report directly to command. At a large incident, several layers of supervision may separate a company officer and command.

To make the ICS work, every fire fighter must know how to function within this system. To fulfill your role as a member of the incident command team, follow the steps in **SKILL DRILL 22-1**.

SKILL DRILL 22-1
Operating Within the ICS Fire Fighter II, NFPA 1001: 5.1.2

1 Verify that the ICS is in use.

2 When given an assignment, repeat that information over the radio to verify it.

3 Assess the scene, hazards, equipment, and PPE to ensure your safety.

4 Account for yourself and for other team members.

5 Update your supervising officer regularly.

6 Provide personnel accountability reports as necessary.

7 Report completion of each assignment to your supervising officer.

© Jones & Bartlett Learning. Photographed by Glen E. Ellman.

Establishing Command

The first fire fighters to arrive at an emergency incident are the foundation of the ICS structure. Rarely is a high-ranking chief sent to a fire to evaluate the situation, design the organization structure, and order the resources that will be needed. Instead, the ICS builds its organization from the bottom up, around the units that take initial action.

The officer in charge of the first-arriving apparatus assumes command, performs an initial size-up, identifies the initial strategy, and requests or returns any additional resources. As the initial IC, this officer is responsible for managing the operation until he or she is relieved by a senior officer. If there is no officer on the first-arriving unit, the fire fighter with the greatest seniority must function as the IC until an officer arrives and assumes command. The position of command must have an unbroken line of succession from the moment the first unit arrives on the scene until the incident is terminated.

Any units arriving after the first unit know that they will be taking their orders from command until a higher-level officer assumes command.

The first-arriving IC must decide whether to take action and directly supervise the initial attack crew or whether to concentrate on managing the incident as a stationary IC. This decision depends on many factors, including the nature of the situation, the resources that are on the scene or expected to arrive quickly, and the ability of the crew to work safely without direct supervision.

If the incident is large and complicated, the best option for command is to establish an ICP and focus on sizing up the situation, directing incoming units, and requesting additional resources. Command's own crew can be assigned to work with an acting officer or to join forces with another company officer and crew. If the situation is less critical, command might be able to function as both a company officer and command simultaneously, at least temporarily.

If the first-arriving unit is a chief officer, the chief officer automatically establishes command and begins executing command responsibilities. If a company officer had previously established command, the company officer would transfer command authority to the first-arriving chief officer. This information would be transmitted over the radio to all units, indicating who has command and the effective time of transfer. The officer relinquishing command should provide an assessment of the situation to the incoming officer, as described in the Transferring Command section found later in this chapter.

Most fire departments have written procedures that specify who will function as command in certain situations. If multiple units respond from the same station and arrive together, there should be an established protocol for which officer establishes command. If a company officer arrives only seconds ahead of a chief officer, the chief officer should establish command from the outset.

Identifying Command

The individual initially establishes command by designating command over the radio. The exact wording varies from department to department, but a common message is as follows: *Dispatch from Pump 15. Pump 15 is on location at 123 Main Street. This is a lightweight Type V two-storey detached single-family residence with flames showing from the second floor. Put in a working fire. We will be in offensive mode laying lines. Pump 15 assuming 123 Main Street Command.* This announcement eliminates any possible confusion over who is in charge. The initial report should include the information in **SKILL DRILL 22-2**.

In smaller, volunteer departments, the first-arriving fire fighter may arrive on the scene in his or her personal vehicle. In this situation, the fire fighter should provide the same information as described, with assignments for incoming units as they arrive. Be sure to follow your local SOPs for establishing the ICS and giving size-up information.

The initial announcement of command confirms that command has been established at an incident and identifies who is the IC. If no one announces that he or she is the IC, all responders should realize that command

has not been established. The announcement of command also reinforces command's personal commitment to the position through a conscious personal act and a standard organizational act.

Identifying the Incident

Individual fire department procedures may vary in terms of the specific protocol they use for naming an incident. The first officer to assume command should establish an identity that clearly identifies the location of the incident, such as "Pump 10 will be Seventh Avenue Command." This announcement reduces confusion on the radio and establishes a continuous identity for command, regardless of who holds that position during the incident.

There can be only one "Seventh Avenue Command" at a time, so there is no confusion about who is in charge of the incident. When anyone needs to talk to command, a call to "Seventh Avenue Command" should be answered by the individual who is in command of the incident. Identifying the name of command also prevents confusion if multiple emergency incidents are operating on the same radio frequency.

Transferring Command

Transfer of command occurs when one person relinquishes responsibility and authority for an incident to another person, usually a superior officer. For example, the first-arriving company officer transfers command to the first-arriving chief officer, who would later transfer command to a higher-ranking officer during a major

SKILL DRILL 22-2
Establishing Command Fire Fighter II, NFPA 1001: 5.1.2

© Jones & Bartlett Learning. Photographed by Glen E. Ellman.

The initial report to establish command should include the following information:

- Arrival announcement verifying address
- Building description and occupancy
- Initial observations and definition of the problem
- Request for or release of resources
- Initial actions and strategy announcement
- Announcement of assumption of command and radio designation for command.

incident. Some departments require the transfer of command when a higher-ranking officer arrives at an incident; others give the higher-ranking officer the option of assuming command or leaving the existing personnel in the command assignment.

When a higher-ranking officer arrives at the scene of an emergency incident and takes charge, that officer assumes legal responsibility of managing the overall operation. Established procedures must be followed when command is transferred. One of the most important requirements of command transfer is the accurate and complete exchange of incident information (**FIGURE 22-18**). The officer who is relinquishing command needs to provide the incoming IC with a current situation status report that includes the following information:

- Strategy and current tactical objectives
- An overview of the incident action plan
- Any safety concerns
- Progress toward achieving tactical objectives
- The establishment of a rapid intervention team, accountability, rehabilitation, and decontamination
- The need for or return of resources
- Status of all fire companies on scene

If transfer of command occurs early in an incident, the transfer of information may be brief. For example, the first-arriving company officer might have been in command for only a few minutes and have little information to report when command is transferred to the first-arriving chief officer. The chief may have heard all of the exchanges over the radio and know what the current situation is. Conversely, if the company officer has been in command for several minutes, there may be a significant amount of information to report, such as the current assignments of all first-alarm companies.

Whether the information is minimal or substantial, the transfer must be accurate and complete.

Most Canadian fire departments have established protocols and SOPs for incident command and incident management. Fire fighters must be well versed in their departmental SOPs. In some cases, the transfer of command may be done via radio. The transfer of command at complex incidents, however, should be done face-to-face to avoid unnecessary radio traffic. This gives both the incoming and outgoing ICs the opportunity to see facial expressions and body language when describing the incident and the situations of which they are about to take command.

The incoming officer should report to the ICP for a situational briefing and the formal transfer of command. Once command has been transferred, the identity of the new IC should be announced to all on scene personnel.

As discussed previously, command is always maintained for the entire duration of an incident. In the later stages of an incident, after the situation is under control, command may be transferred to an officer of lesser rank. A downward transfer of command requires the same type of briefing and exchange of information as an upward transfer of command. The officer in charge of the last company remaining on the scene would be in command. When that company leaves the scene, the command function is terminated.

Command Transfer Rationale

Command may be transferred at different points during an emergency incident for several important reasons. A first-arriving company officer can usually direct the initial operations of two or three additional companies, but as situations become more complex, the problems of maintaining control increase rapidly. A company officer's primary responsibility is to supervise one crew and ensure that the crew members operate safely. When three or more companies are operating at an incident, it is better to have a chief officer assume command.

As more companies are assigned to an incident, the command structure must also expand. That is, the organization must grow to maintain an effective span of control. Additional chief officers may be assigned to the incident, and a higher-ranking officer may assume the command role. A command transfer also may be required if the situation is beyond the training and experience of the current command. A more experienced officer may have to assume command to ensure proper management of the incident.

All fire fighters should know the steps involved in transferring command during an emergency incident (**SKILL DRILL 22-3**).

SKILL DRILL 22-3
Transferring Command Fire Fighter II, NFPA 1001: 5.1.2

1. Establish the ICS.

2. Follow departmental procedures for transferring command.

3. Transfer command in a face-to-face meeting, if possible. If not possible, transfer command over the radio.

4. Communicate to the incoming command officer the tactical priorities, action plans, hazardous conditions, potentially hazardous conditions, accomplishments, effectiveness of operations, status of resources, and need for additional resources.

5. Formally announce the transfer of command over the radio.

© Jones & Bartlett Learning. Photographed by Glen E. Ellman.

FIRE FIGHTER TIP

The goal of the National Incident Management System (NIMS) is to train all emergency responders to be able to operate within this national-scale plan. To make it easier for responders to access this material and to complete this training, a variety of course offerings are available through the U.S. Department of Homeland Security, FEMA's website, and Jones & Bartlett Learning. Check with your training officer to determine which level of incident management training is required by your department.

Incident Action Plan

Based on information gathered during size-up and once the officer establishes command and acts as the IC, the IC develops an IAP that outlines the steps needed to control the situation. The initial IC develops a basic plan for beginning operations. The plan may be revised and expanded as necessary.

Incident Priorities

The IAP should be based on the three incident priorities, which are listed here in order of priority:

1. **Life safety.** This includes keeping fire fighters and other emergency responders safe, as well as rescuing victims.

2. **Incident stabilization.** The incident must be stabilized before the fire can be fully extinguished or the hazardous material can be cleaned up. Incident stabilization may include protecting exposures from impinging fire or containing hazardous fuel spills.

3. **Property conservation.** Once life safety is ensured and the incident is not expanding or becoming more threatening, actions can be taken to prevent further property damage. These actions may include extinguishing the fire, covering furnishings with salvage tarps, and covering ventilation holes in the roof to prevent further water damage.

This system of priorities clearly establishes that the highest priority in any emergency situation is saving lives. All initial emergency scene operations must focus on this priority. If there are no immediate life-safety issues, the other two incident priorities can be addressed.

These priorities are not separate and exclusive, of course. Often, more than one incident priority can be addressed simultaneously, and certain activities help achieve more than one objective. For example, if a direct attack on the fire will bring it under control very quickly, the incident priorities of life safety, incident stabilization, and property conservation can all be addressed simultaneously. Attacking the fire may simultaneously reduce the risk to trapped occupants, limit fire and smoke spread, and prevent further fire damage to the structure.

RECEO-VS

Fire departments and their ICs employ different strategies and tactics to meet the three incident priorities. The acronym RECEO was used in previous decades to identify the primary command objectives. Developed by Chief Lloyd Layman (Parkersburg, WV) in the 1950s, RECEO followed the hierarchy of the three incident priorities, focusing on the most critical to the least critical fire-ground issues. Rescue, Exposure protection, Confinement of the fire, Extinguishment, and Overhaul (RECEO) provided a general guideline for fire-ground commanders to systematically address the incident priorities. Through the years, RECEO was modified by different departments to better meet their incident strategies. A common version used in recent years is RECEO-VS (TABLE 22-2).

Although the acronym provides a suggested order of strategic objectives, the IC needs to determine if and how each of these objectives fits into the IAP. These objectives serve as a guide to help the IC make difficult decisions, particularly if not enough resources are available to address every objective. If a decision must be made between saving lives and saving property, saving lives comes first. After rescue is completed and if the

fire is still spreading, the IC should then place exposure protection ahead of salvage and overhaul.

A similar set of priorities can be established for any emergency situation: Saving lives is always more important than protecting property. The IC must always place a higher priority on bringing the problem under control than on cleaning up after the problem.

Rescue

Life safety is the first incident priority at a fire or any other emergency incident. The need for rescue depends on many circumstances. Notably, the number of people in danger is likely to vary based on the type of occupancy and the time of day. A commercial building that is crowded during the workday might have few, if any, occupants at night. At night, rescue is more likely to be needed in a residential occupancy, such as a house, an apartment building, or a hotel.

The degree of risk to the lives of the building's occupants must also be evaluated. A fire that involves one apartment on the 10th floor of a high-rise apartment building could threaten the lives of both the residents on that floor and the residents who live directly above the fire. In contrast, residents below the fire are probably not in significant danger, and residents several floors above the fire might be safer staying in their apartments until the fire is extinguished instead of walking down smoke-filled stairways.

| TABLE 22-2 RECEO-VS |||
Incident Priorities	RECEO-VS Acronym	Meaning
Life safety	**R**escue	The removal of victims from a life-threatening situation
Stabilize the incident	**E**xposure Protection	The protection of surrounding structures and uninvolved parts of the effected structure from a fire
	Confinement	The containment of a fire to one area or the prevention of further areas from becoming involved in the fire
Property conservation	**E**xtinguishment	The complete extinguishment of the fire
	Overhaul	The process of ensuring that the fire is extinguished completely
	Ventilation	The process of removing smoke, heat, and the products of combustion from a structure
	Salvage	The removal or protection of property that could be damaged during firefighting or overhaul operations

If the size-up suggests that people may be trapped in a burning building, the IC must determine whether it is safe to send fire fighters into the building. The IC should use survivability profiling to make this decision. **Survivability profiling** is a type of **risk/benefit analysis**, an assessment that weighs the risks to be taken versus the benefits of those risks, of the viability and survivability of potential fire victims under the current conditions in the structure. Between 2008 and 2014, 164 fire fighters died in the line of duty while working at structure fires while only nine civilians died at those same structure fires (Shaw, 2015). This means that many fire fighters have died on the fire ground when no victims were in the structure. Recent research also indicates that, due to modern construction methods and materials, victim survivability decreases dramatically with each passing minute after ignition. Survivability profiling helps the IC to decide whether to commit fire fighters to interior search and rescue and fire suppression operations to save live victims. Fire fighters should engage in interior operations only if there is a reasonable probability that trapped civilians might be alive and if interior conditions do not place fire fighters at an unacceptable risk.

Victim survivability declines as room temperatures approach 100°C (212°F). Death occurs at temperatures of 177°C (350°F) in approximately 3 minutes. If an occupant breathes superheated air and gases, burns to the respiratory system may result in death within minutes.

In today's fires, interior temperatures can rise higher than 260°C (500°F) in 3 to 4 minutes. Flashover produces temperatures over 538°C (1000°F) and often occurs before the first fire fighters arrive on the scene. We must always remember that if an environment is not tenable for a fire fighter in full PPE, it is unlikely that trapped victims will survive. It is not acceptable to risk the lives of fire fighters to try to save the lives of victims who have already died or will not survive.

SAFETY TIP

It is not acceptable to risk the lives of fire fighters to try to save the lives of victims who are already dead.

Often, the best way to protect lives is to extinguish the fire quickly. For this reason, efforts to control the fire are usually initiated at the same time as rescue operations. For example, hose lines may be used to protect exit paths and keep the fire away from victims during search and rescue operations. Chapter 12, *Search and Rescue*, discusses specific tactics and techniques for these operations.

Exposure Protection

The second incident priority is to stabilize the incident. Usually, this means mitigating the incident using an outside-to-inside approach and protecting exposures from impinging fire. For structural fire incidents, this means controlling the fire so it does not extend beyond the room, area, or building of origin when the fire department arrives. Many historic conflagrations developed because the original fire quickly spread from the building of origin and involved dozens—or hundreds—of buildings. The IC must start by making sure the fire is not expanding.

In some cases, the IC must look ahead of the fire and identify a place to stop its spread. If flames are extending quickly through the cocklofts in a row of attached buildings, for example, the IC might place hose lines ahead of the fire to stop its progress.

The IC must sometimes weigh potential losses when deciding where to attack a fire. If a fire in a vacant building threatens to spread to an adjacent occupied building, the IC will usually assign companies to protect the exposure before attacking the main body of fire. If, however, a fire in an occupied building might spread to a vacant building, the IC's decision might be to attack the fire first and worry about controlling the spread later.

Confinement

As part of stabilizing the incident and after ensuring that the fire is not actively extending into any exposed areas, the IC will focus on confining the fire to a specific area. If the fire is burning in only one room, the objective should be to ensure that it does not spread beyond that room. If more than one room is involved, the objective might be to contain the fire to one apartment or one floor level. Sometimes the objective is to confine the fire to one building, particularly if multiple buildings are attached or in close proximity to the building of origin.

To accomplish this, the IC or safety officer may define a safety perimeter to prevent injuries and plan operations so that the fire does not expand beyond that area. Fire fighters on the perimeter must be alert to any indications that the fire is spreading to those limits.

Thermal imaging devices can detect hidden sources of heat, and they are valuable tools for finding fires in void spaces. The principles and use of thermal imaging devices are covered more fully in Chapter 12, *Search and Rescue*.

Extinguishment

The third incident priority is property conservation. These actions may include extinguishing the fire, ventilation, and salvage and overhaul operations. Depending on the size of the fire and the risk involved, the IC will mount either an offensive or a defensive operation to extinguish a fire. An offensive operation is used with most small fires. With this approach, fire fighters advance toward a fire building with hose lines or other extinguishing agents to attack a fire. If the fire is not too large and the attacking fire fighters can apply enough extinguishing agent, the fire can usually be extinguished quickly and efficiently. Extinguishing the fire in this way often satisfies several priorities at the same time, including life safety and incident stabilization.

When the fire is too large or too dangerous to extinguish with an offensive operation, the IC will implement a defensive operation. A defensive operation is conducted from the exterior, by directing water streams toward the fire from a safe distance. The IC who adopts a defensive strategy has determined that there are no lives or property left to save or that the potential for saving lives or property does not justify the risk to fire fighters. Sometimes a defensive strategy is effective in extinguishing the fire; at other times, it simply keeps the fire from spreading to exposed properties (**FIGURE 22-19**). For more information about fire suppression operations, see Chapter 17, *Fire Suppression*, and Chapter 23, *Advanced Fire Suppression*.

In some situations, all fire fighters may be withdrawn from the area and the fire allowed to burn itself out. These situations generally involve potentially explosive or hazardous materials that represent an extreme danger to fire fighters.

Each strategy poses some risk to fire fighters, so it is important that the IC consider both the risks and the benefits of each strategy. Sending fire fighters into an unoccupied building may pose a large risk to fire fighters, with the only potential benefit being to save property. Sending fire fighters into an occupied building to save lives has great benefit but only if there is a chance to save the occupants and the operation does not pose an unacceptable risk to the fire fighters. Each IC must evaluate the risk versus the benefit of each action to be taken at an emergency scene. Fire suppression is covered in more detail in Chapter 17, *Fire Suppression*.

Ventilation

Ventilation operations include all activities to remove smoke, heat, superheated gases, and other products of combustion from the interior of the structure. Ventilation contributes to the third incident priority: property conservation. Ventilation can be a stand-alone strategy to control the spread of heat and smoke in the structure, or it may be used in combination with other tactics to support the safe removal of occupants or assist with an interior, offensive fire attack. Ventilation operations are covered in depth in Chapter 13, *Ventilation*.

Salvage and Overhaul

Salvage operations are conducted to protect property by preventing avoidable property damage and losses. Salvage is the removal or protection of property that could be damaged during firefighting or overhaul operations (**FIGURE 22-20**). Salvage operations are often aimed at reducing smoke and water damage to the structure and contents once the fire is under control. However, due to the increase in the number of lightweight structures, many fire services are placing a higher priority on the rapid removal of items of perceived sentimental value because of the effect that fire and water have on degrading lightweight Type V structures. Consequently, reducing the time fire fighters have to fight a fire before collapse becomes an issue.

Depending on the type of building construction, overhaul may be conducted during fire attack operations

FIGURE 22-19 A defensive operation is conducted from the exterior, using heavy water streams.
© Gregory Price, Lewiston Son Journal/AP Images.

FIGURE 22-20 Salvage operations contribute to the third incident priority, property conservation, by preventing avoidable property losses.
© Jones & Bartlett Learning. Photographed by Glen E. Ellman.

or after the fire is under control. The systematic search for hidden fire can be one of the most physically demanding duties fire fighters perform. (**FIGURE 22-21**). The IC is responsible for ensuring that the fire is completely extinguished before terminating operations. These operations are covered in depth in Chapter 19, *Salvage and Overhaul*.

SLICE-RS

Another, more recent tool was developed by the International Society of Fire Service Instructors (ISFSI). The mnemonic **SLICE-RS** incorporates the UL/National Institute of Standards and Technology (UL/NIST) recommendations for initial pump company operations upon arrival at a structure fire scene. While RECEO-VS provides guidance for the ongoing operations and command priorities at the scene of a structure fire, the SLICE-RS model provides the first-arriving officer with a short list of objectives prior to the arrival of additional resources:

- **S**ize up: Gather and analyze information to help develop the IAP.
- **L**ocate: Determine the location and extent of the fire inside a building.
- **I**dentify and control the flow path, if possible.
- **C**ool the space from the safest location: Strategically apply a brief, straight stream of water

through an opening to cool the fire before making entry.
- **E**xtinguish the fire: Fully extinguish the fire, including overhaul of void spaces.
- **R**escue: Conduct search and rescue operations if indicated by a risk/benefit analysis.
- **S**alvage: Protect property from further damage.

Note that the first five objectives (SLICE) are to be done in sequence, while the latter two—rescue and salvage—are actions of opportunity (ISFSI, 2018). After conducting the size-up, the officer attempts to locate the fire and control the flow path of air to the fire. This minimizes fire and smoke spread. Cooling the fire from the exterior with a short burst of water limits the growth of the fire and immediately cools the interior. This transitional attack stabilizes the incident and makes interior conditions more tenable for any victims who might still be alive. Although it is not always possible to cool the fire from an exterior window or door, it should be attempted if it can be accomplished quickly before making entry into the structure. A few seconds of water from a straight stream can dramatically reduce the intensity of the fire. Finally, if conditions justify an interior fire attack, the hose should be advanced into the structure to fully extinguish the fire and begin overhaul operations.

Each of these objectives is directly linked to the three incident priorities (life safety, incident stabilization, and property conservation) and should be attempted if conditions allow. During the first few minutes on the fire ground, fire fighters should look for opportunities to assist trapped occupants or salvage undamaged property. As with other fire-ground mnemonics, SLICE-RS is only as valuable as the fire fighters who thoughtfully implement it.

Each incident must be evaluated and an IAP developed based on the current conditions, priorities, and resources. Other acronyms and mnemonics might be equally valuable, but it is, ultimately, the responsibility of the IC to direct incoming resources in a manner that will best mitigate the emergency.

Incident Reports

In order to track fire occurrences, provinces and territories require fire departments to complete incident reports. Depending on the occurrence, the reports can include but are not limited to the following information:

- Time of call
- Turnout time
- On-location time
- Agent application time
- Benchmark times; all clear, under control, agent applied, loss stopped
- Building identification based on NFPA type

FIGURE 22-21 Overhaul is conducted after a fire is under control; its goal is to completely extinguish any remaining pockets of fire.
© Glen E. Ellman.

- Occupancy
- Estimated dollar loss
- Estimated value of property saved
- Description of actions taken

The incident report is usually saved in the department's Records Management System (RMS) and is submitted to the appropriate regulatory body at regular intervals. The information contained within these reports is often used during postincident briefings. In the United States, the National Fire Incident Reporting System (NFIRS) has a reference guide that identifies standardized codes to use in incident reports (for example, codes that indicate data fields such as incident type or actions taken). These codes are not applicable in Canadian reports. Several different incident-reporting software packages are available, however, and many large fire departments use custom-designed software.

Obtaining the Necessary Information

Property owners and/or occupants are primary sources for the information required to complete reports. Any bystanders or eyewitnesses should also be questioned about what they observed. At a minimum, the name, birthdate, address, and telephone number of any involved persons, the address of the incident, and any relevant information about the specific incident should be recorded. The model number and serial number of any equipment damaged by or involved in the fire should be recorded as well. This information can be shared with the fire investigator or other authorized agencies as required.

Consequences of Incomplete and Inaccurate Reports

As fire department activities and results fall under the scrutiny of the public, all incident records must be completed to place them beyond reproach. Many fire services offer training to their personnel on report writing and note taking. Information must be complete, clear, and concise because these records can become admissible evidence in a court of law. Improper or inadequate documentation can have long-term negative consequences. Careless data entry of statistical fields such as type of property use, type of material first ignited, or floor of origin may undermine the purpose of collecting this data; provincial and territorial governments rely on accurate data to identify national and regional trends in fire cause and loss. Inaccurate coding of data such as fire origin and cause information may also create legal problems for the property owner and for your department. Fire reports are considered to be public records under the Freedom of Information Act, so they may be viewed by an attorney, an insurance company, the news media, or the public. If a fatality or loss occurs, incomplete or inaccurate reports may be used to prove that the fire department was negligent. The department, the fire chief, and others may be held accountable. Careful, accurate documentation of all aspects of the incident is a vital component of NIMS and ICS compliance, and it must be performed with the same level of discipline and professionalism as the emergency response.

Crew Resource Management

The fire service has long operated under the paramilitary chain of command concept; senior-level officers give orders to lower ranks who carry out those orders without question. Although many fire departments now use a more participative management model for routine, daily operations, the strict adherence to orders is still the norm for emergency operations.

Recent fire fighter line-of-duty death investigations have identified several common factors that contribute to injuries and deaths at emergency scenes. Some of these factors are related to the traditional fire-ground model of following orders without providing critical information to the IC. The International Association of Fire Chiefs (IAFC) has adapted the **crew resource management (CRM)** concept from the aviation industry. CRM was first developed to help reduce aircraft accidents caused by human error. Many of the errors identified in aircraft incidents can also be found in fire department line-of-duty deaths. In order to limit these human factors, CRM seeks to enhance emergency scene communication, maintain situational awareness, strengthen decision making, and improve teamwork (IAFC, 2018). While clear lines of authority are maintained on emergency scenes, CRM encourages communication and input from all crew members to assess the situation and to develop the safest and most effective method to complete the assignment. Ultimately, the decision-making authority rests with the officers, but under CRM, crew members share their observations and their suggestions to accomplish the assigned objective.

Implementing CRM requires a change of SOPs, department-wide training for all personnel, and adopting a new culture that promotes a teamwork approach to safety. Safety is everyone's responsibility, and CRM provides the organizational framework to encourage communication and input from all personnel while maintaining clear lines of authority for emergency scene operations. More information about CRM can be found through the IAFC website (www.iafc.org).

After-Action REVIEW

IN SUMMARY

- An incident command system (ICS) is a management tool that identifies a single incident commander (IC) who is responsible for all aspects of the operation. All personnel, apparatus, and resources work under the authority and direction of the IC.

- Key components of an ICS include:
  - Planning
  - Delegation
  - Supervision
  - Communication

- The National Incident Management System (NIMS) model provides a standard approach to management that can be used by many different agencies under a wide range of emergency and nonemergency situations. At the Fire Fighter II level, you will work within the framework of the NIMS ICS model.

- The five major components of an ICS organization are:
  - Command: Responsible for the entire incident. This is the only function that is always staffed.
  - Operations: Responsible for performing tactical operations such as fighting the fire, rescuing any trapped individuals, treating any injured victims, and alleviating the emergency situation.
  - Planning: Responsible for developing the incident action plan by identifying what needs to be done by whom and which resources are needed.
  - Logistics: Responsible for providing supplies, services, facilities, and materials needed to support the incident.
  - Finance/administration: Responsible for tracking expenditures and managing the legal and administrative functions at the incident.

- The command staff assist and report directly to the command at the incident:
  - Incident safety officer
  - Liaison officer
  - Public information officer

- Single resources, such as a pump company or a ladder company, can be combined into task forces or strike teams.

- Other organizational units that can be established under ICS include groups, divisions, branches, task forces, and strike teams.

- The ICS can be expanded infinitely to accommodate any size of incident. Branches can be established to aggregate similar functions, such as suppression, EMS, or hazardous materials.

- As the incident grows or continues, it may be necessary to transfer command to another officer. Transfer of command must be done in a seamless manner to ensure continuity of command.

- When multiple agencies such as police, fire, and EMS work together at an incident, unified command must be established as part of the incident command system. A unified command establishes a single set of incident goals under a single leader and ensures mutual communication and cooperation.

- Unity of command is the concept that each fire fighter answers to only one supervisor.

- Span of control is the number of people whom one person can supervise effectively.

- Size-up is the systematic process of gathering information and evaluating the incident. It is essential for determining the appropriate strategy and tactics to handle the incident.

- Size-up is the first step in making plans to bring the emergency incident under control.
  - The initial size-up is often conducted by the first-arriving company officer, who serves as the IC until a higher-ranking officer arrives at the scene and assumes command.

- The IC uses the initial size-up information to develop an incident action plan. This plan is revised as additional information is gathered.
- Size-up relies on two basic categories of information: facts and probabilities.
  - Facts are data elements that are accurate and based on prior knowledge, a reliable source of information, or an immediate, onsite observation.
  - Probabilities are factors that can be reasonably assumed, predicted, or expected to occur but are not necessarily accurate.
- The preincident plan contains facts that can be essential in creating an IAP. It provides details about a building's construction, layout, contents, special hazards, and fire protection systems. This information can be used in determining how to rescue occupants and attack a fire.
- The IC often operates out of an incident command post outside the area of operation and relies on company officers to provide periodic reports. Progress reports from companies enable an IC to judge whether the IAP is effective or needs to be changed.
- The IAP is based on probabilities, predicting where the fire is likely to spread and anticipating potential problems.
- Based on information gathered during size-up, the IC develops an incident action plan that outlines the steps needed to control the situation. The incident action plan should be based on the three incident priorities:
  - Life safety
  - Incident stabilization
  - Property conservation
- Several acronyms can aid in determining strategic objectives:
  - RECEO-VS (rescue, exposure protection, confinement, extinguishment, overhaul, ventilation, salvage) can guide an IC in accomplishing tactical priorities during ongoing operations.
  - SLICE-RS (size-up, locate, identify and control, cool, extinguish, rescue, and salvage) can guide a first-arriving officer in accomplishing preliminary operations prior to arrival of additional resources.
- An incident report describes where and when the incident occurred, who was involved, and what happened. Incident reports for fires should include details about the origin of the fire, the extent of damage, any injuries or fatalities, and actions taken by fire department personnel.
- Incident reports can be completed and submitted on paper, although many fire departments enter them on computers and store the information in a computerized database.
- The National Fire Incident Reporting System (NFIRS) is used to compile and analyze incident reports to help identify statistical trends, faulty equipment or products, civilian and fire fighter injuries and fatalities, and dollar loss by property type. This information is used to develop programs designed to reduce the loss of life and property by fire.
- From a legal standpoint, records and reports are vital parts of the emergency. Information must be complete, clear, and concise because these records can become admissible evidence in a court of law.
- Crew resource management (CRM) improves emergency scene communication, awareness, decision making, and teamwork by encouraging input from all crew members.

KEY TERMS

Access Navigate for flashcards to test your key term knowledge.

Branch A supervisory level established in either the operations or logistics function to provide a span of control. (NFPA 1561)

Branch director A person in a supervisory level position in either the operations or logistics function to provide a span of control. (NFPA 1561)

Chain of command A rank structure, spanning the fire fighter through the fire chief, for managing a fire department and fire-ground operations.

Command The act of directing and/or controlling resources by virtue of explicit legal, agency, or delegated authority. (NFPA 1026)

Command staff The command staff consists of the public information officer, safety officer, and liaison officer, all of whom report directly to the incident commander and are responsible for functions in the incident management system that are not a part of the function of the line organization. (NFPA 1561)

Company officer The individual responsible for command of a company, a designation not specific to any particular fire department rank (can be a fire fighter, lieutenant, captain, or chief officer, if responsible for command of a single company). (NFPA 1026)

Crew A team of two or more fire fighters. (NFPA 1500)

Crew resource management (CRM) A program focused on improved situational awareness, sound critical decision making, effective communication, proper task allocation, and successful teamwork and leadership. (NFPA 1500)

Division That organizational level having responsibility for operations within a defined geographic location. (NFPA 1026)

Division supervisor A person in a supervisory level position responsible for a specific geographic area of operations at an incident. (NFPA 1561)

Finance/administration section Section responsible for all costs and financial actions of the incident or planned event, including the time unit, procurement unit, compensation/claims unit, and the cost unit. (NFPA 1026)

Finance/administration section chief The general staff position responsible for directing the finance/administrative function. It is generally assigned on large-scale or long-duration incidents that require immediate fiscal management.

Fire Fighter I A person, at the first level of progression as defined in Chapter 4 of NFPA 1001, who has demonstrated the knowledge and skills to function as an integral member of a firefighting team under direct supervision in hazardous conditions. (NFPA 1001)

Fire Fighter II A person, at the second level of progression as defined in Chapter 5 of NFPA 1001, who has demonstrated the skills and depth of knowledge to function under general supervision. (NFPA 1001)

Fire-ground command (FGC) An incident management system developed in the 1970s for day-to-day fire department incidents (generally handled with fewer than 25 units or companies).

FIRESCOPE Fire Resources of California Organized for Potential Emergencies; an organization of agencies established in the early 1970s to develop a standardized system for managing fire resources at large-scale incidents such as wildland fires.

Freelancing Individuals or crews operating independently of the established ICS structure.

General staff A group of incident management personnel organized according to function and reporting to the incident commander, normally consisting of the operations section chief, planning section chief, logistics section chief, and finance/administration section chief. (NFPA 1026)

Group Established to divide the incident management structure into functional assignments of operation. (NFPA 1026)

Group supervisor A person in a supervisory level position responsible for a functional area of operation. (NFPA 1561)

Incident action plan (IAP) A verbal or written plan containing incident objectives reflecting the overall strategy and specific control actions where appropriate for managing an incident or planned event. (NFPA 1026)

Incident commander (IC) The individual responsible for all incident activities, including the development of strategies and tactics and the ordering and release of resources. (NFPA 1026)

Incident command post (ICP) The field location at which the primary tactical-level, on scene incident command functions are performed. (NFPA 1026)

Incident command system (ICS) The combination of facilities, equipment, personnel, procedures, and communications operating within a common organizational structure that has responsibility for the management of assigned resources to effectively accomplish stated objectives pertaining to an incident or training exercise. (NFPA 1670)

Incident report A document prepared by fire department personnel on a particular incident. (NFPA 901)

Incident safety officer (ISO) A member of the command staff responsible for monitoring and assessing safety hazards and unsafe situations and for developing measures for ensuring personnel safety. (NFPA 1500)

Leader The individual responsible for command of a task force, strike team, or functional unit. (NFPA 1026)

Liaison officer A member of the command staff responsible for coordinating with representatives from cooperating and assisting agencies. (NFPA 1561)

Logistics section Section responsible for providing facilities, services, and materials for the incident or planned event, including the communications unit, medical unit, and food unit within the service branch and the supply unit, facilities unit, and ground support unit within the support branch. (NFPA 1026)

Logistics section chief The general staff position responsible for directing the logistics function. It is generally assigned on complex, resource-intensive, or long-duration incidents.

National Fire Incident Reporting System (NFIRS) The standard national reporting system used by U.S. fire departments to report fires and other incidents to which they respond and to maintain records of these incidents in a uniform manner. (FEMA NFIRS)

National Incident Management System (NIMS) A system mandated by Homeland Security Presidential Directive 5 (HSPD-5) that provides a systematic, proactive approach guiding government agencies at all levels, the private sector, and nongovernmental organizations to work seamlessly to prepare for, prevent, respond to, recover from, and mitigate the effects of incidents, regardless of cause, size, location, or complexity, so as to reduce the loss of life or property and harm to the environment. (NFPA 1026)

Operations section Section responsible for all tactical operations at the incident or planned event, including up to 5 branches, 25 divisions/groups, and 125 single resources, task forces, or strike teams. (NFPA 1026)

Operations section chief The general staff position responsible for managing all operations activities. It is usually assigned when complex incidents involve more than 20 single resources or when command staff cannot be involved in all details of the tactical operation.

Planning section Section responsible for the collection, evaluation, dissemination, and use of information related to the incident situation, resource status, and incident forecast. (NFPA 1026)

Planning section chief The general staff position responsible for planning functions and for tracking and logging resources. It is assigned when command staff members need assistance in managing information.

Preincident plan A document developed by gathering general and detailed data that are used by responding personnel in effectively managing emergencies for the protection of occupants, responding personnel, property, and the environment. (NFPA 1620)

Public information officer (PIO) A member of the command staff responsible for interfacing with the public and media or with other agencies with incident-related information requirements. (NFPA 1026)

RECEO-VS An acronym developed for use by the incident commander for accomplishing tactical priorities on the fire ground; stands for rescue, exposure protection, confinement, extinguishment, overhaul, ventilation, and salvage.

Resource management Under the NIMS, includes mutual-aid agreements; the use of special federal, state, local, and tribal teams; and resource mobilization protocols. (NFPA 1026)

Risk/benefit analysis An assessment of the risk to rescuers versus the benefits that can be derived from their intended actions. (NFPA 1008)

Single command A command structure in which a single individual is responsible for all of the strategic objectives of the incident. It is typically used when an incident is within a single jurisdiction and is managed by a single discipline.

Single resource An individual, a piece of equipment and its personnel, or a crew or team of individuals with an identified supervisor that can be used on an incident or planned event. (NFPA 1026)

Size-up The process of gathering and analyzing information to help fire officers make decisions regarding the deployment of resources and the implementation of tactics. (NFPA 1410)

SLICE-RS An acronym intended to be used by the first-arriving company officer to accomplish important strategic goals on the fire ground.

Span of control The maximum number of personnel or activities that can be effectively controlled by one individual (usually three to seven). (NFPA 1006)

Staging area Location established where resources can be placed while they await a tactical assignment. (NFPA 1026)

Strike team Specified combinations of the same kind and type of resources, with common communications and a leader. (NFPA 1026)

Survivability profiling An assessment that weighs the risks likely to be taken versus the benefits of those risks of the viability and survivability of potential fire victims under the current conditions in the structure.

Task force A group of resources with common communications and a leader that can be pre-established and sent to an incident or planned event or formed at an incident or planned event. (NFPA 1026)

Transfer of command The formal procedure for transferring the duties of an incident commander at an incident scene. (NFPA 1026)

Unified command A team effort that allows all agencies with jurisdictional responsibility for an incident or planned event, either geographic or functional, to manage the incident or planned event by establishing a common set of incident objectives and strategies. (NFPA 1026)

Unit The organizational element having functional responsibility for a specific incident operations,

planning, logistics, or finance/administration activity. (NFPA 1026)

Unity of command The concept by which each person within an organization reports to one, and only one, designated person. (NFPA 1026)

REFERENCES

Federal Emergency Management Agency. (n.d.). "ICS Management: Span of Control." Accessed November 19, 2018. https://emilms.fema.gov/IS200b/ICS0102370.htm.

International Association of Fire Chiefs. (n.d.). "Crew Resource Management." Accessed November 19, 2018. https://events.iafc.org/Operations/ResourcesDetail.cfm?ItemNumber=4745.

International Society of Fire Service Instructors (ISFSI). (n.d.). "Principles of Modern Fire Attack: SLICE-RS 8-hour Course." Accessed November 19, 2018. http://www.isfsi.org.

National Fire Protection Association. NFPA 1026, *Standard for Incident Management Personnel Professional Qualifications*, 2018. www.nfpa.org. Accessed November 19, 2018.

National Fire Protection Association. NFPA 1500, *Standard on Fire Department Occupational Safety, Health, and Wellness Program*, 2018. www.nfpa.org. Accessed November 19, 2018.

National Fire Protection Association. NFPA 1521, *Standard for Fire Department Safety Officer Professional Qualifications*, 2015. www.nfpa.org. Accessed November 19, 2018.

National Fire Protection Association. NFPA 1561, *Standard on Emergency Services Incident Management System and Command Safety*, 2014. www.nfpa.org. Accessed November 19, 2018.

Shaw, Steven. 2015. "FHExpo: Survivability Profiling - Size-up On Steroids." Accessed November 19, 2018. https://www.firehouse.com/safety-health/news/12094782/survivability-profiling-starts-with-sizeup.

On Scene

An explosion at an ammunition manufacturing complex has caused a partial collapse of a large, two-storey building. Heavy smoke and fire are coming from the "B" side of the building, and the "B" exposure building is starting to burn. The IC requests multiple fire department units, the local police department, all available ambulances, and a representative from the Bureau of Alcohol, Tobacco, Firearms and Explosives (ATF). As your pump company arrives at the complex, you witness a large, chaotic scene. Your company officer directs you to don your PPE and SCBA and grab as many tools as you can carry. Within a few minutes, the scene is filled with police officers, fire fighters, and EMS personnel, as well as news reporters and a growing crowd of civilians.

1. What is the first action your crew should take upon arrival to this incident?

A. Set up a safety perimeter using cones and barrier tape.

B. Report to the incident command post for assignment.

C. Report to the university president's office.

D. Begin treating the first victims you see.

2. Under NIMS, who should be in charge of this incident?

A. The facility supervisor

B. The ranking local police officer

C. The ranking EMS person

D. Unified command should be established with representatives of all agencies responding to the incident.

3. What is the first incident priority at this incident?

A. Life safety

B. Incident stabilization

C. Property conservation

D. Preservation of the crime scene

(continued)

On Scene Continued

4. Which of the following ICS components will likely be implemented for this emergency?

1. Operations
2. Planning
3. Logistics
4. Finance/administration
5. Command

A. 1, 2, 3, 4, and 5

B. 1 and 5

C. 2, 3, and 5

D. 5

5. Which of the following command staff positions should be filled for this incident?

1. Operations section chief
2. Safety officer
3. Public information officer
4. Liaison officer

A. 1, 2, and 3

B. 1 and 2

C. 2, 3, and 4

D. Unified command can manage all of these responsibilities without appointing a command staff.

6. Command requests multiple pumpers and a supervisor under the area's mutual aid agreement. Typically organized in groups of two or more units, these groups are called:

A. strike teams.

B. task forces.

C. squads.

D. companies.

7. Before assigning personnel to perform search and rescue, the IC should:

A. establish a rehabilitation area.

B. conduct a survivability profile.

C. appoint an operations section chief.

D. request additional resources.

8. At a major incident, who is responsible for releasing information to the media?

A. Incident commander

B. Operations section chief

C. The fire chief

D. Public information officer

Access Navigate to find answers to this On Scene, along with other resources such as an audiobook.

Advanced Fire Suppression

KNOWLEDGE OBJECTIVES

After studying this chapter, you will be able to:

- List the factors that the incident commander evaluates when determining whether to perform a defensive operation or an offensive operation. (**NFPA 1001: 5.3.2**, p. 845)
- Describe the objectives of an indirect attack. (**NFPA 1001: 5.3.2**, p. 848)
- Describe the objectives of a direct attack. (**NFPA 1001: 5.3.2**, p. 848)
- Describe the objectives of a combination attack. (**NFPA 1001: 5.3.2**, p. 848)
- Describe how to coordinate an interior attack. (**NFPA 1001: 5.3.2**, pp. 848, 849)
- List the three incident priorities in structural firefighting operations and how these priorities affect ventilation operations. (**NFPA 1001: 5.3.2**, pp. 850–851)
- Explain how ventilation is coordinated with fire suppression operations. (**NFPA 1001: 5.3.2**, pp. 850–851)
- Describe the characteristics of combustible or flammable liquid fires. (**NFPA 1001: 5.3.1**, p. 851)
- Describe the hazards presented by combustible or flammable liquid fires. (**NFPA 1001: 5.3.1**, p. 851)
- Describe a boiling liquid/expanding vapour explosion (BLEVE). (**NFPA 1001: 5.3.3**, p. 852)
- Describe the characteristics of flammable gas cylinder fires. (**NFPA 1001: 5.3.3**, p. 853)

- Describe the hazards presented by flammable gas cylinder fires. (**NFPA 1001: 5.3.3**, p. 853)
- Describe how to suppress a flammable gas cylinder fire. (**NFPA 1001: 5.3.3**, pp. 853, 855)
- Describe the characteristics of Class A foam. (**NFPA 1001: 5.3.1**, pp. 854, 856)
- Describe the characteristics of Class B foam. (**NFPA 1001: 5.3.1**, p. 856)
- List the major categories of Class B foam concentrate. (**NFPA 1001: 5.3.1**, pp. 857–858)
- Describe the characteristics of protein foam. (**NFPA 1001: 5.3.1**, p. 858)
- Describe the characteristics of fluoroprotein foam. (**NFPA 1001: 5.3.1**, p. 858)
- Describe the characteristics of aqueous film-forming foam. (**NFPA 1001: 5.3.1**, p. 858)
- Describe the characteristics of alcohol-resistant foam concentrate. (**NFPA 1001: 5.3.1**, p. 858)
- Describe the characteristics of compressed air foam. (pp. 858–859)
- Describe how foam proportioner equipment works with foam concentrate to produce foam. (**NFPA 1001: 5.3.1**, pp. 859–861)
- Describe how foam is applied to fires. (**NFPA 1001: 5.3.1**, pp. 861–864)
- Describe how to determine how much foam is required to extinguish fires of certain sizes. (pp. 861–862)

- List three common foam application techniques. (pp. 862–864, 865)
- Describe how to perform a service test on a fire hose. (**NFPA 1001: 5.5.5**, pp. 865–868)
- List the information that should be noted on a hose record. (**NFPA 1001: 5.5.5**, p. 868)

- Suppress a flammable liquid fire by applying foam using the bounce-off method. (**NFPA 1001: 5.3.1**, pp. 864, 865)
- Perform an annual service test on a fire hose. (**NFPA 1001: 5.5.5**, pp. 866–867)

SKILLS OBJECTIVES

After studying this chapter, you will be able to perform the following skills:

- Coordinate an interior attack. (**NFPA 1001: 5.3.2**, p. 849)
- Suppress a flammable gas cylinder fire. (**NFPA 1001: 5.3.3**, p. 855)
- Operate an in-line foam eductor. (**NFPA 1001: 5.3.1**, p. 860)
- Prepare the appropriate type of foam for application to a flammable liquid fire. (**NFPA 1001: 5.3.1**, pp. 854, 856–858)
- Suppress a flammable liquid fire by applying foam using the rain-down method. (**NFPA 1001: 5.3.1**, pp. 862, 863)
- Suppress a flammable liquid fire by applying foam using the roll-in method. (**NFPA 1001: 5.3.1**, pp. 862, 864)

ADDITIONAL STANDARDS

- **NFPA 11**, *Standard for Low-, Medium-, and High-Expansion Foam*
- **NFPA 11A**, *Standard for Medium- and High-Expansion Foam Systems*
- **NFPA 11C**, *Standard for Mobile Foam Apparatus*
- **NFPA 1145**, *Guide for the Use of Class A Foams in Fire Fighting*
- **NFPA 1710**, *Standard for the Organization and Deployment of Fire Suppression Operations, Emergency Medical Operations, and Special Operations to the Public by Career Fire Departments*
- **NFPA 1720**, *Standard for the Organization and Deployment of Fire Suppression Operations, Emergency Medical Operations, and Special Operations to the Public by Volunteer Fire Departments*
- **NFPA 1901**, *Standard for Automotive Fire Apparatus*

Fire Alarm

Your pump company (Pump 22) is responding to a reported residential structure fire in a neighbourhood known for large two-storey houses. You can see smoke on the horizon as you hear the first arriving officer transmit the following radio report: "Dispatch, this is Pump 14. We are on location at 123 Main Street. This is a two-storey, Type V single-family residence with smoke showing from Side D (Side 4). People appear to have exited the structure. Put in a working structure fire. We will be assuming an offensive fire attack. Pump 14 will be Main Street command." Upon arrival, the incident commander (IC) orders your crew to stretch a line for exposure protection on side D (4), while other crews are ordered to complete a primary search and ladder the building. Upon your arrival, your company officer notices the fire intensifying, and embers are landing on the roof of the Delta exposure house. He instructs you to deploy the compressed air foam (CAF) line from the apparatus and apply foam to the roof and Bravo side of the exposure.

1. Why did your officer select a CAF line for this exposure?

2. What are the benefits of CAF in this situation?

3. Could this foam stream also be used as part of a transitional attack on the fire? If so, how and when would you turn the foam stream onto the original fire structure?

 *Access Navigate for more practice activities.*

Introduction

Generally speaking, most fires begin as small events that grow in intensity with time, fuel, and oxygen. It is for this reason that no fire scene should be considered to be routine. Small fires can unexpectedly flash over as they transition from the development stage to the fully developed phase. Basic fire suppression knowledge and skills are learned at the Fire Fighter I level; however, incidents involving rapidly developing structure fires or those involving flammable liquids and gases require advanced fire suppression techniques. In order to safely and effectively manage these incidents, the incident commander (IC) must perform a size-up to evaluate many incident factors, formulate an incident action plan (IAP), and communicate that plan to responding personnel. It is important that fire fighters understand how these tools are used to determine firefighting strategies for difficult incidents.

Command Considerations

As discussed in Chapter 17, *Fire Suppression*, the IC must evaluate many factors to decide whether an offensive operation or a defensive operation should be used at a particular fire. If the risk factors are too great, an exterior attack is the only acceptable option. If the decision is made to launch an interior attack, the IC must determine where and how to attack after considering safety issues and the potential effectiveness of the approach being considered. These factors include, but are not limited to, the following:

- What are the risks versus the potential benefits?
- Is it safe to send fire fighters into the building?
- Are there viable occupants that need to be rescued?
- Are there indicators of potential collapse?
- Is the building made of lightweight construction?
- Do hostile fire event conditions prohibit entry?
- Are enough fire fighters on the scene to mount an interior attack? (Remember the two-in/two-out rule.)
- Is an adequate water supply available?
- Can proper ventilation be carried out to support offensive operations?
- Are hazardous materials present?

Life Safety

Life safety is the most important consideration when evaluating risk. As firefighting is inherently risky, ICs and fire fighters must continually evaluate conditions

FIGURE 23-1 The officer monitors the progress of the fire attack and provides direction to the fire fighter operating the nozzle.
© Jones & Bartlett Learning. Photographed by Glen E. Ellman.

in an effort to mitigate entrapment and injuries. ICs should perform a risk/benefit analysis before identifying a strategy and assigning tactical objectives as part of an IAP. A risk/benefit analysis helps to weigh the risks. If there are known victims in the burning building, the IC must assess the risks associated with aggressive search and rescue operations. Are the victims still alive? Can fire fighters rescue the victims without taking an unacceptable risk? How will the rescue attempt be made? Where will initial handlines be placed? How will ventilation affect interior conditions? Are fire conditions so severe that any interior operation is unsafe?

Structural Collapse

Structural collapse should be a major consideration when developing an IAP. Firefighters must be experts in identifying various types of building construction. Understanding the manner in which buildings are constructed informs decision making. This knowledge empowers officers and ICs with the confidence they need to both evaluate and anticipate collapse potential. The knowledge gained through learning and training combine with experience to form the basis for sound decision making. Many buildings give little or no warning before they collapse. Many newer homes use lightweight engineered structural materials to support large, open floor spaces. These large, open areas may suddenly collapse, especially if the fire is in the basement.

Fire fighters who are assigned to interior operations or vertical ventilation tasks should always be aware of building conditions. Building collapse is also a concern during defensive fire attack operations. Many fire fighters have been injured or killed even while working outside of burning buildings. Fire fighters should immediately retreat and stay out of the collapse zone if they notice any of the following signs:

- Visible indication of sagging of floors or roof supports
- Indication that the roof assembly is separating from the walls, such as the appearance of fire or smoke near the roof edges
- Cracked, crumbling, or bulging walls (**FIGURE 23-2**)
- Smoke or water coming through masonry walls
- Structural failure of any portion of the building, even if it is some distance from the fire fighters
- Heavy fire conditions below any interior operation
- Sudden increase in the intensity of the fire from the roof opening
- High heat indicators on a thermal imaging camera

Other factors that may contribute to building collapse include heavy snow on the roof, master streams flowing a high volume of water over a long period of time, and any building that is under construction or being renovated.

Fire Conditions

Although every fire incident presents a certain level of risk to responding fire fighters, some fire scenes are more dangerous than others. Interior fire and rescue operations are generally more dangerous than exterior operations. Fires above grade and fires in basements make fire-ground operations more difficult, and the risks to fire fighters significantly increase in these situations. As discussed in Chapter 5, *Fire Behaviour*, smoke and fire conditions vary greatly, depending on the size and shape of the compartment, the fuel type and fuel load, ventilation openings, and many other factors. Fire fighters must constantly monitor fire and smoke conditions and be aware of any changes. A gradual increase in heat, smoke, or visible flame may indicate that suppression operations are not effective. A rapid increase in any of these conditions may indicate an imminent **hostile fire event**. Thick, turbulent smoke may indicate a rapid increase in interior temperatures (**FIGURE 23-3**). A roll-over or a rapidly lowering smoke layer may indicate that a flashover is imminent. Notify your crew immediately if you notice any rapidly deteriorating smoke, flame, or heat conditions. All personnel should evacuate the building immediately until it is determined to be safe to resume interior operations.

FIGURE 23-2 A crack indicating serious building instability.
© monbibi/Shutterstock, Inc.

FIGURE 23-3 Turbulent smoke flow.
Courtesy of Keith Muratori.

Water Supply

Before beginning interior operations, a water supply must be established. Initial interior operations may begin with water from an onboard booster tank, but that supply is usually limited to 1893 to 3785 L (500 to 1000 gal). Interior fire suppression operations in large buildings, high-rise buildings, or buildings with heavy fire or smoke should not begin until a continuous water supply is established. In these situations, fire fighters may suddenly find themselves in extreme danger if the water supply is interrupted. It is usually better to take the necessary time to establish a water supply before engaging in interior operations in large buildings or those with substantial fire involvement.

Staffing

In addition to having enough equipment and water resources, NFPA 1710, *Standard for the Organization and Deployment of Fire Suppression Operations, Emergency Medical Operations, and Special Operations to the Public by Career Fire Departments*, and NFPA 1720, *Standard for the Organization and Deployment of Fire Suppression Operations, Emergency Medical Operations, and Special Operations to the Public by Volunteer Fire Departments*, specify the minimum number of people needed for safe and effective interior firefighting operations. Fire fighters should not commit to interior operations without having enough trained fire fighters on the scene to perform all of the essential fire-ground operations. Refer to the 11 Rules of Engagement for Fire Fighter Survival in Chapter 18, *Fire Fighter Survival*, and your department's specific SOPs before beginning any interior fire-ground operations.

SLICE-RS

The mnemonic **SLICE-RS** has been developed to help first-arriving company officers remember the steps to be taken for initial pump company operations upon arrival at a structure fire scene (UL FSRI, 2018). The steps are as follows:

Sequential Actions

S Size-up—gather and analyze information to help develop the IAP.

L Locate the fire—determine the location and extent of the fire inside a building.

I Identify and control the flow path, if possible (**FIGURE 23-4**).

C Cool the space from the safest location—strategically apply a brief, straight stream of water through an opening to cool the fire before making entry.

FIGURE 23-4 The flow path is the movement of heat and smoke from the area of higher pressure within the fire toward the areas of lower pressure accessible via doors, windows, and roof structures.
© Jones & Bartlett Learning. Photographed by Glen E. Ellman.

E Extinguish the fire—fully extinguish the fire, including overhaul of void spaces.

Actions of Opportunity

R Rescue—conduct search and rescue operations if indicated by a risk/benefit analysis.

S Salvage—protect property from further damage.

It is important to remember that SLICE-RS was developed as a tool to assist fire fighters and officers during scene size-up. Your department may prefer another mnemonic that prompts proper evaluations and decision-making processes.

The SLICE-RS steps are intended to be used as a guide to accomplish important strategic goals on the fire ground. When using any acronym, keep in mind the objectives of your actions. It may be necessary to modify your actions to fit a specific situation.

After sizing up the situation, the IC must determine which type of attack is appropriate. *As a new fire fighter, you are not responsible for determining the type of fire attack that will be used.* Nevertheless, you should understand the various factors that go into making these decisions and recognize why the IC orders different types of fire attacks for different fire conditions.

Defensive Operations

As discussed in Chapter 17, *Fire Suppression*, **defensive operations** are typically conducted from the exterior and are implemented when interior fire conditions, building conditions, limited water supply, or other factors make an interior attack unsafe or impractical. If fire conditions are such that there is little or no chance of occupant survival, or the building has sustained substantial damage, the IC will likely determine that a defensive, exterior fire attack is the only option to stabilize the incident. Initial fire streams should protect exposure buildings to keep the fire from spreading. Master stream appliances may be used to apply thousands of gallons of water through doors, windows, or holes in the roof. This is known as an **indirect attack**. Exterior fire attacks may take several hours to fully control and extinguish the fire, especially if part of the building collapses. Fires may continue to burn under piles of concrete, brick, or steel for hours or even days. If the IC determines that an exterior fire attack has substantially improved interior fire conditions, he or she may change the action plan to an offensive operation that uses limited interior handlines for overhaul and extinguishment. Before this occurs, all exterior fire streams, except for those protecting an exposure, must be shut down, safety precautions must be strictly observed, and interior crews must constantly look for signs of standing pools of water, electrical or fuel gas hazards, structural damage, or impending building collapse.

Offensive Operations

Offensive operations typically require fire fighters to enter a building and apply an extinguishing agent (usually water) onto the fire. The objective in an offensive operation is for fire fighters to get close enough to the fire to perform a **direct attack** or to apply extinguishing agents directly onto the burning surfaces.

The larger the fire, the greater the challenge in suppressing it and the more ominous the risks that are involved in fire suppression.

Interior fire attack can be conducted on many different scales. In many cases, an interior attack is used to fight a fire that is burning in only one room; this kind of fire may be controlled quickly by one handline. Larger fires require more water, which could be provided by two or more small handlines working together or by one or more larger handlines. Fires that involve multiple rooms, large spaces, or concealed spaces are more complicated and require more extensive coordination; nevertheless, the basic techniques for attacking these fires are similar to the techniques used when extinguishing smaller fires.

A **combination attack** is performed in two stages, beginning with an indirect attack and continuing with a direct attack. Fire fighters should first use an indirect attack method, from the safest location such as a hallway, adjoining room, exterior, or doorway, to cool the fire gases down to safer temperatures and prevent flashover from occurring. This operation is followed with a direct attack on the main body of fire.

As discussed in Chapter 17, *Fire Suppression*, a trained Fire Fighter II should be able to understand and coordinate an interior fire attack. To coordinate an interior attack, follow the steps in **SKILL DRILL 23-1**.

FIRE FIGHTER TIP

Generally speaking, water applied to a fire through a roof opening using elevated devices tends to be less effective than master streams directed at angles under a roof from below. Quite often, fires are shielded from elevated water application by collapsed roof material, floors, and walls. Fire fighters must remember that water application is only useful if it absorbs heat. One can often tell the effectiveness of elevated water application by the amount of runoff.

Transitional Attack

As discussed in Chapter 17, *Fire Suppression*, a **transitional attack** is an offensive fire attack method intended to cool the fire compartment from a safe location prior to making an interior attack. UL/National Institute of Standards and Technology (UL/NIST) studies have shown that just a few seconds of water from a straight stream applied through an open window or door into the fire compartment can reduce interior temperatures by several hundred degrees (Fahrenheit), improving interior conditions for both fire fighters and occupants (**FIGURE 23-5**) (UL, 2013). This drop in temperature reduces smoke production, which further improves interior conditions. Once the fire is cooled and smoke production is reduced, fire fighters can make entry into the building for an interior attack.

Previously, some departments may have used the term "transitional fire attack" to mean a defensive, exterior fire attack operation that changes to an interior attack once the fire is under control. As a result of the UL/NIST study, however, many departments now consider a transitional attack to be an intentional, brief, offensive application of water from an exterior position in order to cool the fire before making an interior attack. Some departments may use expressions such as "softening the target" or "resetting the fire" instead of transitional fire attack.

For many years, fire fighters believed that applying water through an open window or door could cause the fire to spread further into the structure, causing occupant injuries and greater property damage. Fire spread can be limited by controlling the flow path while cooling the fire from a safe, exterior position. Although it is not always possible to use a transitional attack before entering the structure, fire fighters should always consider using it, especially if the fire has already self-vented through a door or window. Always refer to your department's SOPs when choosing a fire attack strategy.

SKILL DRILL 23-1
Coordinating an Interior Attack Fire Fighter II, NFPA 1001: 5.3.2

1 Don full personal protective equipment (PPE), including self-contained breathing apparatus (SCBA). Report to the IC, check into the personnel accountability system, and proceed to work as a team. Perform size-up, and give an arrival report. Call for additional resources if needed.

2 Ensure that an adequate water supply and appropriate backup resources are available. Select the appropriate attack technique. Communicate the attack technique to the team.

3 Maintain constant crew integrity at all times. Monitor air supply, and notify command of changing fire or smoke conditions.

4 Coordinate fire attack, ventilation, and search and rescue operations.

5 Ensure complete extinguishment of the fire during overhaul. Exit the hazard area, account for all members of the team, and report to incident command.

FIGURE 23-5 A transitional fire attack is an offensive fire attack initiated by an exterior, indirect handline operation into the fire compartment to initiate cooling while transitioning into an interior direct fire attack in coordination with ventilation operations.
© Jones & Bartlett Learning. Photographed by Glen E. Ellman.

Ventilation to Support Incident Priorities

Ventilation is directly related to the three major incident priorities of structural firefighting operations: life safety, incident stabilization, and property conservation. Ventilation must be coordinated with the fire suppression efforts to ensure that both events support the IAP. As explained in Chapter 13, *Ventilation*, ventilation is designed to allow hot gases and smoke to be removed from the building, thereby improving visibility and tenability in the building for any trapped victims and fire fighters. Because many structure fires are ventilation-limited fires, it is important to be sure that charged hose lines are in place before ventilation is initiated. When preparing for ventilation, consider which type of fire attack is appropriate. Remember that uncoordinated or improper ventilation may cause fire conditions to worsen, creating a more hostile environment for both occupants and fire fighters.

Coordination and communication are essential to ensure that hose lines will be ready to attack the fire

when the ventilation openings are made. The officer in charge of ventilation should verify with the fire attack officer prior to performing ventilation. This will ensure that the fire attack team is in place and ready to advance the moment the vent is made. These openings must be located so that the hot smoke and gases will be drawn away from the attack crews.

Venting for Life Safety

Life safety is the primary incident priority of the fire service, whether it involves rescuing civilians or protecting fire department personnel. Good ventilation practices help to clear smoke and toxic gases from the structure, which gives occupants a better chance to survive. It also provides a safer environment for firefighting crews and increases visibility, enabling more rapid searches for victims.

Ventilation, in coordination with an effective fire attack, cools the interior of the building, limits fire spread, and allows fire fighters to advance hose lines more safely and rapidly to attack the fire. A coordinated fire attack and ventilation effort must consider the timing, method, and direction of the attack, as well as the type and location of possible ventilation openings. The swift control of the fire reduces the life-safety risk for fire fighters as well as for building occupants.

Venting for Incident Stabilization

Incident stabilization is an often-forgotten critical incident priority of firefighting. Many times, simply stopping the growth of the fire improves life-safety conditions and reduces property damage. Proper ventilation techniques can help stabilize fire conditions dramatically. Improper ventilation, however, can cause a relatively stable incident to quickly escalate, with catastrophic results.

Because furnishings and finishes in modern construction burn so hot and so rapidly, they produce massive amounts of hot, thick smoke and poisonous gases. These products of combustion expand and create pressure inside the building. Fire fighters often arrive on scene when interior fire conditions are rapidly deteriorating. Temperatures rise, and thick, turbulent smoke may be seen coming from open doors and windows. These conditions may indicate that a hostile fire event is imminent. Hostile fire events include smoke explosions, backdrafts, flashover, or other rapidly deteriorating fire conditions. Fire fighters must be aware of these indicators—both before and after making entry into the structure.

Proper ventilation techniques may prevent or minimize the consequences of these hostile fire events. Horizontal, vertical, or positive-pressure ventilation methods can remove hot, toxic gases from the structure, and this can help stabilize the incident. Recent studies indicate that controlling the flow path until fire attack crews are ready to enter the building may be one of the most important ventilation techniques fire fighters can use on the fire ground (UL, 2014). Opening doors, windows, or ventilating the roof too early can quickly spread fire and smoke throughout the structure. By coordinating fire suppression and ventilation activities under the direction of the IC, interior conditions can be made safer, allowing fire fighters to extinguish the fire.

Venting for Property Conservation

Another priority of the fire service is property conservation. This involves reducing the losses caused by means other than direct involvement in the fire. Because much of the damage from a fire can be smoke damage, ventilation can play a significant role in limiting property damage. If a structure is ventilated rapidly and correctly, the damages associated with smoke, heat, water, and overhaul operations can all be reduced.

Combustible and Flammable Liquid Fires

Combustible and flammable liquid fires can be among the most dangerous and challenging types of fires to extinguish. Because these liquids produce huge amounts of heat energy, they can be dangerous to approach, difficult to extinguish, and they may accelerate fire growth in Class A fuels. Liquid fuel fires can be encountered in almost any type of occupancy. Many fires involving a vehicle (e.g., airplane, train, ship, car, truck) may involve a combustible or flammable liquid. Special tactics must be used when attempting to extinguish a liquid fuel fire, and special extinguishing agents such as foam or dry chemicals may be needed.

Hazards

Fires involving combustible or flammable liquids such as diesel fuel or gasoline require special extinguishing agents. Most liquid fuel fires can be extinguished using either foam or dry chemicals. Class B extinguishing agents are approved for use on Class B (combustible and flammable liquids) fires. Liquid fuel fires can be classified as either two-dimensional or three-dimensional. A two-dimensional fire refers to a spill, pool, or open container of liquid that is burning only on the top surface. A three-dimensional fire refers to a situation in which the burning liquid fuel is dripping, spraying, or flowing over the edges of a container.

A two-dimensional liquid fuel fire can usually be controlled by applying the appropriate Class B foam to the burning surface. Several different formulations of Class B foams are suitable for a variety of liquids and situations. The foam will flow across the surface of the liquid and create a seal that stops the fuel from vapourizing. The foam also isolates the oxygen from the fuel and extinguishes the fire. As the water drains from the foam blanket, the water will cool the liquid fuel and further reduce the possibility of reignition.

When extinguishing liquid fuel fires, fire fighters should look for hot surfaces or open flames that could cause the vapours to reignite after a fire has been extinguished. It is important to determine the identity of the liquid that is involved so as to select the appropriate extinguishing agent and to determine whether the vapours are lighter or heavier than air.

A three-dimensional liquid fuel fire may be more difficult to extinguish with a low-expansion foam stream because the foam cannot establish an effective seal between the fuel and the oxygen. Dry-chemical or gaseous extinguishing agents may be more effective than foam in controlling these kinds of fires. Dry-chemical and gaseous agents can also be used to extinguish two-dimensional fires, although they do not provide long-lasting protection. In some cases, a fire can be extinguished with a dry chemical, and then the surface can be covered with foam to prevent reignition.

Fire fighters should avoid standing in pools of flammable liquids or contaminated runoff, because their PPE will absorb the flammable product and become contaminated. In cases of serious contamination, the PPE itself can become flammable.

Suppression

The skills used in suppressing small flammable liquid fires are presented in Chapter 7, *Portable Fire Extinguishers*. Larger flammable liquid fires may require the use of foam. Foam application techniques are discussed later.

Flammable Gas Containers

Many types of flammable gases are stored in different types and sizes of containers. A variety of flammable gases can be found in big-box stores, supply warehouses, and industrial occupancies. Each of these gases and container types presents unique challenges when it is threatened by fire. It is essential that fire fighters quickly and carefully size-up the scene to determine how, or if, they should approach the flammable gas container.

Propane Gas

The popularity of propane gas for heating and cooking has meant that these cylinders have become commonplace in residential areas and in industrial, commercial, and recreational locations. Dozens of propane cylinders may be stored in front of or behind hardware stores. Propane is increasingly being used as an alternative fuel for vehicles, and it is often stored to power emergency electrical generators. Fire fighters should be familiar with the basic hazards and characteristics of propane as well as with procedures for fighting propane fires.

Propane (LPG) exists as a gas in its natural state at temperatures higher than −42.2°C (−44°F). When the gas is placed into a storage cylinder under pressure, it is changed into a liquid. Storing propane as a liquid is very efficient because it has an expansion ratio of 270:1 (i.e., 0.02 m³ [1 ft³] of liquid propane is converted to 7.64 m³ [270 ft³] of gaseous propane when it is released into the atmosphere). Put simply, a large quantity of propane fuel can be stored in a small container.

Inside a propane container, there is a space filled with propane gas above the level of the liquid propane. As the contents of the cylinder are used, the liquid level drops and the vapour space increases. The internal piping is arranged so as to draw propane gas from the vapour space.

Propane gas containers come in a variety of sizes and shapes, with capacities ranging from a few millilitres to thousands of litres. The cylinder itself is usually made of steel or aluminum. A discharge valve (or service valve) keeps the gas from escaping into the atmosphere and controls the flow of gas into the system where it is used. This valve should be easily visible and accessible. In the event of a fire, closing the valve should stop the flow of the product and extinguish the fire. The valve should be clearly marked to indicate the direction in which it should be turned or moved to reach the closed position.

A connection to a hose, tubing, or piping allows the propane gas to flow from the cylinder to its destination. In the case of portable tanks, this connection is often the most likely place for a leak to occur.

A propane cylinder is always equipped with a pressure relief valve to allow excess pressure to escape, thereby preventing an explosion if the tank becomes overheated. Propane cylinders must be stored in an upright position so that the pressure relief valve remains within the vapour space. If the cylinder is placed on its side, the pressure relief valve could fall below the liquid level. If a fire were then to heat the tank and cause an increase in pressure, the pressure relief valve would release liquid propane, which would expand by the 270:1 ratio and create a huge cloud of potentially explosive propane gas.

Propane Hazards

Propane is highly flammable. Although it is nontoxic, propane can displace oxygen and cause asphyxiation. By itself, propane is odourless, so leaks of pure propane cannot be detected by a human sense of smell. For this reason, ethyl mercaptan is added to propane to give it a distinctive odour.

Propane gas is heavier than air, so it will flow along the ground and accumulate in low areas. Wind and HVAC systems can also cause propane gas to migrate unexpectedly from the source. When using meters to check for LPG, be sure to check storm drains, basements, and other low-lying areas for concentrations of the gas.

When responding to a reported LPG leak, fire fighters and their apparatus should be staged uphill and upwind of the scene. Because an explosion can happen at any time, fire fighters should wear full PPE and SCBA at this type of incident. Life safety should be the highest priority; depending on the type and size of the leak, an evacuation might be necessary.

The greatest danger with propane and similar products is a boiling liquid/expanding vapour explosion (BLEVE). A BLEVE is an explosion that occurs when pressurized liquified materials such as propane or butane inside a closed vessel are exposed to a source of high heat. If an LPG tank is exposed to heat from an external fire, the liquid fuel in the tank absorbs much of the heat. As the liquid fuel is converted to a gaseous form, the vapour pressure inside the tank will increase until propane gas escapes from the pressure relief valve. If the fire is not quickly extinguished, more and more gas escapes, the level of the liquid fuel inside the tank drops, and the surface of the tank begins to weaken due to extreme temperatures. If the internal pressure exceeds the strength of the container, the container can catastrophically rupture (**FIGURE 23-6**). Pressurized propane gas will ignite in an expanding fireball, and the container will be destroyed with explosive force.

The best method to prevent a BLEVE is to direct heavy streams of water onto the tank from a safe distance.

FIGURE 23-6 Small propane cylinder failure due to fire.
Courtesy of Rob Schnepp.

The water should be directed at the area where the tank is being heated. Cool the upper part of the tank to cool the gas vapours. The fire fighters operating these streams should work from shielded positions or use remote-controlled or unmanned master stream appliances. Horizontal tanks are designed to fail at the ends if a catastrophic failure occurs, so fire fighters should operate from the sides of the tank.

Flammable Gas Fire Suppression

Fighting fires involving LPG or other flammable gas cylinders requires careful analysis and logical procedures. If the gas itself is burning because of a pipe or regulator failure, the best way to extinguish the fire is to close the main discharge valve at the cylinder. If the fire is extinguished and the fuel continues to leak, there is a high probability that it will reignite explosively. Do not attempt to extinguish the flames unless the source of the fuel has been shut off or all of the fuel has been consumed. If the fire is heating the storage tank, use hose streams to cool the cylinder, being careful not to extinguish the fire.

Unless a remote discharge valve is available, the flow of propane can be stopped only if it is safe to approach the cylinder. Fire fighters should inspect the integrity of the cylinder from a distance before they make any attempt to approach and close the discharge valve. If the container is damaged or the discharge valve is missing, the fuel should be allowed to burn off, while hose streams continue to cool the tank from a safe distance.

Approach a flammable gas fire with two 45-mm (1¾-in.) hose lines working together. When approaching a horizontal LPG tank, always approach it from the sides. Use a straight stream pattern to cool the tank from a distance. Carefully approach the tank while continuing to adjust the nozzle to a wide fog pattern. The team leader should be located between the two nozzle operators. On the command of the leader, the crew should move forward, remaining together at all times.

Upon reaching the discharge valve, the fire fighter in the center can close the discharge valve, stopping the flow of gas. Any remaining fire may then be extinguished by normal means. Continue the flow of water as a protective curtain and to reduce sources of ignition.

If the fire is extinguished prematurely, the discharge valve should be closed as soon as the team reaches it. Always approach and retreat from these types of fires while facing the objective with water flowing, in case of reignition.

Unmanned master stream appliances should be used to protect flammable gas containers that are exposed to heavy fire. Direct the stream so that it is one-third of the way down the container. This technique will allow some of the water to roll up and over the container, while the remainder projects downward (**FIGURE 23-7**). The objective is to cover as much of the exposed tank as possible.

If the LPG container is located next to a fully involved building or a fire that is too large to control, evacuate the area and do not fight the fire. If there is nothing to save, risk nothing.

Keep in mind that if the pressure relief valve is open, the flammable gas container is under stress. Exercise extreme caution in this scenario. As the gas pressure is relieved, it will sound like the whistle on a teakettle; if the sound is rising in frequency, an explosion could be imminent, and evacuation should be ordered.

Continue to cool liquid fuel containers until either the discharge valve is closed or all fuel escapes from the container. Keep unmanned master streams in place until the incident is declared to be under control by the IC.

To suppress a flammable gas cylinder fire, follow the steps in **SKILL DRILL 23-2**.

Introduction to Firefighting Foam

Because water absorbs heat energy and primarily cools the surface of the burning fuels, it is the extinguishing agent of choice for many fires; however, some fires require the use of special hazard-suppression agents. Among the many hazards with which fire fighters must contend are a wide variety of incidents involving combustible liquids and flammable liquids. Successful control and extinguishment of these incidents require not only the proper application of foam on the fuel surface but also an understanding of physical characteristics of foam. Lack of familiarity with the chemical characteristics of foam and its application can cause severe problems.

This section discusses the types and application methods for firefighting foams used in manual firefighting operations. It also briefly discusses different

FIGURE 23-7 Using a master stream appliance to protect a flammable gas container that is exposed to fire.
© Jones & Bartlett Learning.

types of special hazard fire suppression systems that use firefighting foam. Chapter 26, *Fire Detection, Suppression, and Smoke Control Systems*, discusses most of these special hazard-suppression agents in more detail.

Firefighting foam can be used to fight multiple types of fires and to prevent the ignition of materials that could become involved in a fire. The use of foam is increasing as many new types of foam have become available and efficient systems for applying them have been developed. Foams also have been developed for use in neutralizing hazardous materials and in decontamination. Many fire departments use several different types of foam for a variety of situations.

Firefighting foam is produced by mixing **foam concentrate**, or a concentrated liquid, with water to produce a **foam solution** that can serve as an effective extinguishing agent. Several different types of foam are used for fires involving different types of fuels. Each type of foam requires the appropriate type of concentrate, the proper equipment to mix the concentrate with water in the required proportions, and the proper application equipment and techniques to mix the foam solution with air to generate the appropriate finished foam for the specific application. Fire fighters must become familiar with the specific types of foam used by their fire department and the proper techniques for using them. It is particularly important to learn where and when to use each type of foam that is available.

Foam-Extinguishing Mechanisms

Fire-extinguishing mechanisms do not refer to specific fire protection systems or devices. Instead, fire

extinguishing mechanisms mean that one or more sides of the fire tetrahedron are removed by the application of a particular fire-extinguishing agent. The fire tetrahedron is a geometric shape used to depict the four components required for a fire to occur: fuel, oxygen, heat, and chemical chain reactions. If you remove any one of these elements, the fire will be extinguished. Although foam is a water-based extinguishing agent, its primary extinguishing mechanism is not cooling the surface of burning fuels. Most firefighting foams prevent ignition or extinguish fires by creating a **foam blanket** that isolates the fuel vapours from the heat source and excludes oxygen (in the ambient air) from the burning process. Each class and type of foam extinguishes fire in a unique way, depending on its intended application. It is imperative that fire fighters understand how to select and apply various classes and types of firefighting foam based on the specific hazards they might face.

Foam Classifications

The basic classifications of firefighting foams are either Class A or Class B. Within each classification there are many types of foam.

Class A foam is used to fight fires involving ordinary combustible materials, such as wood, paper, and textiles. It is also effective on organic materials such as hay and straw. Class A foam is particularly useful for protecting buildings in rural areas during forest and brush fires when the supply of water is limited. Because Class A foam is a water-based fire suppression agent, it is effective for cooling ordinary combustibles below their ignition temperatures. In addition to cooling the

SKILL DRILL 23-2
Suppressing a Flammable Gas Cylinder Fire Fire Fighter II, NFPA 1001: 5.3.3

1 Using a straight stream, cool the tank from as far away as possible until the pressure relief valve resets.

2 Wearing full PPE, two teams of fire fighters, using a minimum of two 45-mm (1¾-in.) hose lines, advance toward the side of the tank. Do not approach the tank from either end. The officer should be located between the two nozzle persons. The officer coordinates the advance toward the cylinder.

3 Gradually adjust the nozzles to a wide fog pattern as you approach the side of the tank. Make sure the fog streams overlap as you reach the tank.

4 When the cylinder is reached, the two nozzle teams isolate the discharge valve from the fire with their fog streams while the leader closes the discharge valve, eliminating the fuel source.

5 After the burning gas is extinguished, the fire fighters continue to apply water to the cylinder to cool the metal, with the goal of preventing tank failure and a subsequent BLEVE.

6 As cooling continues, fire fighters slowly back away from the cylinder while adjusting the nozzles to a straight stream as they retreat.

© Jones & Bartlett Learning. Photographed by Glen E. Ellman.

fuel, Class A foam that is applied directly to the burning fuel surface can insulate the fuel from radiant heat, as well as exclude oxygen and reduce vapour production. It can enhance the ability of water to penetrate tightly stacked ordinary combustibles such as large hay bales or deep-seated fires in pallets made from Class A materials. Class A foam contains a solution of water and Class A foam concentrate. Class A foam increases the effectiveness of water as an extinguishing agent by reducing the surface tension (the physical property that causes water to bead or form a puddle on a flat surface) of water. This allows the water to penetrate dense materials instead of running off the surface, and it allows more heat to be absorbed. Because foam tends to stick to both horizontal and vertical surfaces, it also keeps water in contact with unburned fuel to prevent initial ignition or reignition. Class A foam concentrate can be added to water streams through a foam eductor or other proportioner device and applied with several types of nozzles. Refer to the manufacturer's instructions and your department's SOPs for specific application information.

Class B foam is used to fight Class B fires involving flammable and combustible liquids. The various types of Class B foam are formulated to be effective on different types of liquid fuels. Note that some liquids are incompatible with different foam formulations and will destroy the foam before the foam can control the fire.

In a liquid fuel fire, the liquid itself does not burn. Only the ignitable vapours that evaporate from the surface of the liquid and mix with air can burn. Depending on the temperature and the physical properties of the fuel, the amount of vapours released from the surface of a liquid varies. For example, gasoline produces flammable vapours at temperatures as low as −42.7°C (−45°F). This is called the flash point. The flash point is the lowest temperature at which a liquid or solid produces an ignitable vapour. As a result, gasoline is extremely volatile at ambient temperatures and pressures. Some vapours are lighter than air and rise up into the atmosphere. Other vapours are heavier than air and flow across the surface and along the ground, collecting in low spots. Understanding the characteristics of liquid fuels helps fire fighters to select the appropriate type of Class B foam to apply to a spill before ignition or after the fuel has already begun to burn.

Foam extinguishes liquid fuel fires by a combination of fire-extinguishing mechanisms. As the foam is applied to a burning liquid fuel, a blanket of foam bubbles floats on the surface of the fuel. This foam blanket excludes air from the surface of the burning fuel, so the oxygen is removed from the fire tetrahedron. Because the foam blanket separates the fuel from the fire above, the vapours are isolated from the high temperatures, and the fuel component is also removed from the tetrahedron (**FIGURE 23-8**). Additionally, because the foam bubbles contain some water, the surface of the fuel is cooled, further reducing vapour production of some fuels.

Once a foam blanket has been applied, it must not be disturbed. The fuel under the foam blanket is still capable of producing ignitable vapours. If the foam blanket is disturbed by wind, by someone walking through the liquid, or by hose streams breaking up the foam blanket, ignitible vapours may be released and could be reignited easily.

Foam

Leaking flammable liquids

FIGURE 23-8 A foam blanket excludes air and separates the flames from the fuel surface.

© Jones & Bartlett Learning.

When using foam to extinguish a liquid fuel fire, it is critically important to apply enough foam to cover the liquid surface fully. If not enough foam is available to completely cover the surface at one time, the fire will continue to burn, and the suppression efforts will be ineffective. The rate of foam application must also be high enough to cover the surface and maintain a blanket on top of the liquid. All firefighting foam degrades over time, depending on the fuel and certain conditions. If the application rate is too low, the column of hot gases, flames, and smoke rising above the fire (plume) will keep the foam from covering the surface, and the heat of the fire may destroy the foam that has already been applied.

Class B foam can also be applied to a liquid fuel spill in an effort to prevent a fire. A foam blanket floating on the surface of the liquid will inhibit the production of vapours that might otherwise be ignited. In this situation, the rate of foam application is not as critical because there is no fire working to destroy the foam while it is being applied.

Fire fighters must ensure that they use the proper foam for the situation that is encountered. For example, they must use an alcohol-resistant foam if the incident involves a polar solvent. Polar solvents are water-soluble, flammable liquids (such as alcohols, acetone, esters, and ketones) that readily mix with water. An ordinary foam would break down quickly if it came in contact with this type of product. Be sure to use the proper foam for the product involved in the incident.

Class A foams are not designed to resist hydrocarbons or to self-seal on a liquid fuel surface. If Class A foam is used on a liquid fuel fire, the foam blanket may quickly break down or be easily disrupted. In this case, the risk of reignition could create a dangerous situation.

Foam Concentrates

As discussed, foam concentrate is mixed with water in different ratios to produce a foam solution. Air is introduced into the foam solution to produce firefighting foam, which is then applied to extinguish a fire or to cover a spill.

Class A Foam Concentrates

Class A concentrate is designed to mix with water to form foam for application on ordinary combustibles. Class A foam is an excellent extinguishing agent as a means to both cool and coat combustible surfaces. It effectively extinguishes burning material and is designed to prevent reignition. The end product can be adjusted to have different properties by varying the percentage of foam concentrate in the mixture and the application method. It is possible to produce a "wet" foam that has good penetrating properties or a "drier" foam that is more effective for applying as a protective layer of foam to a surface.

Class B Foam Concentrates

Most Class B foam concentrates are designed to be used in either 3% or 6% solutions. Fire fighters must follow the manufacturer's recommended mixing ratios and ensure that the foam applicator is one that is referenced in training materials. Some foams are designed to be used at 3% for ordinary hydrocarbons and at 6% for polar solvents. Fire fighters must determine which type of fuel is involved in the incident so that they can select the correct proportioning rate.

The compatibility of foam agents with other extinguishing agents needs to be considered, as well. For example, some combinations of dry-chemical extinguishing agents and foam agents can cause an adverse reaction. The compatibility data needed are available from the manufacturers of the agents.

Fire fighters must never mix different types of foam concentrate or even different brands of the same type of foam concentrate unless they are known to be compatible. Some concentrates have been known to react with other types of foams, causing a congealing of the concentrate in the storage containers or in the foam delivery system. This type of reaction can plug a foam delivery system and render it useless.

SAFETY TIP

Although most foam concentrates are somewhat corrosive, they pose few health risks to fire fighters if normal precautions are taken. In the past, some fire fighters have experienced minor skin irritations after coming into contact with foam concentrate. Fire fighters must read and follow the manufacturer's directions before using foam. After using foam, thoroughly clean all equipment following the manufacturer's instructions.

Some components in Class B foam concentrates that have been widely used in the past are being phased out because of environmental concerns. Newer concentrates have been developed that are equally effective but lack the undesirable properties. The major categories of Class B foam concentrate are as follows:

- Protein foam
- Fluoroprotein foam
- Film-forming fluoroprotein (FFFP)
- Aqueous film-forming foam (AFFF)
- Alcohol-resistant foam

Standard personal protective clothing provides appropriate protection when working with foam agents. Fire fighters are advised to thoroughly wash their skin after coming into contact with foam and thoroughly clean all equipment with clear water after foam use.

Personal protective equipment should be thoroughly washed and dried after contact with foam. Follow the manufacturer's instructions.

Protein Foam

Protein foams are made from animal by-products. They are effective on Class B hydrocarbon fires (petroleum-based fires) such as gasoline or diesel fuel fires, and they are applied in 3% or 6% delivery rates. Protein foams tend to spread slowly over the surface of burning fuels.

Fluoroprotein Foam

Fluoroprotein foams are made from the same base materials as protein foam but include additional fluorochemical surfactant additives. These additives allow the foam to produce a fast-spreading film. This type of foam is used for hydrocarbon vapour suppression and extinguishment of deep-seated fires.

Film-forming fluoroprotein foam (FFFP) agents are composed of protein plus film-forming fluorinated surface-active agents. This makes them capable of forming water solution films on the surface of most flammable hydrocarbons and of conferring a fuel-shedding property to the foam generated. FFFP foams have a fast-spreading and leveling characteristic. Like other foams, they act as surface barriers that exclude air and prevent vapourization.

Aqueous Film-Forming Foam

Aqueous film-forming foam (AFFF) is a synthetic-based foam that is particularly suitable for spill-related fires involving gasoline and light hydrocarbon fuels. It can form a seal across a surface quickly and has excellent vapour suppression capabilities. It can also be used on actively burning pools of flammable liquids.

Alcohol-Resistant Foam

Alcohol-resistant foam concentrate has properties similar to AFFF; however, it is formulated so that alcohols and other polar solvents do not dissolve the foam. Regular protein foams cannot be used on these types of products.

Foam Equipment

Foam equipment includes the proportioning equipment used to mix foam concentrate and water to produce the foam solution, as well as air-aspirating nozzles and other devices that are used to apply the foam. Many different types of proportioning and application systems are available. Most pump companies carry the necessary equipment to place at least one foam attack line into operation. Structural firefighting apparatus can also be designed with built-in foam-proportioning systems and onboard tanks of foam concentrate to provide greater capabilities. Some fire departments specify new apparatus to include Class A or Class B integrated foam systems that can be used at the discretion of the company officer. Other departments have special foam apparatus available for situations when large quantities of foam are needed.

LISTEN UP!

Never mix different brands or classes of foam. This will cause gelatin-like substances to form, and they will clog the foam system.

Compressed Air Foam Systems

A compressed air foam system (CAFS) is a relatively new method of making class A foam. Compressed air foam (CAF) is produced by injecting compressed air into a stream of water that has been mixed with 0.1% to 1.0% foam. This results in a highly compacted foam with small bubbles. CAF has excellent surface-adherence properties, so it can be a good choice of preventive foam to be applied to exposure buildings. CAF is also an excellent foam for overhaul and for deep-seated Class A fires.

CAF is produced by a combination of a centrifugal fire pump, a foam metering device to provide a 0.1% to 1.0% concentration of foam that is injected on the discharge side of the pump, and an air compressor to inject compressed air bubbles into the foam solution. This solution is said to be "scrubbed" by its movement through the attack line. As it exits the nozzle, the compressed air expands, producing a finished foam product. Hose lines containing CAF are lighter than hose lines that are completely filled with water because the CAF consists of a mixture of water and a significant amount of air. The finished foam product is usually discharged at a rate of between 151 and 473 L (40 and 125 gal) of water per minute as measured before expansion.

Foams of different consistencies can be produced by adjusting the ratios of air to water. Wetter foams have better surface penetration properties and are produced by decreasing the ratio of air to water. Wetter foams will drain more quickly in the presence of heat. Drier foams have longer drain times and produce more durable bubbles. The bubbles do not burst as quickly, so it is a longer-lasting product. Because CAF adheres to most surfaces, it is used to protect exposures during structural fires and to coat the sides and roofs of buildings during a wildland fire as a precautionary measure. When prepared as a drier foam, it will remain in place for a longer period of time than many other foams. When CAF is used for interior structural firefighting, a wetter foam is often used.

In spite of earlier claims that the use of CAF resulted in faster knockdown and more rapid cooling, current research conducted by NIST indicates that a given amount of water applied to a fire produces similar cooling and structural knockdown to using a similar mass flow of compressed air foam (Weinschenk, 2017). In fact, water streams (narrow, fog, and straight streams) were shown to be somewhat more effective than CAF streams for an indirect attack outside of the fire compartment, but both water and CAF were determined to be effective suppression agents for cooling and fire suppression. These tests also indicated that there did not appear to be significant differences in the amount of post-extinguishment smoke between using water and using CAF. Some departments have identified limitations using CAF as the initial interior fire attack stream, but it may be effective as an exterior transitional attack stream before making entry into the structure. If the CAF can be applied directly onto the burning fuel surfaces from a safe exterior location, the fire may be significantly suppressed before making entry for an interior attack for final extinguishment.

The use of CAF requires ongoing classroom and practical training for pump operators, fire fighters, and company officers. This training should include practice with live burn situations using different scenarios.

Foam-Proportioning Equipment

A foam proportioner mixes the foam concentrate into the fire stream in the proper percentage. The two types of proportioners—eductors and injectors—are available in a wide range of sizes and capacities. Foam solution can also be produced by batch mixing or premixing. This is discussed later.

Foam Eductors

A foam eductor draws foam concentrate from a container or storage tank into a moving stream of water or hose line (**FIGURE 23-9**). Eductors operate on the Venturi effect (i.e., suction effect), much like a typical garden hose–end sprayer applies fertilizer or weed killer through the nozzle. As the water stream passes through a narrowing of the eductor, the increase in water velocity (speed) reduces the water's pressure as it flows into the eductor. Foam concentrate is introduced into the eductor using a metering device. A metering device is used to set the percentage of foam concentrate that is educted into the stream (**FIGURE 23-10**). An eductor can be built into the plumbing of a pumper, or a portable eductor can be inserted in an attack hose line (**in-line eductor**). A foam eductor is usually designed to work at a predetermined pressure and flow rate.

Two types of eductors are used by the fire service: in-line eductors and bypass eductors. The most common type is the portable in-line eductor. The portable

FIGURE 23-9 The most common type of eductor used by the fire service is the portable in-line eductor. A portable foam educator draws foam concentrate from a container or storage tank into a moving stream of water.
© Jones & Bartlett Learning. Photographed by Glen E. Ellman.

FIGURE 23-10 The metering device controls the amount of foam concentrate educted into the water.
© Jones & Bartlett Learning. Photographed by Glen E. Ellman.

in-line eductor is sized to work with a 38-mm (1½-in.) or 45-mm (1¾-in.) attack line. This type of eductor uses water pressure to draw foam concentrate from a portable container into the stream. The eductor is placed in the attack line at a coupling. As water flows through the eductor, foam concentrate is drawn into the stream through a pickup tube to produce the proper foam solution. The foam solution is then discharged through a foam nozzle to produce the finished foam. Be sure to follow the manufacturer's instructions or refer to your department's SOPs for proper foam eductor operation.

Bypass eductors are permanently mounted devices that can be used for water or foam application, depending on what is required at the incident scene.

To place an in-line eductor foam line in service, follow the steps in **SKILL DRILL 23-3**.

SKILL DRILL 23-3
Operating an In-Line Foam Eductor Fire Fighter II, NFPA 1001: 5.3.1

1 Don all PPE. Make sure all necessary equipment is available, including an in-line foam eductor and the correct nozzle. Ensure that enough foam concentrate is available to suppress the fire. Deploy an attack line, remove the nozzle, and replace it with the foam nozzle.

2 Place the foam concentrate container next to the eductor, check the percentage at which the foam concentrate should be used (found on container label), and set the metering device on the eductor accordingly.

3 Place the in-line eductor in the hose line according to the manufacturer's instructions and your department's SOPs.

4 Place the pickup tube from the eductor into the foam concentrate, keeping both items at similar elevations to ensure sufficient induction of foam concentrate. Charge the hose line with water per your department SOPs or as directed by the manufacturer.

5 Flow water through the hose line until foam starts to come out of the nozzle. The hose line is now ready to be advanced onto the fuel. Apply foam using one of the three application methods (roll-in method, bounce-off method, or rain-down method—discussed later in this chapter) depending on the situation.

Foam Injectors

Foam injectors add the foam concentrate to the water stream under pressure. Most injector-based proportioning systems can work across a range of flow rates and pressures. A metering system measures the flow rate and pressure of the water and adjusts the injector to add the proper amount of foam concentrate. This type of system is often installed on special foam apparatus.

Batch Mixing

Foam concentrate can be poured directly into an apparatus booster tank to produce foam solution through a technique called a batch mix (**FIGURE 23-11**). If the booster tank has a capacity of 1893 L (500 gal), 57 L (15 gal) of 3% foam concentrate should be added. If 6% foam concentrate is used, 114 L (30 gal) of foam concentrate should be added to the booster tank. Follow the manufacturer's guidelines and your departmental SOP's to ensure that mixing ratios are correct. It might be necessary to drain sufficient water from the tank first to make room for the foam concentrate. After the concentrate has been added, the solution should be mixed by circulating the water through the pump before it is discharged.

LISTEN UP!

Class A foam concentrates that are batch mixed must be used within 24 hours if they are to be effective.

Premixing

Premixed foam solution is commonly used in 9.5-L (2½-gal) portable fire extinguishers. Foam fire extinguishers are filled with premixed foam solution and pressurized with compressed air or nitrogen. Some apparatus are equipped with large tanks holding 189 or 379 L (50 or 100 gal) of premixed foam, which operate in the same manner. Many fire departments use these extinguishers filled with premixed foam for small flammable liquid spills at the scene of a motor vehicle accident.

Foam Application

Foam can be applied to a fire or spill through portable extinguishers, handlines, master stream appliances, or a variety of fixed systems for special applications. Most fire departments apply foam through either handlines or master stream appliances on fire apparatus. Several types of nozzles are used, each with the goal of producing different types of foam:

- Medium- and high-expansion foam generators
- Master stream foam nozzles
- Air-aspirating foam nozzles
- Smooth-bore nozzles
- Fog nozzles

The primary difference between these nozzles is the manner in which air is introduced into the stream of foam solution to produce the desired consistency of foam and air bubbles.

FIRE FIGHTER TIP

Nozzles are an important part of all foam operations. The proper nozzle is needed in order for fire fighters to be able to produce a good-quality foam blanket.

Foam Expansion Ratios

Foam can be applied using a wide range of expansion rates, depending on the amount of air that is mixed into the stream and the size of the bubbles that are produced. NFPA 11, *Standard for Low-, Medium-, and High-Expansion Foam*, identifies three distinct categories of foam based on their expansion rates. The expansion

FIGURE 23-11 Batch mixing is not the most effective way to mix foam concentrate and water.
© Jones & Bartlett Learning. Photographed by Glen E. Ellman.

rate is determined by the volume of foam concentrate compared to the volume of the finished foam as it is applied to the fire or hazard.

Low-expansion foam has an expansion ratio of less than 20:1. This means that 0.02 m³ (1 ft³) of foam concentrate will produce up to 0.56 m³ (20 ft³) of finished foam. Low-expansion foam has little air entrained into the foam solution, which results in smaller bubble size. This type of foam is often produced with standard adjustable fog nozzles, where the air is entrained by the flowing stream and mixed into the foam solution. Although a fog nozzle may be used to apply AFFF or Class A foam, many manufacturers recommend using specific air-aspirating nozzles to produce a higher-quality foam blanket. Most fire departments carry low-expansion foam for use on two-dimensional fuel spills such as gasoline or diesel fuel.

Medium-expansion foam has an expansion ratio between 20:1 and 200:1 and is produced with special air-aspirating nozzles that are designed to introduce more air into the stream and produce a consistent bubble structure (aeration). Although some departments use medium-expansion foam for manual firefighting operations, it is typically used in fixed fire protection systems designed to protect three-dimensional hazards.

High-expansion foam has an expansion ratio between 200:1 and 1000:1 and contains a much higher proportion of air, which produces large bubbles. High-expansion foam is used in fixed fire protection systems to provide a thick layer of foam several feet high. These fixed fire protection systems use a high-expansion foam generator to introduce large quantities of air into the discharge stream. High-expansion foam is typically used in automatic special-hazard fire protection systems that are designed to completely fill entire rooms or a large, enclosed space with foam. The foam covers the liquid fuel hazard and provides insulating protection to high-value equipment. These systems are most likely to be found in aircraft hangars or other large industrial areas. For more information on medium- and high-expansion foam systems, see the text *Fire Protection Systems* (Jones & Bartlett Learning, 2015).

Foam Application Techniques

When applying foam from a handline, the correct application techniques must be used to produce the desired quality of foam and successfully blanket the surface of a burning liquid fuel or liquid spill. Class A foam usually is most effective if it can be applied directly to the surface of the burning solid fuels or to exposure fuels you want to protect from radiant heat. Your department may use specific foam nozzles for Class A foam application.

Some departments, however, use Class A foam in very low percentages (0.25% to 0.5% concentrate) with a standard adjustable nozzle, either as a dedicated foam line or on demand at the company officer's direction. This provides the heat absorption capabilities of a fog stream with the benefits of foam (surface penetration and fuel coating) in one application.

Fire fighters must use caution when using foam to suppress liquid fuel fires. Applying a Class B foam stream directly onto a pool of burning liquid fuel may cause the fuel to splash and spread the fire.

Plunging the stream into an existing foam blanket will allow vapours to escape, which can result in spreading, reignition, or flare-up of the fire. To avoid these problems, three methods are used to apply foam blankets on Class B fires. These techniques can be direct or indirect and include the rain-down method, the roll-in method, and the bounce-off method of application. Whichever method is used, the foam must be applied carefully to avoid disrupting the existing foam blanket or splashing fuel outside of the container or pool. These methods can also be used to apply Class A foam when it is not possible or safe to apply the foam directly to burning objects.

Rain-Down Method

The **rain-down method** is used when there is no object available to deflect the foam onto the burning surface. This method consists of lofting the foam stream into the air above the fire and letting it fall down gently onto the burning surface. This method is also used for applying foam to a large area. Because the stream breaks apart above the fuel surface as it falls, the foam does not disrupt the existing foam blanket or cause the fuel to splash. Carefully observe how the foam blanket is building up, and direct your stream so that the entire surface is covered. If a large spill of combustible liquid is burning, the rain-down method can be used to apply foam to the edge of the fire from a safe distance. The rain-down method is a good choice for initial application if fire conditions make it difficult to approach the fuel surface. Once a portion of the burning liquid is covered with foam, the roll-in or bounce-off method may be used if it is safe to approach the fire.

To perform the rain-down method of applying foam, follow the steps in **SKILL DRILL 23-4**. In this example, foam is being applied as a protective coating.

Roll-In Method

The **roll-in method** is used when a pool of flammable product is located on open ground. The technique is performed by directing the foam stream onto the ground in front of the product involved. The energy of the stream

SKILL DRILL 23-4
Performing the Rain-Down Method of Applying Foam Fire Fighter II, NFPA 1001: 5.3.1

1 Open the nozzle and test to ensure that foam is being produced.

2 Move within a safe range of the fuel product or tank, and open the nozzle.

3 Direct the stream of foam into the air so that the foam gently falls onto the surface of the fuel product or tank.

4 Allow the foam to flow across the surface of the fuel product or tank until it is completely covered.

will push the foam blanket across the surface of the fuel or roll it onto the surface of the fuel. The stream is moved back and forth in a slow, steady horizontal motion to push the foam forward gently until the area is covered. It is important to push the foam slowly and gently so that the foam blanket is not disturbed. Avoid aiming the nozzle down toward the surface of the spill to prevent splashing fuel. The fire fighter might need to move to different positions to be sure that the entire surface of the fuel is covered by the foam blanket.

To perform the roll-in (roll-on or sweep) method of applying foam, follow the steps in **SKILL DRILL 23-5**.

Bounce-Off Method

The bounce-off method is used at fires where the fire fighter can use an object to deflect the foam stream and let it flow down onto the burning surface. This method could be used to apply foam to an open-top storage tank or a rolled-over transport vehicle, for example. The fire fighter should sweep the foam back and forth against the object while the foam flows down and spreads back across the surface. As with all foam application methods, it is important to let the foam blanket flow gently onto the surface of the flammable liquid to form a blanket.

To perform the bounce-off (bank-shot or bank-down) method of applying foam, follow the steps in **SKILL DRILL 23-6**.

> ### FIRE FIGHTER TIP
>
> Using Class A or Class B foam on a brush fire after extinguishment can allow for additional water penetration and reduce the likelihood of rekindle.

Backup Resources

When attempting to use foam to extinguish a flammable liquid fire, it is critically important that enough foam concentrate is available to complete the job. If the flow of foam must be interrupted while additional foam supplies are obtained, the fire will destroy the foam that has already been applied. It is better to wait until an adequate supply of foam concentrate is on hand than to waste the limited supply that is immediately available.

Foam manufacturers provide specific formulas that enable fire fighters to calculate how much foam is required to extinguish fires of a certain size. Most fire departments have contingency plans to deliver large quantities of foam to the scene of a major incident. This foam could be located on designated vehicles or kept in storage at fire stations. Backup sources often include airport crash vehicles or petroleum facilities that have their own foam apparatus and often keep large quantities of foam in storage. Manufacturers of foam products also have emergency programs to deliver large quantities of foam to the scene of exceptionally large-scale incidents.

SKILL DRILL 23-5
Performing the Roll-In Method of Applying Foam Fire Fighter II, NFPA 1001: 5.3.1

1 Open the nozzle and test to ensure that foam is being produced. Move within a safe range of the fuel product or tank, and open the nozzle. Direct the stream of foam onto the ground just in front of the pool of product.

2 Allow the foam to roll across the top of the pool of the fuel product or tank until it is completely covered.

SKILL DRILL 23-6
Performing the Bounce-Off Method of Applying Foam Fire Fighter II, NFPA 1001: 5.3.1

1 Open the nozzle and test to ensure that foam is being produced.

2 Move within a safe range of the fuel product or tank, and open the nozzle. Direct the stream of foam onto a solid structure such as a wall or metal tank so that the foam is directed off the object and onto the pool of product or tank.

3 Allow the foam to flow across the top of the pool of product or tank until it is completely covered. Be aware that the foam may need to be bounced off several areas of the solid object to extinguish the burning product.

© Jones & Bartlett Learning. Photographed by Glen E. Ellman.

Foam Apparatus

Some fire departments operate apparatus that is specifically designed to produce and apply foam. The most common examples are found at airports, where these apparatus are used for aircraft rescue and firefighting. These large vehicles carry the foam concentrate and water onboard and are designed to quickly apply large quantities of foam to a flammable liquid fire. Remote-control monitors can be used to apply foam while the vehicles are in motion. If there is an airport close to your department, you should train with them and know the best way to access them, before an emergency occurs.

Hose Testing and Records

As discussed in Chapter 15, *Fire Hose, Appliances, and Nozzles*, each length of hose should be tested at least annually, according to the procedures listed in NFPA 1962, *Standard for the Care, Use, Inspection, Service Testing, and Replacement of Fire Hose, Couplings, Nozzles, and Fire Hose Appliances*. The Fire Fighter II hose testing procedures are complicated and require special equipment. This equipment must be operated according to the manufacturer's instructions.

To perform an annual service test on fire hose, follow the steps in **SKILL DRILL 23-7**.

SKILL DRILL 23-7
Performing an Annual Service Test on a Fire Hose Fire Fighter II, NFPA 1001: 5.5.5

1 Don turnout gear. Connect up to 91 m (300 ft) of hose to a hose test gate valve on the discharge valve of a fire department pumper or hose tester.

2 Attach a nozzle to the end of each hose. Slowly fill each hose with water at 345 kPa (50 psi), and remove kinks and twists in the hose.

3 Open the nozzles to purge air from the hose, discharging the water away from the test area. Close the nozzles once the air is purged. Measure and record the length of each section of hose.

4 Mark the position of each hose coupling on the hose. This will help determine if slippage occurs during the test (Step 7). Check each coupling for leaks. If leaks are found behind the coupling, remove the hose from service. If the leak is in front of the coupling, tighten the leaking coupling. If the leak continues, replace gaskets if necessary after shutting down the hose line.

5 Once all coupling leaks and any other leaks have been dealt with, open each test gate valve for each hose, in preparation for Step 6.

SKILL DRILL 23-7 Continued

Performing an Annual Service Test on a Fire Hose Fire Fighter II, NFPA 1001: 5.5.5

6 Ensure that all fire fighters are clear of the test area. Increase the pressure on the hose to the pressure required by NFPA 1962, and maintain that pressure for 5 minutes. Monitor the hose and couplings for leaks as the pressure increases during the test. Close the gate valves and open the nozzles to bleed off the pressure. Uncouple and drain the hose.

7 Inspect the marks placed on the hose jacket near the couplings to determine whether slippage occurred.

8 Tag hose that failed.

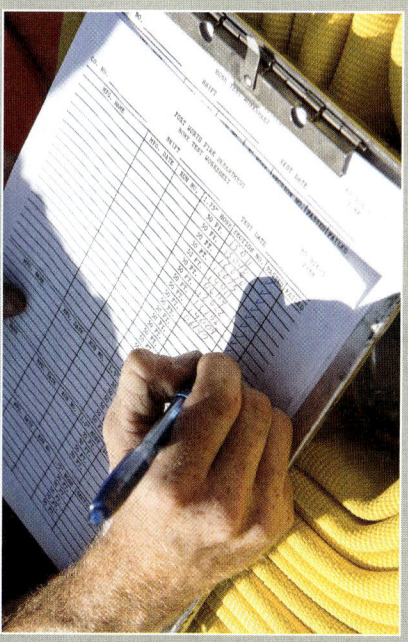

9 Mark hose that passed. Record the results in the departmental logs.

A hose record is a written history of each individual length of fire hose. Each length of hose should be identified with a unique number stenciled or painted on it. A hose record should contain information such as the following:

- Hose size, type, and manufacturer

- Date when the hose was manufactured, purchased, and tested
- Any repairs that have been made to the hose

Some fire departments keep hose records on cards or paper files; others keep the records in a database in a computer system.

After-Action REVIEW

IN SUMMARY

- All fire suppression operations are either offensive or defensive:
  - Offensive—Tasks performed while moving toward the fire or inside the fire structure. Used when the fire is not too large or dangerous to be extinguished using interior handlines.
  - Defensive—Tasks typically performed outside. Used when the fire is too large to be controlled by an offensive attack or when the level of risk is too high.
- The IC evaluates conditions constantly to determine the type of attack that should be used. An interior attack may be switched to a defensive attack if necessary and vice versa. Considerations for the IC include the following:
  - Life safety—A risk/benefit analysis can help to weigh the risks likely to be taken versus the benefits of those risks.
  - Structural collapse—Some buildings give little or no warning before they collapse, so fire fighters should always be aware of building conditions.
  - Fire conditions—A gradual increase in heat, smoke, or visible flame indicates that suppression operations are not effective. A rapid increase in any of these conditions, however, may indicate an imminent hostile fire event. Fire fighters must constantly monitor the fire and smoke conditions and be aware of any changes.
  - Water supply—A water supply must be established before beginning any interior operations.
  - Staffing—Fire fighters should not commit to interior operations without having enough trained fire fighters on the scene to perform all of the essential fire-ground operations.
  - Defensive and offensive operations—Conducted from inside or outside the fire structure.
- Attack types include the following:
  - Interior attack—An offensive operation that requires fire fighters to enter a building and discharge an extinguishing agent onto the fire. In many cases, an interior attack is conducted on a fire that is burning in only one room and may be controlled quickly by one attack hose line. Larger fires require more water, which could be provided by two or more small handlines working together or by one or more larger handlines.
  - Direct attack—The most effective means of fire suppression in most situations. This kind of attack uses a straight or solid hose stream to deliver water directly onto the base of the fire. The water cools the fuel until it is below its ignition temperature.
  - Indirect attack—Used to quickly remove as much heat as possible from the fire atmosphere. Such an approach is particularly effective at preventing flashover from occurring. This method of fire suppression should be used when a fire has produced a layer of hot gases at the ceiling level.
  - Combination attack—Performed in two stages and employs both indirect attack and direct attack methods. This strategy should be used when a room's interior has been heated to the point that it is nearing a flashover condition. Fire fighters first use an indirect attack method to cool the fire gases and then make a direct attack on the main body of fire.
  - Transitional attack—An exterior, indirect method used to cool and control the fire compartment from a safe location before implementing an interior attack.

- Ventilation must be coordinated with the fire suppression efforts to ensure that both events support the IAP.

- Proper ventilation techniques can help to clear smoke and toxic gases from the structure. Improper ventilation can cause a relatively stable incident to quickly escalate, with catastrophic results.

- Hostile fire events can be prevented or minimized through the use of a transitional fire attack that is coordinated with appropriate ventilation.

- Fire fighters who are assigned to vertical ventilation tasks should always be aware of the condition of the roof. They should immediately retreat from the roof if they notice any of the following signs:
  - Visible indication of sagging roof supports
  - Any indication that the roof assembly is separating from the walls, such as the appearance of fire or smoke near the roof edges
  - Structural failure of any portion of the building, even if it is some distance from the ventilation operation
  - Sudden increase in the intensity of the fire from the roof opening
  - High heat indicators on a thermal imaging camera

- Flammable gas cylinders can be found in many places. Many types of flammable gas are stored in many different types and sizes of containers. A variety of flammable gases can be found in industrial occupancies.

- The greatest danger with propane and similar flammable gas products is a BLEVE (boiling liquid/expanding vapour explosion). If an LPG tank is exposed to heat from a fire, the temperature of the liquid inside the container will increase. The fire itself could be fuelled by propane escaping from the tank or from an external source. As the temperature of the product increases, the vapour pressure will increase in tandem. This increasing pressure creates added stress on the container. If this pressure exceeds the strength of the cylinder, the cylinder can rupture catastrophically.

- Foams are either Class A or Class B.
  - Class A foam is used for cooling ordinary combustibles below their ignition temperatures. Ordinary combustible materials include wood, paper, and textiles. Organic combustible materials include hay and straw.
  - Class B foam is used for cooling flammable vapours and liquids below their flash point.

- Foam extinguishes flammable liquid fires by excluding air from the surface of the fuel and by suppressing fuel vapours. When a blanket of foam completely covers the surface of the liquid, the release of flammable vapours stops. Preventing the production of additional vapours eliminates the fuel source for the fire, which extinguishes the fire.

- Foam concentrate is the product that is mixed with water in different ratios to produce foam solution. The foam solution is the product that is applied to extinguish a fire or to cover a spill.

- The major categories of Class B foam concentrate are protein foam, fluoroprotein foam, film-forming fluoroprotein foam, aqueous film-forming foam, and alcohol-resistant foam.

- Compressed air foam systems are a new method of making Class A foam. Compressed air foam is produced by injecting compressed air into a stream of water that has been mixed with 0.1% to 1.0% foam.

- A foam proportioner is the device that mixes the foam concentrate into the fire stream in the proper percentage. The two types of proportioners—eductors and injectors—are available in a wide range of sizes and capacities.

- Foam solution can also be produced by batch mixing or premixing.
  - Batch mixing is a technique where foam concentrate is poured directly into an apparatus booster tank to produce foam solution.
  - Premixed foam is commonly used in 9.5-L (2½-gal) portable fire extinguishers.
  - Class A foam that is batch mixed must be used within 24 hours to be effective.

- When attempting to use foam to extinguish a flammable liquid fire, it is critically important that enough foam concentrate is available to complete the job. If the flow of foam must be interrupted while additional foam supplies are obtained, the fire will destroy the foam that has already been applied.

KEY TERMS

Access Navigate for flashcards to test your key term knowledge.

Alcohol-resistant foam concentrate A concentrate used for fighting fires on water-soluble materials and other fuels destructive to regular, AFFF, or FFFP foams, as well as for fires involving hydrocarbons. (NFPA 11)

Aqueous film-forming foam (AFFF) A concentrate based on fluorinated surfactants plus foam stabilizers to produce a fluid aqueous film for suppressing hydrocarbon fuel vapours and usually diluted with water to a 1 percent, 3 percent, or 6 percent solution. (NFPA 11)

Batch mix The manual addition of foam concentrate to a water storage container or tank to make foam solution. (NFPA 1145)

Bounce-off method A foam application method that applies the stream onto a nearby object, such as a wall, instead of directly onto the surface of the fire. Also referred to as the bank-shot or bank-down method.

Class A foam Foam for use on fires in Class A fuels. (NFPA 1145)

Class B foam Foam intended for use on Class B fires. (NFPA 1901)

Combination attack A type of attack employing both direct attack and indirect attack methods.

Compressed air foam (CAF) A homogenous foam produced by the combination of water, foam concentrate, and air or nitrogen under pressure. (NFPA 11)

Compressed air foam system (CAFS) A foam system that combines air under pressure with foam solution to create foam. (NFPA 1901)

Defensive operations Actions that are intended to control a fire by limiting its spread to a defined area, avoiding the commitment of personnel and equipment to dangerous areas. (NFPA 1500)

Direct attack Firefighting operations involving the application of extinguishing agents directly onto the burning fuel. (NFPA 1145)

Eductor A device that uses the Venturi principle to siphon a liquid into a water stream. The pressure at the throat is below atmospheric pressure, allowing liquid at atmospheric pressure to flow into the water stream. (NFPA 1145)

Film-forming fluoroprotein foam (FFFP) A protein-foam solution that uses fluorinated surfactants to produce a fluid aqueous film for suppressing liquid fuel vapours. (NFPA 10)

Fluoroprotein foam A protein-based foam concentrate to which fluorochemical surfactants have been added. (NFPA 402)

Foam blanket A covering of foam over a surface to insulate, prevent ignition, or extinguish the fire. (NFPA 1145)

Foam concentrate The foaming agent as received from the supplier that, when mixed with water, becomes foam solution. (NFPA 1145)

Foam injector A device installed on a fire pump that meters out foam by pumping or injecting it into the fire stream.

Foam proportioner A device or method to add foam concentrate to water to make foam solution. (NFPA 1901)

Foam solution A homogeneous mixture of foam concentrate and water, in the mix ratio required for the application. (NFPA 1145)

Hostile fire event A general descriptor for hazardous fire conditions, including flashover, backdraft, smoke explosion, flameover, and rapid fire spread. (NFPA 1521)

Indirect attack Firefighting operations involving the application of extinguishing agents to reduce the buildup of heat released from a fire without applying the agent directly onto the burning fuel. (NFPA 1145)

In-line eductor A Venturi-type proportioning device that meters foam concentrate at a fixed or variable concentration into the water stream at a point between the water source and a nozzle or other discharge device. (NFPA 11)

Offensive operations Actions generally performed in the interior of involved structures that involve a direct attack on a fire to directly control and extinguish the fire. (NFPA 1500)

Premixed foam solution Solution produced by introducing a measured amount of foam concentrate into a given amount of water in a storage tank. (NFPA 11)

Protein foam A protein-based foam concentrate that is stabilized with metal salts to make a fire-resistant foam blanket. (NFPA 402)

Rain-down method A foam application method that directs the stream into the air above the fire and allows it to gently fall on the surface.

Roll-in method A foam application method that involves sweeping the stream just in front of the target. Also referred to as the sweep or roll-on method.

SLICE-RS An acronym intended to be used by the first-arriving company officer to accomplish important strategic goals on the fire ground.

Transitional attack An offensive fire attack initiated by an exterior, indirect handline operation into the fire compartment to initiate cooling while transitioning into interior direct fire attack in coordination with ventilation operations.

REFERENCES

National Fire Protection Association. NFPA 11, *Standard for Low-, Medium, and High-Expansion Foam*, 2016. www.nfpa.org. Accessed November 20, 2018.

National Fire Protection Association. NFPA 11A, *Standard for Medium- and High-Expansion Foam Systems*, 1999. www.nfpa.org. Accessed November 20, 2018.

National Fire Protection Association. NFPA 11C, *Standard for Mobile Foam Apparatus*, 1995. www.nfpa.org. Accessed November 20, 2018.

National Fire Protection Association. NFPA 1145, *Guide for the Use of Class A Foams in Fire Fighting*, 2017. www.nfpa.org. Accessed November 20, 2018.

National Fire Protection Association. NFPA 1710, *Standard for the Organization and Deployment of Fire Suppression Operations, Emergency Medical Operations, and Special Operations to the Public by Career Fire Departments*, 2016. www.nfpa.org. Accessed November 20, 2018.

National Fire Protection Association. NFPA 1720, *Standard for the Organization and Deployment of Fire Suppression Operations, Emergency Medical Operations, and Special Operations to the Public by Volunteer Fire Departments*, 2014. www.nfpa.org. Accessed November 20, 2018.

National Fire Protection Association. NFPA 1901, *Standard for Automotive Fire Apparatus*, 2016. www.nfpa.org. Accessed November 20, 2018.

National Fire Protection Association. NFPA 1962, *Standard for the Care, Use, Inspection, Service Testing, and Replacement of Fire Hose, Couplings, Nozzles, and Fire Hose Appliances*, 2018. www.nfpa.org. Accessed November 20, 2018.

UL Firefighter Safety Research Institute. 2018. "SLICE-RS." https://modernfirebehavior.com/s-l-i-c-e-r-s/. Accessed November 20, 2018.

UL. 2013. "Innovating Fire Attack Tactics." https://newscience.ul.com/wp-content/themes/newscience/library/documents/fire-safety/NS_FS_Article_Fire_Attack_Tactics.pdf. Accessed November 20, 2018.

UL. 2014. "Interrupting the Flow Path." https://newscience.ul.com/wp-content/themes/newscience/library/documents/fire-safety/NS_FS_Article_Interrupting_Flow_Path.pdf. Accessed November 20, 2018.

Weinschenk, Craig G., Daniel Madrzykowski, Keith M. Stakes, and Joseph M. Willi. 2017. "Examination of Compressed Air Foam (CAF) for Interior Fire Fighting." NIST Technical Note 1927. https://nvlpubs.nist.gov/nistpubs/Technical Notes/NIST.TN.1927.pdf. Accessed November 20, 2018.

On Scene

You are just returning to the station from a medical emergency when your pump company is dispatched to the report of a car on fire in a garage. The battalion chief arrives first and establishes command. The IC gives a size-up indicating that the structure is a large, single-storey, wood-frame auto repair shop with dark, turbulent smoke coming from the garage area. There are six vehicles in the garage, and large, elevated gasoline storage tanks are close to the building. The IC assigns your crew to initial fire attack.

1. A transitional attack on this fire:

   A. is a defensive operation.

   B. is done as close to the building as possible.

   C. will push the fire.

   D. will improve interior conditions for fire fighters and occupants.

2. Which of the following is *not* a valid reason for ventilating this fire?

   A. Ventilating for water conservation

   B. Ventilating for property conservation

   C. Ventilating for incident stabilization

   D. Ventilating for life safety

(continued)

On Scene Continued

3. If a flammable compressed gas cylinder is exposed to a fire, where should water be applied?

   A. From the end of the cylinder

   B. To the upper part of the cylinder

   C. To the bottom part of the cylinder

   D. To the cylinder valve

4. For an indirect attack outside the fire compartment, which of the following is most effective?

   A. Protein foam

   B. Wide-angle fog stream

   C. CAFS

   D. Straight stream

5. When would you use the roll-in method of applying foam if the gasoline in the storage tanks is ignited?

   A. For a three-dimensional flammable liquid fire

   B. To loft the foam stream into the air above the fire

   C. When a pool of flammable liquid is located on open ground

   D. To deflect a foam stream off an object

6. Which of the following would not be expected to be able to supply additional foam during a large flammable liquid fire?

   A. Airport crash vehicles

   B. Petroleum manufacturers

   C. Highway department

   D. Foam manufacturers

7. Which of the following is *not* part of a hose record?

   A. Hose size and type

   B. Date of manufacture and purchase

   C. Hose tester that was used

   D. Hose manufacturer

Access Navigate to find answers to this On Scene, along with other resources such as an audiobook.

CHAPTER **24**

Fire Fighter II

Vehicle Rescue and Extrication

KNOWLEDGE OBJECTIVES

After studying this chapter, you will be able to:

- Describe a vehicle's anatomy. (**NFPA 1001: 5.4.1**, pp. 874–876)
- List the hazards to look for when arriving on the scene of a vehicle extrication situation. (**NFPA 1001: 5.4.1**, pp. 878–881)
- List the hazards to look for when stabilizing the scene of a vehicle extrication situation. (**NFPA 1001: 5.4.1**, pp. 881–883)
- Describe cribbing. (**NFPA 1001: 5.4.1**, pp. 883–885)
- Describe rescue-lift air bags. (**NFPA 1001: 5.4.1**, pp. 884, 886)
- Describe how to gain access to a victim of a motor vehicle accident. (**NFPA 1001: 5.4.1**, pp. 886–892)
- Describe how to disentangle a victim of a motor vehicle accident. (**NFPA 1001: 5.4.1**, pp. 893–899)
- Describe how to remove and transport victims of a motor vehicle accident. (**NFPA 1001: 5.4.1**, pp. 898, 900)

SKILLS OBJECTIVES

After studying this chapter, you will be able to perform the following skills:

- Disable the electrical system of an electric drive vehicle. (**NFPA 1001: 5.4.1**, p. 877)

- Perform scene size-up at a motor vehicle accident. (**NFPA 1001: 5.4.1**, pp. 880–881)
- Mitigate the hazards at a motor vehicle accident. (**NFPA 1001: 5.4.1**, p. 882)
- Stabilize a vehicle following a motor vehicle accident. (**NFPA 1001: 5.4.1**, p. 885)
- Break tempered glass. (**NFPA 1001: 5.4.1**, p. 888)
- Gain access to a vehicle following a motor vehicle accident. (**NFPA 1001: 5.4.1**, p. 889)
- Force a vehicle door. (**NFPA 1001: 5.4.1**, p. 891)
- Gain access and provide medical care to a victim in a vehicle. (**NFPA 1001: 5.4.1**, p. 892)
- Displace the dashboard of a vehicle by performing the dash roll. (**NFPA 1001: 5.4.1**, p. 896)
- Displace the dashboard of a vehicle by performing the dash lift. (**NFPA 1001: 5.4.1**, p. 897)
- Remove the roof of a vehicle. (**NFPA 1001: 5.4.1**, p. 899)

ADDITIONAL STANDARDS

- **NFPA 1006**, *Standard for Technical Rescue Personnel Professional Qualifications*
- **NFPA 1670**, *Standard on Operations and Training for Technical Search and Rescue Incidents*

Fire Alarm

It has been storming all afternoon. Your station has just finished dinner when the alarm sounds for a motor vehicle accident involving a car that has slid off the road. As you pull out of the station, the rain gets heavier and the wind gets stronger. When you arrive on the scene, you see a passenger car resting on the driver's side in the ditch. One passenger was ejected from the vehicle and is lying motionless by the car. Your officer does a quick size-up and finds two more occupants in the car: The driver is still in the driver's seat and a passenger appears to be unconscious in the back-seat compartment. The officer establishes command, calls for additional resources, and assigns the ambulance crew to assess the ejected victim. He directs you to set up the equipment for an extensive extrication.

1. What safety concerns should you have in this situation? How would you address those concerns?

2. What extrication equipment will you need?

3. How would you stabilize this vehicle?

 *Access Navigate for more practice activities.*

Introduction

Motor vehicle accidents (MVAs), or as some municipalities refer to them, motor vehicle collisions (MVCs), are common. As a fire fighter, you will likely respond to many MVAs during your career, including those that require extrication. In most Canadian jurisdictions, extricating victims from MVAs is the responsibility of the fire service. Arguably, no other incident to which fire personnel respond can be as emotionally challenging. The worst-case scenarios involve critically injured patients, some conscious, some unconscious, and some deceased. Often children and young adults are involved. The best professional approach is to be compassionate without involving yourself emotionally in the event. Focus on your tasks and remain situationally aware. You did not cause the accident. You are there to help. In order to function efficiently you must remain professionally detached while being supportive of the victims. While responding to these incidents, listen to the radio transmissions and mentally prepare yourself.

While this chapter provides the knowledge and skills a fire fighter needs to assist with basic vehicle extrication operations, it will not prepare you to work as part of a special extrication team. Fire fighters who are members of special rescue teams or who may be expected to perform challenging vehicle extrication operations should complete a course in rescue techniques that meets the requirements of NFPA 1006, *Standard for Technical Rescuer Professional Qualifications*, and NFPA 1670, *Standard on Operations and Training for Technical Search and Rescue Incidents*.

Vehicle Anatomy

Parts of a Motor Vehicle

To reduce confusion and minimize the potential for mistakes at extrication scenes, it is important to use standardized terminology when referring to specific parts of vehicles. The left side of a vehicle is on your left as you sit in the vehicle. In the United States and Canada, the driver's seat is on the left side of the vehicle. The right side of a vehicle is where the passenger's seat is located. Always refer to left and right as they relate to the vehicle; do not refer to left and right from where you might be standing.

In general, vehicles consist of three main compartments: the engine compartment, the passenger compartment, and the trunk (cargo area). The front of a vehicle, which normally travels down the road first, is the usual location of the engine compartment. The engine compartment is covered by the hood. The structure that divides the engine compartment from the passenger compartment is called the **bulkhead** or firewall. The passenger compartment includes the front and back seats. This part of a vehicle is also sometimes called the occupant cage or the occupant compartment. The trunk is usually located at the rear of the vehicle. The trunk may be exposed to the back seats, as in a station wagon, or enclosed, as in a sedan.

Vehicles contain vertical **posts** that support the roof and form the upright columns of the passenger compartment. These posts are named alphabetically from the front to the back of the vehicle (**FIGURE 24-1**).

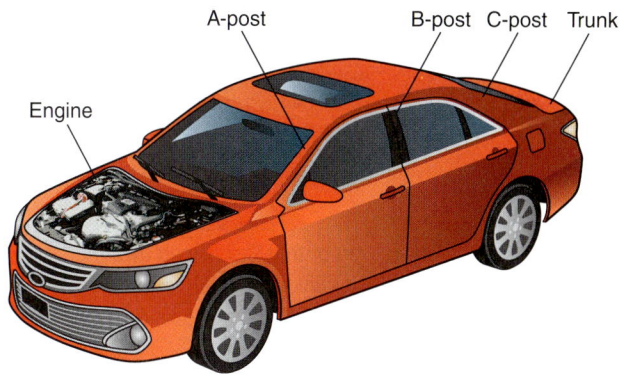

FIGURE 24-1 The anatomy of a vehicle.
© Jones & Bartlett Learning.

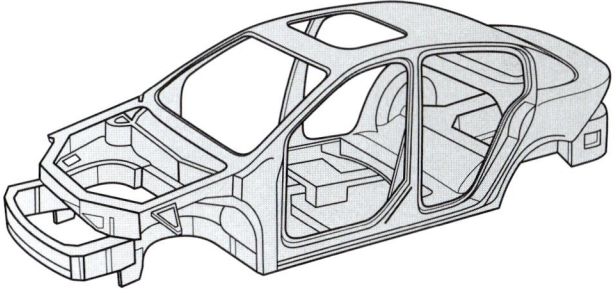

FIGURE 24-3 The design of unibody construction combines the vehicle body and the frame into a single component.
© Jones & Bartlett Learning.

The **A-posts** are located closest to the front of the vehicle; they form the sides of the windshield. In four-door vehicles, the **B-posts** are located between the front and rear doors of a vehicle. In some vehicles, these posts do not reach all of the way to the roof of the vehicle.

In four-door vehicles, the **C-posts** are located behind the rear doors—specifically, behind the rear passenger windows. In two-door vehicles, the C-post is the rear post. Some vehicles have four or more sets of posts that support the roof.

Motor Vehicle Frames

Two types of vehicle frames are commonly found in today's motor vehicles: **body-over-frame construction** and **unibody (unit body) construction**.

Body-over-frame construction consists of two large beams tied together by cross member beams (**FIGURE 24-2**). The beams form the load-bearing frame of the vehicle. The engine, transmission, and body components are attached to this basic frame. This type of frame construction is found primarily in trucks and larger SUVs; it is rarely present in smaller passenger cars.

Unibody construction, which is used for most modern passenger cars, combines the vehicle body and the frame into a single component (**FIGURE 24-3**). By

folding multiple thicknesses of metal together, a column can be formed that is strong enough to serve as the frame for a lightweight vehicle. Unibody construction has the advantage of enabling auto manufacturers to produce lighter-weight vehicles. When extricating a person from a unibody vehicle, remember that these vehicles do not have the frame rails that are present with body-over-frame vehicles. The basic extrication techniques discussed in this chapter must be adapted for each type of vehicle.

Supplemental Restraint Systems

A **supplemental restraint system (SRS)** can be found in most passenger vehicles on the road today. This system commonly includes a series of air bags and seat belts. Identifying the location of these components is critical to avoid the hazards they present.

Air bags can be found in many locations in modern vehicles—even within the seats of the vehicle. Their locations can be identified by markings consisting of acronyms, such as "SRS," "SIR" (supplemental inflatable restraint), "HPS" (head protection system), "SIPS" (side-impact protection system), and "AIR BAG." These markings are generally located in proximity to the inflator and may be embossed, raised, or sewn into the plastic, cloth, or leather material (**FIGURE 24-4**).

The seat belt pretensioning system is designed to reduce the slack in the seat belt when a MVA is detected. The system can be activated in conjunction with the vehicle air bags, or it can act independently. The system can be set up to operate at the belt buckle attachment or to operate at the anchor attachment utilizing the spool. Seat belt assemblies can be housed in any post or column, under the seats, or in the center console.

Roll-over protection systems (ROPS) may be found in some vehicles, specifically convertible vehicles. These systems, which protect occupants in vehicle roll-over incidents, consist of deployable roll bars that are concealed until activated by sensor detection. Roll bars can extend up to 508 mm (20 in.) in some models.

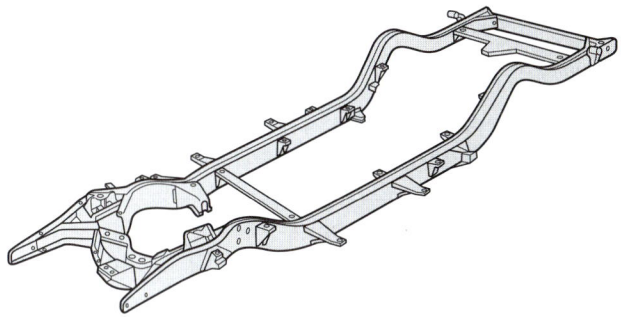

FIGURE 24-2 The basic design of body-over-frame construction consists of two large beams tied together by cross member beams.
© Jones & Bartlett Learning.

FIGURE 24-4 A labelling system using acronyms is generally used to indicate that an air bag system is present. The label shown in this figure indicates there is a head protection system in the vehicle.
Courtesy of David Sweet.

Fire fighters should be familiar with the different types of SRS devices found in modern vehicles. Although these safety devices can provide an extra level of protection during an MVA, unexpected deployment during the extrication process can cause serious injuries to both occupants and rescuers. As part of the size-up procedure, fire fighters should be made aware of any undeployed air bags or ROPS, especially those near trapped occupants or rescuers. Before making any cuts, use a small prying tool to remove interior trim and look for hidden pistons, canisters, or electrical wiring. Rescuers should remove trim around the vehicle's doors and posts to identify the presence of these SRS components.

Alternative-Fuel Vehicles

Most of the vehicles on the road today are **conventional vehicles**, which use internal combustion engines that burn gasoline or diesel fuel to produce power. As concerns about greenhouse gases and limited fossil fuel supplies increase, however, technological advances in alternative fuels will become more common on our roads. An **alternative-fuel vehicle** uses anything other than a petroleum-based motor fuel (gasoline or diesel fuel) to propel a motorized vehicle. These vehicles may be powered by alternative fuels such as propane, natural gas, methanol, or hydrogen, or they may be powered by electricity (battery electric vehicles) or a combination of electricity and fuel (hybrid electric

TABLE 24-1 Types of Alternative-Fuel Vehicles	
Vehicle type	**Description**
Battery electric vehicles	These vehicles use an electric motor that is powered by batteries. Usually these batteries must be recharged from an external source.
Hybrid electric vehicles	These vehicles use both a battery-powered electric motor and a liquid-fuelled engine. The batteries power the electric motor that drives the wheels. The gasoline-powered engine is used for auxiliary power and to recharge the batteries.
Blended liquid fuel–powered vehicles	These vehicles use a blend of liquid fuels. The most common combination is gasoline and another flammable liquid, such as methanol or ethanol. Diesel fuel may be blended with other combustibles to produce a blended diesel fuel, and gasoline may be blended with biofuels, derived from agriculturally produced products such as grains, grasses, or recycled animal oils.
Compressed gas–powered vehicles	These vehicles use one of three forms of compressed gas: compressed natural gas (CNG), liquefied natural gas (LNG), and liquefied petroleum gas (LPG; also called propane).
Fuel cell–powered vehicles	These vehicles use fuel cells that generate electricity through a chemical reaction between hydrogen gas and oxygen gas. The vehicle's electric motor is powered by the electricity generated by the fuel cell.

vehicles) (**TABLE 24-1**). For more information about conventional and alternative-fuel vehicles, see Chapter 17, *Fire Suppression*.

Each of these unique types of alternative-fuel vehicles presents tactical and safety challenges for fire fighters conducting firefighting and extrication operations. Electric drive vehicles, that is **battery electric vehicles** and **hybrid electric vehicles**, are especially hazardous due to their powerful batteries and high-voltage electrical systems. One of the first priorities during extrication operations should be to disable the vehicle's electrical system to prevent accidental starting, electrical shock, or fires. Most of these vehicles are designed so that the high-voltage system is isolated once the 12-volt electrical system is disconnected. Brightly coloured (typically orange) high-voltage cables are routed from a high-voltage battery pack to the inverter and electric motor, which are located in the engine compartment (**FIGURE 24-5**). Avoid touching or cutting these cables and components, as they can cause electrical shock.

To disable the electrical system of an electric drive vehicle, follow the steps in **SKILL DRILL 24-1**.

Alternate disabling methods involve removing fuses. Review the manufacturer's manual regularly to keep up to date with the steps required to disable the electrical systems in electric drive vehicles.

FIGURE 24-5 Electric drive vehicles contain high-voltage electrical systems in the engine compartment.
© Jones & Bartlett Learning. Photographed by Glen E. Ellman.

SKILL DRILL 24-1
Disable the Electrical System of an Electric Drive Vehicle Fire Fighter II, NFPA 1001: 5.4.1

1 Immobilize the vehicle by chocking the wheels. Set the parking brake, and place the vehicle in park, if you can access these controls.

2 Lower all automatic windows, if possible. Disable the low-voltage system by shutting off the vehicle's ignition (power button or conventional key) and disconnecting or cutting the 12-volt battery cables. It is preferable to disconnect the cables as reconnection may be necessary at some point to move the seats. Cut the 12-volt negative cable first, and then cut the positive cable. Double cut each cable to remove a short section. This will prevent the cables from accidentally touching. For vehicles equipped with a proximity key, move the key as far away from the vehicle as possible to prevent the possibility of unintentional restart. It may take up to 10 minutes for the high-voltage system to discharge.

© Jones & Bartlett Learning. Photographed by Glen E. Ellman.

The NFPA offers an *Emergency Field Guide* to assist fire fighters in responding safely to MVAs involving electric drive vehicles and other alternative-fuel vehicles.

A good rule of thumb regarding the management of MVAs is to:

- Stabilize the scene (control traffic, secure dangers around the vehicle, such as leaking fuel).
- Stabilize the vehicle (chock wheels and crib to prevent vehicle motion).
- Stabilize the patient, ensure that trapped victims have an unobstructed airway and they are breathing, control bleeding, and maintain cervical spine control.

If extrication is required, reassure the victim and explain that loud noises and bangs can be expected. Speak to patients in a reassuring voice.

Arrival and Size-Up of the Scene

The first step in the extrication process is arrival and size-up. After arriving at the scene of an MVA, it is important to assess the hazards present and to determine the types of vehicles involved, the scope of the incident, the number and severity of injuries, and the need for additional resources.

> **LISTEN UP!**
>
> Before doing any type of work on a vehicle that has been involved in an MVA, scan the entire area for hazards. Did the vehicle hit an electrical source or rupture a gas line? Hazards are everywhere, and a quick scan might save everyone on the scene from further injury.

Traffic Hazards

Deciding where to position emergency vehicles should take into account the safety of personnel, victims, and motorists travelling along the road. Apparatus should be positioned to protect the scene from traffic. Ideally, traffic flow should not be stopped; however, there may be times when the road will need to be closed. In these incidents the IC may need to confer with police to ensure optimal safety. Sometimes the most important action to take at the scene of an MVA is to slow, stop, or divert the flow of traffic before proceeding with additional actions. These actions are related to scene stabilization in an effort to maximize safety.

Position large emergency vehicles so that they provide a barrier against motorists who fail to recognize or heed emergency warning lights. Many departments place apparatus at an angle to the MVA and pointing away from oncoming traffic (**FIGURE 24-6**). This

FIGURE 24-6 Many fire departments place an apparatus at an angle to the MVA.
© Jones & Bartlett Learning. Photographed by Glen E. Ellman.

position helps to push the apparatus to the side of the accident in the event that the emergency apparatus is struck from behind. Traffic cones or flares can be placed to direct motorists away from the accident (**FIGURE 24-7**). Be sure to look for leaking fuels before using flares. Call for law enforcement to assist in traffic control whenever needed.

> **FIRE FIGHTER TIP**
>
> Extrications are dynamic events, especially on roadways, and fire fighters must be constantly aware of what is going on around them. Maintain situational awareness at all times.

Personal protective equipment (PPE) must be worn at all MVAs. Unless you are exposed to or are likely to be exposed to fire conditions, you must wear high-visibility safety clothing that meets the CSA (Canadian Standards Association) Standard Z96-15 High-Visibility Safety Apparel (CCOHS, 2018) (**FIGURE 24-8**). These break-away vests are designed to provide high visibility to oncoming traffic, but they are easily removed if they become entangled during the extrication process. Remember to follow your department's SOPs regarding MVAs. General

FIGURE 24-7 Traffic cones or flares can be placed to direct motorists away from the MVA.
Courtesy of David Jackson, Saginaw Township Fire Department.

FIGURE 24-8 Break-away vests are designed to provide high visibility to oncoming traffic.
© Jones & Bartlett Learning. Photographed by Glen E. Ellman.

requirements such as vehicle and hose-line placement may vary slightly among fire departments.

Before exiting fire apparatus at an emergency scene, be alert for vehicles that might cause injury to fire fighters. Do not assume that motorists will heed the warning lights. Let law enforcement personnel coordinate traffic control.

The incident commander (IC) will usually perform a size-up of the scene by conducting a 360-degree walk-around of the scene. During this size-up, the IC evaluates the hazards present and determines the number of victims. Using the information obtained from the scene size-up, the IC can create an action plan and call for additional resources, if needed.

Fire Hazards

Look for spilled fuel and other ignitable substances. Motor vehicles use a variety of fuels and lubricants that might pose fire hazards. In addition, vehicles may carry combustible objects and ignitable liquids in the trunk or passenger area. A short in the electrical system or a damaged battery may cause a post-MVA fire by releasing sparks and igniting spilled fuel. These fires may trap the occupants of the vehicle and require rapid fire suppression.

SAFETY TIP

The locations of vehicle batteries can vary widely. If you open the hood to disconnect or cut the battery cables and do not find a battery, look in other less-obvious locations. Batteries may be under or behind the back seat, in the trunk, or inside rear wheel wells. Many vehicle manufacturers provide responder manuals that identify the locations of batteries and other safety devices. These manuals may be available through the manufacturer's website or an emergency response software program.

Electrical Hazards

Downed and low-hanging power lines represent an electrical hazard. Look closely to determine whether the MVA has damaged any electrical power poles. Downed and low-hanging power lines may be difficult to see, and they may energize other objects such as fences, guardrails, and guy wires (cables designed to add stability to a structure). Energized objects create a deadly hazard for rescuers, the victims of the MVA, and bystanders. If power lines are on or near the vehicle, do not approach the scene until electrical power has been disconnected by the utility provider.

As noted earlier, electric drive vehicles contain high-voltage batteries and electrical cables that require special handling because they may pose an electrical or fire hazard. Do not touch any of these components; only specially trained personnel should handle the high-voltage system during an MVA involving an electric drive vehicle.

SAFETY TIP

High-intensity discharge (HID) headlights are becoming more common on vehicles as standard equipment or as after-market modifications. HID headlights discharge several thousand volts of electricity. If the HID bulb is broken or removed, you may be exposed to an electrical hazard.

Other Hazards

Environmental conditions can lead to unique hazards at the scene of an MVA. MVAs that occur in rain, sleet, or snow, for example, present an added hazard for the rescuers and the victims.

Assume that all vehicles are carrying hazardous materials until proven otherwise. Assess the scene for hazardous material placards, unusual odours, or leaking liquids. More information about hazardous materials can be found in the *Emergency Response Guidebook* or in the hazardous materials chapters of this textbook.

Be especially alert for the presence of infectious bodily fluids. Be prepared for the presence of blood, and exercise universal precautions. Specifically, do not let blood or other bodily fluids come in contact with your skin. Wear PPE, including gloves and eye protection, to protect from the contaminated fluids and sharp objects that may be present at the site of an MVA. If you or your clothes become contaminated, report the contamination, document it, and then clean and wash the affected clothes and equipment.

Threats of violence may be present at the scene of some MVAs. In particular, intoxicated people or those who are upset with other motorists may pose a threat to you or to other people present at the scene. Be alert for weapons that are carried in vehicles.

Occasionally, animals become a hazard at the scene of an MVA. Dogs and other family pets may be protective of their owners and threaten rescuers. Farm animals or horses that have been involved in an MVA may need care. You may need to call in specialized resources to assist with this type of incident.

To perform a scene size-up at an MVA, follow the steps in **SKILL DRILL 24-2**.

SKILL DRILL 24-2

Performing a Scene Size-Up at a Motor Vehicle Accident Fire Fighter II, NFPA 1001: 5.4.1

1 Position emergency vehicles to protect the MVA scene and the rescuers. Take any additional actions needed to prevent further MVAs.

2 Perform a quick initial assessment as you arrive on the scene, establish command, and give a brief initial radio report. Establish fire suppression protection if fluids are released or if extrication may be required; a minimum of one 38-mm (1½-in.) hose line should be in place.

3 Perform a 360-degree walk-around to identify potential hazards. Look for hazards above and below the vehicle, and determine the stabilization equipment needed to prevent further movement of the vehicle involved in the incident.

4 Determine the number of patients, the severity of their injuries, and the amount of entrapment. Give an updated report, and call for additional resources if needed.

SKILL DRILL 24-2 Continued
Performing a Scene Size-Up at a Motor Vehicle Accident Fire Fighter II, NFPA 1001: 5.4.1

5 Establish a secure working area and an equipment staging area. Direct personnel to perform initial tasks.

© Jones & Bartlett Learning. Photographed by Glen E. Ellman.

Stabilization of the Scene

After performing a size-up, the scene needs to be stabilized. This step consists of reducing, removing, or mitigating the hazards at the scene. These hazards should have been identified during arrival and scene size-up. The order in which these hazards are addressed will depend on the specific conditions at a scene and the amount of risk that each hazard poses.

Traffic Hazards

The hazards posed by traffic need to be handled quickly before they lead to additional MVAs or injuries. In some cases, the most critical action is slowing or stopping traffic before further damage and injuries can occur. As mentioned previously, emergency vehicles can be used to block traffic from the MVA scene. Traffic cones and flares can slow motorists and direct them in a safe pattern around the MVA scene. Avoid positioning emergency vehicles in a manner that confuses oncoming motorists.

Traffic hazards are best handled by the appropriate law enforcement agency. This is their area of expertise, and giving this duty to law enforcement personnel leaves fire department personnel free to handle extrication tasks. Fire fighters need to work together with law enforcement officials to control traffic in a manner that is safe for emergency workers, victims of the MVA, and motorists. If law enforcement officials are not on the scene when you arrive, verify that they are aware of the incident and that they have been dispatched.

Fire Hazards

If a person is trapped or fuel or other ignitable liquids are spilled at the scene of an MVA, a charged hose line should be advanced to the vehicle. This hose line should be at least 38 mm (1½ in.) in diameter and be staffed by a fire fighter in full firefighting PPE and self-contained breathing apparatus (SCBA). MVAs that pose large fire hazards or actual fires may require additional fire suppression resources, which should be requested as soon as possible. Small fuel spills can be mitigated by using an absorbent material to remove the fuel from the area around the damaged vehicle. Some departments advance a charged hose line on the scene of all MVAs; refer to your department SOPs. See Chapter 16, *Supply Line and Attack Line Evolutions*, for the steps required to advance an attack line. See Chapter 17, *Fire Suppression*, for more information on extinguishing vehicle fires.

To mitigate fire hazards at an MVA, follow the steps in **SKILL DRILL 24-3**.

SKILL DRILL 24-3
Mitigating the Hazards at a Motor Vehicle Accident Fire Fighter II, NFPA 1001: 5.4.1

1 Don PPE, including SCBA. Attach a regulator, if conditions require.

2 Assess the incident scene for hazards. Communicate with other crews and command. Advance a charged hose line to the proximity of the vehicle. Extinguish any fires.

3 Open the hood of the vehicle.

4 Disconnect the battery terminal connects. Avoid cutting these wires as it creates more damage and prevents reconnection if necessary after the incident. Cutting the connections may also create sparking or a short if the cutters make contact with other metal components or the vehicle body.

© Jones & Bartlett Learning. Photographed by Glen E. Ellman.

Electrical Hazards

Disconnecting the vehicle's electrical system can mitigate electrical hazards posed by a damaged vehicle. Follow your department's SOP for disconnecting the electrical system. Fire fighters must weigh the need to disconnect the electrical system against the advantage of being able to operate electrically powered components of the vehicle. Information on mitigating the electrical hazards posed by electric drive vehicles is presented earlier in this chapter.

Stabilizing electrical hazards posed by downed electrical wires is essential before fire fighters attempt to approach the MVA site. At times it may be necessary to instruct victims of an MVA to remain in their vehicles until the power can be turned off. Disconnecting the power to damaged electrical lines should be done by employees of the power company. Do not attempt to approach downed electrical lines until the utility company personnel have turned off the electricity. Direct occupants to stay in the vehicle until the utility company verifies that electrical service has been disconnected. Although it is frustrating to delay extrication and treatment because of downed power lines, this step is essential to avoid potentially life-threatening injuries to both rescuers and victims of the MVA.

Other Hazards

Weather-related issues must be considered for vehicle extrication operations in extreme heat, cold, rain, or snow. Both the victims and the rescuers must be protected from heat and direct sunlight on hot days. Provide shade for victims and rescuers in the vehicles. Watch for signs of heat-related illness. Cover victims with blankets in cold weather. Consider calling for additional resources, so crews can be rotated to a climate-controlled environment for rehabilitation.

Many MVAs occur at night. It is important to provide adequate lighting so that rescuers can work quickly and safely. Chapter 19, *Salvage and Overhaul*, describes the steps required to light an emergency scene. To prevent slips and falls, use caution while walking, working, and moving victims when the roads are wet or icy. Sprinkling sand or an oil absorbent material may help to give solid footing to rescuers. Be alert for oncoming vehicles that may slide into the operations area.

MVAs often leave a variety of sharp objects, such as mangled pieces of metal, plastic, and glass, that pose significant hazards for rescuers and victims. To reduce the chance of injury, rescuers should wear proper personal protective clothing. Sharp edges may be covered with blankets to reduce the possibility of injury to the victims or rescuers. Be especially careful when moving victims,

making sure that they are protected from contact with any sharp objects.

In incidents involving animals, it may be necessary to secure the animals before proceeding with other activities. In particular, dogs can be very protective of their owners. If the dog's owner is injured, remove the dog from the MVA site, and place it with an uninjured family member or other responsible person. If a pet is injured, call the agency that is responsible for transporting animals to a veterinary clinic. As part of preincident planning, you should learn who in your community is responsible for caring for large animals at rescue scenes.

Cribbing

Unstable objects pose a threat to both rescuers and victims of an MVA. Such objects need to be stabilized before fire fighters approach the scene. Most often, the objects that need to be stabilized are the damaged vehicles.

Vehicles that end up on their wheels need to be stabilized with **wheel chocks** to prevent the vehicles from rolling. The vehicles must also be stabilized vertically with **cribbing** (**FIGURE 24-9**). Cribbing consists of short lengths of sturdy lumber, cut to different dimensions. Treated 101.60-mm by 101.60-mm (4-in. by 4-in.) and 457.20-mm by 609.60-mm (18-in. by 24-in.) posts are often used, but other dimensions are available. After cribbing has been placed against the wheels, a vehicle may still be able to move because of the rocking motion the suspension system causes when rescuers get into the vehicle and victims are extricated from the vehicle. This type of instability can cause further injuries to the victims of the MVA.

The suspension system of most vehicles can be stabilized with **step chocks**, which are shaped like stair steps and are placed under both sides of the vehicle.

FIGURE 24-9 The application of cribbing in front and back of the vehicle wheels stabilizes the vehicle.
© Jones & Bartlett Learning. Photographed by Glen E. Ellman.

Place one step chock toward the front of the vehicle and a second step chock toward the rear of the vehicle (**FIGURE 24-10**). Repeat this process on the other side of the vehicle. Once the step chocks are in place, the tires can be deflated by pulling out the valve stems to create a stable vehicle.

If step chocks are not available or are not the right size, rescuers can build a box crib. To do so, place cribbing at right angles to each preceding layer of cribbing (**FIGURE 24-11**).

After an MVA, some vehicles might come to rest on their roofs or sides. Vehicles in these positions are very unstable and present a risk to both victims and rescuers. Placing the slightest amount of weight on them can cause them to move or roll over. Overturned vehicles can be quickly stabilized using box cribs or step chocks on each end of the vehicle. Vehicle stabilization struts can be used for added safety (**FIGURE 24-12**).

Wedges can be used to fill the void between the crib and the object as it is stabilized or raised (**FIGURE 24-13**). Wedges should be the same width as the cribbing, with the tapered end no less than 6 mm (¼ in.) thick. Ends that are less than 6 mm (¼ in.) will commonly fracture under a load.

To stabilize a vehicle following an MVA, follow the steps in **SKILL DRILL 24-4**.

Rescue-Lift Air Bags

A rescue-lift air bag is a pneumatic-filled bladder made of rubber or synthetic material (**FIGURE 24-14**).

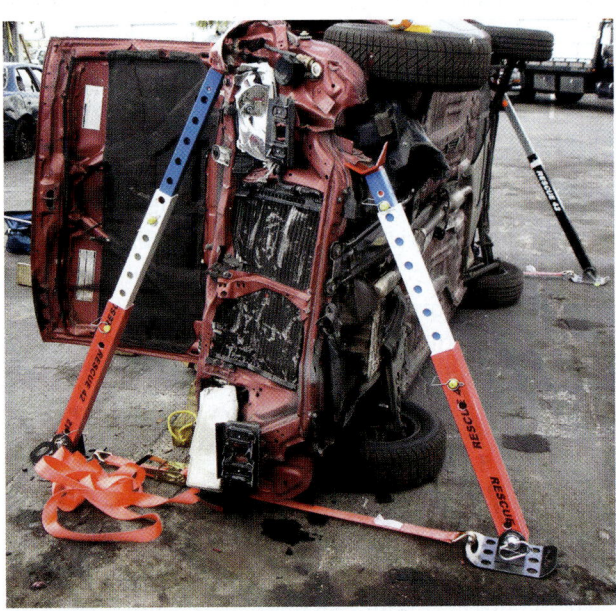

FIGURE 24-12 Struts are used for structural support to stabilize and reinforce a vehicle.
Courtesy of David Sweet.

FIGURE 24-10 Step chocks are used to stabilize a vehicle.
© Jones & Bartlett Learning. Photographed by Glen E. Ellman.

FIGURE 24-11 A box crib is constructed to stabilize a vehicle.
© Jones & Bartlett Learning. Photographed by Glen E. Ellman.

FIGURE 24-13 Wedges are used to snug loose cribbing when the void space is not big enough to accommodate a regular-sized crib.
© Jones & Bartlett Learning. Photographed by Glen E. Ellman.

SKILL DRILL 24-4
Stabilizing a Vehicle Following a Motor Vehicle Accident Fire Fighter II, NFPA 1001: 5.4.1

1 Don PPE, including eye protection. Minimize hazards to rescuers and victims. Chock both sides of one tire to prevent the vehicle from rolling by placing one step chock in front of a wheel and a second step chock in back of the wheel.

2 Use additional step chocks and cribbing, as needed. Consider deflating tires for added stability.

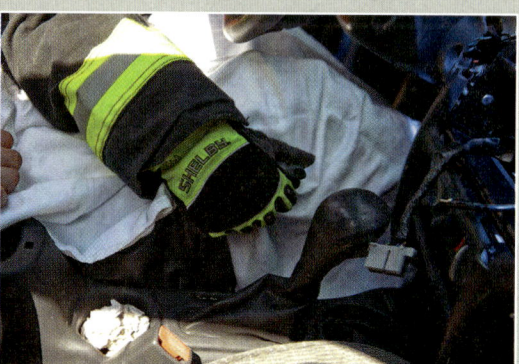

3 Place the gear shift in park, and apply the parking brake, if these controls can be accessed.

4 Lower all automatic windows, if possible. Turn off the ignition, and remove the key or fob.

FIGURE 24-14 A rescue-lift air bag.
© Jones & Bartlett Learning. Photographed by Glen E. Ellman.

They are used to lift an object or to spread one or more objects away from each other to assist in freeing a victim. Rescue-lift air bags are never used to shore or stabilize a vehicle by themselves; cribbing is always required in conjunction with rescue-lift air bags.

Rescue-lift air bags are often used to lift a vehicle or object off a victim. Fire fighters should use extreme caution when applying this technique and should follow all safety precautions outlined in the manufacturer's instructions for the rescue-lift air bag. Cribbing must be used in conjunction with rescue-lift air bags and whenever fire fighters are lifting a load because instability can occur from weight shifts, or the rescue-lift air bags may fail under a load.

Several types of pneumatic rescue-lift air bags are available: low-pressure, medium-pressure, and high-pressure. None of these devices should be used without first properly chocking the wheels opposite of the rescue-lift air bag and cribbing the vehicle or object as it is lifted. These safety precautions will prove invaluable in case the rescue-lift air bag suffers a catastrophic failure.

As they age, rescue-lift air bags become more prone to failure, so they should be tested regularly. Replace rescue-lift air bags per your department's SOPs or the manufacturer's recommendations.

Low-Pressure Rescue-Lift Air Bags. Low-pressure rescue-lift air bags are commonly used for recovery operations and are sometimes used for vehicle rescue operations. These devices come in many sizes and shapes; square air bags, however, offer greater stability. Because of their lightweight construction, low-pressure rescue-lift air bags can be less stable (at least until they are fully inflated) than high-pressure rescue-lift air bags, which have a lower lift height and stronger construction.

Medium-Pressure Rescue-Lift Air Bags. Medium-pressure rescue-lift air bags are designed so that they contain either two or three cells. For general vehicle rescue, these devices are not appropriate. Medium-pressure lift air bags are more suitable for aircraft, medium or heavy truck, or bus rescue and for recovery work.

High-Pressure Rescue-Lift Air Bags. High-pressure rescue-lift air bags are the type used most frequently by fire fighters. These devices have a very sturdy construction and are generally made of vulcanized rubber mats that are reinforced by steel or other material woven into a fibre mat and then covered with rubber. When using this type of air bag, chocking and cribbing remain essential actions to provide a safe working environment.

Gaining Access to the Victim

Open the Door

After stabilizing the vehicle, the simplest way to access a victim of an MVA is to open a door. Try all of the doors first—even if they appear to be badly damaged.

FIRE FIGHTER TIP

- Never stack high-pressure rescue-lift air bags more than two units high.
- Do not use a rescue-lift air bag to pull a steering column.
- Do not use a rescue-lift air bag to stabilize a vehicle; cribbing must be the primary stabilizer.
- Never operate the rescue-lift air bag system without proper training and complete understanding of how the system works.
- When stacking rescue-lift air bags, the largest size should be on the bottom and the smallest on top.
- It may be necessary to place a sheet of protective material (plywood, grip mat, mud flap, etc.) on the ground under the bottom air bag or above the top air bag to protect them from sharp objects. Do not use boards or plywood between rescue-lift air bags.
- Refer to your department's SOPs and the manufacturer's instructions for effective and safe lifting techniques.
- Clean rescue-lift air bags by following the manufacturer's recommendations.
- Test rescue-lift air bags regularly.
- Never store a rescue-lift air bag near gasoline.

Remember the first rule of forcible entry: "Try before you pry." It is an embarrassing waste of time and energy to open a jammed door with heavy rescue equipment, only to discover that another door can be opened easily and without any special equipment.

Attempt to unlock and open the least damaged door first. Make sure the locking mechanism is released. Then try the outside and inside handles at the same time, if possible.

Break Tempered Glass

If a victim's condition is serious enough to require immediate care and you cannot enter through a door, consider breaking a window. Windshields are made of laminated glass, which may be difficult to break. In contrast, the side and rear windows are made of tempered glass, which will break easily into small pieces when hit with a sharp, pointed object such as a spring-loaded center punch, a multipurpose tool, or the point of an axe (**FIGURE 24-15**). Because these windows do not pose as great of a safety threat, they can provide quick access to the victim.

If you must break a window to unlock a door or gain access, cover the victim, if possible, and try to break a window that is farthest away from the victim. If the victim's condition warrants your immediate entry, however, do not hesitate to break the closest window. Small pieces of tempered glass do not usually pose a danger to victims trapped in cars. Advise emergency medical services (EMS) personnel if a victim is covered with broken glass so that they can notify the hospital emergency department.

After breaking the window, use your gloved hands to pull the remaining glass out of the window frame so that it does not fall onto the victims or injure the rescuers. If you use something other than a spring-loaded center punch to break the window, always aim for a low corner.

To break tempered glass using a spring-loaded center punch or a striking tool, follow the steps in **SKILL DRILL 24-5**.

Once you have broken the glass and removed the pieces from the frame, try to unlock the door. Release the locking mechanism, and then use both the inside and outside door handles at the same time. This technique will often force a jammed locking mechanism to open the door, even when the door appears to be badly damaged.

Breaking the rear window will sometimes provide an opening large enough to enable a rescuer to gain access to the victim if there is no other rapid means

A

B

C

FIGURE 24-15 Three tools are used to break tempered glass. **A.** Spring-loaded center punch. **B.** Multipurpose tool. **C.** Axe.

for doing so. The simple techniques of opening a door and breaking the rear window will enable fire fighters to gain access to most MVA victims, even those in an upside-down vehicle.

SKILL DRILL 24-5
Breaking Tempered Glass Fire Fighter II, NFPA 1001: 5.4.1

1 Don PPE, including eye protection. Minimize hazards to rescuers and victims. Ensure stability of the vehicle by using appropriate chocks and cribbing. Select a tool for breaking tempered glass.

2 Ensure that the victim and other fire fighters are as protected as possible.

3 If using a spring-loaded center punch tool, place the tool in the lower corner of the window and sharply apply pressure until the spring is activated. If using a striking tool, strike the lowest corner away from the victim.

4 Remove loose glass around the window opening.

Follow the steps in **SKILL DRILL 24-6** to gain access to a vehicle following an MVA. The steps in this skill drill may have to be modified depending on the situation.

Force the Door

If fire fighters cannot gain access by the previously mentioned methods, they must use heavier extrication tools to gain access to the victim. The most commonly used technique for gaining access to a vehicle following an MVA is door displacement (**FIGURE 24-16**). The opening and displacement of vehicle doors may be difficult and somewhat unpredictable. As mentioned earlier, doors on newer vehicles are sometimes hard to force, and several attempts may be necessary.

When it is necessary to force a door to gain access to the victim, choose a door that will not endanger the safety of the victim. For example, do not try to force a door open if the victim is leaning against it.

To force a door, fire fighters typically use a prying tool to bend the sheet metal at the edge of the door near the locking mechanism. Once this metal has been exposed, a hydraulic spreading tool can be inserted into the space between the door and the post, and it can then pull the hinges or locking mechanism apart. Once a door

SKILL DRILL 24-6
Gaining Access to a Vehicle Following a Motor Vehicle Accident Fire Fighter II, NFPA 1001: 5.4.1

1 Don PPE, including eye protection. Minimize hazards to rescuers and victims. Ensure stability of the vehicle by using appropriate chocks and cribbing. If you can access the passenger compartment, place the gear shift in park, apply the parking brake, lower automatic windows, turn off the ignition, and move the proximity key (if applicable) away from the vehicle. Isolate the power by disconnecting or cutting the battery cables. Determine the best access point, and then open a door, break needed glass, or distort metal.

2 Enter the vehicle, and communicate with the victim.

© Jones & Bartlett Learning. Photographed by Glen E. Ellman.

FIGURE 24-16 The most commonly used technique for gaining access to a vehicle following an MVA is door displacement.
© Jones & Bartlett Learning.

has been removed, it should be placed away from the vehicle where it will not be a safety hazard to rescuers.

Powered hydraulic tools are the most efficient and widely used tools for opening jammed doors. Jammed doors can be opened by releasing the door either from the latch side or the hinge side of the door. The decision regarding which method to use will depend on the structure of the vehicle and the type of damage the door has sustained.

The first step in opening a vehicle door is to create a **purchase point** where the spreader of the hydraulic tool can be inserted. A purchase point can be created by using a pry bar or other hand tool to bend the metal away from the hinges or locking mechanism. A hydraulic spreader tool can also be used in several ways to pinch or spread the metal surface of the door to expose the hinges and locking mechanism. Fire fighters should train with a special rescue team to learn different methods to create a purchase point.

Once rescuers have gained access to the hinges or latch, they can place the spreader in a position so that it is not in the pathway that the door will take when the hinges or latch break. Rescuers should not stand in a position that might put them in danger. To ensure that the door does not swing open violently, rescuers can use a long pry bar or rope to limit the movement of the door when it opens. Activate the hydraulic tool to push apart the outer sheet metal skin of the vehicle, exposing the hinges or the door latch. Once the outer sheet metal has been exposed, close the tips of the spreader and remove them.

Insert the closed tips onto the inner skin of the door and the doorjamb just above the latch or above the hinges. Activate the spreader to spread the tips until the latch or the hinge separates. When separating a door at the latch side, rescuers can place 101.60-mm by 101.60-mm

(4-in. by 4-in.) cribbing under the bottom of the door to hold it up. Once the latch has separated, start to separate the hinges of the door. Some hydraulic tools are capable of cutting door hinges. Check the manufacturer's recommendations for the proper operation of the hydraulic tool.

When separating a door from the hinge side, create a purchase point near the top hinge. Place the spreader tips above the top hinge, and open the tips to separate the door from the hinge. Use the spreader tips or cutting tool to separate the bottom hinge. Once the hinges have separated, place 101.60-mm by 101.60-mm (4-in. by 4-in.) cribbing underneath the door to hold it in place, while separating the latch side of the door.

FIRE FIGHTER TIP

As vehicle manufacturers strive to make new vehicles stronger and lighter, they have begun using increasing quantities of high-strength steel. These stronger steels, in turn, require more robust rescue tools. Tool manufacturers continue to increase the capacity of rescue tools to cut and distort the metals in newer vehicles. By studying individual vehicle manufacturers' specifications, you can determine the easiest places to cut or distort specific vehicle parts.

To force a door following an MVA, follow the steps in **SKILL DRILL 24-7**.

FIRE FIGHTER TIP

The company officer or another Fire Fighter II should coordinate extrication and rescue operations to increase efficiency and help prevent the actions of one team from interfering with another.

Provide Initial Medical Care

As soon as you have secured access to the victim, begin to provide emergency medical care, per your department's EMS protocols. A caregiver should remain with the victim and provide both emotional comfort and physical care until extrication is complete. The IC will coordinate the victim care provided by the medical team with the operations conducted by the extrication team to ensure the safety of everyone on the scene. Although it might be necessary to delay one part of these processes for a short period, all personnel should work toward the goal of getting the victim stabilized and removed from the vehicle as quickly and safely as possible.

To gain access and provide medical care to a victim in a vehicle following an MVA, follow the steps in **SKILL DRILL 24-8**.

SKILL DRILL 24-7
Forcing a Vehicle Door Fire Fighter II, NFPA 1001: 5.4.1

1 Don PPE, including eye protection. Retrieve and set up the required tools. Check the equipment for readiness. Assess the vehicle for stabilization and hazards, including SRS devices.

2 Communicate with the victim; minimize hazards to both the rescuers and the victim.

3 Engage hand tools or a power unit to force the door, using good body mechanics.

4 Remove the door, if possible. If rapid extrication is required, secure the door with ropes or cribbing so it does not shift while removing the victim.

SKILL DRILL 24-8
Gaining Access and Providing Medical Care to a Victim in a Vehicle
Fire Fighter II, NFPA 1001: 5.4.1

1 Don PPE, including eye protection, and minimize the hazards present. Communicate the action plan to the victim.

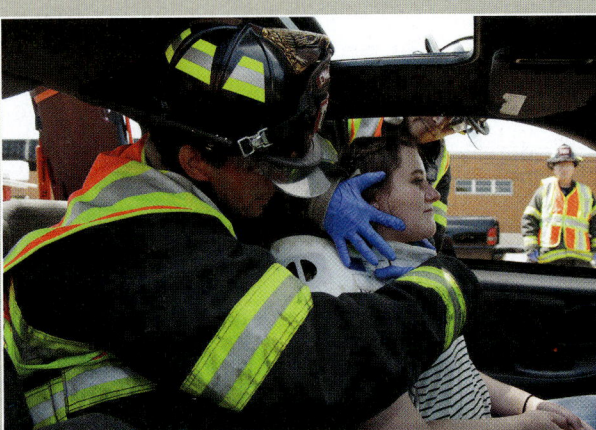

2 Enter the vehicle through the best access point. Begin initial medical care per your department's EMS protocols, and communicate the victim's condition to the outside rescuers.

3 Begin packaging the victim for extrication.

Disentangling the Victim

The next stage in the extrication sequence is disentangling the victim. The purpose of this stage is to remove those parts of the vehicle that are trapping the victim. Note that the goal of this step is to remove the sheet metal and plastic from around the victim—not, to "cut the victim out of the vehicle."

Before discussing the steps of gaining access and disentanglement, it is important to understand the principles involved with these steps of extrication. Although many different techniques and methods are used to gain access and disentangle victims from vehicles involved in MVAs, the objective of these techniques is accomplished by performing four functions:

- Stabilize or hold an object or vehicle. Vehicles are stabilized when cribbing is used to keep them from moving.
- Bend, distort, or displace. An example is bending a vehicle door back to get it out of the way.
- Cut or sever. An example is cutting a roof.
- Disassemble. An example is removing a vehicle door by unbolting the door hinges.

Before beginning disentanglement, study the situation. What is trapping the victim in the vehicle? Perform only those disentanglement procedures that are necessary to remove the victim safely from the vehicle. The order in which these procedures are carried out will be dictated by the conditions at the scene. Many times, it will be necessary to perform one procedure before you can access parts of the vehicle to perform another procedure.

As you work to disentangle the victim from the vehicle, be sure to protect the victim by covering him or her with a blanket or by using a backboard. Be sure that the victim understands what is being done. Provide emotional support because the sounds made by extrication procedures may frighten the victims.

To learn additional methods of disentangling a victim, fire fighters should take an approved extrication course. The five procedures presented here are the ones that are most commonly performed.

Displace the Seat

In frontal and rear-end MVAs, the vehicle may become compressed (**FIGURE 24-17**). As the front of the vehicle collapses, the space between the steering wheel and the seat becomes smaller. In some cases, the driver may be trapped between the steering wheel and the front seat. Displacing the seat can relieve pressure on the driver and give rescuers more space for removal.

If it is necessary to displace a seat, start with the simplest steps. Many times you can gain some room by

FIGURE 24-17 In frontal MVAs, the vehicle may become compressed.
© Jack Dagley Photography/Shutterstock.

moving the seat backward on its track. This maneuver works well with short drivers who have their seat forward. To move a seat back, first make sure that the victim is supported. With manually operated seats, release the seat-adjusting lever and carefully slide the seat back as far as it will go. Attempt to use this simple method first.

For cars that have electrically adjustable seats, use this feature to move the seat back to give the victim more room or lower the seat to help disentangle the victim. Make sure that the victim is adequately supported before attempting these maneuvers.

As a last resort, use a manual hydraulic spreader or a powered hydraulic tool to move the seat back. Place one tip of the hydraulic tool on the bottom of the seat, but avoid pushing on the seat channel that is attached to the floor of the vehicle. Place the other tip of the spreader at the bottom of the A-post doorjamb. Support the victim carefully. Engage the seat adjustment lever on manually operated seats and open the spreader in a careful and controlled fashion.

In some cases, it may be helpful to remove the back of the seat. To accomplish this task, cut the upholstery away from the bottom of the seat back where it joins the main part of the seat. A reciprocating saw or a hydraulic cutter can then be used to cut the supports for the seat back. Be certain that the victim is supported and protected throughout this procedure.

Displacing the seat or removing the seat back will give fire fighters more room to administer treatment and remove the victim from the vehicle.

Remove the Glass

A second technique that is often part of the disentanglement process involves the removal of glass. Removing the rear window, the side glass, or the windshield may improve communication between rescue personnel inside the vehicle and personnel outside the vehicle. Sometimes rescuers can simply roll a window down to gain this advantage. Open windows provide a good route for passing medical care supplies to the inside caregiver. When the roof of a vehicle must be removed, it is necessary to first remove all of the glass from the vehicle.

The side windows and rear window on most vehicles are made of tempered glass, which can be removed by striking the glass in a lower corner with a sharp object, such as a spring-loaded center punch. Whenever this measure becomes necessary, fire fighters should protect the victim as much as possible by covering him or her to prevent the victim from being cut by the small pieces of glass.

Windshields, however, cannot be broken with a spring-loaded center punch. Windshields are made from laminated glass, which consists of a thin layer of flexible plastic between two sheets of glass. When laminated glass is struck by a sharp stone or by a spring-loaded center punch, a small chip or crack appears, but the structure of the glass remains intact. Because of this special construction, it is necessary to remove the windshield in one large piece.

The windshields of most passenger vehicles are glued in place with a strong, plastic-type glue. The windshield must be removed with a special glass saw, reciprocating saw, or by carefully using an axe. Windshield removal can create a significant amount of glass dust, which can be an inhalation hazard to victims and rescuers. Consult your department's SOPs for taking protective measures against this hazard.

To remove a windshield, first make sure that the victim is protected from flying glass. One rescuer makes a purchase point at the top center of the windshield and cuts along the top of the windshield to the nearest A-post. The rescuer continues cutting down the side of the windshield close to the A-post (**FIGURE 24-18**). Finally, the rescuer cuts along the bottom of the windshield to the center. At this point, the first rescuer stabilizes the half of the windshield that has been cut free.

Once this operation is complete, a second rescuer starts on the opposite side of the windshield where the initial cuts were made and cuts the second half of the windshield, following the same sequence of cuts used by the first rescuer. When the second rescuer has completed cutting the windshield, the entire windshield is lifted out of its frame and placed under the vehicle or in another safe place, where it will not present a safety hazard.

Removing a windshield is an essential step prior to removing the roof of a vehicle. It also provides extra space when administering emergency medical care to an injured victim.

FIGURE 24-18 Removing the windshield by cutting close to the A-post.
© Jones & Bartlett Learning. Photographed by Glen E. Ellman.

Remove the Steering Wheel

During normal driving, the steering wheel must be located close to the driver. During an MVA, however, any compression of the front part of the vehicle may push the steering wheel back into the victim's abdomen or chest. Removing the steering wheel can help to disentangle a victim from the vehicle.

Steering wheels can be cut with hand tools such as a hacksaw or bolt cutters. Hydraulic cutters and

SAFETY TIP

- On most vehicles, the steering wheel contains the driver's-side air bag, which is a secondary, life-saving safety feature for the driver of the vehicle. Front-passenger air bags may also be present.
- Air bags may present a hazard after deployment because some air bags deploy in two stages.
- If the air bag did not deploy during the MVA, it presents a hazard for both the occupant of the vehicle and for rescue personnel. An undeployed air bag could deploy if wires are cut or if it becomes activated during the rescue operation.
- If the air bag did not deploy, disconnect the battery and allow the air bag capacitor to discharge. The time required to discharge the capacitor varies for different models of air bags.
- Do not place an object between the victim and an undeployed air bag.
- Do not attempt to cut the steering column if the air bag has not deployed.
- For your safety, never get in front of an undeployed air bag. You could suffer serious injury if it activates unexpectedly.
- Some vehicles contain side-mounted air bags or curtains that provide lateral protection for occupants. Under-dash air bags for knee protection are also common in newer vehicles. Check vehicles for the presence of these devices.

reciprocating saws will also cut steering wheel spokes and the steering wheel rim. One method of removing a steering wheel is to cut the spokes as close to the center hub of the steering wheel as possible. A second method is to cut the rim of the steering wheel. The steering wheel rim can be removed completely, or one section can be cut and removed. A third method is to cut the column (**FIGURE 24-19**).

Note that cutting the steering wheel rim leaves sharp edges that present a safety hazard. These sharp edges must be covered to prevent injury both to the victim and to rescuers.

Displace the Dashboard

During frontal MVAs, the dashboard is often pushed down or toward the seat. When a victim is trapped by the dash, it is necessary to displace this component (**FIGURE 24-20**). Two techniques are commonly used to displace the dash: the dash roll and the dash lift. The objective with these procedures is to displace the dashboard up and away from the victim. These techniques are also sometimes used to displace the steering column up and away from a victim.

Dash displacement requires a cutting tool such as a hacksaw, a reciprocating saw, an air chisel, or a hydraulic cutter. A cutting tool is needed to make one or more cuts on the A-post and other structural members of the vehicle. The dash roll requires fewer cuts than the dash lift and can often be accomplished quicker, but it may not provide as much room to remove the victim. A mechanical high-lift jack, a hydraulic ram, or a hydraulic spreader can be used to push the dash forward and away from the victim, and cribbing is needed to maintain the opening made with these tools.

The first step of dash displacement is to open both front doors. Tie them in the open position so that they will not move as the dash is being displaced. Alternatively, the doors can be removed, as discussed earlier in this chapter in the section on gaining access. Additionally, the roof can be removed.

Place a backboard or other protective device between the victim and the bottom part of the A-post where the relief cut will be made. Do not place a backboard in this position if the driver's-side air bag has not deployed. Next, cut the bottom of the A-post where it meets the sill or floor of the vehicle. It is critical to make the cut perpendicular to the A-post—failure to do so may cut the fuel or power line located in the **rocker panels** (a section of the vehicle's frame located below the doors, between the front and rear wheels).

Place the base of a mechanical high-lift jack or a hydraulic ram at the base of the B-post where the sill and the B-post meet. Place the tip of the jack or ram at the bend in the A-post, which is located toward the top of the A-post. In a controlled manner, extend the jack or ram to push the dash up and off of the victim. This evolution may be difficult if the vehicle contains high-strength steel.

Once the dash has been removed from the victim, cribbing and wedges are utilized to hold the sill in the proper position and to prevent the dash from moving. Only then can the jack or ram be removed.

Performing dash displacement requires careful monitoring of the dashboard's movement to be certain that it is moving away from the victim and is not causing any additional harm to the victim. Throughout this evolution, make sure that the victim is protected and that someone keeps the victim informed about what is happening. Communication is important in this procedure.

FIGURE 24-19 One method of removing a steering wheel is to cut the column of the steering wheel.
© Jones & Bartlett Learning. Photographed by Glen E. Ellman.

FIGURE 24-20 When a victim is trapped by the dash, it is necessary to displace the dashboard.
© Jones & Bartlett Learning. Photographed by Glen E. Ellman.

To perform a dash roll to displace the dashboard of a vehicle following an MVA, follow the steps in **SKILL DRILL 24-9**. To perform a dash lift to displace the dashboard of a vehicle, follow the steps in **SKILL DRILL 24-10**. There are variations to each of these techniques; be sure to train with your department's vehicle extrication personnel to learn best practices.

SKILL DRILL 24-9
Performing a Dash Roll Fire Fighter II, NFPA 1001: 5.4.1

1 Don PPE, including eye protection. Minimize hazards to rescuers and victims. Prevent the vehicle from rolling by placing a chock in front of one tire and a second chock behind the same tire. Use additional step chocks and cribbing, as needed. Consider deflating tires for added stability. Communicate with the victim, and ensure that both the victim and rescuers are protected from hazards. Make a relief cut at the bottom of the A-post.

2 Use a hydraulic ram or spreading tool to roll the dash and bulkhead up and away from the victim. Place cribbing under the rocker panel as the dash is being rolled.

3 Place a wedge in the A-post cut to prevent the dash from returning to its original position.

© Jones & Bartlett Learning. Photographed by Glen E. Ellman.

SKILL DRILL 24-10
Performing a Dash Lift Fire Fighter II, NFPA 1001: 5.4.1

1 Don PPE, including eye protection. Minimize hazards to rescuers and victims. Prevent the vehicle from rolling by placing a chock in front of one tire and a second chock behind the same tire. Use additional step chocks and cribbing, as needed. Consider deflating tires for added stability. Communicate with the victim, and ensure that both the victim and rescuers are protected from hazards. Make necessary relief cuts in the A-post and other structural members, as needed.

2 Use a hydraulic spreading tool in the A-post relief cut to lift the dash and bulkhead up and away from the victim. Place cribbing under the rocker panel as the dash is being lifted.

3 Maintain contact with the hydraulic spreading tool to prevent the tool from shifting while removing the victim.

Displace the Roof

Displacing the roof of a vehicle has several advantages. First, it enables equipment to be easily passed to the emergency medical providers. It also increases the amount of space available to perform medical care and increases the visibility and space for performing disentanglement. Likewise, fresh air supply is improved. The increased space helps to reduce the feeling of panic caused by the confined space of the vehicle. Finally, removing the roof provides a large exit route for the victim (**FIGURE 24-21**).

One method of displacing the roof is to cut the A-posts and fold the roof back toward the rear of the vehicle. This technique provides limited space. Given this fact, it is often preferable to remove the entire roof.

Roof displacement can be accomplished with hand tools such as hacksaws, air chisels, and manual hydraulic cutters. Appropriate power tools include reciprocating saws and powered hydraulic cutters. Be aware that some A-posts are made of high-strength steel and may be difficult to cut, especially with older cutting tools.

Most newer vehicles include additional reinforcement beams for the seat belts. Older hydraulic cutters may not be able to sever these beams, so cuts must be made above the seat belts.

The first step in displacing the roof is to ensure the safety of both the rescuers and the victim inside of the vehicle. In particular, as rescuers cut the posts that support the roof, they must support the roof to keep it from falling on the victim.

To remove the roof, remove the glass from all windows and the windshield to prevent it from falling on the victim. Prior to cutting any materials, be sure to identify the presence and location of any hidden pistons, canisters, electrical wiring, or SRS devices, such as air bags and seat belt pretensioners, and address these hazards accordingly. It may be necessary to remove interior trim around the posts to locate these devices. Cut the vehicle posts farthest away from the victim. Cut each post at a level that will ensure the least amount of post remains in place after roof removal. When cutting the wider rear posts, cut them at the narrowest point of the post. As each post is cut, a rescuer needs to support that post. Cut the post closest to the victim last, because it provides some protection for the victim.

Working together, several rescuers should remove the roof and place it away from the vehicle, where it does not pose a safety hazard. They should then cover the sharp ends of the cut posts with a protective device, thereby ensuring the safety of both rescuers and the victim.

To remove the roof of a vehicle following an MVA, follow the steps in **SKILL DRILL 24-11**.

Removing and Transporting the Victim

The final phase of the extrication process consists of removing the victim from the vehicle. During this phase, the victim needs to be stabilized and packaged in preparation for removal. Follow your department's EMS protocols for victim stabilization and packaging. The definitive treatment of trauma victims will occur at a hospital; therefore, only those steps needed to stabilize and prevent further injury to the victim should be performed.

Develop a plan for the victim's removal, and make sure that a clear exit pathway is available. The actual victim removal should be directed by a designated person—usually the rescuer controlling cervical-spine stabilization—and conducted with clear commands.

FIGURE 24-21 Removing the roof provides a large exit route for the victim.
© Jones & Bartlett Learning. Photographed by Glen E. Ellman.

SKILL DRILL 24-11
Removing the Roof of a Vehicle Fire Fighter II, NFPA 1001: 5.4.1

1 Don PPE, including eye protection. Retrieve and set up the required tools. Check the equipment for readiness. Assess the vehicle for stabilization and hazards. Prior to cutting any materials, be sure to identify the presence and location of any hidden pistons, canisters, electrical wiring, or SRS devices, such as air bags and seat belt pretensioners, and address these hazards accordingly. It may be necessary to remove interior trim around the posts to locate these devices. Communicate with the victim, and minimize hazards to both the rescuers and the victim. Remove any remaining glass.

2 Engage the appropriate tools to remove the roof, while using good body mechanics.

3 Control the vehicle roof and protect against sharp edges at all times.

4 Remove the roof to a safe location. Return the tools to the staging area upon completion of the tasks.

Ensure that an adequate number of rescuers are available to assist with the victim removal, and confirm that everyone involved in this process understands the commands that will be given.

Try to make the removal process as seamless as possible. Ideally, the ambulance cot should be positioned close by so that the victim can be placed on the cot without any delay. The victim should be moved to an ambulance as soon as possible.

Transport the victim to an appropriate medical facility as soon as the initial stabilization has been completed. If your department uses a helicopter service to transport victims, you also need to be familiar with the procedures used to load victims into helicopters.

FIRE FIGHTER TIP

Although the IC or incident safety officer will be responsible for rescuer safety, all personnel on the scene should be aware of hazards that may further injure victims or rescuers.

Terminating an Incident

Terminating an incident includes securing the scene by removing the damaged vehicle and equipment from the scene and ensuring the scene is left in a safe condition. After you have secured the scene and packed your equipment, it is important to return to the station and fully inventory, clean, service, and maintain all of the equipment (per the manufacturer's instructions) to prepare it for the next call. Some items will need repair, but most will need simple maintenance before they are placed back on the fire apparatus and considered in service.

Securing the Scene

Just because the victim has been removed from the wreckage does not mean the scene can be abandoned. Law enforcement will secure the scene by conducting their on scene investigation. A potential crime scene, such as a homicide, should be managed by law enforcement who will preserve and secure any evidence and close off the scene or roadway. Once a scene has been released by law enforcement, the removal of vehicles will be coordinated with a tow agency.

In some cases, a vehicle may need to be turned right-side up. This may cause sparks or a fire. In addition, the high-voltage batteries in electric drive vehicles may cause a delayed fire. Be aware of smoke, sparks, irritating fumes, or a gurgling sound from the battery. These are warning signs of a damaged high-voltage battery that is undergoing a thermal event and is at risk for fire hours or even days later. In addition, there is a risk of toxic off-gassing. Being proactive by standing by with a charged hose line is always a good practice during one of these procedures.

Any fluid hazards that have spilled from the vehicle, such as gasoline, motor oil, transmission fluid, or radiator fluid that was either captured with an absorbent or not, will be removed by the towing agency. Towing agencies must be licensed to transport and dispose of any hazardous substances; fire rescue agencies are typically not licensed to do so. Ask the tow agency representative if you may assist the towing agency by placing any vehicle parts, such as doors, roofs, and fenders, back in a heavily damaged vehicle. The tow agency may have procedures or a policy of their own for stowing and transporting loose objects. Being considerate of their preferences will help to maintain a good working relationship with the tow agency for future endeavors such as acquiring vehicles for use in an extrication class. It is advisable to carry a contact list of various private and public organizations or businesses that can offer a particular resource to be utilized on the incident, such as public works/utilities, Transport Canada, or a heavy equipment company.

FIRE FIGHTER TIP

Battery electric and hybrid electric vehicles should be towed using a flatbed truck. Towing them with the drive wheels on the ground could result in an electrical fire. Law enforcement officers may take command of the scene to conduct an MVA or fatality investigation after the victims have been extricated. In the event that law enforcement investigators take possession of the vehicle, brief them on the hazards associated with electric drive vehicles.

After-Action REVIEW

IN SUMMARY

- The design of motor vehicles is constantly evolving. As a fire fighter, you need to keep up to date on the design of motor vehicles.

- Most vehicles on the road today are conventional vehicles, which use internal combustion engines that burn gasoline or diesel fuel. Alternative-fuel vehicles may be powered by electricity, propane, natural gas, methanol, or hydrogen. The hazards posed by these vehicles include fuel fires.

- Battery electric vehicles use an electric motor that is powered by batteries. These vehicles pose electrical hazards due to the large number of batteries needed to power the engine.

- Hybrid electric vehicles use both a battery-powered electric motor and a liquid-fuelled engine. As a consequence, these vehicles pose both electrical hazards and fuel fire hazards.

- Fire fighters performing vehicle extrication must know the following vehicle components:
  - The A-posts form the sides of the windshield.
  - The B-posts are located between the front and rear doors of a vehicle.
  - The C-posts are located behind the rear passenger windows.
  - The bulkhead divides the engine compartment from the passenger compartment.
  - The passenger compartment (also called the occupant cage) includes the front and back seats.
  - The two common types of vehicle frames are the body-over-frame and the unibody (unit body). Body-over-frame construction is used for SUVs and trucks; it uses beams to form the load-bearing frame of a vehicle. Unibody construction is used in most passenger vehicles; it combines the vehicle body and the frame into a single component.

- Identifying the location of supplemental restraint systems, such as air bags and seat belts, is critical to avoid the hazards they present.

- Extrication should follow a series of logical steps:
  - Arrive and size up the scene.
  - Stabilize the scene.
  - Access the victim.
  - Disentangle the victim.

- After arriving at the scene, assess the hazards present and determine the types of vehicles involved, the scope of the incident, the number and severity of injuries, and the need for additional resources.

- Stabilizing the scene consists of reducing, removing, or mitigating the hazards at the scene.

- Use cribbing and associated materials to stabilize the vehicle and other unstable objects.

- To gain access to the victim, the following techniques are used:
  - Open the door.
  - Break a window.
  - Force open the door.

- To disentangle the victim from the vehicle, the following techniques are used:
  - Displace the seat.
  - Remove the windshield.
  - Remove the steering wheel.
  - Displace the dashboard.
  - Displace the roof.

- The final phase of the extrication process consists of removing and transporting the victim to the hospital. During this phase, the victim is stabilized, packaged, removed from the vehicle, and transported to a medical facility.

- Terminating an incident includes securing the scene by removing the damaged vehicle and equipment and ensuring the scene is left in a safe condition. Once a scene has been released by law enforcement, the removal of vehicles will be coordinated with a tow agency.

KEY TERMS

Access Navigate for flashcards to test your key term knowledge.

Alternative-fuel vehicle A vehicle that uses anything other than a petroleum-based motor fuel (gasoline or diesel fuel) to propel a motorized vehicle.

A-posts Vertical support members that form the sides of the windshield of a motor vehicle.

Battery electric vehicles Vehicles that are powered by electricity.

Body-over-frame construction A type of vehicle frame resembling a ladder, which is made up of two parallel rails joined by a series of cross members. This kind of construction is typically used for luxury vehicles, sport utility vehicles, and all types of trucks.

B-posts Vertical support members located between the front and rear doors of a motor vehicle.

Bulkhead The separation between the passenger compartment and the engine compartment. (NFPA 556)

Conventional vehicle A vehicle that uses an internal combustion engine for power.

C-posts Vertical support members located behind the rear doors of a motor vehicle.

Cribbing Short lengths of timber/composite materials, usually 101.60 mm × 101.60 mm (4 in. × 4 in.) and 457.20 mm × 609.60 mm (18 in. × 24 in.) long that are used in various configurations to stabilize loads in place or while load is moving. (NFPA 1006)

Hybrid electric vehicle A vehicle that uses both a battery-powered electric motor and a liquid-fuelled engine to propel the vehicle.

Laminated glass Safety glass; the lamination process places a thin layer of plastic between two layers of glass so that the glass does not shatter and fall apart when broken.

Post One of the vertical support members or pillars of a vehicle that holds up the roof and forms the upright columns of the passenger compartment.

Purchase point A small opening made to enable better tool access in forcible entry.

Rescue-lift air bag An inflatable device used to lift an object or spread one or more objects away from each other to assist in freeing a victim. Various sizes and types are available.

Rocker panels A section of a vehicle's frame located below the doors, between the front and rear wheels.

Step chocks Specialized cribbing assemblies made of wood or plastic blocks in a step configuration. They are typically used to stabilize vehicles.

Supplemental restraint system (SRS) A system that uses supplemental restraint devices such as air bags to enhance safety in conjunction with properly applied seat belts. Seat belt pretensioning systems are also considered part of an SRS.

Tempered glass A type of safety glass that is heat-treated so that, under stress or fire, it will break into small pieces that are not as dangerous.

Unibody (unit body) construction The frame construction most commonly used in vehicles. The base unit is made of formed sheet metal; structural components are then added to the base to form the passenger compartment. Subframes are attached to each end. This type of construction eliminates the rail beams used in body-over-frame vehicles.

Wedges Material used to tighten or adjust cribbing and shoring systems. (NFPA 1006)

Wheel chocks Wedges or blocks of sturdy materials that are placed against a vehicle's wheels to prevent accidental rolling during extrication.

REFERENCES

CCOHS (Canadian Centre for Occupational Health and Safety). High-visibility safety apparel. https://www.ccohs.ca/oshanswers/prevention/ppe/high_visibility.html. Accessed November 21, 2018.

National Fire Protection Association. NFPA 1006, *Standard for Technical Rescue Personnel Professional Qualifications*, 2017. www.nfpa.org. Accessed November 21, 2018.

National Fire Protection Association. NFPA 1670, *Standard on Operations and Training for Technical Search and Rescue Incidents*, 2017. www.nfpa.org. Accessed November 21, 2018.

On Scene

Your fire department's training division is focusing on different rescue techniques this month. Today, your crew will practise vehicle extrication techniques. When you arrive at the training facility, you see several junked vehicles in the parking lot. One is a hybrid electric passenger vehicle with heavy damage to the front end. Another is a full-size pick-up truck with severe damage to the front and rear passenger doors. A third passenger vehicle has minor damage to the trunk and left-rear side. As you don your PPE, you begin thinking of ways to gain access to the passenger compartments of these vehicles.

1. Which of the following hazards should you expect if performing extrication on the hybrid electric vehicle?

A. Electrical hazards

B. Traffic hazards

C. Sharp glass or metal hazards

D. All of the above

2. Prior to making your first extrication cut, you must first stabilize the vehicle. What tool will you use?

A. Rescue-lift air bag

B. Ropes

C. Cribbing

D. Spreaders

3. What type of glass is generally used in the side and rear windows of motor vehicles?

A. Laminated

B. Tempered

C. Heat-resistant

D. Flat

4. Which of the following is the first step in the extrication process?

A. Disentangle the patient.

B. Stabilize the vehicle.

C. Assess the patient.

D. Arrival and size-up.

5. Which of the following tools is used for cutting metal?

A. Hydraulic spreaders

B. Pry bar

C. Reciprocating saw

D. Rescue-lift air bags

6. After ensuring victim and rescuer safety inside the vehicle, what is the next step in removing a windshield with an axe?

A. Cut along the A-post, starting at the top.

B. Create a purchase point in the top center of the windshield.

C. Cut along both sides of the top of the windshield.

D. Cut along the bottom of the windshield.

(continued)

On Scene Continued

7. When removing the roof of a vehicle, which post or posts should be cut first?

   A. The post(s) closest to the victim

   B. The post(s) farthest from the victim

   C. The B-posts

   D. The C-posts

8. Which of the following is *not* required while conducting extrication operations on a roadway?

   A. SCBA

   B. Gloves

   C. A reflective vest meeting the ANSI 207 standard

   D. Eye protection

 *Access Navigate to find answers to this On Scene, along with other resources such as an audiobook.*

Assisting Special Rescue Teams

KNOWLEDGE OBJECTIVES

After studying this chapter, you will be able to:

- Define the types of special rescues encountered by fire fighters. (**NFPA 1001: 5.4.2**, pp. 907–908)
- Describe the steps of a special rescue. (**NFPA 1001: 5.4.2**, pp. 909–913)
- Explain how tools and equipment are staged for rapid access. (**NFPA 1001: 5.5.4**, pp. 910–911)
- Describe the general procedures at a special rescue scene, including what to do in the case of utility hazards. (**NFPA 1001: 5.4.2**, pp. 914–917)
- Describe how to safely approach and assist at a vehicle or machinery rescue incident. (**NFPA 1001: 5.4.2**, pp. 917–918)
- Describe how to safely approach and assist at a confined-space rescue incident. (**NFPA 1001: 5.4.2**, pp. 919–921)
- List the types of incidents that might require a rope rescue. (pp. 921–923)
- Describe how to safely approach and assist at a rope rescue incident. (**NFPA 1001: 5.4.2**, p. 922)
- Describe the hardware components used during a rope rescue. (**NFPA 1001: 5.4.2**, p. 923)
- Describe how to safely approach and assist at a trench and excavation rescue incident. (**NFPA 1001: 5.4.2**, pp. 924–925)

- Describe how to safely approach and assist at a structural collapse rescue incident. (**NFPA 1001: 5.4.2**, pp. 925–926)
- Describe how to safely approach and assist at a water or ice rescue incident. (**NFPA 1001: 5.4.2**, pp. 926–928)
- Describe how to safely approach and assist at a wilderness search and rescue incident. (**NFPA 1001: 5.4.2**, pp. 928–929)
- Describe how to safely approach and assist at a hazardous materials rescue incident. (**NFPA 1001: 5.4.2**, pp. 929–930)
- Describe how to safely respond to an elevator or escalator rescue. (**NFPA 1001: 5.4.2**, pp. 930–931)
- Describe how to safely assist at an active shooter incident. (p. 931)
- Describe the types of generators used to power lighting equipment. (pp. 932–933)
- Describe how generators operate. (p. 932)
- Describe how to clean and maintain lighting equipment. (**NFPA 1001: 5.5.4**, pp. 932–933)
- Describe how to maintain generators. (**NFPA 1001: 5.5.4**, pp. 932–934)
- Describe how to maintain power equipment and power tools. (**NFPA 1001: 5.5.4**, p. 933)

SKILLS OBJECTIVES

After studying this chapter, you will be able to perform the following skills:

- Establish a barrier. (**NFPA 1001: 5.4.2**, pp. 915–916)
- Identify and retrieve rescue tools. (**NFPA 1001: 5.4.2**, p. 918)
- Conduct a weekly/monthly generator test. (**NFPA 1001: 5.5.4**, p. 934)

ADDITIONAL STANDARDS

- **NFPA 472**, *Standard for Competence of Responders to Hazardous Materials/Weapons of Mass Destruction Incidents*
- **NFPA 1006**, *Standard for Technical Rescue Personnel Professional Qualifications*
- **NFPA 1500**, *Standard on Fire Department Occupational Safety, Health, and Wellness Program*
- **NFPA 1670**, *Standard on Operations and Training for Technical Search and Rescue Incidents*
- **NFPA 1951**, *Standard on Protective Ensembles for Technical Rescue Incidents*
- **NFPA 1983**, *Standard on Life Safety Rope and Equipment for Emergency Services*

Fire Alarm

Your pump company is dispatched on a medical call for a 51-year-old male construction worker who is possibly having a heart attack. While en route, you receive additional information that the patient was working in an underground electrical vault and began to feel ill. Now, his crew can no longer communicate with him. Upon your arrival, witnesses state that two of his crew members entered the vault to retrieve the victim, but they can no longer hear their voices.

1. What must you consider with this type of incident?

2. What are your incident priorities, and in which order will you address them?

3. What actions will you take to safely meet those priorities?

 *Access Navigate for more practice activities.*

Introduction

The services that fire departments provide have expanded dramatically over the last several years. Modern fire services have evolved into multiservice emergency organizations. Fire departments in large Canadian cities provide a wide range of services from CBRN (chemical, biological, radiologic, or nuclear agents) defense to high-angle rescue. Unlike other emergency services, fire departments are extremely adept at responding with large numbers of trained personnel to large-scale incidents. Additionally, fire departments are equipped with vehicles and tools that can be rapidly deployed to a wide variety of specialized rescues.

A **technical rescue incident (TRI)** is a complex rescue incident that requires specially trained personnel and special equipment. TRIs tend to be categorized as low-frequency, high-risk incidents. Due to increased safety requirements, fire departments involved in providing specialized rescue services are required to train their personnel to an exceedingly high standard. Some jurisdictions must be prepared to respond to below-grade rescues in mines, caves, or tunnels, while others train extensively for wilderness search and rescue operations. Because hazardous materials are commonly found in almost every jurisdiction, fire fighters must be prepared to safely respond to any rescue situation where these are present.

This chapter explains how nonspecialized fire fighters can aid technical rescue personnel at the scene of a TRI for information purposes only and is not considered a basis for undertaking any technical rescue. Specialized training is required to acquire true proficiency in TRIs. The more training you receive in these areas, the better you will understand how to conduct yourself at these scenes.

Training in technical rescue areas is conducted at three levels: awareness, operations, and technician (NFPA 1006):

- Awareness level: This level represents the minimum capability of individuals who provide response to technical search and rescue incidents.
- Operations level: This level represents the capability of individuals to respond to technical search and rescue incidents and to identify hazards, use equipment, and apply limited techniques specified in the standard to support and participate in technical search and rescue incidents.
- Technician level: This level represents the capability of individuals to respond to technical search and rescue incidents and to identify hazards, use equipment, and apply advanced techniques specified in this standard necessary to coordinate, perform, and supervise technical search and rescue incidents.

Fire fighter trainees are taught at an awareness level. Experiential knowledge is still a very important part of personal development. You should never be placed in a situation that exceeds your capabilities, training, and experience.

Types of Rescues Encountered by Fire Fighters

Most fire departments respond to a variety of special rescue situations (**FIGURE 25-1**). Special rescue situations include the following:

- Energized electrical line emergencies and other utility hazard incidents
- Vehicle and machinery rescue
- Confined-space rescue
- Rope rescue
- Trench and collapse rescue
- Structural collapse rescue
- Cave and tunnel search and rescue

FIGURE 25-1 Most fire departments respond to a variety of special rescue situations that require additional training for safe operations.
© Jones & Bartlett Learning.

- Water and ice rescue
- Wilderness search and rescue
- Hazardous materials incidents
- Elevator and escalator emergencies
- Industrial incidents
- Active shooter incidents

To become proficient in handling these situations, you must take a formal course to gain specialized knowledge and skills. NFPA 1670, *Standard on Operations and Training for Technical Search and Rescue Incidents*, identifies operational benchmarks and training needed to establish operational capacity for technical search and rescue events.

It is important for awareness-level responders to have an understanding of these special types of rescues. Often, the first emergency unit to arrive at a rescue incident is a fire department pump or ladder company. Consequently, the initial actions taken by that company may determine the safety of both the victims and the rescuers. These initial actions may also determine how efficiently the rescue is completed.

It is important to understand and recognize the tools needed for rescue situations. You may not have to know how to use all of them, but you may be called upon to assist with a tool at any time. Be sure to train

LISTEN UP!

TRIs can be complex and, based on the type of incident, may require a wide variety of tools and equipment to safely and effectively rescue victims. Departments should obtain and train with the tools and equipment carried on their apparatus and used by their special rescue teams.

with your department's special rescue teams to understand how to use the tools and equipment carried on your apparatus.

Guidelines for Operations

When assisting rescue team members, keep in mind the five guidelines that you follow during other firefighting operations:

- Be safe.
- Follow orders, and work within the incident command system (ICS).
- Work as a team.
- Maintain situational awareness and focus.
- Follow the Golden Rule of public service, which emphasizes the ethic of reciprocity: Treat others as you would like to be treated.
- Ensure that you control your emotions. Poor decisions tend to occur when you are anxious or excited.

> **FIRE FIGHTER TIP**
>
> The Fire Fighter II should never attempt to perform technical or dangerous rescue skills without the proper training. Always work within your level of training.

Be Safe

Rescue situations may be full of hidden hazards, including oxygen-deficient atmospheres, toxic agents, weakened floors, electrical hazards, and strong water currents. Knowledge and training are required to recognize signs indicating that a hazardous rescue situation exists. Once the hazards are recognized, determine which actions are necessary to ensure your own safety, as well as the safety of your team members, the victims of the incident, and bystanders. It requires experience and skill to determine that a rescue scene is not safe to enter, but that determination could ultimately save lives.

Follow Orders

When you begin your career as a fire fighter, you will have limited training and experience. By contrast, your officers and the rescue teams with whom you will work on special rescue incidents have received extensive specialized training. They have been chosen for their duties precisely because they have experience and skills in a particular area of rescue. It is critical to follow the orders of those who understand exactly what needs to be done to ensure safety and to mitigate the dangers involved in the rescue situation. Moreover, orders should

be followed exactly as given. If you do not understand what is expected of you, ask for more information. Have the orders clarified so that you will be able to complete your assigned task safely.

To do well as a fire fighter, you must have a strong appreciation for the basic philosophy of many firefighting organizations. Most fire departments are run like military organizations. Grasping the paramilitary structure will enable you to understand the command and control concept of fire departments. Understanding the following principles will enable you to follow the orders of your commanding fire officers quickly and confidently:

- The fire officer's knowledge base and experience are greater than yours.
- Orders come from superiors. Legitimate orders are those given by a fire officer or other designated person.
- Follow rules and procedures. A fire fighter is required to follow rules, procedures, and guidelines regardless of his or her personal opinions.
- You must do your own job.
- Get the job done. In emergency situations, time is critical. Nevertheless, you must not act beyond your own skill and training level, and you must not violate any rules, standard operating procedures (SOPs), or orders of a superior in an effort to get the job done quickly.

Work as a Team

Rescue efforts often require many people to complete a wide variety of tasks. Some personnel have been trained in specific tasks, such as rope rescue or swiftwater rescue (**FIGURE 25-2**). To do their jobs, however,

FIGURE 25-2 Some personnel are trained in specific rescue skills, such as rope rescue.
© Jones & Bartlett Learning.

they need the support and assistance of other fire fighters. Your role is to support rescue operations within the framework of the rescue team.

Situational Awareness and Focus

As you are working on a rescue situation, you must constantly assess and reassess the scene. You might see something that your company officer does not see. If you think your assigned task may be unsafe, bring this matter to the attention of your superior officer. Do not try to reorganize the total rescue effort, as it is being directed by people who are highly trained and experienced—but do not ignore what is going on around you; you must maintain situational awareness.

Observations that a fire fighter should bring to the attention of a superior fire officer include unsafe structures, changing weather conditions, suspicious packages, and broken equipment. In a hazardous materials rescue incident, the wind direction and speed are important to know. For example, an increase in wind speed can cause hazardous vapours to spread more quickly, which can affect the response plan. Anecdotally, a lack of mental focus can result in unexpected injury. It is vital that you maintain both your individual focus and an overall understanding of the situation.

Follow the Golden Rule of Public Service

When you are involved in carrying out a rescue effort, it is easy to concentrate on the technical parts of the rescue and forget that there is a frightened person who needs your emotional support and encouragement. It is helpful to have a rescuer stay with the victim whenever possible. He or she can tell the victim which actions will be performed during the rescue process. This rescuer is often the one the victim remembers and the one who helps to fulfill the Golden Rule.

Maintain Emotional Control

Perhaps one of the most useful skills a firefighter can develop is the ability to maintain a calm demeanor. Allowing a situation to overwhelm your emotional state of mind is dangerous not only for you but for your fellow fire fighters. One of the best ways to overcome mental stress is to be an expert at your job. Even if the tasks you are assigned do not require technical level skills, take pride in your work, do it well, and practise it often. Being a student, throughout your career, is the best protection against emotional stress. If you are diligent at learning your job, the more confident you will become on the scene of an emergency. There can

---

be no corollary substitute for learning; it is as much your individual responsibility as it is your departments to continually train and learn.

Steps of Special Rescue

Although special rescue situations may take many different forms, all rescuers take certain basic steps to perform special rescues in a safe, effective, and efficient manner:

1. Preparation
2. Response
3. Arrival and size-up
4. Stabilization
5. Access
6. Disentanglement
7. Removal
8. Transport
9. Security of the scene and preparation for the next call
10. Postincident analysis

Preparing for the Response

You can prepare for responses to emergency rescue incidents by training with the specialized rescue teams with whom you will be working. This kind of experience will enable you to learn about their techniques and equipment. It will help you to understand what you can do to assist them as well as what you should *not* do. It is also helpful to train with other fire departments in your area, as such training will better enable you to respond to a mutual aid call. In addition, you will be able to learn about the types of rescue equipment to which other departments have access, as well as the training levels of their personnel.

Pay attention to terminology—knowing the terminology used in the field will make communicating with other rescuers easier and more effective. Also, learn to recognize the different types of rescue situations that you could encounter. Prior to a technical rescue call, your department must consider these issues:

- Does the department have the personnel and equipment to handle a technical rescue incident from start to finish?
- Does the department meet National Fire Protection Association (NFPA) and Public Safety Canada standards for technical rescue calls?
- Which equipment and personnel will the department send on a technical rescue call?
- Do members of the department know the hazards located within the department's response area, and have they visited these areas with local representatives?

Responding to the Incident

A TRI should have its own dispatch protocol. If your agency has its own technical rescue team, it will usually respond with a rescue unit supported by other equipment and personnel, such as a medic unit, a pump company, a ladder company, and a chief. The makeup of the response is based on the ability of the agency or department to provide resources for the response. Even if a technical rescue team is available, it may not be trained or equipped to handle all incidents that could be encountered. Your department should make provisions for additional resources and trained personnel to assist in handling incidents when the TRI is beyond the local technical rescue team's capabilities.

It is common for your communications centre to contact other agencies, such as gas and utility companies, to provide support for rescue operations. Many technical rescues involve electricity, sewer pipes, or other factors that may create the need for additional heavy equipment, to which utility companies have ready access.

Arrival and Size-Up

The first company officer to arrive on the scene should assume command. A rapid and accurate size-up is needed to avoid placing would-be rescuers in danger and to identify additional resources that may be needed. Determine how many victims are involved, and note the extent of their injuries. This information will help to determine how many EMS units and other resources are needed.

When responding to a work site or industrial facility, the incident commander (IC) should contact the foreman or supervisor, also known as the responsible party. This individual can often provide valuable information about the work site and the emergency situation. An important part of any rescue is the identification of all hazards. It is also important to determine whether

a TRI is a rescue event or a body recovery operation. After these determinations have been made, the IC can decide whether to call for additional resources and order other actions intended to stabilize the incident.

Do *not* rush into the incident scene until you have adequately assessed the situation because you may inadvertently make the situation worse. For example, a fire fighter approaching a trench collapse may cause further collapse. A fire fighter entering a swiftly flowing river might be quickly knocked down and carried downstream. A fire fighter climbing down into a well to evaluate an unconscious victim may be overcome by an oxygen-deficient atmosphere. Stop and think about the dangers that may be present—do not make yourself part of the problem. Always be alert for invisible dangers such as electrical hazards and oxygen-deficient or poisonous atmospheres. If the incident hazards exceed your level of training, do not attempt to rescue victims or mitigate the incident. Report the size-up information over the radio, request the appropriate resources, and establish a safe perimeter around the incident to prevent other people from becoming victims. When the special rescue team arrives, give updated information to the officer, and wait for instructions to assist within your level of training.

Tool Staging

As part of the size-up, you may be able to identify some of the tools and equipment that will be needed for the TRI. Many fire departments have SOPs for staging necessary equipment nearby during a fire. This procedure often involves placing a tarp or salvage cover on the ground at a designated location and laying out commonly used tools and equipment where they can be accessed readily. A similar procedure may be used for rescue operations, so tools that are likely to be needed can be laid out, ready for use. This kind of tool staging saves valuable time because fire fighters do not have to return to their own apparatus or search several different vehicles to find a particular tool.

A department's SOPs usually specify the types of tools and equipment to be staged. The tool staging area

SAFETY TIP

At many fire and emergency scenes, fireline tape is set up to keep bystanders back. Emergency responders operating at such a scene routinely ignore this barrier by simply ducking under or climbing over the tape. This complacency toward fireline tape as a warning device creates a potential hazard for responders at TRIs and other emergency scenes where the tape is placed to warn not only bystanders but responders as well. All fireline tape should be considered a hazard warning to fire fighters unless and until it is proved to be for bystanders only.

can be located outside the building or, in the case of a high-rise or large building, at a convenient location inside the structure. Additional personnel may be directed to bring particular items to the tool staging location at working fires.

Stabilizing the Incident

Once the needed resources are on the way and the scene is safe to enter, begin to stabilize the incident. Establish an outer perimeter to keep the public and media out of the staging area. When the special rescue team arrives, they will establish a smaller perimeter directly around the rescue site. A rescue area is an area that surrounds the incident site (e.g., a collapsed structure, a collapsed trench, or a hazardous spill area) and whose size is proportional to the hazards that exist. Three controlled zones should be established:

- **Hot zone**: This area is for entry teams and special rescue teams only. The hot zone immediately surrounds the dangers of the site (e.g., hazardous materials releases) and is established to protect personnel outside the zone.
- **Warm zone**: This area is for properly trained and equipped personnel only. The warm zone is where personnel and equipment decontamination and hot zone support take place.
- **Cold zone**: This area is for staging vehicles and equipment. The cold zone contains the command post. The public and the media should be kept clear of the cold zone at all times.

To determine the size and scope of each of these zones, fire fighters should identify and evaluate the hazards that are discovered at the scene, observe the geographic area, note the routes of access and exit, observe weather and wind conditions, and consider evacuation problems and transport distances. Fire fighters should enter only those zones for which they are adequately trained.

The most common method of establishing the control zones for an emergency incident site is to use police or fireline tape. Police or fireline tape is available in a variety of colours. The most commonly used colours are red, for the hot zone; orange, for the warm zone; and yellow, for the cold zone. Refer to your department's SOPs for establishing control zones.

Once the fire department has established the three controlled zones, responders should ensure that the zones of the emergency scene are enforced. Due to the lack of firefighting personnel, scene control activities are sometimes assigned to law enforcement personnel. **Lockout and tagout systems** should be used at this time to ensure an electrically safe environment (**FIGURE 25-3**). Lockout and tagout systems are

FIGURE 25-3 Lockout and tagout systems are intended to ensure that the electricity has been shut down.
© Jones & Bartlett Learning.

intended to ensure that the electricity has been shut off and that electrically powered equipment is not unintentionally turned on.

In an emergency situation, all power must be turned off and locked out. As part of this effort, switches or valves must be tagged with labels to protect personnel and emergency response workers from accidental machine start-up. Lockout and tagout procedures are used to warn personnel and ensure that the electrical power is disconnected. Authorized and trained personnel can disconnect the source of power, lock it out, and tag it. Locks and tags are used for everyone's protection against electrical dangers. For your safety and that of others, never remove or ignore a lock or tag. Always be alert for electrical hazards because they are not necessarily easy to recognize. Vehicle crashes, building collapses, entrapments, and natural disasters, for example, can all create electrical hazards to rescuers and to patients. These scenes require careful overview and continual monitoring.

During stabilization, atmospheric monitoring should be performed to identify any immediately dangerous to life and health (IDLH) environments for rescuers and victims. The next steps involve looking at the type of incident and planning how to rescue victims safely. For example, in a trench rescue, these measures might include setting up ventilation fans for air flow, setting up lights for visibility, or protecting a trench from further collapse.

Gaining Access to the Victim

Once the scene is stabilized, the special rescue team needs to consider how to gain access to the victim. How is the victim trapped? In a trench situation, a dirt pile may be to blame. In a rope rescue, scaffolding may have collapsed. In a confined space, a hazardous atmosphere may have caused the victim to collapse. To reach a victim who is buried or trapped beneath debris, it is sometimes necessary to dig a tunnel as a means of rescue and escape. Identify the actual reason for the rescue, and work toward freeing the victim safely.

Technical rescue personnel are likely to maintain constant contact, if possible, with victims. This communication provides reassurance to the victims and allows rescue personnel to gauge how their actions may be impacting those trapped. Even if they are not injured, all victims need to be reassured that the team is working as quickly as possible to free them.

Emergency medical care should be initiated as soon as safe access is made to the victim. Technical rescue paramedics are vital resources at TRIs: Not only can these responders start intravenous lines and treat medical conditions, but they also know how to deal with the equipment and procedures that are going on around them. It may be necessary to integrate EMS personnel into an ongoing rescue incident. However, EMS involvement depends on the situation and will be determined by the IC. Their main functions are to treat victims and to stand by in case a rescue team member needs medical assistance. If it is determined that emergency medical care is required, as soon as the scene has been secured and stabilized, EMS can be allowed access to the patient. Interagency rivalries can become more acute at these incidents. If concerns arise, communicate them to the IC if they cannot be resolved. If it is not possible to adequately stabilize the scene, special rescue team personnel must move the victims to a safe location for treatment. Some fire departments have crossed-trained paramedics to the technical rescue level so that they can enter hazardous areas and provide direct assistance to the victim. Protocols for patient care will be determined through interagency agreement. Review your departmental SOPs regarding patient care.

The strategy used to gain access to victims depends on the type of incident. For example, in an incident involving a motor vehicle, the location and position of the vehicle, the damage to the vehicle, and the position of the victim are important considerations. The means of gaining access to the victim must also take into account the severity of the victim's injuries.

The chosen means of access may have to be changed during the course of the rescue as the nature or severity of the victim's injuries becomes apparent.

Disentangling the Victim

Once precautions have been taken and the reason for entrapment has been identified, the victim needs to be freed as quickly and as safely as possible. A special rescue team member should remain with the victim to direct the rescuers who are performing the disentanglement. In a trench incident, this step would include digging either with a shovel or by hand to free the victim.

In a vehicle accident, the most important point to remember is that the vehicle is to be removed from around the victim, rather than trying to remove the victim from the wreckage. Many different parts of the vehicle may trap the occupants, including the steering wheel, seats, pedals, and dashboard. In this scenario, disentanglement entails the cutting of a vehicle (and or machinery) away from trapped or injured victims, which is accomplished by using extrication and rescue tools with a variety of extrication methods (see Chapter 24, *Vehicle Rescue and Extrication*). The Fire Fighter II must be ready to maintain perimeter security, provide the correct tools and equipment, or perform other duties outside of the hazard area.

> **FIRE FIGHTER TIP**
>
> Technical rescue incidents may involve longer operational periods than structure fires. Pay attention to your physical state and that of your team members. If you are tired, ask the IC to send your crew to the rehabilitation area.

Removing the Victim

Once the victim has been disentangled, efforts will be redirected toward removing the victim (**FIGURE 25-4**).

FIGURE 25-4 Special rescuers removing a victim from a confined space.
© Jones & Bartlett Learning.

In some instances, this may simply amount to having someone assist the victim up a ladder. In other cases, special rescue teams may use Stokes baskets, backboards, stretchers, or other immobilization devices to remove an injured or unresponsive victim from a trench, a confined space, or an elevated point. Once the victim is out of the hazard area, the Fire Fighter II may assist EMS personnel with preparing victims for transport. **Packaging** is the process of preparing the victim for movement as a unit. Refer to your local EMS protocols for specific patient packaging and treatment before moving victims from special rescue situations.

The overriding objective for victim rescue, transfer, and removal is to complete the process as safely and efficiently as possible. It is important that the rescuer use good body mechanics, victim packaging, removal, and transportation skills.

Transporting the Victim

Once the victim has been removed from the hazard area, transport to an appropriate medical facility is accomplished by EMS personnel. The type of transport will vary depending on the severity of the victim's injuries and the distance to the medical facility. For example, if a victim is critically injured or if the rescue is taking place some distance from the hospital, air transportation may be more appropriate than the use of a ground ambulance.

In rough-terrain rescues, four-wheel drive, high-clearance vehicles may be required to transport victims on stretchers to an awaiting ambulance. Snowmobiles with attached sleds can be used to transport victims down a snowy mountain. Helicopters are increasingly used for quick evacuation from remote areas, but weather conditions sometimes limit their use.

Transporting victims who have been injured in a hazardous materials incident may pose additional problems. Prior to transport, it is extremely important that adequate decontamination occur. Placing a poorly decontaminated victim inside an ambulance or helicopter and then closing the doors put both the victim and the rescue personnel in danger. Any toxic gases or vapours given off by the victim or by the victim's clothing can contaminate the inside of the transport vehicle, causing injury to the EMS personnel. In such an incident, any bags containing contaminated clothing or other personal effects must be properly sealed if they are going with the victim. Receiving hospitals should be notified of the impending arrival of victims who have been involved in a hazardous materials exposure incident. This will alert the hospital's triage teams so that they can take the appropriate precautions once the victims begin to arrive.

Because special rescue incidents are often extended-duration technical events, it is important to address the rehabilitation needs of rescue personnel. During many such situations, it will be necessary to rotate personnel. This course of action reduces the chance of injury or sickness to the rescuers and also provides a safer rescue for the victim. Rehabilitation is especially important when temperatures are very high or very low and when rescuers are subjected to wet conditions. It is better to plan for rehabilitation and not need it than to not have rehabilitation available when it is needed. For more information on rehabilitation, see Chapter 20, *Fire Fighter Rehabilitation*.

Postincident Duties

Security of the Scene and Preparation for the Next Call

Once the rescue is complete, the scene must be stabilized by the rescue crew to ensure that no one else becomes injured. In a trench incident, this may include filling the trench with dirt and roping off the surrounding area.

In a hazardous materials incident, the clean-up of equipment and personnel takes place after the hazardous materials incident has been completely controlled and all victims have been treated and transported. Trained disposal crews should be available to clean the site. All equipment, protective gear, and clothing, as well as the rescue personnel themselves, must be decontaminated.

In a vehicle rescue incident, the vehicle is removed; usually it is transported by a tow truck to a storage lot. The street or roadway is cleaned of all debris from the vehicle. This includes the clean-up of any spilled fluids, including gasoline, from the vehicle that was involved in the incident.

In an industrial setting, the supervisor of the facility is responsible for securing the scene. Nevertheless, the technical rescue team must follow up with the supervisor or the safety coordinator to ensure that further problems are prevented.

Once you have secured the scene and packed up your equipment, it is important to return to the station and fully inventory, clean, service, and maintain all of the equipment to prepare it for the next call. Although some items may need repair, most will require only simple maintenance before being placed back on the apparatus and considered in service.

Back at the station, as you prepare for the next call, complete all reports, and document the rescue incident. Record keeping serves several important purposes. Adequate reporting and the keeping of accurate records ensure the continuity of quality care, guarantee proper transfer of responsibility, and fulfill the administrative needs of the fire department for reporting requirements. In addition, the reports can be used to evaluate response times, equipment usage, and other areas of administrative interest.

Postincident Analysis

As with any type of call, the best way to prepare for the next rescue call is to review the last one and identify the strengths and weaknesses in your response. What did you do well? What could have been done better? Would any other equipment have made the rescue safer or easier? If a death or serious injury occurred during the call, a critical incident stress management session may occur to assist fire fighters. Reviewing a TRI with everyone involved will allow everyone to learn from the call and make the next call even more successful.

General Rescue Scene Procedures

At any scene to which you respond—whether it is a fire, EMS, or technical rescue call—your safety, the safety of your company, and the public's safety are paramount. At a TRI, many issues need to be considered during the response. Many fire fighters have been injured or killed during special rescue incidents because they did not consider their own safety. While the temptation may be to immediately approach the victim or the accident area, it is critically important to slow down and properly evaluate the situation.

In confined-space rescue incidents, for example, potential hazards may include deep or isolated spaces, multiple complicating hazards (such as water or chemicals), failure of essential equipment or service, and environmental conditions (such as snow or rain). In all rescue incidents, fire fighters should consider the potential general hazards and risks posed by utilities, hazardous materials, confined spaces, and environmental conditions, as well as IDLH hazards.

Approaching the Scene

As you approach the scene of a TRI, you will not always know which kind of scene you are entering. Is it a construction scene? Do you see piles of dirt that would indicate a trench? Has a structural collapse occurred in a building? Which actions are the civilians taking? Are they attempting to rescue trapped people, possibly placing themselves at great danger?

Beginning with the initial dispatch of the rescue call, fire fighters should be compiling facts and factors about the call. Size-up begins with the information gained first from the person reporting the incident and then from the bystanders at the scene upon arrival.

The information received when an emergency call is received is important to the success of the rescue operation. It should include the following details:

- The location of the incident
- The nature of the incident (kinds, cause, what is involved, etc.)

- The conditions and positions of victims
- The conditions and positions of vehicles, buildings, structures, terrain, etc.
- The number of people trapped or injured and the types of their injuries
- Any specific or special hazard information
- The name of the person calling and a number where that person can be reached

Many times, the information received on the emergency call will be incomplete. You will have to evaluate the incident with the information available. Once on the scene, you can better identify life-threatening hazards and take the appropriate corrective measures to mitigate those dangers. If additional resources are needed, they should be ordered by the IC.

A size-up should include the initial and continuous evaluation of the following issues:

- Scope and magnitude of the incident
- Risk/benefit analysis
- Number of known and potential victims
- Hazards
- Access to the scene
- Environmental factors
- Available and necessary resources
- Establishment of control perimeters

Careful size-up and coordination of rescue efforts are essential to avoid further injuries to the victims and to provide for the safety of the fire fighters.

Mitigating Utility Hazards

Look for any downed electrical wires near the scene (**FIGURE 25-5**). Is any of the equipment or machinery present electrically charged so that it might present a hazard to the victim or the rescuers? Is a natural gas or propane fuel service connected to the structure? The IC should ensure that the proper procedures have been taken to shut off the utilities in the area where the

FIGURE 25-5 Downed electrical wires present a hazard at many types of rescue scenes.
Courtesy of Lara Shane/FEMA Photo News.

rescuers will be working. Before the utilities are disconnected by the utility provider, fire fighters must assume that gaseous fuels and energized electrical systems and equipment are present.

Rescue operations involving energized electrical lines or other utility hazards require the assistance of trained personnel. Many fire fighters have been injured or killed because they did not recognize the hazards associated with these incidents. Do not rush to assist victims who have come into contact with energized electrical equipment or have been overcome by gaseous fuels. For energized electrical line emergencies, such as downed lines, park at least one utility pole span away. Watch for falling utility poles—a damaged pole may bring other poles down with it. Do not touch any wires, power lines, or other electrical sources until they have been de-energized by a power company representative. It is not just the wires that are hazardous; any metal that they touch is also energized. Metal fences that become energized, for example, are energized for their entire unbroken length. Be careful around running or standing water because it is an excellent conductor of electricity. Once the electrical service has been disconnected, use proper lockout/tagout procedures to prevent electrical shock hazards.

Both natural gas and liquefied petroleum gas are nontoxic but are classified as asphyxiants because they displace breathing air. In addition, both gases are flammable. If a call involves leaking gas, call the gas company immediately. If a victim has been overcome by leaking gas, wear full personal protective equipment (PPE) and positive-pressure self-contained breathing apparatus (SCBA) when making the rescue. Remove the victim from the hazardous atmosphere before beginning treatment.

Providing Scene Security

Has the area been secured to prevent people from entering? Co-workers, family members, and even other rescuers may enter an unsafe scene and become additional victims. The IC should coordinate with law enforcement to help secure the scene and control access. A strict personnel accountability system should be implemented by the fire department to control access to the rescue scene.

You will often be asked to establish barriers to provide scene security at a rescue incident. When emergency incidents occur on streets or highways, vehicle warning lights, traffic cones, and fusees are often used to warn oncoming traffic. When crowd control is an issue, plastic barrier tape marked with the words "Caution" or "Fireline" is often used to keep out nonessential personnel. The steps for establishing a barrier are listed in **SKILL DRILL 25-1**.

Using Protective Equipment

Firefighting gear is designed to protect the body from the high temperatures of fire. It does, however, restrict movement. Technical rescues generally require the ability to move around freely, so most firefighting gear does not work well in a TRI. For this reason, most

SKILL DRILL 25-1
Establishing a Barrier Fire Fighter II, NFPA 1001: 5.4.2

1 Respond safely to the emergency scene. Place the emergency vehicle in a safe position that protects the scene. Don the appropriate PPE.

2 Perform a size-up to assess for hazards. Secure the scene.

(continued)

SKILL DRILL 25-1 Continued
Establishing a Barrier Fire Fighter II, NFPA 1001: 5.4.2

3 Call for needed assistance. Use appropriate devices to establish a barrier, following the orders of the IC.

Note: This is an artist's rendering. The fire apparatus should be positioned away from the building collapse field.

© Jones & Bartlett Learning.

specialized teams also carry items such as harnesses; smaller, lighter helmets; and coveralls that are easier to move in than structural firefighting protective gear (**FIGURE 25-6**).

For personnel operating in a water rescue hazard zone, the minimum PPE includes a Coast Guard–approved **personal flotation device (PFD)**, thermal protection, a helmet appropriate for water rescue, a cutting device, a whistle that works in water, and contamination protection (if necessary).

Other equipment items that are easily carried by fire fighters include binoculars, chalk or spray paint for marking searched areas, a compass for wilderness rescues, first-aid kits, a whistle, a hand-held global positioning system (GPS), and Cyalume-type light sticks. The IC and the technical rescue team will help determine which PPE

FIGURE 25-6 Technical rescue technicians wear different types of protective equipment that allow mobility while still providing an adequate level of protection for the incident.

© Jones & Bartlett Learning.

you will need to wear while assisting in the rescue operation. For more information, see NFPA 1951, *Standard on Protective Ensembles for Technical Rescue Incidents*.

Using the Incident Command System

The first-arriving officer immediately establishes command and starts using the incident command system (ICS). Activating the ICS is critically important because many TRIs will eventually become very complex and require many assisting units. Without the ICS in place, it will be difficult—if not impossible—to ensure effective scene operations and the safety of the rescuers. For more information on ICS, see Chapter 22, *Establishing and Transferring Command*.

Ensuring Accountability

Accountability should be practised at all emergencies, no matter how small. A **personnel accountability system** is important to ensure rescuers' safety. It tracks the personnel on the scene, including their identities, assignments, and locations. This system ensures that only rescuers who have been given specific assignments are operating within the hot zone. By using a personnel accountability system and working within the ICS, an IC can track the resources at the scene, make appropriate assignments, and ensure that every person at the scene operates safely. Review the section on personnel accountability in Chapter 19, *Fire Fighter Survival*.

Making Victim Contact

At any rescue scene, every effort should be made to communicate with the victim, if possible. Technical rescue situations often last for hours, with the victim being left alone for long periods of time. Sometimes making contact with the victim is not possible, but do try to communicate via a radio, by a cell phone, or by yelling. Reassure the victim that everything possible is being done to ensure his or her safety.

For the fire fighter, it is important to stay in constant communication with the victim. If possible, assign someone to talk to the victim while other fire fighters carry out the rescue. Realize that the victim could be sick or injured and is probably frightened. If the fire fighter is calm, his or her reassuring demeanor will calm the victim. To help keep a victim calm, do the following:

- Make and keep eye contact with the victim.
- Tell the truth. Lying destroys trust and confidence. You may not always tell the victim everything, but if the victim asks a specific question, answer it truthfully.
- Communicate at a level the victim can understand.
- Be aware of your own body language.
- Always speak slowly, clearly, and distinctly.
- Use the victim's name.
- If a victim is hard of hearing, speak clearly and directly to the person so that the person can read your lips.
- Allow time for the victim to answer or respond to your questions.
- Try to make the victim comfortable and relaxed whenever possible.

Many victims at TRIs require medical care, but this care should be given only if it can be done safely. Do not become a victim yourself during a rescue attempt. Many fire fighters have been killed while entering a TRI in an attempt to help a deceased victim.

Assisting Rescue Crews

Every TRI is different. If you have the role of assisting a technical rescue team, training with the team is probably the most important thing you can do to prepare for your participation in a TRI. By training with your department's technical rescue teams, you will better understand how the team members operate and how you can assist them. The more knowledge you have, the more you will be able to do.

At any TRI, follow the orders of the company officer who receives direction from the IC. Many of the tasks will involve moving equipment and objects from one place to another. Other tasks will involve protecting the team and victims. Do not take these tasks lightly; they

might not seem important, but they are essential to the effort. Without scene security, traffic control, protective hose lines, and people to maintain equipment, the rescue cannot and will not happen.

You may be asked to retrieve specialized tools and equipment at any type of special rescue incident. You must become familiar with the different tools that are used by the specialized rescue teams in your department. The steps for identifying and retrieving rescue tools are listed in **SKILL DRILL 25-2**.

As you read the information that follows about assisting at specific types of rescue scenes, bear in mind the three factors about safely approaching the TRI scene:

- Approach the scene cautiously.
- Position the apparatus properly.
- Assist the specialized team members as needed.

Vehicles and Machinery

A wide variety of rescue situations involving vehicles and machinery are possible. The actions commonly taken at the scene of motor vehicle accidents are covered in Chapter 24, *Vehicle Rescue and Extrication*. This section focuses on other types of rescues involving vehicles and machinery, including motorcycles, trucks, buses, trains, mass-transit vehicles, aircraft, watercraft, farm machinery, agriculture implements, construction machinery, and industrial machinery. Rescues involving mass-transit vehicles often involve many victims, whereas most other types of rescues involve single victims.

Safe Approach

Vehicle and machinery rescues present a wide variety of problems to rescuers and victims. Hazards present at these scenes may include flammable liquids, electrical hazards, unstable machines or vehicles, and many other types of hazards.

Electricity is an invisible hazard. Consider any machine you encounter to be electrically charged until proven otherwise. Because of this risk, it is important to use lockout and tagout systems with any potential electrical source before approaching the scene. The operator of the machine or a maintenance person who is familiar with the machine can be a valuable resource in this regard.

With vehicle-related calls, traffic control is a crucial issue. Every year, fire fighters and EMS workers are struck and killed at accident scenes by passing drivers who became distracted by the commotion. In the United States, the National Institute for Occupational Safety and Health (NIOSH) independently investigates and issues reports on fire fighter line-of-duty deaths,

SKILL DRILL 25-2
Identifying and Retrieving Rescue Tools Fire Fighter II, NFPA 1001: 5.4.2

Courtesy Bill Larkin.

1. Respond safely to the emergency scene.

2. Place the emergency vehicle in a safe position that protects the scene.

3. Don the appropriate PPE.

4. Perform a size-up to assess for hazards.

5. Secure the scene.

6. Call for needed assistance.

7. Receive the request from the special rescue team to retrieve rescue tools.

8. Identify and retrieve the rescue tools.

including those occurring on roadways. When working in or around traffic, placement of the apparatus is key to the protection of you and those around you. Whenever you are in a traffic-related situation, you must wear a high-visibility vest over your firefighting PPE. Unless you are exposed to or are likely to become exposed to fire conditions, you must wear high-visibility safety clothing that meets the CSA (Canadian Standards Association) Standard Z96-15 High-Visibility Safety Apparel (CCOHS, 2018).

Stabilization of the rescue scene includes making sure that the vehicle or machine cannot move. Gaining access to some victims will be easy. In other situations, access to the victim can occur only after extensive stabilization of the scene.

Unusual and special rescue situations include extrication and disentanglement operations at incidents involving vehicles on their tops or sides, trucks and large commercial vehicles, and vehicles on top of other vehicles. To ensure responder safety, be wary of fuels and fluids spilled at these scenes.

How You Can Assist

At a vehicle or machinery rescue call, you may be called upon to assist in the extrication and treatment of the

victim. Adequate protection of the victim (e.g., with c-spine immobilization or holding of blankets) will allow the technical rescue team to operate hydraulic tools and cut the vehicle or machine apart without causing further injury to the victim. Many hydraulic tools are heavy, so you may need to help support the tools while technicians operate these pieces of equipment.

You may also be called upon to do the following:

- Assist with controlling site security and the perimeter of the rescue incident.
- Obtain information from witnesses.
- Retrieve more rescue tools and equipment from the fire apparatus or rescue trucks.
- Console a family member of a victim.
- Keep bystanders out of the way.
- Assist in moving items that are in the way of the rescue team.
- Service equipment.
- Set up power tools.
- Stand by with a hose line.
- Extinguish a fire.

Tools Used

Many tools—including some inexpensive items—are required for a successful vehicle or machinery rescue situation. Simple tools such as a spring-loaded punch, for example, can safely break out a vehicle window to access a victim. Tools and items that will be required for a vehicle or machinery rescue include the following:

- PPE
- Hydraulic tools (spreaders, cutters, rams)
- Halligan tool
- Cutting torch
- Air chisels
- Cribbing
- Saber saw
- Windshield cutter
- Spring-loaded punch
- Chains
- Rescue-lift air bags (high and low pressure)
- SCBA air cylinders
- Basic hand tools
- Come along
- Portable generator
- Seat belt cutters
- Hand lights and other scene lighting
- Hose lines (protection)
- Blanket

Confined-Space Rescue

A confined space is an enclosed area that is not designed for people to occupy. Confined spaces have limited openings for entrance and exit and may have limited ventilation to provide air circulation and exchange. They can occur in natural, farm, commercial, and industrial settings. For example, grain silos, industrial pits, tanks, and below-ground structures are all confined spaces. Likewise, cisterns, well casings, sewers, and septic tanks are confined spaces that are found in residential settings.

Confined-space incidents are among the most dangerous scenes you may encounter. Confined spaces may be oxygen deficient and contain poisonous gases or combustible atmospheres. There is also the potential for slips, trips, falls, entrapment, encountering deep water, and collapse hazards. NIOSH reports that initial rescuers account for approximately 60% of all confined-space deaths (NIOSH 86-110, 1986).

It is often difficult to extricate an unconscious or injured person from a confined space because of the poor ventilation and limited entry or exit area. For this reason, ropes are often used to remove an injured or unconscious victim (**FIGURE 25-7**).

Be aware that you may be dispatched to another type of incident such as an unresponsive person or a search for a missing person without knowing that a confined-space incident exists. Once you determine that the incident involves a confined space, you must take steps to protect yourself, other responders, and the public.

FIGURE 25-7 Ropes are often used to remove an injured or unconscious victim from a confined space.
© Jones & Bartlett Learning. Photographed by Glen E. Ellman.

SAFETY TIP

The two greatest hazards in a confined space are the lack of oxygen and the presence of poisonous gases or asphyxiants. If rescuers fail to identify a confined space and to ensure that it contains a safe atmosphere, injury and death can occur to rescuers as well as to the original victim. Treat all confined-space incidents with extreme caution. Do not enter the confined space without proper training. Instead, secure the area, gather as much information as possible, and wait for confined-space technical rescuers to arrive on the scene.

All confined spaces should be considered to be IDLH until proven otherwise.

Safe Approach

As you approach a rescue scene, look for a bystander who might have witnessed the emergency. Information gathered prior to the technical rescue team's arrival will save valuable time during the actual rescue. Do not assume that a person in a pit has simply suffered a heart attack; instead, assume that an IDLH atmosphere is present at any confined-space call. An IDLH atmosphere can immediately incapacitate anyone who enters the area without breathing protection. Toxic or explosive gases may be present, or the confined space may not have enough oxygen to support life.

When a rescue involves a confined space, remember that it will take some time for qualified rescuers to arrive on the scene and prepare for a safe entry into the confined space. Given this delay, the victim of the original incident may have died before your arrival. Do not put your life in danger to recover a body.

How You Can Assist

Your main role in confined-space rescues is to secure the scene and prevent other people from entering the confined space until additional rescue resources arrive. If the incident has occurred at an industrial site, ascertain whether an OSHA-required confined-space entry permit was prepared. If so, locate the entry supervisor and attendant listed on the permit and have them stand by at the space. If both of these persons cannot be accounted for, they may have become additional victims inside the confined space. As additional highly trained personnel arrive, your company may provide help by giving the rescuers a situation report.

It is important for the first-arriving company officer to brief incoming crews with regards to all aspects of the situation. Anything that might be important to the response should be noted by the first-arriving unit. Observed conditions should be compared to reported conditions, and a determination should be made as to the relative change over the time period. Whether an incident appears to be stable or has changed greatly since the first report will affect the operation strategy for the rescue. A size-up should be completed quickly and immediately upon arrival, and this information should be relayed to the special rescue team members upon their arrival at the scene. Other items of importance that should be included in a situation report are a description of any rescue attempts that have been made, the exposures, any hazards present, the extinguishment of fires, the facts and probabilities of the scene, the situation and resources of the fire company, the identities of any hazardous materials present, and an evaluation of the progress made so far.

Many pump and ladder companies carry atmospheric or gas detection devices; if you have this capability, obtain readings to determine whether the situation presents a hazard. If you do encounter an oxygen-deficient or other IDLH atmosphere, set up a ventilation fan to help remove toxic gases and improve air flow to the victim. Sometimes you can help a victim without entering the confined space by passing down SCBA, first-aid supplies, or even a ladder that the victim can use to climb out.

FIGURE 25-8 Supplied-air respirator system.
© Jones & Bartlett Learning. Photographed by Glen E. Ellman.

Confined-space rescues can be complex and take a long time to complete. You may be asked to assist by bringing rescue equipment to the scene, maintaining a charged hose line, or providing crowd control. By understanding the hazards associated with confined spaces, you will be better prepared to assist a specialized team that is dealing with an emergency involving a confined space.

Tools Used

One of the key tools used in any confined-space rescue is a supplied-air respirator system (**FIGURE 25-8**). These devices are similar to SCBA except that, instead of carrying the air supply in a cylinder on your back, you are connected by a hose line to an air supply located outside the confined space. This kind of system provides the rescuer with a continuous supply of air that is not limited by the capacity of a back-mounted air cylinder.

Many other tools and pieces of equipment can be used during a confined-space rescue operation:

- Air lines
- Air carts
- Extra SCBA air cylinders
- Tripod and winch for raising and lowering rescuers and victims
- Pry bar or manhole cover remover
- Rescue rope
- Personnel harnesses
- Gas monitors
- Ventilation fans
- Explosion-proof lights
- Radios
- Hard-wired communications equipment
- Protective gear
- Lockout and tagout kits
- Personnel accountability boards
- Victim removal devices such as short backboard devices, Stokes baskets, backboards, and other commercially made devices

- Medical equipment
- Carabiners and locking "D" rings
- Descenders
- Safety goggles
- Hand lights (battery-operated)
- Hearing protection headsets

Rope Rescue

Rope rescue skills are used in many types of TRIs. For example, an injured victim might be lifted from inside a tank using a raising system, or rescuers might need to lower themselves down to the area where victims are trapped in a structural collapse. Rope rescue skills are the most versatile and widely used technical rescue skills. This section outlines the basics of rope rescue and gives you a foundation for learning the more complex parts of rope rescue. An approved rope rescue course is required to attain true proficiency in rope rescue skills.

Rope rescue courses cover the technical skills needed to raise or lower people using mechanical advantage systems and to remove someone from a rock ledge or a confined space (**FIGURE 25-9**). They also cover the equipment and skills needed to accomplish these rescues safely.

Rope Rescue Incidents

Most rope rescue incidents involve people who are trapped in normally inaccessible locations such as a mountainside or the outside of a building (**FIGURE 25-10**).

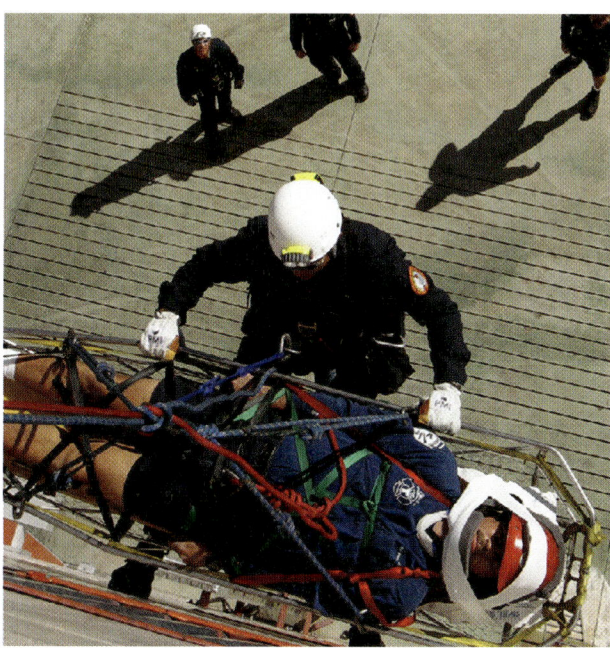

FIGURE 25-9 Rope rescues require intense technical training.
© Jones & Bartlett Learning. Photographed by Glen E. Ellman.

FIGURE 25-10 Ropes are invaluable when a person is trapped in an inaccessible area, such as outside a high building.
© Jones & Bartlett Learning. Photographed by Glen E. Ellman.

Rescuers often must lower themselves using a system of anchors, webbing, ropes, carabiners, and other devices to reach the trapped person. Once rescuers reach the person, they have to stabilize him or her and determine how to get the person to safety. Sometimes the victim has to be lowered or raised to a safe location. Extreme cases could involve more complicated operations, such as transporting the victim in a basket lowered by a helicopter.

The type and number of ropes used in a rope rescue depend on the situation. There is almost always a primary rope that bears the weight of the rescuer (or rescuers) as he or she attempts to reach the victim. The rescuers often have a second line attached to them, known as a belay line, which serves as a backup if the main line fails. Additional lines might be needed to raise or lower the trapped individual, depending on the circumstances.

Rope rescue incidents are divided into low-angle and high-angle operations. **Low-angle operations** are situations where the slope of the ground over which the fire fighters are working is less than 45 degrees, fire fighters are dependent on the ground for their primary support, and the rope system becomes the secondary means of support (**FIGURE 25-11**). An example of a low-angle system is a rope stretched from the top of an embankment and used for support by rescuers who are carrying a victim up an incline. In this scenario, the rope is the secondary means of support for the rescuers; the primary support is the rescuers' contact with the ground.

Low-angle operations are used when the scene requires ropes to be used only as assistance to pull or haul up a victim or rescuer. This step usually becomes necessary when adequate footing is absent—in areas such as a dirt or rock embankment. In such an incident, a rope will be tied to the rescuer's harness, and the rescuer will climb the embankment on his or her own, using the rope only to make sure he or she does not fall. Low-angle operations also include lifelines deployed during ice or water rescues.

Ropes can also be used to assist in carrying a Stokes basket. This use of ropes aids the rescuers and frees them from having to carry all of the weight over rough terrain. Rescuers at the top of the embankment can help to pull up or lower the basket using a rope system.

High-angle operations are situations where the slope of the ground is greater than 45 degrees and rescuers or victims depend on **life safety rope** for support, rather than on a fixed surface such as the ground. High-angle rescue techniques are used to raise or lower a person when other means of raising or lowering are not readily available. Sometimes, a rope rescue is performed to remove a person from a position of peril; at other times, it is needed to remove an ill or injured victim.

In Chapter 9, *Ropes and Knots*, you learned how to tie some basic knots and to use ropes to lift and lower selected tools. Rope rescue training courses will build on the skills you have already learned. There is much more to learn before you are ready to perform high-angle rescues, however, because you need to be able to perform complex tasks quickly and safely in these incidents. Successful rope rescue operations also require a cohesive team effort.

Safe Approach

Rope rescues are among the most time-consuming calls that you will encounter. Extensive setup is necessary, and a significant body of equipment must be assembled prior to initiating any rescue. Protect your safety by remaining away from the area under the victim and away from any loose materials that may fall or slide. Likewise, when you are above the victim, stay back from the edge unless you are securely tied off to keep from falling. Use extreme caution so you do not cause any loose debris to fall on the victim. Work to control the scene by moving the bystanders and friends of the victim to an area where they will not be injured.

You can do a lot to stabilize the scene and prevent further injuries by remaining calm and putting your skills to work. A rope can, for example, be used to secure a rescuer when making a rescue attempt, haul a stretcher or litter up an embankment, or lower a rescuer into a trench or cave.

How You Can Assist

Rope rescues are labour-intensive operations. By remembering the introductory rope skills you have learned, you may be able to assist with many phases of the rescue operation. You should identify and reduce any impediments that may slow down rescue techs from accessing victims. For example, in the event of a high-rise rescue, place the elevators on service and ensure you have keys to that allow for access to all floors. A technical rescue team member might ask you to tie knots and prepare anchors. Do not be offended if the team members check your work—it is protocol to check a system two or three times to ensure its safety.

Avoid stepping on ropes because any damage or friction to the outer layers of a rope can decrease its tensile strength and possibly cause a catastrophic failure. Keep everyone clear of areas where falling objects could cause injuries. Keep in mind the importance of following the ICS. Work to complete your assigned tasks.

Tools Used

Rope rescues rely almost completely on the equipment used. When you are rappelling down to rescue a victim who is 91 m (300 ft) off the ground, you need to have equipment that has been properly designed and maintained to keep you alive. Some of the tools and

FIGURE 25-11 Low-angle rescue operations.
© Keith D. Cullom/www.fire-image.com.

equipment that you will need to recognize and use are listed here:

- PPE (helmet, rescue gloves)
- Personnel harnesses
- Stokes basket
- Harness
- Rescue rope
- Carabiners
- Webbing and prusik cord
- Miscellaneous hardware (racks, pulleys)

In addition to the rope itself, several hardware components might be used. Fire fighters, for example, often use a carabiner (**FIGURE 25-12**). This device is used to connect one rope to another rope or to other hardware such as an anchor plate, swivel, or pulley. Several types of carabiners are available, and you should know how to operate the type used by your department. Only a few are recognized for use in the fire service and rescue. Refer to NFPA 1983, *Standard on Life Safety Rope and Equipment for Emergency Services.*

A harness is a piece of rescue or safety equipment made of webbing and worn by a person. It is used to secure the person to a rope or to a solid object. Two types of harnesses—Class II and Class III—can be used by rescuers, depending on the circumstances encountered:

- Class II harness (seat harness) fastens around the rescuer's waist and legs and has a design load of 272 kg (600 lb). It is used to support a fire fighter, particularly in rescue situations (**FIGURE 25-13**).
- Class III harness (full-body harness) fastens around the rescuer's waist and thighs as well as secures the rescuer's waist and shoulders. It is the most secure type of harness and is often used to support a fire fighter who is being raised or lowered on a life safety rope (**FIGURE 25-14**).

FIGURE 25-13 Class II harness (seat harness).
© Jones & Bartlett Learning. Photographed by Glen E. Ellman.

FIGURE 25-14 Class III harness (full-body harness).
Courtesy of RescueTECH1, Inc.

FIGURE 25-12 Carabiners.
© Jones & Bartlett Learning. Photographed by Glen E. Ellman.

Harnesses need to be cleaned and inspected regularly, just as life safety ropes do. Follow the manufacturer's instructions for cleaning and inspecting harnesses.

Trench and Excavation Collapse Rescue

Rescues in collapsed trenches often are complicated and involve many different skills, such as shoring, air quality monitoring, confined-space operations, and rope rescue. Trench and excavation rescues become necessary when earth has been removed for placement of a utility line or for other construction and the sides of the excavation collapse, trapping a worker (**FIGURE 25-15**). Entrapments can occur when children play around a pile of sand or earth that collapses. Many entrapments occur because the required safety precautions were not taken.

Whenever a collapse has occurred, you need to understand that the collapsed product is unstable and prone to further collapse. Earth and sand are heavy, and a person who is partly entrapped in these materials cannot simply be pulled out. Instead, the victim must be carefully dug out. This step can be taken only after shoring has stabilized the sides of the excavation.

Vibration or additional weight on top of displaced earth will increase the probability of a secondary collapse—that is, a collapse that occurs after the initial collapse. A secondary collapse can be caused by equipment vibration, personnel standing at the edge of the trench, or water eroding the soil. Do not attempt to rescue someone who is trapped by dirt or sand. You may quickly become another victim! Safe removal of trapped persons requires a special rescue team that is trained

and equipped to erect shoring to protect the rescuers and the entrapped person from secondary collapse.

Safe Approach

Safety is critical when approaching a trench or excavation collapse. Walking close to the edge of a collapsed area can trigger a secondary collapse, so stay away from the edge of the collapse, and keep all workers and bystanders at a distance from the site. All equipment within 30 m (98 ft) or more of the scene should be shut down to prevent further collapse through ground vibration. Likewise, vibrations caused by nearby traffic can cause collapse, so it may be necessary to stop or divert traffic away from the scene.

Soil that has been removed from the excavation and placed in a pile is called the spoil pile. This material is unstable and may collapse if placed too close to the excavation. Avoid disturbing the spoil pile.

Make verbal contact with the trapped person, if possible, but do not place yourself in danger while doing so. If you plan to approach the trench at all, do so from the narrow end where the soil will be more stable. It is best not to approach the trench unless absolutely necessary. Stay out of a trench unless you have been properly trained in trench rescue techniques.

Provide reassurance by letting the trapped person know that a trained rescue team is on the way. By moving people away from the edges of the excavation, shutting down machinery, and establishing contact with the victim, you start the rescue process.

You can also size up the scene by looking for evidence that would indicate where the trapped victims may be located. Hand tools are an indicator of where victims may have been working, as are hard hats. By questioning the bystanders, you can also determine where the victims were last seen.

How You Can Assist

As the rescue team starts to work, your company will be assigned to carry out certain tasks, ranging from unloading lumber for shoring to assisting with cutting timbers at a safe distance away from the entrapment. Depending on weather conditions or the length of the incident, a rehabilitation group may need to be established. Early implementation of the ICS will help make this type of rescue go smoothly. If someone in your company has a specialty such as carpentry, he or she should make this fact known because this skill is valuable at a trench rescue. Measuring and cutting timber and shores for a trench operation are an important aspect of the rescue.

Like confined spaces, trenches may have an IDLH atmosphere inside because of the presence of hazardous gases such as carbon monoxide, methane, or sewer

FIGURE 25-15 Trench rescue.
Courtesy of Captain David Jackson, Saginaw Township Fire Department.

gases. Flammable vapours from gas-powered equipment or gas cans may also be present in the trench. Setting up ventilation fans can make a difference in victim and rescuer survivability in these scenarios.

It may be necessary to pump water out of a trench. If the collapse occurred because of a ruptured pipe or during a rainstorm, for example, rising water in the trench may endanger the trapped victim and cause additional soil to collapse into the trench. Removing this water may stabilize the situation.

When extricating victims from the trench, it may be necessary to lift them out using a raising system, just as in a high-angle rope rescue operation. This effort will require that the rescuers have rescue rope, harnesses, pulleys, carabiners, and all of the other associated equipment available. Personnel must be trained in how to use and apply these systems in this type of rescue scenario.

Tools Used

Trench and excavation rescue uses equipment, rescue techniques, and skills similar to those used with both confined-space rescue and structural collapse rescue.

Tools and equipment used in trench and excavation rescue include the following:

- PPE: helmet, gloves, personal protective clothing, harness, work boots, knee pads, elbow pads, eye protection, SCBA, and supplied-air breathing apparatus (SABA)
- Hydraulic, pneumatic, and wood shores
- Lumber and plywood for shoring
- Cribbing
- Power cutting tools and saws
- Carpentry hand tools
- Shovels
- Buckets for moving soil
- Rescue rope, harnesses, webbing, and associated hardware
- Utility rope
- Ventilation fans
- Water pumps
- Lighting and hand lights
- Ladders
- Shovels
- Extrication equipment such as Stokes baskets or backboards
- Harness sets for Stokes baskets
- Medical equipment

Structural Collapse Rescue

Structural collapse is the sudden and unplanned fall of all or part of a building (**FIGURE 25-16**). Such collapses may occur because of fires, removal of supports during construction, renovation or demolition,

FIGURE 25-16 Structural collapse is a sudden and unplanned fall of part or all of a building.
Courtesy of Dave Gatley/FEMA.

vehicle accidents, explosions, rain, wind, snowstorms, earthquakes, and tornadoes.

Chapter 6, *Building Construction*, emphasizes the importance of all building components working together to provide a stable building. Consider the type of building construction when determining the potential for collapse. When any part of a building becomes compromised, the dynamics of the building change. Consequently, fire fighters should always be alert for signs of a possible building collapse. A partial building collapse is extremely hazardous to rescuers because of the potential for secondary collapse during search and rescue operations.

Safe Approach

Many different factors can lead to building collapses, so you must approach the scene carefully. The causes of some collapses—for example, a vehicle crashing through the wall of a house—will be readily evident. Other causes, such as a natural gas explosion, may require extensive investigation before they are definitively identified.

SAFETY TIP

Fires often lead to structural collapses. Be alert for the possibility of a building collapse any time a fire has damaged a building.

As you approach any building collapse, consider the need to shut off utilities. Entering a structure with escaping natural gas or propane is extremely hazardous. Electricity from damaged wiring can also present a deadly hazard.

A prime safety consideration is the stability of the building. Even a well-trained engineer cannot always determine the stability of a building simply by looking at its exterior. The IC must make the decision regarding whether the building is safe for the special rescue team to enter. This is a difficult decision, especially if bystanders cannot tell you whether people were present in the building before the collapse. Unless you are trained in structural collapse rescue, do not attempt to enter any building that has collapsed or is partially collapsed.

How You Can Assist

Rescue operations at a structural collapse vary depending on the size of the building and the amount of damage to the building. In cases of large building collapses, the rescue operation will be sizable. Urban search and rescue teams and structural collapse teams have received special training in dealing with these types of situations. Specifically, they are trained in shoring and specialized techniques for gaining access and extricating victims that enable them to systematically search the affected building. These types of rescue situations take a lot of time and require a lot of personnel.

Personnel who do not have this kind of special training usually will be assigned to support operations. For example, you might be assigned to be part of a bucket brigade that works to remove debris from the building. A substantial amount of manual digging and searching may be necessary, all of which is physically taxing. You may be able to assist with this activity if the work is not in an unstable location.

For a rescue effort of this type to work, there must be teamwork and a well-organized and well-implemented ICS. Without a solid command structure, complicated rescue incidents may become much more difficult, and the likelihood of success diminishes.

Tools Used

Fire fighters should know the tools and equipment designed for structural collapse emergency rescue incidents:

- PPE: helmet, gloves, work boots, harness, elbow pads, knee pads, eye protection, SCBA, SABA, and dust masks
- Shoring equipment
- Lumber for shoring and cribbing
- Power tools
- Hand tools
- Lighting
- Rescue rope, harnesses, webbing, and associated hardware
- Utility ropes
- Buckets
- Shovels

Cave and Tunnel Rescue

Cave and tunnel rescue operations are often lengthy, complicated incidents that require trained rescue teams and specialized equipment. Never attempt to enter a cave or tunnel for rescue operations without having the proper training and equipment.

How You Can Assist

If your department responds to a cave or tunnel rescue incident, you can assist the special rescue team by maintaining a safe perimeter around the operations area, staging tools and equipment in the designated area, and assisting with patient packaging once the victim is removed from the hazard area. Be sure to train with your special rescue team frequently to be familiar with the equipment and procedures used on these incidents.

Tools Used

Many of the tools and equipment used for cave and tunnel rescue will be the same as those used for similar collapse incidents:

- PPE: helmet, gloves, harness, work boots, knee pads, elbow pads, eye protection, SCBA, and SABA
- Hydraulic, pneumatic, and wood shores
- Lumber and plywood for shoring
- Cribbing
- Power cutting tools and saws
- Carpentry hand tools
- Shovels
- Buckets for moving soil
- Rescue rope, harnesses, webbing, and associated hardware
- Tripod and winch for raising and lowering rescuers and victims
- Gas monitors
- Ventilation fans
- Explosion-proof lights
- Radios
- Utility rope
- Ventilation fans
- Water pumps
- Lighting and hand lights
- Ladders
- Shovels
- Extrication equipment such as Stokes baskets or backboards
- Harness sets for Stokes baskets
- Hearing protection headsets
- Medical equipment

Water and Ice Rescue

Almost all fire departments have the potential for being called to perform a water rescue (**FIGURE 25-17**). Water is present in small streams and large rivers, and it

FIGURE 25-17 Water rescue.
© Nate Rawner.

FIGURE 25-18 Ropes ensure rescuer safety during water rescues.
Courtesy of Captain David Jackson, Saginaw Township Fire Department.

fills lakes, oceans, reservoirs, irrigation ditches, canals, and swimming pools. A static source such as a lake may have no current, whereas a whitewater stream or flooded river may have a swift current. During a flood, a dry wash in the desert can temporarily become a raging, swiftly flowing river.

⚠️ **SAFETY TIP**

Wearing structural firefighting protective gear while conducting water rescue is dangerous and can be deadly. Do not wear structural firefighting protective gear within 3 m (10 ft) of water, especially around swiftly moving water.

In North America, a common swiftwater rescue scenario involves a car that has tried to drive through a pool of water created by a storm. When the vehicle stalls because of the depth of the water, it leaves the vehicle occupants stranded in a rising stream with a swift current. If the water is high enough, the vehicle can be swept away. These incidents are especially dangerous for rescue personnel because it is difficult to determine the depth of the water around the vehicle.

Ropes can be used in a variety of ways during water rescue operations. The simplest situation involves a rescuer on the shore throwing a rope to a person in the water and pulling that individual to shore. A more complicated situation could involve a rope stretched across a stream or river (**FIGURE 25-18**). A boat might be tethered

FIRE FIGHTER TIP

Using ropes and knots is challenging in the best of circumstances and should be considered a "perishable skill." To be proficient and to work in an effective and efficient manner when tying knots and moving equipment, you should constantly work on your rope and knot skills. Keep a short piece of rope close at hand so that you can practise your skills during down times.

to the rope, and rescuers on shore might maneuver the boat using a series of ropes and pulleys.

Fire departments should be prepared for all types of water rescue incidents that may occur in their communities. Fixed bodies of water in your community are easy to identify. Identifying the areas in your community where flooding is likely to occur will require preplanning. Other common water rescue scenarios include cars rolling into lakes or streams, cars going off bridges, people falling into bodies of water, and swimmers getting into trouble.

Safe Approach

When responding to water rescue incidents, your safety and the safety of other responders are the primary concern. Your structural firefighting protective gear is designed for structural firefighting, and it provides amazing protection for that purpose. Unfortunately, it is not designed for water rescue activities. Fire fighters who fall into the water while wearing structural firefighting protective gear quickly find that their movements are severely limited by the bulky gear. For this reason, when working at a water rescue scene, you should use PPE designed for water rescue, not structural firefighting. Anytime you are within 3 m (10 ft) of the water, you should be wearing a Coast Guard–approved PFD. Shoes that provide solid traction are preferred to fire boots in this situation.

If you are part of the first-arriving pump or ladder company and the endangered people are in a vehicle or holding on to a tree or other solid object, try to communicate with them. Let them know that additional help is on the way.

Do not exceed your level of training. If you cannot swim, operating around or near water is not recommended. A person who is trained as a lifeguard for still water is not prepared to enter flowing water with a strong current, such as is found in a river, stream, or ocean. Trained swiftwater rescuers may consider using a throw rope, a pike pole, or a ladder to reach a person. Many people who are in trouble in the water are close to shore. Do not use a boat unless you are trained in its use. Ensure that bystanders do not try to rescue the victim and place themselves in a situation where they need to be rescued, as well.

Ice rescues are common in colder climates, and many departments in colder regions have developed specialized equipment for these scenarios. Throwing a rope or flotation device may be helpful, initially. If the body of water is small enough, two fire fighters can slide a rope across the ice to the victim by walking around the shore. It may be necessary to tie ropes together if one is not long enough. Ladders can be extended to the victim or used to distribute the weight of the rescuer on ice-covered water. A fire hose that has been capped at each end and then inflated with an SCBA air cylinder can be used to create a flotation buoy. In addition, many other devices can be used to keep a victim above the ice and pull him or her to safety. Rescuers in this situation need special ice rescue suits to prevent hypothermia and provide flotation. If your department is involved in ice rescues, you need to receive training in these specialized procedures.

FIRE FIGHTER TIP

All fire fighters should be trained in the use of water rescue throw-lines. This rope is designed to be kept in a special throw bag that can be thrown to a struggling victim in water or on ice. The rescuer secures one end of the rope before throwing the bag. Once the victim has grasped the bag, the rescuer can pull the victim to shore.

How You Can Assist

Water and ice rescues are dangerous for rescuers. Do not attempt to go onto the ice to rescue victims without special training and equipment. Do not wear firefighting PPE when operating near ice or water. PFDs should be worn by everyone operating at the scene. Good scene control is needed to prevent additional people from becoming victims. Call for the special rescue team to respond, establish a safe perimeter, and maintain visual and verbal contact with the victim.

At water and ice rescues, you may be called upon to assist by keeping victims in sight, retrieving equipment, assisting with rescue ropes, and changing rescuers' air supplies. You can also assist special rescue teams by relaying communications.

Be alert for changing weather conditions such as an increase in wind speed or a change in temperature that could affect the rescue operations.

Tools Used

Fire fighters should become familiar with the tools used in water rescue operations:

- PFD
- Self-contained underwater breathing apparatus (SCUBA)
- Helmet
- Waterproof whistle
- Water rescue throwline
- Boat
- Rescue rope, webbing, and associated hardware
- Lighting and hand lights
- Gloves
- Goggles or other eye protection
- Suitable footwear for the environmental and rescue conditions

Wilderness Search and Rescue

Wilderness search and rescue (SAR) is conducted by a limited number of fire departments. It is included in NFPA 1670, *Standard on Operations and Training for Technical Search and Rescue Incidents*, as part of technical rescue training. SAR missions consist of two parts: search and rescue. Search is defined as looking for a lost or overdue person. Rescue, in the SAR context, is defined as removing a victim from a hostile environment.

Several types of situations may result in the initiation of SAR missions. For example, small children may wander off and be unable to find their way back to a known place. Older adults who are suffering from conditions such as Alzheimer's disease may fail to remember where they are going and become lost. People who are hiking, hunting, or participating in other wilderness activities may become lost because they do not have the proper training or equipment for that activity, because the weather changes unexpectedly, or because they become sick or injured.

Safe Approach

As a beginning fire fighter, you may respond to a possible SAR mission as part of your fire company. In cases of a lost person, some small fire departments will request all available personnel to respond to assist with a search. It is important to respond to such a call as part of your department and to work together as a team. The IC for such a mission should be a person who is well trained in directing SAR missions. In some communities, SAR is the responsibility of law enforcement personnel. In others, it becomes the responsibility of volunteer search and rescue groups.

The term "wilderness" may be used to describe a variety of environments, such as forests, mountains, deserts, parks, and animal refuges. Depending on the terrain and environmental factors, the wilderness can be as little as a few minutes into the backcountry or a few feet off the roadway. An incident with a short access time could require an extended evacuation and, therefore, qualify as a wilderness incident. Examples of terrain found in wilderness areas include cliffs, steep slopes, rivers, streams, valleys, mountainsides, and beaches. Terrain hazards may include cliffs, caves, wells, mines, avalanches, and rock slides.

When you participate in SAR missions, prepare for the weather conditions by bringing suitable clothing. Make sure that you do not exceed your physical limitations, and do not get in situations that are beyond your ability to handle in the wilderness. Call for a special wilderness rescue team, depending on the needs of the situation and your local protocols.

How You Can Assist

The fire service teaches its members to work in a buddy system; wilderness rescues are another situation in which you never go out alone. By working in teams of at least two and having a radio, teams can be methodically deployed and assigned to a search. The more knowledge you have about the search area, the more vital your role is during the search. Knowing your response area and the places where a child or lost person might hide is beneficial to the search. A well-coordinated team can be an effective SAR force; conversely, an unorganized group of people will produce increasing chaos. Do not enter the search area before the search team arrives. If the team will be using dogs, your scent will distract them.

Tools Used

Each specialized rescue team will bring its own equipment to the scene. Many of the same tools, equipment, and personal protective clothing that you would use for other specialized rescue situations will be used in SAR emergencies:

- Personal clothing appropriate for the environment
- Water
- Food
- Lighting and hand lights
- Thermal imaging device
- Communications equipment, including radios
- Medical equipment
- Maps, compass, and GPS
- Shelter
- Rescue rope, harnesses, webbing, and associated hardware

- Extrication equipment such as a Stokes basket and harness
- Flare gun and flares, whistles, or other signaling devices

Hazardous Materials Incidents

Hazardous materials are defined as any materials or substances that pose a significant risk to the health and safety of persons or to the environment if they are not properly handled during manufacture, processing, packaging, transportation, storage, use, or disposal (**FIGURE 25-19**).

Although hazardous materials incidents often involve petroleum products, many other chemicals found in our communities may have toxic effects when they are not handled properly. We tend to think of hazardous materials incidents as occurring during transportation or at a large industrial setting, but it is important to consider that many retail businesses contain significant quantities of hazardous materials. Home and garden stores, farm cooperatives, hardware stores, and pool supply stores, for example, are places where hazardous materials can be found in significant quantities.

Most fire departments are trained and equipped to recognize these incidents, contain the hazards, and evacuate people from the area if necessary. Refer to your department's hazardous materials SOPs for specific response information.

Safe Approach

Hazardous materials incidents are not always dispatched as hazardous materials incidents. Given this fact, you must be able to recognize the signs indicating that hazardous materials may be present as you approach

FIGURE 25-19 Hazardous materials incident.
Courtesy of Photographer's Mate 2nd Class Daniel R. Mennuto/U.S. Navy.

the scene. Preincident plans may note that a fixed facility stores chemicals, or the use of a specific type of transport container may tip you off to the presence of hazardous materials. You might also see an escaping chemical or smell a suspicious odour. Warning **placards** are required for hazardous materials that are stored or transported.

Once you have recognized the presence of a hazardous material, you must protect yourself by staying out of the area exposed to the material. If you have happened upon a hazardous materials incident, it is important to have the hazardous materials team dispatched as soon as possible. In addition to standard action and precautions, call for a special hazardous materials rescue team immediately upon your arrival at such an incident, implement site control and scene management, and assist specialized personnel after their arrival according to your training level.

How You Can Assist

To assist in operations at a hazardous materials incident, you need formal training in techniques to deal with this kind of threat. Training at the awareness level will provide you with the knowledge and skills you need to be able to recognize the presence of a hazardous material, protect yourself, call for appropriate assistance, and evacuate or secure the affected area. As part of your training, you will learn how to assist other responders to the hazardous materials incident.

The four major objectives of training at the operational level are to analyze the magnitude of the hazardous materials incident, plan an initial response, implement the planned response, and evaluate the progress of the actions taken to mitigate the incident. Being trained and equipped to perform emergency decontamination of victims may help to minimize the effects of chemical exposures while hazardous materials technicians are en route.

Tools Used

The following tools are used in a hazardous materials incident:

- PPE appropriate to the level of the hazard
- Two-way radios
- Lighting
- Gas monitors
- Sensing and monitoring tools to identify the materials involved
- Extensive research material
- Decontamination equipment
- A wide variety of hand tools (e.g., hammers, screwdrivers, axes, wrenches, pliers)
- Devices for sealing breached containers
- Leak-control devices

- Binoculars
- Fireline tape
- Control agents

Elevator and Escalator Rescue

Elevator and escalator rescues are technical responses that require specialized knowledge. Rescue crews should review the Canadian Standards Association Code (CSA B-44), the North American Elevator Code, and the Canadian Centre for Occupational Health and Safety (CCOHS) for lockout and tagout procedures in Canada.

All elevator and escalator responses begin with knowledge of the machinery found in your department's jurisdiction—information that can be gathered only through a preincident analysis of these structures. Specialized training classes are an essential component of safe and successful operations.

Elevator Rescue

As metropolitan areas continue to grow, building heights are also increasing. Because of our aging population, more buildings are being equipped with elevators. In turn, elevator emergencies are increasing in frequency (**FIGURE 25-20**). Fire fighters are to follow their departmental regulations regarding elevator rescues.

Fire fighters should never attempt to move or relocate an elevator under any circumstances. Only professional elevator technicians who are thoroughly trained and authorized to do so should consider this step.

Escalator Rescue

Escalator emergencies can be challenging and complex. Fire fighters should never attempt to move an escalator under any circumstances—only professional escalator technicians who are thoroughly trained and authorized to do so should consider this step. This is particularly true if a victim (usually a child) is caught or entangled in the machinery.

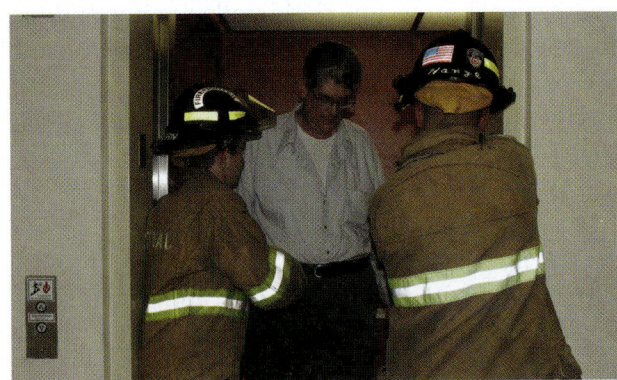

FIGURE 25-20 Elevator emergencies are a common emergency call for many fire departments.
© Dennis Wetherhold, Jr.

Good incident risk management and risk assessment skills are crucial in this type of incident because an error in judgment or a poorly executed reaction on the part of rescue personnel can easily turn an abrasion, laceration, or fractured bone into an amputated limb.

Victims who are caught in machinery must be reassured, calmed, and stabilized medically by the rescuers. Fire fighters are to follow their departmental regulations for an escalator rescue.

Industrial Rescue

Rescue operations at industrial facilities may involve specialized vehicles and equipment, confined spaces, high-voltage electrical equipment, hazardous materials, or high-angle rescue situations. Preincident planning should include possible rescue incidents. Train with your special rescue team to be prepared for emergencies that might occur at these facilities.

Active Shooter Incidents

Responders must be aware of the threats posed by firearms. The FBI reported 50 active shooter situations in the United States in 2016 and 2017 (FBI.gov). These incidents resulted in 943 casualties (221 deaths and 722 wounded). Active shooters are defined as one or more individuals actively engaged in harming, killing, or attempting to kill people in a populated area with the use of a firearm (NFPA, 2018). History shows that many of these incidents occur in buildings where people gather in larger numbers such as shopping malls, churches, concert venues, and schools or during public gatherings during open areas events such as parades, festivals, and sporting events.

In May of 2018, the National Fire Protection Association (NFPA) released the world's first active shooter/hostile event standard. NFPA 3000™ (PS), *Standard for an Active Shooter Hostile Event Response (ASHER) Program*, provides unified planning and response and recovery guidance, as well as civilian and responder safety considerations. NFPA 3000™ (PS) is a provisional standard (PS). Provisional standards are developed in an expedited process to address an emergency situation or other special circumstance. This is only the second time in 122 years that the NFPA has released a provisional standard.

Today, responders must recognize that they are a part of a much larger response mechanism in Canada. Much has changed in the years since the Parliament Hill and Moncton shootings in 2014. Police procedures have changed, and the weaponry carried has been upgraded. It is important for fire fighters to realize that active shooter events are the responsibility of the local police, provincial police, or Royal Canadian Mounted Police, depending on the jurisdiction. Fire and paramedic services provide a support role. IC will reside with the authority having jurisdiction. Any incident involving a hostile event such as an active shooter situation will require the assistance and cooperation of countless resources. The coordination and success of this response mechanism require written and practised protocols. Responders within every jurisdiction and response agency should think about the locations of potential active shooter targets in their response area, maintain good situational awareness, and consider changing conditions and the possibility of additional shooters or planted explosive devices.

In the absence of local active shooter protocols, fire department responders should approach reported active shooter situations cautiously and stage a safe distance away until direction is provided by law enforcement. Responders should don ballistic PPE if provided. Pump and ladder company fire fighters should be prepared to fight fire once the area has been secured by law enforcement. In mass casualty situations, fire fighters may be asked to forego fire suppression tactics to provide immediate rescue/apparatus shielding of victims and to utilize techniques for rapid bleeding control.

In some cases, emergency responders might be dispatched to a typical incident (fire alarm activation, EMS call, or traffic accident) only to discover an active shooter. Emergency responders must take immediate action to protect themselves from the active shooter. Take cover behind buildings, apparatus, or any other object that provides protection. Request immediate law enforcement assistance to secure the area before attempting to rescue victims. Once the area is secure, the goal is to help victims and the uninjured retreat to a protected area. Always follow your local protocols.

Tools and Equipment

Power tools and equipment are used for lighting, ventilation, salvage, and overhaul. For example, powerful emergency scene lighting equipment requires a separate source of electricity, often supplied by a generator. As a Fire Fighter II, you may be responsible for cleaning, maintaining, and testing this equipment to ensure that it works properly at an emergency scene. This section discusses generators, lighting equipment, and power tools, including the use, cleaning, and maintenance of this equipment. As discussed in Chapter 8, *Fire Fighter Tools and Equipment*, all tools and equipment must be properly maintained so that they will be ready for use when they are needed. Use power tools and equipment only after you have received training in their use. Read and follow the instructions supplied by the manufacturer.

Lighting Equipment and Electrical Generators

Powerful lighting equipment requires a separate source of electricity, generally a 120-volt alternating current (AC) delivered through power cords. The electricity for lighting equipment can be supplied by a **generator**, an **inverter**, or a building's electrical system, if the power has not been interrupted.

Power inverters convert 12-volt direct current (DC) from a vehicle's electrical system to 120-volt AC power. An inverter can provide a limited amount of AC current and is typically used to power one or two lights mounted directly on the apparatus. Some vehicles have additional outlets for operating portable lights or small electrically powered tools. Most power inverters, however, do not produce enough power to operate high-intensity lighting equipment, large power tools, or ventilation fans. Be aware that connecting devices that draw too much current to an inverter can seriously damage it.

Electrical generators are powered by gasoline or diesel engines; they can be either portable or permanently mounted on fire apparatus. Portable generators are small enough to be removed from the apparatus and carried to the fire scene. These devices come in various sizes and can produce as much as 6 kilowatts (6000 watts) of power. A portable generator has a small gasoline- or diesel-powered motor with its own fuel tank.

Apparatus-mounted generators have much larger power capacities, sometimes exceeding 20 kilowatts (20,000 watts). The smaller units, which are similar to portable generators, can be permanently mounted in a compartment on the apparatus. By contrast, larger generators are usually mounted directly on the vehicle, often have diesel engines, and draw fuel directly from the vehicle's fuel tank. Permanently mounted generators can also be powered by the apparatus through a power take-off and a hydraulic pump.

Another potential power source for lighting is a building's normal power supply. It is usually not an option if a serious fire has occurred in the building because the power will be disconnected for safety reasons. In contrast, if the fire is relatively minor, the power might not have to be interrupted and can be used for lighting. Alternatively, power can sometimes be obtained from a nearby building. When drawing power from a building, be aware that you might not know the capacity of the circuit from which you are obtaining electricity. Do not overload a household electrical circuit.

Cleaning and Maintenance of Electrical Equipment

It is important to properly maintain lighting equipment and generators to keep them in good operating condition. To clean lighting equipment, follow the manufacturer's instructions for cleaning. In general, the directions specify how to clean the floodlights and search lights with a mild soap or detergent. Always use a soft cloth to clean and polish the lens. Avoid the use of strong solvents, as they may damage plastic or rubber components.

When cleaning and maintaining a generator, follow the manufacturer's instructions. Avoid overfilling the fuel tank on portable generators. When a fuel tank is too full, it can overflow onto the hot engine and ignite. An overfull tank can also leak fuel as the fuel expands, especially if the tank was filled in a cool environment and then brought into a warmer place. Spilled fuel can damage paint under certain circumstances.

Clean the external parts of the generator with a mild soap or detergent as recommended by the manufacturer. Remove any dirt, vegetation, or charred materials on the device. Avoid the use of strong solvents, as they can damage rubber and plastic parts of the generator and can remove paint.

If a solvent is recommended by the manufacturer for any cleaning operation, follow the directions for its use on the product label. Also read the safety data sheet before working with any solvent. Check whether the product is flammable. If it is, be sure to follow the safety recommendations for its use. Always use flammable solvents in a well-ventilated area that is not close to any flame or heat source.

Test and run generators on a weekly or monthly basis to confirm that they will start, run smoothly, and produce power. Run gasoline-powered generators for 15 to 30 minutes to reduce any deposit build-up that could foul the spark plugs and make the generator hard to start.

SAFETY TIP

Never run a portable generator in a confined space or an area without adequate ventilation unless you and everyone else in the area are wearing SCBA. The exhaust from the generator contains carbon monoxide, which can kill you.

SAFETY TIP

Solvents can irritate or burn the skin, eyes, and lungs. Some of these agents are known to cause cancer and should be avoided. Solvents can be absorbed through the lungs as well as through the skin. For solvents that can be absorbed through skin, check whether chemical-resistant gloves are required during their use. Also check whether eye protection or respiratory protection is required.

At the same time, inspect and test other electrical equipment on the apparatus. Take junction boxes out, and check them for cracked covers or broken outlets. Plug them in to make sure the ground-fault circuit interrupter (GFCI) works properly. Check power cords for tears in the protective covering, mechanical damage, fraying, heat damage, or burns. Inspect plugs for loose or bent prongs. Inspect and test the operation of all power tools and equipment on the apparatus. Check lighting fixtures to confirm that they are in good repair and that the bulbs are working.

After shutting down the generator, refill the fuel tank. Follow the manufacturer's recommendations for generator engine maintenance, including regular oil changes. To conduct a weekly/monthly generator test, follow the steps in **SKILL DRILL 25-3**.

Power Tools and Equipment

Some rescue incidents require heavy machinery to move dirt or large sections of collapsed buildings, but special rescue teams often carry smaller tools and equipment for more precise rescue operations. Along with hand tools, powered tools are used to displace, bend, and distort metal to gain access to and to disentangle a victim from a vehicle or machinery. Hydraulic rams and hydraulic spreaders can exert a large amount of force to displace the vehicle or machinery so that the victim can be removed. Powered hydraulic rams and powered hydraulic spreaders use a hydraulic pump powered by an electric motor or a gasoline engine.

A variety of power tools for cutting or severing are also available—some are air powered (pneumatic), some are electrically powered, some are fuel powered, and others are hydraulically powered. Air chisels are powered by pressurized air from portable SCBA air cylinders or from a cascade system mounted on a vehicle.

Several types of powered saws are used for special rescue operations. Some electric saws require power cords, whereas others operate on battery packs. Powered saws produce different types of motion. For example, reciprocating saws move back and forth very quickly, whereas rotary saws move in a circular motion. Whenever you use power and hydraulic tools, cover the victim for protection.

Cleaning and Maintenance of Power Tools and Equipment

Test power tools and equipment frequently, and have these items serviced regularly by a qualified shop. Keep records of all inspections and maintenance performed on power tools and equipment. Preventive maintenance will help to ensure that tools and equipment operate properly when needed. If a power tool or piece of equipment is defective, report this fact to your officer.

Fill each tool with the proper fuel, remembering that some tools operate on gasoline whereas others use various gasoline and oil mixtures. Many small engines operate on gasoline, for example, but others are designed to burn diesel fuel. Fuels have a limited storage life, especially fuels containing ethanol. If the fuel is not used within a certain period of time, it may be necessary to drain and refill equipment with fresh fuel. Check the manufacturer's recommendations before fuelling a particular piece of equipment.

After returning from a fire, clean, inspect, and record maintenance data for all of your overhaul tools to ensure that they are in a "ready state."

All power equipment should be left in a "ready state" for immediate use at the next incident:

- All debris should be removed, and the tools should be clean and dry.
- All fuel tanks should be filled completely with fresh fuel.
- Any dull or damaged blades and chains should be replaced.
- Belts should be inspected to ensure that they are tight and undamaged.
- All guards should be securely in place.
- All hydraulic hose should be cleaned and inspected.
- All power cords should be inspected for damage.
- All hose fittings should be cleaned, inspected, and tested to ensure a tight fit.
- Tools should be started to ensure that they operate properly.
- Tanks on water vacuums should be emptied, washed, cleaned, and dried.
- The hose and nozzles on wet vacuums should be cleaned and dried.

It is essential to read the manufacturer-provided manuals and follow all instructions on the care and inspection of power tools and equipment. Keep all manufacturers' manuals in a safe and easily accessible location, and refer to these documents when cleaning and inspecting the tools and equipment.

Learn the proper procedure for reporting a problem with a power tool and taking it out of service. Remember—your safety depends on the quality of your tools and equipment. Every tool and piece of equipment must be ready for use before you respond to an emergency incident. Always follow the manufacturer's instructions for cleaning and maintaining each piece of equipment.

SKILL DRILL 25-3
Conducting a Weekly/Monthly Generator Test Fire Fighter II, NFPA 1001: 5.5.4

1 Remove the generator from the apparatus compartment, or open all doors as needed for ventilation. Install the grounding rod, if needed.

2 Check the oil and fuel levels, and start the generator. Connect the power cord or junction box to the generator, connect a load such as a fan or lights, and make sure the generator attains the proper speed. Check the voltage and amperage gauges to confirm efficient operation.

3 Run the generator under load for 15 to 30 minutes. Turn off the load, and listen as the generator slows down to idle speed. Allow the generator to idle for approximately 2 minutes before turning it off. Disconnect all power cords and junction boxes; clean all power cords, plugs, adaptors, GFCIs, and tools; and replace them in proper storage areas. Allow the generator to cool for 5 minutes. Refill the generator with fuel and oil, as needed, and return the generator to its compartment. Fill out the appropriate paperwork.

After-Action REVIEW

IN SUMMARY

- A technical rescue incident (TRI) is a complex rescue incident involving vehicles or machinery, water or ice, rope techniques, trench or excavation collapse, confined spaces, structural collapse, cave and tunnel rescue, wilderness rescue, utility hazards, or hazardous materials that requires specially trained personnel and special equipment.

- Training in technical rescue areas is conducted at three levels: awareness, operations, and technician.

- To become proficient in handling these types of situations, you must take a formal course to gain specialized knowledge and skills.

- Tool staging often involves placing a tarp or salvage cover on the ground at a designated location and laying out commonly used tools and equipment so that they can be readily accessed.

- During technical rescue incidents, several hardware components are used in addition to rope to rescue victims. Two common components are:
  - *Carabiner.* Connects one rope to another rope or a harness
  - *Harness.* Webbing that secures a person to a rope or solid object

- Most rope rescue incidents involve people who are trapped in normally inaccessible locations such as trenches, confined spaces, and open water.

- When assisting rescue team members, keep five guidelines in mind:
  - Be safe.
  - Follow orders.
  - Work as a team.
  - Think.
  - Follow the Golden Rule of public service: Treat others as you would like to be treated.

- The basic steps of special rescue operations are outlined here:
  - Preparation—Train with the specialized rescue teams in your area, and become familiar with the terminology used in the field.
  - Response—A technical rescue team will usually respond with a rescue unit supported by other units such as a medic unit, a pump company, a ladder company, and a chief.
  - Arrival and size-up—The first company officer to arrive on the scene should assume command and rapidly assess the situation. Do not rush into the incident scene until the situation has been assessed and the hazards identified.
  - Stabilization—Once the needed resources are on the way and the scene is safe to enter, begin to stabilize the scene by establishing a perimeter around the rescue site.
  - Access—Identify the actual reason for the rescue, and work toward freeing the victim safely. Communicate with the victim at all times, and initiate emergency medical care as soon as safe access is made to the victim.
  - Disentanglement—Once precautions have been taken and the reason for entrapment has been identified, the victim needs to be freed as safely as possible. Disentanglement removes what is confining the victim from around the victim.
  - Removal—This step could be as simple as assisting a victim up a ladder or as complex as packaging the victim in a Stokes basket and lifting him or her out of a trench.
  - Transport—Once the victim has been removed from the hazard area, transport to an appropriate medical facility is accomplished by EMS personnel.
  - Security of the scene and preparation for the next call—Once the victim has been transported, the scene must be stabilized by the rescue crew to ensure that no one else becomes injured.
  - Postincident analysis—Identify what worked well and which procedures could work better.

- Beginning with the initial dispatch of the rescue call, begin compiling facts and identifying factors pertinent to the call. The information acquired when an emergency call is received is important to the success of the rescue operation. The information should include the following details:
  - The location of the incident
  - The nature of the incident (kinds and number of vehicles)
  - The conditions and positions of victims
  - The conditions and positions of vehicles
  - The number of people trapped or injured and the types of their injuries
  - Any specific or special hazard information
  - The name of the person calling and a number where that person can be reached
- A size-up should include the initial and continuous evaluation of the following issues:
  - Scope and magnitude of the incident
  - Risk/benefit analysis
  - Number of known and potential victims
  - Hazards
  - Access to the scene
  - Environmental factors
  - Available and necessary resources
  - Establishment of control perimeters
- After sizing up the scene, take the following precautions to avoid further injuries to the victims and to provide for the safety of other fire fighters:
  - Secure utility hazards.
  - Provide scene security.
  - Use the proper protective equipment for the emergency.
  - Activate the incident command system.
  - Activate the personnel accountability system.
  - Make contact with the victim, and keep the victim calm.
- At any type of TRI, follow the orders of the company officer who receives direction from the IC. Many of the tasks assigned to fire fighters will involve moving equipment and objects from one place to another; others will involve protecting the team and victims.
- Vehicle and machinery rescues occur in many settings. These situations require responders to stabilize the machinery and ensure that the electricity is off.
- With a confined space, rescuers must ensure that a pure and adequate air supply is available before entering the confined spaces.
- Low-angle and high-angle rope rescues require safe equipment, adequate training, and considerable caution.
- Trench and excavation rescues are hazardous and require responders to minimize the chance for secondary collapses.
- A damaged building is prone to structural collapse. When any part of a building becomes compromised, the dynamics of the building change.
- Water rescue training is needed in almost all communities. Training, proper PFDs, and appropriate clothing are important to ensure rescuer safety.
- Wilderness rescue may be called for even when initial access to a lost or stranded individual occurs quickly.
- Hazardous materials incidents are not always dispatched as hazardous materials incidents, so be cautious when approaching any incident. Once the presence of a hazardous material is identified, protect yourself by staying out of the area exposed to the material.

- Never attempt to move or relocate an elevator under any circumstances. Always shut off the power to a malfunctioning elevator, and secure the power supply. Implement incident risk management and risk assessment. Always have additional trained personnel on hand, and anticipate obstacles and problems. Once the incident has been resolved, leave the power supply off.

- Escalator emergencies can be challenging and complex. Never attempt to move an escalator under any circumstances—only professional escalator technicians who are thoroughly trained and authorized to do so should consider this step.

- Electricity for lighting equipment is supplied by a generator, an inverter, or a building's electrical system, if the power has not been interrupted.

- An inverter can provide a limited amount of AC current and is typically used to power one or two lights mounted directly on the apparatus.

- Electrical generators are powered by gasoline or diesel engines; they can be either portable or permanently mounted on fire apparatus.

- Portable electrical equipment should be cleaned and properly maintained to ensure that it will work when needed. Test and run generators on a weekly or monthly basis to confirm that they will start, run smoothly, and produce power. Run gasoline-powered generators for 15 to 30 minutes to reduce any deposit build-up that could foul the spark plugs and make the generator hard to start.

- All tools and equipment must be properly maintained and left in a "ready state" for immediate use at the next incident:
  - Equipment should be cleaned, dried, inspected, and stored properly.
  - Fuel tanks should be filled.
  - Any dull or damaged blades and chains should be replaced.
  - Power equipment and tools should be tested and serviced frequently.

- Read and follow the manufacturer's manuals, and follow all instructions on the care and inspection of power tools and equipment.

KEY TERMS

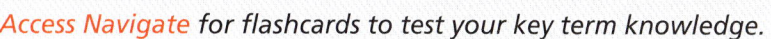

Access Navigate for flashcards to test your key term knowledge.

Carabiner An auxiliary equipment system item; load-bearing connector with a self-closing gate used to join other components of life safety rope. (NFPA 1983)

Cold zone The control zone of an incident that contains the command post and such other support functions as are deemed necessary to control the incident. (NFPA 1500)

Confined space An area large enough and so configured that a member can bodily enter and perform assigned work but which has limited or restricted means for entry and exit and is not designed for continuous human occupancy. (NFPA 1500)

Entrapment A condition in which a victim is trapped by debris, soil, or other material and is unable to extricate himself or herself.

Generator An electromechanical device for the production of electricity. (NFPA 1901)

Harness An equipment item; an arrangement of materials secured about the body used to support a person. (NFPA 1971)

Hazardous material Matter (solid, liquid, or gas) or energy that when released is capable of creating harm to people, the environment, and property, including weapons of mass destruction (WMD) as defined in 18 U.S. Code, Section 2332a, as well as any other criminal use of hazardous materials, such as illicit labs, environmental crimes, or industrial sabotage. (NFPA 1072)

High-angle operation A rope rescue operation where the angle of the slope is greater than 45 degrees. In this scenario, rescuers depend on life safety rope rather than a fixed support surface such as the ground.

Hot zone The control zone immediately surrounding a hazardous area, which extends far enough to prevent adverse effects to personnel outside the zone. (NFPA 1500)

Inverter Equipment that is used to change the voltage level or waveform, or both, of electrical energy. Commonly, an inverter [also known as a power conditioning unit (PCU) or power conversion system (PCS)] is a device that changes DC input to AC output. Inverters may also function as battery chargers that use alternating current from another source and convert it into direct current for charging batteries. (NFPA 70)

Life safety rope Rope dedicated solely for the purpose of supporting people during rescue, firefighting, other emergency operations, or during training evolutions. (NFPA 1983)

Lockout and tagout systems Methods of ensuring that electricity and other utilities have been shut down and switches are "locked" so that they cannot be switched on, so as to prevent flow of power or gases into the area where rescue is being conducted.

Low-angle operation A rope rescue operation on a mildly sloping surface (less than 45 degrees) or flat land. In this scenario, fire fighters depend on the ground for their primary support, and the rope system is a secondary means of support.

Packaging The process of securing a victim in a transfer device, with regard to existing and potential injuries or illness, so as to prevent further harm during movement. (NFPA 1006)

Personal flotation device (PFD) A device manufactured in accordance with U.S. Coast Guard specifications that provides supplemental flotation for persons in the water. (NFPA 1006)

Personnel accountability system A system that readily identifies both the locations and the functions of all members operating at an incident scene. (NFPA 1500)

Placards Signage required to be placed on all four sides of highway transport vehicles, railroad tank cars, and other forms of hazardous materials transportation that identifies the hazardous contents of the vehicle, using a standardized system: 273-mm by 273-mm (10¾-in. by 10¾-in.) diamond-shaped indicators.

Secondary collapse A subsequent collapse in a building or excavation. (NFPA 1006)

Shoring A structure such as a metal hydraulic, pneumatic/mechanical, or timber system that supports the sides of an excavation and is designed to prevent cave-ins.

Spoil pile A pile of excavated soil next to the excavation or trench. (NFPA 1006)

Supplied-air respirator An atmosphere-supplying respirator for which the source of breathing air is not designed to be carried by the user. Also known as an airline respirator. (NFPA 1989)

Technical rescue incident (TRI) Complex rescue incidents requiring specially trained personnel and special equipment to complete the mission. (NFPA 1670)

Technical rescue team A group of rescuers specially trained in the various disciplines of technical rescue.

Warm zone The control zone outside the hot zone where personnel and equipment decontamination and hot zone support take place. (NFPA 1500)

Wilderness search and rescue (SAR) The process of locating and removing a victim from the wilderness.

REFERENCES

Centers for Disease Control and Prevention. 1986, January. "Preventing Occupational Fatalities in Confined Spaces." NIOSH Publication, 86-110. https://www.cdc.gov/niosh/docs/86-110/. Accessed November 26, 2018.

Federal Bureau of Investigation (FBI). *Active Shooter Incidents in the United States in 2016 and 2017.* https://www.fbi.gov/file-repository/active-shooter-incidents-us-2016-2017.pdf/view. Accessed November 26, 2018.

National Fire Protection Association. NFPA 472, *Standard for Competence of Responders to Hazardous Materials/Weapons of Mass Destruction Incidents*, 2018. www.nfpa.org. Accessed November 26, 2018.

National Fire Protection Association. NFPA 1006, *Standard for Technical Rescue Personnel Professional Qualifications*, 2017. www.nfpa.org. Accessed November 26, 2018.

National Fire Protection Association. NFPA 1500, *Standard on Fire Department Occupational Safety and Health Program*, 2018. www.nfpa.org. Accessed November 26, 2018.

National Fire Protection Association. NFPA 1670, *Standard on Operations and Training for Technical Search and Rescue Incidents*, 2017. www.nfpa.org. Accessed November 26, 2018.

National Fire Protection Association. NFPA 1951, *Standard on Protective Ensembles for Technical Rescue Incidents*, 2013. www.nfpa.org. Accessed November 26, 2018.

National Fire Protection Association. NFPA 1983, *Standard on Life Safety Rope and Equipment for Emergency Services*, 2017. www.nfpa.org. Accessed November 26, 2018.

National Fire Protection Association. NFPA 3000, *Standard for an Active Shooter Hostile Event Response (ASHER) Program*, 2018. www.nfpa.org. Accessed November 26, 2018.

National Fire Protection Association. *NFPA releases the world's first active shooter/hostile event standard with guidance on whole community planning, response, and recovery*, 2018. https://www.nfpa.org/News-and-Research/News-and-media/Press-Room/News-releases/2018/NFPA-releases-the-worlds-first-active-shooter-hostile-event-standard. Accessed November 26, 2018.

On Scene

It has been a busy shift for your crew. You have already responded to several medical calls, a motor vehicle accident with three patients, and a small residential fire. You have just cleaned up the equipment and loaded the last section of hose when you are dispatched on another call. A building under renovation has partially collapsed, and there are reports of multiple injuries. When your pumper turns the corner, you see a pile of rubble that was once Side B of a three-storey Type III building. Several workers are outside; some of them have obvious injuries. As you consider the many challenges that your crew might face with this type of incident, the foreman tells you that eight of his workers are missing.

1. Which of the following is not likely to be used at a structural collapse technical rescue incident (TRI)?

- **A.** Shoring equipment
- **B.** Personal flotation device (PFD)
- **C.** Utility rope
- **D.** Shovel

2. What is the major safety concern while working at a partial structural collapse TRI?

- **A.** The collapsed portion, which may burn quickly
- **B.** Deep water in the basement
- **C.** Electrical hazards
- **D.** The potential for a secondary collapse

3. Which TRI area is restricted to entry teams and special rescue teams?

- **A.** The hot zone
- **B.** The warm zone
- **C.** The cold zone
- **D.** The hazard area

4. What is the best way to prevent electrical shock hazards and accidental operation of electrical equipment during rescue operations?

- **A.** Remove the equipment.
- **B.** Wear insulated PPE.
- **C.** Use a lockout/tagout system.
- **D.** Pull the electrical meter on the building.

5. What must be used at all TRI incidents?

- **A.** SCBA
- **B.** Utility rope
- **C.** Lockout/tagout procedures
- **D.** Personnel accountability system

6. Which of the following is not the responsibility of the Fire Fighter II at a TRI?

- **A.** Entering a confined space to remove victims
- **B.** Assisting with controlling site security and the perimeter of the rescue incident
- **C.** Retrieving and setting up power tools
- **D.** Extinguishing any fires that may ignite

(continued)

On Scene Continued

7. The first incident priority for TRIs is:

A. size-up.

B. life safety.

C. incident stabilization.

D. property conservation.

8. The last step of a TRI is:

A. patient packaging and transport.

B. rehab for rescuers.

C. postincident analysis.

D. cleaning and refuelling all equipment.

Access Navigate to find answers to this On Scene, along with other resources such as an audiobook.

Fire Detection, Suppression, and Smoke Control Systems

KNOWLEDGE OBJECTIVES

After studying this chapter, you will be able to:

- Describe the basic components and functions of a fire alarm system. (**NFPA 1001: 5.5.3**, pp. 944–953)
- Describe the fire department's role in resetting fire alarms. (**NFPA 1001: 5.5.3**, pp. 944–945)
- Describe the basic types of fire alarm initiating devices, and indicate where each type is most suitable. (**NFPA 1001: 5.5.3**, pp. 945–953)
- Describe the basic types of alarm notification appliances. (pp. 953–954)
- Describe the basic types of fire alarm annunciator systems. (p. 954)
- Explain the different ways that fire alarms may be transmitted to the fire department. (**NFPA 1001: 5.5.3**, pp. 954–956)
- Identify the four types of sprinkler heads. (**NFPA 1001: 5.5.3**, pp. 957–960)
- Identify the different styles of indicating valves. (**NFPA 1001: 5.5.3**, pp. 961–963)
- Describe the operation and application of the following types of automatic sprinkler systems (**NFPA 1001: 5.5.3**, pp. 965–967):
  - Wet pipe system
  - Dry pipe system
  - Preaction system
  - Deluge system
  - Water mist systems
- Describe the differences between commercial and residential sprinkler systems. (**NFPA 1001: 5.5.3**, pp. 965–969)
- Identify the three classifications of standpipes, and point out the differences among them. (**NFPA 1001: 5.5.3**, pp. 969–970)
- Describe two problems that fire fighters could encounter when using a standpipe in a high-rise building. (**NFPA 1001: 5.5.3**, pp. 970–971)
- Identify the hazards fire fighters might encounter when responding to incidents involving special hazard suppression systems. (**NFPA 1001: 5.5.3**, pp. 971–974)
- Describe the operation and application of smoke control systems. (**NFPA 1001: 5.5.3**, pp. 974–975)

SKILLS OBJECTIVES

- Identify the components of a fire suppression and detection system. (**NFPA 1001: 5.5.3**, pp. 943–974)

ADDITIONAL STANDARDS

- **NFPA 72**: *National Fire Alarm and Signaling Code*
- **NFPA 92**: *Standard for Smoke Control Systems*

Fire Alarm

Your crew is updating the department's preincident plan for a large industrial facility in your city. The building's safety officer is showing you the state-of-the-art fire alarm system and suppression systems. The fire alarm system includes both manual and automatic initiating devices, as well as a comprehensive emergency notification system. The building is fully sprinklered, and a clean agent suppression system protects the large computer room. Class III standpipes are located strategically throughout the building. Although the facility uses a lot of hazardous processes in its operations, you feel that the fire protection systems provide a high level of protection for both the occupants and responding fire fighters if an emergency occurs.

1. What information should be included in the preincident plan?

2. What are the advantages of a Class III standpipe system? Are there any disadvantages?

3. What safety measures should responding fire fighters take if the clean agent system activates?

 *Access Navigate for more practice activities.*

Introduction

Modern fire prevention and building codes require most new structures to have some sort of fire protection system installed. This requirement makes it more important than ever for fire fighters to have a working knowledge of these systems. **Fire protection systems** include fire alarm, automatic fire detection, fire suppression, and smoke control systems. Understanding how these systems operate is important for the fire fighter's safety and necessary to provide effective customer service to the public.

From a safety standpoint, fire fighters need to understand the operational capabilities and limitations of fire protection systems. Fire fighters need to know how to fight a fire in a building with a working fire protection system and how to shut down the system when directed to do so by the incident commander (IC).

The public expects that fire fighters will have more than a basic understanding of how fire protection systems operate. In Canada most jurisdictions have enacted laws making smoke detectors and carbon monoxide detectors mandatory for all dwelling units. Fire prevention programs across the country have made early detection and evacuation a priority for the past several years. Fire fighters play an important role with regard to public fire safety education. As the most visible face of the fire service, fire fighters must maintain consistent fire safety messaging. Every false alarm is an opportunity to explain how false alarms occur and how they can be prevented. Unfortunately, frequent false alarm activations can lead to resident complacency, which can imperil those who choose to ignore evacuation procedures. Most people have no idea how the fire protection systems in their building work, and there are often more nuisance alarms in buildings with fire protection systems than actual fires. Fire fighters can help the owners and occupants determine what activated the system, how they can prevent future nuisance alarms, and what needs to be done to restore a system to service.

The term "false alarm" is frequently used by the public to describe all **fire alarm** activations that are not associated with a true emergency, even though, in some of these situations, the alarm system operated properly. NFPA 72, *National Fire Alarm and Signaling Code*, uses the terms nuisance alarm or unwanted alarm to describe an activation of a fire alarm that is not the result of a potentially hazardous condition. Nuisance alarms may be caused by improperly installed or maintained systems, unintentional conditions that activate the alarm, or through malicious, intentional acts. Occupants may become complacent if the fire alarm sounds frequently, and they may neglect to take appropriate action in a real emergency. Such was the case in January of 2000, when three students died in their dormitory rooms at Seton Hall University. The students ignored the alarm, which signaled a real fire emergency, because they had become used to hearing nuisance alarms on a regular basis. A fire alarm signal should be always treated as a true emergency until someone from the fire department arrives on the scene and indicates otherwise. This chapter discusses the components and basic operation of fire alarm, fire detection, fire suppression, and smoke control systems. Most of North America follows the applicable National Fire Protection Association (NFPA) codes and standards for these systems; however, there are some provincial and local jurisdictions that amend the requirements.

In Canada, the National Building Code establishes fire safety requirements for new construction, while the National Fire Code establishes maintenance of fire safety requirements in existing buildings. Both codes reference applicable NFPA codes and standards for fire detection, suppression, and smoke control systems.

Fire Alarm and Detection Systems

Practically all new construction includes some sort of fire alarm requirement. In most commercial occupancies, the fire detection and fire alarm components are integrated into a single system. A fire detection system recognizes when a fire is occurring and activates the fire alarm system, which alerts the building occupants. Some fire detection systems also automatically activate fire suppression systems to control or extinguish the fire.

Fire alarm and detection systems range from simple, single-station smoke alarms for private homes to complex fire detection and control systems for high-rise buildings. Many fire alarm and detection systems in large buildings also control other fire protection and building systems to help protect occupants and control the spread of fire and smoke. Although these systems can be complex, they generally include the same basic components as the simpler versions.

Residential Fire Alarms

Current building and fire codes require the installation of fire alarm devices in all new residential dwelling units. The most common type of residential fire alarm is the single-station smoke alarm (**FIGURE 26-1**). This life-safety device includes a smoke detection sensor, an automatic control unit, and an audible alarm within a single device. It alerts occupants when a fire occurs. Millions of single-station smoke alarms have been installed in private dwellings and apartments.

Smoke alarms can be either battery-powered, hard-wired to a 120-V electrical system, or both. Current building codes require smoke alarms in all newly constructed dwellings to be hard-wired and powered by the building's electrical system. These smoke alarms must also be interconnected, so all alarms will sound if one alarm detects smoke. Hard-wired smoke alarms must also contain a backup battery in case electrical service is disrupted. Many older residences have single-station battery-operated smoke alarms that are not hard-wired or connected to other smoke alarms. It is imperative that smoke alarm batteries be replaced on a regular basis. Some communities recommend replacing smoke alarm batteries when Daylight Saving Time begins and ends. Some smoke alarms come with a long-life lithium battery that may last up to 10 years. NFPA 72 recommends that all smoke alarms be replaced after 10 years from the date of manufacture.

The model building codes and NFPA 72 require new residences to have at least one smoke alarm on each level of the house, one in the corridor or hallway outside each sleeping area, and one in each bedroom. Older residences built before these requirements were established may not have enough smoke alarms to provide early detection and notification. Fire fighters

FIGURE 26-1 A single-station smoke alarm.
Courtesy of Kidde Residential and Commercial Division.

should recommend that the occupants install additional smoke alarms to improve their protection.

Some home smoke alarms are part of security systems. Like commercial systems, these home fire alarm/security systems have an alarm control unit and require a passcode to set or reset the system. They may also be monitored by contract monitoring companies that receive the signals and initiate the appropriate action or response.

Fire Alarm System Components

A fire alarm system has three basic components: **alarm initiating devices**, **alarm notification appliances**, and a **fire alarm control unit (FACU)**.

The alarm initiating device is either an automatic or manually operated device that, when activated, causes the system to initiate an alarm. The alarm notification appliance is generally an audible device, often accompanied by a visual device that alerts the building occupants when the system is activated. The FACU links the initiating device to the notification appliance and performs other essential functions.

Fire Alarm Control Unit

Most fire alarm systems in commercial or industrial buildings have several types of alarm initiating devices in different areas and use both audible and visible notification appliances to notify the occupants of an alarm. The FACU serves as the "brain" of the system, linking the initiating devices to the notification appliances.

The FACU manages and monitors the proper operation of the system. It can indicate the source of an alarm so that responding fire personnel will know what activated the alarm and where the initial activation occurred. The FACU also manages the primary and backup power supplies for the system. It may perform additional functions, such as notifying the jurisdiction's emergency communications centre or an on- or off-site monitoring centre where trained personnel receive the signals and initiate the appropriate action. The FACU might also interface with other fire protection and building systems, including elevators, stairwell door locks, fans, and dampers in smoke control and management systems.

Control units vary greatly depending on the age of the system and the manufacturer. For example, an older system might simply indicate that a fire alarm has been activated, whereas a newer system might indicate that an alarm occurred in a specific zone within the building (**FIGURE 26-2**). Some modern fire alarm systems specify the exact location and type of the activated initiating device.

Fire alarm control units should always be locked to prevent unauthorized access to the system. Alarms should not be silenced or reset until the source of the

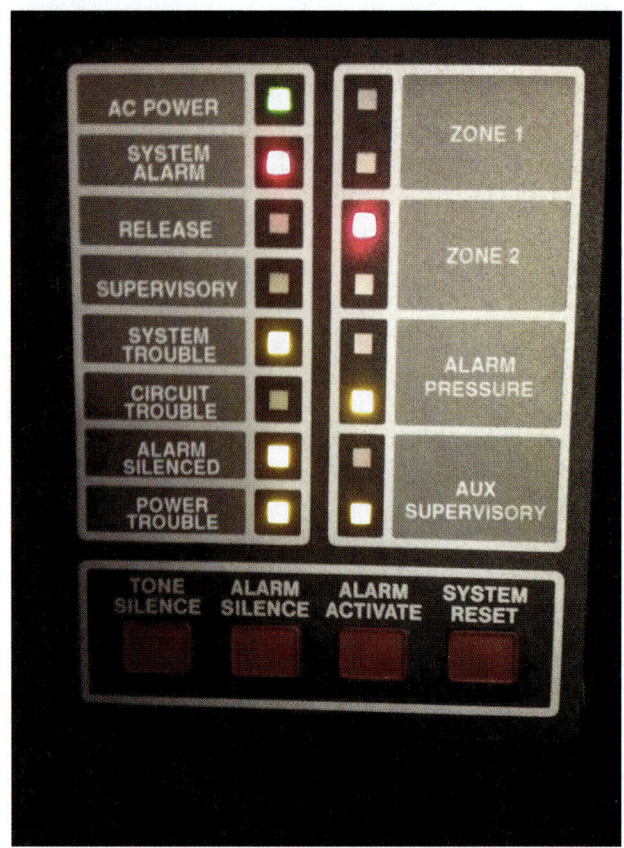

FIGURE 26-2 Most modern fire alarm control units indicate the zone where an alarm was initiated.
Courtesy of A. Maurice Jones, Jr.

alarm has been identified and checked by fire fighters to ensure that the situation is under control. In some systems, the initiating devices must be manually reset before the entire system can be reset; others automatically reset after the problem has been resolved.

Many buildings have one or more remote annunciators located near main entrance doors. A **remote annunciator** enables fire fighters to determine the type and location of the activated alarm device as they enter the building (**FIGURE 26-3**).

The FACU also monitors the condition of the entire alarm system to detect any internal or external faults. Faults within the system are indicated by a **trouble signal**. Trouble signals indicate that a component of the system is not operating properly and requires attention from the building owner. Typical faults include loss of primary power, low or no battery power, a broken wire, or a device that has become inoperable, damaged, or was removed. Changes in the normal ready status of other fire protection systems external to or integrated with the FACU are indicated by a **supervisory signal**. Typical changes in normal ready status include low or high air pressure in a dry pipe sprinkler system, a control valve that has been closed, or low water temperature (below 4.44°C/40°F) in a fire suppression system storage tank.

FIGURE 26-3 A remote annunciator allows fire fighters to quickly determine the type and location of the activated alarm device.
Courtesy of A. Maurice Jones, Jr.

Trouble and supervisory signals do not activate the building's fire alarm, nor should they result in a fire department response. They should, however, make an audible sound and illuminate a special light at the alarm control unit and annunciator unit, if present. They may also transmit a notification to an on- or off-site monitoring location where the signal can be received and acted upon.

Fire alarm systems are usually powered by a dedicated 120-V circuit on the building's electrical system, but the FACU may convert the power to a lower voltage.

In addition to the primary power source, NFPA 72 requires a secondary power source for all alarm systems (**FIGURE 26-4**). Most fire alarm systems use batteries for backup power; however, some buildings use a diesel-powered generator to power essential equipment.

FIGURE 26-4 Fire, building, and life safety codes require that alarm systems have a backup power supply, which becomes activated automatically when the normal electrical power is interrupted. Most systems use batteries as the backup power supply.
Courtesy of A. Maurice Jones, Jr.

The FACU must generate a trouble signal if either the primary or secondary power source fails.

Depending on the building's size and floor plan, a fire alarm may sound throughout the building or only in particular areas. In high-rise buildings, fire alarm systems are often programmed to alert only the occupants on the same floor as the activated alarm as well as those on the floors immediately above and below the affected floor. Alarms on the remaining floors can be manually activated from the system control unit in the fire command centre. Some systems include a public address feature, which enables fire department personnel to provide specific instructions or information for occupants in different areas.

The control unit in a large building may be programmed to perform several additional functions. For example, it may automatically shut down or change the operation of air-handling systems, recall elevators to the ground floor, and unlock stairwell doors so that a person in an exit stairwell can reenter an occupied floor.

Alarm Initiating Devices

Alarm initiating devices begin the fire alarm process either manually or automatically. Manual initiating devices require human activation; automatic devices function without human intervention. Manual fire alarm boxes are the most common type of alarm initiating devices that require human activation. Automatic initiating devices include various types of devices that detect predetermined fire conditions or changes in automatic fire suppression systems.

Manual Initiating Devices. Manual initiating devices are designed so that building occupants can activate the fire alarm system manually by pulling a lever or turning a switch. Many older alarm systems could only be activated manually. The most common manual initiating device is the manual fire alarm box or **manual pull station** (**FIGURE 26-5**). The manual pull station has a switch that closes an electrical circuit to activate the alarm.

Pull stations can be either single-action or double-action devices. A **single-action pull station** requires a person to pull down a lever, toggle, or handle to activate the alarm; the alarm sounds as soon as the pull station is activated. A **double-action pull station** requires a person to perform two steps before the alarm will activate. They are designed to reduce the number of nuisance alarms caused by accidental or malicious activations of the alarm. The person must move a flap, lift a cover, or break a piece of glass to reach the alarm activation device. Although pull stations with breakable glass rods or panes are still in use, most new pull stations use a

FIGURE 26-5 Several types of manual fire alarm boxes (also known as manual pull stations) are available.

Left: © rob casey/Alamy Images; Right: Courtesy of Honeywell/Fire-Lite Alarms.

resettable locking mechanism to indicate that the pull station has been activated. Once activated, a manual pull station should stay locked in the "activated" position until it is manually reset. This status enables responding fire fighters to determine which pull station initiated the alarm. Resetting the pull station requires a special key, screwdriver, or wrench. The pull station must be reset before the building alarm can be reset at the FACU.

FIRE FIGHTER TIP

Troubleshooting Smoke Alarms

Most problems with smoke alarms are caused by a lack of power, dirt in the sensing chamber, or defective equipment.

Power Problems

Smoke alarms require power from a battery or from the building's hard-wired 120-V power source. Some smoke alarms that are hard wired to a 120-V power source are equipped with a battery that serves as a backup power source.

Many battery-powered smoke alarms contain a dead battery. Some people do not change batteries when recommended, resulting in inoperable smoke alarms. People may also remove smoke alarm batteries to use for another purpose or to prevent nuisance alarm activations. Smoke alarm batteries should be replaced twice a year unless they have a long-life lithium battery that may last up to 10 years.

Hard-wired smoke detectors become inoperable if someone turns the power off at the circuit breaker, so the dedicated circuit should remain on at all times.

Most smoke alarms containing batteries will signal a low-battery condition by emitting a chirp every few seconds. This chirp indicates that the battery needs to be replaced to keep the smoke alarm functional. Many hard-wired smoke alarms will chirp to indicate that the backup battery is low, even when the 120-V power source is operational. This feature ensures that the smoke alarm always has an operational backup system.

Dirt Problems

Smoke alarms may generate nuisance alarms or may not activate properly if dust or an insect becomes lodged in the sensing chamber. To prevent this, alarms should be periodically cleaned by removing the cover and gently vacuuming the chamber. In some instances, a puff of air from a can of compressed air might remove the dust or insect; however, it could instead push the contaminant deeper into the detection chamber. Follow the manufacturer's instructions for proper cleaning and maintenance.

Equipment Problems

NFPA 72 recommends that all smoke alarms be replaced after 10 years from the date of manufacture. Older smoke alarms may become overly sensitive and generate frequent nuisance alarms. Others may develop decreased sensitivity and fail to emit an alarm in the event of a fire. The date of manufacture is stamped on each smoke alarm. If a smoke alarm is more than 10 years old, it should be replaced with a new one.

Understanding the basic functions and troubleshooting of simple smoke alarms enables you to respond to citizens who call your fire department when they encounter problems with their smoke alarms. Most important, it enables you to keep smoke alarms operational. Remember, only operational smoke alarms can save lives.

Double-action pull stations may be covered by a plastic tamper cover (**FIGURE 26-6**). The plastic cover is designed to deter malicious alarms, but it also protects the pull station from potential physical damage. These covers are often used in areas where malicious alarms occur frequently, such as schools and college dormitories. The plastic cover must be opened before the pull station can be activated. Lifting the cover triggers a loud tamper alarm at that specific location, but it does not activate the fire alarm system. Snapping the cover back into place resets the tamper alarm.

Automatic Initiating Devices. An automatic initiating device is designed to function without human intervention and will activate the alarm system when it detects predetermined fire conditions, such as smoke, heat, or flowing water. Even if the building is unoccupied, these systems can transmit alarm signals to an emergency communications centre or an on- or off-site monitoring facility, and they can perform other functions when a detector is activated.

There are many different types of automatic initiating devices (detectors). Some detectors are activated by smoke or by the invisible products of combustion. Other detectors react to heat, flame, sparks, or specific gases. Others respond to the activation of fire suppression systems, such as sprinkler systems or Class K suppression systems in commercial kitchens.

Smoke Detectors. A smoke detector is the most common type of initiating device, and it may be a stand-alone device or be part of a comprehensive fire alarm system. Smoke detectors are commonly found in schools, hospitals, businesses, and commercial occupancies that are equipped with fire alarm systems (**FIGURE 26-7**).

Smoke detectors come in a variety of designs and styles designed for different applications. The most commonly installed types are ionization and photo-electric spot-type detectors. Each detector is listed by an independent testing laboratory for use in specific applications based on the size, height, contents, and occupancy type of the protected area.

There are two types of fire detection technology that may be used to detect combustion: ionization and photoelectric smoke detectors:

- An ionization smoke detection device is activated by the smaller, invisible products of combustion.
- A photoelectric smoke detector is activated by the larger, visible products of combustion.

FIGURE 26-6 A variation on the double-action pull station, designed to prevent malicious alarms, has a clear plastic cover and a separate supervisory alarm.
Courtesy of STI-USA.

FIGURE 26-7 Commercial ionization smoke detector.
Courtesy of A. Maurice Jones, Jr.

Ionization smoke detectors work on the principle that burning materials release many different products of combustion, including electrically charged microscopic particles. An ionization detector senses the presence of these invisible charged particles (ions).

An ionization smoke detector contains a minuscule amount of radioactive material inside its inner chamber. This radioactive material releases charged particles into the chamber, and a small electric current flows between two plates (**FIGURE 26-8**). When smoke particles enter the chamber, they neutralize the charged particles and interrupt the current flow. The detector senses this interruption and activates the alarm.

A photoelectric smoke detector uses a light beam and a photocell to detect larger, visible particles of smoke (**FIGURE 26-9**). When visible particles of smoke enter the inner chamber, it reflects some of the light onto the photocell, thereby activating the alarm.

Ionization smoke detectors react more quickly than photoelectric smoke detectors to fast-burning fires, such as a fire in a trash can, which may, at first, produce

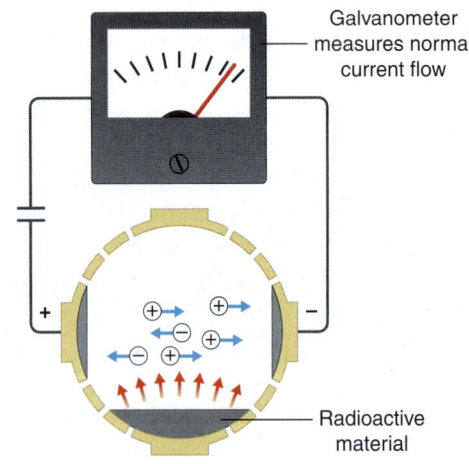

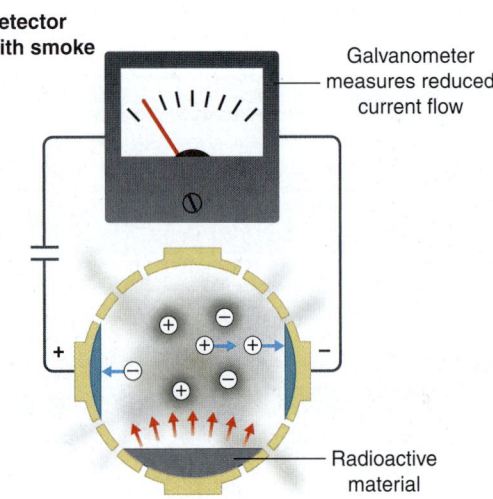

FIGURE 26-8 Principle of operation for an ionization smoke detector.

© Jones & Bartlett Learning.

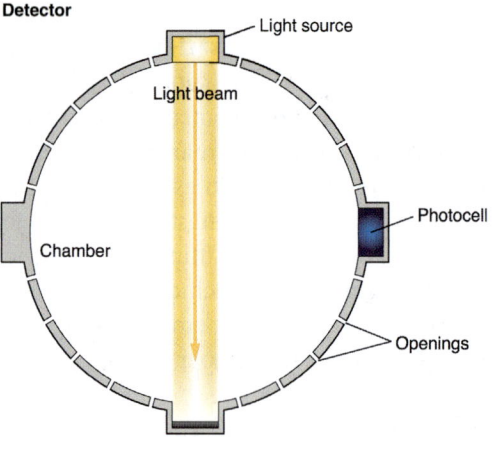

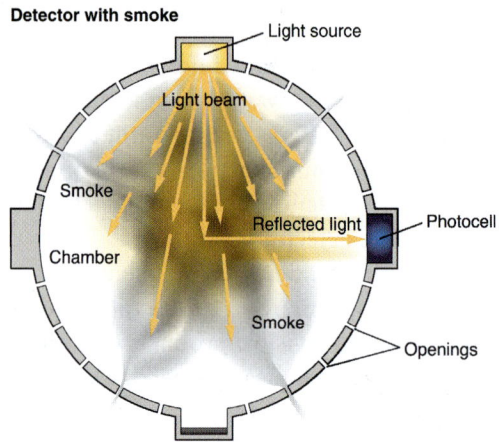

FIGURE 26-9 Principle of operation for a photoelectric smoke detector.
© Jones & Bartlett Learning.

little visible smoke (**FIGURE 26-10**). Ionization detectors are more susceptible to nuisance alarms from common activities, such as light smoke from cooking or steam from a shower.

By comparison, photoelectric smoke detectors are more responsive to slow-burning or smouldering fires, such as a fire caused by a cigarette caught in a couch, which usually produces a large quantity of visible smoke (**FIGURE 26-11**). They are less prone to nuisance alarms from steam than are ionization smoke detectors. Recent studies indicate that both ionization and photoelectric smoke detectors are acceptable life-safety devices. These studies determined that an adequate number of properly spaced detectors of either type detected smoke within acceptable time limits. Combination ionization/photoelectric smoke detectors are also available. These alarms quickly react to both fast-burning and smouldering fires. They are not suitable for use near kitchens or bathrooms because they are prone to the same nuisance alarms as regular ionization smoke detectors.

Most ionization and photoelectric smoke detectors look alike. Refer to the label on each detector to determine

FIGURE 26-10 Ionization smoke detectors react more quickly than photoelectric smoke detectors to a fast-burning fire, such as a fire in a wastepaper basket, which may produce little visible smoke.
© Brendan Byrne/age fotostock.

FIGURE 26-11 A photoelectric smoke detector. These detectors react more quickly than ionization smoke detectors to a slow-burning or smouldering fire, such as a cigarette in furniture, which usually produces a large quantity of visible smoke.
© Jones & Bartlett Learning.

its type and applications. An ionization detector must have a label or engraving stating that it contains a small amount of radioactive material.

Smoke detectors connected to a fire alarm system are powered by a low-voltage circuit, and they send a signal to the FACU when they are activated. After the smoke condition clears, the alarm system can be reset at the control unit. If the detector is damaged by smoke or heat, it must be replaced.

A projected **beam detector** is a type of photoelectric smoke detector used to protect large open areas such as churches, atriums, auditoriums, airport terminals, and indoor sports arenas. Smoke may not rise high enough to reach spot-type smoke detector in these facilities, so one or more beam detectors could be used to provide protection for the entire area (**FIGURE 26-12**).

A beam detector has two components: a sending unit, which projects a narrow beam of light across the open area, and a receiving unit, which measures the intensity of the light when the beam strikes the receiver. When smoke interrupts the light beam, the receiver detects a drop in the light intensity and activates the fire alarm system. Most photoelectric projected beam detectors are set to respond to a certain **obscuration rate**, or percentage of light blocked. If the light is completely blocked, such as when a solid object is moved across the beam, the trouble or supervisory alarm will sound, but the fire alarm will not be activated.

Heat Detectors. A **heat detector** is also commonly used as an automatic alarm initiating device. This device can provide property protection but cannot provide reliable life-safety protection because it does not react quickly enough to incipient fires. It is generally used in areas where smoke alarms or detectors cannot be used, such as dusty environments and areas that experience extreme cold or heat. Heat detectors are often installed in unheated areas, such as attics and storage rooms, as well as in boiler rooms and manufacturing areas.

Heat detectors are generally very reliable and less prone to nuisance alarms than are smoke detectors. You may come across heat detectors that were installed 30 or more years ago and are still in service. Unfortunately, older units lack a visual indicator light that shows which device was activated, so tracking down the cause of an alarm may be very difficult. Newer models have an indicator light that shows which device was activated.

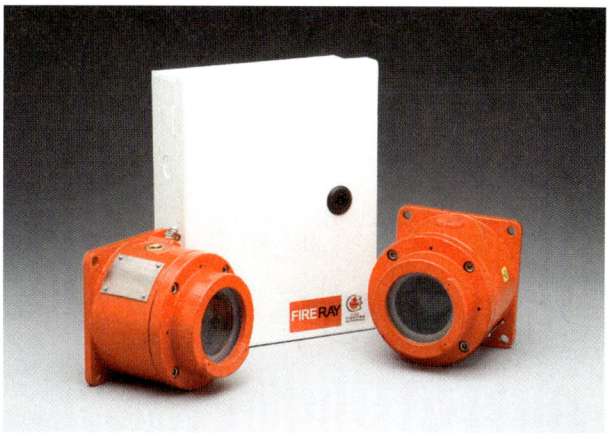

FIGURE 26-12 Beam detectors are used in large open spaces.
Courtesy of Fire Fighting Enterprises Ltd.

Several types of heat detectors are available, each of which is designed for specific situations and applications. Single-station heat detectors are sometimes installed in unoccupied areas of buildings that do not have fire alarm systems, such as attics or storage rooms. A **spot-type detector** is an individual unit that can be spaced throughout an occupancy so that each detector covers a specific floor area. Spot-type detectors are usually installed in light commercial and residential settings; the units may be in individual rooms or spaced at intervals along the ceiling in larger areas.

Heat detectors are designed to operate at a fixed temperature or to react to a rapid increase in temperature. Either fixed-temperature or rate-of-rise devices can be configured as spot-type or line detectors.

A **fixed-temperature heat detector**, as the name implies, is designed to operate at a preset temperature (**FIGURE 26-13**). A typical temperature for a light-hazard occupancy, such as an office building, would be 57°C (135°F). Fixed-temperature detectors typically include a metal alloy that will melt at the preset temperature. This melting alloy releases a lever-and-spring mechanism, which closes or opens a circuit. Most fixed-temperature heat detectors must be replaced after they have been activated, even if the activation was accidental.

A **rate-of-rise heat detector** is activated when the temperature of the surrounding air increases by more than a set amount in a given period of time. A typical rating might be "greater than −11°C (12°F) in 1 minute." If the temperature increase occurs more slowly than this rate, the rate-of-rise heat detector will not activate. By contrast, a temperature increase at greater than this rate will activate the detector and initiate the fire alarm. Rate-of-rise heat detectors should not be located in areas that normally experience rapid changes in temperature, such as near garage doors in heated parking areas.

Some rate-of-rise heat detectors have a **bimetallic strip** made of two metals that respond differently to heat: A rapid increase in temperature causes the strip to bend unevenly, which closes a contact. Another type of rate-of-rise heat detector uses an air chamber and diaphragm mechanism; as air in the chamber heats up, the pressure increases. Gradual increases in pressure are released through a small hole, but a rapid increase in pressure will press upon the diaphragm and activate the alarm. Most rate-of-rise heat detectors are self-restoring, so they do not need to be replaced after an activation unless they were directly exposed to a fire.

Rate-of-rise heat detectors generally respond more rapidly to most fires than do fixed-temperature heat detectors. However, a slow-burning fire, such as a smouldering couch, might not activate a rate-of-rise heat detector until the fire is well established.

Combination rate-of-rise and fixed-temperature heat detectors are available. These devices balance the faster response of the rate-of-rise detector with the reliability of the fixed-temperature heat detector.

A **line detector** uses wire or tubing strung along the ceiling of large open areas to detect an increase in heat. An increase in temperature anywhere along the line will activate the detector. Line detectors are often found in churches, warehouses, and industrial or manufacturing applications. They are also used in data cable trays.

Line detectors use wires or a sealed tube to sense heat. One wire-type model has two wires inside, separated by an insulating material. When heat melts the insulation, the wires short and activate the alarm. The damaged section of insulation must be replaced with a new piece after activation of the detector.

Another wire-type model measures changes in the electrical resistance of a single wire as it heats up. This device is self-restoring and does not need to be replaced after activation unless it is directly exposed to a fire.

The tube-type line heat detector contains a sealed metal tube filled with air or a nonflammable gas. When the tube is heated, the internal pressure increases and activates the alarm. Like the single-wire line heat detector, this device is self-restoring and does not need to be replaced after activation, unless it is directly exposed to a fire.

Flame Detectors. A **flame detector** is a specialized device that detects the infrared or ultraviolet light waves produced by a flame (**FIGURE 26-14**). These devices can quickly recognize even a very small fire.

Typically, flame detectors are found in places such as aircraft hangars, fuel loading racks, or specialized industrial settings where early detection and rapid reaction to a fire are critical. Such detectors are also used in explosion suppression systems where they detect and suppress an explosion as it is occurring.

Flame detectors are both complicated and expensive. Infrared or ultraviolet light sources, such as the sun or a

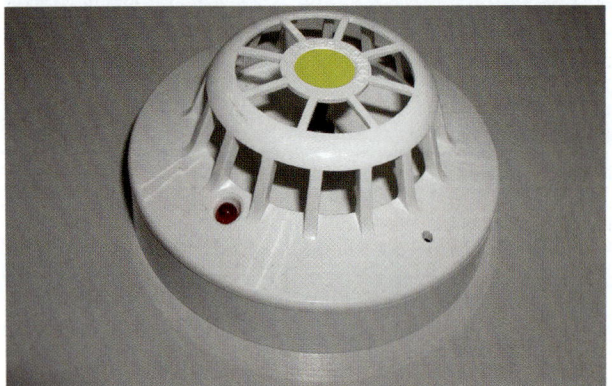

FIGURE 26-13 A fixed-temperature heat detector initiates an alarm at a preset temperature.

Courtesy of Firetronics Pte Ltd.

FIGURE 26-14 Flame detectors are specialized devices that detect the electromagnetic light waves produced by a flame.
© Jones & Bartlett Learning.

FIGURE 26-15 Air sampling detector.
© Jones & Bartlett Learning.

welding operation, can initiate an unwanted alarm. Some flame detectors use combined infrared and ultraviolet technologies to reduce nuisance alarms.

Gas Detectors. A **gas detector** is calibrated to detect the presence of a specific gas that is created by combustion or that is used in the facility. Depending on the system, a gas detector may be programmed to activate either the building's fire alarm system or a separate alarm. Some of these specialized instruments require regular calibration to operate properly. Gas detectors for specific gases are often found in commercial or industrial applications. The most common type of gas detector you will encounter in residential settings is a carbon monoxide detector.

Air Sampling Detectors. An **air sampling detector** continuously captures air samples and measures the concentrations of specific gases or products of combustion. These devices draw air samples through tubing into a sampling unit where they are analyzed using specialized ionization or photoelectric smoke alarm technologies (**FIGURE 26-15**). A **duct detector** is a type of smoke detector installed in the air ducts of buildings to sample the air moving through the air distribution system. It will sound either an alarm or supervisory signal and shut down the air-handling system if it detects smoke.

More complex systems are sometimes installed in special hazard areas to draw air samples from rooms, enclosed spaces, or equipment cabinets. These samples pass through gas analyzers that can identify smoke particles, products of combustion, and concentrations of other gases associated with a dangerous condition much faster than conventional detectors. Air sampling detectors are most often used in areas that hold valuable contents or sensitive equipment.

Residential Carbon Monoxide Detectors. Residential carbon monoxide detectors are designed to sound an audible or visual alarm when the concentration of carbon monoxide becomes high enough to pose a health risk to the occupants of the building. Such an alarm is triggered by a very high concentration of carbon monoxide that accumulates in a short period of time or by a lower concentration of carbon monoxide that occurs over a longer period of time. Many carbon monoxide detectors look like smoke alarms to most people. Some of these units are combined with smoke alarms to provide warning from a fire or from elevated levels of carbon monoxide (**FIGURE 26-16**). These combination alarms have one audible or visible alarm for fire and a different audible or visible alarm for elevated levels of carbon monoxide. As has been mentioned, most jurisdictions in Canada have made CO detectors mandatory

FIGURE 26-16 One type of combination carbon monoxide and smoke alarm.
© kvisel/iStock/Getty Images Plus/Getty Images.

in residential housing equipped with fireplaces and or gas-fired appliances.

Carbon monoxide detector activations need to be investigated by personnel who can use proper gas detection devices to determine the level of the gas and the cause of the alarm activation. Anytime you respond to a carbon monoxide detector, you need to check all building occupants to determine whether they are suffering from the following signs or symptoms of carbon monoxide poisoning:

- Headache
- Nausea
- Disorientation
- Unconsciousness
- Multiple people with flulike symptoms in the same location

Be especially alert for multiple people who are all suffering flulike symptoms, as they may be suffering from carbon monoxide poisoning.

Alarm Initiation by Fire Suppression Systems.

Other fire protection systems in a building may be used to activate the fire alarm system. An **automatic sprinkler system** is usually connected to the fire alarm system and will activate the alarm if water flows through the system (**FIGURE 26-17**). These devices alert the building occupants, on- and off-site monitoring facilities, and the fire department to a possible fire, and they also ensure that someone is made aware that water is flowing, in case of an accidental discharge. Any other fire-extinguishing systems in a building, such as wet-chemical systems found in kitchens or those containing clean agents, should also be connected to the building's fire alarm system.

FIGURE 26-17 Automatic sprinkler systems use an electric flow switch to activate the building's fire alarm system.
© Jones & Bartlett Learning.

Nuisance Alarms. As the number of fire detection and alarm systems increases, so does the number of nuisance alarms. Fire department personnel who understand how fire detection and alarm systems operate can advise building owners when these situations occur.

Three distinct types of nuisance alarms are possible: malicious alarms, alarms due to environmental conditions, and alarms caused by improper installation or maintenance. It is important to distinguish between the three types to determine the cause of the fire alarm activation because the problem must be recognized before it can be corrected.

Regardless of the underlying cause, all three types of nuisance alarms have the same results: They waste fire department resources, and they may delay legitimate emergency responses. Frequent nuisance alarms at the same site can desensitize building occupants to the alarm system so that they might not respond appropriately to a real emergency.

A **malicious alarm** occurs when an individual deliberately activates a fire alarm when there is no fire, causing a disturbance. Manual fire alarm boxes are popular targets for pranksters. Causing a malicious alarm is an illegal act.

Nuisance alarms due to environmental conditions occur when an alarm system is activated by a condition that is not a real emergency. For example, a smoke alarm placed too close to a kitchen may be activated by normal cooking activities. A nuisance alarm could also occur if a person is smoking a cigarette under a smoke detector. The smoke detector functions properly—it detects smoke and activates the alarm system—but there is no real emergency.

Nuisance alarms can also be caused by the malfunction of one or more alarm system components or by an improperly designed and maintained system. Alarm systems are designed and installed to meet the performance objectives established by the fire protection engineer. Once installed, fire alarm systems must be properly maintained on a continual basis. An improperly installed system or a mechanical failure due to the lack of maintenance may cause nuisance alarms, or the system may fail to operate as intended in the event of a real fire emergency.

Preventing Nuisance Alarms.

All nuisance alarms should be investigated to determine the cause of the alarms and to correct the problem. Many nuisance alarms could be avoided by relocating a detector or using a different type of detector in a particular location. Proper design, installation, and system maintenance are essential to prevent nuisance alarms.

If an alarm system is activated whenever it rains, water could be leaking into the wiring or a system component. Detectors subjected to a buildup of dust, dirt, or other debris may become more sensitive and initiate nuisance alarms.

Several methods can be used to reduce nuisance alarms caused by smoke detectors:

- A **cross-zoned system** requires two or more initiating devices to activate before an alarm sounds. If a single device activates, a trouble signal will sound. If a second initiating device activates, the fire alarm will sound, and an alarm signal will be sent to the monitoring station. These devices are sometimes used to activate preaction sprinkler systems or other fire suppression systems.

- Systems with an **alarm verification feature** use a delay of 30 to 60 seconds between the first initiating device signal and sounding the alarm. During this time, the system may show a trouble or "pre-alarm" condition at the system control unit. After the preset interval, the system rechecks the detector. If the condition has cleared, the system returns to normal. If the detector is still sensing smoke or a second detector activates, the fire alarm sounds.

Both cross-zoned and alarm verification systems are designed to prevent nuisance alarms after brief exposures to light smoke or dust. For more information, see the NFPA's Fire Service Guide to Reducing Unwanted Fire Alarms.

Many communities have implemented regulations to control the frequency of nuisance alarms when multiple alarms occur in a short period of time. These regulations encourage occupants to rectify the problems that are causing these types of alarms. Some are system-related problems such as faults in the wiring or sensors, and others are human-related problems such as activating an alarm maliciously. Regulations may set fines or require remedial action when nuisance alarms exceed a specified number in a set time period. Fire fighters should be familiar with the regulations in their jurisdiction.

Alarm Notification Appliances

Audible notification appliances such as bells, horns, and electronic speakers produce an audible signal when the fire alarm is activated. Newer systems also incorporate visual notification appliances. These audible and visual alarms alert building occupants that a fire condition has been detected.

Older systems used a variety of sounds as notification appliances, but this inconsistency often led to confusion about whether the sound was actually a fire alarm. NFPA 72 now requires a distinctive, standardized evacuation signal consisting of a three-pulse temporal pattern. This is often referred to as the **temporal-3 pattern** or T-3

signal. The temporal-3 pattern consists of three tones followed by a short pause. Even single-station smoke alarms designed for residential occupancies are required to use this sound pattern. As a result, people are now more likely to recognize a fire alarm immediately.

Some public buildings also play a recorded evacuation announcement in conjunction with the temporal-3 pattern. This recorded message is played through the fire alarm speakers and provides instructions on safe evacuation of the site (**FIGURE 26-18**). In facilities such as airport terminals, this announcement is recorded in multiple languages. Such a system may also include a public address feature that allows fire department or building security personnel to provide specific instructions, information about the emergency, or notice when the alarm condition is terminated.

Many new fire alarm systems incorporate visual notification devices such as high-intensity strobe lights as well as audio devices (**FIGURE 26-19**). Visual devices alert hearing-impaired occupants to a fire alarm and are useful in noisy environments where an audible alarm might not be heard.

Other Fire Alarm Functions

In addition to alerting occupants and summoning the fire department, fire alarm systems may control other building functions, such as air-handling systems, fire doors, and elevators. To control smoke movement through the building, the system may either shut down or start up air-handling systems. Many modern Type I buildings have systems that will pressurize means of egress and hallways in order to contain smoke to the area of origin. Fire doors that are normally held open

FIGURE 26-18 A speaker fire alarm notification appliance.
Courtesy of Honeywell/Fire-Lite Alarms.

FIGURE 26-19 This alarm notification appliance has both a loud horn and a high-intensity strobe light.
© Jones & Bartlett Learning.

by electromagnets may be released to compartmentalize the building and confine the fire to a specific area. Doors allowing reentry from exit stairwells into occupied areas may be unlocked. Elevators may be summoned to a predetermined floor, usually the main lobby, so they can be used by fire crews.

Responding fire personnel must understand which building functions are being controlled by the fire alarm, for both safety and fire suppression reasons. This information should be gathered during preincident planning surveys and should be available in printed form or on a graphic display at the control unit location.

Fire Alarm Annunciator Systems

Some fire alarm systems provide limited information at the alarm control unit; others specify exactly which initiating device activated the fire alarm. The systems can be further subdivided based on whether they are zoned or **coded systems**.

Many older fire alarm systems simply sounded an audible alarm when the system was activated. Known as **noncoded systems**, these systems provided no information about the type or location of the initiating device that caused the alarm. Responding personnel had to search the entire building to locate the source of the alarm.

Other older fire alarm systems provided limited information through coded signals. These audible or visible signals were typically sounds or flashing lights in coded patterns that indicated the floor or zone from

which the alarm signal originated. A pull station on the fourth floor of an apartment building, for example, may be coded as four bells followed by two bells. Occupants and responding personnel needed to refer to a chart to interpret the coded signals. Coded alarm systems are antiquated, but these systems are still in use in some older buildings.

A **zoned system** divides initiating devices into zones based on the type or location of the devices. Earlier zoned systems had a limited number of zones, so initiating devices were often grouped by floor, area, or type. For example, all pull stations might be on Zone 1, all smoke detectors on Zone 2, and the water flow alarm might be on Zone 3. These zones would be indicated by lights or digital screens on the FACU. As technology improved, some fire alarm systems were capable of identifying dozens of individual zones to provide greater detail to occupants and fire fighters.

In recent years, fire alarm systems have become far more complex and more reliable. Large industrial complexes or high-rise buildings may now be protected by fire alarm systems with hundreds of zones. Addressable systems assign a unique identifier number to each initiating device and notification appliance. The FACU continually monitors each device and provides zone-specific trouble, supervisory, and alarm signals at the FACU and to on- or off-site monitoring facilities. Fire personnel responding to alarms in these buildings can identify the exact location and specific conditions before they enter the building.

In addition to the information displayed on the main FACU or by audible devices, many buildings use remote annunciators near the main entrances. Responding fire fighters can use these remote annunciators to gain important information about the fire alarm without having to find the FACU.

Fire Department Notification

The fire department should always be notified when a fire alarm system is activated. In some cases, a person must make a telephone call to the fire department's emergency communications centre. In other cases, the fire alarm system may be connected directly to the fire department's emergency communications centre or to a remote location where someone on duty calls the fire department. Fire alarm systems can be classified based on how the fire department is notified of an alarm (**TABLE 26-1**).

Protected Premises Fire Alarm Systems

Most **protected premises fire alarm systems** are found in older buildings and do not notify the fire department upon alarm activation. Instead, the alarm sounds only

TABLE 26-1 Fire Department Notification Systems	
Type of System	**Description**
Protected premises fire (local) alarm system	The fire alarm system sounds an alarm only in the building where it was activated. No signal is sent out of the building. Someone must call the fire department to respond.
Public emergency alarm reporting system	The fire alarm system sounds an alarm in the building and transmits a signal to the fire department via a public alarm box system.
Remote supervising station alarm system	The fire alarm system sounds an alarm in the building and transmits a signal to a remote location. The signal may go directly to the fire department or to another location where someone is responsible for calling the fire department.
Proprietary supervising alarm system	The fire alarm system sounds an alarm in the building and transmits a signal to a monitoring location owned and operated by the facility's owner. Depending on the nature of the alarm and arrangements with the local fire department, facility personnel may respond and investigate, or the alarm may be immediately retransmitted to the fire department. These facilities are monitored 24 hours per day.
Central station service alarm system	The fire alarm system sounds an alarm in the building and transmits a signal to an off-premises alarm monitoring facility. The off-premises monitoring facility is then responsible for notifying the fire department to respond.

in the building to notify the occupants. Buildings with this type of system should have notices posted requesting occupants to call the fire department and report the alarm after they exit (**FIGURE 26-20**).

Public Emergency Alarm Reporting Systems

A public emergency alarm reporting system is the name that is now used to describe an auxiliary system. This type

of system can be used in jurisdictions with a public fire alarm box system. With this system, the building's fire alarm system is connected to a master alarm box located outside the building. When the alarm activates, it signals the master box, which transmits the alarm directly to the fire department communications centre. Most cities have phased out public fire alarm boxes because they are expensive to maintain. They have been largely replaced by more modern communication devices.

Supervising Station Alarm Systems

A supervising station alarm system communicates between the protected property and a constantly attended location that will receive and interpret alarm signals and take the appropriate action to handle the alarm condition. There are three types of supervising station alarm systems—remote, proprietary, and central station—and each handles alarms in a different way.

A remote supervising station alarm system sends a signal directly to the fire department or to another monitoring location via a telephone line or a radio signal. This type of direct notification can be installed only in jurisdictions where the fire department is equipped to handle direct alarms. If the signal goes to a monitoring location, that site must be continually staffed by someone who will call the fire department.

A proprietary supervising alarm system is connected directly to a monitoring site that is owned and operated by the building's owner. Proprietary systems are often

FIGURE 26-20 Buildings with a local alarm system should post notices requesting occupants to call the fire department and report the alarm after they exit.

© Jones & Bartlett Learning.

installed at facilities where multiple buildings belong to the same owner, such as universities or industrial complexes. Each building is connected to a monitoring site on the premises (usually the security centre), which is staffed at all times (**FIGURE 26-21**). When an alarm sounds, the staff at the monitoring site report the alarm to the fire department, usually by telephone or a direct line.

A **central station service alarm system** is a third-party, off-site monitoring facility that monitors multiple

FIGURE 26-22 A central station monitors alarm systems at many locations.
Courtesy of www.acimonitoring.com, Doug Beaulieu.

alarm systems. Individual building owners contract with and pay the central station to monitor their facilities (**FIGURE 26-22**). When an activated alarm at a covered building transmits a signal to the central station by telephone or radio, personnel at the central station notify the appropriate fire department of the fire alarm. Other services, including alarm response and investigation, testing and maintenance, and record keeping, may be provided by the central station. The central station facility may be located in the same city as the facility or in a different part of the country.

Usually, fire alarms are connected to the central station through leased or standard telephone lines. The use of either cellular telephone frequencies or radio frequencies, however, is becoming more common. Cellular or radio connections may be used as a secondary method of transmitting alarm signals, or, in remote areas without telephone lines, they may be the primary transmission method.

Fire Suppression Systems

Fire suppression systems include automatic sprinkler systems, standpipe systems, and special hazard suppression systems such as gaseous agent, wet-chemical, dry-chemical, and foam systems. Most newly constructed commercial buildings incorporate at least one of these systems, and increasing numbers of single-family dwellings are being built with residential sprinkler systems as well.

Understanding how these systems work is important because they can affect fire behaviour. In addition, fire fighters should know how to interface with the fire suppression system and how to shut down a system to prevent unnecessary damage.

FIGURE 26-21 In a proprietary system, fire alarms from several buildings are connected to a single monitoring site owned and operated by the buildings' owner.
© Jones & Bartlett Learning.

Automatic Fire Sprinkler Systems

The most common type of fire suppression system is the automatic sprinkler system. Automatic sprinklers are reliable and effective, with a history of more than 100 years of successfully controlling fires. When properly designed, installed, inspected, tested, and maintained, these systems are extremely effective in controlling or extinguishing fires.

Unfortunately, few members of the general public have an accurate understanding of how automatic sprinklers work. In movies and on television, when one sprinkler is activated, the entire system begins to discharge water. This inaccurate portrayal of how automatic sprinkler systems operate has made people hesitant to install automatic sprinklers because they fear the system will cause unnecessary water damage.

The reality is quite different. In most automatic sprinkler systems, the sprinkler heads open one at a time as they are heated to their operating temperature. In most cases, only one or two sprinkler heads are needed to control the fire, producing minimal water damage.

The basic operating principles of an automatic sprinkler system are simple. A system of water pipes is installed throughout a building to deliver water to designated areas. Depending on the design and occupancy of the building, these pipes may be placed above or below the ceiling. Automatic sprinkler heads are located along this system of pipes, such that each sprinkler head covers a particular square footage of floor area. A fire in that area will activate the sprinkler head, which then discharges water on the fire. One of the major advantages of a sprinkler system is that it can function as both a fire detection system and a fire suppression system. An activated sprinkler head not only discharges water on the fire, but it also activates the building's fire alarm system, notifying the occupants and either the fire department or another entity that monitors the system. The building alarm is generated mechanically by a water-motor gong or electrically by a water flow alarm or water pressure switch. This kind of system is so effective that in many instances the sprinklers have often controlled or extinguished the fire by the time the fire fighters arrive.

Automatic Sprinkler System Components

The overall design of automatic sprinkler systems can be complex, especially in large buildings. This complexity is somewhat deceiving, however, because even the largest systems include just a few major components: the automatic sprinkler heads, piping, fittings, various types of valves, and a water supply, which may or may not include a fire pump (**FIGURE 26-23**). Most sprinkler systems are connected directly to a public water supply system.

Automatic Sprinkler Heads. An automatic sprinkler head, which is commonly referred to as a sprinkler head, is the working end of a sprinkler system. In most systems, the heads detect the fire and activate the sprinkler system to apply water to the fire. Sprinkler heads are composed of a frame, which includes the orifice (opening); a heat-sensitive element release mechanism, which holds a cap in place over the orifice; and a deflector, which directs the water in a spray

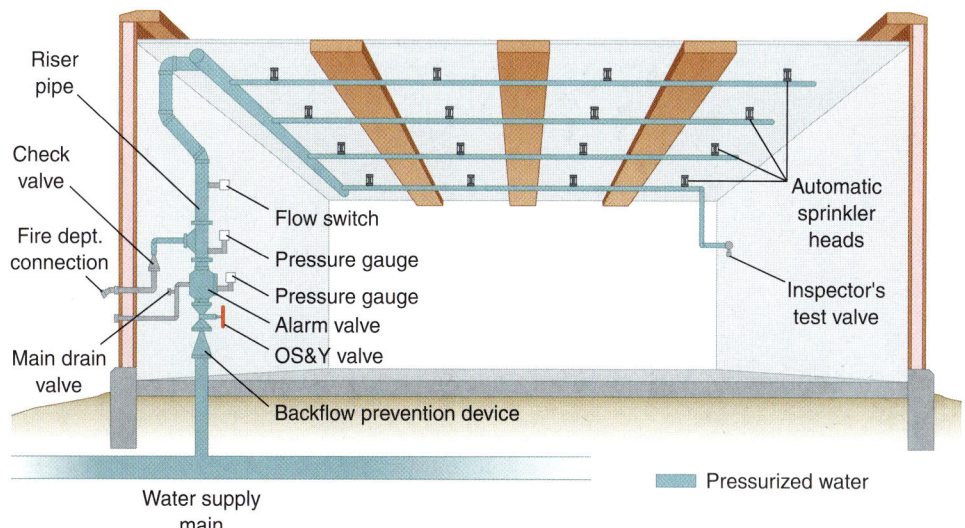

FIGURE 26-23 The basic components of an automatic sprinkler system include sprinkler heads, piping, control valves, and a water supply.
© Jones & Bartlett Learning.

pattern (**FIGURE 26-24**). Standard sprinkler heads have a 13-mm (½-in.) orifice, but several other sizes are available for special applications.

Although sprinkler heads come in several styles, all of them are categorized according to the type of release mechanism used and the intended mounting position—upright, pendent, or horizontal. They are also rated according to their release temperature. The release mechanisms hold the cap in place until the release temperature is reached. At that point, the mechanism is released, and the water pushes the cap out of the way as it discharges onto the fire (**FIGURE 26-25**).

There are three types of release mechanisms for automatic sprinkler heads. A **fusible-link sprinkler head** uses a metal alloy, such as solder, that melts at a specific temperature (**FIGURE 26-26**). This alloy links two pieces of metal that keep the cap in place.

When the designated operating temperature is reached, the solder melts and the link breaks, releasing the cap. Fusible-link sprinkler heads come in a wide range of styles and temperature ratings.

A **frangible-bulb sprinkler head** uses a glass bulb filled with glycerin or alcohol to hold the cap in place (**FIGURE 26-27**). This bulb also contains a small air bubble. As the bulb is heated, the liquid vapourizes, and the air pressure increases until the glass breaks, releasing the cap. The volume and composition of the liquid, the thickness of the glass, and the size of the air bubble determine the temperature at which the head activates as well as the speed with which it responds.

A **chemical-pellet sprinkler head** uses a plunger mechanism and a small chemical pellet to hold the cap in place. The chemical pellet liquefies when the temperature reaches a preset point. When the pellet melts, the liquid compresses the plunger, releasing the cap and allowing water to flow.

FIGURE 26-24 Automatic fire sprinkler heads consist of a frame, heat-sensitive element, orifice, orifice cap, and deflector.
Courtesy of A. Maurice Jones, Jr.

FIGURE 26-26 In fusible-link sprinkler heads, two pieces of metal are linked together by an alloy, such as solder.
© Jones & Bartlett Learning.

FIGURE 26-25 Automatic sprinkler heads are designed to activate at a wide range of temperatures. How quickly they react depends on their temperature rating.
© Jones & Bartlett Learning.

FIGURE 26-27 Frangible-bulb sprinkler heads activate when the liquid in the bulb expands and breaks the glass.
© Jones & Bartlett Learning.

Special Sprinkler Heads. Sprinkler heads can also be designed for special applications, such as covering large areas or discharging the water in extra-large droplets or as a fine mist. Some sprinkler heads have protective coatings to help prevent corrosion. Fire protection engineers and design professionals should consider these characteristics when designing the system and selecting appropriate heads. It is important to ensure that the proper heads are installed and that any replacement heads inserted in subsequent years are of the same type.

An **early-suppression fast-response sprinkler head (ESFR)** has an improved heat collector that speeds up the response and ensures the rapid release of water (**FIGURE 26-28**). ESFRs are used in large warehouses and distribution facilities where early fire suppression is important. These heads often have large orifices to discharge large volumes of water onto a fire.

Deluge Heads. A **deluge head** is easily identifiable by the fact that it has no cap or release mechanism. The orifice is always open (**FIGURE 26-29**). Deluge

FIGURE 26-29 A deluge sprinkler head has no release mechanism.
Courtesy of Tyco Fire & Building Products.

heads are used only in deluge sprinkler systems, which are covered later in this chapter.

Temperature Ratings. Sprinkler heads are rated according to their release temperature. A typical rating for sprinkler heads in a light-hazard occupancy, such as an office building, is 57°C to 79°C (135°F to 175°F). Sprinkler heads that are used in areas with warmer ambient air temperatures would have higher ratings. The rating should be stamped on the body of the sprinkler head. Frangible-bulb sprinkler heads use a colour-coding system to identify the temperature rating. Some fusible-link and chemical-pellet sprinklers also use a colour-coding system by applying paint to the frame of the sprinkler head (**TABLE 26-2**).

The temperature rating of a sprinkler head must match the anticipated ambient air temperatures. If the rating is too low for the ambient air temperature, accidental activations may occur. Conversely, if the rating is too high, the system may react too slowly, and the fire may grow too quickly for the sprinklers to control it.

Spare heads that match those used in the system should always be available onsite. Usually the spare heads are kept in a clearly marked box near the main control valve (**FIGURE 26-30**). Head wrenches specifically made for all of the different heads should also

FIGURE 26-28 The ESFR sprinkler head is larger than a standard sprinkler head and discharges about four to five times the amount of water as a standard head.

TABLE 26-2 Temperature Rating Determined by Colour of Sprinkler Head

Maximum Ceiling Temperature in °C (°F)	Temperature Rating in °C (°F)	Colour Code	Glass Bulb Colours
38 (100)	57–77 (135–170)	Uncoloured or black	Orange or red
66 (150)	79–107 (175–225)	White	Yellow or green
107 (225)	121–149 (250–300)	Blue	Blue
149 (300)	163–191 (325–375)	Red	Purple
191 (375)	204–246 (400–475)	Green	Black
246 (475)	260–302 (500–575)	Orange	Black
329 (625)	343 (650)	Orange	Black

FIGURE 26-30 Spare sprinkler heads should be kept in a special box near the main sprinkler system valve.
© Jones & Bartlett Learning.

be in the head box. Having spare heads and wrenches handy enables sprinkler systems to be returned to full service quickly.

Mounting Position. Each automatic sprinkler head is designed to be mounted in one of three positions, based on the specific application (**FIGURE 26-31**). Sprinkler heads with different mounting positions are not interchangeable because each type is designed to produce an effective water stream down or out toward the fire:

- An **upright sprinkler head** is designed to be mounted on top of the supply piping. Upright heads are usually marked SSU for "standard spray upright."

- A **pendent sprinkler head** is designed to be mounted on the underside of the sprinkler piping, hanging down toward the room. These heads are commonly marked SSP, which stands for "standard spray pendent."
- A **sidewall sprinkler head** is designed for horizontal mounting, projecting the stream out and down from the top of a wall.

Old-Style Versus New-Style Sprinkler Heads. Sprinkler heads manufactured before the mid-1950s were designed to direct part of the water stream up toward the ceiling. At the time, it was believed that this action helped cool the area and extinguish the fire. Sprinkler heads with this design are called old-style sprinklers, and many of them remain in service today.

Automatic sprinklers manufactured after the mid-1950s deflect the entire water stream down toward the fire. These types of heads are referred to as new-style heads or standard spray heads. New-style heads can replace old-style heads, but the reverse is not true. Due to different coverage patterns, old-style heads should not be used to replace any new-style heads.

Sprinkler Piping

Sprinkler piping delivers water to the sprinkler heads, and it includes the main water supply lines, risers, feeder lines, and branch lines. Although sprinkler pipes are usually made of steel, other materials can be used as well (**FIGURE 26-32**). For example, plastic pipe is sometimes used in residential sprinkler systems. The fittings join the pipe together so that the

A

B

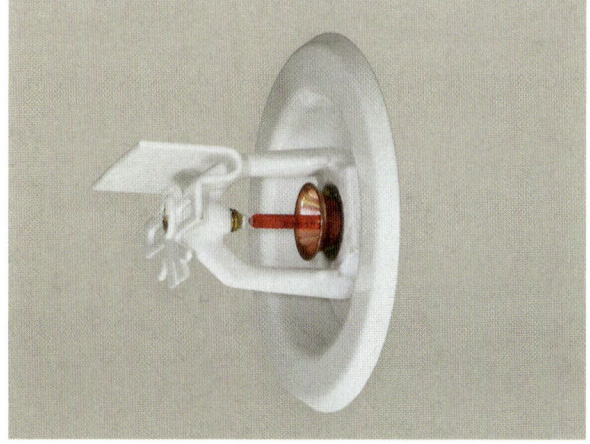

C

FIGURE 26-31 Sprinkler head mounting positions.
A. Upright. **B.** Pendent. **C.** Sidewall.

A and B: © Jones & Bartlett Learning; C: Courtesy of Tyco Fire & Building Products.

FIGURE 26-32 Most new sprinkler systems use steel pipes.

© ImageState/Alamy Images.

piping can change direction or transition between different sizes.

Most new systems are designed using computer software that takes water data and component information to create hydraulic calculations. The calculations determine if the proposed design will deliver the minimum required water at the most hydraulically remote sprinkler heads in the system. By performing hydraulic calculations, the fire protection engineer can design the size and layout of pipe to deliver the minimum water demand needed to control the fire. Near the main control valve, pipes have a large diameter; as the pipes approach the sprinkler heads, the diameter generally decreases. Before hydraulic calculations, older systems were designed using the pipe schedule method.

Valves. A sprinkler system includes several different valves, such as the main water supply control valve, different sprinkler system valves, and other, smaller valves used for testing and service. Many large systems have zone valves, which enable the water supply to different areas to be shut down without turning off the entire system. All of these valves play a critical role in the design and function of the system.

Water Supply Control Valves. Every sprinkler system must have at least one main control valve that allows water to enter the system. This water supply control valve must be of the "indicating" type, meaning that the position of the valve itself indicates whether it is open or closed. Two examples are the **outside screw and yoke (OS&Y) valve** (also called the outside stem and yoke valve) and the **post indicator valve (PIV)**.

An OS&Y valve has a threaded stem that moves in and out as the valve is opened or closed (**FIGURE 26-33**). If the stem is out, the valve is open; if the stem is in, the valve is closed. OS&Y valves are often found in a

FIGURE 26-33 An outside screw and yoke (OS&Y) valve is often used to control the flow of water into a sprinkler system.
© Jones & Bartlett Learning. Photographed by Glen E. Ellman.

FIGURE 26-34 A post indicator valve (PIV) is used to open or close an underground valve.
© Jones & Bartlett Learning.

FIGURE 26-35 A wall post indicator valve controls the flow of water from an underground pipe into a sprinkler system.
© Jones & Bartlett Learning. Courtesy of MIEMSS.

mechanical room in the building, where the water supply for the sprinkler system enters the building. In warmer climates, they may be found outside.

The PIV has an indicator that reads either open or shut depending on its position (**FIGURE 26-34**). A PIV is usually located in an open area outside the building and controls an underground valve. Opening or closing a PIV requires a wrench, which is usually attached to the side of the valve.

A **wall post indicator valve** is similar to a PIV but is designed to be mounted on the outside wall of a building (**FIGURE 26-35**).

The main control valve, whether it is an OS&Y valve or a PIV, should be locked in the open position. This ensures that the water supply to the sprinkler system is never shut off unless the proper people are notified that the system is out of service. It is critically important that the sprinkler system always remain charged with water and ready to operate, if needed.

As an alternative to locking the valves open, the valves may be electronically supervised by a supervisory switch (**FIGURE 26-36**). This device monitors the position of the valve. If someone opens or closes the valve, the supervisory switch sends a supervisory signal to the FACU, indicating a change in the valve position. If the change has not been authorized, the cause of the signal can be investigated, and the problem can be corrected.

Main Sprinkler System Valves. The type of main sprinkler system valve used depends on the type of sprinkler system installed. Options include an **alarm valve**, a **dry pipe valve**, **preaction valve**, or a **deluge valve**. These

FIGURE 26-36 A supervisory switch activates an alarm if someone attempts to close a valve that should remain open.
© Jones & Bartlett Learning.

valves are usually installed on the main riser, above the water supply control valve.

The primary functions of an alarm valve are to signal an alarm when a sprinkler head is activated and to prevent nuisance alarms caused by pressure variations and surges in the water supply to the system. This valve has a clapper mechanism that remains in the closed position until a sprinkler head opens. The closed clapper prevents water from flowing out of the system and back into the public water mains when water pressure drops.

The clapper valve is held closed by [?] high water pressure in the sprinkler piping system. When one or more sprinkler heads activate, there is a commensurate drop in the water pressure allowing the clapper valve to open. The open clapper also allows water to flow to the water-motor gong, sounding an alarm. In many new systems, electrical flow and pressure switches are installed on the valve instead of the water-motor gong to activate the system.

In dry pipe, preaction, and deluge systems, the main valve functions both as an alarm valve and as a dam, holding back the water until the sprinkler system is activated. When the system is activated, the valve opens fully so that water can enter the sprinkler piping. Dry pipe, preaction, and deluge systems are described later in this chapter.

Additional Valves. Sprinkler systems are equipped with a variety of other control valves. Several smaller valves are usually located near the main control valve, and still others are located elsewhere in the building. These smaller valves include drain valves, test valves, and connections to alarm devices. All of these valves should be properly labelled.

In larger facilities, the sprinkler system may be divided into zones, with a specific valve controlling the flow of water to each zone. This design makes maintenance easy

and can prove extremely valuable when a fire occurs. After the fire is extinguished, water flow to the affected area can be shut off so that the activated heads can be replaced. Fire protection in the rest of the building is unaffected by this shutdown.

Sprinkler System Water Supplies

The water used in an automatic sprinkler system may come from a municipal water system, from onsite storage tanks, or from alternative water sources such as storage ponds or rivers. Whatever its source, the water supply must be able to satisfy the demands of the sprinkler system as well as meet the needs of the fire department in the event of a fire.

The preferred water source for a sprinkler system is a municipal water supply, if one is available. If the municipal supply cannot meet the water pressure and volume requirements of the sprinkler system, alternative supplies must be established.

Fire pumps are used when the water comes from a static source. They may also be deployed to boost the pressure in some sprinkler systems, particularly for tall buildings (**FIGURE 26-37**). Because most municipal water supply systems do not provide enough pressure to control a fire on the upper floors of a high-rise building, fire pumps will turn on automatically when the sprinkler system activates or when the pressure drops to a preset level. In high-rise buildings, a series of fire pumps may be needed to provide adequate pressure to the upper floors.

A large industrial complex could have more than one water source, such as a municipal system and a backup storage tank (**FIGURE 26-38**). Multiple fire pumps can provide water to the sprinkler and standpipe systems in different areas through underground pipes.

FIGURE 26-37 In tall buildings, a fire pump may be needed to maintain appropriate pressure in the sprinkler system.
© Jones & Bartlett Learning.

FIGURE 26-38 Municipal water distribution systems often include elevated water storage tanks as a means to maintain consistent water pressure for potable water and firefighting purposes.
© Jones & Bartlett Learning.

Private hydrants may also be connected to the same underground system.

Each sprinkler system should also have a **fire department connection (FDC)**. This connection allows the department's pump company to pump water into the sprinkler system. The FDC may be used as either a supplement or the main source of water to the sprinkler system if the regular supply is interrupted or a fire pump fails.

The FDC usually has two or more 65-mm (2½-in.) female couplings or one large-diameter hose coupling mounted on an outside wall or placed near the building (**FIGURE 26-39**). It connects directly into the sprinkler system after the main control valve or system valve. Each fire department should establish a standard operating procedure for first-arriving companies that specifies how to connect to the FDC and when to charge the system.

In large facilities, a single FDC may be used to deliver water to all fire protection systems in the complex. The water from this connection flows into the private underground water mains instead of into each system. Water pumped into this type of FDC should come from a source that does not service the complex, such as a public hydrant on a different grid.

FIGURE 26-39 A fire department connection provides a point of connection to the fire sprinkler or standpipe system to deliver additional water and boost the pressure in the system.
Top: Courtesy of A. Maurice Jones, Jr.; bottom: © Jones & Bartlett Learning.

Water Flow Alarms. All sprinkler systems should be equipped with a method for sounding an alarm when water begins flowing in the pipes. This type of warning is important both in an actual fire and in an accidental activation. Without these alarms, the occupants or the fire department might not be aware of the sprinkler activation. If a building is unoccupied, the sprinkler system could continue to discharge water long after a fire is extinguished, leading to extensive water damage.

The alarm for many older systems is sounded by a water-motor gong (**FIGURE 26-40**). When the sprinkler system is activated and the main alarm valve opens, some water is fed through a pipe to a water-powered gong located on the outside of the building. This gong alerts people outside the building that there is water flowing. This type of alarm will function even if there is no electricity.

Accidental soundings of water-motor gongs are rare. If a water-motor gong is sounding, water is probably flowing from the sprinkler system somewhere in the building. Fire companies that arrive and hear the distinctive sound of a water-motor gong know that there is a fire or that something else is causing the sprinkler system to flow water.

Newer sprinkler systems are connected to the building's fire alarm system by either an electric **water flow alarm device** or a pressure switch. These initiating devices cause the FACU to alert the building's occupants of its activation; a monitored system will also notify the fire department. Unlike water-motor gongs, water flow alarm devices can be accidentally triggered by water pressure surges in the system. To reduce the risk of accidental activations, these devices usually have a time delay before they sound an alarm.

Types of Automatic Sprinkler Systems

Automatic sprinkler systems are classified into four categories:

- Wet pipe sprinkler systems
- Dry pipe sprinkler systems
- Preaction sprinkler systems
- Deluge sprinkler systems

Although many buildings may use the same type of system to protect the entire facility, it is not uncommon to see two or three systems combined in one building. Some facilities use a wet pipe sprinkler system to protect most of the structure but implement a dry pipe sprinkler or preaction system in a specific area. In some cases, a dry pipe sprinkler or preaction system will branch off from the wet pipe sprinkler system.

Wet Pipe Sprinkler Systems. A **wet pipe sprinkler system** is the most common and least expensive type of automatic sprinkler system. As its name implies, the piping in a wet pipe system is always filled with water. When a sprinkler head activates, water is immediately discharged onto the fire. Although most sprinkler systems are of the wet pipe design, there are potential drawbacks. If a wet pipe system sprinkler head or pipe is damaged by a forklift or other equipment, water will immediately flow from the system. Because water is in the piping at all times, wet pipe systems cannot be used in locations where pipes could freeze, such as on loading docks and in unheated warehouses.

A dry-pendent sprinkler head can be used in very small, unheated areas, such as walk-in freezers (**FIGURE 26-41**). The bottom part of a dry-pendent head, which resembles a standard sprinkler head, is mounted inside the freezer. The head has an elongated neck, usually 152 to 457 mm (6 to 18 in.) long, that extends up and connects to the wet pipe sprinkler piping

FIGURE 26-40 A water-motor gong sounds when water is flowing in a sprinkler system.
© Jones & Bartlett Learning.

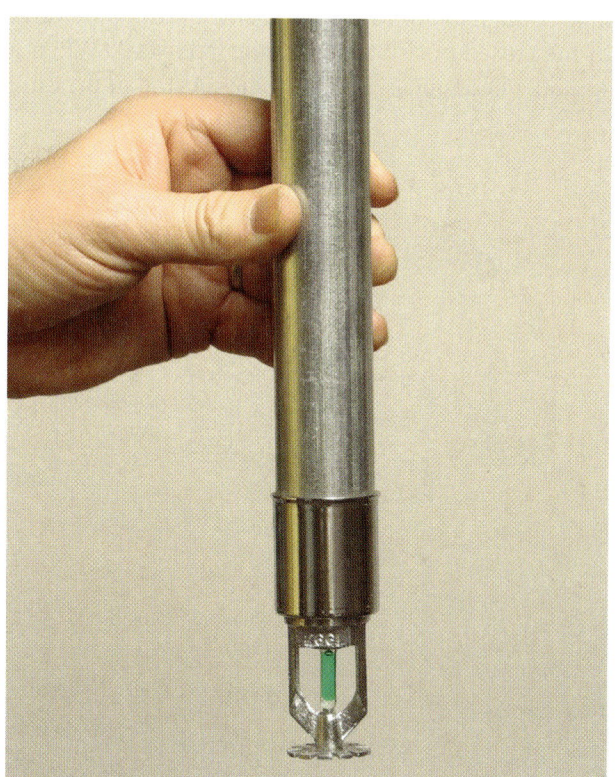

FIGURE 26-41 A dry-pendent sprinkler head can be used to protect a freezer box.
© Jones & Bartlett Learning.

in the heated area above the freezer. The vertical neck section is filled with air and capped at each end. The top cap prevents water from entering the lower section, where it would freeze; the bottom cap acts just like the cap on a standard sprinkler head. When the head is activated and the lower cap drops out, a device inside the neck releases the upper cap, and water flows from the orifice. The entire dry-pendent head assembly must be replaced after it has been activated.

Large, unheated areas, such as a loading dock, can be protected with an antifreeze loop. This small section of the wet pipe sprinkler system is filled with glycol or glycerin instead of water. A check valve separates the antifreeze loop from the rest of the sprinkler system. When a sprinkler head in the unheated area is activated, the antifreeze is discharged first, followed by water. These areas can also be protected by dry pipe sprinkler systems.

Dry Pipe Sprinkler Systems. A dry pipe sprinkler system operates much like a wet pipe sprinkler system, except that the pipes are filled with pressurized air instead of water. A dry pipe valve keeps water from entering the pipes until the air pressure is released. Dry pipe systems are used in facilities that may experience below-freezing temperatures, such as unheated warehouses, garages, or attics.

The air pressure in this kind of sprinkler system is high enough to hold a clapper inside the dry pipe valve in the closed position (**FIGURE 26-42**). When a sprinkler head opens, the air escapes. As the air pressure

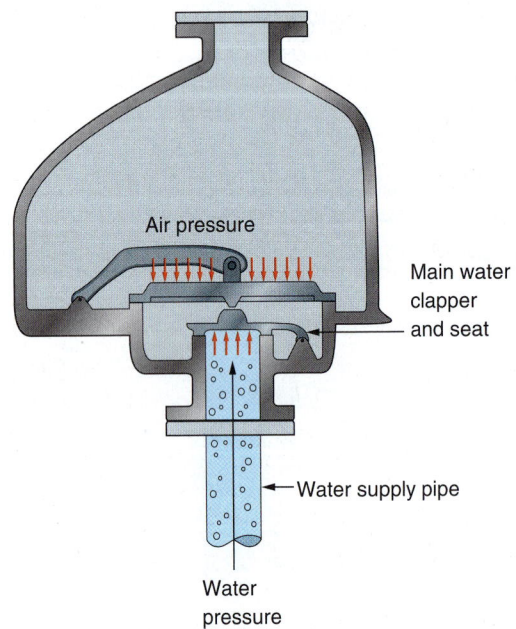

FIGURE 26-42 Water pressure on one side of the dry pipe valve is held down by air pressure on the other side of the valve.

© Jones & Bartlett Learning.

drops, the water pressure on the other side of the clapper forces it open, and water begins to flow into the pipes. When the water reaches the open sprinkler head, it is discharged onto the fire.

Dry pipe sprinkler systems do not eliminate the risk of water damage from accidental activation. If a sprinkler head breaks, the air pressure will drop, and water will flow, just as in a wet pipe sprinkler system.

The action of the clapper assembly located inside most dry pipe valves relies on a pressure differential. The system (or air) side of the clapper has a larger surface area than the supply (or wet) side. As a consequence, a lower air pressure can hold back a higher water pressure. A small compressor is used to maintain the air pressure in the system.

Dry pipe sprinkler systems should have a low air pressure switch connected to the alarm system to alert building management personnel if the air pressure drops. The activation of this alarm could mean one of two things: The compressor is not working, or there is an air leak in the system. If the air pressure in the system is too low, the clapper will open, and the system will fill with water. At that point, the system would essentially function as a wet pipe sprinkler system, which could cause it to freeze in low temperatures. In this scenario, the system would have to be drained and reset to prevent the pipes from freezing.

Dry pipe sprinkler systems must be drained after every activation so that the dry pipe valve can be reset. The clapper also must be reset, and the air pressure must be restored before the water is turned back on. During the warmer months, moisture will develop in the pipes and collect at the drains in the system. Starting in the fall and throughout the winter months, these drains must be purged of water to prevent freeze breaks that will activate the system. Some dry pipe systems use pressurized nitrogen instead of air to reduce the possibility of corrosion from condensed moisture in the piping system.

Accelerators and Exhausters. Because the pipes in dry pipe sprinkler systems are filled with air, there is a delay between the activation of a sprinkler head and the flow of water from the head. The pressurized air that fills the system must escape through the open head before the water can flow. For personal safety and property protection reasons, any delay longer than 90 seconds is unacceptable. Large systems, however, can take several minutes for the air to be purged so water can flow from the heads. To compensate for this problem, two additional devices are used: accelerators and exhausters.

An accelerator is installed at the dry pipe valve. The rapid drop in air pressure caused by an open sprinkler head activates the accelerator, which allows air pressure to flow to the supply side of the clapper valve. This quickly eliminates the pressure differential, opening the dry pipe valve and allowing the water pressure to force the remaining air out of the piping.

An **exhauster** is installed on the system side of the dry pipe valve, often at a remote location in the building. Like an accelerator, the exhauster monitors the air pressure in the piping. If it detects a drop in pressure, it opens a large-diameter portal, allowing the air in the pipes to escape. The exhauster closes when it detects water, diverting the flow to the open sprinkler heads. Large systems may have multiple exhausters located in different sections of the piping.

Preaction Sprinkler Systems.

A **preaction sprinkler system** is used when the accidental discharge of water would cause excessive damage to expensive equipment, archived documents, or artifacts. Depending on the risk of unwanted water damage, these systems may use single interlock or double interlock valves. A single interlock preaction sprinkler system is similar to a standard dry pipe system, but the piping is filled with unpressurized air. An initiating device—such as a smoke detector or a manual pull station—sends an alarm signal to the FACU, which then releases an electric preaction valve, and water enters the piping system. If a sprinkler head opens due to high temperatures, water flows from the head.

Double interlock preaction systems also use pressurized air or nitrogen to keep a preaction valve closed. For water to flow from the system, an alarm-initiating device must notify the FACU, which then opens an electric solenoid valve, and a pneumatic actuator being held shut by the supervisory air pressure holding the preaction valve closed must be released by a sprinkler head. The primary advantage of a preaction sprinkler system is its ability to prevent accidental water discharges. If a sprinkler head is accidentally broken or the pipe is damaged, the preaction valve will prevent water from entering the system. This feature makes single interlock and double interlock preaction sprinkler systems well suited for locations where water damage is a major concern.

Deluge Sprinkler Systems.

A **deluge sprinkler system** allows water to flow from all of the sprinkler heads when the alarm system detects a predetermined fire condition (**FIGURE 26-43**). A deluge system does not have closed heads that open individually at the activation temperature. The FACU opens specific deluge valves when certain initiating devices signal predetermined fire conditions, and multiple sprinkler heads flow in one or more protected areas.

Some deluge systems use a separate dry pipe system instead of the alarm system to open the deluge valve. When a pilot line detector head activates, the air pressure

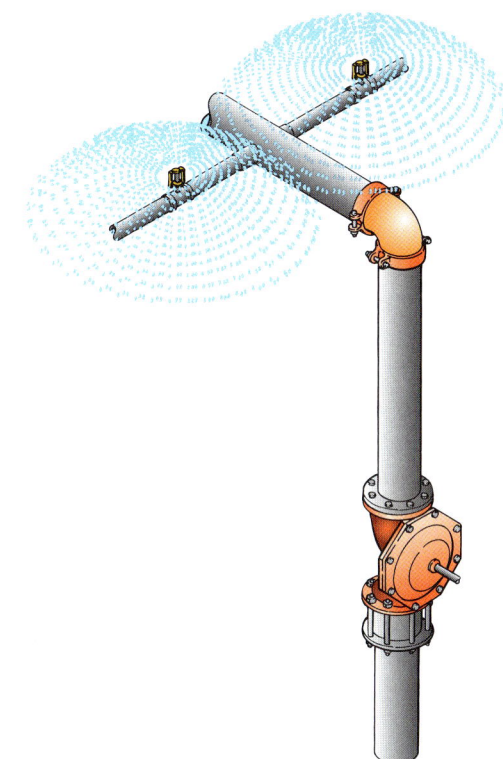

FIGURE 26-43 Water flows from all of the heads in a deluge system as soon as the system is activated.
© Jones & Bartlett Learning.

in the dry pipe system drops, and the water pressure opens the deluge valve. Water then flows through the deluge system piping and from the open deluge heads.

Deluge systems are designed to flow large volumes of water and are used in special applications such as aircraft hangars or industrial processes where rapid fire suppression is critical. In some cases, foam concentrate is added to the water so that the system will discharge a foam blanket over the hazard. Deluge systems are also used for special hazard applications, such as liquid propane gas loading stations. In these situations, a heavy deluge of water is needed to protect exposures from a large fire that ignites rapidly.

LISTEN UP!

Foam suppression systems can be permanently installed and use the same sprinkler valves and components in combination with special equipment to produce a foam solution that protects special hazards such as flammable liquid storage tanks, aircraft hangers, and chemical plants.

Water Mist Systems.

A **water mist system** is a fixed fire protection system that discharges a very fine spray of water particles from specialized sprinkler heads and spray nozzles. The particles extinguish by cooling, displacing the oxygen, or blocking the radiant heat.

Water mist systems can be configured much like any automatic fire sprinkler system: wet, dry, preaction, and deluge. However, these systems operate at high pressures and are classified into three categories based on the maximum working pressure of the system. The *low-pressure system* exposes the system distribution piping to pressures less than 1207 kilopascals (175 psi). The *intermediate-pressure system* exposes the system distribution piping to pressures between 1207 kilopascals (175 psi) and 3447 kilopascals (500 psi), and the *high-pressure system* exposes the system distribution piping to pressures in excess of 3447 kilopascals (500 psi) and, in many installations, actually requires pressures to be over 6895 kilopascals (1000 psi).

The use of water mist systems has become more prevalent because of apparent advantages: A water mist system discharges much less water than a conventional sprinkler system, the water is not hazardous or toxic, the system is self-contained and does not need to rely on a water main, and it is listed and approved for complete fire extinguishment, not only for control over a fire (as is the case for fire sprinkler systems). There are disadvantages as well. These include the finite amount of water, the need for reserve water storage tanks, the need for testing and approval of each specific application, and the need for exceptional water quality to avoid the nozzles clogging from particulate matter floating in the water.

Water mist systems protect many different high-value facilities and equipment, including computer rooms, telecommunication equipment areas, laboratories, archives, museums, and historic buildings. In addition, these systems are widely used to protect marine and offshore facilities and equipment, including ships, offshore platforms, turbine rooms, and personnel accommodation areas. However, water mist systems should not be used with materials that react with water or liquefied gases at cryogenic temperatures. In each case, violent reactions could occur, endangering lives and property.

Residential Sprinkler Systems

Residential sprinkler systems are becoming more common, with many recently constructed homes including them. Some communities require residential sprinkler systems in every newly constructed home.

The design of and theory underlying residential sprinkler systems are similar to those of commercial systems, but with some significant variations. The primary objective of a residential sprinkler system is to protect egress routes so occupants can evacuate safely (**FIGURE 26-44**). Quick response residential sprinkler heads are used in residential applications to control incipient-stage fires.

FIGURE 26-44 Special residential sprinkler heads are used in residential systems. They open more quickly than commercial heads and discharge less water.
Courtesy of A. Maurice Jones, Jr.

Residential systems typically use smaller piping and sprinkler heads with smaller orifices that discharge less water. To control costs, plastic pipe may be used instead of metal pipe. These systems typically protect high-risk areas such as bedrooms, kitchens, living rooms, and corridors; small areas such as closets and bathrooms may not be covered.

Residential systems are usually wet pipe sprinkler systems. They are typically connected to the domestic water supply from the public water main. Residential sprinkler systems do not have a fire department connection, and they are not required to be connected to a fire alarm system, unless required by the local jurisdiction. Fire codes usually require the system to include a water flow alarm.

The responding fire companies in the jurisdiction may not be familiar with every home that has a sprinkler system, but fire fighters should know both how these systems work and how to shut down a system that has been activated. Usually, the main shut-off valve for the sprinklers can be found near the location where water enters the house. If the house has a basement, the shut-off valve may be located near other utility controls. If a separate shut-off valve for the sprinkler system cannot be found, the main water shut-off for the house can be used to deactivate the sprinkler system. Installing working

SAFETY TIP

Do not shut down the water to a sprinkler system until ordered to do so by the IC. The fire must be completely out, and hose lines must be available should the fire reignite. A fire fighter should be stationed at the valve with a portable radio, ready to reopen the valve if necessary.

smoke alarms and a residential sprinkler system reduces the chance of death from a home fire by 82 percent.

Standpipe Systems

A standpipe system consists of a network of inlets, pipes, and outlets for fire hose that are built into a structure to provide water for firefighting purposes. Standpipe systems consist of a water supply, piping to carry the water closer to the fire, and one or more outlets equipped with valves to which fire hose can be connected. Water may be supplied from the municipal water system, fire department connections, or both. Standpipe systems are required in all new high-rise buildings, and they are found in many other structures, as well. Standpipes are also installed to carry water to large bridges and to supply water to limited-access highways that are not equipped with fire hydrants (**FIGURE 26-45**).

Standpipes are found in buildings both with and without sprinkler systems. In many newer buildings, sprinklers and standpipes are combined into a single system. In older buildings, the sprinkler and standpipe systems may be separate systems. Three classes of standpipes—Class I, Class II, and Class III—are distinguished based on their intended use.

Class I Standpipe

A Class I standpipe is designed for use by fire department personnel only. Each outlet has a 65-mm (2½-in.) male coupling and a valve to open the water supply after the attack line is connected (**FIGURE 26-46**). Sometimes the connection is located inside a cabinet, which may or may not be locked. Responding fire personnel carry the hose into the building with them, usually in some sort of roll, bag, or backpack. A Class I standpipe system must be able to supply an adequate volume of water with sufficient pressure to operate fire department attack lines.

Class II Standpipes

A Class II standpipe is designed for use by the building occupants. The outlets are usually equipped with a length of 38-mm (1½-in.) single-jacket hose preconnected to the system (**FIGURE 26-47**). These systems are intended to enable occupants to attack a fire before the fire department arrives; however, most building occupants are not trained to attack fires safely. If a fire cannot be controlled with a portable fire extinguisher, it is usually safer for the occupants to simply evacuate the building

FIGURE 26-45 Standpipe outlets allow fire hose to be connected inside a building.

FIGURE 26-46 A Class I standpipe provides water for fire department hose lines.

FIGURE 26-47 A Class II standpipe is intended to be used by building occupants to attack incipient-stage fires.
Courtesy of Ralph G. Johnson (nyail.com/fsd).

and call the fire department. Class II standpipes may be useful at facilities such as refineries and military bases, where workers are trained as an in-house fire brigade.

Responding fire personnel should not use Class II standpipes for fighting a fire because the hose and nozzles may be of inferior quality and lack maintenance and testing. The water flow may not be adequate to control a fire. Class II standpipe outlets are frequently connected to the domestic water piping system in the building rather than an outside main or a separate system. Instead of using equipment (hose and nozzle) that may not be reliable or adequate, fire fighters should always use department-issued equipment. Some jurisdictions now require existing Class II standpipes to be replaced with Class III systems.

Class III Standpipes

A **Class III standpipe** has the features of both Class I and Class II standpipes in a single system. This kind of system has 65-mm (2½-in.) outlets for fire department use as well as smaller outlets with attached hose for occupant use. The occupant hose may have been removed—either intentionally or by vandalism—in many facilities, so the system functions as a Class I system. Fire fighters should use only the 65-mm (2½-in.) outlets, even if they use a smaller diameter hose. The 38-mm (1½-in.) outlets may have pressure-reducing devices to limit the flow and pressure for use by untrained civilians.

Water Flow in Standpipe Systems

Standpipes are designed to deliver the required water flow at a particular pressure to each floor. The design

requirements depend on the code requirements in effect when the building was constructed.

Flow-restriction devices (**FIGURE 26-48**) or pressure-reducing valves (**FIGURE 26-49**) are often installed at the outlets to limit the pressure and flow. A vertical column of water, such as the water in a standpipe riser, exerts a backpressure (also called head pressure). In a tall building, this backpressure can be dangerously high at lower floor levels. If a hose line is connected to an outlet without a flow restrictor or a pressure-reducing valve, the water pressure could rupture

FIGURE 26-48 A flow-restriction device on a standpipe outlet can cause problems for fire fighters.
Courtesy of Dixon Valve and Coupling.

FIGURE 26-49 A pressure-reducing valve on a standpipe outlet may be necessary on lower floors to avoid problems caused by backpressure and overpressurization of sprinkler systems fed from the standpipe.
Courtesy of Dixon Valve and Coupling.

the hose, and the excessive nozzle pressure could make the line difficult or dangerous to handle.

If flow restrictors or pressure-reducing valves are not properly installed and maintained, these devices can cause problems for fire fighters. An improperly adjusted pressure-reducing valve could severely restrict the pressure to a hose line. Likewise, a flow-restriction device could limit the flow of water to fight a fire.

The flow and pressure capabilities of a standpipe system should be determined during preincident planning. Many older standpipe systems deliver water at a pressure of only 448 kPa (65 psi) or less at the top of the building. The combination fog/straight-stream nozzles used by many fire departments are designed to operate at 689 kPa (100 psi). For this reason, many fire departments use low-pressure combination nozzles for fighting fires in high-rise buildings or require the use of only smooth-bore nozzles when operating from a standpipe system.

> ### LISTEN UP!
>
> It is advisable to treat standpipe systems the same way as one would a fire hydrant and flush it immediately before connecting an attack or supply line. Slag and other debris can flow into hose lines and clog nozzles, especially combination nozzles.

Preincident planning for high-rise buildings should include an evaluation of the building's standpipe system and a determination of the anticipated flows and pressures. This information should be used to make decisions about the appropriate nozzles and tactics for those buildings.

Pump companies that respond to buildings equipped with standpipes should carry a kit that includes the appropriate hose and nozzle, a spanner wrench (hose key), and any required adapters. This kit should also include tools to adjust the settings of pressure-reducing valves or to remove restrictors that are obstructing flows.

Standpipe Water Supplies

Standpipe systems and sprinkler systems are supplied with water in essentially the same way. A wet standpipe system has water in the piping system at all times while a dry standpipe system contains air or nitrogen until it is needed for fire suppression.

Many wet standpipe systems in modern buildings are connected to a public water supply and are equipped with an electric or diesel fire pump to provide additional pressure. Some also have a water storage tank that serves as a backup supply. In these systems, the FDC on the outside of the building can be used to increase the rate of flow, boost the pressure, or provide additional water capacity from an alternative source. An **automatic wet standpipe system** provides adequate volume and pressure at the discharge at all times. The fire department needs only to connect a hose to the discharge and open the valve to provide water for fire suppression.

Dry standpipe systems are found in many older buildings. They are also used if freezing weather is a problem, such as in open parking structures, bridges, and tunnels. A **manual dry standpipe system** does not have a permanent connection to a water supply, so the FDC must be used to pump water into the system. If a fire occurs in a building with manual dry standpipes, connecting the hose lines to the FDC and charging the system with water are high priorities.

An **automatic dry standpipe system** is connected to a water supply, but water is kept out of the piping system by a dry pipe valve, similar to those found in dry pipe sprinkler systems. In such systems, water flows through the system only after the alarm system opens the dry pipe valve.

High-rise buildings often incorporate complex systems of risers, storage tanks, and fire pumps to deliver the needed flows to upper floors. The details of these systems should be obtained during preincident planning surveys. Department procedures should dictate how responding units will supply the standpipes with water as well as how crews should use the standpipes inside the building.

Special Hazard Suppression Systems

Automatic sprinkler systems are used to protect whole buildings or major sections of buildings. In certain situations, more specialized suppression systems are needed. These kinds of systems are often used in areas where water would not be an acceptable extinguishing agent (**FIGURE 26-50**). For example, water is not the extinguishing agent of choice for areas containing sensitive electronic equipment or contents such as computers, valuable books, or documents. Water is also incompatible with materials such as flammable liquids or water-reactive chemicals. Special hazard suppression systems must be used in these applications.

Dry-Chemical and Wet-Chemical Extinguishing Systems

Dry-chemical and wet-chemical extinguishing systems are the most common specialized agent systems. Some

FIGURE 26-50 Specialized extinguishing systems are used in areas where water would not be effective or desirable.
© Jones & Bartlett Learning.

FIGURE 26-51 Dry-chemical extinguishing systems are installed at many self-service gasoline filling stations.
© Jones and Bartlett Learning. Photographed by Christine McKeen.

FIGURE 26-52 Wet-chemical extinguishing systems are used in most new commercial kitchens.
Courtesy of A. Maurice Jones, Jr.

gas stations have dry-chemical systems that protect the dispensing areas. These systems are also installed inside buildings to protect areas where flammable liquids are stored or used. In commercial kitchens, wet-chemical agents are used to protect the cooking areas and exhaust systems.

A **dry-chemical extinguishing system** uses the same types of finely powdered agents as a portable dry-chemical fire extinguisher (**FIGURE 26-51**). The agent is kept in self-pressurized tanks or in tanks with an external supply of pressurized inert gas that provides pressure and fluidizes the dry-chemical agent when the system is activated. The chemical agent flows through the piping system and is discharged by nozzles directed at the target area.

A **wet-chemical extinguishing system** is used in new commercial kitchens (**FIGURE 26-52**). This system uses a liquid Class K extinguishing agent, which is much more effective on cooking fats than are the dry chemicals used in older kitchen systems. Wet-chemical systems are also easier to clean up after a discharge, so

the kitchen can resume operations more quickly after the system has discharged.

Many kitchen systems discharge the extinguishing agent into the ductwork above the exhaust hood as well as onto the cooking surface. This approach helps prevent a fire from igniting any grease build-up inside the ductwork and spreading throughout the system. Although the ductwork should be cleaned regularly, it is not unusual for a kitchen fire to extend into the exhaust system.

With both dry-chemical and wet-chemical extinguishing agent systems, fusible-link or other automatic initiating devices are placed above the target hazard to activate the system (**FIGURE 26-53**). A manual pull station is also provided so that workers can activate the system if they discover a fire (**FIGURE 26-54**). Open nozzles are located over the target areas to discharge the agent directly onto a fire. When the system is activated, the extinguishing agent flows out of all the nozzles.

Most dry- and wet-chemical suppression systems are connected to the building's fire alarm system. Many dry- and wet-chemical systems are designed to shut down electricity, fuel gases, burners, exhaust fans, or other equipment.

Gaseous Suppression Systems

A **gaseous suppression system** is often installed in areas where computers or sensitive electronic equipment is

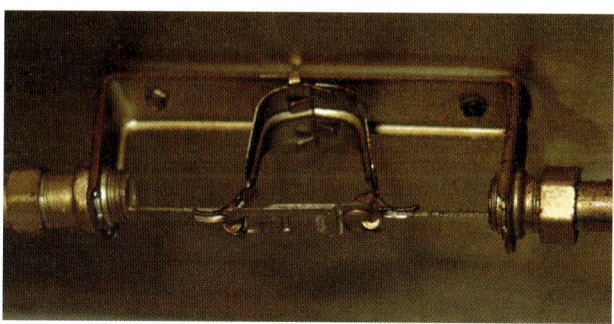

FIGURE 26-53 Fusible links can be used to activate a specialized extinguishing system.
© Jones & Bartlett Learning.

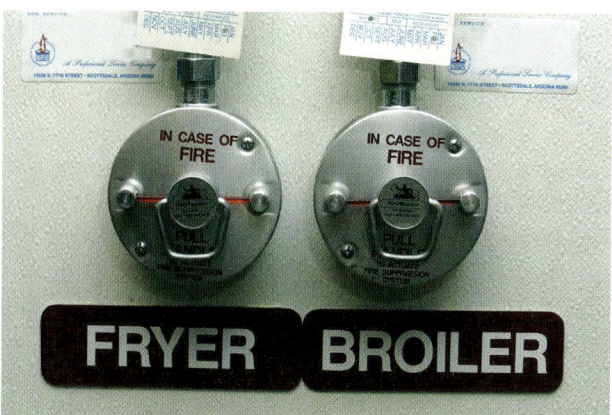

FIGURE 26-54 Most special extinguishing systems can also be manually activated.
© Jones & Bartlett Learning.

used or where valuable documents are stored. Gaseous agents include halogenated agents, halocarbon agents, inert gases, and carbon dioxide. These gases are stored under pressure as liquids, and they are discharged as gases through an engineered piping and nozzle system. Gaseous agents suppress fires by two different suppression mechanisms: They reduce the oxygen level in a room or building by displacing ambient air, or they chemically inhibit the growth of fire. Gaseous agents can be very effective in certain applications because they can extinguish a fire without causing significant damage to the room or contents.

Halogenated Agents

Halogenated agents (also known as Halons) were used to extinguish engine fires in military aircraft in the 1930s. After that time, they were widely used until the 1990s to protect areas such as computer rooms, telecommunications rooms, and other sensitive areas. Halon 1301 was the most common halogenated agent for a total flooding protection system because it is a nontoxic, odourless, colourless gas that leaves behind no residue. It is very effective at extinguishing fires because it interrupts the chemical reaction of fire by sequestering free radicals released during the combustion process. These free radicals are necessary for the chemical chain reaction that allows fires to grow. Unfortunately, the ability of Halons to attach to free radicals also allows these agents to combine with the ozone layer in the earth's upper atmosphere. In 1987, the Montreal Protocol banned the manufacturing and importation of new Halon suppression agents and many other fluorocarbons that have contributed to the ozone depletion. Although new Halon suppression agents cannot be manufactured or imported from other counties, many existing systems are still in use today. Only reclaimed Halon can be used if a system discharges, but eventually, this supply will be depleted. Unfortunately, there currently is no available "drop in" agent that can replace Halon 1301 in an existing system. Once a Halon system is discharged and there is no reclaimed Halon available, the Halon system must be replaced with an entirely new, different type of system. Old systems that are no longer used must be decommissioned by a certified company that captures the existing Halon agent according to strict environmental policies.

Clean Agent Systems

A **clean agent extinguishing system** extinguishes fires by one or more different suppression mechanisms. In some clean agent systems, inert gases are released into a room or building and displace enough ambient air to lower the oxygen content below the level required for combustion. In other systems, halocarbons—not to be confused with the original Halons—extinguish fires by interrupting the chemical chain reaction of the fire or by cooling the fuel. Some halocarbons use both mechanisms. Like Halons, clean agents are stored in pressurized containers as liquids. When the system is activated, the liquid vapourizes and is discharged as a gas. Most clean agents are used in total flooding applications where an entire room or enclosed area is flooded with a specific concentration of the agent, but some systems apply the agent directly to a hazardous process or piece of equipment. Fire alarm initiating devices installed in the hazard area activate the system, although a manual discharge button is also provided with most installations. Discharge is usually delayed 30 to 60 seconds after the detector is activated to allow workers to evacuate the area.

During this delay (the pre-alarm period), a manual override switch can be used to stop the discharge. In some systems, the manual override switch must be pressed until the fire alarm system is reset; releasing the override switch too soon causes the system to discharge. If the system does activate, the clean agent system should be completely discharged within a few seconds, depending on the design specifications. Fire fighters entering the area must use self-contained breathing apparatus (SCBA) until the area has been properly ventilated. Although the gaseous agents used in clean agent extinguishing systems are not considered immediately dangerous to life and health, some gaseous agents release toxic by-products when heated to extreme temperatures. Some gaseous agents reduce the ambient oxygen level to extremely dangerous levels. Fire fighters entering the area must use strict accountability procedures to ensure the safety of all personnel. Clean agent extinguishing systems should be connected to the building's fire alarm system and indicated as a special hazard suppression zone on the control unit. This notification alerts fire fighters that they are responding to a situation where a clean agent has discharged. If the system has a preprogrammed delay, the pre-alarm should activate the building's fire alarm system.

Carbon Dioxide Extinguishing Systems

A **carbon dioxide extinguishing system** is similar in design to an inert gas system. Carbon dioxide (CO_2) extinguishes a fire by displacing the oxygen in the room and smothering the fire. Concentrations of 34 percent carbon dioxide are required to reduce oxygen concentration below 14 percent, so large quantities of the agent are required for total flooding applications (**FIGURE 26-55**).

Fire fighters must use extreme caution when entering a building in which a CO_2 system has discharged. Oxygen levels will be dangerously low; therefore, full

FIGURE 26-55 Carbon dioxide extinguishes a fire by displacing the oxygen in the room and smothering the fire.
Courtesy of Chemetron Fire Systems.

personal protective equipment (PPE) and SCBA are required. Because CO_2 is heavier than air, it tends to settle at the floor or sink to rooms below the fire area. Fire fighters must perform search and rescue operations in the discharge area and on the floors below to look for occupants who may have become incapacitated by the low-oxygen environment. As with other gaseous agents, a strong incident command system (ICS) and personnel accountability system are essential for fire fighter safety.

Carbon dioxide extinguishing systems should be connected to the building's fire alarm system and identified on the FACU or annunciator. Preincident planning can help responding personnel plan for emergency response procedures if the CO_2 system has discharged.

Smoke Control Systems

Fires in buildings often generate large amounts of smoke and superheated toxic gases that can quickly disable, injure, and kill occupants. Smoke can also limit visibility so that occupant evacuation is delayed or prevented. Studies indicate that high temperatures and smoke levels below 1.8 m (6 ft) above the floor significantly reduce the likelihood of occupants escaping from a fire. In 1980, a disastrous fire in the MGM Grand Casino and Hotel claimed the lives of 85 people. Although the fire started in the first-floor casino, most of the deaths occurred in the upper floors of the hotel due to smoke inhalation. More than half of the victims on the upper floors of the building were overcome by smoke that filled the stairways, corridors, and elevator shafts.

Smoke and other by-products of combustion can move throughout a building due to convection, the stack effect, wind, and the flow path. Many buildings now incorporate smoke control systems into their overall fire protection design to minimize the movement of smoke. These systems are designed by fire protection engineers to limit smoke spread and, in some cases, remove the smoke from the building. Coupled with comprehensive fire alarm systems and fire suppression systems, smoke control systems provide a safer environment for both occupants and fire fighters. These systems may also reduce property damage in areas remote from the fire area.

NFPA 92, *Standard for Smoke Control Systems*, provides performance objectives and design fundamentals for various types of passive and active methods to limit the spread of smoke and toxic gases. A **smoke control system** uses a combination of methods that modify smoke movement to provide a safer environment for occupants. Smoke containment systems and smoke management systems, specifically, use natural or mechanical systems to provide safe egress routes or to control the migration of smoke from the fire area to remote areas of the building. Smoke containment systems are primarily designed to keep smoke and other gases in the area or floor of fire origin. Smoke management systems are used to move or exhaust smoke that escapes from the fire area and threatens egress routes, elevator shafts, stairwells, or refuge areas.

There are two ways to contain or control smoke: passive building design and active smoke control systems.

Passive Smoke Control

Passive smoke control uses building features such as fire-rated walls, automatic fire doors, draft curtains, and other smoke barriers that limit the spread of smoke (**FIGURE 26-56**). These building components are designed by fire protection engineers based on the building's size, design, occupancy type, and fire hazards. Passive smoke management uses the building's features to compartmentalize larger spaces to confine the smoke to the room or area of origin. These features require no human intervention because they are built into the design of the building.

Active Smoke Control

Active smoke control uses mechanical systems engineered to prevent the unwanted migration of smoke from the fire area to uninvolved floors or areas of the building. These systems are also used to intentionally move smoke away from occupants or from egress routes. Like all fire protection systems, active smoke control systems are designed by fire protection engineers to meet the performance objectives for each specific building. Active smoke control methods include creating pressure differences between smoke control zones or floors (sometimes called a pressure sandwich) and pressurization of stairwells, refuge areas, and elevators shafts and other vertical shafts. Some systems exhaust smoke from large open areas, such as an atrium, indoor arena, or airport terminal.

Each of these systems is engineered for the specific performance objectives of each building. These systems usually interface with the FACU and are activated when the fire alarm detects a fire condition. Some sophisticated systems allow fire fighters or building personnel to manually control specific functions of the smoke control system from a smoke control panel (**FIGURE 26-57**). Smoke control systems must be inspected and maintained like any other fire protection system to ensure that they will function as designed during a fire emergency.

FIGURE 26-56 The passive design approach uses construction barriers, such this draft curtain, to contain and minimize the amount of smoke leakage from a fire area.
Courtesy of A. Maurice Jones, Jr.

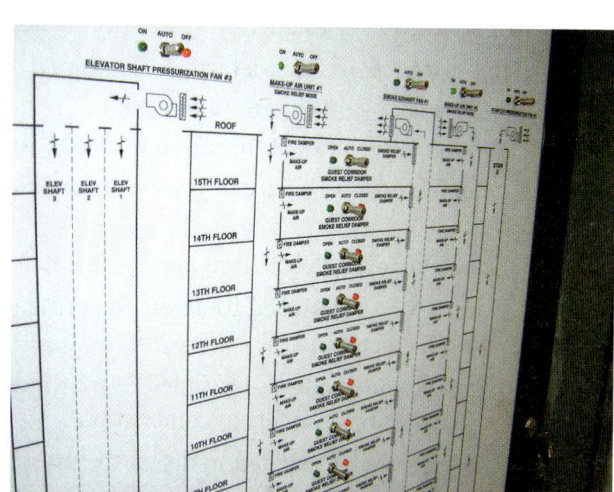

FIGURE 26-57 The smoke control panel must have status indication and control functions for each major smoke control system.
© Jones & Bartlett Learning.

After-Action REVIEW

IN SUMMARY

- Fire protection systems include fire alarms, automatic fire detection, fire suppression, and smoke control systems.

- Fire alarms range from simple, single-station smoke alarms for private homes to complex fire alarm systems for industrial complexes or high-rise buildings. Many fire alarm systems in large buildings also control other systems to help protect occupants and control the spread of fire and smoke.

- A fire alarm system has three basic components:
  - Alarm initiating device—Either an automatic or manually operated device that, when activated, causes the system to indicate an alarm.
  - Alarm notification appliance—Generally an audible device, but it is often accompanied by a visual device, which alerts the building occupants.
  - Fire alarm control unit (FACU)—Links the initiating device to the notification appliance and performs other essential functions.

- The most common type of residential fire alarm system is a single-station smoke alarm. This life-safety device includes a smoke detection sensor, an automatic control unit, and an audible alarm within a single unit. It alerts occupants when a fire occurs.

- NFPA 72 requires new homes to have a smoke alarm in every bedroom, in the corridor outside every bedroom, and on every floor level. It also requires a backup battery for hard-wired smoke alarms in the event of a power failure.

- Two types of fire detection technology may be used in a smoke alarm to detect combustion:
  - Ionization detectors use radioactive material within the device to detect invisible products of combustion.
  - Photoelectric detectors use a light beam to detect the presence of visible particles of smoke.

- Alarm initiating devices begin the fire alarm process either manually or automatically.

- Manual alarm initiating devices include single-action pull stations and double-action pull stations. Automatic initiating devices sense fire using an ionization detector or a photoelectric chamber to detect products of combustion and trigger an audible alarm.

- Various types of automatic initiating devices are available:
  - Smoke detectors—Designed to sense the presence of smoke.
  - Beam detectors—Photoelectric smoke detectors used to protect large open areas.
  - Heat detectors—Designed to sense specific heat conditions.
  - Fixed-temperature heat detectors—Designed to activate at a preset temperature.
  - Rate-of-rise heat detectors—Designed to activate if the temperature of the surrounding air increases by more than a set amount in a given period of time.
  - Line detectors—Use wire or tubing strung along the ceiling of large open areas to detect an increase in heat.
  - Flame detectors—Specialized devices that detect the electromagnetic light waves produced by a flame.
  - Gas detectors—Designed to detect the presence of a specific gas that is created by combustion or that is used in the facility.
  - Air sampling detectors—Designed to continuously capture air samples and measure the concentrations of specific gases or products of combustion.
  - Duct detectors—Designed to detect the presence of smoke in air distribution systems and shut down the air-handling unit and any associated equipment.

- Residential carbon monoxide detectors are designed to sound an audible or visual alarm when the concentration of carbon monoxide is high enough to pose a health risk to the occupants of the building.

- Three distinct types of nuisance alarms are possible: malicious alarms, alarms due to environmental conditions, and alarms caused by improper installation or maintenance.
- Cross-zoned systems and alarm verification features can be used to reduce unwanted and nuisance alarms caused by smoke detection systems.
- Some fire alarm systems give little information at the alarm control unit, while others specify exactly which initiating device activated the fire alarm. Alarm annunciator systems can be subdivided based on whether they are zoned or coded systems:
  - Noncoded—No information is given on the control unit.
  - Coded—The system indicates over the audible warning device which zone has been activated.
  - Zoned—The system divides initiating devices into zones based on the type or location of the devices.
- There are five categories of fire department notification systems:
  - Protected premises fire alarm systems—Sounds an alarm only in the building where it was activated.
  - Remote supervisory station alarm systems—Sounds an alarm in the building and transmits a signal to a remote location.
  - Public emergency alarm reporting systems—Sounds an alarm in the building and transmits a signal to the fire department via a public alarm box system.
  - Proprietary supervising alarm systems—Sounds an alarm in the building and transmits a signal to a monitoring location owned and operated by the facility's owner.
  - Central station service alarm systems—Sounds an alarm in the building and transmits a signal to an off-premises alarm monitoring facility.
- Fire suppression systems include sprinkler systems, standpipe systems, and special hazard suppression systems.
- The most common type of fire suppression system is the automatic sprinkler system.
- The basic operating principles of an automatic sprinkler system are simple. A system of water pipes is installed throughout a building to deliver water to every area where a fire might occur. Depending on the design and occupancy of the building, these pipes may be placed above or below the ceiling. Automatic sprinkler heads are located along this system of pipes, such that each sprinkler head covers a particular floor area. A fire in that area will activate the sprinkler head, which then discharges water on the fire.
- Every automatic sprinkler system includes four major components:
  - Automatic sprinkler heads
  - Piping
  - Control valves
  - Water supply
- Automatic sprinkler systems are divided into four categories:
  - Wet pipe sprinkler systems
  - Dry pipe sprinkler systems
  - Preaction sprinkler systems
  - Deluge sprinkler systems
- A standpipe system consists of a network of inlets, pipes, and outlets for fire hose that are built into a structure to provide water for firefighting purposes. Three classifications of standpipes are distinguished based on their intended use:
  - Class I—Designed for use by fire department personnel only.
  - Class II—Designed for use by the building occupants.
  - Class III—Has the features of both Class I and Class II standpipes in a single system.
- Special hazard suppression systems may be installed to protect areas where water may not be used, such as computer rooms.

- Smoke control systems are designed to contain smoke and gases in the area of origin and provide a tenable environment for relocation and evacuation.
- Smoke control systems use two approaches: passive—the use of building components to form physical barriers, and active—uses mechanical equipment to create air pressure that is introduced into spaces to prevent smoke and gases from leaving the fire area or to remove smoke and gases from egress routes.

KEY TERMS

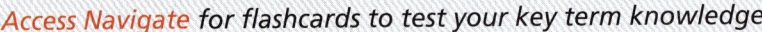 *Access Navigate for flashcards to test your key term knowledge.*

Accelerator A device that speeds up the removal of the air from a dry pipe or preaction sprinkler system. An accelerator reduces the time required for water to start flowing from sprinkler heads.

Air sampling detector A system that captures a sample of air from a room or enclosed space and passes it through a smoke detection or gas analysis device.

Alarm initiating device An automatic or manually operated device in a fire alarm system that, when activated, causes the system to indicate an alarm condition.

Alarm notification appliance An audible and/or visual device in a fire alarm system that makes occupants or other persons aware of an alarm condition.

Alarm valve A valve that signals an alarm when a sprinkler head is activated and prevents nuisance alarms caused by pressure variations.

Alarm verification feature A feature of automatic fire detection and alarm systems to reduce unwanted alarms wherein smoke detectors report alarm conditions for a minimum period of time or confirm alarm conditions within a given time period after being reset, in order to be accepted as a valid alarm initiation signal. (NFPA 72)

Automatic dry standpipe system A standpipe system permanently attached to a water supply capable of supplying the system demand at all times, containing air or nitrogen under pressure, the release of which opens a dry pipe valve to allow water to flow into the piping system and out of the opened hose valve. (NFPA 14)

Automatic sprinkler heads The working ends of a sprinkler system, which serve to activate the system and to apply water to the fire.

Automatic sprinkler system A system of pipes filled with water under pressure that discharges water immediately when a sprinkler head opens.

Automatic wet standpipe system A standpipe system containing water at all times that is attached to a water supply capable of supplying the system demand and that requires no action other than opening a hose valve to provide water at hose connections. (NFPA 14)

Beam detector A smoke detection device that projects a narrow beam of light across a large open area from a sending unit to a receiving unit. When the beam is interrupted by smoke, the receiver detects a reduction in light transmission and activates the fire alarm.

Bimetallic strip A device with components made from two distinct metals that respond differently to heat. When heated, the metals will bend or change shape.

Carbon dioxide extinguishing system A fire suppression system designed to protect either a single room or series of rooms by flooding the area with carbon dioxide.

Central station service alarm system A system or group of systems in which the operations of circuits and devices are transmitted automatically to, recorded in, maintained by, and supervised from a listed central station that is controlled and operated by a person, firm, or corporation whose business is the furnishing, maintaining, or monitoring of supervised alarm systems. (NFPA 72)

Chemical-pellet sprinkler head A sprinkler head activated by a chemical pellet that liquefies at a preset temperature.

Class I standpipe A standpipe system designed for use by fire department personnel only. Each outlet should have a valve to control the flow of water and a 65-mm (2½-in.) male coupling for fire hose.

Class II standpipe A standpipe system designed for use by occupants of a building only. Each outlet is generally equipped with a length of 38-mm (1½-in.) single-jacket hose and a nozzle, which are preconnected to the system.

Class III standpipe A combination system that has features of both Class I and Class II standpipes.

Clean agent extinguishing system A self-contained extinguishing system that expels an electrically non-conducting, volatile, or gaseous fire extinguishant that does not leave a residue upon evaporation.

Coded system A fire alarm system design that divides a building or facility into zones and has audible notification devices that can be used to identify the area where an alarm originated.

Cross-zoned system A fire alarm system that requires activation of two separate detection devices before initiating an alarm condition. If a single detection device is activated, the alarm control unit will usually show a problem or trouble condition.

Deluge head A sprinkler head that has no release mechanism; the orifice is always open.

Deluge sprinkler system A sprinkler system employing open sprinklers or nozzles that are attached to a piping system that is connected to a water supply through a valve that is opened by the operation of a detection system installed in the same areas as the sprinklers or the nozzles. When this valve opens, water flows into the piping system and discharges from all sprinklers or nozzles attached thereto. (NFPA 13)

Deluge valve A water supply control valve intended to be operated by actuation of an automatic detection system that is installed in the same area as the discharge devices. (NFPA 25)

Double-action pull station A manual fire alarm activation device that requires two steps to activate the alarm. The user must push in a flap, lift a cover, or break a piece of glass before activating the alarm.

Dry-chemical extinguishing system An automatic fire-extinguishing system that discharges a dry-chemical agent.

Dry pipe sprinkler system A sprinkler system employing automatic sprinklers that are attached to a piping. (NFPA 13)

Dry pipe valve The valve assembly on a dry pipe sprinkler system that prevents water from entering the system until the air pressure is released.

Duct detector A type of smoke detector that samples the air through the air distribution system ductwork or plenum; upon detecting smoke, the detector sends a signal to shut down the air distribution unit, close any associated smoke damper, or initiate smoke control system operation.

Early-suppression fast-response sprinkler head (ESFR) A sprinkler head designed to react quickly and suppress a fire in its early stages.

Exhauster A device that accelerates the removal of the air from a dry pipe or preaction sprinkler system.

Fire alarm A warning signal that alerts occupants of a fire emergency.

Fire alarm control unit (FACU) A component of the fire alarm system, provided with primary and secondary power sources, which receives signals from initiating devices or other fire alarm control units and processes these signals to determine part of all of the required fire alarm system output function(s). (NFPA 72)

Fire alarm signal A signal that results from the manual or automatic detection of a fire alarm condition. (NFPA 72)

Fire department connection (FDC) A connection through which the fire department can pump supplemental water into the sprinkler system, standpipe, or other system furnishing water for fire extinguishment to supplement existing water supplies. (NFPA 25)

Fire protection system Any fire alarm device or system or fire-extinguishing device or system, or combination thereof, that is designed and installed for detecting, controlling, or extinguishing a fire or otherwise alerting occupants, or the fire department, or both. (NFPA 914)

Fixed-temperature heat detector A device that responds when its operating element becomes heated to a predetermined level. (NFPA 72)

Flame detector A radiant energy-sensing fire detector that detects the radiant energy emitted by a flame. (NFPA 72)

Frangible-bulb sprinkler head A sprinkler head with a liquid-filled bulb. The sprinkler head becomes activated when the liquid is heated and the glass bulb breaks.

Fusible-link sprinkler head A sprinkler head with an activation mechanism that incorporates two pieces of metal held together by low-melting-point solder. When the solder melts, it releases the link, and water begins to flow.

Gas detector A device that detects the presence of a specified gas concentration. Gas detectors can be either spot-type or line-type detectors. (NFPA 72)

Gaseous suppression system A system often installed in areas where computers or sensitive electronic equipment is used or where valuable documents are stored.

Halogenated agents A liquefied gas extinguishing agent that extinguishes fire by chemically interrupting the combustion reaction between fuel and oxygen. Halogenated agents leave no residue. (NFPA 402)

Heat detector A fire detector that detects either abnormally high temperature or rate of temperature rise or both. (NFPA 72)

Ionization smoke detection The principle of using a small amount of radioactive material to ionize the air between two differentially charged electrodes to sense the presence of smoke particles. Smoke particles entering the ionization volume decrease the conductance of the air by reducing ion mobility. The reduced conductance signal is processed and used to convey an alarm condition when it meets preset criteria. (NFPA 72)

Line detector Wire or tubing that can be strung along the ceiling of large open areas to detect an increase in heat.

Malicious alarm An unwanted activation of an alarm initiating device caused by a person acting with malice. (NFPA 72)

Manual dry standpipe system A standpipe system with no permanently attached water supply that relies exclusively on the fire department connection to supply the system demand. (NFPA 14)

Manual pull station A device with a switch that either opens or closes a circuit, activating the fire alarm.

Noncoded system An alarm system that provides no information at the alarm control unit indicating where the activated alarm is located.

Nuisance alarm An unwanted activation of a signaling system or an alarm initiating device in response to a stimulus or condition that is not the result of a potentially hazardous condition. (NFPA 72)

Obscuration rate A measure of the percentage of light transmission that is blocked between a sender and a receiver unit.

Outside screw and yoke (OS&Y) valve A sprinkler control valve with a valve stem that moves in and out as the valve is opened or closed.

Pendent sprinkler head A sprinkler head designed to be mounted on the underside of sprinkler piping so that the water stream is directed in a downward direction.

Photoelectric smoke detector A detector that uses a light beam and a photocell to detect larger visible particles of smoke. When visible particles of smoke enter the inner chamber, they reflect some of the light onto the photocell, thereby activating the alarm.

Post indicator valve (PIV) A sprinkler control valve with an indicator that reads either open or shut depending on its position.

Preaction sprinkler system A sprinkler system employing a piping system that contains air that might or might not be under pressure, with a supplemental detection system installed in the same areas as the sprinklers. (NFPA 13)

Preaction valve A type of sprinkler system valve that holds back water until a manual emergency release is activated, a fire sprinkler head activates, a fire detector activates, or a combination of sprinkler head and initiating device activation takes place.

Proprietary supervising alarm system A fire alarm system that transmits a signal to a monitoring location owned and operated by the facility's owner.

Protected premises fire alarm system A fire alarm system that sounds an alarm only in the building where it was activated; that is, no signal is sent out of the building.

Public emergency alarm reporting system A system of alarm initiating devices, transmitting and receiving equipment, and communication infrastructure (other than a public telephone network) used to communicate with the communications centre to provide any combination of manual or auxiliary alarm service. (NFPA 72)

Rate-of-rise heat detector A device that responds when the temperature rises at a rate exceeding a predetermined value. (NFPA 72)

Remote annunciator A secondary fire alarm notification unit, usually located near the front door of a building, that provides event information and frequently has the same control functionality as the main fire alarm control unit.

Remote supervising station alarm system A fire alarm system that sounds an alarm in the building and transmits a signal to the fire department or an off-premises monitoring location.

Residential sprinkler system A sprinkler system designed to protect dwelling units.

Sidewall sprinkler head A sprinkler that is mounted on a wall and discharges water horizontally into a room.

Single-action pull station A manual fire alarm activation device in which the user takes a single step—such as moving a lever, toggle, or handle—to activate the alarm.

Single-station smoke alarm A detector comprising an assembly that incorporates a sensor, control components, and an alarm notification appliance in one unit operated from a power source either located in the unit or obtained at the point of installation. (NFPA 72)

Smoke control system A mechanical system that can create positive or negative pressure to control, alter, and limit the spread of smoke and gases.

Smoke detector A device that detects visible and invisible products of combustion (smoke) and sends a signal to a fire alarm control unit.

Spot-type detector A single smoke detector or heat detector device. These devices are often spaced throughout an area.

Sprinkler piping The network of piping in a sprinkler system that delivers water to the sprinkler heads.

Standpipe system An arrangement of piping, valves, and hose connections installed in a structure to deliver water for fire hose.

Supervising station alarm system An alarm system that communicates between the protected property and a constantly attended location that will receive, interpret, and take the appropriate action to handle the alarm condition.

Supervisory signal A signal that results from the detection of a supervisory condition that indicates the need for action. (NFPA 72)

Temporal-3 pattern A standard fire alarm audible signal for alerting occupants of a building.

Trouble signal A signal initiated by a dispatch system or device indicative of a fault in a monitored circuit or component. (NFPA 720)

Unwanted alarm Any alarm that occurs that is not the result of a potentially hazardous condition. (NFPA 72)

Upright sprinkler head A sprinkler head designed to be installed on top of the supply piping; it is usually marked SSU ("standard spray upright").

Wall post indicator valve A sprinkler control valve that is mounted on the outside wall of a building. The position of the indicator tells whether the valve is open or shut.

Water flow alarm device An attachment to the sprinkler system that detects a predetermined water flow and is connected to a fire alarm system to initiate an alarm condition. (NFPA 13)

Water mist system A fixed fire protection system that discharges very fine water mist particles through specialized nozzles that extinguish by displacing oxygen, cooling, and blocking radiant heat.

Water-motor gong An audible alarm notification device that is powered by water moving through the sprinkler system.

Wet-chemical extinguishing system A system that discharges a proprietary liquid extinguishing agent. It is often installed over stoves and deep-fat fryers in commercial kitchens.

Wet pipe sprinkler system The most common and least expensive type of automatic sprinkler system. As its name implies, the piping in a wet system is always filled with water. When a sprinkler head activates, water is immediately discharged onto the fire.

Zoned system A fire alarm system design that divides a building or facility into zones so that the area where an alarm originated can be identified.

REFERENCES

National Fire Protection Association. NFPA 72: *National Fire Alarm and Signaling Code*, 2019. www.nfpa.org. Accessed November 27, 2018.

National Fire Protection Association. NFPA 92: *Standard for Smoke Control Systems*, 2018. www.nfpa.org. Accessed November 27, 2018.

On Scene

You are dispatched to the report of an automatic alarm for a 10-storey apartment building. The alarm company reports the activation of a single smoke detector on the fourth floor. Immediately after the alarm was received, calls began coming into the 911 centre advising there is light smoke on the fourth floor and that the occupants are evacuating. You arrive on scene and prepare to go to work.

1. An _____ is an audible and/or visual device in a fire alarm system that makes occupants or other persons aware of an alarm condition.

  A. alarm initiating device

  B. accelerator

  C. auxiliary system

  D. alarm notification appliance

2. A Class _____ standpipe system is designed for use by occupants of a building only.

  A. I

  B. II

  C. III

  D. IV

(continued)

On Scene Continued

3. A(n) _____ fire alarm system transmits a signal to a monitoring location owned and operated by the facility.

   A. proprietary supervising alarm system

   B. public emergency alarm reporting system

   C. central station service alarm system

   D. protected premises fire alarm system

4. A(n) _____ is a fire hose connection through which the fire department can pump water into a sprinkler system or standpipe system.

   A. outside screw and yoke

   B. post indicator valve

   C. dry pipe valve

   D. fire department connection

5. A(n) _____ is a sprinkler head designed to be mounted on the underside of sprinkler piping so that the water stream is directed down.

   A. frangible-bulb sprinkler head

   B. fusible-link sprinkler head

   C. pendent sprinkler head

   D. upright sprinkler head

6. A(n) _____ alarm occurs when an alarm system is activated by a condition that is not really an emergency.

   A. nuisance

   B. malicious

   C. undesired

   D. alarm verification

7. A(n) _____ is the most common fire suppression system.

   A. residential sprinkler system

   B. wet pipe sprinkler system

   C. ionization smoke detector

   D. unwanted

8. A(n) _____ enables fire fighters to determine the type and location of the activated alarm device as they enter the building.

   A. alarm notification appliance

   B. smoke control system

   C. fire alarm control unit

   D. remote enunciator

Access Navigate to find answers to this On Scene, along with other resources such as an audiobook.

Fire and Life Safety Initiatives

KNOWLEDGE OBJECTIVES

After studying this chapter, you will be able to:

- Describe the two Firefighter Life Safety Initiatives that relate specifically to public education and fire prevention. (p. 985)
- Describe the activities that prevent fires and limit their consequences if a fire occurs. (**NFPA 1001: 5.5.2**, pp. 985–987)
- Identify elements of public fire and life safety education programs covering Stop, Drop, and Roll; Exit Drills In The Home (E.D.I.T.H.); and the selection and use of portable fire extinguishers. (**NFPA 1001: 5.5.2**, pp. 987–993)
- Explain the importance of a fire and life safety education program, portable fire extinguishers, smoke alarms, and residential sprinkler systems in preventing residential fire deaths. (**NFPA 1001: 5.5.1**, pp. 987, 990, 992–993)
- Describe the steps in conducting a fire station tour. (**NFPA 1001: 5.5.2**, pp. 993–994)
- Recognize hazards during a fire safety survey of a private dwelling or an occupied structure. (**NFPA 1001: 5.5.1**, pp. 994, 996–998)
- List the typical target hazards that may be found in a community. (pp. 1001–1002)
- Describe why and for which types of properties a preincident survey is created. (**NFPA 1001: 5.5.3**, pp. 1002–1013)

- Describe how to prepare a preincident survey. (**NFPA 1001: 5.5.3**, p. 1002)
- List the information that is gathered during a preincident survey. (**NFPA 1001: 5.5.3**, p. 1002)
- Describe the information included in any sketches or drawings created during the preincident survey. (**NFPA 1001: 5.5.3**, p. 1002)
- Describe the symbols commonly used in preincident plans. (**NFPA 1001: 5.5.3**, pp. 1002, 1004–1005)
- Describe the information that needs to be gathered to assist the incident commander in making a rapid and correct size-up during an emergency incident. (pp. 1000–1001)
- Describe the tactical information that is collected during a preincident survey. (**NFPA 1001: 5.5.3**, pp. 1005–1013)
- Explain how to identify built-in fire detection and suppression systems during a preincident survey. (**NFPA 1001: 5.5.3**, pp. 1007–1010)
- Describe how the sources of water supply for fire suppression operations are identified. (**NFPA 1001: 5.5.3**, pp. 1010–1011)
- Explain why the locations of utilities are noted on the preincident plan. (p. 1012)
- Describe how preincident planning for efficient search and rescue is performed. (pp. 1012–1013)
- Describe how preincident planning for rapid forcible entry is performed. (p. 1013)
- Describe how preincident planning for safe ladder placement is performed. (p. 1013)

- Describe how preincident planning for effective ventilation is performed. (p. 1013)
- List the occupancy considerations to take into account when conducting a preincident survey. (**NFPA 1001: 5.5.3**, pp. 1013–1015)
- List the types of locations that require special considerations in preplanning. (**NFPA 1001: 5.5.3**, pp. 1015–1017)

- Conduct a preincident survey, including sketches, notes, and forms required by your department. (**NFPA 1001: 5.5.3**, p. 1003)

SKILLS OBJECTIVES

After studying this chapter, you will be able to perform the following skills:

- Perform a public fire safety education presentation on Stop, Drop, and Roll. (**NFPA 1001: 5.5.2**, p. 989)
- Perform a public fire safety education presentation on Exit Drills In The Home (E.D.I.T.H.). (**NFPA 1001: 5.5.2**, p. 991)
- Install and maintain a smoke alarm. (**NFPA 1001: 5.5.1, 5.5.2**, p. 992)
- Give a public education tour of a fire station. (**NFPA 1001: 5.5.2**, p. 995)
- Complete a fire safety survey in an occupied structure. (**NFPA 1001: 5.5.1**, p. 999)

ADDITIONAL STANDARDS

- **NFPA 1**, *Fire Code*
- **NFPA 72**, *National Fire Alarm and Signaling Code*
- **NFPA 101**, *Life Safety Code®*
- **NFPA 170**, *Standard for Fire Safety and Emergency Symbols*
- **NFPA 220**, *Standard on Types of Building Construction*
- **NFPA 704**, *Standard System for the Identification of the Hazards of Materials for Emergency Response*
- **NFPA 1031**, *Standard for Professional Qualifications for Fire Inspector and Plan Examiner*
- **NFPA 1035**, *Standard on Fire and Life Safety Educator, Public Information Officer, Youth Firesetter Intervention Specialist and Youth Firesetter Program Manager Professional Qualifications*
- **NFPA 1452**, *Guide for Training Fire Service Personnel to Conduct Community Risk Reduction*
- **NFPA 1620**, *Standard for Pre-incident Planning*

Fire Alarm

You recently entered the fire service to help your community, and you love the excitement of responding to the different types of calls for assistance. You like working with your crew, and even the daily station duties are enjoyable. A group of grade school children will be visiting your station this afternoon, and you and your crew are cleaning the apparatus and equipment before they arrive. As you wipe down the pumper, you think about something your drill school or training school instructor once said: "More lives are saved through fire prevention and code enforcement activities than by fighting fires."

1. What did your instructor mean, and what is your role in fire prevention?

2. What are the key points you will discuss with the group of students?

3. Other than conducting fire station tours, in what ways can you assist your department's fire prevention program?

4. How does fire prevention affect fire fighter safety?

Introduction

As life safety is the highest priority for fire fighters, educating the public about fire risks and other hazards is an important responsibility. As described in Chapter 1, *The Fire Service*, community risk reduction (CRR) is a comprehensive, all-hazard unifying approach that includes programs, actions, and services used by a community to prevent or mitigate loss of life, property, and resources associated with life safety, fire, and other disasters within a community. These programs, actions, and services may also be referred to as fire and life safety initiatives.

Many Canadian fire services are exploring ways to shift resources to proactive fire prevention and risk-reduction programs. Fire services are, by nature, reactionary in their response to achieving their primary service goals. Notwithstanding, it is preferable to prevent fires and develop educational programming through education and code enforcement initiatives.

Regardless of the fire suppression capabilities of your department, once a fire starts damage is immediate. Although North American fire services have made great strides in reducing fire occurrence, loss of life, and property damage, other jurisdictions in the world boast a better overall fire occurrence record. While the reasons for this difference can be complex, one thing is for certain: European and Asian countries have robust public education programs that emphasize safety as an individual responsibility towards society as a whole. In some cultures in Asia, to cause a fire in the home is to be considered irresponsible.

In Canada, fire departments have adopted several initiatives in an effort to prevent fires. Most large fire services have fire prevention divisions that oversee public fire safety education, fire prevention, and fire investigation. Fire investigation is an important part of fire prevention, as it informs the focus for fire education programming. In addition, code enforcement provides an excellent opportunity for fire prevention officers to educate building owners on life safety and fire code requirements. Examples of these strategies include educating the public by delivering public safety presentations and prevention strategies such as conducting inspections to enforce fire codes. This chapter focuses on some of the strategies necessary to participate in fire prevention as a fire fighter.

LISTEN UP!

Fires that are prevented do not cause any deaths, injuries, or property loss. Most fires are caused by actions or situations that could have been avoided. A key element in reducing the number of fire fighter line-of-duty deaths and civilian deaths is proactive fire prevention.

Fire and Life Safety Initiatives

History has taught us that a lack of prevention efforts can result in death and serious injury for civilians and fire fighters. Much can be done to prevent these deaths if fire fighter safety becomes a primary concern of every fire fighter, fire department, and organization. The 16 Firefighter Life Safety Initiatives help reinforce this commitment and describe the steps that need to be taken to change the current culture of the fire service to help make it a safer work environment. The 16 Firefighter Life Safety Initiatives are listed in **TABLE 27-1**.

Initiative 14 states, "Public education must receive more resources and be championed as a critical fire and life safety program." Initiative 15 states, "Advocacy must be strengthened for the enforcement of codes and the installation of home fire sprinklers." In other words, every fire that is prevented reduces the risks to fire fighters and the public.

Although Fire Fighter I and Fire Fighter II personnel may have limited responsibilities for formal fire prevention activities, it is imperative that they understand the objectives of fire prevention, the delineation of responsibilities within their department, and the importance of their role in preventing fires and injuries.

Fire Prevention

Through the use of fire and life safety strategies, such as public education, code enforcement, plan review, and fire investigation, communities will be able to more effectively reduce the outbreak of fires and limit the consequences if a fire does occur. Educating the public by delivering public safety presentations about smoke alarms and the benefits of residential fire sprinklers and conducting inspections to enforce fire codes are just a few of the areas of fire prevention that fire fighters should be able to conduct as part of their job.

Any fire that is prevented is one that creates a safer situation for fire fighters.

Enactment of Fire Codes

Preventing fires and protecting lives and property in the event of a fire are important community concerns and responsibilities. Fire codes—regulations that have been legally adopted by a governmental body with the authority to pass laws and enforce safety regulations—are enacted to ensure a minimum level of fire safety in the home and workplace environments. These codes are adopted by local, provincial, and territorial governments. These codes establish the minimum level of fire protection; local jurisdictions have the authority to adopt more stringent codes.

TABLE 27-1 16 Firefighter Life Safety Initiatives

1. **Cultural change**: Define and advocate the need for a cultural change within the fire service relating to safety and incorporating leadership, management, supervision, accountability, and personal responsibility.
2. **Accountability**: Enhance the personal and organizational accountability for health and safety throughout the fire service.
3. **Risk management**: Focus greater attention on the integration of risk management with incident management at all levels, including strategic, tactical, and planning responsibilities.
4. **Empowerment**: All fire fighters must be empowered to stop unsafe practices.
5. **Training and certification**: Develop and implement national standards for training, qualifications, and certification (including regular recertification) that are equally applicable to all fire fighters based on the duties they are expected to perform.
6. **Medical and physical fitness**: Develop and implement national medical and physical fitness standards that are equally applicable to all fire fighters based on the duties they are expected to perform.
7. **Research agenda**: Create a national research agenda and a data collection system that relates to the 16 Firefighter Life Safety Initiatives.
8. **Technology**: Utilize available technology whenever it can produce higher levels of health and safety.
9. **Fatality, near-miss investigation**: Thoroughly investigate all fire fighter fatalities, injuries, and near-misses.
10. **Grant support**: Grant programs should support the implementation of safe practices and procedures and/or mandate safe practices as an eligibility requirement.
11. **Response policies**: National standards for emergency response policies and procedures should be developed and championed.
12. **Violent incident response**: National protocols for response to violent incidents should be developed and championed.
13. **Psychological support**: Fire fighters and their families must have access to counseling and psychological support.
14. **Public education**: Public education must receive more resources and be championed as a critical fire and life safety program.
15. **Code enforcement and sprinklers**: Advocacy must be strengthened for the enforcement of codes and the installation of home fire sprinklers.
16. **Apparatus design and safety**: Safety must be a primary consideration in the design of apparatus and equipment.

© Jones & Bartlett Learning.

Provinces/territories within Canada have either adopted the National Building Code (NBC) or their own building codes for establishing construction standards. Some provinces, such as Ontario, have enacted their own building codes. Other provinces, such as Manitoba and Prince Edward Island, have adopted the NBC to regulate building construction.

Fire and building codes are created by the Canadian Commission on Building and Fire Codes (CCBFC). In Canada, every jurisdiction has a mechanism for conducting inspections and issuing compliance orders and code enforcement through judiciary means; however, not every jurisdiction performs these activities in the same manner. In large municipalities such as Calgary, Vancouver, Ottawa, Toronto, and Montreal, the fire service has a fire prevention division that is responsible for public education, code enforcement, and fire investigation. In smaller towns and municipalities (fewer than 50,000 inhabitants), the responsibility to conduct inspections typically falls on the chief fire official or a designated person.

Most communities adopt and enforce a full set of codes that cover a range of areas that are designed to establish basic health and safety standards. Generally, these codes include a building code, a fire code, an electrical code, a plumbing code, a mechanical code, and other local rules and regulations. These codes and regulations must be reviewed and coordinated to avoid conflicting requirements and to ensure a safe community. For example, the electrical code includes requirements that are intended to prevent electrical fires as well as requirements intended to reduce the risk of electrocution.

Fire codes are considered "companion documents" to building codes. While the building code has force and effect while a building is under construction, once a building has been deemed habitable and an occupancy permit is granted, the fire code comes into effect and can be enforced. Fire department personnel are usually involved in the enforcement of these codes.

Building codes establish safety requirements that apply to the construction of new buildings and almost always include fire protection requirements intended to minimize the development and spread of fire and smoke. When a new building is constructed, the building code specifies the minimum requirements for the type of construction, allowable heights and areas, distances between firewalls, and necessary building materials. In addition, this code includes requirements for safe exiting and determines when fire alarm systems, automatic sprinklers, and standpipes are required. Chapter 6, *Building Construction*, covers issues that are closely related to building codes.

Generally, fire codes apply to all buildings, new or old, and to many different situations related to fire risks and hazardous conditions. Older buildings may not comply with particular codes if the code has not been retroactively applied. In most cases, only a change of ownership or occupancy type will require compliance with newer codes. In rare circumstances, a building may be required to comply with newer codes due to a major loss-of-life fire in a similar occupancy.

Some code requirements apply to all buildings, such as exit requirements, occupancy load limits, and electrical requirements. Some buildings are required to have fire alarm systems or sprinkler systems. Fire fighters should be aware of the local requirements, because code adoptions and amendments vary from jurisdiction to jurisdiction, even within the same region. In addition, fire codes include regulations for dispensing flammable liquids, for storing them in approved storage containers, and for prohibiting their use or storage in some buildings. Regulations that prohibit parking in a fire lane or obstructing a fire hydrant are usually found in the fire code as well. All of these requirements are intended to promote fire safety and protect both the community and fire fighters.

Inspection and Code Enforcement

After a jurisdiction adopts a fire code, it must be able to enforce it. Citizens have a legal obligation to comply with all rules and regulations set by such an authority. Regular inspections ensure compliance, note any violations, and require their correction. If violations are not corrected, a series of increasing penalties can be imposed, beginning with a verbal warning or a written notice. An individual who fails to correct the violation within a specified time can be fined. In some cases, a jail sentence can be imposed for serious violations.

Fire codes usually specify the types of occupancies that can be inspected and the frequency of the inspections. For example, some fire codes require all schools to be inspected twice each year. In other cases, the schedule of inspections is discretionary. The agency responsible for performing inspections and enforcing codes is usually named in the fire code; it may vary in different communities. Sometimes the responsibility belongs to specific individuals or positions within the fire department, such as fire inspectors or fire marshals. Other fire departments train all fire fighters to conduct inspections and take code enforcement actions. In some communities, building inspectors or special code enforcement personnel are given this responsibility. Generally, newly assigned fire fighters are not directly involved in code enforcement activities, but they should understand how code enforcement is handled in the community.

Fire codes authorize local jurisdictions to conduct fire inspections in all buildings during the construction phase and prior to issuing an occupancy permit. Most jurisdictions inspect commercial occupancies annually or as specified by their regulations. Although fire and building codes apply to both commercial buildings and private residences, most jurisdictions do not inspect single- and two-family residences after they are built and occupied unless there is a reason to believe the residence is unsafe. The occupant can request a voluntary inspection to identify hazardous conditions, however. Many fire departments offer this public service as part of a comprehensive fire prevention program. This type of voluntary inspection is called a home **fire safety survey**, and the legal requirements of the fire code are not applied when it is conducted, and no violations are cited or fines imposed. Voluntary inspections of private dwellings are one way that fire departments can educate citizens about the risks of fire in their homes. Such inspections are an important community service offered by the fire department, and the diplomacy and professionalism of fire fighters conducting home fire safety surveys cannot be underestimated in furthering the mission of the fire department. It is a good idea to offer smoke alarm installations during these survey visits.

Every fire fighter should know how to conduct a home fire safety survey in a single-family residence or other housing—including knowing what to look for during the survey and how to communicate the survey findings. This information is presented in detail later in this chapter.

Fire and Life Safety Education

In order for fire prevention programs to be effective, they must be based on factual analysis. Unfortunately, the Canadian federal government does not have a department whose sole purpose is to collect fire loss statistics. In lieu of this information, vacuum organizations such as the Council of Canadian Fire Marshals and Fire Commissioners collect and publish data on fire loss.

The goals of fire and life safety education, often referred to as *public education*, are to help people understand how to prevent the loss of life, injuries, and property damage from occurring and to teach them how to react in an appropriate manner if an emergency occurs. Educators work to create a change in behaviour that results in greater safety for their community. For example, making people aware of common fire risks and hazards and providing information about reducing or eliminating those dangers can prevent many fires, injuries, and deaths from occurring. Public education programs teach techniques to reduce the risks of death or injury in the event of a fire and prepare for other hazards such as severe weather. The programs are designed to prevent all types of accidents and injury to which the fire service commonly responds.

FIGURE 27-1 The Stop, Drop, and Roll program teaches young children what to do if their clothing catches fire.
Courtesy of Honeywell/Gamewell.

Public fire safety education programs include, but are not limited to, those focused on the following topics:

- Stop, Drop, and Roll (**FIGURE 27-1**)
- Close your door safety initiative
- Crawl low under smoke
- Exit Drills In The Home (E.D.I.T.H.)
- Emergency notification
- Installation and maintenance of smoke alarms
- Advantages of residential sprinkler systems
- Selection and use of portable fire extinguishers
- Learn Not to Burn® preschool program
- Fire safety for special populations
- Fall prevention
- Wildland fire prevention programs

Most fire and life safety programs are presented to groups such as school classes, scout troops, church groups, senior citizen groups, civic organizations, hospital staff, and business employees. Presentations are often made at community events and celebrations, as well as during Fire Prevention Week activities. Five public fire safety education programs—Stop, Drop, and Roll; E.D.I.T.H.; installation and maintenance of smoke alarms; advantages of residential fire sprinklers; and the selection and use of portable fire extinguishers—are described here.

Another popular fire safety education activity is a fire station tour. Both children and adults enjoy the opportunity to tour the local fire station, and a fire station tour is an excellent opportunity to promote fire prevention. Every fire fighter must understand how to conduct an effective fire station tour. The skills and knowledge for conducting a fire station tour are presented later in this chapter.

Stop, Drop, and Roll

The Stop, Drop, and Roll public fire and life safety education program can be taught to children as young as preschoolers, but it is also a valuable educational program for adults. This program is designed to instruct people what to do if their clothing ignites. When clothing catches fire, the normal reaction is to run for help. Unfortunately, running tends to fan the flames and increases the chance for severe burns. Teaching the mnemonic "Stop, Drop, and Roll" and demonstrating each step help to reduce the severity of burn injuries.

When teaching the Stop, Drop, and Roll program to children, ask them to discuss situations that might result in their clothing catching fire. These circumstances might include being too close to a fireplace or kitchen stove or using matches or lighters. Fire personnel should talk about ways to reduce these hazards, such as keeping matches and lighters out of the reach of children or wearing short-sleeved clothing when cooking. Once again, the emphasis should be on preventing the burn by teaching proper behaviours, followed by what to do if their clothes catch on fire. Fire fighters should then stress each step of the sequence:

- **Stop** means just that: Stay in place; do not run. Stress the fact that running will fan the flames and spread the fire.
- **Drop** means getting down on the ground or on the floor. The person should cover his or her face with the hands to help protect the airway and the eyes.
- **Roll** involves encouraging children to tuck in their elbows and keep their legs together and roll over and over, and then roll back in the other direction. This helps to smother the flames.

Remind people that a blanket can be used to smother a fire and that a garden hose both extinguishes the fire and cools the burned area. By teaching ways to prevent clothing from catching fire and steps to smother the fire, you can help to reduce the risk of serious burns.

To teach the Stop, Drop, and Roll program, follow the steps in **SKILL DRILL 27-1**.

Exit Drills in the Home

A second public fire and life safety education program, called **Exit Drills In The Home (E.D.I.T.H.)**, teaches residents how to safely get out of their homes in the event

of a fire or other emergency. Many fire deaths could be prevented if everyone learned a few simple rules about what to do in the event of a fire in the home. As

---

SAFETY TIP

There are many good reasons to have a written plan for fire safety presentations. A written plan offers the following benefits:

- Guides the presentation
- Outlines the main points the presentation will cover
- Ensures consistency when multiple presenters are used

SKILL DRILL 27-1
Teaching Stop, Drop, and Roll Fire Fighter II, NFPA 1001: 5.5.2

1 Explain the purpose of Stop, Drop, and Roll. Instruct students to **STOP**: Stay in place.

2 Instruct students to **DROP** to the floor or ground and cover their faces with their hands.

3 Instruct students to tuck their elbows, keep their legs together, and **ROLL** over and then back to extinguish the flames. Demonstrate the Stop, Drop, and Roll procedure.

4 Have students practise the procedure.

© Jones & Bartlett Learning. Photographed by Glen E. Ellman.

part of this exercise, residents need to understand the importance of having properly installed and working smoke alarms. Working smoke alarms alert residents to a fire and give them more time to escape from a burning building. A public fire and life safety education program about smoke alarms is described later in this chapter. More information about the operation of smoke alarms is also found in Chapter 26, *Fire Detection, Suppression, and Smoke Control Systems*.

As part of your E.D.I.T.H. presentation, stress the importance of keeping bedroom doors closed during sleeping hours, as this step prevents smoke, heat, and flames from reaching bedrooms. Emphasize the need to have two escape routes from each bedroom. If smoke, heat, or flames are present in the primary escape route, use the secondary escape route. The secondary escape route may be through a window or through a different door. Stress the importance of alerting other occupants to the presence of a fire and the benefits of closing the doors as you exit the home (this action may help to smother the fire and limit damage).

When alerted to the presence of a fire, all occupants should roll out of bed, stay low, and crawl toward the designated exit. By staying low, individuals decrease the amount of heat, smoke, and gases they inhale.

Before opening any door, occupants should touch the closed door to see if it is hot. Teach students to use the back of a bare hand to sense the temperature of the door. If the door is hot, they should not open it; instead, they should use a window or another door for escape.

Once outside the home, residents should gather all family members at a pre-established meeting spot. They should not go back into the house for any reason. Stress the importance of making sure that the fire department has been called. If any doubt arises about whether the fire department has been called, they should use a neighbour's phone or a cell phone to call the fire department again.

Once the steps for emergency drills have been taught, families should practise E.D.I.T.H. regularly. Some drills should be held in the daytime, and some should take place at night.

To teach E.D.I.T.H., follow the steps in **SKILL DRILL 27-2**.

LISTEN UP!

When you are working with the public, pay attention to the language that you use. Make sure that it will be understood by your audience. Avoid using firefighting terms and jargon. Speak at a slower pace for young children and older adults.

Smoke Alarms

A third type of public fire and life safety education program covers the proper installation and maintenance of smoke alarms. Smoke alarms can save lives—but only if they are working properly. The goal of this presentation is to help people properly install and maintain approved smoke alarms in their homes. More information about this topic is presented in Chapter 26, *Fire Detection, Suppression, and Smoke Control Systems*.

During your presentation on this topic, stress the importance of having enough working smoke alarms. Interconnected smoke alarms offer the best protection. When one alarm sounds, all alarms sound, notifying occupants in all areas of the home. Older residences that were built before these requirements were established may not have enough smoke alarms to provide early detection and notification. It is recommended that the occupants install additional smoke alarms to improve their protection.

Residents should be sure to install special smoke alarms and alert devices for family members who are deaf or hard of hearing. Remind residents that smoke alarms should not be installed close to kitchens, fireplaces, or garages or near windows, exterior doors, or heating or air-conditioning ducts, as placing smoke alarms in these locations may result in false alarms or delayed activation. Photoelectric smoke alarms can be installed in kitchen areas.

Stress the importance of keeping smoke alarms in working order. Alarms should be maintained and tested according to manufacturer's instructions. It is best practice to ensure smoke alarms are tested at least once a month. Some alarms have a test button that is pressed to test that they are in working order. Residents should identify the type of battery in the smoke alarm. Smoke alarms powered by alkaline batteries should have their batteries changed every 6 months. Many communities participate in the Change Your Clock, Change Your Batteries campaign, which encourages regular battery replacement when Daylight Saving Time (DST) begins and ends. Many new smoke alarms contain an extended-life 10-year battery. All smoke alarms must be replaced 10 years after the manufacture date or if they fail the monthly test. Remind homeowners to keep circuit breakers supplying current to hard-wired smoke alarms on to ensure that these alarms remain operational. Be sure homeowners understand that hard-wired smoke alarms contain backup batteries that require regular replacement. The detector chamber should be vacuumed regularly to prevent the build-up of dust or insects, which can cause false alarms. Some alarms are manufactured to detect the presence of carbon monoxide as well as smoke; it is important to explain the function of each alarm when these dual-purpose alarms are present.

SKILL DRILL 27-2
Teaching E.D.I.T.H. Fire Fighter II, NFPA 1001: 5.5.2

1 **Plan.** Describe the importance of having a sufficient number of working smoke alarms throughout the home and making a quick exit when a fire occurs. There should be a smoke alarm in every sleeping room, outside each sleeping area, and on every level of the home, including the basement (NFPA, EMAC, 2017). Explain why residents should sleep with bedroom doors closed. Have students draw a map of each level of their home, showing all windows and all doors. Have them plan two escape routes from each room, making sure all doors and windows are operable and accessible.

2 **Practise.** Practise the plan with all family members and caretakers, including those with disabilities. Instruct students to roll out of bed and crawl to the exit when they hear a smoke alarm or smell smoke.

3 Instruct students to check exit doors for heat and to not open a door if it is hot.

4 Instruct students to follow the escape route to the pre-established meeting place and to ensure that the fire department has been called. Emphasize the importance of never reentering a burning building.

Teaching these simple facts about the installation and maintenance of smoke alarms will help families keep their homes safe in the event of a fire. To present a public fire safety education program on the installation and maintenance of smoke alarms, follow the steps listed in **SKILL DRILL 27-3**.

LISTEN UP!

Although private residences are usually not inspected by the fire department, it is recommended that homeowners replace their smoke alarms every 10 years or if they fail the monthly test.

Residential Fire Sprinkler Systems

A fourth type of public fire and life safety education program covers the importance of residential fire sprinkler systems in preventing residential fire deaths. Statistics show that the combination of working smoke alarms and residential fire sprinklers can reduce the risk of fire death by approximately 80% (NFPA, 2018). With these statistics in mind, there is a national movement in the fire service to have fire sprinklers installed in all new residential construction. For more information about fire sprinklers, see Chapter 26, *Fire Detection, Suppression, and Smoke Control Systems*.

SKILL DRILL 27-3
Installing and Maintaining Smoke Alarms Fire Fighter II, NFPA 1001: 5.5.1, 5.5.2

1 Explain the importance of properly installing and maintaining smoke alarms.

2 Stress the importance of testing smoke alarms once a month using the test button.

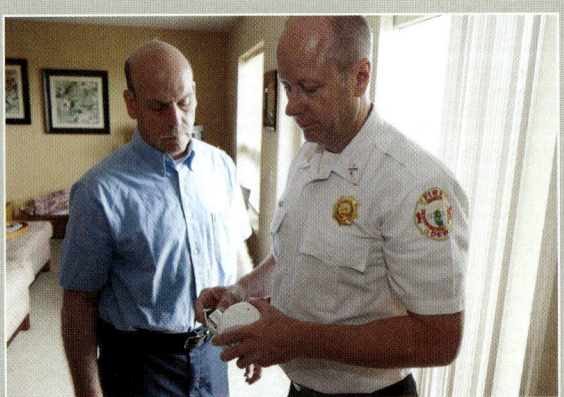

3 Explain that alkaline batteries in smoke alarms should be changed every 6 months and all smoke alarms should be replaced every 10 years.

4 Emphasize the need to clean smoke alarms regularly to prevent false alarms.

Selection and Use of Portable Fire Extinguishers

A fifth type of public fire and life safety education program discusses the selection and use of portable fire extinguishers. Many adults will have questions about which types of fire extinguishers they should have in their home, kitchen, garage, car, or boat. You should be familiar with the information you learned in Chapter 7, *Portable Fire Extinguishers*, so you can answer any questions that arise. When discussing portable fire extinguishers with the public, it should be emphasized that they are only recommended for adult use on small fires. Children should concentrate on knowing the escape routes and getting out immediately.

Conducting Fire Station Tours

Fire station tours present a unique opportunity to help people learn how the fire department operates and how they can help prevent fires. Most fire departments have a set format for conducting fire station tours. Your station officer is in charge of all activities and should teach you how to conduct tours in your station. New fire fighters should be given the opportunity to observe more senior members conducting station tours before being assigned to handle them on their own.

Before tours arrive, check to make sure any areas the visitors will walk through are safe. Check for broken glass, slippery or wet floors, snow and ice on walkways, and tripping hazards. If hazards cannot be removed, cordon them off with caution tape, and advise visitors to avoid them.

Some tours start in the dayroom or a meeting room and provide a general overview of the fire station and the department's expectations for visitors during the tour; this is followed by a tour of the fire apparatus. Alternatively, a tour may start with an inspection of the apparatus and end in a meeting room with a question-and-answer session and information about the department and fire prevention techniques. Showing a fire prevention video or a video about the operation of your department is a good way to start or finish a tour and stimulate questions from your audience.

Remember, when conducting a tour, you are representing your department to the visitors. Your professional appearance and presentation project a positive image of the department to the community.

Explain to the visitors what to expect and what they should do if the station receives an alarm during the tour. Visitors should not wander around the station while fire fighters respond to an emergency call. This is particularly important if the tour guide might have to leave in response to a call.

The tour format will vary depending on the age and interests of the group, which could consist of small children, teenagers, adults, or senior citizens. The tour plan, including your presentation, should be designed to meet the needs of each group. The weather may also be an important factor if you are planning any outdoor demonstrations.

Young children like to see action, so flowing water or raising an aerial ladder can keep their attention (**FIGURE 27-2**). Fire prevention and risk reduction messages for young people include the use of the Stop, Drop, and Roll procedure; learning how to call 911; the importance of E.D.I.T.H.; and the dangers of using lighters and matches.

Teenagers are ready for lessons that they can apply in everyday life. Fire prevention for teenagers should focus on the fact that some may undertake babysitting as a way to make money. Babysitters should be taught proper cooking safety, fire extinguisher use, and methods to avoid risk. Of particular note is the misuse of smoking materials, as Canada has legalized the use of marijuana. Teenagers are also a high-risk group for motor vehicle accidents, so messages about seat belts; proper cell phone use; texting; safe driving; the dangers of driving while under the influence of drugs, alcohol, or marijuana; and actions to take if they see or hear an emergency vehicle while driving are appropriate. Some teenagers might be interested in becoming fire fighters or joining a fire fighter explorer group. This age group

FIGURE 27-2 Young children like to see fire fighters.

Courtesy of Captain David Jackson, Saginaw Township Fire Department.

can be a good source of recruits for both career and volunteer departments.

Tailor your fire prevention and safety messaging to your audience. With a an aging population, the use of in-home oxygen supply systems is becoming more prevalent. Fire personnel must communicate the inherent dangers in the use of oxygen, as it should not be used anywhere near open flame or heat. The possible dangers associated with the use of space heaters, candles, and fireplaces should also be communicated.

Try to leave every tour group with both a message and materials. First, reinforce a simple prevention or safety message that expands their awareness of fire safety and helps them to deal with potential emergency situations. Second, give them printed reference materials (**FIGURE 27-3**). These materials could range

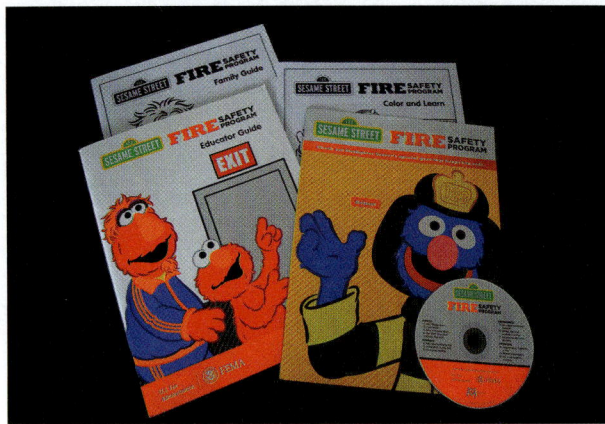

FIGURE 27-3 Always leave visitors with age-appropriate printed fire prevention or safety materials.
Courtesy of Marsha Geisler.

from colouring books for young children to a home fire prevention checklist for adults.

The steps for conducting a fire station tour are listed in **SKILL DRILL 27-4**.

Conducting a Fire Safety Survey for a Private Dwelling

Many fire departments offer to conduct fire safety surveys in private dwellings. These programs are usually based upon a voluntary request. A home fire safety survey helps identify fire and life-safety hazards and provides the occupants with recommendations on making their home safer.

In most jurisdictions, residential fire safety surveys cannot be conducted without the occupant's permission. Fire departments may use public service announcements on the radio and television to increase public awareness of the need for a fire-safe home and the availability of a fire safety survey. In many communities, the fire department offers to check home smoke alarms and to conduct a home fire safety survey at the same time. Some fire departments install smoke alarms free of charge in a dwelling as a community service.

In Ottawa, however, the city's fire service has created the Wake Up fire prevention program, which is part of a smoke alarm bylaw. As part of this program, fire personnel visit every residential dwelling and check to ensure that each dwelling has at least one working smoke alarm. If a residence does not have an alarm, then the fire service provides one with a new battery to that residence. This program involves teaming up with smoke alarm manufacturers and battery companies to supply the needs of the program.

Other municipalities in Canada have similar programs, which are effective because they combine public education, legislation, and code enforcement in an effort to reduce fire loss and injury.

A fire safety survey is a joint effort by the fire department and the occupant to prevent fires and reduce fire injuries and deaths. During the survey, point out hazards, explain the reasons for making recommendations, and answer questions the occupant may have. Many fire departments provide educational materials that address safety issues such as the installation and

SKILL DRILL 27-4
Conducting a Fire Station Tour Fire Fighter II, NFPA 1001: 5.5.2

1 Introduce yourself to the tour participants, and explain the mission of the fire department. Show participants the layout of the fire station and the various types of personal protective equipment (PPE). Consider using an educational video presentation.

2 Explain the purpose of each piece of fire apparatus, keeping all explanations age appropriate. Check for understanding.

3 Conclude the tour in a meeting room. Encourage questions from participants. Hand out appropriate fire safety materials.

maintenance of smoke alarms, the role of residential fire sprinklers in preventing fire deaths, and the selection and use of portable fire extinguishers. These materials can be left with the occupant.

Getting Started

To conduct a fire safety survey, you should follow all department operational guidelines and present a neat, professional image (**FIGURE 27-4**). Identify yourself by name, title, and organization. Inform the occupant of the purpose of your visit and the survey format. Always remember that you are a guest in a private home, and respect the privacy of the people who live there. You are there to help teach how to protect the family and property in the event of a fire.

During the survey, you need to consider several types of hazards and issues. Concentrate on the hazard categories that most often cause residential fires—for example, cooking equipment, heating equipment, electrical wiring and appliances, smoking materials, and candles and other open flames. Some of the hazards you encounter will depend on the local climate. For example, wood-burning stoves and fireplaces are more common in colder climates. Also, consider the occupants. The fire hazards in a residence with young children, occupants with a physical or mental disability, or senior citizens, for example, may be quite different. Ask the homeowner if there are oxygen tanks or concentrators in the home.

Always look for fire protection equipment such as smoke alarms, fire extinguishers, and residential fire sprinkler systems. Make sure that any fire alarm or suppression equipment in the home is properly installed, maintained, and fully operational (**FIGURE 27-5**). Reference your department's policies before making recommendations on fire protection equipment. Most jurisdictions do not allow recommending a specific manufacturer's product.

Some departments address additional safety issues during a fire safety survey such as swimming pool safety, slip-and-fall hazards, and the value of carbon monoxide alarms. Mentioning these additional safety items to the occupant can be very helpful.

Conduct the survey in a systematic fashion for both the inside and the outside of the home. Some fire fighters begin by inspecting the outside of the house, whereas others prefer to start with the interior.

Outside Hazards

On the exterior of a residential occupancy, you need to check several items. Make sure that the house number, or address, is clearly visible from the street so emergency responders can identify the correct house quickly (**FIGURE 27-6**). Look for accumulated trash that could present a fire hazard or obstruct an exit. Note any combustible materials that are located close to the house. Point out shrubs and vegetation that need to be trimmed or removed. Additionally, trees and shrubs should be cleared to allow large fire apparatus access to the residence. Recommend that fireplaces and chimneys be inspected annually by a qualified technician.

FIGURE 27-4 Always look like a professional. Bring plenty of handouts and items you can leave with the occupant that reinforce the concepts that you want to emphasize.

© Jones & Bartlett Learning. Photographed by Glen E. Ellman.

FIGURE 27-5 Ensure that smoke alarms are properly installed, maintained, and fully operational during a home inspection.

Courtesy of Nampa Fire Department - Nampa, Idaho.

FIGURE 27-6 The home's address should be clearly visible from the street.
© LI Cook/Shutterstock.

SAFETY TIP

Security bars and double-cylinder door locks are great deterrents to burglars, but they can prove deadly for occupants who are trying to escape from a house fire. These security measures can also prevent fire fighters from quickly entering homes to rescue the occupants or control a fire. Occupants should always be able to quickly release security bars and open locked doors from the inside.

Inside Hazards

During the interior inspection, explain why different situations are considered potential fire hazards. A person who understands why an overloaded electrical circuit is a fire hazard will be more likely to avoid overloading circuits in the future. Your goal should be to educate the occupant about common fire hazards.

Systematically inspect each room for fire hazards. Each area will have its own hazards. As you walk through the house, help the occupant identify alternate escape routes that can be used in case of fire. Look for primary and alternate exits from each room, and discuss these points of egress with the occupant. Be sure to mention the importance of E.D.I.T.H. for all family members.

Having working smoke alarms on each floor, in the corridor outside each sleeping area, and in each bedroom greatly reduces the risk of death from fire. Take the time to test all smoke alarms, and recommend adding additional alarms if some areas are not protected. Some fire departments install one or more smoke alarms at no cost in any dwelling that lacks this important protection. Give the residents a brochure about smoke alarms. More information about the proper maintenance of smoke alarms is presented in Chapter 26, *Fire Detection, Suppression, and Smoke Control Systems*.

Bedrooms

The most common causes of fires in bedrooms are the improper use of smoking materials (NFPA, 2011). Bedrooms typically have a lot of combustible materials such as bedding, curtains, clothing, books, and toys. If an ignition source—heating devices, smoking materials, or open flame—comes into contact with these combustibles, a fire can ignite and quickly spread. Because occupants are often asleep in these rooms, they may not be able to escape if a fire occurs during the night. Fire fighters conducting fire safety surveys should stress the importance of keeping these ignition sources away from combustible materials, keeping bedroom doors closed to limit fire growth and spread, and having adequate smoke alarms positioned throughout the house. It is especially important to discuss the risk of children playing with lighters and matches.

Kitchens

From 2011 to 2015, almost half (47%) of reported home structure fires were caused by cooking (NFPA, 2017). In particular, these fires are often caused by leaving cooking food on the stove unattended, combustibles stored on or near the burners, and faulty electrical appliances. In the kitchen, look for signs of frayed wires on appliances, combustible towels or potholders stored too close to the stove, and cooking oils stored near the stove. Nothing that can be ignited—such as a plastic food container or paper packaging—should ever be placed on a cooking surface, even if the burners are turned off and the surface is cold. Many fires have resulted from someone inadvertently turning on the wrong burner and igniting something that was left on the stovetop.

Every kitchen should be equipped with an approved ABC-rated fire extinguisher (**FIGURE 27-7**). Recommendations for kitchen safety include the following measures:

- Do not leave anything cooking unattended on a stove.
- Keep all combustible materials, cleaning supplies, cooking oils, and aerosols away from the stove.
- Do not place anything that could ignite on a cooking surface, even when it is turned off.
- Do not place towel racks near the stove.
- Do not overload electrical outlets or extension cords.
- Keep electrical cords properly maintained, and replace any damaged appliances.
- Keep the range exhaust hood clean and in good working order.

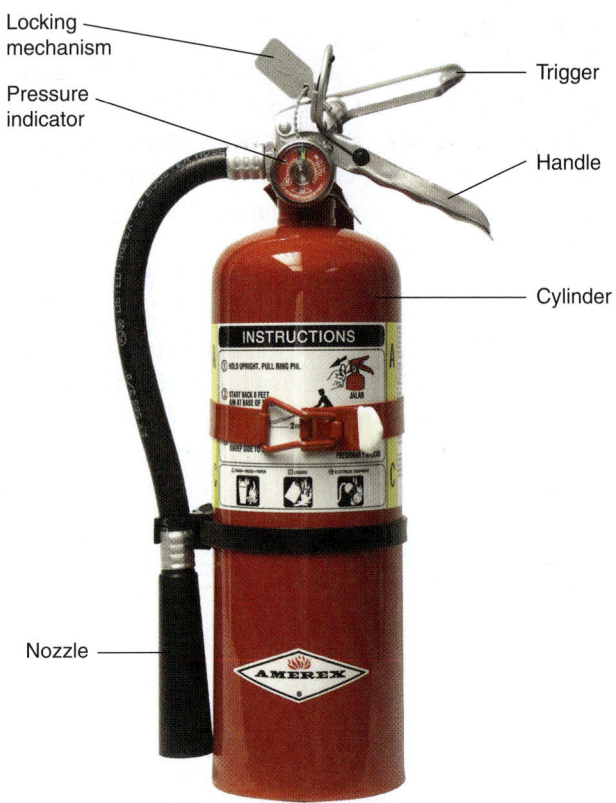

Locking mechanism

Pressure indicator

Trigger

Handle

Cylinder

Nozzle

FIGURE 27-7 Encourage homeowners to purchase an approved ABC-rated fire extinguisher.
Courtesy of Amerex Corporation.

Living Rooms

The "living room" category includes formal living rooms, family rooms, and entertainment rooms. Common causes of living room fires include careless use of smoking materials, fireplaces, overloaded electrical circuits, and portable heating appliances. Look for indications of careless smoking or improperly installed home entertainment equipment, such as multiple devices plugged into a light-duty extension cord. Also look for overloaded electrical outlets and extension cords in home offices with computers, printers, scanners, fax machines, and other devices. Stress the importance of using power strips that contain circuit breakers to prevent overloading of extension cords.

If the room contains a fireplace, a wood stove, or a portable heater, ensure that no combustible materials are stored nearby. A fireplace should have a screen to keep sparks and hot embers from escaping. If solid fuels are used, the chimney or flue pipe should be professionally inspected at least once each year. Stress the importance of placing ashes in metal containers with a tight-fitting lid outside, away from all combustible materials.

Garages, Basements, and Storage Areas

Explain the importance of good housekeeping and the need to clear combustibles out of garages, basements, and storage areas. These areas are often filled with large quantities of unneeded materials. Furnaces and water heaters that operate with an open flame are often located in a basement or garage as well; they can ignite flammable materials that are stored too close to them.

Storage of gasoline and other flammable substances is a major concern because an open flame or pilot light can easily ignite flammable vapours. Gasoline and other flammable liquids should be stored only in approved containers and in outside storage areas or outbuildings. Propane tanks, such as those used in gas grills should also be stored outside or in outbuildings. Small quantities of flammable and combustible liquids (such as paint, thinners, varnishes, and cleaning fluids) should be stored in closed metal containers away from heat sources. Oily or greasy rags should be stored in closed metal containers. It is recommended that homeowners maintain fully charged fire extinguishers in basements and garages.

Other Hazards

During a home fire safety survey, be on the lookout for potential nonfire safety hazards in the home. Poisoning, drowning, burns, falls, toppling furniture, and hoarding are all common causes of injuries in the home. Simple steps can prevent some of these emergencies:

- Pools must be properly enclosed.
- Furniture, televisions, and appliances must be properly secured to prevent tip-overs.
- Medicines and chemicals must be stored where children cannot access them.
- Stairs and balconies need handrails and guards.
- Domestic hot water temperatures should not exceed 49°C (120°F).

Closing Review

During the fire safety survey, listen carefully to any questions from the occupants, because these questions will enable you to address any special concerns or correct inaccurate information. Talk to the occupants, and take the time to answer their questions fully. Bring brochures to leave with the occupants. Emphasize the importance of smoke alarms, home exit plans, and fire drills.

After you have completed a fire safety survey or any other type of inspection, you must file your report according to your department's policies. If you identified hazards that require further action or follow-up, discuss them with your company officer or a designated representative from the fire prevention division. To conduct a fire safety survey, follow the steps in **SKILL DRILL 27-5.**

SKILL DRILL 27-5
Conducting a Fire Safety Survey Fire Fighter II, NFPA 1001: 5.5.1

1 Check for a visible house number. Look outside the house for accumulated garbage, overgrown shrubs, and blocked exits. Explain the purpose of the fire safety survey to the occupant.

2 Check inside the house for properly working smoke alarms, fire extinguishers, and fire sprinkler systems. Stress the importance of proper cooking procedures and the safe storage of cooking oils and flammable objects. Explain the safe use of fireplaces, heating stoves, and portable heaters. Help building occupants identify primary and alternate escape routes. Look for improper wiring, use of candles, and evidence of careless smoking.

3 Complete the inspection form, and review the results of the fire safety survey with the building occupants. Describe the steps that need to be taken to minimize the identified hazards. Give a copy of the inspection form to the occupants.

4 Leave fire and life safety brochures with the building occupants. File your report according to your department's policies.

Preincident Planning

Preincident planning is closely related to other risk reduction activities. While fire inspections, code enforcement, and public education are intended to prevent fire ignition and reduce the consequences if fires occur, preincident planning provides responding fire crews with vital information needed to make fire-ground operations safer and more effective and helps your fire department make better command decisions because important information is assembled before the emergency occurs. The objective is to make valuable information available quickly during an emergency incident that otherwise would not be readily evident or easily determined.

A **preincident plan** is a collection of data that are used by responding personnel to effectively and safely manage an emergency at a particular location. Preincident planning is typically done for properties that are particularly large or that present unusual risks to fire fighters or to the public. Without a preincident plan, you will go into an emergency situation "blind." You will not be familiar with the structure, the location of hydrants, or the potential hazards. By contrast, with a preincident plan, you will know where the hydrants and exits are and which hazards to anticipate. Because a preincident plan identifies potentially hazardous situations before an emergency occurs, fire fighters can be made aware of hidden dangers and prepare for them. A preincident plan puts all of this information at your fingertips, either on paper or on a mobile data terminal (MDT) in the apparatus for use en route to the fire scene (**FIGURE 27-8**).

Preincident planning is performed under the direction of a fire officer. At the emergency scene, the incident commander (IC) can use the information gathered prior to the incident to direct the emergency operations much more effectively (**FIGURE 27-9**). The completed preincident plan can also be used in training activities to help fire fighters become familiar with the properties within their jurisdiction, and they should be made available to all units that might respond to an incident at that location. Invite neighbouring fire departments to train with you and become familiar with the preincident plan.

The amount and the nature of the information provided for specific properties will depend on the size and complexity of the property, the types of risks present, and the particular hazards or challenges likely to be encountered. A preincident plan usually includes one or more diagrams that show details such as the building location and layout, access routes, entry points, lockbox locations, exposures, fire department

FIGURE 27-8 An example of a preincident plan on a mobile data terminal.
Courtesy of David Dodson.

FIGURE 27-9 Preincident plan information is supplied to the IC at the emergency scene. It is intended to help the IC make informed decisions when an emergency incident occurs at the location.
© Jones & Bartlett Learning.

connections (FDCs) for sprinklers and standpipes, and hydrant locations or alternative water supplies. In addition, the location and nature of any special hazards to the public or to fire fighters should be highlighted on the diagrams. Information about the actual building should include its height and overall dimensions, its type of construction, the nature of the occupancy, and the types of contents in different areas of the building. Additional information should include interior floor plans, stairway and elevator locations, utility shut-off locations, and information about built-in fire detection and suppression systems. All of the information must be presented in a uniform and understandable format. For more information, see NFPA 1620, *Standard for Pre-incident Planning*.

The use of modern information technology has greatly enhanced the ability of fire departments to capture, store, organize, update, disseminate, and retrieve preincident planning information. Fire departments using an integrated information technology system to organize data and create preincident plan information can access the information through a network. This network can be accessed in multiple ways—directly, through computer terminals at fire stations, and indirectly, through mobile computer devices (**FIGURE 27-10**). When this approach is used, the preincident plan information for a specific property can be automatically transmitted to the responding fire stations and to responding vehicles.

Accurate and current information such as drawings, maps, satellite and aerial imagery, photographs, descriptive text, lists of hazardous materials, and safety data sheets (SDSs) can be made available instantly to the fire fighters who need it during an emergency incident. In some cases, a geographic information system (GIS) is used to collect data on a geographic environment, such as population characteristics, types of occupancies, water, other utilities, and road maps; organize those data; and present the information digitally to the entire emergency response system, including the fire department.

Updating and disseminating the information are also easier when it is stored electronically on a computer network. As soon as the database is updated, everyone has access to the new information, which eliminates the need to distribute new copies to everyone. In some jurisdictions, the communications centre can send the preincident planning information to a portable fax machine in a command unit at the scene of an incident. If the preincident planning information is stored on paper, copies are distributed to all of the companies and command officers who might respond to the location on an initial alarm. The plans are kept in binders or in a filing system on each vehicle.

No matter which system your department uses, every effort should be made to keep plans current, consistent, and readily available to personnel in the communications centre, in responding vehicles, and to command staff.

Target Hazards

Most fire departments are not able to create a preincident plan for every individual property in their jurisdiction. As discussed, they identify properties that are particularly large or that present unusual risks to fire fighters or to the public. These properties are identified as target hazards. Examples of target hazards may include the following:

- High-rise buildings
- Hospitals
- Nursing homes and assisted-living facilities
- Large apartment buildings
- Hotels and rooming houses
- Schools
- Public-assembly occupancies
- Lumberyards
- Manufacturing plants
- Shopping centres
- Warehouses

A preincident plan should be prepared for every property that poses a high life-safety hazard to its occupants or presents safety risks for responding fire

FIGURE 27-10 Computers allow access to information from any location.
© Jones & Bartlett Learning. Photographed by Glen E. Ellman.

SAFETY TIP

With fire prevention, the goal is to identify hazards and minimize or correct them so that fires do not occur. With preincident planning, the person creating the plan assumes that a fire will occur and compiles information that responding fire fighters would need in order to reduce the risk to everyone involved.

fighters. Preincident planning should also occur for properties that have the potential to create a large fire or conflagration (a large fire involving multiple structures).

Conducting a Preincident Survey for a Building

The information that goes into a preincident plan is gathered during a preincident survey. This survey is usually completed by one of the crews that would respond to an emergency incident at the location. Its performance enables the crew to visit the property and become familiar with the location as they collect information. A preincident survey can often identify the need for a fire inspection. For this reason, many departments conduct preincident surveys simultaneously with fire inspections. If this method is used, different members of the department should be assigned to conduct the inspection and to complete the preincident survey. This separation of duties allows each fire fighter to concentrate on the objectives of one specific assignment.

As with the home fire safety survey, the preincident survey of a building should be conducted with the knowledge and cooperation of the property owner or occupant. Making this initial contact enables the fire department to schedule an acceptable time, to explain the purpose and the importance of the preincident survey, and to clarify that the information is needed to prepare fire fighters in the event that an emergency occurs at the location.

The team members who conduct the survey should dress and conduct themselves in a manner appropriate to the department's mission. A representative of the property should accompany the survey team to answer questions and provide access to different areas. Every effort should be made to obtain complete, accurate, and useful information.

The preincident survey is conducted in a systematic fashion, following a uniform format. Begin with the outside of the building, gathering all of the necessary information about the building's geographic location, external features, and access points. Then, survey the inside of the building to collect information about every interior area. A good systematic approach starts at the roof and works down through the building, covering every level of the structure, including the basement. If the property is large and complicated, it may be necessary to make more than one visit to ensure that all of the required informations is obtained and recorded accurately.

The same set of basic information must be collected for each property that is surveyed. To conduct a preincident survey, follow the steps in **SKILL DRILL 27-6**.

Additional information should be gathered for properties that are unusually large, are complicated, or might be the site of a particularly hazardous situation. Most fire departments use standard forms to record the survey information.

The fire fighters conducting the survey should prepare sketches or drawings to show the building layout and the location of important features such as exits. It takes practise and experience to learn how to sketch the required information while doing the survey and then to convert the information to a final drawing. In some cases, the building owner can provide the survey team with a copy of a plot plan or a floor plan. Some departments use computer software to create and store these diagrams. These systems have the advantage of creating graphics more quickly and accurately than is possible with hand drawings. Many fire departments also use digital cameras to record information during the survey. Drawings and other graphics are usually prepared at the fire station where the fire officer takes the time to organize the information properly and to generate reports and forms that can be used to develop the preincident plan.

The completed drawing should use standard, easily understood symbols (**FIGURE 27-11**).

Preincident Planning for Response and Access

Building layout and access information is particularly important during the response phase of an emergency incident. The preincident plan should provide information that would be valuable to units en route to an incident. For instance, it might identify the most efficient route for the apparatus to take to the fire building and note an alternate route if the time of day and local traffic patterns might affect the primary route. An alternate route should also be identified if the primary route requires crossing railroad tracks, drawbridges, or other potentially blocked routes.

When conducting a preincident survey, fire fighters should ensure that the building address is clearly visible to save time in locating the property during an emergency incident. If the building is part of a complex, the best route to each individual building, section, or apartment should be indicated on a map. The locations of hydrants and FDCs for sprinkler and standpipe systems should always be identified.

Points of access into the building must also be noted, particularly if the building has multiple entrances or if

SKILL DRILL 27-6
Conducting a Preincident Survey Fire Fighter II, NFPA 1001: 5.5.3

1 Schedule the preincident survey in advance.

2 Make contact with a responsible person.

3 Present a neat and professional image.

4 Identify yourself by name, title, and department.

5 Ensure that a representative from the facility accompanies you during the survey.

6 Take notes and pictures as needed, and start outside.

7 Note the building location.

8 Note the size of the building (height, number of stories, length, width).

9 Identify the building construction.

10 Identify the building use and occupancy.

11 Note any life hazards.

12 Note the access points to the interior of the building.

13 Note the location of the fire alarm annunciator panel and the fire department lock box.

14 Note the utility shut-off locations.

15 Assess the apparatus access locations to the building.

16 Note fire hydrant locations, FDCs, and alternative water supplies.

17 Note ventilation concerns.

18 Record information about built-in fire detection, suppression, and alarm systems.

19 Sketch floor plans.

20 Note the elevator and stairway locations.

21 Review exit plans and exit locations.

22 Identify any special hazards and hazardous materials.

23 Note the building exposures.

24 Anticipate the type of incident expected.

25 Identify any special resources needed.

26 Complete and file the preincident survey form.

Preincident Plan Symbols

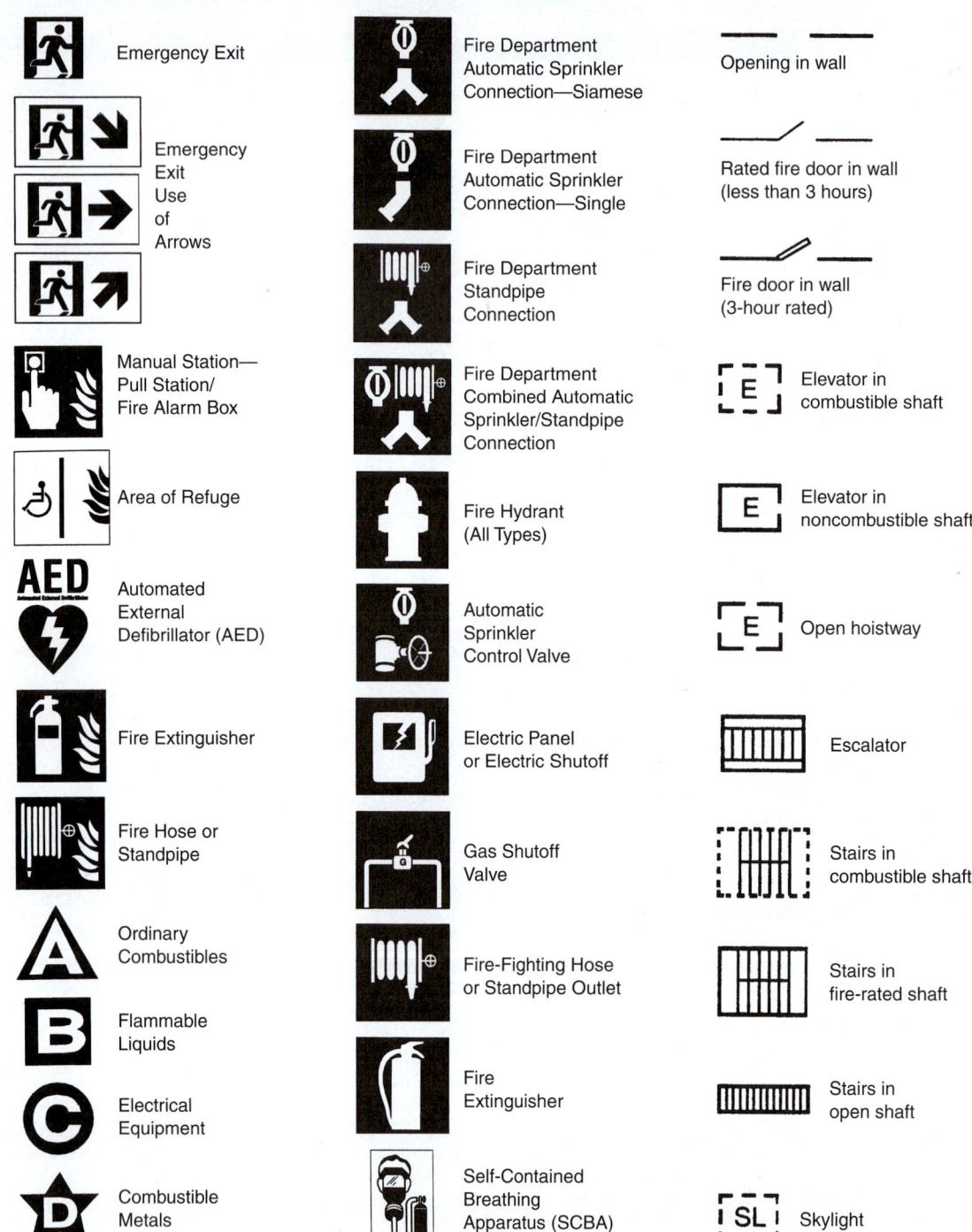

FIGURE 27-11 Preincident plan symbols.

access points are not easily seen from the street. In some cases, different entrances should be used, depending on the location of the emergency within the building. In other cases, the first-due unit should always respond to a designated entrance where the fire alarm annunciator panel is located or to meet a security guard who can act as a guide.

Diagrams should clearly indicate points at which gates, fences, or other barriers block access to parts of a building. The preincident survey should also note whether the topography makes one or more sides of a building inaccessible to fire apparatus or if an underground parking area will not support the weight of fire apparatus.

Access to the Exterior of the Building

The preincident plan should address all possible issues related to access to the exterior of the building. For example, you should ask the following questions during your survey:

- Do several roads lead to the building or just a few?
- Where are the hydrants located?
- Where are the FDCs for automatic sprinkler and standpipe systems located?
- Do any security barriers limit access to the site?
- Are there fire lanes to provide access to specific areas?
- Are any barricades, gates, or other obstructions so narrow or low that they would prevent passage of fire apparatus?
- Are there bridges or underground structures that will not support the weight of fire apparatus?
- Do any gates require keys or a code to gain entry?
- Will it be necessary to cut fences to gain access to the site?
- Does the site include natural barriers such as streams, lakes, or rivers that might limit access?
- Does the topography limit access to any parts of the building?
- Might the landscaping or the presence of snow prevent access to certain parts of the building?

Access to the Interior of the Building

The preincident survey should consider access to the interior of the building. The following questions are helpful in this regard:

- Is there a lockbox containing keys to the building? Do the keys work?
- Where is the lockbox located? Is it clearly visible?
- Are key codes needed to gain access to the building? Who has them? Are they in the lockbox?
- Does the building have security guards? Is a guard always on duty? Does the guard have access to all areas of the building?
- Is a key holder available to respond to the alarm within a reasonable amount of time? How can this person be reached?
- Where is the fire alarm annunciator panel located?
- Is the fire alarm annunciator panel properly programmed so you can quickly determine the exact location of an alarm?

Preincident Planning for Scene Size-Up

The preincident survey must obtain essential information about the building that is important for size-up—that is, the ongoing observation and evaluation of factors that influence the objectives, strategies, and tactics for fire suppression. This information should include the building's construction, height, area, use, and occupancy, as well as the presence of hazardous materials or other risk factors. The locations of other buildings or structures that might be jeopardized by a fire in the building should be recorded as well.

Fire protection system and water supply information is also important for size-up. The preincident survey should identify areas that are protected by automatic sprinklers or other types of fire suppression systems, the locations of standpipes, and the locations of firewalls and other features designed to limit the spread of a fire. Likewise, the survey should note any areas where fire protection is lacking, such as an area without sprinklers in a building that otherwise has a sprinkler system.

Features inside a building that would allow a fire to spread but that are not readily visible from the outside should be noted as well. A common attic space over several occupancies, unprotected openings between floors, or buildings connected by overhead passages or conveyor systems are examples of these features.

Access and egress features, including stairways, exit doors, and elevators, should be diagrammed and labelled. These features are integral in order to identify the easiest access to the upper floors for rescue and fire control activities.

Construction

Properly identifying building construction is an essential factor in completing a incident survey. Although they might have similar appearances, two buildings might nevertheless be constructed differently and, therefore, would react differently during a fire. Five types of construction are defined in NFPA 220, *Standard on Types of Building Construction* (**TABLE 27-2**). (Building construction is discussed in detail in Chapter 6, *Building Construction*.) Preincident planning enables you to take

SAFETY TIP

"Lightweight construction" is a relative term. A structure may be "lightweight" in comparison to other methods of construction, but its components are still heavy enough to cause serious injury or death if they collapse on a fire fighter.

TABLE 27-2 Types of Building Construction

Type I: Fire resistive. Buildings where the structural members are made of noncombustible materials that have a specified fire resistance (**FIGURE 27-12**). Materials include concrete, protected steel beams, and masonry block walls, among others.

FIGURE 27-12 A Type I building.

Reproduced with permission from NFPA 170-2012, *Standard for Fire Safety and Emergency Symbols*, Copyright © 2011, National Fire Protection Association. This reprinted material is not the complete and official position of the NFPA on the referenced subject, which is represented only by the standard in its entirety.

Type II: Noncombustible. Buildings where the structural members are made of noncombustible materials but may not have fire resistance protection (**FIGURE 27-13**). Materials include unprotected steel beams.

FIGURE 27-13 A Type II building.
© Rick's Photography/Shutterstock.

Type III: Ordinary. Buildings where the load-bearing exterior walls are made of noncombustible or limited-combustible materials, but the interior floors and walls are made of combustible materials (**FIGURE 27-14**).

FIGURE 27-14 A Type III building.
© Ken Hammon/USDA.

Type IV: Heavy timber. Buildings where the exterior walls are made of noncombustible or limited-combustible materials, but the interior walls and floors are made of combustible materials (**FIGURE 27-15**). The dimensions of the interior materials are greater than those of ordinary construction (typically with minimum dimensions of 203 by 203 mm. [8 in. by 8 in]).

FIGURE 27-15 A Type IV building.
© Helen Filatova/Shutterstock.

TABLE 27-2 Types of Building Construction (Continued)

Type V: Wood frame. Buildings where the exterior walls, interior walls, floors, and roof are made of combustible wood materials (**FIGURE 27-16**).

FIGURE 27-16 A Type V building.
© Jones & Bartlett Learning.

the time that is needed to determine the exact type of building construction and identify any problem areas. If an emergency occurs, the IC can check the preincident plan for information on construction, instead of hazarding a guess.

Lightweight Construction. It is important to identify whether the building contains lightweight construction (lightweight materials and/or advanced engineering structural members). Lightweight construction uses assemblies of small components, such as manufactured trusses or fabricated beams, as structural support materials (**FIGURE 27-17**). These assemblies are strong and can cost less than traditional construction methods, but they may fail quickly and catastrophically during a fire. This type of construction can be found in newer buildings as well as in older buildings with extensive remodeling. In many cases, the lightweight

FIGURE 27-17 A lightweight wood truss roof assembly.
Courtesy of UL.

components are located in void spaces or concealed above ceilings and are not readily visible.

A wood truss constructed from 51-mm by 102-mm (2-in. by 4-in.) or 51-mm by 152-mm (2-in. by 6-in.) pieces of wood, for example, can be used to span a wide area and support a floor or roof. A truss uses the principle of a triangle to build a structure that can support a great deal of weight with much less supporting material than conventional construction methods (**FIGURE 27-18**).

SAFETY TIP

Two fire fighters died while battling an early-morning blaze in a fast-food restaurant. The fire burned through the lightweight wood truss roof supports, and the roof-mounted air conditioner dropped into the building, trapping and killing the fire fighters. The fire was later determined to be arson.

Remodelled Buildings. Buildings that have been remodelled or renovated may present unique hazards. Remodelling can remove some of the original built-in fire protection and create new hazards. Examples of remodeling-derived hazards include multiple ceilings with void spaces between them, new construction of concrete topping over a wooden floor, and new openings between floors or through walls that originally were fire resistant.

If you can conduct the preincident survey during construction or remodelling, you will be able to see the construction before everything is covered over. It is also important to realize that all buildings under construction—including those that are being remodelled or demolished—are especially vulnerable to fire. Unfinished construction is open, so it lacks many of the

FIGURE 27-18 A lightweight wood truss with gusset plates.
© David Huntley Creative/Shutterstock.

fire-resistant features and fire detection/suppression systems that will be part of the finished structure.

Building Use

The second major consideration during size-up—and hence during a preincident survey—is the building's use and occupancy. Buildings are used for many purposes, such as residences, offices, stores, restaurants, warehouses, factories, schools, and churches. For each use, different types of problems, concerns, and hazards may be present.

For example, knowing the building's use can help fire fighters determine the number of occupants and predict their ability to escape if a fire occurs. It is also a key factor in determining the probable contents of the building. These factors can have a major effect on the problems and hazards encountered by fire fighters responding to an emergency incident at the location.

A building is usually classified by its major use, which identifies the basic characteristics of the building (**TABLE 27-3**). Within the major use classification, various occupancy subclassifications may provide a more specific description of the possible uses of sections of the building and their associated characteristics.

Many large building complexes contain multiple occupancy subcategories under one roof. For example, a shopping mall usually includes both retail stores and restaurants. The mall might also be part of an even larger complex that includes offices, residential condominiums, a hotel, a movie theatre, an underground parking garage, and a subway station. Each area may have very different characteristics and, therefore, different risk factors.

LISTEN UP!

A building can have multiple uses during its lifetime or can incorporate more than one use within the same structure.

TABLE 27-3 Classifications of Buildings

Major Use Classification	Occupancy Subcategories
Public assembly	Theatres, auditoriums, and churches
	Arenas and stadiums
	Convention centres and meeting halls
	Bars and restaurants
Institutional	Hospitals and nursing homes
	Schools
Commercial	Retail stores
	Industrial factories
	Warehouses
	Parking garages
	Offices

© Jones & Bartlett Learning.

Occupancy Changes

Building use may change over time. An outdated factory may be transformed into a residential building, or an unused school may be converted into an office building. A warehouse that once stored concrete blocks may now be filled with combustible foam-plastic insulation or hazardous materials such as swimming pool chemicals. Given this propensity for change, preincident plans should be checked and updated on a regular basis. A building's current occupancy information must always be determined during a preincident survey.

Exposures

An **exposure** is any person or property that could be endangered by fire, smoke, gases, runoff, or other hazardous conditions (NFPA 402). For example, an exposure could be an attached building that is separated from the other structure by a common wall, or it could be a building across an alley or street. Exposures can include other buildings, vehicles, outside storage, or anything else that could be damaged by or involved in a fire. For example, heat from a burning warehouse could ignite adjacent buildings and spread the fire.

A preincident survey identifies potential exposures for the subject property. It should take into account the size, construction, and fire load (the amount of combustible material and the rate of heat release) of the property being evaluated, the distance to the exposure, and the ease of ignition of the exposure by radiation, convection, or conduction of heat.

> **SAFETY TIP**
>
> A component in lightweight building construction is the Trus Joist I beam (or TJI beam). This framing material is made up of three wooden components. Channels may be drilled through TJI beams, thus providing another avenue for fire to spread.

Built-In Fire Protection Systems

The preincident survey should identify built-in fire detection and suppression systems on the property. These systems may include automatic sprinklers, standpipes, fire alarms, and fire detection systems, as well as systems designed to control or extinguish special hazard fires. Some buildings have automatic smoke control or exhaust systems. Most high-rise buildings have systems that control elevator function during an emergency. Each of these systems is covered in detail in Chapter 26, *Fire Detection, Suppression, and Smoke Control Systems*.

Automatic Sprinkler Systems. A properly designed and maintained automatic sprinkler system can help control or extinguish a fire before the arrival of the fire services unit. When properly designed and maintained, such systems are extremely effective and can play a major role in reducing the loss of life and property at an incident. Activated by heat, the system suppresses fire by discharging water from a sprinkler head.

The preincident survey should determine whether the building has a sprinkler system and which parts of the building that system covers. Note the locations of the valves that control water flow to different sections of the system. Normally, these valves are found in the open position.

In addition to the control valves, sprinkler systems should have an FDC outside the building. The FDC is used to supplement the water supply to the sprinklers by pumping water from a hydrant or other water supply through fire hose into this connection. The location of the FDC for the sprinkler system must be marked on the preincident plan.

Standpipe Systems. Standpipe systems are installed in accordance with the requirements of the building code for the purpose of delivering water to fire hose outlets on each floor of a building. Their use eliminates the need to extend hose lines from a pump at the street level up to the fire level. Standpipes may also be used in low-rise buildings, such as convention centres. In such structures, fire fighters can bring attack hose lines inside the building and connect them to a standpipe outlet close to the fire, while the pump delivers water to an FDC outside the building.

The location of the FDC, as well as the locations of the outlets on each floor, should be clearly marked on the preincident survey. The locations of nearby hydrants that will be used to supply water to the pumper that feeds the FDC must also be recorded. In a large building, multiple FDCs may be available to deliver water to different parts of the building. The area or floor levels served by each connection should be carefully noted and labelled at the connection.

Fire Alarm and Fire Detection Systems. The primary role of a fire alarm and detection system is to alert the occupants so that they can evacuate or take appropriate action when an incident occurs. Some fire alarm and detection systems are connected directly to the fire department; others are monitored by a service that calls the fire department when the system is activated.

In some cases, the fire alarm and detection system must be manually activated. In other cases, it is activated automatically. A smoke or heat detection system can activate the alarm, for example, or it may respond to a device that indicates when water is being discharged from the sprinkler system.

The fire alarm control panel serves as the "brain" for the system. Many buildings have an additional display panel near the entrance to the building. The remote annunciator indicates the location and type of device that activated the alarm system. In most cases, the fire fighters who respond to the alarm must check the remote annunciator to determine the actual source of the alarm within the building or complex. The preincident survey should identify which type of system is installed, where the fire alarm control panel and remote annunciator are located, and whether the system is remotely monitored.

Special Hazard Fire Suppression Systems.

Several types of specialized fixed fire suppression systems can be installed to protect areas where water would not be an acceptable extinguishing agent. For example, sophisticated foam or dry-chemical fire suppression systems are found in areas where flammable liquids are stored or used. The type of system and the area that is protected should be identified in the preincident survey.

In addition, details about the method of operation, location of equipment, and foam supplies should be recorded. More information about built-in systems is presented in Chapter 26, *Fire Detection, Suppression, and Smoke Control Systems.*

Considerations for Water Supply

During the preincident survey, use your department's approved method to calculate the fire flow required to fight a fire involving the entire building. The required flow rate, measured in litres per minute or gallons per minute, can be calculated based on the building's size, construction, contents, and exposures. The water supply source should be identified.

In most urban areas, the water will be supplied from municipal hydrants. It is important to locate the hydrants closest to the building. In addition, for large buildings, it will be necessary to locate enough hydrants to supply the volume of water required to control a fire. The ability of the municipal water system to provide the required flow also must be determined. In some cases, it may be necessary to use hydrants that are supplied by different water mains (a generic term for underground water pipes) of sufficient size to achieve the needed water flow (**FIGURE 27-19**).

In areas without municipal water systems, water may have to be obtained from an alternative water supply such as a mobile water supply apparatus (also called tankers or water tenders) or be drafted from a static source such as a lake or dry hydrant. When static water sources are

used, the preincident plan must identify drafting sites (locations where a pump can draft water directly from the static source). It is also important to measure the distance from the water source to the fire building to determine whether large-diameter hose and additional pumpers will be needed for relay pumping. Finally, the preincident plan should outline the operation that would be required to deliver water to the fire.

If mobile water supply apparatus will be used to deliver water, the preincident plan must include several additional details. For instance, sites for filling the mobile water supply apparatus and for discharging their loads must be identified (**FIGURE 27-20**). The preincident plan should also identify how many tankers (or water tenders) will be needed based on the total distance they must travel, the quantity of water each vehicle can transport, and the total time it takes to empty and refill the vehicle.

LISTEN UP!

A private water system may not be able to simultaneously supply both automatic sprinkler systems and fire hose effectively. In one incident, a commercial property in Bluffton, Indiana, was destroyed when the responding fire department hooked its pumpers to private fire hydrants. The subsequent reduction in water flow rendered the building's sprinklers ineffective.

A large industrial or commercial complex may have its own water supply system that provides water for automatic sprinkler systems, standpipes, and private hydrants. These systems usually include storage tanks or reservoirs as well as fixed fire pumps to deliver the water under pressure. The details and arrangement of the private water supply system must be determined during the preincident survey.

The survey should indicate how much water is stored on the property and where the tanks are located. If a fixed fire pump is available, its location, capacity (in litres per minute or gallons per minute), and power source (electricity or a diesel engine) should be noted. It is also important to confirm that these private water systems and fixed fire pumps are being maintained in good operating condition. For example, the private water supply system and fixed fire pump on an abandoned property may be useless if it is no longer properly maintained.

In many cases, the same private water main provides water for both the sprinkler system and the private hydrants. If the fire department uses the private hydrants, the water supply to the sprinklers may be compromised. The preincident plan for these sites should note whether public, off-site hydrants should be used instead of the private hydrants.

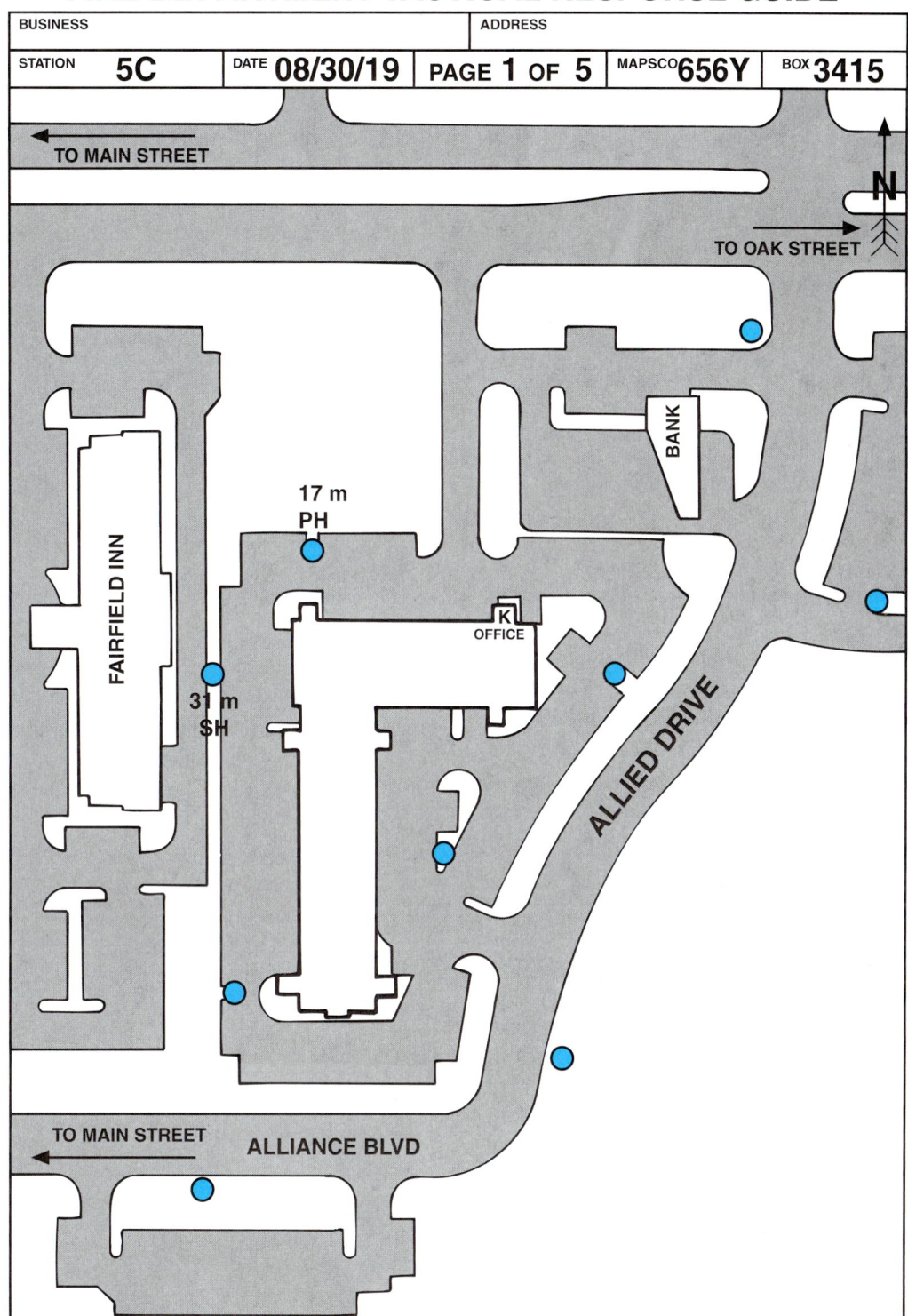

FIGURE 27-19 A diagram showing hydrant locations around a building. PH indicates the primary hydrant for connecting to the FDC. 17 m (55 ft) indicates that the primary hydrant is located 17 m (55 ft) from the FDC. SH indicates the secondary hydrant for connecting to the FDC. 31 m (100 ft) indicates that the secondary hydrant is located 31 m (100 ft) from the FDC.

© Jones & Bartlett Learning

FIGURE 27-20 The points where mobile water supply apparatus can be filled and where they can discharge their loads must be identified in the preincident plan.
© Jim Smalley.

Utilities

During an emergency incident, it may be necessary to turn off utilities such as electricity or natural gas as a safety measure. The preincident survey should note the locations of shut-offs for electricity, water, natural gas, propane gas, fuel oil, and other energy sources. These sites, as well as contact information for the appropriate utility company, should be noted. In some cases, special knowledge or equipment may be required to disconnect utilities. This information must be noted in the preincident plan, along with the procedure for contacting the appropriate individual or organization.

Electrical wires pose particular problems during firefighting operations. Electricity may be supplied by overhead wires or through underground cables. Overhead wires can be deadly if a ground ladder or aerial ladder comes in contact with them. The preincident plan should show the locations of high-voltage electrical lines and equipment that could prove dangerous to fire fighters. If electricity is supplied by underground cables, the shut-off may be located inside the basement of a building or in an underground vault. These locations should be noted during the survey.

If propane gas or fuel oil is stored on the property, the preincident plan should show the location and note the capacity of each tank. The presence of an emergency generator should be noted as well, along with the fuel source and a list of equipment powered by the generator.

Preincident Planning for Search and Rescue

Fire fighters who are conducting search and rescue operations need to know the locations of the occupants

of a building as well as the locations of exits. The preincident survey should identify all entrances and exits to the building, including fire escapes and roof exits.

In addition to conducting a search for occupants who are unable to escape on their own, search and rescue teams may need to assist occupants who are trying to use the exits (**FIGURE 27-21**). For example, fire fighters may need to assist occupants who are descending fire escapes or need to place ladders near windows or other locations for occupants to use.

During the preincident survey, team members should obtain the interior floor plan for the building.

FIGURE 27-21 Fire fighters should plan to assist occupants who are trying to use the building exits.
© Keith D. Cullom/www.fire-image.com.

Information about the floor plan of a building can be life-saving knowledge when the structure is filled with smoke. It is much easier to understand a floor plan when you can tour a building under nonemergency conditions. In large buildings, it may be necessary to plan for the use of ropes during search and rescue to prevent disorientation in conditions of limited visibility.

Preincident Planning for Forcible Entry

As previously noted, the preincident survey should consider both exterior and interior access problems. Locations where forcible entry may be required should be identified and marked on the site diagrams and building floor plan. Noting which tools would be needed to gain entry can save time during the actual emergency. The location of a lockbox and instructions regarding obtaining keys should be identified.

Preincident Planning for Ladder Placement

The preincident survey is an excellent opportunity to identify the best locations for placing ground ladders or using aerial apparatus (**FIGURE 27-22**). The length of ladder needed to reach a roof or entry point should be noted. When planning ladder placement, pay careful attention to any electrical wires and other obstructions that might not be visible at night or in a smoky atmosphere. Plan to place ladders in locations that will not disturb other vital functions. For example, you should not place ladders in front of a door that will be used for entrance to or exit from the fire building.

FIGURE 27-22 Considerations for the use of ladders should include identifying the best locations to place ground ladders and use aerial apparatus.
© Colin Archer/AP Photos.

Preincident Planning for Ventilation

While performing a preincident survey, fire fighters should consider which information would be valuable to the members of a ventilation team during a fire. For example, what would be the best means to provide ventilation? How useful are the existing openings for ventilation? Are there windows and doors that would be suitable for horizontal ventilation? Where could fans be placed? Could the roof be opened to provide vertical ventilation? What is the best way to reach the roof? Are there ventilators or skylights that could be easily removed or bulkhead doors that could be opened easily? Will fire fighters need to use saws and axes to cut through the roof? Would multiple ceilings have to be punctured to allow smoke and heat to escape?

It is also important to determine whether the heating, ventilation, and air-conditioning (HVAC) system can be used to remove smoke without circulating it throughout the building. Many buildings with sealed windows have controls that enable the fire department to set the HVAC system to deliver outside air to some areas and exhaust smoke from other areas. The instructions for controlling the HVAC system should be included in the preincident plan.

Roof construction must also be evaluated to determine whether it would be safe for fire fighters to work on the roof when there is a fire below. If the roof is constructed with lightweight trusses, the risk of collapse is great. Many fire departments do not permit fire fighters on these roofs. The presence of an attic that might allow a fire to spread quickly under the roof should also be noted. In addition, common attics in townhouses and shopping malls can spread fire from one occupancy to another.

Occupancy Considerations

Each type of occupancy involves unique considerations that should be taken into account when preparing a preincident plan. Fire fighters should keep these factors in mind when conducting a survey.

High-Rise Buildings

High-rise buildings present special problems during a fire because of the difficulty in delivering adequate water pressure to the upper floors and because of the large numbers of occupants. As a consequence, major fires in high-rise buildings often result in injuries, fatalities, and millions of dollars in property losses.

A preincident survey should identify both the building construction and any special features that have been installed. It should note all of the systems that are present in a particular building and identify how they are designed to function. This information, in turn, should be incorporated into the preincident plan.

Fire protection features of high-rise buildings vary depending on the age of the building and the specific building and fire code requirements in different jurisdictions. Older high-rise buildings are generally constructed of noncombustible materials and designed to confine a fire within a limited area. Newer Type I buildings generally include automatic sprinklers, smoke detection and alarm systems, emergency generators, elevator control systems, smoke control systems, and building control stations where all of these systems are monitored and controlled. The details of each system must be documented in the preincident plan. In many cases, expert consultants may be contacted to develop an emergency plan for such a building. These plans are intended to be used by the building tenants and management as well as the fire department.

Assembly Occupancies

Public-assembly venues—such as theatres, nightclubs, stadiums, churches, hotel meeting rooms, and arenas—have frequently been the location of major loss of life.

These structures are often very large and complicated, and they may be equipped with complex systems intended to manage emergency situations. Gaining access to the location of the fire or emergency situation may

---

SAFETY TIP

Modern high-rise buildings are often equipped with multiple fire protection systems—but these systems must be maintained properly to be effective. One Meridian Plaza (a modern high-rise office building in Philadelphia, Pennsylvania) had both a standpipe system and a partial sprinkler system on a few floors. When a fire broke out in this building in 1991, fire fighters attached their hose lines to the standpipe system. Because the pressure-reducing valves on the standpipe outlets had been set incorrectly, the fire fighters were unable to obtain sufficient pressure to control the fire. Three fire fighters died, and the fire burned several floors of the building, which eventually had to be demolished. The fire was stopped only when it reached one of the floors that was equipped with a sprinkler system. An NFPA Fire Investigation Report is available on this fire.

---

prove difficult, however, when all of the occupants are trying to evacuate at the same time.

Healthcare Facilities

Healthcare facilities—such as hospitals, nursing homes, assisted-living facilities, surgery centres, and ambulatory healthcare centres—require special preincident planning. Hospitals are often very large and include many different areas, ranging from operating rooms to walk-in clinics. The most challenging problem during an emergency incident at a healthcare facility is protecting nonambulatory patients. **Nonambulatory patients** cannot move themselves to an area of safety during an emergency, owing to their physical condition, medical treatment, or other factors.

A **defend in place** philosophy is used when designing fire protection for these facilities. This approach presumes that patients will not be able to escape from a fire without assistance and that there may not be enough staff present to move all of the patients out of the structure. Instead, the occupants may be relocated to a safe place within the structure. Therefore, the facility itself is designed to protect the patients from the fire. Most healthcare facilities utilize fire-resistant construction, have fire detection systems and sprinkler systems, and are compartmentalized.

If it becomes necessary to move patients from a dangerous area, the preferred approach is often **horizontal evacuation** (moving patients from a dangerous area to a safe area on the same floor). It is much easier to wheel a bed to another area on the same floor level than to carry patients down stairways.

Detention and Correctional Facilities

By their very nature, detention and correctional facilities are designed to ensure that the occupants cannot leave the building, even under normal conditions. The preincident plan must consider the technicalities of removing the inmates from a facility. Additionally, security concerns can make it difficult for fire fighters to gain rapid access to the building or for occupants and rescuers to exit the facility.

Multifamily Residential Occupancies

Preincident plans are usually prepared for multifamily residential properties such as apartment complexes and condominiums. In addition to the standard preincident plan information, the following information will be helpful when developing a preincident plan for residential occupancies:

- Detailed floor plan
- Location of sleeping areas
- Information about any handicapped occupants
- Escape routes and an outside designated meeting place
- Fire hydrant or water source location

As discussed earlier, some homeowners may request an individual fire safety survey of their property. These surveys are intended to identify fire hazards and to educate the homeowner about steps that can be taken to reduce the risk of fire.

If individual homes are not surveyed, many fire departments will survey different neighbourhoods, making special note of addresses, access routes, and hydrant locations. Knowing which styles and types of private residences are present in your jurisdiction will help you prepare for emergency incidents.

Locations Requiring Special Considerations

Preincident planning should extend beyond planning for fires to other types of emergency situations. Specifically, planning should anticipate the types of incidents that could occur at locations such as airports, bridges, and tunnels, as well as incidents along highways or railroad lines or at construction sites (**FIGURE 27-23**).

Detailed preincident plans prepared for airports should note the various types of aircraft that use the airport. In addition, the airport facilities themselves often require extensive preincident planning because of their size, the number of people who may be present, and security issues. Similar planning should be performed for bridges, tunnels, and any other locations where complicated situations could occur.

The following special locations would also require preincident planning:

- Gas or liquid fuel transmission pipelines
- Electrical transmission lines (**FIGURE 27-24**)
- Ships and waterways (**FIGURE 27-25**)
- Subways and tunnels (**FIGURE 27-26**)
- Railroads (**FIGURE 27-27**)

Special Hazards

One of the most important reasons for developing a preincident plan is to identify any special hazards and to provide information that would be valuable during an emergency incident. This information includes chemicals or hazardous materials that are stored or

FIGURE 27-24 Electrical transmission lines.
© Jones & Bartlett Learning.

FIGURE 27-23 Preincident planning should anticipate the types of incidents that are likely to occur at locations other than buildings, such as a tunnel.
© Lucy Autrey Wilson/Shutterstock.

FIGURE 27-25 Ships and waterways.
© GeoStock/Photodisc/Getty Images.

FIGURE 27-26 Subway station.
© Arthur S. Aubry/Photodisc/Getty Images.

FIGURE 27-28 During the preincident survey, look for special hazards, and obtain as much information as possible.
© Jones & Bartlett Learning.

FIGURE 27-27 Railroad.
© JJJ/Shutterstock, Inc.

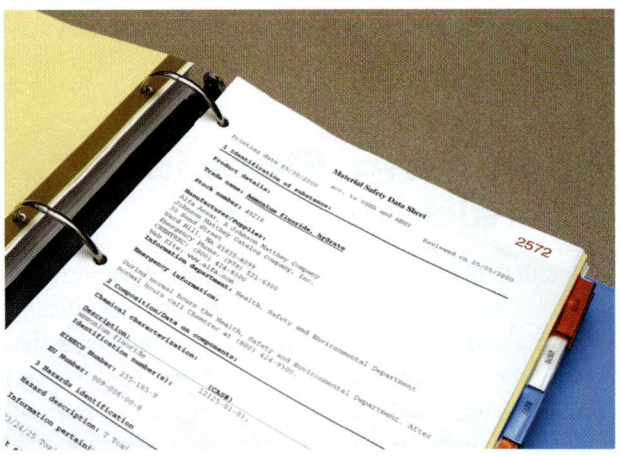

FIGURE 27-29 Safety data sheets (SDSs) are also referred to as material safety data sheets (MSDSs).
© Jones & Bartlett Learning.

used on the premises, structural conditions that could result in a building collapse, industrial processes that pose special hazards, high-voltage electrical equipment, and confined spaces. Preincident plans warn fire fighters of potentially dangerous situations and could include detailed instructions about what to do in those circumstances—information that could ultimately save the lives of fire fighters.

Fire fighters conducting a preincident survey should always look for special hazards and obtain as much information as possible (**FIGURE 27-28**). If the information is not readily available, it may be necessary to contact specialists for advice or conduct further research before completing the preincident plan.

Hazardous Materials

A preincident survey should obtain a complete description of all hazardous materials that are stored, used,

or produced at the property. This assessment includes an inventory of the types and quantities of hazardous materials that are on the premises, as well as information about where they are located, how they are used, and how they are stored. The appropriate actions and precautions for fire fighters in the event of a spill, leak, fire, or other emergency incident should also be listed.

Some jurisdictions may require businesses that store or use hazardous materials to obtain a fire department permit. They may also require that hazardous materials specialists conduct special inspections. If the quantity of hazardous materials on hand exceeds a specified limit, federal and provincial regulations require a property owner to provide the local fire department with current inventories and **safety data sheets (SDSs)** (**FIGURE 27-29**). Fire fighters conducting preincident surveys should learn where the SDSs are stored, ensure that the required information has been provided, and verify that it is up-to-date.

Fire fighters should expect to encounter hazardous materials at certain types of occupancies, such as chemical companies, garden centres, swimming pool supply stores, hardware stores, and laboratories. Of course, hazardous materials can also be found in unexpected locations, so fire fighters should always be alert for their presence.

Some communities put placards on the outside of any building that contains hazardous materials. These placards should use the marking system specified in NFPA 704, *Standard System for the Identification of the Hazards of Materials for Emergency Response.*

In case of fire, some hazardous materials require use of special suppression techniques or extinguishing agents. This information should be noted in the preincident plan. The plan should also contain contact information for individuals or organizations that can provide advice if an incident occurs.

LISTEN UP!

All information in preincident plans, including SDSs, is confidential. The general public is not at liberty to see the information collected in your fire department's preincident plans. Keep all information that you learn about an occupancy during a preincident plan survey confidential.

After-Action REVIEW

IN SUMMARY

- Your highest priority as a fire fighter is to provide education to the public about fire risks and other hazards that would help to limit the loss of life, injuries, and property damage.

- One of the most effective means of avoiding the loss of lives and property is to increase the focus on the prevention of fires, rather than the efforts and resources committed to extinguishing them. Every fire that is prevented reduces the risks to fire fighters and the public.

- Fire prevention includes a range of activities that are intended to prevent the outbreak of fires and to limit the consequences if a fire does occur. Communities will be able to reduce fire outbreak through several fire and life safety strategies:
  - Public education, which helps people understand how to limit the loss of life, injuries, and property damage from occurring and teaches them how to react in an appropriate manner if an emergency occurs
  - Code enforcement, enacted to ensure a minimum level of fire safety in the home and workplace environments
  - Plan review
  - Fire investigation

- Public fire and life safety education programs include those focused on the following topics:
  - Stop, Drop, and Roll
  - Close your door
  - Crawl low under smoke
  - Exit Drills In The Home (E.D.I.T.H.)
  - Emergency notification
  - Installation and maintenance of smoke alarms
  - Advantages of residential sprinkler systems
  - Selection and use of portable fire extinguishers
  - Learn Not to Burn® preschool program
  - Fire safety for special populations

- Fall prevention
- Wildland fire prevention programs

- Most fire departments have a set format for conducting fire station tours. For example, a tour might start with an inspection of the apparatus and end in a meeting room with a question-and-answer session and information about the department and fire prevention techniques.

- Try to leave every tour group with both a fire prevention message and materials.

- A preincident plan is a collection of data that is used by responding personnel to effectively and safely manage an emergency at a particular location. Preincident planning helps your IC and your department to make informed command decisions because important information is assembled before the emergency occurs.

- The use of modern information technology has greatly enhanced the ability of fire departments to capture, store, organize, update, and retrieve preincident plans. Accurate and current information such as drawings, maps, satellite and aerial imagery, photographs, descriptive text, lists of hazardous materials, and SDS information can be made available instantly to the fire fighters who need it during an emergency incident.

- A preincident plan should be prepared for every property that poses a high life-safety hazard to its occupants or presents safety risks for responding fire fighters. Preincident plans should also be prepared for properties that have the potential to create a large fire or conflagration.

- The information that goes into a preincident plan is gathered during a preincident survey.

- A preincident survey should be conducted with the knowledge and cooperation of the property owner or occupant.

- The preincident survey is conducted in a systematic fashion, beginning with the outside of the building and moving inside. A good, systematic approach starts at the roof and works down through the building, covering every level of the structure, including the basement.

- The preincident survey includes, but is not limited to, the following information:
  - Sketches or drawings of building layout and locations of important features such as exits
  - Information that would be valuable en route to an incident
  - Entrances and exits to a building, including fire escapes and roof exits
  - Any exterior and interior access issues and the locations where forcible entry may be required
  - Building construction, height, area, use, and occupancy
  - The presence of hazardous materials
  - Locations of exposed structures
  - Fire protection system information
  - Tactical information about water supply considerations and locations of shut-offs for utilities
  - Best locations for placing ground ladders, aerial apparatus, and ventilation

- Preincident planning should anticipate the types of incidents that could occur at locations such as airports, bridges, and tunnels, as well as incidents along highways or railroad lines or at construction sites.

- The following occupancies involve unique considerations:
  - High-rise buildings—Special issues include difficulty in gaining access and providing adequate water supply to upper floors, as well as assisting the occupants to evacuate.
  - Public assemby venues—These structures are often very large and contain large numbers of people to be evacuated.
  - Healthcare facilities—These structures are very large and contain nonambulatory occupants who need assistance in evacuating.
  - Detention and correctional facilities—Security concerns may make it difficult for fire fighters to gain rapid access to the building or for occupants to exit the facility.
  - Residential occupancies—Include apartment complexes and condominiums. The preincident plan should identify the locations of the sleeping areas and the water supplies.

KEY TERMS

Access Navigate for flashcards to test your key term knowledge.

Defend in place The operational response in which the action is to relocate the affected occupants to a safe place within the structure during an emergency. (NFPA 1620)

Exit Drills In The Home (E.D.I.T.H.) A public fire and life safety education program designed to teach occupants how to safely exit a home in the event of a fire or other emergency. It stresses the importance of having a pre-established meeting place for all family members.

Exposure Any person or property that could be endangered by fire, smoke, gases, runoff, or other hazardous conditions. (NFPA 402)

Fire codes A set of legally adopted rules and regulations designed to prevent fires and protect lives and property in the event of a fire.

Fire prevention Measures taken toward avoiding the inception of fire. (NFPA 801)

Fire safety survey A voluntary inspection of a residence or occupied structure to identify fire and life safety hazards.

Horizontal evacuation Moving occupants from a dangerous area to a safe area on the same floor level.

Nonambulatory patient A term describing individuals who cannot move themselves to an area of safety owing to their physical condition, medical treatment, or other factors.

Preincident plan A document developed by gathering general and detailed data that is used by responding personnel in effectively managing emergencies for the protection of occupants, responding personnel, property, and the environment. (NFPA 1620)

Preincident survey The process used to gather information to develop a preincident plan.

Safety data sheets (SDSs) Formatted information, provided by chemical manufacturers and distributors of hazardous products, about chemical composition, physical and chemical properties, health and safety hazards, emergency response, and waste disposal of the material. (NFPA 1072)

REFERENCES

National Fire Protection Association. 2018. Educational messages desk reference, 2018 edition. https://www.nfpa.org/Public-Education/Resources/Educational-messaging. Accessed November 27, 2018.

National Fire Protection Association. 2018. Home fires involving cooking equipment. https://www.nfpa.org/News-and-Research/Fire-statistics-and-reports/Fire-statistics/Fire-causes/Appliances-and-equipment/Cooking-equipment. Accessed November 27, 2018.

National Fire Protection Association. 2018. Home fire sprinklers. https://www.nfpa.org/Public-Education/By-topic/Home-fire-sprinklers. Accessed November 27, 2018.

National Fire Protection Association. 2011. Home Structure Fires That Began With Mattresses and Bedding Fact Sheet. https://www.nfpa.org/-/media/Files/News-and-Research/Fire-statistics-and-reports/Fact-sheets/beddingfactsheet.ashx. Accessed November 27, 2018.

National Fire Protection Association. 2018. NFPA 1, *Fire Code.* www.nfpa.org. Accessed November 27, 2018.

National Fire Protection Association. NFPA 72, *National Fire Alarm and Signaling Code,* 2019. www.nfpa.org. Accessed November 27, 2018.

National Fire Protection Association. NFPA 101, *Life Safety Code*®, 2018. www.nfpa.org. Accessed November 27, 2018.

National Fire Protection Association. NFPA 170, *Standard for Fire Safety and Emergency Symbols,* 2018. www.nfpa.org. Accessed November 27, 2018.

National Fire Protection Association. NFPA 220, *Standard on Types of Building Construction,* 2018. www.nfpa.org. Accessed November 27, 2018.

National Fire Protection Association. NFPA 704, *Standard System for the Identification of the Hazards of Materials for Emergency Response,* 2017. www.nfpa.org. Accessed November 27, 2018.

National Fire Protection Association. NFPA 1031, *Standard for Professional Qualifications for Fire Inspector and Plan Examiner,* 2014. www.nfpa.org. Accessed November 27, 2018.

National Fire Protection Association. NFPA 1035, *Standard on Fire and Life Safety Educator, Public Information Officer, Youth Firesetter Intervention Specialist and Youth Firesetter Program Manager Professional Qualifications,* 2015. www.nfpa.org. Accessed November 27, 2018.

National Fire Protection Association. NFPA 1452, *Guide for Training Fire Service Personnel to Conduct Community Risk Reduction,* 2015. www.nfpa.org. Accessed November 27, 2018.

National Fire Protection Association. NFPA 1620, *Standard for Pre-incident Planning,* 2015. www.nfpa.org. Accessed November 27, 2018.

National Fire Protection Association. 2018. Top causes of fire. https://www.nfpa.org/Public-Education/By-topic/Top-causes-of-fire. Accessed November 27, 2018.

On Scene

You are a volunteer fire fighter in a small rural community. One night after training, the chief asks if you would be willing to serve as the department's public educator. You know there is a desperate need, and you enjoy talking about the fire department, but you just aren't sure where you would begin. You think back to the programs and concepts you learned in Fire Fighter I and II training.

1. Regulations designed to prevent fires from occurring, provide for safe egress of occupants, eliminate fire hazards, protect lives and property, and limit fire losses are generally called _____ codes.

   A. building

   B. electrical

   C. fire

   D. model

2. Which program focuses on instructing children what to do when their clothes catch fire?

   A. Learn Not to Burn®

   B. Stop, Drop, and Roll

   C. Get low and go

   D. E.D.I.T.H.

3. By approximately what percentage does a working smoke alarm and sprinkler system reduce the risk of fire death?

   A. 23%

   B. 50%

   C. 80%

   D. 94%

4. A _____ is a document developed by gathering general and detailed data that are used by responding personnel in effectively managing emergencies for the protection of occupants, responding personnel, property, and the environment.

   A. preincident plan

   B. preincident survey

   C. fire safety survey

   D. fire station tour

5. A _____ is the process used to gather information to develop a preincident plan.

   A. preincident plan

   B. preincident survey

   C. fire safety survey

   D. fire station tour

6. A _____ is a voluntary inspection of a residence or occupied structure to identify fire and life safety hazards.

   A. preincident plan

   B. preincident survey

   C. fire safety survey

   D. fire station tour

Access Navigate to find answers to this On Scene, along with other resources such as an audiobook.

Fire Origin and Cause

KNOWLEDGE OBJECTIVES

After studying this chapter, you will be able to:

- Explain the reasoning for conducting a fire investigation. (**NFPA 1001: 5.3.4**, pp. 1022–1023)
- Describe the role and relationship of the Fire Fighter II to criminal investigators and insurance investigators. (**NFPA 1001: 5.3.4**, pp. 1023–1024)
- Describe the exigent circumstances rule. (pp. 1024–1025)
- Explain the importance of protecting a fire scene to aid in origin and cause determination. (**NFPA 1001: 5.3.4**, pp. 1025–1028)
- Describe the steps needed to secure a property. (**NFPA 1001: 5.3.4**, pp. 1024–1025)
- Describe how the point of origin of a fire is determined. (**NFPA 1001: 5.3.4**, pp. 1025–1027)
- Describe how the cause of a fire is determined. (**NFPA 1001: 5.3.4**, pp. 1027–1028)
- Describe the four classifications of fire cause. (pp. 1028–1029)
- Describe how to assist fire investigators with processing a fire scene. (**NFPA 1001: 5.3.4**, p. 1029)
- List the types of evidence that may be found at a fire scene. (**NFPA 1001: 5.3.4**, pp. 1029–1030)
- Explain the chain of custody. (**NFPA 1001: 5.3.4**, pp. 1031–1032)
- Describe techniques for preserving fire scene evidence. (**NFPA 1001: 5.3.4**, p. 1032)

- Describe the evidential items and conditions that may be observed during fire-ground operations. (**NFPA 1001: 5.3.4**, pp. 1032–1037)
- Describe the crime of arson. (p. 1037)

SKILLS OBJECTIVES

After studying this chapter, you will be able to perform the following skill:

- Protect evidence. (**NFPA 1001: 5.3.4**, pp. 1030–1031)

ADDITIONAL STANDARDS

- **NFPA 921**, *Guide for Fire and Explosion Investigations*
- **NFPA 1033**, *Standard for Professional Qualifications for Fire Investigator*

Fire Alarm

Your crew is picking up hose after extinguishing a fire in a small clothing store in an old part of town. Other crews are performing overhaul as the department's fire investigator arrives on the scene. As he dons his personal protective equipment and collects his equipment, you think about the things you observed when your crew arrived on the scene. After the investigator photographs the entire scene, he asks your crew to assist with digging through the area of origin.

1. Why should all fires be investigated?

2. What types of fire scene observations should you report to the fire investigator?

3. What tools and equipment will you use to assist the fire investigator in processing the fire's room of origin?

4. How does a fire origin and cause investigation differ from an arson investigation?

 JONES & BARTLETT LEARNING **NAVIGATE 2** *Access Navigate for more practice activities.*

Introduction

Fire fighters usually reach a fire scene before a trained fire investigator arrives. As a consequence, fire fighters are able to observe important fire patterns and evidence that the fire investigator can use to determine how and where the fire started. By identifying and preserving possible evidence, as well as recalling and reporting objective findings, fire fighters provide essential assistance to fire investigators. In some cases, the observations and actions of a fire fighter are a vital part of determining the origin and cause of the fire, a process known as the fire investigation.

Fire cause determination is important because it identifies possible trends or criminal activity and informs the development of fire prevention programming. For example, a fire department might develop an education program to reduce accidental fires, such as those caused when food is left unattended on a hot stove. A series of fires could point to a product defect, such as a design error in a chimney flue or the improper installation of chimney flues. Identifying fires that were intentionally set could lead to the arrest of the person responsible for the crimes.

Fire fighters must understand the basic principles of fire investigation in order to assist with the official investigation. Fire origin and cause determination is often difficult because important evidence may be consumed by the fire or destroyed during fire suppression operations or salvage and overhaul operations. Fire investigators must rely on fire fighters to observe, capture, and retain information, as well as to preserve evidence until it can be documented and examined by the investigator. Evidence that is lost can never be replaced.

Fire fighters should make a mental note of what they witness upon their arrival and during fire-ground operations, such as the fire conditions when they arrive at the scene, who they saw, who opened or closed doors, and tasks they performed. Although the investigation report will be written by a trained fire investigator, the information provided by fire fighters is vital to the investigation and report.

FIRE FIGHTER TIP

Fire departments can use fire cause data to support bylaws or create policies designed to reduce the probability of fires from occurring. For example, these data can lead to the development of bylaws related to removing combustible landscape materials such as pine needles, removing construction materials from the outside of multiple-family buildings, or requiring the use of residential sprinklers.

Why Do We Investigate Fires?

Every fire should be investigated, even if the origin and cause seem obvious. Determining the fire's origin and cause may help prevent future fires; identify who is financially responsible for property damage, injuries, or deaths; and provide evidence to be used in criminal prosecution. Entities such as the National Fire Protection Association (NFPA) and the Council of Canadian Fire Marshals and Fire Commissioners track fire statistics from local fire department incident reports, insurance investigations, and product safety investigation testing facilities. This critical information, which comes from thorough, competent fire investigations, can then be addressed by fire prevention and education programs.

The objective of a fire investigation is to identify the origin and cause of the fire based on the evidence gathered by the investigator. This investigation should not be confused with the criminal investigation of arson. Although the definition and levels of arson vary, the crime generally involves the malicious act of burning property with a criminal intent. Fire investigators must be careful to avoid expectation bias when conducting the investigation. The evidence and the facts alone must determine the origin and cause of the fire. If investigators determine that a crime has been committed, a separate criminal investigation must be conducted by law enforcement personnel.

Who Conducts Fire Investigations?

In most jurisdictions, the chief of the fire department has a legal responsibility to determine fire cause. In very small communities, the fire chief may personally examine the scene to determine the cause. In other small jurisdictions, one or two fire department members may be trained to conduct these investigations. In large fire departments, the fire chief usually delegates this responsibility to the fire prevention division or a special unit staffed with trained fire investigators. Sometimes a special unit conducts fire investigations for several area fire departments under a mutual aid agreement that determines when those unit members should be called to conduct an investigation. For investigations involving serious injuries, deaths, or high financial loss, an investigator from the fire marshal's office or the fire commissioner's office may be called for assistance.

Although the personnel responsible for conducting fire investigations may vary from jurisdiction to jurisdiction, he or she must always meet the job performance requirements (JPRs) established by NFPA 1033, *Standard for Professional Qualifications for Fire Investigator*. This ensures that each fire investigation is conducted in a safe, thorough, and professional manner by trained and qualified personnel.

Depending on the jurisdiction, a fire investigator may be dispatched to all working structure fires, or he or she may be dispatched only when the damage exceeds a predetermined level or when the incident involves injuries or fatalities. In other departments, the incident commander (IC) may be expected to conduct a preliminary investigation and decide whether a fire investigator is needed. If the cause of the fire is evident and accidental, the IC would be responsible for gathering the information and filing the necessary reports. If the cause cannot be determined or appears to be intentional, the IC would then summon a fire investigator.

Many fire larger fire services have their own fire investigation branch as part of their fire prevention division. Some provinces have a fire marshal whose sole responsibility is fire prevention and investigation. In Ontario, the Ontario Fire Marshal's Office investigates any large-loss fire or fires where there have been serious injuries or fatalities. The police authority is usually the primary investigative authority, however, in most provinces.

Although fire fighters are not responsible for conducting the actual fire investigation, they are often the source of valuable information for the investigator. Because fire suppression and overhaul operations usually begin before the fire investigator arrives, fire fighters may see or hear something that could be relevant to the investigation, and they must pass on their observations after extinguishing the fire. Care should be taken to avoid moving, altering, or destroying evidence (spoliation) unless such action is necessary to meet the incident priorities of life safety, incident stabilization, and property conservation. If evidence must be altered in any way before the investigation occurs, fire fighters must report this to the investigator. In some cases, fire fighters may be interviewed or asked to prepare a written report documenting their observations. These tasks are part of the fire fighter's role and responsibility in the fire investigation process.

Law Enforcement Authority

The primary purpose of a fire investigation is to determine the point of origin and the sequence of events that caused fire ignition. As previously mentioned, if the fire investigation determines that a crime may have been committed, the investigation is often turned over to a law enforcement agency, and a secondary, criminal investigation is conducted.

In some jurisdictions, law enforcement personnel are trained as fire investigators, and they conduct the entire investigation process. In other jurisdictions, fire department and law enforcement personnel work together throughout the investigation. Successful fire investigations often depend on good working relationships among different organizations. Fire fighters need to understand how fire investigations are conducted in their community.

Investigation Assistance

Most fire departments and government agencies have limited time, personnel, and resources available for fire investigations. To provide for more extensive investigations, the provincial office of the fire commissioner or fire marshal's office may establish an investigations unit that concentrates on major incidents and supports local investigators on larger fires. Because these investigators cannot always reach the fire scene quickly, the local fire department must be prepared to conduct a thorough preliminary investigation and to protect the scene and preserve evidence.

Private Investigation Entities

Insurance companies often investigate fires to determine the validity of a claim for damages or to identify factors that might help prevent future fires. The cost of an investigation is more than offset by the savings the insurance company would realize by identifying a fraudulent claim. Some insurance companies employ their own fire investigators, whereas others retain the services of independent fire investigators. Product manufacturers and testing laboratories also use their own investigators to conduct destructive analysis investigations that validate or refute claims of faulty equipment. This analysis results in the contamination, spoliation, or destruction of the evidence, so the entire process must occur after the initial fire investigation and should be thoroughly documented by the analyst.

Outside fire investigators often have valuable experience and can provide critical technical support to determine the cause of a fire. If a private investigator finds evidence of criminal activity, he or she must provide that information to the municipal or other governmental fire investigator of record.

Legal Authority to Enter, Secure, and Transfer Property

For the purposes of conducting a fire investigation, a fire investigator may, without warrant, at any time, enter and inspect land or premises where a fire has occurred and, if necessary, land or premises adjoining or near where the fire has occurred.

A fire investigator who enters land or premises under the authority of the Fire Safety Act (Fire Marshals Act) may do any or all of the following: inspect, analyze, measure, sample, or test anything; use or operate anything or require the use or operation of anything, under conditions specified by the fire investigator; take away samples; remove a record or thing from the land or premises; make a record of the land or premises and any person or thing on or in the land or premises. The fire investigator who removes a record or thing must return

the item as soon as practicable. There is no open-ended authority to enter a fire scene and conduct an investigation. The local authority must, within 5 days after the date on which they learned of a fire that destroyed or damaged property or resulted in injury or death, ensure that a fire investigator begins to investigate the cause, origin, and circumstances of the fire. In most circumstances, the fire investigation must be conducted immediately after the fire has been extinguished.

Briefly delaying the investigation due to inclement weather or poor nighttime visibility is generally acceptable, as long as the fire department maintains control of the property until the investigation takes place and all evidence is collected. In the interim, the fire department should limit salvage and overhaul operations to only those that are necessary to stabilize the fire scene. The fire scene should be extensively photographed, starting with the area with the least amount of damage and proceeding toward the area of possible origin. Take several photographs of the point of origin from various angles. Photograph any relevant physical evidence on the premises exactly where they were found. If weather, traffic, or other factors could destroy evidence, take steps to preserve these items in the best way possible.

Maintaining scene integrity during this time is critical to the fire investigation. The building and premises must be properly secured and guarded until the fire investigator has finished gathering evidence and documenting the fire scene. To secure the property, cordon off the area with fire- or police-line tape (**FIGURE 28-1**). A member of the fire department or law enforcement agency should remain at the scene to keep nonessential personnel out of the area and prevent unauthorized persons from entering. Fires, explosions, or other emergency incidents that cause serious injuries, deaths, or substantial property damage may take days, weeks, or even months to fully investigate.

FIGURE 28-1 Security personnel may be needed to monitor access to the investigation site.
© Alex Wong/Staff/Getty Images.

These investigations require round-the-clock security to ensure that the scene is not compromised.

As previously mentioned, the fire department has the authority to deny access to a building for as long as necessary to conduct a thorough investigation and ensure the safety of the public. After receiving permission from the fire investigator, a fire officer or a fire fighter should accompany anyone who enters the premises for any reason until the scene is released. The fire department should keep a log of anyone who enters the property, the times of his or her entry and departure, and a description of any items taken from the scene.

Once the investigator has documented the scene, collected evidence, and interviewed witnesses, the property is released to the owners. Before turning over the property to its owner, make sure that the building is properly secured and that no hazards to public safety exist. Fire departments can secure and protect the premises in several ways. Lock and guard gates, if necessary. Rope off dangerous areas, and mark them with signs. Shut off all utilities, and seal any openings in the roof to prevent additional water damage. Some fire departments have contracts with local companies that provide 24-hour services to board up and secure windows and doors to prevent unauthorized entry.

The fire department's authority ends at this time. Afterward, the fire investigator will need a search warrant or written consent to search the site. Any unguarded evidence left on the scene will likely be ruled as inadmissible if the case goes to court.

If fire fighters or law enforcement officers do not remain on the scene until the fire investigator arrives, the right of entry to conduct fire investigation is generally forfeited if consent or a warrant is not obtained. Except in unusual circumstances, fire fighters may not maintain control of the property for an unreasonable length of time; in other words, the investigation must be conducted in a timely manner, as permitted by safety conditions, weather, daylight, and resources. A search warrant is usually needed for lengthy, complicated investigations that require the scene be secured indefinitely. Additionally, fire fighters must be aware of any provincial or local laws pertaining to right of access by the owners or occupants while the property is under the control of the fire department.

Determining the Point of Origin and Fire Cause

The primary purpose of a fire investigation is to determine the sequence of events that brought together the ignition source and the initial fuel. This is accomplished by first determining the point where ignition occurred and then identifying how and why ignition occurred. For this reason, fire investigations are often referred to as origin and cause investigations.

NFPA 921, *Guide for Fire and Explosion Investigations*, is used by many fire investigators who are conducting fire investigations. NFPA 921 recommends using the **scientific method** to determine the origin and cause of a fire. The scientific method uses a seven-step systematic process to test one or more hypotheses against the evidence collected during the investigation (**FIGURE 28-2**). The investigator must look at the situation objectively to be sure that the evidence is convincing and fully explains the situation. If more than one hypothesis for the observations exists, each

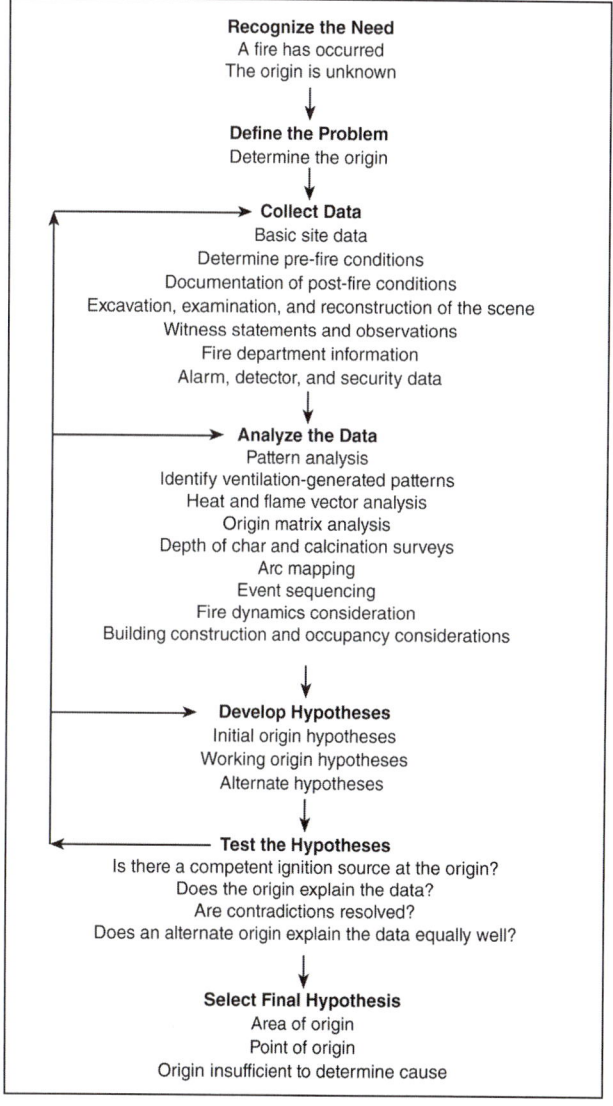

FIGURE 28-2 The scientific method's seven-step process should be used to determine the origin and cause of a fire.

hypothesis must be evaluated against the evidence. The hypothesis with the highest degree of certainty will then be identified as the cause of ignition.

Identifying the Point of Origin

One of the first steps in a fire investigation is identifying the point of origin (**FIGURE 28-3**). At this location, the fire investigator can look for evidence indicating the specific cause of the ignition. Identifying the point of origin requires a careful, systematic analysis of the fire scene. Throughout the entire process, the investigator will use notes, photographs, videos, sketches, diagrams, and other tools to permanently document the investigation. These tools help the investigator to remember the scene, specific evidence, and the process used to conduct the investigation. They may also be used by the investigator if he or she is called to testify in a civil or criminal court case.

When working to identify the point of origin, the fire investigator will usually begin by examining the building's exterior. The fire investigator will look for indications that the fire originated outside the building before looking inside, and he or she will identify important information, such as the overall size, construction, layout, and occupancy of the building. The investigator will also document the extent of damage that is visible from the exterior. He or she will look for any openings that might have created drafts that influenced the fire spread and will examine the condition of outside utilities, such as the electrical power connection and gas meter.

The search for indications of the fire's point of origin continues inside the building, beginning with the area of least damage and moving systematically to the area of heaviest damage. The area of heaviest damage usually

indicates where the fire was burning for the longest time or where it burned the hottest (**FIGURE 28-4**). Areas with less damage were probably not as heavily involved in the fire or were involved for shorter periods.

Fire Effects and Fire Patterns

Fire effects are the observable changes found on objects, surfaces, and construction materials as a result of heat, flame, smoke, and other products of combustion. Fire effects can be as simple as the charring of wood or as complex as the chemical change in a specific material. Each fire effect can be observed and evaluated separately. For example, the analysis of a single piece of melted plastic may provide approximate temperatures of the room during the fire. If multiple objects or larger areas of the room show varying degrees of the same fire effects, fire patterns may be observed (**FIGURE 28-5**).

FIGURE 28-4 The area of heaviest damage usually indicates where the fire burned for the longest time or burned the hottest.
Courtesy of Jamie Novak, Novak Investigations Inc. and the St. Paul Fire Department.

FIGURE 28-3 One of the first steps in a fire investigation is identifying the point of origin.
Courtesy of Captain David Jackson, Saginaw Township Fire Department.

FIGURE 28-5 If multiple objects or larger areas of the room show varying degrees of the same fire effects, fire patterns may be observed.
Courtesy of Robert Schaal.

Fire patterns show the relative degree of fire exposure times and temperatures of the affected objects. For example, several wooden studs may show various degrees of charring. The studs with the heaviest charring were probably exposed to the fire for the longest time or exposed to the highest temperatures.

The **depth of char** is a fire effect that was once believed to indicate exactly how long a material was exposed to fire. Newer research indicates that this method is unreliable (NFPA, 2017). Although charring may be deepest at the point of origin, the presence of **ignitable liquids** and combustible materials, as well as the effects of ventilation, can also influence charring. Still, when considered as a fire pattern, the relative depth of char on similar materials can indicate which areas of a room or which side of a piece of furniture was exposed for the longest time or to the hottest fire (**FIGURE 28-6**). An area that was exposed to a lower intensity of fire or heat impingement or exposed for shorter amounts of time will display shallow charring. Conversely, areas that were exposed to higher intensities of heat and fire or exposed for longer amounts of time will display deeper charring and greater destruction of material.

Because heat rises, the flow of heated gases from a fire will often move up and out from the point of origin. This upward, outward flow can usually be recognized, even when all of the building's contents were involved in the fire. The damage will often spread outward from the room or area where the damage is most severe. Often the point of origin is found directly below the most heavily damaged area on the ceiling, where the heat of the fire was most intensely concentrated. A charred V, U, or hourglass pattern on a wall often indicates that fire spread up and out from something at the base. In this case, the point of origin may be found in a pile of charred debris at the base of the pattern (**FIGURE 28-7**).

Fire patterns can be helpful in identifying the point of origin but are not conclusive evidence when viewed

FIGURE 28-7 Fire patterns often indicate the area or point of origin.
Courtesy of Charles B. Hughes/Unified Investigations & Sciences, Inc.

in isolation to other factors. An experienced fire investigator knows that many factors can influence fire patterns, including ventilation, fire suppression efforts, and the burning materials themselves. As mentioned, a fire pattern on a wall could indicate that an easily ignited and intensely burning fuel source was present at that location. It might also indicate that something fell from a higher level and burned on the floor. Such patterns may be caused by direct flame contact, heat, smoke, soot, or a combination of these elements. The patterns produced by the fire may be apparent on any surface and may indicate movement away from the point of origin. Examples include wall and ceiling surfaces, appliances and furnishings, doors and windows, and even fire victims.

Once the fire investigator identifies the approximate location of the point of origin, the search for indications of a specific cause can begin. This effort involves identifying the source of ignition, the fuels that were first ignited, and the sequence of events that brought the two together.

Determining Fire Cause

Every fire has a cause, which the fire investigator tries to determine. A fire investigation attempts to identify where, how, why, and sometimes when the fire ignited. Some fires have simple causes that are easily identified and understood; others result from a complex sequence of events that must be examined carefully to determine what actually happened. For example, if a fire started in a kitchen, potential causes could include:

- The stove
- Any of the electrical appliances
- Light fixtures
- Smoking materials
- Cleaning supplies
- Other ignition sources, such as candles

FIGURE 28-6 Depth of char.
Courtesy of Charles B. Hughes/Unified Investigations & Sciences, Inc.

In some cases, the cause of a fire will never be determined with a high degree of certainty.

Ignition Source

Once the area of origin is determined, the investigator must search for a competent ignition source. To be considered "competent," the ignition source must have enough energy and be capable of transferring that energy to the first fuel long enough to heat the fuel to its ignition temperature.

In some cases, no obvious ignition source can be found, so the investigator must look for other evidence to establish the cause of the fire. In other cases, one or more possible ignition sources may be found in the area of origin. Each of these possible ignition sources must be evaluated to determine if it could have ignited the fuel under the circumstances.

First Fuel Ignited

In order for the investigator to determine if a heat source is competent, he or she must also identify the first fuel that ignited. Many times, the fuel is obvious, such as in the case of a fire in a recycling bin full of paper. In other cases, many different types of fuel are near or in close proximity to the ignition source. The investigator must try to determine which fuels, if any, could have been ignited by the ignition source.

Oxidizing Agent

Most fires use the oxygen in the ambient air in the combustion process, but some fire scenes may have special oxidizing agents that contribute to fire spread. Medical oxygen cylinders, pool sanitizers, and other chemical oxidizing agents should be reported to the investigator.

Ignition Sequence

As described in Chapter 5, *Fire Behaviour*, the fire tetrahedron is used to explain the chemical reaction that allows a fire to ignite and continue to burn. The fuel and ignition source (heat) must come together, in the presence of oxygen (or an oxidizing agent), to cause ignition and sustained burning. The fire investigator must attempt to determine the exact sequence of events that brought those elements together. For example, if the investigator determines that an unattended candle ignited items on a table, he or she must try to explain how the combustible items came into contact with the candle's flame.

Levels of Certainty

The final step in fire cause determination involves establishing the level of certainty of a hypothesized fire cause. In most cases, it is not possible to determine the cause with 100% certainty unless surveillance video or credible witnesses can attest to the facts. Usually, after all of the evidence has been evaluated, one or more hypotheses are tested to determine which has the highest level of certainty. A hypothesis with less than a 50% certainty level is considered a possible cause. If more than one hypothesis is determined to be possible, the one with the highest level of certainty is listed as the official probable cause. A probable cause must have a level of certainty of 50% or more. If no cause can be determined or if no cause can be considered probable, the fire should be classified as undetermined.

The process of eliminating possible causes and documenting the reasons for rejecting them is just as important as properly documenting the hypothesis that is determined to be the cause of the fire.

Classifications of Fire Cause

Once the point of origin has been identified and the cause of the fire has been determined, the investigator must assign the fire cause to one of four different cause classifications: accidental, natural, incendiary, or undetermined.

Accidental Fire Causes

Accidental fires occur as a result of human action or inaction that does not involve malicious intent. People often do things that result in unwanted fires, but they did not intend for the fire to ignite or become hostile. Forgetting about a pan of grease on the burner, refueling a hot lawn mower, or igniting a trash fire that suddenly spreads to a nearby building are all examples of accidental fires.

Natural Fire Causes

Natural events often cause unwanted fires—some of which can be quite disastrous. Natural fires include those caused by earthquakes and tornadoes that can rupture natural gas pipes and knock down high-voltage power lines. Lightning and high winds cause many natural fires.

Incendiary Fire Causes

Incendiary fires are those ignited by deliberate and intentional acts. In order for a fire to be classified as incendiary, the person who committed the acts must have known or should have known that igniting the fire was wrong. Fires intentionally ignited by small children or by people lacking the ability to understand the consequences of their actions are usually classified as accidental fires. The determining factor for incendiary fires is the intentional commission of acts that were known to be wrong.

Undetermined Fire Causes

Fires that lack a probable cause with more than 50% certainty must be classified as an **undetermined fire**. Even if the investigator feels strongly about a hypothesis, he or she must determine the origin, cause, and cause classification based solely on the evidence collected during the investigation. Under no circumstances should a fire be classified as "suspicious." Fires that have been classified as undetermined may later be reclassified as accidental, natural, or incendiary, if additional evidence is collected.

Assisting the Investigator

After the fire investigator identifies the area of origin, fire fighters could be asked to assist in digging out the fire scene. *Digging out* is a term used to describe the process of carefully looking for evidence within the debris. Sometimes the entire fire scene must be closely examined to gather evidence and determine the cause of the fire.

The fire investigator will take extensive photographs of the fire scene as it first appears. He or she will then begin to remove and inspect the debris, layer by layer, working from the top of the pile down to the bottom. The type of evidence uncovered and its location within the layers of debris can provide important indications of how the fire ignited and progressed.

Removing and inspecting the layers of debris enable the fire investigator to determine the sequence in which items burned, whether an item burned from the top down or from the bottom up, and how long it burned. Did the fire start at a low point and burn up, or did burning items fall down from above and ignite combustible materials below? Did the fire spread along the ceiling or along the floor? Are rags containing the residue of an ignitable liquid found under the furniture? Why are papers that should have been in a metal filing cabinet stacked on the floor and partially burned?

Systematically digging out through the debris often can uncover the exact point of origin and identify the cause of both accidental and deliberate fires. If circumstances or eyewitness accounts indicate that a deliberate fire is a possibility, the fire investigator may request that fire fighters help examine the entire area. The investigator will explain what to look for, how to search, and what to do with any potential evidence.

When you find possible evidence, stop and inform the fire investigator so that he or she can examine it in place. It is the fire investigator's job to document, photograph, and remove any potential evidence, whether or not it supports his or her hypothesis. To determine the point of origin and cause of a fire, the fire investigator must evaluate all evidence gathered at the scene and from other sources of information. Where the fire investigator finds evidence of an incendiary fire or has formed an opinion that the fire was the result of a deliberate act, law enforcement personnel must be notified.

Collecting and Processing Evidence

Evidence refers to all of the information gathered and used by a fire investigator in determining the cause or point of origin of a fire. Evidence can be used in a legal process to establish a fact or prove a point. To be admissible in court, it must be gathered and processed under strict procedures.

Types of Evidence

Physical evidence (also called real evidence) consists of items that can be observed, photographed, measured, collected, examined in a laboratory, and produced in court to prove or demonstrate a point (**FIGURE 28-8**). Fire investigators can gather physical evidence at the fire scene, such as a fire pattern on a wall or an empty gasoline can, to explain how the fire started or to document how it burned.

Demonstrative evidence is anything that can be used to validate a theory or to show how something could have occurred. To demonstrate how a fire could spread, a fire investigator might use photographs, diagrams, charts, or a computer model of the burned building.

FIGURE 28-8 Physical evidence at a fire scene.
Courtesy of Mike Dalton.

Documentary evidence includes any type of written document, record, report, or other information that documents important facts about the investigation. This may include bank records, alarm system activation reports, insurance policies, and credit card receipts. Each of these documents is used by the investigator to support his or her hypothesis about how, when, and why the fire ignited.

Testimonial evidence is the fourth type of evidence gathered during the investigation (**FIGURE 28-9**). Testimonial evidence may be the documented statements given by fire fighters or bystanders. More formal testimonial evidence includes affidavits, interrogatories, and court testimony that documents observations and statements made by lay and expert witnesses.

Evidence used in court can be classified as either direct or circumstantial. **Direct evidence** includes facts that can be observed or reported firsthand. Testimony from an eyewitness who saw a person actually ignite a fire and a video recording from a security camera showing the person starting the fire are examples of direct evidence.

Circumstantial evidence is based on inference and not on personal observation. For example, an investigator may infer that two or more unrelated fires that occurred at the same time and in the same vicinity could be incendiary fires. If a fire victim is found in a chair in the living room, surrounded by ashtrays, the investigator may infer that careless use of smoking materials may be

FIGURE 28-9 Testimonial evidence includes documented statements given by fire fighters or bystanders, affidavits, interrogatories, and court testimony made by lay and expert witnesses.
© Guy Cali/Corbis/Getty Images.

the cause of the fire, if no other probable cause can be determined. Fire investigators often must use circumstantial evidence along with other types of evidence to determine the cause of the fire.

Preservation of Evidence

Fire fighters have a responsibility to preserve evidence that could indicate the cause or point of origin of a fire. They are not, however, responsible for deciding whether the evidence they find is relevant to the investigation. Too much evidence is better than too little, so no piece of potential evidence should be considered insignificant.

Evidence is often found during the salvage and overhaul phases of a fire. Salvage and overhaul should always be performed carefully and can often be delayed until a fire investigator has examined the scene. Fire fighters who discover potential evidence during fire suppression, salvage, or overhaul operations should leave it in place, if possible. Do not move debris any more than is absolutely necessary, and never discard debris until the fire investigator gives his or her approval to do so. As mentioned, it is the fire investigator's job to decide whether the evidence is relevant, not the fire fighter's.

If evidence could be damaged, destroyed, or altered in any way during fire suppression activities, cover it with a salvage cover or some other type of protection. Use barrier tape to keep others from accidentally walking through evidence. These methods may prove to be ineffective, however, if no indication of their purpose is given. For this reason, if evidence is at risk of being damaged or altered by fire suppression activities, it should be attended to by personnel or moved to a secure location. Before moving an object to protect it from damage, be sure that witnesses are present, that a location sketch is drawn, and that a photograph is taken. Immediately notify the IC or investigator of the object's original location and condition.

Evidence should not be cross-contaminated or spoliated in any way. Fire investigators must use clean tools and special containers to store evidence and prevent cross-contamination from any other products. To avoid cross-contaminating evidence, fire investigators must always wash their tools between taking samples. This step ensures that material from one piece of evidence will not be unintentionally transferred to another piece. Fire investigators must also change their gloves each time they take a sample of evidence and place evidence only in clean containers (**FIGURE 28-10**). Before entering the fire scene, fire investigators often wash their boots to keep from transporting any contamination into the fire scene.

FIGURE 28-10 Fire investigators must place evidence in clean containers.
© Jones & Bartlett Learning. Photographed by Glen E. Ellman.

FIGURE 28-11 Do not touch or move evidence until the investigator can document, photograph, and collect it.
© Jones & Bartlett Learning. Photographed by Glen E. Ellman.

Chain of Custody

To be admissible in a court of law, physical evidence must be handled according to certain prescribed standards. Because the cause of the fire may not be known when evidence is first collected, all evidence should be handled according to the same procedure. Evidence relating to the cause of the fire may be presented in court for both criminal and civil cases.

Chain of custody (also known as continuity of evidence, chain of evidence, or chain of possession) is a legal term that describes the process of maintaining continuous possession and control of the evidence from the time it is discovered until it is presented in court. Every step in the collection, movement, storage, and examination of the evidence must be properly documented. For example, if a gasoline can is found at a fire scene, the investigator must record the person who found the can, along with where and when the can was found. Photographs should be taken to show the site where the gasoline can was found and its condition. In court, the fire investigator must be able to show that the gas can presented is the same can that was found at the fire site.

The person who takes initial possession of the evidence must keep it under his or her personal control until the item is turned over to another official. Each successive transfer of possession must be properly recorded. If evidence is stored, documentation must indicate where and when it was placed in storage, whether the storage location was secure, and when the evidence was removed from that location. Often, evidence is maintained in a secured evidence locker to ensure that only authorized personnel have access to it.

Everyone who had possession of the evidence must be able to attest that it has not been contaminated, damaged, or changed in any way. If evidence is examined in a laboratory, the laboratory tests must be documented. The documentation for chain of custody must establish that the evidence was never out of the control of the responsible agency and that no one could have tampered with it.

Fire fighters are frequently the first link in the chain of custody. The fire fighter's responsibility in protecting the integrity of this chain is relatively simple: Report everything to a supervisor, and disturb nothing needlessly. The individual who finds the evidence should remain with it until the material is turned over to a company officer or to the fire investigator (**FIGURE 28-11**). Remember, you could be called as a witness to testify that you are the fire fighter who discovered this particular piece of evidence.

The fire investigator's standard operating procedures (SOPs) for collecting and processing evidence generally include the following steps:

- Take photographs of each piece of evidence as it is found and collected. If possible, photograph the item exactly as it was found, before it is moved or disturbed.
- Sketch, mark, and label the location of the evidence at the fire scene. Sketch the scene as near to scale as possible.
- Place evidence in appropriate containers to ensure its safety and prevent contamination. Unused paint cans with lids that automatically seal when closed are good containers for transporting evidence. Glass mason jars sealed with a sturdy sealing tape are appropriate for transporting smaller quantities of materials. Plastic containers and plastic bags should not be used to hold

evidence containing petroleum products, as these chemicals may lead to deterioration of the plastic. Paper bags can be used for collecting dry clothing or metal articles, matches, or papers. Soak up small quantities of liquids with either a sterile cellulose sponge or sterile cotton batting.

- Tag all evidence at the fire scene. The evidence container should be labelled with the date, time, location, and investigator's name.
- Record the time when the evidence was found, the location where it was found, and the name of the person who found it. Keep a record of each person who handled the evidence.
- Keep a constant watch on the evidence until it can be stored in a secure location. Evidence that must be moved temporarily should be put in a secure place that is only accessible to authorized personnel.
- Preserve the chain of custody in the handling of evidence. A broken chain of custody may result in a court ruling that the evidence is inadmissible.

One person should be responsible for collecting and taking custody of all evidence at a fire scene, no matter who discovers it. If someone other than the assigned evidence collector must seize the evidence, that person must photograph, mark, and contain the evidence properly and turn it over to the evidence collector as soon as possible. The evidence collector must also document all evidence according to department policies. A log of all photographs taken should be recorded at the scene as well.

Identifying Witnesses

Although fire fighters may not interview witnesses, they can identify potential witnesses to the fire investigator. People who were on the scene when fire fighters arrived could have invaluable information about the fire. If a fire fighter learns something that might be related to the cause of the fire, he or she should pass this information on to a supervisor or to the fire investigator.

Interviews with witnesses should be conducted by the fire investigator or by law enforcement personnel. If the fire investigator is not on the scene or does not have the opportunity to interview the witness, the fire fighter should obtain the witness's name, address, and telephone number and give it to the fire investigator. A witness who leaves the scene without providing this information could be difficult or impossible to locate later.

Fire fighters have a primary responsibility to save lives and property. Until the fire is under control, they must concentrate on fighting the fire and not on

FIGURE 28-12 Refer all civilians and news reporters to the PIO for information about the fire.
© Mark C. Ide.

investigating its cause. Even so, fire fighters should pay attention to the scene and make mental notes about their observations. They must tell the fire investigator about any odd or unusual happenings. Information about the fire should be shared only with the fire investigator and only in private.

Do not make any statements of accusation, personal opinion, or probable cause to anyone other than the fire investigator. Comments that are overheard by the property owner, the occupant, a news reporter, or a bystander can impede the efforts of the fire investigator to obtain complete and accurate information. A witness who is trying to be helpful might report an overheard comment as a personal observation. In this way, inaccurate information can generate a rumor, which then becomes a theory, which then turns into a reported "fact" as it passes from person to person.

Never make jesting remarks or jokes at the scene. Careless, unauthorized, or premature remarks could embarrass the fire department. Statements to news reporters about the fire's cause should be made only by the department's public information officer (PIO) after the fire investigator and IC have agreed on their accuracy and validity. Refer all civilians to the IC or PIO for information about the fire (**FIGURE 28-12**).

Observations During Fire-Ground Operations

Although a fire fighter's primary concern is saving lives and property, you will inevitably make observations and gather information as you perform your duties. The fire investigator will usually contact the fire fighters who first arrived on the scene for that information.

Dispatch and Response

During dispatch and response, form a mental image of the scene you expect to encounter. Note the time of day and weather conditions. As the incident progresses, make a mental note of the building, fire and smoke conditions, bystanders, and any other information you observe. This information could help the fire investigator determine the origin and cause of the fire.

Time of Day

Time of day and type of occupancy can indicate the number and type of people at an incident. Offices, stores, and other business places are filled with people during the day but are mostly empty at night. Conversely, restaurants and nightclubs are likely to be crowded after dark. Residential occupancies are more likely to be occupied at night and on weekends and holidays.

As victims evacuate the building, some may stand out, either because they are fully dressed when everyone else is in pajamas or because their behaviour and demeanor are quite different from those of the other victims.

Weather Conditions

Note whether the day is hot, cold, cloudy, or clear and whether conditions in the burning structure match the weather. On a cold day, windows should be closed; on a hot day, the furnace should be off.

Lightning, heavy snow, ice, flooding, fog, or other hazardous conditions may delay the fire department's arrival and make a fire fighter's job more difficult. Because wind direction and velocity help determine the natural path of fire spread, being aware of these conditions will help to determine if the fire behaved in an unnatural way.

Arrival and Size-Up

Size-up operations can provide valuable information for fire investigators. Pay attention to the fire conditions, the building characteristics, positions of doors and windows, and any vehicles and people arriving or leaving the scene.

Description of the Fire

The fire investigator will compare the dispatcher's description with the actual fire conditions. If the fire has intensified dramatically in a short time, an accelerant could have been used. During fire suppression operations, note whether flames are visible or whether only smoke is apparent. Also observe the quantity, colour, and source of the smoke. Does the fire appear to be burning in one place or in multiple locations?

People on the Scene

The appearance and behaviour of people and vehicles at the scene of a fire can provide valuable clues. Note the attitude and dress of the owner and/or the occupants of the building, as well as any other individuals at the scene of a fire. Normally, the owner or occupants of a building will be distressed and wearing clothing appropriate to the time of day. For example, at 4 AM, you would expect to see residents in pajamas.

Notify the investigator if anyone's dress, behaviour, or words do not appear to be appropriate for the situation. Sometimes fire investigators will photograph the crowd at a fire scene, particularly if a series of similar fires has occurred recently, and look for the same faces at different incidents (**FIGURE 28-13**). Fire investigators may also want to know about someone who eagerly volunteered multiple theories or too much information.

Unusual Items or Conditions

Always note any unusual items or conditions about the property, such as whether windows and doors are open or closed upon your arrival or if fire fighters forced entry into the building. Gasoline cans, forcible entry tools, and a damaged hydrant or sprinkler connection should all be reported to the investigator.

Entry

As you prepare to enter the burning structure, note any evidence of prior entry, such as shoeprints leading into or out of the structure or tracks from vehicle tires. Note whether the windows and doors are intact, whether they are locked or unlocked, and whether any unusual barriers limit access to the structure. Also note any evidence of forced entry by others.

FIGURE 28-13 Most persons at the fire scene are intent on watching the fire fighters at work.
© Louis Brems, The Herald-Dispatch/AP Photos.

Forced entry could leave impressions of tools on the windows and doors. Likewise, cut or torn edges of wood, metal, or glass may indicate forced entry. Fire investigators might be able to determine whether the glass had been broken by heat or by mechanical means. Note any evidence of a burglary; if present, the fire may have been set to destroy evidence of another crime.

> **FIRE FIGHTER TIP**
>
> Photographs can document conditions before the arrival of fire investigators, but specific procedures must be followed for these photographs to be admissible in court as evidence. Follow your department's SOPs regarding fire scene photography.

Search and Rescue

As you enter the building to perform search and rescue or interior fire suppression activities, consider the location and extent of the fire. Be especially alert for signs of multiple fires in a building. The presence of multiple fires increases the risk of injury and death both to the building's occupants and to responding fire fighters, and it may indicate that the fire was intentionally ignited.

The location and condition of the building's contents may provide clues for the fire investigator. Unusual building contents or conditions, such as barriers in doors and windows, the absence of inventory in a warehouse, or insurance papers or deeds left out on a desk, should be noted to the fire investigator.

The team responsible for shutting off electrical power to the building should identify whether the circuit breakers were on or off when they arrived. The location of any people found in the building should be noted as well. When turning off electrical power, turn off main breakers, if possible, and do not disturb breakers that are "tripped."

Ventilation

The ventilation crew should note whether the windows and doors of the building were open or closed, locked or unlocked. They should also note smoke conditions and the presence of any unusual odours.

Smoke Conditions

The colour, volume, and velocity of smoke observed upon arrival should be noted and reported to the investigator. This information may help the investigator determine the extent of fire development when fire crews first arrived on the scene.

Unusual Odours

Self-contained breathing apparatus protects fire fighters from hazardous vapours and toxic odours. Nevertheless, sometimes an odour is so strong that it can be detected as you arrive on the scene or lingers after the fire has been extinguished. Cooking fires, overheated light ballasts, and burning electrical equipment produce distinctive, identifiable odours that are familiar to experienced fire fighters. Other common odours familiar to fire fighters include gasoline, kerosene, paint thinner, lacquers, turpentine, linseed oil, furniture polish, or natural gas.

Odours often linger in the soil under a building without a basement, particularly if an accelerant has been used and if the ground is wet. Concrete, brick, and plaster will all retain vapours after a fire has been extinguished. These observations should be passed along to the fire investigator.

Effects of Ventilation

Fire department ventilation operations can dramatically influence the behaviour of a fire and alter fire patterns. Certain ventilation situations can cause a fire to rapidly grow or the flow path to change. Relay information about these operations to the fire investigator so that he or she can correctly interpret fire patterns.

> **FIRE FIGHTER TIP**
>
> Gasoline-powered tools and equipment, including saws and generators, are often used during firefighting operations. Gasoline from such equipment could contaminate the area and lead to an erroneous assumption that accelerants (materials used to initiate or increase the spread of fire) were used in the fire. Limit the use of gasoline-powered equipment inside the structure, particularly in the area of the fire's origin. At a minimum, refuel such equipment completely outside the investigation area and the structure itself to prevent contamination of the site by spilled fuel.

Fire Suppression Operations

During fire suppression operations, note the fire's location and behaviour, evidence of any devices used to initiate the fire, obstacles encountered during forcible entry or fire suppression operations, and charring and fire patterns you observe. This information might help the fire investigator determine the origin and cause of the blaze.

Behaviour of the Fire

During the fire attack, observe the behaviour of the fire, and note how it reacts when an extinguishing agent is applied. Look for unusual flame colours, sounds, or reactions. For example, most ignitable liquids will float, continue to burn, and spread the fire when water is applied to them. Fires that rekindle in the same area or a flare-up when water is applied could indicate the presence of an ignitable liquid. Fires in multiple locations or fuelled by accelerants present an increased risk to you and your fellow fire fighters during suppression activities.

Incendiary Devices, Trailers, and Accelerants

While fighting the fire, be aware of streamers or trailers (combustible materials positioned to spread the fire) (**FIGURE 28-14**). Note any combustible materials such as wood, paper, or rags in unusual locations. Note containers of ignitable liquids that are normally not found in the type of occupancy or are found in an unexpected location. For example, a can of paint thinner in the garage is common. If it is found on the kitchen counter, however, it should be reported to the investigator. Very intense heat or rapid fire spread might indicate the use of an accelerant to increase fire spread.

Incendiary devices (devices used to initiate an incendiary fire or explosion) can include unusual items in unlikely places, such as a packet of matches tied to a bundle of combustible fibres or attached to a mechanical device (**FIGURE 28-15**). Often, incendiary devices fail to ignite, burn out without igniting other materials, or are not completely consumed by the fire.

Condition of Fire Protection Systems

If the building is equipped with a fire alarm or fire suppression system, fire fighters should note whether the system operated properly or was disabled (**FIGURE 28-16**). Note if the outside screw and yoke (OS&Y) valves or post indicator valves (PIVs) were open or closed. Blocked or covered sprinkler heads, missing or disabled smoke detectors or notification appliances, and physical damage to any fire protection system should be reported to the investigator. Note whether the intrusion alarm is sounding when you arrive on the scene.

FIGURE 28-15 Incendiary devices are items used to initiate an incendiary fire or explosion.
© Jones & Bartlett Learning. Photographed by Jessica Elias.

FIGURE 28-14 While fighting the fire, be aware of streamers or trailers.
Courtesy of Joseph Whittaker, Nassau County Fire Marshals.

FIGURE 28-16 Fire setters may sabotage fire protection systems to delay notification of a fire.
© DWImages/Alamy Stock Photo.

Obstacles or Structural Modifications

Use extra caution if you find unusual locks, debris, or other obstacles blocking entrance doors, hallways, stairwells, or access to fire protection systems. These conditions may be the result of renovation or demolition activities, or they may have been intentionally placed to limit fire suppression activities. In these situations, look for unexpected holes in floors, ceilings, and walls that may have been cut to promote fire spread or hinder fire fighters. Report any of these conditions to the fire investigator.

SAFETY TIP

If you encounter an incendiary device that has not ignited, notify your fire officer or fire investigator and exit the structure or area immediately! Do not make any attempt to move or disable the device.

Contents

Fire fighters who are involved in interior fire suppression, like the members of the search and rescue team, should make note of anything unusual about the contents of the building. The absence of personal items in a residence may indicate that they were removed before the incident and that the fire was intentionally set. Empty boxes in a warehouse may belong there, or they may indicate that valuable contents were removed prior to the fire. Sometimes the contents will be removed and replaced by less expensive or nonfunctioning items. Neighbours may report that they observed items being removed or replaced prior to the fire. Check whether appliances are plugged in at the time of the fire. Also note if any contents appear to have been protected from fire, smoke, or water damage.

Charring and Other Fire Patterns

Charring in unusual places, such as open floor space away from any likely accidental ignition source, could indicate that the fire was deliberately set. Char on the underside of doors or on the underside of a low horizontal surface, such as a tabletop, could indicate that a pool of an ignitable liquid was placed below the item.

Salvage and Overhaul

During salvage and overhaul, the smoke and steam should begin to dissipate, enabling both fire fighters and fire investigators to get a better look at the surroundings. Fire fighters should continue to look for the fire patterns and evidence previously discussed while conducting salvage and overhaul in a way that allows that evidence to be identified and preserved. The overhaul process, if not done carefully, can quickly destroy valuable evidence.

If possible, the fire investigator will observe and document an area before overhaul begins. He or she can often quickly identify potential evidence and can help direct or guide the overhaul operation so that this evidence is properly preserved. Evidence located during overhaul should be left where it is found, untouched and undisturbed, until the fire investigator examines it. Evidence that must be removed from the scene should be properly identified, documented, photographed, packaged, and placed in a secure location.

Fire suppression personnel and fire investigators must work as a team to ensure that the fire is completely extinguished and properly overhauled while continually searching for and preserving evidence. If overhaul operations must be done before the arrival of the fire investigator, the IC or company officer must thoroughly document the scene before, during, and after overhaul activities. Personnel should not leave anything at the fire scene that was not there at the beginning of the incident (refrain from leaving a water bottle or power bar wrapper, etc.). Anything that is left must be accounted for by the fire investigator so as to not cast doubt about the integrity of the scene.

In short, be careful to not destroy evidence during salvage and overhaul. Avoid throwing materials into a pile. Use low-velocity hose streams to avoid breaking up and scattering debris needlessly. Thermal imaging devices can be used to find hot spots without tearing apart the interior structure. Avoid damaging electrical wiring in the room of origin. It is possible that faulty electrical wiring caused the fire, but it is also possible that the wiring was damaged by the fire. The investigator may need to remove entire circuits to use them for arc mapping, a process that helps the investigator determine if and where electrical circuits were energized at the time of the fire.

Watch for evidence that was shielded from the fire and is lying beneath burned debris. For example, a wall clock may have fallen during the fire and been covered with debris. If the clock is near the area of origin and stopped approximately when the fire broke out, it could be an important piece of evidence.

FIRE FIGHTER TIP

While searching for hidden fires, avoid altering the area of origin as much as possible. Breach walls from the other side to avoid damaging fire patterns in the room of origin. This will leave valuable fire pattern evidence intact and aid the fire investigator in his or her scene examination.

Nonstructural Fires

Fire fighters are frequently called to fires in automobiles, recreational vehicles, and boats. Many of the investigational techniques used for structural fires apply to these types of fires. Look for fire effects, fire patterns, and any physical evidence that should be reported to the fire investigator. Be careful to avoid destroying evidence or contaminating the fire scene until the investigation is complete.

Wildland fires should be investigated, although the size of the scene and the evidence collected at these fires may be somewhat different from the evidence gathered at a building fire. Follow your department's SOPs for identifying and preserving evidence at a wildland fire.

Fire-Related Injuries and Fatalities

Any fire that results in an injury or fatality must be thoroughly investigated and fully documented. All burn and smoke inhalation injuries must be evaluated and treated at the scene. Patients with more serious injuries must be transported to a hospital for advanced medical treatment. If possible, injuries should be documented and photographed prior to transport to the hospital. Be sure to follow your department's medical privacy policies before photographing any injuries.

While conducting search and rescue operations, always err on the side of victim safety, but do not remove bodies that are found in areas that are obviously not survivable. Notify the investigator if a body was removed during search and rescue operations or because fire suppression or overhaul operations would have caused further damage to the body. The fire investigator must document the original location and position of any victims, especially in relation to the fire and the exits.

Clothing removed from any victim should be preserved as evidence. It may contain traces of ignitable liquids, and fire patterns on the clothing can be used in the investigation. If the clothing is removed in the ambulance or at the hospital, personnel should be assigned to collect it and keep this evidence as intact as possible.

Document anything lying under a victim's body after it is removed; this is often a protected area and may reveal important evidence. Victims who do not escape the fire should not be disturbed, and no attempts should be made to try to identify them. Fire investigators will work with other investigators, including the coroner or medical examiner, to examine the victims and the surrounding area for items of evidentiary value.

The Crime of Arson

The term *arson* has been misused and generalized for many years. Although incendiary fires and arson are often related, they are not the same. Fires classified as incendiary are those determined to have been intentionally ignited by someone who knew that it was wrong. Arson, however, is defined by the Criminal Code of Canada, which is a federal statute and applies to all provinces and territories. There are nine variations of arson-related offenses under the criminal code, each having a separate penalty.

Some arsonists are responsible for multiple fires in a specific location or over a certain time frame. Three repetitive firesetting behaviours have been identified (Douglas, 2013). A serial arsonist is a person who sets three or more fires in succession with an emotional "cooling-off" period between each fire. A spree arsonist sets three or more fires in separate locations with no lengthy cooling-off period between fires. A mass arsonist sets three or more fires at or near the same location during a limited time period.

While all fire investigators attempt to identify the origin and cause of a fire, only those investigators with law enforcement authority can determine if and which crimes have been committed. As previously mentioned, in some jurisdictions, the fire investigator is authorized to conduct both the fire investigation and the criminal investigation; in other jurisdictions, fire investigators determine origin and cause while law enforcement officers investigate criminal activity. Fire fighters should avoid using the term *arson* on the fire scene, even in casual conversation. Your role as a fire fighter is to identify and preserve potential evidence, notify any relevant observations to the investigator, and assist with the investigation as directed by the IC or fire investigator. The determination of origin, cause, cause classification, and any criminal activity is the responsibility of other personnel.

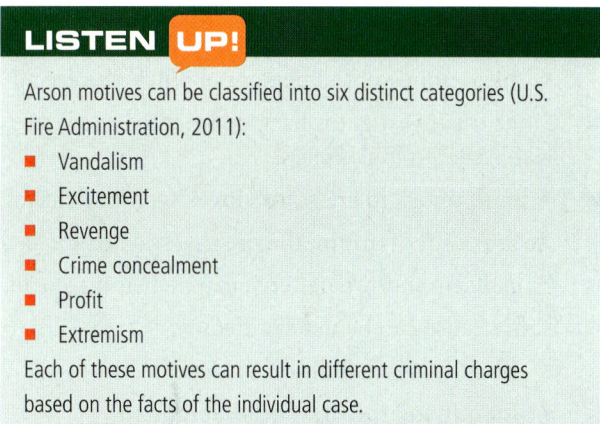

LISTEN UP!

Arson motives can be classified into six distinct categories (U.S. Fire Administration, 2011):

- Vandalism
- Excitement
- Revenge
- Crime concealment
- Profit
- Extremism

Each of these motives can result in different criminal charges based on the facts of the individual case.

Youth Firesetters

Children tend to have a fascination with fire. Most children will quickly learn to respect fire, but some children will become drawn to fire. If this fascination becomes too strong, the child may begin to play with fire or intentionally ignite fires for different reasons. Experts in the field recognize three distinct age categories of youth firesetters.

Child firesetters are children from 2 to 6 years of age. They are often curious about fire. They lack the mental capacity to understand the consequences of their actions, but they often realize that they should not be playing with matches or lighters. As a result, they often play with fire in hidden locations like closets, basements, and garages where many combustible fuels are present. Child firesetters may be injured or killed as a result of playing with fire. Fires ignited by young children are usually reported as accidental fires.

Juvenile firesetters are 7 to 13 years old and often suffer from emotional or psychological problems. Many come from homes with family problems or emotional, physical, or sexual abuse. Juvenile firesetters often ignite fires in or around their own homes or schools.

Adolescent firesetters are typically 14 to 16 years old, and many exhibit serious emotional or psychological symptoms such as extreme stress or anger. These firesetters frequently come from severely dysfunctional homes. Because these firesetters are older, the fires they ignite are often more severe. Target buildings include schools, churches, outbuildings, vacant homes, and vacant lots.

Notify your fire prevention division about a potential youth firesetter so that fire department personnel can contact the parents for preventive education and support. Many youth firesetters need professional counselling to treat the underlying causes of their fire-setting behaviour.

After-Action REVIEW

IN SUMMARY

- Fire fighters usually reach a fire scene before a trained fire investigator and are able to observe important fire patterns and evidence that the fire investigator can use to determine how and where the fire started.
- Determining the fire's origin and cause may help prevent future fires, identify who is financially responsible for property damage, injuries, or deaths, and provide evidence to be used in criminal prosecution.
- The size of the fire department and the size of the fire determine who performs the fire investigation and when a fire is investigated.
- Fire investigation should be performed within a reasonable time following the extinguishment of the fire. Fire fighters must be aware of any provincial or local laws regarding the right of access by the owners or occupants while the property is under the control of the fire department.
- If a fire investigator is not immediately available, the premises should be secured and guarded until the investigation takes place and all evidence is collected. In the interim, take the following steps:
  - Limit salvage and overhaul operations.
  - Photograph the fire scene extensively.
  - If weather, traffic, or other factors could destroy the evidence, take steps to preserve it in the best way possible.
  - Keep nonessential personnel out of the area. Deny entry to all unauthorized and unnecessary persons.
- Once the investigator has documented the scene, collected evidence, and interviewed witnesses, the property is released to the owners.
- A point of origin and cause investigation determines where, how, and why the fire ignited.
- At the point of origin, the fire investigator looks for evidence indicating the specific cause of the fire.
- Fire effects and fire patterns can be helpful in identifying the point of origin but are not conclusive evidence.
- A fire cause can be determined by:
  - Identifying an ignition source
  - Identifying the first fuel that ignited

- Identifying the oxidizing agent
- Identifying the ignition sequence
- Establishing a level of certainty of a possible cause
- The cause of a fire can be classified as accidental, natural, incendiary, or undetermined.
- After the fire investigator identifies the area of origin, fire fighters may be asked to assist in digging out the fire scene to help look for evidence within the debris.
- Evidence refers to all of the information gathered and used by fire investigators in determining the cause of a fire:
  - Physical evidence
  - Demonstrative evidence
  - Documentary evidence
  - Testimonial evidence
  - Direct evidence
  - Circumstantial evidence
- Fire fighters have a responsibility to preserve evidence that could indicate the origin and cause of a fire. Fire fighters who discover potential evidence during fire suppression, salvage, or overhaul operations should leave it in place, if possible, and notify a supervisor or fire investigator immediately.
- Physical evidence must be handled in a manner that protects the integrity of the chain of custody. Every step in the collection, movement, storage, and examination of the evidence must be properly documented.
- The fire investigator's standard operating procedures for collecting and processing evidence generally include the following steps:
  - Take photographs of each piece of evidence as it is found and collected.
  - Photograph, sketch, mark, and label the location of the evidence.
  - Place evidence in appropriate containers to ensure its safety and prevent contamination.
  - Tag all evidence at the fire scene.
  - Keep a constant watch on the evidence until it can be stored in a secure location.
  - Preserve the chain of custody in handling all of the evidence.
- Although fire fighters may not interview witnesses, they can identify potential witnesses to the fire investigator.
- The fire fighter's role in identifying and preserving evidence continues throughout the fire suppression sequence and takes into account the following factors:
  - The time of day and the weather conditions
  - People or vehicles arriving or leaving the scene
  - The extent of the fire and the number of locations in which fire is found
  - Unusual items or conditions
  - The security of the building
  - Any evidence of prior forced entry
  - Indications of unusual fire situations
  - Unusual colour of smoke
  - Unusual odours
  - The position of doors and windows
  - The reaction of the fire during initial attack
  - Any evidence of abnormal behaviour of fire
  - Unusual locks, debris, structural modifications, or other obstacles
  - The condition of the building contents
  - The need to coordinate overhaul and evidence preservation activities

- Repetitive fire setting behaviour includes mass arson, serial arson, and spree arson. Each of these behaviours is defined by the number, frequency, location, and time between fires.
- Experts recognize three categories of youth firesetters: child firesetters, juvenile firesetters, and adolescent firesetters. Each of these categories defines youths of different ages with specific behaviours and psychological characteristics.

KEY TERMS

Access Navigate for flashcards to test your key term knowledge.

Accidental fire Fires for which the cause does not involve a human act with the intent to ignite or spread a fire. (NFPA 556)

Adolescent firesetters Firesetters who are typically 14 to 16 years old and may exhibit serious emotional or psychological symptoms such as extreme stress or anger. Target buildings include schools, churches, outbuildings, vacant homes, and vacant lots.

Arc mapping The systematic evaluation of the electrical circuit configuration, spatial relationship of the circuit components, and identification of electrical arc sites to assist in the identification of the area of origin and analysis of the fire's spread. (NFPA 921)

Arson The crime of maliciously and intentionally, or recklessly, starting a fire or causing an explosion. (NFPA 921)

Chain of custody The trail of accountability that documents the possession of evidence in an investigation.

Child firesetters Firesetters who are typically 2 to 6 years old and are often curious about fire. They lack the mental capacity to understand the consequences of their actions.

Circumstantial evidence Evidence that is based on logical inference rather than personal observation.

Competent ignition source An ignition source that has sufficient energy and is capable of transferring that energy to the fuel long enough to raise the fuel to its ignition temperature. (NFPA 921)

Demonstrative evidence Any type of tangible evidence relevant to a case, such as diagrams, photographs, maps, x-rays, visible tests, and demonstrations.

Depth of char A fire effect that, when evaluated as a pattern on identical fuels, may be used to determine locations within a structure that were exposed longest to a heat source.

Destructive analysis investigation An investigation that uses the methodical deconstruction of evidence to determine specific component conditions, functionality, or failures as they relate to fire investigation.

Direct evidence Testimony of witnesses who observe acts or detect something through their five senses or through surveillance equipment such as CCTV.

Documentary evidence Any type of written record or document that is relevant to the case.

Expectation bias Any preconceived determination or premature conclusions as to the cause of a fire without having examined or considered all relevant evidence.

Fire effects The observable or measurable changes in or on a material as a result of a fire. (NFPA 921)

Fire patterns The visible or measurable physical changes, or identifiable shapes, formed by a fire effect or group of fire effects. (NFPA 921)

Ignitable liquid Any liquid or the liquid phase of any material that is capable of fueling a fire, including a flammable liquid, combustible liquid, or any other material that can be liquefied and burned. (NFPA 556)

Incendiary device A device or mechanism used to initiate an incendiary fire or explosion.

Incendiary fire A fire that is intentionally ignited in an area or under circumstances where and when there should not be a fire. (NFPA 921)

Juvenile firesetters Firesetters who are typically 7 to 13 years old and often suffer from emotional or psychological problems. Juvenile firesetters often ignite fires in or around their own homes or schools.

Mass arsonist A person who sets three or more fires at the same site or location during a limited period of time.

Natural fire A fire caused without direct human intervention or action, such as fire resulting from lightning, an earthquake, or wind.

Physical evidence A physical or tangible item that proves or disproves a particular fact or issue; also referred to as *real evidence*.

Point of origin The exact physical location within the area of origin where a heat source and a fuel first interact, resulting in a fire or explosion. (NFPA 921)

Possible cause A hypothesis determined to have less than 50% probability of being true.

Probable cause A hypothesis determined to have greater than 50% probability of being true.

Scientific method The systematic pursuit of knowledge involving the recognition and definition of a problem; the collection of data though observation and experimentation; analysis of the data; the formulation, evaluation, and testing of hypotheses; and, where possible, the selection of a final hypothesis. (NFPA 921)

Serial arsonist A person who sets three or more fires with a cooling-off period between fires.

Spoliation Loss, destruction, or material alteration of an object or document that is evidence or potential evidence in a legal proceeding by one who has the responsibility for its preservation. (NFPA 921)

Spree arsonist A person who sets three or more fires at separate locations with no emotional cooling-off period between fires.

Testimonial evidence Verbal testimony of a witness given under oath in court or during specific legal proceedings.

Trailers Solid or liquid fuel used to intentionally spread or accelerate the spread of a fire from one area to another. (NFPA 921)

Undetermined fire A classification of fire when the cause cannot be proven to an acceptable level of certainty.

Youth firesetters Recognized classifications of minors who ignite fires for various reasons. Based on age and specific psychological characteristics, youth firesetters include child firesetters, juvenile firesetters, and adolescent firesetters.

REFERENCES

Douglas, J. E., A. W. Burgess, A. G. Burgess, and R. K. Ressler. *Crime Classification Manual,* Third Edition. Hoboken, NJ: John Wiley & Sons, Inc., 2013.

National Fire Protection Association. NFPA 921, *Guide for Fire and Explosion Investigations*, 2017. www.nfpa.org. Accessed November 28, 2018.

National Fire Protection Association. NFPA 1033, *Standard for Professional Qualifications for Fire Investigator*, 2014. www.nfpa.org. Accessed November 28, 2018.

U.S. Fire Administration. 2011. "Working Together to Extinguish Serial Arson." https://www.usfa.fema.gov/downloads/pdf/arson/aaw11_media_kit.pdf. Accessed November 28, 2018.

On Scene

It has been a busy few weeks for your department. There have been several fires in your jurisdiction, and your crew has responded to three of them this week. At the last fire, you overheard the investigator tell the IC that a serial arsonist might be responsible for some of the fires. At this morning's shift meeting, the battalion chief urged each of you to use extra caution on future fire scenes and to watch for any evidence that might be useful to the investigator. Suddenly, the alarm sounds in the bunkroom, and your crew heads to the apparatus. A fire has been reported in a vacant grocery store. As the pump turns the corner, you see smoke rising in the distance.

1. Which of the following must be proven in order to prosecute the crime of arson in Canada?

A. The use of an accelerant such as gasoline or lighter fluid

B. Insurance fraud

C. Criminal intent

D. Three or more fires in succession

2. If this fire was ignited by a very young child, which of the following fire cause classifications would apply?

A. Accidental

B. Natural

C. Incendiary

D. Undetermined

(continued)

On Scene Continued

3. If this fire was ignited by lightning, which of the following fire cause classifications would apply?

A. Accidental

B. Natural

C. Incendiary

D. Undetermined

4. Photographs, maps, charts, and diagrams are found at the scene. Which type of evidence are these?

A. Physical

B. Demonstrative

C. Documentary

D. Testimonial

5. You notice observable changes found on objects, surfaces, and construction materials that are the result of exposure to heat, flame, smoke, and other products of combustion. What are these changes known as?

A. V-patterns

B. Evidence

C. Demonstrative evidence

D. Fire effects

6. Once the fire has been extinguished and the fire investigator releases the property to the owner, which of the following will likely be required to enter the property, search the scene, and seize additional evidence?

A. Owner consent or exigent circumstances

B. Owner consent or administrative warrant

C. Probable cause or administrative warrant

D. The investigator may not return under any condition.

7. Limiting entry and documenting anyone who enters the investigation area is one method of:

A. limiting overhaul.

B. collecting evidence.

C. fire cause determination.

D. securing a fire scene.

Access Navigate to find answers to this On Scene, along with other resources such as an audiobook.

Hazardous Materials Awareness Level

Hazardous Materials Regulations, Standards, and Laws

KNOWLEDGE OBJECTIVES

After studying this chapter, you will be able to:

- Identify the difference between hazardous materials/WMD incidents and other emergencies. (**NFPA 472: 4.2.1**, pp. 1046–1047)

- Identify the location of both the emergency response plan and/or standard operating procedures. (**NFPA 1072: 4.1.3**, p. 1047)

- Define the terms *hazardous materials* (or *dangerous goods*, in Canada) and *weapons of mass destruction*. (**NFPA 1072: 4.2.1; NFPA 472: 4.1.2.2, 4.2.1**, p. 1047)

- Understand the difference(s) between the standards and federal regulations that govern hazardous material response activities. (**NFPA 472: 4.1.1.2, 4.1.2.1**, pp. 1048–1049)

- Describe the different levels of hazardous materials training: awareness, operations, technician, specialist, and incident commander. (**NFPA 1072: 4.1.1, 4.1.2, 4.1.3; NFPA 472: 4.1.1.1, 4.1.1.3, 4.1.2.1, 4.1.2.2**, pp. 1049–1052)

- Explain the need for a planned response to a hazardous materials incident. (p. 1052)

SKILLS OBJECTIVES

This chapter has no skills objectives for awareness level personnel.

Hazardous Materials Alarm

At 13:30 hours, your crew responds to a report of a suspicious odour at a small plastics manufacturing company. When you arrive, you see a cargo delivery truck parked at the loading dock, with the motor still idling. From your vantage point, a few hundred feet away, a liquid is leaking from the back of the truck. On one side of the truck you can see a black and white diamond-shaped placard that reads "corrosive," with the number 8 at the bottom. The facility security guard reports that the truck has been left unattended for at least an hour.

1. Is this an incident? How would you go about analyzing the scene to better understand the problem? What other pieces of information would you want?

2. You are trained to the Hazardous Materials Awareness level. Would this level of training allow you to put on chemical protective equipment, enter the hazardous area, and attempt to determine the nature of the leak and/or fix the problem?

3. What reference source(s) would you want to access for more information on the placard?

 **JONES & BARTLETT LEARNING**
NAVIGATE 2 *Access Navigate for more practice activities.*

Introduction

Fire fighter personnel are called upon to respond to a variety of incidents. These incidents may include structural fires, emergency medical calls, automobile accidents, confined-space rescues, water rescues, or acts of terrorism. Most census metropolitan areas (CMAs) in Canada with fire services that protect more than 500,000 people have dedicated hazardous materials teams. For example, in the province of Ontario there are three fire services that provide hazardous material response in aid of smaller departments. Ottawa provides technical hazardous materials response to the eastern part of the province, whereas Toronto covers the central area and Windsor covers the southwest. These three fire services receive provincial funding to provide services beyond their jurisdiction to other areas within the province. Similar agreements may be in force within your province/territory. Some of these incidents may involve hazardous substances that threaten lives, property, and/or the environment. When an incident clearly involves a hazardous material, or you suspect the presence of a hazardous materials release, you must adjust your approach accordingly (**FIGURE 29-1**).

Hazardous materials incidents are handled in a more deliberate fashion than structural firefighting. There could be unknown substances present, or the setting of the release is such that it may take some time to get a full picture of the incident prior to taking action. For example, train derailments involving hazardous materials will likely cover a large area and require a multiagency response. In any case, it does not mean that hazardous

FIGURE 29-1 The ability to recognize a potential hazardous materials/WMD incident is a critical first step to ensuring your safety.
Courtesy of George Roarty/VDEM.

materials incidents are more or less complicated than fighting fires; it means that responders often take more time to get oriented to the problem and define a rational approach to solving the situation without getting exposed to the released substance.

If a rescue is required during a hazardous materials incident, or if the situation is imminently dangerous in some other way and requires quick action, events may move quickly.

Responders must understand that actions taken at hazardous materials/weapons of mass destruction (WMD) incidents are largely dictated by the chemicals or hazards involved; environmental influences such as wind, rain, and temperature; and the way the chemicals behave during the release. In short, the nature and circumstance of the response as a whole dictate the tactics and strategy.

Additionally, personnel operating at the scene of a hazardous materials/WMD incident must be conscious of the potential or actual law enforcement aspect of the incident. Especially where terrorist or other criminal acts are suspected, responders should be mindful of evidentiary issues associated with the incident. Being mindful of potential evidence at a hazardous materials/WMD event may facilitate later efforts to identify, capture, and prosecute the person responsible for the act. To that end, every responder on the scene must be cognizant of the impact his or her presence and actions will have on potential evidence. Although evidence preservation should not impede the efforts to eliminate the problem or slow life-saving operations, every responder should be diligent in remembering that his or her actions and observations may play a vital role in the successful prosecution of a criminal suspect.

When responding to hazardous materials/WMD incidents, make a conscious effort to change your perspective. Slow down, think about the problem and available resources, and take well-considered actions to solve it.

LISTEN UP!

The goal of the responder is to favorably change the outcome of the hazardous materials/WMD incident. This is accomplished through sound planning and by establishing safe and reasonable response objectives based on the level of training. Don't do it if you're not trained to do it!

Additionally, initial and ongoing actions may be guided by your **authority having jurisdiction (AHJ)** and local or organizational emergency response plans and/or standard operating procedures (SOPs). Every responder should have knowledge of and access to response plans and should have been trained on how to implement the actions in accordance with his or her level of training. The AHJ is the governing body that sets operational policy and procedures for the jurisdiction in which you operate. For example, the AHJ might identify a set of tasks that responders would be expected to perform in their course of duty and match the training competencies to address those tasks.

From a broad perspective, the goal of Sections 3, 4, and 5 of this book is to help personnel learn how to recognize the presence of a hazardous materials/WMD incident; take initial actions, including establishing scene control zones; use basic reference sources such as the *Emergency Response Guidebook (ERG)* (covered in detail in Chapter 30, *Recognizing and Identifying the Hazards*);

select personal protective clothing; implement product control measures when needed; perform appropriate decontamination; and ultimately, understand where you fit into a full-scale hazardous materials/WMD response. The first two chapters are dedicated solely to the awareness level of hazardous materials training. Subsequent chapters focus on the operations level responder. There may be some overlap in content, but the intent is to make a clear delineation between the two levels.

The first points to understand, before diving into the regulatory end of response, are the definitions of *hazardous material* and *weapons of mass destruction*.

What Is a Hazardous Material Anyway?

A **hazardous material** is any substance or material that is capable of posing an unreasonable risk to human health, safety, or the environment when transported in commerce, used incorrectly, or not properly contained or stored. The term *hazardous material* also includes hazardous substances, wastes, marine pollutants, and elevated-temperature materials. This definition also includes illicit laboratories, environmental crimes, or industrial sabotage. It is for this reason that the text will refer to hazardous materials and WMD simultaneously.

As a point of reference, in United Nations model codes and regulations, hazardous materials are called *dangerous goods*. According to the National Fire Protection Association (NFPA) Standard 472, a weapon of mass destruction is:

> (1) Any destructive device, such as any explosive, incendiary, or poison gas bomb, grenade, rocket having a propellant charge of more than 4 oz. (113 g), missile having an explosive or incendiary charge of more than 0.25 oz. (7 g), mine, or similar device; (2) any weapon involving toxic or poisonous chemicals; (3) any weapon involving a disease organism; or (4) any weapon that is designed to release radiation or radioactivity at a level dangerous to human life.

WMD are known by many different abbreviations and acronyms, the most common of which is CBRNE, an acronym for chemical, biological, radiological, nuclear, and explosive. Problems arise when these materials are released as the result of a terrorist attack.

From a responder's perspective, a hazardous material can be almost anything, depending on the situation. Milk, for example, is not routinely regarded as a hazardous substance—but 18,927 litres (5000 gallons) of milk leaking into a creek does, in fact, pose a risk to the

FIGURE 29-2 A hazardous material can be found anywhere.
Courtesy of Rob Schnepp.

environment (**FIGURE 29-2**). A large chlorine gas release also fits the definition, as would any substance used as a terrorist weapon. Regarding the threat of terrorism, it would be unrealistic to assume that your jurisdiction is immune to deliberate criminal acts. Such an event could happen anywhere, at any time, with any type of substance.

Manufacturing processes sometimes generate hazardous wastes. A **hazardous waste** is what remains after a process or manufacturing activity has used a substance and the material is no longer pure. Hazardous waste can be just as dangerous as pure chemicals. It can also comprise mixtures of several chemicals, which may make it difficult to determine how the substance will react when it is released or if it encounters other chemicals. The waste generated by the illegal production of methamphetamine, for example, may produce a dangerous mixture of chemical wastes.

Levels of Training: Regulations and Standards

To understand where you fit in as a responder, you must first recognize some of the regulatory drivers that apply to hazardous materials response, beginning with the difference between a regulation and a standard. **Regulations** are issued and enforced by governmental bodies such as federal, provincial, and territorial **Occupational Health and Safety Acts (OHSAs)**, the Canadian Labour Code, and the **Canadian Environmental Protection Act (CEPA)**. Conversely, **standards** are issued by nongovernmental entities and are generally consensus based. A standard may be voluntary, meaning that an agency such as a fire department may not be required to adopt and follow

the standard completely. For example, organizations such as the **National Fire Protection Association (NFPA)** issue voluntary consensus-based standards that anyone can comment on before committee members agree to adopt them. The technical committee responsible for periodically revising any NFPA standard is required to meet regularly; revise, update, and possibly change a standard; and review and act on any public comments during the revision process. Once the standard is finalized, agencies may *choose* to adopt it. It is important to note that, while the NFPA is based in the United States, NFPA standards are recognized internationally. In Canada, many fire and building codes reference NFPA standards, as do the regulations governing manufacturing processes and laboratory activities.

LISTEN UP!

Responders should understand the relationship between Occupational Health and Safety Act (OHSA) regulations and NFPA standards when it comes to hazardous materials/WMD response. OHSA regulations are the law that governs hazardous materials/WMD responders; NFPA standards are guidelines that agencies choose to adopt. NFPA 472 is clear that responders shall receive any additional training to meet applicable **Transport Canada (TC)**, Canadian Environmental Protection Act (CEPA) and, local, federal, or provincial occupational health and safety regulatory requirements.

There are fourteen jurisdictions in Canada (1 federal, 10 provincial, and 3 territorial), with each one having its own legislated OHSA.

NFPA standards governing hazardous materials/WMD response come from the Technical Committee on Hazardous Materials Response Personnel. This group includes more than 30 members from private industry, the fire service and law enforcement, professional organizations, and governmental agencies. Currently, three published standards are especially important to personnel who may be called upon to respond to hazardous materials/WMD incidents: NFPA 472, *Standard for Competence of Responders to Hazardous Materials/Weapons of Mass Destruction Incidents*; NFPA 1072, *Standard for Hazardous Materials/Weapons of Mass Destruction Emergency Response Personnel Professional Qualifications*; and NFPA 473, *Standard for Competencies for EMS Personnel Responding to Hazardous Materials/Weapons of Mass Destruction Incidents*. Generally speaking, NFPA 472 outlines training competencies for all hazardous materials responders, NFPA 1072 identifies the minimum job performance requirements (JPRs), and NFPA 473 spells out the competencies required for EMS personnel rendering medical care at hazardous materials/WMD incidents.

Similar to NFPA 472, NFPA 1072 applies to the following levels of responder: awareness, operations, mission-specific operations, hazardous materials technician, and incident commander. NFPA 473 also contains mission-specific competencies in a similar manner as NFPA 472 and NFPA 1072. These mission-specific competencies include the following:

- Advanced life support (ALS) responder assigned to a hazardous materials team
- ALS responder assigned to provide clinical interventions at a hazardous materials/WMD incident
- ALS responder assigned to treatment of smoke inhalation victims

Another NFPA standard recently developed and published is NFPA 475, *Recommended Practice for Organizing, Managing, and Sustaining a Hazardous Materials/Weapons of Mass Destruction Emergency Response Program*. The intent of this document is to establish common criteria for the organization, management, programmatic elements, deployment of personnel, and resources for those entities responsible for the hazardous materials/WMD emergency preparedness function.

In Canada, NFPA standards as well as CEPA and federal, provincial, and territorial OHSAs are important to personnel called upon to respond to hazardous material/WMD incidents.

The following descriptions provide a broad overview, as found in NFPA 472, of the different levels of hazardous materials/WMD responders and training competencies. When reading NFPA 472, it is important to understand that the standard is organized to first spell out the *tasks* that a responder (awareness, operations, technician, incident commander) may be called upon to perform on the scene. The subsequent *training competencies* follow. To be clear, NFPA 472 is not a "how to respond" document. It provides no direction on how to plug leaking containers, use detection and monitoring devices, or decide which level of protection to wear in a certain situation. It is intended to provide guidance on the competencies associated with the various training levels.

NFPA 1072, as mentioned earlier, is intended to identify the JPRs that personnel should be able to perform to carry out the job duties. JPRs are to be accomplished in accordance with the requirements of the AHJ. Personnel at the awareness level must meet the requirements defined in Chapter 4 of NFPA 1072. Operations level responders must meet all the requirements defined in Chapter 5 of NFPA 1072. The list of mission-specific duties in NFPA 1072 follows the same list of competencies found in NFPA 472.

To stay focused on the intent of this text—which is to meet and exceed the job performance requirements in the latest edition of NFPA 1001: *Standard for Fire Fighter*

Professional Qualifications—the hazardous materials section of this text will discuss in detail only the competencies, tasks, and JPRs that the awareness level, the core operations level, and select mission-specific operations level personnel and responders may be expected to perform per NFPA 1072 and 472. A general overview of the training competencies for the other response levels—technician, specialist, and incident commander—is provided but is not intended to be definitive or comprehensive.

Awareness Level

Per NFPA 1072 Section 4.1.1, "awareness level personnel are those persons who, during the course of their normal duties, could encounter an emergency involving hazardous materials/weapons of mass destruction (WMD) and who are expected to recognize the presence of the hazardous materials/WMD, protect themselves, call for trained personnel, and secure the area." Per NFPA 472, a person with awareness level training is not considered to be a *responder*. Instead, these individuals are now referred to as awareness level *personnel*. Persons receiving this level of training are not typically called to the scene to respond; rather, awareness level personnel, such as public works employees or fixed facility security personnel, function in support roles.

Tasks that awareness level personnel may be expected to perform on the scene include these duties:

- Analyzing the incident to detect the presence of hazardous materials/WMD
- Identifying the name, United Nations/North American Hazardous Materials Code (UN/NA) identification number, type of placard, or other distinctive marking applied for the hazardous materials/WMD involved
- Collecting information from the current edition of the *ERG* about the hazard
- Initiating and implementing protective actions consistent with the response plan, the SOPs, and the current edition of the *ERG*
- Initiating the notification process

The NFPA 1072 JPRs that apply to awareness level personnel can be found in Chapter 4 of NFPA 1072 and are divided into the following sections. Each section details the knowledge and skills necessary to complete each JPR.

4.2 Recognition and Identification
4.3 Initiate Protective Actions
4.4 Notification

Operations Level

Per NFPA 1072, operations level responders are those persons who are tasked to respond to hazardous materials/WMD incidents for the purpose of taking action to protect nearby persons, the environment, or property from the effects of the release (**FIGURE 29-3**). These persons may also have competencies that are specific to their response mission, expected tasks, and equipment and training as determined by the AHJ—the mission-specific competencies listed earlier in the chapter.

NFPA 472 and NFPA 1072 significantly expand the scope of an operations level responder by separating the operations level suite of competencies into two distinct categories: core competencies and mission-specific competencies. The core competencies are based on those vital tasks that operations level personnel should perform on the scene of a hazardous materials/WMD incident. Some of those tasks are outlined here:

- Analyze the scene of a hazardous materials/WMD incident to determine the scope of the incident
- Survey the scene to identify containers and materials involved
- Collect information from available reference sources
- Predict the likely behaviour of a hazardous material
- Estimate the potential harm the substances might cause
- Plan a response to the release, including selection of the correct level of personal protective clothing
- Perform decontamination
- Preserve evidence
- Evaluate the status and effectiveness of the response

This abbreviated list should serve only as an illustration of the NFPA 472 and NFPA 1072 core competencies. Keep in mind that the core training competencies are designed to provide the skills, knowledge, and abilities to safely accomplish the tasks listed. Core competency training is required of all operations level responders on the scene, no matter what their function. One of the goals of the NFPA 472 standard is to better match the expected tasks that may be required of the responder with the training that the responder should receive.

In addition to undertaking core competency training, an individual AHJ may find the need to do

more training based on an identified or anticipated "mission-specific" need. To that end, those responders who are expected to perform additional missions, beyond the core competencies, must be trained to carry out those mission-specific responsibilities. Remember, both NFPA 472 and NFPA 1072 allow each agency to pick and choose the training program that makes the most sense *for its jurisdiction*. These mission-specific competencies are *nonmandatory* and should be viewed as optional. To that end, operations level responders may end up performing a limited suite of technician level skills but do not have the broader knowledge and abilities of a hazardous materials technician.

NFPA 472 and NFPA 1072 provide a mechanism to ensure that operations level responders, including those with mission-specific training, do not go beyond their level of training and equipment. This is done by using technician level personnel to provide direct guidance to the operations level responder operating on a hazardous materials/WMD incident. Operations level responders are expected to work under the direct control of a hazardous materials technician or allied professional who can continuously assess and/or observe the actions of the operations level responder *and* provide immediate feedback. Guidance by a hazardous materials technician or an allied professional may be provided through direct visual observation or through assessments communicated from the operations level responder(s) to a technician.

Mission-specific competencies as defined by NFPA 1072 (Section 3.4.1) are as follows: "The knowledge, skills, and judgment needed by operations level responders who have completed the operations level competencies and who are designated by the authority having jurisdiction (AHJ) to perform mission-specific tasks, such as decontamination, victim/hostage rescue and recovery, evidence preservation, and sampling." The 11 mission-specific competencies are as follows:

- Personal protective equipment (PPE)
- Mass decontamination
- Technical decontamination

FIGURE 29-3 Operations level responders.
Courtesy of Rob Schnepp.

- Evidence preservation and public safety sampling
- Product control
- Detection, monitoring, and sampling
- Victim rescue/recovery
- Illicit laboratory incidents
- Disablement/disruption of improvised explosive devices (IEDs), improvised WMD dispersal devices, and operations at improvised explosives laboratories
- Diving in contaminated water environment
- Evidence collection

Per the requirements of NFPA 1001, this book covers select mission-specific operations level competencies from NFPA 1072, including those from Section 6.2, Personal Protective Equipment, and Section 6.6, Product Control.

As with the awareness level, the operations level JPRs are listed as Requisite Knowledge and Requisite Skills. The following JPRs are found in Chapter 5 of NFPA 1072:

5.2 Identify Potential Hazards
5.3 Identify Action Options
5.4 Action Plan Implementation
5.5 Emergency Decontamination
5.6 Progress Evaluation and Reporting

As you can see, the JPRs for operations are much more action oriented, reinforcing the fact that at the operations level you are a *responder*.

Technician/Specialist Level

NFPA 1072 Section 7.1.1 defines individuals at the **technician level** as "those persons who respond to hazardous materials/WMD incidents using a risk-based response process by which they analyze a problem involving hazardous materials/WMD, plan a response to the problem, implement the planned response, evaluate progress of the planned response, and assist in terminating the incident." The hazardous materials technician is a person who responds to hazardous materials/WMD incidents using a risk-based response process with the ability to:

- Analyze a problem involving hazardous materials/WMD
- Select appropriate decontamination procedures
- Control a release using specialized protective clothing and control equipment

These persons may have additional competencies that are specific to their response mission, expected tasks, and equipment and training as determined by the AHJ. A number of very detailed training competencies for technician level training are outlined in Chapter 7 of NFPA 472. Technician level personnel are integral to the NFPA 472 standard because they are, in many cases, intended to "supervise" the activities of on-scene operations level responders.

Technicians typically function at a higher level in terms of their cognitive approach to the response. They should be proficient at implementing a comprehensive risk-based approach to solving the problem. They will approach the point of release to plug, patch, or otherwise mitigate the problem (**FIGURE 29-4**).

Incident Commander

Per NFPA 1072, the **incident commander (IC)** is the person responsible for all incident activities, including the development of strategies and tactics and the ordering and release of resources. The IC must receive any additional training necessary to meet applicable governmental occupational health and safety regulations and the specific needs of the jurisdiction.

Individuals trained as ICs should have at least operations level training as well as additional training specific to commanding a hazardous materials incident. ICs, who will assume control of the incident scene beyond the first responder awareness level, must receive at least 24 hours of training equal to the first responder operations level and have competency in the following areas:

- Know and implement the employer's ICS
- Know how to implement the employer's emergency response plan

FIGURE 29-4 Hazardous materials technicians during a product control training exercise.
Courtesy of Rob Schnepp.

- Know and understand the hazards and risks associated with chemical protective clothing
- Know how to implement the local emergency response plan
- Know of the provincial and federal emergency response plans
- Know and understand the importance of decontamination procedures

For more complete information regarding the IC Hazardous Materials position, refer to Chapter 8 of NFPA 1072.

In addition to the initial training requirements for all response levels listed earlier, federal, provincial, and territorial OHSAs require annual refresher training of sufficient content and duration to ensure that responders maintain their competencies or that they demonstrate competency in those areas at least yearly. Consult your local agency for more specific information on refresher training, other hazardous materials laws, regulations, and regulatory agencies.

Other Governmental Agencies

In addition to the federal, provincial, and territorial OHSAs and the CEPA, several other government agencies are concerned with various aspects of hazardous materials/WMD response. Transport Canada (TC), for example, promulgates and publishes laws and regulations that govern the transportation of dangerous goods by highway, rail, pipeline, air, and in some cases, marine transport. TC is responsible for overseeing the regulations for the transportation of hazardous materials throughout the country, regardless of provincial/territorial boundaries.

As discussed earlier in this chapter, the CEPA regulates and governs issues relating to hazardous materials in the environment.

Preincident Planning

It is a mistake to assume that a response to a hazardous materials/WMD incident begins when the alarm sounds. The response really begins with training, learning the regulations and agencies involved, and finding out about the potential hazards in your area. Response agencies should conduct preincident planning activities at **target hazards** and other potential problem areas throughout the jurisdiction (**FIGURE 29-5**). Target hazards include any occupancy type or facility that presents a high potential for loss of life or serious impact to the community resulting from fire, explosion, or chemical release.

Preplanning activities enable agencies to develop logical and appropriate response procedures for anticipated incidents. Planning should focus on the real threats that exist in your community or adjacent communities you could be assisting.

Once the threats have been identified, fire departments, police agencies, public health offices, and other governmental agencies should determine how they will respond and work together in case of a large-scale incident. In many cases, the move toward interoperability before an incident will make the actual response work run smoothly. Remember this important point: People making good decisions solve problems effectively. When people are acquainted with each other before the event happens, they tend to work together better. Get to know your peers around you at whatever level you operate.

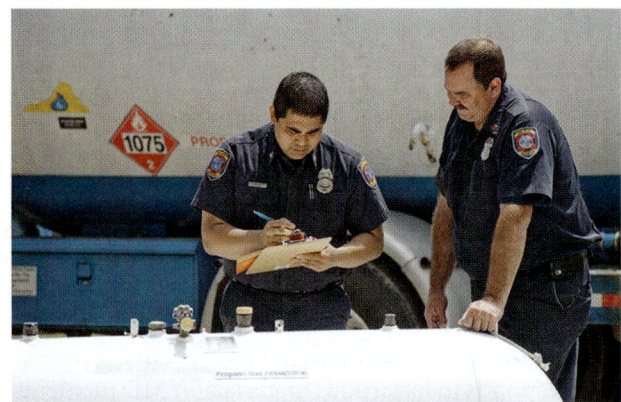

FIGURE 29-5 Conduct preincident planning activities at target hazards throughout the jurisdiction.
© Jones & Bartlett Learning. Photographed by Glen E. Ellman.

After-Action REVIEW

IN SUMMARY

- When an incident clearly involves a hazardous material, or you suspect the presence of a hazardous materials release, the nature of the incident changes and so must the mentality of the responder.
- A hazardous material is any substance or material that can pose an unreasonable risk to human health, safety, or the environment when transported in commerce, used incorrectly, or not properly contained or stored.

- Government entities such as federal, provincial, and territorial Occupational Health and Safety Acts (OHSAs) and the Canadian Environmental Protection Act (CEPA) issue and enforce regulations concerning hazardous materials emergencies.

- Consensus-based NFPA standards relating to hazardous materials/WMD are available for those agencies that choose to adopt them.

- Three NFPA standards are important to responders who may be called upon to respond to hazardous materials/WMD incidents:
  - NFPA 472, *Standard for Competence of Responders to Hazardous Materials/Weapons of Mass Destruction Incidents*
  - NFPA 1072, *Standard for Hazardous Materials/Weapons of Mass Destruction Emergency Response Personnel Professional Qualifications*
  - NFPA 473, *Standard for Competencies for EMS Personnel Responding to Hazardous Materials/Weapons of Mass Destruction Incidents*

- The goals associated with the competencies of awareness level personnel are to recognize a potential hazardous materials emergency, to isolate the area, and to call for assistance. Awareness level personnel take protective actions only.

- NFPA 472 expands the scope of an operations level responder's duties by making a distinction between core competencies and mission-specific competencies.

- For those who choose to adopt the NFPA 472 standard, the core competencies are required for all operations level responders; each agency can then pick and choose to require any or all of the mission-specific responsibilities. The core competencies of operations level responders are defensive actions.

- Hazardous materials technicians will control a hazardous materials release using specialized protective clothing and control equipment.

- The hazardous materials incident commander is responsible for all incident activities.

KEY TERMS

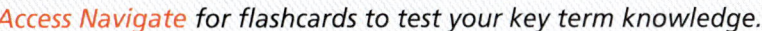

Access Navigate for flashcards to test your key term knowledge.

Authority having jurisdiction (AHJ) An organization, office, or individual responsible for enforcing the requirements of a code or standard or for approving equipment, materials, an installation, or a procedure. (NFPA 1072)

Awareness level personnel Personnel who, in the course of their normal duties, could encounter an emergency involving hazardous materials/weapons of mass destruction (WMD) and who are expected to recognize the presence of the hazardous materials/WMD, protect themselves, call for trained personnel, and secure the scene. (NFPA 1072)

Canadian Environmental Protection Act (CEPA 1999) An important part of Canada's federal environmental legislation aimed at preventing pollution and protecting the environment and human health. CEPA 1999 came into force on March 31, 2000, following an extensive Parliamentary review of the first CEPA.

Hazardous material Matter (solid, liquid, or gas) or energy that when released is capable of creating harm to people, the environment, and property, including weapons of mass destruction (WMD) as defined in 18

U.S. Code, Section 2332a, as well as any other criminal use of hazardous materials, such as illicit labs, environmental crimes, or industrial sabotage. (NFPA 1072)

Hazardous waste A substance that remains after a process or manufacturing plant has used some of the material and the substance is no longer pure.

Incident commander (IC) The individual responsible for all incident activities, including the development of strategies and tactics and the ordering and the release of resources. (NFPA 1072)

National Fire Protection Association (NFPA) The association that develops and maintains nationally recognized minimum consensus standards on many areas of fire safety and specific standards on hazardous materials.

Occupational Health and Safety Act (OHSA) Canadian legislation that regulates workplace health and safety.

Operations level responders Persons who respond to hazardous materials/weapons of mass destruction (WMD) incidents for the purpose of implementing or supporting actions to protect nearby persons, the environment, or property from the effects of the release. (NFPA 1072)

Regulations Mandates issued and enforced by governmental bodies such as the federal, provincial, and territorial Occupational Health and Safety Acts (OHSAs) and the Canadian Environmental Protection Agency (CEPA).

Standards Documents, the main text of which contain only requirements and which are in a form generally suitable for mandatory reference by another standard or code or for adoption into law. Nonmandatory provisions shall be located in an appendix or annex, footnote, or fine-print note and are not to be considered a part of the requirements of a standard. (NFPA 1)

Target hazards Any occupancy types or facilities that present a high potential for loss of life or serious impact to the community resulting from fire, explosion, or chemical release.

Technician level A person who responds to hazardous materials/WMD incidents using a risk-based response process by which he or she analyzes a problem involving hazardous materials/WMD, selects applicable decontamination procedures, and controls a release using specialized protective clothing and control equipment.

Transport Canada (TC) The Canadian government agency that publicizes and enforces rules and regulations that relate to the transportation of many hazardous materials.

On Scene

Your volunteer fire company has just accepted five new members. You are tasked with providing them with core operations level hazardous materials training, so you spend several weeks preparing for the class. As part of the initial instruction, you deliver a section on the laws and regulations that govern hazardous materials response. Your instructional plan is to use this quiz to reinforce the main points of your introductory lecture.

1. Which consensus-based standard describes the hazardous materials training competencies for operations level responders?

  A. NFPA 473, *Standard for Competencies for EMS Personnel Responding to Hazardous Materials/Weapons of Mass Destruction Incidents*

  B. NFPA 472, *Standard for Competence of Responders to Hazardous Materials/Weapons of Mass Destruction Incidents*

  C. Code of Federal Regulations 1910.150(q)

  D. CFR Title 40, *Protection of the Environment, Part 311, Worker Protection*

2. The members of your fire company are tasked to respond to hazardous materials incidents. Which NFPA chapter would apply in order to complete the appropriate JPRs?

  A. Chapter 4 of NFPA 472

  B. Chapter 5 of NFPA 472

  C. Chapter 5 of NFPA 1072

  D. Chapter 7 of NFPA 1072

3. What is the name of the federal document containing the hazardous materials response competencies? This regulation, issued in the late 1980s, standardized training for hazardous materials response and for hazardous waste site operations.

  A. NFPA 471 Paragraph 4, Subsection (q)

  B. NFPA 473 Section 5.2.2.1

  C. NFPA 472; CFR, book number 29, part 1910.120

  D. HAZWOPER; CFR, Title 29, part 1910.120 subpart (q)

Access Navigate to find answers to this On Scene, along with other resources such as an audiobook.

CHAPTER 30

Hazardous Materials Awareness Level

Recognizing and Identifying the Hazards

KNOWLEDGE OBJECTIVES

After studying this chapter, you will be able to:

- Describe how to approach a scene size-up with potential hazardous materials involved. (**NFPA 1072: 4.2.1, 4.3.1, 4.4.1; NFPA 472: 4.2.1, 4.2.2, 4.4.1, 4.4.2, 5.2.2**, pp. 1057–1059)

- Identify and describe the types of containers that are often used to contain hazardous materials. (**NFPA 1072: 4.2.1; NFPA 472: 4.1.2.2, 4.2.1, 5.2.1.1.1, 5.2.1.1.2, 5.2.1.1.3**, pp. 1059–1062)

- Describe the purposes and types of various transportation and facility markings for hazardous materials. (**NFPA 1072: 4.2.1, 4.3.1; NFPA 472: 4.1.2.2, 4.2.1, 4.2.2, 4.2.3, 4.4.1, 5.1.2.2, 5.2.1.2.1, 5.2.1.2.2, 5.2.1.3, 5.2.2, 5.3.1**, pp. 1062–1067)

- Identify and describe the four routes of entry harmful substances take in the human body. (**NFPA 1072: 4.2.1, 4.3.1 NFPA 472: 4.2.3, 4.4.1**, pp. 1078–1081)

SKILLS OBJECTIVES

After studying this chapter, you will be able to:

- Use the *Emergency Response Guidebook (ERG)*. (**NFPA 1072: 4.2.1, 4.3.1; NFPA 472: 4.1.2.2, 4.2.3, 4.4.1, 5.1.2.2**, pp. 1070, 1072–1077)

Hazardous Materials Alarm

During an odour investigation in a light-industrial area of your district, a police officer indicates an abandoned open-head steel 208-litre (55-gallon drum) just off a city street. The officer directs you to a local shopkeeper who made the 911 call. The shopkeeper reports that the drum must have been illegally dropped off overnight, and he smelled an unusual odour when he walked by it. You notice a green label on the side of the drum reading "non-hazardous waste." You and the other members of the pump company are trained to the hazardous materials awareness level, and even though you are not tasked to respond to a hazardous materials incident, you find yourself on the scene of a potential hazardous materials incident.

1. What initial actions would you take based on your level of training?

2. What does the label on the drum signify?

3. How would you go about obtaining more information on the potential contents of the drum? What notifications would you make?

 *Access Navigate for more practice activities.*

Introduction

Transport Canada (TC) estimates that, in terms of tonnage, 70% of dangerous goods are transported by road, 24% by rail, and 6% by marine. A very small quantity of dangerous goods, accounting for less than 1%, is transported by air. The most common dangerous goods commodities transported in Canada are crude petroleum oil, gasoline, and fuel oils, representing 77% of all dangerous goods transported by road. Alberta, because of its oil industry, is the province with the highest volume of dangerous goods movement on the road. Per the **Chemical Abstracts Service (CAS)**, which produces the largest databases on chemical information, including the CAS Registry (https://www .cas.org), millions of organic and inorganic substances are registered for use in commerce, with several thousand new ones being introduced each year. The bulk of the new chemical substances are industrial chemicals, household cleaners, and lawn care products.

Chemicals are used and/or stored in warehouses, hospitals, laboratories, industrial occupancies, residential garages, bowling alleys, home improvement centres, garden supply stores, restaurants, and scores of other facilities or businesses in your response area. So many different chemicals exist in so many different locations that you could encounter almost anything during any type of incident.

Identifying the kinds and quantities of hazardous materials used and stored by local facilities should be an integral part of any comprehensive community response plan. Additionally, the authority having jurisdiction (AHJ) should develop plans and other methods to identify high-hazard occupancies and provide responders with guidance on response procedures and other standard operating procedures to handle hazardous materials incidents at those locations.

Once a chemical incident has occurred at a fixed facility, fire fighters should locate key personnel. Generally speaking, most fixed facilities that use and/or store a significant number of chemicals will have an Environmental Health and Safety (EH&S) department. In many cases, EH&S departments employ certified industrial hygienists, chemists and chemical engineers, and/or certified safety professionals to ensure safe work practices at the site. These industry experts may be a valuable source of information for understanding the chemical inventory of the facility, the ventilation systems, high-hazard chemical storage areas, and specialized areas such as clean rooms or refrigeration systems. Additionally, representatives of the EH&S department can connect the responders with site security, maintenance workers, **safety data sheets (SDSs)**, and other important employees at the facility.

It's important to find people with authorized access to storage locations, site knowledge, and knowledge about the building's design and layout. In many cases, facilities have written response plans with current contact information and/or on-duty representatives that can assist in the event of an emergency. Be prepared, however, to find outdated phone lists, personnel who are no longer in a particular position, inaccurate inventories, and the like that may hamper your ability to get timely and accurate information. Also, you may be speaking with a person who is not familiar with a substance and/or the process, or the area where a release occurred, and may not have all the answers you may be seeking when trying to determine a course of action.

Although it is not possible to know each chemical by name, it is possible to identify many of the more common commodities by way of available identification systems (placards, labels, and other signage) and/or using other methods (detection devices, eyewitness accounts, visible indicators) to identify the presence of a hazardous material. This chapter provides you with guidance on interpreting some of the visual clues that may signal the possible presence of a hazardous materials incident. Fire personnel must realize that hazardous materials incidents require a different approach from fire incidents. The approach taken at hazardous materials calls is slower and more methodical. It requires a great deal of discipline by not immediately taking action before obtaining as much information as possible. Situational awareness and critical thinking are essential factors in properly addressing these incidents.

Scene Size-Up

On any hazardous materials incident, your first action should always be to approach the scene from a safe location and direction. The traditional rules of staying uphill and upwind are a good place to start. If possible, it may make sense to use binoculars and view the scene from a safe distance, looking for labels, placards, container type, or other clues that could help you understand the scene. Be sure to question anyone involved in the incident—a wealth of information may be available to you if you simply ask the right person. Take enough time to assess the scene and interpret other clues, such as dead animals near the release, discoloured pavement, dead grass, visible vapours or puddles, or other indicators that may help identify the presence of a hazardous material. A wet area near unidentified containers on an asphalt parking lot may not initially seem significant, but perhaps it's a hot day, and the "wet" area still looks wet an hour after the containers were noticed, well past the point that water would have evaporated. Perhaps the substance is a hydrocarbon, and the wet-looking pavement is actually a solvent that has permeated into the asphalt (**FIGURE 30-1**).

A thorough and thoughtful size-up will help clarify the problem you are facing. Once you have a basic idea of what happened or have determined that danger may be present, you can begin to formulate a plan for addressing the incident. That plan begins with taking the basic actions known as SIN:

Safety
Isolate
Notify

FIGURE 30-1 Notice the details, and think about what they mean.
Courtesy of Rob Schnepp.

Safety

Scene size-up is important in any incident but especially during hazardous materials incidents. The ability to "read" the scene is a critical skill, and you must interpret the available clues and weave them together to make informed decisions and operate safely. It is not enough to simply scan the incident scene—you must train yourself to stop for a moment and pay attention.

> **LISTEN UP!**
>
> First-responding fire fighters, law enforcement personnel, or representatives of other allied agencies should place a high priority on identifying the released material and finding a reliable source of information about the chemical and physical properties of the released substance. The most appropriate source will depend on the situation; use your best judgment in making your selection.

Looking at something is nothing more than pointing your eyes in the right direction; seeing, by contrast, is taking in the visual clues and piecing them together to form a conclusion. This is the basis for situational awareness (**FIGURE 30-2**). The opening scenario, for example, contains a few clues. An open-head drum (the type where the entire lid is secured by a bolted clasp-type ring that circles the entire head of the drum) typically contains solid materials or those types of substances that cannot be poured or pumped easily. Additionally, steel drums don't usually contain corrosives. The label provides some preliminary information. Also, the shopkeeper mentioned detecting an odour. Each point does not tell the whole story but at least provides a place from which to start your size-up.

FIGURE 30-2 Pay attention. Situational awareness is key.
Courtesy of Rob Schnepp.

To that end, consider these initial first steps to ensure your own safety first:

- Stay upwind, uphill, and out of the problem. Responders must take the time to pay attention to their surroundings.
- Obtain a briefing from those involved in the incident prior to acting. (These individuals may include bystanders, law enforcement personnel, emergency medical services [EMS] responders, facility representatives, or other responders.)
- Understand the nature of the problem and the factors influencing the release. If you cannot understand the problem, then you cannot formulate a proper plan to address it.
- Attempt to make a positive identification of the released substance. If possible, obtain the correct SDS, shipping papers, the **Emergency Response Guidebook (ERG)**, or some other suitable reference source.

Isolate

After you have ensured your safety, the next step is to isolate and deny entry to the scene. This is typically accomplished by establishing a hot zone to identify the area of highest contamination and exclude accidental entry by untrained or unprotected responders or civilians. Your first priority (after ensuring your own safety) is to separate the people from the problem—life safety is always the first consideration. If people are involved (exposed or threatened by the release), do not move past this step until you have addressed the life safety issues. This could include removing affected people from the environment (evacuation); sheltering in place (leaving potential victims inside buildings, vehicles, etc.) until a transient problem, such as a fast-moving vapour cloud or other situation, passes; performing

decontamination; and/or rendering medical care. Isolation and denial of entry to a hazard zone may be accomplished by using law enforcement personnel to provide a physical presence, using scene control identifiers such as barrier tape that reads "Danger" or "Caution," or creating physical barriers with fences or other methods to deny access.

Standard operating procedures (SOPs), the emergency response plan, and the *ERG* will help you identify the various protective actions and notifications that must be made for the types of responses anticipated within the jurisdiction. SOPs should define points of contact for local, provincial, territorial, and federal resources that might be called upon for assistance during hazardous materials emergencies.

Establish a command post in an area where you are protected from the incident and the weather and where you have access to communications and technical reference materials. With hazardous materials incidents, it is important to establish clear and visible command.

Next, determine your response objectives, and begin to formulate a basic incident action plan (IAP). This plan must be carried out as safely as possible, should not involve contacting the released substance in any way, and above all must be well thought out.

Another isolation objective may be to identify and remotely secure potential ignition sources when flammable liquids and gases have been released. Common examples of ignition sources include open flames from pilot lights or other sources, arcs occurring when electrical switches are turned on or off, static electricity, and/or smoking materials.

Notify

Decide whether you need to notify anyone else—for example, other specialized responders, law enforcement, or other technical experts. You may also have to notify regulatory agencies such as Fisheries and Oceans Canada, the provincial or territorial office of emergency management, or county-level agencies. Have a current and comprehensive contact list of local, provincial, territorial, and federal resources available, and understand who the key players are in your jurisdiction. Large-scale incidents usually draw upon the resources of many agencies. Awareness level personnel should know the basic notification procedures for reporting hazardous materials emergencies and requesting assistance from local and regional authorities. The types of approved communications equipment include portable radios, mobile radios, and base station radios. It is not possible in this text to identify all types of communications equipment and procedures for making notifications. Therefore, awareness personnel should be familiar with all communications equipment, radio frequencies, and protocols for using the communications equipment provided by the AHJ.

In some cases, it may be possible to detect the presence of a hazardous materials incident based on information relayed in the initial dispatch, from persons on the scene, or based on your own knowledge of the response area. Clues that are seen or heard may also provide valuable information from a distance, enabling you to take precautionary steps. Vapour clouds at the scene, for example, are a signal to move yourself and others away to a place of safety; the sound of an alarm from a toxic gas sensor in a chemical storage room or laboratory may also serve as a warning to retreat. Some highly vapourous and odourous chemicals—chlorine and ammonia, for example—may be detected by smell a long way from the actual point of release. These and other clues may alert the awareness level hazardous materials personnel to the presence of a hazardous atmosphere.

In other cases, you may have to put on your detective hat and search for clues that may indicate whether a hazardous substance is present. At all times, departmental SOPs and your level of training, along with your information-gathering efforts at the scene, should guide your initial and ongoing actions. Use your senses, but do so carefully to avoid becoming contaminated or exposed. The senses that are typically safe to employ on a regular basis are those of sight and sound. Generally, the farther you are from the incident when you notice a problem, the safer you will be. Using any of your senses that bring you close to the chemical should be done with caution or should be avoided. When it comes to hazardous materials incidents, "leading with your nose" is not a good tactic—using binoculars from a distance is.

Containers

In basic terms, a **container** is any vessel or receptacle that holds a material. Often the container type, size, and material of construction provide important clues about the nature of the substance inside. Responders should not rely solely on the type of container when making a determination about hazardous materials, however, because there are numerous examples of finding substances in the wrong type of container. Red phosphorus from an illicit laboratory, for example, might be found in an unmarked plastic container. In this case, there may be no legitimate markings to alert a responder to the possible contents. Gasoline or waste solvents (from legitimate or illegitimate processes) may be stored in a 208-litre (55-gallon) steel **drum** with two capped openings (51 mm and 19 mm) [2 in. and ¾ in.] on the top (**FIGURE 30-3**). Sulfuric acid, at 97 percent concentration, could be found in a polyethylene drum that might be coloured black, red, white, or blue. In most cases, there is no correlation between the colour of the drum and the possible contents. The same sulfuric acid might also be found in a 4-litre (1-gallon) amber glass container. Hydrofluoric acid, by contrast, is incompatible with silica (glass) and would be stored in a plastic container. Steel or polyethylene drums, bags, high-pressure gas cylinders, railroad tank cars, plastic buckets, aboveground and underground storage tanks, cargo tanks, and pipelines are all representative examples of how hazardous materials are used, stored, and shipped (**FIGURE 30-4**).

Some very recognizable chemical containers, such as 208-litre (55-gallon) drums and compressed gas cylinders, can be found in almost every type of manufacturing facility. Stainless steel containers may hold particularly dangerous chemicals, and cold liquids are

FIGURE 30-3 An abandoned steel drum. Notice the configuration of the holes on top.
Courtesy of Rob Schnepp.

FIGURE 30-4 Drums may be constructed of many different types of materials, including cardboard, polyethylene, or stainless steel.
Courtesy of Rob Schnepp.

kept in a Thermos-like **Dewar container** designed to maintain the appropriate temperature (**FIGURE 30-5**).

In any case, it is important to look closely at a container and form an opinion about the material inside. The following section illustrates a few container types that awareness level personnel should be familiar with. For more information on other types of containers, see Chapter 32, *Understanding the Hazards*, in Section 4 of this book.

Drums

Drums are easily recognizable, barrel-like containers. They are used to store a wide variety of substances, including food-grade materials, corrosives, flammable liquids, and grease. Drums may be constructed of low-carbon steel, polyethylene, cardboard, stainless steel, nickel, or other materials. Generally, the nature of the chemical dictates the construction of the storage drum. Steel utility drums, for example, hold flammable liquids, cleaning fluids, oil, and other noncorrosive chemicals. Polyethylene drums are used for corrosives such as acids, bases, oxidizers, and other materials that cannot be stored in steel containers. Cardboard drums hold solid materials such as soap flakes, sodium hydroxide pellets, and food-grade materials. Stainless steel or other heavy-duty drums generally hold materials too aggressive (i.e., too reactive) for either plain steel or polyethylene.

Closed-head drums have a permanently attached lid with one or more small openings. The opening is called a **bung**. Typically, these openings are threaded holes sealed by caps that can be removed only by using a special tool called a bung wrench (**FIGURE 30-6**).

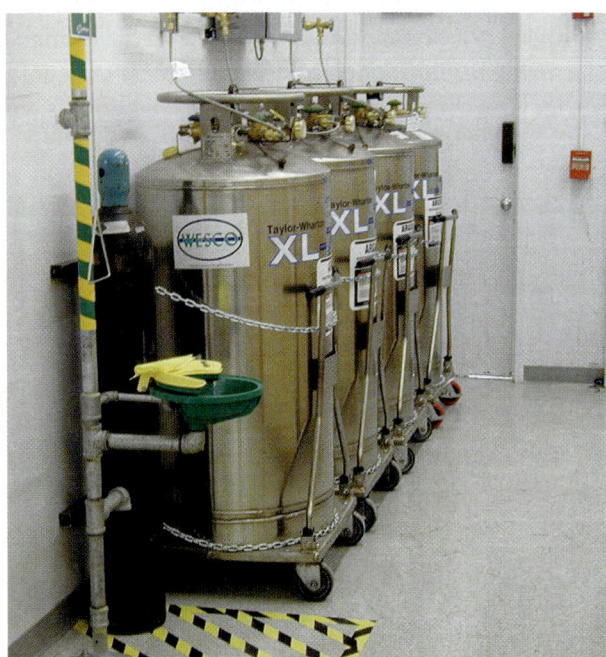

FIGURE 30-5 A series of Dewar containers stored adjacent to a compressed gas cylinder.
Courtesy of Rob Schnepp.

FIGURE 30-6 A bung wrench is used to operate the openings on the top of a closed-head drum.
© Jones and Bartlett Learning. Photographed by Glen E. Ellman.

Closed-head drums usually have one 51-mm (2-in.) bung and one 19-mm (¾-in.) bung. The larger bung is used to pump product from the drum; the smaller bung functions as a vent.

An open-head drum has a removable lid fastened to the drum with a ring (**FIGURE 30-7**). The ring is tightened with a clasp or a threaded nut-and-bolt assembly. These containers typically contain a product in solid form. This is an example of the type of drum described in the opening scenario.

Carboys

Some corrosives and other types of chemicals are transported and stored in a vessel called a **carboy** (**FIGURE 30-8**). A carboy is a glass, plastic, or steel container that holds 19 to 57 litres (5 to 15 gallons) of product. Glass carboys are often placed in a protective wood, foam, fibreglass, or steel box to help prevent breakage. For example, nitric acid, sulfuric acid, and other strong acids are often transported and stored in thick glass carboys protected by a wooden or Styrofoam crate to shield the glass container from damage during normal shipping.

Cylinders

Several types of **cylinders** are used to hold liquids and gases. Uninsulated compressed gas cylinders are used to store substances such as nitrogen, argon, helium, and oxygen (**FIGURE 30-9**). They come in a range

FIGURE 30-8 A carboy is used to transport and store corrosive chemicals.
Courtesy of EMD Chemicals, Inc.

FIGURE 30-9 Compressed gas cylinders in storage—what clues do you see?
Courtesy of Rob Schnepp.

FIGURE 30-7 An open-head drum has a lid that is fastened with a ring that is tightened with a clasp or a nut-and-bolt assembly.
Courtesy of Globalindustrial.com.

of sizes and have variable internal pressures. An oxygen cylinder used for medical purposes, for example, has a pressure reading of approximately 13,790 kPa (2000 psi) when full. By comparison, the very large compressed gas cylinders found at a fixed facility may have pressure readings of 34,474 kPa (5000 psi) or greater.

The high pressures exerted by these cylinders create a potential for danger. If the cylinder is punctured, the valve assembly fails, or the cylinder falls over and damages the valve, causing a rapid release of compressed gas, it will turn the cylinder into an unpredictable missile. In addition, if the cylinder is heated rapidly, it could explode with tremendous force, spewing product and metal fragments over long distances. Compressed gas cylinders have pressure-relief valves, but those valves may not be sufficient to relieve the pressure created during a fast-growing fire (**FIGURE 30-10**).

A propane cylinder is another type of compressed gas cylinder. Propane cylinders have lower pressures (1379–2068 kPa [200–300 psi]) and contain a liquefied gas. Liquefied gases such as propane are subject to the phenomenon known as BLEVE (boiling liquid/expanding vapour explosion). BLEVEs occur when pressurized liquefied materials (propane or butane, for example) inside a closed vessel are exposed to a source of high heat (**FIGURE 30-11**).

The low-pressure Dewar container is another commonly encountered cylinder type. As mentioned previously, Dewars are Thermos-like vessels designed to hold **cryogenic liquids (cryogens)** such as helium, liquid nitrogen, and liquid argon (**FIGURE 30-12**).

Typical cryogens include oxygen, helium, hydrogen, argon, and nitrogen. Under normal atmospheric conditions, each of these substances is a gas. A complex

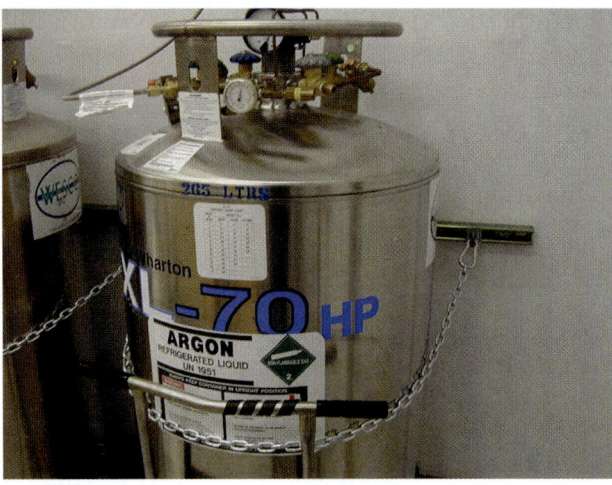

FIGURE 30-12 A small cryogenic Dewar container.
Courtesy of Cryofab, Inc.

FIGURE 30-10 Compressed gas cylinder failure due to heat generated by fire.
Courtesy of Rob Schnepp.

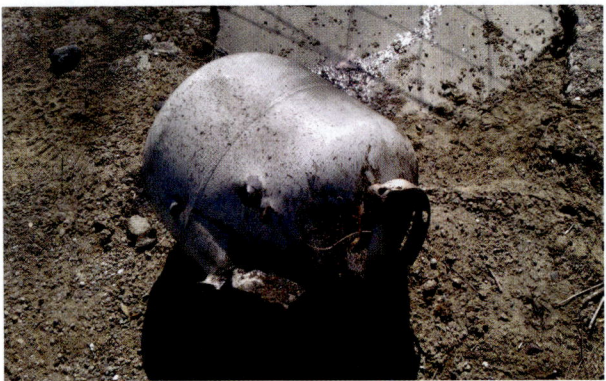

FIGURE 30-11 Small propane cylinder failure due to fire.
Courtesy of Rob Schnepp.

process turns them into liquids that can be stored and used for long periods of time. Nitrogen, for example, becomes a liquid at –160°C (–320°F) and must be kept at that temperature if it is to remain in a liquid state.

Cryogens pose a substantial threat if the Dewar container fails to maintain the low temperature of the cryogenic liquid. Cryogens have large expansion ratios—even larger than the expansion ratio of propane (270:1). Cryogenic helium, for example, has an expansion ratio of approximately 750:1. If one volume of liquid helium is warmed to room temperature and vaporized in a totally enclosed container, it can generate a pressure of more than 99,974 kPa (14,500 psi). To counter this possibility, cryogenic containers usually have two pressure-relief devices: a pressure-relief valve and a frangible (easily broken) metal disk.

SAFETY TIP

Cryogens are stored in a liquid state. Beware of skin exposures! Significant injuries, like those associated with thermal burns, can occur when skin meets one of these liquids.

Transportation and Facility Markings

Markings on buildings, packages, boxes, and containers enable responders to identify a released chemical. When used correctly, marking systems indicate the presence of a hazardous material and provide clues about the substance. Marine pollutants and environmentally hazardous substances, for example, pose a risk to aquatic life and the marine ecosystem. Transportation markings denoting those substances may not be seen often, especially in areas without significant bodies of

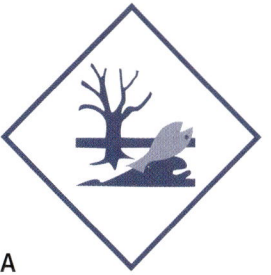

A B C D

FIGURE 30-13 A. Environmental hazard. **B.** Inhalation hazard. **C.** Elevated temperature. **D.** Commodity.
Courtesy of the U.S. Department of Transportation.

water. Other transportation markings such as elevated temperature materials (asphalt and molten sulfur or other liquids transported above 100°C or 240°C [212°F or 464°F] for solids, for example), consumer commodities intended for retail sale, and inhalation hazards are additional markings you may find in your jurisdiction. Examples of those markings can be found in **FIGURE 30-13**.

Safety Data Sheets/Material Safety Data Sheets (MSDSs)

A common source of information about a chemical is the SDS (MSDS) specific to that substance (**FIGURE 30-14**). Essentially, an SDS provides basic information about the chemical make-up of a substance, the potential hazards it presents, appropriate first aid in the event of an exposure, and other pertinent data for safe handling of the material. An SDS will typically include the following details:

- The name of the chemical, including any synonyms
- Physical and chemical characteristics of the material
- Physical hazards of the material
- Health hazards of the material
- Signs and symptoms of exposure
- Routes of entry

LISTEN UP!

The local emergency planning committee (LEPC) is likely to have a substance's SDS. Formerly referred to as a material safety data sheet (MSDS), the term was updated in an effort to standardize terminology worldwide by way of the Globally Harmonized System of Classification and Labelling of Chemicals (GHS). In short, the GHS intends to create a standard methodology to define and classify hazards posed by chemical substances along with a standard method to communicate the hazards—the SDS. The intent of the document remains the same.

- Permissible exposure limits
- Responsible-party contact
- Precautions for safe handling (including hygiene practices, protective measures, and procedures for cleaning up spills or leaks)
- Applicable control measures, including personal protective equipment (PPE)
- Emergency and first-aid procedures
- Appropriate waste disposal

When responding to a hazardous materials incident at a fixed facility, responders should ask the site representative for an SDS for the spilled material. All facilities that use or store chemicals are required by law to have an SDS on file for each chemical used or stored in the facility. Many sites, but especially those that stock many different chemicals, may keep this information archived in a computer database. Although the SDS is not a definitive response tool, it is a key piece of the puzzle. Responders should investigate as many sources as possible (preferably at least three) to gather information about a released substance. An SDS can also be obtained from staffed national resource centres or on the transporting vehicle.

The National Fire Protection Association 704 Marking System

The National Fire Protection Association (NFPA) has developed its own system for identifying hazardous materials. NFPA 704, *Standard System for the Identification of the Hazards of Materials for Emergency Response*, outlines a marking system characterized by a set of diamonds that are found on the outside of buildings, on doorways to chemical storage areas, and on fixed storage tanks. This marking system is designed for fixed-facility use. Responders can use the NFPA diamonds to understand the broad hazards posed by chemicals stored in a building or part of a building.

The **NFPA 704 hazard identification system** uses a diamond-shaped symbol of any size, which is itself broken into four smaller diamonds, each representing

SAFETY DATA SHEET

Section 1. Identification	
Product Name:	**Ammonia, Anhydrous**
Synonyms:	Ammonia
CAS REGISTRY NO:	7664-41-7
Supplier:	Tanner Industries, Inc. 735 Davisville Road, Third Floor Southampton, PA 18966
Website:	www.tannerind.com
Telephone (General):	215-322-1238
Corporate Emergency Telephone Number:	**800-643-6226**
Emergency Telephone Number:	**Chemtrec: 800-424-9300**
Recommended Use:	Various Industrial/Agricultural

Section 2. Hazard(s) Identification

Hazard: Acute Toxicity, Corrosive, Gases Under Pressure, Flammable Gas, Acute Aquatic Toxicity

Classification:
Acute Toxicity, Inhalation (Category 4)   Note: (1 - Most Severe/4 - Least Severe)
Skin Corrosion/Irritation (Category 1B)
Serious Eye Damage/Irritation (Category 1)
Gases Under Pressure (Liquefied gas)
Flammable Gases (Category 2)
Acute Aquatic Toxicity (Category 1)

Pictogram:

Signal word: **Danger**

Hazard statements:
Harmful if inhaled.
Causes severe skin burns and serious eye damage.
Flammable gas.
Contains gas under pressure; may explode if heated.
Very toxic to aquatic life.

Precautionary statements:
Avoid breathing gas/vapors.
Use only outdoors or in well-ventilated area.
Wear protective gloves, protective clothing, eye protection, face protection.
Keep away from heat, sparks, open flames and other ignition sources. No smoking.

FIGURE 30-14 An example of an SDS for anhydrous ammonia (first page only).
Courtesy of Tanner Industries, Inc., Southhampton, PA.

a property or characteristic of a substance or group of substances (**FIGURE 30-15**). The blue, red, and yellow diamonds each contain a numerical rating in the range of 0–4, with 0 being the least hazardous and 4 being the most hazardous (**TABLE 30-1**).

The blue diamond (at the nine o'clock position) indicates the health hazard posed by a material alone or perhaps within a group of other chemicals. Responders must understand that when an NFPA diamond represents a series of hazards posed by several different substances, the most severe characteristic of any of the substances may be used to represent the hazard within any of the four coloured diamonds. For example, if any one of the substances in a grouping of chemicals could be fatally toxic, that single substance causes a 4 to appear in the blue diamond. All other substances could be much less hazardous, but the one causing the 4 represents the health hazard for the group.

The same logic holds true for the flammability and reactivity diamonds. The top red diamond indicates flammability. The yellow diamond (at the three o'clock position) indicates instability. The bottom white diamond

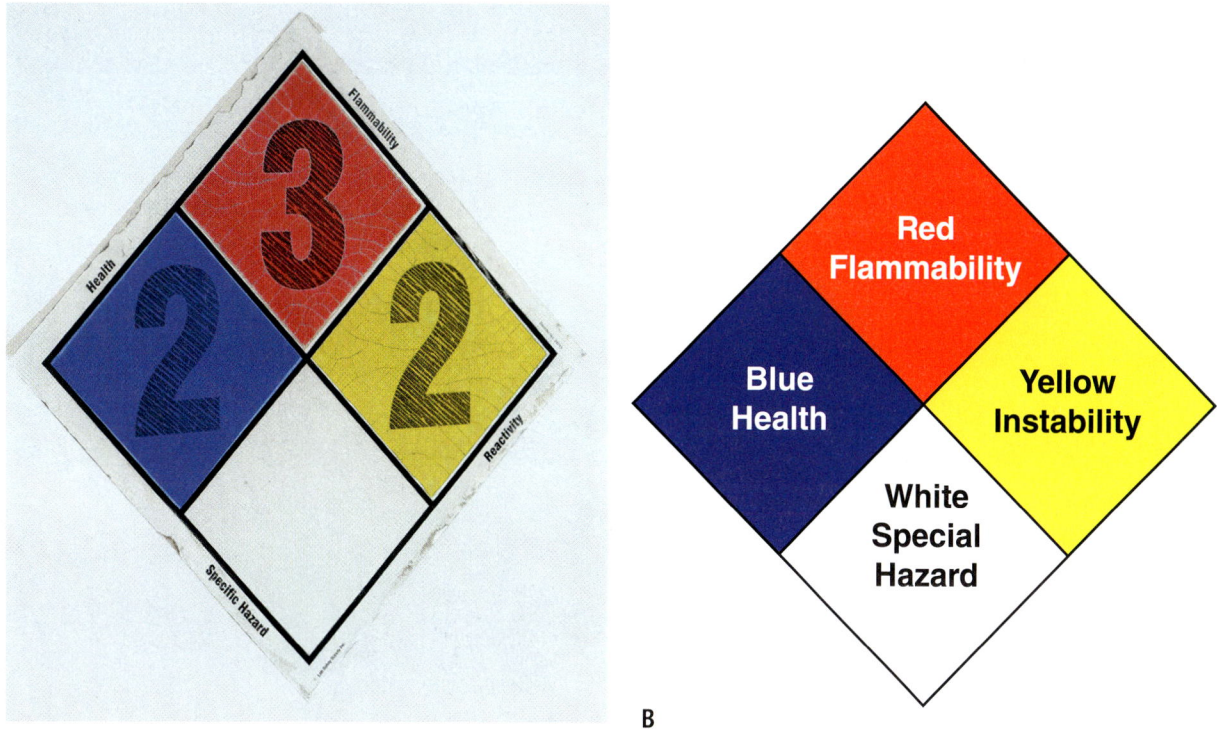

A   **B**

FIGURE 30-15 A. Example of a placard using the NFPA 704 hazard identification system that is for fixed-facility use. **B.** Each colour used in the diamond represents a particular property or characteristic.

A. © Jones & Bartlett Learning. Photographed by Glen E. Ellman; **B.** © Jones & Bartlett Learning.

TABLE 30-1 Hazard Levels in the NFPA Hazard Identification System			
Flammability Hazards (Red Diamond)		**Instability Hazards (Yellow Diamond)**	
4	Materials that will rapidly or completely vaporize at atmospheric pressure and normal ambient temperature or that are readily dispersed in air and that will burn readily. Liquids with a flash point below 22°C (73°F) and a boiling point below 38°C (100°F).	4	Materials that in themselves are readily capable of detonation or of explosive decomposition or reaction at normal temperatures and pressures.
3	Liquids and solids that can be ignited under almost all ambient temperature conditions. Liquids with a flash point below 22°C (73°F) and a boiling point above 38°C (100°F) or liquids with a flash point above 22°C (73°F) but not exceeding 38°C (100°F) and a boiling point below 38°C (100°F).	3	Materials that in themselves are capable of detonation or explosive decomposition or reaction but require a strong initiating source, or that must be heated under confinement before initiation, or that react explosively with water.
2	Materials that must be moderately heated or exposed to relatively high ambient temperatures before ignition can occur. Liquids with a flash point above 38°C (100°F) but not exceeding 93°C (200°F).	2	Materials that readily undergo violent chemical change at elevated temperatures and pressures, or that react violently with water, or that may form explosive mixtures with water.

(continued)

TABLE 30-1 Hazard Levels in the NFPA Hazard Identification System *(continued)*

Flammability Hazards (Red Diamond)

1	Materials that must be preheated before ignition can occur. Liquids that have a flash point above 93°C (200°F).
0	Materials that will not burn.

Instability Hazards (Yellow Diamond)

1	Materials that in themselves are normally stable but can become unstable at elevated temperatures and pressures.
0	Materials that in themselves are normally stable, even under fire exposure conditions, and are not reactive with water.

Health Hazards (Blue Diamond)

4	Materials that on very short exposure could cause death or major residual injury.
3	Materials that on short exposure could cause serious temporary or residual injury.
2	Materials that on intense or continued, but not chronic, exposure could cause incapacitation or possible residual injury.
1	Materials that on exposure would cause irritation but only minor residual injury.
0	Materials that on exposure under fire conditions would offer no hazard beyond that of ordinary combustible material.

Special Hazard (White Diamond)

ACID Acid
ALK Alkali
COR Corrosive
OX Oxidizer
W̶ Reacts with water
☢ Radioactivity

© Jones & Bartlett Learning.

will not have a number but may contain special symbols. Among examples of the symbols used are a burning *OX* (oxidizing capability), *COR* (corrosive), a three-bladed trefoil (radioactivity), and a *W* with a slash through it (water reactive). For complete information on the NFPA 704 system, consult NFPA 704.

Hazardous Materials Information System

Since 1988, the **Workplace Hazardous Materials Information System (WHMIS)** hazard communication program has helped employers comply with health and safety information on hazardous products intended for use, handling, or storage in Canadian workplaces. The WHMIS is like the NFPA 704 marking system and uses a numerical hazard rating with similarly coloured horizontal columns.

The WHMIS is more than just a label; it is a method used by employers to give their personnel necessary information to work safely around chemicals and includes training materials to inform workers of chemical hazards in the workplace. The WHMIS is the law. In addition to describing the chemical hazards posed by a substance, the WHMIS provides guidance about the PPE that employees need to use to protect themselves from workplace hazards. Letters and icons specify the different levels and combinations of protective equipment.

Responders must understand the fundamental difference between the NFPA 704 marking system and the WHMIS. NFPA 704 is intended for responders; the WHMIS is intended for the employees of a facility. Although the WHMIS is not a response information tool, it can give clues about the presence and nature of the hazardous materials found in the facility.

Military Hazardous Materials/ Weapons of Mass Destruction Markings

The Canadian military has developed its own marking system for hazardous materials. The military system serves primarily to identify detonation, fire, and special hazards.

In general, hazardous materials within the military marking system are divided into four categories based on the relative detonation and fire hazards:

- Division 1 materials are considered mass detonation hazards and are identified by a number 1 printed inside an orange octagon (**FIGURE 30-16A**).
- Division 2 materials have explosion-with-fragment hazards and are identified by a number 2 printed inside an orange X (**FIGURE 30-16B**).
- Division 3 materials are mass fire hazards and are identified by a number 3 printed inside an inverted orange triangle (**FIGURE 30-16C**).
- Division 4 materials are moderate fire hazards and are identified by a number 4 printed inside an orange diamond (**FIGURE 30-16D**).

Chemical hazards in the military system are depicted by colours. Toxic agents (such as sarin or mustard) are identified by the colour red. Harassing agents (such as tear gas and smoke producers) are identified by yellow. White phosphorus is identified by white. Specific personal protective gear requirements are identified using pictograms. Military shipments containing hazardous materials/WMD are not required, by exception, to be placarded.

Shipping Papers

Shipping papers are required whenever materials are transported from one place to another. They include the names and addresses of the shipper and the receiver, identify the material being shipped, and specify the quantity and weight of each part of the shipment. Additionally, shipping papers allow the reader to match the chemical name found on the shipping papers with the mode of transportation. Shipping papers for road and highway transportation are called a bill of lading or freight bill and are located in the cab of the vehicle (**FIGURE 30-17**). Drivers transporting chemicals are required by law to have a set of shipping papers on their person or within easy reach inside the cab at all times.

A bill of lading may provide additional information about a hazardous substance, such as its packaging group designation. The packaging group designation is another system used by shippers to identify special handling requirements or hazards. Some TC hazard classes require shippers to assign packaging groups based on the material's flash point and toxicity. A packaging group designation may signal that the material poses a greater hazard than similar materials in a hazard class. There are three packaging group designations:

- *Packaging group I*: High danger
- *Packaging group II*: Medium danger
- *Packaging group III*: Minor danger

Shipping papers for railroad transportation are called waybills (**FIGURE 30-18**). A list of the contents in every car on the train is called the consist or train list. The conductor, engineer, or a designated member of the train crew will have a copy of the consist in mainline use. If the incident happens in a railyard, then you may find a waybill as well.

On a marine vessel, shipping papers are called the dangerous cargo manifest (**FIGURE 30-19**). The manifest is generally kept in a tube-like container in the wheelhouse, in the custody of the captain or master.

For air transport, the air bill is the shipping paper (**FIGURE 30-20**). It is kept in the cockpit and is the pilot's responsibility.

Pipelines

Of all the various methods used to transport hazardous materials, the high-volume pipeline is the one that is most rarely involved in emergencies. A pipeline is a length of

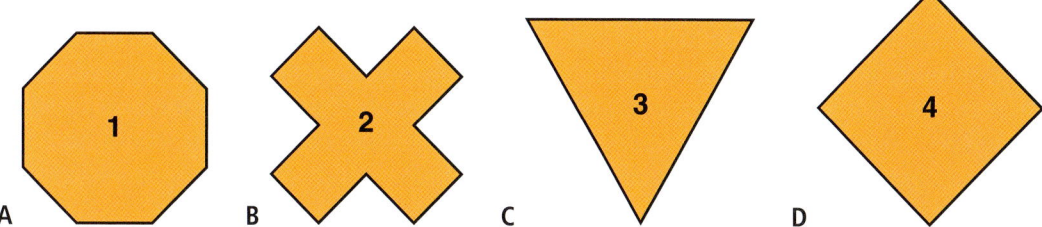

FIGURE 30-16 A. Mass detonation hazards. **B.** Explosion-with-fragment hazards. **C.** Mass fire hazards. **D.** Moderate fire hazards.

STRAIGHT BILL OF LADING
ORIGINAL - NOT NEGOTIABLE

BOL/Reference No.
RSI82715

CARRIER: NORFOLK SOUTHERN   Date: 12/23/2008

Shipper: RSI LOGISTICS, INC (OKEMOS, MI US)

The property described below, in apparent good order, except as noted (contents and condition of packages unknown), marked, consigned, and destined as indicated below, which said carrier (the word carrier being understood throughout this contract as meaning any person or corporation in possession of the property under the contract) agrees to carry to its usual place of delivery at said destination, if on its route, otherwise to deliver to another carrier on the route to said destination. It is mutually agreed, as to each carrier of all or any said property, that every service to be performed hereunder shall be subject to all the terms and conditions of the Uniform Domestic Straight Bill of Lading set forth (1) in Official, Southern, Western and Illinois Freight Classification in effect on the date hereof, if this is a rail or a rail-water shipment, or (2) in the applicable motor carrier classification or tariff if this is a motor carrier shipment

  Shipper hereby certifies that he is familiar with all the terms and conditions or the said bill of lading, including those on the back thereof, set forth in the classification or tariff which governs the transportation of this shipment, and the said terms and conditions are hereby agreed to by the shipper and accepted for himself and his assigns.

Consignee Information: CONSIGNEE DEER PARK, TX Address: City: DEER PARK, TX US	
Route: NS-ESTL-BNSF	
Origin Switch Route:	
Destination Switch Route: HUSTN-PTRA	Rail Car No: GATX290861

For assistance in any transportation emergency involving chemicals, phone CHEMTREC, day or night, Toll Free 1-800-424-9300

DESCRIPTION	*WEIGHT
ONE TANK CAR	Contains: Methyl Esters STCC#2899415 BIODIESEL-15, Biodiesel Sales Order Contract No: RSI82715 Sales Order Contract No: AAT122308-4 Purchase Order Contract No: AAT122308-4	(Sub. To Correction) 204400 Lbs.

SEAL NUMBERS: Gross
 	
                             Tare
 	
                             Net
 	
                             Weighed By: _____	

If charges are to be prepaid, write or stamp here, "To be Prepaid"
Prepaid

Subject to Section 7 of the conditions of applicable bill of lading, if this shipment is to be delivered to the consignee without recourse on the consignor, the consignor shall sign the following statement: *The carrier shall not make delivery of this shipment without payment of freight and all other lawful charges.*

Not In Effect

* This is to certify that the above named materials are properly classified, described, packaged, marked, and labeled, and are in proper condition for transportation, according to the applicable regulations of the Department of Transportation.

FIGURE 30-17 A bill of lading or freight bill.

Courtesy of RSI Logistics, Inc.

pipe—including pumps, valves, flanges, control devices, strainers, and/or similar equipment—for conveying fluids and gases over potentially long distances. Like rail incidents, pipeline incidents may present responders with challenges and hazards not typically encountered at most hazardous materials incidents. In many areas, large-diameter pipelines transport natural gas, gasoline, diesel fuel, and other products from delivery terminals to distribution facilities. Pipelines are often buried underground but may be aboveground in remote areas.

Additionally, subject matter experts from the company that owns the pipeline may be required to assist hazardous materials responders for the local jurisdiction. These incidents, like rail incidents, could have far-reaching implications and present responders with a challenging set of circumstances.

The **pipeline right-of-way** is an area, patch of land, or roadway that extends a certain number of metres on either side of the pipe itself. The company that owns the pipeline maintains this area. The company is also responsible for

WAYBILL
NON-NEGOTIABLE

WAYBILL NO.

SHIP DATE:

SHIPPER NUMBER	SHIPPER REFERENCE NUMBER	CONSIGNEE REFERENCE NUMBER	P.O. NUMBER

SHIPPER	CONSIGNEE
STREET ADDRESS	STREET ADDRESS
STREET ADDRESS	STREET ADDRESS
CITY, STATE AND ZIP CODE	CITY, STATE AND ZIP CODE

CONTACT	PHONE NUMBER	CONTACT	PHONE NUMBER

3rd PARTY NUMBER	
3rd PARTY NAME	SHIPPING CO. LIABILITY IS LIMITED TO $50 PER SHIPMENT OR 50 CENTS PER POUND (U.S. DOLLARS), WHICHEVER IS HIGHER SUBJECT TO A MAXIMUM LIABILITY OF $25,000, UNLESS A HIGHER VALUE IS DECLARED AND APPLICABLE CHARGES (FOR DECLARING A VALUE ON WAYBILL) ARE PAID PRIOR TO SHIPPING (SUBJECT TO THE TERMS AND CONDITIONS ON REVERSE SIDE, AND THE SERVICE CONDITIONS FOUND IN THE SHIPPING CO. SERVICE CONDITIONS POLICY).
STREET ADDRESS	
CITY, STATE AND ZIP CODE	DECLARED VALUE
CONTACT NAME	$

(SUBJECT TO CORRECTION)

NO. OF PIECES	TYPE	HM	KIND OF PACKAGE, DESCRIPTION OF ARTICLES, SPECIAL MARKS & EXCEPTIONS	ACTUAL WEIGHT	LENGTH	WIDTH	HEIGHT

Special Instructions:

☐ SATURDAY DELIVERY ☐ SUNDAY DELIVERY ☐ APPOINTMENT DELIVERY ☐ INSIDE DELIVERY ☐ RESIDENTIAL DELIVERY

Services Requested:

☐ SAME DAY/ NEXT FLIGHT OUT ☐ NEXT DAY AM ☐ NEXT DAY PM ☐ 2ND DAY ☐ ECONOMY DEFERRED (3-5 DAYS)

FOR CHARTER AIR, TIME DEFINITE, OR GUARANTEED SERVICE, PLEASE CALL 1-800-000-XXXX FOR AVAILABILITY.

QUOTE NUMBER	DIM WEIGHT

SHIPPER CERTIFICATION: Shipper certifies by its signature, its agreement to all of the foregoing terms and conditions, and further certifies that the above named materials are properly classified, described, packaged, marked and labeled, and are in proper condition for transportation according to the applicable regulations of the DOT.

SHIPPER REPRESENTATIVE

SIGNATURE X _____ Print Name X _____ Date _____

PICKED UP BY:	RECEIVED BY:	RECEIVED BY CONSIGNEE IN GOOD ORDER UNLESS NOTED BELOW:
DRIVER SIGNATURE _____	CONSIGNEE SIGNATURE _____	# S/W SKIDS DEL'D INTACT _____
PLEASE PRINT _____	PLEASE PRINT _____	# SKIDS DEL'D: _____
COMPANY _____	COMPANY _____	☐ GOOD ORDER ☐ SHORT ☐ OVER ☐ DAMAGED
DATE _____	DATE _____	DESCRIBE EXCEPTIONS:
TIME _____	TIME _____	

All rules as contained in Shipping Co. Services Conditions Policy will apply. Terms and conditions stated on any Bill of Lading used to transfer goods for carriage, other than an Shipping Co. Waybill, will be null and void. Quotes are based on the information provided and are only an estimate. Final charges are based on actual shipment pieces, weight, dimensions, and services performed as a requirement for delivery. Any changes in actual shipment details will affect the final charges.

Shipping Co. is a certified participant in compliance with the Transportation Security Administration Regulations, Part 109, a Federal Security program.

1 - SHIPPER'S COPY

FIGURE 30-18 A waybill.

Courtesy of private source.

CARGO MANIFEST

AIR	AIRCRAFT DATA: CARRIER, A/C NO., A/C MODEL	DEST CODE	REF	DESTINATION	MISSION DATA: NO., SU, DATE	ALW WT	ALW CU	MANIFEST ID: STA, FY, TY, NO.	PAGE NO.	
SUR-FACE	POE	DATE SAILED	VOYAGE DOCUMENT NO.	POD	REF	VESSEL NAME	STATUS, SUST	TRUCK NO.	REMARKS	PAGE NO.
DOC ID	VEHICLE TRAILER OR CNTNR NUMBER	YR, MAKE, CNTNR NUMBER, COMMODTY DESCRIP	COM CODE	CAR-GO EXC	VOYAGE DOC NO., AIRDM, TRNS PT	PORT OF DISCH	TYPE PACK	TRANSPORTATION CONTROL NUMBER	CONSIGNEE, SERV, ACTVTY ADDRESS, PRIORITY	NAME, AMMO LOT NO./NOMEN, DIMENSIONS, RDD, PROJ, STOW LOC, TRANS ACCT

IDENTIFICATION NO. OR REMARKS	PIECES	WEIGHT	CUBE
TOTALS	0	0.00	0.00

ITEMS HAVE BEEN LOADED: DATE | SIGNATURE OF LOADING AGENT

ITEMS HAVE BEEN RECEIVED EXCEPT AS CIRCLED NOTED ON REVERSE SIDE: DATE | SIGNATURE OF UNLOADING AGENT | DATE | SIGNATURE OF RECEIVING AGENT

DD FORM 1385, NOV 78 REPLACES EDITION OF 1 APR 66 WHICH MAY BE USED Adobe Professional 7.0

FIGURE 30-19 A dangerous cargo manifest.

Courtesy of the U.S. Department of Defense.

placing warning signs at regular intervals along the length of the pipeline. Pipeline warning signs include a warning symbol, the pipeline owner's name, and an emergency contact phone number (**FIGURE 30-21**).

Again, pipeline emergencies are complicated events that require specially trained responders. If you suspect an incident involving a pipeline, contact the owner of the line immediately. The company will dispatch a crew to assist with the incident.

Information about the pipe's contents and owner is also often found at the **vent pipes**. These inverted J-shaped tubes provide pressure relief or natural venting during maintenance and repairs. Vent pipes are clearly marked and are located approximately 1 metre (3 feet) above the ground.

Transport Canada Marking System

Transport Canada's marking system employ a system of coded labels and placards similar to the U.S. Department of Transportation (DOT) marking system to identify dangerous goods in transit. The *ERG*—developed jointly by Transport Canada, the DOT, and Mexico's Secretaria de Communicaciones y Transportes—is also part of this system.

The *Emergency Response Guidebook*

The *ERG* offers a certain amount of guidance for responders operating at a hazardous materials incident (**FIGURE 30-22**). This guide, which is intended to help responders decide which preliminary action to take, provides information on several thousand chemicals. (As of the writing of this text, an online version of the *ERG* is also available.) The *ERG* does not list all chemicals that could be shipped by land, sea, air, or rail. In most cases, the package or cargo tank must contain a certain amount of hazardous material before a placard is required. For example, the "1000-pound rule" applies to some explosives, flammable and nonflammable gases, flammable/combustible liquids, flammable solids, air-reactive solids, oxidizers and organic peroxides, poison solids, corrosives, and miscellaneous (class 9) materials. Placards are required for these materials only when the shipment weighs more than 1000 pounds. (More information

Set your tabulator stops here

STAPLE DOCUMENTS ABOVE PERFORATION

← Line-up here →

| Shipper's Name and Address | Shipper's Account Number | Not Negotiable |

Air Waybill

Issued by

Copies 1, 2 and 3 of this Air Waybill are originals and have the same validity.

| Consignee's Name and Address | Consignee's Account Number |

It is agreed that the goods described herein are accepted in apparent good order and condition (except as noted) for carriage SUBJECT TO THE CONDITIONS OF CONTRACT ON THE REVERSE HEREOF. ALL GOODS MAY BE CARRIED BY ANY OTHER MEANS INCLUDING ROAD OR ANY OTHER CARRIER UNLESS SPECIFIC CONTRARY INSTRUCTIONS ARE GIVEN HEREON BY THE SHIPPER, AND SHIPPER AGREES THAT THE SHIPMENT MAY BE CARRIED VIA INTERMEDIATE STOPPING PLACES WHICH THE CARRIER DEEMS APPROPRIATE. THE SHIPPER'S ATTENTION IS DRAWN TO THE NOTICE CONCERNING CARRIER'S LIMITATION OF LIABILITY. Shipper may increase such limitation of liability by declaring a higher value for carriage and paying a supplemental charge if required.

| Issuing Carrier's Agent Name and City | Accounting Information |

| Agent's IATA Code | Account No. |

| Airport of Departure (Addr. of First Carrier) and Requested Routing | Reference Number | Optional Shipping Information |

| To | By First Carrier | Routing and Destination | to | by | to | by | Currency | CHGS Code | WT/VAL PPD COLL | Other PPD COLL | Declared Value for Carriage | Declared Value for Customs |

| Airport of Destination | Requested Flight/Date | Amount of Insurance | INSURANCE - If carrier offers insurance, and such insurance is requested in accordance with the conditions thereof, indicate amount to be insured in figures in box marked "Amount of Insurance". |

Handling Information

These commodities, technology or software were exported from the United States in accordance with the Export Administration Regulations. Ultimate destination

Diversion contrary to U.S. law prohibited.

SCI

No. of Pieces RCP	Gross Weight	kg lb	Rate Class / Commodity Item No.	Chargeable Weight	Rate / Charge	Total	Nature and Quantity of Goods (incl. Dimensions or Volume)

| Prepaid | Weight Charge | Collect | Other Charges |

Valuation Charge

Tax

Total Other Charges Due Agent

Shipper certifies that the particulars on the face hereof are correct and that **insofar as any part of the consignment contains dangerous goods, such part is properly described by name and is in proper condition for carriage by air according to the applicable Dangerous Goods Regulations.**

Total Other Charges Due Carrier

Signature of Shipper or his Agent

| Total Prepaid | Total Collect |

| Currency Conversion Rates | CC Charges in Dest. Currency |

Executed on (date)   at (place)   Signature of Issuing Carrier or its Agent

| For Carriers Use only at Destination | Charges at Destination | Total Collect Charges |

APPERSON K0419 (10/03) WHSE. #05640

FIGURE 30-20 An air bill.

Courtesy of Apperson Print Resources Inc.

FIGURE 30-21 A pipeline warning sign provides information about the pipe's contents, the owner's name, and contact information.

© Photodisc.

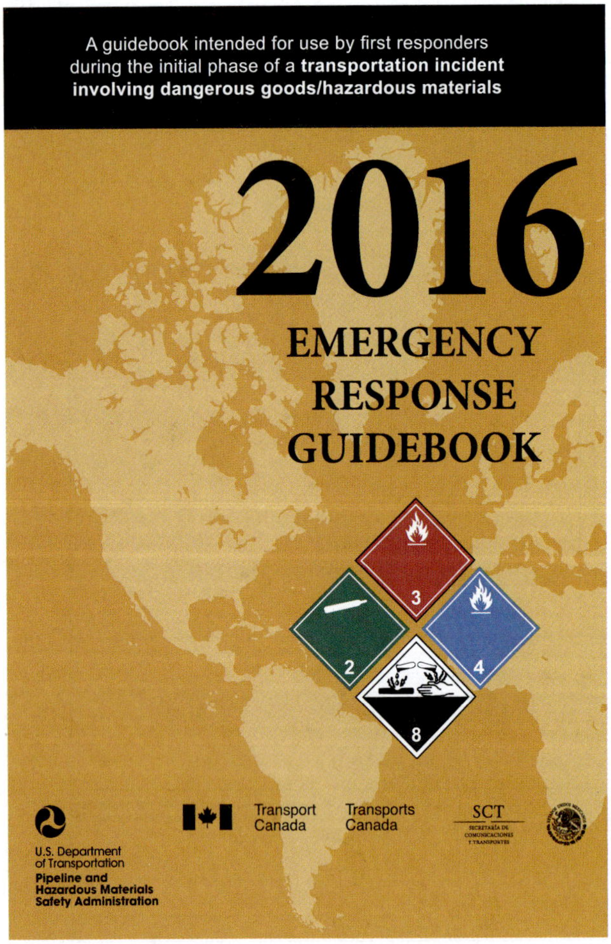

FIGURE 30-22 The *Emergency Response Guidebook* is a reference used as a base for your initial actions at a hazardous materials incident.

Courtesy of the U.S. Department of Transportation.

on the requirements for placards can be found on the Transport Canada website, in the Transportation of Dangerous Goods Act, 1992.)

Conversely, some chemicals are so hazardous that shipping any amount of them requires the use of labels or placards. These materials include some explosives, poison gases, water-reactive solids, and high-level radioactive substances. Responders at the scene should seek additional specifics about any material in question by consulting the appropriate response agency or using the emergency response number on a shipping document, if applicable, to gather more information.

Placards are diamond-shaped indicators (27.3 cm [10¾ in.] on each side) that are placed on all four sides of highway transport vehicles, railroad tank cars, and other forms of transportation carrying hazardous materials (**FIGURE 30-23**). **Labels** are smaller versions (102-mm [4-in.] diamond-shaped indicators) of placards; they are placed on the four sides of individual boxes and smaller packages being transported (**FIGURE 30-24**).

Placards, labels, and markings are intended to give responders a general idea of the hazard inside a container or cargo tank. A placard identifies the broad hazard class (flammable, poison, corrosive) to which the material inside belongs. A label on a box inside a delivery truck, for example, relates only to the potential hazard inside that package. A four-digit United Nations (UN) number may be required on some placards. This number identifies the specific material being shipped; a list of UN numbers is included in the *ERG*. These placards and labels can be viewed from a distance with binoculars, or shipping papers or other reference sources may provide the four-digit UN identification number or the name of the material, or you may see the colours on the placard. All can be used to determine the appropriate guide for a released material.

FIGURE 30-23 A placard is a large diamond-shaped indicator that is placed on all sides of transport vehicles that carry hazardous materials.

© Mark Winfrey/Shutterstock.

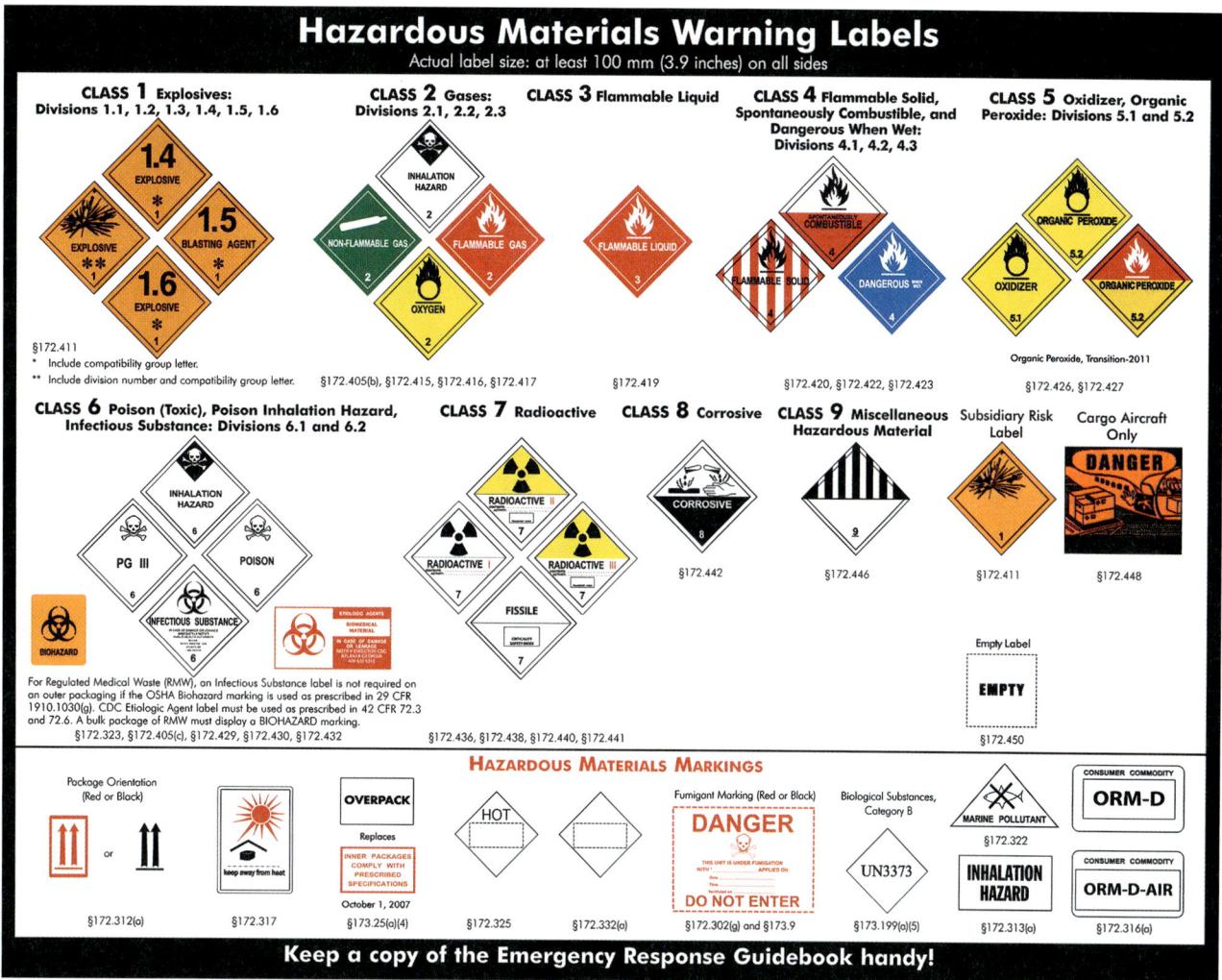

FIGURE 30-24 A label is a smaller version of the placard and is placed on boxes or smaller packages that contain hazardous materials.

Courtesy of the U.S. Department of Transportation.

Using the _ERG_. When the _ERG_ refers to a _small spill_, it means a leak from one small package, a small leak in a large container (up to a 208-litre [55-gallon] drum), a small cylinder leak, or any small leak, even one in a large package. A _large spill_ is a large leak or spill from a larger container or package, a spill from a number of small packages, or anything from a 1-ton (2000-lb) cylinder, tank truck, or railcar.

The _ERG_ is divided into four coloured sections: yellow, blue, orange, and green.

- _Yellow section_: More than 4000 chemicals are found in this section, listed numerically by their four-digit UN number/identification (ID) number. Entry number 1005, for example, identifies "ammonia, anhydrous." Use the yellow section when the UN/ID number is known or can be identified (**FIGURE 30-25**).

   The entries include the name of the chemical and the emergency action guide number.

For example:

ID No.	Guide No.	Name of Material
1005	125	Ammonia, anhydrous

- _Blue section_: The same chemicals listed in the yellow section are found here, listed alphabetically by name. The entry will include the emergency action guide number and the identification number (**FIGURE 30-26**).

Name of Material	Guide No.	ID No.
Ammonia, anhydrous	125	1005

- _Orange section_: This section is organized by guide number (**FIGURE 30-27**). The general hazard class, fire/explosion hazards, health hazards, and basic emergency actions, based on hazard class, are provided.
- _Green section_: This section is organized numerically by UN/ID number and provides the initial isolation distances for certain materials (**FIGURE 30-28**).

ID No.	Guide No.	Name of Material	ID No.	Guide No.	Name of Material
— —	112	Ammonium nitrate - fuel oil mixtures	1014	122	Oxygen and carbon dioxide mixture, compressed
— —	158	Biological agents	1015	126	Carbon dioxide and nitrous oxide mixture
— —	112	Blasting agents, n.o.s.	1015	126	Nitrous oxide and carbon dioxide mixture
— —	112	Explosives, division 1.1, 1.2, 1.3, or 1.5	1016	119	Carbon monoxide
— —	114	Explosives, division 1.4 or 1.6	1016	119	Carbon monoxide, compressed
— —	153	Toxins	1017	124	Chlorine
1001	116	Acetylene, dissolved	1018	126	Chlorodifluoromethane
1002	122	Air, compressed	1018	126	Refrigerant gas R - 22
1003	122	Air, refrigerated liquid (cryogenic liquid)	1020	126	Chloropentafluoroethane
1003	122	Air, refrigerated liquid (cryogenic liquid), non-pressurized	1020	126	Refrigerant gas R - 115
			1021	126	1-Chloro-1,2,2,2-tetrafluoroethane
1005	125	Ammonia, anhydrous	1021	126	Refrigerant gas R - 124
1005	125	Anhydrous ammonia	1022	126	Chlorotrifluoromethane
1006	121	Argon	1022	126	Refrigerant gas R -13
1006	121	Argon, compressed	1023	119	Coal gas
1008	125	Boron trifluoride	1023	119	Coal gas, compressed
1008	125	Boron trifluoride, compressed	1026	119	Cyanogen
1009	126	Bromotrifluoromethane	1027	115	Cyclopropane
1009	126	Refrigerant gas R - 13B1	1028	126	Dichlorodifluoromethane
1010	116 P	Butadienes, stabilized	1028	126	Refrigerant gas R -12
1010	116 P	Butadienes and hydrocarbon mixture, stabilized	1029	126	Dichlorofluoromethane
1010	116 P	Hydrocarbon and butadienes mixture, stabilized	1029	126	Refrigerant gas R - 21
			1030	115	1,1-Difluoroethane
1011	115	Butane	1030	115	Refrigerant gas R - 152a
1012	115	Butylene	1032	118	Dimethylamine, anhydrous
1013	120	Carbon dioxide	1033	115	Dimethyl ether
1013	120	Carbon dioxide, compressed	1035	115	Ethane
1014	122	Carbon dioxide and oxygen mixture, compressed	1035	115	Ethane, compressed
			1036	118	Ethylamine

FIGURE 30-25 Use the yellow section of the *ERG* when the UN/ID number is known or can be identified.
Courtesy of the U.S. Department of Transportation.

Name of Material	Guide No.	ID No.	Name of Material	Guide No.	ID No.
Ammonium hydroxide, with more than 10% but not more than 35% ammonia	154	2672	Ammonium silicofluoride	151	2854
Ammonium metavanadate	154	2859	Ammonium sulfide, solution	132	2683
Ammonium nitrate, liquid (hot concentrated solution)	140	2426	Ammonium sulphide, solution	132	2683
			Ammunition, poisonous, non-explosive	151	2016
Ammonium nitrate with not more than 0.2% combustible substances	140	1942	Ammunition, tear-producing, non-explosive	159	2017
			Ammunition, toxic, non-explosive	151	2016
Ammonium nitrate based fertilizer	140	2067	Amyl acetates	129	1104
Ammonium nitrate based fertilizer	140	2071	Amyl acid phosphate	153	2819
Ammonium nitrate emulsion	140	3375	Amylamine	132	1106
Ammonium nitrate fertilizer, n.o.s.	140	2072	Amyl butyrates	130	2620
			Amyl chloride	129	1107
Ammonium nitrate fertilizers, with ammonium sulfate	140	2069	n-Amylene	128	1108
			Amyl formats	129	1109
Ammonium nitrate fertilizers, with ammonium sulphate	140	2069	Amyl mercaptan	130	1111
			n-Amyl methyl ketone	127	1110
Ammonium nitrate fertilizers, with calcium carbonate	140	2068	Amyl nitrate	140	1112
			Amyl nitrite	129	1113
Ammonium nitrate fertilizers, with phosphate or potash	143	2070	Amyltrichlorosilane	155	1728
Ammonium nitrate-fuel oil mixtures	112	— —	Anhydrous ammonia	125	1005
Ammonium nitrate gel	140	3375	Aniline	153	1547
Ammonium nitrate suspension	140	3375	Aniline hydrochloride	153	1548
Ammonium perchlorate	143	1442	Anisidines	153	2431
Ammonium persulfate	140	1444	Anisidines, liquid	153	2431
Ammonium persulphate	140	1444	Anisidines, solid	153	2431
Ammonium picrate, wetted with not less than 10% water	113	1310	Anisole	128	2222
			Anisoyl chloride	156	1729
Ammonium polysulfide, solution	154	2818	Antimony compound, inorganic, liquid, n.o.s.	157	3141
Ammonium polysulphide, solution	154	2818	Antimony compound, inorganic, solid, n.o.s.	157	1549
Ammonium polyvanadate	151	2861	Antimony lactate	151	1550

FIGURE 30-26 The same chemicals listed in the yellow section of the *ERG* are found in the blue section, listed alphabetically by name.
Courtesy of the U.S. Department of Transportation.

Chemicals included in this section consist of the chemicals highlighted from the yellow and blue sections. The green section includes water-reactive materials that produce toxic gases (calcium phosphide and trichlorosilane, for example); **toxic inhalation hazards (TIH)**, which are gases or volatile liquids that are extremely toxic to humans; chemical warfare agents (CWAs); and dangerous water-reactive materials (WRMs). Examples include such substances as anhydrous ammonia, sarin, and sodium cyanide. These gases or volatile liquids are extremely toxic to humans and pose a hazard to health during transport. Any material listed in the green section is extremely hazardous.

The green section also offers recommendations on the size and shape of protective action zones. This section is useful when it is necessary to protect people from toxic-by-inhalation (TIH) vapours resulting from a release. The initial isolation zones may be used to define an area surrounding an incident where persons may be exposed to potentially dangerous or life-threatening concentrations of the vapour both upwind and downwind from the release. The orange-bordered guides are different from the green section in that the orange-bordered guides are relevant to evacuation distances required to protect against fragmentation hazard from a large container if it should fail due to explosion. The rationale is that if a certain material becomes involved with fire, the TIH hazard may be less than the fire or explosion hazard.

LISTEN UP!

The same information, organized differently, is found in both the blue and yellow sections of the *ERG*.

Any substance highlighted in either the yellow or blue section of the *ERG* is either a toxic inhalation hazard, a chemical warfare agent, or a water-reactive material.

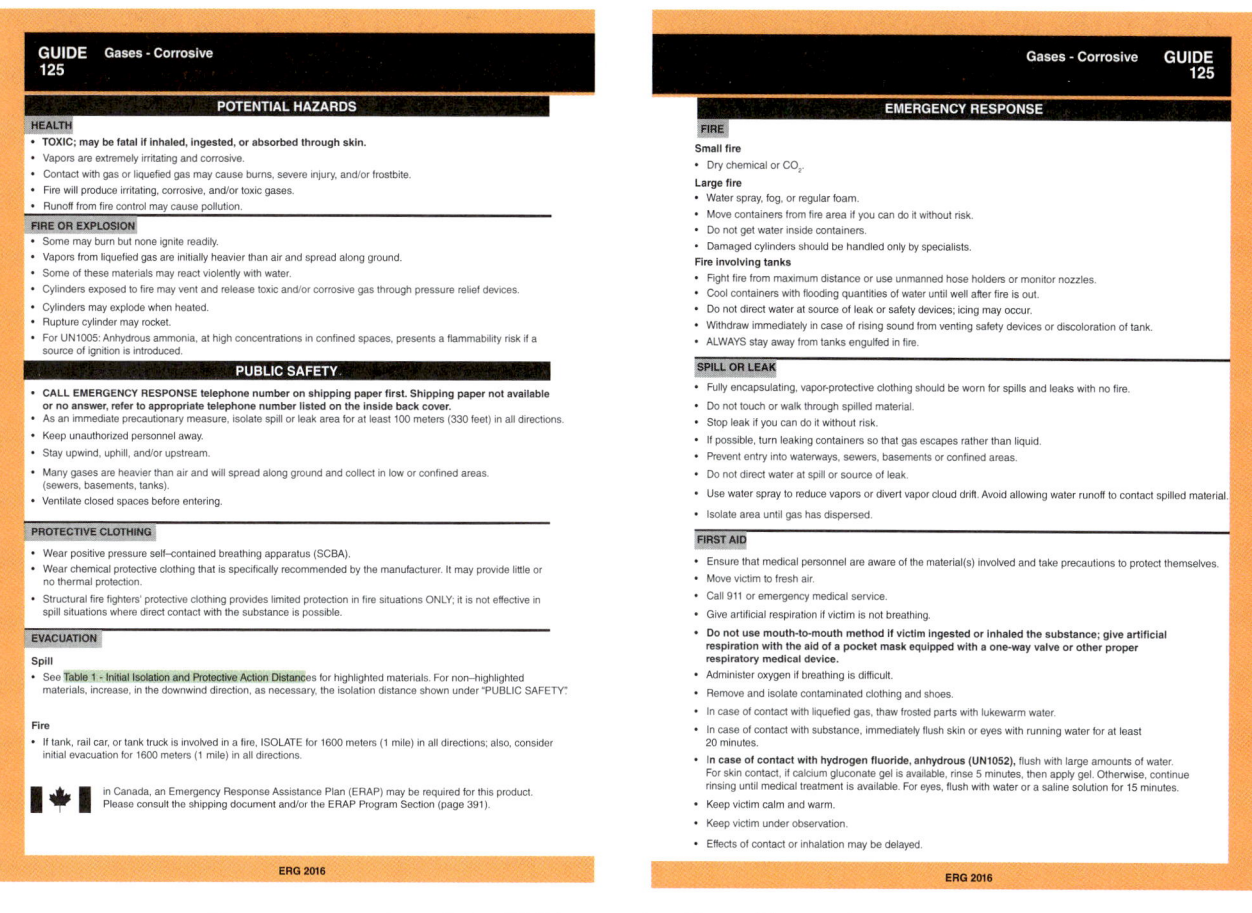

FIGURE 30-27 The orange section of the *ERG* is organized by guide number.

Courtesy of the U.S. Department of Transportation.

The U.S. DOT, the Secretariat of Communications and Transportation (SCT) of Mexico, and Transport Canada jointly developed the *ERG*.

The nine chemical families, and their respective divisions recognized in the *ERG*, are outlined here:

- Class 1—Explosives
    - Division 1.1 Explosives with a mass explosion hazard
    - Division 1.2 Explosives with a projection hazard
    - Division 1.3 Explosives with predominantly a fire hazard
    - Division 1.4 Explosives with no significant blast hazard
    - Division 1.5 Very insensitive explosives with a mass explosion hazard
    - Division 1.6 Extremely insensitive articles
- Class 2—Gases
    - Division 2.1 Flammable gases
    - Division 2.2 Nonflammable, nontoxic gases
    - Division 2.3 Toxic gases
- Class 3—Flammable liquids (and combustible liquids in the United States)
- Class 4—Flammable solids, spontaneously combustible materials, and dangerous-when-wet materials/water-reactive substances
    - Division 4.1 Flammable solids
    - Division 4.2 Spontaneously combustible materials
    - Division 4.3 Water-reactive substances/ dangerous when wet
- Class 5—Oxidizing substances and organic peroxides
    - Division 5.1 Oxidizing substances
    - Division 5.2 Organic peroxides
- Class 6—Toxic substances and infectious substances
    - Division 6.1 Toxic substances
    - Division 6.2 Infectious substances
- Class 7—Radioactive materials
- Class 8—Corrosive substances
- Class 9—Miscellaneous hazardous materials/ products, substances, or organisms

INITIAL ISOLATION AND PROTECTIVE ACTION DISTANCES

ID No.	Guide	NAME OF MATERIAL	SMALL SPILLS (From a small package or small leak from a large package) First ISOLATE in all Directions Meters (Feet)	Then PROTECT persons Downwind during DAY Kilometers (Miles)	NIGHT Kilometers (Miles)	LARGE SPILLS (From a large package or from many small packages) First ISOLATE in all Directions Meters (Feet)	Then PROTECT persons Downwind during DAY Kilometers (Miles)	NIGHT Kilometers (Miles)
1005	125	Ammonia, anhydrous	30 m (100 ft)	0.1 km (0.1 mi)	0.2 km (0.1 mi)	Refer to table 3		
1005	125	Anhydrous ammonia				Refer to table 3		
1008	125	Boron trifluoride	30 m (100 ft)	0.1 km (0.1 mi)	0.7 km (0.4 mi)	400 m (1250 ft)	2.2 km (1.4 mi)	4.8 km (3.0 mi)
1008	125	Boron trifluoride, compressed	30 m (100 ft)	0.1 km (0.1 mi)	0.7 km (0.4 mi)	400 m (1250 ft)	2.2 km (1.4 mi)	4.8 km (3.0 mi)
1016	119	Carbon monoxide	30 m (100 ft)	0.1 km (0.1 mi)	0.2 km (0.1 mi)	200 m (600 ft)	1.2 km (0.7 mi)	4.4 km (2.8 mi)
1016	119	Carbon monoxide, compressed	30 m (100 ft)	0.1 km (0.1 mi)	0.2 km (0.1 mi)	200 m (600 ft)	1.2 km (0.7 mi)	4.4 km (2.8 mi)
1017	124	Chlorine	60 m (200 ft)	0.3 km (0.2 mi)	1.1 km (0.7 mi)	Refer to table 3		
1026	119	Cyanogen	30 m (100 ft)	0.1 km (0.1 mi)	0.4 km (0.3 mi)	60 m (200 ft)	0.3 km (0.2 mi)	1.1 km (0.7 mi)
1040	119P	Ethylene oxide	30 m (100 ft)	0.1 km (0.1 mi)	0.2 km (0.1 mi)	Refer to table 3		
1040	119P	Ethylene oxide with Nitrogen	30 m (100 ft)	0.1 km (0.1 mi)	0.2 km (0.1 mi)	Refer to table 3		
1045	124	Fluorine	30 m (100 ft)	0.1 km (0.1 mi)	0.2 km (0.1 mi)	100 m (300 ft)	0.5 km (0.3 mi)	2.2 km (1.4 mi)
1045	124	Fluorine, compressed	30 m (100 ft)	0.1 km (0.1 mi)	0.2 km (0.1 mi)	100 m (300 ft)	0.5 km (0.3 mi)	2.2 km (1.4 mi)
1048	125	Hydrogen bromide, anhydrous	30 m (100 ft)	0.1 km (0.1 mi)	0.2 km (0.1 mi)	150 m (500 ft)	0.9 km (0.6 mi)	2.6 km (1.6 mi)
1050	125	Hydrogen chloride, anhydrous	30 m (100 ft)	0.1 km (0.1 mi)	0.3 km (0.1 mi)	Refer to table 3		
1051	117	AC (when used as a weapon)	60 m (200 ft)	0.3 km (0.2 mi)	1.0 km (0.6 mi)	1000 m (3000 ft)	3.7 km (2.3 mi)	8.4 km (5.3 mi)
1051	117	Hydrocyanic acid, aqueous solutions, with more than 20% Hydrogen cyanide	60 m (200 ft)	0.2 km (0.2 mi)	0.9 km (0.6 mi)	300 m (1000 ft)	1.1 km (0.7 mi)	2.4 km (1.5 mi)
1051	117	Hydrogen cyanide, anhydrous, stabilized						
1051	117	Hydrogen cyanide, stabilized						

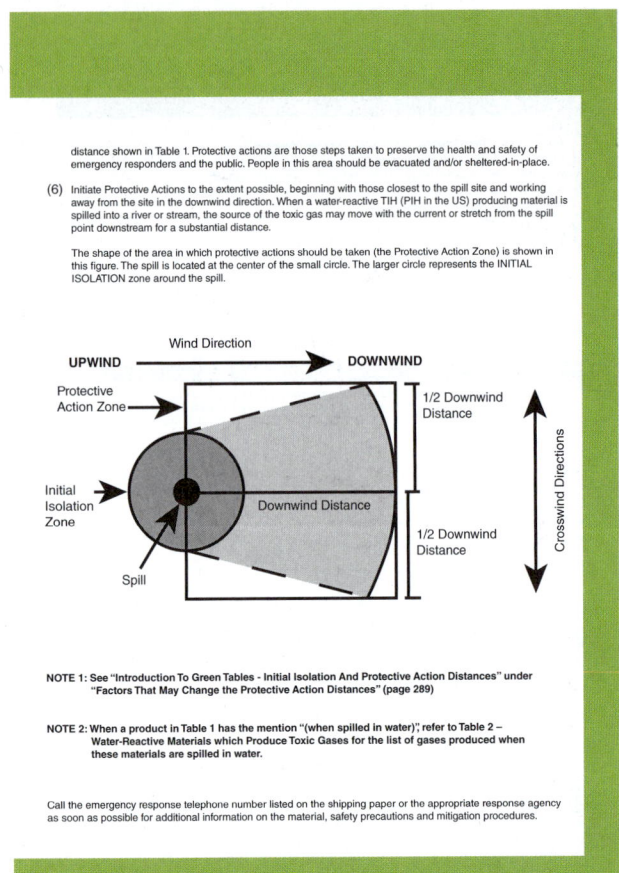

distance shown in Table 1. Protective actions are those steps taken to preserve the health and safety of emergency responders and the public. People in this area should be evacuated and/or sheltered-in-place.

(6) Initiate Protective Actions to the extent possible, beginning with those closest to the spill site and working away from the site in the downwind direction. When a water-reactive TIH (PIH in the US) producing material is spilled into a river or stream, the source of the toxic gas may move with the current or stretch from the spill point downstream for a substantial distance.

The shape of the area in which protective actions should be taken (the Protective Action Zone) is shown in this figure. The spill is located at the center of the small circle. The larger circle represents the INITIAL ISOLATION zone around the spill.

NOTE 1: See "Introduction To Green Tables - Initial Isolation And Protective Action Distances" under "Factors That May Change the Protective Action Distances" (page 289)

NOTE 2: When a product in Table 1 has the mention "(when spilled in water)", refer to Table 2 – Water-Reactive Materials which Produce Toxic Gases for the list of gases produced when these materials are spilled in water.

Call the emergency response telephone number listed on the shipping paper or the appropriate response agency as soon as possible for additional information on the material, safety precautions and mitigation procedures.

FIGURE 30-28 The green section of the *ERG* is organized numerically by UN/ID number and provides the initial isolation distances for certain materials.

Courtesy of the U.S. Department of Transportation.

The *ERG* organizes chemicals into nine basic hazard classes, or families; the members of each family exhibit similar properties. There is also a "Dangerous" placard, which indicates that more than one hazard class is contained in the same load (**FIGURE 30-29**).

To use the *ERG*, follow the steps in **SKILL DRILL 30-1**.

FIRE FIGHTER TIP

To accurately compare the TC marking system and NFPA 704 system, remember this important difference:

- The TC hazardous materials marking system is used when materials are being transported from one location to another.
- The NFPA 704 hazard identification system is designed for fixed-facility use.

FIGURE 30-29 A "Dangerous" placard indicates that more than one hazard is contained within the same load.

Courtesy of Rob Schnepp.

SKILL DRILL 30-1
Using the *Emergency Response Guidebook* NFPA 1072: 4.2.1, 4.3.1; NFPA 472: 4.1.2.2, 4.2.3, 4.4.1, 5.1.2.2

Courtesy of Rob Schnepp.

1 Identify the chemical name and/or the chemical ID number for the placard seen here.

2 Look up the material name in the appropriate section of the *ERG*. Use the yellow section to obtain information based on the UN/ID number. Use the alphabetized blue section to obtain information based on the chemical name. *Note any green highlights, which would indicate the substance will also have an entry and recommendations in the green section of the guide.*

3 Determine the correct emergency action guide to use for the chemical identified.

4 Identify the health hazards, potential fire and explosion hazards, recommended protective clothing, and evacuation recommendations. (For the substance used in this exercise you should find a Table 1 recommendation of initial Isolation and Protective Action Distances as well as isolation distances if the substance is involved in fire. Also, take note of the firefighting recommendation, handling spills or leaks, and first aid measures.)

5 If necessary, identify the isolation distance and the protective actions required for the chemical substance in the green section.

Harmful Substance Routes of Entry into the Human Body

The damage that a hazardous material/WMD will inflict on a human being or the environment is a function of the physical and chemical properties of the released substance, as well as the conditions under which it was released and the duration of the exposure. Among other factors, characteristics such as the concentration of the material, the temperature of the material at the time of its release, and the pressure under which the substance was released affect both the release parameters and the potential health effects on those exposed to the material. Additionally, the age, gender, genetics, and underlying medical conditions of the exposed person will have some bearing on patient outcome. Chemical exposures are complicated events because of the number of variables that may be present.

When it comes to rendering medical care to persons exposed to harmful substances, guidance can be found in NFPA 473, *Standard for Competencies for EMS Personnel Responding to Hazardous Materials/Weapons of Mass Destruction Incidents*. This standard outlines a basic set of hazardous materials response skills that all EMS responders, regardless of their scope of practice, should achieve to work safely on a hazardous materials/WMD scene and deliver effective patient care. In most cases, that care is performed in a safe area (cold zone) away from the hazard, after decontamination. There are very few circumstances where definitive advanced life support must be rendered in the hot zone. EMS responders, however, should understand the nature of the incident and look at the scene with a critical eye to pick up the clues that might assist with defining the nature of the exposure.

For a chemical or other harmful substance to injure a person, it must first get into or onto the individual's body. Throughout the course of his or her service career, a responder will inevitably be bombarded by harmful substances such as diesel exhaust, bloodborne pathogens, smoke, and accidentally and intentionally released chemicals. Fortunately, most of these exposures are not immediately deadly. Unfortunately, repetitive exposures to these materials may have negative health effects after a 20-year career. To protect yourself now and give yourself the best shot at a healthy retirement, it is important to understand some basic concepts about toxicology. Chemical substances can enter the human body in four ways (**FIGURE 30-30**):

- *Inhalation*: Through the lungs
- *Absorption*: By permeating the skin
- *Ingestion*: Via the gastrointestinal tract
- *Injection*: Through cuts or other breaches in the skin

Toxicology is the study of the adverse effects of chemical or physical agents on living organisms. The following sections discuss these agents' potential routes of entry into the human body and methods used to protect against these agents.

Inhalation

Inhalation exposures occur when harmful substances enter the body through the respiratory system. The lungs are a direct point of access to the bloodstream, so they can quickly transfer an airborne substance into the

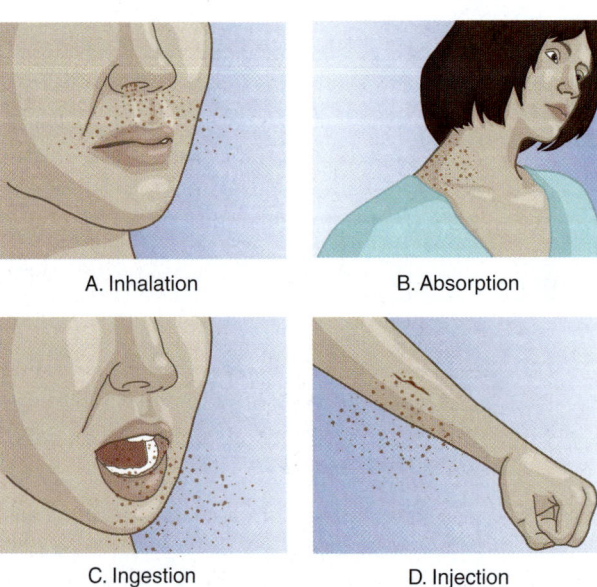

A. Inhalation   B. Absorption

C. Ingestion   D. Injection

FIGURE 30-30 The four ways a chemical substance can enter the body.
© Jones & Bartlett Learning.

circulatory system and onward to the rest of the body. In addition, the lungs cannot be decontaminated, so any exposure will result in some type of harm that cannot be addressed in the same way as an exposure to skin.

The respiratory system is vulnerable to attack from a wide range of substances, from corrosive materials such as chlorine and ammonia to solvent vapours such as gasoline and acetone, or superheated air from a fire, or any other material finding its way into the air. In addition to gases and vapours, small particles of dust, fibreglass insulation, asbestos, or soot from a fire can become lodged in sensitive lung tissue, causing substantial irritation.

Given these facts, it is imperative that responders wear appropriate respiratory protection when operating in the presence of airborne contamination. Fortunately, fire fighters have ready access to an excellent form of respiratory protection—namely, the positive-pressure, open-circuit self-contained breathing apparatus (SCBA). This equipment is by far the single most important piece of PPE that fire fighters have at their disposal. Sometimes fire fighters may need to use other forms of respiratory protection depending on the specific respiratory hazard they are facing while performing a given task. For example, full-face and half-face air-purifying respirators (APRs) offer specific degrees of protection if the chemical hazard present is known and the appropriate filter canister is used (**FIGURE 30-31**).

APRs do not provide oxygen, however. Thus, if the oxygen content of the work area is low, full- or half-face

respirators are not a viable option. Per the CSA (Canadian Standards Group) Standard 94.4-11 (R2016), *Selection, Care and Use of Respirators*, based on the NIOSH standard used in the United States, any work environment containing less than 19.5 percent oxygen in some jurisdictions is considered to be oxygen deficient and is therefore considered an IDLH environment requiring either SCBA or an air-supplying respirator.

Respirators are lighter than SCBA, are more comfortable to wear, and usually allow longer work periods because they are not dependent on a limited source of breathing air. SCBAs certainly offer a higher level of respiratory protection, but in the right circumstances, APRs may represent a viable form of respiratory protection.

When considering protection against airborne contamination, it is important to understand the origin, concentration, and potential impact of the contamination relative to the oxygen levels in the area. In short, the appropriate type of respiratory protection is determined by looking at the overall situation, including the nature of the contaminant. In some cases, the anticipated particle size of the contamination may dictate the level of respiratory protection employed (**TABLE 30-2**).

Anthrax spores offer an excellent example to illustrate this point. Weaponized anthrax spores typically vary in size from 0.5 micron to 1 micron. Based on that size range, a typical full-face APR with a nuisance dust filter, or a surgical-type mask, would not offer sufficient protection against this hazard. Anyone operating in an area contaminated with anthrax should wear SCBA or, at a minimum, a full-face APR with P100 filtration (which filters out greater than 99 percent of 0.3-micron or larger particles).

Anthrax exposures also illustrate another important pair of terms and definitions that responders should

FIGURE 30-31 Air-purifying respirators offer some degree of protection against airborne chemical hazards.
Courtesy of Sperian Respiratory Protection.

TABLE 30-2 Particle Sizes of Common Types of Respiratory Hazards

Fume	< 1 micron
Smoke	≤ 1 micron
Dust	≥ 1 micron
Fog	> 40 microns
Mist	< 40 microns

© Jones & Bartlett Learning.

understand: **infectious** and **contagious**. Anthrax is a pathogenic microorganism capable of causing an illness (infectious). A person with an illness caused by an anthrax exposure, however, is not capable of passing it along to another person (contagious). In other words, anthrax is not contagious. Conversely, smallpox can be both infectious and contagious, which is why this pathogen poses such a high risk in the event of an outbreak.

Now consider this scenario: A container of gaseous helium is leaking inside a poorly ventilated storage room. Helium is nonflammable; the main threat it poses is the possibility of oxygen deficiency. In this case, SCBA is the appropriate type of respiratory protection based on the anticipated hazard. In many cases, when hazardous materials technicians respond to incidents, they will use air-monitoring devices to characterize the work area prior to entry. When this step is taken, choices of respiratory protection are based on more definitive information.

Particle size also determines where the inspired contamination will eventually end up. The larger particles that make up visible mists will be captured in the nose and upper airway, for example, whereas smaller particles may work their way deeper into the lung (**TABLE 30-3**).

Respiratory protection is one of the most important components of any PPE. In all cases where airborne contamination is encountered, take the time to understand the nature of the threat, evaluate the respiratory protection available, and decide if it provides adequate protection. When protecting the lungs, "good enough" is not an option. Chapter 36, *Hazardous Materials*

Responder Personal Protective Equipment, addresses the concept of chemical PPE and describes how respiratory protection fits into the "big picture" of remaining safe at chemical incidents.

Absorption

The skin is the largest organ in the body and is susceptible to the damage inflicted by many substances. In addition to serving as the body's protective shield against heat, light, and infection, the skin helps regulate body temperature, stores water and fat, and serves as a sensory centre for painful and pleasant stimulation. Without this important organ, human beings would not be able to survive.

When discussing chemical exposures, however, absorption is not limited only to the skin. **Absorption** is the process by which substances travel through body tissues until they reach the bloodstream. The eyes, nose, mouth, and, to a certain degree, intestinal tract are also part of the equation. The eyes, for example, will absorb a large amount of liquid and vapour that these sensitive tissues encounter. This absorption is particularly problematic because the eyes connect directly to the optic nerve, which allows the chemical to follow a direct route to the brain and the central nervous system.

Although the skin functions as a shield for the body, that shield can be pierced by many chemicals. Aggressive solvents such as methylene chloride (found in paint stripper), for example, can be readily absorbed through the skin. A secondary hazard associated with this chemical occurs when the body attempts to metabolize the substance after it is absorbed. A by-product of that metabolism reaction is carbon monoxide, a cellular asphyxiant. **Asphyxiants** are substances that prevent the body (at the cellular level) from using oxygen, thereby causing suffocation. In this scenario, the initial chemical is broken down to form another substance that is potentially a greater health hazard than the original chemical. Methylene chloride is also suspected to be a human cancer-causing agent (**carcinogen**).

Absorption hazards are not limited to solvents. Hydrofluoric acid, for example, poses a significant threat to life when it is absorbed through the skin. This unique corrosive can bind with certain substances in the body (predominantly calcium). Secondary health effects occurring after exposure can include muscular pain and potentially lethal cardiac arrhythmias.

TABLE 30-3 Location of Respiratory Trapping by Particle Size	
< 7 microns	Nose
5–7 microns	Larynx
3–5 microns	Trachea and bronchi
2–3 microns	Bronchi
1–2.5 microns	Respiratory bronchioles
0.5–1 micron	Alveoli

© Jones & Bartlett Learning.

In the field, responders must constantly evaluate the possibility of chemical contact with their skin and eyes. In many cases, structural firefighting PPE provides little or no protection against liquid chemicals. Consult the *ERG* or an SDS for response guidance when deciding whether PPE is appropriate for the hazard you are facing. In the event the PPE does not offer adequate protection, responders may have to increase their level of protection to include specialized chemical-protective clothing.

Ingestion

In addition to absorption through the skin, chemicals can be brought into the body through the gastrointestinal tract, by the process of ingestion. The water, nutrients, and vitamins the body requires are predominantly absorbed in this manner. For example, at a structure fire, fire fighters generally have an opportunity to rotate out of the building for rest and refreshment and may not take the time to wash up prior to eating or drinking. This leads to a high probability of spreading contamination from the hands to the food and subsequently to the intestinal tract. If you do not think about every situation where you might become exposed, you may put yourself in harm's way.

Injection

Chemicals brought into the body through open cuts and abrasions qualify as injection exposures. To protect yourself from this route of exposure, begin by realizing when you will work in a compromised state. Any cuts or open wounds should be addressed before reporting for duty. If they are significant, you may be excluded from operating in contaminated environments. Open wounds act as a direct portal to the bloodstream and subsequently to muscles, organs, and other body systems. If a chemical substance encounters this open portal, the health effects could be immediate and pronounced.

After-Action REVIEW

IN SUMMARY

- Millions of organic and inorganic substances are registered for use in commerce, with several thousand new ones being introduced each year.

- Identifying the kinds and quantities of hazardous materials used and stored by local facilities should be an integral part of any comprehensive community response plan.

- All responders must interpret visual clues effectively to improve their ability to safely operate at an incident.

- All responders should be able to recognize the various container profiles and understand the general classifications of materials that may be stored inside each type of container.

- All responders should be able to name, understand, and locate the various types of shipping papers on various modes of transportation.

- When used correctly, various marking systems indicate the presence of a hazardous material and provide clues about the substance. Transport Canada (TC), the National Fire Protection Association (NFPA), the Workplace Hazardous Materials Information System (WHMIS), and the military have all developed marking systems specific to their level of response.

- All responders should be able to demonstrate proficiency when using the *Emergency Response Guidebook*.

- It is important to know how to obtain SDS documentation from various sources, including one's own department, the scene of the incident itself, or the manufacturer of the material.

KEY TERMS

Access Navigate for flashcards to test your key term knowledge.

Absorption The process by which substances travel through body tissues until they reach the bloodstream.

Air bill The shipping papers on an airplane.

Asphyxiants Materials that cause the victim to suffocate.

Bill of lading The shipping papers used for transport of chemicals over roads and highways; also referred to as a *freight bill*.

Bung One or two openings on top of a closed-head drum. Typically sealed with a threaded cap.

Carboy A glass, plastic, or steel storage container, ranging in volume from 19 to 57 litres (5–15 gallons).

Carcinogen A cancer-causing substance that is identified in one of several published lists, including, but not limited to, NIOSH Pocket Guide to Chemical Hazards, Hazardous Chemicals Desk Reference, and the ACGIH 2007 TLVs and BEIs. (NFPA 1851)

Chemical Abstracts Service (CAS) A division of the American Chemical Society. This resource provides hazardous materials responders with access to an enormous collection of chemical substance information—the CAS Registry.

Consist A list of the contents of every car on a train; also called a *train list*.

Contagious Capable of transmitting a disease.

Container A vessel, including cylinders, tanks, portable tanks, and cargo tanks, used for transporting or storing materials. (NFPA 1)

Cryogenic liquids (cryogens) A fluid with a boiling point lower than –90°C (–130°F) at an absolute pressure of 101.3 kPa (14.7 psi). (NFPA 1)

Cylinder A pressure vessel designed for absolute pressures higher than 276 kPa (40 psi) and having a circular cross-section. It does not include a portable tank, multiunit tank, car tank, cargo tank, or tank car. (NFPA 1)

Dangerous cargo manifest The shipping papers on a marine vessel, generally located in a tube-like container.

Dewar container A container designed to preserve the temperature of the cold liquid held inside.

Drum A barrel-like storage vessel used to store a wide variety of substances, including food-grade materials, corrosives, flammable liquids, and grease. Drums may be constructed of low-carbon steel, polyethylene, cardboard, stainless steel, nickel, or other materials.

Emergency Response Guidebook (ERG) The reference book, written in plain language, to guide emergency responders in their initial actions at the incident scene, specifically the *Emergency Response Guidebook* from the U.S. Department of Transportation, Transport Canada, and the Secretariat of Transport and Communications, Mexico. (NFPA 1072)

Freight bill The shipping papers used for transport of chemicals along roads and highways. Also referred to as a *bill of lading*.

Infectious Capable of causing an illness by entry of a pathogenic microorganism.

Ingestion Exposure to a hazardous material by swallowing the substance.

Inhalation Exposure to a hazardous material by breathing the substance into the lungs.

Injection Exposure to a hazardous material by the substance entering cuts or other breaches in the skin.

Labels A visual indication whether in pictorial or word format that provides for the identification of a control, switch, indicator, or gauge or the display of information useful to the operator. (NFPA 1901)

NFPA 704 hazard identification system A hazardous materials marking system designed for fixed-facility use. It uses a diamond-shaped symbol of any size, which is itself broken into four smaller diamonds, each representing a particular property or characteristic of the material.

Pipeline A length of pipe including pumps, valves, flanges, control devices, strainers, and/or similar equipment for conveying fluids. (NFPA 70)

Pipeline right-of-way An area, patch, or roadway that extends a certain number of centimetres or metres on either side of a pipeline and that may contain warning and informational signs about hazardous materials carried in the pipeline.

Placards Signage required to be placed on all four sides of highway transport vehicles, railroad tank cars, and other forms of hazardous materials transportation; the sign identifies the hazardous contents of the vehicle, using a standardization system with 273 mm (10¾-inch) diamond-shaped indicators.

Safety data sheet (SDS) Formatted information, provided by chemical manufacturers and distributors of hazardous products, about chemical composition, physical and chemical properties, health and safety hazards, emergency response, and waste disposal of the material. (NFPA 1072)

Shipping papers A shipping order, bill of lading, manifest, or other shipping document serving a similar purpose and containing the information required by regulations of Transport Canada (TC).

Toxic inhalation hazards (TIH) Any gas or volatile liquid that is extremely toxic to humans.

Toxicology The study of the adverse effects of chemical or physical agents on living organisms.

Train list *See* consist.

Transport Canada (TC) marking system A unique system of labels and placards that is used when materials are being transported from one location to another in Canada. The same marking system is used in the United States by the Department of Transportation (DOT).

Vent pipes Inverted J-shaped tubes that allow for pressure relief or natural venting of a pipeline for maintenance and repairs.

Waybills Shipping papers for railroad transport.

Workplace Hazardous Materials Information System (WHMIS) A colour-coded marking system by which employers give their personnel the necessary information to work safely around chemicals. The Hazardous Materials Information System (HMIS) is the U.S. hazard communication standard.

On Scene

Your crew is dispatched to a biotechnology research company for an odour investigation. Your initial information from the dispatch centre reported an unusual odour in one of the laboratories. Upon arrival, a security guard meets you at the street and relays the location of a spill of an unknown liquid. He reports that a scientist called the security desk to report finding a broken 4-litre (1-gallon) amber-coloured glass container in laboratory #206. He tells you that the lab is evacuated and there are no injuries or exposures, and it is unknown how the container was broken or any other history of the event. You and your crew are trained to the hazardous materials awareness level.

1. Based on the above information, does your crew have the right level of training to enter the lab, identify the liquid, and clean up the spill?

  A. Yes, awareness level responders are trained to take offensive action to clean up unknown chemical spills.

  B. No, you tell the security guard that you don't have adequate training, that it is the responsibility of the lab personnel to clean up the spill, and then you leave the scene.

  C. Yes, you tell the security guard to find the person responsible for the lab and that your crew will enter with him or her to clean up the spill.

  D. No, you are not trained to take offensive action, but you will safely secure the area and request the properly trained personnel to respond.

2. If you are able to get a chemical name of the spilled substance, which reference source listed below would provide the most complete information regarding the scenario above?

  A. The *ERG*

  B. Going to the lab and looking at the spill yourself

  C. Asking a scientist from the site to tell you about the material

  D. An SDS for the material

(continued)

On Scene Continued

3. Initial operational priorities for this scenario are based on the acronym SIN. What does that stand for?

A. Scene, isolate, notify

B. Safety, interview, notify

C. Scene, investigate, notify

D. Safety, isolate, notify

4. Which of the following substances would most likely be found in the amber glass container described in the scenario?

A. Cryogenic nitrogen

B. Grease

C. Hydrofluoric acid

D. Sulfuric acid

Access Navigate to find answers to this On Scene, along with other resources such as an audiobook.

Hazardous Materials Operations Level

CHAPTER **31**

Hazardous Materials Operations Level

Properties and Effects

KNOWLEDGE OBJECTIVES

After studying this chapter, you will be able to:

- Describe states of matter and their physical and chemical changes. (**NFPA 1072: 5.2.1; NFPA 472: 5.1.2.2, 5.2.3**, pp. 1089–1092)
- Discuss the critical characteristics of flammable liquids. (**NFPA 1072: 5.2.1; NFPA 472: 5.1.2.2, 5.2.3**, pp. 1092–1102)
- Discuss a responder's role in working with hazards, exposure, and contamination. (**NFPA 1072: 5.2.1, 5.5.1; NFPA 472: 5.1.1.1, 5.1.2.2, 5.2.3**, pp. 1102–1103)
- Describe how hazardous material exposure can lead to chronic and/or acute health effects. (**NFPA 472: 5.1.2.2, 5.2.3, 5.3.4**, pp. 1103–1105)

SKILLS OBJECTIVES

This chapter has no skills objectives for operations level hazardous materials responders.

Hazardous Materials Alarm

Just after midnight you receive a call for an explosion and fire at a local landscaping company. Upon arrival at the site, you find a working fire in a storage shed located behind the main building. The wooden shed is fully involved, so you begin an indirect attack on the fire from the outside of the shed. As the fire is being knocked down, you receive an order over the radio to shut down the attack and move away from the building. About the same time, you experience an itchy feeling on the back of your neck. Other crew members are also complaining of itching and burning sensations around their wrists and necks. Some lower-floor residents of an adjacent multistorey apartment complex are complaining of eye irritation and asking about a strange odour in the air.

1. Which types of chemicals might be found in this kind of occupancy?

2. Where could you obtain accurate technical information on the products stored in this building?

3. Which actions should be taken to address the complaints of burning and itching skin among fire fighters as well as the complaints of the residents of the apartment complex?

 *Access Navigate for more practice activities.*

Introduction

To safely mitigate hazardous materials incidents, it is important to understand the **chemical and physical properties** of the substances involved. Chemical and physical properties are the characteristics of a substance that are measurable, such as vapour density, flammability, corrosivity, and water reactivity. However, you do not have to be a chemist to safely respond to hazardous materials incidents. In most cases, being an astute observer, referring to your incident response plan and/or standard operating procedures, consulting the appropriate reference sources, and correctly interpreting and understanding the visual clues presented to you will provide enough information to take basic actions at the incident, protecting nearby persons, the environment, or property from the effects of the release.

Pesticide bags are a good example of providing good information when you know what to look for. Pesticide bags must be labelled with specific information, and responders can learn a great deal from the label, including the following details (**FIGURE 31-1**):

- Name of the product
- Statement of ingredients
- Total amount of product in the container
- Manufacturer's name and address

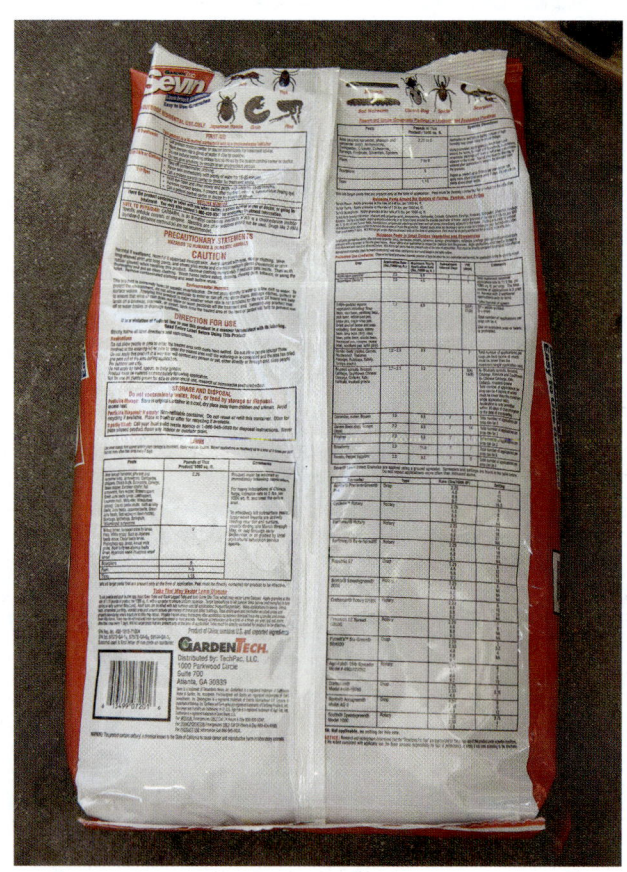

FIGURE 31-1 A pesticide label.
© Jones & Barlett Learning. Photographed by Glen E. Ellman.

- Signal words to indicate the relative toxicity of the material:
  - *Danger—Poison*: Highly toxic by all routes of entry
  - *Danger*: Severe eye damage or skin irritation
  - *Warning*: Moderately toxic
  - *Caution*: Minor toxicity and minor eye damage or skin irritation
- Practical first-aid treatment description
- Directions for use
- Agricultural use requirements
- Precautionary statements such as mixing directions or potential environmental hazards
- Storage and disposal information
- Classification statement on who may use the product

In addition, every pesticide label must carry the statement "Keep out of reach of children."

Materials originating in Canada carry a *Pest Control Products Act* registration number. The Canadian Transport Emergency Centre (CANUTEC), operated by Transport Canada, provides information about these materials when provided the registration number. Canadian products have the same signal words and required information as they would in the United States (e.g., Environmental Protection Agency [EPA] registration number). Registration numbers provide proof that the product was registered.

This chapter provides information pertaining to basic key terminology and how to interpret reference sources such as material safety data sheets (MSDSs). Safety data sheets (SDSs) may be obtained from site facility representatives, found online, or carried by those responsible for shipping hazardous materials. When looking at an SDS during an incident, responders may find a wealth of technical information on the document, such as:

- Identification, including supplier identifier and emergency telephone number
- Hazard identification
- Composition/information on ingredients
- First-aid measures
- Firefighting measures
- Accident release measures
- Handling and storage requirements
- Exposure controls/personal protection
- Physical and chemical properties
- Stability and reactivity
- Toxicological information
- Ecological information (nonmandatory)
- Disposal considerations (nonmandatory)
- Transport information (nonmandatory)
- Regulatory information (nonmandatory)
- Other information

Additionally, making a phone call to resources such as the CANUTEC, (or, in the United States, the Chemical Transportation Emergency Center [CHEMTREC], and, in Mexico, the Emergency Transportation System for the Chemical Industry, Mexico [SETIQ]) may provide the responder with valuable information on a substance. These resources are available 24 hours a day and provide responders with critical information for incidents involving hazardous materials and dangerous goods. A responder can access live information from product specialists or access volumes of technical data from over 6 million SDSs.

Physical and Chemical Changes

An important first step in understanding the **hazard(s)** associated with any chemical involves identifying the **states of matter**, or physical state, of the substance. The state of matter defines the substance as a solid, liquid, or gas (**FIGURE 31-2**).

If you know the state of matter and other physical properties of the chemical, you can begin to predict what the substance will do if it escapes, or has escaped, from its containment vessel. For example, it would be vital to know if a released gas is heavier or lighter than air and how that physical property relates to the environmental factors at the time of the incident, along with the other characteristics of the incident scene. Imagine a release of a heavy gas such as propane in the setting of a trench rescue or other below-grade incident: The physical properties of propane could be a complicating factor to performing a safe rescue in such a scenario.

Another critical part of comprehending the nature of the release comes from identifying the reason(s) why the containment vessel failed. Potential ways that containers could breach include disintegration, runaway cracking, closures opening up, punctures, splits, or tears. In many cases, responders focus on the fact that

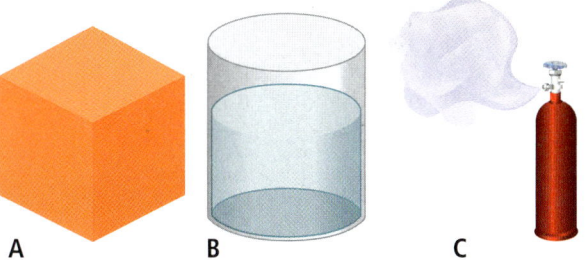

FIGURE 31-2 The state of matter identifies the hazard as a **A.** solid, **B.** liquid, or **C.** gas.
© Jones & Bartlett Learning.

a substance is being released rather than understanding why the product is escaping its container. It's one thing to notice that a container is leaking or generating a cloud from a puncture, crack, split, or tear. It's equally important, however, to figure out what type of stress caused the vessel to fail in the first place. Is the product release caused by a thermal influence—heat from inside or outside the container that may be the root cause of a catastrophic container failure? Maybe an errant forklift driver struck a drum, or a valve failure occurred. In these examples, physical damage allows the release of the container contents. In other cases, a chemical reaction inside a container may cause the entire container to breach (**FIGURE 31-3**).

Generally speaking there are three main types of stresses that can cause a container to fail:

- *Thermal*: Heat created from fire or cold generated by environmental factors or substances such as cryogenics.
- *Chemical*: The interaction of incompatible chemicals and/or the physical and chemical properties of a substance and how those substances interact inside or outside a container may lead to overpressure, disintegration, or other kinds of failures of any type of container.
- *Mechanical*: Falling debris, shrapnel, firearms, explosives, forklift puncture, and the like are all examples of how mechanical means can cause container failure.

These influences often result in predictable types of container failures such as the ones listed in **FIGURE 31-4**.

Responders must determine and/or estimate the duration of the event and link that to the other information gathered. For example, you should think of all parameters of the event as clues and use those clues to form a hypothesis and action plan. What's leaking? What are the properties of the material? How did it get out of its container? How long might the event last, and what happens if we do nothing? If we choose to intervene, are we trained to handle the incident? Do we have the proper personal protective equipment (PPE)? Are

FIGURE 31-3 Four examples of ways containers can release their contents. **A.** Rapid release. **B.** Spill or leak. **C.** Violent rupture. **D.** Detonation.

Courtesy of Rob Schnepp.

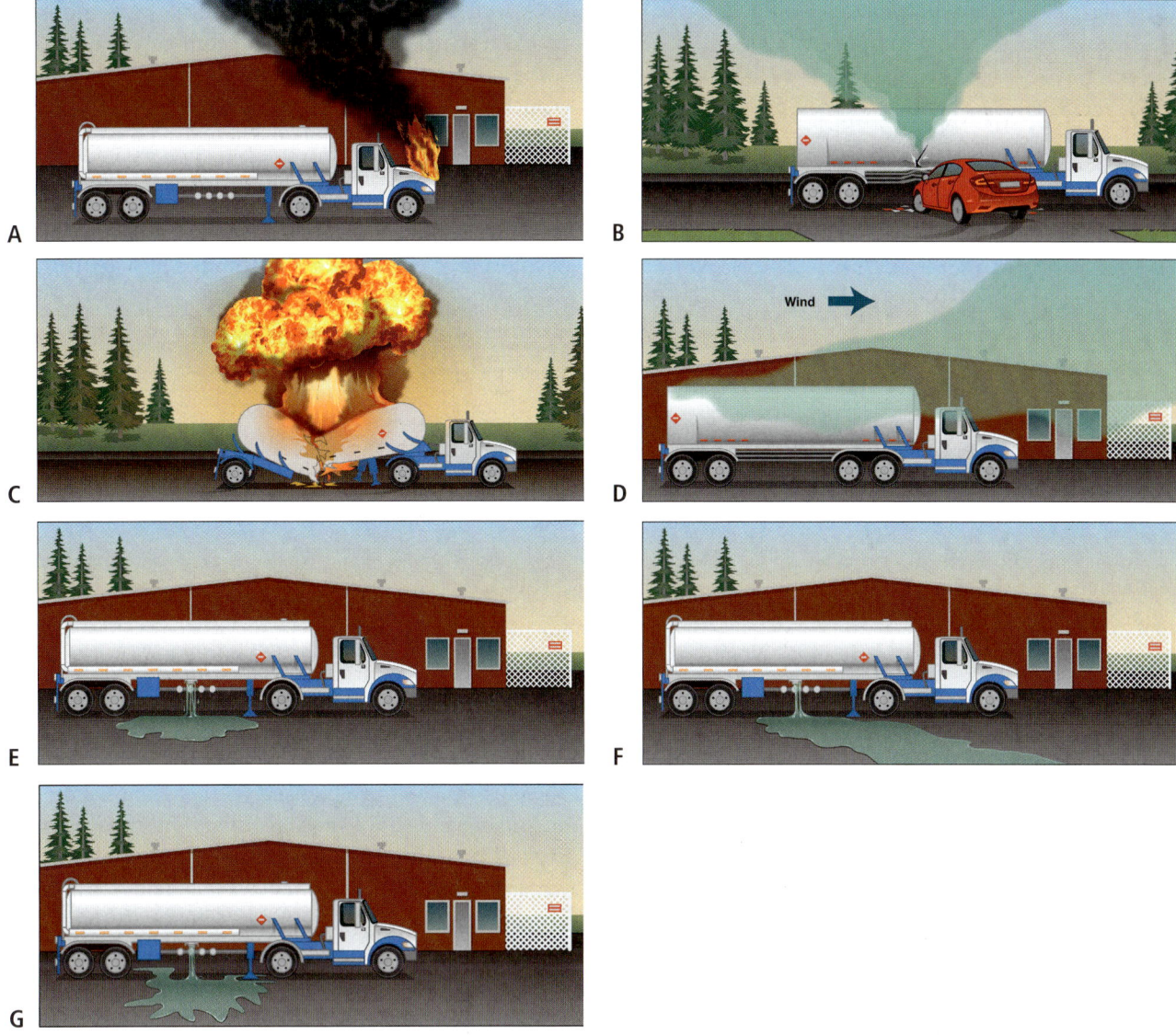

FIGURE 31-4 **A.** Cloud. **B.** Cone. **C.** Hemispheric release. **D.** Plume. **E.** Pool. **F.** Stream. **G.** Irregular dispersion.
© Jones & Bartlett Learning.

additional resources necessary? What is the likelihood of a positive outcome—safely resolving the problem? Incidents may last anywhere from seconds and minutes to several days and, in extreme cases, months or years. Tactics and strategies may vary greatly depending on the projected duration of the event. At any hazardous materials incident, it's important to link together all the bits of information to make an informed decision about how to handle the problem.

Chemicals can undergo a **physical change** when they are subjected to environmental influences such as heat, cold, and pressure. For example, when water is frozen, it undergoes a physical change from a liquid to a solid. The actual chemical make-up of the water (H_2O) is still the same, but the state in which it exists is different.

Like water, propane gas is subject to physical change based on environmental influences such as heat and cold. If a propane cylinder is exposed to heat, the compressed liquefied propane inside changes phase, becoming gaseous propane—which, in turn, increases the pressure inside the vessel. If this uncontrolled expansion takes place faster than the relief valve can vent, a BLEVE could occur. A **BLEVE** (boiling liquid expanding vapour explosion) occurs when pressurized liquefied materials (propane or butane, e.g.) inside a closed vessel are exposed to a source of high heat. In this case, the chemical make-up of propane has not changed; rather, the propane has physically changed state from liquid to gas.

The **expansion ratio** is a description of the volume increase that occurs when a compressed liquefied gas (e.g., propane, chlorine, oxygen) changes to a gas. For example, propane has an expansion ratio of 270 to 1. This means that for every 1 volume of liquid propane, 270 times that amount of propane exists as **vapour** (the gas phase of a substance). If released into the atmosphere,

4 litres (1 gallon) of liquid propane would vaporize to 1022 litres (270 gallons) of propane vapour.

FIGURE 31-5 Responders should take great care when handling incidents involving compressed liquefied gases.
Courtesy of Rob Schnepp.

Chemical reactivity (also known as chemical change) describes the ability of a substance to undergo a transformation at the molecular level, usually with a release of some form of energy. Physical change is essentially a change in state. By contrast, a chemical change results in an alteration of the chemical nature of the material. Steel rusting and wood burning are examples of chemical change. Polymers are good examples of reactive substances. **Polymerization** is a process of reacting monomers together in a chain reaction to form polymers. Many plastics in use in the home and in industry are produced by the process of polymerization.

To relate the concepts of physical and chemical change to a hazardous materials incident, think back to the initial case study. Assume for a moment that the owner of the landscape company wadded up some rags soaked with linseed oil and left them in a corner of the shed. The rags spontaneously ignited and started a fire that ultimately caused the failure of a small propane cylinder (the explosion that occurred prior to your arrival). The fire also melted the fusible plug on a chlorine cylinder (at approximately 71°C [160°F]), causing a release of chlorine gas; chlorine gas is 2.5 times heavier than air.

Looking closely at each event in this sequence, you can see the effect of physical and chemical changes. Linseed oil is an organic material that generates heat as it decomposes (chemical change). That heat, in the presence of oxygen in the surrounding air, ignited the rags, which in turn ignited other **combustibles**. The surrounding heat caused the propane inside the tank to expand (physical change) until it overwhelmed the ability of the relief valve to handle the build-up of pressure. The cylinder ultimately exploded. At the same time, the escaping chlorine gas mixed with the moisture in the air and formed an acidic mist (a chemical change). A slight breeze carried the smoke and mist toward the apartment complex. The fire fighters began to feel itchy and burning sensations where the acidic mist contacted their moist skin. Because chlorine vapours are heavier than air, most of the residents complaining about eye irritation are located on the bottom floor of the complex.

Critical Characteristics of Flammable Liquids

When looking at the fire potential of a flammable liquid, several important aspects must be considered. Among them are flash point, ignition temperature, and flammable range. More important than memorizing the definitions for these terms, however, responders must understand the relationships between these and other physical characteristics of a flammable liquid. Keep in mind that when it comes to combustion (burning), all materials must be in a gaseous or vapour state prior to flaming combustion: Solids do not burn, and liquids do not burn—they give off a gas or vapour that is ultimately ignited. Think of a log burning in a fireplace. If you look closely, the fire is not directly on the log but rather appears slightly above the surface. The reason for this phenomenon is that the wood must be heated to the point where it produces enough "wood gas" to support combustion. The log, then, does not burn—the gas produced by heating the log is what starts and sustains the process of combustion.

Conceptually, liquid fuels such as gasoline, diesel fuel, and other hydrocarbon-based dissolving solutions (**solvents**) behave the same way as the log. One way or another, vapour production must occur before there can be fire. This vapour production must be factored in when estimating the probability of fire during a release.

Flash Point

Flash point is an expression of the minimum temperature at which a liquid or solid gives off sufficient vapours such that, when an ignition source (e.g., a flame, electrical equipment, lightning, or even static electricity) is present, the vapours will result in a flash fire. The flash fire involves only the vapour phase of the liquid (like the example of the log) and will go out once the vapour fuel is consumed.

To illustrate how such an event occurs, consider the flash point of gasoline at –43°C (–45°F). When the temperature of gasoline reaches –43°C (–45°F) because of heat from an external source or from the surrounding environment, it gives off sufficient flammable vapours to ignite but not support combustion. In nearly all circumstances, when gasoline is spilled or otherwise released, the temperature of the external environment is well above the flash point of gasoline, creating the potential for ignition. Diesel fuel, by comparison, has a much higher flash point than gasoline—in the range of approximately 49°C (120°F) to 60°C (140°F), depending on the fuel grade. In either case, once the temperature of the liquid surpasses its flash point, the fuel will give off sufficient flammable vapours to support combustion (**FIGURE 31-6**).

Additionally, liquids with low flash points—such as gasoline, ethyl alcohol, and acetone—typically have higher vapour pressures and higher ignition temperatures. Vapour pressure and ignition temperature are explained in more detail later in this chapter. Consider the case of gasoline, whose flash point is –43°C (–45°F). At a vapour pressure of more than 275 mm Hg (compared to the vapour pressure of ethyl alcohol of 40 mm Hg at approximately 20°C [68°F]), gasoline has an ignition temperature of approximately 246°C (475°F). By contrast, flammable/combustible liquids with high flash points typically have lower ignition temperatures and lower vapour pressures. A grade of diesel—

–43°C (–45°F)
Flash Point
(Gasoline)

FIGURE 31-6 Fire fighters should always be mindful of ignition sources at flammable/combustible liquid incidents.
© Jones & Bartlett Learning.

#2 grade diesel, for example—may have a flash point of 52°C (125°F) and a corresponding ignition temperature of approximately 260°C (500°F); this is clearly a much narrower range than that described in the gasoline example. The vapour pressure of #2 grade diesel fuel is very low—much lower than the vapour pressure of water. **TABLE 31-1** lists several other examples of flash points of flammable/combustible liquids, illustrating the relationships among flash point, ignition temperature, and vapor pressure.

A direct relationship also exists between temperature and vapour production. Simply put, when the temperature increases, the vapour production of any flammable liquid increases, leading to a higher concentration of vapours. Therefore, even liquids with low flash points can be expected to produce a significant amount of flammable vapours at all but the lowest ambient temperatures.

Flash point is merely one aspect to consider when flammable/combustible liquids are released from a container. The fire point is another definition that should be appreciated. **Fire point** is the temperature at which sustained combustion of the vapour will occur. It is usually only slightly higher than the flash point for most materials.

Ignition Temperature

Ignition temperature is another important temperature landmark for flammable/combustible liquids. From a technical perspective, you can think of ignition temperature as the minimum temperature at which a fuel, when heated, will ignite in the presence of air and continue to burn.

From a responder's perspective, it is important to realize that when a liquid fuel is heated beyond its ignition temperature, from any type of heat, it will ignite without an external ignition source. Think of a pan full

TABLE 31-1 Flash Point, Vapour Pressure, and Ignition Temperature

	Flash Point	Vapour Pressure	Ignition Temperature
Water	N/A	25 mm Hg at 20°C (68°F)	N/A
Gasoline	−43°C (−45°F)	275–400 mm Hg at 21°C (70°F)	246°C (475°F)
Acetone	−20°C (−4°F)	400 mm Hg at 40°C (104°F)	465°C (869°F)
#2 grade diesel	52°C (125°F)	< 2 mm Hg at 20°C (68°F)	260°C (500°F)

All readings within the table are closed cup results. These results were created with closed cup testing and are not always accurate within the atmosphere.
© Jones & Bartlett Learning.

of cooking oil on the stove. For illustrative purposes, assume that the ignition temperature of the oil is 148°C (300°F). What would happen if the burner was set on high and left unattended so that the oil was heated past 148°C (300°F)? Once the temperature of the oil exceeds its ignition temperature, it will ignite; there is no need for an external ignition source. In fact, this scenario is a common cause of stove fires.

Flammable Range

Flammable range (explosive limit) is another important term to understand. Defined broadly, flammable range is an expression of a fuel/air mixture, defined by upper and lower limits, that reflects an amount of flammable vapour mixed with a given volume of air. Gasoline will serve as our example. The flammable range for gasoline vapours is 1.4 percent to 7.6 percent. The two percentages, called the **lower explosive limit (LEL)** (1.4 percent) and the **upper explosive limit (UEL)** (7.6 percent), define the boundaries of a fuel/air mixture necessary for gasoline to burn properly. If a given gasoline/air mixture falls between the LEL and the UEL and that mixture encounters an ignition source, a flash fire will occur.

The concept of automobile carburetion capitalizes on the notion of flammable range—the carburetor is the place where gasoline and air are mixed. When the mixture of gasoline vapours and air occurs in the right proportions, the car runs smoothly. When there is too much fuel and not enough air, the mixture is too "fuel rich." When there is too much air and not enough fuel, the carburetion is too "fuel lean." In either of these two cases, optimal combustion is not achieved, and the motor does not run correctly.

Understanding flammable range, as it relates to hazardous materials response, is based on the same line

TABLE 31-2 Flammable Ranges of Common Gases

Gas	Flammable Range
Hydrogen	4.0%–75%
Natural gas	5.0%–15%
Propane	2.5%–9.0%

© Jones & Bartlett Learning.

of thinking. If gasoline is released from a rolled-over cargo tank, the vapours from the spilled liquid will mix with the surrounding air. If the mixture of vapours and air falls between the UEL and the LEL and those vapours reach an ignition source, a flash fire will occur. Ultimately, the entire volume of gasoline will likely burn. As with carburetion, if too much or too little fuel is present, the mixture will not adequately support combustion.

Generally speaking, the wider the flammable range, the more dangerous the material (**TABLE 31-2**). This relationship reflects the fact that the wider the flammable range, the more opportunity there is for an explosive mixture to find an ignition source.

Vapour Pressure

For our purposes, the definition of **vapour pressure** will pertain to liquids held inside any type of closed container. When liquids are held in a closed 208-litre (55-gallon) drum or a 4-litre (1 gallon) glass bottle, for example, some amount of pressure (in the headspace above the liquid) will develop inside (**FIGURE 31-7**).

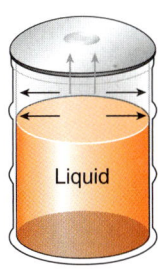

FIGURE 31-7 Vapour pressure. The vapour in the headspace is exerting pressure in all directions above the liquid inside the drum.
© Jones & Bartlett Learning.

All liquids, even water, will develop a certain amount of pressure in the air space between the top of the liquid and the container.

The key point to understanding vapour pressure is this: *The vapours released from the surface of any liquid must be contained if they are to exert pressure.* Essentially, the liquid inside the container will vaporize until the molecules given off by the liquid reach equilibrium with the liquid itself. Equilibrium is a balancing act between the liquid and the vapours—some molecules turn to vapour, whereas others leave the vapour phase and return to the liquid phase.

Carbonated beverages illustrate this concept. Inside the basic cola drink is the formula for the soda and a certain amount of carbon dioxide (CO_2) molecules. The bubbles in the soda are composed of CO_2. When the soda sits in an unopened can or plastic bottle on the shelf, the balancing act of CO_2 is happening: CO_2 from the liquid becomes gaseous CO_2 above the liquid; at the same time, some of the CO_2 dissolves back into the liquid. The pressure inside the bottle can be verified by feeling the rigidity of the container—until you open it. Once the lid is removed, the CO_2 is no longer in a closed vessel, and it escapes into the atmosphere. With the pressure released, the sides of the bottle can be squeezed easily. Ultimately, the soda goes flat when all the CO_2 has escaped.

With the soda example in mind, consider the technical definition of vapour pressure: The vapour pressure of a liquid is the pressure exerted by its vapour until the liquid and the vapour are in equilibrium. Again, this process occurs inside a closed container, and temperature has a direct influence on the vapour pressure. For example, if heat impinges on a drum of acetone, the pressure above the liquid will increase and perhaps cause the drum to fail. If the temperature is dramatically reduced, the vapour pressure will drop. This is true for any liquid held inside a closed container.

What happens if the container is opened or spilled onto the ground to form a puddle? The liquid still has a vapour pressure, but it is no longer confined to a container. *In this case, we can conclude that liquids with high vapour pressures will evaporate much more quickly than will liquids with low vapour pressures.* Vapour pressure directly correlates to the speed with which a material will evaporate once it is released from its container.

For example, motor oil has a low vapour pressure. When it is released, it will stay on the ground a long time. Chemicals such as isopropyl alcohol and diethyl ether exhibit the opposite behaviour: When either of these materials is released and collects on the ground, it will evaporate rapidly. If ambient air temperature or pavement temperatures are elevated, their evaporation rates will increase even further. Wind speed, shade, humidity, and the surface area of the spill also influence how fast the chemical will evaporate.

When consulting reference sources, be aware that the vapour pressure may be expressed in pounds per square inch (psi), atmospheres (atm), torr, or millimetres of mercury (mm Hg); most references give the vapour pressures of substances at a temperature of 20°C (68°F). Each expression of pressure is a valid point of reference in incident response. The term *millimetres of mercury* (mm Hg) is commonly found in reference books; it is defined as the pressure exerted at the base of a column of fluid that is exactly 1 millimetre in height. The *torr* unit is named after the Italian physicist Evangelista Torricelli, who discovered the principle of the mercury barometer in 1644. In his honour, the torr was equated to 1 mm Hg. Another common expression of pressure is bar. The conversion from bar to psi is 1 bar = 14.7 psi.

Certain conversion factors allow calculations from one reference point to another, but some of these factors are very complex. For the purposes of simplicity and incident response, the important point is to have some frame of reference to understand the concept of vapour pressure and to recognize how that concept will affect the release of chemicals into the environment.

To understand the relationships among the various units of pressure, use the following comparison:

101.4 kPa (14.7 psi) = 1 atm = 760 torr = 760 mm Hg = 1 bar

Another method of comparison is to take the values from the reference books and compare those values to the behaviour of substances you may be familiar with. The following example compares three common substances (using mm Hg): water, motor oil, and isopropyl alcohol. The vapour pressure of water at room temperature is approximately 25 mm Hg. Standard 40-weight motor oil has a vapour pressure of less than 0.1 mm Hg at 20°C (68°F)—it is practically vapourless at room temperature. Isopropyl alcohol, by contrast, has a high vapour pressure of 30 mm Hg at room temperature. Again, temperature has a direct correlation to vapour pressure, but all things being equal, these three substances give you a good starting point for making comparisons.

Boiling Point

Boiling point is the temperature at which a liquid will continually give off vapours in sustained amounts and, if held at that temperature long enough, will turn completely into a gas. The boiling point of water, for example, is 100°C (212°F). At this temperature, water molecules have enough kinetic energy (energy in motion) to overcome the downward force of the surrounding atmospheric pressure (**FIGURE 31-8**). (At sea level, a pressure of 101.4 kPa (14.7 psi) is exerted on every surface of every object, including the surface of the water.) At temperatures less than 100°C (212°F), there is insufficient heat to create enough kinetic energy to allow the water molecules to escape.

An illustration of boiling point, which demonstrates how heated liquids inside closed containers might create a deadly situation, can be found by looking at a completely benign and common example—popcorn. Fundamentally, popcorn pops when the trapped water inside the kernel of corn is heated beyond its boiling

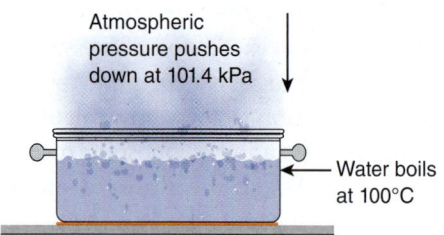

FIGURE 31-8 The concept of boiling point versus atmospheric pressure.

Atmospheric pressure pushes down at 101.4 kPa

Water boils at 100°C

point. When the popcorn kernels are heated, the small amount of water inside each kernel eventually exceeds its boiling point and expands to 1700 times its original water volume. The kernel then "pops." A kernel of corn has no relief valve, so a rapid build-up of pressure will cause it to breach. Unpopped kernels are those with insufficient water inside the kernel to permit this expansion.

Vapour Density

In addition to identifying the flammability of a vapour or gas, responders must determine whether the vapour or gas is heavier or lighter than air. In essence, you must know the **vapour density** of the substance in question. Vapour density is the weight of an airborne concentration of a vapour or gas as compared to an equal volume of dry air (**FIGURE 31-9**).

Basically, vapour density is a question of comparison—what will the gas or vapour do when it is released in the air? Will it collect in low spots in the topography or somewhere inside a building, or will it float upward into the ventilation system or upward into the air? These are important response considerations for hazardous materials incidents and may influence the severity of the incident.

You can find a chemical's vapour density by consulting a good reference source such as an SDS. The vapour density will be expressed in numerical fashion,

FIGURE 31-9 Cylinder **A.** Vapour density less than 1. Cylinder **B.** Vapour density greater than 1.
© Jones & Bartlett Learning.

such as 1.2, 0.59, or 4.0. The vapour density of propane, for example, is 1.55. Air has a set vapour density value of 1.0. Gases such as propane or chlorine are heavier than air; therefore, they have a vapour density value greater than 1.0. Substances such as acetylene, natural gas, and hydrogen are lighter than air and will have a vapour density value of less than 1.0.

In the absence of reliable reference sources in the field, there is a mnemonic you can use to remember many lighter-than-air gases: 4H MEDIC ANNA. This mnemonic translates as follows:

H: Hydrogen
H: Helium
H: Hydrogen cyanide
H: Hydrogen fluoride
M: Methane
E: Ethylene
D: Diborane
I: Illuminating gas (methane/ethane mixture)
C: Carbon monoxide
A: Ammonia
N: Neon
N: Nitrogen
A: Acetylene

If you encounter a leaking propane cylinder, for example, just refer to 4H MEDIC ANNA. Propane is not on that list, so by default, you can assume that propane is heavier than air. You won't find chlorine or butane on the list, either: Both are heavier than air. Again, if the substance is not on this list, it is most likely heavier than air.

Specific Gravity

Specific gravity is to liquids what vapour density is to gases and vapours—namely, a comparison value. In this case, the comparison is between the weight of a liquid chemical and the weight of water. Water is assigned a value of 1.0 as its specific gravity. Any material with a specific gravity value less than 1.0 will float on water (**FIGURE 31-10**). Any material with a specific gravity value greater than 1.0 will sink and remain below the surface of water. A comprehensive reference source will provide the specific gravity of the chemical in question.

Most flammable liquids will float on water. Gasoline, diesel fuel, motor oil, and benzene are all examples of liquids that float on water (**FIGURE 31-11**). Carbon disulfide, by comparison, has a specific gravity of approximately 2.6. Consequently, if water

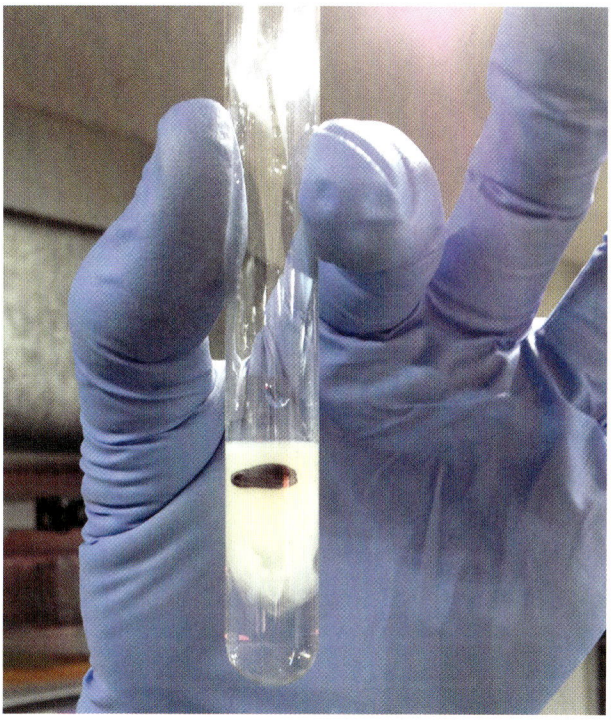

FIGURE 31-10 An example of a substance floating on water.
Courtesy of Rob Schnepp.

FIGURE 31-11 Gasoline floating on water.
Courtesy of Rob Schnepp.

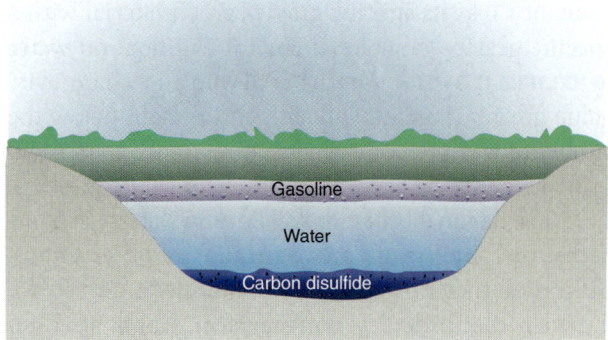

FIGURE 31-12 Gasoline will float on water, whereas carbon disulfide will not.
© Jones & Bartlett Learning.

is gently applied to a puddle of carbon disulfide, the water will rest on top and cover the puddle completely (**FIGURE 31-12**).

LISTEN UP!

It is important to carefully consider the application of water in all cases involving spilled or released chemicals. The results could be something worse than the initial problem. Complications such as chemical reactivity or incompatibility, contaminated runoff, and unwanted increase in the surface area of the spill could occur.

Water Solubility

When discussing the concept of specific gravity, it is also necessary to determine if a chemical will mix with water. **Water solubility** describes the ability of a substance to dissolve in water. Water is the predominant agent used to extinguish fire, but when you are dealing with chemical emergencies, it may not always be the best and safest choice to mitigate the situation. That's because water is a tremendously aggressive solvent and can react violently with certain chemicals.

Concentrated sulfuric acid, metallic sodium, and magnesium are just a few examples of substances that will adversely react with water. If water is applied to burning magnesium, for example, the heat of the fire will break apart the water molecule, creating an explosive reaction. Adding water to sulfuric acid would be like throwing water on a pan full of hot oil—popping and spattering may occur.

In other circumstances, water can be a friendly ally when you are attempting to handle a chemical incident. Depending on the chemical involved, fog streams operated from handlines may knock down vapour clouds. Additionally, heavier-than-water flammable liquids can be extinguished by gently applying water to the surface of the liquid. In some cases, water may be an effective way to dilute a chemical, thereby rendering it less hazardous.

Corrosivity (pH)

Corrosivity is the ability of a material to cause damage (on contact) to skin, eyes, or other parts of the body. The technical definition, as written by the U.S. Department of Transportation (DOT), includes the language, "destruction or irreversible damage to living tissue at the site of contact." Such materials are often also damaging to clothing, rescue equipment, and other physical objects in the environment.

Corrosives are a complex group of chemicals that should not be taken lightly. The tens of thousands of corrosive chemicals used in general industry, semiconductor manufacturing, and biotechnology can be further categorized into two classes: **acid** and **base**.

There are several technical ways to describe the acidity or alkalinity of a solution, but the most common way to define them is by their **pH** (**FIGURE 31-13**). In simple terms, you can think of pH as the "power of hydrogen." Essentially, pH is an expression of the measurement of the presence of dissolved hydrogen ions (H^+) in a substance (technically defined as the potential of hydrogen). When thinking about pH from a field perspective, this value can be viewed as a measurement of corrosive strength, which can loosely be translated into a certain degree of hazard. In simple terms, we can use pH to judge how aggressive a corrosive substance

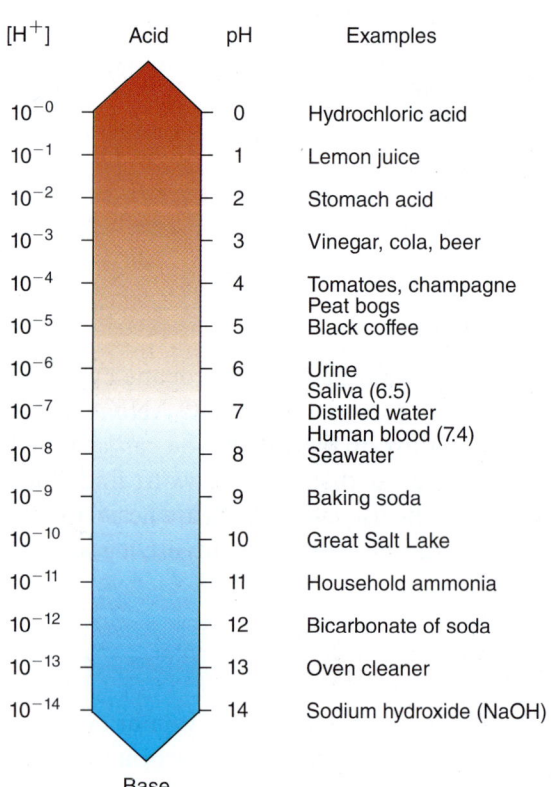

FIGURE 31-13 The pH scale.
© Jones & Bartlett Learning.

might be. To assess that hazard, you must understand the pH scale.

How do we determine pH in the field? One method is to use specialized pH test paper. You can also obtain this information from the SDS or call for a specialized hazardous materials response team to determine the pH level of a chemical. Hazardous materials technicians have more specialized ways of determining pH and may be a useful resource when responders are handling corrosive incidents.

Common acids, such as sulfuric, hydrochloric, phosphoric, nitric, and acetic (vinegar) acids, have a predominant amount of hydrogen ions (H^+) in the solution and, therefore, will have pH values less than 7 (**FIGURE 31-14**). Chemicals that are bases, such as sodium hydroxide, potassium hydroxide, sodium carbonate, and ammonium hydroxide, have a predominant amount of hydroxide (OH^-) ions in the solution and will have pH values greater than 7 (**FIGURE 31-15**).

The middle of the pH scale (7) is where a chemical is considered to be neutral—that is, neither acidic nor basic. A pH of 7 is "neutral" because the concentration of hydrogen ions (H^+) in such a material is exactly equal to the concentration of hydroxide (OH^-) ions produced by dissociation of the water. At this value, a chemical will not harm human tissue.

Generally, pH values of 2.5 or less, and those of 12.5 or greater, are considered to be strong. In practical terms, this designation means that strong corrosives (acids and bases) will react more aggressively with metallic substances such as steel and iron; they will cause more damage to unprotected skin; they will react more adversely when contacting other chemicals; and they may react violently with water.

At the hazardous materials operations level, it is critical to understand that incidents involving corrosives may not be as straightforward as other types of chemical emergencies. The materials themselves are more complicated, and, in fact, the tactics employed to deal with them could be outside your scope of responsibilities.

Toxic Products of Combustion

Toxic products of combustion are the hazardous chemical compounds released when a material decomposes under heat. Recall that the process of combustion is a chemical reaction and, like other reactions, will generate a given amount of gases (many of them toxic) and particulates as by-products of the decomposition of the fuel. Have you thought about what's in the smoke you may be breathing in or taken a moment to think about the toxic gases liberated during a residential structure fire?

An easy way to think about this issue is to apply a phrase long used in the world of chemistry: garbage in, garbage out. This phrase reflects the idea that whatever objects are involved in the fire (chairs, tables, and sofas, e.g.) decompose and release a host of chemical by-products.

Notable substances found in most fire smoke include soot (carcinogen), carbon monoxide, carbon dioxide, polycyclic aromatic hydrocarbons, benzene (carcinogen), water vapour, formaldehyde (carcinogen), cyanide compounds, chlorine compounds, and many oxides of nitrogen. Each of these substances is unique in its chemical make-up, and most are toxic to humans, even in small doses.

For example, carbon monoxide affects the ability of the human body to transport oxygen. When it is present in the body in excessive amounts, the red blood cells cannot get oxygen to the other cells of the body, and, subsequently, a person will die from tissue asphyxiation. Cyanide compounds also adversely affect oxygen uptake in the body and are often a cause or contributing cause of smoke-related illness and death. Formaldehyde is found in many plastics and resins; it is one of the many components of smoke that causes eye and lung irritation. The oxides of nitrogen, which include nitric

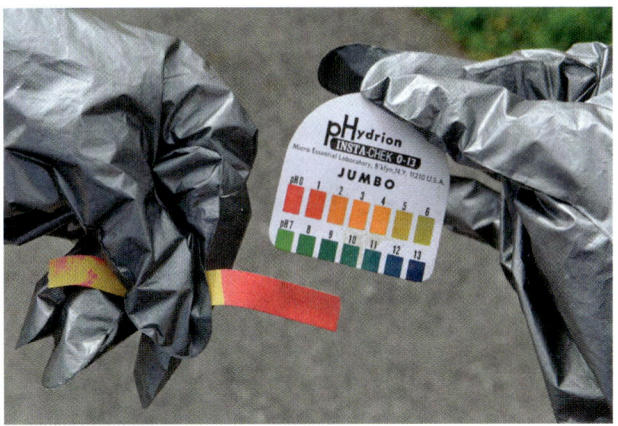

FIGURE 31-14 An acid.
© Jones & Bartlett Learning. Photographed by Glen E. Ellman.

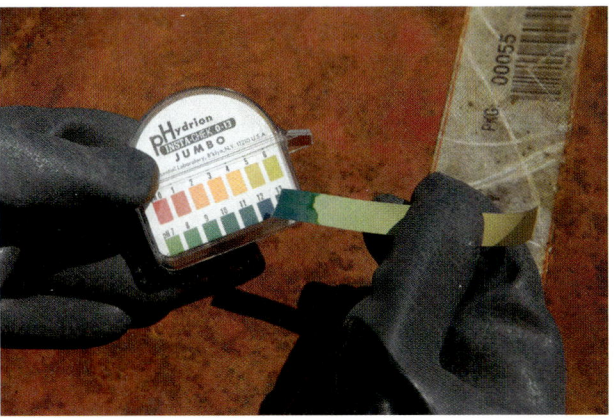

FIGURE 31-15 A base.
© Jones & Bartlett Learning. Photographed by Glen E. Ellman.

oxide, nitrous oxide, and nitrogen dioxide, are deep lung **irritants** that may cause a serious medical condition called **pulmonary edema** (fluid build-up in the lungs). There is more on this topic in Chapter 35, *Hazardous Materials Responder Health and Safety*.

Radiation

Most fire fighters have not been trained on the finer points of handling incidents involving radioactive materials and, therefore, have many misconceptions about radiation. **Radiation** is energy transmitted through space in the form of electromagnetic waves or energetic particles.

Fundamentally, you should understand that you cannot escape radiation. It is all around you, every day. During your life, you will receive radiation from the sun and the soil, when taking a plane ride, or by having an x-ray. Radiation has been around since the beginning of time. Our focus here, however, is not on background radiation—the kind of radiation you receive from the sun or the soil; rather, it is on the occupational exposures encountered in the field. For the most part, the health hazards posed by radiation are a function of two factors:

- The amount of radiation absorbed by your body has a direct relationship to the degree of damage done.
- The exposure time to the radiation will ultimately affect the extent of the injury.

The periodic table of elements illustrates all the known elements that are found in all the chemical compounds on the face of the Earth (**FIGURE 31-16**). Those elements, in turn, are made up of atoms. In the nucleus of those atoms are protons (positive [+] electrical charge) and neutrons (no electrical charge). Orbiting the nucleus are electrons (negative [–] electrical charge).

All stable atoms of any given element will have the same number of protons and neutrons in their nucleus. Those same elements, however, are prone to having an imbalance in the numbers of protons and neutrons. This variation in the number of neutrons creates a **radioactive isotope** of the element. Carbon-14, for example, is a radioactive isotope of carbon because of the imbalance in the numbers of protons and neutrons found in the nucleus of carbon-14. You will notice on the periodic table of elements that stable carbon (C) has an atomic mass of 12. Carbon-14 illustrates a different mass, which shows its imbalance of protons and neutrons. Other examples include sulfur-35 and

> **LISTEN UP!**
>
> Isotopes are atoms of the same element with the *same number* of protons but *different numbers* of neutrons.

PERIODIC TABLE OF THE ELEMENTS

Element	Atomic Number	Symbol	*Atomic Mass
hydrogen	1	H	1.01

Metals / Metalloids / Nonmetals

	1 IA	2 IIA	3 IIIB	4 IVB	5 VB	6 VIB	7 VIIB	8 VIII	9 VIII	10 VIII	11 IB	12 IIB	13 IIIA	14 IVA	15 VA	16 VIA	17 VIIA	18 VIIIA
																		helium 2 He 4.00
2	lithium 3 Li 6.94	beryllium 4 Be 9.01											boron 5 B 10.81	carbon 6 C 12.01	nitrogen 7 N 14.01	oxygen 8 O 16.00	fluorine 9 F 19.00	neon 10 Ne 20.18
3	sodium 11 Na 22.99	magnesium 12 Mg 24.31											aluminum 13 Al 26.98	silicon 14 Si 28.09	phosphorus 15 P 30.97	sulfur 16 S 32.07	chlorine 17 Cl 35.45	argon 18 Ar 39.95
4	potassium 19 K 39.10	calcium 20 Ca 40.08	scandium 21 Sc 44.96	titanium 22 Ti 47.88	vanadium 23 V 50.94	chromium 24 Cr 52.00	manganese 25 Mn 54.94	iron 26 Fe 55.85	cobalt 27 Co 58.93	nickel 28 Ni 58.69	copper 29 Cu 63.55	zinc 30 Zn 65.39	gallium 31 Ga 69.72	germanium 32 Ge 72.61	arsenic 33 As 74.92	selenium 34 Se 78.96	bromine 35 Br 79.90	krypton 36 Kr 83.80
5	rubidium 37 Rb 85.47	strontium 38 Sr 87.62	yttrium 39 Y 88.91	zirconium 40 Zr 91.22	niobium 41 Nb 92.91	molybdenum 42 Mo 95.94	technetium 43 Tc (99)	ruthenium 44 Ru 101.07	rhodium 45 Rh 102.91	palladium 46 Pd 106.42	silver 47 Ag 107.87	cadmium 48 Cd 112.41	indium 49 In 114.82	tin 50 Sn 118.71	antimony 51 Sb 121.75	tellurium 52 Te 127.60	iodine 53 I 126.90	xenon 54 Xe 131.29
6	cesium 55 Cs 132.91	barium 56 Ba 137.33	lanthanum 57 La 138.91	hafnium 72 Hf 178.49	tantalum 73 Ta 180.95	tungsten 74 W 183.85	rhenium 75 Re 186.21	osmium 76 Os 190.2	iridium 77 Ir 192.22	platinum 78 Pt 195.08	gold 79 Au 196.97	mercury 80 Hg 200.59	thallium 81 Tl 204.38	lead 82 Pb 207.2	bismuth 83 Bi 208.98	polonium 84 Po (209)	astatine 85 At (210)	radon 86 Rn (222)
7	francium 87 Fr (223)	radium 88 Ra (226)	actinium 89 Ac (227)	rutherfordium 104 Rf (261)	dubnium 105 Db (262)	seaborgium 106 Sg (263)	bohrium 107 Bh (262)	hassium 108 Hs (265)	meitnerium 109 Mt (266)	ununnilium 110 Uun (269)	unununium 111 Uuu (272)	ununbium 112 Uub (277)	ununtrium 113 Uut (284.18)	flerovium 114 Fl (289.19)	ununpentium 115 Uup (288.19)	livermorium 116 Lv (293)	ununseptium 117 Uus (294)	ununoctium 118 Uuo (294)

Lanthanide Series	cerium 58 Ce 140.12	praseodymium 59 Pr 140.91	neodymium 60 Nd 144.24	promethium 61 Pm (147)	samarium 62 Sm 150.36	europium 63 Eu 151.97	gadolinium 64 Gd 157.25	terbium 65 Tb 158.93	dysprosium 66 Dy 162.50	holmium 67 Ho 164.93	erbium 68 Er 167.26	thulium 69 Tm 168.93	ytterbium 70 Yb 173.04	lutetium 71 Lu 174.97
Actinide Series	thorium 90 Th 232.04	protactinium 91 Pa (231)	uranium 92 U 238.03	neptunium 93 Np (237)	plutonium 94 Pu (244)	americium 95 Am (243)	curium 96 Cm (247)	berkelium 97 Bk (247)	californium 98 Cf (251)	einsteinium 99 Es (252)	fermium 100 Fm (257)	mendelevium 101 Md (258)	nobelium 102 No (259)	lawrencium 103 Lr (260)

*Note: For radioactive elements, the mass number of an important isotope is shown in parentheses; for thorium and uranium, the atomic mass of the naturally occurring radioisotopes is given.

FIGURE 31-16 The periodic table of elements.

phosphorus-32, which are radioactive isotopes of sulfur and phosphorus, respectively.

Radioactivity is the natural and spontaneous process by which unstable atoms (isotopes) of an element decay to a different state and emit or radiate excess energy in the form of particles or waves. Radioactive isotopes give off energy from the nucleus of an unstable atom to reach a stable state. Typically, you will encounter a combination of alpha, beta, and gamma radiation. Each of these forms of energy given off by a radioactive isotope will vary in intensity and consequently determine your efforts to reduce the exposure potential (**FIGURE 31-17**).

Small radiation detectors are available that can be worn on turnout gear. These detectors sound an alarm when dangerous levels of radiation are encountered and alert responders to leave the scene and call for more specialized assistance.

Alpha Particles

As mentioned previously, radiation stems from an imbalance in the number of protons and neutrons in the nucleus of an atom. Alpha radiation is a reflection of that instability. This form of radiation energy is produced when an electrically charged particle is given off by the nucleus of an unstable atom. **Alpha particles** have weight and mass; therefore, they cannot travel very far (less than a few centimetres) from the nucleus of the atom. For comparison, alpha particles are like dust particles and can be stopped by a sheet of paper or a layer of skin. Typical alpha emitters include americium (found in smoke detectors), polonium (identified in cigarette smoke), radium, radon, thorium, and uranium.

You can protect yourself from alpha emitters by staying several feet away from their source and by protecting your respiratory tract with a P100 filter on an air-purifying respirator or a self-contained breathing apparatus (SCBA).

Beta Particles

Beta particles are more energetic than alpha particles and therefore pose a greater health hazard. Essentially, beta particles are like electrons, except that a beta particle is ejected from the nucleus of an unstable atom. Depending on the strength of the source, beta particles can travel 3 to 4.5 metres (10 to 15 feet) in the open air and can typically be stopped by a layer of clothing. The beta particles themselves are not radioactive; rather, the radiation energy is generated by the speed at which the particles are emitted from the nucleus. Due to this phenomenon, beta radiation can break chemical bonds at the molecular level and cause damage to living tissue. This breaking of chemical bonds creates an ion; therefore, beta particles are considered **ionizing radiation**. Ionizing radiation has the capability to cause changes in human cells, which may ultimately lead to a mutation of the cell and become the root cause of cancer. Other examples of ionizing radiation include x-rays and gamma rays.

Non-ionizing radiation comes from electromagnetic waves, which can cause a disturbance of activity at the atomic level but do not have sufficient energy to break bonds and create ions. Typical non-ionizing waves include sound waves, radio waves, and microwaves.

Beta particles can redden (erythema) and burn skin; they can also be inhaled. Beta particles that are inhaled can directly damage the cells of the human body. Most solid objects can stop these particles, and your SCBA should provide adequate respiratory protection against

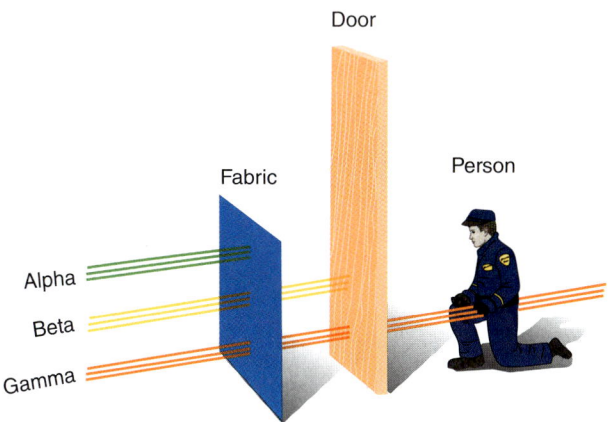

FIGURE 31-17 Alpha, beta, and gamma radiation.
© Jones & Bartlett Learning.

LISTEN UP!

To reduce the effects of a radiation exposure, responders should understand the concept of time–distance–shielding (TDS). Think of TDS in this way: The less time you spend in the sun, the less chance you have of suffering a sunburn (*time*); the closer you stand to a fire, the hotter you will get (*distance*); and if you come inside during a rainstorm, you will stop getting wet (*shielding*). TDS directly correlates to ALARA (**A**s **L**ow **A**s **R**easonably **A**chievable)—a principle of radiation protection philosophy that requires that exposures to ionizing radiation shall be kept as low as reasonably achievable.

them. Common beta emitters include tritium (luminous dials on gauges), iodine (medical treatment), and cesium.

Gamma Rays and Neutrons

Gamma radiation is the most energetic radiation responders may encounter. Gamma radiation differs from alpha and beta radiation in that it is not a particle ejected from the nucleus; rather, it is pure electromagnetic energy. A gamma ray has no mass and no electrical charge and travels at the speed of light. Gamma rays can pass through thick solid objects (including the human body) very easily and generally follow the emission of a beta particle. If the nucleus still has too much energy after ejecting beta particles, it may release a photon (a packet of pure energy) that takes the form of a gamma ray.

Like beta particles, gamma radiation is a form of ionizing radiation and can be deadly. Structural firefighting gear that includes SCBA will not protect you from gamma rays; if fire fighters are near the source of this type of radiation, they will be exposed. Typical sources of gamma radiation include cesium (cancer treatment and soil density testing at construction sites) and cobalt (medical instrument sterilization).

Neutrons are also penetrating particles found in the nucleus of the atom. Neutrons themselves are not radioactive and do not have a positive or negative charge but do have mass. Neutrons may be detected in nuclear facilities or research laboratories. Exposure to neutrons can create radiation, such as gamma radiation.

When it comes to gaining information from labels found on radioactive material, responders should understand the following:

- The type or category of the label
- The contents of the container
- The activity level of the contained substance (rate of disintegration or decay of the substance measured in curies or becquerels)
- The transport index (maximum dose equivalent rate at 1 m from the surface of the package)

- The critical safety index (a number used to provide control over the accumulation of packages or freight containers containing **fissile** material)

These basic pieces of information found on a radiation label will go a long way toward helping you understand the hazards.

Hazard, Exposure, and Contamination

When responders operate at hazardous materials incidents, it is vital to identify the potential hazards and risks to minimize the potential for exposure. This begins by keen observation of the scene to determine whether a hazardous substance is released or has the potential to escape its container. At some point, early in the incident, it is vital to determine whether the pool of responders has the right training, equipment, and protective gear to positively influence the outcome of the incident. In some cases, a "no-go" situation may exist if the problem exceeds the ability of the responders to solve it safely within the boundaries of their training. It is as important to determine what can't be done as it is to decide on what actions to take. Additionally, responders must establish a safe perimeter around the problem, keeping it secure from accidental entrance and possible human exposures. In fact, the most important initial actions taken at a hazardous materials incident revolve around identifying the problem and taking actions to limit the spread of contamination and/or human exposures.

Hazard and Exposure

A *hazardous material* is defined as a material capable of posing an unreasonable risk to health, safety, or the environment—that is, a material capable of causing harm. The same source defines **exposure** as the process by which people, animals, the environment, and equipment are subjected to or come into contact with a hazardous material.

Contamination

When a chemical has been released and physically comes in contact with people, the environment, and everything around it, either intentionally or unintentionally, the residue of that chemical is called **contamination**. In some cases, the process of removing such contaminants is complex and requires a significant effort to complete. In other cases, the contamination is easily eliminated. Chapter 36, *Hazardous Materials Responder Personal Protective Equipment*, covers decontaminating the responders and their gear. For now, simply understand that when a chemical escapes its container, it's possible it will get into or onto something, and you will need a system to safely and efficiently remove that chemical or otherwise reduce the hazard. This could include exposed persons, PPE, or tools and equipment.

Secondary Contamination

Secondary contamination, also known as cross-contamination, occurs when a person or object transfers the contaminant or the source of contamination to another person (responder or civilian) or object by direct contact. Responders may become contaminated and subsequently handle tools and equipment, touch door handles or other responders, and spread the contamination. Additionally, contaminated victims who are improperly handled by unprotected responders also run the risk of spreading contamination (**FIGURE 31-18**).

The possibility of spreading contamination brings up a point that all responders should understand: The cleaner responders stay during the response, the less decontamination they will need later. Many responders have the misperception that PPE is worn to enable you to contact the product. In fact, quite the opposite is true: PPE, including specialized chemical protective gear, is worn to protect you in the event that you cannot avoid product contact. The less contamination encountered or spread, the easier the decontamination will be.

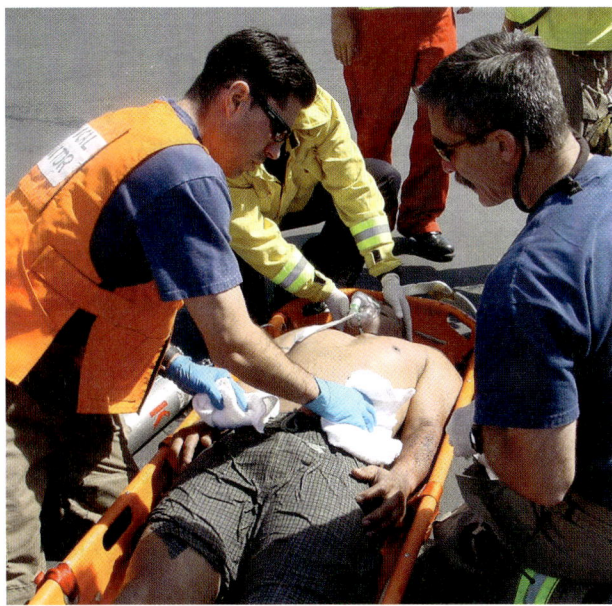

FIGURE 31-18 Responders should ensure contaminated victims are completely decontaminated prior to rendering care.
Courtesy of Rob Schnepp.

Chronic and Acute Health Effects

A **chronic health hazard** (also known as a chronic health effect) is an adverse health effect that occurs gradually over time after long-term exposure to a hazard. Chronic health effects may appear either after long-term or **chronic exposures** or following multiple short-term exposures that occur over a shorter period. The exposures can involve all routes of entry into the body and may result in such health problems as cancer, permanent loss of lung function, or repetitive skin rashes. For example, inhaling asbestos fibres for years without respiratory protection can result in a form of lung cancer called asbestosis.

Chemicals that pose a hazard to health after relatively short exposure periods are said to cause **acute health effects**. Such an exposure period is subchronic, with effects occurring either immediately after a single exposure or as long as several days or weeks after an exposure. Essentially, **acute exposures** are "right now" exposures that produce some observable conditions such as eye irritation, coughing, dizziness, and skin burns.

Skin irritation and burning after a dermal exposure to sulfuric acid would be classified as an acute health effect. Other chemicals, such as formaldehyde, are also capable of causing acute health effects. Among other things, formaldehyde (also classified as a human carcinogen) is a **sensitizer** that can cause an immune system response when it is inhaled or absorbed through the skin. This reaction is like other allergic reactions, in that the initial exposure produces mild health effects, but subsequent exposures cause much more severe reactions such as breathing difficulties and skin irritation. Health Canada's Workplace Hazardous Materials Information System (WHMIS) defines *sensitizer* as a material that can cause severe skin and/or respiratory responses in a sensitized person after exposure to a very small amount of the material. Sensitization develops over time: when a person is first exposed, there may be no obvious reaction; however, repeated exposure can lead to increasingly severe reactions.

It is also important to understand the difference between contagious and infectious as they both relate to exposure. An *infectious* substance is known or reasonably suspected of containing a microorganism or particles that can cause disease in humans or animals. Many biological agents (disease-causing bacteria, viruses, and other agents) we consider as potential terrorist agents are infectious. Biological toxins are not cellular organisms, however, and cannot reproduce. Toxins are poisonous substances produced from the metabolic processes of living plants, animals, or microorganisms. *Contagious*, as defined by the Centers for

Disease Control and Prevention, means that a bacteria or virus can be transmitted from person to person. Malaria serves as a good example in that a person can't "catch" malaria from another person (it is not contagious), but a single bite from an infected mosquito could potentially cause a fatal case of malaria (it is infectious).

Toxicity is a measure of the degree to which something is toxic or poisonous. This term can also refer to the adverse effect(s) a substance may have on a whole organism, such as a human (or a bacterium or a plant), or to a substructure such as a cell or a specific organ such as the liver, kidneys, or lungs. To understand the risks posed by a material's toxicity, responders must consider the physical and chemical properties of the substance causing the illness or injury, as well as the dose–response relationship that exists when a person is exposed to any substance. This important correlation refers to the response (signs, symptoms, illness, or injury) that a specific dose might provoke from the human body. The magnitude of the response depends on several factors, including the concentration of the hazardous substance and the duration of the exposure. The actual dose is the amount taken up by a person through one of the four routes of entry.

Cyanide, for example, is a harmful substance that can enter the body by all routes of entry. If cyanide gets into the body via the lungs, for example, it would be useful to understand at what airborne concentration (dose) the exposure might produce adverse health effects (response). To fully understand the threat, responders must be familiar with several health-related terms and definitions as they relate to airborne and dermal exposures. For example, the **lethal dose (LD)** of a material is a single dose that causes the death of a specified number of the group of test animals exposed by any route other than inhalation. The **lethal concentration (LC)** is defined as the concentration of a material in air that, based on laboratory tests (inhalation route), is expected to kill a specified number of the group of test animals when administered over a specified period of time.

Several subcategories of LD and LC exist, each depicting a benchmark concentration of a particular exposure that will be harmful and/or fatal to a certain percentage of the test population. These values are commonly found in SDS, electronic databases, and printed reference books. In most cases, these values are derived from animal studies, so the results may not exactly correspond to the levels/concentrations that would be harmful to humans.

The following overall descriptions of these benchmark values offer relative guidelines when it comes to determining potential levels of human toxicity. As mentioned earlier, many chemical substances on the market today have established LD or LC values. Typically, if the substance is a vapour or gas, it will be expressed as an LC value. If the substance poses a dermal threat or is harmful when ingested, its toxicity will be expressed as an LD value.

For any given substance, the LD is the lowest dosage per unit of body weight (typically stated in terms of milligrams per kilogram [mg/kg]) of a substance known to have resulted in fatality in an animal species. The median lethal dose (LD_{50}) of a toxic material is the dose required to kill half (50 percent) of the members of a tested population.

For example, the LD_{50} of sodium cyanide (based on testing done on laboratory rats) is approximately 6.4 mg/kg. LD_{50} figures are frequently used as a general indicator of a substance's toxicity. The LD_{hi} or LD_{100} is the absolute dose of a toxic material required to kill all (100 percent) of the members of a tested population.

The LC_{lo} is the lowest lethal concentration of a material reported to cause death in an animal species, when administered via the inhalation route; it is the lowest lethal concentration for gases, dusts, vapours, and mists. The LC_{50} is the concentration of a material in air that is expected to kill 50 percent of the group of an animal species when administered via the inhalation route. Some literature reports the LC_{50} of carbon monoxide to be approximately 3700 parts per million (ppm) for a 1-hour exposure time frame. Carbon monoxide levels have been measured to approach 10,000 ppm—well above the LC_{50}. The LC_{hi} or LC_{100} is the absolute concentration of a toxic material required to kill all (100 percent) of the members of a tested population when the substance is administered via the inhalation route.

The values expressed here figure prominently when it comes to identifying the relative toxicity of a particular substance. The following information reflects exposure levels considered to be toxic and highly toxic.

The label *toxic* is assigned to any chemical that falls into one of these three categories:

- A chemical that has an LD_{50} of more than 50 mg/kg but not more than 500 mg/kg of body weight when administered orally to albino rats weighing between 200 and 300 g each
- A chemical that has an LD_{50} of more than 200 mg/kg but not more than 1000 mg/kg of body weight when administered by continuous contact for 24 hours (or less if death occurs within 24 hours) with the bare skin of albino rabbits weighing between 2 and 3 kg each
- A chemical that has an LC_{50} in air of more than 200 ppm but not more than 2000 ppm by volume of gas or vapour or more than 2 mg/L but not more than 20 mg/L of mist, fume, or dust, when

administered by continuous inhalation for 1 hour (or less if death occurs within 1 hour) to albino rats weighing between 200 and 300 g each

Highly toxic materials are defined as follows:

- A chemical that has an LD_{50} of 50 mg/kg or less of body weight when administered orally to albino rats weighing between 200 and 300 g each
- A chemical that has an LD_{50} of 200 mg/kg or less of body weight when administered by continuous

contact for 24 hours (or less if death occurs within 24 hours) with the bare skin of albino rabbits weighing between 2 and 3 kg each

- A chemical that has an LC_{50} in air of 200 ppm by volume or less of gas or vapour or 2 mg/L or less of mist, fume, or dust, when administered by continuous inhalation for 1 hour (or less if death occurs within 1 hour) to albino rats weighing between 200 and 300 g each

After-Action REVIEW

IN SUMMARY

- The most fundamental of all actions is the ability to observe the scene and understand the problem you are facing. Think before you act—it could save your life!
- An important first step in understanding the hazards of any chemical is identifying the state of matter and defining whether the substance is a solid, liquid, or gas.
- A critical step in comprehending the nature of the release is identifying the reason(s) why the containment vessel failed.
- Chemical change is not the same thing as physical change. Physical change is a change in state; chemical change describes the ability of a substance to undergo a transformation at the molecular level, usually with a release of some form of energy.
- There are many critical characteristics of flammable liquids, and you should be familiar with each of them.
- When responders respond to hazardous materials incidents, they must fully understand the hazards to minimize the potential for exposure.
- Avoid contamination whenever possible—it will reduce the likelihood of harmful exposures.
- Chronic health effects may occur after years of exposure to hazardous materials. Wear all protective gear to minimize the impacts of repeated exposures.

KEY TERMS

Access Navigate for flashcards to test your key term knowledge.

Acid A material with a pH value less than 7.

Acute exposures "Right now" exposures that produce observable signs such as eye irritation, coughing, dizziness, and skin burns.

Acute health effects Health problems caused by relatively short exposure periods to a harmful substance that produces observable conditions such as eye irritation, coughing, dizziness, and skin burns.

Alpha particles Positively charged particles emitted by certain radioactive materials, identical to the nucleus of a helium atom. (NFPA 801)

Base A material with a pH value greater than 7.

Beta particles Elementary particles, emitted from a nucleus during radioactive decay, with a single electrical charge and a mass equal to that of a proton. (NFPA 801)

BLEVE Boiling liquid/expanding vapour explosion; an explosion that occurs when pressurized liquefied materials (e.g., propane or butane) inside a closed vessel are exposed to a source of high heat.

Boiling point The temperature at which the vapour pressure of a liquid equals the surrounding atmospheric pressure. (NFPA 1)

Chemical and physical properties Measurable characteristics of a chemical, such as its vapour density, flammability, corrosivity, and water reactivity.

Chemical reaction Any chemical change or chemical degradation, occurring inside or outside a containment vessel.

Chemical reactivity The ability of a chemical to undergo an alteration in its chemical make-up, usually accompanied by a release of some form of energy.

Chronic exposures Long-term exposures, occurring over the course of many months or years.

Chronic health hazard An adverse health effect occurring after a long-term exposure to a substance.

Combustibles A material that, in the form in which it is used and under the conditions anticipated, will ignite and burn; a material that does not meet the definition of noncombustible or limited-combustible. (NFPA 1)

Contamination The process of transferring a hazardous material, or the hazardous component of a weapon of mass destruction (WMD), from its source to people, animals, the environment, or equipment, which can act as a carrier. (NFPA 1072)

Corrosivity The ability of a material to cause damage (on contact) to skin, eyes, or other parts of the body.

Expansion ratio A description of the volume increase that occurs when a liquid changes to a gas.

Exposure The process by which people, animals, the environment, property, and equipment are subjected to or come in contact with a hazardous material/weapon of mass destruction (WMD). (NFPA 1072)

Fire point The lowest temperature at which a liquid will ignite and achieve sustained burning when exposed to a test flame in accordance with ASTM D 92, Standard Test Method for Flash and Fire Points by Cleveland Open Cup Tester. (NFPA 1)

Fissile Capable of sustaining a chain reaction using neutrons at any level.

Flammable range (explosive limit) The range of concentrations between the lower and upper flammable limits. (NFPA 67)

Flash point The minimum temperature at which a liquid or a solid emits vapour sufficient to form an ignitable mixture with air near the surface of the liquid or the solid. (NFPA 115)

Gamma radiation High-energy short-wavelength electromagnetic radiation. (NFPA 801)

Hazard Capable of causing harm or posing an unreasonable risk to life, health, property, or environment. (NFPA 1072)

Ignition temperature Minimum temperature a substance should attain in order to ignite under specific test conditions. (NFPA 402)

Ionizing radiation Radiation of sufficient energy to alter the atomic structure of materials or cells with which it interacts, including electromagnetic radiation such as x-rays, gamma rays, and microwaves and particulate radiation such as alpha and beta particles. (NFPA 1991)

Irritants Substances (such as mace) that can be dispersed to briefly incapacitate a person or groups of people. Irritants cause pain and a burning sensation to exposed skin, eyes, and mucous membranes.

Lethal concentration (LC) The concentration of a material in air that, based on laboratory tests (inhalation route), is expected to kill a specified number of the group of test animals when administered over a specified period of time.

Lethal dose (LD) A single dose that causes the death of a specified number of the group of test animals exposed by any route other than inhalation.

Lower explosive limit (LEL) The minimum concentration of combustible vapour or combustible gas in a mixture of the vapour or gas and gaseous oxidant above which propagation of flame will occur on contact with an ignition source. (NFPA 115)

Neutrons Penetrating particles found in the nucleus of the atom that are removed through nuclear fusion or fission. Although neutrons are not radioactive, exposure to them can create radiation.

Non-ionizing radiation Electromagnetic waves capable of causing a disturbance of activity at the atomic level but that do not have sufficient energy to break bonds and create ions.

pH An expression of the amount of dissolved hydrogen ions (H^+) in a solution.

Physical change A transformation in which a material changes its state of matter—for instance, from a liquid to a solid.

Polymerization The process of reacting monomers together in a chain reaction to form polymers.

Pulmonary edema Fluid build-up in the lungs.

Radiation The combined process of emission, transmission, and absorption of energy travelling by electromagnetic wave propagation (e.g., infrared radiation) between a region of higher temperature and a region of lower temperature. (NFPA 550)

Radioactive isotope A variation of an element created by an imbalance in the numbers of protons and neutrons in an atom of that element.

Radioactivity The spontaneous decay or disintegration of an unstable atomic nucleus accompanied by the emission of radiation. (NFPA 801)

Secondary contamination The process by which a contaminant is carried out of the hot zone and contaminates people, animals, the environment, or equipment. Also referred to as cross-contamination.

Sensitizer A chemical that causes a large percentage of people or animals to develop an allergic reaction after repeated exposure.

Solvents A substance (usually liquid) capable of dissolving or dispersing another substance; a chemical compound designed and used to convert solidified grease into a liquid or semiliquid state in order to facilitate a cleaning operation. (NFPA 96)

Specific gravity The weight of a liquid as compared to water.

States of matter The physical state of a material—solid, liquid, or gas.

Toxicity The degree to which a substance is harmful to humans. (NFPA 236)

Toxic products of combustion Hazardous chemical compounds that are released when a material decomposes under heat.

Upper explosive limit (UEL) The maximum amount of gaseous fuel that can be present in the air if the air/fuel mixture is to be flammable or explosive.

Vapour The gas phase of a substance, particularly of those that are normally liquids or solids at ordinary temperatures. (NFPA 326)

Vapour density The weight of an airborne concentration (vapour or gas) as compared to an equal volume of dry air.

Vapour pressure The pressure, measured in pounds per square inch, absolute (psia), exerted by a liquid, as determined by ASTM D 323, Standard Test Method for Vapour Pressure of Petroleum Products (Reid Method). (NFPA 1)

Water solubility The ability of a substance to dissolve in water.

On Scene

Your crew is called to respond to a possible leaking container at a nearby biotechnology research facility. Upon arrival, you are met by a member of the site emergency response team (ERT). The person identifies herself as the incident commander (IC) of the ERT and tells you that the white polyethylene drum was knocked over and appears to be leaking. The IC says that the drum contains 190-proof ethyl alcohol that is used in one of their processes. There is an alcohol-like odour in the air, and you move everyone back to a safe distance and isolate the area. You find the information below on an SDS given to you by the IC.

Chemical name: ETHYL ALCOHOL 190 proof
Appearance: Clear, colourless liquid
Flash point: 18.5°C (65.3°F)
Flammable range: LOWER: 3.3% UPPER: 19%
Solubility: Soluble in water
Specific gravity: 0.8
pH: Neutral
Vapour density (air = 1): 1.59
Health effects: Hazardous in case of skin contact (irritant) or eye contact (irritant). Noncorrosive to skin. Noncorrosive to eyes. Noncorrosive to lungs.
Fire hazard: Highly flammable in presence of open flames and sparks or heat. Slightly flammable to flammable in presence of oxidizing materials

(continued)

On Scene Continued

1. At or above what temperature would you be concerned about the risk of fire if an uncontrolled ignition source was nearby?

A. 5.5°C (42°F)

B. 10°C (50°F)

C. 12.7°C (55°F)

D. 29.4°C (85°F)

2. If vapours are being produced, which definition would you use to find out what percentage of air-to-fuel mixture would be required for the vapours to ignite?

A. Flammable range

B. Vapour density

C. Flash point

D. pH

3. You would expect the vapours generated from the release to be _____ than air due to the _____ listed in the data above.

A. lighter; vapour density

B. heavier; vapour density

C. lighter; specific gravity

D. heavier; specific gravity

4. Which of the following best describes the way in which the white drum likely failed?

A. Disintegration

B. Runaway cracking

C. Closures opening up

D. Punctures

Access Navigate to find answers to this On Scene, along with other resources such as an audiobook.

CHAPTER **32**

Hazardous Materials Operations Level

Understanding the Hazards

KNOWLEDGE OBJECTIVES

After studying this chapter, you will be able to:

- Identify and describe common types of hazardous materials containers. (**NFPA 1072: 5.2.1; NFPA 472: 5.2.1, 5.2.1.1, 5.2.1.1.1, 5.2.1.1.2, 5.2.1.1.3, 5.2.1.1.5, 5.2.1.2, 5.2.1.2.1**, pp. 1110–1115)
- Describe the ways in which hazardous materials are transported. (**NFPA 1072: 5.2.1; NFPA 472: 5.2.1, 5.2.1.1, 5.2.1.1.1, 5.2.1.1.2, 5.2.1.1.3, 5.2.1.1.4, 5.2.1.2**, pp. 1115–1123)
- Identify resources for technical chemical information. (**NFPA 1072: 5.2.1; NFPA 472: 5.1.2.2, 5.2.1.2, 5.2.2**, p. 1120)
- Identify the components of potential terrorist incidents. (**NFPA 1072: 5.2.1; NFPA 472: 5.2.1.6, 5.2.1.6.1**, pp. 1124–1137)
- Explain how to respond to terrorist incidents. (**NFPA 1072: 5.2.1; NFPA 472: 5.2.1.1.6, 5.2.1.2, 5.2.1.3, 5.2.1.3.3, 5.2.1.4, 5.2.1.5, 5.2.1.6, 5.2.1.6.1, 5.2.1.6.2, 5.2.1.6.3, 5.2.1.6.4, 5.2.1.6.5, 5.2.1.6.7, 5.2.1.6.8, 5.2.1.6.9, 5.2.4**, pp. 1124–1125)

SKILLS OBJECTIVES

This chapter has no skills objectives for operations level hazardous materials responders.

Hazardous Materials Alarm

Your pump company is dispatched to the scene of a leaking drum at a local lumberyard. It appears to have been hit by a forklift, causing a small tear toward the bottom and a slow leak. The dirt around the drum appears to be wet. The foreman of the lumberyard tells you that the drum holds waste oil, and the forklift driver on duty denies hitting the drum. From a safe distance, you can tell that the drum is made of steel and has a 51-mm (2-in.) bung and 19-mm (¾-in.) bung on the top. You can see a small red label on the side of the drum that says "flammable," with a number 3 at the bottom.

1. What type of material is most likely stored in the drum (e.g., corrosive, solvent, explosive)? Does the description of the drum provide any other useful information? If so, what?

2. What does the label on the drum signify, and does the report of an unusual odour make sense given the information you have?

3. How would you go about obtaining more information on the potential contents of the drum?

 *Access Navigate for more practice activities.*

Introduction

Even the most common chemical substances and the containers they are held in can become powerful weapons. For example, we tend to think of gasoline tankers and railcars as valuable components of transportation and commerce, but in the hands of determined terrorists, these devices can be used as deadly weapons. Designating an incident as an act of terrorism reflects more the intent of the attacker than the use of a certain device, chemical substance, or agent. For example, demolition companies use explosives to purposefully collapse unneeded structures quickly and safely; terrorists could use the same explosives to bring down an occupied building, deliberately killing many people. *Again, it is the intent that distinguishes between an accident and a terrorist event, not the chemical or the container.* This chapter is therefore offered as a combined overview of a wide variety of containers and their physical characteristics, common substances found in those containers, and a broad overview of several classes of terrorism agents.

Containers

Obtaining a basic understanding of storage containers provides personnel with information that can be applied in an effort to mitigate a hazardous materials incident. Some commonly encountered containers include drums, carboys, and compressed gas cylinders. Those examples represent a classification of containers called **non-bulk**

packaging. In contrast, the examples discussed in this chapter represent bulk packaging. **Bulk packaging**, or large-volume containers, is defined by its internal capacity based on the following measures (excluding a vessel or a barge):

- A maximum capacity greater than 450 L (119 gal)
- A maximum net mass greater than 400 kg (882 lb) and a maximum capacity greater than 450 L (119 gal) as a receptacle for a solid
- A water capacity greater than 454 kg (1000 lb) as a receptacle for a gas

To keep it simple, non-bulk packages have volumes less than those listed. Think of bulk packaging as large containers such as fixed tanks, highway cargo tanks, railcars, totes, and intermodal tanks. Essentially, non-bulk packaging includes all types of containers other than bulk containers, including drums, bags, compressed gas cylinders, and cryogenic containers (see Chapter 30, *Recognizing and Identifying the Hazards*).

In general, bulk storage containers are found in occupancies that rely on and need to store large quantities of a substance. Most manufacturing facilities have at least one type of bulk storage container (**TABLE 32-1**). Often these bulk storage containers are surrounded by a supplementary containment system to help control an accidental release. **Secondary containment** is an engineered method to control spilled or released product if the main containment vessel fails. A 18,927-L (5000-gal) vertical storage tank, for example, may be surrounded by a series of short walls commonly

TABLE 32-1 Common Bulk Storage Vessels, Locations, and Contents

Tank Shape	Common Locations	Hazardous Materials Commonly Stored
Underground tanks	Residential, commercial	Fuel oil and combustible liquids
Covered floating roof tanks	Bulk terminal and storage	Highly volatile flammable liquids
Cone roof tanks	Bulk terminal and storage	Combustible liquids
Open floating roof tanks	Bulk terminal and storage	Flammable and combustible liquids
Dome roof tanks	Bulk terminal and storage	Combustible liquids
High-pressure horizontal tanks	Industrial storage and terminal	Flammable gases, chlorine, ammonia
High-pressure spherical tanks	Industrial storage and terminal	Liquid propane gas, liquid nitrogen gas
Cryogenic liquid storage tanks	Industrial and hospital storage	Oxygen, liquid nitrogen gas

© Jones & Bartlett Learning.

referred to as *dikes* or *berms* that form a catch basin around the vessel.

Secondary containment basins typically can hold the entire volume of the tank, along with a percentage of the water flowed from hose lines or sprinkler systems in the event of fire and, in the case of outdoor storage, a certain amount of rainfall. Many storage vessels, including 208-L (55-gal) drums, may have secondary containment systems. If facilities that use bulk storage containers in your response area have secondary containment systems, it will be easier to handle leaks when they occur.

Large-volume storage tanks are also common at fixed facilities. These tanks are usually made of aluminum, steel, or plastic and can be pressurized or nonpressurized. Nonpressurized tanks are usually made of steel or aluminum and commonly hold flammable or combustible materials, such as gasoline, oil, or diesel fuel, but may also hold solids. Examples of large-volume storage tanks include bulk fixed facility pressure containers, bulk fixed facility tanks, 1-ton (approximately 1000-kg) containers for gases such as chlorine, Y cylinders, and bulk fixed facility cryogenic containers (**FIGURE 32-1**).

Pressurized horizontal tanks have rounded ends and large vents or pressure-relief stacks. The most common above-ground pressurized tanks contain liquid propane and liquid ammonia; such containers can hold a few hundred gallons to several thousand gallons of product. These tanks usually have a small vapour space—called the *headspace*—above the liquid. In most cases, 10 to 15 percent of the total container capacity is vapour (**FIGURE 32-2**). Smaller versions of pressure containers may be found in the form of portable propane cylinders and vehicle-mounted pressure containers (**FIGURE 32-3**).

Ton Containers

Ton containers commonly hold compressed liquefied gases such as chlorine and sulfur dioxide, although you may encounter some select refrigerant fluorocarbon gases such as trichlorofluoromethane (Freon-11) and dichlorodifluoromethane (Freon-12) stored in ton containers. Regardless of the material inside, these vessels hold 907 kg (2000 lb) of product (hence the name, *ton container*) and are 2.4 m (8 ft) in length and 1 m (3 ft) in diameter. All ton containers other than those containing phosphine have pressure relief valves called *fusible plugs*. Figure 32-1D shows a ton container after a leak. The Y cylinder seen in Figure 32-1E is used for high-purity gases such as silane and phosphine.

Intermodal Tanks

Intermodal tanks are both shipping and storage vessels. They hold between 18,927 and 22,712 L (5000 and 6000 gal) of product and can be either pressurized or non-pressurized. Intermodal (IM or IMO) tanks can also be used to ship and store gaseous substances that have been chilled until they liquefy (cryogenic liquids), such as liquid nitrogen and liquid argon. In most cases, an IM tank is shipped to a facility, where it is stored and used, and then returned to the shipper for refilling. Intermodal tanks can be shipped by all methods of transportation—air, sea, or land (**FIGURE 32-4**).

Typically, a box-like steel framework, constructed to facilitate efficient stacking and shipping, surrounds an IM tank. Several types of IM tanks are available:

- IM-101 portable tanks (IMO type 1 internationally) have a 23,848-L (6300-gal) capacity, with internal working pressures between 175 and

FIGURE 32-1 **A.** Spherical fixed facility pressure containers. **B.** Bulk fixed facility tanks. **C.** Bulk fixed facility cryogenic container. **D.** A ton chlorine container (US DOT 106A) after a leak. **E.** Y ton cylinders.

A, B, and C: Courtesy of Chris Hawley; **D:** Courtesy of Jason Krusen; **E:** Courtesy of Rob Schnepp.

689 kilopascals (kPa) (25.4 lb per square inch [psi] and 100 psi). These containers typically carry mild corrosives, food-grade products, and flammable liquids (**FIGURE 32-5**).

- IM-102 portable tanks (IMO type 2 internationally) have a 23,848-L (6300-gal) capacity, with internal working pressures between 101 kPa and 175 kPa

(14.7 psi and 25.4 psi). They primarily carry non-hazardous materials but may also contain flammable liquids and corrosives (**FIGURE 32-6**).

- Pressure intermodal tanks (IMO type 5 internationally or DOT Spec 51) are high-pressure vessels with internal pressures in the range of 689–4137 kPa (100–600 psi). These intermodal containers

FIGURE 32-2 In most cases, 10 to 15 percent of the total container capacity is vapour.
Courtesy of Rob Schnepp.

commonly hold liquefied compressed gases such as propane and butane (**FIGURE 32-7**).

- Cryogenic intermodal tanks (IMO type 7 internationally) are low-pressure containers in transport but can be pressurized to 4137 kPa (600 psi). This kind of container commonly carries cryogenic materials that have temperatures less than –101°C (–150°F), such as liquefied oxygen, nitrogen, or helium (**FIGURE 32-8**).
- Tube modules consist of several high-pressure tubes attached to a frame. The tubes are individually specified and have working pressures that range as high as 34,474 kPa (5000 psi). The products commonly carried include hydrogen and oxygen (**FIGURE 32-9**).

FIGURE 32-3 A. Portable propane cylinder in a storage yard. **B.** Vehicle-mounted pressure container (propane) in use on a forklift.
Courtesy of Rob Schnepp.

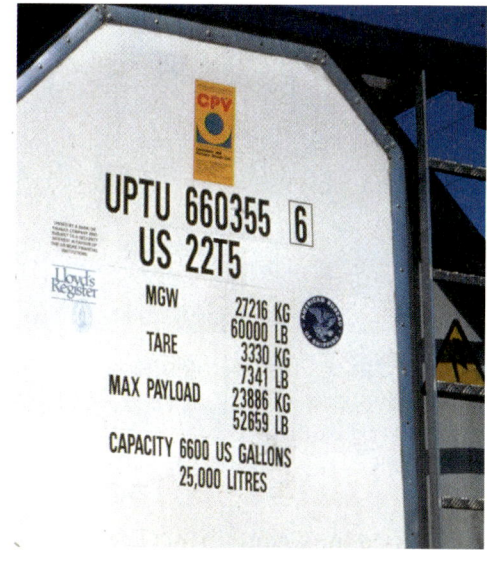

FIGURE 32-4 A. Bulk packaging placed on/in transport vehicles. **B.** Intermodal portable tank markings.
© Jones & Bartlett Learning.

FIGURE 32-5 IM-101 portable tanks (IMO type 1 internationally).
Courtesy of UBH International Ltd.

FIGURE 32-6 IM-102 portable tanks (IMO type 2 internationally).
Courtesy of UBH International Ltd.

FIGURE 32-7 Pressure intermodal tanks (IMO type 5 internationally).
Courtesy of UBH International Ltd.

FIGURE 32-8 Cryogenic intermodal tank (IMO type 7 internationally).
© Jones & Bartlett Learning.

FIGURE 32-9 Tube modules consist of several high-pressure tubes attached to a frame.
Courtesy of Bill Hand.

Intermediate Bulk Containers

Intermediate bulk containers (IBCs) are so named because the volumes stored inside fall between what is typically found in drums or bags and what is in cargo tanks. Their capacities are greater than 450 L (119 gal) but less than 3002 L (793 gal). IBCs are either flexible intermediate bulk containers (FIBCs) or rigid intermediate bulk containers (RIBCs). You may hear RIBCs referred to as "totes" and FIBCs called "sacks" or "super sacks." Super sacks are much larger than a normal bag and may hold solid material weighing anywhere from 227 kg (500 lb) to several thousand kilograms (**FIGURE 32-10**). The fabric of

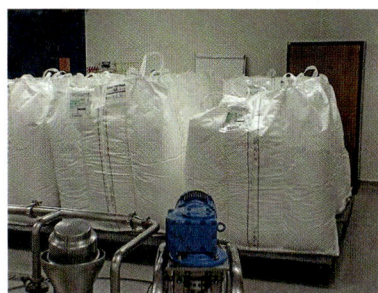

FIGURE 32-10 Super sacks.

Top and bottom: © Jones & Bartlett Learning; Middle: © Courtesy of Hildebrand and Noll Associates, Inc.

FIGURE 32-11 Totes waiting for use at a fixed facility.

Courtesy of Rob Schnepp.

FIGURE 32-12 Flexible bladder.

© Allen Fredrickson/Reuters.

construction ranges from woven cloth to polyethylene, and they may be palletized for transport. Solid materials transported and stored in super sacks may include oxidizers, corrosives, and flammable solids. Totes can be found as a high-density polyethylene tank inside a rigid stainless steel frame or perhaps a square metal tank around 1.2 m (4 ft) wide and 1.8 m (6 ft) tall. Both types have bottom-mounted discharge valves and are commonly found at fixed facilities. Totes may hold a wide variety of substances such as solvents, corrosives, or select oxidizers.

Shipping and storing totes can be hazardous. These containers often are stacked atop one another and moved with a forklift, such that a mishap with the loading or moving process can compromise the tote (**FIGURE 32-11**). Because totes have no secondary containment system, any leak has the potential to create a large puddle. Additionally, the steel webbing around the tote makes it difficult to access and patch leaks.

Another type of unique container used for chemical storage is the flexible bladder. These are typically portable and used for short-term storage and dispensing of low-hazard materials (**FIGURE 32-12**).

Transporting Hazardous Materials

Hazardous materials may be transported by air, sea, or land or a combination of any or all of these modes of transport. Although transportation of hazardous materials occurs most often on the roadway, when another primary mode of transport is used, roadway vehicles often transport the shipments from the rail station, airport, or dock to the point where it will be used. For this reason, responders must become familiar with all types of chemical transport vehicles they might encounter during a transportation emergency.

Roadway Transportation

Per the Transportation of Dangerous Goods (TDG) guidelines, a **cargo tank** (called a highway tank by the TDG) is bulk packaging that is permanently attached to or forms a part of a motor vehicle or is not permanently attached to any motor vehicle and that, because of its size, construction, or attachment to a motor vehicle, is loaded or unloaded without being removed from the motor vehicle. The U.S. Department of Transportation (DOT) does not view tube trailers (which consist of several individual cylinders banded together and affixed to a trailer) as cargo tanks. Transport Canada refers to these vehicles based on the type of cargo being transported (see https://www.tc.gc.ca/eng/tdg/moc-highway-high waysspecs-402.html for more information).

One of the most common and reliable transportation vessels is the **nonpressure liquid cargo tank (MC-306/DOT 406 cargo tank)** (**FIGURE 32-13**). These tanks frequently carry liquid food-grade products, gasoline, or other flammable and combustible liquids. The oval-shaped tank is pulled by a diesel (or liquefied natural gas [LNG] or compressed natural gas [CNG]) tractor and can carry between 22,712 and 37,854 L (6000 and 10,000 gal) of product. The MC-306/DOT 406 is nonpressurized (its working pressure is between 18.3 and 27.5 kPa [2.65 and 4 psi]), usually made of aluminum or stainless steel, and loaded and offloaded through valves at the bottom of the tank. These cargo tanks have several safety features, including full rollover protection and remote emergency shut-off valves (**FIGURE 32-14**).

A vehicle that is like the nonpressure liquid cargo tank is the **low-pressure chemical cargo tank (MC-307/DOT 407 chemical hauler)**. It has a round or horseshoe-shaped tank and can hold 22,712 to 26,498 L (6000 to 7000 gal) of liquid (**FIGURE 32-15**). The low-pressure chemical cargo tank, which is also a tractor-drawn tank, is used to transport flammable liquids,

FIGURE 32-14 The MC-306/DOT 406 cargo tank has a remote emergency shut-off valve as a safety feature.
Courtesy of Glen Rudner.

FIGURE 32-15 The MC-307/DOT 407 chemical hauler carries flammable liquids, mild corrosives, and poisons.
Courtesy of Polar Tank Trailer L.L.C.

mild corrosives, and poisons. This type of cargo tank may be insulated (horseshoe) or uninsulated (round) and may have a higher internal working pressure than the nonpressure liquid cargo tank—in some cases up to 241 kPa (35 psi). Cargo tanks that transport corrosives may have a rubber lining to prevent corrosion of the tank structure.

The chemical cargo tank (**MC-312/DOT 412 corrosive tank**) is commonly used to carry corrosives such as concentrated sulfuric acid, phosphoric acid, and sodium hydroxide (**FIGURE 32-16**). This cargo tank has a smaller diameter than the nonpressure liquid cargo tank or low-pressure chemical cargo tank and is often identifiable by the presence of several heavy-duty reinforcing rings around the tank. The rings provide structural stability during transportation and in the event of a rollover. The inside of this type of chemical cargo tank operates at approximately 103 to 172 kPa (15 to 25 psi) and holds approximately 22,712 L (6000 gal). These cargo tanks have substantial rollover protection to reduce the potential for damage to the top-mounted valves.

FIGURE 32-13 The MC-306/DOT 406 flammable liquid tank typically hauls flammable and combustible liquids.
Courtesy of Polar Tank Trailer, LLC.

FIGURE 32-16 The MC-312/DOT 412 corrosives tanker is commonly used to carry corrosives such as concentrated sulfuric acid, phosphoric acid, and sodium hydroxide.
Courtesy of National Tank Truck Carriers Association.

The **high-pressure cargo tank (MC-331)** carries materials such as ammonia, propane, Freon, and butane (**FIGURE 32-17**). The liquid volume inside the tank varies, ranging from the 3785-L (1000-gal) delivery truck to the full-size 41,640-L (11,000-gal) cargo tank. The high-pressure cargo tank has rounded ends, typical of a pressurized vessel, and is commonly constructed of steel or stainless steel with a single tank compartment. The high-pressure cargo tank operates at approximately 2068 kPa (300 psi), with typical internal working pressures being near 1724 kPa (250 psi). These cargo tanks are equipped with spring-loaded relief valves that traditionally operate at 110 percent of the designated maximum working pressure. A significant explosion hazard arises if a high-pressure cargo tank is impinged on by fire, however. The nature of most materials carried in these tanks means a threat of explosion exists because of the inability of the relief valve to keep up with the rapidly building internal pressure. Responders must use great care when dealing with this type of transportation emergency.

The **cryogenic liquid cargo tank (MC-338)** operates much like the cryogenic intermodal container described earlier and carries many of the same substances (**FIGURE 32-18**). This low-pressure tank relies on tank insulation to maintain the low temperatures required for the cryogens it carries. A box-like structure containing the tank control valves is typically attached to the rear of the tank. Special training is required to operate valves on this and any other tank. An untrained individual who attempts to operate the valves may disrupt the normal operation of the tank, thereby compromising its ability to keep the liquefied gas cold and creating a potential explosion hazard. Cryogenic cargo tanks have a relief valve near the valve control box. From time to time, small puffs of white vapour will be vented from this valve. Responders should understand that this is a normal occurrence—the valve is working to maintain

the proper internal pressure. In most cases, this vapour is not indicative of an emergency.

Compressed gas **tube trailers** carry compressed gases such as hydrogen, oxygen, helium, and methane (**FIGURE 32-19**). Essentially, they are high-volume transportation vehicles that are made up of several individual cylinders banded together and affixed to a trailer. The individual cylinders on the tube trailer are much like the smaller compressed gas cylinders discussed earlier in this chapter. These large-volume cylinders operate

FIGURE 32-17 The MC-331 high-pressure cargo tank carries materials such as ammonia, propane, Freon, and butane.
Courtesy of Rob Schnepp.

FIGURE 32-18 The MC-338 cryogenic tank maintains the low temperatures required for the cryogens it carries.
Courtesy of Jack B. Kelly, Inc.

FIGURE 32-19 A tube trailer.
Courtesy of Jack B. Kelly, Inc.

at working pressures of 20,684 to 34,474 kPa (3000 to 5000 psi). One trailer may carry several different gases in individual tubes. Typically, a valve control box is found toward the rear of the trailer, and each individual cylinder has its own relief valve. These trailers can frequently be seen at construction sites or at facilities that use large quantities of compressed gases.

Dry bulk cargo trailers are commonly seen on the road. They carry dry bulk goods such as powders, pellets, fertilizers, or grain (**FIGURE 32-20**). These tanks are not pressurized but may use pressure to offload the product. Dry bulk cargo tanks are generally V-shaped with rounded sides that funnel the contents to the bottom-mounted valves.

To understand the markings found on cargo tanks, the responder should have a basic understanding of not only the container profiles but also written information found on the tank and tank specification plates (**FIGURE 32-21**).

Railroad Transportation

Railroads move millions of carloads of chemicals and other freight each year in Canada, with relatively few accidents. Even so, responders should recognize that when rail incidents do occur, they can create unique and significant hazards. Railcars include passenger cars, freight cars, and tank cars, some of which can carry more than 113,562 L (30,000 gal) of product. Hazardous materials incidents involving railroad transportation have the potential to be large-scale emergencies.

Operations level hazardous materials responders should recognize the basic types of rail tank cars: low-pressure, pressure, cryogenic liquid, gondolas, and boxcars. Each has a distinctive profile that can be recognized from a distance. Additionally, rail tank cars are usually labelled on both sides with, among other things, the owner of the car, the car's capacity, and the specification. With dedicated haulers, the chemical name is often clearly visible on both sides of the rail tank car.

Low-pressure tank cars (also referred to as general-service rail tank cars) typically carry general industrial chemicals and consumer products such as corn syrup, flammable and combustible liquids, and mild corrosives. Low-pressure tank cars have visible valves and piping without protective housing on top of the car and internal vapour pressures less than 172 kPa (25 psi) (**FIGURE 32-22**). Older low-pressure tank cars may have a dome that covers the top-mounted valves and bottom outlet valves (**FIGURE 32-23**). Low-pressure tank cars may hold volumes ranging from 15,142 to 151,416 L (4000 to 40,000 gal).

Pressure tank cars transport materials such as propane, ammonia, ethylene oxide, and chlorine. These cars have internal working pressures ranging from 689 to 3447 kPa (100 to 500 psi) and are equipped with top-mounted fittings for loading and unloading. These fittings are protected by a sturdy and easily identified protective housing that sits atop the rail tank car (**FIGURE 32-24**). Unfortunately, the high volumes carried in these cars can generate long-duration, high-pressure leaks that may prove difficult to control. For example, a liquid or vapour release from the valve arrangements on a chlorine tank car requires a special kit and specific training to stop the leak. Additionally,

FIGURE 32-20 A dry bulk cargo tank carries dry goods such as powders, pellets, fertilizers, or grain.
Courtesy of Polar Tank Trailer L.L.C.

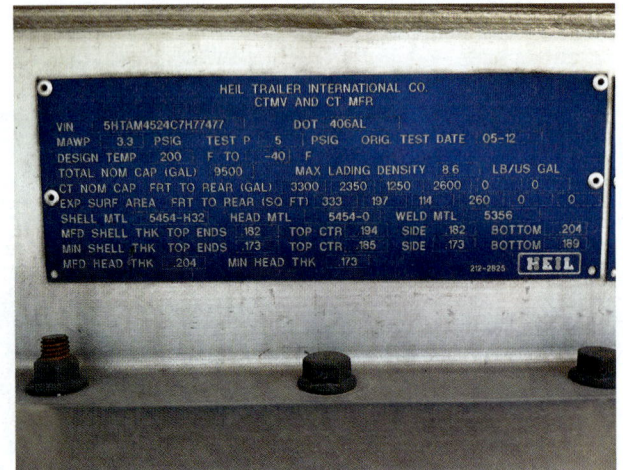

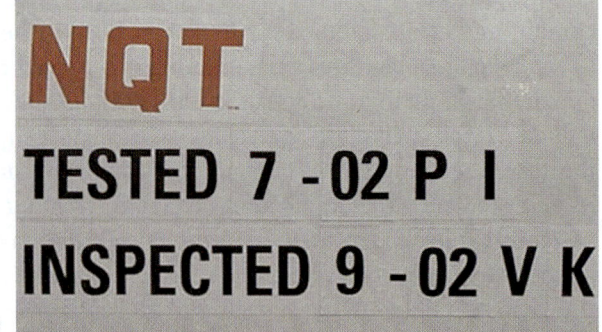

A   B

FIGURE 32-21 Cargo tank trucks. **A.** Specification plate. **B.** Inspection markings.
© Jones & Bartlett Learning.

pressure rail tank cars may be insulated or uninsulated, depending on the material being transported, and chlorine and ammonia railcars have federal security alarms in protective housing to prevent chemicals from being stolen and terrorist activities from occurring.

Cryogenic liquid tank cars are the most common special-use railcars emergency responders may encounter (FIGURE 32-25). Cryogenic liquid cars illustrate an important concept for the emergency responder: The hazard will be unique to the railcar and its contents,

and the responder will need to pay attention to the details and features of each type. Other examples of special-use railcars include railway gondolas used to carry items such as lumber, scrap metal, coal, and pipes and boxcars that carry consumer goods, industrial supplies, and a multitude of boxes and palletized goods (FIGURE 32-26). Freight containers can be used to transport goods on trucks, ships, and railcars and may be challenging for a responder due to the wide variety of goods and materials transported inside (FIGURE 32-27).

The train crew will have information about all the cars on the train. The information is typically found in the caboose with the conductor or at the engine with the engineer. The train crew must retain control of the paperwork at all times and will have knowledge about the availability of a specialized response team from the railroad, access to a 24/7 call centre for the railroad, and information about the location and contents

FIGURE 32-22 Low-pressure tank cars have visible valves and piping.
Courtesy of private source.

FIGURE 32-23 Example of an older-style, out-of-service, low-pressure tank car.
Courtesy of Rob Schnepp.

FIGURE 32-24 Pressure tank car.
Courtesy of private source.

FIGURE 32-25 Cryogenic liquid tank car.
© Jones & Bartlett Learning.

FIGURE 32-26 One example of a special-use railcar is the boxcar, typically used for carrying freight.
© Jones & Bartlett Learning. Photographed by Glen E. Ellman.

FIGURE 32-27 Examples of freight containers in the background.
Courtesy of Chris Hawley.

of the railcars and will be able to help responders understand the scope and impact of the derailment. These types of incidents can have significant effects on communities adjacent to the incident and may pose a significant risk to life, property, or the environment (**FIGURE 32-28**).

Reference Sources

The responsible person in each situation should maintain the information about hazardous cargo and provide it in an emergency. When encountering a chemical release on any of these modes of transportation, take time to find the responsible person to track down these valuable pieces of information. Additionally, the person who finds it for you, or has it in his or her possession, is likely to be a great resource to connect you with subject matter experts from the company or may have some level of knowledge about the material or the mode of transportation that may prove valuable.

CANUTEC

Located in Ottawa, the **Canadian Transport Emergency Centre (CANUTEC)** is operated by the Transportation of Dangerous Goods Directorate of Transport Canada. CANUTEC provides a national bilingual (French and English) advisory service and is staffed by professional scientists experienced and trained in interpreting technical information and providing emergency response advice. CANUTEC can provide responders with technical chemical information via telephone, fax, or another electronic medium. It also offers a phone conferencing service that will put a responder in touch with thousands of shippers, subject matter experts, and chemical manufacturers.

When calling CANUTEC 1-888-CAN-UTEC (1-888-226-8832), 613-996-6666, or *666 (STAR 666) on a cellular phone, be sure to have the following basic information ready:

- Name of the chemical(s) involved in the incident (if known)
- Name of the caller and callback telephone number
- Location of the actual incident or problem
- Shipper or manufacturer of the chemical (if known)
- Container type
- Railcar or vehicle markings or numbers
- Shipping carrier's name
- Recipient of material
- Local conditions and exact description of the situation

When speaking with CANUTEC personnel, be very clear about the name of the substance, and spell out the name if necessary; if using a third party, such as a dispatcher, it is vital that you confirm all spellings to avoid misunderstandings. One number or letter out of place could throw off all subsequent research. When in doubt, be sure to obtain clarification. Be sure to have access to a fax machine or electronic medium for resource information being sent from CANUTEC.

The U.S. equivalent of CANUTEC is the **Chemical Transportation Emergency Center (CHEMTREC)**, which is located in Arlington, Virginia. CHEMTREC is operated by the American Chemistry Council. This organization serves U.S. responders in much the same way that CANUTEC serves responders in Canada. CHEMTREC may be called 24 hours a day for emergency situations.

The Mexican equivalent of CHEMTREC and CANUTEC is the **Emergency Transportation System for the Chemical Industry, Mexico (SETIQ)**. SETIQ also may be called 24 hours a day.

Phone numbers for these agencies can be found in the *Emergency Response Guidebook (ERG)*.

National Response Center

The National Response Center (NRC) is an agency maintained and staffed by the U.S. Coast Guard. The agency should be notified if a hazard discharges into the environment. The NRC has established complex reporting requirements for different chemicals based on the reportable quantity (RQ) for that chemical. The shipper or the owner of the chemical has the ultimate legal responsibility to make this call, but by doing so themselves, response agencies will have their reporting bases covered.

RAILROAD TANK CARS

Railroad Tank Car Nomenclature. The railroad industry uses tank car test pressure as the criterion for differentiating between pressurized and nonpressurized tank cars. Nonpressure tank cars have a test pressure of 100 psig or less, while pressure tank cars have a test pressure greater than 100 psig.

When describing a tank car, the "B-end" is used as the initial reference point. The B-end is where the hand brake wheel is located; numbers 1 through 4 indicate the wheels.

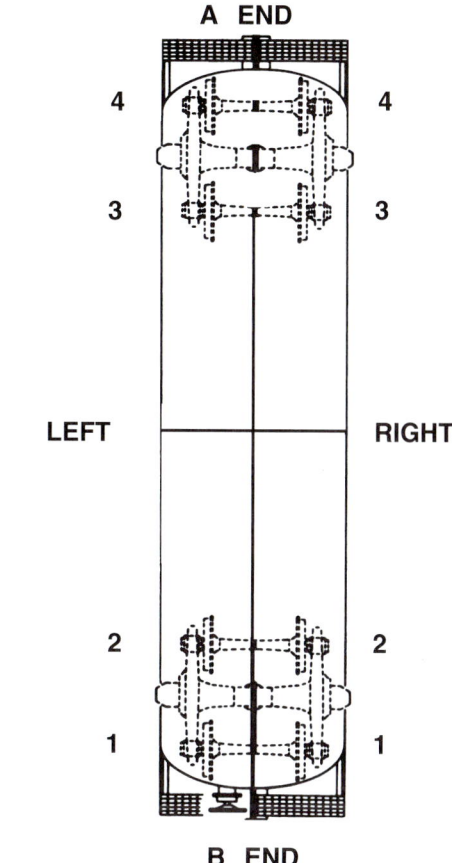

Railroad Tank Cars: Markings

Railroad Tank Car Markings can be used to gain knowledge about the tank itself and its contents. This information would be useful in evaluating the condition of the container. These markings include the following:

- *Commodity Stencil.* Tank cars transporting anhydrous ammonia, ammonia solutions with more than 50% ammonia, Division 2.1 material (flammable gas), or a Division 2.3 material (poison gas) must have the name of the commodity marked on both sides of the tank in 102-mm (4-in.) minimum letters.
- *Reporting Marks and Number.* Railroad cars are marked with a set of initials and a number (e.g., GATX 12345) stenciled on both sides (left end as one faces the tank) and both ends of the car. These markings can be used to obtain information about the contents of the car from the railroad, the shipper, or CANUTEC. The last letter in the reporting marks has special meaning:
  - "X" indicates a railcar is not owned by a railroad (for a railcar, the lack of an "X" indicates railroad ownership).
  - "Z" indicates a trailer.
  - "U" indicates a container.

FIGURE 32-28 Derailments can have significant effects on communities adjacent to the incident and may pose a significant risk to life, property, or the environment. Having access to the information found stenciled on railcars may be valuable to the responders.

Some shippers and car owners also stencil these markings on top of tank cars to assist in identification in an accident or derailment scenario. New tank cars are also stenciled on top for tank car verification during loading operations, as plant personnel can easily verify the car initial and number without climbing down from the rack.

- *Capacity Stencil.* Shows the volume of a tank car in gallons (and sometimes litres), as well as in pounds (and sometimes kilograms). These markings are found on the ends of the car under the reporting marks. For certain tank cars (e.g., DOT-105, DOT-109, DOT-112, DOT-114, and DOT-111A100W4), the water capacity/water weight of the tank car is stenciled near the center of the car.

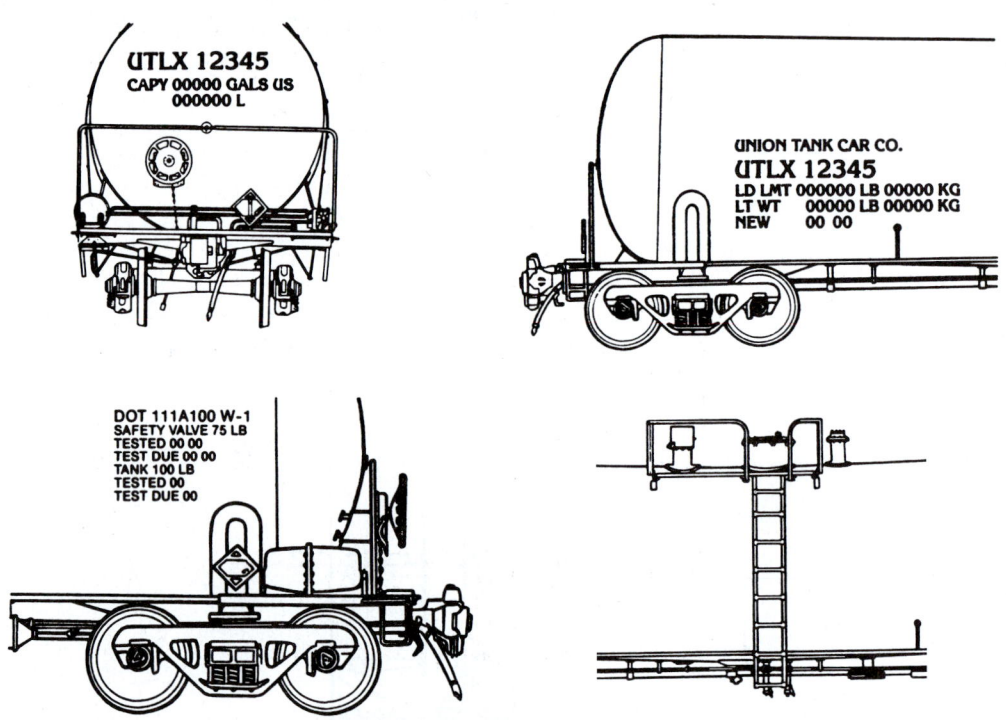

- *Specification Marking.* The specification marking indicates the standards to which a tank car was built. These markings will be on both sides of the tank car (right end as one faces the tank). The specification marking is also stamped into the heads of the tank, where it is not readily visible. The markings provide the following information:
  - Approving Authority (e.g., DOT—Department of Transportation, AAR—Association of American Railroads, ICC—Interstate Commerce Commission (authority to DOT in 1966), CTC—Canadian Transport Commission, TC—Transport Canada)
  - Class Number—three numbers that follow the approving authority designation
  - Separator/Delimiter Character (significant in certain tank cars)
  - Tank test pressure
  - Type of material used in construction—most tank cars are carbon steel and no designation appears. When other construction materials are used (e.g., aluminum, nickel, alloy steel), designations are used.
  - Type of weld used
  - Fittings/material/lining

FIGURE 32-28 *(continued)*

- *Specification Plate.* Tank cars ordered after 2003 will have a plate on the A-end right bolster and the B-end left bolster (i.e., structural cross member that cradles the tank) that provides information about the tank car's characteristics. Although not designed as an emergency response tool, the plate will provide the following information:
  - Car Builder's Name
  - Builder's Serial Number
  - Certificate of Construction/Exemption
  - Tank Specification Tank Shell Material/Head Material
  - Insulation Materials
  - Insulation Thickness
  - Underframe/Stub Sill Type
  - Date Built

DOT SPECIFICATION MARKINGS FOR RAILROAD TANK CARS

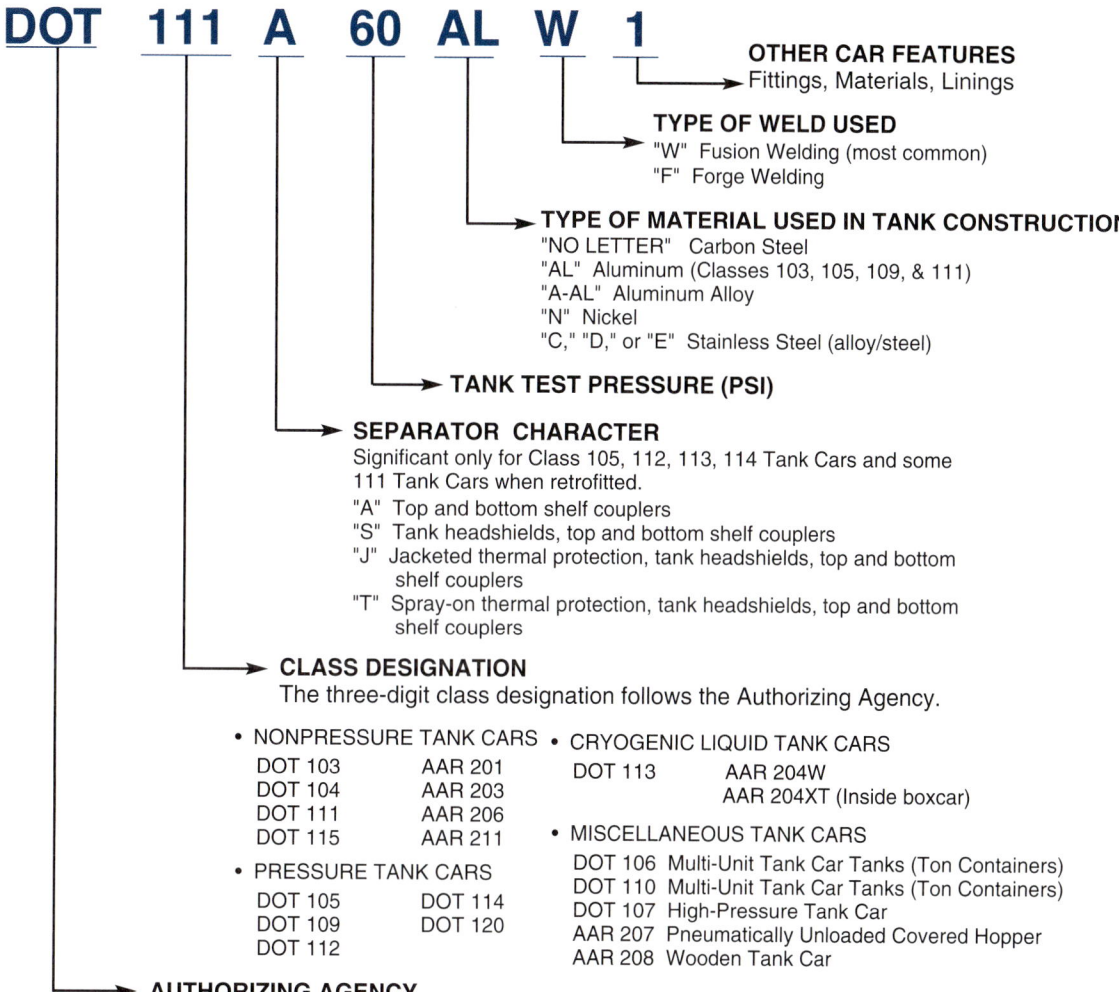

DOT 111 A 60 AL W 1

OTHER CAR FEATURES
Fittings, Materials, Linings

TYPE OF WELD USED
"W" Fusion Welding (most common)
"F" Forge Welding

TYPE OF MATERIAL USED IN TANK CONSTRUCTION
"NO LETTER" Carbon Steel
"AL" Aluminum (Classes 103, 105, 109, & 111)
"A-AL" Aluminum Alloy
"N" Nickel
"C," "D," or "E" Stainless Steel (alloy/steel)

TANK TEST PRESSURE (PSI)

SEPARATOR CHARACTER
Significant only for Class 105, 112, 113, 114 Tank Cars and some 111 Tank Cars when retrofitted.
"A" Top and bottom shelf couplers
"S" Tank headshields, top and bottom shelf couplers
"J" Jacketed thermal protection, tank headshields, top and bottom shelf couplers
"T" Spray-on thermal protection, tank headshields, top and bottom shelf couplers

CLASS DESIGNATION
The three-digit class designation follows the Authorizing Agency.

- NONPRESSURE TANK CARS
  - DOT 103   AAR 201
  - DOT 104   AAR 203
  - DOT 111   AAR 206
  - DOT 115   AAR 211
- PRESSURE TANK CARS
  - DOT 105   DOT 114
  - DOT 109   DOT 120
  - DOT 112

- CRYOGENIC LIQUID TANK CARS
  - DOT 113   AAR 204W
  - AAR 204XT (Inside boxcar)

- MISCELLANEOUS TANK CARS
  - DOT 106 Multi-Unit Tank Car Tanks (Ton Containers)
  - DOT 110 Multi-Unit Tank Car Tanks (Ton Containers)
  - DOT 107 High-Pressure Tank Car
  - AAR 207 Pneumatically Unloaded Covered Hopper
  - AAR 208 Wooden Tank Car

AUTHORIZING AGENCY
Tank Car specifications start with three letters designating the agency under whose authority the specification was issued
- DOT DEPARTMENT OF TRANSPORTATION
- AAR ASSOCIATION OF AMERICAN RAILROADS
- ICC INTERSTATE COMMERCE COMMISSION (Regulatory authority assumed by DOT in 1966)
- CTC CANADIAN TRANSPORT COMMISSION
- TC   TRANSPORT CANADA (replacing CTC)

FIGURE 32-28 *(continued)*

Potential Terrorist Incidents

The threat of terrorism has changed the way public safety agencies operate. In small towns and major metropolitan areas, the possibility exists that on any day, any agency could find itself in the eye of a storm involving intentionally released chemical substances, biological agents (disease-causing bacteria or other viruses that attack the human body), or an attack on buildings or people using explosives. No area of any country is immune to such attacks. Today, responders must recognize that they are a part of a much larger response mechanism in Canada; they must also understand that any incident involving terrorism will require the assistance and cooperation of countless local, provincial, territorial, and federal resources. Responders within every jurisdiction should think about the locations of potential targets for terrorists; the general and specific hazards posed by chemical, biological, and radiological agents (materials that emit radioactivity); possible indicators of illicit laboratories; and basic operational guidelines for dealing with explosive events and identifying the possible indicators of secondary devices. It is vital that responders maintain good situational awareness always, but potential terrorist events require an even higher level of attention. Changing conditions or the possibility of secondary devices or additional attacks should always be considered.

As has been stated, practically any vehicle can be used by criminal elements, fundamentalist activists, or terrorists as a weapon. For those who are intent on disrupting society, even a car can become a lethal weapon. Military-grade explosives or explosives used by demolition companies can prove deadly in the wrong hands.

Although bombings are the most frequent terrorist acts, responders also must be aware of the threats posed by other potential weapons. Shooting into a crowd at a shopping mall or a train station with an automatic weapon could cause devastating carnage. The release of a biological agent into a subway system could make numerous people become ill and die, simultaneously creating a public panic response that could overload the local fire department, EMS, law enforcement, and hospitals. Given the breadth of these threats, planning should consider the full range of possibilities. It is becoming more and more common for federal agencies such as the RCMP, CSIS, and the CSE to share information regarding potential threats with local law enforcement and fire and paramedic services. Canada is a member nation of the "Five Eyes" security alliance between Canada, the United States, Great Britain, Australia, and New Zealand. Through this alliance, federal agencies have access to information that pertains to fundamentalist and political terrorism that may aid local emergency response.

Potential targets for terrorist activities include both natural landmarks and human-made structures. These sites can be classified into three broad categories: infrastructure targets, symbolic targets, and civilian targets (**FIGURE 32-29**). The following sections will help you build a foundation to become a more informed responder when it comes to dealing with potential terrorist incidents.

LISTEN UP!

Preincident planning at infrastructure, symbolic, and civilian targets should consider the possibility of a terrorist attack.

Responding to Terrorist Incidents

The roles of all public safety responders in handling terrorist events involve many of the same functions that responders perform on a day-to-day basis. Regardless of the intent behind the incident, responders are required to make risk-based decisions on tactics and strategy, personal protective equipment (PPE), victim rescue and/or evacuation, decontamination, and information obtained during detection and monitoring activities—the entire gamut of actions required to solve the problem.

What is different during an incident with criminal intent is the landscape upon which that incident is handled and the interagency cooperation that must occur. An incident with criminal intent, causing widespread destruction and/or numerous casualties, will require the involvement of many local, provincial, territorial, and federal law enforcement agencies; emergency management agencies; allied health agencies; and perhaps the military. It is critical that all these agencies train together and work together in a coordinated and cooperative manner. A mass casualty incident involving a weapon of mass destruction (WMD) could quickly overwhelm your agency, neighbouring agencies, and the local healthcare system, in addition to quickly becoming national and international news. To that end, it is important to understand which assets and capabilities of those regional agencies may be available to assist with a large-scale incident. Additionally, preserving potential evidence is an important factor for responders to consider. This may be difficult during an active incident but should be a consideration for all responders. In many cases, high-impact events may require some form of after-action care for the responders in terms of mental health. This will be up to the AHJ to implement, but all responders should at least be aware of the need for these types of services after the incident.

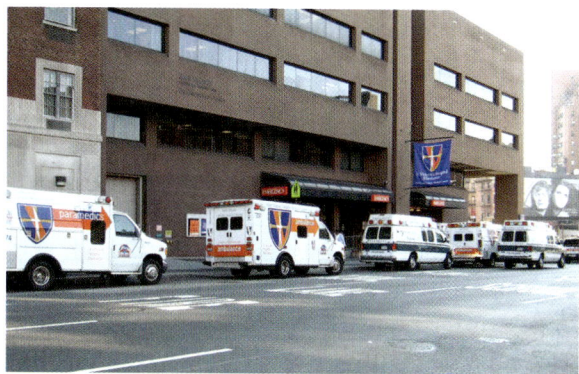

FIGURE 32-29 A. Subways, **B.** airports, **C.** bridges, and **D.** hospitals are all vulnerable to attack by terrorists who seek to interrupt a country's infrastructure.

A: © Arthur S. Aubry/Photodisc/Getty Images; **B:** © Steve Allen/Brand X Pictures/Alamy Images; **C:** © AbleStock; **D:** © Jones & Bartlett Learning. Photographed by Christine McKeen.

In most cases, the first responder emergency units will not be dispatched for a known WMD or terrorist incident. Rather, the initial dispatch report might cite an explosion, a possible hazardous materials incident, a single person with difficulty breathing, or multiple victims with similar symptoms. Emergency responders will usually not know that a terrorist incident has occurred until personnel on the scene begin to piece together information gained from their own observations and from interviews with witnesses.

If appropriate precautions are not taken, the initial responders may find themselves in the middle of a dangerous situation before they realize it. For this reason, initial responders should take note of any factors that suggest the possibility of a terrorist incident and immediately implement appropriate procedures. The possibility of a terrorist incident should be considered when responding to any location that has been identified as a potential terrorist target. It could be difficult to determine the true nature of the situation until a scene size-up is conducted.

Initial Actions

Responders should approach a known or potential terrorist incident just as they would a hazardous materials incident. If possible, apparatus and personnel should approach the scene from a position that is uphill and upwind. Emergency responders should don PPE, including self-contained breathing apparatus (SCBA). Later arriving units should be staged an appropriate distance away from the incident.

The first units to arrive should establish an outer perimeter to control access to and from the scene. They should deny access to all persons except emergency responders, and they should prevent potentially contaminated individuals from leaving the area before they have been decontaminated. The perimeter must surround the affected area, with the goal being to keep people who were not initially involved from becoming additional victims. This operation sounds quite orderly while you are reading this text, but rest assured that a real-world scene would be chaotic. There will be widespread panic, and you will probably be overwhelmed. It will take some time and many more responders before you get a handle on the scene. Expect this kind of tumultuous environment if the incident is significant.

Incident command should be established in a safe location, which could be as far as 914 m (3000 ft) away from the actual incident scene. The incident command post must be set up outside the area of possible contamination and beyond the distance where a

FIRE FIGHTER TIP

As a new fire fighter, responding to a terrorist attack may make you feel psychologically disoriented. An ordinary street can become an unrecognizable war zone in a few seconds. The best way to overcome these feelings is to focus on what you are trained to do. If you find you are becoming panicked or unsure, look to your officer and follow his or her instructions. Focus on smaller aspects of the emergency scene, such as assisting a victim or executing a task. Do not allow yourself to be overwhelmed by the overall entirety of the situation. Remain situationally aware, and maintain your self-control. Remember you are there to bring order to chaos, not become part of the problem. You cannot think clearly when you become emotionally overwhelmed.

secondary device may be planted. The initial task should be to determine the nature of the situation, the types of hazards that could be encountered, and the magnitude of the problems that must be faced.

An initial reconnaissance (recon) team should be sent out to quickly examine the involved area and to determine how many people are involved. Proper use of PPE, including SCBA, is essential for the recon team, and the initial survey must be conducted very cautiously, albeit as rapidly as possible. The possibility that chemical, biological, or radiological agents are involved cannot be ruled out until qualified personnel with appropriate instruments and detection devices have surveyed the area. Responders should begin their reconnaissance mission from a safe distance, working inward toward the scene, while taking care not to touch any liquids or solids or to walk through pools of liquid. Emergency responders who become contaminated must not leave the area until they have been decontaminated.

A process of elimination may be required to determine the nature of the situation. Occupants and witnesses should be asked if they observed any unusual packages or detected any strange odours, mists, or sprays. The presence of many casualties with no outward signs of trauma could indicate a possible chemical agent exposure. In such a case, victims' symptoms might include trouble breathing, skin irritations, or seizures.

A visible vapour cloud would be another strong indicator of a chemical release. Such a release could have occurred as the result of either an accident or a terrorist attack. The presence of dead or dying animals, insects, or plant life might also point to a chemical agent release as the culprit.

When approaching the scene of an explosion, responders should consider the possibility of a terrorist bombing incident and remain vigilant for secondary explosive devices. They should note any suspicious packages

and notify the incident commander (IC) immediately. Responders who have not been specially trained should never approach a suspicious object. Instead, **explosive ordnance disposal (EOD) personnel** should examine any suspicious articles and disable them.

The guidelines presented in this section are general recommendations, of course. You must also rely on your agency's standard operating procedures, your training, and your experience to take the appropriate actions. A terrorist event may never happen in your jurisdiction— but if it does, it will require every bit of skill you have to operate safely and effectively.

Interagency Coordination

If a terrorist incident is suspected, the IC, if he or she is not already a member of a law enforcement agency, should consult immediately with local law enforcement officials. In many cases, a unified command—including law enforcement (local, provincial, territorial, and federal), fire department, and EMS—should be established. If there are casualties, or if a mass casualty situation is evident, the IC should notify area hospitals and activate the mass casualty incident (MCI) medical plan (if one exists). Typically, local hospitals will begin to conduct an open-bed count and communicate with other hospitals about the availability of specialized medical services. Depending on the nature of the event, specialists like tactical law enforcement (SWAT teams) and bomb techs would be involved in the mitigation of the event (**FIGURE 32-30**). Most large municipalities have an Office of Emergency Management that coordinates responses to large-scale disasters and crises.

Provincial or territorial emergency management officials should be notified as soon as possible. This will

FIGURE 32-30 Tactical law enforcement teams may be required to render the area safe for fire and EMS responders to mitigate additional problems and render patient care.
Courtesy of Rob Schnepp.

help ensure a quick response by provincial, territorial, and federal resources to a major incident. Emergency operations centers (EOCs) at the local, provincial, or territorial levels may be established, depending on the severity of the incident. Large-scale search and rescue incidents could require the response of urban search and rescue (USAR) task forces activated through the provincial or territorial Emergency Management Organization (EMO). Medical response teams, such as disaster medical assistance teams (DMATs), may be needed for incidents involving large numbers of people. The Centers for Disease Control and Prevention (CDC) Strategic National Stockpile (SNS) may be requested when large caches of life-saving pharmaceuticals such as antidotes and medical supplies are needed.

An EOC can help coordinate the actions of all involved agencies in a large-scale incident, particularly if terrorism is involved. The EOC is usually set up in a predetermined remote location and is staffed by experienced command and staff personnel (**FIGURE 32-31**). The IC, who remains at the scene of the incident, should provide detailed situation reports to the EOC and request additional resources as needed.

Responders must remember that a terrorist incident is also a crime scene. To avoid destroying important evidence that could lead to a conviction of those responsible for perpetrating the attack, responders should not disturb the scene any more than is necessary. Where possible, law enforcement personnel should be consulted prior to overhaul and before the removal of any material from the scene. Responders should also realize that one or more terrorists could be among the injured. Be alert for threatening behaviour, and make note of anyone who seems determined to leave the scene.

Chemical Agents

Indicators of possible criminal or terrorist activity involving chemical agents may vary depending on the complexity of the operation, and there is no single indicator that may tip you off to the presence of such illicit activities. There may be overt indicators such as chemical-type gloves, chemical suits, respirators, and marked or unmarked containers made of various materials in a variety of shapes and sizes. For example, glass containers are prevalent at such locations but may or may not appear to be obvious hazards or threatening in any way.

The chemicals may provide unexplained odours that are out of character for the surroundings. Residual chemicals (liquid, powder, or gas form) may also be found in the area. Chemistry books or other reference materials may be seen, as well as materials that are used to manufacture chemical weapons (such as scales, thermometers, or torches). There may or may not be some type of easily identifiable signature such as an odour, liquid or solid residue, or dead insects or foliage. The main point is that you must always be on the lookout for items that may appear out of context with the setting—always pay attention to your surroundings. The routine medical or garage fire may quickly bring you up close and personal with a very dangerous situation.

Persons working with illicit materials may become exposed and exhibit symptoms of chemical exposure— for example, irritation to the eyes, nose, and throat; difficulty breathing; tightness in the chest; nausea and vomiting; dizziness; headache; blurred vision; blisters or rashes; disorientation; or even convulsions.

Chemical weapons can be disseminated in several ways. For example, intentionally releasing chlorine gas inside a building or a crowded gathering place could cause many injuries and deaths. To ensure broader distribution of a chemical agent, however, the agent might be added to an explosive device or mechanically dispersed. Crop-dusting aircraft, truck-mounted spraying units, or machine/hand-operated pump tanks could all potentially be used to disperse an agent.

The extent of dissemination of a toxic gas or suspended liquid particles depends on wind direction, wind speed, air temperature, and humidity at the time of the material's release. Because these factors can change quickly, it is difficult to predict the exact direction that might be taken by a chemical release cloud. Hazardous materials teams use computer models to predict the pathway of a toxic cloud. They also have the training and equipment to safely handle these situations.

FIGURE 32-31 An EOC is set up in a predetermined location for large-scale incidents.
© Jones & Bartlett Learning. Courtesy of MIEMSS.

Nerve Agents

Nerve agents are toxic chemical agents that attack the central nervous system. These weapons were first developed in Germany before World War II. Nerve agents are like some pesticides (organophosphates) but are much more toxic—in some cases, 100 to 1000 times more toxic than similar pesticides. Exposure to these substances can result in injury or death within minutes.

In their normal states, most nerve agents are liquids (**FIGURE 32-32**). To be an effective weapon, the liquid must either be dispersed in aerosol form or be broken down into fine droplets so that it can be inhaled or absorbed through the skin.

Pouring a liquid nerve agent, such as **V-agent (VX)**, onto the floor of a crowded building may not affect large numbers of people because at ambient temperature it is not highly vaporous. A more effective method to disperse it would be to aerosolize it in some fashion and distribute it across the widest populated area possible. In such a scenario, the effectiveness of the agent would depend on how long it stayed in the air and how widely it became dispersed throughout the building. VX is persistent because it has a low vapour pressure and will not evaporate quickly. Sarin, on the other hand, is considered nonpersistent because it evaporates at about the same rate as water. Common nerve agents, their method of contamination, and specific characteristics are listed in **TABLE 32-2**.

When a person is exposed to a nerve agent, symptoms of that exposure will become evident within minutes.

FIGURE 32-32 In their normal states, nerve agents are liquids. They must be dispersed in aerosol form if they are to be inhaled or absorbed by the skin.
© AbleStock.

The symptoms may include pinpoint pupils, runny nose, drooling, difficulty breathing, tearing, twitching, diarrhea, convulsions or seizures, and loss of consciousness (**FIGURE 32-33**). The same symptoms are seen in individuals who have been exposed to pesticides.

Several mnemonics can help you remember the symptoms of a nerve agent exposure. The mnemonic used most often in the emergency response community is SLUDGEM (**TABLE 32-3**).

TABLE 32-2 Common Nerve Agents		
Nerve Agent	**Route of Exposure**	**Characteristics**
Tabun (GA)	Skin contact Inhalation	Nonpersistent
Soman (GD)	Skin contact Inhalation	Nonpersistent
Sarin (GB)	Skin contact Inhalation	Nonpersistent
V-agent (VX)	Skin contact	Persistent

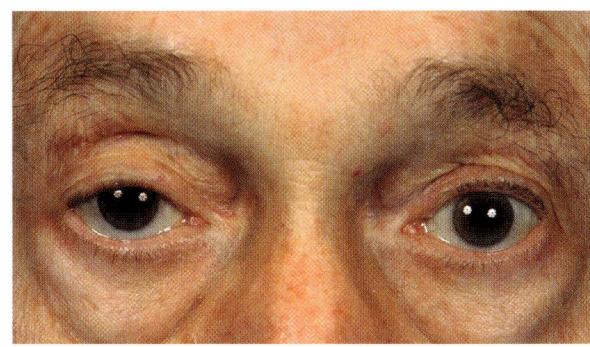

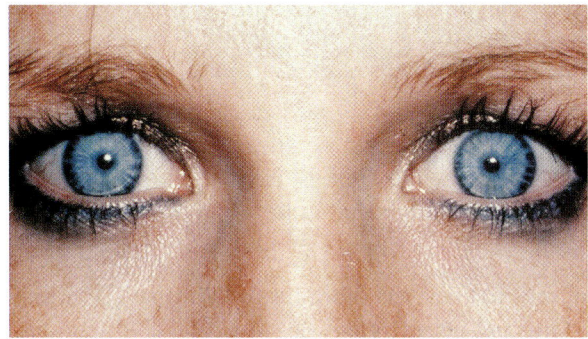

FIGURE 32-33 Pupil responses. **A.** Normal.
B. Pinpoint.
© Jones & Bartlett Learning.

TABLE 32-3 Symptoms of Nerve Agent Exposure

S: Salivation (drooling)

L: Lacrimation (tearing)

U: Urination

D: Defecation

G: Gastric upset (upset stomach, vomiting)

E: Emesis (vomiting)

M: Miosis (pinpoint pupils)

© Jones & Bartlett Learning.

Keep in mind that such mnemonics represent a very basic and limited way to identify a nerve agent exposure. From a medical standpoint, an exposure of this type involves both the sympathetic and parasympathetic nervous systems—each having its own unique set of signs and symptoms. Advanced life support providers (paramedics, e.g.) should have in-depth knowledge when it comes to recognizing the signs and symptoms of a nerve agent exposure. Additionally, many Internet and print resources are available for further study on the subject.

In terms of medical treatment for a nerve agent exposure, the most common field-level treatment is the DuoDote™, which contains 2.1 mg of atropine and 600 mg of pralidoxime (2-PAM), delivered as a single dose through one needle (**FIGURE 32-34**). These kits have been provided to many EMS units around Canada who carry these antidotes in their vehicles.

The antidote medications can be quickly injected into a person who has been contaminated by a nerve agent. Peak atropine levels are reached approximately 5 minutes after administration; peak levels of 2-PAM are achieved in 15 to 20 minutes. For the emergency responder, this delay means that you should not expect to administer this antidote to a person exhibiting serious signs and symptoms of a nerve agent exposure and have the patient recover right away or even at all. Auto-injectors are not an immediate cure-all. In fact, they may be ineffective if the victim has suffered a significant exposure to a nerve agent or pesticide.

Each drug in the auto-injector has a specific target and acts independently to reverse the effects of a nerve

RESPONDER TIP!

The DuoDote™ is intended for patients between 18 and 55 years of age.

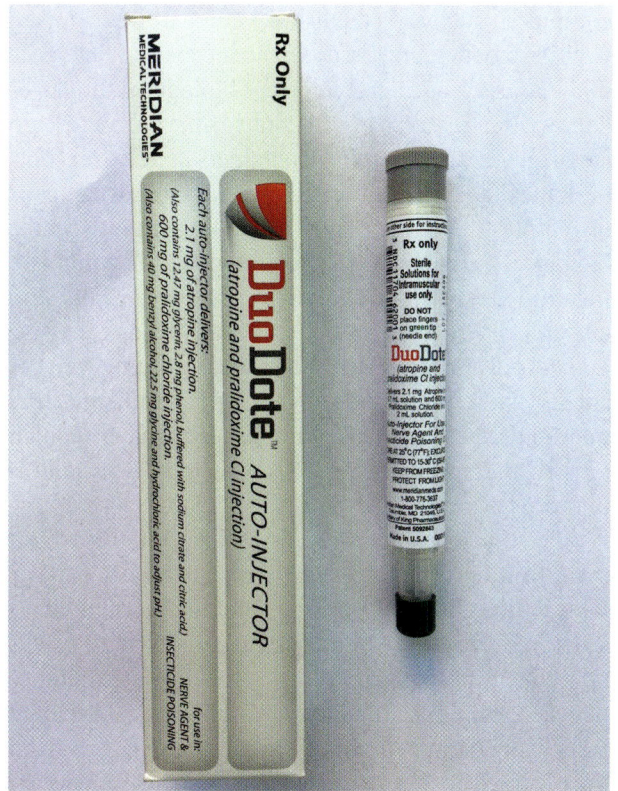

FIGURE 32-34 A DuoDote™ auto-injector.
Courtesy of Rob Schnepp.

agent exposure. Atropine, for example, may reverse **muscarinic effects** such as runny nose, salivation, sweating, bronchoconstriction, bronchial secretions, nausea, vomiting, and diarrhea; essentially, atropine is intended to deal with the SLUDGEM effects of a nerve agent exposure. Atropine dosing is guided by the patient's clinical presentation and should be given until secretions are dry or drying and ventilation becomes less laboured.

2-PAM, by contrast, does not reverse muscarinic effects on glands and smooth muscles. Instead, this medication's main goal is to decrease muscle twitching, improve muscle strength, and allow the patient to breathe better.

When it comes to treating victims of nerve agent exposure, emergency responders should understand that all nerve agents "age" once they are absorbed by the body. Thus, after a certain period of time (which varies depending on the agent), the administration of the DuoDote™ kit may be largely ineffective. Soman, for example, has an aging time of approximately 2 minutes. Sarin's aging time is approximately 3 to 4 hours. The other nerve agents have longer aging times. Follow local protocols before administering any antidote.

Biological Agents

Indicators of incidents that may potentially involve biological agents may include chemicals or production equipment such as Petri dishes, vented hoods, Bunsen burners, pipettes, microscopes, and incubators. Reference manuals, such as microbiology or biology textbooks or handwritten research or directions—maybe in a foreign language—may be present. Containers used to transport biological agents may include metal cylindrical cans or red plastic boxes or bags, probably without specific biological hazard labels. Personal protective equipment, including respirators, chemical or biological suits, and latex gloves, may be on the scene, as well as excessive amounts of antibiotics to protect those working with the agents (**FIGURE 32-35**). Other potential indicators may include abandoned spray devices and unscheduled or unusual sprays being disseminated (especially if outdoors at night).

Persons working in the lab may eventually exhibit symptoms consistent with the biological weapons with which they are working. Biological agents have a delayed onset of symptoms (usually days to weeks after the initial exposure). In fact, the biggest difference between a chemical incident and a biological incident is typically the speed of onset of the health effects from the involved agents. Because most biological agents are odourless and colourless, there are usually no outward indicators that the agents have spread. Again, there may

FIGURE 32-35 Biological agents may be evident in production or containment equipment, such as Petri dishes, vented hoods, Bunsen burners, pipettes, microscopes, and incubators.
Courtesy of Rob Schnepp.

or may not be overt indicators of criminal activities—be aware and attentive!

Biological agents are organisms that cause disease and attack the body. They include bacteria and viruses (**TABLE 32-4**). Other toxins, such as ricin and aflatoxin, are also biological toxins. Some of these organisms, such as anthrax, can live in the ground for years; others are rendered harmless after being exposed to sunlight for only a short period of time. The effects of a biological agent depend on the specific organism, the dose, and the route of entry. Most experts believe that a biological weapon would probably be spread by a device capable of widespread dispersion.

Some of the diseases caused by biological agents, such as smallpox and pneumonic plague, are contagious and can be passed from person to person. Doctors are concerned about the use of contagious diseases as weapons, because the resulting epidemic could overwhelm the healthcare system. Experts have different opinions about how difficult it would be to infect large numbers of people with one of these naturally occurring organisms. Because of their **incubation period**, people would not begin to show signs of being infected until 2 to 17 days after exposure to these organisms.

Anthrax

Anthrax is an infectious disease caused by the bacterium *Bacillus anthracis*. These bacteria are typically found around farm animals such as cows and sheep. For use as a weapon, the bacteria must be cultured to develop anthrax spores. The spores, weaponized into an ultra-fine powder, can then be dispersed in a variety of ways (**FIGURE 32-36**). Approximately 8000 to

TABLE 32-4 Bacteria Versus Virus

	Description	Dispersion	Examples
Bacteria	Single-cell microscopic organisms with a nucleus and cell wall	May form a spore	Anthrax Plague Tularemia
Virus	Submicroscopic agent Protein coated with DNA or RNA	Requires a host to live and reproduce	Smallpox Viral hemorrhagic fever (VHF)

© Jones & Bartlett Learning.

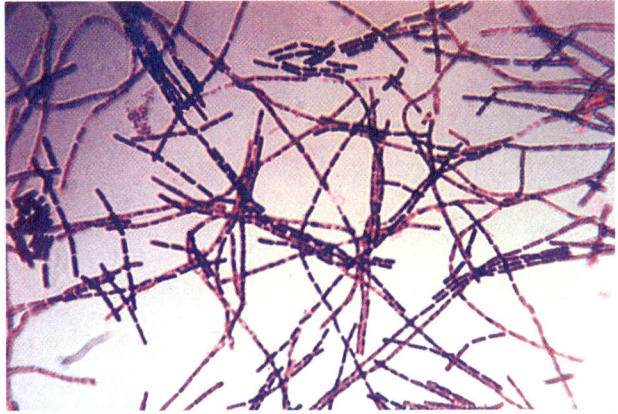

FIGURE 32-36 Anthrax spores can be dispersed in a variety of ways.
Courtesy of CDC.

10,000 spores are typically required to cause an anthrax infection. Spores infecting the skin cause cutaneous anthrax, ingested spores cause gastrointestinal anthrax, and inhaled spores cause inhalational anthrax. Anthrax has an incubation period of 2 to 6 days. The disease can be successfully treated with a variety of antibiotics if it is diagnosed early enough.

The threat posed by anthrax-related terrorism is quite real. In 2001, four letters containing anthrax were mailed to locations in the United States (New York City; Boca Raton, Florida; and Washington, D.C.). Five people died after being exposed to the contents of these letters, including two postal workers who were exposed as the letters passed through postal sorting centres. Several major government buildings had to be shut down for months to be decontaminated. These incidents followed shortly after the terrorist attacks of September 11, 2001, and they caused tremendous public concern. In the wake of these incidents, emergency personnel had to respond to thousands of incidents involving suspicious packages and citizens who believed that they might have been exposed to anthrax.

Today, presumptive field tests are available that hazardous materials teams can use to determine whether the threat of anthrax or other biological agents is legitimate.

The gold standard to positively identify anthrax (and other biological agents), however, is not a field test. Anthrax must be cultured in a lab, by qualified microbiologists, to be positively identified. Your agency should have established procedures for handling suspicious powders and getting samples to a qualified laboratory. Your local Federal Bureau of Investigation (FBI) office can provide guidance on the different types of labs available through the Laboratory Response Network and how to work with law enforcement to get a sample to the appropriate lab.

Plague

Plague is caused by *Yersinia pestis*, a bacterium that is commonly found on rodents. These bacteria are most often transmitted to humans by fleas that feed on infected animals and then bite humans.

The two main forms of plague are bubonic and pneumonic. Individuals who are bitten by fleas generally develop bubonic plague, which attacks the lymph nodes (**FIGURE 32-37**). Pneumonic plague can be contracted by inhaling the bacterium.

Yersinia pestis can survive for weeks in water, moist soil, or grains. These bacteria might also be cultured for distribution as a weapon in aerosol form. Inhalation of the aerosol form would put the target population at risk for pneumonic plague.

The incubation period for plague ranges from 2 to 6 days. This disease can be treated with antibiotics.

Smallpox

Smallpox is a highly infectious and often fatal disease caused by variola, a virus; it kills approximately

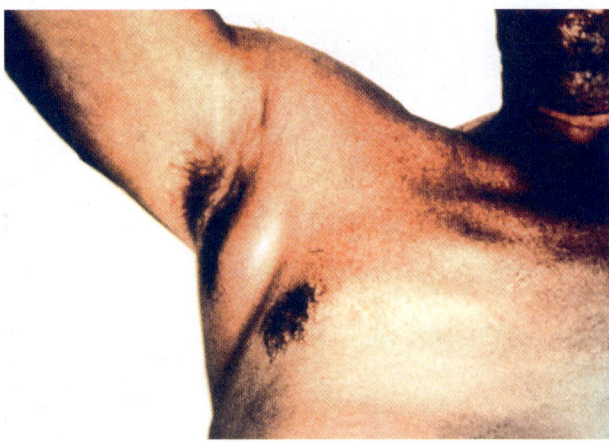

FIGURE 32-37 A bubo—one of the symptoms of plague—consists of a swollen, painful lymph node.
Courtesy of CDC.

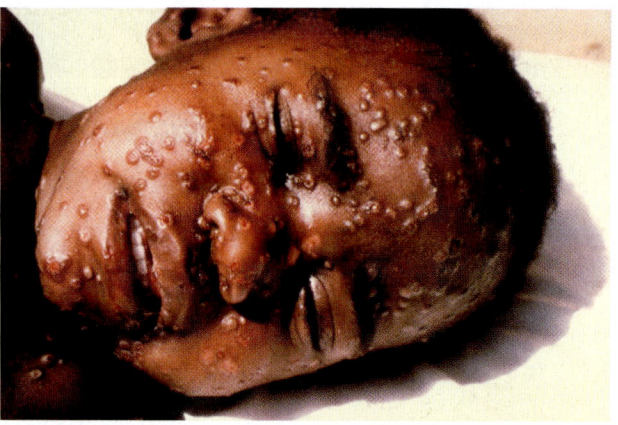

FIGURE 32-38 Smallpox is a highly contagious disease with a mortality rate of approximately 30 percent.
Courtesy of CDC.

30 percent of all persons who become infected with this pathogen. Smallpox first presents with small red spots or as a rash in the mouth. The rash then progresses to the face, followed by the arms and legs, and then farther outward to the hands and feet. Smallpox lesions are unique in that all lesions appear to be in the same stage of development at the same time. In contrast, in chickenpox, the lesions are in different stages of development across the body. Additionally, smallpox lesions can be found on the palms of the hands and the soles of the feet, whereas chickenpox lesions are seldom found on the palms and/or soles.

Although smallpox was once routinely encountered throughout the world, by 1980 it had been successfully eradicated as a public health threat through the use of an extremely effective vaccine. Officially, two countries (the United States and Russia) have maintained cultures of the disease for research purposes. It is possible, however, that international terrorist groups may have acquired the virus.

The smallpox virus could potentially be dispersed over a wide area in an aerosol form; however, widespread broadcasting of this agent may not be necessary to cause a devastating outbreak. Infecting a small number of people could lead to a rapid spread of the disease throughout a targeted population, given smallpox's highly contagious nature: The disease is easily spread by direct contact, droplet, and airborne transmission. Patients are considered highly infectious and should be quarantined until the last scab has fallen off (**FIGURE 32-38**). The incubation period for smallpox is between 4 and 17 days (average = 12 days).

Currently there are millions of people who have never been vaccinated for the disease, and millions more have reduced immunity because decades have passed since their last immunization.

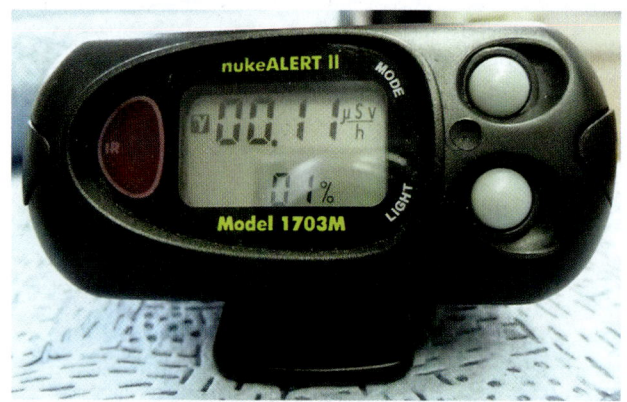

FIGURE 32-39 Only properly trained and equipped personnel, each of whom must carry an approved radiation monitor to measure the amount of radiation present, should make rescue attempts.
Courtesy of Rob Schnepp.

Radiological Agents

Indicators of radiological agents may include production or containment equipment, such as lead or stainless steel containers (with or without labels), and explosives that may be used to disperse the radioactive source, along with containers (e.g., pipes), caps, fuses, gunpowder, timers, wire, and detonators. Personal protective equipment present may include radiological protective suits and respirators. Radiation monitoring equipment such as Geiger counters or radiation pagers may be present, as may similar radiation detection devices used by responders to alert them to the presence of a potential threat (**FIGURE 32-39**). As with other locations where nefarious activities may be present, responders must put the potential threat in context. There may or may not be a valid explanation for the presence of something suspicious.

A major difference between an illicit location where radioactive substances are present and a legitimate operation may be the way the substances are packaged and stored. Legitimate sources are well marked, tracked, and regulated. When radiological agents are shipped, the package type is dictated by the degree of radiation activity inside the package—that is, the labelling is driven by the *amount of radiation that can be measured outside the package*. Three varieties of labels are found on radioactive packages: White I (**FIGURE 32-40**), Yellow II (**FIGURE 32-41**), and Yellow III (**FIGURE 32-42**).

Additionally, shippers of radiological materials are required to include a transport index (TI) number on the package label. This number indicates the highest amount of radiation that can be measured 1 m away from the surface of the package.

Responders must be able to recognize situations where radioactive materials might be encountered. Industries that routinely use radioactive materials include food testing labs, hospitals, medical research centres, biotechnology facilities, construction sites, and medical laboratories. For the most part, there will be some visual indicators (signs or placards) that indicate the presence of radioactive substances, but this is not always the case.

The key is to be able to suspect, recognize, and understand when and where you may encounter radioactive sources. If you suspect a radiation incident at a fixed facility, you should initially consult with the radiation safety officer of the facility. This person is responsible

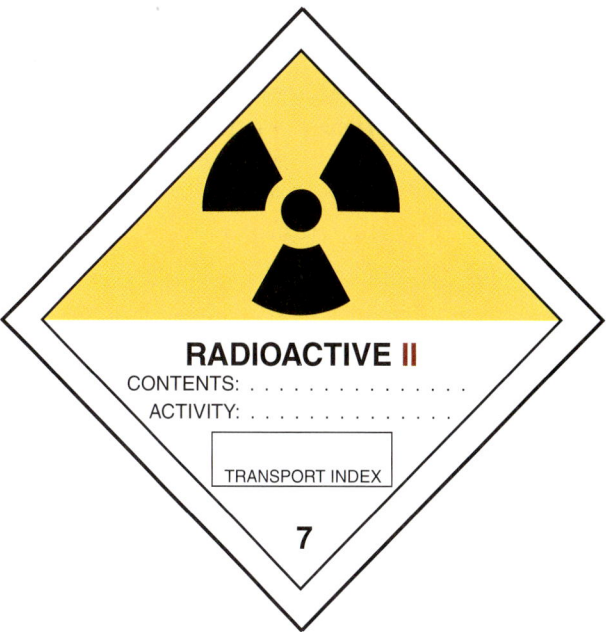

FIGURE 32-41 A Yellow II label.
Courtesy of the U.S. Department of Transportation.

FIGURE 32-42 A Yellow III label.
Courtesy of the U.S. Department of Transportation.

FIGURE 32-40 A White I label.
Courtesy of the U.S. Department of Transportation.

for the use, handling, and storage procedures for all radioactive material at the site. He or she likely will be a tremendous resource to you and will know exactly what is being used at the facility. If the incident is not at a fixed site, the presence of radiation may never be apparent. Radioactive isotopes are not detected by sight, smell, taste, or any of the other senses. Therefore, if you have any suspicion that the incident involves radiation, it will be necessary to call a hazardous materials team or some other resource with radiation detection capabilities.

Significant incidents involving radiation are rare, largely due to the comprehensiveness of the regulations for using, storing, and transporting significant radioactive sources. This is not to say that these incidents will not happen; nevertheless, the regulations have helped considerably in keeping the number of incidents low. Most of the incidents you may encounter will involve low-level radioactive sources and can be handled safely. These low-level sources are typically found in Type A packaging. This packaging method is unique to radioactive substances and contains materials such as radiopharmaceuticals and other low-level emitters.

Radiological Packaging

The most common types of containers and packages used to store radioactive materials are divided into five major categories: excepted range radioactive packaging; industrial radioactive packaging; and Type A (**FIGURE 32-43**), Type B (**FIGURE 32-44**), and Type C packaging (**FIGURE 32-45**).

Excepted packaging is used to transport materials that meet only general design requirements for any hazardous material package. Low-level radioactive substances are commonly shipped in excepted packages, which may be constructed out of heavy cardboard. Excepted packaging is authorized for limited quantities of radioactive material that would pose a very low hazard if released in an accident. Examples of material typically shipped in excepted packaging include consumer goods such as smoke detectors. Excepted packaging is excepted (excluded) from specific packaging, labelling, and shipping paper requirements; however, it is required to have the letters "UN" and the appropriate four-digit UN identification number marked on the outside of the package.

Industrial packaging is used in certain shipments of low-activity material and contaminated objects, which are usually categorized as radioactive waste. Most low-level radioactive waste is shipped in these packages. Transport Canada regulations require that these packages allow no identifiable release of the material to the environment during normal transportation and handling. There are three categories of industrial packages: IP-1, IP-2, and IP-3. The category of package will be marked on the exterior of the package.

Type A packaging is designed to protect the internal radiological contents during normal transportation and in the event of a minor accident. Such packaging is characterized by having an inner containment vessel made of glass, plastic, or metal and external packaging materials

FIGURE 32-43 A Type A package.
A and B: Courtesy of Rob Schnepp.

FIGURE 32-44 Type B packaging.
Courtesy of the U.S. Department of Energy.

made of polyethylene, rubber, or vermiculite. Examples of material typically shipped in Type A packages include nuclear medicines (radiopharmaceuticals), radioactive waste, and radioactive sources used in industrial applications. Type A packaging and its radioactive contents must meet standard testing requirements designed to

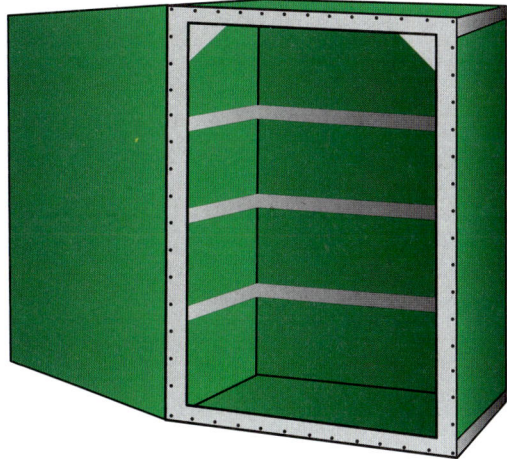

FIGURE 32-45 A Type C package.
© Jones & Bartlett Learning.

ensure that the package retains its containment integrity and shielding under normal transport conditions and may be transported by air. Type A packages must withstand moderate degrees of heat, cold, reduced air pressure, vibration, impact, water spray, drop, penetration, and stacking tests. The consequences of a release of the material in one of these packages would not be significant because the quantity of material in this package is so limited.

Type B packaging is far more durable than Type A packaging and is designed to prevent a release in the case of extreme accidents during transportation. More dangerous radioactive sources might be found in Type B packaging. Some of the tests that Type B containers must undergo include heavy fire, pressure from submersion, and falls onto spikes and unyielding surfaces. Type B packages include small drums and heavily shielded casks weighing more than 100 metric tons. This type of containment vessel contains materials such as spent nuclear fuel, high-level radioactive waste, and high concentrations of other radioactive material such as cesium and cobalt. Type B packages are designed to protect their contents from greater exposure; the amount of protection is based on the potential severity of the hazard. These package designs must withstand all Type A tests and a series of tests that simulate severe or "worst-case" accident conditions. Accident conditions are simulated by performance testing and engineering analysis. Life-endangering amounts of radioactive material are required to be transported in Type B packages. Type B packages are often used for air transportation.

Type C packaging is used for transporting high-activity radioactive substances by air. Dangerous radioactive sources are shipped in Type C packaging. Type C packaging is not certified for use in the United States

and is not part of the transportation regulations. This type of packaging is only referenced in international regulations.

Illicit Laboratories

Many indicators of possible criminal or terrorist activity involving illicit laboratories may be evident to responders. Many of the same materials used to manufacture homemade explosives are used to make illicit drugs. For example, terrorist paraphernalia may include terrorist training manuals, ideological propaganda, and documents indicating affiliation with known terrorist groups. Locations with certain characteristics are also commonly sites of illicit (clandestine) laboratories—for example, basements with unusual or multiple vents, buildings with heavy security, buildings with obscured windows, and buildings with odd or unusual odours. Personnel working in illegal laboratory settings may exhibit a certain degree of unusual or suspicious behaviour; for instance, they may be nervous and have a high level of anxiety. In addition, they may be very protective of the laboratory area and not want to allow anyone to access the area for any reason, or they may rush people out of the area as soon as possible.

Equipment that may be present in illicit laboratory areas includes surveillance materials (such as photographs, maps, blueprints, or time logs of the target hazard locations), non-weapon supplies (such as identification badges, uniforms, and decals that would be used to allow the terrorist to access target hazards), and weapon-related supplies (such as timers, switches, fuses, containers, wires, projectiles, and gunpowder or fuel). Security weapons such as guns, knives, and booby trap systems may also be present.

Drug laboratories are by far the most common type of clandestine laboratory encountered by responders. These laboratories are typically very primitive and can be found in hotel rooms, cars, and in even smaller settings. Materials used to manufacture the drugs often consist of everyday items (jars, bottles, glass cookware, coolers, and tubing) that have been modified to produce the illicit drugs (**FIGURE 32-46**).

Specific chemicals and materials found at the scene may include large quantities of cold tablets (ephedrine or pseudoephedrine), hydrochloric or sulfuric acid, paint thinner, drain cleaners, iodine crystals, table salt, aluminum foil, and batteries. The strong smell of urine or unusual chemical smells such as ether, ammonia, or acetone are very common indicators of clandestine drug manufacturing. Illicit drug laboratories should be considered significant hazardous materials scenes,

FIGURE 32-46 Items typically found at clandestine drug laboratories are everyday items.

© Jones & Bartlett Learning. Photographed by Glen E. Ellman.

FIGURE 32-47 Pipe bombs come in many shapes and sizes.

Courtesy of Captain David Jackson, Saginaw Township Fire Department.

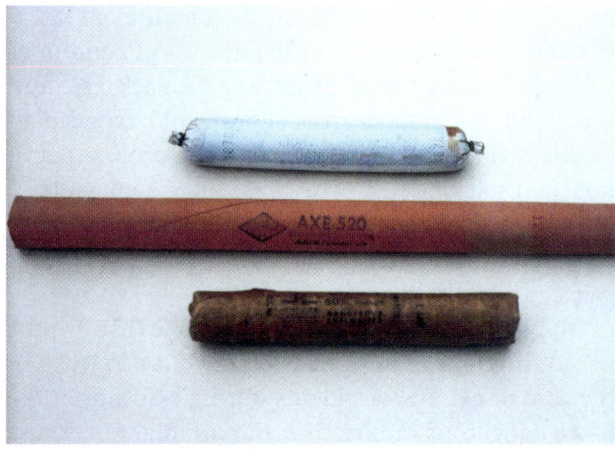

FIGURE 32-48 Every year, thousands of pounds of explosives are stolen from their rightful owners.

Courtesy of Dennis Krebs.

because the inexperienced chemists who run them take many shortcuts and disregard typical safety protocols to increase production.

The examples listed here are intended to give you an idea of some of the things that might tip you off to the presence of suspicious or illicit activity. It is not an exhaustive list, and you may encounter an illicit lab without any of these indicators.

Explosives

Indicators of possible criminal or terrorist activity involving explosives typically include materials that fit into four major categories—protective equipment, production and containment materials, explosive materials, and support materials. Protective equipment may include rubber gloves, goggles and face shields, and maybe even fire extinguishers. Production and containment equipment may include funnels, spoons, threaded pipes, caps, fuses, timers, wires, detonators, and concealment containers such as briefcases, backpacks, or other innocuous-looking packages (**FIGURE 32-47**). Explosive materials may include gunpowder, gasoline, fertilizer, solvents, oxidizers, and similar materials (**FIGURE 32-48**). Support materials may include explosive reference manuals, Internet-based reference materials, and military information. If you encounter a situation where a bomb-making operation is active, it is important to note what you see, get yourself out of danger, secure an appropriate amount of real estate, and call for your local EOD team (**FIGURE 32-49**). This is not a situation in which untrained responders should be investigating (**FIGURE 32-50**).

The Dirty Bomb

In recent years, the **radiation dispersal device (RDD)** or "dirty bomb" has emerged as a source of serious concern in terms of terrorism. Although not considered a weapon of mass destruction, an RDD has been described by the NFPA as "any device that causes the purposeful dissemination of radioactive material across an area without a nuclear detonation." Packing radioactive material around a conventional explosive device could contaminate a wide area, with the size of the affected area ultimately depending on the amount of radioactive material and the power of the explosive device. There are only a select number of radioactive sources that can be used effectively in an RDD. It is also possible for a criminal to construct a nonexplosive RDD, disseminating radioactive material via pressurized sprayers or air-handling systems in buildings. To limit this threat, radioactive materials,

FIGURE 32-49 EOD personnel are trained and have the necessary tools and equipment to evaluate and disable or disrupt a device or suspected device.
Courtesy of Rob Schnepp.

even in small amounts, are kept secure and protected. Such materials are widely used in industry and health care, and a criminal could potentially construct a dirty bomb with just a small quantity of stolen radioactive material.

Other types of radiological devices that may be used as weapons include improvised nuclear devices (INDs) and radiation exposure devices (REDs):

- *Improvised nuclear device (IND)*: A weapon fabricated from fissile material (highly enriched uranium or plutonium) capable of producing a nuclear explosion. A generally accepted successful yield in the 10- to 20-kiloton range—the equivalent to 10,000 to 20,000 tons of dynamite—in addition to high levels of radiation, would make this a very lethal device.
- *Radiation exposure device (RED)*: Radioactive material in a sealed container located where persons nearby would receive a direct exposure. Although not considered a device that would cause widespread death or sickness, it could constantly impact a steady stream of persons over a given period of time. If the container failed, then contamination might occur because the material could come into contact with victims.

FIGURE 32-50 The first responders who arrive at the scene of an explosion should establish a command post in a safe location and begin the process of setting up a unified command with law enforcement.
Courtesy of Captain David Jackson, Saginaw Township Fire Department.

Secondary Devices

A **secondary device** is some form of explosive or incendiary device designed to harm those responders summoned to the scene for some other reason. Terrorists who want to injure responding personnel with a secondary device or attack will typically make the initial attack very dramatic to draw responders into the proximity of the scene. The secondary attack usually takes place as the responders begin to treat victims of the initial attack.

Indicators of potential secondary devices may include "trip devices" such as timers, wires, or switches. Common concealment containers, such as briefcases, backpacks, boxes, or other common packages, may also be present; uncommon concealment containers may include pressure vessels (propane tanks) or industrial chemical containers (chlorine storage containers). Terrorists may watch the site of the primary devices, as part of preparing to manually activate the secondary devices. Responders can use the EVADE acronym to help think critically about the presence of secondary devices. This acronym can be found in NFPA 473, *Standard for Competencies for EMS Personnel Responding to Hazardous Materials/Weapons of Mass Destruction Incidents*:

- **E**valuate the scene for likely areas where secondary devices can be placed.
- **V**isually scan operating areas for a secondary device before providing patient care.
- **A**void touching or moving anything that can conceal an explosive device.
- **D**esignate and enforce scene control zones.
- **E**vacuate victims, other responders, and nonessential personnel as quickly and safely as possible.

After-Action REVIEW

IN SUMMARY

- Lots of information can be gained about an incident by knowing some basic things about containers.

- Although transportation of hazardous materials occurs most often on the roadway, when another primary mode of transport is used, roadway vehicles often transport the shipments from the rail station, airport, or dock to the point where it will be used. Responders must therefore become familiar with all types of chemical transport vehicles they might encounter during a transportation emergency.

- The responsible person in each situation should maintain the information about hazardous cargo and provide it in an emergency. When encountering a chemical release on any mode of transportation, take time to find the responsible person and track down these valuable pieces of information.

- The threat of terrorism has changed the way public safety agencies operate. In small towns and major metropolitan areas, the possibility exists that on any day, any agency could find itself in the eye of a storm involving intentionally released chemical substances, biological agents, or an attack on buildings or people using explosives.

- The roles of all public safety responders in handling terrorist events involve many of the same functions that responders perform on a day-to-day basis, including risk-based decisions on tactics and strategy, personal protective equipment (PPE), victim rescue and/or evacuation, decontamination, and information obtained during detection and monitoring activities. The difference is the landscape upon which that incident is handled and the interagency cooperation that must occur.

KEY TERMS

Access Navigate for flashcards to test your key term knowledge.

Anthrax An infectious disease spread by the bacterium *Bacillus anthracis*; typically found around farms, infecting livestock.

Biological agents Biological materials that are capable of causing acute disease or long-term damage to the human body. (NFPA 1951)

Bulk packaging Any packaging, including transport vehicles, having a liquid capacity of more than 450 L (119 gal), a solids capacity of more than 400 kg (882 lb), or a compressed gas (water) capacity of more than 454 kg (1001 lb). (NFPA 472)

Canadian Transport Emergency Centre (CANUTEC) The Canadian Transport Emergency Centre, operated by Transport Canada, that provides emergency response information and assistance on a 24-hour basis for responders to hazardous materials/weapons of mass destruction (WMD) incidents. (NFPA 1072)

Cargo tank A container used for carrying fuels and mounted permanently or otherwise secured on a tank vehicle. (NFPA 407)

Chemical Transportation Emergency Center (CHEMTREC) A public service of the American Chemistry Council that provides emergency response information

and assistance on a 24-hour basis for responders to hazardous materials/weapons of mass destruction (WMD) incidents. (NFPA 1072)

Cryogenic liquid cargo tank (MC-338) A low-pressure tank designed to maintain the low temperature required by the cryogens it carries. A box-like structure containing the tank control valves is typically attached to the rear of the tanker.

Dry bulk cargo trailers Trailers designed to carry dry bulk goods such as powders, pellets, fertilizers, or grain. Such tanks are generally V-shaped with rounded sides that funnel toward the bottom.

Emergency Transportation System for the Chemical Industry, Mexico (SETIQ) The Emergency Transportation System for the Chemical Industry in Mexico that provides emergency response information and assistance on a 24-hour basis for responders to emergencies involving hazardous materials/weapons of mass destruction (WMD). (NFPA 1072)

Excepted packaging Packaging used to transport materials that meets only general design requirements for any hazardous material. Low-level radioactive substances are commonly shipped in these packages, which may be constructed out of heavy cardboard.

Explosive ordnance disposal (EOD) personnel Personnel trained to detect, identify, evaluate, render safe, recover, and dispose of unexploded explosive devices.

General-service rail tank cars See *low-pressure tank cars.*

High-pressure cargo tank (MC-331) A tank that carries materials such as ammonia, propane, Freon, and butane. This type of tank is commonly constructed of steel and has rounded ends and a single open compartment inside. The liquid volume inside the tank varies, ranging from a 3785-L (1000-gal) delivery truck to a full-size 41,640-L (11,000-gal) cargo tank.

Incubation period The time period between the initial infection by an organism and the development of symptoms by a victim.

Industrial packaging Packaging used to transport materials that present a limited hazard to the public or the environment. Contaminated equipment is an example of such material, because it contains a non-life-endangering amount of radioactivity. It is classified into three categories, based on the strength of the packaging.

Intermodal tanks Bulk containers that serve as both a shipping and storage vessel. Such tanks hold between 18,927 and 22,712 L (5000 and 6000 gal) of product and can be either pressurized or nonpressurized. They can be shipped by all modes of transportation—air, sea, or land.

Low-pressure chemical cargo tank See *MC-307/DOT 407 chemical hauler.*

Low-pressure tank cars Railcars equipped with a tank that typically holds general industrial chemicals and consumer products such as corn syrup, flammable and combustible liquids, and mild corrosives.

MC-306/DOT 406 cargo tank Such a vehicle typically carries between 22,712 and 37,854 L (6000 and 10,000 gal) of a product such as gasoline or other flammable and combustible materials. The tank is nonpressurized; also called *nonpressure liquid cargo tank.*

MC-307/DOT 407 chemical hauler A rounded or horseshoe-shaped tank capable of holding 22,712 to 26,498 L (6000 to 7000 gal) of flammable liquid, mild corrosives, and poisons. The tank has a high internal working pressure; also called *low-pressure chemical cargo tank.*

MC-312/DOT 412 corrosive tank A tank that often carries aggressive (highly reactive) acids such as concentrated sulfuric and nitric acid. It is characterized by several heavy-duty reinforcing rings around the tank and holds approximately 22,712 L (6000 gal) of product.

Muscarinic effects Effects such as runny nose, salivation, sweating, bronchoconstriction, bronchial secretions, nausea, vomiting, and diarrhea.

National Response Center (NRC) An agency maintained and staffed by the U.S. Coast Guard; it should always be notified if a hazard discharges into the environment.

Non-bulk packaging Any packaging having a liquid capacity of 450 L (119 gal) or less, a solids capacity of 400 kg (882 lb) or less, or a compressed gas (water) capacity of 454 kg (1001 lb) or less. (NFPA 472)

Nonpressure liquid cargo tank See *MC-306/DOT 406 cargo tank.*

Plague An infectious disease caused by the bacterium *Yersinia pestis*, which is commonly found on rodents.

Pressure tank cars Railcars used to transport materials such as propane, ammonia, ethylene oxide, and chlorine.

Radiation dispersal device (RDD) A device designed to spread radioactive material through a detonation of conventional explosives or other (non-nuclear) means; also referred to as a "dirty bomb." (NFPA 472)

Radiological agents Radiation associated with x-rays, alpha, beta, and gamma emissions from radioactive isotopes, or other materials in excess of normal background radiation levels. (NFPA 1951)

Secondary containment Any device or structure that prevents environmental contamination when the primary container or its appurtenances fail. Examples of secondary containment mechanisms include dikes, curbing, and double-walled tanks.

Secondary device An explosive or incendiary device designed to harm emergency responders who have responded to an initial event.

Smallpox A highly infectious disease caused by the variola virus.

Soman A nerve gas that is both a contact and a vapour hazard; it has the odour of camphor.

Special-use railcars Boxcars, flat cars, cryogenic tank cars, or corrosive tank cars.

Tabun A nerve agent that disables the chemical connections between nerves and targets organs.

Tube trailers Trucks or semitrailers on which a number of very long compressed gas tubular cylinders have been mounted and manifolded into a common piping system. (NFPA 1)

Type A packaging Packaging that is designed to protect its internal radiological contents during normal transportation and in the event of a minor accident.

Type B packaging Packaging that is far more durable than Type A packaging and is designed to prevent a release of the radiological hazard in the case of extreme accidents during transportation. Type B containers must undergo a battery of tests including those involving heavy fire, pressure from submersion, and falls onto spikes and rocky surfaces.

Type C packaging Packaging used when radioactive substances must be transported by air.

V-agent (VX) A nerve agent, principally a contact hazard; an oily liquid that can persist for several weeks.

On Scene

It is a rainy afternoon when your pump company is dispatched to a motor vehicle accident. Upon arrival, you see a large tractor trailer rig on its side, with a single tank compartment that has rounded ends. The tank is painted white and appears to be a pressurized vessel.

1. Which type of container is this likely to be?

   A. Low-pressure cargo tank

   B. High-pressure cargo tank

   C. Vehicle-mounted pressure container

   D. High-pressure intermodal tank

2. Which of the following designators would be correct for this type of vessel?

   A. MC-407

   B. MC-331

   C. IMO-101

   D. IM-306

3. Which of the following substances would most likely be carried in the cargo tank described in the scenario?

   A. Propane

   B. Liquid nitrogen

   C. Gasoline

   D. Sulfuric acid

Access Navigate to find answers to this On Scene, along with other resources such as an audiobook.

CHAPTER **33**

Hazardous Materials Operations Level

Estimating Potential Harm and Planning a Response

KNOWLEDGE OBJECTIVES

After studying this chapter, you will be able to:

- Explain how to estimate the potential harm or severity of an incident. (**NFPA 1072: 5.3.1; NFPA 472: 5.1.1.2, 5.1.1.3, 5.1.2.1, 5.2.4, 5.3.2**, pp. 1142–1146)
- Explain how exposures might be affected by various types of hazardous materials incidents. (**NFPA 1072: 5.1.1, 5.3.1; NFPA 472: 5.1.2.2, 5.2.4, 5.3.1, 5.3.2**, pp. 1145, 1147–1149)
- Describe how to plan an initial response. (**NFPA 1072: 5.3.1; NFPA 472: 5.1.2.2, 5.2.4, 5.3.1, 5.3.2, 5.3.3**, pp. 1149–1150)
- Describe how to select personal protective equipment for an incident. (**NFPA 1072: 5.3.1, 5.5.1; NFPA 472: 5.3.3, 5.4.4**, pp. 1150–1151)
- Identify and describe the types of personal protective equipment needed for hazardous materials incidents. (**NFPA 1072: 5.3.1, 5.5.1; NFPA 472: 4.4.1, 5.3.3**, pp. 1151–1154)
- Identify and describe the four chemical-protective clothing ratings. (**NFPA 1072: 5.3.1, 5.5.1; NFPA 472: 5.1.2.2, 5.3.3**, pp. 1154–1156)

- Explain the role of respiratory protection. (**NFPA 1072: 5.3.1, 5.5.1; NFPA 472: 5.1.2.2, 5.3.3**, pp. 1156–1159)
- Describe the basic types of decontamination. (**NFPA 1072: 5.3.1, 5.5.1; NFPA 472: 5.1.2.2, 5.3.4, 5.4.1**, pp. 1159–1163)

SKILLS OBJECTIVES

After studying this chapter, you will be able to perform the following skill:

- Perform emergency decontamination. (**NFPA 1072: 5.3.1, 5.5.1; NFPA 472: 5.1.2.2**, p. 1162)

Hazardous Materials Alarm

Your pump company has been called to a semiconductor fabrication facility for a report of shortness of breath. You arrive at the reception area and are led to an employee break room. As you approach the seated patient, you notice an odour of ammonia. The laboratory manager is standing next to the patient—a laboratory technician—who is doubled over in the chair. The laboratory manager tells you that the technician was splashed in the face with approximately 100 mL (0.03 oz) of ammonium hydroxide while pouring chemicals into an instrument. He states that he helped the laboratory technician to an eye-wash station immediately after the incident and then escorted him to the break room. The laboratory manager tells you he has seen this sort of injury before. In his opinion, the laboratory technician doesn't need to go to the hospital—he just needs some oxygen. The paramedic begins an assessment and finds that the laboratory technician has reddened skin over his entire face and is complaining of shortness of breath. You are the officer on the pump company, and your four-person crew is trained to the hazardous materials operations level.

1. What are the initial response objectives for the pump company and ambulance once the ammonium hydroxide exposure is discovered?

2. What are the initial objectives for treating/transporting the patient?

3. Are the ambulance and pump company at risk of being exposed or contaminated to potentially toxic levels of ammonium hydroxide from the patient?

 *Access Navigate for more practice activities.*

Introduction

It is important to have a set of basic priorities to guide your decision making at the scene of a hazardous materials/weapons of mass destruction (WMD) incident. To that end, the first response objective should be to ensure your own safety while operating at the scene. You're no good to anyone if you've become part of the problem! At a minimum, you must arrive at the scene in a safe manner and make sure that you and your crew do not become a liability during the incident.

After ensuring your own safety, your next objective should be to address the potential life safety of those persons affected or potentially affected by the incident. In the chapter-opening scenario, your crew responded for one type of emergency—shortness of breath—and ended up facing a completely different kind of problem—a medical issue related to a chemical exposure. This new set of parameters will require you to quickly adapt your approach. You now have an exposed victim and a potentially exposed laboratory manager. Are additional response personnel and/or equipment required to handle the situation? You can smell the ammonia: Does that mean you and other personnel are in danger, too? Has the patient been adequately decontaminated to

the point you can safely render care? Who performed the decontamination? What type of decontamination was done? Should you back out of the area without treating the victim and call for a hazardous materials/WMD response team? Can the exposed person wait for a team of specially trained hazardous materials responders to perform decontamination? Do you have enough information and a decision-making process to weigh the risks and benefits of safely rendering patient care to a chemically exposed person?

Such real-world challenges present a set of complicated questions. This chapter will help you to think through identifying response options and estimate and plan for the challenges you may encounter at a hazardous materials/WMD incident.

Estimating the Potential Harm or Severity of the Incident

Hazardous materials/WMD incident response objectives should be based on the need to protect and/or reduce the threat to life, property, critical systems, and the environment.

In some cases, the people and the problem are one and the same, as in the chapter-opening scenario. Once the threat to life has been handled, the incident becomes a matter of reducing the impact to the property that may be affected and minimizing environmental complications.

To have some frame of reference for the degree of harm a substance may inflict, fire fighters at the hazardous materials operations level, in addition to having met the job performance requirements and competencies of NFPA 1072 and 472, should have a basic understanding of some commonly used terms and definitions. Two main organizations establish and publish the toxicological data typically used by hazardous materials/WMD responders: the American Conference of Governmental Industrial Hygienists (ACGIH) and the Occupational Safety and Health Administration (OSHA).

For more than 60 years, the ACGIH has been a respected and trusted source for occupational health and industrial hygiene guidelines and information. Its best-known committee, the Threshold Limit Values for Chemical Substances (TLV-CS) Committee, was established in 1941. This committee first introduced the concept of **threshold limit value (TLV)** in 1956. The TLV is the point at which a hazardous material/WMD begins to affect a person. Today, TLVs have been published for more than 600 chemical substances. Any values indicated with a TLV are established by the ACGIH.

OSHA was created in 1971 with three goals: to improve worker safety, to conduct research, and to publish toxicological data. OSHA's **permissible exposure limit (PEL)**, for example, is conceptually the same as ACGIH's TLV term. The PEL is the established standard limit of exposure to a hazardous material. You might see both terms in a reference source. Nevertheless, there is an important distinction between the ACGIH and OSHA standards—ACGIH sets guidelines; OSHA standards are the law.

There are many online resources that may assist in determining the toxicity level of a substance, such as TOXLINE https://envirotoxinfo.nlm.nih.gov/about -nlmenvirotox.html. It is common to see toxicological values expressed in units of parts per million (ppm), parts per billion (ppb), and, in some cases, parts per trillion (ppt). Typically, the amount of airborne contamination

encountered with releases of gases such as arsine, chlorine, and ammonia will be expressed in this manner. Arsine, for example, has an OSHA-established PEL of 0.05 ppm. This is a very small amount compared to the OSHA PEL for chlorine, which is 1 ppm. Comparatively speaking, arsine is a far more toxic substance than chlorine.

Another way to express contamination levels for substances other than gases, such as fibres and dusts, is through units of milligrams per cubic metre (mg/m^3). For simplicity and better understanding of this text, keep in mind that you may see toxicological data expressed in several different ways. Regardless, the lower the value, the more toxic the product.

The **threshold limit value/short-term exposure limit (TLV/STEL)** is the maximum concentration of a hazardous material that a person can be exposed to in 15-minute intervals, up to four times per day, without experiencing irritation or chronic or irreversible tissue damage. A minimum 1-hour rest period should separate any exposures to this concentration of the material. The lower the TLV/STEL concentration, the more toxic the substance. The **threshold limit value/ time-weighted average (TLV/TWA)** is the maximum airborne concentration of material that a worker could be exposed to for 8 hours a day, 40 hours a week, with no ill effects. As with the TLV/STEL, the lower the TLV/ TWA, the more toxic the substance. The **threshold limit value/ceiling (TLV/C)** is the maximum concentration of a hazardous material that a worker should not be exposed to, even for an instant. Again, the lower the TLV/C, the more toxic the substance.

The **threshold limit value/skin** indicates that direct or airborne contact with a material could result in possible and significant exposure from absorption through the skin, mucous membranes, and eyes. This designation is intended to suggest that appropriate measures be taken to minimize skin absorption so that the TLV/ skin is not exceeded.

As mentioned earlier, the PEL is the standard limit of exposure to a hazardous material as established and enforced by OSHA. The **recommended exposure level (REL)** is a value established by the National Institute for Occupational Safety and Health (NIOSH) and is comparable to OSHA's PEL. NIOSH is part of the U.S. Department of Health and Human Services and is charged with ensuring that individuals have a safe and healthy work environment by providing information, training, research, and education in the field of occupational safety and health. NIOSH sets RELs for other chemicals; these limits do not have the force of U.S. federal law, however. In Canada this is often referred to as "Occupational Exposure Limits" and is defined in various provincial and territorial legislation. As an example, in the province of Ontario there are OELs for over 725 substances set out in R.R.O. 1990, Regulation 833 (Control of Exposure to Biological or Chemical Agents) and Ontario Regulation 490/09 (Designated Substances) under Ontario's Occupational Health and Safety Act. The PEL and REL limits are comparable to ACGIH's TLV/TWA. These three terms (PEL, REL, and TLV/TWA) measure the maximum, time-weighted concentration of material to which 95 percent of healthy adults can be exposed without suffering any adverse effects over a 40-hour workweek.

> ### LISTEN UP!
>
> Identifying and measuring the levels of airborne contamination require specific detection and monitoring instruments along with training to interpret the results.

The designation **immediately dangerous to life and health (IDLH)** means that an atmospheric concentration of a toxic, corrosive, or asphyxiant substance poses an immediate or delayed threat to life or could cause irreversible adverse health effects or interfere with an individual's ability to escape unaided from a hazardous environment. Three types of IDLH atmospheres are distinguished: toxic, flammable, and oxygen deficient. Individuals exposed to atmospheric concentrations below the IDLH value (in theory) could escape from the atmosphere without experiencing irreversible damage to their health, even if their respiratory protection fails. Individuals who may be exposed to atmospheric concentrations equal to or higher than the IDLH value must use positive-pressure **self-contained breathing apparatus (SCBA)** or equivalent protection.

With the appropriate equipment, responders will be able to measure concentrations of specific chemicals. Once the exposure values are understood, they can be applied at the scene of a hazardous materials/WMD emergency. For example, exposure guidelines can be used to identify three basic atmospheres or environments that might be encountered at a hazardous materials/WMD emergency. Green environments could be considered low hazard and low risk. These environments would not require any PPE beyond a normal work uniform. Yellow environments are transitional, whereby the hazard is increasing, and some level of PPE is required (at least splash protection and some level of respiratory protection). Red environments are those that pose a high hazard and high risk; therefore, you must wear the highest level of skin and respiratory protection. An exposure to unprotected skin and/or lungs could be fatal. These are above-IDLH environments where toxic chemicals are present. You can certainly work around lethal concentrations of released chemicals safely if the correct type and level of PPE are used. These are rough guidelines and serve only to illustrate an actual or potential level of risk.

In many cases, responders have no control over the hazard—the genie may be out of the bottle upon your arrival. (See Chapter 31, *Properties and Effects*, for information on dispersion patterns, such as hemisphere, cloud, plume, cone, stream, pool, and irregular.) You have some control over your risk by understanding why and determining how you will interact with the problem. How will various levels of PPE change the risk? What is the risk of taking no action? Do you have the proper type and level of protection that allow for safe entry into a contaminated atmosphere?

- *Safe atmosphere:* No harmful hazardous materials effects exist, so personnel can handle routine emergencies without donning specialized PPE.
- *Unsafe atmosphere:* A hazardous material that is no longer contained has created an unsafe condition or atmosphere. A person who is exposed to the material for long enough may experience some form of acute or chronic injury.
- *Dangerous atmosphere:* Serious, irreversible injury or death may occur in the environment without PPE.

All exposure guidelines share a common goal: to ensure the safety and health of people exposed to a hazardous material.

> ### LISTEN UP!
>
> Initial isolation zones and protective actions are just that—initial. You must constantly evaluate the conditions and adjust tactics and strategy accordingly.

Resources for Determining the Size of the Incident

It is vital to understand the incident as a whole. To do so, responders must factor in results obtained from detection and monitoring devices, reference sources, bystander information, the current environmental conditions surrounding the incident, and other information to get a clear picture of what is going on and what is likely to happen next. Sometimes a decision must be made to evacuate or rescue people in danger. In those instances, responders may need to consult printed and electronic reference sources for guidance on evacuation distances and other safety information. Numerous computer programs can be used to model and predict the direction and size of vapour clouds. When used properly, these computer programs can be a valuable source of information for predicting the size, shape, and direction of movement of vapour clouds. Again, it is important to identify the health hazards posed by the substance to accurately set safe parameters around the entire incident.

The *Emergency Response Guidebook (ERG)* is a valuable resource to consult for evacuation distances. This reference outlines predetermined evacuation distances and basic action plans for chemicals, based on spill size estimates. All responders should be equipped with the latest version of the *ERG* and take time to become familiar with it. (Refer to Chapter 30, *Recognizing and Identifying the Hazards*, for specifics on using the guidebook. Figure 30-29 can be reviewed as an illustration of evacuation distance examples in the *ERG*.)

A good way to practice is to imagine a chemical and a credible location or condition in which the chemical might be released. Identify the unique United Nations/North American (UN/NA) Hazardous Materials Code identification number for the chemical, check whether it is highlighted and found in the green section of the *ERG*, and determine the recommended emergency actions and PPE that might be required to handle the incident. Also, it is useful to imagine the release occurring in several different areas of your jurisdiction. Take a few minutes to think about initial isolation distances or other protective actions in each case. To that end, responders should understand that the "initial isolation" distances (the distance at which all persons should be considered for evacuation in all directions) and the "protective action" distances (the downwind distance over which some form of protective actions might be required) are based on the nature of the material, the environmental conditions of the release, and the size of the release. A small spill, for example, means that a spill involves less than 200 L (approximately 52 gal) for liquids or less than 300 kg (approximately 661 lb)

of a solid. The *ERG* also offers suggested stand-off distances for improvised explosive devices (IEDs) and potential boiling liquid expanding vapour explosion (BLEVE) situations (**FIGURE 33-1**).

LISTEN UP!

To reduce the effects of a radiation exposure, responders should understand the concept of time–distance–shielding (TDS). When a radiation source is suspected or confirmed, responders should take action to reduce the amount of time they are exposed to the source, remain as far away as necessary, and place some barrier between themselves and the source. Identifying the presence of a radioactive source may require using a radiation detector. In the event you do not have such a device, or if you have a reasonable suspicion that the incident may involve a radioactive source, employ basic tactics that use the concept of TDS. Think of TDS in this way: The less time you spend in the sun, the less chance you have of suffering a sunburn (*time*); the closer you stand to a fire, the hotter you will get (*distance*); and if you come inside during a rainstorm, you will stop getting wet (*shielding*). These basic illustrations are analogous to the TDS concept for reducing the health effects of a radiation exposure.

Exposures

When considering the potential consequences of a hazardous materials/WMD incident, the on-scene crews must consider how exposures might be affected. In firefighting, the term *exposures* typically applies to those areas adjacent to the fire that might become involved if the fire is left unchecked. At hazardous materials incidents, exposures include any people, property, structures, or environments that are subject to influence, damage, or injury due to contact with a hazardous material/WMD. The number of exposures is determined by the location of the incident, the physical and chemical properties of the released substance, and the amount of progress that has been made in protecting those exposures by isolating the release site or by taking protective actions such as evacuation or sheltering-in-place. Incidents in urban areas are likely to have a greater potential for exposures; consequently, more resources will likely be needed to protect those exposures from the hazardous materials/WMD.

Isolation of the hazard area is one of the first actions responders must take at a hazardous materials/WMD incident. The general philosophy of isolation revolves around the concept of life safety and separating the people from the problem: *Responders and civilians alike must be kept a safe distance from the release site—a vital first step in beginning to establish safe work zones and identify the areas of high hazard.*

Improvised Explosive Device (IED) SAFE STAND-OFF DISTANCE

	Threat Description	Explosives Mass (TNT Equivalent)[1]		Building Evacuation Distance[2]		Outdoor Evacuation Distance[3]	
High Explosives (TNT Equivalent)	Pipe Bomb	5 lbs	2.3 kg	70 ft	21 m	850 ft	259 m
	Suicide Belt	10 lbs	4.5 kg	90 ft	27 m	1080 ft	330 m
	Suicide Vest	20 lbs	9 kg	110 ft	34 m	1360 ft	415 m
	Briefcase/Suitcase Bomb	50 lbs	23 kg	150 ft	46 m	1850 ft	564 m
	Compact Sedan	500 lbs	227 kg	320 ft	98 m	1500 ft	457 m
	Sedan	1000 lbs	454 kg	400 ft	122 m	1750 ft	534 m
	Passenger/Cargo Van	4000 lbs	1814 kg	640 ft	195 m	2750 ft	838 m
	Small Moving Van/ Delivery Truck	10000 lbs	4536 kg	860 ft	263 m	3750 ft	1143 m
	Moving Van/Water Truck	30000 lbs	13608 kg	1240 ft	375 m	6500 ft	1982 m
	Semitrailer	60000 lbs	27216 kg	1570 ft	475 m	7000 ft	2134 m

	Threat Description	LPG Mass/ Volume[1]		Fireball Diameter[4]		Safe Distance[5]	
Liquefied Petroleum Gas (LPG—Butane or Propane)	Small LPG Tank	20 lbs/5 gal	9 kg/19 L	40 ft	12 m	160 ft	48 m
	Large LPG Tank	100 lbs/25 gal	45 kg/95 L	69 ft	21 m	276 ft	84 m
	Commercial/ Residential LPG Tank	2000 lbs/500 gal	907 kg/1893 L	184 ft	56 m	736 ft	224 m
	Small LPG Truck	8000 lbs/2000 gal	3630 kg/7570 L	292 ft	89 m	1168 ft	356 m
	Semitanker LPG	40000 lbs/10000 gal	18144 kg/37850 L	499 ft	152 m	1996 ft	608 m

[1] Based on the maximum amount of material that could reasonably fit into a container or vehicle. Variations possible.

[2] Governed by the ability of an unreinforced building to withstand severe damage or collapse.

[3] Governed by the greater of fragment throw distance or glass breakage/falling glass hazard distance. These distances can be reduced for personnel wearing ballistic protection. Note that the pipe bomb, suicide belt/vest, and briefcase/suitcase bomb are assumed to have a fragmentation characteristic that requires greater stand-off distances than an equal amount of explosives in a vehicle.

[4] Assuming efficient mixing of the flammable gas with ambient air.

[5] Determined by U.S. firefighting practices wherein safe distances are approximately 4 times the flame height. Note that an LPG tank filled with high explosives would require a significantly greater stand-off distance than if it were filled with LPG.

A

BLEVE—SAFETY PRECAUTIONS

Use with caution. The following table gives a summary of tank properties, critical times, critical distances, and cooling water flow rates for various tank sizes. This table is provided to give responders some guidance, but it should be used with caution.

Tank dimensions are approximate and can vary depending on the tank design and application.

Minimum time to failure is based on **severe torch fire impingement** on the vapor space of a tank in good condition and is approximate. Tanks may fail earlier if they are damaged or corroded. Tanks may fail minutes or hours later than these minimum times depending on the conditions. It has been assumed here that the tanks are not equipped with thermal barriers or water spray cooling.

Minimum time to empty is based on an engulfing fire with a properly sized pressure relief valve. If the tank is only partially engulfed then time to empty will increase (i.e., if tank is 50% engulfed then the tanks will take twice as long to empty). Once again, it has been assumed that the tank is not equipped with a thermal barrier or water spray.

Tanks equipped with thermal barriers or water spray cooling significantly increase the times to failure and the times to empty. A thermal barrier can reduce the heat input to a tank by a factor of ten or more. This means it could take ten times as long to empty the tank through the pressure relief valve (PRV).

Fireball radius and emergency response distance are based on mathematical equations and are approximate. They assume spherical fireballs, and this is not always the case.

Two safety distances for public evacuation. The minimum distance is based on tanks that are launched with a small elevation angle (i.e., a few degrees above horizontal). This is most common for horizontal cylinders. The preferred evacuation distance has more margin of safety since it assumes the tanks are launched at a 45 degree angle to the horizontal. This might be more appropriate if a vertical cylinder is involved.

It is understood that these distances are very large and may not be practical in a highly populated area. However, it should be understood that the risks increase rapidly the closer you are to a BLEVE. Keep in mind that the farthest reaching projectiles tend to come off in the zones 45 degrees on each side of the tank ends.

Water flow rate is based on $\sqrt[5]{\text{capacity}}$ **(USgal) = US gal/min needed to cool tank metal.**

Warning: the data given are approximate and should only be used with extreme cautions. For example, where times are given for tank failure or tank emptying through the pressure relief valve—these times are typical but they can vary from situation to situation. Therefore, never risk life based on these times.

B

FIGURE 33-1 The *ERG* offers suggested stand-off distances for **A.** IEDs and **B.** potential BLEVE situations.

Courtesy of the U.S. Department of Transportation.

Imagine arriving on scene as the incident commander and encountering hose lines in service to cool the container. It is the middle of the day in a residential area; there are several blocks of occupied homes. What material is likely to be in the container, and how much? Would you evacuate or shelter-in-place? What resources could you use to determine the evacuation distances? Isolating the hazard may be accomplished in several ways. It's common for law enforcement officers to be posted a safe distance from the release to create a secure perimeter.

Other public safety personnel such as fire fighters may serve the same function, although it is important not to waste the skills of a cache of trained hazardous materials/WMD responders by assigning them to guard doors, other points of ingress or egress from a building, or other contaminated areas. In many cases, responders will stretch a length of barrier tape across roadways, doors, or other access points. Care must be taken, however, not to rely solely on this method of scene control. Quite often, areas marked with barrier tape are not respected by public safety responders or the general public. If barrier tape is used, it should still be backed up by a human presence (**FIGURE 33-2**). Also, keep in mind that the precise type of isolation efforts undertaken will be driven by the nature of the released chemical and the environmental conditions.

Once the hazard is isolated, access is denied to all but a small group of responders who are trained and equipped to enter the contaminated atmosphere. Isolating a contaminated atmosphere is always conjoined in some way with **denial of entry** (i.e., restriction of access) to the site. Practically speaking, one action should not exist without the other. Typically, site access control is established to control the movement of personnel into and out of a contaminated area. Review Chapter 30 for the basic initial actions recommended by the acronym SIN: safety, isolate, notify.

Evacuation is the removal/relocation of those individuals who may be affected by an approaching release of a hazardous material. If the threat will be sustained over a long period of time, it may be advisable to evacuate people from a predicted or anticipated hazard area, making sure to evacuate those in the most danger first.

Evacuation efforts should not require personnel to wear PPE or enter contaminated atmospheres. Think of it this way: If you are wearing PPE to move people from one area to another, you may have shifted gears from conducting an evacuation to performing a rescue. The latter responsibilities may or may not be within the scope of your hazardous materials training.

Sheltering-in-place is a method of safeguarding people located near or in a hazardous area by temporarily keeping them in a cleaner atmosphere, usually inside structures. In some cases—for example, with a transitory problem such as a mobile vapour cloud—it is advisable to use a shelter-in-place strategy. This method is desirable only when the population being protected in place can care for themselves and in a sealed environment with proper ventilation and air supply. Personnel trained to the hazardous materials awareness level, for example, could be expected to initiate some form of protective action such as directing civilians away from the contaminated area or directing certain populations of civilians to follow a shelter-in-place approach. *Remember that NFPA 472 and 1072 do not consider personnel trained to the hazardous materials awareness level to be hazardous materials responders.*

Reporting the Size and Scope of the Incident

Reporting the estimated physical size of the area affected by a hazardous materials/WMD incident is accomplished by using information available at the scene. If a vehicle is transporting a known amount of material, for example, an estimate of the size of the release might be made by subtracting the amount remaining in the container from the maximum capacity of the container. This can be computed by looking at the shipping papers to see whether deliveries have been made. To "see" into containers such as railroad tank cars, steel drums, or cargo tanks and estimate their remaining contents, responders may use thermal imaging devices (**FIGURE 33-3**).

For example, you might use a thermal imaging device to investigate a steel drum discovered in a vacant lot. A quick look at the scene may reveal some wet-looking soil around the base of the drum. By using the thermal imaging device, you might be able to determine the percentage of liquid remaining and make an educated

FIGURE 33-2 Law enforcement can be a great asset to help ensure your restricted area stays restricted.

Courtesy of Rob Schnepp.

FIGURE 33-3 Thermal imaging devices allow a responder to estimate how much material remains in a container. In this photo, the whitish colour at the bottom of the drum is the amount of liquid remaining.
Courtesy of Rob Schnepp.

Concentrated

FIGURE 33-4 A concentrated solution of any kind contains a large amount of solute for a given amount of solution.
© Jones & Bartlett Learning.

Dilute

FIGURE 33-5 A dilute solution contains a small amount of solute for a given solution.
© Jones & Bartlett Learning.

guess about how much could have leaked. Of course, these estimations may be quite rough, especially when it is unknown how much a given vessel may have contained prior to a release; however, it does allow for some "worst-case scenario" estimations.

Depending on its size, the extent of the release may be expressed in units as small as square metres or as large as square kilometres. There are no hard and fast rules here: Be as accurate and as clear as possible when communicating with other responders or assisting agencies or when contacting call centres such as CANUTEC, CHEMTREC, or SETIQ for assistance. Remember that the safety of responders is paramount to maintaining an effective response to any hazardous materials/WMD incident.

Determining the Concentration of a Released Hazardous Material

Concentration, from the perspective of a chemist, refers to the amount of solute in a given amount of solution. From a practical perspective, a concentrated solution of any kind contains a large amount of solute for a given amount of solution (**FIGURE 33-4**). A dilute solution, by contrast, contains a small amount of solute for a given solution (**FIGURE 33-5**).

Consider a natural gas release. If a gas heater were to fail in some way, resulting in a sustained release of natural gas (the solute), a high concentration of gas could build up inside a tightly sealed house (with air being the solution in this instance). The consequence of that accumulation could be an explosion if the gas were to reach the proper proportions and find an ignition source.

If the windows or doors of the house were opened, however, the concentration of natural gas would decrease, perhaps becoming so "dilute" that it would pose no significant fire or health hazard. Typically, concentrations of gases are expressed as percentages—think of flammable range as an example. The flammable range of natural gas is 5 percent to 15 percent, meaning that the concentration of vapours must be between these two values if combustion is to occur.

In the preceding example, responders may be called upon to determine the airborne concentration of natural gas within the house. When this step is necessary, the use of specialized detection and monitoring equipment may be required. Using detection and monitoring equipment properly requires some technical expertise, a lot of common sense, and a commitment to continual training. It is a mistake for responders to believe that they can simply turn on a machine, point it in some direction, and expect it to solve the problem. A reading from a gas detector, taken out of context, may cause an entire response to head off in the wrong direction, leading to an unsafe decision or a series of inefficient tactics. The responder must interpret the information the instrument is providing and make decisions based on the information. Always remember that using a detector/monitor entails more than just reading the screen or waiting for an alarm to sound. For more information on detection and monitoring, see Chapter 19, *Salvage and Overhaul*.

When responding to incidents involving corrosives, it is also important to know the concentration of the released substance. Concentration, when discussing corrosives, is an expression of how much of the acid or the base is dissolved in a solution (usually water). Again, this value is generally expressed as a percentage. Sulfuric acid at a concentration of 97 percent is considered to be "concentrated," whereas the sulfuric acid found in car batteries (approximately 30 percent) is considered to be "dilute." In practical terms, responders should understand whether a released corrosive is concentrated or diluted—but should not confuse *concentration* with the *strength* of the solution. The words "strong" and "weak" do not correspond with "concentrated" and "dilute," respectively. The strength of a corrosive refers to the degree of ionization that occurs in a solution, which is determined by the solution's pH. A strong acid such as hydrochloric acid (HCl) is strong even if it's found in a dilute concentration. Acetic acid (vinegar) is a weaker acid, and it remains a weak acid even when it occurs in high concentrations. In general, strong corrosives will react more vigorously with incompatible materials such as organic substances and will be more aggressive when they contact metallic objects such as metal shelving, shovels, and other items.

To measure pH in the field, hazardous materials/WMD responders can use litmus paper, sometimes referred to as pH paper (**FIGURE 33-6**). Several styles of pH paper are in use today; refer to the tools used in your own jurisdiction to determine the specific styles of pH paper you may be called upon to use. Although specialized laboratory instruments are also used to measure pH, these kinds of tools are rarely deployed in the field.

Skin Contact Hazards

Many hazardous substances on the market today can produce harmful effects on the unprotected or inadequately protected human body. The skin can absorb harmful toxins without any sensation to the skin itself. Given this fact, responders should not rely on pain or irritation as a warning sign of absorption. Some poisons (the nerve agent VX, for example) are so concentrated that just a few drops placed on the skin may result in death.

Personnel must ensure that they take care to protect cuts, abrasions, and moist skin from exposure to any chemicals. This relationship can create critical

FIGURE 33-6 Litmus paper (pH strips) is used to determine the hazardous material's pH.
© Sabine Kappel/Shutterstock.

problems for responders who are working at incidents that involve any form of chemical or biological agents. Responders with large open cuts, rashes, or abrasions should be prohibited from working in areas where they may be exposed to hazardous materials/WMD. Smaller cuts or abrasions should be covered with nonporous dressings.

The rate of absorption can vary depending on the body part that is exposed. For example, chemicals can be absorbed through the skin on the scalp much faster than they are absorbed through the skin on the forearm. The high absorbency rate associated with the eyes makes them one of the fastest means of exposure. For example, a chemical may quickly enter the body through this route when it is splashed directly into the eyes or carried from a fire by toxic smoke particles or when the eyes are exposed to gases or vapours.

Chemicals such as corrosives will immediately damage skin or body tissues upon contact. Acids, for example, have a strong affinity for moisture and can create significant skin and respiratory tract burns. In contrast, alkaline materials dissolve the fats and lipids that make up skin tissue and change solid tissue into a soapy-like liquid. This process is like the way caustic cleaning solutions dissolve grease and other materials in sinks and drains. As a result, alkaline burns are often much deeper and more destructive than are acid burns.

Plan an Initial Response

Planning a response boils down to understanding the nature of an incident and determining a course of action that will favourably change the outcome. On the surface, this seems like a straightforward, uncomplicated task. In truth, the decision to act can be a weighty one, fraught with many pitfalls and dangers. When planning an initial hazardous materials/WMD incident response, it is important to be mindful of the safety of the responding personnel. The responders are there to isolate, contain, and/or remedy the problem—not to become part of it. Proper incident planning will keep responders safe and provide a means to control the incident effectively, preventing further harm to persons or property.

The information obtained from the initial call for help is used to determine the safest, most effective, and fastest route to the hazardous materials/WMD scene. Choose a route that approaches the scene from an upwind and upgrade direction so that natural wind currents blow the hazardous material vapours away from arriving responders (**FIGURE 33-7**). A route that places the responders uphill as well as upwind of

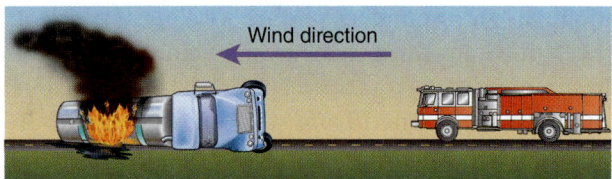

FIGURE 33-7 Approach a hazardous materials incident cautiously; choose a route that approaches from an upwind direction.
© Jones & Bartlett Learning.

the site is also desirable so that a liquid or vapour hazardous material flows away from responders.

Responders need to know as much as possible about the material involved. Is the material a solid, a liquid, or a gas? Is it contained in a drum, a barrel, or a pressurized tank? Is the spill still in progress (dynamic), or has it ceased (static)? The response to a spill of a solid hazardous material will differ from the response to a liquid-release incident or a vapour-release incident. A solid may be easily contained, whereas a released gas can be widespread and constantly moving, depending on the gas characteristics and weather conditions (**FIGURE 33-8**).

The characteristics of the affected area near the location of the spill or leak are also important factors in planning the response to an incident. If an area is heavily populated, evacuation procedures may be established very early in the incident. If the area is sparsely populated and rural, isolating the area from anyone trying to enter the location may be the top priority. A high-traffic area such as a major highway would necessitate immediate rerouting of traffic, especially during rush hours.

Response Objectives

Response objectives should be measurable, flexible, and time sensitive; they should also be based on the chosen strategy. Some examples might include the following:

- A team of three will construct a dirt berm around the drain at the south end of the leaking tanker to protect the adjacent waterway. This task will need to be completed in the next 30 minutes.
- A team of two wearing Level B ensemble (discussed later) will immediately enter the steel door on the east side of the building and shut down the ventilation system.

In some cases, several response objectives may be developed to solve a problem. To be effective and meaningful, however, those objectives need to be tied to the reason you chose to act in the first place. Again, if you don't understand why you are taking action, reevaluate the situation so that you can better understand the problem.

Typically, response objectives fall into one of three main categories: offensive, defensive, and nonintervention.

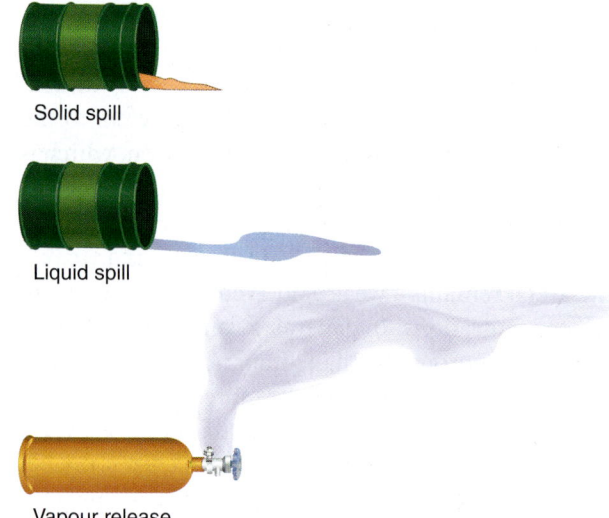

Solid spill

Liquid spill

Vapour release

FIGURE 33-8 The response to a spill of a solid hazardous material will differ from the response to a liquid-release or vapour-release incident.
© Jones & Bartlett Learning.

With offensive actions, responders take action to mitigate the issue. Offensive operations typically take place in the identified hot zone (the area closest to the release, where PPE is required to operate). Defensive actions take place outside of the hot zone or some distance away from the point of release. Examples of defensive actions that can be taken include diking and damming, stopping the flow of a substance remotely from a valve or shut-off, diluting or diverting the material, or suppressing or dispersing vapour. These and other actions are covered in detail in Chapter 37, *Product Control*.

Nonintervention occurs when the hazard of entering the hot zone is too great or for some reason allowing the incident to self-stabilize makes more sense and is much safer. Allowing a compressed gas cylinder to vent off the pressure until empty may be an example of nonintervention. In any case, the mode of operations should be identified in the verbal or written incident action plan (IAP). The main components of an IAP include specific objectives, clear work assignments for the responders, the resources needed to handle the incident, an organizational chart representing the key players on the scene, communications channels or methods, and a medical plan to follow in the event responders or civilians are exposed or injured during the response. Site safety plans should also be considered and are covered in Chapter 34, *Implementing the Planned Response*.

Personal Protective Equipment

The determination of which PPE is needed is based on the hazardous material involved, the specific hazards present, and the physical state of the material, along with a consideration of the tasks to be performed by the operations level hazardous materials responder. (Chapter 36, *Hazardous Materials Responder Personal Protective Equipment*, discusses proper PPE for a hazardous materials/WMD incident in detail.)

In the realm of hazardous materials/WMD response, the selection and use of chemical-protective clothing may have the greatest direct impact on responder health and safety. Without the proper PPE, responders place themselves at risk of suffering harmful exposures. Corrosives such as concentrated sulfuric acid and hydrochloric acid, as well as caustic substances such as sodium hydroxide, for example, can damage the skin. The human body is also susceptible to the adverse effects of solvents such as methylene chloride and toluene, which may penetrate the skin and cause systemic health effects.

Skin protection, however, is just one factor to be considered when discussing PPE. Anyone planning to work in a contaminated atmosphere must also place a high priority on respiratory protection. Respiratory protection is so important that it can be viewed as the defining element of PPE. The head-to-toe ensemble is not complete until the hazards have been identified and the respiratory protection has been properly matched to both the hazard and the garment. Keep in mind that PPE is not intended to function as an impenetrable suit of armor: It has limitations!

The National Fire Protection Association (NFPA) publishes protective clothing standards to provide guidance on the performance of certain types of chemical-protective garments. NFPA 1991, for example, is the *Standard on Vapor-Protective Ensembles for Hazardous Materials Emergencies and CBRN Terrorism Incidents.* NFPA 1992, *Standard on Liquid Splash-Protective Ensembles and Clothing for Hazardous Materials Emergencies*, covers a different type of chemical-protective garment. The NFPA also acknowledges the importance of chemical-protective garments as they relate to WMD response. To obtain guidance in this area, responders may reference NFPA 1994, *Standard on Protective Ensembles for First Responders to Hazardous Materials Emergencies and CBRN Terrorism Incidents.* (**CBRN** stands for chemical, biological, radiological, and nuclear.)

The NFPA does not "certify" any garments. This is a common misperception in the hazardous materials/WMD response industry. Instead, the intent of the NFPA clothing standards is to provide guidance on manufacturing quality and performance standards. A third-party testing laboratory carries out the testing and "certifies" the garment in question. In short, the NFPA publishes performance standards (durability, flammability, chemical resistance, and cold temperature); third-party laboratories then test manufacturers' garments to determine whether they meet the NFPA standards. Much like other NFPA committees, the Technical Committee for Chemical Protective Clothing consists of end users (responders), manufacturers, government representatives, and other recognized experts in the field.

When it comes to the selection and use of PPE, responders must understand how standards and regulations influence their decision making in the field. NFPA protective clothing standards do not explain when, or under which conditions, chemical protection is to be worn. Rather, these standards are performance documents for the garments only—they are not intended to guide responders. For guidance on which level of chemical protection to use under specific conditions, you may consult CANUTEC or other forms of reference materials for suit compatibility.

In addition to performance standards, responders must be aware of the procedures for cleaning, disinfecting, and inspecting PPE. These procedures may vary from manufacturer to manufacturer, so it is important for all responders to understand what is required to maintain the PPE in their jurisdiction.

Some types of PPE—primarily reusable garments—are required to be tested at regular intervals and after each use. Individual manufacturers will have well-defined procedures for the maintenance, testing, and inspection of their equipment. Prior to purchasing any PPE, the authority having jurisdiction (AHJ) should understand what is required in terms of maintenance and upkeep, including the cleaning and disinfection of the PPE.

Most types of chemical-protective garments should be stored in a cool, dry place, free of significant temperature swings and/or high levels of humidity. If repairs are required, consult the manufacturer prior to performing any work. There is a risk that the garment will not perform as expected if it has been modified or repaired incorrectly.

LISTEN UP!

The NFPA does not "certify" chemical-protective garments of any kind but rather sets performance standards for the garments.

Types of PPE for Hazardous Materials

Several types and levels of PPE may be selected for use at a hazardous materials/WMD incident. These levels are spelled out in detail in NFPA 1992, *Standard on Liquid Splash-Protective Ensembles and Clothing for*

Hazardous Materials Emergencies and *NFPA 1994, Standard for Protective Ensembles for First Responders to CBRN Terrorism Incidents.* This section reviews the protective qualities of various ensembles, from the lowest level of protection to the greatest, and discusses the selection criteria for each level.

Street Clothing and Work Uniforms

At the lower end of the PPE spectrum is normal street clothing or work uniforms, which offer the least amount of protection in a hazardous materials/WMD emergency (**FIGURE 33-9**). Work uniforms may prevent a "nuisance" powder from coming into direct contact with the skin but offer no chemical protection. Typically, personnel performing support functions away from the areas of contamination wear normal work uniforms.

Structural Firefighting Protective Clothing

Structural firefighting PPE provides the next level of protection. (**FIGURE 33-10**).

Structural firefighting gear is not recognized as a chemical-protective ensemble, though it does have a place at a hazardous materials/WMD incident. Many support functions can be carried out in structural fire fighter's gear. In some cases (such as during incidents involving chemicals with low toxicity and/or high

FIGURE 33-10 Standard structural firefighting gear.
© Jones & Bartlett Learning. Photographed by Glen E. Ellman.

flammability), structural gear may be safer than traditional types of chemical-protective clothing. A full set of structural fire fighter's gear includes a helmet, structural firefighting protective coat and pants, boots, gloves, a protective hood, SCBA, and a personal alert safety system (PASS) device.

High-Temperature–Protective Clothing and Equipment

High-temperature–protective equipment is a level above the structural fire fighter's ensemble (**FIGURE 33-11**). This type of PPE shields the wearer during short-term exposures to high temperatures. Sometimes referred to as a proximity or entry suit, high-temperature–protective equipment allows the properly trained fire fighter to

FIGURE 33-9 A Nomex jumpsuit.
Courtesy of the DuPont Company.

LISTEN UP!

A fire department should issue PPE that properly fits each fire fighter, regardless of gender, size, or shape.

FIGURE 33-11 High-temperature–protective equipment for short exposures.
© Photodisc.

work in extreme fire conditions. It provides protection against high temperatures only—it is not designed to protect the fire fighter from hazardous materials/WMD.

Chemical-Protective Clothing and Equipment

Chemical-protective clothing is unique in that it is designed to prevent the wearer from contacting chemicals. Not all chemical-protective clothing is the same, and each type/brand/style may offer varying degrees of resistance. There is no single chemical-protective garment on the market that will protect you from everything. Manufacturers supply compatibility charts with all protective equipment; these charts are intended to assist you in choosing the right chemical-protective clothing. You must match the anticipated chemical hazard to these charts to determine the resistance characteristics of the garment. Time; temperature; and resistance to cuts, tears, and abrasions are all factors that affect the chemical resistance of materials. Other requirements include flexibility, temperature resistance, shelf life, and sizing criteria.

Chemical resistance is the ability of the garment to resist damage or become compromised because of direct contact with a chemical. **Chemical-resistant materials** are specifically designed to inhibit or resist the passage of chemicals into and through the material by the processes of penetration, permeation, or degradation.

Penetration is the flow or movement of a hazardous chemical through closures (e.g., zippers), seams, porous materials, pinholes, or other imperfections in the material. Although liquids are most likely to penetrate a material, solids (e.g., asbestos) can also penetrate protective clothing materials.

Permeation is the process by which a substance moves through a given material on the molecular level. It differs from penetration in that permeation occurs through the material itself rather than through openings in the material.

Degradation is the physical destruction or decomposition of a clothing material owing to chemical exposure, general use, or ambient conditions (e.g., storage in sunlight). It may be evidenced by visible signs such as charring, shrinking, swelling, colour changes, or dissolving. Materials can also be tested for weight changes, loss of fabric tensile strength, and other properties to measure degradation.

Chemical-protective clothing can be constructed as a single-piece or multiple-piece garment. For example, a single-piece garment completely encloses the wearer and is referred to as an encapsulated suit. A variety of materials are used to manufacture encapsulated suits and nonencapsulated multiple-piece garments. Some common suit materials include butyl rubber, Tyvek®, Saranex™, polyvinyl chloride, and Viton®, which are used either singly or in multiple layers of several materials. Special chemical-protective clothing is adequate for some chemicals yet useless for other chemicals; no single material provides satisfactory protection from all chemicals.

Fully encapsulating protective clothing offers full body protection from highly contaminated environments and requires supplied-air respiratory protection devices such as SCBA. NFPA 1991 sets the performance standards for these types of garments, more correctly referred to as **vapour-protective clothing** (**FIGURE 33-12**).

FIGURE 33-12 Vapour-protective clothing.
© Jones & Bartlett Learning. Photographed by Glen E. Ellman.

FIGURE 33-13 Liquid splash–protective clothing must be worn when there is the danger of chemical splashes.
© Jones & Bartlett Learning. Photographed by Glen E. Ellman.

NFPA 1991 garments are tested for permeation resistance against several chemicals.

Liquid splash–protective clothing is designed to protect the wearer from chemical splashes (**FIGURE 33-13**). It does not provide total body protection from gases or vapours, however, and should not be used for incidents involving liquids that emit vapours known to affect or be absorbed through the skin. NFPA 1992 is the performance document for liquid-splash garments and ensembles. This type of equipment is tested for penetration resistance against a test battery of several chemicals. The tests include no gases, because this level of protection is not considered to be vapour protection.

Chemical-Protective Clothing Ratings

Chemical-protective clothing is rated for its effectiveness in several different ways. The U.S. Environmental Protection Agency (EPA) defines levels of protection using an alphabetical system.

Level A

The **Level A ensemble** consists of a fully encapsulating garment that completely envelops both the wearer and the respiratory protection, gloves, boots, and communications equipment. The Level A ensemble should be used when the hazardous material identified requires the highest level of protection for the skin, eyes, and respiratory tract. Typically, this level is indicated when the operating environment is above IDLH values for skin absorption. The Level A ensemble is effective against vapours, gases, mists, and even dusts (see Figure 33-12).

Ensembles worn as Level A protection must meet the requirements for vapour-protective clothing as outlined in the most current edition of NFPA 1991.

Recommended PPE of a Level A ensemble includes the following components:

- SCBA or **supplied-air respirator (SAR)**
- Fully encapsulating vapour-protective chemical-resistant suit
- Inner and outer chemical-resistant gloves
- Chemical-resistant safety boots/shoes (including steel shank and toe)
- Two-way radio

Optional PPE for a Level A ensemble includes the following components:

- Coveralls
- Cooling vest
- Long cotton underwear
- Hard hat
- Disposable gloves and boot covers

Level B

The **Level B ensemble** consists of chemical-protective clothing, boots, gloves, and SCBA (see Figure 33-13). This type of PPE should be used when the type and atmospheric concentration of identified substances require a high level of respiratory protection but less skin protection. To that end, the defining piece of equipment with Level B ensembles is an SCBA or some other type of supplied-air respirator (SAR). Garments and ensembles that are worn for Level B protection should comply with the performance standards outlined in NFPA 1992.

Several types of single-piece and multiple-piece garments on the market can be used to create a Level B ensemble. It is the respiratory protection, however, that distinguishes Level B ensembles from the next (lower) level of protection (Level C).

The types of gloves and boots worn depend on the identified chemical. Wrists and ankles must be properly sealed to prevent splashed liquids from contacting skin.

Recommended PPE worn as part of a Level B ensemble includes the following components:

- SCBA or SAR
- Chemical-resistant clothing
- Inner and outer chemical-resistant gloves
- Chemical-resistant safety boots/shoes
- Two-way radio

Optional PPE for a Level B ensemble includes the following components:

- Cooling vest
- Coveralls

- Disposable gloves and boot covers
- Face shield
- Long cotton underwear
- Hard hat

Level C

The **Level C ensemble** consists of standard work clothing plus chemical-protective clothing, chemical-resistant gloves, and a form of respiratory protection. Typically, Level C ensembles are worn with an **air-purifying respirator (APR)** or a **powered air-purifying respirator (PAPR)**; both are discussed in more detail later in this chapter. The APR could be a half-face mask (with eye protection) or a full-face mask. A Level C ensemble is appropriate when the type of airborne substance is known, its concentration is measured, the criteria for using APRs are met (see the respiratory protection section of this chapter), and skin and eye exposure is unlikely (**FIGURE 33-14**). The garments selected must meet the performance requirements outlined in NFPA 1992.

Recommended PPE for a Level C ensemble includes the following components:

- Full-face APR
- Chemical-resistant clothing
- Inner and outer chemical-resistant gloves
- Chemical-resistant safety boots/shoes
- Two-way radio

Optional PPE for a Level C ensemble includes the following components:

- Coveralls
- Disposable gloves and boot covers
- Face shield
- Escape mask
- Long cotton underwear
- Hard hat

Level D

The **Level D ensemble** is the lowest level of protection. This type of ensemble typically comprises coveralls, work shoes, hard hat, gloves, and standard work clothing (**FIGURE 33-15**). It should be used only when the atmosphere contains no known hazard and when work functions preclude splashes, immersion, or the potential for unexpected inhalation of or contact with hazardous levels of chemicals. A Level D ensemble should be used for nuisance contamination (such as dust) only; it should not be worn on any site where respiratory or skin hazards are known to exist.

Recommended PPE for a Level D ensemble includes the following components:

- Coveralls
- Safety boots/shoes
- Safety glasses or chemical-splash goggles
- Hard hat

FIGURE 33-15 The Level D ensemble is primarily a work uniform that includes coveralls and provides minimal protection.
© Jones & Bartlett Learning. Courtesy of MIEMSS.

FIGURE 33-14 A Level C ensemble includes chemical-protective clothing and gloves, as well as respiratory protection.
© Jones & Bartlett Learning. Photographed by Glen E. Ellman.

Optional PPE for a Level D ensemble includes the following components:

- Gloves
- Escape mask
- Face shield

A responder may wear liquid splash–protective clothing over or under structural firefighting protective clothing in some situations. This multiple-PPE approach provides limited chemical-splash and thermal protection. Those trained to the hazardous materials operations level can wear liquid splash–protective clothing when they are assigned to enter the initial site, protect decontamination personnel, or construct isolation barriers such as dikes, diversions, retention areas, or dams.

Respiratory Protection

NFPA 1994 was developed to address the performance of protective ensembles and garments (including respiratory protection) specific to weapons of mass destruction. This standard covers three classes of garments (Classes 2, 3, and 4), which differ from the traditional levels of protection listed in the EPA regulation (Levels A, B, C, and D). The CBRN (chemical, biological, radiological, and nuclear) requirements were added to the performance standards of NFPA 1991, so there is no Class 1 garment in NFPA 1994. The main difference between these classifications is that NFPA 1994 covers the performance of the garments and factors in the performance requirements of the respiratory protection. This consideration is critical when it comes to WMD events because some of the chemicals may cause the components of an APR or SCBA to fail, thereby exposing the responder to a highly toxic environment. Essentially, the NFPA standard acknowledges that the entire ensemble is only as good as the individual components. Based on that criterion, NFPA 1994 requires that all components be certified to perform in a CBRN environment. NFPA 1994 typifies a thought process about PPE that should extend beyond the standard—namely, chemical-protective clothing should be thought of as a *system*.

To help clarify the levels of protection, the NFPA 1994 classes are described here:

- *Class 2*: Liquid-splash garment performance with SCBA. (Class 2 standards for PPE are in line with the CBRN requirements for SCBA.)
- *Class 3*: Liquid-splash garment performance with APR. (Class 3 standards for PPE are in line with the CBRN requirements for APR.)
- *Class 4*: Performance requirements for particles and liquid-borne viral protection.

The CBRN performance requirements for SCBA and APR were born out of the terrorist attacks that occurred on September 11, 2001. In response to the growing threat of terrorism, NIOSH set performance guidelines for SCBA and APR relative to anticipated WMD incidents. In addition to meeting the requirements of NFPA 1981, *Standard on Open-Circuit Self-Contained Breathing Apparatus (SCBA) for Emergency Services*, any SCBA or APR with a NIOSH CBRN certification must have passed a battery of tests that measured their performance against sarin and sulfur mustard. From a practical standpoint, these tests were intended to ensure that the components of the respiratory protection would stand up to the aggressive nature of these chemicals. To reiterate, NFPA 1994 is intended to serve as an integrated performance guideline, factoring in both the garment and the respiratory protection used.

Physical Capability Requirements

Hazardous materials/WMD response operations put a great deal of physiological and psychological stress on responders. During the incident, personnel may be exposed to both chemical and physical hazards. They may face life-threatening emergencies, such as fire and explosions, or they may develop heat or cold stress while wearing protective clothing or working under extreme temperatures. For these reasons, every emergency response organization should have a comprehensive health and safety management program. Briefly, a health and safety program should include the following broad elements: medical surveillance, including pre-employment screening and periodic medical examinations; treatment plans for acute on-scene illness and injury; thorough recordkeeping of all elements of the program; and a mechanism to periodically review the entire process.

The medical surveillance piece of the overall program is the cornerstone of an effective health and safety management system for responders. The two primary objectives of a medical surveillance program are to determine whether an individual can perform his or her assigned duties, including the use of personal protective clothing and equipment, and to detect any changes in body system functions caused by physical or chemical exposures.

As part of a medical surveillance program, responders should be examined by a physician once a year or biennially based on the physician's recommendations. During this examination, the physician may—among other things—perform a routine exam based on the expected tasks the employee may perform, including

LISTEN UP!

The term *CBRN certified*, when used in reference to SCBA and APR, refers to SCBA and APR that are safe to use during a chemical, biological, radiological, or nuclear incident. CBRN-certified SCBA and APR have undergone rigorous testing to ensure their integrity during such incidents.

An SCBA provides a high level of protection at a hazardous materials/WMD incident. To comply with NFPA 1981, positive-pressure CBRN-certified units must maintain an air flow inside the mask at all times. This is a very important feature when responders are operating in an environment characterized by airborne contamination.

The extra weight and reduced visibility are other factors to consider when choosing to wear an SCBA. As with any piece of PPE, there are as many positive benefits as there are negative points to consider when determining whether SCBA is appropriate. Any responder called upon to wear an SCBA should be fully trained by the AHJ prior to operating in a contaminated environment. All responders should follow manufacturers' recommendations for using, cleaning, filling, and servicing the units they use.

FIGURE 33-16 An SCBA carries its own air supply, which limits the amount of air and time the user has to complete the job.
Courtesy of Rob Schnepp.

wearing PPE; conduct a health questionnaire; and take x-rays and perform a respiratory function test to measure lung capacity and function. The physician may also evaluate an individual's fitness for wearing SCBA and other respiratory protection devices. The specific requirements for any responder who is assigned to any duty where any form of respiratory protection will be used can be found in Canadian Standards Association (CSA) 94.4-11 (2016). Additional guidance can be found in NIOSH.

Medical monitoring and support differ in several ways from a medical surveillance program. Medical monitoring is the on-scene evaluation of response personnel who may experience adverse effects because of exposure to heat, cold, stress, or hazardous materials. Such monitoring can quickly identify problems so they can be treated in a timely fashion, thereby preventing severe adverse effects and maintaining the optimal health and safety of on-scene personnel. Medical monitoring, as a support function that takes place on the scene of a hazardous materials/WMD emergency, is covered in depth in Chapter 34, *Implementing the Planned Response*.

Positive-Pressure Self-Contained Breathing Apparatus

Respiratory protection, which in the fire service is commonly provided by an SCBA, is an important consideration at a hazardous materials/WMD incident

(**FIGURE 33-16**). Use of positive-pressure SCBA prevents both inhalation and ingestion exposures (two primary routes of exposure) and should be mandatory for fire service personnel. An SCBA carries its own air supply, a factor that limits the amount of air and time the user has to complete the job.

Supplied-Air Respirators

SARs, also referred to as *positive-pressure air-line respirators* (with escape units), use an external air source such as a compressor or a compressed air cylinder that is not designed to be carried by the user. A hose connects the user to the air source and provides air to the face piece. SARs are useful during extended operations such as decontamination, clean-up, and remedial work.

These units are equipped with a small "escape cylinder" of compressed air. Escape cylinders typically provide the user with approximately 5 minutes of breathing air. SARs may be less bulky and weigh less than SCBA, but the length of the air hose may limit movement, and there is potential for physical damage or perhaps chemical damage to the hose if it were to contact a released product (**FIGURE 33-17**).

FIGURE 33-17 A supplied-air respirator is less bulky than an SCBA but is limited by the length and structural integrity of the air hose.
Courtesy of Rob Schnepp.

Closed-Circuit SCBA

Some hazardous materials/WMD response teams use a form of respiratory protection referred to as a **closed-circuit self-contained breathing apparatus**. Commonly referred to as a "rebreather," this type of unit can be used when long work periods are required. The basic operating principle is different from the SCBA in that exhaled air is scrubbed free of carbon dioxide, supplemented with a small amount of oxygen, and "rebreathed" by the wearer. No exhaled air is released to the outside environment, making this type of unit a closed-circuit system. The earliest rebreathers were developed in the mid-1800s and were used primarily by mine workers.

Air-Purifying Respirators

APRs are filtering devices or particulate respirators that remove particulates, vapours, and contaminants from the air before it is inhaled. They should be worn only in atmospheres where the type and quantity of the contaminant are known and where sufficient oxygen for breathing is available. APRs should not be used when the atmosphere is IDLH. APRs may be appropriate for operations involving volatile solids and for remedial clean-up and recovery operations where the type and concentration of contaminants are verifiable (**FIGURE 33-18**).

FIGURE 33-18 Air-purifying respirators can be used only where there is sufficient oxygen in the atmosphere.
Courtesy of Rob Schnepp.

These devices range from full-face piece, dual-cartridge masks, to half-mask, face piece–mounted cartridges with no eye protection. APRs do not have a separate source of air but rather filter and purify ambient air before it is inhaled. The models used in environments containing hazardous gases or vapours are commonly equipped with an absorbent material that soaks up or reacts with the gas. Consequently, cartridge selection is based on the expected contaminants. Particle-removing respirators use a mechanical filter to separate the contaminants from the air. Both types of devices require that the ambient atmosphere contain a minimum of 19.5 percent oxygen.

APRs are easy to wear, but they do have some drawbacks. Because filtering cartridges are specific to expected contaminants, these devices are ineffective if the contaminant changes suddenly, possibly endangering the lives of responders. The air must also be continually monitored for both the known substance and the ambient oxygen level throughout the incident. For these reasons, APRs should not be employed at hazardous materials/WMD incidents until qualified personnel have tested the ambient atmosphere and determined that the devices can be used safely.

Powered Air-Purifying Respirators

PAPRs are similar in function to the standard APR described earlier but include a small fan to help circulate air into the mask (**FIGURE 33-19**). The fan unit is battery powered and worn around the waist. The fan draws outside air through the filters and into the mask via a low-pressure hose. PAPRs are not considered to be true positive-pressure units like an SCBA because it is possible for the wearer to "outbreathe" the flow of supplied air, thereby creating a negative-pressure situation inside the mask, possibly allowing contaminants to enter the face mask due to a poor seal with the face. The main advantages of the PAPR are that it diminishes the work of breathing of the wearer, helps reduce fogging in the mask, and provides a constant flow of cool air across the face.

FIGURE 33-19 PAPRs in use during a training exercise.
Courtesy of Rob Schnepp.

Decontamination

Even though there should be no intentional contact with the hazardous material involved in a hazardous materials/WMD incident, a procedure or a plan must be established to decontaminate anyone who becomes contaminated (**FIGURE 33-20**). According to NFPA 1072, contamination is "the process of transferring a hazardous material, or the hazardous component of a weapon of mass destruction (WMD), from its source to people, animals, the environment, or equipment, which can act as a carrier" (Section 3.3.11). Decontamination is "the physical and/or chemical process of reducing and preventing the spread and effects of contaminants to people, animals, the environment, or equipment involved at hazardous materials/weapons of mass destruction (WMD) incidents" (Section 3.3.15). There are various types of decontamination that can be performed at a hazardous materials incident, ranging from emergency

FIGURE 33-20 There must be a plan in place for decontamination at every hazardous materials/WMD incident.
© Jones & Bartlett Learning. Photographed by Glen E. Ellman.

LISTEN UP!

The NFPA standards provide guidance on the performance of various types of chemical-protective garments. Performance guidance for responders is provided by NFPA standards.

decontamination to mass decontamination. To understand the difference, the following descriptions provide a general overview of each type:

- **Emergency decontamination**: "The process of immediately reducing contamination of individuals in potentially life-threatening situations with or without the formal establishment of a decontamination corridor" (NFPA 1072, Section 3.3.15.1). The sole purpose is to quickly separate as much of the contaminant as possible from the individual to minimize exposure and injury. This can be due to a chemical exposure or, in the case of fire fighters operating at a fire scene, quickly removing the contaminants from structural firefighting protective gear (with hose lines or other methods) after exposure to soot and other particulates while still at the scene.
- **Gross decontamination**: "A phase of the decontamination process where significant reduction of the amount of surface contamination takes place as soon as possible, most often accomplished by mechanical removal of the contaminant or initial rinsing from handheld hose lines, emergency showers, or other nearby sources of water" (NFPA 1072, Section 3.3.15.2). Gross decontamination is performed on the following:
  - Team members before their technical decontamination
  - Emergency responders before leaving the incident scene
  - Victims during emergency decontamination
  - Persons requiring mass decontamination
  - PPE used by emergency responders before leaving the scene
     Decontamination performed on victims in a hospital setting is generally referred to as *definitive decontamination* but is not covered in this standard.
- **Technical decontamination**: According to NFPA 1072, technical decontamination is "the planned and systematic process of reducing contamination to a level that is as low as reasonably achievable" (NFPA 1072, Section 3.3.15.4). In other words, technical decontamination is the process that occurs after gross decontamination and is designed to remove contaminants from responders, equipment, and victims. It is intended to minimize the spread of contamination and ensure responder safety. Technical decontamination is normally established in support of emergency responder entry operations at a hazardous materials incident, with the scope and level of technical decontamination based on the types and properties of the contaminants involved. In

non–life-threatening contamination incidents, technical decontamination can also be used on victims of the initial release.

- **Mass decontamination**: "The physical process of reducing or removing surface contaminants from large numbers of victims in potentially life-threatening situations in the fastest time possible" (NFPA 1072, Section 3.3.15.3). Viewed from a big-picture perspective, mass decontamination is about making a rapid assessment of the situation and the number of victims present, attempting to identify the contaminant, setting up some form of mass decontamination process approved by the AHJ, selecting and wearing the proper type and level of PPE, and getting the job done.

The overall goal of any hazardous materials decontamination is to make personnel, equipment, and supplies safe by reducing or, in some cases, eliminating the offending substances. Proper decontamination is essential at every hazardous materials/WMD incident to ensure the safety of personnel and property. This section focuses on life safety and emergency decontamination.

Emergency (Field Expedient) Decontamination

Emergency decontamination is the process of quickly reducing or removing the bulk of contaminants from a victim as rapidly as possible. This procedure is undertaken in potentially life-threatening situations without the formal establishment of a **decontamination corridor**, a controlled area located within the warm zone where decontamination is performed. A more formal and detailed decontamination process may follow later. Emergency decontamination usually involves removing contaminated clothing and dousing the victim with large quantities of water (**FIGURE 33-21**).

If an emergency decontamination area has not been designated, responders should isolate the exposed victims in a contained area and establish an appropriate location. If possible, try to prevent the runoff from getting into drains, streams, or ponds; instead, divert the stream of water into an area where it can be treated or disposed of later. Emergency decontamination can easily be accomplished from a pump apparatus. Simply pull a handline, maneuver it into a circle, throw a tarp over the middle of the circle, and pull a booster line or other small handline to accomplish the decontamination. This technique is a quick way to handle a conscious victim of a chemical exposure or a responder who may have come into contact with a chemical and requires PPE to be decontaminated.

FIGURE 33-21 Emergency decontamination involves the immediate removal of contaminated clothing.
© Jones & Bartlett Learning. Photographed by Glen E. Ellman.

If adequate decontamination is not performed, the victim should not be allowed into the transport ambulance or the emergency room at the hospital. When possible, it is important to obtain a safety data sheet (SDS)/material safety data sheet (MSDS) for the substance to which the person was exposed and to make sure the information goes to the hospital with the patient.

Always ensure your own safety first before attempting decontamination of others. Avoid touching contaminated victims and/or entering contaminated environments without the proper level of protection. You will be no help in resolving the incident if you become a victim, too.

Refer to the scenario at the beginning of the chapter. Would the victim be a candidate for emergency decontamination? If so, how could that process be accomplished at a fixed facility? Most fixed facilities that use chemicals will have emergency safety showers and eyewash stations available to the employees. These sites may be the most easily accessible locations in which responders can perform emergency decontamination. The water is clean and readily available; also, employees of such a facility have typically been trained to use the safety showers (**FIGURE 33-22**).

An effective emergency decontamination operation depends on an understanding of the contaminant and its chemical and physical properties. To perform emergency decontamination, follow the steps in **SKILL DRILL 33-1**.

A

B

FIGURE 33-22 A. Indoor safety showers are in rooms that have no drains. Be prepared for a large clean-up process after activating a safety shower. **B.** Some safety showers also have preplumbed eyewash stations.
© Jones & Bartlett Learning. Photographed by Glen E. Ellman.

LISTEN UP!

By removing the appropriate clothing of a contaminated victim, you have significantly reduced the exposure.

SKILL DRILL 33-1
Performing Emergency Decontamination NFPA 1072: 5.3.1, 5.5.1; NFPA 472: 5.1.2.2

1 Confirm that you (the responder) have the appropriate PPE to protect against the contaminant. Stay clear of the product, and avoid physical contact with it if possible. Make an effort to contain runoff by directing the victim out of the hazard zone and into a suitable location for decontamination. It is not imperative to capture the runoff when you are in the process of performing emergency decontamination on an exposed person. Removing the contaminant from the person is more important than capturing the runoff. Instruct the victim in removing contaminated clothing. If the victim is capable of removing his or her own clothing, this is a preferred method. This ensures the assisting responder will stay as clean as possible.

2 Rinse the victim with copious amounts of water. Avoid using water that is too warm or too cold; room temperature water is best. Provide or obtain medical treatment for the victim, and arrange for the victim's transport.

NOTE: The manner in which decontamination occurs will depend on the prevailing weather conditions

Secondary Contamination

Secondary contamination, also known as cross-contamination, is the process of transferring a hazardous material from its source to people, animals, the environment, or equipment, all of which may act as carriers of the contaminant. Secondary contamination occurs when a contaminated person or object comes into direct contact with another person or object. It may happen in several ways:

- A contaminated victim comes into physical contact with another person.
- A bystander or responder encounters a contaminated object from the hot zone (the area immediately around and adjacent to the incident).
- A decontaminated responder reenters the decontamination area and encounters a contaminated person or object.

Because secondary contamination can occur in so many ways, areas at every hazardous materials/WMD incident should be designated as hot, warm, or cold, based on safety considerations and the degree of the hazard found in the area. These areas, known as control zones, should be established, clearly marked, and enforced at hazardous materials/WMD incidents. Establishing control zones is covered in greater detail in Chapter 34, *Implementing the Planned Response.*

After-Action REVIEW

IN SUMMARY

- The first priority for all responders is to ensure their own safety while operating at the scene.
- Hazardous materials/WMD incident response priorities should be based on the need to protect and/or reduce the threat to life, property, critical systems, and the environment.
- To have some frame of reference for the degree of harm a substance may pose, responders should have a basic understanding of some commonly used toxicological terms and definitions.
- Gather information from detection and monitoring devices, bystanders, reference sources, and environmental conditions to obtain a clear picture of the incident.
- Exposures can include people, property, structures, or the environment that is subject to influence, damage, or injury because of contact with a hazardous material/WMD.
- Immediate protective actions include isolation of the hazard area, denial of entry, evacuation, or sheltering-in-place.
- Tactical control objectives include preventing further injury and controlling or containing the spread of the hazardous material.
- Response objectives should be measurable, flexible, and time sensitive; they should also be based on the chosen strategy.
- Defensive actions include diking and damming, absorbing or adsorbing the hazardous material, stopping the flow remotely from a valve or shut-off, diluting or diverting the material, and suppressing or dispersing vapours.
- The decision to act at a hazardous materials/WMD incident should be based on the concept of risk versus benefit.
- The type of PPE required for an incident depends on the material involved, any specific hazards, the physical state of the material, and the tasks to be performed by the operations level responder.
- The selection and use of chemical-protective clothing may have the greatest direct effect on responder health and safety.
- The EPA defines four levels of PPE: Level A (highest level of protection), Level B, Level C, and Level D (lowest level of protection).
- NFPA 1994, *Standard on Protective Ensembles for First Responders to Hazardous Materials Emergencies and CBRN Terrorism Incidents*, was developed to address the performance of protective ensembles specific to weapons of mass destruction. It covers three classes of garments: Class 2, Class 3, and Class 4.
- Respiratory protection is so important that it can be viewed as the defining element of PPE.

- Medical surveillance is the cornerstone of an effective health and safety management system for responders. Before any PPE is worn, hazardous materials response personnel should be aware of the physiological stress those garments create.

- Even though there should be no intentional contact with the hazardous material involved, a procedure or a plan must be established at every hazardous materials/WMD incident to decontaminate anyone who accidentally becomes contaminated.

KEY TERMS

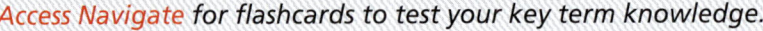

Access Navigate for flashcards to test your key term knowledge.

Air-purifying respirator (APR) A respirator that removes specific air contaminants by passing ambient air through one or more air purification components. (NFPA 1984)

CBRN Chemical, biological, radiological, and nuclear. (NFPA 1991)

Chemical-resistant materials Clothing (suit fabrics) specifically designed to inhibit or resist the passage of chemicals into and through the material by the processes of penetration, permeation, or degradation.

Closed-circuit self-contained breathing apparatus Self-contained breathing apparatus designed to recycle the user's exhaled air. This system removes carbon dioxide and generates fresh oxygen.

Contamination The process of transferring a hazardous material, or the hazardous component of a weapon of mass destruction (WMD), from its source to people, animals, the environment, or equipment, which can act as a carrier. (NFPA 1072)

Decontamination The physical and/or chemical process of reducing and preventing the spread and effects of contaminants to people, animals, the environment, or equipment involved at hazardous materials/weapons of mass destruction (WMD) incidents. (NFPA 1072)

Decontamination corridor The area usually located within the warm zone where decontamination is performed. (NFPA 1072)

Degradation A chemical action involving the molecular breakdown of a protective clothing material or equipment due to contact with a chemical. (NFPA 1072)

Denial of entry A policy under which, once the perimeter around a release site has been identified and marked out, responders limit access to all but essential personnel.

Emergency decontamination The process of immediately reducing contamination of individuals in potentially life-threatening situations with or without the formal establishment of a decontamination corridor. (NFPA 1072)

Evacuation The removal or relocation of those individuals who may be affected by an approaching release of a hazardous material.

Gross decontamination A phase of the decontamination process where significant reduction of the amount of surface contamination takes place as soon as possible, most often accomplished by mechanical removal of the contaminant or initial rinsing from handheld hose lines, emergency showers, or other nearby sources of water. (NFPA 1072)

High-temperature–protective equipment A type of personal protective equipment that shields the wearer during short-term exposures to high temperatures. Sometimes referred to as a *proximity suit*, this type of equipment allows the properly trained fire fighter to work in extreme fire conditions. It is not designed to protect against hazardous materials or weapons of mass destruction.

Immediately dangerous to life and health (IDLH) Any condition that would pose an immediate or delayed threat to life, cause irreversible adverse health effects, or interfere with an individual's ability to escape unaided from a hazardous environment. (NFPA 1670)

Isolation of the hazard area Steps taken to identify a perimeter around a contaminated atmosphere. Isolating an area is driven largely by the nature of the released chemicals and the environmental conditions that exist at the time of the release.

Level A ensemble Personal protective equipment that provides protection against vapours, gases, mists, and even dusts. The highest level of protection, Level A requires a totally encapsulating suit that includes a self-contained breathing apparatus.

Level B ensemble Personal protective equipment that is used when the type and atmospheric concentration of substances require a high level of respiratory protection but less skin protection. The kinds of gloves and boots worn depend on the identified chemical.

Level C ensemble Personal protective equipment that is used when the type of airborne substance is known, the concentration is measured, the criteria for using an air-purifying respirator are met, and skin and eye exposure is unlikely. A Level C ensemble consists of standard work clothing with the addition of chemical-protective clothing, chemical-resistant gloves, and a form of respiratory protection.

Level D ensemble Personal protective equipment that is used when the atmosphere contains no known hazard, and work functions preclude splashes, immersion, or the potential for unexpected inhalation of or contact with hazardous levels of chemicals. A Level D ensemble is primarily a work uniform that includes coveralls and affords minimal protection.

Liquid splash–protective clothing Clothing designed to protect the wearer from chemical splashes. It does not provide total body protection from gases or vapours and should not be used for incidents involving liquids that emit vapours known to affect or be absorbed through the skin. NFPA 1992 is the performance document pertaining to liquid-splash garments and ensembles.

Mass decontamination The physical process of reducing or removing surface contaminants from large numbers of victims in potentially life-threatening situations in the fastest time possible. (NFPA 1072)

Penetration The movement of a material through a suit's closures, such as zippers, buttonholes, seams, flaps, or other design features of chemical-protective clothing, and through punctures, cuts, and tears. (NFPA 1072)

Permeation A chemical action involving the movement of chemicals, on a molecular level, through intact material. (NFPA 1072)

Permissible exposure limit (PEL) The established standard limit of exposure to a hazardous material. It is based on the maximum time-weighted concentration at which 95 percent of exposed, healthy adults suffer no adverse effects over a 40-hour workweek.

Powered air-purifying respirator (PAPR) An air-purifying respirator that uses a powered blower to force the ambient air through one or more air-purifying components to the respiratory inlet covering. (NFPA 1984)

Recommended exposure level (REL) A value established by NIOSH that is comparable to OSHA's permissible exposure limit (PEL) and the threshold limit value/time-weighted average (TLV/TWA). The REL measures the maximum time-weighted concentration of material to which 95 percent of healthy adults can be exposed without suffering any adverse effects over a 40-hour workweek.

Secondary contamination The process by which a contaminant is carried out of the hot zone and contaminates people, animals, the environment, or equipment. Also referred to as *cross-contamination*.

Self-contained breathing apparatus (SCBA) A respirator that supplies a respirable air atmosphere to the user from a breathing air source that is independent of the ambient environment and designed to be carried by the user. (NFPA 1981)

Sheltering-in-place A method of safeguarding people located near or in a hazardous area by keeping them in a safe atmosphere, usually inside structures.

Supplied-air respirator (SAR) An atmosphere-supplying respirator for which the source of breathing air is not designed to be carried by the user. Also known as an air-line respirator. (NFPA 1989)

Technical decontamination The planned and systematic process of reducing contamination to a level that is as low as reasonably achievable. (NFPA 1072)

Threshold limit value (TLV) The point at which a hazardous material or weapon of mass destruction begins to affect a person.

Threshold limit value/ceiling (TLV/C) The maximum concentration of hazardous material to which a worker should not be exposed, even for an instant.

Threshold limit value/short-term exposure limit (TLV/STEL) The maximum concentration of hazardous material to which a worker can sustain a 15-minute exposure not more than four times daily without experiencing irritation or chronic or irreversible tissue damage. There should be a minimum 1-hour rest period between any exposures to this concentration of the material. The lower the TLV/STEL value, the more toxic the substance.

Threshold limit value/skin The concentration at which direct or airborne contact with a material could result in possible and significant exposure from absorption through the skin, mucous membranes, and eyes.

Threshold limit value/time-weighted average (TLV/TWA) The airborne concentration of a material to which a worker can be exposed for 8 hours a day, 40 hours a week and not suffer any ill effects.

Vapour-protective clothing The garment portion of a chemical-protective clothing ensemble that is designed and configured to protect the wearer against chemical vapours or gases. (NFPA 472)

On Scene

It is 9:00 AM when your crew is dispatched to a potential chemical spill at a nearby manufacturing facility. Upon your arrival, the incident commander (IC) of the site emergency response team reports that a worker intentionally mixed sodium cyanide and sulfuric acid in a 19-L (5-gal) container inside a closed 6 X 6-m (20-ft X 20-ft) storage room. The IC gives you a handwritten note from the worker, indicating that the chemical release was an apparent suicide attempt. There is a security camera monitoring the room, and the IC tells you the worker is on the floor next to the bucket, not moving. The entire building has been evacuated. The IC urges you to make an immediate rescue. The IC hands you two SDSs—one for sodium cyanide and one for 97 percent sulfuric acid.

1. After reading the SDSs for both chemicals, you realize the mixture of these two chemicals liberates cyanide gas, which is toxic by all routes of entry. Based on this information, which of the following actions would you take?

   A. Take no action that requires entering the storage room. Call for additional resources, including law enforcement and a hazardous materials/WMD team.

   B. Don full structural firefighting protective gear without an SCBA, and enter the storage area to attempt a rescue.

   C. Instruct a crew member to break out the rear window of the storage room, and then attempt a rescue wearing structural firefighting protective gear and SCBA.

   D. Turn on the ventilation system to the entire building.

2. To safely enter the storage room to assess the worker, which level of protection would you choose to wear?

   A. Level A

   B. Level B

   C. Level C

   D. Level D

3. A Level A ensemble offers full-body protection from highly contaminated environments and requires air-supplied respiratory protection such as an SCBA. Which of the NFPA standards spells out the performance requirements for these types of garments, which are more correctly referred to as vapour-protective ensembles?

   A. NFPA 1994

   B. NFPA 1992

   C. NFPA 1991

   D. NFPA 473

4. Level B ensembles are recommended when the type and atmospheric concentration of a released substance require a high level of respiratory protection but less skin protection. Which of the following items is the defining piece of equipment for a Level B ensemble?

   A. Air-purifying respirator

   B. Tyvek-Saranex chemical-protective garment

   C. Butyl rubber gloves

   D. Self-contained breathing apparatus or supplied-air respirator

Access Navigate to find answers to this On Scene, along with other resources such as an audiobook.

CHAPTER **34**

Hazardous Materials Operations Level

Implementing the Planned Response

KNOWLEDGE OBJECTIVES

After studying this chapter, you will be able to:

- Size up an incident. (**NFPA 1072: 5.1.5, 5.2.1, 5.4.1**, pp. 1168–1169)
- Identify and describe the safety procedures at a hazardous materials incident. (**NFPA 1072: 5.1.5, 5.4.1, 5.6.1; NFPA 472: 5.1.2.2, 5.4.1, 5.4.3, 5.4.4, 5.5.1, 5.5.2**, pp. 1169–1172)
- Describe the protective actions at the hazardous materials operations level. (**NFPA 1072: 5.1.5, 5.4.1; NFPA 472: 5.1.2.2, 5.4.1, 5.4.2, 5.4.4, 5.5.1**, pp. 1172–1178)
- Identify and describe the components of the incident command system. (**NFPA 1072: 5.4.1; NFPA 472: 5.1.2.2, 5.4.3**, pp. 1178–1183)
- Explain the role of the hazardous materials operations level responder in implementing a planned response. (**NFPA 1072: 5.1.1, 5.1.2, 5.1.3, 5.1.4, 5.1.5, 5.4.1; NFPA 472: 5.4.3**, p. 1184)

SKILLS OBJECTIVES

This chapter has no skills objectives for operations level hazardous materials responders.

Hazardous Materials Alarm

Local law enforcement is requesting a response for a suspicious package leaking a liquid. The package is in a small storeroom on the ground floor of a three-storey office building. While en route, the dispatcher announces this update: "Pump 40, be advised that police officers on scene are reporting several people complaining of eye irritation and nausea. There are also reports of an unusual odour on the second and third floors. A full building evacuation is in progress."

1. Based on the initial reports, how would you go about taking control of the scene?

2. Working with the on-scene law enforcement is important. What would you do to facilitate a coordinated approach to managing and jointly commanding this incident?

3. Does your agency have the capabilities to communicate via radio with other public safety agencies (various law enforcement agencies, for example)?

 *Access Navigate for more practice activities.*

Introduction

Scene control is important at all emergencies. At a hazardous materials incident, however, scene control is paramount because it has an influence on both scene security and personnel accountability. Typically, the starting point for implementing any response is sizing up the situation and taking control of the affected area. **Size-up** is the rapid mental process of evaluating the critical visual indicators of the incident (what's happening now), processing that information based on your training and experience (what might happen next), and arriving at a conclusion that will serve as the basis to form and implement a plan of action (what you are going to do about the situation). Sometimes that plan includes an aggressive offensive posture—that is, to address the problem immediately with action. In other cases, a defensive posture is appropriate—that is, isolate the scene, protect exposures, and allow the incident to stabilize on its own. The posture chosen should be driven by prudent decisions based on accurate information, observations, indications, and your level of training.

> **LISTEN UP!**
>
> Think before you act or give a command to others to act. Do not put yourself or others at undue risk or become part of the problem. Your safety is the number one priority!

Size-up is always a work in progress. In other words, as the incident progresses, everyone on the scene should be constantly sizing up his or her own piece of the problem. All responders should maintain situational awareness (SA) and understand the actions they are taking. Remember that responder safety is the number one priority.

> **LISTEN UP!**
>
> The initial actions taken set the tone for the response and are critical to the overall success of the effort.

Emergencies are seldom black-and-white problems; more typically, they involve many shades of gray that require you to *think*. It is always essential that you understand your job and know how to use the tools at your disposal both effectively and appropriately.

> **SAFETY TIP**
>
> Situational awareness (SA) refers to the degree of accuracy to which one's perception of the current operating environment mirrors reality. In essence, SA is the act of processing all the information available to you and understanding how it fits together in the big scheme of things.

It is important to follow a decision-making algorithm when facing a hazardous materials incident (**FIGURE 34-1**). Use it as a loose guide for developing an action plan and to focus your thinking.

> **LISTEN UP!**
>
> Do not let pressure from incoming units push you to make hasty decisions or assignments. Take time to think and formulate a plan of attack.

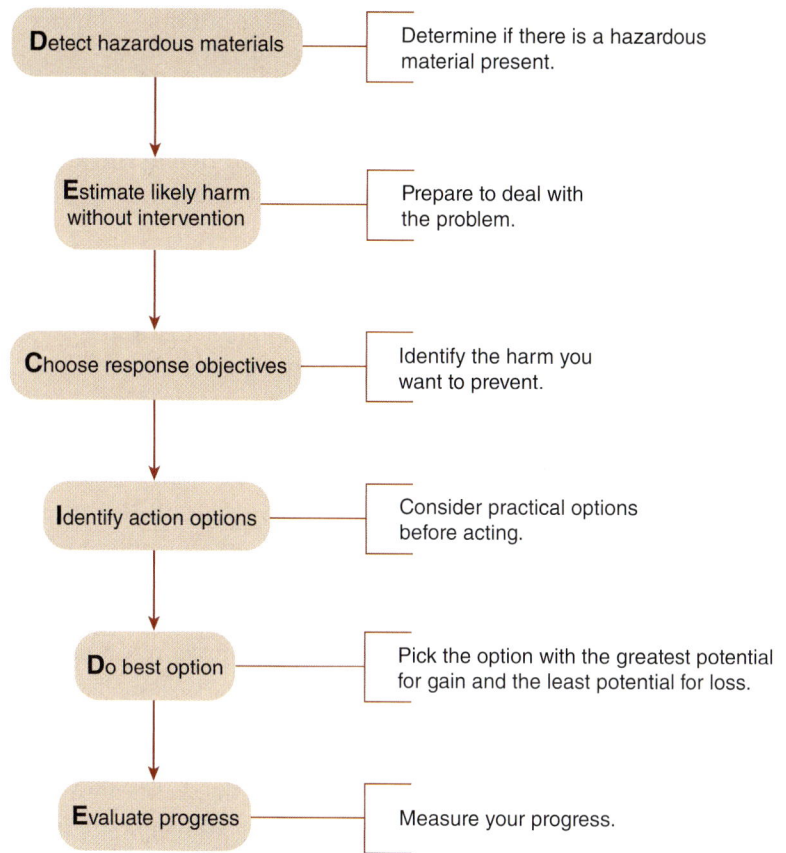

Detect hazardous materials — Determine if there is a hazardous material present.

Estimate likely harm without intervention — Prepare to deal with the problem.

Choose response objectives — Identify the harm you want to prevent.

Identify action options — Consider practical options before acting.

Do best option — Pick the option with the greatest potential for gain and the least potential for loss.

Evaluate progress — Measure your progress.

The DECIDE model, created by Ludwig Benner in the 1970s, is designed to help responders think through a HazMat situation.

The DECIDE acronym represents key decision-making points that occur during a typical HazMat emergency. The model provides a guide for responders to follow in order to understand and update predictions of what's going to happen next and to see how their actions are changing the course of the incident.

The strength of the DECIDE process is this: Even when you don't know what will happen next, you can pinpoint the data gaps that will ultimately allow you to make a prediction.

FIGURE 34-1 The DECIDE model.
Data from: The DECIDE model by Ludwig Benner.

Response Safety Procedures

Isolation of the release area is one of the first measures that operations level hazardous materials responders should implement (**FIGURE 34-2**). For example, access to the scene can be controlled by stretching scene tape across roads to divert traffic away from a spill or to prevent civilians or other responders from entering or reentering a contaminated building or approaching

FIGURE 34-2 Isolating the area of the release is a vital step in gaining control of the incident.
© David Goldman/AP Images.

a rolled-over cargo tank. Law enforcement or site security personnel can be posted in safe areas to prevent bystanders from wandering into a contaminated area. In some instances, scene control actions may be one of your biggest response challenges and could require a coordinated effort among fire, police, emergency medical services (EMS), and other agencies. Be generous with your scene control zones, but *take care that you don't mark off more real estate than you can control!*

Responders should consult the *Emergency Response Guidebook (ERG)* for guidance on protective actions or as a starting point for gathering information about a released substance. Other initial actions may include evacuating others from the immediate area, implementing a sheltering-in-place strategy, rendering emergency medical care in a safe location away from the release, and surveying the scene to detect indicators of a hazardous materials/WMD release.

These basic protective actions should be taken immediately, while maintaining your own safety. For the first several minutes, not even the first responders themselves should enter the area unless properly protected and trained to take action. During this period of time, responders can begin to gather necessary scene information and take initial steps to identify the materials involved (size-up). Again, these measures should be completed from a safe distance, away from potential contamination. Any actions taken should follow the predetermined local response plan or your agency's response guidelines. Additionally, each authority having jurisdiction (AHJ) should have clearly defined policies and procedures for reporting the status of the response through the normal chain of command. This is a vital factor when it comes to ensuring that everyone on the incident is informed of the current conditions and any changes (incident escalating or getting worse, under control, etc.) that might be made to the incident action plan. In concert with passing incident status reports up the chain of command, the AHJ should identify the methods for immediate notification of the **incident commander (IC)** and other response personnel about critical conditions at the incident. Also, if any responders notice an unsafe situation, it should be communicated up the chain of command as soon as possible.

A quick and easy format you can use for communicating your status is the CAN report. CAN is an acronym for Conditions, Actions, and Needs and is an easy way to remember the process for giving a quick, concise briefing or update. *Conditions* refer to your current status or perhaps the status of the incident (such as progress reports; signs or other indicators that the status of the incident is improving, deteriorating, or staying the same; or circumstances where it would be prudent to withdraw from the offensive activities). *Actions* are what you are doing, what is being done by the team, or what is occurring with the incident (largely based on the policies and procedures of the AHJ). *Needs* include any additional resources or actions that are required to accomplish a task. Usually needs include requests for more people or more time.

The CAN report allows you (no matter what job function you have) to distill a message down to its most basic parts. CAN reports are useful to use when:

- You are briefing others about a task or the status of the incident.
- An incident commander is requesting information from individuals or teams engaged in the incident.
- The incident commander, group leader, or any other leadership function needs to broadcast information to all responders within a group or on the scene. Safety messages or general information can be disseminated via a CAN report.

The following is an example of a CAN report from an **entry team** to the entry team leader:

- Conditions
  - "Entry team 1 to entry team group leader. Be advised we have made access to the control

room at the rear of the building and have identified the leaking valve."
- Actions
  - "We have confirmed that this is the valve identified in the safety briefing and are starting to work on stopping the leak."
- Needs
  - "It looks like we need a bigger pipe wrench, too, but we don't have one that size in the toolbox. We will come back to the edge of the hot zone in 5 minutes and pick up a different one."

A CAN report can contain any relevant content, but it's a good format to help keep communications crisp, clear, and concise.

Another important initial safety action might consist of identifying and securing any possible ignition sources—especially when the incident involves the release of flammable materials. So as not to create an unintentional ignition source, responders should use only radios and other electrical devices that have been certified as intrinsically safe (**FIGURE 34-3**). With intrinsically safe electrical devices, the thermal and electrical energy within the device always remains low enough that a flammable atmosphere could not be ignited. Typically, a portable radio that is certified to be intrinsically safe must be specifically wired and used in conjunction with an intrinsically safe battery.

Scene Control Procedures

Managing a hazardous materials incident by setting control zones and limiting access to the incident site helps reduce the potential number of civilians and public service personnel exposed to a released substance. **Control zones** are established at a hazardous materials incident based on the chemical and physical properties of the released material, the environmental factors at the time of the release, and the general layout of the scene. Of course, isolating a city block in a busy downtown

FIGURE 34-3 All intrinsically safe radios and batteries are marked by the factory with a specific label denoting them as such.
Courtesy of Rob Schnepp.

area of a large city presents far different challenges than isolating the area around a rolled-over cargo tank on an interstate highway. Each situation is different, requiring you to be flexible and thoughtful about how you secure the area. Securing access to the incident helps ensure that responders arriving after the first-due units do not accidentally enter a contaminated area.

If the incident takes place inside a structure, the best place to control access is at the normal points of ingress and egress—doors. Once the doors are secured so that no unauthorized personnel can enter, appropriately trained response crews can begin to isolate other areas as appropriate.

The same concept applies to outdoor incidents. The goal is to secure logical access points around the hazard. Begin by controlling intersections, on/off ramps, service roads, or other access points (including pedestrian routes) to the scene. Law enforcement personnel should block off streets, close intersections and sidewalks, or redirect traffic as needed (**FIGURE 34-4**).

During a long-term incident, highway transportation department or public works department employees may be called upon to set up traffic barriers. Whatever methods or devices are used to restrict access, they should not limit or prevent a rapid withdrawal from the area by personnel working inside the hot zone.

It is not uncommon to set large control zones at the onset of an incident only to discover that the zones may have been established too liberally. At the same time, control zones should not be defined too narrowly (**FIGURE 34-5**). As the incident commander gets more information about the specifics of the chemical or material involved, the control zones may be changed. Ideally, the control zones will be established in the right place, geographically, the first time. This may be accomplished by understanding the state of matter or the nature of the material or through visual reconnaissance, or perhaps the control zones will be established based on the results of air monitoring. In all cases, you should be prepared to expand or contract control zones if necessary. Wind shifts are a common reason that control zones are modified during the incident. If there is a prevailing wind pattern or a predicted shift based on time of day in your area, factor that consideration into your decision making when it comes to control zones.

Typically, control zones at hazardous materials incidents are labelled as *hot, warm,* or *cold.* You may also hear other terms used, such as *exclusionary zone* (hot zone), *contamination reduction zone* (warm zone), or *outer perimeter* (cold zone). Both sets of terms are correct and common, so make sure you understand the terminology used in your jurisdiction. Be prepared to discover that different jurisdictions may use terminology and setup procedures unlike the ones used in your agency. If you understand the concepts behind the actions and remember that safety is the main focus, the act of setting up and naming your zones can remain flexible. Avoid confusion by performing regular and consistent training and making sure all responders are on the same page with the terminology.

The **hot zone** is the area immediately surrounding the release, which is also the most contaminated area. All personnel working in the hot zone must wear complete, appropriate protective clothing and equipment. Hot zone boundaries should be set large enough that adverse effects from the released substance will not

FIGURE 34-4 Police officers might assist by diverting vehicle and foot traffic to a safe distance outside the hazard area.
Courtesy of Rob Schnepp.

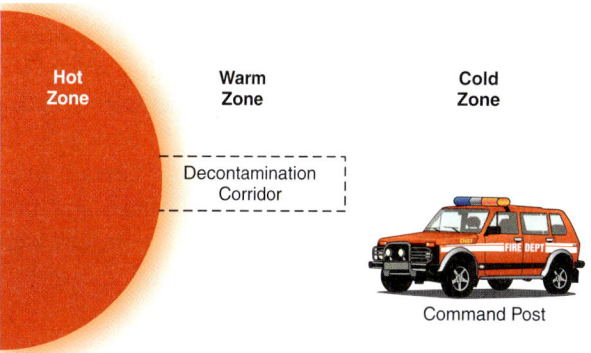

FIGURE 34-5 Control zones spread outward from the hot zone.
© Jones & Bartlett Learning.

affect people outside the hot zone. An incident involving a gaseous substance or a vapour, for example, may require a larger hot zone than one involving a solid or nonvolatile liquid leak. In some cases, atmospheric monitoring, plume modeling, or reference sources such as the *ERG* may prove useful in helping to establish the parameters of a hot zone. Specially trained responders should be tasked with using these tools.

Keep in mind that the physical characteristics of the released substance will significantly affect the size and layout of the hot zone. Additionally, all responders entering the hot zone should avoid contact with the product to the greatest extent possible—an important goal that should be clearly understood by those entering the hot zone. Adhering to this policy makes the job of decontamination easier and reduces the risk of secondary contamination.

LISTEN UP!

It's generally quicker, easier, and safer to *reduce* the size of the control zones than to *enlarge* them.

Personnel accountability is important, so access into the hot zone must be limited to only those persons necessary to control the incident. All personnel, tools, and equipment must be decontaminated when they leave the hot zone. This practice ensures that contamination is not inadvertently spread to "clean" areas of the scene.

The **warm zone** is where personnel and equipment transition into and out of the hot zone. It contains control points for access to the hot zone as well as the decontamination corridor. Only the minimal amount of personnel and equipment necessary to perform decontamination, or support those operating in the hot zone, should be permitted in the warm zone.

Generally, personnel working in the warm zone can use personal protective equipment (PPE) whose protection is rated one level lower than that of the PPE used in the hot zone. For example, if personnel in the hot zone are wearing Level A ensembles, personnel in the warm zone should dress in at least Level B protection. This recommendation is not a hard-and-fast rule, however. You must understand the hazard well enough to make that determination on a case-by-case basis.

Beyond the warm zone is the **cold zone**. The cold zone is a safe area where personnel do not need to wear any special protective clothing for safe operation. Personnel staging; the command post; EMS providers; and the area for medical monitoring, support, and/or treatment after decontamination are all located in the cold zone. Typically, support personnel for the incident operate in this area.

LISTEN UP!

The primary functions of warm zone activities are decontamination and providing ready support to personnel operating in the hot zone.

Protective Actions at the Hazardous Materials Operations Level

Evaluating the threat to life is the number one response priority at a hazardous materials incident. The safety of the responders is always taken into account, but the bottom line is this: If there is no life threat—either immediate or anticipated—the severity of the incident is diminished. This is not to say that property or the environment is unimportant; rather, it is simply an acknowledgment that you as a responder must consider the threat to life before anything else.

Life-safety actions include ensuring your own safety and searching for, and possibly rescuing, those persons who were immediately exposed to the substance or those who may now be in harm's way. For example, an IC might be faced with a decision of whether to evacuate people ahead of an advancing vapour cloud or when an explosive device is discovered or if some other potentially hazardous condition exists. This is not an easy decision, because efforts to relocate people bring up many complications. A risk-based method of decision making should be employed in such cases. Essentially, the IC should weigh the severity of the threat against the potential effort, impact, and/or resource requirements of moving people from one location to another (**FIGURE 34-6**).

FIGURE 34-6 The IC must weigh the severity of the threat against the time, personnel, and other logistical challenges required to carry out an evacuation.

Courtesy of Rob Schnepp.

Evacuating a predominantly ill and elderly population from a care facility, for example, may take a long time, have negative effects on the evacuees, and require many resources. Given these concerns, the IC must decide if the hazard merits such a move or if other methods might be used to protect this type of population.

Evacuation

The IC must consider many factors before making a decision to evacuate. Some factors include the nature and duration of the release; the status of the evacuees, such as their age, underlying health, and mobility; the ability to support the evacuees with basic services such as food, water, and shelter; and transportation challenges. When considering evacuating a significantly large group of people, it may be helpful to consider the host of "shuns" associated with the task. The "shuns" is a word play on the ending of all the following terms:

- Contamination (the trigger for the evacuation)
- Communication
- Transportation
- Nutrition
- Sanitation
- Habitation
- Compassion

You may identify more "shuns" that are specific to your AHJ—be creative and complete when thinking about evacuation.

Once the evacuation order is given, fire fighters and law enforcement personnel may be called upon to assist in the physical relocation of residents to a safe area. Their work may include such actions as travelling to homes and informing residents that they must relocate to a temporary shelter.

Before an evacuation order is given, a safe area with suitable facilities should be established. In many cases, schools, fairgrounds, and sports arenas are used as shelters. The evacuation area should be located close enough to the exposure for the evacuation to be practical but far enough away from the incident to be safe. Depending on the time of day and the season, it may take a considerable amount of time to evacuate even a small residential area. Temporary evacuation areas may be needed to shelter residents until evacuation sites or structures indicated in your community's emergency response plan are open and accessible. Security of the evacuation facility is an important consideration and should be accomplished with the assistance of law enforcement. Access should be monitored and controlled to ensure the maximum amount of security and shielding from the media, onlookers, or those interested in taking advantage of the evacuees in any way.

Transportation to temporary evacuation areas must be arranged for all populations when an evacuation is ordered. As part of this effort, the requirements of elderly, handicapped, and special needs persons should be considered. Accommodations for pets should also be considered. It would be a mistake to assume all evacuees are able and/or motivated to move on their own.

Initial evacuation distances may be derived from the *ERG* (**FIGURE 34-7**). Note that the *ERG* does not give complete detailed information for all conditions that responders might potentially encounter, such as weather extremes, road conditions, or actual conditions encountered at the scene.

In severe weather, evacuation can be challenging at best. Flooded areas, heavy rains, and winds present a danger to the evacuees as well as responders. Even

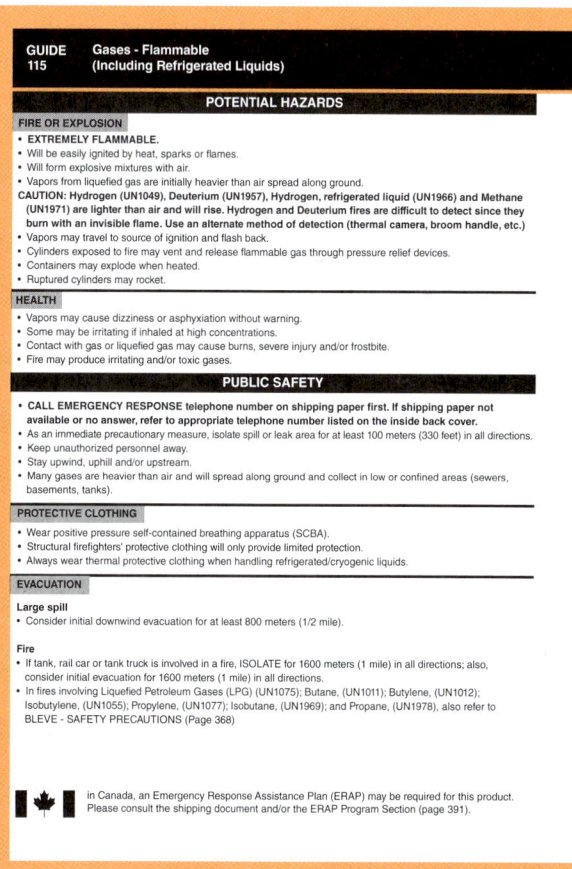

FIGURE 34-7 Sample page from the orange section of the *ERG*.
Courtesy of the U.S. Department of Transportation.

in perfect circumstances, evacuation of residents will be a process that is measured in hours, not minutes. Preplanning for evacuations is key to successful and safe evacuations.

A good way for the IC to determine the area to be evacuated, including how far to extend actual evacuation distances, is to use detection and monitoring devices to identify areas of airborne contamination. Access to local weather reporting conditions is invaluable in this decision-making process.

Sheltering-in-Place

Sheltering-in-place is a method of safeguarding people in a hazardous area by keeping them in an enclosed atmosphere, usually inside structures. Local emergency plans should identify local facilities where vulnerable populations might be found, such as schools, churches, hospitals, nursing homes, and apartment complexes. When residents are sheltered-in-place, they remain indoors with windows and doors closed. All ventilation systems are turned off to prevent outside air from being drawn into the structure.

The toxicity of the hazardous material and the amount of time available to avoid the oncoming threat are major factors in the decision of whether to evacuate or to use a sheltering-in-place strategy. The expected duration of the incident is also a factor in determining whether sheltering-in-place is a viable option. The longer the release is expected to continue, the more time the material has to enter or permeate into protected areas. Short-term events or transient vapour clouds might dictate a sheltering-in-place approach. The IC or unified command should conduct plume-modeling activities or use other reliable expertise or methods to determine whether evacuation or sheltering-in-place makes the most sense. In either case, it is generally a time-sensitive decision.

Search and Rescue

Ensuring your safety is the first priority in any response, and a hazardous materials incident is no exception. The search for and rescue of people can be more complicated in the setting of a hazardous materials incident. In a structure fire, it is understood that smoke and flame

are serious threats to life, and fire fighters are properly protected to conduct reasonable operations to perform search and rescue. In a hazardous materials incident, all response personnel (fire, law enforcement, EMS) must first recognize and identify the released substance. This may take some time and is not as straightforward as a typical structure fire. After this is accomplished, the IC should use a risk-based thought process to understand the hazards as they relate to the possibilities of victim survivability and responder safety. Only then should personnel consider search and rescue.

Ultimately, the IC must determine whether entry into a contaminated environment to perform search and rescue is a worthy endeavor. If the released substance is highly toxic, unprotected victims may not have survived the initial exposure. If it is determined that a search can be safely performed, rescue teams wearing proper PPE may then enter the hot zone to look for or retrieve victims. The victims should be removed to the warm zone, where they can be decontaminated and turned over to EMS providers for transport to a medical facility. In nearly all cases, definitive medical care should be provided to victims who have been decontaminated and removed to the cold zone. NFPA 473, *Standard for Competencies for EMS Personnel Responding to Hazardous Materials/Weapons of Mass Destruction Incidents*, offers additional guidance on the medical management of chemical exposures.

Safety Briefings

Typically, before significant actions are taken at a hazardous materials incident, the IC should ensure that a *written* site safety plan is completed and a *verbal* safety briefing is performed. The purpose of the safety briefing is to inform (at a minimum) all responders of the health hazards that are known or anticipated, the incident objectives, emergency medical procedures, radio frequencies and emergency signals, a description of the site, and the PPE to be worn. Each AHJ should develop templates for site safety plans and verbal safety briefings that should also include the protocols and/or procedures for using approved communications tools and equipment provided by the AHJ and the policies and procedures for contacting and cooperating with outside agencies. In an incident where different entities are operating (law enforcement, fire, and EMS), perhaps on the scene of an incident with criminal intent, all roles and responsibilities among the agencies should be clearly identified. The use of a standardized form means that the safety plan can be completed in a timely manner; standardized formats also keep safety briefings on track and provide a logical format that everyone can follow.

There are many ways to conduct a safety briefing. Chiefly, a good safety briefing should be carried out in the manner that its name implies—it should be brief. It should be complete enough to provide the responders on the scene with relevant information yet not overly detailed with useless or "nice to know" information (**FIGURE 34-8**).

At small incidents where the incident management structure is simple, the IC may be responsible for both putting together a site safety plan and conducting the safety briefing. At larger incidents, where an **incident safety officer** (also referred to as a safety officer) or a **hazardous materials safety officer** is appointed, it is the responsibility of that officer to participate in the preparation and implementation of the site safety plan. The roles of these officers are discussed in more detail later in this chapter. The depth and scope of the safety briefing should have a direct correlation with the severity and complexity of the incident. For significant

incidents, where the released substances are highly toxic or where other significant health hazards are present, the site safety plan and briefing may be comprehensive and documented in writing.

The IC may establish predetermined trigger points, intended to evaluate the status of the planned response, which may lead to withdrawal of responders from the hot zone. Based on a lack of progress toward meeting the incident objectives, the IC may later decide to abandon the current plan of action and withdraw to a safe distance, set a defensive perimeter, and wait for additional resources to arrive—or to allow the hazardous materials incident to run its course. Typically, these decisions are made when the offensive actions are not effective in mitigating the problem, the selected PPE is found to be incorrect or ineffective, the tools and equipment required to solve the problem are ineffective or unavailable, or the problem is simply too complex to handle with the available resources. A reasonable and prudent IC knows when he or she is "outgunned" and does not make a foolhardy decision to risk the health and safety of the personnel. In the event a withdrawal is necessary, the IC should include evacuation signals in the pre-entry briefing.

It is also important, within the context of a safety briefing, to discuss incident communications, including procedures for interacting with the media or other outside entities that provide public information. *It is common for communication to be disrupted or otherwise hampered during an incident.* To avoid undue communications complications, the IC may designate command and tactical channels for the incident. Entry teams may be instructed to communicate on a dedicated radio channel to ensure they are not cut off or excluded from making a critical radio transmission by other radio chatter. In this case, the backup team should also be monitoring the same radio channel for situational awareness. They may learn about the progress the entry team is making on the mitigation effort or be quickly notified in the event the entry team is in trouble and may need rescue assistance.

Radio transmissions are an excellent way to communicate the status of the mitigation efforts and to gauge the success or failure of those efforts (**FIGURE 34-9**).

If things are not going as planned, the entry team may contact the entry team leader, who may have a face-to-face conversation with the safety officer or hazardous materials safety officer (assistant safety officer). That conversation, carried out through the normal chain of command, may result in a change of tactics, modification of the incident objectives, or a complete abandonment of the effort. In any case, by observing the proper chain of command, all parties with management

FIGURE 34-8 All personnel must be briefed before approaching the hazard area or entering the hot zone.
Courtesy of Rob Schnepp.

FIGURE 34-9 Radio communications should be clear, concise, specific, and accurate. Think about what you will say before you push the button to talk!

© Jones & Bartlett Learning. Photographed by Glen E. Ellman.

and safety interests at the scene will be included in the appropriate discussions. This approach will also help to ensure coordinated communication to the public.

In some jurisdictions, the safety officer conducts most of the safety briefing; in others, the IC or hazardous materials group supervisor conducts the safety briefing in conjunction with the safety officer. Regardless of how it's accomplished, a safety plan should be completed (and approved by the IC), and a safety briefing should be conducted before responders attempt to enter the incident site. Make sure that you are familiar with and follow the standard operating procedures in your jurisdiction.

In today's world, the threat of terrorism or other incidents with criminal intent drives the need for complete and meaningful safety briefings. Briefings at these types of incidents may also include procedures for operating at a crime scene or evidence collection procedures. It is imperative to understand the needs of law enforcement, especially as those needs relate to beginning an investigation or collecting evidence for a possible prosecution. Operating at a crime scene can present a complicated set of circumstances, but a good working relationship among all agencies on the scene (ideally, established before the event) will create better operational efficiency during the response. Refer to the standard operating procedures and policies established by your agency to fully understand your role during a hazardous materials incident with criminal intent.

A written safety plan should also include information about excessive heat disorders or cold stress the responders may encounter while engaging in their work. Working in PPE in ideal conditions is challenging enough on its own. When temperature extremes are anticipated, their effects on the responders should be acknowledged and understood by everyone operating at the scene.

The Buddy System and Backup Personnel

It is *not* an accepted practice at a hazardous materials incident to allow only one responder to don a PPE ensemble and enter a contaminated environment alone. The risk of something going wrong is too great, and operating alone should *never* be allowed. To that end, it is an accepted practice to implement the **buddy system** for those personnel entering contaminated areas. The simplest expression of the buddy system is for no fewer than two responders to enter a contaminated area. There should be no deviation from this practice under any circumstance. It is the responsibility of the IC not only to limit the number of responders working in the contaminated area but also to prevent responders from working alone. More than two responders might enter a contaminated area, but never fewer than two. To work alone is to take an undue risk!

On a hazardous materials incident, a **backup team** of at least two properly equipped personnel should be staged in the warm zone and prepared to provide immediate aid should the need arise. Like the use of the buddy system, the use of backup personnel is required. The backup team could also be called upon to remove those individuals working in the hot zone if an emergency occurs and the entry crew members are unable to escape on their own. The concept is much like the practice of establishing a rapid intervention team at a structure fire. *A backup team should never be more than a minute or so from being fully dressed and ready for action.*

Excessive-Heat Disorders

Hazardous materials responders operating in protective clothing should be aware of the signs and symptoms of heat exhaustion, heat stress, dehydration, and heat stroke. If the body is unable to disperse heat because an ensemble of PPE covers it, serious short- and long-term medical issues could result.

Heat exhaustion is a mild form of shock that arises when the circulatory system begins to fail because the body is unable to dissipate excessive heat and becomes overheated. With heat exhaustion, the body's core temperature rises, followed by weakness and sweating. A person suffering from heat exhaustion may become dizzy and have episodes of blurred vision. Other signs and symptoms of heat exhaustion include acute fatigue, headache, and muscle cramps; however, signs and symptoms may vary among individuals. Any individual experiencing heat exhaustion should be removed at once from the heated environment, rehydrated with electrolyte solutions (perhaps by intravenous methods), and kept cool. If not properly treated, heat exhaustion may progress to a potentially fatal condition.

When treating heat-related illnesses, it is important to avoid pouring cold water or otherwise placing a victim in an unusually cold environment. The extreme change in temperature may have an adverse effect on an individual's recovery. Using tepid water for drinking and for cooling the skin is the safest approach. Keep in mind that rehydration by mouth is much slower than rehydration by an intravenous line.

SAFETY TIP

Most heat-related illnesses are typically preceded by dehydration. It is important to stay hydrated so that you can function at your maximum capacity. As a frame of reference, athletes should consume approximately 500 mL (17 oz) of fluid (water) prior to an event and 200–300 mL (6.7–10 oz) at regular intervals. Responders are occupational athletes—so keep up on your fluids!

Heat stroke is a severe and potentially fatal condition resulting from the failure of the temperature-regulating capacity of the body. It is caused by exposure to the sun or working in high temperatures. Reduction or cessation of sweating is an early symptom. The body temperature can rise to 40.5°C (105°F) or higher and be accompanied by a rapid pulse; hot, red-looking skin; headache; confusion; unconsciousness; and possibly seizures. Heat stroke is a true medical emergency that requires immediate transport to a medical facility.

To combat heat stress while their personnel are wearing PPE, many response agencies employ some form of cooling technology under the garment. These technologies include, but are not limited to, air-, ice-, and water-cooled vests, along with phase-change cooling technology. Many studies have been conducted on each form of cooling technology. Each is designed to accomplish the same goal: to reduce the effects of heat stress on the human body. Chapter 36, *Hazardous Materials Responder Personal Protective Equipment*, provides more information on the specific technologies used for cooling.

Cold-Temperature Exposures

Responders at hazardous materials incidents may be exposed to two types of cold temperatures: those caused by the released materials and those caused

by the operating environment, including ambient air temperatures or conditions such as rain, snow, or other adverse cold-weather conditions. Hazardous materials such as liquefied gases and cryogenic liquids may expose responders to the same low-temperature hazards as those created by cold-weather environments. Exposure to severe cold for even a short period of time may cause severe injury to body surfaces, especially to the ears, nose, hands, and feet.

Two environmental factors influence the extent of cold injuries: temperature and wind speed. Because still air is a poor heat conductor, responders working in low temperatures with little wind can endure these conditions for longer periods (if their clothing remains dry). However, when low temperatures are combined with significant winds, wind chill occurs. As an example, if the temperature with no wind is −7°C (20°F), it will feel like −21°C (−5°F) when the wind speed is 24 km (15 miles) per hour.

Regardless of the ambient temperature, personnel will perspire while wearing chemical protective clothing. Wet clothing extracts heat from the body approximately 240 times faster than dry clothing. Wet or damp clothing may also lead to hypothermia, a condition in which the core body temperature falls below 35°C (95°F). Hypothermia is a true medical emergency.

Responders must be aware of the dangers of frostbite, hypothermia, and physical impairment when temperatures are low or when they are working in wet clothing. All personnel should wear layered clothing and be able to warm themselves in heated shelters or vehicles.

The layer of clothing next to the skin, especially socks, should be kept dry. The combination of cold and wet softens skin, causing numbness, tingling, and, in some cases, peeling skin.

Responders should carefully schedule their work and rest periods and monitor their physical working conditions. For example, warm or cool shelters should be available where responders may don and doff their protective clothing.

Personal Protective Equipment: Physical Capability Requirements

Medical monitoring is a process of pre-entry and postentry on-scene evaluation of response personnel who may be experiencing adverse effects because they are wearing PPE or because they have been exposed to heat, cold, stress, or hazardous materials. The goal of medical monitoring is to quickly identify medical problems—while responders are on-scene—so they can be treated in a timely fashion, thereby preventing severe adverse effects and maintaining the optimal health and safety of on-scene personnel. The Canadian Standards Association (CSA) outlines the components of a health and safety management system for hazardous materials responders in CAN/CGSB/CSA-Z1610-11, *Protection of First Responders from Chemical, Biological, Radiological, and Nuclear (CBRN) Events*, and CSA Z1000, *Occupational Health and Safety Management*.

Responders should undergo pre-entry health screening prior to donning any level of chemical protective equipment. (This step may be eliminated if rapid entry is required for rescue or other life-safety reasons.) This screening should include a check of vital signs, including pulse rate, blood pressure, and respiratory rate. Body weight measurements and general health should also be observed. In the event the responder presents with an abnormal reading (determined by the AHJ), that responder may be excluded from wearing PPE or participating in the mitigation phase of the incident.

Once the mission is complete and the responder has been decontaminated and doffed the PPE, a second medical evaluation—similar to the initial evaluation—should be completed. In the event of abnormal findings (again determined by the AHJ), the responder may need to be seen by a physician or transported to an appropriate receiving hospital.

For a more comprehensive set of criteria for medical monitoring/support, consult NFPA 473.

The Incident Command System

As discussed in Chapter 22, *Establishing and Transferring Command*, in the 1970s, a series of devastating wildland fires occurred in Southern California, requiring the services of numerous local and state resources. That rash of fires illustrated the need for a better way to organize and manage large numbers of agencies and resources called to a major incident. Also because of those fires, many of the agencies involved agreed to form a working group, subsequently named FIRESCOPE (Fire Resources of Southern California Organized for Potential Emergencies). The FIRESCOPE group began its work by identifying several problem areas common to almost all major or complex incidents: ineffective communications, span of control challenges, a lack of a common command structure, the inability to track personnel working on the scene, and the inability to effectively coordinate on-scene resources.

Subsequently, the FIRESCOPE consortium developed a standardized yet flexible management system called

LISTEN UP!

Span of control refers to how many responders can be effectively managed by one supervisor. In most situations, one person can effectively supervise only three to seven people or resources. This span of control is flexible and depends on issues such as the criticality of the mission or task being performed, the skill level of the supervisor, and the skill levels of the responders.

the **incident command system (ICS)**. Some of the key benefits of using the ICS are summarized here:

- Common terminology
- Consistent organizational structure
- Consistent position titles
- Common incident facilities

The ICS, which was originally designed for managing wildland fires, subsequently evolved into an all-risk management structure suitable for managing resources at all types of fires, natural disasters, technical rescues, and any other type of incident of virtually any size. When properly used, it provides a strong organizational framework on which to lay the operational goals and objectives. Now more than ever, it's critical that all responders—regardless of the patch on their uniform or the nature of their job or even their geographic location—understand the need for this cohesive incident management tool.

The ICS can be expanded to handle an incident of any size and complexity. Hazardous materials incidents can be extremely complex, such that local, provincial, territorial, and federal responders and agencies may all become involved in many cases of long duration. The basic ICS consists of five functions: command, operations, planning, logistics, and finance/administration (**FIGURE 34-10**).

Command

Command is established when the first unit arrives on the scene and is maintained until the last unit leaves the scene. The concept of incident command is the first main principle of incident response and the cornerstone of the ICS. Without leadership and a process

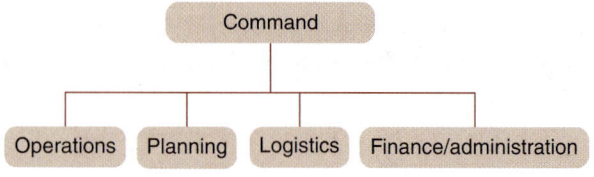

FIGURE 34-10 General staff functions of the ICS.
© Jones & Bartlett Learning.

for organizing and directing personnel and resources, the effort to solve a problem may end up being as chaotic as the problem itself. A certain chain of command must exist from the top to the bottom of the incident organizational structure. Everyone either reports to or is responsible for someone else. Command is directly responsible for at least the following tasks:

- Determining strategy
- Selecting incident tactics
- Establishing the action plan
- Developing the ICS organization as the incident expands
- Managing resources
- Coordinating resource activities
- Providing for scene safety
- Releasing information about the incident
- Coordinating with outside agencies

Unified Command

When multiple agencies with overlapping jurisdictions or legal responsibilities are involved in the same incident, a unified command provides several advantages. Using this approach, representatives from various agencies cooperate to share command authority (**FIGURE 34-11**). Oil spills on navigable waterways; wildfires that threaten or occur in a combination of provincial/territorial lands; hazardous materials incidents on public roadways or at fixed facilities; or perhaps active shooter incidents represent ideal situations for the establishment of a unified command.

Incident Command Post

Regardless of whether there is a single incident command or the incident is run under a unified command, the command function is always located in an **incident command post (ICP)**. The ICP is where the incident commander is located and where coordination, control, and communications are centralized. If additional resources of any kind are required to respond to the incident, the ICP is where those requests originate. The emergency response plan of the AHJ should outline those potential outside agencies that may be called upon to respond to certain circumstances. The plan's command and all direct support staff should be located at the ICP. Ideally, the ICP should be in an area where it is not threatened by the incident and has the necessary infrastructure (e.g., communications, technology support, bathrooms, meeting space) to support sustained operations if required.

FIGURE 34-11 A unified command involves many agencies directly involved in the decision-making process for a large incident.
Courtesy of Rob Schnepp.

> **FIRE FIGHTER TIP**
>
> It is common practice to identify the on-scene incident command post or vehicle with a flashing/strobe green light.

During a hazardous materials incident, the ICP should be established uphill and upwind of the incident, keeping in mind the potential for predicted changes in wind direction based on the time of day.

> **SAFETY TIP**
>
> It is important to consider the prevailing winds of the area when establishing the location of the incident command post.

Command Staff

The IC is the person in charge of the entire incident and should be qualified to be in the position. The on-scene hazardous materials IC, who will assume control of the incident scene beyond the first-responder awareness level, should receive at least 24 hours of training equal to the first-responder operations level and, in addition, have competency in the following areas:

- Know and be able to implement the jurisdiction's ICS

- Know how to implement the jurisdiction's emergency response plan
- Know and understand the hazards and risks associated with responders working in chemical protective clothing
- Know how to implement the local emergency response plan
- Know about the local emergency response plan and the regional or provincial/territorial response teams
- Know and understand the importance of decontamination procedures

The **command staff** consists of the safety officer, the liaison officer, and the public information officer (**FIGURE 34-12**). These job functions report directly to the IC and are critical to the effective management of a hazardous materials/WMD incident.

Safety Officer. The role of the safety officer at a hazardous materials incident is clear:

> The individual in charge of the ICS shall designate a safety officer, who is knowledgeable in the operations being implemented at the emergency response site, with specific responsibility to identify and evaluate hazards and to provide direction with respect to the safety of operations for the incident.

The authority of the safety officer at a hazardous materials incident:

> When activities are judged by the safety officer to be an IDLH [immediate danger to life and health] and/or to involve an imminent danger condition, the safety officer shall have the authority to alter, suspend, or terminate those activities. The safety official shall immediately inform the individual in charge of the ICS of any actions that need to be taken to correct these hazards at the incident scene.

A hazardous materials safety officer, sometimes referred to as the *assistant safety officer (ASO)*, is responsible for the hazardous materials team's safety only. When a hazardous materials branch or hazardous materials group has been established, or when the incident requires a dedicated hazardous materials response, the safety officer may appoint a hazardous materials safety officer to the hazardous materials branch or group. This safety officer reports directly to the incident safety officer, who in turn reports directly to the IC.

FIRE FIGHTER TIP

Most large fire services have created command identification protocols that include arm bands and/or vests to identify specific command functions.

Liaison Officer. The **liaison officer** is command's point of contact for cooperating and assisting agencies. An assisting agency or organization is one that assists in operations and provides personnel or other tactical resources to the agency with jurisdictional authority over the incident. Fire departments, law enforcement agencies, and EMS providers are examples of assisting organizations. Cooperating agencies such as the Canadian Red Cross and the Salvation Army provide supplies and support but do not participate in the actual operations component of incident management.

On a hazardous materials incident, the liaison officer deals with agency representatives (A reps) from provincial or territorial and local resources, and any other outside agency with an interest in the management or outcome of the incident. Ideally, these representatives should have the authority to make decisions on behalf of the agency they represent. It is extraordinarily cumbersome to deal with an agency representative who must constantly go back to a key decision maker within his or her own organization to get approvals. This function is critical when the interests of many jurisdictions are at stake. Liaison officers can find themselves quite busy dealing with questions and providing information to agency representatives during an incident.

Public Information Officer. The **public information officer (PIO)** typically functions as a point of contact for the media or any other entity seeking information about the incident (**FIGURE 34-13**). As with all the other command staff positions, only one person should fill this role. Although many people may work with or for the PIO, they should serve as assistants. This chain of command is necessary to streamline communications and reduce redundancies in the management system.

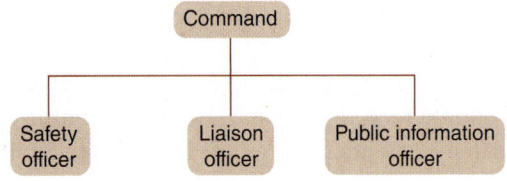

FIGURE 34-12 The command staff members report directly to the incident commander.
© Jones & Bartlett Learning.

FIGURE 34-13 The public information officer functions as a point of contact for the media.
Courtesy of Rob Schnepp.

The value of the PIO should not be underestimated. Keeping the media (and others) well informed is an important piece of successfully managing almost any type of incident. A wise IC will acknowledge that fact and incorporate good media relations into the incident objectives.

General Staff Functions

When the incident is too large or too complex for just one person (the IC) to manage effectively, the IC may assign other individuals to oversee parts of the incident. As discussed in Chapter 22, *Establishing and Transferring Command,* everything that occurs at an incident can be divided among the major functional components within ICS:

- Operations
- Planning
- Logistics
- Finance/Administration

Operations

The operations section, which is typically led by an operations section chief on larger incidents, carries out the objectives developed by the IC and is responsible for all tactical operations at the incident. The operations section chief directs and manages the resources assigned to his or her section. These resources could include fire units of any type, law enforcement resources, emergency medical units, airborne resources such as helicopters and fixed-wing aircraft, and hazardous materials response resources. This position is usually assigned when complex incidents involve more than 20 single resources or when the IC cannot be involved in all details of the tactical operation. When incident operations require many responders, the operations section can be further divided into groups and divisions.

Groups and Divisions. Organizational units such as groups and divisions are established to aggregate single resources or crews under one supervisor. The primary reason for establishing groups and divisions is to maintain an effective span of control.

A hazardous materials group, led by a hazardous materials group supervisor, is often established when companies and crews are working on the same task or objective, albeit not necessarily in the same location. The term "group" is very specific as it applies to the ICS: A *group* is assembled to relieve span of control issues and is considered to consist of functional assignments that may not be tied to any one geographic location. A division usually refers to companies and crews that are working in the same geographic location. Groups and divisions place several single resources under one supervisor, effectively reducing the IC's span of control. **FIGURE 34-14** provides an overview of how divisions and groups are integrated into the overall command structure. The names and functions can be tailored to the specific needs of the incident. The important thing to ensure is that divisions and groups understand their functions and where they fit into the ICS organizational chart.

Hazardous Materials Branches. If necessary during a hazardous materials incident, a special technical group may be developed under the operations section, known as the hazardous materials branch. The hazardous materials branch consists of some or all of the following positions as needed for the safe control of the incident:

- A hazardous materials group supervisor
- An entry team assigned to enter the designated hot zone
- A decontamination team responsible for reducing and preventing the spread of contaminants from persons and equipment
- A technical reference team that gathers information and reports to the IC and the hazardous materials safety officer

A branch represents a higher level of combined resources than either a division or a group. At a major incident, several different activities may occur in separate geographic locations or involve distinct functions. Span of control might still present a problem, even after the establishment of divisions and groups. In these situations, the IC can establish branches to place a higher-level

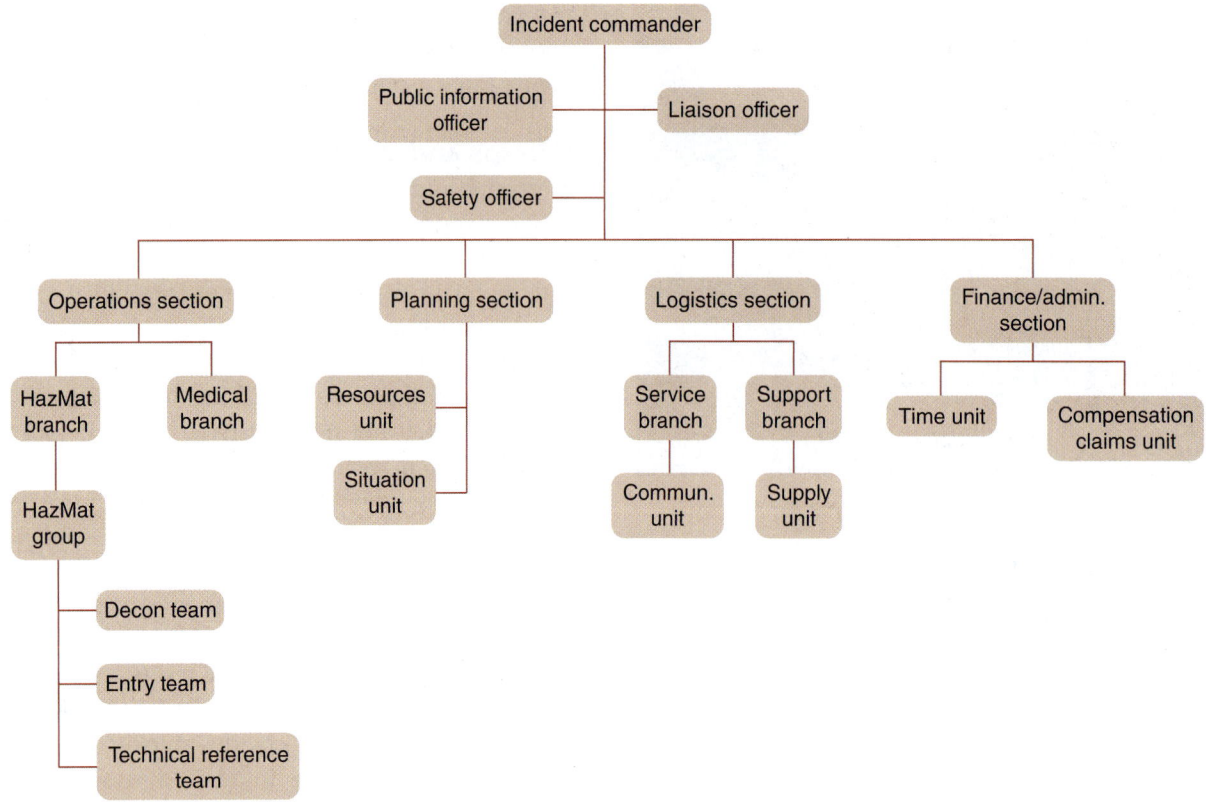

FIGURE 34-14 An overview of the hazardous materials branch.
Modified from: FEMA.

supervisor (a hazardous materials branch director) in charge of several divisions and groups. One incident may require several branches, such as the medical branch, fire branch, or law branch. Keep in mind that each branch would be broken down in a discipline-specific way, based on span of control.

The hazardous materials branch director reports to the operations section chief or directly to the IC if the operations section chief position is not assigned (**FIGURE 34-15**). Responsibilities of the hazardous materials branch director include obtaining a briefing from the IC, staffing the hazardous materials branch functions as required, ensuring that scene control zones are established, ensuring that a site safety plan is developed, maintaining accountability, ensuring that proper PPE is worn, and ensuring that the operational objectives are being met. A hazardous materials group supervisor may be appointed under a hazardous materials branch director to direct and manage positions such as decontamination group supervisor or team leader, entry group supervisor or team leader, or other functions specific to the hazardous materials portions of the response. The management structure and the ICS positions that are ultimately assigned are based on the needs associated with the specific incident.

In some types of incidents, such as a wildland fire or a multistorey structure fire, the safety officer may assign an assistant to each division working the fire. (Remember, the command system is intended to be flexible.) Responsibilities of the hazardous materials safety officer include obtaining a briefing from the IC, the safety officer, and/or the hazardous materials branch director; participating in the preparation of the site safety plan; providing or participating in the safety briefing; altering or suspending any activity that poses an imminent threat to the responders; and maintaining accountability for all resources assigned to the hazardous materials mitigation efforts.

> **LISTEN UP!**
>
> There is only one safety officer (or incident safety officer) for the entire incident, but that position may have many assistants—for example, entry control officers located at various access points to the hot zone.

Planning

The **planning section** is responsible for the collection, evaluation, dissemination, and use of information relevant

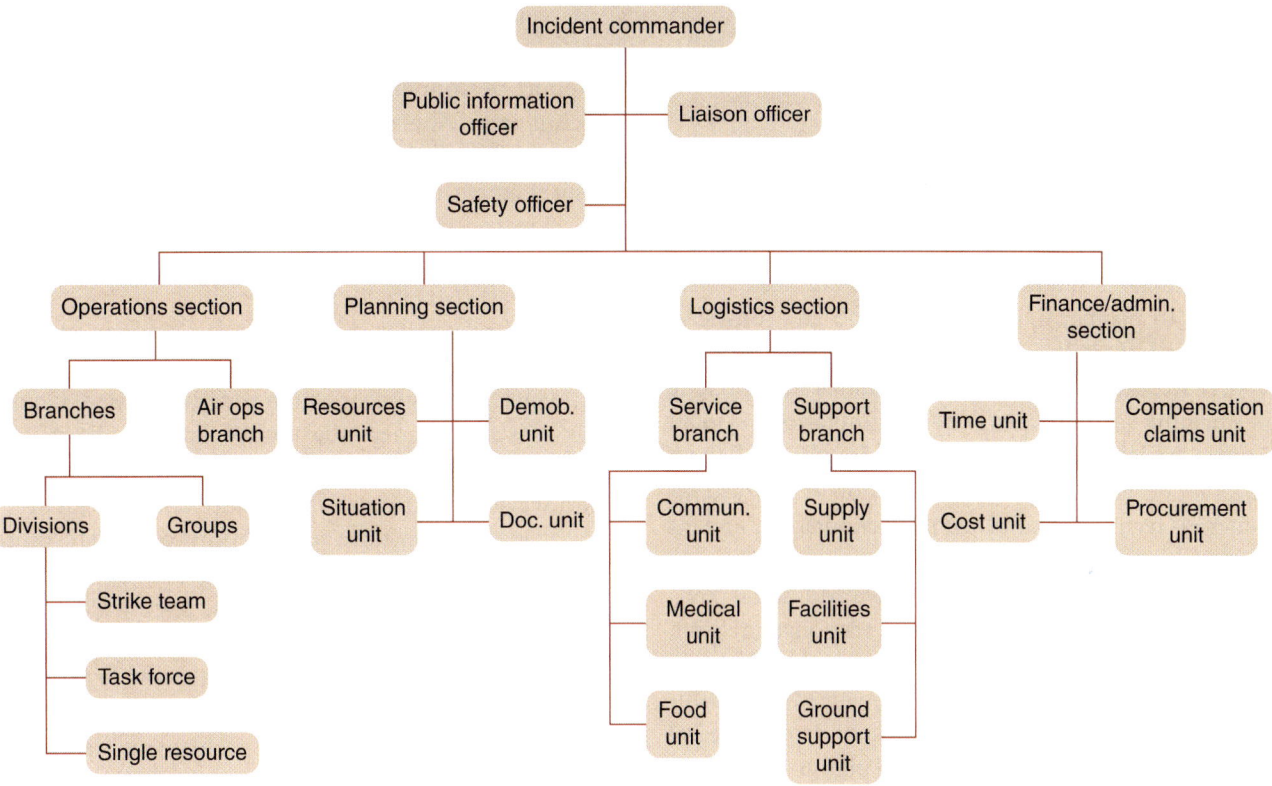

FIGURE 34-15 An example of how to organize a response by function and maintain span of control.
Modified from: FEMA.

to the incident. The planning section, which is led by a **planning section chief**, acts as the central point for collecting information on the situation status (sit-stat) of the event, tracking and logging on-scene resources, and disseminating the written incident action plan. On large-scale incidents, the planning section chief facilitates the incident briefings and planning meetings.

Logistics

The **logistics section** can be viewed as the support side of an incident management structure. This section within the ICS is responsible for providing facilities, services, and materials for the incident. Logistics is headed by a **logistics section chief**, who is responsible for providing the bulk of the support functions for an incident. This position is generally assigned on long-duration or resource-intensive, complex incidents.

Logistical support includes food, sleeping facilities, transportation needs, sanitation facilities, showers, and all other requests for resources needed to manage the incident. If someone on the incident needs a truckload of sand, for example, the logistics section is the group that makes it happen. When resources are needed, the requesting person follows the appropriate chain of command to route the request to logistics. It is then up to logistics personnel to determine whether they will be able to fill the request. Typically, during a planning meeting at a large incident, the operations section chief, planning section chief, safety officer, IC, and a small core of other key management positions meet to discuss the logistical needs of the incident based on the operational objectives. The logistics section then determines whether those needs can be met and, if so, under what time frame.

Finance

The **finance/administration section** tracks the costs related to the incident, handles procurement issues, records the time that responders are on the incident for billing purposes, and keeps a running cost of the incident. Today, cost is certainly a factor in every response, even when it comes to operational objectives. Large-scale incidents can cost millions of dollars to handle, and it is common to see incident objectives tied to financial constraints. Ideally, an incident should be handled in the most cost-effective way possible, to minimize the financial impact to the jurisdiction in which the incident occurs.

Role of the Operations Level Responder

According to NFPA 1072, *Standard for Hazardous Materials/Weapons of Mass Destruction Emergency Response Personnel Professional Qualifications*, operations level hazardous materials responders are tasked with helping to protect nearby people, the environment, or property from the effects of a hazardous materials/WMD release. Examples of these duties may include performing various types of decontamination, product control, or detection and monitoring activities. All these tasks are performed under the direction of a hazardous materials technician, an allied professional, or standard operating procedures.

However, these persons may have additional competencies that are specific to their response mission, expected tasks, and equipment and training as determined by the AHJ. Clearly, this language identifies operations level hazardous materials responders as integral components of an overall response plan to hazardous materials/WMD incidents.

The scope of the operations level hazardous materials responder definition provides a jurisdiction with much more flexibility in its operational procedures than ever before, especially where mission-specific competencies are concerned. An AHJ may, for example, choose to train and use operations level responders strictly as support personnel for a group of hazardous materials technicians. Alternatively, the AHJ could provide its personnel with the training to satisfy operations level requirements, including both core competencies and all mission-specific competencies, thereby ensuring that its responders can handle a wide variety of hazardous materials/WMD incidents.

As a responder, you should be familiar with all emergency response plans for hazardous materials/WMD incidents that may occur within your jurisdiction. If none exists, as a responder you should default to the scope of practice established by your training when you are called into service, regardless of the nature of the incident.

It is common for a jurisdiction to predetermine response levels and response configurations to anticipated hazards. To that end, agencies use a wide variety of methods to identify and denote the severity of certain types of incidents. Establishing these thresholds takes the guesswork out of determining the types and amounts of resources required to handle a problem, as well as the levels of training required for the responding personnel. **TABLE 34-1** illustrates how an agency might choose to denote the different levels of response. For obvious reasons, the role of the operations level responder will differ depending on the severity of the incident.

TABLE 34-1 Hazardous Materials Incident Levels

Level I: Lowest Level of Threat	Level II: Medium Level of Threat	Level III: Highest Level of Threat
<ul><li>A small amount of a low-toxicity/low-hazard substance is involved.</li><li>The incident can usually be handled by a single resource such as a pump company.</li><li>Appropriate level of protection includes turnout gear and SCBA.</li></ul>	<ul><li>An organized hazardous materials team is needed.</li><li>Additional chemical protective clothing will be required.</li><li>Civilian evacuations may be required.</li><li>Decontamination may need to be performed.</li></ul>	<ul><li>Highly toxic chemicals are involved.</li><li>Mitigation efforts may require multiple jurisdictions.</li><li>Large-scale evacuations may be needed.</li><li>Federal agencies will be called in.</li></ul>
Example: A small gasoline spill from a motor vehicle accident.	Example: A tanker carrying sulfuric acid has overturned in a tunnel and is leaking onto the freeway.	Example: A ship in a highly populated harbor catches fire and begins to release chlorine vapours from its cargo area.

© Jones & Bartlett Learning.

After-Action REVIEW

IN SUMMARY

- At a hazardous materials incident, scene control is paramount because it has an influence on both scene security and personnel accountability.

- Isolation of the release area is one of the first measures that operations level hazardous materials responders should implement.

- The CAN report (conditions, actions, needs) allows you to distill a message down to its most basic parts.

- The three levels of control zones around a hazardous materials incident are the cold zone, the warm zone, and the hot zone. The hot zone is the most contaminated; the cold zone is a safe area.

- The IC must consider many factors before making decisions concerning evacuation, such as the nature and duration of the release, the nature of the evacuees, the ability to support the evacuees with basic services, and transportation challenges.

- The toxicity of the hazardous material and the amount of time available to avoid the oncoming threat are major factors in the decision of whether to evacuate or to use a sheltering-in-place strategy.

- Typically, before significant actions are taken at a hazardous materials incident, the IC should ensure that a *written* site safety plan is completed and a *verbal* safety briefing is performed.

- A written safety plan should include information about excessive heat disorders or cold stress the responders may encounter while engaging in their work.

- The incident command system (ICS) has many benefits, including the use of common terminology, consistent organizational structure and position titles, and common incident facilities.

- The ICS can be expanded to handle an incident of any size and complexity.

- The major functional components of the ICS are operations, planning, logistics, and finance/administration.

- The incident commander (IC) is the person in charge of the incident site; he or she is responsible for all decisions relating to the management of the incident.

- The command staff consists of the safety officer, the liaison officer, and the public information officer.

- Groups and divisions are established to aggregate single resources and/or crews under one supervisor.

- A hazardous materials branch consists of some or all of the following staff, as needed: a hazardous materials safety officer (assistant safety officer), an entry team, a decontamination team, a backup entry team, and a technical reference team.

- The operations level responder responds to hazardous materials incidents for the purpose of implementing or supporting actions to protect nearby persons, the environment, or property from the effects of the release.

KEY TERMS

Access Navigate for flashcards to test your key term knowledge.

Backup team Individuals who function as a stand-by rescue crew or relief for those entering the hot zone (entry team). Also referred to as backup personnel.

Buddy system A system in which two responders always work as a team for safety purposes.

Cold zone The control zone of hazardous materials/ weapons of mass destruction (WMD) incidents that contains the incident command post and such other support functions as are deemed necessary to control the incident. (NFPA 1072)

Command staff The command staff consists of the public information officer, safety officer, and liaison officer who report directly to the incident commander and are responsible for functions in the incident management system that are not a part of the function of the line organization. (NFPA 1561)

Control zones The areas at hazardous materials/weapons of mass destruction (WMD) incidents within an established perimeter that are designated based upon safety and the degree of hazard. (NFPA 1072)

Decontamination team The team responsible for reducing and preventing the spread of contaminants from persons and equipment used at a hazardous materials incident. Members of this team establish the decontamination corridor and conduct all phases of decontamination.

Division That organizational level having responsibility for operations within a defined geographic location. (NFPA 1026)

Entry team A team of fully qualified and equipped responders who are assigned to enter the designated hot zone.

Finance/administration section Section responsible for all costs and financial actions of the incident or planned event, including the time unit, procurement unit, compensation/claims unit, and cost unit. (NFPA 1026)

Hazardous materials branch The function within an overall incident management system (IMS) that deals with the mitigation and control of the hazardous materials/weapons of mass destruction (WMD) portion of an incident. (NFPA 1072)

Hazardous materials group See *hazardous materials branch*.

Hazardous materials safety officer The person who works within an incident management system (IMS) (specifically, the hazardous materials branch/group) to ensure that recognized hazardous materials/weapons of mass destruction (WMD) safe practices are followed at hazardous materials/WMD incidents. (NFPA 1072)

Heat exhaustion A mild form of shock that occurs when the circulatory system begins to fail because of the body's inadequate effort to give off excessive heat.

Heat stroke A severe, sometimes fatal condition resulting from the failure of the body's temperature-regulating capacity. Reduction or cessation of sweating is an early symptom; body temperature of 40.5°C (105°F) or higher, rapid pulse, hot and dry skin, headache, confusion, unconsciousness, and convulsions may occur as well.

Hot zone The control zone immediately surrounding hazardous materials/weapons of mass destruction (WMD) incidents, which extends far enough to prevent adverse effects of hazards to personnel outside the zone and where only personnel who are trained, equipped, and authorized to do assigned work are permitted to enter. (NFPA 1072)

Incident commander (IC) The individual responsible for all incident activities, including the development of strategies and tactics and the ordering and the release of resources. (NFPA 1072)

Incident command post (ICP) The field location at which the primary tactical-level, on-scene incident command functions are performed. (NFPA 1026)

Incident command system (ICS) A component of an incident management system (IMS) designed to enable effective and efficient on-scene incident management by integrating organizational functions, tactical operations, incident planning, incident logistics, and administrative tasks within a common organizational structure. (NFPA 1072)

Incident safety officer A member of the command staff responsible for monitoring and assessing safety hazards and unsafe situations and for developing measures for ensuring personnel safety. (NFPA 1500)

Liaison officer A member of the command staff responsible for coordinating with representatives from cooperating and assisting agencies. (NFPA 1561)

Logistics section Section responsible for providing facilities, services, and materials for the incident or planned event, including the communications unit, medical unit, and food unit within the service branch and the supply unit, facilities unit, and ground support unit within the support branch. (NFPA 1026)

Logistics section chief The general staff position responsible for directing the logistics function. It is generally assigned on complex, resource-intensive, or long duration incidents.

Operations section Section responsible for all tactical operations at the incident or planned event, including up to 5 branches, 25 divisions/groups, and 125 single resources, task forces, or strike teams. (NFPA 1026)

Operations section chief The general staff position responsible for managing all operations activities. It is usually assigned when complex incidents involve more than 20 single resources or when command staff cannot be involved in all details of the tactical operation.

Planning section Section responsible for the collection, evaluation, dissemination, and use of information related to the incident situation, resource status, and incident forecast. (NFPA 1026)

Planning section chief The general staff position responsible for planning functions and for tracking and logging resources. It is assigned when command staff members need assistance in managing information.

Public information officer (PIO) A member of the command staff responsible for interfacing with the public and media or with other agencies with incident-related information requirements. (NFPA 1026)

Size-up The process of gathering and analyzing information to help fire officers make decisions regarding the deployment of resources and the implementation of tactics. (NFPA 1410)

Span of control The maximum number of personnel or activities that can be effectively controlled by one individual (usually three to seven). (NFPA 1006)

Technical reference team A team of responders who serve as an information-gathering unit and referral point for both the incident commander and the hazardous materials safety officer (assistant safety officer).

Unified command A team effort that allows all agencies with jurisdictional responsibility for an incident or planned event, either geographic or functional, to manage the incident or planned event by establishing a common set of incident objectives and strategies. (NFPA 1026)

Warm zone The control zone at hazardous materials/weapons of mass destruction (WMD) incidents where personnel and equipment decontamination and hot zone support take place. (NFPA 1072)

On Scene

It is 4:00 AM when your pump company is dispatched to an old manufacturing plant for a report of a 8000-L (2113-gal) above-ground storage tank slowly leaking. The on-scene security guard tells you the storage tank has been there for years, and he thinks it might contain polychlorinated biphenyls (PCBs). Approximately 200 L (53 gal) of liquid have spilled onto the asphalt and are entering a storm drain. You and your crew must establish control zones to limit exposure to the chemical.

1. Which of the following most accurately describes the warm zone?

  A. The area immediately around and adjacent to the incident

  B. The area located between the hot zone and the cold zone. The decontamination corridor is in the warm zone.

  C. An area where personnel do not need to wear any special protective clothing for safe operation

  D. The area where the incident command post is located

(continued)

On Scene Continued

2. After doing some research on PCBs, you learn that PCB is a toxic chemical that was banned in Canada in the late 1970s but can still be found in some older electrical devices. This chemical is very harmful to the environment; it can also cause skin irritation, liver damage, nausea, dizziness, and eye irritation. It is a thick, oily liquid with a low vapour pressure at ambient atmospheric temperature. The leak is occurring approximately 183 m (200 yd) away from an occupied apartment complex. Which of the following actions would be appropriate for this incident?

A. Evacuate the apartment complex.

B. Take no action, because this spill poses no threat of airborne contamination to the apartment complex, and the possibility of the material being PCB is very remote.

C. Notify local law enforcement to shelter-in-place all occupants of the apartment complex.

D. Instruct the public information officer to make an announcement of the spill during the evening news broadcast.

3. After working in PPE for 30 minutes and going through decontamination, one of the responders complains of dizziness and nausea. He is sweating and tells you he feels weak. Which of the following medical conditions most accurately describes his condition?

A. Heat stroke

B. Hypothermia

C. Hypocalcemia, due to the PCB exposure

D. Heat exhaustion

Access Navigate to find answers to this On Scene, along with other resources such as an audiobook.

CHAPTER 35

Hazardous Materials Operations Level

Hazardous Materials Responder Health and Safety

KNOWLEDGE OBJECTIVES

After studying this chapter, you will be able to:

- Discuss the hazards of fire smoke. (**NFPA 1072: 5.2.1; NFPA 472: 5.2.1.7**, pp. 1190–1198)
- Discuss the effects of carbon monoxide and hydrogen cyanide on the body. (**NFPA 1072: 5.2.1; NFPA 472: 5.2.1.7**, pp. 1193–1197)
- Describe methods for treating smoke inhalation. (pp. 1197–1198)
- Discuss postfire detection and monitoring needs. (**NFPA 1072: 5.1.1**, pp. 1198–1203)
- Discuss the purpose of detection devices at fire scenes. (**NFPA 1072: 5.1.1**, pp. 1199–1201)
- Discuss the various technologies available for fire-ground detection and monitoring. (**NFPA 1072: 5.1.1**, pp. 1201–1203)
- Discuss general fire-ground monitoring principles and practices. (**NFPA 1072: 5.1.1**, pp. 1203–1204)

SKILLS OBJECTIVES

This chapter has no skills objectives for operations level hazardous materials responders.

Hazardous Materials Alarm

Just before lunch, your pump company responds to a working fire in a single-family wood frame dwelling. The fire started in a bedroom and extended into the adjacent hallway. Your apparatus is first in, and your crew extends the initial attack line, makes access to the seat of the fire, and quickly knocks it down. The fire is declared under control, and the crew transitions to overhaul. A few minutes later, your officer directs the crew to exit the building. After a short rehab period, your crew reenters to assist with overhaul, wearing protective pants, jacket, helmet, gloves, and eye protection. Your pump company remains on scene for another hour, mopping up and eventually turning the building back over to the homeowner. The crew returns to quarters and places the apparatus back in service. You have a headache, mild nausea, and a feeling of fatigue. It doesn't resolve over the remainder of the shift, and you go home the next morning still not feeling 100 percent.

1. Do you believe the ill feeling you have is related to the smoke exposure? If so, what do you think you were exposed to?

2. Should air monitoring be conducted at the scene prior to reentering the structure for overhaul? At what point would you deem the environment safe to operate without self-contained breathing apparatus?

3. If you began feeling worse either at the firehouse or later at home and sought medical attention at a local hospital, do you think you would receive medical treatment for a smoke exposure?

 *Access Navigate for more practice activities.*

Introduction

Research conducted over the years has proven beyond question that breathing smoke is bad for you. There may be debate about how much of a certain fire gas is present at a given fire scene or what combustion by-products are identified in certain studies, but the fact remains that every time something burns, the combustion process liberates fire gases (many of which are toxic) and particulates.

When asked about fire fatalities, fire fighters typically observe that smoke kills people before the flames ever get to them, and fire death statistics prove likewise. Fire fighters will also acknowledge that after working at a fire and breathing smoke during the firefight, or perhaps during the overhaul phase, they don't feel well in the hours and sometimes days following a fire. Headache, nausea, dizziness, and fatigue are common signs and symptoms that can be traced back to breathing smoke from all types of fires. In Canada, many provinces have enacted "prescriptive compensation" for many fire-related illnesses under their respective Workers Compensation Acts. If a fire fighter is diagnosed with certain forms of cancer, it is considered a work-related illness.

From 2010 to 2014, Statistics Canada reported a marked decrease in fire-related deaths. The specific cause for this reduction at a time when Canada's net population increased by approximately 1.23 million is uncertain. However, the decrease may be the result of a combination of regulatory changes, bylaw enforcement, and increased awareness that are contributing to reductions in loss of life due to fire.

After the fire is knocked down, it is not uncommon to see fire fighters remove their self-contained breathing apparatus (SCBA) and perform overhaul without respiratory protection, breathing smoke at the point in time when smouldering debris is producing volumes of particulates and a variety of gases. It is becoming more common for fire agencies to mandate the use of SCBA for longer periods of time on the fire ground—during and after the active firefighting phase—but there is a long way to go to change the perception of some fire fighters that breathing smoke is just part of the job. Like a SCUBA diver, a fire fighter advancing a hose line during an interior attack at a working structure fire is completely enveloped by the operating environment—the smoke—and if a supply of breathable air is lost or interrupted, the environment invades the body, causing harm (**FIGURE 35-1**).

To underscore this point, consider the case of Captain James Carter from Springdale, Arkansas, as told in this chapter's *Voice of Experience*.

The goal of this chapter is to provide an overview of the hazards of fire smoke and highlight a link between smoke exposure and the resulting acute and chronic

FIGURE 35-1 Think of yourself as a "smoke diver" when you are working at a structure fire.
Courtesy of Rob Schnepp.

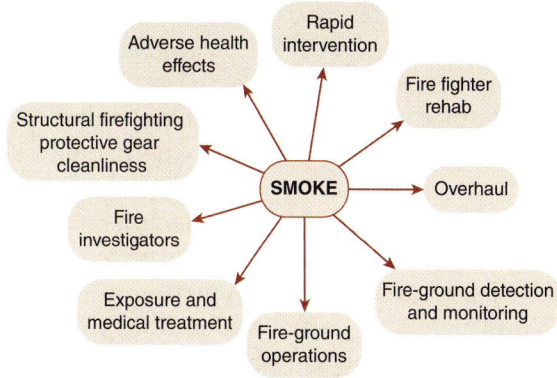

FIGURE 35-2 Think about all the ways that smoke exposures can reach out and affect a fire fighter.
Courtesy of Rob Schnepp.

health effects. The anticipated result is a modification of the fire fighter's attitude toward smoke and an increased interest in better managing repeated dermal and inhalation exposures. This chapter is aimed mostly toward rank and file fire fighters, but other groups of emergency responders such as law enforcement, emergency medical services (EMS) providers, fire investigators, and fire training instructors will find the information useful.

Simply put, if you go into situations where smoke is present you should be informed about the potential health hazards. Envision smoke as reaching out and touching nearly every aspect of the life of anyone who routinely works in or around it. **FIGURE 35-2** represents some of the places and situations in which people interact with particulates and gaseous toxins.

Ideally, the authority having jurisdiction (AHJ) should establish reasonable and safe work practices for structural firefighting, but at the most basic level, it is your responsibility to be informed about the hazards of breathing smoke and the need to take all precautions for reducing your repeated exposure. Keeping structural firefighting protective gear clean is a good example, as is taking a shower after working at a fire scene and ensuring crews are wearing SCBA during active fire-ground operations, even when engaged in

exterior operations or working on a roof. Think about how overhaul is carried out in your organization and whether you are doing everything possible to reduce your dermal and inhalation exposure.

Some questions to ask about your fire scene work practices include:

- Do crews wear SCBA until the building is clear of visible particulates (smoke)?
- If you are doing postfire detection and monitoring, have you established action levels (keeping face mask on) for the gases you are monitoring?
- Are personnel trained to understand the benefits and limitations of detection and monitoring at the fire scene and the operational parameters of the technology used to perform the detection and monitoring?
- Are rapid intervention crews (RICs) located so close to the building that they are breathing smoke prior to being called into service? Is there a better place for those crews to set up and still be available if the need arises?
- Is the command post located in drift smoke?
- Does your prehospital care system have a treatment protocol for smoke inhalation?
- Is the local receiving hospital ready to treat a civilian or fire fighter properly in the event of a significant smoke-related illness?

Think through the potential pathways of a smoke exposure, and identify ways you could reduce or address the places and situations whereby you could be exposed.

Overview of Fire Smoke

Even though smoke is the constant companion of the fire fighter, there is not always a thorough understanding of the link between what's in smoke and the reason people get sick or die when they breathe it or why fire fighters contract heart disease, neurological dysfunction, cancer, or some other illness related to a single or repeated exposure to smoke. Cancer is prevalent in the fire service, more so than most any other profession, and this should be no surprise once the by-products of combustion are understood. Known human carcinogens typically are liberated as fire gases and particulates. (This will be discussed later in this chapter.)

Most people identify carbon monoxide as the main harmful component of fire smoke but struggle to list more than a handful of substances beyond that. Less often acknowledged are compounds such as ammonia, hydrogen chloride, sulfur dioxide, hydrogen sulfide, hydrogen cyanide, carbon dioxide, the oxides of nitrogen, formaldehyde, acrolein, polycyclic aromatic hydrocarbons, and soot (**FIGURE 35-3**).

Smoke production depends on several factors, including the chemical make-up of the burning material, the temperature of the combustion process, and the influence of ventilation (oxygenation). Make no mistake that fire (combustion) is a complex process, and the smoke produced is an intricate collection of particulates, superheated air, and gaseous chemical compounds.

The extensive use of synthetic manufacturing and construction materials (e.g., plastics, nylon, rubber products, fire retardants, laminates, and foams such as

FIGURE 35-3 A different way to look at fire smoke: If this cargo trailer were carrying smoke, this is how it might be placarded. If these placards were on the next house fire you went to, would you think differently about the potential health hazards?
Courtesy of Rob Schnepp.

Styrofoam and polystyrene) has a significant effect on fire behaviour and smoke production. Synthetic substances ignite and burn fast, causing rapidly developing fires and toxic smoke and making structural firefighting more dangerous than ever before.

Polyurethane foam is the most predominant substance in a typical mattress and is made up of many different chemicals including polyol (an organic alcohol molecule and the majority of the polyurethane compound), toluene diisocyanate (TDI), methylene chloride, and ammonia-based catalysts. When polyurethane foam is exposed to heat, the parent substances break down and bond with each other, creating many new compounds. Some of those compounds are irritants, such as hydrogen chloride and ammonia, causing eye irritation or airway problems in smoke exposures. Other compounds, like carbon monoxide and cyanide compounds, are acutely toxic when inhaled.

The thermal decomposition of polyurethane foam in the mattress fire scenario is broadly representative of the way gases are liberated by a working fire. (Keep in mind this is a very limited representation of the process; many more substances are also liberated.) Certainly, each substance is present in varying levels depending on the material(s) involved, the heat of combustion, and the available oxygen (ventilation of the fire). Aside from the toxic gases produced, it is equally important to recognize the visible part of combustion (aside from the flames). These are carbon particles of varying sizes, ranging from large embers to particles you can't see with the naked eye. This collection of particulates—what we traditionally called smoke—is actually soot. Soot is important to acknowledge because it is a known human carcinogen, just like benzene and formaldehyde—also by-products of combustion. With that in mind, you should take a moment and think about your perception of smoke and revise it to this—the combustion process liberates things you can see (particulates) and things you can't see (gases). **TABLE 35-1** breaks "smoke" down into those two main categories to make it easier to understand and separate the hazards. The matrix isn't inclusive of all substances that could fall into the four boxes, but it does illustrate the point that there are many properties in smoke that cause acute and chronic health effects. It also shows that smoke is not only *one thing*; rather, it is a dynamic multifaceted mixture of gases and particulates that changes from minute to minute at any fire.

Another group of compounds commonly found in fire smoke is **polycyclic aromatic hydrocarbons (PAHs)**, classified as probable or possible human carcinogens. Well over 100 PAHs have been identified and categorized by various regulatory agencies. This group of substances occurs naturally in materials such as coal and crude

TABLE 35-1	Smoke Matrix		
		Harm You Now	**Harm You Later**
What you see			Soot, particulates
What you don't see		Carbon monoxide (CO), hydrogen cyanide (HCN), oxides of nitrogen (NO$_x$), sulfur dioxide (SO$_2$), hydrogen chloride (HCl), hydrogen sulfide (H$_2$S)	Aldehydes, benzene

Courtesy of Rob Schnepp.

oil. These substances also are generated during the combustion of organic materials and can be found in vehicle exhaust; tobacco smoke; and the smoke generated from structure fires, vehicle fires, wildland fires, or any other type of fire. PAHs can exist as a particle or a gas. Chances are that you have been exposed to PAHs throughout your life from a common and perhaps surprising activity—grilling food (**FIGURE 35-4**).

When PAHs are generated during a structure fire they may bind with the soot, resulting in dermal and inhalation exposures. PAHs are believed to be immunosuppressants, perhaps contributing to the mechanism by which PAHs are suspected to cause cancer. Some examples of PCHs are:

- Anthracene
- Benzopyrene
- Methylchrysene
- Phenanthrene
- Pyrene

Responders not only have to be concerned about acute exposure to PAHs and other materials but also need to consider this group of chemicals as contaminants to fire fighter structural firefighting protective gear. A 2013 study, "Evaluation of Dermal Exposure to Polycyclic Aromatic Hydrocarbons in Fire Fighters"—released by the Health Hazard Evaluation Program of the U.S. Department of Health and Human Services, Centers for Disease Control and Prevention, and National Institute for Occupational Safety and Health—discusses PAHs and other substances in terms of fire fighter dermal exposure during firefighting, overhaul, and postfire activities. In short, the study illustrates that it is possible to absorb PAHs and other particulates through your skin and that dirty structural firefighting protective gear is unhealthy and is likely to contribute to sustained exposure to the fire, long after you've left the scene. NFPA 1851, *Standard on Selection, Care, and Maintenance of Protective Ensembles for Structural Fire Fighting and Proximity Fire Fighting*, offers guidance on structural firefighting protective gear cleaning and maintenance. More studies are being conducted into the effectiveness of PPE in preventing PAH skin absorption. It is advisable that personnel should wipe down their skin as soon as possible post fire.

Carbon Monoxide and Hydrogen Cyanide: Silent Killers

Smoke is one of the first observable signs of a working fire. Fire fighters note the volume, colour, and force as smoke is exiting a fire building—all good indicators of what the fire is doing inside—and use that information to implement appropriate fire-ground tactics. Fire fighters may aggressively enter smoky buildings to search for victims but rarely perform a conscious evaluation of the toxic substances lurking in the smoke. This could be a significant oversight in terms of treating victims, according to studies performed in Paris, France, and Dallas County, Texas. These studies focused on carbon monoxide (CO) and hydrogen cyanide (HCN) specifically because they are acutely toxic, present to some degree in

FIGURE 35-4 The thermal decomposition of muscle meats such as chicken, beef, and pork generates PAHs and heterocyclic amines (HCAs) when they are cooked over an open flame.
Courtesy of Rob Schnepp.

nearly all fires, and have clinical interventions available to reverse the adverse health effects of the exposure. Clearly many toxic substances are generated during a typical structure fire, and smoke inhalation is a complicated illness. It is also clear that successful medical treatment isn't accomplished by a single drug or single action and that smoke inhalation victims are quite sick.

LISTEN UP!

The discussion of CO and HCN isn't framed in the light that those two substances make smoke toxic. Rather, the important point is that they are toxic substances that are common by-products of combustion (present at some level at most fires) and treatable with clinical interventions.

These studies were done more than 20 years ago, which underscores the fact that the presence of HCN and CO in smoke isn't a new revelation; however, the fire service and other disciplines have recently been connecting the dots to understand that smoke is a bigger issue than previously thought. In short, these studies identify and evaluate the impact of hydrogen cyanide and carbon monoxide on smoke inhalation patients. *Again, there are many other toxins found in smoke—the goal here is to highlight two fire gases found at nearly any kind of fire.*

The Paris study was designed to prospectively assess the role of cyanide in smoke-related morbidity and mortality. Blood samples were drawn from survivors as well as fatalities (at the time of exposure), and cyanide and carbon monoxide levels were measured. In several deaths, cyanide levels were in the lethal range whereas carbon monoxide levels were in nontoxic concentrations. This suggests cyanide toxicity as the primary cause of death. The study also revealed another bit of interesting information: Death occurred in victims with cyanide *and* carbon monoxide levels in the nontoxic range, perhaps revealing a relationship between carbon monoxide and cyanide in smoke inhalation patients. The following summarizes the results of the Paris study:

- Cyanide and carbon monoxide were both important determinants of smoke inhalation–associated morbidity and mortality.
- Cyanide concentrations were directly related to the probability of death.
- Cyanide poisoning may be more predominant than carbon monoxide poisoning as a cause of death in certain fire victims.
- Cyanide and carbon monoxide may potentiate the harmful effects of one another.

The Dallas County, Texas, study measured blood cyanide levels in victims after exposure to fire smoke and in many respects echoed the findings of the Paris study. **TABLE 35-2** shows a summary of the findings. In Dallas County, Texas, over a 2-year period, blood samples were collected from a total of 187 smoke inhalation patients, within 8 hours of exposure. There were 144 viable patients at the University of Texas Health Sciences emergency department; 43 victims were dead on arrival at the Dallas County Medical Examiner's office.

TABLE 35-2 Cyanide Levels After Exposure to Fire Smoke

Patient #	Age, y	% Total Body Surface Area Burn	Blood Cyanide, mg/L	HbCO, %	Outcome
1	47	98%	1.20	18.6%	Died
2	80	3%	1.60	22.6%	Died
3	29	4%	1.40	6.0%	Died
4	22	55%	5.20	35.6%	Lived
5	30	63%	1.40	5.0%	Lived
6	58	32%	2.60	10.9%	Died
7	32	5%	6.00	32.0%	Lived
8	50	25%	2.20	17.2%	Lived

Patient #	Age, y	% Total Body Surface Area Burn	Blood Cyanide, mg/L	HbCO, %	Outcome
9	4	0%	11.50	22.4%	Died
10	36	40%	5.70	3.8%	Died
11	19	90%	1.20	40.0%	Died
12	30	76%	2.72	37.0%	Died

Reproduced from: Silverman SH, et al. Cyanide toxicity in burned patients. *J Trauma* 1988; 28:171–176.

Of the 144 living patients who reached the emergency room, 12 had blood cyanide concentrations exceeding 1 mg/L (see Table 35-2). Of these 12 patients, 8 eventually died. *None had blood carboxyhemoglobin (COHb) concentrations suggesting carbon monoxide as the cause of death (i.e., ≥ 50 percent).* Although some of the patients had extensive burns that may have contributed to their death, three of them (patients 2, 3, and 9) had ≤ 4 percent total body surface area burns. According to the study, blood cyanide levels greater than 1 mg/L had a significant impact on patient outcome. More importantly, the study found that elevated cyanide levels were pervasive in smoke inhalation victims, and cyanide concentrations were directly related to the probability of death.

Lastly, both studies dispel a long-held belief in the fire service—that carbon monoxide is the predominant killer in fire smoke. In fact, it appears that cyanide plays a role in smoke-related death and injury, perhaps more often than we think. Therefore, any victim(s) exposed to significant amounts of smoke or rescued from a closed space structure fire may be suffering from cyanide toxicity. However, remember that there are many properties in smoke that are capable of causing acute illness and leaving an indelible impression on your body.

LISTEN UP!

Morbidity is a disease or the incidence of disease within a population.

Mortality is the incidence of death in a population.

Depending on the dose, hydrogen cyanide has the ability to incapacitate a victim, preventing escape from the fire environment and thereby increasing the exposure to more cyanide, carbon monoxide, and other toxic by-products of combustion. Although this theory is currently unsupported with human data related to smoke exposures, there is information to substantiate the "knock-down" potential of cyanide. In the mid-1980s, studies were conducted on monkeys exposed to the fumes of heated polyacrylonitrile. (When this substance is broken down by **pyrolysis**, cyanide is liberated.) Cyanide-exposed monkeys first hyperventilated and then rapidly lost consciousness at a dose-dependent concentration. A concentration of 200 parts per million (ppm) was associated with rapid incapacitation but not with elevated blood cyanide concentrations measured hours after exposure. The direct correlation to human data is unknown at present but could be interpreted in the following way: Hydrogen cyanide could be partly responsible for rendering fire fighters and civilians incapable of self-rescue when exposed to smoke.

LISTEN UP!

Cyanide exposures can be fatal in the presence of normal oxygen levels.

To appreciate cyanide's mechanism of action, it is first necessary to understand the process of oxygen transportation and use in the body and the basic idea of **aerobic metabolism**. To simplify the concept, imagine the circulatory system as a very efficient public transit system, full of "buses" (**red blood cells**) carrying passengers (oxygen) to and from a multitude of bus stops (the cells). The circulatory system, similar to a network of streets, is loaded with red blood cells (RBCs)—hemoglobin buses—each carrying four oxygen passengers (**FIGURE 35-5**).

During normal cellular respiration, the bus system transports oxygen passengers to the bus stops (cells). At the appropriate stop, four oxygen molecules get off and move through an electron chain, ultimately combining with the final electron acceptor—**cytochrome oxidase** (an enzyme)—before entering the **mitochondria** of each cell.

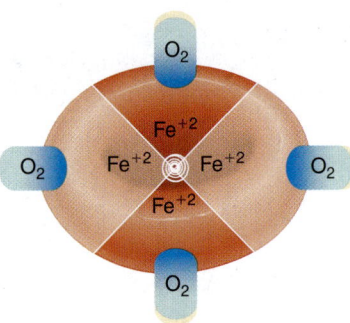

FIGURE 35-5 The circulatory system is loaded with red blood cells (RBCs), each of which carries four oxygen "passengers."
© Jones & Bartlett Learning.

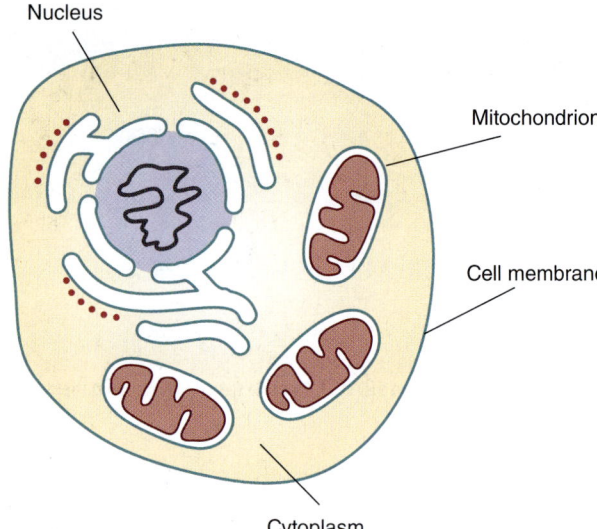

FIGURE 35-6 Mitochondria are responsible for converting nutrients into energy-yielding molecules of adenosine triphosphate (ATP) to fuel the cell's activities.
© Jones & Bartlett Learning.

Mitochondria are responsible for converting nutrients into energy-yielding molecules of adenosine triphosphate (ATP) to fuel the cell's activities (**FIGURE 35-6**). ATP production is highly dependent on oxygen (and glucose); without it, normal aerobic metabolism is impossible. If this process is seriously compromised, death is imminent.

LISTEN UP!

Aerobic metabolism is the creation of energy through the breakdown of nutrients in the presence of oxygen. The by-products are carbon dioxide and water, which the body disposes of by breathing and sweating.

Anaerobic metabolism is the creation of energy through the breakdown of glucose. Without oxygen, the metabolic process results in the production of lactic acid.

Cyanide compounds, once absorbed in the body, "poison" the cytochrome oxidase, barring oxygen from entering the mitochondria and effectively shutting down the process of aerobic metabolism. In short, the buses may be transporting some amount of oxygen passengers, but when they get off the bus, they find their ultimate destination locked. Without oxygen, the cells switch to **anaerobic metabolism**, producing toxic by-products such as lactic acid, ultimately destroying the cell. Therefore, cyanide toxicity is not about the amount of oxygen available to the body; rather, it's about the inability of the body to *use* oxygen for aerobic (life-sustaining) metabolism. Consequently, an elevated lactic acid level is a key indicator of cyanide toxicity.

According to the studies done in Paris and Dallas County, Texas, there is a possible deleterious relationship between carbon monoxide and cyanide in smoke inhalation victims. This relationship could be attributed to the inability of the cells to use oxygen (due to cyanide), coupled with the adverse impact of CO on the RBCs. Carbon monoxide binds in place of oxygen on the RBC,

excluding oxygen from riding on the hemoglobin bus. (The oxygen-carrying capacity of the RBC becomes limited or nonexistent, thereby reducing the amount of oxygen transported to the cells.)

LISTEN UP!

The signs and symptoms of acute cyanide toxicity are similar to those of CO poisoning and mimic the nonspecific signs and symptoms of oxygen deprivation, including headache, dizziness, stupor, anxiety, rapid breathing, and increased heart rate. In extreme cases of cyanide poisoning, patients may present with seizures, a significantly altered level of consciousness (including coma), severe respiratory depression or respiratory arrest, and complete cardiovascular collapse.

Carbon monoxide is one of the most common industrial hazards. It is colourless and odourless and produced during incomplete combustion. As mentioned, CO affects the oxygen-carrying capacity of the red blood cell, causing **hypoxia**. Consequently, the signs and symptoms of exposure will be similar to those caused by cyanide poisoning and consistent with hypoxia—headache, nausea, vomiting, disorientation, and if the exposure is extreme, seizures, coma, and death. **FIGURE 35-7** represents the signs and symptoms of a CO exposure. As the arrow indicates, the severity of health effects is in relation to the severity of the exposure. In summary, you can look at CO and HCN exposures in this way: Carbon monoxide reduces the amount of oxygen carried to the cells; cyanide renders the cells incapable of using whatever oxygen is present.

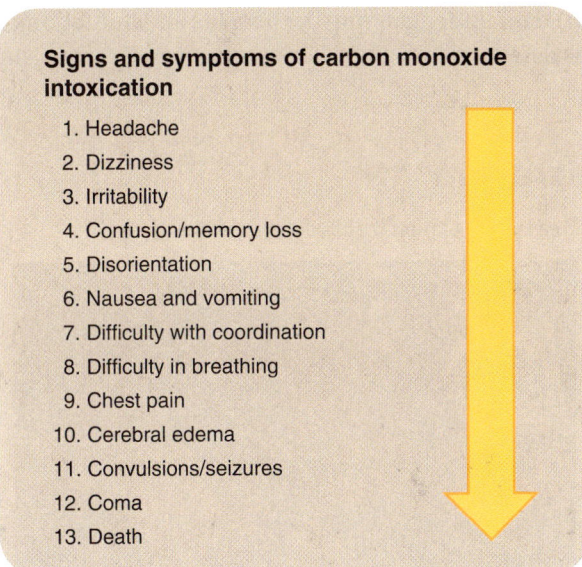

Signs and symptoms of carbon monoxide intoxication

1. Headache
2. Dizziness
3. Irritability
4. Confusion/memory loss
5. Disorientation
6. Nausea and vomiting
7. Difficulty with coordination
8. Difficulty in breathing
9. Chest pain
10. Cerebral edema
11. Convulsions/seizures
12. Coma
13. Death

FIGURE 35-7 As the level of COHb rises, the severity of the signs and symptoms increases. The downward arrow indicates the progression from the most common symptom of CO poisoning, headache, to the most extreme, seizures, coma, and death.
© Jones & Bartlett Learning.

Smoke Inhalation Treatment

Smoke inhalation is one of the most complex and challenging patient presentations faced by medical care providers. Patient outcomes vary greatly, influenced by such factors as the extent and duration of the smoke exposure, amount and nature of toxicants in the smoke, degree of thermal burns to the skin and lungs, quantity/size of inhaled particulates (soot), patient's age, and underlying medical conditions. Nationwide, there is no standard protocol for treating smoke inhalation, leaving paramedics and other prehospital care providers with limited guidance and/or training to properly care for smoke inhalation victims. In many EMS systems, treating smoke inhalation outside the hospital boils down to supportive care: monitoring vital signs, providing high-flow oxygen, establishing intravenous (IV) lines, performing advanced airway management techniques such as endotracheal intubation, cardiac monitoring, and rapid transport.

Referring to the studies done in Paris and Dallas County, Texas, however, it is evident that supportive care alone will not correct the underlying cause of death in smoke inhalation patients—the adverse effects of hydrogen cyanide and carbon monoxide on the body's oxygen transportation and utilization system, causing asphyxia at the cellular level. The bottom line is this: Until the underlying cause of asphyxia is reversed at the cellular level, normal oxygenation is not possible. This requires a clinical intervention—administering an antidote—to restore the body's ability to use oxygen.

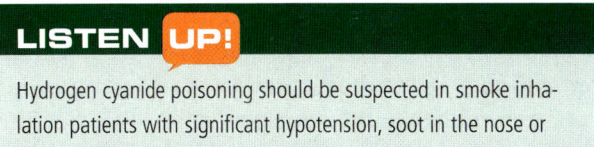

LISTEN UP!

Hydrogen cyanide poisoning should be suspected in smoke inhalation patients with significant hypotension, soot in the nose or mouth, and/or altered level of consciousness.

Oxygen is the natural antidote for CO poisoning, and in all cases of smoke inhalation high-flow oxygen should be administered. For CO poisoning, which primarily targets the heart and brain, the administration of high-flow oxygen reduces the half-life of CO in the body to around 1 hour. If hyperbaric therapy is indicated, the amount of CO in the body can be reduced by one half in approximately 30 minutes. These are approximations, and patient response may vary from case to case. The point is that oxygen is good for CO inhalation victims, whether from fire smoke or some other source.

When it comes to treating the HCN portion of smoke inhalation, a different kind of antidote is indicated along with oxygen administration.

CYANOKIT® (Hydroxocobalamin 5 g Powder for Solution for Infusion) is an emergency treatment (antidote) used in patients with known or suspected cyanide poisoning. The active ingredient in CYANOKIT®, hydroxocobalamin, binds to cyanide that is present in the blood, making it less toxic. The complexed cyanide is then renally excreted from the body. The new CYANOKIT® presentation consists of a single vial containing 5 g hydroxocobalamin powder, which is to be reconstituted in 200 mL of sodium chloride 0.9% solution for intravenous administration. Hydroxocobalamin (a precursor to vitamin B_{12}) is a relatively benign substance, with minimal side effects, making it well suited for use in the prehospital setting.

Hydroxocobalamin has no adverse effect on the oxygen-carrying capacity of the RBCs and no negative effect on the patient's blood pressure—significant benefits when treating victims of smoke inhalation. Surprisingly, the mechanism of action is simple: Hydroxocobalamin binds to cyanide, forming vitamin B_{12} (cyanocobalamin), a nontoxic compound ultimately excreted in the urine.

The CYANOKIT® can be administered to a smoke inhalation patient without first verifying the presence of cyanide in the body, with little fear of making the patient worse. Numerous EMS agencies carry and administer cyanide antidotes to smoke inhalation with success.

It is important that you understand how smoke inhalation victims are treated in your jurisdiction and if a cyanide antidote is carried on advanced life support transport units or stocked in the local hospitals. It is also important that you understand that smoke inhalation is an illness just like congestive heart failure, asthma, or any other medical condition you may encounter in the field and that aggressive clinical intervention is key to the possibility of a good outcome. As a reference source, consult NFPA 473, Section 6.4, "Mission-Specific Competencies Advanced Life Support (ALS) Responder Assigned to Treatment of Smoke Inhalation Victim." This companion document to NFPA 472, *Standard for Competence of Responders to Hazardous Materials/Weapons of Mass Destruction Incidents*, provides background on treating smoke inhalation patients.

Postfire Detection and Monitoring

Although gas detection and atmospheric monitoring are common in hazardous materials response, the typical fire fighter may be unfamiliar with gas detection devices, their methods of use, and procedures for detecting select gases on the fire scene. It is important to note that (at the time of this writing) there is no concrete best practice when it comes to detection and monitoring in the fire environment, specifically during overhaul. To that end, many fire departments are investing in technologies for detecting toxic gases at the fire scene.

Generally, fire fighters wear SCBA during active interior firefighting, but there are still places along the edges of the fire scene, and times during the progression of the incident, where smoke exposures occur. The most common and repeated time and place are during overhaul. The fire service in general is getting better about wearing SCBA during overhaul, but it's still common to see this task being done with no respiratory protection at all. Pump operators working at the panel during a fire, shrouded in drift smoke, are also common victims of exposure, as is a chief officer commanding the fire, standing in the haze at a poorly located command vehicle. Personnel operating exterior lines for extended periods of time are also breathing fire gases and particulates, as are RICs setting up in the front yard of a single-family dwelling fire. Think about all the places on the fire ground, outside of the interior attack crew, where fire personnel are repeatedly exposed to smoke.

Consequently, the increased appetite for education regarding fire gas toxicity and fire fighter safety, coupled with an interest in postfire atmospheric monitoring, adds an entirely new category for detection, outside of the traditional paradigm of detection and monitoring. Over time, a new normal may exist in the fire service

based on understanding the benefits and limitations of gas detection in the hostile environment of a fire scene.

SAFETY TIP

Wearing SCBA means the cylinder *and* the mask!

Courtesy of Springdale FD. Permission granted by James Carter.

To begin the discussion in the right context, consider this before embarking on your quest to perform postfire detection and monitoring: Atmospheric monitoring has a place on the fire scene. The technology and devices selected should be user friendly, durable, cost effective, and easy to maintain, and the benefits and limitations of any instrumentation should be understood. Because there is no single device that will identify all possible toxins on the scene, it is important to know that targeting certain gases (CO and HCN as examples) may be broadly representative of the airborne environment but not an exact indicator of the presence, absence, or concentration of any other gas or particulate in the air. Wearing SCBA is still the gold standard of respiratory protection and is the best way to reduce the possibility of inhalation exposures.

Until recently, there have been three primary uses for detection devices outside of the traditional fire-based hazardous materials response:

- Rescue response including confined space
- Building collapse and trench rescue
- CO detector responses

The intent of this section is to provide some general information about fire scene gas detection, offer descriptions of other types of detection equipment available that could be used at the fire scene, and provide an overview of electrochemical sensor technology (commonly used in fire-ground detection and monitoring). As you will see, the focus is on detecting gases, not particulates. This is important to understand, because the visible part of smoke—the particulates—is hazardous as well and often is overlooked in favour of evaluating the environment for gases like CO and HCN.

Why Use Detection Devices at the Fire Scene?

As stated throughout this chapter, fire smoke is a collection of fire gases and particulates. Most of them can be detected and measured by one type of technology or another, but no single device can detect them all; this is important to understand. Therefore, an initial decision must be made about what substance(s) you want to detect and/or monitor and at what concentrations those substances pose a risk of exposure. To that end, it is vital to understand and apply some basic toxicological terms and definitions. Federal-Provincial-Territorial Ministers of Labour and Part X of the Canada Occupational Health and Safety Regulations (COHSR) for federally regulated workplaces establish safe levels of chemical exposures in the workplace. In addition, the National Institute for Occupational Safety and Health (NIOSH) rules are utilized by most of these federal, provincial, and territorial agencies. These values are for average worker exposure and can be only *estimated* for the rigours of firefighting. Elevated respiratory rate, blood pressure, and heart rate; increased skin temperature resulting in increased permeability; and dehydration are only some of the factors that would add to the intake and effects of these toxic gases on responders. For the purpose of this section, the more conservative NIOSH levels will be used to guide suggested operations and assist in evaluating airborne contamination levels. The exposure levels are as follows:

- Immediately dangerous to life and health (IDLH)
- Short-term exposure limits (STEL)
- Recommended exposure limits (REL)

In short, levels at or above IDLH require use of breathing apparatus or withdrawal from the area if no SCBA is worn. The definition of IDLH as found in Canadian Standards Association (CSA) 94.4-11 (2016) means an atmosphere that poses an immediate threat to life or that will cause irreversible adverse health effects or impair an individual's ability to escape.

A STEL represents a 15-minute exposure no more than four times a day, and a REL is for a 10-hour exposure. The Federal-Provincial-Territorial Ministers of Labor and Part X of the Canada Occupational Health and Safety Regulations (COHSR) also have set comparable levels and, in many cases, use the NIOSH values. **TABLE 35-3** lists the values for the toxic gases commonly present in fire smoke.

TABLE 35-3 The Values for the Toxic Gases Commonly Present in Fire Smoke

Gas	REL	STEL	IDLH	Density
Ammonia (NH_3)	25 ppm	35 ppm	300 ppm	Lighter than air
Carbon dioxide (CO_2)	5000 ppm (0.5% vol.)	30,000 ppm (3% vol.)	40,000 ppm (4% vol.)	Heavier than air
Carbon monoxide (CO)	35 ppm—no higher than 200 ppm allowed (ceiling)	NR	1200 ppm	Lighter than air
Hydrogen chloride (HCl)	No higher than 5 ppm (ceiling)	NR	50 ppm	Heavier than air
Hydrogen cyanide (HCN)	4.7 ppm (15 minutes only)	4.7 ppm—1 time only	50 ppm	Lighter than air
Hydrogen sulfide (H_2S)	10 ppm (10 minutes only)	10 ppm—1 time for 10 minutes	100 ppm	Heavier than air
Oxides of nitrogen: (NO_X, NO_2, NO)	NO_2: 3 ppm; NO: 25 ppm	NO_2: 5 ppm; NO: NR	NO_2: 20 ppm; NO: 100 ppm	Heavier than air
Sulfur dioxide (SO_2)	2 ppm	5 ppm	100 ppm	Heavier than air

NR = Not Reported.

A number of technologies are available to detect fire gases, but the most common method involves the use of single-gas or multigas detection devices. (There is more detail on this technology later in the chapter.) When monitoring for fire gases, responders can reference the levels listed in Table 35-3 to determine if a reading is above or below levels set by NIOSH (for the listed set of substances). As an example, the IDLH for CO, likely the most common and prevalent toxic fire gas, is 1200 ppm. Therefore, if responders obtain meter readings that indicate levels at or above 1200 ppm, the environment or situation should be considered immediately dangerous to life and health (IDLH).

The use of the term "immediately dangerous" imparts a sense of high hazard and is usually interpreted as such, meaning that personnel operating in the environment could suffer serious adverse health effects if not properly protected. Airborne contamination at this level is usually understood to be a high-hazard environment. It's a safe estimation to say that during the active phase of combustion and firefighting, and for some period of time after extinguishment, some gas or the level of aggregate particulates will be above IDLH, therefore making the environment at large above IDLH, requiring the use of SCBA. What can be more difficult to navigate are the situations where the level of a specific gas that you may be looking for, again using CO as an example, is below the IDLH. Consider this question: If CO is below the REL, does that mean any or all other gases are below REL? The answer is unknown unless a host of other gases are being monitored in the same place at the same time. CO readings, then, are more of a rough estimation of the total airborne environment.

The REL is the most conservative level to acknowledge and therefore, a safer end point for detection and monitoring, so it will be used for examples herein. Remember, the REL is an average over a 10-hour period.

things in smoke—because if you are monitoring CO on the scene, you are primarily looking for that gas, but there could be others that the sensor will "see." (**TABLE 35-4** lists other gases that would be picked up by a CO sensor.) Therefore, it isn't possible to draw a straight-line correlation between CO and/or HCN, and the presence of any other gas or particulate that might be present at the scene. Detection of a single gas represents a snapshot of that single gas, at that point in time, at that location. It does not mean the entire building is safe or unsafe to operate in without SCBA or that the entire scene or even a specific location is entirely clear.

With that in mind, it is important to understand the limitations of postfire detection and monitoring when a single gas is being targeted. A common misconception is this: If that single gas is found to be below the REL, the entire airborne environment is safe. The shortcomings of this approach should be clear in that CO levels may be below REL, but something else generated by the combustion process, especially the particulates, may not be in a safe range. Therefore, the philosophy of using the REL of a single gas to determine whether a room, portion of a room, or entire building is safe may be subjective. Again, it's better than nothing but likely not definitive. When the entire amount of materials involved in the combustion process are cooled down past their ignition temperature, and therefore no longer thermally decomposing and off-gassing, the environment is likely safe to occupy without respiratory protection. *CO levels may help ascertain that because CO is always present during the combustion process.*

Because one gas may not be an indicator of the level of airborne contamination across the entire scene, the next logical step is to widen the focus and evaluate the fire environment for multiple gases. A common approach is a multigas meter configured with a variety of sensors—commonly CO and hydrogen

LISTEN UP!

To determine the REL on an emergency scene would involve calculations of the dose of the toxic gas over time. Considering the dynamic nature of fire scenes and that time isn't typically available for emergency responders, this isn't a real possibility. The easiest and safest route is to use the REL level itself as the safe point.

For example, the REL for CO is 35 ppm. Any level above 35 ppm is above the regulated baseline value, but any amount detected below 35 ppm is not necessarily safe—it's below the established regulatory limit for a given time frame. What isn't considered in this example are the other substances present—the other nasty

TABLE 35-4 Examples of Other Substances Picked Up by a CO Sensor

Acetylene	Ethylene	Methyl ethyl ketone
Butane	Hydrogen	Nitric oxide
Chlorine	Hydrogen sulfide	Nitrogen dioxide
Ethylene oxide	Isobutylene	Propane
Ethyl alcohol		Sulfur dioxide

sulfide (H_2S) along with oxygen and lower explosive limit (LEL) sensors.

Certainly, looking for more than a single gas is better because it casts a wider detection and monitoring net, but the practice is still limited in terms of evaluating the particulates and gases other than CO and HCN. Choosing to detect these two gases at the scene is beneficial in that it puts an atmospheric monitoring "stake in the ground" by identifying a couple of nasty players in smoke and drives responders to understand the correlation of feeling bad after breathing smoke and the identification of CO and HCN. Again, looking for these two gases is better than looking for only one, but it is still limited in the view it gives about other substances that might be present.

Another method of postfire detection and monitoring is to use more than one technology and look for multiple gases. Although this is interesting and provides useful information, it still does not address airborne contamination in total (mostly regarding the particulates) and should ultimately lead you to the most conservative conclusion—the safest way to work in smoke is to keep a breathing apparatus on! At the end of the day, each agency must determine how to balance the scales of the impact of wearing SCBA for longer periods of time at the scene versus the attempt to determine if and when the postfire environment is safe enough to work without it.

FIGURE 35-8 Electrochemical sensors ready to be installed and used.
Courtesy of Rob Schnepp.

Common Fire Scene Detection and Monitoring Technologies

This section describes a few common technologies that could be used for detection and monitoring at the fire scene. The list is not definitive in that there are many other instruments/technologies available.

Electrochemical Sensors

One of the most common technologies used in postfire detection and monitoring is electrochemical sensors. Broadly speaking, these sensors are filled with a chemical reagent that reacts with a target gas and results in a meter reading (**FIGURE 35-8**).

For example, a carbon monoxide sensor is filled with a jelly-like substance containing sulfuric acid and has two electrical poles within the sensor. When carbon monoxide enters the sensor, either passively or drawn into the device with a pump, there is a chemical reaction that changes the electrical balance in the sensor. The poles detect the change, which results in a reading seen on a screen. If the amount of gas present is above a preset level, an alarm will sound.

Typical electrochemical sensors used in postfire detection and monitoring include oxygen (O_2), hydrogen

cyanide, carbon monoxide, hydrogen sulfide, ammonia, and chlorine. These sensors are also commonly found in single-gas devices. There are a few important factors to understand about electrochemical sensors, specifically those used to detect O_2, CO, H_2S, and HCN:

- Toxic sensors react to other gases, so you cannot be sure whether you are reading a level of the intended gas or an interfering gas. For example, if you are reading levels of both CO and H_2S there is probably an acidic gas present that is causing the reaction and resulting in the readings displayed. Most of the interfering gases are also toxic, so when readings are found on the detection devices fire fighters should be aware that there are toxic gases present. Some of these gases are also flammable, which means you may also get a reading in your flammable gas sensor. There are ways to determine which gases are present through the use of detection devices your hazardous materials team carries. Nonetheless, the sensor is alerting to the presence of a gas.
- Electrochemical sensors can be easily overwhelmed and will max out with regard to their

readings. For example, most HCN sensors will max out at 50 ppm (IDLH) and will not tell you if you are in levels higher than the maximum. This is largely irrelevant; however, once the readings pass the REL for HCN (4.7 ppm), SCBA should be worn. The same holds true for CO sensors. They usually have a maximum reading of 500 ppm. Remember that CO has an IDLH value of 1200 ppm. High exposures to a gas that causes a reaction will result in the sensor failing sooner than its intended life.

- Electrochemical sensors fail to the 0 (zero) point. With O_2 that's not necessarily a problem because 0 percent oxygen is dangerous, and you would not enter a potentially hazardous environment. But when a CO or HCN sensor fails and reads 0 it may not indicate it has failed, and that can lead responders to believe a toxic gas is not present—a dangerous situation. Proper care and maintenance will help prevent this from occurring.

- Responders should ensure their devices are calibrated according to manufacturers' recommendations and should at least be bump tested prior to use. Some instrument manufacturers recommend that the devices be calibrated before each use, which is not practical in an emergency response environment, but they should be bump tested at a minimum.

- Responders must be mindful of the reaction time of electrochemical sensors because they can take as few as 20 seconds or as many as 200 seconds to react based on the type of sensor.

Photoionization (PID) Sensor

Another type of sensor is a **photoionization detector (PID)**. A PID is available as a stand-alone unit or may be incorporated into a multigas meter (**FIGURE 35-9**). In either case, the PID uses ultraviolet light to ionize the gases that move through the sensor. When gases are ionized, sensors are able to determine the ionization of the molecules. The PID generally detects common materials such as benzene, acetone, toluene, ammonia, ethanol, butane, and many others. It is primarily a sensor that detects organic materials (materials that contain carbon as part of the molecular structure). Ammonia is the most common inorganic chemical detected by a PID. The PID does not identify the material that is present; it only alerts the user of its presence. A PID in a smoke situation will detect parts of the toxic soup that is in the air and alert you that something is present, but it will not identify which toxic gas is in the air. An important thing to remember is that PIDs won't detect the presence of CO or HCN or compounds such as natural gas.

Responders should be aware of differences in the reaction time of the various detection devices: Electrochemical sensors react within 20–200 seconds, and the PID reacts within 1–2 seconds.

Another consideration with any electronic detection device is the dirty nature of the fire environment. There is a lot of particulate matter in the air, and many detection devices have an internal pump that is drawing the gas into the device for analysis, so any particulates in the air will be drawn in as well. To combat this problem, most devices have a protective filter, which keeps out particulate material and will offer some limited protection against liquids (**FIGURE 35-10**).

FIGURE 35-9 A photoionization detector (PID) in service during a session of fire investigation research.
Courtesy of Rob Schnepp.

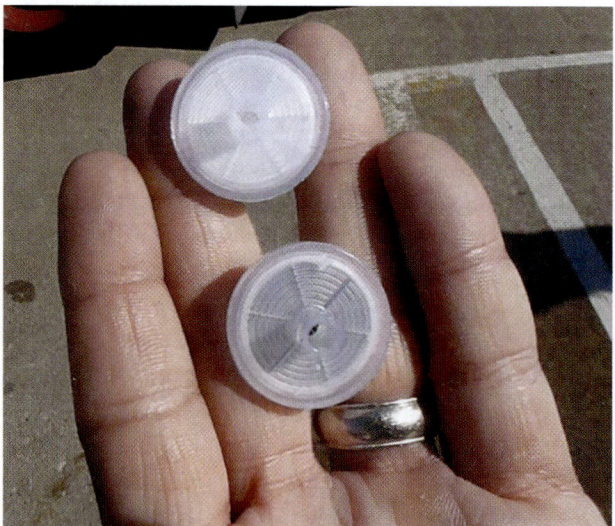

FIGURE 35-10 The filter on the top is clean; the one on the bottom was used during live burns at a training session. The bottom filter was so contaminated the device could not be fresh-air calibrated.
Courtesy of Rob Schnepp.

The devices are typically water resistant, not waterproof, so if there is any chance water can be drawn into the device, it should be shut off as quickly as possible. New filters will be required regularly when the device is used at a fire scene.

FIGURE 35-11 Colorimetric sampling. Keep in mind that colorimetric tubes are a point source sampling tool. This means they only indicate the presence of a material right at the intake of the tube and are not intended to evaluate an area.
Courtesy of Rob Schnepp.

Colorimetric Tubes

Another option for the detection of toxic gases is colorimetric tubes, which can be used to detect specific substances and/or to confirm the readings of the electrochemical sensors or other technologies. The colorimetric tubes are designed to identify the presence and/or levels of a known gas or vapour but can also be used to help determine which unidentified airborne substances may be present. The tubes are set up to detect chemical families, but they can also be used to detect a specific chemical. For example, the ethyl acetate tube is designed to detect ethyl acetate, but if the test is positive it could mean that there is an organic material in the air. The acetone tube detects acetone as well as all the other aromatic hydrocarbon vapours.

The process for colorimetric sampling requires that a certain amount of sample air moves through the tube. To accomplish this, a responder can use a piston-style pump or a bellows pump (**FIGURE 35-11**). In either case, most tubes require a certain number of pump strokes. To accomplish this, a responder usually remains in one location and completes the evaluation of the location before moving on. If a particular tube requires 10–15 pump strokes, it could be quite time consuming to evaluate a single room. If several rooms are to be evaluated during overhaul, it could become tedious and time consuming.

Colorimetric tubes are easy to use and do not require much preparation (other than reading the instructions on each tube) or any calibration prior to use. The tubes are one-time use only, thus creating a cost issue as well as a supply issue. The pump, however, can be reused a number of times. There are standard hazardous materials detection kits that use colorimetric tubes and can help responders detect the range of toxic fire gases. These kits are a bit more complicated and usually are carried by hazardous materials response teams rather than pump, ladder, or rescue companies. They do require additional training and experience to effectively and properly use them.

Fire Scene Detection and Monitoring Practices

The prior section outlined some common technologies used for detecting toxic gases at the fire scene. This section offers some general thoughts on deploying the technology. Regardless of what instrument is selected, hazardous materials responders should develop a defined and systematic strategy for the use of detection devices. There are no hard and fast rules for postfire detection and monitoring, so the AHJ should develop some broad operational guidance. To begin, the use of detection devices during active interior structural firefighting is not necessary. High heat, massive amounts of particulates, and steam or water can be immediately or cumulatively detrimental to the instruments. Additionally, it should be assumed that the environment is IDLH, and firefighting personal protective equipment (PPE), including SCBA, must be worn. Most of the instruments used in fire scene detection can be damaged in temperatures above 37.8°C (100°F). Air monitoring conducted during active firefighting should be limited to exterior operations, such as evaluating downwind exposures, whether an RIC is staged in a safe location, at the command post, and so on—in other words, along the periphery of the fire.

Most commonly, fire scene detection takes place when the fire is declared under control and the operation transitions to overhaul and mop-up. In this setting, typically, detection and monitoring are done in order to determine when personnel may remove their face masks. The discussion earlier in the chapter should serve as a caution against early removal.

Detection and monitoring should start with a general exterior evaluation of the footprint of the fire, working inward toward the areas where crews are operating. It is critical to constantly move through the building or other areas of the fire scene. Fire gas production stops only when all substances involved in the fire are cooled below the point they decompose and off-gas. When many materials are involved, it's nearly impossible to target one culprit as a toxic gas generator. One room or a portion of a room could be producing gases at different times during the overhaul process. A room could be clear, right up until the time a pile of debris is turned over and begins to actively smoulder. There may also be rooms with trapped or unventilated smoke and gases. A good rule of thumb is to initiate interior monitoring once all visible particulate has been ventilated from the structure.

FIGURE 35-12 The best strategy for the protection of the responders is to have air monitoring occur at any location where personnel are operating. Monitoring should be done throughout the building on a continual basis to make sure no toxic hot spots are present.
Courtesy of Rob Schnepp.

> ### LISTEN UP!
>
> Detection and monitoring at the fire scene should be a highly mobile and continuous process.

In particular, fire investigators should evaluate the environment prior to beginning work, because they (and/or canines used for postfire investigation) are even more intimately involved with the smouldering nature of the postfire phase of the incident. It might also make sense to place a detection device near rehab areas or staging areas and near command posts adjacent to the fire building (**FIGURE 35-12**).

Consider also that ventilation practices, or the lack thereof, can change and influence the atmosphere inside buildings. The use of gasoline-powered positive pressure ventilation (PPV) fans may increase airflow throughout the building but could also create a downside—the production of CO. To that end, responders must exercise caution when using gasoline-powered PPV fans. An older fan, or one that is not operating well, can introduce CO levels above the REL in areas that previously had low or no levels of CO. To that end, many agencies are using electric or battery-operated fans for smoke removal and postfire ventilation.

In general, detection at the fire scene is not an exact science due to the limitations of the instrumentation, the dynamic nature of the fire scene, and the narrow view of using only one or two gases as indicators of the airborne environment. Again, a properly functioning instrument, regardless of the technology, only provides you with information. It's up to the humans on the scene to interpret those readings, put them in the context of the overall scene, and make some decisions. When it comes to operating in smoke, and reducing dermal and inhalation exposures, choose the course of action that offers your personnel the highest level of protection.

Certainly, cumulative smoke exposures over the course of a career can be equally dangerous and deadly. Technology is improving so that we are better able to detect the hazards at fires, but it isn't perfect. Most of the toxic gases are odourless and invisible, so even if you inhale a small amount of particulate matter in the air—the part we traditionally called smoke—it does not mean there isn't an invisible toxic soup waiting for you. Wearing your SCBA for longer periods of time at the fire scene will result in reduced exposures. This, in addition to other actions such as being mindful of keeping your structural firefighting protective gear clean, showering after working at a fire, and getting regular health evaluations and cancer screenings, will serve you well in the long run.

After-Action REVIEW

IN SUMMARY

- When asked about fire fatalities, fire fighters typically observe that smoke kills people before the flames ever get to them, and fire death statistics prove likewise.

- Most people identify carbon monoxide as the main harmful component of fire smoke. Less often acknowledged are compounds such as ammonia, hydrogen chloride, sulfur dioxide, hydrogen sulfide, hydrogen cyanide, carbon dioxide, the oxides of nitrogen, formaldehyde, acrolein, polycyclic aromatic hydrocarbons, and soot.

- Studies performed in Paris, France, and Dallas County, Texas, focused on carbon monoxide (CO) and hydrogen cyanide (HCN) specifically because they are acutely toxic, present to some degree in nearly all fires, and have clinical interventions available to reverse the adverse health effects of the exposure.

- Nationwide, there is no standard protocol for treating smoke inhalation, leaving paramedics and other prehospital care providers with limited guidance and/or training to properly care for smoke inhalation victims.

- Many fire agencies are investing in technologies for detecting toxic gases at the fire scene without a clear understanding of the mission, the limitations of the devices, or what it means to check to see if the building is clear and, more commonly, when it is safe to remove SCBA.

- Federal-Provincial-Territorial Ministers of Labor and Part X of the Canada Occupational Health and Safety Regulations (COHSR) and the National Institute for Occupational Safety and Health (NIOSH) establish safe levels for chemical exposures in the workplace; however, the values are for average worker exposure and can be only *estimated* for the rigours of firefighting.

- Equipment such as electrochemical sensors, photoionization sensors, and colorimetric tubes can all be used to monitor conditions at a fire scene.

- Regardless of what instrument is selected, hazardous materials responders should develop a defined and systematic strategy for the use of detection devices.

KEY TERMS

Access Navigate for flashcards to test your key term knowledge.

Aerobic metabolism The creation of energy through the breakdown of nutrients in the presence of oxygen. The by-products are carbon dioxide and water, which the body disposes of by breathing and sweating.

Anaerobic metabolism The creation of energy through the breakdown of glucose. Without oxygen, this metabolic process results in the production of lactic acid.

Colorimetric tubes Reagent-filled tubes designed to draw in a sample of air by way of a manual handheld pump. The reagent will undergo a colour change when exposed to the contaminant it is intended to detect.

Cytochrome oxidase Found in the mitochondria, this is important in cell respiration as an agent of electron transfer from certain cytochrome molecules to oxygen molecules.

Hypoxia A state of inadequate oxygenation of the blood and tissue sufficient to cause impairment of function. (NFPA 99)

Mitochondria Responsible for converting nutrients into energy, yielding molecules of adenosine triphosphate (ATP) to fuel the cell's activities.

Photoionization detector (PID) A sensor that uses ultraviolet light to ionize the gases that move through the sensor; available as a stand-alone unit or may be incorporated into a multi-gas meter.

Polycyclic aromatic hydrocarbons (PAHs) A group of substances that occur naturally in materials such as coal or crude oil. These substances also are generated during the combustion of organic materials and can be found in vehicle exhaust, tobacco smoke, and the smoke generated from structure fires, vehicle fires, wildland fires, or any other type of fire. PAH can exist as a particle or gas.

Pyrolysis A process in which material is decomposed, or broken down, into simpler molecular compounds by the effects of heat alone; pyrolysis often precedes combustion. (NFPA 921)

Red blood cells Oxygen-carrying cells found in mammals. They contain hemoglobin.

Smoke The airborne solid and liquid particulates and gases evolved when a material undergoes pyrolysis or combustion, together with the quantity of air that is entrained or otherwise mixed into the mass. (NFPA 1404)

On Scene

You respond to a working fire with a report of victims trapped. Upon arrival, the homeowner states that all occupants are out of the building. You are directed to a male victim sitting on the ground next to a parked car. The victim is alert and responding to your questions but appears to be disoriented. You suspect the victim is suffering from smoke inhalation.

1. What is the definitive way to confirm cyanide poisoning in the field?

A. Confirm a blood oxygenation level of less than 92%.

B. Use a transcutaneous or in-line cyanide oximeter.

C. Confirm carbon monoxide poisoning—where there's carbon monoxide poisoning, there's cyanide poisoning.

D. There is no detection method for confirming cyanide poisoning in the field.

2. Which of the following organs quickly suffer from the oxygen-deprivation effects of cyanide poisoning?

A. Lungs and kidneys

B. Lungs and heart

C. Heart and brain

D. Heart and kidneys

3. In the face of smoke inhalation, which of the following is an acceptable antidote to administer if you suspect HCN poisoning?

A. Sodium hydroxide

B. Lithium nitrite

C. Methylene blue

D. Hydroxocobalamin

Access Navigate to find answers to this On Scene, along with other resources such as an audiobook.

Hazardous Materials Operations Level: Mission Specific

Hazardous Materials Responder Personal Protective Equipment

KNOWLEDGE OBJECTIVES

After studying this chapter, you will be able to:

- Discuss the similarities and differences in how single-use and reusable personal protective equipment (PPE) is used. (**NFPA 1072: 6.2.1; NFPA 472: 6.2.1.2.1, 6.2.4.1,** pp. 1210–1211)
- Explain how to maintain PPE. (**NFPA 1072: 6.2.1; NFPA 472: 6.2.1.2.1, 6.2.3.1, 6.2.4.1,** p. 1211)
- Explain how PPE needs are determined. (**NFPA 1072: 6.2.1; NFPA 472: 6.2.1.2.1, 6.2.3.1, 6.2.4.1,** pp. 1211–1212)
- Identify and describe specific PPE for hazardous materials response. (**NFPA 1072: 6.2.1; NFPA 472: 6.2.1.2.1, 6.2.3.1, 6.2.4.1,** pp. 1212–1231)
- Explain the safety considerations when wearing PPE. (**NFPA 1072: 6.2.1; NFPA 472: 6.2.1.2.1, 6.2.3.1, 6.2.4.1,** pp. 1231–1235)
- Describe the process of going through emergency decontamination while wearing PPE. (**NFPA 1072: 6.2.1; NFPA 472: 6.2.1.2.1, 6.2.4.1,** pp. 1235–1236)
- Describe the process of going through technical decontamination while wearing PPE. (**NFPA 1072: 6.2.1; NFPA 472: 6.2.1.2.1, 6.2.4.1,** pp. 1235–1236, 1237–1238)

- Explain the inclusion of PPE in reporting and documenting the incident. (**NFPA 1072: 6.2.1; NFPA 472: 6.2.1.2.1, 6.2.4.1, 6.2.6.1,** p. 1236)

SKILLS OBJECTIVES

After studying this chapter, you will be able to perform the following skills:

- Don a Level A ensemble. (**NFPA 1072: 6.2.1; NFPA 472: 6.2.1.2.1, 6.2.4.1,** pp. 1217–1219)
- Doff a Level A ensemble. (**NFPA 1072: 6.2.1; NFPA 472: 6.2.1.2.1, 6.2.4.1,** pp. 1219–1220)
- Don a Level B nonencapsulating chemical-protective clothing ensemble. (**NFPA 1072: 6.2.1; NFPA 472: 6.2.1.2.1, 6.2.4.1,** pp. 1222–1223)
- Doff a Level B nonencapsulating chemical-protective clothing ensemble. (**NFPA 1072: 6.2.1; NFPA 472: 6.2.1.2.1, 6.2.4.1,** pp. 1224–1225)
- Don a Level C chemical-protective clothing ensemble. (**NFPA 1072: 6.2.1; NFPA 472: 6.2.1.2.1, 6.2.4.1,** pp. 1226–1227)
- Doff a Level C chemical-protective clothing ensemble. (**NFPA 1072: 6.2.1; NFPA 472: 6.2.1.2.1, 6.2.4.1,** p. 1228)
- Don a Level D chemical-protective clothing ensemble. (**NFPA 1072: 6.2.1; NFPA 472: 6.2.1.2.1, 6.2.4.1,** p. 1229)

Hazardous Materials Alarm

Your pump company arrives on the scene of a motor vehicle collision involving a small passenger vehicle and a tanker truck carrying 20 tons of anhydrous ammonia. The tanker has rolled onto its side and has slid down the highway for approximately 30.5 m (100 ft). There are no injuries to the three victims in the passenger vehicle, but the driver of the tanker truck is pinned inside the cab. You smell ammonia in the air but see no visible signs of a product release. The regional fire hazardous materials team, fully staffed with technician level responders, is also on scene. Your company officer confers with the hazardous materials team and then directs you and another fire fighter to don your SCBA and full turnout gear and evaluate the driver for injuries.

1. Would full structural fire fighters' protective gear and SCBA offer adequate protection in this situation?

2. Based on your level of hazardous materials training—operations level—would you be qualified to perform this task?

3. Describe the steps you would need to take, including obtaining information about ammonia, to complete your assignment.

Access Navigate for more practice activities.

Introduction

This chapter addresses job performance requirements (JPRs) found in NFPA 1072, *Standard for Hazardous Materials/Weapons of Mass Destruction Emergency Response Personnel Professional Qualifications*, and competencies found in NFPA 472, *Standard for Competence of Responders to Hazardous Materials/Weapons of Mass Destruction Incidents*, for operations level responders assigned mission-specific responsibilities at hazardous materials/weapons of mass destruction (WMD) incidents by the authority having jurisdiction (AHJ) beyond the core responsibilities at the operations level. The operations level responder assigned to use personal protective equipment (PPE) must be trained to meet the operations level and all responsibilities in the Personal Protective Equipment section (Section 6.2 in both NFPA 1072 and 472). PPE at this level refers to chemical-protective equipment. This equipment is used in situations where contact with a hazardous materials/WMD is possible or expected. Standard firefighting PPE may not provide complete protection from the hazards that could be encountered during hazardous materials incidents.

Emergency responders should be familiar with the policies and procedures of the AHJ to ensure a consistent approach to selecting the proper PPE for an expected task or set of tasks. Additionally, all responders charged with responding to hazardous materials/WMD incidents should be proficient with local procedures for technical decontamination as well as the manufacturers'

guidelines for maintenance, testing, inspection, storage, and documentation procedures for the PPE provided by the AHJ. Refer to Chapter 33, *Estimating Potential Harm and Planning a Response*, for the specifics of the National Fire Protection Association (NFPA) standards on protective clothing.

Single-Use Versus Reusable PPE

Much of the chemical-protective equipment on the market today is intended for a single use (i.e., it is disposable) and is usually discarded along with the other hazardous waste generated by the incident. As a consequence, single-use PPE is decontaminated to the point that it is safe for the responder to remove but not so extensively that the garment is completely free of contamination. Single-use PPE is generally less expensive than reusable gear, but it needs to be restocked and/or replenished after the incident. Before the use of any PPE, it should undergo a thorough visual inspection to ensure that piece of equipment is absolutely response ready.

Reusable garments are required to be tested at regular intervals and after each use. Level A suits, for example, are required to be pressure tested—usually upon receipt from the manufacturer, after each use, and annually (**FIGURE 36-1**). Individual manufacturers have well-defined procedures for this activity. Prior to purchasing any type of PPE, the AHJ should understand the maintenance and upkeep requirements. PPE is not

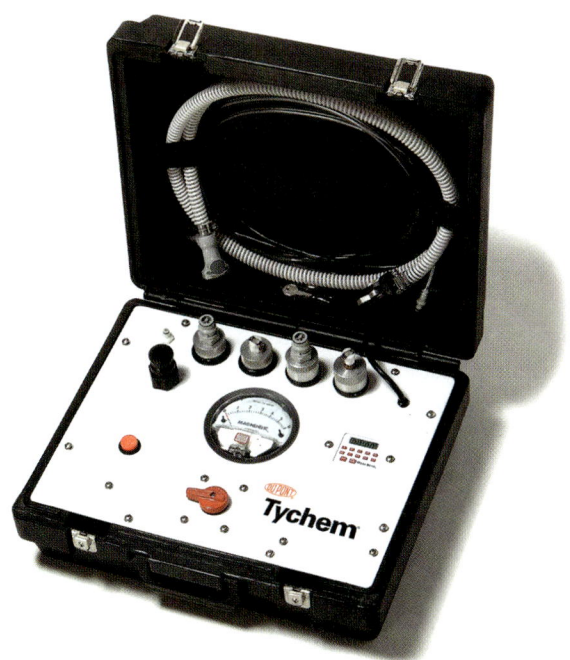

FIGURE 36-1 Level A suit testing kit.
Courtesy of the DuPont Company.

intended to be purchased and completely left alone until it is needed. There must be a level of care and attention devoted to the barrier between you and the released substance. Also, single-use and reusable garments have a shelf life. This time frame should be noted and adhered to. Don't use PPE that is beyond its shelf life—you may not be able to depend on it!

Maintaining PPE

Chemical-protective garments are tested in accordance with the manufacturer's recommendations. Generally speaking, a pressure test is accomplished by using a specially designed kit to pump a certain amount of air into the suit and leaving it pressurized for a specified period of time. At that point, if the garment has lost more than a certain percentage (usually 20 percent) of pressure, it is assumed the suit has a leak. Often, leaks are located by gently spraying or brushing the inflated suit with a solution of soapy water. Small bubbles begin to form in the area of the leak, alerting you to its presence and location. Any garment with a leak should be removed from service until the defect is identified and repaired in accordance with the manufacturer's specifications.

Chemical-protective equipment should be stored in a cool, dry place that is not subject to significant temperature extremes and/or high levels of humidity. Furthermore, the equipment should be kept in a clean location, away from direct sunlight, and should be inspected at regular intervals based on the manufacturer's recommendations. If repairs are required, consult the

manufacturer prior to performing any work—there is a risk that the garment will not perform as expected if it has been modified or repaired incorrectly. Again, individual manufacturers have well-defined procedures for this activity; prior to purchasing any PPE, the AHJ should understand what is required in terms of maintenance and upkeep.

Determining PPE Needs

As the title of this chapter implies, personnel must correlate the mission they are expected to perform with the anticipated hazards. For example, in the ammonia scenario described earlier, the responders should understand that ammonia exposure presents a significant health hazard because it is corrosive to the skin, eyes, and lungs. Ammonia is flammable at concentrations of approximately 15 to 25 percent (by volume) in a mixture with air. Exposure to a concentration of approximately 300 parts per million (ppm) is considered to be immediately dangerous to life and health (IDLH). If the possibility of exposure to a concentration exceeding 300 ppm exists, a **National Institute for Occupational Safety and Health (NIOSH)**-approved **self-contained breathing apparatus (SCBA)** is required. NIOSH sets the design, testing, and certification requirements for SCBA in the United States. Although NIOSH is a U.S. Standard, it is also a referred-to standard in Canada along with NFPA 1852, *Standard on Selection, Care, and Maintenance of Open-Circuit Self-Contained Breathing Apparatus (SCBA)* and Transport Canada for cylinders. Your AHJ will determine the PPE available for use and also the procedures and requirements for selecting and using PPE on the incident scene as part of the incident action plan.

SAFETY TIP

TRACEMP is a common acronym used to sum up a collection of potential types of harms an emergency responder may face.

- **T**hermal
- **R**adiological
- **A**sphyxiating
- **C**hemical
- **E**tiological/biological
- **M**echanical
- **P**sychogenic

Although SCBA will protect the responders from suffering an inhalation exposure, its use is only one piece of the PPE equation. Another question must be answered in this scenario: Will structural fire fighters' protective gear provide sufficient skin protection? Knowing that ammonia presents a flammability hazard is important—and the protective gear would address that potential—but that choice of equipment still does not address the hazard of skin irritation. The release is outside—would you make a different choice if the release were indoors, inside a poorly ventilated room? Responders should always consider the impact of the operating environment as part of the hazard evaluation.

Fire personnel tasked with undertaking a medical reconnaissance mission in the ammonia scenario should balance these hazards and the risk of the mission with the potential gain. This scenario is based on an outdoor release, with unknown variables of wind speed, direction, and ambient air temperature, which may positively or negatively influence the decision to approach the cab of the truck. Unfortunately, it is impossible to decide on the right course of action based on a few sentences in this text—you must make that decision on the street, at the moment the emergency occurs. Not all structural fire fighters' gear is created equal, and not all types are intended to function in an environment that may contain a hazardous material. Structural fire fighters' gear that is 10 years old and well worn will certainly not provide the same level of protection as a new set of gear that meets the certification requirements for the latest edition of NFPA 1971, *Standard on Protective Ensembles for Structural Fire Fighting and Proximity Fire Fighting*.

To that end, it is incumbent on every emergency responder to understand the hazards that may be present on an emergency scene and to appreciate how those hazards may affect the PPE requirements and the mission the responders are tasked with carrying out.

Specific PPE for Hazardous Materials Response

Different levels of PPE may be required at different hazardous materials incidents. This section reviews the protective qualities of various ensembles, from those offering the least protection to those providing the greatest protection.

At the lowest end of the spectrum are street clothing and normal work uniforms, which offer the least amount of protection in a hazardous materials emergency. Normal clothing (or flame-resistant coveralls) may prevent a noncaustic powder from coming into direct contact with the skin, for example, but it offers no significant protection against many other hazardous materials. Such clothing is often used in industrial applications, such as oil refineries, rail yards, or city public works

facilities as a general work uniform (**FIGURE 36-2**). Police officers and emergency medical services (EMS) providers typically wear this level of "protection." Most often, distance from the hazard is the best level of protection with this PPE.

The next higher level of protection is provided by structural firefighting protective equipment (**FIGURE 36-3**). Such an ensemble includes a helmet, a structural firefighting protective coat and pants, boots, gloves, a protective hood, SCBA, and a personal alert safety system (PASS) device. Standard firefighting protective gear is not considered "chemical protection," because the fabric may break down when exposed to chemicals and may not provide complete protection from the harmful gases, vapours, liquids, and dusts that could be encountered during hazardous materials incidents.

Returning to the ammonia scenario, it may be safe and reasonable to carry out the patient assessment mission wearing this level of protection—again, based on a full risk assessment. Keep in mind that structural firefighting gear is primarily intended to protect the wearer from thermal hazards (predominantly encountered during firefighting) and mechanical hazards such as broken glass or other sharp objects. The same gear may be called upon for other reasons, such as for protecting the wearer against alpha and beta radiation sources, but that is not its primary function.

Fire fighters wearing **high temperature–protective clothing** may best address unusually high thermal hazards, such as those posed by aircraft fires. This type of PPE shields the wearer during short-term exposures to high temperatures (**FIGURE 36-4**). Sometimes referred to

FIGURE 36-2 A Nomex jumpsuit.
Courtesy of the DuPont Company.

FIGURE 36-3 Standard structural firefighting PPE.
© Jones & Bartlett Learning. Photographed by Glen E. Ellman.

FIGURE 36-4 High temperature–protective equipment protects the wearer from high temperatures during a short-term exposure.
© Photodisc.

as a proximity suit, high temperature–protective equipment allows the properly trained fire fighter to work in extreme fire conditions. It provides protection against high temperatures only, however; it is not designed to protect the fire fighter from hazardous materials.

Chemical-Protective Clothing and Equipment

Chemical-protective clothing is unique in that it is designed to prevent chemicals from coming in contact with the body. Such equipment is not intended to provide high levels of protection from prolonged exposure to thermal hazards (heat and cold) or to protect the wearer from injuries that may result from torn fabric, chemical damage, or other mechanical damage (tears and abrasion) to the suit. Not all chemical-protective clothing is the same, and each type, brand, and style may offer varying degrees of protection and chemical resistance.

To help you safely estimate the chemical resistance of a particular garment, manufacturers supply compatibility charts with all of their protective equipment (**FIGURE 36-5**). These charts are designed to assist you in choosing the right chemical-protective clothing for the incident at hand. You must match the anticipated chemical hazard to these charts to determine the resistance characteristics of the garment.

Storage conditions, temperature, and resistance to cuts, tears, and abrasions are all factors that affect the

FIGURE 36-5 An example of a compatibility chart.
© Jones & Bartlett Learning.

chemical resistance of materials. Other factors include flexibility, shelf life, and sizing criteria. The bottom line is that **chemical-resistant materials** are specifically designed to inhibit or resist the passage of chemicals into and through the material by the processes of penetration, permeation, or degradation.

Penetration is the flow or movement of a hazardous chemical through closures such as zippers, seams, porous materials, pinholes, or other imperfections in the material. To reduce the threat of a penetration-related suit failure, responders should carefully evaluate their PPE prior to entering a contaminated atmosphere. Checking the garment fully—that is, performing a visual inspection of all its components—before donning the PPE is paramount. Just because the suit or gloves came out of a sealed package, it does not mean they are perfect! Also, a lack of attention to detail could result in zippers not being fully closed and tight. Poorly fit seams around ankles and wrists could allow chemicals to defeat the integrity of the garment. Use of the buddy system is beneficial in this setting because it creates the opportunity to have a trained set of eyes examine parts of the suit that the wearer cannot see. Prior to entering a contaminated atmosphere, or periodically while working, each member of the entry team should quickly scan the PPE of the other member(s) to see if anyone's suit has suffered any damage, discolouration, or other insult that may jeopardize the health and safety of the wearer.

Permeation is the process by which a hazardous chemical moves through a given material on the molecular level. It differs from penetration in that permeation occurs through the material itself rather than through openings in the material. Permeation may be impossible to identify visually, but it is important to note the initial status of the garment and to determine whether any changes have occurred (or are occurring) during the course of an incident. Chemical compatibility charts are based on two properties of the material: its breakthrough time (how long it takes a chemical substance to be absorbed into the suit fabric and detected on the other side) and the permeation rate (how much of the chemical substance makes it through the material) (**FIGURE 36-6**). The concept is similar to water saturating a sponge. Over time, a continuous drip of water will "fill up" the sponge and begin to seep through to the other side. When evaluating the effectiveness of a particular material against a given substance, responders should look for the longest breakthrough time available. For example, a good breakthrough time would be more than 480 minutes—a typical 8-hour workday.

Degradation is the physical destruction or decomposition of a clothing material owing to chemical exposure, general use, or ambient conditions (e.g., storage in sunlight). It may be evidenced by visible signs

Chemical Name	Concentration	Breakthrough Time	Permeation Rate
	(%)	Normalized (min)	(ug/cm2/min)
1,1,2,2-TETRACHLOROETHANE	95+	>480	0.0005
1,1,2-TRICHLOROETHANE	95+	>480	<0.01
1,3-DICHLOROACETONE (40°C)	95+	>480	<0.1
1,4-DIOXANE	95+	>480	<0.05
1,6-HEXAMETHYLENEDIAMINE	95+	>480	<0.01
2,2,2-TRICHLOROETHANOL	95+	>480	<0.01
2,2,2-TRIFLUOROETHANOL	95+	>480	<0.001
2,3-DICHLOROPROPENE	95+	>480	<0.08
2-CHLOROETHANOL	95+	>480	<0.008
2-METHYLGLUTARONITRILE	87	>480	<0.1
2-PICOLINE	95+	46	48
3,4-DICHLOROANILINE	95+	284	2.4
3-PICOLINE	95+	11	22
4,4'METHYLENE BIS(2-CHLOROANILINE)	95+	>480	<0.1
ACETALDEHYDE	95+	>480	<0.01
ACETIC ACID	95+	339	1.3
ACETIC ANHYDRIDE	95+	>480	<0.001
ACETONE	95+	>480	<0.001
ACETONITRILE	95+	>480	<0.01
ACETYL CHLORIDE	95+	181	2
ACROLEIN	95+	>480	<0.02
ACRYLAMIDE	50% in water	>480	<0.1
ACRYLIC ACID	95+	270	1.6
ACRYLONITRILE	95+	>480	<0.0003
ADIPONITRILE	95+	>480	<0.1
ALLYL ALCOHOL	95+	>480	<0.1
ALLYL CHLORIDE	95+	>480	<0.06
AMMONIA GAS	95+	46	0.62
AMMONIUM FLUORIDE	40	>480	<0.01
AMMONIUM HYDROXIDE	28-30	160	4.7
AMYL ACETATE	95+	>480	<0.003
ANILINE	95+	>480	<0.1
ARSINE	95+	>480	<0.01
BENZENE SULFONYL CHLORIDE	95+	>480	<0.1
BENZIDINE	25% in methanol	>480	<0.01
BENZONITRILE	95+	>480	<0.004
BENZONITRILE	95+	>480	<0.004
BENZOYL CHLORIDE	95+	>480	<0.05
BENZYL CHLORIDE	95+	>480	<0.01
BORON TRICHLORIDE	95+	>480	<0.02
BORON TRIFLUORIDE	95+	>480	<0.1

FIGURE 36-6 Breakthrough time is the time it takes a chemical substance to be absorbed into the suit fabric and detected on the other side.
Courtesy of the DuPont Company.

such as charring, shrinking, swelling, colour changes, or dissolving. Materials can also be tested for weight changes, loss of fabric tensile strength, and other properties to measure degradation. Think about the rapid and destructive way in which gasoline dissolves a Styrofoam cup. When chemicals are so aggressive, or when the suit fabric is a poor match for the suspect substance, fabric degradation is possible. If the suit dissolves, the possibility of the wearer's suffering an injury is high.

Types of Chemical-Protective Clothing

Chemical-protective clothing can be constructed as a single- or multi-piece garment. A single-piece garment may or may not completely enclose the wearer and is often found as a coverall-type garment. A multi-piece garment typically has a jacket, pants, an attached or detachable hood, and perhaps attached fabric to cover the feet. Multi-piece garments are found as Level B and Level C protection. (Level A protection is almost always built as an encapsulated one-piece suit with attached gloves and suit fabric that covers the feet.) Chemical-resistant boots should be worn to offer protection from abrasion and mechanical hazards. Chemical-protective equipment suited for law enforcement missions is becoming more popular and finding its way into traditional hazardous materials response. These protective ensembles typically offer protection against liquid and particulate forms of CBRN (chemical, biological, radiological, and nuclear) agents and are much cooler and more comfortable to wear for extended periods of time.

Chemical-protective clothing—both single- and multi-piece—is classified into two major categories: vapour-protective clothing and liquid splash–protective clothing. Both are described in this section. Many different types of materials are manufactured for both categories; it is the AHJ's responsibility to determine which type of suit is appropriate for each situation. Be aware that no single chemical-protective garment (vapour or splash) on the market will protect you from everything.

A **vapour-protective ensemble**, also referred to as *fully encapsulating protective clothing*, offers full body protection from highly toxic environments and requires the wearer to use an air-supplied respiratory device such as SCBA (**FIGURE 36-7**). The wearer is completely zipped inside the protective "envelope,"

leaving no skin (or the lungs) accessible to the outside. If the ammonia scenario described at the beginning of the chapter were occurring in a different location—such as inside a poorly ventilated storage area within an ice-making facility—vapour-protective clothing might be required. Ammonia aggressively attacks skin, eyes, and mucous membranes such as in the eyes and mouth and can cause severe and irreparable damage to the lungs. Hydrogen cyanide would be another example of a chemical substance that would require this level of protection. Hydrogen cyanide can be fatal if inhaled or absorbed through the skin, so the use of a fully encapsulating suit is required to adequately protect the wearer. NFPA 1991, *Standard on Vapour-Protective Ensembles for Hazardous Materials Emergencies and CBRN Terrorism Incidents*, sets the performance standards for vapour-protective garments.

A **liquid splash–protective ensemble** is designed to protect the wearer from chemical splashes (**FIGURE 36-8**). NFPA 1992, *Standard on Liquid Splash-Protective Ensembles and Clothing for Hazardous Materials Emergencies*, is the performance document that governs liquid splash–protective garments and ensembles. Equipment that meets this standard has been tested for penetration against a battery of five chemicals. The tests include no gases, because this level of protection is not considered to be vapour protection.

FIGURE 36-7 Vapour-protective clothing retains body heat, so it also increases the possibility of heat-related emergencies among responders.
© Jones & Bartlett Learning. Photographed by Glen E. Ellman.

FIGURE 36-8 Liquid splash–protective clothing is worn whenever there is the danger of chemical splashes.
© Jones & Bartlett Learning. Photographed by Glen E. Ellman.

Responders may choose to wear liquid splash–protective clothing based on the anticipated hazard posed by a particular substance. Liquid splash–protective clothing does not provide total body protection from gases or vapours, and it should not be used for incidents involving liquids that emit vapours known to affect or be absorbed through the skin. This level of protection may consist of several pieces of clothing and equipment designed to protect the skin and eyes from chemical splashes. Some agencies, depending on the situation, choose to have their personnel wear liquid splash protection over or under structural firefighting clothing.

Hazardous materials fire personnel trained to the operations level often wear liquid splash–protective clothing when they are assigned to enter the initial site, perform decontamination, or construct isolation barriers such as dikes, diversions, retention areas, or dams.

Chemical-Protective Clothing Ratings

A variety of fabrics are used in both vapour-protective and liquid splash–protective garments and ensembles. Commonly used suit fabrics include butyl rubber, Tyvek, Saranex, polyvinyl chloride (PVC), and Viton. Protective clothing materials must offer acceptable resistance to the chemical substances involved, and the garments should be used within the parameters set by their manufacturer. The manufacturer's guidelines and recommendations should be consulted for material compatibility information.

The following guidelines may be used by a responder to assist in determining the appropriate level of protection for a particular hazard. The procedures for the donning and doffing of equipment are described in the following sections.

Level A

A **Level A ensemble** consists of a fully encapsulating garment that completely envelops both the wearer and his or her respiratory protection (**FIGURE 36-9**). Level A equipment should be used when the hazardous material identified requires the highest level of protection for skin, eyes, and lungs. Such an ensemble is effective against vapours, gases, mists, and dusts and is typically indicated when the operating environment exceeds IDLH values for skin absorption.

Level A protection, when worn in accordance with NFPA 1991, will protect the wearer against a transient episode of flash fire. To that end, thermal extremes should be approached with caution. Direct contact between the suit fabric and a cryogenic material, such as liquid nitrogen or liquid helium, may result in immediate suit

FIGURE 36-9 A Level A ensemble envelops the wearer in a totally encapsulating suit.
© Courtesy of Rob Schnepp.

failure. This type of ensemble more than addresses the asphyxiant threat—it's the temperature extreme that must be acknowledged. By contrast, a potentially flammable atmosphere should be considered an extremely dangerous situation. In such circumstances, Level A suits, even with the flash fire component of the suit in place, provide very limited protection. Moreover, it is difficult to see when wearing a Level A suit, which increases the possibility that the person may unknowingly bump into sharp objects or rub against materials that might puncture or abrade the suit's vapour protection. Therefore, some forethought about the operating environment should occur well before entering the contaminated atmosphere. As always, a risk-versus-benefit thought process should prevail. The "best" level of protection is the one that is the most appropriate for the hazard and the mission.

Level A protection is effective against alpha radiation, but because of the lack of fabric thickness (as compared to fire fighters' turnout gear) it may not offer adequate protection against beta radiation and certainly is not a barrier to gamma radiation. Remember—thorough detection and monitoring actions will help you determine the nature of the operating environment.

Ensembles worn as Level A protection must meet the requirements outlined in NFPA 1991. A Level A ensemble also requires open-circuit, positive-pressure SCBA or an SAR for respiratory protection. (Chapter 33 provides a list of the recommended and optional components of Level A protection.)

To don a Level A ensemble, follow the steps in **SKILL DRILL 36-1**.

To doff a Level A ensemble, follow the steps in **SKILL DRILL 36-2**.

SKILL DRILL 36-1
Donning a Level A Ensemble NFPA 1072: 6.2.1; NFPA 472: 6.2.1.2.1, 6.2.4.1

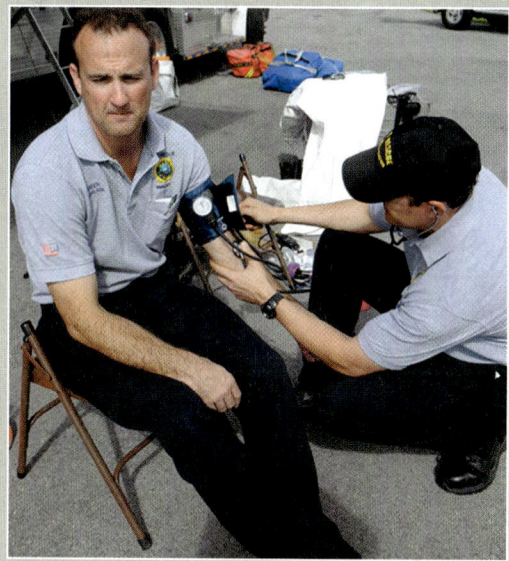

1 Conduct a pre-entry briefing, medical monitoring, and equipment inspection.

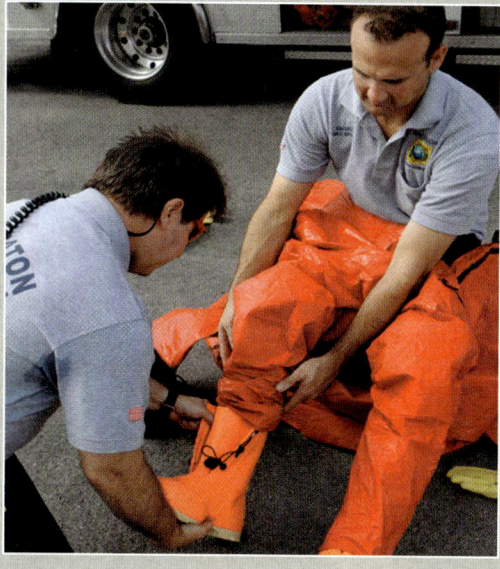

2 While seated, pull on the suit to waist level; pull on the chemical boots over the top of the chemical suit. Fold the suit boot covers over the tops of the boots.

3 Stand up, and don the SCBA frame and SCBA face piece, but do not connect the regulator to the face piece.

4 Place the helmet on your head.

(continued)

SKILL DRILL 36-1 Continued
Donning a Level A Ensemble NFPA 1072: 6.2.1; NFPA 472: 6.2.1.2.1, 6.2.4.1

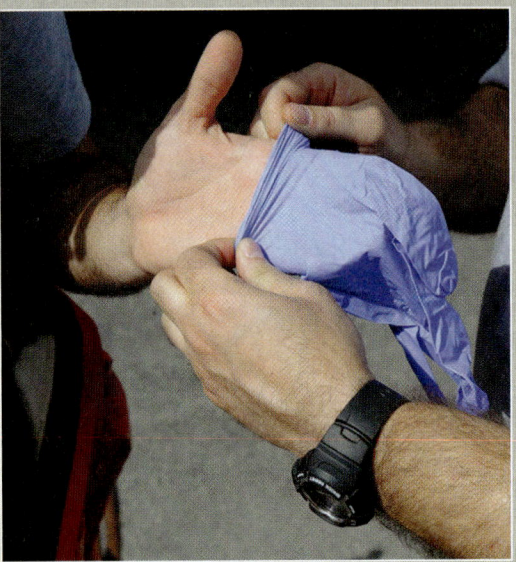

5 Don the inner gloves.

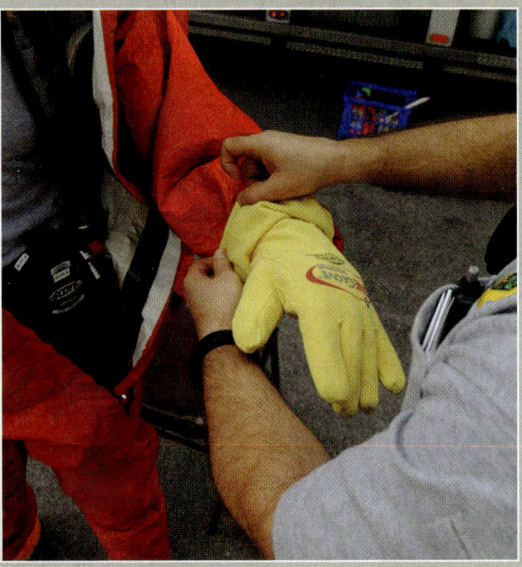

6 With assistance, complete donning the suit by placing both arms in the suit, pulling the expanded back piece over the SCBA, placing the chemical suit over your head, and donning the outer chemical gloves (if required).

7 Instruct the assistant to connect the regulator to the SCBA face piece and ensure air flow.

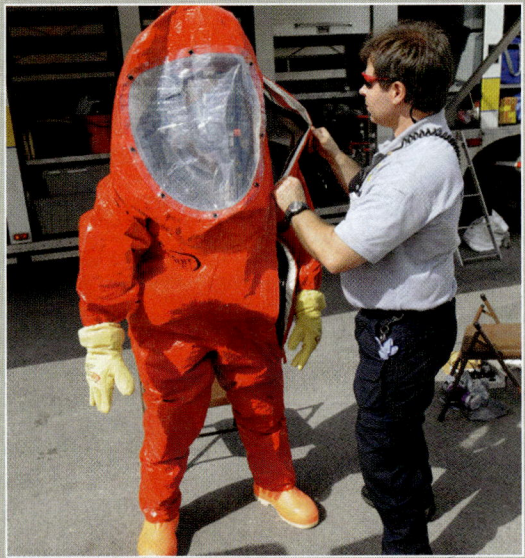

8 Instruct the assistant to close the chemical suit by closing the zipper and sealing the splash flap.

SKILL DRILL 36-1 Continued
Donning a Level A Ensemble NFPA 1072: 6.2.1; NFPA 472: 6.2.1.2.1, 6.2.4.1

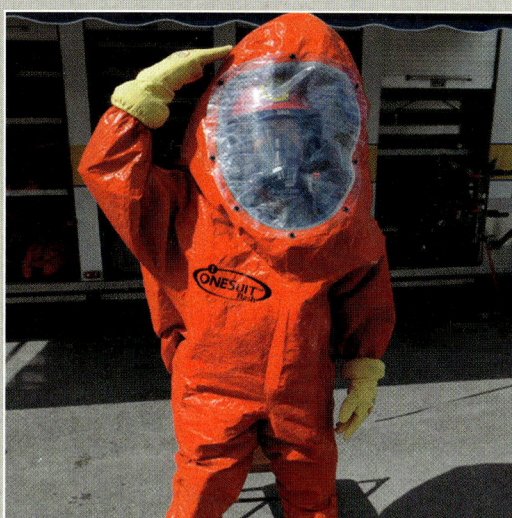

9 Review hand signals, and indicate that you are okay.

© Jones & Bartlett Learning. Photographed by Glen E. Ellman.

SKILL DRILL 36-2
Doffing a Level A Ensemble NFPA 1072: 6.2.1; NFPA 472: 6.2.1.2.1, 6.2.4.1

1 After completing decontamination, proceed to the clean area for suit doffing. Pull your hands out of the outer gloves and arms from the sleeves, and cross your arms in front inside the suit.

2 Instruct the assistant to open the chemical splash flap and suit zipper.

(continued)

SKILL DRILL 36-2 Continued
Doffing a Level A Ensemble NFPA 1072: 6.2.1; NFPA 472: 6.2.1.2.1, 6.2.4.1

3 Instruct the assistant to begin at the head and roll the suit down and away until the suit is below waist level.

4 Instruct the assistant to complete rolling the suit from the waist to the ankles; step out of the attached chemical boots and suit.

5 Doff the SCBA frame. The face piece should be kept in place while the SCBA frame is doffed.

6 Take a deep breath and doff the SCBA face piece; carefully peel off the inner gloves, and walk away from the clean area. Go to the rehabilitation area for medical monitoring, rehydration, and personal decontamination shower.

Level B

A Level B ensemble consists of multi-piece chemical-protective clothing, boots, gloves, and SCBA (**FIGURE 36-10**). This type of protective ensemble should be used when the type and atmospheric concentration of identified substances require a high level of respiratory protection but less skin protection. The kinds of gloves and boots chosen will depend on the physical and chemical properties of the identified chemical. The SCBA components should be considered as well when wearing nonencapsulating PPE.

The Level B protective ensemble is the workhorse of hazardous materials response—it is a very common level of protection and is often chosen for its versatility. Personnel initially processing a clandestine drug laboratory, performing preliminary missions for reconnaissance, or engaging in detection and monitoring duties commonly wear such an ensemble. The typical Level B ensemble provides little or no flash fire protection, however. Thus it should be viewed in the same manner as Level A equipment when it comes to thermal protection and other considerations of use such as protection from mechanical hazards, radiation, or asphyxiants.

Garments and ensembles that are worn for Level B protection should comply with the performance requirements found in NFPA 1992. (Chapter 33, *Estimating Potential Harm and Planning a Response,* provides a list of the recommended and optional components of a Level B protective ensemble.)

You may also encounter single-piece garments that are worn as Level B protection. These suits, referred to in the field as encapsulating Level B garments, are not constructed to be "vapour tight" like Level A garments. Encapsulating Level B garments do not have vapour-tight zippers, seams, or one-way relief valves around the hood like Level A garments. Although the encapsulating Level B suit may look a lot like a Level A garment, it is not constructed similarly and will not offer the same level of protection.

To don and doff a Level B encapsulated chemical-protective clothing ensemble, follow the same steps found in Skill Drill 36-1 and Skill Drill 36-2. Remember, the

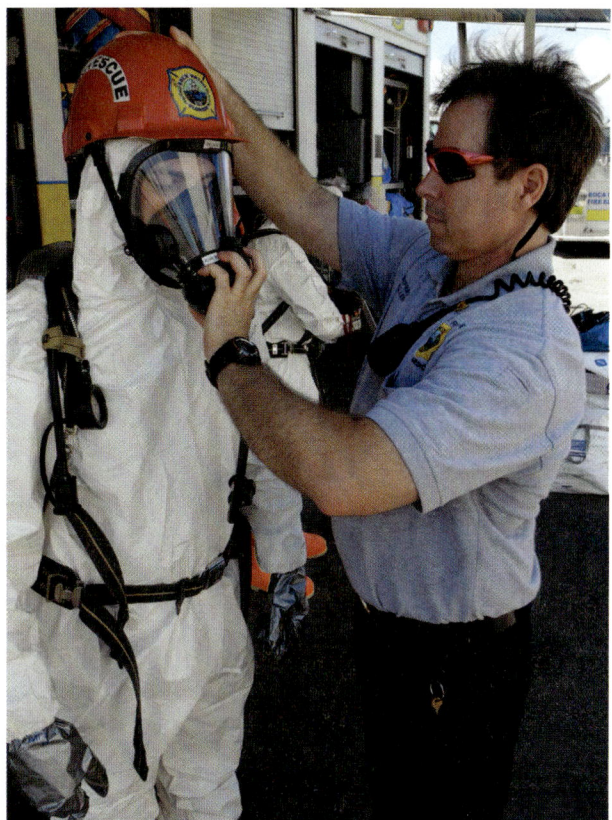

FIGURE 36-10 A Level B protective ensemble provides a high level of respiratory protection but less skin protection.
© Jones & Bartlett Learning. Photographed by Glen E. Ellman.

difference between the Level A ensemble and Level B encapsulating ensemble is not the procedure—it is the construction and performance of the garment.

To don a Level B nonencapsulated chemical-protective clothing ensemble, follow the steps in **SKILL DRILL 36-3**.

To doff a Level B nonencapsulated chemical-protective clothing ensemble, follow the steps in **SKILL DRILL 36-4**.

Level C

A Level C ensemble is appropriate when the type of airborne contamination is known, its concentration is measured, and the criteria for using an air-purifying respirator (APR) are met. Typically, Level C ensembles are worn with an APR or a powered air-purifying respirator (PAPR). The complete ensemble consists of standard work clothing, chemical-protective clothing, chemical-resistant gloves, and a form of respiratory protection other than an SCBA or SAR system. Level C equipment is appropriate when significant skin and eye

---

SAFETY TIP

According to the OSHA HAZWOPER regulation, Level B is the minimum level of protection to be worn when operating in an unknown environment.

SKILL DRILL 36-3
Donning a Level B Nonencapsulated Chemical-Protective Clothing Ensemble NFPA 1072: 6.2.1; NFPA 472: 6.2.1.2.1, 6.2.4.1

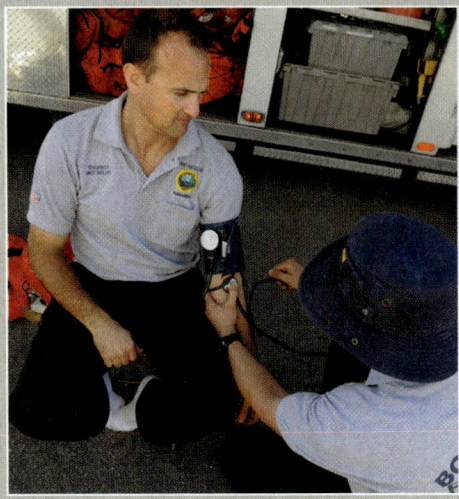

1 Conduct a pre-entry briefing, medical monitoring, and equipment inspection.

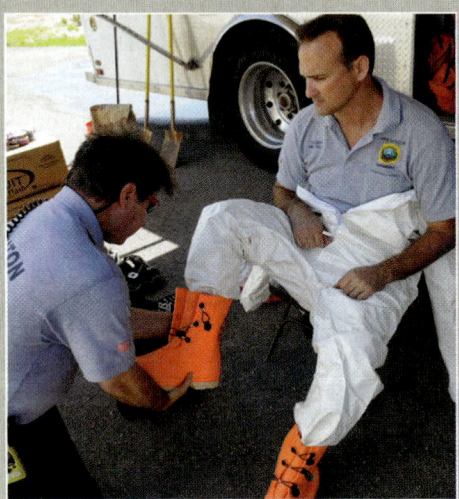

2 Sit down, and pull on the suit to waist level; pull on the chemical boots over the top of the chemical suit.

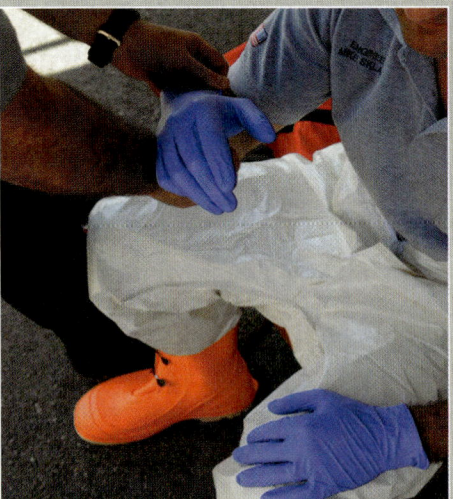

3 Don the inner gloves.

SKILL DRILL 36-3 Continued
Donning a Level B Nonencapsulated Chemical-Protective Clothing Ensemble NFPA 1072: 6.2.1; NFPA 472: 6.2.1.2.1, 6.2.4.1

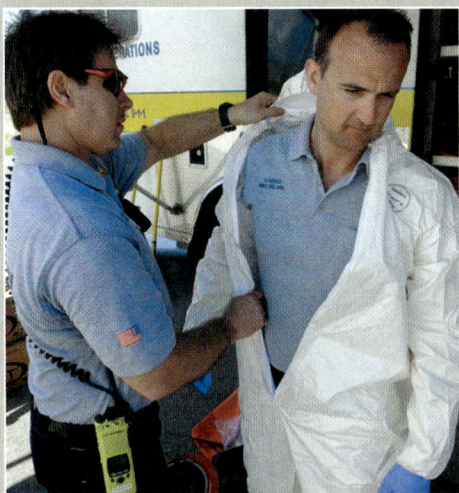

4 With assistance, complete donning the suit by placing both arms in the suit and pulling the suit over your shoulders. Instruct the assistant to close the chemical suit by closing the zipper and sealing the splash flap.

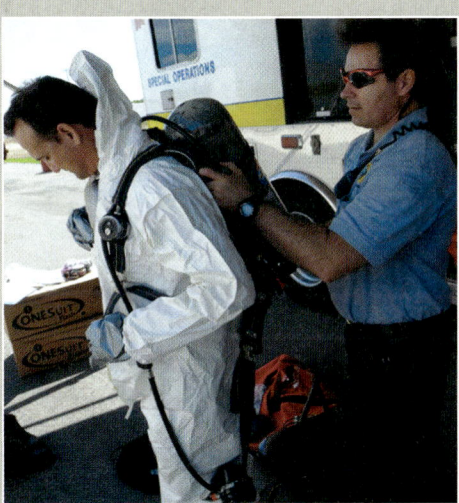

5 Don the SCBA frame and SCBA face piece, but do not connect the regulator to the face piece.

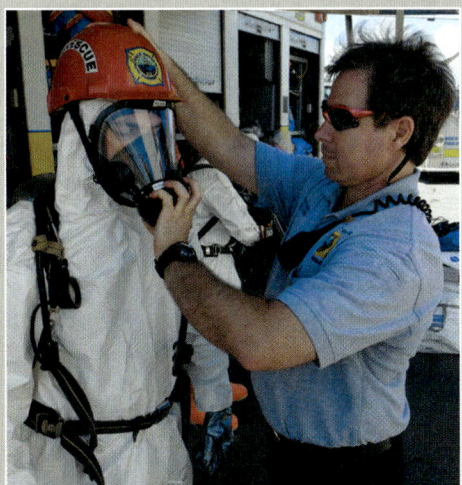

6 With assistance, pull the protective hood over your head and the SCBA face piece. Place the helmet on your head. Put on the outer gloves (over or under the sleeves, depending on the AHJ requirements for the incident). Instruct the assistant to connect the regulator to the SCBA face piece, and ensure you have air flow. Review hand signals, and indicate that you are okay.

SKILL DRILL 36-4
Doffing a Level B Nonencapsulated Chemical-Protective Clothing
Ensemble NFPA 1072: 6.2.1; NFPA 472: 6.2.1.2.1, 6.2.4.1

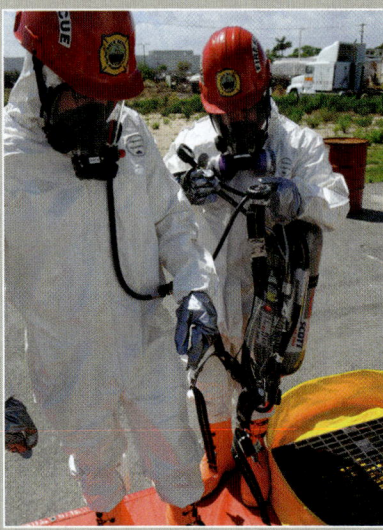

1 After completing the wash/rinse cycle, proceed to the clean area for PPE doffing. The SCBA frame is removed first. The unit may remain attached to the regulator while the assistant helps the responder out of the PPE, or the air supply may be detached from the regulator, leaving the face piece in place to provide for face and eye protection while the rest of the doffing process is completed.

2 Instruct the assistant to open the chemical splash flap and suit zipper.

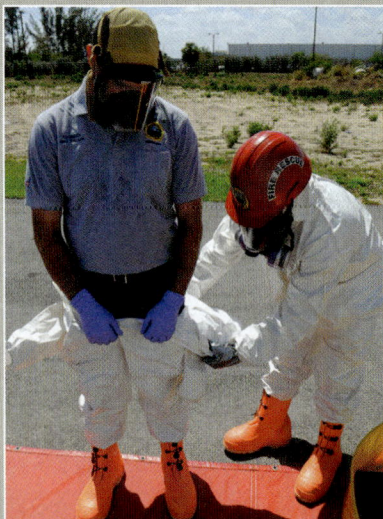

3 Remove your hands from the outer gloves and your arms from the sleeves of the suit. Cross your arms in front inside the suit. Instruct the assistant to begin at the head and roll the suit down and away until the suit is below waist level.

SKILL DRILL 36-4 Continued
Doffing a Level B Nonencapsulated Chemical-Protective Clothing Ensemble NFPA 1072: 6.2.1; NFPA 472: 6.2.1.2.1, 6.2.4.1

4 Sit down, and instruct the assistant to complete rolling down the suit to the ankles; step out of the attached chemical boots and suit.

5 Doff the SCBA face piece and helmet.

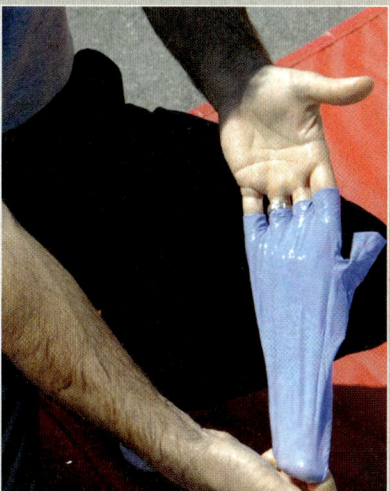

6 Carefully peel off the inner gloves, and go to the rehabilitation area for medical monitoring, rehydration, and personal decontamination shower.

exposure is unlikely (**FIGURE 36-11**). In many cases, Level C ensembles are worn in low-hazard situations such as clean-up activities lasting hours or days, once an area is fully characterized and the hazards are found to be low enough to allow this level of protection, or after responders mitigate the problem to the extent that they can dress down to this lower level to complete the mission. Many law enforcement agencies have provided their officers with Level C ensembles to be carried in the trunk of patrol cars. Based on the mission of perimeter scene control, this may be a prudent level of protection.

Chapter 33, *Estimating Potential Harm and Planning a Response,* provides a list of the recommended and optional components of a Level C protective ensemble and reviews the conditions of use for APRs and PAPRs. The garment selected must meet the performance requirements for NFPA 1992. Respiratory protection may be provided by a half-face (with eye protection) or full-face mask.

To don a Level C chemical-protective clothing ensemble, follow the steps in **SKILL DRILL 36-5**.

To doff a Level C chemical-protective clothing ensemble, follow the steps in **SKILL DRILL 36-6**.

Level D

A **Level D ensemble** offers the lowest level of protection. It typically consists of coveralls, work shoes, hard hat, gloves, and standard work clothing. This type of equipment should be used only when the atmosphere contains no known hazard and when work functions

FIGURE 36-11 A Level C protective ensemble includes chemical-protective clothing and gloves as well as respiratory protection.
© Jones & Bartlett Learning. Photographed by Glen E. Ellman.

SKILL DRILL 36-5
Donning a Level C Chemical-Protective Clothing Ensemble
NFPA 1072: 6.2.1; NFPA 472: 6.2.1.2.1, 6.2.4.1

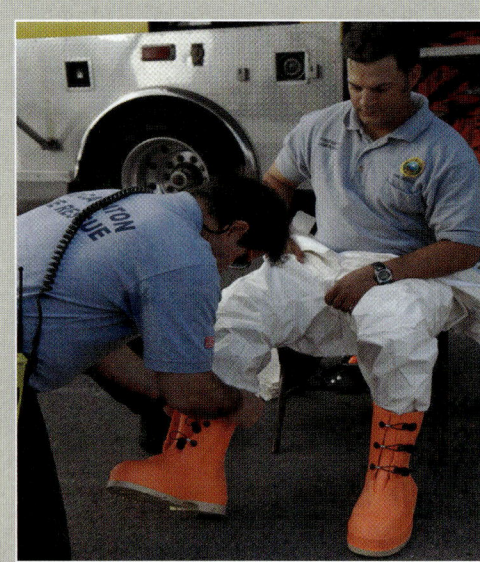

1 Conduct a pre-entry briefing, medical monitoring, and equipment inspection. While seated, pull on the suit to waist level; pull on the chemical boots over the top of the chemical suit.

SKILL DRILL 36-5 Continued
Donning a Level C Chemical-Protective Clothing Ensemble
NFPA 1072: 6.2.1; NFPA 472: 6.2.1.2.1, 6.2.4.1

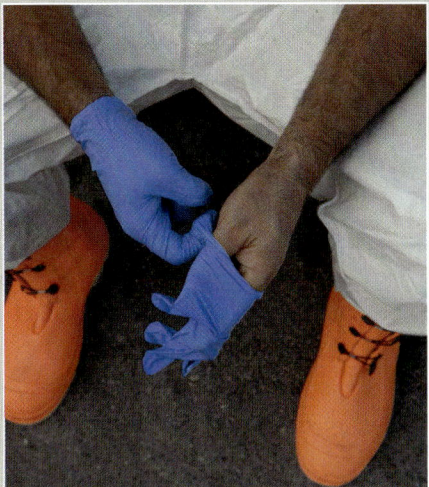

2 Don the inner gloves.

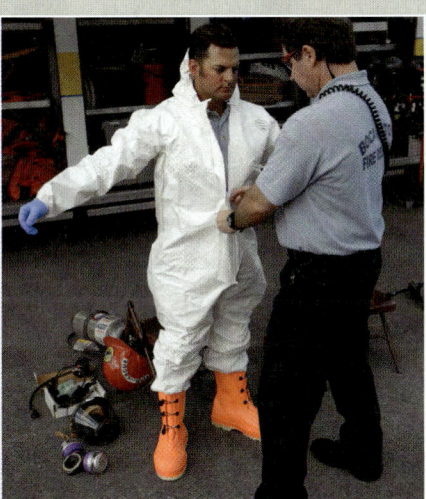

3 With assistance, complete donning the suit by placing both arms in the suit and pulling the suit over your shoulders. Instruct the assistant to close the chemical suit by closing the zipper and sealing the splash flap.

4 Don the APR/PAPR face piece. With assistance, pull the hood over your head and the APR/PAPR face piece. Place the helmet on your head. Pull on the outer gloves. Review hand signals, and indicate that you are okay.

© Jones & Bartlett Learning. Photographed by Glen E. Ellman.

SKILL DRILL 36-6
Doffing a Level C Chemical-Protective Clothing Ensemble
NFPA 1072: 6.2.1; NFPA 472: 6.2.1.2.1, 6.2.4.1

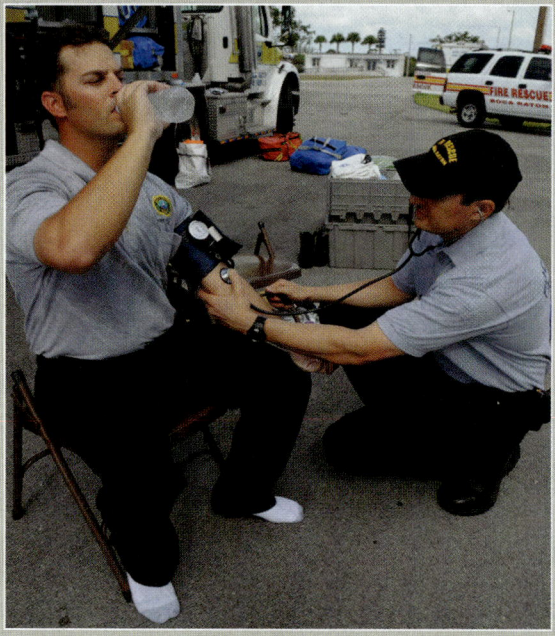

1 After completing decontamination, proceed to the clean area. As with Level B, the assistant opens the chemical splash flap and suit zipper. Remove your hands from the gloves and your arms from the sleeves. Instruct the assistant to begin at the head and roll the suit down below waist level. Instruct the assistant to complete rolling down the suit and to take the outer boots and suit away. The assistant helps remove the inner gloves. Remove the APR/PAPR. Remove the helmet.

2 Go to the rehabilitation area for medical monitoring, rehydration, and personal decontamination shower.

© Jones & Bartlett Learning. Photographed by Glen E. Ellman.

preclude splashes, immersion, or the potential for unexpected inhalation of or contact with hazardous levels of chemicals. Level D protection should be used when the situation involves nuisance contamination (such as dust) only. It should not be worn on any site where respiratory or skin hazards exist. As with the other levels of protection, Chapter 33, *Estimating Potential Harm and Planning a Response,* provides a list of the recommended and optional components of Level D protection.

To don a Level D chemical-protective clothing ensemble, follow the steps in **SKILL DRILL 36-7**. As

for doffing a Level D ensemble, the procedure is simply a reversal of the donning process. Because no chemical contact is expected with Level D, it is a nonhazardous process.

TABLE 36-1 describes the relationships among the NFPA hazardous materials protective clothing standards; Level A, B, and C classifications; and the new NIOSH-certified respirator with CBRN protection standards. The table is intended to clarify the relationship between the NFPA guidelines and to summarize the expected performance of the ensembles.

SKILL DRILL 36-7
Donning a Level D Chemical-Protective Clothing Ensemble NFPA 1072: 6.2.1; NFPA 472: 6.2.1.2.1, 6.2.4.1

■ Conduct a pre-entry briefing, medical monitoring, and equipment inspection. Don the Level D suit. Don boots. Don safety glasses or chemical goggles. Don appropriate head protection. Don gloves, a face shield, and any other required equipment.

Courtesy of Rob Schnepp.

Respiratory Protection

NFPA 1994, *Standard on Protective Ensembles for First Responders to Hazardous Materials Emergencies and CBRN Terrorism Incidents*, was developed to address the performance of protective ensembles and garments (including respiratory protection) specific to WMD. NFPA 1994 covers three classes of equipment: Class 2, Class 3, and Class 4. As discussed later in this chapter, the EPA regulations classify ensemble levels as Level A, Level B,

Level C, and Level D. The main difference between their system and the NFPA 1994 classification is that NFPA 1994 covers the performance of the garment *and* factors in the performance of the respiratory protection.

Simple asphyxiants such as nitrogen, argon, and helium, as well as oxygen-deficient atmospheres, are best handled by using an SCBA or another **supplied-air respirator (SAR)**. SCBA units that comply with the current version of NFPA 1981, *Standard on Open-Circuit Self-Contained Breathing Apparatus (SCBA) for Emergency*

TABLE 36-1 Levels of Protection

NFPA Standard	OSHA/EPA Level	NIOSH-Certified Respirator	NFPA Chemical Barrier Protection Method(s)	Chemical Vapour*	Chemical Liquid*	Particulate	Liquid-Borne Biological (Aerosol)
1991	A	CBRN SCBA (open recruit)	Protection against permeation and penetration*	X	X	X	X
1992	B	Non-CBRN SCBA (or CBRN SCBA)	Protection against permeation and penetration*	X	NA	NA	NA
	C	Non-CBRN APR or PAPR	Protection against penetration	X	NA	NA	NA
1994, Class 1	(Note: The NFPA 1994, Class 1 ensemble was removed in the 2006 edition of the standard because of its redundancy with NFPA 1991.)						
1994, Class 2	B	CBRN SCBA	Protection against permeation	X	X	X	X
1994, Class 3	C	CBRN APR or PAPR	Protection against permeation	X	X	X	X
1994, Class 4	B	CBRN SCBA	Protection against penetration	NA	NA	X	X
	C	CBRN APR or PAPR	Protection against penetration	NA	NA	X	X

Reproduced with permission from NFPA, *Hazardous Materials/Weapons of Mass Destruction Handbook.* Copyright © 2013, National Fire Protection Association. This reprinted material is not the complete and official position of the NFPA on the referenced subject, which is represented only by the standard in its entirety.

* Vapour protection for NFPA 1994 Class 2 and Class 3 is based on challenge concentrations established for NIOSH certification of CBRN open-circuit SCBA and APR/PAPR respiratory equipment. Class 2 and Class 3 do not require the use of totally encapsulating garments.

Services, are positive-pressure, CBRN-certified units that maintain a pressure inside the face piece, in relation to the pressure outside the face piece, such that the pressure is positive during both inhalation and exhalation. This is a very important feature when operating in airborne contamination.

The extra weight and reduced visibility associated with SCBA are factors to consider but are not reason enough to compromise a high level of respiratory protection. If an SCBA is indicated, it should be worn. As with any piece of PPE, there are as many positive benefits as there are negative points to consider in making this decision. Any responder called upon to wear an SCBA should be fully trained by the AHJ prior to operating in a contaminated environment. The OSHA HAZWOPER standard states that all employees engaged in emergency response who are exposed to hazardous substances *shall* wear a positive-pressure SCBA. It is the incident commander's responsibility to ensure that any personnel working within an IDLH environment wear PPE appropriate for the circumstance. This includes but is not limited to the use of SCBA. It is not just a good idea—it's the law. All responders should follow manufacturers' recommendations for using, maintaining, testing, inspecting, cleaning, and filling the SCBA unit. Be sure to document all of these activities so there is a record of what has been done to the unit. Refer back to Chapter 33 for more information about the various types of respiratory protection.

Hand Signals

Wearing PPE not only presents physical challenges but also, in many cases, compromises communications. For this reason, even if radio communication is available, hazardous materials responders often resort to hand signals to communicate. Your safety briefing and the incident action plan should always include a review of hand signals. Everyone should know the basic communication "fall-back" position to ensure that in the event of an in-suit emergency (loss of breathing air, SCBA cylinder failure, or acute illness and injury), anyone having visual contact with the responders will quickly understand that a problem is occurring. **FIGURE 36-12** demonstrates common hand signals used by responders. Your AHJ may have different hand signals; it's the responsibility of all responders to know how to use them correctly.

Safety

There are many hazards associated with wearing PPE. These hazards are best addressed by understanding the NFPA performance requirements for chemical-protective

A

B

C

FIGURE 36-12 A. A hand on top of the head making a tapping motion indicates, "I'm okay." **B.** Hands across the throat indicate "air problems." **C.** Hands over the head in a waving motion or both hands tapping the head is a signal for "trouble." This could indicate a suit problem or that something is amiss with the task or situation on the part of the responder or responders.

Courtesy of Rob Schnepp.

ensembles and the safety considerations taken into account when wearing PPE.

Chemical-Protective Equipment Performance Requirements

Typical chemical-protective equipment performance requirements include tests for durability, barrier integrity after flex and abrasion challenges, cold-temperature flex, and flammability. Essentially, each part of the suit must pass a particular set of challenges prior to receiving certification based on the NFPA testing standards. Users of any garment that meets the performance requirements set forth by the NFPA standard can rest assured that the garment will withstand reasonable insults from most mechanical-type hazards encountered on the scene. Of course, achieving a third-party independent certification that uses the NFPA standard does not mean the suit is invincible and cannot fail; it simply means that the equipment will hold up under "normal" conditions. It is up to the user to be aware of the hazards and avoid situations that may cause the garment to fail.

As described earlier, a variety of materials are used in both vapour- and splash-protective clothing. The most commonly used materials include butyl rubber, Tyvek, Saranex, polyvinyl chloride, and Viton, either singly or in multiple layers consisting of several different materials. Special chemical-protective clothing is adequate for incidents involving some chemicals yet useless for incidents involving other chemicals; no single fabric provides satisfactory protection from all chemicals.

SAFETY TIP

As exemplified by the TRACEMP acronym, many hazards can be encountered during the course of a hazardous materials incident. Given this possibility, multiple layers or multiple types of protection may have to be used in some situations. You should also understand the working environment and match the right garment to the anticipated hazards.

All responders who may be called upon to wear any type of PPE should read and understand the manufacturer's specifications and procedures for the maintenance, testing, inspection, cleaning, and storage of PPE provided by the AHJ. The list of NFPA and NIOSH documents in **TABLE 36-2** offers an overview of the testing and certification standards affecting the PPE currently on the market.

It is important for all hazardous materials responders to remember that some of the mission-specific competencies in this section are taken from competencies required of hazardous materials technicians. *That does*

not mean that operations level hazardous materials responders, with a mission-specific competency in PPE or any other mission-specific competency, are a replacement for a technician.

SAFETY TIP

The AHJ must properly outfit all responders expected to respond to a hazardous materials incident. The current OSHA HAZWOPER regulations [29 CFR 1910.120 (q)(3)(iii)]. (Provinces have established similar regulations that require supervisors to ensure safe work practices.) require the incident commander to ensure that the personal protective clothing worn at a hazardous materials emergency is appropriate for the hazards encountered.

Responder Safety

Working in PPE is a hazardous proposition on two different levels. First, simply by wearing PPE, the responder acknowledges that some degree of danger exists: If there were no hazard, there would be no need for the PPE! Second, wearing the PPE puts an inherent stress on the responder, separate and apart from the stress imposed by the operating environment. Much of this text is devoted to the "safety first" consideration. The next sections are devoted to raising your awareness of the issues that may arise from the very gear used to keep you safe.

Hazardous materials responders commonly experience a variety of heat-related illnesses. Those complications include heat exhaustion, heat cramps, and heat stroke, all of which are usually preceded by dehydration. Given that well-defined relationship, responders should be fully aware that their underlying level of hydration, prior to the response, may have an effect on their safety while they are wearing PPE. The next section, which covers in-suit cooling measures, addresses that fact by revisiting dehydration and the various cooling technologies that may be used to reduce the effects of overheating inside a chemical-protective garment.

Responders should also be aware that their field of vision will be compromised by the face piece of an SCBA or APR and by the encapsulating suit. This factor may result in the responder's slipping in a puddle of spilled chemicals or tripping on something. Moreover, the face piece often fogs up at some point, further limiting the responder's vision. This creates many problems, such as the inability to read labels, see other responders, see the screens on detection and monitoring devices, or quickly find an escape route in the event of an unforeseen problem in the hot zone. Wearing bulky PPE,

TABLE 36-2 PPE Testing and Certification Standards

Agency	Standard Title	Description
NFPA 1994	*Standard on Protective Ensembles for First Responders to Hazardous Materials Emergencies and CBRN Terrorism Incidents*	For chemicals, biological agents, and radioactive particulate hazards. Certifications under NFPA 1994 are issued only for complete ensembles. Individual elements such as garments or boots are not considered certified unless they are used as part of a certified ensemble. Thus, purchasers of PPE certified under NFPA 1994 should plan to purchase complete ensembles (or certified replacement components for existing ensembles).
NFPA 1992	*Standard on Liquid Splash–Protective Ensembles and Clothing for Hazardous Materials Emergencies*	For liquid or liquid splash threats.
NFPA 1991	*Standard on Vapour-Protective Ensembles for Hazardous Materials Emergencies and CBRN Terrorism Incidents*	Includes the now-mandatory requirements for CBRN protection for terrorism incident operations for all vapour-protective ensembles. It also includes the qualifications for the former NFPA 1994 Class 1 protective ensemble.
NFPA 1951	*Standard on Protective Ensembles for Technical Rescue Incidents*	For search and rescue or search and recovery operations where exposure to flame and heat is unlikely or nonexistent.
NFPA 1999	*Standard on Protective Clothing for Emergency Medical Operations*	For protection from blood and body fluid pathogens for persons providing treatment to victims after decontamination.
NFPA 1981	*Standard on Open-Circuit Self-Contained Breathing Apparatus (SCBA) for Emergency Services*	For all responders who may use SCBA; must be certified by NIOSH.
NIOSH	*Chemical, Biological, Radiological, and Nuclear (CBRN) Standard for Open-Circuit Self-Contained Breathing Apparatus*	To protect emergency responders against CBRN agents in terrorist attacks. Compliance with NFPA 1981.
NIOSH	*Standard for Chemical, Biological, Radiological, and Nuclear (CBRN) Full Facepiece Air-Purifying Respirator (APR)*	To protect emergency response workers against CBRN agents.
NIOSH	*Standard for Chemical, Biological, Radiological, and Nuclear (CBRN) Air-Purifying Escape Respirator and CBRN Self-Contained Escape Respirator*	To protect the general worker population against CBRN agents.

such as an encapsulating suit, may inhibit the mobility of the wearer to the point that bending over becomes difficult or reaching for valves above head level is taxing. Furthermore, when gloves become contaminated with chemicals (especially solvents), they become slippery, making it difficult to effectively grip tools, handrails, or ladder rungs. All in all, the environment inside the PPE can be just as challenging as the conditions outside the suit.

To mitigate some of the potential safety considerations that arise when wearing PPE, responders can employ a variety of safety procedures and training. To begin, conducting a pre-entry medical evaluation is important to catch the medical indicators that may signal a responder should not wear PPE.

Further guidance on pre-entry medical evaluation can also be found in NFPA 473, *Standard for Competencies for EMS Personnel Responding to Hazardous Materials/Weapons of Mass Destruction Incidents*, in either the basic life support or advanced life support section. Keep in mind that the medical monitoring station may serve many purposes at the scene of a hazardous materials event. The primary role of the medical monitoring station is to evaluate the medical status of the entry team, the backup team, and those personnel assigned to decontamination duties. On the scene of larger incidents, a medical group or team may be required to obtain basic physiological information from each responder and plan to provide care in the event a responder becomes a patient.

The use of the buddy system is another way that responders can mitigate some of the hazards that may be encountered at the scene of a hazardous materials/WMD incident. Along with the buddy system comes the need to communicate—another potential safety issue on the scene. Prior to entry, all radio communications should be sorted out and tested. To back up that form of communication, all responders on the scene should have a method to communicate by universally accepted hand signals. These hand signals could be used to rapidly share messages about problems with an air supply, a suit problem, or any other problem that might occur in the hot zone. Communications are often problematic on emergency scenes, so take whatever steps you can to minimize problems before anyone enters a contaminated atmosphere.

In-Suit Cooling Technologies

Hazardous materials responders operating in protective clothing should be aware of the signs and symptoms of heat exhaustion, heat stress, heat stroke, dehydration, and illness caused by extreme cold. The most common malady striking anyone wearing PPE is heat related. If the body is unable to disperse heat because an ensemble of PPE covers it, serious short- and long-term medical issues could occur.

Most heat-related illnesses are typically preceded by dehydration. It is important for responders to stay hydrated so that they can function at their maximum capacity. As a frame of reference, athletes should consume approximately 500 mL (16 oz) of fluid (water) prior to an event and 200 to 300 mL (7 to 10 oz) of fluid at regular intervals during the event. Responders can be considered occupational athletes—so keep up on your fluids!

In an effort to combat heat stress while wearing PPE, many response agencies employ some form of cooling technology under the garment. These technologies include but are not limited to air-, ice-, and water-cooled vests, along with phase-change cooling technology. Many studies have been conducted on each form of cooling technology. Each of these approaches is designed to accomplish the same goal—to reduce the impact of heat stress on the human body. As mentioned earlier, the same suit that seals you up against the hazards also seals in the heat, defeating the body's natural cooling mechanisms.

Forced-air cooling systems operate by forcing prechilled air through a hose system worn close to the body. This is similar to the fluid-chilled system described later. As the cooler air passes by the skin, heat is drawn away—by convection—from the body and released into the atmosphere. Forced-air systems are designed to function as the first level of cooling the body would naturally employ. Typically, these systems are lightweight and provide long-term cooling benefits, but mobility is limited because the umbilical is attached to an external, fixed compressor.

Ice-cooled or gel-packed vests are commonly used due to their low cost, unlimited portability, and unlimited "recharging" by refreezing the packs. These garments are vest-like in their design and intended to be worn around the torso. The principle underlying this approach is that the ice-chilled vest absorbs the heat generated by the body. On the downside, this technology is bulkier and heavier than the aforementioned systems, and it may cause discomfort to the wearer due to the nature of the ice-cold vest near the skin. Additionally, the cold temperature near the skin may actually fool the body into thinking it is cold instead of hot, thereby encouraging retention of even more heat.

Fluid-chilled systems operate by pumping ice-chilled liquids (water is often used, so these systems are referred to as "water-cooled") from a reservoir, through a series of tubes held within a vest-like garment, and back to the reservoir (**FIGURE 36-13**). Mobility may be limited with some varieties of this system, because the pump

FIGURE 36-13 A fluid-chilled or water-cooled system.
© Jones & Bartlett Learning. Photographed by Glen E. Ellman.

may be located away from the garment. Some systems incorporate a battery-operated unit worn on the hip, but the additional weight may increase the body's workload and generate more heat, thereby defeating the purpose of the cooling vest.

Phase-change cooling technology operates in a similar fashion to the ice- or gel-packed vests (**FIGURE 36-14**). The main difference between the two approaches is that the temperature of the material in the phase-change packs is chilled to approximately 15.5°C (60°F), and the fabric of the vest is designed to wick perspiration away from the body. The packs typically "recharge" more quickly than those of an ice- or gel-packed vest. Even though the temperature of the phase-change pack is higher than the temperature of an ice- or gel-packed vest, it is sufficient to absorb the heat generated by the body.

SAFETY TIP

Approximately 90 percent of all body heat is generated by the organs and muscles located in your torso.

SAFETY TIP

Remember to take rehabilitation breaks throughout the hazardous materials incident. Wearing any type of PPE requires a great deal of physical energy and mental concentration. Responders should also acknowledge the psychological stress that wearing PPE may present. Claustrophobia is a common problem when wearing chemical-protective equipment, especially encapsulated suits. This is one of the "P" (psychogenic) considerations in TRACEMP and can present a problem for responders.

FIGURE 36-14 Phase-change cooling technology.
Courtesy of Glacier Tek.

Responder Decontamination

As discussed in Chapter 33, *Estimating Potential Harm and Planning a Response*, emergency decontamination is used in potentially life-threatening situations to rapidly remove contaminants and get a responder or victim clean enough to receive medical care from first responders and, if needed, be admitted to a receiving hospital. It is quick and less "formal" than technical decontamination.

Technical decontamination is "the planned and systematic process of reducing contamination to a level that is as low as reasonably achievable" (NFPA 1072, Section 3.3.15.4). It is a much more formal process.

As discussed earlier, all responders charged with responding to hazardous materials/WMD incidents should be proficient with local procedures for emergency and technical decontamination, including the process for being decontaminated while wearing PPE.

Going Through Decontamination in PPE

For emergency decontamination, you may be asked to remove some or all clothing to reduce the harm posed by the contaminant. This ensures the assisting responder will stay as clean as possible. Emergency decontamination usually involves being doused with large quantities of water. A tarp and a booster line or other small handline may be used to accomplish this task. You will be given instructions on the removal of PPE and the process to follow per your local protocols.

The technical decontamination process is intended to minimize the spread of contamination and ensure safety for everyone involved. The pre-entry briefing should include information on the location of the decontamination corridor, the process to be used, and how the decontamination team will communicate with you as you move through the process. Typically, only one contaminated responder is allowed in a wash-and-rinse station at a time and PPE remains in place until it has been thoroughly scrubbed and rinsed. You may be asked to place any contaminated hand tools or other equipment in a tool drop area near the entrance of the decontamination corridor. These items can be cleaned later. Technical decontamination typically involves one to three wash-and-rinse stations, depending on the nature of the expected contamination. The rinser may begin rinsing at your head and thoroughly rinse downward into the collection pool.

> **LISTEN UP!**
>
> There are many ways to accomplish decontamination for the decontamination team. As with many other tasks, the AHJ should have well-defined standard operating procedures for this activity.

After the chemical protective equipment is thoroughly scrubbed and rinsed, it is safe to remove. The SCBA face piece, full- or half-face air-purifying respirator, or powered air-purifying respirator (PAPR) face piece remains in place for as long as possible. (The SCBA harness and cylinder or PAPR fan units can be removed and set off to the side.) The members of the decontamination team typically fold or roll the PPE back and away from you so that the contaminated side of the

garment contacts only itself. Outer chemical gloves (if worn) are carefully peeled away and off both hands. If the procedure is done properly, the contaminated side of the garment and gloves will not touch you.

You will then proceed toward the cold zone end of the decontamination corridor, to an area where helmets, face piece, and any other ancillary equipment are removed and can be placed in a separate area or a plastic bag (or other suitable container). Last, you will be asked to remove and discard inner gloves.

With decontamination complete, you can exit the decontamination area and enter the cold zone, where personal showers are taken and a fresh set of clothes are donned. Afterward, proceed to a medical station for evaluation.

To go through technical decontamination as a responder, follow the steps in **SKILL DRILL 36-8**.

Reporting and Documenting the Incident

As with any other type of incident, documenting the activities carried out during a hazardous materials/WMD incident is an important part of the response. Many responders may pass through the scene, and it could be quite difficult to sort everything out when it comes time to reconstruct the events for an accurate and legally defensible incident report. Good documentation after the incident is directly correlated with how well organized the response was.

Along with the formal written accounts of the event, some agencies require that personnel fill out exposure records that include information such as the name of the substances involved in the incident and the level of protection used. This information, coupled with a comprehensive medical surveillance program (discussed in Chapter 33), provides a method to chronicle the exposure history of the responders over a period of time. Consult your AHJ for the exact details and procedures for reporting and documenting the incident.

SKILL DRILL 36-8
Responder Technical Decontamination NFPA 1072: 6.2.1

1 The contaminated responder drops any tools or equipment into a container or onto a designated tarp.

2 The decontamination team member performs gross decontamination on the contaminated responder, if necessary.

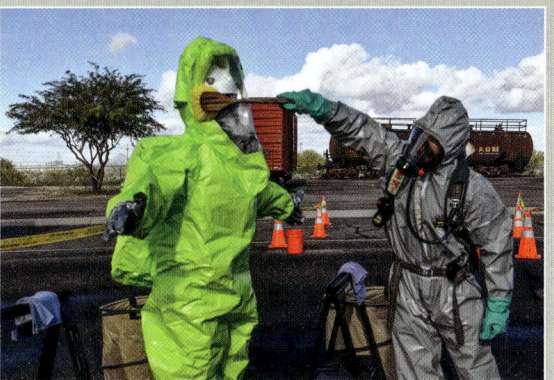

3 For technical decontamination, the decontamination team member washes and rinses the contaminated responder one to three times. The wash–rinse cycle is determined largely by the nature of the contaminant. Each rinse should start at the responder's head, working down to his or her feet. Remember, the goal is to render the PPE safe to remove.

4 The decontamination team member removes the outer hazardous materials–protective clothing from the contaminated responder.

(continued)

SKILL DRILL 36-8 Continued
Responder Technical Decontamination NFPA 1072: 6.2.1

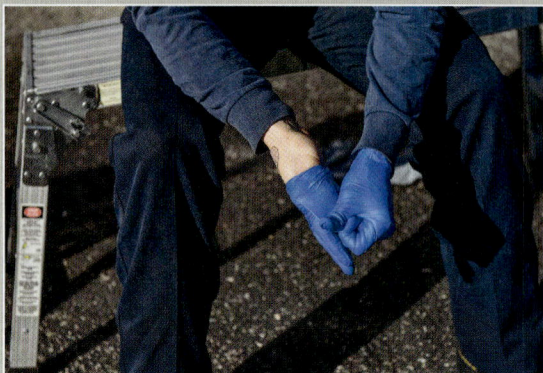

5 The responder removes his or her personal clothing and any respiratory protection and proceeds to the rehabilitation area for medical monitoring, rehydration, and personal decontamination.

© Jones & Bartlett Learning. Photographed by Glen E. Ellman.

After-Action REVIEW

IN SUMMARY

■ The operations level hazardous materials responder assigned to use personal protective equipment (PPE) must be trained to meet all competencies and job performance requirements (JPRs) at the operations level and all competencies and JPRs in the Personal Protective Equipment section in both NFPA 1072 and NFPA 472.

■ Much of the chemical-protective equipment on the market today is intended for a single use (i.e., it is disposable) and is usually discarded along with the other hazardous waste generated by the incident.

■ Chemical-protective garments are tested in accordance with the manufacturer's recommendations.

■ Your AHJ will determine the PPE available for use and also the procedures and requirements for selecting and using PPE on the incident scene as part of the incident action plan.

■ At the lowest end of the spectrum are street clothing and normal work uniforms, which offer the least amount of protection in a hazardous materials emergency.

■ The next higher level of protection is provided by structural firefighting protective equipment.

■ Chemical-protective clothing is unique in that it is designed to prevent chemicals from coming in contact with the body.

■ To help you safely estimate the chemical resistance of a particular garment, manufacturers supply compatibility charts with all of their protective equipment.

■ Typical chemical-protective equipment performance requirements include tests for durability, barrier integrity after flex and abrasion challenges, cold-temperature flex, and flammability.

■ By wearing PPE, the responder acknowledges that some degree of danger exists. Additionally, wearing the PPE puts stress on the responder, separate and apart from the stress imposed by the operating environment.

■ Hazardous materials responders commonly experience a variety of heat-related illnesses. Those complications include heat exhaustion, heat cramps, and heat stroke, all of which are usually preceded by dehydration.

- All responders charged with responding to hazardous materials/WMD incidents should be proficient with local procedures for emergency and technical decontamination, including the process for being decontaminated while wearing PPE.

- Along with the formal written accounts of the event, some agencies require that personnel fill out exposure records that include information such as the name of the substances involved in the incident and the level of protection used.

KEY TERMS

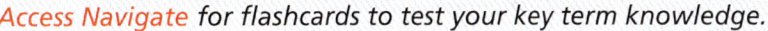

Access Navigate for flashcards to test your key term knowledge.

Air-purifying respirator (APR) A respirator that removes specific air contaminants by passing ambient air through one or more air purification components. (NFPA 1984)

Chemical-resistant materials Clothing (suit fabrics) specifically designed to inhibit or resist the passage of chemicals into and through the material by the process of penetration, permeation, or degradation.

Degradation A chemical action involving the molecular breakdown of a protective clothing material or equipment due to contact with a chemical. (NFPA 1072)

Dehydration An excessive loss of body water. Signs and symptoms of dehydration may include increasing thirst, dry mouth, weakness or dizziness, and a darkening of the urine or a decrease in the frequency of urination.

High temperature–protective clothing Protective clothing designed to protect the wearer for short-term high temperature exposures. (NFPA 1072)

Level A ensemble Personal protective equipment that provides protection against vapours, gases, mists, and even dusts. The highest level of protection, it requires a totally encapsulating suit that includes a self-contained breathing apparatus.

Level B ensemble Personal protective equipment that is used when the type and atmospheric concentration of substances require a high level of respiratory protection but less skin protection. The kinds of gloves and boots worn depend on the identified chemical.

Level C ensemble Personal protective equipment that is used when the type of airborne substance is known, the concentration is measured, the criteria for using an air-purifying respirator are met, and skin and eye exposure are unlikely. A Level C ensemble consists of standard work clothing with the addition of chemical-protective clothing, chemically resistant gloves, and a form of respirator protection.

Level D ensemble Personal protective equipment that is used when the atmosphere contains no known hazard, and work functions preclude splashes, immersion, or the potential for unexpected inhalation of or contact with hazardous levels of chemicals. A Level D ensemble is primarily a work uniform that includes coveralls and affords minimal protection.

Liquid splash–protective ensemble Multiple elements of compliant protective clothing and equipment products that when worn together provide protection from some, but not all, risks of hazardous materials/WMD emergency incident operations involving liquids. (NFPA 1072)

National Institute for Occupational Safety and Health (NIOSH) The U.S. federal agency responsible for research and development on occupational safety and health issues.

Penetration The movement of a material through a suit's closures, such as zippers, buttonholes, seams, flaps, or other design features of chemical-protective clothing, and through punctures, cuts, and tears. (NFPA 1072)

Permeation A chemical action involving the movement of chemicals, on a molecular level, through intact material. (NFPA 1072)

Powered air-purifying respirator (PAPR) An air-purifying respirator that uses a powered blower to force the ambient air through one or more air-purifying components to the respiratory inlet covering. (NFPA 1984)

Self-contained breathing apparatus (SCBA) A respirator worn by the user that supplies a respirable atmosphere that is either carried in or generated by the apparatus and that is independent of the ambient environment. (NFPA 350)

Supplied-air respirator (SAR) An atmosphere-supplying respirator for which the source of the breathing air is not designed to be carried by the user. Also known as an "airline respirator." (NFPA 1989)

Vapour-protective ensemble Multiple elements of compliant protective clothing and equipment that when worn together provide protection from some, but not all, risks of vapour, liquid-splash, and particulate environments during hazardous materials/WMD incident operations. (NFPA 1072)

On Scene

The chief of your fire company asked you to give a brief presentation to the town council about the new Level A suits you are planning to purchase. You decide to use the example of an ammonia release at a local ice-making facility to underscore the reasons why your company needs this particular level of protection.

1. Which of the following NFPA standards would be the proper one to reference regarding your new Level A suits?

A. NFPA 1981, *Standard on Open-Circuit Self-Contained Breathing Apparatus for Emergency Services*

B. NFPA 1951, *Standard on Protective Ensembles for Technical Rescue Incidents*

C. NFPA 1999, *Standard on Protective Clothing for Emergency Medical Operations*

D. NFPA 1991, *Standard on Vapour-Protective Ensembles for Hazardous Materials Emergencies and CBRN Terrorism Incidents*

2. In addition to the garment, you also plan to purchase a forced-air cooling system. Which of the following gives the most accurate description of forced-air cooling technology?

A. Forced-air cooling systems operate by pumping ice-chilled liquids from a reservoir, through a series of tubes held within a vest-like garment, and back to the reservoir.

B. The principle of forced-air cooling systems is that an ice-chilled vest absorbs the heat generated by the body. This technology may cause discomfort to the wearer because the ice-cold vest is placed so close to the skin.

C. Forced-air cooling systems operate by forcing prechilled air through a system of hoses worn close to the body. As the cooler air passes by the skin, it is drawn away from the body and released into the atmosphere.

D. Forced-air cooling technology operates by chilling the hands and feet with ice-chilled liquids in an attempt to increase manual dexterity.

3. Which of the following is a true statement?

A. Forced-air cooling technology is bulky and heavy.

B. With phase-change cooling technology, the temperature of the material in the packs is approximately 15.5°C (60°F).

C. Forced-air cooling technology is loud and interferes with communications.

D. Phase-change cooling technology is created by melting ice packs and placing them on or near the body.

Access Navigate to find answers to this On Scene, along with other resources such as an audiobook.

Hazardous Materials Operations Level: Mission Specific

Product Control

KNOWLEDGE OBJECTIVES

After studying this chapter, you will be able to:

- Describe how to use the following control methods:
  - Absorption and adsorption (**NFPA 1072: 6.6.1; NFPA 472: 6.6.1.2.2, 6.6.3.1**, pp. 1244, 1246)
  - Damming (**NFPA 1072: 6.6.1; NFPA 472: 6.6.1.2.2, 6.6.3.1**, pp. 1245, 1247–1248)
  - Diking (**NFPA 1072: 6.6.1; NFPA 472: 6.6.1.2.2, 6.6.3.1**, pp. 1249–1250)
  - Dilution (**NFPA 1072: 6.6.1; NFPA 472: 6.6.1.2.2, 6.6.3.1**, p. 1250)
  - Diversion (**NFPA 1072: 6.6.1; NFPA 472: 6.6.1.2.2, 6.6.3.1**, pp. 1250–1251)
  - Retention (**NFPA 1072: 6.6.1; NFPA 472: 6.6.1.2.2, 6.6.3.1**, pp. 1250–1251)
  - Remote valve shut-off (**NFPA 1072: 6.6.1; NFPA 472: 6.6.1.2.2, 6.6.3.1, 6.6.4.1**, pp. 1251–1253)
  - Vapour dispersion and suppression (**NFPA 1072: 6.6.1; NFPA 472: 6.6.1.2.2, 6.6.3.1, 6.6.4.2**, pp. 1253–1258)
- Describe the recovery phase of a hazardous materials incident. (**NFPA 1072: 6.6.1; NFPA 472: 6.6.1.2.2, 6.6.3.1, 6.6.6.1**, pp. 1257–1258)

SKILLS OBJECTIVES

After studying this chapter, you will be able to perform the following skills:

- Use absorption/adsorption to manage a hazardous materials incident. (**NFPA 1072: 6.6.1; NFPA 472: 6.6.1.2.2, 6.6.3.1**, p. 1246)
- Construct an overflow dam. (**NFPA 1072: 6.6.1; NFPA 472: 6.6.1.2.2, 6.6.3.1**, p. 1248)
- Construct an underflow dam. (**NFPA 1072: 6.6.1; NFPA 472: 6.6.1.2.2, 6.6.3.1**, p. 1248)
- Construct a dike. (**NFPA 1072: 6.6.1; NFPA 472: 6.6.1.2.2, 6.6.3.1**, p. 1249)
- Use dilution to manage a hazardous materials incident. (**NFPA 1072: 6.6.1; NFPA 472: 6.6.1.2.2, 6.6.3.1**, p. 1250)
- Construct a diversion. (**NFPA 1072: 6.6.1; NFPA 472: 6.6.1.2.2, 6.6.3.1**, p. 1251)
- Use retention to manage a hazardous materials incident. (**NFPA 1072: 6.6.1; NFPA 472: 6.6.1.2.2, 6.6.3.1**, p. 1251)
- Use vapour dispersion to manage a hazardous materials incident. (**NFPA 1072: 6.6.1; NFPA 472: 6.6.1.2.2, 6.6.3.1, 6.6.4.2**, p. 1254)

- Use vapour suppression to manage a hazardous materials incident. (**NFPA 1072: 6.6.1; NFPA 472: 6.6.1.2.2, 6.6.3.1, 6.6.4.2**, p. 1255)
- Perform the rain-down method of applying foam. (**NFPA 1072: 6.6.1; NFPA 472: 6.6.1.2.2, 6.6.3.1, 6.6.4.2**, p. 1256)
- Perform the roll-in method of applying foam. (**NFPA 1072: 6.6.1; NFPA 472: 6.6.1.2.2, 6.6.3.1, 6.6.4.2**, p. 1257)
- Perform the bounce-off method of applying foam. (**NFPA 1072: 6.6.1; NFPA 472: 6.6.1.2.2, 6.6.3.1, 6.6.4.2**, p. 1258)

Hazardous Materials Alarm

Your pump company has been dispatched to assist a rescue company on the scene of a rolled-over diesel truck pulling a flatbed trailer. The trailer was carrying several lengths of large steel pipe, which are now lying on the road. The fuel tanks on the diesel truck are leaking. Upon arrival, your officer receives an assignment to protect a curbside drain located downslope from the leaking tanker. He assigns the task to you and two other fire fighters. There is a slight breeze blowing away from you, back toward the leaking tanker. The product is confirmed to be diesel fuel. The concrete roadway ahead of the spill is completely dry. You have access to three plastic shovels and several bags of loose absorbent.

1. Would full structural fire fighter's protective gear and self-contained breathing apparatus offer adequate protection in this situation?

2. Based on your level of training (i.e., hazardous materials operations core competencies and mission-specific competencies), would you be qualified to perform this task?

3. Describe the steps you would need to take, including obtaining information about diesel fuel, to complete the assignment.

 *Access Navigate for more practice activities.*

Introduction

This chapter addresses the job performance requirements (JPRs) found in NFPA 1072, *Standard for Hazardous Materials/Weapons of Mass Destruction Emergency Response Personnel Professional Qualifications*, and the competencies found in NFPA 472, *Standard for Competence of Responders to Hazardous Materials/Weapons of Mass Destruction Incidents*, for operations level responders assigned mission-specific product control responsibilities at hazardous materials/weapons of mass destruction (WMD) incidents beyond the core responsibilities at the operations level. The operations level responder assigned to perform product control at hazardous materials/WMD incidents must be trained to meet all responsibilities at the awareness level and the operations level, all mission-specific requirements for personal protective equipment (PPE), and all

requirements in product control. Additionally, the operations level responder assigned to perform product control at hazardous materials/WMD incidents must operate under the guidance of a hazardous materials technician, an allied professional, or standard operating procedures (SOPs).

It is not uncommon for a hazardous materials incident to require some form of product control. NFPA 1072 defines **control** as "the procedures, techniques, and methods used in the mitigation of hazardous materials/weapons of mass destruction (WMD) incidents, including containment, extinguishment, and confinement." Scenarios like the one described in the chapter-opening vignette are relatively common, as are incidents involving leaking drums, spills into waterways, and other types of releases involving flammable liquids or gases. Many of these incidents require fire personnel to intervene (control the release) by shutting off valves or applying loose absorbents or various kinds of foam to mitigate the situation.

In most cases, the best course of action is to confine the problem to the smallest area possible. **Confinement** is the process of attempting to keep hazardous materials within the immediate area of the release. This goal can usually be accomplished by damming or diking a material or by suppressing vapour with an appropriate type of foam. **Containment**, by contrast, refers to actions that stop a hazardous material from leaking or escaping its container. Examples of containment include patching or plugging a breached container or righting an overturned container to stop a slow leak. Sometimes it is necessary to first stop the leak (contain) and then confine whatever has been released. In other cases, it may be impossible to stop the leak, and all that can be done is to confine the hazardous material as much as possible. In situations where ignition of a material has occurred, extinguishment (stopping the burning) may be the best course of action for incident containment.

As every hazardous material call is unique, there is no specific sequence for handling each incident. You must size up the situation as you would any other problem and employ the best methods available to handle the incident safely. When considering a control option, certain factors must be evaluated, such as the maximum quantity of material that can be released and the likely duration of the incident without intervention.

It is vital in these situations, as with any other type of release, for responders to understand the nature of the release, wear the appropriate level of PPE for the situation, and have all the tools and equipment—including air monitoring and detection equipment—available to accomplish the task at hand. An incident action plan (IAP) must be developed in accordance with your department's SOPs. This includes safety measures based on the nature of the incident. In many cases, product control measures bring responders near the released product. To work safely in a contaminated (or potentially contaminated) atmosphere, responders must stay informed about their working environment. Readings from monitoring and detection devices should always be interpreted in the context of the specific event. Additionally, all emergency responders should be familiar with the policies and procedures of the local jurisdiction to ensure that they employ a consistent approach to the selected control option.

Control Options

The most challenging aspect of mitigating a hazardous materials emergency is arriving at a solution that can be employed quickly and safely, while minimizing the potential negative effects on people, property, and the environment. If that sounds like a tall order, it is. Handling a hazardous materials incident is a bit like a chess game: One move sets up and (either positively or negatively) influences the next move or series of moves. It is important to consider your departmental risk management principles when developing an IAP. Remember—a solid response objective should be well thought out and realistic, considering both the positive and negative effects of the actions taken. Think creatively, and know what resources are available in your jurisdiction—sometimes you must think outside the box (**FIGURE 37-1**).

Sometimes, no action is the safest course of action. Unfortunately there are instances in which the situation

FIGURE 37-1 Bales of hay procured from a local feed store. In this instance, the bales were used to absorb some of the downstream flow of a gasoline spill (floating on water) from an MC-306/DOT 406 cargo trailer.
Courtesy of Rob Schnepp.

is so extreme; or the responders cannot be properly protected; or the product will evaporate quickly, cool down, or solidify, such that it may be prudent to create a safe perimeter and let the problem stabilize on its own. Choosing not to intervene can be a challenge for fire fighters who are used to taking quick action, but there are times when taking a hands-off approach may be the correct course of action. As an example, a fire in a gasoline tanker may have to burn itself out if not enough foam is available to fight the fire effectively.

Similarly, when incompatible chemicals are mixed, it may be wise to let the reaction run its course before intervening. Imagine a spill of a highly volatile flammable liquid like isopropyl alcohol on a hot summer day. It would likely evaporate before you could make an entry and attempt to clean it up. It is important to understand the chemical and physical properties of any released material and how those properties may be used to your advantage regarding confinement or containment (**FIGURE 37-2**).

As personnel approach a hazardous materials incident, they should be aware of natural control points—areas in the terrain or places in a structure where materials might be contained or confined. Structural barriers that might be used include doors to a room, doors to a building, designated areas for secondary containment, and curbed areas of roadways. A number of control measures may be available if responders are thinking creatively and are aware of their surroundings.

Absorption and Adsorption

Absorption is the process whereby a spongy material (e.g., soil or loose absorbents such as vermiculite, clay, or peat moss) or specially designed spill pads are used to soak up a liquid hazardous material (**FIGURE 37-3**). The contaminated mixture of absorbent material and chemicals is then collected, and the materials are disposed of together. Most states have laws and regulations that dictate how to dispose of used absorbent materials.

Absorption minimizes the spread of liquid spills, but it is effective only on flat surfaces. Although absorbent materials such as soil and sawdust are inexpensive and readily available, they become hazardous materials themselves once they come in contact with the spilled liquid; therefore, they must be disposed of properly. The process of absorption does not change the chemical properties of the involved substance. It is only a method to collect the substance for subsequent containment.

The technique of absorption may prove challenging for personnel because it requires them to be near the spilled material. It also involves the addition of material to a spilled product, which adds volume to the spill. The absorbent material may also react with certain hazardous substances, so it is important to determine whether an absorbent substance is compatible with the hazardous material. Hydrofluoric acid, for example, is not compatible with the silica-based absorbents commonly found in some types of spill booms.

Some absorbent materials repel water while still absorbing a spilled liquid that does not mix with water—an effective property under the right circumstances (such

FIGURE 37-2 Highly volatile flammable liquids like methyl alcohol may be left to evaporate naturally without taking offensive action to clean them up.
Courtesy of Rob Schnepp.

FIGURE 37-3 Spill pads are often used to soak up a liquid hazardous material.
Courtesy of Rob Schnepp.

as when the goal is to mitigate an oil spill on a body of water) (**FIGURE 37-4**). In these situations, spill booms may be deployed as floating barriers used to obstruct passage. These booms are typically constructed of an outer mesh covering and filled with a material such as polypropylene or silica. Many different sizes and types of spill booms are on the market today. Some will float on water, whereas others are designated for use only on dry land (**FIGURE 37-5**). It is up to your AHJ to decide which types of absorbents are appropriate for your jurisdiction.

Another example where the technique of absorption may prove challenging for personnel is an incident involving nitric acid at a concentration higher than 72%. In this situation, the nitric acid becomes such an aggressive oxidizer that using an organic-based absorbent may result in the acid igniting the absorbent material. To reiterate the point made earlier, not all absorbent materials are created equal, and using the wrong one might end up unduly complicating your situation.

The opposite of absorption is **adsorption**. In adsorption, the contaminant *adheres* to the surface of an added material—such as silica or activated carbon—rather than combining with it (as in absorption). In some cases, the process of adsorption can generate heat—a key point you should consider when use of adsorbent materials is proposed. The concept of adsorption is analogous to the mechanism underlying Velcro: Adsorbents are "sticky" and grab on to whatever substance they are designed to be used for. A common adsorbent material is activated carbon, which is found in the filter cartridges used for air-purifying respirators.

To use absorption/adsorption to manage a hazardous materials incident, follow the steps in **SKILL DRILL 37-1**.

A

B

FIGURE 37-5 A. A spill boom deployed by boats to contain a fuel spill in a harbor. **B.** A spill boom can be used to confine a liquid.
A: Courtesy of Rob Schnepp. **B:** © Jones & Bartlett Learning. Photographed by Glen E. Ellman.

FIGURE 37-4 Gasoline floating on water in a creek.
Courtesy of Rob Schnepp.

Damming

Damming is a containment technique that is used when liquid is flowing in a natural channel or depression, and its progress can be stopped by blocking the channel. Three kinds of dams are typically used in such circumstances: a complete dam, an overflow dam, or an underflow dam.

A *complete dam* is placed across a small stream or ditch to completely stop the flow of materials through the channel. This type of dam is used only in areas where the stream or ditch is basically dry and the amount of material that needs to be controlled is relatively small.

SKILL DRILL 37-1
Using Absorption/Adsorption to Manage a Hazardous Materials Incident
NFPA 1072: 6.6.1; NFPA 472: 6.6.1.2.2, 6.6.3.1

1 Decide which material is best suited for use with the spilled product. Assess the location of the spill, and stay clear of any spilled product. Use detection and monitoring devices to determine whether airborne contamination is present. Consult reference sources for the physical and chemical properties of the spilled material.

2 Apply the appropriate material to control the spilled product.

3 Maintain control of the absorbent/adsorbent materials, and take appropriate steps for their disposal.

© Jones & Bartlett Learning. Photographed by Glen E. Ellman.

For hazardous materials spills in streams or ditches with a continuous flow of water, an overflow or underflow dam must be constructed. An *overflow dam* is used to contain materials heavier than water (specific gravity > 1). It is constructed by building a dam base up to a level that holds back the flow of water (**FIGURE 37-6**). Polyvinyl chloride (PVC) pipe or hard suction hose is installed at a slight angle to allow the water to flow "over" the released liquid, thereby trapping the heavier material at a low level at the base of the dam.

An *underflow dam* is used to contain materials lighter than water (specific gravity < 1), and it is basically constructed in the opposite manner from the overflow dam. The piping on the underflow dam is installed near the bottom of the dam so the water flows "under" the dam, thereby allowing the materials floating on the water to accumulate at the top of the dam area (**FIGURE 37-7**). It is critical to have sufficient pipes and hose to allow enough water to flow past the dam without flowing over the top of the dam.

To construct an overflow dam, follow the steps in **SKILL DRILL 37-2**.

To construct an underflow dam, follow the steps in **SKILL DRILL 37-3**.

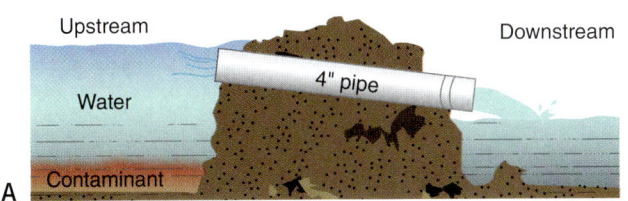

A

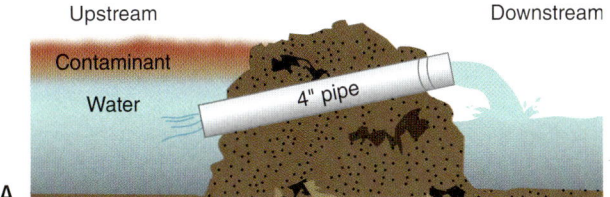

A

B

B

C

C

FIGURE 37-6 A. An overflow dam is used to contain materials that are heavier than water. **B.** The piping at or near the surface of the retained water flow. **C.** A wider view of the overflow dam showing where the water can flow through the dam above the heavier-than-water liquid held back at the base of the dam.

A: © Jones & Bartlett Learning; B: Courtesy of Rob Schnepp; C: © Jones & Bartlett Learning. Photographed by Glen E. Ellman.

FIGURE 37-7 A. An underflow dam is used to contain materials that are lighter than water. **B.** The piping below the surface of the water, allowing the lighter-than-water liquid to float on the surface and be retained by the dam. **C.** Water flowing through the dam.

A: © Jones & Bartlett Learning; B and C: © Jones & Bartlett Learning. Photographed by Glen E. Ellman.

SKILL DRILL 37-2
Constructing an Overflow Dam NFPA 1072: 6.6.1; NFPA 472: 6.6.1.2.2, 6.6.3.1

© Jones & Bartlett Learning.

1 Determine the need for and location of an overflow dam. Build a dam with sandbags or other available materials.

2 Install the appropriate number of 76.2- to 101.6-mm (3- to 4-in.) plastic pipes horizontally on top of the dam, and then add more sandbags on top of the dam. Complete the dam installation, and ensure that the piping allows the proper flow of water without allowing the heavier-than-water material to pass through the pipes.

SKILL DRILL 37-3
Constructing an Underflow Dam NFPA 1072: 6.6.1; NFPA 472: 6.6.1.2.2, 6.6.3.1

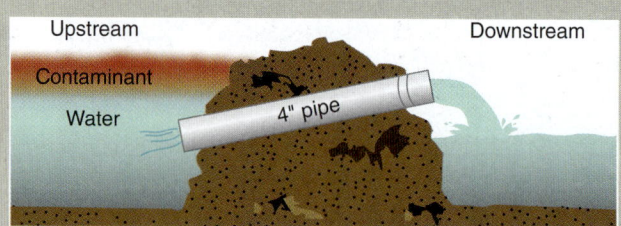

© Jones & Bartlett Learning.

1 Determine the need for and location of an underflow dam. Build a dam with sandbags or other available materials.

2 Install two to three lengths of 76.2- to 101.6-mm (3- to 4-in.) plastic pipes at a 20- to 30-degree angle on top of the dam, and add more sandbags on top of the dam. Complete the dam installation, and ensure that the size will allow the proper flow of water underneath the lighter-than-water liquid.

Diking

Diking is the placement of a selected material such as sand, dirt, loose absorbent, or concrete to form a barrier that will keep a hazardous material (in liquid form) from entering an unwanted area or to hold the material in a specific location. Before contemplating construction of a dike, you must confirm that the material used to control the hazard will not react adversely with the spilled material.

To construct a dike, follow the steps in **SKILL DRILL 37-4**. It may be necessary to construct a

SKILL DRILL 37-4
Constructing a Dike NFPA 1072: 6.6.1; NFPA 472: 6.6.1.2.2, 6.6.3.1

1 Determine the best location for the dike. If necessary, dig a depression in the ground 127 to 203.2 mm (6 to 8 in.) deep. Ensure that plastic will not react adversely with the spilled chemical. Use plastic to line the bottom of the depression, and allow for sufficient plastic to cover the dike wall.

2 Build a short wall with sandbags or other available materials.

3 Complete the dike installation, and ensure that its size will contain the spilled product.

© Jones & Bartlett Learning. Photographed by Glen E. Ellman.

series of dikes if there is a concern about the amount of product being released. A wise incident commander will consider backing up the original structure to ensure that the product will be controlled effectively.

Dilution

Dilution is the addition of water or another substance to weaken the strength or concentration of a hazardous material (typically a corrosive). Dilution can be used only when the identity and properties of the hazardous material are known with certainty. One concern with dilution is that water applied to dilute a hazardous material may simply increase the total volume; if the volume increases too much, it may overwhelm the containment measures implemented by responders. For example, if 3.8 L (1 gal) of water were added to 11.4 L (3 gal) of spilled hydrochloric acid (pH = 3), the result of this action would be the creation of 15.1 L (4 gal) of hydrochloric acid (pH = 3); it takes a *tremendous* amount of water to effectively dilute such a low-pH acid. Dilution should be used with extreme caution and only on the advice of those knowledgeable about the nature of the chemicals involved in the incident.

To use dilution to manage a hazardous materials incident, follow the steps in **SKILL DRILL 37-5**.

Diversion

Diversion techniques in general are intended to redirect the flow of a liquid away from an endangered area to an area where it will have less impact. In many cases, however, responders do not need to build elaborate dikes to divert the flow of a spilled liquid. Instead, existing barriers—such as curbs or the curvature of the roadway—may be effective structures to divert liquids away from storm drains or other unwanted destinations. Additionally, dirt berms; spill booms; and plastic tarps filled with sand, dirt, or clay may provide a rapidly employable diversion mechanism. These diversion methods are not as "permanent" as a dike; however, they can be constructed fairly quickly. Remember—the intent of diversion is to protect sensitive environmental areas or any other areas that would be adversely affected by the spilled liquid.

To employ a diversion technique to manage a spilled liquid, follow the steps in **SKILL DRILL 37-6**.

Retention

Retention is the process of creating a defined area to hold hazardous materials. For instance, it may involve digging a depression in the ground and allowing material to pool in the depression. The material is then held there until a clean-up contractor can recover it. In some cases, some sort of diversion technique may be required to guide the spilled liquid into the retention basin. As an example, a city's public works department or an independent contractor might use a backhoe to create a retention area for runoff at a safe downhill distance from the release.

To use retention, follow the steps in **SKILL DRILL 37-7**.

SKILL DRILL 37-5
Using Dilution to Manage a Hazardous Materials Incident NFPA 1072: 6.6.1; NFPA 472: 6.6.1.2.2, 6.6.3.1

■ Obtain guidance from a hazardous materials technician, specialist, or professional. Determine the viability of a dilution operation. Ensure that the water used will not overflow and affect other product-control activities. Add small amounts of water from a distance to dilute the product. Contact the hazardous materials technician, specialist, or other qualified professional if additional issues arise.

SKILL DRILL 37-6
Constructing a Diversion NFPA 1072: 6.6.1; NFPA 472: 6.6.1.2.2, 6.6.3.1

■ Determine the best location for the diversion. Use sandbags or other materials to divert the product flow to an area with fewer hazards. Stay clear of the product flow. Monitor the diversion channel to ensure the integrity of the system.

© Jones & Bartlett Learning. Photographed by Glen E. Ellman.

SKILL DRILL 37-7
Using Retention to Manage a Hazardous Materials Incident NFPA 1072: 6.6.1; NFPA 472: 6.6.1.2.2, 6.6.3.1

■ Determine the best location for the retention system. Dig a depression in the ground to serve as a retention area. Use plastic to line the bottom of the depression. Use sandbags or other materials to hold the plastic in place. Stay clear of the product flow. Monitor the retention system to ensure its integrity.

© Jones & Bartlett Learning. Photographed by Glen E. Ellman.

Remote Valve Shut-off

A protective action that should always be considered—especially with transportation emergencies or incidents at fixed facilities—is the identification and isolation of the remote valve shut-off. Many chemical processes and many piped systems that carry chemicals provide a way to remotely shut down a system or isolate a leaking fitting or valve. This action may prove to be a much safer way to mitigate the problem (**FIGURE 37-8**).

Fixed ammonia systems provide a good example of the effectiveness of using remote valve shut-offs. In the event of a large-scale ammonia release, most current

FIGURE 37-8 When available, consider using remote shut-off valves to mitigate the problem.
© Jones & Bartlett Learning. Photographed by Glen E. Ellman.

FIGURE 37-9 The remote shut-off valve is typically found near the front of the cab, adjacent to the driver's door, or at the rear of an MC-306/DOT 406 cargo tank.
Courtesy of Glen Rudner.

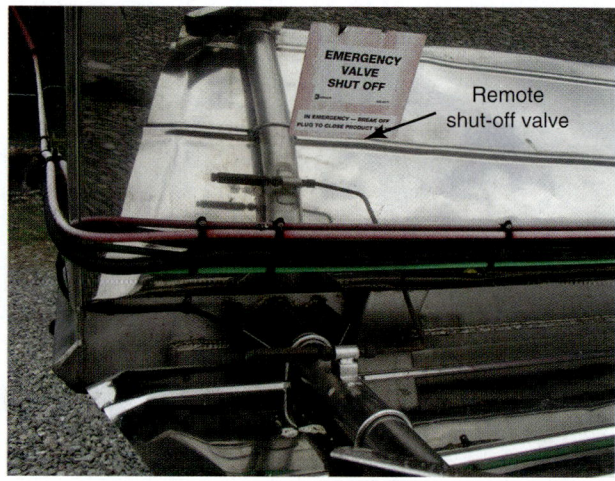

FIGURE 37-10 The remote shut-off valve is typically found near the front of the cab, adjacent to the driver's door, or at the rear of an MC-307/DOT 407 cargo tank.
Courtesy of Glen Rudner.

systems have a main dump valve that will immediately redirect anhydrous ammonia into a water deluge system or diffuser. These systems are designed to knock down the vapours and collect the resulting liquid in a designated holding system.

Many types of cargo tanks have emergency remote shut-off valves. MC-306/DOT 406 cargo tanks, for example, carry flammable and combustible liquids and Class B poisons. These single-shell aluminum tanks may hold as much as 34,826 L (9200 gal) of product at atmospheric pressure. The tanks have an oval/elliptical cross-section and overturn protection that serve to protect the top-mounted fittings in the event of a rollover. In the event of fire, a fusible link is designed to melt at a predetermined temperature (typically around 121°C [250°F]); its collapse causes the closure of the internal product discharge valve. A baffle system within the cargo tank compartment reduces the surge of the liquid contents. Remote shut-off valves are located at the front of the cargo tank on the driver's side or at the rear of the cargo tank on the passenger side (**FIGURE 37-9**).

MC-307/DOT 407 cargo tanks carry chemicals that are transported at low pressure, such as flammable and combustible liquids as well as mild corrosives and poisons. These vehicles may carry as much as 26,498 L (7000 gal) of product and are outfitted with many safety features like those found on the MC-306/DOT 406 cargo tanks (**FIGURE 37-10**).

MC-331 cargo tanks carry compressed liquefied gases such as anhydrous ammonia, propane, butane, and liquefied petroleum gas (LPG). The tanks, which are not insulated, have rounded ends, baffles, and a circular cross-section. MC-331 cargo tanks have a carrying capacity between 9464 and 43,532 L (2500 and 11,500 gal). These types of cargo tanks have remote shut-off valves located at both ends of the tank, internal shut-off valves, a rotary gauge depicting product pressure, and top-mounted vents (**FIGURE 37-11**). Intermodal tanks have emergency shut-offs that close the bottom outlet valve. Typically, the shut-off valves are located on one side of the container near the discharge end.

FIGURE 37-11 The MC-331 cargo tank has remote shut-off valves at both ends of the tank, internal shut-off valves, a rotary gauge depicting product pressure, and two top-mounted vents.
Courtesy of Rob Schnepp.

SAFETY TIP

Participating in product control operations may require a great deal of physical energy and mental concentration. Remember to take rehabilitation breaks throughout the incident to minimize the effects of heat stress and fatigue. Refer to NFPA 473, *Standard for Competencies for EMS Personnel Responding to Hazardous Materials/WMD Incidents*, for more information on heat stress.

Vapour Dispersion and Suppression

When a hazardous material produces a vapour that collects in an area or increases in concentration, vapour dispersion or suppression measures may be required.

Vapour dispersion is the process of lowering the concentration of vapours by spreading the vapours out. Vapours can be dispersed with hose streams set on fog patterns, large displacement fans (being mindful of the fan itself becoming a source of ignition), or other types of mechanical ventilation found in fixed hazardous materials handling systems. Before dispersing vapours, consider the consequences of this action. If the vapours are highly flammable, an attempt to disperse them may ignite the vapours. The dispersed vapours can also contaminate areas outside the hot zone. Vapour dispersion with fans or a fog stream should be attempted only after the hazardous material is safely identified and in weather conditions that will promote the diversion of vapours away from populated areas. Conditions must be constantly monitored during the entire operation.

Vapour suppression is the process of controlling fumes or vapours that are given off by certain materials,

particularly flammable liquids, to prevent their ignition. It is accomplished by covering the hazardous material with foam or some other material or by reducing the temperature of the material. For example, gasoline gives off vapours that present a danger because they are flammable. To control these vapours, a blanket of firefighting or vapour-suppressing foam is layered on the surface of the liquid.

Not all vapour-suppressing foams are appropriate for all types of applications. Much like the concept of choosing the right PPE to protect you from a hazardous substance, responders must choose a type of foam that is designed to work in certain situations. For example, Class A foam is typically used to fight fires involving ordinary combustible materials such as wood and paper. In most situations involving hazardous materials or WMD agents, Class A foam is *not* the most appropriate type of foam to use. Remember, compressed air foam systems (CAFS) use Class A foam. Class B foam, by contrast, is used to fight fires and suppress vapours involving flammable and combustible liquids such as gasoline and diesel fuel. This type of foam seals the surface of a spilled liquid and prevents the vapours from escaping, thereby reducing the danger of vapour ignition. If the vapours are burning, a gently applied foam blanket will also spread out over the surface of the liquid and work to smother the fire.

Responders should be aware of the nature of the materials that are released and the way in which foam concentrates behave when they are applied. Take the time to become familiar with the types of foam concentrates that may be used by your agency.

Reducing the temperature of some hazardous materials will also suppress vapour formation. Unfortunately, there is no easy way to accomplish this goal except in cases of small spills.

Employ the application of foam concentrates in accordance with the policies and procedures of the AHJ and after consulting with a hazardous materials technician or another subject matter expert in the field.

To use vapour dispersion to manage a hazardous materials release, follow the steps in **SKILL DRILL 37-8**.

To perform vapour suppression using foam, follow the steps in **SKILL DRILL 37-9**.

Always remember that as foam is applied, more volume is added to the spill. All of the following foams suppress the ignition of flammable vapours and may be used to extinguish fires in flammable or combustible liquids:

- **Aqueous film-forming foam (AFFF)** can be used at a 1 percent, 3 percent, or 6 percent concentration. This type of foam is designed to form a blanket over spilled flammable liquids to suppress vapours or on actively burning pools of flammable

SKILL DRILL 37-8
Using Vapour Dispersion to Manage a Hazardous Materials Incident
NFPA 1072: 6.6.1; NFPA 472: 6.6.1.2.2, 6.6.3.1, 6.6.4.2

1 Determine the viability of a dispersion operation. Use the appropriate monitoring instrument to determine the boundaries of a safe work area. Ensure that ignition sources in the area have been removed or controlled.

2 Apply water from a distance to disperse vapours. Monitor the environment until the vapours have been adequately dispersed.

© Jones & Bartlett Learning. Photographed by Glen E. Ellman.

SKILL DRILL 37-9
Using Vapour Suppression to Manage a Hazardous Materials Incident
NFPA 1072: 6.6.1; NFPA 472: 6.6.1.2.2, 6.6.3.1, 6.6.4.2

■ Determine the viability of a vapour suppression operation. Use the appropriate instrument to determine a safe work area. Ensure that ignition sources in the area have been removed or controlled. Apply foam from a safe distance to suppress vapours. Monitor the environment until the vapours have been adequately suppressed.

© Jones & Bartlett Learning. Photographed by Glen E. Ellman.

liquids. Foam blankets are designed to prevent a fire from reigniting once extinguished. Most AFFF foam concentrates are biodegradable. AFFF is usually applied by way of in-line foam eductors, foam sprinkler systems, and portable or fixed proportioning systems.

■ **Alcohol-resistant concentrates** have properties like AFFF, except that they are formulated so that alcohols such as methyl alcohol and isopropyl alcohol and other polar solvents will not dissolve the foam. (Regular protein foams cannot be used on these types of products.)

■ **Fluoroprotein foam** contains protein products mixed with synthetic fluorinated surfactants. These foams can be used on fires or spills involving gasoline, oil, or similar products. Fluoroprotein foams rapidly spread over the fuel, ensuring fire knockdown and vapour suppression. Fluoroprotein foams, like many of the synthetic foam concentrates and other types of special-purpose foams, are resistant to polar solvents such as alcohols, ketones, and ethers.

■ **Protein foam** concentrates are made from hydrolyzed proteins (animal by-products), along with stabilizers and preservatives. These types of foams are very stable and possess good expansion properties. They are quite durable and resistant to reignition when used on Class B fires or spills

involving nonpolar substances such as gasoline, toluene, oil, and kerosene.

■ **High-expansion foam** is used when large volumes of foam are required for spills or fires in warehouses, tank farms, and hazardous waste facilities. It is not uncommon to have a yield of more than 3785 L (1000 gal) of finished foam from every litre of foam concentrate used. The expansion is accomplished by pumping large volumes of air through a small screen coated with a foam solution. Because of the large amount of air in the foam, high-expansion foam is referred to as "dry" foam. By excluding oxygen from the fire environment, this agent smothers the fire, leaving very little water residue.

There are several ways to apply foam. Foam concentrates (other than high-expansion foam) should be gently applied or bounced off another adjacent object so that they flow down across the liquid and do not directly upset the burning surface. Foam can also be applied in a rain-down method by directing the stream into the air over the material and letting the foam gently fall onto the surface of the liquid, as rain would.

To apply foam using the rain-down method, follow the steps in **SKILL DRILL 37-10**. The rain-down method is less effective when the fire is creating an intense thermal column. When the heat from the thermal

SKILL DRILL 37-10
Performing the Rain-Down Method of Applying Foam NFPA 1072: 6.6.1; NFPA 472: 6.6.1.2.2, 6.6.3.1, 6.6.4.2

1 Open the nozzle, and test to ensure that foam is being produced. Move within a safe range of the product, and open the nozzle.

2 Direct the stream of foam into the air so that the foam gently falls onto the pool of the product.

3 Allow the foam to flow across the top of the pool of the product until it is completely covered.

© Jones & Bartlett Learning.

column is high, the water content of the foam will turn to steam before it comes in contact with the liquid surface.

Foam can also be applied via the roll-in method by bouncing the stream directly into the front of the spill area and allowing it to gently push forward into the pool rather than directly "splashing" it in.

To apply foam using the roll-in method, follow the steps in **SKILL DRILL 37-11**.

The bounce-off method is used in situations where a fire fighter can use an object to deflect the foam stream and let it flow down onto the burning surface. For example, this method could be used to apply foam to an open-top storage tank or a rolled-over transport vehicle. The foam should be swept back and forth against the object while the foam flows down and spreads back across its surface. As with all foam application methods,

SKILL DRILL 37-11
Performing the Roll-In Method of Applying Foam NFPA 1072: 6.6.1; NFPA 472: 6.6.1.2.2, 6.6.3.1, 6.6.4.2

■ Open the nozzle, and test to ensure that foam is being produced. Move within a safe range of the product, and open the nozzle. Direct the stream of foam onto the ground just in front of the pool of product. Allow the foam to roll across the top of the pool of product until it is completely covered.

© Jones & Bartlett Learning. Photographed by Glen E. Ellman.

it is important to let the foam blanket flow gently on the surface of the flammable liquid to form a blanket.

To perform the bounce-off method of applying foam, follow the steps in **SKILL DRILL 37-12**.

Recovery

The **recovery phase** of a hazardous materials incident occurs when the imminent danger to people, property, and the environment has passed or is controlled and clean-up begins. During the recovery phase, local, provincial, territorial, and federal agencies may become involved in cleaning up the site, determining the responsible party, and implementing cost-recovery methods. The recovery phase in large-scale incidents can go on for days, weeks, or even months and may require large amounts of resources and equipment (**FIGURE 37-12**). As part of recovery, reporting and documentation should be completed in accordance with the AHJ's policies and procedures for incidents requiring product control.

In some situations there is a clear transition between the emergency phase and the clean-up phase. The decision to declare a change from the emergency phase to the recovery phase is typically made by the incident commander. For example, a distinct hand-off between on-scene public safety responders and private-sector commercial clean-up companies would fit this description. It is not uncommon for public-sector responders to hand off the clean-up operations but remain on scene to ensure that the operation is carried out properly. While public safety personnel and responders remain on the scene, they should maintain their vigilance for safety to ensure that no new hazards are created and no new injuries are incurred during the recovery phase.

LISTEN UP!

The ultimate goal of the recovery phase is to return the property or site of the incident to its preincident condition and to return the facility or mode of transportation to the responsible party.

SKILL DRILL 37-12
Performing the Bounce-Off Method of Applying Foam NFPA 1072: 6.6.1; NFPA 472: 6.6.1.2.2, 6.6.3.1, 6.6.4.2

1 Open the nozzle, and test to ensure that foam is being produced.

2 Move within a safe range of the product, and open the nozzle. Direct the stream of foam onto a solid structure such as a wall or metal tank so that the foam is directed off the object and onto the pool of product. Allow the foam to flow across the top of the pool of product until it is completely covered. Be aware that the foam may need to be bounced off several areas of the solid object to extinguish the burning product.

© Jones & Bartlett Learning. Photographed by Glen E. Ellman.

FIGURE 37-12 The recovery phase involves clean-up, determination of the responsible party, and implementation of cost recovery.

Courtesy of Captain David Jackson, Saginaw Township Fire Department.

At other times, the initial responders perform both the emergency response and the clean-up. For example, responders often handle in their entirety small gasoline spills resulting from auto accidents or spills at gas stations.

> **LISTEN UP!**
>
> After performing any product control procedure, follow your local procedures for technical decontamination.

The transition from emergency to clean-up should be carried out in a manner consistent with your agency's SOPs and the guidelines determined by your AHJ. The recovery phase also includes completion of the records necessary for documenting the incident. Additionally, responders should be aware of cost recovery policies and procedures in their AHJ. There are certain instances when the spiller or responsible party for the release may be financially responsible for the costs incurred to mitigate the problem.

After-Action REVIEW

IN SUMMARY

- When considering a control option, a variety of factors must be evaluated, such as the maximum quantity of material that can be released and the likely duration of the incident if no intervention is made.

- Sometimes the situation is so extreme, or the proposed action is so risky, that it may be prudent to create a safe perimeter and let the problem stabilize on its own.

- Control techniques or options generally aim to contain, redirect, or lower the concentration of the hazardous material involved and/or to prevent the ignition of flammable liquids or gases.

- Hazardous materials responders can employ a variety of product-control options, such as absorption, diversion, damming, diking, or isolating a leak with remote shut-off valves.

- Special foams can help both in vapour suppression actions and in extinguishment of fires associated with flammable liquid releases.

- Responders should be aware of the locations of emergency shut-off valves at fixed facilities and on the various types of cargo tanks typically found in their response area.

- The recovery phase may last for days or months and aims to return the exposure area to its original condition and return the facility or mode of transportation to the responsible party.

KEY TERMS

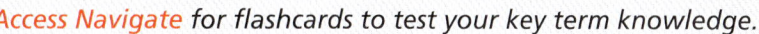 *Access Navigate for flashcards to test your key term knowledge.*

Absorption The process of applying a material that will soak up and hold a hazardous material in a sponge-like manner for collection and subsequent disposal.

Adsorption The process in which a contaminant adheres to the surface of an added material—such as silica or activated carbon—rather than combining with it (as in absorption).

Alcohol-resistant foam concentrate A concentrate used for fighting fires on water-soluble materials and other fuels destructive to regular, AFFF, or FFFP foams, as well as for fires involving hydrocarbons. (NFPA 11)

Aqueous film-forming foam (AFFF) A concentrate based on fluorinated surfactants plus foam stabilizers to produce a fluid aqueous film for suppressing hydrocarbon fuel vapours and usually diluted with water to a 1 percent, 3 percent, or 6 percent solution. (NFPA 11)

Confinement Those procedures taken to keep a material, once released, in a defined or local area. (NFPA 472)

Containment The actions taken to keep a material in its container (e.g., stop a release of the material or reduce the amount being released). (NFPA 472)

Control The procedures, techniques, and methods used in the mitigation of hazardous materials/weapons of mass destruction (WMD) incidents, including containment, extinguishment, and confinement. (NFPA 1072)

Damming The product-control process used when liquid is flowing in a natural channel or depression, and its progress can be stopped by constructing a barrier to block the flow.

Diking The placement of materials to form a barrier that will keep a hazardous material in liquid form from entering an area or that will hold the material in an area.

Dilution The process of adding a substance—usually water—to weaken the concentration of another substance.

Diversion The process of redirecting spilled or leaking material to an area where it will have less impact.

Fluoroprotein foam A protein-based foam concentrate to which fluorochemical surfactants have been added. (NFPA 402)

High-expansion foam A foam created by pumping large volumes of air through a small screen coated with a foam solution. Some high-expansion foams have expansion ratios ranging from 200:1 to approximately 1000:1.

Protein foam A protein-based foam concentrate that is stabilized with metal salts to make a fire-resistant foam blanket. (NFPA 402)

Recovery phase The stage of a hazardous materials incident after imminent danger has passed, when clean-up and the return to normalcy have begun.

Remote valve shut-off A type of valve that may be found at fixed facilities utilizing chemical processes or piped systems that carry chemicals. These remote valves provide a way to remotely shut down a system or isolate a leaking fitting or valve. Remote shut-off valves are also found on many types of cargo containers.

Retention The process of purposefully collecting hazardous materials in a defined area.

Vapour dispersion The process of lowering the concentration of vapours by spreading them out, typically with a water fog from a hose line.

Vapour suppression The process of controlling vapours given off by hazardous materials, thereby preventing vapour ignition, by covering the product with foam or other material or by reducing the temperature of the material.

On Scene

Your pump company has arrived on the scene of an MC-306/DOT 406 flammable liquid cargo tanker parked on the side of the trans-Canada highway. Prior to your arrival, the dispatch center advises you that the tanker is carrying gasoline. Upon arrival, you notice the beginnings of a stream of released liquid flowing from the underside of the tank, away from your location. The driver approaches you and confirms the product to be gasoline and states that he has no idea why the leak is occurring. The air temperature is approximately 32°C (90°F), and the winds are calm. You are considering the options available to control the released liquid.

1. Where would be the most likely places to find the remote shut-off valves on this type of cargo tanker?

   A. Underneath the truck, near the front and rear wheels

   B. Adjacent to the driver's door or at the rear of the cargo tank

   C. Inside the cab next to the safety data sheet (SDS) or under the passenger seat

   D. On top of the cargo tank; one on the front of the tank and the other at the rear

2. If the leak has been stopped and approximately 1893 L (500 gal) of liquid is now retained in a catch basin, which type of foam might be applied effectively to this type of product (gasoline) to suppress the vapours?

   A. High-expansion foam

   B. Foam eduction solution

   C. Surfactant foam concentrate

   D. Aqueous film-forming foam (AFFF)

3. You notice the AFFF foam blanket appears to be breaking down too quickly. Which of the following statements best describes a likely cause?

   A. The pavement temperature is too hot; the foam is breaking down due to heat.

   B. The AFFF is incompatible with the spilled liquid; you should consider switching to alcohol-resistant-type foam.

   C. The pool of spilled liquid is too deep; foam works only on shallow pools of liquid.

   D. You should change nothing; keep applying the same foam.

4. Imagine that the leak worsens, to the point that the flowing liquid has now reached an adjacent 5-m (15-ft)-wide drainage canal with water flowing in it. Your team decides to build a dam to capture the flowing gasoline. Which of the following types of dams would be appropriate given the physical properties of gasoline?

   A. A diversion dam

   B. An underflow dam

   C. An overflow dam

   D. A retention dam

Access Navigate to find answers to this On Scene, along with other resources such as an audiobook.

Appendix A

Extreme Cold Weather: Cautions and Operations

Introduction

Responding to incidents in extreme cold temperatures is a challenging reality that the Canadian fire service faces every winter. Responding to emergency incidents—fires, rescues, hazardous materials spills, and emergency medical calls—is both dangerous and difficult in normal weather conditions; extremely cold temperatures, ice-covered roads, and deep snow combine to make response, arrival, and operations particularly treacherous (**FIGURE A-1**).

In this chapter, we will discuss how to prepare fire apparatus, water supply, and personnel for the special challenges that an incident in extreme cold weather can bring. We will discuss the precautions to take when responding to and arriving at an incident and fire-ground tactics used in extremely cold weather.

Preplanning and Preparing

Prepping the Fire Apparatus

Canadian winters can be extreme events that challenge both firefighters and their equipment. If not managed properly, equipment can become frozen, rendering it useless. In an emergency a frozen fire hydrant or pump can prove life threatening. Most Canadian fire departments have SOPs outlining cold weather operations that include vehicle maintenance and preparation.

The following outlines several useful tips for driver operators. Notwithstanding, follow your individual departmental SOPs and or guidelines.

- Ensure that all drains are in working condition on fire apparatus with pumps.
- Depending on the manufacturers' guidelines, pump mechanisms should be drained and run dry in extreme cold conditions. Some manufacturers may require a pump to be stored wet.
- On fire apparatus with pumps, keep a spray bottle on the fire apparatus that contains alcohol or antifreeze, which can be applied to threads on pump outlets and intakes in case of an ice buildup.
- Before leaving the warmth of the fire station, close all drains on fire apparatus with pumps. If left open, the drains may freeze in the open position and compromise pump performance at the scene.
- Ensure that the seating compartments are dry. Compartment doors may have to be left open at the fire station to allow for complete drying. If they are left closed with moisture still inside, the doors and windows may freeze during an incident.
- Ensure that the cab and compartment heaters are in working order.
- On fire apparatus with pumps, ensure that the heat pans under the pump are in place and that the heaters are working.
- Keep sandbags, salt, or other materials that can be used to spread over ice to provide traction for feet and wheels. Covering icy or snowy spots with such materials will prevent slips and falls and can also help to prevent the fire apparatus from becoming stuck in the snow.
- Keep shovels on the fire apparatus to shovel snow away from fire hydrants and walking paths.

FIGURE A-1 Winter presents a different set of problems for fire fighters.
© Keith Muratori/ShutterStock, Inc.

Prepping Tankers/Tenders

When using tankers/tenders to supply water in extreme-cold-weather conditions, there is a tendency for the valves to freeze or for ice buildup to occur on the tank vents. This can create pressure in the tank and possibly cause a collapse. On fire apparatus with pumps, it is important to always drain water from the pump, valves, and piping after use to prevent freezing and damage.

The turnaround time for the tankers/tenders may also be affected by the cold weather conditions. As discussed in the chapter "Water Supply," when tankers/tenders are used, fire departments establish a dump area where tarps are placed on the ground and portable tanks are placed on top of the tarps. In extreme-cold-weather conditions, the tarps may freeze to the ground and become difficult to remove after the incident. In addition, the portable tanks may become frozen to the tarps and the ground if enough overflowing water creates an icy seal. To prevent this, dumping areas for the tankers/tenders should be well thought out during the preplanning stage to minimize the risk of frozen tarps and tanks.

SAFETY TIP

It is important to ensure that when the incident is over, the water is drained away from the established dump area. This makes it easier for the incident cleanup and reduces the risk of slip-and-fall injuries on any newly formed ice.

Water Supply Cold Weather Maintenance

Fire Hydrants

In Canada, most municipal water distribution systems are maintained through the municipal government. Improperly maintained hydrants can freeze as a result of static water accumulating in the hydrant barrel. The outlet caps may be frozen and be very difficult to remove without first applying heat to thaw the ice.

Fire hydrants may be buried from heavy snowfall and may not be quickly locatable. Some areas identify fire hydrant locations with indicators that are 1.5 m (5 ft) high and not easily buried in snowfall (**FIGURE A-2**). Fire departments in some areas have encouraged citizens to remove snow from the fire hydrants on their properties through public education programs.

Alternative Water Sources

In rural areas, **alternative water sources** can be identified through preincident planning. Many volunteer rural fire services have identified water supply sites and mark them accordingly. In extremely cold conditions,

FIGURE A-2 Some fire hydrants have indicators to help locate them in heavy snowfall.
© Willowpix/iStockphoto.

static sources of water, such as ponds and lakes, will develop a layer of ice, making access to the water supply challenging. Static sources of water may have a layer of ice ranging from 2 cm to 2 m (0.75 in. to 6.6 ft) thick. Because of this, a common practice in many rural areas is to use ice augers or a chainsaw to open a hole in the ice into which a hard suction hose can be inserted. When crews are working on or near the ice, their safety is an added concern. Only properly trained personnel well versed on the fire department's written operating guidelines should be permitted to work on or near the ice. Personnel working on or near the ice should be supplied with personal flotation devices and life lines.

Many fire services in rural areas have installed dry hydrant systems at static water supply locations. These installations allow for ease of access and reliable sources for water. Unfortunately, because of the costs involved, dry hydrants may not be readily available in all rural fire departments.

> ### FIRE FIGHTER TIP
>
> In Canada, the majority of the fire hydrants are dry barrel to prevent freezing in wintry conditions; however, if the dry barrel hydrants are not drained properly, there is a chance that any left-over water may freeze. To prevent this, your fire department may ask the public works department to come to the incident after overhaul operations to ensure that the fire hydrants are drained properly.

Energized Equipment

When operating battery-powered equipment in extreme-cold-weather conditions, battery life is dramatically reduced. When flashlights, radios, atmospheric monitoring devices, self-contained breathing apparatus (SCBA) heads-up displays (HUDs), and personal alert safety system (PASS) devices are used, the batteries can drain very quickly and require replacement often. Ensure that there are extra batteries available for each item of battery-powered equipment. If the batteries are rechargeable, charge them as soon as possible. Equipment must always be in a ready state; in extreme cold conditions, this is critical for incident success.

> ### FIRE FIGHTER TIP
>
> Some atmospheric monitoring equipment may not work properly or give the correct readings under extremely cold conditions.

Structures under Extreme Conditions

Regardless of the time of year or weather conditions understanding building construction is a critical component of being an effective fire fighter. Remember, you should never enter a building without knowing how that building is likely to be constructed. During the planning and building phase, architects and engineers consider the different types of loads the building must endure: buildings must be able to endure stresses such as wind, rain, snow, extreme temperatures, and gravity.

One critical factor that needs to be considered is the load or weight that the more vulnerable structures of the building will have to support. Architects and engineers in many parts of Canada must take into consideration the weight of the heavy, wet snow that will accumulate on the roof areas; however, wet snow, when combined with the additional water added during firefighting operations, can lead to a roof collapse and an early collapse of the structure (**FIGURE A-3**).

Under extremely cold conditions the water used to suppress a fire can accumulate and freeze quickly. This accumulation of ice adds weight to the overall structure, creating additional stress. Fire fighters using a stream that is flowing at a rate of 1020 L per minute (270 gal per minute) can potentially add 1000 kg (2200 lb) of weight onto the building every minute, creating an unengineered load on the building that could cause its collapse. Risk of collapse occurs when a building cannot withstand the weight and force generated by water and ice buildup. When working in these conditions, be aware of any signs that the structure may be compromised and evacuate immediately.

FIGURE A-3 Buildings are vulnerable to overloading during extreme weather conditions due to the extra weight caused by fire suppression activities.

Courtesy of Bernard Dallaire, Fire Chief of Alma Fire Department, Quebec, Canada

Preincident Planning for the Effects of Extreme Cold on Personnel

Factors Affecting Exposure

There are several factors that affect personnel under extreme cold temperatures. This information can be used when personnel are preparing for extreme environmental temperatures.

- Physical condition—People who are in poor physical condition will not be able to tolerate extreme temperatures as well as those with a strong cardiovascular system. A well-conditioned athlete performs much better and is less likely to experience injury due to the cold than the "weekend warrior" who is not in peak physical condition. In addition, it is easier for a well-conditioned athlete to pick up the pace, even in cold weather, and thereby generate more heat.
- Nutrition and hydration—Your body needs calories in order for your metabolism to function. Staying well hydrated provides water to catalyze this metabolism. A decrease in either will aggravate cold stress. Calories provide fuel to burn, which creates heat during the cold.

SAFETY TIP

The temperature does not need to be extreme for a cold injury to occur. Many cases of hypothermia occur at temperatures between −1°C and 10°C (30°F and 50°F).

Cold Exposure

Normal body temperature must be maintained within a very narrow range for the body's chemistry to work efficiently. If the body, or any part of it, is exposed to a cold environment, these mechanisms may be overwhelmed. Cold exposure may cause injury to individual parts of the body, such as the feet, hands, ears, or nose, or to the body as a whole. When the entire body temperature falls, the condition is called hypothermia.

Because heat always travels from a warmer place to a cooler place, the body tends to lose heat to the environment. The body loses heat in the following five ways:

1. **Conduction** is the direct transfer of heat from a part of the body to a colder object by direct contact, such as when a warm hand touches cold metal or ice, or is immersed in water with a temperature of less than 2.8°C (37°F). Heat passes directly from the body to the colder object.
2. **Convection** occurs when heat is transferred to circulating air, such as when cool air moves across the body's surface. A person standing outside in windy winter weather, wearing lightweight clothing, is losing heat to the environment mostly by convection.
3. **Evaporation** is the conversion of any liquid to a gas, a process that requires energy or heat. Evaporation is the natural mechanism by which sweating cools the body. This is why swimmers coming out of the water feel a sensation of cold as the water evaporates from their skin. Individuals who exercise vigorously in a cool environment may sweat and feel warm at first, but later, as their sweat evaporates, they can become exceedingly cooler. This is why it is critical to keep moisture away from the skin in extreme cold conditions.
4. **Radiation** is the transfer of heat by radiant energy. Radiant energy is a type of invisible light that transfers heat. The body can lose heat by radiation, such as when a person stands in a cold room.
5. **Respiration** causes body heat to be lost as the warm air in the lungs is exhaled into the atmosphere and cooler air is inhaled.

The rate and amount of heat loss can be modified in three ways:

1. **Increase heat production.** One way for the body to increase its heat production is to increase the rate of metabolism, which the body can accomplish by shivering. Also, people often have a natural urge to move around when they are cold.
2. **Move to an area where heat loss is decreased.** The most obvious way to decrease heat loss from radiation and convection is to move out of a cold environment and seek shelter from the wind. Just covering the head will minimize radiation heat loss by up to 70 percent.
3. **Wear insulated clothing.** This helps to decrease heat loss in several ways. Insulators, such as specific materials or dry, still air, do not conduct heat; thus, layers of clothing that trap air provide good insulation, as do wool, down, and synthetic fabrics that contain small pockets of trapped air. Protective clothing also traps perspiration and prevents evaporation. Sweating without evaporation will not result in cooling.

Hypothermia

Hypothermia literally means low temperature. It is diagnosed when the core temperature of the body—the temperature of the heart, lungs, and vital organs—falls below 35°C (95°F). The body can usually tolerate a drop in core temperature of a few degrees; however, below this critical point, the body loses the ability to regulate its temperature and to generate body heat. Progressive loss of body heat then begins.

To protect itself against heat loss, the body normally constricts blood vessels in the skin; this results in the characteristic appearance of blue lips and/or fingertips. As a secondary precaution against heat loss, the body tends to create additional heat by shivering, which is the active movement of many muscles to generate heat. As cold exposure worsens and these mechanisms are overwhelmed, many body functions begin to slow down. Eventually, the functioning of key organs, such as the heart, begins to slow, and if no emergency care is given, this can lead to death.

Hypothermia can develop either quickly, as when someone is immersed in cold water, or more gradually, as when a lost person is exposed to a cold environment for several hours or more. The temperature does not have to be below freezing for hypothermia to occur. In winter, homeless people and those whose homes lack heating may develop hypothermia at higher temperatures.

Signs and Symptoms of Hypothermia

The signs and symptoms of hypothermia are as follows:

- Change in mental status—This is one of the first symptoms of developing hypothermia. Examples include disorientation, apathy, and changes in personality, such as unusual aggressiveness.
- Shivering—Shivering is the first, and most important, of the body's defences against a falling body temperature. Shivering can produce more heat than many rewarming methods. As the core temperature continues to fall, shivering decreases and usually stops at about 35°C (95°F).
- Cool abdomen—Place the back of your hand between the clothing and the victim's abdomen to assess the victim's temperature. When the victim's abdominal skin under clothing is cooler than your hand, consider the victim hypothermic until proven otherwise.

There are two types of hypothermia: mild and severe. With severe hypothermia, the victim becomes so cold that shivering stops, which means the victim's body cannot rewarm itself internally and requires external body heat for recovery. In fact, 50–80 percent of all victims of severe hypothermia die.

Signs of mild hypothermia include the following:

- Vigorous, uncontrollable shivering
- Victim has the "umbles."
  - Grumbles—Decreased mental skills
  - Mumbles—Slurred speech
  - Fumbles—Difficulty using fingers or hands
  - Stumbles—Staggers while walking
- Has cool or cold skin on the abdomen, chest, or back.

If you or a crew member has any of these signs, go to the warm area for rehabilitation, discussed later, where you will be treated by an EMS professional.

You may find a victim at an emergency incident with severe hypothermia. The signs of severe hypothermia are:

- No shivering
- Skin feels ice cold and appears blue.
- Muscles can be stiff and rigid, similar to rigor mortis.
- Altered mental status, not alert
- Breathing and pulse are slow.
- Victim might appear to be dead.

Remove the victim from further exposure to the cold and into a warm area. Handle the victim gently. Do not allow the victim to exert himself or herself. Do not rub the victim's arms or legs. Contact EMS to provide emergency medical care to the victim.

> ### FIRE FIGHTER TIP
>
> In certain regions of Canada during the winter, it is a challenge to train and learn when the mind is focused on just keeping warm. It is a great benefit to these fire departments if they have access to indoor training facilities. In Manitoba, all facets of firefighting training (with the exception of live fires) can be performed in an indoor facility.

Localized Cold Injuries

Most injuries from cold are confined to exposed parts of the body. The extremities, particularly the feet, and the exposed ears, nose, and face are especially vulnerable to cold injuries. When exposed parts of the body become very cold but not frozen, that condition is called frostnip, chilblain, or immersion foot (trench foot). When these parts become frozen, the injury is called frostbite.

Frostnip and Immersion Foot

After prolonged exposure to the cold, skin may freeze, whereas the deeper tissues are unaffected. This condition, which often affects the ears, nose, and fingers, is called frostnip. Because frostnip is usually not painful, the victim often is unaware that a cold injury has occurred. Immersion foot, or trench foot, occurs after prolonged exposure to cold water. It is particularly common in hikers or hunters who stand for a long time in a river or lake. Due to the ever-flowing hose lines at an extreme-cold-weather fire ground, fire fighters may be at risk if their feet get wet during an incident.

Frostbite

Frostbite happens only in below-freezing temperatures. Tissue is not composed of water alone, so it will not freeze until it has been cooled to about −2°C (28.4°F). Tissue is damaged in two ways: (1) actual tissue freezing, which results in the formation of ice crystals within the tissues (the ice crystals expand as they freeze, damaging cells), and (2) the obstruction of the blood supply to the tissue, which causes sludgy blood clots and further prevents blood from flowing to the tissues. The second type of tissue damage is more extensive than the first. In severely cold temperatures, flesh can freeze in less than a minute.

Frostbite can be identified by the hard, frozen feel of the affected tissues. Most frostbitten parts are hard and waxy (**FIGURE A-4**).

Signs and Symptoms of Frostnip and Immersion Foot

It is difficult to tell the difference between frostnip and frostbite. Signs and symptoms of frostnip include:

- Skin appears pale (blanched) and cold to the touch.
- Usually painless with no further damage after rewarming.
- Repeated frostnip in the same spot can dry the skin, causing it to crack and become sensitive.

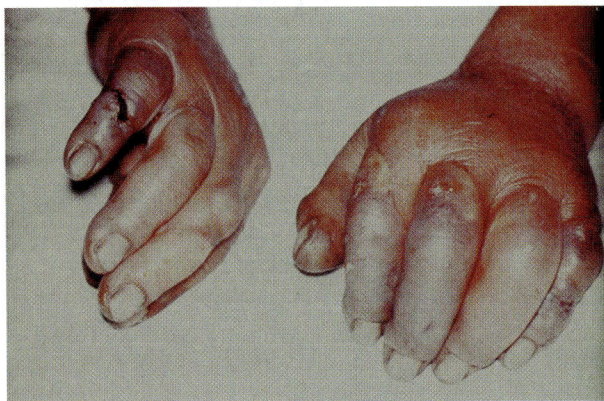

FIGURE A-4 Frostbitten parts are hard and usually waxy to touch.
Courtesy of AAOS.

If you or a crew member has any of these signs, go to a warm area of rehabilitation, discussed later, where you will be treated by an EMS professional. Do not rub the affected area.

Signs and Symptoms of Frostbite

The severity and extent of frostbite are difficult to judge until hours after thawing. Frostbite can be classified as superficial or deep before thawing.

The signs and symptoms of superficial frostbite are as follows:

- The skin is white, waxy, or greyish yellow.
- The affected part feels very cold and numb. There might be a tingling, stinging, or aching sensation.
- The skin's surface feels stiff or crusty, and the underlying tissue is soft when depressed gently and firmly.

The following signs and symptoms indicate deep frostbite:

- The affected part feels cold, hard, and solid and cannot be depressed; it feels like a piece of wood or frozen meat.
- Blisters might appear after rewarming.
- The affected part is cold, with pale, waxy skin.
- A painfully cold part suddenly stops hurting.

After a part has thawed, frostbite can be categorized by degrees:

- First-degree frostbite—The affected part is warm, swollen, and tender.
- Second-degree frostbite—Blisters form minutes to hours after thawing and enlarge over several days (**FIGURE A-5** and **FIGURE A-6**).
- Third-degree frostbite—Blisters are small and contain reddish-blue or purplish fluid. The surrounding skin can be red or blue and might not turn white when pressure is applied.
- Fourth-degree frostbite—No blisters or swelling occur. The part remains numb, cold, and white to dark purple.

If you or a crew member has any of these signs, go to a warm area for rehabilitation, discussed later, where you will be treated by an EMS professional. Frostbite can cause serious damage, especially to fingers and toes.

Preplanning Actions for Extreme Cold

While preparing for the next call, fire fighters must ensure that their personal protective equipment (PPE) is dry, because this will decrease their likelihood of developing

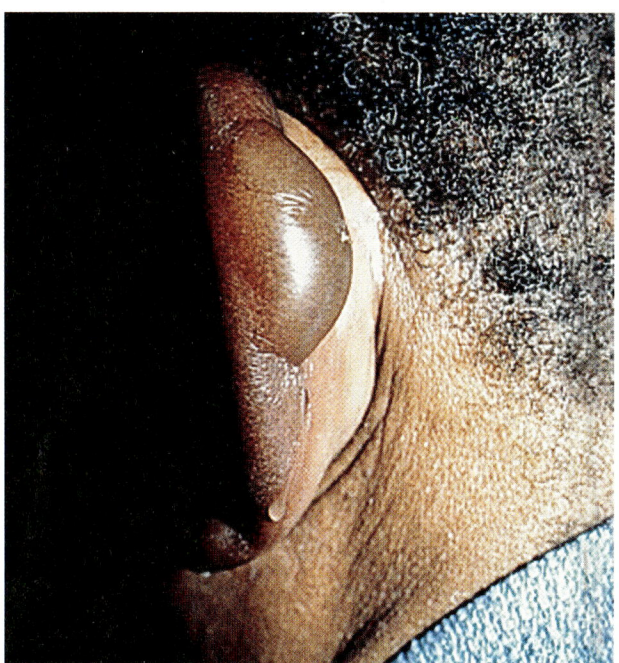

FIGURE A-5 A frostbitten ear.

© Charles Stewart MD, EMDM MPH.

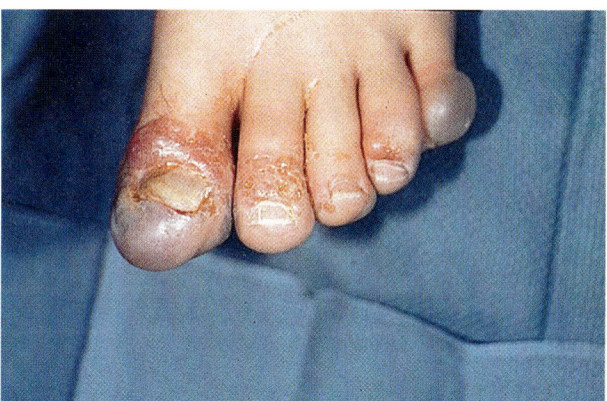

FIGURE A-6 Second-degree frostbite.

Courtesy of AAOS.

hypothermia. Fire fighters must ensure that they have extra clothes to change into during rehabilitation after they get wet. Fire fighters should keep extra gloves, balaclavas, T-shirts, sweatshirts, toques, and extra socks in their PPE cache so they can replace their wet clothes as soon as possible. Wool is one of the best materials a fire fighter can use, because it can keep them warm even when wet.

The effectiveness of fire boots may be reduced in extreme cold temperatures. The rubber in firefighting boots may crack in certain situations. In some extreme cold temperatures, the rubber sole of the boot can become very hard, with decreased ability for traction, which creates a higher risk for slip-and-fall injuries in ice and snow conditions.

Response

Driving in Winter Conditions

The temperature is −30°C (−22°F) with the wind at 60 km per hour (37.3 miles per hour) and a wind chill of −48°C (−54.4°F). It is snowing heavily, and you are dispatched to a structure fire 8 km (5 miles) away. This is a common scenario in many areas of Canada, and fire personnel must ensure that their vehicles and fire apparatus are prepared for the icy conditions and have alternative preplanned routes. Road conditions may be so challenging that some fire departments keep front-end loaders on standby in order to plow their way to the incident or to ensure that the incident area can be cleared of snow during blizzard conditions. Occasionally, snowmobiles have been used to transport medical personnel to an incident, even when the incident is located in an urban area.

Driving on snowy or icy roads requires drivers to reduce their speed (**FIGURE A-7**). Even roads that have been recently plowed can remain slippery. Under these wet conditions, oil slicks often occur and can lead

FIGURE A-7 Driving on snow or icy roads requires the driver to reduce his or her speed.

© Zoran Milich/age fotostock.

to slipping and skidding. Wet roads can cause a loss of tire contact with the road known as **hydroplaning**, which results in loss of control. Depending on road conditions, the stopping distance of the fire apparatus may be reduced, and visibility may also be affected by these conditions.

Fire apparatus that carry water will have increased stopping distances due to liquid surge. **Liquid surge** is the movement of liquid inside a container as the container is moved. As the fire apparatus decelerates or accelerates, the water inside the tank will slosh around. This movement can be very hazardous due to the large amount of water in the tank. If the fire apparatus has to brake quickly, then the liquid surge will try to force the fire apparatus forward, putting an additional strain on the fire apparatus's braking system and increasing the total stopping distance. To reduce the effects of liquid surge, the water tanks are designed with baffles. These plates inside the tank slow down the movement of water and displace some of the energy transferred during the movement of the fire apparatus.

During a snowstorm, visibility may be decreased. In addition to limiting the fire apparatus driver's field of vision, all other drivers on the road will also have a limited field of vision, thus increasing the likelihood of a collision. This is one reason why other drivers on the road have to be taken into consideration during extreme weather conditions, because they can be unpredictable.

Extreme caution is always needed when driving in adverse weather conditions. Extreme weather conditions will affect the response time of departments. Fire personnel should preplan accordingly to ensure that they make it to the incident safely and efficiently.

Arrival

After you have arrived safely at the scene, be careful when dismounting apparatus, because the steps on fire apparatus are often high and can be slippery. Use caution when walking on snow- and ice-covered ground. You cannot help your crew or any victims if you are injured. In order to prevent slips, sand or salt stored on the fire apparatus may be spread underfoot.

Incident Operations

Fire-Ground Tactics

There are many things to consider when dealing with fire-ground tactics in extreme-cold-weather operations. Accessibility to the fire apparatus along with frozen hoses and nozzles adds time and effort to fire-ground tactics. Accessibility to the actual incident site may be an issue if snow drifts are 2 m (6.6 ft) high. Fire apparatus may have to stage and set up at a distance from the scene, which requires hose lays to be stretched farther to reach the fire (**FIGURE A-8**). Because of this, fire apparatus should keep extra lengths of hose to quickly attach to the hose lays. In addition, fire fighters may want to add extra lengths to their preconnected hose lays during the winter months to ensure that the hose will be long enough to cover extended distances.

In extreme-cold-weather conditions, fire fighters need to keep the water flowing through nozzles at all times to prevent them from freezing. This is often called "cracking" the nozzle—that is, opening the nozzle slightly so that a small amount of water will flow through the tip in order to prevent freezing. Cracking the nozzle also prevents freezing because it allows water to keep moving through the entire system—from hydrant to hose. If water becomes static in an extreme cold environment, it will freeze and detrimentally impact the suppression efforts.

FIGURE A-8 Fire apparatus may have to stage and set up at a distance from the scene due to heavy snow.
© Joe McDaniel/iStockphoto.

The pump operator should recirculate the water in the tank on the fire apparatus to keep the water above a freezing temperature. The pump operator also needs to keep an eye on the water temperature and ensure that it does not become too warm. If left to recirculate too long, the water in the tank can actually boil in extreme cold temperatures, which can cause extensive damage to the pump on the fire apparatus.

If unused hose lines are left in place on the ground for a long period of time, they will freeze to the ground. When it is time to clean up the incident scene, fire fighters may cause damage to the hose if they have to use tools to pry the hose from the ground. To minimize this, place unused hose lines in a warmer environment as soon as it is tactically possible.

All water piping must be drained on aerial/tower devices immediately once the water supply has stopped flowing. Leaving a water pipe on a ladder even for a few minutes in extreme cold can cause extensive damage and may prohibit the ladder or boom from retracting or extending. If water spray is an issue, aerial ladder operators should frequently extend and retract the ladder flys to ensure they do not freeze into place.

Use of Ladders

During extreme-cold-weather incidents, using ladders can be a challenge. There may be a large accumulation of snow or ice in the area where the ladders are needed. Fire fighters must ensure that the ladders are stable, even on icy or snow-covered surfaces, in order to perform their duties safely. When roof ladders are required for operations, fire fighters must ensure that the area of placement is free of snow and that the ladder will remain in place with the hooks. This can be very challenging when there is a large amount of snow on the roof and the snow is close to the roof's peak.

Additional threats to safety occur when discharged water accumulates on the ladder rungs and becomes a slip-and-fall hazard. If a fire fighter's gloves are wet, then they may become frozen to the ladder, effectively leaving the fire fighter without the use of his or her hands or arms.

FIRE FIGHTER TIP

Spraying the nozzles with alcohol or antifreeze will aid in freezing prevention.

Rescue Operations

Trying to keep power generators and rescue tools running can be a challenge in extreme cold weather, as the fuel can start to solidify (gel). Hydraulic tools will respond more slowly than usual in extreme cold temperatures, which can affect the time it takes to perform certain tasks, such as vehicle extrication. Keeping hand tools readily available is a common practice to combat this. It is important to have fuel additives in both the fire apparatus and equipment fuel tanks to minimize the risk of fuel gelling and ensure that all fuel is indicated for winter use.

When responding to incidents on roads and highways, all responders must ensure that the apparatus is parked appropriately and safe zones are created. Wintry road conditions cause stopping distances for traffic on roadways to increase, which may put responders at greater risk. There have been many occasions when the traffic driving close to the incident has been involved in secondary incidents that then need to be addressed (**FIGURE A-9**).

Emergency Personnel

Emergency personnel may be dispatched to rescue situations in extreme cold weather. The extreme cold may cause delays in patient removal, creating a risk of cold exposure and injury to responders and to the victim. Responders are typically dressed for the weather, but the victim may not be. To help a patient retain warmth on scene while awaiting removal, responders may deploy extra blankets and cover the victim. Responders may also utilize a tarp as a windbreak to protect themselves and the victim against the wind chill and to help prevent hypothermia. When there is a wind chill of $-50°C$ ($-58°F$), it becomes critical to create protection against the wind.

FIGURE A-9 Winter weather can cause treacherous driving conditions.
© Boris Roessler/EPA/Landov.

Portable heat packs can ensure that the victim's body temperature remains stable. Portable heaters can also be implemented if the victim is in a position where he or she cannot be removed quickly. In certain circumstances, responders may have to perform rapid removal techniques to get the victim to a warm area as soon as possible.

FIGURE A-10 If ice forms on the SCBA, it can affect the low-pressure alarm on the breathing apparatus, making it not as loud as it should be.
© Mark Sauer/Mesabi Daily News/AP Photo.

Self-Contained Breathing Apparatus

In extreme cold weather, SCBA can experience several issues that will complicate its use. For example, when the face piece is donned outside during extreme cold temperatures, it will quickly fog up, and visibility will be reduced. This may lead to trip hazards prior to the fire fighter's entry into the structure. Once the fire fighter enters the structure and the heat from the fire reaches the face piece, condensation will form on the outside of the face piece. This can be alleviated by using a gloved hand to wipe the moisture from the lens on the face piece. This same condition can occur when the fire fighter's helmet visor is lowered.

After the SCBA is donned in the extreme cold, fire fighters must be cognizant of the fact that any condensation in the regulator could lead to its freezing before entry into the structure is even attempted. To reduce the risk of this occurrence, personnel should ensure that the regulator is dry before donning the SCBA. Often, when personnel exit the structure for a full air cylinder, the regulator may freeze. If the secondary regulator attaches to the face piece, avoid allowing it to hang down; always ensure it is clipped into its proper holder when not attached to the face piece itself.

Another issue that can arise is ice forming on the SCBA while the company is working outside (**FIGURE A-10**). This ice can affect the low-pressure alarm (PASS alarm) on the breathing apparatus, effectively muffling it, so it is not as loud as it should be. The ability to use the PASS alarm may also be compromised by a large formation of ice on the PASS alarm. There is also the risk of a fire fighter's gloves freezing to the point where the fire fighter has little dexterity and cannot activate the device.

Warm Area for Rehabilitation

At the incident, command should establish a warm area for the rehabilitation (rehab) of all responders and to ensure that Emergency Medical Services (EMS) are addressing the signs and symptoms of working in the extreme cold. Fire fighters should be trained to identify the signs and symptoms of frostbite or hypothermia not only for themselves but also for all personnel working at an incident during extreme cold temperatures and conditions. Fire fighters need to know when it is time to request rehab, because time is critical in treating cold injuries.

After-Action REVIEW

IN SUMMARY

- Responding to incidents in extreme cold temperatures is a challenging reality that the Canadian fire service faces every winter.
- When preparing the apparatus for cold-weather operations, fire fighters must take preventive action to reduce the risk of the pumps, valves, and piping from freezing.
- Battery life on energized equipment is greatly diminished in extreme cold weather. Ensure that there are extra batteries available for each item of battery-powered equipment.
- In extreme cold weather, building structures have a greater load or weight to support when dealing with wet, heavy snow, or the added weight of firefighting operations, which can lead to roof collapse or collapse of the structure.
- A number of factors affect how a person deals with a cold environment, including:
  - Physical condition
  - Nutrition
  - Hydration
- Cold exposure may cause injury to individual parts of the body, such as hands, feet, ears, or nose, or to the body as a whole.
- Extreme caution is always needed when driving in adverse weather conditions. Extreme weather conditions will affect the response time of departments. Fire personnel should preplan accordingly to ensure that they make it to the incident safely and efficiently.
- After arrival on the scene, fire fighters should use caution when dismounting the apparatus in order to prevent slip-and-fall injuries.
- Accessibility to the apparatus and frozen hoses and nozzles all add time and effort to fire-ground tactics. Accessibility to the actual incident may become an issue if snow drifts are 2 m (6.6 ft) high or greater. Consideration should be taken to keep extra hose lengths for quick attachment to hose lays to ensure the hose will be long enough to cover extended distances.
- Rescue operations become more difficult during extreme weather due to tools responding more slowly than normal because of fuel "gelling," so it is necessary to keep hand tools readily available.
- SCBA use can be complicated by extreme cold weather. Fire fighters should be cognizant of any moisture gathering on their SCBA. Moisture on the SCBA can lead to fogging, freezing, or muffling of the PASS alarm.
- Fire fighters should be trained to identify the signs and symptoms of frostbite and/or hypothermia for themselves and for all personnel working at an incident during extreme cold weather.

KEY TERMS

Alternative water sources Reliable water supply sources found in rural areas, such as ponds or lakes.

Conduction The loss of heat by direct contact (e.g., when a body part comes into contact with a colder object).

Convection The loss of body heat caused by air movement (e.g., a breeze blowing across the body).

Core temperature The temperature of the central parts of the body (i.e., heart, lungs, and vital organs).

Evaporation The loss of body heat caused by the conversion of water or another fluid (often sweat) from a liquid to a gas.

Frostbite Damage to tissues as the result of exposure to cold; frozen body parts.

Hydroplaning The loss of steering or braking ability (and thus, control) because of a layer of water on the road that prevents sufficient contact between the vehicle's tires and the road's surface.

Hypothermia A condition in which the internal body temperature falls below 35°C (95°F), usually as a result of prolonged exposure to cold or freezing temperatures.

Liquid surge The force imposed upon a fire apparatus by the contents of a partially filled water or foam-concentrate tank when the vehicle accelerates, decelerates, or turns. (NFPA 1002)

Radiation The transfer of heat to colder objects in an environment by radiant energy—for example, heat gain from a fire.

Respiration The loss of body heat as warm air in the lungs is exhaled into the atmosphere and cooler air is inhaled.

Appendix B

An Extract from NFPA 1001, *Standard for Fire Fighter Professional Qualifications*, 2019 Edition

Chapter 4 Fire Fighter I

4.1 General. For qualification at Level I, the fire fighter candidate shall meet the general knowledge requirements in 4.1.1; the general skill requirements in 4.1.2; the JPRs defined in Sections 4.2 through 4.5 of this standard, knowledge of the incident management system, and the requirements defined in Chapter 5 as well as mission-specific competencies in Section 6.2, Personal Protective Equipment, and Section 6.6, Product Control, of NFPA 1072.

4.1.1 General Knowledge Requirements. The organization of the fire department; the role of the Fire Fighter I in the organization; the mission of fire service; the fire department's standard operating procedures (SOPs) and rules and regulations as they apply to the Fire Fighter I; the value of fire and life safety initiatives in support of the fire department mission and to reduce fire fighter line-of-duty injuries and fatalities; the role of other agencies as they relate to the fire department; the signs and symptoms of behavioral and emotional distress; aspects of the fire department's member assistance program; the importance of physical fitness and a healthy lifestyle to the performance of the duties of a fire fighter; the critical aspects of NFPA 1500.

4.1.2 General Skill Requirements. The ability to don personal protective clothing, doff personal protective clothing, perform field reduction of contaminants and prepare for reuse, hoist tools and equipment using ropes and the correct knot, and locate information in departmental documents and standard or code materials.

4.2 Fire Department Communications. This duty shall involve initiating responses, receiving telephone calls, and using fire department communications equipment to correctly relay verbal or written information, according to the JPRs in 4.2.1 through 4.2.4.

4.2.1 Initiate the response to a reported emergency, given the report of an emergency, fire department SOPs, and communications equipment, so that all necessary information is obtained, communications equipment is operated correctly, and the information is relayed promptly and accurately to the dispatch center.

(A) Requisite Knowledge. Procedures for reporting an emergency; departmental SOPs for taking and receiving alarms, radio codes, or procedures; and information needs of dispatch center.

(B) Requisite Skills. The ability to operate fire department communications equipment, relay information, and record information.

4.2.2 Receive a telephone call, given a fire department phone, so that procedures for answering the phone are used and the caller's information is relayed.

(A) Requisite Knowledge. Fire department procedures for answering nonemergency telephone calls.

(B) Requisite Skills. The ability to operate fire station telephone and intercom equipment.

4.2.3 Transmit and receive messages via the fire department radio, given a fire department radio and operating procedures, so that the information is accurate, complete, clear, and relayed within the time established by the AHJ.

(A) Requisite Knowledge. Departmental radio procedures and etiquette for routine traffic, emergency traffic, and emergency evacuation signals.

(B) Requisite Skills. The ability to operate radio equipment and discriminate between routine and emergency traffic.

4.2.4 Activate an emergency call for assistance, given vision obscured conditions, PPE, and department SOPs, so that the fire fighter can be located and rescued.

(A) Requisite Knowledge. Personnel accountability systems, emergency communication procedures, and emergency evacuation methods.

(B) Requisite Skills. The ability to initiate an emergency call for assistance in accordance with the AHJ's procedures, the ability to use other methods of emergency calls for assistance.

4.3 Fireground Operations. This duty shall involve performing activities necessary to ensure life safety, fire control, and property conservation, according to the JPRs in 4.3.1 through 4.3.21.

4.3.1 Use self-contained breathing apparatus (SCBA) during emergency operations, given SCBA and other PPE, so that the SCBA is correctly donned, the SCBA is correctly worn, controlled breathing techniques are used, all low-air warnings are recognized, respiratory protection is not intentionally compromised, and hazardous areas are exited prior to air depletion.

(A) Requisite Knowledge. Conditions that require respiratory protection, uses and limitations of SCBA, components of SCBA, donning procedures, breathing techniques, indications for and emergency procedures used with SCBA, and physical requirements of the SCBA wearer.

(B) Requisite Skills. The ability to control breathing, replace SCBA air cylinders, use SCBA to exit through restricted passages, initiate and complete emergency procedures in the event of SCBA failure or air depletion, and complete donning procedures.

4.3.2 Respond on apparatus to an emergency scene, given personal protective clothing and other necessary PPE, so that the apparatus is correctly mounted and dismounted, seat belts are used while the vehicle is in motion, and other personal protective equipment is correctly used.

(A) Requisite Knowledge. Mounting and dismounting procedures for riding fire apparatus, hazards and ways to avoid hazards associated with riding apparatus, prohibited practices, and types of department PPE and the means for usage.

(B) Requisite Skills. The ability to use each piece of provided safety equipment.

4.3.3 Establish and operate in work areas at emergency scenes, given protective equipment, traffic and scene control devices, structure fire and roadway emergency scenes, traffic hazards and downed electrical wires, photovoltaic power systems, battery storage systems, an assignment, and SOPs, so that procedures are followed, protective equipment is worn, protected work areas are established as directed using traffic and scene control devices, and the fire fighter performs assigned tasks only in established, protected work areas.

(A) Requisite Knowledge. Potential hazards involved in operating on emergency scenes including vehicle traffic, utilities, and environmental conditions; proper procedures for dismounting apparatus in traffic; procedures for safe operation at emergency scenes; and the protective equipment available for members' safety on emergency scenes and work zone designations.

(B) Requisite Skills. The ability to use personal protective clothing, deploy traffic and scene control devices, dismount apparatus, and operate in the protected work areas as directed.

4.3.4 Force entry into a structure, given PPE, tools, and an assignment, so that the tools are used as designed, the barrier is removed, and the opening is in a safe condition and ready for entry.

(A) Requisite Knowledge. Basic construction of typical doors, windows, and walls within the department's community or service area; operation of doors, windows, and locks; and the dangers associated with forcing entry through doors, windows, and walls.

(B) Requisite Skills. The ability to transport and operate hand and power tools and to force entry through doors, windows, and walls using assorted methods and tools.

4.3.5 Exit a hazardous area as a team, given vision-obscured conditions, so that a safe haven is found before exhausting the air supply, others are not endangered, and the team integrity is maintained.

(A) Requisite Knowledge. Personnel accountability systems, communication procedures, emergency evacuation methods, what constitutes a safe haven, elements that create or indicate a hazard, and emergency procedures for loss of air supply.

(B) Requisite Skills. The ability to operate as a team member in vision-obscured conditions, locate and follow a guideline, conserve air supply, and evaluate areas for hazards and identify a safe haven.

4.3.6 Set up, mount, ascend, dismount, and descend ground ladders, given single and extension ladders, an assignment, and team members if needed, so that hazards are assessed, the ladder is stable, the angle is correct for climbing, extension ladders are extended to the necessary height with the fly locked, the top is placed against a reliable structural component, and the assignment is accomplished.

(A) Requisite Knowledge. Parts of a ladder, hazards associated with setting up ladders, what constitutes a stable foundation for ladder placement, different angles for various tasks, climbing techniques, safety limits to the degree of angulation, and what constitutes a reliable structural component for top placement.

(B) Requisite Skills. The ability to carry ladders, raise ladders, extend ladders and lock flies, determine that a wall and roof will support the ladder, judge extension ladder height requirements, and place the ladder to avoid obvious hazards, mount, ascend, dismount, and descend the ladder.

4.3.7 Attack a passenger vehicle fire operating as a member of a team, given PPE, an attack line, and hand tools, so that hazards are avoided, leaking flammable liquids are identified and controlled, protection from flash fires is maintained, all vehicle compartments are overhauled, and the fire is extinguished.

(A) Requisite Knowledge. Principles of fire streams as they relate to fighting automobile fires; precautions to be followed when advancing hose lines toward an automobile; observable results that a fire stream has been properly applied; identifying alternative fuels and the hazards associated with them; dangerous conditions created during an automobile fire; common types of accidents or injuries related to fighting automobile fires and how to avoid them; how to access locked passenger, trunk, and engine compartments; and methods for overhauling an automobile.

(B) Requisite Skills. The ability to identify automobile fuel type; assess and control fuel leaks; open, close, and adjust the flow and pattern on nozzles; apply water for maximum effectiveness while maintaining flash fire protection; advance 1½ in. (38 mm) or larger diameter attack lines; and expose hidden fires by opening all automobile compartments.

4.3.8 Extinguish fires in exterior Class A materials, given fires in stacked or piled and small unattached structures or storage containers that can be fought from the exterior, attack lines, hand tools and master stream devices, and an assignment, so that exposures are protected, the spread of fire is stopped, collapse hazards are avoided, water application is effective, the fire is extinguished, and signs of the origin area(s) and arson are preserved.

(A) Requisite Knowledge. Types of attack lines and water streams appropriate for attacking stacked, piled materials and outdoor fires; dangers—such as collapse—associated with stacked and piled materials; various extinguishing agents and their effect on different material configurations; tools and methods to use in breaking up various types of materials; the difficulties related to complete extinguishment of stacked and piled materials; water application methods for exposure protection and fire extinguishment; dangers such as exposure to toxic or hazardous materials associated with storage building and container fires; obvious signs of origin and cause; and techniques for the preservation of fire cause evidence.

(B) Requisite Skills. The ability to recognize inherent hazards related to the material's configuration, operate handlines or master streams, break up material using hand tools and water streams, evaluate for complete extinguishment, operate hose lines and other water application devices, evaluate and modify water application for maximum penetration, search for and expose hidden fires, assess patterns for origin determination, and evaluate for complete extinguishment.

4.3.9 Conduct a search and rescue in a structure operating as a member of a team, given an assignment, obscured vision conditions, personal protective equipment, a flashlight, forcible entry tools, hose lines, and ladders when necessary, so that ladders are correctly placed when used, all assigned areas are searched, all victims are located and removed, team integrity is maintained, and team members' safety—including respiratory protection—is not compromised.

(A) Requisite Knowledge. Use of forcible entry tools during rescue operations, ladder operations for rescue, psychological effects of operating in obscured conditions and ways to manage them, methods to determine if an area is tenable, primary and secondary search techniques, team members' roles and goals, methods to use and indicators of finding victims, victim removal methods (including various carries), and considerations related to respiratory protection.

(B) Requisite Skills. The ability to use SCBA to exit through restricted passages, set up and use different types of ladders for various types of rescue operations, rescue a fire fighter with functioning respiratory protection, rescue a fire fighter whose respiratory protection is not functioning, rescue a person who has no respiratory protection, and assess areas to determine tenability.

4.3.10 Attack an interior structure fire operating as a member of a team, given an attack line, ladders when needed, personal protective equipment, tools, and an assignment, so that team integrity is maintained, the attack line is deployed for advancement, ladders are correctly placed when used, access is gained into the fire area, effective water application practices are used, the fire is approached correctly, attack techniques facilitate suppression given the level of the fire, hidden fires are located and controlled, the correct body posture is maintained, hazards are recognized and managed, and the fire is brought under control.

(A) Requisite Knowledge. Principles of fire streams; types, design, operation, nozzle pressure effects, and flow capabilities of nozzles; precautions to be followed when advancing hose lines to a fire; observable results that a fire stream has been properly applied; dangerous building conditions created by fire; principles of exposure protection; potential long-term consequences of exposure to products of combustion; physical states of matter in which fuels are found; common types of accidents or injuries and their causes; and the application of each size and type of attack line, the role of the backup team in fire attack situations, attack and control techniques for grade level and above and below grade levels, and exposing hidden fires.

(B) Requisite Skills. The ability to prevent water hammers when shutting down nozzles; open, close, and adjust nozzle flow and patterns; apply water using direct, indirect, and combination attacks; advance charged and uncharged 1½ in. (38 mm) diameter or larger hose lines up ladders and up and down interior and exterior stairways; extend hose lines; replace burst hose sections; operate charged hose lines of 1½ in. (38 mm) diameter or larger while secured to a ground ladder; couple and uncouple various handline connections; carry hose; attack fires at grade level and above and below grade levels; and locate and suppress interior wall and subfloor fires.

4.3.11 Perform horizontal ventilation on a structure operating as part of a team, given an assignment, PPE, ventilation tools, equipment, and ladders, so that the ventilation openings are free of obstructions, tools are used as designed, ladders are correctly placed, ventilation devices are correctly placed, and the structure is cleared of smoke.

(A) Requisite Knowledge. The principles, advantages, limitations, and effects of horizontal, mechanical, and hydraulic ventilation; safety considerations when venting a structure; fire behavior in a structure; the products of combustion found in a structure fire; the signs, causes, effects, and prevention of backdrafts; and the relationship of oxygen concentration to life safety and fire growth.

(B) Requisite Skills. The ability to transport and operate ventilation tools and equipment and ladders, and to use safe procedures for breaking window and door glass and removing obstructions.

4.3.12 Perform vertical ventilation on a structure as part of a team, given an assignment, PPE, ground and roof ladders, and tools, so that ladders are positioned for ventilation, a specified opening is created, all ventilation barriers are removed, structural integrity is not compromised, products of combustion are released from the structure, and the team retreats from the area when ventilation is accomplished.

(A) Requisite Knowledge. The methods of heat transfer; the principles of thermal layering within a structure on fire; the techniques and safety precautions for venting flat roofs, pitched roofs, and basements; basic indicators of potential collapse or roof failure; the effects of construction type and elapsed time under fire conditions on structural integrity; and the advantages and disadvantages of vertical and trench/strip ventilation.

(B) Requisite Skills. The ability to transport and operate ventilation tools and equipment; hoist ventilation tools to a roof; cut roofing and flooring materials to vent flat roofs, pitched roofs, and basements; sound a roof for integrity; clear an opening with hand tools; select, carry, deploy, and secure ground ladders for ventilation activities; deploy roof ladders on pitched roofs while secured to a ground ladder; and carry ventilation-related tools and equipment while ascending and descending ladders.

4.3.13 Overhaul a fire scene, given PPE, attack line, hand tools, a flashlight, and an assignment, so that structural integrity is not compromised, all hidden fires are discovered, fire cause evidence is preserved, and the fire is extinguished.

(A) Requisite Knowledge. Types of fire attack lines and water application devices most effective for overhaul, water application methods for extinguishment that limit water damage, types of tools and methods used to expose hidden fire, dangers associated with overhaul, obvious signs of area of origin or signs of arson, and reasons for protection of fire scene.

(B) Requisite Skills. The ability to deploy and operate an attack line; remove flooring, ceiling, and wall components to expose void spaces without compromising structural integrity; apply water for maximum effectiveness; expose and extinguish hidden fires in walls, ceilings, and subfloor spaces; recognize and preserve obvious signs of area of origin and arson; and evaluate for complete extinguishment.

4.3.14 Conserve property as a member of a team, given salvage tools and equipment and an assignment, so that the building and its contents are protected from further damage.

(A) Requisite Knowledge. The purpose of property conservation and its value to the public, methods used to protect property, types of and uses for salvage covers, operations at properties protected with automatic sprinklers, how to stop the flow of water from an automatic sprinkler head, identification of the main control valve on an automatic sprinkler system, forcible entry issues related to salvage, and procedures for protecting possible areas of origin and potential evidence.

(B) Requisite Skills. The ability to cluster furniture; deploy covering materials; roll and fold salvage covers for reuse; construct water chutes and catch-alls; remove water; cover building openings, including doors, windows, floor openings, and roof openings; separate, remove, and relocate charred material to a safe location while protecting the area of origin for cause determination; stop the flow of water from a sprinkler with sprinkler wedges or stoppers; and operate a main control valve on an automatic sprinkler system.

4.3.15 Connect a fire department pumper to a water supply as a member of a team, given supply or intake hose, hose tools, and a fire hydrant or static water source, so that connections are tight and water flow is unobstructed.

(A) Requisite Knowledge. Loading and off-loading procedures for mobile water supply apparatus; fire hydrant operation; and suitable static water supply sources, procedures, and protocol for connecting to various water sources.

(B) Requisite Skills. The ability to hand lay a supply hose, connect and place hard suction hose for drafting operations, deploy portable water tanks as well as the equipment necessary to transfer water between and draft from them, make hydrant-to-pumper hose connections for forward and reverse lays, connect supply hose to a hydrant, and fully open and close the hydrant.

4.3.16 Extinguish incipient Class A, Class B, and Class C fires, given a selection of portable fire extinguishers, so that the correct extinguisher is chosen, the fire is completely extinguished, and correct extinguisher-handling techniques are followed.

(A) Requisite Knowledge. The classifications of fire; the types of, rating systems for, and risks associated with each class of fire; and the operating methods of and limitations of portable extinguishers.

(B) Requisite Skills. The ability to operate portable fire extinguishers, approach fire with portable fire extinguishers, select an appropriate extinguisher based on the size and type of fire, and safely carry portable fire extinguishers.

4.3.17 Operate emergency scene lighting, given fire service lighting equipment, power supply, and an assignment, so that emergency scene lighting equipment is operated within the manufacturer's listed safety precautions.

(A) Requisite Knowledge. Safety principles and practices, power supply capacity and limitations, and light deployment methods.

(B) Requisite Skills. The ability to operate department power supply and lighting equipment, deploy cords and connectors, reset ground-fault interrupter (GFI) devices, and locate lights for best effect.

4.3.18 Turn off building utilities, given tools and an assignment, so that the assignment is safely completed.

(A) Requisite Knowledge. Properties, principles, and safety concerns for electricity, gas, and water systems; utility disconnect methods and associated dangers; and use of required safety equipment.

(B) Requisite Skills. The ability to identify utility control devices, operate control valves or switches, and assess for related hazards.

4.3.19 Combat a ground cover fire operating as a member of a team, given protective clothing, SCBA (if needed), hose lines, extinguishers or hand tools, and an assignment, so that threats to property are reported, threats to personal safety are recognized, retreat is quickly accomplished when warranted, and the assignment is completed.

(A) Requisite Knowledge. Types of ground cover fires, parts of ground cover fires, methods to contain or suppress, and safety principles and practices.

(B) Requisite Skills. The ability to determine exposure threats based on fire spread potential, protect exposures, construct a fire line or extinguish with hand tools, maintain integrity of established fire lines, and suppress ground cover fires using water.

4.3.20 Tie a knot appropriate for hoisting tool, given PPE, tools, ropes, and an assignment, so that the knots used are appropriate for hoisting tools securely and as directed.

(A) Requisite Knowledge. Knot types and usage; the difference between life safety and utility rope; reasons for placing rope out of service; the types of knots to use for given tools, ropes, or situations; hoisting methods for tools and equipment; and using rope to support response activities.

(B) Requisite Skills. The ability to hoist tools using specific knots based on the type of tool.

4.3.21 Air Monitoring. Operate an air-monitoring instrument, given an air monitor and an assignment or task, so that the device is operated and that the fire fighter recognizes the high- or low-level alarms of the air monitor and takes action to mitigate the hazard.

(A) Requisite Knowledge. Knowledge of the various uses for an air monitor, the basic operation of an air monitor, and recognition and emergency actions to be taken upon the activation of the high- or low-level alarms of the air monitor.

(B) Requisite Skills. The ability to operate the air monitor, recognize the alarms, and react to the alarms of the air monitor.

4.4 Rescue Operations. This duty shall involve no requirements for Fire Fighter I.

4.5 Preparedness and Maintenance. This duty shall involve performing activities that reduce the loss of life and property due to fire through response readiness, according to the JPRs in 4.5.1 and 4.5.2.

4.5.1 Clean and check ladders, ventilation equipment, SCBA, ropes, salvage equipment, and hand tools, given cleaning tools, cleaning supplies, and an assignment, so that equipment is clean and maintained according to manufacturer's or departmental guidelines, maintenance is recorded, and equipment is placed in a ready state or reported otherwise.

   (A) Requisite Knowledge. Types of cleaning methods for various tools and equipment, correct use of cleaning solvents, and manufacturer's or departmental guidelines for cleaning equipment and tools.

   (B) Requisite Skills. The ability to select correct tools for various parts and pieces of equipment, follow guidelines, and complete recording and reporting procedures.

4.5.2 Clean, inspect, and return fire hose to service, given washing equipment, water, detergent, tools, and replacement gaskets, so that damage is noted and corrected, the hose is clean, and the equipment is placed in a ready state for service.

   (A) Requisite Knowledge. Departmental procedures for noting a defective hose and removing it from service, cleaning methods, and hose rolls and loads.

   (B) Requisite Skills. The ability to clean different types of hose; operate hose washing and drying equipment; mark defective hose; and replace coupling gaskets, roll hose, and reload hose.

Chapter 5 Fire Fighter II

5.1 General. For qualification at Level II, the Fire Fighter II shall meet the general knowledge requirements in 5.1.1, the general skill requirements in 5.1.2, the JPRs defined in Sections 5.2 through 5.5 of this standard, the requirements defined in Chapter 4, and knowledge of the Incident Management System sections of ICS 200 as described.

5.1.1 General Knowledge Requirements. Responsibilities of the Fire Fighter II in assuming and transferring command within an incident management system, performing assigned duties in conformance with applicable NFPA and other safety regulations and AHJ procedures, and the role of a Fire Fighter II within the organization.

5.1.2 General Skill Requirements. The ability to determine the need for command, organize and coordinate an incident management system until command is transferred, and function within an assigned role in an incident management system.

5.2 Fire Department Communications. This duty shall involve performing activities related to initiating and reporting responses, according to the JPRs in 5.2.1 and 5.2.2.

5.2.1 Complete a basic incident report, given the report forms, guidelines, and information, so that all pertinent information is recorded, the information is accurate, and the report is complete.

   (A) Requisite Knowledge. Content requirements for basic incident reports, the purpose and usefulness of accurate reports, consequences of inaccurate reports, how to obtain necessary information, and required coding procedures.

   (B) Requisite Skills. The ability to determine necessary codes, proof reports, and operate fire department computers or other equipment necessary to complete reports.

5.2.2 Communicate the need for team assistance, given fire department communications equipment, SOPs, and a team, so that the supervisor is consistently informed of team needs, departmental SOPs are followed, and the assignment is accomplished safely.

   (A) Requisite Knowledge. SOPs for alarm assignments and fire department radio communication procedures.

   (B) Requisite Skills. The ability to operate fire department communications equipment.

5.3 Fireground Operations. This duty shall involve performing activities necessary to ensure life safety, fire control, and property conservation, according to the JPRs in 5.3.1 through 5.3.4.

5.3.1 Extinguish an ignitible liquid fire, operating as a member of a team, given an assignment, an attack line, PPE, a foam proportioning device, a nozzle, foam concentrates, and a water supply, so that the correct type of foam concentrate is selected for the given fuel and conditions, a properly proportioned foam stream is applied to the surface of the fuel to create and maintain a foam blanket, fire is extinguished, reignition is prevented, team protection is maintained with a foam stream, and the hazard is faced until retreat to safe haven is reached.

   (A) Requisite Knowledge. Methods by which foam prevents or controls a hazard; principles by which foam is generated; causes for poor foam generation and corrective measures; difference between hydrocarbon and polar solvent fuels and the concentrates that work on each; the characteristics, uses, and limitations of fire-fighting foams; the advantages and disadvantages of using fog nozzles versus foam nozzles for foam application; foam stream application techniques; hazards associated with foam usage; and methods to reduce or avoid hazards.

   (B) Requisite Skills. The ability to prepare a foam concentrate supply for use, assemble foam stream components, master various foam application techniques, and approach and retreat from spills as part of a coordinated team.

5.3.2 Coordinate an interior attack line for a team's accomplishment of an assignment in a structure fire, given attack lines, personnel, PPE, and tools, so that crew integrity is established; attack techniques are selected for the given level of the fire (e.g., attic, grade level, upper levels, or basement); attack techniques are communicated to the attack teams; constant team coordination is maintained; fire growth and development are continuously evaluated; search, rescue, and ventilation requirements are communicated or managed; hazards are reported to the attack teams; and incident command is apprised of changing conditions.

   (A) Requisite Knowledge. Selection of the nozzle and hose for fire attack, given different fire situations; selection of adapters and appliances to be used for specific fireground situations; dangerous building conditions created by fire and fire suppression activities; indicators of building collapse; the effects of fire and fire suppression activities on wood, masonry (brick, block, stone), cast iron, steel, reinforced concrete, gypsum wallboard, glass, and plaster on lath; search and rescue and ventilation procedures; indicators of structural instability; suppression approaches and practices for various types of structural fires; and the association between specific tools and special forcible entry needs.

   (B) Requisite Skills. The ability to assemble a team, choose attack techniques for various levels of a fire (e.g., attic, grade level, upper levels, or basement), evaluate and forecast a fire's growth and development, select tools for forcible entry, incorporate search and rescue procedures and ventilation procedures in the completion of the attack team efforts, and determine developing hazardous building or fire conditions.

5.3.3 Control a flammable gas cylinder fire, operating as a member of a team, given an assignment, a cylinder outside of a structure, an attack line, PPE, and tools, so that crew integrity is maintained, contents are identified, safe havens are identified prior to advancing, open valves are closed, flames are not extinguished unless the leaking gas is eliminated, the cylinder is cooled, cylinder integrity is evaluated, hazardous conditions are recognized and acted upon, and the cylinder is faced during approach and retreat.

(A) Requisite Knowledge. Characteristics of pressurized flammable gases, elements of a gas cylinder, effects of heat and pressure on closed cylinders, boiling liquid expanding vapor explosion (BLEVE) signs and effects, methods for identifying contents, how to identify safe havens before approaching flammable gas cylinder fires, water stream usage and demands for pressurized cylinder fires, what to do if the fire is prematurely extinguished, valve types and their operation, alternative actions related to various hazards, and when to retreat.

(B) Requisite Skills. The ability to execute effective advances and retreats, apply various techniques for water application, assess cylinder integrity and changing cylinder conditions, operate control valves, and choose effective procedures when conditions change.

5.3.4 Protect evidence of fire cause and origin, given a flashlight and overhaul tools, so that the evidence is noted and protected from further disturbance until investigators can arrive on the scene.

(A) Requisite Knowledge. Methods to assess origin and cause; types of evidence; means to protect various types of evidence; the role and relationship of Fire Fighter IIs, criminal investigators, and insurance investigators in fire investigations; and the effects and problems associated with removing property or evidence from the scene.

(B) Requisite Skills. The ability to locate the fire's origin area, recognize possible causes, and protect the evidence.

5.4 Rescue Operations. This duty shall involve performing activities related to accessing and disentangling victims from motor vehicle accidents and helping special rescue teams, according to the JPRs in 5.4.1 and 5.4.2.

5.4.1 Extricate a victim entrapped in a motor vehicle as part of a team, given stabilization and extrication tools, so that the vehicle is stabilized, the victim is disentangled without further injury, and hazards are managed.

(A) Requisite Knowledge. The fire department's role at a vehicle accident, points of strength and weakness in auto body construction, dangers associated with vehicle components and systems, the uses and limitations of hand and power extrication equipment, and safety procedures when using various types of extrication equipment.

(B) Requisite Skills. The ability to operate hand and power tools used for forcible entry and rescue as designed; use cribbing and shoring material; and choose and apply appropriate techniques for moving or removing vehicle roofs, doors, windshields, windows, steering wheels or columns, and the dashboard.

5.4.2 Assist rescue operation teams, given standard operating procedures, necessary rescue equipment, and an assignment, so that procedures are followed, rescue items are recognized

and retrieved in the time as prescribed by the AHJ, and the assignment is completed.

(A) Requisite Knowledge. The fire fighter's role at a technical rescue operation, the hazards associated with technical rescue operations, types and uses for rescue tools, and rescue practices and goals.

(B) Requisite Skills. The ability to identify and retrieve various types of rescue tools, establish public barriers, and assist rescue teams as a member of the team when assigned.

5.5 Fire and Life Safety Initiatives, Preparedness, and Maintenance. This duty shall involve performing activities related to reducing the loss of life and property due to fire through hazard identification, inspection, and response readiness, according to the JPRs in 5.5.1 through 5.5.5.

5.5.1 Perform a fire safety survey in an occupied structure, given survey forms and procedures, so that fire and life safety hazards are identified, recommendations for their correction are made to the occupant, and unresolved issues are referred to the proper authority.

(A) Requisite Knowledge. Organizational policy and procedures, common causes of fire and their prevention, the importance of a fire safety survey and public fire education programs to fire department public relations and the community, and referral procedures.

(B) Requisite Skills. The ability to complete forms, recognize hazards, match findings to preapproved recommendations, and effectively communicate findings to occupants or referrals.

5.5.2 Present fire safety information to station visitors or small groups, given prepared materials, so that all information is presented, the information is accurate, and questions are answered or referred.

(A) Requisite Knowledge. Parts of informational materials and how to use them, basic presentation skills, and departmental standard operating procedures for giving fire station tours.

(B) Requisite Skills. The ability to document presentations and to use prepared materials.

5.5.3 Prepare a preincident survey, given forms, necessary tools, and an assignment, so that all required occupancy information is recorded, items of concern are noted, and accurate sketches or diagrams are prepared.

(A) Requisite Knowledge. The sources of water supply for fire protection; the fundamentals of fire suppression and detection systems; common symbols used in diagramming construction features, utilities, hazards, and fire protection systems; departmental requirements for a preincident survey and form completion; and the importance of accurate diagrams.

(B) Requisite Skills. The ability to identify the components of fire suppression and detection systems; sketch the site, buildings, and special features; detect hazards and special considerations to include in the preincident sketch; and complete all related departmental forms.

5.5.4 Maintain power plants, power tools, and lighting equipment, given tools and manufacturers' instructions, so that equipment is clean and maintained according to manufacturer and departmental guidelines, maintenance is recorded, and equipment is placed in a ready state or reported otherwise.

(A) Requisite Knowledge. Types of cleaning methods, correct use of cleaning solvents, manufacturer and departmental guidelines for maintaining equipment and its documentation, and problem-reporting practices.

(B) Requisite Skills. The ability to select correct tools; follow guidelines; complete recording and reporting procedures; and operate power plants, power tools, and lighting equipment.

5.5.5 Perform an annual service test on fire hose, given a pump, a marking device, pressure gauges, a timer, record sheets, and related equipment, so that procedures are followed, the condition of the hose is evaluated, any damaged hose is removed from service, and the results are recorded.

(A) Requisite Knowledge. Procedures for safely conducting hose service testing, indicators that dictate any hose be removed from service, and recording procedures for hose test results.

(B) Requisite Skills. The ability to operate hose testing equipment and nozzles and to record results.

Appendix C

An Extract from NFPA 1072, *Standard for Hazardous Materials/Weapons of Mass Destruction Emergency Response Personnel Professional Qualifications*, 2017 Edition

Reproduced with permission from NFPA 1072-2017: *Standard for Hazardous Materials/Weapons of Mass Destruction Emergency Response Personnel Professional Qualifications* © 2017, National Fire Protection Association. This reprinted material is not the complete and official position of the NFPA on the referenced subject, which is represented only by the standard in its entirety.

Chapter 4 Awareness

4.1 General.

4.1.1 Awareness personnel are those persons who, in the course of their normal duties, could encounter an emergency involving hazardous materials/weapons of mass destruction (WMD) and who are expected to recognize the presence of the hazardous materials/WMD, protect themselves, call for trained personnel, and secure the area.

4.1.2 Awareness personnel shall meet the job performance requirements defined in Sections 4.2 through 4.4.

4.1.3 General Knowledge Requirements. Role of awareness personnel at a hazardous materials/WMD incident, location and contents of the AHJ emergency response plan, and standard operating procedures for awareness personnel.

4.1.4 General Skills Requirements. (Reserved)

4.2 Recognition and Identification.

4.2.1 Recognize and identify the hazardous materials/WMD and hazards involved in a hazardous materials/WMD incident, given a hazardous materials/WMD incident, and approved reference sources, so that the presence of hazardous materials/WMD is recognized and the materials and their hazards are identified.

(A) Requisite Knowledge. What hazardous materials and WMD are; basic hazards associated with classes and divisions; indicators to the presence of hazardous materials including container shapes, NFPA 704 markings, globally harmonized system (GHS) markings, placards, labels, pipeline markings, other transportation markings, shipping papers with emergency response information, and other indicators; accessing information from the *Emergency Response Guidebook* (ERG) (current edition) using name of the material, UN/NA identification number, placard applied, or container identification charts; and types of hazard information available from the ERG, safety data sheets (SDS), shipping papers with emergency response information, and other approved reference sources.

(B) Requisite Skills. Recognizing indicators to the presence of hazardous materials/WMD; identifying hazardous materials/WMD by name, UN/NA identification number, placard applied, or container identification charts; and using the ERG, SDS, shipping papers with emergency response information, and other approved reference sources to identify hazardous materials/WMD and their potential fire, explosion, and health hazards.

4.3 Initiate Protective Actions.

4.3.1 Isolate the hazard area and deny entry at a hazardous materials/WMD incident, given a hazardous materials/WMD incident, policies and procedures, and approved reference sources, so that the hazard area is isolated and secured, personal safety procedures are followed, hazards are avoided or minimized, and additional people are not exposed to further harm.

(A) Requisite Knowledge. Use of the ERG, SDS, shipping papers with emergency response information, and other approved reference sources to identify precautions to be taken to protect responders and the public; policies and procedures for isolating the hazard area and denying entry; and the purpose of and methods for isolating the hazard area and denying entry.

(B) Requisite Skills. Recognizing precautions for protecting responders and the public; identifying isolation areas, denying entry, and avoiding minimizing hazards.

4.4 Notification.

4.4.1 Initiate required notifications at a hazardous materials/WMD incident, given a hazardous materials/WMD incident, policies and procedures, and approved communications equipment, so that the notification process is initiated and the necessary information is communicated.

(A) Requisite Knowledge. Policies and procedures for notification, reporting, and communications; types of approved communications equipment; and the operation of that equipment.

(B) Requisite Skills. Operating approved communications equipment and communicating in accordance with policies and procedures.

Chapter 5 Operations

5.1 General.

5.1.1 Operations level responders are those persons who respond to hazardous materials/weapons of mass destruction (WMD) incidents for the purpose of implementing or supporting actions to protect nearby persons, the environment, or property from the effects of the release.

5.1.2 Operations level responders shall meet the job performance requirements defined in Sections 4.2 through 4.4.

5.1.3 Operations level responders shall meet the job performance requirements defined in Sections 5.2 through 5.6.

5.1.4 Operations level responders shall have additional competencies that are specific to the response mission and expected tasks as determined by the AHJ.

5.1.5 General Knowledge Requirements. Role of operations level responders at a hazardous materials/WMD incident; location and contents of AHJ emergency response plan and standard operating procedures for operations level responders, including those response operations for hazardous materials/WMD incidents.

5.1.6 General Skills Requirements. (Reserved)

5.2 Identify Potential Hazards.

5.2.1 Identify the scope of the problem at a hazardous materials/WMD incident, given a hazardous materials/WMD incident, an assignment, policies and procedures, and approved reference sources, so that container types, materials, location of any release, and surrounding conditions are identified, hazard information is collected, the potential behavior of a material and its container is identified, and the potential hazards, harm, and outcomes associated with that behavior are identified.

(A) Requisite Knowledge. Definitions of hazard classes and divisions; types of containers; container identification markings, including piping and pipeline markings and contacting information; types of information to be collected during the hazardous materials/WMD incident survey; availability of shipping papers in transportation and of safety data sheets (SDS) at facilities; types of hazard information available from and how to contact CHEMTREC, CANUTEC, and SETIQ, governmental authorities, and manufacturers, shippers, and carriers; how to communicate with carrier representatives to reduce impact of a release; basic physical and chemical properties, including boiling point, chemical reactivity, corrosivity (pH), flammable (explosive) range [LFL (LEL) and UFL (UEL)], flash point, ignition (autoignition) temperature, particle size, persistence, physical state (solid, liquid, gas), radiation (ionizing and nonionizing), specific gravity, toxic products of combustion, vapor density, vapor pressure, and water solubility; how to identify the behavior of a material and its container based on the material's physical and chemical properties and the hazards associated with the identified behavior; examples of potential criminal and terrorist targets; indicators of possible criminal or terrorist activity for each of the following: chemical agents, biological agents, radiological agents, illicit laboratories (i.e., clandestine laboratories, weapons labs, ricin labs), and explosives; additional hazards associated with terrorist or criminal activities, such as secondary devices; and how to determine the likely harm and outcomes associated with the identified behavior and the surrounding conditions.

(B) Requisite Skills. Identifying container types, materials, location of release, and surrounding conditions at a hazardous materials/WMD incident; collecting hazard information; communicating with pipeline operators or carrier representatives; describing the likely behavior of the hazardous materials or WMD and its container; and

describing the potential hazards, harm, and outcomes associated with that behavior and the surrounding conditions.

5.3 Identify Action Options.

5.3.1 Identify the action options for a hazardous materials/WMD incident, given a hazardous materials/WMD incident, an assignment, policies and procedures, approved reference sources, and the scope of the problem, so that response objectives, action options, safety precautions, suitability of approved personal protective equipment (PPE) available, and emergency decontamination needs are identified.

(A) Requisite Knowledge. Policies and procedures for hazardous materials/WMD incident operations; basic components of an incident action plan (IAP); modes of operation (offensive, defensive, and nonintervention); types of response objectives; types of action options; types of response information available from the *Emergency Response Guidebook* (ERG), safety data sheets (SDS), shipping papers with emergency response information, and other resources; types of information available from and how to contact CHEMTREC, CANUTEC, and SETIQ, governmental authorities, and manufacturers, shippers, and carriers (highway, rail, water, air, pipeline); safety procedures; risk analysis concepts; purpose, advantages, limitations, and uses of approved PPE to determine if PPE is suitable for the incident conditions; difference between exposure and contamination; contamination types, including sources and hazards of carcinogens at incident scenes; routes of exposure; types of decontamination (emergency, mass, and technical); purpose, advantages, and limitations of emergency decontamination; and procedures, tools, and equipment for performing emergency decontamination.

(B) Requisite Skills. Identifying response objectives and action options based on the scope of the problem and available resources; identifying whether approved PPE is suitable for the incident conditions; and identifying emergency decontamination needs based on the scope of the problem.

5.4 Action Plan Implementation.

5.4.1 Perform assigned tasks at a hazardous materials/WMD incident, given a hazardous materials/WMD incident; an assignment with limited potential of contact with hazardous materials/WMD, policies and procedures, the scope of the problem, approved tools, equipment, and PPE, so that protective actions and scene control are established and maintained, on-scene incident command is described, evidence is preserved, approved PPE is selected and used in the proper manner; exposures and personnel are protected; safety procedures are followed; hazards are avoided or minimized; assignments are completed; and gross decontamination of personnel, tools, equipment, and PPE is conducted in the field.

(A) Requisite Knowledge. Scene control procedures; procedures for protective actions, including evacuation and sheltering-in-place; procedures for ensuring coordinated communications between responders and to the public; evidence recognition and preservation procedures; incident command organization; purpose, importance, benefits, and organization of incident command at hazardous materials/WMD incidents; policies and procedures for implementing incident command at hazardous materials/WMD incidents; capabilities, limitations, inspection, donning, working in, going through decontamination while wearing, doffing approved PPE; signs and symptoms of thermal stress; safety precautions when working at hazardous materials/WMD incidents; purpose, advantages, and limitations of gross decontamination; the need for gross decontamination in the field based on the task(s) performed and contamination received, including sources and hazards of carcinogens at incident scenes; gross decontamination procedures for personnel, tools, equipment, and PPE; and cleaning, disinfecting, and inspecting tools, equipment, and PPE.

(B) Requisite Skills. Establishing and maintaining scene control; recognizing and preserving evidence; inspecting, donning, working in, going through decontamination while wearing, and doffing approved PPE; isolating contaminated tools, equipment, and PPE; conducting gross decontamination of contaminated personnel, tools, equipment, and PPE in the field; and cleaning, disinfecting, and inspecting approved tools, equipment, and PPE.

5.5 Emergency Decontamination.

5.5.1 Perform emergency decontamination at a hazardous materials/WMD incident, given a hazardous materials/WMD incident that requires emergency decontamination; an assignment; scope of the problem; policies and procedures; and approved tools, equipment, and PPE for emergency decontamination, so that emergency decontamination needs are identified, approved PPE is selected and used, exposures and personnel are protected, safety procedures are followed, hazards are avoided or minimized, emergency decontamination is set up and implemented, and victims and responders are decontaminated.

(A) Requisite Knowledge. Contamination, cross contamination, and exposure; contamination types; routes of exposure; types of decontamination (emergency, mass, and technical); purpose, advantages, and limitations of emergency decontamination; policies and procedures for performing emergency decontamination; approved tools and equipment for emergency decontamination; and hazard avoidance for emergency decontamination.

(B) Requisite Skills. Selecting an emergency decontamination method; setting up emergency decontamination in a safe area; using PPE in the proper manner; implementing emergency decontamination; preventing spread of contamination; and avoiding hazards during emergency decontamination.

5.6 Progress Evaluation and Reporting.

5.6.1 Evaluate and report the progress of the assigned tasks for a hazardous materials/WMD incident, given a hazardous materials/WMD incident, an assignment, policies and procedures, status of assigned tasks, and approved communication tools and equipment, so that the effectiveness of the assigned tasks is evaluated and communicated to the supervisor, who can adjust the IAP as needed.

(A) Requisite Knowledge. Components of progress reports; policies and procedures for evaluating and reporting progress; use of approved communication tools and equipment; signs indicating improving, static, or deteriorating conditions based on the objectives of the action plan; and circumstances under which it would be prudent to withdraw from a hazardous materials/WMD incident.

(B) Requisite Skills. Determining incident status; determining whether the response objectives are being accomplished;

using approved communications tools and equipment; and communicating the status of assigned tasks.

Chapter 6 Operations Mission-Specific
6.2 Personal Protective Equipment.

6.2.1 Select, don, work in, and doff approved PPE at a hazardous materials/WMD incident, given a hazardous materials/WMD incident; a mission-specific assignment in an IAP that requires use of PPE; the scope of the problem; response objectives and options for the incident; access to a hazardous materials technician, an allied professional, an emergency response plan, or standard operating procedures; approved PPE; and policies and procedures, so that under the guidance of a hazardous materials technician, an allied professional, an emergency response plan, or standard operating procedures, approved PPE is selected, inspected, donned, worked in, decontaminated, and donned; exposures and personnel are protected; safety procedures are followed; hazards are avoided or minimized; and all reports and documentation pertaining to PPE use are completed.

(A) Requisite Knowledge. Policies and procedures for PPE selection and use; importance of working under the guidance of a hazardous materials technician, an allied professional, an emergency response plan, or standard operating procedures when selecting and using PPE; the capabilities and limitations of and specialized donning, donning, and usage procedures for approved PPE; components of an incident action plan (IAP); procedures for decontamination, inspection, maintenance, and storage of approved PPE; process for being decontaminated while wearing PPE; and procedures for reporting and documenting the use of PPE.

(B) Requisite Skills. Selecting PPE for the assignment; inspecting, maintaining, storing, donning, working in, and doffing PPE; going through decontamination (emergency and technical) while wearing the PPE; and reporting and documenting the use of PPE.

6.6 Product Control.

6.6.1 Perform product control techniques with a limited risk of personal exposure at a hazardous materials/WMD incident, given a hazardous materials/WMD incident with release of product; an assignment in an IAP; scope of the problem; policies and procedures; approved tools, equipment, control agents, and PPE; and access to a hazardous materials technician, an allied professional, an emergency response plan, or standard operating procedures, so that under the guidance of a hazardous materials technician, an allied professional, an emergency response plan, or standard operating procedures, approved PPE is selected and used; exposures and personnel are protected; safety procedures are followed; hazards are avoided or minimized; a product control technique is selected and implemented; the product is controlled; victims, personnel, tools, and equipment are decontaminated; and product control operations are reported and documented.

(A) Requisite Knowledge. Types of PPE and the hazards for which they are used; importance of working under the guidance of a hazardous materials technician, an allied professional, an emergency response plan, or standard operating procedures; definitions of control, confinement, containment, and extinguishment; policies and procedures; product control methods for controlling a release with limited risk of personal exposure; safety precautions associated with each product control method; location and operation of remote/emergency shutoff devices in cargo tanks and intermodal tanks in transportation and containers at facilities, that contain flammable liquids and flammable gases; characteristics and applicability of approved product control agents; use of approved tools and equipment; and requirements for reporting and documenting product control operations.

(B) Requisite Skills. Selecting and using PPE; selecting and performing product control techniques to confine/contain the release with limited risk of personal exposure; using approved control agents and equipment on a release involving hazardous materials/WMD; using remote control valves and emergency shutoff devices on cargo tanks and intermodal tanks in transportation and containers at fixed facilities; and performing product control techniques.

Annex A: Explanatory Material

A.4.1.2 Awareness personnel include public works employees, maintenance workers, and others who might see or encounter an incident involving hazardous materials/WMD occur while performing their regular assignment.

A.4.2 While the purpose of the JPR is to require the *Emergency Response Guidebook* (ERG) as the minimum reference at the awareness level, other reference sources can be provided as necessary, including an equivalent guide to the ERG; safety data sheets (SDS); manufacturer, shipper, and carrier (highway, rail, water, air, and pipeline) documents (shipping papers) and contacts; and the U.S. DOT *Hazardous Materials Marking, Labeling and Placarding Guide*. If provided, responders should be able to use these sources to accomplish the goals of the JPR.

In transportation, the name, placard applied, or identification number of the material provides access to information in the ERG or an equivalent document.

A.4.2.1(A) Instructors should include indicators of terrorist attacks and other potentials, emphasizing that "if you can smell it, taste it, or feel it, you are now (or might be) part of the problem."

While this is a minimum requirement, the AHJ has the option to select additional information from the operations chapter (Chapter 5) regarding container and hazard information as necessary, based on local conditions and circumstances.

Awareness level personnel should be able to match the hazard classes and divisions with the primary hazards and examples.

Indicators of the presence of hazardous materials include occupancy and locations, including facilities and transportation; container shape (general shape of the container); container owner/operator signage; placards and labels; markings, including NFPA 704 markings, military markings, transportation markings such as identification number marks, marine pollutant marks, elevated temperature marks, commodity markings, inhalation hazard marks, and pipe and pipeline markings and colors; GHS markings; shipping papers and emergency response information and SDS; and sensory clues (dead birds or fish, color of vapors, unusual odors, sheen, hissing noise, dead vegetation, etc.). Other items, such as fume hood exhaust stacks and vents on the exterior of a building, could indicate hazardous materials and can be identified in advance through pre-incident survey activities.

SDS is a component of the Globally Harmonized System of Classification and Labeling of Chemicals (GHS) and replaces the term *material safety data sheet* (MSDS). GHS is an internationally agreed-upon system, created by the United Nations in 1992. It replaces the various classification and labeling standards used in different countries by using consistent criteria on a global level. It supersedes the relevant European Union (EU) system, which has implemented the GHS into EU law as the Classification, Labelling and Packaging (CLP) Regulation and United States Occupational Safety and Health Administration (OSHA) standards. The SDS requires more information than MSDS regulations and provides a standardized structure for presenting the required information.

A.4.2.1(B) These requisite skills can be assessed through cognitive testing.

A.4.3 People not directly involved in emergency response operations should be kept away from the hazard area, and control should be established over the area of operations. Unprotected emergency responders should not be allowed to enter the isolation zone.

At the awareness level, approved reference sources include the current edition of the *Emergency Response Guidebook* (ERG), safety data sheets (SDS), shipping papers with emergency response information, and other approved reference sources.

A.4.3.1(A) Recommended precautions found on numbered guides in the ERG include public safety issues; recommended protective clothing; evacuation; emergency response to fire, spill, and leak; and first aid sections.

Examples of required knowledge include (1) precautions for providing emergency medical care to victims; typical ignition sources; ways hazardous materials/WMD are harmful to people, the environment, and property; general routes of entry for human exposure; emergency action (fire, spill, or leak; first aid); actions recommended not to be performed (e.g., closing of pipeline valves); protective actions (isolation of area and denial of entry, evacuation, shelter-in-place); size and shape of recommended initial isolation and protective action distances; difference between small and large spills; conditions that require the use of the ERG Table of Initial Isolation and Protective Action Distances and the isolation distances in the ERG numbered guide; techniques for isolating the hazard area and denying entry to unauthorized persons; how to recognize and protect evidence; and use of approved tools and equipment; (2) basic personal protective actions: staying clear of vapors, fumes, smoke, and spills; keeping vehicle at a safe distance from the scene; approaching from upwind, uphill, and upstream; and (3) types of protective actions and their purpose (e.g., isolate hazard area and deny entry, evacuation, and shelter-in-place); basic factors involved in the choice of protective actions (e.g., hazardous materials/WMD involved, population threatened, and weather conditions).

A.4.3.1(B) The requisite skills can be assessed through cognitive testing.

A.5.2 At the operations level, approved reference sources should include as a minimum of the *Emergency Response Guidebook* (ERG), safety data sheets (SDS), shipping papers, including emergency response information, and other approved reference sources such as CHEMTREC, CANUTEC, and SETIQ; governmental authorities; and manufacturers, shippers, carriers (highway, rail, water, air, and pipeline), and contacts.

A.5.2.1(A) At the operations level, responders should be able to recognize the following containers and identify them by name: rail tank cars (pressure, nonpressure, and cryogenic tank cars); highway cargo tanks (compressed gas tube trailers, corrosive liquid tanks, cryogenic tanks, dry bulk cargo tanks, high-pressure tanks, low-pressure chemical tanks, and nonpressure liquid tanks); UN portable tanks/intermodal tanks (nonpressure, pressure, cryogenic, and tube modules); storage tanks (nonpressure, pressure, and cryogenic storage tanks); piping and pipelines; intermediate bulk containers (IBC) and ton containers; radioactive materials packages (excepted, industrial, Type A, and Type B packages); and nonbulk containers (bags, carboys, cylinders, drums, and Dewar flasks for cryogenic liquids).

To ensure that operations level personnel also understand how to obtain information pertaining to a pipeline-involved incident, line markers or pipeline markers are added to supplement the list of information sources. In a pipeline incident, the pipeline markers would be the source of information used since no shipping papers, placards, UN numbers, or other information would be available.

Hazardous materials incident survey information. This includes location, weather conditions, topography, populated buildings, bodies of water, other buildings, remedial actions taken, container/package, contents, release, container damage, time of day, and other factors that help determine the scope of the problem.

Physical and chemical properties. Predicting the behavior of hazardous materials/WMD relies on understanding certain characteristics of the material. Information identifying the following characteristics should be collected and interpreted: boiling point, chemical reactivity, corrosivity (pH), flammable (explosive) range [LFL (LEL) and UFL (UEL)], flash point, ignition (autoignition) temperature, particle size, persistence, physical state (solid, liquid, gas), radiation (ionizing and nonionizing), specific gravity, toxic products of combustion, vapor density, vapor pressure, and water solubility.

Identifying hazards. The process for predicting/identifying the behavior of a hazardous material/WMD and its container under emergency conditions is based on the simple concepts that containers of hazardous materials/WMD under stress can open up and allow the contents to escape. The release of contents will vary in type and speed. A dispersion pattern will be formed by the escaping contents, potentially exposing people, the environment, or property to physical and/or health hazards.

This overall concept for identifying the likely behavior of a container and its contents under emergency conditions is often referred to as a general behavior model. The general behavior model considers the type of stress on the container involved and the potential type of breach, release, dispersion pattern, length of contact, and the health and physical hazards associated with the material and its container, as follows:

(1) *Stress.* The three types of stress that could cause a container to release its contents are thermal stress, mechanical stress, and chemical stress.

(2) *Breach.* The five ways in which containers can breach are disintegration, runaway cracking, closures opening up, punctures, and splits or tears.

(3) *Release.* The four ways in which containment systems can release their contents are detonation, violent rupture, rapid relief, and spill or leak.

(4) *Dispersion.* Seven dispersion patterns can be created upon release of agents: hemisphere, cloud, plume, cone, stream, pool, and irregular.

(5) *Contact.* The three general time frames for predicting the length of time that an exposure can be in contact with hazardous materials/WMD in an endangered area are short term (minutes and hours), medium term (days, weeks, and months), and long term (years and generations).

(6) *Hazards.* The seven health and physical hazards that could cause harm in a hazardous materials/WMD incident are thermal, mechanical, poisonous, corrosive, asphyxiating, radiological, and etiologic.

Identifying outcomes. The process for identifying the potential harm and associated outcomes within an endangered area at a hazardous materials/WMD incident includes identifying the size and shape of the endangered area, the number of exposures (people, property, environment, and major systems) within the endangered area, and the physical, health, and safety hazards within the endangered area as determined from approved resources.

Resources for determining the size of an endangered area of a hazardous materials/WMD incident are the current edition of the ERG and plume dispersion modeling results from facility pre-incident plans.

The factors for determining the extent of physical, health, and safety hazards within an endangered area at a hazardous materials/WMD incident are victim presentation (including nonclinical indicators or clues of a material's presence), surrounding conditions, indication of the behavior of the hazardous material and its container, and the degree of hazard.

A.5.2.1(B) The requisite skills can be assessed through cognitive testing.

A.5.3 At the operations level, approved information sources should include a minimum of ERG; SDS; CHEMTREC, CANUTEC, or SETIQ; local, state, and governmental authorities; and manufacturers', shippers', and carriers' documents (shipping papers) and contacts.

A.5.3.1(A) Modes of operation are offensive, defensive, and nonintervention and include the following:

(1) Common response objectives, for example, product control; fire control; protection of people, the environment, and property; identification and isolation; evidence protection; rescue; recovery; and termination

(2) Common response options, for example, spill control, leak control, foam, control exposures, evacuation, isolation, shelter-in-place, and establishment of product control zones

(3) Contamination types: primary, secondary, and tertiary

A.5.3.1(B) The requisite skills can be assessed through cognitive testing.

A.5.4 Operations level responders should be able to identify their role during hazardous materials/WMD incidents as specified in the emergency response plan and/or standard operating procedures; the levels of hazardous materials/WMD incidents as defined in the emergency response plan; the purpose, need, benefits, and elements of the incident command system for hazardous materials/WMD incidents; the duties and responsibilities of the incident safety officer and hazardous materials branch or group; considerations for determining the location of the incident command post; procedures for requesting additional resources; and the role and response objectives of other responding agencies.

Executive Summary – Field Decon

Over the past decade, research has been published linking higher rates of cancer in fire service personnel to repeated, chronic exposure to the by-products of smoke and particulates from structure fires. Various studies have proven that fire fighters are experiencing higher rates of certain types of cancers and that they are more likely to have rare forms of cancers than the general population. See NIOSH Study of Cancer among U.S. Fire Fighters at www.cdc.gov/niosh/firefighters/ffcancerstudy.html.

The fire service has begun to adapt to these findings by changing organizational practices in order to minimize exposures to known and suspected carcinogenic by-products in structure fires. Evolving adaptations include decontamination processes relating to fireground activities. Changes include, but are not limited to, forced air and water decontamination of structural fire-fighting personal protective equipment (PPE), modifying station practices, such as mandating that structural PPE be laundered after exposure to fire contaminants, and personal hygiene changes, such as mandating personnel to shower as soon as possible after interior fire-fighting activities at structure fires. In some instances, fire departments are assigning hazardous materials response assets to structure fire incidents to assist with scene (field) decontamination tasks.

During the recent meeting of the National Fire Protection Association (NFPA) Technical Committee (TC) – Hazardous Materials Response Personnel (HCZ-AAA), lengthy discussions regarding the role of emergency responders during field decontamination practices took place. These discussions led the Technical Committee to a decision that expanded technical language was needed in relation to job performance requirements (JPRs). Secondly, the TC decided that decontamination management does fit within one or more of the technical documents under the purview of the Committee. Of specific focus was NFPA 1072, *Standard for Hazardous Materials/ Weapons of Mass Destruction Emergency Response Personnel Professional Qualifications*. A small task group was formed to further research this subject and develop suggested language for possible inclusion into the upcoming version of NFPA 1072, which is currently in the second draft phase.

On January 19, 2016, the task group met via teleconference and determined that information about the previously referenced decontamination practices does indeed fall within the scope of the JPRs that have been developed as part of NFPA 1072. The task group reached a consensus that additional language should be crafted and inserted into the working copy of the second draft in support of the fire service's efforts to reduce or prevent cancer among fire fighters. The task group believes that the expanded information should be added to the existing language that deals with the use of PPE. The three specific areas include gross decontamination, action plan implementation, and decontamination.

As more information becomes available and this movement gains momentum and as best practices are developed, it is projected that field decontamination of personnel will remain a high priority and the means for minimizing fire fighter exposures to carcinogens. As such, it is incumbent upon the fire service that such practices become standardized and documented to ensure that the goals of supporting fire fighter health and safety

are met by the broadest base of fire service organizations. If the referenced recommendations are accepted by the TC, it will place the NFPA in a position to play an integral role in addressing fire fighter decontamination and cancer concerns.

A.5.4.1(A) *Evidence preservation.* Preservation of evidence is essential to the integrity and credibility of an incident investigation. Preservation techniques must be acceptable to the law enforcement agency having jurisdiction; therefore, it is important to get that agency's input ahead of time on the techniques specified in the AHJ emergency response plan or the organization's standard operating procedures.

General procedures for preserving evidence include the following:

(1) Secure and isolate any incident area where evidence is located. This can include discarded personal protection equipment, specialized packaging (shipping or workplace labels and placards), biohazard containers, glass or metal fragments, containers (e.g., plastic, pipes, cylinders, bottles, fuel containers), and other materials that appear relevant to the occurrence, such as roadway flares, electrical components, fluids, and chemicals.

(2) Leave fatalities and body parts in place and secure the area in which they are located.

(3) Isolate any apparent source location of the event (e.g., blast area, spill release point).

(4) Leave in place any explosive components or housing materials.

(5) Place light-colored tarpaulins on the ground of access and exit corridors, decontamination zones, treatment areas, and rehabilitation sectors to allow possible evidence that might drop during decontamination and doffing of clothes to be spotted and collected.

(6) Secure and isolate all food vending locations in the immediate area. Contaminated food products will qualify as primary or secondary evidence in the event of a chemical or biological incident.

The collection (as opposed to preservation) of evidence is usually conducted by law enforcement personnel, unless other protocols are in place. If law enforcement personnel are not equipped or trained to enter the hot zone, hazardous materials technicians should be trained to collect samples in such a manner as to maintain the integrity of the samples for evidentiary purposes and to document the chain of evidence.

Safety precautions. Safety precautions should include buddy systems, backup systems, accountability systems, safety briefing, and evacuation/escape procedures. The following items should be considered in a safety briefing prior to allowing personnel to work at hazardous materials/WMD incidents:

(1) Preliminary evaluation
(2) Hazard identification
(3) Description of the site
(4) Task(s) to be performed
(5) Length of time for task(s)
(6) Required PPE
(7) Monitoring requirements
(8) Notification of identified risk

A.5.4.1(B) The operations level responder should implement the incident command system as required by the AHJ by completing the following requirements:

(1) Identify the role of the operations level responder during hazardous materials/WMD incidents as specified in the emergency response plan and/or standard operating procedures

(2) Identify the levels of hazardous materials/WMD incidents as defined in the emergency response plan

(3) Identify the purpose, need, benefits, and elements of the incident command system for hazardous materials/WMD incidents

(4) Identify the duties and responsibilities of the following functions within the incident management system:
  (a) Incident safety officer
  (b) Hazardous materials branch or group

(5) Identify the considerations for determining the location of the incident command post for a hazardous materials/WMD incident

(6) Identify the procedures for requesting additional resources at a hazardous materials/WMD incident

(7) Describe the role and response objectives of other agencies that respond to hazardous materials/WMD incidents

A.5.6 All responders should understand why their efforts must be evaluated. If they are not making progress, the plan must be re-evaluated to determine why. The evaluation should include what changes have occurred with the circumstances of the incident (behavior of container or its contents).

To decide whether the actions being taken at an incident are effective and the objectives are being achieved, the responder must determine whether the incident is stabilizing or increasing in intensity. Factors to be considered include reduction of potential impact to persons or the environment and status of resources available to manage the incident. The evaluation should take place upon initiation of the IAP, and the IC/unified command and general staff should constantly monitor the status of the incident. The actions taken should be leading to a desirable outcome, with minimal loss of life and property. Changes in the status of the incident should influence the development of the IAP for the next operational period.

A.5.6.1(A) Remaining in the immediate vicinity of an incident when nothing can be done to mitigate it and the situation is about to deteriorate is pointless. If flames are impinging on an LP-Gas vessel, for example, and providing the necessary volume of water to cool it is impossible, it would be prudent to withdraw to a safe distance. ICs should always evaluate the benefit of operations against the risk. Refer to the ERG or other references to determine appropriate action to be taken under the circumstances.

A.5.6.1(B) The proper methods for communicating the status of the planned response lie within the guidelines of the ICS and are dictated by the incident-specific IAP. The ICS identifies two types of communication at an incident, formal and informal. Formal communication should be used for all policy-related communication, using the ICS principles of unity of command and chain of command, while maintaining span of control. Ideally, all critical information should be communicated face-to-face.

The format for communications within the ICS must be established by the IC/unified command with input from the general staff.

A procedure should be established to allow responders to notify the IC immediately when conditions become critical and personnel are threatened. For example, the notification could take the form of a pre-established emergency radio message or

tone that signifies danger, or it might be repeated blasts on an air horn. The message should not be delayed while responders try to locate a specific person in the chain of command.

A.6.2 At this level, PPE refers to personal protective equipment that would be used in situations where contact with hazardous materials/WMD is possible or expected. Such equipment can include chemical-protective clothing, bomb suits, respirators, or other equipment that typically would not be worn by operations level responders. Specialized PPE also refers to operations level responders' PPE that requires changes to donning, doffing, and usage procedures—for example, taping gaps in fire-fighter protective clothing, doffing in a decontamination corridor, or working in the hot zone as a member of a buddy system. Personnel should be able to describe the types of PPE available and the options for thermal hazards, radiological hazards, asphyxiation hazards, chemical hazards, etiological/biological hazards, and mechanical hazards. *(See also A.6.1.6.)*

A.6.2.1(A) Limitations of PPE include permeation, penetration, and degradation of protective clothing and limitations of respiratory protective equipment, such as air-purifying respirators.

Requisite knowledge includes the ability to describe the types of PPE that are available for response based on NFPA standards and the PPE options for thermal hazards, radiological hazards, asphyxiating hazards, chemical hazards, etiological/biological hazards, and mechanical hazards.

A.6.6 See A.6.1.6.

For the purposes of this section, the intent is to focus on confining or containing the release with limited risk of personal exposure. The applicable techniques include absorption, adsorption, damming, diking, dilution, diversion, remote valve shutoff, retention, vapor dispersion, and vapor suppression. Product control also includes techniques for controlling flammable liquid incidents and flammable gas incidents.

Tools and equipment include such items as Class B foam application equipment, diking equipment, damming equipment, approved absorbent materials and products, shovels and other hand tools, piping, heavy equipment (such as backhoes), floats, and spill booms.

Control agents can include Class B foam, dispersal agents, and so on.

A.6.6.1(A) Product control techniques that focus on confining/containing the release with limited risk of personal exposure include absorption, adsorption, damming, diking, dilution, diversion, remote valve shutoff, retention, vapor dispersion, and vapor suppression. Product control also includes techniques for controlling flammable liquid incidents and flammable gas incidents.

Remote/emergency shutoff devices include those for MC-306/DOT-406, MC-407/DOT-407, MC-331 cargo tanks, and intermodal tanks.

A.6.6.1(B) Product control techniques that focus on confining/containing the release with limited risk of personal exposure include absorption, adsorption, damming, diking, dilution, diversion, remote valve shutoff, retention, vapor dispersion, and vapor suppression. Techniques for controlling flammable liquid incidents and flammable gas incidents (e.g., hose handling, nozzle patterns, and attack operations) can be found in NFPA 1001.

Appendix D

An Extract from NFPA 472, *Standard for Competence of Responders to Hazardous Materials/Weapons of Mass Destruction Incidents*, 2018 Edition

Chapter 4 Competencies for Awareness Level Personnel

4.1 General

4.1.1 Introduction.

4.1.1.1 Awareness level personnel shall be persons who, in the course of their normal duties, could encounter an emergency involving hazardous materials/weapons of mass destruction (WMD) and who are expected to recognize the presence of the hazardous materials/WMD, protect themselves, call for trained personnel, and secure the area.

4.1.1.2 Awareness level personnel shall be trained to meet all competencies defined in Sections 4.2 through 4.4 of this chapter.

4.1.1.3 Awareness level personnel shall receive additional training to meet applicable governmental occupational health and safety regulations.

4.1.2 Goal.

4.1.2.1 The goal of the competencies in this chapter shall be to provide personnel who in the course of normal duties encounter hazardous materials/WMD incidents with the knowledge and skills to perform the tasks in 4.1.2.2 in a safe and effective manner.

4.1.2.2 Given a hazardous materials/WMD incident, policies and procedures, approved reference sources, and approved communications equipment, the awareness level personnel shall be able to perform the following tasks:

(1) Analyze the incident to identify both the hazardous materials/WMD present and the basic hazards for each hazardous materials/WMD agent involved by completing the following tasks:
  (a) Recognize the presence of hazardous materials/WMD.
  (b) Identify the name, UN/NA identification number, type of placard, or other distinctive marking applied for the hazardous materials/WMD involved from a safe location.

  (c) Identify potential hazards from the current edition of the *Emergency Response Guidebook* (ERG), safety data sheets (SDS), shipping papers, and other approved reference sources.

(2) Implement actions consistent with the authority having jurisdiction (AHJ), and the current edition of the ERG or an equivalent document by completing the following tasks:
  (a) Isolate the hazard area
  (b) Initiate required notifications

4.2 Competencies — Analyzing the Incident.

4.2.1 Recognizing the Presence of Hazardous Materials/WMD.

Given a hazardous materials/WMD incident and approved reference sources, awareness level personnel shall recognize those situations where hazardous materials/WMD are present by completing the following requirements:

(1) Define the terms *hazardous material* (or *dangerous goods,* in Canada) and *WMD*

(2) Identify the hazard classes and divisions of hazardous materials/WMD and identify common examples of materials in each hazard class or division

(3) Identify the primary hazards associated with each hazard class and division

(4) Identify the difference(s) between hazardous materials/WMD incidents and other emergencies

(5) Identify typical occupancies and locations in the community where hazardous materials/WMD are manufactured, transported, stored, used, or disposed of

(6) Identify typical container shapes that can indicate the presence of hazardous materials/WMD

(7) Identify facility and transportation markings and colors that indicate hazardous materials/WMD, including the following:
  (a) Transportation markings, including UN/NA identification number marks, marine pollutant mark, elevated temperature (HOT) mark, commodity marking, and inhalation hazard mark
  (b) NFPA 704 markings
  (c) Military hazardous materials/WMD markings
  (d) Special hazard communication markings for each hazard class
  (e) Pipeline markings
  (f) Container markings

(8) Given an NFPA 704 marking, describe the significance of the colors, numbers, and special symbols

(9) Identify placards and labels that indicate hazardous materials/WMD

(10) Identify the following basic information on safety data sheets (SDS) and shipping papers for hazardous materials:

  (a) Identify where to find SDS

  (b) Identify major sections of SDS

(11) Identify the following basic information on shipping papers for hazardous materials:

  (a) Identify the entries on shipping papers that indicate the presence of hazardous materials

  (b) Match the name of the shipping papers found in transportation (air, highway, rail, and water) with the mode of transportation

  (c) Identify the person responsible for having the shipping papers in each mode of transportation

  (d) Identify where the shipping papers are found in each mode of transportation

  (e) Identify where the papers can be found in an emergency in each mode of transportation

(12) Identify examples of other clues, including senses (sight, sound, and odor), that indicate the presence of hazardous materials/WMD

4.2.2 Identifying Hazardous Materials/WMD. Given examples of hazardous materials/WMD incident, awareness level personnel shall, from a safe location, identify the hazardous material(s)/WMD involved in each situation by name, UN/NA identification number, or type placard applied by completing the following requirements:

(1) Identify difficulties encountered in determining the specific names of hazardous materials/WMD at facilities and in transportation

(2) Identify sources for obtaining the names of, UN/NA identification numbers for, or types of placard associated with hazardous materials/WMD in transportation

(3) Identify sources for obtaining the names of hazardous materials/WMD at a facility

4.2.3 Collecting Hazard Information. Given the identity of various hazardous materials/WMD (name, UN/NA identification number, or type placard), awareness level personnel shall identify the basic hazard information for each material by using the current edition of the ERG or equivalent document; safety data sheet (SDS); manufacturer, shipper, and carrier documents (including shipping papers); and contacts by completing the following requirements:

(1) Identify the three methods for determining the guidebook page for a hazardous material/WMD

(2) Identify the two general types of hazards found on each guidebook page

4.3 Competencies — Planning the Response. (Reserved)

4.4 Competencies — Implementing the Planned Response.

4.4.1 Isolate the Hazard Area. Given examples of hazardous materials/WMD incidents, the emergency response plan, the standard operating procedures, and the current edition of the ERG, awareness level personnel shall isolate and deny entry to the hazard area by completing the following requirements:

(1) Identify the location of both the emergency response plan and/or standard operating procedures

(2) Identify the role of the awareness level personnel during hazardous materials/WMD incidents

(3) Identify the following basic precautions to be taken to protect themselves and others in hazardous materials/WMD incidents:

  (a) Identify the precautions necessary when providing emergency medical care to victims of hazardous materials/WMD incidents

  (b) Identify typical ignition sources found at the scene of hazardous materials/WMD incidents

  (c) Identify the ways hazardous materials/WMD are harmful to people, the environment, and property

  (d) Identify the general routes of entry for human exposure to hazardous materials/WMD

(4) Given examples of hazardous materials/WMD and the identity of each hazardous material/WMD (name, UN/NA identification number, or type placard), identify the following response information:

  (a) Emergency action (fire, spill, or leak and first aid)

  (b) Personal protective equipment (PPE) recommended:

    (i) Street clothing and work uniforms

    (ii) Structural fire-fighting protective clothing

    (iii) Positive pressure self-contained breathing apparatus (SCBA)

    (iv) Chemical-protective clothing and equipment

(5) Identify the definitions for each of the following protective actions:

  (a) Isolation of the hazard area and denial of entry

  (b) Evacuation

  (c) Shelter-in-place

(6) Identify the size and shape of recommended initial isolation and protective action zones

(7) Describe the difference(s) between small and large spills as found in the Table of Initial Isolation and Protective Action Distances in the ERG or equivalent document

(8) Identify the circumstances under which the following distances are used at a hazardous materials/WMD incident:

  (a) Table of Initial Isolation and Protective Action Distances

  (b) Isolation distances in the numbered guides

(9) Describe the difference(s) between the isolation distances on the orange-bordered guidebook pages and the protective action distances on the green-bordered ERG pages

(10) Identify the techniques used to isolate the hazard area and deny entry to unauthorized persons at hazardous materials/WMD incidents

4.4.2 Initiating the Notification Process. Given a hazardous materials/WMD incident, policies and procedures, and approved communications equipment, awareness level personnel shall initiate notifications at a hazardous materials/WMD incident, completing the following requirements:

(1) Identify policies and procedures for notification, reporting, and communications

(2) Identify types of approved communications equipment

(3) Describe how to operate approved communications equipment

4.5 Competencies — Evaluating Progress. (Reserved)

4.6 Competencies — Terminating the Incident. (Reserved)

Chapter 5 Competencies for Operations Level Responders

5.1 General.

5.1.1 Introduction.

5.1.1.1 The operations level responder shall be that person who responds to hazardous materials/weapons of mass destruction (WMD) incidents for the purpose of protecting nearby persons, the environment, or property from the effects of the release.

5.1.1.2 The operations level responder shall be trained to meet all competencies at the awareness level *(see Chapter 4)* and the competencies defined in Sections 5.2 through 5.5 of this chapter.

5.1.1.3 The operations level responder shall receive additional training to meet applicable governmental occupational health and safety regulations.

5.1.2 Goal.

5.1.2.1 The goal of the competencies in this chapter shall be to provide operations level responders with the knowledge and skills to perform the competencies in 5.1.2.2 in a safe manner.

5.1.2.2 When responding to hazardous materials/WMD incidents, operations level responders shall be able to perform the following tasks:

(1) Identify the scope of the problem and potential hazards, harm, and outcomes by completing the following tasks:

(a) Survey a hazardous materials/WMD incident to identify the containers and materials involved and to identify the surrounding conditions

(b) Collect hazard and response information from the ERG; SDS; CHEMTREC/CANUTEC/SETIQ; governmental authorities; and shipper/manufacturer/carrier documents, including shipping papers with emergency response information and shipper/manufacturer/carrier contacts

(c) Predict the likely behavior of a hazardous material/WMD and its container, including hazards associated with that behavior

(d) Estimate the potential outcomes harm at a hazardous materials/WMD incident

(2) Plan an initial response to a hazardous materials/WMD incident within the capabilities and competencies of available personnel and personal protective equipment (PPE) by completing the following tasks:

(a) Describe the response objectives for the hazardous materials/WMD incident

(b) Describe the response options available for each objective

(c) Determine whether the PPE provided is suitable for implementing each option

(d) Describe emergency decontamination procedures

(e) Develop a plan of action, including safety considerations

(3) Implement the planned response for a hazardous materials/WMD incident to favorably change the outcomes consistent with the emergency response plan and/or standard operating procedures by completing the following tasks:

(a) Establish and enforce scene control procedures, including control zones, emergency decontamination, and communications

(b) Where criminal or terrorist acts are suspected, establish a means of evidence preservation

(c) Initiate an incident command system (ICS) for hazardous materials/WMD incidents

(d) Perform tasks assigned as identified in the incident action plan

(e) Perform emergency decontamination

(4) Evaluate and report the progress of the assigned tasks taken at a hazardous materials/WMD incident to ensure that the response objectives are met in a safe, effective, and efficient manner by completing the following tasks:

(a) Evaluate the status of the actions taken in accomplishing the response objectives

(b) Communicate the status of the planned response

5.2 Competencies — Analyzing the Incident.

5.2.1 Surveying Hazardous Materials/WMD Incidents. Given scenarios involving hazardous materials/WMD incidents, the operations level responder shall collect information about the incident to identify the containers, the materials involved, leaking containers, and the surrounding conditions released by completing the requirements of 5.2.1.1 through 5.2.1.6.

5.2.1.1 Given examples of the following pressure containers, the operations level responder shall identify each container by type, as follows:

(1) Bulk fixed facility pressure containers
(2) Pressure tank cars
(3) High-pressure cargo tanks
(4) Compressed gas tube trailers
(5) High-pressure intermodal tanks
(6) Ton containers
(7) Y-cylinders
(8) Compressed gas cylinders
(9) Portable and horizontal propane cylinders
(10) Vehicle-mounted pressure containers

5.2.1.1.1 Given examples of the following cryogenic containers, the operations level responder shall identify each container by type, as follows:

(1) Bulk fixed facility cryogenic containers
(2) Cryogenic liquid tank cars
(3) Cryogenic liquid cargo tanks
(4) Intermodal cryogenic containers
(5) Cryogenic cylinders
(6) Dewar flasks

5.2.1.1.2 Given examples of the following liquids-holding containers, the operations level responder shall identify each container by type, as follows:

(1) Bulk fixed facility tanks
(2) Low-pressure tank cars
(3) Nonpressure liquid cargo tanks
(4) Low-pressure chemical cargo tanks
(5) 101 and 102 intermodal tanks
(6) Flexible intermediate bulk containers/rigid intermediate bulk containers (FIBCs/RIBCs)
(7) Flexible bladders
(8) Drums
(9) Bottles, flasks, carboys

5.2.1.1.3 Given examples of the following solids-holding containers, the operations level responder shall identify each container by type, as follows:

(1) Bulk fixed facilities
(2) Railway gondolas, coal cars

(3) Dry bulk cargo trailers
(4) Intermodal tanks (reactive solids)
(5) FIBCs/RIBCs
(6) Drums
(7) Bags, bottles, boxes

5.2.1.1.4 Given examples of the following mixed-load containers, the operations level responder shall identify each container by type, as follows:
(1) Box cars
(2) Mixed cargo trailers
(3) Freight containers

5.2.1.1.5 Given examples of the following containers, the operations level responder shall identify the characteristics of each container by type as follows:
(1) Intermediate bulk container (IBC)
(2) Ton container

5.2.1.1.6 Given examples of the following radioactive material containers, the operations level responder shall identify the characteristics of each container by type, as follows:
(1) Excepted (package)
(2) Industrial (package)
(3) Type A (package)
(4) Type B (package)
(5) Type C (package)

5.2.1.2 Given examples of containers, the operations level responder shall identify the markings that differentiate one container from another.

5.2.1.2.2 Given examples of the following marked transport vehicles and their corresponding shipping papers, the operations level responder shall identify marking used for identifying the specific transport vehicle:
(1) Highway transport vehicles, including cargo tanks
(2) Intermodal equipment, including tank containers
(3) Rail transport vehicles, including tank cars

5.2.1.2.1 Given examples of facility storage tanks, the operations level responder shall identify the markings indicating container size, product contained, and/or site identification numbers.

5.2.1.3 Given examples of hazardous materials incidents, the operations level responder shall identify the name(s) of the hazardous material(s) in 5.2.1.3.1 through 5.2.1.3.3.

5.2.1.3.1 Given a pipeline marker, the operations level responder shall identify the emergency telephone number, owner, and product as applicable.

5.2.1.3.2 Given a pesticide label, the operations level responder shall identify the active ingredient, hazard statement, name of pesticide, and pest control product (CPC) number (in Canada).

5.2.1.3.3 Given a label for a radioactive material, the operations level responder shall identify the type or category of label, contents, activity, transport index, and criticality safety index as applicable.

5.2.1.4 The operations level responder shall identify and list the surrounding conditions that should be noted when surveying a hazardous materials/WMD incident.

5.2.1.5 The operations level responder shall describe ways to verify information obtained from the survey of a hazardous materials/WMD incident.

5.2.1.6 The operations level responder shall identify at least three additional hazards that could be associated with an incident involving terrorist or criminal activities.

5.2.1.6.1 Identify at least four types of locations that could be targets for criminal or terrorist activity using hazardous materials/WMD.

5.2.1.6.2 Describe the difference between a chemical and a biological incident.

5.2.1.6.3 Identify at least four indicators of possible criminal or terrorist activity involving chemical agents.

5.2.1.6.4 Identify at least four indicators of possible criminal or terrorist activity involving biological agents.

5.2.1.6.5 Identify at least four indicators of possible criminal or terrorist activity involving radiological agents.

5.2.1.6.6 Identify at least four indicators of possible criminal or terrorist activity involving illicit laboratories (e.g., clandestine laboratories, weapons lab, explosive lab, or biological lab).

5.2.1.6.7 Identify at least four indicators of possible criminal or terrorist activity involving explosives.

5.2.1.6.8 Identify at least four indicators of secondary devices.

5.2.1.6.9 Identify at least four specific actions necessary when an incident is suspected to involve criminal or terrorist activity.

5.2.1.7 The operations level responder shall describe ways in which emergency responders are exposed to toxic products of combustion.

5.2.2 Collecting Hazard and Response Information. Given scenarios involving known hazardous materials/WMD, the operations level responder shall collect hazard and response information from SDS, CHEMTREC/CANUTEC/SETIQ, governmental authorities, and manufacturers, shippers, and carriers by completing the following requirements:
(1) Match the definitions associated with the hazard classes and divisions of hazardous materials/WMD with the designated class or division.
(2) Identify two ways to obtain an SDS in an emergency.
(3) Using an SDS for a specified material, identify the following hazard and response information:
  (a) Identification, including supplier identifier and emergency telephone number
  (b) Hazard identification
  (c) Composition/information on ingredients
  (d) First aid measures
  (e) Fire-fighting measures
  (f) Accident release measures
  (g) Handling and storage
  (h) Exposure controls/personal protection
  (i) Physical and chemical properties
  (j) Stability and reactivity
  (k) Toxicological information
  (l) Ecological information (nonmandatory)
  (m) Disposal considerations (nonmandatory)
  (n) Transport information (nonmandatory)
  (o) Regulatory information (nonmandatory)
  (p) Other information
(4) Identify the types of assistance provided by, procedure for contacting, and information to be provided to CHEMTREC/CANUTEC/SETIQ and governmental authorities.

(5) Identify two methods of contacting manufacturers, shippers, and carriers (highway, rail, marine, air, and pipeline) to obtain hazard and response information.

(6) Identify the type of assistance provided by governmental authorities with respect to criminal or terrorist activities involving the release or potential release of hazardous materials/WMD.

5.2.3 Predicting the Likely Behavior of a Material and Its Container. Given scenarios involving hazardous materials/WMD incidents, each with a single hazardous material/WMD, the operations level responder shall describe the likely behavior of the material or agent and its container by completing the following requirements:

(1) Use the hazard and response information obtained from the current edition of the ERG, SDS, CHEMTREC/CANUTEC/SETIQ, governmental authorities, and manufacturer, shipper, and carrier contacts, as follows:

  (a) Match the following chemical and physical properties with their significance and impact on the behavior of the container and its contents:

    (i) Boiling point

    (ii) Chemical reactivity

    (iii) Corrosivity (pH)

    (iv) Flammable (explosive) range [lower explosive limit (LEL) and upper explosive limit (UEL)]

    (v) Flash point

    (vi) Ignition (autoignition) temperature

    (vii) Particle size

    (viii) Persistence

    (ix) Physical state (solid, liquid, gas)

    (x) Radiation (ionizing and nonionizing)

    (xi) Specific gravity

    (xii) Toxic products of combustion

    (xiii) Vapor density

    (xiv) Vapor pressure

    (xv) Water solubility

    (xvi) Polymerization

    (xvii) Expansion ratio

    (xviii) Biological agents and toxins

  (b) Identify the differences between the following terms:

    (i) Contamination and secondary contamination

    (ii) Exposure and contamination

    (iii) Exposure and hazard

    (iv) Infectious and contagious

    (v) Acute effects and chronic effects

    (vi) Acute exposures and chronic exposures

(2) Identify types of stress that can cause a container system to release its contents (thermal, mechanical, and chemical).

(3) Identify ways containers can breach (disintegration, runaway cracking, closures open up, punctures, and splits or tears).

(4) Identify ways containers can release their contents (detonation, violent rupture, rapid relief, spill, or leak).

(5) Identify dispersion patterns that can be created upon release of a hazardous material (hemispherical, cloud, plume, cone, stream, pool, and irregular).

(6) Identify the time frames for estimating the duration that hazardous materials/WMD will present an exposure risk (short-term, medium-term, and long-term).

(7) Identify the health and physical hazards that could cause harm.

5.2.4 Estimating Potential Harm. Given scenarios involving hazardous materials/WMD incidents, the operations level responder shall describe the potential harm within the endangered area at each incident by completing the following requirements:

(1) Identify a resource for determining the size of an endangered area of a hazardous materials/WMD incident

(2) Given the dimensions of the endangered area and the surrounding conditions at a hazardous materials/WMD incident, describe the number and type of exposures within that endangered area

(3) Identify resources available for determining the concentrations of a released hazardous materials/WMD within an endangered area

(4) Given the concentrations of the released material, describe the factors for determining the extent of physical, health, and safety hazards within the endangered area of a hazardous materials/WMD incident

(5) Describe the impact that time, distance, and shielding have on exposure to radioactive materials specific to the expected dose rate

(6) Describe the potential for secondary threats and devices at criminal or terrorist events

5.3 Competencies — Planning the Response.

5.3.1 Describing Response Objectives. Given at least two scenarios involving hazardous materials/WMD incidents, the operations level responder shall describe the response objectives for each example by completing the following requirements:

(1) Given an analysis of a hazardous materials/WMD incident and the exposures, describe the number of exposures that could be saved with the resources provided by the AHJ

(2) Given an analysis of a hazardous materials/WMD incident, describe the steps for determining response objectives

(3) Describe how to assess the risk to a responder for each hazard class in rescuing injured persons at a hazardous materials/WMD incident

5.3.2 Identifying Action Options. Given examples of hazardous materials/WMD incidents (facility and transportation), the operations level responder shall identify the action options for each response objective and shall meet the following requirements:

(1) Identify the options to accomplish a given response objective

(2) Describe the prioritization of emergency medical care and removal of victims from the hazard area relative to exposure and contamination concerns

5.3.3 Determining Suitability of Personal Protective Equipment (PPE). Given examples of hazardous materials/WMD incidents, including the names of the hazardous materials/WMD involved and the anticipated type of exposure, the operations level responder shall determine whether available PPE is applicable to performing assigned tasks by completing the following requirements:

(1) Identify the respiratory protection required for a given response option and the following:

  (a) Describe the advantages, limitations, uses, and operational components of the following types of respiratory protection at hazardous materials/WMD incidents:

    (i) Self-contained breathing apparatus (SCBA)

    (ii) Supplied air respirators

    (iii) Powered air-purifying respirators

    (iv) Air-purifying respirators

(b) Identify the required physical capabilities and limitations of personnel working in respiratory protection

(2) Identify the personal protective clothing, required for a given action option and the following:

(a) Identify skin contact hazards encountered at hazardous materials/WMD incidents

(b) Identify the purpose, advantages, and limitations of the following types of protective clothing at hazardous materials/WMD incidents:

(i) Chemical-protective clothing, including liquid splash-protective ensembles and vapor-protective ensembles

(ii) High temperature-protective clothing, including proximity suits and entry suits

(iii) Structural fire-fighting protective clothing

5.3.4 Identifying Emergency Decontamination Issues. Given scenarios involving hazardous materials/WMD incidents, the operations level responder shall identify when emergency decontamination is needed by completing the following requirements:

(1) Identify ways that people, PPE, apparatus, tools, and equipment become contaminated.

(2) Describe how the potential for secondary contamination determines the need for emergency decontamination.

(3) Explain the importance, differences, and limitations of emergency/field expedient, gross, technical, and mass decontamination procedures at hazardous materials incidents.

(4) Identify the purpose of emergency decontamination procedures at hazardous materials incidents.

5.4 Competencies — Implementing the Planned Response.

5.4.1 Establishing Scene Control. Given two scenarios involving hazardous materials/WMD incidents, the operations level responder shall explain how to establish and maintain scene control, including control zones and emergency decontamination, and communications between responders and to the public by completing the following requirements:

(1) Identify the procedures for establishing scene control through control zones

(2) Identify the criteria for determining the locations of the control zones at hazardous materials/WMD incidents

(3) Identify the basic techniques for the following protective actions at hazardous materials/WMD incidents:

(a) Evacuation

(b) Shelter-in-place

(4) Perform emergency decontamination while preventing spread of contamination and avoiding hazards while using PPE

(5) Identify the items to be considered in a safety briefing prior to allowing personnel to work at the following:

(a) Hazardous material incidents

(b) Hazardous materials/WMD incidents involving criminal activities

(6) Identify the procedures for ensuring coordinated communication between responders and to the public

5.4.2 Preserving Evidence. Given two scenarios involving hazardous materials/WMD incidents, the operations level responder shall describe the process to preserve evidence as listed in the emergency response plan and/or standard operating procedures.

5.4.3 Initiating the Incident Command System. Given scenarios involving hazardous materials/WMD incidents, the operations

level responder shall implement the incident command system as required by the AHJ by completing the following requirements:

(1) Identify the role of the operations level responder during hazardous materials/WMD incidents as specified in the emergency response plan and/or standard operating procedures

(2) Identify the levels of hazardous materials/WMD incidents as defined in the emergency response plan

(3) Identify the purpose, need, benefits, and elements of the incident command system for hazardous materials/WMD incidents

(4) Identify the duties and responsibilities of the following functions within the incident management system:

(a) Incident safety officer

(b) Hazardous materials branch or group

(5) Identify the considerations for determining the location of the incident command post for a hazardous materials/WMD incident

(6) Identify the procedures for requesting additional resources at a hazardous materials/WMD incident

(7) Describe the role and response objectives of other agencies that respond to hazardous materials/WMD incidents

5.4.4 Using Personal Protective Equipment (PPE). Given the PPE provided by the AHJ, the operations level responder shall describe considerations for the use of PPE provided by the AHJ by completing the following requirements:

(1) Identify the importance of the buddy system

(2) Identify the importance of the backup personnel

(3) Identify the safety precautions to be observed when approaching and working at hazardous materials/WMD incidents

(4) Identify the signs and symptoms of heat and cold stress and procedures for their control

(5) Identify the capabilities and limitations of personnel working in the PPE provided by the AHJ

(6) Identify the procedures for cleaning, disinfecting, and inspecting PPE provided by the AHJ

(7) Maintain and store PPE following the instructions provided by the manufacturer on the care, use, and maintenance of the protective ensemble elements

5.5 Competencies — Evaluating Progress.

5.5.1 Evaluating the Status of Planned Response. Given two scenarios involving hazardous materials/WMD incidents, including the incident action plan, the operations level responder shall determine the effectiveness of the actions taken in accomplishing the response objectives and shall meet the following requirements:

(1) Identify the factors to be evaluated to determine if actions taken were effective in accomplishing the objectives

(2) Describe the circumstances under which it would be prudent to withdraw from a hazardous materials/WMD incident

5.5.2 Communicating the Status of Planned Response. Given two scenarios involving hazardous materials/WMD incidents, including the incident action plan, the operations level responder shall report the status of the planned response through the normal chain of command by completing the following requirements:

(1) Identify the procedures for reporting the status of the planned response through the normal chain of command

(2) Identify the methods for immediate notification of the incident commander and other response personnel about critical emergency conditions at the incident

5.6 Competencies — Terminating the Incident. (Reserved)

Chapter 6 Competencies for Operations Level Responders Assigned Mission-Specific Responsibilities

6.2 Mission-Specific Competencies: Personal Protective Equipment (PPE).

6.2.1 General.

6.2.1.1 Introduction.

6.2.1.1.1 The operations level responder assigned to use PPE at hazardous materials/WMD incidents shall be that person, competent at the operations level, who is assigned by the AHJ to select, inspect, don, work in, go through decontamination while wearing, and doff PPE at hazardous materials/WMD incidents.

6.2.1.1.2 The operations level responder assigned to use PPE at hazardous materials/WMD incidents shall be trained to meet all competencies at the awareness level (see Chapter 4), all competencies at the operations level (see Chapter 5), and all competencies in this section.

6.2.1.1.3 The operations level responder assigned to use PPE at hazardous materials/WMD incidents shall operate under the guidance of a hazardous materials technician, an allied professional, an emergency response plan, or standard operating procedures.

6.2.1.1.4 The operations level responder assigned to use PPE shall receive the additional training necessary to meet specific needs of the jurisdiction.

6.2.1.2 Goal. The goal of the competencies in this section shall be to provide the operations level responder assigned to select, inspect, don, work in, go through decontamination while wearing, and doff PPE with the knowledge and skills to perform the tasks in a safe and effective manner.

6.2.1.2.1 Given a hazardous materials/WMD incident, a mission-specific assignment in an incident action plan (IAP) that requires use of PPE; the scope of the problem; response objectives and options for the incident; policies and procedures; access to a hazardous materials technician, an allied professional, an emergency response plan, or standard operating procedures; approved PPE; and policies and procedures, the operations level responder assigned to use PPE shall be able to perform the following tasks:

(1) Select PPE provided by the AHJ based on tasks assigned
(2) Inspect, don, work in, go through emergency and technical decontamination while wearing, and doff PPE provided by the AHJ consistent with the AHJ standard operating procedures and the incident site safety and control plan by following safety procedures, avoiding or minimizing hazards, and protecting exposures and personnel
(3) Maintain and store PPE consistent with AHJ policies and procedures
(4) Report and document the use of PPE

6.2.2 Competencies — Analyzing the Incident. (Reserved)

6.2.3 Competencies — Planning the Response.

6.2.3.1 Selecting Personal Protective Equipment (PPE). Given scenarios involving hazardous materials/WMD incidents with known and unknown hazardous materials/WMD and the PPE provided by the AHJ, the operations level responder assigned to use PPE provided by the AHJ shall select the PPE required to support assigned mission-specific tasks at hazardous materials/WMD incidents based on AHJ policies and procedures by completing the following requirements:

(1) Describe the importance of working under the guidance of a hazardous materials technician, an allied professional, an emergency response plan, or standard operating procedures
(2) Describe the purpose of each type of PPE provided by the AHJ for response to hazardous materials/WMD incidents based on NFPA standards and how these items relate to EPA levels of protection
(3) Describe capabilities and limitations of PPE for the following hazards:
   (a) Thermal
   (b) Radiological
   (c) Asphyxiating
   (d) Chemical (corrosive, toxic)
   (e) Etiological/biological
   (f) Mechanical
(4) Select PPE provided by the AHJ for assigned mission-specific tasks at hazardous materials/WMD incidents based on AHJ policies and procedures
   (a) Describe the following terms and explain their impact and significance on the selection of chemical-protective clothing (CPC):
     (i) Degradation
     (ii) Penetration
     (iii) Permeation
   (b) Identify at least three indications of material degradation of CPC
   (c) Identify the different designs of vapor-protective clothing and liquid splash-protective clothing, and describe the advantages and disadvantages of each type
   (d) Identify the advantages and disadvantages of the following cooling measures:
     (i) Air cooled
     (ii) Ice cooled
     (iii) Water cooled
     (iv) Phase change cooling technology
   (e) Identify the physiological and psychological stresses that can affect users of PPE
   (f) Describe AHJ policies and procedures for going through the emergency and technical decontamination process while wearing PPE

6.2.4 Competencies — Implementing the Planned Response.

6.2.4.1 Using Personal Protective Equipment (PPE). Given the PPE provided by the AHJ, the operations level responder assigned to use PPE shall demonstrate the ability to inspect, don, work in, go through decontamination while wearing, and doff the PPE provided to support assigned mission-specific tasks by completing the following requirements:

(1) Describe safety precautions for personnel wearing PPE, including buddy systems, backup systems, accountability systems, safety briefings, and evacuation/escape procedures

(2) Inspect, don, work in, and doff PPE provided by the AHJ following safety procedures, protecting exposures and personnel, and avoiding or minimizing hazards

(3) Go through the process of being decontaminated (emergency and technical) while wearing PPE

(4) Maintain and store PPE according to AHJ policies and procedures

6.2.5 Competencies — Evaluating Progress. (Reserved)

6.2.6 Competencies — Terminating the Incident.

6.2.6.1 Reporting and Documenting Personal Protective Equipment (PPE) Use. Given a scenario involving a hazardous materials/WMD incident and AHJ policies and procedures, the operations level responder assigned to use PPE shall report and document use of the PPE as required by the AHJ by completing the following:

(1) Identify the reports and supporting documentation required by the AHJ pertaining to PPE use

(2) Describe the importance of personnel exposure records

(3) Identify the steps in keeping an activity log and exposure records

(4) Identify the requirements for filing documents and maintaining records

6.6 Mission-Specific Competencies: Product Control.

6.6.1 General.

6.6.1.1 Introduction.

6.6.1.1.1 The operations level responder assigned to perform product control with limited risk of personal exposure shall be that person, competent at the operations level, who is assigned by the AHJ to confine and contain releases of hazardous materials/WMD and control flammable liquid and flammable gas releases at hazardous materials/WMD incidents.

6.6.1.1.2 The operations level responder assigned to perform product control at hazardous materials/WMD incidents shall be trained to meet all competencies at the awareness level (see Chapter 4), all competencies at the operations level (see Chapter 5), all mission-specific competencies for PPE (see Section 6.2), and all competencies in this section.

6.6.1.1.3 The operations level responder assigned to perform product control at hazardous materials/WMD incidents shall operate under the guidance of a hazardous materials technician, an allied professional, an emergency response plan, or standard operating procedures.

6.6.1.1.4 The operations level responder assigned to perform product control at hazardous materials/WMD incidents shall receive the additional training necessary to meet specific needs of the jurisdiction.

6.6.1.2 Goal.

6.6.1.2.1 The goal of the competencies in this section shall be to provide the operations level responder assigned to perform product control, including to confine or contain releases of hazardous materials/WMD and to control flammable liquid and flammable gas releases, with limited risk of personal exposure at hazardous materials/WMD incidents with the knowledge and skills to perform the tasks in 6.6.1.2.2 in a safe and effective manner.

6.6.1.2.2 Given a hazardous materials/WMD incident with release of product; an assignment in an IAP; the scope of the problem; policies and procedures; approved tools, equipment, control agents, and PPE; and access to a hazardous materials technician, an allied professional, an emergency response plan, or standard operating procedures, the operations level responder assigned to perform product control shall be able to perform the following tasks:

(1) Select techniques to control releases with limited risk of personal exposure at hazardous materials/WMD incidents within the capabilities and competencies of available personnel, tools and equipment, control agents, and PPE, in accordance with the AHJ policies and procedures, by completing the following requirements:

(a) Describe control techniques to confine/contain released product with limited risk of personal exposure available to the operations level responder.

(b) Describe the location and operation of remote control/emergency shutoff devices on cargo and intermodal tanks, and containers at fixed facilities containing flammable liquids and gases.

(c) Describe the characteristics and applicability of available control agents and equipment available for controlling flammable liquid and flammable gas releases.

(2) Implement selected techniques for controlling released product with limited risk of personnel exposure at the incident following safety procedures, avoiding or minimizing hazards, and protecting exposures and personnel.

(3) Report and document product control operations.

6.6.2 Competencies — Analyzing the Incident. (Reserved)

6.6.3 Competencies — Planning the Response.

6.6.3.1 Selecting Product Control Techniques. Given examples of hazardous materials/WMD incidents, the operations level responder assigned to perform product control with limited risk of personal exposure shall select techniques to confine or contain releases of hazardous materials/WMD and to control flammable liquid and flammable gas releases within the capabilities and competencies of available personnel, tools and equipment, PPE, and control agents and equipment in accordance with the AHJ's policies and procedures by completing the following requirements:

(1) Explain the importance of working under the guidance of a hazardous materials technician, an allied professional, an emergency response plan, or standard operating procedures.

(2) Explain the difference between control, confinement, containment, and extinguishment.

(3) Describe the product control techniques available to the operations level responder.

(4) Describe the application, necessary tools, equipment, control agents, and safety precautions associated with each of the following control techniques:

(a) Absorption

(b) Adsorption

(c) Damming

(d) Diking

(e) Dilution

(f) Diversion

(g) Remote valve shutoff
(h) Retention
(i) Vapor dispersion
(j) Vapor suppression
(5) Identify and describe the use of tools and equipment provided by the AHJ for product control, including Class B foam application equipment, diking equipment, damming equipment, approved absorbent materials and products, shovels and other hand tools, piping, heavy equipment (such as backhoes), floats, and spill booms and control agents, including Class B foam and dispersal agents.
(6) Identify the characteristics and applicability of the following Class B foams if supplied by the AHJ:
    (a) Aqueous film-forming foam (AFFF)
    (b) Alcohol-resistant concentrates
    (c) Fluoroprotein
    (d) High-expansion foam
(7) Identify the location and describe the operation of remote control/emergency shutoff devices to contain flammable liquid and flammable gas releases on cargo tanks on MC/DOT-306/406, MC/DOT-307/407, and MC-331 cargo tanks, intermodal tanks, and containers at fixed facilities.
(8) Describe the safety precaution associated with each product control technique.

6.6.4 Competencies — Implementing the Planned Response.

6.6.4.1 Performing Product Control Techniques. Given the selected product control technique and the tools and equipment, PPE, and control agents and equipment provided by the AHJ at a hazardous materials/WMD incident, the operations level responder assigned to perform product control shall implement the product control technique to confine/contain the release with limited risk of personal exposure by completing the following requirements:
(1) Using the tools and equipment provided by the AHJ, perform the following product control techniques following safety procedures, protecting exposures and personnel, and avoiding or minimizing hazards:
    (a) Operate remote control/emergency shutoff devices to reduce or stop the flow of hazardous material from MC-306/DOT-406, MC-407/DOT-407, and MC-331 cargo tanks, intermodal tanks, and containers at fixed facilities containing flammable liquids or gases

6.6.4.2 Given the required tools and equipment provided by the AHJ, perform product control techniques following safety procedures, protecting exposures and personnel, and avoiding or minimizing hazards with the following:
(1) Using the equipment provided by the AHJ, control flammable liquid and flammable gas releases using techniques, including hose handling, nozzle patterns, and attack operations, found in NFPA 1001.
(2) Using the Class B foams or agents and equipment provided by the AHJ, control the spill or fire involving flammable liquids by application of the foam(s) or agent(s).

6.6.5 Competencies — Evaluating Progress. (Reserved)

6.6.6 Competencies — Terminating the Incident.

6.6.6.1 Reporting and Documenting Product Control Operations. Given a scenario involving a hazardous materials/ WMD incident involving product control, the operations level responder assigned to perform product control shall document the product control operations as required by the AHJ by completing the following requirement:
(1) Identify the reports and supporting documentation required by the AHJ pertaining to product control operations

Appendix E

NFPA 1001, NFPA 1072, and NFPA 472 Correlation Guide

NFPA 1001: Fire Fighter I

NFPA 1001, *Standard for Fire Fighter Professional Qualifications*, 2019 Edition	Corresponding Chapter(s)	Corresponding Page(s)
4.1 General.	1, 2, 3, 9	See below.
4.1.1	1, 2	5, 8–9, 16–21, 39–46
4.1.2	1, 3, 9	16–17, 74–80, 104, 303–320
4.2 Fire Department Communications.	4, 18	See below.
4.2.1	4	135, 138, 141–144, 148–151
4.2.2	4	143–144, 147–148
4.2.3	4	147–151
4.2.4	18	696–698
4.3 Fireground Operations.	2, 3, 5, 7, 8, 9, 10, 11, 12, 13, 14, 15, 16, 17, 18, 19, 21	See below.
4.3.1	3, 18	72, 81–82, 85–107, 104, 114–119, 703–706, 708–709
4.3.2	2	47–49
4.3.3	2	49–50, 56–59
4.3.4	10	327–364
4.3.5	18	691–696, 699–709
4.3.6	11	373–380, 384–388, 390–414, 416–418

NFPA 1001, *Standard for Fire Fighter Professional Qualifications*, 2019 Edition	Corresponding Chapter(s)	Corresponding Page(s)
4.3.7	17	673–681
4.3.8	17	650–651, 656, 658–663, 667–671
4.3.9	12, 18	427–465, 703–706, 709–718
4.3.10	5, 15, 16, 17	159–160, 551–561, 572, 574–577, 579–583, 610–612, 615, 618, 620–621, 623–635, 644–649, 651–657, 663–667, 672–673
4.3.11	5, 13	161–163, 165, 169–176, 476–491
4.3.12	5, 11, 13	164–166, 171–173, 417–418, 491–513
4.3.13	19	729, 746–747, 749–755
4.3.14	19	728–746, 753–755
4.3.15	14, 15, 16	526–533, 538–544, 551–553, 556, 561–562, 570–571, 574–578, 593–599, 600–601, 607–614
4.3.16	7	229–236, 238–256, 256–257
4.3.17	19	725–728
4.3.18	2, 17	57–59, 682–684
4.3.19	21	778–789, 791–796
4.3.20	9	291–297, 301–320
4.3.21	19	747–749
4.4 Rescue Operations.	General	N/A
4.5 Preparedness and Maintenance.	3, 8, 9, 11, 13, 15, 16, 19	See below.
4.5.1	3, 8, 9, 11, 13, 19	77–80, 107–113, 116, 120–123, 267, 283, 296–301, 380–384, 513–515, 745–746
4.5.2	15, 16	555, 563–573, 599, 601–607, 615–623, 635–636

NFPA 1001: Fire Fighter II

NFPA 1001, *Standard for Fire Fighter Professional Qualifications*, 2019 Edition	Corresponding Chapter(s)	Corresponding Page(s)
5.1 General.	22	See below.
5.1.1	22	807–814, 819–823, 828–831
5.1.2	22	815–819, 823–831
5.2 Fire Department Communications.	22	See below.
5.2.1	22	835–836
5.2.2	22	808, 826–827
5.3 Fireground Operations.	23, 28	See below.
5.3.1	23	851, 854, 856–865
5.3.2	23	845, 848–851
5.3.3	23	852–853, 855
5.3.4	28	1022–1037
5.4 Rescue Operations.	24, 25	See below.
5.4.1	24	874–900
5.4.2	25	907–931
5.5 Fire and Life Safety Initiatives, Preparedness, and Maintenance.	23, 25, 26, 27	See below.
5.5.1	27	987, 990, 992–994, 996–999
5.5.2	27	985–995
5.5.3	26, 27	943–975, 1002–1017
5.5.4	25	910–911, 932–934
5.5.5	23	865–868

NFPA 1072: Core Competencies for Awareness Level Responders

NFPA 1072, *Standard for Hazardous Materials/ Weapons of Mass Destruction Emergency Response Personnel Professional Qualifications*, 2017 Edition	Corresponding Chapter(s)	Corresponding Page(s)
4.1 General.	29	See below.
4.1.1	29	1049–1052
4.1.2	29	1049–1052
4.1.3	29	1047, 1049–1052
4.1.4 General Skills Requirements. (Reserved)	N/A	N/A
4.2 Recognition and Identification.	29, 30	See below.
4.2.1	29, 30	1046–1047, 1057–1067, 1070, 1072–1081
4.3 Initiate Protective Actions.	30	See below.
4.3.1	30	1057–1059, 1062–1067, 1070, 1072–1081
4.4 Notification.	30	See below.
4.4.1	30	1057–1059, 1062–1067, 1070, 1072–1081

NFPA 1072: Core Competencies for Operations Level Responders

NFPA 1072, *Standard for Hazardous Materials/ Weapons of Mass Destruction Emergency Response Personnel Professional Qualifications*, 2017 Edition	Corresponding Chapter(s)	Corresponding Page(s)
5.1 General.	33, 34, 35	See below.
5.1.1	33, 34, 35	1145, 1147–1149, 1184, 1198–1204
5.1.2 (Prerequisites)	34	1184
5.1.3 (Prerequisites)	34	1184
5.1.4	34	1184

NFPA 1072, *Standard for Hazardous Materials/ Weapons of Mass Destruction Emergency Response Personnel Professional Qualifications, 2017 Edition*	Corresponding Chapter(s)	Corresponding Page(s)
5.1.5	34	1168–1178, 1184
5.1.6 General Skills Requirements. (Reserved)	N/A	N/A
5.2 Identify Potential Hazards.	31, 32, 34, 35	See below.
5.2.1	31, 32, 34, 35	1089–1103, 1110–1137, 1168–1169, 1190–1198
5.3 Identify Action Options.	33	See below.
5.3.1	33	1142–1163
5.4 Action Plan Implementation.	34	See below.
5.4.1	34	1168–1184
5.5 Emergency Decontamination.	31, 33	See below.
5.5.1	31, 33	1102–1103, 1150–1163
5.6 Progress Evaluation and Reporting.	34	See below.
5.6.1	34	1169–1172

NFPA 1072: Core Competencies for Operations Level Responders Assigned Mission-Specific Responsibilities

Note: Per NFPA 1001, select mission-specific operations level competencies from NFPA 1072, *Standard for Hazardous Materials/Weapons of Mass Destruction Emergency Response Personnel Professional Qualifications* are required (6.2 Personal Protective Equipment and 6.6 Product Control).

NFPA 1072, *Standard for Hazardous Materials/ Weapons of Mass Destruction Emergency Response Personnel Professional Qualifications, 2017 Edition*	Corresponding Chapter(s)	Corresponding Page(s)
6.2 Personal Protective Equipment.	36	See below.
6.2.1	36	1210–1238

NFPA 1072, *Standard for Hazardous Materials/ Weapons of Mass Destruction Emergency Response Personnel Professional Qualifications, 2017 Edition*	Corresponding Chapter(s)	Corresponding Page(s)
6.6 Product Control.	37	See below.
6.6.1	37	1244–1258

NFPA 472: Core Competencies for Awareness Level Responders

NFPA 472, *Standard for Competence of Responders to Hazardous Materials/Weapons of Mass Destruction Incidents, 2018 Edition*	Corresponding Chapter(s)	Corresponding Page(s)
4.1 General.	29, 30	See below.
4.1.1 Introduction.	N/A	N/A
4.1.1.1	29	1049–1052
4.1.1.2	29	1048–1049
4.1.1.3	29	1049–1052
4.1.2 Goal.	N/A	N/A
4.1.2.1	29	1048–1052
4.1.2.2	29, 30	1047, 1049–1052, 1059–1067, 1070, 1072–1077
4.2 Competencies—Analyzing the Incident.	29, 30	See below.
4.2.1	29, 30	1046–1047, 1057–1067, 1070, 1072–1081
4.2.2	30	1057–1059, 1062–1067
4.2.3	30	1062–1067, 1070, 1072–1081
4.3 Competencies—Planning the Response. (Reserved)	N/A	N/A
4.4 Competencies—Implementing the Planned Response.	30, 33	See below.

NFPA 472, *Standard for Competence of Responders to Hazardous Materials/Weapons of Mass Destruction Incidents*, 2018 Edition	Corresponding Chapter(s)	Corresponding Page(s)
4.4.1	30, 33	1057–1059, 1062–1067, 1070, 1072–1081, 1151–1154
4.4.2	30	1057–1059
4.5 Competencies—Evaluating Progress. (Reserved)	N/A	N/A
4.6 Competencies—Terminating the Incident. (Reserved)	N/A	N/A

NFPA 472: Core Competencies for Operations Level Responders

NFPA 472, *Standard for Competence of Responders to Hazardous Materials/Weapons of Mass Destruction Incidents*, 2018 Edition	Corresponding Chapter(s)	Corresponding Page(s)
5.1 General.	30, 31, 32, 33, 34	See below.
5.1.1 Introduction.	N/A	N/A
5.1.1.1	31	1102–1103
5.1.1.2	33	1142–1146
5.1.1.3	33	1142–1146
5.1.2 Goal.	N/A	N/A
5.1.2.1	33	1142–1146
5.1.2.2	30, 31, 32, 33, 34	1062–1067, 1070, 1072–1077, 1089–1105, 1120, 1145, 1147–1150, 1154–1163, 1169–1183
5.2 Competencies—Analyzing the Incident.	30, 31, 32, 33, 35	See below.
5.2.1	32	1110–1137
5.2.1.1	32	1110–1125
5.2.1.1.1	30, 32	1059–1062, 1110–1123

NFPA 472, Standard for Competence of Responders to Hazardous Materials/Weapons of Mass Destruction Incidents, 2018 Edition	Corresponding Chapter(s)	Corresponding Page(s)
5.2.1.1.2	30, 32	1059–1062, 1110–1123
5.2.1.1.3	30, 32	1059–1062, 1110–1123
5.2.1.1.4	32	1115–1123
5.2.1.1.5	32	1110–1115
5.2.1.1.6	32	1124–1125
5.2.1.2	32	1110–1125
5.2.1.2.1	30, 32	1062–1067, 1110–1115
5.2.1.2.2	30	1062–1067
5.2.1.3	30, 32	1062–1067, 1124–1125
5.2.1.3.1	30	1089–1091
5.2.1.3.2	31	1108–1109
5.2.1.3.3	32	1124–1125
5.2.1.4	32	1124–1125
5.2.1.5	32	1124–1125
5.2.1.6	32	1124–1137
5.2.1.6.1	32	1124–1137
5.2.1.6.2	32	1124–1125
5.2.1.6.3	32	1124–1125
5.2.1.6.4	32	1124–1125
5.2.1.6.5	32	1124–1125
5.2.1.6.6	32	1147–1150
5.2.1.6.7	32	1124–1125
5.2.1.6.8	32	1124–1125
5.2.1.6.9	32	1124–1125

NFPA 472, *Standard for Competence of Responders to Hazardous Materials/Weapons of Mass Destruction Incidents*, 2018 Edition	Corresponding Chapter(s)	Corresponding Page(s)
5.2.1.7	35	1190–1198
5.2.2	30, 32	1057–1059, 1062–1067, 1120
5.2.3	31	1089–1105
5.2.4	32, 33	1124–1125, 1142–1150
5.3 Competencies—Planning the Response.	30, 31, 33	See below.
5.3.1	30, 33	1062–1067, 1142–1163
5.3.2	33	1142–1150
5.3.3	33	1149–1159
5.3.4	31, 33	1103–1105, 1159–1163
5.4 Competencies—Implementing the Planned Response.	33, 34	See below.
5.4.1	33, 34	1159–1163, 1168–1184
5.4.2	34	1172–1178
5.4.3	34	1169–1172, 1178–1184
5.4.4	33, 34	1150–1151, 1169–1178
5.5 Competencies—Evaluating Progress.	34	See below.
5.5.1	34	1169–1178
5.5.2	34	1169–1172
5.6 Competencies—Terminating the Incident. (Reserved)	N/A	N/A

NFPA 472: Core Competencies for Operations Level Responders Assigned Mission-Specific Responsibilities

NFPA 472, *Standard for Competence of Responders to Hazardous Materials/Weapons of Mass Destruction Incidents*, 2018 Edition	Corresponding Chapter(s)	Corresponding Page(s)
6.2 Personal Protective Equipment.		
6.2.1 General.	N/A	N/A
6.2.1.1 Introduction.	N/A	N/A
6.2.1.1.1	36	1237–1268
6.2.1.1.2	36	1237–1268
6.2.1.1.3	36	1237–1268
6.2.1.1.4	36	1237–1268
6.2.1.2	36	1237–1268
6.2.1.2.1	36	1210–1238
6.2.2 Competencies—Analyzing the Incident. (Reserved)	N/A	N/A
6.2.3 Competencies—Planning the Response.	36	See below.
6.2.3.1	36	1211–1235
6.2.4 Competencies—Implementing the Planned Response.	36	See below.
6.2.4.1	36	1210–1238
6.2.5 Competencies—Evaluating Progress. (Reserved)	N/A	N/A
6.2.6 Competencies—Terminating the Incident.	36	See below.
6.2.6.1	36	1236

NFPA 472, *Standard for Competence of Responders to Hazardous Materials/Weapons of Mass Destruction Incidents,* 2018 Edition	Corresponding Chapter(s)	Corresponding Page(s)
6.6 Product Control.		
6.6.1 General.	N/A	N/A
6.6.1.1 Introduction.	N/A	N/A
6.6.1.1.1	37	1271–1290
6.6.1.1.2	37	1271–1290
6.6.1.1.3	37	1271–1290
6.6.1.1.4	37	1271–1290
6.6.1.2 Goal.	N/A	N/A
6.6.1.2.1	37	1271–1290
6.6.1.2.2	37	1244–1258
6.6.2 Competencies—Analyzing the Incident. (Reserved)	N/A	N/A
6.6.3 Competencies—Planning the Response	37	See below.
6.6.3.1	37	1244–1258
6.6.4 Competencies— Implementing the Planned Response.	37	See below.
6.6.4.1	37	1251–1253
6.6.4.2	37	1253–1258
6.6.5 Competencies—Evaluating Progress. (Reserved)	N/A	N/A
6.6.6 Competencies—Terminating the Incident.	37	See below.
6.6.6.1	37	1257–1258

Glossary

911 dispatcher/telecommunicator From the communications center, the dispatcher takes the calls from the public, sends appropriate units to the scene, assists callers with treatment instructions until the EMS unit arrives, and assists the incident commander with needed resources.

Absorption The process of applying a material that will soak up and hold a hazardous material in a sponge-like manner for collection and subsequent disposal.

Accelerator A device that speeds up the removal of the air from a dry pipe or preaction sprinkler system. An accelerator reduces the time required for water to start flowing from sprinkler heads.

Accidental fire Fires for which the cause does not involve a human act with the intent to ignite or spread a fire. (NFPA 556)

Accordion hose load A method of loading hose on a vehicle that results in a hose appearance that resembles accordion sections. This is achieved by standing the hose on its edge and laying it side to side in the hose bed.

Activity logging system A device that keeps a detailed record of every incident and activity that occurs.

Adaptor A device that allows fire hose couplings to be safely interconnected with couplings of different sizes, threads, or mating surfacing or that allows fire hose couplings to be safely connected to other appliances.

Adjustable-gallonage fog nozzle A nozzle that allows the operator to select a desired flow from several settings.

Adolescent firesetters Firesetters who are typically 14 to 16 years old and may exhibit serious emotional or psychological symptoms such as extreme stress or anger. Target buildings include schools, churches, outbuildings, vacant homes, and vacant lots.

Adsorption The process in which a contaminant adheres to the surface of an added material—such as silica or activated carbon—rather than combining with it (as in absorption).

Advanced Emergency Medical Technician (AEMT) A member of EMS who can perform limited procedures that usually fall between those provided by an EMT and those provided by a Paramedic, including IV therapy, interpretation of cardiac rhythms, defibrillation, and airway intubation.

Adze The blade or wedge part of a tool such as the Halligan tool.

Aerial fuels Fuels located more than 2 m (6 ft) off the ground, usually part of or attached to trees.

Aerial ladder A self-supporting, turntable-mounted, power-operated ladder of two or more sections permanently attached to a self-propelled automotive fire apparatus and designed to provide a continuous egress route from an elevated position to the ground. (NFPA 1901)

Aerobic metabolism The creation of energy through the breakdown of nutrients in the presence of oxygen. The by-products are carbon dioxide and water, which the body disposes of by breathing and sweating.

Aircraft/crash rescue fire fighter (ARFF) An individual who takes firefighting actions to prevent, control, or extinguish fire involved or adjacent to an aircraft for the purpose of maintaining maximum escape routes for occupants using normal and emergency routes for egress. (NFPA 414)

Air cylinder The pressure vessel or vessels that are an integral part of the SCBA and that contain the breathing gas supply; can be configured as a single cylinder or other pressure vessel or as multiple cylinders or pressure vessels. (NFPA 1981)

Air line The hose through which air flows, either within an SCBA or from an outside source to a supplied air respirator.

Air management The use of a limited air supply in such a way as to ensure that it will last long enough to enter a hazardous area, accomplish needed tasks, and return safely.

Air-purifying respirator (APR) A respirator that removes specific air contaminants by passing ambient air through one or more air purification components. (NFPA 1984)

Air sampling detector A system that captures a sample of air from a room or enclosed space and passes it through a smoke detection or gas analysis device.

Alarm initiating device An automatic or manually operated device in a fire alarm system that, when activated, causes the system to indicate an alarm condition.

Alarm notification appliance An audible and/or visual device in a fire alarm system that makes occupants or other persons aware of an alarm condition.

Alarm valve A valve that signals an alarm when a sprinkler head is activated and prevents nuisance alarms caused by pressure variations.

Alarm verification feature A feature of automatic fire detection and alarm systems to reduce unwanted alarms wherein smoke detectors report alarm conditions for a minimum period of time or confirm alarm conditions within a given time period after being reset, in order to be accepted as a valid alarm initiation signal. (NFPA 72)

Alcohol-resistant foam concentrate A concentrate used for fighting fires on water-soluble materials and other fuels destructive to regular, AFFF, or FFFP foams, as well as for fires involving hydrocarbons. (NFPA 11)

Alternative-fuel vehicle A vehicle that uses anything other than a petroleum-based motor fuel (gasoline or diesel fuel) to propel a motorized vehicle.

Ammonium phosphate An extinguishing agent used in dry-chemical fire extinguishers that can be used on Class A, B, and C fires.

Anaerobic metabolism The creation of energy through the breakdown of glucose. Without oxygen, this metabolic process results in the production of lactic acid.

Anchor, flank, and pinch attack A direct method of suppressing a wildland or ground cover fire that involves two teams of fire fighters establishing anchor points on each side of the fire and working toward the head of the fire until the fire gets "pinched" between them; also known as the pincer attack.

Anchor point A strategic and safe point from which to start constructing a fire control line. An anchor point is used to reduce the chance of fire fighters being flanked by fire.

Annealed The process of forming standard glass.

Anthrax An infectious disease spread by the bacterium *Bacillus anthracis*; typically found around farms, infecting livestock.

A-posts Vertical support members that form the sides of the windshield of a motor vehicle.

Aqueous film-forming foam (AFFF) A solution based on fluorinated surfactants plus foam stabilizers to produce a fluid aqueous film for suppressing liquid fuel vapors. (NFPA 10 and 11)

Arc mapping The systematic evaluation of the electrical circuit configuration, spatial relationship of the circuit components, and identification of electrical arc sites to assist in the identification of the area of origin and analysis of the fire's spread. (NFPA 921)

Area of origin A structure, part of a structure, or general geographic location within a fire scene, in which the "point of origin" of a fire or explosion is reasonably believed to be located. (NFPA 921)

Arson The crime of maliciously and intentionally, or recklessly, starting a fire or causing an explosion. (NFPA 921)

Assistant or division chief A midlevel chief who often has a functional area of responsibility, such as training, and who answers directly to the fire chief.

Atmosphere-supplying respirator (ASR) A respirator that supplies the respirator user with breathing air from a source independent of the ambient atmosphere and includes self-contained breathing apparatus (SCBA) and supplied air respirators (SAR). (NFPA 1981)

Atom The smallest particle of an element, which can exist alone or in combination.

A tool A cutting tool with a pry bar built into the cutting part of the tool.

Attack hose Hose designed to be used by trained fire fighters and fire brigade members to combat fires beyond the incipient stage. (NFPA 1961)

Attack line operations The delivery of water from an attack pump to a handline, which discharges the water onto the fire.

Authority having jurisdiction (AHJ) An organization, office, or individual responsible for enforcing the requirements of a code or standard or for approving equipment, materials, an installation, or a procedure. (NFPA 1072)

Automatic-adjusting fog nozzle A nozzle that can deliver a wide range of water stream flows. As the pressure at the nozzle increases or decreases, an internal spring-loaded piston moves in or out to adjust the size of the opening.

Automatic dry standpipe system A standpipe system permanently attached to a water supply capable of supplying the system demand at all times, containing air or nitrogen under pressure, the release of which opens a dry pipe valve to allow water to flow into the piping system and out of the opened hose valve. (NFPA 14)

Automatic location identification (ALI) A series of data elements that informs the recipient of the location of the alarm. (NFPA 1221)

Automatic number identification (ANI) A series of alphanumeric characters that informs the recipient of the source of the alarm. (NFPA 1221)

Automatic sprinkler heads The working ends of a sprinkler system that serve to activate the system and to apply water to the fire.

Automatic sprinkler system A system of pipes filled with water under pressure that discharges water immediately when a sprinkler head opens.

Automatic wet standpipe system A standpipe system containing water at all times that is attached to a water supply capable of supplying the system demand and that requires no action other than opening a hose valve to provide water at hose connections. (NFPA 14)

Awareness level personnel Personnel who, in the course of their normal duties, could encounter an emergency involving hazardous materials/weapons of mass destruction (WMD) and who are expected to recognize the presence of the hazardous materials/WMD, protect themselves, call for trained personnel, and secure the scene. (NFPA 1072)

Awning windows Windows that have one large or multiple medium-size panels that do not overlap when they are closed. The window is operated by a hand crank from the corner of the window. The hinge is on the top.

Backdraft A deflagration (explosion) resulting from the sudden introduction of air into a confined space containing oxygen-deficient products of incomplete combustion. (NFPA 1403)

Backfire A fire set along the inner edge of a fire control line to consume the fuel in the path of a wildland fire or change the direction of force of the fire's convection column. (NFPA 901)

Backpack fire extinguisher A portable fire extinguisher usually consisting of a 19-L (5-gal) water tank that is worn on the user's back and features a hand-powered piston pump for discharging the water.

Backup team Individuals who function as a stand-by rescue crew or relief for those entering the hot zone (entry team). Also referred to as backup personnel.

Balloon-frame construction An older type of wood frame construction in which the wall studs extend vertically from the basement of a structure to the roof without any fire stops.

Ball valves Valves used on nozzles, gated wyes, and pumper discharge gates. They consist of a ball with a hole in the middle of the ball.

Bam-bam tool A sliding hammer with a case-hardened screw, which is inserted, secured, and driven into the keyway of a lock to remove the keyway from the lock.

Bangor ladder A ladder equipped with tormentor poles, or staypoles, that stabilize the ladder during raising and lowering operations.

Banked Covering a fire to ensure low burning.

Base (bed) section The lowest or widest section of an extension ladder. (NFPA 1931)

Base station A stationary radio transceiver with an integral AC power supply. (NFPA 1221)

Batch mix The manual addition of foam concentrate to a water storage container or tank to make foam solution. (NFPA 1145)

Battering ram A tool made of hardened steel with handles on the sides used to force doors and to breach walls. Larger versions may be used by as many as four people; smaller versions are made for one or two people.

Battery electric vehicles Vehicles that are powered by electricity.

Beam detector A smoke detection device that projects a narrow beam of light across a large open area from a sending unit to a receiving unit. When the beam is interrupted by smoke, the receiver detects a reduction in light transmission and activates the fire alarm.

Beam raise A ladder raise used to raise a ladder perpendicular to a building.

Beam The main structural side of a ground ladder. (NFPA 1931)

Bend A knot that joins two ropes or webbing pieces together. (NFPA 1670)

Bight The open loop in a rope or piece of webbing formed when it is doubled back on itself. (NFPA 1006)

Bimetallic strip A device with components made from two distinct metals that respond differently to heat. When heated, the metals will bend or change shape.

Biological agents Biological materials that are capable of causing acute disease or long-term damage to the human body. (NFPA 1951)

Black An area that has already been burned.

Black fire A hot, high-volume, high-velocity, turbulent, ultra-dense black smoke that indicates an impending flashover or autoignition.

Block creel construction Rope constructed without knots or splices in the yarns, ply yarns, strands or braids, or rope. (NFPA 1983)

Body-over-frame construction A type of vehicle frame resembling a ladder, which is made up of two parallel rails joined by a series of cross members. This kind of construction is typically used for luxury vehicles, sport utility vehicles, and all types of trucks.

Boiling liquid/expanding vapor explosion (BLEVE) An explosion that occurs when pressurized liquefied materials (e.g., propane or butane) inside a closed vessel are exposed to a source of high heat.

Boiling point The temperature at which the vapor pressure of a liquid equals the surrounding atmospheric pressure. (NFPA 1)

Bolt cutter A cutting tool used to cut through thick metal objects, such as bolts, locks, and wire fences.

Booster hose (booster line) A noncollapsible hose used under positive pressure having an elastomeric or thermoplastic tube, a braided or spiraled reinforcement, and an outer protective cover. (NFPA 1962)

Bounce-off method A foam application method that applies the stream onto a nearby object, such as a wall, instead of directly onto the surface of the fire. Also referred to as the bank-shot or bank-down method.

Bowstring truss A truss that is curved on the top and straight on the bottom.

Box-end wrench A hand tool used to tighten or loosen bolts. The end is enclosed, as opposed to an open-end wrench. Each wrench is a specific size, and most have ratchets for easier use.

B-posts Vertical support members located between the front and rear doors of a motor vehicle.

Braided rope Rope constructed by intertwining strands in the same way that hair is braided.

Branch A supervisory level established in either the operations or logistics function to provide a span of control. (NFPA 1561)

Branch director A person in a supervisory level position in either the operations or logistics function to provide a span of control. (NFPA 1561)

Breakaway type nozzle A nozzle with a tip that can be separated from the shut-off valve.

Bresnan distributor nozzle A nozzle that can be placed in confined spaces such as cellars or basements. The nozzle spins, spreading water over a large area.

Buddy system A system in which two responders always work as a team for safety purposes.

Bulk packaging Any packaging, including transport vehicles, having a liquid capacity of more than 450 L (119 gal), a solids capacity of more than 400 kg (882 lb), or a compressed gas water capacity of more than 454 kg (1001 lb). (NFPA 472)

Bulkhead The separation between the passenger compartment and the engine compartment. (NFPA 556)

Butt The end of the beam that is placed on the ground, or other lower support surface, when ground ladders are in the raised position. (NFPA 1931)

Butt plate An alternative to a simple butt spur; a swiveling plate with both a spur and a cleat or pad that is attached to the butt of the ladder.

Butt spurs That component of ground ladder support that remains in contact with the lower support surface to reduce slippage. (NFPA 1931)

Butterfly valves Valves that are found on the large pump intake connections where the suction hose connects to the suction side of the fire pump.

Call box A system of telephones connected by phone lines, radio equipment, or cellular technology to a communications centre or fire department.

Canadian Environmental Protection Act (CEPA 1999) An important part of Canada's federal environmental legislation aimed at preventing pollution and protecting the environment and human health. CEPA 1999 came into force on March 31, 2000, following an extensive Parliamentary review of the first CEPA.

Canadian Radio-television and Telecommunications Commission (CRTC) The authority that regulates and supervises broadcasting and telecommunications in Canada.

Canadian Standards Association (CSA) The federal agency responsible for the regulation of respirator fit testing, training, and breathing air systems.

Canadian Transport Emergency Centre (CANUTEC) The Canadian Transport Emergency Centre, operated by Transport Canada, provides emergency response information and assistance on a 24-hour basis for responders to hazardous materials/weapons of mass destruction (WMD) incidents. (NFPA 1072)

Captain The second rank of promotion in the fire service, between the lieutenant and the district/battalion chief. Captains are responsible for managing a fire company and for coordinating the activities of that company among the other shifts.

Carabiner An auxiliary equipment system item; load-bearing connector with a self-closing gate used to join other components of life safety rope. (NFPA 1983)

Carbon dioxide A colorless, odorless, electrically nonconductive inert gas that is a suitable medium for extinguishing Class B and Class C fires. (NFPA 10)

Carbon dioxide extinguishing system A fire suppression system designed to protect either a single room or series of rooms by flooding the area with carbon dioxide.

Carbon dioxide (CO$_2$) fire extinguisher A fire extinguisher that uses carbon dioxide gas as the extinguishing agent. It is rated for use on Class B and C fires.

Carbon monoxide (CO) A toxic gas produced through incomplete combustion.

Carcinogen A cancer-causing substance that is identified in one of several published lists, including, but not limited to, *NIOSH Pocket Guide to Chemical Hazards, Hazardous Chemicals Desk Reference*, and the *ACGIH 2007 TLVs* and *BEIs*. (NFPA 1851)

Cargo tank A container used for carrying fuels and mounted permanently or otherwise secured on a tank vehicle. (NFPA 407)

Carpenter's handsaw A saw designed for cutting wood.

Carryall A piece of heavy canvas with handles, which can be used to tote debris, ash, embers, and burning materials out of a structure.

Cartridge/cylinder-operated fire extinguisher A fire extinguisher in which the expellant gas is in a separate container from the agent storage container. (NFPA 10)

Cascade system A method of piping air tanks together to allow air to be supplied to the SCBA fill station using a progressive selection of tanks, each with a higher pressure level. (NFPA 1901)

Case-hardened steel Steel created in a process that uses carbon and nitrogen to harden the outer core of a steel component, while the inner core remains soft. Case-hardened steel can be cut only with specialized tools.

Casement windows Windows in a steel or wood frame that open away from the building via a crank mechanism. These windows have a side hinge.

CBRN Chemical, biological, radiological, and nuclear. (NFPA 1991)

Ceiling hook A tool with a long wooden or fiberglass pole that has a metal point with a spur at right angles at one end. It can be used to probe ceilings and pull down plaster lath material.

Cellar nozzle A nozzle used to fight fires in cellars or basements and other inaccessible places. The device works by spreading water in a wide pattern as the nozzle is lowered through a hole into the cellar.

Central station service alarm system A system or group of systems in which the operations of circuits and devices are transmitted automatically to, recorded in, maintained by, and supervised from a listed central station that is controlled and operated by a person, firm, or corporation whose business is the furnishing, maintaining, or monitoring of supervised alarm systems. (NFPA 72)

Chain of command A rank structure, spanning the fire fighter through the fire chief, for managing a fire department and fireground operations.

Chain of custody The trail of accountability that documents the possession of evidence in an investigation.

Chainsaw A power saw that uses the rotating movement of a chain equipped with sharpened cutting edges. It is typically used to cut through wood.

Channel An assigned frequency or frequencies used to carry voice and/or data communications.

Chase Open space within walls for wires and pipes.

Chassis The basic operating motor vehicle, including the engine, frame, and other essential structural and mechanical parts, but exclusive of the body and all appurtenances for the accommodation of driver, property, passengers, and appliances. Common usage might, but need not, include a cab (or cowl).

Chemical energy Energy that is created or released by the combination or decomposition of chemical compounds.

Chemical-pellet sprinkler head A sprinkler head activated by a chemical pellet that liquefies at a preset temperature.

Chemical-resistant materials Clothing (suit fabrics) specifically designed to inhibit or resist the passage of chemicals into and through the material by the process of penetration, permeation, or degradation.

Chemical Transportation Emergency Center (CHEMTREC) A public service of the American Chemistry Council that provides emergency response information and assistance on a 24-hour basis for responders to hazardous materials/weapons of mass destruction (WMD) incidents. (NFPA 1072)

Chief of the department The top position in the fire department. The fire chief has ultimate responsibility for the fire department and usually answers directly to the mayor or other designated public official.

Chief's trumpet An obsolete amplification device that enabled a chief officer to give orders to fire fighters during an emergency. Also called a bugle, it was a precursor to a bullhorn and portable radios.

Child firesetters Firesetters who are typically 2 to 6 years old and are often curious about fire. They lack the mental capacity to understand the consequences of their actions.

Chisel A metal tool with one sharpened end that is used to break apart material in conjunction with a hammer, mallet, or sledgehammer.

Circumstantial evidence Evidence that is based on logical inference rather than personal observation.

Clapper mechanism A mechanical device installed within a piping system that allows water to flow in only one direction.

Class A fire A fire in ordinary combustible materials, such as wood, cloth, paper, rubber, and many plastics. (NFPA 10)

Class A foam Foam for use on fires in Class A fuels. (NFPA 1145)

Class B fire A fire in flammable liquids, combustible liquids, petroleum greases, tars, oils, oil-based paints, solvents, lacquers, alcohols, and flammable gases. (NFPA 10)

Class B foam Foam intended for use on Class B fires. (NFPA 1901)

Class C fire A fire that involves energized electrical equipment. (NFPA 10)

Class D fire A fire in combustible metals, such as magnesium, titanium, zirconium, sodium, lithium, and potassium. (NFPA 10)

Class I standpipe A standpipe system designed for use by fire department personnel only. Each outlet should have a valve to control the flow of water and a 65-mm (2½-in.) male coupling for fire hose.

Class II standpipe A standpipe system designed for use by occupants of a building only. Each outlet is generally equipped with a length of 38-mm (1½-in.) single-jacket hose and a nozzle, which are preconnected to the system.

Class III standpipe A combination system that has features of both Class I and Class II standpipes.

Class K fire A fire in a cooking appliance that involves combustible cooking media (vegetable or animal oils and fats). (NFPA 10)

Claw bar A tool with a pointed claw-hook on one end and a forked- or flat-chisel pry on the other end. It is often used for forcible entry.

Clean agent Electrically nonconducting, volatile, or gaseous fire extinguishant that does not leave a residue upon evaporation. (NFPA 10)

Clean agent extinguishing system A self-contained extinguishing system that expels an electrically nonconducting, volatile, or gaseous fire extinguishant that does not leave a residue upon evaporation.

Clemens hook A multipurpose tool that can be used for several forcible entry and ventilation applications because of its unique head design.

Closed-circuit self-contained breathing apparatus Self-contained breathing apparatus designed to recycle the user's exhaled air. This system removes carbon dioxide and generates fresh oxygen.

Closet hook A type of pike pole intended for use in tight spaces, commonly 0.6 to 1 m (2 to 4 ft) in length.

Cockloft The concealed space between the top-floor ceiling and the roof of a building.

Coded system A fire alarm system design that divides a building or facility into zones and has audible notification devices that can be used to identify the area where an alarm originated.

Cold zone The control zone of hazardous materials/weapons of mass destruction (WMD) incidents that contains the incident command post and such other support functions as are deemed necessary to control the incident. (NFPA 1072 and 1500)

Colorimetric tubes Reagent-filled tubes designed to draw in a sample of air by way of a manual handheld pump. The reagent will undergo a colour change when exposed to the contaminant it is intended to detect.

Combination attack A type of attack employing both direct attack and indirect attack methods.

Combination hose load A hose loading method used when one long hose line is needed; the end of the last length of a hose in one bed of a split hose bed is coupled to the beginning of the first length of hose in the opposite bed.

Combination ladder A ground ladder that is capable of being used both as a stepladder and as a single or extension ladder. (NFPA 1931)

Combination wrench A hand tool with an open-end wrench on one end and a box-end wrench on the other.

Combustibility The property describing whether a material will burn and how quickly it will burn.

Combustion A chemical process of oxidation that occurs at a rate fast enough to produce heat and usually light in the form of either a glow or a flame. (NFPA 1)

Come along A hand-operated tool used for dragging or lifting heavy objects that uses pulleys and cables or chains to multiply a pulling or lifting force.

Command The act of directing and/or controlling resources by virtue of explicit legal, agency, or delegated authority. (NFPA 1026)

Command staff The command staff consists of the public information officer, safety officer, and liaison officer who report directly to the incident commander and are responsible for functions in the incident management system that are not a part of the function of the line organization. (NFPA 1561)

Community Risk Reduction (CRR) Programs, actions, and services used by a community, which prevent or mitigate the loss of life, property, and resources associated with life safety, fire, and other disasters within a community. (NFPA 1035)

Company officer The individual responsible for command of a company, a designation not specific to any particular fire department rank (can be a fire fighter, lieutenant, captain, or chief officer, if responsible for command of a single company). (NFPA 1026)

Compartment A space completely enclosed by walls and a ceiling. Each wall in the compartment is permitted to have openings to an adjoining space if the openings have a minimum lintel depth of 200 mm (8 in.) from the ceiling and the total width of the openings in each wall does not exceed 2.4 m (8 ft). A single opening of

900 mm (36 in.) or less in width without a lintel is permitted when there are no other openings to adjoining spaces. (NFPA 13)

Competent ignition source An ignition source that has sufficient energy and is capable of transferring that energy to the fuel long enough to raise the fuel to its ignition temperature. (NFPA 921)

Compressed air foam (CAF) A homogenous foam produced by the combination of water, foam concentrate, and air or nitrogen under pressure. (NFPA 11)

Compressed air foam system (CAFS) A foam system that combines air under pressure with foam solution to create foam. (NFPA 1901)

Compressor A device used for increasing the pressure and density of a gas. (NFPA 853)

Computer-aided dispatch (CAD) A combination of hardware and software that provides data entry, makes resource recommendations, and notifies and tracks those resources before, during, and after fire service alarms, preserving records of those alarms and status changes for later analysis. (NFPA 1221)

Conduction Heat transfer to another body or within a body by direct contact. (NFPA 921)

Confined space An area large enough and so configured that a member can bodily enter and perform assigned work but that has limited or restricted means for entry and exit and is not designed for continuous human occupancy. (NFPA 1500)

Confinement Those procedures taken to keep a material, once released, in a defined or local area. (NFPA 472)

Consensus document A code document jointly developed by people representing various organizations and interests. NFPA codes and standards are consensus documents.

Containment The actions taken to keep a material in its container (e.g., stop a release of the material or reduce the amount being released). (NFPA 472)

Contamination The process of transferring a hazardous material, or the hazardous component of a weapon of mass destruction (WMD), from its source to people, animals, the environment, or equipment, which can act as a carrier. (NFPA 1072)

Contemporary construction Buildings constructed since about 1970 that incorporate lightweight construction techniques and engineered wood components. These buildings exhibit less resistance to fire than older buildings.

Control The procedures, techniques, and methods used in the mitigation of hazardous materials/weapons of mass destruction (WMD) incidents, including containment, extinguishment, and confinement. (NFPA 1072)

Control zones The areas at hazardous materials/weapons of mass destruction (WMD) incidents within an established perimeter that are designated based upon safety and the degree of hazard. (NFPA 1072)

Convection Heat transfer by circulation within a medium such as a gas or a liquid. (NFPA 921)

Conventional vehicle A vehicle that uses an internal combustion engine for power.

Coping saw A saw designed to cut curves in wood.

Council rake A long-handled rake constructed with hardened triangular-shaped steel teeth that is used for raking a fire control line down to soil with no subsurface fuel, for digging, for rolling burning logs, and for cutting grass and small brush.

Coupling One set or pair of connection devices attached to a fire hose that allow the hose to be interconnected to additional lengths of hose or adapters and other firefighting appliances. (NFPA 1963)

C-posts Vertical support members located behind the rear doors of a motor vehicle.

Crew A team of two or more fire fighters. (NFPA 1500)

Crew resource management (CRM) A program focused on improved situational awareness, sound critical decision making, effective communication, proper task allocation, and successful teamwork and leadership. (NFPA 1500)

Cribbing Short lengths of timber/composite materials, usually 101.60 mm × 101.60 mm (4 in. × 4 in.) and 457.20 mm × 609.60 mm (18 in. × 24 in.) long that are used in various configurations to stabilize loads in place or while load is moving. (NFPA 1006)

Critical incident stress debriefing (CISD) A post-incident meeting designed to assist rescue personnel in dealing with psychological trauma as the result of an emergency. (NFPA 1006)

Critical incident stress management (CISM) A program designed to reduce acute and chronic effects of stress related to job functions. (NFPA 450)

Cross-zoned system A fire alarm system that requires activation of two separate detection devices before initiating an alarm condition. If a single detection device is activated, the alarm control unit will usually show a problem or trouble condition.

Crowbar A straight bar made of steel or iron with a forked chisel on the working end that is suitable for performing forcible entry.

Cryogenic liquid cargo tank (MC-338) A low-pressure tank designed to maintain the low temperature required by the cryogens it carries. A box-like structure containing the tank control valves is typically attached to the rear of the tanker.

Curtain wall Nonbearing walls that separate the inside and outside of the building but are not part of the support structure for the building.

Curved roof A roof with a curved shape.

Cutting torch A torch that produces a high-temperature flame capable of heating metal to its melting point, thereby cutting through an object. Because of the high temperatures (3149°C [5700°F]) that these torches produce, the operator must be specially trained before using this tool.

Cylinder The body of the fire extinguisher where the extinguishing agent is stored.

Cylindrical locks The most common fixed locks in use today. The locks and handles are placed into a predrilled hole in the door. The outside of the doorknob will usually have a key-in-the-knob lock; the inside will usually have a keyway, a button, or another type of locking/unlocking mechanism.

Cytochrome oxidase Found in the mitochondria, this is important in cell respiration as an agent of electron transfer from certain cytochrome molecules to oxygen molecules.

Damming The product-control process used when liquid is flowing in a natural channel or depression, and its progress can be stopped by constructing a barrier to block the flow.

Dead load Dead loads consist of the weight of all materials of construction incorporated into the building including but not limited to walls, floors, roofs, ceilings, stairways, built-in partitions, finishes, cladding and other similarly incorporated architectural and structural items, and fixed service equipment including the weight of cranes. (NFPA 5000)

Deadbolt Surface- or interior-mounted lock on or in a door with a bolt that provides additional security.

Decay stage The stage of fire development within a structure characterized by either a decrease in the fuel load or available oxygen to support combustion, resulting in lower temperatures and lower pressure in the fire area. (NFPA 1410)

Deck gun A device that is permanently mounted on and operated from a vehicle and equipped with a piping system that delivers water to the gun.

Decontamination The physical and/or chemical process of reducing and preventing the spread and effects of contaminants to people, animals, the environment, or equipment involved at hazardous materials/weapons of mass destruction (WMD) incidents. (NFPA 1072)

Decontamination corridor The area usually located within the warm zone where decontamination is performed. (NFPA 1072)

Decontamination team The team responsible for reducing and preventing the spread of contaminants from persons and equipment used at a hazardous materials incident. Members of this team establish the decontamination corridor and conduct all phases of decontamination.

Defend in place The operational response in which the action is to relocate the affected occupants to a safe place within the structure during an emergency. (NFPA 1620)

Defensible space An area, as defined by the authority having jurisdiction [typically a width of 9 m (30 ft) or more], between an improved property and a potential wildland fire where combustible materials and vegetation have been removed or modified to reduce the potential for fire on improved property spreading to wildland fuels or to provide a safe working area for fire fighters protecting life and improved property from wildland fire. (NFPA 1051)

Defensive operation Actions that are intended to control a fire by limiting its spread to a defined area, avoiding the commitment of personnel and equipment to dangerous areas. (NFPA 1500)

Degradation A chemical action involving the molecular breakdown of a protective clothing material or equipment due to contact with a chemical. (NFPA 1072)

Dehydration A state in which the body's fluid losses are greater than fluid intake. If left untreated, dehydration may lead to shock and even death.

Deluge head A sprinkler head that has no release mechanism; the orifice is always open.

Deluge sprinkler system A sprinkler system employing open sprinklers or nozzles that are attached to a piping system that is connected to a water supply through a valve that is opened by the operation of a detection system installed in the same areas as the sprinklers or the nozzles. When this valve opens, water flows into the piping system and discharges from all sprinklers or nozzles attached thereto. (NFPA 13)

Deluge valve A water supply control valve intended to be operated by actuation of an automatic detection system that is installed in the same area as the discharge devices. (NFPA 25)

Demonstrative evidence Any type of tangible evidence relevant to a case, such as diagrams, photographs, maps, x-rays, visible tests, and demonstrations.

Denial of entry A policy under which, once the perimeter around a release site has been identified and marked out, responders limit access to all but essential personnel.

Depth of char A fire effect that, when evaluated as a pattern on identical fuels, may be used to determine locations within a structure that were exposed longest to a heat source.

Destructive analysis investigation An investigation that uses the methodical deconstruction of evidence to determine specific component conditions, functionality, or failures as they relate to fire investigation.

Digital radio The transmission of information via radio waves using native digital (computer) data or analog (voice) signals that have been converted to a digital signal and compressed.

Diking The placement of materials to form a barrier that will keep a hazardous material in liquid form from entering an area or that will hold the material in an area.

Dilution The process of adding a substance—usually water—to weaken the concentration of another substance.

Direct attack Firefighting operations involving the application of extinguishing agents directly onto the burning fuel. (NFPA 1145)

Direct evidence Testimony of witnesses who observe acts or detect something through their five senses or through surveillance equipment such as CCTV.

Direct line A telephone that connects two predetermined points.

Discipline The guidelines that a department sets for fire fighters to work within.

Dispatch To send out emergency response resources promptly to an address or incident location for a specific purpose. (NFPA 450)

Distributors The smallest diameter underground water main pipes in a water distribution system that deliver water to local users within a neighborhood.

District/battalion chief Usually the first level of fire chief. These chiefs are often in charge of running calls and supervising multiple stations or districts within a city. A district/battalion chief is usually the officer in charge of a single-alarm working fire.

Diversion The process of redirecting spilled or leaking material to an area where it will have less impact.

Division of labor Breaking down an incident or task into a series of smaller, more manageable tasks and assigning personnel to complete those tasks.

Division supervisor A person in a supervisory level position responsible for a specific geographic area of operations at an incident. (NFPA 1561)

Division That organizational level having responsibility for operations within a defined geographic location. (NFPA 1026)

Documentary evidence Any type of written record or document that is relevant to the case.

Doff To take off an item of clothing or equipment.

Don To put on an item of clothing or equipment.

Door An entryway; the primary choice for forcing entry into a vehicle or structure.

Door jamb The upright or vertical parts of a door frame onto which a door is secured.

Double-action pull station A manual fire alarm activation device that requires two steps to activate the alarm. The user must push in a flap, lift a cover, or break a piece of glass before activating the alarm.

Double-female adaptor A hose adaptor that is used to join two male hose couplings.

Double-hung windows Windows that have two movable panels or sashes that can move up and down.

Double-jacket hose A hose constructed with two layers of woven fibres.

Double-male adaptor A hose adaptor that is used to join two female hose couplings.

Double/triple-pane glass A window design that traps air or inert gas between two pieces of glass to help insulate a house.

Drag rescue device (DRD) A component integrated within the protective coat element to aid in the rescue of an incapacitated fire fighter. (NFPA 1851)

Dry-barrel hydrant (frostproof hydrant) A type of hydrant used in areas subject to freezing weather. The valve that allows water to flow into the hydrant is located underground below the frost line, and the barrel of the hydrant is normally dry.

Dry bulk cargo trailers Trailers designed to carry dry bulk goods such as powders, pellets, fertilizers, or grain. Such tanks are generally V-shaped with rounded sides that funnel toward the bottom.

Dry chemical A powder composed of very small particles, usually sodium bicarbonate, potassium bicarbonate, or ammonium phosphate based with added particulate material supplemented by special treatment to provide resistance to packing, resistance to moisture absorption (caking), and the proper flow capabilities. (NFPA 10)

Dry-chemical extinguishing system An automatic fire-extinguishing system that discharges a dry-chemical agent.

Dry-chemical fire extinguisher A fire extinguisher that uses a powder composed of very small particles, usually sodium bicarbonate, potassium bicarbonate, or ammonium phosphate, based with added particulate material supplemented by special treatment to provide resistance to packing, resistance to moisture absorption (caking), and the proper flow capabilities. These fire extinguishers are rated for use on Class B and C fires, although some are also rated for Class A fires.

Dry hydrant An arrangement of pipe permanently connected to a water source other than a piped, pressurized water supply system that provides a ready means of water supply for firefighting purposes and that utilizes the drafting (suction) capability of a fire department pump. (NFPA 1142)

Dry pipe sprinkler system A sprinkler system employing automatic sprinklers that are attached to a piping. (NFPA 13)

Dry pipe valve The valve assembly on a dry pipe sprinkler system that prevents water from entering the system until the air pressure is released.

Dry powder Solid materials in powder or granular form designed to extinguish Class D combustible metal fires by crusting, smothering, or heat-transferring means. (NFPA 10)

Dry-powder fire extinguisher A fire extinguisher that uses solid materials in powder or granular form to extinguish Class D combustible metal fires by crusting, smothering, or heat-transferring means.

Drywall hook A specialized version of a pike pole that can remove drywall more effectively because of its hook design.

Dual-path pressure reducer A feature that automatically provides a backup method for air to be supplied to the regulator of an SCBA if the primary passage malfunctions.

Duck-billed lock breaker A tool with a point that can be inserted into the shackles of a padlock. As the point is driven farther into the lock, it gets larger and forces the shackles apart until they break.

Duct detector A type of smoke detector that samples the air through the air distribution system ductwork or plenum; upon detecting smoke, the detector sends a signal to shut down the air distribution unit, close any associated smoke damper, or initiate smoke control system operation.

Dump valve A large opening from the water tank of a mobile water supply apparatus for unloading purposes. (NFPA 1901)

Duplex channel A radio system that is able to simultaneously use two frequencies per channel; one frequency transmits and the other receives messages. Such a system uses a repeater site to transmit messages over a greater distance than is possible with a simplex system.

Dutchman A short fold placed in a hose when loading the hose into a hose bed; the fold keeps the hose properly oriented and prevents the coupling from turning in the hose bed.

Dynamic rope A rope generally made from synthetic materials that is designed to be elastic and stretch when loaded. Mountain climbers often use dynamic rope.

Early-suppression fast-response sprinkler head (ESFR) A sprinkler head designed to react quickly and suppress a fire in its early stages.

Eductor A device that uses the Venturi principle to siphon a liquid into a water stream. The pressure at the throat is below atmospheric pressure, allowing liquid at atmospheric pressure to flow into the water stream. (NFPA 1145)

Eighteen watch out situations A list of situations published by the National Wildfire Coordinating Group (NWCG) and used to assess whether or not a wildland firefighting assignment is safe to conduct.

Electrical energy Heat that is produced by electricity.

Electrolytes Certain salts and other chemicals that are dissolved in body fluids and cells. Proper levels of electrolytes need to be maintained for good health and strength.

Elevated master stream appliance A nozzle mounted on the end of an aerial device that is capable of delivering large amounts of water into a fire or exposed building from an elevated position.

Elevated platform An apparatus that includes a passenger-carrying platform (bucket) attached to the tip of a boom or ladder.

Elevated water storage tower An above-ground water storage tank that is designed to maintain pressure on a water distribution system.

Elevation pressure The amount of pressure created by gravity. Also known as head pressure.

Emergency breathing safety systems (EBSS) A device on an SCBA that allows users to share their available air supply in an emergency situation. (NFPA 1981)

Emergency decontamination The process of immediately reducing contamination of individuals in potentially life-threatening situations with or without the formal establishment of a decontamination corridor. (NFPA 1072)

Emergency incident rehabilitation A function on the emergency scene that cares for the well-being of the fire fighters. It includes relief from climatic conditions, rest, cooling or warming, rehydration, calorie replacement, medical monitoring, member accountability, and release.

Emergency Medical Responder (EMR) The first trained professional, such as a police officer, fire fighter, lifeguard, or other rescuer, to arrive at the scene of an emergency to provide initial medical assistance. EMRs have basic training and often perform in an assistant role within the ambulance.

Emergency medical services (EMS) company A company that may be made up of medical units and first-response vehicles. Members of this company respond to and assist in the transport of medical and trauma victims to medical facilities. They often have medications, defibrillators, and Paramedics who can stabilize a critical patient.

Emergency medical services (EMS) personnel Personnel who are responsible for administering prehospital care to people who are sick and injured. Prehospital calls make up the majority of responses in most fire departments, and in some organizations, EMS personnel are crossed-trained as fire fighters.

Emergency Medical Technician (EMT) EMS personnel who account for most of the EMS providers in the United States. An EMT has training in basic emergency care skills, including oxygen therapy, bleeding control, CPR, automated external defibrillation, use of basic airway devices, and assisting patients with certain medications.

Emergency traffic An urgent message, such as a call for help or evacuation, transmitted over a radio that takes precedence over all normal radio traffic.

Emergency Transportation System for the Chemical Industry, Mexico (SETIQ) The Emergency Transportation System for the Chemical Industry in Mexico that provides emergency response information and assistance on a 24-hour basis for responders to emergencies involving hazardous materials/weapons of mass destruction (WMD). (NFPA 1072)

Emergency vehicle technician (EVT) The individual who performs maintenance, diagnosis, and repair on emergency vehicles.

Employee assistance programs (EAPs) An employee-sponsored service designed for personal or family problems, including mental health, substance abuse, various addictions, marital problems, parenting problems, emotional problems, or financial or legal concerns. (NFPA 450)

End-of-service-time indicator (EOSTI) A warning device on an SCBA that alerts the user that the reserved air supply is being utilized. (NFPA 1981)

Endothermic Reactions that absorb heat or require heat to be added.

Engine or pump company A group of fire fighters who work as a unit and are equipped with one or more pumping engines that have rated capacities of 2839 L/min (750 gpm) or more. (NFPA 1410)

Entrapment A condition in which a victim is trapped by debris, soil, or other material and is unable to extricate himself or herself.

Entry team A team of fully qualified and equipped responders who are assigned to enter the designated hot zone.

Escape rope A single-purpose, emergency self-escape (self-rescue) rope; not classified as a life safety rope. (NFPA 1983)

Evacuation The removal or relocation of those individuals who may be affected by an approaching release of a hazardous material.

Evacuation signal A distinctive signal intended to be recognized by the occupants as requiring evacuation of the building. (NFPA 72)

Excepted packaging Packaging used to transport materials that meets only general design requirements for any hazardous material. Low-level radioactive substances are commonly shipped in these packages, which may be constructed out of heavy cardboard.

Exhauster A device that accelerates the removal of the air from a dry pipe or preaction sprinkler system.

Exit Drills In The Home (E.D.I.T.H.) A public fire and life safety education program designed to teach occupants how to safely exit a home in the event of a fire or other emergency. It stresses the importance of having a pre-established meeting place for all family members.

Exothermic Reactions that result in the release of energy in the form of heat.

Expectation bias Any preconceived determination or premature conclusions as to the cause of a fire without having examined or considered all relevant evidence.

Explosive ordnance disposal (EOD) personnel Personnel trained to detect, identify, evaluate, render safe, recover, and dispose of unexploded explosive devices.

Exposure Any person or property that could be endangered by fire, smoke, gases, runoff, or other hazardous conditions. (NFPA 402)

Extension ladder A non–self-supporting ground ladder that consists of two or more sections travelling in guides, brackets, or the equivalent arranged so as to allow length adjustment. (NFPA 1931)

Exterior wall A wall—often made of wood, brick, metal, or masonry—that makes up the outer perimeter of a building. Exterior walls are often load bearing.

Extinguishing agent A material used to stop the combustion process. Extinguishing agents may include liquids, gases, dry-chemical compounds, and dry-powder compounds.

Extra (high) hazard locations Occupancies where the total amounts of Class A combustibles and Class B flammables are greater than expected in occupancies classed as ordinary (moderate) hazards. The combustibility and heat release rate of the materials are high.

Face piece Describes both full face pieces that cover the nose, mouth, and eyes and half face pieces that cover the nose and mouth. (NFPA 1404)

Face shield A protective device commonly intended to shield the wearer's face, or portions thereof, in addition to the eyes from certain hazards, depending on face shield type. (NFPA 1500)

Film-forming fluoroprotein (FFFP) foam A protein-foam solution that uses fluorinated surfactants to produce a fluid aqueous film for suppressing liquid fuel vapors. (NFPA 10)

Finance/administration section Section responsible for all costs and financial actions of the incident or planned event, including the time unit, procurement unit, compensation/claims unit, and cost unit. (NFPA 1026)

Finance/administration section chief The general staff position responsible for directing the finance/administrative function. It is generally assigned on large-scale or long-duration incidents that require immediate fiscal management.

Fine fuels Fuels that ignite and burn easily, such as dried twigs, leaves, needles, grass, moss, and light brush.

Finger A narrow point of fire whose extension is created by a shift in wind or a change in topography.

Fire A rapid, persistent chemical reaction that releases both heat and light.

Fire alarm A warning signal that alerts occupants of a fire emergency.

Fire alarm control unit (FACU) A component of the fire alarm system, provided with primary and secondary power sources, which receives signals from initiating devices or other fire alarm control units and processes these signals to determine part of all of the required fire alarm system output function(s). (NFPA 72)

Fire alarm signal A signal that results from the manual or automatic detection of a fire alarm condition. (NFPA 72)

Fire and life-safety educator (FLSE/fire prevention officer [FPO]) The individual who has demonstrated the ability to coordinate, create, administer, prepare, deliver, and evaluate educational programs and information.

Fire apparatus driver/operator A fire department member who is authorized by the authority having jurisdiction to drive, operate, or both drive and operate fire department vehicles. (NFPA 1451)

Fire barrier wall A wall, other than a fire wall, having a fire-resistance rating. (NFPA 5000)

Fire codes A set of legally adopted rules and regulations designed to prevent fires and protect lives and property in the event of a fire.

Fire control line Comprehensive term for all constructed or natural barriers and treated fire edges used to control a fire. (NFPA 901)

Fire department connection (FDC) A connection through which the fire department can pump supplemental water into the sprinkler system, standpipe, or other system furnishing water for fire extinguishment to supplement existing water supplies. (NFPA 25)

Fire department ground ladder Any portable ladder specifically designed for fire department use in rescue, firefighting operations, or training. (NFPA 1931)

Fire door assembly Any combination of a fire door, a frame, hardware, and other accessories that together provide a specific degree of fire protection to the opening. (NFPA 80)

Fire effects The observable or measurable changes in or on a material as a result of a fire. (NFPA 921)

Fire escape rope An emergency self-rescue rope used to escape an immediately hazardous environment involving fire or fire products; not classified as a life safety rope. (NFPA 1983)

Fire Fighter I A person, at the first level of progression as defined in Chapter 4 of NFPA 1001, who has demonstrated the knowledge and skills to function as an integral member of a firefighting team under direct supervision in hazardous conditions. (NFPA 1001)

Fire Fighter II A person, at the second level of progression as defined in Chapter 5 of NFPA 1001, who has demonstrated the skills and depth of knowledge to function under general supervision. (NFPA 1001)

Fire-ground command (FGC) An incident management system developed in the 1970s for day-to-day fire department incidents (generally handled with fewer than 25 units or companies).

Fire helmet Protective head covering worn by fire fighters to protect the head from falling objects, blunt trauma, and heat.

Fire hooks Tools used to pull down burning elements in structures; also called pike poles.

Fire hydraulics The physical science of how water flows through a pipe or hose.

Fire inspector An individual who conducts fire code inspections and applies codes and standards. (NFPA 1037)

Fire investigator An individual who has demonstrated the skills and knowledge necessary to conduct, coordinate, and complete an investigation. (NFPA 1037)

Fire load The total energy content of combustible materials in a building, space, or area including furnishing and contents and combustible building elements expressed in MJ. (NFPA 557)

Fire mark Historically, an identifying symbol on a building informing fire fighters that the building was insured by a company that would pay them for extinguishing the fire.

Fire marshal A person designated to provide delivery, management, and/or administration of fire protection and life-safety–related codes and standards, investigations, education, and/or prevention services for local, county/provincial, federal, tribal, or private sector jurisdictions as adopted or determined by that entity. (NFPA 1037)

Fire patterns The visible or measurable physical changes, or identifiable shapes, formed by a fire effect or group of fire effects. (NFPA 921)

Fireplug Historically speaking, a plug installed to control water accessed from wooden pipes. Today, this is a slang term used to describe a fire hydrant.

Fire point The lowest temperature at which a liquid will ignite and achieve sustained burning when exposed to a test flame in accordance with ASTM 92, Standard Test Method for Flash and Fire Points by Cleveland Open Cup Tester. (NFPA 1)

Fire police officer An individual officially deployed who provides scene security, directs traffic, and conducts other duties as determined by the authority having jurisdiction. (NFPA 1091)

Fire prevention Measures taken toward avoiding the inception of fire. (NFPA 801)

Fire protection engineer A member of the fire department who is responsible for reviewing plans and working with building owners to ensure that the design of and systems for fire detection and suppression will meet applicable codes and function as needed. They also may be employed by an architectural firm to assure that buildings are constructed in a fire-safe manner.

Fire protection system Any fire alarm device or system or fire-extinguishing device or system, or combination thereof, that is designed and installed for detecting, controlling, or extinguishing a fire or otherwise alerting occupants, or the fire department, or both. (NFPA 914)

Fire resistance The measure of the ability of a material, product, or assembly to withstand fire or give protection from it. (NFPA 251)

Fire safety survey A voluntary inspection of a residence or occupied structure to identify fire and life safety hazards.

FIRESCOPE Fire Resources of California Organized for Potential Emergencies; an organization of agencies established in the early 1970s to develop a standardized system for managing fire resources at large-scale incidents such as wildland fires.

Fire separation A horizontal or vertical fire resistance–rated assembly of materials that have protected openings and are designed to restrict the spread of fire. (NFPA 45)

Fire shelter An item of protective equipment configured as an aluminized tent utilized for protection, by means of reflecting radiant heat, in a fire entrapment situation. (NFPA 1500)

Fire streams Streams of water or extinguishing agents.

Fire tetrahedron A geometric shape used to depict the four components required for a fire to occur: fuel, oxygen, heat, and chemical chain reactions.

Fire triangle A geometric shape used to depict the three components of which a fire is composed: fuel, oxygen, and heat.

Fire wall A wall separating buildings or subdividing a building to prevent the spread of fire and having a fire-resistance rating and structural stability. (NFPA 5000)

Fire wardens Individuals who were charged with enforcing fire regulations in the colonial period.

Fire window A window assembly rated in accordance with NFPA 257 and installed in accordance with NFPA 80. (NFPA 5000)

Firing out A wildland firefighting technique that involves setting a fire along the inner edge of a fire control line to consume the fuel between a fire control line and the fire's edge.

Fixed-gallonage fog nozzle A nozzle that delivers a set number of litres per minute (gallons per minute) as per the nozzle's design, no matter what pressure is applied to the nozzle.

Fixed-temperature heat detector A device that responds when its operating element becomes heated to a predetermined level. (NFPA 72)

Flame detector A radiant energy-sensing fire detector that detects the radiant energy emitted by a flame. (NFPA 72)

Flammable range (explosive limits) The range in concentration between the lower and upper flammable limits. (NFPA 67)

Flanking attack A direct method of suppressing a wildland or ground cover fire that involves placing a suppression crew on one flank of a fire.

Flank of the fire The edge between the head and heel of the fire that runs parallel to the direction of the fire spread.

Flashover A transition phase in the development of a compartment fire in which surfaces exposed to thermal radiation reach ignition temperature more or less simultaneously, and fire spreads rapidly throughout the space, resulting in full room involvement or total involvement of the compartment or enclosed space. (NFPA 921)

Flash point The minimum temperature at which a liquid or a solid emits vapor sufficient to form an ignitable mixture with air near the surface of the liquid or the solid. (NFPA 115)

Flat bar A specialized type of prying tool made of flat steel with prying ends suitable for performing forcible entry.

Flat-head axe A tool that has a head with an axe on one side and a flat head on the opposite side.

Flat hose load A hose loading method in which the hose is laid flat and stacked on top of the previous section.

Flat raise (rung raise) A ladder raise used to position a ladder parallel to a building. Also called a rung raise.

Flat roof A horizontal roof; often found on commercial or industrial occupancies.

Floodlight A light that can illuminate a broad area.

Floor runner A piece of canvas or plastic material available in various lengths that is used to protect flooring from dropped debris and dirt from shoes and boots.

Flow path The movement of heat and smoke from the higher pressure within the fire area toward the lower pressure areas accessible via doors, window openings, and roof structures. (NFPA 1410)

Fluoroprotein foam A protein-based foam concentrate to which fluorochemical surfactants have been added. (NFPA 402)

Fly section Any section of an aerial telescoping device beyond the base section. This definition applies to aerial ladder devices and ground ladders. (NFPA 1901)

Foam blanket A covering of foam over a surface to insulate, prevent ignition, or extinguish the fire. (NFPA 1145)

Foam concentrate The foaming agent as received from the supplier that, when mixed with water, becomes foam solution. (NFPA 1145)

Foam injector A device installed on a fire pump that meters out foam by pumping or injecting it into the fire stream.

Foam proportioner A device or method to add foam concentrate to water to make foam solution. (NFPA 1901)

Foam solution A homogeneous mixture of foam concentrate and water, in the mix ratio required for the application. (NFPA 1145)

Fog-stream nozzle A nozzle that is placed at the end of a fire hose and can be adjusted to produce a straight stream or to separate the water into droplets to produce a variety of fog streams. May also be referred to as a combination nozzle, spray nozzle, or adjustable fog-stream nozzle.

Folding ladder A single-section ladder with rungs that can be folded or moved to allow the beams to be brought into a position touching or nearly touching each other. (NFPA 1931)

Forcible entry Techniques used by fire personnel to gain entry into buildings, vehicles, aircraft, or other areas of confinement when normal means of entry are locked or blocked. (NFPA 402)

Forestry fire hose A hose designed to meet specialized requirements for fighting wildland fires. (NFPA 1961)

Fork The fork or claw end of a tool.

Forward hose lay A method of laying a supply line where the hose line starts at the water source and ends at the attack pump. This is also referred to as the straight hose lay.

Four-way hydrant valve A specialized type of valve that can be placed on a hydrant and that allows another pump to increase the supply pressure without interrupting flow.

Frangible-bulb sprinkler head A sprinkler head with a liquid-filled bulb. The sprinkler head becomes activated when the liquid is heated and the glass bulb breaks.

Freelancing Individuals or crews operating independently of the established ICS structure.

Frequency The number of cycles (oscillations) per second of a radio signal.

Fresno ladder A narrow, two-section extension ladder that has no halyard. Because of its limited length, it can be extended manually.

Friction loss The reduction in pressure resulting from the water being in contact with the side of the hose. This contact requires force to overcome the drag that the wall of the hose creates.

Friction points Those places where a hose line encounters resistance, such as corners, doorways, and stairs.

Frostbite A localized condition that occurs when the layers of the skin and deeper tissue freeze. (NFPA 704)

Fuel A material that will maintain combustion under specified environmental conditions. (NFPA 53)

Fuel cells Cells that generate electricity through a chemical reaction between hydrogen gas and oxygen gas to produce water and in the process produce electricity to propel a vehicle.

Fuel compactness The extent to which fuels are tightly packed together.

Fuel continuity The relative closeness of wildland fuels, which affects a fire's ability to spread from one area of fuel to another.

Fuel-limited fire A fire in which the heat release rate and fire growth are controlled by the characteristics of the fuel because there is adequate oxygen available for combustion. (NFPA 1410)

Fuel moisture The amount of moisture present in a fuel, which affects how readily the fuel will ignite and burn.

Fuel orientation The position of a fuel relative to the ground.

Fuel volume The amount of fuel present in a given area.

Fully developed stage The stage of fire development where heat release rate has reached its peak within a compartment. (NFPA 1410)

Fully encapsulated suit A protective suit that completely covers the fire fighter, including the breathing apparatus, and does not let any vapor or fluids enter the suit. It is commonly used in hazardous materials emergencies.

Fusible-link sprinkler head A sprinkler head with an activation mechanism that incorporates two pieces of metal held together by low-melting-point solder. When the solder melts, it releases the link, and water begins to flow.

Gas detector A device that detects the presence of a specified gas concentration. Gas detectors can be either spot-type or line-type detectors. (NFPA 72)

Gaseous suppression system A system often installed in areas where computers or sensitive electronic equipment is used or where valuable documents are stored.

Gas The physical state of a substance that has no shape or volume of its own and will expand to take the shape and volume of the container or enclosure it occupies. (NFPA 921)

Gated wye A valved device that splits a single hose into two separate hose, allowing each hose to be turned on and off independently.

Gate valves Valves found on hydrants and sprinkler systems. Rotating a spindle causes a gate to move slowly across the opening.

General-service rail tank cars See *low-pressure tank cars*.

General staff A group of incident management personnel organized according to function and reporting to the incident commander, normally consisting of the operations section chief, planning section chief, logistics section chief, and finance/administration section chief. (NFPA 1026)

General use life safety rope A life safety rope that is no larger than 16 mm (⅝ in.) and no smaller than 11 mm (⁷⁄₁₆ in.), with a minimum breaking strength of 40 kN (8992 lbf).

Generator An electromechanical device for the production of electricity. (NFPA 1901)

Geographic information systems (GIS) A system of computer software, hardware, data, and personnel to describe information tied to a spatial location. (NFPA 450)

Glass blocks Thick pieces of glass that are similar to bricks or tiles.

Glazing Glass or transparent or translucent plastic sheet used in windows, doors, skylights, or curtain walls. [ASCE/SEI 7:6.2] (NFPA 5000)

Global positioning system (GPS) A satellite-based radio navigation system comprised of three segments: space, control, and user. (NFPA 414)

Glucose The source of energy for the body. One of the basic sugars, it is the body's primary fuel, along with oxygen.

Governance The process by which an organization exercises authority and performs the functions assigned to it.

Gravity-feed system A water distribution system that depends on gravity to provide the required pressure. The system storage is usually located at a higher elevation than the end users of the water.

Green An area of unburned fuels.

Gripping pliers A hand tool with a pincer-like working end that can be used to bend wire or hold smaller objects.

Gross decontamination A phase of the decontamination process where significant reduction of the amount of surface contamination takes place as soon as possible, most often accomplished by mechanical removal of the contaminant or initial rinsing from handheld hose lines, emergency showers, or other nearby sources of water. (NFPA 1072)

Ground cover fire A fire that burns loose debris on the surface of the ground.

Ground duff Partly decomposed organic material on a forest floor; a type of light fuel.

Group Established to divide the incident management structure into functional assignments of operation. (NFPA 1026)

Group supervisor A person in a supervisory level position responsible for a functional area of operation. (NFPA 1561)

Growth stage The stage of fire development where the heat release rate from an incipient fire has increased to the point where heat transferred from the fire and the combustion products are pyrolyzing adjacent fuel sources and the fire begins to spread across the ceiling of the fire compartment (roll-over). (NFPA 1410)

Guideline A rope used for orientation when fire fighters are inside a structure where there is low or no visibility. The line is attached to a fixed object outside the hazardous area.

Guides Strips of metal or wood that serve to guide a fly section during extension. Channels or slots in the bed or fly section may also serve as guides.

Gusset plate Connecting plate made of a thin sheet of steel used to connect the components of a truss. May also be referred to as a gang nail.

Gypsum A naturally occurring material consisting of calcium sulfate and water molecules.

Gypsum board The generic name for a family of sheet products consisting of a noncombustible core primarily of gypsum with paper surfacing. (NFPA 5000)

Hacksaw A cutting tool designed for use on metal. Different blades can be used for cutting different types of metals.

Halligan tool A prying tool that incorporates a sharp tapered pick, a blade (either an adze or wedge), and a fork; it is specifically designed for use in the fire service.

Halocarbon Halocarbon agents include hydrochlorofluorocarbon (HCFC), hydrofluorocarbon (HFC), perfluorocarbon (PFC), fluoroiodocarbon (FIC) types of agents, and other halocarbons that are found acceptable under the Environmental Protection Agency Significant New Alternatives Policy program. (NFPA 10)

Halogenated agent A liquefied gas extinguishing agent that extinguishes fire by chemically interrupting the combustion reaction between fuel and oxygen. Halogenated agents leave no residue. (NFPA 402)

Halogenated-agent fire extinguisher A fire extinguisher that uses a halogenated extinguishing agent; also called a clean agent fire extinguisher.

Halon 1211 A halogenated agent whose chemical name is bromochlorodifluoromethane ($CBrClF_2$) and that is a multipurpose, Class ABC–rated agent effective against flammable liquid fires. (NFPA 408)

Halons Halons include bromochlorodifluoromethane (Halon 1211), bromotrifluoromethane (Halon 1301), and mixtures of Halon 1211 and Halon 1301 (Halon 1211/1301). (NFPA 10)

Halyard Rope used on extension ladders for the purpose of raising a fly section(s). (NFPA 1931)

Hammer A striking tool.

Hand light A small, portable light carried by fire fighters to improve visibility at emergency scenes; it is often powered by rechargeable batteries.

Handle The grip used for holding and carrying a portable fire extinguisher.

Handline A hose and nozzle that can be held and directed by hand. (NFPA 11)

Handline nozzle A nozzle with a rated discharge of less than 350 gpm (1325 L/min).

Handsaw A manually powered saw designed to cut different types of materials. Examples include hacksaws, carpenter's handsaws, keyhole saws, and coping saws.

Hard suction hose A short section of supply hose that is used to draft water from a static source such as a river, lake, or portable drafting basin to the suction side of the fire pump on a fire department pumper or into a portable pump.

Hardware The parts of a door or window that enable it to be locked or opened.

Harness An equipment item; an arrangement of materials secured about the body used to support a person. (NFPA 1971)

Hazardous material Matter (solid, liquid, or gas) or energy that when released is capable of creating harm to people, the environment, and property, including weapons of mass destruction (WMD) as defined in 18 U.S. Code, Section 2332a, as well as any other criminal use of hazardous materials, such as illicit labs, environmental crimes, or industrial sabotage. (NFPA 1072)

Hazardous materials branch The function within an overall incident management system (IMS) that deals with the mitigation and control of the hazardous materials/weapons of mass destruction (WMD) portion of an incident. (NFPA 1072)

Hazardous materials company A fire company that responds to and controls scenes where hazardous materials have spilled or leaked. Responders wear special suits and are trained to deal with most chemicals.

Hazardous materials group See *hazardous materials branch*.

Hazardous materials safety officer The person who works within an incident management system (IMS) (specifically, the hazardous materials branch/group) to ensure that recognized hazardous materials/weapons of mass destruction (WMD) safe practices are followed at hazardous materials/WMD incidents. (NFPA 1072)

Hazardous materials technician A person who responds to hazardous materials/weapons of mass destruction incidents using a risk-based response process by which he or she analyzes the problem at hand, selects applicable decontamination procedures, and controls a release while using specialized protective clothing and control equipment. (NFPA 472)

Hazardous waste A substance that remains after a process or manufacturing plant has used some of the material and the substance is no longer pure.

Hazel hoe A hand tool used to grub out heavy brush to create a fire control line; also known as an adze hoe.

Head of the fire The main or running edge of a fire; the part of the fire that spreads with the greatest speed.

Heads-up display (HUD) Visual display of information and system condition status that is visible to the wearer. (NFPA 1981)

Heat cramps Painful muscle spasms that occur suddenly during or after physical exertion; usually involve muscles in the leg or abdomen.

Heat detector A fire detector that detects either abnormally high temperature or rate of temperature rise or both. (NFPA 72)

Heat exhaustion A mild form of shock that occurs when the circulatory system begins to fail because of the body's inadequate effort to give off excessive heat.

Heat flux The measure of the rate of heat transfer to a surface, typically expressed in kilowatts per metre squared (kW/m^2) or Btu/ft^2. (NFPA 268)

Heat rash Itchy rash on skin that is wet from sweating; seen after prolonged sweating.

Heat release rates (HRR) The rates at which heat energy is generated by burning. (NFPA 921)

Heat sensor label A label that changes colour at a preset temperature to indicate a specific heat exposure. (NFPA 1931)

Heat stroke A severe, sometimes fatal condition resulting from the failure of the body's temperature-regulating capacity. Reduction or cessation of sweating is an early symptom; body temperature of 40.5°C (105°F) or higher, rapid pulse, hot and dry skin, headache, confusion, unconsciousness, and convulsions may occur as well.

Heat transfer The movement of heat energy from a hotter medium to a cooler medium by conduction, convection, or radiation.

Heavy fuels Fuels of a large diameter, such as large brush, heavy timber, snags, stumps, branches, and dead timber on the ground. These fuels ignite and are consumed more slowly than light fuels.

Heel of the fire The side opposite the head of the fire, which is often close to the area of origin.

Higbee indicators Indicators on the male and female threaded couplings that indicate where the threads start. These indicators should be aligned before fire fighters start to thread the couplings together.

High-angle operation A rope rescue operation where the angle of the slope is greater than 45 degrees. In this scenario, rescuers depend on life safety rope rather than a fixed support surface such as the ground.

High-expansion foam A foam created by pumping large volumes of air through a small screen coated with a foam solution. Some high-expansion foams have expansion ratios ranging from 200:1 to approximately 1000:1.

High-pressure cargo tank (MC-331) A tank that carries materials such as ammonia, propane, Freon, and butane. This type of tank is commonly constructed of steel and has rounded ends and a single open compartment inside. The liquid volume inside the tank varies, ranging from a 3785-L (1000-gal) delivery truck to a full-size 41,640-L (11,000-gal) cargo tank.

High temperature–protective clothing Protective clothing designed to protect the wearer for short-term high temperature exposures. (NFPA 1072)

High temperature–protective equipment A type of personal protective equipment that shields the wearer during short-term exposures to high temperatures. Sometimes referred to as a *proximity suit*, this type of equipment allows the properly trained fire fighter to work in extreme fire conditions. It is not designed to protect against hazardous materials or weapons of mass destruction.

Hitch A knot that attaches to or wraps around an object so that when the object is removed, the knot will fall apart. (NFPA 1670)

Hockey puck lock A type of padlock with hidden shackles that cannot be forced open through conventional methods.

Hollow-core door A door made of panels that are honeycombed inside, creating an inexpensive and lightweight design.

Horizontal evacuation Moving occupants from a dangerous area to a safe area on the same floor level.

Horizontal-sliding windows Windows that slide open horizontally.

Horizontal ventilation The opening or removal of windows or doors on any floor of a fire building to create flow paths for fire conditions. (NFPA 1410)

Horn The tapered discharge nozzle of a carbon dioxide fire extinguisher.

Horseshoe hose load A hose loading method in which the hose is laid on its edge around the perimeter of the hose bed so that it resembles a horseshoe.

Hose appliance A piece of hardware (excluding nozzles) generally intended for connection to fire hose to control or convey water. (NFPA 1962)

Hose bridge A device that protects a hose when it is necessary for a vehicle to drive over a hose.

Hose clamp A device used to compress a fire hose to stop water flow.

Hose jacket A device used to stop a leak in a fire hose or to join hose that has damaged couplings.

Hose liner The inside portion of a hose that is in contact with the flowing water; also called the hose inner jacket.

Hose roller A device that is placed on the edge of a roof and is used to protect hose as it is hoisted up and over the roof edge.

Hostile fire event A general descriptor for hazardous fire conditions, including flashover, backdraft, smoke explosion, flameover, and rapid fire spread. (NFPA 1521)

Hot zone The control zone immediately surrounding hazardous materials/weapons of mass destruction (WMD) incidents, which extends far enough to prevent adverse effects of hazards to personnel outside the zone and where only personnel who are trained, equipped, and authorized to do assigned work are permitted to enter. (NFPA 1072 and 1500)

Hux bar A multipurpose tool that can be used for several forcible entry and ventilation applications because of its unique design. It also may be used as a hydrant wrench.

Hybrid building A building that does not fit entirely into any of the five construction types because it incorporates building materials of more than one type.

Hybrid electric vehicle A vehicle that uses both a battery-powered electric motor and a liquid-fuelled engine to propel the vehicle.

Hydrant wrench A hand tool that is used to operate the valves on a hydrant; it also may be used as a spanner wrench. Some models are plain wrenches, whereas others have a ratchet feature.

Hydraulic shears A lightweight, hand-operated tool that can produce up to 4536 kg (10,000 lb) of cutting force.

Hydraulic spreader A lightweight, hand-operated tool that can produce up to 4536 kg (10,000 lb) of prying and spreading force.

Hydraulic ventilation Ventilation that relies on the movement of air caused by a fog stream that is placed 0.6 m to 1.2 m (2 ft to 4 ft) in front of an open window.

Hydrogen cyanide An extremely toxic gas produced by the combustion of many common plastic-based materials. Low-level exposure can cause cyanosis, headache, dizziness, unsteady gait, and nausea.

Hydrostatic testing A test performed by filling pressure-containing components completely with water or other incompressible fluid while expelling all contained air, closing or capping all open ports of the pressure-containing components, and then raising and maintaining the contained pressure to pressurize the pressure-containing components to a prescribed value through an externally supplied pressure-generating device to verify its strength against unwanted rupture. (NFPA 1901 and 10)

Hypothermia A condition in which the internal body temperature falls below 35°C (95°F), usually a result of prolonged exposure to cold or freezing temperatures.

Hypoxia A state of inadequate oxygenation of the blood and tissue sufficient to cause impairment of function. (NFPA 99)

I-beam A ladder beam constructed of one continuous piece of I-shaped metal or fiberglass to which the rungs are attached.

Ignitable liquid Any liquid or the liquid phase of any material that is capable of fueling a fire, including a flammable liquid, combustible liquid, or any other material that can be liquefied and burned. (NFPA 556)

Ignition The action of setting something on fire.

Ignition temperature Minimum temperature a substance should attain in order to ignite under specific test conditions. (NFPA 402)

Immediately dangerous to life and health (IDLH) Any condition that would pose an immediate or delayed threat to life, cause irreversible adverse health effects, or interfere with an individual's ability to escape unaided from a hazardous environment. (NFPA 1670)

Incendiary device A device or mechanism used to initiate an incendiary fire or explosion.

Incendiary fire A fire that is intentionally ignited in an area or under circumstances where and when there should not be a fire. (NFPA 921)

Incident action plan (IAP) A verbal or written plan containing incident objectives reflecting the overall strategy and specific control actions where appropriate for managing an incident or planned event. (NFPA 1026)

Incident commander (IC) The individual responsible for all incident activities, including the development of strategies and tactics and the ordering and release of resources. (NFPA 1026)

Incident command post (ICP) The field location at which the primary tactical level, on-scene incident command functions are performed. (NFPA 1026)

Incident command system (ICS) A component of an incident management system (IMS) designed to enable effective and efficient on-scene incident management by integrating organizational functions, tactical operations, incident planning, incident logistics, and administrative tasks within a common organizational structure. (NFPA 1072 and 1670)

Incident report A document prepared by fire department personnel on a particular incident. (NFPA 901)

Incident Response Pocket Guide (IRPG) A job and training reference for personnel operating at a wildland fire. It may also be used for all-hazard incident response.

Incident safety officer (ISO) A member of the command staff responsible for monitoring and assessing safety hazards and unsafe situations and for developing measures for ensuring personnel safety. (NFPA 1500)

Incipient stage The early stage of fire development where the fire's progression is limited to a fuel source and the thermal hazard is localized to the area of the burning material. (NFPA 1410)

Incomplete combustion A burning process in which the fuel is not completely consumed, usually due to a limited supply of oxygen.

Incubation period The time period between the initial infection by an organism and the development of symptoms by a victim.

Indirect attack Firefighting operations involving the application of extinguishing agents to reduce the build-up of heat released from a fire without applying the agent directly onto the burning fuel. (NFPA 1145)

Industrial packaging Packaging used to transport materials that present a limited hazard to the public or the environment. Contaminated equipment is an example of such material, because it contains a non-life-endangering amount of radioactivity. It is classified into three categories, based on the strength of the packaging.

Information management Fire fighters or civilians who take care of the computer and networking systems that a fire department needs to operate.

Initial attack apparatus Fire apparatus with a fire pump of at least 946 L/min (250 gpm) capacity, water tank, and hose body, whose primary purpose is to initiate a fire suppression attack on structural, vehicular, or vegetation fires and to support associated fire department operations. May also be referred to as quick attack apparatus.

In-line eductor A Venturi-type proportioning device that meters foam concentrate at a fixed or variable concentration into the water stream at a point between the water source and a nozzle or other discharge device. (NFPA 11)

Interior attack The assignment of a team of fire fighters to enter a structure and attempt fire suppression.

Interior finish The exposed surfaces of walls, ceilings, and floors within buildings. (NFPA 5000)

Interior wall A wall inside a building that divides a large space into smaller areas.

Intermodal tanks Bulk containers that serve as both a shipping and storage vessel. Such tanks hold between 18,927 and 22,712 L (5000 and 6000 gal) of product and can be either pressurized or nonpressurized. They can be shipped by all modes of transportation—air, sea, or land.

Inverter Equipment that is used to change the voltage level or waveform, or both, of electrical energy. Commonly, an inverter [also known as a power conditioning unit (PCU) or power conversion system (PCS)] is a device that changes DC input to AC output. Inverters may also function as battery chargers that use alternating current from another source and convert it into direct current for charging batteries. (NFPA 70)

Ionization smoke detection The principle of using a small amount of radioactive material to ionize the air between two differentially charged electrodes to sense the presence of smoke particles. Smoke particles entering the ionization volume decrease the conductance of the air by reducing ion mobility. The reduced conductance signal is processed and used to convey an alarm condition when it meets preset criteria. (NFPA 72)

Irons A combination of tools, usually consisting of a Halligan tool and a flat-head axe, that are commonly used for forcible entry.

Island An unburned area surrounded by fire.

Isolation of the hazard area Steps taken to identify a perimeter around a contaminated atmosphere. Isolating an area is driven largely by the nature of the released chemicals and the environmental conditions that exist at the time of the release.

J tool A tool that is designed to fit between double doors equipped with push bars or panic bars.

Jalousie windows Windows made of small slats of tempered glass that overlap each other when the window is closed. Often found in trailers and mobile homes, jalousie windows are held together by a metal frame and operated by a small hand wheel or crank found in the corner of the window.

Junction box A device that attaches to an electrical cord to provide additional outlets.

Juvenile firesetters Firesetters who are typically 7 to 13 years old and often suffer from emotional or psychological problems. Juvenile firesetters often ignite fires in or around their own homes or schools.

K tool A tool that is used to remove lock cylinders from structural doors so the locking mechanism can be unlocked.

Kelly tool A steel bar with two main features: a large pick and a large chisel or fork.

Kerf cut A cut that is the width and depth of the saw blade. It is used to inspect cockloft spaces from the roof.

Kernmantle rope Rope made of two parts—the kern (interior component) and the mantle (the outside sheath).

Keyhole saw A saw designed to cut keyhole circles in wood and drywall.

Kinetic energy The energy possessed by an object as a result of its motion.

Knot A fastening made by tying rope or webbing in a prescribed way. (NFPA 1670)

Ladder A-frame An A-shaped structure formed with two ladder sections. It can be used as a makeshift lift when raising a trapped person. Sometimes referred to as an A-frame hoist.

Ladder belt A compliant equipment item that is intended for use as a positioning device for a person on a ladder. (NFPA 1983)

Ladder fuels Fuels that provide vertical continuity between the ground and the tops of trees or shrubs, thereby allowing fire to move with relative ease.

Ladder pipe A monitor that attaches to the rungs of a vehicle-mounted aerial ladder. (NFPA 1964)

Laminar smoke flow Smooth or streamlined movement of smoke, which indicates that the pressure in the building is not excessively high.

Laminated glass Safety glass; the lamination process places a thin layer of plastic between two layers of glass so that the glass does not shatter and fall apart when broken.

Laminated wood Pieces of wood that are glued together.

Large-diameter hose (LDH) A hose 89 mm (3.5 in.) or larger that is designed to move large volumes of water to supply master stream appliances, portable hydrants, manifolds, standpipe and sprinkler systems, and fire department pumpers from hydrants and in relay. (NFPA 1410)

Latching device A spring-loaded latch bolt or a gravity-operated steel bar that, after release by physical action, returns to its operating position and automatically engages the strike plate when it is returned to the closed position. (NFPA 80)

Lath Thin strips of wood used to make the supporting structure for roof tiles.

LCES A mnemonic that stands for Lookouts, Communications, Escape routes, and Safety zones. Fire fighters should ensure that the components of LCES are in place before attacking a wildland fire to reduce the risk associated with fighting these types of fires.

Leader The individual responsible for command of a task force, strike team, or functional unit. (NFPA 1026)

Legacy construction An older type of construction that used sawn lumber and was built before about 1970.

Level A ensemble Personal protective equipment that provides protection against vapours, gases, mists, and even dusts. The highest level of protection, it requires a totally encapsulating suit that includes a self-contained breathing apparatus.

Level B ensemble Personal protective equipment that is used when the type and atmospheric concentration of substances require a high level of respiratory protection but less skin protection. The kinds of gloves and boots worn depend on the identified chemical.

Level C ensemble Personal protective equipment that is used when the type of airborne substance is known, the concentration is measured, the criteria for using an air-purifying respirator are met, and skin and eye exposure are unlikely. A Level C ensemble consists of standard work clothing with the addition of chemical-protective clothing, chemically resistant gloves, and a form of respirator protection.

Level D ensemble Personal protective equipment that is used when the atmosphere contains no known hazard, and work functions preclude splashes, immersion, or the potential for unexpected inhalation of or contact with hazardous levels of chemicals. A Level D ensemble is primarily a work uniform that includes coveralls and affords minimal protection.

Liaison officer A member of the command staff responsible for coordinating with representatives from cooperating and assisting agencies. (NFPA 1561)

Lieutenant A company officer who is usually responsible for a single fire company on a single shift; the first in line among company officers.

Life safety rope Rope dedicated solely for the purpose of supporting people during rescue, firefighting, other emergency operations, or during training evolutions. (NFPA 1983)

Light-emitting diodes (LEDs) Electronic semiconductors that emit a single-colour light when activated. LEDs are used for operational displays in SCBA.

Light (low) hazard locations Occupancies where the quantity, combustibility, and heat release of the materials is low, and the majority of materials are arranged so that a fire is not likely to spread.

Line detector Wire or tubing that can be strung along the ceiling of large open areas to detect an increase in heat.

Liquid A fluid (such as water) that has no independent shape but has a definite volume and does not expand indefinitely and that is only slightly compressible.

Liquid splash–protective clothing Clothing designed to protect the wearer from chemical splashes. It does not provide total body protection from gases or vapours and should not be used for incidents involving liquids that emit vapours known to affect or be absorbed through the skin. NFPA 1992 is the performance document pertaining to liquid-splash garments and ensembles.

Liquid splash–protective ensemble Multiple elements of compliant protective clothing and equipment products that when worn together provide protection from some, but not all, risks of hazardous materials/WMD emergency incident operations involving liquids. (NFPA 1072)

Live load The load produced by the use and occupancy of the building or other structure, which does not include construction or environmental loads such as wind load, snow load, rain load, earthquake load, flood load, or dead load. Live loads on a roof are those produced (1) during maintenance by workers, equipment, and materials and (2) during the life of the structure by movable objects such as planters and by people. (NFPA 5000)

Load-bearing wall A wall that supports any vertical load in addition to its own weight or any lateral load.

Loaded-stream fire extinguisher A water-based fire extinguisher that uses an alkali metal salt as a freezing-point depressant.

Lock body The part of a padlock that holds the main locking mechanisms and secures the shackles.

Locking mechanism A device that locks a fire extinguisher's trigger to prevent its accidental discharge.

Lockout and tagout systems Methods of ensuring that electricity and other utilities have been shut down and switches are "locked" so that they cannot be switched on, so as to prevent flow of power or gases into the area where rescue is being conducted.

Logistics section Section responsible for providing facilities, services, and materials for the incident or planned event, including the communications unit, medical unit, and food unit within the service branch and the supply unit, facilities unit, and ground support unit within the support branch. (NFPA 1026)

Logistics section chief The general staff position responsible for directing the logistics function. It is generally assigned on complex, resource-intensive, or long duration incidents.

Loop A piece of rope formed into a circle.

Louver cut A cut that is made using power saws and axes to cut along and between roof supports so that the sections created can be tilted into the opening.

Low-angle operation A rope rescue operation on a mildly sloping surface (less than 45 degrees) or flat land. In this scenario, fire fighters depend on the ground for their primary support, and the rope system is a secondary means of support.

Lower explosive limit (LEL) The minimum concentration of a combustible vapour or combustible gas in a mixture of the vapour or gas and gaseous oxidant, above which propagation of flame will occur on contact with an ignition source. (NFPA 115)

Low-pressure chemical cargo tank See *MC-307/DOT 407 chemical hauler*.

Low-pressure tank cars Railcars equipped with a tank that typically holds general industrial chemicals and consumer products such as corn syrup, flammable and combustible liquids, and mild corrosives.

Low-volume nozzle A nozzle that flows 151 L/min (40 gpm) or less.

Malicious alarm An unwanted activation of an alarm initiating device caused by a person acting with malice. (NFPA 72)

Mallet A short-handled hammer.

Manual dry standpipe system A standpipe system with no permanently attached water supply that relies exclusively on the fire department connection to supply the system demand. (NFPA 14)

Manual pull station A device with a switch that either opens or closes a circuit, activating the fire alarm.

Manufactured (mobile) home A structure, transportable in one or more sections, which, in the travelling mode, is 2.4 m (8 body-ft) or more in width or 12 m (40 body-ft) or more in length or, when erected on site, is 29.7 m^2 (320 ft^2) or more and which is built on a permanent chassis and designed to be used as a dwelling, with or without a permanent foundation, when connected to the required utilities, and includes plumbing, heating, air-conditioning, and electrical systems contained therein. (NFPA 5000)

Masonry Built-up unit of construction or combination of materials such as clay, shale, concrete, glass, gypsum, tile, or stone set in mortar. (NFPA 5000)

Mass arsonist A person who sets three or more fires at the same site or location during a limited period of time.

Mass decontamination The physical process of reducing or removing surface contaminants from large numbers of victims in potentially life-threatening situations in the fastest time possible. (NFPA 1072)

Master stream appliance Devices used to produce high volume water streams for large fires. Most master stream appliances discharge between 1325 L/min (350 gpm) and 5678 L/min (1500 gpm), though much larger capacities are available. These devices include deck guns and portable ground monitors.

Master stream nozzle A nozzle with a rated discharge of 1325 L/min (350 gpm) or greater. (NFPA 1964)

Maul A specialized striking tool, weighing 3 kg (6 lb) or more, with an axe on one side of the head and a sledgehammer on the other side.

Mayday A verbal declaration indicating that a fire fighter is lost, missing, or trapped and requires immediate assistance.

MC-306/DOT 406 cargo tank Such a vehicle typically carries between 22,712 and 37,854 L (6000 and 10,000 gal) of a product such as gasoline or other flammable and combustible materials. The tank is nonpressurized; also called *nonpressure liquid cargo tank*.

MC-307/DOT 407 chemical hauler A rounded or horseshoe-shaped tank capable of holding 22,712 to 26,498 L (6000 to 7000 gal) of flammable liquid, mild corrosives, and poisons. The tank has a high internal working pressure; also called *low-pressure chemical cargo tank*.

MC-312/DOT 412 corrosive tank A tank that often carries aggressive (highly reactive) acids such as concentrated sulfuric and nitric acid. It is characterized by several heavy-duty reinforcing rings around the tank and holds approximately 22,712 L (6000 gal) of product.

McLeod A hand tool used for constructing fire control lines and overhauling wildland fires. One side of the head consists of a five-toothed to seven-toothed fire rake; the other side is a hoe.

Mechanical energy A form of potential energy that can generate heat through friction.

Mechanical saw A saw that usually is powered by an electric motor or a gasoline engine. The three primary types of mechanical saws are chainsaws, rotary saws, and reciprocating saws.

Mechanical ventilation A process of removing heat, smoke, and gases from a fire area by using exhaust fans, blowers, air-conditioning systems, or smoke ejectors. (NFPA 402)

Mildew A fungus that can grow on hose if the hose is stored wet. Mildew can damage the jacket of a hose.

Minuteman hose load A hose loading method that allows a single fire fighter to flake out the necessary amount of hose from the shoulder while advancing toward the fire; this load avoids having to maneuver the hose lines around obstacles and helps to prevent sharp kinks.

Mitochondria Responsible for converting nutrients into energy, yielding molecules of adenosine triphosphate (ATP) to fuel the cell's activities.

Mobile data terminals (MDTs) Technology that allows fire fighters to receive data while in the fire apparatus or at the station.

Mobile radio A two-way radio that is permanently mounted in a fire apparatus.

Mobile water supply apparatus A vehicle designed primarily for transporting (pickup, transporting, and delivering) water to fire emergency scenes to be applied by other vehicles or pumping equipment. (NFPA 1901)

Mortise locks Door locks with both a latch and a bolt built into the same mechanism; the two locking mechanisms operate independently of each other. Mortise locks often are found in hotel rooms.

Multiple-jacket A construction consisting of a combination of two separately woven reinforcements (double jacket) or two or more reinforcements interwoven. (NFPA 1962)

Multiplex channel Simultaneous transmission of multiple data streams, most often voice signals, in either or both directions over the same frequency.

Multipurpose dry-chemical fire extinguisher A fire extinguisher that uses an ammonium phosphate–based extinguishing agent that is effective on fires involving ordinary combustibles, such as wood or paper, and fires involving flammable liquids. It is rated to fight Class A, B, and C fires.

Multipurpose hook A long pole with a wooden or fiberglass handle and a metal hook on one end used for pulling.

Municipal-type water system A system having water pipes servicing fire hydrants and designed to furnish, over and above domestic consumption, a minimum of 950 L/min (250 gpm) at 138 kPa (20 psi) residual pressure for a 2-hour duration. (NFPA 1141)

Muscarinic effects Effects such as runny nose, salivation, sweating, bronchoconstriction, bronchial secretions, nausea, vomiting, and diarrhea.

National Fire Incident Reporting System (NFIRS) The standard national reporting system used by U.S. fire departments to report fires and other incidents to which they respond and to maintain records of these incidents in a uniform manner. (FEMA NFIRS)

National Fire Protection Association (NFPA) The association that develops and maintains nationally recognized minimum consensus standards on many areas of fire safety and specific standards on hazardous materials.

National Incident Management System (NIMS) A system mandated by Homeland Security Presidential Directive 5

(HSPD-5) that provides a systematic, proactive approach guiding government agencies at all levels, the private sector, and nongovernmental organizations to work seamlessly to prepare for, prevent, respond to, recover from, and mitigate the effects of incidents, regardless of cause, size, location, or complexity, so as to reduce the loss of life or property and harm to the environment. (NFPA 1026)

National Institute for Occupational Safety and Health (NIOSH) The U.S. federal agency responsible for research and development on occupational safety and health issues.

National Response Center (NRC) An agency maintained and staffed by the U.S. Coast Guard; it should always be notified if a hazard discharges into the environment.

Natural fire A fire caused without direct human intervention or action, such as fire resulting from lightning, an earthquake, or wind.

Natural ventilation The flow of air or gases created by the difference in the pressures or gas densities between the outside and inside of a vent, room, or space. (NFPA 853)

Negative-pressure ventilation Ventilation that relies on electric fans to pull or draw the air from a structure or area.

Neutral plane The interface at a vent, such as a doorway or a window opening, between the hot gas flowing out of a fire compartment and the cool air flowing into the compartment where the pressure difference between the interior and exterior is equal.

Nonambulatory patient A term describing individuals who cannot move themselves to an area of safety owing to their physical condition, medical treatment, or other factors.

Nonbearing wall Any wall that is not a bearing wall. (NFPA 5000)

Non-bulk packaging Any packaging having a liquid capacity of 450 L (119 gal) or less, a solids capacity of 400 kg (882 lb) or less, or a compressed gas water capacity of 454 kg (1001 lb) or less. (NFPA 472)

Noncoded system An alarm system that provides no information at the alarm control unit indicating where the activated alarm is located.

Nonpressure liquid cargo tank See *MC-306/DOT 406 cargo tank*.

Normal operating pressure The observed static pressure in a water distribution system during a period of normal demand.

Nose cups An insert inside the face piece of an SCBA that fits over the user's mouth and nose.

Nozzle A device for use in applications requiring special water discharge patterns, directional spray, or other unusual discharge characteristics. (NFPA 13)

Nozzle shut-off valve A device that enables the fire fighter at the nozzle to start or stop the flow of water.

Nuisance alarm An unwanted activation of a signaling system or an alarm initiating device in response to a stimulus or condition that is not the result of a potentially hazardous condition. (NFPA 72)

Obscuration rate A measure of the percentage of light transmission that is blocked between a sender and a receiver unit.

Occupancy The purpose for which a building or other structure, or part thereof, is used or intended to be used. (NFPA 5000)

Occupational Health and Safety Act (OHSA) The Canadian act that regulates workplace health and safety.

Occupational Safety and Health Administration (OSHA) The U.S. federal agency that regulates worker safety and, in some cases, responder safety. It is part of the U.S. Department of Labor.

Offensive operation Actions generally performed in the interior of involved structures that involve a direct attack on a fire to directly control and extinguish the fire. (NFPA 1500)

Open-circuit self-contained breathing apparatus An SCBA in which the exhaled air is released into the atmosphere and is not reused.

Open-end wrench A hand tool that is used to tighten or loosen bolts. The end is open, as opposed to a box-end wrench. Each wrench is a specific size.

Operations level responders Persons who respond to hazardous materials/weapons of mass destruction (WMD) incidents for the purpose of implementing or supporting actions to protect nearby persons, the environment, or property from the effects of the release. (NFPA 1072)

Operations section Section responsible for all tactical operations at the incident or planned event, including up to 5 branches, 25 divisions/groups, and 125 single resources, task forces, or strike teams. (NFPA 1026)

Operations section chief The general staff position responsible for managing all operations activities. It is usually assigned when complex incidents involve more than 20 single resources or when command staff cannot be involved in all details of the tactical operation.

Operator lever The handle, doorknob, or keyway of a door that turns the latch to open it.

Ordinary (moderate) hazard locations Occupancies that contain more Class A and Class B materials than are found in light hazard locations. The combustibility and heat release rate of the materials is moderate.

Oriented strand board (OSB) An engineered wood product manufactured from small pieces of wood that are held together with glue or adhesives. The adhesives include urea-formaldehyde resins, phenol-formaldehyde resins, melamine-formaldehyde resins, and polyurethane resins.

Outside screw and yoke (OS&Y) valve A sprinkler control valve with a valve stem that moves in and out as the valve is opened or closed.

Overhaul The process of final extinguishment after the main body of a fire has been knocked down. All traces of fire must be extinguished at this time. (NFPA 402)

Oxidation Reaction with oxygen either in the form of the element or in the form of one of its compounds. (NFPA 53)

Packaging The process of securing a victim in a transfer device, with regard to existing and potential injuries or illness, so as to prevent further harm during movement. (NFPA 1006)

Padlocks One of the most common types of locks on the market today, portable locks built to provide regular-duty or heavy-duty service. Several types of locking mechanisms are available, including keyways, combination wheels, and combination dials.

Parallel attack A method of attack in which the control line is located parallel to the fire edge, at a distance of about 1.5 to 15 m

(5 to 50 ft) from the fire. The intervening fuel usually burns out as the fire control line moves alongside the fire but can also burn out with the main fire.

Parallel chord truss A truss in which the top and bottom chords are parallel.

Paramedic EMS personnel with the highest level of training in EMS, including cardiac monitoring, administering drugs, inserting advanced airways, manual defibrillation, and other advanced assessment and treatment skills.

Parapet The part of a wall entirely above the roofline. (NFPA 5000)

Partition A nonstructural interior wall that spans horizontally or vertically from support to support. The supports may be the basic building frame, subsidiary structural members, or other portions of the partition system. (ASCE/SEI 7:11.2) (NFPA 5000)

Party wall A wall constructed on the line between two properties.

PASS Acronym for the steps involved in operating a portable fire extinguisher: Pull pin, Aim nozzle, Squeeze trigger, Sweep across burning fuel.

Pawls Devices attached to a fly section(s) to engage ladder rungs near the beams of the section below for the purpose of anchoring the fly section(s). Also called locks or dogs. (NFPA 1931)

Peak cut A ventilation opening that runs along the top of a pitched roof.

Pendent sprinkler head A sprinkler head designed to be mounted on the underside of sprinkler piping so that the water stream is directed in a downward direction.

Penetration The movement of a material through a suit's closures, such as zippers, buttonholes, seams, flaps, or other design features of chemical-protective clothing, and through punctures, cuts, and tears. (NFPA 1072)

Permeation A chemical action involving the movement of chemicals, on a molecular level, through intact material. (NFPA 1072)

Permissible exposure limit (PEL) The established standard limit of exposure to a hazardous material. It is based on the maximum time-weighted concentration at which 95 percent of exposed, healthy adults suffer no adverse effects over a 40-hour workweek.

Personal alert safety system (PASS) A device that continually monitors for lack of movement of the wearer and automatically activates an alarm signal, indicating the wearer is in need of assistance; can also be manually activated to trigger the alarm signal. (NFPA 1982)

Personal flotation device (PFD) A device manufactured in accordance with U.S. Coast Guard specifications that provides supplemental flotation for persons in the water. (NFPA 1006)

Personal protective equipment (PPE) Consists of full personal protective clothing, plus a self-contained breathing apparatus (SCBA) and a personal alert safety system (PASS) device. (NFPA 1001)

Personnel accountability report (PAR) Periodic reports verifying the status of responders assigned to an incident or planned event. (NFPA 1026)

Personnel accountability system A system that readily identifies both the location and function of all members operating at an incident scene. (NFPA 1500)

Phosgene A chemical agent that causes severe pulmonary damage; it is a by-product of incomplete combustion.

Photoelectric smoke alarm A detector that uses a light beam and a photocell to detect larger visible particles of smoke. When visible particles of smoke enter the inner chamber they reflect some of the light onto the photocell, thereby activating the alarm.

Photoelectric smoke detector A detector that uses a light beam and a photocell to detect larger visible particles of smoke. When visible particles of smoke enter the inner chamber, they reflect some of the light onto the photocell, thereby activating the alarm.

Photoionization detector (PID) A sensor that uses ultraviolet light to ionize the gases that move through the sensor; available as a stand-alone unit or may be incorporated into a multi-gas meter.

Physical evidence A physical or tangible item that proves or disproves a particular fact or issue; also referred to as *real evidence*.

Pick The pointed end of a tool, which can be used to make a hole or purchase point in a door, floor, or wall.

Pick-head axe A tool that has a head with an axe on one side and a pointed end ("pick") on the opposite side.

Piercing nozzle A nozzle that can be driven through sheet metal or other material to deliver a water stream to that area.

Pike pole A pole with a sharp point ("pike") on one end coupled with a hook. It is used to make openings in ceilings and walls. Pike poles are manufactured in different lengths for use in rooms of different heights.

Pipe wrench A wrench having one fixed grip and one movable grip that can be adjusted to fit securely around pipes and other tubular objects.

Pitched chord truss A type of truss typically used to support a sloping roof.

Pitched roof A roof with sloping or inclined surfaces.

Pitot gauge A type of gauge that is used to measure the velocity pressure of water that is being discharged from an opening. It is used to determine the flow of water from a hydrant or nozzle.

Placards Signage required to be placed on all four sides of highway transport vehicles, railroad tank cars, and other forms of hazardous materials transportation that identifies the hazardous contents of the vehicle, using a standardized system: 273-mm by 273-mm (10¾-in. by 10¾-in.) diamond-shaped indicators.

Plague An infectious disease caused by the bacterium *Yersinia pestis*, which is commonly found on rodents.

Planning section Section responsible for the collection, evaluation, dissemination, and use of information related to the incident situation, resource status, and incident forecast. (NFPA 1026)

Planning section chief The general staff position responsible for planning functions and for tracking and logging resources. It is assigned when command staff members need assistance in managing information.

Plaster hook A long pole with a pointed head and two retractable cutting blades on the side.

Plate glass A type of glass that has additional strength so it can be formed in larger sheets but will still shatter upon impact.

Platform-frame construction Construction technique for building the frame of the structure one floor at a time. Each floor has a top and bottom plate that acts as a firestop.

Platoon chief Manages the on-duty shift. In most career departments there are four platoons working a 24-hour shift. The platoon chief answers to either the assistant deputy chief or the deputy chief of operations.

Plume The column of hot gases, flames, and smoke rising above a fire; also called convection column, thermal updraft, or thermal column. (NFPA 921)

Pocket A deep indentation of unburned fuel along the fire's perimeter, often found between a finger and the head of the fire.

Point of origin The exact physical location within the area of origin where a heat source and a fuel first interact, resulting in a fire or explosion. (NFPA 921)

Polar solvent A water-soluble flammable liquid such as alcohol, acetone, ester, and ketone.

Policies Formal statements that provide guidelines for present and future actions. Policies often require personnel to make judgments.

Polycyclic aromatic hydrocarbons (PAHs) A group of substances that occur naturally in materials such as coal or crude oil. These substances also are generated during the combustion of organic materials and can be found in vehicle exhaust, tobacco smoke, and the smoke generated from structure fires, vehicle fires, wildland fires, or any other type of fire. PAH can exist as a particle or gas.

Portable monitor A monitor that can be lifted from a vehicle-mounted bracket and moved to an operating position on the ground by not more than two people. (NFPA 1965)

Portable radio A battery-operated, hand-held transceiver. (NFPA 1221)

Portable tanks Folding or collapsible tanks that are used at the fire scene to hold water for drafting.

Positive-pressure attack The use of positive-pressure fans to control the flow of products of combustion while fire suppression efforts are under way.

Positive-pressure ventilation Ventilation that relies on fans to push or force clean air into a structure after a structure fire has been controlled.

Possible cause A hypothesis determined to have less than 50% probability of being true.

Post One of the vertical support members or pillars of a vehicle that holds up the roof and forms the upright columns of the passenger compartment.

Post indicator valve (PIV) A sprinkler control valve with an indicator that reads either "open" or "shut" depending on its position.

Potential energy The energy that an object has stored up as a result of its position or condition. A raised weight and a coiled spring have potential energy.

Pounds per square inch (psi) The standard unit for measuring pressure.

Powered air-purifying respirator (PAPR) An air-purifying respirator that uses a powered blower to force the ambient air

through one or more air-purifying components to the respiratory inlet covering. (NFPA 1984)

Power take-off shaft A supplemental mechanism that enables a pump apparatus to operate a pump while the apparatus is still moving.

Preaction sprinkler system A sprinkler system employing a piping system that contains air that might or might not be under pressure, with a supplemental detection system installed in the same areas as the sprinklers. (NFPA 13)

Preaction valve A type of sprinkler system valve that holds back water until a manual emergency release is activated, a fire sprinkler head activates, a fire detector activates, or a combination of sprinkler head and initiating device activation takes place.

Preconnected flat hose load A hose loading method similar to the flat hose load used for supply lines, except the female end of the hose line is preconnected to a pump discharge outlet, and a nozzle is attached to the male end of the hose line.

Preincident plan A document developed by gathering general and detailed data that are used by responding personnel in effectively managing emergencies for the protection of occupants, responding personnel, property, and the environment. (NFPA 1620)

Preincident survey The process used to gather information to develop a preincident plan.

Premixed foam solution Solution produced by introducing a measured amount of foam concentrate into a given amount of water in a storage tank. (NFPA 11)

Pressure gauge A device that measures and displays pressure readings. In an SCBA, the pressure gauges indicate the quantity of breathing air that is available at any time.

Pressure indicator A gauge on a pressurized portable fire extinguisher that indicates the internal pressure of the expellant.

Pressure tank cars Railcars used to transport materials such as propane, ammonia, ethylene oxide, and chlorine.

Primary cut The main ventilation opening made in a roof to allow smoke, heat, and gases to escape.

Primary feeders The largest diameter water main pipes in a water distribution system that carry the greatest amounts of water.

Primary search An immediate and quick search of the structures likely to contain survivors. (NFPA 1670)

Private water system A privately owned water system that operates separately from the municipal water system.

Probable cause A hypothesis determined to have greater than 50% probability of being true.

Projected windows Windows that project inward or outward on a top or bottom hinge; also called factory windows. They are usually found in older warehouses or commercial buildings.

Proprietary supervising alarm system A fire alarm system that transmits a signal to a monitoring location owned and operated by the facility's owner.

Protected premises fire alarm system A fire alarm system that sounds an alarm only in the building where it was activated; that is, no signal is sent out of the building.

Protection plates Reinforcing material placed on a ladder at chafing and contact points to prevent damage from friction and contact with other surfaces.

Protective hood A part of a fire fighter's personal protective equipment that is designed to be worn over the head and under the helmet; it provides thermal protection for the neck and ears.

Protein foam A protein-based foam concentrate that is stabilized with metal salts to make a fire-resistant foam blanket. (NFPA 402)

Pry axe A specially designed hand axe that serves multiple purposes. Similar to a Halligan tool, it can be used to pry, cut, and force doors, windows, and many other types of objects. Also called a multipurpose axe.

Pry bar A specialized prying tool made of a hardened steel rod with a tapered end that can be inserted into a small area.

Public emergency alarm reporting system A system of alarm initiating devices, transmitting and receiving equipment, and communication infrastructure (other than a public telephone network) used to communicate with the communications centre to provide any combination of manual or auxiliary alarm service. (NFPA 72)

Public information officer (PIO) A member of the command staff responsible for interfacing with the public and media or with other agencies with incident-related information requirements. (NFPA 1026 and 1035)

Public safety answering point (PSAP) A facility equipped and staffed to receive emergency and nonemergency calls requesting public safety services via telephone and other communication devices. (NFPA 1061)

Public safety communications centre A building or portion of a building that is specifically configured for the primary purpose of providing emergency communications services or public safety answering point (PSAP) services to one or more public safety agencies under the authority or authorities having jurisdiction. (NFPA 1061)

Pulaski axe A hand tool that combines an adze and an axe for brush removal.

Pulley A device with a free-turning, grooved metal wheel (sheave) used to reduce rope friction. Side plates are available for a carabiner to be attached. (NFPA 1670)

Pump tank fire extinguisher A nonpressurized, manually operated water-type fire extinguisher that is rated for use on Class A fires. Discharge pressure is provided by a hand-operated, double-acting piston pump.

Purchase point A small opening made to enable better tool access in forcible entry.

Pyrolysis A process in which material is decomposed, or broken down, into simpler molecular compounds by the effects of heat alone; pyrolysis often precedes combustion. (NFPA 921)

Quint apparatus Fire apparatus with a permanently mounted fire pump, a water tank, a hose storage area, an aerial ladder or elevating platform with a permanently mounted waterway, and a complement of ground ladders.

Rabbit tool A hydraulic spreading tool designed to pry open doors that swing inward.

Radiation dispersal device (RDD) A device designed to spread radioactive material through a detonation of conventional explosives or other (non-nuclear) means; also referred to as a dirty bomb. (NFPA 472)

Radiation The combined process of emission, transmission, and absorption of energy travelling by electromagnetic wave propagation (e.g., infrared radiation) between a region of higher temperature and a region of lower temperature. (NFPA 550)

Radiological agents Radiation associated with x-rays, alpha, beta, and gamma emissions from radioactive isotopes, or other materials in excess of normal background radiation levels. (NFPA 1951)

Rafters Joists that are mounted in an inclined position to support a roof.

Rail The top or bottom piece of a trussed beam assembly used in the construction of a trussed ladder. Also, the top and bottom surfaces of an I-beam ladder. Each beam has two rails.

Rain-down method A foam application method that directs the stream into the air above the fire and allows it to gently fall on the surface.

Rapid intervention crew/company (RIC) A minimum of two fully equipped personnel on site, in a ready state, for immediate rescue of disoriented, injured, lost, or trapped rescue personnel. (NFPA 1500)

Rapid intervention crew/company universal air connection (RIC UAC) A system that allows emergency replenishment of breathing air to the SCBA of disabled or entrapped fire or emergency services personnel. (NFPA 1407)

Rapid intervention pack A portable air supply that provides an emergency source of breathing air for a single fire fighter who has run out of air or whose air supply is insufficient to safely exit from an IDLH atmosphere.

Rate-of-rise heat detector A device that responds when the temperature rises at a rate exceeding a predetermined value. (NFPA 72)

Rear of the fire The side opposite the head of the fire. Also called the heel of the fire.

RECEO-VS An acronym developed for use by the incident commander for accomplishing tactical priorities on the fire ground; stands for rescue, exposure protection, confinement, extinguishment, overhaul, ventilation, and salvage.

Reciprocating saw A saw that is powered by an electric motor or a battery motor and whose blade moves back and forth.

Recommended exposure level (REL) A value established by NIOSH that is comparable to OSHA's permissible exposure limit (PEL) and the threshold limit value/time-weighted average (TLV/TWA). The REL measures the maximum time-weighted concentration of material to which 95 percent of healthy adults can be exposed without suffering any adverse effects over a 40-hour workweek.

Recovery phase The stage of a hazardous materials incident after imminent danger has passed, when clean-up and the return to normalcy have begun.

Red blood cells Oxygen-carrying cells found in mammals. They contain hemoglobin.

Reducer A fitting used to connect a small hose line or pipe to a larger hose line or pipe. (NFPA 1142)

Regulations Mandates issued and enforced by governmental bodies such as the federal, provincial, and territorial Occupational Health and Safety Acts (OHSAs) and the Canadian Environmental Protection Agency (CEPA), and the Canadian Standards Association.

Regulator purge/bypass valve A device or devices designed to bypass a regulator.

Rehabilitate To restore someone or something to a condition of health or to a state of useful and constructive activity.

Reinhart A hand tool used for constructing fire control lines and overhauling wildland fires. The tool is similar to an oversized garden hoe.

Rekindle A return to flaming combustion after apparent but incomplete extinguishment. (NFPA 921)

Relative humidity The ratio between the amount of water vapour in the gas at the time of measurement and the amount of water vapour that could be in the gas when condensation begins, at a given temperature. (NFPA 79)

Remote annunciator A secondary fire alarm notification unit, usually located near the front door of a building, that provides event information and frequently has the same control functionality as the main fire alarm control unit.

Remote-controlled hydrant valve A valve that is attached to a fire hydrant to allow the operator to turn the hydrant on without flowing water into the hose line.

Remote supervising station alarm system A fire alarm system that sounds an alarm in the building and transmits a signal to the fire department or an off-premises monitoring location.

Remote valve shut-off A type of valve that may be found at fixed facilities utilizing chemical processes or piped systems that carry chemicals. These remote valves provide a way to remotely shut down a system or isolate a leaking fitting or valve. Remote shut-off valves are also found on many types of cargo containers.

Repeater A special base station radio that receives messages and signals on one frequency and then automatically retransmits them on a second frequency.

Rescue Those activities directed at locating endangered persons at an emergency incident, removing those persons from danger, treating the injured, and providing for transport to an appropriate healthcare facility. (NFPA 1500)

Rescue company A group of fire fighters who work as a unit and are equipped with one or more rescue vehicles. (NFPA 1410)

Rescue-lift air bag An inflatable device used to lift an object or spread one or more objects away from each other to assist in freeing a victim. Various sizes and types are available.

Reservoir A water storage facility.

Residential sprinkler system A sprinkler system designed to protect dwelling units.

Residual pressure The pressure that exists in the distribution system, measured at the residual hydrant at the time the flow readings are taken at the flow hydrants. (NFPA 24)

Resource management Under the NIMS, includes mutual-aid agreements; the use of special federal, state, local, and tribal teams; and resource mobilization protocols. (NFPA 1026)

Respirator The complete assembly, including the respiratory inlet covering air purification components, electronics, batteries, harness, cables, and hose where applicable; designed to protect the wearer from inhalation of atmospheres containing harmful gases, vapors, or particulate matter. (NFPA 1994)

Response Immediate and ongoing activities, tasks, programs, and systems to manage the effects of an incident that threatens life, property, operations, or the environment. (NFPA 1600)

Retention The process of purposefully collecting hazardous materials in a defined area.

Reverse hose lay A method of laying a supply line where the supply line starts at the attack pump and ends at the water source.

Rim locks Surface-mounted, interior locks located on or in a door with a bolt that provide additional security.

Risk/benefit analysis An assessment of the risk to rescuers versus the benefits that can be derived from their intended actions. (NFPA 1006 and 1008)

Rocker lugs Fittings on threaded couplings that aid in coupling the hose. Also referred to as rocker pins.

Rocker panels A section of a vehicle's frame located below the doors, between the front and rear wheels.

Roll-in method A foam application method that involves sweeping the stream just in front of the target. Also referred to as the sweep or roll-on method.

Roll-over The condition in which unburned fuel (pyrolysate) from the originating fire has accumulated in the ceiling layer to a sufficient concentration (i.e., at or above the lower flammable limit) so that it ignites and burns. This can occur without ignition of, or prior to the ignition of, other fuels separate from the origin; also known as flameover. (NFPA 921)

Roof covering The membrane, which may also be the roof assembly, that resists fire and provides weather protection to the building against water infiltration, wind, and impact. (NFPA 5000)

Roof decking The rigid portion of roof between the roof supports and the roof covering.

Roof hooks The spring-loaded, retractable, curved metal pieces that allow the tip of a roof ladder to be secured to the peak of a pitched roof. The hooks fold outward from each beam at the top of a roof ladder.

Roof ladder A single ladder equipped with hooks at the top end of the ladder. (NFPA 1931)

Roofman's hook A long pole with a solid metal hook used for pulling.

Rope bag A bag used to protect and store rope so that the rope can be easily and rapidly deployed without kinking.

Rope record A record for each piece of rope that includes a history of when the rope was placed in service, when it was inspected, when and how it was used, and which types of loads were placed on it.

Rotary saw A saw that is powered by an electric motor or a gasoline engine and that uses a large rotating blade to cut through material. The blades can be changed depending on the material being cut.

Round turn A piece of rope looped to form a complete circle with the two ends parallel.

Rubber-covered hose Hose whose outside covering is made of rubber, which is said to be more resistant to damage. Also referred to as a rubber-jacket hose.

Run cards Cards used to determine a predetermined response to an emergency.

Rung The ladder crosspieces, on which a person steps while ascending or descending. (NFPA 1931)

Running end The part of a rope used for lifting or hoisting.

Safe location A location remote or separated from the effects of a fire so that such effects no longer pose a threat. (NFPA 101)

Safety data sheets (SDSs) Formatted information, provided by chemical manufacturers and distributors of hazardous products, about chemical composition, physical and chemical properties, health and safety hazards, emergency response, and waste disposal of the material. (NFPA 1072)

Safety knot A knot used to secure the leftover working end of the rope.

Salvage A fire-fighting procedure for protecting property from further loss following an aircraft accident or fire. (NFPA 402)

Salvage cover A large square or rectangular sheet made of heavy canvas or plastic material that is spread over furniture and other items to protect them from water runoff and falling debris.

San Francisco hook A multipurpose tool that can be used for several forcible entry and ventilation applications because of its unique design, which includes a built-in gas shut-off and directional slot.

Saponification The process of converting the fatty acids in cooking oils or fats to soap or foam; the action caused by a Class K fire extinguisher.

SCBA harness The backpack or frame for mounting the working parts of the SCBA and the straps and fasteners used to attach the SCBA to the fire fighter.

SCBA regulator The part of the SCBA that reduces the high pressure in the cylinder to a usable lower pressure and controls the flow of air to the user.

Scientific method The systematic pursuit of knowledge involving the recognition and definition of a problem; the collection of data though observation and experimentation; analysis of the data; the formulation, evaluation, and testing of hypotheses; and, where possible, the selection of a final hypothesis. (NFPA 921)

Screwdriver A tool used for turning screws.

SCUBA dive rescue technician A responder who is trained to handle water rescues and emergencies, including recovery and search procedures, in both water and under-ice situations. (SCUBA stands for self-contained underwater breathing apparatus.)

Search Land-based efforts to find victims or recover bodies. (NFPA 1951)

Search and rescue The process of searching a building for a victim and extricating the victim from the building.

Search rope A guide rope used by fire fighters that allows them to maintain contact with a fixed point.

Seat belt cutter A specialized cutting device that cuts through seat belts.

Secondary collapse A subsequent collapse in a building or excavation. (NFPA 1006)

Secondary containment Any device or structure that prevents environmental contamination when the primary container or its appurtenances fail. Examples of secondary containment mechanisms include dikes, curbing, and double-walled tanks.

Secondary contamination The process by which a contaminant is carried out of the hot zone and contaminates people, animals, the environment, or equipment. Also referred to as cross-contamination.

Secondary cut An additional ventilation opening made for the purpose of creating a larger opening or limiting fire spread.

Secondary device An explosive or incendiary device designed to harm emergency responders who have responded to an initial event.

Secondary feeders Smaller diameter water main pipes in the water distribution system that connect the primary feeders to the distributors.

Secondary loss Property damage that occurs due to smoke or water or other measures taken to extinguish the fire.

Secondary search A detailed, systematic search of an area that is conducted after the fire has been suppressed. (NFPA 1670)

Self-contained breathing apparatus (SCBA) An atmosphere-supplying respirator that supplies a respirable air atmosphere to the user from a breathing air source that is independent of the ambient environment and designed to be carried by the user. (NFPA 350 and 1981)

Self-contained underwater breathing apparatus (SCUBA) A respirator with an independent air supply that is used by underwater divers.

Self-expelling agent An agent that has sufficient vapour pressure at normal operating temperatures to expel itself from a fire extinguisher.

Self-rescue Escaping or exiting a hazardous area under one's own power. (NFPA 1006)

Serial arsonist A person who sets three or more fires with a cooling-off period between fires.

Seven, nine, eight (7, 9, 8) rectangular cut A ventilation opening that is usually about 1.2 m by 2.4 m (4 ft by 8 ft) in size; it is primarily used for large commercial buildings with flat roofs.

Shackle The U-shaped part of a padlock that runs through a hasp and then is secured back into the lock body.

Sheltering-in-place A method of safeguarding people located near or in a hazardous area by keeping them in a safe atmosphere, usually inside structures.

Shock load An instantaneous load that places a rope under extreme tension, such as when a falling load is suddenly stopped as the rope becomes taut.

Shoring A structure such as a metal hydraulic, pneumatic/mechanical, or timber system that supports the sides of an excavation and is designed to prevent cave-ins.

Shove knife A forcible entry tool used to trip the latch of outward swinging doors.

Shut-off valve Any valve that can be used to shut down water flow to a water user or system.

Siamese connection A hose appliance that allows two hoses to be connected together and flow into a single hose.

Sidewall sprinkler head A sprinkler that is mounted on a wall and discharges water horizontally into a room.

Simplex channel A radio system that uses one frequency to transmit and receive all messages; transmissions can occur in either direction but not simultaneously in both; when one party transmits, the other can only receive, and the party that is transmitting is unable to receive.

Single-action pull station A manual fire alarm activation device in which the user takes a single step—such as moving a lever, toggle, or handle—to activate the alarm.

Single command A command structure in which a single individual is responsible for all of the strategic objectives of the incident. It is typically used when an incident is within a single jurisdiction and is managed by a single discipline.

Single jacket A construction consisting of one woven jacket. (NFPA 1962)

Single resource An individual, a piece of equipment and its personnel, or a crew or team of individuals with an identified supervisor that can be used on an incident or planned event. (NFPA 1026)

Single-station smoke alarm A detector comprising an assembly that incorporates a sensor, control components, and an alarm notification appliance in one unit operated from a power source either located in the unit or obtained at the point of installation. (NFPA 72)

Size-up The process of gathering and analyzing information to help fire officers make decisions regarding the deployment of resources and the implementation of tactics. (NFPA 1410)

Slash Debris resulting from natural events such as wind, fire, snow, or ice breakage or from human activities such as building or road construction, logging, pruning, thinning, or brush cutting. (NFPA 1144)

Sledgehammer A hammer that can be one of a variety of weights and sizes.

SLICE-RS An acronym intended to be used by the first-arriving company officer to accomplish important strategic goals on the fire ground.

Smallpox A highly infectious disease caused by the variola virus.

Smoke The airborne solid and liquid particulates and gases evolved when a material undergoes pyrolysis or combustion, together with the quantity of air that is entrained or otherwise mixed into the mass. (NFPA 1404)

Smoke colour The attribute of smoke that reflects the stage of burning of a fire and the material that is burning in the fire.

Smoke control system A mechanical system that can create positive or negative pressure to control, alter, and limit the spread of smoke and gases.

Smoke density The thickness of smoke. Because it has a high mass per unit volume, smoke is difficult to see through.

Smoke detector A device that detects visible and invisible products of combustion (smoke) and sends a signal to a fire alarm control unit.

Smoke ejectors A mechanical device, similar to a large fan, that can be used to force heat, smoke, and gases from a post-fire environment and draw in fresh air. (NFPA 402)

Smoke explosion A violent release of confined energy that occurs when a mixture of flammable gases and oxygen is present, usually in a void or other area separate from the fire compartment, and comes in contact with a source of ignition. In this situation, there is no change to the ventilation profile, such as an open door or window; rather, it occurs from the travel of smoke within the structure to an ignition source.

Smoke inversion The condition in which smoke hangs low to the ground because of the presence of cold air.

Smoke particles The unburned, partially burned, and completely burned substances found in smoke.

Smoke velocity The speed of smoke leaving a burning building.

Smoke volume The quantity of smoke which indicates how much fuel is being heated.

Smooth-bore nozzle A nozzle that produces a solid stream or a solid column of water.

Smooth-bore tip A nozzle device that is a smooth tube, which is used to deliver a solid column of water. Different-sized tips can be attached to a single nozzle or appliance.

Socket wrench A wrench that fits over a nut or bolt and uses the ratchet action of an attached handle to tighten or loosen the nut or bolt.

Soffit The material covering the gap between the edge of the roof and the exterior wall of the house. The soffits in most modern houses contain a system of vents to maintain air flow to the attic area. Some attics may have sealed soffits to create a conditioned air space for energy conservation.

Soft sleeve hose A short section of large-diameter supply hose that is used to provide water from the large steamer outlet (the large-diameter port) on a fire hydrant or other pressurized water source to the suction side of the fire pump.

Solar panels A general term for thermal collectors or photovoltaic (PV) modules. (NFPA 780)

Solar photovoltaic (PV) systems A power system designed to convert solar energy into electrical energy; may also be referred to as a PV system or a solar power system.

Solid One of the three stages of matter; a material that has three dimensions and is firm in substance.

Solid beam A ladder beam constructed of a solid rectangular piece of material (typically wood), to which the ladder rungs are attached.

Solid-core door A door design that consists of wood filler pieces inside the door. This construction creates a stronger door that may be fire rated.

Solid stream A solid column of water.

Soman A nerve gas that is both a contact and a vapor hazard; it has the odor of camphor.

Sounding The process of striking a roof with a tool to determine where the roof supports are located.

Spalling Chipping or pitting of concrete or masonry surfaces. (NFPA 921)

Spanner wrench (hose key) A type of tool used to couple or uncouple hose by turning the rocker lugs or pin lugs on the connections.

Span of control The maximum number of personnel or activities that can be effectively controlled by one individual (usually three to seven). (NFPA 1006)

Special-use railcars Boxcars, flat cars, cryogenic tank cars, or corrosive tank cars.

Split hose bed A hose bed arranged to enable the pump to lay out either a single supply line or two supply lines simultaneously.

Split hose lay A scenario in which the attack pump forward lays a supply line from an intersection to the fire, and the water supply pump reverse lays a supply line from the hose left by the attack pump to the water source.

Spoil pile A pile of excavated soil next to the excavation or trench. (NFPA 1006)

Spoliation Loss, destruction, or material alteration of an object or document that is evidence or potential evidence in a legal proceeding by one who has the responsibility for its preservation. (NFPA 921)

Spot fire A new fire that starts outside areas of the main fire, usually caused by flying embers and sparks.

Spotlight A light designed to project a narrow, concentrated beam of light.

Spot-type detector A single smoke detector or heat detector device. These devices are often spaced throughout an area.

Spree arsonist A person who sets three or more fires at separate locations with no emotional cooling-off period between fires.

Spring-loaded center punch A spring-loaded punch used to break automobile glass.

Sprinkler piping The network of piping in a sprinkler system that delivers water to the sprinkler heads.

Sprinkler stop A mechanical device inserted between the deflector and the orifice of a sprinkler head to stop the flow of water.

Sprinkler wedge A piece of wedge-shaped wood placed between the deflector and the orifice of a sprinkler head to stop the flow of water.

Stack effect The vertical air flow within buildings caused by the temperature-created density differences between the building interior and exterior or between two interior spaces. (NFPA 92)

Staging area Location established where resources can be placed while they await a tactical assignment. (NFPA 1026)

Standard operating procedure (SOP) A written organizational directive that establishes or prescribes specific operational or administrative methods to be followed routinely for the performance of designated operations or actions. (NFPA 1521)

Standards Documents, the main text of which contain only requirements and which are in a form generally suitable for mandatory reference by another standard or code or for adoption into law. Nonmandatory provisions shall be located in an appendix or annex, footnote, or fine-print note and are not to be considered a part of the requirements of a standard. (NFPA 1)

Standing part The part of a rope between the working end and the running end.

Standpipe system An arrangement of piping, valves, and hose connections installed in a structure to deliver water for fire hose.

States of matter The physical state of a material—solid, liquid, or gas.

Static pressure The pressure that exists at a given point under normal distribution system conditions measured at the residual hydrant with no hydrants flowing. (NFPA 24)

Static rope A rope generally made out of synthetic material that stretches very little under load.

Static sources of water A water source such as a pond, river, stream, or other body of water that is not under pressure.

Staypoles Poles attached to each beam of the base section of extension ladders, which assist in raising the ladder and help provide stability of the raised ladder. (NFPA 1931)

Steamer port The large-diameter port on a fire hydrant.

Step chocks Specialized cribbing assemblies made of wood or plastic blocks in a step configuration. They are typically used to stabilize vehicles.

Stop A piece of material that prevents the fly sections of a ladder from becoming overextended, leading to collapse of the ladder.

Stored-pressure fire extinguisher A fire extinguisher in which both the extinguishing agent and expellant gas are kept in a single container and that includes a pressure indicator or gauge. (NFPA 10)

Stored-pressure water-type fire extinguisher A fire extinguisher in which water or a water-based extinguishing agent is stored under pressure.

Storz-type (nonthreaded) hose coupling A hose coupling that has the property of being both the male and the female coupling. It is connected by engaging the lugs and turning the coupling a one-third turn.

Straight stream A stream made by using an adjustable nozzle to provide a straight stream of water.

Strike team Specified combinations of the same kinds and types of resources, with common communications and a leader. (NFPA 1026)

Strip cut Another term for a trench cut.

Structural firefighting protective coat The protective coat worn by a fire fighter for interior structural firefighting.

Structural firefighting protective pants The protective trousers worn by a fire fighter for interior structural firefighting.

Subsurface fuels Partially decomposed matter that lies beneath the ground, such as roots, moss, duff, and decomposed stumps.

Suction hose A short section of supply hose that is used to supply water to the suction side of the fire pump.

Supervising station alarm system An alarm system that communicates between the protected property and a constantly attended location that will receive, interpret, and take the appropriate action to handle the alarm condition.

Supervisory signal A signal that results from the detection of a supervisory condition that indicates the need for action. (NFPA 72)

Supplemental restraint system (SRS) A system that uses supplemental restraint devices such as air bags to enhance safety in conjunction with properly applied seat belts. Seat belt pretensioning systems are also considered part of an SRS.

Supplied-air respirator (SAR) An atmosphere-supplying respirator for which the source of breathing air is not designed to be carried by the user. Also known as an air-line respirator. (NFPA 1989)

Supply hose Hose designed for the purpose of moving water between a pressurized water source and a pump that is supplying attack lines. (NFPA 1961)

Supply line operations The delivery of water from a water supply source, such as a fire hydrant or a static water source, to an attack pump.

Surface fuels Fuels that are close to the surface of the ground, such as grass, leaves, twigs, needles, small trees, logging slash, and low brush. Also called ground fuels.

Survivability profiling An assessment that weighs the risks likely to be taken versus the benefits of those risks of the viability and survivability of potential fire victims under the current conditions in the structure.

Tabun A nerve agent that disables the chemical connections between nerves and targets organs.

Tag line A separate rope that ground personnel can use to guide an object that is being hoisted or lowered.

Talk-around channel A simplex channel used for onsite communications.

Tamper seal A retaining device that breaks when the locking mechanism is released.

Target hazards Any occupancy types or facilities that present a high potential for loss of life or serious impact to the community resulting from fire, explosion, or chemical release.

Task force A group of resources with common communications and a leader that can be pre-established and sent to an incident or planned event or formed at an incident or planned event. (NFPA 1026)

Technical decontamination The planned and systematic process of reducing contamination to a level that is as low as reasonably achievable. (NFPA 1072)

Technical reference team A team of responders who serve as an information-gathering unit and referral point for both the incident commander and the hazardous materials safety officer (assistant safety officer).

Technical rescue incident (TRI) Complex rescue incidents requiring specially trained personnel and special equipment to complete the mission. (NFPA 1670)

Technical rescuer A person who is trained to perform or direct a technical rescue. (NFPA 1006)

Technical rescue team A group of rescuers specially trained in the various disciplines of technical rescue.

Technical use life safety rope A life safety rope with a diameter that is 9.5 mm (⅜ in.) or greater but is less than 13 mm (½ in.), with a minimum breaking strength of 20 kN (4496 lbf). Used by highly trained rescue teams that deploy to technical environments such as mountainous and/or wilderness terrain.

Technician level A person who responds to hazardous materials/WMD incidents using a risk-based response process by which he or she analyzes a problem involving hazardous materials/WMD, selects applicable decontamination procedures, and controls a release using specialized protective clothing and control equipment.

Telecommunicator An individual whose primary responsibility is to receive, process, or disseminate information of a public safety nature via telecommunication devices. (NFPA 1061)

Telephone interrogation The phase in a 911 call during which the telecommunicator asks questions to obtain vital information such as the location of the emergency.

Temperature The degree of sensible heat as measured by a thermometer or similar instrument.

Tempered glass A type of safety glass that is heat treated so that, under stress or fire, it will break into small pieces that are not as dangerous.

Temporal-3 pattern A standard fire alarm audible signal for alerting occupants of a building.

Ten-codes A system of predetermined coded messages, such as "What is your 10–20?" used by responders over the radio.

Ten standard firefighting orders A set of systematically organized rules developed by the U.S. Department of Agriculture Forest Service task force to reduce danger to firefighting personnel.

Testimonial evidence Verbal testimony of a witness given under oath in court or during specific legal proceedings.

Thermal column A cylindrical area above a fire in which heated air and gases rise and travel upward.

Thermal conductivity A property that describes how quickly a material will conduct heat.

Thermal imaging device An electronic device that detects differences in temperature based on infrared energy and then generates images based on those data. It is commonly used in smoke-filled environments to locate victims as well as to search for hidden fire during size-up and overhaul.

Thermal layering The stratification (heat layers) that occurs in a room as a result of a fire.

Thermal radiation The means by which heat is transferred to other objects.

Thermoplastic material Plastic material capable of being repeatedly softened by heating and hardened by cooling and that, in the softened state, can be repeatedly shaped by molding or forming. (NFPA 5000)

Thermoset material Plastic material that, after having been cured by heat or other means, is substantially infusible and cannot be softened and formed. (NFPA 5000)

Threaded hose coupling A type of coupling that requires a male fitting and a female fitting to be screwed together.

Threshold limit value/ceiling (TLV/C) The maximum concentration of hazardous material to which a worker should not be exposed, even for an instant.

Threshold limit value (TLV) The point at which a hazardous material or weapon of mass destruction begins to affect a person.

Threshold limit value/short-term exposure limit (TLV/STEL) The maximum concentration of hazardous material to which a worker can sustain a 15-minute exposure not more than four times daily without experiencing irritation or chronic or irreversible tissue damage. There should be a minimum 1-hour rest period between any exposures to this concentration of the material. The lower the TLV/STEL value, the more toxic the substance.

Threshold limit value/skin The concentration at which direct or airborne contact with a material could result in possible and significant exposure from absorption through the skin, mucous membranes, and eyes.

Threshold limit value/time-weighted average (TLV/TWA) The airborne concentration of a material to which a worker can be exposed for 8 hours a day, 40 hours a week and not suffer any ill effects.

Throwline A floating rope that is intended to be thrown to a person during water rescues or as a tether for rescuers entering the water. (NFPA 1983)

Tie rod A metal rod that runs from one beam of the ladder to the other to keep the beams from separating. Tie rods are typically found in wood ladders.

Time marks Status updates provided to the communications centre every 10–20 minutes. Such an update should include the type of operation, the progress of the incident, the anticipated actions, and the need for additional resources.

Tip The very top of the ladder.

Topography The land surface configuration. (NFPA 1051)

Trailers Solid or liquid fuel used to intentionally spread or accelerate the spread of a fire from one area to another. (NFPA 921)

Training officer The person designated by the fire chief with authority for overall management and control of the organization's training program. (NFPA 1401)

Transfer of command The formal procedure for transferring the duties of an incident commander at an incident scene. (NFPA 1026)

Transitional attack An offensive fire attack initiated by an exterior, indirect handline operation into the fire compartment to initiate cooling while transitioning into interior direct fire attack in coordination with ventilation operations.

Transport Canada (TC) The Canadian government agency that publicizes and enforces rules and regulations that relate to the transportation of many hazardous materials.

Trench cut A roof cut that is made from one load-bearing wall to another load-bearing wall and that is intended to prevent horizontal fire spread in a building.

Triangular cut A triangle-shaped ventilation cut in the roof decking that is made using a saw or an axe.

Trigger The button or lever used to discharge the agent from a portable fire extinguisher.

Triple-layer hose load A hose loading method in which the hose is folded back onto itself to reduce the overall length to one-third before loading the hose into the hose bed. This load method reduces deployment distances.

Trouble signal A signal initiated by a dispatch system or device indicative of a fault in a monitored circuit or component. (NFPA 720)

Truck or ladder company A group of fire fighters who work as a unit and are equipped with one or more pieces of aerial fire apparatus. (NFPA 1410)

Trunked radios A radio system that uses a computerized shared bank of frequencies to make the most efficient use of radio resources.

Truss A collection of lightweight structural components joined in a triangular configuration that can be used to support either floors or roofs.

Truss block A piece of wood or metal that ties the two rails of a trussed beam ladder together and serves as the attachment point for the rungs.

Trussed beam A ladder beam constructed of top and bottom rails joined by truss blocks that tie the rails together and support the rungs.

TTY/TDD systems User devices that allow speech- and/or hearing-impaired citizens to communicate over a telephone system. TTY stands for teletype, and TDD stands for telecommunications device for the deaf; the displayed text is the equivalent of a verbal conversation between two hearing persons.

Tube trailers Trucks or semitrailers on which a number of very long compressed gas tubular cylinders have been mounted and manifolded into a common piping system. (NFPA 1)

Turbulent smoke flow Agitated, boiling, angry-movement smoke that indicates great heat in the burning building. It is a precursor to flashover.

Twisted rope Rope constructed of fibres twisted into strands, which are then twisted together.

Two-in/two-out rule A safety procedure that requires a minimum of two personnel to enter a hazardous area and a minimum of two backup personnel to remain outside the hazardous area during the initial stages of an incident.

Two-way radios Portable communication devices used by fire fighters. Every firefighting team should carry at least one radio to communicate distress, progress, changes in fire conditions, and other pertinent information.

Type A packaging Packaging that is designed to protect its internal radiological contents during normal transportation and in the event of a minor accident.

Type B packaging Packaging that is far more durable than Type A packaging and is designed to prevent a release of the radiological hazard in the case of extreme accidents during transportation. Type B containers must undergo a battery of tests including those involving heavy fire, pressure from submersion, and falls onto spikes and rocky surfaces.

Type C packaging Packaging used when radioactive substances must be transported by air.

Type I construction (fire resistive) The type of construction in which the fire walls, structural elements, walls, arches, floors, and roofs are of approved noncombustible or limited-combustible materials that have a specified fire resistance.

Type II construction (noncombustible) The type of construction in which the fire walls, structural elements, walls, arches, floors, and roofs are of approved noncombustible or limited-combustible materials without fire resistance.

Type III construction (ordinary) The type of construction in which exterior walls and structural elements that are portions of exterior walls are of approved noncombustible or limited-combustible materials and in which fire walls, interior structural elements, walls, arches, floors, and roofs are entirely or partially of wood of smaller dimensions than required for Type IV construction or are of approved noncombustible, limited-combustible, or other approved combustible materials. (NFPA 14)

Type IV construction (heavy timber) The type of construction in which fire walls, exterior walls, and interior bearing walls and structural elements that are portions of such walls are of approved noncombustible or limited-combustible materials. Other interior structural elements, arches, floors, and roofs are constructed of solid or laminated wood or cross-laminated timber without concealed spaces within allowable dimensions of the building code. (NFPA 14)

Type V construction (wood frame) The type of construction in which structural elements, walls, arches, floors, and roofs are entirely or partially of wood or other approved material. (NFPA 14)

Ultrahigh-frequency (UHF) band Radio frequencies between 300 and 3000 MHz.

Underwriters Laboratory of Canada (ULC) The organization that tests and certifies that fire extinguishers (among many other products) meet established standards.

Undetermined fire A classification of fire when the cause cannot be proven to an acceptable level of certainty.

Unibody (unit body) construction The frame construction most commonly used in vehicles. The base unit is made of formed sheet metal; structural components are then added to the base to form the passenger compartment. Subframes are attached to each end. This type of construction eliminates the rail beams used in body-over-frame vehicles.

Unified command A team effort that allows all agencies with jurisdictional responsibility for an incident or planned event, either geographic or functional, to manage the incident or planned event by establishing a common set of incident objectives and strategies. (NFPA 1026)

Unit The organizational element having functional responsibility for a specific incident's operations, planning, logistics, or finance/administration activity. (NFPA 1026)

Unity of command The concept by which each person within an organization reports to one, and only one, designated person. (NFPA 1026)

Unlocking mechanism A keyway, combination wheel, or combination dial used to open a padlock.

Unwanted alarm Any alarm that occurs that is not the result of a potentially hazardous condition. (NFPA 72)

Upper explosive limit (UEL) The maximum amount of gaseous fuel that can be present in the air if the air/fuel mixture is to be flammable or explosive.

Upright sprinkler head A sprinkler head designed to be installed on top of the supply piping; it is usually marked SSU ("standard spray upright").

Utility rope Rope used for securing objects, for hoisting equipment, or for securing a scene to prevent bystanders from being injured. Utility rope must never be used in life safety operations.

V-agent (VX) A nerve agent, principally a contact hazard; an oily liquid that can persist for several weeks.

Vapour density The weight of an airborne concentration (vapour or gas) compared to an equal volume of dry air.

Vapour dispersion The process of lowering the concentration of vapors by spreading them out, typically with a water fog from a hose line.

Vapour-protective clothing The garment portion of a chemical-protective clothing ensemble that is designed and configured to protect the wearer against chemical vapours or gases. (NFPA 472)

Vapour-protective ensemble Multiple elements of compliant protective clothing and equipment that when worn together provide protection from some, but not all, risks of vapour, liquid-splash, and particulate environments during hazardous materials/WMD incident operations. (NFPA 1072)

Vapour suppression The process of controlling vapours given off by hazardous materials, thereby preventing vapour ignition, by covering the product with foam or other material or by reducing the temperature of the material.

Ventilation The controlled and coordinated removal of heat and smoke from a structure, replacing the escaping gases with fresh air. (NFPA 1410)

Ventilation-limited fire A fire in which the heat release rate and fire growth are regulated by the available oxygen within the space. (NFPA 1410)

Vertical ventilation The vertical venting of structures involving the opening of bulkhead doors, skylights, scuttles, and roof cutting operations to release smoke and heat from inside the fire building. (NFPA 1410)

Very high-frequency (VHF) band Radio frequencies between 30 and 300 MHz; the VHF spectrum is further divided into high and low bands.

Voice over Internet Protocol (VoIP) Technology that converts a person's voice into a digital signal that can be sent via the Internet to another device.

Voice recording system Recording devices or computer equipment connected to telephone lines and radio equipment in a communications centre to record telephone calls and radio traffic.

Volatility The ability of a substance to produce combustible vapours.

Volume The quantity of water flowing; usually measured in litres (gallons) per minute.

Wall post indicator valve A sprinkler control valve that is mounted on the outside wall of a building. The position of the indicator tells whether the valve is open or shut.

Warm zone The control zone outside the hot zone where personnel and equipment decontamination and hot zone support take place. (NFPA 1072 and 1500)

Water catch-all A salvage cover that has been folded to form a container to hold water until it can be removed.

Water chute A salvage cover that has been folded to direct water flow out of a building or away from sensitive items or areas.

Water curtain nozzle A nozzle used to deliver a flat screen of water that forms a protective sheet of water to protect exposures from fire.

Water flow The amount of water flowing through pipes, hose, and fittings, usually expressed in litres (gallons) per minute (L/min or gpm).

Water flow alarm device An attachment to the sprinkler system that detects a predetermined water flow and is connected to a fire alarm system to initiate an alarm condition. (NFPA 13)

Water hammer The surge of pressure that occurs when a high-velocity flow of water is abruptly shut off. The pressure exerted by the flowing water against the closed system can be seven or more times that of the static pressure. (NFPA 1962)

Water knot A knot used to join the ends of webbing together.

Water main A generic term for any underground water pipe.

Water mist fire extinguisher A fire extinguisher containing distilled or de-ionized water and employing a nozzle that discharges the agent in a fine spray. (NFPA 10)

Water mist system A fixed fire protection system that discharges very fine water mist particles through specialized nozzles that extinguish by displacing oxygen, cooling, and blocking radiant heat.

Water-motor gong An audible alarm notification device that is powered by water moving through the sprinkler system.

Water pressure The application of force by one object against another. When water is forced through the distribution system, it creates water pressure.

Water shuttle operations A method of transporting water from a source to a fire scene using a number of mobile water supply apparatus.

Water supply A source of water for firefighting activities. (NFPA 1144)

Water thief A device that has a 65-mm (2½-in.) inlet and a 65-mm (2½-in.) outlet in addition to two 38-mm (1½-in.) outlets. It is used to supply many hoses from one source.

Webbing Woven material of flat or tubular weave in the form of a long strip. (NFPA 1983)

Wedges Material used to tighten or adjust cribbing and shoring systems. (NFPA 1006)

Wet-barrel hydrant A hydrant used in areas that are not susceptible to freezing. The barrel of the hydrant is normally filled with water.

Wet-chemical extinguishing agent Normally an aqueous solution of organic or inorganic salts or a combination thereof that forms an extinguishing agent. (NFPA 10)

Wet-chemical extinguishing system A system that discharges a proprietary liquid extinguishing agent. It is often installed over stoves and deep-fat fryers in commercial kitchens.

Wet-chemical fire extinguisher A fire extinguisher containing a wet-chemical extinguishing agent for use on Class K fires.

Wet pipe sprinkler system The most common and least expensive type of automatic sprinkler system. As its name implies, the piping in a wet system is always filled with water. When a sprinkler head activates, water is immediately discharged onto the fire.

Wetting-agent fire extinguisher A fire extinguisher that expels water combined with a concentrate to reduce the surface tension and increase its ability to penetrate and spread.

Wet vacuum A device similar to a wet/dry shop vacuum cleaner that can pick up liquids. It is used to remove water from buildings.

Wheel chocks Wedges or blocks of sturdy materials that are placed against a vehicle's wheels to prevent accidental rolling during extrication.

Wheeled fire extinguisher A portable fire extinguisher equipped with a carriage and wheels intended to be transported to the fire by one person. (NFPA 10)

Wilderness search and rescue (SAR) The process of locating and removing a victim from the wilderness.

Wildland Land in an uncultivated, more or less natural state and covered by timber, woodland, brush, and/or grass. (NFPA 901)

Wildland/brush company A fire company that is dispatched to vegetation fires where larger pumpers cannot gain access. Wildland/brush companies have four-wheel drive vehicles and special firefighting equipment.

Wildland fire An unplanned fire burning in vegetative fuels. (NFPA 1051)

Wildland/urban interface The line, area, or zone where structures and other human development meet or intermingle with undeveloped wildland or vegetative fuels. (NFPA 5000)

Wildland/urban intermix An area where improved property and wildland fuels meet with no clearly defined boundary. (NFPA 5000)

Wired glass A glazing material with embedded wire mesh.

Wood truss An assembly of small pieces of wood or wood and metal.

Wooden beam Load-bearing member assembled from individual wood components.

Working end The part of the rope used for forming a knot.

Wye A device used to split a single hose into two or more separate lines.

Youth firesetters Recognized classifications of minors who ignite fires for various reasons. Based on age and specific psychological characteristics, youth firesetters include child firesetters, juvenile firesetters, and adolescent firesetters.

Zoned system A fire alarm system design that divides a building or facility into zones so that the area where an alarm originated can be identified.

Index

Note: Page numbers followed by *f* or *t* indicate figures and tables, respectively.